Environmental Profiles

Garland Reference Library of Social Science
(Vol. 736)

Environmental Profiles

A Global Guide to Projects and People

Linda Sobel Katz, Sarah Orrick, and Robert Honig

Foreword by Vice President Al Gore

Illustrations by Jane Svoboda

Garland Publishing, Inc.
New York & London
1993

© 1993 Linda S. Katz, Sarah Orrick, and
Robert Honig
Cover photograph
© The National Audubon Society Collection/PR
All rights reserved

Library of Congress Cataloging-in-Publication Data

Katz, Linda Sobel
    Environmental profiles : a global guide to projects
and people / Linda Sobel Katz, Sarah Orrick, and
Robert Honig ; foreword by Al Gore.
    p. cm. — (Garland reference library of social
science ; vol. 736)
    Includes index.
    ISBN 0–8153–0063–8 (alk. paper)
    1. Environmental sciences—Societies, etc.—
Directories. 2. Environmentalists—Directories.
3. Environmental sciences—Information services—
Directories. I. Katz, Linda Sobel. II. Orrick, Sarah.
III. Honig, Robert. IV. Title. V. Series: Garland
reference library of social science ; v. 736.
GE1.K38   1993
363.7'0601—dc20   92–41800
                                          CIP

Printed on acid-free, 250-year-life, recycled paper
Manufactured in the United States of America

For Hershel, Steven, Simone, and my
parents, Rose and Fred
—*LSK*

For Jesse Orrick
—*SO*

For Rande, Albert, Ilona, Eva, and Frank,
survivors all, whose example brings the
impossible within reach
—*REH*

For our contributors and
environmentalists worldwide

Contents

Foreword

Human civilization has reached a turning point, a defining moment in history, and our response will profoundly affect future generations.

For the past 150 years, industrial expansion has enriched humanity in immeasurable ways, but it has taken a terrible toll on the environment. Resources have been extracted and never replenished, often for short-term use and then disposal. Streams of waste have been formed that flow into nature, poisoning the land, air, and water—everything basic to our very survival.

This has given rise to "pollution control," the technological fix we developed in an attempt to capture and contain at least some of the toxic byproducts of our industrial activities. Some nations have done better than others in this respect, but every nation has produced its share of acute environmental contamination. And as we continue to grow in number, our consumption rates continue to soar, and the wastestreams surge and swell.

Gradually, we are becoming aware that our activities produce some byproducts that are even more insidious than poisonous wastestreams—more insidious both because they are less visible and because they pose threats on a vast scale.

Pesticides, for example, which in small doses seem to provide improvements in agricultural efficiency, can accumulate over time and devastate the ecology of entire regions. Releases of chlorofluorocarbons and related chemicals lead to a destruction of the thin layer of ozone in our upper atmosphere which shields the entire planet from deadly ultraviolet radiation. Excessive fossil fuels consumption has resulted in an accumulation of carbon dioxide in the global atmosphere. This condition elevates temperatures on the Earth's surface and threatens to distort the Earth's ecological balance in ways that can cause the loss of entire species and the disruption of ecosystems and climate patterns upon which our communities and economies are based.

It is now possible for us to recognize these phenomena as symptoms of an underlying malady in our civilization's relationship with nature. Society, as it is presently organized, is on a collision course with the natural environment. In my view, three overarching factors have led to this crisis:

The first is the staggering increase in population that we have witnessed in our lifetimes, placing unprecedented pressures on the natural resources of the world.

Second, scientific and technological advances have magnified our ability to affect the environment and have changed the consequences of exploiting the Earth as profoundly as nuclear weapons changed warfare.

Third, and most important, the change in our way of thinking that began with the scientific revolution 350 years ago was basically benign until our greater numbers and ecological power made it dangerous to the Earth. That change in thinking was based on the misguided assumption that we are separate and apart from the Earth.

In fact, we are part of an intricate web of life, and our own survival is threatened as we continue to disrupt the ecological balance of the planet.

The efforts described in this important volume help us to understand better the complexity and subtlety of the relationship we as humans share with the global environment. Though informed by differing perspectives, these projects are unified by a common purpose— a commitment to the preservation of the diverse natural resources that comprise our world, and to the realign-

ment of our relationship as a civilization with our global environment.

Happily, a new way of thinking about our responsibility to the global environment is gaining momentum. Young people today think this is the number one issue facing the world, and the community of nations is beginning to move toward a commitment of following sustainable patterns of development.

It is my strong belief that the preservation and protection of the environment must and will become the central organizing principle for the post–Cold War world. The solutions we seek will be found in a new faith in the future of life on Earth, a faith in the future that justifies sacrifices in the present. A new moral courage to choose higher values in the conduct of human affairs and a new reverence for absolute principles can serve as stars by which to map the future course of our species and our place within creation.

Al Gore
Washington, DC

Board of Advisors

MICHAEL JEFFREY BALICK is director of the Institute of Economic Botany and philecology curator of economic botany, New York Botanical Garden. He is also a lecturer in tropical studies at the Yale University School of Forestry and Environmental Studies and an adjunct professor at the City University of New York. He has served as assistant curator of the Herbarium at the New York Botanical Garden and as research associate for plant domestication, Botanical Museum of Harvard University; research assistant for Agribusiness Associates, Inc., Wellesley Hills, Massachusetts; and as a research fellow in tropical horticulture and economic botany, Las Cruces Tropical Botanical Garden, Costa Rica. Dr. Balick has conducted extensive fieldwork in the central Caribbean and South America, as well as in Israel, Europe, and Mexico. He is a review editor for the *Journal of Ethnopharmacology*, associate editor of *Advances in Economic Botany*, and chairman of Ix Chel Tropical Research Center in Belize. His memberships include the Conservation Committee of the American Association of Botanical Gardens and Arboreta; the Advisory Committee, Las Cruces Tropical Botanical Garden, Costa Rica; Advisory Committee on Technology Innovation (ACTI) of the Board on Science and Technology for International Development (BOSTID), National Research Council; Scientific Advisory Board, American Society for the Protection of Nature in Israel; and chairman of the Ethnobotany Group and the Palm Specialist Group of the Species Survival Commission, World Conservation Union (IUCN/SSC). Dr. Balick holds a Ph.D. in biology from Harvard University.

LYNNE HARDIE BAPTISTA, a specialist in environmental communications, is the manager of World Wildlife Fund's nationwide environmental education program, Windows on the Wild. Prior to this she was WWF's public information and education specialist and was responsible for developing the organization's speaker's program, writing primary and secondary school education materials, coordinating special events, and curating exhibitions for airports, zoos, and natural history museums. She developed and directed a national consumer awareness campaign on wildlife trade, Buyer Beware, and for six years was editor of *TRAFFIC (USA)*, a technical newsletter on wildlife trade. Baptista is the author or editor of numerous books and articles on wildlife conservation and the environment. Two of her education programs have won national awards for conservation education. She holds a BA degree in English literature from the University of Maryland and has studied advanced writing at George Washington University.

JOAN BAVARIA is the founder of the Coalition for Environmentally Responsible Economies (CERES), a coalition that brings together leading environmentalists and socially concerned investors representing more than 10 million individuals and $150 billion of invested assets. Bavaria has been involved directly in social investing since 1975. She was founding president of the Social Investment Forum, a ground-breaking trade association of over 700 investment professionals, research groups, community loan funds, banks, and socially concerned investors. Bavaria also serves as president and CEO of Franklin Research and Development, a socially responsible investment advisory firm in Boston, Massachusetts, and on the boards of the Council on Economic Priorities, the Industrial Cooperative Association Loan Fund of Sommerville, Green Seal, and Lighthawk.

RICHARD BLOCK is former director of public programs for the World Wildlife Fund (WWF). His responsibilities included developing public and educational projects, coordinating speakers programs, and organizing public exhibits. He also served as WWF's liaison with zoos and aquariums. He has addressed audiences across the USA, Europe, New Zealand, and the Bahamas on subjects ranging from endangered species to individual action for improving the environment. Prior to joining WWF in 1987, he was public relations curator at the Kansas City Zoo and education curator at Zoo Atlanta. Block is now executive director of the Dian Fossey Gorilla Fund, Colorado. He has also taught communications courses at the School of Natural Resources at the University of Michigan. He holds an M.S. degree from the University of Michigan.

JOHN M. FITZGERALD is counsel for wildlife policy for Defenders of Wildlife in Washington, DC, representing the organization before the U.S. Congress and administrative agencies and coordinating litigation. He was a leader of the Endangered Species Act Reauthorization Coalition and presently serves on the steering committee of a coalition seeking to strengthen the Act.

He also represents Defenders of Wildlife at meetings on the Convention on International Trade in Endangered Species of Wild Fauna and Flora (CITES), and has been instrumental in the development of a U.S. "dolphin-safe" tuna policy. He served as counsel to the Subcommittee on Human Resources of the Committee on Post Office and Civil Service of the U.S. House of Representatives and as a legislative aide to a member of Congress. He was formerly communications director of the National Public Interest Research Group (PIRG) Clearing House in Washington, DC.

GARY T. GALLON is president of Environmental Economics International based in Toronto, Canada, specializing in integrating environmental concerns into government and corporate decisionmaking. He previously served as senior policy advisor to the Ontario Minister of the Environment, helping to initiate the government's waste reduction, reuse, and recycling effort, to mediate environmental conflicts; and to establish a public participation policy. Before that, as director of the UN Environment Liaison Centre in Nairobi, Kenya, Gallon helped national governments adopt new environmental laws and incorporate information processes and policies that helped them with issues ranging from pesticides management and water pollution to energy conservation and tropical forest protection. As executive director of the Society for the Promotion of Environmental Conservation in British Columbia, he received the Canadian National Award for Environment for his work on reducing West Coast oil spills, for the Berger Inquiry into the environmental impacts of the proposed MacKenzie Valley natural gas pipeline, and for the campaign to protect the Fraser River farmland and aquatic ecosystems. He was one of the founding members of the board of directors of Greenpeace. Gallon holds a B.A. degree in sociology from California State University, Northridge.

THOMAS E. LOVEJOY is assistant secretary for External Affairs at the Smithsonian Institution. He also serves as chief advisor for the Minimum Critical Size of Ecosystems project, a joint research project of the Smithsonian Institution and Brazil's National Institute for Amazon Research. He serves on numerous scientific and conservation boards, including the New York Botanical Garden, the Academy of Natural Sciences of Philadelphia, Resources for the Future, World Resources Institute, Environmental Defense Fund, World Wildlife Fund, Center for Plant Conservation, Rainforest Alliance, and the Rachel Carson Council, Inc. He is a fellow of the American Association for the Advancement of Science, the New York Zoological Society, the Linnean Society of London, and the American Ornithologists' Union. He is the recipient of the Ibero-

American Award, the Goeldi Museum Certificate of Merit, the Order of Merit of Mato Grosso Commander, the Brazilian National Parks 50th Anniversary Medal, the Order of Rio Branco Commander, the Carr Medal, the Garden Club of America Frances K. Hutchinson Medal, and the UNEP Global 500 Roll of Honour, as well as honorary degrees from Colorado State University, Williams College, and the College of Boca Raton. He was a member of President Bush's Council of Advisors in Science and Technology, and is currently president of the American Institute of Biological Sciences, honorary chairman of Wildlife Preservation Trust International, chairman of the Advisory Board of Earth Communications Office, chairman of the Biosphere II Scientific Advisory Committee, and chairman of the U.S. Man and the Biosphere Program. He is the author of numerous articles and is author or editor of four books. He was previously executive vice president of World Wildlife Fund. Dr. Lovejoy holds a Ph.D. in biology from Yale University.

DANIEL MAGRAW is associate general counsel for international activities, U.S. Environmental Protection Agency (EPA). He has taught public international law, international environmental law, international business transactions, and international development policy and the law at the University of Colorado School of Law. During 1989–1991, Magraw was a visiting scientist at the Environmental and Societal Impacts Group of the National Center for Atmospheric Research in Boulder, Colorado, and an adjunct scientist there. He previously practiced international law and constitutional law at the firm of Covington & Burling in Washington, DC, and served as an economist and business consultant for the Peace Corps in India. Magraw has spoken in the USA and abroad on international law topics ranging from the jurisprudence of the International Court of Justice to international environmental law, nuclear war, and international business law. Recent projects include a compilation of basic international environmental law documents and related references; a coursebook on international environmental law; and studies of the use of reasonableness as a normative element in international law, of the relationship between "black money" and economic development, and of the role of party-appointed arbitrators. He is a member of the American Law Institute and chairman of the International Environmental Law Committee of the American Bar Association's Section of International Law and Practice and of the U.S. Department of State's Advisory Committee on International Business Transactions. He was formerly on the Roster of Experts of the United Nations Centre on Transnational Corporations. He holds a Juris Doctor degree from the University of California, Berkeley.

SHARON MATOLA is director of the Belize Zoo and Tropical Education Center, which she founded in 1983. She also serves as chairperson of the Tapir Specialist Group of the Species Survival Commission, World Conservation Union (IUCN/SSC). She previously was an animal handler and business manager for natural history films and a lion tamer for the Suarez Circus in Mexico. In 1989, Matola was an expedition coordinator for a British Forces scientific expedition to the highest point in Belize, a previously unexplored region, and for the "Dolores Estate Expedition," to a tropical rainforest in southwestern Belize. Matola also participated in rapid ecological assessments (REAs) in Guatemala and Colombia. She is a member of the Turneffe Island and Range Committee, advising the Belize government on conservation strategies for Turneffe Island atoll, as well as of the National Conservation Advisory Board, Belize Audubon Society, Programme for Belize, Belize Center for Environmental Studies, and the International Organization of Women. She holds a B.A. degree in both biology and Russian from New College, Sarasota, Florida, and has served in the United States Air Force.

ALAN S. MILLER has been executive director of the Center for Global Change at the University of Maryland since 1989. He has also taught law at Duke University; Widener University Law School in Wilmington, Delaware; University of Maryland Law School, Washington College of Law at American University; and University of Iowa College of Law. He was a Fulbright scholar from 1977 to 1978, at Macquarie University, Australia, and again in 1987, visiting Tokyo University School of Law and the Japan Energy Law Research Institute. He was an attorney for the Natural Resources Defense Council, assistant director of the National Energy Project of the Special Committee on Energy of the American Bar Association (ABA), and research attorney for the Environmental Law Institute. He has served on Governor William Donald Schaefer's Energy Task Force for the State of Maryland, the Office of Technology Assessment (OTA) Advisory Committee on Energy Efficiency and Global Change Data Needs, the board of the Solar Electric Fund, the board of Environmental Exchange, the Global Climate Committee of the ABA Section on Natural Resources, and the Stratospheric Ozone Advisory Committee of the U.S. Environmental Protection Agency (EPA). He was also a consultant to EPA on climate change and a participant in the Keystone dialogue on electricity transmission issues, the National Academy of Sciences Panel on the Future of Electric Power, and the OTA Advisory Board for study of decentralized power generation. He holds Juris Doctor and a Master of Public Policy degrees from the University of Michigan.

clude the International Council of Environmental Law, American Bar Association, Inter-American Bar Association, International Bar Association, International Law Association, American Society of International Law, Foreign Policy Association, and Council on Foreign Relations. He holds a Juris Doctor degree with distinction from the University of Michigan Law School.

PHILIP SHABECOFF is executive publisher of *Greenwire*, a daily electronic summary of worldwide environmental news coverage. He was a reporter for the *New York Times* from 1959 to 1991. He was a foreign correspondent in West Germany, with responsibilities for covering West and East Germany, Scandinavia, and Czechoslovakia, and later in Tokyo, with responsibilities for covering Japan, Korea, Vietnam, Indonesia, the Philippines, Malaysia, Hong Kong, Singapore, and Thailand. He was subsequently an economics and labor correspondent for the Washington Bureau, White House correspondent covering the Nixon and Ford administrations, and environmental correspondent for the Washington Bureau (1977–1991). He is a contributing author to *American Government* by Charles Hamilton and *The Presidency Reappraised*, Thomas Cronin and Rexford Tugwell, editors. His book *A Fierce Green Fire, Environmentalism in the United States* was published in early 1993. He was an original selectee to the Global 500 Roll of Honour, the United Nations Environment Programme's award for outstanding achievement in protection of the environment. He also received the James Madison Award of the American Library Association, the annual award of the National Environmental Development Association for balance in journalism, and the National Wildlife Foundation's "Connie" Award for achievement in conservation. He holds an M.A. degree from the University of Chicago.

DURWOOD J. ZAELKE is president and founder of the Center for International Environmental Law—US (CIEL US), a Washington, DC, public interest law firm established to develop and use international, comparative, and U.S. environmental law to solve global and transnational environmental problems. He co-founded CIEL at Kings College, London, and served as a senior research fellow there from 1990 to 1991. He is also an adjunct law professor and scholar-in-residence at Washington College of Law, American University, Washington, DC. Zaelke was previously with the Sierra Club Legal Defense Fund, where he served as director first of the Alaska Office, then the Washington, DC, Office, and then the International Program. Prior to that he was a special litigating attorney for the U.S. Department of Justice, Land and Natural Resources Division. Earlier experience included work with the Environmental Law Institute in Washington, DC, and with the Los Angeles law firm of Adams, Duque & Hazeltine. He is a member of the Advisory Panel of the World Resources Institute program "The U.N. System and the Challenges of Global Environmental and Sustainable Development," chair of the Board of Directors for the Environmental Exchange, and a consultant for the United Nations and the Stockholm Environment Institute. He holds a Juris Doctor degree from Duke University School of Law.

Introduction

Tom Lovejoy, who graces the Board of Advisors of *Environmental Profiles*, coined the now famous clarion call for this decade: The race to save our planet from environmental destruction will be won or lost based upon our actions in the 1990s. We have attempted to design this volume as a tool to bring scarce resources and dedicated people together.

Preparing a resource for linking and spotlighting international environmental efforts during such fluid times was a challenge. During the two years it took to plan, compile, and produce *Environmental Profiles*, the face of the world changed dramatically: communism collapsed in the Soviet Union, democracy emerged in Eastern Europe, a Persian Gulf war brought the worst environmental disaster in history, volcanic eruptions of Mt. Pinatubo momentarily cooled the Earth and heated the debate on global warming, and the planet's population grew by about 190 million people!

We believe, however, that it was the best of times for such an undertaking, for along with the upheaval, a new environmental awareness and meaningful action emerged, with nations of widely differing size, stages of development, and forms of government achieving agreements and cooperative relations, most notably at the June 1992 Earth Summit in Brazil.

The most heartening aspect of our project was the extraordinary willingness of so many people (through so many languages and in so many places) to share their accomplishments, travails, needs, and resources—motivated by the common desire to connect with others in similar fields, whether in Venezuela or Vietnam, Poland or Papua New Guinea, Uganda or the USA.

In collecting this material, we came to see that just as the ecosystem may depend as much upon the tiniest insect as the tallest tree, so too is the planet served by the committed efforts of small and large groups alike. Thus, we have tried to treat all organizations equally, with the length of each profile reflecting only the amount and nature of the information submitted to us.

Our goal was to create a book that went beyond the function of a directory—to produce a resource that actually describes how, why, and where environmental efforts have been undertaken and who provided the energy, impetus, and commitment to make them happen.

Sharon Matola, director of the Belize Zoo and Tropical Education Center, deserves much credit for inspiring this book's concept. Sharon brought her views from the grassroots to our advisory board. The example of her own work and accomplishments in Belize, one of the world's least populous and most environmentally rich countries, convinced us that a communications tool that opened doors to cooperation and coordination could have a profound impact. The project then came to life in the hands of Garland Publishing, who joined us in conducting 2 sets of surveys, including the dissemination of questionnaires to more than 5,000 individuals and organizations worldwide.

The descriptions found in this book are gleaned from the responses to those questionnaires and other materials sent to us for review from 115 countries, and so reflect a diversity of philosophy, politics, approach, and interest in environmental concerns. We intentionally set aside personal inclinations to enable each group or project to present itself in its own terms and context. We leave all value judgments about the merits of the organizations to our readers.

In defining "environmental" projects, we cast a wide net to include not only conservation and pollution control, but also health, education, population, and family planning, for instance. Also featured in these pages are zoos, described as the modern Noah's Ark, along with wildlife utilization programs in developing countries, such as Madagascar.

Even with thousands of environmental efforts profiled, we could not have included every group and project making important contributions to the global environment. Faster than we could capture them in print, new projects were being launched. This first edition represents what we hope will be a beginning to building a uniquely useful tool for the kind of progress and change upon which our common future relies. We especially hope the contents will appeal to young people, by arousing their curiosity and awe of the world's remaining bountifulness, empowering their search for meaningful personal and professional roles as adults.

We welcome your ideas. Please send your suggestions and nominations for new inclusions to:

ENVIRONMENTAL PROFILES
Garland Publishing, Inc.
717 Fifth Avenue, Suite 2500
New York, NY 10022 USA

Linda Sobel Katz
Sarah Orrick
Robert Honig

Never doubt that a small group of thoughtful, committed citizens can change the world; indeed, it's the only thing that ever has.

—Margaret Mead

Acknowledgments

We are grateful for the guidance of our notable Board of Advisors: Michael Balick, Lynne Hardie Baptista, Joan Bavaria, Richard Block, John Fitzgerald, Gary Gallon, Tom Lovejoy, Dan Magraw, Sharon Matola, Alan Miller, Gail Osherenko, Frank Potter, Steve Rubin, Philip Shabecoff, Thomas Schoenbaum, and Durwood Zaelke, whose accomplishments are described on pages xi–xv and throughout the *Guide*. We are very honored that Vice President Al Gore, author of *Earth in the Balance*, contributed our book's Foreword prior to leading this country's congressional delegation as U.S. Senator at the 1992 UNCED Earth Summit.

Warmest appreciation is expressed to our enthusiastic editorial and research team: Ancelyn Krivak, Shira Robinson, Pattie Cinelli, Susan Dunn, Karen Grossman, Vince Harriman, Lisa Kamm, Simone and Steven Katz, Mary E. Kepferle, Adina Lack, Faith Noble, Robby Peckerar, Jane Svoboda, and Susan Womeldorf. Ancelyn, Mary, Vince, and Susan were instrumental in translating materials in Spanish, Portuguese, Italian, French, and German. (Materials also arrived all or partly in Icelandic, Polish, Danish, Norwegian, Dutch/Flemish, Russian, Chinese, Arabic, and Turkish, and we attempted to remain as true to the text as our abilities and resources would permit.) Jane enhanced our *Guide* with her original pen-and-ink renderings. Her interpretive drawings appear at random and are not meant to be technical executions of specific species.

The embassies in Washington, DC, were enormously cooperative in identifying appropriate organizations in their respective countries. Statistics and other facts, mostly appearing in the "country introductions," were obtained from various embassies and publications of the World Resources Institute, including *The 1992 Information Please Environmental Almanac* and the *Directory of Country Environmental Studies*. Renew America generously offered much material of use to this *Guide*. Rande Joiner, Tom Gorman, Donald Bogle, Mary Koenig, Steve Raabe, Thomas Gorguissian, and Ram Y. Uppuluri gave invaluable help, as did Garland Publishing's editors and staff.

North America

Africa

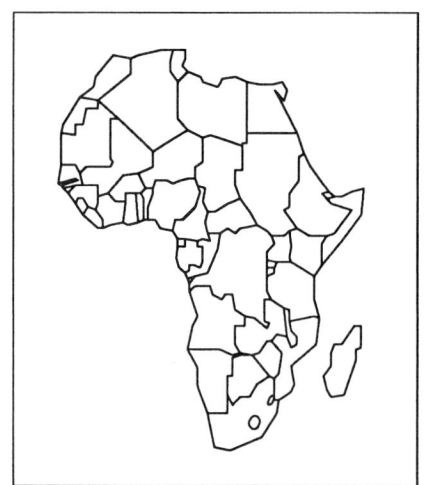

South America

Middle East

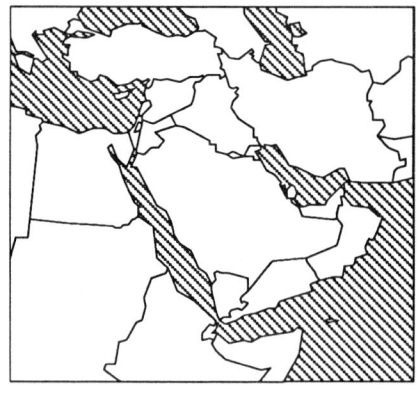

Europe

Asia

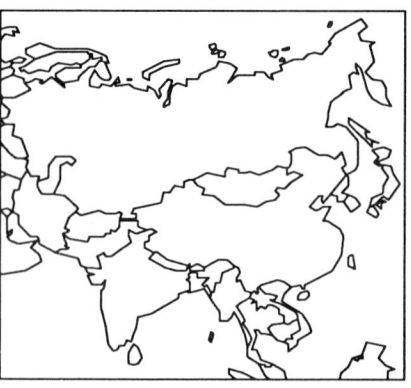

Oceania

How to Use This Guide

Environmental Profiles is arranged in alphabetical order by country. Each country begins with a brief introduction and a map of one of seven regions: North America, South America, Europe, Asia, Africa, Oceania, and the Middle East.

Within each country, the profiles are organized under the following headings: **Government, NGO (non-governmental organization), Private,** and **Universities.** Groups that are ⊕ international (engaged in projects outside their country) and/or ❦ grassroots are identified by these symbols. Headings within each profile include: **Organization, Projects and People,** and **Resources** (both available and needed).

Whenever possible, names of organizations are listed first in their native language, followed by an English translation—both in the profiles and in the index. Because material was collected from such diverse sources, we have taken some liberty in standardizing its presentation, such as in type of organization, address lines, and the use of acronyms in parentheses following the organizational name. Telephone numbers include country and city codes. (Readers and contributors are encouraged to let us know if we can improve on this format.)

A **Global/Regional** section in the front of the *Guide* includes organizations that transcend national boundaries, such as the United Nations Environment Programme (UNEP).

A list of **Abbreviations and Acronyms,** appearing on pages xxiii–xxvi, should be particularly helpful in identifying organizations listed in the **Meetings** subheading.

The **Appendix,** beginning on page 897, lists organizations in alphabetical order, with their country included, according to the following issues: Air Quality/Emission Control, Biodiversity/Species Preservation, Deforestation, Energy, Global Warming, Health, Population Planning, Recycling/Waste Management, Sustainable Development, Transportation, and Water Quality.

Consult the **Index** to locate all profiles relating to a specific country; to research a particular species (such as "leopard, snow"); and to locate individuals, among other uses. Persons interested in ordering publications or other resource materials should ask the organizations about current prices, as these are not included.

Abbreviations and Acronyms

AALS	Association for Arid Land Studies
AAZPA	American Association of Zoological Parks and Aquariums
AFRENA	Agroforestry Research Networks for Africa
AGM	annual general meeting
AGU	American Geophysical Union
AIBS	American Institute of Biological Sciences
AIT	Asian Institute of Technology, Thailand
AMS	American Meteorological Society
ANEN	African NGOs Environment Network
ANZEC	Australia and New Zealand Environment Council
ASEAN	Association of Southeast Asian Nations
ATO	African Timber Organization
AWB	Asian Wetland Bureau
BIR	International Recycling Bureau
BLM	U.S. Bureau of Land Management
CARE	Cooperative for American Relief Everywhere
CATIE	Tropical Agricultural Research and Training Center, Costa Rica
CBSG	Captive Breeding Specialist Group, IUCN
CCAMLR	Convention on Conservation of Antarctic Marine Living Resources
CEAT	Coordinating Office for Friends of the Earth Groups in Europe
CEP	Council on Economic Priorities
CERES	Coalition for Environmentally Responsible Economies
CFCs	chlorofluorocarbons
CI	Conservation International
CIC	International Council for Game and Wildlife Conservation

CIDA	Canadian International Development Agency
CIDIE	Committee of International Development Institutions on the Environment
CITES	Convention on International Trade in Endangered Species of Wild Fauna and Flora
CMS	Convention on the Conservation of Migratory Species of Wild Animals
CNPPA	Commission on National Parks and Protected Areas
CO_2	carbon dioxide
CPC	Center for Plant Conservation
CSG	Crocodile Specialist Group
CSIRO	Commonwealth Scientific and Industrial Research Organization, Australia
DHS	Demographic and Health Surveys
DOE	U.S. Department of Energy
EC	European Community
ECOSOC	UN Economic and Social Council
EDF	Environmental Defense Fund
EEB	European Environmental Bureau
EIA	environmental impact assessment(s)
EIS	environmental impact study
ENDA	Environment and Development Action in the Third World
EPA	U.S. Environmental Protection Agency
ESA	Ecological Society of America
ESCAP	UN Economic and Social Commission for Asia and the Pacific
FAO	UN Food and Agriculture Organization
FDA	U.S. Food and Drug Administration
FOEI	Friends of the Earth International
FUNEP	Friends of United Nations Environment Programme (UNEP)
FWS	U.S. Fish and Wildlife Service
GESAMP	Joint Group of Experts on the Scientific Aspects of Marine Pollution
GEWEX	Global Energy and Water Cycle Experiment
GIS	geographic information system
HELCOM	Baltic Marine Environment Protection Committee (Helsinki Commission)
HHS	U.S. Department of Health and Human Services
IAEA	International Atomic Energy Agency
IAFWA	International Association of Fish and Wildlife Agencies
IATTC	Inter-American Tropical Tuna Commission
IBAMA	Brazilian Institute of the Environment and Renewable Natural Resources
ICBP	International Council for Bird Preservation

ICCAT	International Convention for the Conservation of Atlantic Tuna
ICES	International Council for the Exploration of the Seas
ICLARM	International Center for Living Aquatic Resources Management
ICOLD	International Commission on Large Dams
ICOMOS	International Council on Monuments and Sites
ICRAF	International Council for Research in Agroforestry
ICSU	International Council of Scientific Unions
IDB	Inter-American Development Bank
IDRC	International Development Research Centre, Canada
IFOAM	International Federation of Organic Agriculture Movements
IGBP	International Geosphere-Biosphere Programme
IGS	International Glaciological Society
IGU	International Geographical Union
ILO	International Labor Office
IMATA	International Marine Mammal Trainers Association
IMF	International Monetary Fund
IMO	International Maritime Organization
INTECOL	International Congress of Toxicology
IOBC	International Organization for Biological Control of Noxious Animals and Plants
IOC	International Ornithological Congress
IOPB	International Organization of Plant Biology
IPCC	Intergovernmental Panel on Climate Change
IPM	integrated pest management
ISCO	International Soil Conservation Organization
ISIS	International Species Inventory System
ISO	International Organization for Standardization
ISORMOP	International Symposium on Responses of Marine Organisms to Pollutants
ISS	International Seaweed Symposium
ITTO	International Tropical Timber Organization
IUAPPA	International Union of Air Pollution Prevention Associations
IUBS	International Union of Biological Sciences
IUCN	World Conservation Union (also International Union for the Conservation of Nature)

IUDZG	International Union of Directors of Zoological Gardens
IUFRO	International Union of Forestry Research Organizations
IUGB	International Union of Game Biologists
IUGG	International Union of Geodesy and Geophysics
IUHPS	International Union of the History and Philosophy of Science
IWC	International Whaling Commission
IWRA	International Water Resources Association
IWRB	International Waterfowl and Wetlands Research Bureau
LDC	London Dumping Convention
MAB	Man and the Biosphere Program (UNESCO)
NAAEE	North American Association for Environmental Education
NASA	National Aeronautics and Space Administration, USA
NCSS	National Council for Social Sciences
NGO	non-governmental organization
NIEHS	National Institute of Environmental Health Sciences, USA
NIH	National Institutes of Health, USA
NMFS	National Marine Fisheries Service, USA
NMML	National Marine Mammal Laboratory, USA
NOAA	National Oceanic and Atmospheric Administration, USA
NORAD	Norwegian Agency for International Development
NO$_x$	nitrogen oxides
NRC	National Recycling Congress
NRDC	Natural Resources Defense Council
NSDL	National Soil Dynamics Laboratory, USA
NSF	National Science Foundation
NSTA	National Science Teachers' Association
NWF	National Wildlife Federation
NYZS	New York Zoological Society
NZP	National Zoological Park, Smithsonian Institution
OAS	Organization of American States
OAU	Organization of African Unity
OECD	Organization for Economic Cooperation and Development
OIE	International Office of Epizoites (Animal Parasites)
ORSTOM	Overseas Scientific and Technical Research Office, France
OSCOM	Oslo Convention on the Dumping and Discharge of Toxic Materials
PARCOM	Paris Convention on the Dumping and Discharge of Toxic Materials

PBBs	polybrominated biphenyls
PCBs	polychlorinated biphenyls
PROSEA	Plant Resources of Southeast Asia
PSTC	Program for Studies in Tropical Conservation
PVA	population viability analysis
PVO	private voluntary organization
Ramsar	Convention on Wetlands of International Importance (Ramsar, Iran)
RCRA	Resource Conservation and Recovery Act, USA
REA	rapid ecological assessment
REDES	Regional Network of NGOs for Sustainable Development in Central America
SAARC	South Asian Association for Regional Cooperation
SADCC	Southern African Development Coordination Conference
SCAR	Scientific Committee on Antarctic Research
SCOPE	Scientific Committee on Problems of the Environment, International Council of Scientific Unions (ICSU)
SCS	U.S. Soil Conservation Service
SETAC	Society of Environmental Toxicology and Chemistry
SIL	International Limnological Congress
SO$_2$	sulfur (sulphur) dioxide
SOPAC	South Pacific Applied Geoscience Commission
SPREP	South Pacific Regional Environment Programme
SSC	Species Survival Commission, IUCN
SSP	Species Survival Plan
TFAP	Tropical Forestry Action Plan
TNC	The Nature Conservancy
TRAFFIC	Trade Records Analysis of Fauna and Flora in International Commerce
TVA	Tennessee Valley Authority
UNCED	United Nations Conference on Environment and Development; 1992 Earth Summit
UNDP	United Nations Development Programme
UN ECE	United Nations Economic Commission for Europe
UNEP	United Nations Environment Programme
UNESCAP	United Nations Economic and Social Commission on Asia and the Pacific
UNESCO	United Nations Educational, Scientific, and Cultural Organization
UNFPA	United Nations Population Fund
UNICEF	United Nations Children's Fund

USAID	U.S. Agency for International Development
USDA	U.S. Department of Agriculture
USDA-ARS	USDA–Agricultural Research Service
WCED	World Commission on Environment and Development
WCI	Wildlife Conservation International
WCRP	World Climate Research Programme
WHO	World Health Organization

WMO	World Meteorological Organization
WOCE	World Ocean Circulation Experiment
WRI	World Resources Institute
WRM	World Rainforest Movement
WWF	World Wide Fund for Nature; World Wildlife Fund

Other abbreviations and acronyms appear throughout this *Guide.*

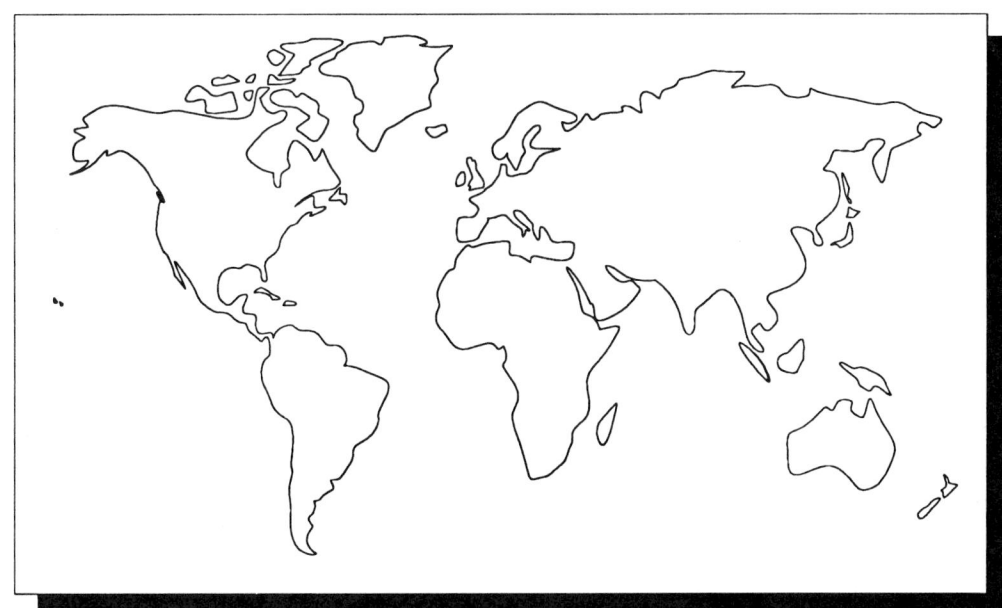

Commission for the Conservation of Antarctic Marine Living Resources (CCAMLR) ⊕

25 Old Wharf
Hobart TAS 7000 Australia

Contact: **D.L. Powell, Ph.D.**, Executive Secretary
Phone: (61) 02 31 0366 • *Fax:* (61) 02 23 2714

Activities: Research • *Issues:* Biodiversity/Species Preservation; Conservation

● Organization

The **Commission for the Conservation of Antarctic Marine Living Resources (CCAMLR)** is at the heart of the Antarctic Treaty System that fosters cooperation among participating nations to do scientific research, prohibits military activity, minimizes harmful human impact, ensures rational uses of resources, and sets up organizations to implement these principles.

In 1980, the guiding Convention was signed by Argentina, Australia, Belgium, Chile, France, Germany, Japan, New Zealand, Norway, Poland, South Africa, UK, USA, and the USSR; it was ratified in 1982. Presently, Brazil, the European Community, India, Italy, Korea, and Spain are Commission members too; also acceding to the Convention are Canada, Finland, Greece, Netherlands, Peru, and Uruguay.

The **Convention on the Conservation of Antarctic Marine Living Resources** stringently states that "exploited populations must not be allowed to fall below a level close to that which ensures their greatest net annual increase; depleted populations must be restored to such levels; ecological relationships between harvested, dependent and related species must be maintained; and risks of changes to the marine ecosystem that are not potentially reversible over two or three decades must be minimized."

The Commission implements the Convention, facilitating research, compiling data, analyzing statistics on harvested populations, identifying conservation needs, adopting conservation measures, and ensuring compliance through inspection and observation systems.

▲ Projects and People

Antarctica and its surrounding ocean, which make up about one-tenth of the Earth's surface, are a great influence on world weather and climate. Mainly an ice-covered land mass year-round, its frozen sea extends some 1,000 kilometers from the coast in wintertime. Here there are no indigenous people or universally recognized owners. Instead, inhabitants are scientists, fishermen, tourists, and adventurers from other countries.

The 2 percent of land that is free from ice and snow and its surrounding ocean is a haven for plentiful birds and mammals that get their food from the sea. They must be protected from exploitation that has nearly led to extinction for some species. For example, the hunting of Antarctic fur seals and elephant seals that yield an oil similar to whale oil nearly depleted their populations. With the protective Convention for the Conservation of Antarctic Seals, these species are increasing in the South Atlantic. The Convention's Scientific Committee on Antarctic Research (SCAR) is an information exchange and management body if commercial sealing were to begin again; however, this is unlikely.

The International Whaling Commission adopted a moratorium on all commercial whaling in 1986, which is still in force. CCAMLR reports that some species are recovering, although slowly. The only whaling permitted today in Antarctica is by Japan, which is able to take several hundred minke whales yearly out of a population of about 760,000 for research purposes.

Fishing in Antarctic waters is regulated with the help of the Working Group on Fish Stock Assessment. A commercial krill fishery with a yearly catch of 400,000 tons is "the largest crustacean fishery in the world, yielding ten percent of all crustaceans taken worldwide." Feeding the estimated 400 million seabirds, 10 million sea mammals, and 850,000 whales takes some 17 million tons of fish, 20 million tons of squid, and 100 million tons of krill each year. Because krill feeds on phytoplankton and is the prey of fish, squid, birds, seals, and great baleen whales, its conservation is "fundamental to the maintenance of the Antarctic marine ecosystem and vital to the recovery of depleted whale populations." A Working Group on Krill evaluates commercial krill fisheries and krill stocks and is prepared to take conservation measures, if necessary.

Also watching krill closely is a scientific group managing the CCAMLR Ecosystem Monitoring Program (CEMP), which detects "significant changes in key components of the Southern Ocean ecosystems and distinguishes between changes due to commercial harvesting and those due to natural causes." Krill predators—fur seals, penguins, petrels, and black-browed albatrosses—are monitored for growth, reproduction, abundance, for example. CEMP is recognized as an "ambitious program" to be ultimately judged on how its results influence decisionmaking within CCAMLR and to what degree member countries appreciate the validity of CEMP's results.

CCAMLR chairman is Ambassador Jorge Berguno [Ministerio de Relaciones Exteriores, Bandera 52, Piso 7, Santiago, Chile, fax (56) 2 699 4202]. Scientific Committee Chairman is Ole-Johan Ostvedt [Institute of Marine Research, P.O. Box 1870, Nordnes 5024, Bergen, Norway, fax 47 5 23 8531]. CEMP convener is Dr. John Bengtson [National Marine Mammal Laboratory, National Marine Fisheries Service, 7600 Sand Point Way, NE, Seattle, WA 98115 USA, fax 1-206-526-6615]. Working Group on Krill convener is D.G.M. Miller [Sea Fisheries Institute, Private Bag X2, Roggebaai 8012, South Africa, fax (27) 21 25 2920]. Working Group on Fish Stock Assessment is Dr. K.H. Kock [Institute fur Seefischerei, Palmaille 9, D-2000 Hamburg 50, Germany, fax (49) 40 38 90 5129].

■ Resources

Published reports of meetings, conservation measures, members' activities, and scientific papers are available in four languages from the Executive Secretary. The *CCAMLR Newsletter* is produced biannually.

Convention on the Conservation of Migratory Species of Wild Animals (CMS) ⊕
United Nations Environment Programme (UNEP)
Postfach 20 1448
W5300 Bonn, Germany

Contact: **Judith Carol Johnson**, Coordinator
Phone: (49) 228 302152 • *Fax:* (49) 228 373237

Activities: Education; International Conservation; Law Enforcement; Political/Legislative • *Issues:* Biodiversity/Species Preservation • *Meetings:* Bonn Convention Standing Committee

● Organization
The UNEP Convention on the Conservation of Migratory Species of Wild Animals (CMS), commonly referred to as the Migratory Species, or Bonn, Convention, aims to improve the conservation status of migratory species through national action and international cooperative agreements. The Convention applies to terrestrial, marine, and avian species over the whole of their normal migratory range. UNEP is the United Nations Environment Programme.

The Bonn Convention came into force in 1983, and now has a membership of 39 parties from an increasingly wide geographic distribution: 16 from Europe, 11 from Africa, 5 from Asia, 3 from the Middle East, 3 from Central/South America, and 1 from Australia. Eleven other countries are signatories.

A small Secretariat under the auspices of UNEP provides general administrative support to the Convention. A Scientific Council consisting of about 30 experts appointed by the Conference of the Parties and by individual member states advises on scientific matters. A six-member Standing Committee provides guidance on matters related to the implementation of the Convention between formal meetings of the parties, which are held every three years.

▲ Projects and People
The Convention incorporates two appendices which list migratory species that would benefit from concerted conservation measures. Endangered species, listed in Appendix I, are accorded full protection. Range states of Appendix I species are to endeavor to conserve their habitat, to counteract factors impeding their migration, and to control other factors which might endanger them. Moreover, range states are obliged to prohibit the taking of animals of these species, with few exceptions. The definition of "taking" includes such activities as hunting, fishing, capturing, harassing, and deliberate killing. The Convention also applies the term "range state" to flag ships engaged in taking migratory species on the high seas, outside national jurisdictional limits.

Appendix II lists migratory species which have a conservation status that requires, or would benefit from, international cooperative agreements. Where appropriate, a species may be listed in both appendices.

The Convention provides for two forms of agreements for Appendix II species. First, there are AGREEMENTS (the capitalization is intentional) intended to benefit migratory species—especially those with an unfavorable conservation status—over their entire range. These AGREEMENTS should be open to accession by all range states of the species concerned, including those that are not parties to the Convention. The text of the Convention spells out very clearly what an AGREEMENT should include. At a minimum, it should incorporate provisions for species research and periodic status assessments; coordinated management plans and exchange of information among Range States; maintenance of suitable habitat; and regulation of factors that are directly harmful to the species concerned or which impede their migration.

A second type of agreement (in lower case) is encouraged for populations of species which periodically cross national jurisdictional boundaries, but which are not necessarily migratory under the definition provided by the Convention. Unlike the instruments mentioned above, the content in such agreements is not stipulated in the text of the Convention.

The third meeting of the Conference of the Parties to the Convention on the Conservation of Migratory Species of Wild Animals was held in Geneva in September 1991. It was attended by 24 of the current 37 parties.

Some of the outcomes of the Convention were as follows:

• Twenty-eight species/populations of small cetaceans were added to Appendix II of the Convention.
• A proposal urging conclusion of agreements for small cetaceans listed in Appendix II, and in particular for those in the Mediterranean and Black seas, was adopted.
• The meeting directed the Scientific Council to: recommend specific conservation and management measures for AGREEMENTs, giving priority to sirenians, albatrosses, and mammals in the Sahalo-Saharan region, the Arabian Peninsula and southern Asia; to provide advice on further species to be added to the Appendices, especially neotropical species; and to conduct a preliminary review of the impact of artificial barriers to migration.
• The Conference appointed the following experts to the Scientific Council: **Pierre Pfeffer** (French Museum of Natural History), Sahalo-Saharan mammals; **Michael Moser** (International Waterfowl and Wetlands Bureau), waterfowl; **William Perrin** (National Oceanic and Atmospheric Administration, USA), small cetaceans; and **Roberto Schlatter** (Southern University of Chile), neotropical fauna.

■ Resources

The following publications are available: proceedings of the first, second, and third meetings of the Conference of the Parties; *Review of the Conservation Status of Small Cetaceans*.

Political support outside of Europe is needed in order to improve the geographic scope and implementation of the Convention.

Plants Committee ⊕
Convention on International Trade in Endangered Species of Wild Fauna and Flora (CITES)
c/o Office of Scientific Authority
Fish and Wildlife Service
Washington, DC 20240 USA

Contact: **Bruce MacBryde, Ph.D.**, Chairman and Botanist
Phone: 1-703-358-1708 • *Fax:* 1-703-358-2202

Activities: Development; Regulatory • *Issues:* Biodiversity/ Species Preservation; Sustainable Development

● Organization

Though not as familiar as its CITES animal committee counterpart, the CITES Plants Committee is equally committed to the conservation of plants through the enforcement of laws and international treaties that protect threatened and potentially threatened flora. Preceded by a CITES Plant Working Group, the committee first met in 1988 in London and organized itself into six geographical regions: Africa; Asia; Europe; North America; Oceania; and South America, Central America, and the Caribbean. Participants are both government and non-government scientists.

The group seeks to develop CITES as an effective treaty for the conservation of plants, "recognizing that its national and international structures offer a unique opportunity to advance knowledge of threatened and potentially threatened taxa, especially in entities involved with planning for the environment and development of natural resources." It is particularly concerned with conserving "qualifying taxa" in world import and export trade "without encumbering trade in specimens that usually are not of conservation concern," writes Dr. Bruce MacBryde, Chairman.

▲ Projects and People

The Plants Committee supports a CITES *Guide to Plants in International Trade* to be published as a book for the general public and regulatory officers; checklists with databases for higher taxa listed in CITES appendices and including IUCN categories on vulnerability; a study on significant international trade in wild orchids; further study for traded non-CITES groups, such as bulbs and timber; and greater education and communications efforts to improve understanding of and cooperation with CITES.

Recently, CITES members rededicated themselves to further conservation of tropical plants in natural settings, parks, and gardens; and to more research in the artificial propagation of orchids, cactus, and other native tropical plants as bromelia. Another goal is to set up a register of plant nurseries with special interest in rare species of orchids, cactus, and bromelia.

Dr. MacBryde belongs to the Species Survival Commission (IUCN/SSC) Cacti and Succulent as well as North American Plant groups.

■ **Resources**

Needed are CITES interns and volunteers as well as information transfer on biodiversity to implement CITES in developing countries.

Inter-American Tropical Tuna Commission (IATTC) ⊕

c/o Scripps Institution of Oceanography
La Jolla, CA 20005 USA

Contact: **William H. Bayliff, Ph.D.**
  Phone: 1-619-546-7100 • *Fax:* 1-619-546-7133

Activities: Research • *Issues:* Fishery Management

● Organization

The Inter-American Tropical Tuna Commission (IATTC) operates under the authority and direction of a convention that was originally entered into by the governments of Costa Rica and the USA in 1950. The convention is open to governments of any countries that participate in tropical tuna fishing in the eastern Pacific Ocean. Member nations now include France, Japan, Nicaragua, Panama, and USA.

The principal duties of the Commission are "(1) to study the biology of the tropical tunas, tuna baitfishes, and other kinds of fish taken by tuna vessels in the eastern Pacific Ocean and the effects of fishing and natural factors upon them and (2) to recommend appropriate conservation measures, when necessary, so that these stocks of fish can be maintained at levels which will afford the maximum sustained catches."

▲ Projects and People

In 1976, the member nations decided to broaden the Commission's duties to include problems arising from the tuna–dolphin relationship in the eastern Pacific Ocean. These include striving to maintain dolphin stocks at levels that assure their survival and making every effort to avoid needless or careless killing of dolphins.

Directing the staff of 61 persons is **James Joseph, Ph.D.** Senior scientists are: **William H. Bayliff, Ph.D.**—Biology; **David A. Bratten**—Biology; **Richard B. Deriso, Ph.D.**—Biology, Population Dynamics; **Eric D. Forsbergh**—Biology; **Martin A. Hall, Ph.D.**—Biology; **Michael G. Hinton**—Biology, Statistics; **Witold L. Klawe**—Biology; **Forrest R. Miller**—Meteorology; **Robert J. Olson**—Biology; **Patrick K. Tomlinson**—Biology, Population Dynamics; and **Alexander Wild, Ph.D.**—Biology.

■ Resources

Research results are promptly published in the Commission's *Bulletin, Special Report, Internal Report* and *Data Report* series. Listings of available publications and prices are furnished upon request.

International Whaling Commission (IWC) ⊕

The Red House
135 Station Road, Histon
Cambridge CB4-4NP, UK

Contact: **Ray Gambell, Ph.D.**, Secretary
  Phone: (44) 223 233971 • *Fax:* (44) 223 232876

Activities: Political/Legislative; Research • *Issues:* Biodiversity/Species Preservation; Whaling Regulation • *Meetings:* CCAMLR; CITES; CMS; IATTC; UNEP

● Organization

The **International Whaling Commission (IWC)** was set up under the 1946 **International Convention for the Regulation of Whaling**. According to IWC information, "the purpose of the Convention is to provide for the proper conservation of whale stocks and thus make possible the orderly development of the whaling industry."

The Schedule to the Convention contains rules "which govern the conduct of whaling throughout the world"—for example, providing for the complete protection of certain species; designating whale sanctuaries; limiting the number and size of whales which may be taken in a season, as well as the time period and location; prohibiting the capture of suckling calves and their mothers; and requiring catch reports and other statistical and biological records. The IWC's primary duty is to review and revise these measures as necessary. The Commission also "encourages, coordinates, and funds whale research, publishes the results of these and other scientific research, and promotes studies into related matters such as the humaneness of the killing operations."

Any country formally accepting the 1946 Convention may become a member, including both whaling and nonwhaling countries. Member nations are: Antigua and Barbuda, Argentina, Australia, Brazil, Chile, China, Costa Rica, Denmark, Ecuador, Finland, France, Germany, Iceland, India, Ireland, Japan, Kenya, Republic of Korea, Mexico, Monaco, Netherlands, New Zealand, Norway, Oman, Peru, Saint Lucia, Saint Vincent and the Grenadines, Senegal, Seychelles, South Africa, Spain, Sweden, Switzerland, former USSR, UK, USA, and Venezuela.

A commissioner represents each member country, with the assistance of experts and advisers. The chairman and vice-chairman are elected from among the commissioners and serve three-year terms. Since 1976, a full-time secretariat has managed the IWC's day-to-day work, based in Cambridge, England. Secretariat personnel include the secretary, executive officer, scientific editor, computing manager, and supporting staff. The Commission's annual meeting is held each spring in a member country.

Three committees—Scientific, Technical, and Finance and Administration—facilitate IWC activity. The **Technical Committee** has two standing subcommittees, dealing with **Aboriginal Subsistence Whaling** and **Infraction**, respectively. Most member countries are represented on the Scientific and Technical committees, at the discretion of their commissioners. The Finance and Administration Committee consists of representatives of countries nominated by the chairman.

The approximately 80 scientists on the Scientific Committee meet immediately prior to the Commission's annual meeting; they may also hold special meetings during the year to consider particular subjects. This committee provides information and advice on the status of the whale stocks, and in turn the Technical Committee uses these findings to develop the Commission's whaling regulations. All regulations require a three-quarters majority of the Commissioners voting; if a member state objects that, for example, "its national interests or sovereignty are unduly affected," the new regulation is not binding on that country. Nationally appointed inspectors enforce the IWC regulations, and international observers may be appointed by and report directly to the Commission.

▲ Projects and People

Many populations of the 12 species of great whales "have been depleted by over-exploitation, . . . both in recent times and in earlier centuries." Because whale stocks, like any other animal populations, naturally increase or decrease, and remain in relative equilibrium when these two factors balance one another, in 1975, the IWC adopted a new management policy based on these characteristics, and "designed to bring all stocks to the levels providing the greatest long-term harvests, by setting catch limits for individual stocks below their sustainable yields." However, because of uncertainties about the precise status of the various whale stocks, the Commission decided in 1982 to issue a moratorium on all commercial whaling beginning in 1985. Although several nations—in particular Japan, Peru, the then USSR, and Norway—objected at first, by 1988 all were participating in the ban.

Such a moratorium would not affect aboriginal subsistence whaling, permitted from Denmark for Greenland, fin, and minke whales; St. Vincent and the Grenadines for humpback whales; the former USSR for Siberian and gray whales; and the USA for Alaska, bowhead, and gray whales. The Scientific Committee is working on a Comprehensive Assessment of whale stocks, evaluating their status in light of the moratorium, and developing new management procedures. By 1991, assessments had been carried out for gray, minke, and fin whales.

Any IWC member government may issue a special permit for the taking of whales for scientific research purposes. Since the moratorium decision, the Commission has established guidelines for the review of such permits, although the decision rests solely with the individual country. The governments of Iceland, Japan, and Norway have permitted limited catches as part of their scientific research programs. In 1989, Iceland completed a four-year research program on catches, part of a long-term project including "vessel and aerial surveys, radio tagging, and theoretical modeling." Norway concluded a five-year study of minke whales in 1990, investigating age determination and energetics as part of a broader ecological project to collect "information for future multi-species management of the Barents Sea." Japan is currently involved in a 12-year study in the Antarctic, catching some 300 minke whales in 1990–1991 in order to investigate the biological parameters of management, especially relating to age, and to learn more about the role of whales in the Antarctic ecosystem.

In addition to research in individual member states, the Commission is sponsoring a second International Decade of Cetacean Research, featuring "a series of ship surveys of the Antarctic minke whale stocks." Other research involves new techniques such as satellite/radio tracking of whales.

■ Resources

The IWC issues annual reports, with special publications on particular species, whaling methods, and behavior, for instance. Other publications include Scientific Committee Documents and *International Whaling Statistics*, a multivolume document. A publications list and order form are available.

North Atlantic Salmon Conservation Organization (NASCO) ⊕
11 Rutland Square
Edinburgh, Scotland EH1 2AS, UK

Contact: Malcolm Windsor, Ph.D., Secretary
Phone: (44) 31 228 2551 • *Fax:* (44) 31 228 4384
Telex: 94011321 nasc g

Activities: Political/Legislative • *Issues:* Biodiversity/Species Preservation; Global Warming; Water Quality

● Organization

The North Atlantic Salmon Conservation Organization (NASCO) was set up in 1984 as a result of the 1983 international Convention for the Conservation of Salmon in the North Atlantic Ocean. NASCO currently involves nine governments as members: Canada, Denmark (with the Faroe Islands and Greenland), European Community, Finland, Iceland, Norway, Sweden, former USSR, and the USA. In addition, the Organization invites the participation of many Non-Government Observers, such as the Atlantic Salmon Trust (UK), the Atlantic Salmon Federation (USA and Canada), the Association of Icelandic Angling Clubs, and others.

Four staff keep NASCO's Edinburgh headquarters running; the Organization itself is divided into a Council and three regional Commissions: the North American, the North-East Atlantic, and the West Greenland. The Commissions hear scientific advice and work on regulatory measures for the fisheries in participating countries; the Council coordinates work of the Commissions and "provides a forum for the study, analysis, and exchange of information and for consultation and cooperation on matters concerning salmon stocks." As of 1990 Allen E. Peterson, Jr., USA, was reelected Council president, and Svein Aage Mehli (Norway) was reelected vice president.

▲ Projects and People

Typical NASCO activities were outlined in a recent report. Having obtained evidence of a salmon fishery in international waters north of the Faroe Islands by vessels reflagged in Panama and Poland, NASCO parties exerted "concerted diplomatic action . . . aimed at ending the development of such fisheries." A second issue involved the rapid growth of salmon fish farms, creating potential threats to wild stocks. More than 150,000 tons of salmon were farmed in 1989 alone. When "large numbers of farmed fish are inadvertently released to the wild," genetic alteration, disease, and environmental damage may result. In 1989 and 1990 the Council held Special Sessions on the topic, and agreed to develop "guidelines for minimizing threats to wild stocks."

NASCO has also established a database of all salmon rivers in the North Atlantic which flow into the Convention area, and in 1990 launched the Tag Return Incentive Scheme, offering prizes of up to $2,500 for the return of scientific tags. Besides regulatory issues, the Organization discussed "problems of comparability of catch statistics, of unreported catches, of the effects of acid rain and of the introduction and transfer of salmonids."

Also in 1990, the NASCO Council adopted a resolution on the use of large-scale pelagic driftnets in response to a communication received from the United Nations.

■ Resources

NASCO regularly publishes a *Report on the Activities of the Organization.*

Department of Regional Development
    and Environment ⊕
Executive Secretariat for Economic
    and Social Affairs
Organization of American States (OAS)
1889 F Street, NW
Washington, DC 20006 USA

Contact: **Richard E. Saunier**, Environmental
        Management Advisor
    Phone: 1-202-458-3228 • *Fax:* 1-202-458-3560

Activities: Development • *Issues:* Regional Planning • *Meetings:* CIDIE; UNEP, Regional

▲ Projects and People

The Department of Regional Development and Environment, Organization of American States (OAS) provides technical cooperation and training in development planning, environmental management, and in the design of specific investment projects to its member states. Emphasis is on maintaining the resource base that makes sustainable development possible. Its environmental management and protection services were valued at more than $7 million during the 1990–1991 biennium, which included the following:

Technical Cooperation in Integrated Regional Development. Projects in subnational regions must be economically and technically feasible as well as financially and institutionally viable. They aim to improve the life quality of populations, assure a resource base for future generations, and minimize conflicts in using natural resources. Recent examples follow.
- *Bolivia:* Amazon Region integrated development; Chaco Region integrated development.
- *Mexico:* Ecological planning of priority geographic regions.
- *Trinidad and Tobago:* Forestry/Agriculture Sector Plan for Eastern Northern Range.

Technical Cooperation in Developing River Basins and Border Areas. River basins and border areas can be shared by more than one country; and the economic, social, and political goals must be balanced between the states. Activities include analyzing and rationally using shared natural resources; facilitating commerce across borders; integrating and jointly using transportation, communication and energy infrastructure and services; and integrating tourism activities into development planning. Recent examples follow.
- *Amazon River Basin:* Pluri-national project for Amazon cooperation.

- *Brazil:* Iguazu River Basin development.
- *Brazil/Colombia:* Tabatinga-Apaporis Axis integrated development.
- *Brazil/Peru:* Development program for border communities.
- *Colombia/Ecuador:* San Miguel-Putumayo River Basin integrated development.
- *Colombia/Peru:* Putumayo River Basin integrated development
- *Dominican Republic:* Multiple resources use of the Artibonito-Macacias River Basin.
- *Haiti/Dominican Republic:* Maribaroux Frontier Zone integral development.

Technical Cooperation in Managing National Parks, Ecologic Reserves and Other Conservation Areas. Efforts are to conserve unique areas for recreation, biodiversity, environmental education, ecologic research, erosion and sediment control, production of fish and wildlife, watershed protection, preservation of cultural and historic sites, and other long-term purposes. Recent examples follow.
- *Costa Rica:* La Amistad International Biosphere Reserve management.
- *Panama:* La Amistad International Biosphere Reserve management.

Technical Cooperation in Preparing National Environmental Action Plans. These plans provide timely information and identify immediate problems regarding a country's "state of the environment." Focus ranges from pollution control to ecosystem rehabilitation to formulating relevant investment projects. Recent activities follow.
- *Suriname:* Natural resources management (including preparation of national environmental action plan).
- *Uruguay:* National environmental action plan.

Technical Cooperation in Formulating Ecotourism Policy and Projects. Assistance to member states includes preparing natural attractions for management and protecting relevant ecosystems, using financing generated by the tourism industry. Recent projects follow.
- *Costa Rica:* Ecotourism components of management of La Amistad International Biosphere Reserve.
- *Dominica:* Tourism and environment.

Preparing Investment Projects for Ecosystem Rehabilitation. Rehabilitating damaged ecosystems with economically feasible programs and investments is key to the Latin American and Caribbean Region where resource potential has suffered degradation: soil erosion, pastureland destruction from overgrazing, salinization and other forms of man-influenced desertification, deforestation, and devastation from natural calamities such as floods, earthquakes, and landslides. Moreover, lack of planning in rapidly growing urban areas has resulted in air and water contamination and improper disposal of solid and toxic wastes. Recent examples of cooperation follow.
- *Brazil:* City of Santos environmental rehabilitation; State of Alagoas rehabilitation and reconstruction program.
- *Guatemala:* Chixoy River Basin natural resources feasibility study.
- *Honduras:* El Cajon Basin natural resources feasibility study; Priority River Basins environmental management program.

Promoting Practical Transfer of Environmentally Sound Technology. This technology minimizes adverse environmental consequences and includes energy conservation, nonpolluting industrial processes, and other environmentally friendly methods. Recent applications follow.

• *Colombia:* Modern technology application of natural resources conservation (CAR) and in the Bogota, Ubate, and Suarez river basins.

• *Uruguay:* Montevideo urban transportation.

Organizing Information for Integrated Development Planning Geographic Information Systems (GISs) is an important development planning tool for evaluating and monitoring natural resources, natural hazards, infrastructure, and human activities as well as for preparing strategies and investment projects. Use of GIS permits efficient mapping and analytical work. The technology is readily transferable through training activities built into technical cooperation programs for integrated development planning and is often accomplished by donation of software programs and peripheral equipment. Recent technical assistance examples follow.

• *Nicaragua:* GIS applications workshop for natural hazard assessment, management, and development planning.

• *St. Lucia:* GIS introduction into Ministry of Planning for natural resources management.

• *Uruguay:* Development of the Natural Environmental Action Plan.

Environmental Education and Public Awareness. Efforts to produce changes in attitudes and public values can yield long-term benefits of improved environmental management, the Department says. A recent project follows.

• *Dominica:* Tourism public attitudes and awareness.

Environmental Management Planning Research. The evaluation of practical experience acquired in the field is essential to providing effective technical assistance, responding to the evolving needs of developing countries, and making training materials available to technical centers. The Department publishes and distributes materials, such as *Disasters and Development: Managing Natural Hazards to Reduce Economic Loss; Primer on Natural Hazard Assessment in Integrated Regional Development Planning;* and *Use of Scientific Information for Sustainable Development: Preparation of a Major Research Proposal for the Scientific Committee on Problems of the Environment of the International Council of Scientific Unions (SCOPE/ICSU).*

Environmental Health Program ⊕
Pan American Health Organization (PAHO)
525 23rd Street, NW
Washington, DC 20037 USA

Contact: **Horst Otterstetter,** Coordinator and
Sanitary Engineer
Phone: 1-202-861-3311 • *Fax:* 1-202-223-5971

Activities: Development; Education • *Issues:* Health

● **Organization**

The **Pan American Health Organization (PAHO)** is the world's oldest international health agency. It includes 35 member countries from the Americas and 3 participating European governments with territories in the Western Hemisphere that have joined forces to improve the health and living standards of their people.

The PAHO headquarters in Washington is also the Regional Office for the Americas of the World Health Organization (WHO) and as such forms part of the United Nations system of specialized agencies. Under an agreement with the Organization of American States, PAHO also functions as the specialized agency for health in the inter-American system.

The Pan American Sanitary Conference is PAHO's highest authority. Composed of delegates from each member government, it meets every four years, defines PAHO's overall policies, and elects the Director of the Pan American Sanitary Bureau.

▲ **Projects and People**

The **Environmental Health Program** is one of six development areas of the Pan American Health Organization, which provides technical cooperation focusing on priority health problems and high-risk population groups.

This program helps solve problems in solid waste management, works to reduce environmental contamination, disseminates information on environmental health, and helps countries train personnel in environmental protection. It also helps countries identify and deal with occupational health problems.

Two centers provide additional resources and cooperation in environmental health:

The Pan American Center for Sanitary Engineering and Environmental Sciences (CEPIS) in Peru works with member countries to improve drinking water supplies and sanitation, provides training, assists in developing and funding projects, promotes research, and disseminates information through a network of 230 collaborating centers. The Center also helps to improve the management and disposal of solid and hazardous wastes.

The Pan American Center for Human Ecology and Health (ECO) in Mexico provides up-to-date information on how to monitor air and water pollution and on the legislative, institutional, and technical methods for preventing contamination.

Metropolitan Environmental Improvement
Programme (MEIP) ⊕
**United Nations Development Programme
(UNDP)**
World Bank/ASTEN
1818 H Street, NW
Washington, DC 20433 USA

Contact: **Stephen R. Stern,** Communications
Consultant/Publications
Phone: 1-202-458 2729 • *Fax:* 1-202-477 7335

Activities: Development; Institutional Strengthening; Investment Lending; Research • *Issues:* Air Quality/Emission Control; Deforestation; Energy; Global Warming; Health; Planning; Sustainable Development; Transportation; Waste Management/Recycling; Water Quality

● **Organization**

The **United Nations Development Program (UNDP)** and the **World Bank** are collaborating in the Metropolitan Environmental Improvement Programme (MEIP), a regional effort designed to support and learn from environmental management efforts of Asia's large, rapidly growing metropolitan areas. Projections are that 1.7 billion people of Asia will live in cities in 2025—that's triple the number of 1985. The objective is to serve governments, industries, and community organizations in reversing the process of environmental degradation around these cities.

The work of MEIP is based on four premises: urban development must not degrade the environment if urban life is to be acceptable; environmentally sensitive development can be cost-effective; coordination among all economic sectors—private businesses, public agencies, civic and non-governmental organizations—is necessary; and the interaction of the cities participating will allow the cities to share knowledge and experience.

The project was begun in late 1989, and in less than a year, five cities were participating: Beijing, China; Bombay, India; Colombo, Sri Lanka; Jakarta, Indonesia; and Manila, Philippines. Each city has a steering committee of senior representatives from the national, provincial, and municipal agencies involved. The steering committee has the power to appoint technical experts to oversee individual projects and provides high-level support and interagency cooperation. MEIP also employs a local environmental professional in each country as a national program coordinator, who communicates with Washington and the other cities involved, and acts as a secretary to the steering committee. MEIP's principal work in each city is the development of an environmental management strategy (EMS). The EMS for the city is based on analysis of the costs of environmental degradation and the benefits of cleanup of the region's natural systems. EMS becomes the basis for future development along conservationist lines.

According to MEIP, "Reliance on governments alone is insufficient for environmental quality management." For this reason, MEIP brings together NGOs, government agencies, industry, and academia to form its steering committees.

MEIP's "central office relationship with the World Bank lending program affords opportunities for immediate action on high-priority pollution problems," it reports, such as feasibility studies for pollution abatement investments. UNDP provides policy guidance and funding for administration and intercountry workshops; the World Bank administers MEIP.

▲ Projects and People

Although the program is young, it is already having a profound influence. Sri Lanka adopted the Colombo Steering Committee which became the National Environmental Steering Committee (NESC), located in the lead cabinet Ministry of Policy, Planning and Implementation. Sri Lanka is beginning a period of rapid industrialization, and the NESC is placed at the highest levels of government to see that this growth is sensitive to environmental concerns. Currently, MEIP is undertaking studies for action plans on solid waste management, joint waste treatment at industrial complexes, and restoration of the city landmark, Beira Lake.

MEIP-Colombo, the Ministry of Housing and Construction, and the National Housing Development Authority held a workshop which has led to the Clean Settlements Project (CSP) and encouraged five low-income settlements to study their environmental situations. The communities are located in degraded areas close to canal banks, garbage dumps, or polluted waterways. They are addressing such problems as priorities of land tenure, stagnant water and drainage, lack of toilet facilities, garbage collection, water supply, and poor housing design. The Clean Settlements Project, in cooperation with the non-government agency SEVANATHA (Technical Services Agency for Community Development), will work with low-income communities to improve the basic amenities of toilets, drains, water supply, and access roads.

MEIP-Jakarta is bringing together city enterprises on community-based composting with the DKI Jakarta Cleansing Department to work to reduce waste volume at the source and to provide income for informal waste collectors. MEIP is working with the Kampung Improvement Program and a river cleanup program to improve and upgrade riverside communities and reduce the effects of domestic wastes on the rivers.

MEIP-Manila has dual-role, large-scale industrial waste abatement and small-scale recycling programs. The Industrial Efficiency and Pollution Control Project is addressing the problems inherent in large-scale industrial waste reduction. The Balikatan Women's Movement privately initiated a recycling project designed to generate income and increase recycling through waste recovery. The program uses junkshop dealers as secondary collectors, who go house to house with carts supplied by the dealers to collect the wastes. The dealers buy the wastes and sell them for reuse. Balikatan estimates that 10 to 35 percent of recyclables are recovered in the community in this way.

In Bombay, the focus is the reconstruction and modernization of the decaying and outdated industrial base, with care for the environment a primary consideration. One MEIP-Bombay project is on environmental management and joint waste collection and treatment in the Chembur and Thane/Belapur industrial areas.

MEIP works in Beijing through the Beijing Environmental Project. Beijing has a well-established institutional structure for environmental protection and monitoring. The Beijing Comprehensive Waste Treatment Plant is located 12 kilometers from Beijing city center and processes about 300 tons a day of municipal garbage. The plant begins by separating the waste-glass, paper, and metals for recycling; the remainder is divided for composting or incineration. Thus, 100 tons of garbage create 30 tons of compost, which is mixed with fertilizers and used by local farmers.

Some members of the MEIP-Beijing Steering Committee come from the Beijing Municipal Environmental Protection Bureau (BMEPB), who run the Beijing Environmental Monitoring Center (BEMC) covering 18 district and county environmental monitoring systems. BMEPB's Main Laboratory and Computer Control Center maintains mobile emission analysis trucks and operates a real-time computer system. The computer system receives data every five minutes on sulfur dioxide (SO_2), carbon dioxide (CO_2), nitrogen oxide (NO_x), ozone, windspeed, wind direction, temperature, and relative humidity from the stations and mobile units. The main task of the center is to monitor the 40 substations linked to the system.

■ Resources

MEIP publishes *Cityscape*, a quarterly newsletter, from its Washington office. Research findings, workshop reports, and case studies are available on request. The World Bank also publishes the annual progress report *The World Bank and the Environment*, regarding its policies and operations worldwide.

Programme on Man and the Biosphere
   (MAB) ⊕
UNESCO Division of Ecological Sciences
7, place de Fontenay
75700 Paris, France

Contact: **Bernd von Droste, Ph.D.**, Director,
   Division of Ecological Sciences
Phone: (33) 1 45 68 40 68
Fax: (33) 1 40 65 98 97

Activities: Research • *Issues:* Biodiversity/Species Preservation; Deforestation; Sustainable Development

● Organization

Dr. **Bernd von Droste**, director of the UNESCO Division of Ecological Sciences, is also secretary of the International Coordinating Council of the **Programme on Man and the Biosphere (MAB)**. MAB was launched in 1971 to encourage interdisciplinary research, demonstration, and training in natural resource management. MAB thus "contributes not only to better understanding of the many factors that affect the environment but to greater involvement of scientists in deciding how resources can be used more wisely."

MAB operates through 110 National Committees established out of the 159 member states of UNESCO, in addition to MAB Committees in the United Kingdom and the United States. Its governing body, the **International Co-ordinating Council (ICC)**, consists of 30 member states elected by UNESCO's biennial General Conference. The MAB Secretariat, which is responsible for day-to-day implementation, is provided by the UNESCO **Division of Ecological Sciences**, whose approximately 35 staff members represent many disciplines.

▲ Projects and People

MAB National Committees use **biosphere reserves** as sites for comparative studies and international pilot projects. These projects act as a national testing ground and demonstration for putting sustainable development principles into practice. Biosphere reserves also hold promise as secure sites for long-term ecological research and monitoring of global change.

By way of definition, biosphere reserves are protected areas of representative ecosystems. In April 1991, there were 300 reserves in 75 countries representing approximately two-thirds of all terrestrial regions of the world. Biosphere reserves are valuable as models for sustainable development because they "help to conserve biological resources, perpetuate traditional forms of land use, monitor nature and social changes, and improve the overall management of natural resources."

MAB also supports international cooperation through various **research networks** that are organized along thematic or geographical lines, or a combination of both. Networks include humid and subhumid tropics; arid, semiarid, and Mediterranean regions; land-use changes in Europe; islands; land/inland water ecotones; high mountains; northern areas; and urban systems.

Networks are linked through joint training programs, newsletters, meetings, and other means of communication. An example is the Pro-**Amazonia project**, designed to strengthen both local and national research capacities for the sustainable development of tropical forests. MAB thus contributes to the testing and application of research results from one country to another, which MAB says is particularly important for South-South cooperation.

MAB attempts to make the most efficient possible use of scarce human and financial resources by **coordinating research** with other international programs within UNESCO and other NGOs. Examples of international cooperation include comparative studies on **Tropical Soil Biology and Fertility** and on **Response of Savannas to Stress and Disturbance** with the International Union of Biological Sciences (IUBS); **Land/Inland Water Ecotones** with the International Limnological Society (SIL); and **High Mountain Research** with the International Center for Integrated Mountain Development (ICIMOD), the International Centre for Alpine Environments (ICALPE), and the African Mountain Association (AMA).

Among the many activities in arid zones, a new project was launched in 1989 to strengthen the scientific capacity of countries in the Sahel for **agro-sylvo-pastoral management**. A Mediterranean **forest research network** has also been developed.

In addition to classical ecological research, MAB now works on practical, site-specific solutions such as **debt-for-nature swaps** in less developed countries with high public debts. MAB also pays attention to **more resourceful urban management**.

During 1990–1991, a new generation of MAB field projects started demonstrating ways of using tropical forests in a sustainable way. The **Co-operative Ecological Research Project (CERP)** is working on problem-oriented research at eight sites in **China**. In **Madagascar**, MAB is helping to protect biological diversity through four sustainable development projects in **biosphere reserves that cover regions suffering from increasing pressure to use their resources**.

■ Resources

MAB publications come in a variety of formats and languages, reflecting the diversity of its international activities.

The *InfoMAB* newsletter is published twice a year in English and Spanish. MAB also produces a *Book Series*, providing state-of-the-art information in various fields of sustainable development. More policy-oriented information is summarized for decisionmakers in the *MAB Digest Series*.

Single copies of MAB publications may be ordered free of charge from MAB Publications, Division of Ecological Sciences.

**UNESCO Regional Office for Science
and Technology for Southeast Asia
(ROSTSEA)** ⊕
Jl. Thamrin 14
P.O. Box 1274/JKT
Jakarta 10012, Indonesia

Contact: **Kuswata Kartawinata, Ph.D.**, Programme
   Specialist in Ecological Sciences
Phone: (62) 21 321 308 • *Fax:* (62) 21 334 498

Activities: Education; Research • *Issues:* Biodiversity/Species Preservation; Deforestation; Global Warming; Marine Environment; Sustainable Development; Water Quality

● Organization

The UNESCO Regional Office for Science and Technology for Southeast Asia (ROSTSEA) strives to develop a broad awareness of and expertise in all fields of the natural sciences and technology through the promotion of higher education, research, and exchange of information. ROSTSEA also "acts as an antenna" for UNESCO in fields such as education, culture, and communications. UNESCO is the United Nations Educational, Scientific, and Cultural Organization.

ROSTSEA contributes to the planning, implementation, and evaluation of UNESCO's programs in science and technology in the region. The general aims of the programs are to bring about the exchange of ideas and experiences, and to identify further courses of action specifically related to development, using regional networks as a mode of operation and communication.

▲ Projects and People

The 10 professional staff work toward the identification and formulation of development-oriented projects in the following fields:

Science and Technology for Development: ROSTSEA works to promote the basic sciences—that is, mathematics, physics, biology, chemistry, genetic engineering, biotechnology, microbiology, informatics, and computer applications—as well as science- and technology-related studies and research. ROSTSEA also promotes engineering sciences and technology, with emphasis on appropriate technology, energy, housing, microelectronics, and management and utilization of wastes.

Environment: ROSTSEA's **Man and the Biosphere Programme** emphasizes environmental sciences involving research, training, demonstration, information diffusion, and biodiversity conservation through the creation of biosphere reserves. The **marine sciences program** concentrates on oceanology, coastal management, and curriculum development. In its global change program, called **Environmentally Sound and Sustainable Development**, ROSTSEA works in close cooperation with UNDP offices to assist developing countries in providing input on issues such as deforestation and climate change at the United Nations Conference on Environment and Development (UNCED).

Science, Technology, and Society: Through its information systems and through identification, formulation, and implementation of projects through its **Science and Technology Policy Asian Network (STEPAN)**, ROSTSEA seeks to popularize and raise public awareness of science and technology.

■ Resources

ROSTSEA disseminates information on a continual basis concerning training activities, especially the UNESCO-supported international postgraduate training courses and workshops.

ROSTSEA also publishes an annual *Calendar of Activities* and an *Annual Report*, as well as many specialized activity reports and research papers.

United Nations Environment Programme (UNEP) ⊕

1889 F Street, NW
Washington, DC 20006 USA

Contact: **Joan Martin-Brown**, Special Advisor to the Executive Director; Chief, Washington Office
Phone: 1-202-289-8456 • *Fax:* 1-202-289-4267

Activities: Policy • *Issues:* Air Quality/Emission Control; Biodiversity/Species Preservation; Deforestation; Energy; Global Warming; Health; Sustainable Development; Waste Management/Recycling; Water Quality

● Organization

"Never has humanity faced so crucial a decade as the 1990s," says executive director of UNEP and Under Secretary General of the United Nations **Mostafa Kamal Tolba**. "The decisions—and, even more important, the actions—taken over these ten short years will determine the shape of the world for centuries. The very fate of life hangs upon them."

Created as a result of the 1972 Stockholm Conference, the **United Nations Environment Programme (UNEP)** works to raise the level of environmental awareness and action in the UN and worldwide. UNEP coordinates all the UN agencies and promotes the participation of governments, scientists and professionals, and non-governmental environmental organizations in environmental action. UNEP's activities include monitoring global environmental quality, implementing an environmental management action plan, and regulating multilateral mechanisms to solve environmental problems. UNEP is also involved with environmental law, public information, education, and training.

▲ Projects and People

Since UNEP's earliest days, the ozone layer has been high among its concerns. In 1977, UNEP convened a meeting of experts that adopted a **World Plan of Action on the Ozone Layer** to assess the problem. As the body of scientific information increased, UNEP pressed on, developing a global convention for the protection of the ozone layer. The convention, a landmark agreement, was signed in Vienna in 1985 by 20 countries and the European Community. The Montreal Protocol followed in 1987. By 1990, 62 countries had signed, committing to halving their consumption and production of chlorofluorocarbons (CFCs). In 1990, in a London conference, this verbal commitment was translated into international law.

The scientific consensus that some degree of global warming is inevitable owes much to research and assessment coordinated by UNEP, through its **Global Environmental Monitoring Systems (GEMS)**. In 1985, UNEP, the World Meteorological Organization (WMO), and the International Council of Scientific Unions (ICSU) covened a climate conference with scientists from 29 countries, urging governments to develop plans for adapting to global warming. In 1990, the world's climatologists, brought together by WMO and UNEP in the **Intergovernmental Panel on Climate Change**, concluded that "the Earth's average surface temperature is likely to rise by one degree Celsius by the year 2025 and by up to three degrees by 2100." (The Earth's average temperature has not varied more than one or two degrees Celsius over the past 10,000 years.)

Today, as manager of the **World Climate Impact Studies Programme**, UNEP is "helping countries to assess how climate change will affect them and to make contingency plans," as well as "setting up an international network so that countries can exchange information gleaned through national climate impact studies."

Since 1981, UNEP's legal department has been working toward an **international agreement to control the movement of wastes**. The **Cairo Guidelines**, on the environmentally sound management of hazardous wastes, were drawn up and approved by UNEP in 1987, and formed the basis for the **Convention on the Control of Transboundary Movements of Hazardous Wastes and Their Disposal**, adopted by 116 states in 1989. The aim is "to encourage countries to cut back on the quantity and toxicity of the wastes they generate, to manage them in an environmentally sound way, and to dispose of them safely and as near to the source of their generation as possible." In addition, UNEP's Industry and Environment Office in Paris has been promoting low- and non-waste technology and launched a network for the exchange of information on cleaner production methods.

Oceans determine climate and supply a vital source of food for half the population of the developing world. However, most wastes created on land end up in the sea, poisoning marine life. Over 120 countries participate in UNEP's **Regional Seas Programmes**, through which countries have overcome ancient conflicts in the interest of protecting their seas. Action Plans for "cooperation on research, monitoring, control of pollution, and the development of coastal and marine resources" have been developed for the Mediterranean, Kuwait, the Caribbean, and the South Pacific, among others.

According to UNEP, every day 25,000 people die as a result of poor water management. Overstrained reserves, industrial wastes, sewage, and agricultural runoff lead to poisoned water supplies. UNEP's **programme for the environmental sound management of inland water (EMINWA)** addresses water management as part of the environmental whole, and seeks to "strike a balance between water's role as a resource for development and its environmental value."

The Sahel drought of 1968–1974 inspired the United Nations to convene the **Conference on Desertification (UNCOD)** in 1977, which adopted a plan of action to combat the problem. In 1982, UNEP's Governing Council adopted a **World Soils Policy**, drawn up with the help of FAO and the United Nations Educational, Scientific, and Cultural Organization (UNESCO), through which member governments agreed to pursue sound soil and land-use management policies. On the global level, UNEP has worked with other international organizations to assess and map global soil degradation and set up a world database. Regionally, UNEP set up the secretariat for the **North African Transnational Green Belt**, a reforestation project that "aims to hedge in the Sahara along its northern edge." Nationally, UNEP has cooperated with UNESCO in two **Integrated Projects on Arid Lands** in Kenya and Tunisia.

Forests and woodlands cover about one-third of the land area of the world; yet, according to UNEP, over two-thirds of Western Europe's natural forests have vanished and 95 percent of the USA's virgin forests have been felled in the last 150 years. Much of this devastation is the result of pollution, including acid rain. In 1979, East and West European states signed the **Convention on Long-Range Transboundary Air Pollution** based on data provided by UNEP through GEMS. In 1985, the UN Conference on Tropical Timber ratified the **International Tropical Timber Agreement**, "regulating tropical timber trade and encouraging cooperation between producing and consuming nations." Today, UNEP continues to work to prepare methodologies for assessing the forest resources and to bring together representatives of nations to discuss these issues.

There are an estimated 50 million species believed to be alive today. Yet about 100 are lost every day. This loss will hurt mankind, because "even the most insignificant species plays a crucial role in the ecosystem to which it belongs." In 1980, UNEP, the World Conservation Union (IUCN) and the World Wildlife Fund (WWF) launched the **World Conservation Strategy**—"the first comprehensive policy statement of the link between living resource conservation and sustainable development." This became "the most influential conservation document of the decade," spurring over 40 national conservation strategies. UNEP supports the World Conservation Monitoring Centre (WCMC), which assesses the distribution and abundance of the world's species, and the International Board for Plant Genetic Resources (IBPGR). UNEP has also worked on- and off-site in a number of countries to help conserve forest genetic resources. The organization also works closely with UNESCO's Man and the Biosphere (MAB) programme, to establish a worldwide network of biosphere reserves.

The following UNEP "Global 500 Roll of Honour" recipients appear in this *Guide*: Anil Agarwal (Center for Science and Environment, India); Dr. Irma Acosta Allen (National Environmental Education Programme of Swaziland); Dr. George Archibald (International Crane Foundation, USA); Asociacion Peruana para la Conservacion de la Naturaleza (APECO, Peru); Frank W. Bohman (Total Resource Conservation Plan, USA); *China Environment News*; Dr. Dhrubajyoti Ghosh (Ashoka, India); Don Henry (WWF- Australia); Prof. Kazi Zaker Husain (University of Dhaka, Bangladesh); Dr. Tom Lovejoy (Smithsonian Institution, USA); Dr. Uri Marinov (Ministry of the Environment, Israel); Dr. Raúl Alberto Montenegro and the Foundation for the Defense of the Environment (FUNAM, Argentina); Norman Myers Consultancy Ltd. (UK); Hideo Obara (Nature Conservation Society of Japan); Dr. Roger Payne (Whale and Dolphin Conservation Society USA); Dr. Nicholas Polunin (Foundation for Environmental Conservation, Switzerland); Dr. M.K. Ranjitsinh (Ministry of Environment and Forests, India); Philip Shabecoff; and Turkish Association for the Conservation of Nature and Natural Resources. Lester Brown (Worldwatch Institute, USA) and the Bureau of the "Three North" Protection Forest System (China) also received UNEP awards.

Honorees of the Friends of the United Nations Environment Programme (FUNEP) are: Council on the Environment of New York City Training Student Organizers Program, Brock Evans (National Audubon Society), Interfaith Coalition on Energy, Mothers and Others for Safe Food, Susan C. Seacrest (Nebraska Groundwater Foundation), Texas Department of Highways and Public Transportation antilitter campaigns, and Textbook Review Project/Sierra Club/Audubon Council/Texas.

Clearing-house Unit for Technical
  Cooperation ⊕
**United Nations Environment Programme
(UNEP)**
P.O. Box 47074
Nairobi, Kenya

Contact: **Mikko O. Pyhala**, Chief
  Phone: (254) 2 520600 • *Fax:* (254) 2 521435
  E-Mail: GN:mikkop

Activities: Development • *Issues:* Air Quality/Emission Control; Biodiversity/Species Preservation; Global Warming; Sustainable Development; Waste Management/Recycling; Water Quality

▲ Projects and People

The **Clearing-house Unit for Technical Cooperation, United Nations Environment Programme** has been UNEP's focal point for the Global Environment Facility (GEF) since 1991. **Mikko Pyhala** is both chief of the Clearing-house Unit and the GEF. Previously, Pyhala was secretary to the Scientific and Technical Advisory Panel (STAP) and an architect of the financing mechanism of the Interim Multilateral Fund of the Montreal Protocol.

Risø National Laboratory ⊕
**UNEP Collaborating Centre on Energy and
  Environment**
P.O. Box 49
DK 4000 Roskilde, Denmark

Contact: **John M. Christensen, Ph.D.**,
    Head of Centre
  Phone: (45) 46 32 22 88 • *Fax:* (45) 46 32 19 99

Activities: Development; Research • *Issues:* Energy; Global Warming; Sustainable Development

● Organization

The **UNEP Collaborating Centre on Energy and Environment** was established in 1990 and is staffed by an international, multidisciplinary team of five. Core funding is provided jointly by United Nations Environment Programme (UNEP), Danish International Development Agency (Danida), and Risø National Laboratory, provisionally for a four-year period.

The Centre collaborates with a number of institutions, mainly in developing countries, and provides direct technical and scientific support to UNEP Headquarters in Nairobi, Kenya. The Centre is also able to undertake project activities on a contract basis for other UN agencies, national and international development organizations, and governments.

The main objective of the Centre is to promote and facilitate the incorporation of environmental aspects into energy policy and planning at the national level in developing countries and in UN agencies and other international organizations. The Centre's activities are concentrated in the four areas listed below.

Risø National Laboratory's Systems Analysis Department hosts the Collaborating Center. According to UNEP, the lab is "one of the main energy research centres in Europe with extensive experience in international cooperation which also includes developing countries." Denmark is a pioneer in the fields of energy efficiency and conservation, and of renewable energy use. Its staff includes energy planners, scientists, and economists.

▲ Projects and People

Environmental Impacts: Included are studies on energy production and use in developing countries; greenhouse gas emission, efficient energy production, and end-use technologies; potentials for energy conservation; and the environmental impacts of specific energy technologies.

Energy Policy: The Centre conducts studies of energy policy in selected countries, with priority given to the major developing countries. The aim is to determine the environmental consequences of current energy plans and policies, and to suggest alternative development strategies that reduce the environmental impacts, while maintaining the same level of energy services.

Information: The Centre has established and maintains databases on energy-related environmental effects and energy planning methods and models. It disseminates information to UNEP and other UN agencies, governments, and other interested institutions and individuals.

Scientific and Technical Support: The Centre provides support to UNEP on energy questions on an ad hoc basis, including participation in conferences and expert groups, preparation of background papers, and collection of information on specific topics.

The Centre is headed by **Dr. John M. Christensen**, Denmark, an engineer with experience in renewable energy technologies and energy planning. He previously was programme officer with the Energy Unit at UNEP Headquarters in Nairobi. Other staff include: **Gordon Mackenzie**, United Kingdom, a physicist working on energy demand modeling, transport energy, integrated energy-environment models, and energy planning in developing countries; **Arturo Villavicencio**, Ecuador, a mathematician with international experience as an energy consultant and planner, particularly in Latin American countries; and **Camilo Lim**, the Philippines, an economist with international experience in energy economics and planning, particularly in Southeast Asia.

■ Resources

The Centre publishes the *C2E2* newsletter, which provides up-to-date information on the activities of the Centre, UNEP, and related events and developments. The newsletter welcomes information on forthcoming conferences, reports, studies, and other activities.

Harmonization of Environmental Measurement
    (HEM) ⊕
**United Nations Environment Programme
    (UNEP)**
c/o GSF
D-8042 Neuherberg, Germany

Contact: **Hartmut Keune, Ph.D.**
    Phone: (49) 89 3187 5488
      Fax: (49) 89 3187 3325
    E-Mail: Internet: keune@gsf.de

Activities: Environmental Monitoring • *Issues:* Harmonization of Environmental Measurement

● Organization

In 1989, the United Nations Environment Programme (UNEP) laid the groundwork for the Harmonization of Environmental Measurement (HEM) project under the auspices of the Global Environmental Monitoring System (GEMS). As data is collected worldwide on the environment, needs exist for such information to be universally compatible—or harmonized—within programs as well as between programs. Otherwise, knowledge is lost, and time and money are wasted due to "duplication of efforts in the collection of data," writes Dr. Hartmut Keune.

For example, as distinct sets and types of data are gathered on varying environmental issues—such as drinking water contaminants or soil degradation or forest depletion—this information cannot be utilized readily by other groups unless it is "harmonized" so that similar matters relate. To overcome the lack of compatibility in environmental measurements and connections between environmental programs, HEM promotes the "availability of appropriate information about the actual condition of the environment at different levels—global, regional, and local." HEM also seeks to upgrade the reliability, quality, and consistency of observations, measurements, and data.

Long-term objectives feature a "central source of information on available environmental data" and "harmonization of taxonomies" or classification systems, as reported in *GeoJournal* (March 1991).

▲ Projects and People

Still in the planning stage, the Information System recommends the following: development of a meta-database, which will be "a source of information on what data is available, where, and the means of access to it, rather than an actual data bank"; compilation of a comprehensive catalog of classification systems used in environmental areas worldwide; a quality guarantee label for data with minimum standards; and "harmonized" environmental specimen banking (ESB), to provide samples of toxic chemicals, for example, for uniform or baseline testing.

HEM points out that "not only can ESBs be extremely useful as a practical approach to a number of aspects of harmonization of environmental measurement, they themselves also require harmonization." With ESBs already conducted in a few cooperating countries, HEM adds, "there is a very good chance for developing internationally accepted and acknowledged guidelines for the establishment and management of ESBs enabling maximum use of their resources in years to come. HEM will act as a centre for promoting the development and application of such guidelines working in close cooperation with UNEP's IRPTC (International Register for Potential Toxic Chemicals)."

Working with other United Nations agencies including World Health Organization (WHO), World Meteorological Organization (WMO), UN Educational, Scientific, and Cultural Organization (UNESCO), and Food and Agriculture Organization (FAO) and with other organizations such as World Conservation Union (IUCN), International Organization for Standardization (ISO), and International Council of Scientific Unions (ICSU), UNEP's GEMS brings together data that has been collected on specific issues and processes this information. HEM exists not only to standardize the GEMS information on an "international and an interdisciplinary" basis, but also to disperse the information.

"Success, however, will depend on the support and cooperation of all those working in the field of environmental measurement," reports HEM.

■ Resources

HEM publishes materials regarding environmental surveys and information management programs of international organizations.

INFOTERRA
**United Nations Environment Programme
    (UNEP)** ⊕
P.O. Box 30552
Nairobi, Kenya

Contact: **Wo Yen Lee,** Director
    Phone: (254) 2 520600 • *Fax:* (254) 2 520711

Activities: Technical Environmental Information • *Issues:* Air Quality/Emission Control; Biodiversity/Species Preservation; Deforestation; Energy; Global Warming; Health; Population Planning; Sustainable Development; Transportation; Waste Management/Recycling; Water Quality • *Meetings:* CITES; UNCED

● Organization

INFOTERRA, directed by Wo Yen Lee, is the technical information service of UNEP—Kenya.

■ Resources

An ad hoc schedule of information services is available upon request.

Deutsches Komitee für das Umweltprogramm der Vereinten Nationen (UNEP) ⊕
German Committee for the United Nations Environment Programme

Adenauerallee 214
5300 Bonn, Germany

Contact: **Wolfgang Burhenne, Ph.D.**
  Phone: (49) 228 269 2216
    Fax: (49) 228 269 2251, 52, 53

Activities: Political/Legislative • *Issues:* Air Quality/Emission Control; Biodiversity/Species Preservation; Deforestation; Energy; Global Warming; Health; Population Planning; Sustainable Development; Transportation; Waste Management/Recycling; Water Quality

● Organization

The United Nations Environment Programme (UNEP) was initiated in 1972 following the **Stockholm Conference**. It is responsible not only for carrying out its own initiatives, but also for coordinating all environment-related or -relevant activities within the framework of the entire UN family, in particular Food and Agriculture Organization (FAO), World Health Organization (WHO), UNESCO, the United Nations Development Programme (UNDP), and regional commissions. At the same time, the UNEP works for closer contact with NGOs, especially with the World Conservation Union (IUCN).

After the UNEP advisory body decided to encourage the development of national committees, the **Deutsche Stiftung für Umweltpolitik (DSU)**, or German Foundation for Environmental Policy, was given the task of assuming responsibility for the German UNEP Committee. With one permanent staff member, the Committee's presidium is comprised of numerous personalities in policy, administration, science and research, business, and manufacturing. Through various projects of the UNEP and environmentally relevant advice to the UN's General Assembly as well as to other organizations, the Foundation informs not only the press and committee members, but others as well, including local and national governments, and interested business groups.

Furthermore, the Committee has as its duty to ensure that German discoveries, new developments and suggestions are brought into international discussion and attention in the decisionmaking process.

▲ Projects and People

Committee members include Federal Minister for Environment, Nature Conservation and Reactor Security Dr. Töpfer, Dr. Günter Hartkopf, Dr. Michael Bothe, and Dr. Eberhard Meller.

■ Resources

The DSU publishes *Umwelt-weltweit* (originally under the UNEP Report title *The World Environment*), 675 pages of solid analysis, answering the most urgent questions as to the "current condition of our environment for progress in our handling of our planet."

U.S. Committee for the United Nations Environment Programme (US/UNEP) ⊕

2013 Q Street, NW
Washington, DC 20009 USA

Contact: **Richard A. Hellman,** President
  Phone: 1-202-234-3600 • *Fax:* 1-202-332-3221

Activities: Education; Political/Legislative • *Issues:* Air Quality/Emission Control; Biodiversity/Species Preservation; Deforestation; Energy; Global Warming; Health; Sustainable Development; Water Quality • *Meetings:* UNCED

● Organization

The **U.S. Committee for UNEP (US/UNEP)** is a private, nonprofit environmental organization established to support UNEP's goals and programs in the United States. US/UNEP seeks to educate the American public concerning the critical environmental challenges UNEP confronts, as well as to raise the awareness of Americans about the importance of a global effort to improve the environment.

▲ Projects and People

The Committee serves as a liaison between UNEP, government, industry, and the public. It also sponsors the annual **World Environment Day,** to commemorate the 1972 Stockholm UN Conference on the Human Environment, as well as other environmental achievements.

US/UNEP seeks to **educate youth about the environment** by distributing information and holding annual poster, essay, and video contests. The Committee also works to arrange connections with other environmental groups worldwide.

US/UNEP, through its director **Milton Kaufman,** has collaborated extensively with the U.S. government and 27 other countries involved with the **Caribbean Environment Programme (CEP)**. The focus has been on the development of the Cartagena Convention Protocol on the **Specially Protected Areas and Wildlife (SPAW)**, and most recently its action arm, the new **CEP Action Plan Regional Program for Protected Areas and Wildlife**. At a Conference of Plenipotentiaries, held in Kingston, Jamaica, in June 1991, three endangered and threatened species annexes to the SPAW Protocol were considered. The annexes will list the species of animals and plants endangered with extinction which the governments are obligated to protect in accordance with the Protocol.

US/UNEP encourages U.S. NGOs to establish liaisons with the Regional Coordinating Unit (CEP Secretariat) to explore the advantages of coordinating the wider Caribbean programs and projects with CEP.

■ Resources

The *US/UNEP Newsletter* is available to members and correspondents at no cost.

Algeria

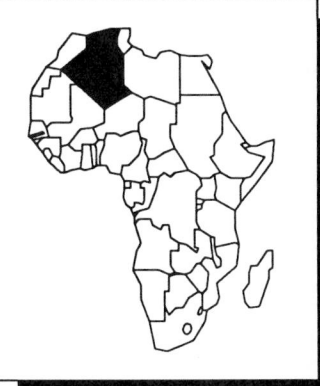

Located on the southern shore of the Mediterranean, this Islamic nation is divided into a fertile coastal plain, a high, largely barren plateau surrounded by the Atlas and Saharan Atlas mountains; and, in the south, part of the Sahara Desert. Most of Algeria's natural wealth lies in its sizable mineral deposits. The arable land comprises only about 10 percent of the total area and is located mainly in the valleys and plains of the coastal region.

The northern sections of the country have suffered from centuries of deforestation and overgrazing, and the relatively sparse vegetation can support only a limited wildlife population. In addition, the rapidly growing population has exacerbated such problems as an inadequate water supply and untreated urban sewage. The large government sector is attempting to come to grips with the country's environmental needs; for instance, substantial reforestation projects have been undertaken in recent years.

🏛 *Government*

Institut National Agronomique (INA)
National Agronomy Institute
B.P. 59 Hacen Badi
16200 El-Harrach
Algiers, Algeria

Contact: **Mahdi Sellami, Ph.D.,** Senior Lecturer and Researcher
Phone: (213) 76 19 87/89
Telex: 64143 DZ

Activities: Development; Education; Political/Legislative • *Issues:* Biodiversity/Species Preservation; Global Warming; Population Planning; Waste Management/Recycling
Meetings: Symposium of Ungulates

● Organization

Dr. Mahdi Sellami of the **Institut National Agronomique (INA)**, National Institute of Agronomy, contrasts the "exceptional natural heritage" of **Algeria** with the disappearance of numerous species, such as the Atlas lion, and the threatened losses of others, such as the monk seal and the dama gazelle.

"Before this alarming affidavit and in view of contributing to the survival of all species, the country has adhered to numerous international conventions," he writes. Among them are the African Convention for the Conservation of Nature and Natural Resources, CITES, and the Ramsar Convention. The previous absence of environmental laws and a lack of a natural resources administration also aggravated efforts to conserve endangered species.

Since 1983, Algeria has three main laws concerning the protection of the environment, hunting, and the Forestry Administration. An official list now defines 70 protected species of birds, 8 reptiles, and 33 mammals—all nondomestic animals. Also during the last nine years, the country established eight national parks, four natural reserves, and four hunting reserves as tools for conservation.

▲ Projects and People

Organizations such as INA are undertaking the scientific and effective capture of Algerian fauna. "Within this framework, we have done research since 1986 on one of the four types of Algerian gazelles, which is the **Cuvier gazelle** in the Mergueb natural reserve (RNM)," writes Dr. Sellami.

The endangered Cuvier gazelle lives in a semidesert habitat characterized by mild temperatures, little rainfall, winter snows, and sometimes violent winds. Its current range is limited to a few small areas in Algeria, Morocco, and Tunisia. A 1988–1989 investigation showed that in Algeria the Cuvier gazelle lives in 29 sites on the Atlas Mountains. A comparison of results with those of the last 150 years reveals that this ungulate is now absent from the Mediterranean coast and the Bécher region. Poaching and intensive raising of pasture livestock contribute to this factor.

INA established a **Vertebrate Laboratory** to study threatened ungulates. Each year, themes dealing with the protection of animal species are proposed to students who are preparing final research projects.

INA accomplishments at the **Mergueb reserve** include a topographical map, vegetation study, inventory of mammalian species, and repatriation of the Cuvier gazelle and related efforts that include a census and studies of this animal's eating and social habits.

Work ahead includes: analysis of botanical specimens and dung specimens; comparison of DNA of various gazelle species (to be carried out in Europe because of lack of equipment); and capture of gazelles which will be outfitted with radio collars in cooperation with a French organization known as IRGM.

■ Resources

Dr. Sellami expresses the need for expanded relations with international environmental organizations as well as aid to facilitate his research and complete his projects. INA is lacking equipment for field studies such as binoculars, telescopes, cameras, and film.

Antigua

An independent island state in the East Caribbean Sea, Antigua is generally low-lying, with a tropical climate, but subject to drought. Antigua and its sister island, Barbuda, were once a single island formed some 30 million years ago from the residue of a volcanic eruption. Antigua is divided into three distinct zones: the southwestern volcanic district, the central plain of mixed terrestrial and marine sediments, and the northeastern limestone upland.

Tourism is the mainstay of the economy, but most of the population is engaged in agriculture, including cotton, fruits, sugarcane, and fishing. Manufacturing is still in the development stages. Antigua is experiencing many of the same environmental problems that affect other Caribbean countries—such as water supply contamination, beach erosion, inadequate waste disposal systems, uncontrolled land-use practices, and fisheries depletion—while lacking the institutional structure for addressing them effectively.

❦ *NGO*

Museum of Antigua and Barbuda
Long Street
P.O. Box 103
St. John's, Antigua, West Indies

Contact: **Desmond V. Nicholson,** Director
Phone: 1-809-462-1469

Activities: Education; Political/Legislative; Research • *Issues:* Historic and Archaeological Sites; Historic/Cultural Resources

● Organization

The **Museum of Antigua and Barbuda** is dedicated to "preserving the past to enrich the future." Geologists report that the islands were formed some 30 million years ago from the residue of a volcano eruption. Coral deposits connected the islands until about 10,000 years ago and make up the Barbuda Bank. Antigua has three zones: "southwestern volcanic district; the central plain of mixed terrestrial and marine sediments; and the northeastern limestone upland. Barbuda consists of low Pleistocene limestones overlapped by modern reef."

Hunters and gatherers are considered to have been the first people to inhabit Antigua—about 6,000 years ago—living on the "abundant fauna and flora of shore and ocean" using tools made from stone and shell. Amerindians first arrived from the Orinoco Delta almost 2,000 years ago and called Antigua "Waladli." Christopher Columbus first sited Antigua in 1493 but never landed there. One family—the Codringtons—leased the sister island of Barbuda for almost two centuries until 1870 for the price of "one fat sheep per year." Today, Barbuda is considered "one of the last pristine ecological havens and truly a treasure to preserve," according to the Museum.

Museum exhibits trace the history of the nation's "geological birth through political independence to the present day" as it strives for self-sufficiency, and describe how colonists, beginning in the seventeenth century, began cutting down forests to grow sugarcane and other crops and to raise livestock with the labor of imported African slaves. Raw sugar was processed into sugar with the use of windmills, locomotives, and other technological advances of the industrial age that are now being restored as funding becomes available.

Particularly geared to children, the Museum instills pride and a sense of identity in this heritage of richness and hardship. Built with locally quarried stones in the eighteenth century, the structure was once the site of both glittery balls and grim court cases; it was refurbished for the Museum with the help of UNESCO, Canadian International Development Agency (CIDA), and private donors.

▲ Projects and People

The Historical and Archeological Society (HAS) of Antigua and Barbuda, which operates the Museum, was founded in 1956 at a pre-Columbian excavation site known as Mill Reef. As HAS expanded its archeological focus, it founded the Museum in 1985. One Amerindian excavation site is at Nonsuch below Emerald Cove; shells including 42 species of mollusc are being analyzed. As people became more aware of the islands' history and ecology, the Environmental Awareness Group (EAG) emerged four years later. In 1990, a separate Trust was set up to restore parts of the old Codrington Plantation—through the Betty's Hope Project and support from friends in the USA. Special children's programs such as Youth Arm, lectures, field trips, and library use are provided. It is expected that the Museums Association of the Caribbean will assist in planning for hurricanes and other disasters.

■ Resources

The Museum provides lectures and tours on requests. Its publications include *the Arawaks of Antigua and Barbuda; Short History of Antigua and Barbuda; Antigua Reef, Rocks and High Roads; Sediment Map*, a quarterly newsletter, and a general brochure, among others. It seeks archaeologists for local explorations.

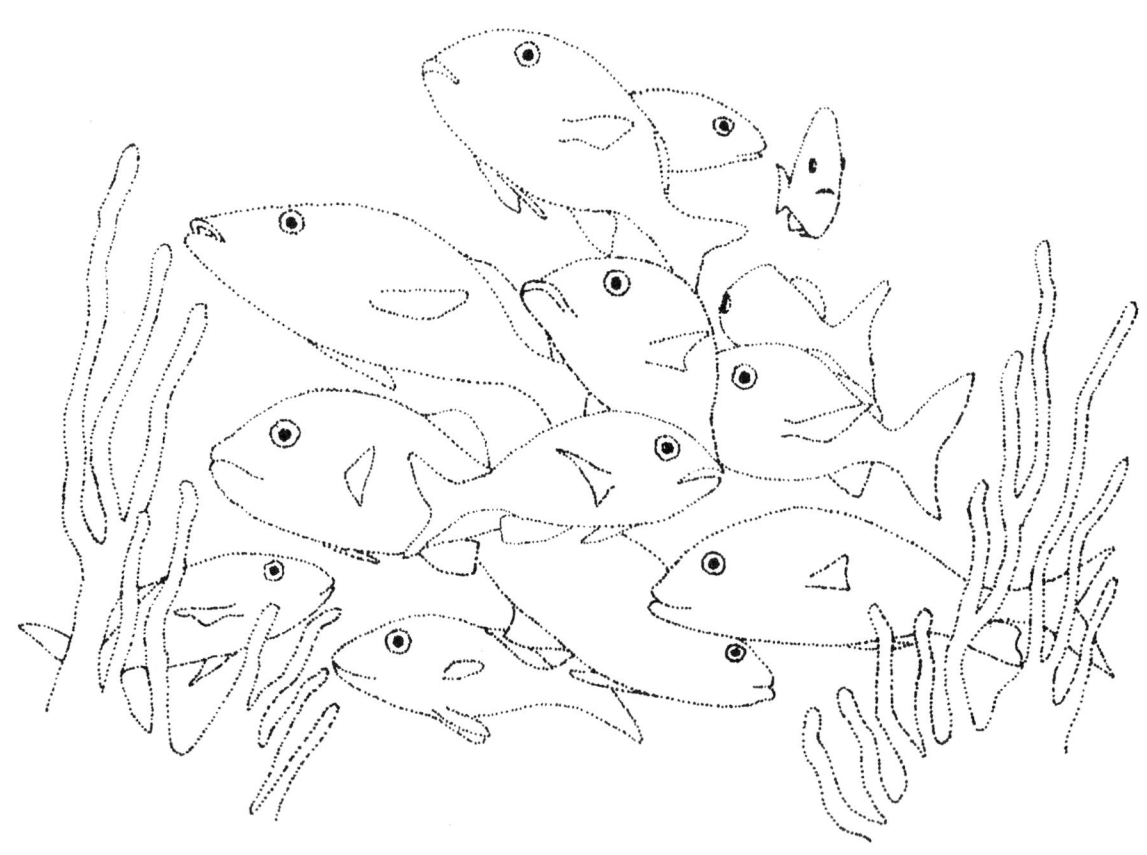

The second largest country in South America, Argentina comprises a diverse territory of mountains, upland areas, and plains. The western boundary falls entirely within the Andes. The traditional wealth of Argentina lies in its vast grasslands and Pampas, which are used for extensive grazing and grain production. The indigenous vegetation varies greatly within different regions of the country.

Despite its size, Argentina's population is heavily concentrated in the cities, especially Buenos Aires, where 50 percent of the people reside. This has led to widespread air and water pollution and waste disposal problems. Heavy flooding is a problem in both urban and rural areas, resulting in soil erosion, especially in the north.

🏛 *Government*

Centro Regional de Investigaciones Científicas y Tecnológicas (CRICYT)

Parque San Martin
5500 Mendoza, Argentina

Contact: **Ricardo A. Ojeda, Ph.D.,** Researcher
Phone: (54) 61 241995 • *Fax:* (54) 61 380370
E-Mail: ntcricyt@criba.edu.ar

Activities: Education; Research • *Issues:* Biodiversity/Species Preservation • *Meetings:* Environmental Education; World Parks Congress

▲ Projects and People

Dr. Ricardo Ojeda encourages Latin American governments and scientists to take responsibility for protecting Latin America's biodiversity from the assault of "burgeoning human populations, uncontrolled habitat conversion, commercial hunting," and other adverse factors. Sharply aware of criticism from developing nations, he says that "such entreaties to Latin American countries to preserve their ecosystems can appear shallow when they are made by scientists living in nations built on the principle of development of 'unused' lands." Leadership must come from Latin America's "front-line soldiers" in battling species loss; adequate funding must be available for long-term research and training.

With Latin America's strong infrastructure of government and private conservation groups, universities, and research centers, such as the **Regional Center for Science and Technology (CRICYT)**, programs are underway to restore the tropical and extratropical ecosystems at the grassroots.

For example, in the **Chacoan scrub forest** of Argentina, private landowners, local inhabitants, and forest managers are working together to recover Fauna and Flora—under assault for centuries by forestry, hunting, ranching, and agricultural activities—and allow some logging and cattle grazing. In the **Andes** of Argentina, Bolivia, Chile, and Peru, local people with ecologists, field biologists, zoologists, botanists, and sociologists are protecting ecosystems. In Costa Rica, **dry tropical forests** are being recovered in similar fashion. **Arid and semiarid ecosystems** and development of national parks are being enhanced in **Man and the Biosphere Preserves**, such as Argentina's Ñacuñán and Mexico's Mapimí.

Overall, **ecological research** is being expanded, **scientific societies** are growing, and quality natural science publications are being produced—despite weak economies. The key to progress is the ability of local environmentalists concerned with reversing species and habitat loss to work with local people with the broad support of government levels, corporations, and banks in Latin America.

Ojeda is encouraging his colleagues to undertake **proper surveys of Fauna and Flora** with time to identify and catalog and to describe distribution patterns.

Dr. Ojeda explains how Latin Americans interpret the push from developed nations to protect their natural resources while repaying "massive and questionable national debts" and employing other economic restrictions: "We (the developed countries) destroyed our ecosystems and adversely affected the global climate and the oceans, and by doing so became rich. Now you must preserve your species and ecosystems for our future use and aesthetic and moralistic principles."

Although global cooperation is vital to solve biodiversity problems, conservation plans need to be developed locally—with a keen understanding of the society that must implement them. **Reducing the exploiting and exporting of fur-bearing animals** is one area where countries can collaborate.

"If Latin America continues to make its natural resources available on an unlimited basis to developed countries, Latin American dependence will continue, and environmental deterioration will not be halted," reports Dr. Ojeda. In a joint commentary with **Dr. Michael A. Mares**, University of Oklahoma, he says preserving nature is a "race that all ecologically concerned people of all countries can win if they work together and act swiftly. Latin Americans, together with people from developed countries, can reduce the rates of environmental degradation."

■ Resources

CRICYT "urgently needs" funding for field research projects and field courses on conservation biology and biodiversity; educational materials such as books, slides, and videos on biodiversity, conservation biology, and wildlife; and a personal computer (PC) with printer.

Materials available are a *Guide to the Mammals of Salta Province, Argentina*; *Mammals of Tucuman, Argentina*; posters on spiders, mammals, and food webs; and books on desert vegetation and Andean flora.

State Government
Chubut Province

Dirección de Intereses Marítimos
9 de Julio 280
9103 Rawson, Argentina

Contact: **Gabriel Punta**, Marine Biologist, Oceanographer
Phone: (54) 965 82604 • *Fax:* (54) 965 82603

Activities: Research • *Issues:* Biodiversity/Species Preservation

● Organization
Dirección de Intereses Marítimos studies the marine environmental problems in the province of Chubut that relate to fisheries, seaweed beds, seabird colonies, and red tides.

▲ Projects and People
Current projects concern the dynamics of populations of **cormorant colonies**, including the interrelationship between seabirds and fish-

eries. Cormorants are generally dark-plumaged diving seabirds having long necks and throat pouches for holding fish. These long-term studies began in the mid-1980s. In one project, efforts are being made to reduce the negative effects of guano exploitation on Patagonian cormorant colonies. Guano is a natural excrement of seabirds or fish, which can be used as fertilizer.

■ Resources

Resources needed include a computer for data analysis and training courses for new work methodologies. **Gabriel Punta**'s publications include *A Newly Discovered Colony of Southern Giant Petrels (Macronectes giganteus) on Isla Gran Robredo* and *Diet and Food Habits of Two Patagonian Cormorants and Relation with Migratory Mouvements of Both Species*. Most other works are published in Spanish.

NGO

Ecobios
Facultad de Ciencias Exactas y Naturales
Ciudad Universitaria Pab:II 4to.piso Lab.76
Catedra de Ecologia.1428
Nunex, Buenos Aires, Argentina

Contact: **Patricia Alejandra Gandini**, Lic. Ciencias Ciologicas
Phone: (54) 1 718 5020/29, int. 204

Activities: Education; Research • *Issues:* Biodiversity/Species Preservation • *Meetings:* CITES; V Congress de Ornitologie Neotropical

● Organization
Ecobios is a nonprofit organization dedicated to biological research, especially the study of environmental problems affecting the seabirds of the Patagonian area in Argentina.

▲ Projects and People
A principal project has been evaluation of the effect of **oil pollution** along the Patagonian coast. Researchers used the Magellanic penguin, a seabird that spends most of its life in water and that is likely to encounter oil, as an indicator of chronic oil pollution. These seabirds are easy to see, and some retreat to land when they become oiled.

Another goal has been to **survey the population size** of different penguin colonies along the Argentine coast. Fourteen colonies have been assigned to the survey, eleven in Santa Cruz province and three in Chubut province.

Since 1987, Ecobios has carried out an intense **study in Cabo Virgenes and Puerto Deseado penguin colonies** in Santa Cruz province. The objectives are to quantify penguin breeding success, detecting changes due to human development; to quantify the importance of habitat quality on breeding success; to determine the increase of kelp gull population and evaluate their potential negative effect on the reproduction of magellanic penguins and other seabirds; and to begin a study of buff-necked ibises in Santa Cruz Province.

Ecological Systems Analysis Group
12 de Octubre, 1915
Casilla de Correo 138, Río Negro
8400 San Carlos de Bariloche, Argentina

Contact: **Gilberto Carlos Gallopín, Ph.D.,** Director
Phone: (54) 944 2 2050 • *Fax:* (54) 944 2 2050
E-Mail: UUCP: uunet!atina!fbl!miguel

Activities: Education; Research; Technical Assistance • *Issues:* Biodiversity/Species Preservation; Deforestation; Global Warming; Sustainable Development; Terrestrial Ecology

▲ Projects and People
Associated with Fundación Bariloche, the **Ecological Systems Analysis Group** with a staff of eight reports the following current environmental activities:

Ecological Prospective for Latin America. This project is anticipating "the probable transformations which the major Latin American ecosystems will suffer under different socio-economic and technological scenarios in a time-horizon of 20 to 50 years." According to **Dr. Gilberto Gallopín,** the "ecological consequences of the diffusion of the new technological wave" are being assessed, and the organization is contributing to the "definition of elements for a scientific-technological strategy adequate to the conditions and needs of Latin America."

Global warming and Latin America: ecological regional impacts, potential contribution of the region to climatic changes and mitigation of the greenhouse effect. This project is examining the major environmental impacts upon Latin America as a result of global climatic change and land-use patterns. Under study are ecological implications of a change in climatic variability and the impact of regional climatic shifts on biodiversity, land use, production potential, and sustainable development in the major life zones. Dr. Gallopín lists the specific work steps regarding Latin America: review of current ecological and land-use situation; selection of climatic change scenario; analysis of current greenhouse gas emissions; preparation of a computerized life-zone map for the present and with the assumed climatic change scenario; analysis of the anticipated ecological problems and opportunities; and discussion of the implications of the ecological changes on present and potential land use as well as major interactions between ecosystems.

"The results will provide valuable information for the development of policy responses to climatic change," writes Dr. Gallopín. The project will also produce valuable data sets.

Sustainable Development in Argentina. This interdisciplinary exercise on the relationship among natural resources exploitation, economic and ecological changes, and the evolution of human settlements in urban and rural communities is leading to a diagnosis of current and potential situations, and the formation of alternative policy proposals for achieving sustainable development.

Dr. Gallopín, who is also affiliated with Argentina's National Council of Scientific and Technical Research, is a professor formerly with the University of Buenos Aires' Department of Biological Sciences as well as its Center for Advanced Studies. Overall research interests are "ecological systems analysis; integrated studies on natural and managed ecosystems, with emphasis on structure and function, including human actions; mathematical models of ecological processes; and environmental impact assessment." He was recently involved in a United Nations Environment Programme (UNEP) project regarding environmental priorities for Latin America and the Caribbean.

Dr. Manuel Winograd is assistant researcher/professor with the Group. **Miguel Rubén Gross** is investigator.

■ Resources
A list of ecological publications is available in Spanish. Dr. Gallopín speaks at conferences worldwide on global climate change as well as on sustainable development in Latin America.

Foundation for the Defense of the Environment (FUNAM) ⊕ ✺
Casilla de Correo 83, Correo Central
5000 Córdoba, Argentina

Contact: **Raúl Alberto Montenegro,** President
Phone: (54) 51 22 62 52 • *Fax:* (54) 51 52 02 60

Activities: Development; Education; Political/Legislative *Issues:* Air Quality/Emission Control; Biodiversity/Species Preservation; Deforestation; Energy; Global Warming; Health; Population; Quality of Life; Sustainable Development; Waste Management/Recycling; Water Quality • *Meetings:* International Environmental Law Conference; UNCED

● Organization
Founded in 1982, **FUNAM**—the Environment Defense Council—is a nonprofit organization operating "independently of government and private business." With an Assembly of Active (500) Members and Founders, Board, Executive Committee, staff of 5, and 20 volunteer program coordinators, FUNAM has headquarters in Córdoba, offices in Buenos Aires and Rio Negro, and representatives in Europe, Africa, North America, and Asia-Pacific.

FUNAM advocates "the creation of self-help groups who can learn to defend their right to a healthier and sustainable environment themselves." With goals of "justice among citizens, justice among countries, and justice among generations," FUNAM supports "freedom of thought and expression and self-determination of individuals and peoples of the world in a framework of respect for human dignity and nature."

Calling for a "new ethic of man in nature and not against nature," FUNAM speaks out against "blind consumption, non-sustainable lifestyles, unnecessary waste production, ecological neocolonialism, and the use of hazardous energy sources." Disapproving of the "foreign debt system" that increases "Southern poverty, social inequality, and environmental destruction," at the same time, FUNAM recognizes a "Northern 'ecological debt.'"

More than 30 national and local programs emphasize networking, workshops, legislation, community action, postgraduate human ecology studies, and environmental education to achieve a nuclear-free future, sustainable development, ecodiversity, and ecozones as parks and reserves. FUNAM also takes part in international programs and conventions regarding children; banks, trade, and foreign debt; climate change; and biodiversity. FUNAM co-organized the 1992 Global Radiation Victims Conference in Berlin. Córdoba's NGO Council and FUNAM are involved in an effort to create a non-governmental United Nations parallel system or "new international institutional order" for existing NGOs to overcome the UN's "bureaucracy" and "operative limitations."

With Greenpeace International, Philippines' Haribon Foundation, Italy's North-South Campaign, Netherlands Committee for the World Conservation Union (IUCN), and the USA's Environmental Defense Fund (EDF), FUNAM is part of an NGO information network and watchdog team on the **Global Environmental Facility (GEF) Program** to reduce negative impacts.

▲ Projects and People

Biologist **Raúl Alberto Montenegro** founded FUNAM and helped to spawn other non-government organizations in Argentina such as Córdoba's Environmental Defense Council and the Argentine Environmental NGOs Confederation. In 1984, he put together the first national meeting of environmental NGOs, Alto Gracia, and that year published the first *Argentine NGOs Directory.*

A past president of the Argentine Association of Ecology, Montenegro was also vice president of the Environmentalists' Argentine Association (ASARDA) and of Greenpeace Argentina. Numerous other organizations have named him an honorary member; recently he was appointed to the board of the Environment Liaison Centre International, Tunisia, to represent Latin America and the Caribbean region.

Using mass media to heighten environmental awareness, Montenegro wrote the ecology pages for *La Voz del Interior* newspaper, helped create **Econews**—a freelance environmental news service—with the Argentina Private Journal Council, broadcasted ecology discussion programs on the University of Córdoba radio station, and promoted a **Phone Emergency Network** to inform and solve environmental crises.

He reaches an audience of 100,000 people with a **children's environmental education program** on the National University television channel. He also created the **Environment's Little Defenders Club** of some 800 active children and wrote the *Environment's Little Defenders Manual,* which is distributed free in all schools. An evolutionary biology professor at the National University of Córdoba, Montenegro also chaired the **Children's Campaign for Peace and Life,** which has trained more than 350,000 children. He is presently co-chairing the **Voice of the Children International Campaign** in 40 countries with the Norwegian Campaign for Environment and Development; the USA Academy for Educational Development cited this campaign in 1990 as "one of the 10 most important educative programs" worldwide.

FUNAM's **Only One Sky** program depicts the ozone depletion plight with a children's program, **The Umbrella Is Broken,** relevant also in Chile, which is nearby Antarctica and Patagonia. FUNAM also shows how to avoid fires and nonorganic farming and how to protect against ultraviolet (UV) diseases and ecological damage.

Montenegro authored the **Environment Protection Codes** for Córdoba—the first of their kind—as well as dozens of laws, acts, and regulations to encourage environmental management and prevent pollution. Mobilizing Argentine public opinion in defense of the environment and against the arms race and nuclear power, he took legal actions against various government agencies and officials that were not protecting the environment or public health.

FUNAM specifically speaks out against uranium mining, sale of hazardous pesticides, illegal wild species trade, chemical pollution of waterways, forest fires, and desertification. It developed a low-cost strategy to deal with cholera disease and sanitation problems; worked on urban air pollution control of automobile and bus exhaust;

spearheaded a "clean lake for us all" recovery program of semiarid basins; and encourages an "environment open fora program" at all government levels to broaden public participation in decision-making.

In addition to chairing **courses** in Córdoba and at the University of Río Cuarto, University of Buenos Aires, and National University of San Luis, Montenegro has published 21 books and other extensive works on environmental protection and politics. He also was editor of the scientific magazine *Ecology.*

■ Resources

FUNAM has available a detailed report on its activities, *Environment's Little Defenders Handbook,* and *Guidelines for the Voice of the Children International Campaign.* Financial resources are needed, along with communication technologies.

Fundación Amigos de la Tierra ⊕
Friends of the Earth Foundation
Zapata 343
1426 Capital Federal
1000 Buenos Aires, Argentina

Contact: **Patricia Gay,** Lawyer
 Phone: (54) 1 553 4318, (54) 1 773 5947
  Fax: (54) 1 331 6720, (54) 1 305 618

Activities: Development; Education; Research • *Issues:* Air Quality/Emission Control; Biodiversity/Species Preservation; Deforestation; Energy; Global Warming; Health; Sustainable Development; Waste Management/Recycling; Water Quality • *Meetings:* AGM; CITES; FOEI; Montreal–Protocol Negotiations; UNCED

● Organization

"Por la restauración, preservación y uso racional de la Biósfera"— Such is the dedication of **Amigos de la Tierra,** the Argentina member of **Friends of the Earth International (FOEI).** Amigos de la Tierra is an independent environmental organization that works with any political party or institution sharing its ideals and goals.

Chapters of Friends of the Earth, which exist in numerous countries, are united by a shared name and a common cause: the conservation, restoration, and rational use of the biosphere; these chapters constitute the federated FOEI. A consultant to the Economic and Social Council and other bodies of the United Nations, FOEI includes an international secretariat, assembly, and a policy-shaping executive committee elected by all member countries. There are also three Regional Committees for Europe, Eastern Europe, and Latin America and the Caribbean.

Amigos de la Tierra sponsors **conferences and public debates** and brings its message to the media to raise "the public's consciousness in respect to environmental problems and their repercussions for the future of the planet." According to its literature, it "creates proposals and acts effectively on them; aids and stimulates collective investigation, acts as a mediator between the public and the authorities, and cooperates with all of the national and international organizations that have objectives similar to ours." Amigos de la Tierra particularly addresses environmental problems that cross geographical and political barriers—bringing together government, non-government, local, national, and international force and resolve.

▲ Projects and People

A national federation of Amigos de la Tierra member groups in Córdoba, Santa Fe, Buenos Aires, and Río Negro concentrate on environmental education, awareness campaigns, national and international politics, environmental health, and sustainable agriculture.

Specific undertakings include a permanent **environmental education project** begun in the mid-1980s; the latest stage is a series of classes that travels throughout the country's interior. Being developed is a course that **trains teachers** in the use of informational materials for environmental activities. The socioeconomic status of **native Argentinean communities and their relationship to the environment** were researched to "counter the celebration of the 500th anniversary of Columbus' discovery of America," reports Amigos de la Tierra; and informational materials were being developed in three languages. To prepare for the UNCED '92 meeting in Brazil, a series of radio and television commercials and print advertisements were prepared about national and global environmental problems. A guide for Buenos Aires residents is being produced to illustrate what individuals can do for the environment.

Regarding future environmental collaboration, **Patricia Gay** writes, "The only way is to cooperate in the projects, interchange experience and information, and train people."

■ Resources

These publications are available in Spanish at U.S. dollars: *Canje Deuda Externa por Naturaleza* about debt swaps; *Qué Hacemos con la Basura* about rubbish; *El Tratado de la Antártida* (Antartica's treaty); and *Pesticdas y Agroquímicos* (Pesticides).

Fundación Cruzada Patagonia (FCP)
Feux San Martin 672, Newquen
8371 Junin de los Andes, Argentina

Contact: **German Pollitzer**, Lawyer
  Phone: 54 944 91286 • *Fax:* 54 944 91262

Activities: Development; Education; Research • *Issues:* Deforestation; Health; Native Americans; Sustainable Development • *Meetings:* National Network of NGOs

● Organization

Marking its tenth anniversary in 1989, **Fundación Cruzada Patagonia (FCP)** is a nonprofit institution founded by young families mostly originating in Buenos Aires who settled in southern Neuquén Province to help improve the living conditions of impoverished local communities, particularly among the indigenous Mapuche people.

FCP work is guided by a conviction which is affirmed daily: "The Mapuche culture encompasses ancestral values which we want to learn about. The interaction necessary to our work must not threaten their cultural institutions. Triumph and progress achieved through the participation of both educated people and the villagers will impart a zealous respect for the indigenous."

▲ Projects and People

Among its activities, FCP has conducted **demographic studies** with varying methodologies in the small urban center of Junin de los Andes as well as in rural areas. The underlying purpose is to investigate infant mortality, where rates are lower in urban areas, but also appear to be declining in rural areas. These studies help bring attention to local **health needs**.

In addition to implementing health services and instituting formal registration of births, deaths, and marriages—the **San Ignacio Education Center** has been established for job training and a locally operated factory is providing work. Scientific research and agricultural projects are being launched.

■ Resources

New energy sources are sought for remote populations. Radios are also needed for adult education outreach. Infant mortality reports are bilingual; brochures and other literature is in Spanish.

Fundación Sirena ⊕
Casilla 1395, Correo Central
1000 Buenos Aires, Argentina

Contact: **Jorge Rabinovich, Ph.D.**, Secretary
  Phone: (54) 1 72 2950 • *Fax:* (54) 1 662 0999
  E-Mail: raminovich@cribab.edu.ar

Activities: Development; Education; Research • *Issues:* Biodiversity/Species Preservation; Deforestation; Environmental Impact Assessment; Health; Sustainable Development; Waste Management/Recycling; Water Quality

● Organization

Founded in 1990, **Fundación Sirena** supports research and teaching on behalf of the environmental sciences. Its activities are: "analysis of scientific policies; environmental impact assessments [EIAs]; organization of graduate programs in ecology and conservation; land reclamation and sanitary engineering; analysis of environmental and ecological components in development planning; environmental aspects of domiciliary and industrial wastes; economic analysis of environmental projects; modeling, databases, and expert systems applicable to renewable natural resources; [and] citizen participation in urban environmental problems." President is **Alberto Larrondobuno** and vice president is **Enrique Oteiza**.

A major undertaking is support for SPIDER (Special Program for the Improvement and Development of Ecological Research), formed in 1986 in Argentina and extending itself throughout Latin America and the Caribbean.

▲ Projects and People

EIAs are concerned about developments of **watersheds, agriculture production, industry,** deterioration of the **railroad structure,** and frontier technologies. Regarding the application of **artificial intelligence** toward environmental problems, the Foundation is investigating the "expert systems" for managing natural resources, developing "inference engines," putting together "interactive programs on environmental information for legislators, administrators, and managers."

Modeling tools are used in predicting environmental effects of development and legislation and helping decisionmaking. **Socio-economic considerations** are evaluated, for example, in studying the relationship between poverty, development projects, and environment; or between population, education, and employment.

The Foundation is involved in **designing university graduate ecology programs**; organizing advanced environmental courses; compiling **environmental bibliography**; surveying "perception and opinions on environmental problems"; publishing books and journals; and creating databases of protected areas, Fauna and Flora, and bibliographies as well as geographic information systems (GIS).

Among SPIDER's undertakings are the development of **a graduate program in wildlife management and conservation** in Latin America; publication of a *Directory of Ecologists and Environmental Scientists of Latin America, the Caribbean, Portugal, and Spain* with entries from 600 institutions in 35 countries; and availability of the Directory, including data from 45 countries, on a "user-friendly system" on IBM-compatible diskettes, called **DECA**. Support for DECA is from UNESCO, Panamerican Health Organization, Higher Council for Scientific Research of Spain, and the Spanish Agency for International Development—along with Fundación Sirena.

Dr. Jorge E. Rabinovich, economics professor, University of Saint Andrew, Buenos Aires, is on expert panels of UNESCO, Food and Agiculture Organization (FAO), World Health Organization, and United Nations Environment Programme (UNEP). Among many publications, he has written two books in Spanish on animal population ecology and compiled two ecological bibliographies.

■ Resources

The *Directory* and DECA are available through the Foundation.

Grupo de Educadores Ambientalistas (GEA) Environmental Educators Group

Laprida 1373

1642 San Isidro, Argentina

Contact: **Horacio E. Rodríguez Moulin**, Agricultural Engineer and Environmental Educator
Phone: (54) 1 743 9467 • *Fax:* (54) 1 812 1350

Activities: Education • *Issues:* Biodiversity/Species Preservation; Deforestation; Nature Interpretation • *Meetings:* GEA Annual; NYZS

● Organization

The philosophy of **GEA**—Grupo de Educadores Ambientalistas (Environmental Educators Group)—is summed up in this translated Baba Dioum poem:

In the end . . .
We will conserve only what we love
We will love only what we are able to understand
We will understand only that which is taught to us.

GEA—a group of professionals, educators, and naturalists from Argentina, UK, and the USA—believes education is the most positive tool for conservation based on individual responsibility. The future is in the hands of our children, the group writes, and we must encourage a familiarity with those factors causing the deterioration

of our planet, awareness of ways to reverse harmful situations, and useful skills in the search for solutions.

▲ Projects and People

Environmental awareness programs are directed countrywide at teachers at the primary and secondary levels, students in both public and private schools at those levels, and interested institutions. GEA develops programs according to the requirements of the users. Included are basic courses, workshops at *Aula Verde*—the "Green Classroom" at a mountaintop in Los Cocos—games, conferences, field trips, outdoor activities, and summer camps.

Goals are to create an understanding of the teacher's responsibility in forming students' environmental consciousness through acquiring knowledge and skills; to build attitudes and practices showing respect for nature; and to develop a sense of urgency in resolving current problems such as pollution, species extinction, deforestation, and desertification. Advanced sessions cope with managing the environment through planning, politics, legislation, and conflict mediation.

Special conferences address the vanishing misionera forest threatened by slash-and-burn and clear-cutting techniques. GEA says the parana forest, of which the misionera is part, is in even greater danger than the Amazon rainforest. Participants also explore the wild Fauna and Flora of Staten Island in the Tierra del Fuego, the exclusive species of the land-locked island Samuncura, the thorny forests and carob trees of Montiel, the cave paintings of Cerro Colorado (red hills), wildlife of the pampas near Buenos Aires, desert and jungle life, and protected reserves and parks.

GEA believes a new era in **environmental education** is emerging to confront these "final risks which must be understood and analyzed." General director is elementary school teacher **Vivian Lavalle de Vignaroli** and educational programs coordinator is chemistry technician **Mónica Ramirez de Spinardi**.

■ Resources

GEA needs environmental experts from other organizations and countries to help train teachers and professionals who would then "act as multipliers." With the Ministry of Education, a special conference could be organized for this purpose.

Species Survival Commission (IUCN/SSC) Veterinary Specialist Group

Boedo 90, Florida

1602 Buenos Aires, Argentina

Contact: **Marcelo D. Beccaceci**, Chairman, South American Section
Phone: (54) 1 797 2251

Activities: Education; Research • *Issues:* Biodiversity/Species Preservation

▲ Projects and People

Dr. Marcelo D. Beccaceci, a veterinarian, founded the Endangered Species Group of the Argentine Wildlife Foundation, which he describes as "the most important non-governmental conservationist group in Argentina." In recent years, he has gathered information about many of the endangered species included in the Red List of the IUCN, the World Conservation Union—a global group made up of some 65 governments, 111 government agencies, and 414 NGOs. The **Species Survival Commission (SSC)** is the largest and most active of the IUCN Commissions with more than 2,500 members in 135 countries who work to stem the loss of the world's biological diversity. The SSC sets the pace for species conservation worldwide, according to Beccaceci.

Launching "the first and only field work in Argentina" about the distribution and ecology of the **maned wolf** (*Chrysocyon brachyurus*), South America's largest canid, Beccaceci updated the population status and compiled other information. The species is listed as vulnerable on the Red List.

He is presently undertaking the country's first research about the status of the endangered **marsh deer** (*Blastocerus dichotomus*), a big South American mammal also designated as vulnerable. Studies are undertaken at the Iberá Natural Reserve, "one of the largest wetlands in the world," writes Beccaceci, "and one of the species' strongholds—the second after Pantanal in Brazil."

Beccaceci also researches the incidence of certain diseases in endangered mammals of the subtropical areas of Argentina and Paraguay and conducts public information programs. The governments of both these countries invited Beccaceci to discuss the impact of the Yacyretá Dam—"largest under construction in the world"—on wild animals inhabiting the area.

The author of *Patagonia Wilderness*, Beccaceci is also an honorary member of the IUCN/SSC Canid Group.

TRAFFIC South America ⊕
Ayacucho 1477 9° piso "B"
1111 Buenos Aires, Argentina

Contact: **Tomás Waller**, National Representative
Phone: (54) 1 424348 • *Fax:* (54) 1 3124104

Activities: Education; Law Enforcement; Political/Legislative; Research • *Issues:* Biodiversity/Species Preservation; Deforestation; Sustainable Development; Wildlife Use and Conservation • *Meetings:* CITES; IUCN/SSC Crocodile Specialist Group; World Congress of Herpetology

● Organization
TRAFFIC South America monitors and controls legal and illegal trade in wildlife. **Director Juan S. Villalba-Macías** (Montevideo, Uruguay) is responsible for international trade; **Juan X. Gruss**, as Argentina consultant, oversees national trade and population surveys. TRAFFIC works closely with the CITES (Convention on International Trade in Endangered Species of Wild Fauna and Flora) management authorities here, the CITES Secretariat in Switzerland, and other global organizations such as the World Wildlife Fund (WWF).

▲ Projects and People
One such collaborative project was AVISA: **Administración de Vida Silvestre en la Argentina**, which examined and published the status of wildlife administration and its failures, along with recommendations for solutions. It discussed exploited animals, such as parrots and parakeets; false otter, foxes, and iguanas for skin or leather; and tortoises, snakes, and flamingos. The report stirred much debate and is now a basic tool for the stimulation of new research, writes co-author **Tomás Waller**, and CITES adapted many recommendations. A subsequent paper, prepared on Argentinean CITES enforcement problems with conclusions and recommendations, also led to positive reforms, he says.

An ongoing project is a **status survey of crocodilian species** in Argentina's northeastern provinces so that management recommendations can be proposed. An earlier survey on the ecology and trade data of *Geochelone chilensis*, a heavily traded land tortoise in Argentina, needs to be continued—but is hampered by lack of funding.

As national representative, Waller's main focus is detecting and stopping illegal trafficking while monitoring and analyzing the legal trade in terms of total specimens involved. **Surveys** are planned on the trade of **Guanaco (camelid) skins and wools, and boas.**

Waller says he worries about "current big-issue campaigns that waste millions in funds that could be better invested in specific conservation programs." Campaigns originating in developed countries and seeking to impose policies elsewhere are not always based on the same facts existing in the developing countries, he writes. Adherence to CITES offers the best way to solve trade problems and conserve wildlife, he urges. He is a member of the IUCN/SSC South American Reptile and Amphibian Group.

(See also TRAFFIC in the Malaysia, USA, and Uruguay NGO sections.)

■ Resources
Radio tracking devices are needed for wildlife research.

The *Alerta* bulletin and reports such as *Diagnostico y Recomendaciones Sobre La Administracion de Recursos Silvestres en Argentina: La Decada Reciente* are published in Spanish; some materials are bilingual.

⚑ *Private*

R. Natalie P. Goodall
Research Biologist
Sarmiento 44
9410 Ushuaia, Tierra del Fuego, Argentina

Phone: (54) 901 22742 • *Fax:* (54) 901 22318

Activities: Research • *Issues:* Biodiversity/Species Preservation; Marine Mammals Biology • *Meetings:* Conference on the Biology of Marine Mammals; IWC; SCAR

▲ Projects and People
Natalie Prosser, a graduate of Kent State University who grew up on an Iowa farm, was first drawn to South America by teaching jobs in Venezuela. While traveling in Argentina on holiday, she read a book, *Uttermost Part of the Earth*, about Tierra del Fuego (TF) and decided to explore the "Fireland." Upon her arrival at the farm that was the setting of the book, she met her husband-to-be, Thomas D. Goodall—descendant of TF's first European settler (at Estancia Harberton). They were married soon afterward and raised two daughters. **Natalie Goodall** remains there today and is recognized as one of the world's foremost marine mammal research biologists.

Private

Goodall's initial research in TF involved studying, collecting, and illustrating the local flora. It was the discovery of "beach-worn dolphin skulls that developed into full-scale research."

Beginning her research on the cetaceans and pinnipeds of Tierra del Fuego in 1976, she became involved in studying the basic biology of 21 species of smaller, mostly exceedingly rare cetaceans, 7 larger cetaceans, and 7 seals. The "size, pigmentation, food habits, organs, parasites, aging, reproduction, and occasionally the behavior" of marine mammals are painstakingly studied. Prof. Goodall and her research team preserve and prepare the animals for future observations and to be used in museum exhibits.

"We have a great deal of information and are preparing papers on each species," writes Goodall. "Ready for publication or in press are several on beaked whales, southern porpoises, and pilot whales. Underway are papers on the general *Lageonorhynchus, Lissodelphis, Grampus,* and *Pseudorca.*"

Goodall began a small publishing company, Ediciones Shanamaiim, to produce works on TF, such as the book *Tierra del Fuego.* She is presently preparing a general book on **wildflowers** based upon the watercolors of her mother-in-law Clarita M. Bridges Goodall, who began painting wildflowers to record those she thought were "disappearing due to sheep grazing." Goodall believes "the book will help local people and visitors identify the Fuegian species and call attention to those that are rare."

"All birds are now protected in TF," notes Goodall, who took part in a community effort first to save the "upland, ashy-headed, ruddy-headed and kelp geese." Goodall reports, "An island belonging to Harberton has a new and growing colony of **Magellanic penguins,** usually accompanied by two **king and two gentoo penguins.** All of these were formerly rare in the eastern Beagle Channel. We manage the islands as **nature reserves** and hope to have them declared legally so."

In 1991, Goodall did "three summer and four winter **ecological surveys of birds and marine mammals,** both on land and by helicopter, of the coasts of Cabo San Sebastián to Cabo Espiritu Santo, for the oil company, Total Austral S.A." The subsequent reports indicate numbers of wildlife, rare species, and habitats in the event of an oil spill. Also discovered were "southern right whales mating in later winter . . . and Commerson's dolphins." Similar surveys were planned for 1992, along with field studies of certain species.

Nature tours are conducted at Estancia Harberton, and Goodall gives short talks on life in TF "from glaciers to present" to groups from large ships who visit the farm. She also lectures on local Fauna and Flora on tourist ships and to schools and local groups.

Prof. Goodall has extensive **scientific collections,** including more than 6,000 plant, 1,600 marine mammal, and 400 bird specimens. Her work is funded by organizations such as the National Geographic Society and Consejo Nacional de Investigación Científica y Técnica (CONICET). For example, she has carried out research on persistent plastics on beaches for the U.S. Marine Mammal Commission. "You might say that none of my work is strictly conservation-oriented, but one cannot conserve if the basic studies of the species in question have not been carried out," she writes.

A member of organizations including the Species Survival Commission (IUCN/SSC) Cetacean Group, Goodall is also an honorary research associate of the following: Centro Austral de Investigaciones Científicas (CADIC); Museo Argentino de Ciencias Naturales Bernadino Rivadavia; Museo Fin del Mundo; University of California, Santa Cruz–Long Marine Lab; National Museum of New Zealand; and CIMMA–Universidad Austral, Chile. She does independent research as well.

■ **Resources**

Research efforts and stronger educational outreach could be helped with funding and assistants. Available publications include *Tierra del Fuego,* area map, and Fauna and Flora map and calendar.

🎓 *Universities*

Instituto Miguel Lillo
Universidad Nacional de Tucumán
Miguel Lillo 205
4000 Tucuman, Argentina

Contact: **Ruben M. Barquez, Ph.D.,** Professor
Phone: (54) 81 237240 • *Fax:* (54) 81 216185

Activities: Education; Research • *Issues:* Biodiversity/Species Preservation; Evolution of Ecology

● **Organization**

The **Instituto Miguel Lillo, National University of Tucumán,** is a nonprofit organization which is centered on developing scientific research and education of the natural sciences, ecology, taxonomy, botany, zoology, and geology. The national government provides funds for developing such research and educational activities.

▲ **Projects and People**

As a vertebrate biology professor, **Dr. Ruben Barquez** conducts undergraduate and graduate students in their biodiversity studies in northern Argentina—especially in the Yungas Forests, Thorn Forests or Chaco, and the Monte Desert. **Field research** is an "integral part of the science of ecology, wildlife management, and conservation," writes Dr. Barquez, who indicates that such studies help train field biologists. "We are currently underway to establish a program in wildlife biology, offering field courses," he reports.

In a current study of the **mammals of the Chaco,** such fauna is analyzed in an east-west gradient. With bat inquiries completed, researchers are concentrating on habitat destruction, by man, of mammals and birds. "Devastation of the gallery forests is moving very fast," writes Dr. Barquez. "Being the main passage for the dispersion of the species, they need to be studied for a better comprehension of the taxonomic, evolutive, and zoogeographical relationships between the two theoretically isolated faunas of the main forests of northern Argentina."

Also being researched are **mammals of the Yungas Forests** in northwestern Argentina, where some 120 species represent half the total number of the country's land mammals. "The Yungas Forests represent one of the richest ecosystems in Argentina and one of the most important from the biodiversity point of view," reports Dr. Barquez. But habitat destruction and over-exploitation are taking their toll. "Our study is an attempt to analyze the biogeography of the mammals and birds of this ecosystem . . . and the pattern of decline of species along about 700 kilometers of the southward extension of the Yungas 'peninsula,'" he also writes. Data already shows a loss of species increasing significantly from the peninsula's base to its tip.

■ **Resources**

A four-wheel-drive vehicle for field research is sought.

Australia is the world's smallest continent and largest island. It is also the driest inhabited continent—with vast arid or semiarid desert. Most of the population lives on the coastal fringe. The country has experienced a general trend of rising temperatures over the past six decades, probably attributable to the greenhouse effect.

Environmental pressures have been created by an expanding population, continuing coastal development for both recreational and industrial purposes, the demand for increased utilization of fisheries resources, and the need to find new oil and gas reserves. Australia has been concentrating on key environmental issues such as soil erosion, water-quality improvement, chemicals control (including hazardous waste handling and control), climatic change, ozone depletion, and assessment of the impact of human activity. The federal government regulates uranium mining and waste dumping at sea, and provides protections for native fauna through state and territorial controls. The government has programs to conserve and manage the kangaroo population and to prevent the killing, capturing, injuring, or interfering with whales, dolphins or porpoises.

🏛 *Government*

Australian Museum ⊕
6-8 College Street
Sydney South NSW 2000 Australia

Contact: **Tim Flannery, Ph.D.**, Research Scientist, Mammalogy
Phone: (61) 2 339 8114 • *Fax:* (61) 2 399 8304

Activities: Research • *Issues:* Biodiversity/Species Preservation

▲ Projects and People

Although the **Solomon Islands** (east of Papua New Guinea in the Pacific Ocean) are one of the most diverse island groups on Earth, little was known about the rare and common mammals who share life with a unique array of plants and other animals. In 1990, **Australian Museum** biologists, working with Ministry of Natural Resources staff from the Solomon Islands, undertook a comprehensive survey mostly of bats and native rats that is expected to take years to analyze.

The findings uncovered 52 species of native mammals, half of which are found *only* in the Solomon Islands. All are either bats—both flying foxes and small insect-eating types—or large, forest-dwelling rats. Presently, 3 of these species are already extinct, 17 are classified as "vulnerable," and 5 species are endangered—representing about half of the mammal species. As human populations expand and current trends of "over-hunting" and forest destruction continue, it is feared that species unique to the Solomon Islands would be lost forever to the world. Animals that have been part of the natural heritage for more than 30,000 years could be destroyed in one generation, warn researchers **Dr. Tim Flannery** and **Dr. Harry Parnaby**.

Among the most threatened species are the giant rats of Guadalcanal and Choiseul which nest in large trees now felled by loggers. Lowland forests are also threatened with destruction. Monkey-faced bats need primary forest to feed in and old hollow trees or large figs for roosting, as do the planet's greatest diversity of flying foxes, which are important plant pollinators and seed dispersers.

Ecological imbalance would be the result of the extinction of small insect-eating bats because an increase in mosquitoes, for example, might cause increases in malaria, encephalitis, and elephantiasis. Insects might also destroy people's food crops. The loss of flying foxes would keep plants from having their flowers pollinated and thus bearing fruit that is now also carried through the forest to create more trees. Owls and eagles that eat giant rats might disappear when their food source vanishes.

It is clear to Drs. Flannery and Parnaby that undisturbed large tracts of forest and caves must be preserved on the island to prevent extinction of mammals and to be protected from logging, mining, gardening, and overhunting. They also recommend bans on the commercial export of flying foxes; tighter controls on logging companies to leave old, hollow trees and younger trees nearby; hunting controls of flying foxes and large rats in some areas; and more research and survey work in the future.

"We feel optimistic," the researchers conclude. "Solomon Islanders generally know their environment well and have a great pride and respect for it. The battle to save the natural heritage will not be lost easily."

Dr. Flannery writes and co-authors books and articles on Australia kangaroos, wilderness heritage, and vanishing mammals; and New Guinea mammals. He says that he is "too busy running conservation programs in the Bush" to attend most meetings. He belongs to the Species Survival Commission (IUCN/SSC) Australasian Marsupial and Monotreme Group.

■ Resources

This project needs educational outreach support.

CSIRO Division of Wildlife and Ecology
Institute of Natural Resources and
Environment (INRE)
Tropical Ecosystems Research Centre (TERC)
PMB 44, Winnellie
Darwin NT 0821 Australia

Contact: **R.W. (Dick) Braithwaite, Ph.D.**, TERC
          Program Leader
   Phone: (61) 89 22 1735 • *Fax:* (61) 89 47 0052

Activities: Research • *Issues:* Biodiversity/Species Preservation

● Organization

The **Commonwealth Scientific and Industrial Research Organization (CSIRO)** is Australia's principal organization of its kind and one of the world's largest such institutions with a staff of 7,500—including 2,500 scientists in more than 100 laboratories and field stations throughout the country. Australia itself is the only developed country with tropics within its borders.

The **CSIRO Division of Wildlife and Ecology** is one of four divisions within the **Tropical Ecosystems Research Centre (TERC)**. TERC is centrally located within the seasonally dry Australian-Asian tropics of the Top End. Headquarters are in Darwin and research is in the field, including three CSIRO operated sites at Kapalga, Katherine, and Manbulloo. The area is especially suited to learning about, designing, and managing biotic resources of the tropics. Research projects are generally collaborative and involve integrated problem solving in identifying biological and ecological principles upon which management and conservation should be based. Projects are grouped into three areas: ecology and management of wet-dry tropical ecosystems; restoration of damaged ecosystems in the wet-dry tropics; and early assessment of weeds and pests in northern Australia.

▲ Projects and People

Understanding **tropical ecosystems** is the focus of research in the Wildlife and Ecology Division. TERC reports that its research areas include "the use of fire as a management tool; the interaction of drought and fire on plants, animals and soil; the relationship between floodplains and woodlands; the management of pests in natural ecosystems; and the restoration of damaged ecosystems."

Dr. Pat A. Werner is the project leader assessing the **biotic effects of fire**, including its interaction with water and vegetation on complex landscapes that can affect the distribution and population of animals. **Dr. Garry D. Cook** is doing a five-year study on the long-term effects of various **fire regimes on soil properties**, including "nutrient uptake and plant growth in the savanna woodlands." **Dr. R.W. Braithwaite** is concluding a study on the distribution and **abundance of small vertebrate animals as a function of seasonal drought**. **Dr. Alan Anderson** is examining **interactions among diverse insects, plants, and vertebrate animals** as a result of fire. As TERC senior research scientist, Dr. Anderson is particularly interested in ants as bio-indicators and in their community ecology, plant population dynamics, insect ecology, and fire ecology. **Dr. Laurie Corbett**, TERC officer-in-charge, is researching **dingoes** in three climatic habitats as part of an overall study of **vertebrate predators** and how they interact with their prey. He also belongs to the Species Survival Commission (IUCN/SSC) Canid Group.

Dr. Werner is the leader of projects concerned with **restoring damaged ecosystems** of the wet-dry tropics. In one project, she is assessing the damage and recovery of landscapes from **introduced water buffalo**. Early results show that buffalo removal increases certain vegetation, which could enhance breeding geese population and gosling survival; that "rodent populations persist longer into the dry season"; and that some small vertebrate populations become greater. Dingoes, owls, kits, and sea eagles are among the 20 predator species feeding on magpie geese.

Again, Dr. Werner leads the research efforts to assess and control weeds and pests in Northern Australia. **Ian Cowie** is identifying known and alien **weed threats** to conservation, tourism, and other land uses in the South Alligator River region.

A conservation project concerning a **wetlands blight** is also underway in the Entomology Division; biological control of the weed *Mimosa pigra* with natural enemies is being tested and the ecology of the weed, including seed longevity, is being researched.

Dr. Richard J. Williams is studying **alpine ecosystems**, important for recreation and nature conservation and also for summer grazing of domestic cattle in Victoria. He is focusing on seed biology in alpine plants as well as in herbs and trees of sub-Antarctic and semiarid regions.

Dr. Braithwaite established the Kakadu National Park and is an honorary member of the Kakadu Board of Management.

■ Resources

TERC library has access to the full CSIRO library network. SIROMATH consultants and CSIRONET computing facilities are available along with a Science Education Centre.

Great Barrier Reef Marine Park Authority (GBRMPA)
GPO Box 791
Canberra ACT 2601 Australia

Contact: **Graeme Kelleher**, Chairman
Phone: (61) 6 247 0211 • *Fax:* (61) 6 257 5761

Activities: Education; Environmental Impact Assessment; Law Enforcement; Planning and Management • *Issues:* Biodiversity/Species Preservation; Global Warming; Managing Marine Protected Areas; Sustainable Development; Water Quality • *Meetings:* UNCED; World Parks Congress

● Organization

The Australian government created the **Great Barrier Reef Marine Park Authority (GBRMPA)** in 1975 so that the people who use and rely on the Great Barrier Reef can do so with care, understanding, enjoyment, and self-management. Although regulations are minimal, the ecological characteristics are in a "multiple-use protected area," and certain perilous activities are forbidden—such as oil exploring, commercial mining, littering, scuba spearfishing, and the taking of large fish of certain species. Some economic development is permitted; sustainable use is the guiding "ethic." A universal treasure, the Great Barrier Reef (along Australia's northeast corner) is on the World Heritage List under the UNESCO Convention.

Broad areas of GBRMPA functions are overall planning, managing the assessment of environmental impact on proposed development, research and monitoring, and education.

Oceanographer **Dr. Donald Kinsey** is executive officer; **Kathleen Shurcliff** is assistant executive officer (AEO) for the environmental impact section; **Simon Woodley** is AEO for research and development; **Dr. Wendy Craik** is AEO for planning and management; and **Raymond Neale** is AEO for education and information.

Senior staff members also consult with international agencies on the conservation and management of marine environments and resources. With UNESCO, they advised the Republic of Maldives on the suitability of Maldivian atolls as biosphere reserves. They aided the Indonesian government and ASEAN/Australian programs in managing marine, coastal, and terrestrial protected areas.

▲ Projects and People

Deciding **zoning** matters is a main means by which the public self-regulates reef use. Here is where conflicts are debated and resolved between different user groups; enlightened scientific knowledge is helping to balance favorably the struggle between ecological protection and human activity. Zoning plans are reevaluated about every five years.

Higher levels of **nitrogen and phosphorus in the Marine Park's waters**—mostly originating from mainland farms—are damaging to some of the coral communities nearby the coast. Government environmental programs are underway to minimize these effects as well as soil erosion and land degradation on the mainland.

Developers are now paying for **monitoring programs** to be sure that environmental effects are not greater than those predicted during the assessment process. Developers must undertake remedial action if environmental effects become excessively hazardous.

Minimizing risks of oil spills is a huge concern due to the amount of traffic within the shipping channels and the damage that

could be caused. The GBRMPA provides the scientific advisor to spill cleanups. A simple computer program with maps shows the location of environmentally sensitive areas and predicts oil spill movement relative to wind and tide conditions; it will also predict the movement of drifting vessels in relation to rescue operations.

The GBRMPA is faced with assessing **environmental impacts** and managing marinas, reef top walkways, pearl farming ventures, clearance of wreckage debris, and tourist facilities. It investigates the infestations of crown of thorns starfish that eat away at live corals, leaving the skeletons behind; detects reef changes caused by cyclones, scales, and other natural phenomena; monitors **marine pollution**; and studies water quality.

Other priority research projects include studying the **effects of trawling and line fishing** on the Marine Park environment, gaining an understanding of **water movements in the Great Barrier Reef lagoon** and the connections with and between reefs, developing the **"Torres Strait Baseline"** to analyze the effects of mining on this marine environment, gauging the ecosystem's natural variations to "differentiate between the effects of human activity and natural variability," and evaluating the impact of social uses.

Graeme Kelleher is marine vice chairman of the Commission on National Parks and Protected Areas (CNPPA) of IUCN, the World Conservation Union, where he is helping to implement a global representative system of Marine Protected Areas (MPAs). Regional working groups are looking for gaps where MPAs in their biogeographic zones ought to be and proposing possible sites to establish them. *Guidelines for Establishing Marine Protected Areas* by Kelleher and Richard Kenchington is available through the GBRMPA. A Member of the Order of Australia, Kelleher also belongs to the Species Survival Commission (IUCN/SSC) Ethnozoology Group.

■ Resources

Developed as part of the Great Barrier Reef Wonderland Bicentennial Commemorative project conceived by Graeme Kelleher, the **Aquarium** opened in 1987 with biological displays and walk-through viewing of a coral reef system—providing educational experiences for hundreds of thousands of visitors yearly.

The GBRMPA publishes a list of research projects in progress. This information is also available in the database **Australian Marine Research in Progress (AMRIP)** on CSIRONET. Publications, staff papers, posters, and audiovisual productions are listed in its Annual Report. Also available from the Aquarium are *Project Reef-Ed*, a resource book describing educational activities; zone maps; and a mail-order catalog. Public seminars are held for recreational fishermen.

Resources needed are remote sensing devices for shallow-water resource assessment.

Office of the Supervising Scientist (OSS)
Alligator Rivers Region (ARR) Research Institute
Private Bag No. 2
Post Office
Jabiru NT 0886 Australia

Contact: **R.M. Fry,** Chief Executive Officer
Phone: (61) 89 79 9711 • *Fax:* (61) 89 79 2076

Activities: Research • *Issues:* Biodiversity/Species Preservation; Health; Water Quality

● Organization
The Ranger Uranium Environmental Inquiry regarding the environmental aspects of mining newly discovered uranium mineralization in the Alligator Rivers Region (ARR) of the Northern Territory led to the creation of the **Office of the Supervising Scientist (OSS)** in 1978. OSS is an Australian government environment protection and research organization. Its purpose is to assure that regulatory and technical measures are sound and adequate to protect this region from the effects of mining operations.

"Alligator Rivers Region is of outstanding interest for its unusual combination of largely uninhabited wilderness areas with attractive wild scenery, abundant Fauna and Flora, . . . and a large concentration of Aboriginal rock art of world significance," reports Mr. **R.M. Fry,** supervising scientist. "The region is rich in natural resources, having a variety of terrestrial and aquatic ecosystems including sandstone heathlands, open woodland, flood plains, seasonal water courses and permanent billabongs, as well as significant mineral reserves including uranium, gold, and platinum metals." It appears in the Convention on Wetlands of International Importance and the World Heritage List under the UNESCO Convention.

▲ Projects and People
The ARR Research Institute is the research arm of the OSS ensuring that accurate information is available to protect the environment and evaluate the level of protection achieved. Scientists weigh whether proposed actions—such as pre-release biological testing, contaminant dispersion studies, and studies of bioaccumulation of radionuclides in components of Aboriginal diet—will be detrimental to the environment. Researchers also determine whether certain environmental phenomena—such as fish kills or high concentrations of radionuclides in certain animals—are a result of natural causes or related to mining. Other possible long- or short-term consequences of mining operations are predicted in order to prevent harm to the region.

Senior scientific manager is **Dr. R.M. Baker,** also a member of the Species Survival Commission (IUCN/SSC) Captive Breeding Group.

 # State Government
New South Wales

Taronga Zoo
Western Plains Zoo
Zoological Parks Board of New South Wales
P.O. Box 20
Mosman NSW 2088 Australia

Contact: **J.R. (Jack) Giles, Ph.D.,** Director of
Scientific Policy and Research
Phone: (61) 2 969 2777 • *Fax:* (61) 2 969 7515

Activities: Education; Research • *Issues:* Biodiversity/Species Preservation; Health of Animals; Sustainable Development; Water Quality • *Meetings:* Directors of International Black Rhino Foundation; International Zoo Directors Conference; IUCN General Assemblies

● Organization
Taronga Zoo, started in 1916, and its sister zoo, **Western Plains Zoo,** which opened in 1977 with an open-range facility for grazing animals and space to breed endangered species, maintain close relationships with animal groups throughout Australia and overseas, including the World Wide Fund for Nature.

▲ Projects and People
The **Conservation and Research Centre** at Taronga is the first of any similar facility in Australasia. Current research programs include **platypus breeding and conservation** with the help of LANDSAT imagery and field surveys; **bat rehabilitation** whereby Gould's long-eared bats are monitored by radio tracking; and **little penguin population studies,** which monitor the health and rehabilitation of Sydney's populations in relation to environmental pollution.

Successful, ongoing **breeding programs** are with Leadbeater's possum, Pilliga mouse, wedge-tailed eagle, victorian crowned pigeon, glossy ibis, nicobar pigeon, epaulette shark, and primates including chimpanzee, giraffe, red panda, Indian antelope, and Przewalski's horse. Collaborative captive breeding with the Conservation Commission of the Northern Territory is continuing for the endangered Rufous hare wallaby and greater bilby.

Launched during "Shark Week" and a "Shark's Down Under Conference" in 1991, a **shark conservation program** is being implemented globally. Experts say that the "commercial fishing catch of sharks has more than doubled during the past five years" —an unsustainable level that can lead to the destruction of young sharks in Australia, as has happened off the Florida Keys, USA. "Sharks are often netted and their fins cut off to supply the lucrative Oriental shark fin markets. In many cases the sharks are returned to the ocean still alive."

Education Centres at both zoos help children develop positive attitudes toward animals, conservation, and concern for the environment, as does the **Zoomobile,** which visits schools and hospitals in more remote areas.

The Zoological Parks Board cooperates with the **IBM Conservation Award** program that honors Australia's outstanding environmentalists. Dr. Giles belongs to the Species Survival Commission (IUCN/SSC) Australasian Marsupial and Monotreme Group.

■ Resources
Zoos need radio tracking for short- and long-range monitoring of land and marine vertebrates.

Proceedings of "Sharks Down Under" international conference are available for a fee.

Northern Territory

Conservation Commission of the Northern Territory
P.O. Box 496
Palmerston NT 0831 Australia

Contact: **William John Freeland, Ph.D.,** Head, Wildlife Division
Phone: (61) 89 89 4400 • *Fax:* (61) 89 89 6526

Activities: Development; Legislative; Research • *Issues:* Biodiversity/Species Preservation; Global Warming; Sustainable Development; Waste Management/Recycling; Water Quality

● Organization
Established in 1980, the **Conservation Commission**'s broad charter is to ensure that the Territory's "economic development is compatible with the use and conservation of its natural resources" now and in the future. With a staff of 50, the Commission protects the environment, establishes and manages parks, trains conservationists, educates the public, assists in soil conservation, helps manage development that impacts on the environment, and aids in environmental impact studies.

▲ Projects and People
The Conservation Commission confronts the world's most pressing environmental issues—the greenhouse effect, ozone depletion, and waste management. It's **Environmental Protection Unit (EPU)** helped develop a national greenhouse strategy and is involved in a regional study on the impact of climate change. Within the Greenhouse Taskforce of the Australian and New Zealand Environment Council (ANZEC), options are being examined to greatly reduce emissions of greenhouse gases. The EPU introduced ozone legislation—a model for Australia—that controls the manufacture, importation, use, and disposal of ozone-depleting substances. Efforts are underway to improve recycling.

A Northern Territory conservation strategy is making the parks more accessible while preserving a rich heritage. The Nitmiluk (Katherine Gorge) National Park was recently returned to the local **Jawoyn Aboriginal people.** And the new Strehlow Research Centre houses a collection of artifacts relating to the Central Australian Aborigines. Progress with preserving the **Nile crocodile** species in the Top End ranching project is recognized globally. A local crocodile farmer, Mr. John Bache, became world president of the Crocodile Farmers Association, and a Territory Crocodile Industry Advisory Committee was set up. Measures to **prevent bushfires** were

undertaken such as firebreaks, aerial burning, and public awareness programs.

Land care issues are in the forefront. Government and private sectors cooperated on the Mary River Wetlands to stem saltwater intrusion. Computerized land resource surveys are underway using geographic information systems (GIS). Rainforest Fauna and Flora surveys are being analyzed. Vegetation management involves protection from browsing feral animals. Erosion stemming and riverbanks restoring occur in the parks. Wildlife programs are progressing on saving the rare and endangered Gouldian finch, mala, and bilby while controlling the less desirable cane toad, rabbit, and feral pests.

For example, the bilby (*Macrotis lagotis*) on the decline since European settlement, is being reintroduced into the Watarrka National Park. Growth of seed-producing plants is being encouraged, along with studies of the effects of ants, birds, and rodents on native seed production so that supplementary seeds for the bilby can be supplied. Meanwhile, Aboriginal people are participating in returning the captive-bred Mala species to their natural habitat. They are also expressing interest is agile wallaby ranching.

Dr. William Freeland's most recent grants are on ghost bat behavior, interactions between freshwater crocodile and cane toad, and modeling of feral water buffalo and bovine tuberculosis. He belongs to the Species Survival Commission (IUCN/SSC) Sirenia Group.

■ Resources
The Conservation Commission publishes a comprehensive annual report that describes these and other projects more fully.

Queensland

Department of Environment and Heritage
Queensland National Parks and Wildlife Service
P.O. Box 155
160 Ann Street, 13th Floor
Brisbane QLD 4000 Australia

Contact: **Mr. N.M. Dawson,** Director
Phone: (61) 7 227 7801 • *Fax:* (61) 7 227 7676

Activities: Development; Education; Law Enforcement; Political/Legislative; Research • *Issues:* Biodiversity/Species Preservation

● Organization
Department of Environment and Heritage officials say that even with Queenland's growing population, the area is not experiencing the "extremes of environmental degradation" occurring elsewhere in the world. Nonetheless, government officials are debating how to share responsibilities for the environment to bring about better cooperation between federal, state, and local levels.

▲ Projects and People
Priority issues are nature conservation, coastal protection, rainforest management, heritage, and recycling. **Maximizing biodiversity** is

the main reason for wanting to double the national park estate. The arid area of the Channel Country—with its endangered animal species—is being studied. Elsewhere, the koala, cassowary, bilby, crocodile, green turtle wading bird habitat, migratory bird habitat, and bitou bush are subjects of conservation research.

The wetlands are undergoing a three-year survey on the western Cape York Peninsula. **Maritime conservation** is essential in Queensland, which contains most of Australia's marine and estuarine areas, including the Great Barrier Reef. The formation of a **Coastal Protection Unit** is in keeping with community concern about changes along the coastlines. "Conservation of mangroves, wetlands, beaches and other coastline features is seen as equal significance to tourism, mining, and other uses as well as an essential buffer for possible sea level rises with the greenhouse effect," says the Department. The **Beach Protection Branch** investigates coastal processes. It helped correct erosion on the southern Gold Coast—one of the world's largest beaches. In another instance, a numerical modeling study of the Noosa River led the way to nourish Noosa Beach with sand dredged from inactive deposits within the river's estuary. Other projects **control wind erosion, revegetate coastal dunes**, and monitor beach nourishment with **wave recording stations** and the compilation of wave data banks.

As the north Queensland **rainforests** are in the World Heritage List (under the UNESCO Convention) of natural resource treasures, a federal/state ministerial council was recently set up with input from wet tropics management experts. Efforts are underway to protect **heritage** sites. Public programs stress proper **disposal of commercial and household wastes** and use of **recycled papers** and other products, as are done within the Department. The nationally known puppet "Agro" broadcasts environmental education messages.

 # NGO

Australian Council of National Trusts

14/71 Constitution Avenue, Campbell
Canberra ACT 2601 Australia

Contact: **Duncan Marshall,** General Secretary
Phone: (61) 6 247 6766 • *Fax:* (61) 6 249 1395

Activities: Education; Political/Legislative; Research • *Issues:* Biodiversity/Species Preservation; Conservation; Cultural Heritage; Deforestation; Sustainable Development • *Meetings:* ICOMOS General Assembly; International National Trusts Conference

● Organization

With more than 80,000 members, the **Australian Council of National Trusts** coordinates Australia's largest voluntary conservation organization, the National Trust. The movement for National Trusts began in 1945 and the Council was formed 20 years later. These are community organizations—independent, autonomous bodies with open membership. The Board is a federal body made up of 17 members representing each state and territory trust; it coordinates policies, programs, and educational projects at these levels and expresses opinions to national and international groups.

With some 442 employees and about 6,500 volunteers, the National Trusts own more than 343 properties. Their purpose is "to acquire, conserve, and present for public benefit lands and buildings of aesthetic, historic, scientific, social, or other special values." A force for conservation among government agencies and other groups, they identify the parts of Australia's natural, Aboriginal, and historic environment which merit preserving.

▲ Projects and People

Environmental emphasis is at the heart of **surveys of landscapes, townscapes, historic buildings, and industrial sites**; with its work on the National Conservation Strategy and in such places as Antarctica, Kakadu National Park, NT; Lemonthyne and Southern Forests, TAS; Shelburne Bay, QLD; and Old Parliament House, ACT.

Lobbying the federal government and promoting public awareness through "**Save the Bush**" projects, **Heritage Week, "Heritage Australia,"** and **conferences on conservation** in the Asia/Pacific Basin are other activities.

■ Resources

Sponsorships are needed for restoration and special projects. Available publication for fee is *Discovering Australia's World Heritage: Lord Howe Island.*

Clean Air Society of Australia and New Zealand (CASANZ) ⊕

P.O. Box 141
Eastwood NSW 2122 Australia

Contact: **Kenneth M. Sullivan, Ph.D.,** President
Phone: (61) 2 858 4663 • *Fax:* (61) 2 858 3854

Activities: Development; Education • *Issues:* Air Quality/Emission Control; Energy; Global Warming; Waste Management/Recycling • *Meetings:* National, International Clean Air Conferences

● Organization

Since 1966, members—now at 600—actively seek solutions to air pollution problems to bring about a better environment here and globally. **CASANZ** is a leader in identifying polluters from "stationary sources" in the early days such as industrial stacks' emitting of smoke, particulates, and chemical fumes to pinpointing more potent mobile sources that are reducing the ozone layer in the stratosphere, increasing radiative gases in the troposphere, and altering the world's climate.

Participants hold conferences and symposia with scientists and researchers to advance the "knowledge and practical experience of air pollution, its abatement, control, measurement, and dispersal." Scholarships and funds promote education in air pollution control. Branches are located in Canberra City, Eastwood, Brisbane, Adelaide, South Yarra, and Perth, Australia, and in Wellington, New Zealand. CASANZ belongs to the International Union of Air Pollution Prevention Associations (IUAPPA).

Dr. Kenneth M. Sullivan was appointed member of the Order of Australia for service to science, particularly in the field of air pollution control, 1984. A former president of IUAPPA, he is a principal in K.M. Sullivan & Associates Pty. Ltd., specialists in coal technology, combustion, air pollution control.

■ Resources

Its quarterly journal, *Clean Air*, is the authoritative publication on air pollution control in these countries. Training course manuals and conference proceedings are also published.

Conservation Council of South Australia, Inc.

120 Wakefield Street
Adelaide SA 5000 Australia

Contact: **Marcus Richard de la Poer Beresford**
   Phone: (61) 8 223 5155
    Fax: (61) 8 232 4782, (61) 8 232 2490

Activities: Education; Political/Legislative; Research • *Issues:* Air Quality/Emission Control; Biodiversity/Species Preservation; Deforestation; Energy; Global Warming; Heritage Preservation Sustainable Development; Transportation; Waste Management/Recycling; Water Quality • *Meetings:* IUCN

● Organization

Representing 60 societies made up of some 50,000 people, the Conservation Council of South Australia is the "peak or umbrella body" for organizations concerned with environment and conservation issues in South Australia. At quarterly meetings of this "parliament" of the conservation movement, policies and individual issues are discussed.

There are also weekly meetings of the most active groups in South Australia—such as the Australian Conservation Foundation, Greenpeace, National Trust, Nature Conservation Society, and The Wilderness Society—at which "current concerns are discussed and courses of action determined."

▲ Projects and People

The **Conservation Centre** includes a bookshop specializing in the environment, library, information services, displays, and a point of contact or office space for most conservation groups. The staff of seven includes **Jo Juchiewicz**, librarian/information officer who answers public enquiries; **Soonpoh Tay**, threatened species network coordinator of campaigns on rare and endangered plants and animals; **Tanya Littlely**, greenhouse information officer; and **Marcus Beresford**, council executive officer.

■ Resources

The newsletter *Environment Conservation News* is published twice monthly. A list of *Environmental and Education Resources*, featuring wildlife and sealife posters, slide kit, issues outlines, special publications, and other services is available. The bookshop also sells T-shirts and other gift items.

Friends of the Earth (FOE) Fitzroy ⊕ ❧

222 Brunswick Street
Melbourne VIC 3065 Australia

Contact: **Linette Harriott,** Coordinator
   Phone: (61) 3 419 8700 • *Fax:* (61) 3 416 2081

Activities: Education; Research • *Issues:* Anti-Uranium; Energy; Global Warming; Health; Stop Consuming; Sustainable Development; Transportation; Waste Management/Recycling

● Organization

With 2,800 members, **FOE Fitzroy** is an "autonomous community-based activist organization" similar to a collective where decisions are made through consensus. It says the way it operates is as important as the issues covered. The group lobbies politicians, organizes public demonstrations and marches, produces educational materials, promotes public debate, and brings together community groups, unions, and councils. It runs a bookshop and gallery and provides "alternatives to mainstream shopping" with its food cooperative.

▲ Projects and People

Bookstore items reflect FOE's most pressing campaigns: the **nuclear cycle, environment, food issues, recycling, pesticides and hazardous chemicals, peace education, nonviolence, alternative technology and energy issues, and organic gardening**. Literature explains, for example, "why Australians are exposed to chemicals banned in the USA and Europe," how teachers can incorporate "peace education" in the classroom, how recycling creates jobs, and what the impacts of the nuclear fuel cycle from uranium mining to waste disposal are. FOE Fitzroy speaks out for renewable or "soft" energy sources such as wind, sun, biomass, and water—in place of "hard" sources such as coal, oil, natural gas, and nuclear energy.

■ Resources

Available are speakers on "consensus decision-making" and environmental topics, and the FOE Bookshop's *Wholesale Catalogue* of publications, posters, stickers, and T-shirts.

National Parks Association (NPA) of New South Wales, Inc.

P.O. Box A96
Sydney South NSW 2000 Australia

Contact: **Rod Bennison,** Executive Officer
   Phone: (61) 62 264 7994 • *Fax:* (61) 62 264 7160

Activities: Education; Political/Legislative; Research • *Issues:* Biodiversity/Species Preservation; Bushwalking; Deforestation; Global Warming; Water Quality • *Meetings:* Australian Committee of IUCN; CNNPA; IUCN; World Parks Congress

● Organization

The **National Parks Association (NPA) of NSW** began with a campaign for a new government body to manage natural areas that resulted in the National Parks and Wildlife Act of 1967 and the establishment of the NSW National Parks and Wildlife Service. It is that government agency's watchdog, organizing lobbying efforts and public demonstrations on behalf of a strong, effective parks and wildlife conservation system.

Formed in 1957, its 8,000 members claim credit for the existence of national parks, such as Barrington Tops, Mount Kaputar, Morton, Oxley Wild Rivers, and Myall Lakes. NPA continues to encourage growth and good management of parks and wildlife estate. Environmental education programs create local awareness of conservation issues.

▲ Projects and People

Recent activities keep watch on more than 100 **estuaries** in NSW and their ecosystems and wetlands, which are vulnerable to pollution, human disturbances, poor planning, and climate change. NPA is pressuring the state government to commit resources to develop environmentally sensitive regional management plans.

NPA alerts its members about **Landcare**—a recent term that describes actions of community farmer groups to tackle resource degradation issues such as declines in crops, trees, water quality, soil structure, and native plants and animals; and increases in soil salinity. A concept is supported of "Total Catchment Management," which coordinates the sustainable use of natural land, water, vegetation, and other natural resources for balanced conservation. Trees planted as windbreaks, for example, may be nesting sites for insect-eating birds that may help prevent soil salinity and stop tree decline, and in turn help improve crop and pasture production. Farmers play a role in protecting koalas and rainforests. Respect for land and water, as the Aboriginal people practiced for some 40,000 years, is coming full circle.

The group grapples with government decisionmakers on coastal policy and the degradation of marine areas. It recommends the "empowerment of communities" so that local citizens and groups can influence such policy. The NPA has fought the government to stop road reopenings in Gap Beach and Myall Lakes in the North Coast national parks, and also to keep auto-free Shelley Headland in the Yuraygir National Park for the sake of the wetlands, heath, and wildlife. Other recent controversial issues involved commercial development at Kosciusko National Park, an International Biosphere Reserve and a proposed World Heritage area; and mining sand and gravel from nontidal rivers—both opposed by the NPA.

■ Resources

National Parks Journal is available through annual subscription. NPA also publishes *Bushwalks in the Sydney Region* and a *Field Guide to National Parks of Northern NSW.*

Pelican Lagoon Research Centre
P.O. Penneshaw
Kangaroo Island SA 5222 Australia

Contact: **Peggy Diane Rismiller, Ph.D.,**
Environmental Physiologist
Phone: (61) 848 33174 • *Fax:* (61) 848 31294

Activities: Education; Environmental Awareness; Research
Issues: Biodiversity/Species Preservation; Ecology

● Organization

Since 1980, **Pelican Lagoon Research Centre** is a facility for marine botany, marine biology, geology, geography, archaeology, terrestrial botany, and wildlife documentation. Scientists, researchers, and educators as well as television media, journalists, wildlife photographers, and artists use the Centre's resources in a setting of unusual marine, estuary, and terrestrial ecosystems.

▲ Projects and People

Dr. Peggy Rismiller has five major research projects here: **Echidna Ecology, Ecology of Island Reptiles, Ecology of the Little Penguin, Island Bird Populations,** and **Feeding Habits and Behavior of the Tammar Wallaby.**

The short-beaked **echidna** (*Tachyglossus aculeatus muliaculeatus*) is a monotreme, one of the Earth's oldest surviving mammals, with heretofore many unanswered questions about its population, reproduction, and other behavioral characteristics. Popularly known as the "spiny anteater," the echidna is undergoing its most intense scrutiny of its ecology and life history in Dr. Rismiller's long-term field-oriented study. "Since echidna population density has never been monitored in Australia, there is no evidence as to the impact of the changing environment on this ancient species," writes Dr. Rismiller. She is finding out what happens when habitats are destroyed or altered, food sources are eliminated, and non-native predators are introduced.

In this five-year or more study, echidnas are being observed in their natural environment on or adjacent to the Research Centre and in every phase of its life cycle, from egg to adult. "Some study echidnas have small radio transmitters attached to the spines on their back," says Dr. Rismiller. "Monitoring involves mapping of each individual's movements, vegetation and habitat, and types of food sources." How they mate and rear their offspring and how they choose their living and nursery burrows will also give clues into how they cope and adapt to rapid changes imposed by man. Understanding their biology and 150-million-year-old preservation may give insight into helping other species survive environmental changes, she writes.

Earthwatch and community volunteers are color coding and radio tracking individual echidnas. Biologist and **photographer Michael W. McKelvey,** who established the Research Centre, is providing photographic documentation.

Dr. Rismiller is working with **Dr. Terry Schwaner** (formerly herpetology curator, South Australian Museum and now at Alabama School of Mathematics and Science) to examine the reproduction, growth rates, and home ranges of the **black tiger snake** (*Notechnis ater niger*) and the **goanna** (*Varanus gouldii rosenbergi*).

"We are planning on measuring daily body temperature rhythms of these reptiles and examining how environmental factors affect seasonal rhythms and adaptations," says Dr. Rismiller. Island populations differ from mainland relatives because they appear to have slower growth and reproduction rates and do not hibernate in the winter. Through road accidents, fear, and lack of human understanding, many are destroyed. Volunteers assist with the project.

Mike McKelvey heads the **Little Penguin** (*Eudyptula minor*) **Project** in conjunction with the Penguin Research Centre, Phillip Island, Victoria. Kangaroo Island lends itself to observing ecology and breeding biology—with its diverse penguin colony and feeding habitats and lack of introduced species. Trained volunteers aid in bird banding and monitoring.

The **Island Bird Populations Project** is a long-term study to determine seasonal movements, habitat use, population dynamics, and distribution. Licensed bird bander McKelvey bands and recaptures the resident populations throughout the island. Collaborator is **Dr. David Paton, University of Adelaide.**

The **Tamar Wallaby Project** is investigating feeding habits and behavior of these marsupials, with McKelvey's coordination. Adaptive behaviors are being researched in the wild to find out why their range and habitat are diminishing—which is environmentally significant. Results are differing markedly with studies conducted with captive populations in pastoral settings, indicates Dr. Rismiller. Observations also show that "this herbivore opportunistically utilizes animal protein in the diet."

Dr. Rismiller believes, "Before environmental matters can be effectively dealt with, we need a sound understanding of . . . whole organism biology of Australian ecosystems, current and past adaptations, and how they are coping with global changes. . . . [At the Centre] we feel that not only scientists should be involved in environmental issues—but that the public should be offered an opportunity to participate in research today. Anyone who is interested can help."

■ Resources

Pelican Lagoon Research Centre is noted for its public and academic lectures and slide presentations both nationally and internationally on current research and environmental topics; photo archive for scientific and commercial use; field research facilities; contract monitoring of studies for nonresident researchers with long-term projects; outdoor education and environmental workshops for small groups; public participation in ongoing field research projects arranged through a docent program; and professional consultants on native Australian Fauna and Flora, radio tracking and biotelemetry techniques in the bush, and general monitoring of wild animal populations.

Needs include biotelemetry monitoring devices for field studies on free-ranging animals; global outreach for outdoor education and greater environmental awareness; and volunteers to search, mark, and radio track echidnas and examine their juvenile development.

Project Jonah ⊕
9 Trelawney Street
Eastwood NSW 2153 Australia

Contact: **Brendon Gooneratne, Ph.D.,** President
Phone: (61) 2 874 4335 • *Fax:* (61) 2 876 8698

Activities: Education; Law Enforcement; Political/Legislative *Issues:* Biodiversity/Species Preservation; Marine Mammal Conservation • *Meetings:* CITES; IWC; Indian Ocean Marine Affairs Cooperation

● Organization

In 1976, Joy Lee founded **Project Jonah,** and shortly thereafter "a team of energetic, intelligent, and good-humoured individuals joined the organization, and waged the lively **Save the Whale** Campaign" to end Australia's participation in commercial whaling.

Young and old were drawn to the initial campaign. School children exhibited projects and art in local libraries to call community attention to the "plight of the gentle giants." Also, "office workers, academics, artists, singers, and musicians pooled their talents, circulated petitions, bombarded their local federal members with information, and made it quite clear that they wanted their nation to end its barbaric treatment of whales."

As a result, the government launched an inquiry, observed the International Whaling Commission's (IWC) deliberations in London, held discussions with experts in the USA and Europe, and eventually recommended that Australia end its participation in commercial whaling and help bring a halt to international whaling. The Whale Protection Act of 1980 was the outcome: it includes a $100,000 fine for causing the death of a whale or a dolphin, on conviction following indictment, as well as the confiscation of boats and other equipment used in the capture and storage of whales. Since then, Project Jonah incorporated and now has more than 3,000 members with an honorary committee of 7.

In continuing its struggles for marine mammal conservation, Project Jonah reports that humpback whales along the Australian coasts have fallen from 10,000 to 200; fin whales dropped from 500,000 to only 2,000, and blue whales have been reduced to a mere 1 percent of their original population. Because minke whales are thriving, they could be a justification for a resumption of commercial whaling.

"There is now still some doubt as to whether blue whales will ever recover," according to Project Jonah. "This would be one of the greatest tragedies enacted by man to destroy the largest creature that has ever lived on this earth—three to four times as large as the biggest dinosaur. The tongue of the blue whale is bigger than an elephant."

One crucial reason why whales are essential to the world's ecology is that they regulate oxygen production as part of the food chain of Antarctic plankton and krill. Moreover, there seems to be "striking similarities" between family life, behavior, brain structure, and physiology with those of the human race, according to Project Jonah literature.

Thus, despite legislation and growing awareness, whales and dolphins are still being slaughtered, says **Dr. Brendon Gooneratne.** So-called "scientific whalings" and the use of gill and purse-seine nettings threaten these species worldwide. "As long as these activities continue," he writes, "and the seas in which the marine mammals live continues to be polluted and deprived of the krill—now being commercially harvested—which constitutes the whales' major Antarctic food supply, the fight on behalf of the cetaceans and their habitat will have to continue."

▲ Projects and People

One major step in this battle is the **Project Jonah Resolution on a permanent whale sanctuary in the Indian Ocean,** which was ratified in 1991 by the standing committee of the Indian Ocean Marine Affairs Co-operation (IOMAC) nations—an organization of governments at ministerial level—and seeks IWC endorsement in 1992. The hope is that this Australian initiative will "set the precedent for future permanent whale sanctuaries and lead to an eventual permanent ban on whaling worldwide," reports Dr. Gooneratne. IWC established such a sanctuary in 1979 where research may be conducted among undisturbed whale populations and for their conservation.

Marine resources in the Indian Ocean are a "major component of the dietary protein for nearly one third of the world's population living in this area," according to Project Jonah. Yet about half of the fish harvested here are caught by non–Indian Ocean nations including France, Spain, Taiwan, Japan, Korea, the former USSR, and Panama at an annual commercial value of about $2 billion; and the need exists for IOMAC to safeguard such resources for Indian Ocean countries, including Australia.

The resolution calls on other international organizations, such as IWC, United Nations Environment Programme (UNEP), and Food and Agricultural Organization (FAO) to cooperate in preserving whale species in the permanent sanctuary.

In addition, Project Jonah **supports the transfer of the whaling issue to the United Nations** if whaling nations fail to honor the IWC Moratorium and more whales are slaughtered for meat, pet food, and fertilizer. It is encouraged by the Japanese and Taiwanese **curtailing of drift nets in the Pacific** that will save marine mammals as well as seabirds and turtles; and also notes the **growing popular attraction of Japanese whalewatch expeditions**. Project Jonah continues to call on members of the World Conservation Union (IUCN) to **ban all oil exploration and mineral activities in the Antarctic continent**.

Dr. Gooneratne is a founder of the Antarctic Society of Australia and a life member of the Wild Life and Nature Protection Society of Sri Lanka, his native land, where he has influenced ecological policy including conservation measures for the Ceylon elephant and the establishment of a Ministry for the Environment. Also a medical doctor, he is the editor and co-author of the book *Lymphography—Clinical and Experimental* and contributes to medical journals worldwide. He publishes in the fields of history and cartography as well.

Project Jonah vice president **Bob McMillan**, also a founder of the Antarctic Society of Australia, is recognized as a "true hero of Australia" by his colleagues. In his eighties, McMillan continues to campaign vigorously to save marine mammals.

■ Resources

The *Save the Whale and Dolphin Campaign* newsletter is available. Project Jonah seeks broader political support and educational outreach for marine mammal conservation and enforcement of laws.

Rainforest Information Centre (RIC) ⊕ ⩊

P.O. Box 368
Lismore NSW 2480 Australia

Contact: **John Revington,** Coordinator and Editor,
    World Rainforest Report
    Phone: (61) 66 218 505 • *Fax:* (61) 66 222 339

Activities: Development; Education; Political/Legislative; Research • *Issues:* Biodiversity/Species Preservation; Deforestation; Global Warming; Sustainable Development • *Meetings:* ITTO; WRM

● Organization

The entry for this publication arrived in a recycled envelope stamped with "Boycott—Don't buy rainforest timber." The **Rainforest Information Centre (RIC)**, which also coordinates the Australia-wide network of Rainforest Action Groups (RAGs), summarized its most recent activities as follows:

▲ Projects and People

Alternatives to Rainforest Timbers. A compiled list includes recommended uses and availability. An established "Good Wood" group includes organizations that promote the use of these alternatives.

Ban Rainforest Imports Campaign. RIC leads the fight to get Australia's federal government to ban the importation of rainforest timbers.

In campaigning for **Environmental Awareness** in Australia's overseas aid program, RIC has spoken out against certain World Bank projects.

Information Service answers rainforest inquiries from students, researchers, activists, and the media.

Reforestation. RIC planted hundreds of trees on the riverbank near its headquarters. RIC lobbies for forest preservation in federal elections. In **Ecuador**, orchards are being established and tree nurseries maintained. In **southern India**, RIC is helping reforest the "sacred mountain of Arunachala."

The **Papua New Guinea** campaign is studying its tropical timber industry. Also in New Guinea and the Solomon Islands, RIC representatives investigated the **Walkabout sawmill**, with which local people can mill timber for their own use—a possible alternative to large-scale, destructive logging. RIC is helping the Zia tribe with an ecoforestry management plan that includes training in sustainable forest management; use of the portable sawmill; and timber treatment, grading, and marketing. Enthusiastic about the program in which the "UK Ecological Trading Company is paying premium price for ecologically produced timber," says RIC spokesman **John Seed**, Zia are denying access to large logging companies and instead are planning for ecotourism and the training of nearby villagers. Similar ecoforestry management plans are underway in the Solomon Islands.

In **Sarawak**, the Penan tribe is being assisted to save their dwindling forests from loggers and themselves from cultural extinction.

Traditional Medicine Project. Through funding and publicity, RIC assists Indian chemist Rajeev Khedkar in revitalizing the Ayurvedic system of traditional medicine, which preserves endangered species—many from the rainforests.

■ Resources

RIC publishes the quarterly magazine *World Rainforest Report*, with readers in 80 countries. Many Third World groups receive free copies. Also available is the "Australia Rainforest Memorandum."

It is looking for "non-timber income sources in tropical forest areas and political and educational support for the campaign to end the tropical timber trade" as well as funding for its projects.

Royal Society for the Prevention of Cruelty to Animals (RSPCA)

201 Rookwood Road, Yagoona
Sydney NSW 2199 Australia

Contact: **David Richard Butcher,** Executive Director
    Phone: (61) 2 709-5433 • *Fax:* (61) 2 796-2258

Activities: Development; Education; Law Enforcement; Political/Legislative • *Issues:* Animal Conservation

● Organization

The **Royal Society for the Prevention of Cruelty to Animals** (RSPCA) is Australia's largest animal welfare organization with 8,000 members and a staff of 160, including 17 veterinarians. Founded in 1873, it is concerned with development of policy both for society and government; enforcement of Prevention of Cruelty to Animals Act; education; provision of veterinary services; and animal welfare generally including consideration of the environment and wild animals. With some 30 branches throughout New South Wales, funding is primarily from the private sector with about $50,000 annually from the government.

In the early 1990s, new policies regarding the environment and wild animals were under review. Generally, these regarded the grazing of introduced animals on arid or semiarid land, which the RSPCA opposes when it is shown to "cause deleterious effects on the environment . . . and the welfare of animals." The RSPCA also "demands the establishment and maintenance of national parks and major conservation zones of a size adequate to retain important habitat systems in the long term . . ." with adequate resources, research, and education.

▲ Projects and People

Specific environmental projects including the **training of veterinarians in management of marine-mammal strandings**, postmortem technique, and information collection. Up till now, the many general-practitioner veterinarians living on the coast have been underutilized in marine-mammal strandings, reports the RSPCA. So the group is beginning two-day training workshops to remedy this situation. Giving instruction are staff from the NSW National Parks and Wildlife Service and two oceanariums as well as veterinary surgeons who are experts in the area of marine mammal management, medicine, and disease.

To help solve a problem whereby thousands of short-billed corella parrots (*Cacatua pastinators*) are being attracted to grain feedlots for dairy animals, the RSPCA recently brought together experts in wildlife, dairy management, and dairy nutrition. The parrots have been known to damage electrical installations and trees with previous remedies being unsuccessful. Current recommendations relate to better feeding practices for the dairy cattle, which are also expected to improve their nutrition. The RSPCA also recommends "manipulating habitat" instead of destroying "pest" species, which "rarely achieves the desired result and in most cases further exacerbates the environmental imbalances," writes veterinary surgeon **David R. Butcher.**

Butcher has been a consultant to numerous Australian zoos and natural parks including a koala sanctuary in Victoria, Hare Krishna World, and feasibility studies for the Aboriginal Development Commission. He was retained by Tokyo's Tama Zoo, Japan, to organize the supply and captive management of glowworms.

■ Resources

A brochure and an *Annual Report* describe the RSPCA.

The Wilderness Society ❦
130 Davey Street
Hobart TAS 7600 Australia

Contact: **Karenne Jurd,** Director
  Phone: (61) 02 34 9799 • *Fax:* (61) 02 235112
E-Mail: Pegasus: twsnat

The Wilderness Society
1A James Lane
Sydney NSW 2000 Australia

Contact: **Andrew Donovan,** Director's Assistant
  Phone: (61) 2 267 7929 • *Fax:* (61) 2 264 2673
E-Mail: Pegasus: twssyd

Activities: Education; Political/Legislative; Research • *Issues:* Biodiversity/Species Preservation; Deforestation; Mining in High-Conservation Areas; Wilderness Protection • *Meetings:* IUCN; UNCED

● Organization

Since the 1970s, **The Wilderness Society** is a leader in the battle to protect Australia's dwindling natural resources "to save what is left." It says that 24 mammal species are now extinct (the WWF Australia reports 18 extinct mammal species) and that mining exploration in the wake of agricultural land clearings are tearing up remote areas, destroying plants, and causing soil erosion. Unregulated tourism also threatens the natural heritage; only 12 percent of South Australia is said to be "high-quality wilderness."

Activists say they must continually pressure state governments to commit to wilderness preservation. Its 16,000 members protest mining in national parks such as the Kakadu wetlands, Australia'a first World Heritage area, and stripping rainforests that contributes to an estimated 25 percent of the carbon dioxide causing the greenhouse effect and leads to plant and animal extinction. Media events and demonstrations generate public awareness and support.

As member **Trish Sykes** reports, "The message coming back to us is that people, young people in particular, value wilderness and want it protected."

▲ Projects and People

The Wilderness Society points to some of this decade's early struggles: fighting to stop imports of tropical timber for unsustainable sources; in **South Australia**—getting a Wilderness Protection Act introduced and gaining passage of the Pastoral Land Management Act affecting arid lands and the Soil Conservation and Landcare Act; in **Victoria**—pushing to get 85,000 hectares added to the Mallee National Parks system, which contain this state's largest wilderness area; in **Queensland**—supporting wilderness legislation and staging a peaceful logging blockade on Fraser Island that eventually won government support; in New South Wales—lobbying to save the southeast wilderness forests; in **Tasmania**—opposing state government policies to endorse the "timber industry's agenda of open slather logging and woodchipping," while celebrating the declaration of Douglas-Apsley National Park and the extension of the World Heritage area by 600,000 hectares; and in **western Australia**—holding dialogues with the local Aboriginal people for their input on the Conservation Plan for the "sacred wilderness" of

Kimberley, where there are "no documented extinctions of native species during the period of European settlement" within the ecosystems richly endowed with rainforests, savannah woodlands, mangroves, and unique plants and animals.

A **National Campaign for Wilderness**, in place since 1984, is largely responsible for expanding protective legislation. One success is that the federal government is funding the completion of a national Wilderness Inventory and studying its options for wilderness protection.

The Wilderness Society opposed "resource security" legislation in 1991 that would permit private timber companies to log in certain zones in the great, public escarpment forests of eastern Australia; instead, it supports the harvesting of privately owned plantations. Also, it helped Australia lead the world in its mining ban in Antarctica.

■ Resources

Education and a gift catalog are available. Members' newsletter is *Wilderness*.

Needs are computers, both hardware and software; "lobbying skills for wilderness legislation, and fundraising skills to raise money," requests **Director Karenne Jurd**, who adds "the Wilderness Society has great spirit—and a vision worth the struggle."

World Wide Fund for Nature—Australia (WWF) ⊕
Level 10, 8-12 Bridge Street
Sydney NSW 2000 Australia

Contact: **Don Henry,** Director
Phone: (61) 2 247 6300 • *Fax:* (61) 2 247 8778
E-Mail: Pegasus: wwfsyd

Activities: Education; Political/Legislative; Research • *Issues:* Air Quality/Emission Control; Biodiversity/Species Preservation; Deforestation; Energy; Global Warming; Health; Population Planning; Sustainable Development; Transportation; Waste Management/Recycling; Water Quality • *Meetings:* CITES; conservation of biological diversity meetings

● Organization

Globally, WWF is one of the largest independent conservation agencies. Founded in 1961, it espouses a universal mission to preserve "genetic, species, and ecosystem diversity;" to sustain renewable natural resources; to reduce pollution and wasteful exploitation and consumption of resources and energy; to reverse the rapid degradation of the planet; and to enable people to live in harmony with nature.

The WWF—Australia Conservation Programme is tailored to face these challenges: conserving native Fauna and Flora, conserving and rehabilitating ecosystems and ecological processes, sustainable use of natural resources, developing a conservation ethic at all society levels, and reducing wasteful consumption and pollution.

Although Australia is an amazingly biologically diverse country—with rainforests and woodland, arid deserts and alpine ecosystems, tropical islands and huge coral reef formations, and a large portion of Antarctica claimed as Australian territory—even with its large land mass and low population, it has suffered "severe environmental

degradation" during the past 200 years. Its 18 extinct mammal species represent about half of all those listed as extinct by the IUCN in 1988. Almost half the forests are gone. Salinity threatens the Murray-Darling river system and agricultural lands. Development is degrading coastal wetlands. These are major threats to biological biodiversity, as are the presence of plant diseases and introduced animals and plants, lack of expertise on managing fires, climate change arising from global air pollution, development of mineral/coal/oil and gas deposits in vulnerable areas, and overexploitation of timber, soils, and fish.

For the past 10 years, WWF has funded research to conserve and manage threatened species and habitats; some wildlife and other conservation programs are now thriving. Presently, WWF is building on these successes and extending its work in Australia, Antarctica, the southwest Pacific, and Indonesia.

Current objectives are the following:

• The **conservation of threatened species and habitats.** WWF is promoting the development of a national strategy and seeks new or improved laws in the Commonwealth and each state for this purpose. It is working on research objectives and support for specific projects that will prevent the decline and extinction of certain species. It is writing action plans and position papers for taxonomic groups addressing species conservation, involving the community and non-government groups in wildlife conservation, and seeking government funding for conservation programs.

• The **conservation of Australian habitats and ecosystems.** WWF is working toward establishing a representative system of **terrestrial and marine nature reserves** that would enhance the conservation of biological diversity. It is encouraging national and regional surveys of habitats and ecosystems—with areas for reservations identified. It seeks **land-use strategies** that protect habitats and ecosystems and promote related educational materials. Likewise, WWF is striving to see **wetland and coastal conservation strategies** and management plans implemented by all governments as well as increased research to conserve the wetlands' Fauna and Flora. Other goals are the **conservation of temperate grasslands, woodlands, and heaths** through land management policies and practices, research, and reservations at key sites, and the **encouragement of native vegetation on rural lands.**

• **International action and cooperation to conserve biological diversity.** Aid and developmental assistance programs, mostly in the South Pacific, as well as international treaties and conventions can be used to save biological diversity, conserve tropical rainforests, and practice sustainable utilization of natural resources. Australia can also contribute to conservation strategies for Antarctica and the southern oceans, and to policies to protect tropical forests in Southeast Asia and the South Pacific and to monitor its role in both the international tropical timber and wildlife trades. WWF International must receive adequate funding for their programs in the South Pacific.

• **Promoting environmentally sustainable use of natural resources.** WWF is helping to prepare and promote public policy that will prevent overexploitation, pollution, and wasteful consumption through sustainable use of natural resources and terrestrial and marine ecosystems. It seeks the "full integration of environmental considerations into economic and social decision making and planning."

Donald James Henry is a trustee of the Rainforest Ecology Trust and an honorary fellow of the Division of Australian Environmental Studies, Griffith University.

■ Resources

Henry welcomes "ideas of ecologically sustainable development that are working." He says, "more international networking is required between conservation and development NGOs."

A quarterly newsletter and *Wildlife Australia Magazine* are published; a publications catalog is available from GPO Box 528, Sydney 2001.

Private

Geo-Processors Pty. Ltd.
Consultants
P.O. Box 195
Holland Park
Brisbane QLD 4121 Australia

Contact: Aharon (Aro) V. Arakel, Ph.D., Director
Phone: (61) 7 343 4876 • *Fax:* (61) 7 849 7787

Activities: Education; Environmental Technology Development; Research • *Issues:* Global Warming; Industrial Minerals Fields; Sustainable Development; Waste Management/Recycling; Water Quality

● Organization

Geo-Processors Pty. Ltd. are consultants in environmental geoscience and technology, offering applied research and related services in cross-disciplinary areas such as minerals and petroleum exploration, sedimentology, geochemistry, site geological and geotechnical investigations, and technical feasibility assessments and testing services for new industrial mineral products. Specialized training programs and project coordination are also offered.

Among environmental services and technology development, baseline surveys relate to soils, landforms, land capability, and water. Other activities include environmental impact assessment and project coordination; environmental auditing and database management; land-use and rehabilitation planning; land degradation monitoring and management; rehabilitation of disturbed mining areas and saline tailings; salt interception in dryland farming, urban, and industrial areas; saline wastewater control technology; and coastal sedimentological and water quality studies for impact assessments.

▲ Projects and People

To discuss coastal zone management approaches, concepts, methodologies, and conflicts, Geo-Processors brings together people of diverse backgrounds and periodically holds workshops on coastal zone management. The proceedings are made available to the public.

To determine the impact of chemical runoff from sugarcane and banana farms on the Great Barrier Reef, a multinational Earthwatch volunteer team with co-investigators Dr. Aro Arakel and Dr. Theodore Loder III, University of New Hampshire, is assessing how hydrology, land use, and climate interact in six watersheds. Studies show that eroded soil loaded with pollutants from farming

and urban growth is being dumped in mangrove estuaries and swept into the ocean during storms. In recent evaluations, researchers are finding that nitrate and phosphate nutrients are mostly being dumped at the mouths of the Johnstone and Moreby river systems and may not be reaching the Great Barrier Reef to destroy coral farther out to sea or to cause abnormal algal growth. Even so, there has been a three-fold increase in the amount of eroded soil flowing from the South Johnstone River since the 1970s.

"Fertilizer from farming, untreated sewage, and logging-induced erosion can choke river estuaries and coastal waters with excess sediment and nutrients," according to Arakel. "These in turn act as unintentional nurseries for vast algal blooms that grow so fast they squeeze out the slower-growing coral polyps that build reefs. Moreover, unnaturally high concentrations of nutrients alone can be toxic to marine organisms," he writes.

An Earthwatch team stationed at each watershed set up river transects for monitoring tidal cycles, collected and analyzed water and sediment samples, mapped distribution of mangrove and seagrass communities, identified fauna and flora, and interviewed residents about land uses. Other principal investigators are Drs. D. McConchie and P. Saenger, University of New England, Dr. C. Pailles, Queensland University of Technology, and Dr. J. Piorewicz, University College of Central Queensland.

In previous studies, Dr. Arakel remarked on the difficulties in making reliable estimates of the "size, types, and quantities of eroded sediments and soils introduced to the Great Barrier Reef shelf. . . . The knowledge of fundamental controls on hydrodynamics and sedimentation pattern of the river-estuaries in the (North Queensland) region needs to be improved significantly."

Earthwatch volunteers also recently assisted Dr. Arakel in his systematic hydrological and sedimentological investigations begun about a decade ago regarding desertification and salinization of arid drainage basins in central Australia. Determining the climatic and hydrological evolution during the past 37,000 years will impact on land degradation, farming and logging practices, and land management today, suggests Dr. Arakel. He is urging government to understand natural processes—such as the interaction of climate, geology, and water—before solutions to complex environmental questions are asserted.

Dr. Arakel is continuing his 20-year-old pioneering efforts to turn an environmental menace into an asset. Science writer Julian Cribb in *The Weekend Australian* describes Dr. Arakel's proposal to convert an accumulation of salt that is "devouring 1.3 million hectares of Australia's farmlands and irrigation areas" and can "poison groundwater, creeks, or arable land" into "valuable minerals worth millions of dollars."

Dr. Arakel believes this salt can be processed and used to make "fertilizers, food additives, glass fibre products, building materials, industrial catalysts and other valuable substances. It could replace chemical imports worth tens of millions of dollars a year, cut farmers' and manufacturers' production costs, and reap substantial export income—from a process which is now a one-way ticket to environmental devastation." Through studies of salt lakes and brines and awareness of the four-century-old Chinese system of extracting industrial chemicals from similar areas, Dr. Arakel is now encouraging the government to pay serious attention to saline water treatment. He has already extracted minerals such as "glaubers and epsom salts, zeolites, potassium, sodium and magnesium salts and many kinds of high-value byproducts all with direct industrial applications," according to Cribb.

Geo-Processors recently introduced a new saline wastewater processing scheme, *Sal-Proc*, to treat such waters generated by agricultural, farming, and industrial activities in Australia's arid and semi-arid regions. Using a combination of natural multiple evaporation and cooling and chemical extraction techniques, the scheme first allows a significant reduction of wastewater volume and then produces economically valuable salt chemicals with industrial and fertilizer application, reports Dr. Arakel.

In other studies with **T. Hungjun** and **W.F. Ridley**, Queensland University of Technology, Dr. Arakel undertook detailed hydrological monitoring and a geochemical survey of a coastal plain near the Brisbane Airport to "assess the role of natural and anthropogenic processes on concentration and dispersion of potentially pollutant metals in the Brisbane coastal environment."

Dr. Arakel, who writes and addresses groups throughout the country on land care and arid zone water management, believes future global cooperation is enhanced through the exchange of ideas and databases between environmentalists "so planning of environmental monitoring and rehabilitation programs can proceed in most time- and cost-effective ways." The public, he believes, deserves "knowledge of qualified and unbiased environmental information so the process of rational decisionmaking can proceed through meaningful public consultation and consensus rather than through emotional outbursts."

■ Resources

Geo-Processors can provide Coastal Zone Management Workshop proceedings. The firm supports the flow of information on major environmental issues, assesses and publicizes the potential on-site and off-site impacts of land degradation, and offers new approaches and technologies for treatment of "wasted resources" in environmentally acceptable ways.

John McEachern
Consultant
c/o The Great Barrier Reef Marine Park
   Authority
P.O. Box 1379
Townsville QLD 4810 Australia

Contact: **Gail McEachern**, Executive Director
          M-ARK Project, Inc.
          P.O. Box 793
          Arkville, NY 12406 USA
Phone: 1-914-586-3500, 1-607-326-7385
Fax: 1-914-586-3044

Activities: Development; Research • *Issues:* Coastal Zone Management; Environmental Rehabilitation; Sustainable Development • *Meetings:* World Parks Congress

▲ Projects and People

John D. McEachern and his associates consulted with **The Great Barrier Reef Marine Park Authority** in 1991–1992 to evaluate the overall impact of the fishing and tourist industries on the resources of the Great Barrier Reef. Special focus was on the "economic benefits generated by these industries and the bio-physical capabil-

ity of the reef to sustain these industries," writes McEachern. His studies also considered "long-term trends in reef utilization, their socio-economic effects, and implications for effective planning and management."

An international consultant, McEachern has also worked with the World Conservation Union (IUCN) as a project manager of the **National Environment Strategy for Jordan** and on missions to **Bangladesh** to evaluate the **Natural Resources Information Centre Project** and the **Environmental Education, Training and Awareness Project**; and to assist the **Chunati Wildlife Sanctuary**, Chittagong, and the **Embankment Conservation and Enhancement Project for the Meghna-Dhonagoda Irrigation Area**. Earlier for IUCN, he surveyed resources of the Arabian Gulf Coast (in 1986) from Dawhat As Salwah to the Kuwait border for **Saudi Arabia's Meteorological and Environmental Protection Administration**. Thus, several years prior to the Persian Gulf War, he investigated resource uses and conflicts while outlining environmental management needs.

For the Canadian International Development Agency (CIDA), McEachern was a coastal resources specialist for an environmental mission to **Thailand**—namely the **Khung Krabane Bay Royal Development Project**, Chantha Buri Province. He also was a coastal zone specialist on a World Bank environmental and resource management mission to the **Philippines**, where he investigated the status of coastal resources including artisanal fisheries, coral reefs, mangroves, and other biomes near Luzon, the Vasayas, and Palawan. As a result, new sustainable resource management recommendations were made.

For industrial and governmental clients, McEachern has undertaken various other studies in Canada, the Caribbean, Southeast Asia, and West Africa.

■ Resources

McEachern can be contacted at his USA address for information about consultant reports and publications.

☜ *Universities*

Department of Prehistory and Anthropology ⊕
Australian National University (ANU)
GPO Box 4
Canberra ACT 2601 Australia

Contact: **Colin Peter Groves, Ph.D.**, University
          Teacher/Reader
Phone: (61) 6 249 4590 • *Fax:* (61) 6 249 2711

Activities: Education; Research • *Issues:* Biodiversity/Species Preservation

▲ Projects and People

Dr. Colin P. Groves, as Food and Agriculture Organization (FAO) consultant in Iran, reports on his environmental work assessing the number of mammal species and making recommendations for the conservation of protected areas.

"Environmental problems in Iran are accelerating, and the government is becoming more aware of them," he writes. "This is therefore a good time to press for new conservation initiatives; the enthusiasm among wardens and game guards is remarkable, and in some areas—particularly Shiraz Province due to the enterprise of the government biologist there—public awareness is increasing." Dr. Groves was expecting a positive response to his recommendations from the Iranian government.

His observations follow:

The **Asian cheetah** (*Acinonyx jubatus venaticus*) is Iran's rarest mammal, a good population of which lived in several regions before the Revolution. Today, in the Bahram-e-Gour Protected Area 250 km east of Shiraz, there are about 20 animals; and it is thought that about the same number may still exist scattered about Iran. The threats are to their habitat, not directly to the animals.

Some 200 to 250 **Persian wild asses** (*Equus homionus onager*) live in 2 protected areas—one of which is Bahram-e-Gour, where the staff's enthusiasm ensures vigilance. However, poaching still occurs, and rangers have been killed the past few years. Dr. Groves' recommendations are: increases in funding; the establishment of a public education facility (such as that in the Bamou National Park, Shiraz, where it successfully increases conservation awareness; restoring of a historic building in the reserve to act as a guard post; and translocation of wild asses to restock other areas.

The **goitered gazelle** (*Gazella subgutturosa*) and **Bennett's gazelle** (*G. bennetti*) appear numerous, but subject to poaching. "The range of Bennett's gazelle is now known, as a result of my visit, to be rather wider than previously thought. . . . I also uncovered evidence of . . . the **Arabian gazelle** (*Gazella gazella*) in a few areas in western Iran, such as the Mooteh Protected Area. In Mooteh, three different species of gazelles thus exist side by side, together with such poorly known species as the felids Pallas's cat (*Otocolobus manul*) and sand cat (*Felis margarita*); this would also be good country for wild ass reintroduction."

Dr. Groves belongs to several groups of the Species Survival Commission (IUCN/SSC), including Asian and African primates, equids, cats, wild cattle, antelopes, pigs and peccaries, and mustelids and viverrids.

Department of Prehistory and Anthropology ⊕
Australian National University (ANU)
GPO Box 4
Canberra ACT 2601 Australia

Contact: **Vern Weitzel,** Physical Anthropologist
  Phone: (61) 6 254 0166 • *Fax:* (61) 6 241 8304

Activities: Research • *Issues:* Biodiversity/Species Preservation; Population Planning; Sustainable Development • *Meetings:* SSC

● Organization

Located at the **Australian National University (ANU)**, Vern Weitzel works within four organizations: the University's Department of Prehistory and Anthropology; Australian Primate Society (APS); Australian Committee for Scientific Cooperation with Vietnam (ACSCV), part of the Australia-Vietnam Society; and the Australasian Society of Human Biology.

▲ Projects and People

With a background in primatology, physical anthropology, computing, science writing, photo-processing, and television producing,

Vern Weitzel has pursued his primate research at ANU and in Java and Sumatra, Indonesia; Singapore; and Vietnam. The evolution of cranial anatomy is one specialty. He belongs to the Species Survival Commission (IUCN/SSC) Primate Group.

Weitzel coordinates APS conservation efforts with environmental specialists **Graeme Crook**, CSIRO Division of Human Nutrition, South Australia, and **Rosemary Markham**, Perth Zoological Gardens, Western Australia. They are concentrating on helping the International Primatological Society with fundraising for its research projects on the **mountain gorilla in Rwanda; primate conservation in Thailand** in cooperation with Mahidol University, Bangkok; **Vietnam conservation** in cooperation with Adelaide Zoological Society; and the **Javan silvery gibbon** in central Java, where the dense human population is encroaching on the small patches of natural habitat.

Researchers report that the conservation community has neglected the reserves and mountains in central Java. The silvery gibbons are an endangered species, with less than 5,000 in the wild and about 60 animals in captivity. Along with the silvery gibbon, a study is proposed of these species inside the remaining reserves: Javan leaf monkey, Indonesian mink, Javan yellow-throated marten, Javan small-toothed palm civet, Javan hawk eagle, Javan laughing thrush, and Javan chestnut-bellied partridge.

The **ACSCV** links Vietnamese environmentalists with their counterparts in Australia. Cooperation so far has focused on migratory birds and mammals. "We recently helped organize a cooperative research and training program between Hanoi University and the University of Canberra," writes Weitzel. "Current areas of cooperation are **sea turtles, land degradation, education,** and graphic information systems (GIs)." He also reports support for hospitals, malaria research, and other medical cooperation.

According to ACSCV, Vietnam is also beginning to address its serious environmental problems of deforestation, soil erosion, and loss of forest cover by implementing the Tropical Forestry Action Plan formulated by United Nations groups, the World Bank, and the World Resources Institute.

The **Australasian Society of Human Biology,** located at the Center for Human Biology, University of Western Australia, is attempting to extend its contacts within the region as far as its limited budget will permit. In preparing anthropological research from Hanoi University, the group encountered problems resulting from "extremely limited contact with colleagues in Hanoi, our lack of relevant Vietnamese publications, and their lack of knowledge of western scientific literature," notes Weitzel.

■ Resources

For his work with varied groups, Weitzel needs these resources:
• Large-scale satellite maps of Indochina to determine forest cover. "Deforestation occurs at such a staggering rate, and ground information is so poor that we cannot be sure whether the land we intend to survey is actually forested."
• Educational materials to teach children in developing nations about environment and conservation. "We also need VCRs and educational tapes; academics sorely need computers and faxes."
• Journals and other promotional materials from relevant organizations and companies.
• Conventional wire-service access to relay newspaper stories internationally and to syndicate popular reports globally.

Ecology and Evolutionary Biology (EEB)
Monash University
Clayton VIC 3168 Australia

Contact: **John Nelson, Ph.D.**, Zoologist
   Phone: (61) 3 565 5659 • *Fax:* (61) 3 565 5613

Activities: Education; Research • *Issues:* Biodiversity/Species
Preservation; Ecology

▲ Projects and People

Dr. John Nelson, Monash University, member of the Species
Survival Commission (IUCN/SSC) Chiroptera Group, is researching the ecology of **dasyurid marsupials**—"feeding, reproduction,
radio tracking, water turnover using double-labeled water, metabolic rates." He is also studying the ecology of **bats** and **flying foxes**,
including "water turnover and oxygen consumption."

Roseworthy Agricultural College
University of Adelaide
Roseworthy Campus
Adelaide SA 5371 Australia

Contact: **Iain Thomas Grierson, Dip.Ed.**, Senior
   Lecturer
   Phone: (61) 85 622 783 • *Fax:* (61) 85 248 007

Activities: Education; Research • *Issues:* Deforestation; Sustainable Development; Waste Management/Recycling

▲ Projects and People

Soil water erosion is commonly associated with agriculture in tropical and semiarid areas, but it can also occur in urban development
and recreational areas.

Iain Grierson has developed a comprehensive computer-based
teaching aid about soil water erosion for teachers and students. The
"Universal Soil Loss Equation Unit" enables researchers to know
"how fast the soil is being eroded" so that conservationists can
compare the estimate of the rate of soil loss with what is acceptable.
He says that this powerful equation can help "analyze the various
relationships between land resources, land use, and land management" eliminating previously slow and difficult calculations.

The National Soil Conservation Funds provided a grant to develop the teaching unit, which includes a software program, comprehensive manual, and 35mm slide set to illustrate soil erosion
problems. Grierson is also involved in a five-year project with the
National Soil Conservation Programme to investigate "**biological
farming for sustainable agricultural production.**" In addition, he
has a grant from the Australian Wheat Research Council to measure
"**biological changes in the soil under various agricultural rotations.**"

Grierson also completed a recent **Earthwatch** study regarding
improved fruit quality that used hot air balloons as instrument
platforms for testing moisture stress and irrigation scheduling for
vineyards. He has done environmental work in Thailand and China.

■ Resources

The Universal Soil Loss Equation kit is available through Grierson.

Deer Research Unit (DRU)
Department of Animal Health
University of Sydney
Private Bag 3
Camden NSW 2570 Australia

Contact: **Anthony William English, Ph.D.**, Director/
   Senior Lecturer
   Phone: (61) 46 55 2300 • *Fax:* (61) 46 55 1212

Activities: Education; Research • *Issues:* Biodiversity/Species
Preservation; Health; Sustainable Development • *Meetings:*
Biology of Deer, Scotland; Wildlife Ranching Symposium,
South Africa

● Organization

Dr. Anthony William English describes the **Deer Research Unit**
(DRU) at Camden as "Australia's premier facility for studies on the
captive management of cervids." Started in 1979 on 2 hectares with
several rusa deer caught in the Royal National Park and the support
of the New South Wales Deer Breeding Association (NSWDBA),
the DRU has grown to a stock of some 244 deer on 23 hectares with
research, teaching, and extension facilities to benefit the Australian
deer farming industry.

Species include fallow, chital, red, and rusa deer. Research studies
have focused on reproduction, artificial breeding, nutrition, diseases, venison production, and antler harvesting. The Australian
Special Rural Research Council supported some projects; CSIRO,
New South Wales (NSW) Department of Agriculture and Fisheries,
and several drug companies also collaborate in certain work.

Final-year veterinary science students work at the DRU and
sometimes on local deer farms. Extension or outreach programs
bring high school and agriculture college students to the unit. Staff
members address seminars in Australia, New Zealand, Canada,
USA, and Great Britain.

The DRU is expected to play a future role in the development of
Australia's deer farming industry.

Dr. English is Australia and New Zealand coordinator of the
Species Survival Commission (IUCN/SSC) Deer Group.

School of Physics ⊕
University of Sydney
Sydney 2006 Australia

Contact: **Harry Messel, Ph.D.**, Professor Emeritus
   Phone: (61) 2 692 2537, (61) 2 328 7261
   Fax: (61) 2 660 2903, (61) 2 328 7213

Activities: Education; Research • *Issues:* Biodiversity/Species
Preservation; Sustainable Development • *Meetings:* CITES

● Organization

Retired as head of the School of Physics since 1987, during his
tenure **Dr. Harry Messel** established and directed the **Science Foundation for Physics**—the first of its kind in the British Commonwealth. Under his 35-year leadership, the **University of Sydney's
School of Physics** grew from a single professorial entity to encompass 6 research departments specializing in theoretical physics, high-

energy nuclear physics (cosmic rays), astronomy, astrophysics, plasma physics, and applied physics which includes solar energy and environmental sections.

The solar energy work, which he initiated, has led to "exciting discoveries which are being commercialized in a number of countries and to excellent cooperative arrangements with industry, both in Australia and overseas," the university reports. The Foundation has contributed to expanding research programs on campus as well as to developing secondary science education—particularly for the International Science Schools for High School Students, which Dr. Messel began, and the publication of integrated science books that have influenced teaching methods worldwide.

▲ Projects and People

In addition to his work as physicist and educator, Dr. Messel is Australia vice chairman of the Species Survival Commission, World Conservation Union (IUCN/SSC), and chairman of the SSC Crocodile Group. A series of his monographs, in conjunction with other researchers and published by Pergamon Press, surveys the navigable portions of tidal rivers and creeks in northern Australia and their crocodile populations. These waterways include Blyth-Cadell river system; Victoria and Fitzmaurice river systems; Alligator Region river systems; Adelaide, Daly, and Moyle rivers; Darwin and Bynoe harbours; and the Liverpool-Tomkinson rivers system.

These studies are considered unique because they present "the first systematic exploration and basic survey data on northern Australian tidal waterways and their crocodile populations since the settlement of the continent." Dr. Messel has also written on the status of crocodiles in the Solomon Islands—in addition to co-authoring and editing books such as on solar and nuclear energy, *Space and the Atom, Our Earth, Australian Animals and Their Environment*, and *The Biological Manipulation of Life*.

Department of Zoology ⊕
University of Western Australia
Nedlands
Perth WA 6009 Australia

Contact: **Gerald Kuchling, Ph.D.**, Research Scientist
   Phone: (61) 9 380 2243 • *Fax:* (61) 9 380 1029

Activities: Education; Research • *Issues:* Biodiversity/Species Preservation • *Meetings:* IUCN; SSC; World Congress of Herpetology

▲ Projects and People

Dr. Gerald Kuchling came to Australia in 1987 from Austria, where he worked with the Herpetological Collection of the Vienna Museum of Natural History and studied reproduction and hibernation of the tortoise *Testudo hermanni* in both a Yugoslavian natural population and captive colony in Vienna's Zoological Gardens. Also, as research fellow, University of Vienna, Dr. Kuchling did field research on the biology and status of threatened tortoises and freshwater turtles in Madagascar. The Austrian government's award of the Erwin Schröder Research Fellowship brought Dr. Kuchling to the University of Western Australia's Zoology Department to work with Prof. S.D. Bradshaw on the reproductive biology of Western Australian freshwater tortoises.

His first project was to launch a **rescue operation for the western swamp tortoise** (*Pseudemydura umbrina*)—considered to be the world's most endangered chelonian species and Australia's most endangered vertebrate—using techniques of habitat rehabilitation, captive breeding, and reintroduction. About 30 years ago, it was discovered that the world population numbered about 200 tortoises. In the early 1960s, scientists had limited success with protecting existing habitats and setting up a captive colony at Perth Zoo; consequently, the decline continued until there were fewer than 50 worldwide in 1987 with 20 to 30 remaining in the wild and some 17 in captivity, where they had not reproduced for many years.

With global, national, and state funding from the World Wide Fund for Nature (WWF) Australia, Australian National Parks and Wildlife Service (ANPWS), and the Western Australian Department of Conservation and Land Management (CALM), Dr. Kuchling—as chief investigator and in collaboration with Prof. Bradshaw and **Dr. A.A. Burbidge** (CALM)—instituted a successful three-year program based at the University's Zoological Department and using tortoises from the Perth Zoo and CALM's Wildlife Research Centre as well as those found on private properties. By 1991, "32 captive hatched tortoises increased the world population by more than 75 percent," writes Dr. Kuchling. The wild population has been secured with a fence from exotic predators such as red foxes, feral dogs, and cats.

A 10-year **management program** for the **western swamp tortoise** has been established with the support of WWF, CALM, and the Species Survival Commission (SSC) of the World Conservation Union (IUCN). The recovery team's goal is that two viable populations will be created in the wild. The Perth Zoo has taken over captive-breeding management, and a program to reintroduce such tortoises into natural habitats has been launched. This project— **restoring viable populations** of the western swamp tortoise—will also continue for several years with plans to control predators, rehabilitate and re-create suitable habitats, study biology and reproduction, and genetically manage and rear hatchlings with success. In addition to the Zoology Department, Perth Zoo, WWF Australia, ANPWS, and CALM, funding is being provided by Aherns Pty. Ltd., the Bundesverband für fachgerechten Natur–und Artenschutz (BNA, Germany), East-West Veterinary Supplies Pty. Ltd., Kailis Bros., and Minerva Air Conditioning.

In a third project, a **conservation strategy** is being established for **a large freshwater turtle of Madagascar** (*Erymnochelys madagascariensis*) which is mainly threatened by its overexploitation as a meat source for local communities. With the help of Conservation International, Dr. Kuchling began in 1991 to "collect information on populations in different regions and habitats with the goal to select study sites for a more comprehensive follow-up study on population structure, reproduction, food and foraging behavior, survival, predation, and the genetic relationships of different populations."

Adds Dr. Kuchling, "Commercial [use] is negligible, but [illegal] subsistence hunting for the cooking-pot remains a serious threat. . . . Due to rapid human population growth and migration of people, the exploitation pressure is constantly increasing." He expects that the conservation plan will include public education, protection of breeding habitats, egg incubation, and hatchlings programs for this species which is the "only living member of the subfamily Podocneminae (family *Pelomedusidae*) to inhabit the old world."

Dr. Kuchling belongs to the SSC Tortoise and Freshwater Turtle Group.

■ Resources

Video films were produced on the western swamp tortoise; T-shirts are also available. Most needed resource is "money" to fund research.

Austria's environmental problems are closely tied to its status as a major tourism country with a strong tradition of social welfare. In addition, because of its Central European location, Austria shares the environmental problems of neighboring nations, especially problems caused by heavy north-south and east-west traffic. Air pollution in particular has contributed to forest decline and destruction.

The country is attempting to solve these problems by encouraging increased use of cleaner alternative fuels. (Austria is opposed to the use of nuclear power.) In addition, waste management laws have been enacted and steps taken to clean the country's rivers and halt forest damage. The Austrian government is encouraging cooperation among European countries to set goals and timetables for emissions levels and air quality standards, and is working to improve traffic management on a local, regional, and international scale.

🏛 *Government*

Internationale Gewässerschutzkommission für den Bodensee (IGKB) ⊕
International Commission for the Protection of Surface Waters at Lake Constance
Stubenring 1
A-1012 Vienna, Austria

Contact: **H. Gerlitz, Ph.D.**, Chairman
Phone: (49) 89 2192 02 (Germany), (49) 89 2192 3350 (Germany)

Activities: Development; Law Enforcement • *Issues:* Waste Management/Recycling; Water Quality

● Organization

The **Bodensee** or **Lake Constance** is a large lake covering 207 square miles. Its 263-kilometer shoreline serves as a shared border for Germany, Austria, and Switzerland. The history of the area is one of shifting national control, from the Holy Roman Empire to the Austrian Empire to the current state system. During the Thirty Years' War, German and Swedish battleships vied for control, and in the Napoleonic Wars, an English flotilla patrolled the waters.

The current battle is one to protect this central point of a watershed vital for millions, stretching into Italy, and members of the **International Commission for the Protection of Surface Waters at Lake Constance (IGKB)** stress that waters have no borders. Organized in 1959 and first meeting in 1960, the Commission is composed of representatives from all the bordering countries, including Liechtenstein; Germany has representatives from both Bavaria and Baden-Württemberg. The chairmanship revolves every two years, from one state or country to the next. As of January 1992, the chairman for two years is **Dr. H. Gerlitz** of Austria; the previous chairman was Bavarian **Karl-Ernst Orbig**.

IGKB's goal is to return the lake and its surrounding areas to a good condition, in the process dealing with both anthropogenic influences and adverse atmospheric conditions such as abnormally mild winters. To this end, in 1967 the Commission came up with a set of guidelines for the improvement of the lake; the guidelines were completely revised in 1987.

▲ Projects and People

With its relatively early start at improving the conditions of the lake, and a commitment to cooperation, IGKB's efforts have been relatively successful.

An important step has been the **reduction of phosphates** in the lake, as high phosphate levels lead to unmanageable masses of algae, thus lowering the oxygen level of the water and endangering fish and other water-dependent life forms. Through a concerted program of studies resulting in strictly enforced legal measures, begun in the 1970s and continuing indefinitely, the IGKB managed to reduce the level of phosphates from 87 mg/cubic meter in 1979 to 43 mg/cubic meter in 1989.

Also receiving close attention is the construction of sewage treatment plants and other methods to reduce or eliminate the amount of sewage going into the lake. As a prime vacation spot, the lake attracts millions of visitors each year; but it also must supply four million area residents with drinking water. Currently underway is a 10-year program (1986–1995) to **improve the wastewater treatment** situation. The Commission is also working to further restrict damaging sport activity, in particular speed boats and other craft that might leak fuel into the water.

Although efforts in the region have borne fruit, IGKB worries that not enough is being done elsewhere in Europe, which in the end will hurt Lake Constance as well. For example, residue from the Sandoz chemical spill in Geneva several years ago eventually made its way to Lake Constance, through various streams and underground leaching. The lake is the final destination of other smaller amounts of spilled toxics as well. In general IGKB-sponsored laws are stricter than many others across the continent. The Commission recommends a **European Community–wide effort** to **regulate environmentally hazardous activity**, from industry to agriculture.

■ Resources

The IGKB publishes annual reports of its activities and of various aspects of Lake Constance's condition. These limnological studies include measurements of phosphate, oxygen, nitrate, and other levels through the year; investigations into the effect of weather patterns on the lake; temperature charts; and many other research results. An important publication is the *1987 Revised Guidelines* for the upkeep of the lake. Some reports have abstracts in English.

 *NGO*

Energieverwertungsagentur
Austrian Energy Efficiency Agency (EVA)
Opernring 1/R/3
A-1010 Vienna, Austria

Contact: **Manfred Heindler, Ph.D.,** Director
Phone: (43) 222 586 1524
Fax: (43) 222 569 488

Activities: Development; Education; Research • *Issues:* Energy; Global Warming; Transportation

● Organization

The **Austrian Energy Efficiency Agency (EVA)** underwent fundamental changes in 1990. Prior to the reorganization, EVA—which was founded in 1977 at the suggestion of then Austrian chancellor Kreisky—functioned as a think tank for energy policy, which was a great concern to business and industry worldwide since the 1972–1973 oil-price crisis.

EVA reports that "for the first time, people had become urgently aware of the fragility of their energy systems." It was necessary to find a way to ensure the long-term prosperity of the Western world. Dependency on oil as a source of energy became thought of as somewhat risky, and people began to look for specific ways of saving energy.

Conceived as a charitable institution dedicated to furthering sensible use of energy, EVA early on posed three questions: How much energy do we need? How much energy do we have? How much energy may we allow ourselves to consume in the future? The agency set a clear and demanding goal: EVA should help bring about an optimum economical procurement and use of energy in Austria. Furthermore, it should seek a long-term stabilization of energy need by maintaining the full protection and safeguarding of economic growth while reducing the need of business and industry for increased energy consumption.

As society came to a consensus on the necessity of rational energy use, EVA began to work with members of important corporations and institutions. The agency came to be recognized nationwide for the significance of its work and its success in raising public awareness.

The Austrian chancellor personally heads the presidium of EVA, and the vice president is the present minister of energy. The presidium also includes representative members of the Austrian legislative body. Since its inception, EVA has tried to create a focus for its activities while fulfilling a number of requests and tasks; from the beginning, its informational and public relations materials have encouraged a responsible energy policy—above all through the formation of nationwide, regional, and local energy awareness and energy-saving programs.

▲ Projects and People

Within the new system of organization, EVA with its staff of 12 will attempt to make **sustainable energy policy** a concrete reality by developing a pool of knowledge about energy issues for the future; assisting in **innovative energy projects and technology**; and constructing a **center which offers the services of qualified energy consultants** for local and global energy policy initiatives, and which provides research, discussion forums, and distribution of information relevant to energy issues.

In the process of being approved, this plan includes important elements of cooperation with the federal government—including a separate department to be established for workers. Already there are work groups for government employees in Vienna Salzburg and Voralberg.

■ Resources

EVA publishes a *Quarterly Report* in German.

Institut für Angewandte Öko-Ethologie
Institute for Comparative Ethology

Forschungsgemeinschaft Wilhelminenberg
Austrian Academy of Science
Glasergasse 20/3
A-1090 Vienna, Austria

Contact: **Hans Peter Kollar, Ph.D.**, Biologist
  Phone: (43) 222 310 98 97 • *Fax:* (43) 2216 2102

Activities: Education; Research • *Issues:* Biodiversity/Species
Preservation • *Meetings:* ICBP

▲ Projects and People

Dr. Hans Peter Kollar is involved in reintroducing the beaver
(*Castor fiber L.*) which once populated Austria's river systems, including the Danube, and "died out in the 19th century, mainly as a
result of hunting and trapping," he writes. "The last known Austrian beaver was probably shot in 1869 near Anthering, Salzburg.
The last **Danube beaver** was killed in 1863 downstream of Vienna
near Fischamend, Lower Austria."

The **Institute for Comparative Ethology, Wilhelminenberg,**
Austrian Academy of Sciences, with the city of Vienna began its
reintroduction program in 1976 with the release of a pair of Swedish
beavers (*Castor f. fiber*) along the River Danube east of Vienna.
Since 1982, the **Institute for Applied Behavioural Ecology** (Institut
für Angewandte Öko-Ethologie) is in charge of the field studies.
Also at Rivers Inn and Salzach that border Germany, Swedish
beavers were released. In subsequent years, 42 beavers including
Canadian (*Castor canadensis*), Scandinavian, and East European or
Russian (*Castor f. belarusicus et osteuropaeus*) animals were released.
Some 32 Swedish wild-caught beavers were also let go on the
Bavarian side of the River Inn; and another pair was released at the
River Salzach in Austria.

Population along the Danube has accelerated, especially from
1984 until now, according to Dr. Kollar, who estimates total stock
to be "130–140 individuals," including those dispersed. Beavers
have been sited in all four tributaries of the Danube downstream of
Vienna and into Hungary and Czechoslovakia. There are also more
than 200 beavers along the Rivers Inn and Salzach.

Riverine forests of poplar and willow trees providing softwood
and herbaceous food are much to the beavers' liking. "Beaver dams
are built only rarely," reports Dr. Kollar, as the stillwaters remain
fairly stable due to longtime, century-old flood regulation measures.
"The banks of the River Danube itself are not settled by beavers
because they are sealed throughout by embankments made of stone."
Dwellings are mainly bank burrows without the use of wood. Along
the Rivers Inn and Salzach, a hydroelectric power plant helps maintain wetlands along the reservoirs and thus favors living conditions
for the beavers.

Mortality has been low and a "strong (European) beaver population in the Austrian-Hungarian-Slovakian Danube basin is to be
expected," says Dr. Kollar. A potential threat is intensified tourism;
cities such as Vienna and Passau already are obstacles to beaver
dispersion. Ways to protect the natural willow stands from a lower
water table that is causing riparian habitats to slowly dry up are
under discussion with the federal forestry commission. Also, as
beavers cut trees, this affects the relevant plant and bird communities.

Measures to preserve the great bustard (*Otis tarda*) and other dry
grassland bird species such as quail, Montagu's harrier, woodlark,
and tawny pipit are underway in the regions of east Austria characterized by "sand steppe habitats of the Marchfeld, the loess steppe of
the *Weinviertel* north of the Marchfeld, and the grass-dominated
xerothermic limestone-hill vegetation of the limestone cliffs in the
Danube basin and in the *Weinviertel.*"

Switching from its primary habitat of grass steppe with ample
herbs that is today impacted upon by afforestation, construction,
farming, fertilization, recreation, and landfills—the great bustard
now seeks wide open expanses of agricultural land, which is causing
their decline. Yet "measures for species conservation must remain
within the prevailing system of land use and adapt to specific local
demands," writes Dr. Kollar.

For this reason, habitat management requires that "special plots
are set up to meet the demands of the species in respect of breeding
sites and resting areas, but which do not try to imitate the original
steppe habitat," writes Dr. Kollar. Plots known as *Trappenäcker* or
bustard fields are tilled with special seeds including rape (*Brassica
napus*), lucerne (*Medicago sativa*), clover (*Trifolium sp.*), high cabbage (*Brassica oleracea convar. acephala var. medullosa*), and other
Leguminosae and *Gramineae* grasses. "The plots are cultivated . . .
with regard to undisturbedness, vegetation structure, plant species
diversity, and food resources, especially, arthropods." Results are
that the great bustard decline is slowing; "each year one or two hens
leading young were observed in the managed plots."

Dr. Kollar is convinced that such habitat management is the only
way "to achieve long-term success in the preservation of the great
bustard in the European cultural landscape." In 1990, an all-Aus-
trian **community of interests** was formed to coordinate efficiently
these conservation initiatives. In so doing, a species-rich green
countryside is restored with natural reserves without human interference. Dr. Kollar, who is co-chair of the Species Survival Commission (IUCN/SSC) Bustard Group, continues to build public awareness on the value of dry habitats—"low compared to wetlands"—to
reduce their destruction and further their preservation.

■ Resources

A *Green Series* of booklets is published in German with English
summaries that describe environmental work in areas such as ecology, preservation of amphibians, water quality, and "classic ethology." A publications list (also in German) is available.

International Institute for Applied Systems
Analysis (IIASA) ⊕

Schlossplatz 1
A-2361 Laxenburg, Austria

Contact: **Yuri Sinyak, Ph.D.**, Principal Investigator
  Phone: (43) 2236 71521-0 • *Fax:* (43) 2236 71313

Activities: Research • *Issues:* Energy; Global Warming; Sustainable Development; Transportation • *Meetings:* UNCED;
World Energy Conference

▲ Projects and People

Dr. Yuri Sinyak of the International Institute for Applied Systems Analysis (IIASA) is engaged in global energy and climate change studies to reduce greenhouse gas (GHG) emissions through the transformation of energy sources from fossil to noncarbon fuels.

Also with the Working Consulting Group for Long-Term Energy Forecasting, (former) USSR Academy of Sciences, Moscow, Professor Sinyak is developing analytical tools for forecasting CO_2 emissions worldwide by region. The study takes into account "the influence on energy demand and supply of changes in social needs, economic and population growth, governmental policies, and technological progress."

He writes that "the model set will include an energy-demand model MEDEE; an energy-supply model LEAP; and an environmental emission model MARS." Scenarios for evolutionary changes and energy-efficiency and saving (or revolutionary) changes are being developed and analyzed.

To reach energy demand-and-supply policymakers and specialists, the study is elaborating on "the consequences of various policy options (to reduce) environmental impacts, for example, through regional interactions such as technology transfer and technological progress," reports Dr. Sinyak. Economic measures, such as energy taxation and noncarbon subsidies, are being emphasized to mitigate or prevent global warming, as are the "institutional steps needed to achieve the desired goals of economic and social development with the least impact on the environment."

An IIASA workshop on Energy/Ecology/Climate Modeling and Projects was expected to review the various modeling approaches in early 1992. Final research results are to be published one year following the study's completion in a book with a working title of *Energy and the Environment*.

Koji Nagano, Central Research Institute of Electric Power Industry (CRIEPI), Japan, is also participating in the research to develop a global, long-term GHG abatement strategy. Other institutions expected to collaborate with the IIASA project include the former USSR's Academy of Sciences' Siberian Energy Institute, and Institute of Global Climate and Environmental Monitoring and Energy Research Institute—both of Moscow; Japan's Institute of Energy Economics and Atomic Energy Research Institute; Germany's Institute of Energy Economy and Efficient Use, and Research Center, Jülich; Finland's Technical Research Center; Netherlands' Energy Research Centre; and USA's Massachusetts Institute of Technology, Electric Power Research Institute, and Gas Research Institute.

■ Resources

Available publications include: *Energy Efficiency and Prospects for the USSR and East Europe*, CRIEPI Report, EY90001; *Global Energy/ CO_2 Projections*, IIASA, WP-90-51; *USSR: Energy Efficiency and Prospects, Energy*, Vol. 16, No. 5; *Energy Scenarios for Eastern Europe (1980–2030)*, ECN-1-90-033. Dr. Sinyak has written 8 books among 150 scientific publications in the field of energy systems planning and forecasting.

Forest Resources Project
International Institute for Applied Systems Analysis (IIASA) ⊕
Schlossplatz 1
A-2361 Laxenburg, Austria

Contact: **Sten Nilsson, Ph.D.**, Project Leader
   Phone: (43) 2236 71521/0 • *Fax:* (43) 2236 71313

Activities: Research • *Issues:* Air Quality/Emission Control; Biodiversity/Species Preservation; Deforestation; Energy; Forest Resources; Global Warming; Population Planning; Sustainable Development; Transportation; Waste Management/Recycling; Water Quality • *Meetings:* 25 annual international meetings

▲ Projects and People

Dr. Sten Nilsson began his commitment to forests worldwide at the Royal College of Forestry, Stockholm, Sweden, more than two decades ago when he was responsible for building a planning system to salvage windthrown forests. Shortly thereafter, he established a new research group for economic planning which later became the **Institute of Forest-Industry-Market Studies** at the Swedish University of Agricultural Sciences. Also a consultant to Canada, Bangladesh, Norway, and the United Kingdom on forest sectors and reforestation, he became principal investigator of the IIASA forest study in 1986 while a professor at the latter university.

Dr. Nilsson has also been a commissioner for both the Swedish government and Sweden's National Board of Forestry. Recently, he was a commissioner for five years to the Food and Agriculture Organization (FAO/ECE) Timber Committee regarding **damage to European forests due to air pollution and forest decline and trade effects**; in 1991 he was consultant to the World Bank on a Polish forestry development project.

Climate change and the greenhouse effect are current topics of publications that also include studies of forest resources, environment, and socioeconomic development of Siberia; forest resources of the European former USSR; options for air pollution control; future development of tropical forest resources; European commercial wood at risk from sulfur and nitrogen (with R.W. Shaw); and future development of the European softwood lumber industry.

For Procter and Gamble Company, he prepared a 1991 report on "forestry resources, forestry management, and diaper fluff pulp versus other wood uses."

Transboundary Air Pollution Project
International Institute for Applied Systems Analysis (IIASA) ⊕
Schlossplatz 1
A-2361 Laxenburg, Austria

Contact: **Markus Amann**, Leader
  Phone: (43) 2236 71521 • *Fax:* (43) 2236 71313

Activities: Education; Research • *Issues:* Air Quality/Emission Control; Biodiversity/Species Preservation; Deforestation; Energy; Global Warming; Population Planning; Sustainable Development; Transportation; Water Quality

▲ Projects and People
With Dr. Markus Amann's direction, the **Transboundary Air Pollution Project, IIASA**, developed the **Regional Acidification Information and Simulation (RAINS)** model to "evaluate alternative strategies to reduce acidification in Europe." More than 30 scientists from 8 nations participated.

According to Dr. Amann, "the RAINS model combines information on energy consumption, emission generation, costs of controlling emissions, [and] the atmospheric long-range transport of pollutants, and simulates environmental impacts of acid deposition on freshwater, forest soils, and vegetation." The model considers pathways of emissions of sulfur dioxide, nitrogen oxides, and ammonia.

The RAINS model can be operated on IBM-PC compatible machines; there are more than 150 users in Europe, North America, and Asia. It is presently providing scientific support to the international negotiations on further emission reductions regarding long-range transboundary air pollution taking place under the framework of the United Nations Economic Commission for Europe (UN ECE) Convention in Geneva, Switzerland.

Research is now focusing on introducing tropospheric ozone formation into the model, the role of economic instruments in stimulating emission reductions, and emission inventory systems for Central and Eastern Europe and for cities in developing countries.

Documentation can be found in the book *The RAINS Model of Acidification: Science and Strategies in Europe*, edited by Alcamo, Shaw, and Hordijk, and published by Kluwer Academic Press, Dordrecht, Netherlands.

International Society for Environmental Protection (ISEP) ⊕
Marxergasse 3/20
1030 Vienna, Austria

Contact: **Werner Pillmann**, Secretary General
  Phone: (43) 1 715 28 28 • *Fax:* (43) 1 715 28 29

Activities: Education; Research • *Issues:* Air Quality/Emission Control; Energy; Global Warming; Health; Sustainable Development; Waste Management/Recycling; Water Quality

● Organization
The philosophy of the International Society for Environmental Protection (ISEP), or Internationale Gesellschaft für Umweltschutz (IGU), is that "environmental problems can only be solved on an international level, in cooperation with experts of all fields." Political, ecological, economic, and social strategies must be pooled to foster a humane environment in harmony with man's dignity.

For these reasons, scientists from all parts of the world got together in 1987 to found ISEP and to coordinate globally the disparate concepts of "ecology and economy" in a neutral country considered to be a "turntable" between east and west, north and south. Members are notable companies or research institutions whose representatives collaborate on integrative solutions independent of government support. Participants are environmental planners and engineers, scientists, regulators, and policymakers from government, industry, and academia who confront environmental issues which are common across political and geographic boundaries.

ISEP is determined to "avoid pollution, . . . promote environmental protection regardless of state borders, . . . recognize and express trends early, . . . provide the general public with objective information, . . . initiate and publish research projects, . . . [and] make its members conscience of their environment."

▲ Projects and People
The **International Forum of Environment** is an annual event which brings together scientists, government representatives, and executives to conceive and discuss such integrative solutions on a global level.

The latest developments and findings in environmental technology are presented at special conferences and seminars, such as the **Envirotech Vienna**, with working languages in German and English. **Industrial waste management**—including new reduction methods, environmentally sound packaging, hazardous waste treatment, recycling, incineration, sanitation of contaminated sites, and wastewater treatment including sludge and purification of sewage and of surface and ground waters—was featured at the 1992 international conference, University of Economics, Vienna. Workshops illustrate computer application to environmental protection.

The first international **Envirotour Vienna**, scheduled for late 1992, is featuring seminars and strategies for reducing the environmental impact of tourism. Key issues are tourism-induced traffic, "environmentally conscious tourism management," regional and landscape planning, psychological and social issues of sports, tourism in mountainous regions, and "ecomodels." French, Italian, and Spanish are being added to the working languages.

National Points of Contact (NPOCs) are individuals or institutions representing ISEP in several countries so that direct contacts can be established quickly—particularly in Eastern European countries.

Working with ministries and NGOs, ISEP takes part in **environmental research projects**.

■ Resources
Published proceedings are available for Envirotech, International Forum of Environment, and computer workshops. Papers and exhibits at upcoming events can be requested. ISEP also provides other conference papers as well as exchanges publications with private and public organizations to make new trends and technolo-

gies available worldwide. Lectures, seminars, and workshops are organized on-site regarding specific and current subjects. Members benefit from the ISEP information service; ISEP is expanding membership and its international contacts to encourage broadened participation in conferences.

Österreichische Gesellschaft für Umwelt und Technik (ÖGUT)
Austrian Society for the Environment and Technology
Floragasse 7
A-1040 Vienna, Austria

Contact: **Hannes Drössler, Ph.D.,** Managing Director
Phone: (43) 222 505 37 60, 61
Fax: (43) 222 505 37 62, 73

Activities: Education; Political/Legislative; Research • *Issues:* Air Quality/Emission Control; Deforestation; Energy; Global Warming; Health; Sustainable Development; Waste Management/Recycling; Water Quality • *Meetings:* UNCED

● Organization
ÖGUT, or the Austrian Society for the Environment and Technology, was founded in 1984 as an independent platform for discussion among industry, government, and environmentalists under the patronage of a partnership between trade unions and industry with the federal government. ÖGUT is based on the principles of "honor, independence, openness, networking, fairness based on partnership, the greatest possible economizing while avoiding public subsidies, and an effort towards a diversity of voices, activities, and perspectives." It provides both public and project-group memberships, political activities, information about nature, and participation of representatives from the federal, provincial, and municipal governments.

▲ Projects and People
Fifteen projects are organized in the following groups: **construction/waste management, transportation, chemistry, packaging/paper, education, energy, media/information, new roads, tourism, agriculture, environmental law, UNCED '92, business/insurance, industry/trade/recycling,** and **Eastern Europe.**

An **Environment Forum** and **Research Institute** are two of the permanent features. The Forum sponsors panel discussions in the provincial capitals in cooperation with local governments, news media, environmental organizations, and representatives of the regional trade unions and industries. The Research Institute offers interdisciplinary work groups in the fields of economics, ecology, and technology—topics of concern to ÖGUT and its members and outside clients. Other functions include a Scientific Advisory Committee, Leadership Seminar, Symposium, Press Club, Information Section, and Video Documentation.

Dr. Hannes Drössler is also a lecturer at the Institut für Politologie, Universität Innsbruck.

■ Resources
ÖGUT has prepared reports on the *Protection of the Environment in Austria, Waste Management,* and the environmental situation in Czechoslovakia. A monthly magazine, *ÖGUT INFO,* is also published.

Welt Natur Fonds
World Wide Fund for Nature (WWF) ⊕
Ottakringer Strasse 114–116
Postfach 1
A-1160 Vienna, Austria

Contacts: **Winfried Walter,** Director
Phone: (43) 222 46 14 63 0
Fax: (43) 222 46 69 29 29

Erhard Kraus, Ph.D., Zoologist
Phone: (43) 222 46 14 81 0
Fax: (43) 222 46 14 81 29

Activities: Education; Lobbying; Nature Conservation; Research • *Issues:* Biodiversity/Species Preservation; Energy; Sustainable Development; Waste Management/Recycling
Meetings: WWF

● Organization
With its motto "the gateway to an understanding of nature"—World Wide Fund for Nature (WWF), also known as **World Wildlife Fund,** has 30,000 members and a staff of 30. The threat to one of the most important European habitats for birds, the Langen Lacke of the Seewinkel in the Austrian state of Burgenland, prompted the founding of WWF in 1963. Secured under a long-term lease, today this area is a national nature preserve. Such **preserves** remain an important focus of WWF. Receiving international recognition are "Marchauen/Marchegg" in the beautiful central European Auwald, "Pürgschachener Moor" in Steiermark, and the Hundsheimer Mountains in Niederösterreich.

▲ Projects and People
WWF had a revolutionary influence on politics because of the **Fight for Hainburg** campaign. In 1984, when the scenic area along the Danube between Vienna and the Czechoslovakian border was threatened by a proposed power station, WWF staff coordinated actions against its construction. As a result, the section of the **Danube River** from Vienna to Hainburg was made a **national park.**

A second focus is on **conservation projects,** large and small, for such animals as the **otter, birds of prey, amphibians, white stork,** and **butterflies.** At the same time, WWF is aware that without the comprehensive preservation of natural habitats, conservation of individual species is almost impossible. For example, trees were purchased for a stork reservation in Burgenland; ponds and shore land is being rented or purchased for otters; and marshland protection is underway in Lower and Upper Austria. Species on the brink of extinction are being helped with breeding and reintroduction programs. Injured bats are being fed and nursed back to health. Of the country's 30,000 different species of animals, one-third are registered on Austria's **Red List** of endangered species. Destruction of habitats and the rapid spread of technical and chemical innovations are responsible for their decline.

After a 130-year absence due to extermination by man, one lone male bear wandered from Yugoslavia into the Alps of Lower Austria in 1973, established residence, and inspired WWF Austria and the Federal Hunting Association (FHA) to begin a plan to **introduce the brown bear (Ursus arctos)** into this area. **Dr. Erhard Kraus** with **Drs. H. Gossow and J. Rauer**, Institute of Wildlife Biology and Management, University of Agronomy, Vienna, are implementing the four-year reintroduction and management plan that initially set as its 1993 goal a population of about 40 bears in the Lower Austrian and Styrian Alps.

The plan was activated with the release of a wild, radio-collared female bear from Yugoslavia named "Wanja" (funded by a Viennese shopping mall) into the home range of the lone male called "Ötscherbär" in an area known as Ötscher-Dürrenstein and characterized by "rocky northern slopes for denning, mixed evergreen forests where bears can forage for berries and insects, and quiet isolated forest tracts where encounters between humans and bears should be minimal."

The subsequent, planned release of up to 10 bears was preceded with discussions among commercial beekeepers, farmers, and FHA representatives to prevent damage to property and people. WWF notes that the government of Lower Austria compensated livestock owners for Ötscherbär's destruction of sheep in the foothills in his early days; but such behavior ceased when he became familiar with the mountain resources. Electric fences protect the beehives from bears; WWF and FHA worked to secure other means to compensate for bear damages and even changed the content of deer feedings that were attracting Wanja.

WWF acknowledges that bear catching in Slovakian parks and other regions for quick transport and release in Lower Austria was physically difficult and complicated by a changing political situation. Too, bears involved in reintroduction cross boundaries of three countries—Yugoslavia, Italy, and Austria; WWF raises the possibility of "an international program aimed at protecting and monitoring these animals [to be] incorporated into the WWF project."

Nevertheless, a third bear wandered into Styria in 1990. And in July 1991, a great success occurred when "three little cubs were seen together with Wanja" on several occasions. "This is a good sign for the ecological quality of this region," reports WWF.

In agricultural regions, new projects, such as one promoting **ecograzing of depleted pastures** in the barren areas of Burgenland, are the first steps in overcoming destruction and reclaiming habitats.

The **National Conservation Programme** also includes the Neusiedlersee Nature Reserve information center; management and education at the Donau-March-Tyaya-Auen National Park; purchase of "Otter-Haven" with ongoing captive breeding and education programs; white stork habitat management; bearded vulture captive breeding and reintroduction into the Alps; rehabilitation of birds of prey and captive breeding of endangered species; education about amphibians to reduce their mortality on roads; pond construction; conservation and education measures at the Riverine Forest, Lech/Tyrol; promotion of landscapes and agricultural conservation on arable lands; captive breeding of the black vulture; habitat management of the great bustard; moorlands management for the black grouse; and game management in Hohe Tauern National Park.

Energy and Transportation policies that are beneficial to people and the environment are also a current focus of WWF's activities.

In its **International European Programme**, WWF reports activities regarding the Gabcikovo-Nagymaros power plant and birds of prey protection.

■ Resources
WWF-Austria brochures, posters, and descriptions of wildlife programs are primarily printed in German.

◊ *Private*

Prof. Johannes Zopp
Arsenal 9A/5/9
1030 Wien, Austria

Phone: (43) 222 78 95 84

Activities: Education • *Issues:* Biodiversity/Species Preservation

▲ Projects and People
Prof. Mag. Johannes Zopp is a free-lance journalist and has been a natural science and geography teacher for some 30 years. He supports the work of herpetologists in Austria as well as of the Cyprus Herpetological Society.

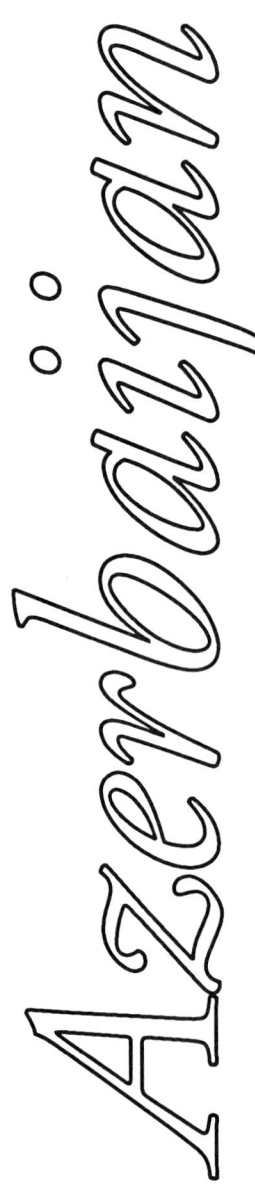

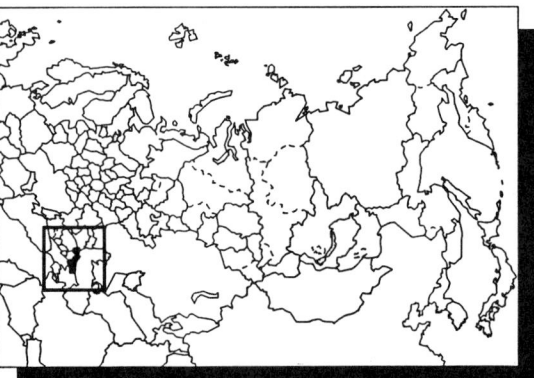

One of the 15 former republics of the USSR, Azerbaijan is characterized by high population density and heavy concentrations of industry and development. Oil refining, petrochemicals, metallurgy, and construction are among the industries which cause serious pollution-related health dangers for the population in the cities, exacerbated by atmospheric inversions. In regions of intensive cotton-growing, the soils are significantly degraded due to the overuse of chemicals and mineral fertilizers.

Nature reserves hold a variety of mammals, reptiles, birds, amphibians, and fish, but no steps are being taken to ensure their continued protection. Much of the fertile soils along the Caspian seacoast have been destroyed.

The country recognizes the gravity of its environmental problems and has called upon the international community for help.

🍁 *NGO*

Azerbaijan Green Movement (GMA)
Zevin Street 4/6
370005 Baku, Azerbaijan

Contact: **Ismail Musa oglu Rustamov**, Board Member
Phone: (7) 8922 930536

Activities: Development; Ecological Business; Education; Law Enforcement; Political/Legislative • *Issues:* Air Quality/Emission Control; Biodiversity/Species Preservation; Deforestation; Energy; Health; Radioactivity Control; Sustainable Development; Transportation; Waste Management/Recycling; Water Quality

● Organization

In 1989 an Ecological Club was founded in Azerbaijan, which later became the **Azerbaijan Green Movement (GMA)**. In addition to the Movement's center, which is located in Baku, the capital of Azerbaijan, there are associated branches in Sumgait and Gandja, the largest industrial cities in Azerbaijan. The organization has approximately 1,000 members, but considers itself to have tens of thousands of supporters. The Azerbaijan Greens are "integrated into the All-Caucasian ecological organization" and are establishing links with other Green parties and similar world conservation groups in Europe and elsewhere.

Holding conferences every two years, the Movement has a Coordinating Council, Government (executive body), Council of Elders (consultants and advisors), and a Revision Committee. The regional organizations are associated with the Movement, but function independently.

Overall aims are "preservation and restoration of the natural environment of Azerbaijan, harmonization of man's relations with [his] surroundings," and making the environmental legislation of Azerbaijan conform to international conventional and agreements. The Movement embraces the ideas of freedom—of speech, of the press, of a free exchange of views, and of conscience. The Movement also advocates nonviolence.

Due to the region's political turmoil, Azerbaijan has been under a state of emergency for several years. The Movement faces other difficulties including severely limited finances, lack of food, poor environmental educational materials, and a population with little knowledge of ecological problems.

▲ Projects and People

Previously, GMA sponsored a joint expedition with the Green Movement of Georgia along the **Cura River** to collect **pollution data** and analyze water composition. They also have collected data about **industrial waste** within Azerbaijan and established a **databank on experts**, pursued the control of **radioactive pollution** of food products, **planted trees** in Baku, and tried to stop the death of **migratory birds** wintering in the oil-polluted lakes at the Apsheron Peninsula.

The Movement is working to get political and moral support to battle the development of the *Azeri* oilfield in the Caspian Sea, a collaborative venture of the (former) Soviet government and a USA oil firm. "The Caspian Sea is a unique lake in the world with unique fauna and flora" already under threat, writes chemical engineer **Ismail Musa oglu Rustamov**. He points to "excessive pollution by old oilfields and by the Volga, Cura, and Ural Rivers." The *Azeri* oilfield "should not be exploited fully, because we have to regard future generations."

GMA's future plans include creating the **Shamkir National Park**, increasing public awareness about the environmental threats of oilfield expansion, organizing ecological camps, investigating alternatives to hydropower plants on the Araks River, and preparing **environmental laws** for the government of Azerbaijan. Research is planned about the health effects on women who work in chemical and industrial enterprises in Sumgait. Most of the women who work in these plants are ill and experience pregnancy complications; and "every fourth child in Sumgait is born with different kinds of deformity. Every third child dies after birth," according to the Movement.

Thus, the Green members have grave concerns about pollution in Azerbaijan resulting from gas, oil, and chemical industries; chemical pesticides and nitrogen fertilizers used in agriculture and aided by the flow of the Cura, Araks, and Volga rivers; and "dust, soot, sulphide gas, carbon monoxide, hydrocarbons, photooxidants, and carcinogenic substances" in cities. "The prolonged ground inversion and calm weather often result in smog, toxic mist." As a result, certain disease rates are higher in Baku populations as compared to other major cities in the former Soviet Union; morbidity rates in agricultural regions also appear to be rising.

Movement activists also include: **Abdullaev Elman Asker oglu, Husseynova Farida Camil gizi, Alieva Leila Aga gizi, Atababaev Elchin Rafic oglu**, and **Gusein-zade Nigar Gusein gizi**. ("Oglu" means "son of"; "gizi" means "daughter of.")

■ Resources

The Azerbaijan Green Movement points out that new technologies are needed for plants and factories locally and throughout the former Soviet Union. The Greens need "public relations and equipment too because we do not have our own technical infrastructure" and welcome connections with ecological business to "help solve financial problems." Training courses are sought for activists "at ecological institutes, to exchange their experiences with [similar] organizations in the West, and to raise professional skills."

Papers prepared in English are "The Effect of Ecological Factors on the Health of Population of Azerbaijan" and "The Present Ecological Situation."

The Bahamas is a sprawling archipelago of about 700 mainly flat, low-lying islands that extend roughly 600 miles from near southern Florida to near Haiti. The entire population of roughly a quarter of a million resides on 14 of the larger islands. Much of the land area lies ten feet or less above mean sea level. While the country is seeking ways to diversify its economy, it is still largely dependent on tourism.

The Bahamas is particularly concerned about the potential effects of unchecked global warming on coral reef systems and marine coastal ecosystems, and about rises in sea level that could damage the coastline. As a small nation, it strongly supports, and is to a great extent dependent upon, international monitoring, evaluation, and conservation efforts. Nevertheless, projects undertaken by the Bahamas National Trust and other Bahamian groups have achieved striking success.

 NGO

Bahamas National Trust (BNT)
P.O. Box N4105
Nassau, Bahamas

Contact: **Susan G. Larson**, Director of Education
Phone: 1-809-393-1317 • *Fax:* 1-809-393-4978

Activities: Development; Education • *Issues:* Biodiversity/Species Preservation; Sustainable Development • *Meetings:* IDB Consultative Meetings on Environment; International Conference of National Trusts; IUCN; UNEP; World Parks Congress

● Organization

Established by an Act of Parliament, the **Bahamas National Trust (BNT)** is "non-government run" and manages all 250,000 acres of the Bahamas' national park system. With 3,500 members from a broad cross-section of the population, the well-known Trust is dedicated to "preserving our lands and submarine areas of beauty or natural or historic interests, and its animal, plant, and marine life for the benefit and enjoyment of our community."

With leadership from former 15-year president **Lynn Holwesko**, the Trust is involved with programs in the "mainstream of international conservation," such as the International Conference of National Trusts and the World Conservation Union (IUCN), where Holwesko's appointment to the Commission on National Parks and Protected Areas is enabling her to "position the Caribbean as an IUCN target area."

▲ Projects and People

Under discussion is a **national conservation strategy** to determine guidelines for development in the Bahamas based on a complete inventory of natural resources.

The country's largest **wildlife conservation survey** was recently undertaken on Grand Bahama, Abaco, New Providence, and Great Inagua by the Trust in cooperation with the Department of Lands and Survey's Division of Forestry and the Department of Agriculture. Follow-up entailed an education campaign for school children, business representatives, government employees, and the general population.

Earlier, the first **marine mammals survey** got underway with the help of whale specialist **Kenneth Balcomb**, Center for Whale Research, and associates **Diane and Rhonda Claridge**, who photographed the whales and recorded their vocalizations under water. During the census of whales, dolphins, and manatees, BNT enlisted the help of members and sea travelers to report any sightings and observations of whale behavior. A photo-identification catalog is one project result, as are school visits with slide presentations and whale-song recordings to urge children to do their part to save these unique mammals from extinction.

The model restoration of the Adelaide Creek and Wetlands has renewed the creek's natural flushing and returned marine life and waterfowl to the waters, it is reported.

Various committees are engaged in bottle and can recycling programs, antilittering and community cleanup campaigns, Earth Week commemorations, outreach to the Bahamas Defense Force and Police College regarding conservation laws, film festivals featuring a National Audubon Society series, live species educational exhibits, and sponsorship of school programs—such as a Protect Our Common Wealth competition featuring short plays about environmental problems and how to solve them. As the BNT describes it, "Mother Nature appeared in a dramatic performance along with earth, sky, and water, telling the story of Man's irreverence for the natural environment and the inevitable consequences of his careless, unthinking ways." A video was produced of the participating plays for use as resource material for environmental studies.

Members also participate in retreat garden workdays on the BNT grounds, recycling Saturdays, and white-crowned pigeon banding. BNT hosts guest speakers from such groups as the World Wildlife Fund—USA (WWF) and the Jersey Wildlife Preservation Trust (JWPT), an international breeding center for some of the world's most endangered species.

■ Resources

BNT publishes the bimonthly newsletter *Currents* and *Field Guides for Children*. Environmental education workshops are held for teachers; other speakers' programs are also available. Resources such as "raw materials" and "subject expertise" are needed to boost educational outreach and public information programs.

(See also Center for Whole Research in the USA NGO section and George Mason University in the USA Universities section.)

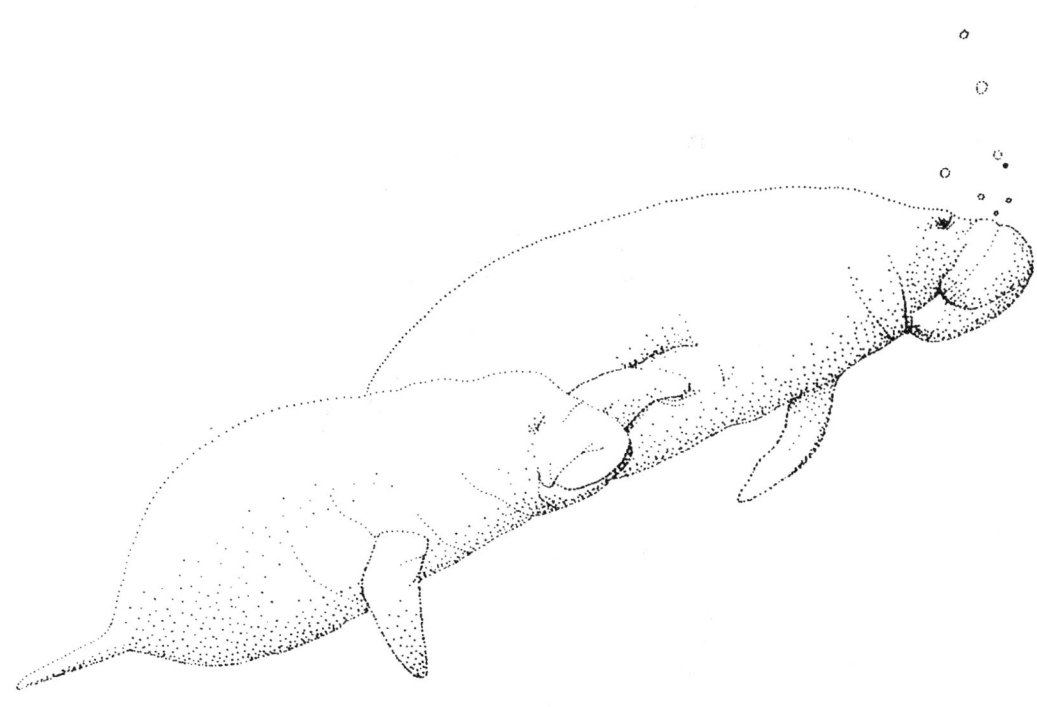

Bahrain

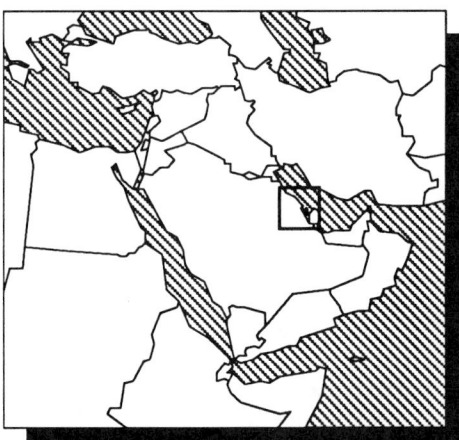

Made up of 33 low-lying islands in the Persian Gulf, between the Qatar Peninsula and Saudi Arabia, Bahrain is rich with petroleum resources. Most of what little arable land exists is used for growing date palms. There are few animals or reptiles in the country, but a variety of birds and fish.

Environmental problems have arisen from oil spills and other discharges, which have caused damage to the coastlines, coral reefs, and sea vegetation, and are posing an added threat to marine life in the Persian Gulf. Desertification and lack of surface water resources are endangering future water supplies. The government and the population in general have been slow to recognize and respond to the country's environmental needs.

🏛 *Government*

Environmental Protection Committee
P.O. Box 26909
Adliya, Bahrain

Contact: **Ekarath Raveendran**, Senior Chemist
Phone: (973) 293693 • *Fax:* (973) 293694

Activities: Law Enforcement; Research • *Issues:* Air Quality/Emission Control; Biodiversity/Species Preservation; Energy; Global Warming; Health; Sustainable Development; Waste Management/Recycling; Water Quality

▲ Projects and People

Ekarath Raveendran describes his research with the State of Bahrain's **Environmental Protection Committee**: "I have set up a laboratory for monitoring air, water, wastewater, soil, and hazardous wastes." Since its 1983 founding, scientists have tested air quality, water quality, and soil, and have answered environmental complaints from the public.

A major project is the testing of **industrial effluent quality in Bahrain**. Eighteen industrial effluents from 13 major industries were monitored on a quarterly basis in a 1-year period, and 38 physical and chemical parameters were measured for each effluent. "The analysis of data . . . indicates that the metal treatment industry is the most polluting among all the major industries on the island, reports Raveendran. "They discharge effluents containing high aluminum, suspended matter, phenols, and heavy metals into the marine environment."

Ongoing **Bahrain studies** are examining heavy metals in dust; lead in air and blood; radioactivity in air, water, and soil; trace metals in drinking water; drainage water quality; and soil amendments. A 1992 report was prepared on the status of the country's air quality following the Kuwait oil fires. Raveendran is also responsible for preparing a **Status of the Environment** report, industrial profiles for each industry, and a **Tubli Bay environmental study**.

Results of a recent investigation into the possibility of **recycling sand blasting grit** generated by a major industry suggested that "copper slag can be used as a replacement of marine sand used in mortars for concrete blockworks," writes Raveendran, who is also studying the fluoride levels in the vicinity of an aluminum reduction plant. He is also involved in global issues such as the reduction of chlorofluorocarbons (CFCs) and the Regional Organization for the Protection of the Marine Environment (ROPME) programs as well as in APELL—Awareness and Preparedness for Emergencies at Local Level—programs.

Future plans are to examine mercury content and hydrocarbons in sediments and biota; nutrients, organic carbon, and heavy metals in marine sediments; soil-water characteristics at the mangrove swamp area; polycyclic aromatic hydrocarbons in sediments and fish; and heavy metals such as vanadium and nickel in air particulate matter as a result of the **Kuwaiti oil fires**.

■ Resources

New air-pollution-control technologies are needed to remove nitrogen oxide and carbon dioxide. Also sought are training assistance to keep abreast of new developments as well as air and water legislation that controls pollution.

Bangladesh

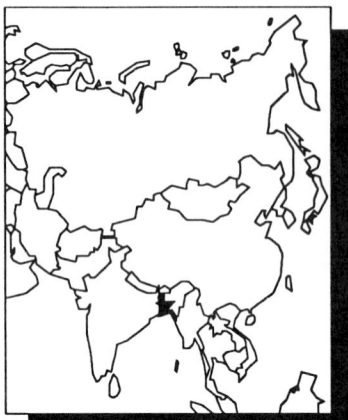

There is much beauty to protect in Bangladesh: because of the abundance of water and sunshine, the country is luxuriant in vegetation and is the home of such majestic animals as the royal bengal tigers. Yet, as one of the world's poorest countries, Bangladesh suffers from rapid population growth, unemployment, low productivity, malnutrition, and illiteracy. All these factors affect the relationship between the people and their environment, with poverty increasing human pressures on natural resources and development often undermining long-term resource sustainability.

Major environmental problems include annual flooding; soil erosion in the hills; soil salinity in the coastal areas; marine pollution from oil and other industries (harmful to mangroves, coastal and estuarine fisheries); loss of wildlife habitats; and deforestation. Although the relevant government agencies have been hampered by lack of training, staffing, and equipment, there is growing awareness within the country of the importance of environmental conservation.

🏛 *Government*

Bangladesh National Herbarium
House 52, Road 8 A, Dhanmandi
Dhaka, Bangladesh

Contact: **M. Salar Khan, Ph.D.,** Honorary Advisor
Phone: (880) 2 311273

Activities: Development; Education; Research • *Issues:* Biodiversity/Species Preservation; Deforestation; Energy; Sustainable Development • *Meetings:* IUCN General Assembly; IUCN/SSC Indian Subcontinent, Palm Specialist groups; SSC Annual Meeting

▲ Projects and People

Dr. M. Salar Khan and principal scientific officer Ahmed Mozaharul Huq, Bangladesh National Herbarium, recently completed a project with M. Kairul Alam, senior research officer, Forest Research Institute on inventorying flora of the Chunati game sanctuary in Chittagong. Dr. Khan reports obstacles from "the sanctuary (being) heavily encroached by local farmers." With the above taxonomists, team leader Dr. Khan plans follow-up of field studies on the threatened plants.

An ongoing project is the biodiversity assessment of Teknaf game reserve using keystone species. With support from the World Wildlife Fund and the Conservation Foundation, Dr. Khan and Herbarium director M. M. Rahman (an agrostologist), Huq, and Herbarium senior scientific officer M.M.K. Mia (an ethnobotanist) are identifying the keystone species of ecological and socioeconomic value from the total plant genetic resources that will be recorded. "The studies would also include the firsthand data on the status of the keystone species in the wild, evaluation of the current use pattern of the resources by the indigenous people, linkages with wildlife, and recommendations for the sustainable development," reports Dr. Khan.

Follow-up includes field studies of wild relatives of mango and banana and the recording of the frequency of indigenous timber trees. "Dissemination of research results through lectures and displays at the local government offices and schools" are planned, writes Dr. Khan.

A retired botany professor at the University of Dhaka, Dr. Khan is also a senior fellow with the Bangladesh Centre for Advanced Studies (BCAS) and a member of both the Species Survival Commission (IUCN/SSC) Palm and Indian Subcontinent Plants groups. He is editor-in-chief of the *Bangladesh Journal of Botany* and national coordinator of the Commonwealth Science Council's Biological Diversity and Genetic Resources project.

Huq, who has an interest in herbal medicine, has pursued plant taxonomic interests at herbariums in India and the United Kingdom; in Turkey, he explored "biomass-energy prospects of fast-growing tree species in Bangladesh." He has written about the 10,000 plant specimens from throughout Bangladesh and has published books on taxonomy, economic botany, and ethnobotany. He is also affiliated with the Bangladesh Association for the Advancement of Science (BAAS).

■ Resources

"Want of a four-wheel drive vehicle is hampering movement (for field studies) especially in wet weather," Dr. Khan reports. Other needs are personal computers for data processing, ethnobotanical studies in aboriginal areas, visits to foreign herbaria for consulting specimens, and financial assistance to produce future *Flora of Bangladesh* publications.

Three available publications are *Aquatic Angiosperms of Bangladesh*, *Flora of Bangladesh* (fascicles 1-43), and *Plant Names of Bangladesh*, priced in U.S. dollars.

🍁 *NGO*

Nature Conservation Movement (NACOM)
29-C-1 North Kamalapur
Dhaka 1217, Bangladesh

Contact: **Md.Anisuzzaman Khan**, President
Phone: (880) 2 418883 • *Fax:* (880) 2 833495

Activities: Development; Education; Research • *Issues:* Biodiversity/Species Preservation; Deforestation; Energy; Population Planning; Sustainable Development • *Meetings:* Wetlands and Waterfowl Conservation in Asia

▲ Projects and People

Dedicated to community-based ecosystem conservation, the **Nature Conservation Movement (NACOM)** is actively involved in inventory building of renewable natural resources of the *Sundarbans*—or beautiful—**mangrove forest**, which **Md.Anisuzzaman Khan** describes as the largest of its kind on the planet. The forest supports 300 species of plants, 50 species of reptiles, 261 species of birds, and 49 species of mammals including tigers, the highly endangered *Batagur baska*, he reports. Total area is 5,800 square kilometers crisscrossed by thousands of canals and tidal creeks.

With a staff of 10, Khan takes aerial surveys of **coastal wetlands** for identification of key potential areas for **conservation**, which are followed by ground surveys. This project is in cooperation with **Abdul Akonda**, Bangladesh Forest Department senior research officer, and consultant **Dr. Derek Scott**, representative of the Ramsar Convention on Wetlands of International Importance, Asian Wetland Bureau (AWB), and International Waterfowl and Wetlands Research Bureau (IWRB).

NACOM also provides employment for jobless environmentalists.

■ Resources

New technologies needed are "alternative energy development against fossil fuel" and "utilization techniques of huge freshwater of our country," reports Khan. Also needed are training on sustainable development economics, grassroots educational material development, and a mobile education unit.

Publications available in U.S. dollars are *Otter and Fishermen: New Techniques of Wildlife Conservation* and *Community-based Ecosystem Conservation*. A speaker makes presentations on "symmetrical development—a new concept in conservation."

🎓 *Universities*

Department of Zoology
University of Dhaka
Dhaka 1000, Bangladesh

Contact: **Kazi Zaker Husain**, Professor
Phone: (880) 2 501306

Activities: Research • *Issues:* Biodiversity/Species Preservation; Deforestation; Sustainable Development

▲ Projects and People

The **white-winged wood duck** of the Chittagong Hill Tracts is a major interest of **Kazi Zaker Husain**, who made recommendations on its protection to the Bangladesh government after his investigations that were sponsored by the University of Dhaka. Also in Chittagong, Husain studied and reported on species in the **Chunati Wildlife Sanctuary** for MARC, a non-government organization, and surveyed wetland fauna at Hail Haor in Sylhet. More recently, Husain conducted surveys of commercially important frogs and of wildlife in the Jamuna Multipurpose Bridge Area for the Bangladesh government. The World Bank sponsored the latter study, which is in preparation.

Husain is the founder and president of the Wildlife Society of Bangladesh and the Bangladesh Bird Preservation Society. Editor-in-chief of the *Bangladesh Journal of Zoology*, he also founded the Zoological Society of Bangladesh. Husain also drafted the *Wildlife and Protected Areas* chapter of the National Conservation Strategy, sponsored jointly by the Bangladesh government and World Conservation Union (IUCN).

Author of three books, *An Introduction to the Wildlife of Bangladesh*, *Birds of Bangladesh*, and *Bagh (Tiger)*, Husain is a former biological sciences faculty dean in Dhaka, where he prepared a wildlife and conservation curriculum for schools and universities throughout Bangladesh. He also introduced the country's first wildlife biology course. "Nearly 150 have come out by now specialized in this branch, some of whom have joined the forest department and other universities," reports Husain.

Husain continues to publicize wildlife and environmental protection through seminars, lectures, writings, and radio and television programs.

■ Resources

Available are the *Wildlife News Letter*, and *Paribesh Parikrama* (Environment), which is published in Bengali. Needs include "video film, editing equipment, binoculars, telescope, and a vehicle like a landrover, which we cannot buy because of lack of funds," reports Husain.

The high population density of Barbados, the most easterly island in the Caribbean, places severe limitations on its soil and land resources. Thus, the rational use of these resources is the country's overriding environmental concern.

Specific problems include threatened coastline erosion due to man-made changes in the land and the natural fragility of the land; coastal pollution from oil slicks; and the potential for groundwater shortages by the end of the century. Like the other countries in the outer chain of the Lesser Antilles, Barbados is low-lying, relatively arid, and of coral formation. The country is also vulnerable to drought, floods, and volcanic activity. The government is actively addressing these problems through legislative and administrative means.

❧ *NGO*

Barbados Environmental Association
P.O. Box 132
Bridgetown, Barbados

Contact: **Stephen Boyce**
Phone: 1-809-427-0619 • *Fax:* 1-809-427-0619

Activities: Education; Research

▲ Projects and People
The **Barbados Environmental Association** reports that it is "committed to preserving the environment through public awareness and education programs, research, and positive action."

Caribbean Conservation Association
Savannah Lodge, The Garrison
St. Michael, Barbados

Contact: **Calvin A. Howell**, Executive Director

Activities: Development; Education • *Issues:* Biodiversity/Species Preservation; Sustainable Development; Waste Management/Recycling

■ Resources
The **Caribbean Conservation Association** publishes *Ecology and the Environment*.

Children's Association for the Protection of the Environment (CAPE)
28 Oxnards
St. James, Barbados

Contact: **Ruth Hope**
Phone: 1-809-424-7232

Activities: Education

▲ Projects and People

The Children's Association for the Protection of the Environment (CAPE) aims to "improve environmental awareness and responsibility among children in Barbados."

🎓 *Universities*

Women and Development Unit (WAND),
   School of Continuing Studies
University of West Indies
The Pine
St. Michael, Barbados

Contact: **Peggy Antrobus**, Coordinator
   Phone: 1-809-436-6312 • *Fax:* 1-809-427-4397

Activities: Education

▲ Projects and People

The **Women and Development Unit (WAND), University of West Indies**, reports that "WAND is a regional program working with women's organizations and other community groups and fellow agencies in the English-speaking Caribbean, using participatory approaches to training and involvement in processes for information and documentation."

Despite its small size and location in the midst of a large industrial center, Belgium has a rich environmental inheritance with much diversity. Yet, the country's dense population, agriculture, and high level of industrialization are the sources of many types of pollution, especially water pollution caused by factories in the steel-making regions, and air pollution from the smokestack industries. The country has moved away from use of oil and coal in meeting its energy needs toward increased use of natural gas and, to a certain extent, nuclear power.

Belgium is also the headquarters of most of the institutions of the European Community.

🏛 *Government*

Institut Royal des Sciences Naturelles Belgique (IRSNB) ⊕
Royal Institute of Natural Sciences of Belgium

29, rue Vautier
B-1040 Brussels, Belgium

Contact: Jacques Verschuren, Ph.D., Biologist
Phone: (32) 2 7 70 86 36

Activities: Conservation • *Issues:* Biodiversity/Species Preservation; Deforestation; National Parks

▲ Projects and People

The Institut Royal des Sciences Naturelles de Belgique (Royal Institute of Natural Sciences of Belgium) is mainly interested in the problems of the environment and the conservation and promotion of national parks, mostly in French-speaking Africa.

Working through the Institute, **Dr. Jacques Verschuren** has won a gold medal from the World Wildlife Fund (WWF) for his work in Zaire and has spent, by his estimation, some "2,500 nights under the tent to approach and study wildlife." Dr. Verschuren also participated in the creation of the Parc Salonga, at "36,000,000 hectares," the largest protected rainforest park in Africa. In addition, he contributed to the development of the national parks of Senegal, Rwanda, Burundi, and Benin, among others. He has written over 160 scientific and popular papers.

Of special interest are "two endangered large mammals—Jara's rhinoceros and white rhinoceros"—as well as the ultrasonics, hibernation, and reproduction of bats.

A regional member of the Species Survival Commission (IUCN/SSC), Dr. Verschuren also belongs to its Antelope Group.

 *NGO*

Les Amis de la Terre/Friends of the Earth ⊕

Place de la Vingeanne
5100 Dave, Belgium

Contact: **Alain Hanssen,** International Secretary
Phone: (32) 81 40 14 78

Activities: Education; Law Enforcement; Research • *Issues:* Deforestation; Energy; Global Warming; Sustainable Development; Town and Country Planning; Waste Management/Recycling; Water Quality • *Meetings:* FOEI

● Organization

With 700 members and a staff of 8, **Les Amis de la Terre—Belgique** is part of the **Friends of the Earth** global network. Established in the mid 1970s, **Les Amis de la Terre** is a federation of about 30 groups spread throughout Belgium to inform and educate the Belgian people. The idea is to increase the power and efficiency of each local organization by including it in a national network of conservationists and educators. The actions of the local groups are coordinated and combined regionally with other groups. Under the umbrella of Les Amis de la Terre, these groups provide a wide range of services and projects throughout Belgium: conferences, nature apprenticeships for the young, publication of a bimonthly review, and the distribution of educational materials.

▲ Projects and People

La Nature dans Mon Jardin (Nature in My Garden) is a project of Les Amis de la Terre that brings conservation into every home. Under the Direction of Le Ministère de la Conservation de la Nature (Minister for the Conservation of Nature), the project attempts to turn every private garden into a mini wildlife preserve. The project is aimed at "tomorrow's adults," that is, today's children, because Les Amis de la Terre believes that with interest and education that begins with children, a lifelong respect for the environment can be instilled.

Beginning with primary school children, Les Amis de la Terre produces films to help different age groups understand the richness and potential of gardens. For the adults, conferences and debates are organized through horticultural associations.

In Rwanda, **Alain Hanssen** has trained teachers and leads the International Scouts Troop summer camp for children. He also writes articles for environmental and Third World newspapers.

■ Resources

Most recent publications, printed in French, are *Nos Dechets Ne Sont Pas Une Fatalite*, *Haies Et Jardins Sauvage*, *Amenager Le Territoire Wallon*, and *Les Bois Tropicaux*.

Bond Beter Leefmilieu ⊕

Overwinningstraat 26
1060 Brussels, Belgium

Contact: **Luc Hens, Ph.D.**, Chairman
Phone: (32) 2 539 2217 • *Fax:* (32) 2 539 0921

Activities: Education; Political/Legislative • *Issues:* Air Quality/Emission Control; Energy; Global Warming; Health; Sustainable Development; Transportation; Waste Management/Recycling; Water Quality

▲ Projects and People

Dr. Luc Hens is also director of the Department of Human Ecology, Vrije Universiteit Brussel, where his research interests are urban ecology and environmental impact assessment. With D. Devuyst, he has recently written on the *Integration of Environmental Education into General University Teaching in Europe*. Other diverse concerns include "environmental education as an element of the quality of life in small towns," monitoring well water from waste deposit

sites, migrants from plastic food- and drug-packaging materials, international networking strategies in human ecology in Europe, and air pollution in Brussels.

European Environmental Bureau (EEB)
20, Luxembourg Street
B1040 Brussels, Belgium

Contact: **R. Van Ermen**
Phone: (32) 2 514 1250 • *Fax:* (32) 2 514-0937

Activities: Education • *Issues:* Sustainable Development

● Organization

A member organization describes the **European Environmental Bureau (EEB)** as the "largest and most representative and efficient international federation of NGOs." EEB is a federation of 140 nongovernmental organizations concerned about the environment in the European Community and representing some 35 million individuals. Established in 1974, the independent, international body has a Brussels-based secretariat and is governed by an Executive Committee with **Dr. Michael Scoullos** as recent chair. Prof. Scoullos, an oceanographer, is also president of Elliniki Etairia (Hellenic Society for the Protection of the Environment and the Cultural Heritage), Athens, Greece. *(See also Elliniki Etairia in the Greece NGO section.)*

The EEB follows closely and comments regularly on European Community environmental policies, with its views widely publicized in its *Memorandum to the Presidency*.

Objectives are to "promote an equitable and sustainable lifestyle," along with advancing environmental conservation and better uses of human and natural resources. In pursuit of these goals, EEB makes recommendations to the appropriate authorities. It also uses educational and other means to increase public awareness of environmental matters.

■ Resources

Memorandum is published semiannually.

Global Legislators for a Balanced Environment (GLOBE) ⊕
50, rue du Taciturne
1040 Brussels, Belgium

Contact: **François Roelants du Vivier**, Director
Phone: (32) 2 230 6589 • *Fax:* (32) 2 230 9530

Activities: Education; Political/Legislative • *Issues:* Biodiversity/Species Preservation; Conservation; Global Warming; Sustainable Development

● Organization

In 1989, parliamentarians from the European Community and the USA created **GLOBE** and were promptly joined by Japanese legislators. Their goal is to improve the state of the global environment through the legislative process, communications, and other means of international cooperation. International and European Community headquarters are in Brussels; GLOBE Japan is located in Tokyo; and GLOBE U.S. is in Washington, DC. U.S. Senator **Albert Gore** was elected president of Globe International in November 1990 for a two-year term. Vice presidents were **Takashi Kosugi**, Japanese Diet, and **Hemmo Muntingh**, European Parliament. Membership is open to English-speaking legislators of those bodies and the U.S. Congress who are committed to the environment. Each member organization has five votes at the biannual General Assemblies. Overall, their intent is to "bequeath a healthy planet to future generations."

GLOBE members pledge "to exchange information on a regular basis, to compare and improve the environmental legislation of the respective countries and regions, to support actions undertaken by individuals or groups working for global conservation, and most importantly, to propose similar new legislation" at home.

▲ Projects and People

Participants are organized into Working Groups: biodiversity, forests, dumping at sea, climate change, toxic waste, and trade and the environment. Earlier, a Working Group helped prepare for the 1992 United Nations Conference on the Environment and Development (UNCED). Their main function is legislation as "one rapporteur and at least one member of each GLOBE organization volunteer to tackle a specific environmental issue of worldwide interest," describes GLOBE. "At the GLOBE biannual General Assembly they report to their colleagues . . . and if necessary submit action plans to the members present for approval. These action plans usually involve guidelines to be worked into legislative texts as well as proposed steps of action." Comparing legislation is essential " . . . to ensure harmonized legislation acceptable to all countries," reports GLOBE.

Information is diffused on three levels: **General Assembly, publications, and daily communications.** Members meet regularly on a regional or national level and twice yearly on the global level to exchange experiences and formulate legislative plans. The *Facts-by-Fax Bulletin* is published bimonthly about proposed or adopted laws, news, and special global conferences. Proposed legislation is frequently transmitted to members by electronic mail.

■ Resources

A brochure describes GLOBE and lists its members.

International Federation of Chemical, Energy and General Workers' Unions (ICEF) ⊕
109, Avenue Emile de Béco
B-1050 Brussels, Belgium

Contact: **Victor E. Thorpe**, Deputy General Secretary
Phone: (32) 2 647 0235 • *Fax:* (32) 2 648 4316
E-Mail: Geonet/Poptel: geo2: icef-two

Activities: Development; Education; Political/Legislative; Research • *Issues:* Energy; Global Warming; Health; Sustainable Development; Waste Management/Recycling; Water Quality

● Organization

The resources of 250 member groups in 75 countries—representing 7,200,000 people—contribute to the International Federation of Chemical, Energy and General Workers' Unions (ICEF), which lobbies and forms trade union policy. Founded in 1907 by unions from several European countries, ICEF has offices in Tokyo, Rio de Janeiro, and Geneva—in addition to its Brussels headquarters. Current target issues are **economic democracy, occupational health, environmental protection, and disarmament.**

ICEF is an industrially based International Trade Secretariat (ITS) which works closely with the International Confederation of Free Trade Unions (ICFTU), International Labour Organization and World Health Organization (WHO) of the United Nations, and the Trade Union Advisory Committee of the Organization of Economic Cooperation and Development (OECD)—among other bodies important to labor.

When ICEF recently surveyed its membership, statistics showed that "general workers" are employed in two main sectors: office-based services, such as typists, clerks, computer operators, sales, and administrative staff; and "a rapidly growing, coalescing and internationalising new sector best described as 'environmental services,'" according to ICEF literature.

ICEF reports that the environment sector "has been recycled from a range of existing occupations, some of them ancient indeed. The new factor is the way in which they are being combined and globalised by multinational companies." ICEF affiliates include environmental workers ranging from "asbestos abaters to zookeepers." As defined by ICEF, main groups in the environmental services industry are **waste** (including management, collection, and disposal; toxic, hazardous, surgical, and nuclear waste disposal; sewage processing; catalyst handling; garbage sorting; energy from waste; and recycling), **pollution control** (including air and ventilation, site remediation, water and toxic waste treatment, landfill management, soil washing, impact assessment, resource recovery, health and safety, and asbestos abatement), **cleaning/maintenance,** and **utility services.**

▲ Projects and People

Chaired by ICEF vice president **Michael Nollet**, national secretary of the Belgian general union Centrale Générale/Algemene Centrale World Conference of General Workers, a recent ICEF World Conference of General Workers recommended these priorities:

• Creation of an ICEF Environmental Services Section.
• Greater exchange of information between unions. To this end, ICEF is issuing a new publication *Environmental Services*.
• Resistance to exploitation of workers, such as the use of the world's Free Trade Zones as sources of cheap labor and dangerous working conditions "particularly in the case of the many 'cowboy companies' operating in this field. This entails compiling existing legislation, urging the adoption and implementation of new laws, [and] lobbying appropriate intergovernmental bodies," reports ICEF.
• Trade union organizing campaigns, "particularly within the big environmental service internationals." ICEF is compiling a global "best practices" manual of organizing techniques; international union networking within such businesses is being promoted.

- Protection of environmental workers' occupational health and safety through development of ICEF's computerized information service on industrial substances and its hazard alert scheme.
- Training for ICEF affiliates on environmental service issues, especially in developing countries and in Central and Eastern Europe. ICEF's first environmental seminar was held in Czechoslovakia in late 1991.

ICEF regularly provides workers with information on handling flammable, hazardous, or poisonous industrial chemicals such as solvents—found in industries from "metal fabrication, petroleum, refineries, plastics, rayon, printing rubber, chemicals, laboratories to arts and crafts." According to ICEF, "solvents rank high in number of exposures and disability costs in terms of industrial hazards."

Vic Thorpe prepared a commissioned report, *The Death of Trees*, on the effects of transboundary air pollution on the pulp and paper industry. Citing the tree as a "very reliable early-warning indicator of life-threatening stresses in the global environment," Thorpe writes that evidence points to acid precipitation and pollution as the underlying and grave threat to temperate and boreal forests of the northern industrialized hemisphere.

"Over half the German and Swiss forests were officially recorded as diseased or damaged in 1986; the vast forest lands of North America and the Nordic countries show increasingly similar signs of stress. Studies indicate a long-term decline in tree growth in many regions," reports Thorpe. The diminishing forest resources and industries that emit "polluting gases now endangering the ecological balance" critically affect ICEF members.

As a result, ICEF's Pulp and Paper Industry Section adopted a resolution in 1985 to gather and disseminate information that will help trade unions on local-to-global levels lessen this pollution. ICEF is also helping its affiliates start action programs that include education "to convince our members that environmental concerns are bargaining issues"; measures to persuade other labor organizations, political parties, and social groups to commit themselves to a better environment; new coalitions to fight pollution; and intensified international cooperation through global agreements.

■ Resources

ICEF offers training and varied publications to trade union members via their sponsoring unions such as health and safety guides; *ICEF Regards* newsletters on health, security, and environment; and *ICEF Info* and newly released *Environmental Services* periodicals.

Royal Zoological Society of Antwerp
Koningin Astridplein 26
2018 Antwerp, Belgium

Contact: Frederick J.E. Daman, Director
Phone: (32) 3 231 16 40 • *Fax:* (32) 3 231 00 18

Activities: Education; Research • *Issues:* Biodiversity/Species Preservation • *Meetings:* CITES

● Organization

Two zoos, one natural reserve, and one natural history museum are part of the Royal Zoological Society of Antwerp. With the support of 20,000 members and a staff of 320, the Zoological Society undertakes a series of endangered species breeding programs and emphasizes its environmental education and research.

A distinguished staff includes veterinary and research department head De Meurichy, conservation biologist Helga De Bois, and Paul Van den Sande, who is general aquariumcurator at the Antwerp Zoo, secretary of the European Union of Aquariumcurators, and member of the Species Survival Commission (IUCN/SSC) Captive Breeding Group. Ethologist Linda Van Elsacker belongs to the Bonobo Advisory Committee; while Bruno Van Puijenbroeck, curator of mammals, is also the International Studbook keeper for the bonobo and okapi. Roland Van Bocxstaele, curator of birds, is vice president of the World Pheasant Association for the Benelux and International Studbook keeper for the Congo peacock. Biologist Chris Struyf, curator at Planckendael, has a special interest in the European otter, stork, and Przewalski's horse.

■ Resources
More funding is needed.

WWF–Belgique C.F. ⊕
Fonds Mondial pour la Nature
World Wildlife Fund—Belgium
Chaussée de Waterloo 608
1060 Brussels, Belgium

Contact: Christiane Linet, President
Phone: (32) 2 347 01 11 • *Fax:* (32) 2 344 05 11

Activities: Development; Education; Law Enforcement • *Issues:* Biodiversity/Species Preservation; Deforestation; Sustainable Development • *Meetings:* CITES; IUCN; UNEP

● Organization

WWF–Belgique C.F. is the French-speaking branch of the Belgian World Wildlife Fund; WWF has 40,000 members in Belgium overall, and this branch has a staff of 46. On the occasion of WWF's twenty-fifth birthday, the organization reiterated its efforts to sensitize the public about nature conservation problems and to raise funds to finance urgent projects.

▲ Projects and People

Educational services are WWF's main endeavors. The Educational Preserves project permits the observation and study of fauna and flora for understanding and verifying ecological processes and for measuring the amplitude of the degradation of the national environment. Such preserves can be created on school playgrounds to teach children the magnitude and variety of problems facing the planet and to enable them to answer questions directly about natural resources. WWF also encourages the integration of environmental subjects in all school disciplines so that students can begin to solve problems of pollution and declining species.

A project entitled La Loutre, The Otter, was begun by a group of zoo guides at Planckendael who are intent in breeding these mustelids in captivity, which is described as "arduous work," and in improving the surface water of their habitats. The guides put together an assembly of people from the private and public sectors who exchange ideas, experiences, and information and conduct activities and further studies. The Royal Zoological Society in Anvers prepared an enclosure for otters; the first candidates to arrive for breeding purposes at Planckendael were a female donated by the Arnhem Zoo, Netherlands, and a male from the Innsbruck Zoo, Austria.

Working with the Conservation Department in an **Investment for Nature** program, WWF-Belgium began a fundraising campaign, **Pro Natura Belgica**, in 1977 to create preserves for the protection of indigenous plants and wildlife. By 1991, 45 nature preserves had been established across the country, with 9 million Belgian francs raised through this WWF endeavor. The Conservation Department also campaigns to protect local endangered species and deals with international issues such as North Sea pollution and the rainforests.

TRAFFIC (Trade Record Analysis of Fauna and Flora in Commerce) is a specialized WWF branch which surveys the exploitation of plants and wild animals. It estimates, for example, that 60 tons of ivory was imported to industrialized countries in a single year, 1989. TRAFFIC EUROPE is headquartered in Brussels and has representatives in France, Italy, Germany, and the Netherlands. Among its current projects in Belgium, TRAFFIC informs customs officials and the Bureau of Water and Forests about CITES regulations, parrot trade in the Netherlands, consumption of tropical woods in European Community (EC) countries, and illegal trade of crocodile skins that are introduced into global commerce. (CITES is the Convention on International Trade in Endangered Species of Wild Fauna and Flora.)

WWF-Belgium is also active in Akagéra National Park in northeast Rwanda and several national parks in Zaire.

A **European bureau of WWF** has been established in Brussels as an environmental voice for EC. Among its concerns are Eastern Europe and the Mediterranean, Baltic, and North seas.

With a quarter-century of "patient devotion to the conservation world," **Christiane Linet** has been associated with the World Wildlife Fund since 1966—launching its magazine of which she is editor-in-chief and serving as "fundraiser, project organizer, consultant to WWF International, trustee of WWF Belgium," and now president of this branch. She contributes to two joint publications of WWF, United Nations Environment Programme (UNEP), and the World Conservation Union (IUCN)—the *World Conservation Strategy* and *Caring for the Earth*. A former journalist, Linet is a recipient of the **Order of the Golden Ark** bestowed by His Royal Highness Bernard, prince of the Netherlands. She belongs to the Species Survival Commission (IUCN/SSC) Otter Group.

■ Resources

Publications printed in French include a list of educational materials *Le WWF au Service de l'Education a l'Environnement*, an informational brochure *WWF Belgique—25 Ans au Service de la Nature*, and the periodicals *Panda*, *Panda Press*, and *Panda Junior*. Teachers can subscribe to group memberships. An informational center features printed materials, slides, and traveling exhibits appealing to children.

🎓 *Universities*

Faculté des Sciences Agronomiques (FSA)
Faculty of Agronomy Sciences
Zoologie Générale et Appliqúee
5030 Gembloux, Belgium

Contact: **Charles Gaspar, Ph.D.,** Professeur
           Ordinaire
Phone: (32) 81 62 2283 • *Fax:* (32) 81 61 4544

Activities: Education; Research • *Issues:* Biodiversity/Species Preservation; Waste Management/Recycling; Water Quality

▲ Projects and People

With the **Faculté des Sciences Agronomiques (FSA) de Gembloux,** Professor Dr. **Charles Gaspar** conducts research in these areas:

Chemical Mediators. Investigating the biochemistry of insects, Dr. Gaspar is centering his studies on nontoxic chemical protective substances of plants from temperate and tropical regions.

Biological Control. Dr. Gaspar belives that the study of predator-prey relationships is expected to promote better management of agroecosystems, and the study of plant-insect relationships will promote better quality control of food products.

A member of the Species Survival Commission (IUCN/SSC) Ant Group, Dr. Gaspar recently researched the relationship between **cereal grains and their insect pests** and sources for improvement of the protection of stored crops.

Pollination. The ecology and ethnology of solitary bees are being analyzed with the goal of using them in fruit growing and truck farming.

Management of **fish farming** in streams, rivers, and ponds is another pursuit.

RUG—Laboratorium voor Ekologie der Dieren ⊕
Institute of Animal Ecology
Rijksuniveristeit—Ghent
State University of Ghent
K. Ledeganckstraat 35
B-9000 Ghent, Belgium

Contact: **H.J. Dumont, Ph.D.,** Professor
Phone: (32) 91 64 54 53 • *Fax:* (32) 91 64 53 43

Activities: Education; Research • *Issues:* Biodiversity/Species Preservation; Water Quality

▲ Projects and People

In the framework of the Species Survival Commission (IUCN/SSC), Prof. Dr. **H.J. Dumont** is monitoring **dragonfly faunas in** Turkey and the Middle Eastern countries, including **Saudi Arabia, Oman,** and **Yemen.**

Professor Dumont is also involved in the following current projects at the **Institute of Animal Ecology, State University of Ghent:** climate reconstruction of the Holocene in the **Middle East, Africa, central Brazil,** and **Easter Island** in the South Pacific from sediment cores taken in selected crater lakes; study of the **aquatic fauna of Easter Island** in a historical perspective including the influence of introduced species and the presence of endemisms; management of **floodplain lakes** in the Niger Delta, Nigeria; and management of **man-made reservoirs** in the State of Sao Paulo, Brazil.

Future research plans include examining the ecological impacts of **Tucurui Dam in Amazonas** as well as the structure and function of freshwater lakes on the **Antarctic Peninsula.**

Dr. Dumont belongs to the IUCN/SSC Odonata Group.

Belize

Bounded by Mexico, Guatemala, and the Caribbean, Belize is a land of mountains, swamps, and tropical jungle. The inner coastal waters are sheltered by the longest barrier reef in the Western Hemisphere. Belize provides a wide range of habitats for its diverse wildlife. Mangrove swamp covers much of the low coastal plain, but the land rises gradually toward the interior. The Maya Mountains and the Cockscomb Range form the backbone of the southern half of the country.

Large mainland forests, which remain largely intact and unspoiled, make up Belize's most distinctive topographical characteristic. The abundant wildlife include jaguar, tapir, deer, crocodile, manatee, and many species of birds, turtles, and fish. There are 16 forest reserves and 8 wildlife sanctuaries, including the world's only jaguar preserve. While industrial projects have threatened Belize's natural heritage in the past, government and both national and international organizations are now working to preserve hundreds of thousands of acres.

NGO

Belize Audubon Society (BAS)
P.O. Box 1001
29 Regent Street
Belize City, Belize

Contact: **Janet Patricia Gibson,** Director
Phone: (501) 2 77369

Activities: Conservation; Development; Education • *Issues:* Biodiversity/Species Preservation • *Meetings:* IUCN

● Organization

After four years as a foreign chapter of the Florida Audubon Society, the **Belize Audubon Society (BAS)** became an independent organization in 1973 and now devotes itself to all aspects of conservation of the country's natural resources, concentrating on the education of children and the cooperation of the general public. A leading environmental group, BAS became the first Belizean member of the World Conservation Union (IUCN), belongs to the Central American network of non-governmental organizations known as its Spanish acronym REDES, is a member of the United Kingdom's Fauna and Flora Preservation Society, and is active in the Belize Tourism Industry Association.

▲ Projects and People

World Wildlife Fund support is helping BAS expand its **environmental education program,** which provides lectures and films free of charge; airs **Audubon Weekly** through the Broadcasting Corporation of Belize; publishes conservation booklets for schools; prepares guidebooks on local flora and fauna; maintains a small library on wildlife for teachers, students, and the public; and develops slide programs. **Protection of reefs and mangroves** is a main theme. The Society is also involved in the environmental education component of the country's coastal zone management project.

Regarding conservation, BAS works to preserve the **marine turtle** and runs a volunteer patrol program at one of the major nesting sites each season. It had the "rare **Jabiru stork** locally known as 'turk' added to the list of fauna protected by law" and conducted studies of its nesting activity. A **Christmas Bird Count** is taken annually and reported to *American Birds* journal. The Society has a "unique opportunity," it says, "in learning from the mistakes of other countries which have allowed unplanned industrial development, incorrect disposal of waste, and inadequate protection of rare and endangered plants and animals to destroy much of their heritage." BAS is determined to "maintain our environmental integrity [such as] our magnificent coral reefs and islands, spectacular wild areas and great diversity of wildlife" that are attracting tourists from around the world.

Since 1984, BAS has assisted the government in financing, developing, and operating protected areas established under the National Parks System. These include the following:

- **Half Moon Caye National Monument,** a marine reserve that supports "the only nesting colony of the red-footed booby in their white-phase adult plumage in the western Caribbean." More than 400 birds nest on this island with its rookery observation platform, educational programs for school children and visitors, and unique "rat control exercise."

- **Crooked Tree Wildlife Sanctuary,** which since 1984 has protected thousands of resident and migrant birds gathering at its several lagoons, swamps, savannahs, and lowland pine ridge during the dry season on their northbound spring migration. The area is prime habitat for the endangered Jabiru stork. Wild Wings and Underhill Foundations of New York are prime funders of this sanctuary, popular among tourists.

- **Cockscomb Basin Forest Reserve,** where hunting is forbidden throughout its 10,000 acres to protect the natural prey of **resident jaguars,** and which in 1990 became a Wildlife Sanctuary (CBWS). With support from the World Wildlife Fund, Cockscomb and its jaguars are internationally famous and attract such renowned visitors as HRH the Duke of Edinburgh, who came here as president of the World Wide Fund for Nature.

- **Guanacaste, a National Park** since 1990, is named for its "very large, old guanacaste or tubroos tree . . . supporting a great variety of epiphytes" and is noted for birdwatching.

- **Blue Hole National Park,** which is rich in undisturbed forests, wildlife, caves, and trails among its 575 acres.

- **Community Baboon Sanctuary,** which protects a healthy black howler monkey population in a volunteer grassroots program. BAS administers funds and works closely with an advisory committee comprised of landowners from the eight villages in the project. *(See also Community Conservation Consultants in the USA NGO section.)*

Scientific research is conducted in the 97,000-acre **Bladen Nature Reserve** in the Maya Mountains to encourage breeding habitat for many species; **Society Hall Nature Reserve** for genetic diversity in the subtropical wet forest; and at 7 small mangrove cayes declared as **Bird Sanctuaries** that provide nesting rookeries for wood storks, great and cattle egrets, boat-billed and tri-colored herons, reddish egrets, white ibis, frigate birds, and anhingas, among others.

■ Resources

These appear above in description of BAS environmental education.

The Belize Zoo and Tropical Education Center (TEC) ⊕ ❦

P.O. Box 474
Belize City, Belize

Contact: **Sharon Matola,** Director and Founder
   Phone: (501) 2 45523 • *Fax:* (501) 2 78808

Activities: Education; Research • *Issues:* Biodiversity/Species Preservation • *Meetings:* CBSG; IUCN, Regional; SSP

● Organization

A small group of Belizean animals once used as movie stars for natural history films were left in 1983 with no financial support for their maintenance. **Sharon Matola,** their caretaker, changed her title to zoo director and started a small zoo now known as **The Belize Zoo and Tropical Education Center (TEC),** which is visited by more than 20,000 persons yearly.

Staffed by young Belizeans, the zoo has grown dramatically and, with help from the U.S. Agency for International Development (USAID) and hundreds of supporters, recently relocated to a 28-acre site, formerly the Parrots Wood Biological Research Station. British Forces Belize and other volunteers, under the supervision of **Steve White,** constructed quality habitats for animals such as tamandua or anteater, gray fox, jaguarundi, ocelot, howler monkey, puma, margay, tayra, kinkajou, crocodile, and a tapir called "April," who was awaiting her mate "Cabel Sylvester," named through a school contest. Zoo animals, including some endangered species, are successfully breeding, reports Matola. The new visitors' center features small invertebrates, a gift shop with local Belizean crafts, and administration offices.

To be powered solely by solar panels from the local Siemens Solar Industries, the zoo and TEC expect to have "the country's largest solar system to help us become energy independent" as funding becomes available, says Matola. The overall purpose is to provide a "dynamic center for environmental awareness as well as a facility for research involving biodiversity and natural history studies on a grassroots and international level," explains Matola. The Center has developed a research network with Dartmouth University, New College, The Peregrine Fund, and the British Museum of Natural History. Belizean teachers and students are attending workshops and training here.

▲ Projects and People

An active education department takes wildlife and conservation education materials and programs into every district of the country in its **Belize Zoo Outreach** program. Director **Myrtle Flowers** previously conducted a conservation program for some 11,000 primary school students in Belize City.

Matola, working with the Belize Center for Environmental Studies for the World Wildlife Fund, led a 12-person expedition to the 103,000-acre, subtropical wet and lower montane wet Columbia River Forest Reserve in the southern Belize district of Toledo. They found uncommon species of **orchids** as well as **zamia,** a rare primitive seed plant that is believed to live up to 1,000 years. Herpetologist **Dr. John Meyer** found a **frog species** in the genus *Eleutherodactylus* never before described. Archaeologist **Paul Francisco** mapped a **cave** that had been used by the ancient Maya Indians and located and reconstructed a polychrome vessel from the cave, which was from A.D. 600 to 900. A new species for Belize's bird list, the common **woodnymph,** *Thalurania furcata,* was recorded. Other scientific expeditions are coordinated, including one to the nearly inaccessible Upper Raspaculo area, where tapir, morelet's crocodile, and flocks of scarlet macaws were sighted.

Matola, a former lion tamer, is also chairperson of the Species Survival Commission (IUCN/SSC) Tapir Group and is on the Board of Advisors of *Environmental Profiles.* She credits British filmmaker **Richard Foster** and the late **Chuck Hettel,** outdoorsman and diver, with creating the "turn of events" that led to the start of the Belize Zoo.

■ Resources

The zoo has an *Animals of Belize* coloring book and *Hoodwink the Owl* children's story. It put together a *Land Use Manual*, an environmental education guide for teachers concerning pollution, prepared in coordination with the U.S. Peace Corps and assistant director Amy Bodwell, formerly with Zoo Atlanta. Computers and funds are needed for educational materials.

Monkey Bay Wildlife Sanctuary
P.O. Box 187
Belmopan, Belize

Contact: Matthew Miller, Director

Activities: Education • *Issues:* Biodiversity/Species Preservation; Deforestation; Ecological/Natural History Tourism; Sustainable Development • *Meetings:* IUCN General Assembly

● Organization

Dedicated in honor of Earth Day 1990, Monkey Bay Wildlife Sanctuary serves as a model for private land stewardship that encourages conservation management, environmental education, and ecotourism. It lies on 1,070 acres of tropical forest and savanna, bordered by the Sibun River, which flows from the Maya Mountains through the coastal savanna on to the Caribbean Sea. Monkey Bay is home to a wide variety of flora and fauna, where species of both northern and southern Belizean hardwoods coexist. Support is derived from members, Rainforest Action Information Network (RAIN), Monkey Bay Wildlife Fund Tokyo, and Belize Center for Environmental Studies. Monkey Bay works closely with the government of Belize to encourage sustainable development and conservation; some 38 percent of the country is managed as parks, sanctuaries, and forest reserves. Even so, tropical forests "are being obliterated at the rate of 3,000 acres per hour," according to the organization. "They are home to 60 percent of the world's plant and animal species, most of which have not yet been documented by our scientists. In addition to the obvious loss of wildlife, the other effects of deforestation are desertification, massive soil erosion, flooding and the likelihood of major climate changes."

The Monkey Bay Wildlife Sanctuary is committed to the long-term environmental goal of preserving the living biological diversity of tropical America while supporting local development. It pays close attention to the wisdom of Chief Seattle of 1863: "Whatever befalls the Earth befalls the sons of the Earth. Man did not weave the web of life; he is merely a strand in it. Whatever he does to the web, he does to himself."

▲ Projects and People

An ideal location for students, researchers, and ecotourists to study at their Field Research Station, Monkey Bay cooperates with the Belize Zoo and Tropical Education Center in sponsoring an environmental education center used by Belizean youth groups and foreign visitors interested in tropical ecology and natural history.

Two aims of Monkey Bay's education message are land stewardship and restoration ecology. An on-site demonstration nursery produces native trees and plants for planting projects and landscap-

ing. Sustainable agriculture is another key component of land stewardship. Monkey Bay's bio-intensive organic garden demonstrates alternative and healthy home gardening practices.

Land next to Monkey Bay Wildlife Sanctuary is a protected wildlands corridor. The goal is to acquire additional land and expand the sanctuary boundaries.

■ Resources

Monkey Bay Wildlife Sanctuary has overnight camping facilities and cabanas for visitors in a "comfortable and beautiful natural setting." Educational tours are offered to secondary schools and university groups. The *Monkey Business* newsletter is published. Contributors can "adopt" one acre of tropical forest to be managed by the Sanctuary. Interns are needed for biodiversity studies and natural resource management planning.

(See also Community Conservation Consultants [CCC] and Programme for Belize in the USA NGO section.)

Wildlife Conservation International (WCI) ⊕
P.O. Box 282
Belize City, Belize 10460

Contact: Janet Patricia Gibson, Conservation Fellow
 Phone: (501) 2 44015 • *Fax:* (501) 2 74819

Activities: Development; Education; Research • *Issues:* Biodiversity/Species Preservation; Deforestation; Global Warming; Sustainable Development; Water Quality • *Meetings:* IUCN; UNCED; World Parks Congress

● Organization

(See also Wildlife Conservation International [WCI] in the USA NGO section.)

▲ Projects and People

Janet Patricia Gibson works with WCI in preparation of a management plan for the Hol Chan Marine Reserve and is a member of its Advisory Board. Gibson also completed a management plan for the Glover's Reef Atoll, which is being declared a marine reserve.

As coordinator of the coastal zone management project for the use and development of the coast and barrier reef complex of Belize, she is preparing an action plan for its use and development. In this capacity, she oversaw the preparation of a special report for Turneffe Islands which emphasized that "tourism development should not take place to the detriment of the fishing industry"; that "mangroves be left intact as far as possible" with ecotrails being acceptable to promote understanding of the mangroves' beauty and the birds and wildlife they support; that provisions be made for dealing with sewage and garbage disposal and water sources for proposed large-scale tourist resorts; that certain areas be designated as reserves to protect the "resident breeding population of crocodiles" and nesting sites of several endangered species of birds. As director and past president of the Belize Audubon Society, Gibson focuses mainly on marine conservation issues. *(See also Belize Audubon Society in this section.)*

■ Resources

In order to increase the understanding of environmental issues and natural heritage appreciation, WCI has programs for schools and user groups. Its center in Belize has brochures on various reserves, posters, a coloring book of reefs, books on *Snakes of Belize* and *Belizean Rainforest*, a *Checklist of Birds of Belize*, and a guide to Hol Chan Marine Reserve.

Wildlife Conservation International (WCI) ⊕
P.O. Box 37
Belize City, Belize

Contacts: **Bruce W. Miller**, Researcher
　　　　　Carolyn M. Miller, Researcher and
　　　　　　　Photographer
Fax: (501) 2 77062

Activities: Research • *Issues:* Biodiversity/Species Preservation; Deforestation

▲ Projects and People

Deep in the tropical forest, Bruce W. Miller, director of the **Belize Biosphere Reserve** and WCI research fellow, studies "bird diversity and relative abundance work of the forest understory and canopy" in the **Chiquibul Forest Project**, as he supervises a team of lepidopterists, or butterfly and moth scientists. Miller and co-director

Carolyn Miller are establishing a 1.4-million-acre Biosphere Reserve in some of the most remote areas of Belize.

Bruce also directs the **Man and Biosphere Program** in Belize to initiate economic development of buffer zones and multiple use areas along the edge of the reserve for local farming communities.

A wildlife photojournalist, Carolyn Miller conducts an avian census of the Vaca Plateau surrounding the archaeological site of Caracol. As part of her studies, she works with her husband in documenting the life history of the endangered **keel-billed motmot** (*Electron carinatum*), black catbird, and neotropical migrants, as well as the diversity of the tropical forests. Photo documenting forest conversion from air and on site, she was a consultant to the Programme for Belize's Rio Bravo Resource Management and Conservation Area, and has taken nature photos for the U.S. Fish and Wildlife Service and *National Geographic* magazine.

Bruce Miller is making conservation recommendations for *Pittidae*, a threatened family of tropical forest birds. As part of avian studies, he is also a vocalist for the Library of Natural Sound, Cornell University. In the Gallon Jug area, he and **Chan Robbins** surveyed wintering birds. Miller's book, the *Annotated Check-list of the Birds of Belize*, is being published in that country.

■ Resources

The **Wildlife Conservation International (WCI)** research center in Belize can provide slide shows and photos on rainforest diversity and adaptations to interested groups.

(See also Community Conservation Consultants [CCC] and Programme for Belize in the USA NGO section.)

Bhutan is a landlocked country located north of Bangladesh and at the eastern end of the Himalayas. The terrain is mostly mountainous and heavily forested. Bhutan is one of the world's poorest countries; more than 90 percent of the workforce is employed in subsistence farming and animal husbandry. However, only about 6 percent of the land can be used for farming or as pasture.

Much of the land is on steep slopes and susceptible to erosion. In addition, it contains high quantities of aluminum, which presents a toxic hazard to certain plants. In rural areas, only about 25 percent of the population has access to safe drinking water, and only about 7 percent has adequate sanitation services. This and a shortage of medical facilities make health problems a continuing concern.

🏛 Government

Department of Forestry
Royal Government of Bhutan
Thimphu, Bhutan

Contact: **Sangay Thinley**, Director
Phone: (975) 22487 • *Fax:* (975) 22395

Activities: Education; Development; Political/Legislative • *Issues:* Biodiversity/Species Preservation; Deforestation; Population Planning; Sustainable Development

● Organization

The philosophy of the **Royal Government of Bhutan** can be summed up in the words of the Lord Buddha: *"The forest is a peculiar organism of unlimited kindness and benevolence that makes no demands for its sustenance and extends protection to all beings, offering shade even to the axeman who destroys it."*

Bhutan is bordered by the "Tibetan autonomous region of China to the north, the Indian states of Sikkim to the west, Assam and West Bengal to the south, and Arunachal Paradesh to the east." Bhutan has a diverse geography and, as a result, has a variety of native species: takin, blue sheep, wolf, lamergier, snow leopard, Asian elephant, greater one-horned rhino, tiger, Asian buffalo, clouded leopard, golden langur (a very rare species of primate now thought only to be found in Bhutan), and the Himalayan black bear and leopard.

"The takin (*Budoreas taxicolor*), the national animal of Bhutan, is found at elevations of 7,000 to 12,000 feet in the dense bamboo thickets and the rhododendron and conifer forests of northern Bhutan," describes the **Department of Forestry**. "The takin has a convex face, heavy, mouth, and tremendously thick neck."

Long closed to the outside world and now beginning to open up, Bhutan is trying to learn from the mistakes of other countries in initiating a nature conservation program. With a population of 1.5 million that is increasing by about 30,000 each year, the country is preparing for development that can place "severe demands on the land for crop cultivation and livestock grazing," its literature reports. Bhutan wants to prevent destruction of underbrush and young forests, forest fire susceptibility, soil erosion, flood damages, and wildlife and habitat depletion.

The Royal Government considers the conservation program so important that His Royal Majesty the King personally helped draft a national policy to "ensure primarily the preservation of the environment, and only thereafter derivation of economic benefits that flow from a rationally managed resource."

▲ Projects and People

Sangay Thinley directs the Department of Forestry, the lead agency responsible for all environmental matters in Bhutan. Under its wise management, the Department of Forestry has saved over 60 percent of Bhutan's land area under forest cover. About 20 percent of the land has been set aside in 10 wildlife sanctuaries, reserved forests, wildlife reserves, and a national park with 2 additional areas proposed as nature reserves.

With the assistance of the World Wildlife Fund (WWF), the Royal Manas National Park, formerly a wildlife sanctuary, was upgraded to a national park. In southern Bhutan, the park is the home of many endangered species including the golden langur, the country's only viable herd of wild Asian buffalo, tiger, clouded leopard, pygmy hog, hispid hare, and greater one-horned rhinoceros, as well as a variety of exotic orchids and rare agar wood.

Also with WWF, the Department of Forestry is developing new ways to protect Manas for future generations by building guard posts, training staff, and providing necessary equipment. They are also working on opening the park to visitors by constructing a research and nature center and initiating conservation education and awareness programs.

The Department of Education is committed to bringing conservation to the schools by revising school curriculum to include more information on the environment and nature. With the consent of this department, nature clubs are opening in different schools. Sherubtse College is planning an undergraduate program in environmental studies and has started "Singye Karm," a nature, ecology, and trekking club.

Joining with such government aims, Dasho Peljore Dorjee leads the Royal Society for Protection of Nature (RSPN), which is the first NGO in Bhutan dedicated to conservation. Its annual art and essay contest draws interest and heightens environmental awareness of young people all over the country. The Society persuaded the government to declare the interior of Phobjikha Valley in western Bhutan a protected area for this wintering ground of the endangered black-necked crane.

Other conservationists in Bhutan are T.B. Moncar, deputy director responsible for planning and policymaking for wildlife and protected areas of the Forestry Department, and Tshering Tashi, head, National Environmental Secretariat, involved in environmental conservation planning under Bhutan's Planning Commission.

■ Resources

An informative brochure and a takin poster have been produced.

Bolivia is located precisely in the center of the South American continent, with the greatest portion of its territory made up of tropical lowlands belonging to the basins of the Amazon and La Plata rivers. Fifty percent of the population lives in rural areas; 67 percent suffers critical poverty. Massive migratory movements from the Andean regions to the tropical regions have resulted in severe pressure on ecologically fragile forests and prairies. This has caused a trend of nonsustainable resource use aggravated by hunger and need.

Flooding and landslides occur frequently during the rainy season in some areas, due largely to forest clearing, caused by logging, agriculture and grazing activities, and the resultant soil erosion. Generally, Bolivia's wildlife is still varied and plentiful; however, some native species are on the verge of extinction. The Bolivian government has initiated efforts to resolve some of these problems, but lack of personnel and funding continue to limit progress. Bolivia was the first country to participate in a debt-for-nature swap, setting the stage for more and more of these transactions between developed and developing countries.

🏛 *Government*

Corporación de Desarrollo (CORDECO)
Casilla 722
Cochabamba, Bolivia

Contact: **Victor Ricaldi Robles**, Director of Natural Resources and the Environment
Phone: (591) 42 25324, (591) 42 28251

Activities: Development; Political/Legislative • *Issues:* Deforestation; Population Planning; River and Soil Erosion Control; Sustainable Development; Water Quality

▲ Projects and People

Corporación de Desarrollo (CORDECO) is involved in projects regarding flood control, basin management, groundwater contamination, and environmental legislation.

Carlos Flores directs the **Flooding Control** project, the purpose of which is to regulate rivers in highland valleys of the **Cochabamba** state as well as agricultural land protection and urban development in flooding areas. Within five years, this project is expected to cover about 50 percent of the largest four basins of the state, reports CORDECO.

Martin Moll Cotesu, Cooperación Tècnica Suiza, directs the pilot **Basin Management** project on control and management of the basins in the Andes Mountains. The aim is to protect soil, forest, and water in the basin. It is a three-year project with large investments for the next ten years.

Leonardo Anaya directs the **Groundwater Contamination** project to "learn and determine the hydrogeological characteristics of groundwater in the central valley of Cochabamba where the capital of the state is established."

Victor Ricaldi directs the **Environmental Natural Legislation** project, with CORDECO acting as a regional environmental commission to assist congressmen in developing future environmental law—which is a "fundamental step [toward] a conservation and preservation policy." CORDECO will also help "implement the law regionally and establish the institutional structure to assess and approve environmental impact studies."

Because many Latin America countries lack technical capabilities to assess such studies, CORDECO says that it wants to help "within the country and within the continent [to] exchange ideas and experiences as well as [dealing with] problems on environmental matters."

■ Resources

RED bimonthly newsletter regarding groundwater contamination is available free of charge to Latin American geoscientists. The quarterly newsletter *Geosciencies* is also available in Latin America.

Herbario Nacional de Bolivia/National Herbarium of Bolivia
Casilla 10077—Correo Central
La Paz, Bolivia

Contact: **Mónica Morales**, Researcher
Phone: (591) 2 792582, (591) 2 792416
Fax: (591) 2 391176

Activities: Education; Research • *Issues:* Biodiversity/Species Preservation • *Meetings:* Congreso Latina Americano Botanica; Etnobotanica

● Organization

Established in 1983 in La Paz, the Herbario Nacional de Bolivia (**National Herbarium of Bolivia**) is a botanical research center created through a cooperative agreement. Participants were Universidad Mayor de San Andrés (major university of San Andrés) and the Nacional de Ciencias de Bolivia (Academy of Sciences in Bolivia) under the sponsorship of the Organization pro Flora Neotropica (Neotropical Flowers) of UNESCO. The Herbario describes its principal goal as "the study of the flora of Bolivia through floristic inventories, building up of botanical collections, and development of basic and applied research projects."

▲ Projects and People

The study of **flora and vegetation** is ongoing in La Paz in an inter-Andean valley; Wildlife Refuge Espiritu in the lowlands savanna; Biosphere Reserve of Beni; serrania Pilón Lajas and Alto Beni in the Yungas region; and the Andean Reserve Eduardo Abaroa.

With researcher **Inés Hinojosa**, the Herbario is conducting **ethnobotanical** studies of the Mosetene (Chimane) culture and other ethnic Bolivian groups regarding their uses of Bolivian plants.

Research with selected plant families features Bolivian palms, compositae, grasses, and mosses. At the Botanical Garden in La Paz, a long-term study of native plants has been established on the university campus.

Emilia Garcia E. is Herbario director. **Mónica Morales** is curator of palms and a member of the Species Survival Commission (IUCN/SSC) Palm Group. Other researchers as **Esther Valenzuela Celis**, whose specialty is "plants that 'fix' soils in the eroded areas of La Paz," and **Elizabeth Vargas**, who also teaches agroecology.

■ Resources

Ecología en Bolivia and *Comunicación MNHN* are available magazines from the Herbario.

 ## NGO

Asociación Boliviana para el Medio Ambiente (ABMA)
Bolivian Association for the Environment
Avinida Camacho 1277, 5th floor, Of. 503
P.O. Box 14174
La Paz, Bolivia

Contact: **Alfredo Ballerstaedt G.**, Secretary
Fax: (591) 2 09322007

Activities: Development; Law Enforcement • *Issues:* Air Quality/Emission Control; Biodiversity/Species Preservation; Deforestation; Energy; Sustainable Development; Water Quality • *Meetings:* CITES; International Tropical Harvester Organization

● Organization

The Asociación Boliviana para el Medio Ambiente (ABMA) (**Bolivian Association for the Environment**) is a new, nonprofit organization interested in the "adequate management" of Bolivian natural resources. The organization is linked to other national and international groups dealing with ecological issues.

▲ Projects and People

ABMA recently started projects concerned with management of watersheds and forests; conservation of Amboró National Park; creation of the Wild Life Sanctuary; and assistance to the indigenous groups.

The **watershed management** project is beginning basic studies of land capability, watershed damage, and current land use—including conflicts. The Association is also undertaking physical work for watershed maintenance, protecting watershed zones, and doing hydrological and meterological monitoring—although funds are lacking to develop and accomplish all of its targets.

The **forest management system** project is analyzing Bolivia's extraction methods in a certain region as well as various forest management strategies. ABMA is implementing a demonstration project on **strip shelterbelts** for sustained forestry production.

Management and conservation of Amboró National Park focuses on protecting the borders of the park to regulate colonization, lumbering, and hunting. The project also intends to quantify ecosystems and floral and faunal resources. As a management plan is developed, the park will be promoted with the aim of attracting national or international funding and developing tourism.

Regarding the creation of a **Wild Life Sanctuary**, whose mission is to establish a protection zone around the **guacharo colony** (*Steatornis caripernis*), a management plan is being developed for its continued maintenance.

The Association is assisting the **indigenous Yuki and Yuracaré groups** to determine the population size and distribution, to assist with integration into Bolivian society with respect to their cultural values, and to help with land titles, health care, education, and agriculture.

With **Alfredo Ballerstaedt Goytia**, ABMA staff includes **Harry P. Brindley, Jose Iporre Bellido, Lincoln Quevedo Hurtado,** and **Adolfo Ernesto Meave Berkhoff.**

■ Resources

ABMA writes that a main problem is "lack of information with regard to the most suitable systems to preserve valuable forestry species from extinction" and requests data on strip shelterbelts or any other systems to achieve this aim.

Help is requested in obtaining *Vetiveria zizanoides* to aid farmers in the valleys and tropics in controling soil erosion. A grass plantlet, "*vetiveria* has already adapted to different Bolivian ecological regions" for this purpose with assistance from the Research Center in Tropical Agriculture, Colombia," reports the Association. Funding is requested from international sources to continue assistance to indigenous groups.

Funding is also sought for studies of humid forest sustainable development in order to hire experts in this field.

The Association provides technical assistance in developing programs and specific projects to stop the destruction of forests and wildlife in Bolivia. Results of field studies and endeavors will eventually be published to enhance global cooperation on environmental matters.

Centro de Datos para la Conservación (CDC-Bolivia)
Conservation Data Center
Cota Cota, Calle 26 y Avinida Muñoz Reyes s/n
Casilla 11250
La Paz, Bolivia

Contact: **María R. de Marconi, Ph.D.**, Executive Director
Phone: (591) 2 797399 • *Fax:* (591) 2 797399

Activities: Education; Political/Legislative; Research • *Issues:* Biodiversity/Species Preservation; Deforestation; Sustainable Development

● Organization

Improper use of Bolivian natural resources resulted in their deterioration, and in some cases, their disappearance. The lack of a scientific and technical basis for aid in planning and executing environmental projects was cited as one crucial cause. Through a technical and financial agreement with The Nature Conservancy, CDC–Bolivia, Conservation Data Center, was formed and legalized in 1987. Its goal is to bridge the gap between the information gatherers and those who use this knowledge. A member of LIDEMA, World Conservation Union (IUCN), International Geographic Union, and the Data Network for Conservation, CDC–Bolivia maintains close ties with all institutions dedicated to the conservation of Bolivia's natural heritage.

▲ Projects and People

In the 1990–1995 strategic plan for the Conservation Data Center, two information systems are designated: "a databank containing current and accurate information about the biological resources of Bolivia [and] a system of data management and distribution which is rapid, useful, and responsive to the needs of the users." The underpinnings are a "highly trained scientific-technical team," evalu-

ation of the country's biological heritage in varying geographic areas, continual updating and reevaluation of computerized information, and a lending services system.

These systems are used to develop regional or national land-use strategies, establish nature preserves, design and monitor management plans for plant and animal species, survey wildlife refuges, prepare legislation and legal arrangements, evaluate environmental impacts, create educational and public awareness programs, and generate similar databases in other institutions or projects. CDC targeted the Amazon, Valleys, Altiplano and Chaco areas to evaluate Bolivia's natural heritage.

■ Resources
A list of publications, printed in Spanish, is available.

Centro Interdisciplinario de Estudios Comunitarios (CIEC)
Interdisciplinary Center for Community Studies
Avinida Ecuador 2459, 2° piso
Casilla 159
La Paz, Bolivia

Contact: **Erick Roth**, Director
Phone: (591) 2 360583 • *Fax:* (591) 2 328933

Activities: Development; Education; Research • *Issues:* Health; Sustainable Development; Environmental Education

● Organization

CIEC, the Interdisciplinary Center for Community Studies in both urban and rural areas, was founded in 1984 and fully legalized by the government four years later. In working on community development, CIEC's approaches to problems involving the environment, health, productivity, and education have been varied and include interdisciplinary training, advice, and information sharing with and for the community. A number of scientists and consultants at CIEC integrate these issues.

CIEC is affiliated or works with the Committee of Man and the Biosphere, Bolivia; Environmental Liaison Centre, Kenya; Andean Subregional Front of NGOs for Environmental Education, Ecuador; World Conservation Union, Switzerland; Social Science and Tropical Disease Network, Venezuela; and Panamerican Network for Information and Documentation on Sanitary Engineering and Environmental Sciences, Peru. Among others, the World Wildlife Fund—USA, Conservation International, U.S. Agency for International Development (USAID), and U.S. Fish and Wildlife Service have provided support.

▲ Projects and People

CIEC's primary goal is the preservation of the health of both the environment and the people, particularly regarding marginalized social groups. CIEC sees the solutions to problems such as high rates of morbidity and mortality as only possible with "the active involvement of those affected. It is they who must, in the end, develop the appropriate framework to fight against those adverse conditions."

Appropriate **educational methods and technology** are developed to help change attitudes and awareness of the community. Offered are national and global health and environmental workshops, including those held on the role of private voluntary organizations (PVOs), and advisors in these fields. Children's environmental literature has been prepared and informal community programs aim at "developing self-sustaining activities."

Environmental projects target, for example, the **area of influence of Reserve of the Biosphere: Biological Station of Beni.** The only bear native to South America—the spectacled bear—is endangered because of deforestation and hunting. Because it is mainly herbivorous, the bear plays "an important role in distributing seeds in the forest," CIEC literature points out.

Courses are taught on traditional medicine and folk remedies, sanitary education, child survival and health, vector biology and control, agroecology, and water and sewage disposal in rural communities. CIEC also conducts research and provides support and feedback through evaluations of projects, reviews of reports, and examinations.

Erick Roth is a founding member of the Bolivian Association of Behavioral Analysis and has been a psychology professor at universities in Bolivia and Mexico. He is ex-director of the School of Public Health and of the Direction of Human Resources of the Ministry of Health of Bolivia.

■ Resources

CIEC has produced more than 25 publications, radio programs/announcements, and at least 8 videos ranging in length from 8 minutes to 2 hours that address 4 main issues—community, health, environment, and drug use. Current materials printed in Spanish are *Manual de Huertos en el Trópico* (Tropical orchards manual), *Manual de Educación Ambiental* (Environmental education manual), *Manual de Organizacion y participación comunitaria* (Community organization and participation manual), *Manual del responsable popular de salud en el trópico* (Tropical people's health manual), and *Actitudes y Patrones de Conducta hacia el Medio Ambiente* (Attitudes and behavioral patterns toward the environment).

CIEC needs instructional aids.

Conservación Internacional—Programa Bolivia ⊕

Avinida Villazón 1958, Edificio Villazon
Piso 10, Of. 10 A, Casilla 5633
La Paz, Bolivia

Contact: **Guillermo Rioja Ballivián,** Project
               Coordinator
Phone: (591) 2 341230, (591) 2 320341
Fax: (591) 2 352279

Activities: Development; Education; Law Enforcement; Research • *Issues:* Biodiversity/Species Preservation; Deforestation; Health; Sustainable Development

● Organization

Ecosystem conservation is the challenge of the 1990s and the goal of **Conservation International (CI).** In Bolivia, **Conservación**

Internacional is particularly working with native communities to help bring about better livelihoods, "more economic autonomy, steps toward greater political influence, and self-determination." *(See also Conservation International in the USA NGO section.)*

▲ Projects and People

Working with Chimane Indian communities in the rainforest near the Beni Biosphere Reserve, for example, CI is helping improve the "harvest and marketing of *jatata* palm" (*Geonoma deversa*), which has been used locally for a **thatch roofing material** for at least 200 years for native and foreign settlers in Bolivia's Amazonian region. Known for their ability to weave roof panels from the palm, the Chimanes, nevertheless, suffer from "unimaginable economic exploitation and inhumane treatment." CI reports that "increases in alcoholism and overt violation of . . . rights are two examples."

Through the **Jatata Project,** a local Committee of Self-Determination was formed to break the monopoly over the market, "presently controlled by 'outsiders' of the 'national society,'" according to CI. Achieving financial autonomy, the Chimane people are finding a way to achieve "considerable pride and dignity." Studies are ongoing in ethnobotanical research and to determine other sustainable uses of the palm, which is also found in Central America and other parts of the Amazon Basin in South America. Conservation Databases (CDC–Bolivia) is now preparing to train a group of Chimanes in biological research. Through a radio broadcasting project, the Biological Station in Beni (EBB)—with its program of **Environmental Education, Communication and the Public Use of the Biosphere Reserve**—also assists in training. CI points out that "the Jatata Project is far from completion."

Generally, projects are oriented toward helping to **integrate Indian groups and peasants into the national socioeconomic system;** to implement sustainable development at the grassroots level; and to create organizations among indigenous people to facilitate their participation in politics, development of autonomy, and ability to communicate needs at the community level and to the government when necessary.

CI projects also aim for the "defense and recovery of traditional lands and resources, recognition of rights to exist as distinct peoples with different cultures and beliefs, [prevention] of all forms of repression and violence against Indians, participation of indigenous groups in the initial planning of large-scale development projects, and preservation of the forest implemented by indigenous peoples who have lived in the rainforest for centuries."

An anthropologist and specialist in ethnomusicology, Guillermo Rioja prepares seminars on this subject as well as lectures on sustainable development and rights of native Bolivians. He directs the Jatata Project.

■ Resources

CI seeks studies of ecology of tropical forests' nontimber resources. CI publishes the quarterly *Tropicus*. Printed in Spanish are *Conservación, Desarrollo y Comunicación* (Conservation, development, and communication), and *Bolivia—Medio Ambiente y Ecología Aplicada* (Environment and ecology efforts).

ENDA–Bolivia ⊕
Casilla 9772, El Alto
La Paz, Bolivia

Contact: **Michel Gregoire,** Coordinator
 Phone: (591) 2 811695 • *Fax:* (591) 1 811446

Activities: Education • *Issues:* Health; Population Planning; Street Children Prevention; Sustainable Development; Waste Management/Recycling • *Meetings:* CHILDHOPE; UN; UNICEF

● Organization
According to *El Riesgo de Ser Menor* (The risk of being young), El Alto, a section of La Paz, is known as "the youngest city in Latin America" and the poorest—with per capita income about $US 4.00 in 1988. Statistics from the UNICEF report note that 3 out of 10 people have potable drinking water at home and of the remaining 7, only 4 have access to water through public fountains. Some 80 percent of buildings lack electric lighting, and barely 4 kilometers of road are paved over.

Founded in 1988 by **Michel Gregoire,** ENDA–Bolivia is part of **Environment and Development in the Third World (ENDA–TM),** an international organization headquartered in Dakar, Senegal, with branches in Africa, Asia, and Latin America and dedicated to a "South–South dialogue" among Third World countries. ENDA–Bolivia seeks a "comprehensive prevention of the marginalization of street children."

With a staff of 30, ENDA proposes "comprehensive alternative development grounded in creativity, self-esteem, solidarity, and mutual respect to **make high-risk children into responsible citizens** and present and future participants in society." This is done through participatory action, research, communication, service, work, and social reintegration. *(See also ENDA–Caribe in the Domincan Republic NGO section.)*

▲ Projects and People
Young workers on the street use donated equipment to **collect waste for recycling** from shops, factories, offices, restaurants, hotels, workshops, and residential areas in La Paz and El Alto. The purpose is to create a dynamic of participation in the community and to reduce the importation of basic materials for local industry. Businesses, organizations, and individuals agree to have their trash picked up by the workers.

In so doing, the project contributes to the improvement of the population's health and hygiene as well as environmental protection. The activities raise community awareness of the production, management, and disposal of solid waste while making solid waste into something of value. Materials are taken to the EMBOL recycling center.

ENDA also participates in drug abuse prevention programs. It reaches young people through active teaching, theater and multifunctional workshops, suburban farm projects, and audio-visual presentations. A library and computer center are among ENDA's facilities.

■ Resources
Printed in Spanish, materials include *El Riesgo de Ser Menor, El Presente en la Ciudad del Futuro* (The present in the city of the future), *Alternativas para una Riqueza Olvidada* (Alternatives for a forgotten wealth) and *Sistema de ahorro para el Menor Trabajador* (Savings plan for the young worker).

Technology needs are solar energy for community showers and help with business administration—particularly for small alternative businesses.

Fundación Amigos de la Naturaleza (FAN) Friends of Nature Foundation
Avinida Irala 421
P.O. Box 2241
Santa Cruz, Bolivia

Contact: **Hermes Justiniano,** Executive Director
 Phone: (591) 33 33806 • *Fax:* (591) 33 41327

Activities: Education; Law Enforcement • *Issues:* Biodiversity/ Species Preservation; Sustainable Development

● Organization
The Fundación Amigos de la Naturaleza (FAN) (Friends of Nature Foundation) is a private, nonprofit Bolivian conservation organization established in 1988 to protect the country's biological diversity. Working with government and non-government groups and international agencies, FAN's goals are to secure proper management of Bolivia's protected areas, identify and promote the establishment of new protected areas, train professionals in conservation, and develop and implement an environmental conservation program. It has a staff of ten.

▲ Projects and People
FAN wrote the first management plan for the **Noel Kempf National Park,** calling for its protection and administration. In December 1990, FAN purchased the 25,000-acre **Flor de Oro,** a strategically situated ranch in the northern section of the park ". . . to serve as a focal point for natural history tours and scientific and research expeditions."

Protecting the southern and western limits of **Amboró National Park,** FAN hired, equipped, and trained guards because of the area's vulnerability to the large number of communities nearby. "Protection and management efforts are being coordinated between FAN, the UTD-CDF (the government agency officially responsible for the protection and management of protected areas), and the **Chimore-Yapacani Project** (supported by the Inter-American Development Bank)," according to a FAN report. FAN works with these guards daily. At the same time, outreach programs are underway with the help of **Edwin Villagomez,** a community extensionist working with families living in or near the park and teaching them how to respect and care for its resources. With the assistance of Peace Corps volunteers, FAN launched another extension program—a tree nursery project in Samaipata.

Both national parks receive support from the Parks in Peril Program, a joint effort of the U.S. Agency for International Development (USAID), The Nature Conservancy, and FAN.

Because of the joint work of FAN and UTD-CDF, the **Rios Blanco y Negro Wildlife Reserve** is now a protected area, where **Dr. Andrew Taber,** a research fellow with **Wildlife Conservation Inter-**

national (WCI), has begun a long-term study of peccaries, tapirs, deer, and otters with the help of local biology students. Dr. Taber's research will enable FAN and UTD-CDF to prepare a management plan for the reserve. FAN regularly provides aerial surveys which help determine immediate and future actions as well as assist UTD-CDF in controlling "illegal lumber extraction, hunting, fishing, and capture of live specimens for the pet trade."

Cristian Vallejos is development director; Nicole Martinez is development coordinator; Hugo Salas is project officer; Luis Marcus helps implement the Parks in Peril Program; and biologist Marielos Peña assists with the environmental education program.

■ Resources

FAN needs educational materials and technical expertise on environmental outreach.

Liga de Defensa del Medio Ambiente (LIDEMA)
League for the Defense of the Environment
Avinida 20 de Octubre, No. 1763
Casilla 11237
La Paz, Bolivia

Contact: Carlos Enrique Arze Landivar, Executive
            Director
Phone: (591) 2 356249, (591) 2 353352
Fax: (591) 2 392321

Activities: Development; Education; Law Enforcement; Research • Issues: Biodiversity/Species Preservation; Deforestation; Energy; Sustainable Development; Water Quality

● Organization

LIDEMA, the League for the Defense of the Environment, was founded in 1985 and legally recognized in 1987. With a staff of 9, it has 16 national member institutions and functions as an interinstitutional organization coordinating multidisciplinary programs, projects, and actions. LIDEMA acts on the belief that "it is important to maintain the genetic potential of all living species and the productive capacity and quality of soils, water and air, for the benefit of humans in all their cultural diversity as well as to protect the natural patrimony."

In recent years, LIDEMA has encouraged the formation of the National General Environmental Secretariat (SEGMA), the National Environmental Fund (FONAMA), and the Congressional Environmental Commission for the Environmental Law Project; LIDEMA has also encouraged the launching of sustainable development projects through Bolivia's Environmental Action Plan (PAAB) and the Tropical Forestry Action Plan (TFAP).

Its members are Bolivian Wildlife Association (PRODENA), Conservation Association of Torotoro (ACT), Ecological Association of the East (ASEO), Data Bank Center for Conservation (CDC), Center for Ecological and Integral Development Studies (CEEDI), Interdisciplinary Center for Community Studies (CIEC), Research Center for the Greater Capacity of Use of the Soil (CUMAT), Bolivian Andean Club (CAB), Biosphere Reserve: Beni Biological Station (EBB), Friends of Nature Foundation (FAN),

Institute of Ecology (IE), National Museum of Natural History (MNHN); "Noel Kempf Mercado" Museum of Natural History (MHNNKM), Multiple Services on Appropriated Technologies (SEMTA), Bolivian Society of Landscape Architects (SAPBOL), and Bolivian Ecological Society (SOBE).

▲ Projects and People

LIDEMA works to raise national consciousness regarding environmental issues while supporting effective conservation. As an informative service, LIDEMA carries out environmental impact assessments. It also instigates agreements between the polluting industries and affected populations, following through with technical advice and assistance in industrial transitions. One such example resulted in dust reduction, as caused by a cement company in one town. It fosters the passage of natural resources legislation and supports such legislation once enacted. LIDEMA provides academic training for professionals in applicable fields and is involved in basic and applied research to gain knowledge about present and future projects.

Among other activities, LIDEMA provides courses and scholarships, monitors illegal activities in forests, organizes Earth Day and World Environment Day celebrations, supported a video production on the spectacled bear and produced other documentaries, prepares public service announcements, evaluates ecologically specific areas and species, attends and sponsors environmental workshops and seminars, and collaborates with global and other national resource groups on common causes.

With the goals of the member institutions intertwined with those of LIDEMA, each supports the other. One example is a joint effort to gain the government's establishment of the Sistema Nacional de Areas Protegidas (SNAP) (National System of Protected Areas), or a "network of parks."

■ Resources

The Annual Report, printed in Spanish and English, describes member organizations, activities, and resources. LIDEMA publishes the monthly newspaper Habitat and other materials.

Protección del Medio Ambiente Tarija (PROMETA)
Environmental Protection Group
Cocon 922
Tarija, Bolivia

Contact: Gabriel Baracatt Sabat, Director
Phone: (591) 66 25865 • Fax: (591) 66 22062

Activities: Education; Research • Issues: Biodiversity/Species Preservation; Deforestation; Sustainable Development

● Organization

Established in 1990 by concerned university students and young professionals in southern Bolivia as Tarija's first conservation organization, Protección del Medio Ambiente Tarija (PROMETA) (Environmental Protection Group) seeks to "protect Tarija's natural environment, raise concerns, and promote adequate uses of the natural resources." This nonprofit, private group is targeting "social and economic causes of environmental degradation and resource management" to develop a long-term strategy for sustainable development. It operates mainly with volunteers, small contributions, and support from the U.S. Park Service, World Wildlife Fund (WWF), and the Farnwood Foundation.

Tarija is Bolivia's southernmost region of contrasting ecosystems—with the Andes Mountains running along its western edge and the valleys to the east, Tarija's first center of grape and wine production. Among the wet cloud forests, both the endangered **guanaco** and **spectacled bear** still roam about.

▲ Projects and People

In 1989, the Bolivian government established the **Tariquia Forest Reserve** to protect nearly 250,000 hectares of such pristine tropical cloud forests. However, the reserve lacks administration, personnel, or unmarked boundaries, currently subjecting Tariquia to illegal logging operations. According to PROMETA, the reserve is an "urgent priority for conservation action" and plans are underway to survey problem areas and to work with small-scale farmers and other local people to preserve the tropical forest.

PROMETA points to Tarija's central valley—a former Pleistocene lake bed that exemplifies "one of the most dramatic cases of soil erosion in Latin America," where about 1,800 hectares of rich agricultural soils are lost every year. As a result, farmers are likely to flee the area and move into the pristine forests. Even though attention is being focused on the technical aspects of land recuperation and erosion control, PROMETA says that social considerations of this problem are being overlooked.

Consequently, PROMETA is embarking on **environmental education** regarding soil erosion dangers and the "need for socially acceptable alternatives as an urgent priority for action." To achieve this conservation effort, PROMETA fostered and created a youth movement, **CERCA**, whose purpose is reforestation and environmental protection in the city of Tarija. CERCA reforested the city's first suburban park and developed an outreach program to sensitize other local youth organizations about the region's main environmental problems.

■ Resources

PROMETA's educational resources include a video describing the natural beauty and abundant wildlife and woods of the Tariquia Forest Reserve. PROMETA gives workshops and photography presentations about conservation.

Botswana's protected area system is one of the world's best; almost 18 percent of the country consists of protected areas, representing all of Botswana's major habitats.

One of the country's greatest environmental problems is the spread of pastoralism and the increase in cattle. Since 1966, the cattle population has more than doubled, leading to overgrazing and land degradation. This has been compounded in the Okavango Delta, Botswana's most important ecological site, by near eradication of the tsetse fly, which previously had deterred cattle from grazing in the area. Deforestation is also occurring as the result of overgrazing, fires, or overconcentration of wildlife (particularly elephants). In addition, collection of fuelwood is very widespread. The country is now in a transition period as it attempts to strengthen its natural resource management system.

 NGO

Forestry Association of Botswana (FAB) 🌿
Lekgotla La Temo Ditlhare
P.O. Box 2088
Gaborone, Botswana

Contact: **Andre Kooiman**, Research Manager
Phone: (267) 351660 • *Fax:* (267) 300316

Activities: Education; Extension; Research • *Issues:* Biodiversity/Species Preservation; Deforestation; Energy; Sustainable Development

● Organization

Founded in 1983 by a group of people troubled by the destruction of the area's woodlands and rangelands, the **Forestry Association of Botswana (FAB)** works throughout Botswana to prevent deforestation and land destruction and to research and educate on related issues, such as overgrazing and overcutting.

Run by 19 staff members, FAB has developed several chief concerns and ambitions. First of all, the organization wishes to "promote public awareness of the importance of trees, forests, tree-planting, and forest destruction by means of extension, education, and social and agroforestry programs." FAB also aims to promote and generate forestry research, and support and cooperate in the efforts of the government and other environmental organizations in the conservation of woodlands and the management of natural resources.

A Board of Trustees meets at least four times a year. The Association is divided into two branches, extension and research, and each branch conducts its own individual and applicable activities.

Supporting FAB are a number of agencies interested in the field of forestry and rural development support, including Humanistic Institution for Development Corporation (HIVOS), Netherlands; International Research Development Centre (IDRC), Canada; Norwegian Agency for International Development (NORAD); and the Netherlands Development Organization (SNV).

▲ Projects and People

Extension activities try to enhance public awareness and "technical know-how" in forestry and other environmental concerns. FAB leads the **Around-the-Home Tree Planting** program, which promotes tree planting in rural areas of Botswana.

FAB visits community schools, and establishes **school tree nurseries** to intensify the environmental curricula and to give children hands-on experience in forestry.

Cooperating with **Dr. Gaositwe K.T. Chiepe**, Minister of External Affairs, FAB co-sponsors **National Tree Planting Day** and also works collectively with the government and organizations on researching **species survival trials** throughout the country and on farms to test the effectiveness of trees in agricultural production.

Research programs include tree seed studies to collect, store, and treat seeds, which are distributed to organizations and the public. Other enterprises include nursery research, tests on the suitability of tree species for production of firewood and fodder, and inventory of available woody biomass and natural woodland management research in southeast Botswana. Occasionally, FAB does consulting on environmental issues that it has studied.

FAB plans to initiate new projects in forestry and in the regulation of wood resources. FAB also has plans for "the use of treated sewage effluent for the irrigation of trees."

Land ecologist and research manager Andre Kooiman is concentrating on "forestry planning studies using remote sensing and geographic information systems (GIS), silvicultural research, and community forestry products." Previously working in Southeast Asia, Haiti, and the Netherlands, Kooiman chairs the research working group of the Steering Committee Forum for Sustainable Agriculture, Botswana.

■ Resources

FAB sells tree seedlings to the public at its headquarters along with FAB T-shirts and brochures. FAB is developing an Information Center—a library of all available materials on forestry and related subjects, including slides, videos, books, and leaflets.

FAB provides its members with a quarterly newsletter and an annual journal about current events, programs, and research. A list of publications, reports, and papers is available.

Kalahara Conservation Society (KCS)
Botsalano House, The Mall
Gaborone, Botswana

Contact: **Eleanor Patterson**, Chief Executive Officer
Phone: (267) 31 314259 • *Fax:* (267) 31 374557

Activities: Education; Research • *Issues:* Biodiversity/Species Preservation; Conservation • *Meetings:* IUCN General Assembly

● Organization

The **Kalahara Conservation Society (KCS)** promotes the conservation of Botswana's natural heritage through education, publicity, and the financing of research projects. It supports the implementation of Botswana's National Strategy, which includes such broad environmental issues as "the conservation of water resources, ensuring proper management of natural pastures, [and] teaching people to live within the constraints of their environment." A member of the World Conservation Union (IUCN), KCS says it is consulted by international organizations.

The KCS has received funds from World Wide Fund for Nature (WWF), United Nations, European Community, Norwegian

Agency for International Development (NORAD), Swedish Society of the Conservation of Nature, and U.S. Agency for International Development (USAID), as well as members' contributions.

▲ Projects and People

Since its inception in 1982, KCS sponsored aerial surveys of elephant populations. In northern Botswana, a study is being conducted to evaluate the elephants' interactions with their habitat. KCS buys equipment and vehicles to help patrol poachers.

Conflicts over land use develop as Botswana's population rises. KCS is promoting several key plans to promote land use suited to the country's varying environmental conditions. "A recent success resulted in the extension of **Nxai Pan National Park** to include the scenic surroundings of Baines baobabs," reports KCS literature.

The Okavango Delta has various, unique ecosystems which necessitate sensitive planning. Some parts of the delta require complete protection because of the animals and plants they harbor, while other areas must be made available for recreational and subsistence use, such as fishing, hunting, and photographic tourism. KCS has conducted a major ecological zoning project to define these areas. KCS also sponsors an ongoing project to monitor the effect on the fish in the delta due to tsetse fly spraying.

Botswana permits hunting of some species in designated areas. A computerized hunting permits process provides valuable information for wildlife management. This successful project prompted the Department of Wildlife and National Parks to use the system nationwide. Among other projects, KCS studies wild dogs, evaluates the value of tourism to the Botswana economy, and examines the effects on the Okavango woodland of increasing use of *mokoros*, or canoes, among other projects.

KCS's education programs are "given high priority." More than 80 clubs, now autonomous, have been established from elementary through college level, including one at the University of Botswana. Students organize tree-planting days, encourage awareness of sensitive nesting areas for Cape vultures and African skimmers, and participate in international conservation conferences. Environmental awareness is extended through a series of teaching aids and workshops to be integrated into the school curriculum. An environmental magazine called *Action*, written for primary school students, supports this classroom curriculum.

KCS lobbies government officials and has personal contact with chiefs, council secretaries, and other government officials. A series of videos, information sheets, and direct mailings are used to reach these individuals. As a result of this contact, a community-based wildlife sanctuary in Nata was formed, "the first of its kind in Botswana," reports KCS.

■ Resources

Members receive a quarterly newsletter, have access to an extensive library of books, periodicals, and videos, and are given discounts at many safari lodges. KCS sells gifts and books at Gaborone and at its branch offices in Maun, Francistown, Serowe, Palapye, and Tsabong.

Brazil

Brazil covers nearly half of South America, and most of the Brazilian Amazon is covered by the world's largest rainforest, with one of the highest indices of biodiversity. It is characterized by a high degree of sustainability; yet, when the topsoil is exposed directly to rainfall and sunlight, nutrients tend to be destroyed rapidly and washed out. In addition, deforestation has occurred as a result of forest fires, flooding, and other causes, leading to loss of biodiversity, topsoil erosion, detrimental effects on watersheds, and regional and possibly global climate changes.

In recent years, federal and state governments have substantially increased efforts to protect Brazil's biological resources, and the pace of deforestation in the Amazon region has decreased. The rise of public concern and the formation of numerous interest groups have made a significant difference. (There are now about 1,000 environmental NGOs in Brazil.) Brazil was the site of the 1992 Earth Summit.

🏛 *Government*

Instituto Brazileiro do Meio Ambiente e dos Recursos Naturais Renováveis (IBAMA)
Brazilian Institute for the Environment and Natural Resources
Alameda Tietê 637
01417 São Paulo, SP, Brazil

Contact: **Antonia Pereira de Ávila Vio**, Coordinator of Ecosystems
   Phone: (55) 11 883 1300, ext. 213

Activities: Education; Law Enforcement; Political/Legislative • *Issues:* Biodiversity/Species Preservation; Deforestation; Sustainable Development

Contact: **Carlos Yamashita**, Biologist

Activities: Research

IBAMA
Rua Bernardino Macieira, 284
37200 Lavras, MG, Brazil

Contact: **Luiz Fernando Macieira de Pádua**, Agronomist
   Phone: (55) 35 821 1582

Activities: Law Enforcement; Research • *Issues:* Air Quality/Emission Control; Biodiversity/Species Preservation; Deforestation; Sustainable Development; Waste Management/Recycling; Water Quality

● Organization
IBAMA is the Brazilian Institute for the Environment and Natural Resources. *(See the following IBAMA profile in this section.)*

■ Resources
Carlos Yamashita has reports available on *Central South America Caiman Study,* the *Status and Distribution on Hyacinth Macaw in Brazil,* and the *Habitat and Status of Spix's Macaw.* A member of the Species Survival Commission (IUCN/SSC) Tortoise and Freshwater Turtle Group, **Luiz Fernando Macieira de Padua** has written numerous reports about his studies with Amazon turtles and their nesting behavior.

Fundação Pró-Tamar, IBAMA

Juracy Margalhães Jr. 608-Rio Vermelho
CX Postal 2219
41911 Salvador, BA, Brazil

Contact: **Maria Angela Guagni dei Marcovaldi,**
Director of Pró-Tamar Foundation
Guy Guagni dei Marcovaldi, Director,
National Center for the Conservation and
Management of Marine Turtles
Phone: (55) 71 835 1150 • *Fax:* (55) 71 876 1045

Activities: Education; Law Enforcement; Research • *Issues:* Biodiversity/Species Preservation • *Meetings:* Annual Workshop on Sea Turtle Biology and Conservation; Global Assembly of Women and the Environment; UNCED

● Organization

The national program for the protection of marine turtles, Projeto Tamar—IBAMA, was established in 1980 to prevent the extinction of these species which are threatened in Brazil as well as throughout the world. IBAMA is the Brazilian Institute for the Environment and Natural Resources.

Before Projeto Tamar—IBAMA, there was "poaching of nearly 100 percent of the eggs as well as the slaughtering of females as they came to shore to lay their eggs." Tamar projects have established nearly 800 kilometers of nesting beaches along the coast of Brazil as well as intermittent monitoring on the oceanic islands. In selected high-priority areas, these hazards are almost completely obliterated. Local fishermen are hired to help protect beaches where they are accustomed to fishing which has provided them with extra income, thereby benefiting the local communities.

▲ Projects and People

Tamar bases are established in 14 locations where principal nesting sites are determined. Each base consists of one field manager in the field of biology or oceanography who remains at the site and patrols the nesting beaches throughout each nesting season. Almost all bases are equipped with an administrative office, a support staff, interns, and a vehicle. A few bases also maintain a visitors' center, hatchery, and holding tanks for public display which are open throughout the year.

Thirty percent of the nests that were once transferred to the hatchery are now able to remain *in situ* without the threat of human predation. Tamar hopes to transfer only those nests that will be inundated by the tide and monitor the remaining nests in their original nesting sites.

Obstacles still exist. In areas of principal nesting sites, illegal acquisition of land by squatters and vacationers continues to occur without any legal action taken to remove them. Unfortunately the beaches remain not only unprotected, but exploited. Administratively, of the 180 staff with Projeto Tamar, only one is a full-time IBAMA employee; the rest are hired seasonally or annually without guarantees for future employment.

Maria Angela and Guy Guagni dei Marcovaldi, both oceanographers, report that once Brazil's high-priority principal nesting sites are secure, "the next step will be to establish protection programs in areas of major feeding grounds where there is a high number of turtle deaths due to drowning in fishing nets."

Guy is also coordinator of Biological Reserves and National Marine Parks in Brazil, as appointed by IBDF—Brazilian Institute for Forest Development. Foundation president Maria Guagni dei Marcovaldi coordinates the Sea Turtle Project in the state of Bahia; she is the Brazilian representative to WIDECAST, and belongs to the Species Survival Commission (IUCN/SSC) Marine Turtle Group. **João Carlos Alciati Thomé** is assistant director of the Marine Turtle Foundation and coordinator of the Tartaruga Marine Turtle Project in Linhares, Espírito Santo.

■ Resources

Funding remains uncertain. The fact that funds are secured for no more than one year at a time makes planning for even the most basic activities difficult and long-term research nearly impossible. A thermocouple thermometer is needed to determine nest temperatures. Promotional gift items include the children's book *Turtle Watch*, T-shirts, stickers, earrings, and posters.

Jardim Botanico do Rio de Janeiro Botanical Garden

Rua Pacheco Leão 915
22460 Rio de Janeiro, RJ, Brazil

Contact: **Tânia Sampaio Pereira,** Seed Laboratory Chief
Phone: (55) 21 294 8696 • *Fax:* (55) 21 294 6590

Activities: Research • *Issues:* Biodiversity/Species Preservation *Meetings:* International Congress for Conservation in Botanic Gardens

● Organization

Originally about 1.2 million square kilometers, Brazil's Atlantic Forest is one of the world's most threatened areas of tropical vegetation. *(See also Fundação SOS Mata Atlantica in the Brazil NGO section and CEPARNIC in the Brazil State Government section.)* Its great diversity of flora includes a large number of unique species, with much yet to be discovered. The Atlantic Forest program at the **Jardim Botanico do Rio de Janeiro (Botanical Garden)** studies the flora in the remaining areas of this ecosystem, organizing it into a databank, and making it accessible for conservation. The genetic inheritance and internal structure of representative species are of economic interest, and such information is being shared with government agencies, NGOs, and businesses. Other supporting groups include Instituto Pro-Natura, Linhas de Ação em Botanica, John D. and Catherine T. MacArthur Foundation, and Shell do Brasil S.A.

▲ Projects and People

The Atlantic Forest program comprises projects in floristics and phytosociology, germplasm bank, vegetal anatomy, and a databank.

Initial studies from the Macaé de Cima district and Nova Friburgo township confirm the presence of unique and bountiful flora. The **Municipal Ecological Reserve of Macaé** was created to integrate the ecosystem's conservation. The Botanical Garden aims to increase the territory of this reserve to include Silva Jardim and Cachoeira de Macaco forests and to manage the forests to conserve threatened species.

Recommendations include developing a management plan, bringing technical equipment into the reserve, promptly installing a meteorological station, and launching research projects on plant groups that are difficult to classify as well as those on reproductive biology and population dynamics of species. The Botanical Garden also advises continuing an **inventory of species through bimonthly collections**, including little-known species, particularly from forests in high altitudes and from rural areas, and reevaluating species in the germplasm bank and vegetal anatomy projects.

Jardim Botanico supports the introduction of **a genetic reserve** in Macaé de Cima and the expansion of the larger reserve to provide greater quantity and variability of seeds for *ex situ* conservation. In addition, cuttings of Atlantic Forest species would contribute to the **creation of a nursery** in the reserve and help improve the Botanical Garden. The **ECOLOG program** stores information and permits its exchange with institutions involved in conserving the Atlantic Forest. A *Flora of the Ecological Reserve of Macaé* publication is designed to make floristic and phytosociological results available to interested groups.

Tânia Sampaio Pereira belongs to the Species Survival Commission (IUCN/SSC) Brazil Plant Group.

■ Resources

New technologies are needed for seed-drying methods for tropical large seeds and fruits. Video equipment is needed for educational programs. Available publications are *Rodriguesia* and *Arquivos do Jardim Botanico*.

Museu Paraense Emílio Goeldi
Avenida Magalhães Barata, 376
66040 Belém, PA, Brazil

Contact: **Manoela Ferreira Fernandes de Silva,**
       Ph.D., Researcher
Phone: (55) 91 228 2341 • *Fax:* (55) 91 241 7384

Activities: Research • *Issues:* Biodiversity/Species Preservation

● Organization

Recognized as the most diverse among known monocotyledons, the orchidaceae family, with approximately 20,000 species, is found mostly in tropical and subtropical zones. The Amazon features the largest diversity, from micro-orchids to huge plants with enormous flowers. Yet indiscriminate collection, for commercial purposes, leads to the destruction of natural habitats. Deforestation is an assault, with unknown species being destroyed forever.

The objectives of the **Museu Paraense Emílio Goeldi** are to survey the orchids of the Brazilian Amazon, especially in the **Solimões and Negro rivers;** to detect new, rare, and endangered species; to create a collection of living plants to produce seeds for replanting in nature and for commercial purposes; to analyze the patterns of geographical distribution of orchid species; to disclose results in publications; to develop books and articles for educational use; to participate in scientific meetings; and to promote the development of human resources.

▲ Projects and People

The **Salvage of Orchids in the Brazilian Amazon** project was begun in 1983 to rescue species in danger of destruction in high risk areas. With the help of **João Batista F. de Silva,** the Museu Goeldi replanted orchids in impacted areas—such as **Carajás,** an area of petroleum exploration. The orchidarium of Carajás was opened in 1985, with about 110 species, and one at **Petrobrás** was opened in 1989, with about 155 species. "The total number of orchid species catalogued in our surveys is presently about 300," reports the Museu Goeldi.

The *Cattleya eldorado,* a species symbolic of the Amazon, is at risk of extinction because of indiscriminate collection in easily accessible areas and the destruction of its natural habitat. Other species are cited only once in literature and then never again seen. It is necessary to **collect and propagate the seeds of orchids for business** so that no species will be lost, urges the Museu Goeldi, which is concentrating on the Solimões River region and its tributaries and other areas suffering high environmental impact. The material collected is brought to the herbarium at the Museu Goeldi where a live collection is being established in cooperation with the Paraense Society of Orchidophiles.

Fearful for the flora in the Serra dos Carajás mining region, the Museu began inventories in 1969 that culminated in a five-year research project in the 1980s and the discovery of new species—as mineral exploration yielded rich sources of iron, manganese, copper, nickel, tin, and gold. Mining is known to contribute to accelerated deforestation and threaten species not yet cataloged, as can related activities, such as construction of roads, towns, and airports.

■ Resources

The Museu Goeldi produces a catalog of publications in Portuguese. A fellowship for training in a foreign country, such as England, is sought.

🏛 *State Government*

Centro de Pesquisas Aplicadas de Recursos
   Naturais da Ilha do Cardoso (CEPARNIC)
Center for Applied Research in Natural
   Resources of Cardoso Island
Avenida Prof. Vladmir Besnard S/N
11990 Cananéia, SP, Brazil

Contact: **Timothy Peter Moulton, Ph.D.,** Research
       Director
Phone: (55) 138 51 1163 • *Fax:* (55) 11 282 1588
E-Mail: Internet: ceparnic@fpsp.fapesp.br
       Alternex: ceparnic@ax.apc.org

Activities: Education; Park Management; Research • *Issues:* Biodiversity/Species Preservation; Deforestation; Sustainable Development; Water Quality • *Meetings:* AIBS; ESA; ISS; SIL

● Organization

Established in 1962, the State Park and Biological Reserve of Cardoso Island, **Ilha do Cardoso**, is located within the largest remnant of Mata Atlantica, the Brazilian coastal rainforest, of which only about 5 percent of its original area remains. *(See also Fundação SOS Mata Atlantica in the Brazil NGO section.)* These 22,500 hectares of "primitive" rainforest and other threatened vegetation in the São Paulo and Paraná states encompass the research center, **Centro de Pesquisas Aplicadas de Recursos Naturais da Ilha do Cardoso (CEPARNIC)**, on Cardoso Island. Cardoso, with its steep, mountainous terrain, sandy dunes, rocky shore, and mangroves, is separated by a narrow strip of water from the mainland—a barrier possibly responsible for the island's survival.

The **Center for Applied Research in Natural Resources of Cardoso Island** began construction in 1976. Facilities comprise 4 large laboratories, 8 smaller laboratories, seminar room, library, museum, separate aquaculture lab complete with outdoor concrete ponds, large covered areas for larval culture, tanks, piped seawater and fresh water, a separate administration building, dormitory complex for 60 people, houses for general workers, and 6 houses for scientists. Unfortunately, CEPARNIC reports that the Center has been both underfunded and underutilized; it experienced problems with its "hydroelectric scheme" and the "expense of producing energy with diesel generators reduced the effectiveness." Mostly, the laboratory complex is used by scientists for brief field collections. Environmental education courses occupy the seminar rooms and dormitories.

CEPARNIC falls under the aegis of the State Environmental Secretariat's Coordinator of Technical Information, Documentation, and Environmental Research. In 1988, research director **Dr. Timothy Peter Moulton**, an Australian, joined the "wonderfully situated" Center to expand its use. He lives on the island with his wife, marine botanist **Maria do Rosário Braga** of the Botanical Institute, São Paulo.

▲ Projects and People

Dr. Moulton's aim is to attract **visiting scientists for research, conservation, and further studies** of the relatively undisturbed ecosystem as a model for other activities, such as recovery of devastated areas and agroforestry. Establishing databases and mapping physical and biological parameters, he is incorporating research information into land-use planning and management of the Ilha do Cardoso island. Moulton is assessing the value of the **natural resources of the mangrove-estuarine system** and, in the future, the forest resources and marine interface system.

"Research is fundamental to conservation," writes Dr. Moulton. "It must be integrated into the broad context of conservation-related activities of the State Park—environmental education, appropriate tourism, incorporation of the local inhabitants, and planning and management of the Park."

A joint study of São Paulo University's Ecology Section, Botanical Institute, and CEPARNIC focuses on **terrestrial productivity, decomposition, nutrient cycling, and structure of Mata Atlantica** and sand-plain, or *restinga,* vegetation. **Aquatic ecology and meteorological data** are collected and analyzed. Others, including **Dr. Yara Schaeffer-Novelli**, São Paulo University's Oceanographic Institute, are focusing on mangrove ecology research, such as the role **of pufferfish and crabs in the distribution of algae**. Braga is studying the "ecology and feasibility of mass culture of the green algae *Monostroma.*"

The Fundação Florestal, with **Rafael Reséndiz**, is carrying out a **marine turtle conservation program**—"looking after turtles that are caught accidentally by nets and traps," then tagging them for further studies. He also has interests in **oyster culture**, fishing, and fish-processing technology. As a result, the Center's aquaculture facilities are being reactivated to provide tanks to study and conserve freshwater and marine organisms.

Vertebratologist **Paulo Maruscelli**, a local resident, is surveying mammals and birds—especially **bats** and the endangered **parrot** *Amazona brasiliensis*—and banding marine and forest birds. His wife, **Miriam Milanelo**, works with community environmental education programs. Research is also underway on ants, "tank fauna" of bromeliads, and stream invertebrates.

With the direction of **Dr. Bill Magnusson**, Instituto Nacional de Pesquisa da Amazônia (INPA), **Earthwatch** volunteers assist with studies of the endangered **broad-nosed caiman**, its nesting patterns, and habitat. Other Earthwatch volunteers study the "biogeography, speciation, and autoecology" of frogs. In 1991, Earthwatch teams participated in three **Island Rain Forest** expeditions to survey water opossum, lizards, snakes, tortoises, and other amphibians, and to assess water and vegetation to gain an understanding of the ecology.

Dr. Moulton says that CEPARNIC has potential for various studies of stream fauna and flora as indicators of chemical and physical environment, relationship of soil and leaf-litter terrestrial ecosystem function, estuarine ecosystem research and modeling, environmental economics, geographic information systems, vegetation analysis, evolutionary biology of hummingbird mites and butterflies, and bird and mammal surveys.

The island also is home to **jaguar, armadillo, raccoon, capivara, agouti, peccary, otter**, and the **monkey** *bugiu* (*Alouatta fusca clamitans*). Marsupial species include the aquatic-adapted *Chironectes minimus,* which is under study. **Dolphins** are common, swimming near observers.

■ Resources

CEPARNIC needs assistance with general awareness and political support to help preserve the world's "number two biodiversity hotspot" at Ilha do Cardoso. CEPARNIC welcomes visiting scientists to work in the program of applied research for conservation. Earthwatch expeditions and environmental education courses are offered.

 ## *NGO*

Associação Brasileira de Caça e Conservação (ABC)
Brazilian Hunting and Conservation Association
R. Mourato Coelho, 1372
05417 São Paulo, SP, Brazil

Contact: Ricardo Freire, Executive Director
Phone: (55) 11 813 8238 • *Fax:* (55) 11 220 7896

Activities: Development; Education; Political/Legislative; Research • *Issues:* Biodiversity/Species Preservation; Deforestation; Energy; Sustainable Development; Wildlife Utilization
Meetings: IUCN

● Organization

ABC—the **Brazilian Hunting and Conservation Association**—was founded in 1977 by a group of 200 hunters. It aims to promote sensible use of natural resources through sustainable development; vigorously advance an appreciation of cultural heritage; cooperate with government and environmental groups on research to assess and solve environmental problems; exchange information and technical, financial, political, and cultural help with other groups; sponsor conferences, lectures, classes, and publications on modern conservation and management techniques; develop conservation projects for recovery of environmentally degraded areas; promote ethics and education in hunting; promote unity among Brazilian amateur hunters; represent its members to the government and public; and provide services, such as issuing hunting licenses, registering weapons, and preparing information on hunting laws and protecting the environment.

A member of the International Hunting Council, World Conservation Union (IUCN), and Fundação Pró Natureza (Pro Nature Foundation), ABC is affiliated with SOS Mata Atlantica and contributes to the World Wildlife Fund (WWF).

ABC believes, "If people could learn to use lasting natural and modified areas [while] improving their lives, getting economic benefits, and making political and social progress, mankind can reach a sustainable way of living—caring for the Earth without threatening essential ecological processes."

■ Resources

ABC publishes *Habitat*, a newsletter with articles on environmental matters in Brazil and worldwide and editorials on disappearing habitats and other problems. Excerpts from scholarship papers are included on wildlife management, with suggestions for further reading.

Hunting and Conservation: Renewable Natural Resources—Concepts, Uses and Hunting Management, published in 1985, is a detailed study of wildlife management techniques in the USA and Brazil, and includes several formal research papers. A newspaper, *Jornal Armaria*, is also published about hunting, historic, and other weapons. So are scientific papers on wildlife research. All materials are printed in Portuguese.

ABC seeks new wildlife management techniques adaptable to Brazil, strengthened lobbying to implement new laws, and understanding of media, public, and non-government groups about its programs. Scientific and financial assistance is needed for sustainable utilization on fauna and flora projects.

Fundação Biodiversitas
Biodiversity Foundation
Rua Bueno Brandão 372
Bairro Santa Teresa
30380 Belo Horizonte, MG, Brazil

Contact: **Gustavo A.B. Fonseca, Ph.D.,** Scientific
          Advisor
Phone: (55) 31 448 1199 • *Fax:* (55) 31 441 1412

Rua Maria Vaz de Melo, 71
Bairro Dona Clara
31250 Belo Horizonte, MG, Brazil

Contact: **Ana Maria Paiva da Fonseca,** Executive
          Director
Phone: (55) 31 443 2119 • *Fax:* (55) 31 441 7037

Activities: Development; Education; Political/Legislative; Research • *Issues:* Biodiversity/Species Preservation; Deforestation; Population Planning; Sustainable Development • *Meetings:* IUCN

● Organization

"There are approximately 1.5 million species of animals and plants known to science but recent projections have estimated that the real number of species existing on earth may lie between 5 and 30 million," according to **Fundação Biodiversitas (Biodiversity Foundation)**, which is concerned that most species on Earth remain unknown and are threatened with extinction. Created in 1988, Biodiversitas works to protect and maintain the natural biodiversity of fauna and flora in Brazil.

Its mission is to develop scientific research in conservation biology; to create, administer, and manage protected areas; to collaborate with public and private institutions for the protection of areas of significant biological value; to create systems for data storage and processing regarding biodiversity conservation; to develop education programs related to threatened species and ecosystems; to support the master's program in ecology, conservation, and wildlife management of the Federal University of Minas Gerais; to offer technical, scientific, and legal consultancy related to the conservation of nature and biodiversity; and to become involved in other organizations with similar aims.

▲ Projects and People

The **three-banded armadillo**, the **giant anteater**, the **red-handed howler**, the **golden lion tamarin**, **Lear's macaw**, and an orchid, *Constantia cipoensis*, are being studied by the Foundation's researchers. The lion tamarin, one of the threatened genera, was the subject of an international workshop held in Belo Horizonte. As a result, a conservation plan for four species is underway that will enable IBAMA (Brazilian Institute for the Environment and Natural Resources) to prepare for their protection and management.

In 1990, IBAMA also assisted Biodiversitas with the publication of Brazil's **Official List of Threatened Species**; the same year, the Foundation established the **Data Centre for Biodiversity Conservation (CDCB)**, whose information will be the basis for strategies to protect Brazilian species and ecosystems.

A first project is the publication of **Red Data Books**, beginning with one on mammals. **Biological inventories** are being undertaken of key ecosystems—beginning with a survey of terrestrial vertebrates of the Atlantic coastal forest with funding from the John D. and Catherine T. MacArthur Foundation. To preserve coastal forest species of southern Bahia, such as the golden-headed lion tamarin, IBAMA now owns part of the **Una Biological Reserve**—with support from World Wildlife Fund (WWF), Conservation International, and the Jersey Wildlife Preservation Trust.

Biodiversitas created and now administers the **Biological Station of Mata do Sossego**, which protects the muriqui, the yellow-faced marmoset, and the brown howler monkey. Also in the Rio Doce Valley in eastern Minas Gerais, Biodiversitas administers the Biological Station of Caratinga, noted for conservation of forest primates. The Biological Station of Vereda Grande in northwestern Minas Gerais is now designated a "Private Natural Heritage Reserve"; *cerrado*'s amphibians are studied among the palm swamp.

Biodiversitas supports enterprises working to create conservation units or develop management plans of Atlantic forest areas such as the Companhia Siderúrgica Belgo Mineira and the Tumbá Reserve, owned by Minerações Brasileiras Reunidas (MBR).

The environmental education program of Biodiversitas aims to reach local population surrounding the Atlantic forest areas. A "mobile unit of environmental education" is used near the National Park of Caparaó and Rio Doce State Park.

A workshop on debt-for-nature swaps emphasized the conversion of the Brazilian international debt for environmental conservation projects and resulted in the official creation of the Consortium of Brazilian NGOs for Conservation to obtain international funds for such projects through the Brazilian external debt.

Among the Biodiversitas staff of 32, Ana Maria Paiva da Fonseca is executive director; Dr. Gustavo A.B. Fonseca is scientific advisor and also zoologist at Federal University of Minas Gerais; Miriam Ester Soares and Aline Tristã Bernardes are biologists; Ilmar Bastos Santos and Célio Murilo de Carvalho Valle are zoologists; and Dr. Angelo Barbosa Monteiro Machado is president.

■ Resources

Biodiversitas created a video and picture library for storing and producing ecological materials for education projects and for use by other institutions with similar objectives; additional equipment is needed, including a vehicle, to expand environmental education programs. Publications include *Biodiversity News, Brazilian Fauna Threatened with Extinction, Ecology and Behavior of Neotropical Primates,* and *La Primatologia en Latino Americano.* A descriptive brochure with membership form is available.

Fundação Pró-Natureza (FUNATURA)/Pro-Nature Foundation
SCLN 107, Bloco B, Salas 201-210
70743 Brasília, DF, Brazil

Contact: **Maria Tereza Jorge Pádua**, President
Phone: (55) 61 274 5449 • *Fax:* (55) 61 274 5324

Activities: Research • *Issues:* Biodiversity/Species Preservation; Deforestation; Energy; Sustainable Development

● Organization

FUNATURA, the Pro-Nature Foundation, is a nonprofit organization created in 1986 to "strengthen the role of the private sector in protecting Brazil's rich biological diversity to improve the quality of life in Brazil." Some 600 members, donors, and international foundations provide support.

FUNATURA works with the government and other conservation groups to "promote the sustainable use of natural resources through research, training, education, management." FUNATURA describes its major accomplishment as the 1989 creation of the Grande Sertão Veredas National Park with the cooperation of

President José Sarney. The 84,000-hectare park protects a wide range of savanna vegetation in central Brazil.

▲ Projects and People

FUNATURA has more than 57 projects throughout Brazil dealing with nature preservation, wildlife sanctuaries, and national parks, particularly in areas experiencing pressures from local populations, agriculture, hydroelectric dam construction, and mining activities.

With the aid of World Wildlife Fund (WWF/USA), FUNATURA made a countrywide selection of species and habitat sites which could be protected through wildlife sanctuaries. Since 1988, six wildlife sanctuaries have been created on private lands in four states and the federal district. Owners are giving "willing support," reports FUNATURA, as the startup and maintenance costs are low and the lands are not appropriated. Biologist Cilulia M. Maury is coordinating the wildlife sanctuary program.

Maria Tereza Jorge Pádua belongs to the Species Survival Commission Primate and Brazil Plant Groups.

■ Resources

FUNATURA welcomes contact with individuals and organizations experienced in similar conservation projects. To build its environmental education program, FUNATURA seeks support for a video cassette system, slide projector, computer and printer, radio transmitter/receptor, four-wheel-drive car, ranger uniforms, satellite-imaging systems, and fire equipment. Also needed is literature on ecological agriculture, agroforestry, and rainforest management, as well as information on working with communities near natural protected areas.

Fundação SOS Mata Atlantica
Rua Manoel da Nóbrega, 456
04001 São Paulo, SP, Brazil

Contact: **João Paulo Ribeiro Capobianco,**
      Superintendent
Phone: (55) 11 887 1195 • *Fax:* (55) 11 885 1680

Activities: Development; Education; Political/Legislative; Research • *Issues:* Biodiversity/Species Preservation; Deforestation; Energy; Sustainable Development; Water Quality

● Organization

Since 1986, SOS Mata Atlantica has had a mission of protecting and preserving what remains of Brazil's Atlantic Rainforest, which once covered 300 million acres. According to the Brazilian Foundation for the Conservation of the Atlantic Rainforest, it is limited to less than 5 percent of its original area and is considered to be the world's second "biodiversity hotspot." The Atlantic Rainforest has an estimated 10,000 plant species (4 percent of the Earth's total) of which 2,500 are believed to be close to extinction. Also endangered are numerous species of primates and birds.

SOS Mata Atlantica's purposes are to protect the biological diversity and ecological processes of its ecosystems; promote the sustainable use of the region's natural resources; advance the comprehensive documentation of the region's ecology, conservation, and people; and establish a strong environmental ethic in the area.

▲ Projects and People

The **Green of Our Earth Is Being Destroyed** media campaign was launched on Brazil's major television and radio stations and its largest circulation newspapers and magazines in order to make Brazilians more aware of the threat of destruction to the remaining Atlantic Rainforest.

The **Federal Constitution Project** enabled the Ecologist Action Front of the Constitutional Assembly to gain approval of a chapter on environmental issues. Coordinating with the Chamber of Deputies' Environmental Committee, SOS Mata Atlantica forwarded proposals of the country's conservation community to presidential candidates. It also **encourages environmental legislation** in municipalities in which rainforest remnants are located.

Employing technology that is accessible to the local population of the Lagamar Region, SOS Mata Atlantica developed a pilot project to demonstrate efficient and productive **oyster farming** on a sustainable basis. Also in this region, SOS Mata Atlantica established **urban bases for environmental education** and other projects in the region. In addition, a **greenhouse** was established at the Technical Agriculture School of Iguape for management of *Caxieta Tabeuia Cassinoides* saplings, a "native wood of great economic and ecological value."

Using **satellite imagery** for the first time in surveying the Brazilian rainforest, the Foundation published an atlas.

The SOS Mata Atlantica has organized **seminars** and workshops on **rational management of tropical forests, remote sensing, and conservation databases,** for example. With other NGOs, the Foundation set up a **forum of private organizations** to encourage countries participating in the UN Conference on Development (UNCED) to make commitments toward improving the world's environment and the quality of life in developing countries.

■ Resources

Publications include proceedings of the *Atlantic Rainforest Workshop,* the *International Seminar on Data Bases for Conservation in Brazil,* and the *International Seminar on Rational Management of Tropical Rainforests.* SOS Mata Atlantica can provide information on a *Minimum Environmental Platform for Presidential Candidates* and a *Minimum Environmental Platform for Candidates for the Governor of São Paulo State.*

Instituto Pró-Ecología (IPE)/Pro-Ecology Institute

Avenida dos Operários 587
13400 Piracicaba, SP, Brazil

Contact: **Claudio V. Pádua**, Director, Biologist
Phone: (USA) 1-904-334-5153
Fax: (USA) 1-904-334-5153
(Brazil) (55) 194 349833
Activities: Education; Research • *Issues:* Biodiversity/Species Preservation; Deforestation; Sustainable Development

● Organization

Instituto Pró-Ecología (IPE) is a small, new group attempting to achieve its mission of **saving four lion-tamarin species** (*Leontopithecus*) whose presently fragmented habitat has been reduced to 2 percent of its original area. **Claudio V. Pádua,** IPE

director, is building his organization with determination, a strong background in captive breeding, linkage with international groups, and little funding.

Having helped prepare a **Population Viability Analysis (PVA)** report for the lion tamarins with the Species Survival Commission (IUCN/SSC) Captive Breeding Specialist Group, Pádua helped develop a conservation plan in 1990 with the sponsorship of Jersey Wildlife Preservation Trust, World Wildlife Fund–USA (WWF-USA), IBAMA (Brazilian Institute for the Environment and Natural Resources), Fundação Biodiversitas, and Conservation International.

To accomplish its goal, Padua writes that IPE is concentrating its focus on "training and education, captivity, and [obtaining] private property." IPE will also undertake studbook keeping and demographic and genetic analyses of endangered species populations.

▲ Projects and People

As reported in the conservation plan, habitat and wild population protection must begin immediately along with an inventory of these species. Captive populations must be scientifically managed so that suitable habitat can later be stocked. The following recommendations were among those made for golden lion tamarin (*Leontopithecus rosalia*), golden-headed lion tamarin (*L. chrysomelas*), black lion tamarin (*L. chrysopygus*), and the newly discovered black-faced lion tamarin (*L. caissara*): "Establish each species in the wild with a total population size of at least 2,000 by the year 2025" and secure sufficient habitat for this purpose; establish subpopulations of each species of 100 breeding-age adults to be monitored and managed; if extinction threatens such species, establish a captive population of sufficient size that retains 90 percent of the genetic heterozygosity of the species for 100 years but do not reintroduce these species until they are "demographically and genetically secure." Also featured are the development of population and metapopulation models and integrative management plans.

What are specifically recommended for South America's most endangered monkeys—**450 black lion tamarins with their golden rumps** found only in São Paulo at the Morro do Diabo State Reserve and Caitetus Reserve—are protection from fire, deforestation, and hunting; increased funding for surveillance guards and park wardens; creation of buffer zones around protected areas; education of local people; and increased funding for the Rio de Janeiro Primate Center, where captive animals are kept. **Prof. Adelmar Coimbra-Filho** started that colony in 1973 by capturing seven animals in the vicinity of Morro do Diabo (hill of the devil) Reserve after hearing that some farmers were using the defoliant Agent Orange nearby. At the time of the mid-1990 report, there were 64 captive animals at 2 locations. Padua hopes to discover more wild populations.

Little is known about the **black-faced lion tamarin**, which mainly lives on Superagui Island; however, its main threats are illegal trade and development. Unlike the black lion tamarins, thus far there is no captive population; this lion-marmoset is recommended for placement on the CITES and Brazilian Lists of Threatened Species. Research on ecology and behavior and environmental education are essential.

It is estimated that present suitable forested habitat for the **golden lion tamarin (GLT)** could support about 844 tamarins—about double today's confirmed population in the wild. To sustain the projected 2,000 animals, areas in and adjacent to the Poço das Antá Reserve require reforestation and protection and other lands must be purchased. Researchers must find ways to prevent diseases from being transmitted, avoid fires, and stop toxic spills from rail cars transported through the reserve.

Some **600 golden-headed lion tamarins** (GHLT) in the wild are also threatened with habitat loss through deforestation and fire as well as illegal capture. Present reserves must be restored, vegetation and wild populations must be surveyed, and forests must be evaluated for regeneration from second growth and abandoned cocoa plantations. Other goals are to "target landowners with proposals for the creation of private reserves" and to "involve [local] Bahian [university] students in conservation research." With 285 captive-bred animals, plans must be made for easy transfer between institutions in Brazil and abroad.

Claudio V. Padua, a member of both the IUCN/SSC Captive Breeding and Primate Specialist groups, received early captive breeding training from the Wildlife Preservation Trust and was assistant director of the Rio de Janeiro Primate Center. The University of Florida's Program for Tropical Forest Studies has supported his tamarin research.

■ Resources

IPE needs about 10 computers to be distributed among zoos and other captive breeding facilities in Brazil to do conservation in captivity. Startup funds for the Institute are sought.

Instituto Sul Mineiro de Estudo e Preservação (ISM)
Sul Mineiro Institute of Studies and Nature Conservation
Faz. Lagoa, Cx. Postal 06
37132 Monte Belo, MG, Brazil

Contact: **Maria Cristina Weyland Vieira**, Director
   Phone: (55) 35 571 1500 • *Fax:* (55) 35 571 1500

Activities: Conservation Campaigns; Education; Research
Issues: Biodiversity/Species Preservation; Deforestation • *Meetings:* ITTO; IUCN; UNCED

● Organization

Forerunner to the **Sul Mineiro Institute of Studies and Nature Conservation** (ISM) was Horto Monte Alegre, founded in 1978 at Monte Alegre Farm, Monte Belo, Minas Gerais, to plant the first seedlings of three native trees for reforesting the area. Horto next embarked on a floristic survey, which in a decade collected 1,200 species; an avifauna survey identified 230 species; and a study of mammals with **Russell Mittermeier** ensued.

With the establishment of ISM at Lagoa Farm, the research was extended, and public education campaigns intensified to reforest with native species. In the Sul de Minas region, ISM works with schools, teachers, farmers, communities, and the media on behalf of nature conservation to "criticize threats to wildlife and publicize ISM events."

ISM is concerned with the extinction of all forms of life. "We must remain alert so that our food, water and atmosphere are not poisoned. When the majority of people are aware of what's happen-

ing right in their own back yard, it's often too late to repair the damage," writes ISM.

▲ Projects and People

Since 1986, ISM has been engaged in **monographical studies on two semideciduous tropical forests** in Fazenda Lagoa to target endangered species and determine the possibility of survival in these forest fragments. In conjunction with this research, a graduate degree is being offered on forest conservation and phytogeography through the Federal University of Rio de Janeiro. Two groups with teenage members are assisting the "pro-forest campaign": Ecological Group Roots of the Future, Monte Belo, and Ecological Association Flowering the Dawn, Areado.

ISM is also involved in the **Fight for the Preservation of the Serra da Mantiqueira (FEDAPAM)**, the highest mountain range in South America. Originally covered by the Atlantic Forest up to 1,500 meters, by the Temperate Forest with its araucária (Brazilian pine) between 1,500 and 2,000 meters, and by fields mixed with temperate forests at higher levels, these isolated ecosystems are considered unique "islands that are separated from tropical forests."

Both the species and the native culture of the Mantiqueira are endangered. "Preserving the significant culture and wildlife of the region, and improving the quality of life and the stability of the population, are our constitutional obligations," states ISM.

A recent campaign **discourages illegal and predatory hunting** of native animals as well as farm invasions and plantation destruction for these purposes.

ISM participated with other NGOs in the following meetings that preceded the United Nations Conference on Environment and Development (UNCED) in 1992.

The **Atlantic Forest Work Group** established legislative, scientific research, environmental education, and informational priorities. Its goals are to **map the remaining areas of the Atlantic Forest**, provide legal incentives for the creation of public and private nature preserves, consolidate existing nature preserves with the goal of founding new ones, inventory the fauna and flora of the Atlantic Forest so that its biodiversity can be known, prioritize **projects for communities where conservation organizations are located**, and encourage cooperation between NGOs, research institutes, and government. A national debate is encouraged so that all opinions on the Atlantic Forest can be heard.

The **Biodiversity and Technology Work Group** recommends increasing inventories of fauna and flora of the remaining Latin American ecosystems before the majority of indigenous species are extinct, and pressuring financial agencies to direct their funds toward **research in alternative technologies**, especially **organic agriculture and anaerobic digestion of organic wastes**. Concerns about biotechnological consequences include the devaluing of genetic material, genetic erosion, and risks to the equilibrium of ecosystems. "Biodiversity should be conserved *in situ*, respecting local populations, and national sovereignty," said the group.

The **Climate Change Work Group** urges that **industrialized countries begin to make lifestyle changes** in consumption and production; that existing forests be preserved and **a network for global reforestation be set up**; and that **nuclear energy be discouraged** because of the high costs, dependency on technical equipment, and environmental impact as well as inevitable military uses and inaccessibility to poorer nations.

Other hazards are agribusinesses with an "enormous use of energy, water, and agrochemicals" that contribute to environmental degradation; livestock ranching at the expense of tropical forests; and methane emissions of cows, rice paddies, and sanitary wastes.

With industrialized countries responsible for more than 90 percent of the emissions that cause global warming, according to the group, the less-developed countries are less able to deal with the effects and will be the ones most harmed by them. Therefore, there must be "fair negotiations on a per capita criteria for emissions." ISM and other NGOs urge the scientific community to play a more responsible and active role explaining the implications of global change.

Serving on ISM's scientific council are representatives of the Brazilian Foundation for Nature Conservation (FBCN), Foundation for Studies on the Seas (FEMAR), Association of the Leme Residents of Rio de Janeiro (AMALEME), Brazilian Movement for Defense of Life (MBDV), "Vertente" grassroots group, Botanical Gardens of Rio de Janeiro, and Association for the Defense of the Environment in Minas Gerais (AMDA).

In 1991, **Maria Cristina Weyland Vieira**, who is also general secretary of FEDAPAM, continued her tropical rainforest studies at University College, London, with a scholarship from the British Council and the Margaret Mee Amazon Trust.

■ Resources

ISM has bulletins and leaflets available, mainly in Portuguese.

ISM seeks new technologies for organic farming/biodynamic agriculture for the production of healthy products and conservation of biodiversity that use alternative sources of energy, watermills, and solar collectors. It also welcomes assistance with educational outreach and plant conservation projects.

Sociedade de Defesa do Pantanal (SODEPAN) Society for the Defense of the Pantanal
Avenida Américo Carlos de Costa No. 320
79020 Campo Grande, MS, Brazil

Contact: **Nilson de Barros**, President
Phone: (55) 67 742 1891 • *Fax:* (55) 67 742 3388

Activities: Development; Education; Law Enforcement; Political/Legislative; Research • *Issues:* Biodiversity/Species Preservation; Deforestation; Health; Sustainable Development; Water Quality • *Meetings:* CHORLAVI (SA IUCN); UNCED

● Organization
The **Sociedade de Defesa do Pantanal (SODEPAN), Society for the Defense of the Pantanal**, promotes the ecological equilibrium of the world's broadest expanse of wetlands and the protection of flora and fauna in the states of Mato Grosso and Mato Grosso do Sul. SODEPAN encourages the self-determination of the region's people integrated with preserving its valuable natural ecosystem; identifies sustainable economic activities; stimulates research, education, and development; and creates protected areas. SODEPAN seeks to adopt technical and legal measures to establish an environmental policy adapted to the region's needs. To strengthen environmental education programs, it encourages the exchange of information with organizations and entities involved in research, defense, funding, development, and preservation of the region.

▲ Projects and People
The **Camelote Project** aims at creating conditions that permit the "implementation of an environmental conservation policy in the Pantanal with community support to develop a sense of unity."

In the **Blue Macaw Project**, scientists survey and identify areas important for feeding, tagging, and studying young macaws and their reproduction.

SODEPAN intends to **survey** the socioeconomic and environmental characteristics of areas with a high concentration of **spotted jaguars** in southern Pantanal, and seeks to protect the **Pantaneiro horse** because of the animal's social, economic, and cultural importance. This horse originated in the vast plain that forms the Pantanal in the two states of Mato Grosso and Mato Grosso do Sul.

Other planned projects include **educating the community** residing in the Hydrographic Basin of the high Paraguay River about their fragile and complex environment; acquiring an area in the Hydrographic Basin of the Pantanal for **research on flora and fauna**; holding **public hearings** with specialists from Brazil and Paraguay regarding the Platina Basin; **introducing courses in recycling** for tourist guides and distributing educational folders in airplanes, hotels, buses, and information centers.

SODEPAN's **Pantanal Image Bank Project** is a photography archive and visual record of both beauty and environmental problems. The **Tapera Project** is concerned with food for woodland fauna; other projects aim to control toxic waste, and involve erecting educational signs in the Parque dos Poderes Ecological Reserve. Rancher **Dayton Hyde** has set up **Operation Stronghold** to demonstrate rational use of natural resources.

■ Resources
Publications include *O Pantanal e O Pantaneiro, Sentir Para Compreender, Um Susto em Piquiri*, and *Jornal da Sodepan*.

🔔 *Private*

Ecopress—Environmental News Agency ⊕
Rua Sampaio Viana, 72 Paraiso
01440 São Paulo, SP, Brazil

Contact: **Sandra Sinicco**, Co-founder, Editor
Phone: (55) 11 288 6515 • *Fax:* (55) 11 289 6621

Activities: News Agency; Research • *Issues:* Biodiversity/Species Preservation; Deforestation; Global Warming; Sustainable Development • *Meetings:* UNCED

● Organization
Cristine Pinheiro Machado and Sandra Sinicco, both journalists, founded Ecopress—Environmental News Agency in 1989. Media and organizational subscribers receive news daily from environmental groups in Brazil and the world. When the news is very important, it is sent free of charge to all newspapers and radio and television stations in Brazil.

Ecopress also edits a bimonthly newsletter relating to environmental matters in Brazil. Ecopress is forming and organizing a databank news service with a MicroIsis software.

▲ Projects and People

In one *Ecopress Newsletter*, the editors assessed the **environmental record of the administration** of the Brazilian President Fernando Collor de Mello during its first 16 months, including debates on decrease of burnings in the Amazon, demarcation of Indian lands, protection of the Atlantic Rainforest, and importance of follow-up to the United Nations Conference on Development (UNCED) in Brazil, known here as ECO-92.

Aware that the conference is linking "poverty and environmental degradation," Ecopress quoted government spokesperson Justice Minister Jarbas Passarinho as expressing concern about pressures of foreign government and NGO groups to create an "internationalization" of the Amazon. Environmental journalist **Randau Marques**, in the *Ecopress* 1991 international issue, said he feared a return of a government mindset that could advance a "pollution of progress against the pollution of misery." Some sources expressed concern that deforestation could accelerate with the country's economic recovery.

Other items reported experiments with microscopic **freshwater crustaceans** found in clean lakes to "monitor the effluents generated by industries all over Brazil and help keep the environment clean." Also, collaborative state government, university, and private efforts are underway to create an **Albatross National Park** in an archipelago along the northeast coast of São Paulo. "With migratory routes of polar birds, Antarctic and sub-Antarctic, the archipelago lives covered in a cloud of birds," the newsletter reports. Reported threats to the nesting area of gannets and albatrosses, or frigate birds (*Fregata magnificens*), are "military maneuvers and predatory activities like hunting, egg pillaging, oil spills, and fires."

Journalists working for Ecopress include **Regina Scomparin, Sonia di Pieri**, freelancer **Karla Brunner, Nira Worcman** in New York, and **Jair Rattner** in Lisbon.

■ Resources

As mentioned above, subscriptions to the Ecopress news service and newsletter are offered. Ecopress wants to expand its efforts in informing the public through independent news sources and disseminating stories about environmental newsmakers.

Ricsel Orquideas ⊕
Rua Jataí, 758
90820 Porto Alegre, RS, Brazil

Contact: **Sergio Inacio Englert, Ph.D.**, Director
  Phone: (55) 512 497 566

Activities: Production and Marketing of Orchids • *Issues:* Biodiversity/Species Preservation • *Meetings:* SSC Orchid Group

● Organization

Within a decade, **Ricsel Orquideas** has collected many orchid species and horticultural varieties. The nursery is registered with the Brazilian Institute for the Environment and Natural Resources (IBAMA) and is regularly inspected by experts from CITES, the Convention on International Trade in Endangered Species of Wild Fauna and Flora. Ricsel Orquideas propagates orchid species from seed, tissue culture, and division with a sense of urgency not only for conservation of orchid species but for other plant and wildlife which are under threat of extinction from deforestation and habitat destruction.

▲ Projects and People

Ricsel works with a laboratory to produce flask bottles of husky seedling and mericlones. The nursery recently bought the complete **seeds and pollen collection** of the late Dr. Walter Haetinger, who worked for nearly four decades on breeding *Laelia purpurata* and *Cattleya intermedia*. Ricsel Orquideas continues his work with the fifth generation of these orchids. A superior *Laelia purpurata*, flâmea "Larissa," won first prize at a state orchid show.

Dr. **Sergio Englert** is a member of the Species Survival Commission (IUCN/SSC) Orchid Group.

■ Resources

Orchids are shipped worldwide and can be ordered in quantity; three weeks should be allowed for delivery. Overseas customers must supply the necessary documents and import permits in accordance with government regulations. A CITES Permit and a Phytosanitary Certificate accompany each order.

🎓 *Universities*

Departamento de Zoologia/Zoology Department
Instituto de Ciencias Biologicas/Biological Sciences Institute
Universidade Federal de Minas Gerais
31270 Belo Horizonte, MG, Brazil

Contact: **Anthony Brome Rylands, Ph.D.**, Professor
  Phone: (55) 31 448 1236 • *Fax:* (55) 31 441 1412

Activities: Education; Research • *Issues:* Biodiversity/Species Preservation • *Meetings:* International Primatological Congress; World Conference on Breeding Endangered Species

▲ Projects and People

Dr. **Anthony Brome Rylands** specializes in primate studies, particularly marmosets and tamarins, as noted in many journals and books. The professor of vertebrate zoology at **Federal University of Minas Gerais** was involved in the rescue of the black lion tamarin in the inundation area of the Rosana Hydroelectric Dam, São Paulo. He was also consultant for faunal surveys for other dam projects. In addition, Dr. Rylands has researched tree porcupines, army ants, howling monkeys, and other neotropical primates.

Dr. Rylands is currently collaborating with **Dr. Russell A. Mittermeier** and artist **Stephen Nash** on a *Field Guide to Neotropical Primates* and with **Dr. Gustavo A.B. da Fonseca** on a *Red Data Book for Brazilian Mammals*. Subjects of other recent publications include an evaluation of the current status of federal conservation areas in the tropical rainforest of the Brazilian Amazon (in conjunction with World Wildlife Fund) and "the elaboration of the legend for a map of priority areas for conservation in the Amazon" with Drs. **Keith Brown, Otto Huber**, and **Ghillean Prance** for the Brazilian Institute for the Environment and Renewable Natural Resources (IBAMA).

A past president of the Brazilian Primatological Society and a member of the Species Survival Commission (IUCN/SSC) Primate Group, Dr. Rylands is an international affiliate scientist with the Wisconsin Regional Primate Center, University of Wisconsin, USA. He advises the Neotropical Taxon Advisory Group in association with the Captive Breeding Specialist Group (CBSG) of the World Conservation Union (IUCN) and the American Association of Zoological Parks and Aquariums (AAZPA).

Instituto de Biologia/Biology Institute
Departamento de Zoologia/Zoology Department
Universidade Estadual de Campinas (UNICAMP)
C.P. 6109
13802 Campinas, SP, Brazil

Contact: **Keith S. Brown, Jr., Ph.D.**, Professor
Phone: (55) 192 39 7022 • *Fax:* (55) 192 39 3124

Activities: Education; Research • *Issues:* Biodiversity/Species Preservation; Chemical Ecology; Sustainable Development

▲ Projects and People
Dr. Keith S. Brown, Jr., UNICAMP, has been teaching ecology and chemistry courses in Brazil since 1964 and has published some 120 papers and books, including a section of *The Conservation of Insects and Their Habitats* and *Tropical Forest Conversion and Species Loss* for the World Conservation Union (IUCN), both in 1991. Dr. Brown belongs to the Species Survival Commission (IUCN/SSC) Lepidoptera Group. To enhance global cooperation, he recommends that environmentalists "check facts before speaking. Speak many languages, like different people. Encourage base democracy. Discourage consumerism, materialism, and imperialism."

Departamento de Ecología/Ecology Department
Universidade Estadual Paulista (UNESP)
Avenida 24A No. 1515 CP 199
13500 Rio Claro, SP, Brazil

Contact: **Miguel Petrere Jr., Ph.D.**, Quantitative Ecology Lecturer
Phone: (55) 195 34 0244, ext. 151
Fax: (55) 195 34 4433

Activities: Development; Research • *Issues:* Air Quality/Emission Control; Biodiversity/Species Preservation; Deforestation; Energy; Health; Sustainable Development; Waste Management/Recycling; Water Quality

▲ Projects and People
Dr. Miguel Petrere, Jr., UNESP Ecology Department, studies life strategies of long-distance migratory **catfish** in relation to **hydroelectric dams** in the Amazon Basin. In a recent project with zoologist **Dr. Ronaldo Borges Barthem**, Museu Paraense Emilio Goeldi, and **Dr. Mauro César Lambert de Brito Ribeiro**, Reserva Ecológica

do IBGE, it was shown that "hydroelectric dams eventually lead to the interruption of the downstream movements of catfish eggs, or young, provided they spawn in the upper tributaries, or obstruct the upstream migrations which yearly restore their stocks up river."

As published in *Biological Conservation* (1991), "synergistic effects of flood control may also prove to be harmful to these species, since their hydrological requirements are drastically modified." The researchers' report continues, "The only hope for preserving migratory catfish stocks above the dams will be for them to spawn upstream from the reservoirs. Otherwise artificial measures such as fish ladders or side channels should be employed." Stocking fish is another alternative. Because many questions remain unanswered about the future level of catches of these catfish in response to artificial changes, Dr. Petrere raises the issue of "balancing present restraints against future benefits to regional economy."

Dr. Petrere's newest project in Central Amazonia, upper Amazon, is also concerned with large catfish migration. He belongs to the Species Survival Commission (IUCN/SSC) Freshwater Fish Group.

Department of Forestry Engineering
Universidade Federal de Viçosa
DEF/UFV
36570 Viçosa, MG, Brazil

Contact: **James Jackson Griffith, Ph.D.**, Associate Professor
Phone: (55) 31 899 2467 • *Fax:* (55) 31 891 2166

Activities: Education; Research • *Issues:* Air Quality/Emission Control; Biodiversity/Species Preservation; Deforestation; Energy; Global Warming; Mining Land Rehabilitation; Sustainable Development; Water Quality

● Organization
Dr. James Jackson Griffith (a recent visiting scholar at the University of Washington) is associated with the **Department of Forestry Engineering** at the Universidade Federal de Viçosa and the cooperative research unit, Sociedade de Investigacões Florestais (Society for Forestry Research).

▲ Projects and People
Dr. Griffith reports on four recent projects at the Federal University of Vicosa.

A five-year **management plan for Rio Doce State Park** in the southeast region of Minas Gerais state encompassed scientific, educational, and recreational uses with the aid of a geographic information system (GIS). Opportunity and restriction maps were generated for each projected use covering 36,000 hectares—the largest remaining area of Mata Atlantica (Atlantic Forest) under government control. The World Bank funded this collaborative effort of eight organizations and agencies. Planning sessions and work seminars featured a "synergy training" technique.

In a second program, "lands drastically altered by mining and related industrial activities at the world's largest operating **open pit mine** were studied to develop **rehabilitation techniques using native species**," writes Dr. Griffith. An objective was to undertake nursery and planting activities at the Itabira, Minas Gerais, operations of the Vale do Rio Doce Company. "A company nursery was planned and implemented to produce 500,000 tree seedlings and transplants per year," reports Griffith. "An environmental education strategy reinforced conservation practices and protection on rehabilitated sites accessible to the public."

With supporting agencies, the Department of Forestry Engineering identified, mapped, and ranked areas for potential natural reserves or parks in Espírito Santo state according to vegetation type and land-use history. Recent vegetation coverage was compared to historic coverage in the survey's analysis.

Regarding abandoned lands following bauxite mining during the 1940s in Poços de Caldas, Minas Gerais, researchers analyzed the phytosociology of natural regeneration of trees and shrubs there to "better understand how drastic land use affects plant communities" and to "provide a database for future mining rehabilitation." Relations between the *cerrado* vegetation appearing naturally and soil and temperature conditions were analyzed. Results will assist rehabilitation of the land which is being mined again, according to Dr. Griffith.

For improved global cooperation, Griffith believes environmentalists need "systematic training in organizational development to improve internal teamwork skills and to work effectively with other environmental groups."

■ Resources

The Department of Forestry Engineering needs computer hardware and software, especially for GIS. Also needed is support in publishing basic forestry textbooks in Portuguese. Publications include *Revista Árvore*, a forestry research periodical, and *Boletim Tecnico da Sociedade de Investigaçoes Florestais*, the research cooperative's technical bulletin.

Instituto de Biociencias/Biosciences Institute
Departamento de Botânica/Botany Department
Universidade de São Paulo
C.P. 11461
05499 São Paulo, SP, Brazil

Contact: **Ana Maria Giulietti, Ph.D.**, Head Professor
Phone: (55) 11 813 6944 • *Fax:* (55) 11 815 4272

Activities: Education; Research • *Issues:* Biodiversity/Species Preservation

▲ Projects and People

Dr. Ana Maria Giulietti, São Paulo University, belongs to the Species Survival Commission (IUCN/SSC) Brazil Plant Group.

Programa de Pesquisa e Conservação de Areas Úmidas/Programme on Research and Management of Wetlands in Brazil
Universidade de São Paulo
Rua do Anfiteatro, 181
Colméia-Favo 6
05508 São Paulo, SP, Brazil

Contact: **Antonio Carlos Santanna Diegues, Ph.D.**, Coordinator
Phone: (55) 11 211 0011, ext. 2307
Fax: (55) 11 814 5050, (55) 11 212 4057

Activities: Research • *Issues:* Biodiversity/Species Preservation; Deforestation; Energy; Global Warming; Population Planning; Sustainable Development; Wetlands Management
Meetings: IUCN

● Organization

The University of São Paulo's Oceanographic Institute began the **Programa de Pesquisa e Conservação de Areas Úmidas (Programme on Research and Management of Wetlands in Brazil)**, in April 1988, with the help of the Ford Foundation and the World Conservation Union (IUCN).

Brazilian wetlands are found in ecosystems such as the undeveloped Pantanal woodlands, areas of the Amazon, swamps, lakes, estuaries, and mangrove swamps. Covered each season by fresh or salt water, these biomes carry out essential ecological functions such as maintaining water balance and water quality. However, such areas are being destroyed by embankments in the mangrove swamps, construction of canals, dredging, and pollution from urban and industrial areas. The wetlands' inability to perform their natural functions has caused grave ecological damage such as invasion of the sea after mangroves are cut, inundations, and extinctions, according to the Programme. Human communities which live off these natural resources are becoming impoverished. Many countries which have destroyed their wetlands are trying to recreate them artificially, at a high cost.

The Programme asserts that "the maintenance of these ecosystems represents a higher economic benefit than the transformation of them into other uses," such as agricultural or industrial. Objectives are to study the relationships between human communities and the wetland ecosystems, to determine their socioeconomic value, and to propose new ways for their sustainable maintenance. The Programme intends to employ researchers from ecology, anthropology, biology, economics, geography, and sociology.

▲ Projects and People

In 1985, the IUCN launched an international program for wetlands to call attention to the importance of wetlands worldwide and the grave dangers facing them as a result of draining, dredging, and urban and industrial development. Some 3,000 scientists from 54 countries, including 124 government institutions, are participating in its scientific and technical committees.

One emphasis is on developing economic evaluation methodologies of wetlands to determine alternatives that are ecologically and socially responsible and fiscally viable. At the same time, management techniques must be devised to protect essential ecological processes and permit sustainable use by local populations.

Ongoing projects include the inventory of wetlands in Brazil; ecological and economic evaluation of Iguape-Cananéia estuary; livelihood strategies of low-income communities in degraded and pristine wetlands in the Cubatão and Iguape-Cananéia estuary; and studies of the mangroves of Itamaracá in Pernambuco, the floodplains of Marituba in Alagoas, the Rondonia Valley in the Amazon, and the Kaigang Indians reserves in Paraná.

Coastal wetlands seminars are held, a computerized information and network system is being established, and a biannual newsletter is planned.

■ Resources

Publications in English include *Biological Diversity and Traditional Cultures in Coastal Wetlands of Brazil*; *Managing Brazil's Coastal Wetlands*; *Management of Wetlands: The Iguape-Cananéia-Paranaguá Estuary*; and *Tradition and Change in the Brazilian Small-Scale Fisheries: A Preliminary Synthesis*. A list of publications in Portuguese and English is available.

Cameroon

The Republic of Cameroon is bordered by the Gulf of Guinea on Africa's west coast and by Chad, the Central African Republic, Congo, Gabon, Equatorial Guinea, and Nigeria. The diverse geography and climate include semiarid hot plains, rainy plateaus and coastal highlands, and mountains.

Cameroon's main environmental problems are forest destruction, range degradation, and waterborne diseases. The range in the northern semiarid region has been harmed severely by drought, the overcrowding of stock animals, and extensive underutilization due to tsetse fly infestation. Most of the country's forest resources may have already been destroyed, thereby endangering wildlife. Although a number of forestry reserves have been set aside, logging concessions and uncontrolled poaching within their boundaries have diminished their effectiveness. Development of a national forest policy is a top priority.

 *NGO*

International Council for Research in Agroforestry (ICRAF) ⊕
IRA/ICRAF Collaborative Agroforestry Project
P.B. 2067
Yaounde, Cameroon

Contact: **Bahiru Duguma, Ph.D.,** Senior Scientist/Project Leader/Forester
Phone: (237) 22 30 22
Telex: 1140 KN-IRA/ICRAF

Activities: Research • *Issues:* Agroforestry • *Meetings:* Africa 2000; ICRAF

● Organization

According to senior scientist **Dr. Bahiru Duguma,** the **International Council for Research in Agroforestry (ICRAF)** is committed to "addressing environmental issues rather indirectly than directly." Dedicated to the advancement and formation of agroforestry research on a global scale, the Council was created to assure the nutritional health of people in developing nations worldwide.

ICRAF researches ways to ensure people's well-being through "the integration of woody components crops and animal management," while land management is also promoted. It seeks to develop a "low-input land use system that ensures improved and sustainable production per unit area." Beginning in Africa, ICRAF plans to extend its research programs to Asia and Latin America.

ICRAF works with **Africa 2000,** which is under the **United Nations Development Programme (UNDP)** and is responsible for funding environmental projects. The two organizations target "environmental pollution, subsistence food production, timber and wildlife exploitation." Projects are tailored to available resources and climatic conditions.

▲ Projects and People

In Cameroon, a humid lowland area, the Council has completed "a baseline data collection, identified production and environmental related constraints, designed appropriate agroforestry technologies to address the problems, conducted field trials to screen tree species that could be used in developing the technologies," and is continuing with field trials, reports ICRAF.

However, sufficient funding is needed elsewhere to attract qualified experts, operational and capital equipment, and training facilities for national scientists who will continue with the projects.

According to ICRAF, what contributes most to worldwide environmental pollution is the "large-scale dependence of people and government of [developing] countries [on] subsistence food production and economic development, timber and wildlife exploitation . . . responsible for the fast disappearance of tropical forests." ICRAF aims to "reduce the pressure" on renewable resources effectively and realistically to satisfy the needs of those who are now destroying their environment in order to survive.

A member of the National Selection Committee of Africa 2000, Dr. Duguma lectures on land-use problems and agroforestry throughout Africa, including Ethiopia, where he was born. He recently wrote a series of articles on liming, phosphorus, and other factors affecting germination of *Leucaena* grown in acid soil.

■ Resources

ICRAF offers articles, books, and publications about agroforestry. Most are free for developing countries or available at a small cost from ICRAF's Information and Communication Division, P.O. Box 30677, Nairobi, Kenya.

🎓 *Universities*

Ecole de Faune/Wildlife School
B.P. 271
Garoua, Cameroon

Contact: **Jean Ngog Nje, Ph.D.**, Director, Animal
            Ecology Professor
Phone: (237) 27 1025

Activities: Wildlife Management • *Issues:* Air Quality/Emission Control; Biodiversity/Species Preservation; Deforestation; Wildlife Conservation • *Meetings:* FAO; IUCN; UNESCO

▲ Projects and People

Ecole de Faune (Wildlife School) has a staff of 13, with **Dr. Jean Ngog Nje** as director. He has written about 12 publications in his field and belongs to the Species Survival Commission (IUCN/SSC) African Elephant and Rhino Group as well as the International Council of Hunting and Conservation of Game.

■ Resources

The School seeks books on wildlife management and field equipment, such as binoculars and compasses.

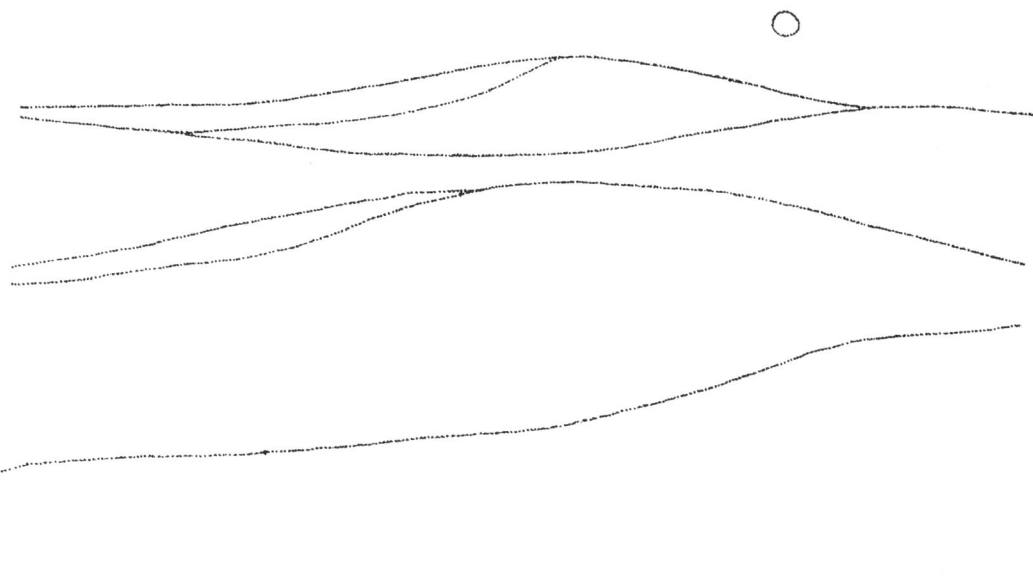

Canada is the trustee of 9 percent of the world's renewable supply of fresh water, and of the world's longest coastline, involving three oceans. It is also home to about 10 percent of the planet's forests. Wildlife populations are under considerable stress, due to loss and degradation of habitat, overharvesting and poaching, disease, and the impact of toxic substances. In addition, the country recycles only about 10 percent of its garbage and over 80 percent of Canadians live in areas with high levels of acid rain–related pollution. Finally, Canada's cold climate and dispersed population make it an energy-intensive country.

The Canadian government has committed funds for a comprehensive set of programs that make up the Green Plan to protect the country's natural resources and reduce waste generation and energy use. These include an environmental strategy for the Arctic, a pollution program for the Great Lakes, an extension of the parks system, wildlife conservation, and measures aimed at such global problems as climate change, ozone depletion, ocean dumping, and deforestation.

🏛 *Government*

Prairie Farm Rehabilitation Administration (PFRA)
Agriculture Canada
1901 Victoria Avenue
Regina, SASK, S4P 0R5 Canada

Contact: **G.M. Luciuk,** Director, Policy and Analysis Service
Phone: 1-306-780-5070 • *Fax:* 1-306-480-5018

Activities: Development • *Issues:* Sustainable Development

● Organization

Established in 1935, the **Prairie Farm Rehabilitation Administration (PFRA)** works with producers, local and provincial governments, and conservation groups to provide a comprehensive approach for managing soil and water resources while promoting sustainable agriculture. With a staff of 830, PFRA concentrates its activities in the semiarid prairie provinces of Alberta, Saskatchewan, and Manitoba, and in the Peace River region of British Columbia. Although aimed at the agricultural sector, many programs also benefit wildlife and recreation.

▲ Projects and People

The **Community Pastures Program** is PFRA's largest and longest running contribution to soil conservation on the Prairies. Started in the 1930s to reclaim badly eroded areas, the program has returned more than 900,000 hectares of poor quality cultivated lands to grass cover.

PFRA currently runs 87 community pastures, described also as "ecological reserves," which provide grazing for some 230,000 head of livestock each year. Each pasture has a full-time manager overseeing its operation. According to PFRA, community pastures allow farmers to supplement their own grazing lands and operate larger, more economical herds of cattle. To help improve the quality of western cattle, a herd of about 3,000 high-quality bulls is kept for breeding purposes. Other services are provided on the pastures as required and include inoculation, spraying, branding, dehorning, and castration.

Environmental projects are also established on behalf of endangered species (such as the **burrowing owl**), **swift fox** reintroduction, key habitat areas, and archeological and historical sites.

Since the prolonged drought of the 1930s, **water development** is crucial in the dry climate. Farmers and others are dependent on the "thousands of man-made water bodies that dot the countryside," such as dams, dugouts, wells, and reservoirs. Federal and provincial governments enter into agreements with "groups of five or more farmers or ranchers and towns or villages with a population of up to 300" to offer engineering, grants, and technical advice in establishing water supplies. Since 1935, PFRA states that it has spent about $84 million for more than 200,000 water supply projects on farms. Such projects also benefit regional parks and enhance habitat for waterfowl and other wildlife as well, reports PFRA.

Major undertakings are the St. Mary Irrigation Project supporting crops such as sugar beets, corn, peas, beans, potatoes, and oil seeds; the Assiniboine River–Shellmouth Dam Project, both a water source and a protection against flooding; the South Saskatchewan River Project with urban and rural uses; the Alberta Irrigation Rehabilitation Program; and the Bow River Irrigation Project of Alberta, which necessitated the resettling of about 436 farm families onto an area of nearly 11,000 hectares. To help in the transition from dryland farming, advice on irrigation practices and techniques was provided as well as assistance with farm layout designs, shelterbelt planting, and the building of on-farm water supplies.

A program entitled Farming for Tomorrow is one of numerous soil conservation efforts, whereby groups of farmers organize locally, prepare workplans, and receive government help for managing crop residues; planting grass, annual barriers, or field shelterbelts; contour stripcropping, extending crop rotations, establishing forage, and seeding waterways. PFRA points out that such activities also "improve air and water quality by reducing airborne particles and silt buildup in waterways." Vegetative cover also provides home for wildlife and farm diversification opportunities.

Protecting against prairie wind, shelterbelts were first introduced in 1902 and also "serve as an energy-efficient shelter to the farm home, trap snow for essential soil and moisture, and provide food and refuge to birds and animals." Each spring, the Shelterbelt Centre at Indian Head, Saskatchewan, "produces and ships more than 10 million tree seedlings." Mostly, shelterbelts help cut the loss of topsoil, improve growing conditions, and enhance crop diversification. Farmer Walt Swedberg, a pioneer in shelterbelt use, writes, "You walk along any part of the shelterbelt we planted [40 years ago] and you will see great horned owls, all sorts of birds' nests, and lots of deer. . . . Even the air seems to be fresher along the trees."

The Shelterbelt Centre also grows trees that benefit wildlife: buffaloberry for waxwings and game birds; bur oak for mule and whitetail deer, chipmunks, jays, flickers, game birds; black-fruited choke cherry for songbirds, deer; red-osier dogwood for rabbits, deer, robins; hawthorn for birds, rabbits, deer; red elder for squirrels, birds; hedge rose; Russian olive; saskatoon; sea-buckthorn; and Siberian crabapple for nesting birds—among other species.

■ Resources

PFRA offers numerous brochures, at no cost, about soil and water conservation, shelterbelts, and permanent cover.

Canadian Wildlife Service (CWS)
4999 98th Avenue
Edmonton, AB, T6B 2X3 Canada

Contact: Ludwig Norbert Carbyn, Ph.D., Research Scientist, Zoologist
Phone: 1-403-435-7357 • *Fax:* 1-403-435-7359

Activities: Research • *Issues:* Biodiversity/Species Preservation; Wildlife Conservation

▲ Projects and People

With the Canadian Wildlife Service (CWS) for nearly 25 years, Dr. Ludwig Norbert Carbyn is dedicated to research on the ecology of canids, namely wolves, coyotes, and swift fox. A major multiagency project emphasizing winter ecology and survival is underway to reintroduce swift fox into northern ranges. Other studies involve wolf predation on ungulates (hoofed mammals) in boreal forest areas within Riding Mountain National Park in Manitoba and in Wood Buffalo National Park in the Northwest Territory, where he is also studying wolf predation on bison.

Dr. Carbyn, who is also associated with the University of Alberta, where he supervises graduate ecosystem studies, organized and convened his second symposium on wolves in 1992.

Dr. Carbyn was one of 15 scientists in North America to be a part of the Delphi Study team which reviewed implications of wolf reintroduction to Yellowstone National Park. He also represents the Canadian perspective in three Species Survival Commission Groups—Canid, Wolf, and Reintroduction—for the World Conservation Union (IUCN/SSC).

International assignments have taken Dr. Carbyn to eastern Poland to assess wolf conservation problems, to Portugal to develop field studies on protecting the Iberian wolf, and to Mexico to assist in a reconnaissance program on the current status of Mexican wolves in the country's remote, northern chaparral areas.

Chairman of both the Recovery Team and Propagation Committee for the swift fox, Carbyn is professionally involved with three zoological gardens and one private game park. He frequently speaks at symposia, colloquia, public meetings, and task forces. Earlier wolf research at Jasper National Park was featured in a 30-minute film; Carbyn is currently involved in an educational film evaluating canids. The World Wildlife Fund has provided some funding for his projects.

Canadian Wildlife Service (CWS)
17th Floor, Place Vincent Massey
Hull, PQ, KIA OH3 Canada

Contact: Fern L. Filion, Chief, Socio-Economic and Marketing Division
Phone: 1-819-997-1360 • *Fax:* 1-819-953-6283

Activities: Education; Research • *Issues:* Biodiversity/Species Preservation; Ecotourism; Sustainable Development; Wildlife Conservation

▲ Projects and People

As chief of his division at the Canadian Wildlife Service (CWS) and a federal-provincial task force chairman, Fern Filion advises senior management on the relationship between wildlife resources and the economy, with implications for sustainable development.

A nationwide authority in his field, his reports range from current and potential demand for wildlife and habitat and wildlife-related tourism to the economic significance of these renewable resources and attitudes toward conservation. "The resulting insights have yielded a strategic advisory framework linking biological and ecological conservation programs in the Department of the Environment with the socio-economic priorities of Cabinet" in Canada's government, his literature notes.

Outside Canada, Filion provides socioeconomic expertise and leadership elsewhere in North America, Latin America and the Caribbean, Europe, and Africa. For example, he advises the World Conservation Union (IUCN) on the significance of wildlife utilization in equatorial Africa. With the Canadian International Development Agency (CIDA), he is designing a multimillion dollar socioeconomic program for the sustainable development of renewable resources in Zimbabwe. For the United Nations Environment Programme (UNEP), Filion is elaborating a proposal to develop and apply socioeconomic databases and expertise to strengthen environmental policies and programs throughout Latin America and the Caribbean. In addition, he represents Canada on the UN Economic Commission of Europe Task Force on Wildlife-Economy Integration.

Filion has organized socioeconomic symposia for the International Council for Bird Preservation (ICBP) on birds and international tourism in New Zealand, and with the International Union of Game Biologists (IUGB) in Hungary regarding the worldwide decline in hunting. He has also been involved in conferences in Poland, Mexico, Brazil, and Norway.

According to Filion's literature, his research findings have helped "justify increased funding for wildlife and habitat management programs, develop renewable resource policies and strategies, and enhance the 'political clout' of conservationists among senior decisionmakers."

Bedford Institute of Oceanography (BIO)/Institut océanographique de Bedford
Department of Fisheries and Oceans (DFO)
P.O. Box 1006
Dartmouth, NS, B2Y 4A2 Canada

Contact: **J.M. Bewers, Ph.D.,** Head, Marine
        Chemistry Division
    Phone: 1-902-426-2371 • *Fax:* 1-902-426-7827

Activities: Research • *Issues:* Marine Chemistry

● Organization

"More than the land, the sea is a metaphor for everything. . . . " Ernest Buckler, *Window on the Sea*

Beginning modestly in 1962, the **Bedford Institute of Oceanography (BIO)** is now among the world's largest ocean research centers. Some 350 scientists, engineers, and hydrographers in specialized fields study the Atlantic and the Arctic oceans. They are supported by a larger group consisting of officers and crews for nine research vessels plus technicians, maintenance personnel, computer programmers, and library staff.

On a 40-acre campus, BIO is administered by the federal Department of Fisheries and Oceans (DFO), Scotia-Fundy region, with scientific work divided into biological sciences, physical and chemical sciences, and hydrography branches. The unit collaborates with scientists from government, universities, and private industries worldwide—assisting with technology transfer, providing scientific advice, and producing atlases, charts, and other publications.

▲ Projects and People

BIO reports that its scientific teams "investigate the physical and chemical properties of the ocean, the life within it, and the geology of the sea floor, as well as the interactions between the atmosphere, the oceans, and the continents." According to BIO literature, "special attention is given to the **effect of the oceans on the global atmosphere**, in particular their abating influence on the greenhouse effect, the relationship between marine **environmental conditions and the abundance of fish**, and the **mineral and energy resource potential** of the continental shelves." BIO describes many scientists as tracing the "influence of the oceans on the well-being of human life itself."

Focusing on **marine chemistry** with "studies of the sources, transport, transformation, and ultimate fate of natural and manmade chemicals in the ocean," BIO examines the incidence and distribution of chemicals, their rates of introduction and removal from the marine environment, their decay and change, and the physical and biological processes that act on them. Chemicals are also used as tracers of oceanic processes to help scientists understand their effects on marine life.

Toxic chemicals derived from human and industrial activities are **researched and monitored.** As part of marine habitat protection management, toxicologists are especially concerned about contaminants on marine organisms and their incidence in seafoods.

Ongoing activities include **evaluating fish species at sea** for their abundance and health; learning about global warming by observing the growth of microscopic plants in the surface layers of the sea to determine **how much carbon dioxide the plants absorb and remove from the atmosphere;** sampling **water properties,** studying **plankton**—vital link in the food chain; investigating the **pollution-vulnerable clam fishery** in the Bay of Fundy with the aid of computer simulation models; and **measuring changes in the oceans' physical environment,** such as temperature, salinity, waves, tides, currents, and ice.

BIO concentrates studies on **ocean currents** "from the enormous Gulf Stream and Labrador Current, to tidal flows in regions close to shore." With sophisticated models and complex measurements, scientists learn "how currents interact with each other and with ice and the atmosphere, how water upwells from the deep ocean, how heat and cold are exchanged in the ocean, how tides flush harbors and estuaries, how pollutants are dispersed, and how fish larvae are transported."

Continental drift and the "deep oil and gas-bearing sedimentary basins" beneath the shelves are measured through seismic techniques and air- and ship-borne surveys. Such basins are also examined for potentially trapped hydrocarbons. Environmental marine geology studies the ocean floor and searches the layers of marine sediments for contamination in order to develop strategies to counter such pollution.

"The sentinels which alert us to the state of an ocean's environmental health," **seabirds** are the subject of a BIO research unit which surveys where these birds feed and breed, whether they are declining—such as the murre, a sea duck—migration routes, impacts of the depletion of capelin—a main food, damage by gillnets, and long-term pollution effects.

BIO maintains 9 research vessels; 50 smaller boats; on-ship, laser, and satellite equipment; and specialized instruments such as the BATFISH, "a towed underwater instrument carrier"; the BIONESS, a fine-mesh net that can "selectively trap" tiny organisms, and OCTUPROBE, which measures "microscale water turbulence."

■ Resources

BIO services include analytical development, quality assurance support, and scientific advice regarding "marine biogeochemistry, toxicology, fisheries contamination and protection, habitat contamination, environmental assessment, and environmental protection and management."

Science Branch
Department of Fisheries and Oceans
P.O. Box 5667
St. John's, NF, A1C 5X1, Canada

Contact: **Garry B. Stenson, Ph.D.**, Research Scientist
   Phone: 1-709-772-5598 • *Fax:* 1-709-772-2156
   E-Mail: stenson@nflorc.nwafc.nf.ca

Activities: Research • *Issues:* Biodiversity/Species Preservation

▲ Projects and People

Dr. Garry B. Stenson, Department of Fisheries and Oceans, specializes in marine mammals research—with attention to the river otter, sea otter, blue whale, right whale, killer whale, hooded seal, harp seal, and leatherback sea turtle. Articles he has authored and co-authored focus, for example, on ice entrapments of whales, "incidental entrapments of marine mammals and birds in an experimental salmon driftnet fishery," and reproduction.

Dr. Stenson is chairman of the Canadian Atlantic Fisheries Scientific Advisory Council's (CAFSAC) marine mammal subcommittee and belongs to the Committee on the Status of Endangered Wildlife in Canada (COSEWIC), International Council for the Exploration of the Seas (ICES) Baltic Seal Working Group, and Species Survival Commission (IUCN/SSC) Otter Group.

■ Resources

Apparatus for remote sensing of free-ranging animals and techniques for estimating populations are sought.

Canada's Green Plan
Canadian Parks Service
Environment Canada
Terrasses de las Chaudière
10 Wellington Street, Room 520
Hull, PQ, KIA OH3 Canada

Contact: **Bill Milliken**, Director of Communications
   Phone: 1-819-997-3736 • *Fax:* 1-819-953-5523

Activities: Development • *Issues:* Air Quality/Emission Control; Biodiversity/Species Preservation; Global Warming; National Parks; Sustainable Development; Transportation; Waste Management/Recycling; Water Quality

▲ Projects and People

Canada's Green Plan is a national action strategy formulated in the early 1990s, with input from every segment, for "reversing the damage caused to the environment today, sustaining development tomorrow, and securing a healthy environment and a prosperous economy." According to **Robert R. de Cotret**, Canada's Minister of the Environment, the Plan for a healthy environment is the federal government's response to Canadians' concerns and ideas for cleaning up and protecting the environment. "We have worked hard to ensure the priorities expressed by Canadians are reflected in the many Green Plan initiatives," he says.

"The resulting Green Plan is the most important environmental action plan ever produced in Canada. It is the source for more than 100 important and well-funded initiatives over the next five years. . . . In addition to the $1.3 billion the government already spends annually on the environment, the Green Plan commits an additional $3 billion in new funds over five years," states de Cotret.

According to the Green Plan summary, the **Health and Environmental Action Plan** is assessing the human health hazards from acid rain, air toxics, ground-level ozone (or smog), climate change, and indoor air contaminants. It deals with the safety of drinking water and the health effects of waste management practices.

The Action Plan is described as focusing "particularly on the problems of those most at risk to environmental stress—our children, pregnant women, and native and northern peoples who live off the land." It is also expected to concentrate on the needs of individuals and communities to learn more about the link between health and the environment. Programs will encourage and support individual and community action in this area.

The overall Green Plan has **compelling goals** to attain clean water, air, and land control of chemical wastes; sustaining renewable resources; protecting wildlife and their habitats; preserving the Arctic; addressing international problems such as global warming, acid rain, and ozone depletion; and preparing for environmental emergencies.

For example, the Green Plan proposes a **Drinking Water Safety Act** with measures supporting **groundwater cleanup** and the acceleration of **water and sewer systems to Indian reserves**. The Great Lakes Pollution Prevention Centre is being set up, and steps are proposed to prevent contamination in inland waters in Alberta and Manitoba that may double fish population in certain areas. On behalf of the St. Lawrence River, Canada and the USA—their governments, industries, and communities—will cooperate. Canada has already begun a five-year **Ocean Dumping Action Plan** against industrial waste, plastics, and other debris; harbors and coastal areas must also be cleaned up.

Regarding **clean air**, Canada has adopted tougher emission standards against leaded gasoline, sulphur dioxide, and carbon monoxide; now it is **targeting smog**, or ground-level ozone, and expects to reduce such emissions by 40 percent by the year 2000. Tighter emission standards for new cars begin in 1994; and 1995 will see a "national agreement on what is ultimately necessary to achieve air quality standards." Again, the USA and Canada must work together.

Canada's goal is to **cut waste by 50 percent by the year 2000**. This means reducing some 30 million tons of garbage a year and managing what is produced; for example, less than half of today's annual 8 million tons of hazardous waste is treated. To reach this goal, the Green Plan is building on the National Waste Reduction Plan and is expanding the National Waste Exchange Program to find markets for reusable materials. Related goals are cleaning up 30 high-risk waste sites by 1995 and destroying all PCBs under federal jurisdiction by 1996. Also set out is a **National Regulatory Action Plan**—accompanied by a database or a **National Toxicology Network** at Canadian Universities—to control toxic substances through their identification, measurement, and periodic assessment. Ultimately Canada intends to eliminate the discharges of "persistent toxic substances."

To Canada, a sustainable renewable resource is its forests that provide one job for every fifteen. The solution includes planting 325 million trees in cities and towns, stemming insects and diseases, setting up a national forest seed and gene bank, establishing model demonstration projects, and intensifying science programs on climate change, acid rain, biodiversity, and fire management. Sustainability is also being emphasized in agriculture and fisheries. For example, the government is developing a **National Sustainable Fisheries Policy and Action Plan** as well as a **Recreational Fisheries Policy**. Fish habitats are being protected and the "race for fish" syndrome is being replaced with a rational "transferable share" system.

The Green Plan also has a **National Wildlife Strategy** to set aside 12 percent of Canada as "protected space"; establish 5 new national parks and possibly 6 new marine parks by the year 2000; safeguard at least 175 species from habitat loss and toxic chemicals; improve wildlife science and health research; introduce laws attacking poaching and illegal trading; and strengthen wildlife law enforcement.

The Green Plan pledges to defend the Arctic with 40 percent of Canada's land mass and two-thirds of its marine coastlines, "unspoiled wilderness," open tundra, great rivers and forests, and native populations who are dependent on these delicate ecosystems. Canada and the peoples of the North are preparing a five-year **Arctic Environmental Strategy** to preserve and enhance the Arctic's "integrity, health, biodiversity, and productivity."

Also called for are new guidelines, regulations, and public education regarding **responses to environmental emergencies** and including upgrades to the natural hazard prediction system with four Doppler radar stations by 1996, a national television environmental warning system, and special devices such as "automated water-level monitoring equipment in flood-prone urban centers."

Building **partnerships** and strengthening decisionmaking are at the heart of the Green Plan. Key elements are the Canadian Council of Ministers of the Environment (CCME) with its Statement on Interjurisdictional Cooperation on Environmental Matters, the involvement of native groups, commitment to Environmental Non-Government Organizations (ENGOs), and enlisting business, youth, women's groups, and other stakeholders.

Canada is adjusting its traditional National Economic Accounts (including the Gross National Product) to reflect the "real costs and benefits" to the country of decisions affecting the environment. Because communication is crucial, a **National State of the Environment Reporting Agency** will be established by 1994. Other aspects are a five-year **Environmental Science and Technology Action Plan**, effective environmental laws with "market-based approaches for environmental protection," and the new Canadian Environmental Assessment Agency with a participant funding program.

■ Resources

Canada's Green Plan is available through Environment Canada.

Canadian Climate Centre (CCC)
Environment Canada
4905 Dufferin Street
North York, ONT, M3H 5T4 Canada

Contact: **Stewart J. Cohen, Ph.D.**, Impacts Climatologist
Phone: 1-416-739-4389 • *Fax:* 1-416-739-4297

Activities: Research • *Issues:* Global Warming; Sustainable Development

▲ Projects and People

The **Mackenzie Basin Impact Study (MBIS)** is an integrated assessment of the potential impacts of global warming scenarios in the Mackenzie Basin, Canada. A six-year project directed by **Dr. Stewart Cohen, Canadian Climate Centre (CCC)**, it has input from researchers and stakeholders representing federal and provincial/territorial government agencies, native organizations, and private industry, which assist in overall planning, including proposal review.

MBIS has targeted these policy issues: "interjurisdictional water management, sustainability of native lifestyles, economic development opportunities, buildings and infrastructure (including transportation), limitation strategies, and sustainability of ecosystems." With the support of the Canadian government's Green Plan, ESSO Resources Ltd., Environment Canada, and others, MBIS is undertaking the following projects, which are active in 1992–1993: forest-wetlands response, **S. Bayley**, University of Alberta; regional accounting framework, **S. Lonergan**, University of Victoria; permafrost in the Mackenzie Valley and Beaufort Sea region, **P. Egginton**, Terrain Sciences Division, Geological Survey of Canada; freshwater fisheries, **D. Hamilton**, University of Manitoba, and **B. Welch**, Freshwater Institute; Peace River ice, **D. Andres**, Alberta Research Council; basin runoff, **R. Soulis**, University of Waterloo; wildlife response to fire, **P. Latour**, GNWT Renewable Resources; agriculture impacts and opportunities, **M. Brklacich**, Agriculture Canada; tourism impacts and opportunities, **G. Wall**, University of Waterloo; and vegetation response at tree line, **R. Wein**, University of Alberta.

Other projects are expected to begin on land-cover data from remote sensing, forest sector impacts, Saskatchewan fisheries, and "Dene traditional knowledge on climatic variability."

CCC is contributing "baseline and scenario climate data sets, analysis of snow cover, sea ice and lake ice, and a framework for assessing potential change in land capability due to climatic change. Atmospheric Environment Service Western Region has also assisted in developing climate data sets." Other agencies, universities, and community groups, such as the Inuvialuit Game Council, are involved.

■ Resources

An interim report was scheduled for late 1992, mid-study report in 1994, and a final report in 1996.

Hydrometeorological Processes Division
Canadian Climate Centre
Environment Canada
11 Innovation Boulevard
Saskatoon, SASK, S7N 3H5 Canada

Contact: **Richard (Rick) G. Lawford,** Chief
Phone: 1-306-975-5775 • *Fax:* 1-306-975-5143

Activities: Research • *Issues:* Global Warming; Sustainable
Development; Waste Management/Recycling • *Meetings:*
AGU; AMS; WCRP/GEWEX; IGBP/AMIGO

● **Organization**

A "science in its infancy," hydrometeorology, as the name implies, is
a combination of "hydrology and meteorology dealing with processes such as precipitation formation, drought, evaporation, tran-
spiration, runoff, ablation, snow accumulation, and snow transport." The **Hydrometeorological Processes Division** is part of the
Canadian Climate Centre's Climate Research Branch. Other divisions in this Branch include the Climate Modeling and Diagnostic
Studies Division, Extended Range Forecast Division, and a special
projects group responsible for remote-sensing activities and the
work of the Boreal Ecological Study. The division has a staff of 10.

Division chief **Richard Lawford** writes in the *Fluxes and Cycles*
quarterly report about opportunities in his field. "Hydrometeorology in Canada like many other interdisciplinary sciences is highly
specialized. . . . No Canadian university and few USA universities
give advanced degrees. . . . Scientists entering the field are embarking on a long-term intensive learning experience because they are
either atmospheric scientists who must learn about hydrology or
hydrologists who must learn about the atmospheric sciences. Furthermore, they often have to leave behind many of the contacts and
to some extent their established reputations with a small scientific
community to reestablish themselves in a new and much larger
community. . . . Major challenges for management are to facilitate
this process of integration into a larger community and to ensure the
relevance of the research programs to a broader user community."
Lawford believes that interdisciplinary science is the way of the
future.

▲ **Projects and People**

According to its *1990–91 Annual Report*, the Division collaborates
internationally on several projects. It participated in the Canada/
USA Symposium on Climate Variability and Change on the Great
Plains. Lawford is involved in an International Geosphere–Biosphere Programme (IGBP) project that compares the responses of
the Northern and Southern hemispheres to global change.

The four major Division areas are: evaporation studies, precipitation studies, hydrometeorological models, and hydroclimatology.
The report states that it acts also as a research broker, managing
contracts which are funded primarily by other agencies. Cooperating are Ducks Unlimited, Canadian Wildlife Service, Prairie Farm
Rehabilitation Administration (PFRA), Western Region of the Inland Waters Directorate, and various provincial and federal offices,
for example.

The **Regional Evaporation Study (RES),** with **Dr. Geoff Strong's**
leadership, is a collaborative effort aiming at "improving our understanding of evaporation at all scales." A portion of this overall study,

entitled the **evaporation and mesoscale research project,** is investigating these processes on the Prairies as well as the "role of local
evaporation in convective precipitation and the mesoscale dynamics
which influence or control the balance between evaporation and
precipitation." A field study concentrating on data collection with
instrumented balloons and other equipment got underway in mid-
June 1991.

Two studies were carried out under contract to the University of
British Columbia: one on **climate variability and change,** the other
on the **hydrometeorology of British Columbia.** As described in
Division literature, the first dealt with the possible effects of climate
change on the water resources, coastal currents, and fisheries of
British Columbia, as well as with the difficulties of carrying out
climate-impact assessments in a region of complex terrain. The
second study investigated the "effects of changing synoptic atmospheric circulation patterns on abnormally moist or dry periods,"
reports the Division.

Another study is the **drought characterization project,** with **Don
Bauer's** leadership, to discover means to describe the "temporal and
spatial variability and trends of meteorological drought, especially
for the North American Prairie." The Division reports, "A method
of gap-filling precipitation data was tested and applied in the creation of a database of precipitation time series for Saskatchewan
crop districts." With **Prof. L. Nkendirim,** University of Calgary, a
project aspect was a study of the "feasibility of using proxy data to
extend the drought time series on the Canadian Prairies."

Other projects focus on **precipitation processes research** using a
weather radar facility with investigator Joe Eley; hydrometeorological modeling and the use of satellite data with **Dr. Geoff W. Kite**
and **Dr. Thian Yew Gan,** formerly of the Asian Institute of Technology (AIT), Thailand; **hydroclimatology** or **CliRed project** in the
Red River Basin with **Les Welsh** and **John Knox** of the Canadian
Climate Centre and **Gregg Wiche** of the U.S. Geological Survey's
(USGS) North Dakota district office; **climate/wetlands study** to
assess climate warming impact on the Prairie's shallow wetlands
with the development of a descriptive model by **Dr. M.K. Woo** and
Bob Rowsell, McMaster University; and collaborative studies following up the impact of the 1988 drought in Saskatchewan and
Manitoba on agriculture, wildlife, forestry, fishery, energy, and
water resources. Stephen Ross Macpherson is also with the Division.

■ **Resources**

The Division offers speakers and consultation services, often free of
charge. It also publishes a quarterly report, *Fluxes and Cycles*, and
can provide a publications list.

Environmental Achievement Awards
Conservation and Protection
Environment Canada
Ottawa, ONT, K1A OH3 Canada

Contact: **Gail Turner**
Phone: 1-819-997-6827 • *Fax:* 1-613-957-2655

Activities: Education; Public Affairs • *Issues:* Biodiversity/
Species Preservation; Conservation; Sustainable Development

▲ Projects and People

In 1989, **Environment Canada**, the Canadian Ministry of the Environment, initiated the annual **Environmental Achievement Awards** to "acknowledge the accomplishments and dedication of individuals and organizations devoted to protecting, conserving, and rehabilitating the environment."

Presented to Canadians "from all walks of life," the awards are divided into six categories: communications, corporate, municipalities, nonprofit, lifetime achievement, and environmental science fair project. An independent body, the Canadian Environmental Advisory Council, selects the winners and other finalists. Award winners follow:

Communications. Graeme Ferguson, IMAX Systems Corporation, Toronto, developed giant-screen technology that is used internationally. The IMAX feature *Blue Planet* has been cited as a major contribution to environmental public awareness. Scientist **David Suzuki**, University of British Columbia's Department of Zoology, was honored for making people aware of the plight of endangered animals through his weekly newspaper column, easy-to-understand books, and articles with the basic message: "We must stop polluting, and must stop it now." *(See also University of British Columbia in the Canada Universities section.)*

The first communications award winner was the **Young Naturalist Foundation**, Toronto, led by **Annabel Slaight**, who reaches millions of children through *Owl* and *Chickadee* magazines as well as through a science and nature television series tailored to young people and broadcast worldwide.

Corporate. INCO Limited of Toronto was recognized for initiating the largest environmental cleanup in Canadian history. *(See also International Nickel Company, Limited [INCO] in the Canada Private section.)*

Quebec's **La Société Laminage Perma Ltée**, with Pierre Matte as owner, was recognized for **industrial sanitation**. La Laminage Perma, in Saint-Leonard, was described as manufacturing a "high-quality product in a clean workplace [using] only wood scraps, thus sparing many trees. . . . Its wood treatment and lamination processes require no toxic glues or sanding. "The material used in manufacturing is recycled, while painting is done in a workshop designed to eliminate fumes." Also, "the company never throws chemical products down its drains. Mr. Matte also gives preference to suppliers that use recycled material. He invests only in companies that demonstrate concern for the environment."

The first corporate award recipient, **Mohawk Oil of Burnaby**, British Columbia, pioneered a way to cut down on the amount of waste oil entering the ecosystem by **recycling used oil** through special treatment at a North Vancouver plant.

Municipalities. The **City of Montréal's Botanical Gardens** designed an award-winning **water purification system** which is used at a public beach on Ile Notre-Dame. The water is purified by passing over 100,000 aquatic plants. Receiving worldwide attention, similar projects may begin in Great Britain, Florida, and California. *(See also Montréal Botanical Garden in the Canada Local Government section.)* The **Regional Municipality of Sudbury** won an award for reclamation of land degraded by the intense coal mining of the region. Such land has been planted and fertilized. *(See also Regional Municipality of Sudbury in the Canada Local Government section)* The **City of Kitchener** won for inaugurating in the mid-1980s the "blue box program" of **curbside recycling** of bottles, cans, and newspapers—reducing the city's landfill by 10 percent.

Nonprofit. Western Canada Wilderness Committee (WCWC), Vancouver, received the prize for helping "give global prominence to the need to save wilderness . . . through its sponsorship of the Wilderness Is the Last Dream (WILD) conferences." WCWC has worked for over a decade to increase public awareness of ecology and sustainable development.

Ducks Unlimited Canada won for protecting the wetlands and prairies of Canada and abroad. Their programs are designed to improve the survival of waterfowl by conserving the birds' natural habitat. The first NGO award winner was **World Wildlife Fund—Canada** for creating the Wildlife Toxicology Fund, various environmental education programs, and the Endangered Species Recovery Fund in conjunction with Environment Canada, which helped save the white pelican and wood bison. WWF–Canada has supported more than 1,000 conservation projects.

Lifetime Achievement. Dr. Louis LaPierre, New Brunswick, was honored for inspiring future environmental work through two decades of developing public and private sector **environmental policy**. A biology professor and broadcaster, Dr. LaPierre has published widely on wildlife ecology and produced a documentary series on Atlantic ecology. **Dr. Andrew Thompson, Ferguson Gifford Barristers and Solicitors**, focuses on Canada's North—the Yukon, Northwest territories, and especially the Arctic. Dr. Thompson organized and chaired several conferences on wildlife and endangered species and has served on government commissions on behalf of environmental groups throughout Canada.

Dr. Pierre Dansereau, professor emeritus at the Université du Québec à Montréal, received the first lifetime achievement award for pioneering **global science ecology**. With 11 honorary doctorates and other worldwide honors, he is known for his work in human ecology, botany, phytosociology (study of plant groupings), and biogeography.

Science Fair Projects. Students who have distinguished themselves are **Samir Gupta and Denis Tsui**, Centennial Regional High School, Brossard, Quebec, for a "pilot-scale bioremediation demonstration of an aerated bacterial process that proved effective in consuming PCBs [polychlorinated biphenyls] and **cleaning up contaminated soil**."

Pascale Charest and Sandra Imbeault, Matane, Quebec, high school students, developed a waste management project that explained how a **disposable diaper** is completely recyclable. The absorbent pad was converted into "pulp" and mixed with sawdust to produce plants "as beautiful as those obtained with the sawdust-and-peatmoss mixture used in hydroponics." The students' research showed that plastic could be converted into rigid plastic for the manufacture of garden tools. According to Environment Canada, throw-away diapers represent about 3 percent of solid waste at municipal disposal sites, and that number is growing.

Environment Canada and the Youth Science Foundation select the winners in the Science Fair category.

Inland Waters Directorate (IWD)
Environment Canada
351 St. Joseph Boulevard, Hull
Ottawa, ONT, KIA OH3 Canada

Contact: **Denis A. Davis**, Director General
Phone: 1-819-997-2019 • *Fax:* 1-819-997-8701

Activities: Research; Development; Political/Legislative • *Issues:* Health of Aquatic Ecosystems; Global Warming; Sustainable Development; Water Quality • *Meetings:* GIS Conference; IWRA; UN Economic Commission

● **Organization**

Promoting the "wise management and protection of water in Canada and globally as a vital contribution to sustainable development" is the mission of the **Inland Waters Directorate (IWD)**, says director general **Denis A. Davis.** IWD also is involved in freshwater/aquatic ecosystem monitoring, forecasting, management, and communications.

To address problems in Canada's waters, such as acid rain and atmospheric change, IWD is enacting a strategy for broadening involvement in multilateral international organizations and agreements, and selected bilateral programs and exchanges; sponsoring training programs in developing countries and creating "indigenous technologies for global sustainable development of waters"; demonstrating Canadian water technology at global forums; participating in international monitoring, research, and emergency preparedness programs; and sharing water-related expertise with the Canadian International Development Agency (CIDA) and other federal departments.

▲ **Projects and People**

At IWD's **National Water Research Institute,** research is underway in the fields of "limnology and aquatic ecology; contaminant properties, fate and effects; exchange of toxic contaminants between air, water, sediment, and biota; acid/toxic rain effects; groundwater contamination; lake rehabilitation; aquatic monitoring; ecotoxicology; large basin studies and modeling; simulation modeling hydraulics and sedimentology; analytical methods development; quality assurance and quality control." **R.J. Daley** is executive director.

IWD also has a **National Hydrology Research Institute** that, according to its literature, is doing experiments in the fields of hydrologic cycle; glacial, alpine, prairie and northern hydrology, hydrological and water quality modeling; glaciology, hydrogeology, groundwater supplies; intergranular and fracture flow; fluvial processes and river norphology; river-ice formation and breakup; ice-jams and floods; groundwater-permafrost processes; contamination transport in surface and groundwaters; pesticide movement and degradation, nutrient dynamics, lake and river ecology, and microbia biology. **T.M. Dick** is director.

The **Water Resources Branch** conducts research in the fields of "development and evaluation of methods for sampling suspended sediment and bedload, for surveying reservoirs and for data computations; evaluation of sampling strategies and sediment survey network designs; analysis and interpretation of sediment data; application of sediment transport equations and mathematical models. This Branch also has a **Hydrology Division.**

A **Water Quality Branch** assesses environmental quality, sets procedures for biological monitoring and testing in northern waters and regarding genetically engineered organisms, researches acid rain and "the long-range transport of toxicants," does computer modeling, and develops guidelines for contaminant substances in fresh and marine water, sediment, and soil. **Dr. D.B. Carlisle** is the contact.

A **Water Planning and Management Branch** defines tools and strategies for integrated planning; does socioeconomic modeling of water demands; undertakes water-use studies; examines water pric-

ing techniques and effectiveness; and investigates socioeconomic aspects of acid rain, climate change, conflict resolution, ecosystem approach to water management, and institutional and legal aspects of water management—among its projects. **H. Foerstel** is the contact.

■ **Resources**

IWD needs equipment and instrumentation to conduct lab analyses. It also needs software and programs to use in field observation and modeling. IWD states that it would like to set up an information exchange to promote awareness and knowledge.

IWD has available a publications booklet in both English and French listing publications including general interest brochures, technical bulletins, scientific series, report series, social science series, data series, water quality reports, flood damage reduction program publications, and western and northern region reports.

Biotechnology Research Institute (BRI)/
Institut de recherche en biotechnologie
**National Research Council Canada (NRC)/
Conseil national de recherches Canada**
6100 Royalmount Avenue
Montreal, PQ, H4P 2R2 Canada

Contact: Peter C.K. Lau, Ph.D., Research Scientist
Phone: 1-514-496-6325 • *Fax:* 1-514-496-5232

Activities: Development; Education; Research • *Issues:* Air Quality/Emission Control; Energy; Health; Pharmaceuticals; Waste Management/Recycling; Water Quality • *Meetings:* International Biohydrometallurgy Symposium; Pseudomonas '93

● **Organization**

The mandate of the **Biotechnology Research Institute (BRI)** is "downstream research and development with industrial relevance." Downstream refers to the "flowchart of scientific development, at the point where biotechnological processes begin building to greater and greater gram masses." Thus, the Institute is focused on "the nature of biochemical change and the behavior of biological molecules." Protein engineering and biochemical engineering are two main fields.

▲ **Projects and People**

Genetic engineering researchers participate in the Protein Engineering Network of Centres of Excellence (PENCE) working to "redesign proteins to meet specific biological needs, particularly those of the health and pharmaceutical industries." Enzymes are studied that are implicated in various medical disorders ranging from degenerative diseases such as muscular dystrophy to parasitic infections and the common cold.

PENCE is investigating cysteine proteases, with **papain,** a plant cysteine protease, as the Network's working model. In a collaborative undertaking, researchers are working to "design inhibitor molecules that block enzymatic activities." They are investigating **cathepins,** which are linked to human degradative diseases, a protozoan-causing enzyme that is responsible for a widespread and dehabilitating tropic disease known as Chagas' disease, and cysteine protease from rhinoviruses that are responsible for the common cold. One goal is to develop antiviral drugs.

Cited as Canada's "most advanced lab" on the molecular genetics of *Bacillus thuringiensis*, the bacterium producing the insecticidal toxin Bt, BRI's Genetic Engineering section is engaged in creating more potent variants—as the country uses "up to half the worldwide production of Bt in protecting trees against insects like the spruce budworm and gypsy moth." Yet, all the bioinsecticide is produced outside Canada. BRI provides the synthezied genes for variants to Forestry Canada's Forest Pest Management Institute for testing.

Biochemical engineering is described as "the middle ground of biotechnology, straddling the fence between lab discovery and product development. . . . The rules governing the growth of living systems come up against the restraints imposed by scale-up. Here, engineering principles come into play." Research is described as both "client-driven" (client calls in with problems) and "technology-pushed" (BRI calls out with new developments or bioprocesses). In this heavily collaborative field, BRI says that about "half the effort is invested in partnerships with industries and public agencies and the other half devoted to work promoting the national interest."

Biosensors are being developed, for example, to detect bacterial contaminants in food and pesticides in water. By optimizing this process, such instruments can be developed at considerable savings for a "large and growing market." Finely tuned biosensors to cell culture environments can sense such things as "the rate of substrate uptake, growth rate, and product formation rate."

One application is in the careful monitoring of the toxic cyanide, which appears in the wastewaters of the chemical industry through the use of "thousands of tons of hydrogen cyanide each year." Another is in the biological assay of the food contaminant *Salmonella typhimurium* whereby biosensors would reduce the microorganism's measurable detection from four days to four to five hours.

In environmental engineering, anaerobic digestors are being developed for "detoxifying pulp and paper mill wastewaters and for treating other industrial wastewaters." Ongoing work includes developing conditions that will ultimately create "faster, tailor-made digestors for specific industrial wastes, particularly the bleach effluents from Kraft pulp mills." The bioaugmentation process includes using soil bacteria to break down chloro-organics, which BRI says, are "ubiquitous, artificial chemicals that pose considerable human health risks." Mixed bacterial cultures are added to contaminated soils to hasten natural remediation.

In another project, BRI environmental engineers are developing an *in situ* soil detoxification process on behalf of Environment Canada to help reclaim contaminated areas in the St. Lawrence River valley.

Emphasizing integrated approaches, BRI reports breakthroughs in research concerned with biodegradation of halogenated aliphatic hydrocarbons, a collaborative venture with Prof. D.T. Gibson, University of Iowa, on the microbial degradation of polycyclic aromatic hydrocarbons (PAHs), and a molecular study of a mineral-leaching organism, *Thiobacillus ferrooxidans*, an economically important bacterium that causes pollution in acid-mine drainage. Research with *T. ferrooxidans* genes indicates a relatedness of certain proteins to those of *E. coli* which has "important ramification about genetic exchange between bacteria of diverse habitats," according to Dr. Peter C.K. Lau. BRI refers to *E coli* as "the workhorse of recombinant–DNA technology."

A recent visiting professor at the University of Iowa's Microbiology Department/Biocatalysis Group, Dr. Lau is also adjunct professor with the Microbiology and Immunology Department, University of Ottawa. Regarding hazardous materials, Dr. Lau believes "a pollutant is a pollutant" and "every effort should be given to curtail the problem regardless of its numerical rank in the EPA [U.S. Environmental Protection Agency] list."

■ Resources

More "financial/industrial support and human resources" are needed for most research projects, writes Dr. Lau. He also recommends joint studies—for example, by USA and Canada government labs working together. As references, he recommends World Resources Institute publications, such as the *Environment Almanac*, and believes that literature, such as this *Guide*, "be read and [provide] a common knowledge to all citizens and not serve as a mere reference to a few concerned."

Biotechnology Research Institute (BRI)/Institut de recherche en biotechnologie
National Research Council Canada (NRC)/ Conseil national de recherches Canada
6100 Royalmount
Montreal, PQ, H4P 2R2 Canada

Contact: **Réjean Samson, Ph.D.,** Group Head, Environmental Engineering
Phone: 1-514-496-6180 • *Fax:* 1-514-496-6265

Activities: Development; Research • *Issues:* Environmental Biotechnology; Waste Management/Recycling; Water Quality • *Meetings:* Air and Waste Management Association; Water Pollution Control Federation

● Organization

In 1984, **Biotechnology Research Institute (BRI)** began collaborating with Canadian industries and universities to function as a "leading edge" toward industrial application in biotechnological research. Five research sectors, including a pilot plant extending bioprocess capabilities, are organized into BRI's research and developmental programs. An incubator program provides laboratory facilities to firms and other research centers for innovative product development. Overall, staff is comprised of some 350 persons including on-site employees and guest researchers.

One example is a **cooperative effort of BRI and the Pulp and Paper Research Institute of Canada (PAPRICAN)** to "reduce the toxic pollutants formed when chlorine and other chemicals are used to bleach wood pulps." With PAPRICAN's discovery that "certain white rot fungi will bleach hardwood pulps," the collaborative undertaking is extending this process to "the bleaching of softwoods [and increasing] the slow reaction rates of the natural organisms, hopefully to acceptable commercial levels," writes Peter E. Wrist, PAPRICAN president.

Another example is the joint undertaking of the Nova Husky Research Corporation (NHRC) and BRI to produce "useful value-added products based on methanol as a cheap chemical feedstock" that can lead to further "product development, downstream processing, and product recovery," according to NHRC scientist **Mike Francis**.

▲ Projects and People

In the late 1980s, projects in applied microbiology, biosensor development, cell culture, and environmental engineering have matured. And, more specifically, environmental protection through bioremediation techniques is the thrust of the environmental engineering group, with **Dr. Réjean Samson** serving as group head with a staff of 25.

In applying biotechnology to various environmental and industrial problems, four specialized labs are used. The **Anaerobic Biotreatment** program uses "upflow anaerobic granular sludge bed reactors" in the treatment of industrial effluents (toxic and nontoxic). Working with the University of Calgary's **Biofilm Group**, the lab team developed a "protocol for industrial processing of efficient granules specific to targeted effluents such as pulp and paper bleacheries, by accretion of selected microorganisms to pre-formed acetate-to-methane nuclei." An economical technology for organochlorides removal appears "promising" for degradation of toxic chemicals, reports BRI.

The **Xenobiotics Biodegradation** lab focuses on characterizing and isolating bacteria degrading organic pollutants, such as petroleum hydrocarbons and chlorinated organics. The application of gene probes and molecular biology that "accelerate the isolation of pollutant degrading bacteria" could benefit industry and government agencies. The techniques are apt in enforcing environmental protection laws and monitoring potentially hazardous environments.

The objective of the **Soil Decontamination Engineering** laboratory is the understanding and development of remedial biotechnological processes to restore contaminated sites. Such contaminants include petroleum hydrocarbons, PAH (polycyclic aromatic hydrocarbons), PCP (pentachlorophenol), PCB (polychlorinated biphenyl), and creosote on wood preservative sites. Experiments with a "genetically modified microorganism" quantified "for the first time" the "effect of competition and inhibition on a microorganism introduced in a wild environment such as soil." Other studies are concerned with UV-radiation technology that could dechlorinate PCB from contaminated soil.

Finally, in the **Environment Analytical Chemistry** lab, new techniques chemically analyze various pollutants in sediments, soil, mud, and pulp and paper effluents. Metabolites are identified from "various degradation pathways of pollutants" and bioreactors are monitored. Working with sunlight and artificial light sources, efforts are underway to develop a "photochemical means of degrading chlorinated aromatic pollutants." The use of a cheap light source or solar radiation is being studied for its effectiveness "to first partially or totally photodechlorinate PCBs. The resulting compounds would then be treated by microbes to complete the mineralization of PCBs. This sequential process could form the basis of a general protocol for soil and effluent detoxification," reports Dr. Samson.

■ Resources

BRI's *Annual Report* and other printed matter describe environmental engineering and other related research.

🏛 *Provincial Government*

Saskatchewan Research Council (SRC)
15 Innovation Boulevard
Saskatoon, SASK, S7N 2X8 Canada

Contact: **Elaine Wheaton**, Climatologist
   Phone: 1-306-933-8179 • *Fax:* 1-306-933-7446

Activities: Research • *Issues:* Air Quality/Emission Control; Biodiversity/Species Preservation; Energy; Global Warming; Sustainable Development; Waste Management/Recycling; Water Quality

▲ Projects and People

Elaine Wheaton, climatologist with the Environment Division, **Saskatchewan Research Council (SRC)**, prepares publications in the areas of climatology, hydrometeorology, and environmental studies. Recent research reports summarized SRC's Climatological Reference Station's annual activities with co-author **V. Wittrock**, reviewed the accomplishments of the *Task Force on Soil-Water-Wetlands Management in Saskatchewan*, and recapped *The Drought of 1988*.

Wheaton has made presentations on **global warming** in relation to prairie agriculture, "the opportunities it may bring," and changing weather patterns as well as on **climatic change** in relation to both dust storms and boreal forest sensitivity. She is an authority on modeling agroclimatic and ecoclimatic impacts of such change and has also written on "estimating impacts on Saskatchewan biomass and wind erosion for several climatic scenarios" with **A.V. Chakravarti**.

■ Resources

A publications list is available.

🏛 *Local Government*

Montréal Botanical Garden ⊕
4101 rue Sherbrooke Est
Montreal, PQ, H1X 2B2 Canada

Contact: **Pierre Bourque**, Director
   Phone: 1-514-872-1454 • *Fax:* 1-514-872-3765

Activities: Architectural Landscape; Education; Research
Issues: Biodiversity/Species Preservation; Waste Management/Recycling; Water Quality

▲ Projects and People

The City of Montreal's overall plan to **revitalize an island park** and provide a larger range of recreational activities was realized under the direction of the **Montréal Botanical Garden**. A staff of 120 worked with the Williams Asselin Ackaoui landscape architectural firm which opened the St. Lawrence River beach park in June 1990.

The Notre-Dame beach park island, reflecting Quebec's natural landscape, was completely man-made for Expo '67. An artificial lake, the Lac des Régates is located in the southern sector with the

600-meter beach at its south end and sheltered from the wind; a pump station feeds the lake from the St. Lawrence River.

To maintain the appearance of a natural lake—as found in the Laurentian Mountains north of Quebec—conventional methods of water purification (such as chlorine treatment) were avoided. A "filter-lakes" system, consisting of four ponds, was adopted instead. Three of the ponds included different submerged and emergent aquatic species of which some 100,000 were planted. According to a report on this subject, "aquatic plants, by their ability to absorb and metabolize dissolved nutrients from the water in which they grow, can be exploited to purify wastewater." Because the organic charge of the water from the beach is relatively low, the system used at the park differs from those currently operating in Europe.

Water in the swimming area is recirculated approximately every five days. The water filtration station uses both sand filters and ultraviolet sterilizers. The UV process leaves "no permanent residue in the water," writes Gilles Vincent of the Garden. The water is then pumped back toward the swimming area "over a rocky cascade which provides final aeration."

There are three functions to the filter-lakes method: degradation of organics by the microorganisms living around and on plant root systems; natural filtration for suspended coliforms and solids; and reduction of the concentration of dissolved pollutants, such as nutrients in excess, trace metal, and pesticides.

Although still considered a new technology at the Canadian latitudes (due to the long cold season which halts plant-growing periods), the use of constructed wetlands on a small and medium scale showed "a good potential during the summer," reports a Garden spokesman. During 1990, its first season, more than 148,000 bathers enjoyed the water.

■ Resources

The Garden has produced a report, in English, on *Natural Processes to Maintain Water Quality, the Beach Park of Ile Notre-Dame* and, in French, *Un Bain de Nature: Le Lac de l'Ile-Notre-Dame*.

Regional Municipality of Sudbury
200 Brady Street, Civic Square
Sudbury, ONT, P3A 5W5 Canada

Contact: **William E. Lautenbach**, Director, Long-Range Planning
Phone: 1-705-673-2171 • *Fax:* 1-705-673-2960

Activities: Development; Political/Legislative; Vegetation Enhancement • *Issues:* Deforestation; Sustainable Development; Waste Management/Recycling

▲ Projects and People

The **Regional Municipality of Sudbury's Land Reclamation Program** is also known as the **Greening of Sudbury**. Since the late 1800s, mining activities in Sudbury, Ontario, have left widespread environmental damage and a negative image to the area itself. Through new public and private sector building projects, community cultural improvements, area promotions, environmental pollution reduction, and landscape improvements, the community of

Sudbury (152,000 inhabitants) began countering its problems in the 1970s.

A multidisciplinary, technical advisory committee was established to examine the local environmental issues. Committee members included: Laurentian University, International Nickel Company Limited (INCO), Falconbridge Nickel Company Limited, and other local and regional governments and conservation authorities. In 1975, the Municipality established the Vegetation Enhancement Technical Advisory Committee (VETAC) charged with the issue of vegetating barren lands. Restraining control orders by the Ontario Ministry of the Environment and voluntary action from mining companies yielded major air quality improvements. However, few changes to the area's visual appearance were seen in the mid-1970s.

When INCO reduced its mining workforce in 1978, the regional planning department offered the idea of recapturing lost student jobs via a major land reclamation program. Funding was secured from the then Ontario Ministry of Northern Affairs, the federal government's Young Canada Works Program, and the Regional Municipality of Sudbury. The summer of 1978 saw 175 students clean up 500 acres of vegetation debris and lime, and fertilize and seed 285 acres of barren land. This program continued for three years.

In addition to students, unemployed individuals and welfare recipients were included in the program during 1982 and 1983, thereby creating 1,740 short-term jobs and reclaiming 3,600 acres of barren land. The regional municipality's first major tree-planting program started in 1983 (230,000 plants) and continued through 1985.

The land reclamation program contains 11 separate aspects—5 pertaining to grassing and greening, site improvement, tree planting, planning and mapping, and monitoring and assessment. The remaining cover native seed collection, reclamation research, trials and forest assessment, experimental transplanting, and wildflower planting. The five-year land reclamation plan begun in 1979 resulted in the rehabilitation of 40,000 hectares of environmentally degraded landscape by 3,200 working individuals.

A second five-year plan (1990–1994) continues the goal of making significant improvements to reverse the negative image of the Sudbury region. VETAC efforts will concentrate on 685 hectares for restoring grassing, fertilizing, and seeding barren land not completed earlier; the liming, seeding, and planting of trees; and the reclamation of 5 lake watersheds as a means to improve lake water quality (930 hectares) and reclaim barren lands. In 1991, for example, 150,000 additional trees were planted and 100 hectares of barren land were grassed. More than 1 million trees and shrubs will be planted including many in urban areas; in all some 1,615 hectares will be restored. Reclamation areas also targeted are road corridors, neighborhoods, and railroad corridors. Support comes from the federal and provincial governments, the private sector, and volunteers.

This example of community partnership has produced about 10 square miles of barren land reclaimed. Some 6,425 acres of barren landscape has been seeded and restored to a more natural setting, 625,000 tree seedlings were planted, and an additional 2,420 acres have been improved with the removal of unsightly vegetational debris. And, 1,670 acres of tailings and 3,090 acres of barren or semibarren lands have been reclaimed by INCO, Limited and Falconbridge Limited. *(See also International Nickel Company, Limited [INCO] in the Canada Private section.)*

■ Resources

Among publications available are *Environmental Degradation and Rehabilitation in the Sudbury Area*, *The Greening of Sudbury*, *Land Reclamation Program 1978–1984*, and the *Five-Year Land Reclamation Plan 1990–1994*. The Municipality conducts study tours for school groups and visiting professionals.

 *NGO*

The Banff Centre for Management ⊕
Box 1020
Banff, AB, T0L 0C0 Canada

Contact: **Felicity N. Edwards**, Program Manager
Phone: 1-403-762-6137 • *Fax:* 1-403-762-6422

Activities: Education • *Issues:* Energy; Global Warming; Health

● Organization

In the heart of the Canadian Rockies, inside the country's first national park, near Calgary, **The Banff Centre for Management** is one of three divisions at **The Banff Centre**. (The other divisions are the Centre for the Arts and the Centre for Conferences.) Its goal is "to help managers function in an efficient, effective, and harmonious manner in organizations" forced to adapt to a rapidly changing environment. Through a variety of intense courses lasting three days to three weeks, managers examine other global programs and new human resource development methods, and focus on creative management leadership.

▲ Projects and People

According to the Centre's annual report, **Dr. Donald G. Simpson's** appointment in early 1990 as vice president and director marks an increased programming effort to develop leadership qualities enabling a more global understanding within organizations striving to respond to the changing economic, political, and social environments. The Centre has a core staff of 20, with an equal number of project associates and 80 part-time faculty drawn from the private and public sectors. Joint ventures with other similar organizations around the world include groups and institutions in Belgium, Germany, Japan, and the USA.

The Centre is a founding member of the **PAC Global Group on Organizational Innovation** and co-chair of the European Foundation for Management Development's **Task Force on Integrating Environment into Business**.

The Centre offered 70 courses in 1990–1991, attracting almost 2,000 participants. Managing in cross-cultural environments, resource and environment management, tools for sustainable development, business and the environment, and municipal waste management are among the Centre's curriculum. Additional areas of consideration are: "business and the environment: the greening of Canadian business, an institute for environmental innovation, and programs in geographic information systems (GIS)." Efforts to strengthen the overall **resource and environment management program** are underway; **Patrick Jackman** is coordinator.

■ Resources

Resource management publications include *Oil Sands/Heavy Oil and the Environment*, *Environmentally Sound Tourism Development in the Caribbean*, *A Policy Dialogue on Toxic Chemical Management in the Prairie Region*, *Sustainable Development through Northern Conservation Strategies*, and *Audit and Evaluation in Environmental Assessment and Management: Canadian and International Experience*. Program schedules are also available.

Calgary Zoological Society (CZS)
1300 Zoo Road
Calgary, AB, T2M 4R8 Canada

Contact: **Peter Karslen**, Director
Phone: 1-403-232-9314 • *Fax:* 1-403-237-7582

Activities: Education; Research • *Issues:* Biodiversity/Species Preservation • *Meetings:* CITES

● Organization

Calgary Zoological Society (CZS), with the help of 78 staff members, currently maintains 1,334 specimens of animals representing 301 species. The zoo's programs include the "Cared for Naturally" initiative, in which **grasses and wildflowers simulate natural habitats** in an unmanicured exhibit; an **animal rescue program for the treatment of injured and orphaned wildlife** at its Animal Health Center; and survival programs for 14 species, most recently including the **palm cockatoo**. CZS also maintains a 65-hectare semiwild preserve for **Przewalski's horses**, part of a program studying the adaptability of captive bred horses native to Russia and Mongolia. The zoo's botanical collection boasts over 100,000 individual plants which serve horticulture lectures, workshops, and tours.

Educational programs include a Summer Interpretive Program with talks, tours, theatre for 150,000 visitors; and a docent corps of 130 volunteers and educational staff who provide on-site outreach programs for children and the community. This includes a Junior Zoologist Program started in 1990.

▲ Projects and People

Ongoing projects include **Site Preparation for the Canadian Wilds**, **Spider Web Activity Pod**, a newly constructed **Education Center**, and **Aspen Woodlands of the Canadian Wilds**.

Canadian Institute of Forestry/Institut Forestier du Canada (CIF/IFC)
151 Slater Street, Suite 1005
Ottawa, ONT, K1P 5H3 Canada

Contact: **C.A. Lee**, Executive Director
Phone: 1-613-234-2242 • *Fax:* 1-613-234-6181

Activities: Development • *Issues:* Forestry; Sustainable Development

● Organization

With projects from Nova Scotia to British Columbia, the **Canadian Institute of Forestry (CIF)** has a "broad professional forestry base including industry, government, consulting, and education." Its three objectives are to "encourage a wider understanding of forestry problems including the need to maintain and/or improve the environment; improve forestry practices throughout Canada; and advance membership in the knowledge and practice of the science of forestry."

▲ Projects and People

To enhance generally the state of the forests, CIF directs several cooperative programs, which serve as model forests throughout Canada. Two principal projects are the St. Mary's River Forestry-Wildlife Project (SMRP) and the Willow River Demonstration Forest.

Located in Antigonish, Nova Scotia, SMRP is a working example of a precisely run forest/wildlife reserve. The project links the cooperative efforts of the federal government, provincial government, industry, and non-government organizations. **Joint study programs** "develop practical and effective methods for managing wildlife and forestry together," reports the Institute. Through workshops and field discussion, "everyone takes an active role to develop realistic and economical methods. . . . This approach encourages a more rapid acceptance of new methods into current forestry practices." Reports, fact sheets, talks, and training videos are produced.

SMRP concentrates on **special management zones (SMZs)** along waterways, wildlife and forest habitat relationships, snags and cavity trees, and field demonstrations.

SMZs are sources of water, food, cover, and travel lanes for wildlife. At sites being harvested, problems occur as small feeder streams become conduits for silt that enters larger waterways during periods of heavy rains and can destroy **trout and salmon spawning beds**; yet these streams do not flow during dry spells. SMRP is looking for practical ways to cross such streams with temporary bridges to help eliminate the buildup of harmful silt resulting from permanent bridges.

Because trout and salmon prefer "places to hide and proper water temperatures," **instream cover** to meet these needs is being provided through "narrowing and deepening the flow during low water conditions, digging pools, and undercutting banks." Staff is also determining how wide vegetation along lakes and streams should be to provide shelter for wildlife.

Regarding **wildlife and forest habitat relationships**, SMRP is examining, for example, "how songbirds, mice, and amphibians respond to vegetation changes on sites where different forestry operations such as cutting, spacing, and planting have occurred." Long-term studies are determining "how the arrangement or spacing of different habitat or forest types affects . . . deer, moose, and other large mammals."

Because **snags and cavity trees** provide nesting, denning, roosting, feeding, and perching for some 250 wildlife species in Nova Scotia, SMRP is preparing recommendations for foresters on how many and where such trees should be left standing—and for how long.

Demonstration sites are being developed as "valuable education tools" to help "forestry contractors, planners, cutters, biologists, foresters," and technicians foster better management of both trees and wildlife.

Canada's **Green Plan** already acknowledges some benefits. "For example, the size of clear-cuts is being reduced to encourage wildlife to remain in the area. . . . Waterways are being monitored to determine the impact of tree harvesting on animal populations along rivers." And ways of stopping soil erosion are being studied.

Across the country in 1985, CIF's **Cariboo Section** created the **Willow River Demonstration Forest** in Prince George, British Columbia, as "an outdoor [public] **classroom of ecological principles and forest management practices**" as well as a recreation site. Once considered limitless in Prince George, timber showed its finite qualities in the 1950s and 1960s when foresters began realizing that "cutover areas failed to regenerate adequately" and a combination of reforestation methods were essential.

The Cariboo Section is made up of 80 foresters, technicians, forestry students, and educators who have a "professional interest in forestry." Tours are sponsored for schools and other interested groups.

A 1.9-kilometer surfaced trail, funded by government and industry, meanders through the Fraser Basin of the Interior Plateau with a "variety of forest ecosystems including mature coniferous forests of white spruce, lodgepole pine, and Douglas-fir." Also found are paper birch, black cottonwood, herbs, and berry-bearing shrubs. Mammals include black bear, red squirrel, porcupine, snowshoe hare, moose, and mule deer. Visitors glimpse hairy woodpecker, yellow-bellied sapsucker, and ruffed grouse, among other birds.

■ Resources

The Nova Scotia Department of Lands and Forests' publication, *Forest/Wildlife Guidelines and Standards for Nova Scotia*, provides more details. *Cutting Up the North: The History of Forestry in the Northern Interior* by Ken Bernsohn is also recommended reading. Other literature is available through CIF.

Canadian Nature Federation (CNF)
453 Sussex Drive
Ottawa, ONT, K1N 6Z4 Canada

Contact: **Paul Griss**, Executive Director
Phone: 1-613-238-6154 • *Fax:* 1-613-230-2054

Activities: Development; Education; Political/Legislative; Research • *Issues:* Biodiversity/Species Preservation; Deforestation; Global Warming; Parks and Protected Areas; Sustainable Development; Waste Management/Recycling

● Organization

Canada's largest naturalist organization is the **Canadian Nature Federation (CNF)**, with over 40,000 supporters and 144 affiliates. During its more than four decades in existence, two main beliefs guide CNF's work—all species have a right to exist and human existence should be led by sound ecological principles.

A core staff of 13, with the support of volunteers, assists in delivering CNF's conservation advocacy and educational programs, which are divided into 10 distinct areas including forests and parks, northern land use, and international and marine conservation.

▲ Projects and People

In 1990 the Canadian Wildlife Minister's Council adopted **A Wildlife Policy for Canada**, for which CNF actively lobbied along with other federal and provincial wildlife agencies. Other CNF roles include the creation of **a marine park** at the Saguenay and St. Lawrence rivers junction **to protect the endangered St. Lawrence beluga whale**, and the inclusion of 20 new species and/or populations to the Committee on the Status of Endangered Wildlife in Canada (COSEWIC).

CNF has taken high-profile and controversial positions, such as **opposing the slaughter of bison** in Wood Buffalo National Park, **challenging the hydroelectric development** of the James Bay II megaproject, and **protecting the boundaries** of the Thelon Game Sanctuary. It also joined the campaign to protect Alaska'a Arctic National Wildlife Refuge. CNF toured cross-country with *The Last Wilderness* photo exhibit by **Freeman Patterson**, depicting Canada's grandeur.

Having helped develop the National Forest Sector Strategy, CNF continues to give priory to such issues. It helped the National Round Table on the Environment and the Economy (NRTEE) produce a **kit on Sustainable Development and the Municipality**. Regarding northern lands, CNF is "the only national conservation organization to contribute to the planning processes of the Mackenzie Delta/Beaufort Sea, North Yukon, and Denendeh land use plans." With support from the Donner Canadian Foundation, it is working to establish a national collaborative **group on marine conservation**, since none presently exists.

CNF is cooperating with the International Council for Bird Preservation (ICBP) on related issues, and with the World Wildlife Fund Canada on campaigning for completing a "network of parks and protected areas representing each of Canada's 350 natural regions," which would preserve about 12 percent of the country.

Continuing CNF target programs focus on the completion of the national parks system by the year 2000 and establishing a 900-square kilometer **Grassland National Park**, home to endangered wildlife, and at least five new parks by 1996. CNF is recognized as "the lead national organization on parks issues for 30 years."

■ Resources

CNF's available publications include *Nature Canada* magazine, *Almanac* bimonthly newsletter, and *The Last Wilderness* photography book. *Birdquest* is an educational program with slide video.

Commission on Developing Countries and Global Change ⊕

c/o The Royal Society of Canada
207 Queen Street
Ottawa, ONT, K1G 5J4 Canada

Contact: **Alvaro Soto, Ph.D.**, Executive Secretary
 Phone: 1-613-992-8480 • *Fax:* 1-613-992-8476

Activities: Development; Education; Political/Legislative • *Issues:* Air Quality/Emission Control; Biodiversity/Species Preservation; Deforestation; Energy; Global Warming; Health; Population Planning; Sustainable Development • *Meetings:* UNCED

● Organization

The **Commission on Developing Countries and Global Change** was created in response to the implications of global warming, other environmental concerns, and the necessity to examine associated social aspects of these problems. Begun in March 1991, initially to run for one year, the charge of the Commission is to determine "what can be done to ensure that research on global change is made as relevant as possible to the needs and interests of both developed and developing countries."

▲ Projects and People

Representatives from East Africa, West Africa, South Asia, Southeast Asia, and Latin America comprise the core group of five social scientists and one natural scientist on the Commission. Due to the limited time frame of the Commission, its substantive work included discussions within meetings, highly selective short-term research, literature and data searches, and correspondence between meetings. **Dr. Alvaro Soto** remains as the contact for the Commission for additional informational requests.

Mostly, the Commission wants to ensure the understanding and participation of Third World representatives in global environmental issues that could have "potentially catastrophic implications for many developing countries." The Commission reports that the "greatest need for research support lies in the field of applied social sciences." It is charged with beginning to evaluate the ability of individuals and institutions within such countries to undertake this research and proposing an appropriate research agenda. The report was to be prepared for the 1992 United Nations Conference on Environment and Development (UNCED).

The International Development Research Centre (IDRC) of Ottawa and the Swedish Agency for Research Cooperation with Development Countries (SAREC) of Stockholm provided support for the Commission.

A Colombian anthropologist, **Dr. Alvaro Soto** is president of the International Center for the Environment in the Tropics (INCENT-NEOTROPICO), head of the Secretariat for the Latin-American Environmental Network, and associate member of the Institute for Research on Environment and Economy, University of Ottawa. He was also director of Colombia's national parks system and has written and implemented "indigenous people's policy" and books.

■ Resources

The *Commission Report on Perspectives from the South in Global Environmental Change* was planned for release in 1992.

Friends of the Earth/Les Amis de la Terre— Canada

701-251 Laurier Avenue West
Ottawa, ONT, K1P 5J6 Canada

Contact: **Julia Langer**, Executive Director
 Phone: 1-613-230-3352 • *Fax:* 1-613-232-4354
 E-Mail: webid:foe

Activities: Education; Political/Legislative; Research • *Issues:* Air Quality/Emission Control; Energy; Global Warming; Ozone; Sustainable Agriculture; Sustainable Development; Transportation

● Organization

Founded in 1978, **Friends of the Earth/Les Amis de la Terre—Canada** has become one of the country's environmental leaders with more than 27,000 members nationwide. As honorary chairman **Bruce Cockburn** claims, "We can stumble into the future as graceless parasites, sucking the life out of our host . . . or we can step forward in reverence, in respect, embracing and being embraced by creation. We have to save our Earth in spite of ourselves."

Through its involvement with businesses, governments, individuals, and organizations, Friends, with a staff of 14, strives to achieve its goal of a sustainable global future. It is also a member of the Friends of the Earth—International, whose one million affiliates address worldwide issues such as rainforest destruction, ozone depletion, air pollution, and marine contamination.

▲ Projects and People

Friends provides Canadians with the educational background and tools to lead an active role in environmental protection. Its input on critical contemporary issues is constantly examined by decision-makers. Four major programs are sponsored:

The **Global ReLeaf Campaign**—an international education, action, and policy effort launched on Earth Day 1990—plants trees in Canadian communities. The campaign seeks to assure the health of the nation's forests and improve the environment by planting more trees. With the support of this program, more than 100,000 trees were planted across the country in one year. Financial support is also offered to community forestry projects twice yearly. *(See also Friends of the Earth, Global ReLeaf Canada in this section.)*

Launched in late 1989, the **Global Warming Campaign** fights against the global warming trend by committing to a 20 percent cut in the carbon dioxide (CO_2) emission levels of 1988 by the year 2005. Fossil-fuel burning accounts for over 80 percent of Canada's carbon dioxide emissions, and over 50 percent of the greenhouse gases released annually are from carbon dioxide, notes Friends' *Annual Report.* In 1990, Canada's first commitment toward this goal was announced by the environmental ministry—to stabilize the CO_2 emissions at 1990 levels by 2000. Friends also was instrumental in the creation of the **Canadian Climate Action Network** (**CANet–Canada**), a group of more than 50 non-governmental environmental organizations pressuring all levels of government to take strong and immediate action on this issue.

Spurred by the 1991 report that ozone depletion over southern Canada is twice as extensive as was earlier thought, the **Ozone Protection Campaign** sprung into a more active mode. Friends says it is the only Canadian national ozone campaign environmental group. Their **Atmosphere Protection Team** (**APT**) is a nationwide volunteer program that brings fresh approaches to protecting the ozone. Canada was among the world leaders to completely ban the ozone-destroying methyl chloroform. The campaign seeks to eliminate the production and import of chlorofluorocarbons (CFCs), methyl chloroform, and halons no later than 1995, with most hydro-chlorofluorocarbons (HCFCs) five years later, and all by 2010.

Canada is "losing ground to soil erosion and facing the degradation of water quality," according to Friends. Hence, the **Sustainable Agriculture Campaign** was begun in 1990 to increase Canadians' understanding, concern, and sense of responsibility for a sound ecological system.

■ Resources

Friends' public outreach program strives to supply the vital information people need to become active participants in environmental protection. Numerous publications, including *Earth Words* newsletter, *No Man Apart* newsmagazine, and *Ozone Protection Campaign* brochure are available. Among the fact sheets, guides, reports, and books are the popular *How to Get Your Lawn & Garden Off Drugs* and *Clean House, Clean Earth.*

Global ReLeaf
Friends of the Earth/Les Amis de la Terre—Canada
251 Laurier Avenue West
Ottawa, ONT, K1P 5J6 Canada

Contact: **Donna Passmore**, Program Director
Phone: 1-613-230-3352 • *Fax:* 1-613-232-4354

Activities: Education; Political/Legislative; Research • *Issues:* Biodiversity/Species Preservation; Deforestation; Global Warming; Sustainable Development • *Meetings:* Climate Institute Conferences; Canadian Forestry Conferences

● Organization

Global ReLeaf is one of the featured projects in the **Friends of the Earth—Canada** operation. Originally created in 1988 by the American Forestry Association, it was brought to this country two years later when Canadians were urged into positive action individually, communally, and nationally. According to a Global ReLeaf report, Canadians are the worst carbon dioxide (CO_2) polluters in the Western World, at five tons per person per year. And, with only 5 percent of the world's population, North Americans produce 25 percent of the annual global CO_2 from burning fossil fuels.

With its slogan "Plant a tree, cool the globe," Global ReLeaf is dedicated to planting more trees in cities, towns, and frequented countrysides where great forests once existed. Today, the Global ReLeaf constituency is made up of groups ranging from school children to seniors, recreation clubs to municipal parks departments, small businesses to environmentalists.

▲ Projects and People

With the planting and maintaining trees and forests, Global ReLeaf strives to reduce the global warming trend. A fact sheet notes that for every ton of new wood grown, about 1.47 tons of carbon dioxide are removed from the air and 1.07 tons of life-giving oxygen is produced. Other Global ReLeaf statistics point out that 25 million hectares, or 6 percent of Canada's forests, burned during the 1980s; and worldwide, trees are destroyed at the rate of 500,000 per hour.

Trees and You education kits aimed at grades 4 to 8 were distributed to 10,000 schools in 1991 and later to summer recreation programs to inspire "wilderness gardens" at schoolyards and campsites. The **Global ReLeaf Fund** supports the planting of larger trees to provide "better energy saving potential and wildlife habitat" in place of seedling giveaways.

Global ReLeaf gains corporate backing from Lever Brothers, Molson Breweries, W.H. Smith booksellers, Aveda Limited, United Distillers Canada, and PineSol; participation of popular magazines as *Reader's Digest, MacLeans, Canadian Living,* and *Outdoor Canada* for free public service messages; and the cooperation of government agencies such as the Ontario Ministry of Natural Resources. Students also helped raise money by selling reusable canvas bags from Canadian Accents.

In addition, reducing dependence on fossil fuels and a shift toward renewable, **sustainable energy sources are targeted** by Global ReLeaf as other methods to stabilize and protect the climate. Pushing for mass transit, upgraded building codes for new houses, and renewable energy sources such as solar electricity and heating, wind power, and biomass energy, the organization is also asking automakers to "double fuel economy in cars to 4.7 litres per 100 kilometers by the year 2000."

■ Resources

Among various brochures, Global ReLeaf offers *Ten Things You Can Do to Fight Global Warming.* The group seeks "lobbyists for legislative change," marketing specialists, and urban foresters to help reach and educate Canadians.

Grand Manan Whale & Seabird Research Station, Inc. (GMWSRS)
P.O. Box 9, North Head
Grand Manan, NB, E0G 2M0 Canada

Contact: **David Edward Gaskin, Ph.D.,** Executive
          Director
Phone: 1-519-824-4120, ext. 2707
  Fax: 1-519-767-1656

Activities: Education; Research • *Issues:* Biodiversity/Species Preservation; Marine Mammals • *Meetings:* IWC Scientific Committee

● Organization
The **Grand Manan Whale & Seabird Research Station (GMWSRS)** operates through direct research grants to the station such as from the Natural Sciences and Engineering Research Council, Department of Fisheries and Oceans, World Wildlife Fund—Canada, and founder/donor **Mary Lou Campbell.** Researchers also receive direct donations from individuals.

▲ Projects and People
Biologists at GMWSRS **monitor annual changes in the Bay of Fundy** and provide basic information on the **biological interactions within the marine community.** Noted for its complex biological and oceanographic events, the Grand Manan Island serves as an ideal base station for researchers working in the mouth of the Bay of Fundy.

Through scientific journals, books, public lectures, museum displays, newsletters, and practical training for visiting scientists, the research station, with its staff of three, expands the base of existing knowledge on the preservation of potentially threatened and endangered species. GMWSRS particularly reports on its investigations of pollutants, such as mercury, lead, and cadmium.

The **harbour porpoise** population, tracked with radiotelemetry, and the **rare North Atlantic right whale** are major station research

projects. **Plankton, Bonaparte gulls, herring,** a variety of shorebirds and seabirds, **red-necked phalaropes,** semi-palmated **sandpipers** and numerous physical and oceanographic features in the Bay of Fundy are also studied.

With GMWSRS since 1969, **Dr. David E. Gaskin** is professor of zoology and marine resources, University of Guelph, and the author of numerous books and papers on marine mammals. He belongs to the Species Survival Commission (IUCN/SSC) Cetacean Group.

■ Resources
Article reprints are available. Funding is needed to further population and habitat ecology studies and for education.

Inuit Circumpolar Conference (ICC) ⊕
650 32nd Avenue, Suite 404
Lachine, PQ, H8T 3K4 Canada

Contact: **Mary Simon,** President
Phone: 1-514-637-3771 • *Fax:* 1-514-637-3146

Activities: Education; Research • *Issues:* Air Quality/Emission Control; Biodiversity/Species Preservation; Global Warming; Health; Political/Social Environment; Sustainable Development; Water Quality • *Meetings:* CITES; ILO; IUCN; UNCED

● Organization
The **Inuit Circumpolar Conference (ICC)** represents approximately 115,000 Inuit of Alaska, Canada, and Greenland. Founded in 1977, the international organization seeks to ensure the continuance of the Inuit culture and the Arctic environment. For thousands of years, the Inuit have lived along the Arctic coastline from "the Chukotka Region of the [former] Soviet Union, across North America, to the east coast of Greenland," developing a unique similarity of language, lifestyle, and culture as a result of the people's deep respect for the land and its natural resources.

▲ Projects and People
The ICC focus includes the **development of an Arctic policy** defining basic principles to govern the range of issues affecting Inuit; implementation of the **Inuit Regional Conservation Strategy** to integrate modern technology with traditional Inuit education; and **protection of the Arctic environment.**

The ICC continues its efforts to establish an **Arctic zone of peace,** secure additional circumpolar cooperation, participate in the drafting of a **Universal Declaration on Indigenous Rights** at the United Nations, and guarantee full membership for Inuit of the former USSR. ICC is a member of IUCN, the World Conservation Union.

■ Resources
The **ICC Foundation** is a charitable organization that accepts donations to further the social, cultural, and environmental activities of Inuit. A descriptive brochure is available.

Union Québécoise for the Conservation of Nature (UQCN)/Union Québécoise pour la conservation de la nature
160 76 rue Est
Charlesbourg, PQ, G1H 7H6 Canada

Contact: **Dr. Pierre Gosselin**, President
Phone: 1-418-628-9600 • *Fax:* 1-418-626-3050

Activities: Development; Education; Political/Legislative; Research • *Issues:* Air Quality/Emission Control; Biodiversity/Species Preservation; Deforestation; Energy; Global Warming; Health; Sustainable Development; Transportation; Waste Management/Recycling; Water Quality • *Meetings:* Canadian Nature Federation; Great Lakes United; IUCN

● Organization
Founded in 1981, the Union Québécoise pour la conservation de la nature (UQCN) is an umbrella group of affiliate organizations representing about 6,000 individuals and more than 100 corporate members. UQCN conservation strategy seeks to maintain essential ecological processes and life-support systems, preserve genetic diversity and the sustainable utilization of species and ecosystems.

▲ Projects and People
With a staff of nine, the primary mission of UQCN under the direction of **Dr. Pierre Gosselin** is educational, although consulting and research are also other avenues followed. At least six times a year, UQCN publishes *Franc-Vert*, a magazine providing readers information on various environmental issues and alerting them to the different ecosystems and species. A field guide for Quebec's wetlands was prepared, as were syntheses of articles on woodlands, farmlands, and aquatic environments. UQCN is working with the transit authority to develop green marketing for mass transit.

Telephonic services for environmental advice, organizing symposia, assisting other science organizations' public relations and communications efforts, and preparing educational materials for school-age children are other areas of UQCN activities. UQCN also provides translations adaptable to Quebec of the World Wildlife Fund's (WWF) *Operation Lifeline* and *The Recycler's Handbook* as well as translating and publishing the *Brundtland Report: Our Common Future*. A documentation room is being established, and liaison is provided among environmental and natural sciences organizations.

In May 1989, UQCN launched Stratégies Saint-Laurent in collaboration with nine regional environmental groups. Its three-year goal is to rehabilitate the public use of the St. Lawrence River area, through increased public awareness and community involvement, and ultimately its restoration and management. For over 15 years prior to this project's announcement, environmental groups have been trying to have the river claimed a national priority. Stratégies is organizing a Beluga and Pollution awards program to acknowledge those who have "distinguished themselves positively or negatively" in saving the river.

With citizen coalitions, UQCN is pushing for "precise laws and regulations that have teeth" along with the changes in the practices of the polluting industries.

According to the UQCN, the St. Lawrence is said to be "polluted from the mouth of the Great Lakes to the Gulf. "Some 60 large companies discard more than 100,000 metric tons of toxic waste into the river annually. In Montreal's vicinity, some 3,900 industrial plants are discarding toxic waste. Such actions are endangering about 27 species of fish, reptiles, birds, and mammals, including the beluga whale; in addition, many fish are contaminated and discouraged for consumption. The drinking water for half of Quebec's population is said to contain "micropollutants, several of which are recognized as toxic, as capable of causing congenital malformations, or as carcinogenic." Vital marshes have been filled in along 70 percent of the river's shoreline, affecting both water quality and wildlife.

■ Resources
As mentioned above, numerous publications, magazines, maps, posters, and other materials—many printed in French—are available from UQCN, which continues to seek philanthropic financial support.

⚑ *Private*

Environmental Economics International (EEI) ⊕
9 Sultan Street, Suite 300
Toronto, ONT, M5S 1L6 Canada

Contact: **Gary T. Gallon**, President
Phone: 1-416-972-7400 • *Fax:* 1-416-972-4660

Activities: Education; Political/Legislative; Research • *Issues:* Deforestation; Energy; Global Warming; Sustainable Development; Waste Management/Recycling; Water Quality

● Organization
A Canadian consulting firm, **Environmental Economics International** (EEI) assists corporations, trade associations, and governments in meeting environmental challenges of the 1990s. Established in 1991, EEI's staff of 5 specialists have a combined total of more than 70 years of experience in the fields of energy, environment, sustainable development, and waste management.

Among the areas in which EEI provides assistance to its clients are planning, policy, and problem-solving capabilities in corporate greening and green community development; energy efficiency and renewable energy; environmental assessment and land-use planning; environmental audits; interpreting and meeting environmental legislation; and reduction, reuse, and recycling of solid and liquid wastes. EEI also specializes in environmental media communications.

▲ Projects and People
As a founding member of **Greenpeace** in 1971, **Gary Gallon** began his emersion into the environmental maze. He realized that "real change takes a long time and to achieve, we had to help come up with some solutions." Gallon's mission was "to provide a voice for wilderness and wildlife, to protect their right to life on earth . . . to speak on behalf of the eagle, the wolf, the dolphin and the tree, in human fora where they could not defend themselves."

In the mid-1970s, Gallon founded CUSEC, the Canadian-US Environmental Council. And, along with **Dr. Ted Mosquin**, **Canadian Nature Federation**, he brought together the major groups in the two countries to discuss emerging boundary environmental issues and coordinated responses.

In 1976, Gallon was named Canada's Environmentalist of the Year for his work on protecting the Fraser River estuary and his efforts to promote energy conservation and solid waste recycling. The following year, he became executive director of the **UN Environment Liaison Centre International** in Nairobi, Kenya, where he remained for four years. Upon his return to Canada in 1982, he helped establish **Probe International** in its efforts to forge connections with the Third World.

The mid-1980s saw Gallon become senior policy advisor to **Jim Bradley**, Ontario's minister of the environment. Here he assisted in **mandating acid rain reductions**, virtually halted garbage incinerator construction, passed chlorofluorocarbon (CFC) phase-out legislation, introduced jail sentences for pollution offenses, and worked within the federal government introducing auto emission standards, according to EEI literature. Gallon is on the Board of Advisors of this *Guide*.

■ Resources

EEI has various papers, pamphlets, articles and special reports available upon request, including *Sustainable Development, the Brundtland Commission and Roundtables in Canada, Green Products Endorsements: The Case of Pollution Probe and Lowlaws;* and *Report on the Nature and Makeup of the Key Environment and Conservation Organizations in Canada.*

Geomatics International ⊕
3370 South Service Road
Burlington, ONT, L7N 3M6 Canada

Contact: **Mark E. Taylor, Ph.D.**, Wildlife Ecologist
Phone: 1-416-632-4259 • *Fax:* 1-416-333-0798

Activities: Education; Research • *Issues:* Biodiversity/Species Preservation; Consultation

● Organization

Geomatics International, an environmental consulting company, specializes in the application and use of geographic information systems (GIS) and remote sensing technologies to provide a full spectrum of environmental services to its clients.

▲ Projects and People

With its staff of 30, its operations include **biophysical and ecological surveys**, facility site selections, **environmental assessments**, resource inventories, **endangered species status and management**, **forest and watershed management**, environmental constraint analysis, environmental resource planning, and **wetland evaluations**.

Geomatics uses such advanced hardware as digitizing tables, high-resolution displays, high-capacity plotters and printers, microcomputer workstations, and several UNIX engineering workstations. ESRI's ARC/INFO GIS, Pamap's PAMAP GIS, PCI's EASI/PACE remote sensing, and Tydac's SPANS GIS are Geomatics' software package options.

Dr. Mark E. Taylor is also adjunct professor, University of Toronto, and a member of the Species Survival Commission (IUCN/SSC) Mustelid and Viverrid Group.

International Nickel Company, Limited (INCO) ⊕
P.O. Box 44, Royal Trust Tower
Toronto-Dominion Centre
Toronto, ONT, M5K 1N4 Canada

Contact: **W.R.O. "Roy" Aitken**, Executive Vice President
Phone: 1-416-361-7809, 1-416-361-7758
Fax: 1-416-361-7736

Activities: Development; Research • *Issues:* Air Quality/Emission Control; Global Warming

● Organization

The Western World's largest producer of nickel, the 100-year-old **International Nickel Company Limited (INCO)** set out to reduce its contributions to acid rain as the continent's "single most visible source" resulting from the high levels of mined sulphur ore at its Sudbury, Ontario, facility. In 1989, INCO announced a five-year, $500 million (Canadian dollar) plan to further reduce sulphur dioxide (SO_2) emissions from the complex. By 1994, INCO plans to have reduced emissions by an additional 60 percent to a level of 265 thousand tons per year—in keeping with the Ontario Ministry of the Environment's "Countdown Acid Rain" regulation.

Used in coinage, stainless steel, and super alloys, nickel is an important ingredient in deep-sea diving equipment, jet engine turbines, and the USA's space shuttle program.

▲ Projects and People

INCO reports that it developed oxygen "flash furnace" smelting technology in the 1950s to improve the capacity to capture sulphur dioxide, and since the 1960s, INCO reduced its emissions by 70 percent, "the largest single tonnage reduction of any organization in North America." Continuing to move away from environmental degradation toward principles of sustainable development, **Roy Aitken** described "what a conservation strategy means to INCO" when serving as vice chairman of Canada's National Task Force on the Environment and the Economy: "Don't exploit the resources at a rate which exceeds your ability to develop another or . . . a substitute product. The fundamental message is: Don't compromise the sustainability of the host environment, the air, the water, and terrestrial resources."

The ongoing, five-year **SO$_2$ Abatement Program** aims to "efficiently extract nickel and copper from the ore [that is, 8 parts sulphur for every one part nickel], yet capture 90 percent of the sulphur before it reaches the atmosphere as sulphur dioxide." With the implementation of key process changes, INCO reports, new milling technology will allow the rejection of additional amounts of the high sulphur-bearing (iron sulphide) portion of ore, called pyrrhotite, prior to smelting. And, rather than separate concentrates, a bulk copper-nickel concentrate will be produced and smelted in more efficient oxygen flash furnaces which will replace current reverberator furnaces.

INCO expects to be able to "contain and safely discard two thirds of the sulphur prior to its ever reaching the smelting stage." Also, the SO_2 offgas can be "captured and later converted into marketable sulphuric acid" for products such as fertilizer. INCO reports its smelting technology is also eliminating the burning of natural gas, as combustion is "being accomplished with pure oxygen from another new plant."

When the project is completed, 90 percent of the sulphur content at the Sudbury facility—or 420,000 tons—will be contained, states Aitken. He further notes that INCO is "dedicated to achieving still further emission reductions" beyond the 1994 target and continues to examine methods to reduce emissions at Sudbury to 175,000 tons per year. "We are convinced that it is cheaper, easier, and better in the long run to build clean plants than to have to clean them up later under governmental edict," says Aitken.

INCO's corporate **environmental impact policy** analyzes every new program while still on the drawing board, for such effects. In so doing, INCO fosters research, applies cost-effective management to advance environmental protection, maintains active monitoring programs, and works with the government and public to enhance occupational health and safety.

At Sudbury, land reclamation programs have been instituted through the **planting of thousands of pine seedlings in abandoned mines**, revegetating some 1,600 acres. Area surveys have identified "86 species of plant life and 92 different types of birds" including Canadian geese that are returning to once barren wastelands. At INCO's Indonesian facility, nursery stock is started for transplanting to mined areas.

■ Resources
Information about INCO's SO_2 Abatement Program is available from the public affairs department.

MacNeill & Associates ⊕
110 Rideau Terrace
Ottawa, ONT, K1M 0Z2 Canada

Contact: **James W. MacNeill**, International
            Consultant
    Phone: 1-613-749-8681 • *Fax:* 1-613-235-8237

Issues: Sustainable Development

▲ Projects and People
James MacNeill's international, national, and urban environmental involvements span more than 20 years, beginning with the initial publication in Canada in 1971 of his *Environmental Management*, which is a text for universities and policy-planning bodies.

Starting life as an "ecological refugee," when his family was forced to migrate from the southern dust belt to the central park belt of Saskatchewan with their one-year-old, MacNeill grew up to become a forceful global spokesperson on behalf of sustainable development. He has previously served as director general of intergovernmental affairs at the Canadian Department of Environment and then with the Canadian Ministry of State for Urban Affairs, initially as assistant secretary and later deputy minister. Early in his professional life, MacNeill chaired a United Nations Committee that addressed environmental pollutants as chemicals.

A highlight of MacNeill's career was his role as secretary general of the independent 22-member World Commission on Environment and Development in Geneva, Switzerland. From 1984 to 1987, he managed and organized the Commission's work and was principal author of the publication popularly known as the *Brundtland Commission Report on Our Common Future* that revolutionized thinking regarding "new strategies for tackling critical global environment and development problems."

Soon to be published in 19 languages, *Our Common Future* continues to influence world leadership, including the Commonwealth Prime Ministers; UN; such regional groups as the Organization for Economic Cooperation and Development (OECD) Association of Southeast Asian Nations (ASEAN); and the World Bank and other multinational institutions. The report also has generated new research programs in publiy policy institutes, graduate programs in universities, and more intense NGO activity, according to MacNeill.

MacNeill was the Canadian commissioner general and ambassador extraordinary and plenipotentiary for the United Nations Conference on Human Settlements in 1976. The next year, he took a sabbatical to examine West European regional, urban, and environmental development and policy. Following this experience, MacNeill became director of environment and urban affairs at OECD in Paris, France. Here he examined issues, trends, and common problems, and proposed policy options dealing with vital environmental issues facing member countries and the world. In the 24 advanced industrialized countries of OECD, MacNeill negotiated and developed international agreement on policies and programs enhancing environmental quality and resources for economic development.

After 12 years abroad, MacNeill joined the Institute for Research on Public Policy as its director of environment and sustainable development. He is on the board of the World Environment Center, New York City, and on the council of the World Resources Institute, Washington, DC.

■ Resources
A list of MacNeill's numerous articles and publications is available upon request.

♠ *Universities*

Environmental Management Development in
  Indonesia (EMDI) ⊕
Dalhousie University
School for Resource and Environmental Studies
  (SRES)
1312 Robie Street
Halifax, NS, B3H 3E2 Canada

Contacts: **Shirley A.M. Conover, Ph.D.**, EMDI
            Project Director
    Phone: 1-902-494-3632

            Barbara Patton, EMDI Project Information
            Officer
    Phone: 1-902-494-1368 • *Fax:* 1-902-494-3728
E-Mail: cosy:emdihfx; envoy:sres

Activities: Development; Education; Research • *Issues:* Air Quality/Emission Control; Biodiversity/Species Preservation; Deforestation; Energy; Global Warming; Sustainable Development; Waste Management/Recycling; Water Quality

▲ Projects and People

Environmental Management Development in Indonesia (EMDI) is a project jointly sponsored by **Dalhousie University**'s School for Resource and Environmental Studies and **Indonesia**'s Ministry of State for Population and Environment (KLH). Through institutional strengthening and human resource development, EMDI strives to upgrade environmental management capabilities. It is funded by the Canadian International Development Agency (CIDA) in addition to contributions received from KLH and the university.

Environmental management—ecological, social, and cultural—is an integral component of all development decisionmaking in Indonesia, according to that government's goal for EMDI. During its third phase (1989–1994), EMDI's agenda includes: spatial planning (allocation of atmospheric space, land, and water), regional environmental management, and the prevention or alleviation of some of the negative impacts of development such as rapid population growth and pollution. Environmental impact assessments (EIAs), the development of a systematic hazardous and toxic waste management system, and the monitoring of environmental standards are used in the program.

Critical to a nation of more than 13,000 islands are **threats to the coastal and marine zone** with its coral reefs, mangroves, and peat and seagrass beds. Attention is being paid to marine pollution in marine ports and terminals with training on spill cleanups. Reports on *The Ecology of the Moluccas and Lesser Sundas* and *The Ecology of the Indonesian Seas* offer baseline information about marine and coastal ecosystems, resources, and human activity and development.

Such support systems as a library, legal courses dealing with environmental law, 21 graduate fellowships, and publication programs are currently in operation. Future initiatives include cooperative work between Canadians and Indonesians in the area of Women in Development and Environment (WIDE) and support for environmental efforts of "self-reliant" organizations at the village level.

EMDI works also with WALHI, the umbrella organization of environmental NGOs in Indonesia, and in the private sector with establishing environmental consulting units through INKINDO, the National Association of Indonesian Consultants, which will provide internships and on-the-job training in environmental management.

■ Resources

Dalhousie University offers programs for both Canadian and Indonesian groups and scholars, including research fellowships and exchanges. Scholarships, seminars, and workshops are also planned. Dalhousie's Southeast Asia Environmental Collection can be accessed through NOVANET at 1-902-494-2500 or by contacting the address and phone/fax numbers listed above. An EMDI brochure, *Environmental Report Series, Ecology of Indonesia Series,* and a publications list are also available.

Information on the University Consortium on the Environment—comprising Indonesia's Universitas Gadjah Mada, Universitas Indonesia, and the Institut Teknologi Bandung as well as Canada's University of Waterloo and York University—can be obtained from UCE, Faculty of Environmental Studies, York Uni-

versity, 4700 Keele Street, North York, Ontario M3J 1PE, phone 1-416-735-5252, ext. 2607. EMDI offers various articles and series and a descriptive brochure.

(See also Environmental Management Development in Indonesia in the Indonesia NGO section.)

International Development Studies (IDS) ⊕
Dalhousie University
Department of Political Science
Halifax, NS, B3H 4H6 Canada

Contact: **Timothy W. Shaw, Ph.D.,** Professor
Phone: 1-902-494-2396, 1-902-494-3769
Fax: 1-902-494-1957, 1-902-494-2319

Activities: Development; Education; Research • *Issues:* Sustainable Development

▲ Projects and People

International Development Studies (IDS) is a joint undergraduate program at both Dalhousie and Saint Mary's universities, whose faculty also serve as consultants to the **Atlantic International Development Associates (AIDA)**. The academic program's purpose is to advance understanding of developing nations which comprise 5 billion people—"some 75 percent of the total world population who live in Asia, Africa, Latin America, and the Caribbean."

According to IDS literature, the Third World "possess 35 percent of the world's critical raw materials and exports enormous amounts of food and agricultural products to industrialized societies." IDS also reports that more capital is being sent to the industrialized world than is received in assistance. "Millions of jobs in Europe, North America, and Japan depend upon export markets" in developing countries where opportunities for tourism and transnational corporations abound.

The interdisciplinary coursework provides knowledge of such dynamics in specific geographic regions and prepares students for employment with NGOs and global development agencies as teachers, journalists, and administrators in the Third World.

AIDA objectives are "to provide multi-disciplinary consulting services in . . . international development based on sound analysis and proven experience in Africa, Asia, the Caribbean and Latin America in cooperation with international, regional, and non-governmental organizations." It is associated with Andante International in France and Sweden.

Dr. Timothy M. Shaw, professor of political science, is a senior partner in AIDA, with 25 years' experience in development—especially in Africa. He is a general editor in a *Macmillan Press Series on International Political Economy.*

■ Resources

Brochures list course offerings.

Philippines Environment and Resource
 Management Project (ERMP) ⊕
Dalhousie University
School for Resource and Environmental Studies
 (SRES)
1312 Robie Street
Halifax, NS, B3H 3E2 Canada

Contact: **Shirley A.M. Conover, Ph.D.**, ERMP
 Project Director
 Phone: 1-902-494-3632 • *Fax:* 1-902-494-3728
 E-Mail: envoy:sres

Activities: Development; Education; Research • *Issues:* Air
Quality/Emission Control; Biodiversity/Species Preservation;
Deforestation; Energy; Global Warming; Sustainable Devel-
opment; Waste Management/Recycling; Water Quality

▲ Projects and People

The **Philippines Environment and Management Project (ERMP)**
is a joint program of **Dalhousie University**'s School for Resource
and Environmental Studies (SRES) and the Institute of Environmental Science and Management (IESAM) at the **University of the
Philippines at Los Banos (UPLB)**. Funded by the Canadian International Development Agency (CIDA) in 1989, the four-year project
aims to develop a greater interaction between non-government
organizations, universities, and government in creating and implementing policies and programs affecting the environment.

Broadening UPLB's expertise, ERMP conducts **community-based
Development Action Research Projects (DAPs)** on managing the
Laguna Lake watershed, dealing with deforestation and indigenous
land tenure in northern Luzon, and addressing marine and coastal
ecosystem concerns in the Visayas. Some **policy areas** likely to be
encountered are environmental law and regulations, taxation policies of key natural resources and industries, macroeconomic pricing,
and hazardous substance management. Technical courses in environmental impact assessment and rapid rural appraisal are offered to
NGO, university, and government staff. Agricultural extension activities are being planned.

With ongoing, mutually beneficial linkage emphasizing training,
teaching, action research, policy development, and environmental
information management systems, ERMP ultimately will support
the Philippine government's environment sector programs and plans.
For example, a database is being expanded at the Department of
Environment and Natural Resources (DENR). A UPLB library is
being developed to support the DAPs and other studies. Women in
Development and Environment (WIDE) is also an important aspect of ERMP.

■ Resources

ERMP fellowships provide Canadian and Filipino students three to
six months in the field for graduate work, faculty exchanges, and
postdoctoral grants. The Southeast Asia Environmental Collection,
located at SRES, includes publications relating to ERMP projects.
An ERMP brochure and related materials are available.

School for Resource and Environmental Studies
 (SRES) ⊕
Dalhousie University
1312 Robie Street
Halifax, NS, B3H 3E2 Canada

Contact: **Peter Stokoe, Ph.D.**, Research Associate
 Phone: 1-902-494-3632 • *Fax:* 1-902-494-3728
 E-Mail: Bitnet: stokes@ac.dal.ca.

Activities: Education; Research • *Issues:* Biodiversity/Species
Preservation; Deforestation; Energy; Global Warming;
Health; Sustainable Development; Waste Management/Re-
cycling; Water Quality

● Organization

Founded in 1973, **Dalhousie University's School for Resource and
Environmental Studies (SRES)** provides an avenue for interdisciplinary research and education in environmental programs and
natural resource management of importance in Canada and abroad.
In its broad-based ecological approach to the identification and
solution of these problems, the School works at the local, national,
and international levels as staff and faculty coordinate with governments, industries, and other research support groups.

For more than 10 years, SRES has been active in international
development, witnessing a burst of expansion in the program area in
the late 1980s. Especially noteworthy during this time was the
beginning of the third phase of the Environmental Management
Development in Indonesia (EMDI), the largest award to a Canadian university from Canadian International Development Agency
(CIDA). In addition, international projects involve the Caribbean,
China, and the Philippines. *(See also Dalhousie University, EMDI
and ERMP profiles in this section.)*

▲ Projects and People

Various SRES research interest themes include the preservation of
the coastal zone, **global climatic change**, and sound economic or
sustainable development. The School continues to expand its interest in applications of knowledge-based systems or "expert systems"
in natural resource and environmental management. These systems
are a subfield of **artificial intelligence** in which "computer systems
are used to solve problems that cannot be managed by conventional
computer models, but require reasoning based on logical relationships."

A 1989 review of the **FMG Information and Mapping System
(Bay of Fundy, Gulf of Maine, Georges Bank)** project led to
appropriate alterations concerning software packaging usage. Climatic impact assessments research and marine ecosystem modeling
and environmental impact assessments (EIAs) were also developed.
A two-year project supported by the Atlantic Canada Opportunities
Agency (ACOA) is surveying knowledge-based systems of small and
medium-sized businesses in the natural resource sector and developing prototypes for a knowledge-based system "to assist in silvicultural decisionmaking for private woodlots in the Maritime provinces" and for an assessment of potential aquaculture sites.

In the international arena, numerous activities have taken place
under CIDA-funded programs, including **Institutional Development for Toxic Chemicals Management** in the Caribbean, and a
project bringing together SRES and the University of the Philip-

pines at Los Banos (UPLB) to strengthen the Philippines' capacity to contribute to environment and resource management in its own country.

■ Resources

SRES prepares literature on its research and academic endeavors, including participation in the interdisciplinary Ph.D. program. An available publications list contains reports on acid rain, fisheries and forest management, fuelwood, biomass energy, herbicides, oceans and coastal zone management, and environmental monitoring, among other topics. It seeks computers/software, developmental support, and broad-based financial assistance for the promotion of its program strategies.

Forest Management and Policy
School of Forestry
Lakehead University
Thunder Bay, ONT, P7B 5E1 Canada

Contact: Peter N. Duinker, Ph.D., Associate
　　　　　Professor
Phone: 1-807-343-8508 • *Fax:* 1-807-343-8116
E-Mail: Netnorth: pduinker@lakehead

Activities: Development; Education; Research • *Issues:* Biodiversity/Species Preservation; Deforestation; Global Warming; Sustainable Development

● Organization

Lakehead University's Forest Management and Policy research program chaired by **Dr. Peter N. Duinker** analyzes land-use policy and user-conflict issues and develops analytical tools for forest managers and policymakers.

▲ Projects and People

With the help of graduate students, Dr. Duinker recently undertook six basic studies to meet these objectives: climate change, community forestry, environmental impact assessment, forest tenure and land-use policy, old growth, and wildlife habitat supply analysis.

At the request of Forestry Canada, Dr. Duinker prepared a synthesis and literature review on the **implications of climate change for the world's managed forests.** Future climatic changes in northern Ontario present "such a strong potential risk and important challenge to forest management that it deserves immediate analytical attention," wrote Duinker. He wants to use geographic information system (GIS)–based simulation "to explore what management strategies ought to be implemented" to curb undesirable impacts.

Renewal of **community forestry** interests show many areas dependent on a single industry—whether mining, forests, or tourism—are exploring ways to "diversify their economies." A study of 20 communities north of Lake Superior tests the theory that when "a wide variety of biophysical and socioeconomic variables are taken into account, some communities will demonstrate favorable traits for successful community forestry, and others will not." Communities biased for success are those that should be selected for government-sponsored community forestry experiments, reports Duinker.

Although the scientific **adequacy of environmental impact assessments** (EIAs) has been studied by Dr. Duinker for over a decade, he feels "technical and scientific quality in EIA have not kept pace" with the regulatory meeting requirements for EIA or the participation in EIA hearings. He continues to evaluate technical specifications.

In examining the numerous forms of **forest-land tenure** in Ontario, the Chair seeks to find the apparent advantages and disadvantages of each form and propose reform when necessary. In addition, a review of Ontario's Crown Forest lands multiuse management is also being investigated. A GIS-based simulation model is being built to test the hypothesis that "the dominant-use approach, in which parcels of land within a region are dedicated to specific uses such as timber, recreation, wildlife, or wilderness, may lead to overall higher regional flows of forest benefits and fewer user conflicts than will implementation of the current multiple-use approach." **Gary Bull** followed through with this study at the University of Toronto's Faculty of Forest.

Another often discussed issue is Ontario's forestry **old growth**, what it is, and whether it should be preserved or harvested. Again, using GIS-based simulation, researchers will develop a variety of forest-level objectives for old growth and test their feasibility.

Lastly, **wildlife habitat carrying capacities** in landscaped forests—where timber is harvested—are being studied with simulated programs, for example, in areas populated by both moose and marten.

Plant Hormones Canada
c/o **McMaster University**
214 Chester New Hall
Hamilton, ONT, L8S 4L9 Canada

Contact: James D. Brasch, Ph.D., Director
Phone: 1-416-525-9140, ext. 4497
Fax: 1-416-527-0100

Activities: Education; Research • *Issues:* Biodiversity/Species Preservation • *Meetings:* CITES; IUCN

▲ Projects and People

Dr. James D. Brasch, associate professor of English, **McMaster University**, also directs **Plant Hormones Canada**, coordinates the New World Orchid Species Collection, Royal Botanical Gardens, Hamilton, and belongs to the Species Survival Commission (IUCN/SSC) Orchid Group. Dr. Brasch uses his knowledge of plant growth regulators to enhance many orchid varieties.

Through the propagation of plants by cloning or asexual methods, **meristem tissue cultures** only a few millimeters in diameter grow in an artificial environment (*in vitro*). According to a report by Drs. Brasch and Ivan Kocsis, many identical plants can develop from a single parent in a relatively short time and generally mature and flower faster than those raised from seed because the meristems are derived from superior stock. With this rapid development of meristem techniques—ferns, cut flowers, and the modern orchid industry are the beneficiaries. Due to the high cost of equipment, the development of sterile techniques, and the necessity to produce from 200 to 2,000 or more clones of one variety, this process mainly benefits large wholesalers in the orchid industry.

However, hormone usage—"meristemming"—makes the reproduction of valuable plants even within the reach of most modest growers. Since hormones serve as a messenger to "trigger" plant functions, they differ from vitamins which add nutrition. Hormones must function in the context of other cultural factors (potting media, genetic makeup, and available nutrients) just as they do in nature. Plant hormones, or phytohormones, are natural compounds produced by one part of a plant, which when moved to another part can control growth or other physiological functions. Differing from animal hormones, phytohormones are not produced in a specific part of the plant, but rather in every cell, thereby offering communications between cells.

"As endangered orchid species become more difficult to move across international borders and already established plants become more expensive, the ability to increase existing collections of these plants becomes more and more important and valuable both to the amateur and professional grower," note Brasch and Kocsis in the *Bulletin of the American Orchid Society* (October 1980). Dr. Brasch has also written about orchid collecting in Nigeria, the species "Chysis," and water quality regarding Canadian orchid culture.

■ Resources

Catalogs regarding *Plant Growth Regulators* and orchids are available. Dr. Brasch seeks additional "expertise on hormone application to propagation and preservation of endangered species, especially orchids."

Canadian Circumpolar Institute (CCI)/Institut circumpolaire canadien (ICC)
University of Alberta
G-213 BioSciences Building
Edmonton, AB, T6G 2E9 Canada

Contact: **Clifford G. Hickey, Ph.D.**, Director
Elaine Maloney-Gignac, Administrator
Phone: 1-403-492-4512 • *Fax:* 1-403-492-1153

Activities: Education; Research • *Issues:* Sustainable Development

● Organization

Established in 1918 in Edmonton—"Gateway to the North"—the **University of Alberta** is committed to northern and native studies and environmental research which was intensified in the 1970s when the oil industry expanded from northern Alberta to the Arctic Islands. It maintains Canada's second largest library with some 3,200,000 volumes.

Research is largely funded through federal councils such as the Natural Sciences and Engineering Research Council (NSERC) as well as various federal government departments as the Canadian Wildlife Service and Indian Affairs and Northern Development. Studies of the international whaling moratorium's impact on Japan's coastal industry; climate change and northern ecosystems; wildlife as reindeer, red squirrel, wolf, coyote, and swift fox; glaciology; dry grassland; and forest stand dynamics are typical.

When budgetary constraints forced the University to close its 30-year-old Boreal Institute for Northern Studies in 1990—just as

Parliament was to establish the Canadian Polar Commission—the **Canadian Circumpolar Institute (CCI)/Institut circumpolaire canadien (ICC)** was set up in its place.

The University reports it is committed to continuing and enhancing, when possible, the Boreal Institute's core functions. Reorganization is designed to foster research continuity with stronger links with northernists on and off campus. CCI is expected to grow, and a separate CCI Library is being maintained.

▲ Projects and People

CCI is sponsoring a large-scale interdisciplinary proposal, **Effects of Climate Change on the Natural Ecosystems of Alberta**, by University of Alberta scientists **Drs. Susanne Bayley, Paul Curtis, Dennis Gignac, Peter Kuhry, Dave Schindler, and Dale Vitt**, with backgrounds as limnologists, wetland ecologists, paleoecologists, and climate modelers. Models—based on "historical climate records, analyses of chemical indicators, and fossils preserved in lake and wetland sediments"—are being proposed to help predict the effects of "increasing greenhouse warming on sensitive natural ecosystems in Alberta." The researchers point out that "early warning of climate-related economic impact is important for long-term planning and development" of agriculture, forestry, wildlife, water quality and quantity, and tourism, among vital resource sectors. *(See also University of Alberta, CCI Belcher Islands Reindeer Project and Project CELIA in this section.)*

Also at the University, **Dr. Ellie E. Prepas** directs the Environmental Research and Studies Centre.

Belcher Islands Reindeer Project ❧
Canadian Circumpolar Institute (CCI)/Institut circumpolaire canadien (ICC)
University of Alberta
Edmonton, AB, T6G 2E9 Canada

Contact: **Milton M. Freeman, Ph.D.**, Senior Research Scholar
Phone: 1-403-492-4682 • *Fax:* 1-403-492-1153

Activities: Education; Research • *Issues:* Sustainable Development • *Meetings:* International Association for the Study of Common Property; International Congress of Arctic Social Sciences

▲ Projects and People

The **Canadian Circumpolar Institute (CCI)** and the Inuit community of Sanikiluaq collaborate in a sustainable community-based research initiative, the **Belcher Islands Adaptive Reindeer Management Project**, which was begun in 1987. Together, they seek to provide a mutual exchange of information concerning population-habitat status and management needs for the reindeer population of the Belcher Islands in the Northwest Territories. It is administered by the Hunters' and Trappers' Association (HTA) in cooperation with project researchers from CCI and a consultant.

Within southeastern Hudson Bay, the largest inland sea in North America, the islands can be snow- and ice-immersed from November to August, leaving only a three-month growing season for plants—mainly Arctic mosses, lichens, sedges, and some grasses.

The approximate 470 Inuit (and 30–40 transient workers) are the only people living year-round at the 1970 government-established center in Sanikiluaq. Formerly known as a *Qikitarmiut* (people of the islands), the natives continue to eat a "highly nourishing and, traditionally, low-disease diet . . . susceptible to food shortages" in synchronization with the annual cycles of marine mammals, sea ducks and other birds, and fish.

Hoping to supplement the Inuit diet, the government of the Northwest Territories brought 60 reindeer to the Belcher Islands in 1978—the first presence of large land-based animals on the islands since the late 1800s when caribou disappeared due to migration or starvation. The reindeer were left alone during their first five years to establish themselves. Community harvests began in 1982, first on an annual and later a seasonal basis. Hunters reported how fast the reindeer population grew, prompting the community's 1987 request for more knowledge about the species and initiation of the reindeer study. In seeking to provide answers for their concerns, CCI was aware that the natives possessed "empirical knowledge" of the wildlife managing process," a benefit of lifelong experiences in the Arctic region.

Even though the understanding of Inuktitut and the Belcher Islands dialect hampered exchanges, CCI learned how to "break the ice," make people feel comfortable, and answer the community's questions about the reindeer's reproduction, herd formations, breeding cycles, and feeding areas; "develop a wildlife management system that incorporates the traditional ecological knowledge of Belcher Island Inuit and scientific techniques of wildlife biologists"; and design a "viable model for community-based management of wildlife resources in the Canadian north," reported Miriam McDonald Fleming, project participant.

As discussions proceeded, they focused on reindeer, Hudson Bay eider duck, and other species. Traditional ecological knowledge is recorded to learn about the natural world, maintain respect of the Inuits' ways, and systemically bring the native Canadians into the management process.

According to Dr. Milton Freeman, "Canadian government agencies and cooperative management boards are assuming an increasingly active role in managing Arctic wildlife species. This is occurring in convergence with the imminent settlement of aboriginal land claims, deterioration of the Arctic marine habitat, and recognition of the importance of wildlife sustaining life in the Arctic."

Methodology includes aerial and "efficient" snowmobile reindeer surveys. HTA also conducts field carcass analysis comparing nutritional conditions faced by reindeer with caribou from other parts of the Arctic, to establish a baseline for evaluating reindeer population management. Researchers are Lucassie Arragutainaq, Robert J. Hudson, and Peter Poole.

CCI points out that the economies and cultures of such rural communities depend upon harvesting wildlife, marine mammals, and fish—activities which are being undermined "by the growing urban-based, anti-harvest movements in Europe and North America." Dr. Freeman writes, "We see sustainable development [associated with] ensuring sustainable livelihoods . . . as involving issues of equity and social justice [among people lacking] economic and political power in decision-making centres." Such communities place few demands on global resources and promote "important environmental and social values that dominant societies are in danger of losing."

■ Resources

The *Documentation of Traditional Knowledge for Adaptive Management of Reindeer, Reindeer Surveys on the Belcher Islands*, and a biodata list are available at no charge. Information is sought on adaptive management elsewhere that combines traditional and scientific expertise. Materials such as video usage and poster displays are requested to advise similar communities on such management procedures.

Project CELIA ⊕
Canadian Circumpolar Institute (CCI)/Institut circumpolaire canadien (ICC)
University of Alberta
G-213 BioSciences Building
Edmonton, AB, T6G 2E9 Canada

Contacts: **Charles E. Scheweger, Ph.D.**
Nat Rutter, Ph.D.
Phone: 1-403-492-4512 • *Fax:* 1-403-492-1153

Activities: Research • *Issues:* Global Warming

▲ Projects and People

According to its literature, **Project CELIA**—Climate and Environment of the Last Interglacial in Arctic and Subarctic North America—is an international cooperative research program developed to "generate and synthesize data on Isotope Stage 5 high latitude terrestrial and near-shore marine environments; test hypotheses generated by general circulation models and other warm earth or interglacial simulations; and, by documenting the past, provide insight into future environments as a result of global change."

Stage 5e of "the marine oxygen isotope record" refers to the last period of time, some 126,000 years ago, characterized by a "lower world ice volume, higher sea-level, and warmer world climate" similar to what is anticipated in the future "as a result of anthropogenic changes to the atmosphere." CELIA scientists believe that this is "an ideal time period for investigating climate change, especially warming, and the attending environmental responses." It is also known that carbon dioxide (CO_2) levels increased during Stage 5e "with the greatest buildup prior to the sharp decline in world ice volumes."

Field researchers will work at high-altitude sites—the first to experience future warming—to investigate fossil pollen, insects, seeds, soil properties, stable isotopes, and other paleoecological indicators. Here is where "trees meet their present-day climatic limits in the Arctic and subarctic and permafrost, or sporadic permafrost is widespread" as well as where "sensitive and accurate paleoenvironmental records can be established."

The U.S. National Geophysical Data Center, Boulder, Colorado, is cooperating in the establishment of data sets for selected field sites and help with geochronology. Dr. John Kutzbach, University of Wisconsin's Center for Climate Research, is helping to test computer-based climate model simulations. CELIA has also helped form LIGA—the Last Interglacial in the Arctic—to coordinate "high-latitude global data sets for climate model testing and global change research."

With fieldwork getting underway in 1991, about three more years of active research are expected.

Friends of the Athabasca Environmental Association (FOTA) 🌱
University of Alberta
Box 672
Athabasca, AB, T0G 0B0 Canada

Contact: **William A. Fuller, Ph.D.**, Professor
   Emeritus
Phone: 1-403-675-2993
E-Mail: Internet: billf.@cs.athabascau.ca

Activities: Education; Political/Legislative; Research • *Issues:* Air Quality/Emission Control; Biodiversity/Species Preservation; Deforestation; Health; Water Quality

▲ Projects and People
Friends of the Athabasca Environmental Association (FOTA) reports that it is "concerned with those actions that may have an adverse effect on the air, land, or water in the Athabasca Basin." With its more than 100 members, FOTA says, "Our primary concern is to delay the Alberta Pacific bleached kraft pulp mill until proper studies have shown that the river can handle the effluent. If studies show the river cannot handle the additional load, then we want the mill stopped," writes **Dr. W.A. Fuller**. An educational action program was instigated at a conference on the Boreal Forest in late 1991.

A zoologist, Dr. Fuller belongs to the Species Survival Commission (IUCN/SSC) Bison Group, is a former member of the World Conservation Union board, and has written books and papers on Canada's circumpolar northern lands and small mammals.

The David Suzuki Foundation
University of British Columbia
2075 West 12th Avenue
Vancouver, BC, V63 2G3 Canada

Contact: **David Suzuki, Ph.D.**, Broadcast Journalist,
   Writer
Phone: 1-604-732-4228 • *Fax:* 1-604-732-0752

Activities: Education; Research • *Issues:* Biodiversity/Species Preservation

▲ Projects and People
A zoology professor since 1969 at the University of British Columbia, **Dr. David Suzuki** also holds 11 honorary doctorate and law degrees from universities in Canada and the USA and is the recipient of numerous awards. Noted for popularizing science, Dr. Suzuki writes a weekly syndicated column for 31 Canadian newspapers. He also has authored or co-authored 15 books (including 7 for children) and produced records, such as *Spacechild* and *Earthwatch;* tapes and radio shows, such as *It's a Matter of Survival (CBC);* films

for the National Film Board of Canada, such as *Telidon* and *This Is an Emergency;* and television shows, for the past 20 years, such as *A Planet for the Taking, The Nature Connection, Voices in the Forest,* and *James Bay: The Wind That Keeps on Blowing.* For his media work, he was named 1990 "Author of the Year" by the Canadian Booksellers Association; received the Children's Literature Roundtables of Canada Award for *Looking at Insects;* and was presented Gold and Grand awards from the Canadian Council for Advancement and Support of Education—among many such honors.

Other publications in his field include the textbook, *An Introduction to Genetic Analysis,* originally written with **Dr. A.J.F. Griffiths** (W.H. Freeman and Co., USA) and other editions co-authored by **Dr. Richard C. Lewontin** and **Dr. Jeffrey H. Miller.** Dr. Suzuki describes this book as "the most widely used genetic textbook in the USA and . . . translated into Italian, Spanish, Greek, Indonesian, Arabic, French, and German." With associates, he has also written more than 80 major publications and some 70 abstracts and short scientific articles.

Long involved in civil rights issues, Dr. Suzuki was once named the "Outstanding Japanese-Canadian of the Year," and is on the board of directors, Canadian Civil Liberties Association.

IUCN/SSC Caprinae Specialist Group ⊕
University of British Columbia
Animal Science Department
Vancouver, BC, V6T 1Z4 Canada

Contact: **David M. Shackleton, Ph.D.**, Deputy
   Chairman
Phone: 1-604-822-6873 • *Fax:* 1-604-822-6394

Activities: Conservation Planning; Education; Research • *Issues:* Conservation; Consultation; Biodiversity/Species Preservation

▲ Projects and People
Dr. David M. Shackleton is both deputy chairman of the Species Survival Commission (IUCN/SSC) Caprinae Specialist Group and associate professor, Animal Science Department, University of British Columbia. With the preparation of action plans as the primary task of the SSC groups, Dr. Shackleton is undertaking such a strategy for the world's wild caprinae—sheep, goats, and their relatives.

Such plans usually assess the species' conservation status, make recommendations to ensure continued survival, and implement a program in keeping with global environmental issues and biodiversity.

Data on the caprinae population and conservation status in 61 countries is currently being gathered. **Dr. Sandro Lovari**, Group chairman, initiated this project in early 1989 with Shackleton as the plan's compiler. Following the plan's drafting and publication, implementation will begin.

At the University, Shackleton is involved with graduate research projects on the ecology of the Roosevelt elk in relation to forestry and recreation. One study focuses on seasonal movements and habitat in a Vancouver Island park setting. Another study is concerned with "elk management in relation to wolf predation and forestry practices."

Also ongoing is a **bioenergetics study of killer whales**. "Captive animals from two facilities have been trained to provide air and urine samples," writes Shackleton. "Free-ranging animals allow data for calculating natural energy costs [such as swimming speeds and diving depths] and physiological measures [such as respiration rates] to be collected." Results will assess the "energy/food requirements for improved management of captive killer whales and to allow fisheries biologists to estimate the impacts of killer whales on fish stocks," adds Shackleton. A related project recently demonstrated that "daily and seasonal movements of the northern killer whale pod along the British Columbia coast are highly related to . . . annual salmon [spawning] migrations," which can impact Pacific salmon stocks.

A 10-year study of **grizzly bears** was carried out in an active **forestry and natural gas development area**. Observing large numbers of radio-collared bears, and using helicopter overflights, it was learned that "through appropriate management of people, industry, and bears—healthy bear populations and human activities are compatible outside protected areas." Former graduate student Dr. Bruce McLellan, research scientist with the BC Forest Service, is continuing the study.

Faculty of Environmental Design
University of Calgary
Calgary, AB, T9N 1N4 Canada

Contact: **Valerius Geist, Ph.D.**, Professor, Environmental Science
Phone: 1-403-220-6601 • *Fax:* 1-403-284-4399

Activities: Education • *Issues:* Air Quality/Emission Control; Biodiversity/Species Preservation; Deforestation; Energy; Global Warming; Health; Sustainable Development; Waste Management/Recycling; Water Quality

● Organization
The **University of Calgary's Faculty of Environmental Design**, established in 1971, offers professional master's degrees in environmental science, industrial design and planning, and architecture. It seeks to provide its students with a better understanding of environmental issues, enabling them to contribute traditional and "new forms of practice" in society. Students and faculty come from Canada and abroad. According to a 1991–1992 report, there are 24 full-time members of the Faculty, 20 part-time and visiting faculty, and 230 students. The Faculty is looking to establish exchange agreements and other cooperative arrangements with numerous other universities and institutions, particularly in developing countries.

▲ Projects and People
The Faculty helped secure a 1,600-hectare park in the City of Calgary and has performed "impact assessment for government decentralization projects and major resource development projects in Alberta," regarding water resources, energy, and historic preservation. One such assessment was of the move of the Alberta Environment Research Laboratory to Vegreville.

The **affordable sustainable community** demonstration project in an urban setting is a cooperative effort of both Alberta industry and public officials. Experimental, interactive community/housing is being designed with attention to innovative site development standards and reduction of infrastructure and construction costs. Model "ecological community" projects in Scandinavia are being investigated.

Ecological studies of both ecosystems and wildlife species are research components that have led to books on mountain sheep, deer, bear, forest growth, and revegetation—and their improved management. Researchers study the design, creation, and operation of natural parks and biosphere reserves; and learn how to restore natural grassland ecosystems—with attention paid to political and financial matters in funding and managing such reserves.

In the **swift fox reintroduction** long-term program begun in 1983, the Faculty collaborates with researchers from Alberta, Saskatchewan, and federal governments, to determine if the fox has a reasonable probability of reestablishing its population in the Canadian prairie.

Integrated resource planning projects address energy development and its land disturbances with the intent to foster environmental protection.

On behalf of four Canadian universities, the Faculty joins the Asian Institute of Technology (AIT) in Bangkok, Thailand, to study **human settlements and natural resources** development in Southeast Asian locations with demonstration projects.

Design of industrial and medical lifesaving equipment is among other noteworthy and interdisciplinary Faculty involvements.

Dr. Valerius Geist, professor of environmental sciences and biology, is particularly interested in maximizing health environmentally and conducting research on large-mammal biology. **Dr. William A. Ross** is also professor of environmental science and involved in numerous energy conservation projects. **Dr. Grant A. Ross** is associate professor of engineering in environmental design. **Dr. J. David Henry** is associate professor of environmental science with a background in managing national parks and ecological reserves; he was president of the Canadian Parks and Wilderness Society.

■ Resources
The Faculty's Environmental Design catalog is available as well as lists of workshop and degree projects.

Department of Zoology
University of Toronto
25 Harbord Street
Toronto, ONT, M5S 1A1 Canada

Contact: **Nicholas Mrosovsky, Ph.D.**
Fax: 1-416-978-8532

Issues: Biodiversity/Species Preservation; Global Warming; Sustainable Development • *Meetings:* CITES

▲ Projects and People
"Sea turtles are beautiful complex creatures, mysterious enough to become addicting for the biologist, absorbing for anyone to watch, and of great value for their eggs, meat, shell, and leather," writes Dr.

Nicholas Mrosovsky, University of Toronto, in his book, *Conserving Sea Turtles*, a British Herpetological Society publication.

In his joint report with **C.L. Yntema**, Upstate Medical Center, New York, he describes research studies with incubated eggs and notes that **higher temperatures produce more females and lower temperatures, more males.** In a range of findings, which appeared in the article "Temperature Dependence of Sexual Differentiation in Sea Turtles: Implications for Conservation Practices" (*Biological Conservation*, 1980), Dr. Mrosovsky showed that the use of styrofoam [incubation] boxes above ground, location of central hatcheries, incubation in reduced clutch sizes, and egg harvesting only during certain seasons could affect the sex ratio of sea turtles. Styrofoam boxes seem to be linked to "masculinizing turtle populations."

Consideration of sand incubation on a natural beach, such as with the absorption of minerals and stimuli for imprinting, may offer turtle eggs additional benefits as well as the correct temperature regime, the article explains.

Dr. Mrosovsky, who belongs to the Species Survival (IUCN/SSC) Marine Turtle Group, urges more research in this area "before unevaluated methods become accepted procedures." He has conducted studies with freshwater snapping, leatherback, loggerhead, green, and Atlantic ridley turtles. Mrosovsky also reported on studies with green and hawksbill turtles in Australia that examined whether nests in shady areas exposed to the wind extend the incubation period.

Environmental Change and Acute Conflict ⊕
University College
University of Toronto
15 King's College Circle
Toronto, ONT, M5S 1A1 Canada

Contact: **Thomas F. Homer-Dixon, Ph.D.,** Co-Director
   Phone: 1-416-978-8148 • *Fax:* 1-416-978-8854
   E-Mail: hdixon@gpu.utcs.utoronto.ca

Activities: Education; Research • *Issues:* Air Quality/Emission Control; Deforestation; Degradation of Agricultural Land; Energy; Fisheries; Global Warming; Health; Ozone Depletion; Population Planning; Sustainable Development; Water Quality

● Organization

The **Peace and Conflict Studies Program**, University of Toronto, and the **American Academy of Arts and Sciences**, Cambridge, Massachusetts, began a two-year project in 1990 on the environmental aspects of international and national security. **Dr. Jeffrey Boutwell** directs the Academy's Committee on International Security Studies in this operation, with **Dr. Thomas Homer-Dixon** serving as his counterpart in Canada. **Professor George Rathjens**, Massachusetts Institute of Technology (MIT), completes the core group members.

▲ Projects and People

The research project concentrates on three specific conditions which could lead to acute conflict in the developing world: **general economic decline, population displacement,** and **water scarcity and degradation.** The causal links between these conditions and previous social and physical variables are examined in Africa, the Far East, and South Asia.

The progressive **impoverishment degradation produced in developing societies** is probably "the most significant yet least studied potential long-term result of environmental degradation," states the project outline. It goes on to note that while logging for hardwood export markets may yield gains in the short-run, it could damage the country's long-term productivity. Also, a lack of finances, materials, and intellectual resources needed to adapt to such climatic changes—created by ozone layer depletion—pose other problems, in addition to civil unrest and insurgency.

A "somewhat misleading" term, "environmental refugees" may surface as a result of environmental degradation, according to this project. However, as the outline explains, it is not so easily distinguished because environmental changes will be only one of several interacting social and physical variables responsible for conflict among various ethnic groups forced together.

This is already illustrated in human **displacement in the West Bengal-Bangladesh-Assam region** caused in part "by a rapidly growing population, unsustainable agricultural practices, and consequent shortages of adequately fertile land." Other environmental problems, such as "rising sea levels coupled with extreme weather events (both induced by climate change), and by flooding arising from deforestation in watersheds upstream on the Ganges and Brahmaputra rivers," can also spur migration that heightens "tribal, cultural, religious, and economic cleavages that . . . aggravate intergroup hostility."

Water scarcity; runoffs resulting from deforesation that "damage roads, dams, irrigation works, and other valuable infrastructure"; extra siltation impeding river transportation; and increases in ultraviolet radiation that can escalate disease in humans and livestock are among other major resource and environmental problems of the next century, notes the project.

Indonesia is cited as a **potential model for other developing countries** for keeping agricultural production well ahead of population growth "as it aggressively works to promote sustainability" and combat deforestation and erosion problems.

Three workshops were held in 1991, followed by the preparation of several project papers. In early 1992, policymakers and environmental analysts in Asia and Africa were called upon for comments and criticisms of the draft summary report prior to issuance of a final version at the Conference on Environment and Development (UNCED) in Brazil.

■ Resources

Project publications are in production. Also, Dr. Homer-Dixon has written a chapter on global change and international security in *Emerging Trends in Global Security*, reports on environmental change, human security, and violent conflict; and "Taxes, Fuel Consumption, and Carbon Dioxide Emissions."

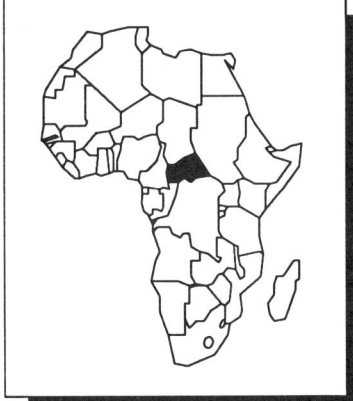

The Central African Republic, in the center of the African continent, is bordered by Zaire and the Congo to the south, Chad to the north, the Sudan to the east, and Cameroon to the west. The country forms a transition between the sub-Saharan savanna and the equatorial West Africa forest.

The land is a vast plateau with rolling hills, mountains in the northwest and northeast, and many rivers with various tributaries that provide fresh water year-round. Most of the country is wooded. About 80 percent of the population lives in rural areas.

Cattle brought in by nomads from Chad and the Sudan have crowded onto the limited grasslands, causing devegetation, soil degradation, and the encroachment of the desert. In addition, hunting has caused a serious threat to wildlife; 75 percent of the elephant population has been destroyed. However, a ban on elephant hunting was imposed in 1985, and projects are underway to create protected areas for wildlife.

 NGO

Association Communautaire de YOBE-SANGHA (ACYS)
B.P. 1053
Bangui, Central African Republic

Contact: **Phil Hunsicker**, Director of Rural Development
Phone: 61 4299 (operator assistance) • *Fax:* 61 4299

Activities: Development; Environmental Conservation/Preservation • *Issues:* Biodiversity/Species Preservation; Sustainable Development

● Organization

The **Association Communautaire de YOBE-SANGHA (ACYS)** grew from the interests of the World Wildlife Fund (WWF) and the local community in this region's flora and fauna, including the country's largest concentration of forest elephants, bountiful bongo, dwarfed buffalo, leopard, golden cat, and 15 species of primates, such as the lowlandplains gorilla, chimpanzee, and colobus guereza. Within the "dense humid forest . . . the most unique vegetal formation in the Central African Republic" (CAR) are said to be "3,600 species of plants and 208 species of mammals."

In early 1988, WWF and the Central African government agreed to establish the National Park and Reserve of DZANGA-SANGHA, DŻANGA-NDOKI, including the entire community of YOBE-SANGHA; and late in 1990, laws were passed that created a "new type of protected area [to] give the Reserve a special usage capacity that will bring in an integrated and controlled manner human and industrial activities: rural development zone, community hunting zone, safari hunting zone, and a forest exploitation zone." For the first time in CAR, a "hunting territory" in a conservation project is "authorized to guarantee a standard of living to the local community."

▲ Projects and People

As ACYS literature describes, the WWF-USA/CAR Project "works with the participation of local communities situated in the Park and Reserve Zones" that must both depend on and manage the area's resources. The Project aims to involve local people in decisionmaking so that they can overcome the present "catastrophic" economic situation of the region. Most residents lost their jobs when a logging company shut down in 1986; many turned to poaching for survival. Today, a newly instituted visitors fee for Park guests is helping to fund community organizational efforts to meet the needs of the population and ultimately develop tourism as a major industry.

Residents have formed "sub-associations by trade," such as fishermen, animal husbandry, bakers, butchers, restaurateurs, gardeners, construction workers, farmers, hoteliers, librarians, traditional hunters (BaAka Pygmies and LINDJOMBO), and artisans. Townspeople are also represented.

Delegates of these sub-associations met in 1991 to form ACYS and to decide how to distribute a foundation grant among themselves. They agreed, for example, to dispense "$425 to the construction sub-association to buy a wheelbarrow and a brick press, . . . $450 to fishermen to buy nets and twine, . . . $293 to the gardeners to buy watering cans, shovels, rakes, nails, and seeds; $92 to a butterfly artist to buy paper, glue, varnish, pins, hooks, tweezers, and black material." As decisionmaking grew, the local population's interests increased.

ACYS is currently waiting for CAR documentation to "unite the population of the YOBE-SANGHA community in the limits of the National Park and Reserve." Its goal remains to "promote the conservation and safeguarding of natural resources in the zones so that the residents of the community can benefit and improve their lives."

■ Resources

ACYS seeks "small business training to give a strong base in determining what businesses are good and how to keep them going."

Chile's large latitudinal area contains conditions ranging from the hot, arid desert areas of the northern plains to the gusting sleets of the southern archipelago. It is also a country prone to natural disasters. Because Chile is situated on one of the most active seismic areas of the world, shocks from earthquakes are a regular occurrence. Chile also has 17 active volcanoes, and forest fires have become more frequent as rural development has accelerated. Floods and drought occur more irregularly.

Overexploitation of forests has led to soil erosion, biodiversity loss, and air pollution. However, the country has instituted incentives for reforestation that may help to reverse this trend.

❦ *NGO*

Colegio de Ingenieros Forestales
Professional Association of Forest Engineers
San Isidro 22, Of. 503
Santiago, Chile

Contact: **Pablo Tiromi,** President
  Phone: (56) 2 393 289 • *Fax:* (56) 2 232 5069

Issues: Deforestation; Sustainable Development; Water Quality

● Organization

Colegio de Ingenieros Forestales, the Professional Association of Forest Engineers, has 435 members and publishes a bimonthly magazine, *Renarre.*

Comité Nacional pro Defensa de la Fauna y Flora (CODEFF)
National Committee for the Defense of Animals and Plants
Santa Filomena 185, Casilla 3675
Santiago, Chile

Contact: **Godofredo Stutzin,** Founder, Honorary President
  Phone: (56) 2 777 0617 • *Fax:* (56) 2 377 290

Activities: Development; Education; Law Enforcement; Research • *Issues:* Antarctic Conservation; Biodiversity/Species Preservation; Deforestation; Sustainable Development; Waste Management/Recycling; Water Quality • *Meetings:* CITES; FOE; UNCED; World Parks Congress

● Organization

Founded in 1968, the **Comité Nacional pro Defensa de la Fauna y Flora (CODEFF) (National Committee for the Defense of Animals and Plants)** and its 1,600 members raise the nation's social consciousness about the protection of endangered Chilean flora and fauna; promote the conservation and rational management of ecosystems and natural resources to achieve sustainable development; research endangered species and propose concrete alternatives for their management; and educate and disseminate information to denounce inadequate resource management.

According to CODEFF, 56 percent of the trees and shrubs that grow in Chile are unique to the area and close to 30 species are in danger of extinction. Each year, between the regions of Maule and Araucanía, 40,000 hectares of native forest are replaced by pine with the following results: reduced water quality and increased fluctuation of levels of river water; intensification of erosion; large-scale demise of animals; destruction of endangered trees and shrubs; and pollution from the use of herbicides and other toxins. If the situation continues, reports CODEFF, the native forests will be gone within 20 years.

Chile, incidentally, is the largest exporter of fish meal. Of the 620 species of fish in the Pacific Ocean off the Chilean coast, the reproductive cycles and biology of only 30 are known. In the Magallanes region, **illegal hunting** of protected species, such as **birds, marine wolves, dolphins, and penguins**, is common; their meat is used as bait for catching shellfish. Illegal traffic in pelts of marine and river nutrias, wildcats, and foxes continue as well.

The *Huemul del Sur*, an indigenous deer (appearing on the country's national coat of arms and as CODEFF's logo), has been reduced to only 1,000 animals through forest fires, hunting, illegal trafficking of its horns, and the introduction of livestock into its territory, which cuts its food supply and increases the spread of disease. Considered a "national monument" in Chile and also found in Argentina, the **larch tree** (which is said to take a thousand years to reach one meter in diameter) is losing its branches to illegal cuttings.

▲ Projects and People

CODEFF, with Greenpeace, previously persuaded the Chilean government to prohibit the capture of whales in its territorial waters; condemned the illegal destruction of native forest in Biobio region with the replacement of pine trees for logging; helped bring a case before the Supreme Court which led to the **cessation of extraction of water from Lake Chungará**—world's highest lake—in a UNESCO biosphere reserve with a **giant coot** population; and introduced the concept of fines on those illegally trafficking in pelts of endangered species.

CODEFF, with WWF–USA, evaluates the impact of exotic species introduction on native forest destruction and seeks **sustainable development alternatives by small landowners in south-central Chile**. CODEFF also studies nutria populations and the impact of illegal hunting. The **habitat of the endangered Andean deer** *Huemul* (*Hippocamelus bisulcus*) is being investigated in the Nevados de Chillán; **Environmental education workshops** are ongoing to raise awareness of unions, community groups, women's groups, and professors about natural resources, ecology, and the environment. A library and environmental documentation center is being developed in Santiago for students, teachers, and the general public.

Attorney **Godofredo Stutzin** is also the founder and honorary president of Unión de Amigos de los Animales (Union of Friends of Animals); Aldeas de Niños SOS (Children's Villages), which emphasizes care of nature and animals by youths; and the Albert Schweitzer Fund for the Animal World. He belongs to the World Conservation Union's International Council of Environmental Law and is a regional member of the Species Survival Commission (IUCN/SSC). Among his achievements, he carried out a worldwide campaign to **stop Chile's massive slaughter of Commerson's and other dolphins for bait** by the crab fishing industry; helped obtain **Chile's ratification of CITES** (Convention on International Trade in Endangered Species of Wild Flora and Fauna) in 1975; and has written hundreds of conservation and animal protection articles for the *El Mercurio* newspaper and other periodicals.

■ Resources

CODEFF needs assistance in developing and carrying out projects, especially to strengthen its institution and to preserve flora and fauna. A publications list, including brochures, bimonthly newsletter *Ecos CODEFF* and the bimonthly journal *Eco Tribuna*, is available. Materials are printed in Spanish.

Fundación Claudio Gay/Claudia Gay Foundation
Alvaro Casanova 613, Peñalolen
Santiago, Chile

Contact: **Adrianna Hoffmann**, Research Officer
Phone: (56) 2 277 2899 • *Fax:* (56) 2 277 5954

Activities: Education; Research • *Issues:* Biodiversity/Species Preservation; Deforestation; Sustainable Development

● Organization

Fundación Claudio Gay (Claudia Gay Foundation) is concerned with environmental education for children, emphasizing conservation, sustainable development, and biodiversity.

Red Nacional De Acción Ecologica (RENACE) National Ecological Action Network
Antonia Lopez de Bellow 024
Santiago, Chile

Contact: **Gabriela Coopman**, Executive Secretary
Phone: (56) 2 274 280, (56) 2 777 5065

Activities: Development; Education • *Issues:* Biodiversity/Species Preservation; Deforestation; Energy; Global Warming; Sustainable Development; Water Quality

● Organization

Founded in 1989, RENACE is the National Ecological Action Network, which groups together ecological organizations and other groups with environmental concerns from throughout Chile. Within the Network, there are presently about 127 such groups representing students, schools, workers' unions, ethnic associations, cultural centers, women's groups, and art collectives. RENACE is comprised of a coordinating secretariat with 11 voluntary members and an executive secretariat with 2 members who carry out actions determined by the coordinating secretariat.

Its mission statement describes RENACE as exchanging information and experiences, promoting actions that contribute to solving environmental problems in Chile, and encouraging the development of an ecological society.

▲ Projects and People

RENACE's activities include: publishing and distributing to members and subscribers the magazine *Ecoprensa*, which culls environmental news from Chile and worldwide; **fundraising** through the sale of T-shirts with pictures of endangered Chilean flora and fauna; holding **annual national meetings** for network members; organizing **area meetings to establish RENACE REGIONAL**; and maintaining correspondence with members, organizations, media, and others seeking information.

Ecoprensa features editorials; contributions from members; sustainable development issues; Network current events; reprints of environmental new articles such as "Latin American Politics and Agroecological Strategies" and "Hundreds of Valleys and Forests Are Threatened by Unnecessary Dams"; and a **Practical Guide to**

Ecology, providing answers to environmental problems, such as noise pollution and nitrates in food.

■ Resources
Besides *Ecoprensa* and T-shirts, RENACE produces a list of member organizations.

 ## *Universities*

Centro de Investigación y Manejo de Mamiferos Marinos (CIMMA)
Universidad Austral de Chile
Casilla 567
Valdivia, Chile

Contact: **Jorge A. Oporto Barria**, President and Researcher
Phone: (56) 63 213911, ext. 1210
Fax: (56) 63 212953
E-Mail: red bit net

Activities: Education; Research • *Issues:* Biodiversity/Species Preservation; Sustainable Development • *Meetings:* Biennial Conference on the Biology of Marine Mammals

● Organization
CIMMA, the **Marine Mammal Center**, conducts research on the biology and ecology of marine mammals, such as whales, dolphins, and seals. Its aim is to decrease the mortality rate of these threatened species by investigating possible causes of their decline.

▲ Projects and People
Currently, **Jorge Oporto**, with **Lila M. Brieva**, is studying the biology of small cetaceans and how they are impacted upon by the local fisheries. Oporto is also assisted by **Claudia Mercado** in an in-depth investigation of the **conflicting interaction between seals and salmon fisheries**, including interviewing salmon farmers and National Fishery Service personnel. In conjunction with **Jorge Olivos**, Oporto is also hoping to complete a study of the **organochloride contamination of the Chilean dolphin**. While much data has been

collected, funding has been the primary obstacle for **gas chromatography analysis**. CIMMA considers this to be an important project due to the lack of research performed in areas which suffer from the "indiscriminate use of toxic substances in agriculture."

CIMMA believes it could play a more vital role in the research, conservation, and education projects at a national level, consequently providing data that would be useful on a global level, with the added staffing of qualified researchers and visiting scientists to update knowledge and techniques. "Conservationist organizations should work [together], improve communication, and give value to local groups or institutions, giving them credit for their activities in their respective regions," writes Oporto, who belongs to the Species Survival Commission (IUCN/SSC) Cetacean Group.

■ Resources
CIMMA needs laboratory equipment: microscope, stereoscope, freezer, digital balance, dissection tools, glass and raw materials, and computer; field equipment: binoculars, zodiac boat and outboard motor, manual counter, camera and telephone lens; assistance with training personnel and mass communications; and publications for setting up a library with collections on aquatic mammal, biological conservation, and International Whaling Commission publications.

Universidad de Concepción—Chillan
Casilla 537
Chillan, Chile

Contact: **Alejandro Valenzuela**, Dean
Phone: (56) 42 276 333 • *Fax:* (56) 42 221 507

Activities: Development; Education; Research • *Issues:* Biodiversity/Species Preservation; Deforestation; Energy; Health; Waste Management/Recycling; Water Quality

● Organization
Universidad de Concepción has a staff of 100, with 60 involved in research. Books and journals are needed for a student and research library.

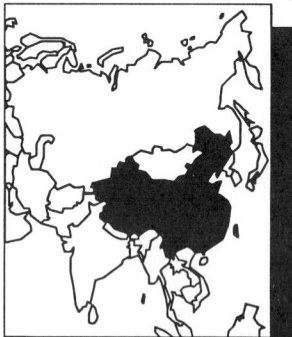

China is characterized by high mountainous regions, rich coastal areas, deserts, and low plains with large river systems. However, an emphasis on industrial production means that the country is increasingly plagued by large-scale deforestation, losses of cropland, reclamation of lakes, soil erosion, and desertification. Some of these problems are attributable to government policies; others, such as deforestation, have resulted from resource exploitation. The cities suffer from air, noise, water, and chemical pollution as well as urban congestion.

Conservation efforts are limited because of the shortage of environmental experts and education. However, environmental awareness is increasing. There is now a national reforestation goal that would increase the current forest area by half. In addition, individual provinces, regions, and municipalities are enacting specific regulations.

🏛 *Government*

Beijing Natural History Museum and
China Ornithological Society
126, Tien Qiao Street
Beijing 100050 China

Contact: **Weishu Hsu, Ph.D.,** Ornithologist
Phone: (86) 1 841 7773, ext. 187

Activities: Education; Research • *Issues:* Biodiversity/Species Preservation • *Meetings:* ICBP; International Ornithological Congress

▲ Projects and People
Weishu Hsu is chief of the Department of Ecology, **Beijing Natural History Museum,** and also vice president of the **China Ornithological Society** and secretary-general of the Peking Zoological Society. Among his 6 books and 40 papers, he has written *Birds of China 1987* and *Rare and Endangered Gamebirds in China 1990.*

■ Resources
The Museum publishes *Nature,* a quarterly periodical, and *Memoirs of Beijing Natural History Museum,* at irregular intervals.

China Environment News
15 A Xiaoxinglongjie
Chongwen District
Beijing 100062 China

Contact: **Chang Xiaoqing,** Editor, Foreign Affairs
Phone: (86) 1 7012039 • *Fax:* (86) 1 7013772

Activities: Media • *Issues:* Sustainable Development • *Meetings:* Global Assembly of Women and the Environment; Enviro Asia

▲ Projects and People
According to foreign affairs editor **Chang Xiaoqing,** the *China Environment News*—sponsored by the Environmental Protection Commission under the State Council—is "the first national newspaper dealing with environmental protection in the world." Beginning in 1984 and with an annual circulation of nearly 420,000 copies, the newspaper provides "significant news both at home and

abroad and various information and knowledge on the environment." The newspaper's influence on society and contributions to environmental protection were recognized twice by the United Nations Environment Programme (UNEP), including its selection for the Global 500 Roll of Honour. It appears three times weekly in a folio, four-page format.

Since mid-1990, the international department of *China Environment News* has published a monthly, 8-page English edition covering 20 countries. "It is aimed at introducing China's environmental protection to the outside world, promoting national and international exchange and cooperation in this specific field and creating environmental awareness," writes Chang.

As she reports, "the Chinese government attaches great importance to environmental protection . . . one of the fundamental national policies." She indicates that the investment devoted to environmental protection more than doubled to "7 billion yuan in 1988" and is increasing from the current investment of 0.7 percent of the GNP—"which provides a vast market for environmental products both at home and abroad."

▲ Projects and People

Chang comments, "According to the Five Principles of Peaceful Coexistence, environmentalists should closely contact each other, discuss strategy for global environmental protection, learn from each other."

"*China Environment News* is playing an important role in introducing Chinese environmental protection experiences to the outside world," she continues. "For example, we have introduced such Chinese environmental protection experiences as Chinese eco-farming, coal technology, desertification, and biological control. Simultaneously, we introduce foreign countries' good experiences into China. Besides, our organization is also contacting foreign organizations to do some specific projects to enhance global cooperation."

A recent special project, Water Resources Management and People's Participation in Sustainable Development in China, was conducted with the support of the Environment Liaison Centre International to arouse enthusiasm for the 1992 United Nations Conference on Environment and Development (UNCED). Other participants include Ren Guangzhao, Yang Cunxing, Deng Gulin, and Ms. Wu Changhua. Also, the International Department organized a trip for 17 NGO representatives to attend a late 1991 global conference in Paris, France, called "Roots of Our Future—Our Agenda." It is now setting up a magazine, *Green Leaves*, with support from the UN Economic and Social Commission on Asia and the Pacific (ESCAP).

■ Resources

China Environment News asks foreign environmental organizations to send newsletters by fax, not just by mail, for translation and publication in the Chinese edition. This gives Chinese writers an opportunity to respond and write articles that are published later in the English edition "in order to awaken world people's consciousness." The newspaper provides most copies of its English edition as complementary and for exchange with international, governmental, and NGO organizations involved in environmental protection such as UNEP, UN ESCAP, World Wildlife Fund (WWF), World Conservation Union (IUCN), Friends of the Earth International (FOEI), Canada Environment, and the Women's Environmental Network. However, because foreign mailing costs are high, free subscriptions are limited

Endangered Species Scientific Commission
19 Zhongguanoun Lu, Haidian
Beijing 100080 China

Contact: **Sung Wang,** Executive Vice Chairman
Phone: (86) 1 256 2717 • *Fax:* (86) 1 7013772

Activities: Law Enforcement; Research • *Issues:* Biodiversity/Species Preservation • *Meetings:* CITES; IUCN; SSC

● Organization
The **Endangered Species Scientific Commission** acts as the scientific authority for CITES, the Convention on International Trade in Endangered Species of Wild Fauna and Flora.

▲ Projects and People
In addition to his role as this Commission's executive vice chairman, **Sung Wang** is the main compiler and organizer of the *Chinese Red Data Book* for endangered species. As research professor with the Institute of Zoology, Chinese Academy of Sciences, Sung Wang has as a major project **Mammalian Taxonomy and Compilation of Fauna Sinica.** Regarding biodiversity, conservation, and wildlife sustainable use, Wang is organizing a project on **Bioinventory, Status Survey, Database, and Action Planning for China;** he also serves as head of the Working Group for Biodiversity Conservation.

Wang is also vice president of both the Mammalogical Society of China and the Chinese Association for Zoological Gardens. For CITES, he is the Asian representative of the Animals Committee and a member of the Nomenclature Committee. For the Species Survival Commission (IUCN/SSC) he is steering committee member, regional vice chairman, and coordinator for northeast Asia's Rodent Specialist Group.

■ Resources
Technologies needed are computer programs on biodiversity conservation or bioinventory. "Video tapes concerning biodiversity conservation would be very helpful," writes Wang.
Available are the *Proceedings of the Symposium on Biological Diversity (1990),* sponsored by the Bureau of Biosciences and Biotechnology, Chinese Academy of Sciences.

Bureau of the "Three North" Protection Forest System
Ministry of Forestry
20, Nanxunxijie
Yinchuan 750001, Ningxia, China

Contact: **Li Jionshu,** Director
Phone: (86) 951 41535 • *Fax:* (86) 951 41684

Activities: Air Quality/Emission Control; Energy; Global Warming; Population Planning; Waste Management/Recycling • *Meetings:* Desert Control Meetings; World Forestry Conference

▲ Projects and People

The "Three North" Protection Forest System, popularly known as The Green Great Wall, got underway in 1978 when China's Central Committee of the Communist Party and the State Council took action to protect from sandstorms and soil erosion an immense region of some 3,470,000 square kilometers that covers 396 counties in 12 provinces in the country's northern regions. Turning "desert into oasis," the Sanbei Shelter is a step-by step program to preserve existing forests and grassland vegetation; to create "shelterbelts" that act as wind breaks and sand fixation to protect farmland and pasture; to encourage soil and water conservation; and to balance forestry, agriculture, and animal husbandry both ecologically and economically.

In these regions are 23 nationalities "that have created the splendid oriental culture and written the brilliant history . . . Han, Hui, Mongolian, Uighur, Kazakh, Khalkhas, Uzbek, Tajik, Tartar, Sibo, Russian, Tu, Sala, Dongxiang, Paoan, Yugu, Tibetan, Korean, Manchu, Daur, Olunchun, Hezhe, and Owenk," indicates the Bureau. Successive dynasties and reigns, lack of rainfall, overgrazing, large-area cultivation, rapid population growth, and wars have taken their toll in destroying the "green natural defense" and opening the region to "serious sandstorms and soil erosion."

Soil became impoverished, and more than a billion tons of silt were carried into the Bohai Sea, carrying along great quantities of nitrogen, phosphate, and potassium; and also harming the Yellow River where the valley was once covered with "dense forests and luxuriant grassland." The "ruthless felling of trees" led to "barren hills and numerous gullies and ravines." Apart from the Gobi, "there are 12 vast stretches of desert and desertificated land within the 'Three North' region; "the drifting sand moved with wind [pollutes] the environment while damaging vast stretches of farmland and grassland," describes the Bureau.

Li Jionshu, Bureau director and forestry senior engineer, writes, "As the great nature has meted out such severe punishment to human beings for the damage of forests, people became painfully aware that in order to live in harmony with nature, it is essential to give before one takes, that man must refrain from taking insatiably from nature while refusing to give anything back to it, and that it is absolutely undesirable to deprive the good earth ruthlessly of its green attirement which nurtures mankind."

During the first eight-year phase, funds were raised mainly from farmers with some government aid, units were organized, and a policy implemented of "those who planted trees own the trees." Using 25,000 specialists, the Bureau assesses that "seven billion man-days were spent by the masses of people and local army units . . . for accomplishing more than 6,060,000 hectares of plantations"— mitigating serious windstorm damage, protecting 8 million hectares of farmland, beginning to bring soil erosion under control, and improving timber, fuel, fertilizer, and fodder supplies. The Bureau says it is on its way to reversing the predicament created when "sand moves forward while people move backward" with this intensive, interdependent system-engineering project.

Now in its second phase (1986–1996), goals are to expand the Green Great Wall to 512 counties; raise the forest cover rate from 5.9 percent to 7.7 percent; and continue large-scale afforestation through air seeding by plane, covering hillsides and deserts with forests and grass, and pinpointing certain areas—such as the Mu Us and Holqin deserts—for sand-fixation shelterbelts. Yet new problems also must be addressed, such as protecting these forests from plant diseases and insect pests. In addition, "fire extinguishing equipment is inadequate and forest machinery is outdated. The forest resources are not fully used."

Lack of investments and a forest pricing system in a commodity economy are also hampering development. Although mineral deposits are abundant, a lower underground water table is resulting from decreased melting of glaciers. According to the Bureau, ". . . it is necessary to speed up the construction of the shelterbelts so as to conserve water and soil and protect the waterheads of rivers. The reckless development of mining and processing industries which would consume huge amounts of water would only cause serious pollution and aggravate the ecological and environmental problems. The consequences would be dreadful to contemplate."

Moreover, the Sanbei project—affecting half the country—is critical for food production, as it is already demonstrated that the grain shortage is eased in areas where newly planted trees and grass are growing.

Expected to take about 73 years to complete, the project presently aims to help bring local farmers out of poverty in both the near- and long-term by gaining from resources including solar energy and fast-growing timber, such as jujube trees. Arbors yield apples, apricots, peaches, pears, persimmons, grapes, dates, and nuts. Seabuckthorn is a base for medicine and cosmetics. Paper, synthetic boards, feed, and fertilizer are made from shrubs. Forage grasses are used in mixed feed. Mushrooms and medicinal herbs are forest byproducts. Wicker weavings find worldwide markets.

Government officials acknowledge that the "Three North" Protection Forest System is the "great but arduous task that takes unremitting and indomitable efforts of several generations to accomplish" and encourage people of varying nationalities to "unite as one" to improve the region's ecological environment.

On World Environment Day in 1987, the United Nations Environment Programme (UNEP) honored this Bureau with a medal and citation as an "advanced unit in the global drive for environmental protection." Various research institutes in the region draw foreign scientists and tourists.

■ Resources

The Rising Green Great Wall is a full-color "photo album" printed in English that beautifully illustrates afforestation on sandy land and the abundant cultures occupying these regions. The title of a smaller brochure explains that *Work on the Project of "Three North" Protection Forest System Is Underway*. Expressed needs include training in forestry resources management as well as afforestry and sand-fixation equipment.

National Environmental Protection Agency (NEPA)
No. 115 Xizhimennei
Nanxiaojie
Beijing 100035 China

Contact: **Qu Geping**, Administrator
Phone: (86) 1 655635

Activities: Political/Legislative • *Issues:* Air Quality/Emission Control; Biodiversity/Species Preservation; Global Warming; Sustainable Development; Waste Management/Recycling; Water Quality • *Meetings:* UNCED

● Organization

"Environmental protection in China has a short history," admits literature from the **National Environmental Protection Agency (NEPA)**. Without formal protection organizations until the early 1970s, the movement's turning point came when a Chinese delegation participated in a United Nations Conference on the Human Environment in 1972. "After this, the government began to put [the] issue on its agenda." With the initial establishment of the Executive Group of Environmental Protection in 1974, regional and local offices followed suit throughout the country. "The state's chief industrial sectors also set up similar organizations," reports the literature.

NEPA became a ministry directly under the State Council in 1988—emerging from the creation of the Ministry of Urban and Rural Construction and Environmental Protection in 1982, and the Environmental Protection Commission in 1984. As a result, NEPA conducts "independent work of management and supervision." According to its literature, "the government has set up environmental protection agencies in 28 provinces, regions and municipalities, and in all big and medium-sized cities. Environmental organizations have been taken into governments at all levels. There are over 60,000 people directly employed in environmental organizations, while the total number of staff involved in environmental protection in China amounts to almost 200,000."

▲ Projects and People

Among the nine departments are **Pollution Management** for water, air solid waste, oceans, and noise, among others; **Nature Protection** drawing up national programs for conservation areas and organizing ecological research; **Development and Supervision** for regulations and technical and economic policies that environmentally manage new construction projects, pollutant discharge and discharge fees, radiation control, and environment monitoring; **Policy and Legislation** for comprehensive, environmental protection strategies; and **Foreign Affairs**, for global cooperation and exchange of environmental protection work.

Qu Geping writes, "China should actively participate in the effort to solve the international environmental issues while paying much attention to the domestic environmental problems. As a member of the international community, China supports the just stand and reasonable demand of the developing countries on one hand, strengthens the relations with the developed countries and develops bilateral or multilateral cooperation on the other hand."

He continues, "The challenge of international environmental issues is both the pressure and the opportunity. It is no doubt that we are under great pressure . . . to solve global environmental issues. To reduce the use of CFCs, the emission of CO_2 and control acid rain, for instance, is really a hard job for us and we have to pay a high price. The appropriate attitude is to face the challenge and turn pressure to momentum by formulating and carrying out policies which encourage and ensure the stable, sustainable, and coordinative development."

Included in the **institutions under NEPA** are the *China Environment News*, Chinese Research Academy of Environmental Science, Nanjing Institute of Environmental Science, and the Xingcheng Environmental Management Research Center. New institutions to be set up are China Environmental Strategy Research Center, China Environmental Information Center, Environmental Publicity and Education Center of China, Haiyang Environmental Supervision Training Center, and the Service Center for NEPA Offices.

Cooperative institutions are the Chinese Society of Environmental Sciences, Chinese Association of Environmental Protection Industry, and the Environmental Section of the China Climate Influence Commission.

Qu Geping has three main posts: NEPA administrator, vice chairman of the Environmental Protection Commission under the State Council, and president of the Chinese Society of Environmental Sciences. As China's first representative to the United Nations Environment Programme (UNEP) in Nairobi, Kenya, 1976, Qu Geping received a UNEP medal in 1987; a year later, he was awarded an honorary doctorate in engineering from Bradford University, UK. Now also a part-time professor with Beijing, Qinghua, Tongji, and Wuhan universities and a guest professor at Oxford University, UK, he has written publications on global environmental issues, China's environmental protection in the year 2000, a population study, and an *Analysis of Environmental Investment in China*.

NEPA deputy administrators include **Professor Zhang Kunmin**, specialist in solid and nuclear wastes and radioactive wastewater; **Wang Yangzu**, senior engineer; **Dr. Jin Jianmin**, in charge of publicity, education, science and technology standards, and nature protection; and **Xie Zhenhua**, radioactive waste expert.

■ Resources

Environmental Management in China is available in U.S. dollars.

Xian Environmental Office
1 East Village, Daxie East Road
Xian 701168, Shaanxi, China

Contact: **Shen Rui-shi**, Engineer
Phone: (86) 29 55660, ext. 262

Activities: Development; Education • *Issues:* Water Beauty

▲ Projects and People

Shen Rui-shi, who recently headed the special project **A New Method of Beautifying Water Body and Its Surface with Organism at Parks**, briefly describes his invention:

"To beautify the environment of water's surface at parks, villas, and sightseeing places, generally it is only in the shallow water that some water plants such as lotus, water lily are planted. But in the deep water, few plants exist. Especially when the water is not suited for some floating water plants, it is often not possible to make use of these plants to beautify the environment of the water.

"Now, however, a new technology and a new method have been developed by our pioneering research work. One is to plant water plants in 'the borrowed water,' the other is to graft dryland plants in water, which escape from the limits to the depth and quality of the water, whether the water is deep or shallow, polluted or not; both of them are available for the surface of the water. And they have no pollution to the original water and are suited for the decorative fish pool as well. A fine picture with water, fish, flowers, and plants would come to our eyes when this technology has come into use.

"Dryland plants have a wide variety, which would make it full of more dazzling beauty for the environment of water's surface at parks, villas, and sightseeing places if they are further beautifully shaped."

Shen, who is primarily engaged in ecology agriculture, applied for a patent at the Patent Office of the People's Republic of China, which, he writes, "has been established after first investigation." Formerly with the Chinese Academy of Sciences, Shen says he is "ready to attend international meetings with this project in order to introduce and collaborate on global environmental activities. Besides, I expect it to be adopted by companies, enterprises, and individuals from other countries."

NGO

Study Group on Regional Hydrological Responses to Global Warming
International Geographical Union (IGU)
Building 917, Datun Road
P.O. Box 771
Beijing 100101 China

Contact: **Changming Liu,** Chairman
Phone: (86) 1 4231539 • *Fax:* (86) 1 4231551

Activities: Research • *Issues:* Global Warming; Sustainable Development

● Organization
Started in 1990, the **Study Group on Regional Hydrological Responses to Global Warming**, or climate change, is providing a foundation on which a "new commission on hydrology can be established" following the 1992 International Geographical Union Congress held in Washington, DC. Objectives of the Study Group are "to foster understanding of the natural processes and human activities that influence the geographical distribution of water; and to study the interface between the hydrosphere, atmosphere, geosphere, and biosphere, including the hydrological responses to global change."

▲ Projects and People
In the near-term, the scientists are examining regional responses of hydrological processes to global warming to help understand its impact on water use, farming, and other human activities, according to IGU.

To determine water resources response to climatic change, the following aspects are being examined: "global precipitation distribution and regional precipitation change; average annual runoff distribution and its maximum and minimum values; water quality change . . . as a critical indicator for regional water resources; sea level rise [and its] magnitude and trend; glacio-eustatism and lake water storage; soil moisture change and its influence on agriculture; groundwater balance change; responses of environment and socio-economic systems to change of the global/regional water cycle; and interface processes among hydrological states."

According to Study Group member Prof. **Isamu Kayane,** Institute of Geosciences, University of Tsukuba, Japan, the activities of hydrologists in IGU need to be strengthened "since water is the most important substance supporting the human and natural environment."

Changming Liu and colleague **Guobin Fu** recommend counter-measures to greenhouse effects, such as ". . . exploring substitute energy sources to reduce the use of fossil fuel, protecting forests, . . . taking proper policy and legal actions, determining such impacts on global scale systematically, and developing some international agreements on this problem."

One research study of concern illustrates "glacier recession and lake shrinkage indicating a climatic warming and drying trend in central Asia," according to **Shi Yafeng and Ren Jiawen,** Kanzhou Institute of Glaciology and Geocryology, Chinese Academy of Sciences. "Since the maximum of the Little Ice Age which occurred in the 18th century, glaciers have decreased in area by about 44 percent in the Urumqi valley," they report.

In addition to Chairman Liu and Prof. Kayane, full members of the Study Group are: **Dr. Mike Bonell,** James Cook University, Australia; **Professor Low Kwai Sim,** University of Malaya, Malaysia; **Dr. Peter Wayien,** University of Florida, USA; **Dr. J.A.A. Jones,** University of Wales, UK; **Dr. O. Martins,** University of Agriculture, Nigeria; and **Professor Ming-ko Woo,** McMaster University, Canada.

Professor Liu is chairman of the Department of Hydrology, Institute of Geography, Chinese Academy of Sciences, and recent project leader of **Regulation Modeling of Water Resources.** He has studied at the University of Arizona and participated in international symposiums on water and the environment in France, Australia, and Germany. A recent book is on *Agricultural Hydrology and Water Resources in the North China Plain;* many of his articles are published worldwide in English, Russian, and Chinese.

■ Resources
Evidence identification is needed for research.

🎓 *Universities and Institutes*

Branch College of Forestry
Guangxi College of Agriculture
Nanning 530001, Guangxi, China

Contact: Li Xinxian
Phone: (86) 771 26713

Activities: Development; Education; Research • *Issues:* Biodiversity/Species Preservation

▲ Projects and People
Li Xinxian is investigator of a forest ecosystem research project in Guangxi. A forest ecology laboratory is a chief facility here. Li is a member of the Species Survival Commission (IUCN/SSC) China Plant Group.

■ Resources
The **Branch College of Forestry** needs technological materials pertaining to geology and more information on forest nutrition elements and geological background. Available publications relate the forest to nutrition element cycling, production efficiency, community population dynamics, water balance, climate, and soil drainage.

Laboratory of Systematic and Evolutionary
    Botany and Herbarium
Institute of Botany
Chinese Academy of Sciences
Xiangshan
Beijing 100093 China

Contact: **Chen Sing-Chi**, Director/Professor
Phone: (86) 1 2591431, (86) 1 2592116

Activities: Research • *Issues:* Biodiversity/Species Preservation

▲ Projects and People

In his report on the **status of the conservation of rare and endangered plants in China,** Chen Sing-Chi quotes American horticulturist E.H. Wilson in his 1929 book *China, Mother of Gardens*: "It is safe to say that there is no garden in this country or in Europe that is without its Chinese representatives and these rank among the finest tree, shrub, and vine." Chen concurs that China with its "wide variation in vegetation and great riches in plant resources" offers as many as 24,357 species of flowering plants belonging to 2,949 genera, 193 species of gymnosperms belonging to 34 genera, 2,600 species of ferns belonging to 204 genera, 2,100 species of mosses belonging to 459 genera, 2,000 species of lichens belonging to 200 genera, 2,500 species of fungi belonging to 600 genera, and no less than 2,500 species of algae known from this country"—as he indicated in *Cathaya* (1989).

However, the destruction of forests due to "rapid population growth and lack of environmental education and regulations" has led to the threatened or endangered condition of about 10 percent of these; some 38 percent of seed plants are also so labeled. This means that about 2,700 species of vascular plants need protection. Examples of two rare trees are the *Ginkgo biloba* and *Abies beshanzuensis*. Listings of such species are in the first *China Red Data Book*, published about 1989; the second volume is expected in about 1994.

Chen notes too that people, even researchers, pay more attention to plants of economic and scientific importance, but that "the main purpose of plant conservation is to preserve the diversity of life forms, regardless of their importance for man." In fact, writes Chen, "the most seriously threatened wild plants are those that can be sold because of their horticultural value." Certain orchids and *Cymbidium* are near extinction due to exploitation and smuggling by foreigners, he points out.

Meanwhile, tropical forests are being destroyed by "rampant wood cutting, slash-burn cultivation and forest fire" as well as "extensive clearing for rubber plantations."

Public education and the creation of nature reserves are making strides in conservation, with the establishment of at least 392 in 30 provinces since 1956, mainly since 1978—covering 24,000,000 hectares or about 2.5 percent of the total area of China. These include six biosphere reserves. Along with reserves for forests that are deciduous, tropical, subtropical, alpine, evergreen, primeval; steppe ecosystems; alpine meadows; and mangroves, there are sanctuaries for golden monkeys; wild horses; and migratory birds, such as hawks and falcons.

Introducing threatened plants into botanical gardens is particularly wise, believes Chen, who points to bamboos, magnolias, camellias, and rhododendrons, for example, in South China, Sun Yat-sen, and Kunming Botanical Gardens and the Plant Conservation Center of Western China.

Chen also recommends a "more comprehensive network" of protected areas with effective management, training of administrators, and coordinated research; control of wildlife and plant trade; expanding on the *China Red Data Book* for further study of taxa; developing economy in backward areas; and assisting with population control.

As director of the **Laboratory of Systematic and Evolutionary Botany and Herbarium,** Chen is devoted to establishing the Orchid Conservation Center of China, Wuxi, Jiangsu Province. He provides training classes for technicians and officers of Chinese nature reserves on orchids and how to protect them.

■ Resources

Financial support of approximately $US200,000 is being sought for the Orchid Conservation Center.

River Dolphin Research Department
Institute of Hydrobiology
Chinese Academy of Sciences
Wuhan 430072, Hubei, China

Contact: **Chen Peixun**, Director/Professor
Phone: (86) 27 813481

Activities: Research • *Issues:* Biodiversity/Species Preservation

▲ Projects and People

Because the **conservation of river dolphin** (*Lipotes vexillifer*), also known as baiji, is of "great concern" to the Chinese government, it is under "first-grade protection—the same category accorded the great panda," according to **Chen Peixun**, director, **River Dolphin Research Department, Institute of Hydrobiology.**

Found only in the middle and lower parts of the Yangtze (or Changjiang) River, surveys assess the population at less than 300. "The main factors causing the dangerous decline . . . are decreases in fish resources and the despoiling of appropriate biotopes for Lipotes due to the harnessing of the river," reports Professor Chen. "Also, increased navigation . . . has more than doubled the numbers of ships since 1949, and because dolphins must depend on echolocation, the engines of ships interfere with high-frequency echo signals and may bring about disorientation" as well as head injuries from collisions. However, the greatest danger appears to come from accidental fishing and the "system of rolling hook fishing in particular." Nearly 50 percent of Lipotes killed between 1950 and 1983 were caught on rolling-hook lines.

Scientists such as Professor Chen want to be able to study the behavior and ecology of Lipotes more closely "to analyze relationships between biotic and environmental factors affecting dolphins and to establish what harmful elements threaten the survival of dolphins." With the 1980 catch of an injured male dolphin which was carefully treated and of a healthy female in 1986, researchers here now have such an opportunity. Affectionately named "Qi Qi" and "Zhen Zhen," these dolphins are the basis of studies in rearing, behavior, biometrics, hematology, acoustics, and hereditary histology.

According to Chen, who has written extensively on Chinese river dolphin, the following steps are necessary to save the species:

- "Intensifying studies on artificial rearing and reproductive biology so as to ensure reproduction. . . . It is necessary to construct a new dolphin pool system including two indoor pools and a special breeding pool with a good water filter system. A series of equipment for researching on behavior, acoustics, and physiology must be prepared."
- "Establishing the semi-natural reserve containing optimal natural habitat for Lipotes—It is necessary to move some individuals or group into a pace under artificial control for careful management so as to protect them and enable them to reproduce under such conditions." The Research Lab recommends a natural ox-bow in a secluded, unpolluted area that is on the main course of the Yangtze River.
- Setting up a conservation area for Lipotes. Professor Chen suggests that a certain section of the Yangtze where dolphins are presently both high in density and mortality could become a protected habitat.

Chen is a member of the Species Survival Commission (IUCN/SSC) Cetacean Group.

■ Resources

Because the research center has fully utilized its resources, Professor Chen asks for "additional support from domestic and foreign institutions and organizations in order to go forward. By helping to finance any portion of this work, you will be joining in the common effort to restore Lipotes to its home waters where it may thrive again," writes Chen. "Such an achievement will stand as a symbol of friendship out of an awareness and appreciation of the one world we inhabit."

Kunming Institute of Botany
Chinese Academy of Sciences
Kunming 650204, Yunnan, China

Contact: **Wu Su-gong**, Associate Professor
Phone: (86) 871 50201, (86) 871 50304

Activities: Research • *Issues:* Biodiversity/Species Preservation

▲ Projects and People

At the **Kunming Institute of Botany**, Prof. **Wu Su-gong** is the captain of the Comprehensive Scientific Expedition to Hohxili region in the Qing-hai-Xizang Plateau. He is also vice-chairman of the Pteridophyte Society of China and a member of the Species Survival Commission (IUCN/SSC) Pteridophyte Group. Pteridophytes are ferns and related plants such as horsetails and club mosses.

Many of Prof. Wu's publications are printed in English or appear in Chinese with English abstracts. Subjects include two new fern species in the limestone area of Yunnan and the "urgent importance" of conserving plants in this region, ferns of The Himalayas, new plants from Tibet, "phytogeographical affinities" between ferns of Japan and China, and mountain flora.

Kunming Institute of Zoology
Chinese Academy of Sciences
Kunming 650107, Yunnan, China

Contact: **Weizhi Ji**, Assistant Director/Professor
          Qi-Kun Zhao, Associate Professor
Phone: (86) 871 82661 • *Fax:* (86) 871 82416

Activities: Research • *Issues:* Biodiversity/Species Preservation
Meetings: International Primatological Society; IUCN

● Organization

Begun in 1959, the **Kunming Institute of Zoology** is known for its "endemic and comprehensive animal research in the tropics and subtropics"—with emphasis on conservation and use of animal resources. It also desires to contribute to the prosperity of southwest China. Activities in the five departments include the following:

Vertebrates—studies of mammals, birds, amphibians, and fishes that concentrate on taxonomy and "faunal succession of vertebrates in Yunnan" and nearby.

Cytogenetics—research on the collection and storage of animal genetic resources and on gene transfer, with the establishment of a cell repository of wild animals from more than 140 endemic and rare species. Evolutional cell biology animal chromosomes are examined, as are new technologies of animal breeding.

Entomology—taxonomy of insects, insect resource use, and pest control.

Animal Biochemistry—such as on venoms and other active substances.

Primate Biology—breeding, disease control, and management of caged primates; and research in ecology, reproduction, behavior, anatomy, and related areas. On a special farm for laboratory animals are rhesus and black snub-nosed monkeys, and black muntjac deer, among others.

Overall, there are over 90,000 vertebrate specimens and some 350,000 insect specimens. An educational museum serves the public. A library houses more than 55,000 books. The Institute is proud of the nearly 200 scientific prizes awarded to researchers here.

▲ Projects and People

Kunming literature mentions some specific areas of concentrated investigations: amphibian and reptile research, animal chromosomes, cultivation of "traditional medicine Chinese insect herb," innovative snakebite treatment, successful reproduction of rare economical birds (*Chrysulophurs amherstiae*), reproduction of albino macaques, and captive breeding of *Rhinopithecus bieti*, among other projects.

Qi-Kun Zhao, founder and head of the Institute's Behavior Research Group, reports on his examination of ethoecological problems resulting from visitors' food handouts to Tibetan macaques (*Macaca thibetana*) at Mt. Emei, a tourist and Buddhist center in China which, he says, "has resulted in disaster for visitors and the monkey." He describes this region with its variable terrain as "one of the best protected in the area." Unique to China, the commensal Tibetan macaque has a large body size, probably related to foliage eating, which can be a "terror for the tourists who first face the wildlife." The monkeys also have a long history of feedings by monks at Buddhist temples, located in China for more than 1,000 years, and are much "respected and tolerated in India."

As popular tourism increased in connection with recent socioeconomic reforms, visitors fed the monkeys for pleasure, to ensure safety, or as an offering to Buddha, reports Zhao. This practice led to begging, robbing, and other aggressive behavior by the monkeys. Consequently, both visitors and macaques have been harmed, disabled, or even killed in tragic accidents on the mountain cliffs.

Following a thorough study, Zhao recommends conservation actions for both humans and monkeys to stem or reverse the "vicious circle" involving the monkeys; tourists as food sources; and natives, "most of whom have not enough land for farming [who] took advantage of interactions between tourists and monkeys for their business." Observations did show that monkeys mostly liked getting food from submissive tourists and brought "trouble to small groups of tourists, females, persons with red things or a camera, persons who did not feed the monkey, or who did not hold in awe and veneration to the monkey." But according to Zhao, such is not the whole story, with food vendors and others fostering some of the folklore. Zhao raises another problem: "I think it is time for us to face the conflict between the urgent need of conservation of biodiversity and publicizing such animal medicine" as that appearing in scientific journals regarding use of organs or tissues of legally protected animals for medicinal treatment.

Presently, Zhao urges more job opportunities for local people, and the posting of signs such as "Do Not Feed the Monkeys" or "Keep the 'Wild' in Wildlife!" He recommends new regulations to keep macaques from being disturbed and education programs, including on television, to guide tourists on how to observe monkeys and approach them with dominance without provoking them. He particularly argues to withhold food reinforcement so the monkeys will return to the wild. "Let us all cooperate," he writes, "to recover the harmony between ourselves and nature."

The National Geographic Society's Committee for Research and Exploration, Wenner-Gren Foundation for Anthropological Research, and the Science Foundation of China have supported Qi-Kun's work.

Weizhi Ji has studied primate reproduction—specifically neurosystem regulating sex hormone secretion—at Oregon Regional Primate Research Center, USA; he was awarded a Smithsonian Institution fellowship to study embryo transfer at the National Zoological Park, Washington, DC; received a Rockefeller Foundation grant to study *in vitro* fertilization and embryo transfer on the rhesus monkey; and participated in global meetings of the International Primatological Society, Species Survival Commission (SSC), and IUCN. An organizer of the International Symposium on Primate Conservation in China (ISPC), Professor Weizhi is deputy chairman of the Chinese Laboratory Animal Society's Primate Council and has co-authored publications on rhesus monkey, tree shrew, gibbon, and Tibetan macaque.

Both Ji and Zhao are members of the Species Survival Commission (IUCN/SSC) Primate Group; Weizhi also belongs to the Captive Breeding Group.

■ Resources

A descriptive brochure is available. Technology support for *in vitro* fertilization is needed to save endangered primates.

Qi-Kun Zhao writes, "For the purpose to recover and keep the harmony between humans and wildlife at . . . tourist places, I need political support to make feeding animals at national park-like sites unlawful in China; I also need materials to increase public awareness of ethoecology in tourist areas or in the country."

Cetacean Research Laboratory
Nanjing Normal University
Department of Biology
122 Ninghai R
Nanjing 210024, Jiangsu, China

Contact: **Zhou Kaiya,** Director/Professor
  Phone: (86) 25 631636, ext. 328
    Fax: (86) 25 307448

Activities: Research • *Issues:* Biodiversity/Species Preservation
Meetings: SSC

▲ Projects and People

Conducting a **World Wildlife Fund** project, **Zhou Kaiya** has entered his second phase of **conservation of the baiji in the Yangtze River,** which is ongoing until 1995. The baiji is a fresh river dolphin "close to extinction," writes Zhou, who is monitoring and studying its social structure and individual movement as well as promoting public awareness and enforcing existing legal protection.

Earlier, surveys were made on board small fishing boats on the Yangtze River as the animals were observed and photographed when sighted. "Seven baiji were recognized by the photographs taken," reports Zhou. "The movement of the animal was recorded for the first time by photographic identification technique." According to Zhou, the "population estimate in 1989 and 1990 showed that about half of the animals were lost in the past decade." As part of a public education campaign, 5,000 brochures and posters were printed and distributed to the fishermen and high schools along the river. A book, *Baiji, the Yangtze River Dolphin and Other Endangered Animals of China,* was published both in China and abroad.

A two-year Toyota Foundation Project studying the ecology and environmental chemistry of **habitat pollution of aquatic mammals** in China was underway through 1992. Zhou is responsible for studying "tissue distribution of heavy metals and organochlorines" in the finless porpoise (*Neophocaena phocaenoides*) to compare the impact of water pollution on the finless porpoise in the Yangtze River and Japanese coastal waters.

A third project is with the China Natural Science Foundation to investigate **morphological differences and genetic variation** in different populations of *Neophocaena.*

Director of the **Cetacean Research Laboratory,** Professor Zhou belongs to the Species Survival Commission Cetacean and Captive Breeding Groups and is active in the Mammalogical Society of China, China Wildlife Conservation Association, China Zoological Society, Herpetological Society of China, Jiangsu Provinces's Wildlife Conservation Association, Zoological Society, Science and Technology Association, and Science and Technology Commission. Zhou is deputy director of the China National Committee for Unifying Zoology Terms and chief editor for Middle School Biology.

■ Resources

Promotional material for the "Save the Baiji" campaign is printed in English and Chinese, as is the above-mentioned *Baiji* book co-authored with Zhang Xingduan. Also in English are certain reports such as *Marine Mammal Studies* and various scientific papers relating to biology and conservation of the river dolphin. The Laboratory needs transponders for individual identification of river dolphin and porpoise.

Botanical Institute
Northeast Forestry University
Harbin 150040, Heilongjiang, China

Contact: **Yiliang Chou**, Professor
   Phone: (86) 451 223443-782 or 622

Activities: Education; Research • *Issues:* Biodiversity/Species
Preservation; Deforestation; Energy; Population Planning;
Water Quality • *Meetings:* CITES; International Botanical
Congress

▲ Projects and People

Yiliang Chou is involved with plant ecology research at the **Botanical Institute, Northeast Forestry University**, and is a member of the
Species Survival Commission (IUCN/SSC) China Plant Group.

■ Resources

The following publications are available and priced in Yuan: *Forests
of China, Vegetation of Da Hinggan Ling in China, Ligneous Flora of
Heilongjiang,* and the quarterly *Bulletin of Botanical Research.*

Pedology and Desert Research ⊕
Xinjiang Institute of Biology
Chinese Academy of Sciences
Urumqi 830011, Xinjiang, China

Contact: **Gao Xingyi**, Head of Research Group/
         Professor
   Phone: (86) 991 337358 • *Fax:* (86) 991 335459

Activities: Education; Research • *Issues:* Biodiversity/Species
Preservation • *Meetings:* International Snow Leopard Symposium; International Symposium of Deer

● Organization

Established in 1961, the **Xinjiang Institute of Biology, Pedology
and Desert Research**, pursues local studies of soil, natural resources,
and other living things—as representative of Chinese arid regions.
Researchers among the staff of 281 observe "important problems in
ecology and environment" and determine how to use and manage
the area's resources. The Institute is organized into 10 research
departments concerned with plant resources, physiology, ecology,
zoology, microbiology, land resources, soil improvement, desert
research, remote sensing, and biotechnology.

Here, more than "70,000 plant and animal specimens and soil,
sand samples" are housed. Experiments include an artificial haloxylon
shelterforest at Muosuowan Desert Research Station, artificial grassland at Bainbuluke Grassland Research Station, planting a "windbreak forest" at Celer Desert Research Station, a building at Turpan
Desert Research Station, and explorations at Fukang Desert Ecology Research Station. As comprehensive desert and mountain studies are underway, subjects include medicinal plants, wild camels,
drift sand, nitrogen fixation, fritillaria plantings, alkali–saline soil
improvements in the Yanji Basin, and tissue culture of seedless
white grapes.

Symposiums are held for scholars from Japan, Soviet Republics,
Australia, and other countries.

Professor Gao Xingyi says that his organization "can be in charge
of international cooperation research" regarding the "status, protection, and utilization of the natural resources of arid regions." He
would like to see launched "science and personnel exchange," cooperative research, and more scientific observations on location. He
has already worked on cooperative ventures in rare wildlife studies
between China and the USA.

Throughout his career, Gao has raised and bred new varieties of
"valuable wildlife," studied birds and mammals, and advanced the
preservation of endangered species. Honored for his investigations
of China's wild horse and wild double hump camel, Gao is presently
working to save the bustard and the goitered gazelle. He has won
prizes for writing about subjects such as Tibetan fox, tiger stoat, and
musked polecat; he has also published on the brown bear of Xinjiang,
"big red fish," "animals for medicines," Swan Lake in Tianshan
Mountain, and the European beaver, among other topics.

■ Resources

Needed are "new research measures, cooperative research and visitation, and enlarged contacts with colleagues in other countries,"
emphasizes Gao.

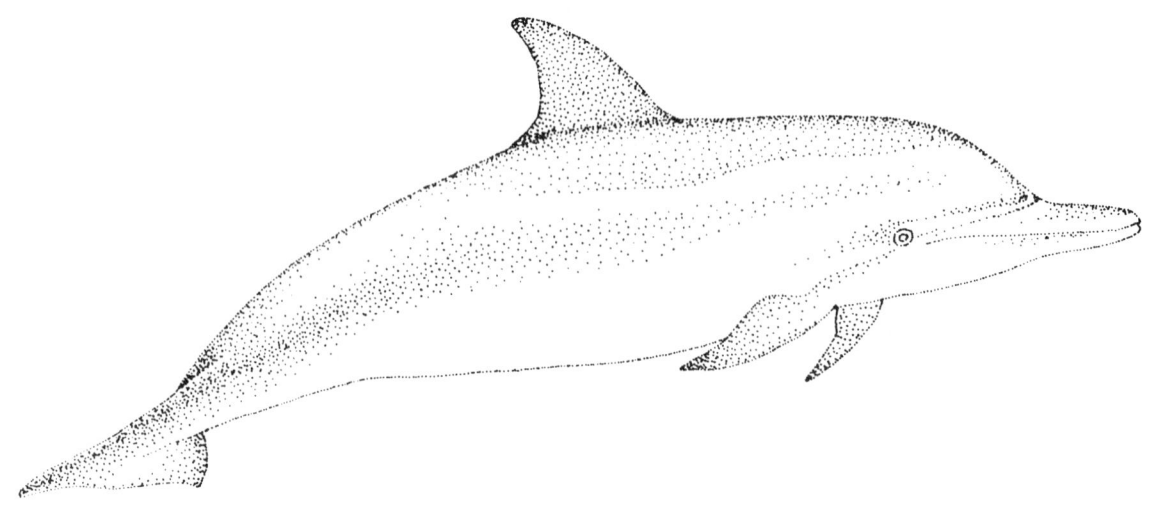

Colombia

Colombia borders Venezuela, Brazil, Peru, Ecuador, Panama, the Pacific Ocean, and the Caribbean Sea. The climate varies from the hot tropical rainforests on the coast and eastern plains to temperate conditions on the rainy plateau. The country is the fourth most populous in South America, with about 70 percent of the people living in cities.

Soil erosion has been caused by clearing for cattle, coffee production, and mining, exacerbated by heavy rainfall. Forest protection has been hurt by exploitation of timber resources and the presence of cocaine traffickers. The Choco region, with one of the world's largest varieties of endemic bird and plant species, is seriously endangered as a result.

♣ *NGO*

Fundación Herencia Verde (FHV)/Green Heritage Foundation
Calle 4 Oeste No. 3A32 El Peñón 32802
Cali, Valle, Colombia

Contact: **Enrique Murgueitio,** President
Phone: (57) 23 813257, (57) 23 808484 • *Fax:* (57) 23 813257

Activities: Education • *Issues:* Biodiversity/Species Preservation; Deforestation; Environmental Education; Sustainable Development; Waste Management/Recycling; Water Quality • *Meetings:* IUCN; Red de Bosques Latinoamericana

● Organization

More than 100 members of the rapidly growing **Fundación Herencia Verde (FHV),** established in 1983, actively work to link conservation and human needs via a series of successful research, environmental education, sustainable development, and preservation programs. FHV's philosophy "strives to" recover the biodiversity which characterizes Colombia, "based on ethical, aesthetic, cultural, socioeconomic and scientific values." FHV receives support from international donors as well as private and public Colombian institutions and individual donations.

▲ Projects and People

A multidisciplinary staff of 26, including 15 professionals, 1 technician, 2 secretaries, 8 caretakers, and a permanent 4-member Advisory Committee make up FHV. While activities grow exponentially, the FHV focus is on two pilot areas of ecological richness in the country: **Alto Quindio** in the Central Andes and **Bajo Anchicaya** on the Pacific coast.

Since 1985, FHV has operated a **private reserve** (part of the buffer zone of the National Park of los Nevados) with facilities for overnight visitors in the municipality of Salento, department of Quindio. Here, 2,600 to 3,000 meters above sea level, research is conducted on the rich flora and fauna. Orchids and birds are inventoried and new species registered, such as the orchid *Telipogon acaimensis* and the bird *Saltatur cinctu.*

Pilot **sustainable agriculture and forestry** work continues with a dozen neighboring farms in the region to instill a conservation ethic. Biodigestors, feedlot cattle raising, crop rotation, tree barriers, and earthworm culture are promoted to minimize the negative impact of agriculture on water, soil, and forest resources. Native tree nurseries provide alternatives to the conventional eucalyptus and pine reforestation. Organic gardens, demonstrations of biodigestors, teach-ins at local schools and training of tourist guides are among FHV's expanding public programs.

Close to the port of Buenaventura, the Bajo Anchicaya region is considered to hold one of the world's highest biological diversities. FHV works with the community to plan a strategy for conservation and management to establish a **natural reserve including 40,000 hectares of tropical rainforest.** Local farmers also receive technical support concerning alternative uses of the forest that would cause minimal damage to the environment, such as production of fruits, plantain, sugarcane, the raising of minor species, and the commercialization and export of baby bananas.

FHV future plans include environmental planning of the municipality of Salento, Quindio, including a community water management system; environmental conservation and sustainable development of the Colombian Pacific, with the advent of appropriate technologies and new community organizations; and integrating conservation of the wax palm forest (*Ceroxylon quindiuense*) in the Central Andes of Tolima, on behalf of the national tree.

■ Resources

FHV publishes the 65-page *Bosque de Niebla, Palma de Cera, Arbol Nacional de Chia*, and a coloring book, *"ACAIME La Reserva del Color.*

FHV needs sustainable agriculture technologies for projects it develops and a strengthened community outreach program.

Fundación Jardín Botánico "Leandro Agreda" Botanical Garden Foundation
Carrera 13, No. 18-50
Sibundoy-Putumayo, Colombia

Contact: Pedro Juajibioy, Director

Activities: Education; Research • *Issues:* Biodiversity/Species Preservation

● Organization

Started in 1983, the **Fundación Jardín Botánico "Leandro Agreda"** (**Botanical Garden Foundation**) is concerned with the recovery and protection of the environment—especially the establishment and management of medicinal plants and the conservation of the scientific and cultural heritage of the indigenous Camentsá people, who are recognized for their knowledge of plants. According to the Foundation, the Camentsá village in the Sibundoy Valley is where "traditional medicine has reached its highest level of refinement."

Community members direct and administer the Foundation and work in cooperation with Camentsá traditional doctors. It is named after the tribal chief and community hero, Leandro Agreda, who defended the interests of the indigenous community before the Spanish crown in the eighteenth century. Foundation support is from grassroots, national, and international agencies, including the Foundation for Higher Education (FES), National Electric Company (FEN), World Wildlife Fund (WWF) International, University of Nariño, Corporación Autónoma (Autonomous Corporation) del Putumayo, **Dr. Richard Schultes** of Harvard University, and the Colombian network of botanical gardens.

▲ Projects and People

Traditional doctors from Camentsá and other regions of Colombia donate **medicinal, hallucinogenic, poisonous, and ritual and magic plants** to the Botanical Garden, where they are maintained on seven hectares. So far, the collection includes 400 plant species, of which 210 are being studied.

Research is conducted on the **preventive and curative health benefits of plants**, which are also studied and being classified for a future **herbarium. Workshops** instruct young people in growing and using medicinal plants and also bring together traditional doctors and Western physicians to exchange knowledge. In the commu-

nities of Cauca and Nariño, **courses** are offered on medicinal plant use.

"Traditional medicine or 'indigenous science' represents an alternative [use of resources] because, in the indigenous world, good health is the result of harmony between man and nature," states the Foundation.

■ Resources

The Foundation seeks funds for its research. An information bulletin is published in Spanish.

Fundación Natura/Nature Foundation
Carrera 12 No. 70-96, A.A. 55402
Bogota, Colombia

Contact: Juan Pablo Ruiz Soto, Executive Director
Phone: (57) 1 310 0097, (57) 1 310 0026
Fax: (57) 1 2104515

Activities: Development; Education; Research • *Issues:* Biodiversity/Species Preservation; Sustainable Development • *Meetings:* IUCN; UNCED

● Organization

Created in 1985, **Fundación Natura** aims to preserve the biological diversity of Colombia with emphasis on human well-being, although according to its *Annual Report*, "Conservation policies designed to provide mankind with a better future sometimes go against the interests of today's generations." With INDERENA, the governmental agency charged with the preservation and management of the country's natural resources, Fundación Natura helps avoid unnecessary duplication and dispersion of the country's human and financial resources. Through the Colombian National Parks System, INDERENA is in charge of 42 conservation units. In 1990, the Fundación developed the National Parks Information System at the request of INDERENA.

▲ Projects and People

Colombia's most densely populated areas are found in the high-Andean ecosystems which are severely threatened. Here, water sources for humans, animals, and industrial use, as well as countless species and natural resources upon which the population depends, are located. At the **Carpanta Biological Reserve**, inventories of birds, small mammals, plants, and soil are underway. Noteworthy is the discovery of the **carpintero real woodpecker** (*Campephilus pollens*), one of the world's largest, nesting in big mature trees, a "sign of the forest's good state of conservation." The **flame-winged parakeet** (*Pirrhura calliptera*) and the **golden-headed quetzal** are found in small numbers.

In the **mammal survey**, 19 species belonging to 10 families in 5 orders, with predominance of the Rodentia order, were identified, indicating "the largest number of small mammal species" in the high Andean region in northern South America. Vegetation survey samples are evaluated at Colombia's National Science Institute and the Missouri Botanical Garden, USA. **Rare forest species**—Colombian pine, cariseco (*Billia colombia*), and cedro rosado (*Cedrella montana*)—were identified. In a reproductive biology study, cariseco, encenillo (*Weimannia sp.*), cuacho (*Hyeronima huilensis*), and granizo (*Hedyosmum sp.*) showed promise in boosting the recovery of certain areas inside and near the Reserve. With cattle raising imposing stress on forests while providing a livelihood to the peasant landowners, a forestry development project will explore other income sources.

Local people are encouraged to attend workshops on local land use, geography, and mapping with field studies in topography, soil use, and plot boundaries. With the 1990 opening of the **Cusumbos trail**, visitors are familiarized with the Páramos and high Andean forest. With the Universidad Externado de Colombia's School of Social Work, students observed "the daily routine of mothers and children in relation to the area's natural resources" and the Reserve.

Within the 54,300-hectare **Utría National Natural Park**, created in late 1986, is the **Utría Inlet**, with its coral reefs, fish populations, nesting sites for sea turtles, and mangroves ecosystem plundered. To overcome damage particularly of the *Pocilloporidos* coral group, colonies are being established at different sites, marked, and observed. Ethnology research among the **Emberá indigenous communities** living in the Park examined their uses and knowledge of wildlife species for food, trade, and magical and religious practices, and inquired about the plants these animals eat. Community environmental education projects continue, and a school is being built for El Valle children.

On land, reductions of **tapir** and **wild hog** populations were dramatic; **spider monkey, red brocket deer, collared peccary,** and **giant anteater** have also declined. Surveys are identifying a potential wildlife sanctuary. A physiographic map and land ownership studies precede consolidation of Utría Park and the essential management plan in which indigenous people are participating through OREWA, the Waunana Emberá Regional Organization. When the Park's northern Pacific coast became the grave site of an estimated 600 **black turtles** (*Chelonia agassizzi*), the Fundación began work in 1991 on the **Colombian Pacific Sea Turtle Conservation and Management Program** to help guarantee the survival of the migratory species.

In the **Santo Domingo river basin**, several groups cooperated to begin **reforestation and erosion control** measures, purchase land to preserve a portion of the Andean forest and 25 water sources, hold teacher training workshops on conservation, and set up pilot sanitation systems. Elsewhere, the 12,000-hectare **Virolin** is being considered as a **wildlife and flora sanctuary within the parks system.**

In the Colombian Amazon region is the **Caparú Biological Reserve**, an Amazon sector accorded a high priority as an international area of conservation. (*See also Fundación Natura, Estación Biológica Caparú in this section.*) With Fundación support, The Nature Conservancy initiated the **Parks in Peril Program** in 1991 in three areas: **La Paya, Chingaza,** and **Utría.**

■ Resources

Fundación Natura offers its *Annual Report, Tropico* bulletin, and the *El Ganzo* newspaper; members receive a quarterly *Nature and Man in the Tropics* newsletter. It seeks a geographic information system (GIS) and management programs.

Estación Biológica Caparú/Biological Station
Fundación Natura/Nature Foundation ⊕
Carrera 12 No. 70-96, A.A. 55402
Bogota, Colombia

Contact: **Thomas R. Defler, Ph.D.,** Director
Phone: (57) 1 249 7590, (57) 1 310 0026
(57) 1 310 0097 • *Fax:* (57) 1 210-4515

Activities: Conservation; Education; Research • *Issues:* Biodiversity/Species Preservation; Deforestation; Tropical Ecosystems • *Meetings:* CITES; IUCN

● Organization

Dr. Thomas R. Defler founded the **Estación Biológica (Biological Station)** at Caparú in 1983 to study lowland tropical forest ecosystems and vertebrates in the northwest Amazon of Colombia and to offer Colombians opportunities to learn about the Amazon. The station is now part of the **Fundación Natura,** a non-government agency devoted to the conservation of natural resources in Colombia which cooperates with INDERENA (Instituto Nacional de los Recursos Naturales Renovables y del Ambiente), the government agency responsible for such protection.

▲ Projects and People

Reserva Caparú, with its "long oxbow lake," is one of the few sites in the world's lowland tropics for which integrated, long-term ecological data is being accrued. Dr. Defler conducts a study of the Colombian Amazon via satellite images and overflights. With a staff of five, he also investigates the ecology and behavior of the reserve's two large, vulnerable primates—the **black-headed uakari** (*Cacajao melanocephalus*) and the **woolly monkey** (*Lagothrix lagothricha*), seeking to develop empirically informed strategies for their preservation.

Called "caparú" by the local Indians, the woolly monkey inspired the reserve's name, and the animal is a vital seed disperser of many key plant species. The wild fruits upon which these animals depend is researched by Dr. Defler's wife, **Dr. Sara Bennett Defler,** who also studies the avian phenology. Conservation International, National Geographic Society, and Wildlife Conservation International support the Deflers' research and the work of their students.

In 1991, the Colombian government approved the Deflers' proposal establishing a 335,000-hectare national park on the lower Apaporis River. Lake Taraira is habitat for the endangered giant otter, freshwater manatee, black caiman, pink and gray river dolphins, and the world's largest freshwater fish. "Commercially valuable fish breed in the flooded forests at the lake's edge," report the Deflers. Biological treasures at Caparú also include six other kinds of monkeys, jaguar, ocelot, puma, tapir, white-lipped and collared peccaries, curassow, guan, eagle, gray-winged trumpeter, and deer.

■ Resources

The Estación Biológica Caparú seeks fundraising, especially in developed countries, for support of the biological station and national park.

International Board for Plant Genetic Resources (IBPGR) ⊕
Ciat, A.A. 6713
Cali, Colombia

Contact: **Katsuo Armando Okada, Ph.D.**,
            Coordinator for South America
Phone: (57) 23 675050, ext. 329
Fax: (57) 23 647243
E-Mail: cgi-157; cgi-301

Activities: Education • *Issues:* Biodiversity/Species Preservation

● Organization
The **International Board for Plant Genetic Resources (IBPGR)** for South America promotes the formulation of national programs of plant genetic resources. As coordinator, Dr. **Katsuo Armando Okada** notes, "We are deeply concerned about issues related to the conservation and utilization of biodiversity and of potential economic value in the South American continent." IBPGR is a center of the Consultative Group on International Agricultural Research. Operating in several South American countries, IBPGR encourages collaboration with the **National Agricultural Research Systems (NARS)** regional organizations and sister international centers in the development of crop genetic resources networks. It also promotes, sponsors, and in some cases provides funds for the following activities: collections, collection/characterization of germ plasm, meetings, conservation research, setting up germ plasm database, and training.

▲ Projects and People
IBPGR plays a fundamental role in the conservation and utilization of plant genetic resources. In attempting to develop a global system in which the NARS are the mainstay, IBPGR considers the safeguard of plant diversity vital for sustainable agriculture and the welfare of humanity as a whole. Due to the continuing genetic erosion affecting most gene pools related to crop plants, this issue is of "utmost importance for a sustainable agriculture," states Okada.

■ Resources
IBPGR has available articles, books, and manuals. Its needs include funds to support training activities in plant genetic resources and assorted pieces of technical equipment. This equipment will be used for *in vitro* and *ex situ* conservation of plant germ plasm for database implementation and germ plasm characterization by izoenzyme and other analyses.

◗ *Private*

Monterrey Forestal Ltda.
Pizano S.A.
Carrera 9, No. 74-08, Piso 9
Bogota, Colombia

Contact: **Miguel A. Rodriguez M.**, Natural Resources
            Director
Phone: (57) 1 211 8356 • *Fax:* (57) 1 255 1709

Activities: Research • *Issues:* Biodiversity/Species Preservation; Deforestation; Sustainable Development • *Meetings:* SSC Crocodile Group

☜ *Universities*

Universidad Industrial de Santander
Ciudad Universitaria
Apartado Aéreo 678
Bucaramanga, Colombia

Contact: **Rafael Serrano Sermiento**, Director
Phone: (57) 73 343655, (57) 73 343656
Fax: (57) 73 350541

Activities: Development; Education • *Issues:* Air Quality/Emission Control; Energy; Health; Waste Management/Recycling; Water Quality

▲ Projects and People
The **Universidad Industrial de Santander (Industrial University of Santander)** is proposing the founding of a **Centro de Investigación para la Creación y Protección de los Recursos Naturales (Research Center for the Creation and Protection of Natural Resources)** which would address environmental problems such as desertification, pollution by the petroleum industry, and the accelerated rate of erosion and contamination of water.

The University proposes to offer nationwide the most contemporary system of education covering Colombia and other Andean countries to further land and water studies.

The Center's objectives include identifying the most critical areas for conservation of natural resources, and using and adapting existing technologies for their protection, management, and recovery. Adverse affects on water and land resulting from inappropriate agricultural practices will be identified and practical methods will be sought for eliminating or minimizing this damage.

Academically, the Center will **improve the quality and focus of environmental education at the University.** Postgraduate programs will emphasize "optimal usage of available logistics and data." Coursework will also promote the interdisciplinary and interinstitutional planning of solutions concerning these problems. Departments of civil and metalurgical engineering, biology and microbiology, and geology would be involved.

Instituto de Ciencias Naturales/Natural Science
Institute
**Universidad Nacional de Colombia/National
University of Colombia**
A.A. 7495
Bogota, Colombia

Contact: **Alberto Cadena G., Ph.D.,** Chairman,
            Zoology Department
Phone: (57) 1 268 2485 • *Fax:* (57) 1 269 2951

Activities: Education; Research • *Issues:* Biodiversity/Species
Preservation • *Meetings:* Congress of Latinamerican Zoology

▲ Projects and People

Dr. Alberto Cadena G., whose doctoral degree in systematics and
ecology was earned at the University of Kansas, USA, was chairman
of the Zoology Department, Universidad Nacional de Colombia
(National University of Colombia) until his appointment in 1991–
1992 as visiting professor at Kyoto University's Primate Research
Institute. A major research interest is South American bats. He is a
consultant member of the Species Survival Commission (IUCN/
SSC) Chiroptera Group. Bat nets are a research need.

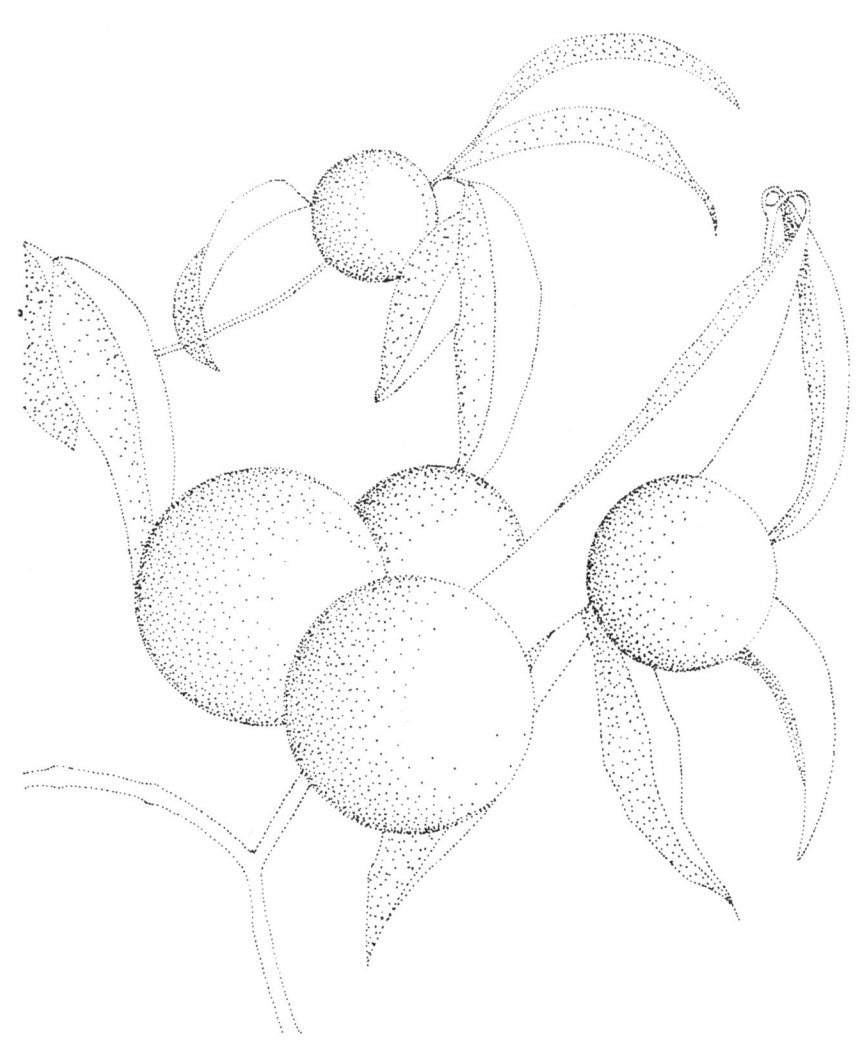

Congo

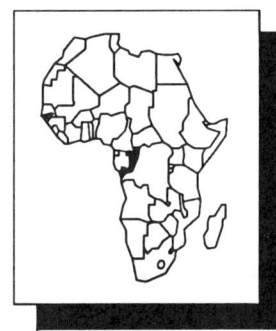

Located on the west central coast of Africa, the Congo consists of treeless plains, well-forested escarpment, savannahs, and swampy lowlands. The country boasts a rich variety of plant and animal species, with about half the area covered by rainforest. Most of the population lives in the four major cities, and agriculture is stagnant. This has resulted in a heavy dependency on imports.

Deforestation has proceeded at an alarming rate in recent years, and urban growth has brought congestion and air pollution. In addition, there is a need to improve the Congo's current system of protected area reserves with better equipment, personnel, and enforcement.

🏛 *Government*

ORSTOM ⊕
Centre ORSTOM
B.P. 181
Brazzaville, Congo

Contact: **Alain Brauman**, Head, Microbiology Laboratory
  Phone: (242) 81 2680 • *Fax:* (242) 81 0322

Activities: Research • *Issues:* Air Quality/Emission Control; Deforestation; Global Warming; Sustainable Development

● Organization

ORSTOM is the **French Scientific Research Institute for Development and Cooperation**. To find information and solutions for environmental problems such as the **greenhouse effect**, ORSTOM conducts in-depth and extensive experiments on integral parts of its surroundings. It seeks to heighten the information exchanged with international environmental scientists.

▲ Projects and People

A series of studies focuses on the relationship between termites of the **African tropical rainforest** and the production of greenhouse gases. Also explored are nesting habits of the termites in the **Congo's Mayombe Forest**, and the **Ivory Coast's Thaï Forest**. The study searches for the link between the termite nests and the CO_2 (carbon dioxide) and CH_4 (methane) gases which are naturally recycled and produced by the termites.

In one study aspect, the guts of two **wood-feeding termite species** showed "significant rates of CO_2 fixation." However, soil-feeding termites (*Cubitermes speciosus*) did not. Rather, the latter was described as "among the most productive methanogenic species measured so far." Also investigated were "endosymbiosis and exosymbiosis of **fungus-growing termites** to understand the digestive tract and the actions of enzymes from different origins—termite, fungus, microflora."

In another undertaking in the Mayombe Forest of Central Africa, Drs. **J.P. Tathy** and **R.A. Dalmas** produced preliminary results to "differentiate between methane production and oxidation in termite mounds." Their findings noted that "a considerable large part of newly generated methane is oxidized microbiologically and never reaches atmosphere." Experiments compared termites *in situ* with direct measurement of gas at mounds and *in vivo* where the insects' gas emission was measured in glass flasks with rubber stoppers. Although calculations were uncertain, observations seemed to bear out that "the forest globally acts as a sink rather than a source for atmospheric methane."

In Equatorial Africa's mountain forest, experiments concerned with **sources and sinks of methane and carbon dioxide exchanges** continued testing earlier findings of other researchers that "flooded forest and floodplains in the tropics are major sources of methane" whereas "dry forest soils absorb atmospheric methane." This study "identified strong sources of methane in flooded shallows" while CH_4 emissions by termites appeared to be a "minor source" possibly due to "methane oxidizing bacteria living in the termite nest which would act as a biofilter absorbing the methane produced by termites." Once again, the experiments showed that the tropical evergreen forest appears as "a net sink for atmospheric methane."

Similar investigations continue with **Dr. B. Cross**, Brazzaville University of Science, in both the Mayombe and Thaï forests in savanna and mountain regions. **Dr. Edouard Miambi**, also a researcher with the Laboratory of Microbiology, conducts training courses on "soils microbiology and fertility."

ORSTOM's research has been featured in international scientific journals, among them *Science* magazine, *Chamrousse*, and the *Journal of Geophysical Research. (See also ORSTOM in the Mexico and Senegal Government sections.)*

■ Resources

ORSTOM offers scientific reports, such as *Preliminary Studies on the Gut Microbiota of the Soil-feeding Termite: Cubitermes Speciosus; Sources and Sinks of Methane and Carbon Dioxide Exchanges in Mountain Forest in Equatorial Africa; Endosymbiosis and Exosymbiosis in the Fungus-Growing Termites; Preliminary Results to Differentiate between Methane Production and Oxidation in Termite Mounds,* and reprints of other studies, published in French and English. ORSTOM encourages the "scientific exchange of students or research assistants" with Europe and the USA and welcomes grant proposals to this effect. Workshop and training assistance is also sought.

Projet Inventaire et Amenagement de la Faune (PIAF)

Inventory and Management of Fauna Project

Parcs Nationaux et les Aires Protégées/National Parks and Protected Areas

B.P. 2153
Brazzaville, Congo

Contact: **N'Sosso Dominique**, Director
Phone: (242) 83 17 18 • *Fax:* (242) 83 24 58

Activities: Development; Education; Law Enforcement; Political/Legislative • *Issues:* Biodiversity/Species Preservation; Deforestation; Natural Ecosystems; Personnel Management; Sustainable Development; Waste Management/Recycling
Meetings: CITES

● Organization

Director **N'Sosso Dominique** lists six primary interests of the National Parks and Protected Areas: inventories of plants and animals, management, personnel training classes, environmental education for the public, publication of journals and manuals, and defining laws and participating in politics concerning parks, plants, and animals. Currently, the work of the Parks Department is somewhat hampered by lack of supplies and personnel.

▲ Projects and People

N'Sosso and the National Parks are conducting a **census of elephants, gorillas, and chimpanzees** of the Congo. A preliminary census has been completed, and the financing has been assured by the European Community (EC) for the remainder.

Companies such as British Petroleum, Chevron International, and Conoco are financing several **micro-projects** in collaboration with the World Conservation Union (IUCN) on the **Conkouati Reserve** and its periphery.

Another current environmental project is a **Regional Park for Central Africa**, comprised of seven countries: **Congo, Cameroon, Gabon, Equatorial Guinea, Central Africa, Sao Tome and Principe,** and **Zaire**, with some financing from EC. In cooperation with each country, National Parks and Protected Areas, and several foreign consultants, IUCN has conducted the preliminary studies. The plans concerning the Congo are being realized, and the **Odzala National Park** has begun to take shape with the construction of personnel buildings, construction of trails and bridges, personnel training, and education programs.

■ Resources

The National Parks and Protected Areas are in need of both staff and materials. In order to carry out the inventory, a bulldozer and leveler, all-terrain vehicles, walkie-talkies, citizens' band (CB) radios, and materials for trapping and keeping animals are needed. Also sought are civil engineers, builders, biologists, and botanists. Simple items are always needed, such as camping supplies, veterinary equipment, first-aid kits, tents and beds, boots, and compasses. Forestry and ecological studies are available, printed in French.

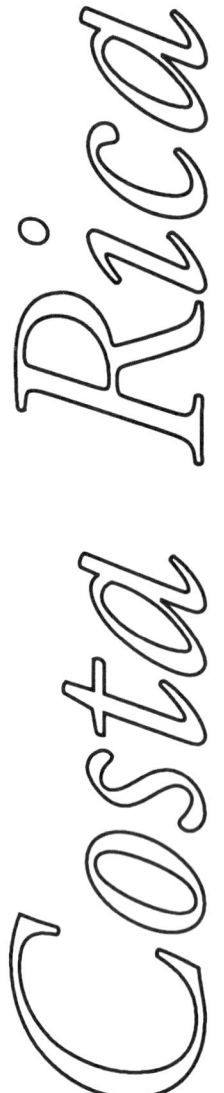

Costa Rica

Costa Rica is one of the smallest Central American republics, located in the isthmus between Nicaragua and Panama. In spite of its small size, Costa Rica is endowed with a rich variety of natural resources. Nonetheless, environmental deterioration is a serious problem. The best agricultural lands are being degraded by soil erosion or eliminated by urbanization. There has been insufficient reforestation, due in part to lack of incentives. Overused areas also suffer from flooding, landslides, and general erosion.

At the same time, progress has been made in such areas as population growth, the establishment of an extensive park system, land-use reform, and pollution control.

✤ *NGO*

Asociación Talamanqueña de Ecotourismo y Conservación (ATEC)
Puerto Viejo de Talamanca
San José, Limón, Costa Rica

Contact: **Mauricio Salazar**, President
Phone: (506) 583844, (506) 246090 • *Fax:* (506) 537524

Activities: Education; Research • *Issues:* Biodiversity/Species Preservation; Deforestation; Health; Sustainable Development; Waste Management/Recycling; Water Quality

● Organization

"Along Costa Rica's Caribbean coast, natural resources and traditional lifestyles based on cacao farming and artesanal fishing are threatened by tourism development and the expansion of the banana industry." So reports the **Asociación Talamanqueña de Ecotourismo y Conservación (ATEC) (Talamanca Association for Ecotourism and Conservation)**, which is devoted to the "development of socially responsible ecological tourism in Talamanca."

Founded in 1990 by coastal residents, this NGO seeks to provide "information and educational activities for local residents and tourists concerning tropical ecology and regional environmental problems, and training to local residents to assist them in developing environmentally and culturally appropriate activities and facilities for tourists." Among the region's resources are its rainforests, wildlife, coral reefs, and beaches.

According to ATEC, these programs aim to "generate action at local, national, and international levels to protect Talamanca's forest and marine resources and to strengthen the position of local black and indigenous peoples vis-a-vis 'outside' developers." The Talamancan community also wants to maintain a sense of ethnic and natural pride among the native people, as the area's cultural flair is one of the "most attractive" features and therefore needs protection, reports ATEC.

▲ Projects and People

Among ATEC's recent accomplishments is its publication of **educational brochures**, in English, to provide tourists with knowledge of the area's ecology, history, and activities.

With the support of the World Wildlife Fund (WWF) and the Inter-American Foundation, ATEC provides **training for Talamancan residents** to become **nature guides** who "interpret indigenous as well as scientific knowledge of tropical ecology." ATEC has organized community projects, such as voluntary **beach cleanup campaigns** and **garbage separation for recycling**.

The nonprofit group continually researches **appropriate wastewater management** and banana industry expansion in Talamanca and its potential social and environmental impacts. **Tourists and tourism managers** are surveyed "to learn their opinions, complaints, needs, interests, and suggestions concerning tourism in Talamanca."

An important aspect is bringing the Talamanca people into the decisionmaking process regarding resource management and development.

President **Mauricio Salazar**, who owns a small nature lodge, leads tours into the region's protected areas and indigenous reserves. **Benson Venegas**, vice president and marine biologist, formerly administered the Manzanillo-Gandoca Wildlife Refuge and now coordinates conservation projects for Asociación ANAI. **Paula Palmer**, secretary, sociologist, and oral historian, has written books including *What Happen: A Folk History of Costa Rica's Talamanca Coast* and *Taking Care of Sibo's Gifts: An Environmental Treatise from Costa Rica's Kéköldi Indigenous Reserve.* **Monica Donley**, treasurer, is designing a mountain conservation-tourism project with her husband.

Other members include **Enrique Brown**, native farmer, fisherman, and carpenter; **Florentino Grenald**, native farmer, fisherman, and forest ranger in the Manzanillo-Gandoca Wildlife Refuge; **Pamela Carpenter**, educator and tropical plant specialist who owns a small nature lodge and nursery; and **Willis Rankin**, coordinator of environmental education and health programs.

▉ Resources

ATEC organizes activities for ecotourist groups, such as jungle hikes, horse-riding trips, and discussions with Talamanca residents about medicinal plants, traditional lifestyles, and cultural and natural history. A small library of materials about the region, people, and resources has been established. Information is also provided on dealing with threats to the environment.

ATEC needs Spanish-speaking volunteers with backgrounds as environmental educators for a minimum three-month commitment. The organization also desires environmental education materials, or funding for their purchase, as well as television and VCR equipment, video cassettes, and binoculars. ATEC also needs support for constructing an environmental activities center.

Biomass Users Network (BUN) ⊕
200 mts. West and 50 mts. South Saint Francis Chaptel
P.O. Box 1800–2100, Moravia
San José, Costa Rica

Contact: **Jose Blanco**, Regional Director
Phone: (506) 40 8997 • *Fax:* (506) 40 8998

Activities: Development • *Issues:* Deforestation; Energy; Global Warming; Sustainable Development; Waste Management/Recycling

● Organization

The term "biomass" refers to the "organic products of the agriculture and forestry systems developed to provide **food, fuel,** and **fiber** and the **organic waste** captured by sewage and waste treatment facilities," writes the **Biomass Users Network (BUN).** For certain countries such as Ethiopia, Nepal, Kenya, and Brazil, biomass is an important energy source.

BUN enables developing countries to assist one another in managing biomass to produce fuel; conserve soil and water; and improve sanitation, waste management, and public health while creating jobs and stimulating growth in rural economies.

For example, treating sewage to produce methane also generates water suitable for irrigation. Also, bioenergy crops, such as trees, can be grown on marginal lands unsuitable for food crops. Traditional agriculture can become more profitable and foreign exchange positions can be enhanced "by reducing oil imports," according to BUN.

The Network's overall goal is to "integrate the appropriate technology into an efficient bioenergy system, the bioenergy system into effective biomass management, and biomass management into a successful economic development program." In so doing, valuable information is exchanged, technology is evaluated, and vendors and consultants are referred.

BUN was inaugurated in Bangkok, Thailand, following two years of planning by representatives from Costa Rica, Indonesia, Jamaica, Philippines, and Sudan. Assisting were the Canadian Industrial Development Agency through the International Institute for Environment and Development, Rockefeller Brothers Fund, Swedish International Development Authority through the Beijer Institute, U.S. Agency for International Development (USAID), U.S. Agricultural Resources Development Foundation, and World Resources Institute. About 20 countries are involved.

▲ Projects and People

Numerous programs and projects have been undertaken worldwide. For instance, **Brazil** launched an **alcohol fuels program to reduce oil imports** that also succeeded in creating many more jobs than would have resulted from processing and refining imported crude oil. In São Paulo, the urban population is decreasing as people move from the city to rural areas and new jobs in the alcohol fuels industry.

Coffee-producing countries discard some 17 million tons of coffee waste yearly. Improper disposal of waste creates pollution and health hazards. Technical innovations developed in **Costa Rica** convert coffee pulp residues into animal feed, ethanol, pectin, and tannin.

Also in Costa Rica, BUN provided technical assistance as the government developed the **National Conservation Strategy for Sustainable Development,** with watershed protection and management as the focal point. BUN also co-sponsored an International Conference on Peace and Sustainable Development held under the aegis of **President Oscar Arias** and focusing on improving the utilization of natural resources to promote national development and political stability. In 1990, BUN co-designed and implemented a **workshop for journalists and NGOs** to promote awareness of the interdependence and tradeoffs between environment and economic development; and also assisted the Federation of Livestock Producers (FLP) to develop ecological and sustainable methods.

In **Guatemala,** BUN worked to establish the **Central American Commission for Environment and Development (CACED)** to support regional efforts to devise "indigenous approaches and solutions to environmental degradation and natural resource depletion." In **Sri Lanka,** with the support of the Commission of European Communities (CEC), BUN designed a project using agricultural residues for energy to **reduce the need for fuelwood,** a major cause of deforestation in watershed areas.

BUN provided technical assistance to the **Philippines** government to design the diversification of the **sugar industry** and conduct a symposium on sugarcane diversification. The Asian Development Bank subsequently made a grant of $250,000 to the government to finalize diversification planning.

An **integrated energy program** was implemented in **Jamaica** with a co-generation feasibility study, energy-efficiency demonstration projects, and a wastewater component including design of a sewage discharge and high-nitrate system that can irrigate 14,000 acres of energy crops such as sugarcane. According to BUN, "using organic pollutants as irrigating nutrients . . . will help decrease pollution of Kingston Harbour and also reduce $1.5 million spent annually for fertilizer imports." BUN, Jamaican government, Rockefeller Foundation, and the Conservation Law Foundation of Boston, Massachusetts, cooperated.

In Zimbabwe, BUN is implementing a South-South Knowledge Transfer Research Program with funding from the African Development Foundation, USA, to help rural communities adopt "more productive and ecologically sound farming techniques." Demonstration projects and handbooks on composting, vermiculture, and enzymatic fermentation are field testing organic fertilizer production using crop residues, animal waste, and rock minerals to lessen animal waste uses; integrated livestock production to reduce natural resource degradation; and production of botanical pesticides from native plants "to help reduce environmental and health hazards of petro-chemical pesticides."

Also in Zimbabwe, a case study documents "one of the most successful [alcohol] programs of petroleum substitution in developing countries."

In Kenya, the Baringo Fuel and Fodder Project demonstrates how soil erosion in this watershed of a vital fishing lake area can be prevented through economical revegetation with biomass crops that can produce "directly marketable products, or raw material for value-added processing" in agri-industry.

With the cooperation of the Conservation Foundation, and using low-cost technologies developed by a Peruvian NGO, BUN is undertaking a case study of value-added products, such as organic fertilizer, from municipal or urban waste.

Rootfuels: Non-traditional Biomass Fuel Production is a case study with application in arid and semiarid zones whereby "this energy source is derived from the *Cucurbitaceae* species which grow relatively well in water-stressed areas and provide a renewable and low-cost source of domestic fuel." A handbook is planned for rural communities and local NGOs.

■ Resources

BUN publishes a bimonthly newsletter called *Network News* about institutions, projects, and initiatives concerning biomass resources. It has a circulation of some 5,000 in over 70 countries. Technical information sheets are prepared on biomass plant species such as cassava, sugarcane, neem (*Azadirachta indica*), and ipil-ipil. A series of *Biomass Management* fact sheets is also available.

Centro de Derecho Ambiental y de los Recursos Naturales (CEDARENA)/ Environmental and Natural Resources Law Center ⊕

APDO 134-2050
San Pedro, Costa Rica

Contact: **Rodrigo Barahona Israel, Ph.D.**, President
Phone: (506) 25 1019, (506) 24 8239
Fax: (506) 25 5111

Activities: Development; Education; Political/Legislative; Research • *Issues:* Biodiversity/Species Preservation; Deforesta-

tion; Energy; Legal Issues on Natural Resources; Sustainable Development • *Meetings:* World Parks Congress

● Organization

In 1989, a group of attorneys, professors, and law students founded the **Environmental and Natural Resources Law Center (CEDARENA)** to function as a clearinghouse for legal information and expertise, help governmental and NGO groups draft model laws, promote cooperation, and work with local communities and organizations to ensure sustainable development. CEDARENA concentrates on deforestation, watershed protection, Indian rights, creation and protection of public and private conservation areas, management of coastal and wetland zones, and promotion of sustainable forestry and agriculture.

▲ Projects and People

With Ford Foundation support, CEDARENA undertook an 18-month study of **land tenure and use** to identify the impacts in rural areas of real property law, conservation laws, and economic incentive programs focusing on cattle raising, agriculture, and forestry. The existing legal regime contributes to environmental degradation and poverty, believes CEDARENA, which is fostering cooperative projects to begin to improve laws affecting land use, titling, and property registration. Workshops on technical assistance, education, and model legislation were begun in 1992 with **Rodrigo Barahona, Robert Wells**, and **Steve Mack**.

Since **establishing a protected biological corridor** between Tortuguero National Park and the Barra del Colorado Wildlife Refuge in 1991 with the help of the Caribbean Conservation Corporation and Costa Rica's Ministry of Natural Resources, Energy and Mines, CEDARENA is working with the Fundación Neotropica to map the area, purchase lands, and determine allowable uses for the zone.

The ongoing **Boscosa project** is a long-term conservation and sustainable development program for the Osa Peninsula jointly with the World Wildlife Fund (WWF), Fundación Tropica, and Costa Rican government agencies. CEDARENA attorney **Sylvia Chavez** is an advisor in helping local residents secure land tenure and the government protect other lands. Early support was obtained from WWF, Rainforest Action Group, Parks in Peril of the U.S. Agency for International Development (USAID), and The Nature Conservancy.

The **Talamanca Indigenous Law** project is helping the Bribri, Cabecar, and Kekoldi Indians protect their lands and resource rights from squatter invasions and other exploitive projects, mainly in coal, lumber, and oil enterprises.

Farmers and townspeople joined together to **protect the Santa Cruz watershed** with CEDARENA's legal advice. Local residents established the Fundación Gran Chorotega to legally represent them; the project includes an inventory of coastal lands and environmental impact assessments of proposed developments as tourism and the construction of second homes expand. Further hotel construction threatens the nesting site of the leatherback turtle.

The Costa Rican government has initiated feasibility studies for the **Pacuare hydroelectric project** to meet the country's energy demands and avoid using oil-based energy sources. Meanwhile, the Association for the Protection of Costa Rica's Rivers asked CEDARENA to help halt the dam's construction because of the area's biodiversity, aesthetic value, and ecotourism potential that draws white-water rafters from around the world. Investigations of alternative energy sources and smaller hydroelectric projects are underway.

CEDARENA is designing a project to set up a land trust—Fincas Conservacionistas—for private nature reserves with the help of Fundación Neotropica, The Nature Conservancy, and others.

To protect the Pacific and Caribbean coastlines and estuarine ecosystems, such as mangroves, from destructive agricultural and sewage treatment practices and unrestricted tourism development, CEDARENA will propose recommendations for improving the existing legal framework.

■ Resources

Dr. Barahona has written a natural resource and agrarian law textbook, essays, and articles on environmental, development, and related subjects.

Tropical Agricultural Research and Training Center (CATIE)
P.O. Box 90
Turrialba 7170, Costa Rica

Contact: **Tomas Schlichter, Ph.D.**, Project Leader
Phone: (506) 56 1712 • *Fax:* (506) 56 1533

Activities: Development; Education; Research • *Issues:* Biodiversity/Species Preservation; Deforestation; Sustainable Development; Water Quality

● Organization

Central America's "rich combination of climate and terrain that has produced the astonishing biodiversity of the region is also the reason for the limited capacity of many of its ecosystems to withstand substantial modifications without suffering irreparable degradation," according to CATIE (Tropical Agricultural Research and Training Center). Tree cover removal has led to intense erosion, soil depletion, and at the same time, rapid population growth in the area and "incessant expansion of the farm frontier toward the areas of greatest ecological fragility," it reports.

CATIE and the World Conservation Union (IUCN) are attempting to develop rural areas through the **Conservation for Sustained Development Project (OLAFO)** in Central America. One goal is to identify forest nontimber products that can be used for the development of communities living in or near the forest. Another aim is to provide a guide for rational use of resources from mangroves such as wood, fish, shrimp, tannins, and shells. Overall, the purpose is to improve the quality of life for the 26 million people of Central America—a region "producing less food per inhabitant than it did 10 years ago, [where] the deforestation rate continues high, and migration is on the rise."

CATIE, which receives funding from development agencies of Norway and Sweden, intends to involve the local population in extracting products from the ecosystem such as **lumber, bark for tannin, shellfish, and fish** while conserving the environment and ensuring the equitable distribution of wealth. At the same time, the ecosystem must be kept intact: soils must not suffer erosion, species must be protected, genetic diversity must be fostered in reserves, nutrient and hydrological cycles must be maintained, and pollution must be minimized by refraining from introducing any solid, liquid, or gaseous chemical compounds that take a long time to decompose.

▲ Projects and People

Launched in 1989, the program for sustainable development features four components: field demonstration projects, training and technical exchange, support for regional NGOs, and wetlands research. Demonstration sites included in the following:

• The **Bocas del Toro** area in **Panama**, with 9,000 square kilometers of wet tropical forest undisturbed by humans. Coastal and marine resources make it compatible with developing a sustainable economy by OLAFO and with launching an overall management of the region's coastal wetlands.

• **Talamanca**, which borders Bocas del Toro Province on the eastern side of Costa Rica and is a 2,800-square-kilometer canton of indigenous reserves and protected wildlife areas. More populated than its neighboring province, this fragile wetland offers sustainable use of forest plants for medicinal, ornamental, or tinctorial purposes, in keeping with traditional uses by local people.

• The region of **Heroes y Martires, Veracruz, Nicaragua**, situated on the Pacific coast near Leon and Corinto, which relies on resources extracted from the mangroves, such as tannin from tree bark and timber for fuel and construction. Shrimp, crabs, clams, and fish from the mangroves are a subsistence food source and provide cash income as well. These "practices hold no prospect for sustainable production due to high intensity of extraction," according to CATIE, which supervises proposals to bring sustainable development to the region.

• **Choluteca, Honduras**, in the southeast, with a "high population density and several ecological limitations to agricultural development." Rapid degradation of the forests, soils, and water results from a combination of the region's steep slopes, light precipitation, six-month dry season, and low technological application. CATIE is working to implement a conservation-for-development project involving ecological and socioeconomic assessments.

• **Barra de Santiago, El Salvador**, with a largely rural population of five million and "the highest demographic concentration in Central America," where the vegetation is significantly altered and mangroves and other forest cover have been reduced to 7 percent. Here, peasants use timber for fuelwood and construction and hunt aquatic animals. CENREN, a Natural Resources Center which is successfully managing one mangrove cover, is able to implement sustainable development efforts.

• **El Petén, Guatemala**, a subtropical rainforest area with a "very distinct dry season" where the population has risen nearly tenfold in almost 3 decades to 230,000. The fragile ecosystem is marked by "undeveloped" soils with drainage limitations. A conservation-for-development plan was designed.

Training has included international mobile seminars on conservation for sustainable development; graduate scholarships for training in the field; regional workshops whereby staff was exchanged among the different demonstration projects and opportunities were broadened for women; and a wetlands management workshop to identify the many resources for providing livelihoods.

The plan also features a documentation and publications center with references on the region's wetlands; bibliography on conservation for development; and preparation and distribution of teaching materials and case studies based on demonstration project activities.

Networks are being created to strengthen NGOs and reduce duplicative efforts. To counter the destruction of wetlands in Central America, methodology is being developed for "economically identifying and evaluating the goods and services provided by tropical wetlands . . . in Guatemala and Nicaragua." A manual on this economic assessment will be produced.

Donald P. Masterson is training coordinator, and Dr. Tania Ammour is agricultural economist.

■ Resources

Assistance is needed in managing and processing secondary forest products. Project leader Tomas Schlichter has prepared 17 scientific publications on sustainable development and gives presentations on his experiences.

(See also Center for Environmental Study/Centro de Estudio Ambiental in the USA NGO section.)

Tropical Science Center (TSC) ⊕

P.O. Box 8-3870
San José 1000, Costa Rica

Contact: **Raúl Solórzano Soto**, Agricultural
　　　　Economist
Phone: (506) 53 3267 • Fax: (506) 53 4963

Activities: Education; Research • Issues: Biodiversity/Species Preservation; Deforestation; Global Warming; Sustainable Development

● Organization

Founded in 1962, the **Tropical Science Center (TSC)** is a professional association of about 45 members who conduct and support scientific research and education in the fields of basic and applied ecology, forest resource management, and agricultural development. Projects have been undertaken in 15 countries in tropical areas: Belize, Colombia, Costa Rica, Dominican Republic, Ecuador, El Salvador, Guatemala, Honduras, Nicaragua, Panama, Paraguay, Peru, Venezuela, Thailand, and Tanzania.

Facilities at the San José headquarters include a specialized library; laboratory, cartographic, drafting, photo interpretation room; computer room, and four-wheel-drive vehicles for the support of ongoing projects.

▲ Projects and People

The Tropical Science Center carries out national and international ecological classifications, using the **World Life Zone System of Ecological Classification** in Latin American countries, Africa, southern Asia, and Australia.

Ecological research focuses on tropical forest environments, climate, soils, landforms, vegetation, and wildlife. Environmental impact assessments for economic and infrastructural development projects are researched in tropical countries. Activities also include land-use capability studies, watershed planning projects, wildlife management, soil conservation, relocation of communities, and agricultural diversification for regional and global development programs.

The Tropical Science Center owns and administers the 28,000-acre **Monteverde Cloud Forest Biological Preserve** located high in

northern Costa Rica's Tilaran Mountains and extending down the Caribbean and Pacific slopes. In addition, the Preserve is located between North and South America, so there is a mixture of species from both hemispheres abounding in "cloud-enveloped forests," rainforests, and drier habitats.

The Preserve is so described: "Wind-sculptured elfin woodlands on the exposed ridges are spectacularly dwarfed, whereas protected cove forests have majestically tall trees festooned with orchids, bromeliads, ferns, vines, and mosses. Poorly drained areas support swamp forests, while parts of the Preserve, dissected by deep gorges, have numerous crystal clear streams tumbling over rapids and waterfalls."

Wildlife includes the **golden toad** and its entire habitat—**jaguar, ocelot, Baird's tapir, three-wattled bellbird, barenecked umbrellabird, and resplendent quetzal** and its important nesting sites. Altogether, more than 100 species of mammals, 400 species of birds, 120 species of amphibians and reptiles, 2,500 species of plants, and thousands of insect species live here. Because large tracts of forests are essential for the survival of large mammals as well as migrating birds and butterflies, TSC continues efforts to expanding the Preserve.

■ Resources

The Tropical Science Center can provide a list of publications. TSC offers the *Land Capability Classification System* and *Ecological Maps* for each Central American country and Bolivia, Brazil, Colombia, Ecuador, Peru, and Venezuela. Also published are an *Occasional Paper Series* covering controversial or lengthy scientific articles, a *Facsimile Series* including articles of outstanding interest to resources conservation, and *Economic and Technical Notes.*

TSC wants to establish an Endowment Fund to carry out research, such as modeling, which could benefit the standard of living for Costa Rica's population.

The Monteverde Cloud Forest Biological Preserve includes a visitors' center, field station, laboratory facilities, dormitory-style lodging for up to 30 people, a small research outpost, a trail system, and backpacking shelters.

Voluntarios en Investigación y Desarrollo Ambientalista (VIDA)

Calle Central Avinida 9 y 11, casa No. 928
APDO 7-350-1000
San José, Costa Rica

Contact: Mario León Ch., Executive Director
Phone: (506) 211411, (506) 364274
Fax:　(506) 572273, (506) 223620

Activities: Development; Education; Law Enforcement; Political/Legislative; Research • Issues: Biodiversity/Species Preservation; Community Development; Deforestation; Health; Sustainable Development; Community Development with International Assistance • Meetings: REDES; UNCED

● Organization

In 1988, VIDA, Volunteers for Environmental Development, was created primarily by young students and professionals concerned about pressing environmental problems, such as burgeoning population growth, and desiring to find solutions. "Sustainable development," says VIDA, "depends on the stabilization of the worldwide population on a level compatible with its resources, but it is very improbable that this will occur if there is no improvement in the living conditions of poorer countries [requiring] positive action [by] more opulent industrialized nations . . . to reduce 'hyperconsumerism.'" Global environmental cooperation is required to "guarantee a better tomorrow for all human beings."

Yet, VIDA's 200 members believe that diverse natural resource preservation actually is the only process whereby sound sustainable development is achieved. VIDA is "fighting a battle whose only arms are the raising of consciousness." At 1992 UNCED, VIDA was an officially accredited organization representing Costa Rica.

▲ Projects and People

VIDA works to provide support to Costa Rica's environmental politics and programs through activities emphasizing sustainable development, conservation, ecotourism, and youth leadership skills.

Adhering to the principles of Global Strategy for Conservation (WSC) and the National Strategy for Sustainable Development (ENCODE), VIDA focuses on **environmental education** to foster global biodiversity and preserve ecosystems that make such development possible. VIDA gives **technical advice** to teaching organizations regarding the conservation of plants and animals.

Projects recently undertaken at the Palo Verde Wildlife Refuge include **constructing nesting boxes for the endangered black-bellied whistling duck,** restoring and painting two observation towers used by scientists to study migratory birds, burning of fire lanes to **control numerous forest fires** in the dry season, patrolling the refuge to **prevent illegal hunting,** and building an **innovative septic system** for the park station. At Guayabo National Monument, a petroglyph of a bird was discovered during their excavation of a drainage system and two quadrants on an ancient road.

For the Caribbean Conservation Corporation (CCC), Tortuguero, a marking **program for the endangered green sea turtle** was launched to study nesting and migratory habits, beaches were patrolled to prevent poachers from stealing turtle eggs, research station and housing for volunteers were renovated, tourists were taught how to observe the turtles, **a beach cleanup inventory was** sent to the Center for Marine Conservation, Washington, DC, to be included in an international survey, and a local school was painted.

VIDA intends to **involve youth directly into the national political environmental movement.** Cooperative conservation and rural development programs are developed with young volunteers from Costa Rica and other countries, focusing on long-term quality of life. To these ends, VIDA collaborates with the International Youth Federation for Environment (IYE) headquartered in Denmark and India, the Latin American Federation of Young Environmentalists (FLAJA), Network of NGO Conservationist Organizations of Central America (REDES-CA), Costa Rican Federation for the Conservation of the Environment (FECON), Youth Challenge International in Canada, and Youth Service International in North Carolina, USA.

■ Resources

VIDA's youth programs and summer camps teach conservation, low-impact agricultural techniques, and how to form environmental groups. For example, YSI expedition participants recently con-

structed a research and education center at the Campanario coastal and forest reserve, home to toucans, scarlet macaws, and hundreds of species; repaired schools; worked with indigenous people at Chirripo; and built a school and health center and installed a water pump in the remote area of Bocas. They also hiked, swam, and snorkeled. VIDA needs funds, information, and training about management of protected areas, environmental education, and sustained development in the tropics.

Paseo Pantera Project ⊕
Wildlife Conservation International (WCI)
APDO Postal 246
2050 San Pedro, Costa Rica

Contact: **James R. Barborak,** Technical Advisor
Phone: (506) 249215 • *Fax:* (506) 341061

Activities: Education; Research • *Issues:* Biodiversity/Species Preservation • *Meetings:* CITES

● Organization

"Whatever else divides the human inhabitants of the Western Hemisphere, the Paseo Pantera silently unites us," writes **Dr. Archie F. Carr,** regional director, Wildlife Conservation International (WCI).

"By thinking in terms that reach beyond the cramped political boundaries of the modern-day Central America," says Dr. Carr, "we may intelligently address the challenge of biodiversity conservation in the entire region. Paseo Pantera originates from a phenomenon of nature, but its successful completion will perhaps breach a human phenomenon in the region—the partitioning of the isthmus into seven small nations, whose isolation and independence from one another is considered by economists and historians to be a major factor contributing to the chronic underdevelopment of the region."

The Paseo Pantera Project was "proposed in response to the 'Regional Wildlands Management Programs' component of the Regional Environmental and Natural Resources Management Project (RENARM). The RENARM project was initiated by the Agency for International Development's (USAID) Regional Office for Central American Programs (ROCAP) in direct response to urgent and widespread conservation needs in Central America." *(See also Agency for International Development [USAID] in the USA Government section.)* Implementing the project is a consortium of "WCI (a division of the New York Zoological Society) and the Caribbean Conservation Corporation (CCC) in collaboration with Tropical Research and Development, Water and Air Research, University of Florida Program for Tropical Studies, and a number of Central American governmental and NGO organizations." *(See also Wildlife Conservation International [WCI] and Caribbean Conservation Corporation [CCC] in the USA NGO section.)*

▲ Projects and People

The Paseo Pantera Project was created in response to the alarming devastation and overexploitation of the natural resources on Central America. Paseo Pantera, meaning "Path of the Panther," was named for the area of Central America which is the route of many species that travel throughout North and South America. The destruction of the Central American natural environment poses a threat not only to "the wild flora and fauna, but also to the 30 million people living in the region today," reports Jim Barborak, technical advisor.

The extreme degradation and destruction of natural wildlife habitats has left nothing but "islands in a sea of civilization." This destruction results from many forces, especially deforestation that destroys entire ecosystems as well as watersheds which provide the water for irrigation and domestic consumption.

Paseo Pantera is dedicated to preserving the prodigious species and natural resources so threatened as well as to contributing to the region's sustainable development. The Project's goal is to "provide the methods, tools, and knowledge to the nations of Central America to work toward conservation."

To protect and maintain the diversity of Central America, Paseo Pantera has set forth a five-year plan which encompasses both regional and site-specific activities.

The regional activities include buffer zone management, which will place an emphasis on "applying ecological principles of island biogeography and wildlife corridors." This project includes an international training workshop on the subject of buffer zone management, which is organized by the University for Peace in Costa Rica.

An ecotourism program is expected to provide the seriously needed funds to acquire and manage protected areas and to "enhance economic development of local communities, and promote environmental education of park visitors" throughout Central America. A handbook will be produced.

A main feature of the regional environmental program is the production of two books so that people can learn about Central America's unique and diverse environment and the importance of conservation. Together, the books cover the history of the area and list Paseo Pantera's initiatives in conservation, ecotourism, and a guide to the outstanding geographic features of the region. Regional conservation strategies will be continually updated.

Pilot projects are also being planned in specific areas:

In Honduras, the Paseo Pantera works in conjunction with the government and NGOs to improve the coastal management and protection of the Bay Islands and also in the forested interior on behalf of the Rio Platano Biosphere Reserve, Central America's oldest such reserve and one of the largest protected areas. To save the region from colonization and illegal harvesting, the plan is to "establish a protected area linking the Rio Platano with Nicaragua's Bosawas Biosphere Reserve and to provide secure land tenure for indigenous peoples of the area."

In Belize, the Project is working with the government and NGO conservation groups to establish new biosphere reserves—one including the entire Belize Barrier Reef, the largest reef in the Western Hemisphere, and the other in the Maya Mountains, which encompass one-fifth of the country. Underway are biodiversity inventories in the rainforest and coral reef species research.

With WCI and CCC, Paseo Pantera plans to assist the Costa Rican government in the expansion of the Caribbean lowland rainforest park, Tortuguero, and the linkage between that site and the rainforest across the Nicaraguan border.

In Panama, with the help of a national NGO, an effort is underway to protect the coastal bays of Bocas del Toro. The area is home to "endangered manatees, four species of sea turtles, marine bird rookeries, lobster fisheries, and homelands of the Guaymi Indians."

In Guatemala, field biologists are conducting "species specific" research and preparing for vertebrates censusing and training of local persons as Project leaders plan the design and operation of the 1.4 million hectare Maya Biosphere Reserve—with its "fabled Mayan ruins of Tikal" and "continuous sweep of forest into Mexico and Belize."

Ecological restoration is the mission of war-beleaguered El Salvador. And in Nicaragua, Paseo Pantera is supporting proposals for "three huge protected areas": one million hectares of rainforests at Bosawas; the green sea turtle's "most important feeding ground" in the Caribbean—Miskito Cays, a 5,000-square-mile coastal site that is also home to the Miskito Indians; and a binational project with Costa Rica known as SI-A-PAZ in the San Juan River valley.

World Wildlife Fund (WWF-USA) ⊕
7170 CATIE
Turrialba, Costa Rica

Contact: **Miguel Cifuentes**, Regional Coordinator–
  Central America
Phone: (506) 56 1712, (506) 56 6188
Fax: (506) 56 1421

Activities: Development; Education; Law Enforcement; Political/Legislative; Research • *Issues:* Biodiversity/Species Preservation; Deforestation; Sustainable Development • *Meetings:* UNCED; World Parks Congress

▲ Projects and People
For schools in Costa Rica—as well as in Colombia, Brazil, and Argentina—World Wildlife Fund (WWF–USA) has "developed materials and trained educators to teach basic environmental and ecological concepts to young students." *(See also World Wildlife Fund [WWF] in the USA NGO section.)*

WWF's Central American regional coordinator Miguel Cifuentes also teaches protected areas management and planning courses at CATIE—Centro Agronómico Tropical de Investigación y Enseñanza (Tropical Agricultural Research and Training Center). He has written about planning and management at Galapagos National Park, Ecuador, among other subjects.

(See also Center for Environmental Study/Centro de Estudio Ambiental in the USA NGO section.)

🎓 *Universities*
Center for Sustainable Development Studies ⊕
School for Field Studies
Universidad Nacional
APDO 1350
Heredia, Costa Rica

Contact: **Christopher Vaughan**, Director
Phone: (506) 37 7039 • *Fax:* (506) 37 7036

Activities: Education • *Issues:* Biodiversity/Species Preservation; Deforestation; Sustainable Development

▲ Projects and People

In 1991, at the **Center for Sustainable Development Studies, School for Field Studies,** director **Christopher Vaughan** organized a new concept in environmental education for students to learn under field conditions about sustainable development projects. The innovative staff consists of a cultural anthropologist, an agroecologist, a natural resource economist, a forester, a farmer, and a land-use planner in addition to Vaughn, the biological conservationist/wildlife manager.

During the program's first semester, students developed case studies around important conservation issues: **land utilization in the Coto Brus area;** the relationship between **transnational and sustainable development;** and the attempts of both the **La Amistad Biosphere Reserve** and the **Boscosa-Corcovado National Park** attempt toward sustainable development.

Vaughn, who also supervises the first graduate training program in wildlife management in Latin America, organizes the study sites, logistical support, case studies, and an agreement with the Universidade Nacional for bringing to Costa Rica approximately 80 North American undergraduate and graduate students for semester and summer courses every year.

In 1974, after serving in the U.S. Peace Corps, Vaughan set up a self-sustaining farm, with several hundred fruit and forest trees, where he built a house, became a beekeeper, began a dairy herd, and planted a vegetable garden using natural fertilizers. Because the zone was not electrified, he built a solar water heater, solar stove, and a solar fruit dryer. A member of the Species Survival Commission (IUCN/SSC) Cat and Tapir groups, Vaughan is a past president of the Wildlife Society of Costa Rica.

■ Resources

Assistance is needed in acquiring educational staff.

Cuba consists of a number of islets and cays located about 140 miles south of Florida. The main island is the largest in the West Indies and is characterized by plains, hills, and mountains as high as 6,000 feet.

Cuba has the world's fourth largest deposits of nickel, which it exports along with shellfish and sugar. The country has the greatest species diversity and the most endemic species in the West Indies. Its vast wetlands and mangroves are home to a rich variety of marine life. Nevertheless, over 100 species have become extinct and others are endangered due to changes in natural habitats, hunting, and the introduction of exotic plants and animals.

Only 14 percent of the land is now forested (down from 90 percent a century and a half ago); however, reforestation efforts have protected watershed and helped to prevent soil erosion. In addition, shortages of lumber and other construction materials have encouraged efforts to prevent waste.

🎓 *Universities*

Jardín Botánico Nacional/National Botanical Garden
University of Havana
Calabazar C.P. 19230
Havana, Cuba

Contact: **Carlos Sánchez Villaverde**, Assistant Professor
Phone: 44 5525 (operator assistance)

Activities: Education; Research • *Issues:* Biodiversity/Species Preservation

▲ Projects and People

Carlos Sánchez Villaverde is a professor and researcher at the **National Botanical Garden** of Cuba, which is affiliated with the **Department of Botany** at the **University of Havana**. Much of his research focuses on ferns and other vegetation found in mountainous regions of Cuba, including at the Sierra del Rosario Ecological Station. He is also a member of the Species Survival Commission (IUCN/SSC) Pteridophyte Group.

■ Resources

Sánchez is the author of over 20 scientific publications in Spanish.

Cyprus

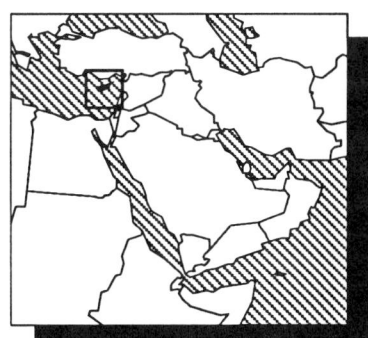

Cyprus is an island nation located in the northeastern corner of the Mediterranean Sea. It encompasses a fertile central plain, a mostly barren mountain range along the northern coast, and a forest range in the southwest. Like all Mediterranean countries, Cyprus suffers from coastal development problems, fueled by industrialization, tourism, urbanization, and intensive agricultural modernization. Transboundary concerns include urban and industrial waste and oil pollution.

In Cyprus's case, the country's rapid economic growth and transformation from a basically rural and agrarian society into a predominately urban and mobile one has strained the ecosystem. The government has established programs for natural resource management, but their effectiveness has been limited by modernization and scarce resources.

NGO

Cyprus Herpetological Society (CHS)
P.O. Box 2133
Paphos, Cyprus

Contact: **Hans-Jorg Wiedl**, Director-Manager
Phone: (357) 61 238160

Activities: Education; Research • *Issues:* Species Preservation

● Organization

In Cyprus, the local people are extremely fearful of snakes, and kill them whether they are harmless or not. Consequently, this slaughter—as well as loss of habitat to agriculture and tourism—has endangered the island's herpetofauna. In 1990, the volunteer **Cyprus Herpetological Society (CHS)** was founded to teach people about the value of reptiles and amphibians and encourage them to coexist with the island's species.

Nevertheless, CHS encounters negative responses by those who believe that "only a dead snake is a good one," and the practice of killing reptiles continues. However, CHS remains steadfast in its goal in stressing the importance of biodiversity, and CHS sees no reason why reptiles and amphibians should have to clash with "economic development and public well-being."

Hans-Jorg Wiedl, director-manager, writes, "With the establishment of a Red Listing of endangered species . . . together with a concept for a representative system of protected habitats, we expect to stimulate interest and discussion." At the same time, he is working with other groups to stem pollution of soil and water, curb urbanization, and stop "uninhibited destruction of animal and plant life."

▲ Projects and People

The Cyprus Herpetological Society initiated an **educational program** on the value of snakes, in conjunction with other Cypriot agencies that are also committed to the protection of the environment.

CHS publicly protests the butchery of the island's serpents, by issuing media statements and news releases. CHS also prepares **exhibitions**, which include the showing of live animals as an attempt to shatter unfounded myths about snakes, which CHS believes, lead to their inevitable slaughter. Such exhibitions are being extended and improved, and CHS plans to create an outdoor landscaped terraria for the snakes so that the public may see them in their natural environment.

Seeking advice from European scientists, the organization is presently **establishing research objectives** and a plan to carry them out. A **publication on herpetofauna of Cyprus** was recently produced, with "advice to people venturing into the countryside." Versions are expected to be published in English, German, and possibly other languages.

Wiedl is recognized internationally for his herpetological efforts, lecturing in the USA, South America, Asia, Africa, and Europe, and cooperating with various universities. According to Prof. Johannes Zopp, a journalist, "Wiedl fights a brave fight against the hard-core prejudices against reptiles in Cyprus," and deserves praise.

■ Resources

A brochure is available. CHS welcomes scientific expertise on herpetology and environmental education.

Laona Project

P.O. Box 257
Limassol, Cyprus

Contact: **Artemis Yiordamlis**, Project Manager
    Phone: (357) 51 58632 • *Fax:* (357) 51 52657

Activities: Development; Political Legislative; Research • *Issues:* Sustainable Development

● Organization

The **Laona Project** is a program begun by the Friends of the Earth in Cyprus to help maintain and fund the Laona rural villages on the Akamas Peninsula, an ecologically important coastal region of Cyprus. The Project is funded through the Coordinating Office for Friends of the Earth (FOE) Groups in Europe (CEAT) and by Mediterranean Specially Protected Areas (MEDSPA) programme that supports NGOs in environmental and conservation projects. A matching sum is provided by the Cypriot A.G. Leventis Foundation, a local trust, and other local donors.

The Project has the cooperation of government from national to local levels as well as from the Greek Tourism Organization (EOT), Greek Ministry for the Environment, UK Countryside Commission, European Federation of Nature and National Parks, and Sunvil Holidays, a Cypriot tour operator.

The Project's main goal is to bring rural, or **soft, tourism** to the provincial towns along the Laona coastline. The five Laona villages include: Kathikas, Kritou Terra, Pano Akourdalia, Kato Akourdalia, and Miliou. The Project seeks to preserve the land with funds gathered from tourism and interest groups. The goal is to provide the Laona villages with enough tourism to sustain the communities, but not mass tourism which would be more of a burden. In turn, visitors are treated to a countryside vacation with the lure of wildlife and antiquities.

▲ Projects and People

The Laona Project began work to attract tourism by studying the tourist potential in the towns and their benefits. The Project also began restoring old Laona buildings to show the communities' innate beauty. Other efforts include assisting the government in providing low-interest loans and providing grants to builders who wish to restore homes and buildings.

The second phase of the Laona Project involved conducting a door-to-door survey in the Laona villages in order to obtain descriptions and records of all of the buildings in the region. The survey "also produced information on the inhabitants' occupations, their potential for developing handicrafts, and their views on the desired form of tourism development." These results provided the Laona Project with invaluable information on the inception of the quickly developing tourist industry.

The third phase of the Laona Project consisted of overseeing restoration loans, studying new agricultural methods and introducing organic farming, recording typical Laona cuisine, marketing the area in Europe, and developing a Nature Study Centre of particular appeal to children. The first houses were planned for 1991. Overall coordinator **Artemis Yiordamlis** worked with **Adrian Akers Douglas**, FOE Cyprus; many professionals and students collaborated.

■ Resources

The Project offers a color brochure of the Laona villages, printed in French and English. The Project also provides a final report of phases one and two, detailing the progress in the communities. Help is needed with new technologies in agriculture for areas with poor soil and insufficient water and with promotion in potentially new markets.

In 1993, the Czech Republic and the Slovak Republic became two separate nations. Czechoslovakia's economic growth from the Second World War through the early 1980s took place at the expense of severe degradation of its natural resources and environment—especially in terms of health problems and losses in agricultural and forestry production. The extensive use of low-quality brown coal contributes to a level of air pollution that is among the highest in the world, and as much as one-third of the forests may have been irreversibly damaged. Other environmental problems include water shortages due to surface and ground water contamination, and pollution from municipal wastes. Policy changes that are currently being implemented should have a positive effect on the environment, through reduced, cleaner, and more efficient energy use; restrictions on agricultural chemical use; and vehicle emissions standards.

🏛 *Government*

Cesky Ústav Ochrany Příody
Czech Institute for Nature Conservation
Sleská 9
12029 Prague 2, Czech Republic

Contact: **Jaroslav Hromas, Ph.D.**, Director
    Phone: (42) 2 215 2800, (42) 2 215 2986 • *Fax:* (42) 2 215 2810

Activities: Education; Law Enforcement; Political/Legislative; Research • *Issues:* Biodiversity/Species Preservation; Protected Areas Management; Sustainable Development
Meetings: Centre Naturopa; Council of Europe; IUCN

● Organization

The Czech Republic (CR) and the Slovak Republic (SR) each have separate conservation laws, with State Nature Conservancy Acts passed, respectively, in 1955 and 1956. Supplemented with guidelines since then, the Acts define nature conservation as "the preservation, renewal, enhancement, and use of natural wealth and the special protection of important areas and natural features." The objective is to "integrate conservation and use of natural resources and to apply principles of ecosystem conservation." In addition, governments of the Republics establish national parks; district councils declare protected natural monuments and nature features.

Established in 1990, the **Federal Committee for the Environment** evolved from the Trust for Enhancement and Protection of Native Country of 1904. In 1989, a **Czech Ministry** was set up to cover water, air, nature, agriculture, forestry, and mineral resource protection; a **Slovak Commission for the Environment** was created in 1990 with similar jurisdiction, including "fuel-wastage management."

The **Czech Institute for Nature Conservation** and the **Slovak Institute for Nature Conservation** originated from the State Institutes for Protection of Monuments and Conservation of Nature in both Prague and Bratislava. These Institutes currently conduct research on threatened species, select protected areas, and prepare management plans. Their work is complemented by three NGOs: Taxum (TIS), the Czech Union of Nature Conservation, and the Slovek Union of Nature and Landscape Conservationists.

Two 1990 documents—*The Environment in Czechoslovakia* and *Ecological Programmes and Projects, Czech and Slovak Federative Republic*—detail major environmental threats such as "acid precipitation and agricultural mechanization" and are directed toward "integrating ecological and economic activities." The Czech Ministry's *Blue Book* and *Rainbow Book* contains plans for agricultural and water management and energy projects. A **species preservation strategy** has also been outlined for the CR.

▲ Projects and People

With 12,121 square kilometers in CR and 8,810 square kilometers in SR, these protected areas include **8 national parks**, open to visitors, with "controlled area" zones for villages, holiday homes, campsites, and sanitaria. Some logging is "strictly controlled," and hunting is regulated. Forty **protected landscape areas (CHKOs)** encourage "rational use of natural resources" with projects dealing with water and forest management, agriculture, industry, transportation, building, tourism, recreation, and regulated mining. **State nature reserves (SPRs)** manage ecosystems, where scientific research is conducted. **Protected habitats** are concerned with the preservation of one or more plant or animal species. **Protected parks and gardens (CHPZs), study areas (CHSPs),** and **natural monuments (CHPPs)** as well as **protected natural features (CHPVs)** in karst areas to preserve selected trees are also part of the system. Such areas date from 1838 when two forest reserves were created in South Bohemia.

Biosphere reserves, including wetlands, have been designated for inclusion in the World Heritage List and presented to UNESCO for this purpose.

Emphasizing **environmental education,** new universities have been founded in CR to address such issues, including the Forestry Faculty of the Prague Agricultural College. A Training and Education Nature Conservation Centre is also active, and a School of Nature Conservation is beginning near Zilina in SR. Certain protected areas, with visitors' centers and nature trails, are widely used for educational purposes.

■ Resources

The Federal Committee for the Environment can be reached at the above address. The CR Ministry of Environment is at Vrsovická, CS 10000, Prague 10. The Slovak Commission for the Environment is at Hlboká 2, CS 81235, Bratislava; and the Slovak State Centre of Nature Conservation is at Hejrovského 1, CS 84103, Bratislava-Lamac. Both Republics publish *Pamiátky a príroda* (Monuments and nature) 10 times a year and contribute to the Red Data Book for birds and vertebrates. The independent journal *Nature Conservation* was renewed in 1992. SR also publishes the annual *Chránená územi Slovenska* (Protected areas of Slovakia).

Institute of Landscape Ecology (ILE)

Czechoslovak Academy of Sciences
Na sádkách 7
37005 C. Budejovice, Czech Republic

Contact: Jaroslav Bohác, Ph.D., Scientific Secretary
 Phone: (42) 38 40240 • *Fax:* (42) 38 45719

Activities: Education; Research • *Issues:* Air Quality/Emission Control; Biodiversity/Species Preservation; Deforestation; Waste Management/Recycling; Water Quality • *Meetings:* International Colloquium on Soil Zoology; International Symposium on Bioindicators and Biomonitoring

● Organization

The interdisciplinary scientific research teams of the Institute of Landscape Ecology (ILE) are concerned with better land manage-

ment; control and repair of air pollution damage; restoration of land damaged by industry; rejuvenation of habitats and lost species; studies of population biology; and environmental microbiology's application to forestry, agriculture, and environmental protection.

Director is RNDr. Ing. Václav Mejstrík, Deputy Director is Ing. Jan Tesitel, and Scientific Council Chairman is Dr. Miroslav Gottlieb.

Scientific achievements include developing a concept of man's role in landscape ecology, finding a method for stabilizing landscape units and their structures, gaining insights into evaluating ecotoxological criteria in forest systems, and gathering data on species sensitivity to different chemical forms. For example, ILE studies are determining the relationship between the degree of forest damage and the extent that short roots of Norway spruce trees are dying.

Technical achievements include assessing the impact of air pollution in the North Bohemia Brown Coal Mining District; proposing "ecological optimization" of the Bohemian Mesozoic Protected Landscape Region; predicting and maintaining optimal agricultural and forest landscape subsystems; and developing monitoring stations and equipment for testing organic and inorganic pollutants, microorganisms, radionuclides, and meteorological parameters.

ILE organizes regional conferences concerned with biomonitoring, heavy metals in the environment, and landscape ecology. It also provides an international training course in landscape ecology with the cooperation of UNESCO and UNIDO.

▲ Projects and People

Ten projects receive grant money from the Czechoslovak Academy of Sciences:

Rural Settlements as Biocentres in Agricultural Landscape studies the relationship of natural and social structures to rural settlements. Scientists are gathering data on the effect of historical development, location, and function within the settlement system on the diversity of plant and animal communities. Higher plants, birds, small mammals, and beetles are being used in the study.

Principal investigator is biologist/zoologist **Dr. Jaroslav Bohác,** who for 20 years has studied the environmental interrelationships among communities of organisms—focusing on the role of the beetle as a bioindicator of change in ecosystems. Author of articles in international journals and scientific bulletins, Bohác began publishing monographs while still a student at Charles University, Prague. He lectures on general ecology at the University of Agriculture, Ceske Budejovice.

Interrelationship between Photosynthetic Apparatus and Root System of Norway Spruce Forest Stands with Different Features of Foliage Disturbances is using 25- to 30-year-old trees in 6 permanent plots in Beskydy Mountains to determine mechanisms of foliage damage and impact upon speed of defoliation. Results should facilitate more precise prediction of forest stand development.

Principal investigator is biologist/geobotanist **RNDr. Pavel Cudlín,** with ILE since 1974 and on the Faculty of Natural Sciences, Charles University, Prague. His focus includes mycorrhizal development during forest decline, the role of the success of mycorrhizal symbiosis in plant succession on spoil banks, and protection and management of terrestrial orchids.

Interrelationship between Photosynthetic Apparatus and Root System of Norway Spruce Seedlings under SO$_2$ and Acid Rain Stress is about the response of fine roots in seedlings and cuttings to these factors. Cultures of seedlings are being established in mist nutrient solutions with added acids and in plastic growth chambers with natural soils, fumigated by SO_2. The project's findings are expected to protect forests from environmental aspects. Cudlín is in charge.

Significance of Private Farming in the Ecology of the Landscape is concerned with whether private land ownership in Bohemia stimulates the best ecological behavior—as it is assumed—compared to practices on collective and state farms. Social ecologist Dr. Miroslav Lapka is principal investigator. He also heads a multidisciplinary team of anthropoecologists working on issues regarding man in the biosphere within the context of culture and tradition. He has conducted studies in Poland, former USSR, Bulgaria, and Germany.

Why Is the Mountain Spruce Forest of Sumava Mountains Dying Back? Results from this study on the synergic effect of extreme weather conditions, acid rain, and nutrient deficiencies are expected to answer this question and to evaluate the liming effect on spruce forest ecosystems. Anna Lepsova is the principal investigator.

The Common Vole as Indicator of Emission Load in the North Bohemian Coal Basin is a study of free-living rodents and how they mirror the contamination of food chains. Results are expected to suggest how this process may be reversed. RNDr. J. Paukert is in charge.

Biometal Investigation of Interactions between Bacteria and Heavy Metals and Their Biotechnological Application to Decontamination examines the frequency of bacteria resistant to toxic metals in the soil and the relationship to biologically available forms of metals, using chromatographic, electrophoretic, and spectrophotometric methods. RNDr. Ing. Vladimir Riha is in charge.

Role of Feeding and Habitat Factor in Abundance: Changes of Select Waterfowl Species in South Bohemian Fishpond Regions is developing information about interaction of waterfowl and food, habitat, pollution and pathogenic agents to improve wetland conservation. RNDr. Zuzana Svecova is principal investigator.

Characterization and Deposition of Air Pollution to Severely Impacted Mountain Plateau of Northern Bohemia is an interdisciplinary undertaking with University of East Anglia, UK, on characterizing chemical pollution sources and examining their impact on an adjacent mountain range. Engineer Z. Zeman is in charge.

Model for an Ecological Information System for Use in Sustainable Land Management of Landscape is developing the Landscape Ecological Information System (LEIS) to monitor landscape changes and prepare ecological simulation models to assess environmental impact. Zemek Frantisek is the principal investigator.

Financial support is sought for the following pending projects:

Biocentres and Biocorridors in Agricultural Landscape: The Influence of Different Forms of Land Ownership on Their Biological and Social Functions is a study of how changes in land ownership can affect woodlands, streams, and other features of the agriculture landscape that do not directly contribute to agricultural production. RNDr. J. Kubes and RNDr. J. Sadlo are principal investigators.

Effect of Atmospheric Inputs on Changes in Forest Ecosystems of the Sumava Mountains/Bohemia Forest is a quantitative study of the effects of levels of air pollution on forest ecology. Dr. Jana Kubiznakova is principal investigator.

Radionuclide Transfer through the Components of the Environment and Assessment of Accident Situation in the Surroundings of Nuclear Power Plants. Its goal is to develop a simple model to examine impact on surroundings of Temelin nuclear power plant and synergies of air pollutants and ionizing radiation. Chief scientist is J. Knocuyt.

Model of Artificial Radionuclide Transport Through Some Components of Water Ecosystems would develop a method to use water as a bioindicator of surface contamination. M. Svadlenkova is principal investigator.

Some Effects of Stress in Several Mammal Species including Man by Biological Effective Fluorides in North Bohemia will track congenital abnormalities associated with increased concentration of SO_2—a pollution marker. RNDr. Roman Valach is principal investigator.

■ Resources

A bibliography of ILE publications is available. Featured are: Proceedings of the Fifth International Conference Bioindicatores Deteriorisationis Regionis, 1989; and Proceedings of the Sixth International Conference Bioindicatores Deteriorisationis Regionis, 1991.

Institute of Landscape Ecology (ILE)
Czechoslovak Academy of Sciences
Na Sadkach 7
37005 Ceske Budejovice, Czech Republic

Contact: Jirí Konečný, Ph.D., Research Scientist
Phone: (42) 38 40240 • *Fax:* (42) 38 45719

Activities: Research • *Issues:* Air Quality/Emission Control; Energy; Health; Transportation

▲ Projects and People

One of the Institute's research scientists is Dr. Jirí Konečný, a specialist in radiochemistry and radiation chemistry. Formerly chief of dosimetric laboratories at the nuclear power plant Jaslovske Bohunice, Dr. Konečný researched radiation protection for plant staff members and citizens, monitored plant surroundings, and performed risk assessment studies. Later, on a team preparing the Dukovany nuclear power plant, he served as chief of the dosimetry division; in that capacity, he monitored the natural and working environment in the whole complex and developed an education program and strategies for possible accidents. Later, at the central directorate of the Slovak Power Generation Board, Dr. Konečný was involved in the field of "radioactive waste treatment, transport, and dumping"; and comparing nuclear energy with coal burning and water energy "from the economical point of view." Dr. Konečný subsequently continued studies to estimate the "consequences of coal burning and mining in the vicinity of uranium mines."

Dr. Konečný has since 1985 worked at the Institute of Landscape Ecology as a radioecologist, performing research that emphasizes food chains and biomonitoring "the fate of radionuclides from the Chernobyl fallout." He has also helped organize national seminars and international conferences in the field of nuclear power plant operation and nuclear safety.

■ Resources

Dr. Konečný has published more than 33 studies and articles, including "Occurrence of Radioactivity in Basidomycete Fruiting Bodies in South Bohemia after Chernobyl Accident" (1991) and "Siting of Nuclear Power Plant in CSR from Point of View of Effect of Effluents on Soil" (1989). A publications list includes other titles in English.

Institute of Landscape Ecology (ILE)
Slovak Academy of Sciences
Stefanikova 3
81434 Bratislava, Slovak Republic

Contact: **Zuzana Kasanická**, Engineer
  Phone: (42) 7 330167 • *Fax:* (42) 7 332560

Activities: Education; Research • *Issues:* Biodiversity/Species Preservation; Optimal Landscape Ecological Planning; Sustainable Development • *Meetings:* CITES; International Symposia on Problems of Landscape Ecological Research

● Organization

The **Institute of Landscape Ecology (ILE)** of the **Slovak Academy of Sciences** was established in 1990 as a continuation of the Institute of Landscape Biology (1965–1974). With 78 staff members, including 41 researchers and 7 individuals involved in educational outreach, the Institute has offices in several locations. In Bratislava, the focus is on agroecology and urbanized landscapes; in Nitra, structural and dynamic landscape ecology are studied; and in Kosice, members work on agricultural landscape, soil ecology, and water ecology.

A member describes the Institute as "an interdisciplinary scientific workplace" oriented toward basic research, application of landscape ecology knowledge to decisionmaking, management and planning processes, and linking the science with teaching at universities, technical colleges, and other educational levels.

▲ Projects and People

Institute researchers analyze, synthesize, and interpret "the abiotical, biotical and socio-economical elements of landscape for . . . optimal space, organization and utilization" of the area. In Bratislava, they study the **ecological problems of an urbanized landscape**, including the "impact of anthropogenic factors [and] optimization of urbanized space," as well as the "protection and nutrition of crop plants" and "movement of chemicals in landscape." Nitra-based Institute staff work to interpret the various components of a landscape for ecological syntheses, studying their structure and dynamics, ecological stability, connectivity, and bearing on a particular landscape.

Realizing the importance of international research and cooperation, the Institute has since 1967 held a **triennial international symposium** on the ecological problems of landscape. Attendees at the 6th Symposium held at Pestany in 1982 founded the **International Association of Landscape Ecology (IALE)**. The Secretariat for the East European region of IALE is now housed at the Institute. Institute researchers are also involved in **ecological evaluations for the Danube Park** on the Czechoslovak/Austrian border.

Institute studies integrate economic development with the need for ecological stability and conservation of natural resources. Through special **summer courses** and other opportunities, the Institute also works to educate young scientists.

■ Resources

The Institute seeks new information on "ecological problems and optimal utilization of landscape and urbanized space."

Institute publications include proceedings of international symposia, and two journals: *Ecology of CSFR* in English, and *Zivotne prostredie* (Environment) in Slovak with English abstracts. In addition, the Institute holds international symposia on Problems of Landscape Ecological Research.

Institute of Systematic and Ecological Biology (ISEB)
Czechoslovak Academy of Science
Kvetna 8
60365 Brno, Czech Republic

Contact: **Jiri Gaisler, Ph.D.**, Professor of Zoology
  Phone: (42) 5 331143 • *Fax:* (42) 5 338979

Activities: Education; Research • *Issues:* Biodiversity/Species Preservation; Deforestation; Ecological Monitoring; Population Planning • *Meetings:* International and European Bat Research Conferences; International and European Congresses of Mammalogy and Ornithology

● Organization

The **Institute of Systematic and Ecological Biology (ISEB)** of the **Czechoslovak Academy of Science**, with its nine staff members, gathers researchers from various Czechoslovak universities to cooperate on projects investigating the current environmental situation of the country.

▲ Projects and People

An ongoing project (1991–1994) involves monitoring **terrestrial vertebrates in the Palava Biosphere Reserve and the surrounding** Pavlovské vrchy Hills, located in southern Moravia, south-central Czechoslovakia. Palava was named to UNESCO's list of biosphere reserves in 1986. Its 75 square kilometers include a variety of habitat types, including an unusual Upper Jurassic limestone cliff. Research focuses on "fluctuations in population densities of mammals, birds, reptiles, and amphibians" in the region. The project goal is to collect data that will be used for future landscape management and protection, particularly in a planned bilateral national park on the Czechoslovak/Austrian border. Among Palava's species are orchid varieties, *Iris pumila*, the endangered green lizard (*Lacerta viridis*), eagle owl, bezoar goat, and saxicolous birds.

Scientists at the Institute point to the growing urgency of collecting data on the environment of the country, especially following the signing of the Danube Charter, and with the ongoing construction of the Nové Mlyny waterworks, in the face of vigorous protests by environmentalists—"another grave problem requiring scientific support."

The Palava project is a cooperative effort involving the Institute, the Masaryk University Department of Plant Biology, Shippensburg University in Pennsylvania, USA, and Universität für Bodenkultur, Austria. Scientists collaborating in the effort include **Marta Heroldová**, **Jirina Nesvadbová**, **Jitka Pellantová**, and **Jan Zejda**, a "leading scientist at the ISEB, . . . acknowledged worldwide as an expert in mammal ecology."

One of the lessons learned at ISEB is that although "vineyards on terraces may be beneficial to the landscape," shrubs and trees must also be planted to prevent erosion. Another regards the "importance of windbreaks for ecological diversity in agricultural landscape." Research of Dr. **Jiri Gaisler** and others acknowledged that windbreaks, or shelterbelts, provide refuges and biocorridors for wildlife, helping animals to survive. Earthworms, ground beetles, birds, and small terrestrial mammals were investigated.

In another study, five-year surveys of hibernating bats in the Moravian Karst indicated that numbers of certain species, such as *Myotis myotis*, were increasing and that other populations were stable—unlike those in areas of Western Europe or Bohemia. Generally, abrupt changes in numbers of bats are caused by "human interference."

Designer of the Palava project, Dr. Gaisler has since 1991 split his time between ISEB and the Faculty of Science at Masaryk (formerly Purkyne) University in Brno. He is also general secretary of the National Committee for the International Union of Biological Sciences, a member of the Species Survival Commission (IUCN/SSC) Chiroptera Group, and a member of the Scientific Advisory Board of Bat Conservation International. Dr. Gaisler emphasizes the need for "more cooperation between and among ecologists," in the form of internationally organized projects. For example, he suggests the "monitoring of certain animal or plant groups in various countries using the same methods within the same time span."

■ Resources

Resources needed include radiotelemetry equipment to study small mammals, including bats.

The Institute sponsors five-day workshops on the ecological monitoring of mammals. Dr. Gaisler has an extensive list of publications. A report on Palava appears in four languages: Czech, German, Russian, and English.

Muzeum Vychodních Cech V Hradci Králové
Museum of East Bohemia
Eliscino nábrezí 465
50039 Hradec Králové, Czech Republic
Contact: **Petr Rybář, Ph.D.,** Director
Phone: (42) 49 23416
Activities: Education; Research • *Issues:* Biodiversity/Species Preservation; Sustainable Development

● Organization

The Muzeum Vychodních Cech V Hradci Králové (Museum of East Bohemia) describes itself as "one of the most important regional museums in Czechoslovakia." Founded in 1880, and in its present location since 1912, the Museum houses more than one million items relating to archaeology, history, and natural history.

Its staff of 50 includes 6 naturalists and 16 Department of Natural History specialists in paleontology, botany, and zoology. These experts work on regional investigations of nature, sharing their findings for the purpose of environmental management, including developing projects for protected areas, endangered species, and environmental education. In addition, various sections of the volunteer Naturalists Club work in close connection with the Museum.

▲ Projects and People

Dr. Petr Rybář, whose specialty is bat aging, was from 1967 to 1990 head of the Department of Nature Reserves and Protected Species at the Center of Monuments and Nature Conservation in Pardubice. Since 1990, he has directed the Museum of East Bohemia in Hradec Králové. Dr. Rybář is the author of numerous scientific and popular papers on nature conservation, ecology and protection of fauna, especially bats, published both in Czechoslovakia and elsewhere. He belongs to the Species Survival Commission (IUCN/SSC) Chiroptera Group.

A traveler and photographer, Dr. Rybář has written books on "ecological" travels, including *The Diadem of Mother Earth* (on The Himalayas–1980) and *Trails to the Wilderness* (1990); he has also published popular books about nature, and is the editor of an encyclopedia article "Through the Nature between the Krkonose Mountains and Ceskomoravska Vrchovina Highlands" (1989).

■ Resources

The Museum needs assistance completing a computer network for an information sharing system, and is also seeking wide international public awareness of its activities, as well as an international exchange of exhibitions.

The Museum publishes *Acta Musei Reginaehradecensis, Sci. Nat.*

◢ *Private*

EKOCONSULT
Hurbanovo námestie 9
81103 Bratislava, Slovak Republic

Contact: **Mikulás Janovsky, Ph.D.,** Director
Phone: (42) 7 332468 • *Fax:* (42) 7 332468

Activities: Development; Education; Research • *Issues:* Air Quality/Emission Control; Energy; Waste Management/Recycling • *Meetings:* CITES

● Organization

An engineering and consulting firm, **EKOCONSULT** works in research and development, providing expert advice, mainly to industry, on how to take the environment into account. The organization has a staff of 20, as well as about 70 outside consultants who work cooperatively on an as-needed basis. The firm describes its focus as the "ecologization of industrial processes [and] biotechnology."

▲ Projects and People

Many of the firm's projects involve feasibility studies and technical/economic studies of governmental or private industry activity. Subjects include the possibility of storing hazardous wastes in defunct coal mines, done for the East Bohemian Coal Mines District; a system for separating and collecting municipal and complex wastes for the city of Bratislava; an assessment of location for the **construction of a solid municipal waste disposal site**, EKO-INPROS, in Brno; technical solution for liquidating bad-smelling sulfur compounds produced by sulphate way of pulp production, Slovak Pulp and Paper Mill, Zilina; **waste separation strategies**, on the basis of analyses and toxicology tests and in accordance with the laws and regulations, for INORGA, in Ostrava; and a proposed **slag and ash disposal site** in Devinska Nova Ves, and incineration technologies for municipal waste, both for the Municipal Authority of Bratislava.

Technical and economic studies have covered the **decontamination and reclamation of land around industrial plants**, for the Hradec Králové road construction department; **elimination of hazardous waste** produced for trucks, in Jablonec; and the production of tissue paper of nonbleached magnesium-bisulphite pulp, for the Chemicellulose company in Zilina.

Several other studies involve more general environmental protection opportunities in various regions of the country.

Dr. Mikulás Janovsky has directed EKOCONSULT since 1990. Prior to that, he was head of the research department and deputy director at the Czechoslovak Institute for the Environment. At EKOCONSULT, Dr. Janovsky "deals with conceptual and prognostic questions of environmental problems and waste management." Environmentalists can enhance cooperation "in the framework of the European Community (EC), between individual firms, with the USA," and through global programs.

Dr. Matús Povanaj, deputy director of EKOCONSULT, is a specialist in microbiology, genetics, biotechnology, and bioengineering, with special emphasis on the taxonomy, physiology, and genetics of industrial microorganisms. Previously, he was a researcher at the Institute of Fermentation Technology of the Czechoslovak Institute for the Environment, serving as head of the applied research department. He currently addresses the application of biotechnological processes in environmental protection, as well as conceptual questions of environmental problems and waste management.

■ Resources

EKOCONSULT seeks environmental technologies for industry and specialized experience with hazardous wastes.

Denmark

Denmark was the first industrialized country to establish a ministry to deal exclusively with environmental matters (in 1971). In 1988, new policies were developed. These include a stringent plan to combat the greenhouse effect by lowering carbon dioxide emissions; the Rube Biogas plant, located in the flat marshlands in southwest Denmark, which converts farm fertilizer and slaughterhouse wastes into a clean energy source (with the goal of using organic kitchen wastes as well); and an expansion of windpower and other alternative energy forms. The Danish government is also working with other Nordic countries and international organizations on limiting marine pollution and other transborder environmental problems.

🏛 *Government*

Division of Wildlife Ecology, Kalø
National Environmental Research Institute (NERI)
Ministry of the Environment
Grenåvej 12
DK-8410 Rønde, Denmark

Contact: **Karsten Laursen**, Director of Research
Phone: (45) 89 201400 • *Fax:* (45) 89 201514

Activities: Research • *Issues:* Biodiversity/Species Preservation

● Organization

According to **Ellen Margrethe Basse** and **Henrik Sandbach**, respective board chairman and director general of the **National Environmental Research Institute (NERI)**, **Ministry of the Environment**, the goal of NERI is to "ensure reliable environmental data and professional data to the environmental administration." Established in 1989 "through the amalgamation of a number of smaller institutions," NERI also plans and executes research related to NERI's themes.

"What is new," they say, "is that environmental policy is not only being discussed in international forums; more and more decisions are being moved from the national to the international level; and both nationally and internationally the decisions are changing from selective single decisions to collecting many partial decisions which affect each other."

The task of this "sector-research institute" is to "increase background knowledge necessary for suitable regulation of various conditions in society." NERI, with 100 researchers, works on the "long-range collection of information and expertise" and develops new methods and theories on the area it studies.

NERI's research is conducted in seven divisions: Freshwater Ecology, Wildlife Ecology, Emissions and Air Pollution, Marine Ecology and Microbiology, Environmental Chemistry, Policy Analysis, and Terrestrial Ecology. Each division monitors national plans and programs which relate to its interests. It reports back to the Institute, which relays information to the Minister of the Environment.

▲ Projects and People

When engineer **Hans Schrøder** wrote a short book, *A Piece of Ecological Advice,* he influenced new directions in society, "new moral and cultural values for our civilization," that have an important bearing on **systems analysis**, NERI's *1990 Annual Report* states. As the Danish Ministry of the Environment recently published, "It is really a paradox that, especially during the last three–four decades, we have succeeded in turning the idea of economy almost upside down: to many people— at least in the industrialized part of the world—'economy' is almost synonymous with the unlimited use of natural resources resulting in pollution, instead of suitable management and economical administration and development of our resources."

Others such as writer **Erik Christensen** in his book *New Values in Politics and Society: Change of Paradigm and Cultural Convulsion*, physicist **Fritjov Capras**, and physicist **Poul Davies** express a movement of thinking toward an "organic, ecological and holistic world picture." NERI says, "when we move from simplicity towards complexity new things are added, a new form of order, which at the same time maintains simplicity and transgresses it."

Consequently, EVE is "the chaotic way" of a systems **analysis environmental model** moving from simplicity to complexity in the fields of science and mathematics. The EVE model combines "known models and tools from the single disciplines; and interdisciplinary analyses which integrate the sub-descriptions and proper system-oriented analyses." This model is dependent upon broad cooperation with other national and global research groups.

A contrasting systems analysis tool, ADAM, analyzes and forecasts economic factors, as NERI begins research activity within environmental economy. "It would be wise to acknowledge the contrasts [of ADAM and EVE] and accept that the confrontation between them enables the creation of a new development, a new life," writes NERI. "When we have to handle the conflict between the environment and society, it is necessary to develop a public spirit which will remove them from their present destructive collision course."

NERI also conducts **ecological bird research** in the Wadden Sea, Denmark's only nature area with pronounced tides influencing an abundance of fauna, including large numbers of birds flocking to the seabed. In studying migratory birds, aerial censuses occur monthly in Danish waters. NERI has been able to track the "distribution, circadian rhythm, and foraging of the individual species," of birds such as the brent goose, shelduck, wigeon, eider, common scoter, oystercatcher, bar-tailed godwit, knot, dunlin, and herring gull.

These investigations have "provided comprehensive insight into the ecological importance of the Wadden Sea to these migratory species." Other studies in the Wadden Bay also focus on mapping human activities, including hunting and mussel fishing, and noting bird displacement and land conditions. The impact of a sea dike built in 1979 on migratory birds was also investigated.

Mathematical models and various tools are being developed by NERI to calculate the **concentration and deposition of air pollution**. These include: OML (Operational Meteorological Air Quality) model, which evaluates emissions from industry and in urban areas; advanced long-range transport models, which calculate transboundary transport of air pollution across Europe; and the OSPM (Operational Street Pollution) model, which calculates traffic pollution in urban areas.

In the field of **biotechnology**, NERI focuses on ecology of microbes and experiments with microcosms (contained models). Regarding genetic engineering, NERI conducts experiments for the Danish Environmental Protection Agency in order to evaluate risks and formulate regulations in this area.

■ Resources

NERI's *Annual Report* features a publications list as well as detailed explanations of its programs.

Ecological Division
Miljø/ministeriet—Skov- og Naturstyrelsen ⊕
Ministry of the Environment, National Forest and Nature Agency
Slotsmarken 13
2970 Horsholm, Denmark

Contact: **Veit Koester**, Head of Division
Phone: (45) 4 576 5376 • *Fax:* (45) 4 576 5477

Activities: Law Enforcement; Political/Legislative • *Issues:* Biodiversity/Species Preservation; Sustainable Development *Meetings:* CITES; global inter-governmental

● Organization

The Danish Ministry of the Environment's National Forest and Nature Agency has 12 divisions, with a staff of approximately 260; the Ecological Division has a staff of 27. The Ecological Division performs administrative functions, rather than project direction.

▲ Projects and People

The Ecological Division is "the relevant governmental body responsible for the international inter-governmental cooperation [global and regional conventions, for example] in the field of nature protections and certain forest issues."

Lawyer and division head **Veit Koester** was a main contributor to the current *Danish Conservation of Nature Act*, which he describes as an "advanced legal system whereby many nature [habitat] types in Denmark (lakes, bogs, salt meadows, heathland . . .) are protected." Besides working on Danish legislation, Koester has contributed to the World Heritage Convention, the Bonn Convention on the Conservation of Migratory Species of Wild Animals, the Convention on the Conservation of European Wildlife and Natural Habitats (Bern Convention), the European Community Directive on the Conservation of Wild Birds, and Community Regulations on the Implementation of the Washington Convention (CITES) within EC, and has worked on "Danish implementation of these international legal instruments." In the process, he has also worked closely with UNESCO, World Conservation Union (IUCN), and other international bodies.

■ Resources

The Division's and Ministry's numerous publications are primarily in Danish. Koester has authored or co-authored numerous articles and publications (in Danish) in the fields of nature conservation and international environmental law. His works in English include the IUCN Environmental Policy and Law Paper No. 14: "Nordic Countries Legislation on the Environment"; No. 23: "The Ramsar Convention"; and *From Stockholm to Brundtland in Environmental Policy and Law*, Vol. 20, 1990, on international environmental customary law.

Nature and Wildlife Reserve Division
Miljø/ministeriet, Skov- og Naturstyrelsen
Ministry of the Environment, National Forest
and Nature Agency
Ålholtvej 1
6840 Oksbol, Denmark

Contact: **Palle Uhd Jepsen**, Senior Conservation
Officer, Division Head
Phone: (45) 7 527 2088 • *Fax:* (45) 7 527 2514

Activities: Development • *Issues:* Biodiversity/Species Conservation; Habitat Restoration; Wildlife Management • *Meetings:* CITES; Marine Mammals Conference; Wetlands Conference

● Organization
In 1989 the Danish Ministry of Agriculture became the Ministry of the Environment; the Nature and Wildlife Reserve Division is a subsector of the Ministry's National Forest and Nature Agency. The Agency employs 250 staff members, 5 of whom work in the Division. Primary tasks of the Division include conservation and wildlife management, planning and habitat management, enforcement coordination, disseminating information, and wildlife monitoring.

▲ Projects and People
The Division is involved in activities of the **Trilateral Wadden Sea Cooperation**, the Ramsar Convention, the European Cetacean Society, and the World Conservation Union (IUCN). Plans are currently underway for a project to **restore the wetlands associated with the Danish Ramsar site Fiil-So** (a freshwater lake). Restoration will involve **improving sewage treatment** in the surrounding area, constructing "a traditional lock in the outstream" to enlarge the lake, clearing the adjacent area, and providing grazing cattle to maintain the habitat. The project is being planned in conjunction with the County Council of Ribe, the Danish Heathland Company, the local State Forest District, and landowners.

As part of a joint project between the Ministry of the Environment and the Ministry of Fisheries, the Division is studying the "problems concerning incidental by-catch of [cetaceans] mainly **harbor porpoise** in Danish waters." The multilevel project, which continued in 1992, included collecting information from fishermen; registering stranded cetaceans; analyzing fishing methods with the most harmful impact; and analyzing possibilities of minimizing incidental by-catch by modifying fishing gear or using preventive measures. Other contributing organizations are the Danish Fisheries Society, Danish Organization of Deep-Sea Fisheries, and National Environmental Institute.

Palle Uhd Jepsen has been working in the conservation field since 1964, when he began monitoring waterfowl at the Game Biology Station in Kalo, researching the goldeneye (*Bucephala clangula*); in 1976, he was named Senior Conservation Officer and head of the Wildlife Reserve Section at the Wildlife Administration. As coordinator for the International Waterfowl and Wetlands Research Bureau (IWRB), he reviewed a **waterfowl management plan** for the Strangford Lough Wildlife Scheme in Northern Ireland. Jepson reports that the IWRB Wetlands Division, with World

Wildlife Fund International, is preparing a wetlands strategy for the Lower Volga, former USSR. Jepson has been in his present position since 1990, with more than 23 publications in Danish, German, and/or English.

■ Resources
The Division publishes brochures on specific areas or reserves, disseminated to the public free of charge.

NGO

Danmarks Naturfredningsforening (DN)
Danish Society for the Conservation
of Nature
Norregade 2
1165 Copenhagen, Denmark

Contact: **David Rehling**, Director
Phone: (45) 3 332 2021 • *Fax:* (45) 3 332 2202

Activities: Development; Education; Law Enforcement; Political/Legislative • *Issues:* Air Quality/Emission Control; Biodiversity/Species Preservation; Energy; Global Warming; Sustainable Development; Transportation; Waste Management/Recycling; Water Quality

● Organization
Danmarks Naturfredningsforening (DN), the **Danish Society for the Conservation of Nature**, with a staff of 40 and 273,000 members, describes itself as "Scandinavia's largest nature conservation organization."

Global cooperation is growing as the Society works with "sister organizations" in other Nordic countries, such as Norway, Sweden, Finland, Aland Islands, Iceland, and the Faroe Islands. Representing more than 600,000 members, these societies hold biannual meetings to share ideas and strategies. DN has a seat on the executive board of the European Environmental Bureau (EEB) and works through the World Conservation Union (IUCN) to improve working conditions for NGOs in developing countries, and thus help ensure "the long term success of efforts toward sustainable development in Africa, Asia, Latin America, and Oceania."

Founded in 1911, the Society now has more than 200 local chapters, each of which is self-governing. Scientists from Danish universities and research centers serve on special committees to provide technical expertise. Shortly after its founding, the Society drafted, and persuaded Parliament to pass (in 1917), the **Nature Conservation Act**, "the first major Danish land use legislation." Twenty years later, another law gave the Society special status: "the right to appeal decisions made by local or regional authorities if those decisions do not sufficiently take into account environmental considerations."

▲ Projects and People
A major part of DN's effort involves **proposing areas** to the National Superior Nature Conservation Board **for preservation easements**. More than 160,000 hectares, about 4 percent of Denmark, have been preserved in this manner. In addition, the Society uses its **right of appeal** to question a Board decision; for example, allowing

development along a previously restricted beach, or to reexamine a county council's permit "for development within wetlands, heaths, moors and beach meadows." The arbiter in these cases is the National Agency for Forestry and Nature Protection. Each year, the DN reviews approximately 4,000 permits, and appeals about 75; "half of the appeals lead to decisions that are more favorable to the environment."

The Society acknowledges that Denmark has an excellent planning system—and sees its job as ensuring that the public makes use of it. DN promotes "active citizen involvement in the planning process at all stages, with the aim that the plans respect nature and the environment. And that afterward, the plans, too, are respected." This requires a major lobbying effort, providing the public with the necessary information, and urging industry, public works, and agriculture interests not to abuse the public's trust—or the environment.

David Rehling spent nearly 10 years working under the Danish minister of the environment, in nature conservation and physical planning. In 1986 he became vice-president of the EEB, which represents 20 million Europeans. He was chairman of the 1989–1990 international "Bridging the Gap" initiative, which brought together representatives from 350 voluntary environmental organizations in Western and Eastern Europe, the former Soviet Union, Canada, and the USA to participate in the UN Conference on Environment and Development.

■ Resources

The Society publishes *Natur og Miljø*, a quarterly membership magazine with a circulation of 300,000.

International Association of Zoo Educators (IZE) ⊕
Copenhagen Zoo
Sdr. Fasanvej 79
2000 Frederiksberg, Denmark

Contact: **Lars Lunding Andersen**, Curator/President
Phone: (45) 3 630 2555 • *Fax:* (45) 3 644 2455

Activities: Education • *Issues:* Biodiversity/Species Preservation; Nature Conservation

● Organization

The **International Association of Zoo Educators (IZE)** grew out of a 1972 meeting of zoo educators at Frankfurt. Any professional educator working in a zoo, aquarium, or related institution may join, and can vote on IZE policy and in the election of the Executive Committee. The current membership is about 250, from more than 24 countries. The IZE's goal, emphasizing conservation education, is "to promote a greater use of zoos, aquaria and other collections of living animals for educational purposes." To this end, the Association "encourages communication and co-operation on an international level among zoo/aquarium educators and other professionals interested in the field."

▲ Projects and People

An example of recent IZE activities is the six-week **Zoo Educator Training Course** held in 1991, with the cooperation of the Jersey Wildlife Preservation Trust, as well as the Wildlife Preservation Trust International offices in the USA and Canadian offices. With the growing awareness of zoos' "potential role for environmental and conservation education . . . the goal of the course is to teach participants how to create or improve conservation education programmes in their zoos."

A year earlier, in conjunction with the New York Zoological Society (NYZS), Wildlife Conservation International (WCI), and the Bronx Zoo Education Department, IZE organized the **First Pan American Congress on Conservation of Wildlife through Education** in Caracas. The aims were to help zoo educators "develop more effective strategies for fostering conservation actions among the audiences they serve," and to strengthen the existing network among institutions, encouraging further cooperative efforts.

■ Resources

The IZE publishes the biannual *Journal of the International Association of Zoo Educators.* International conferences take place every two years, with regional conferences held in alternate years.

ICBP/IWRB Grebe Research Group ⊕
c/o Zoological Museum
University of Copenhagen
Universitetsparken 15
DK 2100 Copenhagen, Denmark

Contact: **Jon Fjeldså, Ph.D.**, Chairman
Phone: (45) 3 135 4111 • *Fax:* (45) 3 139 8155

Activities: Education; Research • *Issues:* Biodiversity/Species Preservation • *Meetings:* CITES Scientific Working Groups

● Organization

The **ICBP/IWRB Grebe Research Group** was organized in 1985 as a joint effort of the **International Council for Bird Preservation** and the **International Waterfowl Research Bureau**. Since then, and including some of their earlier research, the 12 group members have researched all 20 species of grebes.

▲ Projects and People

The Group members have **researched the history of threats to grebes**. In the last century, they note, the great crested grebe faced extinction "because of the use of 'grebe-fur' in women's fashion." Other grebes have been victimized "on suspicion of harming freshwater fisheries." Now they are losing their wetland habitats to agriculture and other environmental changes, which have caused the extinction of the giant pied-billed grebe of Lake Atitlán, Guatemala, the Columbian grebe in the eastern Andes, and possibly Delacour's grebe of Madagascar. Other grebe species are endangered, including the Junín flightless grebe in central Peru, and the hooded grebe of Patagonia, discovered only in 1974. Larger grebe populations are also at risk and should be monitored.

Besides advocating action to protect specific grebe species and habitats, the Group points to the **value of the birds "as indicators of** wetland quality." For example, grebes will stop reproducing if a lake undergoes excessive eutrophication, making it more suitable for plant than for animal life. At the top of a lake's food chain, piscivorous grebes "are also good indicators of environmental pollutants." Group members continue to research "the relationships between grebe numbers and wetland quality."

Chairman of the Grebe Research Group Dr. Jon Fjeldså has extensively studied waterbirds—primarily in South America. In 1990, he co-published, with Niels Krabbe, an 880-page study of Andean birds, *Birds of the High Andes*, based on 9 years of field research. Realizing the need to conserve high-elevation woodlands, "a once widespread habitat," Dr. Fjeldså began a survey in 1987, in order to lay "a sound scientific basis for . . . conservation of this habitat, and studying evolutionary processes that operate in tiny relict bird populations." With colleague P. Arctander, Dr. Fjeldså, who is also chairman of the Species Survival Commission (IUCN/SSC) Grebe Group, worked on techniques to obtain DNA samples from live birds. The Group is working on an extensive collection of DNA, which they intend to use "for 'case studies' of importance for general speciation theory and for biogeographic analysis."

In 1991, Dr. Fjeldså began a long-term program, with both professors and students, to study biodiversity and conservation of tropical high-elevation forests. Current projects are in Ecuador, Peru, Bolivia, Tanzania, and the Philippines as part of a program of the Zoological Museum, University of Copenhagen (ZMUC), where Dr. Fjeldså is curator of the bird collections, chairman of the biodiversity program, and advisor to some 13 master's students and Ph.D. candidates.

Dr. Fjeldså holds positions on boards or other decisionmaking bodies of the ICBP, the Danish Ornithological Society, the Scandinavian Ornithologists' Union, the International Ornithological Commission, and the Danish advisory committee for CITES, among other groups.

■ Resources

Drs. J.J. Vlug and Fjeldså recently published a *Working Bibliography of Grebes of the World*, through ZMUC. According to Dr. Fjeldså, "The main intention behind the bibliography is to stimulate the further development of grebe research and to strengthen the focus on grebes in conservation biology." Dr. Fjeldså has published more than 210 articles, books, and chapters, and has illustrated some 55 publications.

(See also International Council for Bird Preservation [ICBP] and International Waterfowl and Wetlands Research Bureau [IWRB] in the UK NGO section.)

International Solid Wastes and Public Cleansing Association (ISWA) ⊕
Vester Farimagsgade 29
DK-1780 Copenhagen, Denmark

Contact: **Susan McCarty**
Phone: (45) 3 315 6565, ext. 281
Fax: (45) 3 393 7171

Activities: Education • *Issues:* Energy; Health; Sustainable Development; Transportation; Waste Management/Recycling

● Organization

The **International Solid Wastes Association (ISWA)** emphasizes information exchange to improve systems for handling solid wastes. Besides having members in 60 countries, the Association also has, in its words, "a very close working relationship" with the World Health Organization (WHO), United Nations Environment Programme (UNEP), UN Development Program (UNDP), and other international organizations.

International officers include Han den Dulk of the Netherlands (president), John H. Skinner of the Office of Research and Development, U.S. Environmental Protection Agency (EPA) (vice president), and William S. Forester, head of the Washington subsecretariat. In Denmark, Jeanne Møller heads the General Secretariat, and Niels Jørn Hahn is chairman of ISWA's Recycling Working Group. The Danish Waste Management Association is the Danish National Member of the ISWA.

▲ Projects and People

The Association has six active working groups, the most recently formed being the Sewage and Water Works Sludge Working Group, founded in 1991 and chaired by Erik Bundgaard. Members join a working group depending on their interests, and each group organizes about one conference or seminar each year, sometimes in close cooperation with a national member or another organization. The Hazardous Wastes Working Group, for example, organized a conference on the Environmental Impacts of Hazardous Wastes held in Poland in late 1991.

A three-day conference on Waste Minimization and Clean Technology, a cooperative effort of the Working Group on Hazardous Wastes and the British Institute of Wastes Management in conjunction with the UNEP Industry and Environment Office was held in 1992. The goal of ISWA members is to reduce or eliminate wastes at their source(s). The Working Groups on Sanitary Landfill, Incineration, Collection and Transport, and Recycling organized or participated in 1992 conferences as well.

The Association as a whole also holds annual and quadrennial conferences; the 1991 Toronto conference covered "An Integrated Approach to Solid Waste Management," featuring current trends and waste management activities in North America, Eastern and Western Europe, and Japan. The quadrennial Sixth International Solid Wastes Congress and Exhibition was held in Madrid in 1992 on the theme "Towards Future Practice."

ISWA has sponsored the 1992 International Symposium on Anaerobic Digestion of Solid Waste, and the first International Conference and Exhibition of Biological Waste Treatment in 1992 provided an opportunity to share current global information "on the possibilities of biological waste treatment from research to practical application." At a packed conference sponsored by the University of Denmark's Environmental Engineering Laboratory, "wastewater sludge dewatering" was the topic. The 1993 Annual Conference is slated to take place in Jönköping, Sweden, on the theme "East meets West—Problems and Solutions."

■ Resources

ISWA members' research contributes to the International Solid Waste Professional Library Series, which includes such titles as *International Perspectives on Municipal Solid Waste and Sanitary Landfilling* and *Nitrogen in Organic Wastes Applied to Soils*, edited by Jens A. Hansen and Kaj Henriksen. ISWA conference proceedings are regularly published, and *Waste Management and Research: An International Journal for an International Problem* appears bimonthly.

International Youth Federation for Environmental Studies and Conservation (IYF) ⊕

Astridsvej 7-5-1
8330 Brabrand, Denmark

Contact: **M. Mohan Mathews**, Biologist
Phone: (45) 8 625 0190

Activities: Education; Research • *Issues:* Biodiversity/Species Preservation; Deforestation; Energy; Global Warming; Health; Sustainable Development • *Meetings:* CITES; UNCED; World Youth Summit 1992

● Organization

The **International Youth Federation** (IYF) was founded in 1956 under the auspices of the World Conservation Union (IUCN). The Federation is now a worldwide network of independent environmental youth groups, with 130 member organizations in 54 countries. The Danish chapter has one regular staff member and 50 to 100 volunteers, depending on the needs. The Federation is composed of four independent regional organizations, whose representatives meet every other year at a general assembly. There they elect a new board and discuss the activities and policies of the organization.

Although originally tilted toward Europe, the IYF has expanded in the last decade to include many organizations in the developing world, where the focus is on environmental issues. In its policy statement, the organization states that it "does not divorce the study and conservation of nature from social and political awareness and action. . . . We resolve to promote . . . the conservation and proper disposition of the world's resources, and the creation of better living conditions in harmony with the environment."

▲ Projects and People

Using networks to link their activities with those of other groups working on development, consumer, and peace issues, IYF tries to "involve young people . . . in environmental activities and to bring them in contact with nature," as well as to spread public awareness of environmental problems. To this end, the IYF offers **training courses** for youth leaders from different countries, who gather to "study issues [such] as **tree-planting, urban pollution, soil erosion,**" and the like, via field trips, lectures, and discussions. A former attendee notes the value of gathering "people who have different backgrounds but who share the same goals. . . . At the courses we acquire not only the knowledge, but also the enthusiasm we need . . . to carry out" the lessons learned.

In its capacity as "a **world forum** of young environmentalists," the IYF concentrates on such environment and development issues as tropical forests, agriculture, pesticides export to developing countries, ocean pollution, and acid rain.

While studying at the University of Madras, India, M. Mohan Mathews was involved in numerous public service environmental activities, such as a rodent control and safe grain storage project,

rural health camps, and a tree-planting campaign. His M.S. thesis on "Conservation of Nature and Natural Resources—Priorities and Prospects" was one of the first academic treatments of that subject in India. Mathews founded the **Madras Environmental Group** in 1980; organized the first Youth Leader Training Course in Environmental Education in the state of Tamil Nadu, which resulted in the formation of two environmental groups by participants; and was project coordinator for two IYF projects—one on **technology transfer** and one on pesticides exports, then later was IYF's secretary general. Mathews focuses on the environmental problems of developing nations, many of which have roots in developed nations; the Union Carbide disaster in Bhopal is one terrible example of the linkage between "increasing transfer of pesticide technologies to less-industrialized countries," he writes.

Mathews has prepared and updated several directories of European and Indian environmental NGOs.

■ Resources

Resources needed include fax and E-Mail facilities; funds for research and for establishing regional offices; slide projectors, screens, and other equipment to enhance networking and educational efforts.

IYF publishes a magazine for its members, *Taraxacum,* as well as booklets such as *Lessons from Bhopal* and *Environment and War.* The organization arranges camps and training courses, youth exchange opportunities, and "public actions tours," often in cooperation with other youth groups.

World Assembly of Youth (WAY) ⊕

4 Ved Bellahoj, 2700 Brnshoj
Copenhagen, Denmark

Contact: **Shiv Khare**, Secretary General
Phone: (45) 3 160 7770 • *Fax:* (45) 3 160 5797

Activities: Development; Education • *Issues:* Deforestation; Global Warming; Health; Population Planning; Sustainable Development; Water Quality • *Meetings:* Costa Rica Youth Workshop; UN; UNCED

● Organization

The **World Assembly of Youth** (WAY) grew out of an international youth council meeting in 1949. Its original goals were to provide information services and coordinate member activities. However, at WAY's first General Assembly, held in 1951, members "formulated the initial action programmes which were to lead to WAY's steady development." There are currently WAY chapters in at least 90 countries; Denmark's has a staff of 10. Members focus on youth participation in development, youth education, environmental education, and other causes. Education is paramount in "non-violent, peaceful, and democratic means" to achieve change and address grievances.

Participating countries form national committees made up of the main youth organizations in each country. According to WAY, a typical committee includes "youth sections of the major political parties, rural youth, young workers' organizations, religious youth groups, service organizations, student and educational groups, and cultural organizations." Each of these committees has a representative in WAY's General Assembly, which meets every four years. The Assembly in turn elects an executive committee of 16 members, divided into 5 subcommittees, which meets more often and is responsible for the organization's administration.

▲ Projects and People

A large part of WAY's effort is concentrated on developing countries, where the organization seeks to "provide constructive leadership and an effective framework" for the hopes and energies of youth. The organization works to promote the "economic and social advancement of all people" and to uphold the Universal Declaration of Human Rights.

In recent years, WAY has expanded its programs to include population, drug abuse control, adult education, and the environment, among others. Activities with rural youth "have dealt with the pressing problems of the agrarian world today, the need to transform social structures and attitudes, to reform land tenure systems, and to revolutionize agricultural methods in order to feed the rapidly growing world population."

■ Resources

WAY seeks an "innovative program" of environmental education to enrich its current program.

The organization conducts workshops, offers consultations, and does educational networking. WAY has several publications, including *Rural Youth Information Service*.

WWF Verdensnaturfonden—Denmark
World Wide Fund for Nature (WWF)—
Denmark ⊕
Ryesgade 3 F
2200 Copenhagen, Denmark

Contact: **Lene Witte**, Secretary General
Phone: (45) 3 536 3635 • *Fax:* (45) 3 139 2062

Activities: Development; Education; Research • *Issues:* Air Quality/Emission Control; Biodiversity/Species Preservation; Deforestation; Energy; Global Warming; Sustainable Development; Waste Management/Recycling; Water Quality
Meetings: CITES

● Organization

The WWF Verdensnaturfonden—Denmark (World Wide Fund for Nature) (WWF) was founded in Denmark in 1972 under president Hans Kongelige Hojhed Prinsen. With 42,000 members and 17 staff, the organization participates in international WWF projects, engages in fundraising to save rainforests, pandas, and other endangered species or habitats, and works on specifically Danish conservation projects as well.

▲ Projects and People

The object of **Project Vadehavet i Jylland** is to save a 950-square-kilometer area on the North Sea that is rich in flora and fauna.

Internationally, the Danish WWF is working to save the rainforest of Pacaya-Samiria in northeastern Peru. This involves a five-year fundraising campaign, "Junglevagt for Amazonas," in collaboration with the Arbejderbevaegelsens Internationale Forum (AIF) and several other national organizations, and the efforts of Lillian Knudsen and Niels Hausgaard. Another international project concentrates on the Tai National Park in West Africa, a UNESCO-sponsored preserve, where WWF helps fund park rangers and basic conservation and antipoaching efforts.

Since 1991, Lene Witte, trained as a lawyer, has been secretary general of WWF—Denmark. Before that, she worked for 12 years in the Ministry of the Environment.

■ Resources

WWF-Denmark produces numerous pamphlets, brochures, and other publications. These include the quarterly magazine *Levende Natur* (Living nature); many of WWF's international publications, in Danish version; many Denmark-specific materials, including teaching materials; and the *Whale Report*, which provides updates on the struggle to limit commercial whaling and prevent the indiscriminate slaughter of whales. The Panda Club is WWF's children-oriented division.

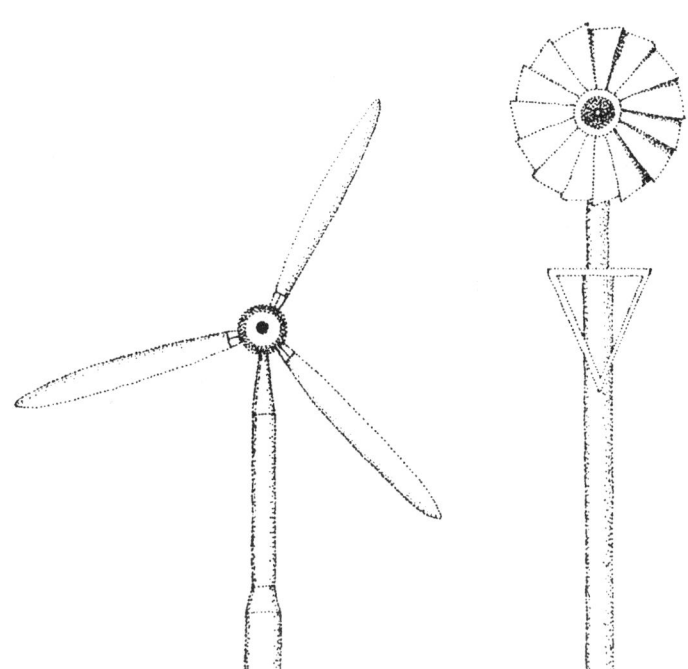

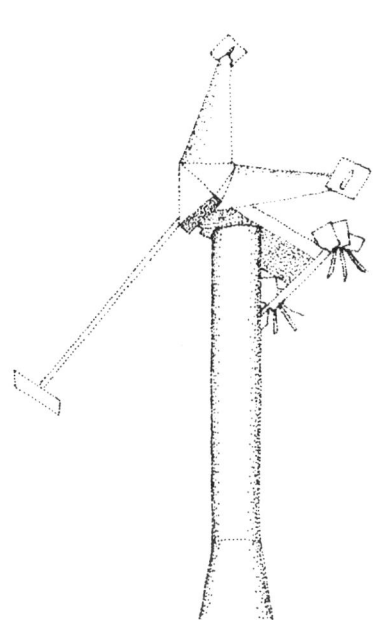

Dominica is the Caribbean's most mountainous and wettest island, receiving as much as 300 inches of rainfall each year. The climate along with the mostly rocky coastline has prevented the establishment of a significant tourism industry. Agriculture forms the backbone of the economy. The mountains shield one of the world's last oceanic rainforests, where rich flora and fauna abound, including 162 bird species, many of them threatened due to poaching.

🏛 *Government*

Forestry and Wildlife Division
Botanical Gardens
Roseau, Dominica

Contact: **Felix W. Gregoire**, Director of Forests, Wildlife, Parks
Phone: 1-809-448-2401, ext. 415 • *Fax:* 1-809-448-5200

Activities: Development; Education; Law Enforcement; Research • *Issues:* Biodiversity/Species Preservation; Deforestation; Sustainable Development; Water Quality • *Meetings:* UNCED

▲ Projects and People

The Forestry and Wildlife Division encourages Dominicans to be a tourist for a day on their Caribbean "Nature Isle" and explore Emerald Pool, River Blanc, Trafalgar Falls, Bwa Nef Falls, Boiling Lake, Pan Lake, the historic Cabrits, Morne Diablotin, and the Valley of Desolation, for instance. Forestry Week promotes the "wise use of our natural resources for . . . economic and social advancement."

The Division fosters ecotourism, agroforestry, and watershed management as components of a national development strategy. "As planners, developers, and environmental collaborate to ensure sustainable development," writes director **Felix W. Gregoire**, "the Division has been placed center-stage in the move towards integrating conservation and development."

Division activities include completing a 10-year **Forest Sector Plan** (1989–1999) and a 10-year **Morne Trois Pitons National Park Management Plan** (1990–2000), both sponsored by the Organization of American States (OAS). With Food and Agriculture Organization (FAO) assistance, the Division completed a **Forest Management Plan**; with help from the International Council for Bird Preservation (ICBP), a management plan is being developed for the proposed Morne Diablotin National Park/Reserve that includes the forests of Syndicate. Within Dominica, the Division is helping to implement the Tropical Forestry Action Plan (TFAP) of the FAO and the World Bank. It is involved in "plans for Cabrits National Park, the proposed Soufriere/Scotts Head Development, and the upgrading of tourist sites island-wide."

Including Caribbean buckeye, flambeau, Dominica snout butterfly, red rim, and white peacock, some 55 butterfly species abound from the coastline to the woodlands. The following passage appears in Division literature: "Finally, while strolling down memory lane, we may sometimes see the child in us, chasing a butterfly, and long for the freedom that surrounds such a beautiful creature. But in describing our relationship with the lonely butterfly, H.A. Giles notes, 'I do not know whether I was then a man dreaming I was a butterfly, or whether I am now a butterfly dreaming I am a man.'"

During the egg-laying season, conservation efforts are underway for "Dominica's dragons" or the "peaceful, harmless," **Lesser Antillean iguana**—threatened by "loss of habitat and over-hunting."

To satisfy growing ecotourism and a demand for indigenous products, the Division is researching ways to increase the propagation of the **larouman plant**, which is used for Carib crafts. Other economically important plants are **bamboo, verti vert, roseau, bakwa,** and **tamarind.** About 20 native species of **bromeliads** are also found. Epiphytes, such as orchids, gain support, but not nutrients, from other trees. Their leaves catch rain water, conserve water when temperatures rise, and are "an intricate part of the rainforest ecology."

Line, loop, and horsehoe trails are maintained in the national park system, known for their "soufriere areas" of "hot rocks, steam, and hot sulphurous gases." While Dominica's natural forests are protected, commercial logging is accepted as a viable economic alternative through "multiple-use management" and "within the guidelines of environmental safety and sound scientific principles."

The Division owes a debt of gratitude to local naturalist **Daniel Green**, whose wildlife collection ranges from parrots and doves to land turtles, or *morocoy*.

■ Resources

The Forestry Division offers numerous brochures, *The New Forester* and other booklets, maps, posters, and slide sets regarding conservation, birds and other wildlife, and the Cabrits Peninsula and National Park.

🎓 *Universities*

(See Clemson University, Archbold Tropical Research Center [ATRC] in the USA Universities section.)

The Dominican Republic shares the island of Hispaniola with Haiti. Its environment and economy have been devastated by 139 hurricanes over the past 100 years, the effects of which have been compounded by flooding, deforestation and erosion, and forest fires.

The most serious of the Dominican Republic's environmental problems is soil erosion, brought about by poor hillside farming practices, slash-and-burn agriculture, poor drainage techniques, and land exploitation. As a result, virtually all native plant and animal species are threatened. Government efforts to enforce conservation legislation and manage protected areas have been weak.

🏛 *Government*

Parque Zoological Nacional—Zoodom ⊕
P.O. Box 2449
Santo Domingo, Dominican Republic

Contact: **Jose A. Ottenwalder, Ph.D.**, Research and Conservation Director
Phone: 1-809-562-2080 • *Fax:* 1-809-562-2070

Activities: Conservation Biology; Research • *Issues:* Biodiversity/Species Preservation; Captive Breeding; Environmental Education • *Meetings:* AAZPA; IUCN/SSC

▲ Projects and People

In 1991, **Dr. Jose A. Ottenwalder**, research and conservation director, **Parque Zoological Nacional—Zoodom**, worked on a number of research projects at the University of Florida, Gainesville, USA. With support from Wildlife Conservation International (WCI) and the Brehm Funds for International Bird Conservation, Dr. Ottenwalder's ongoing research projects include "population status and conservation biology of **sea turtles, crocodiles, flamingo, native endemic mammals,** and other threatened species in the **Greater Antilles.**"

On behalf of the Species Survival Commission (IUCN/SSC) Insectivore, Tree Shrew and Elephant Shrew Group, he prepared an Action Plan for the conservation of two extant species of the **West Indian mammalian insectivore** genus *Solenodon* in **Cuba, Dominican Republic,** and **Haiti.**

Ottenwalder is at work on a long-term captive breeding program (which began at the Dominican Republic's National Zoological Park in 1978) of Greater Antillean threatened native fauna including *Amazona, Aratinga,* and *Dendrocygna* birds; *Cyclura* iguanas, *Epicrates* snakes, and *Trachemys* slider turtles. Captive bred progeny are reintroduced into protected areas in the wild.

In cooperation with **Dr. Charles A. Woods,** University of Florida, and with support from the MacArthur Foundation, a conservation strategy and management plan is being developed for the **endangered Hispaniolan solenodon** (*Solenodon paradoxus*) in the Pic Macaya Biosphere Reserve, Haiti; and speciation and extinction of the endemic **West Indian shrews** of the genus *Nesophontes* are being studied as are fossil and subfossil bats from southwestern Haiti. A book is being prepared on the birds of Haiti.

Dr. Ottenwalder also developed a captive breeding program for native **Iberic fauna** of the Balearic Islands, Minorca, **Spain.** He also participated in the 1991 CITES mission to **Cuba** to "evaluate the local hawksbill sea turtle (*Bretmochelys imbricata*) fishery, including the turtle farm in Isla de la Juventud; evaluate possibility to conduct a CITES project to assess the population status of the Cuban (*Crocodylus rhombifer*) and American (*C. acutus*) crocodiles in Cuba, concerning government plans for the developing of intensive farming and eventual marketing of hides of the endangered Cuban crocodile (*C. rhombifer*)."

An advisor in the preparation of the Dominican Republic's **Biological Diversity Assessment Project,** carried out by the local wildlife authority and funded by World Wildlife Fund (WWF), Dr. Ottenwalder also conducts faunal inventories of mammals, bird, and reptiles in his country's National Park Armando Bermudez and National Park Jose del Carmen Ramirez.

Dr. Ottenwalder is a Caribbean and North American Regional Member of the World Conservation Union's Species Survival Commission as well as a member of the Sea Turtle, Crocodile, and Captive Breeding Specialist Groups. He also belongs to the International Waterfowl and Wetlands Research Bureau (IWRB)/International Council for Bird Preservation (ICBP) Flamingo Research Group, and advises Lizard, Crocodile, and Chelonian Groups of the American Association of Zoological Parks and Aquariums (AAZPA). He is a consultant to WIDECAST—Caribbean Sea Turtle Recovery Team and Conservation Network.

NGO

ENDA–Caribe
APDO 3370
Santo Domingo, Dominican Republic

Contact: **Lionel Robineau, M.D.,** Director
   Phone: 1-809-566-8321 • *Fax:* 1-809-541-3259

Activities: Development; Education • *Issues:* Biodiversity/Species Preservation; Deforestation; Health; Sustainable Development • *Meetings:* CITES

● Organization

Enda–Caribe, which works for environment and development in the Caribbean that serves community needs, began its activities in 1978, mainly in the Lesser Antilles. Signing an accord with the Dominican Republic government, ENDA–Caribe in 1982 permanently located in Santo Domingo, with space in the National Botanical Garden. The ENDA–Caribe team is comprised of 28 professionals: artisans, technicians, agronomists, architects, sociologists, and medical doctors of Latin American and European background.

Interest areas are agroforestry, alternative technology, environmental health, community development, health and medicine, water, communications, and information. Goals include increasing the level of community organization, participation, and education through linking technical and social development; encouraging self-reliance; combining ideas and training into action; and exchanging experiences and techniques with others in the Third World. Other ENDA offices are in Bogota, Colombia; La Paz, Bolivia; and Dakar, Senegal. *(See also ENDA–Bolivia in the Bolivia NGO section.)*

▲ Projects and People

In 1983, ENDA–Caribe developed the DESIZ (Comprehensive Rural Development in Zambrana) program. A DESIZ Agroforestry Project is experimenting with techniques to benefit small farms. An extension project is also being developed to involve **farm women in agricultural production** by cultivating family vegetable gardens which helps improve the community's diet. A DESIZ **Social-Technical Project** is researching and experimenting with alternative technologies for construction, especially with bamboo, tiles, and blocks of earth.

A DESIZ **Environmental Health Project** implements experimental techniques in rebuilding dilapidated houses and building latrines, kitchens, stoves, and community centers. Workshops and seminars use audio-visual materials to train in low-cost construction techniques. **Utilitarian Craft Projects** help to organize small community businesses for construction and repair of simple farm tools and production of traditional utilitarian crafts, such as products woven from local natural fibers.

Collaborating with the NGO Inter-Aide, a DESIZ **Rural Hydraulics Project** attempts to improve the quality of rural drinking water through constructing wells and raising community awareness. **Agroforestry research**, reaching farmers in different areas, involves soil conservation and cultivation of rapidly growing trees, plants along terraces and streets, and family vegetable gardens.

Started in 1984 with the cooperation of anthropologists, botanists, pharmacists, and physicians in Haiti and the Dominican Republic, **TRAMIL** is a program of applied research in **popular uses of medicinal plants in the Caribbean.** Studying ways in which families use plants to alleviate illness, TRAMIL holds international scientific and public seminars on reevaluating traditional therapies. An urban program in La Chorrera, a section of Santo Domingo, aims to **improve sanitary, economic, and social conditions** especially through activities geared to women and children and that focus on environmental health, flood control, neighborhood cleanups, small business promotion, and health education, for instance.

For public use, ENDA–Caribe's **documentation center** contains at least 2,500 publications and audio-visual materials organized according to the SATIS system. Information is processed with the aid of a computer program developed by UNESCO. In the planning stages, with UNICEF's cooperation, is REDESC—a National Network for Documentation of Community Development. ENDA–Caribe also edits scientific and popular literature, primarily about agroforestry and medicinal plants.

■ Resources

Available in Spanish, English, and French is *Hacia una Farmacopea Caribeña.* Materials printed in Spanish include *Construcción de Viviendas con Bambu* and *El Arbol al Servicio del Agricultura.*

Ecuador is a mountainous country on the northwest coast of South America and the home of the ecologically unique Galapagos Islands. The country may have the greatest number of plant species of any South American country.

The country faces three main environmental problems: natural erosion caused by rivers and aggravated by overgrazing, uncontrolled cultivation, and deforestation; deforestation, accelerated by colonization efforts and petroleum exploration and logging; and desertification, mostly along both the southern coast and the border with Peru. A successful tourism trade in the Galapagos Islands and a debt-for-nature swap that channeled money toward environmental protection have been positive factors in encouraging conservation efforts.

 *NGO*

Amigos de la Naturaleza de Mindo/Friends of Nature
Manosca 1011, Residencias Altamira, Casa 07
Quito, Ecuador

Contact: **Maria Elena Garzon J.**, Vice President
Phone: (593) 2 455 344

Activities: Development; Education; Law Enforcement; Research • *Issues:* Biodiversity/Species Preservation; Deforestation; Health; Sustainable Development; Waste Management/Recycling

● Organization

The ecological corporation **Amigos de la Naturaleza de Mindo (Friends of the Nature of Mindo)** was founded in 1986 at the initiative of local residents to promote the message of preservation and respect toward the natural habitat. The Friends are volunteers working in approximately 48,000 acres of protected tropical forest—including the Páramo, Andean Forest, and Lower Mountain Rainforest. Here are found 500 bird species of "incomparable beauty such as toucans and hummingbirds, and mammals, such as the spectacled bear, puma, deer, wild boar, raccoon, and agouti." Support is provided from the World Wildlife Fund, Ecuadorian Center for Community Development (CEDECO), and the German Service of Social/Technical Cooperation (DED).

▲ Projects and People

The Friends submit **monthly reports** on the status of the forest and wildlife to appropriate government officials to engender support for their work. A plant nursery helps keep native species alive. **Environmental education** is provided in the local schools; talks and slide shows are presented to adults. Promotion of **sanitation** and assistance in "making **organic farming** a viable alternative for local families" are offered. **Nature tours** in the forest are also planned.

■ Resources

These include the Ecological Education Center and cabins for students or others in the protected tropical forest. Amigos de la Naturaleza de Mindo would like assistance with developing ecotourism in a nondestructive manner and exploring other sustainable uses of the forest. The Friends request audiovisual equipment and educational materials regarding tourism.

EcoCiencia
Ecuadorian Foundation for Ecological Studies
Ave. 12 de Octubre 959 y Roca. Off. 701
P.O. Box 17-12-00257
Quito, Ecuador

Contacts: **Danilo Silva**, Executive Director
　　　Luis Suárez, Research Fellow
　Phone: (593) 2 502409 • *Fax:* (593) 2 565759

Activities: Education; Environmental Communications; Protected Area Planning/Management; Research • *Issues:* Biodiversity/Species Preservation; Communications; Deforestation; Sustainable Development • *Meetings:* IUCN; Ramsar; World Parks Congress

● Organization
Founded in 1989, EcoCiencia, the Fundación Ecuatoriana de Estudios Ecológicos (Ecuadorian Foundation for Ecological Studies) is a private, nonprofit, scientific organization dedicated to communication, education, and research for the conservation of wildlife species and their habitats. EcoCiencia's programs seek alternatives for use and management of ecosystems that allow the preservation of biological diversity and natural resources while continuing to fulfill human needs. EcoCiencia serves as the coordinator for the Ecuadorian committee to the World Conservation Union (IUCN).

▲ Projects and People
EcoCiencia's activities are divided into communications and outreach, education and training, planning and management of nature reserves, and research. It views the relationship between research, education, and communication as imperative when presenting data and study results to the public. Some researchers and professors are with the Pontifical Catholic University's Biological Sciences Department, Quito.

The Foundation has a micro-isis computational system documentation center containing 2,000 books, 2,500 journals, and over 5,000 articles on botany, conservation, ecology, and zoology. However, due to limited access and distribution, many investigations remain unpublished. EcoCiencia plans a conservation publication fund allowing national and foreign researchers the chance to circulate study results. A series of books geared toward children about the country's history and environmental education is also on its agenda. EcoCiencia scientists have been associated with the television series *Ecuador, Tierradentro* (Inside Ecuador), developed by Fundación Natura and recognized worldwide for its originality and novelty.

At the school level, the Foundation conducts awareness campaigns and courses on the correct use of natural resources, holds conferences on ad hoc issues, and provides technical services. The EcoCiencia's Center for Environmental Documentation develops various audiovisual programs on Ecuador's natural history. With the assistance of various other international organizations, EcoCiencia provides conferences and courses for visiting and local scientists. Training classes have also been offered to naturalist guides in Ecuadorian Amazonia, other tourism operators, and members of the police and armed forces.

Debt-for-nature swap beneficiary La Perla Protected Forest, with the country's last remnants of wet forest, and the dry area in the Jerusalem Protected Forest near Quito are subjects for management plans.

Universities have almost exclusively conducted research in Ecuador, but "do not have a historical tradition towards scientific research" resulting in "scant" studies with inappropriate personnel for their "correct accomplishment," reports EcoCiencia. Therefore, the Foundation began working with postgraduate students in conservation biology, research, and environmental education. As a result, EcoCiencia has executed national, global, private, and public collaboration studies geared toward its research programs. One international course, with sponsorship of the Smithsonian Institution and Wildlife Conservation International (WCI), led to a "management plan for the Antisana reserve and preliminary research in Zancudococha, a highly diverse area." Dr. Peter Feinsinger, University of Florida, instructed on plant-animal interactions, and Dr. Craig Downer, also a USA scientist, taught radiotelemetry for the wooly tapir.

A project identifying the status of tropical forest on western Ecuador, funded by Fundación Natura, was conducted earlier. Another study, financed by WCI, examines the impact of mining on the flora and fauna in the Podocarpus National Park, and similar research is being conducted in the Cayambe-Coca Ecological Reserve. WCI is also supporting a study on the economic use of wildlife. With the New York Botanical Garden, EcoCiencia is involved in a two-year project examining nontimber forest products in extractive reserves.

An annual grants program, sponsored by WCI, offers five students the opportunity to earn their degree and provide financial assistance for specific conservation studies.

■ Resources
Technical assistance and reports, audiovisuals, photographs, and a speakers program are available from EcoCiencia, which also seeks additional field equipment and funding for its programs.

Fundación Charles Darwin para las Islas Galápagos ⊕
Avinida 6 de Diciembre 4757 y Pasaje California
APDO 17-01-3891
Quito, Ecuador

Contact: **Alfredo Carrasco V.**, Secretary General
　Phone: (593) 2 241 573, (593) 2 244 803
　Fax: (593) 2 443 935

Casilla 09-01-10355
Guayaquil, Ecuador

Contact: **Mario Hurtado**, Biologist
　Phone: (593) 4 310 617, (593) 4 307 558
　Fax: (593) 4 565 049

Activities: Education; Research • *Issues:* Biodiversity/Species Preservation; Sustainable Development • *Meetings:* CITES

● Organization

In 1959, the Fundación Charles Darwin para las Islas Galápagos (Charles Darwin Foundation for the Galapagos Islands) existed only on paper. Today, the Foundation includes an **established research station** where scientists from throughout the world take advantage of unique environmental opportunities. The small and relatively isolated islands allow one to investigate the "fragile relationship between the fauna, flora, and habitats of the terrestrial and the marine areas." The "enchanted islands," a volcanic area, are graced with flightless cormorants, marine iguanas, Darwin finches, fur seals, sea lions, nearly extinct giant tortoises, green turtles, blue-footed boobies, albatrosses, and feral goats, among others. However, threats to this ecosystem are many and preservation is a major priority.

▲ Projects and People

A history of human interference, including land encroachment and the impact of tourism, has endangered many unique species. In addition to hunting various animals, many islanders wished to open up protected areas for raising cattle. By providing **education in natural history, biology, and conservation**, the Foundation gained support and understanding for their struggle for the preservation of native species.

However, significant damage has occurred through introducing certain exotic species such as rats, dogs, pigs, and goats that threaten native species through predation and depletion of resources. "Some exotics would eat the young, and some would eat their food," it is pointed out. In 1965, **Roger Perry**, former Foundation director, initiated a **captive breeding program** with a group of **Pinzon** tortoises, 40 years or older, "as the rats were killing off every hatchling with no young reaching breeding age." Tortoise eggs were dug up and incubated in converted bird cages, and eventually 100 young were raised. With the help of the San Diego Zoological Society, the captive breeding program was expanded to include other native species.

A national interest in the islands' preservation was insured in the late 1960s with the formation of the **Galapagos National Park Service (GNPS)**. The GNPS and the Foundation, while not officially connected, have a history of mutual cooperation. A great victory for both was the declaration of all the internal waters of the archipelago as a reserve in 1986.

In addition to research, the Foundation continues to provide educational opportunities for local students and to help the environment cope with damage produced by fire, flooding, and exotic species.

■ Resources

Noticias de Galápagos describes the Foundation's history in English. Help with fundraising is sought.

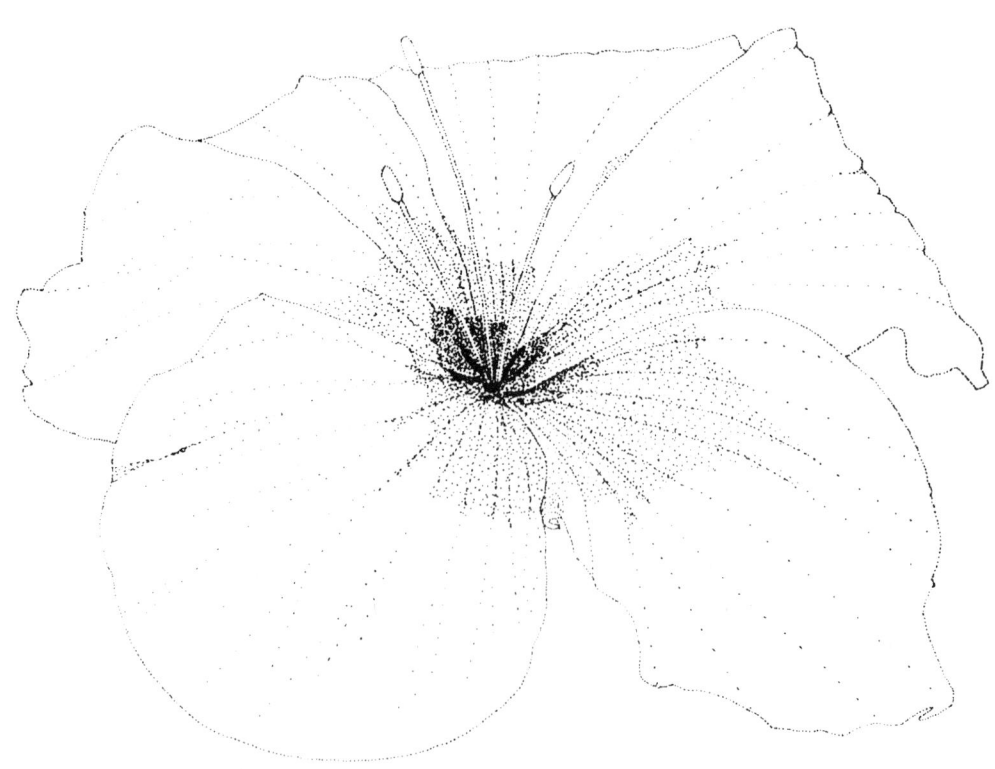

Egypt

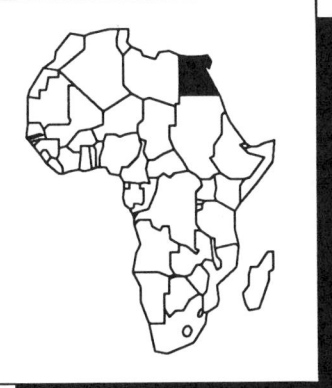

Egypt, with the highest population in the Arab world, is bordered by Libya, Sudan, and Israel, with the Suez Canal to the north, and coasts on the Mediterranean and Red seas. The entire country is a vast desert area except for the abundance of soil and water carried by the Nile River, which supports virtually the entire population.

Many of Egypt's environmental problems relate to the combined effects of intensive irrigation, extreme aridity, and high population densities. Soil damage, water pollution, and waterborne diseases are particularly prevalent. Soil damage, along with waterlogging, has reduced agricultural output by 30 percent in damaged areas. Water pollution has reached dangerous proportions in the highly populated areas north of Cairo, caused by salinized drainage water, agricultural pesticides, industrial effluents, sewage disposal, and overpumping of acquifiers. Poor sanitation contributes to the spread of waterborne diseases.

Oil pollution, caused by petroleum residue from ships, offshore oil facilities, and pipelines also presents a significant threat to coral reefs, commercial fisheries, and other coastal resources. Tourism is threatening to cause habitat degradation.

NGO

Egyptian Association for the Conservation of Natural Resources
Zoological Gardens
P.O. Box 318, Dokki
Giza, Egypt

Contact: **A. Maher Ali, Ph.D.**, General Secretary
Phone: (20) 2 703988 • *Fax:* (20) 2 3462029

Activities: Education; Research • *Issues:* Biodiversity/Species Preservation; Sustainable Development

● Organization

The Egyptian Association for the Conservation of Natural Resources is a non-government organization with 575 members "mostly devoted to wildlife conservation, conservation of nature and natural resources." Members are associated with Egyptian universities, Academy of Science and Technology, and various research centers. And a main aim is to foster **environmental awareness** through trips, lectures, and annual symposium, and through discussing conservation problems with government authorities.

Located at the Giza Zoo, its facilities include a wildlife library, conference hall, lecture room, and audio-visual materials. The Association maintains contacts with the World Conservation Union (IUCN), U.S. Fish and Wildlife Service, United Nations Environment Programme (UNEP), and United Nations Development Programme (UNDP).

▲ Projects and People

Dr. A. Maher Ali, professor emeritus, Assiut University's Plant Protection Department, who is general secretary of both the Association and the Egyptian Zoological Society, expresses his organization's concern about the human population increase that is contributing to "problems of pollution and of wilderness and natural resources."

Dr. Maher Ali is a scientific advisor of **Wadi El Assiuti Protected Area**, one of 12 such areas in Egypt—with **Ras Mohammed Protected Area** south of Sinai as the most important one," he writes. Wadi El Assiuti is devoted to "breeding and **propagation of wild animals** of the Eastern Desert that are now endangered species," he adds. "There is an additional breeding and propagation of **wild plants** which are of **medical importance**."

Author of a textbook on *Pest Control*, Dr. Maher Ali is also the recipient of a conservation merit award from the World Wildlife Fund (WWF) and other citations from the Egyptian government.

■ Resources

The Association publishes an annual scientific journal, *Egyptian Journal for Conservation of Wildlife and Natural Resources*, for which Dr. Maher Ali is chief editor. The Association wants to join "any meeting, conference, study, or project concerning diversity and/or conservation."

🎓 *Universities*

Solar and Space Research Branch
National Research Institute of Astronomy and Geophysics
Helwan-Cairo, Egypt

Contact: **Mosalam Ahmed Shaltout, Ph.D.**, Vice Dean and Professor
Phone: (20) 2 780046, (20) 2 780645
Fax: (20) 2 782683

Activities: Research • *Issues:* Air Quality/Emission Control; Energy; Sustainable Development • *Meetings:* UNCED

▲ Projects and People

With Egypt facing an energy shortage, Dr. Mosalam Ahmed Shaltout is taking a leadership role among the country's scientists in the field of clean, renewable energy. A professor of solar energy physics and vice dean of the **Solar and Space Research Branch, National Research Institute of Astronomy and Geophysics**, Dr. Shaltout is also the vice president of the Egyptian Solar Energy Society (ESES) and Egyptian secretary of the International Solar Energy Society (ISES).

"The new world order, so often referred to in the aftermath of the Gulf War, gives us the chance to solve our real problems of survival on this planet," writes Dr. Shaltout. "There is no doubt that the time of polluting, risky energies is coming to an end. Clean energies are needed for our survival."

Dr. Shaltout points out that "if Egypt continues to consume energy at its present annual rate of 10 percent and no large petroleum finds are made, then Egypt will be forced to import oil early in the next century," writes Dr. Shaltout, also draining Egypt of hard currency.

"Of the new [energy] sources," says Dr. Shaltout, "solar energy is the one from which all other renewable energies branch out. Every year the sun drenches Egypt with. . .one of the highest rates in the world."

Working with **solar radiation and solar energy projects**, he is dedicating his research to overcoming technical and economic problems. One such obstacle is adequate storage of solar and renewable energy to "provide power that could be generated during intermittent clouds, overcast sky, at night or during summer doldrums." Another problem is lack of government funds to support such research. A third hurdle is lack of public awareness. "People are always afraid to use something they don't know," he says. "As long as fossil fuels were plentiful and easy to get, it was considerably more profitable to collect and sell these energy sources than to capture the current energy emissions."

Arguing for the wide use of renewable energy before the end of the century, Dr. Shaltout is urging the government to take these steps: "improvement in technology, demonstrations of its practicality in different regions of the country, reform of building codes and zoning laws, and incentives such as tax breaks, low-interest loans, and outright subsidies."

With such initiatives, "renewable energy can supply Egyptians with five percent of their energy needs by the year 2005," he predicts. According to Dr. Shaltout, the Egyptian Ministry of Electricity and Energy has formulated a "national strategy for the development and widespread use of new and renewable energy systems . . . such as solar thermal, photovoltaic, wind, and biomass technologies."

In April 1992, ESES and the Egyptian Society of ISES co-sponsored the **International Conference on Applications of Solar and Renewable Energy** in Cairo, which reminded the participants of "how volatile and unpredictable the oil market is as well as its global warming effect due to carbon dioxide emission." Dr. Shaltout and ESES chairman **Dr. H. El-Agamawy** also believe that nuclear energy "is not the option for the nineties" due to problems with waste disposal and accidents such as Chernobyl. The conference covered topics such as passive solar architecture, solar pond, solar desalinization, photovoltaics, wind energy, bioconversion, rural and desert development, and industrial applications.

With the Academy of Scientific Research and Technology, Dr. Shaltout published the first Egypt Solar Radiation Atlas entitled *Solar Energy Input to Egypt* in 1981. Among his 48 published papers, 26 are in the field of solar energy physics; the remainder are in the field of solar-terrestrial physics.

■ Resources

Dr. Shaltout has reprints of scientific papers, among them: "Solar Radiation and Air Pollution in Cairo"; "Solar Energy Characteristics and Some Photovoltaic Testing Results in Jeddah"; "Estimation of the Different Components of the Solar Radiation over Egypt from the Meteorological Data"; "Solar Energy Distribution over Egypt Using Cloudiness from Meteosat Photos," with **Dr. A.H. Hassen**; "Variations of the Solar Activity and Irradiance and the Influence on the Flooding of the River Nile," with **Dr. M.T.Y. Tadros**; and "Photovoltaic Activities in Egypt: Applications and Research" with **Dr. Reda Botros**.

Focal Research and Training Programme Support Unit (PSU)
Supreme Council of Universities
P.O. Box 982
Cairo, Egypt

Contact: **Salwa Rasmy, Ph.D.**, National Project Director
Phone: (20) 2 342 2330 • *Fax:* (20) 2 341 4255

Activities: Education; Research • *Issues:* Health; Pollution; Sustainable Development

▲ Projects and People

In 1991, the **Focal Research and Training Programme Support Unit (PSU)** was established under the aegis of the **Supreme Council of Universities**, with headquarters at the Faculty of Medicine, Cairo University; overall coordination from **Dr. Khairy Samra**, Dean of Faculty; and assistance from the United Nations Development Programme (UNDP).

"The main objectives of the PSU are to mobilize technical assistance resources for training and research in priority areas for national development where activities focus on environment and health-related issues during the first stage," writes **Dr. Salwa Rasmy**, PSU director and microbiology professor at Cairo University.

Such assistance can be made available from "national, bilateral, multilateral, or international sources," says Dr. Rasmy. PSU objectives conform to Egypt's "national plan of economic recovery," aim to improve career development prospects, and are "deemed desirable by both donors and recipients." With pilot projects in health and environment areas, PSU intends to "bridge the gap" and "marry the potential supply to the pressing demand" between those who can and want to give and are "faced by stumbling blocks on the part of local procedures and red tape" and those who "need the assistance [who] are not aware that it exists or could be obtained," writes Dr. Rasmy.

Recently returned from a World Bank consulting assignment on **environmental protection in the Arabian Gulf**, Dr. Rasmy also heads a research team at the Biotechnology Center for Pollution Control, also funded by UNDP. She recently coordinated an **Energy and Environment workshop** in Tripoli, which was sponsored by the International Energy Foundation and other UN agencies. "We have also proposed convening a regional/interregional symposium on **Women and the Environment** in association with the UN International Institute on Training and Research for the Advancement of Women (INSTRAW)" in Cairo.

El Salvador is the smallest and most densely populated Central American country. It is bordered by Guatemala, Honduras, and the Pacific Ocean. The country has few natural resources, aside from hydropower and geothermal energy potential. Agriculture is the main economic activity.

El Salvador's most serious environmental problem is deforestation, brought on by cultural traditions, inappropriate land tenure laws, a lack of economic incentives, population growth, inequitable land ownership, and dependence on fuelwood as an energy source. This in turn has led to soil degradation, loss of water quality, and the extinction of many native species of flora and fauna. Subsistence hunting is widespread. Pollution, caused by urban waste and toxic chemicals, is also a serious problem.

❦ *NGO*

Asociación Audubon Capitolo El Salvador
Residencial Palermo, No. 8-C, Pasaje No. 1, Autopista Sur
San Salvador, El Salvador

Contact: **Zoila Esperanza Peréz Molina**, Honorary President
Phone: (503) 74 1877, (503) 98 0338 • *Fax:* (503) 23 5267

Activities: Education • *Issues:* Biodiversity/Species Preservation

▲ Projects and People

Staffed with 200 volunteers, the **Asociación Audubon Capitolo El Salvador (Audubon Society)** is dedicated "para la defensa de las aves y su habitat," which means "for the defense of the birds and their habitat." With commitment to both native and visiting birds, 10 "brigades," or groups of student volunteers, have been formed for this purpose.

Devoted to the education of the local communities, the Audubon Society is organizing an **environmental education** office with formal and informal programs for adults and children. The office will have a library, educational materials for sale or rent, and essential equipment and space.

In addition, the Audubon Society hopes to begin bird breeding programs, such as **creating a bird habitat in the Pargue Infantil** with constructed nests and feeders to increase the number of species. Research on the number of resident and visiting birds is being published in a formal report. Park visitors are provided environmental education programs including discussions, films, and guided tours.

In another project, **15 breeding coops are planned for the common pigeon** to provide food for villagers. Petitioning the local community is one step toward realizing this plan.

■ Resources

A poster and script for environmental theater are available. The Society needs assistance with publishing a pamphlet and advertising through radio and television.

Centro Salvadoreño de Tecnología Apropiada (CESTA)
33 Calle Poniente 316, APDO 3065
San Salvador, El Salvador

Contact: **Ricardo A. Navarro, Ph.D.**, President
Phone: (503) 25 6746 • *Fax:* (503) 26 6903

Activities: Development • *Issues:* Energy; Global Warming; Sustainable Development; Transportation; Waste Management/Recycling

● Organization

Founded in 1980, CESTA, the Salvadorean Center for Appropriate Technology, carries out projects of both social and ecological benefit. CESTA defines "appropriate technology" as that which is necessary to reach and maintain a process of sustained development with equity in El Salvador.

CESTA is one of the 19 institutions comprising the Unidad Ecológica Salvadoreña (UNES), a national organization formed to work on a program of national ecological recovery that would affect every facet of society.

▲ Projects and People

CESTA describes the "Salvadorean reality" as one of economical, political, social, and environmental hardships. Some of the environmental deterioration issues involve the elimination of 80 percent of the natural vegetation, the destruction of the forest until only 6 percent remains in its original state, and the threat of extinction for many species of birds, reptiles, mammals, and freshwater fish. High rates of human malnutrition, infant mortality, death from parasites and infections, and illiteracy compound this situation.

CESTA works to preserve renewable natural resources to improve the quality of air, water, and land, and therefore the quality of life, especially for those with fewer resources. It intends to decrease technological and economical dependence on other countries by training local scientists and technologists who can increase benefits to more people—in a way that is less harmful to the environment.

This includes promoting bicycle and cart usage, manufacturing low-impact and sanitary latrines, forming ecological communities in both urban and rural areas, initiating recycling programs, and launching an environmental education campaign with presentations for professional associations, laborers, community groups, and schools.

■ Resources

CESTA publishes in Spanish *El Pensamiento Ecologista* (Ecological thoughts), *La Bicicletas y los Triciclos* (Bicycles and carts) and *Tecnología Apropiada*. The Unidad Ecológica Salvadoreña has prepared a comprehensive publication, *Propuesta del Cerro Verde* (Green hills proposal).

Estonia is located on the Baltic Sea in northern Europe. Ten percent of the country consists of the Baltic Islands. The terrain is chiefly low-lying plain; farmland covers 40 percent of the total area; forests cover 30 percent; and swamps cover 20 percent. The sandy western coast is resort area. The climate is unusually mild for being so far north.

Estonia enjoys a relatively high average income, with 70 percent of the population living in cities and towns. Industrial development was introduced during the Soviet occupation, bringing with it shortages and poor quality goods and services, as well as industrial pollution.

🏛 *Government*

Tallinn Zoo
Paldiski Road 145
Tallinn 20035, Estonia

Contact: **Tilt Maran**, Curator
 Phone: (7) 0142 55 99 44 (operator assistance)

Activities: Research • *Issues:* Biodiversity/Species Preservation • *Meetings:* All-Union Conference of Zoo Culture; All-Union Theriological Conference; Central-European Mustelid Symposium; First Baltic Theriological Conference; International Conference for Breeding of Endangered Species

▲ Projects and People

Curator **Tilt Maran** launched at the **Tallinn Zoo** a captive breeding program of the **European mink**, "one of the most endangered mammal species in Europe," he writes. An ecologist and nature conservationist, Maran studies minks in the wild as well as those in captivity. He also is conducting field studies "concerning the relationship between the European mink and the American mink."

Successful in breeding the European mink, he has 27 animals in captivity at the zoo. In addition, he plans to **organize an international committee for preserving the European mink**. "The committee will consist of the zoos from Scandinavia and other parts of Europe," he reports. "The main purpose is to coordinate the captive breeding program of the European mink and to monitor the wild population of this species in Estonia."

Working in the Tallinn Zoo's Theriological Laboratory, Maran is also involved in the activities of the Estonian Fund for Nature, a new organization with close ties to World Wide Fund for Nature (WWF) national organizations in Europe—especially in Scandinavia. He belongs to the Species Survival Commission (IUCN/SSC) Mustelid and Viverrid Group.

Ethiopia

One of the world's oldest countries, Ethiopia has been devastated by political strife. Accelerated population growth has led to serious land degradation, partly from adherence to traditional land-use practices and partly from recurrent drought conditions in many parts of the highlands. Deforestation is proceeding rapidly, along with the associated problem of soil erosion. A contributing factor is the collection of fuelwood to meet growing energy needs (current demand is much greater than the sustainable supply). Encroachment by people and grazing cattle into protected areas presents another major threat to biodiversity.

The Soil and Water Conservation Department has attempted to rehabilitate the ravaged lands, but has had difficulty enforcing laws and functioning effectively with sparse technical assistance and training.

🏛 *Government*

Ethiopian Valleys Development Studies Authority (EVDSA)
P.O. Box 1086
Addis Ababa, Ethiopia

Contact: **Zewdie Abate, Ph.D.**, General Manager
  Phone: (251) 1 513111, (251) 1 129084 • *Telex:* 21278 ADESV ET

Activities: Development • *Issues:* Air Quality/Emission Control; Biodiversity/Species Preservation; Deforestation; Energy; Global Warming; Health; Sustainable Development; Water Quality

● Organization

Because of the increasing pressures of an escalating birth rate, industrialization, and urbanization, Ethiopia's natural resources are suffering. According to **Dr. Zewdie Abate**, general manager of the **Ethiopian Valleys Development Studies Authority (EVDSA)**, "water is the most critical of these resources, and per capita consumption is rising even faster than population growth . . . due to the requirements of improving drinking water supply and sanitation, supplying water to new industries, and developing irrigation schemes to increase agricultural production." Dr. Abate describes the Ethiopian river valleys as "a national heritage," as they supply not only food and water, but also housing, clothing, firewood, and jobs.

Earlier valley development schemes lacked "overall coordination, and . . . a full examination of all the implications, especially the impact on the environment. This . . . has resulted in . . . soil erosion, deforestation, pollution, poaching of wildlife, drought, and sometimes flooding." In contrast, the tasks of the EVDSA involve careful planning and management of sustainable river valley development, "having due regard for their delicate environmental balance." This involves conducting studies of the effects of irrigation, and monitoring the environmental conditions in the valleys. The semi-autonomous Authority reports directly to the Council of Ministers.

▲ Projects and People

The EVDSA carries out its responsibilities in six basic steps. The first is to **inventory the natural resources** within the **14 river valleys** or basins under their jurisdiction, including land, air, water, and the corresponding flora and fauna, as well as mineral resources. After assessing the prospects for resource development, the EVDSA prepares "comprehensive development plans and strategies." The Authority then identifies and chooses "specific projects which make efficient and sustainable use of natural resources," and prepares to implement them. The final responsibility involves working "towards **equitable sharing of boundary and transboundary rivers**, lakes, and groundwater aquifers."

Dr. Abate, who has managed EVDSA since 1987, is a specialist in water resource management, sustainable development of renewable and nonrenewable natural resources, systems engineering, and environmental economics. Among other projects, he has directed a study of the "environmental impact of industrial effluents in the highly urbanized centre of Ethiopia." It is his responsibility to ensure "wise use of natural resources, . . . assessing and replacing resource reserves, protecting the destruction of forests, and combating the advance of **desertification**."

From 1975 to 1984, Dr. Abate was executive director of the Amibara Irrigation Project in Ethiopia, involving irrigation, environmental conservation, and settlement of a 25,000-acre area. In the course of the project, funded by the World Bank, African Development Bank, and European Community, Dr. Abate designed master plans for water utilization drainage, "to enhance sustainable development of the area." At Harvard University's John F. Kennedy School of Government, Dr. Abate researched sustainable development policy choices for the drone-prone African Sahel.

In 1992, Dr. Abate wrote *Observed Context of Environment and Development in Ethiopia: An Approach That Needs Development*, which was published by the Centre of Near and Middle Eastern Studies and the Centre of African Studies, School of Oriental and African Studies, University of London. One conclusion is that Ethiopians must plant about 5 million hectares of fuelwood by the year 2000 "to begin to contain deforestation" resulting from "clearing for agriculture and scavenging for firewood." Other recommendations are offered.

■ Resources

Dr. Abate has written and/or presented numerous publications, studies, and papers, including several on the need for integrated and cooperative Nile Basin development.

Ethiopian Wildlife Conservation Organization (EWCO)
P.O. Box 386
Addis Ababa, Ethiopia

Contact: **John Osborne**
Phone: (251) 1 15 44 36

Activities: Development; Education; Law Enforcement; Political/Legislative; Research • *Issues:* Biodiversity/Species Preservation; Deforestation; Sustainable Development; Wildlife Utilization • *Meetings:* CITES

■ Resources

The Ethiopian Wildlife Conservation Organization (EWCO) publishes *National Park Guides* and *Information Sheets, Environmental Education Magazine* for schools, *Wildlife Clubs Magazine* for members, and various other educational materials and audiovisuals. EWCO seeks educational materials and training.

 *Universities*

(See Washington University, School of Medicine, Awash National Baboon Project in the USA Universities section.)

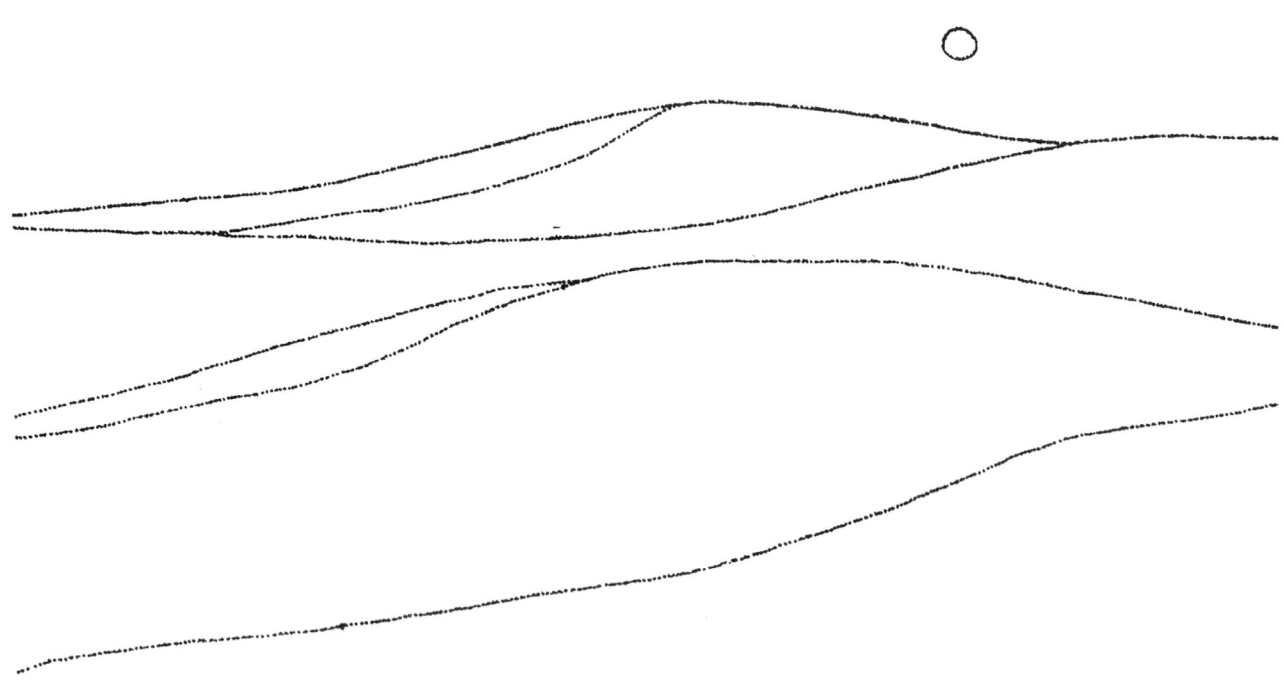

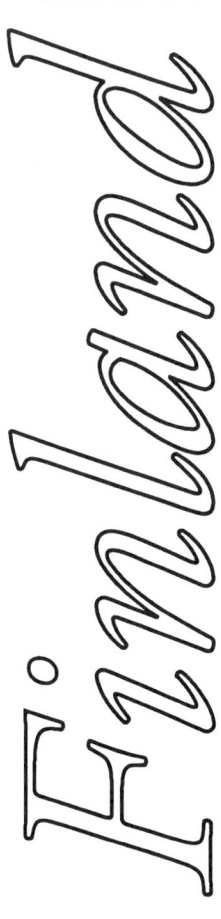

Finland

Thousands of lakes dot Finland's landscape, and thick forests cover more than 70 percent of the land—a higher percentage than in any other European country. Most of the population is concentrated in the southern part of the country, where the climate is milder.

Finland depends heavily on other countries for trade and for environmental cooperation. The Finnish government has developed an Action Programme for participation with countries in Eastern Europe, mainly on environmental protection measures—including the provision of training and research, institution building, and environmental impact assessment and monitoring. Priority is being given to the problem of airborne and waterborne pollutants reaching Finland and the Baltic. In 1990, a cooperative agreement was reached by Finland and Poland, based on the "debt-for-nature" principle.

🏛 *Local Government*

Urban Planning Office of the Town Espoo
Kirkkojärventie 6
02770 Espoo, Finland

Contact: **Heikki Wuorenrinne**, Environmental Researcher
Phone: (358) 90 869 3223 • *Fax:* (358) 90 869 2895

Activities: Research • *Issues:* Biodiversity/Species Preservation; Population Planning; Town Planning; Transportation • *Meetings:* Leipziger Symposium Urbane Ökologie

● Organization
The **Espoo Urban Planning Office** is involved in applying biological knowledge to the practice of urban planning.

▲ Projects and People
Research projects include **Isolated Woodlots as Ecological Inseln**, conducted by **Margareta Ehnberg** under the leadership of **Heikki Wuorenrinne**.

Another project is the **habitats diversity of town parks and greeneries in semiurbanized South-Espoo** by Wuorenrinne with students from the University of Helsinki's Botanical Institute. In preparation since 1988, this project will describe and classify the habitats of the town parks and greeneries—of which there are approximately 500 in South-Espoo. It will also attempt the cleanup of the parks and study the environmental effects of urban pressures. The project will estimate the future of the natural development of the parks, and the future availability of the parks for people.

Wuorenrinne's previous research with ant pupas underscored the **importance of ant colonies to forest ecosystems**. Although ant pupas have been collected in Central Europe since the 1700s, she writes that commercial exploitation, which removes the whole mound, can destroy the colonies. Mounds collected for scientific purposes can best be found in cattle-grazing forests or "burn-over areas." Preferring a light, 40- to 80-year-old spruce-birch mixed forest, ants seem to control certain insects that damage trees, according to scientists. Before World War II, farmers used the mound material for "manure, as litter in cowsheds, and in storing potatoes."

■ Resources
The Urban Planning Office has a number of publications in German, English, Swedish, and Finnish. Those in English include "The Influence of Collection of Ant Pupas upon Ant Populations in Finland," "Effects of urban pressure on colonies of *Formica rufa* group [red wood ants]," and a paper on the metal content of ants "collected in the vicinity of spruces showing different degrees of needle loss."

 *NGO*

Confederation of Finnish Industries (CFI)
Eteläranta 10, P.O. Box 220
00130 (1) Helsinki, Finland

Contact: **Esa Tommila**, Director, Environment and
          Transport Affairs
Phone: (358) 0 18091 • *Fax:* (358) 0 1809 209

Activities: Political/Legislative • *Issues:* Air Quality/Emission
Control; Global Warming; Industrial Environmental Policies; Industrial Transport Policies; Sustainable Development;
Transportation; Waste Management/Recycling; Water Quality

● Organization
The **Confederation of Finnish Industries (CFI)** is a national NGO
with a staff of 10.

▲ Projects and People
The Confederation has published several pamphlets to encourage
and support environmentalism within Finnish industry. According
to *Finnish Industry and the Environment*, Finland has a per capita
GNP of $US21,000, which puts Finland in the Organization for
Economic Cooperation and Development's (OECD) top 10. Industry and agriculture are responsible for 35 percent of the overall
product. "Because of its northerly location and energy-intensive
industrial products, Finland is a significant energy consumer. . . .
Nevertheless its energy efficiency is high by international standards."

    Environmental standards in Finland, which industry must re-
spect, are strict. Water pollution control and the use of new technology have led to a decrease in effluents coinciding with an increase in
production, according to CFI. Since the 1983 Air Pollution Control
Act, air quality has improved. "The remarkable progress in various
emissions reductions is largely due to two factors: the modern
legislation allowing individual control solutions and long term positive dialogue and program cooperation between industry and the
authorities," reports the Confederation.

    Finland has also been the source of a number of **pollution control
innovations** including the flash smelting process for copper, nickel,
and lead; advanced pressurized grinding for mechanical pulp production that saves water, wood, and paper; a method of flue gas
desulphurization used in retrofitting power plants; and the circulating fluidized bed boiler that efficiently binds sulphur and prevents
"the production of nitric oxides in power generation." CFI reports
that Finland has "long made use of its own recycled raw materials
and waste," and these levels are increasing."

    CFI also published *Guidelines for Environmental Care and Protec-
tion in Finnish Industry* and *Guidelines for Environmental Arguments
in Marketing.* According to the former, "Minimizing the harmful
effects on the environment is an indispensable part of industrial
operations. . . . Industry has the important duty of insuring the **safe
manufacture, treatment, transportation, use, and disposal** of its
product." **Guidelines for member companies** cover the following
areas: holistic thinking, recognizing responsibility, duties of corporate management in ensuring that environmental standards can be
met, personnel know-how, exceptional situations, interaction with

authorities, clients and consumers, and communication. The guidelines for marketing include ensuring that a package is truthfully
labeled with information about the environmental effects of the
product.

■ Resources
The above pamphlets are available in English.

Helsinki Zoo ⊕
Korkeasaari
SF-00570 Helsinki, Finland

Contact: **Leif Blomqvist**, Zoologist
Phone: (358) 0 199 8224 • *Fax:* (358) 0 135 4551

Activities: Education; Research • *Issues:* Biodiversity/Species
Preservation; Health; Population Planning

▲ Projects and People
At the Helsinki Zoo, Leif Blomqvist is International Studbook
Keeper of Snow Leopards (*Panthera uncia*). The Studbook contains
resolutions of the International Snow Leopard Symposium and
describes the distribution, status, and studies of snow leopard populations as well as management of captive populations.

    In 1989, the Symposium urged the **establishment of additional
protected areas, including national parks**, with attention paid to
reserves on international boundaries such as the Western Tien Shan
in Kirgizia adjoining the Tomur Reserve in China and "in the
Pamirs where the Taxkorgan Reserve of . . . China and the
Khunjerab Park of Pakistan meet Tajikistan [Tadzhikistan] and
Afghanistan." Also designated for further study are habitat and
reserve areas such as those in the Altai Mountains where Kazakstan,
China, and Mongolia meet. Another resolution encouraged corridors between large reserves that could be "designated for multiple
use, including grazing lands and timber harvest so that both wildlife
and indigenous peoples may benefit."

    It was recommended that in the former USSR, zoos strengthen
their captive breeding efforts with a coordinated plan; that the
Academy of Sciences "create an informational working group" regarding snow leopards in the wild and captivity; and that study
programs be coordinated with the International Snow Leopard
Trust. The Alma-Ata Zoo was designated coordinator between the
former Soviet groups and the Trust.

    In Siberia, the snow leopard is regarded as "one of the rarest
animal species," found primarily in "mountain massifs predominately in the subalpine and alpine belts." Of 38 leopards sighted in
a 30-year period, half were shot. Hunting has been prohibited in
Tadzhikistan since 1968, where there are an estimated 200 to 300
animals living in areas where there are no trees or bushes. "In very
hard and snowy winters, they may descend to forested mountain
areas," writes A.I. Sokov, Institute of Zoology and Parasitology.

    The snow leopard preys mainly on hoofed mammals, particularly
the Siberian ibex, which shares its habitat. Hunting by hiding in its
surroundings, the snow leopard also eats rodents, mainly marmots,
and less often squirrels and moles. Sometimes, it catches birds. As
the deer population has increased, the snow leopard has started
hunting primarily adult males in wintertime. Males appear more
vulnerable as they are not as quick as females and the young; they are

also restricted by their big, heavy horns. On rare occasions, the snow leopard attacks domestic animals, such as calves, sheep, dogs, and swine. "The snow leopard has no natural enemies except human beings," it is reported.

Nepal is working on conserving the snow leopard, found on the country's northern border with Tibet. In the mountains are seven existing protected natural areas to preserve unique ecosystems and endangered species in general. However, snow leopard ranges are linear and often narrow; and large ranges are needed. Again, it is difficult to estimate the population in Nepal because the snow leopards are rare, shy, and inhabit some of the most "remote and forbidding terrain on earth," report **Rodney Jackson** and **Gary Ahlborn**, California Institute of Environmental Studies and the Trust. They report that the best way to "engender a conservation ethic is to improve the standard of living of the people who share the land with wildlife."

In other words, the snow leopard's habitat quality would be enhanced with "improved livestock management, reduction of hunting, and the promotion of income-generating jobs." Moreover, local communities need to become more involved in making land management decisions, write Jackson and Ahlborn. A "Conservation Area" concept would also place emphasis on "sustainable tourism, cottage industries, handicrafts, or other income-generating activities."

Hemis National Park, located in India's northernmost district of Ladakh, is perhaps the best protected area for snow leopards in India, according to **Joseph L. Fox**, University of Tromsø, Norway, and **Chering Nurbu**, Department of Wildlife Protection, Ladakh, Jammu, and Kashmir. Opened to tourism in 1974, the number of visitors is growing; northern areas of the park have become popular trekking routes. Currently one of the largest protected areas in the Himalayan region, conservation measures are underway to prevent human–snow leopard conflicts, as there are an estimated 50 to 75 snow leopards there. **Conservation education programs** in schools and programs within the park help to insure the protection of wildlife. The government of India's **Project Snow Leopard** aids local efforts.

Captive breeding efforts are ongoing at the Bronx Zoo, New York Zoological Society, USA; Woodland Park Zoo, Seattle, Washington; Novosibirsk Zoo, West Siberia; Alma-Ata Zoo, Kazakh; and Moscow Zoo, among others. Blomqvist emphasizes that today "no snow leopards should be caught for zoo purposes." Successful captive breeding programs are urged to transfer stock to China and the former USSR, where wild animals are still captured.

Blomqvist belongs to the Species Survival Commission (IUCN/SSC) Cat Group, among other groups.

Maailman Luonnon Säätiö Suomen Rahasto
World Wide Fund for Nature—Finland ⊕
Uudenmaankatu 40
SF-00120 Helsinki, Finland

Contact: **Mauri Rautkari**, Secretary General
Phone: (358) 0 644 411 • *Fax:* (358) 0 602 239

Activities: Nature Conservation • *Issues:* Biodiversity/Species Preservation • *Meetings:* CITES; Europe Nature Conservation 2000; HELCOM; IUCN; UNCED

☏ *Universities*

Department of Physiology
University of Kuopio
P.B. 1627
SF-70211 Kuopio, Finland

Contact: **Pirjo Lindström-Seppä, Ph.D.**, Junior Fellow, Academy of Finland/Research Council for the Environmental Sciences
Phone: (358) 71 163 095 • *Fax:* (358) 71 163 410

Activities: Education; Research • *Issues:* Health; Water Quality • *Meetings:* Congresses of the European Society of Toxicology; International Conferences on Biochemistry and Biophysics of Cytochrome P-450; International Symposium on Responses of Marine Organisms to Pollutants; Nordic Symposiums on Organic Environmental Chemicals; SETAC

▲ Projects and People

The **Biomonitoring Group** of the **University of Kuopio's Department of Physiology** has been studying the **biotransformation of different organisms** for almost 20 years. A major three-year program to study "xenobiotic biotransformation and the adaptation of organisms to chemical load" was designed to map out the biotransformation reactions and capabilities in different organisms "from microbe to man."

Another major project was work on **xenobiotic metabolism in fish.** This has included applied research on the **biomonitoring of pulp and papermill effluents.** The wood processing industry is a major source of pollution in Finland, and this project has monitored the effects on fish exposed to nonfatal doses of effluents.

One hope was that this study would provide ways of monitoring both the pollutants and their effects. Measuring "PSMO enzymes" and biotransformation products in fish has had some success, but it is still not clear which specific effluents relate to enzyme responses. "Despite this fact, the transformation responses studied can function well as early warning biomarkers of bleached pulp and paper mill effluents," reports Dr. Lindström-Seppä. She has participated in various international symposia on environmental biochemistry and toxicology as well as on responses of marine organisms to pollutants.

■ Resources

Research papers with English titles discuss biomonitoring the aquatic environment, including oil spills, and examining the responses of rainbow trout and feral fish to pulp mill effluents.

Department of Zoology
University of Oulu
Linnanmaa
90570 Oulu, Finland

Contact: **Erkki Pulliainen, Ph.D.**, Professor of
Zoology; Member of Parliament
Phone: (358) 981 511 227
Fax: (358) 981 511 827

Activities: Education; Political/Legislative; Research • *Issues:*
Air Quality/Emissions Control; Biodiversity/Species Preservation; Sustainable Development; Waste Management/Recycling; Water Quality

▲ Projects and People

Both a zoology professor at the University of Oulu and a Member of Parliament, **Dr. Erkki Pulliainen** is engaged in a variety of activities with a number of organizations. He is **promoting green policy**, and believes that the rights of nature take precedence over the rights of man. His publications include approximately 350 scientific publications and several books.

In addition to his work at the University of Oulu, Pulliainen is the director of the **Värriö Subarctic Research Station** in Finnish Lapland, where researchers are constructing a system to study the **effects of air pollutants on the assimilation of pine trees.** "We record the fallout rates of the pollutants coming especially from the nickel mines and smelts of the Kola Peninsula, [former] USSR," reports Dr. Pulliainen. "We have also recorded heavy metals in the soils and organic matter in that area."

At Bothnia Bay, projects concern the **lowered rate of reproduction in earthworms,** and scientists are studying the effects of heavy metals, such as PCBs (polychlorinated biphenyls), on this phenomenon.

Pulliainen is also studying endangered animal species including Arctic fox, wolf, grizzly bear, wolverine, lynx, and pine marten. In Parliament, he is known for his expertise on matters of agriculture, forestry, nature conservation, and the environment.

■ Resources

A Zoology Department publications list is comprised of numerous reports with English titles on the above-named animals as well as the Baltic ringed seal, reindeer, roe deer, shorebirds, wild boar, squirrel marten.

France

The largest country in Western Europe, France has wide differences in geography, from the snow-capped Alps to the beaches and steep cliffs along the Mediterranean and the wooded Loire Valley and the vineyards and orchards of the countryside. Rich soils are the country's most important natural resource.

Within the European Community, France plays a major role in environment and development. It is ecologically one of the most well-preserved regions of Europe, mostly because of its geographical location and the diversity of its natural environment. The French government has set goals for reducing emissions, placed strict controls on transborder shipments of hazardous materials, successfully banned the sale of ivory to protect the African elephant, and supported an extension of the commercial whaling moratorium.

🏛 Government

Directeur des Parcs et Reserves/Parks and Reserves
Manoir de Ker Amel

22810 Plounevez, Bretagne, France

Contact: **André Dupuy, Ph.D., Biologist**
   Phone: (33) 98 9638 6575

Activities: Conservation and Management • *Issues:* Biodiversity/Species Preservation; Forestry Protection; Ornithology; Waste Management/Recycling; Water Quality

■ Resources
National Parks and Reserves offers *Les Parcs Nationaux en Senegal* (National Parks in Senegal), among other publications printed in French. **Dr. André Dupuy** belongs to the Species Survival Commission (IUCN/SSC) Antelope Group.

Institut d'Elevage et de Médécine des Pays Tropicaux (IEMVT/ CIRAD) ⊕
Institute of Breeding and Veterinary Medicine for Tropical Countries
10, rue Pierre Curie

94704 Maisons-Alfort, France

Contact: **Philippe Chardonnet, Ph.D., Veterinarian/Wildlife Specialist**
   Phone: (33) 1 43 68 88 73 • *Fax:* (33) 1 43 75 23 00

Activities: Development; Research • *Issues:* Biodiversity/Species Preservation; Sustainable Development • *Meetings:* IUCN/SSC; OIE; wildlife ranching; deer biology; ungulates

▲ Projects and People
Currently with the **Institut d'Elevage and de Médécine des Pays Tropicaux (IEMVT/CIRAD)** (Institute of Breeding and Veterinary Medicine for Tropical Countries), **Dr. Philippe Chardonnet** has undertaken development project management, such as: a deer project in New Caledonia; a "traditional farming project" in Muramvya province of Burundi with 50,000 cattle, 150,000

sheep and goats, and 450 employees on 150,000 hectares; "various assignments and inspection of breeding development projects in Togo, Zaire, Rwanda, and Senegal"; and veterinary responsibilities in Guinea-Bissau's eastern region covering 200,000 hectares and 150,000 cattle.

Dr. Chardonnet recently evaluated rusa deer ranching in Réunion (a French island near Madagascar) and Brazil, as well as deer ranching in Canada, USA, and Malaysia, among other countries. He also studies the economic importance of wildlife in Tanzania, Ivory Coast, Burkina Faso, and Ethiopia; prepared a bibliography for the Food and Agriculture Organization (FAO) on rational exploitation of wildlife worldwide; and has knowledge of protected areas throughout Africa, North and South America, and the Pacific.

He lists his special skills in using dart guns, ground nets, and net guns from helicopters in capturing wildlife; counting wildlife; installing game fences; commercial deer ranching; and organizing and management hunting; and experience in equine medicine and surgery. Dr. Chardonnet belongs to the Species Survival Commission (IUCN/SSC) Antelope, Ethnozoology, and Veterinary groups.

■ Resources

Dr. Chardonnet's publications list includes literature on rural development in Africa, and wildlife, with French and English titles.

Centre National de la Recherche Scientifique ⊕
National Center for Scientific Research
Laboratoire d'Ecologie
46, rue d'Ulm
75005 Paris, cedex 05 France

Contact: **Claude Grenot, Ph.D.,** Director of Research
Phone: (33) 1 43 29 12 25
Fax: (33) 1 43 25 70 85

Activities: Education; Research • *Issues:* Biodiversity/Species Preservation; Energy

▲ Projects and People

In addition to serving as research director for the Centre National de la Recherche Scientifique, Dr. Claude Grenot is chairman of the vertebrate ecophysiology unit at École Normale Supérieure's Department of Ecology. Comparative ecology and ecophysiology of vertebrates, especially in arid zones, are his major topics.

Much of his work deals with "aspects of the ecophysiology of organisms relating to the particular problems imposed by their environment and way of life." He pursues factors affecting the water and energy balance of terrestrial vertebrates and other osmoregulatory considerations, such as electrolyte intake and excretion. Dr. Grenot reports he was "one of the first in France to initiate the use of double-labelled water for measuring water turnover and field metabolic rates on free animals." Since 1960, Dr. Grenot has conducted comparative research studies on ecological and physiological adaption in reptiles and mammals in the arid regions of the Sahara in Algeria and Tunisia; Baja California, and the Chihuahuan Desert, Mexico; and Venezuela.

Dr. Grenot has written 3 books and more than 30 abstracts, articles, journal studies, reports and reviews, in addition to co-

authoring another 80 publications. Topics include desert biology, grazing land ecosystem of the African Sahel, desert lizards, and European common lizards. On the international scene, he has spoken at numerous seminars and symposia. Among other societies, he belongs to the Species Survival Commission (IUCN/SSC) Tortoise and Freshwater Turtle Group.

■ Resources

A list of publications is available, some with English titles.

Muséum National d'Histoire Naturelle (MNHN)
National Museum of Natural History
55, rue Buffon
75005 Paris, France

Contacts: **Philippe Bouchet, Ph.D.,** Curator of Recent Molluscs
Phone: (33) 1 4079 3103 • *Fax:* (33) 1 4079 3484

Jean Lescure, Research Director
Phone: (33) 1 4079 3495

Activities: Research • *Issues:* Biodiversity/Species Preservation; Deforestation

▲ Projects and People

Dr. Philippe Bouchet, Muséum National d'Histoire Naturelle (National Museum of Natural History), is especially interested in "taxonomy of marine gastropods, with emphasis on the deep-sea faunas, larval ecology of marine invertebrates, and conservation of endangered non-marine molluscs." His research is conducted on oceanographic expeditions in the Atlantic, Indian, and Pacific oceans and in land-based projects from Corsica to New Caledonia, Mauritania to the Philippines.

Dr. Bouchet is editor of MNHN's *Memoires,* chairman of the Species Survival Commission (IUCN/SSC) Mollusc Group, fellow of the Royal Geographical Society, and board member of Unitas Malacologica.

Jean Lescure is research director of NMHN's Reptiles and Amphibians Laboratory.

Parliamentary Office for Scientific and Technical Assessment
National Assembly
233 boulevard Saint Germain
75 Paris, cedex 07 France

Contact: **Michel Destot, Ph.D.,** Deputy
Phone: (33) 1 40 63 88 09
Fax: (33) 1 40 63 88 08

Activities: Political/Legislative • *Issues:* "Any scientific-based political-related question"

● Organization

Established in 1983, the **Parliamentary Office for the Assessment of Scientific and Technological Options** officially began operations two years later with the goal of providing sufficient technical and unbiased information to members of Parliament, regardless of party affiliation. This enables legislators to obtain significant information prior to formulating policies.

The Office is a purely parliamentary body, independent of the Government of France and the Administration, according to Deputy **Dr. Michel Estot**. The Parliament and Senate are each represented by eight members. Each member also has an appointed deputy. According to procedure, "recourse to the Office is thus reserved for parliamentarians and they alone have the right to decide on the desirability of initiating a study programme on a given subject."

A rapporteur, appointed with each case presented to the Office, conducts a feasibility study to determine the manner in which information gathering is to be achieved. Upon his recommendations, the Office then decides whether or not to initiate a study program. Next, a working party—comprised of a scientific council of 15 senior experts, independent consultants, and representatives from trade unions, professional organizations, associations, or consumers' and users' interests—is convened. Upon receipt of the findings of this group, the rapporteur submits a draft report to the scientific council and members of the Office, who in turn decide the merits of publishing the findings.

▲ Projects and People

Among the Office's completed studies are those dealing with **acid rain**, the consequences of the **Chernobyl nuclear power** station accident, the effects of **chlorofluorocarbons (CFCs)** on the environment, and **mineral extraction in the Antarctic**. Other completed reports include the preservation of **water quality**, management of long-term **radioactive waste**, **industrial waste treatment**, **biotechnology** in agriculture, and the supervision of the **safety and security** of **nuclear installations**.

A nuclear engineer, Dr. Destot is also CEO of Corys, a simulation and training software company; deputy of the Isère department; and member of the Parliamentary Office for Scientific and Technological Assessment.

■ Resources

For wider distribution, copies of all parliamentary studies are being published by a commercial printer.

🏛 *Local Government*

Conseil Général de l'Isère

7, rue Fantin-Latour
38000 Grenoble, France

Contact: **Jean-François Noblet**, Technical Advisor on Environment
Phone: (33) 76 60 38 38 • *Fax:* (33) 76 60 38 35

Activities: Political • *Issues:* Air Quality/Emission Control; Biodiversity/Species Preservation; Ecological Journalism; European Bats; Protection of Natural Spaces

▲ Projects and People

Jean-François Noblet began his commitment to the environment in his early twenties, when with five friends, he founded the **Rhône-Alps Federation for the Protection of Nature (FRAPNA)**. Since then, he has worked with both FRAPNA and the **Conseil Général de l'Isère**. He says, "I am unhappy about the towers on the horizon, invasive billboards, anarchy and urbanization. A plastic bag or a 4x4 vehicle in a pretty forest takes away my pleasure." These feelings keep him active in the local government and concerned about the environment. The question he wants everyone to ask is simple: "Am I happy? Can I be happier?" Questions like this, he believes, help lead people to a better understanding of their needs and the needs of the environment. These questions help lead to what he calls "a politics of goodness." He advises the Department of Isere in environmental issues, as well as on recycling, sewage, pollution, and quality of life. He is also FRAPNA's education commission president.

He writes, "I look to the elected to give priority to the elements necessary for daily survival, to drive with unleaded gasoline and catalytic converters, make use of public transportation, write on recycled paper, and to interest themselves, not only in their electors, but also in the little animals who live in their sectors, and work for the defense of the air, for clean drinking water, and silence." Noblet expects to attain his goal by urging others to activism.

A member of the Species Survival Commission (IUCN/SSC) Chiroptera Group, Noblet is an expert on **bats**, organizing projects to protect bats and their habitats and publishing his results in French scientific periodicals.

🍁 *NGO*

Association pour la Sauvegarde de la Nature Néo-Calédonienne (ASNNC)
Association to Safeguard New Caledonian Nature

37, rue Georges Clemenceau, B.P. 1772
Noumea, New Caledonia

Contact: **Jean-Louis d'Auzon**, President
Phone: (687) 28 32 75 • *Fax:* (687) 28 32 75

Activities: Education; Law Enforcement • *Issues:* Air Quality/Emission Control; Biodiversity/Species Preservation; Deforestation; Energy; Waste Management/Recycling; Water Quality • *Meetings:* ICBP; IUCN; UNEP

● Organization

Created in 1971 and located on the island of **New Caledonia**, the **Association pour la Sauvegarde de la Nature Néo-Calédonienne (ASNNC)** (Association to Safeguard New Caledonian Nature) has worked to preserve the local flora and fauna. ASNNC reports that even though New Caledonia is a small French protectorate island (about 1,500 kilometers east of Australia, 1,700 kilometers north of New Zealand, and just north of the Tropic of Capricorn), it is faced with all the problems and pressures of its larger continental cousins. It also says that the "picture-perfect image" emanating from post cards and travel posters is somewhat false.

Among its problems, the ASNNC writes, New Caledonia is faced with "excessive fishing and hunting, boaters who pollute beaches and islets, picnickers who cut branches from very rare trees for their barbecues; one cannot put a gendarme behind each individual." ASNNC realizes that it is more than individuals and tourists who are polluting the island. New Caledonia is reported to be the world's second largest reserve for nickel, and mining of nickel and other minerals has done extensive damage. Also, New Caledonia loses tracts of vegetation to bushfires each year.

With over 3,000 members registered and nearly 500 active members, ASNNC is working for change. ASNNC directs substantial energy to its young, and has a special youth section. ASNNC writes, "The protection of nature needs to be an act of faith. Faith in the future. . . . Faith in a reversal of the tendencies that tomorrow will make the protectors of nature the majority and the destructors a very weak minority."

▲ Projects and People

An indigenous flightless bird, the cagou (*Rhynochetos jubatus*), was classified by the World Conservation Union (IUCN) in 1984 as one of the planet's 12 most endangered species. ASNNC has the tasks of increasing public awareness for the bird and insuring that it has a future habitat as well as safeguarding its present survival.

ASNNC also has a reforestation project, planting pines and plants with special attention to the needs of the birds that live in the trees. The project is conducted with the help of the youth section. So far, some 50,000 trees have been planted; with the help of the army, 8 large palm trees were transplanted.

New Caledonia is also an important breeding ground for marine turtles and has an international roster enlisted to insure the preservation of turtles and of their breeding grounds. ASNNC previously sponsored an expedition to Surprise and Huon islands to count and tag the turtles.

ASNNC directs its educational program through publications, press releases to radio and television outlets, "Protection of Nature" postage stamps, nature expos, guided walks, games, contests, and other activities. ASNNC is a participant in many international organizations, including the World Wildlife Fund (WWF) and Conservation International.

Other projects include "collecting and destroying 800,000 introduced snails (*Achatina fulica*) and numerous *Acanthaster*, a dangerous starfish"; cleaning up beaches and roads; campaigning against bushfires; and collecting used batteries containing mercury.

ASNNC labors extensively for the creation of many kinds of reserves on New Caledonia. These include Integral Reserves such as Montagne des Sources and Reserve Marine Yves Merlet; Special Reserves of Fauna and Flora such as Haute-Yaté, ilôt Leprédour, ile Pam, and l'étang de Koumac; and Territorial Parks such as Rivière Bleue, le Ouen-Toro, and the South Lagoon. ASNNC is working to create reserves specifically to house the cagou.

■ Resources

Lack of funds prevents ASNNC from continuing to publish its magazine *Nature Calédonienne*. However, it does produce posters, stickers, and leaflets on marine turtles, trees, and bushfires free of charge.

Confédération Mondiale des Activités Subaquatiques (CMAS) ⊕
World Underwater Federation
47, rue de Commerce
75015 Paris, France

Contact: **Marcel Bibas, Ph.D.**, General Secretary
Phone: (33) 1 45 75 4275
Fax: (33) 1 45 77 1104

Activities: Development • *Issues:* Protection of the Sea • *Meetings:* CITES; International Maritime Environmental Award

● Organization

The World Underwater Federation (CMAS) was created in 1959 under the care of Commander Jacques-Yves Cousteau in Monaco. What began as 15 countries participating has grown to 73, as well as millions of permanent and active divers and thousands of diving centers and clubs. The only organization representing international divers, CMAS calls itself the "true United Nations of diving."

An independent, nonprofit organization, CMAS sets as its objective the development of underwater activities and the coordination of diving instruction around the world. CMAS is affiliated with many other international organizations, including the United Nations Educational Scientific and Cultural Organization (UNESCO), International Olympic Committee (IOC), World Conservation Union (IUCN), General Association of International Sports Federations (GAISF), and the Organization of World Games.

▲ Projects and People

CMAS has many ongoing projects: technical instruction in all aspects of diving; scientific endeavors, such as underwater archeology, geology, and biology; and medical studies on diving accident prevention and treatment. CMAS has also established a system of international certificates allowing the standardization of diving instruction around the world.

The CMAS funds and directs the International Marine Environment Award, with the involvement of the French Department of State for the Environment, IUCN, World Wildlife Fund (WWF), and several other global organizations. The award is given each year in January at UNESCO headquarters to the "individual, organization or company whose work has made a major contribution to the protection of the marine environment." The winner is awarded 50,000 French francs to be used in furthering his or her work in environmental protection.

According to its new publication, *Subacmas* (May 1991), and Prof. Maurice Auber, director of CERBOM (Centre d'Etudes et de Recherches de Biologie et d'Oceanographie) (Study and Research Center of Biology and Oceanography), the sea has been tapped as a rich source of medicines since the days of Homer and Virgil: roe and fish liver that benefit premature babies; anti-inflammatory substances from coral; iodine to soothe rheumatic illness; "extracts of marine species, sponges, cod, and bonito" used in endocrinology for thyroid conditions; antidepressants from sea-sponge; "agar-agar, an extract of red algae," which acts as a regulator of the intestines; other antibacterial and antiparasitic substances extracted from algae; antiviral substances that are effective against herpes and polio; and other extractions being tested to combat cancers and sarcomas.

According to Prof. Auber, research with phytoplanktonic species began in 1960 in both France and the USA to examine antibacterial properties. "The great variability in species capable of producing antibiotics, and the large number of chemical substances of phytoplanktonic origin which have anti-bacterial properties are therefore apparent," he writes, with substances used in therapeutics and in dermatology for treating "infectious skin diseases and metabolic lesions."

Marine microorganisms are also being researched for their antibiotic properties. "One of the most recent discoveries in this field is Cephalosporine, an extract of the sea-mushroom (*Cephalosporium acremonium*), which offers a wide range of antibiotic properties," Prof. Auber reports. "A whole new pharmacopoeia of marine origin is in the process of emerging and in this respect the future looks most hopeful."

On the downside, *Subacmas* also describes the dumping of 20 billion tons of waste annually into the sea, as quoted from the Barner parliamentary report (April 11, 1990). "Ninety percent of waste ends up stagnating near the coastline, . . . habitat for the richest diversity of animal and plant life. . . . Two thirds of the world's population live within 80 kilometers of the coast and half of the largest towns and cities are built on or near to estuaries. . . . These eject their sewage water into the sea, 80 percent of which has not been treated. . . . Each year, one thousandth of the world's oil production, five million tons, finds its way into the oceans. . . . Rain pours a similar amount of nitrates into the seas as do rivers. Sixty percent of the lead circulating in the atmosphere in the northern half of the planet ends up in the North Atlantic, due to the great concentrations of industry in the northern hemisphere."

■ Resources

Subacmas is a bilingual publication in French and English.

Regarding the International Marine Environment Award, candidates must submit documentation (résumé, text, photos, and films) to the headquarters of CMAS at least 30 days before the panel begins its deliberations. All individuals, companies, and organizations are welcome to enter, or to be nominated by others.

Conservatoire Botanique National de Brest
National Botany Conservatory
52 Allée du Bot
29200 Brest, Brittany, France

Contact: **Jean Yves Lesouëf**, Curator
Phone: (33) 98 41 88 95 • *Fax:* (33) 98 41 57 21

Activities: Education; Law Enforcement; Research • *Issues:* Biodiversity/Species Preservation

● Organization

Founded in 1975, the **Conservatoire Botanique National de Brest** describes itself as the first structure devoted entirely to the conservation of vascular plants in both *ex situ* and *in situ*. Situated on a 22-hectare varied landscape, the Conservatory features extensive greenhouses, laboratory, library, seedbank, and information office. It has more than some 1,200 threatened species in cultivation, mainly from Europe and oceanic islands.

▲ Projects and People

The Conservatory collections include the **threatened flora of the Brittany and Armorican chain**, most threatened endemics of Europe, and flora species listed in the French Red Data Book. Mass reintroduction of an extinct species, *Lysimachia minoricensis*, is also present. A Mallorca conservation project is being conducted in conjunction with Gobierno Autonomo de las Baleares and the Botanical Garden of Cordoba.

"An almost complete collection of the threatened endemic flora of the Azores has been realized," reports the Conservatory. "These plants have found in Brest a climate very similar to their own climate. . . . Even the coastal ones are unaffected by our frosts. These plants reproduce freely in [their] area of the Conservatoire." Endangered species from Madeira and Canarias are cultivated in the open or in the greenhouse for the lowland species.

Threatened endemics of the Mascarenes also are cultivated here. Regarding *Ruizia cordata*, the number of living specimens has grown from 2 survivors to 2,000 plants produced in Brest and reintroduced in Réunion, a small French island near Madagascar. A special "Reserve Garden" for the surviving flora of the dry region of Réunion has been set up.

■ Resources

Occasional papers are available from the Conservatory, which seeks political support and new technologies for its programs.

Fondation Internationale pour la Sauvegarde du Gibier (IGF)
International Foundation for the Conservation of Game ⊕
15, rue de Téhéran
75008 Paris, France

Contact: **Bertrand des Clers, Ph.D.**, Director
Phone: (33) 1 45 63 51 33
Fax: (33) 1 45 63 32 94

Activities: Development; Education; Political/Legislative; Research • *Issues:* Biodiversity/Species Preservation; Deforestation; Ethnozoology; Natural Habitats; Sustainable Development; Wildlife Conservation/Management • *Meetings:* CIC; CITES; FAO Wildlife Working Group; IUCN; UNCED

● Organization

Founded in 1976, the nonprofit **International Foundation for the Conservation of Game (IGF)** set the aim of "conserving wildlife in a developing world." To reach that goal, "the value of natural habitats and of their produce must be highlighted and developed, for the benefit of rural people."

IGF is dedicated to "conserving, managing, and rehabilitating natural habitats, stopping the extinction of animal species, reintroducing, for the benefit of local people, those wildlife species which had been driven to extinction, and leaving to our children, as a heritage, a friendly environment, a strong and lively agriculture, rich and diverse forests, hosting a wealth of wildlife."

▲ Projects and People

Ongoing projects include the organization of **international workshops and symposia** in Europe, Asia, Africa, and Latin America; creation of **protected areas** for threatened species; reestablishment of **hedges in cereal-growing plains** for bird nesting habitats; initiation of international agreements for habitat conservation and management; assistance to **fight elephant poaching** in **Central Africa**; moving **black rhinos** to safe areas; studies on **wildlife utilization in Africa**; and wildlife conservation, development, and management in **Mongolia**.

IGF director **Dr. Bertrand des Clers** is also Tropical Game Commission chairman, International Council for Game and Wildlife Conservation (CIC); president of Institute for World Conservation and Development (IWCD) and of European Bureau for Conservation and Development (EBCD); chairman of the Species Survival Commission (IUCN/SSC) Ethnozoology Group; and member of SSC's Commission on Environmental Policy, Law, and Administration.

■ Resources

IGF needs political support at CITES—as well as broader educational outreach to gain backing of governments and the general public—to "give greatest value to wildlife products and keep markets even more open to them, as an incentive to sustainable management [and] income to local people," writes Dr. Clers. (CITES is the Convention on International Trade in Endangered Species of Wild Fauna and Flora.)

Technical publications on *Wildlife Conservation/Development* are marketed through CIC.

Nature & Entreprises/Nature & Enterprises
Fonds Français pour la Nature et l'Environnement (FFNE)
French Fund for Nature and Environment
45, rue de Lisbonne
75008 Paris, France

Contact: **André Szybowicz**, Director
Phone: (33) 1 42 25 60 19
Fax: (33) 1 43 59 00 51

Issues: Air Quality/Emission Control; Biodiversity/Species Preservation; Deforestation; Energy; Global Warming; Health; Population Planning; Sustainable Development; Transportation; Waste Management/Recycling; Water Quality

● Organization

Nature & Entreprises was created in 1988 by André Szybowicz as a special program within **Fonds Français pour la Nature et l'Environnement (FFNE)** (French Fund for Nature and Environment). Szybowicz had been studying the feasibility of corporate-sponsored patronage of the environment. Traveling extensively throughout Europe, Szybowicz found the idea to be well advanced and that France was somewhat behind. He then wrote a book on sponsoring and patronage, and Nature & Enterprises began to

grow. Today Nature & Enterprises has 15 firms sponsoring 140 projects, ranging from a 16,000-franc billboard on pollution to a program costing over 11 million francs to renovate public gardens on the Mediterranean coast.

According to Nature & Enterprises, the purpose it fulfills is twofold: financing for projects and an outlet for money that corporations would like to invest in the environment. Many corporations want to make this kind of investment, but do not know how to find or create a project to suit their needs. Nature & Enterprises has 140 projects from which firms may choose, and with its growing network of environmental care providers, it can create projects to fit any need.

▲ Projects and People

Nature & Enterprises participates in all manifestations of conservation and preservation, working on dozens of **reforestation** projects as well as **cleanup** projects and construction.

Nature & Enterprises is funding the development of **wild bird surgery and infirmary**. If a wild bird is injured by a hunter or an automobile, or by some natural cause, it will be brought to the bird hospital for treatment. The hospital will consist of two parts—an "emergency room" for immediate care and a recovery room, with 12 cages to contain the birds until they can be released.

Nature & Enterprises is directing the **reconstruction of two churches** whose sites share one of the oldest trees—possibly 1,000 years old—in Europe. The project is being completed by the Regional Parks of Upper Normandy and the Ministry of the Environment. Space will be made for the tree to receive more air and light, and the chapels and steeples will be restored.

In lower Normandy, Nature & Enterprises is constructing a rangers' station equipped as an **Educational Center on the Environment**. The station will welcome adults and children, with programs in plants and the sea, and will lead nature walks in the surrounding country.

In northwestern France, the Grand Marais is a lake of 50 hectares, fed directly by streams, with clean, clear water, abundant vegetation, and a wide diversity of fish and birds. Nature & Enterprises converted the **hunting blinds into observatories** and has opened the lake for public and scholarly use.

Along six kilometers of the **Loire River** banks, the organization seeks to **protect the local animal and vegetable species**. Welcome structures, trails, buildings, and educational posters are being erected.

Nature & Enterprises is developing a new, unique environmental education project. The objective is to raise money to buy an 18-meter, multi-hulled **sailboat**. The boat (christened the *Soyons Nature*) will be outfitted for a **media campaign**, and will be available for lease to firms. With a 6.5 million franc price tag for the boat, and another 9.75 million francs to bring together the package, this is one of Nature & Enterprises biggest projects.

The Tower of Ruynes in Auvergne welcomes around 3,000 visitors a year to an exhibition, which Nature & Enterprises is replacing with an **exposition on medicinal plants**, their usage and growth. The Botanical Garden will be enlarged and extended.

The Deux-Sèvres department of France is host to an **ornithological film festival**. The rural canton of 9,000 inhabitants welcomes 20,000 images and 40 films on ornithology from 10 to 15 countries each October and November. Prizes of 5,000 to 20,000 francs are given for the best quality films in the following categories: cinematography, originality, portrayal of the relationship between man and bird, and educational interest. Also organized are a **Hall of Animal**

Art by French and foreign artists, a forum of 20 to 30 regional and national associations, and bird watches and hikes.

Nature & Enterprises has guided funds in Normandy to create a **glossary of environmental and ecological terms**. The specialized vocabulary used to describe the environment by professionals in the field and scientists is often unknown to the average person, and inhibits communication and understanding. The concise glossary in a small, pocket-sized format will facilitate communication and contact on local to global levels between environmental organizations and the general public.

One of the richest fossil sites in France is in Languedoc-Roussillon, in the Hérault department. More than 800 fossils, up to 255 million years old, contain dinosaur ancestors. The project seeks to **protect the site and fossils from damage** and construct a building in which to house and display the fossils. When completed, the building will include educational materials and a public welcome center.

On Corsica, a **tortoise village** has been established for the protection of the **Hermann tortoise, France's last wild land tortoise**. The Mediterranean forests, which are the normal habitat of the tortoise, are threatened by the yearly fires that strike all over France and Corsica. The "village" will contain breeding enclosures, nurseries for the young, and space for adults. The tortoises will be released to their natural environment at the age of five.

In Lorraine, a **farm for children** is being set up. According to Nature & Enterprises, in some towns and cities, "cows, sheep and ducks are as foreign as the elephant, the giraffe or the toucan." Nature & Enterprises believes that "contact with domestic animals is an excellent introduction to nature." Children will care for the daily needs of the animals, learning about the economics of farms and the responsibility they have to wild and domestic animals.

Situated between Cavalaire and Lavandou, the **Domaine du Rayol** recently became **public lands**. This historic site contains 20 hectares of botanical gardens and buildings bordering the Mediterranean. In addition to the rehabilitation of the gardens, the buildings will be restored, and information booths will be installed on the **creation and care of the botanical gardens**. The grounds will be prepared to host seminars, colloquia, and training courses.

The European Community (EC) is collaborating with Nature & Enterprises on an important project in Champagne-Ardenne. Seventy percent of the **European crane** population passes through Champagne-Ardenne on its migratory route. This phenomenon attracts visitors from all over Europe, and the great volume of birds causes some problems for local farmers. The project's goal was to acquire a farm of 74 hectares on which to plant food for the cranes and provide cover for them. In the winter of 1990, 1,400 cranes stayed at the farm. The project also includes a bird museum, which is in progress, and nature walks along local ponds and rivers.

■ Resources

An information kit, in French, describes memberships and programs. *Mecenat d'Entreprise* can be ordered. Szybowicz produced a handbook, *Mécénaturel*, for the Ministry of Environment on how businesses are protecting nature. The group writes, "We only work on a national level and we'd like to develop our network on an international level. We need clues to this development" so that relations can be improved between "two worlds"—industry and ecology.

Friends of Animals (FoA) ⊕
16, rue Beaugrenelle
75015 Paris, France

Contact: **Bill Clark, Ph.D.**, International Program
Director
Phone: (33) 1 40 79 30 69
Fax: (33) 1 45 77 49 71

Activities: Education; Law Enforcement; Political/Legislative; Research • *Issues:* Biodiversity/Species Preservation; Deforestation; Population Planning of Pets • *Meetings:* CITES

▲ Projects and People

The international **Friends of Animals (FoA)** reports a staff of 30 and overall membership of about 103,000. Some of its recent environmental projects include cooperative efforts with park departments in Chad, Kenya, Gambia, Ghana, and Senegal; organizing a microlight aircraft reconnaissance squadron for the Tanzania Wildlife Department; and providing U.S. Army surplus vehicles to wildlife departments in eight African countries.

The organization is also establishing **animal refuges protecting elephants**; working to "knock out the trade in **wild-caught birds**," and setting up **antipoaching** programs. A wildlife orphanage and rehabilitation in Liberia was temporarily shut down due to the civil war.

Dr. Bill Clark, international director and eco-ethologist, recently evacuated 80 wolves from Hungary to a refuge in the south of France. A member of the Israel delegation to CITES (Convention on International Trade in Endangered Species of Wild Fauna and Flora), Dr. Clark also belongs to the Species Survival Commission (IUCN/SSC) Antelope and Equid groups.

Priscilla Feral, FoA president, can be reached at national headquarters, P.O. Box 1244, Norwalk, CT 06856, phone 1-203-866-5223, fax 1-203-853-9102. Corporate headquarters are in New York City.

Institut Méditerranéen de la Communication (IMCOM) ⊕
Mediterranean Institute of Communication
Palais du Luxembourg—15, rue de Vaugirard
75006 Paris, France

Contact: **Éric Naulleau**, Editor
Phone: (33) 40 22 91 41 • *Fax:* (33) 40 22 91 27

Activities: Development; Political/Legislative

● Organization

The **Mediterranean Institute of Communication (IMCOM)** has a simple ethic to combat a complex task: "To aid the Mediterranean people to know and comprehend better, and to favor the social, economic, and cultural development, to create a bridge of communication between the two banks of the Mediterranean." Spurring development between two vastly different cultures is the task at hand, with a sea dividing people that requires the cooperation of all for its preservation. IMCOM was founded in the early 1980s by Senator Louis Perrein.

▲ Projects and People

IMCOM sponsors colloquia and conferences on a variety of topics. Some recent topics include "Cooperation and Communication in the Mediterranean," "Communication in Direct Relation to the Environment," and "Developing Communications between Men, Enterprises, and Nations, in Order to Better Control Their Environments."

In *Imcom*, the Institute's publication, overcrowding was cited as one of the greatest dangers facing the sea. The result of technology, economics, and sociology—overcrowding is said to lead to degradation of the countryside, overexploitation of resources, and pollution. For example, the kinds of companies and development that occur most rapidly on the coasts are usually the least environmentally friendly: chemical and petrochemical factories, petroleum refineries, steel factories, fertilizer factories, and electrical centers.

The rapid increases in population in both Europe and northern Africa have had great impact on the sea. The number of large cities (700,000–4,000,000 inhabitants) on the coasts has increased to include Athens, Alexandria, Marseille, Beirut, Tripoli, Istanbul, Algiers, Cairo, and Barcelona, among the 17. The stimulation of coastal economies and the urban explosion brings with it overpopulation, chaotic construction without zoning, absence of proper hygiene, lack of urban planning, and lack of public service.

Moreover, the introduction of "paid holidays, the increasing standard of living, the proliferation of the automobile, the need of relaxation and evasion, and the habit of vacationing have thrown the crowds toward the beaches." The coastal populations double in the summer months. The sea and coasts are unable to accommodate the impact.

One of the greatest barriers to progress, reports *Imcom*, is the "egoism" of the countries involved. Each country would like to retain the rights to decide how it is treating its coasts, and none wants a third party or international organization to tell it what to do. So even if all countries regularly participate in and confirm or sign international conferences and treaties, most still regard their neighbors with suspicion and fear.

According to the *Imcom* article, European countries have been loathe to spend great amounts on cleanup efforts, and African countries are unwilling to sacrifice industrialization and development for the defense of the environment. The problem is getting the countries to honor the treaties and accords. The London Convention of 1954 was ignored by 14 Mediterranean countries, and many countries fail to ratify the accords within their own countries.

In February 1976, all the Mediterranean countries except Albania signed a comprehensive accord, the Mediterranean Action Plan, which went into effect two years later. The Convention enumerated general principles for the states to prevent, reduce, and combat pollution. Since that time, there has been a renewed interest and commitment on behalf of the Mediterranean states, and IMCOM looks to the future with optimism, it reports.

■ Resources

IMCOM materials are prepared in French, Spanish, and Arabic. *Imcom* magazine is published quarterly.

Scientific Committee on Problems of the Environment (SCOPE) ⊕
International Council of Scientific Unions (ICSU)

51 boulevard de Montmorency
75016 Paris, France

Contact: **V. Plocq**, Executive Secretary
Phone: (33) 1 45 25 04 98
Fax: (33) 1 42 88 94 31
E-Mail: Omnet: scope.paris

Activities: Education; Development; Research • *Issues:* Air Quality/Emission Control; Biodiversity/Species Preservation; Global Warming; Health; Sustainable Development; Water Quality

● Organization

The Scientific Committee on Problems of the Environment (SCOPE) of the International Council of Scientific Unions (ICSU) was established in 1969 as an international, non-governmental, nonprofit organization. Its mandate is to advance the knowledge of the role humans play in their environment, and the effects of alterations in the environment on people's health and welfare, especially those influences on a global or multinational dimension.

As an interdisciplinary scientific council, SCOPE is a source of advice to both governmental and NGO groups. Scientists from various disciplines are invited to workshops and ad hoc meetings, available knowledge is examined and information gaps are identified. "Carefully reviewed information . . . is published in books which have been widely used as 'state of the art' assessments of environmental problems," indicates SCOPE. The reports also "focus attention on controversial questions, emphasize the stimulation of new approaches, identify research needs, and encourage the adoption of sound environmental practices."

▲ Projects and People

With about 23 projects underway, the 4 major themes deemed critical according to SCOPE's 1990–1992 scientific program are: sustainable development, biogeochemical cycles, global change and ecosystems, and health and ecotoxicology. Each cluster has a major coordinator with specific chairpersons representing the individual subtopics.

The sustainable development cluster considers coastal subsidence, groundwater contamination, phosphorus, and small catchments in its program chaired by Prof. J.K. Syers, UK.

Prof. C.V. Cole, USA, is undertaking a project in understanding **Phosphorus Cycles in Terrestrial and Aquatic Ecosystems and Their Interaction with the Cycles of Other Elements.** Workshops are being held in Africa, Asia, Europe, and South America regarding "worldwide phosphorus requirements for sustainable food and fiber production while minimizing adverse effects on the environment." At the same time, conceptual and simulation models are being developed of phosphorus cycling in major ecosystems.

A project in **Groundwater Contamination** is aimed at integrating "understanding of the chemical, biological, and physical processes" affecting such contaminants and evaluating "the vulnerability of aquifers to contamination and the capacity of aquifers to eventually detoxify pollutants." In some parts of the world, ground-water provides almost all drinking water. Chairpersons are **Profs. C.A. Shoemaker**, USA, and **A.J.B. Zehnder**, Netherlands.

The **Use of Scientific Information for Sustainable Development** is being investigated, especially in confronting climate change, desertification, and deforestation as well as hazardous waste management and biodiversity loss. A series of meetings were held in Latin America, USA, India, and Kenya—with a final synthesis conference in Europe in mid-1993.

The **Biogeochemistry of Small Catchments** applies to surface areas from one to ten square kilometers in undisturbed, agricultural, and urban landscapes. With the leadership of **Drs. B. Moldan** and **J. Cerney**, Czechoslovakia, data is being collected regarding "the flux of elements, minerals, organic matter, gases, and water entering or leaving the system" in subarctic and temperate zones of North and Central Europe and North America. The project is expected to provide guidelines for environmental monitoring programs and "encourage the creation of a worldwide network of small catchments yielding scientifically reliable information."

Dr. J. Milliman, USA, is preparing a "user-friendly" volume to accompany its scientific synthesis report which is focusing on the "local impact of natural and accelerated subsidence on low lying coastal areas" and to recommend action such as engineering and environmental management.

Nitrogen Transport and Transformation is a regional and global analysis regarding eutrophication of coastal marine ecosystems, effects on the open ocean, and causes of forest dieback, possibly a result of "nitrogen as air-borne ammonium." The acidification of soils in agro-ecosystems is also being examined.

Sustainable Agriculture is a new study pinpointing the need for research and technology transfer in semiarid regions, including the strengthening of global information networks and international institutions for these purposes.

The **biogeochemical cycles** cluster examines trace gas exchange, radionuclide pathways, sulphur, metals, and particle flux in the ocean and interactions in continental seas. Coordinator is **Dr. A.G. Ryaboshapko**, of the former USSR. A four-year project on **Trace Gas Exchange between Biosphere and Atmosphere** was chaired by **Prof. M.O. Andreae**, Germany, to gain greater understanding of the processes controlling "the flux rates of biogenic trace gases, such as sulphur gases, methane, ammonia, and nitrogen oxides"—leading to greater interaction among scientists dealing with global change issues. Emphasis is on the tropics "where changing land use is leading to profound changes in the interactions between biosphere and atmosphere," and in the high-latitude regions, "where the impact of climate change on the biota will be felt earliest and most pronouncedly."

Environmental Pathways of Artificial Radionuclides, chaired by **Prof. Sir Frederick Warner**, UK, is an outgrowth of an earlier project on "environmental consequences of nuclear war," with Chernobyl as a case study. Focus is on the abundant and toxic isotopes in this study which aims to "identify pathways which are poorly understood or quantified." According to SCOPE, this study can be a "valuable tool in advancing knowledge generally of biogeochemical cycling."

Sulphur Cycle in Terrestrial and Aquatic Systems, led by academician **M.V. Ivanov**, former USSR, is a joint effort of SCOPE and the United Nations Environment Programme (UNEP) examining contaminated Great Lakes, other freshwater lakes, wetlands, and inland marine basins—and the impact of industrial, agricultural, and domestic sewage, and of atmospheric deposition.

The objectives of **Metals Cycling**, chaired by **Prof. T.C. Hutchinson**, Canada, is to understand cycles of certain heavy metals, such as mercury, and other elements toxic to human health and the environment. **Particle Flux in the Ocean**, chaired by **Dr. S. Kempe**, Germany, also was started by SCOPE and UNEP; the project formerly concentrated on river transport, lakes, and estuaries and now is focusing on oceans and the open sea. In the early planning stages, **Interaction of the Main Elements in Marginal and Intercontinental Seas** is linked to the above project and will assess the "interactions of nitrogen, phosphorus, carbon, sulphur, oxygen, . . . heavy metals, . . . and the impact of human activities on these interactions."

Within the **global change and ecosystems group**, ecotones, genetically modified organisms, ecosystem experiments, long-term ecological research, organic matter budgets and ultraviolet (UV) radiation effects are scrutinized. Coordinator **Dr. J.M. Melillo**, USA, co-chairs **Organic Matter Budgets** with **Prof. A. Breymeyer**, Poland, regarding "whole plant response to simultaneous changes in atmospheric CO_2 content, temperature region, and water availability." Documentation of environmental changes on ecosystem processes are to be documented and ecosystem models linking plant and soil response to climate change are to be developed.

Ecotones in a Changing Environment, chaired by **Prof. P.G. Risser**, USA, probes the decrease in the predictability of globalized environmental problems as related to "unprecedented human disturbance" and the resulting modifications of the shapes of the landscapes. "Most interactions between the various components of the landscapes occur in the boundaries, usually called ecotones," explains SCOPE. "The ecotone is a zone of transition between adjacent ecological systems."

Dr. Risser is also chairing **Long-Term Ecological Research**, now initiating international coordinating and networking of such collaborative efforts at specific sites with the newest technologies. Working groups are being identified for three selected biomes.

Ecosystem Experiments, chaired by **Prof. H.A. Mooney**, USA, raises these questions, among others: What are the most critical needs in the understanding of ecosystem functioning that can best (or only) be addressed by new experimentation? In view of the high cost and long-term nature of ecosystem manipulations, what mechanisms can be utilized to ensure the involvement of a wide group of scientists, both nationally and internationally? Prof. Mooney is also chairing the **Biodiversity** project, which is an "integrated study program on the role of biodiversity on ecosystem functioning." To be monitored are "demography/biogeography of species; interactions between species; sources and sinks for water, minerals, and gases; and productivity."

Effects of Increased UV Radiations on Biological Systems, chaired by **Dr. E. DeFabo**, USA, in cooperation with the International Union of Pure and Applied Biophysics (IUPAB) and Scientific Committee on Antarctic Research (SCAR), is examining such effects on polar systems, changes in primary productivity, and impacts on agriculture, forestry, and fisheries. Other focuses are the "range of defense mechanisms against UV damage" and carcinogenic and other effects on mammals and their immune system, cataract, and hormonal responses, as examples.

Genetically Designed Organisms in the Environment, undertaken with ICSU's Committee on Genetic Experimentation (COGENE), is examining "genetically designed organisms" and biotechnologies to solve environmental problems and to cut down on the levels of chemicals already in the environment.

Prof. C.R. Krishna Murti, India, coordinates the health and ecotoxicology cluster. With research areas defined previously, SCOPE will begin a new ecotoxicology project to "contribute to . . . existing information on the mechanisms of adaptation to chemical stress of ecosystems thriving in diverse climate zones, on the buildup of toxic residues in individual species, and on their impact on food chain dynamics."

Prof. Ph. Bourdeau, Belgium, is chairing Safety of Chemicals, known as SGOMSEC 8 (Scientific Group on Methodologies for the Safety Evaluation of Chemicals), to assess new methodologies for safety evaluation with respect to human health, DNA damage and repair, and pharmacokinetic mechanisms. One goal is to identify "characteristic similarities and differences among species, so that methods for estimating genetic damage may be developed through extrapolation (or interpolation) from one species to another."

Another new study, Health Effects of Climate Change, is expected to result in a "state-of-the-art report" regarding outcomes of water and food scarcity, population migration, and interaction between man and animal vectors as responses to "diverse external and internal stresses."

■ Resources

SCOPE's publications are listed in its *1990–1992 Scientific Programme Report*, which also includes project descriptions as well as directories of SCOPE and ICSU contact persons.

Commission on the Coastal Environment (CCE) ⊕
International Geographical Union (IGU)
10 Square Saint Florentin
78150 Le Chesnay, France

Contact: **Roland Paskoff, Ph.D.**, Chairman
Phone: (33) 39 55 60 02

Activities: Development; Research • *Issues:* Sustainable Development

● Organization
Representatives from 12 nations serve on the board of the Commission on the Coastal Environment (CCE) of the International Geographical Union (IGU). Dr. Roland Paskoff chairs its more than 500 members.

▲ Projects and People
With a worldwide membership, latest projects include reconstruction of coastal dunes, convened by Norbert P. Psuty, Rutgers—The State University of New Jersey, USA; coastal processes and instrumentation, with Joost H.J. Terwindt, State University of Utrecht, Netherlands; Southern Hemisphere dune systems, with J.F. Araya Vergara, Universidad de Chile, Chile; onshore impacts of the petroleum industry with Donald W. Davis, Nicholls State University, Louisiana, and Ian Manners, University of Texas at Austin, Texas; Australian beach systems authored by Andrew D. Short, University of Sydney, NSW, Australia; coastal tourism in developing countries, convened by P.P. Wong, National University of Singapore, Singapore; Tombolos, with C.L. So, University of

Hong Kong, Hong Kong; experiences in planning in the coastal zone, with Paolo Fabbri, Università di Bologna, Italy; physical and human response to sea-level rise, with Eric Bird, University of Melbourne, Australia; sediment dynamics, deposition, and erosion in temperate salt marshes, with Denise Reed, Louisiana University Marine Consortium, USA; geographical coastal zonality, with Dieter Kelletat, Universitat Essen GHS, Germany; and coastal applications of geographic information systems (GIS), with Darius Bartlett, University College, Ireland.

Bartlett is compiling an annotated bibliography of literature relating to coastal GIS, and contributions are welcome from CCE members and nonmembers, sent to him at Ireland's University College, Department of Geography, phone 353 21 276871, ext. 2835; and fax 353 21 275948.

Ongoing projects are international glossary on coastal geomorphology compiled by Vitautas Gudelis, Lithuanian Academy of Sciences, Lithuania, and Maria Eugénia Moreira, Universidade de Lisboa, Portugal; and international bibliography on coastal geomorphology, with Charles W. Finkle, Jr., editor-in-chief, *Journal of Coastal Research*, and Douglas Sherman, University of Southern California, USA.

■ Resources
A newsletter provides the international community with brief project descriptions of its members' activities and special reports on meetings and seminars.

Mouvement National de Lutte pour l'Environnement (MNLE)
National Movement Fighting for the Environment
B.P. 79
93505 Pantin, France

Contacts: **Jean-Yves Guezenec**, Secretary General
        Henri-Jacques Ghendrih, Official
Phone: (33) 1 48 46 04 14
Fax: (33) 1 48 46 44 53

Activities: Education • *Issues:* Biodiversity/Species Preservation; Deforestation; Energy; Sustainable Development; Transportation; Waste Management/Recycling; Water Quality

● Organization
MNLE fights for environmental protection, is a working partner of the Centre for our Common Future on behalf of sustainable development, and publishes *Naturellement*.

Office pour l'Information Eco-Entomologique (OPIE)
Office for Entomological Ecological Information

B.P. 9

78280 Guyancourt, France

Contact: **Robert Guilbot**, General Secretary
  Phone: (33) 1 30 44 13 43
    Fax: (33) 1 30 83 36 29

Activities: Education; Research • *Issues:* Biodiversity/Species Preservation; Deforestation • *Meetings:* Berne Convention; Conseil National Protection de la Nature (Ministry of the Environment)

● Organization

The **Office for Entomological Ecological Information (OPIE)** is a national association created in 1969 by professional entomologists from the **National Institute for Agronomic Research (INRA)** along with a group of amateurs. The OPIE finds itself in a preeminent position as the meeting point of entomological activities in France. With 1,500 members, the OPIE collaborates with many groups, including INRA, Pasteur Institute, France's Museum of Natural History, and Overseas Scientific and Technical Research Office (ORSTOM), among others, and obtains financial support for its activities through regional councils and private companies. The OPIE is a scientific advisor to the World Wildlife Foundation (WWF) and a member of the World Conservation Union (IUCN).

▲ Projects and People

The OPIE educates the general public about **insects and their roles in the life of the planet**. A permanent exhibition at the OPIE headquarters features guided visits, slide shows, and live insects, with ample facilities for school groups and adults. OPIE also organizes special courses for teachers, nature group leaders, and amateur entomologists.

In order to meet the demands of natural science teachers and research laboratories, OPIE has set up **permanent insect rearing units**. These units have 10 species always available, with an annual output of over 20,000 insects.

According to OPIE, the insect population in France is in decline because of several factors: improper use of pesticides and herbicides, changing techniques in agriculture and forestry management, urbanization and changes in land use, uncontrolled tourism, and overcollecting by some entomologists of insect populations already at risk.

OPIE also attributes the decline of insects to an inappropriate and outdated legal framework. In France, the protection of insects is limited to the publication of a national list of 33 species and subspecies for which capture or destruction is prohibited. There is some small protection provided by the national parks system, nature reserves, and some environmental decrees, but these cover less than 4 percent of the total land area.

OPIE has set up a national coordinating structure, the **National Research and Information Group for the Conservation of Insects and Their Environment (GNERCIM)**, with an eight-point mission to provide guidance for the preservation and management of insects in France: "organize concern, carry out field studies, protect environments in order to protect insects, provide information for greater awareness, harmonize development and protection, manage natural resources, reveal the importance of insects, and set an example" by putting forward a code of ethics for French entomologists.

The goal is for GNERCIM to become a national advisory council, recognized throughout France by ministries and local governments as an authority on insects. OPIE would like GNERCIM to represent France in Europe as the concerned body on insects.

■ Resources

OPIE publishes two quarterly journals, *Imago* and *Insectes*, with a distribution of 5,000 copies. It needs help with educational outreach and biotope protection.

Société Nationale de Protection de la Nature (SNPN)
National Society for the Protection of Nature

57, rue Cuvier, B.P. 405

75221 Paris, France

Contact: **Marc Gallois**, Editor
  Phone: (33) 1 47 07 31 95

Activities: Education; Law Enforcement; Research • *Issues:* Biodiversity/Species Preservation • *Meetings:* IUCN

● Organization

One of the oldest organizations active today, **La Société Nationale de Protection de la Nature (SNPN)** (National Society for the Protection of Nature) was founded in 1854 as the Imperial Zoological Society. Started by **Isidore Geoffroy Saint-Hilaire**, professor of the Museum of Natural History, SNPN was soon recognized by the French government as existing for the public good. The Society was later renamed, as its goals and interests grew.

SNPN began in response to what it thought was the abuse of nature in the nineteenth century: "In the eyes of the men of the 19th century, in all the euphoria of the conquest of the Americas and the discovery of Africa, nature was thought of as an inexhaustible capital and its resources could be exploited without limit." SNPN has been fighting this kind of cause for almost 140 years with conservation projects and educational programs. SNPN is particularly interested in preserving wild animal and plant species in France, and works hard to preserve their natural habitats.

▲ Projects and People

SNPN has many types of conservation and educational programs, as featured in its **year-round calendar** filled with excursions and nature walks. It also has ongoing **work and construction projects** in the Camargue Park and the Grand-Lieu Reserve. These projects typically take two weeks, fall on scholastic vacations, and allow participants to take advantage of the park resources. Observation blinds, public information stations, posters, and expositions are made.

SNPN has a wide variety of what it calls "nature rendez-vous" available: one-day trips to local forests, multiday trips to regional parks, and extended trips to volcanoes, mountains, and on rivers.

In 1987, SNPN launched an Amnesty for Elephants campaign. SNPN cites some alarming statistics: between 1971 and 1985 it is estimated that between 1.6 million and 2 million elephants were killed for their ivory, with only 400,000 to 700,000 remaining today. The illegal poaching still continues, and SNPN estimates that at the present rate of destruction, around 100,000 a year, the elephants have very little time remaining. The Amnesty campaign wants increased commitment and policing from CITES (Convention on International Trade in Endangered Species of Wild Fauna and Flora) and a total boycott of ivory and ivory products. Since 1987, SNPN has collected over 350,000 names petitioning the end of ivory trade. Today, with some African countries calling for a lift on the ivory trade ban, the Amnesty campaign has more work. SNPN would like to see the ban continue for another 20 years to allow the elephant population to grow and stabilize.

■ Resources

SNPN wants to expand its educational outreach. It publishes several periodicals in French, including: *Le Courrier de la Nature* (Nature courier), issued every two months, with articles on nature and ecology; and the quarterly *La Terre et la Vie* (Earth and life) edited by SNPN since 1931, described as the only scientific journal in France dedicated to ecology and containing original articles (some in English) on the relations between man and the environment, biological equilibrium, population dynamics, and inventories of flora and fauna. A publications list is available, also printed in French.

Société Réunionnaise pour l'Étude et la Protection de l'Environnement (SREPEN) Réunion Society for Environmental Study and Protection
4, rue Jacob or B.P. 1109
97482 Saint Denis, Réunion

Contact: **Bernadette Ardon**, President
Phone: (262) 20 30 30 • *Fax:* (262) 42 29 32

Activities: Development; Law Enforcement; Research • *Issues:* Deforestation; Energy; Health; Waste Management/Recycling; Water Quality • *Meetings:* CITES

● Organization

Located on the tiny French island of Réunion near Madagascar, Société Réunionnaise pour l'Étude et la Protection de l'Environnement (SREPEN) (Réunion Society for Environmental Study and Protection) has 22 unpaid volunteers working on various projects. The most visible and symbolic of the projects is the goal to save the tuit-tuit (*Coracina newtoni*), a small bird found only on Réunion and thought to number less than 100. According to SREPEN, more than 25 species of birds have already disappeared or left the island. SREPEN has launched an educational campaign, maintaining that "today, we have no more excuse for the ignorance" that leads to the extinction of animals.

SREPEN lists four goals or interests: to promote scientific studies on the environment, to safeguard the natural riches, to participate in projects of management on the island, and to educate the public on the protection of the environment. SREPEN exists only with the help of volunteers, and is constantly seeking new volunteers.

▲ Projects and People

Indigenous to Réunion, the tuit-tuit has been in steady decline since 1972 when a census found 120 couples on the island. Réunion has already lost a number of animal and plant species, including the Réunion tortoise and the solitaire, also known as the dodo bird. For this reason, SREPEN has made the tuit-tuit symbolic. SREPEN hopes that by raising public awareness for the tuit-tuit and saving its habitat, it will save the habitat of numerous other rare and endangered species. SREPEN identifies three keys to rescuing the tuit-tuit: protecting the bird from poachers and scourges, studying its way of life and food supply, and identifying a protected zone for the bird. The natural habitat of the bird at present is only a few square kilometers.

SREPEN designs laws for the protection of lagoons and marine life, as well as creating forest reserves for the protection of rare types of vegetation. SREPEN took a leading role in the reintroduction of an endangered plant, *Ruizia cordata*.

■ Resources

SREPEN produces bulletins, leaflets, and stickers about its work, printed in French. A portfolio on *Fleurs Rares des Iles Mascareignes* (Rare Flowers of the Mascarene Islands) can be ordered. SREPEN seeks help in protecting the forest.

Tour du Valat Foundation ⊕
Le Sambuc
13200 Arles, France

Contact: **Alan Roy Johnson, Ph.D.**, Biologist
Phone: (33) 90 97 20 13 • *Fax:* (33) 90 97 20 19

Activities: Conservation; Research • *Issues:* Biodiversity/Species Preservation • *Meetings:* ICBP; IWRB; WWF

● Organization

The conservation, management, and study of Mediterranean wetlands in cooperation with NGOs in eight other countries, numerous universities, European Community, World Bank, and the French government is conducted by the Tour du Valat Foundation.

Information gathered since 1954, when the Foundation's Station Biologique began studying wetlands in southern France's Camargue area, is now used throughout the Mediterranean to protect and manage valuable threatened habitat and encourage responsible authorities to do the same. Well-known scientists from the international community serve on the Foundation's board. It has a staff of more than two dozen postdoctoral, graduate, and undergraduate scientists.

The Ramsar Convention, or convention on Wetlands of International Importance especially as Waterfowl Habitat, defines wetlands as "areas of marsh, fen, peatland or water, whether natural or artificial, permanent or temporary, with water that is static or flowing, fresh, brackish or salt, including areas of marine water, the depth of which at low tide does not exceed six meters."

▲ Projects and People

Although **integrated wetlands management is economically and technically feasible**, it is also "politically difficult because it implies restrictions on use in order to restore productivity," reports the Foundation. Its mission is to show those in and around the wetlands that "they will derive far greater long-term benefits from wetlands if they temporarily accept restrictive rules to prevent overuse."

Because water resources are limited in the Mediterranean basin, it is crucial to study the water's quantity and quality which is constantly being affected by human activities. Birds, fish, and plants in the area that rely on the water sources are also under threat from man.

For 30 years, **Dr. Alan R. Johnson** has served as ornithologist at the Station, responsible for handling **band recoveries**, research of **water migration patterns and monitoring of aquatic birds** via nest countings. Since 1965, he has managed a research project on the once threatened **flamingos** in the Camargue and western Mediterranean. The project involves creation of an island for breeding and using artificial nest-mounts to attract other birds to the site. The population is estimated at 60,000 to 80,000 birds. Annually since 1977, Dr. Johnson has initiated and organized the **capture and ringing of 500 to 800 flamingo chicks** with the assistance of 120 to 140 persons. And he collects data on the movements and breeding of these marked birds throughout the Mediterranean and West Africa.

Cooperating with colleagues in Andalucia, "identical observations at the only other European breeding site for flamingos are carried out, "including ringing of chicks, underway since 1986," he reports. "One unforeseen aspect involved working with farmers in order to find a solution [of bird scaring] to keep flamingos out of ricefields where crop damage was being caused."

As a major wetland species, **herons** are particularly important because they serve as an **indicator of the wetland's biological richness**. A successful artificial breeding habitat was built for little egrets, cattle egrets, squacco herons, and night herons—among the 150 species of birds in the Mediterranean wetlands.

To **stop lead poisoning of waterfowl**, the Foundation asks that hunters use pellets of nontoxic metals, as lead pellets contaminate the bottom of many marshes and can kill birds when "taken as grit."

Another goal of Tour du Valat is finding effective sustainable management and manipulation techniques, known as **biomanipulation**, techniques for the **fish population**, a vital link in the wetland's food chain. Researchers conclude that "overfishing and deoxygenation due to algal blooms" are responsible for the problems of traditional freshwater and lagoon fisheries, particularly the dwindling of at least 15 endemic species in the Mediterranean region. Habitats are also destroyed through "dam and canal construction, agricultural and industrial pollution, . . . introduction of exotic species and aquaculture, and leisure activities," reports Tour du Valat.

The Foundation's main goal in plant ecology is to increase understanding of **how plant communities function**, showing the effects of domestic (cattle, horses) and wild herbivores (wild boar) on the plants' structure and species composition. Grazing plays a significant role in Mediterranean wetlands management, because overgrazing leads to degradation of both habitat structure for fauna and productivity, says the Foundation. On the other hand, controlled grazing efforts and "flooding of temporary marshes in early autumn

can enrich plant communities and increase food resources for wintering ducks."

Dr. Johnson is the (Old World) Chairman of the International Council for Bird Preservation and International Waterfowl and Wetlands Research Bureau (ICBP-IWRB) Flamingo Specialist Group and editor of its annual newsletter. Also a member of the Species Survival Commission (IUCN/SSC) Flamingo Group, he was scientific director on two films about flamingos with the Service du Film de Recherche Scientifique, Paris.

■ Resources

The Tour du Valat offers a full list of publications available upon request.

The World Conservation Union (IUCN) ⊕
2/4, rue de Bel Air
37400 Amboise, France

Contact: **Gèrard Sournia,** Administrator, French-
        speaking countries
Phone: (33) 16 4757 3023

Activities: Development; Education • *Issues:* Biodiversity/Species Preservation; Protected Areas

(See also World Conservation Union [IUCN] in the Niger, Pakistan, Switzerland, USA, and Zimbabwe NGO sections.)

▲ Private

Jean-Pierre Sclavo ⊕
Villa "La Finca," Plateau du Mont Boron
06300 Nice, Alpes Maritimes, France

Phone: (33) 93 54 67 90 • *Fax:* (33) 93 54 71 34

Activities: Research • *Issues:* Biodiversity/Species Preservation
Meetings: CYCAD

▲ Projects and People

Jean-Pierre Sclavo, managing director of a construction company, is an international researcher into cycads of the family Cycadales—large nonflowering seed plants, somewhere between ferns and palm trees, and considered rare in the wild, in cultivation, and in herbaria. Sclavo has been honored for his work in cycadales by having an example named for him—*Encephalartos sclavoi*, which he discovered on a research trip to Tanzania, near the Kenyan border. Sclavo organizes research expeditions to Africa and Central and South America and has research projects with the Naples Botanical Garden, New York Botanical Garden, and at Xalapa, Mexico.

Among Sclavo's main projects are the triannual **International Conferences on Cycad Biology.** Sclavo inspired the first of these conferences, which took place in 1987 in Ville de Beaulieu sur Mer, France. The second occurred in 1990 in Townsville, Queensland, Australia. CYCAD 93 is being held in Pretoria, South Africa. Sclavo is presently planning CYCAD 96 in mainland China and CYCAD

99 in Florida, USA. Sclavo belongs to the Species Survival Commission (IUCN/SSC) Cycad Group.

Resources

Sclavo can provide the following publications, printed in English: "*Encephalartos sclavoi*, De Luca, Stevenson, and Moretti (Zamiaceae), a New Species from Tanzania," *Delpinoa*; "*Encephalartos voiensis* (Zamiaceae), a New East Central African Species in the *E. hildebrandtii* Complex," *Annals of the Missouri Botanical Garden*; "The Biology, Structure and Systematics of the Cycadales," *Memoirs of the New York Botanical Garden*, edited by Dennis W. Stevenson, Volume 57, 1990; and "Zamiceae, a New Family for Zambia," Malaisse, Sclavo, and Turner, also from *Memoirs of the New York Botanical Garden*.

Sclavo seeks contact with specialized local botanists to determine areas where new species might be growing.

🎓 Universities

Laboratoire de Zoogéographie/Zoogeography Laboratory
Université Montpellier III
B.P. 5043
34032 Montpellier, France

Contact: **Charles P. Blanc, Ph.D.**, Director
Phone: (33) 67 14 23 15 • *Fax:* (33) 67 14 20 52

Activities: Education; Research • *Issues:* Biodiversity/Species Preservation; Deforestation; Zoogeography

● Organization

The objective of the **Laboratoire de Zoogéographie** at Université Montpellier III is "to promote the development of **zoogeography** in a fundamental plan and its applications to the management of species and natural spaces." The Laboratory has eight permanent staff, headed by **Prof. Charles P. Blanc**, to specifically study animal populations. The research includes taxonomy, phylogenetics, paleoecology, and biological rhythms.

Under the guidance of Prof. Blanc, the Laboratory conducts research in France, Corsica, Spain, Mauritania, Madagascar, and Polynesia, among other countries, with results forwarded to museums worldwide and appearing in international and French scientific and popular presses. According to the Laboratory, numerous qualified doctors and scientists working with the environment have emerged through these efforts. The Laboratory participates in several national commissions and 17 colloquia.

▲ Projects and People

The Laboratory has created a bank of bibliographical and factual information on zoogeography (Zoogéco), organized international scientific **colloquia** on terrestrial vertebrates, and published the Edition of Documents for a *Zoogeographical Atlas* (from L to R). The Laboratory has also published **four volumes on the fauna of Madagascar**, with two additional volumes in preparation.

The Laboratory has ongoing studies to determine the **impact and** survivability of two different kinds of animals: **partridges**, which are hunted throughout France, and **sea molluscs**, such as oysters, which are harvested for their meat and pearls. The genetic characteristics and the ability of these animals to repopulate are being investigated.

The Laboratory is currently studying ways to recycle biological materials with the help of macro- and micro-particles to consume the waste products of animals in a sylvo-pastoral setting. This undertaking is linked to the **shore management project**, which includes **inventories of shore life**, as well as a study of the compatibility of different species.

Resources

A list of publications, printed in French, is available. Many papers deal with ecology, ecological cartography, population genetics, zoogeography, and conservation.

Institut d'Embryologie/Institute of Embryology
Université Louis Pasteur (ULP)
11, rue Humann
67085 Strasbourg, France

Contact: **Yves Rumpler, Ph.D.**, Director
Phone: (33) 88 35 87 77 • *Fax:* (33) 88 24 20 05

Activities: Education; Research • *Issues:* Biodiversity/Species Preservation • *Meetings:* IPS

▲ Projects and People

Since 1979, chromosomal evolution and specification in lemurs has been a major research orientation at the **Institut d'Embryology**, Université Louis Pasteur (ULP). A breeding colony was initiated to produce hybrids for the study of chromosomal reproductive barriers with the Universities of Strasbourg and Madagascar cooperating, and later joined by the University of Besançon and the Paris Museum of National History.

In the early 1980s, **Dr. Yves Rumpler** realized the threat of extinction of the Malagasy primates and reoriented a portion of the research toward the protection and conservation of lemurs, with assistance from the World Wildlife Fund (WWF) and the World Conservation Union (IUCN). The Institut maintains a genetic reserve through captive breeding and a bank of gametes and embryos.

"We have organized our colony in that we have reduced the number of individuals of more common species such as *L.f. mayottensis* and *L. catta*, replacing them with species under great threat in the wild, such as *L. rubriventer*, *L. coronatus*, and *L. flavifrons*," writes Dr. Rumpler.

Exchange programs at zoos in Köln, Mulhouse, and Saarbrücken provide important common stock for species of *Hapalemur*, *Lemur*, and *Varecia*. Deep freezing of gametes and embryos, *in vitro* fertilization, and artificial insemination have been applied in the cryoconservation of lemur material. A new technique is being used for obtaining sperm from other centers or the wild for artificial insemination.

Among the Institute's field programs are a systematic cytogenetical study of different lemurs from Madagascar and prosimians from other countries. Some experiments share "differences large enough to predict sterility of expected hybrids," writes Prof. Rumpler.

In other field investigations in **Madagascar**, "the repartition area of *E.m. macaco* and *E.m. flavifrons* is not well known and recent observations of a hybridization zone between them has complicated the situation," reports Dr. Rumpler. One purpose of this project is to detect regions allowing the "creation of a protected area involving pure *E.m. flavifrons* as well as the natural hybrids."

Also in Madagascar, **genetic studies** of different *Eulemur macaco* populations are underway at Nosy-Komba Insula and on the Ambato peninsula—using **molecular biologic markers** (DNA-RFLP) to study erythrocyte enzymes and compare reproductive rates.

Dr. Rumpler belongs to the Species Survival Commission (IUCN/SSC) Madagascan Primate Group. ULP and Ruhr-Universität Bochum, Germany, offer a **certificate in primatology** which is supported by the Inter-University Cooperation Programme (ICP) ERASMUS of the European Community (EC) which facilitates the exchange of European graduate students.

■ Resources

The Institut is interested in new DNA fingerprinting technologies to aid captive-breeding programs and the creation of protected areas. It has available more than 20 publications, most with English titles, concerning lemur conservation and cytogenetics of primates.

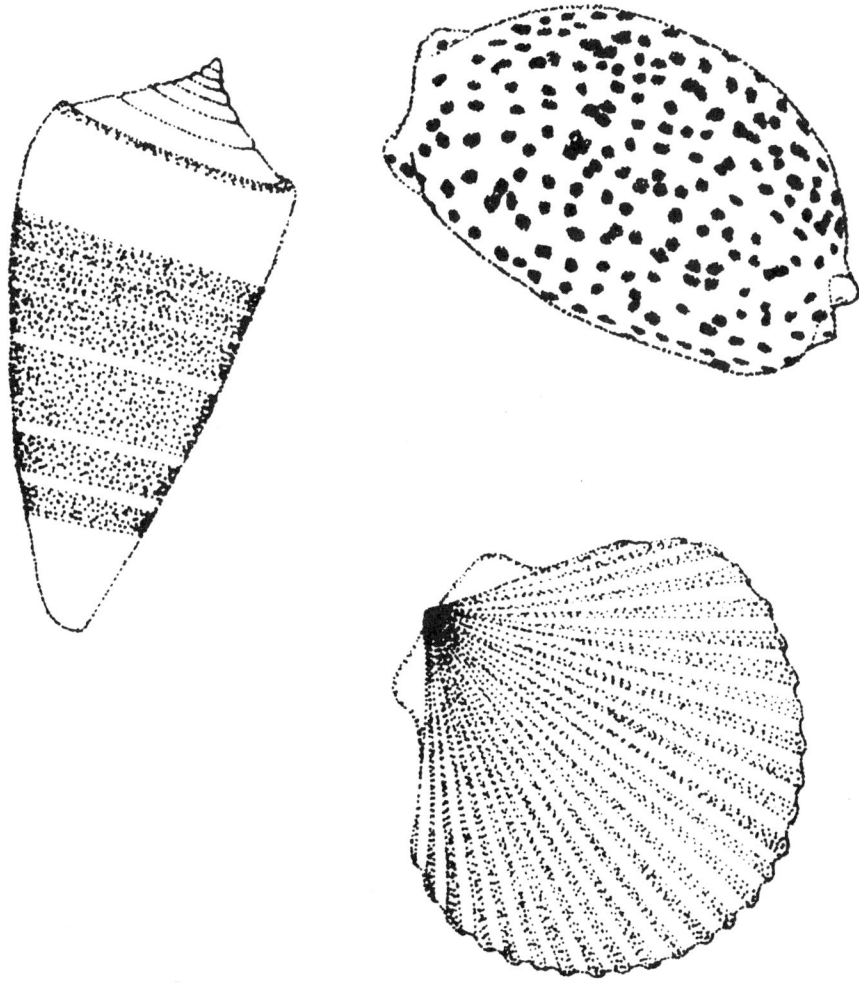

Germany is one of the most densely populated countries in the world. It is also one of the most modern, urbanized, and industrialized. The country's geographical location in the center of Europe, and its closely meshed traffic network, are other key environmental factors. In addition, much of the country's pollution is due to the production, conversion, transport, and utilization of energy sources.

East Germany became subject to the environmental laws of West Germany and the European Community in 1990. The EC's waste management, water quality, and air pollution provisions will be enforced in eastern Germany in 1996. In the meantime, steps are being taken to address protection of the ozone layer, reduction of the greenhouse effect, and protection of tropical rainforests.

🏛 *Government*

IOBC Working Group ⊕

Institut für Biologischen Pflanzenschutz/Institute for Biological Pest Control

Biologische Bundesanstalt für Land-und Forstwirtschaft
Federal Biological Research Centre for Agriculture and Forestry
Heinrichstrasse 243
6100 Darmstadt, Germany

Contact: **Sherif Ali Hassan, Ph.D.**
Phone: (49) 6151 44061 • *Fax:* (49) 6151 422502

Activities: Development; Research • *Issues:* Plant Protection

▲ Projects and People

The IOBC Working Group at the Institute for Biological Pest Control, Federal Biological Research Centre for Agriculture and Forestry, consists of "125 research workers from 36 countries and aims to promote research on this group of beneficial arthropods," writes biologist **Dr. Sherif Ali Hassan**. Particularly interested in **trichogramma and other parasites**, the group researches "biosystematics and genetics; host relation and biology; physiology and behavior; ecology and population dynamics; rearing (*in vivo* and *in vitro*), production and release; compatibility (environmental, biological, chemical); effectiveness and assessment."

Dr. Hassan, convenor of the "mass production utilization behavior subgroup," is researching "the mass-culturing and utilization of entomophagous insects and mites to control agricultural pests as well as the effects thereon of the use of chemical pesticides." He previously investigated the biology and control of the **cabbage root fly** (*Erioischia brassicae*), **asparagus fly** (*Platyparea poeciloptera*), and **asparagus minor fly**. **Dr. Wajnberg**, France, is convenor of systematic genetic ecology. World symposia are held every four years; the last occurred in San Antonio, Texas, USA, in 1990. IOBC is the International Organization of Biological Control.

■ Resources

Symposia proceedings are available from Service des Publications, CNRA, route de Saint-Cyr, 78026 Versailles Cedex, France. Dr. Hassan can provide back issues of *Trichogramma News.*

Bundesamt für Ernährung und Forstwirtschaft (BEF) ⊕
Federal Office for Food and Forestry
Adickesallee 40
6000 Frankfurt/Main 1, Germany

Contact: **Dietrich Jelden, Ph.D.**, Biologist
  Phone: (49) 69 1564 930 • *Fax:* (49) 69 1564 445

Activities: Education; Political/Legislative; Research • *Issues:* Species Preservation • *Meetings:* CITES; IUCN; Societas Europea Herpetologica; SSC; World Congress of Herpetology

● Organization
The **German Scientific Authority to CITES** was established as an office of the **Federal Office for Food and Forestry (BEF)** in 1976 following the decision of the German government to become signatory to the **Convention on International Trade in Endangered Species**, drafted in Washington, DC, in 1973. The Authority is thus part of a comprehensive international control system for trade of endangered plants and animals, and works in close cooperation with the German Customs Authority and other nations' watchdog groups.

▲ Projects and People
Biologist and geographer **Dr. Dietrich Jelden** has served as deputy head of the Authority since 1985. Previously, he had worked with the United Nations' Food and Agriculture Organization (FAO) in Papua New Guinea.

In the process of monitoring which species cross the borders of Germany, either entering or exiting, Authority staff produce reports on, for example, "the impact of international trade on the prehensile-tailed skink," with Dr. Jelden as project leader, and the "impact on the species *rana* of the international frog legs trade" with researcher **H. Martens**. From 1984 to 1990, Jelden conducted a study on European trade in **live herpetofauna**.

More recent investigations involve international trade in **live coral reef fish**, particularly for aquariums and including *in situ* studies, and in **Madagascar succulent plants** and the impact of their survival in the wild. Project leader is U. Schippmann.

Jeldon is involved in both the Species Survival Commission (IUCN/SSC) Crocodile and Cacti/Succulent groups. He has conducted research in Thailand, India, Sri Lanka, USA, and Ecuador.

■ Resources
The Authority publishes Germany's *Annual CITES Report*, as well as the German version of the CITES manual for identifying protected and endangered species, in three volumes.

Bundesforschungsanstalt für Naturschutz und Landschaftsökologie (BANL) ⊕
Federal Research Centre for Nature Conservation and Landscape Ecology
Konstantinstrasse 110
W-5300 Bonn 2, Germany

Contact: **Wolfgang Erz, Ph.D.**
  Phone: (49) 228 8491 0 • *Fax:* (49) 228 8491 200

Activities: Political/Legislative; Research • *Issues:* Biodiversity/Species Preservation • *Meetings:* CITES

● Organization
Noting that "almost 50 percent of all species of vertebrates and 30 percent of all superior species of plant life are endangered in [Germany] today," the **Federal Research Centre for Nature Conservation and Landscape Ecology (BANL)** works to fulfill the need for "specialized nature conservation and landscape management bodies . . . to provide scientific advice to politicians and administrative bodies." As such, the BANL works as a linking organization between the Federal Ministry for Environment, Nature Conservation and Nuclear Safety, and state institutes, under the auspices of the Ministry.

The Centre grew out of the 1906 establishment, in Danzig, of a State Centre for the Care of Natural Monuments. That first office became, in 1953, a Federal Centre for Nature Conservation, and in 1962 joined with an already existing Vegetation Office to form the current BANL.

Even earlier than 1830, citizens banded together on behalf of nature conservation. One objective was to preserve the "Drachenfels" scenic landscape along the Rhine River. A "vigorous" movement began evolving around 1870, and many conservation societies were formed. Parliament discussed the issue extensively at the turn of the century.

BANL's staff of 85, of which 37 are scientists (as of 1990), are in constant contact with scientific institutions, and with foreign and international organizations which conduct research in the field of bio-ecological environmental protection. The information collected serves "as a basis in the decisionmaking process for new legislation, regional planning, spatial planning, administration, practical work for nature conservation and landscape management as well as for international commitments."

On behalf of the German government, BANL cooperates with global programs such as the United Nations Environment Programme (UNEP), World Conservation Union (IUCN), World Wide Fund for Nature (WWF), and the International Council for Bird Preservation (ICBP).

The Research Center consists of the **Institute for Vegetation Ecology**, the **Institute for Nature Conservation and Animal Ecology**, the **Institute for Landscape Management and Landscape Ecology**, and a General Service Department. Projects are generally divided among the three institutes.

▲ Projects and People
The Vegetation Ecology section investigates endangered plants and plant communities as a basis for regulations and programs; maps vegetation of Germany to set up a system of protected areas; tests

plant species for riverbank erosion projects and for highway turf seeding. Ongoing projects involve protection of Germany's distinctive moors and finding viable species to introduce to riverbank areas to halt erosion.

The Nature Conservation and Animal Ecology section gathers data for legislation and international conventions, and performs applied basic research. Its main activities involve analyzing ecosystems and compiling criteria for the establishment of a system of reserves.

Since 1990, the Centre has housed the office of the German National Committee for the United Nations program Man and the Biosphere (MAB), and carries out Germany's contribution to that program. Also since 1990, the Centre has been involved with the International Nature Protection Academy of the 100-hectare Vilm Island on the East Sea coast.

Wider-ranging projects include ecological research on the Taimyr Peninsula in northern Siberia; bird tracking using satellite telemetry; and joint work with 25 European nations on a map of natural vegetation of Europe.

■ Resources

BANL publishes a series on landscape management and nature conservation, on vegetation ecology (results of research), a monthly journal *Nature and Landscape*, a quarterly review *Documentation on Environmental Conservation*, bibliographies, and an annual report. A few materials are printed in English.

The reference library, part of the General Service Department, combines two previous library collections, one from the original organization founded in 1906, and the other from the previous Vegetation Office. The library houses more than 70,000 books, some 1,150 periodicals, a cartographic section, information on more than 2,000 nature reserves and 18 internationally important wetlands in Germany, information on endangered species, and many indices.

Nationales Global Change Sekretariat
National Global Change Secretariat

Alfred-Wegener Institute
Postfach 12 01 61
2850 Bremerhaven, Germany

Contact: **Manfred Lange, Ph.D.**
  Phone: (49) 471 4831 217
  Fax: (49) 471 4831 149

Activities: Science Management • *Issues:* Biodiversity/Species Preservation; Deforestation; Global Warming; Sustainable Development

● Organization

The **National Global Change Secretariat** was established in 1990 by the **German Ministry for Research and Technology (MRT)**, organized concurrently with several other groups. Located at the Alfred-Wegener Institute for Polar and Marine Research, the staff of four works to fulfill the Secretariat's goals: to serve as a liaison between the scientific community and political decisionmakers; to help plan and implement national and international global change

research programs; to inform the public; and to contribute to the **Cabinet report,** *Research on Global Environmental Change.*

The Secretariat is committed to both a "national plan for the advancement of global change research" and to a future international global change program—possibly based upon the USA's Global Change Research Program (US/GCRP)—for collaboration beyond national borders. Among other organizations, the Secretariat works with the International Geosphere-Biosphere Programme (IGBP) and the World Climate Research Programme (WCRP)—considered the field's "building blocks." *(See also World Ocean Circulation Experiment [WOCE] in the UK Government section.)* In 1990, government officials organized the International Group of Funding Agencies (IGFA) for global change research and began discussing the implementation of an International Global Change Programme. Also, the Commission of the European Communities recently drafted an environmental research program.

▲ Projects and People

The Secretariat does not so much focus on special projects as try to serve as a link between or provide an overview of many projects; the staff is especially interested in the human threat to the global environment, and in the social, economic, and human dimensions of the issue. Research concentrates on climate, stratospheric ozone layer, oceans, tropical forests, and sensitive ecosystems. Population growth and its impact on land use, desertification, urbanization, CO_2 emissions, and acid rain are underscored.

According to Dr. Manfred Lange, the Secretariat's "basic strategy [is] measuring, understanding, predicting." Much of the Secretariat's work involves monitoring the progress of other, longer-standing governmental efforts. The first climate research program was initiated by the German government through the MRT in 1982. This program was replaced by a major research program on the greenhouse effect in 1989. Those who participate in such programs include universities, the Max Planck Institute of Meteorology in Hamburg, the Fraunhofer Association Institute, and other major research centers. University research in particular is carrying out numerous single projects, supported by the German Research Foundation through special focus programs.

To fulfill a need for a national center for climate modeling, entailing large, high-performance computers, the Deutsches Klimarechenzentrum GmbH (German Climate Computer Center) was founded in Hamburg in 1987 with the substantial support of the MRT. The partners involved are the state of Hamburg, the Max Planck Society, and the GKSS Forschungszentrum Geesthacht GmbH (GKSS Research Center). Preparations are underway for the Alfred-Wegener Institute to become a member. The German Climate Computer Center has consolidated the status of German climate research scientists as international leaders.

In 1988, the German Ozone Research Program joined in the activities with the aim of carrying out long-term observations of the ozone concentration in the stratosphere, in particular in the North Polar Regions. There is a permanently manned observatory located in Svalbard, and numerous research outposts can be found in both polar regions.

Other activities include the 1991 workshop on Global Change Data to assess existing data collection and management systems.

■ Resources

The Secretariat publishes the report *Global Change: Our World in Transition*, in both English and German; *Global Change Prisma*, a

16-page newsletter, appears several times yearly. The Secretariat urges curricula development and public information programs on behalf of global environmental change.

 ## State Government

Staatliche Vogelschutzwarte für Hessen, Rheinland-Pfalz und Saarland—Institut für Angewandte Vogelkunde (SV-IAV)
State Bird Protection Warden for Hessen, Rheinland-Pfalz and Saarland—Institute for Applied Ornithology
Steinauer Strasse 44
6000 Frankfurt/Main 60, Hessen, Germany

Contact: **Klaus Richarz, Ph.D.**, Chief
Phone: (49) 69 411538
Fax: (49) 69 425152

Activities: Education; Research • *Issues:* Biodiversity/Species Preservation

● Organization

In 1937 the Southwest German Bird Wardens group was established; in 1957 it became the **State Bird Protection Warden for Hessen, Rheinland-Pfalz and Saarland**, with its offshoot the Institute for Applied Ornithology (SV-IAV). Since 1973 the organization has been under the direction of the **Hessen Ministry for Agriculture, Forests and Nature Conservation**. The staff of eight offers expert advice and administers various programs for the protections of bird species and habitats in the three-state region, aided by volunteers and civil servants.

▲ Projects and People

One of the primary duties of the SV-IAV is to advise the three state governments on matters of **general bird conservation**, including using birds to fight biological damage in agricultural and wooded regions, and working to prevent bird/plane accidents. In general, the organization works on habitat and species protection, maintains preserve areas, and researches bird biology and ecology. Members stress the need to turn theoretical research into practical efforts, and are constantly experimenting with new ways to carry out conservation goals.

With more than 50 percent of the regional bird species threatened, such as the **white stork**, the SV-IAV has since 1973 conducted an ongoing effort to monitor birds on the **Red List** of endangered species.

A recent successful project involved the reintroduction of the **peregrine falcon** into the area after several decades of rapid decline. In 1950, there were 320 to 380 breeding pairs of the species in the region; by 1965, the number had dropped to about 70 to 90, and the low point came in 1975, when only 40 pairs were left. Causes included expansion of urban areas, industrial and tourist activity (such as rock climbing and hang gliding), as well as poaching. Encouraged by the success of American experiments with breeding the falcons in captivity, the SV-IAV's **Dr. Werner Keil**, working with other area conservation groups, such as the Hessen chapter of

the DBV (German Union for Bird Protection) and the Hessen Union for Ornithology and Nature Conservation, bred and released several pairs of the banded birds beginning in 1978. By 1987 there were six breeding pairs in the state of Hessen, including one nest in the television tower of Frankfurt.

Dr. Klaus Richarz succeeded in transplanting a nursery colony of the lesser horseshoe bat (*Rhinolophus hipposideros*) when it was rescued from a Peissenberg hotel, Upper Bavaria, about to be torn down. The species has been declining rapidly in Germany with a loss of its terrain. The municipality assisted in offering a substitute roost in a nearby "mining museum" building, which was simulated with niches, chambers, and shelters constructed of old untreated timber, equipped with heating, and monitored by an infrared barrier.

During the move, the animals were "picked from their hanging position and were temporarily stored in small linen bags" in cool deep tunnels. Horseshoe bat feces were moved into the "substitute roost" beforehand to provide an added "odor of the home roost." A male *Myotis myotis* joined about 11 horseshoe bats in the migration. After a series of flights back to the "traditional roost" and a final return to the new home, some 12 to 13 individuals eventually adapted—which was considered "a great success."

Biologist Dr. Richarz has been chief of the SV-IAV since 1991. For more than 10 years prior to that, Dr. Richarz served as officer for species conservation for the government of Upper Bavaria; he continues to serve on the advisory board of the Regional Group for Bird Protection in Bavaria, and is a member of the Species Survival Commission (IUCN/SSC) Chiroptera Group as well as of Bat Conservation International. His more than 70 publications range from scientific to general to children's works, and focus primarily on bats.

■ Resources

In addition to pamphlets and guides, the SV-IAV publishes a magazine for ornithology and environmental protection in Hessen titled *Bird and Environment*. The organization regularly updates the *Red List of Currently Endangered Bird Species in Hessen*, in conjunction with the Hessen Union for Ornithology and Nature Conservation. The Institute's library is open to students and members of other conservation groups. A few papers are printed in English.

Local Government

Botanischer Garten und Botanisches Museum Berlin-Dahlem
Botanic Garden and Botanic Museum of Berlin-Dahlem
Königin-Luise-Strasse 6-8
1000 Berlin 3, Germany

Contact: **Isolde Hagemann, Ph.D.**, Head Curator
Phone: (49) 30 8300 6129
Fax: (49) 30 8300 6218

Activities: Research • *Issues:* Biodiversity/Species Preservation

● Organization

Like many European botanic gardens, the **Botanic Garden and Botanic Museum of Berlin-Dahlem** was originally a royal kitchen garden for herbs and vegetables; with colonization and travels, the garden grew in size and variety of species as its owners received samples of plants from around the world. These days, botanic gardens have graduated from being curiosities to providing a practical method of conserving threatened plant species. Of an estimated 250,000 vascular plant species worldwide, about 60,000 will probably become extinct or at great risk in the next 60 years.

The Botanic Garden in Berlin currently covers 43 hectares of land, including 15 hectares of extended meadow areas which alone support 392 species, including 88 on West Berlin's list of threatened plants (for example, moonwort). In Berlin-Dahlem as a whole, a staff of 10, including 5 scientists and 3 gardeners, works to maintain about 600 globally endangered plant species, such as *Freylinia Visseri*, found only in the Capetown, South Africa area and *Sophoro toromiro* of the Easter Islands. Both of these plants no longer survive in their original habitat; botanic gardens such as the one in Berlin are trying to breed such species for reintroduction to their native areas.

▲ Projects and People

Currently, the Berlin-Dahlem Botanic Garden is working to carry out the resolution of the 1975 Kew Conference on the Conservation of Threatened Plants, which urges botanic gardens to give **priority to the conservation of their local flora.** The most effective method of maintaining flora is through the establishment of nature reserves; however, that is not always sufficient. Because of environmental damage, some need "additional supportive measures . . . such as the maintenance of selected species in botanic gardens." In doing so, however, gardeners must be aware that such cultivation might produce a different variety of the plant in question.

Following a recent decision of Berlin's Department of Environmental Protection, the Botanic Garden has been working to **collect seeds from endangered species in the city's natural environment,** multiply them in culture, return plants to their original site, and **reintroduce** them to support the on-site threatened population. Equally important is the elimination or reduction of threats to extinction at those sites; if these are insurmountable, another place must be selected for reintroduction.

The Botanic Garden's scientists have experimented with different methods of **artificial propagation and replanting,** including different techniques for perennials and annuals. In all cases, they try to reintroduce the plants as soon as possible, because older plants have less chance of survival after replanting. Since 1989, workers have multiplied and reintroduced several extant species; the new plants remain under care of the city parks department until fully established and, as environmental reserves are created, human trampling is minimized.

Alyssum alyssoides were transplanted as seedlings in "jiffi-pots" in **dry grassland habitats.** However, "it is much more difficult," reports the Garden, "to conserve threatened plants in damp or wet areas" because of changes in the water table in the biotopes. The Garden is attempting to rescue such threatened plants as *Dianthus suberbus l., Iris sibirica l.,* or *Euphorbia palustris l.* in Berlin's wetlands. "It remains to be seen whether these reintroductions will become established permanently and continue to produce offspring by self-seeding, and respond successfully to natural competition when the care of the sites by gardeners ends in due course."

Scientific supervision receives support from Berlin's Development and Environmental Protection Department. Workers carefully document site details such as growth, type of on-site reproduction, and rate of multiplication.

The Botanic Garden plans to enlarge its program in the coming years; with reunification, it can now work to **conserve plants in the county of Brandenburg,** not just Berlin. It also offers **guided tours** for the general public on plant conservation, and has erected information panels at sites in the botanic garden where endangered species are located.

Dr. Isolde Hagemann pursued her studies in gardening and botany in former East Germany, primarily at Martin-Luther University (MLU) in Halle-Wittenberg; after working as a scientific assistant at MLU, she then worked as a scientific assistant at the Institute for Hormone Research at the East German Academy of Sciences in Berlin from 1974 to 1976. In 1979, she fled to the West and has been at Berlin-Dahlem since then, becoming head curator in 1990.

■ Resources

Dr. Hagemann is the author of numerous scientific publications, including several on the Berlin Garden's efforts at plant breeding, as well as general articles on such topics as aromatic plants, and children's activity guides to fruits and leaves. Printed in English is literature on mountain flora, sedum growth forms, *Hypericum,* and descriptions of local flora.

With the help of the Technical University of Berlin's information department, the Botanic Garden began work on a **database/computer program** in 1987. The goal is to compile all of Berlin-Dahlem's data, and to share or exchange data with other botanic gardens and organizations.

Krefelder Zoo
Uerdinger Strasse 377
4150 Krefeld 1 Germany

Contact: **Paul Vogt, Ph.D.,** Assistant Director
Phone: (49) 2151 5804 3
Fax: (49) 2151 5908 87

Activities: Education; Research • *Issues:* Biodiversity/Species Preservation • *Meetings:* EEP

▲ Projects and People

Dr. Paul Vogt, assistant director of the **Krefelder Zoo** since 1971, is currently *Studbook* keeper for *Lutra l. lutra,* the European otter, his specialty. He has also published an article on breeding otters in the *International Zoo Yearbook* (1987).

In 1980, under Dr. Vogt's direction, the Krefeld Zoo began planning an **exhibit for otters,** which he describes as "attractive but delicate" creatures. The goal was "to build a natural-looking enclosure in which breeding and rearing could occur without the need for separation of the male at any stage." By 1981, the zoo had completed the exhibit, had obtained two European otters bred in captivity from the Norfolk Wildlife Park, and released them into the enclosure.

After that, zoo workers carefully monitored the activities of the otters; one of the biggest dangers is that fighting might break out between male and female of the species, or between adult and young, especially when the female is ready to give birth or is rearing her young. In such cases, separation would be necessary. In 1982 the two successfully mated for the first time. By 1986 they had produced 14 offspring, of which all but 3 had survived to adulthood. At least six otters have been sent to other collections.

■ Resources

Dr. Vogt seeks funding to attend global research conferences.

Staatliches Museum für Naturkunde Karlsruhe
Karlsruhe Natural History Museum

Erbprinzenstrasse 13

Postfach 6209

D-7500 Karlsruhe 1, Germany

Contact: **Monika Braun**, Biologist; Norbaden Bat
Protection Coordinator
Phone: (49) 721 175165
Fax: (49) 721 175110

Activities: Education; Research • *Issues:* Biodiversity/Species Preservation • *Meetings:* European Bat Research Symposium; International Bat Research Conference

● Organization

At the **Karlsruhe Natural History Museum**, a staff of 4 and about 20 volunteers work under the leadership of **Monika Braun** to protect German bats. More than 22 native bat species in Germany are on the Red List of endangered species; the bat protection division at the Museum works to save and protect species through preserving their habitat and educating the public.

▲ Projects and People

At the Museum, Monika Braun is responsible for public relations and exhibitions, working to increase people's—especially children's—understanding of importance of the environment in general. However, her main interest is bats. Braun first grew interested in bats after seeing a few youths at an evening rock festival throwing stones at bats flying overhead. She has been working with bats since 1979, and has been since 1985 leader of the **Fledermausschutz Norbaden (Nordbaden Bat Protection Organization)**, which began in 1980.

Current projects include working with the bats themselves, performing body, skull, and dental measurements, investigations of their residues, and hand-rearing young or injured bats. But because taking care of individual creatures will not alone save a species, Braun and the organization focus on wider conservation measures as well.

Significant effort goes into **investigating and mapping bat habitats**. From April to September each year, Braun spends her afternoons and evenings investigating old churches, schools, town halls, water towers, and castles, searching for bat colonies. She monitors and examines 150 to 200 locales each season, identifying, counting,

mapping, and documenting her findings. After determining that there is a viable bat population in a particular area, she and her co-workers, many of them volunteers, work to ensure that the area is preserved.

Braun has conducted investigations on the bat fauna and hunting habitat in the Rhine valley, and has initiated special bat conservation projects in other cities. Cooperating with the municipal government on roost-conservation, she serves on advisory boards for architects, public works organizations of church and state, natural protection organizations, forest rangers, and other public offices.

Braun's forte is bat publicity: through articles, interviews, stickers, and the like, she has been very successful at establishing volunteer working groups for bat protection. She teaches, explaining her methodology, and works to dispel anti-bat prejudice, based on a misguided fear of "vampires." Her motto is "überzeugen—nicht anordnen" (persuade, dont push), and in this way she has managed to acquire plenty of extra help. She belongs to the Species Survival Commission (IUCN/SSC) Chiroptera Group.

■ Resources

Resources needed include infrared technology for bat detection.

Monika Braun has contributed many individual and joint publications, mainly on bats. In 1989, she began a project to produce a handbook on *Wild Living Mammals of Baden-Württemberg*, financed by the state ministry. Some publications are printed in English, including *Foraging Areas as an Important Factor in Bat Conservation* with co-author E. Kalko.

 *NGO*

Aquarium–Zoologischer Garten Berlin AG
Zoo-Aquarium Berlin

Budapester Strasse 32

D-1000 Berlin 30, Germany

Contact: **Jürgen Lange, Ph.D.**, General Curator and
Assistant Director
Phone: (49) 30 2540 1216
Fax: (49) 30 2540 1255

Activities: Education; Research • *Issues:* Biodiversity/Species Preservation

● Organization

The original **Aquarium of the Berlin Zoo** opened in 1913. After World War I, from 1923 on, the Aquarium became world-famous for the number of species it maintained. Late in 1943, World War II bombing almost totally destroyed the structure; however, with the generous support of *Stiftung Deutsche Klassenlotterie Berlin*, the Aquarium, including Terrarium and Insectarium, was completely rebuilt by 1959. By 1968, the Aquarium again had the greatest variety of species in the world, according to its literature, including many rare animals maintained in Europe for the first time. The most recent renovation and expansion project was completed in 1983, overseen by curator **Dr. Jürgen Lange**. The reopened Aquarium included a new building extension housing five landscape aquariums (two coral seas, a South American river, an African lake, and Southeast Asia) and a turtle tank.

▲ Projects and People

As in most zoos and aquariums, the primary activity at the Berlin Aquarium involves maintaining and propagating diverse species. At the Berlin Aquarium this involves "not only aquatic animals, but also reptiles, amphibians, insects and spiders." A great deal of effort goes into imitating as closely as possible the native environments of the species housed. In this pursuit, "technology . . . and the experience gained from keeping numerous delicate species in the Zoological Gardens over the past two decades" plays a vital role. No matter what efforts are taken, however, Dr. Lange notes that "successful breeding of various difficult species . . . can only create an approximate equivalent home for the animals."

A popular exhibit is the dogfish, a shark whose eggs are attached to seaweed or coconut ropes by the female. Because the eggshells are transparent, growth of the embryo can be observed easily. Anemone fishes are among the few coral fishes that can be bred in aquariums. The paddlefish is one of the species found only in the Mississippi River, USA, and China. The pair of Komodo monitor lizards is a gift from the Indonesian president; only 10 others are found in zoos outside that country.

Important considerations include water quality (ensured by a complex system of filters), temperature, salt or fresh water supply, type of piping needed (plastic for the salt water, copper for the fresh), aeration (maintained by five air compressors), and lighting. The whole system is computer controlled and supported by a backup generator in the event of power outage.

However, the fish and the water tanks are only the tip of the iceberg. An entire range of secondary activity revolves around feeding and breeding the fish and other aquarium denizens. For example, the aquarium must maintain reserve tanks for quarantining new offspring or sick animals; because an adequate food supply is vital, rats, mice, and rabbits are reared as food, along with plankton, insects, and plants. Almost all the food needs are met through on-site planting and breeding to ensure that feed-animals are also properly fed, to avoid the toxic chain whereby a fish might consume, for example, a worm poisoned by pesticides.

In 1985 the Berlin Aquarium began a project to breed coral fish and has successfully bred 11 species. The process begins with cultivating plankton to nourish the prospective fish; this phase alone demands a great deal of space. Other concerns include knowing when to shift larvae into special breeding tanks, when to separate the territorial species during their aggressive phases, and what type of currents and counter-currents various species prefer for breeding purposes. In general, it is difficult to breed coral fish, but breeding avoids the alternative: environmentally damaging removal of species from their native sites.

Dr. Lange has been general curator for the aquarium, terrarium, and insect section at the Berlin Aquarium since 1978, and since 1988 has served as assistant director of the Zoo. From 1970 to 1977, he was aquarium curator and assistant at Stuttgart Zoological and Botanical Garden, Wilhelma. Internationally, Dr. Lange is a founding member of EUAC (European Union of Aquarium Curators), and was scientific advisor for the construction of new aquarium houses in Assisi, Italy; Saudi Arabia; and elsewhere in Germany. He belongs to the Species Survival Commission (IUCN/SSC) Freshwater Fish Group.

■ Resources

The Berlin Aquarium has several booklets and pamphlets, some printed in both German and English, explaining its layout and the various species of plants and animals maintained there. In addition, Dr. Lange has published several articles on fish breeding and other aquarium activities.

Arbeitskreis für Umweltrecht (AKUR) Working Group for Environmental Law
Adenauerallee 214
5300 Bonn, Germany

Contact: **Rainer Wahl, Ph.D.**, Chairman
Phone: (49) 228 269 2216
Fax: (49) 228 269 2250

Activities: Political/Legislative • *Issues:* Environmental Law

● Organization

The organization Arbeitskreis für Umweltrecht (AKUR), or the Working Group for Environmental Law, began in 1972 to fill a gap left by the federal government. Earlier, the government had established a special advisory body to assess periodically the environmental situation, but had stipulated that the body would not deal with matters of law and administration. AKUR deals specifically with legal and administrative policy relating to the environment, following all developments, both positive and negative, in the environmental situation, and sharing its information with the appropriate bodies.

AKUR has a maximum of 12 members, who cannot be working for the government, for any law-giving body, nor for a trade association.

▲ Projects and People

The organization's main activity involves advising and offering commentary on environmental issues.

Members of the working group include: **Dr. Everhardt Franssen**, **Dr. Eckart Rehbinder** of University of Frankfurt, **Ernst Kutscheidt**, Bonn attorney **Dr. Dieter Sellner**, **Dr. Rudolf Stich** of University of Kaiserslautern, and **Dr. Jürgen Salzwedel** of University of Bonn.

■ Resources

Some 740 research reports support the suggestions AKUR makes. Special attention goes to drawing up, in a widely shared effort, the publication *Fundamentals of Environmental Law*.

Brehm-Fonds für Internationalen Vogelschutz ⊕
Brehm Fund for International Bird Conservation
Adenauerallee 214
5300 Bonn, Germany

Contact: **Wolfgang Burhenne, Ph.D.**
Phone: (49) 228 269 2223
Fax: (49) 228 269 2251

Activities: Development; Research • *Issues:* Biodiversity/Species Preservation

● Organization

With 4 staff members and 6,000 members, the **Brehm-Fonds für Internationalen Vogelschutz**, or **Brehm Fund for International Bird Conservation**, is an institution originating from the **Walsrode Birdpark**. The administration works from the shared headquarters in Bonn, in order to save costs and work with other groups with common interests. The Fund, a member of the World Conservation Union (IUCN), is currently working with the International Crane Foundation and the International Council for Bird Protection (ICBP). Leadership is in the hands of the Fund's board of directors, including film director **Henry Makowski** and chairman **Wolf W. Brehm**.

▲ Projects and People

Projects which serve the practical application of scientific results are supported by a circle of patrons and by donations of visitors to the **Birdpark**.

■ Resources

The fund regularly publishes a newsletter, *Zum Fliegen geboren* (Flying free).

Bund für Umwelt und Naturschutz Deutschland e.V. (BUND) ⊕
Coalition for Environment and Nature Protection of Germany

Im Rheingarten 7
5300 Bonn 3, Germany

Contact: **Arno Behlau**, International Coordinator
  Phone: (49) 228 40097-0
    Fax: (49) 228 40097-40
  E-Mail: Green Net: bundbn

Activities: Education; Law Enforcement; Political/Legislative *Issues:* Air Quality/Emission Control; Biodiversity/Species Preservation; Deforestation; Energy; Global Warming; Health; Sustainable Development; Transportation; Waste Management/Recycling; Water Quality • *Meetings:* IMO; IPCC; UNCED

● Organization

Bund für Umwelt und Naturschutz Deutschland (BUND), or the Coalition for Environment and Nature Protection of Germany, was founded as a national organization from the union of various regional environmental groups in 1975. BUND now covers all states of Germany; membership has grown from 40,000 in 1978, to the current 200,000 members, and a staff of 50. In addition to hundreds of regional and local groups, BUND has 23 "Working Circles" on specific topics, each headed and staffed by experts or "speakers" in such fields as waste disposal/recycling, bio-/gene-technology, agriculture, and legal questions. BUND is a member of Friends of the Earth International (FOEI), and worries that, despite

some progress on environmental issues, "all too often, business interests are placed before the protection of our environment."

BUNDjugend is the youth organization of BUND, but operates independently, with 26,000 members in 700 groups. BUNDjugend's goal is to protect the environment through creative and spontaneous actions, from the youth perspective.

▲ Projects and People

Describing itself as "Nature's lobby in Bonn," BUND has staffers working actively on laws, ordinances, and planning, and is in frequent contact with various ministries. Of particular interest is the recent reunification of Germany, and the planned United Europe. These events require special action, special consideration, and the drawing up of comprehensive environmental plans, and laws that address the reorganized political entities.

BUNDjugend works to conserve areas, plant trees, hold protest actions; it also engages in street theater and other means of informing and mobilizing the public to act. The annual New Year's Eve youth congress draws hundreds. Like its parent organization, BUNDjugend has working circles (AKs); the AK-international organizes an **annual bike tour** with the **European Youth Forest Action**—riding in 1990 from Bergen to Budapest, a distance of 4,000 kilometers.

BUND campaigns include **Europe Without Poison**, symbolized by the endangered crane; Save the Streams; and a heavy focus on forests, both those of Germany and the Northern Hemisphere, where acid rain is the main threat, and the tropical rainforests, threatened by overexploitation, and slash-and-burn farming techniques. BUND demands that immediate measures be taken to protect forests from further destruction.

■ Resources

The main resource needed is help with translation from German into other languages.

Through BUNDinfoservice, the organization makes available a large number of publications. For parents and teachers of young children, *Tips in Season—Experience the Environment with Children* appears quarterly, covering such topics as hibernation; puddles, ponds, and lakes; and night. *BUNDarguments* covers a single environmental issue in depth. For example, one on excess packaging points out the literal mountains of garbage being produced in Germany; notes that fast food equals faster garbage; warns that plastic bags contain 2 percent lead and that most packaging is full of chemicals; and reminds readers that most packaging can be neither reused nor recycled, benefiting only the chemical industry. *BUNDfacts* explores ways in which the individual can make a difference in environmental care. In *Environmental Protection in Your Home*, the newsletter discusses recycling, composting, dealing with the car (changing oil, speed limits, and asbestos-free brake lining), laundry habits, and other household areas for action.

In addition, BUND publishes the quarterly *Nature and Environment*, maps, high school–directed environment periodicals such as *Not for School—For Survival: An Action Brochure*, position papers, posters, stickers, and books to provoke action, like *How the World Bank Makes the World Sick: Environmental Destruction via World Bank Projects*.

Bund Naturschutz in Bayern e.V. (BN)
Coalition for Nature Conservation in Bavaria
**Bund für Umwelt und Naturschutz
  Deutschland (BUND)**
**Coalition for Environment and Nature
  Protection of Germany**
Kirchenstrasse 88
8000 Munich 80, Germany

Contact: **Christoph Markl**, Press and Information
  Phone: (49) 89 4599180
    Fax: (49) 89 485866

Activities: Education; Political/Legislative • *Issues:* Air Quality/Emission Control; Biodiversity/Species Preservation; Deforestation; Energy; Global Warming; Health; Sustainable Development; Transportation; Waste Management/Recycling

● Organization

The **Coalition for Nature Conservation in Bavaria (BN)**, a regional chapter of the national organization BUND, describes itself as Bavaria's "green conscience and ecological early warning system." The oldest and largest conservation organization in Bavaria, BN was founded in 1913 in Munich, and now has more than 100,000 members in 77 regional and 650 local chapters, including a Youth Organization with 15,000 members. The organization, with its various subgroups, requires a staff of 55.

▲ Projects and People

The Coalition works on influencing **legislation for environmental protection**, primarily through publicizing the issues. In 1990, BN successfully pushed for new Bavarian laws governing industrial waste and restricting the use of incinerators to reduce dioxin emissions. The group is also working to pass laws related to excess packaging, highway construction, agriculture, and other environmentally related issues.

Much of BN's work involves **conserving species and habitats.** Through purchase, grants, easements, and other methods, BN by the end of 1990 had protected some 1,600 hectares of land in Bavaria. Projects to protect endangered plants and animals have included programs to replenish the **wildcat** and **beaver** populations in Bavaria.

Many of the local BN chapters work intensively on recycling projects and biological farming/gardening efforts. Under an **Adopt-A-Tree** program, members "sponsor" fruit trees and are then entitled to reap the trees' harvest.

The Coalition's **museum programs**, like "Between Heaven and Earth" and "Waters Around Us," have educated thousands of visitors, both adults and children, regarding environmental issues and environmentally friendly behavior. The group has also sponsored an **Ecohouse** at the Bavarian Garden Show, providing a walk-through example of recycling, composting, energy conservation, and other actions individuals can take in their own homes.

■ Resources

The Coalition offers numerous publications on the issues it tackles, including pamphlets, position papers, and educational materials; and sponsors seminars, field trips, and other educational forums. *(See also previous BUND profile.)* BN recently joined the Bavarian Film Services organization (LFD) to expand its media options.

Bund für Umweltschutz—Reutlingen
Reutlingen Coalition for Environment Protection
**Bund für Umwelt und Naturschutz
  Deutschland (BUND)**
**Coalition for Environment and Nature
  Protection of Germany**
Lederstrasse 86
7410 Reutlingen, Germany

Contact: **Wolfgang Zielke**
  Phone: (49) 7121 3209 93

Activities: Education; Political/Legislative; Research • *Issues:* Air Quality/Emission Control; Biodiversity/Species Preservation; Energy; Global Warming; Health; Transportation; Waste Management/Recycling; Water Quality

● Organization

The **Bund für Umweltschutz—Reutlingen** is a regional chapter of BUND, Germany's largest environmental organization. The national organization has about 200,000 members, organized in over 2,000 local groups throughout the country. The Reutlingen chapter has about 600 members active in such specialized areas as "energy, traffic, waste problems, and agriculture." A special subgroup exists for young people, called **BUNDjugend (BUNDyouth)**. With a staff of 2 and more than 680 members, as well as volunteers, the Reutlingen BUND is working to stop the destruction of nature and to reduce "radically" the pollution of air, water, and soil.

▲ Projects and People

The local BUND organization pursues many of the same goals and activities as the national group: promoting an **environmentally viable traffic system** with fewer cars, trucks, and streets but more public transportation and bicycles; working to change the country's energy system, especially through **alternatives to nuclear energy** such as solar power, to stop inefficient use and waste of energy; supporting **biological, non-chemically dependent agriculture;** and developing ideas to avoid waste and systems to **recycle** waste.

BUND–Reutlingen works "to achieve these goals by dispersing public information, publishing books and brochures, lobbying, and providing information on problems of conservation."

■ Resources

BUND–Reutlingen acknowledges the need for nations to cooperate in protecting the environment; therefore, it is seeking to further its contacts with European environmental groups to encourage information exchange, the setting of common goals, and to encourage broad-based, unified action.

Examples of publications of BUND–Reutlingen include *Gardening without Artificial Fertilizers and Pesticides,* and *Nature without Borders.*

Conseil Européen de Droit de l'Environnement European Council of Environmental Law ⊕
Adenauerallee 214

5300 Bonn, Germany

> *Phone:* (49) 228 269 2216
> *Fax:* (49) 228 269 2250

Activities: Political/Legislative • *Issues:* Environmental Law

● Organization
Founded in 1974 within the International Council for Environmental Law (ICEL), the Conseil Européen de Droit de l'Environnement (European Council of Environmental Law) fosters the development and research within Europe in the area of environmental law. To achieve its aims, the Council works with the European Community's (EC) Parliament and advisory bodies. Every member state of the EC sends a "jurist" to collaborate with the Council on specific questions.

■ Resources
The Council's printed *Resolutions* cover nearly all questions of importance to the sphere of "supranational" environmental law. In addition, commentary is available on the World Charter and on the environmental sections of the Rome Treaty.

Deutsche Delegation des Internationalen Jagdrates zur Erhaltung ⊕ International Council for Game and Wildlife Conservation—German Delegation
Adenauerallee 214

5300 Bonn, Germany

> *Contact:* Johann-Daniel Gerstein, Ph.D., Chairman and Leader of German Delegation
> *Phone:* (49) 228 269 2216
> *Fax:* (49) 228 269 2251

Activities: Political/Legislative • *Issues:* Biodiversity/Species Preservation

● Organization
With a staff of one and 105 members, the Deutsche Delegation des Internationalen Jagdrates zur Erhaltung, or German delegation to the International Council for Game and Wildlife Conservation, is a tax-supported union for the support of scientific measures regarding conservation. In close contact with the international body, based in Paris, the German delegation regulates the activities of its commissions and proceedings, and supports national and international projects which maintain free-living wildlife.

▲ Projects and People
The delegation offers specialized advice and consulting for local and national governmental authorities, in addition to more generalized explanations for the public, regarding issues related to wildlife protection.

Besides chairman and German delegation leader Dr. Johann-Daniel Gerstein, CIC members include treasurer Emil Underberg, and Peter C. von Siemens.

Deutsche Stiftung für Umweltpolitik (DSU) German Foundation for Environmental Policy
Adenauerallee 214

5300 Bonn, Germany

> *Contact:* Wolfgang Burhenne, Ph.D.
> *Phone:* (49) 228 269 2216
> *Fax:* (49) 228 269 2251

Activities: Political/Legislative • *Issues:* Air Quality/Emission Control; Biodiversity/Species Preservation; Deforestation; Energy; Global Warming; Health; Population Planning; Sustainable Development; Transportation; Waste Management/Recycling; Water Quality

● Organization
The Deutsche Stiftung für Umweltpolitik (DSU), or German Foundation for Environmental Policy, was established as a means of advising on and carrying out an environmental policy based on scientific grounds. Its governing body, or board of trustees, includes representatives from political, business, and scientific circles. The office has a staff of three.

▲ Projects and People
Foundation members include: Dr. Wolfgang E. Burhenne, Dr. Martin Uppenbrink of the Free University of Berlin and director of the Federal Environment Office in Berlin, Dr. Heinz Markmann, Martin Hirsch, state secretary Dr. Günter Hartkopf, Dr. Klaus Boisserée, and Dr. Michael Bothe of the University of Frankfurt.

■ Resources
The DSU is known for its **Environmental Policy Colloquia,** on such topics as the problem of economic instruments for environmental policy, or on the World Commission's work on perspectives for the year 2000. The organization also publishes reports on current questions of environmental policy, and such books as *Umwelt Weltweit* (Environment worldwide).

Each year the Foundation awards the **Ernst-Haeckel Medal** to worthy individuals, and the **Beatrice-Flad-Schnorrenberg Prize** to a publicist of merit.

Deutscher Naturschutzring (DNR)/German Nature Conservation Group
Kalkuhlstr. 24
53 Bonn 3, Germany

Contact: **Helmut Roescheisen**, Managing Director
Phone: (49) 228 441505
Fax: (49) 228 442277

Activities: Development; Political/Legislative • *Issues:* Air Quality/Emission Control; Biodiversity/Species Preservation; Energy; Global Warming; Nature Conservation; Transportation

Environmental Law Center der IUCN (ELC) ⊕
Adenauerallee 214
Bonn 5300, Germany

Contact: **Francoise Burhenne-Guilmin, Ph.D.,** Head
Phone: (49) 228 269 2216
Fax: (49) 228 269 2250

● Organization
The **World Conservation Union (IUCN),** founded in 1948, comprises both governments and international and national NGOs from 119 countries. More than 3,200 specialists from around the world work in 6 special commissions: ecology, development, environmental planning, endangered species, protected areas, and environmental policy, law, and administration. *(See also World Conservation Union [IUCN] in the France, Niger, Pakistan, Switzerland, USA, and Zimbabwe NGO sections.)* The Environmental Law Center (ELC) of the IUCN has offices in Bonn.

▲ People and Projects
Chairman of the IUCN Commission on Environmental Policy, Law and Administration (CEPLA) is **Dr. Wolfgang E. Burhenne;** vice chairs are **Dr. Oleg S. Kolbasov** of the Moscow Academy of Sciences, **Malcolm J. Forster** of the University of Southampton, **Nicholas Robinson,** and **Charles Okidi.** Head of the ELC is **Dr. Francoise Burhenne-Guilmin.**

■ Resources
The ELC's **Environmental Law Information System (ELIS)** was first proposed at a 1972 UN Environment Conference. It now contains some 37,000 environmentally relevant law reports from around the world. ELIS works closely with the UN and with the German federal government.

ELC is also devoting a great deal of effort to developing **ENLEX,** an information system of the European Community.

In addition to a relevant literature databank containing some 45,000 titles, ELC has a special databank for more than 10,000 animal species named in law reports. The Center is currently building up a similar databank for plant species.

Fonds für Umweltstudien (FUST) Fund for Environmental Studies
Adenauerallee 214
5300 Bonn, Germany

Contact: **Martin Hirsch,** Chairman
Phone: (49) 228-269 2216
Fax: (49) 228 269 2251

Activities: Education • *Issues:* Environmental Education

● Organization
Fonds für Umweltstudien (FUST), or the **Fund for Environmental Studies,** active since 1970, has sponsored more than 100 projects related to environmental education and understanding. Although it does not receive a great deal of attention from the government, many individual parliamentary deputies are very interested in the Fund's work.

▲ Projects and People
FUST is headed by **Martin Hirsch,** with **Helga Haub** as chairman of the board.

■ Resources
In 1989 the Fund published Volume 110 in its series *Contributions to Environmental Formation.* Two other series of note are the six-volume *Multilateral Treaties* and the four-volume *European Environmental Law.*

Förderungsverein für Umweltstudien—Tirol Association for the Promotion of Environmental Studies
Adenauerallee 214
5300 Bonn, Germany

Contact: **Johann Astner, Ph.D.,** Chairman
Phone: (49) 228 269 2216
Fax: (49) 228 269 2251

Activities: Research • *Issues:* Air Quality/Emission Control; Biodiversity/Species Preservation; Deforestation; Sustainable Development; Transportation; Water Quality; Alpine Environment

● Organization
The **Förderungsverein für Umweltstudien (Association for the Promotion of Environmental Studies)** is the Tirol-based sister organization to the German **Fond für Umweltstudien (FUST)** (Fund for Environmental Studies).

▲ Projects and People
The Tirol Association, with eight staff members, is active in the development project **Alpine Environment in Achenkirch,** Tirol. In general, the Association carries out basic research and practical model experiments for the protective use of high mountain regions,

covering such activities as forestry, hunting, animal husbandry, tourism, and human settlement, and working to reduce the overuse of land and excessive emissions levels.

The prime goal of the Association is maintaining the forests and their natural reproductive ability, as forests are the most important habitat for man and for free-living animals.

Head is **Botho, Prince of Sayn-Wittgenstein,** who also serves as president of the German Red Cross. Others active include chairman **Dr. Johann Astner,** and **Emil Underberg.**

▪ Resources

The scientific results and practical successes of the Tirol Association have been published in a series entitled *Contributions to Environmental Formation*. The organization also offers excursions, lectures, and other educational experiences.

Elisabeth Haub Foundation ⊕

Adenauerallee 214

5300 Bonn, Germany

Contact: **Helga Haub,** Chair
Phone: (49) 228 269 2216
Fax: (49) 228 269 2250

Activities: Political/Legislative • *Issues:* Sustainable Development

● Organization

The **Elisabeth Haub Foundation** is the sister organization of the **Karl-Schmitz-Scholl Fund.** Its aim is to develop and improve legal groundwork, especially in the developing nations, for sensible and protective use of natural resources.

▲ Projects and People

The Foundation took part in work on the United Nations World Charter for Nature. Chaired by **Helga Haub,** Foundation membership includes former U.S. Congressman **Morris K. Udall** and World Wide Fund for Nature (WWF) president **Russell E. Train.**

International Council for Environmental Law (ICEL) (Genf/Geneva) ⊕

Adenauerallee 214

5300 Bonn, Germany

Contact: **Marlene Jahnke**
Phone: (49) 228 269 2240
Fax: (49) 228 269 2250

Activities: Education; Political/Legislative • *Issues:* Environmental Law

● Organization

The **International Council for Environmental Law (ICEL)** was established in New Delhi in 1969, and includes 300 environmental

jurists and policy experts, who represent 10 world regions and speak for those regions' leaders.

▲ Projects and People

Executive governor of ICEL is **Dr. Wolfgang E. Burhenne;** governor is **Taslim Elias,** professor at the University of Lagos.

▪ Resources

The ICEL's greatest resource is its constantly expanding databank, bound with that of the Environmental Law Center (ELC) of the World Conservation Union (IUCN), Gland, Switzerland. The comprehensive library contains some 45,000 books, articles, and other documents on environmental law and policy.

Each year the Council, in conjunction with the Free University of Brussels, awards the Elisabeth Haub Prize for outstanding contributions in the field of environmental law.

The resource journal *References* appears 16 times annually, with some 2,000 references to environmental law and policy studies worldwide. In addition, the Council publishes the magazine *Environmental Policy and Law* eight times each year.

Interparlamentarische Arbeitsgemeinschaft (IPA) ⊕

Interparliamentary Work Union

Adenauerallee 214

5300 Bonn, Germany

Contact: **Hermann Leeb,** Chairman
Phone: (49) 228 269 2216
Fax: (49) 228 269 2250

Activities: Political/Legislative • *Issues:* Environmental Law

● Organization

The **Interparlamentarische Arbeitsgemeinschaft (IPA),** or Interparliamentary Work Union, has since 1952 strived for environmental progress through democratic means. A venue for cooperation among various German political parties, regional councils, the federal parliament, and the European Parliament, the IPA has worked on numerous pieces of environmental legislation; indeed, the first important environmental laws stemmed from IPA initiatives.

Among IPA's goals are cooperation among members of various representative elected bodies; initiatives in areas that offer clear opportunities for cooperation; environmentally responsible use and maintenance of natural resources to ensure survival of the animal and plant worlds as well as of human beings; and improved exchange of specialized support and experience between parliaments.

▲ Projects and People

Members of the IPA include **Hermann Leeb; Dr. Johannes Rau,** the minister-president; **Dr. Heinz Peter Volkert,** president of the Rheinland-Pfalz State Congress; **Dr. Rose Götte** and **Klaus Beckmann,** Bundestag members; and **Dr. Wolfgang E. Burhenne.**

Karl-Schmitz-Scholl-Fonds ⊕
Karl-Schmitz-Scholl-Fund for Environmental Policy and Law
Adenauerallee 214
5300 Bonn, Germany

Contact: **Helga Haub**, Chairman of the Board
Phone: (49) 228 269 2216
Fax: (49) 228 269 2251

Activities: Funding; Political/Legislative • *Issues:* Biodiversity/Species Preservation; Environmental Law

● Organization
The **Karl-Schmitt-Scholl-Fonds** (KSS-Fund) was founded in 1968 by **Elisabeth Haub**, who clearly understood that animals, plants, air, and water could be protected only if the necessary legal mechanisms were improved and implemented. The Fund is expressly concerned with national and international efforts to this end.

▲ Projects and People
The Fund has given financial support to **Africa Convention** and other organizations, and has provided technical assistance on many projects.

Personalities like **Ulrike Nasse-Meyfarth** and **Paul Breitner** have joined the effort to publicize KSS-Fund's goals. In addition to the board chairwoman **Helga Haub**, of the Atlantic and Pacific Tea Company (A&P), Tengelmann department stores, and Kaiser's Kaffee, Fund members include Dr. Francoise Burhenne-Guilmin, Walter Gerling, Dr. Volkmar Mair, and Dr. Herbert P. Koening.

Landesbund für Vogelschutz in Bayern e.V., Verband für Arten-und Biotopschutz
Regional Alliance for Bird Protection in Bavaria, Union for Species and Habitat Protection
Kirchenstrasse 8
8543 Hilpoltstein, Germany

Contact: **Gabriele Kappes**, Species Protection Consultant
Phone: (49) 9174 9085
Fax: (49) 9174 1251

Activities: Education; Law Enforcement; Research • *Issues:* Biodiversity/Species Preservation

● Organization
The **Regional Alliance for Bird Protection in Bavaria** is the Bavarian chapter of the German Alliance for Bird Protection (DBV), which has 35,000 members spread over some 350 subgroups. With 41 staff members, approximately 70 workers fulfilling their civil service requirement, and many volunteers, the Regional Alliance has purchased 500 hectares of land in Bavaria and protected 300, for a total of 800 hectares. It is currently "developing management plans for nature reserves and national parks, . . . conducting guided tours within these areas."

The Alliance focuses on birds, but stresses that in protecting a particular bird species, it is also protecting habitat for numerous other species.

▲ Projects and People
The Regional Alliance subscribes to the national organization's **Bird of the Year** program. Each year, the Alliance singles out a particular species for special attention, public awareness campaigns, and protection. In 1989, it was the reed warbler; in 1992, the robin. Bird of the Year in 1984 was the white stork, and through **Projekt Weissstorch**, a white stork protection project with the support of the European Community, that protection campaign has continued. In Bavaria in 1965, there were some 200 breeding white stork pairs; by 1980, there were 104, and in 1988, the population hit an all time low of 60 pairs. The main cause has been loss of wetlands, which the Alliance is working to reclaim, and overdevelopment, which the Alliance is trying to slow or halt. Conservation efforts focus not only on Germany, but on **Africa** as well, where the Alliance helps efforts to **maintain white storks' winter sites**. Other international efforts of the Alliance include the support of environmental organizations and research programs in Poland, Czechoslovakia, and elsewhere in Eastern Europe.

On a local level, the Alliance carries out multifaceted projects such as **mapping locations and conducting inventories, management, and development of animal and plant species**. The group works politically to inform voters and legislators of pressing concerns, tries to bring "scofflaws" (flouters of the law) to justice, and looks for "acceptable political solutions for problems concerning nature."

Habitat protection efforts are important to the Alliance. In 1984 with help of the WWF, the group purchased **Roetelseeweiher**, saving it from becoming farmland; the wetland is home to many endangered species including the white stork and **black throated diver**, and is a **kingfisher breeding area**. The **Schwarzach River** watershed area is another key locale. The Alliance is working toward ensuring the free course of the river, returning its natural curves and meanderings that are vital to maintaining marshy areas. With just minimal attention, the organization notes, species like the king quail/land rail (wachtelkonig) return to their habitat; thus there is all the more reason for greater efforts.

The Alliance's **Youth Organization** works on its own projects, such as "campaigns against fast-food companies like McDonald's or against waste production in our society, [and] tree planting actions." Indeed, in one recent year more than 100,000 trees and shrubs were planted by Alliance members.

■ Resources
The Alliance sells books, brochures, tapes, posters, figurines, clothing, buttons, and stickers—printed in German. Successful educational series include *Learning Naturally* and *Environment Tips for Students*. The organization offers seminars and programs for children and young adults, works with teachers/schools, and holds several annual events: the Green Scene student theater festival, in conjunction with Bavarian Broadcasting; Encounter Nature Day; and Experience Spring, a nature observation game in which 100,000 students participate each year.

Naturfreunde/Friends of Nature ⊕ Bundesgruppe Deutschland/The German Coalition

P.B. 60 04 41

7000 Stuttgart 60, Germany

Phone: (49) 711 337 687
Fax: (49) 711 337 310

Activities: Education; Political/Legislative • *Issues:* Biodiversity/Species Preservation; Ecotourism; Waste Management/Recycling

● Organization

With 120,000 members, **Naturfreunde** is an umbrella organization with international headquarters in Vienna, Austria. Activities in local groups are political, cultural, environmental, and sports related. *Wandern and Bergsteigen* is its main publication. Other resources are a "Manifesto for a New Europe" and a "Landscape of the Year" program.

Schutzgemeinschaft Deutscher Wald Association for the Conservation of German Forests

Kernerplatz 10

7000 Stuttgart 1, Germany

Contact: **Rainer Deuschel**, Geschaftsfuhrer
Phone: (49) 711 292 099
Fax: (49) 722 226 2455

Activities: Education; Law Enforcement; Political/Legislative; Research • *Issues:* Air Quality/Emission Control; Biodiversity/Species Preservation; Deforestation; Energy; Forest Conservation; Global Warming; Waste Management/Recycling; Water Quality

● Organization

Schutzgemeinschaft Deutscher Wald has 2,000 members.

Schutzgemeinschaft Deutsches Wild Association for the Conservation of German Wildlife

Adenauerallee 214

5300 Bonn, Germany

Contact: **Count Joachim Schönburg**, Chairman
Phone: (49) 228 269 2217
Fax: (49) 228 269 2251

Activities: Political/Legislative • *Issues:* Biodiversity/Species Preservation

● Organization

The **Schutzgemeinschaft Deutsches Wild (Association for the Conservation of German Wildlife)** was founded after World War II, and was for a time considered worthy of invitation to international summits and meetings with chancellors, minister-presidents, and high commissioners. The Association was the first German organization active in the World Conservation Union (IUCN). Today, with so many competing environmental organizations, the Schutzgemeinschaft is that much more significant because of its work toward consensus among supporters of hunting, on one hand, and conservationists, on the other.

▲ Projects and People

The Association has supported or carried out numerous projects, both completed and ongoing, to protect endangered species such as the **rock eagle, black stork, butterflies, peregrine falcon, bat,** and **sea eagle**. In addition, the organization carries out educational work, both in schools and outdoors, where its informative placards provide clear explanations.

Chairman is **Count Joachim Schönburg**; also on the board is state secretary **Hans-Jürgen Rohr**.

■ Resources

The Association publishes the journal *Für unsere freilebende Tierwelt* (For our free-living animal world). The book *Wild hinter unsichtbaren Gittern* (Wildlife behind invisible bars) provides comprehensive coverage of the Schutzgemeinschaft's history and activities.

Stiftung zur Internationalen Erhaltung der Genetischen Pflanzenvielfalt (Zürich) ⊕ Foundation for the International Conservation of Plant Genetic Diversity

Adenauerallee 214

5300 Bonn, Germany

Contact: **Emil Underberg**, Chairman
Phone: (49) 228 269 2216
Fax: (49) 228 269 2252

Activities: Political/Legislative; Research • *Issues:* Biodiversity/Species Preservation

● Organization

The **Stiftung zur internationalen Erhaltung der genetischen Pflanzenvielfalt**, or **Foundation for the International Conservation of Plant Genetic Diversity**, sees as its main task taking measures for the preservation of genetic diversity of wild-growing plants in their natural habitats. Special focus is on **wild plants** considered by many to be weeds, for their preservation, regeneration, and repopulation. Also receiving intense attention are **threatened plants and their biotopes**, biological plant protection measures, and maintenance of **wild relatives of useful domesticated crops**.

The Foundation focuses primarily on the practical application of theoretical findings, especially those relating to the food chain of wild and domestic animal species, and to commerce, particularly in agriculture, forestry, food, and pharmaceuticals.

Umweltstiftung WWF—Deutschland ⊕
World Wildlife Fund (WWF)—Germany

Hedderichstrasse 110
D-6000 Frankfurt (M) 70, Germany

Contact: **Arnd Wünschmann, Ph.D.**, Managing
               Director
  Phone: (49) 69 605003-34
    Fax: (49) 69 617221

Activities: Development; Education; Law Enforcement; Political/Legislative; Research • *Issues:* Biodiversity/Species Preservation; Deforestation; Global Warming; Protected Areas; Sustainable Development; Water Quality • *Meetings:* CITES; IUCN; WWF

● Organization

Established in 1963, Umweltstiftung WWF—Deutschland (WWF-Germany) became the fifth national affiliate organization of the World Wildlife Fund (now the World Wide Fund for Nature). Early activities involved land purchases, various national projects, and international WWF projects with the World Conservation Union (IUCN) that brought about the establishment of the IUCN Environmental Law Centre in Bonn. In 1978, after the election of a new executive board of trustees, the organization became more active. Within 10 years, WWF-Germany became one of the leading national conservation organizations, with a current membership of more than 90,000. Current foundation president is **Casimir, Prince of Sayn-Wittgenstein-Berleburg.**

WWF-Germany recognizes a "highly developed consciousness and public awareness of environmental issues" throughout Germany. In order not to waste the opportunity, WWF is working on "joint lobbying efforts and coordination" with other environmental groups. Many WWF-funded projects (mainly in species conservation) are managed by other NGOs or small local groups. WWF estimates there are some 1,500 conservation-related NGOs in Germany with about 5 million members. With WWF-Germany and Greenpeace as the leading international NGOs, WWF is hoping to coalesce these groups.

In programs and projects of national or international dimensions (such as Wadden Sea and Floodplains Conservation or TRAFFIC-Germany), in which WWF has a key position, "cooperation and coordination with other organizations are working well," it reports. In all its endeavors, whether with like-minded groups or with government and industry, WWF-Germany seeks coordination and cooperation—"constructive dialogue," not confrontation.

▲ Projects and People

In addition to the WWF-Conservation Programme and a long-term Action Plan, WWF-Germany has numerous project priorities, such as habitat protection. Since 1980, WWF has been working on the **Wadden Sea, North Sea, and East Sea,** often in cooperation with the Netherlands and Denmark, to protect endangered tidal marshes, wetlands, and other fragile coastal areas threatened by oil/gas drilling, development, military maneuvers, water sports, hunting, excessive fishing, and other human activity. WWF sees the Wadden Sea as a test case for European environmental politics: it asks whether countries will live up to their legislation.

In 1990, the organization launched a new strategy, dubbed WWF 2000. An important element of the strategy is political involvement. The WWF plans to take political positions on most environmental issues, with a major focus on German government policies toward developing nations. WWF is increasing its financial support of conservation work in the Third World, and striving to influence the funding policy of German development aid agencies and authorities. The organization is also pursuing **debt-for-nature swaps** with German banks, working for sustainable development, and combatting poverty through a "help for self-help" strategy.

Important new efforts are going on in the five new states (formerly East Germany), where there are numerous practically undisturbed areas for rare plant and animal species; in 1990, **five new national parks** and eight other reserves were set aside by the government. "With these new areas being finally confirmed under German conservation law, nearly 10 percent of the former territory will be under nature protection—a remarkable step forward for the whole country," especially when compared to the barely one percent of the total area of former West Germany that is so designated.

One specific east German WWF project, in conjunction with the Berlin Zoological Garden, centers on the **Untere Havel/Gülper Lake area,** a breeding ground for black tern, grey goose, and many other bird species. The area has been designated as "FIB"—a wetland of international significance. Along the German/Polish border, a shared national park area **Unteres Odertal** is being developed along the Oder River; 60 kilometers long, and 2.5 kilometers on both sides of the river, it involves an unprecedented cooperative effort between the two countries. The area includes many different habitats—forest, wetland, meadow, and moor; up to 100,000 geese and ducks at one time rest there during migration each spring, as well as over 100 types of nesting birds, including 25 under special protection like black storks, sea eagles, and cranes.

Following the European Community regulation on trade of endangered species, IUCN and WWF have begun **TRAFFIC—Trade Records Analysis of Fauna and Flora in Commerce.** The groups collect and analyze data and facts about the international trade of wild animals and plants, monitoring compliance with CITES (Convention on International Trade in Endangered Species of Wild Fauna and Flora), drafted in 1973 in Washington, DC. WWF-Germany monitors what comes into Germany.

Youth Protects Nature is a countrywide WWF action effort to foster concrete activities by young people on behalf of the environment. WWF encourages students and teachers to work together to plan and implement environmental protection projects such as tree planting, and recycling. The **Panda Club** for younger children builds awareness of environmental needs as well.

■ Resources

WWF-Germany publishes an *Information* newsletter, *Report* series on current activities, and *WWF Journal.* Periodicals appear in German.

Wildbiologische Gesellschaft München (WGM)
Wildlife Society of Munich
Postfach 170
D-8103 Oberammergau, Germany

Contact: **Goetz Kerger**, Forester
Phone: (49) 88 22 6363
Fax: (49) 89 53 89171 (c/o A. Scholz)

Activities: Development; Education; Research • *Issues:* Biodiversity/Species Preservation; Deforestation; Water Quality; Wildlife Management • *Meetings:* International Union of Game Biologists

● Organization

With a current membership of about 150, the **Wildlife Society of Munich (WGM)** was founded in 1977 by a handful of people who got together to work toward solving problems of nature and environment problems in the light of ecological knowledge and in an interdisciplinary way. The group works in cooperation with universities, institutes, and individual experts and is advised by a board of trustees and supported by members. Current co-chairmen of the board of directors are **Dr. Wolfgang Schröder** and **Wolfgang Dietzen**; **Horst Stern** is chairman of the board of trustees.

The Society's mission is to support and encourage all life forms and methods that serve the sustaining of nature and a dignified life: "It is our duty to maintain species and their habitats; we are duty-bound to use our experience and knowledge to take concrete measure and action." This action includes a wide range of activities from research and planning to carrying out projects, as well as publicity work. The projects seek to unite the needs of humans and nature.

▲ Projects and People

Several projects have involved constructing nature information centers, such as the one in the **Bavarian Forest National Park**. A more unusual construction project is the innovative "Green Island," built on the site of a shut-down gravel pit in the Rhein-Main area. At the heart of **Project Weilbach** will be an information center; the place aims to give people in the large cities nearby more understanding of ecology and the need for greening of land. Similarly, **Project Ecology Education** invites people to learn from nature. In a national study, WGM found that the German people cared intensely for animals, but concluded that "love alone is not enough"; thus, they teach ecological responsibility and build up an understanding of nature as a whole through experience-oriented nature instruction in schools and information centers.

Other WGM efforts involve individual species. **Project Black Grouse**, begun in 1982, can claim a contributing role in preserving habitat of the black grouse in the Langen Rhön area—perhaps the last non-alpine colony with a chance of survival. The area is now under control of a nature warden after this coalition effort of hunting groups and bird protection groups.

Project Ibex focuses on this lucky animal—nearly eradicated in the last century, ironically because of its value as a "living apothecary" for its horn, blood, and organs. Early efforts to repopulate often failed because of introducing the ibex into wrong land types. The WGM used satellite photos to plan out a wide enough area for the ibex, to protect that area and thus the ibex. Project Red Deer focuses on the largest free-living wild animal in Middle Europe; although very adaptable, the red deer still finds it difficult to survive the encroachment of constant development and tourism on its lands. WGM performed radiotelemetric observations, followed annual migrations, and observed the influence of human activity on red deer. The organization is now working in several German states on red deer protection plans. In all cases, the goal is to try to find common ground among the often conflicting interests of tourists, hunters, environmentalists, and agricultural and forestry groups.

In another land protection project, WGM is working to save **Chiemsee, a wetland and delta breeding area for local birds**, as well as a stopping point for more than 250 **species of migratory birds**. It has also been overrun by tourists and developers, ironically because of its natural beauty. WGM developed a land protection concept with the goal of better managing the tourist traffic, informing the visitors of the fragility of the land. A WGM warden has this duty, and is turning the area into a **model breeding program**. Other WGM ecotourism projects have been launched in the Black Forest and with cross-country ski routes.

Several years ago the town of Sindelfingen commissioned WGM to come up with a plan to salvage as much of the nearby forest as possible from a through-road project. The result of **Project Sindelfingen** was a unique biobridge—really a 110-meter-long tunnel under the forest—permitting a corridor for animals and people to move safely through the entire forest, without fear of being run over. The related **Project Traffic** researched thousands of wildlife accidents in order to come up with legal measures to influence street planning. WGM developed and tested numerous warning devices for wildlife, such as sonic and ultrasonic signals, and developed an exhibition that traveled to many cities of Germany and Austria to inform people of the need for awareness; about 200,000 people visited the exhibition in total.

Goetz Kerger founded a student magazine which is published and sold all over Germany; he first began to work with the WGM while a student at Ludwig Maximilian Universität in Munich. WGM supported his thesis project, an analysis of reintroduction of species in Europe. Kerger became a member of the Species Survival Commission (IUCN/SSC) Reintroduction Group in 1989; in 1990, he reported with another WGM scientist about lynx reintroductions in Europe.

■ Resources

WGM publishes monthly its *News from Nature Research* as well as various research reports and student theses. WGM has sold over 100,000 copies of its book *Save the Wild Animals.* A publications and materials list is available from WGM, Postfach 170, D-8103 Oberammergau, Germany. (Although Kerger is fluent in English, literature reviewed for this *Guide* was printed in German.)

🎓 *Universities*

Fachhochschule Darmstadt ⊕
Darmstadt Technical University
Wimpfenerstrasse 12
Ü6800 Mannheim, Germany

Contact: Peter Hennicke, Ph.D.
Phone: (49) 621 7989 73

Activities: Education; Research • *Issues:* Energy; Global Warming

▲ Projects and People

Dr. Peter Hennicke has been a tenured professor at the **Darmstadt Technical University** since 1988. His areas of specialization include business administration, industrial management, and the economics of energy; a major focus has been **economically and environmentally viable energy alternatives.**

Beginning in the 1980s, Dr. Hennicke has participated in numerous government- or university-sponsored projects, panels, and advisory boards. In 1984–1985, for example, he took part in a research project on local and regional systems of **rational energy use;** the following year, he served as a member of the scientific advisory board of a government-sponsored research project in Saarbrücken, and in 1986–1988 was a consultant on basic questions of energy policy for the Ministry for Environment and Energy in the German state of Hesse; he was also a member of the energy advisory board of the Senate (City Council) in the City of Bremen. He has twice been called as an expert witness to the official inquiry commission (Enquete Commission) of the German Bundestag (Parliament) investigating **provisions for the protection of the Earth's atmosphere.**

In 1989, Dr. Hennicke founded and became an advisory board member of the **Institute for Communal Economics and Environmental Planning (IKU),** in conjunction with the Technical University of Hesse. The same year, he participated in planning, establishing, and following through with a **course on Energy Economics** at the Technical University of Darmstadt.

More recently, Dr. Hennicke was appointed to the advisory circle of the minister president of the state of Baden-Württemberg on questions of ecology and environmental protection in 1991. He is currently director of a project for the development of a methodical instrument for a local/regional least-cost planning model, including **strategies to reduce carbon dioxide (CO_2),** with the cities of Hannover and Berlin (1991–1993).

■ Resources

Dr. Hennicke has authored numerous publications, including the book *Conserving Energy: A Handbook for Rational Energy Use in Municipalities and Industries* (1991) and articles such as "Energy Policy as Destroyer of the Environment," "Atomic Energy and Coal Policy," "Urban Energy Strategies for CO_2 Reduction in the FRG," and "A German Approach to Prevent Global Warming." Materials are printed in German.

Botanical Institute and Botanical Garden
Friedrich-Wilhelms-Universität Bonn
University of Bonn
Meckenheimer Allee 170
D-5300 Bonn 1, Germany

Contact: **Wilhelm Barthlott, Ph.D.,** Department
Head
Phone: 49-228-732526 • *Fax:* 49-228-733120

Activities: Education; Research • *Issues:* Biodiversity/Species Preservation

● Organization

The Botanical Institute and Botanical Garden of the University of Bonn is essentially a department within a governmental university, with a staff of 25.

▲ Projects and People

The Institute conducts research in two major fields. Of particular interest is the biodiversity of plants, especially **tropical epiphytes and cacti.** The Institute continues to research "**inselbergs**" (island-mountains)—unusual rock outcrops, often found in tropical areas, which are "highly differentiated extreme habitats" containing a large variety of plant species.

A second pursuit involves "**high resolution scanning electron microscopy (SEM)** of cuticular surfaces of plants." Botany professor **Dr. Wilhelm Barthlott** writes, "We are particularly interested in the context between wetability of surfaces and ability to contaminate."

The extremely detailed views of plants made possible through SEM enable Institute scientists to more fully investigate the connection between the form and function of plants. For example, the high number of **ultramicroscopic stomata** found on tropical plants, as compared to the low number on plants from arid regions, corresponds to their varying water and temperature control needs. Other micromorphological investigations have shown the engineering details of plants—for example, the particular design that ensures the mechanical stability of seeds so that they fall to the ground right side up.

Further Institute efforts go toward **propagating endangered plants,** such as the *Sophoro toromiro,* or "lost tree," of Easter Island, now extinct on this 150-square-kilometer "isolated volcanic outcrop west of Chile." The origin of the Garden's "beautiful small tree" is a mystery. It is hoped that seeds gathered at flowering and rooted cuttings will yield specimens that can be exchanged with other botanic gardens.

Dr. Barthlott has been at the Institute/Garden since 1985; previous posts were at the University of Heidelberg and the Free University of Berlin. He is currently working on the Institute's 20-year SEM **survey of 10,000 species of mosses and vascular plants,** mainly angiosperms, for their taxonomic significance. He has participated in a research project in the Ivory Coast for the German Ministry of Economic Cooperation, and in other research visits around the globe. Also in the Ivory Coast, he is undertaking an Inselberg study from 1991 to 1993 with support from the German Research Community (Deutsche Forschungs Gemeinschaft—DFG). The study focuses on plant colonies, viability, biogeography, and other aspects of the Inselberg formations. Future Inselberg studies will be conducted in East Africa, Madagascar, Brazil, Venezuela, and Australia.

■ Resources

Dr. Barthlott's publications list is comprised of literature in both German and English. His paper on "Scanning Electron Miscroscopy of the Epidermal Surface in Plants" appears in the *Systematics Association Special Volume 41* publication, Clarendon Press, Oxford, 1990.

Zoologisches Institut/Zoology Institute
University of Heidelberg
Im Neuenheimer Feld 230
D-6900 Heidelberg, Germany

Contact: **Arnd Schreiber, Ph.D.,** Biologist
 Phone: (49) 6221 424 607
  Fax: (49) 6221 49 665

Activities: Education; Research • *Issues:* Biodiversity/Species
Preservation • *Meetings:* IUCN, Central Europe

▲ **Projects and People**

Dr. Arnd Schreiber, University of Heidelberg, has co-authored *Weasels, Citets, Mongooses and Their Relatives: An Action Plan for the Conservation of Mustelids and Viverrids* for the World Conservation Union (IUCN), and "MHC Polymorphisms and the Conservation of Endangered Species" in *Biotechnology and the Conservation of Genetic Diversity,* Oxford University Press, Zoological Society of London. **Dr. Schreiber** belongs to the Species Survival Commission (IUCN/SSC) Mustelid and Viverrid Group. He seeks assistance with research funding in genetics and ecology.

Greece's extensive coastline comprises a third of the length of the Mediterranean Sea. Steps have been taken to combat one of the country's greatest environmental problems—the *nefos,* or air pollution, covering the capital city of Athens, caused primarily by emissions from industrial boilers and processes, space heating furnaces, and vehicles, all exacerbated by the city's rapid and uncontrolled expansion.

Other environmental concerns are deforestation and ensuing soil erosion (only 19 percent of the country's 51,000 square miles are forested, and forest fires are common), overdevelopment of many fragile areas, and the dumping of untreated sewage into the seas. In 1976, seventeen countries surrounding the Mediterranean pledged to fight pollution in a program coordinated by the United Nations.

NGO

Elliniki Etairia (EE)
Hellenic Society for the Protection of the Environment and the Cultural Heritage
28 Tripodon Street
10558 Athens, Greece

Contact: **Michael Scoullos, Ph.D.**, President
Phone: (30) 1 322 5245

Activities: Education; Political/Legislative • *Issues:* Sustainable Development; Waste Management/Recycling

● Organization

Elliniki Etairia (EE), or the **Hellenic Society for the Protection of the Environment and the Cultural Heritage**, has 3,000 members, 2 permanent staff, 3 project staff, and numerous volunteers. The Society is a member of the World Conservation Union (IUCN) and Europa Nostra, and houses the Mediterranean Information Office (MIO) for joint projects with the European Environmental Bureau (EEB). Activities of EE include public awareness campaigns, pilot projects on restoring monuments and buildings, and management of biotopes.

▲ Projects and People

The Society's efforts involve restoring and protecting both manmade and natural places. Indeed, the neoclassical building now housing the Society is the product of a major EE restoration effort, only recently completed. It is now "a center for permanent environment work, information, and documentation" open to all environmental NGOs and other concerned organizations. A second restoration project involved the Convent of Ossios Loukas, with a permanent exhibition of the sculptures and frescoes in the refectory.

The EE recently began acting as a coordinator in Greece for the project **1000 Communities for the Protection of the Environment**; in this capacity, the Society encourages Greek communities to finance local conservation projects, and will eventually select the best of these for membership in the 1000 Communities throughout Europe.

At the request of the Greek government, EE leads research and manages projects related to the **wetlands** in the Ambracian Gulf, the Northern Sporades, Cephalonia, and Lake Metrikon. In the case of Lake Prespa, an important wetlands area, the EE and other Greek and European environmental organizations are trying to establish a foundation to assure its continued environmental protection. Since 1976, EE has been managing the **model Prespa Station for field environmental work** in the area.

Society president since 1983, **Dr. Michael Scoullos** teaches environmental chemistry and chemical oceanography at the University of Athens. He was president of the EEB from 1986 to

1991, and has been president of the Greek National Committee of UNESCO's Man and the Biosphere (MAB) program. Scoullos is also a member of the EC Scientific Advisory Committee on Toxicology and Ecotoxicology. An active researcher, he is also the author of several books on oceanography and the environment and of more than 150 articles. He has acted as advisor to UNESCO, the EC, United Nations Environment Programme (UNEP), IUCN, and the Greek government, among others.

■ Resources

EE offers a series of publications, all in Greek, including *Sinasos in Cappadocia*, a book on the monuments and architecture of Cappadocia, and a *Guide for Visitors to Prespes National Park*. The Mediterranean Information Office (MIO) provides important information for all Mediterranean NGOs—including addresses, a general databank, library, publications, and workshops.

Hellenic Society for the Protection of Nature (HSPN)
24 Nikis Street
10557 Athens, Greece

Contact: **Grigoris Tsounis**, General Secretary
Phone: (30) 1 3224944

Activities: Education; Political/Legislative; Research • *Issues:* Biodiversity/Species Preservation • *Meetings:* EC Youth; International Conference on Brown Bear; International Conference on Wetlands; Naturopa Meetings, Council of Europe

● Organization

The **Hellenic Society for the Protection of Nature (HSPN)** was founded in 1951 by "a group of Greek scientists and other individuals in collaboration with the Hellenic Alpine Club and the National Foundation . . . motivated by their deep faith in the uniqueness of Greek Nature and in the need for protection of its natural resources." Its more than 2,000 members, some of them non-Greek, and a staff of 2, promote wildlife and habitat conservation in Greece and advise the government on environmental legislation and programs. In addition, HSPN serves as the Greek National Agency with the European Information Centre for Nature Conservation of the Council of Europe, and is a member of both the World Conservation Union (IUCN) and the International Council for Bird Preservation (ICBP).

Greece boasts more than 6,000 plant species, approximately 650 of them endemic, as well as numerous animal species. Although many of the species, such as lions, present in ancient Greece—as seen in period artwork such as vases—have since disappeared, the HSPN is working to ensure that those remaining, including the northern bear, lynx, wild mountain goat, and otter, do not face the same fate. Many are on the endangered species list.

The Society emphasizes, however, that even more important than conservation projects "is a long-term campaign to make the local people and the visitors aware of the need to protect and preserve the rare natural resources of Greece, . . . which constitute a universal heritage." Thus, the HSPN stresses "educational and informational goals," and works "to provide a networking informational centre which can be useful for Greeks and foreigners alike."

An important element of HSPN's strategy is collaboration with other Greek and international organizations. The Society works closely with the IUCN, World Wide Fund for Nature (WWF) (since 1969), Council of Europe, ICBP, International Waterfowl Research Bureau (IWRB), European Environmental Bureau (EEB), European Community (EC), Organization for Economic Cooperation and Development (OECD), and United Nations Environment Programme (UNEP)—with the latter's Mediterranean Action Plan and the "Blue Plan."

In addition, the HSPN has played an important role in updating Greek environmental legislation. The government often calls on the Society for advice and recommendations related to conservation matters.

▲ Projects and People

An ongoing project of HSPN, with EC funding, involves protection of the **brown bear**. In the first phase, biologist **Giorgios Mertzanis** surveyed the bears' habitats and numbers in Greece in 1987–1988. Then the Society worked on "measures to protect beekeepers and their beehives from the depredation of bears," including the erection of electrified fences around beekeeping areas in some 40 villages in northern Greece, under the guidance of HSPN secretary **Grigoris Tsounis**. The final phase involves studying the biology of the bear and considering the need for further specific protective measures; once these are determined, the Society will submit its proposals to the relevant governmental ministries.

An earlier project, in collaboration with both the EC and WWF, involved protection of the **Mediterranean monk seal**, which has significant colonies in Greece. The WWF and IUCN also worked with HSPN in a project in Zante to protect the **Mediterranean sea turtle**, whose breeding places are endangered by touristic development. Bird protection projects have concentrated on **raptors** such as the eagle (as of 1985, Greece still had about 10 breeding pairs of **imperial eagles**), ammergeier, and vulture, but have also worked to protect the pelican, ibis, spoonbill, and the beautiful bee-eater.

Needing urgent protection are Greek flora, "known from the time of Aristotle," including some 650 endemics, such as 9 species of tulips, orchids (*Ophrys delphinensis*), Cretan dittany, and a "spectacular wildflower," *Paeonia peregrina.*

Besides focusing on individual species, the HSPN acknowledges the need to preserve habitat. To this end, the organization has worked to establish and/or preserve Greece's 10 **national parks**, including Olympus (1938), Parnassus (1938), the White Mountains of Crete (1962), Vikos-Aoos (1973), Prespa (1974), and others. **Prespa**, a vital wetland area, is the object of some of the most intense efforts. HSPN started working to preserve it in 1959, and has archives on it dating to 1917. HSPN notes that "of the 28 species of birds which in Europe have been declared as threatened, 13 breed in this area."

A second wetlands project is located in Evros; development began there in 1961, and since then HSPN and IUCN have worked to protect and preserve this way station for migrating birds. In 1976, HSPN organized a symposium on Evros under the auspices of the Ministry for Northern Greece, and in 1975 the Society established in the Evros Delta the first **biological station** in Greece, open to both Greek and foreign students and observers.

■ Resources

To further its educational and networking goals, the HSPN needs equipment such as computers, videos, and televisions. In addition, the Society seeks "contact with other educational groups abroad so as to learn about other methods and materials used, and other initiatives being undertaken."

Available resources include many publications by the various members of the Society, mostly printed in Greek but with some English translations. The brochure, *35 Years of Struggle for the Protection of Nature in Greece*, is published in English. The Society also publishes a quarterly periodical with articles and news mostly about Greek conservation matters. Books published by HSPN members include *Flowers of Greece* by **G. Sfikas**; *Birds of Greece* by G. Tsounis; *Mammals of Greece* by **G. Tsounis**; *Raptors of Greece* by G. Handrinos; and *Nature Around Us* by **P. Broussalis**. Members offer lectures on various topics related to the environment; HSPN also possesses an archives, with an up-to-date library.

Guatemala

Guatemala's ecosystems, though small, are among the most diverse in Latin America. Yet the country's natural resources, on which the majority of the population depends, have been strained by an expanding population and the demand for wood, cultivable land, and sewage disposal facilities. Watershed deforestation, especially in the Central and Western Himalayas, has caused severe soil erosion; many animal populations are near extinction; water quality is deteriorating; and health problems are extensive in the cotton=growing areas due to pesticide use.

🏛 *Government*

Consejo Nacional de Areas Protegidas (CONAP)
National Council for Protected Areas
2a. Avenida 0-69, Zona 3, Colonia Bran
Guatemala City, Guatemala

Contact: **Arturo Duarte**, Secretary General
   Phone: (502) 2 51895 • *Fax:* (502) 2 518951

Activities: Law Enforcement; Political/Legislative • *Issues:* Biodiversity/Species Preservation; Deforestation; Sustainable Development • *Meetings:* CITES; Ramsar

● Organization

The **Consejo Nacional de Areas Protegidas (CONAP)** (National Council for Protected Areas) organizes and coordinates the governmental, NGO, and private institutions administering Guatemala's protected areas. CONAP manages the Maya Biosphere and the Sierra de las Minas Biosphere Reserve, does field studies to determine potential protected areas in the Izabal region, and conducts a workshop on the condition of protected areas throughout the country.

CONAP registers the commercial reproductive facilities of wild flora and fauna, the scientific wildlife collections at zoos and museums, and researchers and their current studies on natural resources in protected areas; controls and regulates the exportation and importation of wild flora and fauna; and administers the CITES and the Ramsar Conventions.

▲ Projects and People

Currently, CONAP is developing a system to regulate hunting activities and is studying the population status of various commercial *Tillandsia* species. CONAP is preparing the Guatemalan flora and fauna Red Lists for publication.

"The first step to global cooperation is the establishment of communication between organizations with the same interests," writes **Arturo Duarte**, former head of the National Institute of Indian Affairs (IIN) and former director of the Guatemalan chapter of the National Audubon Society. "CONAP . . . may serve as the focal point for distribution of information to all other related institutions."

Other staff members are **María José González**, Flora and Fauna Department advisor; **Gerda María Huertas**, Wildlife Section botanist; **Erick Toledo**, phytotechnician; **Juan Carlos Villagran Colon**, researcher; and **Mario Roberto Garcia Aldana**, agricultural engineer.

■ Resources

CONAP can benefit from new technologies for sustained development to increase the living standard of the rural communities. "This technology must be developed specifically. . . . It does not work to just apply or imitate previous activities without adequate evaluation," reports CONAP. Also needed are materials such as posters, exhibitions, and books for environmental education among rural populations.

Museo Nacional de Historia Natural "Jorge A. Ibarra"

6a Calle 7-30, Zona 13
P.O. Box 987
Guatemala City, Guatemala

Contact: **Jorge A. Ibarra**, Founder and Director
Phone: (502) 2 720468

Activities: Education; Research; Writings for the Media
Issues: Biodiversity/Species Preservation

▲ Projects and People

Jorge A. Ibarra founded the National Museum of Natural History of Guatemala in 1950 and the Museum of Natural History of Quetzaltenango two years later—both dedicated to the preservation of endangered species. Even earlier, Ibarra began writing articles in the press to lament the loss of species and the dangers of insecticides to his countrymen. He has authored almost 100 scientific articles.

Five plants and a toad (*Bufo ibarrae*) are named in honor of Ibarra; one flower genus, *Ibarraea*, appears in southern Mexico, Guatemala, Belize, Honduras, and El Salvador. The son of a naturalist and grandson of an ornithologist, Ibarra is also the great-grandson of the co-founder of the first museum of natural history, in 1797, in Guatemala City, and a signer of the Act of Independence of Central America. His 12 or more books include *Fables and True Tales Pro Natura*, *True Stories and Fables*, *Ecological Tales*, and *Mammals of Guatemala.*

In his charming *Ecological Tales*, Ibarra describes the fate of some species threatened with extinction: quetzal (Guatemala's national bird), whale, toad, swallow-tailed kite, chimpanzee—"the most sociable monkey," Bengal tiger, buffalo, koala bear, kiwi bird, panda, and Kodiak bear.

In one story, a quetzal talks to the endangered Atitlán grebe (*Podilymbus gigas*), a diving bird known as *poc* by the local Indians: ". . . you were hunted for food . . . then the people who make fans and mats cut the reeds where you nest and, lastly, my poor friend . . . I heard that an exotic voracious fish [the black bass] was planted in a lake to make your life even harder. . . . Your friends have built you a meshed-in refuge to help you defend yourself against the fierce attacks of bass." The grebe answers, "The refuge was built and then some evil men cut holes in the mesh so that our voracious enemies can again pursue us. . . .The water we drink is not nearly as pure as before, since insecticides and herbicides are being used on the farms, on the surrounding hills and when it rains the streams and brooks feeding the lake bring this contamination with them."

The quetzal continues, "I came to quench my thirst in your lake and find the same story. Far away, near the streams in my mountains, man has arrived to plant coffee, for which he first felled and burned the trees, compelling us to venture out to look for new virgin forests. . . . He kills us with guns and blowpipes and burns our nests in forest fires. However, . . . we can speedily fly to those forests where man has not yet arrived with his destructive weapons."

A project to establish a hydroelectric plant at the Lake of Atitlán actually was disapproved because Jorge A. Ibarra had initiated the creation of a water haven to protect the rare Atitlán grebe. "Lake Atitlán's waters are pure and refreshing," writes Ibarra.

Ibarra asks environmentalists to join together against "industry responsible for the harmful gases warming the planet" and to "introduce new techniques which will cause less problems for human survival. Forests are vital to take in the carbon monoxide expelled by engines," he writes.

"My organization will intensify the environmental education which we started in the National Museum of Natural History through courses and articles in the press to make people aware of the seriousness of pollution," including efforts to reach native peoples in their languages. Believing that museums today must not only exhibit species but defend the environment, Ibarra has raised the issue of "immoderate use of insecticides by peasants" for 30 years, as reported in his paper *El Hombre ante Natura* (Man before nature).

On marine mammal destruction, he reports that "in the Pacific . . . fishermen fishing for tuna have placed immense nets in which dolphins . . . are dying in vast numbers and I believe that we should ask national and international organizations to bring their influence to bear to prevent this slaughter."

Ibarra founded the Central American Association of Natural History, Committee Operation Quetzal, and Guatemalan Circle of Scientific Press, and co-founded the Association of Authors and Friends of the National Book of Guatemala and the Latin American Committee of National Parks in Ecuador. He is director of the Guatemala section of the International Council for Bird Preservation and is an honorary councilor with the World Conservation Union.

■ Resources

Materials published in Spanish include a list of activities. Some publications are printed in both Spanish and English, such as *Conservation in Guatemala.*

NGO

Asociación Amigos del Bosque
Friends of the Forest Association

9a Calle 2-23, Zona 1
C.P. 01001
Guatemala City, Guatemala

Contact: **Josefina Chavarria**, Executive Director
Phone: (502) 2 83486 • *Fax:* (502) 2 513478

Issues: Biodiversity/Species Preservation; Deforestation

Defensores de la Naturaleza/Defenders of Nature

7a Avenida y 13 Calle, Zona 9
01009 Guatemala City, Guatemala

Contact: **Andreas C. Lehnhoff**, Executive Director
Phone: (502) 2 325064 • *Fax:* (502) 2 322671

Activities: Development; Education; Law Enforcement; Political Legislative • *Issues:* Biodiversity/Species Preservation; Deforestation; Population Planning; Sustainable Development • *Meetings:* ECO 92; UNCED

● Organization

Founded in 1983, **Defensores de la Naturaleza (Defenders of Nature)** is a nonprofit conservation group of business persons, artists, students, and volunteers collaborating to preserve Guatemala's natural resources. With a staff of 27, it works to "foster respect for nature and the sustainable use of natural resources; promote the establishment of protected wildlife areas, their administration, and adequate management; encourage the enactment of both legislation that protects Guatemala's biodiversity and to publicize this legislation; create a reference library with environmental materials available to educators and students; and encourage other organizations and institutions" in its mission.

▲ Projects and People

Major focus is **Sierra de las Minas**, a "spectacular mountain chain in Central America" that is home to at least "885 species of mammals, birds, amphibians and reptiles," including Guatemala's national symbol—the resplendent quetzal, 100 bat types, jaguarundi and other wildcats, howler monkeys, horned guan, and harpy eagle. Defensores writes, "Forestry experts already consider Sierra de las Minas to be an important tropical gene bank of conifer endoplasm" with 17 distinct species, "an irreplaceable seed source for reforestation and agroforestry" worldwide. As the country's biggest water source, some 62 rivers provide fresh water to farms and villages.

A biosphere reserve since 1990, Sierra de las Minas is now managed by Defensores—with funding from the Guatemalan government, The Nature Conservancy, and World Wildlife Fund (WWF)—to protect against "timber extraction, expansion of agriculture on the steep and fragile slopes, and inappropriate development." Unproductive soil and loss of biodiversity are other threats.

Defensores is attempting to buy land, whenever possible, to protect it from logging. For example, the Swedish Children's Rainforest assisted with the purchase of cloud forest, bringing total land owned by Defensores to 26,000 acres.

"Our park guards are marking reserve boundaries with signs, building control posts, patrolling and communicating with locals to share the importance of the reserve." Also, Defensores is working "directly with Kerchí Indians, helping them stabilize and improve agricultural production through sustainable-use techniques. This will reduce their migration into the forest where their slash-and-burn activities have caused much damage." Means to better serve agricultural populations in the Motagua Valley, Central America's driest region, are being investigated. WWF also assists Defensores with environmental education programs.

Andreas Hehnhoff was previously executive secretary for CONAP, Council for Protected Areas. **Margaret Kohring** is Guatemala in-country director, The Nature Conservancy.

■ Resources

Defensores prepares a newsletter, in English, and an information packet, Spanish and English.

Wildlife Conservation International (WCI)
APDO 2703
Guatemala City, Guatemala

Contact: **María José González**, Research Fellow
Phone: (502) 2 690157 • *Fax:* (502) 2 365829

Activities: Education; Research • *Issues:* Biodiversity/Species Preservation

▲ Projects and People

With worldwide activities, **Wildlife Conservation International (WCI)** in Guatemala has undertaken a project on the **Ecology and Conservation of Vertebrate Populations in the Maya Biosphere Reserve.** The purpose is to develop management guidelines that may be useful to the agencies that administer the protected areas in this country. *(See also Wildlife Conservation International in the USA NGO section and throughout this Guide.)*

A guide on local wildlife is being developed to train naturalist tour guides.

María José González is also wildlife advisor to the **National Council for Protected Areas (CONAP)** as well as WCI research fellow working on the **Paseo Pantera Project.**

■ Resources

New technologies are needed for censusing and trapping tropical mammals and birds in order to "develop the optimum methodology for habitat conditions in El Petén that will permit comparison of results with those of other studies."

Wildlife Preservation Trust International (WPTI) ⊕
Parque Zoológico Nacional "La Aurora," Zona 13
Guatemala City, Guatemala

Contact: **Lorena Calvo**, ITC Regional Coordinator for Central America
Phone: (502) 2 720507 • *Fax:* (502) 2 715286

Activities: Education; Development; Research • *Issues:* Biodiversity/Species Preservation

▲ Projects and People

Long Live the Quetzal! You and the quetzal need the forest in order to live . . . don't destroy it!

In rainforests, different species of trees make use of the humidity in the atmosphere, which retains water and forms clouds and fogs. The rainforests also distribute water to the surrounding areas, by the evaporation of water or by small streams, giving rise to a large variety of flora and fauna.

Living high on mountain peaks in the rainforests where the annual precipitation is 2,000 to 6,000 millimeters, the quetzal is vital to these forests, which have few ways of distributing seeds from its trees in order its assure survival. The quetzal plays an important role in disbursing seeds in the forest; therefore, the flora and fauna depend on the quetzal. A few hours after eating the fruit, the quetzal eliminates the seeds far from the "mother tree," giving them the opportunity to germinate and therefore participate in the natural cycle of reproduction in the forest. The quetzal, in turn, depends on the forest because it constructs its nests in hollow tree trunks. Unfortunately, its habitat is being severely compromised because of deforestation.

If the rainforest is devastated, the quetzal—the national symbol of Guatemala—will also be destroyed, reminds **Lorena Calvo, Wildlife Preservation Trust International (WPTI)**. The root system of the rainforest trees is similar to a sponge in that it keeps the Earth from being washed away during heavy rains. When a tree is cut down, the roots die, and the water that the roots retain can form streams and small waterfalls, carrying away soil, crops, and eventually causing landslides. This process of erosion is irreversible, yet it can be alleviated by planting more trees and halting indiscriminate deforestation, reports WPTI.

In Guatemala, the quetzal is found in Alta Verapaz, Sierra de Chamá, Sierra de las Minas, the Cuchumatanes region, and the Sierra de Chelem-ha. The quetzal also makes its home in Mexico, Honduras, and El Salvador with a subspecies in Costa Rica and Panama. But because there are fewer rainforests today, fewer quetzals can survive. If the rainforests continue to be deforested at the present rate, they will disappear in about 70 years, predicts WPTI.

Published in cooperation with WPTI, Asociacion Mesoamericana de Zoologicos (zoological association), and the Pew Charitable Trusts, an educational poster proclaims, "You can help save the rainforest and the quetzal. Don't destroy what Mother Nature took so long to create." *(See also Wildlife Preservation Trust International in the USA NGO section.)*

Lorena Calvo Martinez has attended wildlife conservation and management training programs of the National Zoological Park, Smithsonian Institution, and American Association of Zoological Parks and Aquariums (AAZPA) in the USA. She also took part in a captive raptor propagation workshop sponsored by the Peregrine Fund and Boise State University as well as captive breeding and conservation programs of the Jersey Wildlife Preservation Trust, Channel Islands, UK.

■ Resources

Posters and pamphlets about the quetzal are available, printed in Spanish.

(See also RARE Center in tthe USA NGO section and Frostburg State University in the USA Universities section.)

Guyana

Bordered by the Atlantic Ocean, Suriname, Brazil, and Venezuela, Guyana is in the unique position of having exceptionally rich resources, most of which remain unexploited. The population lives on a small strip along the Caribbean Sea; the interior forests and savannas are largely unexplored.

Compared to other countries, Guyana's environment problems are minor, mostly because of lack of development. What problems do exist are mostly confined to the coastal areas, where 10 to 15 percent of the land supports 90 percent of the population. Especially needed are alternative irrigation sources to protect the coastal arterial wells that supply most of the country's water and to maintain the natural forest soils.

🏛 *Government*

Guyana Agency for Health Sciences Education, Environment and Food Policy (GAHEF)

Liliendaal, Greater Georgetown

Georgetown, Guyana

Contact: **Gurudatt Naraine**, Environmental Officer
Phone: (592) 2 57523 • *Fax:* (592) 2 57524

Activities: Development; Education; Law Enforcement; Political/Legislative • *Issues:* Biodiversity/Species Preservation; Global Warming; Health; Sustainable Development; Water Quality • *Meetings:* UNCED

● Organization

Overall, the **Guyana Agency for Health Sciences Education, Environment and Food Policy** (GAHEF), with a staff of seven, is involved in environmental education, World Environment Day activities, Kaiteur National Park, protecting biological diversity, environmental impact assessment of gold mining and logging activities, curbing industrial pollution and pesticides, environmental health, cleanup campaigns, legislation, and regional and international seminars, conferences, and workshops.

▲ Projects and People

The government approved the **expansion of the Kaiteur National Park** in response to the recommendations of **Dr. Goetz Schuerholz**, consultant, who was obtained through the World Wildlife Fund (WWF) to assess the park. GAHEF is working on a management plan.

Seminars for teachers focus on basic environmental concepts and local to global environmental problems including water resources, proper sanitation, global warming, forest resources, resource conservation, and disaster preparedness. Handbooks on *Man and the Environment* and *Environmentally Sound Agriculture* are distributed. In preparation is a special booklet for young students, *Learning About Your Environment*. Guyana's sustainable development, environmental management, and social equity are also discussed.

Guyana officials prepared a brief in anticipation of the country's position on the International Convention on Biological Diversity. GAHEF is involved in the Tropenbos Programme on behalf of sustainable tropical forestry. Regarding environmental impact assessments, guidelines were prepared for gold mining and logging in tropical forests.

Inspections are made at factories, for example, to "assess the effect of noise and fumes on the nearby residents and factory workers," reports GAHEF. One recent visit to a bag factory noted that "the generator room was lined with egg trays, a sound absorbent material," and that "generator exhaust was directed into an underground pit . . . to reduce noise level." Another visit at a tannery necessitated corrections as "waste solutions were being discharged into a ditch whilst the solid waste (skins, hairs, and flesh) was dumped at the back of the tannery."

Questionnaires are distributed to farmers to determine the quantity, types, and misuse of pesticides; to industries regarding industrial waste; and to petrol stations in the Georgetown area. GAHEF collaborates with the **Caribbean Environmental Health Institute (CEHI)** on extraction techniques of certain pesticides, with the help of CEHI scientist **Dr. Naresh Singh**. Environmental health personnel recently showed community health workers how properly to construct **ventilated improved pit (VIP) latrines** out of indigenous materials.

Among other activities, GAHEF comments on national environmental protection legislation and participates in various conferences such as the Caribbean Seminar on Human Ecology, Environmental Management, and Education; Inter-American Bank meetings; and Latin American and Caribbean preparation for the United Nations Conference on Environment and Development (UNCED).

■ Resources

GAHEF needs wastewater treatment technology and—for pollution monitoring—"analytical and sampling equipment, bauxite dust-health effects and air pollution control technologies, and biological indicators of pollution in tropical waters." GAHEF also seeks to encourage sustainable use of mangroves; produce environmental educational materials such as posters, pamphlets, videos, and films; and gain local expertise in carrying out environmental impact assessment of development projects.

🎓 *Universities*

University of Guyana
Turkeyen, ECD, Guyana

Contact: **Father Malcolm Rodrigues**, Deputy Vice-Chancellor
Phone: (592) 2 54841-6 • *Fax:* (592) 2 65407

Activities: Development; Education; Research • *Issues:* Biodiversity/Species Preservation; Deforestation; Energy; Health; Sustainable Development; Water Quality

▲ Projects and People

The Natural Sciences Department at the **University of Guyana** is establishing a **Teaching and Research Environmental Unit** that links the major faculties and their departments.

The University of Guyana and the Smithsonian Institution are involved in a long-term joint undertaking in collecting, documenting, and preserving the flora of Guyana.

The World Wildlife Fund (WWF), the Smithsonian Institution, and the University of Guyana together have formed a **Biodiversity Center**, responsible for research, training, and public education. The university's core programs are used to enhance its work in gathering and maintaining the region's technical information and biological collections.

The **Amerindian Unit** is concerned with the life, culture, and general well-being of native Amerindian communities. The Environmental and Amerindian units will work closely with seven tribes in Guyana to develop **impact assessment in mining and forestry**, for example.

■ Resources

These seminar papers may be of interest: "The Greenhouse Effect: Nature's Thermostat"; "Sustainable Development and Our Forests"; and various studies on the Amerindian communities.

Located on the western third of the island of Hispaniola in the West Indies, Haiti is the most densely populated and the poorest country in the Western Hemisphere. Its environmental degradation is due to interrelated problems of overpopulation, deforestation, and soil erosion, caused by farming erodible lands and the stripping of forests for fuel. Less than 2 percent of the country remains forested. Strong rains wash nutrient soils off farmed hillsides and into dams and irrigation works, rendering them inoperable. Outside Port-au-Prince, barely 30 percent of the population has access to safe water, and waterborne diseases are widespread. Fish populations are also reduced by the high alkaline levels in the water.

🍁 *NGO*

Agroforestry Extension Project
Pan American Development Foundation (PADF) ⊕
Delmas 31 #27 Port-au-Prince
B.P. 15574
Petionville, Haiti

Contact: **Arlin Hunsberger**, Director, Project Trees
  Phone: (509) 46 3286 • *Fax:* (509) 46 4616

Activities: Development; Education; Research • *Issues:* Biodiversity/Species Preservation; Crop Improvement; Deforestation; Soil and Water Conservation; Sustainable Development

● Organization

Founded in 1962 and affiliated with the Organization of American States, the **Pan American Development Foundation (PADF)** for Latin America and the Caribbean strengthens "the private sector's ability to help the needy and enterprising poor" in Latin America and the Caribbean. *(See also Organization of American States in the Global/Regional section.)* With headquarters in Washington, DC, PADF's overall mission is income generation and institution building, health services and training, and disaster relief programs. Among its accomplishments, the private voluntary organization (PVO) created surgical care units in Honduras and Nicaragua, a Hurricane Hugo relief program for Nevis and Montserrat islands, an advisory group to promote and improve cocoa production in Caribbean countries, and tree plantings to help save the environment and increase incomes.

▲ Projects and People

Suffering from "rapid desertification," Haiti has received help from PADF for more than a decade through Pwojè Pyebwa (Project Trees)—a large-scale agroforestry program reaching small farmers with the early help of the U.S. Agency for International Development (USAID), Belgium, Canada, Switzerland, and private-sector donors. By 1988, some 100,000 farmers planted more than 20 million fast-growing trees with the help of about 80 local PVOs and church groups each spring and fall rainy season.

About 83 persons are employed to supervise tree planting, soil and water conservation, and crop improvement programs. Some 1,200 local animators—persons who train farmers—are first taught agroforestry techniques and how to set up PVO nurseries. During 1990, over 6.5 million tree seedlings were distributed to farmers in this manner. As a result, 16,000 meters of living terraces were established and almost 60,000 farmers learned conservation methods. This network succeeds in changing agricultural practices because, says PADF, "it is delivered by local groups which have been long established and recognized by [farmers]."

Director Arlin Hunsberger reports, "Originally planned to meet acute fuelwood and charcoal shortages leading to denuding of the mountains, the project now develops exotic and indigenous species seedlings for many uses. Only a few are fruit and food trees. Prunings become animal fodder, or green manure. Thinnings become poles for simple houses or fences. Larger trees are saved for windbreaks, lumber, or to improve land values." Planted on mountain contours, "living hedges" enhance intercropping with food crops, retain eroding slopes, accumulate agricultural soils, and enrich slopes with nitrogen-fixing tree species.

Farmers value trees which they may plant only on their small plots or leaseholds and can harvest them as they please. The most popular fruit trees are chadek (*Citrus maxima*), coffee, cocoa, papaya, breadfruit, Jamaican plum, Key lime, tamarind, mango, and sweet orange.

An **environmental education program** is another aspect implemented in 73 schools in 5 Haitian regions. Students learn about basic ecology, seedling and compost production, plantation management, and soil conservation.

Agroforestry research is undertaken at the Research and Documentation Center, which has a library, field data, topographic maps,

and computer. The center analyzes harvest case studies and evaluates species adaptation, pruning effects, and impacts of hedgerows and grass strips.

The Agroforestry Project's staff includes **Louis H. Bergner**, assistant director for finance and administration; **Scott J. Josiah**, assistant director for agroforestry; **Michael Bannister**, research and documentation coordinator; **Daniele Mangones Dejean**, training coordinator; **Jean Michel Tessono**, nursery specialist; and **Stuart North** and **Adrien Joseph**, regional team leaders.

■ Resources

Expertise on agroforestry information is needed. A 244-page tree nursery manual in Haitian Creole, *Gid Pepinyeris*, describes the aspects of growing seedlings. The *Pwojè Pyebwa Annual Report* describes the program in detail.

Honduras

Honduras is bounded by Nicaragua, El Salvador, and Guatemala, in the middle of the Central American isthmus. The fertile coastal regions make up 20 percent of the land; the remaining area is mountainous.

The Honduran economy is based for the most part on its natural resources—including timber and minerals—which are in the process of being degraded. This makes Honduras especially vulnerable to deforestation and related forms of environmental damage. Uncontrolled fires and wasteful logging practices are considered the principal reasons for forest depletion. Other problems include flooding and water contamination leading to waterborne diseases, over-exploitation of wildlife, and urban industrial expansion. Ecotourism presents one hope for sustainable future development.

❁ NGO

Comité para la Defensa y Desarollo de la Flora y Fauna del Golfo de Fonseca (CODDEFFAGOLF)
Committee for the Defense and Development of the Flora and Fauna of the Gulf of Fonseca
APDO 3663
Tegucigalpa, Honduras

Contact: **Mauricio Alvarado**, President
Phone: (504) 37 2655 • *Fax:* (504) 37 2655

Activities: Development; Education; Law Enforcement • *Issues:* Air Quality/Emission Control; Biodiversity/Species Preservation; Coastal Resources; Sustainable Development; Water Quality • *Meetings:* UNCED

▲ Projects and People

CODDEFFAGOLF's work is concentrated on the southern coast of Honduras, a region known as the "Mosquito Coast." Here lagoons, mangrove forests, and even the El Quebrachal Wildlife Reserve are being destroyed by development projects, says CODDEFFAGOLF. Many small-scale fishermen are no longer able to support themselves because large-scale shrimp farming is crowding them out. At the local level, CODDEFFAGOLF organizes **environmental education** programs for children in public schools, training workshops for fishermen, and seminars on sustainable development for workers in small villages. Campaign activities include **defending the mangrove forests** and opposing the Stone Container Corporation's destruction of the forests along the Mosquito Coast.

To focus attention on the region's problems, CODDEFFAGOLF stages demonstrations in the capital, Tegucigalpa, and meets with important people and organizations in Honduras and abroad. Some such organizations are: Canadian International Development Agency (CIDA), Red Latinoamericana de Bosques Tropicales (Latin American Tropical Forests Network), CATIE, Instituto Nicaragüense de Pesca (Nicaraguan Fishing Institute), and World Wildlife Fund (WWF).

CODDEFFAGOLF is also a member of the newly formed **FED AMBIENTE**, the Fedaración de Organizaciones No Gubernamentales Ambientalistas de Honduras (Federation of Honduran Non-Government Environmental Organizations). A project to **produce salt through solar radiation** is underway in cooperation with FIA, the Fundación Inter Americana (InterAmerican Foundation). This unusual activity is designed to provide jobs for unemployed workers along the coast and to avoid the deforestation and air pollution associated with traditional methods of salt production.

In an editorial, CODDEFFAGOLF noted that people in Honduras and around the world have gradually begun to accept the organization's ideals of conservation and sustainable development. When CODDEFFAGOLF first spoke out in 1987 against dumping municipal wastes from other parts of the country on the southern coast, they were laughed at, the group reports. Now its positions are supported at least in part by the U.S. Agency for International Development (USAID), among others.

■ Resources

CODDEFFAGOLF's newsletter, *Boletín Informativo*, is written primarily for people of the Gulf of Fonseca area. In addition to editorials on local environmental issues, articles explain how cholera is transmitted and how to obtain potable water. CODDEFFAGOLF also sells bumper stickers and will provide photocopies of leaflets entitled *Recursos Costeros* (Coastal resources), *Plan de Manejo de Humedales* (Wetland management plan), *Sal Solar* (Solar salt), and *Problema de Manglares* (The mangrove problem).

Proyecto Ambiental para Centro América (PACA)
Environmental Protection for Central America
APDO 87
San Pedro Sula, Honduras

Contact: **Sergio José Midence L.**, Coordinator·

Activities: Development; Education; Law Enforcement; Research • *Issues:* Biodiversity/Species Preservation; Deforestation; Environmental Education; Sustainable Development

▲ Projects and People

"At this moment the project is at the initial level, but in the near future, there will be an environmental magazine focused on the selected site for the [conservation] project," writes Sergio José Midence, coordinator, Proyecto Ambiental para Centro América (PACA) (Environmental Protection for Central America). "In the case of Honduras, the site is the Cusveo National Park and Merendon Watershed." A consortium of CARE, The Nature Conservancy, and Conservation International support this effort.

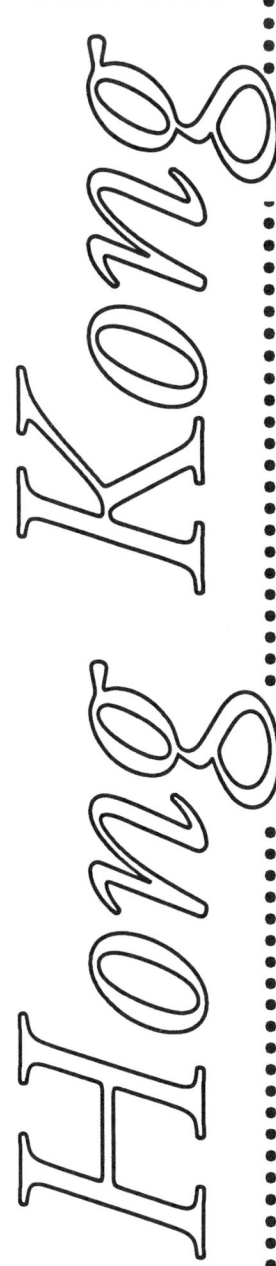

Hong Kong

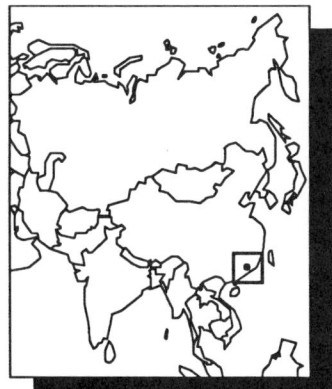

Hong Kong is located to the southeast of China, adjoining the province of Guangdong. Geographically, Hong Kong consists largely of steep, unproductive hillside scattered among more than 230 islands and islets. Developed areas account for about 13.3 percent of the total land area; farming is limited to only 8 percent. Hong Kong is one of the most densely populated places in the world, with a population density of 5,390 per square kilometer.

Despite the dense population and small area, the hilly topography has constrained building and allowed large expanses of the countryside to be preserved, and with it a surprisingly rich variety of indigenous animal and plant life. The major environmental concerns are pollution of streams by livestock wastes, smoke emissions from diesel-engined vehicles, and sewage and waste management. Steps have been taken by the Environmental Protection Department to reduce emissions of air polluting chemicals, to regulate collection and disposal of wastes, and to control water and noise pollution, among other environmental measures.

🏛 Government

Agriculture and Fisheries Department
393 Canton Road
Kowloon, Hong Kong

Contact: Director of Agriculture and Fisheries, Attn: **Gregory NG**
   Phone: (852) 3 733 2211 • *Fax:* (852) 3 311 3731

Activities: Development; Education; Law Enforcement; Research • *Issues:* Biodiversity/Species Preservation; Health; Sustainable Development; Waste Management/Recycling; Water Quality • *Meetings:* CITES; International Commission on Irrigation and Drainage Congress; International Conference on Toxic Marine Phytoplankton; International Society of Arboriculture; World Parks Congress

● Organization

The **Agriculture and Fisheries Department** is "directed to enhancing the productivity of the local farming and fishing industries through increased technical and economic efficiency, improved stability of production and maintenance of orderly and efficient marketing." Policy is also to "protect consumers from unnecessarily high food prices by ensuring that local produce of acceptable standards is marketed efficiently and to maintain a reliable supply of fresh primary products to the community."

▲ Projects and People

The Agriculture and Fisheries Department is studying an "effluent-free" pig-raising technique called the **pig-on-litter** method, which involves rearing pigs on sawdust litter bedding with the pig waste decomposed *in situ.* "The main studies include the performance measurement of pigs reared under the method and the biochemical changes of the in-situ composting process." It is a joint project with **Christina Y.Y. Lo**, Agriculture and Fisheries Department; **Prof. D.K.O. Chan**, Hong Kong University; and **Dr. Nora Tam**, City Polytechnic of Hong Kong. Obstacles are that pig buildings "need to be efficiently ventilated and the bedding needs to be regularly turned for good fermentation." In addition, "specially designed machines" for this purpose are "essential." Pigs, poultry, and dairy cows are the major livestock animals.

The Department is also **studying the establishment of marine parks and reserves** in Hong Kong. It has opened to the public a 16.5-hectare **Nature Education Center** for nature education and recreation purposes, featuring exhibition halls on Hong Kong's agriculture, fisheries, and countryside resources as well as an arboretum, orchards, field crops, Chinese herbal garden, and rock and mineral corner.

With the Environmental Protection Department, the Department is commissioning a study entitled *An assessment of the environmental impact of marine fish culture in Hong Kong.*

The Department acts as the center of a red-tide **reporting network**, monitoring the occurrence of red tide and its possible impact in fisheries and human health by prompt identification of causative species and toxicity bioassay, if necessary.

In collaboration with the Baptist College of Hong Kong, a baseline study was conducted of the level of **heavy metals** and **tributyl tin**, an active ingredient of an antifouling paint, in fish culture zones.

The Department is undertaking the demolition and realignment of certain weirs on selected streams in the New Territories to increase stream flow, a procedure which **directs stagnant pools of water at pollution black spots** to improve water quality and reduce flooding of stream courses.

A **Conservation Policy** is in effect, in keeping with CITES (Convention on International Trade in Endangered Species of Fauna and Flora) and to protect and manage the "landscape, vegetation and wildlife for present and future generations."

Park staff educates visitors on the proper use of countryside resources and wages antilitter and antifire campaigns.

■ Resources

The Agriculture and Fisheries Department is in need of house and ventilation designs in humid and hot areas for higher ventilation rate in open-sided buildings; mechanization for litter management in a pig-on-litter system; and law enforcement for waste control.

The Department has demonstration units available showing systems at government and private farms. Seminars and farm visits are offered to farmers. International symposium proceedings are available on the treatment and recycling of livestock wastes.

 *NGO*

Beidaihe Bird Conservation Society ❦
c/o I/F 15 Siu Kwai Won
Cheung Chau Island, Hong Kong

Contact: **Martin Williams, Ph.D.**, Vice President
Phone: (852) 5 981 3523

Activities: Development; Education; Research • *Issues:* Biodiversity/Species Preservation; Deforestation

● Organization

The **Beidaihe Bird Conservation Society** was founded in 1989 with the help of the local government of **Beidaihe**, a resort on the China coast about 280 kilometers east of Beijing, where bird migration is observed. President is **Lee Fu**, former chief of local government, and vice president is **Professor Hsu Weishu**, head of the Department of Zoology, Beijing Natural History Museum. Dr. **Martin Williams** is also a vice president.

Dr. Williams is setting up the Beidaihe Bird Society in Hong Kong, which will be a sister organization to the Bird Conservation Society. The Society will operate like an overseas branch and will take a broader interest in east Asian migrants.

▲ Projects and People

The Society conducted its first spring migration survey of Beidaihe in 1985, which was also the first systematic survey of the town since the 1940s.

Dr. Williams writes that "while substantial numbers of birds [are] to be seen, populations of several species have fallen, or crashed, since earlier this century . . . thus [showing] a need for conservation." Birds in substantial numbers are Oriental white stork, Siberian hooded and red-crowned cranes, great bustard, and pied harrier.

The Society backs the establishment of the **Beidaihe Bird Reserve**, currently consisting of fields and marshes that will be transformed into a lagoon with a visitor center.

"We have also made visits to coastal estuaries south of Beidaihe," writes Dr. Williams. "Observations suggest they are of international importance for wetlands. We have discussed ideas for conservation with [government officials]; interest [was] expressed but their priorities lie with economic development."

According to Williams, "Ecotourism has played a useful role" in attracting a small amount of bird watchers to Beidaihe from several countries, including Britain, Denmark, Finland, and Japan. Presently, the main help to the town is the tourists' presence—"the fact that people will come to China mainly to visit Beidaihe for birdwatching impresses locals."

■ Resources

The China Flyway is published twice yearly for members. Beidaihe Bird Conservation Society seeks funds to help continue its work and set up the Beidaihe Bird Society in Hong Kong.

Buddhist Perception of Nature ⊕
5 H Bowen Road, First Floor
Hong Kong

Contact: **Nancy Nash**, International Coordinator
Phone: (852) 523 3464 • *Fax:* (852) 869 1619

Activities: Education; Research • *Issues:* Biodiversity/Species Preservation; Deforestation; Energy; Global Warming; Health; Population Planning; Public Affairs; Sustainable Development • *Meetings:* UNCED

● Organization

The philosophy of the **Buddhist Perception of Nature** can be summed up by the words of His Holiness the XIVth Dalai Lama: *Our ancestors viewed the earth as rich and bountiful, which it is. Many people in the past also saw nature as inexhaustibly sustainable, which we now know is the case only if we care for it.* (From *An Ethical Approach to Environmental Protection*)

The Buddhist Perception of Nature was founded in 1985 as a research project where Buddhist scholars and administrators came together to define the task of reading Buddhist literature and assembling those passages that spoke of forestry, wildlife, and conservation. The group seeks ways to make this literature more interesting to modern readers while maintaining its original integrity. The research team has included scholars from both branches of Buddhism: Theravada, from Southeast Asia, and Mahayana, or the northern school.

"Buddhist Perception of Nature, while a response to the crisis of disappearing nature, is also an education project based on cultural, social, and perceptual factors. Human thinking, attitudes, and behavior are recognized as inherent both in the problems and in any long term solutions," reports the group.

The organization has received endorsements from Wildlife Conservation International (WCI), New York Zoological Society; World Wildlife Fund (WWF-USA); WWF-Hong Kong; Wildlife Fund Thailand (WFT); Beldon Fund; and the Sacharuna Foundation, among others.

▲ Projects and People

"Centuries before contamination of the Earth's water would be the widespread threat to human health that it is today, the Buddha set down rules forbidding pollution of water resources," said the project's chief scholar, **Dr. Chatsumarn Kabilsingh**, Thammasat University, Bangkok, at the beginning of the undertaking. Scholars working on the project commented, "For the first time, this resource of teachings **was being compiled** in an original way to highlight what the philosophy has to teach us about both our interdependence with other living things and the protection of our natural environment."

The organization prepares **teaching materials** for use by educators, in public awareness programs, and as models for similar projects involving other faiths and cultural traditions. Materials are featured in conferences and in press, radio, and television coverage of environmental issues. **Environmental ethics** is emphasized.

With a new approach, Buddhist Perception of Nature confronts the crisis of vanishing species and natural resources and the "awakening of people to new potential solutions." Offices are in Hong Kong, India, and Thailand.

Originator and international coordinator **Nancy Nash** was the "negotiator for the first foreign cooperation in the Chinese Great Panda conservation program." She also worked on the inauguration of WWF-Hong Kong, wildlife conservation exhibitions for Hong Kong and Peking governments, and World Conservation Strategy with WWF, World Conservation Union (IUCN), and the United Nations Environment Programme (UNEP) in Asia.

■ Resources

Tree of Life: Buddhism and Protection of Nature is a book (printed in several languages) produced from part of the research the group has completed that is used for college-level courses on ecology and/or ethics. *A Cry in the Forest*, in Thai and English, incorporates work from *Tree of Life* and expands material on Buddhist teachings about protection of the environment. *Buddhist Perspectives on the Ecocrisis*, produced by the Buddhist Publication Society, contains a chapter on the Buddhist Perception of Nature, and is available from P.O. Box 54, Sangharaja, Mawata, Kandy, Sri Lanka.

🎓 *Universities*

Center of Urban Planning and Environmental
  Management
University of Hong Kong
Pokfulam Road
Hong Kong

Contact: **Peter Hills, Ph.D.**, Acting Director and
               Reader
Phone: (852) 859 2721 • *Fax:* (852) 559 0468

Activities: Education; Research • *Issues:* Air Quality/Emission Control; Deforestation; Energy; Global Warming; Sustainable Development; Transportation

▲ Projects and People

Dr. Peter Hills' project, **Energy Consumption in the Household Sector in Hong Kong**, investigated such patterns in the territory focusing on overall levels of fuel and electricity usage, interfuel substitution, and appliance ownership and use. It also examines "household exposure to air pollutants."

The survey of 500 households was part of a larger comparative analysis regarding household energy use in China, India, Philippines, and Thailand. The Canadian International Development Research Centre provided funding.

Dr. Hills, in conjunction with **William F. Barron** and **K.V. Ramani**, have developed a *Handbook for Environmental Impact Assessment for Energy*. This is a general reference and training document for energy planners providing methodological guidance, information, and training materials. Funding is from the Asian and Pacific Development Centre's Energy Programme, Malaysia.

■ Resources

A publications list of conference proceedings and working papers is available. Sample titles are *Greenhouse Gas Emissions: Issues and Possible Responses in Six Large Economy, Lower Income Asian Nations*, and *Responding to Training Needs in the Energy Sector: An Asian Perspective*.

Department of Botany
University of Hong Kong
Hui Oi Chow Science Building
Pokfulam Road
Hong Kong

Contact: **Maureen Ann Weatherhead, Ph.D.**,
               Lecturer
Phone: (852) 5 859 2820 • *Fax:* (852) 5 858 3477

Activities: Education; Law Enforcement; Research • *Issues:* Biodiversity/Species Preservation; Deforestation; Energy; Health; Population Planning; Transportation; Waste Management/Recycling; Water Quality • *Meetings:* Asia Pacific Orchid Conference; World Orchid Congress

▲ Projects and People

Dr. Maureen Weatherhead's research, conducted at four institutions, is concerned with "plant breeding and conservation biology being directed towards the preservation of genetic diversity in food plants and in endangered species. The work is principally involved micropropagation and *in vitro* techniques which are the first steps in any biotechnological exploitation of plants and their commercially useful secondary products."

At the Scottish Society for Research in Plant Breeding, Dr. Weatherhead was responsible for the potato breeding program. At the University of Birmingham (UK), she conducted research into the "application of tissue culture techniques to haploid induction in potatoes." During the course of her studies, she and her colleagues were able to "demonstrate the successful storage of potato pollen in liquid nitrogen and to elucidate some hitherto unknown effects of activated charcoal in plant tissue culture media."

While at the University of Leeds, Dr. Weatherhead investigated tissue culture methods in rapid plant propagation. Her research stimulated interest by commercial horticultural companies, and this led to the funding of various research-based tissue culture projects on a number of plant species in popular demand. Dr. Weatherhead says that "such plants, which may be rare or difficult to propagate in a conventional manner, included new orchid hybrids, aquatic species, hardy nursery stock, and ornamentals." With Dr. D.J. Harberd, she worked for the conservation of wild species of endangered British orchids, a project initially funded by the World Wildlife Fund (WWF) and then by The Nature Conservancy Council.

Dr. Weatherhead is currently collaborating with the Kadoorie Agricultural Air Association on research on the conservation of the endangered wild orchids of Hong Kong. She writes, "Some 120 species are known in the territory and of these, 33 have yet to be recorded elsewhere." She explains, "The vulnerability of many of the local orchid species indicated that the use of conventional methods of micropropagation utilizing the meristem as an explant might be suitable. I have, therefore, developed non-destructive methods which utilize plant organs such as leaf and root tips as starting material for rapid multiplication."

She also initiated an orchid seed bank, unique in Southeast Asia.

Dr. Weatherhead is cooperating with the Food and Agriculture Organization (FAO), United Nations, and Professor D.A. Griffiths in a biotechnological project concerned with the eradication of citrus greening disease. She is also working on developing micropropagation systems for local rhododendrons, hibiscus, and several other species.

■ Resources

Memoirs of Hong Kong Natural History Society can be ordered. The University of Hong Kong publishes an annual catalog on *Current Research*, including publications; additional information on listed projects can be obtained from the Secretary of the Committee on Research and Conference Grants. Dr. Weatherhead seeks tissue cultures for the conservation of wild plants.

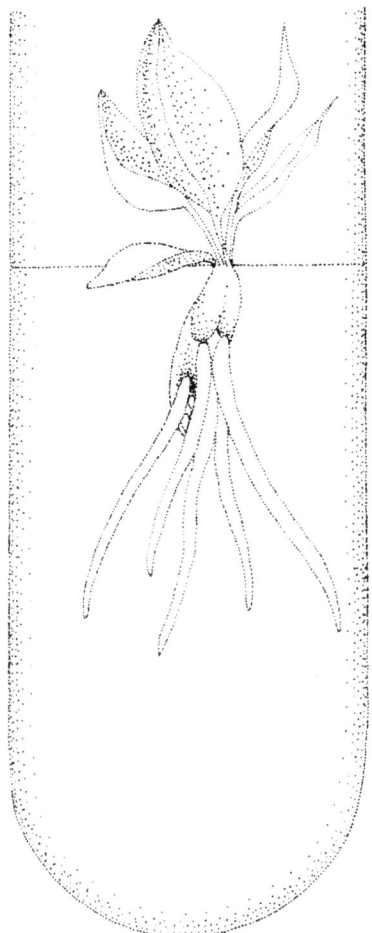

Hungary

Hungary has undergone radical changes in its social and economic structure over the past four decades, including rapid industrialization and urbanization. Unchecked during most of this period were the environmental consequences.

Today, the country is confronting environmental challenges stemming from circumstances common to Eastern European countries generally: the need to rebuild a market-oriented economy while coping with pollution stemming from an outmoded industrial infrastructure. Hungary suffers from a serious air pollution problem, especially in the industrial regions. Groundwater resources are also at risk, particularly from mining activities, and hazardous waste has increased in proportion to the growth of heavy industry. In recent years, however, environmental considerations have been incorporated more and more into industrial policy and physical planning.

🏛 *Government*

Forest Research Institute ⊕
Department of Forest Protection
2100 Gödöllö, P.O. 49
Gödöllö, Hungary 2100

Contact: **György Csóka, Ph.D.**, Forestry Engineer/Researcher
Phone: (36) 28 30 688 • *Fax:* (36) 28 10 856

Activities: Development; Research • *Issues:* Air Quality/Emission Control; Biodiversity/Species Preservation; Forest Damage/Protection • *Meetings:* IUFRO World Congress; World Congress of Forestry

● Organization

In actuality, the Forest Research Institute can claim its beginning in 1899 when the Hungarian Royal Forestry Experiment Station began operations. A few years earlier, the Hungarian Royal Experiment Station came into existence at the Mining and Forestry College. Its objective was to organize forestry science based on specific Hungarian experiences.

Since 1907, the Gödöllö Experiment Station of Conifers has been under the guidance of the Central Experiment Station, renamed the Forest Research Institute in 1933.

Technological, ecological, phytocenological, and phytogeographical afforestation research in addition to hydrology, soil physiology, experiments in natural and thinning regeneration, and investigations on yield and stand structure were conducted. Following the interruption of World War II, the Institute developed a wider organizational and technical structure. According to the Institute, its economic reform began in the late 1960s and accelerated in the mid-1980s.

Six scientific departments, two groups, and seven Experiment Stations comprise the Institute's organizational units throughout the country. The departments are breeding and propagation material production, ecology, forest economics, forest protection, forest utilization, silviculture, and yield science. The groups are the black locust and oak research, and environmental protection. Researchers at the departments and groups also provide expert and technical advice.

▲ Projects and People

Some examples of the types of research are utilization of common walnut in forestry; breeding of poplar, willow, Norway spruce, Austrian black pine, oak, and larch trees; forest ecosystems; biochemical investigations; computer-based forestry methods; harvesting techniques; water management; sandy sites; nutrient circulation and other ecological studies; microfungi; killing of weeds and shrubs; oak decline; mechanization of biomass utilization, chip production, and silviculture; entomology; and economic regulation of industry.

Researchers participate in study tours abroad; likewise about 50 to 60 visitors are accepted at the Institute annually. Scientists also exchange propagation materials, such as cuttings and seeds. The Institute works closely with "foreign partner institutes" in Central and Eastern Europe, former USSR, Turkey, Scandinavia, and the USA.

Dr. György Csóka's ongoing research is concerned with three insect groups: dragonflies, cynipid wasps, and caterpillars. Dr. Csóka studies the life history, behavior, and prey of dragonflies and "what their presence or absence mean in a biotop."

He is investigating the spread of cynipid wasps in Hungary and Central Europe, and whether this phenomenon is related to climate change. Other aspects are their host-seeking ability, native enemies, and the "forest protectional importance of the gall wasps."

Regarding caterpillars, Dr. Csóka is examining the species living on forest trees, their economic importance, and the circumstances of caterpillar outbreaks—and what the differences of such outbreaks would be in native, artificial, or mixed forests, for example. He is artificially rearing and breeding caterpillars to investigate their connection to the host plant and to determine host preference and host-seeking ability as well as to observe overall behavior.

■ Resources

A comprehensive booklet (dated 1990) on the Forest Research Institute is available; it includes a list of recently published books. A central library contains about 13,000 volumes; another 12,000 volumes are at various Experiment Station libraries. *Library Information*, published biannually, describes new acquisitions and translations.

 *NGO*

Zöld Akció
Green Action
Kossuth 13
3525 Miskolc, Hungary

Contact: Iván Gyulai, Ph.D., Founder
Phone: (36) 46 26436

Activities: Development; Education; Research • *Issues:* Biodiversity/Species Preservation; Deforestation; Energy; Health; Sustainable Development; Waste Management/Recycling; Water Quality • *Meetings:* Northern People Alliance

● Organization

Founding member and co-president of the Hungarian Green Party, Dr. Iván Gyulai started Green Action in 1990 to give the public "trustworthy information about the real state of the environment as well as the seriousness of dangers." Green Action also seeks to provide ways to cope with these problems in an ecologically supportable manner. Green Action saw a need for action because it believed that "most official bodies responsible for the preservation of nature have rather vague ideas about the degree of the pollution. . . . Data published by them have very often proven unreliable and there is a striking discrepancy between the results they produce and the data measured by independent organizations."

Green Action built a high-tech laboratory for independent research, to inspect and control governmental studies, and assist other organizations by supplying them with otherwise inaccessible, reliable information based on controlled measuring.

Generally, Hungary's technologies are outdated both economically and ecologically, points out Green Action. Agricultural practices drained wetlands and cleared forests, taking with them numerous species and the "traditional flood-plain economy."

▲ Projects and People

Green Action believes that Hungary's "general backwardness" is due to insufficient ecological awareness and lack of a comprehensive educational program. Therefore, it initiated the integration of ecological studies in the country's school curriculum by a new, global educational dual-language project at the Avasi Grammar School in Miskolc.

A three-year project, started in August 1991, investigates the heavy metal pollution in the Sajó Valley. This area is significant because of its heavy industries—such as coal mining, power plants, and steel manufacturing—which represent the backbone of the urban and industrial agglomeration in northern Hungary. Researchers discovered that the Sajó River was strongly contaminated with heavy metals in the 1970s, and now regard the river as "dead."

The Borsod Chemical Combine (BVK) represented the main mercury source and the North-Hungarian Chemical Works (EMV) discharged cadmium, according to Green Action. Some 500 tons of mercury contaminated the soil and threatens drinking water supplies. Also, as nitrogen accumulates, the oak is becoming extinct. "Nickel rains" are believed to be contributing to hair loss among the region's people. Sludge with high concentrations of these pollutants as well as lead are swept into Tisza-lake.

The Sajó Valley Heavy Metal project aims to identify and locate the heavy metal releasing sources; assess the volumes, types, and impacts of the metals thus far; forecast the long-term potential harmful impact; compile an action plan to avoid the impacts of the impurities; apply heavy metal bioindicators and biomonitoring systems; form a heavy metal database; and put the background and experience from this project into other pollution areas with heavy metals.

The reconstruction of the habitat on Nagykopasz Hill in Tokaj was also begun by Green Action in 1991. This five-year program's ultimate goal is to save *in situ* a gene bank and preserve the existing biodiversity by reconstructing the habitat. By regenerating and restoring those plant communities which foster the most valuable wildlife, a much needed buffer zone also results. Five years earlier, this region became part of the Tokaj-Bodrogzug Landscape Protected Area. Since the ninth century, the hill's wildlife history is inextricably linked to the development of grape cultivation and the Tokaj wine. The decreasing or eventual destruction of the habitat here is due to the spread of alien species of trees such as acacia, the deterioration of the microclimate, and the afforestation of original grassland.

To date, researchers have identified species to be saved, mapped existing vegetation, and acquired the technology to clear the area of the introduced acacia, black pine, and Scotch fir trees. Still needed are knowledge about the soil's structure and chemical conditions, further map elaboration, detailed microclimate maps, and concrete methods for creating "distinct stages of reconstruction." Proposals are also underway to improve the "eco-potential of mine waters in the areas of gravel pits."

In a third major project, Green Action reports that BEFÅG (the state forestry department's local office) is undertaking the afforestation of 12,000 hectares of plowed fields and pastures within 10 years, increasing biodiversity where some 30,000 possible species form the basis of a Central European gene bank, and favorably changing forest structure in Borsod-Abaúj-Zemplén County, where 34 percent of the forest is protected.

Green Action estimates that long-term silviculture needs a 30 percent decrease in clear fellings and "a replacement of large arable lands with forests at the same time." It sees economic benefits from moving away from "the dominance of heavy industry . . . and unemployment derived from it" toward a reliability on a "small and medium company sphere depending on local resources."

Dr. Gyulai, an ecologist, is also a member of the Environmental Committee of the Miskolc Senate.

■ Resources

Green Action seeks new contacts with other independent organizations worldwide. Its detailed reports, many in English, are on sustainable development, global ecology, recultivating peatbogs, reconstructing indigenous forests, and regional national treasures, for instance. A publications list is available.

Action Group on Air Pollution
Talento Foundation
Budaörs, Pf. 102, H-2041 Hungary

Contact: **András Lukács**, General Secretary
Phone: (36) 1 153 7154

Activities: Education; Political/Legislative • *Issues:* Air Quality/Emission Control; Transportation

● Organization

The **Economists' Group of the Hungarian Esperanto Association** established the **Talento Foundation** in 1989 in an effort to protect the environment. It utilizes the principle "think globally, act locally" in its aim to increase children's environmental awareness through an educational system "from the moment they start school" and further improve international communications via teaching and meetings.

▲ Projects and People

The first experimental child development class began a year later in Budaörs, near Budapest. The Foundation plans additional classes to achieve its goal of a 12-year program of study in which students will be fluent in at least 3 foreign languages, including Esperanto, and achieve an internationally recognized level of excellence in a certain field. Computer courses are also taught.

The **Action Group on Air Pollution** (Levegő Munkacsoport), an alternative social organization dealing with environmental pollution problems, receives financial, organizational, and technical assistance from the Foundation. It was initiated in November 1988 by Budapest Technical University's Nature Conservation Club and the Esperantists' Nature Protection Organization. Representatives from 20 various social groups, regardless of political affiliation or membership, meet 2 or 3 times monthly discussing significant environmental issues, making decisions, and taking action.

Its strategic goal requires careful consideration of the technical, economic, and social aspects of **motor vehicle transportation**—a main source of Budapest's serious air pollution, where, for example, lead concentrations are 27 times higher and the cancer-causing benzopyrene is 54 times higher than established health norms. With the population's support, the Group strives to **reduce traffic** and thereby environmental pollution and health hazards. Technical measures regarding motor vehicle adaptation must also be considered.

In 1990, the Action Group developed the **country's first air pollution prevention program** and included a major press campaign to educate the public while pressuring political and economic leaders into action. Its three-phase approach includes dissemination of detailed informational materials, paid advertisements, and demonstrations. Among its activities are national conferences, Earth Day events, cooperation with the National Traffic Safety Council, local demonstrations, calling for decreased traffic patterns in certain areas, pressuring the country's largest petrol company for less damaging activities, advocating a citywide smog alarm system, studying the feasibility of a cyclers' path network, and developing a **public transportation system**.

The Group formulated a 25-point plan to reduce environmental pollution caused by motor vehicles. It includes raising the social consciousness, initiating partially or totally **car-free zones**, annual **vehicle checkups**, tightening of **emission regulations**, **reduced gas lead content**, and **recycling motor oils**.

Medical lectures inform the public about the carcinogenic effects of auto emissions. Among the harmful substances are carbon monoxide, which affects the heart and respiratory system; nitrogen oxides and sulphur dioxide, which irritate mucous membranes; hydrocarbons which reduces the body's immunity; ozone, which harms animal and plant life; lead, which damages the liver, kidney, and blood production; and asbestos (from brake use), which is a known carcinogen, reports the Group. In addition, pollution adversely affects tourism and other economic benefits.

Regarding Budapest's consideration as the site of the 1996 World's Fair, the Action Group arranged meetings between officials and environmentalists to discuss the event's ecological impact. It has campaigned against large commercial developments and a new highway in southern Hungary.

■ Resources

Talento Foundation seeks financial support. It recently published a book, *Transport and Environment,* and began a biweekly newsletter, *LÉLEGZET* (Breath).

Few countries depend as heavily on the marine environment for their sustenance and growth as does Iceland. The largest single environmental issue in Iceland is the conservation of the North Atlantic and Arctic oceans. Recent studies have revealed increasing pollution in the waters north of the country, mainly from transport of contaminants from remote places, including airborne pollutants from the European continent. Situated in a gateway to the North Atlantic, Iceland is also highly sensitive to conditions in the Arctic.

The country is particularly concerned about potential consequences of accidents involving seaborne nuclear reactors and about underground nuclear weapons testing. Another major concern is soil erosion, due mostly to the fragile climate. Terrestrial, air, and water pollution are comparatively low because of the sparsity of people and of polluting industries, and the government has taken steps to prevent such problems from developing.

🏛 Government

Directorate of Shipping
P.O. Box 7200
IS-127 Reykjavik, Iceland

Contact: **Magnús Jóhannesson,** Director of Shipping
Phone: (354) 1 25844 • *Fax:* (354) 1 29835

Activities: Law Enforcement • *Issues:* Air Quality/Emission Control; Transportation; Waste Management/Recycling; Water Quality

Ministry for the Environment
Sölvhóll
IS-150 Reykjavik, Iceland

Contact: **Eiður Guðnason,** Minister
Phone: (354) 1 609000 • *Fax:* (354) 1 624566

Activities: Administration; Education; Research • *Issues:* Air Quality/Emission Control; Biodiversity/Species Preservation; Deforestation; Global Warming; Land Use Planning; Sustainable Development; Waste Management/Recycling; Wildlife Management

● Organization

The **Ministry for the Environment** has five main areas: nature conservation; pollution control; land-use planning; education, information, and research; and international relations. Nature conservation comprises wild animal and vegetation protection, including land reclamation, and general nature conservation, including administration of national parks and other protected areas. Pollution control is carried out on land and air, fresh water, and ocean; there is also a National Centre for Hygiene. Geodetic surveys, international cooperation on environmental affairs, environmental education, and meteorological research are other activities. The Ministry also oversees the Icelandic Nature Conservation Council and the Icelandic Museum of Natural History, among other facets.

■ Resouces

Prepared in 1992, the *Iceland National Report to UNCED* is a 159-page book that covers the country's "environmental endowment," marine pollution, management of ocean resources, afforestation, environmental education, global cooperation, and NGOs, for example. As Eiður

Guðnason quotes English poet John Milton in the preface, "Nature has done her part, Do thou but thine."

Soil Conservation Service (SCS)
Gunnarsholt
850 Hella, Iceland

Contact: **Sveinn Runólfsson**, Director
Phone: (354) 8 75088 • *Fax:* (354) 8 75899

Activities: Development; Education; Law Enforcement; Research • *Issues:* Soil Conservation

● Organization
The most serious environmental hazard in Iceland today is the continued **degradation of vegetation and extensive soil erosion.** In the year 874, at least 60 percent of the island was covered with vegetation—but it is now reduced to about 25 percent. Woodland remnants cover only about 1 percent. The interaction of livestock grazing with weak soil structure, volcanic eruptions, and severe winds cause soil degradation, according to the **Soil Conservation Service (SCS),** which was founded in 1907 to fight erosion.

▲ Projects and People
Reclamation is more difficult in Iceland than in most other countries due to volcanic loessial soil that has been blown away leaving glacial pavement or old lava flows. About 2 percent of Iceland has been enclosed by reclamation fences to exclude livestock grazing. At the turn of the last century, many farms were "decimated by erosion and abandoned," especially in areas where "soils are characterized by ash from volcanic eruptions," Sveinn Runólfsson wrote in a paper reprinted in *Arctic and Alpine Research* (1987). Other soil deficiencies that restrict plant growth are "low water-holding capacities and low levels of nitrogen and available phosphorus."

Today, the Icelandic *Elymus arenarius* plant is used to **stabilize moving sand dunes,** a continued concern of farmers. It is "usually seeded in strips perpendicular to the prevailing wind direction and fertilized until the most serious sand drift has been stopped," according to Runólfsson. Seeding then follows with grasses such as Alaskan *Lupinus nootkatensis*, which was introduced in 1945 for use following reduced grazing intensities; *Festuca rubra*; and *Poa pratensis*. Where land protected from grazing is reseeded, willows and low-growing birches gain a "foothold."

In 1958, the SCS acquired a single-engine airplane for seed and fertilizer dispersion. In 1973, Icelandair donated a Douglas DC3 to expand dispersion. Airline pilots operate it "free of charge demonstrating the widespread interest in the reclamation work," says SCS.

In 1974, conservation efforts intensified in commemoration of Iceland's first settlement 1,100 years earlier, with Parliament beginning a **series of reclamation and range improvement programs.** With grazing as the only major controllable factor, careful ecosystem management is crucial; thus **studies on livestock capacity, causes of erosion, and revegetation** are ongoing at the Agricultural Research Institute. Vegetation maps have been made of about three-quarters of the country.

SCS works in cooperation with the State Forest Service and the Icelandic Agricultural Society as well as with citizens' groups such as *Landvernd* (National Reclamation and Conservation Federation),

Icelandic Forestry Commission, Icelandic Youth Association, Lions Clubs, and Rotary International, among others.

Runólfsson reports that "sheep numbers are currently decreasing and management of both sheep and horses is improving. More may now be gained than lost in the constant fight against erosion. However, an enormous task is ahead for the nation."

■ Resources
Available are a brochure on *Soil Conservation in Iceland* and a reprint of Runólfsson's *Land Reclamation in Iceland.*

Wildlife Management Unit
Hlemmi 3, P.O. Box 5032
Reykjavik 125, Iceland

Contact: **Pall Hersteinsson, Ph.D.,** Director
Phone: (354) 1 626990 • *Fax:* (354) 1 627790

Activities: Research; Wildlife Management • *Issues:* Animal Damage Control; Biodiversity/Species Preservation

● Organization
Under the Ministry of the Environment, the **Wildlife Management Unit** is an institute stationed in the same building as the Natural History Museum (NHM) and the Nature Conservation Council (NCC) of Iceland—thus accessing the NHM library with its zoology collection, other facilities of both groups, and laboratories of the University of Iceland's Institute of Experimental Pathology. The Unit keeps kennels for hunting dogs outside Reykjavik. The staff of seven full- and part-time employees comprise scientific, technical, and office staff. Up to 10 students, research assistants, and hunters join the staff during the summer.

▲ Projects and People
The Wildlife Management Unit lists its two current environmental projects as wildlife management and animal damage control and wildlife management research. In the former, the Unit coordinates "hunting and other means of prevention of damage caused by birds and terrestrial mammals, in particular Arctic foxes, American mink, reindeer, ravens, and gulls." In the latter, the Unit uses "electric fences to exclude Arctic foxes from eider (*Somateria mollissima*) colonies" and cover to "alleviate avian predation of eider eggs. The Unit is studying the "breeding distribution and population size of the lesser black-backed gull (*Larus fuscus*), including daily movements of breeding gulls, which are undergoing "a population explosion" in Iceland. Also being investigated is the "prevalence of *Salmonella* in gulls and ravens and its relevance to salmonellosis in farm animals." Reindeer, their population, sex/age ratios, seasonal movements, and favored vegetation are being monitored.

Dr. **Pall Hersteinsson** is a member of the Species Survival Commission (IUCN/SSC) Canid Group.

■ Resources
The Wildlife Management Unit seeks satellite radio-tracking technology.

 # NGO

Landvernd—Icelandic Environmental Union
Skólavörðustígur 25
Reykjavik 101, Iceland

Contact: **Svanhildur Skaftadðttir**, General Manager
Phone: (354) 1 25242 • *Fax:* (354) 1 625242

Activities: Education • *Issues:* Energy; Sustainable Development; Waste Management/Recycling • *Meetings:* Scandinavian Associations of Environmental Issues and Protection

▲ Projects and People
Landvernd is a non-government organization that attempts to **draw public attention** to environmental problems, such as waste management. For several weeks, Landvernd held an **educational exhibition,** entitled **Pollution-Garbage-Recycling**, at a large shopping center in Reykjavik. Then the exhibition was sent to schools in northern and eastern Iceland; it will also be shown in other parts of the country. "The exhibition has been a great inspiration for children and their work and was put up at the Scandinavian Conference on Environmental Education–MILJÖ 91 in Reykjavik," writes **Svanhildur Skaftadðttir.**

The organization also **produces educational material.** In association with the Nature Conservation of Iceland, the Consumers' Association, and the Environmental and Food Agency, Landvernd prepared posters and pamphlets entitled "The Future Lies in Your Shopping Basket."

■ Resources
The Icelandic Environmental Union produces a series of books, reports, posters, postcards, and various items such as shopping bags and pins—in Icelandic.

India's large population, expected to exceed one billion people by the year 2000, is the source of many threats to the country's biodiversity, which include agricultural expansion, shortened cultivation cycles, increased collection of fuelwood and other forest products, cattle grazing, destructive logging practices, deforestation resulting from dam construction, and conversion of mangroves for commercial purposes. Marine ecosystems are threatened by destruction of coral reefs, siltation, and chemical and oil pollution; as a main thoroughfare for oil tankers, the northern Indian Ocean is particularly vulnerable to oil pollution.

In recent years, the Indian government has become increasingly attentive to these problems, creating new institutions and laws to preserve the environment.

🏛 *Government*

Botanical Survey of India
Cryptogamic Unit, INCHARGE
Post Office Botanic Garden
Howrah 711103, West Bengal, India

Contact: **R.D. Dixit, Ph.D.**, Scientist, SE
  Phone: (91) 60 3231 (operator assistance)
  E-Mail: botsurvey:calcutta

Activities: Research • *Issues:* Biodiversity/Species Preservation

▲ Projects and People

Dr. R.D. Dixit, Botanical Survey of India, is involved in "taxonomy, rare and endangered species and conservation strategies, environmental studies, floristics, and ecology of cryptogams." A cryptogam is a plant without true flowers or seeds that reproduces by spores, including pteridophytes, ferns, mosses, fungi, and algae. A member of the Species Survival Commission (IUCN/SSC) Pteridophyte Group, Dr. Dixit seeks conservation of these plants for "academic and biodiversity reasons." He would like to see a "common expertise survey exploration team [formed] at the international level for extensive collection of cryptogams in tropical countries."

■ Resources

A publications list and reprints are available. The books *A Census of the Indian Pteridophytes* and *A Dictionary of Indian Pteridophytes* can be ordered from the Botanical Survey, P-8 Brabourne Road. *Lycopodiaceae of India* and *Selaginellaceae of India* can be ordered from Bishen Singh Mahendra Palsingh, 23A Connought Place, Dehradun 1, India.

Central Marine Fisheries Research Institute
West Hill
Calicut 673005, Kerala, India

Contact: **R.S. Lal Mohan, Ph.D.**, Principal Scientist
  Phone: (91) 495 52769

Activities: Education; Research • *Issues:* Biodiversity/Species Preservation • *Meetings:* IUCN

▲ Projects and People

Along the coast of India, about 500 dolphins are killed in gillnets each year. The **Central Marine Fisheries Research Institute** educates fishermen and others about biodiversity and conservation of nature in a vigorous awareness program through newspapers, radio, and television. This has reduced the number of dolphins sold in markets.

River dolphins are among the most endangered animals of the world, notes expert **Dr. R.S. Lal Mohan**. The Indian gangetic dolphins in the Ganges and Brahmaputra rivers are killed for their oil in order to lure the catfish *Clupisoma garua*. Dr. Mohan recently discovered that shark oil and sardine body oil are good substitutes and should bring down the mortality of river dolphins. He advises international groups about protecting the river dolphin, taking part, for example, in workshops in China on the world's **platanistoid dolphins** and in the USA on mortality of dolphins in gillnets. A member of the Species Survival Commission (IUCN/SSC) Cetacean Group, Dr. Mohan also observes and writes about **marine mammals of Antarctica**.

To stimulate interest, Dr. Mohan **designed the artwork for two Indian postal stamps**, one on river dolphins and another on dugong, a sirenian found along the Indian coast. The dugong is being studied along the **southeast Indian coast** and around the Andaman Islands, as is the **stranding of whales** along the Indian coast to estimate the number of whales washed ashore annually.

An organizer of **national seminars** on the environmental **impacts of the Gulf War** and science, Dr. Mohan is founder and president for the CSI Conservation of Nature Society in Nagercoil and in Calicut, and the Eco-awareness Committee in Nagercoil, Tamil Nadu.

Three new species of sciaenid fishes—*Johnius macrorhynus* Mohan, *Johnius elongatus* Mohan, and *Johnius mannarensis* Mohan—were discovered by Dr. Mohan along the Indian coast. Among other works, he completed a **bibliography of the Indian Ocean** with 25,000 references, surveyed along Indian coasts for prawn and fish seeds, "developed methods of culture of fishes and prawns in sandy seashore by using polythene film lining in the ponds," and "classified the Indian species of dolphins based on their skulls."

■ Resources

The Institute seeks education outreach support. Dr. Mohan offers a video tape and color slides on dolphins of India.

Ministry of Environment and Forests

B Block, Paryavaran Bhavan, CGO Complex, Lodi Road
New Delhi 110003, Delhi, India

Contact: **M.K. Ranjitsinh, Ph.D.**, Additional Secretary
Phone: (91) 11 362281 • *Fax:* (91) 11 360678

Activities: Education; Law Enforcement; Legislative; Research
Issues: Air Quality/Emission Control; Biodiversity/Species Preservation; Deforestation; Energy; Global Warming; Health; Sustainable Development; Transportation; Waste Management/Recycling; Water Quality; Nature Conservation

▲ Projects and People

The Central Indian **barasingha**, or the Manipur brow-antlered deer, was rescued by Dr. M.K. Ranjitsinh when he saved its habitat in the Kanha National Park. Once down to 66 animals, the barasingha now numbers over 500. As a result of such contributions to conservation of world fauna and flora, Dr. Ranjitsinh was awarded the Order of the Golden Ark in 1979 and the United Nations Environmental Programme (UNEP) "Global 500" Award in 1991. In 1983, scientist **Colin Groves** named the third subspecies of the Indian swamp deer *Cervus duvauceli ranjitsinhi* also to honor the achievements of Dr. Ranjitsinh, who was the first to point out the variation in this eastern race of barasingha.

At Kanha Park, he also undertook the **first translocation of a village outside a national park** in India—leading the way for future such translocations outside protected areas. The park has also been extended. In 1972, Dr. Ranjitsinh **drafted the Wildlife Preservation Act**, the first comprehensive wildlife legislation applicable throughout India, and which remains the legal basis of wildlife conservation and for the establishment of national parks and sanctuaries. Under this Act, Dr. Ranjitsinh was named first director of wildlife preservation of India.

When Dr. Ranjitsinh made the first aerial survey of the Keibul Lamjao Sanctuary, he discovered that the total population of brow-antlered deer there was 14, "the lowest wild population of any large mammalian taxa in the world then." As a result of upgrading Keibul Lamjao to a national park, the population has increased to over 80 deer.

Dr. Ranjitsinh persuaded the state governments to establish a number of national parks for endangered fauna, including the **Eravikulam National Park**, which holds the world's highest population of the Nilgiri tahr. He assisted in the identification of the Tiger Conservation areas, **initiated Project Tiger**, and was involved in the conservation of the **snow leopard**. A number of new protected areas have been established especially in The Himalayas and the Andaman and Nicobar islands.

The final ban of export of **snake and lizard skins, furs, and rhesus monkeys** was the work of Dr. Ranjitsinh, who outlawed the trade in endangered species and banned the export of **frog legs** from India.

At the same time, Dr. Ranjitsinh launched the captive breeding and rehabilitation program for three species of endangered crocodiles in India, including the **gharial**, which are now safe in the wild.

As chairman of the Standing Committee of CITES, chairman of the technical committee of the First Conference of Parties of the International Convention on Migratory Species of Wildlife, and India's commissioner on the International Whaling Commission, he has contributed significantly to conservation. He previously was regional adviser in nature conservation for UNEP in Bangkok, Thailand, helping to prepare a national conservation plan for that country, and later for Bangladesh.

On his second tenure as India's director of wildlife preservation and as additional secretary, he deals with **pollution control**, biodiversity preservation, environmental research, **eco-regeneration**, conservation, project evaluation and approval, and public education. He is in charge of the **Ganga Action Plan** for "pollution detachment of River Ganga and its tributaries and the restoration of the biotic diversity and productivity of this most important river system."

■ Resources

Dr. Ranjitsinh authored the *Indian Blackbuck* and *Beyond the Tiger* book, and these publications: *Keibul Lamjao Sanctuary and the Brow-antlered Deer, Conservation of the Tiger in India, Wildlife Legislation in ESCAP Region, The Manipur Brow-antlered Deer,* and *Forest Destruction in Asia and the South Pacific.*

Zoological Survey of India

Southern Regional Station
100 Santhome High Road
Madras 600028, India

Contact: T.S.N. Murthy, Scientist 'SD'

Activities: Research • *Issues:* Biodiversity/Species Preservation

▲ Projects and People

With 32 years of field and museum experience with Indian reptiles, T.S.N. Murthy, Zoological Survey of India, is engaged in faunistic surveys, "studies on the reptiles of Nilgiris," and upkeep of national herpetological collections. Murthy belongs to the Species Survival Commission (IUCN/SSC) Indian Subcontinent Reptile and Amphibian Group.

■ Resources

The following can be ordered: a set of 50 herpetological reprints, *The Snake Book of India, A Field Guide to the Lizards of Western Ghats, India,* and *A Field Book of Indian Lizards.*

Zoological Survey of India

Western Ghat Regional Station
Department of Environment
2/355, Evenhepalam,
Kozhikkode 673006, Kerala, India

Contact: G.U. Kurup, Ph.D., Regional (Deputy)
        Director
Phone: (91) 495 53148

Activities: Development; Education; Research • *Issues:* Biodiversity/Species Preservation; Sustainable Development

● Organization

Established in 1980, the Western Ghat Regional Station, Zoological Survey of India, investigates and inventories fauna in a region of nearly 1,400,000 square kilometers of rich biological diversity. Located on the west coast peninsular part of India from the Tapti River in the north to Kanyakumari in the south, the Western Ghats comprise high mountains, hilly terrain with tropical evergreen and deciduous forests, and flat hinterland running to the seacoast. A remarkable quality is that the slopes on the windward western side receive two to four times as much rain as the leeward eastern side—resulting in distinctly different species of both plants and animals.

▲ Projects and People

The bonnet macaque, Hanuman langur, Nilgiri langur, and lion-tailed macaques are the main subjects of ecological and behavioral studies. An extensive census of primates of a feral nature conducted at the Regional Station was "hailed as the most comprehensive work of its nature conducted anywhere in the world," according to a 1991 report released during the Platinum Jubilee (75th year) of India's Zoological Survey.

The Malabar civet and the small Travancore flying squirrel have been rediscovered after they had been believed to be extinct. Three other rare animals, the Indian pangolin, the masked booby, and the *Dravida nilamburensis,* the largest species of earthworm found in India, were confirmed by scientists of the Regional Station.

Relict species, or species exclusively confined to the region, found in the Western Ghats are the slender and slow lorises in the family *Lorisidae,* the rodent subfamily *platacanthomyinae* consisting of two spiny mice genera, the flying squirrel, the marten, the tahr, and mouse deer, for example. "Ten species of butterflies, nine genera of fishes, four genera of lizards, six genera of snakes, the entire relict reptile family of *uropeltidae,* and about 41 percent of the genera of amphibia are endemic to Western Ghats," according to the report.

Western Ghats is also home for the elephant, flying fox, false vampires, sambar, four-horned antelope, scale anteater, civet, hyena, tiger, panther, mongoose, giant squirrel, and others. Of the 85 snake species, 14 are poisonous, such as pit vipers, coral snakes, king cobra, and sand boa. Among the 300 butterfly species are the large tree nymph and common birdwing as well as the diminutive grass jewel. "The Paris peacock (*Papillio paris*) and the common banded peacock (*P. crino*) both found here are two of the most beautiful of Indian butterflies," writes Dr. G.U. Kurup. Of the 150 species of birds are the flycatchers (15 species), woodpeckers (7 species) and cuckoos, hornbills, and babblers (each with 4 or more species). Eagles, owls, laughing thrushes, and orange minivet are seen there as well as about 38 species of frugivorous and insectivorous bats. Invertebrates, of which there could be thousands of species, are yet to be described.

Among previous research, "freshwater sources like rivers, streams, lakes, reservoirs, and temporary waterholes in . . . Western Ghats were surveyed for its protozoan and planktonic . . . elements," and cladoceran crustacea and freshwater fishes were also investigated. Some 22 species of protozoa were observed in the city's household wells. During a three-year period, samplings were undertaken of the Chaliyar River to examine the effects of pollutants and effluents on fauna.

The Regional Station's national zoological displays feature 870 named collections and 8,114 unnamed collections of insects and other invertebrates, fishes, amphibians, reptiles, birds, and mammals. The library contains 1,520 scientific books and 72 Indian and foreign journals. Plans are underway to establish a National History Museum, with exhibits open to the public. Previous ecology exhibits have drawn as many as 5,000 people.

Dr. Kurup, who has produced more than 50 research papers, 2 video films, and radio broadcasts, belongs to the Species Survival Commission (IUCN/SSC) Primate and Mustelid/Viverrid groups.

■ Resources

In the 1991 *Zoological Survey of India Report,* 42 scientific papers are listed. Books and periodicals are available. Individual queries will be answered. The Western Ghat Regional Station seeks exchange of scientists, visiting fellowships, and seminar support.

🏛 *Regional Government*

Forest Department, West Bengal
Parks and Gardens Wing, Eden Gardens
Calcutta 700021, West Bengal, India

Contact: **Pranabesh Sanyal**, Special Officer
Phone: (91) 33 287944

Activities: Development; Education; Research • *Issues:* Biodiversity/Species Preservation • *Meetings:* CITES; International Wetlands Conference

● Organization
With a staff of 200, the Forest Department of West Bengal works to manage its forests and save the tiger and the ridley turtle in the mangrove Sunderbans Tiger Reserve.

▲ Projects and People
Pranabesh Sanyal is also the president of the Calcutta Wildlife Society, which has launched an awareness campaign on "Global Environmental Concern" particularly for the school children to "catch them young" with audiovisual talks, "Sit and Draw Competitions," and "Poster Competitions."

West Bengal's Forest Directorate celebrates "Forest Week" during monsoon and "Wildlife Week" during October in order to make local people more aware of issues of wildlife and the environment. Sanyal belongs to the Species Survival Commission (IUCN/SSC) Cat Group.

■ Resources
Among Sanyal's 15 principal publications are: *Managing the Maneaters in the Sunderbans Tiger Reserve of India*, *Saving the Ridley Turtle in Sunderbans of India*, *The Mangrove Otter of Sunderbans*, and *Dancing Mangals of Indian Sunderbans*.

🏛 *State Government*

Wildlife Wing, Government of Orissa
315 Kharavel Nagar
Bhubaneswar 751001, Orissa, India

Contact: **Sudhakar Kar, Ph.D.**, Research Officer, Crocodile Expert
Phone: (91) 674 403390

Activities: Education; Development; Research • *Issues:* Biodiversity/Species Preservation

■ Resources
Dr. Sudhakar Kar, who is writing a book on crocodiles, has published 90 scientific papers worldwide on crocodiles, sea turtles, and India's largest lizards and snakes, along with popular articles. He wishes to visit the USA and African countries to "study and acquire knowledge on crocodile management programs and to interact with scientists." Dr. Kar belongs to the Species Survival Commission (IUCN/SSC) Crocodile Group.

 *NGO*

Asian Elephant Conservation Center (AECC) ⊕
Center for Ecological Sciences
Indian Institute of Science
Bangalore 560012, Karnataka, India

Contact: **R. Sukumar, Ph.D**, Scientist, Coordinator
Phone: (91) 812 340985 • *Fax:* (91) 812 341683

Activities: Development; Education; Research • *Issues:* Biodiversity/Species Preservation; Deforestation

● Organization
The **Asian elephant** once inhabited vast regions through most of Asia but has now disappeared entirely from west Asia, Persia, and most parts of China. "Its sparse distribution over the Indian subcontinent and southeast Asia is now confined largely to fragmented, forested hilly tracts." Total Asian elephant population in the wild is between 34,000 and 54,000, with animals occurring in 13 countries: Bangladesh, Burma, India, Malaysia, Nepal, Sri Lanka, Thailand, Bhutan, Cambodia, China, Indonesia, Laos, and Vietnam. The **Asian Elephant Conservation Center (AECC)** was set up to assist the **Asian Elephant Specialist Group (AESG)** of the World Conservation Union's Species Survival Commission (IUCN/SSC) to promote conservation of elephants in the wild and in captivity.

The AESG has representatives from Germany, USA, France, Switzerland, UK, and Singapore as well as from the above Asian countries. Due to the increased pressure on wild animals to replenish stocks of working elephants, the AESG has drawn on the expertise of several zoos which support **Species Survival Plans** in their countries and have bred Asian elephants in captivity over a considerable period of time.

▲ Projects and People
One of the main objectives of the Asian Elephant Conservation Center is to consolidate the work by the IUCN/SSC and to assist the AESG through the promotion and implementation of the **Asian Elephant Action Plan**. A bibliography of some 600 references on Asian elephants has been **computerized** and is accessible. Information has been collected on status, distribution, biology, and conservation. **Field surveys** are also carried out to check for habitat continuity and to identify crucial corridors between habitats for elephant movement.

Work on **Population Viability Analysis (PVA)** of the Asian elephant is being carried out. Since many elephant populations are small and isolated, the probability of survival of herds of different sizes are being studied using a demographic model in order to develop management strategies for small elephant populations.

The AECC provided input to the government of India's plan for **Project Elephant**. Approved by the Planning Commission, the project is being launched.

■ Resources
The AECC conducts workshops and training programs in censusing techniques and elephant management for both administrators and researchers from various Asian countries where elephants are found.

Technical reports are prepared for researchers and administrators: *Translocation of Wild Elephants*, *Censusing Elephants in Forests*, International Workshop proceedings, and a compilation of elephant bibliography.

Assam Valley Wildlife Society
Pertabghur Tea Estate
Chariali 784176, Assam, India

Contact: **A.M. Khan**, Chairman
Phone: (91) Chariali 74, Sootea 28 (operator assistance)

Activities: Development; Education • *Issues:* Biodiversity/Species Preservation; Deforestation; Energy; Health; Population Planning; Water Quality

● Organization
The **Assam Valley Wildlife Society** describes itself as the only private organization in the state dedicated to educating people on conservation and preservation of wildlife and the environment. India's rich diversity of wildlife is said to include more than 30,000 species in its sanctuaries, parks, and other forest areas—including those that are threatened due to shrinking forests, human habitation, and poaching. Wild tigers, for example, decreased from 40,000 at the turn of the century to less than 2,000 in the 1970s. The Society bears the message, "There is enough in the world for everyone's need, but not enough for everyone's greed."

▲ Projects and People
The Society is locating and protecting the declining **white-winged wood duck** (*Cairina Scutulata*) in Doom Dooma and Dangri reserve forests. It is involved as well in successfully **captive breeding** these ducks at Bordubi Tea Estate in Doom Dooma and Namdang Tea Estate, Margherita, Assam State's Dibrugarh district.

The creation of a 25-hectare safe haven for **capped langurs** is underway at the Mijicajan Tea Estate in Sonitpur district. Also in this district, the Society is creating a **deer park** at Bargang Tea Estate.

Assam has 8 sanctuaries preserving over 33 animals, among them the **great pied hornbill, elephant, python, leopard, wild buffalo, and barking deer**. Over 100 hectares of barren land in Assam State are being brought into **afforestation**.

The Society also educates local people about the dwindling **Asian rhino**, and the myths speeding its destruction, such that as its "head boiled in coconut oil [will] cure toothache, stomachache, and deafness; urine preserves youth when drunk, chases away evil spirits and ghosts if it is hung outside the door in a container; umbilical cord [made into soup] cures rheumatism and arthritis; bone . . . stitched to upper arm gives strength; and tail hung outside the door of a pregnant woman assures painless childbirth."

Additional sanctuaries and the end to degrading practices are called for in assisting the battle of the **mahseer fish** against poachers, floods caused from logging and road building, electrocution by hydroelectric power plants, and explosions by dynamite as a fishing means; barricading of streams; and chemical dumping. In clear waters, the mahseer can grow to 80 to 90 pounds.

Past folklore and superstition that once saved the **hornbill** from persecution due to its habits thought to symbolize "purity and marital fidelity" cannot stop its large nesting trees from being felled today. But the description of the nesting behavior, as described in *Preserve* magazine, is unique: The mother and father work to feed the nestlings. After the mother has left her nest full of chicks, the chicks rebuild the nest wall. Then, as each chick is ready, it pecks its way out through the mud. The remaining chicks repair the damage after each chick leaves. When the last one has flown, the parents abandon the nest.

■ Resources
The Society seeks new technologies for a **dolphin preservation** project under consideration and assistance with public outreach efforts regarding wildlife preservation. *Preserve* magazine, with enlightening tales and descriptions of animals' lives in full color, is published biannually for a limited circulation.

Bombay Natural History Society (BNHS) ⊕
Hornbill House, Shaheed Bhagat Singh Road
Bombay 400023, Maharashtra, India

Contact: **Aloysius G. Sekar**, Herpetologist

Activities: Research • *Issues:* Biodiversity/Species Preservation

● Organization
In 1883, eight people started a society to discuss natural history and met at the Victoria and Albert Museum to "exchange notes and exhibit their collections of butterflies, shells, beetles, birds, and birds' eggs." The **Bombay Natural History Society (BNHS)** evolved into "one of Asia's premier conservation groups." It houses a vast collection of specimens of birds, mammals, reptiles, and invertebrates, some of which are now extinct. Its library holds rare books and contemporary natural history publications from around the world.

▲ Projects and People
The BNHS initiated the creation of a number of parks and sanctuaries including the **Keoladeo Ghana Sanctuary** in Bharatpur, which is a winter haven for **Siberian cranes**—benefitting from "detailed hydrobiological studies." Long-term scientific field research carried out by the BNHS contributed to the survival of a number of endangered species including the **Asiatic lion, great Indian bustard, black-necked crane, flamingo, and lion-tailed macaque**. The **golden gecko** was rediscovered in the Eastern Ghats of Andhra Pradesh.

BNHS has made a study of **bird hazards at Indian airports**, the ecology of endangered species of wildlife and their habitats, hydrobiological research at Bharatpur, and the movement and population structure of Indian avifauna. BNHS seeks to understand the regions of the **tiger**, as this "great cat stands at the apex of the food chain in varying habitats" which must be understood if the tiger is to survive.

Bombay University authorizes only the BNHS to give postgraduate guidance in field biology. **Nature education programs** are provided to school children. BNHS sponsored and drafted the **Bombay Wild Birds Protection Act, 1951**, which formed a model for subsequent wildlife protection legislation in India.

"Over a million species of arthropods [such as insects and centipedes] have been identified by man. Possibly over two or three times this number exist on earth today." The BNHS emphasizes entomology and has in its collection a vast number of invertebrate specimens.

The BNHS together with the Species Survival Commission (IUCN/SSC) formed the **Asian Elephant Specialist Group** to study ways to preserve the "severely threatened" elephant life and habitat. *(See also Asian Elephant Conservation Center in this section.)*

Herpetologist **Aloysius Sekar** belongs to the IUCN/SSC's Indian Subcontinent Reptile and Amphibian Group. He has written about the gliding frog, golden gecko, skink, garden lizard, and other amphibians.

■ Resources

BNHS publishes a *Journal*, noted to be "one of the finest scientific natural history source-books for the Oriental Region." The newsletter *Hornbill* keeps its members informed of BNHS activities. BNHS needs assistance in training to implement new research technologies.

Center for Science and Environment (CSE) ⊕
F-6. Kailash Colony
New Delhi 110048, India

Contact: **Anil Agarwal**, Founder and Director
Phone: (91) 11 643 3394 • *Fax:* (91) 11 644 1711

Activities: Education; Research • *Issues:* Air Quality/Emission Control; Biodiversity/Species Preservation; Deforestation; Energy; Global Warming; Health; Population Planning; Sustainable Development; Waste Management/Recycling; Water Quality • *Meetings:* UNCED

● Organization

An important role of the **Center for Science and Environment** (CSE) during its 10-year existence has been in information dissemination. Reports entitled *The State of India's Environment Series* target specific topics throughout the year, such as floods, green revolution, or forest conservation. The CSE uses over 30 newspapers and 400 magazines to analyze and disseminate information. Some 10,000 color slides and black-and-white pictures, and over 400 maps and posters are included in the CSE, holding "one of the largest collections on the environment in India." The *Green File*, a monthly clipping service, is produced during the year.

▲ Projects and People

In recent years, CSE has also launched an integrated research program stressing "ecosystem-specific development" regarding the balance between "croplands, grazing lands, and forest or tree lands." **Ecologically sound land-use systems** have been studied in the Thar Desert area, the eastern Himalayan area, the northeastern states of Meghalaya, Mizoram, Tripura, Nagaland, Manipur, and the hill regions of Assam.

The **village ecosystems of Bemru** in Uttarakhand Himalaya region and Bhadrajun in arid Rajasthan were selected as models for developing such methodology. Bemru is home to the Chipko Move-

ment and an active women's group that has undertaken afforestation. Also in Rajasthan and other communities, traditional knowledge systems—considered vital "because of [their] ecological rationality"—are being examined before they are destroyed by modern practices or "scientific progress." According to CSE, "cultural diversity is not an historical accident. It is the direct outcome of the country's extraordinary biological diversity" as people learned to use natural resources in harmony with the immediate environment. CSE reports it has seen "several outstanding examples of traditional systems in land use, water use and water harvesting, agriculture, animal care, food preservation, and herbal medicine."

Water harvesting, conservation, and management has been examined in areas where water storage tanks, some 2,000 years old, have served for drinking and irrigation. Piped water supply, large irrigation canals, and tubewells have led to a decline of these traditional systems and have created erosion and a need for dams. The "monsoon runoff now becomes a flood and the use of tubewells during the dry season leads to a rapid lowering of the groundwater table as is now occurring in several parts of India," reports CSE.

Groundwater pollution in Bicchri was so severe "that the color of the water had turned into a dark color like Coca Cola." With the Department of Environment's financial support and help of scientists from the Universities of Roorkee and Aligarh, CSE organized a detailed survey—confirming that wells were unfit for human or cattle consumption or for irrigation. Several farmers lost entire orchards of fruit trees. CSE requested that the government provide for immediate and safe disposal of the **toxic sludge** from a local factory. The supreme court passed the order and the district administration assessed damages for compensation to the villagers.

Throughout India, as CSE researchers **search for successful grassroots efforts** in greening the environment, the group notes the need to develop national strategies for "ensuring people's participation in village-level environmental management and improvement." To this end, CSE began studies on "effective and innovative village-level institutions," a review of laws to determine their positive or negative impact on people's participation in natural resources management, and reviews of India's experience in watershed development and management and in linking "rural employment generation with ecological regeneration." CSE believes that "the stupendous task of holistic planning for every Indian village can be achieved . . . only if it is participatory." There must also be "equity in the distribution of the biomass resources generated," states CSE.

CSE awards short-term **fellowships to Indian journalists** for their consciousness-raising environmental articles and has a **training workshop** in environmental issues for the general public and for teachers, as part of its new Environmental Education unit. An exhibition is being planned for use in schools which includes satellite imageries. In a series of 10 booklets, a *State of India's Environment Report* for children is being developed focusing on land, water, forests, habitat, energy, health, and people's action for change.

A comprehensive database regarding land, water, and forests is being built from the grassroots level to include information from government, universities, technical institutions, and voluntary groups.

Cooperating with the World Conservation Union's IUCN-Pakistan, CSE recently organized an **India-Pakistan Conference on the Environment** to encourage better understanding among these countries on such issues. Other seminars have been held on sustainable use of forests and wasteland development.

Anil Agarwal, an engineer and former press correspondent addressed the Indian Parliament on the impact of environmental destruction on floods and droughts. He has also lectured on "Gandhi, Ecology, and the Last Person," "Human-Nature Interactions in a Third World Country," and "Beyond Pretty Trees and Tigers: The Role of Increasing Poverty and People's Protests."

■ Resources
Publications, posters, photographs, and addresses of environmentalists and institutions are needed as a resource for students, policymakers, researchers, and professionals seeking information on environmental and development issues.

CSE's publications include *Towards Green Villages—A Strategy for Environmentally Sound and Participatory Rural Development; Fight for Survival—People's Action for Environment; The Wrath of Nature—Impact of Environmental Destruction on Floods and Droughts;* and *Temples or Tombs? Decisions in Three Environment Controversies.* The *Annual Report* describes ongoing projects more fully. The Resource Centre also loans video films, tapes, and posters.

Centre for Wildlife Studies ⊕
499, Chitrabhanu Road
Mysore 570023, Karnataka, India

Contact: **K. Ullas Karanth**, Wildlife Biologist
Phone: (91) 812 30364

Activities: Education; Research • *Issues:* Biodiversity/Species Preservation; Deforestation

● Organization
The Centre for Wildlife Studies is working on a study of large predators in Nagarahole National Park in south India focusing on the tiger, leopard, and dhole (*Cuon alpinus*). Using radio-transmitter collars and markings, the Centre is determining the animals' activity and movement patterns, densities, home range sizes, habitat utilization patterns, food habits, and social interaction patterns. The conflicts between human interests and the needs of large carnivores are assessed to better understand how to manage these mammals, as is the "impact of predation . . . on wild prey populations and domestic livestock." The project also "trains Indian researchers and park managers in capture, immobilization, and radio-tracking techniques," writes wildlife biologist K. Ullas Karanth.

Principal investigators are Karanth and researcher Dr. Melvin E. Sunquist, who is with the University of Florida's Wildlife and Range Sciences Department and Museum of Natural History. They are collaborating with the Karnataka Forest Department's chief wildlife warden and other local staff in carrying out fieldwork.

Early data is being analyzed at the Centre's computer facilities; final analysis is being carried out in the USA, with results being published in scientific journals. "The information on prey, predators, and the interactions among them will be integrated to develop management strategies for Nagarahole and other similar reserves," reports Karanth.

Honorary wildlife warden in Karnataka State and worldwide speaker on wildlife management, Karanth previously headed a project to prepare a conservation plan for the lion-tailed macaque in its rainforest habitat; investigated the impact of afforestation on

blackbuck (*Antelope cervicapra*); for Bhadra Wildlife Sanctuary, completed an ecological survey and prepared a management plan; worked to establish the Nilgiri Biosphere; preliminarily assessed a proposed dam's impact on the wildlife and forestry of a Karnataka region; and, also with Dr. Sunquist, examined "ecological relations and resource use in the carnivore-herbivore community" of Nagarahole National Park. Funding sources for various projects include Karnataka government, U.S. Fish and Wildlife Service, National Geographic Society, Wildlife Conservation International (WCI), and US-India Fund.

Karanth, who is a doctoral candidate in zoology, Mangalore University, belongs to the Species Survival Commission (IUCN/SSC) Cat, Mustelid/Viverrid, and Antelope Specialist groups, Wildlife Society (USA), Indian Board for Wildlife, and Bombay Natural History Society's advisory committee. He received training in capture techniques and radio telemetry from the Smithsonian Institution's Conservation and Research Center. *(See also Smithsonian Institution, National Zoological Park, Conservation and Research Center in the USA Government section.)*

■ Resources
Some publications by K. Ullas Karanth are *How Many Tigers Can There Be?, Conservation Prospects for the Lion-tailed Macaque in Karnataka,* and *Dry-Zone Afforestation and Its Impact on Blackbuck Populations.*

Data Center for Natural Resources (DCNR)
143, Infantry Road, Kamala Mansions
Bangalore 560001, Karnataka, India

Contact: **Johnson E. David**, Administrator
Phone: (91) 812 573206

Activities: Development; Education; Law Enforcement; Political/Legislative; Research • *Issues:* Air Quality/Emission Control; Biodiversity/Species Preservation; Deforestation; Energy; Global Warming; Health; Population Planning; Sustainable Development; Transportation; Waste Management/Recycling; Water Quality

● Organization
The Data Center for Natural Resources (DCNR) compiles recent literature on pollution, genetics, conservation and management, sustainable agriculture, renewable energy, forestry and fishery, fauna and flora, and ecology and disseminates it in the form of the periodical *Environmental Resources Abstracts* (*ERA*), which is issued quarterly, in collaboration with World Wide Fund for Nature—India. The issues each year "contain over 1,000 scientific abstracts and 600 significant news items as reported in the print media." Papers are also prepared occasionally for publication in journals or for presentation in seminars or workshops.

▲ Projects and People
The DCNR held a National Seminar on Natural Farming, organized jointly by the Rajasthan Agriculture University and the Department of Agriculture Production and Cooperation, at the Rajasthan College of Agriculture Campus in Udaipur, with the aid

of Johnson E. David. Discussions included "effects of economic policies on agricultural productivity and the environment" and "relative nutritional efficiency and toxicity of food produced with and without chemical inputs." David is involved in other activities regarding **organic and low-external input farming**, speaking on such topics before the Indian Institute of World Culture, for example. DCNR is also active in **national environment awareness campaigns**.

Mapmaker and cartographer David, deputy director of the Southern Circle Survey of India, has been DCNR administrator since 1981. Staffers include **Christina Sathyendra** and **Clara A. Raj**.

■ Resources

David presented or published the following: *Transfer of River Waters, Pricing and Marketing of Organic Foods in India, Role of Wildlife Cropping/Ranching/Farming and Fee Hunting in Conservation, Genetic Engineering: Benefits and Dangers*, and *Sustainable Industrialization*. The Center seeks audiovisuals, books, and periodicals for disseminating environmental information. It fears that its quarterly *Environmental Resources Abstracts* will have to be discontinued for lack of funding. A publications list is available.

Ecosystem Research Group ❦
Rechana, A/33, Elankom Gardens
Trivandrum 695010, Kerala, India

Contact: **S. Usha Kumari**, Scientist
Phone: (91) 67755

Activities: Development; Education; Research • *Issues:* Biodiversity/Species Preservation; Deforestation; Energy; Global Warming; Health; Natural Farming; Sustainable Development

● Organization

The informal **Ecosystem Research Group** is comprised of individuals working in different parts of the state who meet approximately every two months to "exchange ideas and give moral support." Their interests include forest related issues, agriculture, irrigation, education, health, energy, and sustainable development.

"Ultimately everything is connected to one another," writes **S. Usha Kumari**. "And we hope to continue our work and living without separating the two [as] personal and professional life should be the same. . . . People are changing internally to build a new world order in which every life-form will have its own place with a mutual respect and love. . . . This decade is very important, to think, to reflect, and to act sanely to create small self-sustaining societies, like the one many of our forefathers thought and dreamed of.

"We all should cooperate, taking little from this tortured Mother Nature and give back as much as possible to reduce destruction and plan a model for sustainable development," believes Kumari. The Research Group is creating environmental awareness among children.

■ Resources

The Ecosystem Research Group need books, booklets, slides, and films for educational outreach and materials, books, and informa-

tion to instruct the activists on specific issues. Such material could help the Group continue research in pesticide toxicity, organic farming, philosophy for new life-value–based development models, appropriate technology, ocean-wetland situations, water use in agriculture, and irrigation ecology, for example. The Research Group has a few printed booklets in the regional Malayalam language and contributes to magazines and newspapers. Several books/booklets will be published shortly.

Indian Herpetological Society (IHS)
"Usant" Poona Satara Road
Pune 411009, Maharashtra, India

Contact: **Anil Khaire**, Treasurer
Phone: (91) 212 424154

Activities: Education; Research • *Issues:* Biodiversity/Species Preservation; Eradication of Misbeliefs about Wildlife

● Organization

Formed about 1986, the **Indian Herpetological Society (IHS)** describes itself as "a group of young scientists, snake lovers, and social workers" with a fast-growing membership from throughout the country. Among its aims are to eradicate the fear and misunderstanding about snakes, update taxonomical information about reptiles through periodic survey tours through the Western Ghats, plan and implement various conservation programs, conduct research and breeding programs, and collect snakes in populated areas for release in natural habitats.

▲ Projects and People

To prevent commercial exploitation of the **saw-scaled viper** (*Echis carinatus*), IHS surveyed the Konkan region and reported conservation steps to India's Forest Department. It also **protested mass killing of snakes and monkeys** near Bangalore and posted for the first time in India **information boards in braille** for the use of the blind at Poona Snake Park. "Sarpa Rath Yatra" projects were organized in both India and the USA to "respect the sentiments and convey scientific information about snakes." The Yatra uses various placards to explain the role of snakes in nature and ask for their preservation.

IHS also rescued a trapped marsh crocodile, which now lives at Poona Snake Park. Joining other conservation groups, IHS participates in World Environment Day and in a new concept, **Orphanage for Wildlife**, to help injured animals, birds, and reptiles in and around Pune. "These creatures are usually cared for till recovery and released back to their natural habitat." To augment round-the-clock care, IHS is currently raising funds for proper medical and surgical facilities, quarantine and hygienic conditions, and food and rehabilitation.

Anil Khaire, director of a developing Snake Park and Aviary, belongs to the Bombay Natural History Society, Ethological Society of India, World Wide Fund for Nature (WWF), and Species Survival Commission (IUCN/SSC) Reptile and Amphibian Group. He has written scientific papers on, for example, the Indian cobra, checkered keelback water snake, brown whip snake, slender coral snake, common sand boa, and venomous snake bites.

■ Resources

IHS offers a lever-operated snake-catching stick and copies of the scientific journal *Herpeton*. It plans to publish pamphlets, books, and journals in Marathi and English. It seeks guidance and material for artificial incubation of reptilian eggs and veterinarian assistance.

Madras Crocodile Bank ✹
Post Bag 4
Mamallapuram 603104, Tamil Nadu, South
      India

Contact: **Romulus Earl Whitaker**, Director
      Fax: (91) 44 491 0910

Irula Tribal Women's Welfare Society ✹
Post Bag 4
Mamallapuram 603104, Tamil Nadu, South
      India

Contact: **Zai Whitaker**, President

Activities: Development; Education; Research • *Issues:* Biodiversity/Species Preservation; Deforestation; Sustainable Development

▲ Projects and People

The **Madras Crocodile Bank and Herpetological Centre** began in 1975 as a breeding center to "conserve India's three crocodilians, then all endangered." Starting with 50 animals of these species, the Croc Bank, as it is called, has grown to "a population of 10,000 crocodiles of 10 species."

Director **Romulus Whitaker** reports, "The programs of the state and central governments have, with inputs from the Crocodile Bank, brought the mugger (*Crocodylus palustris*), gharial (*Gavialis gangeticus*), and saltwater crocodile (*Crocodylus porosus*) back from the brink of extinction. The 'salty' still has to worry about its dwindling mangrove habitats and the gharial has river damming as only a few of their original habitats remain intact. But the mugger has really made a comeback with 10,000 in the wild and another 20,000 in captivity. So well that commercial farming of mugger to benefit tribal people is being proposed to the government."

The Croc Bank, a popular educational attraction for some 750,000 annual visitors, half of whom are children, with its "standard signboards, guides hired from the nearby village, and articles and television programs will soon be supplemented by a project proposal that will convert part of the Croc Bank into an interpretative center" also featuring turtles, tortoises, and lizards, according to Whitaker. Donors are being sought for this enterprise. In the past, support has come from the World Wide Fund for Nature (WWF), Wildlife Preservation Trust International, New York Zoological Society, and India's Tourism Department, among other groups.

A related undertaking is the **Irula Tribal Women's Welfare Society**, founded in 1986, which plants trees in "thousands of hectares of barren wasteland" in the surrounding Chingleput District. Crocodiles, natives, and tourists benefit, as sandy beach land is transformed with evergreen neem, Australian acacias, and other drought-resistant trees that are also a source of berries, fodder, fuelwood, and timber.

"With the help of the [Indian government's] National Wasteland Development Board, the Indo-German Social Service Society, and the Norwegian Agency for International Development (NORAD), large nurseries of forest trees are established and animators hired to take the message of land regeneration to the land-holding farmers and landless, hunter-gathering tribals alike," according to Whitaker.

He maintains that "tree planting can save India from its rapid downhill environmental slide. . . . the main hurdle is getting access to the land." As the Society is offered land and funds, Whitaker describes the results as "one of the most gratifying sights in the world—the soft evening light on a couple of hundred acres of saplings that have made it through their first hot season and are well on their way to becoming trees."

Zai Whitaker remarks that the Women's Society demonstrates that tribal technology is "ideally suited for sustainable use projects using wild plants and animals. They have a vast botanical and zoological knowledge." However, "Tribal people are losing ground now just as badly as they were in the days of colonization of the large continents. . . . We are losing the priceless knowledge. . . . By employing and encouraging the use of [tribal] technologies to help us solve environmental problems and teach us about the quality of life, we will help ensure that their skills, pride, and will to survive are preserved."

The **Irula Snake Catcher's Cooperative** is another tribal self-help program that "taps Irula tribal technology in the catching of snakes for venom extraction and sale." Based on "generations-old skills," the virtually self-supporting cooperative got underway in 1982 to produce and supply venom to major laboratories which make "life-saving anti-venom serum." Snakes, whose skins once supplied a main source of income, are now "happily released back into the wild after three extractions of venom," Whitaker told the *Illustrated Weekly of India* (August 29, 1989).

Living in a rapidly urbanized area outside Madras, the Irula are traditionally forest dwellers and hunters, who "mainly survive by capturing and selling frogs and snakes," as the Whitakers wrote in *CS Quarterly* and *International Wildlife* (March/April 1979). "The villagers subsist on a diet that includes rats, termites, monitor lizards, turtles, and crayfish" as well as roots and tubers. They are also adept at using medicinal plants to treat muscle and joint pains, allergies, and swellings. The Cooperative demonstrates the principle of "finding appropriate avocations for tribal groups without a drastic wrenching from traditional pursuits," write the Whitakers. "The tribal identity and pride in doing a skilled and dangerous occupation better than anyone else may, in the long run, be as helpful as the cash income the Cooperative provides for Irula members."

The Cooperative's 350,000 annual visitors see and learn about the four dangerous snakes of India, uses of snakes for rodent control, and medically valuable venom. "An innovative pest control program, using Irula skills at rat and termite capture without pesticides, has been started," advises Whitaker. "RATS—**Rodent and Termite Squad**—has been funded by Oxfam Trust (India) and [India's] Department of Science and Technology and is now preparing to form a tribal owned and operated pest control company." A serious problem, rats are believed to destroy 15 to 20 percent of the country's stored and standing grain. Without harming the environment, the Irula hunters physically destroy the burrows and kill whole families of rats at a time.

Whitaker reports on a project that serves as a "logistics base in the remarkable rainforest clad Andaman and Nicobar islands. The forests, mangroves, and coral reefs of the islands are teeming with endemic plants and animals, but the settlers most from the Indian mainland have little 'feel' for the natural ecosystems or pride in being islanders now." The Croc Bank is working with the islands' Environmental Trust, WWF-India, and the government's National Environmental Awareness Campaign, among others, to introduce environmental education into the schools and to establish a Nature Centre adjacent to the first National Park, Wandoor, South Andaman.

"We have close to a billion people in India and precious little resources and time to waste," reminds Romulus Whitaker. "When we get involved in almost any new project we automatically include the environmental awareness component, with a particular slant toward the village grassroots level."

The Whitakers are honorary consultants to the World Conservation Union's Species Survival Commission (IUCN/SSC). Romulus Whitaker, who advises the Food and Agriculture Organization (FAO) and the Bombay Natural History Society, is also an executive committee member of the World Congress of Herpetology and vice chairman of the SSC's Crocodile Group. Regarded as "India's best-known snakesman," according to the *Illustrated Weekly of India*, Whitaker has written books, scientific papers, and numerous articles. Zai Whitaker, who directs Whitaker Films Private Limited with her husband, has written film scripts and more than 100 articles on Indian natural history. Her recent books include *Snakeman*, published by Penguin Books, and for the National Book Trust, *The Snakes Around Us* and *The World of Turtles and Crocodiles*.

■ Resources

Two films, *Seeds of Hope* and *Cooperative for Snakecatchers*, were produced. On behalf of the Irulas, the Whitakers seek new technologies on preparing medicines from plants and marketing such products as well as help with educational outreach on health and literacy.

National Women's Welfare Centre 🌿
Vanitha Bhavan
Ariyancode, Ottasekharamangalam P.O.
Trivandrum 695125, Kerala, India

Contact: **Lilly Genet**, Social Work Secretary

Activities: Development; Education • *Issues:* Deforestation; Energy; Global Warming; Health; Population Planning; Sustainable Development; Water Quality

▲ Projects and People

The **National Women's Welfare Centre** of Kerala has a staff of 21, most working in education activities, and 200 members. Most of the Centre's projects involve educating and training local young women in skills for income generation, and in general awareness of issues affecting their lives. Many activities are in cooperation with other institutions and organizations in the area.

Seminar topics include **agricultural development** (with Kerala Agricultural University); **cattle rearing and milk production** arranged with the cooperation of *Mitranikethan* Farmers Training Center; on world peace; World Environment Day, in collaboration with the Kerala Forestry Board. The Centre also holds **training classes** in tailoring for 35 educated young girls, including National Women's Welfare Centre members.

The Centre has also organized a **family planning campaign** for 100 participants in collaboration with the Family Planning Association of India; distributed **seedlings** received from the Social Forestry Board; collected books and journals for the **Women's Peace Library**, reading room, and research center; conducted a three-day **National Environmental Awareness Campaign** in Kollam District; and conducted a one-month **drought survey** of Pathanamthitta District.

■ Resources

The Centre needs new energy technologies for rural development, improved village-level education, and help with mass awareness campaigns.

Centre publications include books on paralegal training for women's social workers; *Youth Leadership Training Course for Women's Social Workers*, National Science Day reports; World Earth Day reports; series on *Legal Literacy Training Camps for Women* on topics related to violence against women; *International Directory of NGOs and Youth Organizations and Agencies, 1990*; *Youth Leadership Training Camp*; and *Voluntary Action*.

Peace and Disarmament Society of Kerala 🌿
Peace Bhavan
Perumkadavila P.O.
Trivandrum 695124, Kerala, India

Contact: **M. Rajayyan**, Social Work Secretary

Activities: Development; Education; Political/Legislative • *Issues:* Energy; Global Warming; Health; Population Planning; Transportation; Waste Management/Recycling; Water Quality

▲ Projects and People

The **Peace and Disarmament Society of Kerala** is a grassroots NGO with a staff of seven, promoting a better life for the people of the region, primarily through education.

In recent years, staff conducted a day-long **seminar** for 50 youths on **agricultural development** with the cooperation of the Kerala Agricultural University, as well as other seminars on topics including **cattle rearing and milk production** (with the cooperation of *Mitranikethan* Farmers Training Center), world peace, adult education (in cooperation with KANFED/SRC), and **World Environment Day** (in collaboration with the Forestry Board). The Society has also sponsored, alone or jointly, **training in various fields**, such as leadership (in cooperation with the District Health and Social Welfare Departments), weaving for local women, and the installation of **smokeless chulas** (with assistance from ANERT and the Forestry Board).

Other activities include a **family planning campaign** in collaboration with the Family Planning Association of India; distribution of **seedlings** received from the Social Forestry Board; securing **loans** for individuals from the Indian Bank under the **self-employment** program; sponsoring 75 people to receive buildings from the Kerala Housing Board and Banks; collecting books and journals for a **Peace Library**; and celebrating National Science Day.

■ Resources

Publications include souvenir pamphlets for National Science Days; National Environmental Awareness Jatha Programme Souvenirs; Drought Hit Survey 1990, for Kollam and Kasargode District, two books on the Anti-Poverty Programme 1990; and a Youth Leadership training course book, 1991.

Ranthambhore Foundation (RF)

19 Kautilya Marg, Chanakyapuri
New Delhi 110021, India

Contact: **Valmik Thapar**, Naturalist
Phone: (91) 11 301 6261

Activities: Development; Education; Research • *Issues:* Biodiversity/Species Preservation; Deforestation; Energy; Health; Population Planning; Sustainable Development; Waste Management/Recycling; Water Quality

● Organization

The **Ranthambhore Foundation (RF)** describes itself as an "integrated development programme for the conservation of **Ranthambhore National Park**, . . . the smallest **Project Tiger (PT)** reserve." The Park covers 640 square kilometers in Rajasthan, and the surrounding area holds 60 villages with a total population of nearly 200,000.

The tiger has generated the most concern because of its position at the top of the food chain. Its dwindling numbers signal the demise of hundreds of other species of animals and plants lower on the chain. Ranthambhore is planned as the first conservation project, which will then expand to cover other forests. Although the situation of the tiger population has greatly improved since the 1970s because of better management, the "ever increasing clash between man and nature" keeps the tigers and the reserve as a whole at risk.

Wildlife and plants are not the only victims, however. In creating the Park, the government relocated 14 villages; Kailashpuri and Gopalpura, for example, were resettled in 1977. Besides being forced out of their home areas, villagers became *de facto* criminals when they tried to continue their vital subsistence activities of collecting timber or fodder, banned on the reserve. The Foundation began as an effort "to reach a compromise solution": preventing the destruction of the forest and the species living there by offering alternatives to the neighboring rural communities.

The Foundation works to create "a natural integration between man, nature, and wildlife" so that those living near the Park will have a stake in its existence. By providing services in such fields as health, education, dairy development, and income generation, the Foundation's strategy is to "create an atmosphere of trust so as to

finally initiate the more complicated activities of afforestation and alternative energy."

The Foundation's staff of 25 works in close cooperation with 4 other organizations and with government agencies in some areas, including "advising and assisting local Park authorities in the management of the area." The current ecodevelopment plan will continue through 1996.

▲ Projects and People

Initially, the Foundation surveyed the needs of the communities bordering the Park. On the basis of their research, they began several projects "with the help of a series of NGOs already specialists in these fields."

RF funds a **mobile primary health service** run by the NGO **Parivar Seva Sanstha**. In addition to providing basic services and health education, there are plans for building a static clinic.

The Foundation's **dairy development program**, funded in part by nationalized banks, involves the purchase of indigenous Murrah cattle, stall fed with high milk yield, to replace existing low milk yield, free-roaming cattle which put "tremendous pressure on the Park." The program includes a cattle development center for researching breeding, yields, and other relevant topics. In addition, RF sends selected villagers to the national dairy center for more specialized training.

The dairy and energy efforts go hand in hand. The Foundation promotes use of **alternative energy** with help of **DANFOSS INDIA LIMITED** "in order to relieve pressure on wood as fuel." However, biogas for use as fuel, for example, needs stall-fed cattle to provide the manure. The Foundation is also encouraging use of "smokeless chulas, which reduce health hazards and reduce by half firewood consumption."

The RF has an extensive **education** program, including Saturday **nature trips** for children, a mobile audiovisual center and library, posted signs regarding the history of the area, and restoration of historic sites such as an old fortress in the Park, implemented by the Indian National Trust for Art and Cultural Heritage (INTACH). According to RF, "At least 600 children around the Park are in regular and frequent involvement in our eco-education projects."

In the area of **afforestation**, the Foundation works with the Society for the Promotion of Wasteland Development (SPWD) "to plan the most efficient use for the land around the Park," creating a buffer zone which will provide villagers with sufficient resources to prevent incursions into the Park itself. A **fruit sapling nursery** in Sherpur has 25,000 saplings, and the villages of Sherpur and Shampura now have nurseries which distributed 5,000 saplings each during monsoon season of 1991. The Foundation plans further nurseries and a **seed bank** for biogenetics. Staffers are also working to "record the traditional knowledge and the **medicinal value** of botanical species and demonstrate their growth potentials."

The Foundation is also involved in arts and crafts, seeking to **revive traditional handicrafts** to provide alternative sources of income to villagers. In programs implemented by **DASTKAR**, the women of Sherpur village have a training center and the city of Sawai Madhopur has an art school. RF stresses that "no conservation effort acquires significance without income generation measures."

Project Tiger has begun working with the Foundation to **plant barren areas** and **dig ditches** to keep cattle out of the reserve. To overcome villagers' reluctance to work with PT, the Foundation "managed to persuade Project Tiger to rebore a well for the resettled village of Kailashpuri, and to reinforce a traditional reservoir for Sherpur village"—in effect bartering services for conservation efforts. The RF is also engaged in wildlife research, studying the behavior of tigers and leopards and their interactions with domestic livestock.

■ Resources

Resources needed include alternative energy technologies to relieve pressures on natural resources.

The Foundation offers three annual fellowships to graduate students who wish to research "topics that may contribute to the knowledge, understanding, and approach" of the various issues related to conservation and sustainable development, and which may promote "active and useful means for improving the life and environment . . . of local people."

Publications include an *Activities Update* and *Annual Report*.

Centre for Research on Sustainable Agricultural and Rural Development (CRSARD)

M.S. Swaminathan Research Foundation
14, II Main Road, Kottur Gardens
Madras 600085, Tamil Nadu, India

Contact: **M.S. Swaminathan, Ph.D.**, Director
Phone: (91) 44 416923 • *Fax:* (91) 44 478148
Telex: 91-41 5119 SARD IN

Activities: Education; Research • *Issues:* Biodiversity/Species Preservation; Deforestation; Global Warming; Sustainable Development; Waste Management/Recycling

● Organization

When Dr. M.S. Swaminathan received the first World Food Prize in 1987, he used the funds from that award to found the M.S. Swaminathan Research Foundation a year later. The Foundation's staff of 20 focuses primarily on research, with a few working in education.

The Foundation's major aims are "to integrate the principles of ecological sustainability with those of economic efficiency and social equity in the development and dissemination of farm technologies"; to blend traditional and new technologies in such a way as to improve opportunities for skilled jobs in the farm and nonfarm sectors in rural areas; and "to develop and introduce technology, knowledge and input delivery and management systems which will enable disadvantaged sections of rural communities, particularly women, to derive full benefit from technological progress."

The Foundation's main concern is sustainable development based on a belief that attaining "scientific goals should lead to a concurrent strengthening of the ecological security of rural areas and the livelihood security of rural families." Acceptable levels of constant improvement "will be possible only through a *participatory research* mode involving scientists and rural families." Threatened problems coming with global warming, technological changes, and other disturbances or advances require, in addition, what the Foundation terms "*anticipatory research* programmes." In pursuing these goals, the Foundation relies on close partnership with institutions and individuals working toward the same ends. All Foundation endeavors involve extensive collaboration with researchers in the area's numerous institutes and universities.

▲ Projects and People

In 1990, the Foundation established the **Centre for Research on Sustainable Agricultural and Rural Development (CRSARD)** in Madras City. Centre facilities include 50 hectares in **Pichavaram mangrove area**, 200 kilometers south of Madras City, designated as a Genetic Resources Centre. Support has come from the Tamil Nadu state government and the Ministry of Science and Technology of the government of India. The Centre researches such areas as coastal systems, biological diversity, biovillages, education, training, and communication. The Centre's general Research Council makes overall decisions; each program, with Project Advisory Committees, has subprograms, each with a Project Leader.

The **coastal systems research** project focuses on India's 7,500-kilometer coastline, 1,000 kilometers of which is in Tamil Nadu. Coastal communities, which depend on fisheries, forestry, animal husbandry, and tourism, are "in terms of their biological richness . . . the marine equivalent of tropical rain forests." CRSARD gives special attention to the coastal region, which is large enough to "provide a reasonable quality of life to large numbers of people provided economic development is based on the foundation of ecological security." Using its **Coastal Systems Research (CSR) Methodology**, CRSARD surveys, analyzes, plans, introduces technologies, mobilizes support, and spreads its message by example. As in all Foundation projects, the "key is participation of local peoples at all stages."

One example of a coastal development project involves **shrimp farming** for income generation. Currently, "shrimp seed production through hatcheries and wild collection can meet only about 10 percent of the actual seed requirements." Thus, CRSARD researchers are investigating ways to **increase seed production** to popularize shrimp farming, make it more viable as a source of income, and at the same time encourage people to maintain mangrove areas necessary for good shrimp farming. A subprogram involves marketing schemes, without which the farming would be useless.

According to the Foundation, "Only opportunities for assured and remunerative marketing can help to stimulate and sustain the interest of rural families in new technologies." The United India Insurance Co. Ltd. has agreed to insure those shrimp farmers on the project to encourage acceptance of new technologies. The Foundation offers **training programs in coastal systems research**: training local people in prawn farming, shrimp hatchery, prawn feed production, and agroforestry.

Biodiversity projects are another major focus of the Foundation and CRSARD, because "biological diversity is essential for achieving sustainable gains in productivity per units of land, water, energy and time." Such productivity is essential in India, whose 844 million people suffer endemic hunger because of limited productivity on limited land. To investigate areas related to biodiversity, the Foundation and CRSARD created in 1991 the **N.I. Vavilov Research and Training Centre for the Sustainable Management of Biological Diversity**. Working with the Ministry of Environment and Forests, government of India, and State Forest Department of Tamil Nadu, the Vavilov Centre encourages community involvement in such projects as **conservation monitoring** through bioindicators such as insects, algae, lichens, and **earthworms**. The Centre has an "information center and computerized databank on grassroots level conservation activities and on genetic resources for sustainable agriculture and for adaptation to climate change."

The Foundation/Centre's long-term planning for **global warming**, and the projected concomitant rise in sea level, centers on the potential damage to **mangroves** in India, which host 59 species in habitats ranging from untouched sacred groves to the biosphere reserves. At the Genetic Resources Centre, researchers work to preserve genetic diversity through *in situ* and *ex situ*, *in vitro* and *in vivo* conservation. Not only do mangrove ecosystems support a large variety of wildlife, but "in the densely inhabited areas, where species diversity of the mangrove ecosystem is high, people [have learned] to make use of many different products." Researchers have at their disposal the **Mangrove Ecosystems Information System** (MEIS), two databases on mangrove experts, and a bibliography. The Foundation stresses that "meeting the needs of the rural poor for fuelwood and fodder should be an integral part of any strategy designed to conserve mangrove ecosystems."

"Unless the ecological security of the farm and the economic well-being of the farm family are linked in a symbiotic manner, sustained advances in agricultural productivity and family welfare cannot be achieved." **Biovillages** are the Foundation's answer to this problem, integrating traditional and new technologies "to promote the efficient and sustainable use of natural resources, and to achieve a continuous and steady growth of agricultural production while protecting and improving the environmental capital stocks of the village."

In each village, Foundation staff first survey, analyze, and identify the best-suited technologies; they then share their findings and the possible options with the village community and together develop a plan, practicing the principle: "before you teach the farmers, listen to them." In 1991, biovillage projects were launched in the Union Territory of Pondicherry, with support of the Asian Development Bank, with **Mr. P.P. Zacharia**, director of Department of Agriculture, as liaison. The biovillages are the site for **bioenergy projects** such as finding viable nitrogen sources to **improve yields of rice crops.**

The Foundation promotes "ecological horticulture—holistic resource management of farms"—which encourages participation and involvement from the farming community. A 0.4-hectare farm is used to produce high-quality seeds and to multiply fruit plants which are in heavy demand. In the process, the Foundation teaches local people "improved nursery techniques, micro-propagation, water management, and ecological pest-proofing." In Ninnakarai village, after analyzing the needs of school children, the Foundation has embarked on "carefully planned programmes of health intervention, nutrition education, and income generation through **Nutrition Gardens**, ecological horticulture, and producer-oriented marketing."

Reaching the Unreached is a program intended "to enhance the efficiency of organizational and delivery systems" in order to increase the access of "the population normally bypassed by ongoing programmes . . . to available technology, services, and information." The program is tackling child and maternal mortality rates, malnutrition, illiteracy, child protection, and "universal access" to education, safe drinking water, and sanitation.

For many of its projects, the Foundation uses traditional communications media, especially *Koothu* folk theater, to impart its development message. Rural audiences are more receptive to such traditional forms; in addition, this provides jobs for experienced artists and training for new ones. The folk theater project is funded by the Kalaimanram Foundation of the Netherlands.

■ Resources

The Foundation's first *Annual Report* (1990–1991) comprehensively describes its programs. In addition, RF sponsors *Dialogues* and publishes proceedings on such topics as plant genetic resources; genesis and spread of wheat revolution in India; global network of Genetic Resources Centres in mangroves; biotechnology; and information sciences and technology.

Tata Energy Research Institute (TERI) ⊕
9 Jorbagh
New Delhi 110003, India

Contact: **Rajendra K. Pachauri, Ph.D.**, Director
Phone: (91) 11 462 7651 • *Fax:* (91) 11 462 1770

Activities: Development; Research • *Issues:* Air Quality/Emission Control; Biodiversity/Species Preservation; Deforestation; Energy; Global Warming; Sustainable Development; Transportation; Waste Management/Recycling; Water Quality

● Organization

The **Tata Energy Research Institute** (TERI) was organized in 1974, funded by Tata Chemicals and other Tata companies. TERI is working to find solutions to problems caused by the gradual depletion of nonrenewable energy sources, and by current methods of energy use leading to increasing pollution. The Institute focuses on eight major areas: Energy Policy and Planning; Rural Energy; Forestry; Renewable Energy Technologies; Energy Conservation; Biotechnology; Computer Applications; and Training and Information Dissemination. Activities range from low-level studies, for example on the effect of excess smoke in rural kitchens, to the development of mathematical models simulating India's energy economy into the next century.

With both local and international support, including that of the World Bank, Ford Foundation, UNESCO, and the Food and Agriculture Organization (FAO), TERI's 240 research and support staff work to bridge "the gap between lab and land." In India, TERI serves as advisor to the **Ministry of Environment and Forests** on **global warming** policy issues. TERI has also had an office in Washington, DC, since 1990, and also hosts the secretariat of the **Asian Energy Institute**, "a network of energy institutions" from about 12 countries of the Asia-Pacific region. As "the largest research institute in the world focusing on energy and environmental policy issues in developing countries," TERI can provide insight and hard facts as to how a developing country deals with the "field of energy-environment interface," reports the Institute.

▲ Projects and People

In 1990, TERI established a **Tissue Culture Pilot Plant** (TCPP), using tissue culture and cloning techniques for the production of planting material, which is then distributed to forest departments for planting in a number of states. By focusing on scarce and/or easily propagated tree species, TERI hopes to "bring about major increases in the supply of superior quality planting material such that wastelands can be afforested." A top priority is to develop cost-effective production methods.

In a related series of projects, TERI is comparing the strengths of both exotic and indigenous acacia trees as fast-growing sources of biomass energy. Another TERI project involves research into ways of increasing crop yield of the oilseed *Brassica*, including mustard and rapeseed using hybrids and genetic manipulation. Afforestation on degraded lands is ongoing, with bamboo, acacia, *Casuarina equisetifolia*, and other varieties.

More specifically in the energy field, TERI has developed a successful biomass gasifier. The TERI prototype machine requires 3 hours to make enough briquettes to run the alternator for 20 to 30 hours. The Institute hopes to find alternatives for the biogas plants which produce methane from cattle dung; as dung is not always in adequate supply, TERI researchers are investigating substances other than cow dung, "concentrating on wheat straw and rice straw as a possible alternative." However, researchers are currently trying to find bacteria that can break down the lignin contained in such straw, which binds cellulose and hemicellulose of straw to inhibit gasification.

In the area of renewable energy technology, TERI is working to find ways to make solar or wind energy "available in a convenient form and at the point where it is needed." This entails concentration on rural energy research and extension, as more than 75 percent of India's population is rural.

Many of TERI's studies focus on actual energy use, encouraging energy audits in the field of industry (which accounted for 57 percent of energy consumed in India in 1987–1988). Power policy and planning projects search for ways to deal with and ease the burdens on India's already "erratic power supply," as well as to increase domestic energy efficiency and conservation. In 1989–1990, India consumed 53 million tons of fossil fuels, compared to 46.4 million the previous year, according to TERI. The Institute recently completed a study for the World Bank of India's commercial energy demand, analyzing "factors that have determined energy demands in the past [and] various policies/measures adopted by the government to manage energy demand."

Other specific projects include: the Centre for Research and Information on Global Warming; the India Habitat Centre to be solar powered; a database on renewable energy technologies; a study of energy use and environmental effects in Garhwal watershed area; and development of national standards on waste disposal in soils, sponsored by the Central Pollution Control Board.

TERI director since 1981, Dr. Rajendra K. Pachauri has doctorates in both economics and industrial engineering from North Carolina State University, USA. He is a member of both the World Energy Council (WEC) in London and the governing council of Asian Energy Institute, on the steering committee for the Global Warming Assessment Project of the Stockholm Environment Institute, and former president and chairman of the Washington-based International Association for Energy Economics (IAEE). In 1990, Dr. Pachauri was a visiting fellow at the World Bank and formerly served as a member of the Advisory Board on Energy for the Government of India, reporting directly to the Prime Minister.

Amar Nath Chaturvedi, who specializes in forestry, silviculture, and biometrics, is senior fellow at TERI. He is a member of the UNESCO Man and the Biosphere (MAB) Committee of the Indian Department of Environment and the Species Survival Commission (IUCN/SSC) Indian Subcontinent Plant Group. Chaturvedi was

previously Conservator of Forests throughout India, among other government forestry assignments since 1955. He has published nearly 20 Indian Forest Records (IFR)—*General Standard Volume Tables* for various species. Of his more than 85 other publications since 1965, the most recent include *Social Forestry: Concepts and Practices* and *Reforestation in India*. He also co-authored (with L.S. Khanna) the college textbooks *Forest Mensuration*, and *Van Mapiki* in Hindi.

Another senior fellow is Dr. Prodipto Ghosh, whose Carnegie-Mellon doctoral thesis was entitled "Simulating 'Greenhouse Gases' Emissions Due to Energy Use by a Computable General Equilibrium Model of a National Economy." At TERI, Dr. Ghosh oversees the "development of national economic models focused on energy and the environment, economic appraisal of energy technologies and projects (including environmental impacts), and policy analysis of adaptation and regulatory approaches to climate change." He also has a permanent position with the Indian Administrative Service (IAS). Among other organizations, he is a member of the Association of Environmental and Resource Economists, Washington, DC, and has served as consultant to the World Resources Institute in Washington and the Food and Agriculture Organization (FAO) in Rome.

After receiving his master's degree in forestry from Yale, senior fellow O.N. Kaul went on to spend 36 years with the Indian Forest Service before joining TERI in 1989. A life member of the Society of Indian Foresters, including positions as general secretary, vice-president, and executive committee member, Kaul edits the Society's journal *Van Vigyan* and is associated with the International Union of Forestry Research Organizations (IUFRO). His forestry research focuses on environmental impact assessment (EIA), energy conservation, global warming, and forest management programs that involve the local population. He has contributed to national documents, including the latest one, for the World Bank, on an integrated forestry development project in Maharashtra.

Dr. Veena Joshi has been a TERI fellow since 1984, specializing in energy environment interface, rural energy research and extension, renewable and efficient energy technologies, and application software development. Dr. Joshi has about 60 publications, primarily papers and research reports. Dr. Damyant Luthra, industrial engineer, has been a visiting research associate of TERI since 1989, focusing on energy modeling, energy policy, and environmental impacts. Dr. Vikram Madhao Pattarkine is an environmental engineer with expertise in the role of metals in removing phosphorus from wastewater.

Among TERI's research associates are Sumeet Saksena, V. Srinivasachary, Ajay Sharma, P. Venkata Ramana, Priyamvada Zutshi, Dr. Maria Ligia Noronha, Atul Kansal, Sujata Gupta, and Amrita N. Achanta.

■ Resources

Resources needed include new technologies in sophisticated chemical analysis equipment and training in its use for pollution measurement; public relations assistance to "create awareness in developed countries regarding environmental problems of less developed countries—[such as] biomass stress, health hazards of biofuel combustion, and displacement of people."

TERI has a wide range of publications "to complement its research and training activities," and a list is available. These include: *Annual Report*; films on biogas and global warming; *Abstracts of Selected Solar Energy Technology (ASSET)*, six to eight times per year;

semiannual *Energy Environment Monitor, Energy Policy Issues,* annual workshop proceedings; quarterly *Indian Energy Abstracts, Industrial Energy Conservation,* currently numbering five case studies; quarterly *Mycorrhiza News,* semiannual *Pacific and Asian Journal of Energy (PAJE);* semiannual *SESI Journal* published for the Solar Energy Society of India (SESI); quarterly *SESI Newsletter, Proceedings of the International Conference on Global Warming and Climate Change,* New Delhi (27 papers); *TERI Energy Data Directory and Yearbook (TEDDY); TERI Information Service on Global Warming (TISGLOW);* Discussion Papers; and the semiannual *Teriscope* newsletter.

In addition, TERI boasts a well-equipped computer center and a library of energy literature, which by 1990 held more than 8,500 technical books, 550 current journals and 3,500 back volumes. The resources also include a Mycorrhiza Information Centre (MIC); a Documentation and Information Centre (DIC); and INFOTERRA, under the auspices of the United Nations Environment Programme (UNEP).

WILD

Society for Conservation of Forest and Wildlife
21/A 'Asmita,' Swanand Society, Sahakarnager
Pune 411009, Maharashtra, India

Contact: **Chandrashekhar Devidas Nanajkar,**
Founder

Activities: Development; Education; Law Enforcement; Research • *Issues:* Air Quality/Emission Control; Biodiversity/Species Preservation; Deforestation; Global Warming

● Organization

In 1985, several engineering students from Pune, including **Chandrashekhar Devidas Nanajkar,** were on one of their regular treks when they became trapped in the Koyna Wildlife Sanctuary for two nights. The experience was an eye-opener for them, as they were able to view many animals in their natural habitat—and also saw the "negligible efforts taken to conserve this forest and its wildlife." WILD was formed by Nanajkar and his colleagues, initially to conserve Koyna, on Indian Independence Day, August 15, 1989, with "conservation of forest and wildlife" as its main aim.

An independent group supported by its members, WILD has divisions for finance, study, membership, projects, and college outreach. Its headquarters are in Pune, with one branch in Pimpri-Chinchwad Pradhikaran City and three in Bombay and its suburbs of Sahakarnagar and Dattawadi. By 1991, WILD had 1,000 members at its main office, and 700 in the branches; in addition, 400 volunteers are about evenly divided between the main office and the branches.

▲ Projects and People

The first project, to preserve the **Koyna Wildlife Sanctuary,** began with a detailed survey of the 800-square-kilometer forest and inlying villages. Having surveyed for two years, WILD members "prepared a proposal on conservation of this forest including all the alteration and modification and solutions [to] the problem of conservation," which they then submitted to the Forest Department.

Through audiovisual presentations, WILD built public awareness and support, and attracted the government's attention. In response, the government has agreed to many of WILD's suggestions, including converting the status of Koyna from "sanctuary" to "national park," thus strengthening the available conservation measures, according to Nanajkar. WILD volunteers erected signs along the trails encouraging responsible behavior of hikers, many of whom make the trek to the old fort in the forest.

In another endeavor, WILD members installed 60 boards containing information on the various animals housed at **Peshwe Park Zoo** in Pune, which receives some 4,000 visitors daily. According to WILD, the boards play a role "in reducing the misunderstandings and fear of many wild animals" that zoo visitors might have. The Bombay WILD branch has a similar project underway at their local **Ranibaugh Zoo.** WILD plans to expand the **zoo education** project even further, throughout the state of Maharashtra and eventually all of India.

In the area of **social forestry,** the Nigadi Township Development area has provided WILD with 20 acres of land for an **afforestation project,** planting mainly fruit trees. Currently, WILD employs 20 disabled students from the Suhrud Mandal school on the project, and plans to hire up to 100 as the project expands.

A project still in the planning stages involves **reintroducing wildlife** into the **Panchgaon Parvati Sanctuary** outside Pune. Negotiations with the Forest Department are currently underway regarding the reintroduction of such species as the **spotted deer, sambar, jackal, porcupine,** and **rabbit.** WILD has assisted the Forest Department in taking a **census** of animals in the **Bhima-Shankar Wildlife Sanctuary.**

Each month, WILD hosts programs for large groups of people including exhibitions on forest conservation, pollution, and other environmental issues, with sufficient staff to answer questions individually; well-attended **Wildlife Training Camps;** and five-day **jungle expeditions** to train members and volunteers, preceded by classes in track recognition, animal habits, and responsible behavior. Members also gather statistical data on wildlife behavior and translate technical information for everyday use. To commemorate Wildlife Week, the Society sponsors **essay contests** in the local schools.

■ Resources

In addition to funding, WILD needs equipment (video cameras, cameras, projectors), tools (binoculars, tents, compasses), raw material (chemicals for preserving animal droppings) for wildlife study; political support for conservation of Koyana Forest in Maharashtra; and video cassettes, wildlife literature, and books for research and education purposes.

The organization has handbills, booklets, and folders on wildlife; publishes monthly articles and other literature on wildlife, often on specific species; and holds slide and film shows each week for members and volunteers.

World Wide Fund for Nature—India (WWF-India) ⊕
c/o Tollygunge Club
120 D.P. Sasmal Road
Calcutta 700033, West Bengal, India

Contact: **Anne Wright**, Trustee
Phone: (91) 33 741923

Activities: Development; Education; Law Enforcement; Political/Legislative; Research • *Issues:* Biodiversity/Species Preservation; Deforestation; Energy; Global Warming; Health; Population Planning; Sustainable Development; Waste Management/Recycling; Water Quality • *Meetings:* CITES (when held locally)

▲ Projects and People

Anne Wright, MBE, was born in England to a member of the Indian Civil Service, and lived until age 12 "in the jungles of Panchmarhi and the Melghat Range in Central Provinces (now Madhya Pradesh)." She has spent the last 25 years working energetically for the cause of species and habitat conservation in India.

Wright's first involvement in the Indian environmental movement came during the 1967 Bihar drought, "worst of its kind for a century," when she initiated a fundraising campaign, the **Save the Wildlife Fund**, to dig water holes in the forest area of Bihar, and persuaded the government of Bihar to carry out the work. She toured the region extensively with the then Chief Conservator of Forests, **S.P. Shahi**, during 1967 and 1968.

Having reared an orphaned tiger, leopard, and Indian lion cub, Wright has firsthand experience in such protection efforts. In 1969, she became a founder trustee of the newly formed World Wildlife Fund—India, and the next year initiated a campaign to publicize the extensive Calcutta-based **tiger skin trade**, writing articles for the *Statesman* (republished in the *New York Times*) and other publications. As a result, "large hauls of illegal tiger skins were exposed, and Customs persuaded to arrest a renowned Italian smuggler." In 1970, she joined the 10-member **Tiger Task Force** that worked out the location and layout of tiger reserves for funding and management under **Project Tiger**; nine reserves were the result. As part of her Task Force work, the government later asked Wright to survey and submit a report and recommendations for the management and enlargement of the Manas Reserve in Assam and the Palamau Reserve in Bihar.

Wright has not limited her activities to the large cats. In 1970, she requested the Kenya Polo team, then playing in Calcutta, to send her a copy of the Kenya Wildlife Act. After modifying this legislation, Wright sent it through West Bengal government channels to the central government, and it served as a model for the existing **Wildlife Protection Act** that became law in 1972.

Recent efforts include an **ecodevelopment project to preserve the Manas National Park/Biosphere Reserve**, a World Heritage site, and work to **restrict** the extensive **Indian rhino poaching** at Kaziranga. Projects recently completed have involved the **preservation** of such species, along with their habitats, as the **threatened olive ridley turtle** on the east coasts of India; **lesser cats in eastern India**; and the **Asian elephant** in east and northeast India—involving the **creation of corridors** for unhindered movement of elephants

along traditional routes, which in turn helps preserve many other animals who also use these routes.

In addition, Wright has been instrumental in the government's declaration of Balphakram, Neora Valley, and Buxa as **Protected Areas**, and in the **captive breeding** of endangered species such as the **brow-antlered deer** and the **white winged wood duck**. Also, she once organized an **International Crane Workshop** at Bharatpur, Rajasthan, attended by representatives from 26 countries.

She continues to monitor the above-mentioned projects, and is involved in new conservation efforts centering on the **Himalayan salamander** (*Tylototriton verrucosus*) and the **greater adjutant stork** (*Leptoptilus dubius*), Assam, and their habitats. In her ongoing work to **stop illegal trade in animal products**, Wright, with the help of government and NGOs, promotes sustainable ecodevelopment. She has introduced a **mobile health clinic** in the fringe villages of the Palamau Tiger Reserve and helped create a **buffer zone** of friendly villages around the Manas Biosphere Reserve. Another project involves **conservation of the threatened wetlands** of Manipur.

In addition to her membership on the Indian government's **Wildlife Advisory Board** and roles as trustee and Eastern Region representative of WWF-India, Wright belongs to the Species Survival Commission (IUCN/SSC) Asian Elephant and Cat groups.

■ Resources

Resources needed include advice on institutions or individuals whose expertise might help ongoing projects, and names of establishments "willing to give free training to young Indian conservationists of our region specializing in habitat preservation, ecodevelopment and species preservation." She also calls for assistance in "educating politicians worldwide, including such institutions as the World Bank, as to the actual needs of rural population in fringe areas and forest land," and requests funds for "conservation projects, especially those being carried out by small village organizations."

Wright has written several articles for *Oryx*, published by the Fauna Preservation Society; for the *WWF Newsletter*; and for other magazines and journals.

▲ Private

E.R.C. Davidar
Advocate
David Nagar, Padappai
Madras 601301, Tamil Nadu, India

Phone: (91) 44 33 (Padappai exchange)

Activities: Education • *Issues:* Species Preservation • *Meetings:* SSC

▲ Projects and People

E.R.C. Davidar is an advocate who describes his interests as natural history and conservation, wildlife photography, authoring children's books, and writing natural history papers and popular articles. Ongoing projects include work to establish **movement corridors for the Asiatic wild dog and the striped hyena**. Earlier, Davidar studied **elephant migration** in the Nilgiris and Anamalais, and has carried out other studies of Nilgiri Tahr.

He serves as honorary consultant to the World Conservation Union's Species Survival Commission (IUCN/SSC) and as a member of its Asian Elephant and Hyena groups. He is also trustee of the Madras Crocodile Bank, and a life member of the Nilgiri Wildlife and Environment Association, World Wide Fund for Nature (WWF)—India, and the Bombay Natural History Society, as well as a member of the Tamil Nadu State Wildlife Advisory Board.

■ Resources
Davidar has published two recent works on the wildlife sanctuaries of Mukerti and Anamalai, supported by the government of Tamil Nadu; he has also written children's books, including *Tales from the Jungle*, *The Runaway Elephant Calf*, and *Adventures of a Wildlife Warden*.

Balachander Ganesan
Independent Researcher/Consultant
2 Poes Garden
Madras 600086, India

Phone: (91) 44 453250

Activities: Education; Research • *Issues:* Biodiversity/Species Preservation; Deforestation; Sustainable Development

▲ Projects and People
With a background in corporate finance, including an MBA from Carnegie-Mellon University, USA. **Balachander Ganesan** is pursuing a career in ecology as he completes a Rutgers University, USA, doctoral program, while doing fieldwork in the Nilgiris Biosphere Reserve in South India. His experience as a vice president at both Citibank and Marine Midland Bank allows him to take "a multipronged approach" to environmental economics, writes Ganesan.

Until mid-1992, he was stationed at **Mudumalai Wildlife Sanctuary** on a project funded by Wildlife Conservation International, World Wildlife Fund—USA, and a food and health foundation. Here he investigated "the role of **fire and grazing** in modifying **vegetation composition** at three sites," Ganesan reports. "There is intense cattle grazing in parts of the Mudumalai Wildlife Sanctuary (within the biosphere reserve) and outside" that tramples the soil, affecting the ability of seedlings to prosper. Moreover, "periodic man-made fires cause considerable seedling mortality." An additional problem is the overcutting of forest for fuelwood, which "exacerbates soil erosion and nutrient leaching, besides opening up the canopy cover."

Besides evaluating government forestry policy and surveying the use of forest resources, Mr. Ganesan is also studying the **economics of cattle grazing**, including permitting free ranging, export of nutrients in the form of organic manure, milk, meat, and other issues.

Ganesan advocates "projects that cut across artificial political boundaries," and emphasizes the need for pragmatism. In particular, he suggests that governments' conservation plans "incorporate economic development of the local people of critical habitats so that they will have a stake in preserving their habitat as a heritage for all of mankind."

In anticipation of setting up an NGO, Ganesan attended a wildlife management and conservation training course, with Smithsonian Institution support, and visited various research centers in Africa.

Mohammed Abdul Rashid
Consultant
Ketan Apartments, Flat 103
Fatehganj Camp
Baroda 390002, Gujarat, India

Phone: (91) 265 21370

Activities: Education; Research • *Issues:* Biodiversity/Species Preservation; Deforestation; Global Warming; Human Ecology; Sustainable Development • *Meetings:* CITES; UNCED

▲ Projects and People
Before retiring in 1982, **Mohammed Abdul Rashid** was forestry professor and chief conservator of forests, wildlife and chief wildlife warden, Gujarat State, in the **Indian Forest Service**. Vice President of the Wildlife Preservation Society of India, Dehra Dun, Abdul Rashid belongs to various other organizations, including the Species Survival Commission (IUCN/SSC) Cat Group; Commonwealth Human Ecology Council (CHEC), UK; and the Bombay Natural History Society (BNHS). Rashid is now involved in "short and long term projects pertaining to . . . wildlife conservation, social forestry, ecodevelopment, human ecology, and nature education."

His main current activity involves writing an illustrated reference book on the **Asiatic lion** (*Panthera leo persica*), assumed to be the first book on this subject and a UNESCO Man and the Biosphere (MAB) project financed by the Indian Department of the Environment. The Asiatic lion is a rare species now threatened with extinction; Rashid has a long history with the conservation of this species, working for over a decade as manager of the **Gir Lion Sanctuary** in Gujarat.

Another current project involves examining the causes behind the decline of **raptors in Gujarat State** for the Bombay Natural History Society. Completed projects include a study of **urban forestry** around Baroda for the World Wide Fund for Nature (WWF)—India's Central Gujarat Committee; and a technical evaluation of Gujarat's **Social Forestry** Program (under the auspices of the United Nations Food and Agriculture Organization (FAO).

A firm advocate of **sustainable development**, Abdul Rashid writes: "The need for living in harmony with nature, and not in confrontation with it as we have done hitherto, has to be brought home forcefully to all sections of our society, ranging from the up and coming generation . . . to our policy and decisionmakers, politicians, technocrats, and industrialists." Abdul Rashid envisions all environmental organizations pulling together to create a global body that can speedily and effectively implement a "World Conservation Strategy."

■ Resources
Abdul Rashid has a publications list; he has an extensive personal library, color slides and photographs, a projector, and a camera, and can undertake consultancy assignments abroad.

🎓 *Universities*

Turtle Research and Conservation Centre ⊕
Department of Zoology
K.S. Saket P.G. College
Avadh University
3/16 Murai Tola, Ayodhya
Faizabad 224123, India

Contact: M.R. Yadav, Ph.D., Research Director

Activities: Development; Education; Law Enforcement; Research • *Issues:* Air Quality/Emission Control; Biodiversity/Species Preservation; Deforestation; Energy; Global Warming; Health; Population Planning; Sustainable Development; Waste Management/Recycling; Water Quality • *Meetings:* International Symposium on Tortoises, Terrapins and Turtles

▲ Projects and People

Dr. M.R. Yadav has taught in the Department of Zoology at K.S. Saket P.G. College, Avadh University, since 1970, specializing in turtles. Among his research projects are a recent study of feeding strategies and behavior of soft-shelled turtles; examination of the ecology and nesting habit of freshwater turtles inhabiting the Ghaghra River system; and, currently, investigation of the life history management and conservation of Indian freshwater turtles. He participated in a training program on Snakes and Human Welfare, 1990, and in a previous All-India Symposium on Aquatic Organisms.

Dr. Yadav is concerned by the sharp decline in Indian wildlife due to "overexploitation and habitat destruction." Therefore, his work, particularly in northern India, involves "safeguarding the environment and promoting tourism." He stresses the need for developed countries and existing strong environmental organizations to help developing countries' efforts at safeguarding the environment as well as fostering ongoing collaborative efforts on global research. "I offer my services for any such program and for subject-oriented training courses in the future," he writes.

■ Resources

Dr. Yadav is in need of funding, and would like the opportunity to attend an intensive course in conservation training. He has written numerous articles, such as on the Indian tropical pond turtle's reproductive potential.

Centre for Ecological Sciences
Indian Institute of Science
Bangalore 560012, India

Contact: Madhav Gadgil, Ph.D., Astra Professor of Biology, Chairman
Phone: (91) 812 340985 • *Fax:* (91) 812 341683

Activities: Development; Education; Research • *Issues:* Biodiversity/Species Preservation; Deforestation

● Organization

The Indian Institute of Science in Bangalore has been described as "the premier institution of scientific research and graduate level training in India." Its Centre for Ecological Sciences, founded in 1972 by Dr. Madhav Gadgil, has a staff of 40.

▲ Projects and People

According to the Indian Ministry of Environment and Forests, Dr. Gadgil's dissertation for Harvard in 1969, on the "mathematical theory of evolution of life history strategies, [is] a piece of work that has since been recognized as a citation classic."

In India, his research interests range from ecology, evolutionary biology, and social behavior to the evolution of human cultural behavior, in particular, human use of natural resources. Besides his basic scientific research in these areas, Dr. Gadgil has pursued "applied research, policy research, and field action in the areas of sustainable development and conservation of biodiversity."

Among his many contributions to Indian environmental research, Dr. Gadgil has pursued a "theoretical exploration of evolutionary forces moulding the evolution of repeated versus big-bang patterns of reproduction." This included field studies of the reproductive behavior of bamboo, many species of which show evidence of big-bang reproduction, he indicates.

More applied research includes an "analysis of the distribution of elephant populations of south India and the context in which elephant–man conflicts arise" and an investigation of the differences in the foraging and crop-raiding behavior of male and female Asiatic elephants. His studies of queen-worker caste differentiation among paper wasps (*Ropalidia*) have prompted further investigations on "insect sociality." He has also documented traditional systems of nature conservation, focusing on "sacred groves and their relevance in the modern context," and has pioneered investigations into the human use of natural resources.

In the realm of public policy, Dr. Gadgil initiated "the first ever public assessment of a river valley project in India, the Bedthi hydroelectric project," and was in charge of field studies which led to the abandonment of the Silent Valley hydroelectric project, because of its "unacceptably high costs in terms of loss of biodiversity." He also pioneered the biosphere reserve concept in India, preparing the project for such a reserve in the Nilgiris, according to his literature.

To create public awareness of environmental issues, Dr. Gadgil has worked in the Western Ghats, organizing a "Save the Western Ghats march," and with the farmers and villagers of the Uttara Kannada District of Karnataka Western Ghats to involve local people in ecodevelopment efforts. He has also conducted a winter school in Conservation Biology, with the participation of scientists and managers from nine Asian countries.

Dr. Gadgil has lent his expertise to various councils and commissions, including the Scientific Advisory Council to the Prime Minister; Scientific Committee on Problems of Environment of the Indian National Science Academy; World Conservation Union (IUCN); UNESCO's Scientific Advisory Committee on the Biosphere Reserves; and the Biodiversity Program Advisory Board of IUBS/SCOPE/UNESCO. Since 1985, he has chaired SCOPE (Scientific Committee on Problems of the Environment).

Other positions held include: Fellow of the Indian National Science Academy; Foreign Associate of the U.S. National Academy of Sciences; member of Southern Indian Regional Committee of World Wildlife Fund (WWF); Project Tiger National Steering

Committee; IUCN Ecology Commission Steering Committee; Indian Committee for the Man and the Biosphere Program (MAB-UNESCO); Bombay Natural History Society (BNHS) advisory committee, among others.

■ Resources

Dr. Gadgil "has written and broadcast extensively in the two Indian languages, Marathi and Kannada, as well as in English. Many leading newspapers and magazines in these three languages have regularly carried his contributions." He has also held editorial positions on at least eight publications, including *Surveys in Evolutionary Biology, Conservation Biology,* and *Biodiversity and Conservation.*

In addition, Dr. Gadgil has authored or co-authored more than 162 publications in fields of theoretical population biology, plant population biology, animal population biology, human communities, productivity, zoogeography, human geography, sociobiology theory, animal sociobiology, history of resource use, water resources, hunting-gathering, animal husbandry, forestry, conservation of diversity, perceptions of resource use, and ecodevelopment.

School of Zoology
Jiwaji University
Gwalior 474011, Madhya Pradesh, India

Contact: **Rayavarapu Jagannadha Rao, Ph.D.,** Lecturer

Activities: Education; Research • *Issues:* Biodiversity/Species Preservation; Deforestation; Water Quality • *Meetings:* IUCN/SSC CSG; World Congress of Herpetology

● Organization

According to **Dr. R.J. Rao,** "The Jiwaji University is the pioneer organization . . . promoting environmental awareness and education." The University is the center for studies on pollution, toxicology, wildlife conservation, and environmental biology, providing a basis for further action in the region and the country as a whole.

Dr. Rao also notes that "India is one country which is deeply involved in environmental protection. The government as well as NGO organizations and even general public are highly concerned."

▲ Projects and People

A zoologist, Dr. Rao has studied primarily the reproductive physiology and ecology of turtles, crocodiles and aquatic mammals, and management of wildlife sanctuaries. Before becoming a lecturer at Jiwaji University in 1989, Dr. Rao worked at the Wildlife Institute of India (WII), Dehradun. Since 1983, he has pursued field-oriented research activities at the National Chambal Sanctuary's Deori Gharial Rearing Centre, including a turtle ecology project, which continued the work of **Dr. L.A.K. Singh,** currently a research officer in the Simlipal Tiger Reserve.

In his efforts to conserve and manage aquatic animals, in particular the indicator species, Dr. Rao is currently monitoring crocodile populations and studying turtle bioecology in the Chambal River. With the support of WII, his research in the National Chambal Sanctuary covers habitat utilization by crocodiles and turtles in different areas. He has also studied the management of the Great

Indian Bustard Sanctuary, and proposed studies to monitor water quality in the Ganga River, and to investigate the status of conservation management of lizards.

Dr. Rao is a member of the Indian Herpetological Society, Bombay Natural History Society, International Society of Tropical Ecology, World Wide Fund for Nature (WWF)—India, and the Species Survival Commission (IUCN/SSC) Turtle and Tortoise, Crocodile, and Indian Subcontinent Herpetological groups.

■ Resources

Needs include radio-tracking and remote-sensing technology for wildlife management, and field training to develop better understanding of sustainable use of wildlife.

Dr. Rao has prepared some 60 publications since 1969, mainly on turtles and crocodiles. Among the most recent are *Conservation Status of Gharial in the National Chambal Sanctuary: A 10 Year Review, Management of Wildlife Resources in the Chambal River,* and *Status and Conservation of Sea Dolphins along East Coast of Andhra Pradesh, India.*

College of Veterinary and Animal Sciences
Kerala Agricultural University
Mannuthy
Trichur 680651, India

Contact: **Jacob V. Cheeran, D.V.M.,** Professor
Phone: (91) 487 22344

Activities: Education; Research • *Issues:* Biodiversity/Species Preservation

▲ Projects and People

Research experiences of veterinarian **Prof. Dr. Jacob V. Cheeran,** Kerala Agricultural University, range from the "immobilization of Indian elephants with nicotine" and "acupuncture in animals" to the treatment and surgery of domestic and wild species, including bovines, broiler chickens, dogs, lions, and bonnet monkeys. Major undertakings are the "restraint of animals by pharmacological means using projectile syringes" and, with the government of India, "toxic effects of industrial effluents on animals."

Particularly interested in "elephants in the wild and captivity, chemical immobilization of wild animals, and environmental toxicology," Dr. Cheeran has written book chapters (Malayalam language) on animal pneumonia and antibiotics as well as numerous papers in English. He belongs to the Species Survival Commission (IUCN/SSC) Veterinary and Asian Elephant groups, Wildlife Preservation Society of India, Society Against Cruelty to Elephants, and the Indian Society of Zoo Veterinarians, among other groups. He also trains personnel in his fields.

■ Resources

Dr. Cheeran seeks "new drugs for chemical immobilization and other consumables."

Department of Animal Behavior and Physiology
School of Biological Sciences
Madurai Kamaraj University
Madurai 625021, Tamil Nadu, India

Contact: **Maroli K. Chandrashekaran, Ph.D.,**
Department Head
Phone: (91) 452 85216 • *Fax:* (91) 452 85205
E-Mail: grams:university

Activities: Education; Research • *Issues:* Biodiversity/Species
Preservation

▲ Projects and People

Dr. Maroli K. Chandrashekaran has been associated with **Madurai Kamaraj University** since 1975, and a professor there since 1980. His main focus is chronobiology, especially of **bats**. Dr. Chandrashekaran studies circadian rhythms—the tendency to 24-hour cycles—and other ways in which these animals' lives depend on very careful schedules.

Serving as a spokesman for bats in India, he is working against prevalent prejudices and misconceptions. As one publication for an Indian audience puts it: "Bats are man's friends. Their study has contributed to the development of **navigational aids for the blind, birth control and artificial insemination techniques, vaccine production,** and **drug testing.**

"The bat populations are decreasing at an alarming rate due to slaughter for restaurants, use of pesticides, human interference, human consumption for imagined medicinal and aphrodisiac properties. There is a great need for bats to be studied in greater details and for the public to be informed of the myths and realities about bats." *(See also Bat Conservation International in the USA NGO section.)*

Dr. Chandrashekaran has also studied extensively at University of Tübingen, Germany, and the University of California at Berkeley, USA. He is a fellow of the Indian Academy of Science; president of the Indian Society of Chronobiology; and member of the Guha Research Conference, Species Survival Commission (IUCN/SSC) Chiroptera Group; Scientific Advisory Board of Bat Conservation International, USA; Bombay Natural History Society's research advisory panel; Nilgiri Biosphere Reserve's research committee.

■ Resources

Dr. Chandrashekaran is the author of *Biological Rhythms*; he is the co-author with G. Fleissner and G. Neuweiler of *Animal Behavior: Basic Experiments in Neurophysiology.* He is on the editorial boards of 6 journals, and has written more than 88 publications, primarily on chronobiology, especially of bats. Among the most recent, with P. Kumarasamy, are *The Role of Natural Dawn and Dusk Twilight in Entraining the Circadian Activity of a Nocturnal Rodent Mus Booduga* and *A Controversy in Chronobiology.*

Rabindra Bharati University
c/o 45 Suhasini Ganguli Sarani
Calcutta 700025, India

Contact: **Dhriti Kanta Lahiri-Choudhury, Ph.D.,**
Professor
Phone: (91) 33 481144 • *Fax:* (91) 33 282070

Activities: Education; Research • *Issues:* Biodiversity/Species
Preservation; Deforestation

▲ Projects and People

Dr. Dhriti Kanta Lahiri-Choudhury, professor and head of the English Department, **Rabindra Bharati University**, believes environmentalists worldwide need to preserve "tropical and subtropical forests of South and Southeast Asia by launching Project Elephant and other such projects, using elephant as the flagship species."

A member of the Species Survival Commission (IUCN/SSC) Asian Elephant Group and principal investigator of the World Conservation Union's (IUCN)/World Wildlife Fund (WWF) project on "the status and distribution of elephant in northeast India, Dr. Lahiri-Choudhury belongs to numerous such national and global groups, including the Indian government's Project Elephant Task Force. His presentations include "Status and Distribution of Wild Elephants in India," "The Indian Elephant in a Changing World," and "Translocation of Wild Elephants."

■ Resources

Dr. Lahiri-Choudhury has reprints of about 31 articles that can be ordered, including "Preventing Confrontation between Man and Elephant," "Saving Elephants for Posterity," "India: Changing Times in a Wildlife Sanctuary," and "The Endangered Wild Buffalo." He seeks political support and strengthened educational outreach to "preserve and manage tropical forest ecosystems."

Department of Anthropology ⊕
University of Delhi
Delhi 110007, India

Contact: **Praveen Kumar Seth, Ph.D.,** Reader in
Anthropology
Phone: (91) 11 251 5329

Activities: Education; Research • *Issues:* Biodiversity/Species
Preservation; Ecobiology; Health; Population Planning
Meetings: International Primatological Society

▲ Projects and People

With research and teaching interests in primate biology for nearly three decades, Dr. Praveen Kumar Seth, anthropology reader (equivalent of professor), University of Delhi, has edited five books in his field, including *Primates—The New Revolution*, and published some 74 articles. Asia Member of the Species Survival Commission (IUCN/SSC) Primate Group, Dr. Seth is also a member of the *International Journal of Primatology*'s editorial board, Indian Anthropological Society, and India's Society of Cytologists and Geneticists as well as Life Member of the Indian Association of

Experimental Psychologists and Indian Society of Human Genetics—among other groups.

■ Resources
Dr. Seth seeks audiovisual aids for research and offers consultations and technical assistance.

Department of Geography
University of Delhi
Delhi 110007, India

Contact: **R.B. Singh, Ph.D.**, Research Scientist
  Phone: (91) 11 252 1521, ext 215
    Fax: (91) 11 672427

Activities: Development; Research • *Issues:* Air Quality/Emission Control; Deforestation; Energy; Sustainable Development • *Meetings:* CITES; IGU; INTECOL; UNEP; UNESCO/MAB

▲ Projects and People
An assistant professor in the Department of Geography, University of Delhi, Dr. R.B. Singh leads seminars on the monitoring of geosystems and attends scientific conferences throughout the world. According to Dr. Singh, "The main task of geographical monitoring is to observe the state of land use and environmental change in natural and anthropogenically changed geosystems at regional and local level for improvement, development, and amelioration of the environment."

A member of the International Geographical Union (IGU) Commission on Mountain Geoecology and Resource Management, Dr. Singh writes, "The major geosystems in India differ greatly in natural environment, population, local customs, traditions, and level of development." He notes that "human impact takes the form of over-cultivation, overgrazing, over-clearing of woodlands, and overuse of mountain resources."

Dr. Singh utilizes LANDSAT imagery and digital image processing techniques to **study environmental degradation in The Himalayas**, which he considers to be "ecologically very fragile," in order to promote "an integrated and rational strategy for sustainable regional development."

He **locates soil erosion hazards** by comparing rainfall erosivity, slope, soil erodibility, and vegetation cover with elevation, water availability and nutrient availability. This data allows him to classify land into four levels of vulnerability. The level of vulnerability suggests possible land uses or alerts scientists to the need for conservation and preservation measures.

■ Resources
Dr. Singh is lacking facilities to do geographic information system (GIS) monitoring and modeling work. His publications include: *Geography of Rural Development, Environmental Geography, Changing Frontiers of Indian Village Ecology, Dynamics of Mountain Geosystems,* and *Environmental Monitoring.*

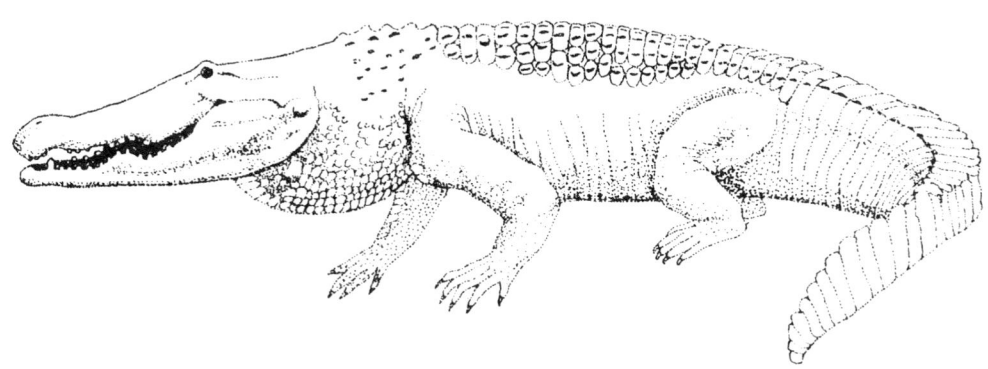

Indonesia is the world's largest archipelago, with more than 13,700 islands. It is located between mainland southeastern Asia and Australia. More than half the country is covered by tropical rainforest with diverse and highly endemic flora and fauna. Oil, gas, and mineral resources are abundant, and only partially explored.

The country has made great strides in alleviating poverty. However, as economic and population growth have occurred, many coastal waters have lost their rejuvenation capacities; reef mining has damaged coral regeneration, for example. In addition, wood processing and paper industries have caused selective logging of much virgin forest territory and, along with the pressures of subsistence agriculture, have led in some cases to irreversible deforestation.

🏛 Government

Badan Koordinasi Survey dan Pemetaan Nasional (BAKOSURTANAL)

Jl. Raya Jakarta–Bogor Km. 46
Cibinong, Bogor, Indonesia

Contact: **Dr. Rubini Atmawidjaja**, Deputy Chairman
Phone: (62) 21 80732/3 • *Fax:* (62) 219 83067

Activities: Development; Education; Law Enforcement; Research • *Issues:* Biodiversity/Species Preservation; Deforestation; Energy; Global Warming; Population Planning; Sustainable Development; Water Quality

▲ Projects and People

Since 1990, **Dr. Rubini Atmawidjaja** has been president of Pakuan University in Bogor and deputy chairman of the **Badan Koordinasi Survey dan Pemetaan Nasional (BAKOSURTANAL)**, the national coordinating agency for survey and mapping. With a staff of 600, BAKOSURTANAL is committed to providing information on the current conditions of the **Indonesian natural resources** via maps, aerial photographs, satellite images, and ground data, using computer processing.

Dr. Atmawidjaja, who received his Ph.D. degree in forest engineering from the State University of New York (SUNY), was dean of Bogor Agricultural University's School of Forestry before becoming president of the University of Cenderawasih, cultural attaché of Indonesia in West Germany in the early 1980s, and director general of Forest Protection and Natural Conservation for six years. He is presently a member of the National Research Council on Natural Resources and Energy, and belongs to the Species Survival Commission (IUCN/SSC) Asian Elephant and Rhino Group.

In establishing **elephant training schools** in Wai Kambas (Lampung), Aceh, and Riau, Dr. Atmawidjaja hopes to resolve the "elephant raid problems." His other conservation efforts include encouraging local citizens to foster breeding programs for crocodiles, sea turtles, orangutan, and Sumatran rhinos, for example; developing conservation specified areas in addition to existing national parks; advancing "sustainable utilization of . . . land, forest, water, and minerals by performing balance accounting system"; and developing conservation-minded NGOs.

His publications, written mostly in Indonesian, concern remote sensing, sampling techniques, national park management, tropical rainforests, and ecosystems.

**Pursat Studi Reboisasi dan Rehabilitasi Hutan
    Tropis Basah (PUSREHUT)**
Tropical Rain Forest Research Center
Kampus Gunung Kelua,
Samarinda, Kalimantan-Timur
East Kalimantan 75113, Indonesia

Contact: **Shigeki Yasuma, Ph.D.**, Animal Ecologist
  Phone: (62) 541 21421, ext. 27
    Fax: (62) 541 21840

Activities: Development; Education; Research • *Issues:* Biodiversity/Species Preservation; Deforestation

● **Organization**

"As one of the largest in the world, the tropical rainforest of Indonesia can be considered as the lungs of the world. It is therefore our duty to conserve this precious property and to make it last forever." (Dr. Jajah Koswara)

As tropical rainforest resources began decreasing drastically, the government of Indonesia realized the need to develop rehabilitation techniques to preserve them. It asked the government of Japan for assistance. In return, the **Japan International Cooperative Agency (JICA)** initiated the **Tropical Rain Forest Research Phase 1**, to run a five-year program that started January 1985. Five areas of research included: agroforestry, forest land use classification and planning, forest site classification, manmade forest management, and natural forest management.

Upon Phase 1 completion, JICA renewed its program in Indonesia with a second five-year phase running through December 31, 1994. Four research areas were selected this time: evaluation of forest site environment, inter-aerial studies, inventory of forest ecosystem environment, and rehabilitation techniques of forest ecosystem.

Through a 1979 Japanese grant, the **Tropical Rain Forest Research Center (PUSREHUT)** was built at Mulawarman University, serving as its basic scientific mission. Two other universities, Bogor Agricultural University and Gadjah Mada University, work with the Indonesian Ministry of Education and Culture in cooperation with the Japanese Ministry of Agriculture, Forestry and Fisheries and the Ministry of Education.

▲ **Projects and People**

Five long-term experts and a liaison officer from Japan help support those objectives adopted annually at the joint committee meetings. **Dr. Shigeki Yasuma** serves as one of the Japanese experts. Dr. Koswara is director of Research and Community Service Development, Directorate General of Higher Education.

Six PUSREHUT research laboratories focus on forest ecology, forest inventory and remoteness, protection, silviculture, socioeconomics, and soil. The Center also has a greenhouse, nursery, research stations at Bukit Soeharto and Lempake, and a 5,000-hectare research forest area in Bukit Soeharto.

East Kalimantan, Indonesia's second largest province, serves as the country's tropical rainforest barometer, says PUSREHUT. Its natural forest resources are more than 80 percent of the total land. As logging activities increase here, so do the number of people living in the surrounding area—creating a "people-forest interaction" prob-

lem. The need for replanting techniques after logging and shifting cultivation following forest fires is evident.

There are currently four research projects in East Kalimantan. The **Strek Project** evaluates Indonesia's selective cutting system in the "logged over rainforest"; tests and develops simple silvicultural methods of thinning and logging; observes seed/seedling adaptation; and raises productivity. In the burned forest area, the **Forest Fire Project** inventories and suggests rehabilitation techniques. A third project, the **Stimulation Programme for Research in Humid Tropical**, Sambodja, seeks to develop operational knowledge on multipurpose forest management for conservation and utilization of the dipterocarp forest. The **German Forestry Project** in Samarinda focues on education, training, and research contributions.

In 1990, PUSREHUT published the 88-page book *Mammals of Bukit Soeharto Protection Forest* co-authored by Dr. Yasuma and **Dr. Hadi S. Alikodra** of Bogor Agricultural University. Classifying about 70 percent of the mammals of Kalimantan states, their research details animal behavior and points out **how wildlife helps forest regeneration**. A few species described are sun bear, pygmy squirrel, flying lemur, fruit bat, bearded pig, Bornean yellow muntjac, clouded leopard, and banded palm civet. Those in the Bukit Soeharto Forest region know that "many animals eat seeds and fruits, and their high reproductive rate is predicted. Some of the fruits are . . . very sweet. . . . People who live in the forest are also happy."

■ **Resources**

PUSREHUT has special publications and reports available at cost.

 *NGO*

**Environmental Management Development in
    Indonesia (EMDI)**
BDN, 11th Floor, Jl. Kebon Sirih #83
Jakarta, Indonesia

Contact: **Gerry Glazier**, Project Leader Jakarta
  Phone: (62) 21 380 2182 • *Fax:* (62) 21 380 2183

Activities: Education; Development; Research • *Issues:* Global Warming; Health; Water Quality

● **Organization**

Environmental Management Development in Indonesia (EMDI) seeks to stimulate human resource development through a joint program sponsored by its country's Ministry of State for Population and Environment (KLH) and Canada's School for Resource and Environmental Studies at Dalhousie University. *(See also Dalhousie University, EMDI in the Canada Universities section.)* The government of Indonesia's goal is to have environmental management aspects (cultural, ecological, and social) firmly entrenched in all its country's development decisionmaking processes.

▲ **Projects and People**

With financial support from the Canadian International Development Agency (CIDA), EMDI is now in its third operational phase, 1989–1994. The two previous phases each lasted three years, 1983–1986 and 1986–1989.

Causing increasing concern in many areas of Indonesia is **environmental pollution**. Helping alleviate or prevent some of the related problems is an aim of EMDI. Working with KLH, it provides environmental impact assessments (EIAs), environmental standards and guidelines for industry, and a systematic hazardous/toxic waste management plan resulting in a computerized National Register of Potentially Toxic Chemicals information system.

During its current phase, EMDI funds also are specified for the publication of *The Ecology of the Moluccas and Lesser Sundas* and *The Ecology of the Indonesian Seas*, which offers baseline information on coastal and marine ecosystems, resources, and human activity and development. A geographic information system (GIS) devoted to spatial planning, an environmental statistics program within the Central Statistics Bureau, and the Coordinating Board for Surveys and Mapping are other elements of the EMDI program.

Realizing the necessity of projecting strong environmental law to protect its nation of more than 13,000 islands, Indonesian lawyers learn **legal strategies for pollution management** through training visits to Canada. And, EMDI gives 21 graduate fellowships in environmental management and environmental law.

EMDI also supports the umbrella environmental organization of NGOs in Indonesia, aids in the **establishment of consulting groups** through the National Association of Indonesian Consultants, offers **internships** to strengthen Canada–Indonesia ties in the private sector and cooperates in WIDE, Women in Development and Environment.

Through EMDI programs, Dalhousie University expects its Canadian citizens to gain a better understanding of Indonesia's developmental needs; in turn, the Indonesians will learn from Canadian applications regarding environmental management experience.

Federation of Indonesian Speleological Activities (FINSPAC)
Jln.Ir H Juanda 30/P.O. Box 55
Bogor, West Java, Indonesia

Contact: **Robby K.T. Ko, Ph.D.**, President
Phone: (62) 251 4376 • *Fax:* (62) 251 5343
E-Mail: dr ko cisarua bogor indonesia

Activities: Development; Education; Research • *Issues:* Biodiversity/Species Preservation; Deforestation; Health; Sustainable Development; Tourism; Water Quality • *Meetings:* karst hydrology and conservation; International Congress of Speleology; International Union of Speleology

● Organization
The Federation of Indonesian Speleological Activities (FINSPAC) is the only officially recognized professional organization dealing with caves and the cave environment in Indonesia. With its staff of 15, FINSPAC sponsors expeditions, courses, slide presentations, and cave conservation practices. As an affiliate of the Union Internationale de Speleologie, it is associated with UNESCO and in close contact with similar groups in France, Netherlands, UK, and USA.

▲ Projects and People
In rural Indonesia, cave-supplied spring wells and rivers are an important water supply source. Scientists study caves and their animals, hydrology, minerals, microenvironment, and microecology. Bat droppings, called *guano*, are "highly prized as fertilizer." *(See also Bat Conservation International in the USA NGO section.)* Because a moldlike bacteria, called Actinomycetes, produces valuable antibiotics, "caves are being searched for new miracle drugs," reports FINSPAC. Caves also attract "speleotourism" and give evidence of past human and animal activities.

In reminding explorers that the underground is fragile, Dr. Robby K.T. Ko notes that wild caves are most sensitive. Visitors are cautioned about the presence of bats, which require a dark and quiet environment, and are the only nighttime predator of insects. Threats to caves include water pollution from raw sewage, industrial waste, pesticides from farming, and landfill contamination; vandalism; dam construction, whereby flooding "kills rare cave animals, halts the ongoing geological processes, and makes caves inaccessible"; limestone quarrying that cause karst spring wells to dry up; logging, whereby sawdust and other byproducts enter caves in surface streams that then become clogged and rob the cave of water—its "lifeblood"; and deforestation.

With tropical cave biology "in its infancy," new communities of animals are being discovered in Papua New Guinea, Sarawak, and in lava in Hawaii, reports Dr. Ko, who urges "detailed, systematic studies of whole cave faunas and their relationships to the cave environment."

Moreover, carbonic rock aquifers in karst areas can be "very efficient [clean] water-bearing geological formations, but need protection from pollution," reports FINSPAC.

Dr. Ko and FINSPAC are asking the Indonesian government to promote the ecological management of caves and to carefully examine proposals to "exploit karst areas for commercial reasons" among the competing interests in agriculture, limestone quarrying, chalk production, and cement factories. Working within the system, FINSPAC invites government officials to sponsor speleological courses, lecture, and participate in field trips—encouraging sound relations.

Among FINSPAC's other recommendations are the compilation and computerization of data on caves and karst; recognition of caves as underground laboratories for interdisciplinary study; and the education of the Indonesian public on the biological, cultural, ecological, and physical values of caves.

Dr. Ko belongs to the Species Survival Commission (IUCN/SSC) Chiroptera Group and represents Indonesia in the International Union of Speleology.

■ Resources
FINSPAC lacks adequate funds and personnel for its various projects and needs assistance from other NGOs to halt destruction of Indonesia's karst, caves, bats, and swiftlets, as well as to identify rare cave animals and *guano* microecosystems, karst vegetation, karst water sources, and ages of sediments. It has available *Warta Speleo* quarterly magazine, reports, a speakers' program, and reprints from *Biotrop*.

🎓 *Universities*

Marine Fisheries Resources Management ⊕
Bogor Agricultural University
Jalan Rasamala, Bogor-Dramaga Campus
Bogor, West Java, Indonesia

Contact: **I Nyoman S. Nuitja, Ph.D.**, Professor,
   Faculty of Fisheries
Phone: (62) 251 312635 • *Fax:* (62) 251 322167

Activities: Development; Education; Law Enforcement; Political/Legislative; Research • *Issues:* Air Quality/Emission Control; Biodiversity/Species Preservation; Deforestation; Global Warming; Health; Population Planning; Sustainable Development; Water Quality

▲ Projects and People

Dr. I Nyoman S. Nuitja, while head of Bogor Agricultural University's Marine Fisheries Resources Management, is also president of the Indonesian Herpetological Foundation and a consultant on sea turtles for the Species Survival Commission (IUCN/SSC) Marine Turtle Group. Previously, he was a consultant to the country's Ministries of Research and of Finance on seaweed culture and marine turtles.

Since 1970, Dr. Nuitja, who is also a fisheries engineer, has written numerous national and international scientific papers and publications on the management and study of marine turtles, including fisheries management, habitat studies, ecology of the leaterhback turtle, nesting site of hawksbill turtle, bioecology of seaweeds, ranching for food and drug production from the sea, and associations with coral reefs.

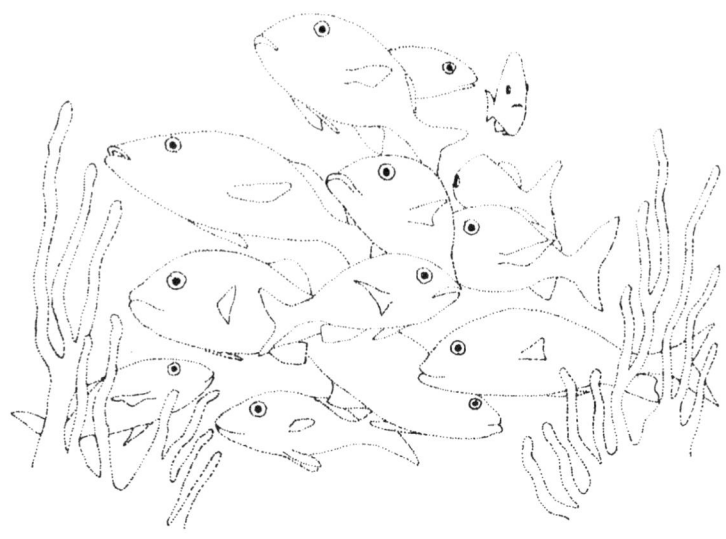

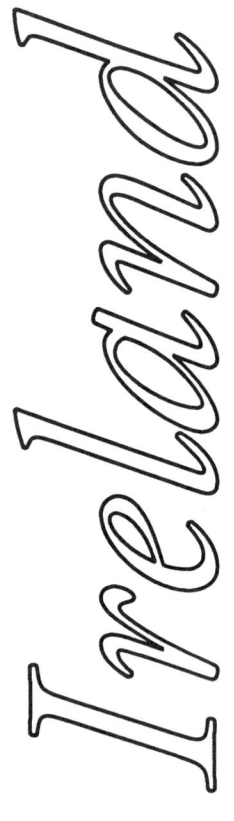

Ireland

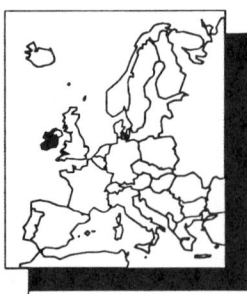

Ireland is a predominantly rural country, with most of its area used for agriculture. Yet, accelerating urbanization, farm development techniques, and energy demands may change the landscape significantly in the near future. The use of fertilizers and pesticides have caused runoff problems, endangering wildlife and plant diversity. A serious pollution problem in Dublin, resulting from smog caused by the burning of bituminous coal, led to a government ban on the coal in parts of the city.

❧ NGO

An Taisce, The National Trust for Ireland
The Tailors' Hall, Back Lane
Dublin 8, Ireland

Phone: (353) 1 541786 • *Fax:* (353) 1 533255

Activities: Heritage Conservation • *Issues:* Air Quality/Emission Control; Architectural Planning; Global Warming; Sustainable Development; Transportation; Waste Management/Recycling; Water Quality • *Meetings:* EEB; Europe Nostra; IUCN

● Organization

An Taisce, The National Trust for Ireland, founded in 1948, is a voluntary membership organization concerned with the conservation of Ireland's heritage and developmental aspects as they relate to the environment.

According to its literature, An Taisce holds unique status as the only non-governmental body allowed to receive and comment on planning proposals. It is nationally recognized and accepted internationally as the "most influential voluntary and independent environmental organization in Ireland." With a staff of 6 and approximately 6,000 members, it examines all aspects of the environment.

An Taisce's first president, **Robert Lloyd Praeger**, called for a solid body of public opinion "to safeguard our treasures, both of the past and the present, . . . for the benefit of all the people." Today, An Taisce continues its mission by practicing consultation first, to prevent development confrontation later. It also promotes legislative changes where needed and encourages use of older buildings of quality, while recognizing modern structures and landscaping with special awards. An Taisce holds and maintains numerous properties for public enjoyment and research.

■ Resources

An Taisce seeks additional financial support for core and project funding activities. It publishes *Living Heritage*, a reference source on environmental matters as well as other specific guides, books, and pamphlets. A publications list is available.

Earthwatch, Ltd.
The Irish Environmental Organization
Harbour View
Bantry, County Cork, Ireland

Contact: **Jeremy Wates**, Coordinator
Phone: (353) 27 50968 • *Fax:* (353) 27 50545

Activities: Education; Political/Legislative • *Issues:* Air Quality; Deforestation; Energy; Water Management/Recycling • *Meetings:* AGM; CEAT

● Organization

Earthwatch, Ltd., the Irish Environmental Organization and member of **Friends of the Earth International (FOEI)** and the **European Environmental Bureau (EEB)**, was established in 1986. Its mission is to meet the increasing demand for information and action on a wide range of environmental issues. Approximately 2,000 members help with Earthwatch's in-depth campaigns on major environmental problems while promoting general community awareness.

According to Earthwatch, Ireland's "recycling rates are among the lowest in the European Community." In addition, there is no renewable energy program "despite having greater wind and wave power potential than any other EC country." It is also reported that Ireland's east coast is "bounded by one of the most radioactive seas in the world," and that more than one-fourth of its rivers are polluted.

▲ Projects and People

Earthwatch's major campaigns focus on an immediate **ban on chlorofluorocarbons (CFCs)**, greater energy efficiency, alternatives to the proposed national incinerator and a shift to **clean technologies**, full **emission controls** to be fitted at Moneypoint power station, a **ban on nuclear submarines** around coastlines and a closing of a nuclear plant, **ban on imports of tropical hardwood**, and ending the discharge of **raw sewage into coastal waters**.

Earthwatch lobbied the government and local authorities in 1991 on the European Community (EC) pesticides directive and started a **Dirty Dozen** campaign calling for a **ban on a group of highly toxic agents**. The same year, it also launched a recycling campaign, partially sponsored by the EC, and set up **Minewatch Ireland**, a network of local groups concerned about mining issues. Earthwatch continues its lobbying efforts for an **Environmental Protection Agency and a Radiological Protection Institute**.

Extensive support and advice on environmental problems at local levels are provided by Earthwatch, which also presents informational data, reports, documents, and brochures to key decisionmakers and the public.

■ Resources

Earthwatch publishes three magazines annually; special reports as on acid rain, gold mining, and energy alternatives; and a *Green Living Guide*. Gift items are listed in a merchandise order form.

Kerry Recycling Co-op ❦
Chutes Lane, John Joe Sheehy Road, Boherbue
Tralee, County Kerry, Ireland

Contact: **Daniel Sheehy**, Chairman
Phone: (353) 6 626260

Activities: Education; Recycling; Research • *Issues:* Environmental Education; Waste Management/Recycling • *Meetings:* Worldwide Women Forums

● Organization

A group of nine unemployed Irish people set up **Kerry Recycling Co-op** in July 1989, to create work for themselves and others and establish a trend for the rest of the country which they considered on the "bottom rung of the ladder in relation to recycling in Europe." After overcoming many obstacles, such as convincing local government of the project's potential, Kerry Recycling showed that the unemployed could make a valuable contribution to society.

Now, with a membership of over 40, the Co-op counts among its supporters the Kerry County Council, the Tralee and Killarney Urban District Council, Tralee's Regional Technical College, and the Tralee Chamber of Commerce. The project is mainly funded by the Department of Environment in the form of grants.

▲ Projects and People

From the Co-op's drop-off warehouse (for households and industries), cardboard, paper, drinks cans, and glass are presorted and processed, then sent to respective recycling plants. A confidential shredding service recycles the paper for animal bedding. The Kerry County Council pays the Co-op a rebate for every ton of landfill it saves at the dump through the recycling process at its depots in Castleisland, Killarney, and Tralee.

Other aspects of Co-op's activities involve informational and consultant services on recycling issues to community groups, businesses, and industry. Its **environmental information office** offers leaflets, posters, tapes, and videos on numerous ecological issues and a lending library of educational materials (such as slides). Kerry Recycling representatives **visit schools and local groups**, conducting workshops and discussing relevant subjects. They operate Ireland's **Cash for Cans** program in schools and take an **eco-info caravan** to local festivals in summer months.

"We plan to get involved in the manufacture of products using recyclable materials, insulation, and aids to packaging materials as an alternative to styrofoam pellets which are both harmful to the environment during manufacture and unrecyclable thereafter," writes **Daniel Sheehy**.

In an effort to expand its scope, the Co-op added a resource warehouse where buttons, cloth, and plastic offcuts are recycled for schools and play groups. They participated with the Belfast "play resources warehouse" in a **Cooperation North exchange program**.

"The spinoff in jobs from recycling operations could be significant," believes Sheehy. "Recycling is labor-intensive and any increase in recycling levels will increase the potential for sustainable jobs."

Various local and national awards have been presented to the Kerry Co-op, which reminds its countrymen to to "recycle, reuse, repair, resell, reclaim and refill."

■ Resources

The Co-op needs recyclable materials and educational tools to carry out its community efforts. In return, it has available videos and recycling guides on its program.

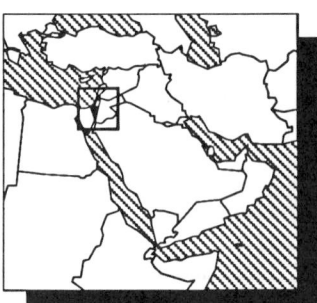

Israel is characterized by a very wide range of physical conditions within a relatively small area, with desert, tropical, and alpine environments in close proximity to one another. The country's enormous population growth during the first two and a half decades of its existence led to the development of roads, railways, airports, and other infrastructure on a scale and at a rate that has had far-reaching environmental implications. More recently, the immigration of Soviet and Ethiopian Jews has brought another wave of population growth. The air quality has deteriorated; the coastal strip is subject to many stresses that affect the 75 percent of the population that is concentrated in the coastal plain; solid waste disposal has been exacerbated by consumer habits and industry growth; and water quality has declined through improper sewage disposal. A Ministry of the Environment was created in 1988 to focus heightened attention on these problems.

Government

Ministry of the Environment
P.O. Box 6234
Jerusalem 91061, Israel

Contact: Uri Marinov, D.V.M., Director General
Phone: (972) 2 701 606 • *Fax:* (972) 2 513 945

Activities: Development; Education; Law Enforcement; Political/Legislative; Research
Issues: Air Quality/Emission Control; Biodiversity/Species Preservation; Energy; Health; Sustainable Development; Waste Management/Recycling; Water Quality

● Organization

The Ministry of the Environment was established in December 1988 to carry on the work of the Israel Environment Protection Service, Ministry of the Interior, formed in 1973. Earlier, the Life Sciences Division of the National Council for Research and Development (NCRD) set up committees to create a national environmental policy—in light of the country's population growth and intensive development. During the more recent reorganization, departments from the Ministries of Health, Interior, Defense, Agriculture, Industry, and Commerce, and the Prime Minister's Office were transferred to the new Ministry.

With a staff of about 170 professionals including scientists, engineers, lawyers, economists, educators, and planners, the Ministry has three levels: national, regional, and local. At the national level, policy is defined and strategies are developed on air quality; noise; waste disposal; marine pollution; land-use planning; preparation of environmental standards; creation and operation of a system of monitoring, follow-up, and control; raising awareness of environmental concerns within the community; support for environmental research; and forming a national focal point for contacts with international organizations and nations elsewhere.

On the regional level, the Ministry serves as a focal point for inspection and enforcement activities. On the local level, environmental units within local authorities carry out the policy of the Ministry in all fields. These include air and water pollution, noise prevention, waste disposal, sewage treatment, advice to local building and planning committees, establishment of centers for environmental education, and assistance with special environmental projects.

▲ Projects and People

With data obtained from monitoring stations, new **air pollution standards are determined and directives issued** regarding "industrial activities, emissions from power stations and refineries, and a rise in the number of cars."

With 75 percent of the population living along the Mediterranean coast with its beaches, nature reserves, and national parks, **marine pollution prevention** is a key activity. As a signatory to the Barcelona Convention of 1976, Israel cooperates with other countries to protect the Mediterranean Sea and enacts laws accordingly. Mostly through the framework of the United Nations

Environment Programme (UNEP), Israel participates in workshops on guidelines for municipal wastewater reuse, managing and maintaining wastewater purification plants in Mediterranean coastal cities, and renewable energies.

Education centers are featured at Israel's 373 nature reserves covering about 1,250,000 acres. Also attracting millions of visitors, the 102 national parks extending over 80,000 acres feature archeological and historical sites as well as recreation areas.

A **Cleanliness Fund** has been established from fines and penalties collected by designated "cleanliness trustees" to benefit cleanup campaigns, recycling research, education, and surveys. Some 25,000 volunteers participate. Regarding a growing solid waste problem, Israel is investigating possibilities for recycling waste paper, glass, plastic, and automobile scrap.

The **Environmental Services Company (Ramat Hovav) Ltd.** is responsible for administering a **hazardous waste disposal** site for "supervision of disposal at all stages, . . . transport . . . neutralization, and storage of waste." Monitoring hazardous chemicals is the **Research Institute for Health and Environment.**

The most serious environmental problem is the "deterioration in water quality," reports the Ministry. "Groundwater in the coastal aquifer is being contaminated by seawater, garbage dumps, wastewater, uncontrolled effluent irrigation, hazardous industrial sewage, pesticides, fertilizer, and fuels," according to the *Israel Environment Bulletin* (Spring 1991). "Groundwater in the mountain aquifer is exposed to pollution and is threatened by salination, bacterial pollution, and contamination by fertilizers and industrial and municipal wastewater." The Ministry's publication reports that "the problem of decreasing [water] quantities, exacerbated by three years of drought, is compounded by deteriorating quality as a result of demographic, industrial, and agricultural pressures on the country's scant water resources."

To alleviate such conditions, the Ministry urges passage of new water pollution prevention regulations and amendments, recommends immediate steps to improve sewage treatment, and proposes more efficient administrative reorganization. A proposed **integrated management plan** would require the "implementation of a balanced production regime in which pumping does not exceed replenishment, prevention of seawater or saline water intrusion throughout the raising of groundwater levels, and transition to reclaimed wastewater for agricultural use in designated areas." Most important, the Ministry is emphasizing a new "water ethic" among citizens to view water "as a precious asset to be carefully protected, managed, and allocated."

Dr. Uri Marinov was an Israeli delegate to numerous international environmental conferences, temporary advisor and consultant to the World Health Organization (WHO), and author of UNEP's *An Approach to Environmental Impact Assessment for Projects Affecting the Coastal and Marine Environment.*

■ Resources

The Ministry of the Environment publishes in English the quarterly *Israel Environment Bulletin; The Environment in Israel,* a 294-page report; and a general brochure. Published in Hebrew are the *Annual Report on the Environment in Israel* and *The Biosphere,* a monthly journal. All materials are free to the public.

🝑 *Private*

Amnir Recycling Industries Ltd.
P.O. Box 142
Hadera 38101, Israel

Contact: **Ofer Dressler, Ph.D.**, Business Development Manager
Phone: (972) 6 349 582 • *Fax:* (972) 6 33 3104

Activities: Development • *Issues:* Energy; Waste Management/Recycling • *Meetings:* BIR

● Organization

Founded in 1968, **Amnir** is the main subsidiary of the American Israeli Paper Mills Ltd.'s Recycling Division (AIPM). "The leading company of its kind in Israel, Amnir collects and recycles waste for use in industry," its literature reports. "About 100,000 tons of waste paper and cardboard are collected each year, over 50 percent of the total quantity of fibers used by the Israeli paper industry.

"Amnir set up a plant in Hadera, Israel to recycle thermoplastic waste into a wide variety of high-quality low-cost materials. The plant includes a very modern washing system that handles plastic waste from industry agriculture and garbage collection," according to the company. "Amnir collects plastic waste and it also offers to granulate it on site under customer supervision. Other recycling services include cleaning, separating, and sorting operations."

▲ Projects and People

An article in *The Israel Economist* (February 1989) describes some of Amnir's processes: "The company has 20,000 paper collection points throughout Israel. . . . In the collection centers the first task is sorting out different types of paper—such as paper, newspaper, and cardboard—and eliminating garbage and unusable paper. A relatively small amount of paper will be finely grated and then sold as padding. The majority of the paper, however, is transferred to the company's plant in Hadera for processing into paper pulp, the primary component of paper production."

According to 1989 statistics, Israel ranked eleventh in the world in recycling paper (24 percent). Holland topped the list, recycling 53 percent of its paper, Japan was second at 50 percent, Austria was third at 46 percent, and the USA was tenth at 29 percent. Such recycling is crucial to Israel, which lacks land for trash burial in Tel Aviv and Haifa, particularly.

In addition, Amnir opened a new plant in Afula to recycle plastic goods, organic matter, glass, and paper. "Plastic waste is recycled into plastic pellets, which are the building blocks of plastic production," the magazine reported. "Compost fertilizer is produced from the organic materials."

Illan Freund, economics and development manager for Amnir, says the key to successful recycling is "increased public and governmental awareness of both the need for and the potential of recycling." Also associated with Amnir are economist **Joram Shetrit**; engineer **Ehud Bar David**; **Uri Sapir**, who "developed recycling and waste management in Israel" and is now new business developing manager for AIPM; and founder and consultant **Dario Navarra**, who inspired the country to be "waste paper and recycling minded."

■ Resources

Amnir seeks information regarding new technologies for landfill design and operation. Amnir can provide an information kit and consulting services on the collection, removal, and recycling of municipal solid waste and industrial waste. A reprint is available of *Recycling Municipal Solid Waste (MSW): A Common Interest of Commercial Ventures and Environmentalists.*

Laster and Gouldman

10 Hanassi Street
Jerusalem 92188, Israel

Contacts: **Richard Laster, M. Dennis Gouldman,**
  Attorneys at Law
  Phone: (972) 2 635224 • *Fax:* (972) 2 636926

Activities: Law Enforcement; Political/Legislative; Research
Issues: Air Quality/Emission Control; Energy; Health; Town Planning; Waste Management/Recycling; Water Quality

▲ Projects and People

At their law firm established in 1980, attorneys Richard Laster and M. Dennis Gouldman specialize in matters to improve Israel's quality of the environment. Laster's major field is environmental law, and Gouldman's area of expertise is in planning and building law.

Believing that "improving the quality of the environment must begin at the grassroots level," Laster represents the Association of Towns for environmental quality for the areas of Hadera, Ashkelon, Southern Judea, and Ashod. "The Associations of Towns are attempting to manage the new area of regional sewage control," he writes. He also represents the city of Arad, which is "contesting the creation of a large phosphate quarry near the city's borders" as well as workers and their families exposed to asbestos in the Nahariya area.

Since Israel has a "tremendous shortage of water," comments Laster, "the reuse and proper distribution of purified sewage will be a major environmental challenge in which the Associations of Towns, Drainage Boards, and the Yarkon River Authority will all be required to handle."

Working on behalf of "those trying to protect the environment," Laster drafted an act to enable class action. "I believe that the public has the right to know about all phases of environmental activities, especially in the field of dangerous substances," he writes. "To this end, I have begun representing labor organizations and individual workers who are exposed to dangerous substances in order to protect the rights of the worker and his right to know to which products he and his family are exposed."

Laster is chairman of the Environmental Law Association in Israel, monograph author on environmental law in Israel for the International Encyclopedia of Laws, and environmental law and policy instructor at Hebrew University's Faculty of Law and the Department of Environmental Sciences. The University is attempting to "create a center for environmental sciences," Laster reports, which will be "multidisciplinary and composed of representatives of practically all the faculties and schools." At the start of Laster's law career, he was a "raider" with Ralph Nader's staff in Washington, DC.

■ Resources

With the firm since 1989, Gouldman is the author of a book on *Legal Aspects of Town Planning in Israel,* articles on planning and building law, *The Role of Law in Solving Urban Problems,* and the *Legal Framework for the Preservation of Buildings*—among other publications.

🎓 Universities

Mitrani Center for Desert Ecology (MCDE) ⊕
Jacob Blaustein Institute for Desert Research
Ben-Gurion University of the Negev
Sede Boqer Campus 84990, Israel

Contact: **Moshe Shachak, Ph.D.,** Senior Desert
  Ecologist
  Phone: (972) 57 565824 • *Fax:* (972) 57 555058
  E-Mail: dcab100@bgunve

Activities: Research • *Issues:* Biodiversity/Species Preservation; Deforestation; Global Warming

● Organization

The Mitrani Center for Desert Ecology (MCDE), a research unit of the Jacob Blaustein Institute for Desert Research, aims to "study deserts as model ecosystems for advancing ecological knowledge" and to specifically investigate the "ecological properties of Israeli deserts," making this information available to government, industry, and the scientific community for "conservation and prudent development of desert regions."

At the Mitrani Center, ongoing research deals with relationships between "precipitation, soil runoff, and animal and plant life; the population dynamics and community structure of desert plants and animals; and the behavioral and physiological ecology of a variety of desert animals."

The Blaustein Institute, an integral part of Ben-Gurion University, was so named in 1980. It was earlier established at Midreshet Sede Boqer by Israel's government as a national center for desert research which focused on the Negev to seek "new insights into how to balance the utilization of the desert's resources with the protection of its delicate ecology." With the Negev comprising more than 60 percent of Israel's landmass, the University was established in 1969.

As a result of its studies on climate, water resources, natural energy sources, and habitat for plants and animals as well as development of suitable biotechnology, the Blaustein International Center attracts desert researchers from throughout the world. In Third World countries, Blaustein scientists from its Applied Solar Calculations Unit, for example, have lectured in China, Thailand, India, Kenya, and Egypt and have hosted colleagues from Nigeria, Pakistan, and India. "The Water Resources Center has ties with scientists in Senegal and Mexico, while the Microalgal Biotechnical Laboratory collaborates with projects on colleagues in Thailand, the Philippines, and Guatemala," reports the Institute.

Among other participating countries are Portugal, Nepal, Chile, Ivory Coast, Cameroon, and Ghana. Native Americans such as Navajos worked with Experimental Farm manager David Mazigh both in Avdat and Arizona to learn how to "achieve far higher yields of tomatoes, potatoes, onions, and corn despite the arid conditions."

In addition to MCDE, the Institute comprises the following units:

Algal Biotechnology, where such water plants "paradoxically" are grown in brackish or sea water as a potential renewable resource for food, feed, and energy. Biofertilizer is one byproduct.

Controlled Environment Desert Agriculture (CEDA), which is putting "high levels of solar radiation" to use on behalf of agriculture. "CEDA requires so little water that it makes desalination economically feasible," it reports. Artificial seeding beds eliminate the need for naturally fertile soil and in the self-enclosed system, "fertilization by carbon dioxide is continuous, resulting in more than double normal yields."

Energy and Environmental Physics, regarding the development and application of solar energy, such as in photovoltaic batteries, water diffusion in the soil, and heating buildings and water.

Desert Architecture, for energy conservation and other suitable human habitat conditions in the Negev and worldwide.

Desert Meteorology, which aims to foresee changes related to global warming, surveys potential for wind and solar energy, and examines hazards from the region's pollutants or dusts.

Desert Hydrology, which estimates water flow and transport systems and estimates better ways to locate surface and subsurface waters in arid lands.

Fish Laboratory, whereby biotechnologies are being developed to raise fish in "closed systems" of recycled brackish water—providing an "important source of protein [that] can be produced in the desert with considerable economic advantages."

Environmental Applied Microbiology, regarding "water management, groundwater, agricultural drainage systems, wastewater treatment and reuse, and industrial wastewater treatment." Ongoing projects evaluate hazards of accumulated heavy metals in soils and crops irrigated with domestic wastewater and "aim to decontaminate Israel's nitrate-polluted aquifers," for example.

Ecophysiology and Desert Plant Introduction, which has placed about 1,350 plant species—many with medicinal, ornamental, or other economical value—from 16 deserts worldwide into the **Plant Introduction Garden** and **Desert Research Botanical Garden** to assess their growth and further research here and abroad.

Center of Desert Agrobiology, which investigates plant responses to desert conditions and develops agrotechniques that can advance agriculture in such regions worldwide. Some research examples include **saline water irrigation** to enhance peanut, wheat, and maize production; **plant productivity in sand** and primary productivity of **fast-growing agroforestry trees**; and studies of ancient and modern methods of **water harvesting** in desert areas. Internationally, the Center is involved in studies on "crop production under saline conditions in the Philippines; modernization of carob and cork oak production in Portugal; foliar fertilization in Thailand and Mexico; desert agriculture in Chile; runoff wheat production in the desert of Chad; and forest tree productivity with Peru.

Social Studies Center, which deals with urban development problems in the Negev region, including the growth of a new Bedouin town.

▲ Projects and People

Dr. Moshe Shachak, MCDE's senior desert ecologist is currently working on two major projects: the study of the structure and function of rocky desert ecosystems, and "savannization," or the study of methods to reverse the trend of decreased productivity of arid and semiarid regions.

In the first project, Dr. Shachak tracks the population fluctuations of a number of desert organisms to "understand the different strategies adopted by the various organisms" in order to confront the difficult desert conditions. "Understanding the importance of biotic and abiotic constraints . . . may enable us to predict the effects of variation in the environment on . . . survivorship and fecundity," he writes. Because of water's scarcity in the desert, it is relatively easy to track. The flux of nitrogen, also "a limiting factor" under certain circumstances, is being studied too.

Initiated with the Jewish National Fund (JNF), the Savannization Project was launched to develop methods of fighting desertification. The objective is to "foster development of hydrological and ecological landscape engineering for increasing biotic productivity, while conserving biodiversity in arid lands," writes Dr. Shachak. The project aims to reverse four negative trends: "depletion of available water, decrease in biological productivity, inefficient land use, and decreasing biological diversity."

Patch cultivation is the project's essence, whereby "only a small portion of the landscape is altered artificially in order to create a resource-enriched patch within the poorer surrounding matrix," reports Dr. Shachak. This method results from two constraints: not enough water for large-scale cultivation and "runoff water harvesting" as the only major source for increasing water availability for the biota.

"The methods being developed make use of surface runoff water for increasing plant production," explains Dr. Shachak. "Runoff is collected into catchments" or patches varying in size to promote water infiltration into the soil and diminish evaporation.

The "patch cultivation" method derives from water-harvesting techniques adopted by the ancient people of the Middle East, now being investigated from the ecological perspective. It is presently incorporated into the JNF's modern "liman," and its "environmental soundness and technological simplicity" have practical application in deserts of industrialized and developing countries, believes Dr. Shachak, who has written extensively on desert ecology.

He recognizes three "important scientific problems related to bioresources and biodiversity" that have global impact: "the relationship between productivity and diversity; heterogeneity, diversity relationship; and relationship between disturbance and diversity."

Among other comments, Dr. Shachak says that "we must study the relative importance of the biotic and abiotic variables in controlling the productivity, diversity relationship." Regarding abiotic variables, "we must study the relative importance of resources [such as nutrients, water] as well as the climatic variables [such as temperature, humidity]. Similar ideas on this subject are expressed in *Ecological Heterogeneity* (Kolasa and Pickett, 1991)," he writes. He also encourages global research to test mechanisms controlling manmade or natural disturbances to ecosystems and to pursue ecological knowledge regarding "the tendency of biotic communities to change through time" and the "explicit dependence of ecological systems on spatial heterogeneity."

Dr. Shachak also recommends the **creation of an international biodiversity organization** be established with an information center containing publications from throughout the world dealing with related issues-and research. Such a center would fund international biodiversity studies, function as an education center and foster the development of curricula on "ecological and social aspects of bioresources and biodiversity," hold global workshops and conferences on such matters, and be the nucleus for an "international network of biodiversity research stations representing the diversity of biomes."

■ Resources

Published is *A Selected Bibliography on Desert Ecology*, edited by Dr. Uriel N. Safriel, who can provide additional information about the MCDE. Article reprints and informational brochures about Ben-Gurion University and the Jacob Blaustein Institute are also available.

Department of Zoology
Tel Aviv University
Ramat-Aviv
Tel Aviv 69978, Israel

Contact: **Heinrich Mendelssohn,** Professor of
Zoology (Emeritus)
Phone: (972) 3 545 8096 • *Fax:* (972) 3 642 5518

Activities: Education; Nature Conservation; Research • *Issues:* Biodiversity/Species Preservation • *Meetings:* Congresses of Herpetology, Ornithology, Theriology; IUCN/SSC groups

▲ Projects and People

Prof. Heinrich Mendelssohn, Tel Aviv University, belongs to the Species Survival Commission (IUCN/SSC) Birds of Prey, Bustard, Cat, Hyena, Hyrax groups. He offers materials on his University's Research Zoo for local fauna and Zoological Museum.

Italy

Italy has been slow to enforce its many environmental laws for a number of reasons, including the recession of the 1980s and transfer of environmental authority to regional jurisdictions. New treatment plants are being built to reduce pollution, however, and sulfur dioxide emissions are declining. In addition, political groups that emphasize conservation have raised public awareness.

In addition to threatening natural diversity, Italy's environmental problems, especially air pollution, endanger historic monuments. Other concerns are industrial and agricultural pollution in the coastal waters and inland rivers, and acid rain pollution in the northern lakes.

Government

Centro Documentazione Internazionale Parchi (CEDIP)
International Park Documentation Center
Villa Demidoff
50036 Protolino Firenze, Italy

Contact: **Giovanni Valdré**, Scientific Committee Coordinator
Phone: (39) 55 409051 • *Fax:* (39) 55 476116

Activities: Education; Research • *Issues:* Sustainable Development • *Meetings:* Federation of National Parks of Europe; IUCN

● Organization

Founded in 1986, on the occasion of the "European Year of the Environment," **Centro Documentazione Internazionale Parchi (CEDIP)** started its documentary archive on the world's protected areas. It offers scholars and scientists access to these informational materials.

▲ Projects and People

Through the cooperation of thousands of agencies and parks, CEDIP provides researchers and scholars a vast, well-organized **library on environmental issues**. Materials include thousands of "printed matter, manuscripts, videocassette and films." Files are computerized and cataloged according to subject, title, and author in a range of languages. Subjects range from the greenhouse effect to environmental education, waste disposal to tourism and biodiversity.

CEDIP is an active participant in numerous local, national, and international conferences and meetings. In 1991, CEDIP organized the **Conference on Legislation for Protected Areas** (Saint-Vincent); joined Pordenone Province in the Fourth Study Sojourn for Minor Parks; created a working group for an analysis of issues presented at the **Workshop Conference for Coordination of European Regional Parks**; and held meetings rendering guidelines for classifying protected areas compatible with current European situations in certain instances.

An organization committee got underway to create an **International Court for the Environment at the United Nations** which would "prosecute cases of crimes against the environment and will protect rights to and of the environment."

Previously, **Professor Giovanni Valdré** and other CEDIP representatives went to the former Soviet Union to meet the directors of ministries of culture and the environment and of scientific institutions. While seeking to establish **cooperative relations for cultural activities in protected areas**, they visited the Prioksko Terrasnyi Biosphere Reserve—famous for a surviving herd of European bison.

In 1990, CEDIP organized a meeting suggesting greater cooperation of protected areas/parks bordering East and West Europe using the **"twinning"** process to exchange experiences for common use. Three suggestions came out of this meeting: removal of heretofore closed and de facto subjects to special protection; **creation of new parks considering border areas of river systems, mountain chains**, lacustrine basins, uninhabited and coastal regions as an ecological unit; and the **involvement of local people** in all planning aspects.

The same year, CEDIP cooperated with the National Coordination for Parks and Reserves in planning the first organizational meeting for the **Creation of the European Regional Parks and Reserves Coordination** to expand "homogenous management" of Europe's protected natural areas.

Italian and foreign students and researchers at Italian universities, institutes, and university-level schools have an opportunity to compete in CEDIP's **Lufthansa Awards for Ecology**, regarding studies of specified parks. Entries must be submitted in Italian.

Prof. Valdré teaches sociol psychology at Pisa University and is a member of the World Conservation Union's Commission on National Parks and Protected Areas.

▲ Resources

CEDIP offers various free bulletins and lists of parks and protected areas.

 *NGO*

Amici della Terra—Italia (AdT)/Friends of the Earth

Via del Sudario 35
00186 Rome, Italy

Contact: **Laura Radiconcini**, Co-Founder
   Phone: (39) 6 686 8289 • *Fax:* (39) 6 654 8610

Activities: Political/Legislative; Research • *Issues:* Biodiversity/Species Preservation; Deforestation; Energy; Waste Management/Recycling

● Organization

With demonstrations, scientific studies, and proposals to Parliament and political parties, **Amici della Terra—Italia (AdT)** (Friends of the Earth—Italy) began in 1977 with a national campaign against nuclear energy and a call for a new energy policy. In the early days, it introduced a "smiling sun" symbol to Italy as well as the first antinuclear referendum. Today, there are some 10,000 full members and 9,000 junior members and supporters in 72 local groups.

Following the Milan-based Altamira meeting in 1989, AdT organized "the first international meeting with natives of the rainforest and the first big public demonstration in defense of the Amazon rainforest." Through Friends of the Earth International (FOEI), it cooperates with environmentalists from Eastern Europe. One recent initiative is a study of the national waste problem, undertaken with the help of the Ministry of the Environment.

"All of our initiatives are directed towards raising public awareness and producing concrete proposals for solutions to problems. We're not only about protests; we're also responsible for contributing to large-scale environmental reforms," reports AdT, which aims to build a consensus among political forces.

▲ Projects and People

Ongoing campaigns include: Defend the Indians, Save the Forests, of which AdT describes itself as the world's leader; **waste reduction,** for collection and recycling of garbage; **East and West Environment,** to promote cooperation between environmentalists in Eastern and Western Europe; energy saving project; Friends of the

Trees, an initiative to safeguard green areas in urban environments; **Animals and Us,** to protect domestic strays and hunted, wild, and zoo animals; **Ciao Farfalla (Hello Butterfly),** to create butterfly areas in villas or public parks; **Non-nuclear Europe** initiatives against such civil and military technology; **Open Sesame,** which is pushing for the right to access government information—a central theme in environmental and consumer protection. Also fighting for a "better functioning public administration" is the **Green League for Reforms;** AdT coordinator is **Enzo Di Calogero.**

In conjunction with the **Body Shop,** which markets natural skin and hair products worldwide, AdT is **distributing environmental information in stores** in Italy to raise public awareness.

In 1991, the group worked to strengthen FOE's presence in the Brazilian rainforests, in light of the 1992 United Nations Conference on Environment and Development (UNCED). A new endeavor is **Radio Amazonia,** in cooperation with Brazilian NGOs, to reach isolated, indigenous communities through the delivery of shortwave, solar battery-powered radios. In the Philippines, AdT's **Dario Novellino** is working with the Filipino Association Haribon on **Project Palawan** to preserve the surviving indigenous communities by establishing protected areas on this island and to encourage renewable use of forest products.

With FOEI and FOE-USA, Amica della Terra, through **Operation Desert Blue,** is urging coordinated cleanup efforts of the Persian Gulf by the Italian government and other countries following the war. "In particular, we are evaluating the juridical feasibility of legal action against Saddam Hussein for 'ecological crimes' and the possibility of producing a photo exhibit on the environmental disaster."

Project Butterfly continues, with cooperation of the Silkworm Institute of Padua, and the **banning of the pesticide FENOXICARB** in some Italian regions; **Enzo Moretto** is coordinator. AdT is formulating a **database on laws governing waste** in the European Community (EC), researching the **recovery and recycling of plastic containers,** and pushing for completion of the second phase of the National Plan for Waste. On energy, AdT is lobbying for **new laws to strengthen the National Energy Plan,** monitoring the behavior of regional governments, and organizing training courses for the *Italgas* utility staff; **Pierluigi-Tito Lombard** is coordinator.

Two educational packets were produced: **Viva Amazonia**—at school in the forest, which includes a photo exhibit, video, forest handbook, booklet *Paradise Lost?*, and teacher's kit; this program reached 5,000 students in 150 Italian schools in 1991. **Ciao Farfalla, a class on the environment,** allows children to grow silkworms at school, teaching them how to set up a butterfly garden in the playground or in their town. So far, the *Let's Raise the Silkworm Together* and *The Butterfly Garden* handbooks have been distributed with silkworm eggs and mulberry plants to 50 schools; 2,000 children have joined the **Butterfly Club.**

Co-founder **Laura Radiconcini** has been a member of AdT's National Secretariat since 1981 and has also served as a member of the National Council for the Environment, an independent body assisting the Minister of the Environment. AdT president is **Mario Signorino;** National Secretariat head is **Walter Baldassarri.**

■ Resources

Educational materials are described above.

Centro per un Futuro Sostenibile ⊕
Center for a Sustainable Future

Viale Giulio Cesare 49
00192 Rome, Italy

Contact: **Roberto Giachetti**, Director
  Phone: (39) 6 321 5491 • *Fax:* (39) 6 321 5493

Activities: Political/Legislative • *Issues:* Air Quality/Emission Control; Deforestation; Energy; Global Warming; Sustainable Development; Transportation; Waste Management/Recycling; Water Quality

● Organization

Twenty Italian political leaders and members of Parliament from both the majority and opposition parties founded the **Center for a Sustainable Future**. Later a scientific committee consisting of economists, jurists, scientists, and experts was formed in conjunction with the Center's activities to resolve global environmental problems.

According to **Roberto Giachetti**, the Center's director, its purpose is to "provide a forum on selected environmental issues for people seeking to promote action in Italy, especially in the institutional framework, thereby contributing to the solution of fundamental problems regarding the global environment."

To carry out its mission, the Center seeks to establish ongoing relationships with environmental experts and associations. It plans yearly **international seminars** on topical issues, the first of which was held in Rome, December 1990, and dealt with the **greenhouse effect and climate changes**.

▲ Projects and People

A set of recommendations emerging from the Center's first seminar were to influence Italian government policymaking as the country prepared for participation in the 1992 United Nations Conference on Environment and Development (UNCED).

"Italy can surely shake off the image it has acquired of a country that proclaims an environmentalist credo without drawing its (sometimes costly) practical conclusions," reported the Center, which is urging Italy to adhere to the decision of the European Community's Council of Environment and Energy Ministers to "stabilize carbon dioxide (CO_2) emissions . . . at the 1990 levels within the end of this century."

The Center is asking Italy to "examine its energy policy to estimate if by 2000 more ambitious goals may be reached, as other European countries plan to do." The Center proposes a cut in emissions by 20 percent by 2005 and more substantial reductions of up to 50 to 60 percent in the following decades. At the same time, with the decline in oil prices, Italy's National Energy Plan was projecting an "11 percent increase in emissions within 2000," according to the Center.

Global warming, acid rain, city smog, and "heavy dependency on the instable international oil market" are reasons for the government to begin to **exploit the Italian "energy conservation reservoir"** and develop large-scale use of renewable sources, such as "solar, wind, biomass, geothermic, and hydro." In addition, chlorofluorocarbons (CFCs) production must be "rapidly stopped" and "reforestation and a careful agricultural policy must be promoted over the next 40 or 50 years."

For example, the Center estimates that replacing incandescent light bulbs with compact, efficient **fluorescent bulbs** would save 15 billion kilowatt hours per year and reduce 10.4 million tons of CO_2 emissions by the year 2005. Regarding agriculture, "Better management of the organic content of farm soil and . . . rational reuse of organic waste could absorb a part of CO_2." In implementing energy conservation, Denmark is cutting heat demands in new buildings 50 percent by 2000, wind turbine capacity is being expanded, and other "cost-effective, low-energy technological solutions" are being encouraged.

Consumer education and "vigorous and integrated policies" that promote research and spread new technologies worldwide are essential, emphasizes the Center.

■ Resources

The Center publishes quarterly monographs with themes such as "environmental aspects of North-South cooperation," with particular reference to **technology transfer**. Copies of conference proceedings, also in English, can be ordered as well.

Lega Per L'Abolizione Della Caccia ⊕
European Federation Against Hunting (EFAH)

Carlo Alberto 39
00185 Rome, Italy

Contact: **Carlo Consiglio**, Professor
  Phone: (39) 6 482 0857

Activities: Education; Political/Legislative • *Issues:* Biodiversity/Species Preservation

● Organization

The **Lega Per L'Abolizione Della Caccia** (European Federation Against Hunting) was founded in 1984 to abolish hunting in developed countries because "hunting is no more necessary for subsistence, very dangerous owing to the greater number of hunters and the modern weapons that are at their disposal, and unjustifiable since it is made for fun." Reasons for opposition are "maintenance of ecological systems, respect of life, and fight against unnecessary animal suffering." EFAH does not oppose subsistence hunting in undeveloped countries.

Representing more than 75,000 members in 11 European countries—Austria, Belgium, France, Czechoslovakia, Germany, Greece, Italy, Luxembourg, Netherlands, Switzerland, and UK—EFAH intervenes on behalf of its cause to European governments, Council of Europe, and European Parliament, and in the Justice Court in Luxembourg regarding violations of the European Community (EC) Directive on wild bird conservation.

▲ Projects and People

EFAH secretary **Carlo Consiglio**, professor of zoology, University of Rome "La Sapienza," researches Italian and Ethiopian **dragonflies**, Italian and Corsican **stoneflies**, and the "polymorphism and ecology of **isopods** of the genus *Sphaeroma* in the Mediterranean Sea. The author of some 116 publications in his field, including books against hunting and about zoos (printed in Italian), Prof. Consiglio is also president of the Italian League Against Hunting and represented the Italian National Research Council in the International Waterfowl and Wetlands Research Bureau (IWRB).

■ **Resources**

The bimonthly *EFAH NEWS* bulletin is printed in English.

Lega Italiana Protezione Uccelli (LIPU) 🕊
Italian League for the Protection of Birds
Vicolo San Tiburzio 5
43100 Parma, Italy

Contact: **Marco Lambertini, Pharm.D.**,
               Conservation Director
   Phone: (39) 521 233414 • *Fax:* (39) 521 287116

Activities: Education; Law Enforcement; Political/Legislative; Research • *Issues:* Biodiversity/Species Preservation; Nature Conservation; Sustainable Development • *Meetings:* ICBP; IWRB

● **Organization**

Originally formed as the League Against the Destruction of Birds in 1966, the League for the Protection of Birds (LIPU) presently has a staff of 20 and membership of 25,000 in 110 branches in Italy and one in the United Kingdom. LIPU seeks the protection of birds by encouraging public awareness and lobbying for more adequate laws.

▲ **Projects and People**

LIPU manages 20 nature reserves, some featuring visitors facilities. It also has four separate centers for the reintroduction of the white stork and white-headed duck; treatment and rehabilitation of injured raptors and wounded aquatic and sea birds; and reproduction.

A major LIPU campaign in 1991 was the restocking of the Griffon vulture colony in Sardinia, which was nearly extinct. Reintroduction of the white-headed duck, which had disappeared for two decades, and the white stork (thought to be extinct as a breeding species for centuries), and the protection of raptor migration in the Straits of Messina were also undertaken in 1991. Among other LIPU projects are antipoaching activities in the northern valleys of Lombardy, Veneto, and Friuli Venezia Giulia—"where trapping and netting songbirds during the fall is still widespread."

LIPU also conducts bird species and habitat research in collaboration with other bird societies, the Italian government, and the European Environmental Bureau (EEB).

Conservation director Dr. Marco Lambertini, who holds a pharmacy degree, contributes regularly to *Giornale Nuovo*, a nature conservation daily newspaper published in Milan, and to *Airone*, a monthly nature magazine. The head of Montepulciano Lake Museum's Ornithology Department, Lambertini's studies include "long-term monitoring of pollutants in eggs of yellow-legged herring gull from Capraia Island," and "pollutant levels and their effects on Mediterranean seabirds."

Also at LIPU are Giuliano Tallone, research head; Ugo Faralli, reserves head; Roberto Spacone, nature centers head; and Barbara Lombatti, international department head. Tallone also collaborates

with the University of Turin on "vocalization and display of night herons, little egret, and white stork" and belongs to collaborative stork, ibis, spoonbill, and heron study groups of the World Conservation Union (IUCN), International Council for Bird Preservation (ICBP), and International Waterfowl and Wetlands Research Bureau (IWRB).

■ **Resources**

LIPU publishes *Ali Notizie* for adult members and *Ali Giovani* for youth members. In English are *Bird Killing in Italy* and abstracts to *Important Bird Areas in Italy*. It seeks political support to "enforce and implement environmental legislation in Italy" and needs promotional and marketing materials for bird conservation campaigns. LIPU consults on treatment of injured birds, land restoration, and wetland creation.

🎓 *Universities*

Department of Geography
Padova University
Via del Santo 26
35123 Padova, Italy

Contact: **Ugo Sauro, Ph.D.**, Professor
   Phone: (39) 49 875 6333 • *Fax:* (39) 49 875 4213

Activities: Education; Research • *Issues:* Landscape; Water Quality • *Meetings:* IGU; International Geomorphological Association; International Speleology Union

▲ **Projects and People**

At the University of Padova since 1971, **Dr. Ugo Sauro** has more than a decade of research in the "morphoneotectonics" and regional and karst geomorphology in the Southern Alps. He described for the first time the "surface faulting phenomena" and searched for relative chronology of these forms, some of which were from the Holocene age. He also serves as chairman of the International Geographical Union (IGU) study group on environmental changes in karst areas.

Dr. Sauro, with doctoral degrees in pharmacology and natural science, continues his work on karst morphology and man's impact on karst areas. Presently, Dr. Sauro is studying the relative chronology of limestone alpine landslides by means of the development of small corrosion forms on the block surfaces; the high mountain karst in the Alps; and the human ecology and impact in some alpine and Mediterranean karst.

■ **Resources**

A publications list is available of Dr. Sauro's books and scientific papers, some with English titles.

Institute of Zoology
University of Perugia
06100 Perugia, Italy

Contact: **Dr. Bernardino Ragni**, Senior Researcher
  Phone: (39) 75 585 5721 • *Fax:* (39) 75 585 2067

Activities: Education; Law Enforcement; Research • *Issues:* Biodiversity/Species Preservation; Wildlife Management

● **Organization**

Perugia University's Institute of Zoology has three main subjects for study and research: vertebrates zoology, aquatic or freshwater invertebrates, and ecology of freshwater animal communities. **Dr. Bernardino Ragni** is responsible for the vertebrates zoology division, with laboratories for macroscopical analysis, dissection, and necroscopy; personal computers "with packages for treatment of biological data," live-traps and radios for field studies, and video cameras for behaviorial studies and didactic purposes. Ongoing projects incorporate naturalistic, telemetric, radio-tracking, and capture-recapture methods.

▲ **Projects and People**

A five-year plan, started in 1989, involves the status and biological conservation of the wild cat in the Mediterranean Islands. With insular populations of *Felis silvestris libyca* "nearly unknown" and feared to be on the "verge of extinction because of direct human persecution and habitat destruction or alteration," the project's purpose is to define the "systematic-genetic status, basic habitat requirements, and true conservation status of the [wild cat] populations," writes Dr. Ragni.

The **reintroduction of European wild cats** in the Maremma Regional Park (Tuscany) got underway in 1988 and is scheduled to be completed in 1993. Again, due to human persecution, the wild cat is a rare and threatened subspecies which is dwindling in the park area. Seeking to reestablish a new population of this felid will serve as a model for similar initiatives in other sections of the wild cat range, reports the Institute.

A recolonization of the **European lynx** in the Italian Alps is the third five-year project. Started in 1991, field research on the presence, distribution, activity, pattern, food, habits and abundance of the recolonizing lynx is determining the adaption of the "new" lynx population after its "extinction about a century ago."

Most important, writes Dr. Ragni, "Wild cats and lynxes carry out a role of ecological indexes in the inhabited ecosystems, and the initiatives for the conservation of their populations must be comprehensive of intervention on the entire biocommunity." Dr. Ragni belongs to the Species Survival Commission (IUCN/SSC) Cat Group.

Department of Evolutionary Biology
University of Siena
Via P.A. Mattioli 4
53100 Siena, Italy

Contact: **Sandro Lovari, Ph.D.**, Professor of Animal Behavior
  Phone: (39) 577 280436 • *Fax:* (39) 577 298898

Activities: Education; Research • *Issues:* Biodiversity/Species Preservation

▲ **Projects and People**

Dr. Sandro Lovari collaborates with doctoral students at the University of Siena on the behavioral ecology of urban stone martens in the town of Siena; a three-year study on tourism in the Gran Paradiso National Park and its effects on habitat, feeding patterns, and activity rhythms of the Alpine ibex; the reintroduction of the Apennine chamois to the Gran Sasso Massif, Central Apennines; and a study on the reproductive behavior of the Himalayan tahr in the Sagarmatha National Park, Nepal.

Earlier, Prof. Lovari conducted research on the behavioral ecology of the crested porcupine, red fox, and stone marten in designated areas of central Italy. Another program was conducted with the goral (goat antelope) in the Majathal Harsang Wildlife Sanctuary, Himachal Pradesh, India, regarding the rutting behavior and habitat use of a wild caprin with management implications," in collaboration with Dr. Marco Apollonio, University of Pisa. A year earlier, Dr. Lovari surveyed wild ungulates of Sagarmatha (Mt. Everest) National Park, Nepal, concentrating on the foot of high-altitude wild ungulates.

Chairman of the Species Survival Commission (IUCN/SSC) Caprinae Group, Dr. Lovari has authored three books in his field and more than 100 scientific publications. His research interests are in the behavioral ecology, behavior, and management of higher vertebrates, especially ungulates and carnivores. He lectures and gives seminars at universities worldwide, serves as a consultant for national parks, and has led numerous research expeditions in Asia. In 1983 and 1989, he organized two international conferences on the biology and management of mountain ungulates.

■ **Resources**

Dr. Lovari seeks software equipment and public outreach assistance. A selected scientific publications list is printed in Italian and English.

The third largest island in the Caribbean, Jamaica faces the challenge of balancing its overwhelming need for economic development with the equally important need to manage and conserve the natural resource base on which this development depends: fisheries, forestry, minerals, and mining, energy, and agricultural resources.

The country's most pressing environmental problems include air pollution, surface and groundwater pollution, deforestation, soil and shoreline erosion, destruction of wildlife and their habitats, and urban infrastructure deficiencies. Particular threats are the serious pollution and eutrophication caused by inadequate waste disposal systems in the coastal waters, especially Kingston Harbour, and the inadequate water supply in rural areas.

❧ *NGO*

Portland Environment Protection Association (PEPA) ❧
P.O. Box 199
Port Antonio, Portland, Jamaica

Contact: **Tom Wirth**, Natural Resources Planner; **Maurgarite Gaoran**
Phone: 1-809-993-2720 • *Fax:* 1-809-993-3407

Activities: Education; Political/Legislative • *Issues:* Air Quality/Emission Control; Biodiversity/Species Preservation; Deforestation; Energy; Global Warming; Health; Marine Environment Protection; Population Planning; Sustainable Development; Transportation; Waste Management/Recycling; Water Quality

● Organization

The **Portland Environmental Protection Association (PEPA)** is a regional environmental support agency that uses **PEP Clubs** to inform the public about the environment and natural surroundings. PEPA organizes events and activities and collaborates with other environmental groups throughout Jamaica, including the U.S. Peace Corps and local youth camps.

▲ Projects and People

With many activities geared toward children, projects include **school cleanup and beautification** efforts, for which students and their teachers were given awards for their work; **tree plantings** at local primary schools; assemblies with songs and skits about the environment; library display on deforestation, soil erosion, reef and mangrove destruction, overfishing and other pollution problems; and a **World Environmental Week** celebration that included an exhibition at the Hope Zoo Environmental Fair. PEP Club members also appear on *Vibrations*, a Jamaican television show which airs weekly. At Buff Bay National Youth Camp, PEPA members speak to **campers** about **environmental awareness.**

Through PEPA, adults as well took part in a **beach clean-up** with help from the Jamaican Defense Force and local fire department; sponsored an exhibit at the Manchioneal Fair, an exposition held by the community's 4-H Club, to call "attention to environmental problems in Portland and solutions to the problems"; purchased "NO DUMPING" signs; supported the work of the Port Antonio Marine Park; created an area ecological map showing potential land for ecotourism development; worked with the newly created Blue Mountain National Park to disseminate information; encouraged the use or recyclables and stopping use of harmful detergents; and promoted organic gardening.

PEPA is also investigating local problems such as "pollution in the Rio Grande by banana bags impregnated with pesticides," destruction of yellow snakes and other wildlife, sand mining, construction in the coastal zone, and deforestation.

With **Tom Wirth** as a Peace Corps worker, PEPA holds a seminar for other volunteers on initiating environmental education programs and starting environmental organizations like PEPA on other Peace Corps sites.

■ Resources

PEPA needs assistance with environmental education for people of all ages and with marine and terrestrial park formation. PEPA provides speakers to interested groups.

 *Universities*

Department of Zoology
University of the West Indies (UWI)
Kingston 7, Jamaica

Contact: **Dr. Peter Vogel Ph.D., Zoology Lecturer**
Phone: 1-809-927-1202 • *Fax:* 1-809-927-1640

Activities: Development; Education; Research • *Issues:* Biodiversity/Species Preservation; Deforestation; Sustainable Development

▲ Projects and People

Dr. Peter Vogel, a lecturer at the University of the West Indies (UWI), **Department of Zoology**, brings to the public's attention the state of Jamaica's diverse natural habitat through several projects.

Working with the 50-year-old Natural History Society of Jamaica, Dr. Vogel is the editor-in-chief of the biannual *Jamaica Naturalist* magazine, started in 1991. Support is from UWI's Zoology Department, Panos Institute, and the World Wildlife Fund (WWF). The official Society publication reports on recent research regarding, for example, **mongooses** on the increase in the West Indies and whether the introduced animal is capable of controlling the arboreal black rat and house mouse as effectively as it does the Norway rat; the endangered **West Indian manatee** and how its digestion of aquatic plants produces fatty acids (as well as carbon dioxide and methane) that are absorbed into the bloodstream to provide energy for growth; and the abundant **bananaquits** and the birds' propensity to nest close to wasp nests.

The magazine points to the disappearance of the **Jamaican hutia** or Indian coney (*Geocapromys brownii*), which Laurie Wilkins, University of Florida's State Museum, describes as "the only surviving endemic terrestrial mammal found in Jamaica." Wilkins—who believes that deforestation, habitat destruction, other land-use pattern changes, and illegal hunting are accelerating—participated in a captive-breeding reintroduction program sponsored by the Jersey Wildlife and Preservation Trust with inclusive results.

With some 3,000 flowering plants, Jamaica has at least "822 endemic flowering plant species," and almost one-third are threatened, according to botanist **Daniel Kelly**, University of Dublin's Trinity College. His published report on *The Threatened Flowering Plants of Jamaica* also points to deforestation as a major menace, along with overcollecting—especially of orchids. When commercial collectors strip whole trees, or whole hillsides, "orchids [which] are slow-growing plants . . . may never recover," Kelly points out. Other endemic plants have horticultural value as perfumes or medicinal drugs. "It was the folk medicine of Jamaica that first brought the **Madagascar periwinkle** [*Catharanthus roseus*] to the attention of the pharmacologists," writes Kelly. "This herb is now the commercial source of several anti-cancer drugs and the basis of a multi-million dollar industry." Kelly urges the cultivation of endangered species in the Botanic Gardens of Jamaica as well as education programs to heighten public awareness.

L. Alan Eyre, physical geography reader, UWI, describes the country's crisis in forestry and watershed management and urges sustainable development to replace the destruction of forests by exploitive agriculture; to stem flood, drought, and other disasters; to end "praedial larceny" or trespassing into forest reserves for "fuelwood to honey to rare parrots and butterflies to yam sticks to fully mature lumber trees"; to reduce the population; to stop the misuse of fire and steep slopes; and to prevent soil erosion and degradation.

"There is no hope for the Jamaican environment, or for long-term economic well being of Jamaicans, unless consistent application of adequate laws on forests, land use, watershed protection, national parks and wildlife is made a national priority," Eyre emphasizes.

With 50 rivers having lost "perennial flow since 1950," rainfall declining, and other water resources diminishing; with temperatures rising, some 409,000 tons of carbon emitted yearly from forest burning, and charcoal (whereby dry woodland is stripped down) constituting a key industry; crop yields falling, landslides posing disaster, and forest squatters increasing—Eyre proposes an immediate plan of action.

He urges a national media campaign of environmental education, massive and voluntary tree plantings, reestablishing multi-species forest nurseries, properly setting up national parks, acquiring critical watersheds for sustainable forestry, relocating forest squatters, controlling housing sites, and guarding against environmental crime. Eyre included these points in a 1990 "sectoral presentation" in the **National Conservation Strategy** of the Jamaican Ministry of Development, Planning, and Production.

What protects certain habitats are "remoteness and inaccessibility," writes Dr. Vogel. He agrees that even though "ecologically important forests are owned by the government," laws that are not enforced fail to stop slash-and-burn agriculture and destructive practices pointed out by Eyre. However, Vogel reports he is encouraged by the emergence of new environmental groups, increased media coverage, interest by the private sector and politicians, and the "promising" Protected Areas Resource Conservation (PARC) project for the establishment of national parks.

Dr. Vogel heads the **Jamaican Iguana Research and Conservation Group** made up of five representatives from UWI, Hope Zoo, Natural Resources and Conservation Authority of Jamaica (NRCA), and the Institute of Jamaica. Believed to be extinct, the native iguana's presence was confirmed in 1990 when a live specimen was retrieved. As a result, the group organized and within a year embarked on a 137-day research mission in the Hellshire Hills, with support from WWF-USA.

Initially, at least 15 different iguanas (*Cyclura collei*) were sighted. At a later point, some were videotaped. When egg clutches were sighted, fearing the mongoose—the researchers removed one clutch to the UWI's Zoology Department, where 11 hatchlings survived. In addition, 20 hatchlings were collected in the field; all were brought to Hope Zoo, where they are being reared in a specially built nursery, with help from the Jersey Wildlife and Preservation Trust. According to Dr. Vogel, "These animals will be used as a stock for captive breeding and for re-release into the wild, after reaching a size at which they are no longer vulnerable to mongoose predation." Other threats are feral pigs, dogs, charcoal burning, and planned developments.

To help the iguana, the Group's long-term plan is to establish viable iguana populations in protected natural habitats; set up captive breeding programs at Hope Zoo; continue scientific study, including aerial surveys; conserve certain biological communities; emphasize sustainable forest use practices; and encourage public appreciation of natural heritage through the production of films, brochures, slide shows, and other educational material.

■ Resources

Society members receive *Jamaica Naturalist*.

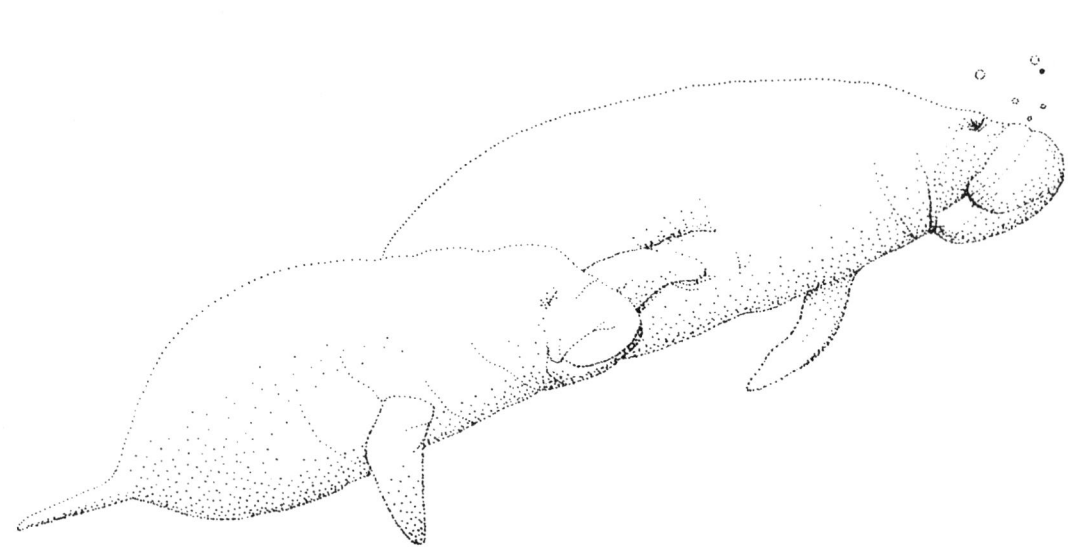

Japan

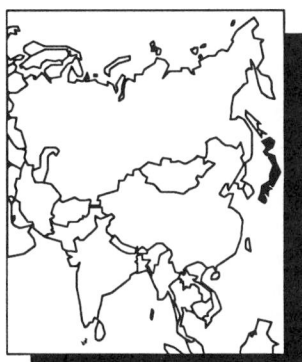

Japan is made up of four large and many small islands in the western Pacific, near the Asian continent. The country's geographic and geologic characteristics give it a limited natural resource base; most of the country is mountainous, and less than 12 percent of the land is arable. As one of the world's leading industrial and financial powers, Japan is one of the most efficient users of energy, yet air quality is still a major concern.

While measures have been taken to improve Japan's air quality, pollution levels are still too high in some areas and acid rain precipitation is as much as or greater than that recorded in the USA and Europe. Other serious problems result from noise, caused especially by automobiles, and inadequate waste disposal. Forests cover 68 percent of the land in Japan, one of the highest levels in the world, but natural forests cover only 18 percent; the number of species in danger of extinction exceeds 600 for all wildlife inhabiting the country.

🏛 *Government*

Nature Conservation Bureau
Environment Agency ⊕
1-2-2, Kasumigaseki, Chiyoda-ku 100
Tokyo 100, Japan

Contact: **Mitsuo Usuki**, Coordinator for International Affairs
Phone: (81) 3 581 3351, ext. 6413 • *Fax:* (81) 3 595 0029

Activities: Political/Legislative • *Issues:* Biodiversity/Species Preservation; Deforestation; Global Warming; Sustainable Development • *Meetings:* CITES; ESCAP; Ramsar; UNCED; UNEP

▲ Projects and People

Established in 1971, the **Environment Agency** focuses on conservation and management of the natural environment for the Japanese Archipelago, consisting of more than 3,900 islands stretching 3,000 kilometers long. The Agency reports that "land use is very complicated due mainly to natural conditions. . . . Most area of mountains, hills, and volcanoes are covered with forests, and some area is used for pastures and orchards. Flatlands including plateaus, terraces, and plains are used for agriculture activities or dwelling areas. Plains are mainly utilized as paddy fields excepting the areas of urban sprawling."

Development and urbanization "coupled with economic growth continuously put pressure on natural landscapes as well as biosphere," states the Agency. Consequently, since 1972, in accordance with the Nature Conservation Law, it has conducted national surveys on the state of the natural environment every five years. Through these *Green Census* reports, Japan's fauna, flora, geology, natural landscapes, and topography are examined. Placed on magnetic tapes, the survey results provide essential data utilized in national planning for land use, development, and park management. Local governments, universities, and research institutes have access to the databank.

The first survey was carried out in 1973, the second in 1978–1979, and the third in 1983–1987, at which time approximately 100,000 volunteers participated. A similar method using volunteers is currently being employed in the fourth survey.

Results of these surveys include the creation of a detailed, nationwide vegetation map, with some 1,000 experts on plant ecology and plant sociology participating in the second and third surveys. In its methodology, the Environment Agency "selected 5,085 specific plant communities covering . . . three percent of the national land area." Of these communities, 643, or 12 percent, were categorized as "extremely rare." Categorized as "planted forest, not to be felled for a long term" were 177 communities, or 3 percent. "Locally endangered plant community and population . . . by human activities such as indiscriminate collecting" were 525 communities, or 10 percent—however, certain communities are categorized under more than one heading. About 60,000 big trees were surveyed, the largest being the camphor tree of Kamou with a "24.2 meter trunk periphery at the height of human breast."

With details of 8,118 species collected and classified, the Environment Agency indicates it has "a plan to carry out nationwide distribution surveys on each plant species."

Among surveys of selected species of animals, eight species of large and medium-sized mammals were investigated: Japanese monkey (*Macaca fuscata*), Shika deer (*Cervus nippon*), Asiatic black bear (*Selenarctos thibetanus*) brown bear (*Urus arctos*), wild boar (*Sus scrofa*), red fox (*Vulpes vulpes*), raccoon-dog (*Nyctercutes procyohoides*), and badger (*Meles melese*). The surveys showed, for example, a relationship between Asiatic black bear and beech forests, the distribution of wild boar and bamboo partridge in areas lacking much snow, the decline of the Japanese firefly, and the appearance of the Japanese marsh crab in the freshwater Hokkaido area for the first time.

Regarding the Survey of Surface Water, the Environment Agency notes, "Most of the rivers, lakes, and marshes are to some extent modified artificially for the purpose of irrigation or flood control works. . . . The conditions of biological habitat . . . are affected, so that this survey aims at revealing the artificial modification extent and the state of inhabiting biome including fish species around and in rivers, lakes, and marshes" throughout Japan.

It was noted, for example, that "construction of dam and weir usually prevents seasonal wandering fishes from migration between upper and lower course of a stream. Many weirs have fishway to ease their migration, but some . . . [are] not always well designed for fish migration and is causing adverse effect to their habitat environment." In the Lake and Marsh Survey covering 483 such bodies of water, it is noted that imported fish species—generally introduced for popular game fishing—are increasing, sometimes "oppress[ing] the native environment of indigenous fish."

The **Nature Conservation Bureau** submitted in 1990 a report on Efforts for Conservation of Tropical Ecosystems, focusing mainly on tropical forests, indicating that "15 percent of the currently existing species on the Earth will become extinct by the year 2000—more than one half [occurring] in the tropical zones." The afforestation rate was noted at one tenth the rate of deforestation between 1980 and 1985. The report outlines the function of tropical forests as providing timber; charcoal; fuelwood; products such as food, plant fiber, cosmetics, rubber, resin, and wax; pasturage for livestock; cropland; wildlife habitat; and maintaining genetic resources contributing to pharmaceuticals, industrial materials, and agriculture.

The report indicates, "Through a great amount of evapo-transpiration, tropical forest have considerable effects on water circulation. Heat consumption associated with the evapo-transpiration, at the same time, maintain low temperature near the ground surface. A large scale of deforestation will, therefore, affect the global meteorological conditions as well as local microclimate." The report also points out, "In the large accumulation of biomass in tropical forest, there is a significant amount of carbon fixed as organic matters absorbing carbon dioxide in the atmosphere. Conserving tropical forest, therefore, could contribute to tapering off the global warming trend."

According to the report, international collaboration to cope with "deforestation and degradation of tropical forest environment" includes a World Commission on Environment and Development (WCED) report advocating "environmentally sound and sustainable development and efforts of International Tropical Timber Organization (ITTO), Tropical Forestry Action Plan (TFAP) of the

UN Food and Agricultural Organization (FAO), United Nations Environment Programme (UNEP) through CITES (Convention on International Trade in Endangered Species of Wild Fauna and Flora) and other efforts, recommendations of the Bellagio, Italy, Conference (1987 and 1988) regarding tropical forest issues, and debt-for-nature swaps.

The report indicates the role of the Environment Agency should support "environmentally sound utilization of nature resources in protected areas [and] efforts to achieve recovery and restoration of disturbed tropical ecosystems as well." The Agency also needs to "be actively involved in basic research on tropical forest ecosystems to build a firm foundation for the applied technology in designation of protected areas, environmental impact assessment, environmentally sound engineering," for example.

Ecotourism should be encouraged in developing nations for markets such as Japan, with surveys of ecosystems and ecotourism resources, design of facilities, training and hiring of park employees, and other means of conservation aid, the report notes.

■ Resources

Green Census and other reports are published by the Printing Bureau of the Ministry of Finance. *Aspects of Nature* summarizes the census data. Vegetation and animal distribution maps are purchased through the Japan Wildlife Research Center. The Ad Hoc Committee Report on Tropical Ecosystems is available through the Nature Conservation Bureau.

Laboratory of Plant Chromosome and Gene Stock
Hiroshima University
Faculty of Science
1-3 Kagamiyama
Higashi-Hiroshima 724, Japan

Contact: **Katsuhiko Kondo, Ph.D.**, Director
Phone: (81) 824 22 7111

Activities: Research • *Issues:* Biodiversity/Species Preservation
Meetings: IUCN

● Organization
The **Laboratory of Plant Chromosome and Gene Stock** has a permanent staff of four along with assistance from graduate students.

▲ Projects and People
The Laboratory is involved in "population recovery and maintenance of species diversity in certain species of carnivorous plants." It has developed its own "shoot primordium method" of tissue culture to maintain genetically stable plant resources. This method allows genetic material to be preserved using less space and with less maintenance than would be needed to care for live plants in the laboratory setting. "Cultured shoot primordia can be maintained free from viruses, have high regeneration rates, and show no genetic change or mutation," points out **Dr. Katsuhiko Kondo**, Laboratory director.

However, the Laboratory needs permission from CITES for transporting live plants for research purposes, which was yet to be obtained by early 1992.

Dr. Kondo, who belongs to the Species Survival Commission (IUCN/SSC) Carnivorous Plant Group, writes that habitats of such plants "can be easily destroyed or deeply disturbed by heavy land uses and water pollution. An example is the free-floating aquatic *Aldrovanda vesiculosa*, which in 30 years was reduced from three critical populations to one because of "floods, grazing by waterfowl, and water pollution." Dr. Kondo advocates that tissue cultures be removed from endangered carnivorous plants to save their genetic resources.

"My Laboratory has preserved cultured shoot primordia of approximately 100 strains of 90 varieties in 80 species of higher plants, including some carnivorous plants," writes Dr. Kondo, who is on the editorial board of the Orchid Society of India. "Moreover, my laboratory has been developing a new technique of cryopreservation of cultured shoot primordia."

■ Resources

Dr. Kondo's publications list is available.

National Research Institute of Far Seas Fisheries ⊕
5-7-1, Orido
Shimizu 424, Shizouka, Japan

Contact: **Toshio Kasuya, Ph.D.**, Head, Offshore Resources Division
Phone: (81) 543 34 0715 • *Fax:* (81) 543 35 9642

Activities: Research • *Issues:* Biodiversity/Species Preservation; Sustainable Development; Water Quality • *Meetings:* Inter North Pacific Fisheries Committee Meeting; IWC, Scientific Committee Meeting

● Organization

The **National Research Institute of Far Sea Fisheries** conducts research on the conservation and management of cetaceans in the North Pacific and Southern Hemisphere. The Institute particularly focuses on those "species and stocks which have been affected by Japanese fishing industry, or have [the] potential of being affected by Japanese fishery in [the] near future." The Institute has five researchers.

▲ Projects and People

Dr. S. Hayase and **Dr. Y. Watanabe** are involved in an observer program to estimate "incidental take of cetaceans by Japanese driftnet fisheries."

Mr. T. Miyashita and **Dr. H. Shimada** are conducting a sighting study "of cetaceans in the Japanese coastal waters and offshore waters where Japanese driftnet fishery operates."

The Institute is also involved in coordinating the International Whaling Commission (IWC) whale sighting surveys in Antarctica. This project is being performed by **Dr. H. Kato**.

Dr. T. Kasuya, the head of the Offshore Resources Division, Dr. Kato, and Dr. Miyashita are all involved in a project that monitors the "direct take of cetaceans by Japanese porpoise fisheries and small-type commercial whaling."

The Institute is also conducting studies which focus on **life history and population dynamics** of the following species: Baird's beaked whale, short-finned pilot whale, false killer whale, Risso's dolphin, bottlenose dolphin, Pacific white-sided dolphin, northern right whale dolphin, Dall's porpoise, and the minke whale. These studies are being conducted by Dr. Kasuya, Dr. Kato, **T. Kishiro**, and **Dr. T. Iwasaki**.

■ Resources

Dr. Kasuya can provide a list of more than 100 publications in English on whales and dolphins.

🏛 *Local Government*

Port of Nagoya Public Aquarium
1-3, Minato-Machi, Minato-ku
Nagoya City 455, Japan

Contact: **Itaru Uchida, Ph.D.**, Director
Phone: (81) 52 654 7000 • *Fax:* (81) 52 654 7001

Activities: Development; Education; Research • *Issues:* Biodiversity/Species Preservation; Global Warming; Population Planning; Water Quality • *Meetings:* IUCN/SSC; IUDZG

▲ Projects and People

Dr. Itaru Uchida of the Port of Nagoya Public Aquarium, Nagoya City, Aichi, is conducting several projects with certain **marine animals** at the aquarium, such as studies of the "status of nesting loggerhead sea turtles, *Caretta caretta*." He is checking the conditions of stranded or incidentally caught sea turtles in Japan.

Dr. Uchida is also doing **joint research** in Malayan waters with the Malaysian Fisheries Department on the **recovery of leatherback turtles** (*Dermochelys coriacea*). Since 1990, Dr. Uchida has been studying population dynamics of the **hawksbill turtle** in the **Cuban Archipelago** with Cuban and Mexican marine scientists. Earlier, he undertook conservation studies of the hawksbill and **green turtles** in **Indonesia** and **Malaysia**.

Dr. Uchida belongs to the Species Survival Commission (IUCN/SSC) Marine Turtle and Captive Breeding Specialist groups and is a member of the Scientific Committee of CITES (Convention on International Trade in Endangered Species of Wild Fauna and Flora) in Japan. He has written at least 8 books and 80 scientific papers in his field.

■ Resources

The Aquarium has a staff of 70; the Chelonian Institute opened in May 1992, and publications will be prepared in the future. The Aquarium seeks information about physiological and ecological research of sea turtles, equipment such as depth and body temperature recorders, and support from university and national research institutions.

 # NGO

Elsa Nature Conservancy (ENC) ⊕
Tsukuba-Gakuen, P.O. Box 2
Tsukba 305, Ibaraki, Japan

Contact: **Eiji Fujiwara**, President
Fax: (81) 298 51 1637

Activities: Development; Education • *Issues:* Biodiversity/Species Preservation; Whale and Dolphin Issues • *Meetings:* IWC

● Organization
Elsa Nature Conservancy (ENC) was established in 1976 in Tokyo as a voluntary membership organization, which now has three domestic chapters and three overseas offices in France, Kenya, and the USA. ENC is a branch of **Elsa Wild Animal Appeal** initiated by Joy Adamson, author of *Born Free—The Story of the Lioness Named Elsa.*

ENC's has five main goals: "conserving all living creatures on the Earth; associating with all nature conservation activity worldwide; educating the younger generation to work for nature conservation; campaigning for nature conservation through mass media, lectures, movies, and fundraising projects; and uniting with all other activities which are of help to conserve as well as to use wisely our natural resources and environment."

▲ Projects and People
ENC is concerned with "protection of **dolphins, whaling issues,** rare and endangered species, zoo and safari parks, and **anti-vivisection and animal experiments.**"

ENC director **Eiji Fujiwara** is an author of 31 books and translator of publications on nature and animals; he translated *Born Free* and the other books written by Joy and George Adamson. A founder of the World Wildlife Fund (WWF)—Japan, Fujiwara is known as: Nature Conservation Society director; president of the Institute of Environmental Science and Culture; and a member of the World Conservation Union (IUCN) Environmental Education Committee, International Council Against Bullfighting (UK), and Japan Science Council's Nature Conservation Committee.

■ Resources
Membership and publications are available, including *Activities of ENC on the Whaling Issue, Japan's Scientific Whaling,* and *Investigation of Japanese Coastal Whaling.*

Industrial Pollution Control Association of Japan (IPCAJ)
Hirokohji NDK Building
17-6 Ueno 1-chome, Taitoh-ku
Tokyo 110, Japan

Contact: **Tatsuo Hiratani**, General Manager, International Affairs Department
Phone: (81) 3 3832 7084 • *Fax:* (81) 3 3832 7021

Activities: Development; Education; Law Enforcement; Political/Legislative; Research • *Issues:* Air Quality/Emission Control; International Cooperation; Policy Planning; Waste Management/Recycling; Water Quality

● Organization
Established in 1962, the **Industrial Pollution Control Association of Japan (IPCAJ)** is a nonprofit, non-government organization described as "the acting end of the Ministry of International Trade and Industry" with experience as an independent body lending services to both government and industries. IPCAJ has a full-time staff of 63, 125 consultants, and a membership of 6,000 firms. With headquarters in Tokyo, there are eight regional offices and a liaison office in Washington, DC. The goal of IPCAJ is to encourage simultaneously industrial growth and sound environmental management.

Within Japan, IPCAJ is involved in such work as formulating environmental criteria in industrial organizations, conducting environmental impact assessments, environmental auditing of existing industry, "granting subsidies to small and medium enterprises for pollution mitigating technical developments," and information services including publications and the consolidation of a database.

Its purpose is to "sensitize the industries successfully to environmental concerns so that negative impacts of their projects and activities on the environment could be addressed through process design, resources management, plant operation, and disciplinary pollutant depletion efforts."

IPCAJ is especially concerned about "changes in the global atmosphere due to fossil fuel combustion, massive ozone depletion over the Antarctic, large-scale destruction of tropical forests, acid rainfall damage," among other global changes.

▲ Projects and People
Human resources is one area of industrial pollution control in which IPCAJ is involved. With a range of training and educational programs, these involve monitoring and evaluation of "environmental pollutant concentrations, quality of public water areas, ecosystems, . . . product environmental impact assessment, and other areas." As of March 1991, IPCAJ hosted 658 foreign trainees from 22 countries in the fields of industrial pollution, and dispatched 37 members to 8 countries to lecture on air pollution and related problems.

IPCAJ works internationally as well as domestically, participating in a project of the Agency of Science and Technology to estimate gaseous emissions in 25 Asian countries. It is also involved in pollution control efforts in "Japanese affiliated overseas ventures in ASEAN countries." It provided secretariat services for the International Organization for Standardization (ISO), the Air Quality Meeting in Tokyo, International Union of Air Pollution Prevention Associations, and the Study Conference on Global Environmental Issues.

IPCAJ also undertakes country assessments, such as the **study on air quality management planning** for the Samut Prakarn Industrial District, Thailand, and the study on **environmental impact of coal-fired power plants** and integrated steelworks in Singapore. IPCAJ also provides other diagnostic studies, lectures, and workshops.

A **database** is maintained regarding environmental pollution and control status. **Tatsuo Hiratani** was previously head of Kawasaki Steel Corporation, San Francisco, California, office. His most recent study is on "concerted environmental conservation efforts for Asian and Pacific Region."

■ Resources

IPCAJ publishes *Environmental Control Regulations in Japan, Industrial Pollution Control (Air and Water), Industrial Pollution Control (Noise and Vibration)*, a monthly journal, textbooks, and other consolidated reports.

Institute of Cetacean Research

Tokyo Suisan Building, 5th Floor
4-18, Toyomi-cho, Chuo-ku
Tokyo 104, Japan

Contact: **Seiji Ohsumi, D.Agr.**, Executive Director
Phone: (81) 3 3536 6521 • *Fax:* (81) 3 3536 6522

Activities: Research • *Issues:* Biodiversity/Species Preservation; Sustainable Development • *Meetings:* IWC

▲ Projects and People

The **Institute of Cetacean Research** is an NGO with a staff of 25 involved in a long-term project that began in 1986 to research **whale resources in the Antarctic**. The objectives of this study are estimation of the biological parameters required for the stock management of the Southern Hemisphere minke whales; and elucidation of the role of whales in the Antarctic marine ecosystem.

A "systematic combination of lethal and non-lethal methods" is used, reports the Institute. This project has met with opposition from "anti-whaling countries in the International Whaling Commission (IWC) and disturbance by a radical anti-whaling group." The Institute hopes to continue its research for "scientific purpose" with the "understanding and support" of the IWC.

Institute director **Dr. Shoychi Tanaka** and the late **Dr. Ikuo Ikeda**, former Institute president, both have been involved in this project.

Nature Conservation Society of Japan (NACS-J)

2-8-1 Toranomom, Minato-ku
Tokyo 105, Japan

Contacts: **Makoto Numata, Ph.D.**, President
         Hideo Obara, Director General
Phone: (81) 3 3503 4896 • *Fax:* (81) 3 3592 0496

Activities: Development; Education; Law Enforcement; Political/Legislative; Research • *Issues:* Biodiversity/Species Preservation; Deforestation; Sustainable Development

● Organization

The work of the **Nature Conservation Society of Japan (NACS-J)** began in 1949 to block the construction of a dam which it said would have destroyed the Oze Moor, the largest moor in Japan. NACS-J is now involved in many Japanese conservation issues by organizing individual funds for different conservation projects. Currently it has a staff of 25 and a membership of 17,000.

▲ Projects and People

The **Primeval Forest Fund** fights logging in the beech forests of Mt. Shiragami, and struggled against the construction of a logging road through the region. As the result of long-term efforts of the conservation movement, this region has been designated as Forest Ecosystem Reserves; NACS-J wants to have the region declared a UNESCO World Heritage Region.

The construction plan for the New Ishigaka Airport would endanger the Shiraho coral reef, believes NACS-J. The **Coral Reef Fund** was involved in assisting the World Conservation Union's (IUCN) survey of the area, and in 1987 the IUCN adopted the recommendations for reef conservation. The first plan for the airport has been rejected, but NACS-J is working to stop the second plan in order to protect the whole reef ecosystem.

Many Japanese rivers have multiple-use dams in the upper river, shore protection works with concrete blocks or river mouth dams which obstruct sea water. The state of these rivers has reached the point that there are only a few rivers left that **salmon and other migratory fish** can use for breeding. The **Wild River Fund** is involved in conservation efforts pertaining to this problem, and NACS-J has "established a special research committee for the conservation of freshwater ecosystems."

The **Wildlife Protection Fund** sponsors activity and scientific research to protect the **Japanese serow, the Kuril seal**, and other wildlife in Japan. NACS-J sees the destruction of habitat by logging as one of the major threats to wildlife. It also acknowledges the conflicting interests of agriculture, forestry, fisheries, and wildlife conservation.

NACS-J is involved in a number of **scientific research** projects that include **monitoring of endangered plant species and vegetation**. A Red Data Book list of Japanese plant species was produced. Also published are *Fundamental Studies for Establishing Yaku-shima Wilderness Area* and *Scientific Studies of Proposed Forest Ecosystem Reserve in Mt. Shiragami*.

Conservation Education is another aspect of the work of NACS-J. In addition to **publications**, the Society runs **workshops** in which they train people to serve as conservation education leaders in schools, local groups, and national parks throughout Japan.

A member of IUCN, NACS-J houses the Japan Committee for IUCN office. It works with Pro Natura Fund and Makita Fund on international projects.

Professor Emeritus of Chiba University's Laboratory of Ecology, **Dr. Makoto Numata** has conducted expeditions in Brazil and The Himalayas, and is currently director of the Natural History Museum and Institute (Chiba) and of World Wildlife Fund—Japan as well as president of the Japanese Society of Environmental Education.

Hideo Obara is a professor of humanology and zoology at Joshi-Eiyo University (Kagawa Nutrition College), whose special interests are "etho-ecological study of man and animals in the urban ecosystem" and conservation of wildlife. Prof. Obara has prepared more than 60 books and 80 papers in his field.

■ Resources

NACS-J publishes the magazine *Conservation of Nature*, textbooks, and field guides such as *A Nature Observation Handbook* and *Indicator Plants and Animals for Nature Observation*.

Wild Bird Society of Japan (WBSJ)
Aoyama Flower Building
1-1-4, Shibuya, Shibuya-ku
Tokyo 150, Japan

Contact: **Noritaka Ichida**, Director
Phone: (81) 3 3406 7141 • *Fax:* (81) 3 3406 7144

Activities: Education; Law Enforcement; Publication; Research • *Issues:* Biodiversity/Species Preservation; Deforestation; Sustainable Development • *Meetings:* East Asian Bird Protection Conference; ICBP World Conference and Asian Section Meeting

● Organization
The **Wild Bird Society of Japan** (WBSJ) is an international NGO with a staff of 70 and a membership of 35,000. Since the Society was founded in 1934, it has grown to be the largest nonprofit wildlife conservation organization in Japan.

▲ Projects and People
One of WBSJ's primary goals is to protect birds and their habitats. This has involved conservation activities, "such as **protecting the wetlands and mudflats** and in being aware of any decision made that may affect birds or other animals." Protection campaigns in the past have included **geese** (which are now on a protected birds list, thanks to the efforts of the Society), **Pryer's woodpecker**, and **raptors**.

Another step that the Society has taken to protect birds is the establishment of the **Wild Bird Protection Fund** and the **Japanese Crane Protection Fund** to set up sanctuaries and to preserve bird habitats. Called *Tancho*, this largest crane in Japan is the country's symbol of happiness, and makes its home in Japan's largest wetland, the Kushiro-Nemuro area of Hokkaido. The World Conservation Union (IUCN) lists Tancho as a vulnerable species in its Red Data Book.

Established on the crane's behalf in 1987 is the **Tsurui-Ito Tancho Sanctuary**, "the first private operation in Japan designated to preserve this particular species, with full-time rangers attending." Prior to its opening, the local villagers kept the cranes from extinction by feeding them every winter for 36 years. With 383 cranes counted in 1986–1987, members of the Sanctuary are to equal that amount so that each may foster one individual for a five-year period. The Sanctuary reports that is has not yet reached its target figure.

Working to increase the number of sanctuaries, such areas that WBSJ maintains include Lake Utonai Sanctuary in Hokkaido, Fukusima-city Sanctuary in Fukushima, Oi Bird Park in Tokyo, Kaga-city Kamoike Sanctuary in Ishikawa, Keep Kiyosato Sanctuary in Yamanashi, and Yokohama Nature Sanctuary in Kanagawa.

Research projects of the Society include conducting the **Nationwide Waterfowl, Wader, and Steller's Sea-Eagle Count**, to study the habitat and distribution of birds and to investigate poaching.

WBSJ works with international conservation organizations "to protect important areas for migratory birds and tries to prevent the import or export of endangered birds and animals," it reports.

■ Resources
The Society publishes an English-language field guide entitled *Birds of Japan* and *Yacho*, a monthly magazine for members. Gift items and other bird books are also available.

WBSJ seeks satellite tracking for migration bird studies and help with sanctuary planning and public education.

World Resources Institute (WRI)
ABS Building, 2-4-16 Kudan Minami
Chiyoda-Ku
Tokyo 102, Japan

Contact: **Miwako Kurosaka**, Senior Associate
Phone: (81) 3 3221 9767 • *Fax:* (81) 3 3239 2817

Activities: Research • *Issues:* Biodiversity/Species Preservation; Deforestation; Energy; Global Warming; Sustainable Development

▲ Projects and People
Miwako Kurosaka is involved in a number of environmental projects, including his work with the **World Resources Institute** (WRI). Kurosaka served on the Advisory Committee in Global Environment regarding participation on the 1992 United Nations Conference on Environment and Development (UNCED). He also is on the Consultative Committee on Environmental Problems, Development Bureau of the Overseas Economic Fund (OECF); the Yokohama Green Preservation Planning Council for the city government; and the Study Group on Environmental Issues in International Development, Ministry of Finance. "Technology transfer" was the subject of a study group with the Research Institute of Innovative Technology for the Earth.

In the private sector, Kurosaka is in a Think Tank Study Group of the Sasakawa Peace Foundation and involved in the Global Environmental Forum and the Japan Center for Human Environmental Problems. With a background in occupational health and safety, Kurosaka also assisted in research on "Conceptions of Nature and Wildlife in Japan" with Dr. Stephen Kellert, Yale University's School of Forestry and Environmental Studies.

Kurosaka also served as the chief editor of the three-volume set *Resonance in Nature* published by Shisakusha Publishing. The volumes are *Nurturing Creative Imagination in Childhood*, *Reflections on Japanese Relationships between Man and His Environments*, and *Towards Creative Relations with Nature*. He has also prepared reports on environmental education in urban areas. *(See also World Resources Institute in the USA NGO section.)*

♜ *Universities*

Graduate School of Environmental Science
Hokkaido University
N10, W5
Sapporo 060, Japan

Contact: **Seigo Higashi**, Associate Professor
　Phone: (81) 11 716 2111, ext. 2265
　　Fax: (81) 11 747 9780

Activities: Research • *Issues:* Biodiversity/Species Preservation

▲ Projects and People
Prof. Seigo Higashi, Hokkaido University, belongs to the Species Survival Commission (IUCN/SSC) Ant Group.

Department of Obstetrics and Gynecology ⊕
Tokyo Medical College
1–7, Nishishinjuku 6
Tokyo 160, Shinjuku, Japan

Contact: **Hiroaki Soma, Ph.D.**, Professor Emeritus
　Phone: (81) 3 342 6111 • *Fax:* (81) 3 3349 8277

Activities: Education; Research • *Issues:* Biodiversity/Species Preservation; Health; Population Planning

▲ Projects and People
With an interest in medical aid for developing countries, Dr. Hiroaki Soma is professor emeritus, Tokyo Medical College; visiting professor at Saitama Medical School and Nepal's Tribhuvan University, Institute of Medicine; and honorary professor at West China University of Medical Sciences.

Involved in preserving and breeding endangered species in the Nepal Zoological Garden, Dr. Soma is a member of the Species Survival Commission (IUCN/SSC) Caprinae Group and a Fellow of the Zoological Society of San Diego, USA. He is co-editor of *The Biology and Management of Capricornis and Related Mountain Antelopes.*

Tokyo University of Fisheries (TUF)
4-5-7, Konan, Minato-Ku
Tokyo 108, Japan

Contact: **Takashi Okutani, Ph.D.**, Invertebrate
　　　　　　Zoology Professor
　Phone: (81) 3 3471 1251, ext. 316
　　Fax: (81) 3 3471 5794

Activities: Education • *Issues:* Fishery Science

● Organization
Tokyo University of Fisheries (TUF) has four major departments: Marine Science and Technology, Aquatic Biosciences, Fisheries Resources Management, and Food Science and Technology. "Every department concerns environmental studies from its standpoint and expertise, such as physico-chemical oceanography, safety food, biological oceanography, and marine biology," writes Dr. Takashi Okutani, invertebrate zoology professor.

▲ Projects and People
Dr. Okutani reports he is working on the "biological process of concentration of harmful substances into the molluscan body." He is also working on the Red Data Book of the Japanese Environmental Agency, CITES consultation with the Japanese Fisheries Agency, and is "temporal advisor" in biological aspects of environmental changes. (CITES is the Convention on International Trade in Endangered Species of Wild Fauna and Flora.) Dr. Okutani was senior curator of the National Science Museum's Zoology Department, Tokyo, and did postdoctoral studies at Scripps Institution of Oceanography, USA. He belongs to the Species Survival Commission (IUCN/SSC) Mollusc Group.

Botanical Gardens
University of Tokyo
3-7-1 Hakusan, Bunkyo-ku
Tokyo 112, Japan

Contact: **Kunio Iwatsuki, Ph.D.**, Director
　Phone: (81) 3 3814 0318 • *Fax:* (81) 3 3814 0319

Activities: Education; Research • *Issues:* Biodiversity/Species Preservation

● Organization
The University of Tokyo has two botanical gardens, the main Koishikawa Botanical Garden and the branch Nikko Botanical Garden. The primary functions of both of these gardens are education and research.

Founded in 1684 as "a garden for medicinal plants in service to the Yedo Era Tokugawa Shogunate," the Koishikawa Garden was transferred to the University in 1877 and cultivates 4,000 taxa of plants primarily from Japan and adjacent Asia. "They are cultivated in compartments such as the systemic garden, medicinal plant garden reserve section, broad-leaved and coniferous forests, montane plant garden, Japanese-style garden, fern garden, nurseries and greenhouses," according to the Gardens' literature. The Nikko Garden has "2,200 taxa mainly indigenous of temperate and alpine regions planted in deciduous broad-leafed and coniferous forests on hilly grounds, rock gardens, and bog gardens. Of particular interest are the collections of indigenous wild species of *Acer, Prunus, Rhododendron,* and conifers."

▲ Projects and People
Plant taxonomy and systematic botany are the main subjects of research in the Botanical Gardens. Studies on a variety of vascular plants are in progress. Biosystemic studies are also being conducted.

Koishikawa Botanical Garden is the site of the bimonthly pteridological meeting and a monthly meeting on evolutionary ecology.

The Botanical Gardens have much international contact. Staff members conduct research overseas, such as Sino-Japanese joint botanical research in Yunnan, and USA-Japan research on the comparison of vegetation between East Asia and eastern North America. Garden scientists participate in global conferences and host researchers from abroad. Staff members have conducted overseas field research in China, Indonesia, East Asia, and eastern North America. They have hosted botanists from China, India, Russia, UK, USA, and Singapore.

One of the most important activities of the Gardens is the storage and exchange of seeds and spores, "particularly those of wild plants." Collected every season, the current stock "is contained in 4,523 bags/tins of seeds of 37 families, 83 genera, and 111 species."

A special effort is being made to conserve endangered plants such as *Melastoma tetramerum* Hayata, *Metrosideros boninensis* (Hayata) Tuyama, and *Rhododendron boninense* Nakai of the Bonin Islands.

Seedling cultivation of two endangered orchid species, *Calanthe hoshii* S. Kobayashi and *C. hattorii* Schlecter, is being attempted on the Bonin Islands.

The Gardens run a herbarium, which was founded in 1877. "The number of vascular plant specimens represented totals 1,500,000," which have been collected from around the world. "Loan of herbarium specimens is available in request from recognized botanical institutes," report the Gardens. Exchange of specimens from Japan, The Himalayas, and Southeast Asia is also available.

Staff also participates in undergraduate and postgraduate teaching and research.

■ Resources

The Journal of the Faculty of Science, the University of Tokyo Section III is published yearly by the Botanical Gardens. An annual *Report on the Botanical Gardens* also lists publications and hours open to the public.

Department of Veterinary Anatomy
Faculty of Agriculture
University of Tokyo
1–1 Yayoi, Bunkyo-ku
Tokyo 113, Japan

Contact: **Yoshihiro Hayashi, D.V.M., Ph.D.,**
            Professor
Phone: (81) 3 3812 2111, ext. 5383
Fax: (81) 3 5800 6896
Activities: Research • *Issues:* Biodiversity/Species Preservation
Meetings: IUCN/SSC

▲ Projects and People

Dr. Yoshihiro Hayashi, veterinary anatomy professor, University of Tokyo, is currently engaged in protecting "endangered animal species in Nansei Islands—Ryukyu and Amami Islands—in Japan. We have organized a special committee for Nansei Islands as part of World Wide Fund for Nature—Japan (WWF) activity."

Dr. Hayashi, who has also researched "morphological studies on the skull of Japanese wild boars," belongs to the Species Survival Commission (IUCN/SSC) Lagomorph Group. He seeks new technologies in "reproductive biology to rescue endangered mammals."

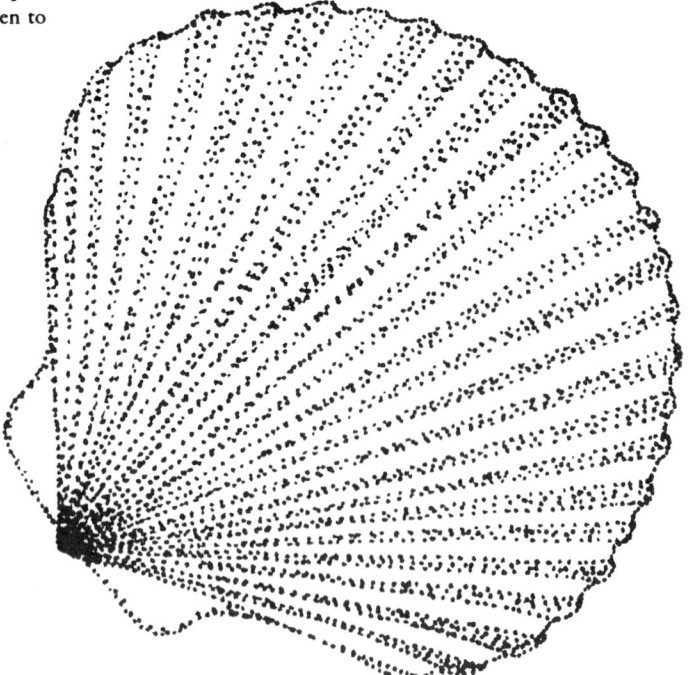

Jordan

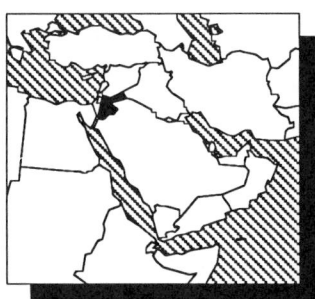

Jordan is an almost landlocked country of rocky deserts, mountains, and plains, bordered by Syria, Saudi Arabia, and Israel. The country does not have abundant natural resources: only 9 percent of the land is cultivable, and forests cover less than 1 percent of the country. Wildlife has been decreased by overhunting, and there are few mineral resources. Environmental problems include a shortage of water resources for human consumption and irrigation, which leads to disease and infection; deforestation due to agricultural clearing, fuelwood exploitation, and overgrazing; soil erosion, which leads to reduced land productivity and siltation of reservoirs. The government has responded to these problems, however, with the passage of major environmental laws.

🏛 *Government*

Environmental Research Centre (ERC)
Royal Scientific Society (RSS)
P.O. Box 925819
Amman, Jordan

Contact: **Dr. Murad Jabay Bino**, Director
Phone: (962) 6 840373 • *Fax:* (962) 6 844806

Activities: Development; Research • *Issues:* Air Quality/Emission Control; Deforestation; Sustainable Development; Waste Management/Recycling; Water Quality

● Organization

In 1970, the **Royal Scientific Society (RSS)** was established as a "nonprofit national institution" involved in scientific and technological research and development work in Jordan. When the Higher Council for Science and Technology was established in 1987 under the chairmanship of HRH Crown Prince El-Hassan, RSS became one of its scientific and technological centers. RSS has a president, 10 directors, and 551 employees, many of whom have advanced degrees. It also consists of 10 centers and departments, including the **Environmental Research Centre (ERC)** and the **Renewable Energy Research Centre**. ERC was inaugurated in late 1989, during Sweden's King Carl XVI Gustaf's official visit.

▲ Projects and People

ERC conducts studies and applied research; provides technical services, consultations, and training in the environmental field for the private and public sectors in Jordan and neighboring Arab countries; designs and operates water and wastewater treatment plants; maintains hazardous chemicals records; and proposes suitable storage and handling methods. Its staff of 30 is divided into three divisions: air and hazardous chemicals, ecology, and water and soil, with additional specialized laboratories and units also cooperating with RSS, which are considered the country's reference laboratories.

Several ERC projects focus on water quality. These include the **National Project**, the first comprehensive water quality surveillance in Jordan; the **Assamra Waste Stabilization Ponds Project** testing water quality at one of the area's largest treatment plants at a point where wastewater is collected; the **Wadi Arab Reservoir Project**, which assesses pollutants and eutrophication, for example; the **King Talal Reservoir Project**, evaluating seasonal and annual water and studying potential water reuse for drinking and agriculture; and the **Environmental Control at Aqaba Region**, monitoring drinking water, industrial and domestic treatment plants' effluent, thermal pollution of sea water, and water quality for bathing.

Other ERC activities involve industrial wastewater treatment plants, the design and construction of a cyanide treatment plant, and a transformer oil project to identify locations and determine concentrations of PCBs (polychlorinated biphenyls) in Greater Amman to take necessary steps to protect the environment.

Air pollution monitoring in the city of Amman was conducted by ERC in a project from 1985 to 1990 to discover the impact of industry and automobiles on air and establish a national air quality standard in Jordan. A similar program at selected populated sites in Zarka, around the Jordanian Oil Refinery, determined the concentration of hydrogen sulfide and compared its results with previous studies. These findings are expected to provide a parameter of safety for the future and discourage pollution emitters.

Three other ERC projects delved into the role of environmental pollution. The comprehensive hazardous chemicals survey evaluated handling, transporting, and storing of chemicals, as well as emergency procedures in care of accidents. A safety directory and databank including 3,000 dangerous chemicals was established. Phosphate dust in the Port of Aqaba and at the Jordan Phosphate Mines Company was monitored at four stations to test for phosphorus penta-oxide. And, as a result of the medical gases project in 1987 to 1988, a fully equipped laboratory specializing in medical and industrial gases analysis was developed, and standards and regulations for these gases were issued.

Dr. Murad Jabay Bino, chemical engineer, is project leader for the Jordan National Water Pollution Study, among other duties as ERC director. He recently wrote articles on "The Possible Effects of the Gulf War on Marine Resources Pollution" and "Environmental Issues in Water Resources and Coastal Management." Dr. Ali Elkarmi, ecology division manager, is responsible for creating a countrywide ecology map pinpointing pollution, biotechnology research, industrial pollution surveys of Arab States, studies regarding solid waste recycling, and investigations of solar water disinfection systems.

Dr. Nageh Y. Akeel, air pollution and hazardous material division manager, is also project manager for monitoring PCBs and hydrogen sulfide in transformer oils. Subhi A. Ramadan is water and soil division manager as well as advisor to United Nations Environment Programme (UNEP), UN, and World Health Organization (WHO) on agricultural related pollutants, eutrophication, water quality, and globally significant, harmful chemical substances. Mohamad Saidam is water technology unit manager and author of *Anaerobic and Facultative Ponds in the Middle East.* Ayman Afif Al-Hassan is head of the industrial chemical department's air pollution laboratory and project manager of monitoring of phosphate dust in Aqaba as well as of Jordanian-Canadian air pollution monitoring undertaking.

■ Resources

ERC seeks lab equipment and any other assistance to help continue its programs.

 *NGO*

Royal Society for the Conservation of Nature (RSCN)
P.O. Box 6354
Amman, Jordan

Contacts: **Maher Abu Jafar,** Director General
Phone: (962) 6 811689 • *Fax:* (962) 6 628258

Anis Mouasher, President
Phone: (962) 6 624907

Activities: Education; Law Enforcement; Political/Legislative; Research • *Issues:* Biodiversity/Species Preservation; Conservation Education; Deforestation; Energy; Sustainable Development; Water Quality • *Meetings:* CITES; IUCN; Ramsar; WWF

● Organization

Established in 1966, the **Royal Society for the Conservation of Nature (RSCN)** is a private, voluntary organization with public service status and noted as the only non-governmental organization in Jordan dedicated to the conservation of nature and natural resources. Its role expanded in the mid-1970s to include law enforcement of wildlife protection. Acting in conjunction with official Jordanian institutions, and Arab and international organizations with similar interests, RSCN conducts scientific and technological research and developmental work and disseminates its services to the public and private sectors.

Among RSCN's objectives are protection of indigenous species and their natural habitat; reintroduction of endangered and locally extinct species; prevention of environmental pollution, particularly in the Gulf of Aqaba; supervision and enforcement of hunting regulations; establishment of national parks and nature reserves; promotion of public awareness on wildlife issues; coordination with Ministry of Education on environmental curricula in school system; and the protection of national heritage sites.

▲ Projects and People

With a membership of 450, a staff of 52 helps the Society conduct its environmental agenda, which is divided into three divisions—information/public awareness, nature reserves, and protection. There are also nine separate committees which further examine individual Jordanian areas of concern. These include the Aqaba Gulf, olive farming, hunting control, and nature reserves committees.

In cooperation with the Ministry of Education, RSCN's information/public awareness division works in schools, setting up 150 wildlife conservation clubs. RSCN staff presents films, lectures, and audiovisual programs periodically.

The nature reserves division seeks to protect indigenous fauna and flora species and their natural habitats, while implementing management plans and reintroducing endangered or locally extinct species. In 1975, RSCN established its first reserve at Shaumari and offered a visitors information center, another first. Here at the completely fenced location, reintroduction programs of several species began in 1979 with the Arabian oryx, gazelle, ostrich, and the Syrian wild ass. Today, the oryx population is about 100. Some 134 bird species, 130 plant varieties, and mammals—such as red fox, cape hare, and wild cat found—are here.

Other nature reserves include the **Azraq Wetland Reserve (AWR)**, with a "flat area of pools, marshes, water meadows, and salt dunes lying at the heart of the oasis from which it takes its name." AWR is recognized as a wetland of international importance for its **migratory waterfowl**. **Dana Wildlife Reserve** is known for its breathtaking canyons and gorges where the **mountain gazelle, ibex, leopard,** and **striped hyena** are protected. Vegetation includes natural cyprus, pine, oak, juniper, and pistacia.

The **Wadi Mujib Wildlife Reserve**, ranging from the Dead Sea shore up to the eastern highlands, is important for the protection of the small remaining **Nubian ibex** population and rare orchid species. With its rocky mounts separated by flat, sandy corridors, the **Wadi Rum Wildlife Reserve** is a site of special archaeological findings of prehistoric dwellings and rock drawings. It is also inhabited by ibex, gazelle, hedgehog, hyena, and porcupine. The **Zubiya Wildlife Reserve**, one of the few remaining relatively undisturbed oak woodlands in Jordan, is awaiting the reintroduction of the **Persian fallow deer** and the **roe deer**, once locally extinct. Brought from Turkey, two pair of roe deer are breeding successfully in a special enclosure.

An RSCN proposal calls for **establishing additional wildlife reserves** in desert regions, plateaus, and mountains for the purpose of protecting and reintroducing endangered species at Abu Rukbah, Bayir, Burqu, Jarba, Jebel Mas'ada, and Rajil.

Daily patrols are organized by RSCN's protection division to **supervise hunting rules** and the application of legislation. Special hunting licenses and schedules limiting bagging of animals during season are also among the division's assignments.

Another daily patrol activity is done by RSCN boats in the **Aqaba Gulf** to **control pollution** and keep the area free of sewage, oil, and dumping to protect marine life including coral reefs.

RSCN initiated the **olive harvest campaign** when the olive trees were being neglected due to the high cost of labor wages for harvesting. Thousands of students voluntarily participate to help farmers preserve the valuable trees, which have provided sustenance to the area's people for centuries.

RSCN also joins various ministries, universities, and the army in a national "**Clean-Up Jordan**" campaign; and in a plan to tackle waste disposal by encouraging recycling and alternative, reusable grocery bags, for instance. Students are encouraged to take part in **conservation clubs**, and RSCN offers training courses to teachers. Through its **Green Cover** program, the Society each year plants a new forest of indigenous tree species in cooperation with the government.

Future plans include the designation of an ecological **Jordan Nature Year** in which "all laws, legislation, and programs involving the protection of Jordan's nature are completely adopted in order to insure man's safety, health, and happiness."

Maher Abu Jafar belongs to the Species Survival Commission (IUCN/SSC) Equid and Re-Introduction groups. A pharmaceutical chemist and businessman, RSCN president **Anis Mouasher** is also a member of World Wildlife Fund's (WWF) international advisory council, and holds other leadership posts.

■ Resources

RSCN offers informational materials in both Arabic and English, such as *The Reintroduction of the Arabian Oryx to Its Habitat in Jordan, The Azraq Wetland Reserve*, and a bird guide to 343 species. The quarterly magazine *El-Reem* is published in Arabic with brief English summaries. A publications list is included in the RSCN handbook.

🎓 *Universities*

Science & Environmental Education
Yarmouk University
Irbid, Jordan

Contact: Moh'd Said Subbarini, Ph.D., Associate Professor
Phone: (962) 2 271100, ext. 2503
Fax: (962) 2 274725

Activities: Education; Research • *Issues:* Biodiversity/Species Preservation; Deforestation; Energy; Health; Sustainable Development

▲ Projects and People

At Yarmouk University, Dr. Moh'd Said Subbarini is both associate professor of science and environmental education as well as Dean of Faculty of Education and Fine Arts. In addition, Dr. Subbarini is a consultant in environmental awareness and environmental issues for the Regional Organization for the Protection of Marine Environment (ROPME/Kuwait) and Arab Towns Organization (ATO/Kuwait).

He holds membership in the Royal Society for the Conservation of Nature in Jordan; the Kuwait Environmental Protection Society; and the Commission for Biological Education of the International Union of Biological Science (CBE/IUBS). Fourteen published books and 38 research papers in local, regional, and international journals are to his credit in the fields of environmental education, science education, and general education.

Kenya is a land of considerable environmental diversity, including forests, woodlands, swamps, grasslands, and more than 7,800 plant and animal species. Its resplendent natural resources include the rolling savannah, the desert regions of the north, the glaciated peaks of Mt. Kenya, Rift Valley, and the coral beaches. Yet, forests cover no more than 2 percent of Kenya's land area.

The country's environmental and political goals have been closely connected for many decades. These include: the reduction of poverty, which has contributed to the exploitation of natural resources and prevented long-term investment in conservation; food security, currently hindered by land degradation; biological diversity maintenance; the securing of water and energy sources; and environmentally sound industrial planning. Non-governmental organizations have been most effective in working for environmental protection.

🏛 *Government*

Environment and Natural Resource Management ⊕

International Development Research Centre (IDRC) Canada and Kenyatta University
Social Science Division, IDRC
P.O. Box 62084
Nairobi, Kenya

Contact: **Daya U. Hewapathirane, Ph.D.,** Professor/Director of Research
Phone: (254) 2 330850 • *Fax:* (254) 2 214583

Activities: Development; Research • *Issues:* Biodiversity/Species Preservation; Deforestation; Flood Hazard Management; Sustainable Development • *Meetings:* International Symposium on Environmental Change; Kenya National Symposium on Natural Disasters and Their Management

▲ Projects and People

Environmental changes and instabilities in the Kano Plain region of Kenya prompted **Dr. Daya U. Hewapathirane,** professor at **Kenyatta University** and researcher at the **International Development Resource Centre (IDRC) Canada,** to study natural conditions so affected. Previously, Dr. Hewapathirane researched in the fields of disaster management, floods and related human adjustments, water and related land resource development, and environmental impact assessment (EIA).

Funded by the IDRC's Regional Office in Nairobi, Dr. Hewapathirane began an intense examination of the Kano Plain, located to the east of Lake Victoria, which is subject to flooding and other dangers. Three interdisciplinary projects are analyzing the situation: **assessment of flood effects and identification of nonstructural measures to minimize losses; study of indigenous adaptations and adjustment to floods; and swamps: conservation versus development,** including a videotape production.

The projects utilize the assistance of researchers in the fields of geography, sociology, anthropology, economics, engineering, hydrology, climatology, biology, ecology, computer science, and remote sensing.

Objectives include compiling a description of the Kano Plain's people—mainly rural farmers—in terms of their "demographic structure, cultural traits, and socio-economic characteristics"; examination of the attributes of floods and their related environmental and socioeconomic impacts; investigation of the "nature, costs, and organizational aspects (including the role of public agencies) of the strategies and measures adopted to cope with floods"; and finally, the identification and characterization of feasible "non-structural responses to floods."

Results are expected to have "direct implications for planning against flood losses and planning flood plain resources use and management for sustainable development," writes Dr. Hewapathirane, "in terms of [the region's] atmosphere, hydrosphere, biosphere and lithosphere. The social implications of these are of extreme importance particularly because of the special characteristics of the culture of communities inhabiting the area."

All government agencies in the Kano Plain region are involved with the research, as are local grassroots organizations regarding the village-level study. There also is participation in international symposiums, seminars, and congresses which deal with the subject of environmental change, hazards, and hazard effects.

Since 1987, Dr. Hewapathirane has been an associate of the World University Service of Canada, which is affiliated with Kenyatta University's Geography Department.

■ Resources

Project reports, speakers, and a 45-minute video are available on various aspects of Kano Plain research.

Kenya Wildlife Service ⊕
P.O. Box 40241
Nairobi, Kenya

Contacts: **Richard E. Leakey, Ph.D.**, Director
          Fred K. Waweru, Ph.D., Senior Ecologist
Phone: (254) 2 501081 • *Fax:* (254) 2 505866

Activities: Development; Education; Research • *Issues:* Biodiversity/Species Preservation; Deforestation; Global Warming
Meetings: CITES

▲ Projects and People

Dr. Richard E. Leakey, Kenya Wildlife Service, is a world renowned conservationist. Dr. Fredrick K. Waweru, also a wildlife management lecturer at Moi University, where he develops field activities for plant ecology and vertebrate population dynamics, has a background as an aerial surveys biologist.

Research interests and consultancies include "the black rhinoceros population status and ecology in Nairobi National Park, establishing **ecological monitoring** programmes in all rhino sanctuaries within the country, status of the **Rothschild's giraffe** translocated to Lake Nakuru National Park, above-ground **browse assessment** in Oljogi Game Ranch, [and] checklist of fauna and flora of Olambwe Valley National Park." He has also published articles on such topics.

Dr. **Waweru** represents the East African Wild Life Society (EAWLS) in the World Conservation Union's (IUCN) Species Survival Commission (SSC) and belongs to the IUCN/SSC African Elephant and Rhino Group.

 *NGO*

African Centre for Technology Studies (ACTS)
P.O. Box 45917
Nairobi, Kenya

Contact: **Calestous Juma, Ph.D.**, Executive Director
Phone: (254) 2 744047 • *Fax:* (254) 2 743995

Activities: Development; Education; Legislative; Research
Issues: Biodiversity/Species Preservation; Deforestation; En-

ergy; Global Warming; Health; Sustainable Development; Waste Management/Recycling; Water Quality

● Organization

The **African Centre for Technology Studies (ACTS)** is a nonprofit institution which conducts policy research and training and disseminates information on the environment, science, and technology.

▲ Projects and People

Founded in 1988 by **Dr. Calestous Juma**, ACTS has a staff of 20. In 1991, Dr. Juma received the **Pew Scholars Program in Conservation and the Environment** grant award for $150,000 to support professional endeavors. The Pew Charitable Trusts award helps rally "African talent and scholarship" toward problems facing the continent, specifically those emphasizing biological diversity, Juma noted. The funds, administered over a three-year period, enable ACTS to explore alternative management techniques.

Founded by ACTS in the Netherlands, the **Biopolicy Institute** at Maastricht draws maximum benefits from the award. The Institute promotes collaborative research in biological diversity and biotechnology with industrialized countries' organizations. According to *Innovation Magazine of Technology and Sustainable Development* (August 1991), it is the first time an institution in a developing country established a research arm in the industrialized world.

"Juma wants to see ACTS grow into a world class policy research institution, a dream he has nursed ever since its inception three years ago," reports *Innovation.* "He regards his award as part of Africa's institutional evolution and a demonstration of the viability of public policy research—'an activity that has often been viewed as either impossible or irrelevant,' he says."

ACTS also receives support from the Mennonite Central Committee (MCC) and the Lutheran World Relief.

■ Resources

ACTS needs educational outreach literature and materials, political advocacy support, and environmental policy studies to continue its programs. It publishes *Innovation,* which also features notices of upcoming events.

African NGOs Environment Network (ANEN) ⊕ 🍃
P.O. Box 53844
Nairobi, Kenya

Contact: **Simon M. Muchiru**, Executive Director
Phone: (254) 2 28138 • *Fax:* (254) 2 33518

Activities: Environmental Management • *Issues:* Biodiversity/Species Preservation; Deforestation; Energy; Global Warming; Sustainable Development • *Meetings:* CITES

● Organization

"In pursuit of sustainable development," representatives of African indigenous NGOs from nine African nations founded the **African NGOs Environment Network (ANEN)** in 1982. They realized environmental problems affecting the continent transcend national boundaries and were associated with "lack of development or the mismanagement of the environment and natural resources."

▲ Projects and People

ANEN's main objective is to respond to the needs of grassroots community groups while strengthening their capacity and technical competence. Through its technical assistance program, ANEN organizes workshops for training, monitoring, evaluation, and implementation of environmental and development projects. Seed monies for grassroots projects are also available from ANEN. And an African expert and consultant exchange program allows such persons to devote 3 to 12 months in a particular community assisting where needed.

With a staff of 6 and more than 430 members from 45 African countries, ANEN also strives to ensure that the concerns of women and youth are considered, that local people are fully involved, and that the environmental dimensions are incorporated initially in the planning stage. A network of African environmental journalists helps promote environmental awareness and information exchanges. An Environmental Resource Centre houses information on African NGO activities.

Globally, ANEN works with the Cairo AMCEN Programme of Action for Africa Cooperation, OAU Africa Priority Programme for Economic Recovery (APPER), Africa 2000 Network, and the United Nations Programme of Action for Africa Recovery and Economic Development (UNPAARED).

Simon M. Muchiru is also coordinator of the African Rainforest Network, member of the Earth Day International Advisory Board, vice president of the Forum of Africa Voluntary Development Organizations (FAVDO), editor of *Rainforest News* magazine, and founder and editor of *Pesticide Digest Newsletter* and *EcoAfrica*. Topics of his publications and presentations range from Lake Nakuru's fragile ecosystem to sustainable agriculture and Third World farmers, from energy planning and pesticides use to developing youth environmental citizenship. For children's magazines, Muchiru writes about "hippos, the floating giants," and "giraffes, the gentle giants," among other wildlife.

■ Resources

EcoAfrica, a bimonthly English/French magazine that documents grassroots programs and features emerging issues, is available from ANEN in addition to other issue papers, case studies, action guides, directories, and environmental reports. ANEN needs new technologies, educational outreach, and public relations to continue its programs.

Amboseli Elephant Research Project
African Wildlife Foundation ⊕
P.O. Box 48177
Nairobi, Kenya

Contact: **Cynthia J. Moss,** Director
Phone: (254) 2 223235 • *Fax:* (254) 2 332294

Activities: Education; Research • *Issues:* Biodiversity/Species Preservation

▲ Projects and People

Formerly a journalist, **Cynthia J. Moss** is director of the **Amboseli Elephant Research Project** and senior associate of the **African Wildlife Foundation.** Early in her career, Moss participated in research projects concerning the feeding ecology of elephants at Tsavo National Park, and the environmental physiology of elands, hartebeest, zebra, wildebeest, and ostrich.

Starting in 1982, she began supervising research on the Amboseli elephant as well as directing fundraising and writing books, scientific papers, and popular articles. She also lectures on elephant behavior to students and wildlife groups throughout Kenya, USA, and UK. Recently, she was scientific director for a BBC documentary on Amboseli elephants. Previously, *Portraits in the Wild* was nominated for an American Book Award for best science paperback of the year. *Echo of the Elephants* is being published by BBC Books in late 1992 and *Elephants* is in preparation for Whittet Books. Moss, who is a recipient of the Smith College medal for alumnae achievement, has also written about the black rhino and the hyena.

A member of the Species Survival Commission (IUCN/SSC) African Elephant Group, Moss participates in elephant surveys and trains researchers in Tanzania, Zaire, and Botswana.

East African Wild Life Society (EAWLS) ❦
P.O. Box 20110
Nairobi, Kenya 254

Contact: **Nehemiah K. arap Rotich,** Executive Director
Phone: (254) 2 748170 • *Fax:* (254) 2 746868

Activities: Education; Research; Law Enforcement • *Issues:* Biodiversity/Species Preservation; Deforestation; Global Warming; Health; Waste Management/Recycling; Water Quality • *Meetings:* CITES; IUCN; New England Environmental Network Conference; UNCED

● Organization

The East African Wild Life Society (EAWLS) originated in 1961 when the Wild Life Societies of Kenya (founded in 1956) and Tanzania joined wildlife enthusiasts in Uganda with a goal of conserving wildlife in East Africa and "to safeguard wildlife and its habitat in all its forms, as a national and international resource." The Society has a worldwide membership of 12,000, a staff of 30, and representatives from over 30 countries serve on its various committees.

Once, the Society's main goal was to save large wildlife—cheetah, lion, leopard, and elephant—from the game hunter. It soon became apparent that smaller species, such as butterflies and birds, were vital to habitats now regarded as national and international resources. Today, the Society's biggest challenge is increasing public awareness and wildlife education. "Without this understanding, all efforts by conservationists and government, however financially well supported, are doomed," writes the Society.

▲ Projects and People

There are five categories in which the Society's work is performed: animal rescue, antipoaching support, conservation education, research, and specific projects. Because many developing countries lack funds for conservation and development, the role of NGOs is welcomed. "In this manner the Society takes a great deal of pressure off the governments of East Africa," states the Society.

Animal rescue or "**translocation**" activities funded by the Society relocate threatened animals to safer areas, thereby helping ensure the species. For example, the **black rhinos** have been provided with both safe breeding areas and protection from poachers by transfers to rhino sanctuaries.

Society funds are also used to train personnel for the government's **antipoaching** units, supplying radio communications equipment, and servicing of aircraft and other vehicles.

A large portion of the Society's budget is geared to the promotion of education and wildlife awareness because, notes the Society, the people themselves must first realize the value and need for cooperation conservation if the programs are to be effective. However, the Society reports, "The man in the bush does not appreciate being told that hunting is wrong, for he and his ancestors have always hunted. The man in the bush . . . does not read the daily paper." Films, publications, and wildlife activities are financed by the Society to help educate **200,000 young people** in wildlife clubs in Kenya, Tanzania, and Uganda.

Many East African **university students** train with local wildlife managers and ecologists using allocated Society funds for species surveys and establishing conservation strategies.

Among special activities sponsored by the Society were the efforts of **Michael Werikhe**, also known as "Rhino Man," whose walks in Africa, Nairobi to Mombasa, Kampala to Mombasa, and through Europe increased public awareness of the plight of the black rhino.

Some of the wildlife projects in Kenya funded by the Society include: the relationship between **flamingoes and phytoplankton**, with aquatic vegetation mapping and estimations of pollution in Lake Nakuru; effects of **water stress and grazing intensity** on two forage grasses, *Panicum maximum* and *Eragrostis Superba*, being tested at the Kiboko National Range Research Station; best **grass establishments methods** and irrigation impacts in the Uaso Nyiro River area; overcoming slow growth and low productivity of the indigenous *Acacia albida* in arid and semiarid areas through manipulating the tree's microsymbionts and other **genetic improvement**; fence extension at the **Ngulia Rhino Sanctuary**, Tsavo West National Park, and camera surveillance at the **Aberdare Rhino Sanctuary**; common edible **mushrooms** study; and manpower training in **monitoring wetlands** and counting waterfowl.

In a 3-year program to prevent "illegal incursions" into a 25-kilometer area of Mau Forest, the Society is providing vehicles, radios, and technical support. A comparative study of structure composition and distribution of plant communities in Lake Nakuru National Park is ongoing, and so is a taxonomic and ecological survey of the **orchid flora** of the Ngong-Hills, with a herbarium collection and potential orchid house for public education. A mollusk **survey** will determine effects of shell collection. **Aquatic ecosystems** are being assessed for physicochemical factors causing the spread of certain weeds, and the **biomass of algal species** in Lake Victoria's Winam Gulf is being measured. **Crowned cranes** and their role as indicators of habitat quality in some of Kenya's wetlands is being valued. The Society participates in the **annual Kenya Winterfowl Counts** with the International Waterfowl and Wetlands Research Bureau (IWRB).

As the deforestation and decimation of plants and animals in Emkwen, Kericho District is being investigated, "attitudes and perceptions" of the local people are being identified that would "reinforce or go against the conventional conservation measures for soil, natural forests, and endangered species." With the University of Nairobi, **medicinal plants** are being surveyed and with the National Museums of Kenya, aloes are undergoing phytochemical studies. A clear boundary, also serving as a forest break, is delineating the Mount Kulal Biosphere Reserve. At Ruma National Park, **roan antelope** are being counted and their relationship with other herbivores observed.

■ Resources

The Society needs computers, materials for schools and colleges, political support, media publicity, and leadership training to continue its variety of programs. It produces the scientific publication *African Journal of Ecology; Swara* (meaning antelope in Swahili) bimonthly magazine with 16,000 subscribers worldwide; *Swara Ranger* newsletter; research studies; and slides, films, and lectures.

Eastern Africa Environmental Network (EAEN) ⊕
P.O. Box 20110
Nairobi, Kenya

Contact: **Nathaniel Kiprugut arap Chumo**, Director
Phone: (254) 2 748170 3 • *Fax:* (254) 2 746868

Activities: Education • *Issues:* Air Quality/Emission Control; Biodiversity/Species Preservation; Deforestation; Energy; Global Warming; Health; Population Planning; Sustainable Development

● Organization

A coalition of governmental and non-governmental organizations formed the **Eastern Africa Environmental Network (EAEN)** in 1990. Its goal is to promote environmental education, intensify public awareness, and find lasting solutions toward achieving sustainable development in Ethiopia, Kenya, Somalia, Sudan, Tanzania, Uganda, and Djibouti. The Network uses the **New England Environmental Conference** as a model. *(See also Tufts University, Environmental Citizenship Program in the USA Universities section.)*

▲ Projects and People

With its 60 environmental organizations and over 400 members, EAEN's **networking** allows others to know they are not working in isolation, prevents duplications of efforts, and creates a forum for discussions on common East African issues.

More than 350 persons representing 4 nations participated in EAEN's first annual conference in June 1991. Issues addressed included biodiversity, national conservation strategies, toxic waste management, harmful effects of chemical residues in agricultural lands, human population growth challenges, lobbying for environmental protection, rural space management, indigenous forest conservation, and the future of Lake Victoria's Nile perch.

EAEN director **Nathaniel Kiprugut arap Chumo** has a varied background as a biology teacher, ecologist with the Ministry of Environment and Natural Resources, Kenya organizer of Wildlife

Clubs for nearly 20 years for 11- to 19-year-old members, editor, creator of radio scripts and film shows, workshop and leadership course arranger, and designer of peaceful rallies and other antipoaching activities. For his outstanding efforts in "duty and service to wildlife," Chumo received conservation awards from the African Wildlife Foundation and Mzuri Safari Foundation. The latter honor commemorates the late David Sheldrick, who was warden of the huge Tsavo East National Park for 28 years. With a World Wildlife Fund (WWF) fellowship, Chumo studied environmental education at Michigan State University, USA, later receiving his master's degree at the University of Nairobi.

Charles J. Kara, EAEN executive coordinator, launched in 1991 a yearlong horseback Ride for Sudan through Zimbabwe, Zambia, Tanzania, and Kenya to generate funds and public awareness regarding a "wildlife conservation strategy as a basis for economic infrastructure." He is coordinator of the Jebel Kujur Rehabilitation Project, which is a tree-planting program to combat desertification, financed with the Netherlands' help. Kara is also founder of Wildlife Clubs of South Sudan, founder and editor of *Sabra Magazine* and *Trees Newsletter*, and editor of *East Africa Environmental News*.

■ Resources

EAEN has various publications, including a directory of environmental organizations in Eastern Africa, and materials for workshops, conferences, consultations, and speakers programs available. It seeks additional networking in the eastern Africa region and financing for staff travel to attend conferences.

Global Coalition for Environment and Development ⊕

Environment Liaison Centre International (ELCI)

P.O. Box 72461
Nairobi, Kenya

Phone: (254) 2 562015 • *Fax:* (254) 2 562175

Activities: Development • *Issues:* Biodiversity/Species Preservation; Deforestation; Energy; Global Warming; Institutional Building; Sustainable Development; Waste Management/Recycling • *Meetings:* UNCED

● Organization

Established in 1974, the Global Coalition for Environment and Development (ELCI) created an international network of NGOs—linking up with the United Nations Environment Programme (UNEP) in Nairobi, Kenya. Today, ELCI has over 526 member NGOs in 100 countries and contact with around 10,000 organizations and community groups worldwide. The ELCI uses its wide base and large membership to protect the world's ecosystem and promote sustainable development.

ELCI believes that solutions to environmental and resource development problems will be reached through a combination of modern science and grassroots activism. Its aims are to empower grassroots environment and development organizations, to increase NGO involvement in government environmental policy, and to promote networking among regional and worldwide NGOs.

To help NGOs at the local level, ELCI carries out most of its work through five main programs: Women, Environment, and Development; Food Security and Forestry; Energy for Sustainable Development; Industrialization and Human Settlements; and International Environment and Economic Relations.

ELCI's staff includes French and Spanish liaisons to strengthen their networking capacity, as global environmental and development challenges grow. Also, ELCI is a founding member of several issue-oriented and regional networks, such as Pesticide Action Network (PAN), Seeds Action Network (SAN), Kenya Energy Non-Governmental Organizations (KENGO), African NGOs Environmental Network (ANEN), African Water Network (AWN), and the International Toxic Waste Action Network (ITWAN).

▲ Projects and People

With many NGOs or community groups lacking necessary professional, administrative, and management skills, ELCI assists in facilitating processes whereby such organizations can learn from one another through networking and skills exchanges. ELCI also provides grants ranging from $100 to $10,000 to NGOs and community groups for practical field projects.

The program objective of ECLI's internship for visiting NGO staff is to build and strengthen links between NGOs from Africa, Asia, and Latin America. ELCI also accepts volunteers who wish to learn about global environmental and development issues, giving them a chance to work on substantive projects in the Secretariat's Nairobi offices.

■ Resources

ELCI maintains an extensive, updated database of global activities of NGOs, available to NGOs, donor agencies, international organizations, and other interested groups. Clippings of environmental articles from newspapers and journals worldwide are compiled and sent to members on a quarterly basis. *Ecoforum*—published bimonthly in Arabic, English, French, and Spanish—serves as a clearinghouse for information, ideas, and strategies on the environment and development issues. *Ecoprobe* is a series of briefing papers on pressing issues. ELCI also produces practical guides and "how to" books, directories, seminar reports, occasional and discussion papers. A publications list is available.

International Council for Research in Agroforestry (ICRAF) ⊕

P.O. Box 30677
Nairobi, Kenya

Contact: Ester Zulberti, Ph.D., Director, Training and Information
Phone: (254) 2 521450 • *Fax:* (254) 2 521001

Activities: Research • *Issues:* Biodiversity/Species Preservation; Deforestation; Energy; Sustainable Development • *Meetings:* AFRENA Zonal Conference

● Organization

The International Council for Research in Agroforestry (ICRAF), established in 1978, is an autonomous, nonprofit international research council representing developed and developing countries. Headquartered in Kenya, ICRAF initiates, stimulates, and supports research for added sustainable and productive land use. Through the introduction or better management of trees in farming and land-use systems, ICRAF offers a three-fold strategy for the 1990s:

- "Encourage and conduct, jointly with national institutions, applied and adaptive research leading to the development of appropriate **agroforestry technologies** through a rational selection of research priorities based on the identified needs and potentials of selected land-use systems in the **major agroecological zones in Africa.**
- "Conduct **strategic research** on selected topics of **global importance** in which a need has been recognized, arising out of collaborative applied research. It will also encourage its partners and others to undertake strategic research in areas outside its own comparative advantage.
- "Take a leading role in strengthening national capacities to conduct agroforestry research by **encouraging inter-institutional collaboration** and by promoting the dissemination of information on agroforestry through **training** and other activities."

▲ Projects and People

Since many countries lack any national institution with a mandate for agroforestry and few scientists are trained in the multicommodity methodologies required for such research, ICRAF uses an inter-institutional coordinated approach within and across countries. Each collaborating country's goal is the development of its own scientific capacity and research experience, bringing together its own inter-institutional mechanisms and linkages, and allocating necessary levels of resources to conduct its own programs.

Through a memorandum of understanding with ICRAF, eight countries in Africa made **collaborative** arrangements for **agroforestry research programs** in 1989. They are Burundi, Cameroon, Kenya, Malawi, Rwanda, Tanzania, Uganda, and Zambia. After collaborative agreements are made, multidisciplinary teams of national scientists join ICRAF staff in analysis of land-use systems. This includes a macro diagnosis and design exercise covering the entire ecological zone, followed by regional or national workshops analyzing common problems and potential land use. Finally, a detailed analysis, or micro diagnosis and design, of a land-use system selected in accordance with national priorities and institutional capabilities, is conducted.

ICRAF set up four **Agroforestry Research Networks for Africa (AFRENAs)** while recognizing the potential benefits of project work in similar ecological zones. These are the Bimodal Highlands of Eastern and Central Africa; the Humid Lowlands of West Africa; the Semiarid Lowlands of West Africa; and the Upland Plateau of Southern Africa. The ongoing research here emphasizes the importance of soil fertility, fodder, and fuelwood production, and soil loss through erosion.

Launched in 1986 by the International Institute of Tropical Agriculture (IITA) of Nigeria and the International Livestock Centre for Africa (ILCA) of Ethiopia, the **Alley Farming Network for Tropical Africa (AFNETA)** was formed to promote research on hedgerow intercropping or "alley cropping."

In **Kenya**, experiments with dryland agroforestry showed that hedgerow species *Cassia siamea* and *Gliricidia sepium* were well suited to the environment, although farmers expressed concern about labor required to establish and maintain hedgerows. None-

theless, farmers showed an interest in growing citrus, papaya, banana, and mango trees interspersed with maize or fodder for oxen.

At ICRAF's **Maseno Agroforestry Research Centre (MARC)** in western Kenya, **Arne M. Heineman** is especially concerned with sustainable agroforestry land-use alternatives for small-scale farmers on the highlands of east and central Africa. Also at Maseno, scientist **Barrack O. Owuor** is undertaking a tree/shrubs germ plasm project; earlier he researched Arabica coffee breeding. And agricultural economist **Rob Swinkels** in 1990 began pursuing experiments with hedgerows of fast-growing leguminous trees, showing that maize yields are higher when grown between hedges. Prunings of such trees check soil loss through mulching, improve soil fertility, and provide fodder and firewood. The scientists communicated their findings to local "women groups" (which also include men) to gain their cooperation for continuing similar experiments on selected farms, using *Leucaena leucocephala* tree seedlings. Twenty-two farmers volunteered and planted about 550 seedlings per farm; in 1991, participation more than doubled. MARC research officer **Willy Mwangi Muturi** designs, plans, and implements MARC's socioeconomic studies, and has co-authored papers on **economic evaluation of hedgerow intercropping** for AFRENA and other conferences.

Researchers in **Zambia** and **Malawi**, for instance, interviewed farmers who identified multipurpose indigenous trees useful for fruit, fuelwood, medicines, timber and construction poles, shade, soil fertility, and living fences. As of 1993, agroforestry is a national commodity program in **Ethiopia**.

In **Cameroon**, the emphasis is on increasing food crop production on small farms, as the urban population is expected to increase by 70 percent by the year 2000; yet soil is fragile.

In **Ghana**, where crop yields are declining, hedgerow intercropping is also recommended with fodder or other shade-tolerant intercrops introduced on oil palm and coconut plots. In **Mali**, "farmers grow trees scattered in their croplands, giving a 'parkland' appearance." Agroforestry technologies in a variety of settings are being tested in **Niger** to help bring about "self-sufficiency in food production, environmental protection, improved wood production, and the development of intensive livestock production."

Overall, positive results of ICRAF's efforts to establish strong institutional bases for agroforestry research and development at national levels are shown by the governments of Ghana, Malawi, Niger, Senegal, Tanzania, and the Kenya Forestry Research Institute. Governments of Benin, Botswana, Lesotho, Mozambique, and Zaire have also requested ICRAF assistance.

Meanwhile, in **South Asia**, ICRAF works in India—the world's first country to "create a national institute for agroforestry research"—to support its All-India Coordinated Research Project on Agroforestry. ICRAF scientists, for example, designed experiments and demonstrations about multipurpose tree screening and management for home gardens, hedgerow intercropping, fruit trees in cropland, boundary planting, and improved pastures at the site of the new Indian Council of Agricultural Research (ICAR) near the National Research Center for Agroforestry in Jhansi. Several technologies were proposed to meet the different groups of land users' needs. In addition, surveys were conducted on the availability of literature and other library materials within the country. In **Bangladesh**, ICRAF collaborates with the National Agroforestry Task Force under the auspices of the Bangladesh Agricultural Research Council (BARC).

■ Resources

ICRAF has an extensive *Annual Report* with program summaries and publications and staff lists, technical research reports, newsletters, individual project articles, and other materials, including an AFRENA staff list. ICRAF welcomes any developed support materials on agroforestry education and scholarships in this field.

(See also International Council for Research in Agroforestry [ICRAF] in the Uganda Government and Zambia NGO sections.)

Teachers Environmental Association (TEA) ❧
P.O. Box 119, KISII
Nairobi, Nyanza, Kenya

Contact: **Christopher A. Nyakiti**, Founder/Coordinator

Activities: Development; Education; Research • *Issues:* Biodiversity/Species Preservation; Deforestation; Energy; Global Warming; Population Planning; Sustainable Development; Water Quality • *Meetings:* ANEN

▲ Projects and People

"The bell is ringing now for the environmentalists to enhance future global cooperation," writes Christopher A. Nyakiti, founder and coordinator of **Teachers Environmental Association (TEA)**, who believes that teachers are the best channel through which to disseminate environmental information. "After the Gulf crisis, time is ripe . . . to realize that man is the most endangered species [who needs] to develop a positive attitude towards the resources around him."

With 500 TEA members, a leading project is the **Teachers Trainees and Environmental Conservation Task Force (ECTF)**, a branch of TEA in colleges. The outreach mission of ECTF is to reduce community resistance to sound strategies of sustainable development by capitalizing on early relationships between teachers and young students. Teachers can influence youth to practice soil and water conservation and appreciate ecological balance, as Nyakiti did at a primary school in the mid-1980s. "Today, the school looks so beautiful with its green carpet of blossomings [nearby] Lake Victoria," he remarks.

TEA is encouraging the cultivation of fruit and tree nurseries at colleges, secondary and primary schools, and by women's groups and individual farmers. At Asumbi Teachers College, students collect seeds and raise them in a central nursery run by members of Wildlife and Young Farmers Clubs. Seedlings are provided free of charge or sold to certain groups.

The **woodlot project** encourages farmers and institutions to set aside small plots to plant stands of trees. TEA notes that "trees planted together have higher chances of survival than those scattered in gardens. The woodlots are more protected and beautiful."

To conserve energy, TEA urges its members to use bicycles. The group participates in environmental youth weeks, Earth Days, and Friends of the Earth (FOE) activities. Nyakiti belongs to the African NGOs Environment Network (ANEN) and the East African Wildlife Society.

■ Resources

TEA needs donors to sponsor its seminars and financial assistance, such as revolving loans for educational programs. TEA has available for cost: workshops, research studies, technical assistance and teaching tools, texts, and slides.

Eastern Africa Regional Office
The World Conservation Union (IUCN) ⊕
P.O. Box 68200
Nairobi, Kenya

Contact: **Robert Malpas, Ph.D.**, Regional Representative
Phone: (254) 2 502650 • *Fax:* (254) 2 503511

Activities: Development; Education; Political/Legislative; Research • *Issues:* Biodiversity/Species Preservation; Deforestation; Sustainable Development

● Organization

An international network of government agencies, governments, NGOs, and scientists, the **World Conservation Union (IUCN)** began in 1948 with a goal to preserve the Earth's rapidly diminishing natural resources. Today, its 663 members include more than 60 governments, 110 agencies, and over 450 NGOs associated with at least 3,000 scientific, technical, and legal experts. With its headquarters in Gland, Switzerland, there are 14 regional and country offices active in 119 of the world's 149 countries, according to IUCN's 1988–1990 triennial report. *(See also World Conservation Union in the USA, Switzerland, and Zimbabwe NGO sections.)*

As eastern Africa regional director, **Dr. Robert Malpas** observes that, ". . . despite overwhelming pressures from competing and immediate issues, governments and local communities continue to accord high priority to conservation concerns. . . . For those of us in the office who have come from developed countries, . . . we are only too aware that, as members of the 20 percent of society responsible for consuming 80 percent of the world's resources, our efforts do not compare well with those of eastern Africa's population. In comparison with the rural communities who are making major changes in their lifestyles to find improved ways of managing resources, western populations still have much to learn and a long way to go."

▲ Projects and People

Since its establishment in 1986, this office has launched more than 30 projects in the 7 countries where it is active: Ethiopia, Djibouti, Kenya, Somalia, Sudan, Tanzania, and Uganda. The underlining theme of IUCN's actions is the inextricable link between the "wise management of natural resources and the well-being of the people who depend on the food, water, heat, and income." All projects strive to reconcile legitimate human development needs with those of natural resource conservation.

IUCN is at the forefront trying to ensure that sustainable development is applied throughout the world's conservation programs. This is particularly significant in eastern Africa, where millions of people depend directly on resources within their immediate vicinity. Three main theories are incorporated here: ecological sustainability seeks development in compatible ways with vital ecological processes, diversity, and resources; economic sustainability checks development management in efficient ways ensuring future generations will be supported; and social and cultural sustainability "ensures that development increases people's control over their lives, is compatible with the culture and values of the people affected by it, and maintains and strengthens community identity."

The IUCN adheres to the National Conservation Strategy (NCS) concept, since it is the national level where major policies are made concerning the integration of sustainable development. Donors, industrialists, local communities, and resources managers join government ministers and departments in NCS proposal efforts in Ethiopia, Kenya, and Tanzania.

A two-year national wetlands conservation and management program was launched in 1989 by the Uganda government in conjunction with the IUCN. Uganda, in fact, became the first member of Ramsar to tackle this issue in an integrated, comprehensive manner. A year later, the World Wide Fund for Nature (WWF) and the Union joined the National Environment Management Council (NEMC) in exploring the need for a national wetlands program in Tanzania.

Another regional initiative is the IUCN Sahel Studies regarding climate, population growth, food production and pricing, land use, and natural resources management. The "socio-political impacts of resource management" are also being studied jointly with the United Nations Environment Programme (UNEP) in the Sahelian countries of Ethiopia, Somalia, and Sudan.

Those particularly threatened areas of the region—coastal zones, tropical forests, and protected areas of global significance—are targeted by IUCN and its members and partners in a wide range of field projects. In Tanzania, where the coastal strip forests and the Eastern Arc Mountains are some of the most threatened ecosystems, the East Usambaras Agricultural Development and Environmental Conservation Project is underway to improve the standard of living, protect forest functions such as water catchment, and preserve biological diversity. Playing a lead role are village coordinators, who assist with day-to-day management. One undertaking has been to switch from cardamom—a spice crop requiring a thinning of the forest canopy, that depletes topsoil, and that is only sustainable for 8 to 10 years—to cloves, black pepper, and coffee in forest land and to investigate the feasibility of growing cardamom on agricultural land. While a detailed inventory is determining the scope of logging exploitation, forest boundaries are being defined and teak, eucalyptus, and cedrella are being used to reforest logging gaps. The European Community and Tanzanian government are among the funders.

Forest reserves are being depleted by illegal agricultural encroachment in Uganda, according to IUCN Eastern Africa Regional Office *Triennial Report, 1988–1990*. A four-year conservation project started in 1987 in the East Usambaras Mountains (part of the Eastern Arc Mountain chain) and a five-year forest development and conservation plan was put in place in Mount Elgon, Uganda, in 1988, with help from NORAD (Norwegian Agency for International Development).

The Tanzania government also joined IUCN in two protected areas projects: the Serengeti Regional Conservation Strategy and the Ngorongoro Conservation and Development Project (NCDP).

The Serengeti-Mara ecosystem, probably the world's "best known wildlife region," has been the site of increased human/conservation conflicts. Antipoaching and protectionist methods, for example,

have alienated local communities and furthered the rift between parks and people. Moreover, "wildlife are seen as a source of livestock diseases, as competition for grazing, and a threat to their crops," reports IUCN. Following a regional workshop, the Serengeti strategy set up a framework for integrating activities across the various sectors, thereby ensuring planned and sustainable use of the natural resources.

Assessing the extent to which the Ngorongoro Conservation Area (NCA) could accommodate change and development without compromising its conservation value is the mission of this protected area project started in 1987. Well known for its natural splendor and archaeological significance—such as the Olduvai Gorge, site of the discovery of the fossil remains of early man—the NCA, with its "innovative and multiple use approaches," is generally succeeding. The local communities and government are also working toward a "comprehensive land-use zoning scheme" and new recommendations for "food security" and protection of elephant, rhinoceros, and other dwindling species. What happens here may provide a global answer as to whether "it is possible for people and wildlife to live together in harmony," states IUCN.

The Tanga Coastal Zone Conservation and Development Project in Tanzania started its three-year program in 1991 to protect rich resources in eastern Africa, such as coral reef, mangrove forests, green turtle, and dugong. Here, most of the 150,000 inhabitants are dependent on coastal resources. "Artisanal fisheries, production of building poles, boat building, salt manufacture . . . are all supported by the ecosystem," reports IUCN. A National Environment Management Council (NEMC) survey revealed the results of years of dynamiting reefs to obtain shells and fish, of exploiting the mangrove forests for charcoal, of overfishing to meet tourism demands, and of pollution of estuaries, lagoons, and beach areas. According to the NORAD-funded study, fishing catches are declining, "the cover of live coral is now less than 20 percent" on most reefs, and commercial operators are major threats to the mangroves. The project aims to promote sustainable use and integrated management of the coastal resources with innovations, such as "solar salt production methods as an alternative to boiling methods." A small research facility and village-level extension programs are expected to be established.

With arid and semiarid lands constituting more than "50 percent of tropical Africa and support [for] over 35 percent (116 million) of the population, IUCN is joining Sudan officials in exploring ways to halt further environmental degradation in the Darfur region— blessed by "well watered mountains of the Jebel Marra" that permits wadi (alluvial soil) agriculture in the desert but nonetheless has been recently hit by famine and conflict over natural resources.

In 1991, IUCN started a technical program department supporting existing activities and developing new priorities within the country programming framework while providing direct technical assistance to its members and supporters.

That same year, IUCN mounted its environmental NGO support program, with financial support from the Swedish International Development Authority (SIDA) to promote the development of the NGO sector. Other areas of IUCN participation include the Tanzania National Park Planning Project, Mweka College Support Program, Tanzania, Commission on European Community's Uganda National Parks Project, Zaire's Garamba National Park Rehabilitation Project, and the Tanzania Wildlife Conservation Project.

■ Resources

IUCN has a detailed *Triennial Report* and other materials available.

▌ *Private*

Baobab Farm Ltd.

P.O. Box 81995
Mombasa, Kenya

Contact: **Rene Daniel Haller**, Managing Director
   Phone: (254) 11 486155 • *Fax:* (254) 11 485151

Activities: Development • *Issues:* Deforestation; Waste Management/Recycling • *Meetings:* CSG; IUCN

● Organization

In the 1950s, a major Kenyan enterprise, the Bamburi Portland Cement Company, which belongs to the Zurich-based concern Cementia, mined prehistoric beds of coral for limestone to manufacture cement, a process that devastated the landscape. The Company reportedly strip mines up to 1.8 million tons of fossil coral limestone for cement annually.

Yet today, where there was once wasteland, 600,000 trees are home to 120 species of birds and wildlife. Banana plants and vegetable crops flourish. An 18-acre enclosure features buffalo, hippos, mongoose, civets, duiker, and warthogs, among other species. At least seven acres a year are regenerated through natural pollination. And fish farms in the desert are a model for other tropical areas.

Swiss agronomist **Rene Daniel Haller** arrived in 1959, was entrusted to look after the company's "Garden Department," and set out to create a natural ecosystem that could eventually maintain and renew itself and yield economic benefits as well. In 1971, Bamburi started a rehabilitation project which led to the formation of **Baobab Farm** as a limited company in 1977 and a "fully owned subsidiary of the cement company."

▲ Projects and People

Due to bulldozers digging down to almost groundwater level and the ocean's proximity, the seeping salt water created a seemingly insoluble regeneration problem. Hence, the only way to generate topsoil there was in a natural way, on the spot. The Farm's aim was a one million tree planting project over a five-year period. Quick-growing varieties of trees were used, with the *Casuarina equisetifolia* serving as the mainstay among the bushes and grasses. Resembling the drooping feathers of the cassowary bird, which shares the Malay name *Kasuari*, the tree has needles that do not decay easily. So the redfooted millipedes were introduced into the area, serving a two-fold purpose: eating the needles and helping in the formation of humus with their fecal matter.

New tree-planting techniques were developed by the Farm, when the brackish water close to the surface problem arose in the South Quarry. It was discovered that the root system of the *Casuarina* trees, unable to anchor themselves sufficiently in the rock beds, were top-heavy and capable of falling in strong overnight winds and high water tables. This was not the same situation in the North Quarry. **Clare V. Wood's** Baobab Farm handbook, *Trees for Wastelands*, details these projects.

From the *Casuarina*, also called She-oak and *Mvinje* (Swahili), the long-lived tree is harvested as a source of poles for houses and boat masts in coastal areas, fuelwood, tannin, livestock fodder, and mulch—thus sparing the destruction of some nonrenewable indigenous forest resources.

It is the Farm's practice to fight pests via natural means, not using synthetic fertilizers or insecticides. The farming of chickens in the 1960s and sheep and goat breeding were other areas of Farm experimentation and expansion.

In the 1970s, the reclamation of the limestone quarries "inhospitable moonscape" was initiated, resulting in nature trails attracting 80,000 visitors worldwide. In 1972, the first fish ponds were dug to make use of the ample supplies of groundwater. However, when tilapia, a genus of fish food able to thrive in both salt and fresh water, failed to grow to marketable size, small concrete tanks fed with a constant supply of flowing water provided the answer.

Within 10 years, the commercial fish farm sold more than 30 tons of fish. The problem of feeding the fish was solved with local farm vegetable refuse, water plants, and cottonseed oil cakes. This fish breeding system is recognized worldwide. The lung fish (*Protopterus amphibius*), another significant fish for the Kenya population, shows artificial breeding success on the Farm.

The breeding of prawns and crocodiles is another ecosystem activity of the Farm. In 1986, some 700 young crocs were fed on fish waste at zero cost. A profit will be realized from their skins sales, and their meat in turn will be fed back to the fish.

More than 100 acres of former stone quarry have been totally reclaimed. The low cost Baobab Farm is essentially self-financing and, according to a report in 1990, showed a sizable annual Farm profit return from a combined 45 sources. With the parent company's cooperation, research at Baobab Farm results in advanced degrees and scientific publications.

News media accounts have described Haller's work as having "great significance for Africa's developing countries, such as those of the Sahel region." His ideas, such as a 10-square-yard fish pond, appear to be "easy to duplicate lessons for local people." Such a pond could yield "more animal protein in a week than many Africans get in a month," according to *International Wildlife* (November 1986). For his efforts, Haller has been honored by the Kenyan government.

■ Resources

Baobab can provide publications lists of scientific papers, reprints, and handouts, such as *Turning the Desert into a Forest of a Million Trees*, a *Baobab Farm News* series, and the handbook *Trees for Wastelands*.

Game Ranching Ltd.

P.O. Box 47272
Nairobi, Kenya

Contact: **Malte Sommerlatte, Ph.D.**, Wildlife
      Manager
   Phone: (254) 2 20467 • *Fax:* (254) 2 725316

Activities: Consultancy; Education; Development; Research
Issues: Biodiversity/Species Preservation; Sustainable Development; Wildlife Management • *Meetings:* Game Ranching

and Utilization Conferences; Training & Education Conferences

● Organization

Dr. David Hopcraft established Game Ranching Ltd. in 1980 as a means to develop ways to manage and utilize wildlife and 2,800 head of cattle together in semiarid environments. A 2.5-meter-high fence surrounding the 8,100-hectare ranch enables the staff of 10 to control the animals. Game cropping is the only form of wildlife utilization on the ranch, with venison sold on the local Nairobi market.

The ranch includes a research camp, laboratory, slaughterhouse, and refrigeration unit. Grasslands dominate the ranch, which has 15 species of plains game as well as predators—hyena, jackal, cheetah, and an occasional lion and wild dog. Thomson's gazelle, Grant's gazelle, wildebeest, and kongoni are some of commercial species comprising approximately 80 percent of the Ranch's wildlife.

Game ranch policy requires a game population at 2,500, allowing an offtake of 400 to 450 animals per year. In 1990, 362 animals or 20 tons of dressed game meat was produced for government meat inspectors.

▲ Projects and People

Visiting wildlife researchers take advantage of the unique collection of biological data collected at the ranch. The Ranch ecologist developed a program that is management oriented and provides answers to questions in wildlife utilization. Two or three wildlife students (mostly from Kenya, although other African, European, and American countries also have representatives) collect field data for their master's or Ph.D. thesis at the Ranch.

Courses in game counting, wildlife and ranch management, and harvesting techniques are offered, including lectures at the Universities of Nairobi, Moi, and Makerere on involving "the private rancher and the village community in wildlife conservation and management."

Senior representatives from nearly all wildlife NGOs and international aid agencies have come to the region. Annually, the Ranch receives approximately 100 official visitors. According to ecologist Dr. Malte Sommerlatte, a group of visiting American Indians expressed interest in a similar operation in the USA.

Various field days, wildlife meetings, cooperative management programs, position and policy papers, courses, and consultancies have developed at the initiation of the Ranch. Government wildlife departments and international aid agencies have also sought advice from the Ranch regarding wildlife management and utilization.

Realizing the need to diversify, the future of the Ranch may include tourism (foot and horse safaris), game capture, bird shooting, and the establishment of a local wildlife consultancy company.

Recently, Dr. Sommerlatte was senior author of Tanzania's Tropical Forestry Action Plan, Food and Agriculture Organization (FAO), to reconcile "wildlife conservation and the needs of rural people living next to national parks and game reserves." His follow-up policy paper is now a government document of Tanzania's Ministry of Natural Resources. Sommerlatte is presently working on a similar policy paper for the Kenya Wildlife Service, outlining "the economics of wildlife utilization and how the private sector can participate in this new form of land use." He recommends regulations and the creation of a Wildlife Producers Association.

A College of African Wildlife Management course, which Sommerlatte presented to park wardens and wildlife extension officers, shows how local people in Zambia and Zimbabwe benefit from wildlife utilization. With Hilary Sommerlatte, he published a field guide on the important *Trees and Shrubs of the Imatong Mountains, Southern Sudan*, which is especially useful to botanists and plant collectors working in East and Central Africa.

■ Resources

The Ranch needs radio telemetry and other wildlife equipment and a wildlife management consultant. Its representatives are available for game captures, consultancies, and wildlife training.

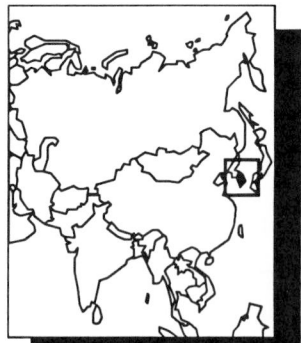

Korea occupies a mountainous peninsula northeast of China that separates the Yellow Sea from the Sea of Japan. The country has limited natural resources and is highly forested. Farming is a major activity. A strong industrial sector is the mainstay of the economy.

South Korea, during the past quarter-century, has experienced significant economic and industrial growth, but at the cost of environmental damage. There is a shortage of piped water due to a dearth of storage and purification facilities. Industrial plants and pesticides and fertilizers have contaminated many water supplies. In addition, the increased use of coal has led to high air pollution levels, exacerbated by the growing number of automobiles. More than 100 animal species have disappeared from rural areas. The Naktong River delta is a particular area of concern.

 # *NGO*

Korean Central Council for Nature Preservation (KCCNP)
44-2, Chu ja-dong Chung-gu
Seoul 100-240, Korea

Contact: **Yung Ho Chung**, President
Phone: (82) 2 267 8451 • *Fax:* (82) 2 277 5726

Activities: Education; Research • *Issues:* Biodiversity/Species Preservation; Deforestation
Meetings: IUCN

● Organization

"Man is born a part of nature, lives on its provisions, and eventually returns thereto. All things in the sky, the earth and the seas have been provided as resources for our life. . . .

"However, . . . Man's indiscriminate plundering has destroyed the balance of nature. The living environment has deteriorated, threatening the survival of man and all other living creatures. It is imperative, therefore, that all people renew their appreciation of nature, take loving care of it, eliminate all sources of pollution, and endeavor to restore and maintain the order and harmony therein."

This passage is from the constitution of the **Korean Central Council for Nature Preservation (KCCNP)**. Founded in 1977 as a non-governmental, independent, and international organization, KCCNP devotes itself to the responsibilities set forth by its constitution.

KCCNP aims to achieve its six main objectives, which are to "recognize the value of environmental preservation and develop its practice among the general public; promote public awareness towards a better environment through available educational systems; attain integrated and scientifically surveyed results with the widest range of beneficial uses of the environment without degradation; enhance the recommended uses of the environment and the quality of renewable resources; assure for all Koreans a safe, healthy, productive, and culturally pleasing natural environment; and participate, communicate, and exchange ideas, with all levels of conservation, governmental, private, and regional establishments."

▲ Projects and People

KCCNP plans and organizes **conservation policies** to be enacted into national practices. The Council also campaigns for **conservation movements** and helps other conservation organizations in the country.

To heighten public awareness, KCCNP leads many **educational activities**. Using the public education system and private forums, the KCCNP lectures, tours, and produces teaching aides on conservation. KCCNP holds **photographic exhibitions**, seminars, and contests concerning environmental issues. For younger children, KCCNP has bird-watching trips and environmental cleanup contests.

■ Resources

For children, KCCNP produces teaching guides of slides, fliers, and literature on the environment. Other publications include a bi-monthly magazine, *The Preservation of Nature*, *Conservation Song Book*, *Collection of Literary Works on Conservation*, and scientific works, such as technical surveys of the natural environment in Korea with specific surveys of Wando and its adjacent island, Dokjuk Island, and the Judo and Kuhmoondo Area of the Dadohae Marine National Park.

KCCNP offers consulting services on environmental impact and waste disposal. For its publicity campaigns, it distributes posters, stickers, banners, waste bags, and related paraphernalia.

This tiny European principality is located between Switzerland and Austria. The eastern two-thirds of the land is comprised of rugged foothills whose lower slopes are covered with evergreen forest and alpine flowers. The mountains contain three major valleys and are drained by the Samina River; the western section is occupied by the Rhine River floodplain. A drainage channel built in the 1930s has made the area's rich soil suitable for agriculture.

Liechtenstein has a remarkable variety of vegetation, and all of the forests are protected to maintain the ecology of the mountain slopes and guard against erosion. The country has no heavy industry, and virtually all raw materials, including wood, are imported. Thus, Liechtenstein has less industrial pollution than other developed areas, though it suffers from many of the same transboundary pollution problems that affect its neighbors.

Government

Landesverwaltung des Fürstentums/National Office of Forests
Landesforstamt
9490 Vaduz, Liechtenstein

Contact: **Felix Näscher, Ph.D.,** Minister
Phone: (41) 75 66400 • *Fax:* (41) 75 66411

Activities: Conservation Management; Development; Law Enforcement; Political/Legislative • *Issues:* Air Quality/Emission Control; Biodiversity/Species Preservation; Deforestation; Global Warming; Nature and Landscape Protection; Sustainable Development *Meetings:* CITES

🍁 NGO

Commission Internationale pour la Protection des Alpes
International Commission for the Protection of the Alps (CIPRA) ⊕
Heiligkeutz 52
FL-9490 Vaduz, Liechtenstein

Contacts: **Mario Broggi, Ph.D.,** President
Ulf Tödter, Director
Phone: (41) 75 81166 • *Fax:* (41) 75 82819

Activities: Information/Experience Exchange; Law Enforcement; Political/Legislative; Research • *Issues:* Air Quality/Emission Control; Biodiversity/Species Preservation; Deforestation; Energy; Sustainable Development; Transportation; Waste Management/Recycling; Water Quality • *Meetings:* CIPRA

● Organization

The **International Commission for the Protection of the Alps (CIPRA)** was founded in 1952, with the initiation of the World Conservation Union (IUCN). Since then, CIPRA has become a group of about 70 associations in Austria, France, Germany, Italy, Slovenia, Liechtenstein, and Switzerland, with over 2.5 million members collectively. Its philosophy is based on preserving "the Alps as a mosaic of human living space with its historic cultural landscapes and natural environments."

The main focus of CIPRA is the exchange of information and experience throughout the Alpine states in an effort to overcome the political, linguistic, and regional boundaries and to help enhance public awareness of the environmental threats facing the Alps.

CIPRA concentrates on the "network of reserves and protected zones, traffic and transport, future perspectives for a sustainable tourism, sport activities and environment, energy-producing facilities, water supply and quality, clean air and waste management, mountain forests and farming, and regional planning" throughout the Alpine region.

▲ Projects and People

CIPRA's main activity is the organization of the **Alpine Convention**, in which the Ministers of the Environment from the seven Alpine states discuss and make environmental policy. CIPRA consulted with over 300 government authorities, institutions, and experts to devise and plan the convention, which was first held in Bavaria in 1988.

In addition, CIPRA holds an **annual congress** for its international members and representatives on topics such as the ecological situation of Alpine rivers, traditional mountain farming in the Alps, and the Alpine Convention. It also sponsors **scientific workshops** on environmental issues, such as hydroelectric power plants, ecotourism, and infrastructure.

■ Resources

CIPRA produces publications in the three official languages of the Commission—French, German, and Italian. These include a quarterly bulletin entitled *CIPRA-Inform*, a series of documents which discusses specific environmental problems in the Alps, and *CIPRA-Documentation*, which comprises the results of the annual CIPRA congress.

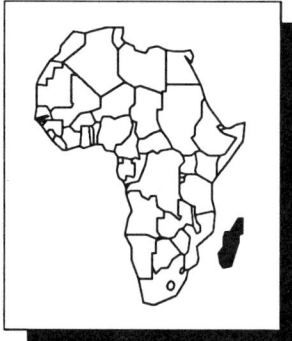

Separated from the African mainland for 100 million years, Madagascar has an exceptionally rich collection of plant species. Some 80 percent of the approximately 10,000 plant species are thought to be endemic, and bird and mammal species, while few in number, are for the most part unique to the island. There are 194 national parks, protected reserves, and classified forests; yet, species extinction rates are growing, and the environment is seriously threatened by deforestation due to fire and slash-and-burn agriculture.

A national conservation strategy is emphasizing biodiversity protection, creation of an environmental fund, land management, education and training, and institutional support.

🏛 *Government*

Parc Botanique et Zoologique de Tsimbazaza
Botanical and Zoological Park of Tsimbazaza
P.O. Box 4096
Antananarivo 101, Madagascar

Contact: **Felix Rakotondraparany,** Head, Fauna Department
 Phone: (261) 2 31149

Activities: Research • *Issues:* Biodiversity/Species Preservation; Deforestation • *Meetings:* SSC

▲ Projects and People

Felix Rakotondraparany of the **Botanical and Zoological Park of Tsimbazaza** assesses the impact of forest destruction on biodiversity/species preservation, particularly small mammals, in the central forests of Madagascar—a region he describes as "disregarded nowadays, though human population density is growing."

The Fauna Department assists in captive breeding the "dwindling wildlife population" and "educates the public about the existence and importance of such biodiversity."

With increasing international attention focused on Madagascar, Rakotondraparany is hopeful about his country's environmental future.

■ Resources

Research assistance and a computer for satellite imagery are needed.

Crocodile Conservation Projects
c/o Lot XV 18, Andrefandrova
Antananarivo, Madagascar

Contact: **Olivier Behra,** Consultant
 Phone: (261) 2 28651 • *Fax:* (261) 2 28651

Activities: Development • *Issues:* Biodiversity/Species Preservation; Deforestation; Sustainable Development • *Meetings:* CITES; IUCN/SSC Crocodile Group

▲ Projects and People

A crocodile research consultant, **Olivier Behra** recently ran the **United Nations Development Programme (UNDP)** crocodile farming project in Madagascar to train rural populations in sustainable crocodile management. Earlier, he managed a similar project for the **Food and Agriculture Organization (FAO)**, including inventorying existing crocodile populations, and also involving the cooperation of local people in collecting crocodile eggs, which can only survive in suitable conditions. When the crocodile ranch organized an egg collection, about 2,800 were spotted by local residents, who were then paid for their services.

Among its beneficial results, the ranching program trains government technicians and potential farmers and provides a source of income. According to Behra, the population is poor while the biodiversity remains rich. However, "80 percent of the forests . . . are considered to have disappeared and only 1.8 percent of the territory is classified as protected areas, whereas the country still has about 10,000,000 hectares of natural forests."

Behra also produced a report on **Sustainable Utilization of Natural Resources in Madagascar**, in which he reiterates the need to alleviate rural poverty while preserving biological diversity. Even with Madagascar's ratification of CITES (Convention on International Trade in Endangered Species of Wild Fauna and Flora), the report expresses concern about wildlife exploitation for commercial international purposes, beginning with the early trade in crocodile skins and medicinal and ornamental plants, and presently live animals—with tourists' greater access of the island.

In urging the involvement of rural populations in sustainable management, Behra points out that "many people in developed countries and particularly conservationists are not in favor of the idea of using wildlife. It is probably because in these countries, the populations are no longer depending on nature for their food and energy needs." Behra adds, "However, in developing countries, the utilization of nature was and is still an important element of the battle to survive and a source of considerable income for many indigenous groups. Their situation needs our collective respect."

Also working with the National Office of Environment on proposals on sustainable development, Behra is deputy vice chairman of the Species Survival Commission (IUCN/SSC) Crocodile Group; president of the conservation committee, International Society for the Study and Conservation of Amphibians; and vice chairman of GERRIC, a research group on reptiles of commercial interests. He is also surveying crocodiles in central Africa and French Guyana.

■ Resources

Behra has published 15 research reports.

 NGO

Earth Preservation Fund (EPF) ✉
4D Immeuble "Vitasoa"—Analakely
Antananarivo 101, Madagascar

Contact: **Sylvie Rebesahala**, Representative
Phone: (261) 2 21893

Activities: Education • *Issues:* Deforestation; Population Planning; Species Preservation; Sustainable Development

● Organization

The **Earth Preservation Fund** (EPF) is a nonprofit affiliate organization of **Journeys International, Inc.**, which promotes worldwide nature and culture exploration and encourages **environmentally responsible travel.** Through Journeys or EPF, travelers can assist and encourage better conservation and environmental practices at a local level worldwide. *(See also Earth Preservation Fund in the USA NGO section.)*

▲ Projects and People

Madagascar's resident staff of one, **Sylvie Rebesahala,** contributes environmental conservation services in the area. Major concerns include deforestation and the lack of awareness among the local population. Projects include **tree-planting** programs, a scholarship **or award** program for young naturalists, and acquisition of a piece of land in order to **create a private natural reserve.**

■ Resources

The EPF seeks funding, greater involvement, and trip participants. In addition, writes Rebesahala, "being very eager to contribute to the environmental preservation, I would need the maximum of technical advice and any form of aid on how to direct such an enterprise." EPF welcomes any participation in terms of ideas or financing.

⚱ *Private*

Jonah A. Andrianarivo, Ph.D.
Madagascar Forestry Consultant
c/o ILRAD
P.O. Box 30709
Nairobi, Kenya

Phone: (254) 2 593075 (Kenya)
Fax: (254) 2 593499

Activities: Education; Research • *Issues:* Deforestation; Sustainable Development

▲ Projects and People

Dr. **Jonah A. Andrianarivo** works as a forestry consultant for several environmental projects in his native Madagascar. While working on his doctorate in forestry at Duke University in Durham, North Carolina, USA, Dr. Andrianarivo was research assistant at Duke's Wetland Center, responsible for setting up from start to finish a geographic information system (GIS). He is currently "developing course and sample materials for a project to **promote the use of GIS in Madagascar.**" In addition, he is working on an **ecotourism** project, conducting ecological analysis for a new national park project, and "assisting a group of local artisans in promoting nature-related handcrafts."

Dr. Andrianarivo has also consulted on several international projects in Madagascar, including a Canadian/Malagasy joint venture on ilmenite (titanium ore) mining and a West German/Malagasy forestry program for integrated agricultural development. From 1978 to 1981, he was head of Tsimbazaza Park under Madagascar's Ministry of Scientific Research.

He has received grants and awards from Fulbright, World Wildlife Fund—USA, Duke University, and other organizations, and is a member of the American Society for Photogrammetry and Remote Sensing, as well as of the International Society of Tropical Foresters.

■ Resources

Dr. Andrianarivo seeks more work assignments from or in Madagascar. His bibliography lists such research articles as: "Integration of Digital Analysis of Landsat MSS Imagery with Field Investigation for Estimating Lemur Population in Madagascar"; "Standardization of Malagasy Names for Animal and Plant Species"; and "Field Check for Laser Remote Sensing in Tropical Forest"; among other titles in French and English, including an article on a critically endangered tortoise (*Geochelonia yniphora*) in northwestern Madagascar.

(See also Washington University, Department of Anthropology, Beza Mehafaly Reserve Project in the USA Universities section.)

Malaysia

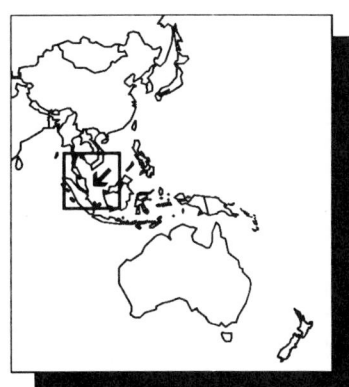

Malaysia has paid a high environmental price for its economic progress in terms of the effect on national parks, wildlife, agriculture, fisheries, and urban areas. The country is one of the largest timber-exporting nations in the world. While standards of living have improved, environmental problems, including erosion, siltation of rivers, flash flooding, heat waves and drought, deforestation, and air and water pollution, have increased. Land conversion, logging, and poaching have posed particular threats. Fortunately, the country is making significant inroads in preserving its considerable natural beauty and resources, especially as public awareness has grown and the environmental activities of government and non-governmental organizations, citizens groups, businesses, and universities have expanded.

🏛 *Government*

Department of Wildlife and National Parks

KM 10, Jalan Cheras
56100 Kuala Lumpur, Malaysia

Contacts: **Mohd Khan Momin Khan**, Director General
Mohd Tajuddin Abdullah, Wildlife Officer
Phone: (60) 3 905 2872 • *Fax:* (60) 3 905 2873

Activities: Development; Education; Research • *Issues:* Biodiversity/Species Preservation; Wildlife Management • *Meetings:* CITES; IUCN/SSC Asian Rhino, CBSG

■ Resources

Available publications include *Birth of a Sumatran Rhinoceros at Malacca Zoo* and *The Husbandry and Veterinary Care of Captive Sumatran Rhinoceros at Zoo Melaka*; and articles, such as "Movement of a Herd of Elephants in Upper Perak Area," "Population and Distribution Studies of Perak Elephants," and "Population and Distribution of the Malayan Elephant in Peninsular Malaysia."

The Department seeks information on embryo transfer and genetics, particularly for Sumatran rhino captive breeding, and biotelemetry for study on gaur (a large wild ox).

Forest Research Institute of Malaysia (FRIM)

Kepong
52109 Kuala Lumpur, Malaysia

Contact: **Kong How Kooi**, Librarian
Phone: (60) 3 634 2633 • *Fax:* (60) 3 636 7753
E-Mail: c:malaysia; a:stm.telemail; d:frim.ill; un:frim

Activities: Research • *Issues:* Air Quality/Emission Control; Biodiversity/Species Preservation; Deforestation; Energy; Global Warming; Sustainable Development; Waste Management/Recycling; Water Quality

● Organization

Forestry research was formally organized in **Malaysia** in 1918 when **Dr. F.W. Foxworthy** was appointed the first forest research officer within what was then the Forest Department. In 1929, the Research Branch was moved to its present premises at Kepong and renamed the Forest Research Institute (FRI). In 1985, the government restructured FRI with the Malaysian Forestry Research and Development Board (MFRDB), renaming it the **Forestry Research Institute of**

Malaysia (FRIM), known locally as Institut Penyelidikan Perhutanan, and transferred its direction from the Forest Department to the Ministry of Primary Industries.

All of its forestry-related functions are governed by the MFRDB, which identifies three primary areas of interest: to conduct and promote research; to collect, collate, and disseminate information relating to forestry; and to coordinate activities within Malaysia relating to forestry. To carry these out, FRIM has 499 employees.

For its research mission, FRIM has extensive facilities. When FRI moved to Kepong in 1929, it opened 600 hectares of experimental plantations, arboreta, and natural forest. Today, the institute grounds flourish, with more than 1,000 species of indigenous and exotic trees for biological and silvicultural studies. In addition to the arboretum begun in 1929, FRIM now has five distinct arboreta, containing dipterocarp, nondipterocarp, gymnosperm, monocot, and fruit trees. A herbarium contains over 125,000 specimens, including all of the dipterocarp species of Malaysia. FRIM has a collection of wood samples numbering over 10,000 for the study of wood structures. FRIM's library, one of the oldest in Malaysia, now has over 120,000 volumes and 1,000 journal titles.

With a commitment to the Langkawi Declaration on the Environment of 1989, Malaysia's government realizes that its tropical forest resources must be conserved and sustainably managed. An accepted strategy is the establishment of "tree plantations with exotic and indigenous species" to supplement timber production from the natural forests.

▲ Projects and People

FRIM researches basic climactic and hydrological parameters to understand their relationships to forest degradation and deforestation. It has gathered considerable data on stream flow, sediment load, soil erosion, and microclimates. By comparing data from forests before and after logging, FRIM can assess damage and make recommendations to minimize logging damage. FRIM has projects to study watershed management, soil management, logging impact, forest biology, pest and disease management, and forest ecology.

FRIM is developing guidelines on improving urban forests, promoting new indigenous species for urban planning, testing for pests and diseases, and developing a directory of urban trees.

Another series of projects involves developing techniques for optimizing seed harvest, improving germination and storage capabilities of forest seeds, assessing genetic variability, initiating a certification system for all collected seeds, and improving techniques for seedling growth.

In addition, FRIM provides technical consultancy and advisory services, training courses in forest products and forestry, library services, publications, and forestry extension services.

■ Resources

FRIM publications include: *Journal of Tropical Forest Science, Malayan Forest Records, Timber Digest, Urban Forestry Bulletin*, in-house *FRIMA*, as well as FRIM *Annual Reports, Research Reports*, technical information, occasional papers, and other pamphlets on research.

Local Government Department
Ministry of Housing and Local Government
Block K Damansara Town Center
50782 Kuala Lumpur, Malaysia

Contact: **Lim Cheng Tatt**, Director General
 Phone: (60) 3 254 1818 • *Fax:* (60) 3 255 4066

Activities: Education; Political/Legislative; Research • *Issues:* Air Quality/Emission Control; Deforestation; Global Warming; Health; Waste Management/Recycling; Water Quality *Meetings:* Solid Waste Management

▲ Projects and People

Headed by **Lim Cheng Tatt**, the **Local Government Department** of the **Ministry of Housing and Local Government** serves such governments with advice on how to manage the environment. Current countrywide projects are solid waste management, sanitary landfills, sewage management, and water pollution from domestic sources.

Tatt has worked for over a decade on housing problems and solutions, producing a book, *The Malaysian Housing Scenario*. He has traveled widely throughout Europe and Asia to study housing and sewage treatments, lecturing frequently. In 1991, he studied firsthand the sewage works sector of the Japanese government and systems of local government and housing in Denmark, UK, and Mauritius.

■ Resources

Training materials on sanitary landfills are produced.

🏛 *State Government*

Sabah Foundation
Box 11623
88817 Kota Kinabalu, Sabah, Malaysia

Contact: **Clive Wallis Marsh, Ph.D.**, Principal Forest
 Officer (Conservation)
 Phone: (60) 88 35496 • *Fax:* (60) 88 422410

Activities: Development; Education • *Issues:* Biodiversity/Species Preservation; Deforestation • *Meetings:* World Parks Congress

● Organization

Borneo, the third largest island in the world after Greenland and New Zealand, is shared by tiny Brunei, Indonesia, and the Malaysian state of Sabah. Here are found some 420 bird species, 92 bat varieties, and 120 other land mammal species—of which one-third exists nowhere else.

Visiting Royal Society scientist Dr. **Tony Greer** described some observations in 1990: "Orangutan, Bornean gibbon, red leaf monkey, pig-tailed and long-tailed macaques are regularly encountered. . . . On night drives along the road, leopard cat and flat-headed cat are often seen and less commonly, clouded leopard and marbled cat. The tangalung, or common Malay civet, appears to enjoy visiting the Centre. . . . Giant red flying squirrels perform almost nightly. . . . Seen regularly are sambar deer and Bornean red and yellow muntjac. . . . The oriental small-clawed otter may be seen in the smaller rivers. . . . Four sun bear were recorded in December . . . and increased elephant activity [is noted] along the logging road."

The **Sabah Foundation (Yayasan Sabah)**, one of Malaysia's top timber producers, realizes the value of sustainable development and conservation. In Sabah's Danum Valley, "a magnet for researchers from all over the world," according to *Time* magazine, the Sabah Foundation has set up the **Danum Valley Field Centre (Pusat Luar Lembah Danum)**, with a project staff of 19.

In 1975, WWF-Malaysia (World Wide Fund for Nature) recommended National Park status for the Danum Valley; in 1981, a 438-square-kilometer area was designated as the Danum Valley Conservation Area within the Yayasan Sabah Forest Concession, and it officially opened in 1986. Principal users are scientists and course participants; however, others can use the center, with its more than 30 kilometers of marked trails, on advance booking when space is available.

The **Kelab Pencinta Alam Sabah (Sabah Nature Club)** was launched in 1988 as a joint effort of Yayasan Sabah and the Education Department. Directed at school children, the Club offers a weeklong nature orientation course at the Field Centre several times a year for up to 30 students per course. The Club now boasts 4,500 members in 50 schools around the state.

▲ Projects and People

Through "Operation Raleigh" held in 1987 and 1991, youth from around the world come to work on Centre facilities. In 1987, for example, the program resulted in 21 km of new trail and construction of a 6-bed cabin.

The Danum Valley Field Centre began a long-term research program in 1984, as a cooperative effort of Yayasan Sabah, the Forest Department, University of Malaya in Sabah (UKMS), and the Royal Society, London. By 1990, nearly 80 studies had been initiated; although some were halted for various reasons, most were completed, and some are ongoing as new projects are proposed. There were 20 active projects in 1990.

Most recent studies have focused on forest dynamics, in particular **comparing flora and fauna in primary and logged forests**, including freshwater fish, birds, and dung beetles, and on hydrology, including studying the **effects of logging on water quality**, is a long-term interest. A recently completed study by Smithsonian's **Dr. Louise H. Emmons**, which focused on five species of treeshrews, comparing two habitats (logged vs. primary), "will help to predict the future of many of the extant Bornean populations and species, and to understand the mechanisms of animal-forest interactions." Other research subjects include the ecology of Borean frogs, fine root dynamics, latent extinction of pyralid moths, and delayed greening and red coloration in young leaves.

In 1990, the Sabah Foundation hosted the International Conference on Forest Biology and Conservation in Borneo for participants from 13 countries to foster Bornean rainforest ecology and better management practices to replace "unsustainable logging and conversion to intensive agriculture." Regeneration of the dipterocarp forests, orangutan rehabilitation, and uses of traditional plant medicines were among issues discussed. Other goals are the expansion of transfrontier reserves and protected areas in general, mangrove protection, realistic valuation of renewable natural resources, and caution in resettlement programs of indigenous peoples.

Dr. Clive Wallis Marsh joined the Sabah Foundation in 1982 as wildlife officer, and since 1989 has been principal forest officer, responsible for conservation planning of the million-hectare timber concession. Dr. Marsh planned and developed the Danum Valley Field Centre, organized Sabah Nature Club, and co-edits the Sabah Society journal. Among other positions, he is a fellow on the Borneo Research Council, a member of the Primate Society of Great Britain and the International Primatological Society, and a member of the Species Survival Commission (IUCN/SSC) Primate Group.

■ Resources

Besides its *Annual Report*, containing a cumulative project list, the Sabah Foundation has two periodicals: the Sabah Nature Club magazine, *Majalah*, and the *Journal of the Sabah Society*. In addition, the Nature Centre has pamphlets on rules, regulations, and how to get there.

Dr. Marsh has published numerous reports and scientific papers, including the recent "Primates, Forest Conservation, and Development" and "Malaysian Forests: A Case Study in Problems of Conservation and Development." He has also collaborated on three books on tropical rainforests, particularly in Sabah.

NGO

Asian Wetland Bureau (AWB) ⊕
Institute of Advanced Studies
IPT-AWB, University of Malaya
59100 Kuala Lumpur, Malaysia

Contact: **Faizal (Duncan) Parish**, Executive Director
Phone: (60) 3 756 6624 • *Fax:* (60) 3 757 1225

Activities: Research • *Issues:* Biodiversity/Species Preservation; Sustainable Development; Water Quality

● Organization

According to the **Asian Wetland Bureau (AWB)**, up to 85 percent of the 120 million hectares of wetlands in East and South Asia "are threatened by degradation or destruction . . . [causing] severe declines in fishery and forestry resources." With offices in the Philippines, UK, Malaysia, and Indonesia, AWB aims to promote the protection and sustainable utilization of wetland resources in Asia. Working through governmental and non-governmental bodies in 13 nations in the region, AWB "promotes the development of national centers of wetland expertise, staffed by local experts."

The Malaysian office serves as the regional coordination center and runs a national program in conjunction with the **Institute of Advanced Studies**, "a multidisciplinary applied research center" at the University of Malaya. *(See also University of Malaya, Institute of Advanced Studies in the Malaysia Universities section.)* The British office coordinates technical and financial support from Europe for project work. The office currently has a staff of 35, including 18 project workers; counterparts from government agencies also work on projects. Primary funding comes from the Asian Development Bank, governments of the Netherlands and New Zealand, U.S. Agency for International Development (USAID), United Nations Environment Programme (UNEP), U.S. Fish and Wildlife Service, World Wide Fund for Nature (WWF), and other groups.

▲ Projects and People

The AWB works in four major areas: biological diversity, water resources, institutional strengthening and awareness, and environmental management and policy. Projects finished in 1991 included environmental impact assessments of Tumboh Swamp and Sg. Golok, and work on Muar flood mitigation. A bird-strike assessment project for airports was slated for completion in 1992. Ongoing projects include developing the Sg. Buloh Sanctuary in Singapore; studying the socioeconomic value of wetland plants; and biodiversity projects at key wetlands. Slated for completion in 1994 is a study of freshwater fish status on the peninsula.

To date, the AWB has completed more than 60 projects and reports on many aspects of wetlands in Asia, including national wetland inventories in Malaysia, Indonesia, and the Philippines, peat swamp and mangrove studies with guidelines for conservation, identification of key sites for migratory waterbirds and providing assistance for their protection, pesticide impact and water quality in Malaysian ricefields, and a major program with the Indonesian Ministry of Forestry to develop management recommendations for 35 million hectares of wetlands throughout Indonesia. The Bureau has also conducted more than 22 training courses in 8 countries and monitored development projects funded by international agencies which affect wetlands in Asia.

■ Resources

The AWB has training materials on wetland assessment and management, available from Mrs. Norzedah Ali, publications officer. In addition, AWB publishes the biannual *Asian Wetland News*, and a long list of publications including books, articles, pamphlets, and reports, many of them authored or co-authored by Duncan Parish. Updated publications lists are available.

International Center for Conservation Biology (ICCB) ⊕

Department of Wildlife and National Parks
Km. 10 Jalan Cheras
56100 Kuala Lumpur, Malaysia

Contact: **Raleigh A. Blouch,** Course Coordinator
Phone: (60) 3 905 2872 • *Fax:* (60) 3 905 2873

Activities: Education • *Issues:* Biodiversity/Species Preservation; Deforestation; Sustainable Development

● Organization

The International Center for Conservation Biology (ICCB) was established in 1991 as a joint effort of the Smithsonian Institution of the USA, Malaysian Department of Wildlife and National Parks (DWNP), and the National University of Malaysia (UKM). Its objective is to train biologists, resource managers, environmental educators, and zoo personnel from developing countries in the conservation of biological resources. The ICCB supports research that leads to better sustainable management of wild flora and fauna and promotes information exchange among nations in tropical zones.

The Smithsonian's Wildlife Conservation and Management Training Program (WCMTP) of the National Zoological Park in Washington, DC, since 1981 has held courses at its base in Front Royal, Virginia, and in developing countries. To date, more than 650 conservationists from 46 nations have studied with the Smithsonian's program. *(See also Smithsonian Institution, National Zoological Park, Department of Zoological Research in the USA Government section.)*

The WCMTP and the DWNP began working together in 1986, when the first Smithsonian course for resource managers was held in Malaysia. In 1988, the Association of Southeast Asian Nations (ASEAN) sanctioned the course, which also received the enthusiastic support of the Malaysian Minister for Science, Technology, and the Environment.

The Smithsonian provides instructors, with one permanently stationed in Malaysia as ICCB coordinator, funds for travel and other expenses of course participants, and support and scientific advice to researchers. The DWNP provides instructors, offices, and other administrative support, transportation, and a training center in the Krau Wildlife Reserve. In turn, the Department is able to conduct research on particularly relevant topics. In addition, about 20 UKM faculty members work to develop ICCB academic and research programs.

Other agencies and organizations that have worked with the ICCB include the Asian Wetland Bureau (AWB), North American Association for Environmental Education, U.S. Fish and Wildlife Service (USFWS), Singapore Zoo, World Wildlife Fund (WWF)—USA, University of Tennessee, and Sumatran Rhino Trust.

▲ Projects and People

The ICCB offers several options. The core course, in conservation biology and wildlife management, runs annually for seven weeks. Participants have the chance to observe firsthand "ongoing DWNP management programs such as elephant translocation and captive breeding of endangered species." Similar courses are conducted in other Asian countries at the request of local sponsoring agencies.

The Zoo Biology Training Program aims "to improve the management of the captive animals," which have a "vast potential for public education, research, and captive breeding of endangered species." In cooperation with the Singapore Zoo, the ICCB trains zoo professionals from around the world. In 1990, ICCB taught a special course in Environmental Education; it was so popular that the Bureau instituted it as a separate four-week course in 1991. The ICCB offers other short specialized courses, on such topics as the use of computers in wildlife research, the techniques of capturing and handling wild elephants, population genetics, and survey techniques for otters and migratory shorebirds. In addition, ICCB offers instructor training for stellar participants.

Raleigh A. Blouch began working in 1988 with the Smithsonian's WCMTP, helping to organize and conduct six to ten training courses per year for wildlife professionals in developing countries. Since 1989, he has been in Malaysia working with the DWNP to establish the ICCB. Previously with WWF-International and WWF-Indonesia, Blouch worked on various projects in Indonesia and Malaysia, including capturing Sumatran rhinos to start captive breeding colonies in England and Indonesia.

He has worked with elephants and other large mammals, developed a buffer zone using agroforestry in Kerinci-Seblat National Park in Indonesia, studied the Javan warty pig (*Sus verrucosus*), the rare Bawean deer (*Axis kuhli*) endemic to Bawean Island in Java Sea, and other endangered species. Blouch, who is a Malayan Nature Society member, also belongs to the Species Survival Commission (IUCN/SSC) Deer Group and is SSC's Southeast Asia regional coordinator for the Pigs and Peccaries Group.

■ Resources

Researchers interested in practical management implications, such as habitat evaluation, wildlife surveys, and environmental monitoring, can benefit from the combined resources of DWNP, Smithsonian, and UKM: accommodations, assistance with logistics, scientific advice, administrative support, lab facilities, university credits and degrees, and some funding. ICCB's high priority is funding for long-term program development.

TRAFFIC Southeast Asia ⊕
Locked Bag No. 911
Jln. Sultan P.O.
46990 Petaling Jaya, Selangor, Malaysia

Contact: **Stephen V. Nash,** Director
   Phone: (60) 3 757 9192 • *Fax:* (60) 3 756 5594

Activities: Education; Law Enforcement; Research • *Issues:* Biodiversity/Species Preservation • *Meetings:* CITES

● Organization

The international TRAFFIC Network, begun in 1976 and based in Cambridge, England, describes itself as "the world's largest wildlife trade monitoring program." TRAFFIC now has offices in South America, Europe, East/Southern Africa, and Oceania, and national offices in the USA, Japan, Taiwan, and India. The Southeast Asia office, which includes 10 countries—Myanmar, Thailand, Malaysia, Singapore, Indonesia, Brunei, Laos, Cambodia, Vietnam, and the Philippines—is relatively recent.

TRAFFIC aims to enhance wildlife conservation by monitoring and reporting on trade in wild animals and plants and derivative products such as ivory, skins, or herbal medicines; identifying particularly detrimental areas of trade; and assisting the Convention on International Trade in Endangered Species of Wild Fauna and Flora (CITES), government agencies, and other bodies in controling trade and possible threats to species created by trade.

Over 110 countries have signed on to CITES, which aims "to encourage rational and sustainable utilization of living resources for human development, in accordance with the principles of the World Conservation Strategy." The main activities of TRAFFIC involve "collecting, storing, and analyzing data and information from as wide a variety of sources as possible." The organization then publishes that data in the form of bulletins and special reports to a variety of receivers, "especially those in a position to influence or make decisions affecting wildlife conservation."

▲ Projects and People

The primary jobs of TRAFFIC Southeast Asia are to "operate a regional wildlife/plant trade monitoring program," to reinforce the presence of regional CITES authorities in those countries that have signed the Convention, and to assist nonsignatories in enforcing trade restrictions. When necessary and appropriate, the resources of

the TRAFFIC Network, the World Wide Fund for Nature (WWF), and the World Conservation Union (IUCN) are called upon to provide assistance. Investigative priorities include trade in birds, oriental medicines, mammals, and reptiles throughout the region.

Stephen V. Nash established the TRAFFIC office in Malaysia and is director for 1991–1993. Specific investigations that Nash has overseen include trade of passerines and reptile skin in Southeast Asia. He has studied the cultural and social factors affecting the regulation of plant and animal trade in Southeast Asia, and worked to identify individual ASEAN nations' concerns regarding CITES implementation. He is also recruiting and training a counterpart from the region to replace him as office director, is working on the establishment of a national office in Indonesia, and is recruiting a national representative there.

From 1988 to 1991, Nash served as WWF-International Program Coordinator for Irian Jaya Province in Indonesia. The program dealt with "human needs-focused conservation" involving community development initiatives such as agroforestry, agriculture, fisheries, livestock, transportation, and small business. In his position he worked on projects in the Cyclops and Arfak montane reserves, the Teluk Cenderawasih Marine Reserve, and Wasur and Lorentz national parks.

Besides conducting a national seminar on buffer zones, Nash has compiled computer databases on marine biodiversity, habitats, and animal trade. Earlier studies for WWF involved rainforest birds in the Tanjung Puting National Park, Borneo; the Padang-Sugihan Sumatran elephant herd; and estuarine and New Guinea crocodiles in southern Irian Jaya.

■ Resources

Steven Nash is co-author, with D.A. Holmes, of *The Birds of Java and Bali, The Birds of Sumatra and Kalimantan,* and *The Birds of Sulawesi;* and author of more than 23 scientific journal articles, and more than 50 technical reports for WWF and other organizations.

(See also TRAFFIC in the Argentina, UK, USA, and Uruguay NGO sections.)

⚑ Private

Ooi Leng Sun Orchid Nursery and Laboratory
Bukit Jambul Orchid and Hibiscus Garden
873 Sungei Dua
13800 Penang, Butterworth, Malaysia

Contact: **Michael H.C. Ooi,** Horticulturist
   Phone: (60) 4 842248 • *Fax:* (60) 4 842238

Activities: Development; Education; Research • *Issues:* Biodiversity/Species Preservation; Tourism • *Meetings:* CITES

● Organization

A pamphlet describes the **Bukit Jambul Orchid and Hibiscus Garden** as a "one-stop nature lovers' watering hole." A formal garden and nursery, with traditional Malay architecture, the Garden frequently holds "orchid shows, song-bird get-togethers, talks and gatherings for orchidists and horticulturists." Other attractions include exotic birds and an ornamental Japanese pond. Bukit Jambul "aims to bring nature and our rich heritage of flora closer to the people."

▲ Projects and People

The gardens' and nursery's main attraction are the orchids, and they offer a wide range of some of the finest hybrids and species, from Vanda and Dendrobiums to rare varieties like *Paraphalaenopsis serpentilingua* and the albo form of *Phalaeopsis violacea*. Bukit Jambul sells its flowers, live or cut, as well as gardening needs like fertilizer, pesticides, and tools. In addition, the garden offers "free advice and information on orchid cultivation." Hibiscus, the national flower, is also popular at the garden, in a wide range of colors, combinations, and forms. Rubber trees and fruit trees, such as the star fruit (*Averrhoa carambola*), miniature banana trees, and the cashew nut tree (*Anacardium occidentale*), are also found there.

Michael Ooi belongs to the Species Survival Commission (IUCN/SSC) Orchid Group.

■ Resources

The Gardens feature monthly orchid shows, aviaries, deer park, cactus garden, pottery demonstrations, tea kiosk, and gift items.

 *Universities*

Fisheries and Marine Science Center (FMSC)
Universiti Pertanian Malaysia
21030 Kuala, Terengganu, Malaysia

Contact: **Chan Eng-Heng**, Lecturer
  Phone: (60) 9 896411 • *Fax:* (60) 9 896441

Activities: Education; Research • *Issues:* Biodiversity/Species Preservation; Sustainable Development; Water Quality *Meetings:* International Symposium on Biotelemetry; Sea Turtle Workshop, USA

▲ Projects and People

The Fisheries and Marine Science Center (FMSC) of the Universiti Pertanian Malaysia, with a staff of 16, is particularly interested in the leatherback turtle, which nests at the east coast rookery of Rantau Abang, in Terengganu. The Malaysian government controls the behavior of those who observe the nesting season, when the turtles crawl from the sea to their nests to lay eggs; according to the Turtles Enactment of 1987, "Any person who disturbs any turtle in any way . . . during its passage up to the nesting place" faces fines and/or imprisonment. In addition, egg collection for breeding purposes is strictly licensed.

Despite such restrictions, the species which has "survived for at least 150 million years" is threatened with extinction due to human ignorance. All too often, because of the crowds, turtles return to the sea without laying their eggs; other dangers include plastic bags in the water—which the turtle eats because it thinks they are jelly fish—as well as oil, tar, and other forms of pollution.

According to the FMSC's Chan Eng-Heng, "In developing nations, many scientists and environmentalists still work in isolation. . . . The conservation of sea turtles require regional cooperation since these animals are known to feed in the waters of one nation, and then migrate thousands of kilometers away to nest on the beaches of another nation."

She continues, "International conservation and environmental organizations must provide the forum and funding for scientists from developing nations to enable them to come together to bring about and promote conservation on a regional/global basis."

Lecturer since 1976, she is also a member of both the Turtle Sanctuary Advisory Council of State Government of Terengganu, and the Species Survival Commission (IUCN/SSC) Marine Turtle Group. Chan Eng-Heng has authored more than 25 publications, 10 educational pamphlets, and numerous conference presentations on her specialties: green turtles, sea turtles, and leatherback turtles.

Another FMSC lecturer is **Liew Hock-Chark**. Besides tracking leatherback turtles, he has been involved since 1986 in coastal oceanographic studies in the Straits of Malacca and the South China Sea. He is vice chairman of the Terengganu Branch of the Malayan Nature Society, and has written numerous reports and articles.

In a recent series of biotelemetric studies on the green turtles of Pulau Redang (Redang Island), Eng-Heng and Hock-Chark collaborated to locate the internesting habitats to determine if current management provided sufficient offshore protection. Using ultrasonic and radio transmitters, hydrophones, antennas, and receivers, the team monitored daily the movement patterns and behavior of the turtles in the sea, until they returned to lay their next clutch of eggs.

The studies, performed with the cooperation of Terengganu State authorities, the Fisheries Department, Pulau Redang Marine Park personnel, and licensed egg collectors, indicated that current regulation provides effective offshore protection. However, the researchers stress that effective enforcement of the Marine Park regulations is vital to the continued safety of the turtles, and advocate a uniform national law to guarantee their protection. The pair hopes to employ satellite telemetry techniques to map out the turtles' migration paths, after which they can push for regional cooperation.

Another joint Eng-Heng/Hock-Chark project involved studying the relationship between variable temperatures of green turtle nesting sites and a balanced sex-ratio. The team hopes to convince the Fisheries Department of Malaysia to replace their current beach hatcheries practice of high temperature incubation, which results in more female than male hatchlings.

Along with Scott A. Eckert and Karen L. Eckert, they have worked to locate the turtles' internesting habitats using radio telemetry, radio transmitters, and time-depth recorders. The team has successfully mapped the movement patterns of 11 turtles. Their proposed offshore sanctuary, covering a shore distance of 30 kilometers and stretching 5 nautical miles offshore, is now being established by the Turtle Sanctuary Advisory Council of Terengganu and the Fisheries Department. In addition, activities identified as harmful to turtles will be prohibited from April to September each year.

Eng-Heng and Law Ah Theem have studied hydrocarbon and tar ball pollution on the nesting beaches of the leatherback turtle, using fluorospectrometric methods to measure petroleum hydrocarbon concentrations. Future studies will determine the vulnerability of sea turtles to oil.

■ Resources

Resources needed include satellite tracking technology to determine international migration routes of sea turtles, and political support from Asian-Pacific governments to cooperate in regional efforts to save the sea turtles which migrate across national sea boundaries.

Among their many publications are *The Leatherback Turtle: A Malaysian Heritage*, by Chan Eng-Heng and Liew Hock-Chark, 1989; *Sea Turtles in Malaysia*, a coloring book in Bahasa Malaysia and English; and an educational brochure, *Why Our Sea Turtles Are Disappearing.*

Department of Zoology
University of Malaya
59100 Kuala Lumpur, Malaysia

Contact: **Bong-Heang Kiew, Ph.D.,** Associate Professor
Phone: (60) 3 755 5466, ext. 368

Activities: Education • *Issues:* Biodiversity/Species Preservation; Deforestation; Sustainable Development; Water Quality

▲ Projects and People
Dr. **Bong-Heang Kiew**, specialist in frogs, turtles, and tortoises, has been working at the University of Malaya's Department of Zoology since 1969, and as associate professor since 1984. Also co-chairman of the Malayan Nature Society's Conservation Committee, he has served as scientific consultant to the World Wide Fund for Nature (WWF)—Malaysia, leading WWF surveys of the Danum Valley in Sabah State, of the Samunsam Wildlife Reserve in Sarawak, and elsewhere. He has also consulted for the World Conservation Union's (IUCN) Species Survival Commission (SSC)—of which he is a member emeritus for the Malaysian Postal Service's stamp series on the protected wildlife of Malaysia, and for several other governmental projects.

In addition, Dr. Kiew has worked on environmental impact assessments (EIAs) for numerous projects, including a **feasibility study on flood mitigation and agricultural development projects** in the Krian River Basin; the **Linggiu Reservoir** project; the **East-West Highway**, Phase II; the **Upper Krian Swamp Drainage** scheme; the **Sungai Buloh Reservoir** construction; and fourth phase development of **Johore Port**.

Dr. Kiew has written several books, including *Malaysian Frogs and Toads* and *Malaysian Mammals* (with Ruth Kiew); in addition, he has authored reports, scientific studies, and a weekly newspaper column on Malaysian environmental issues, interesting species, and other topics for the *Sunday Star.*

■ Resources
Resources needed include technological, political, educational, and public relations support.

Institute of Advanced Studies (IAS)
University of Malaya
59100 Kuala Lumpur, Malaysia

Contact: **Yap Siaw Yang, Ph.D.,** Lecturer
Phone: (60) 3 757 7000 • *Fax:* (60) 3 757 3661

Activities: Development; Education; Research • *Issues:* Air Quality/Emission Control; Deforestation; Sustainable Development; Waste Management/Recycling; Water Quality
Meetings: CITES

● Organization
The University of Malaya's Institut Pengajian Tinggi (IPT), or the Institute of Advanced Studies (IAS), was founded in 1979. It serves to complement and expand the university's research activities, particularly on projects relating to national development. In addition, through working with other universities and research institutions, the IAS provides research and training opportunities to meet the needs of the University and the country as a whole. Present facilities include 2 buildings, a library with more than 9,000 titles, a computer unit with desktop publishing capabilities, laboratories, and a 5-hectare farm.

There are six focus areas for IPT research: environment and natural resources; human development; Malaysian land use; health and medical sciences; technology and industrial development; and biotechnology.

▲ Projects and People
IPT runs numerous environmental research programs. For example, to investigate ways of best **treating agro-industrial waste**, the Institute is studying microbial utilization, algal biomass production—a bioconversion system whereby nutrients in the wastes are recycled into protein-rich algae which can, in turn, be used for animal feed or for extraction of certain biochemicals—and composting, so that indigenous microorganisms can bring about a natural breakdown of the higher polymers.

Under its **Environmental Assessment and Management** program, the IPT carries out environmental impact assessments (EIAs) studying **coastal and marine ecosystems**, as well as pollution studies regarding pesticides, organic chemicals, heavy metals, and many other toxins. A joint project with the Asian Wetland Bureau is gathering information on "the value of wetland ecosystems such as mangroves, swamp forests, and lakes to promote their sustainable management." A **survey of freshwater and marine algae** from various Malaysian habitats is underway, and those with "potential economic importance are screened for valuable biochemicals which may be extracted for use," reports the Institute.

Other IPT environmental projects relate to urban or rural living. For example, the **Kuala Lumpur Ecoville** carries out research on the effects of **urbanization in the Third World** and assesses the alternatives. The IPT **integrated biological farming** system promotes small-scale farming for optimum use of farm land, labor, and capital. Tests of borers living in forests nearby the sea at Lumut aim to **identify wood that is resistant to attack** by these and other organisms. A **Save Our Slopes** project is investigating "stability and erodibility," particularly along a proposed north-south highway. In addition, a **mapping and survey of urban green areas** in the Federal Territory is being conducted to identify sites important for indigenous flora and fauna, wildlife, and education and recreation benefits. Solar energy experiments are being tested as an alternative source.

Dr. **Yap Siaw Yang**, IPT lecturer, is business manager of *WALLACEANA*, a global newsletter for tropical ecology. He is a member of the International Center for Living Aquatic Resources Management's (ICLARM) Network of Tropical Fisheries Scientists, Species Survival Commission (IUCN/SSC) Freshwater Fish Group, and UN's Food and Agriculture Organization (FAO) working party of experts on inland fisheries.

Dr. Yang has worked on more than 23 projects; the most recent include a "feasibility study on **fish-cum-prawn culture** at Ramuan Cina Besar" and a study of the ecology and fisheries of Subang

Reservoir. Lecturing on natural history and environmental management activities in Malaysia, he has authored more than 55 publications, mainly related to fisheries.

■ Resources

The Institute needs telecommunications and planning assistance. Lectures are offered, with slides and video, of Malaysian natural history and environmental management.

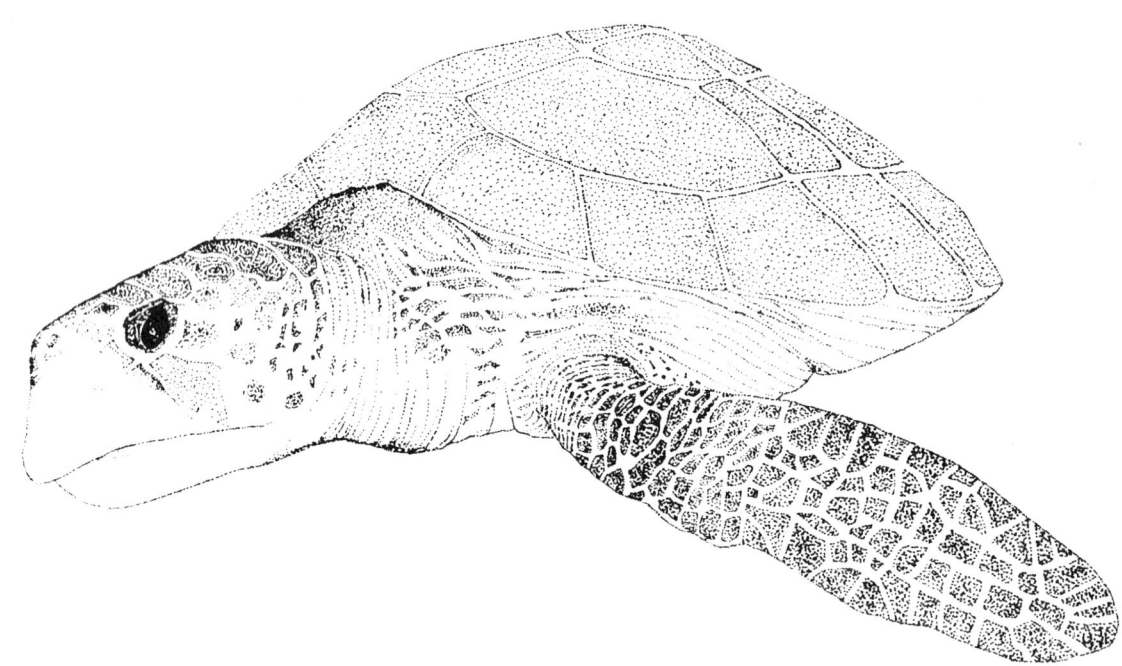

Maldives

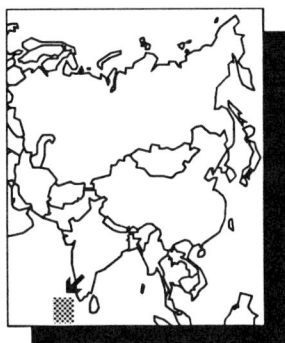

The Maldive archipelago, resting on a submerged mountain range stretching over one million square miles of the north central Indian Ocean, is the largest coral atoll in the world. It is the habitat of a vast array of shellfish and sponge life and a number of marine turtle species. The limited land area and high population density have made pollution from domestic sources a serious concern. The poor and unsafe water supply, coastal erosion, lagoon reclamation, and over-fishing are all major problems.

🏛 Government

Ministry of Planning and Environment (MPE)
Malé, Maldives

Contact: **Hussain Shihab**, Director of Environmental Affairs
Phone: (960) 323825 • *Fax:* (960) 327351

Activities: Development; Education; Law Enforcement; Political/Legislative; Research
Issues: Air Quality/Emission Control; Biodiversity/Species Preservation; Deforestation; Energy; Global Warming; Health; Marine Environment; Sustainable Development; Waste Management/Recycling • *Meetings:* SAARC; UNCED; UNEP

▲ Projects and People

Ongoing projects of the **Ministry of Planning and Environment (MPE)** include **environmental training** to bring about a "qualified and experienced staff" in the Ministry's Environment Section. Funding is through the United Nations Development Programme (UNDP). **Nonformal education materials** are prepared to "to create public awareness on environmental issues"; a newsletter and popular booklets are being produced and distributed. Funding is through United Nations Environment Programme (UNEP).

The Ministry wants to implement an "environmental management and planning capability" to strengthen MPE and to "provide scientific guidance in the execution of various activities." It intends to establish a **sea-level monitoring network** in Maldives; its purpose is to "monitor sea level and wave pattern of the Indian Ocean. " The Australian International Development Assistance Bureau (AIDAB) is providing funding.

The Ministry also wants to undertake a **solid waste management assessment and control project** for urban centers and tourist resorts, possibly with the help of the Canadian government. And it is planning to analyze **demographic trends and migration in Maldives.** Hussain Shihab directs all projects and coordinates other environmental activities. He is co-author of a report on *Sustainable Development, an Imperative for Environmental Protection* and has presented a paper on "Environmental Impact Assessment Procedures in the Maldives" at a UNEP Regional Housing Seminar in Indonesia. He has reported on disaster management, climate change, and coastal management problems in Maldives for other global groups.

■ Resources

The Ministry seeks a desk computer for data processing and short-term training for staff development.

 *NGO*

**South Huvadhu Association for Health
  Education Environmental Development**
H. Gissage
Malé, Maldives

Contact: **Ahamed Didi**
 Phone: (960) 324620

Activities: Development; Education • *Issues:* Global Warming; Health; Population Planning; Sustainable Development; Transportation; Waste Management/Recycling; Water Quality

Malta

Malta is a small archipelago in the central Mediterranean consisting of three inhabited islands: Malta, Gozo, and Comino. The island of Malta is in essence a 600 to 800 foot coralline limestone upland surrounded by blue clay slopes. The summers are dry and hot, and the winters cool but with no snow or frost. The condition of the land is extremely favorable to underground water storage. Soil formation is slow, and vegetation sparse. Forty-four percent of the total land area is arable; most of the cultivation is conducted on small strips of land. Malta is agriculturally self-sufficient.

While industrial development has brought some pollution problems, the per capita income is high—growing at a rate faster than the population and infant mortality has been greatly reduced. Tourism is a major industry, and the infrastructure is highly developed.

🏛 *Government*

Secretariat for the Environment
Floriana, Malta

Contact: **Stanley Zammit, Ph.D.**, Parliamentary Secretary
Phone: (356) 221401 • *Fax:* (356) 243759

Activities: Political/Legislative • *Issues:* Air Quality/Emission Control; Biodiversity/Species Preservation; Sustainable Development; Waste Management/Recycling; Water Quality • *Meetings:* CITES

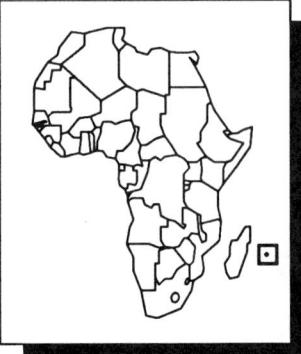

The island of Mauritius lies 500 miles east of Madagascar in the Indian Ocean. It is of volcanic origin and is surrounded by coral reefs. Sugarcane is the most stable crop of the natural resource base. The well-educated populace and rich cultural heritage help to give Mauritius a stable economy. Tourism is a major industry.

At the same time, rapid economic development has led to a proliferation of potentially polluting industries; sewage and solid waste systems are inadequate; and the overuse of pesticides and fertilizers threaten soil fertility and surface and groundwater pollution. In addition, coastal pollution and overfishing have led to the loss of marine ecosystems.

🏛 Government

Department of the Environment
Ministry of Environment and Land Use
Line Barracks Street
Port Louis, Mauritius

Contact: **Raj H. Prayag**, Director
Phone: (230) 212 6080 • *Fax:* (230) 212 6671

Activities: Development; Education; Law Enforcement; Political/Legislative; Research
Issues: Air Quality/Emission Control; Biodiversity/Species Preservation; Deforestation; Energy; Global Warming; Health; Population Planning; Sustainable Development; Transportation; Waste Management/Recycling; Water Quality

▲ Projects and People

Mainly serving as an environmental watchdog in Mauritius, the **Department of the Environment** has numerous functions. Besides formulating the goals and strategies for environmental protection, evaluating the environmental implications of economic and sectoral policies, and conducting studies of environmental planning, the Department examines environmental impact assessment (EIA) reports on all governmental and major commercial and business enterprises; coordinates and monitors the national **Environmental Investment Program (EIP)**; researches and publicizes contaminants, pollution, waste management, waste and litter disposal; and investigates and acts on pollution, waste management, and disposal problems.

It provides local and international organizations with environmental information in the public and global interest, and promotes environmental awareness, education, and research. It is also an enforcer, establishing standards and criteria for air, water, noise, and marine pollution, ensuring the enforcement of and compliance with environmental protection legislation.

The Department prepares and publishes an annual report, and provides assistance and advice to other ministries in developing new policies or modifying existing policies on environmental quality and sustainable development. It is also responsible for calling conferences and conducting seminars, workshops, and training programs relating to the protection of the natural environment.

According to the Department, "By bringing international cooperation through regional cooperation first, the effort of all developing and developed countries in the field of environment can be geared towards a fruitful sharing of experience. This in turn will enable us all to develop a code of conduct vis a vis the degrading environment and approach the requirements and demands in a sound and sustainable way."

■ Resources

The Department has available brochures on recycling programs for used mercury batteries and plastics, as well as other pamphlets on the environment.

Mauritius

Mexico

Bordered by the United States, Guatemala, and Belize, Mexico has a climate with both desert and tropical zones. Mexico's biodiversity ranks fourth in the world: the rainforests in the southeastern states contain exceptional biological richness. Yet Mexico's great natural abundance and diversity has been increasingly threatened by air and water pollution and deforestation. High population growth rates, urban migration, and economic expansion have all contributed to the worsening environmental situation. Deforestation is caused mainly by cash cropping, shifting cultivation, and cattle ranching. Mexico City, in particular, suffers from atmospheric inversions that trap smog and high levels of toxic materials in the air. There is a crucial need in much of the country for a safe water supply.

The government has begun to impose stricter regulations on industrial facilities and has shut down a state-owned refinery for environmental reasons. Controls on auto emissions also have been recently established.

🏛 *Government*

Instituto de Ecología, A.C./Ecology Institute
APDO Postal 63
91000 Xalapa, Veracruz, Mexico

Contact: **Andrew Peter Vovides, Ph.D.,** Director, Botanic Garden
   Phone: (52) 281 86000 • *Fax:* (52) 281 86910

Activities: Research • *Issues:* Biodiversity/Species Preservation; Sustainable Development
Meetings: AIBS; CITES-London; Kew Conservation Conference

● Organization

The mission of **Instituto de Ecología (Ecology Institute)** is to "promote scientific and technological development in the area of ecology and conservation of natural resources, linking its research to national ecological problems."

This mission is accomplished through basic research in the areas of ecology, animal and plant taxonomy, biogeography, conservation of germ plasm, and the structure and dynamics of ecosystems; applied research and development of technology using renewable natural resources to aid commercial production; contributions to conservation and rational use of ecosystems and animal and plant species; publication and dissemination of research; collaboration with local, national, and international institutions within the context of the mission; and education and training of human resources through research.

▲ Projects and People

The Institute is engaged in various basic projects. Of these, seven are strictly concerned with basic and applied research, three involve development, and endeavors on **reserves** focus on the conservation and rational use of ecosystems and animal and plant species. Six projects comprise **vegetation and flora,** regarding the taxonomy, distribution, ecology, and use of plant and fungus species and also regional studies of the physical environment, uses of soil, and location of biological resources. **Ecology and biosystematics of animals** represent a research area.

The **general ecology and conservation of germ plasm** projects include "conservation of ecosystems, relationship between soil, water, and vegetation in the Chihuahuan desert, use of new products of biological origin, and recycling of industrial and urban waste," according to the Institute. Another project, in conjunction with the Museum of Natural History of Mexico City, **publishes and disseminates research results.**

Specific projects have taken place at the Mapimí and La Michilía biosphere reserves in the desert, Durango State; El Pinacate Reserve, Sonara; and El Cielo Biosphere Reserve, Tamaulipas; with other ecological studies in the Valle de México and on interactions between livestock raising and pasture areas; biosystematics, ecology, and biogeography of diverse groups of insects; soil biology; and flora. Also at Mapimí, while the desert laboratory was being constructed, Mexican biologists—in cooperation with ecologists from the Ecole Normale Supérieure in Paris—researched reptiles, birds, and mammals, including the endangered desert turtle. Later, researchers from ORSTOM in France developed projects in hydrology and in the area's primary ecological productivity. *(See also French Institute of Research for Development [ORSTOM] in this section.)*

With the direction of the Institute's senior research entomologist, **Dr. Miguel-Angel Morón**, a 1992–1994 project on **protection and diversity of shine leaf beetles** (*Coleoptera: Scarabaeidae, Rutelinae*) in Mexico and Central America is a cooperative venture with Sociedad Mexicana de Entomología (Mexican Entomological Society). "Insect species are rarely included in international lists of endangered species," writes Dr. Moron. "However, around 500,000 species of insects have very specialized habitats and a restricted distribution [representing] a substantial loss in biodiversity." Moreover, certain insects as *Lepidoptera Papilionoidea* and *Coleoptera Scarabaeidae* and *Cerambycidae* are sought by collectors and traders.

About 80 species in Dr. Moron's research populate the cloudy rainforests, which are being decimated. In the project, at least 20 species—some collected from rotting tree stumps—are being reared to study reproductive behavior, repopulate certain protected areas, and make some available to collectors. Potential ecological reserves' areas will be noted. Guatemalan and Costa Rican institutions are expected to cooperate, and scientific and technical papers and a guide brochure will be produced.

Among other organizations, research botanist **Dr. Andrew P. Vovides** belongs to the Species Survival Commission (IUCN/SSC) Cycad Specialist Group and is scientific secretary of the Mexican Association of Botanic Gardens.

National Sea Turtles Program
Instituto Nacional de la Pesca/National Fish Institute
CRIP/Manzanillo, A.P. 591, P. Ventanas S/N
28200 Manzanillo, Colima, Mexico

Contact: **Rene Marquez M., Ph.D.**, Coordinator, Biologist
Phone: (52) 333 23750 • *Fax:* (52) 333 22102

Activities: Research • *Issues:* Biodiversity/Species Preservation; Sea Turtle Research and Conservation • *Meetings:* Sea Turtles (national, global)

■ Resources
Dr. Rene Marquez, coordinator, National Sea Turtles Program, reports that technical articles can be requested free of charge. He seeks "vehicles, equipment, voluntary work in turtle camps." Dr. Marquez belongs to the Species Survival Commission (IUCN/SSC) Marine Turtle Group.

Anaerobic Treatment of Wastewater in Mexico Program
French Institute of Research for Development in Cooperation (ORSTOM)
Calle Homero 1804-1002, Colonia Los Morales
11510 Mexico City, DF, Mexico

Contact: **Jean-Pierre Guyot, Ph.D.**, Program Head
Phone: (52) 5 395 1085 • *Fax:* (52) 5 395 4227

Activities: Development; Education; Research • *Issues:* Biodiversity/Species Preservation; Deforestation; Energy; Global Warming; Health; Population Planning; Sustainable Development; Water Quality • *Meetings:* International Symposia on Anaerobic Digestion

● Organization
The **French Institute of Research for Development in Cooperation** (ORSTOM) is involved in environmental projects worldwide, such as reforestation using symbiotic trees, global warming, and health. In Mexico, ORSTOM is conducting a program for the **Anaerobic Treatment of Wastewater**. Its purpose is "to promote and develop in Mexico and Latin America the technology of second-generation processes of anaerobic treatment of wastewaters," a technology which ORSTOM sees as matching the economic and technical background of the region.

Launched in 1986, the program involves a group of academics and researchers (including members of ORSTOM's Agricultural Environment and Activity Department). Besides promoting anaerobic wastewater treatment, ORSTOM is working toward greater regional cooperation, providing pilot demonstrations, professional training, and local conferences. In 1990, the first regional symposium dedicated to the implantation and development of anaerobic processes was held. With a staff of 10, ORSTOM uses the following techniques: manipulating and isolating strict anaerobes, biodegradation tests, lab and pilot studies using UASB and fixed film reactors, and gas chromatography.

▲ Projects and People
Mexico's traditional aerobic recycling plants, in which "wastewaters are oxygenated through the ventilation of retention tanks," are described as "investment and energy intensive," with high maintenance costs. Eight of Mexico City's 11 water-recycling plants use the ineffective aerobic sludge blanket (oxygenated) process, while Brazil, India, and China have switched to anaerobic (unoxygenated) plants, according to ORSTOM. The latter are easy to operate and maintain, require technology better suited to the countries' socioeconomic realities, and are effective for a wide range of effluents. ORSTOM also argues that the methane gas released during anaerobic treatment is "ecologically more appealing than the sludge produced by aerobic recycling and which requires additional treatment for stabilization."

In 1982, **Dr. Jean-Pierre Guyot**, molecular biologist/microbiologist, joined ORSTOM, which is headquartered in Paris, France, with **Dr. Gérard Winter** as director-general. *(See also ORSTOM in the Congo Government section and Institut Français de Recherche Scientifique pour le Developpement en Coopération [ORSTOM] in the Senegal Government section.)* Guyot, who began work in Mexico in 1986, writes, "ORSTOM . . . is trying to bring some answers to a very particular way to solve environmental problems [through] fighting by means of biotechnological process against the pollution before it can reach the environment." Other contributors on the project are Hervé Macarie, Oscar Monroy, Rocio Torres de Caña.

■ Resources

Program staff offer lectures on the microbiology of anaerobic treatment; technical assistance on start-up and operation of anaerobic digesters; and consultations on industrial wastewater treatment. ORSTOM has a list of publications, including *ORSTOM Actualités*, a bimonthly news magazine printed in French. ORSTOM director-general Dr. Gérard Winter is located at 213 rue La Fayette, 75480 Paris CEDEX 10, France, fax (31) 1 48 03 08 29.

Secretaria de Desarrollo Urbany y Ecología
Secretary of Urban and Ecology Development
(SEDUE)

Av. Constituyentes 947, EFIF. "A" P.B.

Col. Belen de las Flores

11110 Mexico City, DF, Mexico

Contact: **Sergio Estrada Orihuela, Ph.D.**, Director General, Regulations
Phone: (52) 5 271 2640 • *Fax:* (52) 5 271 1270

Activities: Law Enforcement; Political/Legislative • *Issues:* Environmental Impact Assessments (EIAs); Risk Assessment; Sustainable Development

🏛 *State Government*

Secretary of Rural Development and Ecology, Chiapas

APDO 182

29000 Tuxtla Gutierrez, Chiapas, Mexico

Contact: **Froilán Esquinca Cano**, Biologist
Phone: (52) 961 35542 • *Fax:* (52) 961 32615

Activities: Development; Education; Law Enforcement; Political/Legislative; Research • *Issues:* Agroforestry; Biodiversity/Species Preservation; Deforestation; Global Warming; Sustainable Development; Waste Management/Recycling; Water Quality

▲ Projects and People

Biologist **Froilán Esquinca Cano** reports that Chiapas State's Secretary of Rural Development and Ecology is embarking on new conservation and rural sustainable development projects, in keeping with the state's new ecology law, the Secretary's new Direction (Department) of Ecology and Forestry, and the commitment of the Chiapas governor and the president of Mexico.

A major goal is to "eradicate the extremely poor localities. Chiapas, in the extreme southeast of Mexico, [has] many indigenous and cultural types, archeological value, touristic importance, and agri-

culture, fisheries, cattle ranching . . . important to the development of the state." Chiapas also offers "richness in natural resources," but has a long way to go to attain agroecology and rural sustainable development with no wastes, indicates Esquinca. A huge need exists to train conservationists, technicians, and other professionals to "transform [rural] activities with low-impact and [low] cost to the government and [benefit from] the experience and conviction of the poor and [new involvements] of the state."

■ Resources

Formerly general director of FUNDAMAT, a Chiapas-based NGO, Esquinca now produces government technical papers and books about environmental laws and related matters. He seeks new sustainable agriculture technologies regarding soil conservation and biodiversity; political support to encourage change in the use of soil and land; educational materials on conservation, including technical books on such topics and tropical rainforests; and exchanges with others in these fields, especially to get feedback on projects.

NGO

Amigos de Sian Ka'an A.C. (ASK)/Friends of Sian Ka'an

APDO Postal 770

77500 Cancun, Quintana Roo, Mexico

Contact: **Juan E. Bezaury Creel**, Executive Director
Phone: (52) 988 49583 • *Fax:* (52) 988 73080

Activities: Development; Education; Research • *Issues:* Biodiversity/Species Preservation; Deforestation; Sustainable Development; Protected Areas

● Organization

Sian Ka'an is Mayan for "Where the Sky Is Born"—the name given to the central-southern area of the state of Quintana Roo, Mexico, by the Maya in the fifth century A.D. The Sian Ka'an Biosphere Reserve was created by presidential decree in 1986, covering 1.3 million acres along the central coast of Quintana Roo. It is a complex watershed comprising both land and marine habitat, including "a variety of wetlands and coastal habitats such as seasonally flooded forest, savannas, mangroves, freshwater canals, lagoons, bays, . . . sinkholes, and coral reefs," including 70 miles of the second longest barrier reef in the world.

These habitats house numerous indigenous mammal species, such as jaguar, puma, ocelot, margay, spider and howler monkeys, tapir, and crocodiles, and the beaches are nesting sites of the endangered green, loggerhead, hawksbill, and leatherback sea turtles. In addition, the Sian Ka'an provides habitat for some 336 species of birds, including ocellated turkey, parrots, toucans, trogons, white ibis, roseate spoonbill, flamingo, and 15 species of herons, egrets, and bitterns.

Sian Ka'an is a part of the International Network of Biosphere Reserves, and in 1987 was included in the UNESCO list of World Heritage Sites. In a biosphere reserve, "the goals of preserving the flora, fauna, and ecosystems are integrated with the needs of the local inhabitants." Conservation is seen as "the rational and long-term sustainable use of resources." At Sian Ka'an, fishermen and farmers who live within the reserve and on adjacent lands are viewed as "the inheritors of the ancient Mayan culture that lived in this area for centuries."

Amigos de Sian Ka'an (ASK) (Friends of Sian Ka'an) promotes the conservation of the reserve, with authorization from Mexico's Secretary of Urban Development and Ecology (SEDUE). The organization has 500 members and a staff of 33.

▲ Projects and People

The majority of ASK's current 12 research projects are carried out with the help of other institutions such as the Regional Fisheries Research Center of the National Institute of Fisheries, the Puerto Morelos Station of the Institute of Marine Science and Limnology of the National Autonomous University of Mexico, and Biocenosis, A.C.—with SEDUE authorization. In addition, inhabitants of the reserve, "in particular the fishing cooperative at Punta Allen," participate.

These projects include four to study spiny lobster in Ascension Bay since 1986 to enhance production and develop new models of artifical refuges; three further projects focus on diversification of fisheries, especially for red snapper and stone crab. ASK is also studying the local—relatively pristine—coral reef to adopt norms for maintaining and improving other, damaged reefs. There are two sustainable harvest studies, one for chicle, to avoid overexploitation of the high demand resin of the chicozapote tree (*Manilkara achras*); and one of currently overexploited vines for natural handcrafts. In the latter study, ASK is examining regeneration possibilities and alternative vines.

Four adjacent communities are being encouraged to organize and improve their living and working conditions, substituting organic agricultural techniques, and reducing slash-and-burn agriculture. ASK is also working on an educational outreach program involving 10,000 primary school students and 200 teachers studying tropical forest, marine habitat, and wetlands of the biosphere reserve.

■ Resources

ASK seeks environmental education materials. Publications include the *Sian Ka'an Bulletin*, *Sian Ka'an Coloring Book*, and *100 Birds of the Yucatan Peninsula*. ASK offers Sian Ka'an Nature Tours, as day trips.

Asociación Nacional de Ganaderos Diversifacados (ANGADI)
National Association of Animal Breeders
3639 Toluca
88280 Nuevo Laredo, Tamaulipas, Mexico

Contact: **Juan Francisco Flores Alvarado,** President
  Phone: (52) 871 56782 • *Fax:* (52) 871 49377

Activities: Development; Research • *Issues:* Biodiversity/Species Preservation; Deforestation; Sustainable Development of Wildlife

● Organization

ANGADI (National Association of Animal Breeders) was founded in 1987 by a group of breeders in the northern regions of three states

to legalize hunting of the fauna they raise. Its objective is to "promote the ecological conservation of natural resources and increase the productivity of rural groups and individuals in a comprehensive and sustainable manner with little harm to the environment."

The organization of 320 members, who are devoted to management of domestic cattle and wildlife species, hope their example of conservation on 600,000 hectares will transform other farmers and breeders into ecologists and provide an opportunity to avoid an environmental catastrophe. According to ANGADI, it has saved the white-tailed deer from near extinction; now there are more than 200,000—2,000 of which may be hunted each year.

▲ Projects and People

Goals are to introduce adequate legislation to protect the fauna that breeders' livelihoods depend on; repopulate all areas of Mexico with fauna with the active participation of farmers and ranchers; provide technical advice with the help of universities and NGOs, such as Ducks Unlimited (DUMAC); maintain a publicity campaign to increase tourist activity in hunting and fishing; encourage techniques of sustainable agriculture among animal breeders; maintain vigilance against illegal hunting; provide a registry of breeders of fauna; and conduct population studies of wildlife species.

■ Resources

Brochures for potential members, hunters, and other materials describing the organization are available, printed in Spanish. ANGADI seeks new wildlife management technologies to conserve and improve habitat.

Biocenosis, A.C.
Calle 49, #523A x 68 Centro
97000 Mérida, Yucatan, Mexico

Contact: **Enrique Duhne Backhauss,** Director
  Phone: (52) 99 248417 • *Fax:* (52) 99 247102

Activities: Development; Education; Law Enforcement; Research • *Issues:* Biodiversity/Species Preservation; Deforestation; Sustainable Development

● Organization

Founded in 1983 by a group of professionals from various disciplines, Biocenosis is an NGO named and working for the "community of organisms, plants, and animals that occupy a certain habitat" throughout Mexico. Objectives are the recovery of species in danger of extinction, adequate management of those considered threatened, preservation of natural habitats, and conservation of ecosystems, along with research and communications.

Regarding the reproduction of economically valuable wild species, Biocenosis notes, "The cultivation of species, such as deer, crocodiles, snakes, armadillos, cacti, and orchids, in moderation will not only help to lessen the pressure on wildlife populations, but will also represent a way of making money in areas where the traditional industries are limited by environmental conditions." Biocenosis also encourages the formation of organizations that regulate and organize hunting.

▲ Projects and People

Biocenosis's projects are numerous. Since June 1991, Biocenosis has been directing the management plan of **El Palmar State Reserve**. With Europe Conservation (an NGO based in Milan, Italy), Biocenosis organizes and operates **ecotourism activities** with volunteers through **Proyecto Yucatan (Project Yucatan)**. For the state government of Oaxaca, it is formulating an analytical document and other proposals to serve as a legal basis for environmental protection. Also in Oaxaca, it is establishing a management strategy in the **Chimalapas Zone**, "capable of reconciling the local population's development aims, with those searching for long-term conservation . . . considered one of North America's richest as to biological diversity."

For the **Calakmul Biosphere Reserve**, Campeche, Biocenosis is gathering data to prepare management guidelines in a three-year project supported by Conservation International, MacArthur Foundation, Universidad Autónoma de Campeche, and the federal government's forestry pilot plan (SARH). General guidelines are also being prepared regarding the aquatic portion of the Sian Ka'an Biosphere Reserve, which features the world's second largest **coral barrier reef**. A study of the *Tres Carabelas* tourist resort, Playa del Carmen, Quintana Roo, is the subject of an **environmental impact study** so that mitigation measures can be established.

With **moths and butterflies** regarded as indicators of regeneration processes, research on the status of *Lepidoptera* in areas burnt out after Hurricane Gilbert was recently concluded, at the request of the federal government's Secretary of Urban Development and Ecology (SEDUE).

A series of **television programs** and environmental workshops relating to the conservation of Yucatan's natural resources is being produced, at the request of the state government's Ministry of Ecology. With the Maya Studies Center, National University of Mexico (UNAM), a publication is being produced on the jaguar in Mayan mythology.

Concluded projects include an environmental impact study for a 14-kilometer-long electricity transmission line; in Yucatan, drafting a state environmental law and working with the government on guidelines for **protected areas** and for establishing a state reserve, with management plan, in the Dzilam de Bravo area; recovery and conservation plan for Sonora's endangered pronghorn antelope (*Antelocapra americana*) with SEDUE, Pronatura, Centro Ecológica de Sonora (Ecology Center), Arizona Nature Conservancy, and Arizona Game and Fish Department—with plans to publish a series of monographs; ecological overview of states of Oaxaca, Baja California, Baja California Sur, Sonora, and Chihuahua; translating the *CITES: Identification Manual* and the World Conservation Union (IUCN) book *Managing Protected Areas in the Tropics* (compiled by Mackinnon, Child, and Thorsell) into Spanish; publishing *Geography of Life* books for high school students; and distributing the Spanish version of World Wildlife Fund's (WWF) *Wildlife Management Techniques Manual*.

Enrique Duhne Backhauss has made numerous presentations in North America on conservation and rational management of natural resources. He is a founding member of the Yucatan School of Biology and president/founder of the Yucatan Association for Subterranean Exploration, among leadership in various groups.

■ Resources

Biocenosis writes that its "greatest lack lies in vehicles, computers, and field equipment, and any help along these lines will be welcome." Special requirements include four-wheel drive vehicles, "all-surface" motor tricycles for beach monitoring, and photographic equipment. Publications on Biocenosis and conservation are generally printed in Spanish.

Biósfera Jalisco—Colima (BIOJACO)
Av. Laureles 130, Colonia Tepeyac
1545 Zapopan, Jalisco, Mexico

Contact: **Enrique Estrada Faudon, M.D.**, President
Phone: (52) 36 331914 • *Fax:* (52) 36 213814

Activities: Education • *Issues:* Biodiversity/Species Preservation

▲ Projects and People.

Dr. Enrique Estrada Faudon, medical surgeon and president of BIOJACO, which promotes reforestation and biological conservation, is also director of the Geography and Statistics Institute, University of Guadalajara.

Ducks Unlimited of Mexico (DUMAC) ⊕
APDO 776
64000 Monterrey, Nuevo León, Mexico

Contact: **Eric W. Gustafson, Ph.D.**, National Vice President
Phone: (52) 83 786335 • *Fax:* (52) 83 786439

Activities: Development; Education; Research • *Issues:* Biodiversity/Species Preservation; Deforestation; Sustainable Development; Water Quality; Wildlife Habitat • *Meetings:* Ducks Unlimited International; North American Wildlife; UNCED

● Organization

Ducks Unlimited of Mexico (DUMAC) is a sister organization of the U.S. and Canada Ducks Unlimited organizations. *(See also Ducks Unlimited in the USA NGO section.)* According to DUMAC, its main mission is wetland rehabilitation; however, the organization's efforts are directed toward restoring and protecting a wide variety of habitats and species.

In 1982, 90 percent of DUMAC's funding was from DU, Inc.; that amount is now down to about 50 percent; through other donors and debt-for-nature swaps, DUMAC hopes to reduce its dependence on its northern sibling even further. Also in 1982, DUMAC could boast of 3 projects covering 15,000 acres. By 1990, the organization, with 12,000 members and a staff of 30, had "successfully completed over 93 projects encompassing 910,000 acres."

▲ Projects and People

DUMAC's current president is Don Adrian Sada Treviño; vice president since 1982 is Dr. Eric W. Gustafson, who states: "At DUMAC, we . . . educate and train, we make local communities participants. We look for commitment. When a project is working as wetland habitat and we witness the beautiful wildlife that results, it all seems worth it."

Many of the DUMAC projects involve specific locations. At the **Colorado River Delta**, south of Mexicali, DUMAC is using **runoff irrigation water** to recuperate 10,000 acres of marsh. DUMAC is working on this project together with the Secretariats of Ecology, Agriculture, Fishing and Land, the National Water Commission, and the Baja California Norte State Government. The **Estero San Jose Del Cabo** is a freshwater lagoon which shelters migratory and resident waterfowl. DUMAC is coordinating five government secretariats and agencies in the search for solutions to the problem of its deterioration.

DUMAC also conducts **harvest surveys** in several locations. In **Sinaloa**, where about 40 percent of duck hunting in Mexico takes place, the organization is documenting migratory bird concentration and harvest data funded by the U.S. Fish and Wildlife Service (USFWS). By means of aerial surveys and land observations, 34,500 ducks and geese were tallied in a recent harvest. Another survey in **Yucatan**, also supported by the USFWS, will "provide networking and communication for [DUMAC's] Research Center in Celestun and habitat projects."

DUMAC's **Hunting Ethics in Mexico** project, with the cooperation of DU, Inc. and other organizations, "asks outfitters, hunters and guides to subscribe to an international code of ethics." In voluntarily signing the code, hunters agree to bag limits, seasons, no motorized vehicles, firearm safety, and harvest statistics. There are plans to expand this program to Central and South America.

Since 1988, the **Reserva Program**, also with USFWS help, features annual 90-day programs with participation of more than 11 countries of Central and South America. Instructing professionals from Latin America in the management of parks and natural resources, this program is also backed by SEDUE (Secretariat of Ecology), ITESM (Monterrey Institute of Technology), and the U.S. government.

The **Ranch Wetland Administration Program** provides "technical expertise to those private lands that have an agreement with DUMAC to help provide for the wildlife on their property." The **Nesting Box Program** is "evaluating and replacing our nest boxes, some of which have been destroyed by hurricanes."

Research Stations give graduate students in U.S. and Mexican universities a chance to do research, such as on wildlife diseases affecting waterfowl and specific species such as black-bellied whistling duck and Caribbean flamingos.

Funding for the **Laguna Madre Project**, a proposed waterfowl reserve plan for 250 square kilometers of the "highest priority wetland in Mexico" is being sought from Mexican and U.S. governments, DU, Inc., other groups, and individuals—with the U.S. Senate's Migratory Bird Commission approving matching funds. This area holds 15 percent of Mexico's wetlands.

The Institute for Wetland and Waterfowl Research (IWWR) was recently formed to "sustain research and education and help . . . develop highly skilled professionals, . . . enhance communications, . . . and broaden support for wetland wildlife conservation."

■ Resources

DUMAC publishes the bimonthly *DUMAC Magazine*.

ECOSFERA

5 de Mayo #21
APDO Postal 219
29200 San Cristobal de las Casas, Chiapas, Mexico

Contact: **Ignacio J. March,** Vice President
Phone: (52) 967 8 06 97 • *Fax:* (52) 967 8 06 97

Activities: Development; Education; Research • *Issues:* Biodiversity/Species Preservation; Deforestation; Sustainable Development

● Organization

ECOSFERA, the Centro de Estudios para la Conservacion de los Recursos Naturales (Center for Studies of the Conservation of Natural Resources) is a nonprofit non-governmental organization created to promote integration of natural resource conservation and economic development in Mexico. The philosophy of ECOSFERA is that it is possible to conserve biological diversity and maintain the ecological integrity of natural resources without compromising their potential social benefits. The Executive Board includes Romeo Dominquez-Barradas, president; Ignacio J. March, vice president; Gerardo Garcia-Gil, secretary; Antonio Muñoz Alonso, treasurer; and Board members Marco A. Lazcano-Barrero, Miguel Angel Vásquez, and Eduardo E. Inigo Elias.

ECOSFERA staff members specialize in the following fields: biology and ecology of vertebrates, conservation of endangered species, wildlife utilization, habitat analysis, design and management of natural areas, environmental impact assessment, environmental education, and wildlife legislation. Their expertise has been developed in a variety of regions and habitats in Mexico, particularly tropical environments.

ECOSFERA's activities are divided into four areas: Research and Conservation of Threatened and Endangered Species; Research, Management and Utilization of Wildlife; Research, Planning and Management of Wild Areas; and Training and Academic Development.

▲ Projects and People

Current projects include the following:

Habitat Evaluation and Current Status of the White-Lipped Peccary in Southern Mexico. This is an endangered rainforest indicator species found in the neotropics of Mexico, Central and South America. The peccary is an important source of protein for rural and indigenous communities but one that is hunted throughout its range. The project is evaluating the nature, extent, and condition of the habitat currently available to this species in Mexico. The information obtained will be used to propose to the government ways in which to conserve this mammal and its habitat.

Evaluation of the Current Situation of Cloud Forest in the Meseta Central of Chiapas. The purpose of this project is to determine the current condition of the endangered cloud forest in northern Chiapas, Mexico, and promote its conservation. The few remaining areas of cloud forest will be identified from area photographs and then surveyed for their current extension, presence of bird species, degree of disturbance, and intensity of human pressure. Project leaders then will suggest priority sites for conservation measures.

Diagnosis and Evaluation of the Ocote Ecological Reserve, Chiapas. Biological surveys and evaluation of impacts of development, such as deforestation, are assessing this Reserve's current conditions. With support from the World Wildlife Fund—USA, the project team—including a geographer, physician, environmental planner, mammalogist, botanist, herpetologist, and ornithologist—is working with local people to see how their activities are affecting the Reserve and to develop ways to alleviate pressures on natural resources here.

Cartographic and Geographic Data Base of Calakmul Biosphere Reserve, Campeche. The project's goal is to expand this Reserve's geographical database, providing baseline information which will guide future decisions and management plans. It is part of a wider program developed by the Mexican Association for the Conservation of Nature (Pronatura) and supported by WWF.

Study for the Self-Regulation of the Subsistence Hunting in the Calakmul Biosphere Reserve, Campeche. This project is focused on developing and implementing hunting regulations with Calakmul communities dependent on wildlife as an important source of food. The goal is to ensure long-term sustained use of these resources among the local, rural people. The Secretaria de Desarrollo Urbano y Ecologia (SEDUE) is cooperating with this project.

Preliminary Analysis of the Protected Natural Areas in the Mayan Region. This is a multicountry effort to produce an inventory and analysis of the human and environmental characteristics in the area that incudes southern Mexico, Guatemala, Belize, El Salvador, and Honduras. Research organizations from each country are cooperating to compile the information. The University of California, Riverside, which is supporting funding, will incorporate the project's results in its "Mayan Sustainability Project."

Effects of Forest Fragmentation on the Tropical Raptor Community in the Montes Azules Biosphere Reserve, Chiapas. How does forest disturbance affect the falconiform bird community—both resident and migratory species—in this Reserve? The project is examining population levels and diversity, habitat use, perch use, and prey capture among the birds of prey. Sources of funding are Wildlife Conservation International, American Museum of Natural History, Sigma Xi Scientific Research Society, Program for Studies in Tropical Conservation (PSTC), and the WWF-USA.

Ecology and Current Status of the Aplomado Falcon in Southern Mexico. Once familiar in the southwestern United States to southern Mexico, the northern aplomado falcon is now rare. It breeds regularly in only one locality in this region; only 40 active nests were sighted in the past 40 years. This project is determining the falcon's fate along Mexico's east coast, detecting human activities and factors such as pollution which may be affecting its population and distribution. Expected outcome will be policies, regulations, and actions to protect this species and its habitat.

Reproductive Biology and Population Dynamics of Crocodilians in Sian Ka'an, Quintana Roo. This second largest protected area in the Mexican tropics is the only region where both the endangered American crocodile and Morelet's crocodile are said to co-exist. Plans are to initiate an experimental, economically based conservation program for crocodiles from which local communities may benefit. The potential is in these wetlands to link species

research with ecosystem conservation and rural development. WWF, the University of Florida, and SEDUE are providing funding.

Ignacio J. March, a specialist in wildlife management, conducted research for the Mexican Wildlife Program (Programa Fauna de Mexico) in the Instituto Nacional de Investigaciones sobre Recursos Bioticos (INIREB) of Mexico. Among other activities, he was founder and curator of the INIREB mammal collection. His current ECOSFERA research projects are focused on the Biosphere Reserve of Calakmul, Campeche; the Reserve El Ocote, Chiapas; and the natural areas of the Mayan zone.

Marco Antonio Lazcano-Barrero created the Department of Herpetology within the Wildlife Research and Conservation program at INIREB (National Research Institute on Biotic Resources), specializing in the development of strategies for the protection, recuperation, and management of endangered amphibian and reptile species. He also consolidated INIREB's Herpetological Collection, located in San Cristobal de las Casas, Chiapas, and containing 700 cataloged specimens, including the largest crocodile collection in the country.

▪ Resources

ECOSFERA needs equipment and software for research and education, academic scholarships, and funds for institutional support. Publications awaiting funding are *The Wild Areas of the Mayan Region* and the *Field Guide to the Protected Areas of Southern Mexico.*

Fundación Miguel Alvarez del Toro para la Protección de la Naturaleza (FUNDAMAT)
APDO 970
29000 Tuxtla Gutierrez, Chiapas, Mexico

Contact: **Ramón Perezgil S.**, Board President, Biologist
Phone: (52) 961 33362 • *Fax:* (52) 961 10750

Activities: Education; Fundraising; Research • *Issues:* Biodiversity/Species Preservation; Deforestation; Protected Areas; Waste Management/Recycling • *Meetings:* IUCN; Park Congress

● Organization
With more than 850 members and a staff of 9, FUNDAMAT aims to protect biological diversity in Chiapas and throughout the American tropics. It produces educational materials in both Spanish and English.

Fundación Universo Veintiuno, A.C.
21 Universe Foundation
APDO Postal 44
54600 Tepotzotlán, Edo. México, Mexico

Contact: **Rodolfo Ogarrio**, Director
Phone: (52) 5 874 0222 • *Fax:* (52) 5 876 0214

Activities: Education • *Issues:* Air Quality/Emission Control; Biodiversity/Species Preservation; Deforestation; Energy;

Global Warming; Sustainable Development; Waste Management/Recycling; Water Quality

● Organization

Fundación Universo Veintiuno (21 Universe Foundation) is "dedicated to the protection of nature and the improvement of the environment's quality." Its staff of 40 works toward participation of all organizations and individuals, locally and globally, through information exchange, discussion, and research, "to solve the most pressing problems related to the environment."

▲ Projects and People

Many of the Foundation's activities involve school children. In morning visits, children learn basic ecological concepts and begin to understand the importance of environmental protection. In weekend retreats, students become "familiar with regional ecosystems, natural places and man made processes in the metropolitan area," and learn how they can participate in solving environmental problems. The five-day green weeks expand children's understanding of rural and urban ecology. Summer camps, generally for two weeks, offer field trips with on-site ecological instruction whereby children familiarize themselves with basic ecological concepts, horticultural practices, and wildlife protection techniques. All school programs "are carried out through special agreements with the main schools in the Mexico City metropolitan area."

The Foundation has also worked with the University of New Mexico to monitor the air in Mexico City and "evaluate the major risk groups exposed to these pollutants." In addition, the Foundation has worked on rural studies of such subjects as the consumption and production of wood in Mexico. The most recent project is a demonstration of state-of-the-art equipment for wastewater treatment, atmospheric emissions control, waste recycling, and alternative energy. The Foundation is working to develop a guide on effective environmental auditing.

Many of the Foundation's educational activities take place at the Naturetum, a combination arboretum and gardens "dedicated to research, education and recreation." Containing both evergreens and deciduous trees, both native and exotic species, the Naturetum allows firsthand visitors' experience of "different plant species, environmental problems, horticultural techniques and terrestrial ecology." Plans include a traditional pre-Hispanic Mexican garden, a fragrance garden, a desert garden, and an ethnobotanic garden.

■ Resources

Resources needed include the latest information on new technologies for the demonstration center on environmental control. In addition, the Foundation can use the following: educational materials on environmental topics, including videos, films, and posters of endangered species; advisors to give conferences, workshops, and seminars; and a landscape architect and irrigation systems for reforestation projects.

Foundation publications include an "environmental series" consisting of a summary book and eight volumes on topics like *Mexican Environmental Law*, *Environmental Health in Mexico*, and *Rural Production in Mexico: Ecological Alternatives*. The Foundation offers seminars and workshops, both national and international, covering, for example, voluntary environmental auditing, hazardous industrial waste management, and reforestation and consumption of wood in Mexico.

Grupo Ecologista del Mayab, A.C. (GEMA)
Ecology Group of the Mayab
Plaza Cancun Galerias A.V. Tulum Local No. 13
APDO 479
77500 Cancun, Quintana Roo, Mexico

Contact: Leticia Rubello, Secretary
Phone: (52) 988 45365

Activities: Education; Law Enforcement; Political/Legislative
Issues: Biodiversity/Species Preservation; Deforestation; Health; Population Planning; Sustainable Development; Waste Management/Recycling; Water Quality

● Organization

For six years, GEMA (Ecology Group of the Mayab), based in Cancun, has been attempting to keep further development from causing harm to the environment. Working with Amigos of Sian Ka'an, which organized earlier on behalf of the 1.3-million-acre biosphere reserve that is now Mexico's largest protected area reserve in southern Quintana Roo, GEMA is trying to gain the cooperation of local government officials to safeguard the Cancun area from pollution and overdevelopment. *(See also Amigos de Sian Ka'an in this section.)*

"We know our state has to be developed for tourism," writes Leticia Rubello, GEMA secretary, "but we feel, from our experience of Cancun, that development plans are not respected. Our main concern is to create awareness amongst the developers [regarding] sustainable development and to respect the laws and regulations already established."

▲ Projects and People

With audiovisual presentations, GEMA undertakes environmental education campaigns in schools, through mass media. Subjects include sea turtles, contamination of lagoons, reforestation, coral reefs, Monarch butterfly, and other threatened species. GEMA is accumulating books for an ecology library and cooperates with state Departments of Education and Agriculture and Water Resources.

Efforts to establish a natural reserve encompassing the wetlands of the northern portion of the state include meetings with CALICA, a huge gravel mining and port operation, to gain a "seed money trust fund to buy lands and support such a reserve" as well as with originators of this idea—the Yucatan branch of Pronatura conservation group, and CINVESTAV, a government investigative institution. Such a trust fund would need the immediate support of "industry and developers in other parts of the state to purchase lands and administer the reserve," says GEMA. The state government must also be convinced of the value of making its lands part of the reserve, and technical studies of the reserve accompanied with a public education program must begin.

GEMA is also promoting a state sustainable development plan and, to protect forests, has launched a forest fire prevention program in the city schools, related public outreach programs, and tree planting in Cancun. Rubello points out that "a massive and realistic campaign to eliminate the custom of burning wood areas to clear for agriculture, . . . power lines, road building and maintenance" must be launched. Studies need to demonstrate "the futility of clearing land for cattle growing in this type of terrain," reports GEMA.

Increased monitoring of lagoons and other technical assessment is urged, especially for Cancun's hotel zone sewage treatment plants. Rubello, who left smog-filled Mexico City in the 1980s with her husband to raise a family in Cancun, founded GEMA with other women who discovered that "a sewage treatment plant wasn't operating properly and convinced authorities to fix it," according to the *Mesquite News*, Austin, Texas (June 27, 1990). Now the group works from a donated office in a shopping center, distributing environmental material to residents and thousands of tourists.

■ Resources

GEMA seeks videos and printed materials on environmental education, including endangered species. Resources include audiovisuals and a book, *Medio Ambiente y Desarrollo en Quintana Roo* (Environment and development).

Instituto Indigenista Interamericano
Inter-American Indian Institute

Insurgentes Sur 1690, Col. Florida
01030 Mexico City, DF, Mexico

Contact: **Jose Matos Mar, Ph.D.**, Anthropologist
Phone: (52) 5 660 0007 • *Fax:* (52) 5 534 8090

Activities: Development; Education; Political/Legislative
Issues: Population Planning

■ Resources

The Inter-American Indian Institute offers a publications catalog and publishes the *América Indígena* (Native America) review and the quarterly *Anuario Indigenista*.

Instituto para la Naturaleza y la Sociedad de Oaxaca (INSO)

APDO 6-39
68020 Fovissste, Oaxaca, Mexico

Contact: **Juan Jose Consejo Dueñas**, Coordinator
Phone: (52) 951 30253

Activities: Development; Education; Research • *Issues:* Biodiversity/Species Preservation; Deforestation; Sustainable Development; Waste Management/Recycling

● Organization

With exceptional biological, ecological, and cultural richness, the state of Oaxaca is, nevertheless, scarce in its research and actions to protect wildlife. To this end, Instituto para la Naturaleza y la Sociedad de Oaxaca (INSO) (Institute for Nature and Society) is being formed—particularly to understand regional ecosystems and how they interact with human communities. Its main areas of action are environmental documentation, protection of natural areas, alternative technology, environmental education possibly combined with a botanical garden and zoo, and municipal environmental law.

As of mid-1991, some 30 NGOs and research centers, along with federal and state governments, agreed to support INSO. Project plans are developing, such as an environmental documentation center, Oaxaca State system of natural protected areas, and environmental ordinances in the Sierra de San Felipe. As funds are being raised, World Wildlife Fund (WWF) and Conservation International (CI) are among sponsors.

INSO coordinator Juan Jose Consejo Dueñas is a CI consultant for natural protected areas and also Oaxaca representative to Biocenosis. *(See also Biocenosis in this section.)*

Pronatura, Chiapas Chapter
Mexican Association for the Conservation of Nature

Ave. 5 de Mayo No. 21
San Cristobal de las Casas, Chiapas 29200, Mexico

Contacts: **Rosa Maria Vidal Rodriguez**, Executive Director
Pablo Farias Campero, President
Phone: (52) 967 82723 • *Fax:* (52) 967 80697

Activities: Development; Education; Research • *Issues:* Biodiversity/Species Preservation; Deforestation; Sustainable Development

● Organization

Founded in 1981, **Pronatura** describes itself as "the first and, to date only private, general membership conservation organization in Mexico dedicated to the protection of biological diversity and endangered species of flora and fauna through concrete programs for habitat protection, research, and environmental education."

Pronatura believes that "environmental problems are a result of several causes and complex situations; the solutions are rarely simple and require a diversity of actions at different levels." Based on that principle, Pronatura initiates innovative and feasible projects in different regional locations and with separate areas of focus.

"In developing countries, conservation and development must go hand by hand," writes Pronatura, which works closely with scientific groups to create alternative programs in agriculture, forestry, ecotourism, and others for sustainable development. Social and economic studies are incorporated in projects "to aid local communities and education programs."

Among Pronatura's state chapters are those in Chiapas, Veracruz, Michoacan, and Yucatan Peninsula; its central office is in Mexico City. Friends of Pronatura is an international sister organization.

Started in 1989, the **Chiapas Branch**—with scientists, business leaders, and volunteers working together—is situated among both tropical and temperate forests contributing to biological richness. Here, "some 60 percent of the bird species, 55 percent of the mammals and 33 percent of the amphibians and reptiles that exist in Mexico are found." Some 8,248 plant species of the estimated 12,000 species have been recorded. Altogether, "446 wildlife species endemic to the Mesoamerican region occur in Chiapas."

In contrast, human poverty in Chiapas is acute, especially in rural areas where "60 percent of its population . . . are at, or barely above, subsistence levels." With agricultural lands developed and overcrowding in many areas, the remaining fragile lands are being colonized. Collaboratively, Pronatura-Chiapas assists in land protection and restoration projects.

▲ Projects and People

Pronatura helps preserve and maintain the Huitepec Ecological Reserve, also originated by Pronatura in 1988 in an effort to "find a legal way for this kind of land tenure in Mexico." Situated on a cloud forest, the "most fragile type of forest," the vegetation is considered endangered. Huitepec contains over 700 plant species, some indigenous to Chiapas. The reserve also contains a bird sanctuary of over 100 bird varieties—of which the saw-whet owl, bearded screech owl, blue-throated motmot, black-throated jay, and golden cheeked warbler are endangered species. Medicinal plants, bromeliads, and mushrooms are found in the forest interior. Research is underway on natural forest regeneration and bird and insect ecology studies. Huitepec also features educational activities including a summer training course for children, school visits, and interpretative activities.

A reforestation program called Un Solo Bosque (One Forest) aims at preventing poor land management resulting in deforestation by helping local communities benefit from trees' "fuelwood, traditional medicines, fruits, construction materials, shelter and shade." Emphasizing the planting of local species, including rare ones, Un Solo Bosque is a grassroots effort in which community nurseries are being established, supported by training and environmental education. Participants and visitors are reminded how trees "reduce soil erosion and help sustain soil fertility, water quality, and wildlife habitat." The project also "creates a space for public participation and a focus for talk about other environmental issues in the state." The Canadian Embassy in Mexico and a Canadian company, Brinckman and Associates, support the project, which began in 1991.

In an area where seven languages or dialects are commonly used, Pronatura-Chiapas is beginning an environmental education program to reach rural communities and indigenous populations. Working with Sna Jtzi Bajom (The Writer's House) to develop slide shows, Pronatura hopes its outreach program will gain the support of people living on the borders of large reserves to safeguard the land.

Informal environmental education programs also target children and teachers in the urban area of San Cristobal de las Casas. A teachers' manual and training course were developed, summer courses are offered at the Huitepec Reserve for 7- to 11-year-old youngsters, and a children's book on Mexico endangered species and other materials are being prepared.

As the number of migratory birds is diminishing—especially warblers, thrushes, and flycatchers—the Migratory Bird Center of the Smithsonian Institution, USA, and Pronatura-Chiapas launched a project to determine the impact of tropical rainforest fragmentation on birds and featuring censusing in different habitats as grasslands, corn fields, and second- and old-growth forests. Dr. Russell Greenberg is undertaking the study in the highlands; a Smithsonian team is censusing the Lacandon Rainforest lowlands. Located along the migratory pathways crossing the Gulf of Mexico and passing through the isthmus of Tehuantepec, Chiapas supports more than 190 species of birds that breed in North America. Participating local students benefit from training.

A project to install nest boxes, study feeding requirements, and collect prey is underway on behalf of the southern saw-whet owl and bearded screech owl.

Working with ECOSFERA, the chapter is inventorying the region's species to develop a conservation strategy.

■ Resources

Pronatura needs books and journals to build information on development in tropical rural areas; videos and video-making equipment to expand production of educational materials; training for its participants; and development of a membership program for fundraising.

Pronatura offers coloring books about tropical animals and ecosystems. Literature includes *Cloud Forest in Chiapas, Current Status*; and *Strategy for Conservation of Wildlife in Chiapas*.

🎓 Universities

Instituto de Biologia/Biology Institute
APDO Postal 70-233
04510, Mexico City, DF, Mexico

Contact: **Antonio Lot H., Ph.D.**, Director
Phone: (52) 5 550 5002 • *Fax:* (52) 5 548 8207

Activities: Research • *Issues:* Biodiversity/Species Preservation

▲ Projects and People

Dr. Antonio Lot H., director, **Instituto de Biologia (Biology Institute)** specializes in ecology and plant life of water ecosystems, particularly in Mexican wetlands.

Laboratorio Natural Las Joyas de la Sierra de Manantlán (LNLJ)/Nature Laboratory
Universidad de Guadalajara
Niños Héroes 53
48740 El Grullo, Jalisco, Mexico

Contact: **Enrique J. Jardel P.**, Director
Phone: (52) 338 72748 • *Fax:* (52) 338 72749

Activities: Development; Education; Research • *Issues:* Biodiversity/Species Preservation; Deforestation; Resource Management; Sustainable Development

● Organization

When the University of Guadalajara formed an institution in 1985 to conduct research and management activities, its action preceded the creation of the Sierra de Manantlán Biosphere Reserve (SMBR) by two years. Part of the International Network of Biosphere Reserves, SMBR encompasses approximately 140,000 hectares with altitudes as high as 2,960 meters in the cloud forests and tropical dry forest at lower summits with important oak, pine, and fir resources. For local people, the forests yield mushrooms, blackberries, orchids, bamboo, freshwater shrimps and fish, wildlife, lumber and other building materials, medicinal plants, and a newly found perennial corn, *Zea diploperennis*, resistant to many diseases.

According to **Enrique Jardel** and other researchers at the **Laboratorio Natural Las Joyas (LNLJ)**, such a discovery underscores "the need for *in situ* preservation of genetic diversity in wilderness areas." With this purpose, the Laboratory is carrying out "small-scale sustainable development projects" in which the government, university, and communities within the reserve participate. At workshops, the "local inhabitants identify their problems and suggest their solutions through a feedback research process," Jardel reported in a joint paper with **C. Eduardo Santana** and **M. Rafael Guzman**, entitled "The Sierra de Manantlán Reserve: The Difficult Task of Becoming a Catalyst for Regional Sustained Development."

With 5,000 people living on the reserve, the high rates of illiteracy, malnutrition, and ill health are compounded by conservation problems such as "slash-and-burn agriculture, forest fires, overgrazing, logging activities, poaching and unsustainable levels of firewood consumption." Other symptoms are soil erosion and towns stripped of life by transient logging companies, whose roads scarred landscapes and activities stirred "corruption and community strive."

▲ Projects and People

Thus, LNLJ was started to promote the creation and development of SMBR as well as manage it; conduct scientific research and monitoring activities of ecosystems; train technicians and scientists in ecology and natural resources management; and develop sustainable uses of such resources. With headquarters at El Grullo, LNLJ has a 1,245-hectare scientific station in the mountains. Programs include ethnoecology; environmental education; ecodevelopment; forestry; soils and watersheds; flora; *Zea diploperennis*; fauna; cartography and photointerpretation; information and data processing; publicity; public relations; field station management; and administration.

As the local people realize the "ecological and economical importance of their natural resources," they begin to protect and defend them—a "first step" in gaining control and benefit from these resources. Intensive **summer ecology camps** are held for children of both Guadalajara city and the rural reserve. Special **school programs**, **radio broadcasts**, theater, puppet shows, games, and songs bear a conservation message.

At Ayotitlán, the local people **halted a mining project** and, instead, established a group for the defense of natural resources. In Guzalapa, workers devised a successful plan to prevent and **control forest fires**. In El Terrero, residents gained expertise to manage a business in sync with **sustainable forestry**, also learning how to renegotiate their bank debt.

Also to foster **sustainable agriculture** and raise the standard of living, projects started include a fruit-tree grove, aquaculture and mushroom production, reforestation in town plazas, a nursery of local tree species at the Science Station, and ongoing advice on production of corn, vegetables, firewood, and honey. The **Flora Program** undertook a five-year inventory of plants at SMBR; a herbarium and orchid house were established; and studies reviewed algae, medicinal plants, ferns, and trees. Similarly, the **Fauna Program** inventoried species beginning with insects, fish, birds, and bats. **Vegetation mapmaking and aerial surveys** are ongoing as are an SMBR databank and an LNLJ library. Active in outreach programs, LNLJ staff makes presentations to professional, business, civic, children's government, and media groups.

A costly project, LNLJ sees only benefits to be derived. "The national government benefits . . . because the project insures the sustained use of resources, encourages local development, increases the standard of living in the region, and produces trained professionals. . . . The international community benefits. . . . Migratory birds do not respect international boundaries and their conservation is the responsibility of all the countries involved. Genetic resources could also benefit the world at large."

■ Resources

Brochures are available on the Biosphere Reserve and the Laboratory, printed in Spanish.

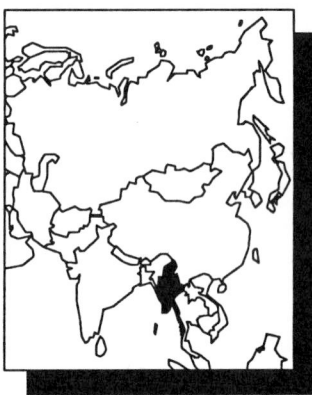

Myanmar (formerly Burma) has long coastlines, high mountains, and more than 8,000 miles of waterways; however, its natural resources are among the least protected in the Indo-Malaysian region. The country suffers from recurring natural disasters, such as cyclones, which cause harm to property, soils, and crops, as well as to people and animals; flooding caused by cyclones and excess precipitation, which causes erosion and sediment loading and helps spread infectious disease; earthquakes; and natural fires, which, along with timber exploitation and shifting cultivation, have destroyed as much as two-thirds of Myanmar's tropical forests. The country's isolationist policy has hampered environmental and resource management.

🏛 *Government*

Forest Department of Myanmar
West Gyogone
Insein, Yangon, Myanmar

Contact: **Maung Ohn**, Advisor
Phone: 66405 (overseas operator)

Activities: Development • *Issues:* Sustainable Development

● Organization

The 261,228 square miles of the Union of Myanmar, formerly Burma, is about half covered by hardwood forests which the **Forest Department** manages. The Department oversees sustained development of hundreds of tree species from eight forest types including evergreen and deciduous forests to dry, hill moist, tidal, and beach and dune forests. It is charged with supervising the cutting, regrowth, and ecological soundness of the forests.

▲ Projects and People

In 1990, the Department produced an in-depth analysis of the stand structure of the **Bago Yoma Forests**, including descriptions, ecological statistics, data, and recommendations on the forests.

Valuable **teak wood** occurs in about 48,000 square miles of mixed deciduous biomes with climatic changes (including a dry season) rather than in the evergreen tropical rainforest. It grows mostly on slopes, under 3,000 feet in altitude, with good drainage without requiring particularly rich soil. Other advantages are its fire-hardiness and ability to regenerate naturally in the same stand for more than a century. In the moister forests, teak and bamboos grow together.

Deciduous forests where teak, pyinkado, and padauk grow together seem to produce the best quality teak in quantity, according to **Maung Ohn and Saw Han**. Teak is best harvested in a 30-year felling cycle when trees reach a girth of 90 inches. "Now with the change in market trend, exploitable trees are greenfelled" as well, reports the Department.

The study's main purpose was to find out if nonvaluable tree species were increasing at the expense of teak and whether "the management system is creaming the forests or not."

Regarding the valuable pyinkado, the study showed that these trees are declining and require protection from illegal cutting and further regulation. Other species are also suffering from "illegal encroachment into the reserves and timber poaching." Yet teak appears to be thriving, according to the Forest Department.

One observation was that the tropical rainforests—while fostering biodiversity for both plants and animals—were yielding fewer "merchantable trees" than similar forests in Malaya, the Far East, and Central America. According to the study, the Bago Yoma deciduous forests were also lower yielding than expected.

It was recommended to research the Malayan forests with regeneration samplings to determine sustainable production levels of teak, study the "overall ecological consequences on the increase of teak stocking at the expense of all other hardwood species," and stop all illegal activities.

Ohm is an honorary consultant to the Species Survival Commission (SSC). Earlier, he helped develop wildlife and marine parks and submitted 12 research papers to his country's research congress on the subjects of teak plantations and other Myanmar national forests.

■ Resources

The Myanmar Forest Department needs help with educational programs that promote conservation and sustainable development.

The report on *Stand Structure Analysis of Bago Yoma Forests* was submitted in 1990 to a United Nations Development Programme (UNDP) seminar of the Economic and Social Commission for Asia and the Pacific (ESCAP) in Myanmar in 1990.

Namibia

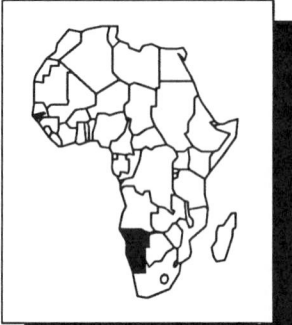

Namibia, formerly known as South-West Africa, is an international territory on the Atlantic coast of southern Africa. The country is mostly high plateau, where most of the population is concentrated. The Namib Desert, stretching along the entire coast, is extremely arid and uninhabited. Most of eastern Namibia forms part of the Kalahari Desert, portions of which are grazed. Rivers flow on the north and south boundaries. Thirty percent of the land is considered arable.

While urbanization and industrialization have increased, most of the population in the relatively wet and wooded northwest are settled farmers and herders. The country's chief income (59 percent) is from mining, mainly diamonds, and fishing (25 percent).

🏛 *Government*

Ministry of Wildlife, Conservation, and Tourism
Private Bag 13306
Windhoek 9000, Namibia

Contact: **Jan H. Joubert**, Chief Public Relations Officer
Phone: (264) 61 63131 • *Fax:* (264) 61 63195

Issues: Biodiversity/Species Conservation • *Meetings:* CITES

● Organization

In Namibia—a country encompassing the western coastal plains, or Namib Desert; central north-south eroded escarpment; and interior plateau to the east dominated by the Kalahari semi-desert—the national conservation strategy aims at fostering biological diversity and sustainable utilization of natural resources to benefit Namibians and the international community in the present and future.

Referred to as "Africa's gem," Namibia entices, from its 200-million-year-old Petrified Forest to its Atlantic Ocean beaches, its "apricot glow of desert dunes, the silvery shimmer of Etosha Pan; the amethyst of distant mountains outlined against the pale sky . . . a wealth of birds and animals."

▲ Projects and People

From marshland to desert landscape, there are 21 major Conservation Areas, many with hiking trails; descriptions of most follow. **Hot Springs Ai-Ais**, meaning "burning hole" or "steaming water," is near **Fish River Canyon**—second to the Grand Canyon in size—and home to **red rock hare**, **Kaokoland ground squirrel**, **dassie rat**, and other indigenous animals and birds in the dense reeds and rushes along the river beds. "Large **yellow fish** and **barbel** abound in the permanent water pools," reports the **Ministry of Wildlife, Conservation, and Tourism**.

As many as 100,000 **Cape fur seals** draw visitors to **Cape Cross Seal Reserve**, where black-backed jackal and brown hyena are also present. Known for its bird life, **Daan Viljoen Game Park** is mainly a highland savanna featuring an endemic aloe (*Aloe viridiflora*) and **red hartebeest**, **wildebeest**, **kudu**, **gemsbok**, **springbok**, **eland**, and the **Hartmann's mountain zebra**. Etosha National Game Park claims 325 bird species and mammals such as **roan antelope**, **black rhinoceros**, **elephant**, **giraffe**, **klipspringer**, **black-faced impala**, **Burchell's zebra**, **Hartmann's zebra**, **lion**, and **Damara dik-dik**. Among the "fascinating plants" is the *Moringa ovalifolia* tree, which forms a thicket called the Haunted Forest. Wild figs, tamboti, maroela, palms, wild dates, acacia species, and shrubs also grow here.

At Gross Barmen Hot Springs are 150 bird species and highland savanna vegetation interspersed with thornbush savanna along the river. More than 800 pelicans are found at **Hardap Recreational Resort** in addition to flamingo, fish eagle, Goliath heron, black rhino, mountain zebra, steenbok, gemsbok, and springbok. Among the dry woodland-savanna and dry bushveld grass plains at **Kaudom Game Park** are **tsessebe**, **Cape hunting dog**, **reedbuck**, and **side-striped jackal**. In the marshes at **Mahango Game Park** are endangered crocodile and hippopotamus as well as endemic water plants. Some 280 bird species are noted in the "dense, riverine forest and adjacent wooded areas," as are other endangered mammals, such as impala, elephant, buffalo, and water-buck. Proclaimed in 1990 were the Mamili and Mudumo national parks. Mamili is Namibia's largest conserved wetland and "the last stronghold on the remnant population of puku"; wattled cranes also breed here. As many as 620 bird species have been recorded at Mudumo; sitatunga, lechwe, and Cape hunting dogs are also found here.

Namibia's largest conservation area, and the world's fourth largest, is Namib-Naukluft Park. Wild horses frequent the southern section. The dunes at the Sossusvlei Area are regarded as Earth's highest, rare game frequents the Naukluft Region, and the western section is noted for the unusual plant *Welwitschia mirabilis* as well as lichens. The Namib is regarded as "the only real desert in southern Africa and is reputed to be the oldest in the world." Known for its crayfish and desert roses is the National Diamond Coast Recreational Area. The rare Damara tern breeds at the National West Coast Recreation Area, also noted for lichen fields and the Cape Cross Seal Reserve.

Namib Desert elephants are found in riverbeds at Skeleton Coast Park, also populated with giraffe, springbok, and gemsbok. The hilly Von Bach Recreation Resort is covered with thornbush savanna and home to kudu, steenbok, and klipspringer.

"Sandstone formations and plateau, the rare Cape vulture, and dinosaur footprints" draw thousands of annual visitors to Waterberg Plateau Park. Bradfield's hornbill, mouse-colored flycatcher, and Alpine swift are located here, as are black and white rhino, sable antelope, tsessebe, roan antelope, and dwarf and rock pythons. Western Caprivi offers fishing and habitat for kudu, steenbok, and klipspringer.

Other protected species include the ostrich, penguin, Kori bustard, martial eagle, and bank cormorant. Seeking shelter in the sand are the "little shovel-nosed lizard, translucent palmato gecko, sidewinder snake, golden mole, [and] 'fog-basking' beetles." Hunting is strictly controlled, advises the Ministry.

■ Resources

The Ministry seeks appropriate (low-tech) technology for rural people and funding for environmental education centers.

Directorate of Wildlife, Conservation, and Research
Ministry of Wildlife, Conservation, and Tourism
P.O. Box 1204, Walvis Bay
Gobabeb 9190, Namibia

Contact: Hermanus H. Berry, Ph.D., Chief Wildlife Researcher
Phone: (264) 642 3581 • *Fax:* (264) 641 2796

Activities: Research • *Issues:* Biodiversity/Species Preservation; Sustainable Development

▲ Projects and People

"In view of the dramatic increase in the number of tourists to the [Namib-Naukluft Park (NNP)] and in view of the sensitivity of the Namib [Desert] to uncontrolled usage and development," guidelines are being established for the Park management, according to Dr. H.H. Berry, Directorate of Wildlife, Conservation, and Research. A strategy is being based on threats to the ecosystems, practical problems involved in countering threats, human carrying capacity for tourism in various zones, sustainable utilization of zones, and management practices to be adopted in the Park's "day-to-day running."

Of particular interest are the control of alien plants and animals, such as *Prosopis* pods growing abundantly in the Swakop River. In one instance, an exotic beetle species has been introduced to control *Prosopis*. An aspect of the plan is to educate farmers as to the dangers of alien plants to the natural environment.

With Hartmann's zebra present in high numbers in small, rather inaccessible areas mainly in the Naukluft Mountains, results are "over-utilization of the food resource and trampling" with the potential to "detrimentally affect the zebra themselves as well as other animal and plant species," reports Dr. Berry. Conservationists have been unable to capture and translocate the zebra in the mountainous terrain, nor has culling been effective. Management staff continues to observe the zebra's distribution, group size, and age-sex throughout NNP and to "investigate the possibility of applying artificial methods to control the population, such as sterilization of stallions and contraception in mares."

As black rhinoceros are reintroduced into the Naukluft section of the Park, monitoring is essential to determine "their behavior and adaptation to a different environment." Researchers are examining reproductive behavior, drinking habits and food preferences, climate effects, and evidences of disease and parasites. They also want to protect the rhinos from wandering onto neighboring farmland or entering unfamiliar mountainous terrain.

The Gobabeb 2000 Project, a proposal of Berry, N. Bessinger, M. Seely, and S. Kooitjie, has the goal of creating the Namib Environmental Complex—a center for desert environmental education and tourism on the Kuiseb River to benefit all Namibians. Three distinct and integrated centers would be incorporated: Namib Visitor's Centre, providing information to local and foreign tourists about the "Namib Desert and other environments and its sensitivity to human pressures"; Kuiseb Education Centre for Namibian educators, students, and pupils; and the Gobabeb Research Centre for visiting scientists. Current funding derives from the Namibian government, Desert Research Foundation of Namibia with the Friends of Gobabeb, and corporations and research funding bodies from Germany, USA, UK, Israel, South Africa, and elsewhere.

The rationale is that the Namib Desert, which "contains the greatest diversity of plant and animal life of any desert in the world . . . is not arable . . . and is largely uninhabited, desolate, uncultivated and barren . . . must be developed and managed with utmost circumspection." According to the proposal, expected consequences of the Environmental Centre are: increased foreign exchange for Namibia, awareness of the Namibian environment and arid environments worldwide, educational opportunities, and human impact on the sensitive desert environment; loss of isolated desert research area to the international research community; and permanent settlement of subsistence farmers in NNP's lower Kuiseb Valley.

The proposed plan calls for zoning the Kuiseb River above Gobabeb as a wilderness area with "wilderness campsites being the only human habitation." Subsistence farming would continue downriver with a staff village near the proposed Visitor's Centre.

Dr. Berry has expertise in entomology through testing pesticides and pest control, ornithology, mammalogy, aerial and ground censuses, and arid savanna animal communities. A regional member for Africa, he belongs to the Species Survival Commission (IUCN/SSC) Equid and Cat groups.

■ Resources

A Directorate publications list features Dr. Berry's works on wildlife, such as lion, wildebeest, pelican, flamingo, reed cormorant, Egyptian vulture, and black tern. A recent publication is *Large-scale Commercial Wildlife Utilization: Hunting, Tourism, and Animal Production in Namibia*.

 NGO

Integrated Rural Development and Nature Conservation (IRDNC) ❦
c/o Palmwag, P.O. Box 339
Swakopmund 9000, Namibia

Contacts: **Garth Owen-Smith**, Director, Conservation
Margaret Jacobsohn, Director, Community Development
Phone: (264) 642 2547

Activities: Development; Education; Research • *Issues:* Biodiversity/Species Preservation; Sustainable Development

● Organization

Integrated Rural Development and Nature Conservation (IRDNC) was established in the late 1980s with support from the World Wide Fund for Nature (WWF)—International, Namibia Nature Foundation, Endangered Wildlife Trust, Rossing Uranium, and De Beers, among others to "involve local communities in nature conservation and to integrate nature conservation into rural development." With a staff of 6 field officers and 42 community game guards, work is concentrated mainly in Namibia's northwest region.

▲ Projects and People

More than an antipoaching unit, the **Community (Auxiliary) Game Guard (CGG)** system provides the structure for rural communities to "actively participate in the conservation of their own natural resources"—considered a key factor in the recovery of wildlife populations since the 1980s. The main objective is to stop poaching—not catch poachers.

With supervision by the IRDNC's field officers, CGGs are responsible to their community leaders and are perceived to be working directly for the local people—rather than the government or the NGO. "We believe that unless rural communities have a real say in decisionmaking regarding the wild animals that live around them and often have a negative effect on their economic development, they will become passive bystanders or even hostile opponents of nature conservation," writes IRDNC. CGGs enable the community and government to enter into a partnership. "In view of our

colonial and apartheid history, this changed relationship is a vital first step towards weaving nature conservation, in its broadest sense, back into Namibia's social fabric," believes IRDNC. A similar project is now in East Caprivi.

Started in 1987, the **Purros Project** links "tourism and wildlife to the economic and social development of an impoverished rural community" in southwestern Kaokoland. Goals are to foster grassroots conservation and a "more dignified relationship" between tourists and rural people as well as to reduce tourism's "negative impact" on the area's traditional seminomadic livestock grazing patterns. Tour operators now pay a levy to Purros residents as wildlife caretakers. A Purros Conservation Committee was set up to decide how the tourist levy is distributed and to discuss problems related to tourism. Underway are a craft market and "ad hoc employment" for local community members in conservation, such as monitoring the local rhino population or the use of trees for crafts.

Because of its strides, Purros was selected for a government wildlife restocking program—gaining 29 gemsbock and 10 giraffe; the economic base is broadened beyond subsistence farming; relationships with tour groups have improved, although problems remain with private, non–levy-paying tourists; and the levy is distributed to all community members, with positive effects on livestock watering patterns. This project is being expanded.

IRDNC also supports the northwest region's **Wereldsend Environmental Centre**, opened in 1990 by the Ministry of Wildlife, Conservation, and Tourism, for students, teachers, and local residents to enhance environmental awareness.

Conservation director **Garth Owen-Smith**, who has worked in Namibia for both the government and NGOs since 1967, and community development director **Margaret Jacobsohn**, consulting social scientist and former news reporter and editor, offer some observations on projects relating to wildlife that contribute to community reempowerment and encourage local resource management:

The goal should be self-sufficiency; "unrealistic dependence on outside funding, technology, and supervision" should be avoided; "development should build on sustainable human and economic resources." As new community institutions are formed, adult males and females—"basic unit of all modern societies, rural and urban"—should be included in economic decisionmaking, instead of relying on "family heads or village elders." Vital opinionmakers, women must be brought into conservation and development programs.

Experience in Africa shows that progress is best achieved from the bottom up. Avoid imposing "masterplans, blueprints, or models onto local communities"; development workers need to avoid the "real temptation to take over decisionmaking from the community when problems arise." Benefits must go directly to local people, who should determine their use. To stop negative social impacts, local people need to decide how tourism takes place.

In Namibia, it is recommended that the Ministry initiate national and regional programs that lay the foundation of sustainable rural development through identifying key communities with valuable wildlife resources, gathering information and facilitating the creation of appropriate new institutions at the local level, and changing laws to accommodate community-based projects.

■ Resources

IRDNC has publications on conservation and development in Africa.

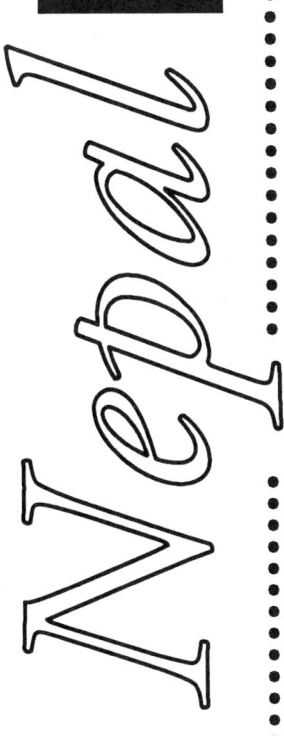

Nepal

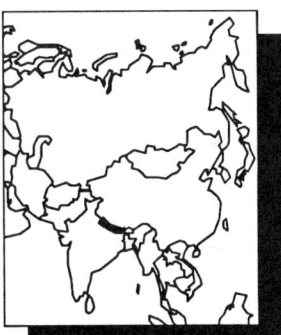

This small, landlocked country, located between northern India and Tibet, suffers from a number of environmental hazards. Nepal's delicate mountain environment is threatened by deforestation resulting from clearing for farming, firewood, and fodder. This, along with grazing livestock, has exacerbated the natural soil erosion process, and has made Nepal more vulnerable to landslides and flooding. Earthquakes, droughts, fires, and epidemics are additional problems, and the continually expanding population (with an annual growth rate of 2.5 percent) puts a severe strain on the country's diminishing resources.

Despite these conditions, conservation awareness dates back many centuries in Nepal, and the current agenda emphasizes improving conservation management and reducing demand.

🏛 Government

Central Zoo
Department of National Parks and Wildlife Conservation
P.O. Box 860
Kathmandu, Nepal

Contact: **Sunder Prasad Shreshta, D.V.M.,** Assistant Superintendent and Curator
Phone: (977) 1 522479 • *Fax:* (977) 1 526570

Activities: Conservation; Development; Education; Propagation; Recreation • *Issues:* Biodiversity/Species Preservation • *Meetings:* CITES

▲ Projects and People

Dr. Sunder Prasad Shreshta studied veterinary physiology at Texas A&M University, USA; undertook a wildlife management course at the Conservation and Research Center, Smithsonian Institution's National Zoo, USA; and was trained in zoo medicine at the St. Louis (Missouri) Zoological Garden and in endangered captive species breeding at the Jersey Wildlife Preservation Trust, UK.

As a veterinary officer at Nepal's Central Zoo, he has studied the effects of "chemical immobilization, particularly in big cats; successfully extracted musk from a live musk deer; captured and treated wild rhinoceroses in Chitawan National Park; and was responsible for the health care of elephants in five government stables" countrywide. Earlier, he was also involved in the breeding of the rare black buck antelope as well as big and small cats. For ecological studies, he has captured wild tigers and greater one-horned rhinoceroses.

Dr. Shreshta belongs to the Species Survival Commission (IUCN/SSC) Veterinary and Antelope groups, American Association of Zoo Veterinarians, and Nepal Veterinary Association.

On behalf of the Central Zoo, he seeks laboratory equipment and information on building new zoo enclosures.

 NGO

Environmental Camps for Conservation Awareness (ECCA) ⊕ ❦
P.O. Box 3923, Jawalakhel
Kathmandu, Nepal

Contact: **Anil Chitrakar,** Project Coordinator
Phone: (977) 1 521506 • *Fax:* (977) 1 226820

Activities: Education • *Issues:* Biodiversity/Species Preservation; Deforestation; Energy; Global Warming; Health; Sustainable Development; Traditional Heritage Maintenance; Waste Management/Recycling; Water Quality • *Meetings:* IUCN

● Organization

Environmental Camps for Conservation Awareness (ECCA) was founded in 1987, and aims "to provide school children a broad environmental program aimed at raising their level of understanding of nature and the need for sound resource management and conservation." Its operations are decided by a board of eight members, in addition to a project coordinator, program director, program officer, and office assistant. ECCA runs more than 40 camps for young people in Nepal and Bhutan (since 1990) each year, including 3 for the disabled, and has a network of counselors in 8 towns in Nepal. Altogether, its 5 offices employ 3 full-time and more than 150 part-time staff.

ECCA is a member of the World Conservation Union (IUCN) and also works with a wide range of NGOs at all levels to carry out its work; these "range from village schools [and] youth clubs, to United Nations agencies, the World Wide Fund for Nature (WWF) and the IUCN."

A typical ECCA camp last 5 days, with 10 camp staff assigning 20 campers representing both sexes and various ethnic groups to activities in "technology, health and sanitation, nature hikes, natural science, arts and crafts, and music and culture." The aim is to foster environmental responsibility, train young people to monitor the environment, and help protect and improve it. The camps are sponsored by the International Trust for Nature Conservation (ITNC), an organization founded in 1980 with the goal of protecting "parts of the world where wildlife is threatened by human activities," in particular Nepal and India. ITNC encourages tree planting and soil improvement activities, as well as "schemes which provide alternative sources of income for local communities."

▲ Projects and People

ECCA often chooses for its campsites nature reserves where local populations are resentful that they can no longer move freely, gathering wood or other resources; they also often blame reserve wildlife for damaging crops and otherwise creating problems. ECCA camps try to teach the children in these areas that the reserves are necessary and good.

Near the **Koshi Tappu Wildlife Reserve** near Nepal's border with India, for example, villagers "complain about the destruction of their crops by the wild buffalo and wild boar that makes their way across the river at night and after filling up on paddy, are back in the reserve by daylight." However, the reserve protects a vital wetlands area, habitat to 114 species of waterfowl, including migratory ducks and shorebirds, the last surviving population of **wild water buffalo**, and the **Ganges dolphin**. The camp, funded by Panos Institute, teaches villagers how to appreciate the reserve.

In 1973, **Meghauli (Royal) Chitwan National Park**, a World Heritage Site, became Nepal's first national park after years of degradation. A U.S. Agency for International Development (USAID)–sponsored malaria eradication scheme in 1950 was so successful that the population of Chitwan increased from 36,000 in 1950 to about 100,000 by 1960, now reaching a density of 117 per square kilometer. As population increased, so did **poaching** of the indigenous rhino, swamp deer, and water buffalo, among others. However, the growing number of people face an acute fuel shortage, and are allowed into the park only 15 days a year to harvest elephant grass. Grazing land is disappearing, and the indigenous Tharu people have been reduced to a minority, with the possible extinction of a popular and unique cultural heritage. According to ECCA, "ignorance is the real villain. . . . There is a need to present the park as an opportunity for better livelihood, rather than a threat [to] their very existence."

A similar situation exists at the **Shivapuri Reserve** in the Shivapuri Watershed area, 144 square kilometers just north of Kathmandu, which is home to sloth bear, leopard, barking deer, langur, the threatened Himalayan dragonfly, and many bird species. The reserve was intended to "improve the quality and quantity of water coming from this area, to conserve soil, flora and fauna," and to meet the needs of the local people. Local populations, however, view the reserve as a threat to their livelihood because they no longer have access to sufficient fuel wood, fodder, and building material.

Children aged 11 to 14 who have attended an ECCA camp "are encouraged to form a catalytic group within the school system," to organize regular activities such as nature hikes, "school projects to revive their local heritage and culture, . . . seed collection, energy data collection," and similar projects.

In addition to the camps, ECCA runs 8 counselor training camps each year, and 20 to 30 **Adventure Times**, one-day activities for urban children, such as cycling, leaf collection, caving, river pollution study, museum visits, or historical site visits, which take place about every 2 weeks. ECCA also offers special training in **food technology** and public forums with other NGOs; trains U.S. Peace Corp volunteers, trekking guides, and school teachers; and provides "input to mass media on environment and children related issues." There are plans to create an ECCA in every village that now hosts a U.S. Peace Corps volunteer.

Anil Chitrakar, renewable energy engineer and project coordinator since ECCA's inception, writes, "Environmental problems are truly global in nature. We must all ask,'Do we have a solution or are we part of the problem?' I plan on contributing to children's understanding of their environment globally." While launching ECCA, Chitrakar was a recipient of an Ashoka fellowship. *(See also Ashoka: Innovators for the Public in the USA NGO section.)*

A member of Nepal Heritage Society, Chitrakar is also IUCN's program officer for Nepal and member of the IUCN Commission on Education and Training. He worked to implement Nepal's National Conservation Strategy, Environmental Impact Assessment (EIA) Program, State of the Environment Project, National EIA Guideline Preparation, National Environmental Planning Guidelines, and Environmental Law Review.

In addition, he is organizer of a special **radio program** over Radio Nepal and of a special environmental camp to mark Earth Day 1990. Chitrakar helped found **TREES Group** of the Godavari Alumni Association, and helped establish *HIMAL* magazine, for which he remains a regular contributor. *(See also HIMAL Magazine in this section.)*

■ Resources

ECCA sees its greatest obstacles as meeting the growing request for camps and carrying out follow-up activities after a camp—especially since villages where camps are conducted are so remote. In addition, the organization needs audiovisual equipment; new teaching methods; means to "assess local needs, problems, and potentials"; educational material reproduction; and a permanent training site. One of Chitrakar's publications is *Towards Decentralized Energy Management in the Nepal Himalaya.*

HIMAL Magazine ⊕
P.O. Box 42
Lalitpur, Nepal

Contact: **Kanak Mani Dixit**, Editor
Phone: (977) 1 523845 • *Fax:* (977) 1 521013

Activities: Development; Education; Research • *Issues:* Air Quality/Emission Control; Biodiversity/Species Preservation; Deforestation; Energy; Health; Population Planning; Sustainable Development; Transportation; Waste Management/ Recycling; Water Quality • *Meetings:* CITES

▲ Projects and People

HIMAL is a bimonthly, nonprofit magazine of The Himalayas, described by its editor, **Kanak Mani Dixit**, as "an open-minded, independent periodical . . . a citizen's forum for Bangladeshis, Bhutanese, Indians, Nepalis, Pakistanis, and Tibetans [which] seeks sustained dialogue and debate across ridgelines, frontiers, and divisions of the mind . . . an *alternative* journal that tackles issues the mainstream media does not." Through conscientious reporting on regional issues ranging from culture to politics, *HIMAL* seeks to raise consciousness and provide a voice to those who believe "that the modernization process must not be allowed to destroy lives and obliterate cultures." While most readers are from the region, the magazine has subscribers around the globe.

Editorial policy screens advertising; cigarette and liquor ads are not accepted, and other advertisements "are scrutinized for possible conflict with HIMAL's world view."

Regarding future global cooperation, Dixit writes, "We believe that environmentalism must be based more on understanding of the physical and social sciences, and less on emotionalism."

■ Resources

HIMAL magazine seeks comments, criticism, advice, subscribers, and assistance in expanding its reach throughout Southeast Asia. It can also be reached through Manoj Basnet/*HIMAL*, 4 South Pinehurst Avenue, #6A, New York, NY 10033 USA, 1-212-928-3761.

Resources Nepal
G.P.O. Box 2448
Kathmandu, Nepal

Contact: **Pralad B. Yonzon, Ph.D.**, Team Leader
Phone: (977) 1 523002 • *Fax:* (977) 1 227132

Activities: Education; Research • *Issues:* Biodiversity/Species Preservation; Deforestation; Energy; Sustainable Development

▲ Projects and People

Dr. Pralad B. Yonzon founded **Resources Nepal** in 1985 and is also president of the Nepal Zoological Society (through 1992), which hosted seminars in 1991 on gharial (crocodilian) conservation, red panda survival, and other topics.

Dr. Yonzon publishes with **Narendra Khadka** the *Agenda Survival* newsletter on behalf of Resources Nepal, which describes both community ecological and nonsustainable activities. "Effects of grazing disturbance on large mammals are particularly significant in the mountain environments," reports their publication. "The Langtang Valley has been subjected to grazing pressures from wild herbivores and livestock for many years."

Cheese factories in the Langtang Valley are said to be adversely affecting the area's forests as *chauri,* large domestic milk herds, trample vegetation and deprive wild animals of habitat, writes Yonzon. "In Langtang, alpine forests are more important to support sustainable agriculture and wildlife than to produce cheese to cater to the tourism industry."

With a doctorate in wildlife resources from University of Maine, USA, where he taught a graduate seminar on "Conservation Issues in Developing Nations," in the 1970s, Dr. Yonzon was a field biologist on the **Nepal Tiger Ecology** project for the Smithsonian Institution; from 1979 to 1982, he was investigator on a **Himalayan Pheasant Ecology Research** project for the World Pheasant Association (WPA); from 1985 to 1987, he was principal investigator for WWF-Nepal's **Himalayas Red Panda Project**. Dr. Yonzon is currently the WWF consultant/advisor for the Ministry of Forests in Bhutan. He serves as instructor and principal coordinator of a **workshop on wildlife habitat and population analysis,** a joint effort of the WWF, the International Center for Integrated Mountain Development (ICIMOD), and Resources Nepal in 1986. He was also the founding editor, in 1988, of *Wildlife Nepal,* for the Department of National Parks and Wildlife Conservation. Dr. Yonzon has authored numerous research publications on pheasants, barking deer, red panda, and other indigenous wildlife species, as well as an article entitled "Cheese, Tourists, and Red Pandas in the Nepal Himalayas" with M.L. Hunter in *Conservation Biology* journal.

■ Resources

Resources needed include a plotter and scanner for a geographical information system (GIS); and desktop publishing equipment for materials encouraging environmental awareness in schools.

South Asian Association for Regional Cooperation (SAARC) ⊕
P.O. Box 4222
Kathmandu, Nepal

Contacts: **Ahmed Saleem**, Director, Environment
Phone: (977) 1 221785 • *Fax:* (977) 1 227033

Activities: Development; Education • *Issues:* Deforestation; Global Warming; Health; Population Planning; Sustainable Development; Transportation

● Organization

The South Asian Association for Regional Cooperation (SAARC), which includes Bangladesh, Bhutan, India, Maldives, Nepal, Pakistan, and Sri Lanka, aims to solve shared problems, and "to accelerate the process of economic and social development in the Member States through joint action in certain agreed areas of cooperation."

The process to establish SAARC began in 1980, and a Charter was adopted at the Dhaka Summit in 1985. SAARC's stated objectives are "to promote the welfare of the peoples of South Asia and to improve their quality of life; to accelerate economic growth, social progress and cultural development in the region, and to provide all individuals the opportunity to live in dignity and to realize their full potentials." The Association has identified 12 areas of cooperation under the Integrated Programme of Action (IPA): agriculture; education; health and population activities; meteorology; postal services; prevention of drug trafficking and drug abuse; rural development; science and technology; sports, arts, and culture; telecommunications; transport; and women in development.

SAARC's highest authorities are the heads of state or government of the member countries, who meet at least once a year. During the period 1985–1990, five such meetings were held in Dhaka, Bangalore, Kathmandu, Islamabad, and Male. The Council of the Foreign Ministers of member states formulates policies, meeting twice a year and occasionally in extraordinary session. The standing committee is composed of the foreign secretaries of member states, who monitor and coordinate the Association's cooperative programs, approve of projects, and submit periodic reports to the Council of Ministers. The standing committee meets as often as necessary. In addition, 12 technical committees are responsible for the implementation, coordination, and monitoring of programs in their respective fields.

The Secretariat, inaugurated in Kathmandu, Nepal, in 1987, bears the role of coordinating and monitoring the implementation of SAARC activities. Secretary-General in 1989–1991 was Kant Kishore Bhargava of India; Ahmed Saleem is one of seven directors in the Secretariat, all of whom serve three-year terms. Regional offices include the SAARC Agriculture Information Center (SAIC) in Dhaka since 1988; the Center for Human Resource Development in Pakistan; and the Tuberculosis Center in Nepal, the Documentation Center in India. Plans for a Meteorological Research Center are underway.

▲ Projects and People

At the Kathmandu Summit in 1987, SAARC commissioned a study for the protection and preservation of the environment and the causes and consequences of natural disasters. By 1991, the Association had completed seven country studies, to be developed into a regional study. The Islamabad Summit in 1988 initiated a joint study on the greenhouse effect and its impact on the region. At the Male Summit in 1990, the Secretary-General was directed to convene a "meeting of experts" to decide on the specifics of the study. SAARC named 1992 the Year of the Environment; of growing concern is the subject of women and environment, under the Association's Women in Development (WID) umbrella.

SAARC's agriculture efforts cover seminars, workshops, field trips, and exchange of germ plasms and prototypes. The organization acknowledges the need to improve utilization of existing regional knowledge, and is planning long-term programs for testing strains of rice, potatoes, and similar staples, as well as improving livestock breeds. Various committees are pursuing natural resources

surveys, organic farming techniques, ways to reclaim problem soils, water management, forestry, seed production, and artificial insemination. Sustainable agriculture is an important SAARC goal.

The Association encourages cooperation between various technical committees. For example, the committees on meteorology and agriculture are working to link meteorological activities with climatic changes, environment, and natural disasters as related to farming. Similarly, the Technical Committee on Science and Technology has plans for cooperation in micro-electronics, forestry development and erosion control, mining and ore beneficiation, genetic engineering, appropriate technology, and the use of agrowaste for energy production. As a whole, SAARC is increasing its studies on the environment and the greenhouse effect, especially in light of the 1992 United Nations Conference on Environment and Development (UNCED).

■ Resources

SAARC publishes in-house reports of meetings, activities, and findings.

Universities

Dolphin Research Foundation (DOPHONE) ⊕
Tribhuvan University
Kirtipur Campus
Kathmandu, Nepal

Contact: **Tej Kumar Shrestha, Ph.D.**, Associate
              Professor
Phone: (977) 2 13748

Activities: Education; Research • *Issues:* Biodiversity/Species Preservation; Water Quality

▲ Projects and People

"By harnessing the symbolic value of dolphin as a mark of river quality and holy symbol of river Ganga," the **Dolphin Research Foundation, Tribhuvan University,** intends to conserve dolphin and its habitat throughout Nepal. **Dr. Tej Kumar Shrestha,** Department of Zoology, coordinates the small, international NGO known as DOPHONE, which was founded in 1987 with the purpose of ending extinction of all river wildlife through "research, conservation, and education programs involving dolphins, otters, crocodiles, fishes and other aquatic life."

DOPHONE describes dolphins as "first classified by Aristotle, often [adorning] the art of ancient Greece and Rome." Freshwater dolphins are "sleek and streamlined with a long beak, . . . active swimmers, often leaping in and out of the water, splashing, pounding on the water with their chins or flukes or zigzagging just below the surface. . . . Curious, they often swim alongside the fisherman's boat."

Rare in the waters of Nepal, India, Bangladesh, and Pakistan where it has been killed for food, cosmetics, and medicinal oil, the dolphin remains a casualty of fishing—creating a pressing problem in Gangetic waterfronts to conserve and manage this "graceful and intelligent animal." Even governmental protection does not guarantee the dolphin's rapid recovery, according to DOPHONE.

Yet the dolphin is considered "an ideal indicator species for monitoring the health of the Ganga river system, as they are the top predators and very sensitive to the environmental perturbation."

During the 1990s, DOPHONE seeks to "study status, distribution, ecology, behavior; identify and map feeding, breeding, and resting habits" through river and feeder stream surveys; "identify areas suitable for habitat enhancement, threats, and problems," such as pollution; launch public awareness campaigns in mass media and schools; encourage related science and research programs at universities; and reduce dolphin/fishermen conflicts. A river park with fish farm is also planned.

Among Dr. Shrestha's recent projects are a study of the ecology of the Koshi River and impact of a dam on the river dolphin, partially funded by the Whale and Dolphin Conservation Society; and a project for ranching mahseer—a migratory fish—in the Himalayan waters, partially funded by the U.S. Agency for International Development (USAID). Also with USAID help, Dr. Shrestha conducts a successful mahseer propagation and ecology program, as part of innovative fisheries research in collaboration with the U.S. Fish and Wildlife Service and University of Arkansas, USA. Interested in the Asian otter, he recently undertook a study of the impact of deforestation, habitat fragmentation, and road building on the bird and animal life of Kathmandu Valley, partially funded by British Ecological Society.

The environmentalist/ecologist has had training in statistical indicators in Japan, as well as in environmental impact assessment (EIA) sponsored by the World Health Organization (WHO) and the Center for Environmental Management and Planning, University of Aberdeen, Scotland.

With more than 17 years' experience in aquatic ecology, EIA, wildlife ecology, and similar fields, Dr. Shrestha is a member of the Linnean Society of London, the Whale and Dolphin Conservation Society, and the Species Survival Commission (IUCN/SSC) Cetacean and Indian Subcontinent Reptile and Amphibian groups. He is foreign secretary of the Zoological Society of India, and editor of Tribhuvan University's *Journal of Science and Technology*.

■ Resources

DOPHONE contributors receive *The River Dolphin Bulletin*, *Annual Report*, "Save Nepal's Dolphin" T-shirt, and opportunities for special dolphin-watching expeditions and internship programs. Dr. Shrestha's numerous publications include: *Environment of Suklaphata* with Dr. D.D. Bhatta; *Wildlife of Nepal*, *Fish Catching Methods of Nepal*, *Dolphin and Whale—Nature's Intellectual Design*, *Dolphins and Whales*; *Fish Culture in Nepal*; *Resource Ecology of Himalayan Waters*, and others.

While the Netherlands enjoys a healthy economy, its natural resources are limited, so that foreign trade is important to its domestic well-being. In addition, given the country's dense population, the land must be used wisely. Dairy farming, mining (particularly natural gas), flower exports, and fishing (most importantly, mussels) are major industries.

The Netherlands has adopted a national environmental policy for the 1990s, which includes stabilizing carbon dioxide, CO_2, emissions by 1995 by taking steps in the areas of fuel use, energy conservation, traffic and transport, waste prevention, and afforestation. Goals also have been set for reducing the level of emissions of substances responsible for acidification, which has damaged forests and eliminated all indigenous species of fish and amphibians in acidified lakes. The Netherlands is especially conscious of the need to consolidate international strategy in meeting environmental goals—60 percent of acid deposits, for example, are from other countries.

🏛 *Government*

Commission on Environmental Impact Assessment ⊕
P.O. Box 2345
3500 GH Utrecht, Netherlands

Contact: **Hans Herman Cohen, M.D.,** Chairman
Phone: (31) 30 331443 • *Fax:* (31) 30 331295

Activities: Independent Environmental Expertise; Law Enforcement • *Issues:* Air Quality/Emission Control; Biodiversity/Species Preservation; Deforestation; Energy; Health; Sustainable Development; Transportation; Waste Management/Recycling; Water Quality

● Organization

The **Commission on Environmental Impact Assessment** concerns itself with the preparation and submission of materials to governmental authorities on environmental aspects which may have an adverse effect. The Dutch government credits the "environmental impact assessment (EIA)" idea from the USA; it is now considered a product of the second generation of environmental legislation within the Netherlands. Since 1987, the Commission has served as part of the formal decisionmaking process—having reviewed more than 300 projects. Some 50 to 60 new projects are submitted annually.

In addition, "at the request of the Ministry for Environment, the Commission will soon start to build up an international advisory environmental capacity for Dutch bilateral assistance activities," it reports. Chair Dr. Hans H. Cohen has 4 deputies and a staff of 25, with an additional 300 advisors and experts. An EIA directive has also been adopted by the European Community (EC).

▲ Projects and People

According to one of its publications, EIA "provides a set of instruments which enable us to visualize the environmental impact of major projects and wide-scale policy plans and of alternatives to these. It takes full account of the environmental interest at stake in the course of the decision-making process."

There are five main parties associated with an EIA. They are the initiator of the activity, the competent authority which receives the decision on the project, the advisors, the Commission on EIA, and the public. A project's impact on the air, flora and fauna, noise, soil, water, microclimate, public health, waste pollution, and cultural heritage and landscape are reviewed by EIA.

Some of the compulsory activities considered by EIA include energy suppliers (coal, electrical, nuclear, wind); defense (air bases, naval ports, training areas); harbors, artificial islands, airports; housing (all major urban extensions); hydraulic works (dams, dikes, reclamation projects); industrial sources (chemical, coal, refineries, steel); mineral extraction; pipelines; railways, roads, waterways; recreational and tourists' amenities (parks, marinas); storage and trans-shipment of fuels; waste disposal; and water extraction and water supplies.

Jules J. Scholten is secretary-general of the Commission; A.A.M.F. Staatsen is vice chair; Dr. Jacques T. de Smidt, Jacobus Van Dixhoorn, and K.H. Veldhuis are deputy chairs.

■ Resources

Summaries of the Commission's annual reports are available in English, as is *Environmental Impact Assessment, The Netherlands—Fit for Future Life*. Most other publications are in Dutch.

International Soil Reference and Information Centre (ISRIC) ⊕

P.O. Box 353 Duivendaal 9
Wageningen 6700 AJ, Netherlands

Contact: **Roel Olderman, Ph.D.**, Head, Programmes and Projects, World Soils and Terrain Digital Database (SOTER)
Phone: (31) 8370 19063 • *Fax:* (31) 8370 24460
Telex: 45888 intas nl

Activities: Education; Research • *Issues:* Global Warming; Soil Geography; Sustainable Development • *Meetings:* CITES; international conferences on global change, land use, soil conservation

● Organization

A documentation center on the world's land resources, emphasizing developing countries, the **International Soil Reference and Information Centre (ISRIC)** houses a large collection of soil monoliths, books, reports, and maps. It was founded in 1966 by the Netherlands government, and until 1983, it was formally known as the International Soil Museum (ISM). The Dutch Ministry of Education and Sciences provides most of its working funds, although it is technically an NGO.

ISRIC aims to improve soil analysis and compilation methods through international correlation and research with particular attention given to soil characterization and classification. By advising, consulting, lecturing, publishing its findings, and training, ISRIC transfers the crucial data gathered.

In 1986, ISRIC's Soil Information System (ISIS) became operational, thereby storing all data collected about climate, landform, land use, soils, vegetation, and other relevant ecological data on a computerized database.

▲ Projects and People

Six sections, three programs, and other miscellaneous and consulting projects comprise ISRIC activities. The sections are documentation, laboratory, micromorphology, soil classification, correlation and mapping, soil monolith collection, and transfer of knowledge to the various visitors to ISRIC facilities.

Its **soil monolith collection** has approximately 800 profiles from 60 countries. Every two years, a **UNESCO-ISRIC short course** is given to help establish national soil reference collections which review land use, extension, training, and research. Some 100 labora-

tories in 70 countries offer a laboratory exchange program to compare and possibly standardize analytical methods and procedures and overcome the problems of polluted and contaminated soils.

The programs of ISRIC are the Laboratory Methods and Data Exchange Programme (LABEX), the National Soil Reference Collection Programme (NASREC), and the Preparation of a World Soils and Terrain Digital Database (SOTER). **Dr. Roel Oldeman** heads the SOTER unit.

LABEX seeks to cross-check, correlate, and standardize analytical methods for soil characterization in order to improve and facilitate correlation studies and international soil classification. One of its objectives is to reduce the "within laboratory" and "between laboratory" variability.

Initiated in 1986, **NASREC** received funds for a 30-month period to strengthen national soil collections in Brazil, Ecuador, Ghana, Indonesia, Mali, Sri Lanka, and Venezuela.

SOTER examines the need to correlate and collate regional and national geographic soil databases and bring them under a common denominator which can serve as a legend for a new world soil map. It's objective is to improve the capability to deliver accurate, timely, and useful information on soils and terrain resources to decisionmakers. In "most developing countries," reports SOTER, "there is a pressing need . . . for a system [in which] each combination of land, water, vegetation, and population which exists within the country can be rapidly analyzed and classified from the point of view . . . of food requirements, socioeconomic factors, and environmental impact or conservation." Such a system would include detailed database, geographical information system (GIS), crop yield models, "table of user-decided dietary levels" for calculating population to be supported in a given area, economical analysis capabilities, and various environmental impact models on erosion and other factors.

This project is divided into three phases with the topic of **Global Assessment of Soil Degradation (GLASOD)**, with the United Nations Environment Programme (UNEP), serving as the pilot phase. According to an ISRIC annual report, GLASOD increases decisionmakers' and policymakers' awareness of the dangers from inappropriate land and soil management to the global well-being. It also aims to improve the capability in regional and national institutions to deliver accurate information on quantitative and qualitative soil degradation processes for regional and national agricultural planning purposes.

SOTER's second phase refines and tests all elements in the database system, and during years 6 through 15, the final phase will transfer the operational input to and output from the database technology. Other related SOTER developmental projects include work in portions of Argentina, Benin, Burkina Faso, Canada, Central America, Czechoslovakia, the Middle East, Niger, Togo, and the USA.

The 1989 international conference on **Soils and the Greenhouse Effect**, held in the Netherlands, was one of ISRIC's major consulting projects. Participants of the numerous working groups presented papers and conclusions which were published a year later.

Recent ISRIC head **Dr. Wim G. Sombroek** is now director of the Land and Water Development and Conservation Division, Food and Agriculture Organization (FAO), Rome, Italy.

■ Resources

ISRIC needs financing to strengthen its outreach programs in the Third World. Cartographic materials, reports, and other technical publications are available upon request. The *Annual Report* and descriptive brochure are published in English.

Nature Policy Plan (NPP) ⊕
Nature Management and Fisheries
Department for Nature Conservation, Forestry, Landscape Planning, and Wildlife Management
Ministry of Agriculture
73 Bezuidenhoutseweg
P.O. Box 20401
2500 EK The Hague, Netherlands

Contact: **Ferdinand H.J. von der Assen**, Head, International Affairs
Phone: (31) 70 3792921 • *Fax:* (31) 70 3478228

Activities: Education; Political/Legislative; Research • *Issues:* Biodiversity/Species Preservation; Deforestation; Landscape Planning; Sustainable Development • *Meetings:* Antarctic Treaty Consultative Meetings; Berne Convention; CITES; CMS; IWC; Ramsar; UNEP Biodiversity Convention

▲ Projects and People

The **Nature Policy Plan (NPP)** is the Dutch government's main nature conservation program. It covers sustainable conservation, development, and rehabilitation of nature and landscape, as well as an international nature conservation policy. Specifically, it focuses on Antarctica, coastal zones, large animal species (whales), migratory birds, North and Wadden seas, tropical forests, and wetlands. The strategic plan offers a long-term approach covering 30 years, with 37 projects and 42 action issues slated for implementation during the first 8 years.

The Nature Policy Plan is one of three national documents that comprise the Netherlands essential policy on the environment. The other two are the **National Environmental Policy Plan (NEPP)** and the Third National Policy Document on Water Management. *(See also Ministry of Housing, Physical Planning and Environment in this section.)* Each has its own special features and character. The NPP is area-oriented toward the conservation of certain species and areas which contribute to the implementation of town and country planning. The NEPP is more general in targeting environmental quality dealing with reduction and prevention of pollution. The Third National Policy Document on Water Management sets down water management system rules.

The NPP plays a significant role in the preparation, assessment, and implementation of governmental policies by assuring that its objectives for landscape and the natural environment and their interrelationships are taken into consideration. The general public, industry, and other groups are also called upon to help improve and protect the quality of the country's landscape and nature. The **Nature Conservation Act** provides the legal basis for NPP regulations and coordination of its activities.

There are four criteria from which the NPP sets its priorities: cultural heritage, ecological value, geomorphological and geological structure, and scenic value.

Archaeology or historical geography significance defines **cultural areas** such as those characteristic of old, man-made features like embankments, livestock watering places or duck decoys, old marine clay polders with flood-refuge mounds, old country estates and parks, and reclaimed old dunes, river terraces, ridges, and river basins.

In selecting those **ecologically valuable sites**, three categories are considered: the characteristics of how natural vegetation or a population associates to its surroundings; the diversity in the number of different species and how they relate to rarity at regional, national or global scope; and naturalness, or how well it fits into the larger ecosystems. Wetlands in the Netherlands, for example, are globally important to waterfowl—spoonbill, cormorant, shoveler, grey heron; sandwich tern and common tern shorebirds; waders—black-tailed godwit, oyster-catcher, avocet, and lapwing; and marshland birds—hen harrier and bluethroat.

Of the 180 breeding birds, however, about 60 have declined since 1950, according to NPP. Some 21 mammal species are at stake, and "recently two species of bats died out, while most probably the otter has to be considered as extinct in the Netherlands too." NPP also notes that "all seven reptile species are dwindling"; 10 fish species are considered very rare; "more than 50 percent of the butterfly species have become less in number, whereas seven species died out this century and eight species have no fixed populations in the Netherlands any more"; 71 species of flowering plants have died out since 1930, and 497 species are declining; also declining are lichens, mosses, and fungi. Acidification, eutrophication, and contamination, falling water tables, desalinization, human disturbances, neglect, and fragmentation and habitat loss are all contributing causes.

Among other species, NPP is focusing on saving mammals such as bat, badger, otter, seal, common porpoise, and tundra vole; various birds; yellow-bellied toad, midwife toad, tree frog, and grass snake; invertebrates as butterfly, dragonfly, and crayfish; wildflowers, orchids, marsh marigold, and other plants.

While evaluating **geomorphological and geological structures** the NPP also follows three criteria: its nonreplaceability, rarity, and topicality. Among those sites given priority are bogs and parts of peatland areas, landscapes structures formed during the glacials, places created under the influence of tectonic movement, or directly/indirectly by the sea, the dissolution of bedrock, or by the wind.

The NPP reports that its country's landscape is "becoming increasingly uniform; the spatial contrasts are becoming less pronounced." Therefore, the government strives to maintain those areas where traditional small- and large-scale open spatial patterns still remain intact.

In addition, the Dutch ecological network increases sustainability by intensifying its relations across the neighboring borders of Belgium and Germany. Ecological corridors and buffers are yet other examples of the total program. The government also implements an international policy by using international cooperation in the field of nature conservation policy proper and introducing ecological considerations into other relevant foreign policies of its governmental operations.

Linkups with various conservation organizations, NGOs, and private/public partnerships create a total environmental package for the Netherlands. And, through the **Policy Document on Nature Studies and Environmental Education**, the general public acquires vital information. The Dutch government realizes that, though it may not have immediate results, **environmental education** will in the long run enhance its country's vital nature awareness. **Sustainable tourism** policies are also being developed, especially in protecting the Alps.

■ Resources

The Nature Policy Plan of the Netherlands, various brochures on nature conservation within its country, and a governmental memorandum on tropic rainforests are available at no cost.

Directorate-General for Environmental Protection
Ministry of Housing, Physical Planning and Environment ⊕
P.O. Box 450
2260 MB Leidschendam, Netherlands

Contacts: **Marius E. Enthoven,** Director-General
Phone: (31) 70 317 4015 • *Fax:* (31) 70 317 4020
Kees Zoeteman, Ph.D., Deputy Director
Phone: (31) 70 317 4017

Activities: Law Enforcement; Political Legislative; Research
Issues: Air Quality/Emission Control; Energy; Global Warming; Health; Sustainable Development; Waste Management/Recycling; Water Quality

● Organization

The **Ministry of Housing, Physical Planning and Environment** of the **Directorate-General for Environmental Protection** is one of four cooperating ministries implementing the Dutch government's National Environmental Policy Plan (NEPP) begun in 1989. It has a staff of 1,000.

"Unless we set a different course quickly and resolutely, we are heading for an environmental catastrophe. The only way to avoid it is to lay a basis now for sustainable development," reports the Dutch government, which has committed itself to reversing present trends. It has concrete measures contained in the short-term environmental program (1991–1994) and long-term courses planned reaching until 2010. However, realizing that all aspects of such an extended program require constant updating, NEPP revises its program every four years.

The policy initially deals with environmental hazards caused by traffic, curbing carbon dioxide (CO_2) emissions, remedial action in cases of soil contamination, tackling acidification, and gaining better control of the entire waste chain. It provides major goals for each level of the environmental scale (continentally, fluvially (rivers), globally, locally, and regionally). Included in these goals are an 80 or 90 percent reduction in emissions of acidifying substances on the continent, regionally a 70 to 90 percent reduction in emissions of acidifying, eutrophying, and poorly degradable substances and 70 to 90 percent decrease in the quantity of waste, and globally a carbon dioxide emissions reduction by 90 percent in the industrial nations.

The environmental policy contains a number of themes with separate measures directed toward each. These are acidification, climate changes, dehydration, diffusion, disposal, disturbance, and euthrophication.

A call for emission-oriented measures to abate acidification include **three-way catalytic converters** on cars, working manure di-

rectly under the soil, large-scale **manure processing**, and extra fuel gas desulfuring at coal-fired power plants. Expanded conservation methods will also be needed in forests and nature areas.

A "fertilizer balance" is sought by NEPP to handle the problems caused by agriculture and industry. **Groundwater** must meet similar **standards** as those of drinking water; no more phosphate and nitrate may enter water and soil than can be absorbed through the natural process. Diffusion of cadmium and solvents are also to be checked.

By 2000, NEPP sets several goals to be met, among them, only a maximum of 12 million tons dumped or incinerated, and that waste produced within the country is also processed there; the number of people experiencing **noise nuisance** be no higher than it was in 1985; creation of an **antisquandering policy** for raw materials, clean water, energy, and fertile soils.

In addition to consumers who are being targeted particularly to save energy, recycle and reuse products, and use mass transit—other groups "will be asked to make large efforts." Among these specialized sectors are agriculture, building and construction, energy, industry, product manufacturers, advertisers, societal organizations, traffic and transport, and waste processing firms. Some of the specific NEPP's requests to these target groups are a 50 percent drop in use of pesticides in agriculture; balanced fertilization with phosphorus and nitrogen; construction of manure processing plants; doubling the recycling of construction and demolition waste; a 25 percent energy conservation in heating systems; establish internal environmental concern systems in energy sector; an 80 percent reduction (relative to 1980) in sulfur dioxide emissions; expand and improve bicycle routes and public transportation; and reduce by 75 percent auto emissions of nitrogen oxides and hydrocarbons and 35 percent from trucks.

NEPP reports that public support continues to grow as numerous organizational groups and specific municipalities units are established. The Directorate-General for Environmental Protection coordinates and monitors the progress and activities of these groups. Instruments, time, economic support, and measures are basic needs in achieving the NEPP, and so are legislation, regulations, and enforcement.

■ Resources

Numerous NEPP literature is available, including *Environmental Programme 1991–1994*, the quarterly *Environmental News, Highlights of the Dutch National Environmental Policy Plan*, and a publications list—all printed in English.

Raad voor het Milieu-en Natuuronderzoek (RMNO)
Advisory Council for Research on Nature and Environment
P.O. Box 5306
2280 HH Rijswijk, Netherlands

Contact: **Johannes B. Opschoor, Ph.D.,** Chairman
Phone: (31) 70 398 5880 • *Fax:* (31) 70 398 5837

Activities: Research • *Issues:* Air Quality/Emission Control; Biodiversity/Species Preservation; Deforestation; Energy; Global Warming; Sustainable Development; Waste Management/Recycling; Water Quality

● Organization

Founded in 1981, the Advisory Council for Research on Nature and Environment (RMNO) is an independent entity which provides information to the government either upon request or at its own direction with a goal to promote and coordinate research. NGOs, policymakers, and researchers make up the Council's tripartite structure.

RMNO's reports to the various governmental ministries on medium- and long-term policies in the areas of economics, social sciences, and public administration that can benefit the environment.

▲ Projects and People

Dr. Johannes B. Opschoor chairs the Council with its 12 members and 4 advisory members from the participating ministries, aided by a staff of 10. Elected for three-year terms (and eligible for reappointment once), the members come from engineering consultancies, environmental organizations, government, large technical research institutes, private enterprises, and universities.

Previously, Dr. Opschoor served as director at the Institute for Environmental Studies at the Free University of Amsterdam, where he is currently a professor of environmental economics.

Research planning studies are commissioned by RMNO to find any gaps and/or areas of overlap within a specific field, and then a coherent research program is formed. A project proposal is formulated when a subject is found suitable for research planning study and presented to the Council to decide whether or not it merits a study. These studies also give early warning signals on long-term environmental problems and serve as discussion platforms between research users and producers. The six areas that RMNO research planning and study groups examine are air pollution, economics and the environment, landscape ecology, recreation and the natural environment, safety and the environment, and water and soil.

Every four years, RMNO releases a long-term perspective advising the government on research policies to be pursued regarding the environment and nature, such as agriculture and fisheries, extraction of raw materials, nature conservation and development, and water supply.

RMNO also conducts workshops and symposia to stimulate new ideas and suggestions for further research, collaborates with other research organizations and advisory councils, and publishes numerous articles on its activities.

■ Resources

RMNO has numerous publications of which summaries are available upon request; subjects range from landscape fragmentation to preventive waste treatment technologies to water management. It publishes the newsletter *Milieu en Economie* (Environment and economics) six times yearly. Dr. Opschoor also has a listing of 8 books, 20 reports, and more than 70 other articles which he has written.

🏛 *Local Government*

OMEGAM Environmental Research Institute
1075 XJ Amsterdam, Netherlands

Contact: **Ronald van der Oost, Ph.D.,**
          Ecotoxicologist
Phone: (31) 20 589 6143 • *Fax:* (31) 20 160327

Activities: Political/Legislative; Research • *Issues:* Air Quality/Emission Control; Biodiversity/Species Preservation; Ecotoxicology; Waste Management/Recycling • *Meetings:* ISORMOP

● Organization

The **OMEGAM Environmental Research Institute** examines the effects and fate of organic trace pollutants, such as polychlorinated biphenyls (PCBs), polycyclic aromatic hydrocarbons (PAHs), polychlorinated dibenzodioxins (PCDDs), organochlorine pesticides (OCPs), and polychlorinated dibenzofurans (PCDFs), in the aquatic environment.

▲ Projects and People

OMEGAM's initial study (1991–1992) concentrates on Lake Nieuwe Meer, contaminated by dredge materials discharged from the Amsterdam city canals. A set of biochemical parameters investigates fish from five sites with various pollution levels. And the bioaccumulation of total toxic equivalents through the freshwater food chain are tested via four toxicity examinations. Field data is further verified with laboratory studies results of experiments with caged fish, as "biomarkers for environmental pollution," with the cooperation of Dr. Antoon Opperhuizen, environmental chemist, State University of Utrecht.

Eels and sediment samples involving water, plankton, molluscs, and crustaceans were collected from each sampling site in 1991. Muscle tissue specimens and livers of eels were removed for analysis and portions stored for future histological examinations. Hans Schoon, water research head, is supervising a subsequent research phase involving "bioaccumulation and effects of heavy metals."

■ Resources

OMEGAM needs antibodies and RNA probes.

🍁 *NGO*

Allard Blom ⊕
Conservation Adviser
Krommestraat 89
4711 NB St. Willebrord, Netherlands

Phone: (31) 1653 3291 • *Fax:* (31) 5753 2788

Activities: Development; Education; Research • *Issues:* Biodiversity/Species Preservation; Deforestation; Sustainable Development; Transportation; Water Quality • *Meetings:* CITES

▲ Projects and People

Biologist **Allard Blom** recently competed work for the World Wide Fund for Nature (WWF) on the **Ituri National Park Project**, hampered by civil strife in **Zaire**. Blom's assignment was to advise local staff of IZCN—Institut Zairois pour la Conservation de la Nature (Zaire Institute for Nature Conservation), help develop a management plan, train park staff, and build support for the creation of protected area status for the Reserve/Park.

Also on behalf of WWF, Blom coordinated with other organizations to help bring about sustainable development practices for local communities. Previously for WWF, in cooperation with the European Community (EC) and Wildlife Conservation International (WCI), he participated in an ongoing **Forest Elephant and Big Ape Survey** to determine the number and distribution of such animals in Central Africa. He also collected **elephant census** data in **Gabon**, where he helped prepare a country report on the utilization of forest ecosystems.

Other specialties include ecology study, with radio tracking and scatanalysis, of the pardellynx (*Lynx pardina*) and the fallow deer (*Dama dama*) in Donana National Park, Spain; game management of the roe deer (*Capreolus capreolus*) in the Netherlands' Oude Venen Reserve; and "conservation of a savanna area in relation with the African buffalo (*Syncerus caffer*) in the Lopé Reserve, Gabon," including setting up a herbarium collection. Blom also consulted in **Ivory Coast** on the **impact of a coastal road on large mammals** and trained **Kenya's** Wildlife Conservation and Management Department (WCMD) staff of Aberdare National Park on forest survey techniques.

Friends of the Earth International (FOEI) ⊕
P.O. Box 19199
1000 GD Amsterdam, Netherlands
Contacts: **Bert van Pinxteren**, International
                Coordinator
          Eka Morgan, International Officer
Phone: (31) 20 622 1369 • *Fax:* (31) 20 627 5287
E-Mail: GreenNet id:foeintsecr

Activities: Development • *Issues:* Air Quality/Emission Control; Biodiversity/Species Preservation; Deforestation; Global Warming; Marine/River Pollution; Sustainable Development; Water Quality

● Organization

Founded in 1971, **Friends of the Earth International (FOEI)** says it is the largest international network of organizations striving to protect the environment. It operates on a three-tier approach at the local, national, and international levels. More than 40 nations now have affiliate Friends of the Earth groups. In 1986, FOEI's activities expanded with the creation of CEAT (Coordination Européene des Amis de La Terre), the first regional network joining European Community member groups. *(See also Friends of the Earth in the USA NGO section and throughout the Guide.)*

As a United Nations accredited NGO, FOEI reports that it has "observer status" at the Food and Agriculture Organization (FAO), International Maritime Organization, London Dumping Convention, International Whaling Commission, Ramsar Convention on Wetlands, and International Tropical Timber Organization (ITTO); as well as consultative status with the UN Economic and Social Council. FOEI belongs to the World Conservation Union (IUCN), Environment Liaison Centre, European Environment Bureau, and the Asian Pacific People's Environment Network.

▲ Projects and People

The conservation, restoration, and rational use of the planet's resources is the common cause of FOE's national groups. Each group is autonomous, politically independent, with its own methods of operations, legal structure, and funding base. However, all groups work jointly on international campaigns, participation in annual meetings, sharing and exchanging pertinent environmental information and action, and in producing the newsletter, *FOE-Link.*

FOEI's mission notes that environmental problems do not respect geographical and political boundaries; economic, political, social, and environmental concerns are interdependent; cooperation with other world organizations is vital; and there are positive alternatives to policies and practices that cause ecological degradation. A leader in developing "soft energy paths," for example, FOEI brings together activists, researchers, and policymakers from many countries to design and implement **renewable and safe energy strategies**.

"FOEI does not just publicly condemn the causes of environmental destruction—we develop environmentally alternatives," its literature notes.

National member groups, elected at the annual FOEI meeting lead international campaigns, as well as coordinate citizen action initiatives, the flow of information within the network, and political lobbying. These national member groups often assume an international role on behalf of the entire FOEI network. Some of FOEI's campaigns include conserving **tropical rainforests**, protecting, preserving, and restoring the Earth's **rivers and watersheds**, and tackling **global warming** and stopping air pollution, especially **ozone depletion** and **acid rain**.

Other potential projects which may be included in the FOEI agenda are monitoring the development of **biotechnology** and its impact on the environment, promoting **East-West cooperation** among the European environmental groups via information exchanges and assistance, and developing an international program to examine commercial and social changes.

■ Resources

FOEI regularly seeks grants enabling continuation of its numerous activities. *ECO* is a conference newspaper which "reports and comments on proceedings and policy matters" at various global meetings.

Greenpeace International ⊕
Keizersgracht 176
1016 DW Amsterdam, Netherlands

Contact: **Roger Wilson**, Director, Treaties and
Conventions Division
Phone: (31) 20 523 6555 • *Fax:* (31) 20 523 6500

Activities: Political/Legislative; Research • *Issues:* Air Quality/
Emission Control; Biodiversity/Species Preservation; Defor-
estation; Disarmament; Energy; Global Warming; Nuclear
Testing; Sustainable Development; Transportation; Waste
Management/Recycling; Water Quality • *Meetings:*
CCAMLR; CITES; FAO; IAEA; IATTC; IPCC; IWC; LDC;
OSCOM; PARCOM; SPREP; UNCED; UNEP

● Organization
Greenpeace is an "independent and non-political international organization dedicated to the protection of the environment through
peaceful means." Started in 1971, it has expanded its operations to
24 countries with more than 5 million members and supporters
worldwide. *(See also Greenpeace USA and Greenpeace, International Pesticides Campaign in the USA NGO section.)*

▲ Projects and People
A multilevel approach to each of its environmental campaigns is
coordinated by a member of the international Greenpeace staff or at
an appropriate national office. Campaigns are divided into five
specific areas: atmosphere and energy, nuclear power and weapons,
ocean ecology, toxics, and tropical rainforests.

Greenpeace calls for **drastic reductions in chemical and gas
emissions** from the major atmospheric polluters while promoting
alternative renewable energy sources and stressing the need for
greater energy efficiency and conservation. Efforts to stop the transport of radioactive materials, the disposal of nuclear waste into the
environment, reprocessing of spent nuclear fuel, and uranium mining are the focus of its **nuclear campaign.**

The **ocean ecology activities** strive to end waste dumping and
pollution into the oceans; more sensitive management of fisheries;
the protection of dolphins, seals, and turtles entangled in hunters'
nets; and a continued **ban on commercial hunting of great whales.**
Greenpeace protests against ships dumping waste at sea. It also is
concerned about the transfer of hazardous toxic waste from industrialized countries to less developed nations.

The newest Greenpeace campaign deals with the tropical
rainforests of Africa, Latin America, and Southeast Asia, which
contain half of all the world's species of animals and plants in only
7 percent of the land surface they cover, according to Greenpeace.
The destruction of this area and the burning of thousands of hectares further add to the global warming trend. Greenpeace works
with various groups to help save the rainforests.

■ Resources
Although Greenpeace International has no publications list, the
NGO's literature is available in its 25 national offices. Needed
resources vary with each campaign and from country to country.

Natuurmonumenten
Society for the Preservation of Nature
Schaep en Burgh, Noordereinde 60
1243-JJ's Graveland, Netherlands

Contact: **Ch. M. van Schaik**
Phone: (31) 35 62004

Activities: Conservation; Education • *Issues:* Biodiversity/Spe-
cies Preservation

● Organization
With a staff of 400 and some 350,000 members, **Natuurmonumenten** owns and operates more than 200 varying reserves including "woodland, heatherland and dunes, lakes and marshland,
country estates, mudflats, peatlands and grassland-areas which form
the habitat of breeding waders like the black-tailed godwit, for
which the Netherlands are famous." In 1905, the Society began
purchasing land, now totaling more than 60,000 hectares. On
broad environmental issues, Natuurmonumenten cooperates with
Stichting Natuur en Milieu (Foundation for Nature and the Environment).

▲ Projects and People
Every 10 years a **management plan** is designed for each site. This
enables the Society to increase , for example,"the number of grazing
animals such as sheep, cattle or horses, to avoid scrub encroachment
on the dunes and heath-lands." The Society notes that "grazing is
also used in forests as a way of achieving diversity in the vegetation
which in turn will create better conditions for animal life."

Hunting is mainly banned, except where it is necessary to control
populations or when animals damage crops outside the reserve.
Nature walks, hides, and boat tours are available for birdwatching—
purple heron, spoonbill, and a cormorant colony of 4,000 pairs.

■ Resources
The Society maintains visitor information centers, and fundraising
campaigns are ongoing activities. *Natuurbehoud* is published quarterly. A *Handbook* is presented to new members. Other books and
publications are issued periodically.

Netherlands Economic Institute (NEI) ⊕
K.P. van der Mandelelaan 11
3062 AD Rotterdam, Netherlands

Contact: **Jan G.D. Hoogland**, Senior Economist
Phone: (31) 10 453 8800 • *Fax:* (31) 10 452 3660

Activities: Development; Law Enforcement; Research • *Issues:*
Air Quality/Emission Control; Deforestation; Economic
Planning; Energy; Health; Population Planning; Sustainable
Development; Transportation; Waste Management/Recy-
cling; Water Quality

● Organization

The **Netherlands Economic Institute (NEI)** links fundamental and applied research to find solutions to pressing socioeconomic and business problems. Prominent in the Netherlands and Europe since 1929, NEI is an independent, nonprofit organization with an ever-expanding circle of clients and contacts covering all levels of economic activity.

▲ Projects and People

Its staff of 150 includes specialized researchers drawn from universities, government, private enterprise, and other research facilities. NEI cooperates with the European Economic Research and Advisory Consortium (ERECO) and independent research groups in France, Italy, Germany, and the United Kingdom. Close ties are also maintained with Wageningen Agricultural University and Erasmus University.

Since **development economics and planning** are major NEI components, it has become a leader in this field over the last 30 years with projects undertaken in more than 70 countries. These are commissioned by the Netherlands Directorate-General for International Cooperation, the European Development Fund, World Bank, regional development banks, United Nations agencies, and bilateral groups.

Presently, agriculture and rural development, energy and industry, international trade, macro-economics, regional and sector development, and transport and port economics are major NEI focus programs. However, through continued theoretical studies, NEI says it is able to stay at the forefront in economic thought—enabling it to meet future challenges.

Ongoing NEI projects in the Netherlands include a **comparative study on the influence of environmental laws as location factors in seaports**, and an appraisal of the integration of costs and benefits of waste processing in market prices of products. In **Hungary**, NEI is providing technical assistance covering an **environmental sector program** from nature conservation to hazardous waste disposal.

Since 1986, NEI has been responsible for more than two dozen studies and programs. Some of the NEI subjects within its own country were the implications of **waste prevention** on the management of environment, natural resources and energy; nontechnical impediments on the **reuse of waste**; implication on noise pollution at **shipyards**; forecasting **air pollution** by car traffic; **economic consequences of a nuclear plant accident**; relation between the price of wastepaper and reuse; economic consequences of added environmental measures for the petroleum refining sector in Rijnmond; **recycling** of metallic raw materials; and economic problems of **Dutch forestry**.

NEI's operations have also included activities in Colombia, Guatemala, Indonesia, the Ivory Coast, Morocco, Sri Lanka, and the USA. NEI's international programs have involved establishing an **environmental policy evaluation**, wastewater treatment technologies, and biomass gasifier pre-investment study. One study for the Probe Pollution Foundation Canada indicated "socio-economic methods to evaluate alternative solutions to clear, neutralize, or isolate **chemical dumpsites near Niagara Falls**" in the USA. Another, in the **Ivory Coast**, identified options for reusing agro-industrial waste from "coffee decortication, rice peeling-mills, wood processing, and palm oil factories" for **biomass energy**.

The development and adaptation of methods and techniques to meet the specific needs of applied research is the key to all NEI activities. **Computer models and software packages** for data man-

agement and statistical analysis are used. Qualiflex, NEI original multicriteria analytical approach initiated in 1975, helps select projects and is regularly refined for this purpose. DATONEI, NEI's comprehensive data bank, contains information from numerous research activities.

Jan G.D. Hoogland is senior economist with the Regional and Sector Studies Head Department, Development Economics and Planning Division. Michiel J.F. van Pelt is economist with the same Division, and Willem Keddeman is senior economist with that Division's Department of Agriculture and Rural Development. Teunis H. Botterweg is research coordinator and senior project economist to the Development Economics and Planning Directorate.

■ Resources

NEI publishes the weekly *Economisch Statistische Berichten*, described as the Netherlands' leading journal for economics. An extensive publications list is printed in Dutch. An information kit, printed in English, describes NEI's "methomatics"; in Development and Economics and Planning, the thrust of research activities in "Analytical Methods, Macro-Economics and International Trade; Regional and Sector Studies' Agriculture and Rural Development; Transport and Port Economics; and Human Settlements Economics"; and in the Netherlands and Europe, research aspects of the "Labour Market; Business and Markets; Society and Policy; Region, Housing and Environment; Sectors and Spatial Patterns; Health and Health Care; Transport and Logistics; and Marketing and Strategy."

Royal Rotterdam Zoological Gardens ⊕
P.O. Box 532
Rotterdam 3000 AM, Netherlands

Contact: **Angela R. Glatston, Ph.D.**, Research Biologist
Phone: (31) 10 654 3333 • *Fax:* (31) 10 467 7811

Activities: Education; Research • *Issues:* Biodiversity/Species Preservation • *Meetings:* Captive Breeding Endangered Species

▲ Projects and People

Since 1987, **Dr. Angela R. Glatston** has been head of the biological section of the Biological/Veterinary Department at the **Royal Rotterdam Zoological Gardens**. She chairs the Species Survival Commission (IUCN/SSC) Procyonid Group and is deputy chair of the SSC **Mustelid and Viverrid Group**. Dr. Glatston is the international studbook keeper for the **red panda** as well as the species' international breeding program coordinator. She belongs to the Association for the Study of Animal Behavior, Primate Society of Great Britain, and the Universities Federation for Animal Welfare as well as a fellow of the Zoological Society of London.

■ Resources

Dr. Glatston needs software to help interpret possible red panda habitat and additional political support to create more national parks and to have the red panda protected under CITES I.

Stichting Milieu-Educatie (SME)
Institute of Environmental Education
P.O. Box 13030
3507 LA Utrecht, Netherlands

Contact: **Frits J. Hesselink,** Director
   Phone: (31) 30 713734 • *Fax:* (31) 30 714868

Activities: Education • *Issues:* Environmental Education; Sustainable Development

● Organization

Founded in 1975, the **Stichting Milieu-Educatie (SME), Institute of Environmental Education,** works in the field of communication, formal and nonformal education, and information dissemination. With a staff of 27, it seeks to alter public attitudes toward nature and the environment by targeting educators, journalists, and policymakers in industry, trade, government, and non-government areas. It also offers consultancy services and takes on projects from private or public commissions, trade unions, women's organizations, government ministries, and nature conservation groups. SME's basic funding is provided by the Dutch Ministry of the Environment. Individual campaigns are financed either through private institutions or the government. The Institute also is a member of the Dutch National Platform for the Environment.

▲ Projects and People

SME cooperates on numerous **environmental educational programs** with the Central Examination Institute, the Centre of Energy Conservation and Appropriate Technology, and the National Institute for Curriculum Development. These various programs serve adults, students, and workers. Working with commercial education publishers, it promotes the use of **environmental text in schoolbooks.**

   The Netherlands' first **national conference** on environmental education in secondary schools was organized by SME in 1986, with its report serving as an important reference point. Four years later, SME joined two major Dutch advertising companies in presenting an environmental education and information conference. The United Nations' **Day of the Environment** is coordinated by SME within its country.

   A future joint project is a **national environmental park** in Amsterdam, focusing on the wonders of nature, the evolution of the Earth's ecosystems, the origins of environmental programs and their solutions, and options for sustainable development.

■ Resources

SME publishes a monthly magazine, *Milieu-Educatie,* for teachers and those in the educational field. It also has a publications list containing 100 titles, in Dutch, on environmental education topics.

Project Modern Economics (PROMODECO) ⊕
Stichting Milieu-Educatie (SME)
Institute of Environmental Education
P.O. Box 13030
3507 LA Utrecht, Netherlands

Contact: **Willem Hoogendijk**
   Phone: (31) 30 713734 • *Fax:* (31) 30 714868

Activities: Education • *Issues:* Economics; Sustainable Development

● Organization

Stichting Milieu-Educatie (SME), Institute of Environmental Education "specializes in working for the Dutch educational system, both formal and non-formal. The objectives of SME are to change public attitudes towards nature and the environment, aiming in particular at teachers, teacher-trainers, journalists, educationalists, and local and national policymakers." *(See also previous Stichting Milieu-Educatie profile in this section.)*

▲ Projects and People

Willem Hoogendijk is a co-founder of SME, where he first worked as a volunteer for five years and then became a paid staff member. Today, as one of the Institute's policymakers, Hoogendijk works on his economic theories, such as "the money-system" in his spare time, because "it is not directly educational and moreover, controversial."

   In his recent book, *The Economic Revolution Towards a Sustainable Future by Freeing the Economy from Money Making,* Hoogendijk writes that "only a U-turn can save us" from the destruction of the environment, which contemporary production puts at risk. To slow down would be positive and real growth, he believes. "The 'money-must-grow system,' however, dictates continuous production, continuous (old-style) growth. It also gives rise to a narrow-minded, lop-sided kind of production, mainly of commodities that yield quick profits. Moreover, the negative 'spin-off' from this production—environmental degradation and human suffering—constantly creates new needs which we try in vain to repair with new production, that is by technical means."

   Hoogendijk calls for a two-track strategy: Reduce, convert, or halt current production modes; and construct "a more intelligent form of economy, tuned more to basic than derived needs and building on the many alternative solutions already being practiced all over the world." His book describes roads to take—beyond recycling, biking, and "installing a filter on a factory chimney"—to make the "great U-turn" and contribute to a "more intelligent and convivial world."

   With a background in art, politics, media, and environmental education, the author is a co-founder of the Dutch Association for the Integrity of the Creation and participant in the Critical Farmers Platform, National Environmental Platform, and Alliance for Sustainable Development. Hoogendijk is also a member of the National Council of Refuse Collectors, Council for the Environment of Utrecht, and the European ecological group ECOROPA.

■ Resources

The Economic Revolution is available through its publishers—Green Print (London), and Jan Van Arkel (Utrecht). Hoogendijk can provide a paper on environmental education, without charge. Jan Van Arkel and its International Books division publish other books on the environment, energy, and development issues; a catalog can be obtained from publisher Lin Pugh, Alexander Numankade 17, 3572 Utrecht, Netherlands, phone (31) 30 731840.

Stichting Natuur en Milieu
Society for Nature and Environmental Conservation
Donkerstraat 17
3511 KB Utrecht, Netherlands

Contact: **Lucas Reÿnders,** Professor
Phone: (31) 30 331328 • *Fax:* (31) 30 331311

Activities: Education; Political/Legislative; Research • *Issues:* Air Quality/Emission Control; Energy; Global Warming; Sustainable Development; Transportation; Waste Management/Recycling; Water Quality

● Organization
Stichting Natuur en Milieu (Society for Nature and Environmental Conservation) cooperates with local action groups from the 12 provincial federations and has links with 7 national sister organizations. Its staff of 50, plus 10 to 15 volunteers, run the 5 separate divisions: administrative, environment, information and education, nature and landscape, and policy analyses. Government grants from the Ministries of Agriculture, Nature Conservation and Fishery, and of Housing, Physical Planning and Environment, along with private contributions, help finance the Society's activities. With its 20 affiliates—nature conservation and environmental organizations—some 700,000 individuals are involved. The Society belongs to the European Environmental Bureau (EEB), World Conservation Union (IUCN), Environmental Liaison Centre, and Antarctic and Southern Ocean Coalition and is active in the Pesticide Action Network and the No More Bhopal Network.

▲ Projects and People
In this highly industrialized country, the Society projects focus on achieving **clean air, water and unpolluted soil,** sustainable **land use, durable energy sources,** preservation of nature, and a good **public transportation** system. In order to achieve its goals, the Society **lobbies government** officials and politicians at the local, regional, and national levels, as well as the European Community (EC) in general on the interest of nature and the environment.

When necessary, the Society initiates appropriate studies of these issues, publishes books and research reports, serves as an informational resource, holds monthly press conferences, stimulates public awareness, plans communications programs, and takes legal action in administrative and judicial proceedings.

Lucas Reÿnders is a professor of environmental science, University of Amsterdam.

■ Resources
The Society publishes *Natuur en Milieu* (Nature and environment) monthly magazine, *Jaarverslag* (Annual report), and other literature listed in a publications brochure, printed in Dutch.

Stichting Reservaten Przewalski–paard
Przewalski Horse Reserves Foundation ⊕
Mathenesserstraat 101a
3027 PD Rotterdam, Netherlands

Contact: **J.G. Bouman, Ph.D.,** Board Chairman
Phone: (31) 10 437 0447 • *Fax:* (31) 10 415 3740

Activities: Development; Education; Research • *Issues:* Air Quality/Emission Control; Biodiversity/Species Preservation *Meetings:* IUCN

▲ Projects and People
Started in 1982, the **Przewalski Horse Reserves Foundation** aims to breed a population of healthy horses on semi-reserves and also to reintroduce them into the steppes of Mongolia. Designated as the "only truly wild horse still living," the Przewalski horse—named after its "rediscoverer" Colonel Przewalski—was last seen in the wild in the late 1960s in southwest Mongolia and China and is now assumed to be extinct in the wild. Smaller than a domestic variety, the Przewalski horse has "upright manes, a mealy nose and dark brown legs mostly with zebra stripes."

Altogether, there are more than 700 Przewalski horses, mainly in 100 zoos and animal parks worldwide and said to be descended from about 13 original ancestors caught around the turn of the century. "It was not easy to breed these rare wild horses in captivity," according to Foundation literature. "The narrow founder base, and the increased inbreeding had detrimental effects on reproduction and life span. The loss of the wild environment, adaption to man, and the artificial surrounds in captivity leading to creeping domestication and a loss of hardiness has become so acute that something had to happen."

In a cooperative effort with the World Wide Fund for Nature (WWF) and the World Conservation Union (IUCN), the Foundation maintains 72 Przewalski horses in 4 semi-reserves in the Netherlands and another in Germany, where the sister organization is called Deutsche Stiftung Urwildpferd. In order to set the horses free into the wild, semi-reserves are necessary to help the animals readjust to natural conditions and to a "minimum of human intervention." An earlier project on a semi-reserve in France failed, so additional measures are planned prior to reintroduction into the Parc National des Cévennes.

The project initiated with one stallion and four to six mares living on three of the Dutch semi-reserves, with a bachelor group living apart. The first colt born in 1984 in the Lelystad Nature Park joined a one-year-old male colt imported from Russia in 1985. The imported colt will become a stallion of a group of mares born in semi-reserves.

In 1992, the first Przewalski horses from the semi-reserves were scheduled to be sent to the Mongolian steppe reserve Hustain Nuruu.

The Foundation keeps an extensive record system documenting the ancestry of each horse and its breeding lines. The studbook is edited at the Prague Zoo.

■ Resources
A list of publications comprises English titles and features articles authored and co-authored by **Dr. J.G. Bouman.** The *Przewalski Horse* magazine and study packets are also produced.

The Foundation needs funds to continue its long-term project as well as equipment such as a landrover and high-frequency radio communications systems.

Nicholaas Jan van Strien, Ph.D. ⊕
Biologist, Consultant
Julianaweg 3
3941 DM Doorn, Netherlands

Activities: Political/Legislative; Research • *Issues:* Biodiversity/Species Preservation • *Meetings:* CITES

▲ Projects and People
As a lecturer of vertebrate zoology, morphology, and mammal biology, Dr. Nicholaas van Strien most recently was with Chancellor College of the University of Malawi for the Netherlands' Directorate-General for International Cooperation (DGIS). Earlier, he served as a consultant for the School of Environmental Conservation Management, Bogor, Indonesia and for the World Conservation Union (IUCN), helping develop the Sumatra rhino conservation master plan. He belongs to the Species Survival Commission Asian Rhino and Tapir groups.

Some of his research activities include: project leader and specialist for the World Wide Fund for Nature's (WWF) program in the Gunung Leuser National Park and assistant researcher on the ecology and distribution of the Sumatran Rhinoceros in Indonesia.

He also participated in numerous international conferences and workshops as a delegate, and drew 35 wild plants for sign boards for the State Forestry Service for use along nature trails in the Netherlands.

Currently in preparation by Dr. van Strien are a comprehensive taxonomic atlas of the mammals of the Australian archipelago, to be published and on diskette through the Fauna Malaysiana Foundation, Leiden and Amsterdam; a checklist of the vertebrates of Zomba Plateau, Malawi; and several short publications on Malawi bats and barn owl pellets. He is the principal writer and editor of the recent *Inventory of Flora and Fauna of Mulunguzi Marsh and Upper Mulunguzi River,* a Mulunguzi Marsh Dam Environmental Impact Assessment report.

■ Resources
Dr. van Strien has available a publications list, including major works, articles, notes, and reports on various subjects.

Werkgroep Noordzee ⊕
North Sea Work Group
Vossiusstraat 20-111
1071 AD Amsterdam, Netherlands
   Phone: (31) 20 676 1477 • *Fax:* (31) 20 675 3806

Activities: Education; Political/Legislative • *Issues:* Air Quality/Emission Control; Biodiversity/Species Preservation; Energy; Sustainable Development; Transportation; Water Quality

■ Resources
The following publications in English can be ordered from **Werkgroep Noordzee (North Sea Work Group)**: *North Sea Monitor* quarterly magazine; *The North Sea and Its Environment: Uses and Conflicts,* first North Sea Seminar proceedings, 1980; *Reasons for Concern,* second North Sea Seminar proceedings, Volumes I and II, 1986; and *Distress Signals,* third North Sea Seminar proceedings.

Beached Bird Surveys in the Netherlands 1915–1988, "seabird mortality in the southern North Sea since the early days of oil pollution," can be ordered only from Vogelbescherming (Protection of Birds), Driebergseweg 16c, NO-3708 JB Zeist, Netherlands.

♦ *Private*

Twijnstra Gudde Management Consultants
P.O. Box 907
3800 AX Amersfoort, Netherlands

Contact: **Dr. T. Van Der Tak**, Managing Partner
Phone: (31) 33 677777 • *Fax:* (31) 33 677666
E-Mail: telex 43689

Activities: Research • *Issues:* Air Quality/Emission Control; Energy; Environmental Management; Sustainable Development; Transportation; Waste Management/Recycling; Water Quality

● Organization
Twijnstra Gudde Management Consultants, with its 400 employees in offices in the Netherlands and Brussels, advises business, industry, commercial operations, government, the trades, and nonprofits on environmental issues. Among its areas of consultancy are education and training, improving relations, research and analysis, trouble-shooting, and organizational programming.

▲ Projects and People
According to the consultants, environmental concerns offer excellent business opportunities for companies prepared to take them on. Advising on the specific positive aspects of preparing an environmental business plan, Twijnstra Gudde emphasizes four points: "anticipating increasingly stringent government regulations, avoiding the expense of dealing with (potentially grave) environmental problems in the future, maximizing opportunities by anticipating new ecological trends, and maintaining or improving the company's image as an environmentally responsible firm."

An environmentally aware business can also be the first to develop environment-friendly products and gain an edge over the competition as well as receive government assistance, such as grants and tax benefits, say the consultants. Such aspects will increase the company's consumer image as a responsible member of society.

Twijnstra Gudde suggests developing an environmental protection policy with priorities involving all company members, from the directors down. Twijnstra Gudde's criteria for such a sound plan is:

"The environment should be given a proper place in the company's overall policy. This also means that sufficient funds should be made available.

"Environment protection should not become an end in itself. The costs of a given measure should always be weighed against the contribution it makes towards reducing actual or potential environmental problems.

"The implementation of environment protection measures calls for a systematic approach, namely the establishment of priorities, phased planning and organization, all of which should be weighed against and integrated with other areas of management.

"An integrated approach to all aspects of the environment is needed if the policy is to succeed. Solving one set of environmental problems should not generate another."

■ Resources

A brochure on *Company Policy and Environment Protection* is available in English.

 *Universities*

Institute for Environmental Studies (IES)
Free University of Amsterdam
Provisorium 3
De Boelelaan 1087A
1081 HV Amsterdam, Netherlands

Contact: **J.B. Opschoor, Ph.D.**, Director, Professor
Phone: (31) 20 548 3827 • *Fax:* (31) 20 445056
E-Mail: Earn/Bitnet: aiking@sara.nl

Activities: Education; Research • *Issues:* Air Quality/Emission Control; Conservation Management; Energy; Health; Water Quality

● Organization

Founded in 1971, the **Institute for Environmental Studies (IES)**, Free University of Amsterdam, is the oldest academic institute for basic and applied environmental research in that country. With 30 to 40 researchers from backgrounds including "chemistry, ecology, hydrology, geography, geochemistry, economics, ecometrics, psychology and sociology," the nonprofit IES mainly works on externally funded contract research projects. It produces some 70 to 80 annual publications as well.

IES collaborates with the World Conservation Union (IUCN), U.S. Environmental Protection Agency (EPA), and other organizations globally. Its funding agencies include the EC, Norwegian Organization for Research and Development (NORAD), Organization for Economic Cooperation and Development (OECD), United Nations Environment Programme (UNEP), and UNESCO. Nationally, support comes from Ministries, Dutch National Science Foundation (NWO) and NGOs.

▲ Projects and People

IES has eight research areas:
- Environmental analytical chemistry, including the study of heavy metals; polycyclic aromatic hydrocarbons (PAH) in sediments, soil, and biota; pesticides in water; and polychlorinated biphenyls (PCBs) in paper, waste oils, and biota.
- Environmental chemistry, toxicology, and health, such as "biomonitoring air pollution by means of mosses and moles;

prediction of toxicity of meterocyclic PAH and heavy metals to fish and aquatic invertebrates; risk evaluation of polluted sediments."
- **Pollution Assessment**, particularly of past, present, and future emissions; environmental policy; "damage to cultural property by air pollution"; nonpoint sources of water pollution; and soil pollution.
- **Conservation and Development of Natural Ecosystems**, with management strategies design; assessment of impacts of forestry, groundwater management, agriculture, and outdoor recreation; national park feasibility analysis.
- **Integrated Environmental Modeling**, such as a "regional outdoor recreation impact mode; a national air pollution control model generating economic cost and ecological effectiveness of abatement strategies; risk assessment for range management strategies and theoretical and practical studies to make the concept of **Ecologically Sustainable Economic Development** operational.
- **Social and Behaviorial Aspects of the Environment**, to improve communications between the public and government and bring about changes in attitudes to "enhance the efficiency of environmental conservation programs"; and research public reactions to environmental hazards, including noise, soil, and air pollution.
- **Policy Studies and Program Evaluation** of environmental effects of alternative energy strategies and cost-benefit analysis, for example.
- **Environment and Third World Issues**, involving the "environmental impact assessment [EIA] of north-south trade flows" and general interactions between environmental and development issues, particularly in semiarid areas.

Centre for Environmental Studies (CML) ⊕
University of Leiden
P.O. Box 9518
2300 RA Leiden, Netherlands

Contact: **Wouter T. de Groot**, Deputy Scientific Director
Phone: (31) 71 277486 • *Fax:* (31) 71 277496

Activities: Education; Research • *Issues:* Biodiversity/Species Preservation; Deforestation; Health; Sustainable Development

● Organization

Education and research into the causes and solutions of environmental problems are the primary objectives of the **Centre for Environmental Studies (CML)**, an interdisciplinary department of **Leiden University**. Its activities are oriented toward environment and policy, with emphasis on the Netherlands and Europe, and program development in developing nations. A large portion of CML's research is contractual from industry, government, and social groups.

▲ Projects and People

Wouter T. de Groot chairs the management team of the Programme Environment and Development unit, which collaborates with universities in **Cameroon** and the **Philippines** at field stations that

CML help set up. Dutch students conduct research, for example, in soil and water management in the agrarian North Cameroon and in forest exploitation in the Cagayan Valley, Philippines.

Four themes are examined within the Programme Environment and Development unit: analysis and explanation of environmental problems, assessment of environmental or carrying capacity in an area, environmental planning and projects, and peoples' backgrounds or traditions with respect to the environment.

CML's general environmental science research is conducted within the framework of two Ph.D. research projects. One focuses on the basic concepts of environmental science as a problem-oriented discipline, the other is geared toward a historical analysis in decisionmaking.

Ecosystems and environmental quality and substances and products research are the two sub-units within the environment and policy division. Each has three focal points of concern. In the first sub-unit are the ecological basis of the policy instruments for environmentally dangerous substances, empirical research concerned with environmental quality, and hierarchical ecosystem classification. The second sub-unit consists of environmental assessment of products, market-conformity policy instruments, and substance-flow analysis.

Senior staff members are deputy scientific director and civil engineer de Groot, agricultural engineer and environmental scientist Carel Drijver, ecologist Hendrik Huibert De Iongh, and anthropologist and environmental scientist Gerard Persoon.

■ Resources

Leiden University has an available publications list (printed in English, French, and Dutch), and copies of reports and reprints of articles. Topics include wetlands, water resources, and floodplain management; women's role in the environment; role of indigenous people; land reclamation; health; and sustainable development of mangroves.

Optimization of Environmental Data Collection Programs
Environmental Sciences
University of Utrecht
Popelenburg 15
2641 MX Pijnacker, Netherlands

Contact: Louis A. Clarenburg, Ph.D., Professor
Phone: (31) 1736 93144 • *Fax:* (31) 1736 93144

Activities: Development; Education; Political/Legislative *Issues:* Air Quality/Emission Control; Biodiversity/Species Preservation; Energy; Environmental Management; Global Warming; Sustainable Development; Water Quality • *Meetings:* IUAPPA; World Conference on Clean Air

▲ Projects and People

As the Netherlands administered its environmental policies at the provincial and national levels, a need arose to establish an optimal data collection package based on fixed budgets. As a result of the research of Dr. L.A. Clarenburg and colleagues, a new process can enable individual programs to be judged and reviewed to select those with the highest utility. The method requires more than 12 scientists to determine the priority of any programs using the standardized system. In the studies on **Optimization of Environmental Data Collection Programs**, associates are Drs. M.A.T. Maenhout, Environmental Studies Department, Agricultural College, Delft, and P. Eilers, Environmental Service Rijnmond, Schiedam.

Dr. Clarenburg explains that "the usefulness of a program is found by multiplying its utility at a given performance level and the overall priority. . . . Careful time phasing can avoid frictions in the transition period from the present set of monitoring programs to an optimized set."

Details of the study report, using this system for the Province of South-Holland, were presented to the International Union of Air Pollution Prevention Associations (IUAPPA) 1991 regional conference on air pollution, Seoul, Korea, and also at the Ninth Clean Air World Congress, Montreal, Canada, in 1992.

Dr. Clarenburg is chairman of the Clean Air Society, Netherlands, and a former IUAPPA president, now serving on its executive committee. Previously honored by the National Society of Statistics and Biometrics for "statistical innovation," he is also a member of the National Council of Environmental and Nature Research, the National Council for Physical Planning, and the National Health Council.

■ Resources

Most materials are printed in Dutch, German, and French. A reprint is available in English of *Optimization of Environmental Data Collection Programs.*

Department of Entomology ⊕
Wageningen Agricultural University
P.O. Box 8031
6700 EH Wageningen, Netherlands

Contact: Arnold van Huis, Ph.D., Tropical Entomologist
Phone: (31) 8370 84653 • *Fax:* (31) 8370 94921

Activities: Research • *Issues:* Pest Management

▲ Projects and People

A tropical entomologist at Wageningen Agricultural University, Dr. Arnold van Huis is responsible for all contacts with developing nations and participates in international meetings, seminars, and workshops. He also chairs a working group to organize a graduate-level English course in crop protection for students from developing countries.

His research, in cooperation with the University of Tours, France, and the University of Niamey, Niger, focuses on the biological control of two stored-product cowpea pests, *Callosobruchus maculatus* and *Bruchidius atrolineatus*, by means of the egg parasitoid *Uscana lariophaga* in West Africa. In his host selection studies, Dr. van Huis examines the effect of abiotic factors, olfaction studies, functional response, biology, and ecology. He formulates, evaluates, and inspects projects for international groups such as the Food and Agriculture Organization (FAO) and the United Nations Development Programme (UNDP).

Dr. van Huis also chairs the biological control of bruchids group, International Organization for Biological Control of Noxious Animals and Plants (IOBC), as well as the **Commission for Tropical Crop Protection and Vector Control**, Crop Protection Departments of the Wageningen Agricultural University, Research Institute for Crop Protection, and Plant Protection Service of the Netherlands. IOBC is affiliated with the International Council of Scientific Unions (ICSU).

Biological controls are being sought for the **pea weevil** (*Bruchus pisorum*), which damages pea seeds, wheat, and barley and causes yield losses in Australia and Chile, for instance.

Earlier, van Huis researched the development of **integrated pest management (IPM) technology in cotton, maize, sorghum, and beans** in Nicaragua, and the behavior of the **harvesting termite** (*Hodotermes mossambicus*) in Kenya. More than two dozen publications contain van Huis's various studies and results.

■ Resources

The IOBC Bruchids Working Group produces a newsletter that lists research projects worldwide. IOBC also publishes a newsletter for its members.

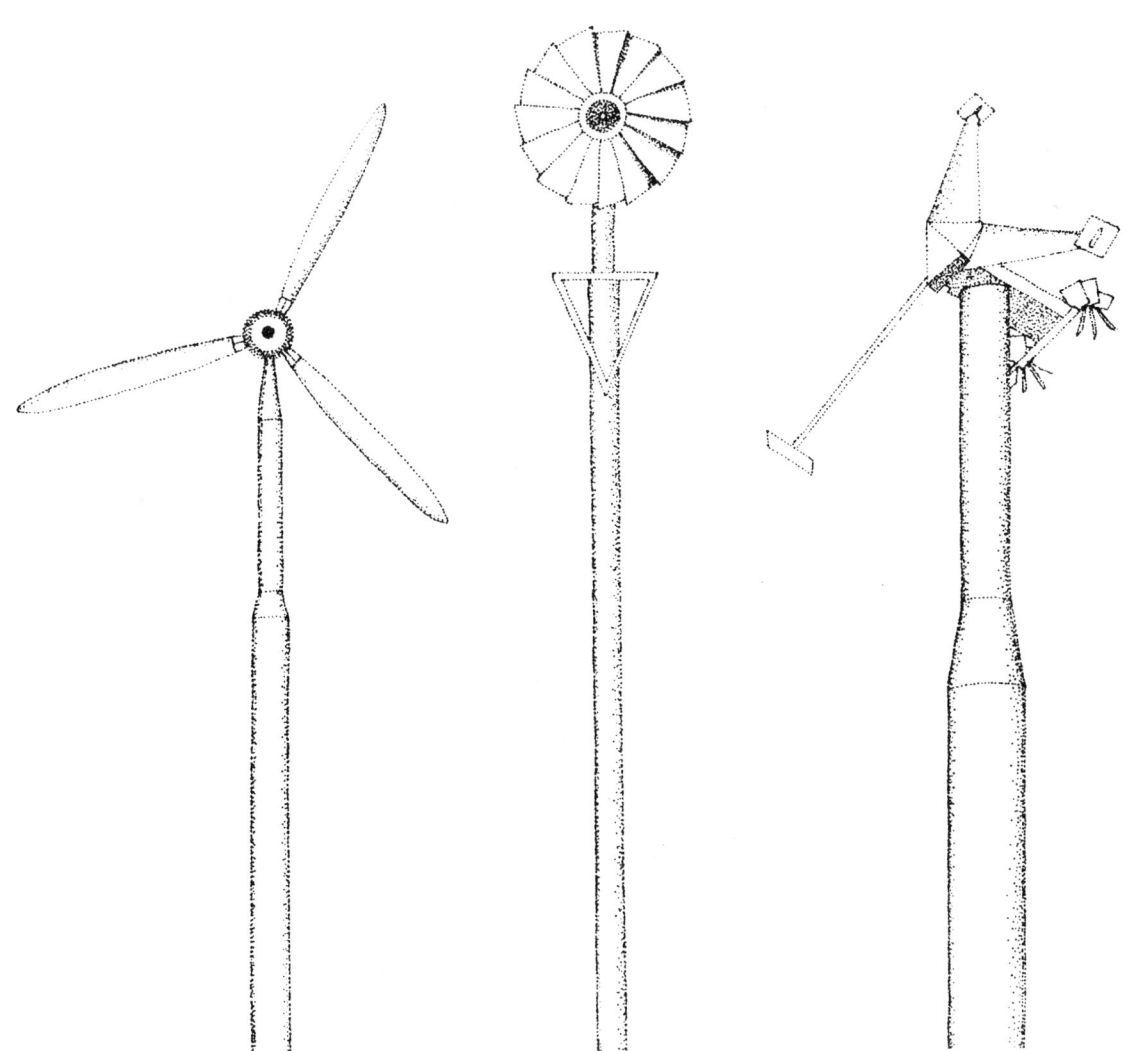

These two widely separated groups of Caribbean islands, covering a total of only 383 square miles, have been an autonomous part of the Netherlands since 1954. Rainfall and vegetation are heavier in the northern islands. The proximity of the southern islands to Venezuelan oil fields, and their location on well-established shipping routes, have made refining storage, and transshipment of oil the mainstay of the economy—particularly on the island of Curaçao. It is also the primary determinant of the country's GNP. Tourism is the other major industry, with some offshore fishing. Little farming is possible because of the rocky terrain. Thus, most of the food must be imported. A stable government has brought steadily improving living conditions and increased attention to conservation concerns, especially marine resources.

 *NGO*

Saba Marine Park (SMP)

Fort Bay, P.O. Box 18
The Bottom
Saba, Netherlands Antilles

Contact: **Susan Walker**, Park Manager
Phone: (599) 46 3295 • *Fax:* (599) 46 3435

Activities: Development; Education; Facility Maintenance; Law Enforcement; Political/Legislative; Research • *Issues:* Biodiversity/Species Preservation; Sustainable Development • *Meetings:* World Parks Congress

● Organization

Established in 1987, the **Saba Marine Park (SMP)** aims to preserve and manage marine resources surrounding the island—an extinct volcano—to a depth of 200 feet as well as two offshore seamounts, with support from the World Wildlife Fund, Prince Bernhard Fund, and the Dutch and Saban governments. The Saba Conservation Foundation administers the park, now self-sustaining for operational expenses through "dollar-a dive" and other user fees, souvenir sales, and donations.

Reminding that it was not set up "in an attempt to help repair our damaged environment," SMP safeguards its extraordinary resources through prohibiting spearfishing and conch and turtle collecting except by Saban residents whose free-diving catches are also regulated; banning the use of poisons, chemicals, or explosives; prohibiting the taking of coral or other bottom-dwelling marine animals or plants; and restricting boating, fishing, and littering.

▲ Projects and People

SMP manager **Susan Walker** outlines current environmental projects as education, research and monitoring, cleanup, and coast resources management.

Biweekly slide shows are presented to visitors and locals, with annual classes provided to school children. "Emphasis is on passive interaction with the marine environment," writes Walker. An audiovisual slide presentation is being developed.

Monitoring, which provides SMP with baseline data, includes fish censusing, diver impact monitoring, and benthic invertebrate mapping. "Major obstacles include financing the programs and getting the technical expertise to continue ongoing monitoring," reports Walker. Time restraints of personnel is another problem. Researchers include **Joan Boomsma**, **Julie Hawkins**, and Drs. **George Dennis III**, **Joe Kimmel**, **Callum Roberts**, **Nick Polunin**, and **Carole McIvor**.

SMP is island coordinator for the **International Coastal Clean-Up**, sponsored by the Center for Marine Conservation.

With the direction of **Dr. Tom Van't Hof**, a project is underway to develop a multiple-use zoned protected area for SMP for "sustainable use of the resources so that they are utilized in perpetuity."

▪ Resources

SMP seeks coral reef and coastal zone management educational materials. It publishes a *Saba Marine Park* brochure, *SMP: Anchorages and Mooring Buoys*, and *Guide to the SMP*. Contributors or "Friends" receive SMP logo pins.

New Zealand

No place in New Zealand, with its snow-capped mountains, green lowlands, beaches, lakes, and waterfalls, is more than 80 miles from the coast or out of sight of mountains or hills. Over three-quarters of the land is above sea level. With one of the world's highest standards of living, the economy depends mainly on farming and foreign trade.

In 1991, New Zealand finalized major revisions of natural resource management laws, with the aim of sustainable use. Concerns over ozone depletion and climate change have also promoted laws and policies for reducing damaging atmospheric pollutants. Intensive conservation work has succeeded in saving some endemic species, though others have been harmed by introduced animals. In addition, only one-third of the indigenous forest remains, due to extensive clearing for pastureland.

🏛 Government

Department of Conservation (DOC)/Te Papa Atawhai
59 Boulcott Street
P.O. Box 104-20
Wellington, New Zealand

Contact: **Bill Mansfield**, Director General
   Phone: (64) 4 471 0726 • *Fax:* (64) 4 471 1082

Activities: Conservation Advocacy; Conservation Management • *Issues:* Biodiversity/ Species Preservation; Conservation Advocacy; Conservation Land Management • *Meetings:* CITES; IWC; World Parks Congress

● Organization

The mission of the **Department of Conservation (DOC)** is "to conserve the natural and historic heritage of New Zealand for the benefit of present and future generations."

Established in 1987, the DOC carries out its "hands-on" activities through a network of field offices and 14 conservancy offices located throughout New Zealand. The Department works closely with government agencies including the New Zealand Conservation Authority, 17 regional conservation boards, the Historic Places Trust, and the Queen Elizabeth II National Trust. Each group represents a broad range of community interests, and each organization plays a role in the task of conservation management.

"A special partnership is developing between DOC and the tangata whenua," its literature reports. A goal is the protection of Maori cultural and spiritual values. In the name **Te Papa Atawhai**, *Te Papa* means a "treasure chest," and *Atawhai* means "to care for." With the New Zealand Historic Places Trust, DOC helps protect the country's historic heritage, including Maori rock drawings, spiritual (*wahi tapu*) and archaeological sites, and other historic buildings.

▲ Projects and People

DOC's endangered species program aims to preserve some of the world's oldest flora and fauna that are found in New Zealand. These include the "southern beech forests, kauri trees, tuatara, native frogs, kiwi, kakapo, giant weta, and land snails"—all at risk, as are some 75 bird species.

Special breeding programs are underway as well as efforts to protect threatened ecosystems of interdependent plants, insects, and animals. Certain endangered species are being transferred to offshore islands without predators or "controlling cats, rats, possums, and other introduced animals."

The coastal projects aim is to protect the fragile marine ecosystem from growing pollution and overuse problems. Activities include examining ways to reduce such pollution, caring for marine mammals in New Zealand's waters, inventorying the coastal waters and identifying areas needing protection, and working with the iwi community and other interest groups to set up a marine reserves network.

Through DOC's Kaupapa Atawhai Unit, a cooperative partnership with iwi and tribal authorities is being developed so that indigenous people will have a vital role in decisionmaking regarding conservation. Maori require access to "traditional cultural materials, such as timber, flax, pingao (a sand-dune plant), and bird feathers."

With visitors using outdoor parks for recreation, DOC provides facilities and services, such as guided walks, white-water rafting, and sea kayaking. To safeguard the environment, DOC oversees commercial activities in protected areas, such as mining, grazing, whitebaiting, and sphagnum moss harvesting. Also in protected areas, DOC controls pests and weeds; deals with threats of fire, erosion, and pollution; organizes community beach cleanups; and works with local authorities to manage local reserves.

Through the Natural Areas Program, an ecological survey project is being carried out to find and protect examples, such as tussock grasslands and peat swamps. While promoting the conservation of wild and scenic rivers, DOC is working on ways to prevent water pollution and manage the land to minimize erosion and flooding. It also runs environmental education and promotion campaigns for the community and school groups.

A research division covers social and biological science on land and sea, often with contracts from scientists in universities and other agencies. The division also manages scientific databases that list wetlands, endangered species, wildlife habitats, and archaeological sites.

On behalf of the global community, DOC works with international groups to oppose driftnet fishing, protect migratory birds, participate in CITES (Convention on International Trade in Endangered Species of Wild Fauna and Flora), and work with neighboring countries to promote conservation and sustainable development in the South Pacific.

Ministry of Agriculture and Fisheries (MAF)
Invermay Agricultural Centre
Private Bag
Mosgiel, New Zealand

Contact: **James M. Suttie, Ph.D.**, Program Leader,
Growth Physiology/Behavior

Activities: Education; Research • *Issues:* Biodiversity/Species Preservation

▲ Projects and People
Dr. James M. Suttie researches the "reproductive and nutrition biology of ruminants, particularly the seasonal, developmental and endocrine aspects." His particular interest is in the "seasonal control of growth and appetite in animals and the relationship between these rhythms and the annual cycle of reproduction and the neuroendocrine control of these systems." Another interest is the "physiological control of body growth and antler development in deer and body growth in sheep." Suttie investigates "developmental endocrinology, particularly the role of and relevance of changes in the pulsatile secretion of growth hormone in determining patterns of body growth and nutrient partitioning." A third pursuit is the "immunization of animals against their endogenous hormones for livestock production or investigative endocrine research purposes."

Dr. Suttie also studies captive animal behavior with special reference to stress and welfare.

On various study tours, Dr. Suttie analyzed the growth endocrinology of reindeer at the Institute of Arctic Biology, University of Alaska, USA; observed puberty in sheep at the University of Michigan, USA; investigated the relationship of metabolism to rhythms of growth and food intake in Australia; and blood sampled free-living Weddell seals in Antarctica.

■ Resources
Dr. Suttie has an extensive publications list.

Botany Institute
New Zealand Department of Scientific and Industrial Research (DSIR)
Private Bag
Christchurch, New Zealand

Contacts: **William R. Sykes**, Taxonomic Botanist
David R. Given, Ph.D., Conservation
Biologist
Phone: (64) 3 325 2511 • *Fax:* (64) 3 325 2074

Activities: Development; Research • *Issues:* Biodiversity/Species Preservation; Deforestation; Global Warming • *Meetings:* CITES; IUCN/SSC; Pacific Science Congress

● Organization
The **Botany Institute** was part of the former DSIR Land Resources, which has been the National Institute of Land Environments since mid-1992.

▲ Projects and People
The primary research interest of biologist **William Sykes** is the taxonomy of cultivated and naturalized plants in the tropical and subtropical South Pacific Islands, including New Zealand. He especially advises on the selections of species for various sites and studies the relationships of plant species' distribution with climate—both in New Zealand and where the species originate.

Monitoring the spread of noxious weeds in New Zealand, according to the country's weed law, Sykes evaluates the suitability of exotic plant species to prevent the risk of importing weeds and insect pest/disease hosts. He is also researching native and naturalized weeds of tropical origin in New Zealand and their spread into the warm temperate areas of northern New Zealand.

Sykes studies, identifies, and registers poisonous plants, preparing educational materials especially about plants toxic to children.

Among other projects, he compiled a taxonomy of seed plants for *Flora of New Zealand, Volume IV* in an eight-year undertaking; researched conifers in Guangxi Province of China and stored 850 specimens in the DSIR herbarium, including descriptions on computerized database; investigated rare and endangered plants of Norfolk Island; is preparing a book on the vegetation of Tonga; and is part of a multidisciplinary team studying the ecology of the Kermadec Islands.

A writer of floras, consultant on ornamental plants for horticulture, and Lincoln University instructor of plant taxonomy courses,

Sykes has 40 years experience as a botanist—having published more than 30 papers and 50 reports.

Dr. David Given has published widely about naturalized flowering plants, ferns, rare and threatened plant taxa, plant conservation, conservation ethics, and Antarctic plants. With the Species Survival Commission (IUCN/SSC), he is a regional member and chairman of the Pteridophyte Group.

Indonesia–New Zealand Land Resources Mapping Project (INZLRMP) ⊕
New Zealand Department of Scientific and Industrial Research (DSIR)
Private Bag
Lower Hutt, New Zealand

Contact: **Nick C. Lambrechtsen, Ph.D.**, Project Manager
Phone: (64) 4 795469 • *Fax:* (64) 4 673114

Activities: Development; Education; Research • *Issues:* Deforestation; Global Warming; Sustainable Development

▲ Projects and People
Jointly funded and managed by the Indonesian and New Zealand governments, the **Indonesia–New Zealand Land Resources Mapping Project (INZLRMP)** was executed by **DSIR Land Resources** until mid-1992, and now by the **National Institute of Land Environments**. The goal is to "demonstrate and evaluate an integrated multi-factor land resources mapping system (developed and applied in New Zealand) in the Wiroko and Keduang sub-watersheds of the Wonogiri watershed, Central Java. Information will be used for watershed management and the planning of sustainable land use."

Multifactor land resources mapping and computerized data storage and retrieval via geographic information systems (GISs) will enable the Ministry of Forestry to carry out its assigned tasks under the "Indonesian Land Development Plan Pelita V." Presently, Indonesia lacks a "uniform mapping system or method of synthesizing data to provide a good basis for soil conservation planning" that is essential for "regreening" in upper areas of some 36 priority watersheds, reports INZLRMP.

Among other goals, the project will identify the "influence of rock type, soil, slope, vegetation, climate, erosion" and other important physical factors as well as develop a "uniform standard of classification of these physical factors" and provide counterpart training for Watershed Management Technology Centre staff in Indonesia.

All the inventory and interpretative data will be stored in a computer; information from other agencies and point data, such as hydrological and climatic data, can be added to the database to further enhance the usefulness of the GIS for soil conservation planning. Maps will be produced either as single-factor or multifactor overlays, or as complete maps incorporating all necessary base information.

Dr. Nick C. Lambrechtsen, INZLRMP project manager with the overall responsibility for his government agency's "commercial and future opportunities" in Indonesia, has a background in DSIR soil conservation research program planning and review of botany and ecology divisions.

He writes that DSIR contributes to global environmental cooperation by "information exchange; promotion of scientific relationships between New Zealand and overseas agencies, collaboration in multi-national research programmes; inputting into international conferences; and technology transfer."

■ Resources
INZLRMP publications are: Report 1: *Land Resource Survey Handbook for Soil Conservation Planning in Indonesia*; Report 2: *Pedoman Survai Sumberdaya Lahan untuk Perencanaan Konservasi Tanah di Indonesia* (translation of Report 1); Report 3: *Land Resource Survey of the Wiroko Subwatershed, Upper Solo Watershed, Central Java, Indonesia*; Project 4: *DSIR ARC/Manger User's Guide for Indonesian Ministry of Forestry*.

National Institute of Land Environments
New Zealand Department of Scientific and Industrial Research (DSIR) ⊕
Private Bag
Christchurch, New Zealand

Contact: **Colin D. Meurk, Ph.D.**, Plant Ecologist
Phone: (64) 3 325 2511 • *Fax:* (64) 3 325 2074
E-Mail: meurkc@chpc.dsir.govt.nz

Activities: Education; Research • *Issues:* Biodiversity/Species Preservation; Habitat/Land Restoration; Subantarctic Ecology; Sustainable Development

● Organization
The newly named **National Institute of Land Environments**, similar to its recent predecessor, **DSIR Land Resources**, conducts research, develops technology, and provides related services which "deliver a scientific basis for the management of New Zealand's land environments through sustainable land use strategies and analyses and prediction of the impacts of major resource utilization." *(See also other profiles on National Institute of Land Environments in this section.)*

The overall mission of the Department of Scientific and Industrial Research (DSIR), the country's "largest scientist and technology organization," is to "make science work for New Zealand." DSIR, with its breadth of scientific talents, reaches out to industry, resulting in innovative partnerships, training programs, and research and development applicable to medicine and engineering, agriculture and biotechnology.

For example, DSIR has developed 40 drought-tolerant and pest-resistant cultivars for pasture that are marketed worldwide. The clover varieties save in nitrogen fertilizer costs and avoid groundwater pollution, reports DSIR. Its "Grasslands Huia dominates the world's white clover seed market." Effects of climate change on subtropical grasses is under study. The Department experiments with pollination techniques and new varieties of asparagus, onion, apple, and grape, which expand export markets—such as for "pinos, babaco, feijoas, tamarillos, lentils, and durum wheat."

DSIR experiments with organic, or chemical-free, systems of livestock farming, and a hormone microcapsule, melatonin, fed to cashmere-producing goats to encourage "an extra fleece growth in spring, ready for summer shearing." Intending to reinforce its "green" image worldwide, DSIR is using a "holistic approach" to reduce or eliminate chemicals in pest and disease management. Biological control methods focus on the parasitic wasp *Aphidus rhopalosiphi,*which attacks the rose-grain aphid that feeds on barley and oat crops. Sex pheromone technology, "the attraction of one sex of a species by the sex attractant produced by the other," holds hope for "monitoring pest populations thus reducing the need for pesticide spraying [and controlling] insect populations by disrupting mating to prevent egg laying." Forensic scientists determine the identification of the origin of illegal plants (by identifying insects in the foliage) as well as contribute to other crime solving.

Among DSIR's advanced computer capabilities, its image processing software, EPIC, is used here as well as is in the USA, Europe, and China. Benefits from such technology vary from chromosome analyses in medical laboratories to cutting industrial energy costs by 50 percent through monitoring for cheaper electricity rates. Engineering processes result in techniques to protect buildings from earthquake damage as well as to create an artificial hip joint. A National Pilot Plant Centre is unique in Australasia, leading to new products and industrial processes. New materials and equipment are being developed, such as superconductors that "will lower the present 30 percent of electricity lost in transmission to users."

In **Antarctic ozone studies,** DSIR monitors trace level gases and measures flavonoid levels in mosses. "Flavonoids are chemicals produced by many plants to protect them from UV [ultraviolet] radiation." Algae and plankton are also analyzed, as are the mercury levels in snow. In New Zealand, where UV radiation levels are "among the highest in the world," DSIR researches the impacts of ozone depletion and the greenhouse effect as well as monitors carbon dioxide (CO_2) levels in the ocean and atmosphere and other gases, such as oxides of nitrogen. It is also developing a "high temperature plasma furnace for destroying used CFCs [chlorofluorocarbons] safely."

The environmental impact of hazards, such as earthquakes, volcanos, and landslides, are evaluated; air and water pollution are measured; and rehabilitation efforts are recommended. DSIR also claims to be a global leader in multidisciplinary research to evaluate geothermal resources, such as gas, oil, coal, gold, and silver. With the "world's fourth largest Exclusive Economic Zone" and the DSIR research ship *Rapuhia,* New Zealand highly values the seas and its rivers and lakes. Groundwater surveys, studies of catchments, and research into algal problems are ongoing.

▲ Projects and People

Dr. Colin D. Meurk is a plant ecologist "researching grassland and subantarctic vegetation, environmental relationships and vegetation change." He undertakes research and consultancies on **impacts of burning, grazing, and land uses** in tussock grassland, shrublands, forest, wetlands and urban environments. As such, he has designed and maintained long-term monitoring programs, developed and implemented habitat restoration programs for wet and dry environments, and set up field trials to refine the principles of habitat restoration.

An authority on alpine plant ecology, **Dr. Meurk** observes vegetation and land use in the USA, Europe, Japan, and Australia, particularly regarding "rehabilitation, land use impacts, and the management of natural areas." In the Ross Dependency section of Antarctica, he conducted a "vegetation-environment analysis" and examined habitat recovery following vegetation disturbance. Locally, he is observing the effects of "sheep and rabbit grazing on short tussock grasslands in the Mackenzie Basin." In the Canterbury drylands, he

is assessing the performance of indigenous trees, shrubs, and tussocks for water, soil, and habitat conservation purposes. Recently, he reviewed the evergreen broad-leaved forests of New Zealand and the "effects of burning on soil properties and vegetation."

■ Resources

In cooperation with the World Wide Fund for Nature (WWF) and the Plant Advisory Group of the World Conservation Union (IUCN), the former DSIR Land Resources is producing a comprehensive guidebook on *Plant Conservation* for an international audience interested in managing both *in situ* (seed bank) and *ex situ* (wild) resources. Books presently available include *Flowering Plants of New Zealand; Poisonous Plants; Flora of New Zealand Series;* and *Threatened Plants of New Zealand,* which replaces the Red Data Book of New Zealand and is authored by DSIR botanists Dr. David R. Given, also a Species Survival Commission (SSC) member, and Catherine M. Wilson. A publications list also includes *Economic Native Plants of New Zealand; Nga Mahi Maori O Te Wao Nui A Tane,* proceedings of an International Workshop on Ethnobotany; *Wetland Plants in New Zealand;* and *Forgotten Fauna.*

DSIR facilities include a large Library Centre for print materials and more than 500 overseas databases, a publishing house, a DSIR ASEAN Centre in Singapore, and headquarters at P.O. Box 1578, Wellington, NZ, phone (64) 4 729979, fax (64) 4 724025.

National Institute of Land Environments (formerly Land Resources Division) ⊕
New Zealand Department of Scientific and Industrial Research (DSIR)
Private Bag
Palmerston North, New Zealand

Contacts: **Noel A. Trustrum,** Program Leader,
            Catchment Processes and Natural Hazards
  Phone: (64) 6 356 7154 • *Fax:* (64) 6 355 9230

         Garth O. Eyles, Scientist

Activities: Education; Research • *Issues:* Biodiversity/Species Preservation; Deforestation; Global Warming; Soil Conservation; Sustainable Development; Waste Management/Recycling

● Organization

DSIR Land Resources (DLR), formed in 1990 from three prior divisions, is New Zealand's "foremost provider of expert scientific information and advice on the land and its biota." It is one of ten divisions of the New Zealand Department of Scientific and Industrial Research (DSIR). With a staff of nearly 2,700, DLR has some 250 scientists and technicians engaged in "resource assessment and evaluation, plant and animal ecology, botany, soil science, data processing, information systems, land use planning, and soil engineering," with experience in some 40 countries in the Pacific, Southeast Asia, Africa, and South America. With other divisions, DLR forms interdisciplinary teams to work on environmental issues and run training programs.

In mid-1992, DSIR Land Resources fully integrated with the Forest Wildland Ecosystems Group at the Forest Research Institute (FRI), Ministry of Forestry; Rabbit and Land Management Programme of another Ministry; Remote Sensing Unit of DSIR Physical Sciences; Plant Materials Unit of DSIR Fruit and Trees; and taxonomists from DSIR Plant Protection. The **National Institute of Land Environments** is the name of the new organization. It has a staff of 410.

Even though it is a government agency, DLR's services and expertise can be accessed by the private sector in New Zealand as well as by international clients. It consults, for example, on using "effective, low-cost options in soil/vegetation surveying and mapping" together with advice on applying cost-effective computer hardware and software, and training local people who can continue such "development" in their country. One commercial undertaking is **Sirtrack Electronics**, which designs and manufactures innovative wildlife radio-tracking equipment—including transmitters, receivers, and antennas—for use in the field.

With modeling, multispectral photography, satellite imagery, and a range of geographic information systems (GISs) and remote sensing tools, DLR pursues soil conservation to minimize soil erosion, investigate soil compaction, prevent future losses in pasture productivity, and verify landslide damage and other natural hazards. Soil classification is a particular strength worldwide.

With the South Pacific Regional Environment Programme (SPREP), staff is working on a bird conservation strategy. It also inventories forests and wildlife habitats, evaluates land resources, maintains terrestrial resource databases, manages pests such as birds and mammals, restores land degraded by mining activities, assesses land for effluent irrigation, plans an ethnobotanical garden—among many other capabilities.

▲ Projects and People

Team leader of the **Catchment Processes and Natural Hazards** research programs Noel A. Trustrum uses "geomorphic and remote sensing techniques to date land forms and measure changes in erosion and soil resources associated with land use changes," such as deforestation.

Trustrum worked on **sustainability of pastoral agriculture, Taranaki hill country, New Zealand**. This project analyzed the deforestation of steep, soft-rock hill slopes in western North Island, New Zealand, which has accelerated the landsliding. "Studies by DSIR Land Resources has demonstrated that soil depths are now considerably less than they were prior to deforestation," he notes. "Landsliding has removed soil and soil formation. . . . This means that productivity on a watershed basis has decreased markedly since deforestation and continues to decline."

An ongoing project is the **sediment budgets for cyclone-induced erosion, Lake Tutira, New Zealand**. Starting in 1988, it includes "detailed field investigation, aerial photographic interpretation, and coring of floodplain and lake floors" to estimate accurately the "amount of material moved by landsliding in a large cyclone in 1988, the amount of material remaining on slopes, the amount put into storage on floodplains and in terraces, and the amount deposited in the lake at the base of the 30 square kilometer watershed." Trustrum, with colleagues Dr. Paul Blaschke and Ron De Rose, is able to examine erosion from the past 6,000 years, "particularly as it relates to land use changes, with the view to predicting the sustainability of present and potential land uses in the watershed."

Regarding erosion control, Trustrum and colleagues have developed "expert systems" that are being applied in more than 100 small watersheds in "highly erodible terrain" to appraise the effectiveness of tree planting and check dams in controlling earthflow, soil slip, and gully erosion."

As a consultant, Trustrum has conducted mining rehabilitation and watershed management workshops in Indonesia for the Food and Agriculture Organization (FAO). He was team leader for project mapping of flood and landslide hazards on Guadalcanal, Solomon Islands, and is a resource person for the East-West Center on sustainable land use, natural hazards, and watershed research.

Garth Eyles, active in the New Zealand Association of Soil and Water Conservation, developed and maintained the country's "first national multi-factor GIS for land resource planning, including erosion control," known as the **New Zealand Land Resource Inventory**. His team produced the **New Zealand Erosion Map** and a **Vegetative Cover Map of New Zealand**. Also an international consultant on land mapping and rehabilitation programs in Cook Islands, Fiji, Samoa, Brazil, and Indonesia, Eyles continues to develop methods of monitoring erosion control works, ways to rehabilitate mined land using indigenous plants, and new soil conservation programs for New Zealand.

■ Resources

Background information on DLR is available. New materials, including publications lists, are being prepared as the National Institute of Land Environment is established. According to Eyles, "The major restriction on our work is underfunding!"

National Institute of Land Environments ⊕
New Zealand Department of Scientific and Industrial Research (DSIR)
Private Bag
Lower Hutt, New Zealand

Contact: John E.C. Flux, Ph.D., Scientist
Phone: (64) 4 673119 • *Fax:* (64) 4 673114

Activities: Research • *Issues:* Biodiversity/Species Preservation; Deforestation; GIS Mapping; Global Warming; Soil Erosion; Waste Management/Recycling • *Meetings:* International Theriological Congress

▲ Projects and People

In addition to mammal and bird research at DSIR's former Land Resources Division, Dr. John Flux is deputy chairman of the Species Survival Commission (IUCN/SSC) Lagomorph Group and has compiled the *Status Survey and Conservation Action Plan for Rabbits, Hares, and Pikas* with co-editor and former Lagomorph Group chair Joseph A. Chapman, Dean of the College of Natural Resources, Utah State University, USA.

At DSIR, Dr. Flux has been concentrating on three projects and their resultant publications: helping to prevent the introduction of **myxomatosis** (a viral disease of rabbits), researching what controls rabbit numbers, and calculating rates of evolution.

Myxomatosis, he writes, disestablishes "the very efficient control of rabbits by predation over most of New Zealand." One work in press says that "watching a rabbit population decline for three years indicated some factors were not involved (floods, drought, diseases) while others could be (predation, habitat change). Dr. Flux has also written that "a single pair of rabbits could produce millions in three years, so the problem is what controls survival, not reproduction."

Flux's evolution project at DSIR is focused on the "effect of selection pressure and gene flow on a wild population of starlings," which helps to explain the "development of bait-shy rabbits in Central Otago," he says. "The cause is over-reliance on a single killing method, poison, because it was cheapest; an interesting example when the drive for efficiency has made a problem worse." Other research on how fitness affects subsequent survival in starlings is expected to be relevant to other species, such as rabbits.

As described in the *Conservation Action Plan*, the successful, worldwide lagomorphs include some 78 living species, including 25 small and rodentlike pikas, 24 rabbits, and 25 large hares. In comparison, however, *Rodentia* has some 1,685 species. Ecologically, lagomorphs are mammalian herbivores low in the food chain requiring habitat that provides forage and an adequate escape cover as they are the source of prey of weasels, foxes, coyotes, cats, civets, and large birds, for instance. Scientists remark that lagomorphs are "ideally suited as models for ecological research" as they inhabit diverse environments and have high population turnovers. With habitats imperiled, five genera are endangered: Amami rabbit (*Pentalagus furnessi*), volcano rabbit (*Romerolagus diazi*), hispid hare (*Caprolagus hispidus*), riverine rabbit (*Bunolagus monticularis*), and the Sumatran rabbit (*Nesolagus netscheri*). Also regarded as threatened or endangered are Tehuantepec jackrabbit (*Lepus flavigularis*), black jackrabbit (*L. insularis*), Tres Marias cottontail (*Sylvilagus graysoni*), Omilteme cottontail (*S. insonus*), Koslov's pika (*Ochotona koslowi*), and Muli pika (*O. muliensis*). Mainly, these lagomorphs are in China, Mexico, USA, India, Indonesia, Nepal, Japan, and South Africa.

The Survey and Plan lists lagomorph classifications; describes the genera and species, including economic importance, conservation, management, and education programs; and includes numerous references.

■ Resources

IUCN/SSC Lagomorph Group members receive a *Lagomorph Newsletter*. The 168-page *Conservation Action Plan* for rabbits, hares, and pikas can be ordered.

What is needed, writes Dr. Flux, is "international pressure on New Zealand government to do more environmental research."

NGO

World Wide Fund for Nature—New Zealand (WWF-NZ) ⊕
Education and Environment Centre
Botanic Garden, Glenmore Street
P.O. Box 6237
Wellington, New Zealand

Contact: **Dave Boardman**, Marketing Manager
Phone: (64) 4 499-2930 • *Fax:* (64) 4 499-2954

Activities: Development; Education; Research • *Issues:* Biodiversity/Species Preservation

● Organization

World Wide Fund for Nature—New Zealand (WWF-NZ), incorporated in 1975, is an "independent part of the international WWF network." Taking part in formulating and following general WWF global conservation strategies, it "determines its own local priorities and controls the expenditure of its own funds," which it is responsible for raising.

▲ Projects and People

Conservation and environmental education are WWF-NZ's primary activities. It helped fund surveys of the **Tuatara** lizard on Cook Strait Island and of Hector's dolphin, the country's only native dolphin. It purchased land to preserve and enhance the habitat of the **yellow-eyed penguin** on Otago Peninsula, supported a study of the **endangered water plants** in New Zealand's lakes, sponsored research into **whale strandings**, and is involved in re-planting of native vegetation on Tiritiri Matangi Island, Hauraki Gulf.

With the Wellington City Council, WWF-NZ recently established an **Education and Environment Centre** in the **Wellington Botanic Garden**—by constructing a new building and renovating an existing one—to create two classrooms, library, public display and information areas in conjunction with the National Museum and Victoria University, and resource and operation areas for school and public environmental projects.

"The Centre will add to the community's enjoyment of the Garden but particularly to their understanding of the inter-relationship between habitat and species survival," reports WWF. The Centre is free to visitors, easily accessible, and open year-round.

WWF-NZ's environmental education strategy is "designed to reach the widest possible audience cost effectively, and to ensure that its school programs bring lasting value within the stated aims of the school curriculum," it reports. For this purpose, its first target group in 1991–1992 were 8- to 12-year-olds from 240 primary and intermediate schools.

WWF-NZ's executive director is **Ian Higgins**, education officer is **Judith Benson**, and chairman is **John Anderson**, also chief executive of the Naitonal Bank of New Zealand.

■ Resources

WWF-NZ notes a "lack of leadership, training, and materials" for environmental education. Consequently, the Centre is undertaking teacher training, preparing and distributing literature for both teachers and students, building an international collection of environmental reference materials, and introducing a national environmental awards program with appropriate materials.

 Universities

Centre for Resource Management ⊕
Lincoln University
P.O. Box 56
Canterbury, New Zealand

Contacts: **Janice C. Wright,** Senior Research Officer
**Jonet Carr Ward, Ph.D., Janet Dalton
Gough, Carolyn Blackford,** Research
Officers
Phone: (64) 3 325 2811 • *Fax:* (64) 3 325 2156

Activities: Education; Natural Resource Policy Analysis; Political/Legislative; Research • *Issues:* Biodiversity/Species Preservation; Energy; Environmental Monitoring; Global Warming; Mediation; Natural Resource Economics; Population Planning; Sustainable Development; Waste Management/Recycling; Water Quality

● Organization

The Maori name of the Centre for Resource Management—*Te Whare Taoka mo ka Kaiwhakahaere*—means "The House of Learning for Managers of Resources." The Centre is a postgraduate teaching and research activity of the University of Canterbury and Lincoln University committed to "searching for methods of using natural resources that will lead to a sustainable future; [and] the ability to deal with the many dimensions and implications of natural resources management."

According to Centre director **John A. Hayward,** "Natural resources management and the rights of indigenous peoples are two international issues whose time has come." The Centre aims to provide the leadership needed to deal adequately with Maori cultural aspects of natural resources use.

The Centre's activities encompass studies of risk management, economic instruments applied to environmental problems, sustainable land use, sustainable management of the energy industry, environmental quality monitoring, depletion policies, natural resource use, bicultural resource management and mediation, and understanding of Maori values.

▲ Projects and People

Janice Wright researches energy demand management, intergenerational equity, and natural resource accounting. With a background at Lawrence Berkeley Laboratory, California, USA, she and a colleague developed a methodology known as "supply curves of conserved energy" which assesses the "technical and economic potential of energy conservation." Her concept was applied in California, Oregon, Washington, Alaska, and Michigan, and later adopted by the Energy Management Directorate, Ministry of Energy, New Zealand.

With New Zealand's ownership and control of energy resources recently shifting from the public sector to state-owned corporations and the private sector, Wright is analyzing the impact of this shift and trying to retrieve inaccessible energy data.

Committed to intergenerational equity, Wright is helping the Ministry for the Environment develop a public policy to manage natural and physical resources in such a way that the needs of future generations will not be jeopardized—in accordance with New Zealand's Environment Act of 1986. Recently, she has given seminars and talks on "societal values and ecological sustainability" before the World Bank's Environment Department, Free University of Amsterdam's Institute for Environmental Studies, and Australian National University's Centre for Resource and Environmental Studies.

Wright works in the area of "natural resource accounting" to help scope out such methodologies such as "environmental accounting" elsewhere in the world to see whether they would help the Ministry of the Environment better formulate sustainable policy. Her conclusion is that "New Zealand is better directed at developing useful sets of information" rather than modifying its present System of National Accounts. Consequently, Wright is shifting her interests to environmental statistics as linked to economic statistics.

Dr. Jonet C. Ward monitored the quality of the environment through measuring different aspects to provide decisionmakers with valuable information that allows them to design new environmental policies or review existing policies.

On behalf of the Ministry of Environment and its proposal to contribute to a "global understanding of environmental quality," the project examined overseas monitoring efforts as well as surveyed monitoring agencies in New Zealand. "The important issue of people's perceptions of environmental quality and how they may be appropriately incorporated into assessments of environmental quality was also examined," reports Dr. Ward.

Janet Gough measures people's perceptions of risk to set priorities for environmental management. Disagreements over types and levels or risks "leads to conflict over how we should use our resources," reports Gough, who analyzed literature on the subject and conducted pilot surveys regarding the controversy surrounding sewage disposal in Lyttelton Harbour and a pipeline between Lyttelton and Woolston.

She has also worked on a **Visitor Survey** at Mt. Cook to help estimate a value for visits to the National Park, and therefore to "impute a value" for the Park; undertaken water allocation studies in the Upper Waitaki basin; and examined approaches to estimate the cost of "non-supply of electricity."

Carolyn Blackford assesses the social impacts of flooding and floodplain management options. In another project, Blackford explored "when mediation might be an appropriate alternative for resolving resource management problems." As the structure of New Zealand local government and resource management legislation has undergone reform, it must be ensured that "no one set of values held by community groups overrides any other, that the rights of individuals and the welfare of the public in general are balanced." A project on **principles of defining problems** covered issues such as whether a "problem is a private or public one, the complexity of the problem, the involvement of facts and values, and the political processes that shape problem definition and resolution."

The Centre's communications manager, **Tracy Williams,** writes that "global cooperation will only happen when people with political power realize there is something to be gained for their particular country. Environmentalists need to raise their credibility and talk the 'talk' of politicians if they are to have an impact, while staying true to their positions." She points out that ECOPOLITICS in Australia "is a wonderful conference in bringing 'greenies' and politicians together." But she also expresses concern about the diminishing funding for environmental research from the national government.

■ Resources

The Centre publishes a *Biannual Report* and *Biannual Research Proposal* and can provide a publications list. It seeks printed materials, such as policy statements, State-of-the-Environment reports, legislation, and any material on environmental management. It will reciprocate by providing newsletters from New Zealand agencies, policy documents, and other information on contemporary issues not generally found in libraries and databases.

Centre for Maori Studies and Research
Centre for Resource Management
Lincoln University
P.O. Box 56
Canterbury, New Zealand

Contacts: **Rev. Maurice Manawaroa Gray**, Director
Heather Jonson, Manager, Research Unit
Phone: (64) 3 325 2811, ext. 8098
Fax: (64) 3 325 2156

Activities: Development; Education; Research • *Issues:* Biodiversity/Species Preservation; Energy; Health; Mana Maori Motuhake; Population Planning; Sustainable Development; Waste Management/Recycling; Water Quality

▲ Projects and People

With the Reverend Maurice Manawaroa Gray as director of the Centre for Maori Studies and Research and Heather Jonson as research unit manager, four goals are pursued: "Working cooperatively with other tribal groups within Maoridom on environmental concerns and the management of resources; encouraging cooperation between Maori and non-Maori groups over environmental issues and resource management; developing and promoting networking and cooperation between indigenous peoples, sharing issues and area of concerns plus strategies (successful or otherwise) for environmental protection and sustainable development; and imparting the knowledge and skills gained through the network of indigenous peoples, to enrich and lend weight to the work of other environmentally concerned peoples and communities in New Zealand and elsewhere."

Maori people are concerned about environmental matters that relate to their use of traditional environmental resources, such as water pollution and quality, waste management, energy development—especially hydro and geothermal, fishery management and control, preservation and harvesting of endangered bird populations, management and ownership of *pounamu* or greenstone, and sustainable development of land-based resources.

"The Maori are significantly disadvantaged in virtually all areas of society in comparison to the non-Maori population of New Zealand," writes Jonson. During the next decade, "the Treaty of Waitangi Maori will continue to regain control over more of their traditional resource base." While many Maori wish to continue the age-old conservation practices, there are others who support the commercial development of natural resources as the primary goal of the future. They are anxious to be able once more to provide for their own people and not be dependent on handouts from (and thus be so beholden to) a predominately non-Maori state.

"However," Jonson continues, "the wish for economic independence and a rise in living standards will need to be balanced by [preventing] further exploitation and despoliation of natural resources. Achieving this balance is likely to be one of the major challenges faced by Maori over the next decade and beyond."

■ Resources

Information on other indigenous peoples, including contact lists, and resource management strategies are sought.

Niger

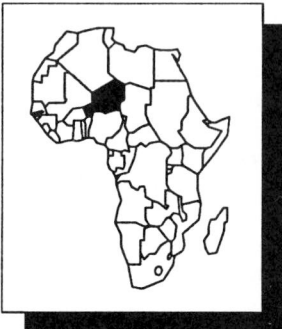

Many of Niger's environmental conditions are related to its location within the semiarid zones of West Africa. Over half the country is desert or semidesert and thus uninhabitable, and only 20 percent is suitable for sedentary agriculture. Problems include droughts, which have typically lasted several years and have caused losses in nomadic livestock; devegetation, caused by clearing, burn techniques, overgrazing, and fuelwood collection; soil erosion; and surface water shortages. The country has developed a comprehensive national strategy to combat desertification.

❧ *NGO*

Sahel Program
The World Conservation Union—Niger (IUCN) ⊕
P.O. Box 10933
Niamey, Niger

Contact: **Anada Tiega**, Technical Advisor
Phone: (227) 73 33 38 • *Fax:* (227) 73 22 15

Activities: Education; Development; Research • *Issues:* Biodiversity/Species Preservation; Deforestation; Energy; Global Warming; Sustainable Development; Water Quality

● Organization

Niger, a member state of the **World Conservation Union (IUCN)**, is presently the greatest beneficiary of the **Sahel Program**, which became operational in 1988 with two principal objectives: for the present generation, to research the development of renewable resources, and for future generations, to research the development and maintenance of the potential productivity of regional natural resources and the ability of those resources to withstand change. IUCN plays an important role in the development of environmental policy and in the direction of environmental efforts, but wants to do so in a complementary manner, specific to each country or region.

▲ Projects and People

Presently, IUCN is turning its efforts to enlarging the juridical principles designated by the Rural Code concerning the place of the environment and rules governing access to natural resources. To this end, IUCN has formed a collaboration with the National Committee for the Rural Code to devise a **National Conservation Strategy**. Together, they work with many other national organizations to revise the following: the **National Plan to Combat Desertification**, the Tropical Forestry Action Plan, and the **Integrated Program for the Management of Natural Resources**. IUCN also works with the Practical Institute of Rural Development on the subject of reinforcing the programs of ecological planning within the Institute.

According to IUCN, the fauna of Niger has not been studied with the attention it merits. For that reason, it has several studies in progress regarding fauna, fish, and pisciculture. IUCN will continue its various collaborations with the goal of helping the environment and ameliorating present damage.

With a background in forestry, wildlife management and agronomy in arid and tropical zones, environmental education, fisheries, and renewable resources, **Anada Tiega** has studied at the University of Arizona, George Mason University, University of Idaho, USA, University of Niamey, and in Cameroon and the Ivory Coast.

(See also World Conservation Union [IUCN] in the France, Pakistan, Switzerland, USA, and Zimbabwe NGO sections.)

Nigeria

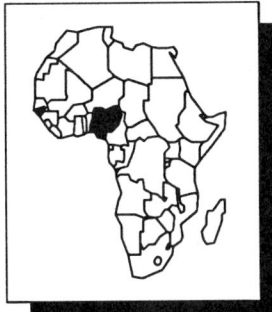

Nigeria, with more population than any other African country, is also the home of extraordinary biological diversity, including arid areas, swamps, many different forest types, and a vast array of plant and animal species. About three-quarters of the people live in rural areas (most Nigerians make their livelihoods by farming, fishing, or herding), but there are several large, overcrowded cities. The country's rich natural resources are threatened by logging, forest clearing for plantation development, fuelwood demand, hunting, wetlands exploitation, and harmful agricultural practices. Fortunately, government and non-government conservation organizations have recognized the need for carefully designed land-use management plans to maintain ecological processes and protect diversity.

🏛 *Government*

Afforestation Programme Co-ordinating Unit (APCU)
223K Hadejia Road, P.M.B. 3441
Kano, Nigeria

Contact: **Lawan B. Marguba,** Programme Head
  Telex: 77330 KN BTH NG ATTN. TDS 0070
E-Mail: apcu kano, ng

Activities: Development; Education; Law Enforcement; Political/Legislative; Research *Issues:* Biodiversity/Species Preservation; Deforestation; Energy; Soil Conservation; Sustainable Development; Waste Management/Recycling • *Meetings:* ATO

● Organization

Started in 1987, the Afforestation Programme Co-ordinating Unit (APCU) concentrates on pursuing "global, regional, and sub-regional cooperation on matters of drought, desertification, and general environmental degradation" with a locally and internationally recruited staff, consultants, and other experts. APCU aims to keep its staff up to date on "latest developments in the art and science of environmental protection management."

State by state, APCU's achievements so far include the contacting of some 72,706 farmers, production of 17 million tree seedlings, and planting and establishing of 1,256 kilometers of shelterbelts.

▲ Projects and People

APCU offers "technical assistance free of charge to the public on the establishment and development of woodlots, orchards, farm forestry, [and] nurseries," for example. In its incentive package for target communities, seeds, polypots, fertilizer, and fencing wires are provided to individuals, associations, schools, and communities. "Already, Foresters Clubs and Associations have been established in some schools, and communities within our areas of operation," reports APCU.

Workshops, seminars, and short-duration courses are conducted on various forestry topics, leadership, and social development for the public, students, women's groups, government personnel, and staffs of collaborating groups. National and regional workshops on conservation, regional planning, and environmental assessment emphasize, for example, strategies for drought and desertification control in Nigeria's arid zone and an integrated approach to agriculture and forestry development.

Ecologist **Lawan B. Marguba,** APCU head, recommends that the United Nations create a UN "Fund for Global Environmental Protection," similar to permanent funds for health and population, to set "global standards on the funding and care of the environment," with cooperation, training, and sharing of ideas encouraged on local to worldwide levels. He also suggests that UN "certificates of honor" could be presented to those who are contributing to the care of the environment.

■ Resources

APCU produces publications of particular value in arid zones on tree planting, shelterbelts, integration of fruit trees into farming, raising seedlings, role of women in afforestation, checking desertification in other African countries, social forestry in India, and forestry revenue studies as well as posters, greeting cards, and school materials. A publications list is available.

APCU seeks consultancy services and technologies on agroforestry, silvi-pastural and agri-silvi-pastural development, forest conservation, social forestry, forestry and rural development, alternative energy sources, and conservation education, for example. Training needs are in forestry extension, research, monitoring field activity, human resources development, and teaching trainers.

Equipment sought are vehicles, film production units, teaching and audiovisual aids, literature and stationery, and materials for distribution to the public, such as "fencing wire, polypots, fertilizers, chemicals and insecticides." Raw materials needed are "fast-growing tree seeds or seedlings with high adaptability with agricultural crops, fruit seeds or seedlings." APCU also seeks national and regional political support for afforestation and a strengthened, viable, and dynamic "legal and institutional framework" to attain conservation and environmental protection.

Forestry Management, Evaluation, and Co-ordinating Unit (FORMECU)
Federal Ministry of Agriculture
P.M.B. 5040, Jericho Reservation
Ibadan, Nigeria

Contact: **Luke Ifeanyichukwu Umeh, Ph.D.**, Head of Unit
Phone: (234) 22 411249
Telex: 31137 FORMECU NG

Activities: Development; Education • *Issues:* Biodiversity/Species Preservation; Deforestation; Energy • *Meetings:* Commonwealth Forestry Association; UNCED

● Organization

The **Forestry Management, Evaluation, and Co-ordinating Unit (FORMECU), Federal Ministry of Agriculture**, is a World Bank and African Development Bank (ADB) assisted Forestry Development and Environmental Programme that has achieved the following accomplishments: supplying 41 million seedlings to cover an estimated area of 750,000 hectares and establishing 1,700 kilometers of shelterbelts that protect about 34,000 hectares of farmlands in 6 Nigerian states; improving the management of some 37,600 existing forest plantations and establishing new ones covering some 15,150 hectares; implementing the Nigerian Tropical Forestry Action Plan/Programme (TFAP); and improving the "forest management capability of the Nigerian Forestry Sector through training, policy analysis, research, and studies" as well the development of 2 Forestry Manpower and Vocational Development Centres.

▲ Projects and People

The **Arid Zone Afforestation Project** aims at checking drought and desertification in Nigeria's north with the World Bank, European Community (EC), and Nigerian government. **Dr. Luke I. Umeh** coordinates the International Bank for Reconstruction and Development (IBRD) (World Bank) and Nigerian government efforts.

The **Erosion and Watershed Management Project** aims at checking erosion siltation of rivers and streams. As president of the Forestry Association of Nigeria, an NGO, Dr. Umeh helps educate the government and the public on the adverse affects of deforestation.

The **Biodiversity Conservation Project** is helping to preserve gene resources in some Nigerian reserves with the assistance of the Nigerian Conservation Foundation, of which Dr. Umeh is a life founder member.

Dr. Umeh reports that all three ongoing projects employ the technique of mass mobilization. "Future plans are to have an integrated management of the environment by involving all sectors concerned." Obstacles are lack of knowledge for a stable environment and of financial and material resources to bring it about.

To enhance global cooperation, Dr. Umeh recommends that environmentalists form regional associations with the help of the UN, EC, and Organization of African Unity; create a "regional trust fund" to assist with activities and attendance at conferences; and stimulate more publications and mass media programs.

Publishing extensively on agroforestry, agri-silviculture, tree seeds, pulpwood, and timber resources, Dr. Umeh formerly co-managed projects of the United Nations Development Programme (UNDP) and Food and Agriculture Organization (FAO). He also produces news media scripts on deforestation and environmental management.

■ Resources

FORMECU seeks computer programs and remote sensing for forestry work, visual aids and books, training and extension programs, help with public outreach, and funding.

Sokoto Energy Research Centre
Usmanu Danfodiyo University
P.M.B. 2346
Sokoto, Nigeria

Contact: **A.S. Sambo, Ph.D.**, Director
Phone: (234) 60 234052

Activities: Development; Research • *Issues:* Energy; Sustainable Development • *Meetings:* CITES; World Renewable Energy Congress, UK

● Organization

Started in 1982, the **Sokoto Energy Research Centre** is one of four such facilities of the Nigerian government which aims to cover all forms of "renewable energy applications over the range of technical, economic, environmental, social, and legal issues, so as to galvanize Nigerians" to its use. A main focus is on rural communities where suitable energy systems are being developed and tested for commercialization.

▲ Projects and People

Such energy systems include: a **solar cooker**, which is a portable box adaptable to both rural and urban areas and used for cooking meals. It can be easily built of native materials by local craftsmen. A **solar water heater** was designed to provide hot water in the 70–90°C temperature range using average sunshine conditions. A **solar still** can provide drinking water for rural communities; the Research Centre describes "a prototype of a 'double-slope solar still coupled with a hot water heater' that has a higher yield than the conventional still."

A **solar dryer**, to be used for fast and hygienic drying of some agricultural products, can be easily and cheaply constructed. A **solar-powered vaccine and medicine storage refrigerator** can be vital in rural areas without electricity for delivery of effective health care, including immunization and epidemic control. A **solar-powered water pumping system** is being tested in community pilot demonstrations. A **solar-powered air cooler** is a substitute for conventional air conditioning in rural areas lacking electricity and has the advantage of increasing humidity in dry atmospheric conditions.

The Sokoto Energy Research Centre has also developed a series of **wood-burning stoves** made of clay, rice husk, and other locally available materials. According to the Centre, this stove is less hazardous than traditional cooking methods, uses about half the wood, and appears to be "much accepted" by villagers living near the demonstration project. Also being tested is a one-pot **sawdust cooking stove** constructed with sheet metal that appears to be comparable to the improved wood-burning stoves.

Four kinds of **biogas digesters** produce cooking gas from waste materials, such as cow dung and poultry droppings. "The gas produced can also be used for lighting and electricity generation," reports the Centre. "In addition, the digested organic materials have been found to be better fertilizer." Again, these can be constructed with local materials.

Under study is a **solar cooker with oil-storage unit**, designed to be used indoors during no-sunshine periods as well as on sunny days. A vertical-axis **wind water pumping system** is the result of Sokoto experiments with wind energy. "The system was constructed using empty oil drums, vertically cut in half, to serve as the rotor which is the source of power to drive a simple acting reciprocating water pump." A horizontal-axis **wind-electricity conversion system** with a three-blade propeller and induction generator is also being observed; its "wind alignment unit enables the whole system to be independent of the direction of the wind."

At the Centre, three **laboratories are powered with photovoltaic (solar cells) panels** to power fluorescent lamps, ceiling fans, refrigerators, and small appliances. The system is being monitored.

A mechanical engineer, **Dr. A.S. Sambo** is a member of the International Energy Foundation and International Solar Energy Society, vice president of the Solar Energy Society of Nigeria (SESN), recipient of numerous academic awards, and deputy vice chancellor of Usmanu Danfodiyo University. He participate in alternative energy forums worldwide and publishes extensively in his field.

■ Resources

The Centre also offers consultancy services to government, private organizations, and industries regarding energy and renewable energy systems, energy policy, planning, efficiency, and management. Available are the *Nigerian Journal of Renewable Energy* and the *Directory of Renewable Energy Research and Development Activities in*

Nigeria. The Centre seeks new technologies with solar meters, data loggers, and computers for more accurate recording data; it also welcomes linkage with similar organizations.

 *NGO*

Nigerian Conservation Foundation (NCF)
Plot 5, Moseley Road, Ikoyi
P.O. Box 74638, Victoria Island
Lagos, Nigeria

Contact: **Pius A. Anadu, Ph.D.**, Executive Director
Phone: (234) 1 684717 • *Fax:* (234) 1 685378

Activities: Development; Education; Nature Protection • *Issues:* Biodiversity/Species Preservation; Deforestation; Global Warming • *Meetings:* WWF

● Organization

The **Nigerian Conservation Foundation (NCF)**, an associate member of the World Wide Fund for Nature (WWF), is described as the "largest private conservation organization in Nigeria" with chapters in 3 states, a staff of 43, and some 1,000 members. Its mission is to protect genetic, species, and ecosystem diversity; promote sustainable use of renewable natural resources; and promote actions to reduce pollution, wasteful exploitation, and consumption of resources and energy. Ultimately, NCF wants to "stop, and eventually reverse, the accelerating degradation of Nigeria's natural environment and to help build a future in which humans live in harmony with nature." It maintains strong relations with the government, educational institutions, and other NGOs to achieve its ends.

▲ Projects and People

Activities range from "complex development projects" such as **Cross River National Park, Hadejia-Nguru Wetlands**, and **Okomu Wildlife Sanctuary** with international NGO partners to small grants for financing **species inventory** or **baseline data surveys**, for instance. A main approach is to "promote **alternative income generating activities** in the villages surrounding a protected area." NCF offers advice and **technical assistance** on environmental rehabilitation as well as on conservation action plans or strategies. **Evaluation and training** are also undertaken.

"All nations, rich and poor, must share in the task of caring for the Earth because we have nowhere else to go," reminds NCF, which in 1992 co-hosted with WWF a **meeting of religious leaders** to discuss the role of religion in promoting the conservation ethic. "For a nation like Nigeria, where religious differences have sometimes led to social tensions, such a conference would hopefully bring about a better understanding of the need for mutual cooperation in tackling a common threat—irrespective of ethnic, religious, or racial differences," hopes NCF.

Regarding global cooperation, executive director **Dr. Pius Amaeze Anadu** writes, "Environmentalists have insisted that unless rich industrialized nations address the underlying causes of environmental degradation and loss of biodiversity in the developing world—poverty, ignorance, heavy debt burden, equitable sharing of the earth's resources, and acquisition of appropriate technology—there would be no consensus on how to tackle global environmental problems."

An animal ecologist with research interests in both large and small mammals, Dr. Anadu was acting head of the University of Benin's Department of Forestry and Wildlife with numerous scientific publications to his credit. He belongs to the Species Survival Commission (IUCN/SSC) Primate Group and Nigeria's National Advisory Committee on Conservation of Renewable Resources and is a regional member for Africa, World Conservation Union (IUCN).

Nigerian Environmental Study/Action Team (NEST)

18B Aare Avenue, New Bodija
P.M.B. 5297
Ibadan, Nigeria

Contact: **Ademola Tokunboh Salau, Ph.D.,**
            Chairman, Executive Committee
Phone: (234) 22 31589
Telex: 3112 CUSO NG

Activities: Development; Education; Research • *Issues:* Air Quality/Emission Control; Biodiversity/Species Preservation; Deforestation; Energy; Global Warming; Health; Sustainable Development • *Meetings:* UNCED

▲ Projects and People

The **Nigerian Environmental Study/Action Team (NEST)** says it published in 1991 "the first-ever environmental profile on Nigeria," entitled *Nigeria's Threatened Environment*, which is described in *Haramata* (No. 13, 1991) as documenting "environmental issues relating to land, water resources, atmosphere, vegetation, wildlife, human habitat, and culture, emphasizing the need to adopt sustainable development pathways."

More than a database, the report provides editorial comment, for example, on pastoralism, conflicts over land, and the inability of the government to enforce curbs on tree cutting and bush burning. As *Haramata* commented, "The case studies in the book clearly show that a regulatory approach to environmental protection inherited from a top-down colonial model is deficient. Sadly, the profile does not move on to discuss what forms of incentive approaches could be encouraged that would allow local communities, with state support, to engage in environmental management."

For its efforts in launching the profile, however, NEST—a nonprofit advocacy group working "to increase understanding and awareness of the Nigerian environment"—received messages of thanks from groups such as Earth Search, Canadian International Development Agency (CIDA), and the Nigerian Association of Non-governmental Development Agencies (NANDA). Such groups express concern that while NGOs promote organic farming, for example, some government officials continue to push "massive use of chemical fertilizers without previous soil tests, not to speak of post-use soil tests." NANDA reports that "agrochemicals already banned in the countries of manufacture get advertised in our newspapers or taken round farms by agricultural extension workers, without the consumers realizing how much danger they contain."

The profile also characterizes Nigeria's desertification as a permanent degradation of vegetation with the genetic loss of valuable plant species, loss of soil's organic matter and its ability to retain water, increased wind erosion and local sandstorms, increased runoff and sheet erosion, declining crop yields and failures, "leading to a constant state of famine," insufficient forage and browse resources, dune formation in certain areas, and "breakdown of traditional community relations as well as socially and economically accepted farming systems."

NEST spokespersons, such as Prof. Uzo M. Igbozurike, executive director, participate in symposia on socioeconomic implications of soil erosion and sustainable uses of tropical forests with representatives from the Nigerian Conservation Foundation (NCF); Forestry Management, Evaluation, and Co-ordinating Unit (FORMECU); Tropical Forestry Action Plan (TFAP); and the Forest Research Institute of Nigeria (FRIN).

Dr. Ademola T. Salau is professor and deputy vice chancellor at the University of Port Harcourt; consultant on urban housing, tourism, industrialization, and land degradation; coordinator of the Multinational Working Group on Environment for CODESRIA (Council for the Development of Economic and Social Research in Africa in Dakar, Senegal), and author of numerous books and articles in his field.

■ Resources

Available publications are *Nigeria's Threatened Environment: A National Profile*; *Sustainable Development in Nigeria's Dry Belt: Problems and Prospects*; *NGO Country Report (Nigeria) for UNCED*; and *The Nigerian Environment: "Non-Governmental Action."*

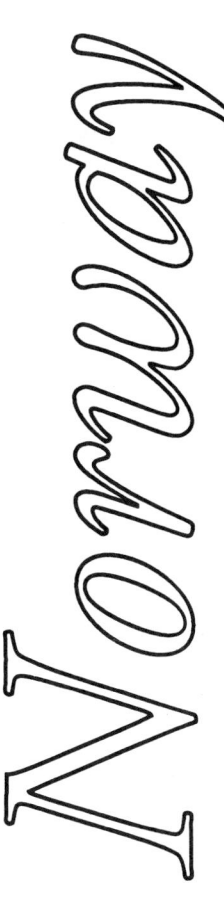

Norway

Known as "Europe's last wilderness," Norway has large areas of almost untouched natural habitats. The countryside is home to a tremendous population of wild reindeer; bears and wolverines inhabit the country's wild mountain forests; and seals, whales, and otters reside undisturbed among the more than 50,000 islands along the coast of the North Sea. During the last few decades, Norway has seen a transformation of its economy from one based primarily on agriculture, forestry, and fishing, to one with a more industrial focus. This change, along with the country's proximity to the pollution of its more industrialized and urbanized European neighbors, has taken an environmental toll.

While the country has made significant progress in reducing emissions from some sources, a rise in oil consumption has resulted in a steep increase in nitrogen oxide (NO_x) and carbon dioxide emissions. In addition, atmospheric pollution from other countries has caused increased damage by acidification. The government has long-range plans to increase energy efficiency, improve transport, control marine oil pollution, clean up the North Sea, and improve waste disposal and recycling.

 # NGO

Norges Naturvernforbund (NNV)
The Norwegian Society for Conservation of Nature
Postboks 2113, Grünerløkka
0505 Oslo 5, Norway

Contact: **Dag Hareide,** Secretary General
Phone: (47) 2 715520 • *Fax:* (47) 2 715640

Activities: Development; Education; Law Enforcement; Political/Legislative • *Issues:* Air Quality/Emission Control; Biodiversity/Species Preservation; Deforestation; Energy; Global Warming; Sustainable Development; Transportation; Waste Management/Recycling; Water Quality

● Organization

Founded in 1914, Norges Naturvernforbund (NNV) (The Norwegian Society for Conservation of Nature) is Norway's "oldest and biggest" environmental organization, with 60,000 members; it includes Inky Arms, a children's club, and Nature and Youth, for young adults. With nearly 200 groups throughout the country, NNV says its numbers and political influence are growing. "People trust us—a survey shows that 70 percent of the voters have confidence in the environmentalists," says NNV literature. NNV is preparing to win the environmental battles into the "next millennium."

▲ Projects and People

Ensuring environmentally friendly Olympic Games in 1994 is a pertinent new project. NNV spoke up successfully on behalf of a swampy wildlife reserve area with 200 bird species which faced encroachment by an ice rink and supermarket for the Games outside Lillehammer. NNV is also trying to "change the Charter of the Olympic Games, to make it more environmentally concerned" for all future Games.

The **Environmental Home Guard** is an information network appealing to consumers, workers, and organizations. Those who join promise to fulfill 10 of 16 different activities listed on an Action Paper. The program is being tested in counties throughout Norway. NNV also analyzes various consumer goods and publishes an "environmentally friendly" guidebook, *Miljøboka.* The Inky Arms Clubs encourage some 15,000 children to become "eco-detectives" and search for solutions to environmental obstacles.

NNV's **Rainforest Foundation,** with the Development Fund and Future in Our Hands—groups originated by Sting, the entertainer—launched in 1992 a rainforest awareness and fundraising campaign that targets Brazil. It also aims to defend the rights of rainforest inhabitants, such as the Yanomami Indians, boycotts furniture and flooring of tropical woods, supports the purchase of sustainable rainforest products, and gives evidence of NNV's commitment to global issues.

With participation in the **North Sea Conference**, through the organization **Seas at Risk**, NNV was part of the effort to gain the pledge of countries to "reduce sewage discharges to the North Sea by 50 percent within 1995. All use of PCB [polychlorinated biphenyl] shall stop, and the discharge of dioxin, lead, cadmium, and mercury shall be reduced by 70 percent within 1995," according to NNV literature.

Through its **Barents Sea Board**, NNV is against "negotiable quotas," which could mean "the richest companies will gain access to the fish. This implies fewer fishermen, bigger boats, and less sustainable use of resources as a result of depletion of the fish stocks"—northern Norway's most important industry. Viewing the conflict between coastal fisherman and big trawlers as a global situation, NNV is building a network of organizations to work to "preserve maritime resources." Yet, after examining the environmental policy of the European Community (EC), NNV does not support Norwegian membership. "The EC of today is not protecting our nature," it says.

Fighting to reduce carbon dioxide (CO_2) emissions, NNV helped win a CO_2 **tax on petrol**, or gasoline for cars. NNV is lobbying for a **high-speed railway network** that could reduce energy consumption "by one billion kilowatt hours" and employ some 19,000 people for a decade. It also wants government transportation plans revised to reduce auto traffic.

Campaigning against chlorofluorocarbons (CFCs), NNV is helping bring about the **reduced discharge of chlorine**, such as in removing polyvinyl chloride (PVC) in packaging and encouraging the use of recycled and chlorine-free paper—and is asking European NGOs to work toward "blacklisting PVC on the international market." One ally is the Norwegian Federation of Trade Unions.

Pressing for **expanded protection of conifer forests**, NNV says Norwegian leaders, including Gro Harlem Brundtland of the Labour government, are only safeguarding one-third of the size that conservationists and scientists recommend. NNV is also urging that new national parks and preservation areas be established, and seeks specific protection for the **wolf**.

With each Norwegian generating about a ton of rubbish each year, NNV seeks **mandatory rubbish sorting** in every municipality and special **plants to treat hazardous wastes**. One goal is to use trash as a resource; another is to keep hazardous wastes from being produced in the first place. Many Norwegian households now sort paper, plastics, and batteries—at NNV's urging.

While NNV is saying "no" to new gas power stations and hydroelectric plants, the group is convinced that **renewable energy sources**—wind, solar, biomass, and perhaps tidal power—can generate power for Norway and its neighboring countries. Heat pumps also offer potential. According to NNV, the Water Resources and Energy Board also says that the country can comfortably "reduce our energy consumption by about 30 percent."

■ Resources

NNV publishes the widely distributed *Natur&Miljø* magazine and *N&M Bulletin*. It encourages broad press coverage through its media relations office.

 *Private*

Norsk Institutt for Naturforskning (NINA) ⊕
Norwegian Institute for Nature Research
Tungasletta 2
N-7004 Trondheim, Norway

Contact: **Thrine Moen Heggberget,** Research
           Ecologist
Phone: (47) 7 91 30 20 • *Fax:* (47) 7 91 54 33

Activities: Research • *Issues:* Biodiversity/Species Preservation; Deforestation; Global Warming; Sustainable Development; Water Quality • *Meetings:* Congress on Regulated Rivers; International Association of Vegetation Science Symposia; IUGB; Ornithological Congress; Theriological Congress; World Fisheries Congress

● Organization

Established in 1988, the **Norwegian Institute for Nature Research (NINA)** is a continuation of the former Research Division of the Directorate for Nature Management (headquartered at the Directorate's address) and its program for applied ecological research (Økoforsk). With a staff of 150, NINA is "intended as a service center for all persons and institutions needing information about ecology." Co-workers are at the Universities of Oslo, Bergen, and Tromsø and at the Norwegian Agricultural University at Ås. NINA describes itself as "among the largest centers for ecological research in Europe," with work consisting of "long-term research programs, commissioned research, consultation, training, and information services." Within Norway, it provides services for these government offices: Department of the Environment, Directorate for Nature Management, State Pollution Control Agency, and County Environmental Administration.

▲ Projects and People

NINA works in the following areas: pollution, environmental encroachment, conservation of natural habitat, game ecology, fish ecology, coastal ecology, species preservation, and outdoor life.

Regarding **environmental pollution**, NINA measures the effects of acid rain and heavy metals on soil, vegetation, and aquatic and terrestrial ecosystems; studies the effects of liming acid rivers; investigates possible ecological results of climatic changes; examines the "spread of radioactivity in natural food webs"; and cooperates with the former Soviet Union on pollution studies in northern regions. Similarly, NINA's **environmental monitoring** includes effects of acid rain and other long-range pollutants on ecosystems, "distribution and population size of seabirds," "population of large carnivores," and certain fauna and flora monitoring.

Work on **river encroachment** includes "effects of hydropower development on fish and fisheries in several major river systems, . . . changes in water chemistry in rivers and lakes due to transfers of drainage areas in regulated systems; studies of transmission wire collision hazards to game birds; vegetation changes in river deltas due to lowering of water levels; and effects of changes in waterflow and silting on plankton and other invertebrates in the Dokka watershed.

Nature conservation work includes designing conservation plans for natural or virgin coniferous forests, rivers, geological and geomorphological conditions; registering coastal beaches and yew locations; protecting coastal heather vegetation; and evaluating human impacts. In **landscape management**, NINA researches "effects of changes in agricultural practice and management on vegetation and animal occurrence and distribution; roe deer in cultural landscapes; changes [relating to] fragmentation and monocultures in spruce forest; and climatic changes and land management."

Genetic resources work encompasses studies of genetic diversity of plants and animals, including insects; surveying the "natural status and ecological characteristics of vulnerable plant and animal species"; and simulating the introduction of genetic modified organisms in nature. Material is preserved in a gene bank.

Under study in NINA's **wildlife ecology** work are large herbivores, such as moose, roe deer, and wild reindeer; brown bear and conflicts with domestic animals, such as sheep in pasture; lynx; badger in conflict with an urban environment; and yew vegetation in subalpine areas and the affect of its removal on plant and animal life. NINA points out that although hunting yields some five million kilos of venison annually, it must be conducted "in a manner which safeguards population and long-term sustainable yields."

Coastal ecology work centers on "the role of sea urchins in the decimation of kelp forest; seabird, otter, and coastal seal abundance and treatment due to oil exploitation on the Norwegian continental shelf; seabird mortality from oil pollution and net fisheries; abundance and environmental requirement of harbor porpoise and coastal seals and interactions with marine fish population; [and] structural changes in marine hard bottom fauna following toxic algae bloom in Skagerrak." **Fish ecology** includes "sea ranching with Atlantic salmon and anadromous Arctic char; fish enhancement in hydroelectric reservoirs; infections of the monogenous fluke *Gyrodactylus salaris* and other parasites on wild Atlantic salmon stocks; effects on wild salmon by escaped salmon from fish farms; and biomanipulation of fish stocks in entropic lakes."

In **outdoor recreation**, NINA undertakes studies of nature tourism and nature management, impacts of hydropower development and oil exploitation, recreational fisheries and boating, and effects of state and local actions in improving such recreation in suburban areas.

■ Resources

NINA's yearbook, *Årsmelding*, contains a publications list with Norwegian and English titles. The *Nina Report* is published in Norwegian or English. Other series, such as *NINA Oppdragsmelding, NINA Utredning, NINA Notat,* and *NINA Temahefter,* have English summaries.

🎓 *Universities*

Laboratory of Marine Molecular Biology ⊕
University of Bergen
Thormøhlensgt. 55, H1B
N-5008 Bergen, Norway

Contact: **Anders Goksøyr, Ph.D.,** Associate Professor
   Phone: (47) 5 544000

Activities: Education; Research • *Issues:* Environmental Health
Meetings: International Symposium on Responses of Marine Organisms to Pollutants

▲ Projects and People

Dr. Anders Goksøyr works in the field of molecular marine toxicology at the Laboratory of Marine Molecular Biology, University of Bergen. Subjects of his research include Atlantic cod, rainbow trout, salmon, seal, and minke whale—with studies examining effects of oil pollution and characteristics of cytochrome P450 in environmental monitoring.

Dr. Goksøyr indicates that since 1986 his Laboratory has "developed an indirect ELISA for studying the P450 1A1 induction response in fish, and have applied this technique to characterize the induction response to different environmental xenobiotics in different fish species at different life stages. . . . We have been able to show induction by oil exposure in cod larvae, where no catalytic activity could be measured.

"We have also shown that certain inducers (e.g. PCBs) [polychlorinated biphenyls], and endogenous factors (e.g. estradiol) are able to inhibit catalytic activity, which will affect induction studies which are performed with catalytic assays only. We are currently working with the production of various immunoreagents, and are developing polyclonal antibodies against a synthetic P450 1A1-peptide (from rainbow trout), monoclonal antibodies against code P450 1A1, and polyclonal antibodies against cod cytochrome b₅."

He reports that this research has a "high relevance for industrial applications" and expects that the P450 induction response will form "an important and integrated part of a biomarker-based monitoring system for environmental stress."

For example, the "induction response of the fish cytochrome P450 system to certain classes of organic pollutants (oil compounds, aromatic hydrocarbons, chlorinated hydrocarbons, dioxins, pesticides) has been shown to reflect the increased expression of a single P450 gene, called P450 1A1 (or CYP 1A1). This induction response has been studied with both catalytic and immunochemical methods.

"The catalytic system has recently been incorporated into several national and international monitoring programmes, such as the North Sea Task Force Monitoring Master Plan of ICES (International Council for Exploration of the Sea) and the Oslo/Paris Commission, and the National Status and Trends Program of NOAA [National Oceanic and Atmospheric Administration] in the USA. It was also apart of the biological effect studies carried out by NOAA after the Exxon *Valdez* oil spill in Alaska, and has been used in regulatory actions against industrial effluents," such as in Sweden.

The Laboratory also points out that "sampling for catalytic determinations demands fresh processing and/or cryopreservation, conditions that may not be easily met in remote areas, far offshore, or in developing regions of the world. Immunoassays, especially simple ones such as ELISA and dip-stick tests, are not critically dependent on sampling, and may be more easily adapted to low-cost laboratories."

Funding from industrial partners is being sought for a three-year research plan, which also will include studies of the regulation by environmental and endogenous factors of certain genes in the cytochrome P450-system in selected fish species; links between pollution exposure and lowered reproductive capability as linked to ongoing fish reproduction, biochemistry, and endocrinology experiments; and immunotoxicology.

■ Resources

Dr. Goksøyr's publications list includes more than 100 titles in English.

Department of Ecology ⊕
University of Tromsø
N-9000 Tromsø, Norway

Contact: **Joseph L. Fox, Ph.D.**, Associate Professor
  Phone: (47) 83 44386 • *Fax:* (47) 83 71961
  E-Mail: joseph@mack.uit.no

Activities: Education; Research • *Issues:* Biodiversity/Species Preservation; Sustainable Development

▲ Projects and People

Dr. Joseph L. Fox, associate professor with **University of Tromø's Department of Ecology**, conducts wildlife research on large mammal communities in the Indian Himalaya on behalf of his University as well as with the U.S. Fish and Wildlife Service and the Wildlife Institute of India.

As the research director of the International Snow Leopard Trust, Dr. Fox is involved in several studies of these mammals in India. *(See also International Snow Leopard Trust in the USA NGO section.)* In an international effort with **Satya P. Sinha, Raghunandan S. Chundawat,** and **Pallav K. Das**, Wildlife Institute of India, Dr. Fox conducted a nine-month study on foot along remote valley bottoms between high mountain deserts to track *Panthera uncia* in areas of three northwestern Himalayan states, including Hemis National Park, Shang Wildlife Reserve, other points in central and southern Ladakh; near the vicinity of Kulu-Manali in Himachal Pzradesh; and the Govind Pashu Vihar Wildlife Sanctuary, Uttar Pradesh. Also encountered were blue sheep, ibex, and other wild ungulate prey; (*Panthera pardus*); and evidence of the lynx.

Most plentiful in Ladakh where parks and reserves are being established, it is estimated that about 400 snow leopards occur throughout northwest India. The Trust estimates that there are only about 1,500 wild snow leopards in their historic range through Mongolia, China, former USSR, Afghanistan, Pakistan, India, and Bhutan.

With Hemis as a prime habitat for about 50 to 75 snow leopards, one goal is to set up corridors linking this national park to other reserves so that a viable population can be maintained. The snow leopard is a protected endangered species both under India law and CITES (Convention on International Trade in Endangered Species of Wild Fauna and Flora). Yet the snow leopard is threatened by hunters who retaliate for the killing of their livestock—as well as by its shrinking habitat. In 1990, India set in motion a five-year **Project Snow Leopard**, similar to Project Tiger. The Trust supports the expansion of Hemis from its present 600 square kilometers to several thousand square kilometers through its **Nurture a Nature Reserve** program that supports both the needs of local people and preservation of ecosystems.

Hemis is also known for such small mammals as fox, otter, Himalayan weasel, and marmot as well as such large birds as the golden eagle and griffon vulture. The abundant wildlife, it is believed, is fostered by the presence of Buddhists at the Hemis Monastery and elsewhere who inspire other villagers to revere life. With tourism in Ladakh on the increase since 1974, this impact can affect wildlife conservation, for example, as exotic livestock are introduced that could alter pastoral patterns. The popularity of mountain trekking also stimulates the demand for transport services, food, and firewood sources. Park management must takes these factors into account and also foster sustainable development alternatives for local people.

Dr. Fox and other colleagues have also observed **six endangered or threatened wild ungulate species** in Ladakh: Tibetan wild ass or kiang (*Equus hemionus kiang*), Tibetan antelope (*Pantholops hodgsoni*), Tibetan gazelle (*Procapra picticaudata*), wild yak (*Bos grunniens*), Tibetan argali (*Ovis ammon hodgsoni*), and Ladakh urial (*Ovis vignei vignei*).

Dr. Fox's past research focused on the ecology of the wolverine in Norway and on mountain goat habitat in Alaska. He belongs to the Species Survival Commission (IUCN/SSC) Cat and Caprinae groups for which Action Plans are currently in preparation and was an organizer and proceedings editor for the Seventh International Snow Leopard Symposium, China, in 1992.

■ Resources

Publications and information on education programs about the snow leopard can be obtained from the Trust, 4649 Sunnyside Avenue North, Seattle, Washington 98103, USA. Dr. Fox also has an extensive publications list in his field.

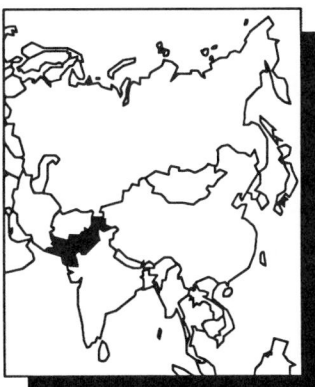

The landscape in Pakistan changes from coastal beaches, lagoons, and mangrove swamps to sandy deserts, desolate plateaus, fertile plains, dissected uplands and, stretching from east to west, a series of mountain ranges that include The Himalayas, the Karakoram Range, and the Hindu Kush mountains. The Indus River, which traverses the country, is the world's largest irrigation system. Pakistan is rich in wildlife, including over 6,000 species of plants, 2,000 of which are medicinal plants. The country has extensive natural gas reserves. Arable land comprises about 36 percent of the area.

The country's extremely high population growth rate, exceeding 3 percent, intensifies and contributes to the environmental problems. Water pollution—due to inadequate sewage facilities, chemical pollution from industry, and pesticide runoff—is a major concern, and only about half the people have access to safe water. Overgrazing, overharvesting, and erosion caused by wind and water are responsible for land degradation; and the limited forest resources are shrinking by more than 1 percent a year due to fuelwood collection and clearing for agriculture. Coastal pollution, due to oil washing ashore from other countries, is another major concern.

🏛 Government

Pakistan Meteorological Department ⊕
Meteorological Complex, P.O. Box 8454
University Road
Karachi 75270, Sindh, Pakistan

Contact: **F.M. Qasim Malik**, Director-General
Phone: (92) 21 476661 • *Fax:* (92) 21 470527

Activities: Research • *Issues:* Global Warming

▲ Projects and People

The Pakistan Meteorological Department collects long-term temperature and rainfall data of the country's various meteorological observatories. According to director-general F.M. Qasim Malik, "Pakistan is rich in long-term meteorological data," including data covering 100 to 130 years from 16 to 20 observatories.

Malik reports that "the long-term temperature and rainfall variations are being studied using trend-line method as well as decadal averages method. Temperature variation of five principal cities of Pakistan have been studied and compiled into a paper, 'Climatic Change and Temperature Rise in Pakistan,' a summary of which has been produced in *Climate System Monitoring (CSM) Monthly Bulletin* (No. 4, April 1990, World Health Organization, Geneva)."

Malik writes that, based on his Department's data of more than a century, a computerized database is being created "which when completed will be helpful in the study of man-caused changes in temperature and rainfall and [in assessing] future effects and make future plans to stop or decrease such atmospheric changes which are capable of creating havoc in the life of man." The World Health Organization (WHO) will be the recipient of these studies.

An "underdeveloped Third-World country with limited financial resources," Pakistan can nonetheless contribute to the benefits of mankind through publication of its long-term studies, says Malik.

■ Resources

Available publications include *Rainfall Variability and incidence of Drought over Pakistan, 1989*; *Climate Change and Temperature Rise in Pakistan, 1989*; and *Rainfall Seasonality and Incidence of Droughts over Pakistan, 1990*.

 # NGO

The Scientific and Cultural Society of Pakistan
B-7, Sheet No. 25 Model Colony
Karachi 75100, Sindh, Pakistan

Contact: **Muhammad Zaheer Khan**, President, Founder
Phone: (92) 21 409336 • *Fax:* (92) 21 409336

Activities: Development; Education; Research • *Issues:* Biodiversity/Species Preservation; Global Warming; Remote Sensing; Waste Management/Recycling; Water Quality

● Organization

Established in 1977, **The Scientific and Cultural Society of Pakistan** is an independent NGO with memberships in the World Conservation Union (IUCN) and the International Federation of Organic Agriculture Movements (IFOAM). Among its activities, the Society, with a membership of 4,000, arranges seminars, conferences, and workshops in its fields; promotes education, research, and awareness on the conservation of nature and natural resources; interests the younger generation in science and technology; and maintains youth welfare and cultural centers and a library.

Among its eight branches are a Wildlife and Environmental Science Division and a Biotechnological Division. Included in its specialties are youth and public environmental education, conservation and wildlife education, and conservation ethics.

▲ Projects and People

With only 5 percent of Pakistan's land forested, the Society launched a seedling program during Earth Day 1990, bringing its message to schools and colleges. Its environmental education theme on posters and other materials proclaimed, "Who Says You Can't Change the World." It also organized a **Natural and Cultural Exhibition** for young people and a **national seminar** on "Universal Problems and Environmental Management" with the Pakistani government. On Earth Day 1991, seeds and plants were distributed throughout Pakistan, among other activities.

To stem pollution, the Society notes that "25 percent of any country should comprise forests." **Muhammad Zaheer Khan** also points to marine and freshwater pollution that is increasing due to industrial waste dumping. Khan wants to see a "functional" Institute of Environmental Studies, Karachi University; a "functional" Regional Transport Authority to control auto emissions; and more cooperation among NGOs, government, and the people.

At a Society-sponsored scientific session, **S.H. Niaz Rizvi**, National Institute of Oceanography (NIO), described the consequences of "sea level rise and coastal environmental problems in South Asian Seas (SAS) region," which are related to global warming. This rise could cause coastal inundation, such as "loss of land, destruction of wetlands, coral reefs, and mangrove swamps; . . . flooding of low-lying areas"; accelerated coastal erosion, including the loss of recreational beaches and resort islands; saltwater intrusion contaminating freshwater aquifers and upstream waters; drainage and irrigation system problems; coastal water production changes; and threats to port facilities, storm barriers, and other coastal structures.

The session underscored that more than "one billion people live in SAS region." Of these, some 10 percent live along the coasts and deltas of the Ganges, Brahamputra, Iravady, Indus, and other major rivers—with this population largely dependent on coastal resources and amenities. With Maldives and Bangladesh particularly vulnerable in the SAS region, scientists are urging programs of research and environmental data collection to formulate measures for better management and mitigation of sea-level rise impacts.

Remote sensing technology as an environmental monitoring tool, organic pest control, and wildlife conservation and management are the subjects of other Society workshops and seminars. The purpose of a 1991 trip to the American Museum of Natural History, New York, USA, was to discuss the proposed **Scientific Museum of Natural History, Karachi.**

Khan, who writes that environmental problems in Pakistan are "very serious," believes they can be solved by "sharing the expedient views with other environmentalists. The international scientific cooperation can make the earth an Eden." Khan, a doctoral research scholar, is working on a toxicology project with mosquitoes and is co-investigator of studies on ecobiology and population dynamics of Marco Polo sheep and snow leopard in Khunjerab National Park. Principal investigator of the latter project is **Dr. S.N.H. Naqvi**, zoology professor, University of Karachi, who is the Society's chairman of the board of advisors.

■ Resources

The Society seeks technologies in molecular biology to train young scientists; broader contact with environmentalists, toxicologists, and wildlife scientists worldwide; and expanded environmental education and training for the country's scientists.

Pakistan Programme ⊕
The World Conservation Union (IUCN)
1 Bath Island Road
Karachi, Sindh, Pakistan

Contact: **Aban Marker Kabraji**, Country Representative
Phone: (92) 21 573079 • *Fax:* (92) 21 530976

Activities: Development; Education • *Issues:* Biodiversity/Species Preservation; Deforestation; Sustainable Development
Meetings: CITES

● Organization

Headquartered in Gland, Switzerland, the **World Conservation Union (IUCN)** was founded in 1948 and is the "world's largest and most experienced alliance of active conservation authorities, agencies and interest groups. Its 636 members include states, government departments, and most of the world's leading independent conservation organizations in some 120 countries," according to its literature. *(See also World Conservation Union [IUCN] in the France, Kenya, Niger, Switzerland, USA, and Zimbabwe NGO sections.)*

To implement its World Conservation Strategy, published in 1980, IUCN established the Conservation for Development Centre (CDC) a year later—now integrated with the IUCN Secretariat as its Field Operations Division. Its keystone is "Sustainable utilization of natural resources yields greater benefits than exploitation for short-term gain." The Division operates with a staff of fewer than 20 and "coordinates a worldwide network of consultants and experts as well as the activities of Regional Offices in Costa Rica, Kenya, Pakistan, Senegal, and Zimbabwe, and programmes in the Caribbean, and with the South Pacific Regional Environment Programme (SPREP)."

An independent country since 1947, Pakistan is the second largest land area of the countries that resulted from the break-up of the British Empire in South Asia. In the early 1970s, IUCN began working with Pakistan; in 1975, the country became a state member with the Ministry of Food and Agriculture as the focal point. Since then, various NGOs and other government institutions have become IUCN members. A project office that opened in 1985 gained the status of a country office three years later; Mrs. Aban Marker Kabraji is its head. Here, Pakistan and IUCN work as partners. Among closely working NGOs are World Wide Fund for Nature (WWF)—Pakistan and Shirkat Gah, a women's development group. The Journalists' Resource Centre for the Environment is an "autonomous, in-house media unit."

▲ Projects and People

One of a number of countries preparing a National Conservation Strategy (NCS), Pakistan has as its underlying theme: "Though careless, often destructive use, creates a problem or degrades a resource, thoughtful, appropriate, sustainable use can rehabilitate the resource, turn destruction into reconstruction and frequently generate high levels of productivity on a sustainable basis." Following the five-year planning process that featured draft reviews, workshops, public hearings, and village meetings—with support from the Canadian International Development Agency (CIDA) and the United Nations Development Programme (UNDP)—implementation was expected to begin after federal Cabinet approval in 1991.

With a population of 107 million that is expected to reach 200 million in about the year 2015, Pakistan is concerned that "our supporting capacity is not sufficient to take care of this rapid" growth. Consequently, a future population-environment program in Pakistan is being incorporated into the NCS.

A Department of Environmental Planning and Management (EPMP) has been established at the University of Peshawar, with IUCN support and start-up funding from the Norwegian Agency for International Development (NORAD), to design a two-year course to train environmental managers and planners, link up with foreign universities and teaching programs in fellowship exchanges, set up a library, and prepare a short-term training package for EPMP faculty.

A proposal for the establishment of a National Wildlife Institute for Pakistan is awaiting approval. The country's diversity includes some "5,700 wild plants, 188 mammals, over 666 birds and 166 reptiles, a number of which are not found elsewhere." Yet, habitat destruction such as deforestation and wetlands drainage, hunting, and competition from livestock are taking their toll—with numerous species endangered and some already extinct. Even with Pakistan's 8 national parks, 74 wildlife sanctuaries, and 72 game reserves, the Institute is needed to educate wildlife managers at undergraduate and graduate levels and sustain in-service training for professionals. The government's National Council for Conservation of Wildlife (NCCW) is a partner in this request.

With the Pakistan Forest Institute (PFI), the Pakistan Programme is promoting courses to train women in forestry, particularly short courses for women trainers involved in Women in Development (WID) portions of rural development projects. Funding is sought to implement the training course curriculum.

In 1986, IUCN began working with the Aga Khan Rural Support Programme (AKRSP) in northern Pakistan to help Village Organizations (VOs) build "forestry management capability" through training, extension, and research. This Sustainable Forestry Programme, expected to extend to 1995, has a goal of reaching some 1,350 village organizations with special attention to training Village Forestry Specialists (VFSs) for providing support to "VO tree planting and management and applied research to enhance village forest productivity." Even though the present demand for firewood and other forestry products outstrips production, successful irrigation and steps to conserve and develop natural resources in watersheds can bring about perennial cropping. Future training materials for VOs will include economic analyses and other information on forestry investment in the form of leaflets, calendars, and manuals.

Also with AKRSP, a pilot pasture management project for three degrading pasture areas in the Hunza Valley, northern Pakistan, is investigating "adapted grazing systems, winter fodder production, hydrological research, support to research in breed improvement, and integration of these into an appropriate socioeconomic framework."

Under pressure from pollution, sprawling urbanization, and industrial development, the Korangi–Phitti Creek Coastal Area is the subject of a Management Plan to protect the ecosystems. "Here, one finds power plants adjacent to villages, factories discharging effluent to fishing grounds, port developments next to camel feeding areas, and stands of mangrove timber within reach of shanties," reports IUCN. With the cooperation of various government agencies and NGOs, studies assessed pollution sources and produced surveys on a marine pollution baseline, public health of creekside villagers, and socioeconomics of villages. An inventory of natural resources is being compiled and "mangrove conservation/restoration work" is being initiated.

Other coastal surveys are the basis for a proposed management plan for the endangered green turtle (*Chelonia mydas*) and threatened olive ridley turtle (*Lepidochelys olivacae*). Sea turtle populations "of international importance" are said to exist along the Sindh-Baluchistan coast, with nesting beaches used by an estimated "2,000–3,000 females annually." The Sindh Wildlife Management Board (SWMB), WWF, and local government offices are cooperating in this effort, which would be incorporated into an overall Coastal Zone Management Plan for Pakistan and develop trained personnel.

The islands of Bhaba, Bhit, and Shams Pir, west of Karachi, were the subject of a Shirkat Gah–sponsored feasibility study to protect mangroves through the proposed establishment of village-run mangrove cooperatives.

As soon as funds are raised, an Integrated Resource Management program of the Ziarat, Loralaie, and Sibi Juniper (*Juniperus excelsa*) forests in Baluchistan is expected to be underway to repair this damaged natural resource and promote sustainable development of the watershed. The rare juniper, which is very slow growing, provides "watershed protection, construction timber, thatching, poles for fencing and other uses, firewood being the only source for cooking, heating, and grazing." A significant vegetation cover, "its disappearance will have severely adverse consequences such as changes in microclimate, shortage of water, which is already a scarce commodity, and on all other socioeconomic benefits," reports the Pakistan Programme.

IUCN is cooperating with the Orangi Pilot Project (OPP), an NGO concerned with development and welfare issues in Pakistan's largest squatter settlement, the *katchi abadi*, with a population of one million people. OPP is cited as a model worldwide in addressing sanitation, health, housing, and income problems in such slum development. At OPP's request, IUCN initiated a feasibility study for a Social Forestry program to grow trees in urban surroundings such as in courtyards, banks of creeks ("open sewers"), or on communal property. Results were incorporated into a detailed proposal about tree-related activities including research, training, and the launching of an OPP-run nursery with local input on choice of trees.

Subject to recent drought, the Sindh Arid zone, which has supported a livestock population of more than 4 million animals, is "under severe pressure and its productivity is declining." Working with the Sindh Arid Zone Development Authority (SAZDA), a series of studies, including an environmental impact assessment (EIA), is being carried out to help restore and sustainably develop the desert ecosystem. Also, a demonstration project is being proposed through the NCS to show farmers how to reclaim waterlogged and salt-affected land—estimated to be about 40,000 hectares of cultivated areas.

Improved management planning is underway for the Khunjerab National Park in northern Pakistan, particularly to address tourism pressures; endangered species suchas Marco Polo sheep, snow leopard, Tibetan wild ass, and their habitat; and livestock rearing practices of local people.

A rural development project—underway in the district of Swabi, Mardan—is integrating the concepts and workings of social forestry—at the request of local villagers who expressed their need for "greater access" to trees, fuelwood, fodder, and other tree products. Collaborators are the Pakistan–German Integrated Rural Development Programme (IRDP) and the government of the North West Frontier Province (NWFP). The training of women forestry extensionists is a component.

To encourage local media coverage, public awareness, and government support, the Journalists' Resource Centre for the Environment (JRC) has been created. The documentation and publication resource center features lectures and journalists' workshops and encompasses a library of print materials, press clippings databank, photographs, video films, educational posters, and teaching kits. Reporters are commissioned to investigate environmental problems. JRC is also working to place environmental films on prime-time television, with help from the nonprofit Television Trust for the Environment, UK.

Working with the Teachers' Resource Centre (TRC) and Aga Khan Education Service (AKES), an Environmental Awareness and Education Programme is underway to define "environmental education" and to work through adult literacy projects, health and family planning extension programs, teacher training, improved teaching material, and the media. JRC and the Allama Iqbal Open University (AIOU) are also involved.

Pakistan and Zambia are two countries selected by IUCN to pilot a series of popular publications to reach decisionmakers and the educated public who could help implement the NCS. *The Nature of Pakistan*, a 72-page publication, is one result.

The Pakistan Natural Resources Expertise Profile (NREP) identifies institutions and people who are in-country experts in sustainable development, environmental matters, and natural resource management. Pakistan's publication also demonstrates to the development assistance community that such expertise is available to promote sustainable development. IUCN has also helped produce NREPs for Kenya, Zambia, Zimbabwe, Botswana, Tanzania, and Sri Lanka.

The Pakistan Programme also sponsored an India-Pakistan Conference on the Environment preceded by a Pakistani Women's Study Group Tour to observe grassroots environmental movements in the Indian Garhwal-Himalaya region, birthplace of the "Chipko" movement, for transfer of such knowledge in their country. The purpose of the conference, co-hosted by the Centre for Science and Environment (CSE) of India, was to "establish a communication system on the environment between India and Pakistan; provide a forum which would encourage joint ventures; and initiate the establishment of a regional network."

IUCN's Environmental Law Services was introduced in 1990 in Pakistan to initiate and undertake EIAs of existing and future development projects and to help the government draft environmental laws and review and update existing environmental legislation.

■ Resources

IUCN produces an information kit on *The Pakistan Programme*, including a publications list. The Journalists' Resource Centre for the Environment produces media materials.

World Wide Fund for Nature—Pakistan (WWF) ⊕

Ali Industrial Technical Institute
Ferozepur Road, P.O. Box 5180
Lahore, Punjab, Pakistan

Contact: **Dawood N. Ghaznavi,** Director
Phone: (92) 42 856177 • *Fax:* (92) 42 370429

Activities: Education; Development; Research • *Issues:* Air Quality/Emission Control; Biodiversity/Species Preservation; Deforestation; Health; Protected Areas; Sustainable Development; Water Quality; Waste Management/Recycling *Meetings:* UNCED; World Bank; WWF

▲ Projects and People

Dawood N. Ghaznavi, director, World Wide Fund for Nature—Pakistan (WWF), writes, "The road to future global cooperation lies in NGOs throughout the world cooperating on joint projects, exchanging relevant information, showing skills and methods developed and successfully used in one region with others. To me, living in the industrializing world, the New World Order is really ultimately an NGO-centered system of human social organization."

■ Resources

WWF-Pakistan, with a staff of 26 and 1,000 members, publishes *Natura*, quarterly English magazine; *Qudrat*, quarterly Urdu magazine; *Mithoo Begum*, quarterly Urdo comic book; *Tree Kit*, a teacher's guide; and *Alphabet Book/Coloring Book* on local birds and mammals. It seeks geographic information system (GIS) technologies to improve protected area planning techniques, political support for lobbying and public action campaigns, and training in environmental impact assessment (EIA) techniques and CITES monitoring methods.

 *Universities*

Herp Laboratory
Talimul Island College
15/6 Darul Saddar North
Rabwah 35460, Pakistan

Contact: **Muhammad Sharif Khan,** Associate
            Professor of Zoology
Phone: (92) 452 4919

Activities: Education; Research • *Issues:* Biodiversity/Species Preservation

▲ Projects and People

Prof. M.S. Khan reports that the **Herp Laboratory** specializes in the study of herpetofauna, presently concentrating on Pakistan "since much of the land is yet herpetologically unknown." The aim is to "record herpetofauna as accurately as possible," region by region. Surveyed so far are the Potwar Plateau and Baluchistan. Dr. Khan is now collecting herps in the D.G. Khan and Rajanpur districts; a future three-year field survey will take place in northern Pakistan. The Laboratory also encourages students in other universities and colleges to learn about herps.

A member of the Species Survival Commission (IUCN/SSC) Indian Subcontinent Reptile and Amphibian Group, Khan is writing a book in Urdu on Pakistani snakes and studying the taxonomy and distribution of geckos. He has prepared checklists on snakes and reptiles and published articles in journals worldwide on toads, lizards, tiger frogs, amphibian tadpoles, pit vipers, cobras, and snakes of medical importance.

■ Resources

Herp Laboratory seeks funding to augment Prof. Khan's underwriting of projects and to attend herpetological meetings. Offered free of charge are information and literature on Pakistani herpetofauna.

Institute of Public Health Engineering and
  Research (IPHER)
University of Engineering and Technology
Lahore 54890, Pakistan

Contact: **Mohammad Nawaz Tariq, Ph.D.,** Director
Phone: (92) 42 339248

Activities: Education; Research • *Issues:* Air Quality/Emission Control; Global Warming; Sustainable Development; Waste Management/Recycling; Water Quality; Water/Wastewater Treatment/Disposal

● Organization

Twenty-five years following Pakistan's creation, the **Institute of Public Health Engineering and Research (IPHER)** was brought into being in 1972 with support from the government, United Nations Development Programme (UNDP), and World Health Organization (WHO). Later, the World Bank, UNICEF, and the U.S. Agency for International Development (USAID), among others, collaborated in IPHER's development. An outgrowth of West Pakistan's Department of Public Health Engineering and the Division of Public Health Engineering in the Civil Engineering Department, **University of Engineering and Technology**, the Institute was the consequence of the need for qualified engineers to deliver water supply and sanitation as the country experienced rapid urbanization and industrialization.

With a staff of 29, the Institute offers a postgraduate program in water and wastewater engineering, solid waste management, air pollution control, and related fields. The library contains at least 2,000 works on environmental and public health engineering. Laboratories are used for sanitary microbiology, sanitary chemistry, unit processes, air pollution, and solid waste studies. In addition, commercial testing of water and wastewater samples is undertaken.

▲ Projects and People

The Institute is presently participating in a three-year joint research project on **air-quality monitoring** in Pakistan with funding from the Commission of the European Communities (CEC) and the participation of the University of Birmingham, UK, and University of Aveiro, Portugal.

Also underway are "coagulation studies in relation to **water filtration** as well as pre-treatment studies in relation to **slow sand filtration**." As part of the **Global Environmental Monitoring Systems (GEMS) Water/Air Project**, water quality is monitored at seven sites with information transmitted to WHO.

A 10-year **Oxidation Pond Research study**, with USAID support, resulted in various research reports and helped in "developing design guidelines for this low-cost wastewater treatment technology under local conditions," according to the Institute. Working with the Water and Sanitation Agency of Lahore, the Institute undertook a two-year study on **Ravi River pollution**. A model now indicates the "relationship of the assimilative capacity of the river and the degree of treatment needed for the wastewater from the city before disposal."

Another two-year study, in conjunction with UNICEF, developed a "**deep well handpump model** suitable for village level operation and maintenance in rural areas." Working with the World Bank, the Institute helped develop "low-cost water supply and sanitation modules for use as training materials in developing countries."

The Institute holds **seminars and symposia**, sponsors short-term **training programs and workshops** for professionals, and **advises national organizations on environmental engineering**. Its reports have influenced the World Conservation Union's (IUCN) National Conservation Strategy for Pakistan, for instance. The Institute expects that it must develop further to meet the country's needs, to learn about recent advances in developed countries, and to expand its library.

Dr. Mohammad Nawaz Tariq, also professor of Public Health Engineering, helped to establish the Institute. He is an advisor to Pakistan's National Planning Commission; Ministry of Environment and Urban Affairs; and Lahore's Water and Sanitation Agency. He has also served as consultant to the World Bank in Indonesia and Korea and to the Asian Development Bank.

■ Resources

The Institute publishes *Public Health Engineer* journal. Its brochure includes a publications list.

Panama's location, which provides a path between the Atlantic and Pacific oceans, is intricately tied to its destiny. The Panama Canal, which opened in 1914, brought a degree of prosperity to the Canal Zone and Panama's larger cities.

The country is home to a variety of habitats, including tropical rainforests, upland forests, savannahs, and coastal mangroves—providing a high degree of biodiversity. Unfortunately, these habitats are endangered by colonization projects intended to accommodate the country's rapid population growth. Increased production of bananas, coffee, sugar, and other products for export is resulting in deforestation and wetlands destruction, which in turn is causing major soil erosion.

In addition, agricultural runoff, construction of commercial shrimp ponds, and recreational development are threatening marine fishery habitats. A lack of public awareness and understanding, as well as the absence of a central conservation authority, are hampering efforts to address these problems.

🏛 *Government*

Comision Nacional del Medio Ambiente (CONAMA)
National Environmental Commission
Ministerio de Planificación y Politica Economica
Ministry of Economic Planning and Policy
APDO 10120, Zona 4
Panama, Panama

Contacts: **Jorge A. Carles**, Director
Maria Daisy de Gracia, Librarian
Phone: (507) 32 6055, ext. 13 • *Fax:* (507) 69 6822

Activities: Development; Education; Political/Legislative • *Issues:* Air Quality/Emission Control; Sustainable Development

● Organization

CONAMA was created in 1985 and a year later was incorporated into the **Ministerio de Planificación y Politica Economica (Ministry of Economic Planning and Policy)**. Its four principal functions are: advising the executive branch of the Panamanian government on matters concerning the protection, improvement, and defense of the environment and natural heritage of Panama, including the environmental impact of state projects; developing environmental strategies and projects in cooperation with other governmental institutions; coordinating the execution of those strategies and projects approved by Panama's executive body; and obtaining technical cooperation from international groups for environmental projects.

Currently, CONAMA has 12 senior staff members; their specialty is environmental management—an activity that is multisectorial in character; consequently some environmental functions are assigned to preexisting organizations, including regional ones. A catalyst, CONAMA promotes interinstitutional cooperation at various levels in response to specific environmental problems. At the national level, CONAMA is engaged in planning, legislation, education, statistics, and cleanups.

CONAMA, the Marine Resources Directorate, and the Renewable Natural Resources Institute are the government's major environmental management organizations. Likewise, NGOs are growing and cooperating with these agencies on matters of environmental protection and conservation.

Despite the recent activity, according to CONAMA, a lack of coordination and cooperation among some institutions has had a "grave impact" on Panama's natural and human resources—due in part to the lack of an adequate legal framework for environmental management and protection.

Often, environmental considerations are not incorporated into social and economic development planning. Also, environmental management and conservation are too often directed toward immediate human well-being; CONAMA says the "ultimate goal must be to assure the sustainability of our natural resources."

▲ Projects and People

In an effort to correct these problems, CONAMA has prepared a National Plan for Environmental Protection and Improvement, which is an integral part of the National Development Plan. It advocates the increased use of legal measures and environmental impact planning and focuses on 19 environmental problems, including **saltwater and freshwater pollution, solid waste disposal,** management of **hazardous substances, population growth, deforestation** (particularly in the Panama Canal basin), **soil erosion,** and protection of natural heritage.

CONAMA research is concerned with **climate change, pollution sources, landfills, oil spills, marine pollution, transport of hazardous materials, wetlands,** and improving the environment.

■ Resources

CONAMA's Information Center contains materials on the ozone layer, environmental law and education, tropical ecosystem, biosphere reserves wetlands, desertification, conservation of natural resources, endangered species, pesticide use, and pollution, for example.

Marine Resources Directorate
Ministry of Commerce and Industries
Recursos Marinos, APDO 3318
Panama 4, Panama

Contacts: **Gusavo Justines Aragón,** Deputy Director, Biologist
Epimenides Manuel Diaz Cedeno, Coordinator, Oil Pollution Studies, Scientific Observer, Inter-American Tropical Tuna Commission
Phone: (507) 27 4691 • *Fax:* (507) 27 3104

Activities: Development; Education; Law Enforcement; Political/Legislative; Research • *Issues:* Biodiversity/Species Preservation; Sustainable Development • *Meetings:* Fisheries/Environment Regional

■ Resources

The **Marine Resources Directorate** offers yearly subscription to the *Fishery Marketing Bulletin* (*Info Pesca*) and limited numbers of the *Fishery Statistical Bulletin,* printed in Spanish. The Directorate seeks to expand its educational outreach on coastal zone management and integrated management of coastal resources.

NGO

Asociación Nacional para la Conservación de la Naturaleza (ANCON) ⊕
National Association for Nature Conservation
APDO 1387
Panama 1, Panama

Contact: **Juan Carlos Navarro Q.**
Phone: (507) 64 8100 • *Fax:* (507) 64 1836

Activities: Development; Education; Research • *Issues:* Biodiversity/Species Preservation; Deforestation; National Parks Conservation and Management; Sustainable Development

● Organization

A group of prominent scientists and business and community leaders founded the nonprofit ANCON in 1985 to "protect and conserve Panama's natural resources and biodiversity for present and future generations of Panamanians." To this end, ANCON, with its 2,000 members and staff of 41, works closely with many other public and private organizations.

▲ Projects and People

At the 50,000-acre Soberanía National Park, where some 400 bird species and 1,000 plant varieties thrive, ANCON helped "fence the park's 40 kilometer eastern boundary [to halt encroachment and deforestation], **posted signs** along park boundaries, **trained and equipped the park's rangers,** and launched an **environmental education campaign** to enlist local communities in the park's protection." Noted for helping protect the Panama Canal Watershed, the Soberanía is also a key site for scientific research, tourism, and recreation.

At the 300,000-acre Chagres National Park that protects the Alajuela Lake Watershed and upper Chagres River—major sources of drinking water and about half the fresh water needed to operate the Panama Canal—ANCON has planted 100,000 trees in launching its **reforestation project** along the border of Chagres and Soberanía, trained park rangers, provided equipment, and "reinforced the limits of both parks, while providing **employment** for over 700 local inhabitants."

The **Rio Cabuya Agroforestry Farm Project** also helps preserve the Panama Canal Watershed through a long-term reforestation, conservation, and protection program in which neighboring communities are trained in agroforestry techniques.

To make way for 100,000 seedlings of teak, albicia, melina, acacia, eucalyptus, cedar, mahogany, oak, nin, and fruit trees, ANCON cleared 64 hectares of elephant grass–covered land in 1990. Seeds from these trees will be used to reforest other areas and to create a genetic bank of native species. At the same time, data from experiments with the growing conditions of various tree species is expected to improve agroforestry techniques throughout the Panama Canal Watershed.

In addition, some 600 temporary jobs were generated; work and training continues in agroforestry, fertilization, and natural pest control. **Sustainable agriculture** techniques are also used in planting corn, rice, guandu, and other crops. Demonstration projects at Rio Cabuya include captive breeding of the **green iguana** (*Iguana iguana*) and the spotted paca (*Aguoti paca*) are aimed at developing economic and food production alternatives for local families. The U.S. Agency for International Development (USAID) provides support.

For Darién National Park, 1.5-million-acre virgin rainforest preserve that is also a World Heritage Site and a UNESCO's Man and the Biosphere Reserve, ANCON helped to prepare a management plan, train and equip park rangers and maintain their stations, fence boundaries, and post signs.

According to ANCON, it was the catalyst for the creation of La Amistad (Friendship) International Park—a binational initiative between Panama and Costa Rica. At the Park, covering more than two million acres of highlands, ANCON constructed the first sentry posts, marked the boundaries, and equipped the park rangers.

ANCON also helped create Bastimentos National Marine Park, composed of "30,000 acres of forested islands, coastal lagoons, mangroves, sandy beaches, and coral reefs, including primary nesting sites for four species of endangered marine turtles." Here, ANCON conducts studies to monitor and protect marine turtles and environmental education and community support programs.

ANCON's Science Division maintains a geographical database of Panama's flora and fauna that helps improve park management.

With Juan Carlos Navarro Q. as executive director, other staff members are Alexis Pineda, administration director; Carlos Brandaris M., conservation director; Madeleine Lescure, environmental education and development director; Graciela Palacios S., science director; Raúl Fletcher, deputy director, conservation/sciences; and Oscar Vallarino B., deputy director, environmental education.

Asociación Centro de Estudios y Accion Social Panameño (CEASPA)
Panamanian Center for Research and Social Action
APDO 6-133, El Dorado
Panama, Panama

Contact: **Charlotte Elton**, Research Economist
Phone: (507) 26 6602 • *Fax:* (507) 26 5320
E-Mail: Internet: uunet!huracan!ceaspa

Activities: Development; Education; Research • *Issues:* Health; Human Rights; Sustainable Development • *Meetings:* Latin American Studies Association; UNCED

▲ Projects and People
CEASPA's new project, Political Economy of Sustainable Development in Central America, aims to "identify and promote development alternatives in Central America and Panama that will both satisfy basic human needs of the impoverished majorities of the region and stem the ongoing degradation of the natural resource base."

Building on advances in analysis concerning "current structural adjustment policies and how they affect the popular sector as well as experiences existing in the region, the research and action project is being implemented under the auspices of CRIES, Coordinadora Regional de Investigaciones Económicas y Sociales (Regional Coordinator of Economic and Social Research), Managua, Nicaragua, to

which CEASPA belongs. Activities will include "inventory and multiplication of alternative sustainable agricultural experiences in the region, in addition to policy analysis," according to Charlotte Elton, economist and research coordinator.

In general, CEASPA is concerned with the quality of human life, human rights, poverty, and social organization. CEASPA played a national role in preparing for the United Nations Conference on Environment and Development in 1992.

It also participated with organized groups in "fighting the exploitation of natural resources in Indian lands and national park areas of Panama." Elton writes, "We lobby for the rights of local people to be consulted and to be the beneficiaries of projects, and that environmental safeguards be included from the conception stage of the project."

One such "unsustainable project" that CEASPA fought was oil exploration and exploitation by Texaco in the province of Bocas del Toro in the Guaymi Comarca and La Amistad National Park, a World Heritage Site, and frontier park with Costa Rica, candidate to be part of a biosphere reserve. "The contract became effective in 1991," reports Elton. Earlier, CEASPA worked with the Guaymi Indians and allies concerning the Cerro Colorado copper mine development planned in Panama.

Elton, who was an Advanced Research Fellow of Harvard University's Program on USA–Japan Relations, has worked with the United Nations Development Programme (UNDP), Organization of American States (OAS), University of Panama, Panama Canal Commission, and Natural Renewable Resources Institute of Panama (INRENARE). She is currently a research associate with the University of Panama's Institute of National Studies (IDEN).

■ Resources
El Pueblo Guaymi y Su futuro (The Guaymi people and their future) is an available publication. Elton's papers include "Japan and Panama: Who Is Setting the Agenda?" and "Japan's Natural Resource Strategies in Latin America and Their Environmental Impact."

CEASPA seeks political support and international solidarity regarding "unsustainable" foreign investments and a bibliography on natural resource and/or sustainable development economics.

Center for Propagation of Endangered Panamanian Species
Box 2026
Balboa, Panama

Contact: **N.B. Gale, D.V.M.**, Director
Phone: (507) 526614

Activities: Research • *Issues:* Biodiversity/Species Preservation; Endangered Species Propagation

Fundación Dobbo-Yala/Dobbo-Yala Foundation 🌿
APDO 83-0308, Zona 3
Panama, Panama

Contact: **Geodisio Castillo Diaz,** Executive President
Phone: (507) 276022, ext. 330
Fax: (507) 625942

Activities: Development; Education; Research • *Issues:* Biodiversity/Species Preservation; Deforestation; Indigenous People; Population Planning; Sustainable Development

● Organization
Fundación Dobbo-Yala was formed in 1990 by professionals of indigenous descent to conserve and make use of the environment while promoting cooperation and self-sufficiency among the indigenous communities of Panama. *Dobbo* (in the Guaymi language) and *Yala* (in the Kuna language) stand for the land, natural resources, and the environment. With its motto, "Through the development of indigenous peoples and conservation of the environment," the Foundation focuses on activities of environmental empowerment and the development of sustainable agriculture. It receives support from World Wildlife Fund—USA (WWF) and Cultural Survival.

▲ Projects and People
Projects in Sustainable Agroforestry Extension (in the community of Ukubseni in Kuna Yala District) and Environmental Education (also in Kuna Yala) have been developed to this end. Both projects are an attempt to "start small" with modest technological innovations or small-scale experiments. Another significant project is the Strategy for the Legalization of the Kuna District because it focuses on the sustainable development and conservation of a river basin.

The **Environmental Education Project** informs instructors about environmental problems in the Kuna Yala area and teaching techniques in this field. Teaching materials for elementary schools are being prepared so that a program about marine and terrestrial resources can be carried out. According to the Foundation, the program's success depends upon the cooperation and participation of institutions involved in formal and nonformal education, and that of traditional leaders through teaching seminars and interpersonal, group, and community communication.

The Environmental Education Project forms an integral part of the **Sustainable Agroforestry Extension Project** because it centers on the "practical empowerment of educators within the context of the realities of the Kuna District." The Extension Project attempts to revive the traditional Kuna knowledge of the use and management of natural resources: medicinal plants, food, crafts, and conservation of the environment, among other aspects. Additionally, discussion groups are being held, with slides and films on environmental problems and alternatives for school children and others involved in the project.

A main objective is to "reclaim agriculture as a valid and attractive alternative" that can improve the quality of life through cooperation with farmers and religious and political leaders. For the Kuna community, cultural development and economic development go hand in hand, reports the Foundation, as efforts are made to raise the local people's self-esteem and their awareness of their role in history and in current society. Thus, the Extension Project is attempting to reintroduce indigenous techniques of soil conservation, such as the use of "green" fertilizers like mulch, and the cultivation of bananas and plantains interspersed with corn and sugarcane to prevent soil erosion on parcels of land.

Another important aspect is strengthening the administrative structure of farming organizations—without which sustainable agricultural development will fail. The Extension Project also includes the cooperation of women on matters of nutrition and hygiene, promoting basic knowledge such as the necessity to boil water to prevent the spread of cholera. The project also teaches school-age children how to run a tree nursery.

Regarding the **Legalization of the Kuna District of Madungandi,** the Foundation proposes that this present reserve, established in 1934, be elevated to the status of a district. It is located on 82,000 hectares in the eastern part of the country, and comprises Bayano Ascanio Villalaz Hydroelectric Dam, which provides energy for Panama City and other adjacent towns. Ironically, the energy produced so close to the Kuna does not reach their communities. It is reported that "development has made orphans out of the Kuna; they are left on a few infertile hillsides, where once they grew plantains and bananas in the rich soil of the Madungandi riverbanks." According to the Foundation, the construction of the dam was followed by that of the PanAmerican Highway and a government policy that extended the agricultural frontier. This led to the building of new towns, cutting of trees, and conversion of the land into pastures for livestock raising in areas close to and within the Kuna "reserve."

This project's goal is to define a strategy for legalizing the Kuna de Madungandi District to guarantee the conservation of the cultural and natural resources of the area and their sustainable development. A document is being prepared to outline viable alternatives for the area's development and to list steps leading to the district's legalization. In the meantime, cultural and legal information related to the sociopolitical status of the region is being compiled, and field research is being conducted on the natural resources and biodiversity of the Kuna territory. Success of this project depends on the participation of the Kuna people and recognition of the cultural and ecological richness of indigenous territories on national and global levels. "Management of natural resources and forest areas always has a cultural component, because it is here that human cultures are found," writes the Foundation.

According to executive president **Geodisio Castillo,** an agricultural engineer, the "Kuna people view the Fundación Dobbo-Yala as an organization which sooner or later must be successful. The improved quality of life of the people who participate in its projects is the greatest incentive to continue these activities on their own." Lessons learned can easily be transferred to other indigenous communities in Panama. "Ecological knowledge, creativity, and global vision" of indigenous people "can go a long way towards solving the world's environmental problems," believes Castillo.

Fundación para el Desarrollo de los Jóvenes (FUNDEJOVEN)
Foundation for the Development of Youth
APDO 87-1740, Zona 7

Panama, Panama

Contact: **Julio Iván Rovi Sanchez,** Area Coordinator
Phone: (507) 61 1520 • *Fax:* (507) 29 1051

Activities: Development; Education; Research • *Issues:* Deforestation; Energy; Health; Sustainable Development; Waste Management/Recycling; Water Quality

● Organization

Founded in 1990, **Fundación para el Desarrollo de los Jóvenes (FUNDEJOVEN)** (Foundation for the Development of Youth) is aptly named. Objectives are to promote programs that satisfy the needs of young people, particularly in the areas of productive work, leisure and free time, sports, and culture; foster the protection and improvement of the environment as a means of guaranteeing the future; provide training for the people involved in the Foundation's projects; research the problems faced by Panamanian youth and possible solutions; and encourage participation of young people in economic, social, and cultural life.

FUNDEJOVEN is divided into four main bodies: the board of trustees funds the organization and is responsible for suggesting and maintaining the direction of the programs; the board of advisors, which is made up of people distinguished by their service to the community; the board of directors, which is responsible for the Foundation's administration; and the area directors and program promoters, who run the Foundation's programs.

▲ Projects and People

Among its areas of interest are **ecotourism**—to promote the enjoyment of Panama's ecological riches; **agroforestry**—reforestation for sustainable use of the tropical forest; and **recycling**—to make use of resources and reduce waste.

FUNDEJOVEN aims to "decrease the high unemployment among urban youth in Panama City by creating jobs in recycling programs." The goal of the **Recycling Project** is to train young people to be "self-employed recyclers." As the program gets underway, youths will collect recyclable materials from homes and businesses to sell to factories and workshops. Young people themselves participate in all aspects from project design and administration to building recycling carts, forming recycling centers, and teaching community members to recycle.

While there is no official recycling program in Panama, FUNDEJOVEN has targeted the San Miguelito District, with its dense population of 243,000—about one-fourth the population of Panama City—and an unemployment rate of 15.6 percent. FUNDEJOVEN hopes to create 160 permanent jobs for youths as recyclers and more jobs at recycling centers and workshops in the future. Emphasis is on teaching people to run their own businesses and learning transferable job skills, such as carpentry, welding, financial management, and community relations—along with separation of recyclables. FUNDEJOVEN also encourages leadership skills and traits such as self-esteem, responsibility, and independence.

Improved sanitation, slowed deforestation, and reduced landfills are three community benefits. FUNDEJOVEN estimates that $50 million of materials are buried annually in landfills. Moreover, the government spends about $20 million a year to dispose of garbage in landfills. The Recycling Project plans to collect "40 tons of recyclables daily or 12,000 tons a year." Natural resources will be saved as recycled products are substituted for virgin or imported products. FUNDEJOVEN foresees that "metals are recast, scrap pieces of fabric are made into curtains and quilts, used containers are used as wall fillers in the construction of new buildings."

FUNDEJOVEN sponsors a **university internship program** for students who research topics related to recycling and volunteer to help the project. "FUNDEJOVEN welcomes international scholars and volunteers to study and assist in the success of this program and future FUNDEJOVEN projects," it reports.

Julio I. Rovi is area coordinator; **Carmen Ortíz** is on the executive board; **Ronda Mosley** is on the board of advisors; and **Lourdes Alvarado, Vasco Torres, Jazmina Rovi,** and **Julio Rovi Fong** are on the board of directors.

The independent state of Papua New Guinea is located north of Australia in the southwestern Pacific. It includes the eastern half of the island of New Guinea, as well as several smaller islands. The terrain includes a high mountain range, three large river basins, and 14 active and 22 dormant volcanoes. The high degree of rainfall, rich soils, moderate temperatures, and low population densities give the country much agricultural and economic potential. Earthquakes are another constant danger.

The country's extremely rich flora and fauna (including over 11,000 plant species, the world's largest and smallest parrots, and largest lizard, doves, and butterflies) are found in forests that cover 85 percent of the land mass. Yet many of the forests are falling victim to unsustainable logging, resulting from insufficient government management. The government, however, has initiated a new conservation management program that includes a national forest action plan. Large-scale mining projects also have caused serious environmental damage.

Government

Department of Environment and Conservation
P.O. Box 6601, Boroko
Port Moresby, NCD, Papua New Guinea

Contact: **John-Mark G. Genolagani,** Acting Officer in Charge
Phone: (675) 271791 • *Fax:* (675) 253589

Activities: Development; Education; Law Enforcement; Political/Legislative; Research
Issues: Air Quality/Emission Control; Biodiversity/Species Preservation; Environmental Assessments; Health; Pollution Monitoring; Sustainable Development; Water Quality
Meetings: CITES

NGO

Foundation for the Peoples of the South Pacific (FSP) ⊕
P.O. Box 1119, Boroko
Port Moresby, NCD, Papua New Guinea

Contact: **Joe Kelly Bik,** Program Director, Awareness Community Theater
Phone: (675) 258 470 • *Fax:* (675) 252 670

Activities: Development; Education; Research • *Issues:* Biodiversity/Species Preservation; Deforestation; Ecotourism; Fisheries; Global Warming; Health; Population Planning; Sustainable Development

● Organization
The **Foundation for the Peoples of the South Pacific (FSP),** a nonprofit organization founded in 1965, has innovative programs in Melanesian, Polynesian, and Micronesian nations that emphasize environmental issues such as ecoforestry, health and nutrition, child survival, income generation, and family food production. A partner with the government of Papua New Guinea and with World Wildlife Fund (WWF), Conservation International, Greenpeace, and the Rainforest Information Center in Australia, FSP is helping to preserve rainforest habitats.

▲ Projects and People
Helping indigenous people to maximize timber resources, the **Village Development Trust (VDT)** trains villagers in the use of portable sawmills (Wokabaut Somils), "light enough to mill a log where it falls, eliminating the need for bulldozers, roads and trucks which cause irreparable damage to the rainforests." With 500 locally owned and operated mills in PNG, the **Mobile Sawmill**

Program helps people gain jobs, maximum profits, and timber for their own buildings. VDT aims to discourage clear-cutting and nonsustainable harvesting methods by large foreign companies, who receive as much as $500 for a tree, while paying a local villager 62¢ for a tree.

"Village people sell their 'timber rights,' or forest resources, because they do not understand the full implication of deforestation." Through the sawmill training program offered at the Timber Institute Training College, Lae, and environmental awareness campaigns, villagers begin decisionmaking on sustainable use of their forest resources. One market opportunity is the Ecological Trading Company Ltd., UK; other markets are planned for the USA, Canada, Australia, and New Zealand.

The Grassroots Opportunities for Work (GROW) Project helps "farmers to establish low-input renewable agriculture systems designed to produce surplus food crops suitable for marketing." GROW combines land-use management plans with subsistence agriculture, pest and disease control using native plant and animal materials, and cash cropping for honey production, insect and butterfly farming, and agroforestry. A holistic community development program, GROW is designed to strengthen the family unit and role of women as well as emphasize nutrition and health. For example, training programs mandate that 50 percent participation is by women; some field workshops are geared to couples and youth. Banana sip is the result of one women's program, whereby "dried, processed banana chips are packaged and sold commercially in PNG." Vocational education helps overcome the alienation of rural youth. Training materials, with pictures and diagrams, are being prepared on nutrition, food processing, regenerative agriculture, and sustainable agriculture. Nutrition is also taught at community schools through puppet shows and recipes that use home-grown vegetables.

With an increased use of chemical fertilizer in some areas, GROW is designing an integrated pest management (IPM) approach. However, "we believe that when farmers are informed of the complexity of making dilutions, and the expense and discomfort of using adequate safety equipment, many will find the use of pesticides to be inappropriate," reports GROW. Moreover, many chemical pesticides exported to PNG are "strictly controlled or are outright illegal in their countries of origin."

GROW is testing various leguminous species in nitrogen-fixing hedgerows for long-term fertility of soil and water-conserving techniques, and various repellant and attractant plants in managing pest and beneficial insect populations, with some plants being processed as insecticidal or repellant sprays. Methods of alley cropping, mulching, water control, and composting enable farmers to permanently cultivate a single plot, easing pressure on rainforest resources. Fruit and legume seedlings uncommon to PNG are being made available to selected families. Volunteers from the U.S. Peace Corps, German Development Service, and United Nations are helping to expand GROW.

The Flora Conservation Project seeks funding to establish a gene bank and sterile laboratory for artificial propagation of endangered species of orchids and other flora. "By exporting flasked orchids, the only type which can be legally traded under the Convention on the International Trade in Endangered Species (CITES), and by distributing easily and inexpensively lab-propagated flower cuttings to local villages, FCP will be [central to establishing] a PNG floral industry involving the orchid export and domestic cut-flower markets." The Foundation expects that this project will be self-sufficient

in five years with profits beyond operational costs being "reinvested in other conservation and research projects and ancillary village development programs."

Some 20,000 flora species are said to be found in Papua New Guinea; yet these are being threatened by commercial mining, agriculture, large-scale logging, and other threats to the habitat, which is disappearing at a rate of more than 40,000 hectares annually. Plants of potential "commercial scientific and medicinal value" are at risk. And unregulated orchid collectors pose problems too. Since 1990, a government ban has been in place on the export of wild-collected orchids for an indefinite period; presently, only artificially propagated plants may be exported.

Wildlife Conservation International (WCI) and Research Conservation Foundation (RCF) of PNG are joining with FSP and the Department of Environment and Conservation to expand a WCI-launched wildlife and forest reserve in the Eastern Highlands Province (EHP) for ecotourism. "The project is supported by local people because they understand that there may be a long-term economic payoff from the protection of large tracts of forest land," reports FSP. At Crater Mountain Wildlife Management Area, "rather than locking up the land as a strict park, we aspire to create a 'living' reserve in which the landowners steadily increase their material well-being without sacrificing the natural resources."

To enhance ecotourism, FSP wants to construct a small airfield with village labor and global volunteers; improve a lodge ideally located for observation of birds of species; conduct a two-year analysis of the effects of tourism on the region; with GROW, create and stabilize sustainable agriculture, targeting the lodge as a major market for local garden products; and market forest products such as rattan cane (used in building and handicrafts), and okari and koroka nuts.

Focusing on the interactions between indigenous people and the forest, for more than 20,000 years, FSP plans to undertake doctoral research programs on forest dynamics and ethnobiology in mid-montane and sub-montane settings. Dr. Michael Hopkins, University of Papua New Guinea, will direct a training program of talented local student interns at Varirata National Park. *(See also Wildlife Conservation International, Dr. Hopkins, in this section.)* An ecotourism strategy will be prepared, which could include the development of a lowland jungle reserve, where no forests are presently protected. Dealing with the situation of some 400 language groups in PNG, a conservation awareness campaign will be launched to discourage, for example, the annual burning of grasslands and open habitats. FSP wants to create small-scale grants to fund local projects as a means to generate income and to encourage Papuans to foster biological diversity.

Through local performances, the popular Awareness Community Theatre promotes, for example, AIDS awareness and also the preservation of the Queen Alexandra butterfly, described by FSP as the world's rarest, and found only in PNG's Oro Province. More than theater, the program encourages dialogue between audience and actors, identifies needs of landowners, and provides information on development and villagers' democratic rights over resources. Following an Earth Day performance, for instance, villagers decided to organize to better protect their lands.

With the inaccessibility of radio and television among the 3.7 million population mainly living in isolated settlements and with only about a 5 percent literacy rate, the Community Theater reaches rural communities; informs landowners about possible development alternatives; trains provincial and local theater groups to "multiply the effect of conservation awareness in their language"; is creating a network of local theater groups; and is raising the importance of conservation to the national level where policymaking needs to be geared toward more environmental planning.

A playwright and environmentalist, program director **Joe Kelly Bik** underscores the importance of reaching rural landowners, the "principal stewards of their forests, reefs, and other habitats" whose decisions often have large environmental impacts. On a global scale, Bik urges environmentalists to encourage conservation and sound socioeconomic development in such developing countries as PNG. FSP country director is **David G. Vosseler**, a former U.S. Peace Corps volunteer and staff trainer, who is a liaison between the PNG government and local and international NGOs.

FSP also stresses potential **alternative energy sources**, such as biomass, geothermal, solar, wind, and ocean, that could be effectively harnessed by the year 2000.

FSP is also expanding its role in **child survival intervention** programs, such as oral rehydration therapy (ORT). "ORT is a simple sugar, salt, and water solution that enables a dehydrated child's body to absorb up to 25 times more water than it could before," reports FSP. With the government of the **Solomon Islands**, FSP is undertaking a program to **prevent and control diarrheal** disease, which accounts for a high percentage of infant deaths. In Kiribati, FSP is collaborating with the Ministry of Health and UNICEF to combat vitamin A deficiency that needlessly causes blindness. Vitamin A capsules are being distributed to pregnant mothers and children under six years of age; foods rich in vitamin A are being encouraged in this region with the "world's highest rate" of the deficiency, according to FSP.

■ Resources

FSP publishes *FSP News* and various materials outlining ecoforestry and other projects.

Research Conservation Foundation of Papua New Guinea (RCF of PNG)

P.O. Box 495
Port Moresby, Papua New Guinea
c/o Wildlife Conservation International (WCI)
Bronx Zoo
Bronx, NY 10468 USA

Contact: **David Gillison**, Founder Member
Phone: 1-718-220-5159
E-Mail: Bitnet: daglc@cunyvm; Bitnet:
        daglc@cunyvm.cuny.edu

Activities: Development; Education; Political/Legislative; Research • *Issues:* Biodiversity/Species Preservation; Population Planning; Sustainable Development • *Meetings:* CITES

● Organization

The **Research Conservation Foundation of Papua New Guinea** (RCF of PNG) is a nonprofit, tax-exempt membership organization that was incorporated in Port Moresby, PNG, in 1986. Originally, the Foundation's main goal was to preserve birds of paradise in their natural environment, and New Guinea's "lush forests provide shelter to the greatest variety of birds of paradise and bowerbirds." The Foundation has expanded its goals, and now also seeks to protect other flora and fauna within selected conservation areas.

Membership is open to residents of PNG, and also to interested people living overseas.

Other support is from Wildlife Conservation International, New York Zoological Society; Research Foundation of the City University of New York; Zoological Society of San Diego; and Pacific Helicopter Group, Trans Niugini Tours, University of Papua New Guinea, and Department of Environment and Conservation, PNG.

▲ Projects and People

The Foundation's primary project is the **Crater Mountain Wildlife Management Area**, which was "established as a cooperative venture between its owners, the Gimi and Pawaians people, the Department of Environment and Conservation in Port Moresby, and the founder members" of the Foundation. The reserve is approximately 2,600 square kilometers, and 98 percent of the land is primary forest or alpine scrub. The area is rainforest, subject to monsoons year-round; Crater Mountain is actually a series of detached peaks.

The program "is designed to generate employment through research support, tourism, and small-scale business development compatible with the national government's wildlife conservation strategy." The program hopes to show landowners the benefits of conservation.

Mary LeCroy and **William S. Peckover** have compiled an initial bird list which includes 21 species of birds of paradise, ratpors such as harpy eagle and peregrine falcon; dwarf cassowary; and palm cockatoo and Pesquet's parrot, among other rare parrots. Also noted are placental mammals and pouched marsupials; fruit bats; and reptiles including pythons and rainbow-colored Bolen's snake—considered by the Gimi to be "the source of the rainbow." Researchers are still needed to help carry out a thorough resource survey. Simon Finlay of the Royal Melbourne Zoo has made an initial botanical survey of the mid-montane forest, but intends to continue to inventory plant life—"rich in beeches, oaks, conifers, rhododendrons, podocarps, nutmegs, a wide variety of aroids, and epiphytes, which include some rare orchids and scheffleras."

RCF assists researchers who are interested in doing work in the wildlife management area. They provide information on how to reach Crater, tour guides, possible places to stay, and how to set up on arrival. They also address health issues. Much information is available on the culture of the Gimi and the Pawaians and on ways to work with them.

Gimi, for example, use **medical plants** to cure migraines, malaria, diarrhea, and hepatitis-type illnesses. **David Gillison**, who is also a WCI fellow, reports, "Since Gimi doctors practice their own form of psychiatry in conjunction with all medical treatment, we are unable to ascertain just how much of a treatment's success is due to pharmacology and how much occurs as a placebo effect. A detailed ethno-medical study is clearly indicated."

Future projects include providing assistance to the Environment and Conservation Department's plans to provide parks to protect all flora and fauna unique to PNG; establishing field stations for use by researchers; and using these stations to provide "hands-on" experience for biology students at the University of Papua New Guinea. RCF also wants to promote scientific studies aimed at increasing knowledge about a specific species or problem. Solar electric panels provide lighting.

■ Resources

In addition to international support for species preservation in PNG, the Foundation is in need of researchers who would be prepared to carry out resource surveys at Crater. They also need help to disseminate information about Crater and its potential use as a research site, assistance with fundraising and funding proposals, and the help of a typographer to prepare their newsletter. They also are looking for materials such as posters and books to be placed in village schools.

The Foundation is interested in establishing an international electronic bulletin board to help spread information about Crater Mountain and other wildlife management areas. RCF has access to computer facilities at the City University of New York, but needs help writing the software and maintaining the system.

The Foundation publishes *An Introduction to Living and Working in the Crater Mountain Wildlife Management Area* by Gillison. It has information from previous resource surveys of the Crater Mountain area.

Wildlife Conservation International (WCI) ⊕
c/o University of Papua New Guinea
Biology Department
P.O. Box 320
University, Papua New Guinea

Contacts: **Eleanor Brown, Ph.D.,** Research Fellow
       Michael J.G. Hopkins, Ph.D., Research
       Fellow and Lecturer
   Phone: (675) 267 154 • *Fax:* (675) 267 187

Activities: Research • *Issues:* Biodiversity/Species Preservation

▲ Projects and People

Recent joint research studies of Drs. Eleanor Brown and Mike Hopkins, Wildlife Conservation International (WCI) concentrate on collecting data on the **Behavior of Birds at Rainforest Trees in Papua New Guinea.** Focusing on the interactions between birds and their food plants, they seek understanding in "pollination and seed dispersal of rainforest trees and ecological requirements of endemic bird species." Data set components include "comparison of avian visitors to flowers and fruits; identification of flower types visited and quantification of flower resources available during different seasons; forest inventory information; and identification of individual forages and their movements from color-bands."

The scientists write, "We focus on flowering and fruiting tree species of various ecological types, ranging from 'big bang' (a large amount of resource during a short period) to 'steady state' (a small amount of resource during a long period)."

Four observed bird groups are birds of paradise (*Paradisaeidae*), fruit-doves (*Columbidae*), parrots (*Psittidae*), and honeyeaters (*Meliphagidae*). Among the preliminary results, even though it was noted that "the highest diversity of bird visitors was found at some fruiting trees rather than at flowering trees," there were "notable exceptions." For example, "*Schefflera*, a plant whose flowers and fruit are both visited by birds, had a far more restricted set of fruit visitors than flower visitors"—especially the raggiana birds of paradise and the great cuckoo-doves. "This suggests a more specialized

relationship between the plant and its seed dispersers versus its pollinators, a result that runs counter to current ecological theories," report Brown and Hopkins.

Continually monitoring some 2,000 marked flowering trees at their forest inventory site, the scientists note that generally "flowers with long peduncles (*Dysoxylum, Sloanea*) are visited by honeyeaters such as the Papuan black myzomela (*Myzomela nigrita*), which have thin elongated beaks and tongues specialized for probing.... Other flowers structured so that nectar or pollen is accessible on or near an outer surface are visited not only by honeyeaters; but also by parrots, the lories, and lorikeets with specialized brush tongues for lapping nectar or pollen from flowers; and even by typically frugivorous/insectivorous species such as black-fronted white eyes (*Zosterops atrifrons*)."

As the two researchers calculate the percentage of trees used as food resources by birds and also probably pollinated by birds, they believe the data being gathered is important because of the lack of previous, similar studies in the New Guinea forests "where birds play a more important part than in other tropical regions.

Overall, the six-hectare forest inventory contains about "2,500 tagged, mapped, measured, and identified trees." With specimens from each tree, a flora of Varirata collection is being established at the University of Papua New Guinea (UPNG) herbarium.

At 10 permanent net sites, hundreds of birds have been banded in the ongoing color-banding program. "Resightings of banded birds will allow us to infer feeding strategies of individuals and species, and movement of potential pollinators and seed dispersers," write Brown and Hopkins.

UPNG students are employed as research assistants to survey, map, and tag trees, and identify and observe birds. The scientists built a research hut at Varirata National Park, with support from Conservation International and the consent of the Departments of Environment and Conservation and of National Parks. Dr. Hopkins prepared an illustrated guide to the park in both English and Melanesian pidgin. A series of "mini-proposals" outlines ideas for future conservation action.

A UPNG lecturer, Dr. Hopkins has also studied seed beetles (*Bruchidae*) feeding on the seeds of *Parkia* in Brazilian rainforests and, at the same time, pollination, mostly by bats, and the seed-dispersal ecology of *Parkia* with Dr. Helen C. Fortune Hopkins. With Varirata serving as the only "functional national park" in Papua New Guinea," Dr. Hopkins urges the establishment of more such parks "as centres for research and education, and not merely as recreational areas."

Dr. Brown is also a research affiliate of UPNG, research associate of the Smithsonian Institution's National Museum of Natural History, and former researcher at the University of Maryland, USA. Along with WCI, the National Geographic Society helps fund her present work. Her previous research has centered on the "acoustic communication and social behavior of birds," such as the American crow (*Corvus brachyrhynchos*), whose songs she taped and analyzed; the endangered Hawaiian crow in its cloud-forest habitat; and the Australian magpie (*Gymnorhina tibicen*). About magpies, she says, "Both sexes advertise territory ownership with loud communal chorus or duet songs called carols. Juveniles also participate."

Of the Hawaiian crow, she writes, "Fewer than 15 crows survive, of which nine are in captivity, so that captive breeding and reintroduction are probably the only hope for this species." An ethologist, Dr. Brown writes, "Observing and understanding an animal's way of life in its natural habitat is a source of aesthetic, intellectual, and cultural enrichment, as well as an essential link with the natural world. . . . Preservation and restoration of the diversity of species and their habitats is the central issue of our time."

■ Resources

A list of U.S. graduate programs in conservation biology/tropical biology and of scholarships and other financial support opportunities are needed for UPNG biology students and others from developing countries.

⚗ *Private*

Ambunti Lodge and Sepik Adventure Tours
P.O. Box 248
Wewak, ESP, Papua New Guinea

Contact: **Alois Mateos,** Managing Director
  Phone: (675) 862525 • *Fax:* (675) 862525

Activities: Development • *Issues:* Ecotourism; Transportation

▲ Projects and People

In a desire to draw public attention to water pollution and waste dumping, **Alois Mateos** expresses his concern about mining companies' exploration and extraction of minerals and oil that affects the living standards of landowners and surroundings in his country. Mateos is worried about "forest exploitation and destruction of trees, plants, insects, and animals" as well as "dumping of wastes by industrialized countries" and the "destruction of cultural systems."

While encouraging organizations to "fight for the rights of preservation" of native cultures and resources, he is fearful that "government and [large] companies through political leaders are discouraging environmental groups to educate the landowners about what will happen to their natural environment." Mateos is a local coordinator of the Earth Preservation Fund.

🎓 *Universities*

Papua New Guinea Forestry College
P.O. Box 92
Bulolo, Morobe, Papua New Guinea

Contact: **Lawong Balun,** Botany/Ecology Lecturer
  Phone: (675) 445226 • *Fax:* (675) 445470

Activities: Education • *Issues:* Biodiversity/Species Preservation; Sustainable Development; Water Quality

▲ Projects and People

Previously a medicinal plants curator, **Lawong Balun,** who teaches botany, ecology, and soil and wildlife conservation at **Papua New Guinea Forestry College,** is researching **soil seed banks of four vegetation types**—from three old-growth forests and one disturbed savanna grassland—to determine their natural regeneration after disturbance, among other factors. He observes his samples "under glasshouse conditions."

Balun writes, "The soil seed bank is the seed component of a flora deposited in a soil over a period of time." Lying there for varying lengths of time, "they either germinate or die at varying rates depending on the nature of the species." Balun continues, "Forest floors after the fruiting seasons are often carpeted with seeds and seedlings," ranging from 39 to 4,500 seeds per square meter, according to prior research. Shade tolerance is a factor in the growth of pioneers.

Balun has contributed to publications on **ethnomedicine** and **ethnobotany** studies.

■ Resources

Balun needs equipment for research, education, and communications outreach.

Paraguay

This landlocked country in the middle of South America has two distinct environments, separated by the Paraguay River: east is subtropical climate with grassy plains, wooded hills, and tropical forests; west, in the Chaco region, is a variable climate, with low, flat plains. About 39 percent of the land is forested; 96 percent of the population lives in the eastern region.

The country's economy relies mostly on agriculture, with about half the land used for pasture. Paraguay has reduced its dependence on imported oil by developing hydroelectric resources and an alcohol fuel industry.

Despite attempts at reforestation, forests are disappearing at an alarming rate. The primary cause is slash-and-burn clearing by farmers, which also releases carbon dioxide into the atmosphere and contributes to soil erosion. Water pollution, from industry and poor sewage systems, and unsanitary solid waste collection are other problems.

🏛 Government

ITAIPU Binacional/ITAIPU Binational
De La Residenta 1075
Asunción, Paraguay

Contact: **Dario Perez Chena**, Superintendent of the Environment
Phone: (595) 21 207 161 • *Fax:* (595) 21 207 176

Activities: Development; Education; Research • *Issues:* Air Quality/Emission Control; Biodiversity/Species Preservation; Energy; Water Quality • *Meetings:* UNCED

● Organization

ITAIPU Binacional (ITAIPU Binational) is a government agency in charge of the world's largest hydroelectric project, according to the organization. Created in 1974 through an agreement between the governments of Paraguay and Brazil, ITAIPU oversees every aspect of the hydroelectric dam and reservoir located on the Paraná River, which forms a border between the two countries.

The scale of this project is astonishingly huge; according to ITAIPU, the amount of concrete used during construction would be enough to rebuild every building in Rio de Janeiro. As a result, ITAIPU has tried to incorporate ecological planning in the construction and operation of the project. It describes its guiding principles as concern for "the balance of the environment as a whole, . . . special attention to social consequences, . . . protection of essential ecological processes," such as food chains, and "preservation of ecological diversity."

▲ Projects and People

Over 20 environmental projects have been developed in the ITAIPU Reservoir area. These include **environmental education** in regional schools and an **environmental museum**, the Ecomuseu or Museo de Historia Natural y Antropología; development of **recreational and tourist facilities**; control of **water quality and sedimentation**; reforestation along the banks of tributary rivers in Paraguay; **waste management**; soil conservation; **health** measures to control human and animal disease; analysis of regional weather data; **aquaculture**; and creation of **wildlife refuges** and repopulation of animal species.

The ITAIPU Aquiculture Station, founded in 1988, is the center for **artificial reproduction of fish species** endangered by the environmental impact of the hydroelectric project. The pacú (*Piaractus mesopotamicus*) is one species that was greatly affected because it must swim long distances to reproduce and prefers to discharge its eggs in the river's headwaters. It is also of great commercial importance because its flesh is highly valued in that region. The scientists at ITAIPU fertilize and incubate the pacú's eggs entirely *in vitro*, then transfer the larvae to enclosed tanks, where they develop into juvenile fish.

Another species protected through the efforts of ITAIPU is the bush dog (*Speothos venaticus*), a singular mammal once found throughout the continent but now registered in the Red Book *of South American Mammals*. In 1979, scientists conducting an inventory of fauna for ITAIPU captured two specimens in Paraguay. Bush dogs are generally found in forest areas near river banks, where they like to swim and dive. They typically prey on agoutis, pacas, rats, and birds, and their footprints are often confused with those of domestic dogs. The ITAIPU Zoo currently possesses 10 bush dogs in captivity, as a result of its successful breeding project.

■ Resources

ITAIPU's publications available upon request include: the texts of their First (in Spanish) and Second (in Spanish and Portuguese) Seminars on the Environment; brochures on the ITAIPU project, the bush dog, and artificial reproduction of the Pacú in Spanish; a bilingual (Spanish and Portuguese) brochure on ITAIPU's environmental projects; and a book in English, *ITAIPU Hydroelectric Development: The Project of the Century*.

Peru

Peru has a long Pacific coastline and borders Ecuador, Colombia, Bolivia, and Chile. Protection of the country's great beauty and diversity is hampered by political and economic problems. Most of the population lives on the coastline, behind which the Andes Mountains reach a height of 22,000 feet. East of the mountains the tropical rainforests begin, covering 60 percent of the country and stretching across to the mouth of the Amazon River at the Atlantic Ocean.

Peru has extraordinary natural resource wealth, but political corruption and conflicts and the accompanying economic decline have taken a toll on the environment. Much of the infrastructure has become dysfunctional, leading to the spread of disease. Soil degradation has been caused by centuries of grazing and recent logging; and overfishing and coastal water pollution are threatening the anchovy fishery. At the same time, active environmental groups and researchers are keeping conservation efforts alive.

🏛 *Government*

Consejo Nacional de Ciencia y Tecnologia (CONCYTEC)
National Council of Science and Technology
Avenida Paseo de la República No. 3505, San Isidro
Lima 27, Peru

Contact: **Carlos Chirinos Villanueva,** President
Phone: (51) 14 422580 • *Fax:* (51) 14 522580

Activities: Development; Education • *Issues:* Biodiversity/Species Preservation; Deforestation; Energy; Health; Sustainable Development

● Organization
CONCYTEC—the National Council of Science and Technology—describes itself as a "state institution devoted to national policy, promotion, support and coordination of the scientific and technological research in Peru." Originally called the National Research Council in 1968, CONCYTEC includes environmental concerns in its scope, although primary emphasis is on research and development in science and technology, according to **Carlos Chirinos V.**, president. In international cooperation activities, CONCYTEC is the country's leading institution in science and technology.

▲ Projects and People
CONCYTEC supports funds for research at academic centers, and distributes information on postgraduate scholarships granted by government and global institutions, such as UNESCO, U.S. Agency for International Development (USAID), and others. It also provides financial support to students at the graduate and postgraduate levels as well as for the publication of scientific and technological books.

Current environmental projects are the evaluation of geological data due to climatic evolution and the Niño current in Peruvian Northwestern Coast, with José Machare O.; geodynamics and disaster risks in Payhua Narrow Valley—Matucana; hydromorphologic data processing on Ucayali River within Pucalipa Area, with Jorge Brousset B.; climatic evaluation of possible location of astronomic observatory within Arequipa, with Alfonso Maguiña L.; and meteorological databank for forecast and follow-up of the Niño Phenomena, with Rodolfo Rodriguez A.

Executive director is Ruben F. Inga. Scientific specialist is Eduardo Neptali Perez Sandoval.

 *NGO*

Asociación para la Conservación de la Naturaleza Amazonica (ACONA-Iquitos)
Association for the Conservation of Amazonian Wildlife

Avenida Mariscal Cáceres no. 666

Iquitos, Loretto, Peru

Contact: **Filomeno Encarnacion,** Wildlife Field
        Researcher
Phone: (51) 94 237408 • *Fax:* (51) 94 235256

Activities: Development; Education; Research • *Issues:* Biodiversity/Species Preservation; Deforestation; Environmental Education; Global Warming; Population Planning; Sustainable Development; Waste Management/Recycling

▲ Projects and People

The Asociación para la Conservación de la Naturaleza Amazonica (ACONA-Iquitos) (Association for the Conservation of Amazonian Wildlife), with an executive council and membership of 65, has objectives of **environmental education, training and information on conservation,** and **research and management of biodiversity.** Committees are set up for fundraising, membership, planning, and organization. Recent activities include developing a **directory** of similar national and global organizations; acquiring a new headquarters and personnel; building institutional links; strengthening ACONA's recognition; setting up **public meetings;** broadening outreach to potential members; expanding environmental education; creating a calendar of activities; establishing a **database;** putting together **training classes,** seminars, symposia, conferences, panels, workshops; and evaluating these plans every month.

Filomeno Encarnacion writes, "Developing countries often have a limited number of project proposals. International agencies should recommend that local conservationists use their creativity within the environment where they live. It's no use introducing techniques and procedures foreign to an area's ecosystem." Encarnacion also believes that "easy, flexible training in the administration and management of natural resources" will aid global cooperation among conservationists. "Every project should emphasize the inventory and evaluation of natural resources." Also, rural projects should be promoted in such areas only "to avoid conflicts between city and country dwellers."

■ Resources

On behalf of Amazonia, ACONA-Iquitos seeks environmental education training and equipment, such as literature, audiovisuals, and information exchanges. ACONA offers *Guia de Educación ambiental para el Maestro en la Amazonia* (Guide to environmental education for teachers in Amazonia) and *Libro de Lectura Selvita, 1er. Grade* (A first grade forest reader).

Asociación Peruana para la Conservación de la Naturaleza (APECO)
Peruvian Association for the Conservation of Nature

Parque José de Acosta 187

Lima 17, Peru

Contacts: **Silvia Sanchez,** President
Phone: (51) 14 625410 • *Fax:* (51) 14 633048

        Mariella Leo, Research and Management
        Coordinator
Phone: (51) 14 616316

Activities: Development; Education; Natural Resources Management; Research • *Issues:* Biodiversity/Species Preservation; Deforestation; Sustainable Development • *Meetings:* IUCN; World Parks Congress

● Organization

Founded in 1982, APECO—the **Peruvian Association for the Conservation of Nature**—is committed to the "conservation and management of wildlife and other renewable natural resources in Peru." Four main goals of the 350 members are to "increase public understanding and support for natural resource conservation issues; develop and conduct applied research projects that lead to improved management . . . ; promote and assist the establishment of regional conservation organizations . . . in response to local initiatives and need; and provide technical training to researchers and conservationists working towards shared goals of wise resource management."

APECO participated in Peru's National Conservation Strategy and the National Program for Forestry Action.

▲ Projects and People

APECO is working on **Integrated Conservation Programs (ICPs),** including lobbying, environmental education, research, and resource management—in specific geographical localities, such as the 90,000-square-kilometer area of northeastern Peru and the Lake Titicaca area. Strategic planners are economist **Victor Merino** and World Wildlife Fund (WWF) advisors **Rusty Davenport** and **Laura Campobasso.**

When two construction projects—a highway and a canal in the isthmus of Fitzcarrald—threatened what is "without doubt the best representative of Amazonia," APECO launched an **educational campaign** on behalf of **El Manu National Park**—a UNESCO-declared Biosphere Reserve and World Heritage Site—in the mid 1980s. With the support of WWF, literature, audiovisuals, media clippings, and presentations were directed to schools to enlighten young people about the park's thousands of species of flowering plants, 1,000 bird types, 200 mammal varieties, and more than a million insects and other invertebrates. Endangered species in the park include the **black lizard, jaguar, river wolf,** and **spectacled bear.** Indigenous groups living in the Park are the Machiguengas, Nahuas, Amahuacas, and Mashco-Piros. Among the archeological remains are the **petroglyphs** of Pusharo and the ruins of Mameria.

The **white-winged guan** (*Penelope albipennis*) is the symbol of APECO. Thought to be extinct, it was rediscovered in 1977 by

Peruvian and North American scientists and remains "highly endangered due to hunting and habitat destruction." APECO works for its survival.

The Humboldt penguin has been a research subject since the "El Niño current" of 1982 with help from the WWF and the Portland Audubon Society. Effects of selective logging and poaching on populations of the endangered yellow-tailed woolly monkey have been investigated. A public awareness campaign about the woolly monkey and its Andean cloud forest habitat accompanied this effort.

A three-year inventory of the Río Abiseo National Park (ARNP), with Mariella Leo now as main researcher, has identified species of mammals, birds, reptiles, anurans (frogs and toads), fish, and butterflies. APECO wants to survey the Park's southern watershed and launch species- and community-oriented research there, but points out that "it is very difficult for a Latin American NGO to obtain the amount of money required [for an inventory]. Logistics are hard and expensive since the ARNP is far away from the main roads, and the local support (guides, mules, porters, food) is expensive for Peruvian standards." Other participants are Enrique Ortiz, Lily Rodriquez, Hernan Ortega, Miriam Medina, and Patricia Moore.

Leo is also leading APECO's efforts to lower the rate of deforestation of the Amazon montane forest in northeastern Peru as well as help improve "the quality of life of local people and to preserve the forest biodiversity." Joining the effort are Victor Merino, president Silvia Sanchez, Juan Carlos Riveros, Aura Murrieta, director Augusto Lainez of the Center for Studies of the Amazon Environment in San Martin, and director Carlos Chavez of the Association for Conservation of the Natural Resources of Amazonas. Adequate funding is being sought.

Biologist Susana Moller Hergt and environmental education coordinator Aura Murrieta have embarked on a National Information Campaign on Nature Conservation. They seek to spread knowledge through training; involvement of local authorities and grassroots leaders; and producing popular booklets, posters, audiovisual aids, and related materials specifically targeted for children, youths, adults, and rural and urban people to help them "face the complex geography and ethnic aspects of Peru in the highlands, coast, and jungle ecoregions." APECO-NET is underway as a means to link grassroots groups. The campaign aims to convince regional authorities to include environmental education in its programs. Obstacles are a lack of funds to support volunteers and meet local expectations as well as the "lack of efficient communication means in a geographically complex territory [and] inefficient mail service."

An Environmental Education Program for Peruvian Teachers is also ongoing, but "the bureaucratic organization of the education sector and their traditional points of view were barriers hard to open," reports APECO. The program teaches about the "heterogeneity of the nation ecologically and its socio-cultural aspects."

■ Resources

A publications list is available, with titles in both Spanish and English, including *Huasceran National Park* and *Manu*. Expertise is sought in cloud forests and hydrology for project design as well as facilitators to achieve consensus, teacher training in environmental education, and print and audiovisual media materials.

Association for the Conservation and Development of the Environment (ACODEMAC)
Avenida Progreso H-15, Urb. R. Castilla
Cajamarca, Peru

Contact: **Homero Bazán Zurita**, Executive Director
Phone: (51) 44 923356 • *Fax:* (51) 44 923356

Activities: Development; Education; Research • *Issues:* Biodiversity/Species Preservation; Deforestation; Environmental Education; Sustainable Development; Water Quality • *Meetings:* International Conference on Environmental Education

▲ Projects and People

The **Association for the Conservation and Development of the Environment**, with a membership of 45, has 2 principal projects. Regarding the **conservation and management of the National Park of Cutervo**, Cajamarca, the Association is evaluating the "status of natural resources within the park and conservation-related problems and [is proposing] a conservation and management plan for the park." However, the Association reports it has "no specific financial resources" to carry out the project. Overseeing the effort are **Isidoro Sánchez**, **Homero Bazán**, **Pablo Sánchez**, **Alfonso Miranda**, and **Gabriel Sánchez**.

Also in Cajamarca, an **environmental education** program is designed to promote an "ecological conscience among the population—especially young people and teachers—in order to respect and conserve the environment." **Isabel Bardales** joins the abovementioned staffers in running this program.

Bazán is also assistant director of the Cajamarca Integrated Rural Development Project and recently was professor of biology, zoology, and ecology at the National University of Cajamarca.

■ Resources

For its environmental education program, the Association seeks audiovisual equipment and raw materials for printing. Other equipment, such as a truck, is needed to evaluate protected areas.

Centro para el Desarrollo de Campesino y del Poblador Urbano (CEDECUM)
Center for the Development of Rural and Urban People
APDO 542, Calle Libertad 137
Puno, Peru

Contact: **Arturo Vasquez Salazar, Ph.D.**, Agricultural Engineer
Phone: (51) 54 353461

Activities: Development • *Issues:* Sustainable Development
Meetings: CITES

■ Resources

CEDECUM seeks confirmation and circulation of Andean technologies.

Centro para el Desarrollo del Indigena Amazónico (CEDIA)
Center for the Development of the Amazonian Indian
Pasaje Bonifacio 166, La Perla
Lima, Callao 4, Peru

Contact: **Leslis Rivera Chavez**, President, Anthropologist
Phone: (51) 14 650708

Activities: Development; Education; Law Enforcement; Political/Legislative; Research • *Issues:* Biodiversity/Species Preservation; Deforestation; Health; Indigenous People's Land Ownership; Population Planning; Protected Areas; Sustainable Development

▲ Projects and People
"It is our institution's position that wild areas are being used indiscriminately and irresponsibly and that the best way to protect them is to entrust most of the Amazonian areas to the indigenous people who have maintained them in a state of equilibrium until today," writes Leslis Rivera Chavez, president, Centro para el Desarrollo del Indigena Amazónico (CEDIA) (Center for the Development of the Amazonian Indian).

Rivera, who founded CEDIA in 1985 and is also a founding member of Seminario de Estudios Antropológicos de Selva (Seminary of Anthropological Studies of Forests), Universidad Nacional Mayor de San Marcos, continues, "Conservation has come to be synonymous with protection of and respect for the Amazon Indian. Therefore, we ask that ecologists all over the world support the defense of indigenous peoples and the struggle for their territories as a direct adjunct to conservation."

CEDIA focused studies on the 443,887-hectare **Kugapakori-Nahua Reserve,** so declared in 1990, where the Kuga-pakori and the Nahua indigenous people develop their basic activities of hunting, fishing, gathering, and participate in "itinerant agriculture."

At the Megantoni **Sanctuary** (or Pongo de Mainigue), research was conducted from 1988 to 1991 to justify its establishment, which includes the famous Pongo de Mainigue canyon formed by the Urubamba River in the Cordillera de Vilcabamba. Four rivers rise there: Yuyato, Ticumpnía, Siguaniro, and Timpía; the mountains reach heights of 2,300 meters above sea level; it will be considered a refuge for flora and fauna and is the only untouched part of the Alto Urubamba forest, where indiscriminate cutting of trees has resulted in 450,000 hectares being lost, according to CEDIA. The Machiguenga, who have lived in the area "from time immemorial" consider it a sacred location. The government is now considering CEDIA's study.

At the Vilcabamba **Community Reserve,** CEDIA is studying the zone made up of the east slope of the Cordillera de Vilcabamba, north of the Megantoni Sanctuary, among forming rivers. The Community Reserve is being established in an area of forests used exclusively by indigenous people who live in neighboring zones, where they hunt, fish, and gather forest products. The CEDIA studies will be completed in 1993.

■ Resources
CEDIA publishes forest management textbooks, maps, and other literature that is distributed free of charge to the indigenous population. A publications list is available. CEDIA seeks new technologies in topographical mapping and communications, international support to consolidate territory gained, and techniques of environmental education for indigenous people of the rainforests.

Centro de Estudios y Prevención de Diastres (PREDES)
Center for the Study and Prevention of Disasters
Avenida San Felipe N° 370, Jesus Maria
Lima, Peru

Contact: **Gilberto Romero Zeballos,** Sociologist
Phone: (51) 14 633240 • *Fax:* (51) 14 633240

Activities: Development; Education; Research • *Issues:* Deforestation; Soil Conservation; Sustainable Development; Water Conservation • *Meetings:* CITES

■ Resources
PREDES's publications list, printed in Spanish, is available.

Centro de Estudios y Promoción del Desarrollo (DESCO)
Center for Research and Promotion of Development
León de la Fuente 110
Lima 17, Peru

Contact: **Charles de Weck Pendavis,** Technical Assistant
Phone: (51) 14 627193 • *Fax:* (51) 14 617309

Activities: Development; Education; Law Enforcement; Research • *Issues:* Biodiversity/Species Preservation; Deforestation; Population Planning; Sustainable Development; Waste Management/Recycling

▲ Projects and People
Founded in 1965, **Centro de Estudios y Promoción del Desarrollo (DESCO) (Center for Research and Promotion of Development)** takes environmental protection or ecological preservation into account in all of the projects it undertakes. With a staff of 110, **Marcial Rubio Correa** is president.

DESCO is currently researching collection and disposal of solid waste in Cusco and metropolitan Lima, under the direction of **Gustavo Riofrío** and **Luis Olivera.**

In the rural valley of Chincha on the Peruvian coast, DESCO is developing a project on **biological control of pests** in cotton, citrus, corn, olive, apple, and pear seedlings—through the use of Trichogramma and other natural predators.

In the mountains, 3,500 meters above the sea, DESCO has 2 important projects in the Colca Valley. In terms of livestock breeding, researchers are trying to improve the health of alpaca, which suffer from congenital defects due to inbreeding. In agriculture, DESCO is trying to preserve the germ plasm of native varieties of corn and also is working to improve the quality of the corn, beans, and potatoes grown in the area through new methods of cultivation and selection of the best seeds.

Plans for development of the river basin call for recovery of grazing areas, protection against erosion, and reforestation with trees and shrubs native to the region. Oscar Toro and Aquilino Mejía direct this effort.

■ Resources

DESCO seeks information on new technologies, funding for future projects, and opportunities to exchange information.

A publications list is available of materials from DESCO's Documentation Center, including *DESCO, 25 Años de Quehacer Institucional* (25 years of institutional work). Also published are the newsletter *Resumen Semanal* (Weekly summary) and *Revista Quehacer* (Task magazine).

Centro de Investigación Documentación Educación y Asesoramiento (IDEAS)
Center for Research, Documentation, Education, and Consultation
Avenida Arenales 651

Lima 11-0170, Peru

Contact: **Julio Chávez Achong,** Director
Phone: (51) 14 237773 • *Fax:* (51) 14 230645

Activities: Development; Education; Research • *Issues:* Agroforestry; Deforestation; Health; Small Business; Sustainable Agriculture; Sustainable Development; Water Quality • *Meetings:* ECO 1992; Women and the Environment

● Organization

The IDEAS Center is an NGO founded in 1978 to research and promote national development in order to "help reverse the problems of poverty and marginalization and to construct new structures and social relationships based on justice, liberty, peace and progress for all—contributing to a strengthened society on the local and national level."

IDEAS aspires to sustainable development "principally through mobilization of the natural resources of Peru, recognizing the interdependency of the countries of the world, and valuing the possibility of a better integration of Latin America in the Andean and Amazonian areas."

The Center emphasizes, "Peru needs sustained economic growth that includes the rational and productive utilization of sources of energy and the conservation and improvement of the environment." Among its objectives are the "promotion of development in urban and rural areas characterized by poverty," the stimulation of international convergence and debate, and the "building of the Center as a dynamic agent of local, regional, and national development."

▲ Projects and People

Its long-term plan (1990–1993) features an urban program in Lima, Piura program on the North Coast, Cajamarca program in the North Sierra, and the agroindustry program in the Central Sierra and Lima. A research and organization unit coordinates such activities, together with the Center for Documentation and Central Administration.

In Lima, activities are aimed at women and at strengthening local government's capacity for management and planning at the grassroots level. IDEAS focuses attention on health, nutrition, and environmental cleanliness.

In areas along the coast, IDEAS is organizing small-scale farmers and helping them to develop management skills to increase their income as well as improve the quality of life and of their natural resources. At the same time, the project "values the help of farm women and rural youth." IDEAS undertakes activities in training, research, and experimentation to "generate appropriate technologies, improve the economic and social infrastructure, health, and nutrition, and to help design policy."

Aiming to restore natural resources in mountain areas and to increase production at low cost, IDEAS is involved in research and experimentation in organic agriculture, training, and investment in infrastructure as well as complementary actions in health and nutrition.

Throughout the country, IDEAS is developing support for agroindustries through research and aid to management, technology, and marketing.

Julio Chávez Achong, IDEAS director, says the Center "works with poor sectors of society that have demonstrated economic and social vitality, to encourage their development and ensure their survival." It does so while incorporating ecological perspectives and valuing the cultural identity of native peoples and of women. It tends to "forge permanent links for financial and technical cooperation between the north and south on matters of development."

■ Resources

An IDEAS brochure and publications list are available, printed in Spanish.

Comisión de Coordinación de Tecnología Andina (CCTA)
Commission for Coordination of Andean Technology
Javier Prado 595, Magdalena del Mar

Lima 17, Peru

Contact: **Gricelda Rodríquez T.,** Secretary
Phone: (51) 14 617353
Fax: c/o CCTA (51) 14 421766

Activities: Development; Education; Research • *Issues:* Sustainable Development

● Organization

Founded in 1984 to coordinate and promote actions oriented toward technological development in the Andes in Peru, **Comisión de**

Coordinación de Tecnología Andina (CCTA) (Commission for Coordination of Andean Technology) is made up of NGOs with a similar purpose mainly located in the Peruvian mountains. CCTA also coordinates activities with rural development teams, farmers' associations, and universities across the country.

Its executive committee develops and administers projects, maintains relations with other groups, prepares documents, and sets institutional policy. Another work area is CENDOTEC, which helps coordinate teams, institutions, and projects to facilitate the exchange of documents and provide information to researchers and technicians and for promotion. A databank has been installed with relevant bibliographic information.

▲ Projects and People

Working toward sustainable development, CCTA helps manage soil, pastures, water, and forests within the country's system of river basins. CCTA gives priority to the transformation of farm products in the rural environment, while aiming to create work alternatives for rural youth and to overcome "agriculture's dependency on industry and rural areas' dependency on the city."

With the participation of local farmers and the technical assistance of member institutions, CCTA is building a germ plasm bank for Andean crops—including potatoes and grains. It is also helping to adapt new technologies to alternative and sustainable uses.

■ Resources

CCTA publishes the *Hoja Informativa* (Informative page) newsletter, the annual magazine *Cuadernos Informativos* (Informative notebooks), technical manuals and project reports.

ECOTEC S.A.
Independencia 461, Miraflores
Lima 18, Peru

Contact: **Alejandro Camino**, General Director
Phone: (51) 14 47 4310 • *Fax:* (51) 14 42 4365

Activities: Development; Education; Political/Legislative; Research • *Issues:* Biodiversity/Species Preservation; Deforestation; Energy; Population Planning; Sustainable Development
Meetings: International Conference on Ethnobiology

▲ Projects and People

Alejandro Camino is the director of both ECOTEC S.A., an environmental consulting company, and PRODIMA, an international, non-governmental development project, and is also an associate professor at the Universidad Católica de Peru (Catholic University of Peru).

ECOTEC, Ecología y Tecnología Ambiental S.A. (Ecology and Environmental Technology Inc.), is staffed by professionals in the social and natural sciences, engineering, and environmental technology who offer their expertise to government entities, research institutions, and private businesses concerned with the environmental impacts of development projects in Peru. The company's services include design and supervision of projects to improve environmental health and use natural resources responsibly, monitoring of pollution and other environmental impacts, and advice on environ-

mental legislation and management of protected areas. More specifically, past projects have included recycling, the use of bacteria to refine mineral products, organic agriculture, and the development of ecotourism as an "industry without smokestacks."

In the countries of the Andean region, PRODIMA, the **Programa para el Desarollo Integrado de Montañas Andinas (Program for Integrated Development of the Andean Mountains)**, coordinates environmental projects and serves as a storehouse of information on the region's ecology. Its goal is to improve the health and productivity of the Andean ecosystem, currently threatened by deforestation, overgrazing by livestock, and inappropriate agricultural technologies which have resulted in widespread soil erosion and the continued poverty of the Andes' indigenous inhabitants. To this end, sustainable development projects are carried out in cooperation with the Peruvian government and other local entities as well as with ICIMOD, the International Center for Integrated Mountain Development, in Nepal, and the International Society for Mountains in Bern, Switzerland.

■ Resources

Mr. Camino notes that new technologies need to be developed for use by rural communities in mountainous areas so that they may sustain their livelihood now and in the future, and also expresses a need for audiovisual materials for use in environmental education.

Instituto Natura/Nature Institute
Jr. Francisco Bolognesi N° 1144
Chimbote, Santa, Peru

Contacts: **Maria Elena Foronda Farro**, Coordinator, Sociologist
Francisco Rogger Carruitero Lecca, Lawyer, Sociologist

Activities: Education • *Issues:* Air Quality/Emission Control; Health; Waste Management/Recycling; Water Quality
Meetings: CITES

▲ Projects and People

Instituto Natura (Nature Institute), "dedicated to preservation of the environment and natural resources," carries out research, public relations, and training in Chimbote, which it describes as one of Peru's "three most polluted cities," exemplified in fishing practices, iron and steel industry, and the poor sanitary conditions in which the population lives.

Examples of the negative environmental impact on the ecosystem and the people include air pollution—particles of grain meal and silicon, carbon monoxide, sulfur dioxide; water pollution—dumping of industrial waste from more than 30 flour mills, discharge of untreated wastewater, lack of dissolved oxygen, and elimination of marine flora and fauna; soil pollution; and absence of sanitary water and poor garbage collection. Because of the lack of potable water, Chimbote is known as the "Capital of Cholera," reports the Institute.

Consequently, Instituto Natura is forming an **Environmental Health Plan** which will include direct participation by the people. It is proposed that the industrial sector improve their currently obsolete and polluting technology. Both formal and informal **environmental education programs** are being formed.

The Institute participates in local and regional environmental forums such as **Forum Regional del Ambiente**, which proposed the creation of a **Regional Institute for Environmental Management** in Chavín (IRGARCH).

■ Resources

Available reports, in Spanish, are *Forum Regional del Ambiente* (Regional environmental forum), *Forum Medio Ambiente y Desarrollo Regional* (Regional environment and development forum), and *Estudio: Sondeo Epidemiologico en La Ciudad de Chimbote* (Study: epidemiological survey of Chimbote City).

Instituto Regional de Ecología Andina (IRINEA)
Andean Regional Ecological Institute
Jr. Arequipa N° 470
Hyancayo, Peru

Contact: **Raul Palomino Vila**, Director, Agronomist
  Phone: (51) 64 237169 • *Fax:* (51) 64 231111

Activities: Development; Education; Research • *Issues:* Biodiversity/Species Preservation; Deforestation; Sustainable Development • *Meetings:* CITES

■ Resources

IRINEA seeks new agroecology technologies and educational resources for a landscape school for children. Included in its publications list are materials on farming, rural development, and Andean ecological projects, such as the following: *Diagnost. Ecológico de una Microcuenca Andina* (Ecological diagnosis of an Andean river basin); *Organización Andina, Drama y Posibilidad* (Andean organization: drama and possibility); *Método de Diagnóstico y Planificación de Sistemas Agrarios* (Diagnostic and planning method for agrarian systems); *Mejoremos la Agricultura y la Vida Campesina* (Let's improve agriculture and country life); *Nuestra Comunidad puede lograr su Desarrollo* (Our community can achieve development); and *Aporte para el Balance de las Ciencias Sociales en la Sierra Central* (Contribution to the balance of social sciences in the central sierra).

Naturaleza, Ciencia y Tecnología Local para el Servicio Social (NCTL)
Nature, Science, and Local Technology for the Social Services
Conquistadores 1220, San Isidro
Lima 27, Peru

Contact: **Luis Masson Meiss**, Director
  Phone: (51) 14 229545

Activities: Development; Education; Research • *Issues:* Environmental Education; Sustainable Development • *Meetings:* Environmental Meetings, International Development Banks

● Organization

Founded in 1979, **NCTL—Nature, Science, and Local Technology for the Social Services**—has a main strategy: "development from inside to outside." NCTL favors local initiatives, especially those dealing with irrigation, potable water supply, and sanitation. Environmental projects are based mainly in rural areas, preferably in mountain ecosystems to encourage ecodevelopment or sustainable development.

NCTL takes into account the "special needs of the inhabitants of such areas" as well as their "know-how" about land and water management, including early technologies, such as terracing and ancient irrigation systems. As projects are planned, NCTL considers "folk traditions and habits, as a local pattern of happiness." It encourages art in relation to folklore and nature—music, painting, theater, ceramics—as an alternative way to reach children and raise "the scale of values" of the population in the future, avoiding alcoholism and other vices. Believing in the long-term approach, NCTL states, "450 years of economic, social, cultural, political, and environmental impacts in the native population cannot be arranged in the short term."

▲ Projects and People

Since 1983 in three native communities, NCTL has been undertaking **ecodevelopment in Andean rural villages**, including San Pedro de Casta, that are in the Santa Eulalia River basin in the Andes Mountains near Lima.

NCTL participates in the **Programa Pirámide de Educación Ambiental (Pyramid Program of Environmental Education)**. Workshops on ecodevelopment are offered to government officials, students, teachers, and journalists. In 1991, NCTL participated in an international workshop in La Paz on development for government decisionmakers from Peru and Bolivia. The subject was Titicaca Lake basin development.

Research projects of agronomy students at the Universidad Nacional Agraria, La Molina, are encouraged concerning "erosion control, rational use of irrigation water, available wild vegetation, and collection of wild species of feeding plants for breeding purposes." NCTL also undertakes its own research projects that assess the evolution of Peru's "small mangle vegetation ecosystem" on 5,000 hectares during a 44-year period (1943–1987), using a series of aerial photographs.

Luis Masson Meiss, agronomist and former government administrator, has special expertise in soils and ancient technologies for steep slopes. **Cesar Quiroz Peralta**, advisor in educational aspects of UNESCO, is NCTL's scientific and environmental education advisor. **Dr. Carmen Felipe Morales**, also an NCTL scientific advisor, is agronomy faculty dean at the Universidad Nacional Agraria.

■ Resources

NCTL provides annual and other reports to sponsors and interested persons. It seeks financing for its Pyramid Program workshops and for fieldwork in developing rural areas.

PROTERRA ⊕
Avenida Esteban Campodónico 208, Urb. Santa
   Catalina, La Victoria
Lima 13, Peru

Contacts: **Antonio Andaluz,** President
   Carlos Andaluz, Administrative Director,
   Environmental Lawyer
   Phone: (51) 14 703930 • *Fax:* (51) 14 417001

Activities: Development; Law Enforcement; Political/Legislative • *Issues:* Biodiversity/Species Preservation; Deforestation; Sustainable Development • *Meetings:* International Environmental Law Conference; International Environmental Education Conference; UNCED

▲ Projects and People
PROTERRA's administrative director, Carlos Andaluz, writes, "Environmentalists can enhance future global cooperation by consolidating their environmental work within their own countries and promoting the establishment of regional and global networks."

He believes that the "problems of degradation of the development countries' non-renewable natural resources [are] worth much more attention. It should be linked to the international exchange terms with developed countries."

The Yanesha Project, which is based on principles of agroecology, could serve as a pilot project for disseminating its accomplishments at a subregional level in countries of the Amazonian basin, writes Andaluz, who represents PROTERRA in forums on watershed and forest management and on the "crisis and alternatives [of] Lima's environmental resources." With the National Office for the Evaluation of Natural Resources (ONERN), he is participating in efforts to develop a **National Action Plan for Fighting Against Desertification in Peru.** An environmental lawyer, Andaluz has co-authored books on the *Code on Environment and Natural Resources, Concordances and Summaries,* and *Local Governments and Environmental Code, General Guidelines to Put into Practice.* A recent research paper is on the "Regime on Tenancy and Ownership of Rural Properties in Peru."

■ Resources
Proterra-Informa is a bimonthly bulletin launched in 1991 that focused on preparation for the 1992 Earth Summit in Brazil, as PROTERRA was the regional coordinator for Latin American NGOs participating in the United Nations Conference on Environment and Development (UNCED), as well as the coordinator for the Andean Sub-Region—Bolivia, Colombia, Ecuador, Peru, and Venezuela. The bulletin covers progress made by member nations that attended the Conference as well as other environmental meetings, new environmental publications, Peruvian environmental campaigns as the National Forest Action program, and the formation of new organizations.

Other literature includes an *Environmental Network Directory, Derecho Ambiental—Propuestas y Ensayos* (Environmental law—proposals and essays), and *Derecho Ecológico Peruano—Inventario Normativo* (Peruvian environmental law).

RAIZ—Centro para el Desarrollo Regional (CDR)
Center for Regional Development
Calle Melgar 223, Cercado
Arequipa, Peru

Contact: **Carlos Machicao Pereyra,** Civil Engineer
   Phone: (51) 54 222972 • *Fax:* (51) 54 245970

Activities: Development • *Issues:* Sustainable Development; Waste Management/Recycling; Water Quality

● Organization
RAIZ identifies and interprets environmental impacts generated by civil works in the city of Arequipa, formulating recommendations for avoiding negative effects in subsequent infrastructure projects.

▲ Projects and People
Specifically, RAIZ seeks to determine the degree of **water, soil, and air pollution** in the city; propose corrective measures; and provide a foundation for adopting the proposal that the National Construction Regulations include a chapter on **Analysis of Environmental Impact.**

RAIZ researchers **Darwin A. Romero Alvarez** and **Natalia Vizcarra Noriega,** with advisor **Carlos Machicao Pereyra,** are establishing parameters and indicators of environmental impact to evaluate pollution in the city and in semirural areas elsewhere. Confronting the mistaken notion that natural resources are "inexhaustible sources of raw materials for modern society," RAIZ notes that the lack of environmental impact assessment (EIA) has resulted in "no clear predictions of the current environmental degradation." Obstacles are a lack of knowledge of practical methods of EIA, insufficient bibliographical information available in Arequipa, and inadequate contacts with international organizations working in the same area.

RAIZ intends to publish its findings, motivate civil engineering students to carry out research and future professional activities with a view to preserving the environment, and incorporate aspects of ecology into their teaching curriculum. The latter is "already being done in irrigation and physical hydrology classes," reports Pereyra, who teaches hydraulic engineering at Universidad Nacional San Agustín.

■ Resources
RAIZ seeks software equipment, teaching experience, magazines, and other educational materials.

Sociedad Peruana de Derecho Ambiental (SPDA)
Peruvian Environmental Law Society
Plaza Arrospide, No. 9, San Isidro
Lima 27, Peru

Contact: **Jorge Caillaux Zazzali,** President, Founder
   Phone: (51) 14 400549 • *Fax:* (51) 14 424365

Activities: Political/Legislative • *Issues:* Air Quality/Emission Control; Biodiversity/Species Preservation; Deforestation; Energy; Environmental Legislation; Global Warming; Health; Population Planning; Sustainable Development; Transportation; Waste Management/Recycling; Water Quality • *Meetings:* International Environmental Law Conference; IUCN; UNEP; World Parks Congress

● Organization

In 1987, Dr. Jorge Caillaux founded **Sociedad Peruana de Derecho Ambiental (SPDA) (Peruvian Environmental Law Society)**, the "only NGO in Peru dedicated entirely to environmental law and policy." With a present staff of 13, SPDA plans to expand and aims to create the "professionals, institutions, and legal instruments necessary to conserve our natural resources."

According to SPDA's literature, "We write environmental legislation, train young lawyers, provide legal counsel on environmental issues, do policy analysis, and provide consulting services for the Senate and Congress of Peru, the Executive, the Ministries of Agriculture, Mining, Tourism, and Fishing as well as local NGOs, National Chamber of Forestry, United Nations, Canadian International Development Agency (CIDA), U.S. Agency for International Development (USAID), Conservation International, The Nature Conservancy, World Wildlife Fund, and many others." SPDA belongs to the World Conservation Union (IUCN).

In 1990, with much SPDA involvement and Dr. Caillaux's leadership, Peru enacted its first comprehensive National Environmental Code—considered "a great victory for the Peruvian conservation movement." Following this action, the Peruvian Prime Minister assigned Dr. Caillaux as president of the Multisectorial Commission to write the regulations and norms to complement the Environmental Code. An SPDA staffer advises various Senate and congressional committees on environmentally relevant matters, including 16 bills.

▲ Projects and People

SPDA has worked for the **systematization of all existing environmental laws and regulations**, and is embarking on setting up a database—including the new regulations. As SPDA continues to try cases in court, public and political awareness is heightened. An ongoing public education program will entail sharing the SPDA database and "publishing and overview of legal and political issues facing Peru today." SPDA also intends to "maintain a media and public relations campaign to guarantee that our legal activities take their full and intended effect." An Environmental Law Gazette is also planned.

With adequate funding, SPDA says it will be prepared to "maintain a sustained presence in Congress, [otherwise] the code may fail to take any real effect in Peru." It hopes to appoint a **support team** to follow the application of the code, assist Dr. Caillaux, and introduce pertinent new laws.

Consulting work keeps SPDA "on the cutting-edge of day-to-day applications," guaranteeing its lawyers that they will "remain realistic in their policy advocation." It also enables the group to strengthen the environmental coalition between the public and private sectors and NGOs.

With a shortage of trained environmental lawyers, SPDA has developed three projects: the **Friday Workshop and Internship Program**, whereby 10 law students or recent graduates meet weekly and undertake their own environmental projects. Student interns also work with SPDA for up to 25 hours a week. Following two years of negotiations, Catholic University of Peru pioneered an environmental law course in 1992 with SPDA-developed materials. SPDA, which also organized short informal seminars at the University, expects that similar courses will soon be offered at other universities.

SPDA wants to begin publishing theses and other legal articles, such as "The Environment and Its Legal Protection in Peru: A First Analysis," "A Study of the Legal Constraints and Possibilities of Debt-for-Nature Swaps," and "A Legal History of the Development of the Peruvian Amazon."

As the Peruvian government undergoes decentralization and democratization, SPDA is continuing to strengthen its regional networks and plans to build its internship program and hold informal meetings and short seminars with lawyers, judges, and in universities in other cities. Small regional SPDA offices are anticipated in the future.

Enrique Ferrando Gamarra is executive director. **Fiorella Ceruti**, environmental education project director, has produced a teacher's guide on *The Amazon and Me.*

■ Resources

New technologies sought are computers, photocopiers, and office equipment. Environmental materials are needed for its legal library; SPDA hopes to disseminate its legal literature worldwide. SPDA seeks expanded contact with U.S. environmental agencies, Congress, United Nations, and other government institutions; media support; and information exchanges with other NGOs and foundations. It welcomes access to funding sources.

Resources available include national workshops, legal environmental articles, public access to its library, and the following papers, which are to be published: "Compilation and Systemization of Relevant Peruvian Environmental Law since 1904"; "Study of the Legal Norms, Institutional Structure and Public Sector Politics for the Management of the Peruvian Environment and Natural Resources"; and legal analyses of management at Rio Abiseo National Park, Huascarán National Park, and Pacaya Simiria Natural Reserve.

🏛 *Private*

Equipo de Desarrollo Agropecuario de Cajamarca (EDAC)
Agricultural Development Team of Cajamarca
Las Casuarinos F-3, 340 Urb. El Ingenio
Cajamarca, Peru

Contact: **Luis B. Guerrero Figueroa**, Executive Director
Phone: (51) 44 922924 • *Fax:* (51) 44 923429

Activities: Education; Promotion; Research; Rural Development • *Issues:* Agricultural Management; Biodiversity/Species Preservation; Population Planning; Sustainable Development; Water Quality • *Meetings:* Coordinating Committee for Andean Technology; Latin American Consortium on Agroecology and Development

● Organization

"The short-term role of our institution is to propose methods of regional development, focusing on the management of natural resources, and to help the community play a role in policy decisions at the regional level," writes **Equipo de Desarrollo Agropecuario de Cajamarca (EDAC) (Agricultural Development Team of Cajamarca)**. EDAC is oriented toward recovering its ecosystem's equilibrium through management of water, soil, plants, and animals in the area called "sub cuenca," or lower river basin. Its Central Development Committee was formed to represent this area's small settlements and take responsibility for resource management.

▲ Projects and People

Among the Committee's areas of activity are **agroforestry** regarding native forest species, **soil conservation** on farming terraces, breeding of alpaca, textile crafts using **vegetable dyes**, and growing of **crops** of Andean species. Crop rotation and strip planting are employed as well as **organic fertilizers** for compost—together with modern techniques such as mechanization to develop **sustainable agriculture** in the river basin. Through sharing experiences and developing proposals, the Committee attempts to influence the environmental policies of political bodies at all levels.

■ Resources

EDAC seeks information on world agroecological advances, support for exchange of experiences and professionals, and teams for chemical analysis of soil and selection of alpaca fibers.

Publications printed in Spanish include *Encounter Magazine* and brochures and research reports on *El Agua* (Water), *Lucha Contra el Cólera* (The fight against cholera), *Construcción de Letrinas* (Latrine construction), *Sistematización Promoción Forestales Nativos* (A system for promoting native forests), *Construcción Sistema de Agua Portable* (Construction of a drinking water system), and *De Nuestras Chacras Salen Las Comidas* (Our food comes from farms).

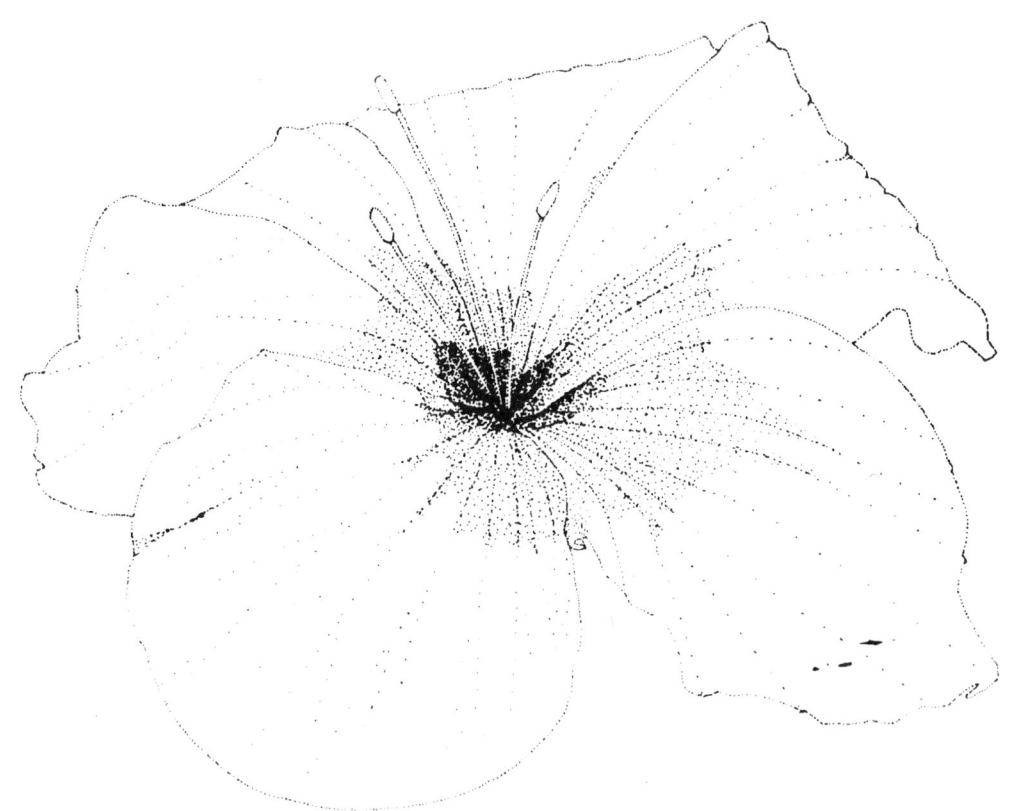

The Republic of the Philippines is an archipelago composed of over 7,000 islands along the southwest rim of Asia. Sixty percent of the total land mass is on the islands of Luzon and Mindanao. It is one of the most disaster prone areas in the world, subject to typhoons, flooding, landslides, earthquakes, volcanic eruptions, tsunamis, drought, and fires. The large and growing population and comparatively weak national economy have led to widespread poverty which is exacerbated by concentrated ownership of the land.

The country's major environmental problems are: rapid deforestation; water pollution caused by mining operations and other industry; air pollution, especially in urban areas, again due to industry and to some extent motor vehicles; and the loss of natural habitats. The coastal reefs, in particular, are expiring due to damage by eroded river silt and dynamiting for fish. Mangrove swamps also have been dramatically reduced due to cutting for forest products.

🏛 Government

Botany Division
National Museum
Old Congress Building, P. Burgos Street
Manila, Philippines

Contact: **Domingo A. Madulid, Ph.D.**, Officer-in-Charge
Phone: (63) 2 476887 • *Fax:* (63) 2 461969

Activities: Research • *Issues:* Biodiversity/Species Preservation; Deforestation • *Meetings:* CITES; IUCN; SSC

▲ Projects and People

The **National Museum's Botany Division** has four environmental projects underway. A **Survey of Critical Plant Sites in the Philippines** is financed by the World Wildlife Fund (WWF), and a debt-for-nature swap is financed through the Haribon Foundation. **Dr. Domingo A. Madulid** is the principal investigator.

The **Flora of the Philippines** is a joint project of the National Museum and the Bishop Museum, Honolulu, Hawaii, USA. Funding is from the National Science Foundation and U.S. Agency for International Development (USAID). Dr. Madulid serves as the co-principal investigator. He is also a Regional Member of the Species Survival Commission (IUCN/SSC) and belongs to its Palm Group.

The **Ecological Resource Assessment of Manila Bay** is being conducted in collaboration with the Bureau of Fisheries and Aquatic Resources. **Vegetation Mapping of Critical Terrestrial Habitats**, concerning the integrated protected areas system in the Philippines, is a project of the Department of Environment and Natural Resources.

■ Resources

The Museum needs funding to publish research results and other information for circulation in the Philippines and elsewhere.

Publications include: *A Pictorial Guide to Philippine Ornamental Plants; A Dictionary of Philippine Plant Names; A Bibliography on Philippine Flora, Vegetation and Biodiversity;* and *A Bibliography on Philippine Ethnobotany and Ethnopharmacology.*

Environmental Research and Development
   Program (ERDP)
**Philippine Council for Agriculture, Forestry,
and Natural Resources Research and
Development (PCARRD)**
Los Baños 4030, Laguna, Philippines

Contact: **Rogelio C. Serrano, Ph.D.,** Coordinator
Phone: (63) 94 50014 • *Fax:* (63) 94 50016

Activities: Development; Research • *Issues:* Biodiversity/Species Preservation; Deforestation; Energy; Sustainable Development; Upland Development; Waste Management/Recycling; Water Quality

▲ Projects and People

Within the Environmental Research and Development Program (ERDP), Dr. Rogelio C. Serrano, among a staff of 320, is undertaking a project on the Ecological Rehabilitation of Volcanic Ash–Laden Mt. Pinatubo and Adjoining Mountains. With Evangeline Castillo, Grassland and Degraded Areas Research Division, Ecosystems Research and Development Bureau, ERDP is developing a scheme to revegetate the mountainous areas that were devastated by the eruption of Mt. Pinatubo. "Seedlings are raised in the nursery and treated with mycorrhiza to promote fixation of nitrogen," reports Dr. Serrano. "In field planting, soil amendment treatment employed to improve survival and establishment include composted materials from mixed grass/straw with manure, forest soil, and pure volcanic ash (control). The results have yet to be compared." He notes that "introduction of grass and other cover crop species in the volcanic ash–laden areas will also be explored later."

According to Dr. Serrano, ERDP contributes to the global environmental movement through coordinating and monitoring research on global warming, deforestation, and air and water pollution. For environmental groups, ERDP suggests making published information more widely available, attending international conferences/symposia/workshops on specific environmental issues, and developing bilateral and multinational projects and networks for information gathering. "Our organization can serve as organizer and host international conferences, Dr. Serrano writes." We can also arrange training and study tours."

He has published works on conserving Philippine mangroves, soil erosion, agroforestry, social forestry, sand dunes, fire as an obstacle to forest renewal, paper and pulp industry, pineapple farming, bamboo, rattan, sago palm, logging, and ipil-ipil as hedgerow species, for example.

■ Resources

ERDP is in need of sloping land agricultural technology to promote sustainable food production in the uplands as well as technical publications on the environment and natural resources, with research results. Its publications list includes semitechnical packages, bibliographies, abstracts, books, data, and proceedings.

🏛 *Local Government*

**Highland Agriculture and Resources Research
and Development Consortium (HARRDEC)**
Benguet State University, La Trinidad
Benguet 2601, Philippines

Contact: **Mary Ann Pollisco-Botengan, Ph.D.,**
            Program Officer, Regional Applied
            Communication
Phone: (63) 74 22176

Activities: Development; Research • *Issues:* Biodiversity/Species Preservation; Deforestation; Sustainable Development
Meetings: World Parks Congress

● Organization

According to Dr. Mary Ann Pollisco-Botengan, Highland Agriculture and Resources Research and Development Consortium (HARRDEC) is "the lone highland government organization in the Philippines concerned with the development of usable, practical, adaptable, profitable and environmentally acceptable technologies or information on various agriculture and natural resource related commodities."

Believing that the manner by which "society treats its highland areas directly affects the nation as a whole, especially outlying lowland areas," HARRDEC is committed to "establishing linkages in efforts of conserving whatever is left of highland resources to assist in sustainable development programs."

One of several consortia organized by the Philippine Council for Agriculture, Forestry, and Natural Resources Research and Development (PCARRD), it is responsible for "planning, reviewing, coordinating, integration, and monitoring" such programs in the highlands of northern Philippines."

HARRDEC also disseminates applicable technologies and useful information developed or generated by its 13 member agencies: Abra State Institute of Sciences and Technology (ASIST); Agricultural Training Institute (ATI), Department of Agriculture (DA); Benguet State University (BSU); Bureau of Agricultural Research (BAR/DA); Department of Agriculture (DA), Cordillera Administrative Region (CAR); Department of Environment and Natural Resources (DENR/CAR); Department of Science and Technology (DOST), Region I/CAR; Ecosystems Research and Development Bureau (ERDB/DENR); Ifugao State College of Agriculture and Forestry (ISCAF); National Economic and Development Authority (NEDA), Region I/CAR; Philippine Council for Agriculture, Forestry, and Natural Resources Research and Development (PCARRD); Philippine Council for Aquatic and Marine Resources Research and Development (PCAMRD); and Philippine Textile Research Institute (PTRI).

The Consortium Secretariat has a staff of 15, including executive director Dr. Percival B. Alipit. Total staff of the member agencies is more than 500.

▲ Projects and People

HARRDEC has three regular programs, and a **Special Concerns** Unit which is responsible for the management of special projects under the aegis of the Consortium Secretariat.

The Research and Development Management Support Program (RDMSP) is involved in Consortium research and development, program and project planning, monitoring, review, and evaluation. The **Regional Applied Communications Program (RACP)** focuses on **technology transfer** and works to promote "the widespread adoption of useful research-based technology among the greater number of small farmer-producers so that they may benefit from the regional investment in science and technology."

The **Management Information Systems Program (MISC)** handles data processing activities.

Recent highlands projects include the use of indigenous farm wastes for **mushroom production**, improving farming systems, studying **vegetative terracing in agroforestry** area, preventing crop insect pests and diseases, **biomass** production, "effectiveness of *trichoderma* fungus in the decomposition of various substrates," fuelbreak construction as an aid for **fire control, soil conservation in abandoned mine areas,** technology transfer agroforestry demonstration farm, leguminous cover crops to **minimize soil erosion,** reforestation, "tiger grass–based upland community development project," production of **pharmaceutic/medicinal and aromatic plants** in the Cordillera region, seed production, wildfood production, survey of forestry-related problems of the Ifugao **woodcarvers,** improving rice farming, and Benquet pine forest studies. Research stations are concerned with development of national crops, dairy farms, root crops, silk industry, and ecosystems.

■ Resources

HARRDEC publishes the quarterly *Highland Express* newsletter. A list of abstracts of completed research is available. HARRDEC welcomes "information exchanges regarding technological breakthroughs, state-of-the-art research and development efforts, and documentation of events as they transpire across boundaries," writes Dr. Pollisco-Botengan, who reminds, "Information is power."

NGO

Asian NGO Coalition for Agrarian Reform and Rural Development (ANGOC) ⊕

47 Matrinco Building
2178 Pasong Tamo Street, Makati
Metro Manila 1200, Philippines

Contacts: **Antonio B. Quizon,** Executive Director
Roel R. Ravanera, Program Officer
Phone: (63) 2 816 3033 • *Fax:* (63) 2 815 1198
E-Mail: Econet:angoc; MCI:angoc pn

Activities: Development • *Issues:* Deforestation; Sustainable Development • *Meetings:* UNCED

● Organization

The **Asian NGO Coalition for Agrarian Reform and Rural Development (ANGOC)** is an "autonomous, nonprofit, Asian regional association of development NGOs" who operate locally, regionally, and nationally within the framework and perspectives of the World Conference on Agrarian Reform and Rural Development (WCARRD).

It aims to "provide a continuing forum for exchange of experiences, ideas, knowledge, and information; promote solidarity and partnership among NGOs so that they become effective democratic pressure groups for people-centered development; serve as a liaison and voice to articulate the perspectives and concerns of Asian development communities at various national regional and international fora; promote South-South and South-North dialogues as well as technical and financial cooperation; and act as a catalyst of selected programs related to training/human resource development/capacity building of NGOs, policy research and advocacy of selected issues, documentation and publications for development education."

ANGOC members include: **Bangladesh**—Association of Development Agencies in Bangladesh (ADAB), **Dr. Khawja Shamsul Huda,** executive director; **India**—Asian Institute for Rural Development (AIRD), **M.V. Rajasekharan,** executive trustee and coordinator; Association of Voluntary Agencies in Rural Development (AVARD), **A.C. Sen,** president; Center for Agrarian Research, Training, and Education (CARTE), **Amar N. Seth,** director; Consortium on Rural Technology (CORT), **Y.K. Sharma,** secretary; Gandhi Peace Foundation (GPF), **Shri Radhakrishna,** secretary; Indian Institute for Regional Development Studies (IIRDS), **K. Matthew Kurian,** director; MYRADA, **Aloysius Fernandez,** executive director; South Asia Rural Reconstruction Association (SARRA), **Dr. G.N. Reddi,** chairman; South Gujarat Rural Labor Trust (SGRLAT), **J.D. Chauhan,** chairman.

Also, **Indonesia**—Bina Desa, **Mr. Kartjono,** executive director; Bina Swadaya, **Bambang Ismawan,** director; Wahana Lingkungan Hidup, (WALHI), **M.S. Zulkarnaen,** executive director; **Malaysia**—Management Institute for Social Change (MINSOC), **Bishan Singh Bahadur,** executive director; **Pakistan**—Rural Development Foundation of Pakistan (RDFP), **Dr. M. Sadiq Malik,** executive president; **Philippines**—Center for Community Services (CCS), **Enrico Garde,** acting director; Philippine Partnership for the Development of Human Resources in Rural Areas (PHILDHRRA), Philippines, **Cresente Paez,** national coordinator; Southeast Asia Rural Social Leadership Institute (SEARSOLIN), **Fr. Antonio Ledesma,** director; **Sri Lanka**—Agroskills, **Ravi Abayagunawardana,** managing director; National NGO Council of Sri Lanka (NNGOC), **L.M. Samarasinghe,** chairman; Sarvodaya Shramadana Movement, **Dr. A.T. Ariyaratne,** president; UCAGRAM Foundation, **Chandra de Fonseka,** chairman; and **Thailand**—THAIDHRRA Foundation, **Paiboon Chareonsap,** executive director.

ANGOC supports a "universal code of environmental conduct" that emphasizes economic and social justice, care of ecosystems, respect for nature, and initiating a process to establish a "global convention on environmental rights and duties."

▲ Projects and People

A major ANGOC undertaking is the **Sustainable Development and Environmental Action Program** to alleviate poverty and promote grassroots initiatives. As part of this effort, ANGOC participates in global conferences, such as **Our Common Future: Making It Happen;** "International Citizens Conference on the World Bank, Environment, and Indigenous Peoples" to bring about "policy reform within multilateral development banks"; and the United Nations Conference on Environment and Development (UNCED).

It later collaborated with other regional and national NGO networks representing 500 groups to conduct the **Southeast Asia Regional Consultation on People's Participation in Environmentally Sustainable Development.**

ANGOC continues with its international citizens campaign to bring about a "fundamental shift in policies and practices of bilateral and multilateral development institutions," particularly banks (MDBs) that influence Third World development. According to ANGOC, NGOs now participate in the World Bank project sector.

Another major program is NGO Institutional and Capability Strengthening to develop managerial and technical capabilities. ANGOC collaborates on the *NGO Management* newsletter. Workshops are held on strategic networking, and study tours regarding natural resource development are arranged.

Influencing public policy on strategic development issues and increasing the recognition of the NGO role in development activities are main goals pursued through the Public Policy Dialogue, Development Education and Cooperation program.

Rural development projects in the member countries focus, for example, on social forestry (in cooperation with United Nations Development Programme [UNDP] and Food and Agriculture Organization [FAO]), training women leaders, land reform, freshwater and fishpond improvements, community-based mushroom production, alternative technologies, expansion of rice trading operations, and "economic empowerment for farmers, fisherfolk, and small vendors."

■ Resources

ANGOC publishes the bimonthly *Alternatives* newsletter and the quarterly *Lok-Niti* journal; and provides special reports, occasional papers, NGO database, and consultancy services. Proceedings are available from the Southeast Asia Regional Consultation on People's Participation in Environmentally Sustainable Development.

Ecological Society of the Philippines
53 Tamarind Road
Forbes Park, Makati
Metro Manila, Philippines

Contact: **Antonio M. Claparols**, President
Phone: (63) 2 810 9962 • *Fax:* (631 2 631 7357

Activities: Education; Law Enforcement; Political/Legislative
Issues: Biodiversity/Species Preservation; Deforestation; Health; Sustainable Development; Transportation; Water Quality • *Meetings:* IUCN

● Organization

The Ecological Society of the Philippines, with a staff of 25 and a nationwide membership of 500, is primarily concerned with issues affecting its nation as well as Southeast Asia. It is affiliated with the World Conservation Union (IUCN).

"It is time that the value of our dwindling natural resources, the protection of our ecological balance, the relationship between man and nature be incorporated in the [Philippines]' new constitution," President Antonio Claparols wrote earlier. He points to the fact that the Philippines once had "one of the richest tropical rainforests in the world. . . . Today, as reported, we have less than two million hectares left out of 12 million hectares and our population is over 54 million. Our water and air are polluted, our rich topsoil continues to erode as our forest continues to be cut."

Claparols also expresses concern about depleted and exploited ocean resources; and points to Malaysia's successful marine turtle protection program to safeguard breeding and nesting grounds. He has urged his government to explore new polyculture techniques to raise prawns sustainably and overcome pond viruses and bacteria, for example.

He also writes articles for newspapers and magazines on corals, mangroves, soil erosion, water conservation, typhoons and volcanoes among other natural disasters, world hunger, ozone hole, and Earth Day. According to Claparols, ecologists and industrialists need more dialogue.

■ Resources

Ecological articles from *Impact Magazine* are available. The Society seeks political support from Congress, and expanded educational and media outreach for greater environmental awareness.

International Center for Living Aquatic Resources Management (ICLARM) ⊕
MC P.O. Box 1501, Makati
Metro Manila 1299, Philippines

Contact: **Roger S.V. Pullin, Ph.D.**, Director, Aquaculture Program
Phone: (63) 2 818 0466 • *Fax:* (63) 2 816 3183
E-Mail: Cgnet:iclarm; Sciencenet:iclarm.manila

Activities: Research • *Issues:* Biodiversity/Species Preservation; Sustainable Development; Water Quality

● Organization

"Aquaculture is relatively new and strange for many organizations concerned with Third-World development and may be misconceived from impressions [of] intensive commercial aquaculture systems topical of the 'North,'" writes Dr. Roger S.V. Pullin, aquaculture program director, International Center for Living Aquatic Resources Management (ICLARM). "Most agriculturists are unaware of the merits of aquatic food production, particularly its high efficiency and scope for integration with Third-World agriculture."

Active in more than 20 countries, ICLARM operates an international aquatic resources center, where research is conducted on fish and other aquatic organisms. It aims to "improve the efficiency and productivity of culture and capture fisheries, . . . upgrade the social, economic, and nutritional status of peoples in the less-developed areas of the world through improvement of small-scale rural and subsistence market fisheries, [and] work toward the development of labor-intensive systems to aid employment and of low-energy systems to minimize capital and cost requirements." ICLARM also holds conferences at local to global levels.

With populations in Asian developing countries dependent upon fish as both a protein source and livelihood, as prices skyrocket while stocks dwindle and the environment degrades, and with less than 1 percent of farmers and coastal dwellers engaged in aquaculture or farm fishing in Asia, Dr. Pullin believes the time is ripe for a "blue revolution," which could be "largely based on low-technology systems by resource-poor farmers and fishermen."

According to Dr. Pullin, ICLARM has been working for more than 15 years to build scientific foundations for a blue revolution, with its research network involving some 2,000 scientists throughout the developing world. Without a sound technical base, he writes, "a frontier situation [is perpetuated] in which the fishing 'cowboys' take their short-term gains and the ensuing boom-and-bust cycles leave communities and the natural environment in ever worse shape. Fishing has always been risky but poor fisheries science and bad management made it more so."

Yet, "aquaculture is also risky," Dr. Pullin continues. "Whereas crops and livestock have been raised for centuries with confidence, most new fish farms are experimental. Farmed fish are scarcely domesticated. Pedigree fish are a future dream. Aquatic food webs are too poorly understood to be well managed. Aquaculture science is really only just beginning. Its further development is essential to attract new entrants."

▲ Projects and People

The **Coastal Areas Management Program (CAMP)** is an integrated approach in influencing regional political leadership and local communities to "push for concrete action towards the sustainable utilization of renewable coastal resources."

Four goals are: "long-term sustainability of economic activities and the use of natural resources which are directly or indirectly linked to the coastal environment; reduction of resource use conflicts through proper planning; human resources development through training; and creation of public awareness through information dissemination, workshops, and seminars."

Biogeographical research studies are undertaken in numerous countries regarding resources of mangroves, coral reefs, fish and shrimp, and aquaculture. Also investigated are artificial reefs, environmental quality, wildlife resources, land use capability, and socioeconomic and legal and institutional studies. Programs try to integrate "population, health and nutrition, education, alternative livelihood, and protected area management."

The **Capture Fisheries Management Program (CFMP)** was started in 1988 "to focus on research relevant to the management of tropical and subtropical capture fisheries. . . . By developing management tools and methods based on its research, ICLARM works to equip developing country fisheries' scientists and managers with the means to . . . manage the fisheries resources for which they are responsible."

In 1990, CFMP launched new projects, FISHBASE, and Global Comparisons of Aquatic Ecosystems. FISHBASE is the "elaboration of a computerized database on fisheries. . . . It can be confidently expected that the project's aims will be reached by providing key information on the identity, biology, ecology, and status of basically all commercially important tropical and subtropical fish species of the world." The **Global Comparisons of Aquatic Ecosystems** project is designed to "encourage and support ecosystem modeling by researchers in developing countries [and] . . . to conduct comparative studies on the functioning of aquatic ecosystems with special reference to their sustained exploitation and management."

The **Aquaculture Program** stresses "research for the development of better breeds and better farming systems for organisms best suited to low-income producers, raising produce mainly for low-income consumers. This requires systems with high biological and resource use efficiency." The focus is on "African tilapias—which we call the 'aquatic chicken,'" says Dr. Pullin, "and carps to the giant clams of the South Pacific; these clams are the world's only truly self-feeding animals, having green algal cells in their living tissues that 'feed' on sunlight like plants."

ICLARM also has an **Information Program** which includes the ICLARM Library and their publications. The library contains books, monographs, dissertation and other hard-copy materials, and computerized databases. A computerized system, using Compleat ELEFAN software, assesses the health of fish stocks and is used as a training tool. ECOPATH II now measures "the steady-state flow of nutrients through an ecosystem" and their ecological impact. A simple geographic information system (GIS) is also employed for biological and economic analyses. In concert with state-of-the-art technology, ICLARM takes a "holistic approach" by bringing together "senior politicians, policymakers, planners, local community leaders, farmers, fishermen, and scientists" to help solve coastal zone problems.

■ Resources

ICLARM publications include conference proceedings, technical and annual reports, studies and reviews, bibliographies, posters, translations, directories, software series, newsletters, education series, magazines, and the Asian Fisheries Society literature. A catalog of publications is available.

(See also International Marinelife Alliance in the USA NGO section.)

World Ecologists (WE) Foundation ⊕
Gold Building, 15 Annapolis
Greenhills, San Juan
Metro Manila, Philippines

Contact: **Charley Barretto**, Founder/President
Phone: (63) 2 722 4016 • *Fax:* (63) 2 721 3356

Activities: Education; Political/Legislative • *Issues:* Air Quality/Emission Control; Biodiversity/Species Preservation; Deforestation; Global Warming; Waste Management/Recycling
Meetings: UNCED

● Organization

The **World Ecologists (WE) Foundation** is comprised of "individuals dedicated to the conservation of our natural resources, to the improvement of the environment; and to the maintenance of ecological balance in the Philippines. Realizing that environmental problems know no political boundaries and have repercussions all around the globe, this ecological movement is a World movement."

Founded in 1988 by **Charley Barretto**, WE says it has grown to more than four million members and is affiliated with Greenpeace International and the World Conservation Union (IUCN).

Organized as a citizen's action group, WE informs Filipinos about ecological issues, and mobilizes them to take action toward their solution. WE believes that adequate legislation and research programs are in place to protect natural resources. It remains for the citizenry to abide by and carry them out, says WE. Each member pledges to plant one fruit tree each month in a public place.

▲ Projects and People

"Imagine all the streets, highways and public parks filled with fruit-bearing trees. People can just reach up and pick a fruit anytime they were hungry, at the same time benefitting from improved environment." Such is the concept of the **Fruitopia Project**, which, so far, has planted more than 1.2 million fruit trees "nationwide in all public places such as highways, parks, [and] schools" to produce food, cool the air, filter pollutants, provide a habitat for birds and other species that eat flies and mosquitoes, add beauty and shade, and prevent soil erosion and flooding.

Seedling banks are being established, in cooperation with the military at Villamor Air Base, Camp Capinin in Tanay, Fort Bonificio Naval Station, and Camp Crame. The **Punong Yaman** livelihood project provides materials for seedling banks to orphanages, homes for older residents, and other institutions. When ready for planting, the seedlings are bought back. So far, some 50 banks are in operation and "purchases have reached 500,000 pesos." **Corporate seedling banks** involve "home-raising of seedlings by private business—management and labor—as its contribution to the improvement of our ecological balance," reports Barretto. Seeds are also collected from restaurants, schools, and fruit processors.

On Fridays WE observes Filipino Day—*Iba Na Ang Pilipino*—by "wearing Filipino attire, speaking Pilipino, eating Filipino food, listening to Filipino music and buying Filipino products."

WE introduces educational, awarding-winning television shows, such as *The Living Planet, The Man Who Planted Trees, First Eden, Commandos of Conservation, Towers of Wax and Paper Palaces, A Killing Rain, The State of the Planet, Secret Weapons, Through Animal Eyes, Supersense*, and the *Karvonen* series.

■ Resources

WE needs technologies and information to assist in growing city trees, television programs, and educational materials. It publishes the quarterly *WE Newsletter*.

▓ *Private*

Office of the Environment ⊕
Asian Development Bank (ADB)
No. 6 ADB Avenue, Mandaluyong
Metro Manila 1501, Philippines

Contact: **Bindu N. Lohani,** Assistant Chief
   Phone: (63) 2 632 6882 • *Fax:* (63) 2 741 7961

Activities: Development • *Issues:* Sustainable Development

● Organization

The Asian Development Bank (ADB) writes that it is "committed to the idea that development need not come at the expense of the environment" and is challenged as well by the implementation of sustainable development. Involved in environmental programs for more than two decades, the Bank established an Environmental Unit in 1987, and upgraded it to the Office of the Environment in 1990. "Bank environmental officers work closely with other Bank staff, with national line agencies, with regional and subregional institutions, and with nongovernmental organizations (NGOs). The Bank has acted as a catalyst for communication and consultation between donors, planners, and developing member countries (DMCs)."

ADB hopes that it can stimulate sustainable development. The population of the Asia-Pacific region is expected to increase by 40 percent in the next 15 years, and the urban population is expected to increase by twice that amount. This will create major environmental stresses and challenges. It is for this reason that ADB now integrates environmental considerations into "program and project cycles." This includes providing technical advice, reviewing and monitoring programs, and performing impact assessment. The Bank helps mobilize resources for countries to encourage them to undertake environmental projects. In 1991, ADB lent $1.14 billion for environment-oriented projects, according to its literature.

The Bank works to forge links with the relevant government officials in the member countries, regional and subregional groups such as the ASEAN (Association of South East Asian Nations) Environmental Program, and with NGOs. Some projects are administered by indigenous NGOs.

▲ Projects and People

The Bank has identified major action areas, such as tropical forest management, biodiversity conservation, agriculture development, fisheries development, poverty alleviation, industrial efficiency, integrated economic and environmental planning, energy sector development, urban environmental improvement, social and economic analysis, and institution building and human resource development.

One example is **Forest Rehabilitation in Nepal.** In 1990, the Bank reports that it gave an interest-free loan of roughly $40 million and a technical assistance grant of $698,000 to Nepal. This loan will "help the government reforest 29,000 hectares of degraded forest, improve management of 75,000 hectares of natural forests, restore major watersheds in 13 of the country's 75 districts, establish and maintain 7,000 hectares of nature reserves, set up 5,000 biogas plants, and encourage the cultivation of medicinal and aromatic plants." In addition, the loan will enable government to formulate environmental policies consistently based on the country's development needs.

Another Bank project involved studying **energy sector development** in Bangladesh, Indonesia, Myanmar, Vietnam, India, and the People's Republic of China. "The regional study provides the Bank and DMC governments with the framework for formulating medium- and long-term energy strategies," ADB states.

An interest-free loan in 1990 went to **Bangladesh** to help upgrade the urban environment. The $43 million is allowing "10 towns to improve roads and drainage systems, provide low-cost sanitation and solid waste facilities, upgrade slums and markets, and rehabilitate water supply systems."

Thailand and the Philippines are recipients of environmental loans and/or environmental assistance grants. The **Songkhla Lake Basin, Thailand,** gained funding for feasibility studies and other environmental planning to foster the region's economic and social development. Innovative **fisheries'** research projects in both the Philippines and Malaysia receive support as do small-scale **aquaculture** operations that benefit the poor. In 1988, the Philippines gained nearly $122 million in loans (with a maturity of 35 years with a 10-year grace period and a 1 percent per year service charge) to help restore its ravaged forests that are causing soil erosion, degradation of watersheds, and sedimentation in rivers, reservoirs, and canals. Components of this forestry sector's investment program include reforesting 358,000 hectares of denuded lands, "reestablishing forest cover over 50,000 hectares of major watershed areas," improving timber stands, and rehabilitating 120,000 settler families.

A Sri Lanka loan supports "rehabilitation of degraded irrigation and drainage systems." ADB has given technical assistance grants "to strengthen environmental management capabilities in the Pacific Islands." The grant of $600,000 was given to the South Pacific Regional Environment Programme (SPREP) to help the nine member countries prepare country reports for the United Nations Conference on Environment and Development (UNCED). A further grant of $900,000 "will support environmental training activities, and will assist governments in developing environmental management strategies."

■ Resources

ADB literature includes guidelines, handbooks, information brochures, paper series, project cycle review reports, proceedings, profiles of selected developing member countries, studies and reports. A publications list, including the booklet *The Environmental Program of the Asian Development Bank*, is available on request.

Universities

Public Education and Awareness Campaign for the Environment (PEACE)

Miriam College Foundation

Katipunan Road

Diliman, Quezon City, Philippines

Contact: Angelina P. Galang, Ph.D., Coordinator
Phone: (63) 982421 • *Fax:* (63) 996233

Activities: Developmental; Education • *Issues:* Air Quality/ Emission Control; Deforestation; Environmental Education; Sustainable Development; Waste Management/Recycling

● Organization

The Public Education and Awareness Campaign for the Environment (PEACE), part of Miriam College Foundation's Environmental Education and Research Center (EERC), "seeks to spread awareness of the need for individuals, groups, and people to link arms around the globe and care for it as their common home [sounding] the call for change in values, attitudes, and behavior."

Initiated in 1986, Miriam-PEACE believes that it is important to instill environmental ideals in children in order to preserve the environment. "The wants, cultivated by media and the modern lifestyles that children are reared in, must be counteracted by the development of an environmental ethic early in their life."

Miriam-PEACE develops curriculum-forming projects and other activities for different levels, from elementary to secondary, and in the public and private sectors, in its attempt to educate the public and to enhance awareness.

▲ Projects and People

In the public education sector, Miriam-PEACE has created a National Curriculum Development Project for elementary and secondary schools, with funding from the World Wildlife Fund (WWF). These projects focus on environmental principles, con-

cepts, skills, and values, which are worked into everyday subjects. The project is done in collaboration with the other departments at Miriam, including the Department of Education, Culture, Sports, and the Environmental Management Bureau (EMB).

Miriam-Peace has begun a Solid Waste Management Program (SWMP) which focuses on the growing and worsening refuse problem in the metropolitan Manila area. After a meeting of government and private sector representatives, Miriam-Peace formed several proposals on waste management treatment.

Each of the resolutions are being tested, with Miriam-PEACE receiving feedback on each. Response has been encouraging to one proposal of separating recyclable materials in homes, offices, and organizations. A second proposal involves pilot testing a recycling and composting materials program in the Loyola Heights community of both high-income and low-income groups. Though the program had its setbacks, the Homeowners' Association has developed an agreement with the Metro-Manila Authority that unseparated garbage will not be picked up. Through courses and hands-on demonstrations, Miriam-PEACE educated the public on problems and solutions of waste management.

In an attempt to test their effectiveness in networking, Miriam-PEACE organized a group of NGOs and the EMB in 1988 to take action against local problems, beginning with the abundance of smoke-belching automobiles in the metro Manila area. With the Philippines Department of Energy and Natural Resources (DENR), the group assisted apprehension and fine collection regarding smoke-belching automobiles.

Another project in which Miriam-PEACE worked with the government is the reforestation of uplands which had been stripped bare. PEACE contracted the DENR to reforest 50 hectares of land in Bulacan. Area residents were informed about deforestation, and citizen involvement in reforestation was encouraged. The community's commitment surfaced when "a fire broke out in the neighboring area and the citizens rushed to our site to ensure the affectivity of our fire breaks and to do what had to be done to protect our sites."

Miriam-PEACE sponsors a radio program which "aims to deliver environmental education to the general public." The show features interviews with environmental workers and audience participation contests. Listeners can join the Radyo Kalikasan Club, "whose aim is to provide a venue for environmental involvement."

Educational sessions, which range in length and age groups, are held by PEACE. At one- or two-day ecology camps, participants spend time outdoors and "view environmental films, laboratory species, and the heavens through the Manila Observatory Telescope." Seminars and workshops span from two hours to five days for grassroots organizations, communities, industries, and religious congregations.

Miriam-PEACE is also a lobbying organization. Through open letters to government officials, it alerts the public, media, and officials about environmental problems, such as commercial logging.

Commissioned by DENR, a youth mobilization program evaluation "monitors and gives follow-up environmental education as needed to college groups and schools who have been given training by the DENR."

■ Resources

Miriam-PEACE seeks information and training on biogas systems for different types of institutions and also needs computer hardware and software for the development of environmental education materials.

Marine Laboratory
Silliman University
Dumaguete City 6200, Philippines

Contact: **Angel C. Alcala, Ph.D.**, President
   Phone: (63) 3522 4251 • *Fax:* (63) 3522 3200

Activities: Development; Education; Research • *Issues:* Biodiversity/Species Preservation; Community Development; Deforestation; Health; Marine Resources Management; Population Planning; Sustainable Development • *Meetings:* Coral Reef, Fisheries Congresses

■ Resources
Silliman University's president, **Dr. Angel C. Alcala**—who belongs to the Species Survival Commission (IUCN/SSC) Crocodile, Chiroptera, Marine Turtle, and Coral Reef Fish groups—offers a list of publications on marine biology as well as vertebrate ecology and systematics. Sample titles include "Comparative Growth of Hard Corals on Natural and Artificial Substrata in the Central Philippines," "A Direct Test of the Effects of Protective Management on Abundance and Yield of Tropical Marine Resources," and "An Annotated Checklist of the Taxonomic and Conservation Status of Land Mammals in the Philippines."

The **Marine Laboratory** seeks new technologists to test for cyanide in fish and to aid in enforcement of regulations. Suggesting "cooperative research involving specialists" to enhance global environmental cooperation, the Laboratory welcomes research grants for conservation projects, especially for education and community outreach.

Project Planning and Development Office
Research and Development Foundation
University of Eastern Philippines (UEP)
University Town, Calarman
Northern Samar 6400, Philippines

Contact: **Rolando A. Delorino**, Head, Assistant Professor
Activities: Research • *Issues:* Biodiversity/Species Preservation; Deforestation; Sustainable Development • *Meetings:* EIA; wetlands management

▲ Projects and People
The **University of Eastern Philippines' Research and Development Foundation**, with support from the International Development Research Centre (IDRC), Canada, is involved in a project regarding the Socioeconomic Impacts of Mangrove Overexploitation to ensure the "long-term efficient use and management of mangrove resources." Ultimately, such programs aim to improve the well-being of the local people and to set up policies that will stem resource depletion and environmental degradation.

Project leader **Rolando A. Delorino**, with his research team, is developing a socioeconomic profile of those using mangroves for fuel to "determine the pattern and rate of land use transformation" occurring as a result of this and other destructive activities. Delorino and principal investigator **Imelda E. Gelera**, with research assistants **Pedro V. Destura, Jr., Mercedes G. Sosa**, and **Josephine N. Toleran**, will recommend policies for sustainable use of mangrove areas.

Dr. Enrique P. Pacardo and Delorino also undertook a project entitled **An Evaluation of Reservoir Sedimentation: The Case of Macagtas Water Impounding Project, Catarman, Northern Samar, Philippines** to study the rate of sedimentation in the Macagtas reservoir during a one-year period. Analyzing the relationship between water discharge and sediment inflow, they then projected the cost of sedimentation over the next 30 years both with and without watershed management, and found that watershed management would provide more than 26 million pesos' value than the same reservoir without watershed management. The researchers also determined that, if the present rate of sedimentation and water impoundment continues, "the dead storage volume capacity will be filled up" by the year 2001.

(See also Dalhousie University, Philippines Environment and Resource Management Project in the Canada Universities section.)

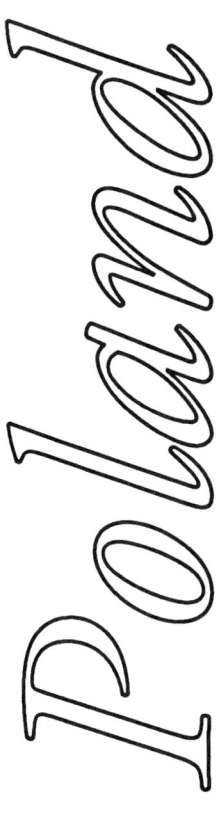

Poland

Poland shares borders with the former USSR, Germany, and Czechoslovakia, with the Baltic Sea lying to the north. The land along the southern boundary is mostly flat, while the southern boundary is mountainous. The country has abundant natural resources, including coal and natural gas, and nearly half the area is used for growing crops.

Poland is one of the most polluted industrialized countries in the world. The concentration on industrial production since World War II has resulted in a great deal of energy use (especially coal) and the resulting air and water pollutants; however, the current transition to a market economy should bode well for the environment, especially with the help of Western countries and international organizations. In the meantime, pollution has measurably affected human health, especially in the provinces of Krakow and Katowice, where about 35 percent of the population live. The forests also have suffered severe damage: about three-fourths of the trees show some injury from air pollution.

NGO

Fundacja Biblioteka Ekologiczna
Ecological Library Foundation
ul. Kościuszki 79
61-715 Poznań, Poland
Contact: **Jarek Fiszer,** Director
  Phone: (48) 61 521325 • *Fax:* (48) 61 528276
Activities: Education • *Issues:* Air Quality/Emission Control; Biodiversity/Species Preservation; Energy; Global Warming; Waste Management/Recycling; Water Quality

● Organization

The **Ecological Library Foundation** was created in 1990 to "provide information about the natural environment, its endangered state, and its protection in the form of an accessible bibliography in biological sciences, natural protection, health protection."

Individuals, foundations, and state and private firms in Poland and abroad support the Library, which also gets an assist from volunteers of the "Polish Ecological Club, Biology Studies, and Poznań high schools." The collection already has 20,000 volumes, and 15,000 issues from 400 various journals. Material is being collected from 31 different fields connected with environmental protection, such as "medicine, energy, economy, agriculture, law sciences, business and environmental technologies." All library material is publicly accessible; the staff, with head librarian Przemysław Biłozor, help users find required information.

In partnership with Poznań University—and of particular benefit to its students—the Library also distributes books to other universities, research institutions, and hospitals, for instance. In an effort to make important Western environmental information useful in Poland, the Library is translating key titles into Polish. *(See also Green Library in the USA NGO section.)*

■ Resources

The Library seeks both printed materials and popular and documentary movies for its collections.

Commission of Human Ecology (CHE) ⊕
International Union of Anthropological and Ethnological Sciences (IUAES)
00-330 Warsaw, Nowy Swiat 72, Poland

Contact: **Napoleon Wolański, Ph.D., D.Sc.,** Chairman
Phone: (48) 22 268312

Activities: Education; Research • *Issues:* Air Quality/Emission Control; Health; Human Ecology; Sustainable Development
Meetings: World Academic Conference on Human Ecology

● Organization

The Commission of Human Ecology, International Union of Anthropological and Ethnological Sciences (CHE/IUAES), is an "umbrella organization in human ecology on the world scale, related to the International Social Sciences Council (ISSC) and UNESCO," according to **Prof. Dr. Napoleon Wolański**, chairman. CHE/IUAES has up to 40 permanent members and an unlimited number of observers.

Prof. Dr. Wolański is also head of the **Department of Human Ecology, Polish Academy of Sciences (DHE/PAS)**, which conducts "field research on problems of environmental factors in human development and structure of human populations on conceptual human ecology." DHE/PAN has a staff of 11, with funding from the Polish government.

With some 750 papers and books about human ecology and auxolgy to his credit, Dr. Wolański contributed to the Commission's study—"Biosocial Status of Human Populations as an Indicator of Global Environmental Change"—as part of the program Human Dimensions of Global Environmental Change (HDGCP). This document influenced the World Academic Conference on Human Ecology (WACHE) in 1990, which emphasized "strong interdisciplinary cooperation between bio-medical and social sciences, [to promote a] systemic [concept] of ecosystems."

CHE/IUAES defines human ecology as "the transdisciplinary problem and scientific discipline about man and his culture and a dynamic component of ecosystems." CHE/IUAES supports greater attention to this field "as a scientific subject," which it believes has been lacking among other institutions, societies, and organizations in human ecology." CHE/IUAES is working on a "synthesis of problems of man and his culture as a dynamic component of ecosystems."

Dr. Anna Siniarska, associate professor, Department of Human Ecology, has written widely in her field, such as about "family socioeconomic conditions and the biological development of children" and "genetics and the motor development of man."

■ Resources

CHE/IUAES publishes the annual *Studies in Human Ecology* and *CHE Newsletter* and organizes WACHE every several years, publishing conference results. A July 1993 meeting is planned in Merida, Yucatan, Mexico.

Polish Association of Environmental Law
ul. Kuznlcza 46/46
50-138 Wroclaw, Poland

Contact: **J. Sommer,** President
Phone: (48) 71 444747

Activities: Education; Political/Legislative; Research • *Issues:* Air Quality/Emission Control; Biodiversity/Species Preservation; Deforestation; Sustainable Development; Waste Management/Recycling; Water Quality

▲ Projects and People

J. Sommer has been involved in environmental law since the 1970s. As president of the **Polish Association of Environmental Law** and a member of the National Council on Environmental Protection, Sommer is particularly interested in halting industrial pollution, advancing nature conservation, the "integration of environmental law" in such matters, and the inclusion of environmental protection in the political decisionmaking process.

■ Resources

Sommer's publications include material on air pollution, mining, environmental planning, and protection of recreational and health resorts, all printed in Polish.

The Association needs the cooperation of lawyers from both Eastern and Western Europe to exchange ideas and to prepare publications and reports. Access to databases on environmental information in as many countries as possible is especially sought. To launch a journal, financial support is needed.

Polski Klub Ekologiczny/Polish Ecological Club (PKE)
ul. Garbarska 9
31-131 Krakow, Poland
Contact: **Maria Gumińska, M.D., Ph.D.,** President, Professor
Phone: (48) 12 222098
Telex: PL 325424

Activities: Development; Education; Law Enforcement • *Issues:* Air Quality/Emission Control; Biodiversity/Species Preservation; Deforestation; Energy; Global Warming; Health; Sustainable Development; Transportation; Waste Management/Recycling; Water Quality • *Meetings:* FOE; IUCN; Regional Environmental Center, Budapest

● Organization

The Polish Ecological Club (PKE) was founded in 1980 as the "first independent, nongovernmental organization in Poland based on the principle of respect for human and natural resources." Organizer **Dr. Maria Gumińska** writes, "The scientists, physicians, journalists, and other concerned parties who gathered in Krakow to establish PKE were the first to break the barrier of silence on environmental problems in Poland."

A year later, the Club forced the closure of the Skawina aluminum plant's electrolysis division, a major source of toxic fluorides. The Club expanded and established 17 branches, but some were forced to close during the period of martial law. By 1990, the group had grown again to include 16 branches and approximately 5,000 members.

PKE encourages **ecological education** and contributes to that process through seminars, publications, bulletins, letters of protest, and books. Introducing the concept of sustainable development, it has been "particularly active in making eco-development a reality in Poland." Past activities include "preserving and extending Poland's national parks, evaluating public health, introducing biodynamic methods into ecological agriculture, protecting the air, water and soil against pollution, organizing 'days without cars,' creating the ecological library, opposing nuclear power stations, stimulating regulation of environmental law, saving energy, and protecting the urban and rural environment and the nation's cultural heritage."

Dr. Gumińska points out, "The historical hazards of the outdated Eastern European model of industrialization, as well as the new threats posed by a free-market transfer of dangerous technologies, have created serious ecological problems in Poland. Environmental conditions have been further aggravated by transboundary emissions. In light of these challenges, PKE aims to create conditions of sustainable development and a healthier environment for the country and for the region."

PKE sites the Constitution of Poland, the Stockholm Declaration of Human Rights, and the Geneva Convention for the basis of its legal rights.

▲ Projects and People

Even with Poland's ecological problems, Dr. Gumińska, PKE president, says her group "would like to help other organizations in their campaigns against dams on big rivers and against destruction of [the] rainforests. We have only one Earth and emissions do not stop at any border." PKE organized a joint "Polish-German-Czech-Slovakia" conference in early 1992 to work on solving the problems in the "Black Triangle" area in southern Poland. This conference took into consideration the consequences of pollution and a "program of pro-ecological government action," such as energy conservation, industry reorganization, and heating system changes. People involved include Wojciech Belbo, Piotr Poborski, Aureliusz Mikłaszewski, Katarzyna Klich, and Aleksandra Chodasewicz.

Aleksander Kalmus, Kazimierz Rabsztyn, Mieczysław Górny, Stanisław Wróbel, and Zbigniew Krysiński are working on a project to include ecological agriculture into the Polish government's agriculture program.

Helena Przybyła and Stanisław Maluty are involved in a project of communal waste segregation, collection, and recycling.

Cooperating with the Clean Air over Krakow Foundation, another project involves reducing low emissions from communal sources in Krakow, with the help of Ryszard Geyer, Józef Gęga, and Danuta Szymońska.

PKE is also concerned about clean water in Poland, so Wojciech Dąbrowski, Włodzimierz Wójcik, Juliusz Chojnacki, Jacek Lendzion, and Andrzej Baranowski are working on such problems in the main Polish rivers and the Baltic Sea.

"Global problems are also our own problems," says Dr. Gumińska. Consequently, PKE sponsors Earth Day celebrations for school children, including national ecological poster contests. "We are especially interested in the greenhouse effect, the problem of ozone layer, and threats for biodiversity."

■ Resources

PKE is in need of office equipment, including photocopiers, computers, and audiovisual installations.

Publications include an ecological book for school children and a newsletter entitled *To Be or Not To Be*. Polish translations are planned for *Water Is Life* and *Nature Protection*. Costs are being covered in part by the Polish Ministry of the Environment and with help from NGOs abroad. PKE's publication list is available including a reprint of Dr. Gumińska's *Air Pollution and Health Hazards in Some Regions of Poland*, which describes her nation as "one of the most coal-dependent countries in the world."

★ Universities and Institutes

Institute of Animal Systematics and Evolution
Polish Academy of Sciences
Slawkowska 17
31-016 Krakow, Poland

Contact: **Adam Krzanowski, Ph.D.**, Professor Emeritus
Phone: (48) 12 227513, ext. 219
Fax: (48) 12 222791

Activities: Research • *Issues:* Species Preservation; Systematics of Animals • *Meetings:* International Bat Research Conference

▲ Projects and People

Prof. Dr. Adam Krzanowski, Institute of Animal Systematics and Evolution—who has conducted bat studies for more than 30 years, was the "official bat conservation expert to the Polish Nuclear Agency," and has "contributed to the preservation of the famous Nietoperek undergrounds as a bat hibernaculum reservation"—can provide a list of publications concerned with **bat conservation**. He is currently working on the bibliography of bats, *Mammalia: Chiroptera, 1968–1977*. A corresponding bibliography covering the period 1958–1967 was published earlier in Bonn, Germany. Dr. Krzanowski belongs to the Species Survival Commission (IUCN/SSC) Chiroptera Group.

Chiropterological Information Center (CIC)
Institute of Animal Systematics and Evolution
Polish Academy of Sciences
Slawkowska 17
31-016 Krakow, Poland

Contact: **Bronisław W. Wołoszyn Ph.D., D.Sc.**, Associate Professor
Phone: (48) 12 221901

Activities: Education; Research • *Issues:* Biodiversity/Species Preservation • *Meetings:* European Bat Conferences; International Congress of Theriology

● Organization

Dr. Bronisław W. Wołoszyn founded the Chiropterological Information Center (CIC) in 1978, and in 1988 it was incorporated into the Institute of Systematic and Experimental Zoology (now Institute of Animal Systematics and Evolution) of the Polish Academy of Sciences. CIC is a small organization, with Dr. Wołoszyn and Danuta Wołoszyn comprising the staff. Cooperating with bat-banding problems is Dr. Wincenty Harmata, Jagiellonian University, Krakow.

CIC's primary goals are to create a database of all information on bats in Poland, to "promote systematic and biographical study on bats," and serve as a resource for the government and for scientific groups concerned with the protection of bats.

▲ Projects and People

CIC's major project is its **annual bat count** each February conducted with the help of more than 60 professional and amateur chiropterologists. Counted are bats hibernating in caves, mines, cellars, and tunnels. In 1991, the census found 34,511 bats in 143 hibernation roosts. CIC wants the yearly census to be a start to the constant monitoring of the bat population hibernating in Poland. "We are planning to prepare a monograph concerning the results of the bat counts after five years' investigation period," writes Dr. Wołoszyn.

CIC is also involved in the **EMMA project**—an atlas of European mammals—coordinated be **Dr. Francois de Baufort**, National Museum of Natural History, Paris. Wołoszyn is Poland's national coordinator for this atlas, which is to be completed in 1995 and is to include changes in species distribution over the past 25 years.

Dr. Wołoszyn, a member of the Species Survival Commission (IUCN/SSC) Chiroptera Group, is also involved in yearly **international expeditions to the Nietoperek Reserve**, an underground system of corridors and chambers that were created by the Germans before and during World War II. Some 20,000 bats now hibernate within this system every winter. These international expeditions are designed to show "the famous bat hibernaculum—the biggest in Central Europe—to the chiropterologists from Western Europe and also to establish contacts between chiropterologists from different countries." Assisting is **Dr. Peter Lina** of Holland.

■ Resources

The Center publishes the *Chiropterological Information Center Bulletin* and *Wszechświat Nietoperzy* (World of bats), which is a quarterly supplement to the monthly, popular science magazine *Wszechświat.*

Dr. Wołoszyn has an extensive bibliography on fossils and recent mammals of Central and North America, Europe, Middle East, and Far East, mainly with Spanish, Polish, and English titles.

The Center is in need of items difficult to obtain in Poland, such as recent publications on mammals, mist-nets for collecting bats, Sherman live-traps for small mammals, ultrasonic detectors for bats, and a computer.

Department of Vertebrate Ecology
Institute of Ecology
Polish Academy of Sciences
Dziekanow L. (near Warsaw)
05-092 Lomianki, Poland

Contact: Jerzy Romanowski

Activities: Research • *Issues:* Biodiversity/Species Preservation; Ecology

▲ Projects and People

Wildlife biologist and ecologist Jerzy Romanowski is a research assistant with the Department of Vertebrate Ecology, Institute of Ecology, Polish Academy of Sciences, whose current project is The Status and Conservation of Mustelids in Eastern Europe and [former] USSR. The Species Survival Commission (IUCN/SSC) Mustelid and Viverrid Group, of which Romanowski is a member, is cooperating with this study to complete and review information on mustelids, their status and protective legislation, and conservation and ecological studies.

A member of the Polish Zoological Society, Romanowski is concentrating on the **diet of the stone marten in Poland** in doctoral studies. He has also written the Polish book *Tracks of Animals,* and presented papers on the **effects of heavy metals on wildlife** and on ecological studies of otters in Eastern Europe and the former USSR. In the *Polish Ecological Studies* volume, which evaluates "ecological aspects of the transformation of a farmland into a suburban area," Romanowski contributed to sections on the **domestic cat** (*Felis silvestris f. catus*) and on the tropic ecology of *Asio otus* and *Athene noctua.*

■ Resources

Romanowski offers these publications for exchange: *Suburban Environment and Evaluation of Its Transformations, Part II, Biotic Environment* and *The Effect of Microorganisms, Heavy Metals, Pesticides, and Predators on Egg and Nestling Mortality of Granivorous Birds.*

Mammals Research Institute
Polish Academy of Sciences
17-230 Białowieża, Poland

Contact: Zdzisław K. Pucek, Director, Professor
Phone: (48) 12278 (operator assistance)

Activities: Research • *Issues:* Biodiversity/Species Preservation
Meetings: International Congress of Theriology; INTECOL Congresses

● Organization

In 1952, Professor A. Dehnel founded the **Mammals Research Institute** as a field station of the Department of Comparative Anatomy of Vertebrates, M. Curie-Skłodowska University. Five years later, the Institute became a research unit of the Division of Biological Sciences, Polish Academy of Sciences.

Benefiting the Institute's development was the presence of a large collection of small mammals that had originated in Białowieża National Park, described as "unique" and the "best preserved forest of lowland Europe."

▲ Projects and People

The Institute consists of 45 people who work in four teams: Biomorphology and Fauna, Ungulate Ecology, Ecology of Small Mammals, and Evolutionary Ecology. **Professor Zdzisław Pucek** heads the **Biomorphology and Fauna team**, which investigates the problems of morphological variability in populations of mammals, taxonomy, and fauna. **Dr. A.L. Ruprecht** is the curator of the continued effort to collect animals, and **A. Świerszcz** conducts a rodent breeding program.

Dr. M. Krasińska leads the **Ungulate Ecology team** that conducts studies on **European bison** ecology. Topics of study include "reproduction, population dynamics, age and social structure of the population, strategy of habitat use, activity rhythms, and some other aspects of behavior," according to Institute literature. It is expected that the bison project will "yield the rationales for the management of this protected species which now re-occupies the human-altered forests of Central and Eastern Europe."

The **Ecology of Small Mammals team**, with investigator **Dr. E. Rajska-Jurgiel**, concentrates on "the functioning of population of rodents in relation to the mosaic of habitats and diversified predator pressure."

Dr. J. Wójcik and the **Evolutionary Ecology team** study "chromosome and enzyme polymorphism, cyto-taxonomy, and population genetics." Cyto-taxonomic studies have been conducted on several species of bats and the common shrew.

Researchers prepare "extensive monographs" of Poland's mammal fauna and the chapters about certain endangered mammals for the **Polish Red Data Book**. Altogether, the staff has prepared about 650 scientific publications. In its collections, the Institute reports to have "173,000 specimens (plus 400,000 from owl pellets) and represents 271 species from 11 orders of mammals. The most valuable are the large series of shrews and small rodents from the Białowieża Forest . . . as well as the series of skulls of European hare, souslik, root vole, birch mouse, fox, polecat, marten, wild boar, red deer, and European bison."

Considered the "central mammalogical library in Poland," the Institute's library contains about 6,000 books, 9,000 reprints, and more than 10,000 volumes of over 800 various journals.

■ Resources

The Institute needs telemetry equipment, chemicals, computers, and software for its research. It publishes *Acta Theriologica: International Journal of Mammalogy*.

Museum of Natural History
Wrocław University
ul. Sienkiewicza 21
50355 Wrocław, Poland

Contact: **Andrzej Wiktor, D.re.nat.**, Professor
Phone: (48) 71 225041

Activities: Education; Research • *Issues:* Biodiversity/Species Preservation

▲ Projects and People

As a member of the Species Survival Commission (IUCN/SSC) Mollusc Group, **Prof. Dr. Andrzej Wiktor** is the specialist in **terrestrial slugs** of the Northern Hemisphere—with field studies in Poland, Bulgaria, Yugoslavia, Greece, Italy, Spain, Tadzhikistan, and Papua New Guinea. He reports that he has "accumulated probably the largest collection of *Gastropoda Pulmonata nuda*," which are at the University of Wrocław, where Prof. Dr. Wiktor has been director of the **Museum of Natural History** since 1980. Immediate past president of the Polish Zoological Society, he is co-author with **A. Riedel** of the Red Book paper entitled "List of Threatened Terrestrial Snails in Poland."

Portugal's mainland is divided into two regions, separated by the Tagus River. The area to the north is mountainous, rainy and cool; the area to the south is warmer and characterized by rolling plains. The country shares the Iberian peninsula with Spain. Natural resources include fisheries, tungsten, iron, uranium, and marble. Almost 40 percent of the land is forested; 30 percent is cropland. The country depends mostly on foreign sources for its energy supply.

The soil in Portugal is not of high quality and has been overworked, resulting in erosion of the topsoil. In the coastal areas, heavy discharge is a serious threat to wetlands, and is affecting underground water. Air pollution is caused mostly by heavy traffic and industry, which is also responsible for the production of 180,000 tons of toxic waste each year. The country is addressing the waste issue, however, through construction of a new system.

🏛 *Government*

Direção-Geral da Qualidade do Ambiente (DGQA)
General Directorate of Environmental Quality
Ministério do Ambiente e Recursos Naturais (MARN)
Ministry of the Environment and Natural Resources
Rua do Século, 51
1200 Lisbon, Portugal

Contact: **João Vila Lobas,** General Director
Phone: (351) 1 346 2751 • *Fax:* (351) 1 346 8469

Activities: Political • *Issues:* Air Quality/Emission Control; Biodiversity/Species Preservation; Consumer Protection; Deforestation; Global Warming; Sustainable Development; Waste Management/Recycling; Water Quality

● Organization

The **Ministério do Ambiente e Recursos Naturais (MARN) (Ministry of the Environment and Natural Resources)** is responsible for the promotion and coordination of aid to and participation in national environmental policy. Created because of growing environmental concern linked to Portugal's accelerated process of economic development, MARN is also committed to consumer protection and maintenance of the quality of life. The Ministry works with regional coordination committees (CCRs) and is involved in international cooperation. It carries out a national strategy for nature conservation, prevention of natural and industrial risks to the environment, and related scientific and technological research.

Among MARN's divisions are the General Directorate of Environmental Quality, General Directorate of Natural Resources, Cabinet of Nuclear Security and Protection, Environmental Consulting Commission (CCA), National Commission Against Marine Pollution (CNPM), National Institute for the Environment, National Institute for Consumer Protection, National Institute of Meteorology and Geophysics, National Institute of Water, Cabinet of European Affairs, Cabinet of Basic Health of Costa do Estoril, and the National Parks, Reserves, and Nature Conservation Service.

■ Resources

Paginas Verdes, or *Green Pages*, is a collaborative directory of the Direção-Geral da Qualidade do Ambiente (DGQA) (**General Directorate of Environmental Quality**) and the Portuguese Association of Environmental Engineers. It lists environmental organizations in Portugal, including government directorates, institutes, and agencies at the federal and regional levels; university departments; research centers and laboratories; businesses, including environmental consultants, industrial manufacturers, and for-profit "think tanks"; NGOs (including student, professional, and scientific associations) concerned with defending the environment, consumer protection, and renewable energy; and the Portuguese Green Party.

Areas of activity include: sustainable agriculture; environmental impact assessment (EIA); laboratory analysis; technical assistance; ecology; nature conservation; environmental education; consumer protection; environmental economics; renewable energy; geology and mining; protection of marine, water, and forestry resources; information services; urban planning; politics; water, air, soil, and noise pollution; recycling; and treatment of liquid, solid, and gaseous wastes.

This "Who's Who in the Environment" is printed in Portuguese. Information can be obtained from (351) 1 346 3241 or 346 3562 or fax 346 0150.

 *NGO*

Associação Portuguesa de Ecologistas/Amigos da Terra
Portuguese Association of Ecologists/Friends of the Earth (FOE)

Travessa da Laranjeira 1-A
1200 Lisbon, Portugal

Contact: **Antonio Eloy,** Director
Phone: (351) 1 347 0788 • *Fax:* (351) 1 347 3586

Activities: Development; Education; Political/Legislative • *Issues:* Biodiversity/Species Preservation; Energy; Global Warming; Sustainable Development; Transportation; Waste Management/Recycling • *Meetings:* EEB-AGM; FOEI-AGM; UNCED

▲ Projects and People

"The beating of a butterfly's wings in California can cause a typhoon in the China Sea."

According to **Amigos da Terra (Friends of the Earth) (FOE)**, environmental devastation and some agricultural and gardening practices are endangering butterflies. Like other plants and animals, the butterfly plays the role of biological indicator because it suffers or disappears completely when its habitat is seriously altered. They also play an important role within their ecosystem by pollinating flowers."

The **Project Butterfly—Olá Borboleta/Ciao Farfalla—campaign** began in 1987 at the initiative of the Swiss, German, Austrian, Luxembourg, and Italian chapters of FOE. On Earth Day 1990, a European Alliance for the protection of butterflies was formed, and Amigos da Terra is preparing a program of classroom activities for teachers and students.

Butterflies are food for many birds and bats, and their eggs and larvae are eaten by spiders, lizards, frogs, and wasps, among others, reports Amigos da Terra. Yet the role of insects is often overlooked in helping maintain nature's equilibrium. Causes of butterflies' extinction include variations in climate, diseases, parasite, and human factors that destroy habitats of meadows, marshes, riverbanks, forests, mountains, and gardens; civil and industrial construction; releases of chemical substances, such as pesticides, herbicides, fertilizers, and industrial and auto emissions; raising of pasture livestock; fires; draining of swamps and marshes; and heavy tourism.

Amigos da Terra writes that some butterflies that eat cultivated plants as fruit trees and vegetables could benefit from human activities. But often gardens and parks are treated with chemicals to discourage the presence of insects and green meadows are mowed to discourage the growth of wildflowers.

To save the butterflies, Amigos espouses research to map out their territories and habits, legal recognition of endangered species and protected areas, and the creation of new nature reserves. "The most direct and immediate way . . . is through educating both the general public and farmers (on sustainable and organic agricultural techniques) so that industry and tourism do not devastate rural areas and urban areas are looked upon as homes for both people and butterflies."

With thousands of butterfly species in tropical forests including those with wing spans greater than many birds and life spans a year long, Amigos da Terra takes part in the international campaign to save these forests.

Amigos da Terra offers these steps: Avoid polluting by using biodegradable, nontoxic, natural and recycled/recyclable products. Avoid littering, especially with cigarette butts and smoke, and practice energy efficiency, such as using public transportation. Do not capture butterflies—take photos or draw them instead. Garden with as few chemicals as possible. Encourage green spaces in the city. Lobby for government-maintained protected areas; preserve meadows, forests, swamps, and marshes. Buy organically grown food. Learn more about environmental problems.

According to Amigos da Terra literature, butterflies were first studied in Portugal by the Jesuits, and their research continues. With about 140 species in continental Portugal and many unique to Madeira and the Azores, endangered species appear in the *Red List of Endangered Lepidoptera*. As protected areas are created, some butterfly populations are experiencing a "boom."

Other projects being developed are **biotechnology** information and related problems, creation of an environmental **database** for Portuguese-speaking countries, coordination of environmental information of Mediterranean countries; promotion of **ozone-friendly** projects, **alternative energy** studies, **inventory of plant and animal species in cities**, and educational **colloquiums.**

■ Resources

Amigos seeks information on new technologies to develop projects and contact with potential sponsors. Available materials, printed in Portuguese, are *Olá Borboleta* (Hello, butterfly), *Ecologia Política* (Political ecology), *Nuclear em Portugal e no Mundo* (Nuclear in Portugal and the world), *As Florestas Tropicais* (The tropical forests), T-shirts, bumperstickers, calendars, and postcards.

Romania is bordered by Bulgaria, Yugoslavia, Hungary, the former Soviet Union, and the Black Sea. The Carpathian Mountains cross the country in the northeast; the Transylvanian Alps cross the center. About 46 percent of the land is used to grow crops; about half the population lives in rural areas.

Like other Eastern European nations, Romania has relied on heavy industry, including metallurgy, oil refining, and petrochemicals. Energy has been used inefficiently and environmental controls have been avoided, making industrial pollution a severe problem in some areas. The town of Copsa Mica in central Romania, with two highly polluting factories, is considered one of the most contaminated places in Europe. Water pollution and soil degradation, from particulate emissions, mining, and chemical fertilizer use, are other concerns. One encouraging development is that environmentalists have significant representation in the newly elected government.

🏛 *Government*

Romanian Marine Research Institute (RMRI)
Bd. Mamaia 300
8700 Constantza, Romania

Contacts: **Alexandru S. Bologa, Ph.D.**, Scientific Deputy Director
Cornelia G. Maxim, Ph.D., Senior Researcher
Radu Mihnea, Biologist, Senior Scientist
Simion Nicolaev, Ph.D., Engineer
Constantin Vlad, Engineer, Senior Scientist
Phone: (40) 16 50870

Activities: Research • *Issues:* Biodiversity/Species Preservation; Water Quality • *Meetings:* CITES; European Society of Radiation Biology; International Conference on Toxic Marine Phytoplankton; IUHPS

● Organization

Established in 1970, the **Romanian Marine Research Institute (RMRI)** is the country's preeminent body of its kind, with research efforts in the Black Sea, Mediterranean Sea, and Atlantic Ocean. Enduring a series of government reorganizations, the Institute is now affiliated with the Ministry of the Environment since 1990, following the Romanian Revolution. RMRI's staff of about 240 are located in a "main building with research laboratories and annexes, radiobiology unit, computer unit, workshop, [and] research library." The vessel STEAUA DE MARE 1 is used for research and experimental fishing, and the vessel PALAMIDA is used for inshore research. EXPLORATORUL is a multipurpose motor boat.

With marine research, RMRI assists other government agencies, organizations, educational institutions, and private companies in Romania and abroad. It maintains a database on oceanographic stations, sea level, waves, chemistry, shore dynamics, and fish stock assessment, and provides diving services and international symposia. "Research projects are performed in the Laboratories for Marine Hydrology, Ecology and Radiobiology; Biochemistry, Extraction and Utilization of Aquatic Living Resources; Aquaculture; Fishery Resources; and Marine Technology," according to RMRI literature.

As biologist **Radu Mihnea** points out, "Marine pollution cannot be confined in the national boundaries. The western part of the Black Sea is a good example. . . . Because of the general north-south circulation, polluting agents carried into the Black Sea by the tributary rivers from the northwestern part influence the [former] Soviet Union, Romanian, Bulgarian, and Turkish coastal waters [and] the same happens with the wastewaters." He believes that long-term cooperation between these countries is necessary and that "common programs for pollution monitoring can be a good example of regional or global cooperation."

Mihnea, who implemented the "monitoring system of the Romanian coastal waters quality," points out that a convention to protect the Black Sea against pollution was negotiated between Romania, former USSR, Bulgaria, and Turkey but was delayed in signing because of political changes in the Soviet Union. "This is another example of how environmental problems should be more important than the political ones," he writes.

Globally, RMRI also participates in UNESCO's Intergovernmental Oceanographic Commission (IOC), Council of Mutual Economic Assistance (CMEA) Convention on ocean research, Fishery Committee of the Food and Agriculture Organization (FAO), Tripartite Agreement (with Bulgaria and former USSR) for Black Sea Fisheries, General Fisheries Council for the Mediterranean (GFCM), International Commission for the Southeast Atlantic Fisheries (ICSEAF), European Society for Radiation Biology (ESRB), and International Union of the History and Philosophy of Science (IUHPS), among other groups.

▲ Projects and People

National programs include HYDROMAR, for "hydrological, chemical, and morphological research"; POLMAR, research to protect the marine environment; RADMAR, for "radioactivity monitoring in the Romanian Black Sea sector and marine radioecological research"; BIOMAR, for biological resources and marine aquaculture research and management; PESCAMAR and ROMOCEAN for fishing resources utilization research in the Black Sea and Atlantic Ocean; research in the sea's "biologically active compounds . . . for pharmaceutical, alimentary, and other economic purposes"; and TECHNOMAR, for researching gear and technical means for exploring using marine resources.

For example, the Biochemistry Lab experiments with "emulsifiers obtained from trash fish and fish processing wastes"; the Aquaculture Laboratory researches cultivation of the Black Sea mussel and shrimp and the "acclimatization of alien commercial species" there; and the Laboratory of Marine Technology investigates oil spill management—among other activities.

RMRI operates a marine coastal water quality monitoring system, which tests and evaluates the following: physico-chemical— "water temperature, transparence, pH, suspended matters (mineral and organic), salinity, dissolved oxygen and saturability, BOD_5, dissolved organic substances, nitrogen (NO_2, NO_3, NH_4, N-ureic), phosphorus (PO_4 organic), silicium-SiO_4, biota, pesticides in water and biota, marine activity"; and biological—"bacterial indicators of fecal and organic pollution, parasitic fungi, fitoplankton [phytoplankton], chlorophyll *a*, zooplankton, zoobenthos, parameters describing the state of ecosystems."

Experiments are also undertaken on "potential biological effects of long-term exposure to low, sublethal, concentration of toxic substances as well as the transfer through the food chain of heavy metals" along the Romanian coast, reports senior scientist Mihnea, who is a contributing author to *Health of the Oceans*, a publication of the United Nations Environment Programme (UNEP) Joint Group of Experts on the Scientific Aspects of Marine Pollution (GESAMP)

Senior scientist Constantin Vlad, who designed and engineered diving systems and the first Romanian-built scientific submersible, supervised the building of two research ships in Poland, and devised a recent plan for offshore oil cleanup technology, is involved in means and methods for marine environment protection and antipollution fighting. Three ongoing programs are bacterial cataborism biochemistry on polluted marine environment compounds, led by Dr. Maria Mîrza; research, design, and testing of oil spill cleanup technology, led by Vlad; and research "in purge of residual waters from food industry," led by engineer and senior scientist Angela Butu.

Vlad reiterates that research regarding environmental protection and pollution controls is "very weak." Because of the country's economic difficulties, "real and adequate financial support does not exist; there are many research programs, but few people are involved, with insignificant technical or scientific tools." Romania also lacks national organizations to struggle against pollution. "The Ministry of Environment hasn't enough money to support the whole national environment protection system—after many years of bad administration of natural resources." Vlad urges too that "all environmentalists must cooperate, because the environment has no borders." RMRI scientists are trying to "confer a greater role for our institute, both by our own efforts and international cooperation."

■ Resources

A *Romanian Marine Research Institute* brochure describes activities and includes a publications list. Dr. Bologa and Mihnea each have publications lists, including English and French titles. The annual journal *Recherches Marine* (Marine research) is offered in exchange with other publications. *Divers and Underwater Vehicles* (printed in Romanian) *Oceanographic Atlas, Halieutica*, and the *Handbook of Fishery* are also offered, as are "trawling fishery with small boats in the Black Sea [and] technical assistance for design and construction of fishing gear."

RMRI seeks new research and development methods of oil spill cleanups and a gamma spectrometry detector. As Mihnea points out, "Romania is a developing country just implementing the market economy" and needing, for example, laboratory equipment, reagents, and computer facilities. "We are ready to discuss our need" with any organizations willing to help.

 # Universities

Department of Zoology
University of Cluj
R-3400 Cluj, Romania

Contact: **Bogdan Stugren, Ph.D.,** Professor
Phone: (40) 51 60311

Activities: Education • *Issues:* Biodiversity/Species Preservation

▲ Projects and People

Since 1951, **Dr. Bogdan Stugren** has been associated with Cluj University's Department of Zoology, where first he worked with amphibians and reptiles in the vertebrate zoology laboratory. As of 1961, Dr. Stugren began teaching ecology and subsequently published a general ecology textbook, printed in Romanian, of which several editions were later published in Germany. Recently, Dr. Cluj submitted a new version of *Fundamentals of General Ecology* for translation and publication in Germany. Among other works, Dr. Stugren has published more than 50 scientific articles on taxonomy and zoogeography of amphibians and reptiles from Romania, Ukraine, the Balkans, and the islands of Crete and Cyprus.

Russia's vast area features a wide variety of climates and terrains—including mountain ranges, tundra, deserts, and forests. The country also has abundant natural resources, with a large percentage of the world's reserves of iron ore and manganese, standing timber, and energy resources—including coal, oil, and natural gas. Petroleum and petroleum products make up a significant portion of exports.

The rapid industrialization of the former Soviet Union occurred with cheap energy and little attention to environmental concerns. In some urban areas, air pollution was as much as 50 percent higher than the national standards. Inadequate sewage treatment and soil degradation are other major problems. Additionally, the Chernobyl accident appears to have had more long-term health consequences than was originally thought, and concern about the country's other nuclear reactors continues.

🏛 *Government*

Ministry of Natural Resources Management and Environmental Protection ⊕
11, Nezhdanovoi Street
Moscow K-9, Russia

Contact: **Vladimir B. Sakharov**, Director, Department of International Cooperation
Phone: (7) 095 229 65 60 • *Fax:* (7) 095 230 27 92

Activities: Political/Legislative • *Issues:* International Cooperation • *Meetings:* UNCED

▲ Projects and People

Dr. Vladimir B. Sakharov and his Ministry were involved in the following projects under the former USSR government:

• A proposal for the establishment of a **Global Center of Ecological Emergency Assistance.** The idea is adopted; this project was to be implemented in 1992.
• Implementation of joint international projects with United Nations Environment Programme (UNEP) on the **Aral Sea and Chernobyl nuclear accident.**
• Development of a joint project with International Waterfowl and Wetlands Research Bureau (IWRB) and World Wildlife Fund (WWF) on the **protection of the Volga River delta.**
• Collaboration on a joint Soviet-American project to create an **international park in Beringia.**
• Implementation of a project on USSR **participation in the Global Resources Information Database (GRID)** under the auspices of UNEP.
• Implementation at national level on various global and regional **conventions and agreements,** such as CITES, European Convention on Transboundary Air Pollution, and the Convention on Environmental Impact Assessment in Transboundary Context.

The Ministry is detailing the global **Convention on the Protection of Biodiversity** as well as developing and improving the national network of natural reserves or "zapovedniks"—strictly protected areas—as its part in global efforts.

The Ministry is the principal counterpart of numerous intergovernmental bilateral agreements in the environmental field with countries including Canada, Finland, France, Germany, Hungary, Italy, Japan, Poland, United Kingdom, and the USA. Its staff of 40 prepared for the 1992 UN Conference on Environment and Development (UNCED) .

Dr. Sakharov began environmental work in 1971 when he was national consultant for the UN Conference on Human Environment and worked on a USSR-USA intergovernmental cooperation agreement in environmental protection. He then joined the USSR State Committee for Science and Technology, Department of International Cooperation, Environmental Section, for six years.

Working at the UN ECE (Economic Commission for Europe) Secretariat on the interrelationships of natural resources policy development and environmental management, Sakharov also carried out comprehensive studies of ecological, economic, organizational, and institutional aspects of government policies. He reported in 1991 that he was still active in the UN ECE as chairman of an international group dealing with the concerns of the European Convention on Transboundary Water Pollution.

His role is to give guidance on international cooperation matters on multilateral and bilateral levels as well as to develop intergovernmental policies regarding nature conservation and sustainable development. He formulates and presents environmental proposals to the government relating to international organizations and agreements, conventions, and programs.

"To enhance future global cooperation, it is essential to demonstrate that environment and development are complementary to each other and not contradictory," he writes. "An important and difficult issue is resolving the complicated problem of different interests in developed and developing countries, such as those of Central and East European states in their period of transition to a market economy."

Dr. Sakharov was nominated deputy chairman of the USSR National Commission for the World Conservation Union (IUCN). He is also national coordinator for implementing the Soviet-Dutch Memorandum of Understanding on environmental protection.

Other memberships include USSR Commissions for UNEP and UNESCO, Soviet Committee for UNESCO's Man and the Biosphere Program, Scientific Council of USSR Academy of Sciences on Problems of Biosphere, and All-Russia Society for Nature Protection.

■ Resources
Needed is "convertible currency to enable national organizations to be more closely involved in international efforts."

Institute of Nature Conservation and Reserves
Ministry of Nature Conservation and Rational Usage
Sadki-Znamenskoe
113628 Moscow, VILAR, Russia

Contact: **Olga Pereladova Ph.D.**, Senior Scientist
Phone: (7) 95 423 2144 • Fax: (7) 95 432 2322

Activities: Research • *Issues:* Air Quality/Emission Control; Biodiversity/Species Preservation; Water Quality

▲ Projects and People
The Institute of Nature Conservation and Reserves is involved in the monitoring of ecosystems of different regions as well as the monitoring, reintroduction, and restoration of rare species. In addition, it has worked on the Red Data Book of the (former) USSR.

Senior scientist **Dr. Olga Pereladova**, also a member of the Species Survival Commission (IUCN/SSC) Deer Group, conducts research on the **ethology and acoustic communication of marmots, deer, wolves, jackals, and rare species of ungulates** at both nature reserves and breeding centers.

Even with its "very wide and interesting role," problems are presently "connected with the whole situation in our country; we have no needed currency, this leads to great difficulties in contacts,

in membership in different environmental organizations and programs," reports Dr. Pereladova. As of early 1992, it remained uncertain as to how its programs would work in the former Republics.

■ Resources
The Institute lacks equipment for marking of individual animals of different species and other materials and medicines to conduct night behavioral observations and to tranquilize ungulates for transport purposes. Literature describing its programs is printed in Russian.

Committee on Ecology
(former) Supreme Soviet of the USSR
Kremlin
Moscow, Russia

Contact: **Alexey V. Yablokov, Sci.D., Bio.D.,** Deputy Chairman, Professor
Phone: (7) 95 203 5952 • Fax: (7) 95 203 5941

Issues: Biodiversity/Species Preservation

▲ Projects and People
Prof. **Alexey V. Yablokov** fills several roles in both the scientific world and the environmental movement. After receiving degrees in vertebrate zoology and animal morphology, Yablokov served as the head of a laboratory for N.K. Koltzoff's Institute of Developmental Biology. In 1976, he joined the Species Survival Commission (SSC) of the World Conservation Union (IUCN), now serving on its steering committee, as his country's regional member and as a member of the Cetacean and Seal groups. He later was elected into the (then) USSR Academy of Science. He has served as president of the Moscow Society for the Protection of Animals and as chairman of Greenpeace USSR. In 1989, he was appointed to the position of deputy chairman of the Committee of Ecology of the Supreme Soviet.

Yablokov's main fields of interest are "biology and protection of **whales, dolphins,** and other **marine mammals;** population and **conservation biology;** theory of evolution, and **ecopolitics.**"

🏛 *State Government*
Caucasus State Biosfere Reserve ⊕
Sovietskaya 187
352700 Maikop, Russia

Contact: **Aleksandr S. Nemcev,** Senior Research Officer
Phone: (7) 21697 (operator assistance)

Activities: Research • *Issues:* Biodiversity/Species Preservation; Deforestation

▲ Projects and People

The **Caucasus State Biosfere Reserve** has a staff of 20. Biologist **Aleksandr S. Nemcev** belongs to the Species Survival Commission (IUCN/SSC) Bison Group.

 *NGO*

Ecology and Peace Association
19, Kuznetskij Most
103031 Moscow, Russia

Contact: **Sergei P. Zalygin**, President, Editor-in-Chief, *Novy Mir*
Phone: (7) 95 209 5702
E-Mail: Econet: sfmt:ecopeace

Activities: Development; Research • *Issues:* Energy; Hydrologic Problems; Sustainable Development; Water Quality

● Organization

The **Ecology and Peace Association**, with a staff of 5 and 120 members, is an NGO comprised of "prominent scientists and publicists" providing an "independent scientific examination" of hazardous water and other environmental projects and their ecological impact. Their findings are published to help create public awareness.

The Association was founded in 1987 by a group of scientists and experts who fought to stop the Siberian and European river diversion projects. Their success both protected the environment in that region and saved huge costs. Since then, Ecology and Peace also stopped projects on the Volga-Chograi, Volga-Don 2, Danube-Dnieper, and Volga-Ural canals.

▲ Projects and People

Sergei Pavlovich Zalygin heads the Association and serves as the editor-in-chief of *Novy Mir*, which publishes his group's views on ecological problems. His articles appear in other Russian papers, such as *Trud* (July 5, 1991), in which he called for the creation of an "international environmental school, international environmental inspection, and an international environmental court."

Ecology and Peace sponsors meetings and press conferences. When the results of a 1988 conference on ecology and agriculture were published, for example, they were also used as the basis for a report by the Russian Supreme Soviet Ecological Committee.

In 1990, a conference on the **experience of independent scientific examination of the Leningrad Dike** resulted in a study by Dr. **A.S. Mishchenko**, Moscow University professor and vice chairman of Ecology and Peace. The dike, a system of dams and watergates, was designed to protect St. Petersburg from flooding, but Mishchenko believes that the risk of flooding was overestimated by those who wanted to build the system. The dike brought about drastic ecological changes in the Neva Bay, according to Dr. Mishchenko. One such change is the total desalinization of the region resulting in the loss of salt species of planktons and a massive buildup of contamination. Mishchenko also points out that the same people who are responsible for much of the environmental degradation have been awarded the right to assess the damage.

After Ecology and Peace examined the **Katun Valley Water project** and presented its findings to the Parliamentary Sessions of the Russian Supreme Soviet Ecological Committee, a decision was made to discontinue the project. The Association also analyzed the **Tehri Dam project** in which the former USSR was participating. The results of these studies were presented to the (then) USSR Ministry of Foreign Affairs and the Indian Embassy.

In February 1991, the Association held a major press conference, **Ecological Disasters in the USSR: Facts, Causes, Consequences**, to discuss their findings, mainly on three major water management problems—the Aral Basin, the Lower Volta, and Caspian region, and the Neva Bay region mentioned above.

The **Aral Sea** is already in a state of ecological calamity, and other parts of the Aral Basin are close to crisis, according to Ecology and Peace. This is primarily the result of improper irrigation and drainage construction in Central Asia during the past 30 years. The Aral Sea is actually drying out, and this is something that was forecast and ignored more than 20 years ago. The same organizations that created this plan to drain the Aral are now receiving funding to attempt a "rescue" of the area, reports the Association.

To alleviate the Aral crisis, Ecology and Peace recommends "conservation of soil covering and reclamative afforestation in the mountains with putting mountainous farming in good order; [and] restoration, conservation, and amelioration (predominantly phytoimprovement) of pastures as well as newly desertified areas in lowlands"—with beneficial studies to be tapped at the Institute for Deserts, and the Institute for Karakul Sheep Breeding, for instance.

Dr. **N.G. Minashina**, Ecology and Peace board member, also writes, "A century-old experience in arable farming should be made use of, the reconstruction of the drainage-accumulation network should be carried out with great care, maintaining the principles of contour arable farming; it is necessary to [reduce cotton production and] start to grow again traditional crops in the oases, to diminish water intake per unit of the irrigated land, to improve techniques, and the irrigation regime. Discharge of irrigation waters into the drainage accumulation network should be completely excluded."

Dr. **M.I. Zelikin**, board member, with Moscow University colleagues **M.V. Lomonosov** and **A.S. Demidov**, also notes that the agricultural economy in the Aral region must be restructured, toxic chemicals must be banned, irrigation methods that conserve water must be implemented, the "development of a scientifically grounded economic scheme for the use of waters from the Amu-Darya and Syr-Darya Rivers" must take place with the help of an independent, "international, all-union" commission—with resulting laws strictly observed.

The **Volga River** is also threatened. Eleven dams with hydroelectric power stations have partitioned off the Volga and its main tributary, according to research by **E.M. Podolsky**, board member. In addition, it is the victim of contaminants including effluents from cities, waste from industrial enterprises, and runoff from fields. The Association has some ideas "to put an end to the losses suffered in fishery and farming in the Lower Volga." In publicizing the plight of the Volga, Ecology and Peace hopes that action can be taken to protect "the great Russian river."

Regarding the "ecological crisis in the northwestern Caspian area" of both semi-desert rangelands and melon and crop fields, wildlife reserves, and hunting localities, Ecology and Peace board member Dr. **B.V. Vinogradov** writes about the spreading desertification. "The moving sands here [in Kalmykia] have increased by 15 times over the last 25 years, while the pasture productivity has decreased by three times." According to Vinogradov, the government permitted ecologically dangerous land-use projects to go forward. "The plough-up of sandy soils has resulted in soil deflation. . . . Tens of thousands of resalted and overwetted soils have emerged as a result of the improper water engineering constructions."

Scientists supporting alternative approaches recommend "large scale afforestation of sandy soils, rangeland improvement, hay land restoration, bringing the cattle population into accord with the feeding capacity of ranges, the development of a local water supply system, revival of the conditions for Kalmyk people's traditional mode of life."

Other problems in the area include "noxious industrial wastes discharged by the Volgograd plants, underflooding the areas with irrigation runoffs from the Stavropol district, air and soil pollution by the Astrakhan group of enterprises engaged in gas condensations, illegally ploughing the sandy ranges to grow melons, overgrazing resulted from additional hundred thousand cattle introduced from Dagestan."

Also in Astrakhan, "The Lower Volga is the site of concentration of harmful flow from the entire river basin and this has inflicted an irreparable damage on fishery and is posing a threat to sturgeon species." Vinogradov also reports, "Gas deposit exploitation in the Astrakhan region create a particular ecologically grave danger. . . . Technological imperfections of gas prospecting and extraction has resulted in pasture, surface, and [groundwater] contamination, the drop in farm produce quality." Agricultural production has also been adversely affected by Volga-Akhtuba floodplain regulation. Vinogradov and colleagues recommend an alternative agroecological program to rehabilitate "traditional fishery and vegetable growing as well as highly performing livestock production, afforestation, recreation areas, hunting farms."

Independent technical and economical corroboration of environmental projects, such as offered by Ecology and Peace and the Kalmyk Steppe conservation movement, as well as press coverage, are viewed as significant in solving such ecological programs on the local and state level. A "new nature conservation strategy" must also be developed to guard against harmful "joint venture and cost accounting establishments" as the country's "economic and organizational alterations" occur, says Vinogradov, who is also professor of the Institute of Animal Evolution, Morphology, and Ecology, Academy of Sciences.

Zalygin, who ran a commission on environmental transformation problems in Western Siberia and was head of a hydrotechnical land-reclamation chair at Omst Agricultural Institute, also studies literature. He has written some 200 books published in more than 30 languages, receiving literary prizes.

■ Resources

The Association has published more than 200 papers in the (former) Soviet and foreign press, with some available in English. It seeks increased means of communications and computers.

Pacific Oceanological Institute ⊕
7 Radio Street
690032 Vladivostok, Russia

Contact: **Boris I. Vasiliev, D.Sci.,** Head, Submarine Investigations Laboratory
Phone: (7) 239187 (operator assistance)

Activities: Research • *Issues:* Environment/Geology of Ocean Floor • *Meetings:* International Geological Congress

● Organization

Dr. Boris Vasiliev is engaged in a **Pacific Geological Map Project,** described as being part of the "World Ocean" complex program that includes "collecting, generalization and studying of a geology-geophysical data on the Pacific floor and then compiling the first geological map of this part of the world in scale 1:10,000,000."

According to Dr. Vasiliev, marine geology department head, who is conducting his long-term studies with the support of **Earthwatch,** "Our Project envisages the compilation of the ocean floor geological map using the same method [as] for land geology mapping." Such a map is the "most visual and effective methods of generalization and presentation of geological data," particularly for "tectonic, palaeogeographic, and metallogenetic" constructions. Several marine cruises are being conducted as geological and geophysical data is compiled through 1994. The map and explanatory note are expected to be published in 1995. *(See also Earthwatch in the USA and UK NGO sections.)*

▲ Projects and People

Dr. Vasiliev—whose specialties are researching the origins of the Earth, moon, Pacific, and "tectonics of the deep-sea trenches and marginal seas" in the Asia-Pacific transition zone—describes the methods and equipment he is using:

"Relief investigation by echo-sounding; magnetometry survey by proton magnetometer; gravimetry survey by marine aboard gravimeter; continuous seismic profiling with electric spark and pneumatic radiator; rock dredging by cilinric dredge, bottom sampling by cores and scoopers, and submarine photography."

Aboard the research vessel, the following data will be processed: "Making and studying petrographic thin sections, paleontological determinations, physics-chemical analyses of rocks and sediments, and magnetic and paleomagnetic measurement."

Yevlanov Yurii Borisovich, also of the Pacific Oceanological Institute and collaborating on the project, specializes in the study of volcanic rocks on the Pacific floor, Japan and Philippine seas, and deep-sea trenches of the western part of the Pacific. His scientific works are mainly in Russian.

✿ *Universities and Institutes*

All-Union Research Institute of Nature Conservation

Znamenskoe-Sadki
113628 Moscow, Russia

Contact: **Valery Orlov, Ph.D.**, Science Secretary
Phone: (7) 95 423 0322 • Fax: (7) 95 423 2322

Activities: Development; Education; Law Enforcement; Political/Legislative; Research • *Issues:* Air Quality/Emission Control; Biodiversity/Species Preservation; Deforestation; Ecological Mapping; Energy; Global Warming; Health; Sustainable Development; Water Quality • *Meetings:* CITES; Ramsar

● Organization

Established in 1979, the All-Union Research Institute has a staff of 280, including scientists and technicians in Moscow, and regional affiliates in St. Petersburg, Brjansk, Vladivostok, and Kishinjov.

In research and coordinating scientific activity, the Institute's priorities include developing a scientific basis for establishing protected areas networks, preserving biodiversity, compiling the country's Red Data Books, making up standards on protected areas, assessing and forecasting ecosystem changes, developing strategies for sustainable use of resources, breeding rare and endangered species in nurseries while recovering their natural habitats, and cooperating with international nature conservation.

Also among its achievements are programs on preserving and recovery of the endangered **European bison, Przewalski's wild horse, cheetah,** and **white Siberian crane;** carrying out research to establish **marine reserves in the Arctic and Far East seas;** and participation in international nature reserves.

▲ Projects and People

The Institute's divisions are: sustainable use of natural resources and environmental standards; ecosystem conservation and recovery (aquatic, forest, soil, and marine ecosystems); protected areas; plant conservation (seed plants, scope plants and fungi, and rare plant communities); animal conservation; coordination center on transboundary air impact assessment; wildlife health control center; information and ecological forecast; international sector; and environmental education and interpretation.

The sustainable use division is involved in regional problems of nature conservation, **biochemical assessment of human impacts** and environmental standards, technogenic emission impact assessment, and environmental demands of natural resource utilization. The protected areas division is concerned with biodiversity in already protected areas and well as developing a **network of protected areas** and monitoring research within protected areas. General wildlife conservation activities includes work with **rare vertebrates, wildlife resources,** and **invertebrates.**

Dr. Valery Orlov writes, "The Institute is active on working out the theoretical approaches in ecological mapping, [and] environmental standards . . . as well as in practical implementation of both

all-Union and regional environmental programs. It is busy building up the ecological forecast for natural resource use in the [former] USSR up to 1995 and beyond." It also helps draft environmental legislation and has implemented more than 470 practical regulations and methodological guides.

The Institute has many international ties, including links with groups such as World Conservation Union (IUCN) and the World Wildlife Fund (WWF). It also serves as a scientific body for international conventions and agreements such as CITES and the Natural Heritage Convention. With the authority to arrange for field projects within the (former) Soviet Union, it "will welcome any proposals on mutually beneficial cooperation."

Current international undertakings include the **Siberian Crane** project, conducted by Dr. V. Flint; **Birds of Prey (Falcons)** and **Sea Colonial Birds** projects by Dr. A. Sorokin; **Waterfowl and Wetland** by Dr. V. Krivenko; Soviet–Finnish shared National Park by Dr. V. Kuleshova; Soviet–American **Bering Bridge National Park** by Dr. V Krassilov and others; **Arctic Ecosystem Conservation** project by S. Belikov; **Environmental Impact Assessment of Pollution in the Aquatic and Terrestrial Ecosystems** by Dr. Pastukhova; and the **Volga Delta Project.** Dr. Orlov belongs to the Species Survival Commission (IUCN/SSC) Chiroptera Group.

A co-publisher of the popular scientific journal *The Green Cross— Nature Protection,* the Institute and its staff have produced some 2,500 publications also including *National Park, Waterfowls, Rare Bird Species Breeding in Captivity, The White Bear,* and *Orchids of the USSR,* among others.

■ Resources

The Institute seeks to exchange knowledge and training, including creating more links with NGOs, public groups, and publishers. With limited funding and means of communication, the Institute also is in need of lab and office equipment to improve and make research more efficient. Scientific publications are available on request. Computer systems are being developed for the "visualization of economical information."

Institute for Lake Research

Academy of Sciences
Sevastynov Street 9
196105 Saint Petersburg, Russia

Contact: **Kirill Yakovlevich Kondratyev, Sc.D.**, Counsellor for Directorate, Academician
Phone: (7) 812 231 7773 • Fax: (7) 812 218 4172

Activities: Research • *Issues:* Deforestation; Global Change; Global Warming; Sustainable Development; Water Quality

▲ Projects and People

An internationally known expert in **environmental physics,** Dr. Kirill Yakovlevich Kondratyev researches atmospheric radiation problems relevant to the physical basis of the climate and has been involved in the development of remote-sensing techniques for environmental study. Kondratyev has written more than 800 scientific papers, of which over 100 have been published in international journals. One recent example is *Climactic Shocks: Natural and Anthropogenic*, which was published by Wiley in 1988.

In addition to his work with the **Academy of Sciences** of the former USSR, Kondratyev spent many years on the faculty of Leningrad State University and headed the Main Geophysical Observatory in (then) Leningrad. He serves on the editorial boards of a number of former Soviet and foreign scientific journals. Other monoliths include *Meteorological Satellites, Global Climate, Key Problems of Global Ecology*, and as co-author—*Remote Sensing of Soils and Vegetation*.

A member of the International Academy of Astronautics and an honorary member of the American Academy of Arts and Sciences, Royal Meteorological Society of Great Britain, and German Natural Science Academy "Leopoldina," he has a doctorate in physico-mathematical science as well as honorary doctorates from the Universities of Lille (France) and Budapest (Hungary).

Institute of Animal Evolution, Morphology, and Ecology

Academy of Sciences
Leninskyi Prospect 33
117071 Moscow, Russia, CIS

Contact: **Dmitry I. Bibikov**, Professor
Phone: (7) 95 232 2088

Activities: Research • *Issues:* Biodiversity/Species Preservation
Meetings: CITES

▲ Projects and People

Prof. Dmitry I. Bibikov, Institute of Animal Evolution, Morphology, and Ecology, Academy of Sciences, belongs to the Species Survival Commission (IUCN/SSC) Wolf Group and can provide *Der Wolf*, printed in German, as well as some books in Russian.

Leningrad State Pedagogical University

Moyka 48
191189 St. Petersburg, Russia

Contact: **Lev A. Kuznetsov**, Professor of Ecology
Phone: (7) 106684 (operator assistance)

Activities: Education; Research • *Issues:* Biodiversity/Species Preservation; Deforestation; Desertification

■ Resources

Prof. Lev. A Kuznetsov, ecologist, Leningrad State Pedagogical University, reports that equipment for field environmental studies is needed and exchanges are sought with other scientists. The book *Stationary Examinations of the Aral Region* can be ordered.

Pacific Institute of Geography

Academy of Sciences
7 Radio Street
690032 Vladivostok, Primorye, Russia

Contact: **Anatoley P. Bragin**, Wildlife Research Biologist

Activities: Research • *Issues:* Biodiversity/Species Preservation

▲ Projects and People

In his work for the **Pacific Institute of Geography** and as the tiger and leopard representative of the Species Survival Commission (IUCN/SSC) Cat Group, **Anatoley P. Bragin** largely collects "information in the size, area of distribution, degree of isolation and the number of breeding animals within the populations of tigers and leopards in the Russian Far East. Most research involves the strategic management of their wild populations in the [former] USSR. In particular, I am now developing an action plan for restoration of the Sikhote-Alin wild population of Amur leopards."

According to Bragin, "The **Amur leopard** is now on the brink of extinction, with no more than 30 surviving in the Russian Far East and adjoining northeast China." While there are 82 of the leopards in captivity "74 have originated from just two founders, one of which is of suspect origin." That leopard is suspected to have been a Tibetan leopard, not an Amur.

Bragin proposes removing 10 to 12 Amur leopards from the wild and breeding them with leopards in captivity. This could produce "250 second-generation and third-generation animals to retain the inbreeding rate below the one percent threshold." He believes that the "potentially rapid expansion of a captive population could help to mitigate the 'founder effect' in comparison with wild populations of the same size."

Such a breeding center would be based on a "self-repaying sika-deer-breeding farm, from which the cats will be provided with fresh meat at the expense of female and young deers, while stag products [such as young antlers and scut] will be subject to sale."

The captive-breeding program will enable the introduction of "genetically remote" individuals into the wild, particularly in the southern Sikhote-Alin region, which it is believed could support more than 100 such leopards.

In addition, Bragin plans to feature "acclimation facilities where the cats can become familiar with the types of foods they will encounter in the wild and develop predatory skills for self sufficiency." These facilities are expected to be sited in two remote locations with an abundance of prey, where young cats—following weaning—would be transported and exposed to whole carcasses at an early age, then live prey. Because monitoring is essential to the program's success, "a team of wildlife biologists with radio telemetry experience is another aspect of the program," he writes.

■ Resources

Available in English are Bragin's papers, "Population Characteristics and Social-Spatial Patterns of the Tiger (*Panthera Tigris*) on the Eastern Macroslope of the Sikhote-Alin Mountain Range" and "Problems of the Amur Tiger," co-authored with **Victor V. Gaponov**, Primorye Agricultural Institute (*Hunting* 1989).

Rwanda

Located deep in the heart of central Africa, Rwanda's landscape is green and mountainous, earning it the name "Land of a Thousand Hills." In the northwest is the National Park of the Volcanoes, established for the protection of the last mountain gorillas.

While Rwanda is rich in wildlife and vegetation, it is the most densely populated country in Africa, with a high fertility rate. Almost all the people live in the countryside, so urban centers remain small, and the rural population is encroaching upon the 10 percent of the land that is comprised of protected areas. Forests provide the fuelwood and charcoal that supplies 90 percent of energy consumption, so land clearing is pervasive. The arable land has been overused to meet the subsistence needs of the massive population. On a more positive note, the recent widespread attention to the encroachment into the natural habitat of the mountain gorillas by poachers and farmers have served to reverse the trend toward extinction.

Government

Ministry of Agriculture
P.B. 621
Kigali, Rwanda

Contact: **Jacques Muberuka, D.V.M.,** Director of Animal Health
Phone: (250) 76845 • Fax: (250) 82103

Activities: Development; Research • *Issues:* Biodiversity/Species Preservation; Health; Sustainable Development

▲ Projects and People

Tsetse fly control is a major undertaking of the **Ministry of Agriculture, Livestock, and Forestry** with **Dr. Jacques Muberuka,** a veterinarian and director of animal health, as coordinator of a team for tsetse and trypanosomiasis control. "Four members are permanently in the field, based in RUSUMO near the boundaries with Tanzania," writes Dr. Muberuka. "They perform a daily survey, collect flies, count and classify them, and examine salivary glands and blood smears to detect infection."

According to Dr. Muberuka, tsetse fly control started in Rwanda in the 1960s as the government allocated arid and semiarid lands in the country's eastern and southeastern portions. This area is recognized as a "fly belt neighboring Burundi, Tanzania, and Uganda," writes Dr. Muberuka. "It had to be developed as far as agriculture and livestock are concerned—given the growth rate of the population [at] more than 3.7 percent per annum."

Control methods have evolved from air spraying of dieldrin in the early 1960s to its selective application until 1969. Since 1986, methods have included "trapping by biconical, Zambabwean, and NGURUMAN models combined with odour baits like acetone and cow urine; [and] electrical screens in limited peri-domestic areas." Dr. Muberuka notes that screens "impregnated with Decis (R) have been abandoned."

"Environmental aspects of the project," he writes, are "to use harmless sound methods and techniques, to avoid desertification, and to protect wildlife and aquatic non-target animals." A method currently in use in the 1990s is "trapping combined with attractants."

Reinvasion of the tsetse fly remains a threat, even though certain pastoral zones have been cleaned up. In areas such as "across the Akagera River from Tanzania and also from the Akagera National Park . . . a high density of *glossina morsitans, brevipalpis,* and *pallidipes*" have been found. Dr. Muberuka also notes that "massive colonization by humans contributed a lot to the prevention of the reinfestation" in an area known as BUGESERA.

Team leader at RUSUMO is **Jean Baptiste Nyiligira.** Involved technicians are **Milton Ngendahimana,** and **J. Bosco Sebakiga.**

Dr. Muberuka is also national coordinator of the Panafrican Rinder Pest Campaign and involved with rabies and African swine fever control. In 1990, he participated in Rwanda's first national seminar on bovine artificial insemination. His scientific papers on vaccines and other control methods are published in French.

"Environmentalists can enhance future global cooperation by first creating a network and a newsletter to exchange experiences and information; secondly by strengthening training capabilities through seminars, training courses, and workshops," he writes.

■ Resources

The Director of Animal Health seeks help with educational outreach for entomology, immunology, and protozoology. It is also in need of financing for drugs, a vehicle, and other equipment.

🏛 *Regional Government*

Kagera Basin Organization (KBO) ⊕
P.B. 297
Kigali, Rwanda

Contact: **Charles Ruzindana**, Research Agronomist
Phone: (250) 84665 • Fax: (25) 82172

Activities: Development • *Issues:* Deforestation; Energy; Human Resources Development; Sustainable Development; Transportation

● Organization

In 1977, the governments of **Burundi, Rwanda,** and **Tanzania** established a regional institution, the Organization for the Management and Development of the Kagera River Basin—now known as **Kagera Basin Organization (KBO)**—to advance integrated development. **Uganda** joined later. Its name is derived from the Kagera River, which crosses the member countries, and the basin is the "principal contributor of the waters of Lake Victoria and commonly regarded as the source of the White Nile," according to KBO.

KBO's *1990 Information Document* describes the basin as "among the most impoverished regions of the world, with an average per capita income of US$250. About 90 percent of the population resides in rural areas where they are engaged mainly in subsistence agriculture. The principal food crops are bananas, beans, rice, maize, cassava, and potatoes, while only coffee and tea [are] the main cash crops."

While food production appears to be presently adequate, as the population increases at about 3 percent per year, so must the food supply. Because of overcultivation and overgrazing, however, yields and soil fertility are declining and erosion is increasing. Cash crops expand at the expense of food crops, "improved seed varieties are largely unavailable . . . and beyond the means of the typical peasant farmer whose annual income is below US$50," and storage facilities and markets for food surpluses are lacking. Labor is abundant, but health services are "very poor" with malaria "endemic in all the lowlands" and human trypanosomiasis occurring in large areas where the tsetse fly is not eradicated or controlled.

In this land-locked region with poor roads, transportation alternatives are being sought to establish links with overseas import/export markets, particularly to the Indian Ocean. With wood as a major source for heating and cooking, forests are being depleted. However, there are potential energy sources, such as hydropower, peat, and methane gas in Lake Kivu.

Therefore, goals of KBO's natural resources development program concern water and hydropower resources; furnishing of water and water-related services for mining and industrial operations, potable water supplies for other needs; forestry and land reclamation, agriculture and livestock; mineral exploration of iron ore and nickel, tin and tungsten mining; disease and pest control; transportation and communications; trade; tourism; wildlife conservation; fisheries and aquaculture; industrialization, including fertilizer production and exploration and use of peat; and environmental protection.

With limited resources, the present focus is on "transport and communications, energy, agriculture, and information and training . . . to disenclave the region of the basin, supply cheap and renewable locally available power, protect the environment, control diseases, and increase the food production."

▲ Projects and People

Among ongoing projects, **control of the trypanosomiasis in the Kagera Basin** aims at controlling the tsetse fly in a 10,000-kilometer region, using environmentally safe "soft methods." Traps and screens are the preferred methods, instead of ground and aerial spraying of insecticides. Charles Ruzindana is team leader. (*See also Ministry of Agriculture in the Rwanda Government section.*)

With economic and technical studies completed, KBO plans to build the **RUSUMO hydropower plant** on the Kagera River between Rwanda and Tanzania. "This project comprises an important environmental aspect," with an artificial lake created upstream, the location of Akagera National Park downstream, and the necessity of environmental impact assessments (EIAs) before implementation, reports KBO.

Afforestation in southwestern Uganda aims at planting 10,000 hectares of firewood and timber trees to protect overgrazed and badly eroded areas. These trees would also be a sustainable source of fuelwood and construction timber for rural and urban uses. KBO is also considering valley drainage.

KBO's *Information Document* also lists proposed projects such as a railway to the Indian Ocean, various road improvements, Kagera River navigability for a route to the Indian Ocean to enhance commerce and tourism, rehabilitation of the Haziba small hydroelectric power plant, Mulindi Valley drainage, Rubaare ranching in southwestern Uganda, and establishment of the Kagera Polytechnic Institute (KPI).

■ Resources

KBO seeks advice from environmental specialists for input on regional development programs—particularly material (in French or English) on environmental land use and training people.

KBO has project documentation on trypanosomiasis control, afforestation in southwestern Uganda, and RUSUMO hydropower as well as its *1990 Information Document.*

 *NGO*

Projet Conservation de la Foret de Nyungwe (PCFN)
Nyungwe Forest Conservation Project
Wildlife Conservation International (WCI)
B.P. 363
Cyangugu, Rwanda

Contact: **Liz Williamson, Ph.D.,** Director
Phone: (250) 37193 • *Fax:* (250) 72128

Activities: Education; Research • *Issues:* Biodiversity/Species Preservation; Deforestation; Ecotourism • *Meetings:* International Primatological Society

● Organization

Established in 1988, PCFN—the **Nyungwe Forest Conservation Project**—is a "collaborative effort between the Rwandan Office of Tourism and National Parks (ORTPN) and Wildlife Conservation International (WCI) with major support from the U.S. Agency for International Development (USAID) and the U.S. Peace Corps." PCFN's purpose is to promote ecotourism, education programs for the local population, and ecological research regarding the tropical forest and human impacts on the ecosystem. **Amy Vedder,** WCI, conceived and started PCFN.

The rugged Nyungwe Forest, contiguous with Burundi's Kibira National Park, is one of Africa's "largest montane (mountainous) rainforests." Covering some 970 square kilometers—East Africa's "largest single forest block"—it features more than 190 tree species, 100 orchid varieties, some 275 bird species with 24 endemic to the Central African highlands, 13 types of primates representing 20 percent of all African primate species, and terrestrial mammals including leopard, elephant, golden cat, serval, and duiker. Rare golden and owl-faced monkeys romp about; "black and white colobus monkeys travel in groups as large as 400 individuals"—considered an "extraordinary" number. Some 500 to 1,000 chimpanzees are vocal but elusive. Prosimians—eastern needle-clawed galago and greater bush-baby—are seen here, along with six species of social guenons.

For its human population, Nyungwe provides "vital ecological services such as watershed protection, stream-flow regulation, and erosion control." The forest products include "fuel and construction wood, honey, bamboo, natural ropes, medicinal plants, and thatching materials." Bamboo, for example, is used in weaving more than 50,000 baskets a year. Beekeeping results in a profitable honey and "beer" industry.

▲ Projects and People

PCFN has established a **tourist reception center, organized tours** on which tree primate species can be viewed, **driving tour, camping facilities,** and extensive **hiking trails.** An orchid garden has also been established. The Scientific Institute of Agricultural Research (ISAR) maintains **enrichment plantings**—mainly of *Podocarpus* tree species used for wood to build houses, furniture, and beehives—which are sighted on tours.

PCFN reports that the Rwandan government has launched an "ambitious" conservation program—with 40 percent of the Nyungwe Forest Reserve designated an undisturbed and fully protected **natural reserve,** where scientific research and controlled tourism is permitted. Two **multiple-use zones** border the natural reserve and are set aside for "controlled harvest of forest products." **Buffer plantations of trees,** eventually comprising the remaining 10 percent of the forest reserve, will encircle these areas. Pines, eucalyptus, and cypress, among other exotic trees, mark the reserve borders and will provide a source of wood.

Rwanda's Forestry Department implements the **forestry action plan** that includes a forest inventory. ORTPN provides park guards, and PCFN contributes to conservation through its "tourism program, ecological research, rural environmental education initiatives, and training of Rwandan professionals in the field of conservation."

Among medicinal plant uses of forest vegetation, bark of the *Syzygium parvifolium* (umugote) is used for cough medicine, and the bark of the *Hagenia abyssinica* (umugeti) treats worms.

Resembling the coconut, fruit of the *Carapa grandiflora* (umushwati) is eaten by people and monkeys; its seeds have been reported to have floated across the Atlantic Ocean. Wood of the *Olea hochstelleri* (intobo), a member of the olive family, is used in sculpture, jewelry, and heavy construction. Rwandans call the *Macaranga neomildbraediana* tree that is easy to cut and split for fuelwood *umusekera*—meaning "the tree that makes you laugh."

Bird species include brightly colored and fruit-eating turacos, hornbills, sunbirds, crowned eagle, cinnamon-breasted bee-eater, olive pigeon, white-headed wood hoopoes, and black-headed waxbill, among others. Other animals are squirrels, the heavily poached and now rare forest hog and bush pig, 120 species of butterflies, 5 chameleon species, viper, and **driver ants**—predators considered "the most important carnivores in the forest." The ants stir up prey for birds and devour "many herbivorous insects which would otherwise threaten forest vegetation if their numbers were left unchecked."

Prior to joining PCFN in 1991, **Dr. Liz Williamson,** a psychologist, studied chimpanzees in Liberia and gorillas and chimps in Gabon. **Dr. Samuel Kanyamibwa** is PCFN assistant director with special interests in sustainable development, biological conservation, and ornithology, particularly the white stork.

■ Resources

PCFN publishes the *Nyungwe Guide Book.*

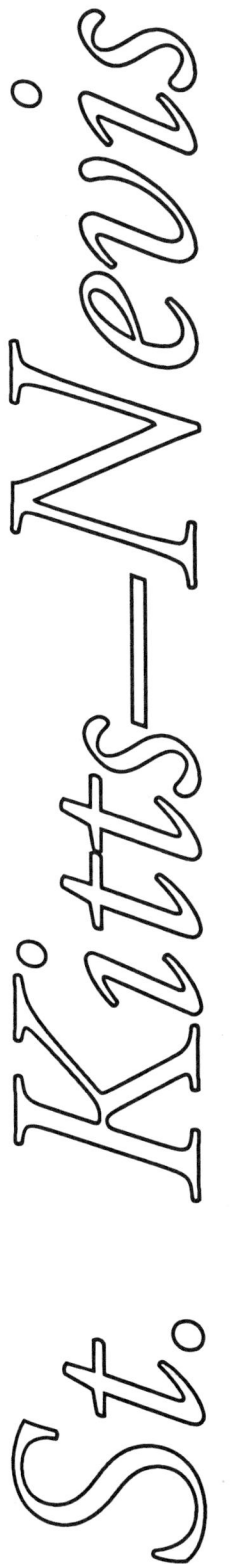

St. Kitts–Nevis

The islands of St. Kitts and Nevis became the independent nation of St. Kitts and Nevis in 1983. St. Kitts, part of the Leeward Islands group, is situated 200 miles southeast of Puerto Rico. The island is volcanic with a high peak, Mount Liamiuga, dominating the central mountain range. St. Kitts is 23 miles long, with an area of 65 square miles. Nevis is to the south of St. Kitts, separated by a two-mile wide channel. Its central mountain range forms a cone about 3,300 feet high. The islands boast a relatively healthy climate, with little tropical disease. The major environmental problem is soil erosion.

Traditionally, sugar was the most important industry, dominating the economy for three and a half centuries. The government is now seeking to develop further the agricultural sector, diversify the economy, and encourage the relatively new tourism business, while avoiding some of the accompanying environmental problems that have plagued other Caribbean countries.

🍁 NGO

Nevis Historical and Conservation Society (NHCS)
Alexander Hamilton Museum
Charlestown, St. Kitts–Nevis

Contact: **Joan and David Robinson**, Curators
Phone: 1-809-469-5786 • *Fax:* 1-809-469-0274

Activities: Development; Education; Research • *Issues:* Architectural Preservation; Biodiversity/Species Preservation; Conservation of Cultural and Historic Sites; Deforestation; Energy; Global Warming; Sustainable Development; Waste Management/Recycling; Water Quality • *Meetings:* Caribbean Conservation Association; Museum of the Caribbean

● Organization

In the eastern Caribbean, Nevis, with a population of about 12,000, is part of the two-island nation of St. Kitts–Nevis. Here, the **Nevis Historical and Conservation Society (NHCS)** is a nongovernmental organization dedicated to "conserving the natural, historical (including prehistorical) architectural and cultural aspects."

▲ Projects and People

With eight ongoing environmental projects, NHCS is working under a World Wildlife Fund—USA (WWF) grant to **monitor six ecological study areas** on the island; promoting **environmental awareness**; and participating with the Nevis Island government to **interpret the recommendations of the St. Kitts–Nevis Country Environmental Profile**, developed under the direction of the Caribbean Conservation Association with technical help from the Island Resources Foundation and grant assistance from the U.S. Agency for International Development (USAID).

NHCS is also conducting a study for a **marine park**, helping the Nevis Environmental Education Committee, and providing technical support to U.S. Peace Corps volunteers who teach environmental education in local schools, including publishing *ECO News for Children* in grades three to six. NHCS also publishes a bimonthly newsletter for public distribution.

■ Resources

NHCS seeks technologies for solar, wind, and other renewable energy sources and support in "making environmental protection popular." It also welcomes ideas for children's newsletters and "pre-written" materials for radio, television, or newspaper spots. It publishes several newsletters, *Birds of Nevis* book, historic information, and other literature.

St. Lucia is an island state in the East Caribbean south of Martinique. The land is of volcanic origin, with mountains running from north to south and many streams flowing from them through fertile valleys. What remains of St. Lucia's lush natural rainforest now exists only in the relatively inaccessible highlands. At lower levels, the forest has been cleared for farmland. Half the island is used for agricultural purposes.

St. Lucia's developing free enterprise economy is based mostly on agriculture—dominated by bananas, which make up 45 percent of exports—and tourism. The country's infrastructure, crops, and tourism industry was heavily damaged in 1980 by a hurricane.

🏛 Government

Caribbean Environmental Health Institute (CEHI) ⊕
P.O. Box 1111
Castries, St. Lucia

Contact: **Naresh C. Singh, Ph.D.**, Executive Director
Phone: 1-809-452-2501 • *Fax:* 1-809-453-2721

Activities: Development; Education; Monitoring; Technical Advice • *Issues:* Air Quality/Emission Control; Deforestation; Global Warming; Health; Sustainable Development; Waste Management/Recycling; Water Quality • *Meetings:* UNCED

● Organization

The **Caribbean Environmental Health Institute (CEHI)** was founded by the member governments of the Caribbean Community (CARICOM) to provide "scientific, technical and advisory services in all areas of environmental management, research and monitoring." The member countries and territories of CARICOM (all English-speaking) are: Antigua and Barbuda, Bahamas, Barbados, Belize, Dominica, Grenada, Guyana, Jamaica, Montserrat, St. Kitts–Nevis, St. Lucia, St. Vincent and the Grenadines, Trinidad and Tobago, the Turks and Caicos islands, and the British Virgin Islands. CEHI began its operations in 1982 on St. Lucia and is still expanding its activities and functions.

Envisioned as a comprehensive environmental management institute, CEHI has the following functions: controlling water, beach, and air pollution; managing waste; conserving natural resources; monitoring pesticide use in agriculture and occupational health in industry; and preventing destruction from natural disasters, such as hurricanes.

CEHI's mandate also includes **environmental education** through courses, seminars, symposia, workshops, on-the-job training, and special scholarships. CEHI receives funding from CARICOM and other international organizations such as the Canadian International Development Agency (CIDA), the German government, the Pan American Health Organization (PAHO), and the United Nations Environment Programme (UNEP). Research is done in cooperation with the Caribbean Conservation Association, Dalhousie University (Canada), the University of the West Indies, and the U.S. Environmental Protection Agency (EPA), among others.

Currently, CEHI's laboratory includes equipment for **monitoring chemical and bacterial water pollution and pesticide residues** in agriculture, managing **industrial chemicals**, providing sanitary engineering and conducting environmental impact studies. Since 1982, scientists have been monitoring coastal and marine pollution throughout the region, testing sediment samples for bacteria and chemicals and keeping records of the incidence of tar balls on beaches.

CEHI deals with new environmental problems as they arise, providing sanitary engineering services to St. Kitts–Nevis and Montserrat after Hurricane Hugo, assessing the damage from an **oil spill** in Castries Harbour, St. Lucia, and analyzing mysterious episodes of **fish kills** in the Caribbean region. Its program is expanding to include training in **toxic chemicals management** and development of an environmental information system to serve member countries.

Dr. Naresh C. Singh, CEHI's executive director and author of more than 40 scientific publications, notes that more NGOs in the Caribbean area need to be created to increase environmental awareness at the grassroots level. He is an adjunct professor of the University of Waterloo's Department of Civil Engineering, Canada, and a member of the Joint WHO/FAO/UNEP Panel of Experts on Environmental Management for Vector Control (PEEM), United Nations.

 *NGO*

St. Lucia National Trust
P.O. Box 595
Castries, St. Lucia

Contact: **Carleen Jules**, Education/Information Officer
Phone: 1-809-452-5005 • *Fax:* 1-809-453-2791

Activities: Development; Education; Research • *Issues:* Biodiversity/Species Preservation; Cultural Preservation; Deforestation; Population Planning; Sustainable Development *Meetings:* Caribbean Conservation Association; UNEP

▲ Projects and People

The St. Lucia National Trust is a grassroots organization of approximately 450 members working to conserve native wildlife in the Caribbean island nation of St. Lucia. It is directly involved with three protected natural areas in particular: the Savannes Bay Marine Reserve, the Fregate Islands Nature Reserve, and the Maria Islands Nature Reserve. These are part of a National Parks and Protected Areas Systems Plan sponsored by the U.S. Agency for International Development (USAID).

Savannes Bay, on the southeast coast of St. Lucia, supports three different ecosystems, each one vital to the continued health of the country's fishing industry. The **mangrove forests** along the bay's coastline serve as a breeding ground and nursery for fish, crabs, shrimps, and oysters. Mangrove trees have an extensive system of roots; the red mangrove's roots extend into the water to protect fish and shellfish, and the white and black mangroves' roots trap particles of silt, preventing erosion of the shoreline. Within the bay, seagrass beds trap algae, providing food for young fish, crustaceans, and molluscs. Coral reefs separate the bay from the ocean, ensuring calm waters for the mangroves and seagrasses to thrive in. The reefs are made up of soft corals such as sea fans and sea whips interspersed with hard corals and green, brown, and red algae.

The twin **Fregate Islands** off the east coast of St. Lucia provide a nesting and roosting place for the rare **fregate bird** each year from May to July. The **boa constrictor** is also found on the island, as are the St. Lucian oriole, trembler, and ramier, birds which nest in the tall bay trees. A nature trail has been engineered so that visitors can enjoy the scenic beauty of the islands, which were home to the Arawak and Carib Indians before the time of Columbus. People from the nearby village of Praslin help manage the Reserve.

The tiny **Maria Islands**, off the southern coast of St. Lucia near the town of Vieux Fort, are home to a variety of wildlife. Two species of reptiles, the turquoise-tailed Maria Islands **ground lizard**, and **grass snake**, said to be the "world's rarest," are found only on these islands. Never inhabited by man, the islands feature undisturbed forest and coral reef areas, as well as many cacti. From mid-May to the end of July, no visitors are allowed, so that hundreds of seabirds may nest undisturbed. The rest of the year, the Maria Islands Nature Centre on the mainland is the starting point for public tours.

■ Resources

The Trust publishes a newsletter, *Conservation News*, which is free to members, and also offers several books for sale, including a guide to Maria Islands Nature Reserve, *The Serpents' Tale: Reptiles and Amphibians of St. Lucia*, and *A Recognition Guide to the Insects of St. Lucia*. National Trust T-shirts are also available.

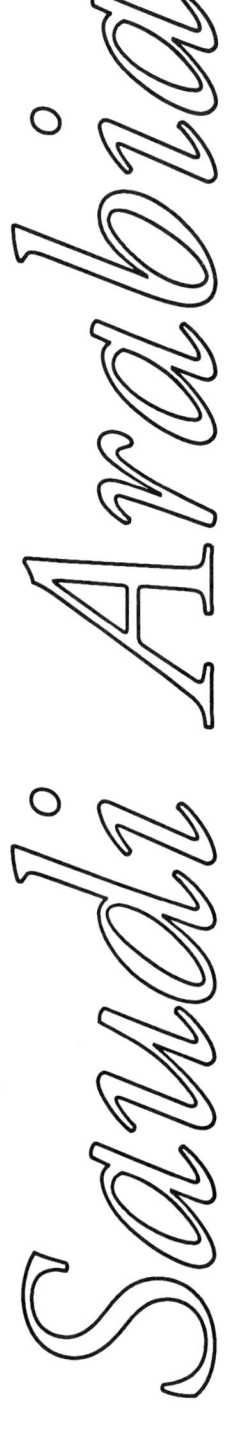

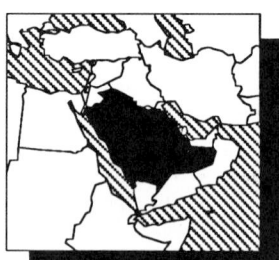

Saudi Arabia, occupying about 80 percent of the Arabian Peninsula, consists mostly of desert. It is the world's largest producer of petroleum products and the largest holder of proven oil reserves. The oil fields have transformed the Saudi society over the last 20 years, bringing high per capita income. Until the 1960s, most of the population was nomadic, but now 95 percent of the people live in cities. Nearly one-third of the population is resident foreigners.

Heavy oil tanker traffic in the Persian Gulf and spillage from coastal refineries have resulted in many oil spills. This has caused severe harm to fisheries and wildlife. The Persian Gulf war contributed greatly to the damage. Another problem is desertification, which is occurring at an alarming rate. Attempts have been made to check this trend through massive tree planting. There is also a danger that Saudi Arabia's aquifers could dry up within the next generation.

🏛 *Government*

National Commission for Wildlife Conservation and Development (NCWCD)

P.O. Box 61681
Riyadh 11575, Saudi Arabia

Contact: **Abdulaziz H. Abuzinada, Ph.D.**, Secretary General, Professor
Phone: (966) 1 441 8700 • *Fax:* (966) 1 441 0797

King Khalid Wildlife Research Centre (KKWRC)

Thumamah

Contact: **Nick Lindsay**, Curator, Zoologist
Phone: (966) 1 404 2527 • *Fax:* (966) 1 404 4412

Activities: Education; Law Enforcement; Research • *Issues:* Biodiversity/Species Preservation; Conservation of Wildlife Habitat; Deforestation Control; Desertification; Sustainable Development • *Meetings:* CITES

● Organization

The **National Commission for Wildlife Conservation and Development (NCWCD)** was established in 1986 by the Saudi Arabian government to initiate conservation programs and to research the wildlife of the area. NCWCD continues to maintain and begin new conservation programs including the establishment of reserves.

The projects that NCWCD has completed in the past has led to the partial restoration of plant cover and an increase in wildlife populations, it reports.

▲ Projects and People

NCWCD leads field programs throughout Saudi Arabia to record the numbers of the fauna and to survey the land used to determine conservation requirements.

Birds were recorded in four categories: breeding residents, residents, migratory, and flying visitors. In areas where the plant density and diversity most improved, the numbers of **Arabian bustard** increased, for example, from 81 in 1987 to 968 in 1990 in Wadi Juwwa and Al Dahna. In other ornithological expeditions, 120 bird species were recorded in the Farasian Islands. A large part of work has been completed on the ongoing *Atlas of Breeding Birds of Arabia* project.

A national wetland survey was begun both inland and along the coast. "For instance, the karstic ponds at Layla were found to be overexploited for irrigation water," writes NCWCD. "Many other wetlands such as Al Hair and Wadi Gizan dam have been proved to be nationally important sites for breeding birds or of international significance for migratory birds.

In studying problems caused by **commensal baboons**, the Commission came up with proposals on fertility regulation, public awareness campaign, garbage management, the use of trained dogs, and further research and practical actions.

At Ras Baridi, NCWCD maintains a turtle monitoring program at the country's most important mainland green turtle rookery on the Red Sea coastline. Of 100 egg-laying females, 70 were tagged.

In pursuing **animal reproduction**, NCWCD has maintained a fertility rate of 45 to 65 percent in the **artificial insemination of houbara bustards**, which increased its population from 70 to 256 in one year. With an Arabian oryx herd continuing to increase, a reintroduction program was launched at Mahazat As Say'd Reserve. Also in "pre-release enclosures for acclimatization and scientific research" are the rheem (sand gazelle), ostrich, and houbara bustard.

Since 1987, NCWCD has been succeeding in its captive breeding efforts at the **King Khalid Wildlife Research Centre (KKWRC)**, Thumamah, with the management of the Zoological Society of London. Originally a counting place for the late King Khalid's private animal collection, KKWRC has become the breeding center for Arabian animals, mainly different species of gazelle such as the sand gazelle and the mountain gazelle, or *idmi*. Nearly 1,000 Arabian gazelle have been numbered. Other animals which have been successfully bred include the nubian ibex, ostrich, and onager. The reintroduction into the wild of the Arabian oryx, ostrich, and houbara bustard has been initiated.

The **Saudi gazelle** had been considered extinct in the wild, but with the acquisition of four of the rare species from Bahrain, a recovery program is being established. Other KKWRC projects are the **classification of gazelles** that involved the establishment of a karyotyping laboratory.

Gazelle studies include examinations of the gazelles' endurance to harsh desert conditions. "Telemetry implants are used to monitor the body temperatures during different conditions, both climatic and dietary," reports KKWRC. "The laboratory is continually monitoring the physiological aspects of gazelles, giving baseline data as well as diagnostic work." Under study is the control of tuberculosis in the gazelle herds, which is significant to farming techniques.

"All aspects of animal management are constantly monitored with the aim of developing successful working strategies for these species," writes KKWRC.

KKWRC staff includes veterinarians Frank E. Rietkerk, Saeed Mubarak Mohammad, Osama Badri Mohammed, Edgardo C. Delima, and Mohammed Abdulkader Sandoka. Also on staff are wildlife consultant Dr. Douglas Williamson, zoologist Nick Lindsay, biologists Khaled Mohammed Al-Basri and Kevin Dunham, and laboratory manager William Flavell.

With the help of the World Conservation Union (IUCN), NCWCD took "a major step forward" in completing a **comprehensive plan** identifying 58 terrestrial and 47 marine and coastal sites to be protected within a decade. Areas are designated as special nature reserves, natural reserves, biological reserves, resource use reserves, and controlled hunting reserves. Proposals were also made to preserve 12 new protected areas each year, and legislation is expected to strengthen conservation measures. A **species conservation strategy** was also recently executed, denoting species needing urgent attention.

To enhance public awareness, NCWCD is creating a visitors' center at its NCWCD headquarters, which will serve as the base for the future natural history museum in Riyadh. **Databases** are also being set up, and communication links were made with the King Abdulaziz City for Science and Technology and with the National Computer Centre.

■ Resources

NCWCD seeks to expand its national expertise in new conservation technologies. KKWRC needs help with genetic research, especially with gazelles.

The NCWCD can provide a number of research publications, for exchange, including studies on the fauna of Saudi Arabia, the proceedings of the first international symposium on wildlife conservation in Saudi Arabia, the captive breeding program of the oryx, and the biology of gazelles. NCWCD also offers various brochures, booklets, annual reports, and slide or video programs.

Senegal

The Republic of Senegal, the most biologically diverse country in the Sahel, is the westernmost nation in Africa, bounded by Mauritania, Mali, Guinea, Guinea-Bissau, and the Atlantic Ocean. The economy is mainly agricultural; peanuts, a major crop, is grown in the mid-coastal section known as the "peanut basin."

The greatest threat to the country's health and well-being is the drought that has persisted since the 1960s: a quarter of the land is arid and 70 percent is semiarid. Senegal has serious desertification and deforestation problems. Fuelwood harvesting, charcoal production, and overgrazing have damaged much of the landscape. The government has protected over 10 percent of the parks and reserves, preserving much of the abundant wildlife. Yet poaching and harvesting of birds and game are concerns. Senegal is the world's largest exporter of exotic birds, going way beyond the official quotas. Water supplies are limited and polluted, and have led to water-related diseases.

🏛 *Government*

Institut Français de Recherche Scientifique pour le Développement en Coopération (ORSTOM) ⊕
French Institute of Scientific Research for Development and Cooperation
B.P. 1386
Dakar, Senegal

Contact: **Géraut Galat, Ph.D.,** Director of Research
Phone: (221) 326746 • *Fax:* (221) 324307

Anh Galat-Luong, Ph.D., Researcher
Phone: (221) 321846

Activities: Development; Research • *Issues:* Biodiversity/Species Preservation; Biotechnology; Botany; Climatology; Ecology; Economy; Global Warming; Health; Human Sciences; Oceanography; Sustainable Development; Water Quality; Zoology • *Meetings:* IUCN

● Organization

Since 1949, ORSTOM, Institut Français de Recherche Scientifique pour le Développement en Coopération (French Institute of Scientific Research for Development and Cooperation), has been conducting inventories and surveys of the natural resources of Senegal, with its work greatly diversified over the years. Currently, the Institute can be divided into five scientific departments: Earth, Ocean, Atmosphere; Continental Water; Agricultural Activity; Health; and Society, Development, Urbanization.

Some 45 research programs in collaboration with Senegalese authorities or organizations—such as the Pasteur Institute of Senegal as well as international groups—concern national problems as degradation and regeneration of the natural heritage, management of natural resources and seas, amelioration of cultures with reference to food, health, urbanization, and demographic growth. Overall, ORSTOM has a staff of 2,000. *(See also ORSTOM in the Congo and Mexico Government sections.)*

▲ Projects and People

ORSTOM runs **seismic observatories** in M'Bour and Kédougou that record data and fluctuations in the electro-magnetic field surrounding the Earth. They are part of a worldwide network of seismic research, gathering data on electro-conductivity of the Earth's core and outer shell, the nature and frequency of earthquakes, and the seismic activity that precedes and follows them.

The water and earth project was begun in 1950 as an inventory of available agricultural land. Today, the project attends to its task with a new sense of global community, and attempts to plan its future with a global commitment to the environment in mind. It features a cartographic survey to locate various land types and determine their best uses. Other aspects include: fighting the salinization of farmland and waterways, creating educational maps, and studying the effect of polluted rain and water on soils.

The sea and its resources projects, conducted with many collaborators, include: fishing industry—understanding the biology of certain fish such as tuna and sardines and the methods used to catch them; socioeconomics of fishing—how the Senegalese economy is affected by and reacts to the fishing industry; and carbon dioxide (CO_2) surveillance—studying the effects of atmospheric CO_2 on the ocean.

Among several biotechnology studies, genetic engineering techniques have been applied very successfully to a variety of plants: palm dates, palm oil, potatoes, papayas, and sweet potatoes. These projects also include hydroponics, molecular biology, and bacterial production of ethanol.

Projects regarding trees and rivers often require a multidisciplinary approach. Because the damming of rivers for flood control and hydroelectricity has changed the nature of some rivers, their future management rests on understanding the effects that present dams have on the ecology and biology of wildlife. ORSTOM observes that runoff from farms is harming wildlife in some rivers, and ichthyological studies are in progress. Senegal has a concerted effort to fight desertification, with the Institute. Important projects include an attempt to reforest ground damaged and desertified by salt and acids with trees that are tolerant to such soil conditions.

Among health programs, Project Vaccination seeks to study the effects of vaccines on increasing the health of the population. The vaccines are given to children under age five to fight such diseases as measles and whooping cough. ORSTOM is conducting studies of AIDS and its spread, as well as educational campaigns on sanitation and hygiene. There are programs on malnutrition and growth, water, and health. A number of studies focus on animal diseases, and the transmission of those diseases to humans, such as monkey AIDS.

■ Resources

ORSTOM's information kit is available, printed in French.

 NGO

Satis ⊕
B.P. 2664
Dakar, Senegal

Contact: Martine Abraham, Provisional Executive Secretary
Phone: (221) 217595 • *Fax:* (221) 222695

Activities: Development; Education • *Issues:* Sustainable Development

● Organization

Satis, made up of organizations ranging from grassroots to global groups and funding agencies to research institutes, is a "global union for technologies for environment and sustainable development" founded in 1982 in the Netherlands and headquartered in Senegal since 1991. Its more than 150 members in some 70 countries are said to create "137 new permanent jobs, give technical advice to 3,400 clients, publish five practical handbooks, [and] distribute 8,800 publications" in fields including "agriculture, construction and shelter, food processing, forestry, energy, . . . health research, enterprise development, . . . management, training, communication, [and] environmental assessments."

Satis sustains itself through a "modest annual membership fee," grants from selected funding agencies, and the sales of consulting services and publications.

▲ Projects and People

Used worldwide, the Satis Information System comprises Satis Classification, a "structured listing of thousands of topics" accessed by more than 700 organizations in various languages and in its third revision; Satis Thesaurus with 3,300 terms; SatisFile software package for information management and compatibility with CDS-ISIS; regional training programs; and Satis Databases, planned for the 1990s, which will feature "selected references of all member organizations, and will appear in printed form, as well as different electronic media, such as diskette, CD-ROM, and online database hosts."

The Satis Communications program "supports members in publishing and disseminating their publications and audiovisual materials." Also featured are an annual Satis Catalog of members publications; quarterly SatisNews; resource guides; electronic publications; and training in publishing and distribution. SatisNet is a network for electronic mail, fax, telex, graphics exchange, file transfers, and databases.

Satis also helps its members develop, transfer, and promote technologies across international boundaries, with sensitivity, such as North-South and South-North exchanges.

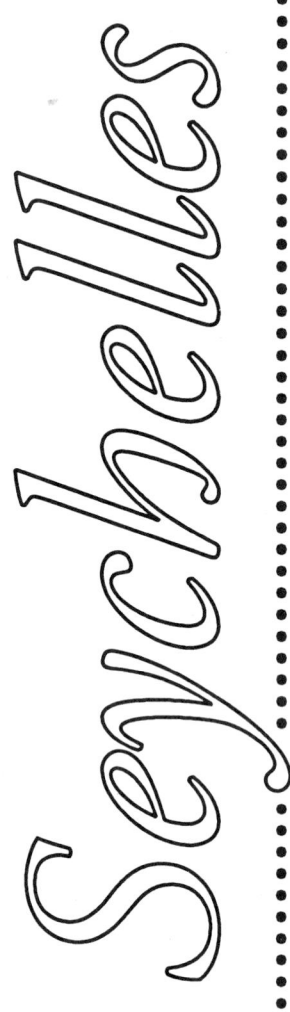

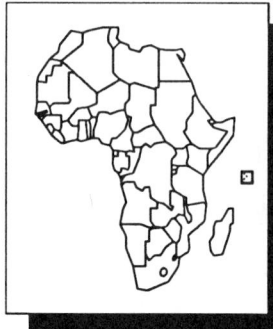

Seychelles is made up of 115 islands, divided into two main groups: one rocky, with a narrow coastal strip and lush vegetation; and the other consisting of elevated coral reefs. Wildlife include giant tortoises, green sea turtles, few insects, and no animals dangerous to man. Emigration has helped to keep the population growth rate at a minimum. The country's mixed developing economy is heavily dependent upon tourism.

🏛 Government

Conservation and National Parks Division
Department of Environment (DOE)
President's Office, P.B. 445, Victoria
Mahé, Seychelles

Contact: **Nirmal Jivan Shah,** Director
Phone: (248) 24644, ext. 18 • *Fax:* (248) 24500

Activities: Development; Education; Enforcement; Legislative; Research • *Issues:* Biodiversity/Species Preservation; Deforestation; Global Warming; Pollution Assessment and Control; Sustainable Development • *Meetings:* CITES; IUCN; IWC; UNCED; UNDP; UNEP

● Organization

"Every island in the Seychelles is like a spaceship, a small and remote unit with finite resources which need to be managed on a sustainable basis for us to survive," writes Seychelles President **France-Albert René.**

In its post-independence days since 1977, the country first focused on health, education, and other essential services for the Seychellois. "Economic goals took priority in our 1984–89 National Development Plan which concentrated on creating wealth and income by encouraging private enterprise and appropriate industries in order to diversify our economy," reports René. Sustainable development, locally and globally, is the cornerstone of the 1990–1994 National Development Plan, of which the **Environmental Management Plan of the Seychelles (EMPS) (1990–2000)** is an integral part.

The Seychelles government prepared EMPS "with the advice and assistance of the United Nations Environment Programme (UNEP), United Nations Development Programme (UNDP) and the World Bank" and presented it to the people at a joint celebration of Liberation Day and World Environment Day, June 5, 1990.

The Plan has six policy goals: to protect health and life quality; "to ensure that future economic development proceeds on an equitable and sustainable basis"; to preserve biological diversity and natural heritage; to improve decisionmaking, laws, and institutional framework for sustainable development; to increase public information and understanding of links between environment and development; and to strengthen global cooperation on these matters.

▲ Projects and People

The Plan contains more than 60 environmental and natural resource management projects, categorized either for action immediately or by 1995. The **Department of Environment (DOE),** with 240 employees, is the executing agency for 24 projects; the remainder are being carried out by other Ministries in cooperation with the DOE. "Priorities include state of the environment reporting [and] new environmental impact assessment procedures: making annual sustainable development audits, assessing the impact of climate warming and sea level rise," according to the Plan's Executive Summary.

Priority projects comprise a new **Pollution Control and Advisory Service**, including plans to combat oil spills and ocean dumping; a **Waste Management Programme**, including improved sewerage for Greater Victoria and other areas, the Mahé solid waste plan and treatment plant, and a regional management plan for hazardous wastes; a **Land Management Programme** to improve soil fertility; **Energy Policy and Conservation Programme**, with a feasibility study for lead-free gas; **National Parks and Conservation Programme** to protect endangered species, upgrade Curieuse National Park, the Aldabra World Heritage Site, and marine parks; pesticide controls; **Forests Management Programme** to preserve endemic tree species and intensify fire prevention measures; **Coastal Environmental Management Programme** to control beach erosion, perform a baseline study, and reduce sand use in construction; **Marine Resources Management Programme** to protect endangered sea turtles and improve fisheries surveys and assessment; **Environmental Law and Enforcement Programme**; and **Environmental Information, Education, and Training Programme**, with new public information campaigns and developing a new school and in-service training for sustainable development.

Government interagency committees on environmental education and chemicals management were also added.

The Plan will require an estimated total of about $47 million, of which 77 percent is for infrastructure for sewage, solid waste management, and water supply and distribution. Spain, Norway, and Finland are helping to finance several projects. "We have taken special care to ensure that this proposed investment in support of sustainable development does not pass on an unsustainable debt service burden to our children," says the Executive Summary.

Among other environmental activities on Seychelles, about nine pairs of paradise flycatcher birds at the La Digue Veuve Reserve are protected with a new conservation law. Similarly, the Curieuse Marine National Park—a "sanctuary of tropical vegetation, giant tortoises, and marine life"—has a new series of protective laws that include the banning of shelling, fishing, and taking of coral and the outlawing of weapons, traps, or fishing gear. National Youth Service (NYS) environmental associations and projects were recently organized in the villages of Port Launay and Cap Ternay. Seychelles Breweries is marketing distilled water for use in car batteries.

With Seychelles at the center of the tuna fishing industry in the western Indian Ocean, a computer system of ORSTOM (French Institute of Scientific Research for Development and Cooperation) is being used for tuna research. The Tourism for Tomorrow awards honored Seychelles' Ministry of Tourism's beach cleaning project. School children learn about medicinal plans used in traditional medicine. New park rangers' uniforms and recruits call attention to the mission of conservation and national parks. The only ozone-measuring station in the Indian Ocean is at the Seychelles international airport; according to the World Meteorological Organization (WMO), tropical meteorology has a high priority.

■ Resources
Copies of the EMPS and the EMPS–90 Investment Plan can be ordered.

Sierra Leone

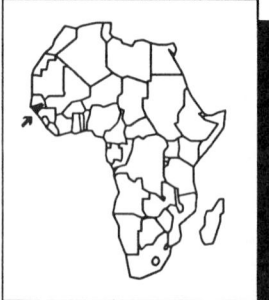

One of West Africa's smallest countries, Sierra Leone is also the site of its highest point—the 6,390-foot Bintimane in the Loma Mountains of the western peninsula. The country is rich in minerals, and has relied on minerals for its economic base—especially diamonds, which account for half the export income.

Much of the land is used for subsistence agriculture, upon which 75 percent of the population depends. Malnutrition is widespread. The "bush fallow" agricultural system is a central influence in the lives of the people. Trees are cut to grow crops for a few seasons, then allowed to regrow, so that most of the original forest has been replaced. This system is breaking down, however, and the soil is becoming exhausted because of the growing population and demand for food. Overharvesting of timber is causing additional pressures, and habitat loss has reduced wildlife populations.

 NGO

Concerned Environmentalists Action Group (CEAG) ❧
22A Hindowa Street, Kenema
Freetown, Sierra Leone

Contact: Cyril Edward Wilson, Coordinator

Activities: Development; Education • *Issues:* Air Quality/Emission Control; Biodiversity/Species Preservation; Deforestation; Global Warming; Health; Population Planning; Sustainable Development; Waste Management/Recycling; Water Quality

▲ Projects and People

Started in 1989 by Friends of the Earth representative M.K. Sei, Concerned Environmentalists Action Group (CEAG) is a "subset" of FOE, with a membership of 10. The grassroots group formed when "farmers persistently reported that the fertility of the soil on which they cultivate is dwindling in an alarming rate to its lowest ebb," reports project coordinator Cyril E. Wilson. Lacking a tree-planting plan and also concerned about the extinction of certain animal species, the group pioneered an **awareness campaign** to educate local people about the dangers of deforestation in tropical rainforests.

CEAG arranged village meetings in areas with large expanses of dense vegetation, and discussions were held in local languages. When funds allow, CEAG plans to prepare charts, maps, photographs, and other illustrative materials.

Involving the entire Tunkia Chiefdom community, a five-year **afforestation project** got underway in 1992 in the southeastern Kenema District. CEAG intends to replant two-thirds of the deforested land, beginning on a 60-acre plot. As the project was launched, "Holes were dug by community people, and seedlings were transplanted, and fertilizer was put at the base of the plants to ensure rapid growth." The plantings occur every six months. "In the future, farmers will be given seedlings for replanting in those areas where trees have been cut down for commercial and domestic purposes," writes Wilson, a teacher.

Wilson explains the importance of gaining the support of local chiefs: "The practical aspect of the work is done at chiefdom levels where the chiefs are custodians of the land. In addition, the chiefs are delegated to oversee the progress of the work and are the immediate contact persons for evaluation."

CEAG plans to undertake drawing and essay competitions and produce a quarterly magazine on environmental issues with the participation of other groups.

"Environmental issues are such significant aspects to man and his environs that it deserves critical analysis not only within the community but globally," writes Wilson. "Therefore, it is quite necessary to create an opportunity wherein environmentalists meet on a common platform such as seminars and workshops to discuss views which could be important for implementation with various localities." Wilson also believes that groups such as FOE should analyze different aspects of environmental issues, such as pollution.

Other CEAG participants are, from the Mende tribe, teacher Frank Coker and banker Patrick A. Bundu; teacher Irene Carew, from the Temne tribe; and technician Moses Pratt, from the Sherbro tribe.

■ Resources

CEAG needs financial aid, tools, and teaching materials to carry out its awareness and afforestation programs. It seeks new technologies in applying chemical fertilizers, making compost, improving planting methods, and introducing plant species that will thrive under tropical climatic conditions.

Friends of the Earth (FOE) ⊕ ⚘
P.M. Bag 950, 33 Robert Street
Freetown, Sierra Leone

Contact: **Olatunde Johnson,** Chairman
Phone: (232) 22 225223

Activities: Development; Education; Human Rights; Political/Legislative; Research • *Issues:* Air Quality/Emission Control; Biodiversity/Species Preservation; Deforestation; Energy; Global Warming; Health; Population Planning; Sustainable Development; Transportation; Waste Management/Recycling; Water Quality • *Meetings:* CITES; FOE AGM; UNCED

■ Resources

Friends of the Earth (FOE), with 500 members, is engaged in education, communication, and "human rights peace and progress" in Sierra Leone. Available materials are *Cotton Tree* (the FOE newsletter), *Sanitation Worksheet, Global Warming, Tropical Rainforest, Energy Worksheet,* and *Pollution.*

FOE needs telex and fax equipment, training and communication tools and other resources for schools, research centers, community development, and transportation.

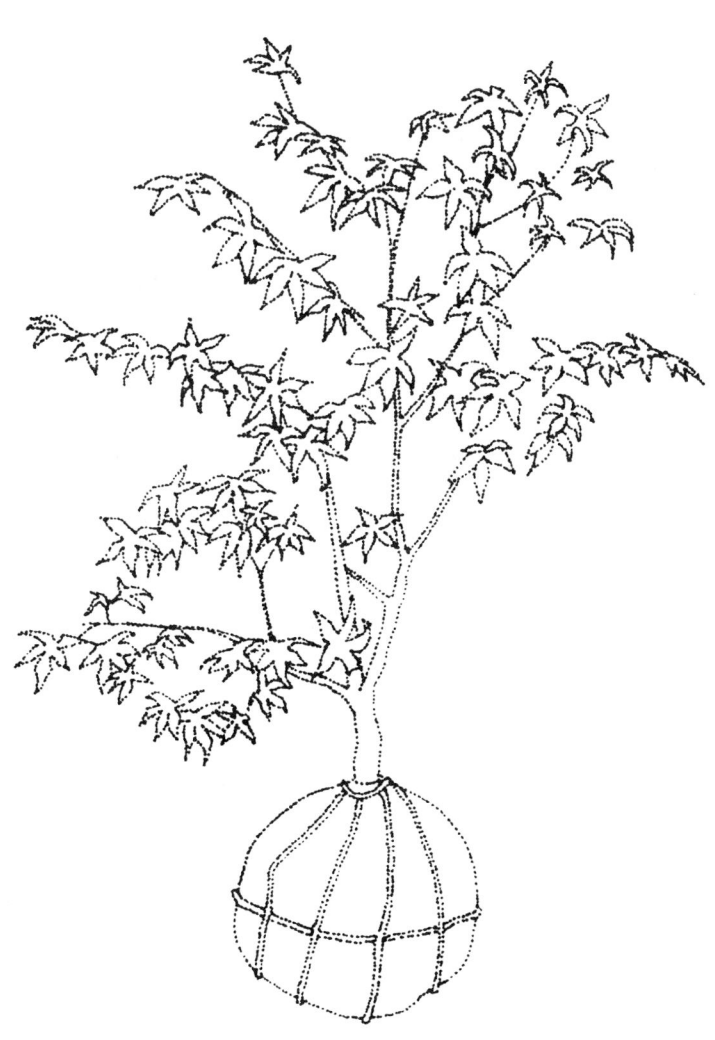

Singapore is a tropical island located at the southern tip of the Malay Peninsula. A lowland area that was once comprised of jungles and swamps, the highest point on the island is only 581 feet above sea level. Despite the high population (one of the most dense in the world), Singapore enjoys a relatively high standard of living—second only to Japan among Asian countries. The island's natural forests have been completely replaced by urban high-rise buildings. While there is some limited intensive agriculture, much of the country's food is imported.

Singapore's hot and humid climate with high annual rainfall provides ideal ecological conditions for rapid bacterial growth, organic decomposition, and the propagation of disease-bearing insects. In addition, urbanization has brought with it congestion, air pollution, and waste disposal and water supply problems. A combination of public, private, and grassroots efforts have created a strong solid waste management and recycling program.

🏛 *Government*

Public Affairs Department
Ministry of the Environment
19th Storey Environment Building
40, Scotts Road
Singapore 0922, Singapore

Contact: **Donald Goh**, Head
Phone: (65) 732 7733 • *Fax:* (65) 731 9866

Activities: Development; Law Enforcement • *Issues:* Air Quality/Emission Control; Energy; Global Warming; Sustainable Development; Water Quality; Waste Management/Recycling

▲ Projects and People

Singapore aims to become a "model Environmental City," according to its draft Green Plan proposal, to accommodate comfortably a population of four million people and ensure "high standards of public health . . . clean air, clean land, clean water, and a quiet living environment."

A major undertaking of the Ministry of the Environment has been cleaning up the Singapore River and Kallang Basin (draining five other rivers) that "once lay hidden and obscured in the murky waters, debris, litter, and the stench of rotting matter." According to **Dr. Ahmad Mattar**, Minister for the Environment, the 10-year task beginning in 1977 involved "the phasing out of some 21,000 unsewered premises, the relocation of some 5,000 street hawkers, the resiting of the lighterage industry to Pasir Panjang, the phasing out of pig farms and wet duck farming, the relocation of the street vegetable wholesalers and pollutive marine and backyard industries, the clearance of large squatter tracts, and the physical improvement works."

The Singapore River is now considered to be a "prime national asset." Today, boatyards provide pollution control measures, riverbanks with new walkways are turfed and landscaped, debris has been cleared, sandy beaches are in place, and food centers feature "lighting, water supply, and wash areas connected to the sewer." The water is "clean and free of offensive smell," and aquatic life is returning, according to National University of Singapore studies.

Ongoing projects include deepening the Kallang River and allowing for flood protection of the catchment, thereby creating a "permanent waterbody." The landscaped riverbanks will feature jogging and cycling tracks. Debris is being removed from the Geylang River and parks are being installed at the attractive mechanical debris removal facility. At the northeastern **Api Api River**, drainage and landscaping are also being improved and **mangrove plants** are expected to be regenerated.

Elsewhere, leacheate treatment facilities are being constructed. The import and sale of mercury oxide batteries are being prohibited (except for hearing aids); the sale of certain zinc carbon and alkaline batteries are also being banned. A recycling scheme for nickel-cadmium (Ni-Cd) batteries is in place. Hospitals and primary health care clinics must each appoint a waste manager to write procedures for biohazardous wastes—to be disposed of in specially designed incinerators. A telemetry system is being set up to monitor air pollution throughout Singapore. The government expects to lessen foul odors from wastewater treatment sites so that buffer zones can be reduced and "better land use" realized. Sewerage infrastructures are being rebuilt and added to serve private and public development.

■ Resources

The draft proposal *Singapore's Green Plan, Clean Rivers*, and the Ministry of the Environment's *Annual Report* are published.

Singapore Botanic Gardens
National Parks Board
Cluny Road
Singapore 0409

Contact: **Tan Wee Kiat, Ph.D.**, Executive Director
Phone: (65) 470 9915 • *Fax:* (65) 475 4295

Activities: Education; Research • *Issues:* Biodiversity; Botany; Species Preservation • *Meetings:* World Parks Congress

● Organization

The first Botanic Gardens in Singapore was created in 1822 as a place to introduce new, economically important crops into Singapore. These crops included timber, spices, and the all important rubber plant. In 1859, the gardens were refounded by an agri-horticultural society and were subsequently handed over to the government to be maintained.

The century-old national parks system came under new leadership and management when in June 1990 the **National Parks Board** was created under the Singapore Ministry of National Development in an effort to take care of all of the republic's natural reserves.

The National Parks Board is in charge of developing, managing, and promoting the **Singapore Botanic Gardens**, the Fort Canning Park, and the nature reserves including the Labrador and Bukit Timah reserves.

▲ Projects and People

The National Parks Board's first task was to continue the previous work of the Parks and Recreation Department in the **Garden City Programme**, which called for the "greening" of Singapore. The Board was to develop and maintain the entire green region of Singapore. The project was extremely successful, and according to the Board executive director **Dr. Tan Wee Kiat**, "Today, visitors from around the world can view the successful achievements of the Department from the moment their plane touches down at Changi Airport."

The formation of the National Parks Board enhanced the public amenities that the Botanic Gardens had provided in the past. With the success of the Garden City project, the Board was able to start horticultural research programs. The older School of Ornamental Horticulture, living plant collection, extensive herbarium and reference library at the Botanic Gardens were augmented as well.

At the Botanic Gardens, an orchid breeding and hybridization program was started throughout the region. The Gardens continue to introduce plants of botanic and horticultural value into the area.

At the Bukit Timah Nature Reserve, a visitors' center was opened to add to the public's appreciation of the park's plants and animals, such as gingers, ferns, fishtail palm, rattans, and small mammals—macaques, treeshrews, flying lemurs, lizards. Darting about are woodpeckers, tit-babblers, cuckoos, and the noisy racquet-tail drongo.

Sponsored by the Hongkong Bank, the National Parks Board and the Ministry of Environment organized a **Nature Trek** in conjunction with the festivities of **Clean and Green Week**. The Trek drew 125 students to the Bukit Timah Reserve, many of whom journeyed through the reserve for the first time. Volunteers participated from the National University and the Malayan Nature Society. Similar **Care for Nature Programmes** continue.

Among research endeavors, the National Parks Board may be able to identify and possibly cure an old tree disease. As far back as 1875, there have been recorded outbreaks of a disease that hits supposedly healthy angsana trees, causing them to wilt and die after about five weeks of infection. The disease is now thought to be caused by a naturally occurring fungus, *fusarium solani*, which exists in the soil. The researchers are still trying to control the quickly spreading fungus. Attempts are being made to inoculate young angsana trees.

■ Resources

The National Parks Board seeks help with a public awareness program to promote conservation and biodiversity.

The Board publishes *GardenWise* newsletter, brochures, booklets, and maps of its nature reserves and gardens. *Visions of Delight* is a pictorial guide and book.

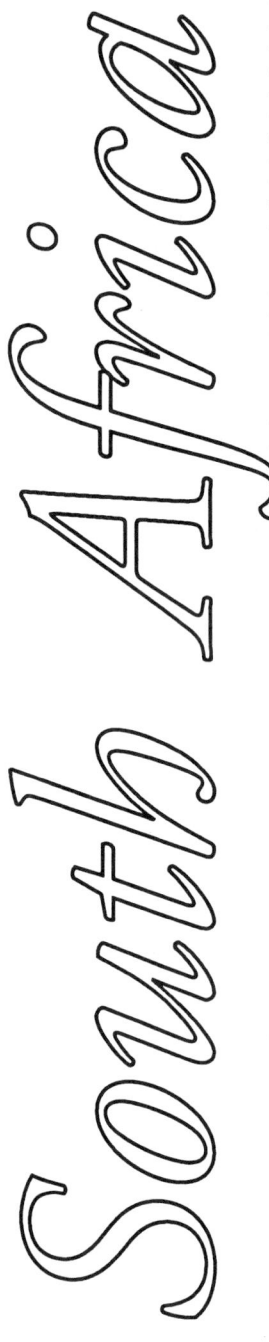

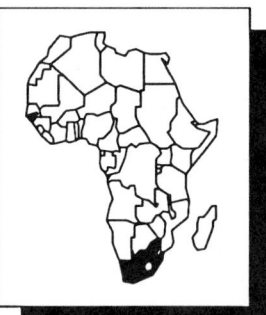

South Africa is the home of 25,000 native plant species, mountains, desert, wildlife, magnificent seacoasts, rangeland, and modern cities. Its supply of gold, of which it is the largest producer, makes up almost half the world's total known reserves.

As the country has emerged from apartheid, it is also emerging from an era of rapid development from which the environmental damage is only now being assessed and for which policies are being developed. Of particular concern is the region known as the Highveld, where 80 percent of the country's electricity is generated by coal-fired power stations. This and other plants and industries contribute to high levels of pollution which build up in stagnant air masses causing severe acid rain. Soil erosion is occurring in areas overpopulated with people and livestock, and desertification is a serious concern in the Karoo, where the grasslands on which South Africa's sheep are stocked are deteriorating.

🏛 *Government*

Department of Environment Affairs
Private Bag 447
Pretoria 0001, Transvaal, South Africa

Contacts: **C.M. Cameron,** Deputy Director General
   Phone: (27) 12 310 3666 • *Fax:* (27) 12 32 2682

   Sydney Albert Gerber, Chief Director, Environmental Conservation
   Phone: (27) 12 310 3695

Activities: Development; Education; Law Enforcement; Political/Legislative; Research
Issues: Biodiversity/Species Preservation; Global Warming; Sustainable Development;
Waste Management/Recycling • *Meetings:* CITES; IUCN; Ramsar

● Organization

Among the Department of Environment Affairs' four components are management of the Antarctic program; Environmental Conservation, which includes "planning, impact studies and advice, education, pollution control, conservation and the management of some 60 research projects; Weather Bureau; and Sea Fisheries with regulatory and research functions.

Directorate of Nature and Environmental Conservation
Private Bag 5014
Stellenbosch 7600, Western Cape, South Africa

Contact: **Neil Fairall, D.Sc.,** Senior Specialist Scientist
   Phone: (27) 2231 70111 • *Fax:* (27) 2231 71606

Activities: Research • *Issues:* Biodiversity/Species Preservation; Conservation • *Meetings:*
CITES; IUCN

▲ Projects and People

Current research interests of Dr. Neil Fairall, Directorate of Nature and Environmental Conservation, are "monitoring requirements for conservation; plant/animal interaction, chiefly in arid environments; population dynamics and population genetics in the management of wildlife populations on small reserves; the theory of non-equilibrium (dissipative) structures in biology; and expert systems modeling."

Dr. Fairall writes about African wildlife, including springbok farming, comparison of blue wildebeest and red hartebeest, dassie management, foraging behavior of impala and blesbok, hyrax, and wild antelope, for examples. He has researched zoology at Israel's Tel Aviv University and visited the Institute of Arctic Biology, University of Fairbanks, Alaska, as well as the University of Maryland, USA, and Zoology Departments at Oxford, London, and Exeter Universities to discuss conservation programs.

Kruger National Park
Private Bag X402
Skukuza 1350, Transvaal, South Africa

Contact: **Darryl R. Mason, D.Sc.**
 Phone: (27) 1311 65611, ext. 2238

Activities: Research • *Issues:* Biodiversity/Species Preservation
Meetings: IUCN/SSC

▲ Projects and People
At the **Kruger National Park**, Dr. **Darryl R. Mason** monitors ungulate population dynamics and the natural processes that regulate them. Field classifications of sex and age classes are undertaken to establish baseline data on long-term population dynamics. And correlations between sex and age ratios and rainfall patterns are investigated.

"Data on populations of ungulates (excluding buffalo and hippopotamus) have already been collected over an eight-year period (1983–1990), by means of ground surveys conducted in the months of August, September, and October," reports Dr. Mason.

Dr. Mason would like to see the formation of a "Federation of African National Parks and Game Reserves" which would provide a vehicle for exchange of staff to gain perspective of a variety of research and management problems in different habitats and to assist special conservation projects where the means are limited." A member of the Species Survival Commission (IUCN/SSC) Antelope and Pigs and Peccaries groups, Dr. Mason has published articles and papers on the warthog, albino bushbuck, springbok, impala, mountain reedbuck, wildebeest, and buffalo.

Natal Parks Board
P.O. Box 662
Pietermaritzburg 3200, Natal, South Africa

Contacts: **George R. Hughes, Ph.D.**, Chief Director
 Martin Brooks, D.Sc., Head, Scientific
 Services
 Phone: (27) 331 471961 • *Fax:* (27) 331 471037

Activities: Conservation Management; Development; Research • *Issues:* Biodiversity/Species Preservation; Conservation Education; Utilization of Natural Resources; Visitor Facilities and Opportunities • *Meetings:* CITES; IUCN; Ramsar; World Parks Congress

● Organization
The **Natal Parks Board** had its beginnings at the end of World War II when its goal was to save what little wildlife was left for future children. Since that time, the Park, now with a staff of 3,700, has grown and changed. During the 1960s, the Board stabilized the population of large mammals and soon looked toward the farmlands of Natal as potential habitat. By the 1970s, the farm community had developed a conservancy system, then a Biosphere Reserve system, in partnership with the Board.

"At the same time, the Zululand and KwaZulu Bureau of Natural Resources began to involve the rural black communities in nature conservation," reports the Board. "Today, over 10 percent of all KwaZulu and Natal is formally protected in parks and nature reserves."

Game reserves are traced to the last century when such areas were set up to protect "dwindling wildlife populations and their habitats." For local populations, they have also been a source of controlled harvesting of thatching grass, reeds, woods, and "Ncema grass" (*Juncus krausii*) used in crafts, as well as marine resources, such as mussels. The Board says that it committs to "ensure that reserves become recognized as a benefit to the community for an educational, employment, and recreational point of view. . . . More formal neighborhood programs are being prepared."

▲ Projects and People
Cape buffalo are in great demand by private game ranchers and conservation bodies. Many currently reside in Zululand; however, the Directorate of Animal Health has refused to allow them to be moved to other destinations, as they are carriers of corridor disease, which, when transmitted to cattle, is fatal. Ticks, particularly the brown ear tick, carry the disease, which is transmitted from buffalo to buffalo and also to cattle. So far, attempts to obtain disease-free calves by transplanting buffalo embryos into domestic cows have not succeeded. The Natal Parks Board is continuing to seek methods of developing buffalo herds which are corridor-disease free.

The elephant is being reintroduced into the **Itala Game Reserve** that it inhabited in the nineteenth century and earlier, until driven extinct by ivory hunters. Its presence is expected to slow down the woody plant encroachment and become a valuable tourist attraction. "The elephant was accepted very well by other game in the Ngubhu basin," reports the Board, "and could even be seen browsing together with the giraffe. They have been feeding selectively and do not tend to over-utilize a particular area before moving on."

Managed by the Board since 1988, the **Sodwana State Forest** is part of the Greater St. Lucia Park, and boundaries are largely unfenced with large numbers of cattle grazing in the area. Negotiations are in progress with local tribal authorities about proposed fencing. Game animals translocated here are oribi, bushbuck, impala, nyala, zebra, warthog, and waterbuck—with plans to introduce wildebeest, white and black rhino, and eland, once the fence has been constructed. Species such as reedbuck, hippo, crocodile, duiker, and suni already dwell in the forest.

Chief director Dr. **George R. Hughes**—with expertise in coastal marine and freshwater systems and especially sea turtles—is the African regional member to the World Conservation Union (IUCN) and South African coordinator to its Species Survival Commission (SSC); he is also a member of the SSC Marine Turtle Group and research associate with Oceanographic Research Institute (ORI), Durham.

Dr. **O. Bourquin** reports on recent projects, among several hundred, including developmental behavior ecology of infant baboons; distribution ecology of *Acacia* species; erosion in the Mkuzi Game Reserve; ecology of four-toed elephant shrews; social transmissions of food selection by vervet monkeys; social behavior of adult male samango monkeys; small mammal surveys; grasslands monitoring in Coleford Nature Reserve; geohydrological study of the Zululand coastal plain; mating behavior of ants; reedbuck reproduction; large herbivore dynamics; spotted hyena study; parasites of the bushpig; importance of Lake St. Lucia grasslands to birdlife; grazing impact of white rhinoceros in Umfolozi Game Reserve; monitoring vultures; blindness in red duiker; analyses of riverine and savanna vegetation; cave sandstone weathering; fish utilization in Zululand inland waters; design of a crocodile monitoring program; mangrove crab study; biomass of mangroves; marine molluscs and seaweeds of Natal; ecology of black parrotfish; ragged tooth shark ecology; estuarine restoration demonstration; and effects of prawn farming on biota of Mlalazi estuary.

Recently completed projects on the conservation of wetlands in Natal and the Mkuzi Swamp System pointed to rehabilitation of the riparian forest and controls on excessive cattle grazing and harvesting of plant species. Elsewhere, new insights into the foam-nest frog's reproductive biology were made. A monitoring program now tests the effects of burning regimes.

Dr. Martin Brooks oversees a staff of 70 in the Ecosystems Conservation, Species Conservation, and Planning divisions with emphasis on understanding terrestrial, freshwater, estuarine, and marine ecosystems and providing management advice. He chairs both the Rhino Management Group (RMG), which is coordinating a conservation plan for the black rhinoceros, and the Species Survival Commission (IUCN/SSC) African Rhino Group.

■ Resources

Reports and publications are available in the Natal Parks Board library, Pietermaritzburg. Drs. Hughes, Brooks, and Bourquin each have lists of published works and projects.

Dr. Brooks writes that the Parks Board is in need of "implantable transponders for remote identification of black and white rhinos, conservation biology expertise [concerning] genetics of small populations in particular, and resource economics expertise."

Natal Parks Board
P.O. Box 456
Mtubatuba 3935, Natal, South Africa

Contact: Jacques R.B. Flamand, Veterinarian
Phone: (27) 35 550 0666 • *Fax:* (27) 35 550 0641

Activities: Research • *Issues:* Biodiversity/Species Preservation; Conservation Biology; Game Capture • *Meetings:* CITES

▲ Projects and People

Dr. Jacques R.B. Flamand, as Natal Parks Board veterinarian, has expertise in "African wild game animal captures, adaptation, holding and transportation of such game species, parasitism of game species, and general wildlife veterinary ecology."

His current projects in Natal are "parasites of game species; population genetics of small populations, currently black rhinoceros, white rhinoceros, and lion; and chemical immobilization procedures of larger game species."

In Saudi Arabia, Dr. Flamand recently conducted research including "diagnosis and combatting tuberculosis in a population of Arabian oryx and handrearing of Arabian oryx calves." He was also director of the National Wildlife Research Center, Taif.

Dr. Flamand belongs to the Species Survival Commission (IUCN/SSC) Veterinary Group.

National Botanical Institute (NBI)
Private Bag X101
Pretoria 001, Transvaal, South Africa

Contact: Jacobus N. Eloff, D.Sc., Director of Research
Phone: (27) 12 804 3200 • *Fax:* (27) 12 804 3211

Activities: Education; Research • *Issues:* Biodiversity/Species Preservation; Global Warming; Sustainable Development

● Organization

With the goal of becoming "a world leader in the development of botanic gardens, in plant conservation and education, and in botanical research by the year 2000," the National Botanic Gardens and the Botanical Research Institute merged in 1989 to become the National Botanical Institute. Its mission is to provide the "facilities, knowledge and expertise necessary to ensure the conservation, sustained use, appreciation and enjoyment of South Africa's exceptionally rich flora and vegetation."

NBI is now a network of eight National Botanic Gardens with research groups at Kirstenbosch and Pretoria, additional facilities at the Universities of Cape Town and Stellenbosch, Albany Museum, and a Natal Herbarium adjacent to the Durban Botanic Gardens.

▲ Projects and People

With research activities concentrated in the fields of taxonomy and ecology, achievements in developing a "pre-eminent herbaria" and in mapping and describing vegetation are noted. In the computerization of herbarium data, the research directorate is considered a world leader, according to NBI literature.

The Herbarium Collections and Services Program is the "backbone to all present and future taxonomic research on the flora of southern Africa." Plans are to expand the herbaria database and prepare a manual on managing herbaria in southern Africa.

The Systematics Research Program is concerned with the Flora of southern Africa (FSA) project, although "only 14 percent of the region's 24,000 taxa have been included." To speed up the process, NBI planned to launch in 1992—with the cooperation of the Foundation for Research Development (FRD)—a development program in which three Ph.D. graduates in plant systematics would be produced annually for five years.

The Stress Ecology Research Program deals with the intensity of impacts on the natural environment. "Southern Africa is confronting the problems of invasive alien plants, desertification, acid rain, and the pervasive threat of global warming," reports NBI, which has "a small but strong team of experimental ecologists which have made important contributions in relation to invasive plants, commercial flower picking, and water stress."

The Conservation Biology Research Program plans to help revise the *South African Red Data Book for Plants* and develop an "integrated conservation biology research program" with universities. It is concerned with the "preservation of genetic resources, either *in situ* in parks or *ex situ* in botanical gardens [and] . . . changes in both gene pool size and in selective forces acting upon breeding populations." With NBI's Endangered Plants Laboratory, undertakings are underway in tissue culture and seed banking—which must now be applied to "endangered species, species of medicinal value, and species with potential for the horticultural trade."

As the botanic gardens are developed, **indigenous flora** is emphasized, **and collections of threatened plants** are being established.

Dr. Jacobus N. Eloff is also professor of botany at the University of Cape Town, worldwide lecturer, author of numerous publications, scientific editor of *South African Journal of Botany*, and radio series commentator on nature conservation.

■ Resources

Annals of Kirstenbosch is a supplementary series to the *South African Journal of Botany*. The NBI is expanding its library and educational services, has botanical information on an extensive database of southern African plants, and is willing to collect plant materials.

Kirstenbosch Botanical Gardens
National Botanical Institute (NBI)
Private Bag X7, Claremont 7735
Cape Town 8000, Cape Province, South Africa

Contact: **Ernst van Jaarsveld**, Horticulturist
Phone: (27) 21 762 1166, ext. 2018
Fax: (27) 21 762 3229

Activities: Education; Research • *Issues:* Biodviersity/Species Preservation • *Meetings:* CITES

▲ Projects and People

Horticulturist **Ernst van Jaarsveld** describes the mission of the **Kirstenbosch Botanical Gardens' Succulent Section** as the "conservation and display of southern African succulent plant diversity through cultivation and to provide information (propagation and other), and propagules, to other institutions or for reintroduction to the wild."

Van Jaarsveld's projects include "the growing of rare and endangered **southern African xerophytes**, some for reintroduction to their native habitat. It was extinct from the wild, but was reintroduced in 1989 to its native habitat in the western Cape." He is studying the rare *Gasteria baylissiana*, whose "numbers are being increased for subsequent reintroduction to the Zuurberg in the eastern Cape."

Coordinator of the Alooideae section of the International Organization of Succulent Plants (ISO) study, van Jaarsveld is also writing three books on *Tylecodon, Plectranthus,* and *Gasteria,* among some 80 publications. He belongs to the Species Survival Commission (IUCN/SSC) Cacti and Succulent Group.

Transvaal Museum
P.O. Box 413
Pretoria 0001, Transvaal, South Africa

Contact: **Ignatius L. Rautenbach, Ph.D.**, Director
Phone: (27) 12 322 7632 • *Fax:* (27) 12 322 7939

Activities: Education; Research • *Issues:* Basic Research; Biodiversity/Species Preservation

▲ Projects and People

With the **Transvaal Museum** since 1967, **Dr. Ignatius L. Rautenbach** is involved in **biosystematic research** and research on **bat biology.** He writes, "73 species of bats occur in the southern African subregion, which constitute 22 percent of the total mammal fauna of the subcontinent. In spite of the fact that bats are, second to the rodents, the most diverse order of our local fauna, they have until recently received virtually no attention from researchers." One reason is that "free-living bats are difficult to study." Consequently, Dr. Rautenbach, southern African representative to the Species Survival Commission (IUCN/SSC) Chiroptera Group, has established a laboratory for bat research and is supervising a group of specialists in investigating the "reproduction, ecology, behavior, and population dynamics of bats."

One collaborator is **G. van Urk**, Potchefstroom University, with whom Dr. Rautenbach has developed "versatile and affordable locally produced radiotelemetry equipment, *inter alia*, to study the biology of free-living bats and their predators." Dr. Rautenbach is focusing current research on **predation on bats by raptors** and a comparative **study between aerial hawking bats and birds of flight** mechanics, agility, and foraging patterns.

Commenting on southern African mammal taxonomy, Dr. Rautenbach notes that research in southern Africa "not only fails to keep pace with that in North America and Europe, but is in fact falling progressively further behind, not only in tempo but also in sophistication." Therefore, he is establishing a "vigorous research program that continues to gain momentum." Along with ongoing collaboration from "First World colleagues," Dr. Rautenbach reports that interest is being expressed by scientists in Malawi, Zimbabwe, Zambia, and Zaire in joining his program. An active fieldwork program is maintained to accrue the necessary specimens, tissues and information, as is an expanded laboratory, with updated computer equipment.

■ Resources

Dr. Rautenbach has an extensive publications list mainly on small mammals including bats. Research results published in in-house journals are available at nominal costs. Reprints of research results published in other journals are available free of charge.

🏛 *State Government*

KaNgwane Parks Board
P.O. Box 1990
Nelspruit 1200, Transvaal, South Africa

Contact: **Jeremy Anderson, Ph.D.**, Director
Phone: (27) 1311 53931 • *Fax:* (27) 1311 23153

Activities: Development; Education; Law Enforcement; Political/Legislative; Research • *Issues:* Biodiversity/Species Preservation; Conservation Training • *Meetings:* World Parks Congress

▲ Projects and People

Formed in 1986, **KaNgwane Parks Board** is run by **Dr. Jeremy Anderson**, who, as director of the Environmental Affairs Board, is also "responsible for the overall development of conservation in the Swazi Homeland," he writes. "There are four conservation areas to be developed, the largest—Songimvelo—will be the third largest conservation area in southern Africa."

Recruitment, development, and training of Swazi staff are high priorities, as Swazis are to occupy all the executive positions within the organization by the year 2000.

The Board is responsible for conservation issues outside the proclaimed areas too. It emphasizes **environmental education** among school children, teachers, and policymakers.

With tourism a key industry in KaNgwane, the Board encourages the development parks with visitor facilities as well as "safari hunting." Visitor facility operations are managed by the private sector on behalf of the tribe in whose area the Reserve is situated. **Tribal Resource Areas** are being created whereby such land is "developed and managed for the sustained utilization by the Tribe of its wild natural resources."

A training center was recently built to provide "modular courses" for personnel from the Board as well as from other conservation groups.

Dr. Anderson advised on proposals for the development and management of Keran National Park, Togo, and on the expansion of Fosse-aux-Leons, a national park for the forest elephant. He also surveyed elephant, buffalo, waterbuck, zebra, sable, and hartebeest in the 2,500-square-kilometer Marromeu Complex, Mozambique, "to evaluate the potentials of managing these populations so that they can provide a resource base for the regions," including establishment of a safari hunting operation. Previously, he was first director of the Pilanesberg Game Reserve, a World Wildlife Fund (WWF) project, where 6,000 animals of 10 species—"from elephant, rhino and giraffe to impala and springbok"—were reintroduced in the 1980s. He is a consultant to the Species Survival Commission (IUCN/SSC) Antelope and African Elephant and Rhino groups.

 *NGO*

Cycad Society of South Africa ⊕
P.O. Box 1537, Lynn East
Pretoria 0039, Transvaal, South Africa

Contact: **Nat. Grobbelaar**, Professor of Botany
Phone: (27) 12 808 0995 • *Fax:* (27) 12 432185

Activities: Education; Research • *Issues:* Biodiversity/Species Preservation

● Organization

Cycad Society of South Africa is a nonprofit organization with 800 members "devoted to promoting interest in and conservation of a cycads" with these goals: "To encourage the cultivation and propagation of cycads; to disseminate information on cycads by various means *inter alia* through the regular publication of a magazine; to arrange the legal exchange of plants, seedlings, seed, and pollen of difficult cycad species between members; to encourage scientific research on cycads; to promote all aspects of cycad conservation;

and to foster and maintain links with organizations having similar aims on an international basis."

■ Resources

The quarterly magazine *Encephalartos* is free to members.

Endangered Wildlife Trust (EWT) ⊕
Private Bag X11, Parkview
Johannesburg 2122, Transvaal, South Africa

Contact: **John A. Ledger, Ph.D.**, Director
Phone: (27) 11 486 1102 • *Fax:* (27) 11 486 1506

Activities: Education; Research • *Issues:* Biodiversity/Species Preservation • *Meetings:* CNPPA

● Organization

Founded in 1973 by artist and writer **Clive Walker**, the **Endangered Wildlife Trust (EWT)** is a nonprofit citizen organization in southern and central Africa that is dedicated to conservation of species diversity. EWT is a non-government member of the World Conservation Union (IUCN).

"With the rapid social and political changes taking place in southern Africa," writes **Dr. John Ledger**, " . . . there is an urgent need for all . . . to agree about the value of the environment that must sustain this and future generations."

▲ Projects and People

EWT's three-point plan focuses on research, awareness, and conservation action. The **research program** is aimed at **identifying threatened species**. The **awareness program** draws the attention of the community and statutory bodies to the problems facing threatened species. The **conservation action program** is directed toward reversing the decline of threatened species and the deterioration of their life-support systems both locally and globally.

The EWT is currently involved in 17 research projects, including those in rural communities in Namibia and Mozambique aimed at sustainable resource use. The emphasis is on preserving biodiversity in southern Africa, where "there are an estimated 23,200 plant species, . . . 80 percent [of which] are endemic to the region," reports Dr. Ledger. Among the African countries with the highest numbers of species for selected groups, South Africa is "sixth for mammals (279 species), ninth for birds (725 species), sixth for amphibians (93 species), and first for reptiles (281 species)."

One such vulnerable species is the **pangolin** (also called scaly anteater) threatened by habitat loss, susceptibility to insecticide poisoning, electrocution on game fences, and overutilization for traditional magical and medicinal purposes. The Mammal Research Institute, University of Pretoria, is undertaking this three-year project.

For the declining **humpback dolphin**, the rarest found on the northern Natal coast, the project is establishing population size and trends and making conservation recommendations with the supervision of **Dr. Vic Cockcroft**, Port Elizabeth Museum.

The **Indian Ocean marine mammal** project is being set up on Bazaruto Island, Mozambique, with the guidance of **Dr. Harry Richards** to pursue studies of the threatened dugong and several species of dolphins and small whales.

In Kruger National Park, Dr. Tony Bowland and his wife Jane Bowland are studying the cheetah with the sponsorship of, among others, Carlton Paper, through a Kleenex tissues campaign. Dr. Gus Mills is continuing radio-telemetry tracking of the "only viable wild dog group in South Africa." And in Botswana, John and Nell McNutt are investigating a healthy wild dog population in preparation for a conservation plan.

Dr. Paul Bartels is pioneering work on "embryo transplantation, artificial insemination, and other advanced techniques" to ultimately benefit threatened species.

Three annual black rhino surveys were undertaken at Pilanesberg National Park. Also, Rhino Management Program (RMG), representing all agencies responsible for populations of black rhinos in southern Africa, targets 2,000 animals each of the 2 subspecies *Diceros b. bicornis* and *D.b. minor*.

Because of the increasing popularity of recreational diving, the future of coral reefs at Sodwana is in jeopardy. "Virtually nothing is known about the species diversity of the reefs, or the impacts of human activities upon them," reports EWT, which is jointly sponsoring this research project with the University of Cape Town.

An ethnobotany research project in Namibia underscores the need to preserve indigenous peoples' knowledge of medicinal and other valuable plants as well as species in danger of extinction. Botanist Pat Craven, University of Cape Town, is in charge.

Other ongoing research projects focus on the Southern African Bird Atlas, spotted thrush, blue swallow, sand dollar, Maluti minnow, and gaboon adder.

With the Foundation for Research Development (FRD), EWT will produce future South African Red Data Books (SARDBs) and is seeking sponsorship from the business sector so that companies can "express their concern for environmental issues." With the Swaziland Conservation Trust, EWT produce Red Data Books for that country.

Other materials in preparation include an information booklet on predators and farmers, booklet and poster on cranes and farmers, and a series of posters on threatened birds, mammals, reptiles, amphibians, and fishes.

With help from World Wildlife Fund International (WWF), the Integrated Rural Development and Nature Conservation Project in Namibia is EWT's largest undertaking. As project leaders, Garth Owen-Smith and Margaret Jacobsohn head a team of 50, mostly community game guards. A companion project is establishing a rural environmental education program with courses on conservation and desert ecology.

Also involving local people, the Community Game Guard Project in the Bazaruto Archipelago, Mozambique has a large potential for ecotourism, while providing livelihood for those living on the islands for centuries. The game guards protect renewable resources against overexploitation.

Other conservation programs aim to protect eagles from electrocution on powerlines, halt poisoning of nontarget animals, safeguard vultures—including captive breeding and reintroducing the Egyptian vulture, provide training in species preservation, and monitor—with TRAFFIC—legal and illegal trade in endangered species. Computers for Conservation is an ongoing program to provide personal computers for wildlife management agencies.

■ Resources

Sponsorship is sought for various projects, such as echo parakeet research. EWT publishes the quarterly *Endangered Wildlife Journal* and a gift catalog.

National Association for Clean Air (NACA)
P.O. Box 5777
Johannesburg 2000, Transvaal, South Africa

Contact: **Gordon H. Grange**, Director
Phone: (27) 11 728 2418

Activities: Research • *Issues:* Air Quality/Emission Control; Energy; Global Warming; Health • *Meetings:* IUAPPA; World Clean Air Congress

● Organization

Founded in 1969, the National Association for Clean Air (NACA) is an independent, interdisciplinary organization and a corporate member of the International Union of Air Pollution Prevention Associations (IUAPPA).

With 450 members, NACA is dedicated to "advancing the science and art of air pollution management and therefore secure the maximum of natural light and air as pollution free in every form as may be practicable."

NACA coordinates air pollution prevention activities and serves as a communication link between the Department of Health and Welfare, local authorities, and the population on matters regarding South Africa's Air Pollution Prevention Act of 1965. It disseminates information and data and encourages dialogue among those in industry, local government, and education concerned with air pollution. And it holds regular meetings and annual seminars.

■ Resources

Members receive the semiannual *Clean Air Journal* and the *NACA Newsletter*. The *Source Book on Air Pollution Topics, Parts I and II* can be ordered; fact sheets are free of charge.

Share-Net ⊕
Box 394
Howick 3290, Natal, South Africa

Contact: **Jim Taylor**, Coordinator
Phone: (27) 332 305721

Activities: Development; Education; Research • *Issues:* Biodiversity/Species Preservation; Deforestation; Energy; Environmental Literacy; Global Warming; Health; Population Planning; Sustainable Development; Water Quality

● Organization

Share-Net is a "collaborative network for environmental education" initiated through the Wildlife Society of Southern Africa, Natal Parks Board, Southern African Nature Foundation, and the Environmental Education Association of Southern Africa.

Neither a "computer retrieval system nor a clearing house for resource materials," Share-Net, with 250 members and a staff of 4, encourages "grassroots resource development by teacher groups and local communities, and fosters joint resource development activities between conservation and environmental education agencies in southern Africa."

▲ Projects and People

Share-Net currently has five main projects. The **Water and Water Pollution** project has the support of the Global Rivers Environmental Education Network (GREEN, USA) to develop water quality test kits for South African schools. *(See also University of Michigan, GREEN Project in the USA Universities section.)* This international linkup program, receiving television coverage, "now has prototype test kits for biotic indicators, chemical analysis as well as the ability to test for harmful bacteria in rivers." Such kits make the science syllabus more relevant to people's everyday living.

The nationwide **Environmental Literacy** project uses "Story Teller group comics" developed in South Africa for black children not attending school and unable to read—provoking interest in Britain and the USA.

With the help of teachers, the **Soil and Soil Erosion** project produced a booklet, *Soil Is Life*, by Mathilda Roos, to encourage schools to participate in soil erosion prevention projects. The project features soil activities, including the "unique compost column and soil food chain puzzles."

The **Household Life** project is ecology for the home. "People unnecessarily kill animals like spiders and other useful insects through shear ignorance of their ecological importance," informs Share-Net. Mba Manqele, with the help of Jason Londt, Natal Museum, produced a booklet on common household organisms.

The **Enviro-Facts** project involves many conservation groups including the Environmental Education Association of Southern Africa, to produce one-page fact sheets on topical environmental issues that are distributed at supermarkets and are available on computer disk for schools and other groups.

■ Resources

Share-Net offers *Hands on Field Guides* on stream and pond life, soil and compost life, grasses and grassland life, and some common spiders, among resource booklets, pamphlets, and puzzles as well as research and reference journals to support resource development for teacher and community projects.

Share-Net needs a desktop publishing system to produce educational materials and funding to continue educational outreach.

South African Association for Marine Biological Research (SAAMBR)
P.O. Box 10712
Marine Parade 4056, Natal, South Africa

Contact: **Antonio J. De Freitas, Ph.D,** Director
Phone: (27) 31 373536 • *Fax:* (27) 31 372132

Activities: Education; Research • *Issues:* Marine Resource Management • *Meetings:* IMATA; IUDZG; Various sport fisheries conferences

● Organization

Founded in 1951 by the Natal community, the South African Association for Marine Biological Research (SAAMBR), with a staff of 100, investigates and disseminates knowledge relating to the biology of animals and plants found on the coasts of southern Africa and in waters surrounding such coasts. SAAMBR also promotes scientific research regarding the food habits and conditions of sea fishes to aid both the fisherman and the government in assuring a food supply from the sea.

SAAMBR also incorporates **Sea World Durban**, including the Aquarium, Dolphinarium, restaurant, and Sea World souvenirs; as well as the **Oceanographic Research Institute (ORI)**.

▲ Projects and People

Sea World Durban breeds dolphins and seals and rehabilitates stranded or damaged marine animals. It claims to have one of the world's finest shark displays and the largest captive brindle base. In its collection of some 2,500 species overall, Sea World exhibits more than 200 species of bony fish, 4 species of turtle, and 10 species of sharks and rays.

Through ORI's sea turtle research, such species are recovering from near extinction and "increasing healthily."

Likewise, once shad catches and sizes were deteriorating. ORI reports that through its research, shad is once again providing a valuable food and recreation source. ORI also conducts linefish research, leading to "innovative legislation providing for sustained use of this resource."

Along the Natal coast are many organisms valuable to man, such as mussels, rock lobster, pink prawn, langoustine, mole crab, prawn, red bait, mud crabs, and oysters. ORI scientific studies are directed toward their rational management and sustained use. Involved in the prevention and treatment of shark attacks, ORI helps to save lives.

Prof. Antonio J. De Freitas points out that although the seas cover over 70 percent of the planet, it is hard to raise funds for marine research. "Yet, the oceans govern the climate of the world; the vast areas of phytoplankton harness a sizeable proportion of the sun's energy needed to drive the world ecosystems; the finite fish resources supply a considerable amount of protein to the populations of this planet; and the sea seems to be expected to act as a limitless dump for the waste of our industrial sector."

Dr. De Freitas adds, "For the seas to live . . . much research is needed into the impact of man on populations which are being utilized, including the 'minor' organism such as oysters, mussels. . . . The man on the street must get involved and children must be made aware. . . . To achieve this, powerful and meaningful educational programs are needed.

"In developing countries, and I include here South Africa of the future, adequate national funds will not be available as they are needed for more 'important' issues such as housing, health, and formal education. Monies to look after the environment must be found elsewhere," he writes.

Dr. Michael Henry Schleyer, ORI assistant director, researches in fields of "fish aging, zooplankton, microbial ecology, and invertebrate ecology." Rudolf P. Van Der Elst, ORI deputy director, chairs the National Marine Linefish Research Groups and researches and manages migratory gamefishes in the southwest Indian Ocean.

■ Resources

SAAMBR has an extensive research publications list, including investigational reports on sharks, batoid fishes, spiny and rock lobsters, sea turtles, and game fishes; reprints from scientific journals on dugongs, seals, coastal zone management, algae, and water samples; data reports; posters; and books, among other literature. SAAMBR seeks funds for research projects, equipment, and marine environmental education.

Southern African Nature Foundation (SANF) ⊕
116 Dorp Street
456 Stellenbosch, 7600, Cape Province, South Africa

Contact: **John Hanks, Ph.D., Chief Executive**
   Phone: (27) 2231 72801 • *Fax:* (27) 2231 79517

Activities: Fundraising • *Issues:* Biodiversity/Species Preservation; Deforestation; Land Purchase for Conservation; Sustainable Development; Waste Management/Recycling; Water Quality • *Meetings:* CITES

● Organization
The **Southern African Nature Foundation (SANF)**, local representative of World Wide Fund for Nature (WWF), was established in 1968 with the help of the business community to channel public concern about the destruction of the living world into effective action. It acts in three main areas: "preserving genetic species and ecosystems diversity, ensuring that the use of renewable natural resources is sustainable, and fighting wasteful consumption of resources and energy and reducing pollution."

▲ Projects and People
SANF's projects are numerous. The following are examples: The **Black Rhino Strategy Project**, to establish new breeding grounds in safe areas, is jointly funded by SANF, Natal Parks Board, and Endangered Wildlife Trust. Many black rhinos are killed by poachers throughout Africa north of Limpopo. SANF hopes to provide secure refuge for about one-third of Africa's remaining 3,400 black rhinos.

As Africa's rangelands face destruction caused by cattle, the **Zimbabwe Multispecies Project** demonstrates and tests the viability of multispecies land-use options. Such a system, based on using a mix of indigenous species for meat production, tourism, and live capture and resale, can lead to the rehabilitation of degraded lands and help reduce starvation in many parts of Africa, SANF believes.

SANF supports the **Ivory Sourcing Project**, whereby University of Cape Town researchers have developed techniques to identify accurately the area of origin of ivory within Africa, and to date it. These techniques are "important to help prevent the sale of poached ivory and to enable southern African countries to establish a mechanism to sell genuinely culled ivory as a sustainable use of thriving elephant populations."

The **Teacher Training Network** program carries out a worldwide environmental education (EE) priority, with the support of Liberty Life. The pilot project has produced a lecture module for trainee teachers at Stellenbosch University, and a teacher training college is developing a full EE curriculum. Environmental teaching material as primary school resources is being prepared about the marine environment, for example.

As indigenous plants are used extensively by certain communities for medicines and purposes such as basket weaving, overuse is leading to shortages and even extension of certain species. The **Ethnobotany Project**, with support from Southern Life for the Institute of Natural Resources, researches human use of plants and how to protect rare species for future use.

SANF has worked to expand the Addo Elephant and West Coast national parks. Wetlands awareness campaigns, agroforestry support, and a conservation management plan for the Bazaruto Archipelago (rich with coral reefs and tropical forests) are underway.

Other projects aim to control alien vegetation, add data on Botswana birds to the *Southern African Bird Atlas*, develop an urban nature reserve in Johannesburg, protect nesting beaches of loggerhead and leatherback turtles, establish the blue swallow reserve, acquire land to protect crucial wetlands and expand protected areas, conduct whale research and viewings, research the endangered and income-generating fynbos wildflower, monitor illegal wildlife trade, study ecological impact of elephants in northern Botswana, prepare a natural history of bottlenose and humpback dolphins, produce comic books as an environmental education method, compile an atlas of more than 800 species of southern African fish, and sponsor environmental awards with the M-Net television company, among others.

■ Resources
WWF-SANF publishes *Our Living World* magazine, *Enviro* fact sheets, books, and other environmental education materials. Additionally they have for sale several gift items including books, posters, and paintings.

Umgeni Valley Project
P.O. Box 394
Howick 3290, Natal, South Africa

Contact: **Tim Wright, Principal Education Officer**
   Phone: (27) 332 303931

Activities: Education • *Issues:* Biodiversity/Species Preservation; Deforestation; Energy; Global Warming; Health; Sustainable Development; Waste Management/Recycling; Water Quality

● Organization
"I respect the birds, trees, and plant much more now because I have learnt," writes Olivia, a student. "I feel warmed up inside by the stillness all around me," notes classmate Raedene.

Started in 1976, **Umgeni Valley Project** is an environmental education endeavor of the Wildlife Society of Southern Africa. "The programs are designed to enable all students, irrespective of age or cultural background, as well as the general public, to experience nature . . . not just for the sheer joy, exhilaration and fun of it, but to use nature as an extension of the class or lecture room." Consequently, mathematics, language, and history are incorporated into the program, along with biology, geography, and conservation. With more than 10,000 students participating annually, it is the largest such NGO project in southern Africa.

▲ Projects and People
The Resource Center is at the Umgeni Valley Nature Reserve, situated in the Umgeni River valley, where giraffe, zebra, wildebeest, eland, nyala, and impala are the inhabitants. More than 200 species of birds range from tiny warblers to black crowned and martial eagles. A network of footpaths allow participants to experience a

hands-on approach to nature to experience the richness of the geographical terrain as well as the varieties of the flora and fauna.

Most courses begin with lectures, films, slides, and displays emphasizing man's role in the total environment. Trails lead to campsites along rivers, streams, and waterfalls, inside cliffs and among fig, celtis, combretum, dombeya, and acacia trees. Topics include "energy flow, adaptations, natural resources, soil and water studies, population explosion, mapwork, animal and plant behavior and inter-relationships."

Other activities are the Wildlife Adventure Trails, Wildlife Clubs for children, a Treasure Beach project emphasizing coastal conservation, in-service teacher training, community-based conservation programs, field excursions for schools, syllabus advisory service, resource material distribution, guidance on setting up environmental education centers, and lodge facilities for casual visitors.

■ Resources

Umgeni Valley Project welcomes the exchange of ideas, resources, and methodology regarding environmental education.

Rhino and Elephant Foundation ⊕
Lapalala Wilderness School
Wilderness Trust of Southern Africa
P.O. Box 645
Bedfordview 2008, Transvaal, South Africa

Contact: **Clive Walker**, Director and Chairman
Phone: (27) 11 4537648 • *Fax:* (27) 11 4537649

Activities: Education; Project Development; Project Funding; Research • *Issues:* Biodiversity/Species Preservation; Sustainable Development; Water Quality • *Meetings:* CITES

● Organization

Noted conservationist, author, and artist Clive Walker is a mainstay of the Rhino and Elephant Foundation, Chobe Wildlife Trust, Lapalala Wilderness School, Educational Wildlife Expeditions, and the Wilderness Trust of Southern Africa, among similar endeavors in Africa.

▲ Projects and People

To explore new concepts in conserving rhinos and elephants on both publicly and privately owned land, the Rhino and Elephant Foundation works closely with wildlife management agency policymakers in the countries where it operates. Goals are to translocate animals to safe new areas, make protection more effective in existing sanctuaries, support conservation research, and educate the public—particularly youth—through media and publications on saving these species and their habitats in Africa, Asia, and elsewhere.

While poaching for rhino horn and ivory is the "immediate cause for decline" and expanding human populations the greater long-term threat, the Foundation believes that conservation of these species are "better served through rational utilization of the animals and their products than the emotional application of trade bans and impractical regulations." Thus, another Foundation goal is to monitor aspects of wildlife utilization industry, such as the ivory and rhino trade and the hunting industry.

President is Dr. Mangosutho G. Buthelezi, chief minister of KwaZulu. Founders Walker, Anthony Hall-Martin, and Peter Hitchins belong to the Species Survival Commission (IUCN/SSC) African Elephant and Rhino Group.

Among Rhino and Elephant Foundation undertakings: the Botswana black rhino projects are the result of a rhino sanctuary in eastern Botswana. In the Itala Game Reserve, elephants have been reintroduced. Assistance with feed has been given to the Maputo Zoo, which has the only white rhino living in Mozambique's zoo. Monies have been given to the Save the Rhino Trust for the antipoaching unit via the Namibia Nature Foundation. The Rhino horn sourcing project is being carried out by the Department of Archaeology at the University of Cape Town. An aerial census is being undertaken of the Ndumo Game Reserve's black rhino population with the help of the KwaZulu Bureau of Natural Resources. A white rhino survey was conducted in Kruger National Park. The Zimbabwe Department of Wildlife received funds to assist them with antipoaching equipment and vehicle maintenance.

In Botswana, the Chobe Wildlife Trust is newly established to provide support for "ecological and wildlife research projects relevant to the conservation of the Chobe National Park." Working with the Department of Wildlife and National Parks and affiliated with the Rhino and Elephant Foundation, among other groups, the Trust intends to reintroduce species, set up a conservation awareness school in the Kasane village, expose international poaching networks active in the region, affiliate with global conservation groups, and investigate new water points for future use.

Already, the Trust helped bring about one of the largest seizures in southern Africa of ivory and rhino horn, valued at $4.5 million. Plans for a black rhino breeding sanctuary in the Serowe area are underway, as the African population of this species is presently 3,500, compared to 65,000 in 1970.

In 1985, Walker founded the Lapalala Wilderness School as a field study center for environmental education, which is administered by the Wilderness Trust of Southern Africa. Bringing young people into contact with nature, the school is an outdoor classroom in conservation, ecology, protection and appreciation of wild creatures and habitats, human impacts and pollution, and the natural resources of the southern Africa environment. Facilities also include a rustic stone and thatch building for up to 60 students and 2 teachers. The Molope Camp has provisions for 10, situated along the Palala River.

"Education cannot be left to the next generation," writes Walker. "Conservation must have a revolution in conservationist thinking. It must cease to be the exclusive cult of the privileged. It must be an exercise in common sense and the domain of all people. And what better way than to start with young people."

Clive Walker Trails are incorporated into the nonprofit Educational Wildlife Expeditions as a "wildlife experience" that uses the bush as an "environmental unit" for ecology as well as for archaeology, anthropology, history, and geology. Along the Fish Eagle Trail to the Okavango Delta of Botswana are the "red lechwe, Mokoro dugouts, palms, papyrus, . . . and the cry of the haunting fish eagle." The Ivory Trail meanders through the Mashatu Game Reserve in eastern Botswana, an area rich in elephant herds, other game, and birdlife. The Klaserie Lowveld Trail offers a chance to observe the "humble dung beetle" as well as buffalo, lion, rhino, giraffe, antelope, hyena, and other wildlife. Giant ebony trees shade the campsite along the Klaserie River.

■ Resources

A one-hour video documentary, *Genesis: The Resurrection of the African Elephant*, on rhino and elephant conservation in southern Africa can be ordered. Foundation members receive the semiannual *Rhino and Elephant Journal* and newsletters. Educational manuals for field guide training are needed.

The Wilderness Trust of Southern Africa publishes *Sanctuary*. Lapalala provides to groups talks, films, and slide shows on conservation, endangered species, and ecology.

The Chobe Wildlife Trust recently produced *Trial of the Giants* on elephants for television distribution and video resale. Members receive the semiannual *Serondella News*.

Wildlife Society of Southern Africa
100 Brand Road
Durban 4001, Natal, South Africa

Contact: **Keith H. Cooper**, Director of Conservation
Phone: (27) 31 210909 • *Fax:* (27) 31 219525

Activities: Development; Education; Political/Legislative; Research • *Issues:* Biodiversity/Species Preservation; Deforestation; Global Warming; Sustainable Development; Urban Conservation; Waste Management/Recycling; Water Quality • *Meetings:* IUCN

● Organization

Environmental education, conservation, and research are the main thrusts of the Wildlife Society of Southern Africa, founded in 1926 as a "body concerned with conservation of air, water, land, and all forms of life." With 30,000 members and a staff of 60, the Society aims to save threatened plants and animals through its urban and rural network, liaison with universities and government agencies, and participation in the World Conservation Union (IUCN).

More than 40,000 children participate in the Society's environmental education program run by the professionally trained staff at its education centers and nature reserves, such as Umgeni Valley, Abe Bailey, and Ben Lavin. A unique marine environmental education center, the first in South Africa, is now underway at Treasure Beach, Durban, to explore the values of grassland, coastal fores, rocky shore, sand dunes, and mangroves, estuaries, and birdlife of the wetlands.

Helping to establish the Kruger National Park, the Society is engaged in starting new and defending existing nature reserves, conserving endangered and threatened species and their habitats, launching antipollution campaigns and reviewing and proposing environmental legislation.

▲ Projects and People

In the Leopard Conservation Project, this large and elusive carnivore's status and long-term survival in areas outside national parks are being investigated, with new methods developed for monitoring changes. Conservationists are concerned that "the loss of top predators, such as the leopard, could give rise to a cascade effect that could result in the extinction of numerous lesser species."

The Society, with the Southern African Nature Foundation, launched a Wetlands Awareness Campaign throughout southern Africa to stem their loss. Estimating that "in some catchments, 90 percent of wetlands have been destroyed," the Society reminds that "wetlands are vital for storing and cleansing water naturally for the benefit of man. Simultaneously, they provide an indispensable home for important and threatened forms of wildlife such as wattled cranes, flufftails, crakes, frogs, fish, reptiles, and mammals."

The Blouberg Mountain Project in the northwestern Transvaal is concerned with the region's protection of rare and endangered species and ecosystems, sustainable resource management, protection of cultural and historical sites, preservation of cultural knowledge and community development of the Hananwa people, and the development of aesthetic values and recreational potential. A conservation action plan has been prepared, with the help of the Lebowa government.

The Orange Free State Conservation and Education Project is setting up a network comprising 18 conservancies with a total extent of 500,000 hectares. This project involves 200 farmers who employ 22 conservancy guards. Its purpose is to provide environmental education for all residents, promote conservancy concepts among farmers, promote a Metropolitan Open Space System (MOSS) concept in all cities and towns, and establish hiking trails.

With Dirk Uys and the Eksteenfontein community, the Richtersveld Conservation Project is underway to conserve natural resources and set up camping facilities that would employ local people. The Society's conservation division is surveying Transkei's indigenous forests; brochures, posters, and audiovisual material will be produced.

To highlight South Africa's statutory protected area network—and to show its gaps—the Society is producing a Protected Areas Map of South Africa. A public awareness campaign on behalf of cycads was begun. Whale research and conservation and resulting in an increase of the once threatened southern right whales in South African waters. A campaign to reintroduce the beneficial oxpecker is aimed at farmers, so they will halt the use of toxic insecticides. The oxpecker removes ticks, mites, and bloodsucking flies from game.

■ Resources

The Society produces a conservation journal *African Wildlife*, and for its junior members, *Toktokkie*—a magazine with environmental puzzles, games, competitions, and cartoons. Literature emphasizes recycling, water and energy conservation, public transportation, substitutions for hazardous products, organic foods, wetlands, and nature preserves. Branches also produce posters, newsletters, and information booklets.

⚗ *Private*

Manyane Game Lodge and Crocodile Farm ⊕
P.O. Box 3
Buhrmannsdrif 2867, South Africa

Contact: Johan Marais, Managing Director
Phone: (27) 140 32144
Fax: (27) 140 32144, ext. 203

Activities: Commercial Crocodile Farming; Education; Research • *Issues:* Biodiversity/Species Preservation • *Meetings:* IUCN/SSC

● Organization

The Manyane Game Lodge and Crocodile Farm is a private for-profit business whose main activity is the commercial farming of the Nile crocodile (*Crocodylus niloticus*) for the exotic skin trade.

Johan Marais writes, "We also have a game farm with nine white rhino, 23 Cape buffalo, zebra, giraffe, lion, kudu, eland, hartebeest, gemsbok, springbok, impala. . . . Excess animals are sold off to other institutions."

▲ Projects and People

Marais developed *CROCLIT*, a computerized bibliographical database on crocodile literature to include references pertaining to the study of crocodilians such as taxonomy, behavior, biology, physiology, farming, and nutrition. "An IBM-compatible PC(AT) is used and several keyword terms are used for each reference," says Marais. With the database "growing at a phenomenal rate," it will "eventually be made available to scientists and farmers throughout the world."

A member of the Species Survival Commission (IUCN/SSC) Crocodile Group, Marais coordinates the Crocodilian Study Group of Southern Africa and is past chair of the Herpetological Association of Africa.

■ Resources

Marais is the author two books, *Snake vs. Man—A Guide to Dangerous and Common Harmless Snakes of Southern Africa* and *A Complete Guide to Snakes of Southern Africa.*

Marais seeks assistance with DNA work on crocodile broodstock and hatchlings to "investigate the possibilities of genetic advantages in captive breeding." Marais also seeks school materials and projects to help educate Third World students in environmental issues. He needs public relations help in marketing the concept of utilizing a renewable resource, such as crocodiles.

🎓 *Universities*

Department of Zoology and Entomology
University of Natal
Pietermaritzburg 3200, Natal, South Africa

Contact: **Michael J. Samways, Ph.D.**, Professor of Entomology
Phone: (27) 331 955328 • *Fax:* (27) 331 955105

Activities: Education; Research • *Issues:* Biodiversity/Species Preservation • *Meetings:* International Entomological Congress

▲ Projects and People

With the University of Natal's Department and Zoology and Entomology, Dr. Michael J. Samways' research has been "principally entomological, both pure and applied" and concentrated on insect pest management "for abundance" and insect conservation biology "for scarcity"—addressing the interface between the two areas.

"Much of the pest management research relates to citrus insects," writes Prof. Samways. "Results have been put into practice on a large commercial scale and cover the monitoring of citrus thrips, citrus psylla, and parasitoids." He also researches the control of ants and red scale.

"Conservation of insects has important ethical and practical reasons for consideration," writes Prof. Samways, who has thus written papers "of a more philosophical nature." Related studies pertain to the landscape and how it may be managed for conservation.

Prof. Samways is an honorary consultant to the Food and Agriculture Organization (FAO) and belongs to the Species Survival Commission (IUCN/SSC) Odonata and Lepidoptera groups and was invited to chair the recently formed Orthopteroidea Group. Also, with the World Conservation Union, he has helped to prepare a Landscape Ecology Working Group. He serves on the editorial boards of *Biodiversity and Conservation* and *South African Journal of Ecology* and is associated with the International Dragonfly Society Council. The Entomology Chair has an extensive list of publications, conference papers, and major presentations.

Institute of Natural Resources
University of Natal
67 St. Patricks Rd., P.O. Box 375
Pietermaritzburg 3200, Natal, South Africa

Contact: **Charles Breen, Ph.D.**, Director
Phone: (27) 331 68317 • *Fax:* (27) 331 68891

Activities: Development; Education; Research • *Issues:* Deforestation; Energy; Environmental Impact Assessments; Global Warming; Health; Population Planning; Sustainable Development; Waste Management/Recycling; Water Quality

● Organization

The Institute of Natural Resources aims to "contribute to the socioeconomic advancement of rural people and management of natural resources in southern Africa through the integration of development and conservation initiatives."

To these ends, the Institute assesses rural development and resource management needs; analyzes environmental and social impact of development initiatives; designs and tests innovative rural development and resource management systems on a pilot scale; designs and tests appropriate education and training programs to support their implementation; and formulates policy and procedural guidelines for adoption and implementation of these systems by development agencies and organizations.

■ Resources

The Institute offers publications for order regarding research and development activities in agroforestry, cattle raising, coastal planning, community gardens, wetlands, wildlife conservation, and land use, for example. A list is available of monographs, investigational reports, working papers, and reprints.

Centre for Wildlife Research ⊕
University of Pretoria
Pretoria 0002, Transvaal, South Africa

Contact: **J. du P. Bothma, Ph.D.,** Professor and
          Head
Phone: (27) 12 420 2627 • *Fax:* (27) 12 420 2569

Activities: Education; Research • *Issues:* Biodiversity/Species
Preservation; Sustainable Development; Wildlife Management

▲ Projects and People

In addition to heading the University of Pretoria's Centre for
Wildlife Research, Prof. J. du P. Bothma is a co-director of Bothma
and Van Hoven Wildlife Consultants with Associate Prof. Wouter
van Hoven. Among Dr. Bothma's current posts are advisory board
member of the Rhino and Elephant Foundation, member of the
Species Survival Commission (IUCN/SSC) Cat Group, trustee for
the Transvaal Museum, and board chairman of the Tourism and
Nature Conservation Corporation of Qwaqwa. Topics of publications include the Kalahari Desert leopard, shepherd's tree, elephant,
food habits of certain carnivores, game range management, and
conservation of wild animals as renewable resources.

  With special interests in nutritional ecology and digestive physiology of herbivores, Dr. Wouter van Hoven is also involved in **game
research**—making presentations, for example, on giraffe nutrition
and high blood pressure, nutrient value of certain bushveld plants,
wildlife management, and the economics of game ranching.

  As consultants, Bothma and van Hoven do baseline studies for
management of nature reserves and game farms. One study was on
the "habitat preferences and flying patterns of birds in the Jan
Smuts Airport area"; another was on **wildlife conservation in
Botswana.**

Centre for African Ecology
Department of Zoology
University of Witwatersrand
Johannesburg 2050, Transvaal, South Africa

Contact: **Norman Owen-Smith, Ph.D.,** Director,
          Professor
Phone: (27) 11 716 4193 • *Fax:* (27) 11 339 3607

Activities: Research • *Issues:* Biodiversity/Species Preservation

● Organization

The University of Witwatersrand's **Department of Zoology** an-
nounces the formation of the **Centre for African Ecology,** a resource ecology group. Its mission is to "foster the development and
maintenance of a centre of excellence in ecology and related biological disciplines, and their application to conserving and managing
biological resources, with special reference to African environments."

  The Centre is offering a one-year Master of Science course in
qualitative conservation biology, with research in "large herbivore
ecology and plant-herbivore interactions, ecology of soil-plant-at-
mosphere relations, plant population biology, ecological biogeography of ectothermic vertebrates, and sustainable utilization of natural
resources."

  Dr. Norman Owen-Smith belongs to the Species Survival Commission (IUCN/SSC) African Elephant and Rhino Group. Group
members are also associated with Decade of the Tropics—Tropical
Soil Biology and Fertility and Responses of Savannas to Stress and
Disturbance and with the International Geosphere-Biosphere Global Change Programme.

Spain

The second largest country in Europe, Spain's climate varies from rainy in the north to arid in the central plains. About 41 percent of the country is cropland and 5 percent is protected.

The urban population is growing rapidly, especially in the capital of Madrid, and the level of urban air pollution is above average for Western Europe. Tourism has taken its toll on the Mediterranean coast, and poor sewage and water treatment facilities have contributed to severe pollution in the sea. A new reforestation plan is underway to increase wood production and contain soil erosion, but forest fires have slowed the process. Spain's energy is provided by its rivers and by its 10 nuclear reactors.

🏛 *Government*

Estación Biológica de Doñana/Doñana Biological Station
Avenida de María Luisa s/n. Pabellón del Perú
41013 Seville, Spain

Contact: **Miguel Delibes de Castro, Ph.D.**, Director
Phone: (34) 54 423 2340 • *Fax:* (34) 54 462 1125

Activities: Research • *Issues:* Biodiversity/Species Preservation

● Organization

Estación Biológica de Doñana (Doñana Biological Station) is a government research facility, staff of 68, belonging to the Consejo Superior de Investigaciones Científicas (CSIC) (Spanish Council for Scientific Research). Biological research is mainly carried out at the Parque Nacional de Doñana (Doñana National Park) and falls into two basic areas: evolutionary biology and applied biology. In the first area, scientists conduct studies in ethology, evolutionary ecology, and population biology, and in the second area, scientists conduct studies in conservation biology and game and pest species biology.

▲ Projects and People

Dr. Miguel Delibes is on the research staff of Doñana Biological Station since 1979 and director since 1988. He is the author of more than 100 scientific papers and several books dealing primarily with zoology and animal ecology, particularly that of mammals, birds, and other vertebrates, and is a member of the Species Survival Commission (IUCN/SSC) Cat Group.

■ Resources

Dr. Delibes edits a scientific review entitled *Doñana Acta Vertebrata* (in Spanish), which features research done at the Doñana Station. The Station also has a library with a large number of publications and an important scientific collection of vertebrates which may be consulted by interested professionals and students.

State Government
Baleares

Conselleria d'Agricultura i Pesca
Department of Agriculture and Fishing
Servei de Conservación de la Natura
Division of Wildlife Conservation
Ed. Sena. Els Geranis
07002 Palma, Spain

Contact: **Joan Mayol,** Chief of Wildlife Department
Phone: (34) 71 725840

Activities: Education; Law Enforcement; Political/Legislative; Wildlife and Nature Management • *Issues:* Biodiversity/Species Preservation • *Meetings:* EUROSITE

● Organization

The **Servei de Conservación de la Natura (Division of Wildlife Conservation)** is dedicated to the conservation and administration of the fauna and flora of the Balearic Islands, with the exception of hunting- and forestry-related aspects, which are dealt with by separate administrative divisions. It also has environmental education programs. The Division's mandate is to ensure the maintenance of biodiversity in the Balearics, which means the protection of specific species and their habitats. It also carries out research when existing studies are insufficient.

▲ Projects and People

The *Red List of Balearic Vertebrates* lists endangered vertebrates and related studies. As of 1990, 53 **black vultures** have been raised in captivity and reintroduced in the wild, where their reproduction is being supervised. The seven nesting **fishing eagles** on Mallorca are being monitored. **Bats** appear to be making a modest recovery, according to the Division. There are 70 to 80 pairs of **red hawks** at Dragonera and 175 at Cabrera. Researchers are also watching the nests of **peregrine falcon. Marine turtles** are being tagged to study accidental capture by fishermen; a map shows distribution of land and sea turtle population to help analyze causes of possible extinction. The National Science Museum at Sòller grows **endangered plant** species and is overseeing the **germination of their seeds** in the wild. A project to protect **Balearic orchids** is planned, beginning at the Parc de S'Albufera.

When a virus began affecting the health of **Balearic dolphins** in the summer of 1990, the Wildlife Division immediately began collecting dead dolphins for analysis, with the help of the University of Barcelona. An **Action Butterfly** campaign is being planned. The eating habits and distribution of **martens** on Minorca are being studied with the cooperation of the University of León.

Regarding **environmental education,** nature walks for student, cultural, garden, and natural history groups are scheduled. School exhibits, teaching materials, and camp activities are prepared. Researchers lead public tours to England and Scotland to study nature reserves and protected species.

Forestry agents watch for the illegal presence of goats on Porros Island that threatened endangered species, collect exotic species, and have begun a campaign to end trade in illegally hunted animals.

■ Resources

Nature guides, posters, slides, videos, and postcards of protected fauna are produced. Books about vultures, seals, and other wildlife are prepared for children. Monographic technical documents about conservation include a minimum of six titles, printed in Spanish.

Catalonia

Departament de Medi Ambient de la
**   Generalitat de Catalunya (DMA)**
Catalonian Regional Department of the
**   Environment**
Passeig de Gracia 94
08008 Barcelona, Spain

Contacts: **Albert Vilalta i Gonzalez, Ph.D.,** Minister
        of the Environment
        Kenty Richardson, International
        Coordinator
Phone: (34) 3 487 2234 • *Fax:* (34) 3 487 3180

Activities: Education; Law Enforcement; Political/Legislative; Research • *Issues:* Air Quality/Emission Control; Waste Management/Recycling; Water Quality

● Organization

Founded in 1991, the **Departament de Medi Ambient (DMA)** **(Department of the Environment)** is the executive environmental body of the Autonomous Government of the Region of Catalonia. The staff includes 50 at the Industrial Waste Agency, 110 at the Water Treatment Agency, and 160 at the General Directorates of the Environment, previously under the control of the Ministry of Public Works. DMA plans to hire an additional 450 people in 1992 to work at the General Directorates. The Department's activities are organized into three basic areas: air and water pollution control and treatment of waste, educational and promotional campaigns, and protection of natural areas.

DMA, through the General Directorate of Environmental Quality, headed by **Agustí Comadran i Monso,** the Industrial Waste Agency, directed by **Ferran Relea i Gines,** and the Water Treatment Agency, led by **Antoni Garcia i Coma,** is in charge of preparing regional laws and decrees and **implementing** European Community (EC), national, and regional legislation. It also **monitors air emissions and water and waste production by industries and oversees water treatment stations.**

DMA works to keep **Catalonia's beaches and Mediterranean waters clean, to restore contaminated sites and quarries,** and to **reduce noise pollution.** Additionally, DMA conducts environmental impact assessments (EIAs) and **promotes waste reduction and clean industrial technologies.**

Joan Puigdollers i Fargas heads the General Directorate of Environmental Education and Promotion, which coordinates environmental education programs at schools and universities, including environmental training and research. DMA also promotes TV and radio programs on the environment and does public relations work promoting clean industrial technologies, green products, waste reduction, and, "in general, . . . a new ethical society more protective of the environment."

Two new General Directorates of Physical Planning and Nature Protection are planned. A Regional Plan of Areas of Natural Interest is being prepared which will protect approximately 20 percent of Catalonia's territory.

The DMA maintains contact with the EC and the World Conservation Union (IUCN) and also cooperates on environmental projects with industrial regions of France, Germany, and Italy.

■ Resources

DMA's publications (in Catalan) include Catalonian, Spanish, and EC environmental legislation, a guide to reducing air pollution, a guide to minimizing waste in industry, and a technical plan for restoring quarry sites. A Documentation Center and additional publications are planned. DMA's major needs are additional qualified staff members and links to environmental administrations in other countries that would like to exchange technical information and experiences.

 ## NGO

Asociación Amigos de Doñana ⊕
Friends of Doñana Association
APDO Correos 2182
41080 Seville, Spain

Contact: **Javier Castroviejo Bolibar, Ph.D.**, President
   Phone: (34) 54 423 6551 • *Fax:* (34) 54 492 2253

Activities: Development; Education; Law Enforcement; Political/Legislative; Research • *Issues:* Biodiversity/Species Preservation; Deforestation; Sustainable Development

● Organization

Asociación Amigos de Doñana was founded in 1986 to promote and organize activities and research relating to conservation and management of the Earth's natural heritage and resources. The 22 staff members are particularly concerned with educating the public and training students in environmental research skills. The Association's work was originally centered on the area near Doñana National Park in Spain, but it now also includes projects in Africa and South America.

▲ Projects and People

One of the Association's most important projects is the Doñana Workshop, funded by the National Employment Institute and the European Social Fund, which trains young unemployed workers in skills for the recovery of Spain's cultural and ecological heritage. Among the subprojects carried out by workers are the restoration of traditional huts in Doñana National Park, replanting of native vegetation in Andalusia, studies in wolf ecology, and studies in cattle-raising techniques traditional to Spain.

In Doñana National Park, students build and maintain inhabited huts out of rushes, wood, and mud and provide other community services, while studying for a graduate degree. The subproject is supported by the Andalusian and national governments and the patrons of Doñana Park. Also in southern Spain, the Center for Research on Wild Flora and Fauna, developed in cooperation with

the Andalusian Environment Agency, is located on 20 hectares of Mediterranean forest donated by the town council of Huevar. Here students plant oak and wild olive trees.

In northwestern Spain, students study ecological aspects of endangered wolf populations through radio-tracking methods, in a subproject developed with the Castille-León government and World Wildlife Fund's (WWF) Spanish chapter. Students also research the imperial eagle and work with the zoological collections at Doñana Biological Station. Cattle management and herding methods are studied at Doñana–Las Marismas, where cattle are raised using traditional methods exported to the Americas by Spanish settlers.

Approximately 50 additional staff members teach and provide assistance to around 40 students involved in the Doñana Workshop project each year. The small size and inverse student-faculty ratio apparently makes for a successful project; according to the Association, over 75 percent of the Workshop's graduates find work using their newly learned skills.

The Association's international projects are cooperative ventures in nature research and conservation in South America and sub-Saharan Africa. Friends of Doñana also maintains international ties as a member of the World Conservation Union (IUCN) and the European Environmental Bureau (EEB).

In the plains of Venezuela's Apure State, the Association supports El Frío Biological Station in cooperation with the La Salle Fund for the Natural Sciences. This region is relatively unknown scientifically, so research on its ecosystem is of the utmost importance.

An ecotourism project, with the help of the Ivan Maria Maldonado Fund, has been underway on the 80,000 hectares of Hato El Frío since 1988. A tourism center is visited by 1,000–1,500 people yearly, who explore the area on horseback and by all-terrain vehicle and boat, accompanied by a naturalist as their guide.

The Association has also worked with the Ibero-American Cooperation Institute (ICI) to protect endangered aquatic and wetland species in Venezuela. Thanks to the efforts of this organization, the Ministry of the Environment and Restoration of Natural Resources of Venezuela created the Caño Guaritico Fauna Refuge.

Also with the support of ICI, the Association collaborates with the University of Santa Cruz de la Sierra in Bolivia to train future environmental specialists, and is promoting the creation of Noel Kempf National Park and El Beni Biological Station in the same country (in cooperation with the Development Corporation of Santa Cruz de la Sierra and the National Science Academy of Bolivia, respectively).

The Program for Research and Nature Conservation in Equatorial Guinea (a former Spanish colony) includes personnel training and environmental education, inventories of native flora and fauna, and technical and legal advice on environmental matters. Two important results of this program are a law establishing 15 percent of the country as protected natural areas and a planned UNESCO Man and the Biosphere committee.

In 1989, the governments of Spain and Angola signed a series of agreements pledging cooperation in a number of areas including the development of natural resources. The Spanish Ministry of Foreign Affairs commissioned Friends of Doñana to draw up a four-year plan entitled Conservation, Management and Tourism Development of Quissama National Park. The recently completed plan includes details on needed infrastructure development and personnel management and proposes the installation of a crocodile farm near the Park that would contribute to repopulating the animals as well as provide needed funds.

Asociación Malagueña para la Protección de la Vida Silvestre (SILVEMA) ⊕
Malaga Association for the Protection of Wildlife

Mariblanca 21, Bajo Izqda
29012 Malaga, Spain

Contact: **Andres Florencio Alcantara Valero,**
Secretary General
Phone: (34) 52 227641 • *Fax:* (34) 52 229595

Activities: Development; Education; Law Enforcement; Research • *Issues:* Air Quality/Emission Control; Biodiversity/Species Preservation; Deforestation; Energy; Global Warming; Sustainable Development; Transportation; Waste Management/Recycling; Water Quality; Wildlife Protection

● Organization

SILVEMA was founded in 1980 with the basic objectives of defending, studying, and enjoying the environment in rural and urban areas. Its principal activities are in the province of Malaga, but it does cooperate with other ecological organizations at the regional, national, and even global levels, including the World Conservation Union (IUCN), International Council for Bird Preservation (ICBP), Coordinator of Environmental Defense Organizations (CODA), and Iberian Council for the Defense of Wildlife (CIDN).

▲ Projects and People

A 10-year program to protect the **tawny vulture in Andalusia** remains in progress. SILVEMA **tracks the dumping of toxic substances** in the sea, pollutant emissions of contaminating industries, and dumping of urban waste. It researches **effects of mining** in Malaga, proposes protected areas, undertakes environmental impact studies, and **disseminates news** stories regarding urbanization, development, roads, and mining. Campaigns champion birds of prey, such as the endangered **ashen eagle**, which nests in southern Spain and winters in Africa and performs a beneficial service by eating grasshoppers, lizards, and rats that threaten wheat and barley crops.

SILVEMA promotes the care of riverine forests and Mediterranean scrub and forests, while **cleaning rivers** and repopulating their banks and mountains with native species. It is working to **recover public paths** in wild areas, speaks out against the catching and consumption of immature fish, and **defends the dunes of Marbella** and other coastal areas. SILVEMA sponsored efforts to clean deposits of combustible fuels off Malaga's urban center and studied the incidence of food poisoning in wild fauna in Serrania de Ronda.

The group also sponsors public seminars on the environment, trips to areas of ecological interest, and other environmental education programs. It maintains a **green telephone** line for the public on environmental issues.

Coordinadora de Organizaciones de Defensa Ambiental (CODA) ⊕
Coordinator of Environmental Defense Organizations

Plaza Santo Domingo, 7 7° B
28013 Madrid, Spain

Contact: **Juan Gallego Luque,** Information and Promotion
Phone: (34) 1 559 6025 • *Fax:* (34) 1 559 7897

Activities: Education; Political/Legislative • *Issues:* Biodiversity/Species Preservation; Energy; Global Warming; Sustainable Development; Waste Management/Recycling • *Meetings:* CITES

● Organization

CODA is known as the "super-organization of the ecology movement in Spain." Founded in 1978, it is largely a federation of local and regional ecology organizations, but also includes FAPAS, responsible for the protection of the last bears in Western Europe; and the Pro-Amazonia Commission, committed to saving the South American rainforests. CODA's campaigns feature the preservation of natural areas in Spain and of specific species such as the bear, imperial eagle, and black vulture. CODA is also concerned with global problems, such as global warming, the disappearing ozone layer, and deforestation.

Once known primarily for protecting bird species, CODA transferred these functions in 1986 to Sociedad Española de Ornitología (SEO) (Spanish Ornithology Society) and established new links with Grupo Ecologista de la Escuela de Ingenieros Agrónomos (GEDEA) (Ecologists' Group of the School of Agricultural Engineers). Consequently, CODA enlarged its activities to include not only wildlife preservation but also the fight against pollution and nuclear power, among other issues. It has a staff of 37 and 150 member groups representing more than 35,000 people.

▲ Projects and People

CODA is organized into the following working committees: The **International Committee** represents member groups at organizations such as the World Conservation Union (IUCN), European Environmental Bureau (EEB), and the International Council for Bird Preservation (ICBP)—a task that individual groups could not easily accomplish. CODA representatives work in the national chapters of global organizations and takes on campaigns against international environmental problems, such as global warming, acid rain, deforestation, erosion, and overpopulation. It pressed for approval of the **European Habitats Directive** of the European Community (EC).

The **Pollution and Waste Committee,** cooperating with other groups, deals with industrial and urban waste, air pollution, and recycling. The **Legal Committee** assists CODA and its member groups with statements, publications, and conferences that are legally precise. It also launches environmental law campaigns such as promoting legislation to protect meadows, forests, and the Mediterranean scrub in the Madrid area and to protect *carballeiras,* Atlantic forests unique to Galicia. The Committee also organized a national environmental law symposium, in cooperation with the Secretary General for the Environment.

The **Natural Areas Committee** is concerned with activities that might alter or threaten natural areas—public works, urbanization, forest exploitation—and pays special attention to their protection and the sensible use of their natural resources. Campaigns were aimed against construction of urban projects in Isla Canela and Alcaraz, irrigation plans for the Saladas de Alcañiz, and development plans for the Canary Island of La Gomera.

The **Energy Committee** is linked with the Asociación Ecologista de Defensa de la Naturaleza (AEDENAT) (Ecologists' Association for the Defense of Nature) and looks for alternatives to the current rate of energy consumption and to sources of energy harmful to the environment. These have included public campaigns against wasting electricity and research and proposals for the rational use of energy resources. In cooperation with 200 ecological groups from throughout Spain, the Committee promoted an antinuclear law that would be established by popular initiative.

The **Species Committee** carries out campaigns for the protection of endangered species, in cooperation with other groups, through research, meetings, and informational materials. Protection of colonies and nesting areas are emphasized. The Committee also focuses on the increase of animal mortality on highways and on the control of hunting activities—fighting for better enforcement of hunting seasons and against nonselective and illegal hunting.

■ Resources

Quercus is a monthly scientific magazine dedicated to the study of wildlife. *El Cárabo Magazine,* named after the tawny owl, is also a wildlife publication with three issues yearly. The *Linneo Book Catalog* lists more than 2,000 titles on wildlife and the environment. *El Movimiento Ecologista en Europa Occidental* describes the European ecological movement and lists 200 such organizations. *Panorama Ambiental de las Comunicades Europeas* describes the institutions, legislation, and political environment of the EC as they relate to environmental issues.

Grup Balear d'Ornitologia i Defensa de la Naturalesa (GOB)
Balearic Ornithology Group in Defense of Nature
Veri 1, 3 er

07001 Palma de Mallorca, Balearic Islands, Spain

Phone: (34) 71 721105 • *Fax:* (34) 71 211375

Activities: Education • *Issues:* Biodiversity/Species Preservation; Energy; Sustainable Development; Waste Management/Recycling; Water Quality

■ Resources

GOB, the Balearic Ornithology Group in Defense of Nature, maintains a library on ornithology and conservation and a collection of videotapes.

◢ *Private*

Biosfera/Biosphere ⊕

Fundación Enciclopedia Catalana/Catalan Encyclopedia Foundation
Diputació, 250

08007 Barcelona, Spain

Contact: **Ramon Folch i Guillén, Ph.D.**, Secretary General, UNESCO MAB Spanish Committee
Phone: (34) 3 301 7118, (34) 3 323 5691 (home)
Fax: (34) 3 412 0172

Activities: Education; Research • *Issues:* Biodiversity/Species Preservation; Deforestation • *Meetings:* EuroMAB

▲ Projects and People

Dr. Ramon Folch, Secretary General of the UNESCO Man and the Biosphere (MAB) Spanish Committee since 1989, has embarked on the related project *Biosfera (Biosphere),* a multimedia series which focuses on the characteristics of the world's 15 biomes, or bioclimatic units, and some of the UNESCO Biosphere Reserves designed to manage them. The project includes a television series, a series of reference books, an interactive compact disc (ICD), and other materials such as home video cassettes, video games, and magazine articles.

There are 15 biomes on Earth: tropical rainforests; rainy tropical savannas (or grasslands); dry tropical savannas; tropical (warm) deserts; subtropical (cool) deserts; Mediterranean woodlands; temperate (coniferous) rainforests; temperate broadleaf forests; temperate grasslands and steppes; taiga (dry subarctic coniferous forests); tundra; highland and mountain systems; island systems; wetlands; and marine systems.

Most of these biomes are found in more than one location around the world. Human activity is an integral part of, and a transforming influence on, each of these biomes' ecosystems. The 285 Biosphere Reserves created by UNESCO in 72 countries function as examples of ecosystems where human activities such as agriculture, fishing, and forest management are carried out in ways that do not disrupt the balance of nature. Research and education are also vital activities in the Biosphere Reserves, so that they are an important resource for the global community of scientists studying the Earth's biodiversity.

The *Biosphere* television program is being produced initially by SONO TV and Televisió de Catalunya, although other European and American television networks are considering participating in the production. Thirteen fifty-five-minute programs are planned, each one focusing on one or two Biosphere Reserves which are representative of a specific biome. All of the programs will be shot entirely on location, and interviews with special guests are planned. The first broadcast is scheduled for Spanish television in 1993.

The published version of *Biosphere* is being put out in Catalan by Fundació Enciclopèdia Catalana, a Barcelona publishing company specializing in reference books. The books will be published as co-editions in other languages by other European and American publishers. *Biosphere* will consist of 10 450- to 500-page volumes dealing with the 15 biomes (and Antarctica) in greater detail than is possible in a television show. One of the book's unique features will be approximately 900 pages of inserts presenting facts on flora and fauna species as well as other supplementary information. The first volume, an introductory book about the whole of the Earth's biosphere, will be published in 1993. Subsequent volumes are scheduled through 1997.

The *Biosphere* ICD will make use of the latest technology by allowing viewers to choose information on the biome and/or Biosphere Reserve he/she is interested in through a menu on the screen. ICDs are played back onto an ordinary television set, and it is hoped that educational institutions, science museums, and the general public will make use of this format, which will also be available to visitors to the Reserves.

Dr. Folch's other recent projects include the television series *Mediterrània* and the reference set *Història Natural dels Països Catalans* (Natural History of the Catalan Lands). *Mediterrània, The Continuing Story of the Mediterranean Ecosystems*, is a documentary series of 50 half-hour programs which focuses on "the annual cycle of nature and human activity," not only in countries bordering the Mediterranean Sea but also in parts of California, Chile, South Africa, and Australia that share the same climate and many of the same natural resources. The shows were produced by Televisió de Catalunya.

Natural History of the Catalan Lands, published by Enciclopèdia Catalana, is a 15-volume set published from 1984 to 1988 which provides a comprehensive look at the geology, flora, and fauna of the Catalan regions of Spain and France in a format that is accessible to professionals and lay people.

Dr. Folch is also the president of the Consell de Protecció de la Natura (Catalonian Nature Protection Council), a government organization, and a member of the Institut d'Estudis Catalans (Catalan Studies Institute). He serves as a UNESCO consultant for environmental management projects all over the world, and he was Conceptual Director of the Environment Pavilion for EXPO '92. According to his literature, *Biosphere* has been admitted by the Spanish Committee of the UNESCO's MAB program as an authorized, homologous work on ambient activity. This inclusion by the Committee gives the work the UNESCO seal of approval, moral guarantee, and conceptual support.

Biologist **Dr. J.M. Camarasa**, member of the UNESCO MAB Spanish Committee, will help direct the authorship of the book series.

🎓 *Universities*

Department of Animal Biology ⊕
Faculty of Biology
University of Barcelona
08071 Barcelona, Spain

Contact: **Alejandro (Alex) Aguilar Vila, Ph.D.,**
          Professor
Phone: (34) 3 402 1453 • *Fax:* (34) 3 411 0887

Activities: Education; Research • *Issues:* Biodiversity/Species Preservation; Water Quality • *Meetings:* IWC

▲ Projects and People

Main research interests of **Dr. Alex Aguilar**, animal biology professor at the **University of Barcelona**, are the effects of **water pollution** and the **fishing industry on marine mammals**, particularly in the Mediterranean Sea. He is currently studying the recent **striped dolphin die-off** in the western Mediterranean Sea, factors in the mortality of other **small cetaceans** in that area, the effects of chlorine-based pollutants such as PCBs (polychlorinated biphenyls) on the reproduction and growth of these animals, and conservation measures for Mediterranean cetaceans and **pinnipeds**. Dr. Aguilar belongs to the Species Survival Commission (IUCN/SSC) Cetacean Group and frequently presents the findings of his research at meetings of the Scientific Committee of the International Whaling Commission (IWC) and other global meetings.

■ Resources
Dr. Aguilar is the author of approximately 45 scientific and popular articles and a co-editor of the book *European Research on Cetaceans*.

Department of Ethology and Primatology ⊕
Faculty of Psychology
University of Barcelona
08028 Barcelona, Spain

Contact: **Jordi Sabater Pi, Ph.D.,** Chairman of
          Ethology and Primatology
Phone: (34) 3 240 4179 • *Fax:* (34) 3 334 3456

Activities: Education; Research • *Issues:* Biodiversity/Species Preservation

▲ Projects and People

For more than 45 years, **Dr. Jordi Sabater Pi**, chairman of Ethology and Primatology at the **University of Barcelona**, has studied **primate ethology** in western and central Africa, notably the **gorillas and chimpanzees** of Río Muni, **Equatorial Guinea** and the mountain gorillas of Rwanda, in collaboration with the late **Dian Fossey**. His most recent projects are a study of the behavioral ecology of the **Bonobo**, or **pygmy chimpanzee**, in **Zaire** and research on the ecoethology of forest monkeys on Bioko Island, Equatorial Guinea.

The pygmy chimpanzee "has fascinated human evolutionary theorists since the 1930's," according to Dr. Sabater Pi. His research on this species (*Pan paniscus*) is concentrated in an area of tropical rainforest about one degree south of the Equator in Zaire and focuses on population density and ecological characteristics of the species' habitat. He notes that "human influence on the population is indirect and is based principally on competition for space and the ecological resources of the region." The chimpanzees are not hunted or captured by humans in the area because of their magical and religious beliefs.

However, Dr. Sabater Pi concludes that "it would be of enormous interest to protect this small population of bonobos because their survival outlook . . . may change for cultural reasons including nutritional imperative." The people of this region of Zaire appear to consume little animal protein, which may contribute to high levels of mortality. It is possible that these people may decide to hunt chimpanzees for food in the near future.

The southern part of Bioko Island, Equatorial Guinea, is "an economically undeveloped area virtually unknown to scientists." The forest monkeys of this region include several unique species which have never been studied before. The Species Survival Commission (IUCN/SSC) Primate Group (of which Dr. Sabater Pi is a member) and the International Primatological Society "consider Bioko of maximum interest, giving this biotope a priority rating for conservation and preservation of the primate communities." Dr. Sabater Pi's research team is studying population density and distribution, possible predators, feeding habits, and social structures of these species. This type of information is essential to determining what types of measures need to be taken to preserve the monkeys and conserve their natural habitat.

■ Resources

Dr. Sabater Pi is the author of approximately 120 published articles and 7 books, including *Comparative Ecology of the Gorillas and Chimpanzees of Río Muni, West Africa.*

Jardin Botanico de Córdoba/Cordoba Botanic Garden
University of Córdoba
Box 3048
14080 Cordoba, Spain

Contact: **J. Esteban Hernández Bermejo, D.Ag.Eng.,**
Director, Botany Professor
Phone: (34) 57 200077 • *Fax:* (34) 57 295333

Activities: Education; Research • *Issues:* Biodiversity/Species Preservation • *Meetings:* CITES

▲ Projects and People

The **Cordoba Botanic Garden** includes the following research programs: taxonomic and ecological knowledge of endangered species, conservation of endangered species, genetic prospecting, economic potential evaluation, current and historical ethnobotany and the history of agriculture, recovery of forgotten agricultural heritage, transfer of threatened species of economic interest, collection of threatened flora and experiments in propagation and adaptation, promotion of *in situ* methods of conservation, reintroduction, and restoration, search for "repressed" species, and protection and increase of relative biodiversity of Iberian phytogenic resources—including the promotion of new crops and publication of results.

Prof. Dr. J. Esteban Hernández Bermejo, director of the Botanic Garden, is a principal researcher of the Spanish team working on the European Community (EC) project **Conservation of Genes of Aromatic and Condimentary Plants of the Mediterranean Area.** He is also researching **Conservation of Natural and Vegetal Heritage of Ethnobotanical Interest in Andalusia** and is reviewing the works of Andalusian agronomists **Ibn al-Beithar** and **Ibn al-Awann** with colleagues.

■ Resources

Dr. Hernández Bermejo, who has published widely in his field, has a publications list with mostly Spanish and some English titles.

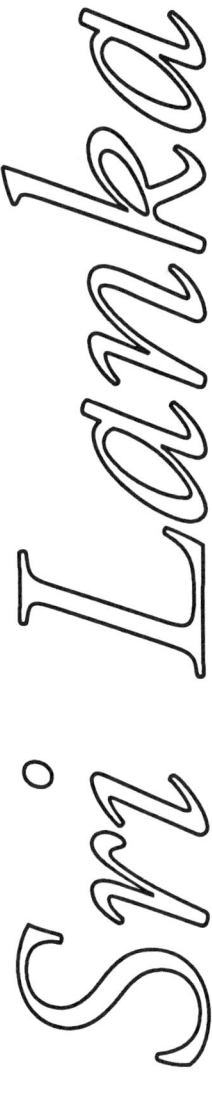

Sri Lanka

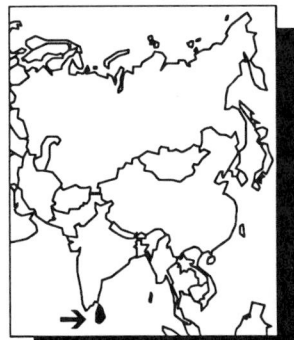

Sri Lanka is a tropical island nation in the Indian Ocean. The south-central part of the island is mountainous, while the northern and southern regions are characterized by a lowland plain. Two-thirds of the country is in the dry zone, with many dry evergreen forests; the rest is in the wet zone, with tropical rainforests.

Sri Lanka is endowed with more than 700 indigenous species of wildlife, and has a centuries-old tradition of conservation. Over 40 percent of the land has protected status, although poaching is a continuing threat, and the government is encouraging the achievement of sustainability at the community level. Loss of forest cover caused by the collection of timber for export or fuelwood and for agricultural expansion is a serious problem. Coastal degradation has been caused by the mining of coral for lime production; and Sri Lanka's freshwater resources are dwindling, due to drought and sewage and irrigation system problems.

🏛 Government

National Herbarium
Department of Agriculture
P.O. Box 15
Peradenia, Sri Lanka

Contact: **Anthony Harold Magdon Jayasuriya, Ph.D.,** Curator
Phone: (94) 8 88053

Activities: Education; Research • *Issues:* Biodiversity/Species Preservation • *Meetings:* International Botanical Congress; International Society of Tropical Root Crops; PROSEA

▲ Projects and People

Since 1977, **Dr. A.H.M. Jayasuriya** has been curator of the **National Herbarium.** With a staff of 10, the Herbarium maintains approximately 100,000 specimens. Researching his country's flora, Dr. Jayasuriya's specializations include plant systematics, forest botany, "assessment of biotic and abiotic effects on vegetation," germ plasm and biological conservation, medicinal plants, and ethnobotany.

Dr. Jayasuriya is continuing a **herbaceous plant inventory** of the Sinharaja MAB (Man and the Biosphere) Reserve in conjunction with UNESCO, World Wildlife Fund (WWF), and the Natural Resources, Energy, and Science Authority of Sri Lanka (NARESA). Also with NARESA, WWF, and the Ministry of Indigenous Medicine, an **inventory of the medicinal plants** of the Sinharaja MAB Reserve is ongoing. In other projects, Dr. Jayasuriya is revising the **yam family** for Sri Lanka, **assessing biodiversity** in the wet zone, and **assessing soil-vegetation relationships** in the Mahaweli River system. Previously, he collected germ plasm of economic plants and their wild relatives, and investigated "two newly recognized forest types" in middle elevations and mountains. As a consultant he reviews forest management plans.

A fellow of the Linnean Society of London (LSL), Dr. Jayasuriya belongs to the International Palm Society, Species Survival Commission (IUCN/SSC) Indian Subcontinent Plant Group, and Sri Lanka Association for the Advancement of Science (SLAAS). He serves on both the Committee of the Combined Indigenous and Allopathic Medical Research and the Technical Committee on the Conservation of Genetic Resources, and chairs the Committee of the Botanical Society of Sri Lanka.

■ Resources

The National Herbarium is in need of computers, microscopes, books and journals on topics related to botany, other herbarium supplies, and training opportunities. Dr. Jayasuriya can provide a publications list.

 # NGO

Sarvodaya Movement of Sri Lanka ⊕ ❧
98 Rawatawatte Road
Moratuwa, Sri Lanka

Contact: **Ari T. Ariyaratne, Doctor of Humanities,
Doctor of Letters,** President
Phone: (94) 1 507159 • *Fax:* (94) 1 507084

Activities: Development; Education; Research • *Issues:* Health;
Provision of Needs to the Poor; Sustainable Development;
Water Quality

● Organization

With groups in 8,600 villages throughout Sri Lanka, **Sarvodaya
Shramadana** is a pathway to answer the basic needs of more than 4
million people who live in poverty in Sri Lanka's rural areas. Described as one of the Third World's largest NGOs, Sarvodaya is
dedicated to helping more than half the country's estimated population of 16 million people who live "below the poverty line" in Sri
Lanka. Literally, "Sarvodaya Shramadana" means "Awakening of
All by voluntarily sharing peoples, resources, especially their time,
thoughts, and efforts."

Dr. A.T. Ariyaratne founded Sarvodaya in 1958 when he gathered together high school teachers and students to help the "socially
depressed and economically backward" to survive. He took the
name of the movement from an expression used by Mahatma
Gandhi to mean "the awakening of all." The organization is based
upon Gandhi's philosophies and Buddhist thought in its desire to
promote a kind of spiritual awakening. The organization's motto is
"First the awakening of the individual, then the community, the
country, and the world."

Another who has influenced Sarvodaya is Chinese philosopher,
missionary, and scholar Yang Chu James Yen, who said: "Don't take
the villagers out of their homes. Go where they are." According to
Sarvodaya, Pearl S. Buck, who wrote about Yen's revolutionary
mission, observed: "We must first think and plan how to remove
oppression, hunger, and ignorance. Where people are oppressed by
bad governments and by ignorance and by hunger, there can be no
contentment and therefore no peace."

Sarvodaya wishes to help arrange and maintain a state of "no
poverty–no affluence" based on plans developed by and for the
people. Related goals are "the release of developmental processes;
mass movement of mutual help; objectives and plans agreed upon
by the masses; and masses depending on their own resources with
other assistance being complimentary."

▲ Projects and People

Sarvodaya follows a set schedule of developmental phases. From
1958 to 1968, the movement focused solely on **Shramadana camps,**
which were organized throughout Sri Lanka to include hundreds of
people to live and work in village reconstruction projects. In these
10 intensive years, Sarvodaya created 300 camps—helping to make
the village inhabitants more aware of their personal resources as
laborers and of their land. What is being accomplished is the
"building of a social infrastructure," in which villagers organize
themselves into groups based on age or occupation—setting up
preschool programs and community centers and taking steps toward

"cooperative utilization of natural resources, agricultural implements, and irrigation facilities." The project also initiated the formation of grassroots movements throughout the country which
helped to enhance pride in cultural values.

The second 10 years highlighted **integrated rural development,**
which, although only planned to include 100, encompassed the
efforts of some 1,000 villages. The period emphasized people's
participation in self-development programs such as the Shramadana
camps, international village-to-village link-up program, self-employment schemes, development education programs, and continued formation of grassroots groups. Traditional cultural values and
happenings surfaced, such as renewed interests in the dance, songs,
and self-expression of individual villages. Sarvodaya also formed
peer groups, which emphasized "spiritual, moral, cultural, educational, social, and economic elements" of village and individual life.
Also at this time, Development Education Institutes were established in the villages.

When Sri Lanka was established in 1972, the parliament passed
an act which recognized the organization and "gave a considerable
impetus to the movement and contributed significantly to its quantitative expansion." (Now both a movement and a legal organization, Dr. Ariyaratne reflects with **D.A. Perera** that the growth of one
entity was not always parallel with the growth of the other.)

The next phase was the change in the structure of the whole
movement. With the formation of a government-recognized program, Sarvodaya felt it necessary to transfer decisionmaking power
from the "centre" to village leaders, or district coordinators. The
theme of this decade was "non-violent total social change from the
village up," with the purpose of getting villages to be legally recognized as independent societies. This action enabled villages to develop the community as they wished, "acting on their own, but
united by a common philosophy and thus being a considerable force
for national development."

Full-time professionals are now hired to assist both with the
recently created Suwa Setha Society, which undertakes welfare activities, and with the Sarvodaya Economic Enterprises Development Services to improve the economic sector. "People should be
made economically literate," the movement believes. The "no poverty–no affluence" society encourages cooperative purchasing of
agricultural needs, for instance. "The Lanka Jathika Sarvodaya
Shramadana Sangamaya is dedicated to its own liquidation," writes
Dr. Ariyaratne. "If the movement succeeds, the Sangamaya has
achieved its mission and is no longer needed."

Today, Sarvodaya remains committed to human service and
sustainable development. It advocates "a clean and beautiful environment" and "a national awareness of the socioeconomic realities
in the country and their probable causes." In 1983, a time of civil
strife, Sarvodaya workers "called a conference of civic leaders of all
races and religions. A historic document named 'A People's Declaration for Mutual Peace and Harmony' was adopted," highlighting
the symptom's of Sri Lanka's woes and organizing Peace Walks,
Shanti Sena (Peace Brigade), and the People's Peace Offensive
(PPO), in which peace-loving people intervene in armed conflict.

In 1990, when Dr. Ariyaratne received an international award for
promoting Gandhian values outside India from the Jamnalal Bajaj
Foundation (inaugurated in 1977 by the Prime Minister of India for
constructive work, service to women, and rural development), it was
noted that the purpose of Sarvodaya is "to both heal human relationships and empower people to stand together in the midst of fear,
hate, and mistrust." And its most "invaluable contribution to Sri

Lanka [is its] distinctive capacity to unite people of all ages, ethnic groups, and political affiliations from the grassroots up. . . . In addition, through its network of centres. . . . Sarvodaya distributes food, medicine, and temporary housing material to those who have fled their homes and villages. Through activities as these, the Sarvodaya Movement led by Dr. Ariyaratne is playing a dynamic role in the transformation of Sri Lanka."

Dr. Ariyaratne is board chairman, Asian Institute for Rural Development, Bangalore, India; chairman, Approtech Asia, Manila, Philippines; chairman, Asian and Pacific Bureau of Adult Education, Region 1; and affiliated with Brandeis University, Massachusetts, USA; International Commission on Peace and Food, California, USA; and the University of Universal Wisdom on Earth, Tokyo, Japan.

■ Resources

Sarvodaya seeks training and financial support for the effective implementation of future projects to benefit the poor and to "further develop the pre-school structure and other facilities for the welfare arm of Sarvodaya–Suwa Setha where handicapped children are looked after and vocational training for young children is given."

As Sarvodaya is a social service organization, its resources include a network of preschools, rural technical services, Sarvodaya Economic Enterprises, development services, poverty alleviation programs, and relief for empowerment of the poor. "Rehabilitation, reconstruction, reconciliation, reawakening, and early childhood development programs are in operation." Literature includes *Sarvodaya as a Movement* and *Sarvodaya as an Organization. Dana* (Sharing) is the International Journal of the Sarvodaya Movement; certain back issues are available.

Wildlife and Nature Protection Society of Sri Lanka (WNPS) ✦
Chaitiya Road, Marine Drive
Colombo 1, Sri Lanka

Contact: **Ranjen Fernando, Ph.D.,** President
Phone: (94) 1 25248 • *Fax:* (94) 1 580721

Activities: Education; Political/Legislative; Research • *Issues:* Biodiversity/Species Preservation; Deforestation; Natural Resource Conservation; Water Quality • *Meetings:* IUCN

● Organization

The **Wildlife and Nature Protection Society of Sri Lanka (WNPS),** with 3,500 members, was founded in 1894, and for almost a century has worked both alone and with other groups on behalf of its mission.

Funded by a variety of sources, including Norwegian Agency for International Development (NORAD), U.S. Agency for International Development (USAID), the Netherlands government, and private corporations such as the Ceylon Tobacco Company, the Society leads a number of projects to better the environment and to educate the public.

▲ Projects and People

WNPS develops **School Nature Club Programmes** to bring about environmental evaluations, nature camps, excursions, exhibitions, and other naturalistic activities. The program has elicited a "positive response from the education authorities of the districts," reports WNPS. With 125 School Nature Clubs throughout Sri Lanka, WNPS encourages its members to start more clubs in their neighborhoods. WNPS conducts **teacher seminars** for leaders of Nature Club activities.

The Society takes part in the **Sri Lankan Fauna and Flora Advisory Committee** which is held by the nation's Ministry of Lands, Irrigation, and Mahaweli (river) Development.

WNPS sends representatives to national and international congresses and events, such as the World Heritage Convention and CITES Convention. For the Sri Lankan's Department of Wildlife, WNPS makes recommendations on such matters as encroachments into Hakgala Strict Nature Reserve and Flood Plains National Park, and amendments to the Fauna and Flora Protection Ordinance. WNPS is lobbying and helping formulate policy for the recently formed Ministry of the Environment.

It pushes for **studies of forest damage** caused by "illicit timber fellings from clearing of land for agriculture," or terrorists, and by the armed forces. WNPS quotes from **Bernard Nietschmann's** article "Battle Fields of Ashes and Mud" (*Natural History,* November 1990): "Degraded land and resources are as much a reason for taking up arms as are repression, invasion, and ideology."

Dr. Ranjen Fernando, WNPS president and honorary consultant to the World Conservation Union's (IUCN) Species Survival Commission (SSC), is particularly concerned about the survival of **elephants** in Sri Lanka, which is linked to the restoration of forests. With regret, WNPS reports that "although in the late 1960s, it was possible to say that reports of elephant deaths and elephant damage were few, today they seem to be a common occurrence. With the enormous reduction in forest cover, a greater part of our forests may have little chance of survival because of our failure to provide and popularize alternatives for use in place of hard timber, with species like bamboo."

On the Mahaweli, Sri Lanka's largest river, the Society leads the **Mahaweli Environmental Project (MEP)** to maintain the river's biodiversity and to monitor building programs and research projects carried out in the area.

In 1978, a proposal was made for the wholesale development of the Muthurajawela Swamp. Since then, the Society has served as a watchdog in monitoring activity, gaining international assistance to prevent any ecological damage to the marsh and to "carve out a conservation area." To this end, it recently conducted a botanical and avifaunal survey.

Centenary celebrations are planned for 1994, when WNPS launches its second century.

■ Resources

WNPS has completed the reorganization of its library. However, funds are needed to complete the project; books and collections on conservation are welcome. WNPS also seeks collections of biological specimens, along with inventories. For educational purposes, the Society needs audiovisual aids for workshops and seminars. *Loris Magazine* and *Warana Magazine* are published.

Swaziland

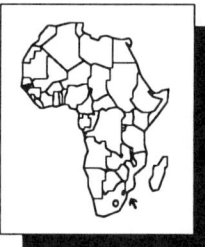

A small, landlocked country in southern Africa, Swaziland's terrain ranges from mountains to plateau to lowland. The country is one of the most prosperous in Africa, with small farmers expanding operations to include more cattle and dairy production.

Swaziland has abundant natural resources, but serious environmental problems. These include deterioration of grazing lands, through overstocking of livestock and soil erosion; the high incidence of waterborne diseases; and the pressures of a growing population and rapid urbanization, leading to poor sanitation and water facilities. The government is working to alleviate these conditions by establishing new nature reserves and public health/family planning programs, among other efforts.

❦ NGO

National Environmental Education Programme of Swaziland
P.O. Box 100
Lobamba, Swaziland

Contact: **Irma Acosta Allen, Ph.D.,** Coordinator
Phone: (268) 61151 • *Fax:* (268) 54328

Activities: Community Development; Environmental Education • *Issues:* Waste Management/Recycling • *Meetings:* Environmental Education Association of Southern Africa

▲ Projects and People

Responding to problems of insufficient solid waste disposal leading to water pollution, gastrointestinal disease, and death of cows from ingestion of plastic bags, in 1986 **Dr. Irma Allen** launched a nationwide cleanup and beautification campaign that brought together 40 organizations, such as the Boy Scouts, women's groups, and Environmental Health Association; government departments; and town councils.

The participants of the "Clean and Beautiful Swaziland" campaign organized themselves into special-interest subgroups: recycling, education, decisionmaking, media, and business; planned their activities; and reported their achievements to the main group at subsequent meetings. What started as a short-term project is now ongoing—attracting more individuals and organizations.

"The campaign turned out to be a resounding success," reports Dr. Allen. "Whole stretches of roadside were 'adopted' and kept clean by various schools. . . . The recycling group, with mostly women members, carried out a paper and glass recycling project. The group has found the means to purchase a mobile can baler and will soon be recycling cans. Two rivers which run past the towns of Mbabane and Manzini have been the objects of a cleanup program by the Rotary Group. School groups have organized tree-planting days, thereby helping to halt soil erosion. This year, the campaign is focusing on human waste disposal and providing adequate pit latrines for residents."

Dr. Allen has nearly 30 years' work in education, conservation, management, and development—including food production and processing—in rural and urban environments in Africa. Recently, she directed in-service education and training, Ministry of Education, Swaziland; worked with Ohio University and the U.S. Agency for International Development (USAID) to improve teacher education at the University of Swaziland and three colleges; and holds her present position as coordinator of the **National Environmental Education Programme** "without pay, as a service to Swaziland." She is a commissioner of the National Trust Commission; and takes part in international conferences, such as the World Congress in Environmental Education, UNICEF, and UNESCO.

■ Resources

Dr. Allen has co-authored teachers' resource books on environmental science and agriculture, among other publications. Materials are used by local conservation clubs, rural residents, other Swaziland NGOs, and local teachers. She seeks "appropriate technologies for sustainable development [and] tools and/or financial assistance to help school conservation clubs and communities carry out small conservation projects, such as water protection, erosion control, tree-planting."

Sweden

Located in the heart of Scandinavia, Sweden is the fourth largest country in Europe. Almost half the land surface is covered with forest; less than 10 percent is farmland. There is a long mountain chain in the northwest, and almost 100,000 lakes dot the countryside. Over 85 percent of the population, which is fairly low-density, live in the southern, more urban half of the country.

Sweden is rich in minerals, but manufacturing has evolved to advanced industries, such as specialty steel production. Energy is mainly derived from hydroelectric or nuclear plants; however, a decision was made to phase out nuclear power in the wake of the Chernobyl accident. Sweden has a comprehensive and ambitious environmental program: great strides have been made in reducing sources of acid rain, which has seriously affected the soil and many lakes. However, much air and water pollution in Sweden originates in nearby countries, and international cooperation is an important element of the government's environmental policy.

❧ *NGO*

International Geosphere-Biosphere Programme (IGBP) ⊕
The Royal Swedish Academy of Sciences
Box 50005
S-104 05 Stockholm, Sweden

Contact: **Thomas Rosswall**, Executive Director, Professor
Phone: (46) 8 166448 • *Fax:* (46) 8 166405
E-Mail: Omet: T.Rosswall

Activities: Research • *Issues:* Biodiversity/Species Preservation; Deforestation; Energy; Global Warming

● Organization

The **International Geosphere-Biosphere Programme (IGBP)** was founded by the International Council of Scientific Unions (ICSU) in 1986 to create in-depth studies of the Earth's changes brought about by both natural and man-made forces, serious ramifications for the environment. IGBP works in conjunction with other international agencies to conduct research and exchange and relay information.

With concern about the Earth's capability to support "the human need for food, fuel, and fiber," in light of degradation of air, water, and soil, IGBP is launching an unprecedented "worldwide research effort" to understand these changes. As data is gathered and interpreted on greenhouse gases, ozone deterioration, and alterations in both ocean and terrestrial ecosystems, for example, results will be communicated to scientists globally.

With advancements in measurement technology, environmental monitoring on a global scale is possible, reports IGBP. With the use of satellite-borne remote sensing, computer synthesized data sets, and numerical models, IGBP can link physical, chemical, and biological operations to create "an integrated and interdisciplinary approach to Earth System Studies." IGBP aims to develop a predictive knowledge of the Earth, with an emphasis on the biosphere.

▲ Projects and People

In designing its research program, IGBP poses four questions: "How is the chemistry of the global atmosphere regulated and what is the role of terrestrial processes producing and consuming trace gases? How do ocean biogeochemical processes influence and respond to climate change? How does vegetation interact with physical processes in the hydrological cycle? How will climate change affect terrestrial ecosystems?"

To find answers, IGBP has set up several core projects. To begin, IGBP established the **International Global Atmospheric Chemistry (IGAC) Project** and is proposing to establish a study of **Stratosphere-Troposphere Interactions and the Biosphere (STIB)**.

Other projects which answer basic core questions are the establishment of the **Joint Global Ocean Flux Study**, and a study of **Biospheric Aspects of the Hydrological Cycle**.

IGBP studies atmospheric concentrations of several trace gases or greenhouse gases. IGBP is also trying to understand key processes that regulate the atmosphere's composition and feedback mechanisms affecting the behavior of the atmosphere. With the joint efforts of the World Meteorological Organization (WMO), United Nations Environment Programme (UNEP), and Intergovernmental Panel on Climate Change (IPCC), a **Global Climate Observing System (GCOS)** is being established. GCOS will monitor the climate system, detect climate change, and give data to be applied to national economic development and for research on atmospheric conditions.

To study climatic regions, IGBP has set up **Regional Research Networks (RRNs)** and **Regional Research Centers (RRCs)** with the help of the World Climate Research Programme (WCRP).

■ Resources

IGBP produces the *Global Change Newsletter*, *Global Change Reports* of meetings, and brochures such as *A Study of Global Change*.

Världsnaturfonden WWF ⊕
World Wide Fund for Nature (WWF)—
Sweden
Ulriksdals Slott
S-17171 Solna, Sweden

Contact: **Jens Wahlstedt**, Secretary General
  Phone: (46) 8 850120 • *Fax:* (46) 8 851329

Activities: Education • *Issues:* Biodiversity/Species Preservation; Deforestation; Global Warming • *Meetings:* CITES

● Organization

The **World Wide Fund for Nature (WWF)—Sweden** runs hundreds of different projects, divided into four different categories: national projects; pan-European projects in the North Sea, in the Arctic, and the Baltic Republics/Poland; international projects in the Third World, including Brazil, Ecuador, East Africa, Madagascar; and international projects in the Third World managed through WWF-International.

With 150,000 supporting members and 300,000 regular donors, WWF-Sweden raised $12 million in 1990. Of the money, one-third goes to Swedish projects and two-thirds goes to international ones. The King of Sweden, **Carl XVI Gustaf**, heads WWF-Sweden; board chairman is county governor **Ingemar Öhrn**. Forty-six members on the council of trustees represent all of Sweden.

▲ Projects and People

More than 160 projects, based and operated in Sweden, cover six topics: **living mountains, living forests, living seas, living wetlands, living agricultural landscapes, and conservation awareness.** In each of these groups, the field projects involve research and conservation of nature, ecosystems, plants, and animals.

Ornithologist **Jens Wahlstedt**, secretary general since 1983, has published numerous books and articles and produced television programs on conservation. Earlier, Wahlstedt served on the board of the Swedish Ornithologists Union and the Swedish Society for the Protection of Nature.

Switzerland

Green and mountainous, Switzerland is the great watershed of Europe: the Rhine flows to the North Sea, the Inn feeds the Danube, the Rhone goes into the Mediterranean, and the Ticino runs through Lake Maggiore. The Alps, which cover about 60 percent of the country, run east and west through the southern region; the Jura Mountains stretch southwest to northwest, covering 10 percent; and the rest is plateau, where urban and industrial centers are concentrated. Switzerland's small population exports the vast majority of what it produces.

Though air pollution is a problem, low emission ceilings have been established and leaded gasoline is prohibited. Switzerland is also a leader in wastewater treatment; however, water pollution from agricultural sources has increased. A high percentage of Switzerland's plant and animal life is now vulnerable to habitat destruction.

🏛 *Government*

Federal Office of Environment, Forests, and Landscape (BUWAL)

Hallwylstrasse 4

3003 Berne, Switzerland

Contact: **B. Böhlen,** Director
Phone: (41) 31 619311 • *Fax:* (41) 31 619981

Activities: Education; Law Enforcement; Political/Legislative; Research • *Issues:* Air Quality/Emission Control; Deforestation; Energy; Global Warming; Waste Management/Recycling; Water Quality

● Organization

In 1885, when avalanches and floods caused by deforestation aroused widespread concern for the environment, the Swiss Federal Institute of Forestry Research went to work. The first federal law protecting Swiss forests—passed in 1902, establishing the superintendence of the federal authorities over the forest police—sought to maintain the quality and quantity of the nation's forests. Responding to increasing water pollution, the Swiss Federal Institute for Water Resources and Water Pollution control was set up in 1936; vital laws to protect water were passed in 1955 and 1971. A **comprehensive law on environmental protection** went into effect in 1985. This law covers six aspects previously not covered or covered insufficiently: air protection, noise control, protection against vibrations and radiations, hazardous substances, waste materials, and soil pollution. Enforcement and jurisdiction mainly falls to the **Federal Office of Environment, Forests, and Landscape (BUWAL)**.

Along with soil, air, and water protection, forest preservation measures and laws to protect habitats and landscapes are now under the auspices of BUWAL, a result of a 1989 merger of the former federal offices for water and environmental protection.

Also in 1989, the **ordinance on air protection (OAP)** went into effect with several provisions for reducing air pollution. Exhaust gas from road traffic is being reduced to the full extent "technically and industrially possible and economically tolerable." Rules govern light motor vehicles, heavy motor vehicles, and motorcycles. Private cars are subject to yearly exhaust checks, and speed limits are changing. With a host of regulations concerning industrial air pollution, OAP contains emission value limits for approximately 150 pollutants as well as for about 40 types of installations. The sulfur content for heating oils will be reduced in order to cut the amounts of sulfur dioxide in the atmosphere. One of Switzerland's more striking moves is legislation renouncing further development of the national road network.

According to law, **noise control** rests primarily with the originators of the emissions, such as factories and transportation devices. Noise reduction is achieved through effective measures taken during construction—walls or dikes, soundproof windows, sound-sensitive rooms—as well as operational and traffic restrictions. The government sets standards and time schedules for improving noisy areas. As noise reduction statutes are implemented, Switzerland examines for noise impact national road sections, railway lines, airports, and military installations. Efforts are made to educate government and private engineering offices in the field of noise control.

Switzerland's ordinance on environmentally hazardous substances provides protection for persons, plants, and animals as well as their habitats. It also stipulates that all chemicals should be submitted for impact assessment. CFCs (chlorofluorocarbons) have been banned, with the exception of medicinal sprays for which there are no known CFC substitutes, according to the Federal Office.

The ordinance on the content of hazardous substances in the soil defines the concept of soil fertility, containing guide levels for the highest admissible levels of pollutants in the soil. To observe soil pollution, there are currently 98 locations for sample takings in the national Swiss reference measuring network for the observation of soil pollution. The protection of soil is fairly critical, as there are presently no known practicable methods of improving polluted soils.

For years, waste disposal was unmanaged and carried out in the cheapest, easiest ways. Now Switzerland has made this disposal of wastes a national concern, seeking to dispose in an environmentally safe way. One new concept is the incineration and mineralization of organic and organic-chemical wastes before dumping. Regarding incinerator flue ashes, new methods are being developed to remove heavy metals and render the ashes inert.

Ordinances stipulate packaging of nonreturnables. Through voluntary household actions, about half of all paper and glass are recycled. Batteries are collected separately. Switzerland exports about one-third of its special wastes, with 95 percent going to European Community (EC) countries, where about half is recycled in the recipient country. Switzerland does not export special waste to Third World countries.

Swiss water protection entails central sewage purification and prevention of oil accidents. Recent important measures include the removal of phosphates from textile washing agents. Switzerland is working to revise all aspects, including provisions to secure safe drinking water supplies in cases of national emergencies. The revisions offer better protection to waterways and water sources.

Mainly, environmental research is conducted at state-run universities, such as the Swiss Federal Institutes of Technology in Zurich and Lausanne, and the cantonal universities of Basle, Berne, Geneva, Lausanne, Neuchâtel, and Zurich. The various research institutes—Water Resources and Water Pollution Control, Forest, Snow and Landscape Research, Materials, Testing and Research—together account for a staff numbering over 1,000 and a budget of about SwFr100 million.

"In view of the mostly global dimension of environmental pollutions," reports BUWAL, "it is important to develop and encourage international collaboration and cooperation in the field of environmental research." BUWAL cooperates with the International Commission for the Protection of the Rhine Against Pollution, International Commission for the Protection of Border Waters Against Pollution, Organization for Economic Cooperation and Development (OECD), United Nations Environment Programme (UNEP), United Nations Economic Commission for Europe (ECE), and EC.

Swiss Federal Laboratories for Materials Testing and Research (EMPA)
Ueberlandstrasse 129
8600 Dübendorf, Switzerland

Contact: **Sabine Voser-Mübus,** Marketing
Phone: (41) 823 5511 • *Fax:* (41) 821 6244

Activities: Development; Research • *Issues:* Air Quality/Emission Control; Energy; Waste Management/Recycling

● Organization
The Swiss Federal Laboratories for Materials Testing and Research (EMPA) was originally founded in 1880 to address the needs of a rapidly industrializing country in materials testing and technology evaluation. The basic goals remain, but today it has broadened its scope to promote safety for the population and the environment. Administratively attached to the Federal Department for Home Affairs (EDI), the "board of the Swiss Federal Institutes of Technology" acts as directing authority and reports to the Swiss Federal Council.

Along with EMPA, staff of 570, directed by **Dr. Fritz Eggimann,** are the Federal Research Institute for Forest, Snow, and Landscape (WSL), staff of 270, directed by **Prof. Dr. Rodolphe Schlaepfer;** Swiss Federal Institute for Water Resources and Water Pollution (EAWAG), staff of 220, directed by **Prof. Dr. Werner Stumm;** Paul Scherrer Institute (PSI), staff of 1,100, directed by **Prof. Dr. Anton Menth;** and the Swiss Federal Institutes of Technology—ETH Zurich, with some 11,000 students and president **Prof. Dr. Jakob Nüesch;** and EPF Lausanne, with some 3,400 students and president **Prof. Dr. Bernard Vittoz.**

▲ Projects and People
EMPA's main activities are "official tests and consulting; research and development in the areas of advanced materials and ecology; publication of research results, . . . courses and seminars; preparation of audiovisual media; teaching activities at the Swiss Federal Institutes of Technology, University of St. Gall for Economic and Social Sciences and other schools; and collaboration in the formulations of regulations, recommendations, ordinances."

In its materials research, EMPA is "convinced that the advantages of fibre-reinforced materials will bring innovations in many areas of machinery and vehicle production as well as the building industry and medical technology, for example." Investigations use the latest "destructive and non-destructive examination methods," particularly in relation to acoustic emission methods. A surface analysis center has been set up for researching new interface technologies to improve resistance to wear and corrosion. Experiments are underway, for example, with laser-modified surfaces and in areas such as high-strength ceramics; composite materials; non-destructive examination (NDE) with equipment, such as an x-ray computerograph and x-ray microscope; and behavior of metallic materials under shock loading, low temperatures, and multi-axis fatigue.

Dr. Hans Peter Tobler, Sewage/Waste Disposal Department, specializes in biodegradation. EMPA chemists are **Dr. Heinz Vonmont,** Inorganic Chemistry Department; **Dr. Max Wolfensberger,** Organic Chemistry Department; and **Dr. Peter Hofer,** Air Pollution Department. Section head is **Mark Zimmermann.**

International Traffic Division ⊕
Swiss Federal Veterinary Office
Schwarzenburgstrasse 161
3097 Liebefeld-Berne, Switzerland

Contact: **Peter Dollinger, D.V.M.,** Head
   Phone: (41) 31 598503 • *Fax:* (41) 31 598522

Activities: Law Enforcement; Political/Legislative • *Issues:*
Animal Health; Biodiversity/Species Preservation; Health;
Transportation • *Meetings:* CITES

● Organization

The **Swiss Federal Veterinary Office** was developed in several stages.
In 1872, the Federal Council created a special division, the Department of Commerce, Industry and Agriculture, to combat animal
diseases. The direction of the new division was assumed by the
Federal Commissar of Animal Disease, who was at the same time
also the Veterinarian in Chief of the Army. In 1886, the Federal
Council created a veterinary service for the borders. The chief task
of the border veterinarians was to inspect imported meat, assuring it
was of sanitary condition and free of disease.

▲ Projects and People

Working for the Federal Veterinary Office in various functions
since 1974, **Dr. Peter Dollinger** now presides as the head of the
International Traffic Division. Formerly with the Mulhouse Zoo,
France, and then the Zurich Zoo, Dr. Dollinger now directs the
work of 8 full-time border veterinary officers and about 70 part-
time officers. As a member of CITES (Convention on International
Trade in Endangered Species of Wild Fauna and Flora) Management Authority for Switzerland and Liechtenstein, Dr. Dollinger
represents Switzerland at the CITES committees and conferences.

   In 1974, the federal assembly approved the bans imposed by
CITES a year earlier. The Federal Veterinary Office took authority
for the enforcement of the ordinances in Switzerland. This country
is deeply involved with the annual CITES meeting—particularly as
a depository nation for the documents and declarations of the
member states of CITES. The **control of traffic of protected animals** is considerable, and the Federal Veterinary Office created a
special section to handle the work.

   Some 2,500 species are protected under the CITES convention,
and a CITES "species protection document" must be procured for
any such animal or product to allow the issuance of an import
permit. Animals from "strictly protected species" can be imported
only under special circumstances, such as, "If the animals have been
born in captivity or if their destination is a zoological garden or a
qualified breeder." The Swiss Traffic Office is developing a *CITES
Identification Manual* and helps fund CITES species surveys.

   Switzerland considers itself to be a **pioneer in the care, treatment, and conservation of animals.** In 1978, it passed comprehensive laws in this field: Animals must be treated in a manner which
best takes account of their needs; all persons who keep animals
must, as much as circumstances permit, attend to their well-being;
and persons must not impose on animals pain, injury, or harm, nor
put the animal in a state of anxiety in an unjustified fashion.

   The **reduction of animals for experimentation,** for example,
came about after these laws took effect. Encouraging laboratories to
use fewer animals in testing, Switzerland points out that between

1983 and 1988 there was a 40 percent decrease in animals used in
experiments. The Veterinary Office also notes that over 90 percent
of the animal testing occurs on small animals like mice, rats, and
hamsters.

■ Resources

The Federal Veterinary Office publishes brochures in English,
French, German, and Italian on import and export regulations, as
well as information on travel permits, endangered species lists, and
banned products lists. The office can also give up-to-date advice on
the current epizootic situation in various countries, including what
meat and animals should be avoided.

🏛 *State Government*

Museum of Zoology
Palais de Rumine, P.O. Box 448
1000 Lausanne, Switzerland

Contact: **Daniel Cherix, Ph.D.,** Curator
   Phone: (41) 21 312 8336 • *Fax:* (41) 21 236 840

Activities: Education; Research • *Issues:* Biodiversity/Species
Preservation • *Meetings:* CITES

● Organization

Since 1982, **Dr. Daniel Cherix** has been curator of the **Museum of
Zoology** in Lausanne, where his current research projects include
ecology of red wood ant; biology and control of pest ants such as
Argentine ant, fire ant, and pharao ant; and control of the little fire
ant (*Wasmannia auropunctata*) on the **Galapagos Islands.**

   He is also an associate professor of entomology, Institute of
Zoology and Animal Ecology, University of Lausanne. Previously,
Dr. Cherix was chairman of the Species Survival Commission
(IUCN/SSC) Ant Group.

■ Resources

Dr. Cherix publishes scientific papers on ants, and has collaborated
on a film presentation on red wood ants.

 *NGO*

Centre for Our Common Future ⊕
52, rue des Pâquis
1201 Geneva, Switzerland

Contact: **Warren H. Lindner,** Executive Director
   Phone: (41) 22 732 7117 • *Fax:* (41) 22 738 5046

Activities: Information Clearinghouse • *Issues:* Sustainable
Development • *Meetings:* Chautauqua Conference on Global
Ethics and the Environment; Global Forum; International
Facilitating Committee, Intersectoral Dialogue; UNCED

● Organization

The **Centre for Our Common Future** was established in 1988 to
ensure the dissemination of information in the report *Our Common
Future*, which was issued by the World Commission on Environ-

ment and Development. Published in 1987, the *Brundtland Report* is now available worldwide in over 20 languages. It has been called "one of the most important documents of our age on the future of the world."

The *Brundtland Report* and the work of the Centre concerns the politics of the future. With 21 scientists and politicians contributing to the *Brundtland Report*—a combination of research, consultation, analysis, and argumentation—it was endorsed by the United Nations General Assembly and has become a "basis of future environment and development policy for many governments and most major international institutions and non-governmental organizations."

An independent, nonprofit organization with a staff of 10, the Centre director is **Warren H. Lindner**, former Secretary of the World Commission. Of the Centre, Mr. Lindner says, "We were determined that the momentum created by *Our Common Future* not be lost or allowed to slow. Working to keep its principal message of sustainable development and its vital recommendations visible in all areas of public activity is the task we have undertaken. This is the most important contribution any of us can make if we are to secure the world's future." The Centre works with a small budget, relying on voluntary contributions to provide follow-up and support for the ideas presented in the *Brundtland Report*.

The Centre provides information, advice, and encouragement to NGOs, youth, women's groups, industry, trade unions, the scientific community, and individuals on the implementation of the report's suggestions. The goal of the Centre can be stated quite simply: the Centre is to promote the recommendations and vision of the World Commission's report. The woman for whom the *Brundtland Report* is named, **Gro Harlem Brundtland**, former Prime Minister, Norway, and Chairman of the World Commission on Environment and Development, says, "We must set forth on a worldwide campaign to inform and to educate. We must secure a constructive debate and persuade public opinion, governments and policymakers of the overriding goal of sustainable development."

To facilitate its goal, the Centre has over 120 "working partners" drawn from key environmental, development, media, and industrial organizations in some 60 countries. This constituency worked closely with the Centre in the formation and organization of the 1992 United Nations Conference on Environment and Development (UNCED). Two groups in particular are credited with support for the Centre: the World Conservation Union (IUCN) and the International Institute for Environment and Development (IIED).

▲ Projects and People

The Centre for Our Common Future calls to attention the following organizations "for their deep concern about environmental and development problems and for their imaginative way in which they try to raise international consciousness about them."

Founded in New Delhi in 1983, **Development Alternatives** is a nonprofit corporate organization designed to help foster the relationship needed between people and the environment. Its objectives are to "design options and promote sustainable development through programs which are economically efficient, socially equitable and just, environmentally harmonious, and lead to self reliance." Through its three branches—Technology Systems, Environmental Systems, and Institutional Systems—Development Alternatives works on the design of innovative and appropriate technologies, institutions, and societal paradigms. Recent actions include a study of common property resources (CPRs) and community-based re-

source management systems (CBRMSs) in India. With support from the World Bank, the project appraised eight highly degraded micro-watershed systems in varying eco-zones to identify a more "productive partnership of decisionmaking and resource sharing" and to preserve and restore the watershed.

The **Green Belt Movement** of Kenya is an indigenous grassroots campaign with tree planting as its basic activity. The project was developed under the auspices of the National Council of Women of Kenya, and has developed into a program that approaches issues of development holistically and campaigns against desertification. One of its major accomplishments is the establishment of over **1,000 tree nurseries**, which have produced millions of tree seedlings that have been issued to Kenyans by small-scale farmers, schools, and churches. Its origins in the National Council of Women is still apparent, as over 50,000 women cultivate trees at the nurseries.

According to the Centre for Our Common Future, the Green Belt Movement is spreading to other countries and has produced two booklets and several films. Over 7 million trees are recorded as having been planted at a survival rate between 70 and 80 percent in a 10-year period. More than 1 million children planted trees on some 3,000 schoolgrounds.

The Movement's short-term goals include educating the public about desertification, encouraging tree planting in order to reforest Kenya, creating jobs in rural areas, and encouraging soil conservation and soil rehabilitation. Long-term goals include promoting sustainable development, fostering a positive image of women in leadership roles in national development, and encouraging indigenous people's initiatives to restore self-confidence in a population overwhelmed by foreign experts.

The **Communal Areas Management Programme for Indigenous Resources (CAMPFIRE)** is Zimbabwe's approach to the management of natural resources in common property-type situations. *(See also Zimbabwe National Conservation Trust in the Zimbabwe NGO section.)* To advance the concept of "local people managing local resources," the countrywide CAMPFIRE program sponsors 26 project areas in 13 districts. At the same time, Zimbabwe legislation allows local landholders to manage the wildlife and utilize it—if not in a protected area. CAMPFIRE seeks to use this legislation to advise and monitor the local resident user groups.

The Ministry of Environment chairs a consortium on NGO organizations to facilitate the implementation of CAMPFIRE. The three collaborating agencies are: Zimbabwe Trust, World Wide Fund for Nature (WWF), and University of Zimbabwe, Centre for Applied Social Sciences. According to CAMPFIRE, "The critical issue involved at present is the integration of natural resource management and economic issues to ensure the decentralization of authority over wildlife resources from national to district level, leading to dynamic natural resource cooperatives that can be spatially determined with given resource user groups instead of furthering bureaucratic systems."

The **Bangladesh Rural Advancement Committee (BRAC)**, an NGO founded and managed by Bangladeshis, began in 1972 as a reconstruction effort after the War of Liberation, and continues now to encourage infrastructure development and long-term sustainable growth. In 1976, BRAC realized that the very structure of Bangladesh society was an impediment to the development of resources and adjusted its strategies accordingly, aiming at the bottom 40 percent of the population, the landless poor. By 1986, BRAC had grown into an organization of 2,500 workers with over 2,500 self-reliant groups in 1,500 villages. Its principal health program—

teaching **oral rehydration therapy to stem diarrhea, for instance**—has reached over seven million households.

Of its goals, BRAC says, "A fundamental intention is to help individuals and communities to become self-reliant so they can function effectively without external assistance. The ultimate objective is to end the longstanding exploitative relationships that dominate rural life in Bangladesh."

■ Resources

The Centre publishes the quarterly *Brundtland Bulletin* and the monthly *Network*.

Foundation for Environmental Conservation (FEC) ⊕

7 Chemin Taverney, 1218 Grand-Saconnex
Geneva, Switzerland

Contact: **Nicholas Polunin, Ph.D.**, President, Professor
Phone: (41) 22 798 2383/4
Fax: (41) 22 798 2344

Issues: Air Quality/Emission Control; Biodiversity/Species Preservation; Deforestation; Population Planning; Waste Management/Recycling • *Meetings:* International Conference on Environmental Future

● Organization

Based in Geneva, Switzerland, the **Foundation for Environmental Conservation (FEC)** is an international non-governmental organization founded in 1975 by **Dr. Nicholas Polunin** in collaboration with the World Conservation Union (IUCN) and the World Wildlife Fund (WWF).

▲ Projects and People

Ongoing activities of founder/president Dr. Polunin and his organization include: *Environmental Conservation (EC)*, an international quarterly journal, for which supplements to index the titles and authors are also being published; **International Conferences on Environmental Future (ICEFs)**, an "independent forum for global survival"; environmental **monographs and symposia; Cambridge Studies in Environmental Policy; World Council for the Biosphere;** *World Who's Who in Environment and Conservation*, a list of leading specialists, administrators, and benefactors; and the **World Campaign for Holistic Thinking and Concomitant Action towards Global Survival.** Other endeavors include: a "best paper prize" awarded annually, sponsorship of the Baer-Huxley Memorial Lectures; **International Conferences on Waste Minimization and Clean Technology;** efforts to encourage environmental awareness; and since 1982, the **World Campaign for the Biosphere,** which is now a joint function of the World Council for the Biosphere and the International Society for Environmental Education.

Affiliated with FEC, the World Council for the Biosphere promotes the idea of a global biosphere and with Dr. Polunin, its founder and president, established in 1991 **Biosphere Day** on September 21, which marks the autumnal equinox in the Northern Hemisphere and the spring equinox in the Southern Hemisphere.

Dr. Polunin writes, "The biosphere is that favored layer of our planet's periphery (solid, liquid, and gaseous) in which any form of life exists naturally. It is our sole home and living heritage whose protection should be our first human imperative. Energized in virtual eternity by the sun, the biosphere is our only proven life-support—as the world is at last beginning to realize, while concomitantly recognizing the immense implications of this most basic fact. It is also the only proven venue of organized life in our Universe, hence multiplying our gargantuan responsibilities to safeguard it in every possible way. . . .

" . . . such resolve to save the biosphere, and reorientation of our lives accordingly, will be necessary if our world is to survive, and that the most effective means of attaining this most fundamental objective would be the emergence of a dominant category of humans based on this resolve."

This declaration also included the World Council's motto and guiding principle: "Every day a Biosphere Day." As Dr. Polunin explains, "It is the biosphere which is so gravely threatened by human overpopulation and profligacy—not the inert and more solid planet Earth."

With degrees from Oxford, Yale, and Harvard universities, Dr. Polunin has been a researcher and writer for more than 50 years as well as a professor in universities from Canada to Nigeria, Iraq to Switzerland. Explorations have taken him to the Arctic, Mount Cameroon in West Africa, and Yakutsk, Siberia. In 1936, he discovered descendants of plants introduced to Greenland from North America by Vikings 900 years earlier. Later, he uncovered the existence of microbial life in the North Pole and "correlated spora with air-mass provenance." Dr. Polunin has written and edited dozens of books and journals and published more than 500 research and scientific papers.

■ Resources

The 400-page *Environmental Conservation* has been published seasonally since 1974, with advisory editors on animal ecology, climatic variation, microbiological implications, and population impact. Advocating timely action for the protection and amelioration of the environment worldwide, topics cover pertinent case histories, rational use of resources, foreseeing ecological consequences, enlightened environmental policy, antipollution measures, low-impact development, environmental education and law, and ecologically sound management "for the lasting future of Earth's fragile biosphere."

International Association Against Noise (AICB) ⊕

Hirschenplatz 7
CH-6004 Luzerne, Switzerland

Contact: **Willy Aecherli, Dr.**, Lawyer
Phone: (41) 41 513013

Activities: Development; Law Enforcement; Political/Legislative • *Issues:* Noise • *Meetings:* AICB

International Organization for Standardization (ISO) ⊕

1, rue de Varembé
1204 Geneva, Switzerland

Contact: **Lawrence D. Eicher,** Secretary-General
Phone: (41) 22 749 0111 • *Fax:* (41) 22 733 3430

Activities: Education; Political/Legislative • *Issues:* Air Quality/Emission Control; Energy; Health; Population Planning; Sustainable Development; Transportation; Waste Management/Recycling; Water Quality • *Meetings:* UNCED

● Organization

The International Organization for Standardization (ISO) is a worldwide federation of national standards organizations, presently comprising 90 member groups. ISO covers standardization in all fields except electrical and electronic engineering. ISO members are the national bodies "most representative of standardization in its country," mostly governmental institutions or organizations incorporated by public law. ISO and the International Electrotechnical Commission (IEC) together make up the world's largest non-governmental system for voluntary industrial and technical collaboration at the international level.

With international standardization launched more than 80 years ago in the electrotechnical field, ISO was the first group whose aim was devoted to international industrial standardization. Delegates from 25 countries met in London in 1946 and created a standards organization, "the object of which would be to facilitate the international coordination and unification of industrial standards," which began functioning in early 1947. Now ISO work is carried out by 2,600 technical bodies, with more than 20,000 experts participating in the technical work. ISO is responsible for publishing 7,778 standards.

New standards are constantly circulating among the committee members, sometimes as many as 10,000 working documents in a single year. Because of the incredible volume, committee members conduct most of their work through correspondence, and meetings are convened only when justified. Most standards require periodic revision, and the ISO has a rule that all standards are reviewed at least every five years. On occasion, as with developing technologies, it is necessary to revise the standards sooner. Various ISO technical committees are working on roughly 500 environmentally related standards projects.

ISO has recently started the **Strategic Advisory Group on Environment (SAGE).** Its responsibilities include assessing the need for international standardization of key elements of sustainable industrial development with respect to nonrenewable resources, energy consumption and alternative sources, environmental effects during production or use of products, disposal and recycling, consumer information, and eco-labeling. Working within the ISO, SAGE will recommend the overall strategy for environmental performance standardization.

■ Resources

ISO publishes a list of publications on the environment in these categories: air quality, exhaust emissions, water quality, acoustics, vibration and shock, and radiation protection. A members' list is also available.

Swiss League for Nature Protection (SLNP)

Centre LSPN de Champ-Pittet
1400 Cheseaux-Noréaz/Yverdon, Switzerland

Phone: (41) 24 231341

Activities: Education • *Issues:* Biodiversity/Species Preservation; Nature Protection

● Organization

The **Swiss League for Nature Protection (SLNP)** was created in 1908 in response to the need for a body to raise money to create a Swiss National Park. Within four years, SLNP grew to 25,000 members and raised the funds to create Europe's first national park. Early on, SLNP tackled important conservation projects and soon realized the need for "purposeful education" to ensure that future generations would value nature, and attempted to have relevant legislation passed by the Swiss parliament. The legislation was not at first successful, but the educational campaign flourished and continues today.

As Switzerland experienced a development boom in the decade prior to World War II, SLNP swung into action as "protector of the landscape." SLNP chairman Adolf Nadig was appointed head of a new federal commission for the protection of nature and the preservation of national heritage. World War II curbed SLNP's efforts, as some reserves were converted to farmland and water projects proliferated.

Since the war, SLNP intensified its educational campaign, called for "comprehensive national planning," focused on the protection of rare plant species, and increased nature reserves and protected areas. Remaining faithful to its mission as a "reserves society," today SLNP owns, leases, or supports more than 500 reserves. SLNP's protected areas cover about 500 square kilometers and its membership of more than 100,000 comprises almost 2 percent of the Swiss population.

▲ Projects and People

SLNP runs two **Field Study Centers** where adults and young people can study nature and enroll in courses, seminars, and conferences. The **Villa Cassel,** built in 1902 by Sir Ernest Cassel, looks down on the Aletsch forest and glacier. In 1976, SLNP converted the castle, which now draws 12,000 visitors daily in the summer to its lectures, library, exhibitions, excursions, and alpine garden.

The Champ-Pittet Castle, situated at the edge of the **Grande Cariçaie** reserve on the bank of Lake Neuchâtel, was built in the eighteenth century. On the lake's southern shore, SLNP, World Wide Fund for Nature (WWF), and cantons of Fribourg and Vaud are managing 550 hectares of marsh and 300 hectares of woodland with a range of habitats including littoral ponds and dunes, reedbeds and reedswamps, sedge meadows, riparian woods, hillside forest, and hinterland.

A "much valued environment of international importance," the area features amphibians such as the threatened **agile frog and warty and smooth newts,** and declining **natterjack and common tree frog.** Some 70,000 waterbirds winter on the lake; marshland birds include **grebes, warblers, purple heron, cuckoo, bearded tit,** and **coot.** Among the rare plants are **orchids, flowering rush,** and **arrowhead.** Thirty species of **dragonflies and damselflies** and 400 species of **butterflies and moths** breed here; 34 mammal species include polecat, beaver, otter, and pygmy shrew; 7 fish species reproduce in the marsh; 20 more are in the lake and tributaries.

Management activities at Grande Cariçaie include checking lakeshore erosion; revitalizing or creating new ponds, mudflats, sandbanks, and other habitats; restraining the forest's advance into the open marshland; and clearing the marsh of accumulated dead plant material.

■ Resources

SLNP produces books, pamphlets, posters, exhibitions, slide shows, and audiovisual presentations. A 30-page catalog of materials is available.

The World Conservation Union (IUCN) ⊕
Avenue du Mont Blanc
1196 Gland, Switzerland

Contact: **Martin Wyatt Holdgate, Ph.D.**, Director
General
Jeffrey A. McNeely, Chief Conservation
Officer
Phone: (41) 22 64 9114 • *Fax:* (41) 22 64 2926

Activities: Development; Education; Political/Legislative; Research • *Issues:* Biodiversity/Species Preservation; Deforestation; Global Warming; Population Planning; Sustainable Development

● Organization

The **World Conservation Union (IUCN)**, formerly known as the International Union for the Conservation of Nature and Natural Resources, was founded in 1948 at an international conference in Fountainbleau, France, sponsored by the government of France, Swiss League for the Protection of Nature, and the United Nations Educational, Scientific, and Cultural Organization (UNESCO).

Today, IUCN is described as the world's "largest professional body" that is working for the environment, natural resources, and all life on the planet. Active in 120 countries, IUCN's members include 61 states and 121 government agencies—with a total membership of more than 600 organizations, an additional 37 international NGOs, and some 700 nonvoting individual and organizational sponsors. For example, these include the World Wide Fund for Nature (WWF), Sierra Club, Royal Society for the Protection of Birds, Fédération Française des Sociétés de Protection de la Nature, and the Instituto de Recursos Naturales Renovables.

IUCN aims to be scientifically sound, economically realistic, and socially constructive. It provides knowledge and leadership worldwide for the application and implementation of sustainable development principles. It guides governments in writing sound environmental policy and legislation. IUCN helped initiate such well-known international conventions as the Convention Concerning the Protection of the World Cultural and Natural Heritage, Convention on International Trade in Endangered Species of Wild Fauna and Flora (CITES), and Convention on Wetlands of International Importance (Ramsar).

IUCN is a principal author of the World Conservation Strategy (WCS), published in 1980 with the United Nations Environment Programme (UNEP), WWF, UNESCO, and the Food and Agri-

culture Organization (FAO). Known to many as the "most important conservation manifesto of the 1980s," the strategy is based on "maintenance of essential ecological processes and life support systems, preservation of genetic diversity, [and] sustainable utilization of species and ecosystems." WCS maintains that "conservation and sustainable development are not enemies, but are inseparably one."

To implement these strategies, IUCN has a four-step approach: Information—IUCN collects scientifically based data about species and ecosystems; Analysis—IUCN investigates causes of environmental change and determines options for solutions; Planning—specialists examine information data and analysis and decide how to reverse destructive trends and affirm sustainable development options; and Action—IUCN designs projects and advises and helps others to fruition.

IUCN has offices in Nairobi, Kenya; Harare, Zimbabwe; Dakar, Senegal; Karachi, Pakistan; San Jose, Costa Rica; Washington, DC, USA; Cambridge and Kew, UK; and Bonn, Germany. In addition, IUCN sponsors the **World Conservation Monitoring Centre** (with UNEP and WWF) and the **Environmental Law Centre**—repositories of information including records on more than "15,000 protected areas, 80,000 species, 860,000 details of trade in wild species, and 31,000 laws and administrative instruments." Each year, it participates in over 500 international, national, and regional meetings, scientific conferences, symposia, and conferences.

Formed in 1956, IUCN's **Species Survival Commission (SSC)** is a worldwide network of some 2,500 qualified, volunteer authorities in species conservation who prepare Action Plans, assess the status of species through "population viability analysis" (PVA), issue Red Data Books, and advise CITES and other global groups. *(Species Survival Commission [SSC] projects and people are listed throughout this Guide.)*

▲ Projects and People

IUCN helps countries prepare **National Conservation Strategies**, based on WCS, as well as government policies based on sustainable use of natural resources and respect for human dignity. A **Global Plan of Action**, with UNEP and FAO, helps protect marine animals and other resources of the sea; as do coastal management efforts along the Red, Wadden, and Mediterranean seas, and the Persian Gulf. The **Action Plan for the Alps** addresses "dying forests and excessive tourist development." African elephants and rhinos have been saved through bans on illegal ivory trading, as have whales, through commercial bans. IUCN encourages countries to end the conflict "between commercial logging and the rights and lives of village peoples." It works to stop the "export of persistent pesticides from European countries which no longer use them at home, and have been dumping them in the developing world." IUCN helps countries secure sites on the World Heritage List and management plans for national parks. China's giant panda as well as gorilla, dugong, green iguana, tree ducks, and rainforests are beneficiaries in IUCN's several hundred field projects, in cooperation with other groups.

Dr. Martin Wyatt Holdgate, director general since 1988, was chief environment scientist and deputy environment protection secretary with UK's Department of the Environment and was also involved in UNEP's establishment—serving as president of its governing council. Among other achievements, Dr. Holdgate conducted oceanic biology research in South America and established a biological research program in Antarctica. He was editor of *Antarctic Ecology*, based on the Scientific Committee on Antarctic Research (SCAR) symposium that he organized.

Jeffrey A. McNeely, chief conservation officer, previously worked with WWF and undertook missions or fieldwork in dozens of countries, such as Burma, China, Republic of Ireland, Mexico, Tanzania, and Venezuela. An accomplished photographer, his work has been published in *Audubon, Nature*, and *International Wildlife*, among others. He has authored or co-authored dozens of publications and reports.

■ Resources

IUCN publishes Red Data Books on species conservation, directories of protected areas, handbooks on managing protected areas, guides for conservation in developing countries, and guidelines on managing buffer zones, resettlement of transmigrants, and tropical forests, for example. A *Publications Catalogue* gives a complete listing, including books on sustainable development and literature on wetlands, Action Plans, ecological science, environmental law, and membership information. *IUCN Bulletin* and *TRAFFIC Bulletin* are published quarterly; *Species*, the SSC newsletter, is published semiannually.

(See also World Conservation Union [IUCN] in the France, Kenya, Niger, Pakistan, USA, and Zimbabwe NGO sections.)

World Meteorological Organization (WMO) ⊕
CP 2300
1211 Geneva 2, Switzerland

Contact: **Dr. Pierre Morel**, Director, World Climate Research Program
Phone: (41) 22 730 8246 • *Fax:* (41) 22 734 2326

Activities: Research • *Issues:* Climate Science; Global Warming

● Organization

The World Meteorological Organization (WMO) and the International Council of Scientific Unions (ICSU) have collaborated to form the World Climate Research Programme (WCRP). The need for scientific information on climate and the oceans has never been greater, they say, and the program seeks to study both. In *Global Climate Change*, a joint publication of WMO and ICSU, the foreword states: "To a large extent, human societies have become attuned to current climate; crops flourish within an expected range of climatic variations, our dams and dykes are built on assumptions based upon the climatological record of the last 30 to 50 years. Yet, the earth's climate has never been static and, for the first time in history, human activities have now reached the stage where they affect climate on regional and global scales."

The World Climate Research Programme has many contributing partners. Founded in 1979 by WMO and ICSU, the Programme is currently affiliated with many national and international organizations, such as UNESCO (United Nations Educational, Scientific and Cultural Organization), Intergovernmental Oceanographic Commission, ICSU's Scientific Committee on Oceanic Research, and their joint Committee on Climatic Change and the Ocean.

WCRP relies on the understanding of four major components to achieve its objectives: the global atmosphere, which is the most rapidly varying component; the world ocean, which interacts greatly with the atmosphere; the cryospheres, which is made up of the continental ice sheets and ice caps, mountain glaciers, and sea ice; and the land surface of the continents, with all surface water and groundwater systems.

▲ Projects and People

WCRP has instituted three major projects to investigate climate change. The **Tropical Ocean and Global Atmosphere (TOGA)** studies the interactions of the tropical oceans and the global and regional atmospheres, in particular the Southern Oscillation and the associated El Niño phenomenon. The purpose of **Global Energy and Water Cycle Experiment (GEWEX)** is to study, model, and predict the transport and exchanges of radiation, heat, and water with the atmosphere and at the Earth's surface. The **World Ocean Circulation Experiment (WOCE)** is designed to provide an observation of all oceans simultaneously as the basis of mathematical models of global ocean circulation. WCRP has other projects that deal with more specific aspects of the world climate, such as the **Joint Global Ocean Flux Study, International Satellite Cloud Climatology Project**, and the **International Geosphere-Biosphere Programme**.

WCRP director Dr. Pierre Morel began his career as the Scientific Attaché for the French Embassy in the USA, and helped establish the French Space Agency, where he served from 1975 to 1982 as its deputy director general and assisted in the development of future space programs, such as Ariane-5. Dr. Morel initiated the European meteorological satellite program METEOSAT. He assumed leadership of the WMO/ICSU Joint Planning Staff for WCRP in 1982.

■ Resources

The publication *Global Climate Change* describes WMO and its programs. The *WCRP Report Series* is available to scientists and scientific institutions. Multinational support is sought for climate research.

▌ *Private*

Applied Environmental Sciences (AES) ⊕
Sonnenbergstrasse 3
8280 Kreuzlingen, Switzerland

Contact: **Hendrik N. Hoeck, D.rer.nat.**, Ecologist, Consultant
Phone: (41) 72 752878 • *Fax:* (41) 72 752878

Activities: Education; Law Enforcement; Political/Legislative; Research • *Issues:* Biodiversity/Species Preservation; Deforestation; Sustainable Development; Water Quality • *Meetings:* IUCN

● Organization

Applied Environmental Sciences (AES) is a joint venture formed by Ambio AG in Zurich, and BiCon AG in Kreuzlingen, Switzerland, which offers consulting services on marine, freshwater, soil, grassland, and forest environments. Using a wide range of techniques, its impact studies provide baseline information in engineering biology, conservation, ecotoxicology, ecotourism, forestry, agriculture, wildlife ecology, and project evaluation.

AES also conducts special courses and lectures, organizes conferences and symposia, and compiles literature reviews, providing such services in English, French, German, Italian, and Spanish.

▲ Projects and People

Dr. Hendrik Hoeck—with a research and teaching background in ecology, ethology, wildlife management, and ecodevelopment at the Max-Planck Institute für Verhaltenphysiologie and Universities of Munich and Konstanz of Germany; Serengeti Research Institute of Tanzania; and the World Health Organization (WHO) malaria survey in North Syria—is an AES consultant and co-partner of BiCon AG. He is a director of the Charles Darwin Research Station, Galapagos, Ecuador; chairman of the Species Survival Commission (IUCN/SSC) Hyrax Group; and involved with Tucán Travel, which provides tropical tours for ecological and educational purposes.

Among AES senior personnel are Dr. Andreas Bally, Bettina Bally, Uell Busin, Dr. Gabriele Gerlach, Markus Haberthür, Dr. Regula Huber, Dr. Annette Johnson, Dr. Marianne Knecht, Dr. Bruno M. Oberle, and Dr. François Schnider.

■ Resources

AES has a large body of written material available from environmental impact studies, as well as several films.

F. Hoffmann–La Roche Ltd. ⊕
4002 Basel, Switzerland

Contact: Albert E. Fischli, Ph.D., Pharma Research
        Department Head
Phone: (41) 61 688 1111 • *Fax:* (41) 61 691 9391

Activities: Research • *Issues:* Health; Pharmaceutical Research
Meetings: CITES

● Organization

Hoffmann–La Roche is a chemical/pharmaceutical company employing 47,000 people in more than 50 countries worldwide and producing "drugs, vitamins, diagnostic aids, and plant protection agents." According to Hoffmann–La Roche, "The production of pharmaceutical specialties does require considerable amounts of organic solvents. Recycling of such solvents and the disposal of leftover solvents are therefore central to Roche's programme of environmental protection."

In its literature, Hoffmann–La Roche describes its three-pronged approach to protecting the environment: utilizing organizational measures such as educating employees and establishing clear safety guidelines; personnel measures such as making clear each employee's responsibility for conservation; and technical measures, such as choosing the chemicals and chemical processes to undertake, along with safety mechanisms to prevent accidents. Its literature also addresses environmental accidents.

The company describes its surface water retention device that "holds back water used in firefighting or other accidents and which would otherwise pollute the groundwater and the Rhine," wastewater treatment plant that is a cooperative venture with local government and other industries, steps to comply with Switzerland's clean air ordinance and reduce emissions, storage of chemicals, disposal of hazardous and other wastes, and impact of vitamin C manufacture.

A pilot biotechnology plant is producing and testing interferons through the genetic manipulation of microorganisms. "The interferons protect non-infected cells against a second viral infection for a certain time," reports Hoffmann–La Roche. They also have an "inhibiting effect on cancer."

Prof. Dr. Albert E. Fischli, who heads the Pharma Research Department, is vice president of the Swiss Academy of Sciences and secretary of the International Union of Pure and Applied Chemistry (IUPAC), Committee on Chemistry and Industry.

■ Resources

Hoffmann–La Roche publishes a 64-page booklet on *Environmental Protection*, printed in English and German.

🎓 *Universities*

Département de Génie Rural/Rural Engineering
    Department (DGR)
Ecole Polytechnique Fédéral de Lausanne
    (EPFL)
Federal Polytechnical School of Lausanne
1015 Lausanne, Switzerland

Contact: Lucien Yves Maystre, Professor
Phone: (41) 21 693 2715 • *Fax:* (41) 21 693 2727

Activities: Education; Research • *Issues:* Air Quality/Emission Control; Biotechnology; Ecotoxicology; Health; Sustainable Development; Waste Management/Recycling; Water Quality

▲ Projects and People

The Federal Polytechnical School of Lausanne's Rural Engineering Department (DGR), with its staff of 38, identifies as its motivating force the patrimonial management of the territory-ground. The "territory-ground" is the two-fold use of land: the spaces and territories occupied and used by humans, and the ground used for farming and agriculture. DGR sees its patrimonial duty to Switzerland as part of a larger commitment to global conservation objectives. DGR also divides the work of the school into fundamental research and methodology in the evolution of natural phenomenon, and applied research and experimentation. The research is often multi- or interdisciplinary. DGR works closely with the Department of Civil and Chemical Engineering within the school, as well as the Universities of Lausanne, Geneva, and Neuchâtel in Switzerland, and schools and institutions in France and Canada.

DGR comprises the Institute of Measurement, including geodesy (land survey) and photogrammetry; Institute of Management of Earth and Water, with hydrology and pedology; and the Institute of Environmental Engineering, featuring sanitary engineering and ecotoxicology, biological engineering, and atmospheric pollution.

Its goal is to educate engineers capable of understanding the problems of contemporary society and mastering the techniques necessary to contribute to the knowledge, management, and protection of the environment. With nearly 200 students, DGR reports that "record numbers of students" are applying to the department.

■ Resources

Based on current research projects, DGR publishes on all aspects of the environment, rural engineering, and development, and holds conferences and symposia.

Zoologisches Institut/Zoological Institute ⊕
Universität Zürich/Zurich University
Winterthurerstr. 190
CH-8057 Zurich, Switzerland

Contact: **Donat Agosti, Ph.D.**, Entomology Research
Phone: (41) 1 257 4830 • *Fax:* (41) 1 361 3185

Activities: Research • *Issues:* Ant Systematics; Biodiversity/Species Preservation

▲ Projects and People

Dr. Donat Agosti is involved in the study of **ants** worldwide. In a collaborative project of Zurich University's Zoological Institute and the Natural History Museum's Department of Entomology, UK, and with the help of a Swiss Science Foundation grant, he investigated *Cataglyphis* ants in 1991. Dr. Agosti recently researched ants in Australia and South America. He also has embarked in the late 1980s and early 1990s on collecting expeditions to the Canary Islands, Turkey, Tunisia, Alps, Pyrenees, Balkans, Scandinavia, Pakistan, Syria, Jordan, Iraq, Indonesia, and Malaysia.

Dr. Agosti has authored or co-authored numerous studies, reports, and other papers in his field—as noted in his publications list with English and German titles.

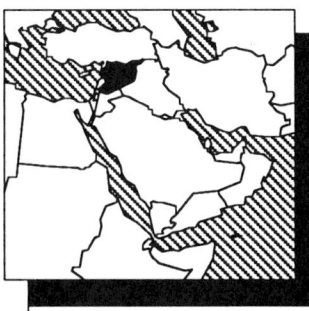

Syria is located on the northeastern corner of the Mediterranean and is bordered by Turkey, Iraq, Jordan, Lebanon, and Israel. The Syrian Desert dominates the country in the southeast, while the east is characterized by a narrow coastal plain backed by a mountainous region in the west. Only 2 percent of the land is arable; 31 percent is covered by forests. Oil production and mining are major industries, but most of the population works in agriculture, animal herding, or service jobs. Half the people live in cities.

Much of the water is unsafe for drinking, due to lack of adequate sewage systems and waste dumping from petroleum processing and other industries. Land degradation has occurred as a result of poor irrigation practices. Overgrazing has contributed to desertification. Oil pollution is a problem along the Mediterranean coast.

🏛 *Government*

Arab Center for the Studies of Arid Zones and Dry Lands (ACSAD) ⊕
P.O. Box 2440
Damascus, Syria

Contacts: **Gilani M. Abdelgawad, Ph.D.**, Director, Soil Sciences Division
Nabil Hosny Naguib Rofail, Ph.D., Deputy Director, Water Resource Division
Phone: (963) 11 755713

Activities: Education; Research • *Issues:* Deforestation; Sustainable Development; Waste Management/Recycling; Water Quality • *Meetings:* CITES

● Organization

In 1971, the Arab Center for the Studies of Arid Zones and Dry Lands (ACSAD) was founded as an intergovernmental research agency by the Arab League comprising Algeria, Egypt, Iraq, Jordan, Kuwait, Lebanon, Libya, Mauritania, Morocco, Oman, Palestine, Qatar, Saudi Arabia, Sudan, Syria, and United Arab Emirates (UAE).

In addition to fostering regional research regarding "arid zones such as water resources, soils, plants, and animal production," ACSAD trains Arab scientists, exchanges knowledge among Arab nations, and cooperates with other Arab and international organizations. A country representative from each participating country serves on the Administration Board, with the director general as the executive authority.

ACSAD has eight research branches: Water Resources; Soil; Plant; Livestock; Climatologic; Social, Economic, and Statistical Studies; Scientific Documentation and Publications and Regional and International Cooperation; and Administrative and Financial Affairs.

With arid land "comprising the greater part of the agricultural resources," ACSAD aims to strengthen joint endeavors at the regional level; "develop Arab capability in the areas of science, technology, and self-sufficient productivity; establish a comprehensive database; conserve agricultural resources, thereby "preventing desertification; and determining the causes and best means of treating deterioration of land, water, plant and animal resources." ACSAD carries out its mission, while developing and applying modern technology and training and developing scientists and the workforce.

▲ Projects and People

Research in the Water Resources Studies is applied to "geophysical and other groundwater investigation methods, hydrogeological mapping and water research cartography, and other methods of water resource studies including remote sensing and other ways of water monitoring, development techniques and techniques for rational management including mathematical models."

Using hydrogeographical and geophysical investigations, the Water Resource Division appraises and surveys available water supplies. The division also determines the quantity and quality of these water resources and how the water should be utilized. The Division utilizes hydrological and hydrogeological mapping, such as a project detailing the water resources of the entire Arab region. In other cartographical studies, the Division designs criteria and methodology for map representation.

The Water Resource Division also processes information and documentation in ACSAD's Water Resources Databank and the Water Resource Documentation unit. The Division is also preparing an inventory of Arab water technology and uses at regional levels, with technical and economic evaluation of these technologies. Projects also in progress include programs for water resource legislation to help realize ACSAD's principles for the management of water resources and the environment. Other programs help improve the efficiency of water usage by means of maintenance and irrigation systems and for the standardization of terminology used in water resource mapping. A major project is the completion of the water resources survey on the Hamad Basin, involving Saudi Arabia, Iraq, Jordan, and Syria.

In the Soil Studies Division, an Arab Soil Resources Map is being established to encompass all land resources in the Arab region. To be universally understood, the map will use "uniform concepts and terminology" and will be "an essential tool in studies on land uses and in formulating plans and programs dealing with the deterioration of land resources and desertification in the Arab World."

Other development projects include an assessment of economical uses of irrigation water and water requirements of principal crops to determine the best means of distributing water in irrigation systems. Researching ways to increase crop production and to improve soil quality, the Division investigates fertilizers and tests saline soil in order to reclaim, manage, and improve its condition. The Division is also analyzing the use of saline water in irrigation. It has determined "the water and fertilizer requirements for crops of wheat, cotton, sugar beets, and clover.

The Plant Studies Division maintains a cereals program, involving the "breeding and selection of drought and saline tolerant variety of wheat and barley," which are being tested throughout the Arab countries in differing environmental climates. The agricultural rotation program, determining the ideal sequence for crop rotation, involves the selecting and breeding of drought-tolerant, quick-yielding varieties of legumes and sorghum. Favorable varieties of chick peas and lentils have been developed. Research is also underway on millet.

A project in progress to improve the varieties of fruit trees in rainfed areas and to find a suitable way to grow these trees in arid areas. The Division also conducts range studies including "environmental studies on range plants; studies on salt-indicator plants in arid and semi-arid regions; determining the requirements, efficiency, and nutritional value of range plants; standardizing range measurement criteria; and establishing a genetic range-plant reservation and central Arab Herbarium." A book was compiled on standardizing methods of forage and range-plant classification; a second book on "methodologies of chemical analyses and nutritional evaluations" of these resources is being prepared.

Clones of high-level producing pistachio nut species are being are being surveyed in genetic, grafting, and pollination studies. Highly productive walnuts and olives are also being selected.

Aiming to expand the animal population in arid zones, the Livestock Studies Division is compiling abstracts of Arab sheep species and other livestock to assess their distribution and vulnerability under different environmental conditions. The Division works genetically and nutritionally to improve sheep, goats, and camels for more efficient production. The Awassi sheep, found in Arab countries east of the Mediterranean, is being bred to increase its production of milk, wool, and meat. The "first volumes of an encyclopedia on livestock resources in the Arab World" have been completed.

The Agro-Climatic Studies Division conducts a "comprehensive program for agricultural observation in rainfall regions, ways of benefitting from climactic observation data from the agricultural standpoint, and studies on agricultural rotation in rainfall regions."

ACSAD's Statistics Division accumulates and organizes such data to create a "unified, scientific survey of agriculture in Arab countries" and to publish an annual journal of agricultural statistics.

Dr. Gilani M. Abdelgawad, Soil Sciences Division director and executive committee member of the International Center of Inorganic Fertilizer, has lectured—at the invitation of Food and Agriculture Organization (FAO)—on "treated water for agriculture use"; was a member of study teams on agriculture development in Algeria and Mauritania, along the Senegal River; and consulted on Libyan national agriculture studies, surveying for grazing and forestation. Recently, he has published papers on heavy metal concentration in soils as well as grains and plants irrigated with sewage treated water. He is an editor of the Ninth World Fertilizer Congress proceedings on *Fight Against Hunger Through Improved Plant Nutrition*, three volumes (1986).

Dr. Nabil Rofail, Water Resource Division deputy director, works to evaluate and manage groundwater resources by designing mathematical models, which "represent various hydraulic and hydrogeologic conditions." Used in varying regions to "define hydraulic parameters and the water balance," his models are helping to create comprehensive plans to manage and develop water resources.

Rofail has supervised programs throughout the Middle East, including projects to manage water resources in Western Noubareya, Egypt; Syria; and UAE. He also helped to prepare water resource databanks unifying the Gulf states and the Arabian Peninsula, and form national databanks throughout the Arab region. Rofail has served as a consultant to the World Meteorological Organization (WMO) in the field of groundwater networks; he also has supervised training courses for Arab specialists and engineers sponsored by ACSAD, UNESCO, United Nations Development Programme (UNDP), and WMO, among others, on solving hydrological problems with mathematical modeling.

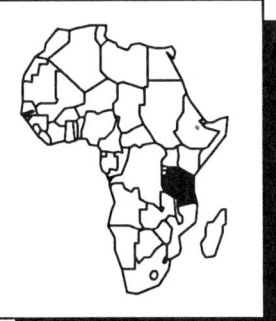

Ninety percent of the people in this East African country earn a living directly from the soil, and the dense population and traditional farming systems (characterized by overgrazing and bush fires) have endangered the regenerative capacity of soils, vegetation, forests, and water.

Tanzania is determined to protect its natural resources, however, and protected areas make up a quarter of the land. Yet forests are being reduced at unsustainable rates—by 55 percent in the Kilimanjaro region and by 30 percent in the Mwanza and Mara regions due to commercial timber extraction and firewood gathering. Industrial pollution, marine pollution, and water contamination have resulted from rapid urban growth. The country's world-famous wildlife is threatened by poaching, shrinking forests, and population pressures on fragile ecosystems.

🏛 *Government*

Hado Soil Conservation Project

P.O. Box 144
Kondoa, Dodoma, Tanzania

Contact: **Alfred Cyril Mbegu,** Project Manager
Phone: (255) 61 102

Activities: Development; Education; Law Enforcement; Political/Legislative; Research
Issues: Deforestation; Energy; Health; Sustainable Development; Transportation • *Meetings:* International Conference on Sustainable Land Management; ISCO

■ Resources

The **Hado Soil Conservation Project,** with a staff of 100, seeks new technologies and outreach means for rapid rural appraisal methods, agroforestry in semiarid areas, and mass communication workshops.

Moyowosi Game Reserve Project

P.O. Box 77
Kibondo, Tanzania

Contact: **Issai Swai,** Ecology Project Manager

Activities: Law Enforcement • *Issues:* Sustainable Development; Wildlife Management
Meetings: CITES

▲ Projects and People

The **Moyowosi Game Reserve Project,** with a staff of 60, is concerned with sustainable conservation and utilization of wildlife through **tourist hunting.** Issai Swai belongs to the Species Survival Commission (IUCN/SSC) Antelope Group.

■ Resources

Available publications are *Wildlife Conservation Status in Zanzibar* and *The Status of Antelopes in Tanzania, Antelopes Global Survey and Regional Action Plan, Part I, East and Northeast Africa,* compiled by **Rod East.** Radio telephones and walkie talkies are needed for patrols.

National Environmental Management Council (NEMC)
P.O. Box 63154
Dar es Salaam, Tanzania

Contact: **Bernard Lutta Mwombeki Bakobi,**
        Principal Environmental Education Officer
        Buzika N. Muheto, Principal Natural
        Resources Officer
Phone: (255) 51 34603

Activities: Development; Documentation; Education; Law Enforcement; Research • *Issues:* Air Quality/Emission Control; Biodiversity/Species Preservation; Deforestation; Energy; Global Warming; Health; Pollution; Sustainable Development; Waste Management/Recycling; Water Quality

● Organization
A fledgling government organization, the **National Environmental Management Council (NEMC)** aspires to create a united global effort to help combat environmental hazards. NEMC feels that this can be achieved if politicians throughout the world can be made to realize that "environmental issues have no boundary," and that international forums can help to educate the public to cooperate in order to constrain environmental problems.

Involved with many global environmental issues, NEMC is a member of the Southern African Sub-Regional Environmental Group (SASREG), East African Sub-Regional Environmental Group (EASREG), UNESCO Man and the Biosphere (MAB), INFOTERRA, and a united effort against desertification in the Sahelean region.

▲ Projects and People
Projects aim to **control beach erosion, disposal of hazardous waste**—including waste oil—and the **storage of agrochemicals.**

Monitoring programs supervise and regulate **urban air quality, desertification,** and the conflict between **mining** and the environment.

NEMC introduced **environmental education** into the school curricula as well as general programs for the public. Utilizing local knowledge in solving environmental problems, NEMC is forming **research teams** to analyze and propose possible answers. Groups include a **Documentation and Geographic Information System (GIS) Center** and teams on **water quality, industrial environmental development,** and **marine research.** NEMC is also contributing to the **National Conservation Strategy.**

■ Resources
Understaffed and short of equipment and funds, NEMC needs help in training various environmental workers in various disciplines. Financial assistance will permit NEMC to continue its work in environmental education, launch pilot projects, research pollution and desertification, and produce scientific and educational publications.

 *NGO*

Wildlife Conservation Society of Tanzania
P.O. Box 70919
Dar es Salaam, Tanzania

Contact: **Paul Y. Nnyiti,** Conservation Officer
Phone: (255) 51 33592

Activities: Education • *Issues:* Biodiversity/Species Preservation; Sustainable Development • *Meetings:* CITES

● Organization
The **Wildlife Conservation Society of Tanzania** is a nonprofit organization seeking to preserve the country's diverse wildlife and focusing its projects on restoring the environment in the wake of human destruction.

Due to overexploitation in Tanzania, several species of animals are on the verge of extinction. In the waters of Kunduchi, Tanzania, **turtles** once bred and nested. Now, "they are caught and drowned in fishermen's nets. They are taken for their meat. Their eggs are excavated and eaten. They suffer from polluted water. The shell is used to make a variety of combs and ornaments." The sea turtles have also been killed by eating plastic bags. Now, very few turtles live in the waters of Tanzania.

The bird trade takes an estimated 96,000 of the native **Fischer's lovebird** and the yellow-collared lovebird from the wild every year; 60 percent of these birds are due to die in transit, from disease, and other problems in a caged environment.

▲ Projects and People
The Society educates the community at large with the production of the monthly newsletter *MIOMBO*, containing articles which describe animals of Tanzania, often with illustrations, photographs, maps, and charts giving more details. These articles introduce new **educational campaigns.**

In an article highlighting the **green sea turtle,** *MIOMBO* initiated an educational drive on the turtles' plight, giving tips on how not to support the turtle shell market or litter the water with dangerous materials. Messages are targeted to varying audiences, as fishermen and local coastal residents.

With 30 species of chameleons in Tanzania, and 15 of these endemic, the Society is concerned with their perpetuation. The specialized habitats, such as rainforests, of these shape- and color-changing reptiles are constantly being threatened. Because they do not thrive well in captivity, translocation is another hazard.

Among other concerns is the devastation of the **Pande Forest Reserve** outside of Dar es Salaam, which is in the process of being cut down. Also, in conjunction with the International Primate Protection League, the Society is requesting Tanzanians to write their president, insisting that the **sale of chimpanzees** be stopped, as according to CITES (Convention on International Trade in Endangered Species of Wild Fauna and Flora).

The Society works under the guidelines of the African Wildlife Foundation (AWF) and the World Wide Fund for Nature (WWF) in establishing a project to "conserve Tanzania's rich wildlife resource by promoting sustainable economic development of the country's wildlife sector as part of Tanzania's economic recovery program." The plan has two objectives: To enhance the **planning** and development of wildlife management in the Wildlife Division of the Tanzanian government, and to facilitate sound development in the **wildlife viewing** industry, **tourist hunting industry,** and **sustainable hunting.** The U.S. Agency for International Development (USAID) is funding the three-year project.

To raise needed funds for conservation efforts and wildlife research, the Society has organized a **Charity Walk.** To celebrate **Gombe 30 Wildlife Week** and a 30-year commitment to wildlife through the Arusha Manifesto, in conjunction with the Jane Goodhall Institute and the Wildlife Division of the Tanzanian government, the Society sponsored train trips through wildlife reserves, exhibitions in the National Museum of Tanzania in Dar es Salaam, and radio programs.

■ Resources

MIOMBO is printed in both English and Swahili. The Society seeks to strengthen the public's conservation awareness.

Universities

University of Dar es Salaam, Wildlife Ecology Programme ⊕
P.O. Box 35064
Dar es Salaam, Tanzania

Contact: **Dr. R.B.M. Senzota,** Senior Lecturer

Activities: Education; Research • *Issues:* Biodiversity/Species Preservation; Deforestation

▲ Projects and People

Dr. R.B.M. Senzota is senior lecturer and coordinator of wildlife ecology at the University of Dar es Salaam. He has supervised research on lion-prey relationships in Lake Manyara National Park, ecology in Serengeti National Park, and ecology of red colobus monkeys in Jozani Forest Reserve and has conducted studies with rodents, such as grass and house rats and plains gerbils. Related interests include "effect of intercropping sorghum and cowpea on their lepidopteran stem and pod borer population buildup"; tropical pests; "dry season food refuge of Thomson's gazelles," rodent-ungulate resource partitioning, and regulation of Serengeti zebra, wildebeest, and gazelle. Dr. Senzota serves on the National Pastures and Forages Research Coordinating Committee, Ministry of Agriculture and Livestock Development.

■ Resources

Dr. Senzota seeks equipment for research and teaching ecology as well as journals and other reference books on ecology.

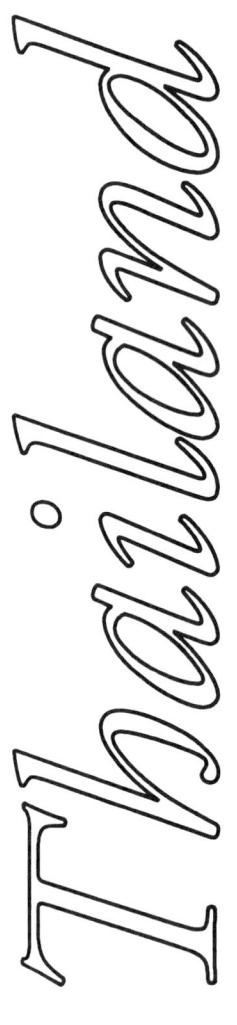

Thailand

Located in the middle of the Indochinese Peninsula, Thailand straddles two mountain systems, with large alluvial plains in between. Southern Thailand, part of the Malay Peninsula, borders the Gulf of Thailand and the Andaman Sea.

Natural resources include timber, rubber, and natural gas, and the country is rich in flora and fauna, with as many as 15,000 flowering plant species, over 900 bird species, and about 265 species of mammals. Nearly half the population lives in rural areas, and nearly half the land is cultivated. The country is rapidly urbanizing, bringing problems of industrial pollution, air pollution, inadequate sewage systems, and traffic congestion. Other environmental problems are deforestation, primarily due to logging, which also results in wildlife habitat eradication; water scarcity, due to forest watershed destruction and poor irrigation system management; mangrove destruction; and overfishing.

🏛 *Government*

Watershed Management Division
Royal Forest Department
61 Phahol Yothin Road
Bangkok 10900, Thailand

Contact: **Anan Nalampoon,** Chief, Research Section
 Phone: (66) 2 579 7587 • *Fax:* (66) 2 579 2811

Activities: Research • *Issues:* Biodiversity/Species Preservation; Catchment Area Rehabilitation; Deforestation; Water Quality

▲ Projects and People

Anan Nalampoon, former chief of the ASEAN (Association of South East Asian Nations) Forestry Cooperation Liaison Office, now is in charge of 18 field research units in the **Watershed Management Division, Royal Forest Department.** In addition to developing and monitoring these activities, he gives technical advice regarding forestry and watershed management and cooperates with international agencies on such matters as well as on nature conservation.

🔥 *Private*

Samutprakan Crocodile Farm and Zoo Company, Inc.
555 Mou 7 Taiban Road
Samutprakan 10280, Thailand

Contact: **Charoon Youngprapakorn,** General Manager
 Phone: (66) 2 387 1166 • *Fax:* (66) 2 387 0060

Activities: Development; Education; Research • *Issues:* Biodiversity/Species Preservation; Sustainable Development • *Meetings:* CITES

▲ Projects and People

The **Crocodile Farm** has a staff of 300. **Charoon Youngprapakorn** belongs to the Species Survival Commission (IUCN/SSC) Crocodile Group.

🎓 *Universities and Institutes*

Asian Institute of Technology (AIT)
GPO Box 2754
10501 Bangkok, Thailand

Contact: **Robert H.B. Excell, Ph.D.,** Chairman,
Division of Energy Technology
Phone: (66) 2 516 0110 • *Fax:* (66) 2 516 2126

Activities: Education; Research • *Issues:* Energy; Water Quality

● Organization
"The Asian Institute of Technology (AIT) is an autonomous international post-graduate technological institute.... The primary objective is to help serve the technological requirements of the people of Asia by providing advanced education in engineering, science, and allied fields," according to its literature. Some 800 students, mainly from Asia, and 200 faculty and international staff are at the 160-hectare campus.

AIT has academic programs which lead to doctoral and master's degrees, regional outreach activities "which contribute to the continuing career development of practicing professionals," and research programs.

Academic programs are offered in agriculture and food engineering, agriculture land and water development, computer science, environmental engineering, energy technology, geotechnical and transportation engineering, human settlements development, industrial engineering and management, natural resources development and management, structural engineering and construction, technology management, telecommunications, and water resources engineering.

Research facilities include environmental engineering laboratories, a regional research and development center, and a regional environmental center. An Energy Park entails "more than 60 devices using renewable sources of energy: solar, thermal, and photovoltaic, and biomass." Here, commercially available products are tested under tropical climatic and environmental conditions.

Founded in 1979, the Division of Energy Technology trains "energy specialists in response to regional needs." Three fields of study are Renewable Sources of Energy, Energy Planning and Policy, and Rational Use of Energy. The Renewable Resource program trains "graduates who can work in research and development departments of industries, energy utilities, and national or international institutions." Research is on "solar thermal and photovoltaic processes, biogas plants, and biomass conversion technology." Students of Energy Planning and Policy are prepared "for positions in economics or planning departments of energy ministries, energy industries, utilities or consulting firms. It comprises fundamental techniques in global and sectoral energy planning, economic appraisal of energy projects, energy policy analysis, and electricity economics." The Rational Use of Energy coursework, which trains students "in the theory and practice of efficient energy use," focuses on research in cogeneration potential assessment and energy management in buildings.

The Regional Energy Resources Information Center (RERIC) reports that it "collects and repackages information in the fields of energy planning, energy conservation, solar, wind and biomass energy, and small-scale hydropower for dissemination to target audiences."

■ Resources
The AIT library contains more than 180,000 volumes and subscribes to more than 1,000 journals in science, engineering, technology, and management.

RERIC publishes the quarterly *RERIC News*, *RERIC International Energy Journal*, annual *RERIC Holdings List*, annual *Abstracts of AIT Reports and Publications on Energy*, and the *RERIC Membership Directory*. Proceedings of AIT conferences and workshops are also published. Brochures describes RERIC computerized services and the master program in the Division of Energy Technology.

Environmental Sanitation Information Center (ENSIC) ⊕
Asian Institute of Technology
P.O. Box 2754
10501 Bangkok, Thailand

Contact: **Marta Miyashiro,** Senior Information Scientist
Phone: (66) 2 524 5863 • *Fax:* (66) 2 516 2126

Activities: Education; Information Services; Research • *Issues:* Global Warming; Waste Management/Recycling; Water Quality

● Organization
Founded in 1978 at the Asian Institute of Technology (AIT), Thailand, the Environmental Sanitation Information Center (ENSIC) is a joint endeavor of the Environmental Engineering Division and the Library and Regional Documentation Center (LRDC). ENSIC initially received funds from the government of Australia and the International Development Research Centre (IDRC) of Canada. It currently receives funds from the Canadian International Development Agency (CIDA), Asian Development Bank, AIT, and from earnings through memberships and services. Among its 450 members are "engineers, technical workers, administrators, educators and other professionals working in the fields of water supply, waste management and reuse, and sanitation."

▲ Projects and People
Its mission in urban and rural sanitation and water supply and use is "to collect, repackage and disseminate information; make information and technology available to users; provide assistance, facilities, and training in establishing local information centers to institutions within the ENSIC regional network; and collaborate with similar information centers worldwide."

ENSIC has created a computerized database, updated from over 300 journals, plus abstracts, indexes, and research reports. Information is also stored on microfiche. Library and Regional Documentation Center director H.A. Vespry notes that ENSIC members are entitled to three free database searches annually, along with "the first 50 citations from each search also free."

With technology transfer a main activity, ENSIC produces publications and information services; runs short training courses, workshops, seminars, and conferences; and offers consultancy services.

With a technical assistance grant from the Asian Development Bank, ENSIC is developing ENSICNETS—focal information centers in China, Indonesia, Nepal, Pakistan, Philippines, and Vietnam. ENSIC is the center for "exchanging information among the members of the network providing technical and information support for water supply and sanitation information development." This project will also allow ENSIC to gather and translate information that until now has not been available in English.

■ Resources

ENSIC publishes the quarterly *ENFO Newsletter*, a semiannual journal, environmental sanitation abstracts and reviews, and occasional papers. *Solid Aspects of Solid Waste Recovery in Asian Cities* is a recent review. A publications list with order form is available.

School of Energy and Materials ⊕
King Mongkuts Institute of Technology Thonburi (KMITT) ⊕
Bangmod, Rasburana
Bangkok 10140, Thailand

Contact: **Prida Wibulswas, Ph.D.**, Professor and Environmental Programme Chairman
Phone: (66) 2 427 8094 • *Fax:* (66) 2 427 8077

Activities: Development; Education; Research • *Issues:* Air Quality/Emission Control; Energy; Global Warming; Transportation; Waste Management/Recycling; Water Quality *Meetings:* Asia Energy; ESCAP

● Organization

The graduate School of Energy and Materials at King Mongkuts Institute of Technology Thonburi (KMITT) was established in 1977 to help Thailand develop efficient energy sources that are alternatives to imported oil, on which this country relies.

Since the late 1980s, as Thailand's economy and industrial sector have undergone significant growth and changes, "new sets of needs and problems arise," according to the School, which has also grown to meet such challenges. The five School divisions are Energy Technology, Energy Management Technology, Biotechnology, Materials Technology, and Environmental Technology.

In addition to the graduate degree programs, the School offers short courses, training, and seminars for industry and other organizations.

▲ Projects and People

With Prof. **Prida Wibulswas** as chairman, the Division of Environmental Technology specializes in research in air, water, and noise pollution; waste treatment and utilization; and environmental impact of energy systems. Selected projects include: "SO_2 and CO_2 emission-form combustion of fuel; gaseous pollution from diesel engines; methane emission from agriculture areas; utilization of water hyacinth to improve the quality of domestic effluents; role of microorganism action on Bangkok metropolitan compost; and absorption of toxic gases by activated carbon."

A fellow of the Royal Institute of Thailand, Dr. Wibulswas is also a member of the National Research Council of Thailand, National Committee on Energy Conservation and Alternative Sources, and Engineering Profession Board, as well as chairman of the Thai Subcommittee on Non-Conventional Energy Research, ASEAN (Association of South East Asian Nations), among other groups. **Dr. Pojanie Khummongkol**, a lecturer in the Environmental Technology Division, specializes in "control and design of air pollution equipments; and thermo-chemical reaction of solid fuel by pyrolysis, gasification, and combustion techniques." **Dr. Sirintornthep Towprayoon**, assistant professor, is concerned with "utilization of solid waste, microbial degradation of waste, and microorganisms effected to the ecosystem."

■ Resources

The School seeks information on "clean coal technology" for industrial applications and the development of a "continuing education centre" for training and seminars. Research publications of the ASEAN Subcommittee on Non-Conventional Energy Research are available from KMITT.

Center for Conservation Biology
Department of Biology
Mahidol University
Rama 6 Road
10400 Bangkok, Thailand

Contact: **Warren Y. Brockelman, Ph.D.**, Director, Center for Conservation Biology, Professor, Mahidol University
Phone: (66) 2 246 1358, ext. 431
Fax: (66) 2 247 7051

Activities: Education; Research • *Issues:* Biodiversity/Species Preservation

▲ Projects and People

Dr. Warren Y. Brockelman directs the Center for Conservation Biology, a subsidiary to the Department of Biology, Mahidol University. An American, Brockelman has spent much of the last 24 years in Thailand, and has been a visiting professor at Mahidol University since 1973.

A major project is the Conservation Data Center, which is supported by the World Wide Fund for Nature (WWF). The Center for Conservation Biology collaborates with the Office of the National Environment Board, Royal Forest Department, and Thailand Institute of Scientific and Technological Research in its efforts to collect and store information necessary for species conservation.

The Data Center comprises "files and a computerized database, established in order to compile reliable and comprehensive information on the status of wildlife resources throughout Thailand. . . . Its function is to serve as a coordinated data bank for interested users, amateur and professional; governmental and non-governmental; for the conservation or development agency, the biological researcher, and the wildlife tourist," according to University literature.

Data are currently available for all bird and mammal species in Thailand. This is being expanded to other groups of terrestrial vertebrates. The information available includes a habitat summary, a biounit summary, a reserve summary, and a species summary. The computerized bibliography contains references "pertaining to species, ecology, and protected areas in Thailand . . . [and] national and international legal protected status of species are also stored." The Center welcomes "reliable records from anyone who studies wildlife in Thailand for research or recreation. Special forms have been printed to aid compilers of data."

Along with wildlife survey and inventory, Dr. Brockelman's research interests include "social structure, communication, and interspecific relations of the gibbons *Hylobates lar* and *H. pileatus* in

Khao Yai National Park," with the support of Wildlife Conservation International (WCI); human transmission of the liver fluke (*Opisthorchis viverrini*), in collaboration with **Dr. Suchart S. Upatham**, and of other intestinal parasites; and "developing a method for rapid assessment of forest cover and height." A United Nations Environment Programme (UNEP) consultant, he also belongs to the Species Survival Commission (IUCN/SSC) Cat, Chiroptera, and Primate groups.

■ Resources

Publications include *Birds of Khao Yai National Park*, *Birds of Doi Inthanon National Park*, and *Birds of Khao Sam Roi Yot National Park*. The Data Center provides information on its use.

Scholarship funds are needed for Master of Science candidates in conservation and wildlife biology.

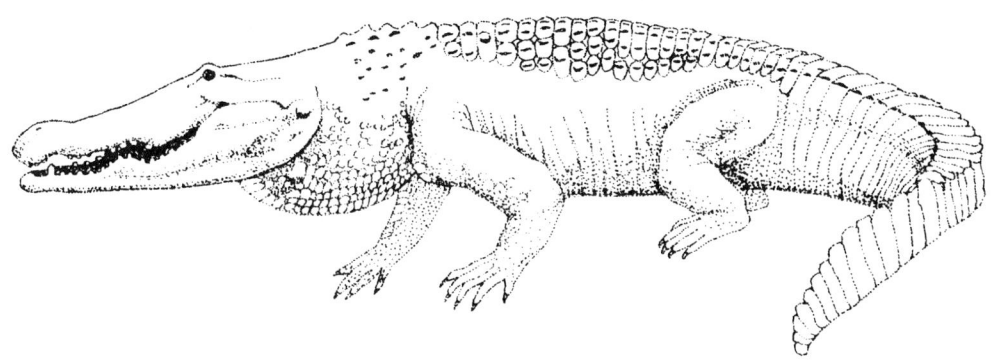

Trinidad & Tobago

Trinidad and Tobago, near the coast of Venezuela, is the southernmost island country of the Lesser Antilles. In the 1970s, the country achieved the highest gross national product per capita after the United States and Canada through the export of oil, natural gas, and oil-based fertilizers. In the 1980s, however, the country went into debt as oil prices fell. Human services remain at a high level, however, and the country is attempting to diversify its economy. Trinidad and Tobago boasts one of the few examples of sustainable forestry products—the Arena Reserve.

A combination of poverty and affluence have contributed to the country's environmental problems—through urban concentration, agricultural modernization, and industrial growth. Some of the greatest concerns are deforestation, soil loss and land degradation, pollution and mismanagement of water resources and sewage systems, and oil pollution.

🏛 *Government*

Institute of Marine Affairs (IMA)
Hill Top Lane, P.O. Box 3160
Carenage Post Office
Chaquaramas, Trinidad

Contact: Director
Phone: 1-809-634-4291 • *Fax:* 1-809-634-4433

Activities: Education • *Issues:* Water Quality

▲ Projects and People

The **Institute of Marine Affairs (IMA)** is a "statutory body which is working to promote and encourage a broader understanding of the marine environment." It makes available information and expertise on problems of the marine and coastal environment to the governments of the Caribbean region.

🍁 *NGO*

Zoological Society of Trinidad and Tobago
Emperor Valley Zoo
Port of Spain, Trinidad and Tobago

Contact: **Hans E.A. Boos**, Curator
Phone: 1-809-622-3530

Activities: Education; Research • *Issues:* Biodiversity/Species Preservation

● Organization

Founded in 1947 by the Field Naturalists Club, the **Zoological Society of Trinidad and Tobago** was formed as the guide and organizer of the Emperor Valley Zoo, which is named for the emperor, or morpho, butterfly that once lived in the valley. The Zoo opened its doors to the public at the end of 1952 with an emphasis on local flora and fauna and continues today in this spirit. The Zoo, with a staff of 45, is located in the Botanical Gardens north of the Queen's Park Savannah in Port of Spain, Trinidad, on 8 acres and welcomes a quarter-million people a year.

The Zoo not only displays local and foreign animals, it breeds them as well. The Zoo has a pride of **African lions**, as well as the **local ocelot**, or tiger cat, and **jaguars**, **Bengal tigers**, **mountain lions**, **peccaries**, and **red brocket deer**. The Zoo is proud of its large primate collection, with **mandrills**, **chimpanzees**, and a variety of African, Asian, and South American **monkeys**.

The "twin islands" Trinidad and Tobago are known as a bird watcher's paradise, and the Zoo has a large collection of **indigenous and foreign birds** on display and actively breeding. Visitors watch for the national birds, **scarlet ibis and cocrico**. A **large reptile house** features venomous **vipers, tortoises and turtles, iguanas, caimans, and crocodiles.** About 30 tanks and some outdoor ponds are stocked with a variety of **fish and marine life**, such as **coscorob, cascadura, guabine, and mama-teta.**

The Emperor Valley Zoo is run with financial support provided in part by the Government of Trinidad and Tobago. The Zoo has begun an education program that will allow visitors to leave the Zoo more informed about the animals.

With the Emperor Valley Zoo since 1973, **Hans E.A. Boos** worked for the Taronga Park Zoo, in Sydney, Australia. Subjects of his writings include geckos, skinks, legless lizards, dangerous snakes, and fauna from "agouti to zandoli." He belongs to the Societas Europaea Herpetologica, American Association of Zoological Parks and Aquariums (AAZPA), and Audubon Society, among other groups.

Tunisia is a small Arab nation on the southern coast of the Mediterranean, bordered by Libya and Algeria. The land along the coast is fertile, while the land to the south is desert. Tunisia's literacy rate and public health services are high compared to other North African countries.

The major environmental problem, intensified by a rapidly growing population, is an eroding landbase—as much as 76 percent may be threatened—caused by farmland expansion, overgrazing, and depletion of vegetation. Water resources are also being depleted by the use of groundwater for irrigated areas and for industry and household needs. Untreated urban sewage is another concern, as it contaminates water supplies and contributes to contamination of the Mediterranean. The government has taken steps to protect the country's wide variety of ecosystems through conservation farming and managed rangeland, among other programs.

 NGO

L'Association Tunisienne pour la Protection de la Nature et de l'Environnement (ATPNE)
Tunisian Association for the Protection of Nature and the Environment
12 rue Tantaoui El Jawhari
1005 El Omrane, Tunis, Tunisia

Contact: **Mohamed Ali Abrougui, D.V.M.**, President
Phone: (216) 1 288 141

Activities: Development; Education; Law Enforcement; Political/Legislative; Research
Issues: Air Quality/Emission Control; Deforestation; Energy; Global Warming; Health; Population Planning; Sustainable Development; Water Quality

● Organization

Regarding its mission, which has resulted in positive government measures and public recognition, **L'Association Tunisienne pour la Protection de la Nature et de l'Environnement (ATPNE)** **(Tunisian Association for the Protection of Nature and the Environment)** writes, "The road is still long and the challenges are important. Each one of us should participate and take up these challenges to associate with the action of protection."

Since 1971, ATPNE has been defending the environment on political and public levels, seeing its aim as a "'rendez-vous' with the history of humanity and of the entire earth; a 'rendez-vous' that must not be missed." The Association, with 4,000 members, was one of the first to be a part of the Tunisian ecological movement, which is growing stronger. ATPNE is a member of Amis de la Terre (Friends of the Earth) and works closely with the Association of Tunisian Journalists (AJT).

As Dr. Mohamed Ali Abrougui, ATPNE president, notes in his organization's *Internal Bulletin*, "Our ancestors always lived in symbiosis with their environment until the last centuries. The first environmental deterioration could perhaps be dated from the invasion by tribes of Beni Helal who ravaged the vegetation. However, the most serious attack on the ecosystems and on environment did not really begin until the last century when the colonials took over Tunisian lands, provoking not only social disorder due to lost lands but also causing a rural exodus.

"At the time of independence, the policy of development adopted was based on a large consumption of natural resources. The damage caused to nature at that time triggered the beginnings of the ecological movement in Tunisia."

Today, the Association is concerned with "the fight against pollution; the protection of nature, soil, and our natural heritage; and water and non-regenerating resources management." According to its charter, ATPNE is involved in "defending conservation of vegetation and fauna as well as protection of natural and archeological sites; defending the preservation of natural resources and the good management of these resources; opposing all projects endangering conservation and environment; fighting against all forms of pollution; awareness of the population about questions on environment and forming partners concerned with protection and conservation; helping to insure environmental education at all school and university levels; organizing training courses to allow a better comprehension of environmental questions and management of natural resources; and collaborating with officials and organizations who work for protection of environment and for development of natural resources."

▲ Projects and People

The ATPNE has projects based in four areas: education, public awareness campaign, environmental impact assessment (EIA) studies, and the realization of development projects closely related to environmental protection.

When the Tunisian Ministry of Education undertook reforms in 1989, it expressed to ATPNE its commitment to environmental education. A special commission was created to draft an environmental education curriculum to be used in elementary and secondary schools. The commission drew up themes to be used in the curriculum, but uncovered some problems in the final drafting of the curriculum. It found, for example, that teachers were poorly trained on the environment and therefore could not adequately impart messages to the students. Also, there were not enough proper materials for teaching including books, slides, films, and brochures. To alleviate these problems, ATPNE began organizing "periodic theoretical and practical training sessions of the teachers; preparing handbooks and audio-visuals; and providing didactic support that allows children to assimilate the information presented in class."

"When children take over the environment, everything is transformed: greenery, gardens, cleanliness, decorated classrooms. . . . The adult of tomorrow teaches the adult of today," according to ATPNE literature, which further describes, "Before planting a tree, each child sings about the event. From now on the tree will become a living being that must be cherished and respected."

Such actions are based on the Association's Strategy for Environmental Education—involving a national network, United Nations input, easy-to-apply ecological concepts, and changes in attitudes and behavior among school children, professionals, and ordinary citizens—to ensure sustainable development.

Likewise, the awareness campaign is founded on the belief that the best way to conserve natural resources is to educate the public. The campaign targets all sectors of Tunisian people including political and administrative officers, industrial and farm workers, students, the media, and the rest of the general public. ATPNE has plans to organize seminars on precise themes and to create a periodical for wide distribution to targeted groups and enterprise leaders.

The third type of project is environmental impact assessment (EIA). Among such ATPNE studies is an examination of how the environment is affected by the tourism industry. This study incorporated an analysis of wastewater treatment, the degradation of the coastline and of Mediterranean forests, and the development of a plan to improve the coastline. Other EIA studies are on the impacts of tanneries on the neighboring environment, and of the textile industry and the related waste problems and consumption of scarce water.

ATPNE intends to create new conservation projects involving local people to make everyday life more environmentally beneficial, such as improved cattle raising, managing parks for threatened species, new uses of agricultural and agro-industrial byproducts, inclusion of rural women in local development projects, implementing model wastewater treatment stations, and beginning a national center for the environment.

Dr. Abrougui, who also is a cabinet environmental attaché to the Mayor of Tunis and former Director of Animal Production and Health, has developed breeding programs and policies for livestock, poultry, and bees; he has planned zoological and national parks in Tunisia, Algeria, and northern Libya. In 1992, he was scheduled to give talks both on the "survival of the child and its rights to education and health" organized by UNICEF with the Ministry of Children and Youth, Tunis, and on urban environment organized by the International Association of Mayors of French-speaking Towns, Niamey, Niger. He is a participant in numerous African and Mediterranean conferences concerning the environment, including those sponsored by the United Nations Environment Programme (UNEP), World Bank, and World Conservation Union (IUCN).

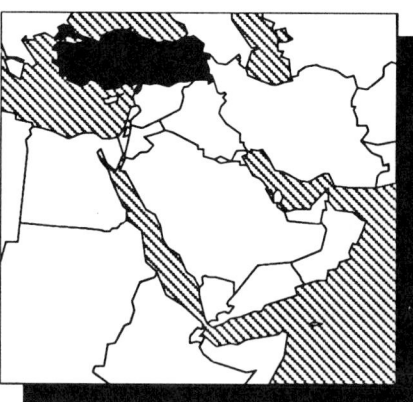

Located in west-central Asia, with a small portion in southeastern Europe, Turkey has an extensive coastline that fronts various international bodies of water, including the Mediterranean and Black seas. The land is characterized by narrow coastal plains and inland plateau that becomes increasingly mountainous. Mining—of coal, chromium, and copper, among other resources—is a chief economic activity, along with a growing industrial sector. Much of the population works on farms, and the country is generally self-sufficient in food. In recent years, however, there has been significant migration to urban areas, and the manufacturing and industrial sectors are growing. Earthquakes pose the most serious natural threat; environmental pressures resulting from modernization include air pollution, water pollution, pesticide use, traffic, and noise.

 NGO

Turkish Association for the Conservation of Nature and Natural Resources
Menekse Sokak 29/4 Kizilay
Ankara, Turkey

Contact: Ekrem Y. Demetçi, Ph.D.
Phone: (90) 4 125 1944 • *Fax:* (90) 4 117 9552

Activities: Education; Law Enforcement; Research • *Issues:* Air Quality/Emission Control; Biodiversity/Species Preservation; Deforestation; Energy; Global Warming; Health; Sustainable Development; Waste Management/Recycling; Water Quality

● Organization

Founded in 1955, the nonprofit **Turkish Association for the Conservation of Nature and Natural Resources** has some 2,500 volunteer members and is affiliated with the World Conservation Union (IUCN). Environmental education is a major aim, especially in reaching young people. To this end, the Association publishes literature; promotes trips, meetings, seminars, exhibitions, and films; encourages the **planting of trees** by making available seeds, saplings, and hedges; and emphasizes sustainable development policies.

The Association also encourages "de-desertification" and conversion of arable regions into fertile fields and productive pastures, preparing reports and proposals for government agencies and others regarding this goal, and conducts ecological research projects and studies.

Also examined are **national laws** relating to the nature conservation and environmental problems. Again, the Association makes proposals on policy to the relevant government ministries and examines and reports on their actions. Stimulating public **environmental awareness** on auto emissions, industrial and domestic waste, forest fires, habitat destruction, soil erosion, Black Sea and eastern Mediterranean coastal and other environmental problems in Turkey—the Association disseminates findings through press releases, radio, or television. To reach children at every level—preschool to universities—the Association holds poster and writing contests and prepares special publications.

Presently, the Association seeks financial support to start an **Environmental Education Center.** In the meantime, it sponsors research and projects, such as the **Environmental Education Project,** the **Wildlife Inventory Project,** and the **Inventory of Wetlands in Central Anatolia.** It also cooperates on projects with universities, government, and global groups, such as IUCN, United Nations Environment Programme (UNEP), Food and Agriculture Organization (FAO), International Council for Bird Preservation (ICBP), and International Waterfowl and Wetlands Research Bureau (IWRB). And it takes part in worldwide conferences and symposia, while sponsoring dozens of meetings, colloquia, panels, and exhibitions at home.

The Association awards the Honor Prize of the Conservation of Turkish Nature as well as other citations.

■ Resources

An *Introductory Bulletin* is prepared in Turkish and English.

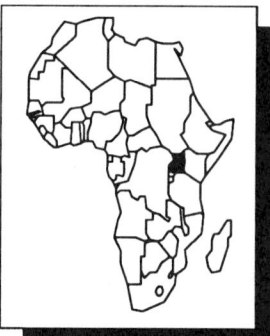

All of Uganda is situated on top of the east-central African plateau at an average height of 2,600 to 6,500 feet; it contains Lake Victoria, the source of the Nile River system. The country is 18 percent rivers, lakes, and wetlands, and contains extensive wildlife, including elephants, lions, giraffes, zebras, antelopes, and a rhinoceros population that was reduced almost to extinction in the political instability of the 1970s. Agricultural development, transport facilities, medical care, research, and training were also casualties of the period. The economy has been rebuilt through the export of coffee.

Because wetlands are a source of malarial mosquitoes, they are being drained indiscriminately and used in agriculture. Conflicts also have arisen over the little remaining forest, with the government attempting to protect these areas from agriculture and other development.

🏛 *Government*

ICRAF/Uganda AFRENA Project ⊕

P.O. Box 311
Kabale, Uganda

Contact: **John Okorio**, Agroforestry Scientist
Phone: (256) 360 Kabale (operator assistance)

Activities: Development; Research • *Issues:* Biodiversity/Species Preservation; Deforestation; Energy; Sustainable Development

● Organization

ICRAF, the International Council for Research in Agroforestry, is a research network for the highlands of Eastern and Central Africa and links the research of institutions in Kenya, Uganda, Rwanda, Burundi, and Ethiopia. Participants meet annually to review progress as well as to plan future research work. The AFRENA (Agroforestry Research Networks for Africa) Project, staff of 80, has been conducting agroforestry research in Uganda since 1987 with funding from the U.S. Agency for International Development (USAID).

▲ Projects and People

AFRENA began its work in Uganda by conducting "diagnostic and design land use surveys" to identify a number of potential technologies that could contribute to **farm income** through increased **farm production**. These surveys were similar to ones conducted in the other AFRENA nations (Kenya, Rwanda, Burundi, Ethiopia), with the understanding that the different information would be collected and shared by all nations.

AFRENA has been conducting screening trials in an effort to find the trees that best suit the needs of people and **agroforestry**. Several trees are under investigation, and the final choices will depend on factors such as growth rate, compatibility with certain kinds of farming, and timber production. Some trials are too recent to yield promising results, but several trees seem likely candidates. There are currently eight trials in progress at five different sites. In three trials, hedgerows are planted in combinations with trees. Several types of hedgerows have considerable potential for use as fodder and as barriers against soil erosion and water loss.

In 1990, over 200 farms benefited from agroforestry research, with hedges and trees introduced to the farms. **Women's groups** have proven to be very effective recipients of agroforestry concepts, with their demand for trees and agroforestry very high. Although the projects have had some problems, notably uncontrolled grazing by domestic animals and theft of seedlings, the project has promise.

AFRENA trains personnel through sponsoring studies at the ICRAF headquarters in Nairobi, Kenya, and by sending scientists to international meetings to present papers and exchange information on agroforestry.

With assistance from Ugandan television, AFRENA has produced television shows on agroforestry. Scientific staff give lectures and presentations to government agencies, NGOs, and individuals. The AFRENA technical coordinator is developing an appropriate curriculum at the university level for implementation at Makerere University of Kampala, Uganda. AFRENA also welcomes visitors to its testing and trial sites to witness progress on its various projects.

Scientist **John Okorio**, AFRENA team leader in Uganda, is also doing **root studies** with a grant from the International Foundation for Science of Stockholm, Sweden. Authored or co-author on scientific papers on agroforestry, including several for the *AFRENA Report Series*, he believes his program contributes to global cooperation "in conserving trees, checking soil erosion, and sustaining food production at a national and regional level."

■ Resources

Okorio, currently conducting root studies of various trees planted together with crops in agroforestry systems, seeks advice from those with experience in this field. He also needs equipment for nitrogen-fixation and mycorrhiza sites, and welcomes much needed financial support.

The *AFRENA Report Series* includes available publications on agroforestry and related topics.
(See also International Council for Research in Agroforestry [ICRAF] in the Kenya and Zambia NGO sections.)

Uganda National Parks
P.O. Box 3530
Kampala, Uganda

Contact: **Eric L. Edroma, Ph.D.**, Director
    Phone: (256) 41 256534 • *Fax:* (256) 41 241247

Activities: Development; Education; Law Enforcement; Political/Legislative; Research • *Issues:* Biodiversity/Species Preservation; Wildlife • *Meetings:* CITES; UNCED; World Parks Congress

■ Resources

Uganda National Park, with a staff of 500, seeks "equipment, tools, experience [for] proper management of wildlife" as well as training of wildlife biologists and resource managers. Its director, **Dr. Eric L. Edroma**, is a regional member of the Species Survival Commission (IUCN/SSC) and belongs to its Antelope Group.

Water Development Department
P.O. Box 19
Entebbe, Uganda

Contact: **John Karundu**, Head of Hydrogeology
        Section
    Phone: (256) 42 20863

Activities: Development; Research • *Issues:* Health; Sustainable Development; Water Quality

● Organization

John Karundu, Hydrogeology Section head, **Water Development Department**, writes of the need to strengthen links among environmental researchers. "Dissemination of information is very important," he writes. "In countries like mine, information facilities are limited and the percentage [of persons] aware, for example, of the depletion of the ozone layer is very small." He does point out that Uganda now has an Environmental Information Centre in its cabinet as well as a geographical information system (GIS) and a ministry that houses an environment department.

With support from Canada's International Development Research Centre (IDRC) and the University of Toronto, Karundu is the principal investigator of an ongoing **Hydrogeology Uganda project** through 1994 to **investigate groundwater** in mantle rock (or regolith), interaction of surface water and groundwater, and groundwater quality and recharge. "The project has a socioeconomic aspect," says Karundo, who expects that a countrywide groundwater management policy will evolve from it. The public is particularly concerned about "numerous boreholes being drilled . . . by the government and donor agencies," he writes. The research will determine if "water tables are falling and springs are dwindling." Groundwater models will also be developed for sustainable development.

■ Resources

The Department seeks new technologies in groundwater surveying and monitoring in crystalline rocks, help with modeling groundwater in hard rocks, and information about groundwater recharge in hard rocks, especially Africa. Available publications are *Preliminary Assessment of Water Resources in Uganda* and *Hydrogeology of Fractured Bedrock Systems in Southwest Uganda*.

 NGO

Impenetrable Forest Conservation Project ⊕
World Wildlife Fund (WWF)
P.O. Box 7487
Kampala, Uganda

Contact: **Thomas M. Butynski, Ph.D.**, Director/
        Wildlife Ecologist
    Fax: (256) 41 245597

Activities: Development; Law Enforcement; Research • *Issues:* Biodiversity/Species Preservation; Deforestation; Sustainable Development • *Meetings:* American Zoological Society; East/West Center; Pan African Ornithological Congress

● Organization

The **Impenetrable Forest Conservation Project** is sponsored primarily by the **World Wildlife Fund (WWF)** and U.S. Agency for International Development (USAID) with more than 10 other organizations, institutions, and agencies.

Objectives are to preserve the biological diversity and ecological well-being of Uganda's forests, enhance the environmental quality of life for Ugandans, establish teams who can conduct biological inventories and assessments in Uganda and throughout Africa, increase productivity and sustainability in rural areas near tropical rainforests, stimulate ecotourism, and train sorely needed biologists to undertake "effective tropical forest research, training, conservation, management, and tourism development."

Dr. Thomas M. Butynski, project director, with a background in wildlife ecology—especially the blue monkey—at Michigan State University and with Wildlife Conservation International (WCI), among other organizations, is especially interested in primates, rodents, and raptors. He is also the co-director of the Institute of Tropical Forest Conservation.

Among current assignments, Dr. Butynski is writing a recovery plan for "Uganda's two mountain gorilla populations," supervising graduate students with tropical forest biology projects, and helping the government develop comprehensive management plans for three tropical forests as well as for gorilla-based tourism. He belongs to the Species Survival Commission (IUCN/SSC) Primate Group, International Primatological Society, and Ecological Society of America, among other groups.

■ Resources

Dr. Butynski has an extensive publications list based on his worldwide research.

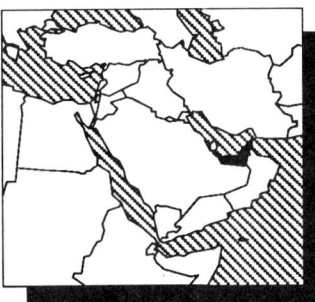

This nation is a union of seven emirates lying along the oil-rich eastern coast of the Arabian Peninsula. The only landscape feature that breaks the low-lying desert plain is Oman's Hajar Mountains along the Musandam Peninsula in the east. Rainfall in the UAE is only three to four inches a year, and less than 2 percent of the land is arable. Date palms are the major crop. The reserves of petroleum, which are among the world's largest, and oil and natural gas extraction are the principal industries. Most of the population and industry is concentrated in Abu Dhabi.

The gulf waters are home to schools of mackerel, grouper, tuna, and occasional sharks and whales. Inland, animal life is mostly restricted to domestic goats, sheep, and camels.

🏛 *Local Government*

Dubai Zoo
P.O. Box 67
Dubai, United Arab Emirates

Contact: **Mohammad Ali Reza Khan, Ph.D.,** In Charge
Phone: (971) 4 440462 • *Fax:* (971) 4 231795

Activities: Conservation; Education; Research • *Issues:* Biodiversity/Species Preservation
Meetings: IUCN/SSC

▲ Projects and People

Dr. Mohammad Ali Reza Khan has conducted field research on various species of wildlife in the tropical and subtropical forests of Bangladesh, India, and Sri Lanka as well as in the desert and mountains of the United Arab Emirates (UAE) and Oman. Previous projects on **primates of Bangladesh** were funded by the World Wildlife Fund—USA. **Asian elephants, birds of Bangladesh,** and **frogs** were subjects of other projects.

In UAE and Bangladesh, Dr. Khan helped lay the "foundation for the movement in wildlife conservation" through television and radio programs and publications, including 7 books and some 50 scientific articles.

A member of the Species Survival Commission (IUCN/SSC) Cat, Crocodile, Chelonian, Elephant, Otter, Primate, and Stork/Ibis groups, Dr. Khan is also active in the World Conservation Union's Commission on National Parks and Protected Areas and Commission on Education; Bombay Natural History Society; Zoological Society of London; and Oriental Bird Club, London, among other groups.

■ Resources

A zoo guide is being prepared. A zoo design is sought for a biopark.

The terrain of the United Kingdom includes mountains, green hills, and plains sloping to cliffs down to the sea. Only 9.5 percent of the land is forested, and 29 percent is devoted to cropland. While some areas are densely populated, there is still much uncrowded countryside. Natural resources include coal and North Sea oil and gas.

As the cradle of the Industrial Revolution, the United Kingdom was once much more polluted than it is today. Much progress is being made in achieving cleaner air, and salmon have returned to the Thames. The country has a complex series of environmental laws and regulations covering land use, resource management, and pollution control. It now works closely with other members of the European Community on limiting emissions of greenhouse gases and other pollutants. Water pollution from waste dumping is a continuing problem, causing serious harm to the North Sea. Radon, seeping up from the soil and into houses, has produced a high level of radiation exposure.

🏛 *Government*

English Nature ⊕
Northminster House
Peterborough PE1 1UA, UK

Contact: **Lynne Farrell**, Botanist/Plant Ecologist
Phone: (44) 733 340345 • *Fax:* (44) 733 68834

Activities: Education; Development; Law Enforcement; Political/Legislative; Research
Issues: Air Quality/Emission Control; Biodiversity/Species Preservation; Global Warming; Sustainable Development; Water Quality • *Meetings:* CITES

● Organization

"The ancient countryside of England survives still," reports **English Nature**. "From our mild, wet Atlantic coasts with their wooded valleys rich in mosses, liverworts and lichen, across our estuaries teeming with invertebrates and birdlife, by our trout streams shaded by willows and fringed with bulrush and loosestrife, to the windswept emptiness of the moors of north Yorkshire, there is a tremendous variety, . . . thousands of years of history amongst the anemones and primroses, . . . old hedges, ancient woods, small fields, . . . commons, hamlets, and villages that date back to medieval times."

English Nature advises the British government on nature conservation and promotes England's wildlife and natural features in its international responsibilities. Formerly known as the **Nature Conservancy Council**, it implemented an extensive research program on a wide variety of nature conservation topics for more than 40 years. English Nature was one of three agencies created as a result of the Environmental Protection Act of 1990, which eliminated the Council. The other two were the Countryside Council for Wales and the Countryside Commission.

▲ Projects and People

"Our natural heritage matters to us all. We are committed to helping people make choices and decisions that favor nature conservation. This heritage is ours to care for, now and in the future," states the English Nature mission. In this endeavor, it establishes, selects, and manages **National Nature Reserves**, noting special scientific interests, and conducts relevant research. English Nature also works through the Joint Nature Conservation Committee with sister organizations in Scotland and Wales on UK and international nature conservation issues.

Lynne Farrell is species recovery project officer and plant ecologist, Science Directorate, who has contributed to the British Red Data Books on vascular plants, updated the third edition of the *Atlas of British Flora*, and written on Chinese water deer, heathland management, orchids, and ferns, among other topics. Farrell belongs to Species Survival Commission (IUCN/SSC) Orchid

Group, British Ecological Society, several Wildlife Trusts, and is chairman of the Berne Committee for Threatened Plants' Cypripedium Committee.

■ Resources

A *Books and Journals Catalogue* is available.

Global Environmental Research (GER) Office⊕
Polaris House, North Star Avenue
Swindon, Wilts SN2 1EU, UK

Contact: **David A. Brown,** Head
Phone: (44) 793 411734 • *Fax:* (44) 793 411691

Activities: Education; Development; Research • *Issues:* Air Quality/Emission Control; Biodiversity/Species Preservation; Deforestation; Energy; Global Warming; Health; Water Quality

● Organization

The **Global Environmental Research (GER) Office** came into existence in 1990 when the five Great Britain research councils saw the need to coordinate responses, research opportunities, and challenges within its country and internationally on those issues related to global change. GER is jointly funded by the Agricultural and Food Research Council, Economic and Social Research Council, Medical Research Council, Natural Environment Research Council (NERC), and the Science and Engineering Research Council.

GER's primary function is to establish contact points and to provide and disseminate information on UK and international science and policy developments and initiatives. It describes itself as reaching beyond the issues of global environmental change (GEC) to incorporate "the technologies and remedial measures aimed at mitigating or removing undesirable influences."

▲ Projects and People

The Globe quarterly **factsheet** is one vehicle for dispensing information. A **database** of UK's global environmental research—inherited from NERC, based on a 1990 survey, and to be updated approximately yearly—allows GER to structure the material on natural sciences and supplement it with equivalent information in the human dimensions—thereby making it accessible to the research community. This in-depth description of the UK's role in global environmental research includes activities in higher education institutes, government agencies, research facilities, and from socioeconomic studies.

GER also provides the **secretariat for the UK Inter-Agency Committee on Global Environmental Change (IACGEC),** which represents all central government departments, the British National Space Centre, the Meteorological Office, and all five research councils. IACGEC reviews the country's global environmental change research interests. And, together with GER, it prepared a 1991 report on those changes occurring at regional levels with consequences to the total Earth system. "Modification of the climate system as a result of the buildup of 'greenhouse gases,' depletion of the stratospheric ozone layer, changing sea levels, loss of biodiversity, desertification, and loss of rainforest are all examples of global change," states the publication.

According to GER, understanding the global environment requires the concentration and organization of research, international cooperation, progress in basic theoretical ideas (those involving chaos and predictability), and modern facilities (satellites and supercomputers). Emphasis is on seven interdisciplinary themes: climatology and hydrology, stratospheric processes, biogeochemical dynamics, ecological systems and dynamics, past environmental change, socioeconomic considerations, and data and facilities.

GER points out that UK's strengths in global change research are in the "discovery of the Antarctic ozone 'hole' by British scientists in 1984," which led to worldwide investigations and, eventually, the adoption of the Montreal Protocol. The UK, with the USA, is a leader in developing three-dimensional climate modeling and related predictions. And UK has strong ties with the International Planning Office for the World Ocean Circulation Experiment (WOCE), located in Britain—as is the associate office for the Global Change and Terrestrial Ecosystems project of the International Geosphere-Biosphere Programme (IGBP).

Current research priorities include "the problem of **cloud radiation feedback,** the **ocean/cryosphere/atmosphere interaction;** understanding the **carbon,** and other elemental, **cycles;** impacts on ecosystems and the biosphere (including human health); and socioeconomic dimensions of global change."

Although the Earth's environment is the result of complex and powerful interactions of biological, chemical, and physical processes, the impact of human activities is also responsible. Hence, GER asks, "Can the planet cope with these changes without life being made intolerable for us? How can we reduce our impact on the planet but at the same time sustain our development?"

■ Resources

GER offers its *Global Environmental Change, The UK Research Framework* publication free of charge. Its quarterly *Globe* and an edited breakdown of the *UK GER Database* are also available on request.

House of Commons
London SW1A 0AA, UK

Contact: **Tam Dalyell,** Member of Parliament, Labour Party
Phone: (44) 71 219 4343

Activities: Political/Legislative • *Issues:* Biodiversity/Species Preservation; Deforestation; Transportation

▲ Projects and People

Tam Dalyell, M.P., makes frequent contributions to official reports in the **House of Commons** on environmental subjects.

Directorate of Fisheries Research (DFR)
Ministry of Agriculture, Fisheries, and Food (MAFF)
Burnham-on-Crouch
Essex CM0 8HA, UK

Contact: **John E. Portmann, Ph.D.**, Head of
            Laboratory
    Phone: (44) 621 782658 • *Fax:* (44) 621 784989

Activities: Law Enforcement; Research • *Issues:* Global Warming; Waste Management/Recycling; Water Quality

▲ Projects and People

Through the **Directorate of Fisheries Research (DFR)**, the **Ministry of Agriculture, Fisheries, and Food (MAFF)** conducts fisheries research in the waters worked by British fishermen. DFR has been operating in this capacity since 1910 for MAFF.

The Directorate **monitors some 40 species and stock of marine and freshwater fish and shellfish** regarding appropriate fish conservation and management policies. It also supports governmental policies protecting the quality of the marine environment including regulatory work under the Radioactive Substance Act, 1960, and the Food and Environment Protection Act, 1985.

There are four laboratories, two units, and two research vessels operated by the Directorate. **Dr. John E. Portmann** heads the Burnham-on-Crouch Laboratory with a staff of 55. This Laboratory has been in operation for more than 25 years. Here, the Aquatic Environment Protection Division 2 (AEP2) offers advice on all aspects of **nonradioactive pollution control** that may harm fish or shellfish and the public who eats them. AEP2 also works to save fisheries in fresh water from the effect of pesticides and advises those governmental departments with no specialists on staff. The Laboratory has established a worldwide reputation, and the AEP2 staff often cooperates with the European Community and other countries on projects such as the Oslo Commission, the International Council for the Exploration of the Sea, and the North Sea Task Force, and with the United Nations Environment Programme (UNEP), UNESCO, and Food and Agriculture Organization (FAO).

The Laboratory studies the **effects of chemicals on aquatic species** and for chemical analysis for a wide variety of substances at trace levels in environmental samples. It has general programs of **toxicological research** and surveillance of pollutant levels in water, sediments, and biota around the coasts—particularly for disposal of industrial wastes and sewage sludge—and in freshwater rivers.

Some of the Laboratory's successes include development of a **new generation of oil dispersants** replacing highly toxic materials; halting production and **strict control of PCBs** (polychlorinated biphenyls) use; **reduction of mercury inputs**; and imposed **controls** of the "highly **toxic compound TBT** in anti-fouling paints, on diesel oil in oil-based drilling muds on the phasing out of the remaining few uses of dieldrin."

MAFF published its first reports on nonradioactive contaminants monitoring covering 1984–1987, with a revised version released in 1990 for the period 1988–1989. DFR's AEP2 monitored a bulk of the activities.

■ Resources

DFR publishes *Aquatic Environment Monitoring Reports*, and has various publications lists available.

Natural History Museum ⊕
Cromwell Road
London SW7 58D, UK

Contact: **Neil R. Chalmers, Ph.D.**, Director
    Phone: (44) 71 938 9123 • *Fax:* (44) 71 938 8799

Activities: Education; Research • *Issues:* Biodiversity/Species Preservation; Deforestation; Health; Human Origins; Mineral Resources; Water Quality

● Organization

As one of the world's foremost institutes for the study and classification of animals, minerals, and plants, the **Natural History Museum** houses a wealth of informational material against which ecological and geological changes are measured.

The Victorian-era museum continues to display "the works of the Creator" since its beginnings in the early 1880s. The Sir Hans Sloane collection, probably the largest of any private person at that time, actually served as the basis of the Museum. During the Museum's early years, the collections grew as a result of British interest overseas—most notably those of Captain John Cook and Charles Darwin. With Cook on his first voyage around the world was naturalist Joseph Banks and skilled natural history illustrator and artist Sydney Parkinson.

Today, more than 350 scientists and librarians are behind the scenes along with about 67 million specimens and 1 million books and manuscripts that create the most complete source of natural history material worldwide. The Museum acquires approximately half a million specimens each year while loaning out over 75,000 worldwide. Described as the ultimate source of reference in this field, the Museum receives thousands of national and international inquiries and requests annually from national educational institutes, broadcasting organizations, commercial and industrial companies, government departments, the press, and public citizens. The Museum also offers interpretation and management advice regarding its various activities, environmental impact studies, and conservation reports. In 1989, the centralized functions of acquisition and cataloging were computerized using the McDonnell Douglas URICA system.

Training opportunities and facilities for undergraduate and postgraduate students in biology, geology, and social sciences are also available. And fully equipped, experienced registered scientific divers are prepared for fieldwork when needed in aquatic habitats and terrestrial locations around the world.

▲ Projects and People

The Museum's major research focus is **taxonomy**; realizing that "no institution can carry out research across the whole range of animal and plant life," it therefore now concentrates on issues of contemporary human concern. There are six programs in the Museum's series: biodiversity, environmental quality, human health, human origins, living resources, and mineral resources. These areas are further divided into earth sciences, environment, living resources, and parasite and health issues. At the root of much endeavors are **molecular biology studies involving DNA**. A frozen tissue collection is being undertaken for long-term storage so that, "ultimately, we will be able to examine genes just as we now examine skeletons or preserved organisms," reports the Museum.

Diversity at local and global levels in **marine benthic communities** and **tropical forests** are studied to test theories, investigate, and prepare taxonomic revisions relevant to present issues. It forms the basis for future monitoring and designation of natural habitat for conservation. "Because many living things are "sensitive 'indicators' or 'predictors' of change in the environment," the Museum observes these as a "barometer of environmental pressure" regarding the extinction of species and changes in diversity.

Identification and **analysis of rocks, minerals, fossils, and artifacts** are also conducted by the Museum, stressing the finite mineral resources. "At present the rates of use of the known reserves of some will result in their exhaustion in a few decades."

Through the use of lichens, nematodes, protozoa, and other organisms, the Museum **measures soil, sediment, and water quality** to find biological approaches to environmental problems. Scientists work on biological control, fish and shellfish, genetic resources, and pests and weeds.

A valuable resource for animal and human studies comes from the Museum's **libraries and database collections** identifying **parasites, pests, and their vectors**. DNA probes and other cytological and biochemical techniques are used. Veterinary work studies insects, ticks, mites, and paretic worms. The Museum operates the **World Health Organization (WHO) collaborating centers** for three tropical diseases: **leishmaniasis**—transmitted by certain sandflies and affecting 12 million people; **onchocerciasis**, or river blindness; and **schistosomiasis**, a disease caused by parasitic worms and causing disease in over 200 million people in the tropics and subtropics.

The significance of environmental changes on man's evolution and the separation of humans from apes are also under review by the Museum. These studies involve research into changes in the brain, the origin and diversity of modern people, toolmaking, and walking upright.

The 1980s saw many alterations in the Museum's operations. With the Geological Museum becoming fully integrated into the Natural History Museum complex, the expanded scope of the Museum stretched. The Museum also realized that "a key to the strategic thinking for science must be an improvement in funding, particularly support from the government," and that additional commercial and fundraising methods were deemed necessary. A developmental trust was established, encouraging individuals and companies to support specific projects. In addition, HRH The Princess of Wales became Patron of the Natural History Museum in 1989, the first time a national museum has had a royal patron.

Among the scientific diversity of the Museum's work are the Spitalfields project, the Flora Mesoamericana initiative, and the "Coming Clean in the Mersey" pollution effort.

The **Spitalfields project** involves **analysis of nearly 1,000 skeletons** (396 of known age within two years) initiated in 1984. "A peculiar pattern emerged in which young adults were overaged and the old underaged. . . . The bones had aged rapidly in those who died young and were eternally young in those who lived to a great age [for both] men and women," notes the Museum's triennial report, 1987–1989. Exploring this paradox, researchers examine the role of environmental, lifestyle, and family inheritance in determin-

ing aspects of skeletal changes. It is hoped this information will be useful in the identification of individuals at risk from premature aging of bones and osteoporosis.

A global priority for botanists is the richness of species and rate of their loss as a result of deforestation in the tropics. The Natural History Museum particularly targets the Central America region between the Isthmus of Tehuantepec in southern Mexico, south to the Colombia and Panama borders in its 18-year **Flora Mesoamericana program**. This is the first time there will be a general means of identifying Central American plants. Several volumes include descriptions and aids to identification of 15,000 to 20,000 species of flowering plants and ferns. The Museum collaborates with the Missouri Botanical Garden, USA, and the Instituto de Biología (Biology Institute) of the Universidad Nacional Autónoma de Mexico, Mexico, in this endeavor.

Closer to home, "Coming Clean in the **Mersey**" refers to one of Europe's most polluted waterways. The **25-year program** commissioned in 1991 and sponsored by the Museum along with the Water Research Center (WRC) strives **to rescue this important waterway**. This survey will be one of the most comprehensive meiofaunal (microscopic organisms) data sets ever obtained from a polluted estuary.

From the Museum's five science departments are a wide variety of research work and projects. The following illustrate some of these examples: structural studies of minerals using synchrotron radiation; gold-bearing mineral assemblages: their relevance to hydrothermal precipitation and growth mechanism; cretaceous echinoid faunas of the Atlantic marginal basins of Brazil; evolution of terrestrial mammals during the Miocene in Turkey; the classification and relationships of pharyngognath fishes; systematics, ecology, and biogeography of British lichens; a manual of tropical seedlings; the taxonomy, biology, and community structure of tropical forest insects; *Crustacea* copepod evolution; biodiversity and ecology of North Atlantic deep demersal fishes; pollution in Hamilton Harbour, Bermuda; DNA phylogeny of the extinct marsupial wolf; and antimicrobial activities associated with amphibian skin function.

With numerous collaborating institutions, some selected tropical forest biodiversity research activities include faunisitic studies on Costa Rican insects and of smaller moths of Southeast Asia, estimates of local and global species-richness, estimates of values of areas for conservation planning using phylogenetic techniques, field guide to the plants of Belize, wasps as important biological control agents in the Orient, reference text on ants of the world, Fern Red Data Book, diatom diversity studies in Central America, resource fragmentation investigation, *Malesian Mosses Handbook*, *Fern Flora of Borneo*, guides to aphids on world trees and to leafhoppers of the Pacific region, and spittlebug studies.

Still more biodiversity studies worldwide focus on terrestrial environments such as "nematodes of European grasslands in relation to soil quality," computerized database on European plants, coastal marine habitat microfauna, European water macrofauna and flora, and algae; and freshwater studies on groundwater protozoa.

An array of permanent Museum exhibitions include such subjects as Britain's **offshore oil and gas**, "creep crawlies," human biology, and a Discovery Center geared toward children, allowing them to handle specimens, do experiments, and make observations under supervised conditions. Numerous temporary exhibits also provide additional information sources. The Museum has a strong working relationship and commitment to work with teachers and their students.

■ Resources

The Natural History Museum Publications lists its numerous books, guides, catalogs, magazines, special reports, and bulletins. Of particular note are the *Triennial Report, Natural History Museum* guide, and *Science for Your Needs*. The Museum has a teachers center, members club, and museum shops to promote its activities.

Natural Resources Institute (NRI) ⊕
Central Avenue, Chatham Maritime
Chatham, Kent ME4 4TB, UK

Contacts: **Andrew Willson**, Executive Officer, Project Office
Stephen H. Walker, Team Leader
Phone: (44) 634 880088 • *Fax:* (44) 634 880066

Activities: Development; Research • *Issues:* Biodiversity/Species Preservation; Deforestation; Energy; Sustainable Development; Waste Management/Recycling

● Organization

The **Natural Resources Institute (NRI)** became an independent, government-funded agency in April 1990. Previously, it was the scientific branch of Great Britain's Overseas Development Administration (ODA), originally formed by the merger of the Land Resources Centre (LRDC), the former Tropical Products Institute (TPI), and the Centre for Overseas Pest Research, with over 100 years of service in natural resources starting with the Imperial Institute. Although a majority of its works is conducted for ODA, other external contracts are made with industry, Asian Development Bank, European Community (EC), and World Bank.

▲ Projects and People

Among some 240 current environmental projects, NRI's major goal is alleviating hardship and poverty in developing countries through increased productivity of renewable natural resources via science and technological applications. Its three target areas are food science and crop utilization, integrated pest management, and resource assessment and farming systems. Short- and long-term projects overseas are conducted by its staff of 500, often in multidisciplinary terms in more than 60 countries.

The **food science/crop utilization section** offers the vital connection between the producer and consumer and tries to reduce losses, improve food supplies (fresh and processed), and quality value at both ends. Its five branches concentrate on cereals and pulses (leguminous plants producing edible seeds) food security; fisheries; forest products; horticultural and tree crops; and oilseeds, edible nuts, and fibers.

As cereals and pulses are 75 percent of the basic food that developing countries produce, accounting for more than 60 percent of per capita energy intake, significant strategies are implored by NRI. Its program covers **biodeterioration control, grain distribution regulation, management of insect pests of stored grain/products, post-harvest consequences of breeding,** and **processing improvements** for cereal and grain legumes.

Fish are also an important food source in developing nations; therefore, NRI improves management of existing stocks, **aquaculture development,** and **reduction of post-harvest losses**.

In the forest products program, NRI's efforts deal with **forest fuels** and multipurpose use of *Eucalyptus* and *Pinus*. Developing nations produce more than 60 percent of the world's supply of fruit, vegetable, root, and tuber crops, making it another major focus of NRI activities. The nutritional value, dietary importance, and potential for agro-industrial development of **oilseeds** make them an important **foreign exchange commodity** under NRI's watchful actions.

With the **integrated pest management (IPM)** strategy, NRI develops robust solutions to management problems that are cost effective and adaptable to local criteria, emphasizing the role of natural regulatory mechanisms in containing pest damage. The five divisions of this program are biological identification and variability, component technologies for pest management, epidemiology of plant disease and windborne pests, pest control for annual crops, and pest management for perennial and industrial crops.

Pests from flowers, fruits, seeds, soil, and vegetative shoots are examined in this program. NRI also works on pre- and post-harvest **diseases of tropic tree crops,** especially avocado, banana, cocoa, cotton, and mango. It develops forecasting and control strategies for windborne pests and plant diseases. And advances in biochemical, numerical, immunological, and molecular genetic techniques aid in the rapid identification and diagnosis of variation within the pest populations.

The overall aim of NRI's **Resource Assessment and Farming Systems** program is increasing agricultural productivity and the farmers income via more effective methods. This program strives to increase knowledge of the natural resource base and the constraints on its development and improve understanding of the farming systems and man's interactions with livestock and natural resources. Agronomy and cropping systems, forestry, livestock production and protection, and resources assessment are considered in this endeavor.

NRI's **remote sensing and mapping** section plays a major role in resource assessment via satellite imagery or aerial photography. With this information, NRI monitors drought, seasonal vegetation, rainfall estimates and the effect of weather on migrant insect pests, such as African armyworm and desert locust.

Recent NRI programs concerned the following countries: **Indonesia**—to tabulate land resource data, provide new topographical maps, and train forestry officers; **India, Brazil, Costa Rica,** and **Colombia**—to evaluate **phosphate fertilizer residues in tropical soils; Ghana**—to determine sustainable yields for timber extraction, training, and remote sensing; **Kenya** and **Nigeria**—to evaluate "open access and common property land tenure systems" and their contributions to land degradation; and also **Honduras, Sri Lanka,** and **West Africa,** for examples.

Stephen H. Walker, as team leader of the **Regional Physical Planning Programme for Transmigration (RePProT)** and of **Map Improvement and Training (RePPMIT),** provides the training program and support for a national geographic information system (GIS) in Indonesia.

Other staff members are **Reginald Allsopp,** ecologist and livestock protection manager; **Barry Francis Blake,** fisheries and aquatic sciences researcher concerned with pre- and post-harvest trends regarding Lake Victoria's Nile perch; **Robert John Douthwaite,** environmental biologist, who recently assessed tsetse and trypanosomiasis control in **Zimbabwe;** and **Lincoln David Charles**

Fishpool, who studies the whitefly *Bemisia tabaci* as a vector of African cassava mosaic disease in Ivory Coast; **Michael John Jeger**, plant pathologist and mathematical modeling authority; **John Michael Thresh**, tropical and temperate plant virologist; **Henry Wainwright**, horticulturist; and **Thomas George Wood**, agricultural zoologist with expertise in IPM worldwide.

NRI also offers on-the-job training opportunities in Great Britain and developing countries, a soil chemistry unit, engineering workshops, and a library recognized as a major source of information on tropical agriculture utilizing a computerized in-house database.

■ Resources

A publications catalog booklet lists NRI's informational materials that can be ordered. Two descriptive ODA publications emphasizing sustainable development are *The Environment and the British Aid Programme* and *NRI—Development Through Science*.

Peak National Park
Aldern House, Baslow Road, Bakewell
Derbyshire DE4 1AE, UK

Contact: **Michael Dower**, Park Officer
    Phone: (44) 629 814321 • *Fax:* (44) 629 812659

Activities: Development; Management/Protection • *Issues:* Biodiversity/Species Preservation; Heritage Conservation; Sustainable Development; Tourism • *Meetings:* Annual Conference of European Federation of National Nature Parks; IUCN; World Parks Congress

● Organization

Peak National Park is one of 10 national parks in Great Britain, created under the National Parks and Access to the Countryside Act. It receives funding through the government's National Parks Supplementary Grant.

▲ Projects and People

Each year, demands increase on the limited Peak Park resources, creating a "vicious circle" which makes it a "victim of its own success," according to an annual report. From improvements at the Park's headquarters at Aldern House to the traffic, Peak Park continues to expand with its record-keeping technology, computer operations, and youth-training programs. Traffic flow alone has increased over 50 percent since 1980.

The country's first major **review of the national parks** using its major land-use policy document, the Structure Plan, and the National Park Plan Review, began in the 1990s. It hopes to enable parks to face the future "with a sound, agreed [upon], and complete package of relevant policies."

The three main considerations of Peak Park are conservation, recreation, and rural development, all of which depend upon increased budgets, doubling "in real terms, within five years," if the Park is to conduct its duties.

Peak Park's national officer is **Michael Dower**, also senior vice president of ECOVAST, the European Council for the Village and Small Town. With members in 25 European countries, ECOVAST

is concerned with the **well-being of rural people** and the rural heritage throughout Europe. *(See also European Council for the Village and Small Town (ECOVAST) in the UK NGO section.)*

The Peak District Rural Housing Association provides **housing for local people** in the national park under an Interim Housing Policy. Housing Corporation grants assist these efforts.

As development pressures prevail, Peak Park sought an extra planner, approved a major supermarket, shops, flats, a business park, a new factory, and the conversion of a high-tech facility. Applications for tourist facilities continue, along with the call for environmental impact statements.

The 1989 Water Act gives the Park safeguards over the privatization of the 15 percent of area owned by three water companies. The companies must meet strong conservation and recreation obligations and consult with Park authorities on any proposed disposal of property.

Starting in 1990, the Farm Conservation Scheme gives **financial incentives to farmers** to conduct work contributing to conservation of the landscape, wildlife, or historic character of the Park. Some 300 farmers or owners showed interest in this program, which initially approved 88 new schemes.

Other Peak Park successes include: the Integrated Rural Development Project (1982–1988) which operated in three distinctly different trial areas; the Upper Don Management Plan with its car park and wheelchair accessible routes; the reconstruction of the Dovedale Stepping Stones and Hartington footpath; and the listing of some 70 buildings within the Conservation Areas. Still under examination are additional improvements in the 4,000 miles of **public path networks** and better dog control by their owners. As in all areas, volunteers continue to play significant roles.

■ Resources

A publications list is available. Peak Park's *Annual Report* describes its activities and projects in detail.

Royal Botanic Gardens (RBG), Kew ⊕
Richmond, Surrey TW9 3AE, UK

Contacts: **John Dransfield**, Ph.D., Research Botanist
       David Pegler, Ph.D., Head of Mycology
       Thomas Laessø, Senior Scientific Officer
  Phone: (44) 81 940 1171 • *Fax:* (44) 81 332 0920

Activities: Education; Political/Legislative; Research • *Issues:* Air Quality/Emission Control; Biodiversity/Species Preservation; Deforestation; Health • *Meetings:* CITES; Congress of European Mycologists; European Council for the Protection of Fungi; International Mycological Congress

▲ Projects and People

Dr. John Dransfield is senior principal scientific officer at the **Royal Botanic Gardens**, chair of the Species Survival Commission (IUCN/SSC) Palm Group, and recent consultant on rattan silviculture in Malaysia on behalf of the Food and Agriculture Organization (FAO)/United Nations Development Programme (UNDP). Also editorial board member for *Flora of Thailand* and *Kew Bulletin* and member of the Royal Society's Committee for Southeast Asian Rainforest Research, Dr. Dransfield has lectured worldwide.

He has written widely on palms of Borneo, Sumatra, Malaya, and Indonesia, for example. Recent literature includes "Raising Cane: The Natural History of the Rattan Race," "Prospects for Rattan Cultivation," "The Palms of Africa and Their Relationships," "Outstanding Problems in Malesian Palms," and proceedings from a 1987 symposium on the "Conservation Status of Rattans: A Cause for Great Concern."

The fungi collections at the **Royal Botanic Gardens (RBG)** represent one of the major worldwide collections in terms of authenticated material—possibly the most important—states **Dr. David Pegler**, head of the mycology section and senior principal scientific officer. He has been responsible for initiating, planning, and implementing the **complete recuration of the fungus collections**, the first such task attempted this century. Since becoming section head in 1987, Dr. Pegler has rearranged the estimated 600,000 collections into 95 orders, 382 families, and 6,660 genera.

Current RBG projects include the **preparation of an Agaric Flora of Sao Paulo State**, Brazil—comparable to those made for Antilles, East Africa, and Sri Lanka, which can offer a "pantropical overview of the agaric orders not currently existing." Dr. Pegler also continues **spore ultrastructural investigations**, which he began in 1969, to provide a new aspect of taxonomic research now considered a standard technique. And a two-year monograph preparation of the **British hypogeous fungi** involved research and revisionary studies utilizing light and electron microscopy.

Senior scientific officer **Thomas Laessø** is engaged in numerous taxonomy and floristic studies of fungi worldwide, including pyrenomycetes and European macromycetes as well as hypocrealean fungi in Denmark. *Rhodocybe stangliana* are also being investigated.

As chair of the Species Survival Commission (IUCN/SSC) Fungi Group, Dr. Pegler reports that this organization is helping to ensure "full representation internationally" of some 1,500,000 species— "one of the largest kingdoms of organisms." Laessø is the Group's secretary. Its *Fungi and Conservation Newsletter* details and updates significant activities worldwide.

For example, the European Committee for the Protection of Fungi was renamed the **European Council for Conservation of Fungi (ECCF)**, which, among other tasks, is preparing a European Red Data List of threatened fungi. And regional committees for fungi conservation are planned. Efforts are underway to increase the number of protected "lower plants (mosses, lichens, algae, fungi) [which] have tended to be badly neglected by conservationists" through the Berne Convention on the Conservation of European Wildlife.

In **North America**, a **database** of literature on the distribution of North American mushroom species is being prepared by **Dr. S.A. Redhead**, Biosystematics Research Centre, Ottawa, Canada. Two volumes on North American **polypores** have been produced, as are publications on *Harvesting Edible Wild Mushrooms* in British Columbia, and commercial regulations on picking wild mushrooms in Washington State, USA. According to Dr. J. Ginns, also of the Biosystematics Research Center, there is no project comparable to Red Data Lists for North American mycoflora. "Obviously, a considerable amount of data is needed before the fungal flora can be subdivided into a rare versus the endangered species."

Fungi are also at risk in Central America from deforestation, squatters, and, in some instances, the military. According to D. Jean Lodge, Belize has a threatened area with a "unique ectomycorrhizal flora associated with oak and pine savannahs," and funds are being raised for a nature reserve. Some conservation efforts are underway

in the **Dominican Republic**, considered a "critical area" for mycorrhizal fungi."

Needs in **Africa** include collection of information, "broad-based mycology curriculum in African universities, training of mycologists, overcoming the isolation in which the few African mycologists are working, and public enlightenment." At present, **Prof. A. Peerally**, University of Mauritius, chairs the new Committee for the Development of Mycology of Africa, International Mycological Association.

■ Resources

The Mycology Section produced a series of at least 20 information leaflets, useful to agriculture, horticulture, and forestry fields, police, pathology, medical, and forensic laboratories; and authorities and the general public. The IUCN/SSC *Fungi and Conservation Newsletter* is available. Dr. Pegler also has a publications list.

Scottish Natural Heritage (SNH)
12 Hope Terrace
Edinburgh, Scotland EH9 2AS, UK

Contact: **Magnus Magnusson, KBE**, Chairman
Phone: (44) 31 447 4784 • *Fax:* (44) 31 447 0055

Activities: Education; Research • *Issues:* Biodiversity/Species Preservation

● Organization

In April 1992, the **Scottish Natural Heritage (SNH)** came into existence with the merger of the Countryside Commission for Scotland (CCS) and the Nature Conservancy Council for Scotland (NCCS). SNH acts as a single Scottish agency while continuing to build upon the work of the two other bodies. It encourages sustainable use of natural resources while promoting understanding and enjoyment of the environment, and considers Scotland's cultural, economic, and social well-being.

"Scotland's natural environment has never before been under such severe pressure. We want to use the countryside to grow things, to dig things out of, to walk over, to live in, to drive through, to preserve for our children, to visit, to 'get away from it all' . . . , [but] we are beginning to endanger the very environment we cherish," writes chairman **Magnus Magnusson**.

As an independent body, SNH expresses its views publicly and advises the government on its country's natural heritage. It is accountable to Parliament and funded by grant-in-aid from the Scottish Office. It also approves grants which follow its goals.

▲ Projects and People

SNH divides its programs into four categories: **communication**, the **countryside**, **research**, and **urban areas**. By developing working relationships with numerous groups and individuals, SNH practices an open approach to communications in raising awareness and stimulating commitment to the country's environment.

Inherited from its predecessor bodies, are the regional parks, long distance footpaths, National Nature Reserves, national scenic areas, and other sites of special scientific interest. SNH projects in the countryside include **conservation of endangered species of animals** and plants, of special habitat, and classic scenic landscape, regeneration of native woodland, and renewal of run-down industrial landscape.

Relevant research topics to the SNH mission are commissioned or undertaken by the staff to create an in-depth knowledge of changes and patterns on the local scene and its interaction between man and nature. In so doing, SNH works closely with the Joint Nature Conservation Committee and other British environmental agencies on those numerous environmental issues which know no political boundaries.

Since 80 percent of Scots live in cities and towns, SNH has a strong commitment to the enhancement of urban areas. It involves local participation, including schools, to help develop more green spaces and wildlife gardens, footpaths, bridle paths, recreation areas, and renewal of derelict landscapes in outlining areas.

Other activities being proposed are a demonstration project on native pinewood regeneration, monitoring countryside changes with aerial photography, developing national and regional strategies for forestry, fish farming, and ski development; training rangers and wardens; and environmental education through media use and publications.

Magnusson was previously NCCS chairman and narrator of popular television programs on archaeology. SNH's first chief executive is Roger Crofts. Dr. Michael Usher continues his NCCS post as chief scientific advisor within the new SNH.

■ Resources
SNH is producing a new publications list.

World Ocean Circulation Experiment (WOCE)⊕
IOS Deacon Laboratory, Brook Road
Wormley, Godalming Surrey GU8 5UB, UK

Contact: N.P. Fofonoff, Ph.D., Director
Phone: (44) 428 68 4141, ext. 311
Fax: (44) 428 68 3066
E-Mail: Omnet: (Sciencenet) woce.ipo

Activities: Research

▲ Projects and People
"Increasing concentrations of greenhouse gases in the atmosphere and threats of global warming and sea level rises have given urgency to research on the world ocean, which plays a critical role in determining climate change."

So reports the World Ocean Circulation Experiment (WOCE), which in the late 1970s began planning the scientific programs being carried out from 1990 to 1997. According to Dr. R. Allyn Clarke, WOCE co-chair and Bedford Institute of Oceanography (BIO) scientist from Canada, the goal is to "develop and test global ocean models that then can be used to better predict or model climate change both natural and anthropogenic." The largest such ocean study "ever attempted," scientists from governments, universities, and organizations in more than 40 nations are participating.

Part of the World Climate Research Programme (WCRP), WOCE is sponsored by intergovernmental organizations, including the World Meteorological Organization (WMO) and the Intergovernmental Oceanographic Commission (IOC); and global organizations, such as the International Council of Scientific Unions (ICSU) and the Scientific Committee on Ocean Research (SCOR).

WOCE's work in climate change is complementing global programs such as the Tropical Ocean and Global Atmosphere (TOGA), which focuses on the El Niño phenomenon and other interactions; the Joint Global Ocean Flux Study (JGOFS), regarding "processes controlling marine biogeochemical cycling on time scales appropriate to carbon dioxide problems"; the Global Energy and Water Cycle Experiment (GEWEX); and the International Geosphere-Biosphere Programme (IGBP), a sister program to WCRP that investigates the Earth's "interactive physical, chemical and biological processes."

WOCE is using a "handful of satellites, dozens of ships, and thousands of instruments to take a comprehensive global 'snapshot' of the physical properties of the ocean." According to WOCE literature, the goal is to "develop a data set of unprecedented magnitude and quality for use in developing numerical models of ocean circulation and physical processes. Then, as these ocean models are improved, WOCE says they will be coupled with atmospheric models to simulate, and perhaps predict, how the ocean influences the global and regional climate over periods of decades."

The ocean, in fact, absorbs "more than half of the incoming solar radiation. . . . The temperature falls or rises much less over the ocean than over land." According to WOCE, greenhouse gases, such as carbon dioxide, methane, nitrogen oxide, and chlorofluorocarbons—and resulting from human activities as deforestation and fossil fuel burning—will reach a level "early in the next century that will be equivalent to doubling the pre-industrial concentration of CO_2." As the ocean absorbs much of these gases, its capacity influences climate.

Meanwhile, sun energy drives "large-scale wind systems" which affect the ocean as does "the sinking of cold water masses in the polar and subpolar oceans." As ocean currents—such as the Gulf Stream and Kuroshio currents—redistribute energy absorbed from the sun, climates in various regions are affected. Basic to comprehending these processes is the understanding of world ocean circulation.

In 1990, WOCE began its field phase with hydrographic cruises in the Atlantic Ocean and Southern Hemisphere. In early 1991, the European Space Agency launched the satellite ERS-1 carrying equipment that measures global sea-surface topography and wind stress. The "high-precision altimetric mission, TOPEX/POSEIDON," was planned for 1992.

WOCE research should ultimately benefit marine life; create conditions to further regional economics, agriculture, fisheries, shipping, and offshore drilling and mining; and determine how and where to establish long-term systematic ocean observations for global monitoring and predicting climate change.

However, funding is not ensured to completely carry out the modeling, field, and analysis phases. "Without the firm commitment of the necessary resources over the next few years, the overall objective of predicting global change will be in jeopardy from lack of adequate ocean models," WOCE believes. With oceanographic sensors on satellites and other data-gathering equipment in hand, WOCE is now "pressing for the design and the international funding of a global climate observing system."

█ Resources

WOCE literature, such as the *WOCE Handbook*, *Summary of Resource Commitments*, and general brochure, is available from the international office. Interested persons may also receive newsletters published by national affiliate groups. WOCE presents scientific information at workshops, conferences, and panels dealing with climate issues.

 *NGO*

Patricia C. Almada-Villela, Ph.D. ⊕
Conservation and Aquatic Biology Consultant
60 Newington
Willingham, Cambridge CB4 5JE, UK

Phone: (44) 954 60520 • *Fax:* (44) 954 60520
E-Mail: Patty@wcnc.uuep

Activities: Development; Education; Research • *Issues:* Biodiversity/Species Preservation • *Meetings:* Coral Reef Symposium

▲ Projects and People

Dr. Patricia C. Almada-Villela, who co-chairs the Species Survival Commission (IUCN/SSC) Coral Reef Fish Group with Dr. Don E. McAllister, Canadian Museum of Nature, Centre for Biodiversity, reports that Ocean Voice International is the new name of International Marinelife Alliance Canada. (International Marinelife Alliance [IMA] USA retains its name.) *(See also International Marinelife Alliance, Inc. in the USA NGO section.)*

Ocean Voice sponsors projects that "train people to use environmentally sound marine resource harvesting methods," emphasizing sustainable harvest, enhancing the "quality of life and income of coastal fisherfolk." For example, Ocean Voice provides a "green checklist" to schools to measure "environmental friendliness." Other Ocean Voice projects include publication of *Sea Wind*, *Guide to Selected Fishes of the Maldives*, *Coral Reef Conservation Manual*, three-volume bibliography on the *Rare Fishes of Quebec*, Netsman Project, Philippines, with the Haribon Foundation, to train "aquarium fish collectors . . . to use small environmentally friendly monofilament nets"; newly developed world Coral GIS (geographic information system) Map; and a study of a proposed airport at the site of Shiraho Coral Reef, Ishigaki Island, Japan. Ocean Voice determined that the reef was a "world-class heritage site, had a very high level of biodiversity of corals and fishes, had the biggest blue coral site in the world and largest known individual coral colonies—up to 3 metres high and six centuries old." As a result, says Ocean Voice, the Japanese government selected another location for its airport.

Also with McAllister, Almada-Villela authored the "World Coral Fish Programme Proposal" for the Species Survival Commission, which states, "Coral reefs are one of the richest environments on earth. . . . They are the most species-rich marine ecosystem, . . . the only aquatic or marine ecosystem comparable in species biodiversity to the tropical rainforests. . . . Coral reefs are amongst the most productive of marine environments, producing up to 35 metric tons of fish per square kilometer each year."

McAllister and Almada-Villela estimate that "coral reef fishes comprise 25 percent of all marine fishes, some 4,000 species, even though coral reefs constitute less than one percent of the world's oceans." Such fishes are an important source of animal protein. Meanwhile, they are subject to overfishing and habitat loss, particularly in Japan, Philippines, and Costa Rica, where 70 percent of the coral reefs are "in poor or fair condition."

Unfortunately, information is lacking on the status of coral fishes; in fact, none was listed in the "700 plus species of concern in the 1990 [World Conservation Union's] IUCN Red List of Threatened Animals.

The World Coral Fish Programme aims to prepare a World Coral Fish Status Report and a Conservation Action Plan. Dr. McAllister, as overall coordinator, has embarked on a three-year computerized listing of coral fish species on the GIS and conservation manual with Ocean Voice and on a marine aquarium industry study.

Dr. Almada-Villela is project leader on the country survey studies and of the threatened fishes of Mexico, and conserving diversity of Mexican marine fish, with Dr. Salvador Contreras-Balderas, Mexican Ichthyological Society, Bioconservación, and the Universidad Autónoma de Nuevo León.

Dr. Almada-Villela also belongs to the SSC Freshwater Fish Group, chaired by Dr. Christopher Andrews, National Aquarium in Baltimore, USA; SSC Shark Group chaired by Dr. Samuel H. Gruber, University of Miami, USA *(See also University of Miami in the USA Universities section)*, of which Sarah L. Fowler, Nature Conservation Bureau, UK, is deputy chair; and SSC Sustainable Use of Wildlife Group, co-chaired by Robert and Christine Prescott-Allen, Canada. Conservation Action Plans are being prepared for the SSC specialist groups.

█ Resources

The *Green School Checklist*, a 16-page publication, and *Sea Wind*, a quarterly marine environmental bulletin, can be ordered. The Coral Reef Specialist Group requires a full-page scanner for its word processing system.

British Herpetological Society (BHS)
c/o Zoological Society of London
Regents Park
London NW1 4RY, UK

Contacts: Trevor John Clark Beebee, Ph.D., Chairman
Michael R.K. Lambert, Ph.D., Former Chairman
Phone: (44) 634 880088 • *Fax:* (44) 634 880066

Activities: Education; Political/Legislative; Research • *Issues:* Biodiversity/Species Preservation • *Meetings:* Societas Europaea Herpetologica; World Congress of Herpetology

▲ Projects and People

Current chairman of the British Herpetological Society (BHS), with 1,100 members and an all-volunteer staff, is Dr. Trevor John Clark Beebee. His activities, over the past 20 years, include conservation of habitat, rare reptiles, and amphibians of UK heathlands; working on the biology, conservation management, and a national site register for the **natterjack toad** (*bufo calamita*); and giving talks and seminars for schools, colleges, natural history societies, and other educational groups.

Immediate past chairman and BHS Council member since 1968, **Dr. Michael R.K. Lambert** also belongs to the Species Survival Commission (IUCN/SSC) Tortoise and Freshwater Turtle Specialist Group. A pest ecologist at Natural Resources Institute (NRI), Kent, he mainly studies control of insects such as locusts, grasshoppers, and bollworms, primarily in tropical Africa. Earlier, Dr. Lambert was consultant to the Fauna and Flora Preservation Society of London and the World Wide Fund for Nature (WWF) to investigate the effects of the pet trade on wild tortoises in Morocco.

Dr. Lambert has studied the effects of DDT ground-spraying against tsetse flies on **lizards in Zimbabwe**, and monitored the effects of using deltamethrin as an alternative to DDT. Dr. Lambert notes that amphibians and reptiles "are useful indicators of habitat condition and the health of marshy and terrestrial environments since they have poor powers of dispersal, are vulnerable through being unable to adapt to rapid changes in their habitats and . . . are at the end of insecticide-sensitive food chains." He has authored more than 50 publications since 1969, including notes, book reviews, and research reports.

■ Resources
BHS publications include *Surveying for Amphibians, Save Our Reptiles,* and *Garden Ponds as Amphibian Sanctuaries.*

British Organic Farmers/Organic Growers Association (BOF/OGA) 🌾
86 Colston Street
Bristol, Avon BS1 5BB, UK

Contact: **Patrick Holden**, Director
Phone: (44) 272 299666 • *Fax:* (44) 272 252504

Activities: Development; Education; Political/Legislative • *Issues:* Sustainable Agriculture • *Meetings:* IFOAM

● Organization
The **British Organic Farmers/Organic Growers Association (BOF/OGA)**, with staff of 7 and some 1,100 members, serves primarily as an information clearinghouse for small- and large-scale organic farmers and gardeners, or those who wish to embark on organic growing.

▲ Projects and People
According to director **Patrick Holden**, from his book *Organic Farming, An Option for the 90s:* "After many years of government policies encouraging increased output by specialization and intensification, it is now widely recognized that the adverse consequences of this approach are so serious that drastic action must be taken. To date, organic farming has not been seen as having a strategic potential in solving these problems. Instead, overproduction, nitrate pollution, the damage to our landscape and wildlife habitat, and pesticide residues in the food chain have each been tackled by a series of

piecemeal measures which lack any overall strategic integrity. . . . Farmers have always been motivated by the concept of good stewardship and protection of the countryside. Now that the economic environment for organic farming is becoming more attractive, the opportunities for taking up this option are greater than ever."

■ Resources
BOF/OGA's book catalog, like its address, is shared with the **Soil Association**. Among recent BOF/OGA publications is *Organic Farming, An Option for the 90s,* which "provides the ideal introduction for farmers considering conversion to organic methods [and] illustrates the viability and new found status of organic farming." The publication *20% of Britain Organic by the Year 2000* is described as "a policy document for the organic movement of the 1990s." The goal of its title can be achieved by "using existing legislation in Britain and the EEC [European Community]," and would have immense impact on the quality of life in the country." *The Case for Organic Agriculture* is the proceedings of the Sixth National Conference on Organic Food Production, and *Conservation and Wildlife Guidelines* is the proceedings of a 1990 Conference on "Conservation Wildlife and Organic Farming," which gave organic farmers and growers, and environmental groups in the Midlands area a chance to exchange ideas. *New Farmer and Grower* is a quarterly journal "for professional organic producers, with news, policy developments and research papers."

Care for the Wild ⊕
1 Ashfolds Horsham Road, Rusper
West Sussex RH12 4QX, UK

Contact: **William J. Jordan**, MVSc., Chairman, Veterinarian
Phone: (44) 293 871596 • *Fax:* (44) 293 871022

Activities: Education; Research • *Issues:* Biodiversity/Species Preservation; Protection against Cruelty and Exploitation
Meetings: CITES; IUCN; IWC; SSC

● Organization
Care for the Wild tackles the problems and sufferings of wildlife caused by man and those of exploited species which are not endangered. "As a wildlife vet, I know the problems and can help make the public aware. . . . These are problems which are ignored by nearly all conservation organizations," states **Dr. William J. Jordan**, Care chairman. He also cites the need to "deal with the ethics and the serious problems arising from the mistaken attitude that man can manage nature."

▲ Projects and People
Care's staff of seven focuses its commitment on badgers, tawny owls, dolphins, elephants, rhinos, turtles, whales, among others. In 1991, Care provided grants to more than 20 volunteer- operated **badger protection groups**, for instance.

Care for the Wild belongs to a consortium of groups responsible for **returning three dolphins to the wild**. The dolphins—"Missy," "Rocky," and "Silver"—lived in a 100-acre fenced-in bay in the Turks and Caicos Islands, where they first learned to catch fish prior to being set free to roam the ocean.

The **Rhino Ark Sanctuary**, in Aberdare, has an electric generator powered by a water wheel that furnishes electricity for the fence on days when the solar panels do not receive enough sunlight. Nearby villages also receive electricity from it. A goal of 220 kilometers of fence is anticipated by Care, pending sufficient funding.

Concerns about the **trade in freshwater turtles** from Bangladesh to Hong Kong and other Far East countries led to banning the importation of the endangered *Trionyx hurum*. After a reported decline in the turtle population on the beaches of Cephalonia, Greece, in 1990, Care sent in a team of 34 scientists and volunteers to help reverse the trend. And, two **loggerhead** turtles (removed to Britain from Cephalonia in their egg stage) were prepared to be returned back to the wild, near the beach on which they were originally deposited as eggs.

"If cetaceans and their marine environment are to be conserved successfully into the 21st century and beyond, a more modern environmental ethic to supersede the humanistic approach must be developed and adopted, ... acceptable by human communities worldwide," states Care in its booklet *Whaling? An Ethical Approach*. Care concludes that all whaling of small and large cetaceans is incompatible with an environment-centered ethic and not in the international public interest. Specifically, it notes that cetaceans have little or no capability to evade capture by modern whaling methods; some species remain with an injured animal thereby being vulnerable themselves to be killed; humane killing cannot be guaranteed; individual animals are indiscriminately killed; there is an absence of understanding the boundaries and structures of whale populations worldwide; and killing of cetaceans is crucially different from the killing of domestic animals for food.

Care representatives also correspond to various worldwide governmental officials regarding specific animal treatment and fund missions to collect data and report on activities.

■ Resources

Care for the Wild needs additional addresses for educational outreach, public relations, and consultant advice to further its fundraising efforts. An order form is available for gift items and publications, including *Care for the Wild News* newsletter and *Whaling? An Ethical Approach*.

Centre for Alternative Technology (CAT) ⊕
Canolfan Y Dechnoleg Amgen
Machynlleth, Powys SY20 9AZ, UK

Contact: **Brian Horne**, Information Officer
  Phone: (44) 654 702400 • *Fax:* (44) 654 702872

Activities: Education • *Issues:* Energy; Environmental Building; Sustainable Development; Transportation; Waste Management/Recycling; Water Quality

● Organization

The **Centre for Alternative Technology** (CAT) was founded in 1974 as an educational charity, and "exists to promote and demonstrate technologies which can help the West to live more in harmony with the planet."

On 40 acres of a former slate quarry in Wales, with 12 resident staff, CAT "is a living and working community which has pioneered

the practical application of sustainable technologies" such as solar, wind, and water power, low energy buildings, organic agriculture, and conservation. There is no main electricity, and the Centre also "provides its own water, treats its own sewage, and produces a good proportion of its own food." Continuously open to the public, the Centre receives about 55,000 visitors each year, who pay a modest admission. The staff of 40 "work[s] as a cooperative, with low, needs-based salaries and a non-hierarchical structure." Technical director is **Dr. Robert Todd**.

According to its literature, "The Centre combines a successful business enterprise, an educational charity, a major tourist attraction, a research laboratory, a databank, a wildlife refuge, and a place of pilgrimage for committed environmentalists."

▲ Projects and People

The primary activity at the Centre is educational. Since 1979, CAT has offered to the public **courses** taught by its expert staff, and in 1981 the Centre began teaching for academic institutions as well. Courses can be tailored according to the length of time desired, class size, and level of experience—whether for school children, teachers, or university students or faculty members. Recent courses include "green living," waterpower (how to install a small water turbine), "green teaching" in or out of the classroom, windpower, solar collectors and systems, blacksmithing, organic gardening and permaculture, and self-building. Most courses last about five days.

Today, CAT also offers "practical training courses for people going to Third World countries as volunteers." The 10-day **Development Courses** provide practical, hands-on training in building and construction, water supply and sanitation, energy, agriculture, and basic manufacturing.

The Centre's "renewable energy hardware and gardens are ideal for practical sessions, and the holistic approach of the Centre enables topics to be examined in a broad environmental context. . . . It is a place where you can experience, with all your senses, the workings of a model green future. You can see energy efficient buildings and feel their warmth, hear wind and water turbines at work, touch animals reared in a natural environment, smell herbs, flowers and vegetables grown organically, taste wholefood cooking at its best and talk to men and women working cooperatively."

The 3,500 members of the **Alternative Technology Association** (ATA) support the Centre's efforts to effect "genuine, deep down change, both politically and in the way we all live our lives." Like CAT, the Association "promotes Earth-friendly practices like organic farming and renewable energy technology," spreading awareness that "an ecological way of life is viable." The ATA network is growing across the UK, working toward the development of a sustainable society.

The Centre also houses **Dulas Engineering**, a small company "involved in innovative work in electronic monitoring and control systems for renewable energy technologies."

■ Resources

CAT produces numerous publications, including *Information Sheets* which comprehensively introduce various aspects of alternative technology, such as biofuels, organic growing, and solar water heating. *Resource Lists* provide a reference for "useful contacts, equipment suppliers, and further reading" on such topics as *Buildings—Energy and Environmental Factors* or *Components of Small Renewable Electricity Systems*. *Do-It-Yourself Plans* explain how to put together bicycle wheel wind generators, haybox ovens, clip-fin solar panels, and other relatively simple environmentally friendly devices. A *Buy Green by Mail* gift catalog is also offered.

The Centre also provides educational material for students at various levels, and for teachers, on solar, wind, and water power. Other publications include *Technical Reports* and *Development Course Notes* on such topics as Agriculture in Sub-Saharan Africa, beekeeping, and building design for tropical climate. The ATA's quarterly magazine is *Clean Slate*.

CAT's physical facilities include a lecture room, two teaching workshops, a covered work area, outdoor space for project work, an extensive library, a bookshop, accomodation space, and a vegetarian restaurant. CAT also offers various information services, "answering questions on all aspects of alternative technology from both the general public and from specialist organizations," by phone, by mail, or on the spot. If more in-depth help is needed, the Centre provides a list of professional consultants.

Council for Posterity
20 Heber Road
London NW2 6AA, UK

Contact: **Nicholas Albery**, General Secretary
    Phone: (44) 81 208 2853 • *Fax:* (44) 81 452 6434

Activities: Education; Political/Legislative; Research • *Issues:* Advocacy Interests of Future Generations

● Organization
A group of self-described "advocates for the interests of future generations," the **Council for Posterity** was launched with Earth Day 1990 "to protect and represent the interests of generations to be born even a thousand years from now." Members include **Anita Roddick** of the Body Shop, writer **Sir William Golding**, international lawyer **Maxwell Bruce**, ecologists and others from organizations such as **Ecoropa**, **Gaia Foundation**, **Sustainability Ltd**, and **Survival International**. The Council is working to transform "the way people think about the future," emphasizing the need for more responsible consideration of future generations. Their *Declaration of the Rights of Posterity* stresses: "Those who live after us have no voice amongst us. We therefore declare and determine their right to inherit a planet which has been treated by us with respect for its richness, its beauty and its diversity; a planet whose atmosphere is life-giving and good, and can remain so for aeons to come; a planet whose resources have been carefully maintained and whose forms of life retain their diversity; a planet whose soil has been preserved from erosion with both soil and water unpoisoned by the waste of our living; a planet whose people apply their technologies cautiously with consideration for the long-term consequences; a planet whose people live in human-scale societies unravaged by population excess; a planet whose future generations have interests which are represented and protected in the decision-making councils of those alive today."

▲ Projects and People
The Council's major activities involve publicizing their cause. One such project was the launch of an **annual award** for the best UK-published article on the theme of "posterity."

On Earth Day, the Council began its **Adopt-a-Planet** program, a competition encouraging school classes to adopt and improve a part of their local environment—from a schoolyard corner to a nearby stream—on a permanent basis. Students then become **Planetary Guardians**, with the Council hoping that each adoption will be permanent. The first prize was offered in 1991, and, given appropriate funding, the Council plans to make it an annual event. Teachers may pass a project on to each year's new class, unless students prefer to choose a new area; the same spot can be reentered with new plans.

The competition is not just a matter of submitting before and after pictures; according to the Council, a "class should also say what their successes, difficulties, and failures have been, how it all compares with what they hoped to achieve at the outset, and [add] any good ideas they have for combatting vandalism or graffiti." Judging criteria include imagination, perseverance, success in gaining local support and/or funding, ability to view the project in a wider context, local media use, and quality of material submitted.

■ Resources
The Council and the Academic Inn have published a paper by ecologist and filmmaker **Herbie Girardet** entitled *Being Fair to Posterity*, funded via the Fourth World Educational and Research Association Trust.

Earthwatch, Europe ⊕
Belsyre Court, 57 Woodstock Road
Oxford OX2 6HU, UK

Contact: **Andrew W. Mitchell**, Deputy Director, Europe
    Phone: (44) 865 311600 • *Fax:* (44) 865 311383

Activities: Education; Research • *Issues:* Biodiversity/Species Preservation; Deforestation; Global Warming; Sustainable Development; Water Quality • *Meetings:* IUCN

● Organization
Earthwatch, Europe is the European affiliate of the USA-based **Earthwatch**, "an international organization offering members of the public the opportunity to join research expeditions to help scientists of all disciplines to accomplish their field work." The organization has about 5,500 European members, 60,000 in the USA, and 3,000 in Australia. The European office has a staff of 12; other affiliate offices are located in Los Angeles, Sydney, and Moscow.

Earthwatch was founded in 1971 by its president, **Brian Rosborough**. Since then, it has supported more than 1,300 field seasons for about 800 projects around the world, using more than 30,000 **EarthCorps** volunteers' contributions of millions of dollars and man-hours in field research. The European office opened in 1990. Earthwatch sees its mission as improving "human understanding of the planet, the diversity of its inhabitants, and the processes which affect the quality of life on earth." *(See also Earthwatch in the USA NGO section and Earthwatch projects throughout this Guide.)*

▲ Projects and People
Earthwatch serves as a bridge between the public and the scientific community. The organization supports more than 135 projects in

50 countries annually, sending thousands of volunteers to assist hundreds of scientists. The **Centre for Field Research** receives more than 400 research proposals each year; the Centre, with its **scientific advisory board**, then does a peer review, screening and selecting projects for Earthwatch support. EW Europe emphasizes environmental research, allocating funds for such research programs as Understanding the Earth, Strategies for Survival, Threatened Habitats, The Human Factor, and Managing the Planet.

Projects are divided into two- to three-week periods so that company employees, teachers, and students can participate. EarthCorps volunteers range in age from 16 to 85, with the average age 45. Through its **Education Awards** scheme, Earthwatch raises funds to sponsor teachers and students on projects, a particularly successful program in Australia and America, where more than 3,000 students and 1,500 teachers "have received career training on EW projects, an investment which is greatly multiplied on their return."

Here are a sampling of Earthwatch 1992 projects: exploring **volcanoes** in New Zealand, Mexico, Japan, and USA; probing the **ozone hole** from Kamchatka; understanding Florida's **aquifers**; rehabituating **orangutans** in the rainforest; compiling a "comprehensive global **atlas of temperate trees and shrubs**"; witnessing **dancing birds** in Costa Rica; **tagging trees** in Puerto Rico; researching **exotic plants**; inspecting **United Nations Biosphere Reserves**; studying **dry forests** in Central America; radio tracking **caiman** in a Brazil rainforest; monitoring **giant clams**, **butterfish**, and other indicator species along **Tubbataha Reef**, Philippines; hiking through **old-growth forests** in Ontario; capturing and releasing **dolphin**; observing **baboon**, **macaque**, and **Bali's temple monkeys**; sampling pollen on the snouts of **honey possum**; checking out the **caribou and muskoxen** at Alaska's North Slope; sighting **whales**; saving **sea turtles**; identifying **black rhino**; looking for **loons**; rescuing **cranes** in Vietnam; recording **bird migrations** in China; tagging **Tasmanian hens**; monitoring **elephants** in Botswana; and **snow tracking coyotes** in Yellowstone Park—among many others.

Earthwatch, Europe hopes to gain the support of private donors, trusts, and corporations. Earthwatch team members share the costs of research expeditions. Over the course of a project, members "learn to excavate, map, photograph, gather data, make collections, assist diving operation, and share all other field chores associated with professional expedition research."

The European office is chaired by **Sir Crispin Tickell**, former Ambassador to the United Nations. Zoologist and vice chancellor of Oxford University, **Sir Richard Southwood** is chairman of the **Science Advisors Group**, comprised of 15 eminent scientists and archaeologists from the UK, Poland, Sweden, France, Kuwait, India, and elsewhere.

Executive Director **Brian Walker** spent 25 years in the corporate sector as manager of a major engineering plant, after which he became director general of Oxfam, then served as president of the Band Aid–Live Aid Projects Committee until 1991. A leading expert on environmental policy, Walker was also president of the International Institute for Environment and Development and is a member of the World Commission on Food and Peace.

Deputy Director **Andrew W. Mitchell** is a zoologist, author, and BBC broadcaster with 15 years' experience of organizing research projects overseas. In the 1970s and 1980s, he was scientific coordinator of the Scientific Exploration Society (Operation Drake), and led pioneering studies of tropical rainforest canopies. Mitchell has written seven books on natural history, including rainforests.

■ Resources

Earthwatch constantly seeks "volunteers in all areas to work alongside scientists on conservation and research projects worldwide."

The organization publishes the bimonthly *Earthwatch Magazine*, and *Earthwatch Expedition Briefings*. Each year, the magazine's January/February edition describes that year's upcoming Earthwatch projects.

Elm Farm Research Centre (EFRC)
Hamstead Marshall, Near Newbury
Berkshire RG15 0HR, UK

Contact: **Lawrence Woodward**, Coordinator
Phone: (44) 488 58298 • *Fax:* (44) 488 58503

Activities: Development; Education; Political/Legislative; Research • *Issues:* Organic Agriculture; Sustainable Development; Waste Management/Recycling; Water Quality

● Organization

Elm Farm Research Centre (EFRC) was established in 1980, with the aim of developing and promoting "organic agriculture as the most environmentally sound way of producing health food." The Centre, with a staff of 14, operates from the 232-acre Elm Farm, in Berkshire. Once a conventional intensive dairy farm, Elm Farm is now a fully organic, mixed farm with a Jersey dairy herd, a flock of Poll Dorset ewes producing early lambs, and long crop rotation including grass/clover leys, cereals, and green manures. The Farm demonstrates new developments in organic methods and techniques, and provides a site for research.

According to EFRC, "The underlying principle of organic agriculture is to produce food of optimum quality and quantity, using methods which seek to co-exist with, and not dominate, natural systems." Organic farming avoids or minimizes the use of synthetic fertilizers, pesticides, growth hormones, and livestock feed additives. Soil fertility is maintained via "crop rotations, crop residues, animal manures, legumes, green manures, off-farm organic wastes, mechanical cultivation, and mineral-bearing rocks." Pest control uses biological methods.

EFRC is a member of the **International Federation of Organic Agriculture Movements (IFOAM)**, the umbrella research coordinating organization. However, according to coordinator **Lawrence Woodward**, "IFOAM at the moment is not a particularly efficient vehicle for ensuring global cooperation in that it has grown very big, very rapidly. . . . There seems to me to be a major lack of understanding between environmentalists, conservationists, and those of us pursuing the development of an environmentally resource-conserving agriculture—organic farming." Woodward stresses the need for more dialogue among these groups, and sees a role for IFOAM in coordinating communication.

Indeed, EFRC maintains close ties with organic research institutions in Europe and the USA. The Centre also works closely with **The Soil Association and British Organic Farmers/Organic Growers Association (BOF/OGA)** and contributes regularly to the latter's journal, *New Farmer and Grower*. The Centre has worked to develop organic standards as a member of the **British Organic Standards Committee**, and on the Board and advisory committees of the **United Kingdom Register of Organic Food Standards (UKROFS)**.

▲ Projects and People

Noting that the goals of UK's agricultural policy since World War II—maximum production and food self-sufficiency—have been so well met that the country and Europe now face a problem of surplus produce, EFRC stresses that "agriculture cannot continue along its present course." Increased production has led to "loss of habitat and species, pollution of ground and surface water, and an increasing incidence and risk of soil erosion," as well as health risks to consumers in the form of chemical residues on produce.

More than 1,000 organic farms in the UK, including EFRC, are proving that "organic methods are the most effective way of tackling these problems and are the best way of producing a reasonable quantity of healthy food without damaging the environment." The Centre's research program seeks to further improve existing organic farming methods.

Expert German and British researchers are working at the Centre, and are using a three-tiered approach to their investigations. In the first phase, the researchers **monitor and evaluate** "whole farms, both organic and in the process of conversion." The second involves pursuing "specific **research projects** concentrating on key technical aspects of organic farming methods." Then, at the Centre and elsewhere, **on-farm trials and demonstrations** can be carried out.

Studies of mixed farms, which deal with both livestock and crops, have focused on "nitrogen accumulation, conservation and losses; the management of complex grass/clover swards; manure management; organic livestock husbandry systems; the use of veterinary homeopathy; wheat variety trials; [and] mechanical and biological weed control." For primarily crop-based farms, in addition to nitrogen studies, the researchers investigate crop rotation; green manures; "recycling of vegetable residues; treatment and use of municipal waste; inter-cropping of grain, legumes and cereals; mechanical and thermal weed control; potato and vegetable variety trials; [and] monitoring of the soil ecosystem. Other studies center on the quality of the food produced, environmental impact comparisons, and cost-benefit analyses.

Emphasizing the need for improved communication among those interested, the Centre's **Organic Advisory Service** provides "advice to existing and potential organic producers." The Farm's full-time advisor is supported by a team of six regionally based advisors, who are on call to visit farms requesting help, and provide "full conversion planning and followup." The **Soil Analysis Service** provides more specific assistance. In addition, EFRC staff and associates have worked as consultants for "a variety of clients from the corporate sector, government, European Community, . . . individuals, and charitable organizations."

Beginning in 1982, EFRC has organized a series of international **Research Colloquia**, as well as farmer meetings. The Centre has also worked closely with BOF/OGA to organize biennial conferences on organic agriculture. Three times yearly, the Farm is open to visits by groups and individuals from the general public.

EFRC advises Ph.D. projects at Reading and Aberystwyth universities, and is also working on research contracts for the Department of the Environment and the Ministry of Agriculture, Fisheries, and Food (MAFF). Other collaboration has involved the Scottish Crops Research Institute, National Institute for Agricultural Botany, Potato Marketing Board, and the University of Lancaster.

■ Resources

The Centre seeks support in biotechnology, genetic engineering, and agricultural engineering technologies.

The Centre describes its **J.A. Pye Research Laboratory** for general research and soil analysis as well equipped. EFRC's small library is open to the public, and contains journals in several languages, reports dealing with alternative agriculture, and an updated collection of papers and books on organic agriculture.

The Centre regularly publishes the results of its research in the form of *Report Series*—reviews of literature or proceedings of meetings and seminars; *Practical Handbook Series* covering issues and techniques of an applied agriculture; *Research Note Series* reporting specific project results; and *General Interest Papers* on a range of topics relating to organic agriculture. Reading lists are provided on general topics.

Among EFRC publications listed in the Soil Association Organic Book Catalog are *The Effects of Mechanical Aeration on Cow Slurry, Green Manures, Converting to Organic Farming,* and *Nitrate in Vegetables.*

Environmental Investigation Agency (EIA) ⊕
208/209 Upper Street
London N1 IRL, UK

Contacts: **Allan Thornton,** Co-founder, Chairman
Dave Currey, Executive Director
Jennifer Longsdale, Director
Phone: (44) 71 704 9441 • *Fax:* (44) 71 226 2888

Activities: Development; Education; Political/Legislative; Research • *Issues:* Biodiversity/Species Preservation; Habitat Preservation • *Meetings:* CITES; IWC; UNCED

Environmental Investigation Agency USA
1506 19th Street, NW
Washington, DC 20036 USA

Phone: 1-202-483-6621 • *Fax:* 1-202-483-6625

● Organization

The **Environmental Investigation Agency (EIA)** was established in 1984 "to protect the natural environment and the species that inhabit it." With 5,000 members, a staff of 12, and volunteers, the EIA's "Eco-Detectives" have a reputation for no-holds-barred "investigations into environmental problems and abuses." The Agency shares its findings with other conservation and animal welfare organizations, governments, and multilateral organizations, working "to develop workable and effective solutions to environmental abuses, and to implement changes which will protect exploited species."

▲ Projects and People

EIA chairman **Allan Thornton,** executive director and photographer **Dave Currey,** and director **Jennifer Lonsdale** co-founded the agency after working together with Greenpeace. In fact, Thornton helped found Greenpeace, set up the UK office, and has twice been its executive director—in 1978–1981 and 1986–1988. Quoted in a March 1991 article in the *Evening Standard* magazine, Thornton commented on attempts to block the Agency's work: "Most of the opposition you meet is from people who are very good at it. I am not intimidated."

The Agency's most effective weapon is evidence; "it gets the facts and the pictures and then invites others to join the cause."

Bird Friends is an eye-opening campaign. Launched in 1991 jointly with the **Royal Societies for the Prevention of Cruelty to Animals (RSPCA)** and **for the Protection of Birds (RSPB)** to end the trade in wild-caught birds, the campaign has met with some success. As a result of EIA-gathered film and documentation reporting on transport conditions of birds, there have been changes in air-travel regulations for the transport of live animals, and several international **airlines, including Lufthansa, have stopped transporting wild-caught birds.** The agency is "currently lobbying the U.S. and Commission [of the European Communities] to ban the import of wild-caught birds" altogether.

Existing controls, including those put in place by the 109 signatories to CITES, are clearly ineffective, according to the EIA, which notes that "for each bird that is bought from a pet shop up to three others have died." Birds that can live for up to 70 years in the wild are destined for early deaths when trapped for international trade. Nearly one-third of all parrot species are threatened, with 77 on the edge of extinction. For example, the blue-fronted Amazon parrot, a popular pet in the USA and Europe, suffers from the destruction of 150-year-old Quebracho trees, cut down by trappers who remove the chicks from the nests.

Agency workers posing as buyers in Argentina "were offered some of the rarest birds in the world, including **hyacinth macaws**—of which there are less than 5,000 left in the wild." EIA reports that some Argentinian parrot traders push wires down the throats of baby parrots before selling them. Their gizzards are punctured, causing infection and eventual death—about six weeks later. "The disappointed pet owner then buys another parrot—and so this vile trade continues." Ironically, the EIA emphasizes that "alternatives are already available: many species can be successfully bred in captivity avoiding most of the problems."

More than five million birds are traded each year, and the absence of complete records makes it difficult to enforce any controls, according to EIA. The European Community (EC) alone imports one to three million birds every year; dropping border controls after 1992 will make "enforcement as weak as its least effective member." The EIA asks members and others to join a letter writing campaign, to British and European Commission officials, airlines, and local pet shops, requesting that they stop the transport and trade of wild-caught birds.

The EIA's **Save the Elephants** campaign involves intensive research and evidence gathering to combat a move by certain South African countries to reopen the ivory trade, planned for the 1992 meeting of CITES. The original trade ban, which came into effect in January 1990, was a result of Agency efforts to expose the extent of damage to elephant populations in the area. In addition to monitoring and countering South African press coverage, EIA plans to attend the CITES meeting "with forceful new film and documentary evidence . . . maintaining the pressure to keep traditional ivory markets closed, and will continue to monitor and expose the illegal trade." The campaign has been funded mainly by the **Animal Welfare Institute** in the USA.

In a 3-year period, over 70 percent of all Dall's porpoises were killed off the coast of Japan, writes EIA, which also reports that thousands of pilot whales and dolphins have been killed in the Faroe Islands. Such fishing techniques as driftnets—"Walls of Death"—lead to the death of "hundreds of thousands of dolphins" each year, according to EIA. In its **Dolphin Friends** campaign, which began in 1990, the Agency has now expanded beyond dolphins to "campaign for increased protection for all cetaceans." Besides working to convince the International Whaling Commission (IWC) to regulate all whale, dolphin, and porpoise kills, the Agency is also working with the United Nations and is "pushing for Antarctica to be made into a cetacean sanctuary."

■ Resources

The Agency has published several studies including two reports on *Global War Against Small Cetaceans* and *To Save an Elephant* by Currey and Thornton.

European Council for the Village and Small Town (ECOVAST) ⊕
c/o Peak National Park
Aldern House, Baslow Road, Bakewell
Derbyshire DE4 1AE, UK

Contact: **Michael Dower**, Vice President
Phone: (44) 629 814321 • *Fax:* (44) 629 812659

Activities: Development; Education; Lobbying • *Issues:* Biodiversity/Species Preservation; Sustainable Development; Heritage Protection

● Organization

More than one-third of all Europeans live in rural areas, including small villages and towns, reports the **European Council for the Village and Small Town (ECOVAST).** However, their lifestyle and environment are threatened, both by migration of the young to urban areas, and by encroaching urban blight, modern farming methods, and pollution. In some areas, including northern Scandinavia, Portugal, Spain, southern Italy, and Greece, "the rural population has been falling continuously for a century, and the people who remain are poor and elderly." Eastern Europe faces similar problems, with the added burdens of poverty, "lack of resources, [and] official neglect." *(See also Peak National Park in the UK Government section.)*

ECOVAST describes itself as "a pan-European association of organizations and individuals deeply concerned about these developments." Founded in 1984, the organization has grown rapidly despite a staff of only 3; current membership numbers more than 300, in 24 countries throughout Europe, including more than 80 organizations—both government and NGO bodies—representing about 5 million people. Enjoying close working relationships with other European organizations, including consultative status with the Council of Europe, ECOVAST sees itself "as a bridge between decisionmakers and those who are active at local level, between experts and practitioners." National committees in Belgium, France, Germany, Hungary, Poland, and the United Kingdom "provide a focus for exchange and activity within each country, in the interests of its rural communities and rural heritage."

The aims of ECOVAST are "to foster the economic, social and cultural vitality and the administrative identity of rural communities throughout Europe; and to safeguard, and to promote the sensitive and imaginative renewal of, the built and natural environments of such communities."

▲ Projects and People

Recent activities include a major role in the **European Campaign for the Countryside** launched by the **Council of Europe**; the protest against Ceaucescu's systematization program in Romania; forming an active working group on rural architecture; and conferences in 1990 on rural development and rural tourism. Working with **Telecottages International**, ECOVAST is using modern information technology to stimulate rural enterprise throughout Europe, and is developing plans for a **House of Rural Europe** at Hoxter in North Rhine-Westphalia, Germany. ECOVAST co-founded EUROTER, promoting tourism in rural Europe.

About ECOVAST's **Strategy for Rural Europe**, president **Jean-Pierre Dichter** says its intent is to "stimulate discussion and action to benefit rural communities, and the rural heritage, throughout Europe." Among other points, the strategy emphasizes "long-term stewardship of Europe's heritage of wildlife, landscape, and culture"; good housing; forestry policy; economic revival; sustained farming; quality food products; diversified farm incomes; and decent schools, mass transit, and other public services.

Vice President **Michael Dower**, also ECOVAST former president, is now national park officer of the Peak District. Earlier, he served on the United Nations Special Fund team supporting An Foras Forbartha (The National Institute of Physical Planning and Construction Research) in Dublin, where he helped to prepare the model Amenity and Tourism Plan for County Donegal, as well as its first County Development Plan. He also wrote the report on the **Protection of Ireland's National Heritage** and researched environment and rural development issues at the Dartington Institute in Devon. In 1980, Dower was founder/chairman of **Rural Voice**, "the alliance of national organizations representing rural communities in England."

ECOVAST stresses that "government alone cannot do the job" of preserving rural Europe; the people themselves must become involved in their future. ECOVAST is trying to provide the means for rural communities "to express their needs and aspirations, and to apply their energies to meeting these . . . through the exchange of practical ideas and living experience. We wish to strengthen the voice, and illuminate the action, of rural people everywhere."

■ Resources

The organization needs assistance in the areas of political support, educational outreach, public relations, financing, and increasing membership.

ECOVAST publishes a newsletter in three languages (English, French, and German), produced twice each year. In addition, ECOVAST holds biennial conferences and runs seminars, working groups, and specialist exchanges.

Falklands Conservation
21 Regent Terrace
Edinburgh, Scotland EH7 5BT, UK

Contact: **Katherine Russell Thompson, Ph.D.,** Consultant Ecologist/Secretary
Phone: (44) 31 556 6226 • *Fax:* (44) 31 556 6226

Activities: Education; Research • *Issues:* Biodiversity/Species Preservation; Habitat Protection; Sustainable Development

● Organization

The Falkland Islands, discovered in 1592 by the English navigator John Davis, were settled in 1764 by the French at Port Louis, and two years later by the British at Port Egmont on Saunders Island; the Islands came under British rule in 1833, and their population of around 2,000 is largely of British descent. There are several hundred islands in the group, many uninhabited; the total land area is about 4,700 square miles.

After a 1979 meeting, at which conservationists, including the late **Sir Peter Scott**, "agreed on the need to protect the unique natural and historic heritage of the Falklands," the **Falkland Islands Foundation** was established "to promote the conservation of wildlife, wrecks and places of historic interest in the Falklands." Two years later, the **Falkland Islands Trust** was set up in Stanley with the same aims. In 1991, two groups merged to form **Falklands Conservation**, which acts locally and internationally to achieve its goals. The organization has 300 members and a staff of 2.

▲ Projects and People

Since the early 1980s, the Foundation, the Trust, and now Falklands Conservation, have pursued a broad agenda of conservation projects. Those ongoing or being developed include a long-term program of **seabird monitoring**, in which local people are trained to observe trends, in particular watching for possible declines, in seabird populations. They also learn to oversee **prey stocks of fish and squid** which, when overexploited by fishermen, could cause decline in populations of, for example, the **blackbrowed albatross, rockhopper** and **Magellanic penguin**, and **thin-billed prion**.

Another project, co-sponsored by the **World Wide Fund for Nature (WWF)**, involves **sea lion research**. Over the past 60 years, this population has declined 99 percent, from 300,000 to 3,000 animals. **Tussac grass**, a vital and unique local habitat, which grows to a height of over 10 feet, is now restricted to tiny uninhabited offshore islands; Falklands Conservation is working to monitor, protect, and investigate the possibility of reintroducing the grass on main islands. The organization is also performing **surveys of breeding birds** and other species, as well as of habitats. For example, the coast is ringed by thick beds of kelp which provide a valuable habitat for a wealth of marine life; these **kelp beds** need the attention of conservationists to prevent their degradation.

Falklands Conservation owns or leases several **nature reserves**, and is working to acquire more. It is pursuing a program of **conservation education**, and has ongoing projects related to the **preservation of historical sites** around the islands and the stabilization of the hulks of ships in the harbor. A **video** is being produced to educate visitors.

Dr. Kate Thompson joined the Falkland Islands Foundation in 1989, after serving as a consultant, "to assess and report upon the potential impact of commercial fisheries on Falkland Islands seabirds."

■ Resources

Falklands Conservation seeks general publicity, especially in the USA; larger membership; and more funds.

The organization regularly publishes a newsletter, and has also produced *Wildflowers of the Falkland Islands; Those Were the Days*; a report on seabirds and fisheries; and a report on tussac grass.

Fauna and Flora Preservation Society (FFPS)
c/o 34 Steele's Road
London NWE 4RG, UK

Contact: **Charles Leofric Boyle**, Ex-Secretary
Phone: (44) 71 722 5716

Issues: Biodiversity/Species Preservation

▲ Projects and People

Lieut. Col. Charles Leofric Boyle, retired, was the chairman of the World Conservation Union's (IUCN) Species Survival Commission (SSC); he is now one of six Members of Honor. He suggests that persons interested in the **Fauna and Flora Preservation Society (FFPS)** write to 1 Kensington Gore, London SW7 2AR, UK.

■ Resources

Both scientists underline the need for funding. In particular, the fledgling SSC Social Insect Group wants to finance "a full-time coordinator for . . . three years"—in order to establish a firm base of operations.

Dr. Morris has published more than 40 papers, articles, and other reports in the last 10 years. Dr. Elmes has authored or co-authored more than 40 publications. The Societas publishes the quarterly *Nota Lepidopterologica*, as well as *SEL News* and the proceedings of biennial lepidoptera congresses.

International Council for Bird Preservation (ICBP) ⊕
32 Cambridge Road, Girton
Cambridge CB3 0PJ, UNK

Contact: **Christoph Imboden, Ph.D.**, Director-General
Phone: (44) 223 277318 • *Fax:* (44) 223 277200

Activities: Development; Education; Research • *Issues:* Biodiversity/Species Preservation; Deforestation; Habitat Conservation; Sustainable Development • *Meetings:* CITES; ICBP; UNEP; World Parks Congress

● Organization

The **International Council for Bird Preservation (ICBP)** is a federation of 360 member organizations, with its directorate in Cam-

bridge. Work requires the efforts of 35 permanent staff and 35 to 50 temporary staff. Projects other than world conferences are generally carried out by member organizations.

■ Resources

ICBP has a technical publications series, currently listing 12, some of which are out of print. In 1991 the Council published *Seabird Status and Conservation: A Supplement* and *Conserving Migratory Birds*. The 1991 ICBP Monograph, joining five others listed, was on *Conservation of the Slender-billed Curlew*. In addition to a series of *Bustard Studies* in four volumes, ICBP also publishes regular Bulletins and Study Reports (48); the 1991 report was on the *Ornithological Importance of Coastal Wetlands in Guinea*. A Red Data Book is *Threatened Birds of Africa and Related Islands*, by N.J. Collar and S.N. Stuart.

(See also International Council for Bird Preservation [ICBP] in the Denmark and USA NGO sections.)

International Tree Planting Committee (ITPC) ⊕
International Tree Foundation
3 Hawthorn Crescent
Caddington, Luton LU1 4EQ, UK

Contact: **George William Ford**, Committee Chairman
Phone: (44) 582 31307

Activities: Development • *Issues:* Deforestation; Sustainable Development

● Organization

The **International Tree Foundation** (formerly Men of the Trees) was founded by **Dr. Richard St. Barbe Baker**; its patron is the **Prince of Wales**, and its president the **Earl of Bessborough**. In the UK, Foundation members are seeing the fruits of their efforts, working with schools and community groups, planting trees in urban and degraded rural areas. The organization's international arm, the **International Tree Planting Committee (ITPC)**, works with branches in 47 other countries "in the vanguard of the fight against deforestation all over the world." With an area of tropical rainforest larger than the UK being destroyed annually by slash-and-burn agriculture aggravated by high population growth rates, the Committee's staff of 10 and 4,000 members have their work cut out for them. According to Foundation data, "For every 10 hectares cleared, less than one hectare is replanted."

▲ Projects and People

The Committee focuses most of its work on the poorest Third World countries, especially those experiencing loss of tropical rainforest areas, needing to update agroforestry techniques, or trying to **halt desertification** with buffer zones of plant growth. ITPC efforts help reduce erosion, improve agriculture and thus the food base, and increase the amount of water available—for example, "in arid land, the root system of certain trees will improve the water table."

ITPC's limited funds are carefully distributed among a few projects each year. The Committee has supported tree planting work in, among other places, the Highlands area of Gursam, Ethiopia; northern **Kenya**, including establishing **tree nurseries** and training women and children; **Swaziland**, where **species trials** have been held; and West Mirriah, **Niger**, site of an **agroforestry** project. In all cases, "the Committee ensures that the projects take account of the needs of local people and the environment."

In 1991, the theme was "Tropical Rainforests," with projects complementing work already carried out, including forest conservation research in Maquipacuna, **Ecuador**; establishment of tree plantations in the Cape Verde Islands; Rondonia, **Brazil**, in the Amazon Basin; and rainforest support in **Belize**. **Reforestation** is particularly important because, as Foundation founder St. Barbe Baker put it: "If a man loses one third of his skin he dies. If a tree loses one third of its bark it dies. . . . If the Earth loses one third of its green mantle it, too, will die."

The chairman of ITPC, **George Ford**, began the Foundation's Bedfordshire Branch and is a founding member of the Tree 2000 Committee. Professionally, he has worked for the United Nations Food and Agriculture Organization (FAO), was chairman of the International Coffee Organization, and was for a time in charge of agriculture and forestry in **Malta**.

Tony Rule joined the Foundation in 1975; after working 20 years in advertising, he was inspired by work on desert reclamation in the Sahara to give up his career and go back to school to study arboriculture. After obtaining his degree, he worked on **tree planting projects** in the Negev of **Israel** and in the deserts of **Abu Dhabi** and **Saudi Arabia**, where no trees had grown for 30,000 years. He is now in charge of publicity and fundraising for ITPC, jointly with **Derrick Dray**.

Mike McCoy-Hill, former chairman and now ITPC technical adviser, worked with the Colonial Forestry Service in Tanganyika and Kenya on "timber grading, export control and improvement of the mangrove swamps." **Robin Hanbury-Tenison**, "explorer, geographer, naturalist, environmentalist and TV broadcaster" as well as president of Survival International, is ITPC's honorary technical adviser.

Betty Saunders, who joined the Foundation in the 1950s, is ITPC secretary. **John Moreland** joined MOT in the 1960s. St. Barbe Baker's books and idea of a "green front" led Moreland to create the exhibition *Green Belt Around the Sahara* following the 1980s drought and famine. **Donald Palmer**, another member of the Tree 2000 Committee, is a surveyor and land agent. **Knyvet Carr** serves as the Foundation's overseas liaison officer.

■ Resources

The Foundation and ITPC need fundraising and publicity assistance; with a broader base of members and sponsors, "more funds can be devoted to tree planting."

Their newsletter, *Trees Are News*, appears three times yearly; the organization also publishes a yearbook.

International Waterfowl and Wetlands Research Bureau (IWRB) ⊕
Slimbridge, Gloucester GL2 7BX, UK

Contact: **Michael Moser, Ph.D.**, Director
Phone: (44) 453 890624 • *Fax:* (44) 453 890697

Activities: Education; Monitoring; Research • *Issues:* Biodiversity/Species Preservation; Sustainable Development; Wetland Conservation and Management

● Organization

Wetlands "are among the most productive habitats known, ranging from estuaries and mangrove swamps—the nurseries for fish, to floodplains and swamps with their phenomenal production of vegetation, to arid zone ephemeral lakes teeming with waterfowl during intermittently occurring floods," writes the **International Waterfowl and Wetlands Research Bureau (IWRB)**. They provide food, drinking water, agricultural area, flood control, and water quality. Nevertheless, "wetlands are often . . . drained, built upon, or polluted." Because of the rapidity of their disappearance, "there is an urgent need to stem the losses of wetland habitats worldwide."

Perhaps the most visible feature of wetlands are the migrating waterfowl, which often travel thousands of miles before and after resting there. Besides "providing a rich and renewable resource for recreation, education and research," waterfowl serve as indicator species, a key to the condition of the entire wetland habitat. As the wetlands vanish and overhunting continues, "populations of ducks are at their lowest recorded levels in North America, and similar trends are apparent around the Mediterranean Basin and in Asia."

Since 1954, IWRB has worked for "international cooperation for the conservation of migratory waterfowl and their wetland habitats." IWRB's staff of 14 operates out of its secretariat at Slimbridge in the UK, shared by **The Wildfowl and Wetlands Trust**, fostering "international partnerships between scientists, organizations and institutes." IWRB's executive board, including governmental and non-governmental delegates from each member country, meets every three years to review its triennial program and to approve budgets. Research groups, comprising experts from around the world, coordinate conservation activities.

The Bureau encourages membership of governments or their agencies, asking an annual contribution on a scale related to a given member country's gross national product (GNP).

▲ Projects and People

IWRB played a significant role in initiating the **Ramsar Convention**, "an intergovernmental treaty for the conservation of wetlands." The more than 50 signatories to the Convention, or Contracting Parties, accept general obligations regarding conservation of wetlands within their borders, and special obligations to those in a "List of Wetlands of International Importance." IWRB views the Convention as a valuable legal tool for wetland conservation and provides technical support to the Convention secretariat, the **Ramsar Bureau**. When IWRB learns that a particularly important wetland is endangered, it contacts the appropriate national agencies to make them aware of the environmental implications. IWRB can also assist "in assessing the potential impact of threats and in identifying solutions to user conflicts."

IWRB has also worked to produce **inventories** (directories) of wetlands in **Europe**, **North Africa**, **Latin America**, and **Asia**, identifying "priority sites for conservation activities and for listing under the Ramsar Convention." More recent inventory work concerns the wetlands of **Oceania** and **Australasia**. In addition, IWRB and hundreds of volunteers have compiled an **International Waterfowl Census** covering over 70 countries in Europe, Africa, Asia, and Latin America. The results of the census help to define conservation priorities. When it becomes clear that key species are declining, IWRB initiates **Waterfowl recovery plans** which identify "habitat and population objectives for the recovery of the species." At the same time, IWRB promotes responsible hunting as "a traditional activity on wetlands, which can be compatible with conservation." The results of two recent studies—demonstrating lead poisoning in waterfowl caused by the ingestion of shotgun pellets and the need for refuges from hunting during periods of severe weather—have led the IWRB to seek changes in **hunting legislation**.

IWRB's **Wetland Management Group** encourages research in management techniques through training courses and workshops. IWRB also organizes "numerous **scientific symposia** and workshops to stimulate international cooperation for waterfowl and wetland conservation."

Dr. Michael Moser, IWRB director since 1988, received his doctorate in 1984 from the University of Durham for his work on the ecology of herons in the Camargue, southern France. He later worked as estuaries officer and development director for the British Trust for Ornithology. Dr. Moser has written more than 30 scientific papers and recently completed as joint editor, with **Max Finlayson**, a book entitled *Wetlands*. He has also written many popular articles and given radio and television interviews. Dr. Moser is a member of the Scientific Council of the Bonn Convention and of the Board of the Biological Station of the Tour du Valat.

■ Resources

IWRB publications include a newsletter, *IWRB News*; the proceedings of symposia, for example *Western Palearctic Geese*, edited by **Tony Fox, Jesper Madsen**, and **Johan Van Rhijn**; and a series of *Special Publications* such as an information pack on *Lead Poisoning in Waterfowl*, with video and booklet.

Oceania Wetland Inventory (OWI) ☘
c/o **International Waterfowl and Wetlands Research Bureau (IWRB)**
Slimbridge, Gloucester GL2 7BX, UK

Contact: **Derek A. Scott, Ph.D.**, International Coordinator
Phone: (44) 453 860062 • *Fax:* (44) 453 890697

Activities: Development; Education; Research • *Issues:* Biodiversity/Species Preservation; Deforestation; Sustainable Development; Water Quality • *Meetings:* ICBP; IWRB

● Organization

The **Oceania Wetland Inventory (OWI)**, 1989–1992, covers 24 nations "from Palau, Guam, and the Solomon Islands in the west to Easter Island and Sala-y-Gomez in the east, and from the Mariana and Hawaiian Islands in the north to New Caledonia and French Polynesia in the south."

The project has brought together many conservation organizations, including the Ramsar Bureau, International Waterfowl and Wetlands Research Bureau (IWRB), Asian Wetland Bureau (AWB), South Pacific Regional Environment Programme (SPREP), World Conservation Monitoring Centre (WCMC), International Council

for Bird Preservation (ICBP), World Conservation Union (IUCN), World Wide Fund for Nature (WWF), and national agencies.

In addition to providing an inventory of the actual areas and their flora and fauna, the OWI seeks to promote information sharing and awareness in the countries where it is being carried out, as well as to encourage "active participation in . . . the Ramsar Convention" by those countries."

The term "wetland" covers areas of marine water less than six meters deep at low tide, and for this inventory, coral reefs are being excluded—as the World Conservation Monitoring Centre completed such a study.

▲ Projects and People

Although ornithological consultant **Derek Scott** writes that he has "no official links to any organization," he works closely with IWRB and other conservation groups in an advisory capacity, most recently as international coordinator of the Oceania Wetland Inventory project. He periodically consults with the IUCN on **designing wetlands programs** for Vietnam and Tanzania, for example. He is also a World Bank consultant on projects such as "preparing a series of maps of Asian wetlands" and on an environment component of a forestry loan to **Bangladesh**. He previously was an advisor in ornithology to **Iran's** Department of the Environment; IWRB consultant in the West Palearctic; team member for British Ornithologists' Union/World Wildlife Fund (BOU/WWF) doing **avifaunal surveys** in the forests of southeastern **Brazil**; WWF consultant to **China** on a proposed nature reserve and marsh survey; and international coordinator of the **neotropical wetlands project**, including neotropical waterfowl census report, and the **Asian Wetlands Inventory**, including directory. Early studies concentrated on the **storm petrel**.

■ Resources

The Inventory culminates in the publication of a *Directory of Wetlands in Oceania* and a database. Scott has authored, singly or jointly, more than 30 publications including several directories and inventories of wetlands.

International Union of Air Pollution Prevention Associations (IUAPPA) ⊕
136 North Street
Brighton BN1 1RG, UK

Contact: **John Langston**, Director-General
Phone: (44) 273 26313 • *Fax:* (44) 273 735802

Activities: Education; Political/Legislative • *Issues:* Air Quality/Emission Control; Energy; Health; Waste Management/Recycling • *Meetings:* World Parks Congress

● Organization

The **International Union of Air Pollution Prevention Associations** (IUAPPA) describes itself as "a union of national professional or voluntary, nonprofit making associations concerned with the maintenance of clean air." Since its founding by 7 national associations in 1964, the Union has grown to include 28 members and observers representing 30 countries, working "to promote the public education, worldwide, in all matters relating to the value and importance of clean air and methods and consequences of air pollution control." Central to the organization is the Secretariat, which serves as the communications link for member organizations, the administrative center, and the information office.

The IUAPPA deals with, among other organizations, the United Nations (UN), Organization for Economic Cooperation and Development (OECD), European Community (EC), and groups and individuals representing national governments, professional bodies, and industry worldwide.

▲ Projects and People

A major IUAPPA activity involves convening triennial World Congresses on air quality, which are hosted by member associations in rotation. The host member is responsible for all aspects of the Congress, subject to Secretariat approval, and for publishing the proceedings. Typical Congress participants include representatives from national and local governments, international organizations, industry, professional bodies, academia, and the press, as well as private individuals. Those attending have "the opportunity to exchange valuable experience and research, and to establish and maintain contact with key individuals in all facets of air pollution control."

At the Conferences, and at every other opportunity, IUAPPA members pursue an agenda of information exchange—discussing air pollution legislation and control techniques; encouraging the use of uniform terminology, monitoring, and measuring methods; and networking.

In 1987, the IUAPPA Fund providing Educational Assistance for Developing Countries was established. Fund money may go to air pollution control associations that wish to establish their own journals, publish and disseminate information in the national language, or fund/expand libraries.

IUAPPA officers as of 1990 include President G. Steve Hart, USA/Canada; Vice Presidents Martin Rivers, USA/Canada, and Eija Lumme, Finland. The director-general of permanent secretariat is John Langston, also secretary general of the National Society for Clean Air and Environmental Protection, UK, and a fellow in both the Royal Society of Arts and the British Institute of Management.

■ Resources

The quarterly *IUAPPA Newsletter* appears in English and French; it informs members of clean air activities and developments in legislation and control worldwide. The *Members' Handbook*, published every three years, lists the addresses, leading officials, aims, and activities of IUAPPA members, and includes the Union's Reports and Accounts, and other information. *World Congress Proceedings* are published every three years by the host member association. *Clean Air Around the World, the Law and Practice of Air Pollution Control in 14 Countries in Five Continents* was published in English in 1988. The reference library at the Secretariat Headquarters contains an international collection of journals, conference and congress proceedings, and information on air pollution legislation and standards worldwide.

World Congresses are also occasions to bestow triennial awards, including the Christopher Ernest Barteh, Jr. Award "to an individual whose contributions of a civic nature . . . have aided substan-

tially in the abatement of air pollution." The IUAPPA first World Clean Air Congress Award to an individual or group making "a contribution of outstanding significance internationally to the progress of science or technology pertaining to air pollution prevention" was scheduled to be presented in 1992 at the ninth Congress, Montreal.

Jersey Wildlife Preservation Trust (JWPT) ⊕
Les Augrès Manor, Trinity
Jersey, Channel Islands JE3 5BF, U2K

Contact: **Phillip Coffey**, Education Officer
Phone: (44) 534 64666 • *Fax:* (44) 534 65161

Activities: Education; Research (Captive Breeding) • *Issues:* Biodiversity/Species Preservation

● Organization

Gerald Durrell founded the Jersey Wildlife Preservation Trust (JWPT) in 1963, for the express purpose of breeding rare and endangered species of animals in captivity. The USA-based Wildlife Preservation Trust International (WPTI) and Wildlife Preservation Trust Canada are supporting organizations; altogether there are 16,000 members. At Les Augrès Manor, parkland, waterfalls, and gardens with magnolia, climbing rose, iris, acacia, bamboo, and other plants of the country origin of the animals grace the Zoological Park. *(See also Wildlife Preservation Trust International [WPTI] in the USA NGO section.)*

▲ Projects and People

In JWPT's specific breeding projects at its headquarters, the ultimate goal is eventually to release the captive-bred animals into their natural habitat; when this is not possible, as is often the case, the Trust loans animals to zoos around the world, or exchanges animals with them. Because of limited resources, the Trust concentrates on a few areas, in particular northern India and Nepal, southeastern Brazil, the Caribbean, Madagascar, the Mascarene Islands, and Mexico. In most cases, the Trust works closely with governments, other conservation organizations, and native populations.

With the National Zoo of Washington, DC, the Trust has worked on a project to conserve the golden lion tamarin; in 1987 the first group of captive-bred tamarins were introduced into natural habitat in Brazil, with radio collars to enable scientists to monitor their condition and the success of the program. Other successful reintroductions have involved the thick-billed parrot of Mexico (that once also inhabited the pine forests of Arizona and New Mexico) and the St. Lucia parrots of the Caribbean. In many cases, commercial airlines donate cargo space and care to the transported animals.

In 1978 the Trust initiated its International Training Centre, and has since then taught more than 250 trainees from around the world specialized captive breeding and conservation techniques. The courses last between six and sixteen weeks; in addition, the Trust offers a three-week summer school course on the breeding and conservation of endangered species.

The Trust's breeding program has had numerous successes. For example, the first pair of **babirusa**—or "pig-deer"—of Sulawesi and several off-shore islands in **Indonesia**, came to the Trust in 1984. According to the Trust, "The natives of one island hunted them for wedding feasts, and now due to their scarcity, weddings are having to be postponed." JWPT has successfully bred the species, and has sent pairs to other zoos for further breeding efforts. Other successful mammal breeding programs have involved the **spectacled bear, snow leopard**, various species of **marmosets, lemurs** at **Madagascar's Parc Tsimbazaza, gorillas,** and **orangutans**, among others. Also in Madagascar, breeding programs are underway for the **radiated tortoise** and the very rare **ploughshare tortoise**.

The government of **Morocco** recently agreed to an international cooperative breeding program with the Trust for the **Waldrapp ibis**, which now breeds only in one colony in Turkey and in Morocco. The first of birds arrived in Jersey in 1986, joining several already bred in captivity there. In addition, since 1984, the Trust has been involved in the **Bali Starling Project Indonesia**; since then, two pairs of Jersey-bred starlings have returned to Indonesia, there breeding with mates arriving from other zoos around the world. Eventually, the birds will be released into the National Park of Bali Barat. The program includes a significant educational component, teaching the local population of the importance of conserving this bird and its habitat.

At the Trust's **Gaherty Reptile Breeding Centre**, among the most important species in the collection, are on loan from Mauritius's 375-acre **Round Island**—the skink, gecko, and boa. Jersey has bred more than 200 geckos from its original 16 since 1976, as well as more than 150 skink and several boas. JWPT reptiles section head **Simon Tonge** and senior herpetology curator **Quentin M.C. Bloxam** both belong to the Species Survival Commission (IUCN/SSC) Tortoise and Freshwater Turtle Group.

Also regarding Mauritius, after years of captive breeding and study, the rare **pink pigeon** is being returned to its native forest. Also being released is the "world's rarest kestrel, *Falco punctatus*." Captive breeding programs are also ongoing on the country's **Rodriques fody** and **Rodriques fruit bat**.

The Trust's zoological director is **Jeremy Mallinson**.

JWPT communicates with an "expanding network of zoos" through ARKS (Animal Records Keeping System).

■ Resources

The Trust's Education Centre receives some 8,000 pupils annually. The *Dodo Dispatch* is the Trust's educational newsletter for school children. *Before Another Song Ends* is an audiovisual program that tells the Trust's story to visitors.

The Trust's journal, *Dodo*, contains updates of the animals in the current collection. Since 1963, the Trust has published hundreds of articles covering the results of fieldwork from projects around the world. Also, the Trust has been keeper of the International Studbook for the **golden-headed lion tamarin** and of Regional Studbooks for the Goeldi's monkey, gorilla, and thick-billed parrot.

Landscape Institute (LI)
6/7 Barnard Mews
London SW11 1QU, UK

Contact: **Peter Broadbent**, Registrar
Phone: (44) 71 738 9166

Activities: Education • *Issues:* Habitat Preservation; Sustainable Development

● Organization

Founded in 1929 as the Institute of Landscape Architects, the **Landscape Institute (LI)**, as of 1978, "serves the three divisions of professional landscape work—design, sciences, and management." With a secretariat based in London, the Institute's activities are organized through a central standing committee system of local chapters. The Institute has strong links with professional bodies in many other countries, particularly in Europe, the Commonwealth, and the USA; graduates from many of these countries can become members. The Institute represents British landscape professionals, registering their practices and classifying design firms in terms of the number of landscape professionals employed, and the type of landscape dealt with—arid, tropical, semiarid, subtropical, temperate, or Mediterranean.

LI helped found the **Joint Council of Landscape Industries**, responsible for developing formal contracts for implementing landscape works. It also has an important role in the **International Federation of Landscape Architects** and helped create the **European Foundation for Landscape Architects** in preparation for the single European market of 1992. Other organizations with whom LI deals include the Tree Council, National Trust, and Royal Town Planning Institute.

The Institute is the primary standard setting body for entry, professional qualification, and methods of operation for the three landscape practitioner professions, according to its literature. Associate of the Landscape Institute (ALI) is "a qualification of equivalent status to professionally qualified architects, civil engineers and quantity surveyors."

Through LI officers, council, committees, regional chapters, and secretariat, the Institute represents and promotes the landscape professions "with national and local government, other official bodies and related professions and organizations." In addition, it nominates landscape practices for commissions at the request of clients, and gives career information and advice.

The size and impact of new developments such as housing, roads, factories, and power stations, and even parks, gardens, and reclamation projects, "can pose a significant threat to the environment." The Landscape Institute advocates careful site selection and landscape design, conservation, and rehabilitation to bring livability and sustainability into development projects. According to the Institute, "Landscape science and landscape management have become disciplines central to conservation of the world's resources. Changes in our environment are inevitable. They must be planned and managed with vision and expertise."

Institute presidents serve two-year terms. The most recent past president was **Andrew Bannister**; **Hugh Clamp** was named president in 1991.

▲ Projects and People

Clamp became an architect some 40 years ago, and in the mid-1970s, he joined the LI, acquiring a diploma in landscape architecture. On the LI Council for several years, Clamp, as president, is working to ensure that agricultural land set aside by the government is put to good use, warding off the prospect "of coast-to-coast golf courses." Clamp believes that landscape architects should not just *design* new landscapes, but also *train* farmers "in alternative uses for their land." He hopes to work with the National Farmers' Union toward this end. Other priorities include preserving "the sanctity of the Green Belts" around cities, and helping inner cities "realize their potential as complete living environments rather than continuing to exist as grey, hostile places from which the inhabitants escape at weekends if they have the means to do so."

As an example of LI involvement in projects, Northern Ireland businessman Roy Baillie of W and G Baird enlisted the Institute's help "after being told by a client that the attractive grounds of his offices had been a factor in their choice of his company as a supplier." With the local LI chapter and the local small business agency, as well as other official support, Baillie launched the Business Environmental Endeavour Awards in 1989, with 77 entrants to the competition in its first year. The Institute itself bestows annual Maintenance Awards.

Currently, LI is trying to work more closely with the Royal Town Planning Institute (RTPI) on questions of urban renewal, energy, and other relevant development topics. The LI acknowledges the importance of environmental legislation, and stresses areas that need special effort: increasing the size of its Sciences Division; careful preparation for a United Europe; addressing landscape issues in the Third World; and dealing with recent educational legislation and developments.

As member Jay Appleton wrote in a recent issue of *Landscape Design*: "We all have to live in a continuum whose poles are represented respectively by total wilderness and total urbanization. . . . We have little option but to find compromises somewhere in between."

■ Resources

The Institute's journal, *Landscape Design*, appears ten times each year, between issues of the monthly periodical *Landscape Design Extra*—both free to members. *Plant User* is enclosed with *Landscape Design* three times a year, and "aims to improve communication between plant producers and professional plant users." The Institute also operates a reference library.

Men of the Trees (MOT)

(See International Tree Foundation in this section.)

National Association for Environmental Education (NAEE)

Wolverhampton Polytechnic, Walsall Campus, Gorway
Walsall WS1 3BD, UK

Contact: **Philip Neal**, General Secretary
Phone: (44) 922 31200

Activities: Education • *Issues:* Environment

● Organization

The National Association for Environmental Education (NAEE) is "the association concerned with all aspects of environmental education," or EE, which it describes as "the recognition of values, clarification of concepts and development of the skills and attitudes needed to understand the inter-relatedness of man, his culture and his biophysical surrounding, practice in decisionmaking, concern for environmental quality and the adoption of a code of behavior."

Membership, now approximately 2,500, is open to anyone involved with EE in educational institutions. Student membership is open to those beginning teacher training. With a staff of three, the Association, either solo or through the Council for Environmental Education (CEE), lobbies for financial and other support for environmental education in schools. It focuses its efforts on the Department of Education and Science, Examination Boards and local authorities, working toward, for example, teacher training in environmental education.

Because of fairly recent developments in the UK's national curriculum, environmental education forms part of the curriculum in all schools and institutes of higher education. Specifics are discussed in the National Curriculum Council's publication *Curriculum Guidelines 7*. NAEE strongly endorses using the environment "as a source of motivation and experiences" for primary school students to develop literacy and awareness through nature observation, examination, and recording. In high schools and colleges, more formal courses in environmental studies or science are available.

▲ Projects and People

The Association serves primarily to channel information and material from outside bodies to teachers, and serves on many national bodies and examination boards. Its "members are linked with others . . . who believe that . . . problems of the environment can only be solved long term through education and that future citizens should be taught to respect, and to learn from, the world around them." Some NAEE funding comes from the Department of the Environment.

Unfortunately, because of economic constraints, some districts in the UK are finding it necessary to cut EE facilities such as learning farms where children can get hands-on experience. According to general secretary Philip Neal: "At the behest of the NAEE, the CEE is obtaining the facts concerning the fate of environmental study centres, rural and urban, nationwide so that combined action can be taken."

Through its *Statement of Aims* and through regular national conferences, NAEE clarifies and pursues its objectives of environmental education at all levels in detail. In addition, "members are encouraged to help in study conferences or working parties to carry out research, construct syllabuses or suggest practical teaching methods."

Patron of the NAEE is the honorable **Lady Bowes Lyon**; president is **Dr. David Bellamy**, who is also president of WATCH, the Wildlife Trust's junior wing.

■ Resources

The NAEE is seeking more educational outreach, including information and networking with details of activities.

Members receive *Environmental Education*, published three times yearly and, on special request, *Connect*, the UNESCO-UNEP (United Nations Environment Programme) education newsletter. NAEE produces its own newsletters, a series of practical teachers guides, and other publications.

National Society for Clean Air and Environmental Protection (NSCA)
136 North Street
Brighton BN1 1RG, UK

Contact: **Kenneth S. Dunn**, Secretary General
Phone: (44) 273 26313 • *Fax:* (44) 273 735802

Activities: Education; Political/Legislative; Research • *Issues:* Air Quality/Emission Control; Deforestation; Energy; Global Warming; Health; Noise; Sustainable Development; Transportation; Waste Management/Recycling; Water Quality

● Organization

The **National Society for Clean Air and Environmental Protection** (NSCA) has its origins in the Coal Smoke Abatement Society founded in 1899. A founding member of the International Union of Air Pollution Prevention Associations (IUAPPA), and UK's member of the **International Association Against Noise**, the Society "brings together pollution expertise from industry, local and central government, technical, academic, and institutional bodies." The NSCA describes itself as a respected consultant for national legislation and regulation, and campaigns on "international pollution problems, noise, energy, waste disposal and other environmental issues."

The Society is organized into a governing council, five specialist committees, administrative headquarters, and twelve U.K. geographical divisions, which arrange meetings and activities and consider local issues. Besides electing their local councils, the Divisions elect representatives to serve on the Society's national council. Divisional Councils are headed by an elected chairman (president in Scotland), with day-to-day matters being handled by the honorary secretary. The various Divisions, annual Conference attendees, and the Society's governing national Council can initiate policy discussion and formulation. Standing committees—technical, noise, and parliamentary and local government—comprise nominated Council members in addition to co-opted specialists, and are responsible for working out policy details. A steering group deals with urgent matters, and policy changes are ratified by the Council.

The Society has a long history of established relations with central and local government representatives and agencies, industry, professional bodies, environmental groups, the media, and the European Parliament. Thus, membership "offers the opportunity of influencing environmental decisionmaking at the very highest levels."

▲ Projects and People

NSCA objectives include "environmental improvement by promoting clean air through the reduction of air, water and land pollution, noise and other contaminants while having due regard for other aspects of the environment." Current efforts are being aided by a grant from the Department of the Environment.

Regular NSCA-sponsored conferences and workshops, especially the **annual National Conference**, involve specialists and professionals from a wide range of areas; through these events members can "establish and maintain contacts with key individuals in all facets of environmental protection." NSCA officials brief the government and media on aspects of environmental policy; "the Society also comments on government consultation papers and gives evidence to select committees and other commissions of enquiry when appropriate."

Issues range from acid rain and agricultural pollution to waste disposal and incineration—and also include auto polution, asbestos, diesel smoke, heavy metals, global warming, nuclear power, motorcycle noise, lead in gasoline or petrol, and energy efficiency, for example.

■ Resources

The NSCA press publishes conferences and workshop proceedings, technical papers, books, reports, and education material such as readings on acid rain and the greenhouse effect, teaching packs, and *The Junior Guide to Air Pollution*. The annual *NSCA Pollution Handbook* is a standard reference book on the law and practice of environmental protection, and the journal *Clean Air* is published quarterly. Many of the publications are *Briefings*, with about 25 published each year giving the "NSCA view on matters of immediate environmental concern, pointing out developments in policy, or giving information on environmental matters."

The NSCA Information Department provides members with current information on national and international legislation and standards, advice on controls, and referral to other expert bodies.

The National Trust for Places of Historic Interest or Natural Beauty
Estates Advisers Office, Spitalgate Lane
Cirencester GL7 2DE, UK

Press Office
36 Queen Anne's Gate
London SW1H 9AS, UK

Phone: (44) 71 222 9251
Fax: (44) 71 222 5097

Contact: **H. John Harvey, Ph.D.**, Chief Adviser on Nature Conservation
Phone: (44) 285 651818 • *Fax:* (44) 285 657935

Activities: Development (Conservation) • *Issues:* Biodiversity/Species Preservation

● Organization

In 1881, activists **Robert Hunter, Octavia Hill,** and **Hardwick Rawnsley** fought together successfully to maintain public access in the Lake District—the beginnings of the **The National Trust for Places of Historic Interest or Natural Beauty.** Formally approved in 1894, the Trust had its first meeting in 1895, when it was given its first property, 4.5 acres of cliffland overlooking Cardigan Bay, called Dinas Oleu. In 1907, Parliament passed the **National Trust Act,** so that only Parliamentary permission could allow Trust land to be taken away from it. The first National Trust Centre, founded in Manchester in 1948, was the precursor to the current 167 supporting associations.

In nearly a century, the Trust has grown substantially, increasing from 100 members in 1895, to 1 million in 1981, and now more than 2 million. The staff of 2,500 oversees the approximately 570,000 acres owned and 78,000 acres covenanted by the Trust. The Project Neptune Appeal alone covers nearly 116,000 acres of coastal land along 521 miles of coast. The acreage includes more than 300 properties open to the public; 18 national nature reserves; 54 nature reserves leased to county and local trusts; woodland, heath, and pasture; archaeological sites including Stonehenge and "a substantial length" of Hadrian's Wall; dozens of houses, castles, and other structures; and even 59 "villages and hamlets wholly or largely Trust-owned."

In the realm of education, the Trust rolls include 3,750 schools as "corporate members." An educational adviser was appointed in 1978, along with the launching of the **Young National Trust Theatre Company** and **Acorn Camp** scheme—"week long working holidays for young people at Trust properties."

The Trust's nature conservation department employs a staff of nine including a chief, two advisers, four biological surveyors, and two technicians.

▲ Projects and People

Primary Trust activities involve conservation of Trust-controlled landscape, wildlife, gardens, and buildings and their contents, as well as fundraising to support such conservation.

Recent or ongoing fundraising appeals related to nature conservation on a national scale target the following areas or needs: In the **Lake District,** Trust funding "has already paid for extra workers, footpaths, repair teams, woodmen, and others who can maintain Lakeland's magnificent landscape . . . and respect the local agricultural community which is the very basis of the character of the lakes." The **Trees and Gardens Storm Disaster Appeal** in February 1990 "sought to . . . cover the cost of clearing, preparing, and replanting trees and gardens, damaged or destroyed during the previous month by some of the worst series of storms this century. **Enterprise Neptune** has since 1965, when it was launched by the **Duke of Edinburgh,** funded the Trust's acquisition of more than 520 of the 900 miles of coastline threatened by housing, industrial development, oil and gas exploration, marinas, new roads, and other encroachments. Efforts have focused on Cornwall, northeast England, and Wales—where a parallel effort called **Apel Glannau Cymru (Wales Coastline Appeal)** was launched in 1990.

On a regional level, the Trust currently owns more than 23,000 acres of countryside in Cornwall, Devon, Yorkshire, and other key spots. Special fundraising efforts include the **Snowdonia Appeal,** led by actor **Anthony Hopkins,** for the conservation of nature in Wales. Its goal is to collect 2 million pounds by 1995—the year of the Trust's Centennial—in order to "finance the protection, preservation and care of . . . 50,000 acres of beautiful Snowdonia." Included in the plan is a work program, which will greatly aid the local economy by providing training and job opportunities.

The Trust's **Shropshire Hills Appeal** has increased fundraising efforts in order "to buy more land at risk, to give more public access, to improve educational and visitor facilities and to provide further help to the warden, tenants and volunteers in caring for this superb countryside." The **Goring Gap Appeal,** launched in 1989, aims to reclaim 114 acres of chalk downs and woodland called The Holies, healing the marks of dirt bikes and skeet shooting.

Activities in 1990 included opposing a British Aerospace proposal to construct a huge repair depot to the west of Liverpool's Speke Hall, one of Britain's most important half-timbered buildings. The Trust defeated another proposal, to extract sand and gravel near the landscape garden at Stowe, Buckinghamshire. In addition, the Trust pressed for amendments to the proposed **Environmental Protection Bill,** so that it covers landscape and access issues. In conjunction with the National Trust for Scotland, the Trust hosted the first conference of European conservation organizations, **Europe Preserved for Europe,** at York University in 1990; 40 groups from 15 countries were represented. Now, planning is underway for the Trust's centenary celebrations in 1995.

Dr. Henry John Harvey, a specialist in plant ecology and former Cambridge educator, has been the National Trust's chief adviser on nature conservation since 1986. A member of the British Ecological Society and the Recreation Ecology Research Group, Dr. Harvey also belongs to numerous ornithological and nature conservation clubs and trusts in Cambridge, Cornwall, and all England, and has authored 20 major and 50 minor publications, jointly or singly.

Previous Trust Chairman was **Dame Jennifer Jenkins,** from 1986 to 1991; since then **Lord Chorley** has served in the position. Trust Director-General is **Angus Stirling,** and Director of Public Affairs is **Brian Lang.**

■ Resources

The Trust is in need of help in habitat management of all types, for improved nature conservation performance; in addition, improved public relations with members would encourage greater involvement.

The Trust produces a large number of publications, "ranging from national focus to individual property and covering all aspects of organizations work for a variety of age groups." These include nature trail guides, a young people's newsletter, members' newsletters, property guidebooks, and area leaflets. Besides *The National Trust Guide,* the *National Trust Handbook* is published annually. In 1990, the Trust launched two educational initiatives: *Cross Currents,* sponsored by Midland Bank, is a coastal studies handbook for the classroom; and the *National Trust Educational Supplement,* published twice a year, provides news and comments on matters relating to education at National Trust properties and is sent free to all 3,511 schools that are corporate members. A list of Trust books is available.

Institute of Terrestrial Ecology (ITE)
Natural Environment Research Council (NERC)
Brathens
Banchory AB31 4BY, UK

Contact: **Hans Kruuk, Ph.D.**, Senior Principal
Scientific Officer
Phone: (44) 3302 3434 • *Fax:* (44) 3302 3303

Activities: Research • *Issues:* Biodiversity/Species Preservation; Global Warming; Water Quality

● Organization

The **Institute of Terrestrial Ecology (ITE)** was established in 1973 within the **Natural Environment Research Council's (NERC)** Terrestrial and Freshwater Sciences Directorate. With north headquarters at Edinburgh and south at Monks Wood, ITE's 180 scientific staff work at 6 research stations around the UK. ITE also collaborates with sister institutes within NERC, drawing upon an extensive range of scientific specialties.

ITE Edinburgh is part of a consortium comprising NERC, University of Edinburgh, Royal Botanic Gardens of Edinburgh, and the International Forest Science Consultancy, forming the new Edinburgh Centre for Tropical Forests, reports ITE.

Research is financed by the government, as well as by private and public sector clients and international organizations. Current projects include joint research programs in Europe, the Cameroons, Kenya, Nigeria, the Sudan, China, and South America.

▲ Projects and People

ITE research focuses on "wise management of natural resources," including several long-term projects investigating the factors controlling species abundance. The Institute stresses that "ecological knowledge is as important in agriculture and forestry as it is fundamental to wildlife management." Projects in 1990 included work on a **comprehensive survey of Britain**, with data obtained by monitoring "land use, land cover and vegetation in 197 representative sites in Scotland."

The survey work complemented development of a **satellite imagery–based map** of the complete land cover of **Scotland** by the Environmental Information Centre at ITE Monks Wood. Other Monks Wood facilities include the Biological Records Center, British National Space Centre's Remote Sensing Applications Development Unit, and the Eastern Rivers Group of the Institute of Freshwater Ecology. Station head is **Dr. M.D. Hooper**, and southern headquarters director is **Dr. T.M. Roberts.**

ITE scientists work to develop **mathematical models** simulating natural conditions, for example, "nutrient cycling in **forests**, the population dynamics of **swans and gamebirds**, and the effects of changes in **land use and climate**." These are then used to predict systemic reactions to environmental changes caused by pollution—especially greenhouse gases, erosion, and other threats.

At the Edinburgh station, headed by **Dr. M.G.R. Cannell**, the focus is on trees, especially **tropical forestry**. The station also houses ITE's north headquarters, under director **Dr. O.W. Heal**. The Merlewood research station, under **Dr. M. Hornung**, covers mainly **woodland ecology, soil microbiology,** and **nutrient cycling**—in general, land use; it is also the base for the ITE **Chemical Service**. **Dr. J.E.G. Good** heads the Bangor station in Wales, which specializes in **geochemical cycling** and land use, montane ecology, and the effects of **pollutants on ecosystems.**

At Furzebrook, headed by **Dr. Michael George Morris**, special topics include feeding and population ecology of **migrant and resident vertebrates**, populations and **genetic ecology**, and community ecology, with particular expertise in environmental impact assessments (EIAs).

Banchory, the northernmost station, under **Dr. B.W. Staines**, specializes in **upland ecosystems**, including population ecology of vertebrates, environmental impact, and habitat reinstatement. An ongoing project involves the study of the predator-prey relationship between fish and various birds and mammals, "with the ultimate aim of providing data for careful management of the **aquatic habitats** in Britain." Study topics include the relationship between the quantity of **salmon** and **trout** eaten by **sawbill ducks, goosanders,** and **mergansers**, and later fish harvests; and the relation between the location of fish and that of **predator birds** (for example, red-breasted mergansers do not settle upstream, where young salmon exist in large populations, but downstream, where although there are fewer fish they are more easily caught because of lack of cover).

Banchory's 20-year study of **puffins** is "one of the few that has spanned the period of change in sufficient detail to understand the demographic processes that have been involved." Another study, on **otter ecology**, examines the relation between the decline in otter numbers and the "decrease in prey populations, increasing environmental pollution and disturbance, and major changes in the habitat . . . [including] fish productivity."

Dr. Hans Kruuk, senior principal scientific officer at ITE Banchory, has been with the Institute since 1975, studying mainly tropical and temperate mammals and birds, and wetland ecosystems. Dr. Kruuk has been involved in other projects, with UNESCO, World Wildlife Fund (WWF), and the Frankfurt Zoological Society; earlier, he co-founded and was deputy director of the Serengeti Research Institute in **Tanzania**. He has authored alone and jointly more than 85 reports, articles, monographs, and books, many focusing on gulls, hyenas, otters, and badgers.

■ Resources

ITE facilities comprise "analytical chemistry laboratories; the Environmental Information Center, which includes remote sensing, aerial photo interpretation, geographical information systems and the Biological Records Centre; pollution exposure chambers; a radiochemistry laboratory; tropical glasshouses; stable isotope unit; statistical analysis and networked interactive computing; engineering workshops; library services and information retrieval."

ITE research results appear in a wide range of scientific journals. Two series of ITE publications are published as well as an *Annual Report* and descriptive brochures. The Biological Records Centre holds national distribution maps of a wide range of species.

Furzebrook Research Station
Institute of Terrestrial Ecology (ITE) ⊕
Natural Environment Research Council (NEEC)
Wareham, Dorset BH20 5AS, UK

Contact: **Michael George Morris, Ph.D.**, Station
      Head
Phone: (44) 929 551518 • *Fax:* (44) 929 551087

Activities: Education; Political; Research • *Issues:* Biodiversity/Species Preservation • *Meetings:* International Congress of Entomology; International Union for the Study of Social Insects

▲ Projects and People

Institute of Terrestrial Ecology scientists **Dr. Michael George Morris** and **Dr. Graham W. Elmes** at ITE's Furzebrook Research Station describe other environmental involvements. *(See also Natural Environment Research Council in this section.)*

Dr. Morris—while head of both Furzebrook Sation and its subdivision of invertebrate ecology as well as terrestrial and freshwater science program leader, Natural Environment Research Council—is a member of the **Societas Europea Lepidopterologica (SEL)** (Europe Lepidoptera Society), a scientific society of about 800 amateur and professional lepidopterists. In 1990, he became secretary of the UK Committee for UNESCO's **Man and the Biosphere (MAB)** program. As consultant to the World Wildlife Fund (WWF) and other organizations, Dr. Morris has investigated **butterfly farming**, utilization, and conservation, focusing on Indonesia, Papua New Guinea, Solomon Islands, Singapore, and Malaysia.

Dr. Elmes, whose main focus is **ants**, and **ant-butterfly interactions**, is chairman of the Species Survival Commission (IUCN/SSC) Social Insect Group. With 10 to 20 members, the Social Insect Group grew out of the old Ant Group in 1990, in collaboration with the International Union for the Study of Social Insects (IUSSI). According to Dr. Elmes, "The aim over the next few years is to identify 'at risk' social insect species, providing lists and priorities for conservation efforts, [and] to further direct species conservation in those parts of the world where this is practicable and to identify conservation projects that would benefit from consideration of social insects in those parts of the world where the main problem is habitat destruction."

Royal Horticultural Society ⊕
Vincent Square
London SWP 2PE, UK

Contact: **Alasdair Andrew Orr Morrison**, Chairman,
      Orchid Committee
Phone: (44) 452 521747 • *Fax:* (44) 452 302612

Activities: Education; Research • *Issues:* Biodiversity/Species Preservation; Deforestation • *Meetings:* CITES; orchid conferences

▲ Projects and People

The Honorable **Alasdair Morrison**, who chairs the orchid committee, **Royal Horticultural Society** with its 150,000 members, comments: "I am very skeptical about both the aims and methods of CITES as it applies to plants—my direct familiarity is with its impacts on **orchids** in particular. The system, originally designed for animals but then as a pious afterthought applied to plants also, is seriously flawed—both unworkable and ineffective at the same time. I would like to see a concerted international effort made to devise a conservation regime specifically for plants (which took, for example, the importance of hybrids in commerce into account) and I am trying to get the Royal Horticultural Society to take a more positive role in this. The present system, at least so far as orchids are concerned, is an expensive and burdensome shambles."

The Royal Society (RS) ⊕
6 Carlton House Terrace
London SW1Y 5AG, UK

Contact: **Amanda Dale**, Press Office
Phone: (44) 71 839 5561 • *Fax:* (44) 71 930 2170
E-Mail: Telecom gold 01:ynk 139

Activities: Education; Research • *Issues:* Biodiversity/Species Preservation; Deforestation; Global Warming; Water Quality

● Organization

Formally known as the **Royal Society of London for Improving Natural Knowledge**, The Royal Society (RS) notes that "only through a better understanding of the complex mechanisms of, and interaction between, all parts of the physical, geological and biological worlds can we begin to predict the consequences of Man's activity." Covering a broad range of scientific disciplines, The Royal Society "works by offering **grants and fellowships for research**, done in other research establishments," as well as through scientific symposia and publications, participation in international **collaborative research programs**, and providing "expert scientific advice on issues of public concern." The Society currently has some 1,200 elected fellows.

▲ Projects and People

A recently completed project was the five-year (1985–1990) **Surface Water Acidification Programme (SWAP)**, on which The Royal Society collaborated with the Norwegian Academy of Science and Letters and the Royal Swedish Academy of Science. Directed by Sir John Mason and under a Management Group chaired by Sir Richard Southwood, SWAP investigated "acidification processes in lakes and rivers, the effects on fish and the role of acid deposition."

In the **South-east Asia Rainforest Programme**, launched in 1985, The Royal Society is funding UK scientists studying the rainforest ecology in the Danum Valley of Sabah, northern **Borneo**. The program focuses on the **effects of logging** and other disturbances on the dynamics of tropical rainforests. It is a cooperative effort with the Danum Valley Management Committee.

Another ongoing project is the **Baikal International Centre for Ecological Research**, in Irkutsk, Siberia, of which the RS is a founding member. The Society has already provided funding for at least five research groups, with more to follow. Lake Baikal, described as the Earth's largest body of fresh water, provides "opportunities to study plankton, water chemistry, speciation and evolution of animals and plants, some unique to the lake, and geological fault systems."

As the UK national member of the International Council of Scientific Unions (ICSU), the Society participates in ICSU activities as well, underlining two in particular. The World Climate Research Programme (WCRP), a joint effort of the World Meteorological Organization (WMO) and ICSU, "aims to determine to what extent climate can be predicted and the extent of man's influence on it." Begun in 1980 and slated to run for 20 years, the program focuses on ozone depletion in its two major programs, the Tropical Oceans and Global Atmosphere (TOGA) project and the World Ocean Circulation Experiment (WOCE), and has proposed a third, the Global Energy and Water Cycles Experiment (GEWEX).

As part of the International Decade of Natural Disaster Reduction (IDNDR) 1991–2000, the RS and the Fellowship of Engineering are sponsoring a UK Science, Technology and Engineering Committee, which "has initiated steps to identify UK scientific, technological, engineering and medical expertise relevant to mitigating the effects of natural disasters." The Committee has also stressed the need to train scientists and technicians in the developing countries who are assigned to hazard minimization and disaster reduction. In addition, the Committee is responsible for the UK reactions to the recommendations of the ICSU special Committee for IDNDR, prioritizing those projects in the areas of volcanoes, earthquakes, and tropical cyclones which are best suited to UK expertise.

Approximately 180 of the RS Research Fellows and Professors are at UK universities, investigating a range of topics, from mathematics to medicine, and including "population biology and ecology, air-sea interaction and the dynamics of waves, land management techniques and conservation, fuel combustion and engine design, and resistance to insecticides." In addition, the Society subsidizes studies through Overseas Field Research grants and marine science grants, and runs exchange programs with foreign scientific academies and universities, encouraging both individual and collaborative projects to be pursued. Pointing out that "no one in 1949 thought that the seemingly academic study of free radicals in chemistry would be crucial to our understanding of CFCs (chlorofluorocarbons) and the ozone hole. . . . The Society supports fundamental research in all areas to ensure our deeper understanding, to provide a base to solve the problems of the future."

Committed to the principle that "a basic understanding of science in all disciplines is necessary if individuals are to be able to appreciate . . . environmental issues and to make informed decisions," the Society has joined the British Association for the Advancement of Science and the Royal Institution in the Committee on the Public Understanding of Science (COPUS), working on activities that will improve the public's understanding of science and the activities of the scientific community. The Society advocates the inclusion of science in the school curriculum, on a general basis until age 16, and a more specific agenda for ages 16 to 19.

■ Resources

Fellows of the Society have contributed to submissions to parliamentary or governmental inquiries, including: *The Nitrogen Cycle*; *Coal and the Environment*; *The Greenhouse Effect*; *Marine Science in the UK*; and *The Effects of Oil Pollution*. A study on pollutant control priorities was also slated for release.

Royal Zoological Society of Scotland ⊕
Edinburgh Zoo, Murrayfield
Edinburgh, Scotland EH12 6TS, UK

Contact: **Roger J. Wheater**, Director
Miranda F. Stevenson, Zoo Curator
Phone: (44) 31 334 9171 • *Fax:* (44 31 316 4050

Activities: Education • *Issues:* Biodiversity/Species Preservation; Captive Breeding • *Meetings:* ISIS; IUCN/SSC CBSG; IUDZG

● Organization

The Edinburgh Zoo has a staff of 120, including mainly zookeepers and wardens, as well as caterers, works/training staff, and administrators. The Royal Zoological Society of Scotland has 12,000 members.

▲ Projects and People

Director of the Royal Zoological Society of Scotland since 1972, **Roger J. Wheater** was also president of the 140-member International Union of Directors of Zoological Gardens (IUDZG) from 1988 to 1992. Much of his early career was spent in Uganda, where from 1961 to 1970 he was chief warden of the Murchison (Kabalega) Falls National Park, and from 1970 to 1972, director of Uganda National Parks. Among his many positions and memberships, Wheater has been vice chairman of the National Federation of Zoological Gardens of Great Britain and Ireland since 1980 and member of the UK Committee of the World Conservation Union (IUCN) since 1988. He has served as consultant to the World Bank since 1974 and to the United Nations World Tourist Organization since 1980. Zoos in Indianapolis (USA), Johannesburg, and Oman have sought Wheater's expertise. He has authored more than 20 articles and reports, gives some 35 lectures each year, and speaks frequently on radio and television.

Dr. Miranda F. Stevenson has been curator of animals at the Edinburgh Zoo since 1978, having obtained her doctorate degree in animal behavior from University College of Wales, Aberystwyth, studying playful interactions of the marmoset. Among other organizations, Dr. Stevenson is a member of the Association for the Study of Animal Behavior, International Primatological Society, Jersey Wildlife Preservation Trust, and Fauna and Flora Preservation Society. In charge of the zoo's animal collection of mammals, birds, reptiles, and amphibians, she oversees a staff of 35 in dealing with matters of "animal husbandry and diets, breeding projects, some research and animal records," as well as medical treatment in cooperation with the zoo's veterinary surgeon. Her work with the animals also entails knowledge of national and international import/export laws. Dr. Stevenson is both UK regional and international Studbook Keeper for the endangered diana monkey species, and a member of the Species Survival Commission (IUCN/SSC) Primate and Captive Breeding Specialist groups. Author of more than 35 research publications, since 1985 she has taught conservation and socioecology to honors students in primate ecology at the University of Edinburgh.

■ Resources

The Edinburgh Zoo publishes an *Annual Report, Park Guide Edinburgh, Park Guide Highland Wildlife Park,* and *Arkfile.* Publication of a *World Zoo Conservation Strategy* is in progress.

Societas Internatonalis Odonatologica ⊕
International Odonatalogy Society

The Farm House, Swavesey
Cambridge CB4 5RA, UK

Contact: **Norman W. Moore, Ph.D.,** Retired
      Scientist
Phone: (44) 954 30233

Activities: Research • *Issues:* Biodiversity/Species Preservation
Meetings: International Symposia, Societas Internationalis Odonatologica

● Organization

The world's rainforests provide habitat for most of the 5,500 known species of dragonfly (*Odonata*) in the world. Members of the **Societas Internationalis Odonatologica (International Odonatalogy Society)** set research and conservation priorities for conserving the species, supporting each other and conservation organizations in the effort to conserve dragonfly habitat, in particular the rainforest.

▲ Projects and People

As a member of the World Conservation Union's (IUCN) Invertebrate Conservation Task Force, a group of some 14 "leading professional scientists" representing all inhabited continents, **Dr. Norman W. Moore** is also chairman of the Species Survival Commission (IUCN/SSC) Odonata Group. In his long career, Dr. Moore has served as lecturer at the University of Bristol; regional officer for the Southwest England Nature Conservancy Council (NCC); head of NCC's toxic chemicals and wildlife division, and later NCC's chief advisory officer; and visiting professor for Wye College, University of London.

■ Resources

Funding is needed to cover basic administrative and travel costs.

Dr. Moore's publications include reports of the SSC Odonata Group; one or two reports per year published by Societas Internationalis Odonatologica; and numerous scientific papers and books on such topics as the effects of pesticides on wildlife, human impact on habitats, and dragonfly behavior and ecology.

TRAFFIC International ⊕

219c Huntingdon Road
Cambridge CB3 ODL, UK

Contact: **Jørgen B. Thomsen,** Director
Phone: (44) 223 277427 • *Fax:* (44) 223 277237
Telex: 817036 SCMUG

Activities: Consulting; Coordination; Research • *Issues:* Biodiversity/Species Preservation; Sustainable Development; Monitoring Trade in Wildlife • *Meetings:* CITES

● Organization

Trade Records Analysis of Fauna and Flora in Commerce, or **TRAFFIC,** describes itself as "the world's largest wildlife trade monitoring programme." The TRAFFIC Network, established in 1976, is an initiative of the World Wide Fund for Nature (WWF) and the World Conservation Union (IUCN). The Cambridge-based **TRAFFIC International** coordinates the Network's activities and monitors global wildlife trade issues, from headquarters shared with the **World Conservation Monitoring Centre.** In addition, there are regional offices in Europe, South America, Australia, and more recently East/Southern Africa and South East Asia. National offices operate in the USA, several European countries, and Japan. *(See also TRAFFIC in the Argentina, Malaysia, USA, and Uruguay NGO sections.)*

▲ Projects and People

TRAFFIC's primary activity involves monitoring for compliance with CITES (the Convention on International Trade in Endangered Species of Wild Fauna and Flora) to which more than 110 member countries are signatory. In order "to prevent international trade from threatening species survival in the wild . . . CITES establishes a regulatory system for controlling wildlife trade that is implemented by domestic legislation in the member countries." With the help of staff like **Steven Broad,** senior investigations officer, TRAFFIC monitors both illegal and legal trade, working to prevent the former and assess the impact of the latter. TRAFFIC will publicize instances of excessive trade in a particular species and work to tighten regulations. For example, "until the EEC [European Economic Community] banned the import of three types of **tortoises** in 1984, tens of thousands of these creatures came as pets from the **Mediterranean** each year. Eight percent were dead within one year of arrival, thus perpetuating the demand, and wild populations declined steeply."

TRAFFIC also is working to control the international trade in **plants,** helping to save species from extinction. Along with other organizations, TRAFFIC put pressure on companies marketing **bulbs of snowdrop, narcissus,** and **cyclamen,** many of which "are dug up from the wild in Southern Europe." As a result, the bulb companies now "differentiate between wild and propagated bulbs on their labeling." In 1989, TRAFFIC began to focus more intensively on the **tropical timber trade.** Current projects "include a detailed analysis of the tropical timber trade in **Papua New Guinea** and the **Solomon Islands.**" TRAFFIC is also working on studies of trade in reptile skins, and live birds and reptiles.

In order to counter public ignorance of the threats to the various species and of existing regulations, TRAFFIC works hard at **public awareness** projects. Through "exhibitions on wildlife trade problems, the presentation of lectures, production of information for the press, and publication of posters and literature," TRAFFIC spreads its conservation message.

■ Resources

In addition to occasional reports, TRAFFIC publishes an international journal, the quarterly *TRAFFIC Bulletin,* including "original studies of the trade in animals and plants, as well as news items on trade problems, changes in regulations and implementation of CITES."

UK2000 Scotland
c/o Scottishpower
201 Drakemire Drive
Glasgow, Scotland G45 9TD, UK

Contact: **David Westwood**, Executive Director
Phone: (44) 41 634 2155
Fax: (44) 41 631 1221, ext. 246

Activities: Development; Education; Project and Training
Issues: Energy; Habitat Creation; Sustainable Development; Waste Management/Recycling

● Organization

UK2000 Scotland was organized in 1987 with the **Scottish Development Department** and is a collaborative effort involving government, local authorities, leading environmental groups, and the business community, working together on "high quality environmental projects in cities, towns and the rural areas throughout Scotland." With a staff of 10 and multitudes of volunteers, UK2000, with 4 other main organizations—Community Service Volunteers, Keep Scotland Beautiful, Scottish Conservation Projects Trust (SCPT), and Scottish Wildlife Trust (SWT)—focuses on several priority areas: **inner cities, new environmental business,** and **alternative land use** in rural economies.

Sam Toy, named UK2000 chairman in 1990, notes that despite the welcome involvement and support of, for example, Scottish Minister for the Environment James Douglas Hamilton and the Rural Affairs Division of the Scottish Development Department, the work "cannot all be left to national and local government.... Business and commerce must acknowledge that the environment is just as important to our well being as the arts or sport."

▲ Projects and People

In 1990, UK2000 maintained "a core portfolio of 50 projects with a further 20 in various stages of development." By summer 1991, over 200 major projects had been developed and supported, as well as hundreds of event- and site-based activities.

One project centered on **Bluther Burn**, which flows into the Firth of Forth, and "has suffered years of neglect and abuse, both from the communities that live along its banks and by old mineworks and other sources of silt." UK2000's **Ron Macraw** has provided "manpower, coordination, and tools," and the Natural Resources Division of Lothian Regional Council's Planning Department, Dunfermline District Council's Planning Department, and the Forth River Purification Boards have worked together "to conserve and improve the Burn and to provide a recreational resource, ... encourage trout numbers, involve local angling groups, and gradually educate the local communities."

With UK2000 support, volunteers on the Isle of Arran have organized a **bottle recycling system** where there was none. The Arran Recycling Company now collects glass throughout the island each week, using earnings to purchase bottle banks and support future projects. The group recently received 500 pounds' Sterling

worth of plastic recycling boxes from British Telecom, making the collection process more efficient. Another recycling effort, **Recycling City**, is a joint Friends of the Earth/UK2000 initiative to set up **four model recycling bases** in the UK. Dundee, Cardiff, and Sheffield are currently active, and Devon is slated to join them as a "recycling county" in 1992.

Yet another UK2000-supported project began when the Buchan Countryside Group set up a **tree nursery**. By October 1989, the extremely successful nursery had become a trading company, **Buchan Woodlands Ltd.**, and planted more than 200,000 trees in its first season. In another success story, the Scottish Conservation Projects Trust (SCPT) created a **Footpath Team** in 1988 "to repair, maintain, and improve Scotland's ever more popular highland and lowland footpaths." Demand was so great that, in January 1990, it became an independent company, **Pathcraft**, "employing around 20 people with good prospects for further expansion in the near future."

The award-winning **Landwise Glasgow** project involves local communities in transforming unimproved backcourts or lots into gardens and play areas.

UK2000 support has also helped SCPT develop a **Training Centre** and curbside paper recycling project in Falkirk. The latter, a partnership with Stirling Fibre Ltd. and supported by Falkirk District Council, is a model project "to establish 'good practice' for . . . recycling in urban areas." By 1990, the project covered some 25,000 households, producing 75 tons of material per month, and planned to expand by 3,000 more households, possibly including other recyclable materials.

The **Scottish Wildlife Trust (SWT) UK2000 Unit**, under the Government's Employment Training program, has taught conservation and administrative skills to more than 150 trainees, preparing them for entry into municipal and guild employment. In 1990, the unit received the "Center of Excellence" award for its **environmental training program** from the Department of Employment. The unit includes **wildlife management projects** in six major towns including Edinburgh, and a "habitat creation project on formerly derelict land on the Montrose Basin." As an example of their activities, members of the Dumfries Habitat Survey Team learn skills like "surveying and recording techniques, **habitat mapping**, and identification of animal and plant life, while producing valuable data to add to the Trust's already extensive knowledge of Scotland's wildlife habitats." Thus, besides helping the environment, the trainees learn increasingly marketable skills.

In another project, the unit established a **hospital wildlife garden** in central Edinburgh. The goal is to provide "a variety of **therapy opportunities** for patients undergoing rehabilitation." Initial support came from SWT's UK2000 **Scotland Green Machine**—"a small but highly effective team who help set up urban greening projects, which eventually become self supporting." Also at Britain's first AIDS hospice, a wildlife garden was established, both as therapy, and as a natural habitat to attract "birds, small mammals, insects, and butterflies."

David Westwood was elected an associate of the **Landscape Institute** in 1983, and served as secretary of Landscape Institute, Scotland, 1986–1987. In 1987, he began working for the national UK2000 as development officer for Scotland, and "contributed regularly to strategic planning of UK2000 at both a Scottish and UK level particularly in the areas of training and marketing/media." Since 1989, Westwood has served as operations manager, leading to his election as chairman of the **Scottish Environmental Standards**

Group (SESG), formed in 1989 by "40 national agencies representing the main employment interest in environmental conservation in Scotland." A joint initiative of UK2000 Scotland and the Training Agency in Scotland, SESG is the official Scottish liaison group with the recently formed COSQUEC (Council for Occupational Standards and Qualifications in Environmental Conservation).

■ Resources

UK2000 Scotland publishes two quarterly newsletters, COSQUEC's *Green Side Up* and *Scotland's Environment Matters*.

Wildfowl and Wetlands Trust (WWT) ⊕
Slimbridge, Gloucester GL2 7BT, UK

Contact: **Brian Bertram, Ph.D.**, Director-General
Phone: (44) 453 890333 • *Fax:* (44) 453 890827

Activities: Education; Research • *Issues:* Biodiversity/Species Preservation; Wetland Conservation • *Meetings:* CBSG; IOC; IUCN

● Organization

With 45,000 members and a staff of 220, including a research team of 35, the **Wildfowl and Wetlands Trust (WWT)**, founded in 1946, "coordinates the efforts of 1,500 volunteers who contribute by collecting information about wildfowl and wetlands." The Trust also pursues research in conjunction with university departments.

▲ Projects and People

Sir Peter Scott (1909–1989), who has been described by Sir David Attenborough as the "patron saint of conservation," founded the Trust. As chairman of the World Conservation Union's Species Survival Commission (IUCN/SSC), he "pioneered development of IUCN's Red Data Books—the world's first comprehensive catalogue of endangered species." In 1961, Sir Peter co-founded the World Wildlife Fund (WWF).

In 1947, the Trust began its **National Wildfowl Counts**; it has been working to collect data on **waterfowl**, and particular species. In 1966, a study of **white-winged wood ducks** began with 12 ducks from Assam; in 1990 in **India**, the Trust was working on a **public awareness program** for the conservation of the **threatened duck** and its shrinking **tropical rainforest habitat**. The program, in close collaboration with Indian zoo and wildlife societies, distributes teachers' packs for primary and secondary schools and press packs. In 1980, the **whooper swan monitoring project** began, and in 1990 Trust researchers were in **Iceland** to study the swan and the **pink-footed goose**. The Trust has pursued long-term studies of **Bewick's swans** and **barnacle geese**; the latter, begun in 1970, is "the longest running research project into the habitat needs of any wild bird." Research in **Norway's** Spitsbergen and Helgeland monitor "distribution and habitat requirements of barnacle geese on migration."

An ongoing project focuses on "the factors affecting flamingos' erratic **breeding** success both in captivity and in the wild." Breeding programs have recently been established for endangered species such as the **Hawaiian goose** and **white-headed duck** of Southern Europe and West Asia. By maintaining stocks of these birds at its Centres, the Trust has been able to reintroduce them into the wild when the right conditions are created. Once reintroduced, the birds become "a powerful living resource in the campaign for greater public awareness in their country of origin," possibly reducing threats. The breeding program entails "careful planning and monitoring," including ringing, computerized record-keeping, studbooks for endangered species, and disease screening and treatment programs.

Other Trust projects are broader. The **International Waterfowl Census** began in 1967, widening the scope of the long-running national census. Indeed, "a major part of the Trust's research activity is concerned with monitoring the numbers and breeding success of wildfowl in the U.K." with the financial support of the Nature Conservancy Council (NCC) and the Department of the Environment (Northern Ireland). Every winter, some 1,500 volunteers count the wildfowl on wetlands throughout the country and send the results to the Trust. The resulting ability to "monitor trends in the numbers of individual species and to identify specific sites which are particularly important for wetland birds" has helped the Trust earmark hundreds of specific sites for protection.

The Trust's **database** on wetlands and waterfowl "is used by practical conservation managers to identify the potential of local areas in relation to particular threats or management practices." For example, the Trust has studied the impact of water-based recreation, shooting, and cold weather on the wildfowl.

In another cooperative effort necessary to the monitoring, the Trust joins with the British Trust for Ornithology to coordinate all **ringing of wildfowl in Britain**, marking about 6,000 birds each year. This aids in "establishing links between breeding and wintering areas and in identifying important staging areas for birds to refuel on the migration journey."

Studies of captive collections of wildfowl provide further vital information which can be applied to the wild populations. The Trust is currently studying **flamingo breeding**, among others, and embarking upon careful reintroduction schemes. Technology is also an aid; for example, **electronic eggs** transmit "information on temperature and humidity which enables researchers to understand natural nest conditions more clearly."

The Trust is branching out, in terms of both activities and locations. From **working with developers on a commercial development** around Cardiff Bay that would remain wildlife friendly "as an alternative to an ecologically unsound barrage proposal"; to collaborating with the Thames Water Authority on a **wildfowl reserve and visitor center** at the out-of-use Barn Elms Reservoir in London; to consulting with the London Wildlife Trust, to advising the London Docklands Development Corporation on how it can include "a wildlife component" in its development project; to **advising mining companies** in planning "a beneficial future use for excavated areas after extraction is completed"—the Trust emphasizes environmentally sound, sustainable development.

With a united Europe at hand, the Trust is expanding its consultancy work for "production of wetland management plans, and especially in the development of visitors' sites" like the Trust's own centers. To this end, the Trust has established **The Wetlands Advisory Service** to streamline its consulting work.

In 1990, the Trust began work in Northern Ireland, in a partnership to manage the **Castle Espie Conservation Centre** near Belfast as the ninth Wildfowl and Wetlands Centre, noting that more than 6,000 **light-bellied brent geese** wintered on the shore nearby. Work is underway on a new Centre at the 100-acre Barn Elms reservoirs near Hammersmith Bridge in London. The Trust received funding, to improve the visitor facilities at the Martin Mere Centre. The

Llanelli Centre also recently opened, involving the construction of hides and scrapes, very popular among wild ducks and waders. To boost the wildfowl population at Llanelli at the beginning, the other centers contributed eggs to a rearing program. The Caerlaverock Centre opened in summer for the Wetland Explorer Programme for school children, organized by the Centre's new Education Officer. Youngsters learn, for example, "that most of the fish we eat spawn on coastal wetlands and that two-thirds of the world's human population live on land bordering the sea."

In fact, the Trust's Education Department has emphasized the need for environmental education in the UK. Through the Education Working Group of Wildlife Link—the umbrella organization of British conservation groups—the Trust has helped to draw up the Charter for Environmental Education presented to the Department of the Environment. Many Trust programs are planned in conjunction with the new National Curriculum and established syllabuses. Various centers offer regular teachers' courses, and the Trust provides numerous resources such as teachers' packs, guides, slide sets and videos, and a fun activity pack. Each year, about 120,000 children visit Trust centers, many with school groups; several centers have braille trails and other facilities for the disabled, and many provide hands-on activity stations.

The Trust's Pondwatch outreach program involves campaigning for pond conservation and study, encouraging membership of thousands of schools, community groups, families, and individuals through award schemes, factsheet distribution, and other means. In 1991, the Trust began a pilot project for an extension of Pondwatch called Wetland Watch International, linking schools internationally, with funding through 1993.

International cooperation plays a significant role in Trust activities. In 1969, the International Waterfowl and Wetlands Research Bureau (IWRB) joined the Trust at Slimbridge; nine years later, the Ramsar International Secretariat began to share the offices too. According to the Trust, "The necessarily close contact is of considerable benefit to all parties." With funding through 1993, the Trust launched Wetland Link International, "an initiative . . . to bring together for mutual assistance wetland sites and organizations all around the world."

Dr. Brian Bertram, director-general, was curator of mammals, aquarium, and insect house for the Zoological Society of London. Previous research fellowships took him on field studies of the Indian Hill mynah in Assam; of lions and leopards at the Serengeti Research Institute in Tanzania; and of ostrich breeding in Kenya. Dr. Bertram has written more than 45 publications and given some 90 lectures; he is a member of the Association for the Study of Animal Behavior, Association of British Wild Animal Keepers, Mammal Society, World Wildlife Fund, Fauna and Flora Preservation Society, and Species Survival Commission (IUCN/SSC) Cat Group.

■ Resources

The Trust needs increased funding to finance its many activities.

Available resources include the consultancy group Wetlands Advisory Service, Ltd.; *Wildfowl* annual scientific journal; *Wildfowl and Wetlands* magazine, twice each year; teachers' packs; and *Pondwatch Pack*, a guide to conservation action in the local community.

Wildlife Conservation International (WCI) ⊕
University of Edinburgh
ICAPG, Zoology Building
West Mains Road
Edinburgh, Scotland EH9 3JT, UK

Contact: **Lee J.T. White**, Research Fellow
Phone: (44) 31 650 5510 • *Fax:* (44) 31 667 3210

Activities: Research • *Issues:* Biodiversity/Species Preservation; Deforestation

▲ Projects and People

Presently analyzing data at the University of Edinburgh following three years in the field, Lee J.T. White writes that his most recent study was undertaken in the Lopé Reserve, Gabon, where he "investigated the effects of mechanized commercial timber extraction on forest structure and composition, and attempted to relate observed changes in mammal densities to these changes." White was based at the Station des Etudes des Gorilles et Chimpanzees (SEGC) (Station for the Studies of Gorillas and Chimpanzees), co-directed by Dr. Caroline Tutin. He says his data "once analyzed . . . will enable me to make recommendations aimed at minimizing the damage caused by logging." While in Gabon, he worked in the areas of wildlife censusing, forest botany, and forestry; performed surveys for the World Wildlife Fund (WWF); and with a Conoco Ltd. commission undertook an "environmental impact study of on-shore oil exploration in southwest Gabon." Support came from WCI and Royal Society of London grants and a Leverhulme Trust Study Abroad Studentship.

Earlier, on behalf of the Nigerian Conservation Foundation, White set up a project in the Okomu Forest Reserve, "the location of Nigeria's first rainforest wildlife sanctuary." It was his role to physically protect the 70-square-mile sanctuary, conduct research on which to base Okomu's management plan, and develop education programs for local people. One survey resulted in the relocation of Sclater's guenon (*Cercopithecus sclateri*) in the Niger Delta and also discovered "Africa's most westerly population of western lowland gorillas (*Gorilla g. gorilla*)," which was highly publicized. As an honorary research assistant of the Zoology Department, University College London, he linked that school's Ecology and Conservation Unit (ECU) with the University of Benin to "improve conservation courses at both centres."

In Sierra Leone, White worked on the Tiwai Primate Project and surveyed both villages and wildlife, including a population census of the pygmy hippopotamus.

While completing his project and studying primate ecology, White is currently lecturing at the University of Edinburgh. *(See also Wildlife Conservation International [WCI] profiles in the USA NGO section and throughout this Guide)*

■ Resources

White seeks "reliable and efficient identification of botanical specimens. Could some herbariums provide such a service for field biologists?"

The Wildlife Trust of Bedfordshire and Cambridgeshire
5 Fulbourn Manor, Fulbourn
Cambridge CB1 5BN, UK

Contact: **Mark Rose**, Director
Phone: (44) 223 880788 • *Fax:* (44) 223 881807

Activities: Education; Political/Legislative; Research • *Issues:* Biodiversity/Species Preservation; Habitat Management; Water Quality

● Organization
The **Royal Society for Nature Conservation (RSNC)**—Wildlife Trust Partnership has been described as "the largest voluntary organization in the UK concerned with all aspects of wildlife protection," and is a coalition of 47 Wildlife Trusts, 50 Urban **Wildlife Groups** and **WATCH**, the junior wing under **Dr. David Bellamy**. The combined Trust of **Bedfordshire** and **Cambridgeshire** united in 1991, with 20 staff and a membership of 8,000 serving a community of 750,000 people, owns, manages, or leases about 86 reserves totaling more than 2,200 acres in both counties, including woodland, grassland, and water—from the Ouse Washes in north Cambridgeshire, to the flower-rich chalk downland at Totternhoe Knolls in south Bedfordshire.

▲ Projects and People
The **Wildlife Trust of Bedfordshire and Cambridgeshire** manages nature reserves where staff "teach children and adults about wildlife and work with all members of the local community seeking to unite them in the campaign for a richer countryside." In addition, the Trust works to save threatened habitats and advises on all aspects of conservation.

In the two counties, 40 percent of "traditionally managed ancient woodland has been lost since 1945," threatening the wildlife that depend on it. Through the **Ancient Woodland Appeal**, launched in 1990, the Trust is raising funds to buy or otherwise control the woods "and restore traditional management practices." Recent acquisitions include the 23-acre **Lower Wood** and the 118-acre **Gamlingay Wood**, "one of the best documented of all woods," as well as Kings Wood Heath and Reach, the most important wood in Bedfordshire. The latter was subdivided into one hundred one- and two-acre **leisure plots** in the 1960s, twelve of which the Trust had acquired by 1991. However, the Trust stresses that "buying a wood is not enough." At least as important is the long-term, "careful and constant management [including] mowing, path clearing, and coppicing."

Many of the Trust's activities, such as **guided walks**, are organized by local volunteer groups. Through **WATCH**, the Trust involves schools and youth groups in nature activities and provides them with literature. Independently, the Trust sets up **courses and lectures on wildlife and environmental issues** for adults. The **volunteer training program** covers habitat workshops on woodland, grassland, wetland, and other types, throwing work parties and teaching skills from management planning to the safe use of chain saws and brush cutters. **Jenny Ryle** is the Trust's education officer.

According to **Neil Burgess**, honorary warden of Cherry Hinton West Pit, the **Cambridge City Trust Group** has been helping the Nature Conservancy Council (NCC)—the government's conserva-

tion agency—conserve three nationally rare plants, the **great pignut**, **moon carrot**, and **grape hyacinth**. All are now found on the road verge and other highly vulnerable positions outside the Trust Reserve. Using techniques including "regeneration from seeds still present in the soil but awaiting suitable conditions to germinate," the Group hopes to recolonize West Pit, "establishing additional new colonies in safe locations."

Other ongoing projects include a campaign to save **Flitwick Moor**, an important wetland now suffering from neglect and reduced water levels. By 1991, the Trust owned 26 acres and was in the process of buying another 54, thus saving at least 100 species of **moss**, 200 species of **flowering plants**, **grass snakes**, **frogs**, **toads**, and other wildlife. In 1990, RSNC and other conservation groups began a **Peat Campaign** to save valuable peatlands from agriculture, forestry, and—the most serious threat—commercial extraction. Flitwick Moor is now a peatland nature reserve. In addition, the Trust has lobbied local authorities regarding peat use, encouraging the public to use non-peat composts for garden use: "They may cost a few pence more, but environmentally they are much less expensive."

Through the **Barn Owl Conservation Programme**, the Trust provides **nest boxes** for the birds; by 1991 a total of 70 boxes had been supplied in Cambridgeshire, and many have been used as roost sites, some for breeding. The **WATCH Ozone Project** boasts some 1,400 members in Bedfordshire and Cambridgeshire, aged 7 to 18.

Mark Rose, wildlife conservationist, began his work in **Papua New Guinea**, first with the United Nations Development Programme (UNDP) and the Food and Agriculture Organization (FAO), later as research manager and director, Mainland Reptiles. With the Trust since 1988, Rose also belongs to the Species Survival Commission (IUCN/SSC) Crocodile and Freshwater Chelonian groups as well as the RSNC's UK2000 Steering Group. He is also founder and director of Wild Horizons, a commercial countryside initiative, and author of reports and other publications on **crocodiles**, **pitted shelled turtle** (*Carretochelys insculpta*), and other species.

■ Resources
In addition to its most valuable resource, the nature reserves, the Trust publishes *Conservation Advice* and *Planning Advice*. *Wildlife* magazine appears three times yearly. Educational training is provided for a per-day fee, as are other special courses. Also available through the Trust are *Natural World*, the RSNC magazine, and *Wetlands: An Exploration of the Lost Wilderness of East Anglia*, a book by David Bellamy and Brendan Quayle. It also offers forums on birdwatching, recycling, and "greening your village."

World Association of Soil and Water Conservation—Europe (WASWC) ⊕
40 Church View
Freeland OX7 2HT, UK

Contact: **Martin J. Haigh**, Ph.D., Vice President for Europe
Phone: (44) 865 819785

Activities: Development; Education; Political/Legislative; Research • *Issues:* Deforestation; Landscape Reconstruction;

Sustainable Development; Soil Conservation • *Meetings:* International Conference on Headwater Control; ISCO

● Organization

The **World Association of Soil and Water Conservation (WASWC)** was founded in 1983, and grew to more than 500 members in 63 countries by 1991. The president, currently **Hans Hurni,** serves for a three-year term supported by ten regional vice presidents; the UK office has a staff of two. The Association "is dedicated to encouraging the wise use and conservation of soil and water resources," and places primary responsibility for this conservation on scientists, conservation professionals, and policymakers. The WASWC views itself as a forum for these people in particular, to assist them "in assessing soil and water conservation needs worldwide."

▲ Projects and People

The Association works to identify problems and solutions related to soil and water conservation in part by creating an **information exchange.** WASWC informs members "of conservation meetings, professional opportunities, and new publications and educational materials"; provides a forum for members to share their problems and possible solutions to those problems; and sponsors a biennial conference. WASWC works closely with other worldwide conservation organizations, and as part of its information exchange **publicizes programs of multilateral and national agencies** around the world that sponsor soil and water research, such as the U.S. Agency for International Development (USAID), Food and Agriculture Organization (FAO), and others. According to past president **R. Lal,** "There are about two billion hectares of worldwide degraded land whose productivity can be restored," and through interdisciplinary cooperation that goal can be achieved.

Other activities include **policy studies** in the fields of soil and water use around the world; suggesting courses of action; working "for adoption of long-range soil and water policies"; meeting with policymakers throughout the world to present the Association's views, findings, and suggestions; and encouraging "projects that will help people better understand the means by which more effective **soil and water conservation methods** can be developed and put into practice."

Specifically, WASWC has organized several **workshops on steeplands** and on **soil management,** as well as six international conferences in collaboration with the International Soil Conservation Organization (ISCO) and national organizations.

In the field of **environmental reconstruction,** WASWC scientific teams funded by **Earthwatch** are working to **rebuild landscapes damaged by industry, and strip mining** in Britain, Bulgaria, and other parts of Europe. At a research center in Blaenavon, South Wales, WASWC, Oxford Polytechnic, and the Bulgarian Academy of Forest Engineering are researching ways to deal with a "universal" problem—symptoms of degradation of land that has been "reclaimed" from coal mining, such as "accelerated runoff, erosion, and the decline of soil and vegetation covers."

The goal of the ecologists and soil scientists on the project is to improve the land sufficiently so that it "can look after itself." WASWC notes that this stands in marked contrast to the philosophy of engineers and agriculturalists, which leads to imposed structures and land control systems that do not **allow the land to truly heal.** Similar experiments are taking place at the Pernik Coal Basin, western Bulgaria.

Dr. Martin J. Haigh, WASWC-Europe vice president since 1988, is principal lecturer and chair, Geography Unit of the Environment Faculty, Oxford Polytechnic University. Earlier, he taught at the Universities of Oklahoma and Chicago, covering environmental systems, soil conservation, soils, physical geography, cartography, and applied geomorphology. He belongs to the Certified Professionals in Erosion and Sediment Control, Gandhi Foundation Council, Indian Association of Soil Conservationists, and Himalayan Research Group, among others. Dr. Haigh has pursued research from Arizona to South Asia, and including Indian Himalaya, Bulgaria, and Wales.

According to WASWC founder and executive secretary **Dr. William C. Moldenhauer,** Dr. Haigh developed cooperation and WASWC participation in the International Symposium on Water Erosion held in Varna, Bulgaria; Soil Erosion of Agricultural Land Workshop in Coventry; and the 1992 International Conference on Headwater Control, Prague, with **Dr. Josef Krecek.** Proper management will lead to "sustainable land husbandry in the vulnerable highland and steepland headwater regions that lie on the margins of expanding economies," believes WASWC. *(See also World Association of Soil and Water Conservation [WASWC] in the USA NGO section.)*

■ Resources

WASWC-Europe seeks help to promote soil conservation issues and involve schools/colleges in soil conservation work, secretarial support for the membership network, and sponsorship for young conservationists.

Dr. Haigh has published more than 100 papers and reports, including the books *Evolution of Slopes on Artificial Landforms, S. Wales; Nepal Himalaya: Geoecological Perspectives* (with four co-editors); and *Dynamic Systems Approach to Natural Hazards* (with one co-editor). Another WASWC member book is the 1989 *Policy and Practice in the Management of Tropical Watersheds* by H. **Charles Pereira,** who suggests that "land misuse is due to poor administrative organization and political guidance rather than a lack of well established technologies."

WASWC publications include *Soil Erosion and Conservation,* the proceedings of the third International Soil Conservation conference; a manual on *Land Husbandry;* and the quarterly *Newsletter.*

Membership information is available from the Soil and Water Conservation Society, 7515 Northeast Ankeny Road, Ankeny, IA 50021, USA, 1-515-289-2331.

World Conservation Monitoring Centre (WCMC) ⊕
219c Huntingdon Road
Cambridge CB3 0DL, UK

Contact: **Richard Luxmoore, Ph.D.,** Wildlife Trade Monitoring Unit
Phone: (44) 223 277314 • *Fax:* (44) 223 277136

Activities: Research • *Issues:* Biodiversity/Species Preservation; Deforestation; Sustainable Development • *Meetings:* CITES

● Organization

The **World Conservation Monitoring Centre (WCMC)**, which shares its offices with **TRAFFIC International**, is an independent charity established jointly by the **World Conservation Union (IUCN)**, **World Wide Fund for Nature (WWF)**, and **United Nations Environment Programme (UNEP)**. WCMC serves as "an information service on global conservation," using its network of contacts to gather, manage, and disseminate information on species, habitats, and sites such as national parks throughout the world.

The Centre notes that "the demand for reliable up-to-date information on the conservation of species and ecosystems is growing as the consequences of environmental change become more widely appreciated." Accordingly, its goal is to provide conservation organization around the world, including those in developing countries, with the information necessary for them to operate.

The kinds of information sought ranges from names and locations of endangered plants and animals to "important sites to conserve tropical forests or coral reefs," to threatened parks, and the extent of the illegal trade in elephant ivory, reptile skins, or tropical hardwood.

Other areas covered by the databank include wildlife utilization and a bibliography of conservation. WCMC is currently involved in a **five-year program to promote and expand these networks**.

■ Resources

WCMC makes its **information service** available to "users ranging from governments, development agencies, non-governmental organizations and multinational corporations to individual scientists, journalists and conservationists." Although a charge may be made for its services based on staff time involved and ability to pay, WCMC encourages a free two-way exchange of data with conservation agencies and research scientists.

(See also Zimbabwe Trust in the Zimbabwe NGO section.)

Zoological Society of Glasgow and West of Scotland (ZSGWS)

Glasgow Zoo, Uddingston
Glasgow, Scotland G71 7RZ, UK

Contact: **David George Hughes**, Research Assistant
  Phone: (44) 41 771 1185 • *Fax:* (44) 41 771 2615

Activities: Education; Research • *Issues:* Biodiversity/Species Preservation • *Meetings:* IUCN/SSC; international zoo

● Organization

The **Zoological Society of Glasgow and West of Scotland (ZSGWS)**, most commonly known as the **Glasgow Zoo**, was founded in 1936 and belongs to the World Conservation Union (IUCN). Director/secretary is **R.J.P. O'Grady**.

▲ Projects and People

A major project of the Glasgow Zoo was the **Hugh Fraser Tropicarium** at the Glasgow Garden Festival. This huge **butterfly and orchid house** was a cooperative effort of the Garden Festival Company, University of Glasgow Botany Department, Botanic Gardens of Glasgow District Council, and other groups, under the supervision of the Zoo's **David Hughes** and **Lynne Collins**. The Zoo flew in butterfly species from South America, Malaysia, and Thailand, and also had to "obtain the [food] plants to match the butterflies." Butterflies arrived in pupae form from butterfly ranches—"regarded in the World Conservation Strategy as acceptable because they are operating a 'sustainable, renewable resource' and are also providing a livelihood for the native people who ranch them."

Recent developments at the Zoo itself include construction of the 3.5-acre Alloa Brewery **Himalayan Black Bear** Enclosures and House, and an educationally valuable **Urban Wildlife Garden** containing 10 separate habitats and a large pond.

As a founding member of the **Joint Management of Species Group (JMSG)**, the Zoo is heavily involved in computer linkages with other zoos around the UK and the world, through such programs as the **ARKS (Animal Record Keeping System)** database, which records "movements, births, acquisitions, illness," and many daily activities for every single animal in the collection. The ARKS information is eventually entered into a main database called **NOAH (National On-Line Animal History)**, and the NOAH data are sent to the USA to be registered in a third database called **ISIS (International Species Inventory System)**, which has been accumulating records on individual animals for more than 15 years. (As of 1989, it held data from 343 institutions in 34 countries.)

Through the computer links, using available data, the Glasgow Zoo can, for example, seek breeding partners, find space available for excess animal stock, or obtain complete parentage reports of prospective mates and thus avoid the genetic deterioration caused by inbreeding, formerly a strong possibility in zoos. Such deterioration would preclude reintroduction into the wild. The Glasgow Zoo has responsibility for maintaining two **studbooks** on the **white throated capuchin monkey** (in preparation), **heloderm lizard**, and **clouded leopard**, and notes that, because most zoos now generate their collections from captive breeding rather than from collecting animals in the wild, studbooks are more important than ever.

Animals in captive breeding programs require patience and encouragement. The Zoo's 24-year old **Indian elephant**, "Kirsty," on breeding loan to Chester Zoo, has taken several years to adjust to her new surroundings and has yet to begin her breeding cycle. A similar problem is faced by the **white rhino pair**, both on loan from other zoos; although the out-of-shape female was put on a special diet and exercise program, she could not seem to get in cycle.

A more successful effort involved cooperating with the West of Scotland Branch of the **Parrot Society** to form a joint informal breeding pool for the **African grey parrot**. From former pet parrots which had been handed in to the Edinburgh and Glasgow zoos and to several private aviculturalists, seven pairs were eventually assembled, and within two years five chicks had been successfully reared, with more expected. Because the Zoo lacks a large aviary, all of the birds are being housed in aviaries in the gardens of private aviculturalists.

Private individuals also play a part in supporting the Zoo's activities through the **Animal Support Scheme**, "whereby individuals or groups support an animal by contributing towards the cost of its upkeep." Junior Zoo members, aged eight and above, meet biweekly for entertaining and educational activities.

Hughes started with the Zoo in 1974 as an animal keeper, then worked with collections, and later coordinated the Tropicarium. As research assistant, Hughes is responsible for the Zoo's ARKS database and studbooks; watching changes in legislation that could affect the zoo with regard to the movement of animals; and assisting in the development of the Zoo's development/master plan, among other assignments.

Hughes serves with the JMSG and IUCN-UK on behalf of the Society, chairs the Invertebrate Working Group of the National Federation of Zoos, and co-chairs the Species Survival Commission (IUCN/SSC) Invertebrate Captive Breeding Specialist Group (CBSG). "Education is the key to international conservation, " he believes.

■ Resources

With the National Federation of Zoos, Invertebrate Working Group, the Glasgow Zoo has prepared *Notes for Inspectors of Invertebrate Collections* and a *Code of Practice on the Keeping of Hazardous Invertebrate in Zoos*. An article on the "Management of Terrestrial Invertebrate Displays in Zoological Collections" was also prepared. Using **Dialog**, an American system with over 600 databases covering subjects from the sciences to government and business, the Zoo compiled a **Polar Bear Bibliography** for custodians' groups, searching the Zoological Record from 1864. The bibliography is being updated annually. The Society's *Annual Report* describes programs and animals in the collection.

The Zoological Society of London (ZSL) ⊕
Regent's Park
London NW1 4RY, UK

Contact: **Alexandra M. Dixon**, Conservation Officer
Phone: (44) 71 722 3333 • *Fax:* (44) 71 483 4436

Activities: Education; Research • *Issues:* Biodiversity/Species Conservation • *Meetings:* CITES; IUCN; IUDZG; SSC CBSG

● Organization

In 1826, **Sir Stamford Raffles, Sir Humphry Davy** (Royal Society president), and other naturalists founded the **The Zoological Society of London (ZSL)** as a scientific society. A Royal Charter in 1829 incorporated it "for the advancement of Zoology and Animal Physiology and the introduction of new and curious subjects of the Animal Kingdom."

The Society's various activities, with the help of some 400 staff members, cover 3 main areas:
• Practical and theoretical zoological research is carried out at the **Institute of Zoology**, formed in 1977 by joining the Wellcome Institute of Comparative Physiology, the Nuffield Institute of Comparative Medicine, the Veterinary Hospital, and the Curators' research units. Since 1989, the Institute has been funded through the University of London.
• The Society increases public understanding of animals through its two main animal exhibits. The **London Zoo** opened in 1828, in what had been the Society's Gardens in Regents Park. A century later, the Society acquired **Whipsnade Park**, some 500 acres of farm and downland, which opened in 1931. Together, the Zoo and the Park "house one of the finest and most comprehensive collections of wild animals in the world."

• Finally, through the activities of the **Learned Society** itself— discussion meetings, publication of research results, and the library—the ZSL encourages the sharing of knowledge. As of 1990, membership included some 2,250 Fellows and nearly 3,000 Associates.

Current Society president **Dr. Avrion Mitchison** began his term in 1990.

▲ Projects and People

The Zoo's **breeding activities** are quite extensive. Within the past several years, London became the first zoo outside Australia to breed the endangered small marsupial **Leadbeater's possum**—so successfully that offspring have been sent to other collections in Europe and Australia. Another success involved the hatching of a **spectacled owl**, last successfully bred in the Zoo in 1971. The event belied the apparent incompatibility of the mating pair. Demonstrating its ongoing cooperation with zoos around the world, the London Zoo sent to the Bronx Zoo (New York) three **Chinese alligators** (two female) from the Peking Zoo; the Bronx Zoo, which has a specific breeding program for this species, will in return send four captive-bred juvenile alligators to London.

Two pairs of **Rodrigues fody**, "insectivorious, weaver-like birds," were recently sent to the new incubation and rearing unit on breeding loan from Jersey Wildlife Preservation Trust. The birds, of which less than 200 remain in the wild on Rodrigues Island, remain the property of the Mauritius government. Because the wild birds "are considered to be extremely vulnerable to the effects of cyclones, habitat destruction and drought," the **cooperative management project** is working "to establish a self-sustaining captive population, with the possibility of future re-introduction to another island."

In 1989, the Zoo began "a **coordinated activities program** for visitors." For example, in one series, keepers introduce the animals for which they are responsible and take questions from the public. The Zoo also held an **exhibit** detailing the **illegal trade in reptiles**, noting that in 1989–1990, "more than one third of our acquisitions have come from . . . Customs and Excise seizures." On a related note, the Park lost several birds, including scarlet macaws, a Leadbeater's cockatoo and a large number of roseate cockatoos, to thieves. The Society suggests that this is one clear "detrimental effect that the high prices in the trade (legal or illegal) in these species is having on their conservation whether in the wild or captivity."

Whipsnade Park has been involved in several international **reintroduction efforts**. For example, some **yaks** were sent to reestablish a herd for a tribe of **Kirghiz refugees** who had lost their own herd on their flight to Turkey from northern Afghanistan. According to the Society, "The yak had been at the core of their social and cultural activities for centuries." In another project, an **Arabian oryx** was sent to support local **reintroduction** projects for the species in Saudi Arabia, where the Society manages the King Khalid Wildlife Research Centre (KKWRC) near Riyadh for the Saudi Arabian National Commission for Wildlife Conservation and Development. KKWRC work concentrates on **gazelles**, studying capture techniques, assessing gazelle populations in the wild, and advising on the reintroduction of Arabian and goitered gazelles. In another reintroduction program, two **Przewalski's horses** were sent to Oberwil, Switzerland.

Both the Zoo and the Park have special **educational programs** for younger visitors—for example, E is for Elephant. In addition, the Society is considering the National Curriculum's guidelines when planning exhibits and activities, "the relevance of programmes being clearly indicated." Another popular program is **Adopt an Animal**, in which individual contributions can be earmarked for specific Zoo inhabitants. For example, celebrity adopter and actor **Anthony Hopkins** adopted a penguin in 1989. **Lifewatch**, a membership program which replaced the Friends of the Zoo scheme includes the junior and senior explorer categories for younger members—whose ranks include the Royal Family's HRH Prince William of Wales and HRH Prince Henry of Wales as honorary explorers.

In 1989, the Institute of Zoology established the **Conservation Biology Research Group** under **Dr. Georgina Mace**. The Group includes units for **molecular genetics** and **population genetics**, working with the World Conservation Union's (IUCN) Species Survival Commission on "population analyses of captive and reintroduced populations [and] scientific management of captive populations, through . . . studbooks and in collaboration with population geneticists." Other research activities have dealt with artificial breeding of **Sumatran tigers**, behavioral physiology, endocrinology, comparative medicine, immunology, microbiology, and biochemistry.

Dr. Joshua Ross Ginsberg, research fellow in ecology at the Zoo's **Institute of Zoology**, beginning in 1991, and previously a conservation fellow of Wildlife Conservation International (WCI), has conducted major research efforts relating to canids, equids, and ungulates (hoofed mammals). He began work on the **African Wild Dog Research Project** in **Zimbabwe** in 1988; since 1988 he has been deputy chairman of the Species Survival Commission (IUCN/SSC) Canid Group, and holds membership in the SSC Equid and Reintroduction groups.

Dr. Ginsberg has also worked on projects for the World Bank, National Geographic, and World Wildlife Fund (WWF). His investigation of conservation of the **painted wolf** in Zimbabwe is described as "one of the first studies which simultaneously addresses questions of genetics, disease, and the influence of a species' social behaviour on its interaction with game and domestic stock outside the park [and] will have a great impact on implementing a general strategy for wildlife conservation."

■ Resources

The ZSL's library, nearly as old as the Society itself, is "one of the major zoological libraries in the world." In addition, the Society publishes extensively. Publications include the *Journal of Zoology*, issued in monthly parts; *Symposia* series of books, each containing papers presented at a symposium on a particular topic; *International Zoo Yearbook*; *Zoological Record*, an annual bibliography of zoological literature; *Nomenclator Zoologicus*, updates and zoological nomenclature (published at intervals); and for members, *Lifewatch* magazine, which replaced *Zoo News* in 1989.

In addition to some 20 publications—including a text on *Conservation Biology: Patterns and Processes* and a study on why zebras have stripes—articles, and reports, Dr. Ginsberg has published *Foxes, Wolves, Jackals and Dogs—An Action Plan for the Conservation of Canids*. Funding is crucial to the survival of the London Zoo.

▌ *Private*

Aquarist and Pondkeeper Magazine ⊕
28 Kidston Way, Rudloe Park
Corsham SN13 0JZ, UK

Contact: **John Dawes**, Editor
   Phone: (44) 225 810084 • *Fax:* (44) 225 811461

Activities: Education; Consultancy/Writing • *Issues:* Biodiversity/Species Preservation; Sustainable Development • *Meetings:* pet/aquatic trade and consumer shows and conferences

▲ Projects and People

John Dawes, who edits the monthly magazine *Aquarist and Pondkeeper*, comments: "I encourage well-informed people to write about conservation and animal welfare in the hope that balanced, well-argued contributions will help educate some readers and allow them to formulate sensible opinions."

In one issue (May 1991), author **Dr. Elizabeth Wood**, Marine Conservation Society, describes her "eco-labeling scheme" to get hobbyists, conservationists, and the aquatic trade to cooperate by rating fish that are "unacceptable" for home aquariums because their chances of survival are low and their collection damages reef habitats or specimens, causes a decline in population, and "contravenes local management plans and legislation." Dr. Wood is currently gathering information on such species in the hopes of preparing guidelines and labeling on behalf of "environmentally friendly marines."

A fellow of both the Zoological Society of London and the Linnean Society, Dawes belongs to the Institute of Biology and Species Survival Commission (IUCN/SSC) Freshwater Fish Group. Fascinated with aquatics since the age of seven, "when he started keeping mosquito fish in a small battery tank," today Dawes judges fish at international shows, produces television documentary and video scripts, and arranges consultancy expeditions.

■ Resources

Dawes has written more than 500 articles for educational, scientific, and popular journals. Among his books on aquaria, ponds, and fishkeeping is *Book of Water Gardens*—which describes how to construct a "natural" pond and guard against diseases and enemies of fishes.

Julian Oliver Caldecott, Ph.D. ⊕
Conservation Consultant
79 Windsor Road
Cambridge CB4 3JL, UK

   Phone: (44) 223 353828

Activities: Development; Education; Research • *Issues:* Biodiversity/Species Preservation; Deforestation; Sustainable Development • *Meetings:* International Primatological Society Congress; World Parks Congress

▲ Projects and People

Independent consultant **Dr. Julian Oliver Caldecott**, with a doctorate in rainforest primate ecology from the University of Cambridge carries out projects in rainforests worldwide. In Washington State, USA, Dr. Caldecott "evaluated **environmental education activities and opportunities** . . . in relation to reform of the timber industry" for the Living Earth Foundation. Also in 1991, the Commission of the European Communities sent him to work with the World Bank and the government of **Nigeria** to help develop a **Tropical Forestry Action Plan**. In developing the plan, Dr. Caldecott had to consider the needs of the various habitats, including savannah, river delta, and near-desert, and to design "an environmental management system applicable at the village level throughout Nigeria." The same year, on behalf of Environmental Resources Ltd, Dr. Caldecott traveled to **China** to help plan an **environmentally sustainable socioeconomic development master plan** for Hainan Province, in conjunction with the provincial government and the Asian Development Bank.

Previously, Dr. Caldecott was a fellow of the Environment and Policy Institute, East-West Centre, and consultant to the World Wildlife Fund (WWF), World Conservation Union (IUCN), and various other organizations. He led feasibility study work to develop two divisions of **Cross River National Park** in Nigeria. Other projects have taken him to the **Philippines, Thailand, Indonesia, Venezuela,** and **Peru,** among other countries. In 1986, he led the Bioresources/Sarawak Forest Department team which rediscovered the **Sumatran rhinoceros** in Sarawak, where he also designed a model **rattan nursery** development project and studied **hunting patterns** and "economic contribution of wild meat and the ecology of hunted species." Primates, such as the **pig-tailed macaque, Asian elephant,** and **gibbon,** are research pursuits.

Dr. Caldecott is a fellow of the Borneo Research Council and a member of the International Society of Tropical Foresters, Institute of Biology, and the Species Survival Commission (IUCN/SSC) Ethnozoology and Pigs and Peccaries groups.

■ Resources

Dr. Caldecott has authored numerous scientific papers, books, and technical reports on his research; a full bibliography and selected reprints are available upon request.

Chester Zoo
North of England Zoological Society
Upton
Chester CH2 1LH, UK

Contact: **Gordon McGregor Reid, Ph.D.,** Curator in Chief
Phone: (44) 244 380280 • *Fax:* (44) 81 2915506

Activities: Education • *Issues:* Biodiversity/Species Preservation

▲ Projects and People

As of 1992, **Dr. Gordon McGregor Reid** became curator-in-chief of the nonprofit **Chester Zoo,** under the auspices of the **North of**

England Zoological Society. Earlier, he was natural history keeper at London's Horniman Museum, responsible for such international biological/geological collections and exhibitions as **For the Love of Birds,** featuring the Royal Society for the Protection of Birds (RSPB), and **Wildlife, the Law, and You.** He also established the **Living Waters Conservation Centre,** which recreates the habitats of numerous endangered fish and reptile species, providing them a chance to breed and scientists a chance to study them. Visitors can view a wide variety of water habitats, from African tropical rainforest to British pond, and the species that live in them. The project was supported in part by the Worldwide Fund for Nature (WWF), and includes a biological records computer center for conservation monitoring.

With a background as teacher and volunteer in Nigeria and Botswana, Dr. Reid's research specialties are tropical freshwater fish, or cichlids, as well as the potential for controlling the debilitating snail-borne disease **bilharzia.**

A consultant editor for various scientific publications, including Time-Life books on maritime science and for such journals as *International Zoo News, Journal of Fish Biology,* and *Journal of the Zoological Society of London,* Dr. Reid belongs to the Linnean Society, International Committee of Natural History Museums, International Council of Museums (ICOM), and Species Survival Commission (IUCN/SSC) Fish Group, and chairs a woodland management committee of London Wildlife Trust.

■ Resources

Dr. Reid stresses the need for more captive breeding programs for endangered species. Published articles are about bilharzia, cichlids, natural history museum curatorship, among other interests.

Environmental Solutions (ES) ⊕
Meunier House, Main Street, Caldecote
Cambridge CB3 7NU, UK

Contact: **Martyn Greer Murray, Ph.D.,** Director
Phone: (44) 223 262761 • *Fax:* (44) 223 333840
E-Mail: mgm2@uk.ac.cambridge.phoenix

Activities: Development; Education; Research; Management Consultancy for Conservation and Environmental Groups; Project Evaluation • *Issues:* Biodiversity/Species Preservation; Deforestation; Energy; Sustainable Development

● Organization

The newly formed **Environmental Solutions (ES)** advocates "the sustainable use of natural resources and the maintenance of biological diversity. We also believe that developments and schemes which create jobs can be advanced in such a way as to enhance rather than damage our natural environment. Man can exist in harmony with his environment." ES describes itself as "a versatile company," with a nucleus of specialists as well as a large body of consultants with expertise in environmental fields such as species and habitat conservation, marine and terrestrial ecology, alternative energy, and education policy, planning, and legislation.

▲ Projects and People

With co-directors **Dr. Martyn Greer Murray and Linda Phillips,** ES consultants evaluate planned, ongoing, and past projects for clients; this stage "is designed to help organizations overcome immediate difficulties with ongoing projects and to provide information that will improve project design. At the strategic planning level, it enables organizations to assess the overall effectiveness of their environmental programs."

Environmental Solutions also provides advice "on strategic planning, promotion, fundraising, internal organization and monitoring, and other aspects affecting viability. The main purpose . . . is not to streamline management (although this is a usual consequence) but to remove barriers to expansion and productivity."

Some ES projects are **a management strategy review** for the Bedfordshire and Cambridgeshire Wildlife Trust, an evaluation of a Cambridge University research project investigating **relationships between livestock and wildlife** on **Kenyan ranches,** and several ongoing collaborative research programs at the University of Cambridge.

Dr. Murray, who is a research group member in **mammalian ecology and reproduction** at Cambridge, stresses his "particular interest in the fate of the remaining **long-distance migrations of ungulates in Africa,**" including studies of migrating **wildebeest** in the Serengeti ecosystem of Tanzania and Kenya. Such migratory systems, according to Dr. Murray, "are the focal points for ecosystems and constitute the highest drama of life as it was in the Pleistocene. . . . They are exceptionally vulnerable to changes in land use and deserve much more attention than they are currently receiving."

In his views on global environmental cooperation, Dr. Murray writes, "A major problem lies in the quality of communication between representatives of environmental bodies based in developed countries and government representatives and other involved parties in under-developed countries." In his experience, many projects "start well with good will on all sides, then either during the lifetime of the project, or frequently a few years after the project has been completed, the good work is undone."

He blames this on a failure on the part of the environmentalists to communicate properly with those involved—"the politicians, businessmen and rural communities of underdeveloped countries." Thus the projects fail to generate long-term local support. One of the specific goals of ES is "to help other environmental/conservation/ aid-giving organizations to increase the long-term success rate of their projects."

During his tenure at Cambridge, which began in 1979, Dr. Murray was seconded to the University of Malaya, where he "established a research program that investigated the interaction between wild figs, their pollinating insects, and frugivorous predators." While a student, Dr. Murray founded the **Charles Darwin Journal Club,** "devoted to the presentation and debate of new and controversial ideas in evolutionary biology." He has also consulted for the World Conservation Union (IUCN), and established a research program in the Serengeti National Park, **Tanzania,** regarding "the migration of antelope [and] interaction of grazing herbivores with natural grasslands; energy metabolism and locomotion of **antelope;** ecological separation of herbivores; [and] diet selection of tame antelope." Dr. Murray also advised the Department of National Parks and Wildlife of Tanzania on a management plan for the Serengeti National Park.

■ Resources

In Dr. Murray's article published in *Biological Conservation* (1990), entitled "Conservation of Tropical Rain Forests: Arguments, Beliefs and Convictions," he analyzed the results of a questionnaire which showed that conservationists distinguished between arguments that would best convince others to save the rainforests—primarily focusing on the sustainable exploitation of hardwoods, medicinal plants, for instance—and those which they most strongly believed—mainly arguments regarding indigenous peoples, ethical responsibility toward the planet, and the like. Dr. Murray concludes: "The paradox is that deeply held beliefs which carry the full force of personal conviction and commitment can be immensely persuasive. A more open approach about strongly held personal beliefs might . . . strengthen the case put for long-term conservation."

Dr. Murray has authored or co-authored numerous other publications, both theoretical and empirical, some of which focus on impala, antelope, **wildebeest,** and **figs and fig wasps.**

Fish Conservation Centre ⊕

Easter Cringate
Stirling FK7 9QX, UK

Contact: **Peter S. Maitland, Ph.D.,** Ecological Consultant
Phone: (44) 786 51312

Activities: Education; Research • *Issues:* Biodiversity/Species Preservation; Global Warming; Water Quality

▲ Projects and People

Dr. Peter S. Maitland has been a freelance ecological research consultant since 1986, working with such clients as the European Community (EC), Council of Europe, Nature Conservancy Council (NCC), British Broadcasting Company (BBC), U.S. Environmental Protection Agency (EPA) (at a workshop on introduced species to the Great Lakes), and local groups. With a background in freshwater ecology at the University of Glasgow and scientific posts at both The Nature Conservancy and the Institute of Terrestrial Ecology, recently Dr. Maitland has concentrated on "**conservation of freshwater fish,** impact of **salmon farming** on the environment, impact of **afforestation** on fresh waters, **acidification of Loch Doon,** [and] motorway impact assessments."

In 1968, he founded the **Scottish Freshwater Group,** a liaison organization of freshwater scientists in Scotland, with a current membership of 280, and semiannual meetings; he is secretary. Dr. Maitland chairs the Species Survival Commission (IUCN/SSC) Fish Conservation Group, coordinating 55 member fish biologists from 45 countries.

■ Resources

Funding for research and travel is sought—activities, along with exchange visits, that Dr. Maitland believes enhance global environmental cooperation. His extensive publications list includes *The Natural History of the Freshwater Fishes of the British Isles* (co-authored with **R.N. Campbell**), and articles on ecology and conservation of **arctic char** in Loch Doon, and on **nonnative fish species** in Loch Lomond.

Herpetofauna Consultants International (HCI) ⊕
P.O. Box 1
Halesworth, Suffolk IP19 9AW, UK

Contact: **Thomas Edward Salatheil Langton,**
Environmental Consultant
Phone: (44) 98 684518 • *Fax:* (44) 98 684579

Activities: Education; Law Enforcement; Political/Legislative; Research • *Issues:* Biodiversity/Species Preservation; Environmental Impact Assessment Research; Species and Habitat Survey

● Organization
Founded in 1989, **Herpetofauna Consultants International (HCI)** works with public, private, and non-governmental organizations to assess and advise on environmental issues, particularly focusing on amphibians and reptiles—which comprise about "one quarter of all land vertebrate species"—in light of rapidly changing legislation. HCI's staff of 10 includes trained biologists; survey and mapping facilities are also available.

▲ Projects and People
The U.K. Wildlife and Countryside Act of 1981 requires public and private entities to consider wildlife habitat and species when planning potentially disruptive construction or activity. The Act provides for special protection of such reptiles and amphibians as the **natterjack toad, crested newt, and sand lizard.** HCI has served as a liaison between the Nature Conservancy Council (NCC)—"the government's statutory body for nature conservation in England"—and clients, helping with field assessments of various species, surveys of habitat, engineering, and budgeting suggestions. HCI "can prepare management plans for the establishment and maintenance of protected areas, and . . . [can] provide a coordinated approach to consultation and after-use."

Recent consulting efforts of HCI include a survey and site management scheme to **protect crested newts during a road widening project** in the London Borough of Hillingdon. The consulting firm also "provided specialist advice and action to meet the statutory regulations" regarding **lizard protection** at the site of a **proposed Channel Tunnel** facility in another London borough. When a neighborhood association "needed representation at a public inquiry" to protect an ecologically valuable area, HCI supplied the necessary assistance.

■ Resources
Herpetofauna collects and publishes information regarding the management of reptiles and amphibians and their habitats. Its advisory services are wordwide.

Kudu Publishing ⊕
The Coach House, Notton
Lacock SN15 2NF, UK

Contact: **Russell Kyle, D.V.M.,** Publisher
Phone: (44) 249 730729

Activities: Development; Wildlife Management • *Issues:* Biodiversity/Species Preservation; Economic Management of Wildlife; Sustainable Development

● Organization
Kudu Publishing "was founded in order to fill a gap in the natural history market" between coffee table books and academic books. The publisher's goal is to produce academically respectable books that are still a pleasure to read. "The future of wild animal management and conservation will depend on a number of different interests working together: agriculture, zoology, tourism, sport, education and politics, as well as the great body of amateur naturalists. Kudu will encourage books with this eclectic approach," writes veterinary surgeon and publisher **Dr. Russell Kyle.**

▲ Projects and People
A state veterinarian with the Ministry of Agriculture, Fisheries, and Food (MAFF), earlier Dr. Kyle spent several months in East Africa "studying schemes to domesticate the **eland.**" Later world travels also focused on projects "where the aim was to **manage the wild species for food production.**" According to Dr. Kyle, "The best way to conserve the animals and their environment is to show that they can have an economic value."

■ Resources
Dr. Kyle's book on economic management of wildlife, *A Feast in the Wild,* "opens with a question: although there are 200 species of wild, large herbivore in the world, why have less than two dozen been domesticated for food?" A result of 15 years of research, the book investigates various projects which have had precisely the goal of expanding the types of domesticated food animals.

Dr. Kyle writes: "In some areas the meat can be used to improve the diet of the malnourished, local people. In other cases the meat can be sold for a high price to the luxury market, and the revenue can be used to aid the management of the wild animals. Often a properly planned exploitation of the local, wild species can improve the conservation of the animals themselves, and will maintain better land conditions in the wild areas which they inhabit, and these things bring in their train other benefits such as tourism, recreation, and local employment."

The Nature Conservation Bureau (NCB) Limited ⊕
36 Kingfisher Court, Hambridge Road
Newbury RG14 5SJ, UK

Contact: **Paul Goriup,** Director
Phone: (44) 635 550380 • *Fax:* (44) 635 550230
Telex: 849125 GOTRAV

Activities: Education; Political/Legislative; Research • *Issues:* Biodiversity/Species Preservation • *Meetings:* Bonn Convention; World Parks Congress

● Organization
Founded in 1986, **The Nature Conservation Bureau (NCB)** Limited's main goals are "to provide reliable and professional sup-

port to official and voluntary conservation organizations"; to help smaller conservation organizations with administration, fundraising, and project implementation; to serve as coordinator for projects with more than one participant; and to provide a consultancy service. Current clients include local, national, and international government agencies, NGOs, and private individuals or companies. The Bureau has associate offices in Rome and Vienna.

Sharing NCB offices is the **Institute of Ecology and Environmental Management**, established in 1991 under the sponsorship of the British Ecological Society, the British Association of Nature Conservationists, the Institute of Biology, and the Royal Geographical Society. The Institute, governed by a council of 15, serves as a representative body for ecologists and environmental managers, setting work and qualifications standards, providing career development opportunities, and in general raising the profile of the profession. The Institute requires that members accept the Code of Professional Conduct, a commitment to global stewardship and all that it entails.

▲ Projects and People

Recent education-related projects of the NCB include providing **trail signs** or **interpretative boards** for clients such as the Sussex Wildlife Trust, English Nature, and Oxford City Council, among others. The Bureau has also organized numerous **seminars** and **conferences**, such as a three-day seminar on the "conservation of lowland dry grassland birds in Europe" for the Nature Conservancy Council, and a postgraduate seminar on International Organization of Nature Conservation at Reading University.

Specific conservation projects include **coastal surveys** for the Nature Conservancy Council of Lochs Gairloch, Ewe and Sunart; advice to the Scottish Salmon Growers Association on **fish farm monitoring programs**; and a review of implications for conservation efforts of **seabed damage** by commercial fishing. In addition, the NCB assesses **wildlife habitats, flora,** and **fauna** for private companies and develops management plans, such as for the Aston Rowant National Nature Reserve, Oxfordshire.

International efforts of the NCB include assisting in the World Conservation Union's (IUCN) **East European** program for **conservation** initiatives; devising for the International Council for Bird Preservation (ICBP) a management plan for international conservation measures to protect the **white stork**; and developing **satellite biotelemetry** equipment to track migratory species, with the National Commission for Wildlife Conservation and Development of Saudi Arabia.

NCB director **Paul David Goriup** was also executive director of the Institute of Ecology and Environmental Management in 1991. He is secretary of ICBP Dry Grassland Birds Group and was principal investigator for World Wildlife Fund/World Conservation Union (WWF/IUCN) bird projects. Memberships include the Anglo-Arab Association, Ornithological Society of the Middle East, and Oriental Bird Club. He is also UK group leader for the Bombay Natural History Society, British Ornithologists Union, the Royal Society for the Protection of Birds, and Zoological Society of London, among others.

Author of at least 20 articles on bustards worldwide, Goriup has edited *Parks*, the journal of the IUCN Commission on National Parks and Protected Areas, as well as *Bustard Studies*, the journal of the ICBP Bustard Specialist Group (now Dry Grassland Birds).

Other NCB directors are **Dr. Philip Bacon, Tom Clark,** artist **Peter Creed,** researcher **Sue Everett, Sarah Fowler,** and **Dr. David** Macdonald. Dr. Bacon has extensive overseas experience in agriculture methods, including germ plasm evaluation. Clark, a member of the British Association of Nature Conservation, has conducted numerous environmental audits for various corporations. Fowler, the Bureau's marine services director, was marine ecologist and diving officer with the Nature Conservancy Council for more than 8 years; she is deputy chair of the Species Survival Commission (IUCN/SSC) Shark Group. Dr. Macdonald is also head of the Wildlife Conservation Research Unit at Oxford University, and is a specialist in animal behavior.

■ Resources

Describing itself as "Britain's leading agency for preparing, editing, and producing publications for conservation organizations," NCB offers *Parks* and *Reef Encounter*, the newsletter of the International Society for Reef Studies, among other literature, environmental services, and aspects of international consultancy.

Norman Myers Consultancy Ltd. ⊕
Upper Meadow, Old Road, Headington
Oxford OX3 8SZ, UK

Contact: **Norman Myers, Ph.D.,** Professor, Scientist, Consultant
Phone: (44) 865 750387 • *Fax:* (44) 865 741538

Activities: Development • *Issues:* Biodiversity/Species Preservation; Deforestation; Energy; Global Warming; Population Planning; Sustainable Development

▲ Projects and People

Since 1981, **Dr. Norman Myers** has consulted on "projects for development organizations and research bodies, including the World Bank, United Nations (UN) agencies, Organization for Economic Cooperation and Development (OECD), the Commission [of the European Communities], U.S. Departments of State and Energy, the World Commission on Environment and Development, and the U.S. National Research Council." His latest work for the **World Bank** involved revising the **Forestry Sector Policy Paper**; for OECD he has studied "interdependency relationships between the developed world and the developing world, especially in terms of critical resource endowments such as tropical forests and gene reservoirs." His consultancy has a staff of four.

Previous work included such clients as the "Rockefeller Brothers Fund **Project on Threatened Species and Genetic Resources,** survey for U.S. National Academy of Sciences on **Conversion of Tropical Forests,** Wildlife and Parks Officer for Africa under FAO (UN Food and Agriculture Organization), and director of **UNEP [United Nations Environment Programme] Conference on Tropical Forests.**" For the Intergovernmental Panel on Climate Change (IPCC), he has analyzed biotic sources of greenhouse gases and economic impacts.

Dr. Myers' assignments include a visiting professorship in International Environment at the University of Utrecht, professor-at-large position at Cornell University, senior fellow of World Wildlife Fund (WWF-USA), and special advisor to **Maurice Strong,** the Secretary-General for the 1992 UN Conference on Environment and Development (UNCED) in Brazil. His work and studies have taken him to more than 80 countries. He is also a visiting fellow with the World Resources Institute and an advisor to the environment and population programs under the MacArthur Foundation, Chicago.

■ Resources

Dr. Myers has published more than 250 papers in journals such as *Science, Scientific American*, and *Environmental Affairs*, as well as hundreds of articles in popular magazines and newspapers. He has written seven books, from *The Long African Day* (1972) to *An Atlas of Future Worlds* (1990).

Trendrine Press ⊕
Trendrine, Zennor
St. Ives, Cornwall TR26 3BW, UK

Contact: **William Frank Harding Ansell**, Farmer and Proprietor
Phone: (44) 736 796926

Activities: Education (Publishing) • *Issues:* Biodiversity/Species Preservation

▲ Projects and People

William F.H. Ansell founded the **Trendrine Press** "mainly to publish books on **African mammals**." Working independently, Ansell seeks to fill a gap left by larger, profit-oriented publishing houses and "provide up-to-date authoritative accounts of the wild mammals for those countries where basic literature is either out-of-date, inadequate, or even non-existent." The book *Mammals of Ghana, Sierra Leone and The Gambia* (Grubb et al.), scheduled for publication in 1992, may be followed by a similar work on the mammals of **Angola**—of which the last study was published in 1941.

After serving in the British and Indian armies in World War II, including a stint with the Gurkhas, Ansell joined the **National Parks and Wildlife Service of Zambia** (formerly Northern Rhodesia) as a game ranger, subsequently becoming chief ranger, chief wildlife research officer, and, later, deputy director until his retirement in 1974. While in Zambia, he researched and wrote two books, *Mammals of Northern Rhodesia* (1960) and *Mammals of Zambia* (1978).

He has spent his "retirement" farming in Cornwall and more recently with publication work. Former secretary of the **Cornwall Farming and Wildlife Advisory Group**, he is a member of the Species Survival Commission (IUCN/SSC) Antelope Group, Wildlife Conservation Society of Zambia, and Cornwall Trust for Nature Conservation, for which he served as administrative officer.

■ Resources

In 1989, Ansell published his work *African Mammals 1938–1988*, using G.M. Allen's 1939 *A Checklist of African Mammals* as primary reference, updating and expanding the taxonomy based on more recent descriptions, and supplying geographic coordinates where lacking in the original version. In 1988, he published *Mammals of Malawi*, which he co-wrote with **R.J. Dowsett**. A previous taxonomy, which was published in 1959, contained 30 fewer species

than this version. Ansell notes that **Malawi** has missed being included in recent taxonomies of both East African and Southern African mammals.

Susan M. Wells ⊕
Environmental Consultant
56 Oxford Road
Cambridge CB4 3PW, UK

Phone: (44) 223 350409

Activities: Development; Research • *Issues:* Biodiversity/Species Preservation; Water Quality • *Meetings:* International Society for Reef Studies; IUCN/SSC Coral Reef Fish and Mollusc groups

▲ Projects and People

Susan M. Wells is an independent environmental consultant also furthering the Species Survival Commission (IUCN/SSC) Coral Reef Fish Group, with **Don McAllister** in Ontario, Canada, and the fledgling Mollusc Group. Much of Wells' work concerns **coral reef conservation**.

Wells has prepared World Conservation Union (IUCN) background documents for the United Nations Environment Programme's (UNEP) **Regional Seas Program for South Asia**, and for establishing an SSC **Task Force on Invertebrate Conservation**. For IUCN's Commission on National Parks and Protected Areas, she researched and co-authored the audiovisual program "A Global Overview of **Tourism** Activities in Coastal and Marine Parks," co-produced by the East-West Center. Working for the World Wildlife Fund (WWF-USA), Wells prepared a proposal to add **stony corals** to CITES (Convention on International Trade in Endangered Species of Wild Fauna and Flora) and investigated the conservation requirements of **Mediterranean precious coral**. For WWF-UK and the Council of Europe, she reported "on threatened **non-marine molluscs** of Europe" and advised on their inclusion under the Bern Convention and the European Community's Habitats Directive. Other projects have been in conjunction with Greenpeace, World Conservation Monitoring Centre, and UNEP, among others.

Earlier, Wells spent several years working with the International Council for Bird Preservation (ICBP), and with the IUCN Conservation Monitoring Centre. She has studied **reefs and marine life** in the Philippines, Papua New Guinea, the Bahamas, and Maldives.

■ Resources

Wells gives radio and television interviews "on conservation issues relating to coral reefs, invertebrates and birds." She has also served as scientific advisor for the television programs *World About Us* and *Natural World*. She edits the biannual newsletters of the International Society for Reef Studies, *Reef Encounter*, and of the SSC Mollusc Group, *Tentacle*; in addition she was senior editor of the three-volume study *Coral Reefs of the World*.

Wells has recently written on the **ornamental shell trade**, status of **snails** in French Polynesia, and **effectiveness of CITES**. She is currently working on *The Atlas of the Oceans*.

George Kenneth Whitehead ⊕
Author
Old House, Withnell Fold
Chorley, Lancashire PR6 8AZ, UK

Phone: (44) 254 830444

Activities: Education; Research • *Issues:* Biodiversity/Species Preservation • *Meetings:* CIC

▲ Projects and People
George Kenneth Whitehead, a former paper mill manager, has been writing books about deer, goats, and cattle, both in the UK and around the world, since 1950. He belongs to the Species Survival Commission (IUCN/SSC) Deer Group.

■ Resources
Whitehead's works include *Deer and Their Management, Deer of Great Britain and Ireland, Deer Stalking in Scotland, Wild Goats of Great Britain and Ireland, Ancient White Cattle of Britain and Their Descendants,* and *Hunting and Stalking Deer Throughout the World.* Whitehead has also written numerous articles for such magazines as *The Field* and *Country Life.*

 *Universities*

Department of Biological Sciences ⊕
Manchester Polytechnic
Chester Street
Manchester M1 5GD, UK

Contact: **David P. Mallon**
   Phone: (44) 61 224 2553 • *Fax:* (44) 61 236 7383

Activities: Education; Research • *Issues:* Biodiversity/Species Preservation; Deforestation; Sustainable Development • *Meetings:* International Snow Leopard Symposium; IUCN/SSC Caprinae and Cat groups

▲ Projects and People
Presently, **David Mallon** is compiling a detailed report on the **lagomorphs of Ladakh**, northwest Himalaya, and working on three projects concerned with protected areas. These are updating a 1989 report on Hemis National Park located in Ladakh, with a significant **snow leopard** population; revising an *Ecological Survey of the Protected Area Network in Ladakh;* and preparing a *Survey of the Kanji Wildlife Sanctuary.* Mallon belongs to the Species Survival Commission (IUCN/SSC) Cat and Caprinae groups and is country author for Mongolia on *Caprinae Action Plan* of 1992. He remains interested in "ecology of **ungulates** in Ladakh" and a "wider interest in the ecology and conservation of Ladakh, the Himalaya, and Mongolia."

Mallon describes Ladakh as "one of the most sparsely populated parts of India; settlements have an oasis-like character, and crop growing is everywhere dependent on irrigation." Barley, peas, and wheat are grown at lower altitudes, and "apricots are cultivated up to around 3,500 meters. Livestock consists of sheep, goats, cattle, yaks, dzo (a cattle-yak hybrid), horses, donkeys, and mules."

■ Resources
Mallon's *Biodiversity Guide to Pakistan* for the World Conservation Union (IUCN) was newly published by the World Conservation Monitoring Centre. He has authored some 15 publications, on snow leopards, **wild sheep, wallabies,** and other mammals. In 1991, Mallon published "Status and Conservation of Large Mammals in Ladakh" in *Biological Conservation;* in it he concludes: "Any management strategy for the protection of wildlife and the environment in Ladakh will have to take into account the needs of the people and their development, and ensure an equitable sharing out of resources between their needs and those of the wildlife."

Department of Zoology ⊕
University of Cambridge
Cambridge CB2 3EJ, UK

Contact: **Andrew Laurie, Ph.D.**, Biologist
   Phone: (44) 223 336600 • *Fax:* (44) 223 352618

Activities: Education; Research • *Issues:* Biodiversity/Species Preservation; Deforestation; Sustainable Development • *Meetings:* Rhino Horn Trade

▲ Projects and People
By March 1992, **Dr. Andrew Laurie** had completed a year-long first phase of the **Tanzania Rhino project**, surveying rhinos in the **Selous Game Reserve**. He estimates that some 95 percent of Tanzania's rhinos were killed by poachers between 1975 and 1989. Despite intense governmental efforts since then, poaching continues because "the rewards . . . are so high." In addition, Dr. Laurie points out that "the conditions of employment of the rangers are so poor that their commitment to the arduous and dangerous work of patroling is severely tested." Besides poachers, the rhinos face threats from a planned stock route which would herd some 20,000 head of cattle annually across part of the Selous Reserve, and from other such encroachments.

The purpose of the survey phase of the project was to improve knowledge of "distribution, numbers and reproduction of rhinos, or whether the populations are viable under present levels of security," given the vastness of the reserve. Despite reports of scattered rhino populations in several other reserves and parks in Tanzania, there is little concrete data.

Originally conceived as "an emergency rescue operation . . . when poaching was at its height, the plan was to capture rhinos from all over Tanzania and airlift them to . . . sanctuary areas . . . on the grounds that anti-poaching operations alone were doomed to eventual failures and the loss of the species in Tanzania." Although the poaching problem is less severe now, the lack of background data for the original plan also slowed the process of carrying it out.

As Dr. Laurie reports, "There was no guarantee that the rhinos would be any safer from poachers in the release areas than where they were taken from, and the risks of death during capture, translocation and after release are so great that accurate information is needed first, in order to decide whether taking those risks is justified." Another significant factor was the cost of relocation.

Following his "detailed survey of the remaining populations and individuals, their reproductive status, the threats to their survival, and the ecology and security of the proposed release areas," Dr.

Laurie will prepare a rhino conservation plan for Tanzania based on the findings. The project will also share information with more established rhino conservation programs in South Africa, Namibia, Zimbabwe, Zambia, and Kenya. The project is funded by the Faith Foundation and the Frankfurt Zoological Society, with the Ministry of Tourism, Natural Resources and Environment.

Dr. Laurie's early studies at Cambridge were on the "ecology and behavior of the greater one-horned rhinoceros." He has studied wildlife in Serengeti Research Institute, Tanzania; crown of thorns starfish in Red Sea off Sudan; gorillas in Rwanda; wild goats in Pakistan (information used in establishment of a national park in the Kirthar Range); and rhinos in Nepal, India, Thailand, Malaysia, and Indonesia. He has served as "consultant biologist for production of the British Broadcasting Company/*Reader's Digest* expanded edition of Sir David Attenborough's book *Life on Earth*," and has worked as a World Wildlife Fund (WWF) consultant in Indonesia, where, working with Operation Drake, he prepared a management plan for Morowali Nature Reserve.

Dr. Laurie has been involved in a "long-term study of population dynamics, ecology and social organization of marine iguanas in Galapagos Islands, Ecuador," turning over the results to the Galapagos National Park Service. A particularly bad year for the iguanas was "1983, when 70 percent of the iguanas died of starvation as a result of the disappearance of their preferred food species during the 1982–83 El Niño sea warming, which was the most severe for at least 100 years." Dr. Laurie has also surveyed the Fijian crested iguana, and served as WWF consultant on a project for conservation and management of the giant panda and its habitat in China, gathering data and teaching a UNESCO training course for panda reserve managers.

Dr. Laurie has been a research associate at the Max-Planck Institut für Verhaltenphysiologie in Germany; fellow of the Zoological Society of London; and member of the British Ecological Society, Association for the Study of Animal Behavior, WWF-UK, Fauna and Flora Preservation Society, and Species Survival Commission (IUCN/SSC) Asian Rhinoceros and Hippopotamus groups.

■ Resources

Dr. Laurie has authored some 35 scientific publications, and another 15 in popular magazines.

Scott Polar Research Institute (SPRI) ⊕
University of Cambridge
Lensfield Road
Cambridge CB2 1ER, UK

Contact: **Peter Wadhams, Ph.D.,** Director
  Phone: (44) 223 336542 • *Fax:* (44) 223 336549
  E-Mail: Omnet: p.wadhams

Activities: Education; Research • *Issues:* Biodiversity/Species Preservation; Global Warming; Sustainable Development; Transportation; Waste Management/Recycling • *Meetings:* IGS; IUGG; SCAR

● Organization

In its sixty-seventh year (1993), the Scott Polar Research Institute (SPRI), affiliated with the University of Cambridge, has staff of 75. The Institute is supported by Friends of the SPRI, numbering about 450 members. It also houses the Scientific Committee on Antarctic Research (SCAR) of the International Council of Scientific Unions (ICSU), and the International Glaciological Society, with Hilda Richardson as secretary general.

▲ Projects and People

Recent projects concentrate on climate change and increasing Arctic research. SPRI scientists monitor the polar ice levels, as an increased rate of melting would lead to a rise in global sea level, with potentially drastic results. The Institute is working on sea ice studies with the recently opened Hadley Centre for Climate Prediction and Research, and has helped plan the new Interdisciplinary Environmental Centre at Cambridge. Four SPRI scientists involved in climate change research are director Dr. Peter Wadhams, Dr. Michael Hambrey, Michael Gorman, and Dr. Charles Swithinbank.

Arctic research, which SPRI states "has always had a lower priority than Antarctic research" in terms of government funding, has gained ground. SPRI has been involved with various new government initiatives, including creation of a Britain in the Arctic program, establishment of a small research facility in Svalbard, and joining the International Arctic Science Committee, "a newly established body of circumpolar nations which seeks to coordinate and encourage Arctic research." All of these efforts were launched by the Natural Environment Research Council (NERC).

In 1990, SPRI began a program of collaborative research with scientific institutes in the then Soviet Union. Along with the Fridtjof Nansen Institute of Norway and the Woods Hole Oceanographic Institute, USA, SPRI—represented by Dr. Terence Armstrong—is participating in a project to study the Northern Sea Route. If the Route could be opened to year-round operation, shipping trade time between Europe and the Far East would shrink dramatically. The SPRI Social Sciences and Soviet Northern Affairs group, led by Dr. Piers Vitebsky and Dr. Armstrong, investigates such issues as ethnic kinship among indigenous Arctic peoples and economic viability of the region. Specific research topics cover reindeer breeding, resettlement programs, and Dr. Vitebsky's focus, shamanism. In another project, Dr. Vitebsky is studying the affect of current historic changes on environmental management in what was Soviet Arctic territory.

Recent SPRI projects also include working with the British National Space Centre to interpret data on ice sheets and sea ice received from the ERS-1 satellite launched in 1991. In this effort and several other projects, SPRI scientists work with a variety of technologies, such as remote sensing, image analysis, and geographic information systems (GIS), to measure ice thickness and roughness. Previously, this data could be obtained only "by use of upward sonar from nuclear submarines." Ice thickness and movement studies in the Antarctic "seek to establish the role of sea ice in the dynamics and thermodynamics of the Antarctic Ocean."

The ongoing NERC-funded Weddell Ice Dynamics Experiment (WIDE) also involves efforts of scientists from the Alfred-Wegener Institute in Germany, World Meteorological Organization, and the U.S. National Oceanographic and Atmospheric Administration

(NOAA). Sea ice studies involve SPRI researchers **Drs. Robin Williams, Norman Davis,** and **Wadhams.** The **Remote Sensing Group** is headed by **Dr. W. Gareth Rees.**

SPRI's **Glacier Geophysics Group** studies ice mass behavior, and the relations between glaciers, marine environment, and climate. Researchers include **Drs. Julian A. Dowdeswell,** Hambrey, **Gordon Robin,** and Swithinbank. **Dr. Bernard Stonehouse** leads the **Polar Ecology and Management Group's** "research on Antarctic marine bird and mammal predator-prey relations," as well as the effect of tourism on the Antarctic. Underway is the establishment of a tourist monitoring unit in the **South Shetland Islands.**

■ Resources

SPRI has the "world's largest polar library," with everything from a copying service for library and archival material, to polar photographs and paintings. **William Mills** is librarian and information officer. SPRI also sells postal and greeting cards, posters, maps, and pamphlets. The Institute publishes two quarterlies: *Polar Record* journal and *Polar and Glaciological Abstracts* bibliography. The thrice-yearly *Journal of Glaciology* and the *Ice* news bulletin are literature of the International Glaciological Society. Scientific books, novels, and handbooks are available for both young and old, including *Scott's Last Expedition,* the 1910–1912 journals of Robert Falcon Scott, with a foreword by his son, the late conservationist Sir Peter Scott. A publications list is available.

Alisa Macqueen manages SPRI's **World Data Centre "C"** for **Glaciology,** responsible for "acquisition, cataloging, and abstracting of published glaciological literature." WDC contributes to SPRI's *Polar Digest* newsletter, launched in 1989 at the request of then Prime Minister Margaret Thatcher. The SPRI Museum and Archives, headed by **R.K. Headland,** contain a wealth of material from two centuries of polar expeditions: diaries, sketches, films, and even scrimshaw.

SPRI offers a graduate course in Polar Studies, directed by **Peter Speak.** In 1990 the Institute began the annual David Sexton Memorial Lecture Series, in honor of a research student killed in the Soviet Tien Shan Mountains in 1989.

Anthropology Department ⊕
University College London
Gower Street
London WC1E 6BT, UK

Contact: **Anthony Glyn Davies, Ph.D.,** Tropical Forest Ecologist

Activities: Development; Ecological Surveys; Research • *Issues:* Biodiversity/Species Preservation; Deforestation; Sustainable Development

▲ Projects and People

Field research of **Dr. Anthony Glyn Davies** focuses on the **interaction between fauna and tree flora** in Malaysia, Borneo, West Africa, and South India. His projects have involved surveys to identify areas for **wildlife management** and **timber resources;** ecological studies of **tree abundance,** their flower cycles, fruit and leaf production, and chemical composition; and primatological studies "relating features

of the tree flora with feeding, ranging and social behavior of **colobine monkeys.**"

Dr. Davies has worked to relate human activities with **rainforest conservation,** surveying the "bushmeat" trade, interviewing hunters, and monitoring market sales; examining the **trade in commercial timber** species and timber company activities; and **assessing farming systems** regarding the "farmbush-fallow agricultural system." He has also managed projects ranging from arranging governmental contracts and terms of reference, to developing project descriptions and organizing counter-part staff, and liaising between NGOs and government. Project sponsors range from international agencies like the Food and Agriculture Organization (FAO), to governments and NGOs such as World Wide Fund for Nature (WWF-Malaysia), and the UK's Royal Society for the Protection of Birds (RSPB).

Dr. Davies belongs to the Species Survival Commission (IUCN/SSC) Antelope, Mustelid and Viverrid, and Primate groups.

Department of Biological Sciences ⊕
University of Durham
South Road
Durham DH1 3LE, UK

Contact: **Nigel Dunstone, Ph.D.,** Lecturer, Ecotourism Consultant
Phone: (44) 91 374 3348 • *Fax:* (44) 91 374 3741
E-Mail: zzk2@dur.mts; Greennet: ecodurham

Activities: Education; Research • *Issues:* Biodiversity/Species Preservation; Deforestation; Sustainable Development • *Meetings:* IUCN; World Parks Congress

▲ Projects and People

At the **University of Durham, Dr. Nigel Dunstone** focuses on the behavioral ecology of mammals, especially endangered and pest species such as **mink,** small **rodents,** and **carnivores.** In South America, Dr. Dunstone has investigated the status, distribution, and behavior ecology of **mountain tapir** in **Ecuador's** Sangay National Park, and of the **spectacled bear** in Ecuador; **biodiversity** in the Podocarpus Cloud Forest National Park, Ecuador, and in the Manu National Park, Peru; behavioral ecology of rainforest carnivores, especially **ocelot, jaguar,** and other **spotted cats;** and the impact of **ecotourism** development on behavior of animals, particularly in the **Madre de Dios** region of Peru. The latter study "involves species inventories, estimation of animal abundance and telemetric recording of indicator species."

Dr. Dunstone is a member of the UK Committee for the World Conservation Union (IUCN) and Mammal Society; scientific advisor for television documentary films, especially on **neotropical biology** and environments on South American and European mammals; ecotourism consultant; and **expert witness** for Crown Prosecution Service in cases concerning conservation, management, and wildlife. He has consulted on "Compassion in World Farming" regarding **farmed mink and foxes** and on environmental impact assessment (EIA) projects of mammals. Dr. Dunstone belongs to the Association for the Study of Animal Behavior, Fauna and Flora Preservation Society, and Otter Trust; he is a founding member of the Durham Rainforest Action Group.

■ Resources

Dr. Dunstone offers lectures and workshops on biodiversity, the impact of ecotourism, guidelines for ecotourism development, species conservation, and development of species status monitoring techniques; he maintains an extensive South American slide library. In addition, Dr. Dunstone has authored some 25 publications, singly and jointly, primarily on various species of mink.

School of Education ⊕
University of Durham
Leazes Road
Durham DH1 IT4, UK

Contact: **Joy A. Palmer, Ph.D.**, Director, B.A.
   Degree Program
   Phone: (44) 91 374 3540 • *Fax:* (44) 91 374 3740

Activities: Education • *Issues:* Environmental Education

▲ Projects and People

Director of the B.A. Education Degree Program at the University of Durham, Dr. Joy Annette Palmer is also lecturer in education and environmental issues. Among her courses are **environmental education**, and study of the environment: ethics and global issues.

Previously a primary school teacher, she led a science/environmental education senior management team and later became advisory head of the Primary Centre in Birmingham Education Department. A fellow of the Royal Geographical Society and various professional organizations, she belongs to the National Curriculum Council's Working Group and advisory subgroup for Environmental Education. Dr. Palmer is member and former chair of the **National Association for Environmental Education**.

Recently, Dr. Palmer advised the University of Thessaloniki on **promoting environmental education in Greece**, with the cooperation of the University of Athens. Other consultancies include advising Greenpeace International, USA, on an environmental education handbook; educational films, and the British Broadcasting Company (BBC) *Watch* natural science program for primary students in school; a science and health series for Central Television; and book reviews for *Environmental Education*, the journal of the NAEE. *(See also National Association for Environmental Education in the UK NGO section.)*

■ Resources

Dr. Palmer has edited, co-authored, or authored numerous books and articles on environmental education, such as books for children on specific environmental issues and activities, plants, and animals. Planned for 1992 publication, for example, are *Environmental Education: History and Development with International Perspectives*, and *Handbook of the Environment*, both with P.D. Neal; the First Starts Series, *Sun, Rain, Wind, and Snow*, with Franklin Watts; and *Tropical Rain Forests*.

Durrell Institute of Conservation and Ecology (DICE) ⊕
University of Kent
Canterbury CT2 7NX, UK

Contact: **Ian R. Swingland, Ph.D.**, Founding
   Director
   Phone: (44) 227 475480 • *Fax:* (44) 227 475481
   E-Mail: irs@ukc.ac.uk

Activities: Education; Research • *Issues:* Biodiversity/Species Preservation; Global Warming; Sustainable Development
Meetings: CITES

● Organization

The **Durrell Institute of Conservation and Ecology (DICE)**, established in 1991 at the **University of Kent**, is a research and postgraduate training institute. With a staff of 12, DICE "centers on conservation biology as a full academic discipline and works in ecology, environmental management, and genetics to conserve species and their habitats. The Institute's mission is to assist peoples to maintain their own ecosystems through education, training, research and field implementation."

Research covers the areas of wildlife ecology and conservation, environmental management, conservation genetics, and evolutionary ecology and behavior. The Training Programs offer higher degrees in **conservation biology, ecology, environmental management in business**—introduced in 1992 in conjunction with the Canterbury Business School, and other areas. In particular, the **diploma program in endangered species management**, in conjunction with the Jersey Wildlife Preservation Trust (JWPT), is an "international course . . . aimed at those already working in species conservation in tropical or temperate countries or those who want to enter conservation or captive species management. The course is either full-time or part-time for a minimum of 10 weeks training in Jersey, followed by a research project in Jersey or in the home country." An international course is also offered in **raptor biology**, focusing on conservation of birds of prey, including **owls**. Special services include short or in-service training courses, conference administration, and consultancies.

▲ Projects and People

"When he isn't traipsing around the world," Dr. **Ian R. Swingland** teaches at the University of Kent (since 1979), or "can generally be found in a medieval village in Kent." His research interests are "the evolutionary aspects of animal behavior and ecology, particularly the evolution of environmental sex determination and of intraspecific differences in behavior, [and] conservation and management, especially international conservation of endangered species and habitats, and protected area design." Dr. Swingland is also an honorary member of the Smithsonian Institution and the University of Michigan.

In 1980, the late **Sir Peter Scott** asked Dr. Swingland to found the Species Survival Commission (IUCN/SSC) Tortoise and Freshwater Turtle Group, which he currently chairs. This led to increased involvement in global research projects, particularly in tropical area efforts at "species conservation, land-use management and resource conservation." As a result of these activities, Dr. Swingland worked to establish the University of Kent's Ecology Research Group, which he later chaired.

Dr. Swingland carried out research in Africa, first as director of research and management at **Kafue National Park in Zambia**, "one of the oldest in Africa," and then for two years on the uninhabited coral **Aldabra Atoll** in the Indian Ocean, where he studied the **giant tortoises**. Chelonia, particularly flying lizards, and the "location on memory in vertebrates," communal roosting of corvids, and large mammal movement patterns were early interests. **Bats, New England herpetofauna, wood ants, and global-warming effects on vertebrates** are more current pursuits.

Dr. Swingland belongs to JWPT's Scientific Advisory Committee, Fauna and Flora Preservation Society, Natural Environment Research Council, British Chelonia Group, among others.

■ Resources

Needs include software/DNA probes, and conservation course material.

Dr. Swingland has authored *The Ecology of Animal Movement, Living in a Patchy Environment,* and *Evolutionary Ecology of Giant Tortoises.* In addition to publishing more than 150 papers and reports, Dr. Swingland founded the journal *Biodiversity and Conservation.*

Department of Plant Sciences ⊕
University of Oxford
South Parks Road
Oxford OX1 3PS, UK

Contact: **Q.C.B. Cronk, Ph.D.,** Botanist
Phone: (44) 865 275000 • *Fax:* (44) 865 274144

Activities: Research • *Issues:* Biodiversity/Species Preservation

▲ Projects and People

Dr. Q.C.B. Cronk, Department of Plant Sciences, University of Oxford, belongs to the Species Survival Commission (IUCN/SSC) Pteridophyte Group.

Department of Zoology ⊕
University of Oxford
South Parks Road
Oxford OX1 3PS, UK

Contact: **Raphael Ben-Shahar, Ph.D.,** Research Assistant

Activities: Research • *Issues:* Biodiversity/Species Preservation

▲ Projects and People

Elephants and their habitats in northern Botswana is the subject of **Dr. Raphael Ben-Shahar's** current Earthwatch-assisted research, which aims to describe feeding patterns of vegetation and damage in the 80,000-square-kilometer range of the region's 50,000 elephants. The study is under the academic auspices of the University of

Oxford, in coordination with Botswana's **Department of Wildlife Conservation and National Parks.** Funding comes from the David Shepherd Foundation for Wildlife and the Environment through World Wide Fund for Nature (WWF-International) and the South African Nature Foundation, administered by the Kalahari Conservation Society of Botswana.

Further stages will "concentrate on aspects of vegetation dynamics and feeding patterns of elephants in the region." Dr. Ben-Shahar writes that "global issues involved in the project include the aridification of semi-arid savannas, the change in biodiversity, and conservation versus development for agriculture."

He offers "consultations on habitat reclamation and erosion control, wildlife management, [and] introduction of rare antelope species." He also conducts speakers' programs on "conservation and wildlife management in southern Africa with particular reference to elephants."

Earlier, Dr. Ben-Shahar was a member and teacher in the Society for the Protection of Nature in Israel; caretaker in the Zoological Centre of Tel-Aviv and Ramat-Gan, attending elephants and carnivores; involved in radio tracking **leopards** in the Judean Desert; and **habitat manager** at the Sabi-Sabi Game Reserve, Transvaal, South Africa.

■ Resources

Funding is needed for aerial surveys, which would expand the area covered during vegetation and elephant distribution transects, and for computerizing the data for implementation of a geographic information system (GIS).

Dr. Ben-Shahar has written some 14 publications; recent ones discuss "the effects of bush clearance on African ungulate populations." He has provided articles for the *Journal of Vegetation Science, Ecological Applications, African Journal of Ecology, Vegetation, Biological Conservation, Ecology, South African Journal of Botany,* and others.

Animal Ecology Research Group ⊕
Department of Zoology
University of Oxford
South Parks Road
Oxford OX1 3PS, UK

Contact: **Malcolm James Coe, Ph.D.,** Tropical Ecologist
Phone: (44) 865 271174 • *Fax:* (44) 865 310447

Activities: Education; Research • *Issues:* Biodiversity/Species Preservation

▲ Projects and People

The latest project of **Dr. Malcolm Coe** is the **Mkomazi Ecological Research Programme,** a joint effort between the Royal Geographical Society and the University of Oxford's **Department of Zoology** and the University of Dar es Salaam's Department of Zoology, **Tanzania.**

In the fall of 1991, with a Memorandum of Understanding in hand, Dr. Coe reported that he and his colleagues "spent two years trying to get [such an] agreement with the Tanzanian government."

At that time, he was "approaching a number of aid bodies to enable us to raise sufficient funds to have people in the field for between three to five years." In the combined research and training program, senior ecologists from the UK will work in the field with Tanzanian biologists, some of whom are earning advanced degrees, he reports.

"Mkomazi lies on the northern Tanzanian border adjacent to the Tsavo National Park in Kenya," writes Dr. Coe. "The Mkomazi National Reserve has been damaged by pastoral intrusion for many years and recent very high elephant poaching has reduced the number of these animals dramatically. It is our intention to engage in a research programme that will provide a complete physical and biological description of the area in order that the biological information may be placed at the disposal of the [Tanzania] Department of Wildlife to plan the area's future management neutralization."

Along with these major scientific efforts, continues Dr. Coe, "we will also be attempting to model plant and animal diversity in the area in relation to environmental parameters."
A small team is scheduled to carry out a reconnaissance in the field in 1992. Dr. Coe also belongs to the Species Survival Commission (IUCN/SSC) Tortoise and Freshwater Turtle Group.

In the past year, Dr. Coe has also been on research missions in Seychelles, Namibia, and South Africa; the Department of Zoology also has assignments in India. Such ecological research programs must involve local people. "If we leave nothing behind, we might as well not bother," he says.

Animal Ecology Research Group ⊕
Department of Zoology
University of Oxford
South Parks Road
Oxford OX1 3PS, UK

Contact: **Indraneil Das, Ph.D.**
   Phone: (44) 865 271173 • *Fax:* (44) 865 310447

Activities: Education; Political/Legislative; Research • *Issues:* Biodiversity/Species Preservation; Deforestation; Sustainable Development • *Meetings:* World Congress of Herpetology

▲ Projects and People

Conservation interests of **Dr. Indraneil Das, Animal Ecology Research Group, University of Oxford**, range from environmental education and management at various **tiger reserves** in **India** to the conservation of **turtles and tortoises** in **Bangladesh** and other Asian countries to the trade status of **monitor lizards**.

Co-chairman of the Species Survival Commission (IUCN/SSC) Indian Subcontinent Reptile and Amphibian Group and member of the Tortoise and Freshwater Turtle Group, Dr. Das researches "community ecology, functional morphology, [and] herpetology." He has also been involved in numerous nature/environmental education camps, sponsored by the World Wildlife Fund (WWF) and the U.S. Fish and Wildlife Service (USFWS). Dr. Das has conducted **Nature Orientation Camps** at Tiger Reserves of Sunderbans, Simlipal, Kanha, and Palamau, as well as at Madhya Pradesh, Bakkhali, and Jaldapar Wildlife Sanctuary; he has also conducted several National Environmental Education Workshops for WWF-India and USFWS.

■ Resources

Dr. Das has authored or co-authored several books, including *Indian Turtles: A Field Guide* (1985) and *Colour Guide to the Turtles and Tortoises of the Indian Subcontinent* (1991). He is associate editor of *Hamadryad,* for Madras Crocodile Bank Trust, India, and on the editorial board for *Asiatic Herpetological Research,* published by the University of California. He has written some 50 articles, with another 10 or so in press, mainly on turtles and crocodiles—both living and fossilized.

Department of Psychology ⊕
Scottish Primate Research Group
University of Stirling
Stirling FK9 4LA, UK

Contact: **William Clement McGrew, Ph.D.**, Reader
            in Behavioural Primatology
   Phone: (44) 786 67640 • *Fax:* (44) 786 67641
   E-Mail: wcm1@uk.ac.stir.forth

Activities: Education; Research • *Issues:* Biodiversity/Species Preservation • *Meetings:* International Primatological Society

▲ Projects and People

With a background in both psychology (focusing on social behavior of preschool children) and anthropology (focusing on chimpanzee material culture), **Dr. William Clement McGrew**, until 1991, was director of the **University of Stirling's** Primate Unit. He continues to research the behavioral ecology of **African monkeys**, apes, and other large mammals. Since 1980, he has been involved in a long-term study of the socio-ecology of **gorillas and chimpanzees** in Lope Reserve, Gabon, with a permanent field station, the Station d'Etudes des Gorilles et Chimpanzes. Other projects are in Liberia, Ivory Coast, Rwanda, and Bolivia. Associated principal investigators are **Dr. Caroline E.G. Tutin, Dr. Elizabeth A. Williamson**, and **Michel Fernandez-Puente**. Project funding comes from scientific and conservation organizations such as the World Wildlife Fund—USA.

Dr. McGrew has taught and researched at universities and research centers worldwide and organized at least seven conferences, including, in 1990, **Origins of Monogamy** for the International Primatological Society in Kyoto and, in 1989, **Behavioural Ecology of Neotropical Primates** for the Tropical Ecology Group of British Ecological Society, London. He is affiliated with numerous national and international associations related to primatology, anthropology, ecology, zoology, and human ethology, and belongs to the Species Survival Commission (IUCN/SSC) Primate Group.

He sits on editorial boards of at least seven journals, and is review co-editor of *Human Ethology Newsletter,* and European editor of *Ethology and Sociobiology.* In addition, he reviews manuscripts for some 20 periodicals.

■ Resources

Dr. McGrew's long list of publications includes reports, book reviews, and recent articles, such as on **cotton-top tamarins**. Much literature describes his investigations of **the use of tools by chimpanzees**—using hammers or other tools to crack nuts, forage for termites, or engage in dental grooming. An example of a 1991 publication is "Brains, Hands, and Minds: Puzzling Incongruences in Ape Tool-Use" in *The Use of Tools in Primates*, edited by J. Chevaillon.

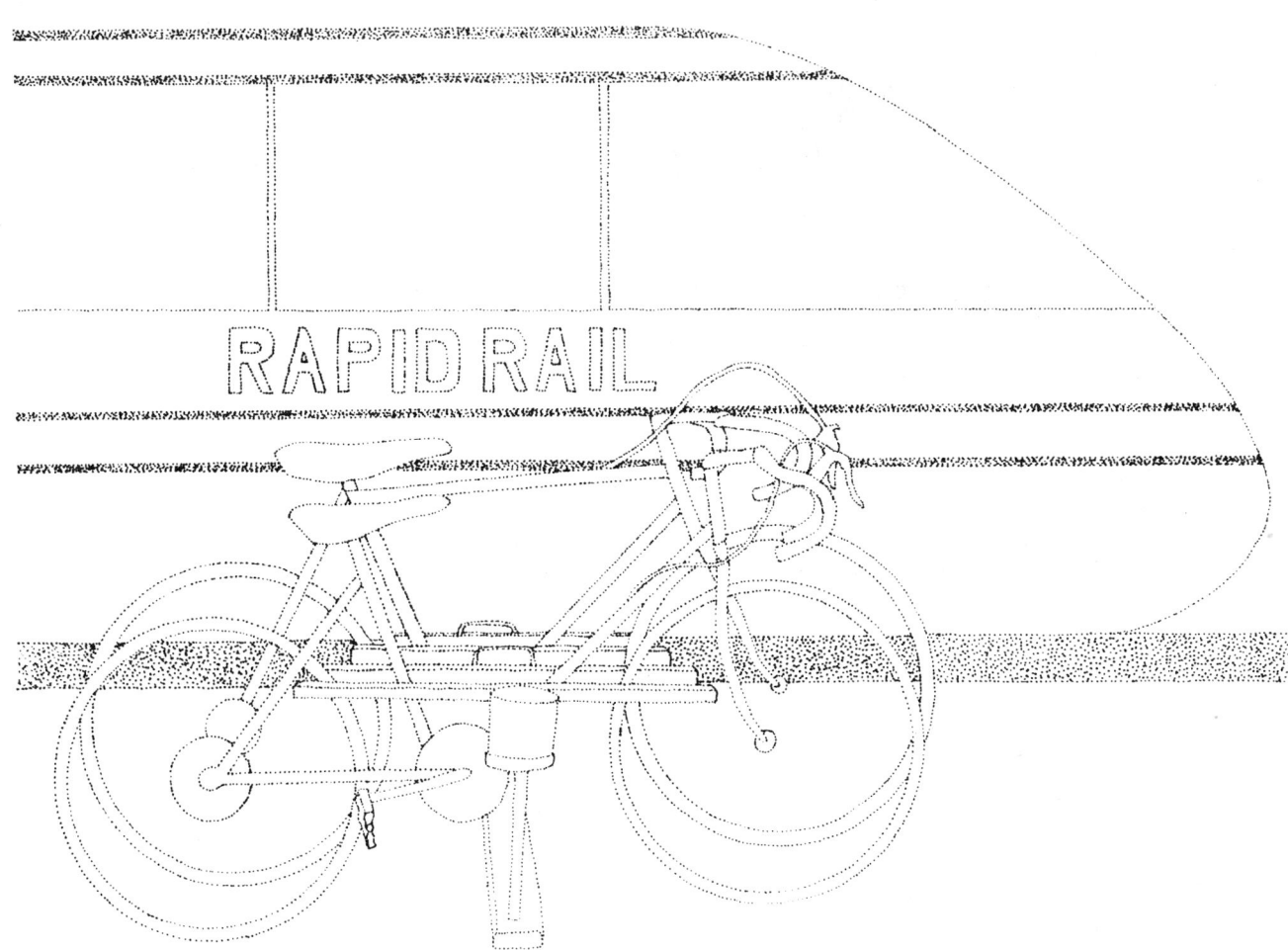

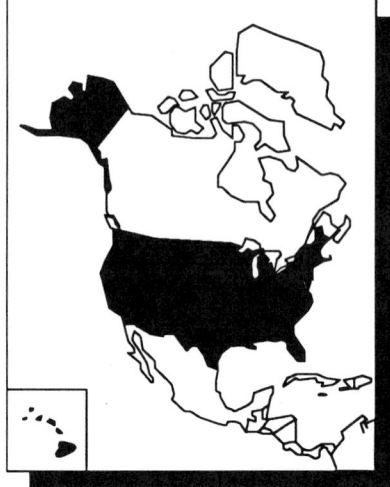

The USA is the world's fourth largest country in both population and area. While it is one of the most plentiful of all nations—in the diversity of its ecology, the abundance of its resources, and the productivity of its economy—it is also one of the most extravagant.

The USA produces 15 percent of the world's oil, 20 percent of the coal, and 25 percent of the natural gas. It is the leading producer of nuclear power and the largest producer and exporter of grains. At the same time, the USA consumes even more natural resources than it generates, especially on a per capita basis: it uses more energy than any other country and is the second largest consumer of nine metals. It is the world's second largest user of pesticides, and it operates more than twice as much farm machinery as any other country.

In June 1992, the USA was criticized for its failure to take a leadership role at the Earth Summit in Rio de Janeiro. What made this missed opportunity particularly disappointing to many was the fact that the USA has been a pioneer in such areas as national parks and wilderness preservation, urban air pollution controls, water quality improvement, endangered species protection, and recycling efforts.

The U.S. environmental record has been particularly tarnished by the nation's contribution to global warming and an official response to the problem that many view as unenlightened and inadequate. The country is the largest single emitter of greenhouse gases, with more than 20 percent of the carbon dioxide released from burning fossil fuels worldwide originating from the USA. Yet, it was the only industrial country at the UN conference to oppose emissions targets.

The country faces its own set of environmental watersheds at home. A broad spectrum of U.S. environmental laws have come to the point of reassessment and reauthorization. These cover water quality, waste management, and endangered species, among other issues. As 1993 begins, however, there is much hope that the USA will meet its environmental challenges by recognizing the need to respect, rather than dominate, the natural environment, and by redefining the present situation in terms of technological and spiritual growth, rather than ideological conflicts and economic trade-offs.

🏛 *Government*

Agency for International Development (USAID) ⊕
U.S. Department of State
Office of Environment and Natural Resources
Washington, DC 20523 USA

Contact: **Maria Beebe**, Senior Program Adviser
Phone: 1-703-875-4255 • *Fax:* 1-703-875-4639

Activities: Development • *Issues:* Air Quality/Emission Control; Biodiversity/Species Preservation; Deforestation; Economic Development; Energy; Global Warming; Health; Population Planning; Sustainable Agriculture; Sustainable Development; Waste Management/Recycling; Water Quality

● Organization

With an approximate $400 million budget that has tripled since the mid-1980s and some 25 new environmental projects in 1992, the **Agency for International Development's (USAID)** environmental strategy for the next decade is to "build an awareness at all levels in AID-assisted countries of the critical link between environmental protection and economic/social development; assist developing countries in improving the management of their natural resources for sustained economic growth and social equity while conserving and improving their environment; help developing countries contribute to resolving key global and regional environmental problems while they work to meet their development objectives; and integrate environmental considerations into all AID-supported development activities."

To USAID, "sustainable development" means economic, social, political, and environmental development that "meets the needs of the present without compromising the ability of future generations to meet their own needs"—as originally defined in *Our Common Future*, the Brundtland Commission Report, 1987.

The Agency focuses on five "critical and environment and natural resource areas: tropical forests and biological diversity, sustainable agricultural practices, environmentally sound energy production and use, urban and industrial pollution, and management of water and coastal resources." Within these areas, USAID pays attention to global climate change impacts.

Three "cross-cutting" and often integrative approaches are identified to implement programs and projects:

Strengthening human and institutional capacity while building public awareness. To help host countries analyze problems and develop appropriate solutions; overcome lack of awareness of policy options and alternatives to unsound practices; and address weak enforcement of environmental laws and regulations and lack of coordination among relevant government agencies, USAID intends to help strengthen institutions at the grassroots level. "The empowerment of local people, particularly traditionally disadvantaged groups, is an effective means of building an on-site constituency with a long-term vested interest in protecting the environment as well as being a necessary precursor for democratic pluralism," writes USAID.

Increased **training programs and advisory services** from federal agencies, universities, and contractors are being planned to reach individuals and groups needing better environmental management skills as well as to strengthen NGOs and community-based organizations working to conserve local natural resources. Public **outreach** techniques, such as the USAID population program, need to be expanded for "increased awareness and education on the consequences of environmental mismanagement." "Missions should work with indigenous NGOs to place environmental problems in a local development context," writes USAID. "Changes in school curricula should also be considered."

Reforming macro- and sectoral economic and environmental policies. Heavily contributing to environmental degradation, reports USAID, are ". . . pricing policies and economic accounting methods that undervalue the benefits provided by the natural resources area; ill-advised subsidies and tax policies that lead to inefficiency and environmentally harmful land resource use [such as] subsidies for ranching and timber extraction; and insecure land and resource tenure that encourages unsustainable resource exploitation and indifference towards environmental protection."

To help bring about policy reform, USAID's strategy is to "pursue discussions with government and the private sector" that will "foster practices that maintain and, if possible, enhance stocks of natural resources." USAID intends to help improve the host country's capacity to conduct environmental impact assessments (EIAs) and economic analyses; "identify and change those policies, laws, and regulations that lead to environmental degradation and natural resource depletion; develop policies with incentives for envi-

ronmentally sound production; and work towards securing the rights to land and other natural resources for the less privileged."

Seeking private sector solutions. Often considered a "major cause of environmental problems," the private sector is now being sought as an active participant on behalf of the environment in endeavors that "link trade, investment, and economic development to mutually benefit developing nations and the USA." One example is the USAID **Partnership for Business and Development Initiative** that targets environment and energy.

To encourage such partnerships, USAID plans to "identify and develop **local business opportunities** which are both environmentally sound and commercially viable [with] local private industry, local populations, NGOs, and the local private banking sector; encourage **technology transfer/assistance and increased trade and investment** by U.S. businesses [such as] non-timber forest products, ecotourism, extractive reserves, 'smart-wood' certificates, and in natural forest management; collaborate on **innovative financing** of environmental projects with NGOs and the private banking sector [such as] debt for equity swaps [and] environmental bonds; [and] encourage policy reform to provide **market incentives** that will improve the trade and investment climate for environmentally beneficial energy technologies as well as environmental services, systems, and technologies."

USAID will help host governments "establish policies and laws that stimulate the development of environmentally pro-active enterprises" and ensure that "private sector activities claiming to be environmentally responsible and/or environmentally pro-active remain so." Periodic EIA monitoring is strongly recommended.

Office of Agriculture ⊕
Bureau for Research and Development
Agency for International Development (USAID)
R&D/AGR, SA-18/408, AID
Washington, DC 20523-1809 USA

Contact: **Raymond E. Meyer, Ph.D.**, Soil and Water Resources Management Specialist
Phone: 1-703-875-4122 • *Fax:* 1-703-875-4186
E-Mail: Telemail: st/agr

Activities: Development; Research • *Issues:* Biodiversity/Species Preservation; Deforestation; Global Warming; Sustainable Development; Waste Management/Recycling; Water Quality

▲ Projects and People

TropSoils is a partnership of the Agency for International Development (USAID), U.S. universities, and cooperating institutions in developing countries. Initiated in 1981, it is one of the Collaborative Research Support Programs (CRSPs) created to implement Title XII of the United States Foreign Assistance Act.

Its goal is to "develop and adapt improved soil-management technologies that are agronomically, ecologically, and economically sound for developing countries in the tropics" through three global activities, or *thrusts*.

Research thrusts are Natural-Resource Management and Sustainable Agriculture, where soil-management constraints and problems are identified, projects are developed, and results are assessed. In the Outreach thrust, research results are incorporated into communications, soil management networks, decision support systems, and training.

Natural-resource management thrusts tackle the following problems:

- **Land-clearing pressure.** As populations swell and farms multiply, the soil recovery period shortens, soils rapidly degrade to nonproductive levels, and farmers clear more land to grow the same amount of food. TropSoils has identified production constraints and demonstrated that "43 consecutive crops can be grown—without yield reduction—on a fragile soil in the humid tropics."

- **Inadequate resource inventory.** TropSoils works collaboratively on collecting reliable information on the production capability of soil systems, effects of forest conversion on soil dynamics, and indigenous technology.

- **Landscape restrictions.** TropSoils production technologies are integrated with indigenous methods to improve watershed management—particularly where arable tropical landscapes are often rolling or steep and susceptible to water and wind erosion.

- **Climate variability.** Research for sustainable agriculture integrates fertility with water-use efficiency for semiarid regions and nutrient management practices to minimize losses for the humid tropics.

- **Sustainable agricultural production** thrusts maintaining or enhancing the natural-resource base while avoiding environmental degradation deal with the following:

- **Soil acidity.** TropSoils research provides options for managing the very acid tropical soils, using crop selection and soil amendments where available. Recommendations on liming and economic evaluation methods are also provided through a computer-based decision-support system.

- **Nutrient deficiencies and losses.** When soils low in nutrients are continuously cultivated, this practice results in low yields, soil degradation leading to land abandonment, rainforest destruction, and encroachment onto marginal semiarid lands—which can end in desertification. TropSoils research quantifies nutrient requirements for rejuvenating soils and sustaining production. It also provides insight into soil biology and nutrient cycling—both of which are necessary for low- or high-input management of such tropical soil systems.

- **Physical limitations.** Degradation of soil that is highly weathered and low in organic matter can result in poor water infiltration, rainwater runoff, and accelerated erosion. Research on the use of legume fallows and green manures offers management options to prevent soil degradation and rejuvenate land already degraded.

- **Topographic limitations.** An example of this is Texas A&M's interdisciplinary approach to topographical limitations in Niger which involves integrating natural-resource management components with the socioeconomic structure, human requirements, indigenous technologies, and physical characteristics of the landscape—into an agricultural watershed designed to meet its area's needs.

- **Water stress.** Too much or too little water can cause natural or imposed moisture stress; a constraint that can be relieved through proper crop selection. TropSoils develops techniques for water harvesting and for helping roots reach subsoil moisture. The practice of placing lime deep in acid Indonesian soils increased

food yields and decreased the risk of crop failures as cropping intensity rose from two to three plantings and harvests annually.

- The outreach thrust enables technology to be extended from a limited number of research sites to many users' fields, particularly when Soil Taxonomy or Fertility Capability Classification (FCC) that inventory soil diversity are used as a basis for transfer. TropSoils outreach overcomes other problems:

- **Lack of local expertise.** More than 100 soil scientists from Central and South American developing countries earned graduate degrees with the help of TropSoils, and are now part of a network of collaborating scientists.

- **Information-knowledge gap.** TropSoils develops decision-support systems and adapts geographic information systems (GIS) to capture new technologies that can be integrated with indigenous knowledge to hasten technology transfer to individual farmers in diverse situations.

- **Lack of skills and research capabilities.** TropSoils provides training programs, backstopping, and educational support to overcome these constraints.

Since 1981, USAID, North Carolina State University, University of Hawaii, Cornell University, and Texas A&M University have been providing worldwide research and technical assistance in the humid tropics, wet/dry acid tropical savannas, and semiarid tropics. Sustainable production technologies as alternatives to rainforest destruction are being developed.

"High-input systems in Peru have produced 41 crops in 18 years with no yield decline," according to TropSoils. "Low-input systems by maximizing nutrient recycling will sustain three crops, followed with one year legume fallow." In Indonesia, farmers accepted rotation by matching ethnic preference; deep lime placement through the use of an indigenous specially manufactured plow encouraged deeper rooting of maize and lowered risks. An integrated watershed management program got underway in the Sahelian region with improved soil management and water use to reduce risks of millet production. Dr. Roger G. Hanson, North Carolina State University, is project director.

A worldwide project in conjunction with the U.S. Department of Agriculture (USDA) provides technical assistance in developing dryland and rainfed agriculture to improve soil and water resources as well as output and income in crop and livestock production systems. So far, databases of current in-country research information were created for Jordan and northeast Thailand; regional workshops give recommendations for soil and water management; and agroclimatological analyses of improved cropping strategies in the Sahel are ongoing. Involved USDA agencies are the Agricultural Research Service (ARS) and Economic Research Service (ERS). Dr. James F. Parr, USDA-ARS, Beltsville, Maryland, is principal investigator.

The USDA's Soil Conservation Service (SCS) is researching improved soil classification systems for more effective technology transfer; training developing country scientists in soil resource inventories, soil analysis and management, and soil survey methodology; and providing technical assistance to more than 50 countries. So far, a common language for soil-based agrotechnology transfer is accepted by more than 30 countries; soil research and survey capability is being strengthened in at least one country of every region; and 900 developing country scientists were provided short-term training. Dr. Richard Arnold, USDA-SCS, Washington, DC, is principal investigator; Dr. Raymond E. Meyer is project officer for the USDA-ARS/ERS/SCS related endeavors.

Socioeconomic components—to assess how crop and livestock growers and communities choose their farming systems and resource management practices, and how improved alternatives can be encouraged—are basic to projects in Benin, Burkina Faso, Cameroon, and Mali in collaboration with numerous U.S. universities.

■ Resources

Lists of publications on soil management support services and technology of soil moisture management are available through USAID. So is additional information on TropSoils projects.

Office of U.S. Foreign Disaster Assistance
(OFDA) ⊕
Food and Humanitarian Assistance Bureau
**Agency for International Development
(USAID)**
Room 1262A, N.S.
Washington, DC 20523-0008 USA

Contact: **Gudrun H. Huden,** Senior Program Officer;
Prevention, Mitigation & Preparedness
(PMP) Division
Phone: 1-202-647-7530 • *Fax:* 1-202-647-5269

Activities: Development • *Issues:* Air Quality/Emission Control; Sustainable Development; Waste Management/Recycling; Water Quality

▲ Projects and People

The Office of U.S. Foreign Disaster Assistance (OFDA), now part of the Food and Humanitarian Assistance Bureau in the Agency for International Development (USAID) is designing a five-year program in industrial accident prevention through a cooperative agreement with the World Environment Center (WEC). Up to six high-risk, urban industrial sites are being selected from a current list which includes Brazil, Colombia, Egypt, India, Indonesia, Mexico, Philippines, Poland, Thailand, and Turkey.

Local Accident Mitigation and Prevention (LAMP) programs are being developed to analyze local chemical and technological hazards, implement disaster response plans, and to enhance preparedness and prevention capabilities. OFDA is working with local and visiting experts from public and private sectors to determine goals, form community advisory boards, assign staff, and create programs in Awareness and Preparedness for Emergencies at the Local Level (APELL). Examples of program support to be developed are Chemtrec-style hazardous materials transport accident response systems and poison control centers.

LAMP programs, with their community involvement, are expected to contribute to overall host country disaster preparedness as well as to improved readiness for natural calamities. Improved land use and zoning legislation are likely results of risk analysis; environmentally improved technologies are other benefits from hazards analysis and public participation.

Gudrun Huden summarizes the OFDA pesticide management and disposal activities, from fiscal years 1992–1997. "Accumulation of surplus and out-dated pesticides presents a growing problem

in developing countries worldwide, but especially in Africa," she explains. "The risk to human and environmental health is both acute and constant"; communities surrounding these pesticide build-ups are in danger.

A yearly $200,000 is being set aside for the following activities:

- **Cement kiln disposal of surplus and overage pesticides.** This program is already initiated in Morocco, with cooperation from the government of Morocco and the German Technical Cooperation Agency. It is based upon an earlier pilot pesticide disposal project that OFDA sponsored and executed in Pakistan, where 17,000 liters of liquid pesticides were burned and emissions sampled and analyzed. "In Morocco, it will be shown that this is not only an environmentally sound and most cost-effective disposal technique for liquid pesticides, but is also the method of choice for power and bran formulations," writes Huden. A similar project may begin in Zimbabwe.
- **Pesticide drum reconditioning plants.** This is being based on technology appropriate to developing countries.
- **Fluidized bed incineration pilot project.** This is a pesticide disposal program for situations in which materials cannot be transported or in which significant quantities of contaminated soil must be treated.
- **Controlled decomposition** environmental study and demonstration in limited access sites such as the desert or tropical environment.
- **Prototype pesticide storage facility.** With technology already in place, this program is based on a model developed by the Netherlands.
- **Model systems and training programs** for managing pesticide spills, transporting and consolidating surplus pesticides, and monitoring agriculturists for exposure to certain chemicals.
- **Microbial decomposition** of pesticides, which is incorporated into pesticide disposal procedures.

Since 1978, Huden has undertaken global responsibilities with OFDA, including grasshopper/locust campaigns in Africa, where she led the USA fight against the use of dieldrin and other chlorinated hydrocarbon pesticides. She also fielded an international study team to observe typical problems of pesticide wastes and evaluate disposal options in the East African countries of Ethiopia, Kenya, and Somalia; and supported environmental assessments of the effects of pesticides on non-target organisms and the development of protocols for ecological monitoring.

Section of Energy and Environment (SEE)
Office of Economics
Interstate Commerce Commission (ICC)
12th Street and Constitution Avenue, NW
Washington, DC 20423 USA

Contact: **Elaine Kaiser,** Chief
Phone: 1-202-927-7684 • *Fax:* 1-202-927-6225

Activities: Environment Analysis of Transportation Licensing
Issues: Transportation

● Organization

The Section of Energy and Environment (SEE) conducts independent environmental reviews of transportation cases, prepares environmental assessments and environmental impact statements, and advises the Commission on environmental matters. In conducting its environmental review, SEE examines the impacts of proposed actions on the environment and recommends conditions to mitigate environmental effects.

When evaluating cases—generally involving railroad abandonments, acquisitions, and construction—SEE considers the protection of water resources, historic sites and structures, threatened or endangered species, public safety, land use, and other pertinent environmental areas. For example, several environmental issues may be identified in one situation such as the proposal of an extension of an authorized 82-mile rail line in Montana, where construction impacts must be considered on adjacent farming and ranching activities, wildlife, water quality of the nearby Tongue River, and significant sites of Native Americans. SEE works closely with various federal and state agencies to ensure that environmental impacts are fully considered and effective mitigation is imposed.

In 1991, the Commission revised and clarified its environmental rules so as to more effectively consider environmental issues. Diverse groups who commented on the proposed rules included the Council on Environmental Quality (CEQ), National Trust for Historic Preservation, Association of American Railroads (AAR), San Francisco Bay Conservation and Development Commission, U.S. Environmental Protection Agency (EPA), and National Oceanic and Atmospheric Administration (NOAA). These new rules are designed to allow applicants, interested parties, and SEE to better identify and more expeditiously resolve environmental concerns.

■ Resources

Environmental impact statements and environmental assessments are announced in the *Federal Register*; they are served on all parties to the proceeding; and copies are made available to the public. Other public documents are available through SEE.

Upper Atmosphere Research Program (UARP)
**National Aeronautics and Space
  Administration (NASA)** ⊕
Code SEP-04
Washington, DC 20546 USA

Contact: **Michael J. Kurylo, Ph.D.,** Manager
Phone: 1-202-358-0237 • *Fax:* 1-202-358-3098

Activities: Research • *Issues:* Global Warming; Stratospheric Ozone Depletion

● Organization

The **Upper Atmosphere Research Program (UARP)** of the National Aeronautics and Space Administration (NASA) was created by a congressional directive in June 1975 and by the Clean Air Act Amendments of 1977. UARP was to "develop and carry out a comprehensive program of research, technology, and monitoring of the phenomena of the upper atmosphere so as to provide for an understanding of and to maintain the chemical and physical integrity of the Earth's upper atmosphere."

Since then, NASA has launched dozens of programs and projects to study the upper atmosphere, with a special emphasis placed on the depletion of the ozone layer. NASA emphasizes that the information and data gathered by its many projects is not limited to American concerns, but is a global enterprise. The data comes from every corner of the world, and the insights and information gained are shared internationally with scientists and other countries.

▲ Projects and People

The **Total Ozone Mapping Spectrometer,** or **TOMS,** was launched aboard NASA's Nimbus-7 in 1978. Since then it has provided daily high-resolution mapping of global total ozone. TOMS is managed by the Goddard Space Flight Center (GSFC), Greenbelt, Maryland, and is the primary source for high-quality global maps about the total content of ozone in the atmosphere. TOMS data has traced the development of the Antarctic "ozone hole," and an analysis of past TOMS data indicates that the hole has existed since at least 1979. Recent studies of over a decade of TOMS data has shown that reductions in ozone over the mid-northern latitudes are approximately twice as severe as previously believed. In order to insure the uninterrupted flow of TOMS data, NASA will launch several more TOMS satellites. On August 15, 1991, the (former) Soviet Union launched a Meteor-3 satellite carrying a TOMS instrument. A third TOMS will be launched by NASA in 1993, and the Japanese Advanced Earth Observations Satellite will carry a TOMS when launched in 1995.

In September 1991, NASA launched the **Upper Atmosphere Research Satellite (UARS),** the first satellite dedicated to studying stratospheric science. The satellite will focus on the processes that lead to ozone depletion, and will complement the information and measurements made by TOMS. UARS is equipped with 10 different instruments and will provide data on energy inputs, winds, and the chemical composition of the atmosphere. The satellite was developed and is managed by the Goddard Space Flight Center.

The Goddard Space Flight Center has developed an instrument for periodic flight in the Space Shuttle, the **Shuttle Solar Backscatter Experiment (SSBUV).** The SSBUV determines ozone level by measuring reflected ultraviolet (UV) light. The SSBUV is sensitive enough to measure the total amount and height distribution of ozone and use that data to calibrate ozone-measuring instruments on other satellites. The SSBUV has flown several times and has regular flights scheduled throughout the 1990s.

In March 1992, NASA used the *Atlantis* space shuttle to launch ATLAS-1, the **Atmospheric Laboratory for Applications and Science.** ATLAS-1 was flown with projects and instruments supplied by the USA, France, Belgium, Germany, Netherlands, Japan, and UK. The ATLAS satellite is designed to study the effects of the sun on climate, and the effects of industrial complexes and agricultural activities. The ATLAS-1 is the first in an 11-year series of missions to study the interactions between the atmosphere and the sun. The ATLAS missions are managed by the Marshall Space Flight Center, Huntsville, Alabama.

In 1984, NASA began the **Global Tropospheric Experiment (GTE)** in collaboration with the Global Tropospheric Chemistry Program. The first project was to increase the capabilities of existing measurement technology. The Chemical Instrumentation Test and Evaluation (CITE) compared existing technology and provided a basis for improvement. GTE is primarily an aircraft-based program supplemented by ground-based measurements. The first GTE expeditions, the Atmospheric Boundary Layer Experiment (ABLE), measured the interactions of the biosphere and the atmosphere. ABLE-2 was carried out in collaboration with the government of Brazil and measured carbon monoxide (CO) concentrations above the Amazon rainforest. At their peak, the ABLE-2 expeditions involved 60 USA and over 100 Brazilian scientists.

Although the Arctic seems "remote from human habitation and industry," its proximity to certain "sources of trace gases and aerosols"—such as industrial emissions from the USA and the former USSR—cause "episodes of pollution" in the spring and fall, according to initial findings of the ABLE-3 project, which measured the northern tundra and boreal forests as well as methane emissions from Alaskan wetlands.

Preliminary conclusions suggest a "role for Arctic tundra in controlling global tropospheric ozone levels." As NASA reports, "Enhanced input of nitrogen to the atmosphere over tropical rainforests and arctic tundra provides a photochemical source for ozone production, thereby reducing net ozone destruction in these regions . . . providing potentially important feedback to global warming."

With Dr. Robert J. McNeal's management, upcoming GTE projects are investigating tropospheric chemistry especially over the tropical Atlantic Ocean and the "economically burgeoning Pacific Rim region."

NASA has sponsored ground-based airborne expeditions in addition to its satellites. In collaboration with industry, the National Oceanic and Atmospheric Administration (NOAA), and the National Science Foundation (NSF), NASA has conducted two airborne campaigns over the polar caps. In 1987 planes loaded with instruments crossed the Antarctic, and in 1989 the Arctic. These expeditions have led NASA to conclude that chemical reactions involving human-produced chlorine are a main cause of ozone depletion. The Ames Research Center, Mountain View, California, is the managing center for the airborne expeditions.

The Langley Research Center, Hampton, Virginia, manages the Stratospheric Aerosol and Gas Experiment (SAGE), first launched in 1979. The original SAGE provided ozone measurements for two years. SAGE II began operation in 1984 with the Earth Radiation Budget Satellite, and has made important observations of the Antarctic ozone hole.

Mission to Planet Earth is an example of NASA's future direction. The centerpiece of Mission to Planet Earth is the Earth Observing System (EOS), a series of environmental research satellites planned to begin launches in 1998. The EOS program will combine and continue the measurements begun by TOMS, ATLAS, UARS, and SSBUV. The EOS program will bring together data on the atmosphere, oceans, land surfaces, and the biosphere.

Working with dozens of foreign countries, agencies, universities, and organizations, NASA seeks to "understand the physics, chemistry, and transport processes of the upper atmosphere, and to accurately assess possible perturbations of the upper atmosphere caused by human activities as well as natural phenomena."

■ Resources

NASA periodically publishes a large guide, *Research Summaries,* outlining results and data obtained from its programs. A brochure on *Global Tropospheric Experiment* and *Reference Publications* are also available.

Global Environmental Studies Center ⊕
Oak Ridge National Laboratory (ORNL)
P.O. Box 2008, MS-6206
Oak Ridge, TN 37831-6206 USA

Contact: **Michael P. Farrell, Ph.D.,** Director
Phone: 1-615-576-7785 • *Fax:* 1-615-576-9977

Activities: Research • *Issues:* Air Quality/Emission Control; Biodiversity/Species Preservation; Deforestation; Energy; Global Warming; Health; Sustainable Development; Transportation; Waste Management/Recycling; Water Resources

● Organization

Government owned and contractor operated, ORNL established its **Center for Global Environmental Studies** in 1989 with three goals: improving the understanding of global workings of environments in the air, water, and on land; developing capabilities to anticipate the long-term, large-scale effects that human actions have on the biosphere; and identifying appropriate options for technological and society's responses.

Looking ahead—not just to the twenty-first century but to the next millennium—the Center is confronting the most complex and global issues comprehensively: greenhouse gases, climate change, ozone breakdown, deforestation, desertification, resource depletion, and the spread of pollution.

Director **Michael Farrell** says the Center is looking toward a "new view of the kinds of research needed in this formidable but exciting field" that is collaborative, politically practical, economical, and of planetary scope.

Global systems analysis provides the framework for the Center's work. Sophisticated models are being developed to "reflect the dynamic interactions of . . . global vegetation; human cultures and behaviors; and Earth systems such as atmospheric chemistry, ocean composition and circulation, and the links between air, land, and sea." These models eventually will reflect the "interplay of demographics, land-use patterns, economics, ecological relationships" and other factors influencing global environment.

▲ Projects and People

The Center is concentrating in four areas—measurement science and instrumentation, data systems, large-scale environmental studies, and policy, energy, and human-systems analysis.

With an instrumentation background in high-energy and health physics, pollution monitoring, nuclear reactor technology, and nuclear and chemical waste, the Center is directing this expertise in atmospheric, terrestrial, and aquatic research. Its foundation is strong in laser technology that could be used to measure trace gases, temperature, and pressure; remote sensing and fiber optics—applicable to climate studies; spectrometry and isotopic analysis; and automation, miniaturization, and portability.

Environmental data collection is already staggering. The Center says "in a few years NASA's Earth Observing System will begin transmitting enough data to fill all the books in the Library of Congress—every three weeks." Inasmuch as data being gathered today may not be pertinent for decades to come, information systems must be user-friendly and clear.

To examine the aggregate processes that are invisible at close range, ORNL is developing tools and techniques for **large-scale** environmental studies that require a longer perspective than traditional ecological research.

The role of people in global environmental change is central to studies here. Anthropologists, economists, political scientists, sociologists, planners, geographers, climatologists, and ecologists are examining decisionmaking processes that affect global risks and resources.

The Center says it "stands in a unique position to serve the needs of federal agencies and international efforts and to make a significant contribution to our understanding of the delicately balanced biosphere we call home."

■ Resources

Publications on global environmental issues, including carbon cycle, energy technologies, environmental analyses, policy studies, and international relations, are available free of charge . Expanded government funds are needed for "human dimensions research."

Smithsonian Institution ⊕
1000 Jefferson Drive, SW
Washington, DC 20560 USA

Contact: **Robert S. Hoffmann, Ph.D.,** Assistant
Secretary for Research
Phone: 1-202-357-2939 • *Fax:* 1-202-357-4482

Activities: Education; Research • *Issues:* Biodiversity/Species Preservation; Deforestation; Energy; Global Warming; Sustainable Development

● Organization

The Smithsonian Institution—which is the nation's center for the study of science, art, and history—is involved in research relating to environmental issues as well as educational efforts for the public and professionals. With an operating budget of about $360 million and some 7,000 employees, the Smithsonian encompasses 14 museums in Washington, DC, and research stations worldwide. The seven research bureaus most concerned with the conservation of the environment are the **National Air and Space Museum (NASM);** the **National Museum of American History (NMAH);** the **National Museum of Natural History (NMNH);** the **National Zoological Park (NZP),** which is described below; the **Smithsonian Astrophysical Observatory (SAO);** the **Smithsonian Environmental Research Center (SERC);** and the **Smithsonian Tropical Research Institute (STRI).**

▲ Projects and People

The following research bureaus are of special interest: the SERC, which is primarily concerned with the preservation of estaurine, wetland, and upland habitats in temperate zones, is located on a subestuary of the Chesapeake Bay near Annapolis, Maryland. SERC strives to understand how ecosystems and their components function and interact. The STRI focuses on marine and terrestrial ecosystems in and near the Isthmus of Panama with laboratories and facilities among lowland tropical forests at the Barro Colorado Nature Monument and at Tivoli and Ancon; marine laboratories are at Galeta (Atlantic), Naos Islands (Pacific), and San Blas (Carib-

bean). STRI also is involved in comparative studies in the Old World Tropics of India, Malaysia, and Papua New Guinea. Scientists are convinced that baseline data resulting from long-term studies of ecosystems is the best means to detect changes due to human-induced or natural causes. Moreover, studies that provide information about the normal functioning of ecosystems can help protect habitats and ecosystems from deleterious changes.

With less than 2 million of the estimated 30 million species of organisms identified on the planet, the NMNH—the world's largest research museum—is at the forefront of discovering, describing, cataloging, and studying the world's plants and animals. In collaborative projects, biological inventories of the Amazon Basin have been undertaken, such as the Neotropical Lowland Research Program and Biodiversity in Latin America (BIOLAT). Databases are being built on threatened plant species in the Western Hemisphere in cooperation with the IUCN, for example. Specific inventories range from flora and/or fauna surveys in North and South America and the Caribbean to India's medicinal plants.

The remainder of the Smithsonian Institution's (SI) environmental activities are grouped as follows: biogeochemical and atmospheric processes, earth system history, remote sensing, human interactions, site management, *ex situ* and *in situ* species conservation, restoration of degraded ecosystems, professional training, primary and secondary education, public education and exhibits, symposia and programs for professionals and the public, and print and electronic media.

Dr. Robert Hoffmann recently launched a cooperative enterprise, the Smithsonian's **Conservation Training Council,** comprised of SI scientists who head training programs in separate SI bureaus but with similar conservation goals. **Dr. Francisco Dallmeier** represents the SI Man and Biosphere Biological Diversity Program; **Dr. Don Wilson** heads the SI Biolat Program; **Dr. George Angehr** leads the SI Tropical Research Institute Training Program; **Dr. Rudy Rudran** oversees the Wildlife Conservation and Management Training Program; and **Dr. Chris Wemmer** is in charge of the Zoo Biology Training Program. Hoffman and **Dr. Tom Lovejoy** are *ex officio* members. The Council's purpose is to address conservation training needs in the developing world.

Internationally, Hoffmann is an honorary member of the All-Union (former USSR) Theriological Society. He was on the organizing committee of the 1st Symposium on Asian-Pacific Mammalogy, Beijing, China; and a U.S. leader in the USA/USSR workshop on Pan-Arctic Biota Project, Moscow, 1991. He belongs to the Lagomorph and Insectivore, Tree Shrew, and Elephant Shrew groups of the Species Survival Commission (IUCN/SSC).

■ Resources

The Smithsonian also has an independent **Office of Environmental Awareness (OEA),** which works with SI bureaus to gather environmental information and reach public and professional groups through exhibitions, publications, conferences, and workshops. **Judith Gradwohl** is director. Supported by private grants instead of federal funding, the OEA recently sponsored forums and exhibits on oceans (including issues on pollution, coral reef destruction, natural-resources exploitation, and coastal development), the disappearing Latin American rainforests, and media and environmental reporting. A "Better World Starts at Home" poster on air, water, and energy conservation is available to the public.

External Affairs
Smithsonian Institution ⊕
Castle 317
1000 Jefferson Drive, SW
Washington, DC 20560 USA

Contact: **Thomas E. Lovejoy, Ph.D.,** Assistant
Secretary
Phone: 1-202-786-2263 • *Fax:* 1-202-786-2304

Contact: **Martha Hays Cooper,** Research Assistant
Phone: 1-202-786-2287

Activities: Development; Education; Environmental Awareness; International Affairs; Legislative; Scientific Research
Issues: Biodiversity/Species Preservation; Deforestation; Energy; Ethnobiology; Global Warming; Sustainable Development • *Meetings:* AIBS; SCOPE; Society for Conservation Biology; UNCED

▲ Projects and People

Thomas E. Lovejoy is one of the world's chief spokespersons for the environment. He holds many concurrent leadership posts, including chief advisor of the **Minimum Critical Size of Ecosystems Project,** a joint effort of WWF–USA and the Instituto Nacional de Pesquisas da Amazonia (Brazil's National Institute for Amazon Research); scientific advisory council member, FUNATURA, The Foundation for Nature, Brazil; advisory and technical board member, Fundacion Neotropica and Fundacion de Parques Nacionales, Costa Rica; board member, Fundacion Maquipucuna, Foundation for Ecodevelopment and the Study of Biodiversity in Ecuador; scientific council member, FPCN (Fundacion Peruana para la Conservacion de la Naturaleza); advisory board member, American Society for the Protection of Nature in Israel; honorary chairman, Wildlife Preservation Trust International; director, Henry Foundation for Botanical Research; member, Species Survival Commission, IUCN (World Conservation Union); founder and advisor, WNET (Channel 13, New York) "Nature" series; advisor, Environmental Assessment Council; scientific advisory board member, Center for Conservation Biology, Stanford University, California; chairman, U.S. Man and the Biosphere Program; member, National Research Council Committee on Biodiversity; past treasurer, SCOPE (Scientific Committee on Problems of the Environment); director, Rainforest Alliance; council member, Jersey Wildlife Preservation Trust; member of board of directors for Resources for the Future and World Resources Institute; president, American Institute of Biological Sciences; and member, White House Council of Advisers in Science and Technology

Dr. Lovejoy is a fellow in the American Association for the Advancement of Science and the American Ornithologists' Union and a recipient of the Florida Museum of Natural History Carr Medal for outstanding contributions to natural heritage knowledge.

Subjects of his publications range from saving Amazonia to the white stork's migration in Egypt, from primates in the tropical forest to changing Third World environments. He testifies in Congress on areas of international affairs, the International Environmental Assistance Act, sustainable development in the Third World,

the National Biological Diversity Conservation Act, and other timely legislation.

Tom Lovejoy joined the Smithsonian in 1987, following 14 years with World Wildlife Fund. He is also on the Board of Advisors of this publication.

Conservation and Research Center (CRC) ⊕
National Zoological Park
Smithsonian Institution
Route 522 South
Front Royal, VA 22630 USA

Contact: **Chris M. Wemmer, Ph.D.,** Associate
Director for Conservation
Phone: 1-703-635-6522 • *Fax:* 1-703-635-6551

E-Mail: Bitnet: nzpcrco1@sivm

Activities: Research • *Issues:* Biodiversity • *Meetings:* ICBP; IUCN; International Ornithological Congress; International zoo meetings; Neotropical Ornithological Congress

▲ Projects and People

"Zoos in the developing world are popular recreational institutions, yet their potential for education, conservation, and research is rarely realized," writes **Chris Wemmer.** "This is regrettable because few other public institutions are as numerous, as firmly established in the fabric of society, or as suitable for stimulating public awareness of the vital conservation issues facing these nations."

In response to this need, the Smithsonian's National Zoo developed the **Zoo Biology and Animal Management Training Course,** which has been taught worldwide in Brazil, China, Morocco, Malaysia, Guatemala, Mexico, Thailand, and Indonesia. Wemmer describes one popular class in taming and crate-training Malayan tapirs where the "Tapir Massage Team" provided a good rubdown. "After two weeks of daily attention . . . tapirs eagerly awaited their daily massage and willingly entered the shipping crate to receive a bucket of veggies." Advantages: the keeper gains intimate knowledge of the animals' health, reproductive condition, and well-being; the animal receives better care; and scientific information is more easily collected.

In classes on immobilization and handling, veterinarians showed how to collect blood samples for genetic analyses of maned wolves and Magellanic penguins. Identified for studbook development were birds such as the *guaroba* (Brazil's national bird), the curassow, the harpy eagle; mammals such as the golden-rumped tamarin and the giant Brazilian otter; and reptiles such as the broad-snouted caiman and Amazon River turtle. Class projects in Indonesia focused on monitoring the sexual behavior in the Sumatran rhino and providing private nesting space for bonded pairs of black palm cockatoos. In Rabat, recovery plans for the declining bald ibis and the endangered Cuvier's gazelle were class projects; and plans to conserve a herring gull rookery off the coast of Eassaouira are underway.

Zoos are playing a growing role in preserving the world's vanishing creatures as they shift their roles from exhibition and recreation to propagating selected species through scientific management. The

Species Survival Plan (SSP) of the American Association of Zoological Parks and Aquariums (AAZPA) is the blueprint to propagating and maintaining demographically and genetically healthy populations of certain endangered species through cooperative captive management as a supplement to preserving animals in nature.

The National Zoo has actively participated in more than 30 SSP programs, from the Aruba Island rattlesnake to the gorilla. Among the recent inventory of species at the CRC are red pandas, Siberian polecats, clouded leopards, Przewalski's wild horses, Arabian oryxes, black-footed ferrets, blue-headed parrots, and Rothschild's mynahs.

The computer is a vital partner in species preservation. CRC microcomputer workshops, such as in Bombay and with Indonesian researchers, help conservationists evaluate complex ecological data to plan protective strategies. A computer distribution center was set up at CRC to collect donated software and hardware for training and conservation projects overseas. CRC staff encourages the development of coordinated and compatible technology worldwide.

CONSLINK is a CRC service providing an electronic medium to discuss biological conservation and a bulletin board with files on topics including updated lists of conferences, symposia, and workshops; biological field stations in tropical countries worldwide; newsletters of IUCN's Captive Breeding Specialist Group and the Indonesian/Malaysian Faunal Interest Group; and the Rainforest Network.

Dr. Christen Marcher Wemmer became assistant director for conservation and captive breeding in 1984. A wildlife biologist with the CRC since 1974, Wemmer has field experience in 10 countries of Latin America, Asia, and Africa. Recent grants were for "International Conservation Training" in Third World countries and for "Population Biology and Genetics of the Asian Elephant" from the U.S. Agency for International Development. He belongs to several Species Survival Commission (IUCN/SSC) groups, including Captive Breeding and Re-Introduction.

Dr. Scott Richard Derrickson is deputy associate director for conservation and captive breeding as well as curator of ornithology. He is an authority on the captive breeding and reintroduction of the endangered Guam rail and whooping cranes.

Dr. John H. Rappole is research coordinator and uses remote sensing and GIS software to assess changes in populations of migratory birds. A recent research grant from the World Wildlife Fund was for tropical forest conservation in Veracruz. He is co-author of books on Texas birds and mammals and Nearctic avian migrants in the Neotropics.

Dr. Michael Stüwe is research associate with international research and teaching experience in Sri Lanka, India, Malaysia, Italy, Switzerland, and Venezuela. His expertise is in developing and teaching microcomputer applications to wildlife management, particularly on behalf of deer and Himalayan ibex species and protected area design. He established the international biological conservation network CONSLINK. For these efforts, Dr Stüwe has grants from the World Wildlife Fund—USA, the Wildlife Institute of India, the Atherton-Seidell Foundation, and the James Smithson Society.

Dr. Martha S. Fujuita is international conservation officer and liaison for Smithsonian projects in Thailand, Malaysia, Indonesia, and Washington, DC. She is on a leave of absence to serve as director of the Indonesia Field Office, Pacific Region Program of The Nature Conservancy, where she is designing and implementing a protected areas program.

■ Resources

The *CRC Newsletter* and publications list are produced. A major need is Remote Sensing GIS (geographic information system) for conservation and management.

Department of Mammalogy ⊕
National Zoological Park
Smithsonian Institution
Mann Building
Washington, DC 20008 USA

Contact: **John Christian Seidensticker IV, Ph.D.,**
       Curator of Mammals
   Phone: 1-202-673-4779 • *Fax:* 1-202-673-4766

Activities: Education; Research • *Issues:* Biodiversity/Species Preservation

▲ Projects and People

Dr. John Christian Seidensticker focuses his research in conservation biology and wildland management on large mammals and the consequences of habitat insularization in Asia. Curator of mammals since 1989, Dr. Seidensticker is responsible for the Mann lion-tiger exhibit, forest carnivores exhibit, and the hoofed stock unit. Working with ten to thirteen animal keepers, two graduate students, and one postdoctoral fellow, he manages **programs for bongo and Sumatran tigers**, oversees the **Carnivore Stereotypic Behavior Project**, and studies **genetic variations of leopards in captivity.** His latest project concentrates on **Behavioral and Physiological Response to Confined Environments in Domestic and Non-domestic Felids.**

Seidensticker has 25 years experience working, traveling, and living in the wildland areas of Asia and North America—including the Rocky and Appalachian mountain ranges in the United States; the *terai* forests and *dun* valleys of India and Nepal; monsoon forest types of Sri Lanka, Thailand, Java and Bali; lowland and hill rainforests of Java and Sumatra; and mangrove forests of Bangladesh. Eagles, owls, grizzly bears, mountain lions, and tigers—and their survival in their natural habitats—are the subjects of his surveys and management plans. Seidensticker belongs to the Species Survival Commission (IUCN/SSC) Cat Group.

■ Resources

Needs include wildlife radio tracking.

Department of Zoological Research ⊕
National Zoological Park
Smithsonian Institution
Washington, DC 20008 USA

Contact: **Rudy Rudran, Ph.D.,** Conservation Officer
   Phone: 1-202-673-4826 • *Fax:* 1-202-673-4686

Activities: Education; Research • *Issues:* Biodiversity/Species Preservation; Sustainable Development • *Meetings:* CITES

▲ Projects and People

The **Wildlife Conservation and Management Training Program** (WCMTP) provides scientific training to citizens of developing countries. Since 1981, participants from some 44 countries and the Caribbean learn how to collect, analyze, and effectively present data on wildlife populations so that informed, intelligent management and conservation decisions can be made. Field courses are from 5 to 10 weeks long and taught at the Conservation and Research Center as well as in tropical nations.

An important objective is to reach wildlife professionals who can immediately put the training to use in their home countries. In Argentina, the training helped to launch several long-term wildlife research projects. The course in Indonesia initiated systematic animal censuses within a heavily exploited nature reserve.

Trained personnel are the essential catalyst for establishing protected areas and implementing public education programs, writes Rudy Rudran. "Given the urgency of the task of averting major ecological disaster, personnel training is a critical component of tropical ecosystem conservation."

Trainees benefit from interacting with other professionals from other regions. They are exposed to new ideas and approaches to conservation problems faced by many developing countries from China to the Caribbean and realize they are part of a global network of conservationists. Coursework promotes environmental education and biological diversity conservation, particularly in the tropics. Likewise, instructors gain insights and offer solutions to the challenges confronting wild animals and their habitats in Asia, Africa, and the Americas.

Dr. Rudy Rudran holds three major posts with the Smithsonian: National Zoo conservation officer, WCMTP coordinator, and Venezuela Wildlife Research Project director. A 1970 Fulbright scholar from Sri Lanka, Rudran later advised the World Bank and the U.S. Agency for International Development on Sri Lanka water and wildlife projects. Rudran belongs to the Species Survival Commission (IUCN/SSC) Primate Group.

Dr. Devra Kleiman is National Zoo research director. **Dr. Katherine Ralls** is research zoologist concerned with conservation biology, especially the genetic problems of small mammal populations and field studies of threatened and endangered mammals. She is also adjunct professor of environmental studies at the University of California, Santa Cruz, and a fellow in the American Association for the Advancement of Science. Her articles on captive breeding programs; marine mammals, such as sea otters, harbor seals, baleen whales, and river dolphins; and ferrets, weasels, horses, and gazelles appear in science and research journals. Ralls belongs to the Species Survival Commission (IUCN/SSC) Otter Group.

NOAHS Center ⊕
National Zoological Park
Smithsonian Institution
Washington, DC 20008 USA

Contact: **David E. Wildt, Ph.D.,** Head, Reproductive Physiology Program
Phone: 1-202-673-4793 • *Fax:* 1-202-673-4733

Activities: Development; Education; Research • *Issues:* Biodiversity/Species Preservation • *Meetings:* IUCN/SSC CBSG

● Organization

The **Center for New Opportunities in Animal Health Sciences** (NOAHS Center) is a cooperative venture between the Smithsonian's National Zoological Park and the National Institutes of Health (NIH). Among the staff of 15, research scientists in genetics, reproductive physiology, veterinary medicine, and infectious disease focus on the health, management, and reproduction of endangered species and on the maintenance of genetic diversity in zoo and free-living populations.

Graduate and postgraduate students are involved in global training programs, working with NOAHS scientists who provide to conservationists information that will encourage species survival. Biomedical analyses and training occur at the National Zoo, Washington, DC; Conservation and Research Center, Virginia; NIH Animal Research Center, Poolesville, Maryland; and National Cancer Institutes' Section of Genetics, Frederick, Maryland.

Global interaction with field workers is essential to wildlife preservation. NOAHS researchers travel worldwide to collect biological materials including blood samples and sperm from healthy, free-living animals. In the laboratory, endocrine tests of urine or stool samples detect hormones indicating stress, breeding cycle, or time of an expected birth. With these new tests, wild animals can be studied regularly without anesthesia.

▲ Projects and People

Why is the cheetah, the world's fastest land animal, vulnerable to predators? What accounts for the male's low sperm count? The **National Cheetah Research Master Plan** is a result of ongoing biomedical research which pinpointed the cheetah's long history of inbreeding that reduces genetic diversity while promoting infertility and poor reproduction. NOAHS Center also records the births and deaths of cheetahs in captivity through its **International Cheetah Studbook**.

NOAHS' projects are geared to sustaining sufficient **genetic diversity** to prevent inbreeding and problems such as birth defects, juvenile mortality, infertility, and increased susceptibility to disease. Genetic variability is studied in endangered species such as the orangutan, Asian lion, African wild dog, black-footed ferret, Florida panther, Alaskan sea otter, and humpback whale. DNA is investigated, along with electrophoretic surveys of enzymes and other gene products. Molecular biology techniques and genetic tests help identify those at risk, such as the cheetah, Florida panther, and the Ngorongoro Crater lion.

As life-threatening diseases are identified, **vaccines** and treatments are being developed to protect wildlife such as the African wild dog from rabies and parvovirus. **Fertility** problems of giant pandas and Asian Gir lions are being addressed. NOAHS scientists were the first to produce tiger cubs by *in vitro* fertilization. New approaches to breeding the Florida panther and black-footed ferret are being tested. Genetic studies of the humpback whale are underway. Methods of freezing sperm keep gene pools alive and can yield young from rare animals years after their deaths.

Based upon its wildlife studies, the Center's clinical database is growing; ongoing clinical research involving medicines, antibiotics, and anesthetics will help it expand further.

"Rescuing a species from extinction does not simply mean confining it to a safe place," writes **Dr. David Wildt**. "It means propagating a healthy population so that viable offspring can be returned to the wild where the species can live, roam, hunt, and thrive in freedom."

Dr. Wildt began collaborating with the National Zoo in reproductive physiology in 1976. His work on freezing sperm and embryos of laboratory animals led to his interest in adapting these procedures for exotic species conservation efforts. He joined the National Zoo staff as head reproductive physiologist in 1983 and leads a team of graduate and postgraduate students in state-of-the-art biotechnologies to save endangered species.

Water Resources Division
Tennessee Valley Authority (TVA)
Evans Building 1A
Knoxville, TN 37902-1499 USA

Contact: **Ralph H. Brooks, Ph.D.,** Manager
   Phone: 1-615-632-6770 • *Fax:* 1-615-632-6137

Water Resources Division
Tennessee Valley Authority (TVA)
1101 Market Street
Chattanooga, TX 37402 USA

Contact: **Ronald Pasch,** Manager, Biomonitoring
            and Habitat Assessment
   Phone: 1-615-751-7309 • *Fax:* 1-615-751-7479

Activities: Education; Environmental Monitoring/Testing; Research • *Issues:* Biodiversity/Species Preservation; Energy; Water Quality

▲ Projects and People

The TVA's Water Resources Division describes some of its recent activities, such as its first-time, long-term sampling of 12 facilities in the **Reservoir Monitoring Program** to study aquatic ecosystems and determine if each is meeting the "swimmable and fishable goals" of the Clean Water Act.

Physical, chemical, and biological indicators are examined in three areas of each reservoir, such as the upstream of the dam, the transition zone where water changes from free flowing to more quiescent, and the inflow or headwater region. Certain results noted that "more animals and taxa were found in downstream Tennessee River reservoirs than in upstream or tributary reservoirs," and that threadfin shad dominated the mainstream and tributary reservoirs, bluegill was prominent in the forebay, and white crappie was seen most in the transition zone of the Douglas tributary reservoir. More fish were seen in tributary than mainstem reservoirs.

Overall, the initial monitoring showed the reservoirs to be in good health; where problems existed in certain reservoirs, these were "low hypolimnetic" dissolved oxygen (DO), high nutrient loads and algal productivity, low fish density and biomass, sediment toxicity including elevated mercury, and fewer benthic species. Participating aquatic and fisheries biologists include Ronald W. Pasch, Dr. John J. Jenkinson, Robert Wallus, and Donald L. Dycus.

Dr. Jenkinson, a malacologist, is in charge of the **Large River Mussel Restoration Demonstration** to restore such populations, including endangered species, to a sizable river segment in the Tennessee system where they once occurred. Of the 37 endangered North American mussel species, "21 once occurred in the larger

streams within the Tennessee River drainage," Dr. Jenkinson reports. The TVA, which is the nation's leader in freshwater mussel propagation, is interested in "improving the quality of dam tailwaters and managing the species these controlled stream reaches could support," he continues.

Dr. Neil M. Woomer, manager of Biological Assessment, is experienced in monitoring aquatic programs at nuclear power plants, such as the channel catfish study at Sequoyah Nuclear Plant.

A **Reservoir Releases Improvements** program is working to improve water quality discharged from TVA dams, particularly "enhancing DO levels and sustaining minimum flows" which—according to the TVA—"have recovered over 170 miles of aquatic habitat lost from intermittent drying of the river bed."

Because dissolved oxygen is vital to the entire aquatic community—fish, plants, insects, and microorganisms—technologies are being employed at dams to improve DO levels. Air compressors, turbine hub baffles, surface water pumps, draft tube aeration, surface water pumps, autoventing turbines, oxygen injection, and aeration are being used for this purpose.

The **Water Quality Department (WQD)** is concerned with water quality in new or modified TVA facilities as well as with surface and groundwater issues within the Tennessee Valley. Working with other state and federal agencies, WQD undertakes projects such as assessing watershed land use to determine "nonpoint source pollution potential," helping landowners implement sound management practices, evaluating water resources to meet current and projected needs, doing hydrogeochemical studies, and helping local governments and small industries in designing and operating potable water supply, wastewater treatment, and solid waste disposal or hazardous waste management facilities. Its scientists are world leaders in "investigating the potential of constructed wetlands for wastewater treatment," WQD reports.

Water quality and solid waste research is ongoing here and at the **TVA Environmental Engineering Laboratory (EEL),** which seeks to develop cost-effective solutions to environmental problems involving wastewater and pollutants. Dr. Richard D. Urban, Water Quality Department, is leading the effort to search for environmentally sensitive solutions to water, air, aquatic, biological, and waste problems.

With a staff of 28 scientists, the **TVA Environmental Chemistry Laboratory (ECHE)** is this agency's primary environmental analytical lab and performs most of the regulatory analyses to support TVA's power plant system activities as well as the reservoir monitoring programs. Equipped with state-of-the-art instruments and quality-control programs, the ECHE is also able to do duplicate analysis of 10 percent of samples to get precise results. Environmental chemical analyses are performed on "biological, soil, sediment, and water matrices for nitrogen and phosphorus parameters, pesticides, pesticide metabolites, heavy metals, PCBs, and other priority pollutants" as well as on wastewater, airborne particulates, coal ash, and hazardous wastes. C. Wayne Holley is environmental chemistry manager.

The **Watershed Water Quality Program,** with leadership from environmental engineer Dr. John M. Higgins, is a historic effort seeking to "restore and preserve surface and ground water resources in the Tennessee Valley . . . to protect public health and maximize beneficial uses."

As early as 1936, the TVA reported to the U.S. Congress that "water control on the Tennessee River in a system of unified development can best begin where the rain falls . . ." and foresaw that "many problems such as flood control, navigation, the prevention of stream pollution and the silting of streams, the improvement of domestic and industrial water supplies, and the development of power, are not only inter-related, but are interstate problems, and require some agency to bring about and to facilitate cooperation."

Land- and water-use considerations are integrated in addressing watershed needs; while federal, state, and local efforts are mobilized to control point and nonpoint source pollution to protect surface and ground waters. For example, the Bear Creek Floataway, northwest Alabama, is a national model for restoring water quality impaired by nonpoint pollution sources—called the USA's "single most significant remaining water-quality problem" by the U.S. Environmental Protection Agency (EPA).

Aerial photography and stream sampling here and in other areas show the close relationship between water quality and land-based activities and target pollution control needs now being met with special projects. Satellite remote sensing identifies pollution loads in tributary watersheds so that priorities can be established; plans are to use this technique for forestry and TVA lands management, and facility siting studies.

Aerially applied land reclamation techniques have reduced erosion from 200 tons to 8 tons per acre per year and reclamation costs from $3,000 to $600 per acre, according to the TVA. Other innovative management practices include the use of solar electric fencing to limit cattle access to streams and a synthetic fabric to stabilize cattle crossings and prevent streambank erosion.

With the region's livestock generating "seven times more waste than people" and discharged directly into tributary streams, according to the TVA, demonstration projects at four watersheds at Duck River, Middle Fork Holston River, Boone Reservoir, and Bear Creek are removing "animal wastes equivalent to 70,000 people . . . at a cost of 10 to 20 percent" less than for treating human waste. Another significant pollution problem being addressed is sediment from abandoned mineland that fills reservoirs.

Cooperative **Watershed Water Quality Projects** with local citizens and groups are underway in each of the seven valley states—Mississippi, Alabama, Georgia, Kentucky, North Carolina, Tennessee, and Virginia.

Land and Water 201 brings together the TVA with the U.S. Department of Agriculture (USDA), EPA, and the valley states for coordinated watershed management activities.

The TVA is building a **watershed database** for economically assessing priorities and needs, while taking steps to improve its data collection, satellite imagery, geographic information systems (GIS), and interactive computer analyses.

■ Resources

A *Catalogue of Water Quality Educational Materials* is available, along with an extensive bibliography of published works on topics ranging from dissolved oxygen to reservoir vital signs monitoring to wastewater treatment, and including fact sheets on aquatic plants. A five-volume clothbound set of *Reproductive Biology and Early Life History of Fishes in the Ohio River Drainage* is obtainable through Dr. Wallus. The TVA works closely with local communities on matters of mutual interest.

Aquatic Biology Department
Water Resources Division
Tennessee Valley Authority (TVA)
Office Service Annex 1S 122B
Muscle Shoals, AL 35660 USA

Contact: **Joseph C. Cooney, Ph.D.,** Program
Manager
Phone: 1-205-386-2277 • *Fax:* 1-205-386-2087

Activities: Control Operations; Education • *Issues:* Health; Resource Management

▲ Projects and People

The **TVA Aquatic Plant Management Program** supports in an "environmentally and economically responsible manner" the balanced multiple uses of the Tennessee Valley's water resources in an integrated approach to stopping the introduction and spread of noxious species. The TVA reports that it controls undesirable vegetation with "water level fluctuations and supplemental herbicide applications" and that other alternative chemical, biological, and mechanical techniques are being investigated for more "environmentally compatible and cost-effective management operations."

An abundance of exotic, invasive species such as hydrilla, spiny-leaf naiad (*Najas minor all.*), and Eurasian water milfoil (*Myriophyllum spicatum l.*) is found at Guntersville Reservoir, northeastern Alabama, and is capable of spreading among small private lakes and shallow ponds throughout the region. Stimulated and spread by changing weather conditions, noxious non-native species in excessive amounts interfere with recreational activities such as swimming, fishing, and boating; mess up shorelines with plant debris; attract mosquitoes, which can cause disease; use up dissolved oxygen (DO) during decay and respiration; reduce water storage capacity; degrade water quality; and clog water intakes. Hydrilla, one of the most noxious aquatic plants in North America, according to the TVA, at one point was scattered through 45 river miles and is expected to be the Tennessee River system's dominant submersed aquatic plant before the year 2000, if it is not quelled.

Led by **Dr. Joseph C. Cooney**, the **TVA Vector and Aquatic Plant Management Program** is a cooperative venture with the U.S. Army Corps of Engineers (HQUSACE), specifically the Waterways Experiment Station (WES) under the Corps of Engineers' Aquatic Plant Control Research Program (APCRP). A five-year aquatic management plan was begun in 1988 to identify vegetation and to test controls locally for their effectiveness as well as environmental and economic impacts. Without compromising the reservoir's biological integrity, efforts are being made to accommodate a diversity of users such as sport fishermen who want little or no control and recreationists and bank fishermen who want maximum control, for example.

One demonstration features the stocking of 100,000 sterile triploid grass carp to reduce the biomass of hydrilla and stop its spread. In certain areas, benthic barriers have been anchored to the reservoir substrate to keep submersed vegetation from rooting. The use of insects and fungal pathogens as biological controls and the introduction of competing plants that permit a free exchange of water and support for aquatic organisms are also being evaluated. The effects of herbicides are being studied in soil, water, fish, molluscs, plants, and water supplies to check dissipation rates and monitor residues.

"Overall findings from this project are expected to have wide applicability to reservoir projects throughout the nation; thus, technology transfer is a major element of this joint project," reports A.L. Bates, TVA Aquatic Biology Department, an author of the *Joint Agency Plan, Aquatic Plant Management on Guntersville Reservoir.*

A crucial element of the Vector Control Program is the creation of an environment relatively free of pests, such as mosquitoes and ticks, often with the use of nonchemicals. Mosquito control is difficult in organically enriched water; however, few mosquitoes are found in wastewater-treating wetlands receiving inorganic, acid waters, it was determined.

■ Resources

Technical assistance to communities and individuals along with public presentations and workshops are available at no charge; also available is a publications list. Consultative services are provided at negotiated contract charges.

Oceans and Environment Program
Office of Technology and Assessment (OTA)
U.S. Congress
Washington, DC 22101 USA

Contact: **Robert W. Niblock,** Program Manager
  Phone: 1-202-228-6850 • *Fax:* 1-202-228-6098

Activities: Political/Legislative • *Issues:* Air Quality/Emission Control; Deforestation; Global Warming; Sustainable Development; Waste Management/Recycling; Water Quality

● Organization

The Office of Technology Assessment (OTA) is an "analytical support agency of the U.S. Congress," working directly with and for congressional committees to provide them with "objective, thorough analysis of technological issues."

Governed by a congressional board, OTA operates with a multidisciplinary staff that draws on the technical and professional resources of "universities, industry, public interest and citizen groups, state and local officials, and individuals to ensure excellence in quality and impartial presentation of all views." In so doing, OTA examines the impacts of technology, including possible adverse affects; identifies and evaluates alternative methods or programs; notes areas where additional research is essential; and reports these findings back to the appropriate legislative committees.

▲ Projects and People

Robert Niblock directs the Oceans and Environment Program, which is responsible for all ocean-related questions, including ocean resources and maritime policy, and for large-scale environmental issues, such as climate change and water pollution. The Oceans and Environment Program is composed of five categories: federal services, natural resources, pollution control, marine industry, and large-scale environmental issues. Current undertakings include the following:

The **Assessment of Chemical Weapons Destruction** project assesses how to destroy chemical weapons that are currently stockpiled at army bases at eight locations in the USA and on one island territory in the Pacific Ocean. OTA is conducting a technical evaluation of current and proposed systems to destroy these weapons.

Several years ago, when the army decided to incinerate the weapons, a plant was built and operated for this purpose. However, concerns have been raised by some groups that "more effective disposal means are available and should be evaluated for future use." OTA is reviewing the status of technologies for disposal of chemical weapons, with particular attention to alternatives to incineration. **Peter Johnson** is project director.

A paper was recently prepared on **Alaskan Water for California? The Subsea Pipeline Option** with **William Westermeyer** as the project director. The background paper examines the "potential of building a subsea pipeline to transport fresh water from southeast Alaska to California." Studies are being conducted about the engineering feasibility and the costs involved. Related research is concentrating on Alaska's water availability, California's projected water demand, and other alternatives for meeting future water needs in California.

Howard Levinson, staff contact for **Industrial Waste Issues,** has been preparing a paper to address issues associated with industrial waste not regulated as hazardous waste.

Cleanup Worker Health Risks is a paper describing the "current status of standards and programs to protect workers engaged in the cleanup of the nuclear weapons facilities under the U.S. Department of Energy's environmental restoration program." Data is being analyzed to determine the risks to workers from various cleanup technologies and approaches. **Tara O'Toole** is project director.

Peter Johnson directs the **Dioxin Treatment Technologies** project and is reporting on the "status of national cleanup efforts and the programs underway to research, design, test and evaluate dioxin treatment technologies." OTA has been asked to evaluate current proposed technologies and viable alternatives that could be used in the cleanup work. Johnson's background paper covers thermal and nonthermal approaches and techniques such as stabilization and storage—including the state of development of the various technologies and the advantages and disadvantages of their use.

Rosina Bierbaum is the project director for the **Systems at Risk from Climate Change** project. OTA is taking a "strategic look at the interplay between the natural and engineered systems potentially at risk from climate change, the timing of information needed for planning these systems, and how well coordinated the federal research program is to provide such answers." OTA recognizes the "great uncertainty" regarding such potential effects while policy decisions need to be addressed. It points out that it will be at least 10 years before the General Circulation Models (GCMs) "offer the kind of temporal and regional detail desirable" along with other long-term, worldwide research results.

■ Resources

A detailed publications list is available from OTA, Washington, DC 20510-8025, attn: Publications; or from the request line at 1-202-224-8996.

National Program Staff
Agricultural Research Service (ARS) ⊕
U.S. Department of Agriculture (USDA)
Building 005, BARC-West
Beltsville, MD 20705 USA

Contacts: **R.D. Child, Ph.D.,** National Program
        Leader, Global Change
Phone: 1-301-504-5618

        C.R. Amerman, Ph.D., National Program
        Leader, Soil Erosion and Models
Phone: 1-301-504-6441

Activities: Development; Research • *Issues:* Air Quality/Emission Control; Global Warming; Waste Management/Recycling; Water Quality

● Organization

The **Agricultural Research Service (ARS)** plans, develops, and implements national research that is designed to produce the new knowledge and technologies required to assure the continuing vitality of the USA's food and agriculture enterprise. The ARS is equipped to respond to regional or national emergency situations and is "international in nature" as it supports foreign policy initiatives of the U.S. government. Its unique collections and repositories of information are also used by other public and private research organizations. These include the National Seed Storage Laboratory; disease-free seed stock; fruit and nut crop repositories; ARS Culture Collection; and other taxonomic collections of plants, microbes, and insects.

A network of research centers and worksites in the USA and abroad nourishes a scientific environment and encourages the coordination of resources with state agricultural experiment stations and other institutions. The **Beltsville Agricultural Research Center (BARC)** with 450 full-time scientists on 7,000 acres is the USA's largest national agricultural research center. The National Agricultural Library, Energy Metabolism Laboratory, Human Nutrition Research Center, and vital collections of insects, fungi, seeds, nematodes, animal parasites, and rhizobia are located here. Regional research centers are in Pennsylvania, Illinois, Louisiana, and California; the Richard B. Russell Agricultural Research Center is in Georgia.

Crop sciences facilities throughout the country do research with aquatic weeds, barley and malting quality, honey bees, biological control of pests, cereal diseases, vegetables, sugarcane, cotton, insect control, forest/range/livestock, metabolism and radiation, ornamental plants, peanuts, rice, foreign plant germplasm, and tree fruits and nuts. Special facilities are the Subtropical Horticultural Research Station, Miami, Florida, with unique germplasm collections of coffee, cacao, sugarcane, and other tropical and subtropical plants; and the Mayaguez Institute of Tropical Agriculture, Puerto Rico, where year-round, outdoor experiments aid agriculture in the tropics.

Researchers at livestock and veterinary sciences facilities investigate domestic and foreign animal diseases, insects affecting livestock, internal parasites, and poultry production.

Postharvest science and technology facilities help improve animal products, evaluate cotton quality, increase efficiency in marketing agricultural goods in Europe, study the economic problems of families, test textiles and clothing, assess fruit and vegetable quality, control insects in stored commodities through unique vacuum and atmospheric test conditions, and improve grain handling. Research on metabolic safety determines the fate of pesticides, agricultural chemicals, and drugs used to prevent and control diseases or improve efficiency of animal production; prevents losses from poisonous plants and other naturally occurring toxicants; and undertakes toxicological and nutritional studies of food and feed.

Also, there are laboratories for erosion and sedimentation studies, National Tillage Machinery Laboratory, and Soil and Water Conservation Research Centers. Water management laboratories, watershed research centers, and the U.S. Plant, Soil, and Nutrition Laboratory studies are on geochemical and chemical processes involved in the cycling of nutritionally important substances in rocks, soils, plants, and animals.

▲ Projects and People

Within the ARS Global Change Research Program, programs in climate and hydrology, biogeochemical dynamics, and ecosystem and dynamics are contributing to the development of **TEAM** (Terrestrial Ecosystem Assessment Model).

Climate and hydrology deals primarily with energy and water fluxes between the land surface and the atmosphere. **Biogeochemical dynamics** is concerned with trace gas fluxes between the land surface and the atmosphere. **Ecosystem and dynamics** is about the responses of the biosphere to the environment. TEAM scientists are from various ARS facilities. Ongoing projects include the following:

• **Assessing biogeochemical fluxes from agriculture and rangelands** with the goal of developing the capability for long-term monitoring, this research includes modeling the water balance of selected ecosystems; modeling carbon, nitrogen, and phosphorus cycles; and incorporating state-of-the-art remote-sensing technology into these models as a primary data source and at a scale that is compatible with global processes.

• **Ecosystem modeling.** Modular, quantitative simulation models are being developed to provide the basic framework for ARS ecological systems research and the ability to assess critical knowledge gaps and set program priorities. Building on current research in cooperation with the U.S. Department of Energy (DOE), which is assessing increased carbon dioxide concentrations on terrestrial ecosystems, this project aims to produce models that will also predict the performance, health, and stability of agricultural and rangeland species resulting from changes in water stress, temperature, ultraviolet (UV)–B radiation intensity, and tropospheric ozone levels.

• **Ozone effects.** "Tropospheric ozone and other air pollutants could substantially alter the biosphere response to global change variables such as temperature, carbon dioxide concentration, water stress, and UV-B radiation intensity," reports Dr. Rawlins, the former national program leader for soil erosion, models, and global change. "The goal of this research is to measure thresholds and sensitivities of crop and rangeland species to ozone exposure, discover the fundamental physiological processes affected, and integrate these response functions into ecosystem models that include the full range of environmental variables subject to global change."

- Biological response to increased UV-B radiation. "Increased levels of UV-B radiation could substantially alter the biosphere response to global change variables such as temperature, carbon dioxide concentration, water stress, and tropospheric ozone concentration," writes Dr. Rawlins. This research is measuring thresholds and sensitivities of crop and rangeland species to UV-B radiation intensities—with the same goals as for ozone effects.
- Scale effects of hydrological processes in support of the National Aeronautics and Space Administration (NASA/Eos) Program. One of the most important unresolved issues in hydrology, according to the ARS, is understanding the hydrologic responses over the range of scales represented in this research endeavor—including spatial and temporal distribution of precipitation, soil moisture, evapotranspiration, runoff, and erosion. Newly developed quantitative simulation models will help determine feedback mechanisms between the land and atmosphere. This research addresses the "critical problem of scale in modeling hydrologic processes" and the interaction with atmosphere processes. "It includes field experimentation and data collection with existing ARS research watersheds, water and energy balance models that are merged with mesoscale meteorological models and data, and use of current and future (Eos) remote sensing data."
- Predicting the impacts of global change on sustainable water supply. Various ARS hydrologic models are being evaluated for applicability, limitations, and performance under different conditions. Models will be modified or developed where necessary to predict adequately global change effects on water supply, rainfall regime, and snowpack accumulation. New technique such as extended weather and quantitative precipitation forecasts will be incorporated into the models.
- Temporal changes in soil carbon inventory. This research is testing the hypothesis that the 25 percent carbon dioxide enrichment of the atmosphere since pre-industrial times has increased the amount of carbon stored within the soils underlying cropland, rangeland, and forests. Measurement techniques including carbon isotope ratios are being developed to determine temporal trends in soil carbon storage. Models will be modified or developed where necessary to predict the carbon sink strength of major components of the terrestrial ecosystems; they will draw upon soil and climate databases being developed to support ecosystem models.

Dr. Rawlins is a recognized international authority in the field of soil and plant physics and in the analysis of holistic agricultural systems. Two conclusions of his collaborative environmental control experiments are that "elevated humidity can greatly increase crop yields" and "salinity tends to protect plants from ozone damage—raising serious questions about the applicability of salt-tolerance data obtained in high ozone areas."

He helped establish Systems Science within ARS as a means of coordinating research to deliver new technologies that are transferable to users in solving important national problems. Dr. Rawlins was appointed as the USA representative to the United Nations' Intergovernmental Panel on Climate Change to develop policies related to social and economic impacts of global environmental change.

Cropping Systems Research Laboratory
Agricultural Research Service (ARS) (Southern Plains Area)
U.S. Department of Agriculture (USDA)
P.O. Box 909
Big Spring, TX 79720 USA

Contact: **Donald W. Fryear,** Research Leader/ Agricultural Engineer
Phone: 1-915-263-0293 • *Fax:* 1-915-263-3154

Activities: Research • *Issues:* Air Quality; Sustainable Development; Water Quality

▲ Projects and People

Donald Fryear's research is conducted on the movement of soil material by wind. "The research is designed to expand our understanding of the basic wind erosion process," he says, "and to develop the technology necessary (for) wind erosion control systems for all farmland in the USA." The material being eroded is sampled and the potential is determined for transport to other states or regions downwind. "The resulting impact on global weather and air quality will be addressed with additional analysis of the samples collected."

Fryear states that "concern for the environment is a major factor in U.S. agriculture maintaining the high level of productivity we all enjoy." But he is cautious about environmental controls. "When people that do not make their living from the land tell production agriculture what they can and cannot do—as a society we (will be) seeing weeds growing in the streets of every city."

National Soil Dynamics Laboratory (NSDL)
Agricultural Research Service (ARS)
U.S. Department of Agriculture (USDA)
P.O. Box 3439
Auburn, AL 36831-3439 USA

Contact: **Hugo H. Rogers, Ph.D.,** Plant Physiologist
Phone: 1-205-887-8596

Activities: Research • *Issues:* Air Quality/Emission Control; Soil/Plants

● Organization

The National Soil Dynamics Laboratory (NSDL) on the Auburn University campus began as the Farm Tillage Machinery Laboratory in 1933 and became the National Tillage Machinery Laboratory in 1957; its research programs led to the concept of soil dynamics—the relation between forces applied to soil and the resulting reactions in the soil. Since 1985, this Laboratory pursues the interrelationship of soil, crops, and machines and addresses problems arising within highly mechanized agricultural practices. It is located within the Agricultural Research Service (ARS), which develops technical information and products that help the nation manage soil, water, air, and climatic resources and improve the environment among other objectives relating to providing high-quality, nutritious agricultural commodities at reasonable prices and improving the standard of living in rural communities.

▲ Projects and People

Increasing crop production while managing soil compaction are NSDL missions. Research is conducted within a system of nine outdoor and two indoor soil bins containing samples of U.S. agricultural soils as well as within a rhizotron—a plant root observatory of 20 containment units where root growth and development are observed through glass fronts. Soil/machine reactions are studied, as are the effects of wheel traffic on soil, which impacts on the conservation of soil, water, and energy. Scientists also explore the relationships between **atmospheric variables and below-ground processes**, especially crop roots and their rhizospheres.

Dr. Hugo H. Rogers is a plant physiologist concerned with environmental sciences and engineering. Research leader is Dr. Jerrel B. Powell.

■ Resources

"NMR imaging" is needed for plant roots/soil studies, writes Dr. Rogers.

Southwest Watershed Research Center
Aridland Watershed Management Research Unit
Agricultural Research Service (ARS)
U.S. Department of Agriculture (USDA)
2000 East Allen Road
Tucson, AZ 85719 USA

Contact: **Leonard J. Lane, Ph.D.**, Research Leader
Phone: 1-602-670-6481 • *Fax:* 1-602-670-6493

Activities: Development; Education; Research • *Issues:* Global Change; Global Warming; Sustainable Development; Waste Management/Recycling; Water Quality

● Organization

The **Southwest Watershed Research Center** develops computerized "decision-support systems" for environmental science as well as for agriculture and natural resource management. This set of computer programs strives to bring the most up-to-date databases together with the best computer simulation models to help farmers, land managers, and other decisionmakers evaluate the environmental and economic consequences of alternative farming practices. Its goal is to arrive at the "best management practice," which the Research Center defines as that which is "environmentally and economically sustainable in the long term." A main objective is to produce food and fiber without having expensive fertilizers and pesticides go to waste or become pollutants.

▲ Projects and People

One of its first projects is a decision-support system that helps protect the nation's groundwaters and surface waters from potential contamination by fertilizers and pesticides used in agriculture. The optimal design of **Shallow Land Burial (SLB) Systems** regarding waste is in cooperation with Los Alamos National Laboratory and

Purdue University. A USDA Water Quality Initiative project is underway with the Soil Conservation Service and universities. Dr. Leonard Lane, hydrologist and research leader, believes that this technology's future application will help develop the best management practices in soil and water conservation; improve pollution assessment for environmental planning; improve management, conservation, and protection of watershed resources; and improve overall sustainable land use.

■ Resources

The Aridland Watershed Management Research Unit publishes an annually updated bibliography listing related works from 1959 to the present. Reprints are available.

Biological Assessment and Taxonomic Support ⊕
Animal and Plant Health Inspection Service (APHIS)
U.S. Department of Agriculture (USDA)
APHIS-PPQ-OS
Federal Building, Room 626
6505 Belcrest Road
Hyattsville, MD 20782 USA

Contact: **Matthew H. Royer, Ph.D.**, Chief Operations Officer
Phone: 1-301-436-8896 • *Fax:* 1-301-436-8700
E-Mail: os.bats

Activities: Political/Legislative • *Issues:* Plant Protection and Quarantine

● Organization

The **Animal and Plant Health Inspection Service**, known as **APHIS**, acts to ensure the health and care of animals and plants, to improve agricultural productivity and competitiveness, and to contribute to the national economy and public health through excluding exotic agricultural pests and diseases; detecting, monitoring, and managing agricultural pests and diseases; protecting endangered species and the welfare of animals; facilitating agricultural exports; providing scientific and technical services; and collecting, analyzing, and disseminating information.

Plant Protection and Quarantine (PPQ) operates on the premise that it's cheaper to keep alien pests out of the country than to combat these problems once they are here and capable of damaging some $40 billion worth of agricultural products destined for foreign markets annually as well as those intended for domestic consumption. The 1980s threat of the medfly to California citrus crops is one example of how an unwanted insect can harm international trade. Other damaging invaders are the San Jose scale, cotton boll weevil, apple ermine moth, Egyptian cotton worm, gypsy moth, and the "killer bee," or the Africanized honeybee—first sighted in 1985 in California as an apparent stowaway on oil-drilling equipment from South America, then exterminated, although its natural northward migration through Mexico continues. PPQ uses x-ray equipment, dog teams, and other methods to check for destructive pests—often in food—at points of entry.

▲ Projects and People

Biological control is a means to fight unwanted insects and weeds—without pesticides. A century ago, the U.S. Department of Agriculture (USDA) successfully used the Australian lady beetle to curb an infestation of cottony cushion scale. APHIS and its cooperators rear beneficial biological agents in large numbers, then carefully ship, release, and monitor them in infested fields. More recently, APHIS used four species of parasitic wasps in the 1960s to defeat the cereal leaf beetle before grain fields were seriously damaged from New England and Maryland west to Wisconsin and Missouri.

Today, a technical review group with members from inside and outside USDA advises APHIS on promising target pests for biological control. Cooperation comes from state agencies, agricultural colleges, industry groups and other USDA agencies such as the Agricultural Research Service (ARS), Economic Research Service (ERS), and Federal Extension Service (FES).

Predators, pathogens, and parasites are the three categories of beneficial agents that must "prove their merit" in experimental trials. Predators are hunters who feed on and quickly kill multiple individuals of a target pest. American lady beetles, or lady bugs, are being fortified with sevenspotted lady beetles and directed against aphids or plant lice. A single beetle can devour as many as 5,000 aphids, and farmers successfully use them to protect vegetables in France, cotton in China, alfalfa in Germany, and fruit orchards in England. APHIS has spread the sevenspotted beetle from New Jersey to states west of the Mississippi River. "Teams are organized to introduce the beetles to new areas and give them a good start," reports APHIS. "Surveyors monitor their progress in colonizing new territories."

A parasite can attack only one individual of a target pest; even so, they are effective weapons against insects and weeds. The seed-head fly is one example of a parasite that feeds on the injurious rangeland plant, the knapweed. The use of the parasitic wasp against the alfalfa weevil has saved farmers millions of dollars annually through reduced use of chemical insecticides. Pathogens kill their targets with infections or diseases. A fungus that is a natural enemy of rush skeletonweed is the subject of research in the western USA.

At APHIS biological control laboratories in Michigan, Texas, and Montana, beneficial agents are kept in proper environmental conditions with appropriate diets, lighting, and ventilation. Cooperators also establish experimental colonies. "Under the right circumstances, beneficial agents employed on a large scale can control pests effectively at relatively low cost and at no risk to man, animals, and the environment," according to APHIS.

Dr. Matthew H. Royer is particularly interested in the integration of agricultural databases and models—such as the World Plant Pathogen Database, expert systems, and geographically referenced data to assist in pest-risk analysis. This data combined with expert opinion would "evaluate plant pests epidemiologically and economically, predict yield loss, and determine the feasibility of control measures and need for quarantine," he writes.

Forest Fire and Atmospheric Sciences Research (FFASR)
Forest Service
U.S. Department of Agriculture (USDA)
P.O. Box 96090
Washington, DC 20090-6090 USA

Contacts: **William T. Sommers, Ph.D.,** Director
Phone: 1-202-205-1561
E-Mail: Omnet: w.sommers;

Elvia E. Niebla, Ph.D., National Coordinator; Global Change Research Program
Phone: 1-202-453-9561;

Michael A. Fosberg, Ph.D.
Phone: 1-202-205-1561 • *Fax:* 1-202-205-1551
E-Mail: Omnet: m.fosberg;

Marla Lacayo–Emery, Writer/Editor, Forest Fire and Atmospheric Sciences Research
Phone: 1-202-205-1561 • *Fax:* 1-202-205-1610

Activities: Research • *Issues:* Air Quality/Emission Control; Deforestation; Global Warming; Sustainable Development

● Organization

Every 10 years, the Secretary of Agriculture updates its Renewable Resources Assessment, according to the Forest and Rangeland Renewable Resources Planning Act of 1974. In response to this federal legislation, one of its most recent documents summarized "Climate Change and America's Forests." Even though the Forest Service has been carrying out resource analyses in the United States for more than 100 years, the emphasis today is on reporting on all renewable resources from forests and rangelands—not just timber resources.

To complicate matters, the validity of knowledge collected during the past century is being challenged, as climate and the chemical environment are changing. These changes appear to vary from region to region. And it's estimated that the increased concentrations of greenhouse gases, carbon dioxide, methane, and nitrogen oxides in the atmosphere in the next century could result in a global warming of 1°C to 4°C—according to researchers. These projections are based on forecasts of energy consumption, energy efficiency, and population growth.

The warming trend raises unanswered questions regarding environmental stresses due to adverse climate, mainly: "How will future physical and chemical changes alter the structure, function, and productivity of forest and range ecosystems?" "What are the implications for forest and range management, and how must forest and range management activities be altered to sustain forest and range health and productivity?"

Moreover, predictions about future global change are uncertain because of events such as fire, insect, or disease outbreaks and stresses of air pollution and ultraviolet radiation related to ozone depletion that cannot be anticipated.

▲ Projects and People

To develop strategies for managing forests and related ecosystems in the next century, Forest Service scientists say models are essential for describing basic processes such as photosynthesis, respiration, and carbon allocation to various organs. Models also must be developed which "characterize and predict whole-plant response to climate and atmospheric change." Ecosystem modeling is limited by knowledge of plant processes and interactions. And current global change modeling (GCM) is described as "very coarse" with a need to resolve difference in scale and resolution. Yet, Forest Service scientist Michael Fosberg says it is necessary to couple ecosystem models to GCMs to "provide for the exchange of energy and matter." He continues, "The most useful output from the models of the physical and biological worlds are probability statements on the outcomes"—which more accurately reflect the real world. Developing economic and risk-analysis models are also essential for managing natural ecosystems in a changing environment. "Political decisions regarding resource management issues are ultimately based on analysis of socioeconomic and biophysical factors," writes Dr. Fosberg.

With Forest Service scientists and cooperators in government and universities, its **priority research program** has four focal points: processes of atmospheric effects on forests, effects of forest management on the atmosphere, long-term assessment effects, and development of management options and strategies.

As a few examples, scientists need to determine the following: sensitivity of aquatic ecosystems and fisheries, wildlife, and wood to changes in the climate and physical or chemical atmosphere; effects of large-scale forest removal on carbon, oxygen, and water budgets and their relative composition of the atmosphere; ways to assess ecosystem response to climate change; and how to predict the impact of projected physical and chemical climate change on ecosystems through effective economic and risk-analysis techniques.

The Forest Service identifies three options in preparing for **future greenhouse warming**: conserving current forest resources, developing a global strategy to curtail climate change through the conservation of energy or use of nonfossil fuels, or adapting to change through developing new technologies and industries and relocating activities. A combination of all three options are expected to occur as the Forest Service continues to "carry out research on the effects of multiple stresses on our forests in order to assure their health and productivity in a changing atmospheric environment," the agency reports.

Dr. Sommers is responsible for forest fire research, air pollution and acid rain effects research, global change research related to forest ecosystems, forest health monitoring, and natural disaster reduction research.

Institute of Tropical Forestry (ITF) ⊕
Southern Forest Experiment Station
Forest Service
U.S. Department of Agriculture (USDA)
Agricultural Experiment Station Grounds
Call Box 25000
Rio Piedras, Puerto Rico 00928-2500 USA

Contact: **Ariel E. Lugo, Ph.D.**, Director and Project Leader
Phone: 1-809-766-5335 • Fax: 1-809-250-6924

E-Mail: Bitnet: a_gillespie@upr1

Activities: Research • *Issues:* Biodiversity/Species Preservation; Deforestation; Global Warming; Sustainable Development

● Organization

Surprisingly little is known about the tropical and subtropical forests that account for half of the world's forests. Tropical forests also harbor about half of the world's species. These facts coupled with economic, social, and environmental problems are the challenges of the Institute of Tropical Forestry (ITF), which since 1939 is charged with providing scientific information to resolve these diverse factors.

Despite the complexities, ITF—cooperating with the University of Puerto Rico and the Caribbean National Forest—has made gains in maximizing production of tropical forest lands and improving tropical forest management, such as screening more than 350 tropical trees species for plantation use; describing wood properties of hundreds of tree species; documenting plantation and natural forest yields in tropical America; performing international species trials; training foresters, resource managers, and specialists from throughout the tropical world; establishing and studying permanent tree growth plots throughout the islands; and saving the Puerto Rican parrot from extinction.

Today, ITF does research in tropical forest ecology and management, plantation forestry, wildlife, watershed management, and global climate change. One global research program is the Tropical Soil Biology and Fertility program of the International Union of Biological Sciences and the Man and the Biosphere Program. International research involves collaboration with local and U.S. universities and government agencies including other Forest Service programs as well as with Caribbean and Latin American governments and institutions—particularly from Argentina, Brazil, Costa Rica, Dominica, Saint Lucia, and Venezuela.

▲ Projects and People

The **tropical forest and ecology management** research program addresses how to manage complex tropical forests without endangering their biodiversity and complexity, with the help of ecologists, silviculturists, ecological economists, biometricians, and geomorphologists. Its purpose is to optimize contributions of tropical rainforests to both the world at large and to tropical peoples. This program impacts on all other ITF research endeavors, such as the study of diverse tropical ecosystems as cloud forests, dry forests, palm wetlands, mangroves, rivers, and estuaries. It also deals with the global status of tropical forests, ecosystems, management of natural succession in secondary forests, and regenerative and productive responses of forests to natural disasters and exploitation.

Plantation forestry research in Puerto Rico is being expanded to the Amazon Basin, where it will document the natural process of reforestation, test plantation establishment techniques for marketable timber species, and provide a bilingual manual of Brazilian species. Test plantations are constantly being evaluated; one discovery is that the Caribbean pine is capable of yields three to five times greater than usually obtained on good sites in southern USA.

Endangered wildlife research includes studies on North American birds that overwinter in the Caribbean's Guanica dry forest, coffee plantations, and other localities. Effects of hurricanes on wildlife are being examined. Under study are black-throated blue warblers, pearly-eyed thrashers, and the Puerto Rican parrots' habitat along with the seasonal production of fruits and seeds.

Three watersheds in the Luquillo Experimental Forest are equipped to measure climatologic, hydrologic, and water quality data to understand the impacts and benefits of management in humid tropical forests. This is a centerpiece of a long-term U.S. National Science Foundation ecological research program. Some 50 scientists from mainland universities, ITF, and the University of Puerto Rico's Center for Energy and Environment Research are collaborating.

Global change research shows that the Luquillo Mountains affect weather systems that originate as far as the Azores, East Africa, North America, the Amazon Basin, and the Pacific. University and government scientists are operating a network of sites to find out the ecological effects of climate and land use change in the tropics. The role of tropical forests in the world's carbon balance is also being explored through global models.

Chief scientists among the staff of 42 are Drs. Ariel Lugo, Peter L. Weaver, and Frank H. Wadsworth.

■ Resources

Its 40-year-old library contains 6,000 books, 20,000 unbound volumes, 300 maps, and 2,000 slides and photos used worldwide. Summary documents on tropical forestry are prepared for federal agencies. ITF is expanding its use and teaching of state-of-the-art telecommunication devices, computers, field data recorders, geographic information systems (GISs), and global positioning systems (GPS). The Institute trains U.S. Peace Corps volunteers, junior forestry officers from Caribbean islands, and colleagues from Latin America and the Caribbean. Plant material is collected to aid botany and biochemistry researchers worldwide. Seeds of local trees are also sent to foresters abroad.

Resources Inventory Division
Soil Conservation Service (SCS)
U.S. Department of Agriculture (USDA)
P.O. Box 2890
Washington, DC 20013 USA

Contact: **Gale TeSelle**, Acting Director
   Phone: 1-202-720-5420 • *Fax:* 1-202-256-2019

Activities: Resource Data Collection • *Issues:* Air Quality/ Emission Control; Biodiversity/Species Preservation; Deforestation; Global Warming; Status, Conditions, Trends of Nation's Resources

● Organization

The **Resources Inventory Division** provides technical and executive leadership for collecting data needed to support the programs and activities of the USDA, SCS, and other public and private groups, organizations, and individuals. This includes collecting, interpreting, and reporting, soil, water, and related resources data on a nationwide basis and operating a network of snow courses and data collection sites in the western USA.

▲ Projects and People

Data collection was recently undertaken for the **1992 National Resources Inventory (NRI).** This is the fourth in a series of inventories conducted to determine the status condition and trend of the USA's soil, water, and related resources on private lands. This inventory provides data that will be a basis for detection of changes and trends in resource conditions between 1982, 1987, and 1992. Data elements include land use, land cover, "sheet and rill" erosion, wind erosion, wetlands, irrigation, rangeland conditions and species, tillage, conservation treatment needs, and wildlife habitat. The data, which in the previous inventory was drawn from 300,000 scientifically selected sample sites in all counties in the continental USA and Hawaii, is compiled into an extensive database.

Wind Erosion Monitoring Inventories are conducted each year, principally in the Great Plains states. Reports are issued in January, April, and June. Information collected includes acres of land damaged, acres of crops or cover destroyed, acres protected by emergency tillage, and acres of land in condition to blow.

In the **Global Change Pilot Project,** 20 sites are being installed throughout the USA to measure soil moisture, soil temperature, and climatic data related to global change. These sites will report daily information through the SCS's Meteor-burst Telemetry System. The data will also be used to monitor drought and calibrate soil genesis models.

Snow Surveys and Water Supply Forecasting are conducted at 900 manual snow courses and 500 SNOTEL (snow telemetry program) sites in the western USA. Data compiled is on snowpack depth and water equivalent, precipitation, air temperature, and related climate data. This information is used to derive and save forecasts of seasonal streamflow and water supply on most of the major users and streams in the western states.

■ Resources

The *National Resources Inventory* and *Water Supply Forecasts* are available to the public.

National Oceanic and Atmospheric Administration (NOAA) ⊕

U.S. Department of Commerce
14th Street and Constitution Avenue, NW
Washington, DC 20230 USA

Contact: **John A. Knauss, Ph.D.,** Under Secretary of
Commerce for Oceans and Atmosphere
Phone: 1-202-482-3436 • *Fax:* 1-202-408-9674
E-Mail: Omnet: j.knauss

Activities: Development; Education; Law Enforcement; Political/Legislative; Research • *Issues:* Air Quality/Emission Control; Biodiversity/Species Preservation; Deforestation; Energy; Global Warming; Health; Sustainable Development; Transportation; Water Quality • *Meetings:* CITES

● Organization

NOAA—as the National Oceanic and Atmospheric Administration is called—is dedicated to "better protection of life and property from natural hazards and a better understanding of the total environment." Since its 1970 founding, its 12,000-plus men and women have carried out a variety of environmental science and research ranging literally from the "seafloor to the sun." Best known among Americans for its National Weather Service, NOAA with its advanced satellites and supercomputers can accurately predict 10-day forecasts and is at the forefront of anticipating hurricanes and floods to "mesoscale" or small-area storms which can cause greater death and destruction than more spectacular counterparts. The Climate Analysis Center is a global leader in spotting precursors of El Niño–Southern Oscillation (ENSO), a Pacific Ocean phenomenon capable of vicious weather damage around the world.

NOAA research today is focusing on the future of Earth's climate and its effect on planetary life. A recent discovery is the process that produces the Antarctic ozone hole.

▲ Projects and People

Every earthly environmental activity is impacted by NOAA's 24 operational satellites, which have orbited more than 10 billion miles. The **Geostationary Operational Environmental Satellite** (GOES) monitors severe storms and jet streams worldwide, saving the economy billions of dollars. GOES takes pictures—now more than 530,000—to collect daily information on rainfall, stream levels, and wave heights from more than 6,000 platforms. GOES sensors monitor high-energy particles and x-rays emitted from solar flares to warn of impending magnetic storms. Instruments aboard NOAA's **polar-orbiting satellites** that measure ozone amounts, snow and ice cover, vegetation, and precise air temperatures across the world's oceans are vital to the study of global climate change. The **National Environmental Satellite, Data, and Information Service** (NESDIS), the world's fourth largest satellite service in monitoring Earth from space, operates global data centers housing the world's largest environmental data bank.

This nation's oldest scientific and technical agency, the National Ocean Service (NOS) is nautical and aeronautical chartmaker and protector of marine sanctuaries that safeguard open-water areas and nearshore ecosystems. NOS has designated Estuarine Research Reserves (NERR) as field laboratories for studying natural and human

processes where rivers and oceans meet; more than 300,000 acres of estuarine land and water are in this system. NOS monitors the effects of pollutants and toxins, assesses the environmental activities of humans, and responds to hazardous materials in the marine environment through HAZMAT, which assists in thousands of cleanup operations for oil spills and other resource-endangering accidents. Among other activities, NOS provides technical staff to help run the NAVY/NOAA Joint Ice Center (JIC), the world's only organization that provides global sea ice analyses and forecasts; and collects global ocean data that contributes to the understanding of worldwide environmental changes.

Environmental studies of NOAA's **Office of Oceanic and Atmospheric Research** (OAR) at laboratories and universities include the following:

- **The National Undersea Research Program,** which provides public and private sector scientists and engineers with funding for peer reviewed research, leadership, and access to state-of-the-science underwater equipment and facilities needed to learn more about oceanic processes. One discovery is that certain seafloors are steaming with mineral-rich geysers barreling upward through cracks; this may have meaning for seabed mining or climate change.
- **The Sea Grant Program,** which encompasses more than 300 universities and affiliated institutions—now the leader in identifying and developing national marine research goals and priorities.
- **Equatorial Pacific Ocean Climate Study** (EPOCS), which observes El Niño–Southern Oscillation (ENSO) events occurring after prolonged, powerful trade winds along the Pacific's equator blow warm surface water westward—a process that can slow or reverse, carrying water back to the east, which throws global patterns out of place and kills marine life.
- **Environmental Research Laboratories** (ERL), with remote sensing instruments and techniques applicable to both seas and skies, are responsible for modernization of the National Weather Service.
- **Geophysical Fluid Dynamics Laboratory,** where super-speed computers help scientists with mathematical modeling of Earth's oceans and atmosphere, a boon to America's population—half of which lives along the coasts.
- **Mauna Loa Observatory,** which is part of a global network of atmospheric monitoring stations alerting investigators to buildup of undesirable chemicals, gases, and aerosols that may be eroding the ozone layer and contributing to the greenhouse effect.

"NOAA faces an era of supercharged environmental demands," writes oceanographer John A. Knauss, under secretary for oceans and atmosphere and NOAA administrator. "We must help meet such basic human needs as safety from the elements, clean air and water, healthy coastal zones and oceans, and wholesome and abundant seafood." Dr. Knauss has served on two presidential commissions on marine affairs and was a professor and dean of the Graduate School of Oceanography, University of Rhode Island.

Jennifer Joy Wilson, assistant secretary of commerce for oceans and atmosphere, is the first woman to hold a presidential-rank position at NOAA. She was assistant administrator for external affairs, U.S. Environmental Protection Agency. **Carmen J. Blondin** is NOAA's deputy assistant secretary for international interests as well as U.S. Commissioner to the International Commission for the Conservation of Atlantic Tunas.

Marine biologist and aquanaut Dr. Sylvia Earle is NOAA's chief scientist whose life work is exploring the ocean at record-breaking depths. President of Deep Ocean Technology and Deep Ocean Engineering, Dr. Earle has logged more than 5,000 hours underwater exploring Earth's last frontier. Her goal at NOAA is to push for more marine sanctuaries and to protect the sea against oil spills and other toxic wastes. Dr. Elbert W. Friday, Jr., became assistant administrator of the National Weather Service after 20 years with the Air Weather Service, where he developed computer models for predicting clouds and weather conditions. Louis J. Boezi, deputy assistant administrator for modernization, National Weather Service, oversees a $2 billion program that includes deployment of new technologies for space and atmospheric observation, information processing, and communications.

Thomas N. Pyke, Jr., as assistant administrator for Satellite and Information Services, is in charge of this nation's civil operational environmental satellite program. He directed computer science programs at the National Bureau of Standards. Dr. Ned A. Ostenso directs the National Sea Grant College Program. He did geophysical research in Antarctica, where a mountain is named for him, and in the Arctic Ocean, Africa, and the Great Lakes. Marine biologist Dr. William W. Fox, Jr., heads the National Marine Fisheries Service and directed a joint NOAA/university research institute at the University of Miami's Rosenstiel School of Marine and Atmospheric Service.

National Marine Mammal Laboratory (NMML)
National Marine Fisheries Service (NMFS)
National Oceanic and Atmospheric
Administration (NOAA)
U.S. Department of Commerce
7600 Sand Point Way, NE
Seattle, WA 90115 USA

Contact: **Charles W. Fowler, Ph.D.**
   Phone: 1-206-526-4031 • *Fax:* 1-206-526-6615

Activities: Research • *Issues:* Biodiversity/Species Preservation
Meetings: ICES; IWC

● Organization

The National Marine Mammal Laboratory is engaged in a five-year research plan (through 1993) in the Antarctic, Arctic, Bering Sea and Gulf of Alaska, and California current. It is housed within the National Marine Fisheries Service (NMFS), which studies and manages living resources within 200 miles of U.S. shores; among its accomplishments, NMFS called public attention to the large killing of dolphins during yellowfin tuna catches in the eastern tropical Pacific; cut overall the incidence of Pacific driftnet fishing, which trapped and killed thousands of fish, sea animals, and birds; is increasing the survival of female Hawaiian monk seal pups; reversed the decline of striped bass; helped end commercial whaling; stopped sewage dumping close to New York; analyzed effects of contaminants as trace metals and synthetic organic materials on winter flounder, hard shell clam, and the American oyster.

▲ Projects and People

Just hours after the Exxon *Valdez* dumped 11.2 million gallons of North Slope Alaska crude oil in Prince William Sound, March 1989, scientists surveyed by air and sea the marine mammals' encounter with oil, searched for stranded sea mammals, rehabilitated seal pups and sea lions, and began assessing damage to killer whales and other cetaceans. Within a year, harbor seals declined 45 percent in oiled areas; one oiled seal had severe damage to the central nervous sytem, liver, and trachea; others assimilated petroleum hydrocarbons; inhaling oil fumes was also toxic. An early conclusion was that humpback whales may have been disturbed by the oil cleanup activities and increased vessel traffic. Researchers were Kathy Frost, Lloyd Lowry, Don Calkins, Craig Matkin, Charles Jurasz, Dan McSweeney, and Jan Straley.

Elsewhere in the western Gulf of Alaska to the western Aleutian Islands, a survey of Steller sea lions counted almost 25,000—as compared with nearly 68,000 four years earlier, representing a 63 percent decline. Pup production declined 66 percent at the five major rookeries studied. Decline of adults and juveniles has been noted since 1960; in the eastern Aleutian Islands, levels have fallen 94 percent. The study was carried out under the USA-USSR Environmental Protection Agreement.

Northern fur seal adult males and pups are being monitored on the Pribilof and Bogoslof islands of Alaska and San Miguel Island, California; radio tagging seals—led by Scripps investigator Tim Regan—shows declining population. Researchers also look for debris entanglement and how seals use rookery space. In the Bering Sea, scientists learned aquatic copulations occur in fur seals; nonbreeding fur seals do not avoid ground vibrations or airborne construction sounds. A conservation plan is underway to pinpoint causes of decline, threats to the population and habitat, and pup mortality around the rookeries, and recommend to management actions.

Chinstrap and macaroni penguins, crabeater seals, and Antarctic fur seals were fitted with radio transmitters and time-depth recorders that tell their deepest dive, where they get krill, and how far they swim for food.

More than 10,000 low-altitude aerial photographs of bowhead whales were shot to identify whales and help determine length, calving production and intervals, and juvenile abundance. Scientists estimate that about 40 percent of the bowhead population is sexually mature.

The migratory and feeding behavior of northern elephant seals is being studied in California currents. Working with the Makah Indian Tribe, the incidental take of harbor porpoise in the coastal salmon set-net fishery in Washington State is being examined. Sea lions are being recaptured, radio tagged, and relocated mainly to prevent escapement of steelhead trout. Most sea lions return to Puget Sound within 15 days; some have been recaptured 3 times.

Ecologist Charles W. Fowler is the manager of NMML's Population and Ecosystem Assessment Program and affiliate associate professor at the University of Washington, where he received his Ph.D. at the Center for Quantitative Science in Forestry, Fisheries, and Wildlife and College of Fisheries. He is a U.S. delegate to the Scientific Committee of the International Whaling Commission and a member of the Species Survival Commission (IUCN/SSC) Seal Group. Dr. Fowler's specialties are studies of large marine mammal population dynamics and of community and ecosystem structure and function.

Office of Law Enforcement
Southeast USA Region
National Marine Fisheries Service (NMFS)
National Oceanic and Atmospheric
  Administration (NOAA)
U.S. Department of Commerce
1A Max Brewer Parkway
Titusville, FL 32796 USA

Contact: **Paul W. Raymond,** Special Agent
  Phone: 1-407-269-0004 • *Fax:* 1-407-269-2558

Activities: Law Enforcement • *Issues:* Biodiversity/Species Preservation; Sustainable Development • *Meetings:* SE USA Sea Turtle Conservation Meeting

▲ Projects and People

Paul Raymond of the Office of Law Enforcement is charged with enforcing the Endangered Species Act, the Marine Mammal Protection Act, the Magnuson Fishery Conservation and Management Act, the National Marine Santuary Program, the Atlantic Tunas Convention Act, and the U.S. Lacey Wildlife Protection Act.

On Florida's east coast, he enforces the Turtle Excluder Device (TED) laws—training the U.S. Coast Guard and state conservation agencies on TED regulations and how they apply to the Southeast shrimp fleet. When inserted into shrimp nets, these devices prevent thousands of turtles from drowning. Other environmental activities include coral protection, enforcement of illegal importation of endangered species and their products, and enforcement of federal laws on domestic commercial fisherman and foreign fishing in USA waters. Raymond belongs to the Species Survival Commission (IUCN/SSC) Marine Turtle Group.

Southeast Fisheries Science Center
National Marine Fisheries Service (NMFS)
National Oceanic and Atmospheric
  Administration (NOAA)
U.S. Department of Commerce
75 Virginia Beach Drive
Miami, FL 33149 USA

Contact: **Nancy B. Thompson, Ph.D.,** Chief, Coastal
    Resources Division
  Phone: 1-305-361-4487 • *Fax:* 1-305-361-4515

Issues: Biodiversity/Species Preservation • *Meetings:* ICCAT; ICES; IWC

▲ Projects and People

Marine turtles are the focus of endangered species research in the southeastern United States. The Southeast Fisheries Center developed the **Marine Turtle Habitat Plan** to protect four endangered species in southeast U.S. waters from the Atlantic to the Gulf of Mexico, including riverbeds, keys, sounds, and bays. These are the Kemp's ridley, leatherback, hawksbill, and the Florida breeding populations of green turtles as well as the threatened loggerhead turtle and green turtle, outside Florida. The NMFS safeguards these species in the water; the U.S. Fish and Wildlife Service has jurisdiction of turtles on land.

Because marine turtles use different habitats during various life history stages, this plan focuses on characterizing inshore and offshore habitats, then determines distribution of species by size, age, and sex within the habitats. A combination of radio/sonic tagging to note short range movements and satellite tagging to determine long-range movements are recommended (so long as these do not negatively alter behavior) along with aerial surveys and sampling such as within coral reef areas, including marine parks and sanctuaries. This approach also identifies potential interactions between turtles and human activities and measures the resulting impact on the turtles and their localities. Risks to marine turtles include accumulation of debris and toxin; oil exploration and related activities; fisheries' incidental catch with longlines, shrimp and finfish trawling, gill nets, dredges, traps and pots, nets, and hooks and lines; entanglement in ghost fishing gear; military bombing; power plant entrainment; dredging; upland drainage; and stream diversion and channelization.

Cooperation with other government agencies and industry is essential. The U.S. Army Corps of Engineers (COE), which maintains the nation's navigable waters; the U.S. Navy; and the Minerals Management Service (MMS) are obliged to provide information as to whether their activities can jeopardize the recovery of endangered or threatened species. These federal agencies have funded surveys of channels and offshore platforms as well as other protected species research. Fishermen help by reporting sightings of turtles or "hot spots."

As chief of the Coastal Resources Division, Nancy B. **Thompson** manages diverse research and provides timely assessments of coastal fisheries and endangered species stocks to the appropriate management agency. Dr. Thompson heads a staff of 14 research scientists and 14 scientific technicians. She is program coordinator of the Endangered Species Program and senior stock assessment scientist for endangered species. Honored by her agency and author of peer review and technical articles, she belongs to the Species Survival Commission (IUCN/SSC) Marine Turtle Group.

Atmospheric and Geophysical Sciences
  Division ⊕
Lawrence Livermore National Laboratory—
  University of California
U.S. Department of Energy (DOE)
P.O. Box 808 (L-262)
Livermore, CA 94550 USA

Contact: **Michael C. MacCracken, Ph.D.,** Division
    Leader
  Phone: 1-510-422-1800 • *Fax:* 1-510-422-5844
  E-Mail: Omnet: m.maccracken;
    mmaccracken@llnl.gov

Activities: Research • *Issues:* Air Quality/Emission Control; Global Warming

● Organization

The **Atmospheric and Geophysical Sciences Division (G-Division)** of the **Lawrence Livermore National Laboratory (LLNL)**, staff of 90, uses carefully formulated and verified numerical models of the atmosphere-geosphere system to advance understanding of these science issues with local-to-global significance. Scientists study the effects of energy- and defense-related emissions on the environment and apply emergency response models to stem the effects of high-impact, technological accidents involving releases of nuclear or toxic materials into the air.

Most research is on the perturbing effects of carbon dioxide, chlorofluorocarbons and other trace gases, and aerosol emissions on climate and long-term atmospheric composition; studies also explore the role of the atmosphere-geosphere system in dispersing, transforming, and depositing radionuclides, particles, trace gases and toxic and heavier-than-air gases.

Models allow scientists to test how the atmosphere and oceans behave, become the framework for interpreting and understanding observations, assist in developing strategies for investigating issues, help in estimating accuracy for measuring instruments, and provide the basis for projecting future conditions. Primary support for G-Division activities comes from the U.S. Departments of Energy and Defense, National Aeronautics and Space Administration, Environmental Protection Agency, and within the Laboratory.

▲ Projects and People

The **Atmospheric Release Advisory Capability (ARAC)** is the designated national response center when potential or actual releases of radionuclides occur in the environment. A key emergency preparedness training resource for federal agencies, ARAC also consults with other countries and the Commission of the European Communities (CEC) to help improve their emergency response systems as well as through the International Atomic Energy Agency (IAEA). "Safety Series" guides, prepared for developing countries, recommend emergency response procedures for nuclear facilities.

ARAC calculated smoke dispersal from the Kuwait oil-field fires and forecast the scattering of volcanic ash from the Mt. Pinatubo eruption, Philippines. ARAC director is atmospheric physicist **Dr. Thomas J. Sullivan**, who has also published papers on hazardous materials dispersal from ground-based and spacecraft reentry events as well as on radioactivity from Chernobyl.

The **Program for Climate Model Diagnosis and Intercomparison (PCMDI)** is helping to bring improvements, coordination, cooperation, and advanced standards in assessing climate change with more sensitive climate models. Two long-range goals are innovative software for visualization and diagnosis and a comprehensive model-oriented database. **Dr. W. Lawrence Gates** is director and chief scientist.

The **Atmospheric Microphysics and Chemistry Group** is developing coupled climate models and chemistry-transport models to study tropospheric trace gas and aerosol distributions and their effects on climate. Cloud models are also being used to study the dynamics of certain clouds that interact with aerosols. Three-dimensional models are examining changes in the gaseous elements in the atmosphere as well as studying processes occurring above large, postnuclear-exchange fires. Global ozone level predictions are an expected result. Group leader is **Dr. Joyce E. Penner**, who concentrates research in "tropospheric chemistry and budgets, cloud and aerosol interaction and cloud microphysics, urban photochemistry, and model development and interpretation."

The **Global Radiative, Chemical, and Dynamical Interactions Group** advances understanding of these processes that determine the state of the global atmosphere, particularly the troposphere and stratosphere. Trace gases, including chlorofluorocarbons, emitted into the air are assessed for their impact on global distributions of ozone and temperature. The group is using data from NASA's Upper Atmospheric Research Satellite (UARS) to gain insights into stratospheric and lower mesospheric processes. Group leader is **Dr. Donald J. Wuebbles**, atmospheric scientist. On several international working groups, he explores the influence of environmental change on climate with scientists of the former USSR.

Goals of the **Mesoscale Modeling and Heavy Gas Dispersion Group** include developing new and better fluid mechanics models and applying existing models to areas of concern, including studies of toxic releases. Physicist and deputy division leader **Dr. Marvin H. Dickerson** is concerned with the atmospheric dispersion process and its applications to emergency response as in the Chernobyl reactor accident. Dr. Dickerson is a consultant to the IAEA, Vienna, Austria.

The **Model Applications and Nuclear Effects Group** develops and applies computer models that simulate the release, dispersion, and depositing of radionuclides. Models have been applied to simulating the global-scale radioactivity dispersion resulting from the Chernobyl reactor accident; radioactive material dispersion over Europe from underground nuclear weapons tests in the former Soviet Union; and assessing risks of operating nuclear facilities. **Dr. Paul H. Gudiksen**, chemist and group leader, does atmospheric field experiments and research in environmental radiation measurements and air quality impact of developing geothermal energy.

The **Climate and Climate Change Group** wants to reduce the uncertainty in projecting future climate change that can result from fluctuations in makeup of the atmosphere. Scientists are pursuing the effects of clouds on radiation and developing three-dimensional ocean models to determine how quickly the climate responds to changing atmospheric composition. A collaborative project on ocean-atmosphere modeling involves five University of California campuses. **Dr. Michael C. MacCracken** is group leader as well as division leader of LLNL's G-Division at the University of California. His chief research interests are climate change, climate modeling, and the greenhouse effect.

Environmentalists can enhance future cooperation by "clearly conducting and reporting on research results without being unnecessarily alarmist or sanguine," writes Dr. MacCracken. "[While] it is appropriate to be strenuous in protection of the environment, it is also appropriate to be cautious about projecting dire and adverse economic and societal impacts when scientific results are uncertain. With nearly 6 billion people on Earth and the prospect of a doubling of the population in the next several decades . . . limiting population and improving technology are essential steps to limiting environmental damage."

National Institute of Environmental Health
   Sciences (NIEHS) ⊕
National Institutes of Health (NIH)
Public Health Service (PHS)
**U.S. Department of Health and Human
Services (HHS)**
P.O. Box 12233
Research Triangle Park, NC 27709 USA

Contact: **Kenneth Olden, Ph.D.,** Director
Phone: 1-919-541-3201

Activities: Education; Research • *Issues:* Global Warming;
Health; Health Effects of Environmental Agents

● Organization

The **National Institute of Environmental Health Sciences (NIEHS)**
is a world research center which frequently hosts major scientific
conferences that focus on reducing environmental health risks for
people in the next century. Scientists here are determining how the
human body maintains health to fight disease and what makes
people succumb to exposure to harmful environmental agents.

Advances in technology such as computer enhanced magnetic
resonance spectroscopy and imaging, rapid progress in the genetics
field, and sophisticated application of x-ray and scanning electron
microscopy, for instance, offer researchers new tools for insights and
discoveries. Results of investigation are published widely; regulatory
agencies and the medical and scientific communities use the information to initiate improved public health programs.

Superfund support provides basic training and research at universities as well as health and safety training for workers at hazardous
waste sites and for emergency response personnel.

NIEHS has four divisions: Intramural Research, Extramural Research, Biometry and Risk Assessment, and Toxicology Research
and Testing—the core of the National Toxicology Program (NTP),
a cooperative venture of federal agencies also including the National
Cancer Institute, Food and Drug Administration's National Center
for Toxicological Research, and the Centers for Disease Control's
National Institute for Occupational Safety and Health. NTP testing
identifies the main toxic effects of each chemical studied for genetic
mutations; damage to critical organs such as the lungs, liver, and
nervous system; birth defects; or cancer.

Scientists now know that potentially toxic chemicals can affect
the entire biological system; many effects are subtle and much more
far-reaching than originally thought. It appears that when chemical
toxins interrupt any one of the normal body processes, they create
problems elsewhere in the system that are not yet fully understood.

One discovery is that prenatal exposure to lead can be associated
with low birth weight and poorer mental development of children
from all socioeconomic backgrounds. Another is that aging seems to
enhance vulnerability to toxic substances. Other findings from an
ongoing six-city study of air pollution and respiratory illness provide
data for setting air quality standards.

▲ Projects and People

NIEHS conducts scientific exchanges and collaborative research
through international programs. Cooperating with the World Health

Organization (WHO), it acts as a **Collaborating Center for Environmental Health Effects** and helps formulate research programs
about the biomedical aspects of environmental pollution. NIEHS is
involved with WHO, the United Nations Environment Programme
(UNEP), and the International Labor Organization in the International Programme on Chemical Safety (IPCS), which evaluates the
biological effects of chemicals, assesses health hazards, and reviews
testing for cancer and toxicity that affects neurobehavior and reproduction. NIEHS scientists exchange information on IPCS committees, conferences, and technical working groups. They also take part
in global meetings of the Scientific Group on Methodologies for the
Safety Evaluation of Chemicals (SGOMSEC) and the working
groups of WHO's International Agency for Research on Cancer.

USA–China (Mainland) Cooperation exchange visits focus on
reproductive and developmental toxicology, short-term testing methods for carcinogens and mutagens in the environment, and extrapo-
lating laboratory animal data to man. A joint reach project on
oncogene activation in human liver cancer is being initiated.

In **USA–China (Taiwan) Cooperation,** NIEHS scientists collaborate with Taiwanese students to study the effects of accidental
human exposure to polychlorinated biphenyls (PCBs) in Taiwan.
Early investigations with children show abnormalities of pigment,
nails, teeth, and skin and some developmental delay.

USA–Egypt Cooperation is carried out through its Joint Working Group on Health Cooperation (JWGHC), Cairo, supported by
the U.S. Agency for International Development (USAID). Egyptian
government, groups, and universities gain access to U.S. information on environmental and occupational health hazards.

USA–Finland Cooperation features information exchanges on
pharmacokinetics; reproductive, neurobehavioral, and genetic toxicology; and epidemiology and risk assessment.

In **USA–India Cooperation,** NIEHS and NTP scientists are
conducting experimental animal studies on the toxicity of methyl
isocyanate, the chemical responsible for the world's worst industrial
disaster in Bhopal, India (1984).

USA–Italy Cooperation in environmental health exchanges information on chemical contamination of drinking water studies, for
example.

USA–Japan Cooperation takes place under two formal agreements on medical sciences and science/technology research and
development. Joint research focuses on the detection and testing of
mutagenic and carcinogenic chemicals and on monitoring human
populations for exposure to these.

USA–Soviet Cooperation is carried out under two agreements.
Joint research is studying the effects of physical and chemical environmental agents on human health. Experiments determine the
biological effects of electromagnetic fields on the nervous system.
Another working group is involved with the biological and genetic
effect of pollution.

USA–Yugoslavia Cooperation is advancing studies that evaluate
genetic effects of low levels of environmental chemical mutagens in
bacterial systems, and comparison with eukaryotic cells.

Dr. Kenneth Olden directs both NIEHS and the National Toxicology Program. In 1991, President George Bush appointed him to
the National Cancer Advisory Board. Division directors are **Dr.
John A. McLachlan,** Intramural Research; **Dr. Anne Sassaman,**
Extramural Research and Training; **Dr. David G. Hoel,** Biometry
and Risk Assessment; and **Dr. Richard A. Griesemer,** Toxicology
Research and Testing.

Dr. Terri Damastra is in charge of International Programs. Dr. John Dement heads the Office of Occupational Health and Technical Services, where Davenport Robertson is chief of the Library and Information Services Branch.

■ Resources

The NIEHS library maintains one of the world's most advanced computerized literature searching capabilities, with access to more than 600 databases, including TOXLINE, MEDLINE, and Hazardous Substances Data Bank. Books are catalogued in an automated system, OCLC, used by more than 3,500 libraries nationwide.

TOXNET
National Library of Medicine (NLM)
National Institutes of Health (NIH)
Public Health Service (PHS)
**U.S. Department of Health and Human
  Services (HHS)**
8600 Rockville Pike
Bethesda, MD 20894 USA

Contact: **Bruno M. Vasta,** Administrator
  Phone: 1-301-496-6531 • *Fax:* 1-301-480-3537

Activities: Environmental Online Information • *Issues:* Health

▲ Projects and People

The **National Library of Medicine (NLM)** describes **TOXNET** as a toxicology data network, which is a computerized system of files oriented to toxicology and related areas. NLM's Toxicology Information Program manages TOXNET, which is run on a series of microcomputers in a "networked client/server architecture." Its integrated system contains modules for building and reviewing records as well as providing sophisticated search and retrieval features for NLM online users.

Available 24 hours a day and designed to be user-friendly, TOXNET enables researchers to extract data on known or unknown chemicals through free-text searching. Cross-file capability—with MEDLINE or TOXLINE files, for example—permits simultaneous searching and printing from multiple files. Menus also allow inexperienced computer users to access easily certain files. Other assistance is also available. Registered NLM users access TOXNET by direct dial or through COMPUSERVE, SprintNet (formerly TELENET), or TYMNET telecommunication networks.

The TOXNET files include the following:

- **HSDB (Hazardous Substances Data Bank).** The toxicology of more than 4,300 potentially hazardous chemicals is featured as well as information on emergency handling procedures, environmental fate, human exposure, detection methods, and regulatory requirements. A Scientific Review Panel of expert toxicologists and other scientists reviews the file, which is co-supported by the Agency for Toxic Substance and Disease Registry (ATSDR).
- **RTECS (Registry for Toxic Effects of Chemical Substances).** Data is included on more than 105,000 chemicals with acute and chronic effects such as skin and eye irritation, carcinogenic-

ity, mutagenicity, and reproductive consequences. The National Institute for Occupational Safety and Health (NIOSH) built and maintains the file.
- **CCRIS (Chemical Carcinogenesis Research Information System).** Scientifically evaluated data is derived from carcinogenicity, mutagenicity, tumor promotion, and tumor inhibition tests on more than 2,500 chemicals. The National Cancer Institute (NCI) is the sponsor.
- **IRIS (Integrated Risk Information System).** An online database built by the U.S. Environmental Protection Agency (EPA), IRIS contains EPA carcinogenic and noncarcinogenic health risk and regulatory information on more than 450 chemicals. EPA scientists review and agree to the risk-assessment data. IRIS also contains EPA Drinking Water Health Advisories, chemical and physical properties, and references. Plans are to add Aquatic Toxicity Assessments.
- **TRI (Toxics Release Inventory).** This series of files contains information on the "annual estimated releases of toxic chemicals to the environment." Mandated by the Emergency Planning Community Right-to-Know Act, TRI is based upon data submitted to the EPA from industrial facilities throughout the USA. Included are facilities' names and addresses and the amounts of certain toxic chemicals that are released into the air, water, or land, or transferred to waste sites. To date, there is information on more than 300 chemicals and chemical categories.
- **DBIR (Directory of Biotechnology Information Resources).** Included are online database, networks, publications, organizations, collections and repositories of cells and subcellular elements, and other resources—all dealing with biotechnology.
- **GENE-TOX (Genetic Toxicology).** This online databank, created by EPA, features genetic toxicology results on more than 4,000 chemicals. For each test system under evaluation, work panels of scientific experts review the genetic toxicology, select mutagenicity assay systems, and the source literature; their data review activities are included in GENE-TOX.
- **ETICBACK (Environmental Teratology Information Center Backfile).** A bibliographic database covering teratology literature, ETICBACK contains more than 49,000 citations to literature published between 1950 and 1989. Produced by the Oak Ridge National Laboratory, it is funded by ATSDR, EPA, and the National Institute of Environmental Health Science (NIEHS).
- **DART (Developmental and Reproductive Toxicology).** Contains nearly 7,000 citations to literature published since 1989, DART is a continuation of ETICBACK and also covers literature on teratology and other aspects of developmental toxicology. Produced by NLM, it is funded by EPA and NIEHS.
- **EMICBACK (Environmental Mutagen Information Center Backfile).** The backfile for EMIC, this bibliographic database on chemical, biological, and physical agents have been tested for genotoxic activity and contains more than 70,000 citations to literature published since 1950. Produced by the Environmental Mutagen, Carcinogen, and Teratogen Information Program, Oak Ridge National Laboratory, EMICBACK is funded by EPA and NIEHS.

Bruno Vasta is primarily responsible for the development of TOXNET and many of its data files, including HSDB. He is also NLM's chief of Biomedical Files implementation activities, chairman of the Division of Chemical Information, and councilor, American Chemical Society.

■ Resources

Vasta expresses resource needs for increased efficiency in computer hardware and software; greater educational outreach to "all medical schools, public health, and environmental curriculums"; and planned exhibits for all relevant technical meetings.

Office of Environment and Energy

U.S. Department of Housing and Urban Development (HUD)

451 Seventh Street, SW
Washington, DC 20410-7000 USA

Contact: **Richard H. Broun,** Director
  Phone: 1-202-708-2894 • *Fax:* 1-202-708-3363

Activities: Administration; Education • *Issues:* Air Quality; Energy; Water Quality

● Organization

Human habitats and the environment were first legislatively linked in 1949 with the passage of the Housing Act mandating "a suitable living environment for every American family." Twenty years later, the National Environmental Policy Act authorized review requirements for major federal actions "significantly affecting the quality of the human environment." Title I objectives were next broadened in 1980–1981 to include energy conservation, energy efficiency improvements, and provisions of alternative and renewable energy supply sources.

Today, the **U.S. Department of Housing and Urban Development (HUD)** is also the lead agency for administering certain specialized laws or their provisions regarding historic preservation, air quality, noise, water quality, aquifers, and endangered species in addition to its HUD regulations and numerous federal Executive Orders.

▲ Projects and People

Richard H. Broun of the **Office of Environment and Energy** oversees and evaluates program environmental reviews and originates performance standards for noise, explosives, clear zones, hazards, and land use. He stimulates **energy efficiency** through district heating, weatherization, and energy planning, and prepares handbooks for implementing environmental laws.

Opportunities are developed for **energy feasibility studies** and to promote links to public housing modernization, neighborhood revitalization, jobs, and economic development. Certain localities are assisted with **waste-to-energy systems** development and operations. Workshops are held for mayors and public works directors regarding incineration equipment. **Cogeneration and alternative fuel opportunities** are explored. Other **regional energy workshops** and seminars are convened on energy and economic development, and local action is promoted through **American Energy Awareness Week** and related efforts.

To encourage energy efficiency through **building retrofit** and to understand the air quality consequences of superinsulation, analyses and technical guides on energy assistance are prepared.

A leader in his field in international collaborations, Broun was U.S. delegate to the Urban Energy Management Conference in Paris and Berlin; served on the World Bank Advisory Review Committee for the United Nations Habitat Environmental Guidelines; was co-chairman of the USA/Mexico Bilateral Agreement; and was urban environment coordinator of the USA-USSR Environmental Protection Agreement in the 1970s.

Endangered Species Program
  Bureau of Land Management (BLM)

U.S. Department of Interior

Arizona State Office
3787 North Seventh Street
P.O. Box 16563
Pheonix, AZ 85011

Contact: **William H. Radtkey,** Coordinator
  Phone: 1-602-640-5501 • *Fax:* 1-602-640-2398

Activities: Development; Education; Political/Legislative; Research • *Issues:* Biodiversity/Species Preservation; Energy; Global Warming; Sustainable Development; Water Quality

■ Resources

Desert system management technologies are sought to stabilize desert geosystems.

Cal McCluskey, BLM's Division of Wildlife and Fisheries (Washington, DC 20240-0001, phone 1-202-653-9502), reports on "Bring Back the Natives!—it's new national joint effort with the U.S. Department of Agriculture's Forest Service, National Fish and Wildlife Foundation, and volunteers. Goals for "Fish and Wildlife 2000," a plan to improve habitats on public lands by the year 2000, are described in the literature, as are the new "Riparian-Wetland Initiative for the 1990s," the "Rise to the Future" program to enhance fisheries and aquatic resources on national forests and grasslands, and other conservation measures to reintroduce endangered and threatened species. Also available is a list of special status plant species on the 270 million acres of BLM lands.

Endangered Species Division ⊕
Fish and Wildlife Service

U.S. Department of Interior

4401 North Fairfax Drive, MS 452
Washington, DC 22203 USA

Contact: **John J. Fay, Ph.D.,** Biologist/Botanist
  Phone: 1-702-358-2171, 1-703-358-1735

Activities: Conservation • *Issues:* Air Quality/Emission Control; Biodiversity/Species Preservation; Deforestation; Global Warming; Sustainable Development; Waste Management/Recycling; Water Quality • *Meetings:* CITES; Ramsar; SSC

● Organization

The Fish and Wildlife Service describes itself as leading "the federal effort to protect and restore animals and plants that are in danger of extinction both in the United States and worldwide." As scientifically as possible, species are identified that seem to be endangered or threatened. "After review by scientists and opportunities for public comment, species that meet the criteria of the Endangered Species Act are placed on the Interior Department's official 'List of Endangered and Threatened Wildlife and Plants.'"

Next, recovery plans are developed with scientists from government, universities, and the private sector. "The Service also consults with other federal agencies and renders 'biological opinions' on the effects of proposed government projects on endangered species." Its research laboratories and field stations are located throughout the country.

Through the Convention on International Trade on Endangered Species of Wild Fauna and Flora (CITES), the Fish and Wildlife Service works with more than 90 countries to preserve wildlife and keep rare species from being exploited in commercial trade. The agency also participates in more than 40 global treaties and statutes, gives technical assistance to other countries, and trains environmentalists worldwide.

With authorization for the Endangered Species Conservation Act of 1973 expiring in 1992, this law that originated in 1969 as the Endangered Species Preservation Act also faces amendments—as it has in previous years. Growing through such amendments and interpretations, "the Act embodies a fairly comprehensive approach to maintaining species diversity in the USA and elsewhere," writes Dr. John J. Fay, who participates in policy development with the Endangered Species Division. "It is doubtful that anyone anticipated in 1973 what a large, complicated, and sometimes contentious job this would become." However, there has been a "continuing underlying commitment . . . that the goal of protecting species is important for the nation."

Fay explains this legislation, beginning with Section 2 of the Act, as a way to identify human activities causing "the extinction of some species and [putting] the survival of other species at risk"; he attributes the USA's commitment through various treaties to the "conservation of endangered and threatened species and the ecosystems upon which they depend."

Section 3 provides the following definitions, according to Fay:

Species—Any species or subspecies of plant or animal and, in the case of vertebrate life forms, includes any distinct population segment as well.

Endangered Species—A species in danger of extinction through all or a significant portion of its range.

Threatened Species—A species likely to become endangered in the foreseeable future throughout all or a significant portion of its range.

Critical Habitat—Specific areas within the geographical area occupied by a species at the time of listing, within which are found those physical or biological features essential to the conservation of the species and which may require special management considerations or protection, as well as areas outside the range of the species which are essential to its conservation.

Conservation—The use of all methods and procedures necessary to bring a species to the point at which the protective measures of the Act are no longer necessary. Lately, the Fish and Wildlife Service does not designate critical habitat for listed species. As economic impacts must be evaluated "somewhat independently" from eco-

logical criteria for critical habitat listing, yet considered in the same rulemaking, "keeping faith with these requirements is a complex undertaking," writes Fay.

According to Section 4, both the Secretaries of Interior and Commerce ascertain "those species in need of attention"; operational authority is delegated to both the Fish and Wildlife Service and the National Marine Fisheries Service. Lists of protected species are part of the Code of Federal Regulations, and changes are accomplished through "a rulemaking process involving proposal, public comment, and adoption of a final rule." Also, the public may petition for a species to be listed. A "boll weevil clause" excludes certain insect species considered to be pests. Section 4 also provides for preparation of a recovery plan—a set of recommendations that are *not* self-implementing—and appointment of a recovery team.

Section 5 grants authority to acquire land for the benefit of endangered and threatened species; Section 6 "establishes a federal cost-sharing grant program for state governments undertaking endangered species conservation," informs Fay.

Section 7 enlists all federal agencies in support of conserving endangered species and "requires them to ensure that actions they authorize, fund, or carry out are not likely to jeopardize the continued existence of listed species or to destroy or adversely modify critical habitat." One "little-used" process allows a "cabinet level committee to exempt an agency's action from the duty to avoid jeopardy to a species or adverse modification of critical habitat."

Section 8 encourages global cooperation such as implementing CITES.

Section 9 prohibits killing, wounding, harassing, or harming endangered animals, and prohibits maliciously destroying, reducing, or removing endangered plants on federal land or as a result of state law. "For reasons that certainly are not based in science," comments Fay, "the current law treats plants differently and less protectively than it does animals."

Section 10 grants permits to be issued for the "taking, import and export" of species for scientific research and other special circumstances.

Section 11 sets penalties of civil fines up to $25,000 and criminal damages up to $50,000 and up to one-year imprisonment. Vehicles, vessels, and other equipment may be seized.

Section 12, regarding endangered plants, is based on a Smithsonian Institution list of some 3,000 endangered or threatened plant species native to the USA.

In Section 18, the Endangered Species Act defers to the Marine Mammal Protection Act, whenever it is more restrictive in species protection. Other sections relate to housekeeping provisions, authorization of appropriations, and annual cost analysis.

From his perspective, Fay sees endangered species conservation issues as becoming "very litigious" and points to the difficulty of analyzing economic impacts in designating critical habitat. He believes that whole ecosystems are served through the protection of individual species; "It is very doubtful that we now know enough about the working of ecosystems to allow reversal of the paradigm."

Fay summarizes that "Putting a species on the endangered or threatened species list is only a first step, and an uncertain one at that, toward ensuring its survival. One need only call to mind the California condor and the black-footed ferret, or *Kokia cookai*, an endangered Hawaiian tree that has existed only as grafts on the roots of a related [endangered] species . . . to know that the conservation of species, and ecosystems, is long-term, hard work."

▲ Projects and People

In 1991–1992, Dr. Fay was on special assignment to help prepare a recovery plan and organize a recovery team—on which he serves—for the northern spotted owl in the Northwest USA.

Wetland conservation under the North American Waterfowl Plan is a multilevel, public/private partnership plan in Canada, USA, and Mexico. The Plan covers the 64,000-acre ACE Basin Project, South Carolina, Atlantic Coast Joint Venture; 12,000 acres of private lands in the Chase Lake Project, North Dakota, Prairie Pothole Joint Venture; more than 100,000 acres in San Joaquin Basin Project, California, Central Valley Joint Venture; and key waterfowl nesting areas in the Prairie and Habitat Joint Venture of Canada and USA—among an estimated 6 million acres of protected, restored, or enhanced, ponds, lakes, and marshes to be conserved by the year 2000. Implementation in Mexico is just getting underway.

Also in 1992, the Fish and Wildlife Service submitted the USA's species listing proposals and implementation resolutions to CITES for its biennial meeting, held in Japan.

Recommended for inclusion in CITES Appendix I—for species threatened with extinction that either are or may be affected by trade—are: Goffin's cockatoo, blue-streaked lory bird, bog turtle, blue-fronted Amazon parrot, and paddlefish, whose roe is a main source of exported U.S. caviar.

Recommended for Appendix II—CITES-regulated animals and plants that may become threatened unless trade control is practiced are: wood turtle, high-jumping African Goliath frog, Venus flytrap, queen conch, American mahogany, and a small eastern Caribbean tree known as *lignum vitae*.

The Service also submitted a resolution to suspend commerce in heavily traded wild bird species while scientific studies are underway to "ensure that any future trade . . . will be sustainable." The Service is also asking that all crocodile skins in international trade be appropriately marked as are legal American alligator skins, which bear tags.

Dr. Fay—who specialized in plant systematics, cytology, ecology, and phytogeography while earning his Ph.D. degree at City University of New York—was formerly associated with the Pacific Tropical Botanical Garden in Hawaii and the Field Museum of Natural History, Chicago, Illinois, before joining the Fish and Wildlife Service in 1978.

■ Resources

The Service produces a series of publications on endangered and threatened species including the 400-page *Endangered and Threatened Species Recovery Program* available from the U.S. Government Printing Office, Washington, DC 20402, 1-202-783-3238.

Brochures on wetlands, waterfowl, sea turtles, volunteer programs from conducting surveys to photography and caring for animals, sport fish restoration, guidelines for purchasing souvenirs abroad, the National Wildlife Refuge System, and the Fish and Wildlife Forensics Laboratory that identifies species and mortality cause are available from the Service's publications unit, 130 ARLSQ 1849 C Street, NW, Washington, DC 20240, 1-703-358-1711. The Service also publishes an in-house *Fish and Wildlife News* bimonthly magazine which describes advances in protecting species, among other publications.

Fish and Wildlife Service ⊕
U.S. Department of Interior
National Museum of Natural History
Washington, DC 20560 USA

Contact: **Roy W. McDiarmid, Ph.D.**, Research
          Zoologist/Curator
Phone: 1-202-357-2780 • *Fax:* 1-202-357-1932

Activities: Research • *Issues:* Biodiversity/Species Preservation
Meetings: CITES; World Congress of Herpetology

▲ Projects and People

Standard methods for measuring biological diversity of amphibians is the subject of a manual published in 1992 by Smithsonian Press and prepared by Dr. Roy W. McDiarmid, curator of amphibians and reptiles, National Museum of Natural History, Fish and Wildlife Service, with Maureen Donnelly, W. Ronald Heyer, and Lee-Ann Hayek.

According to McDiarmid, ". . . growing awareness of the impact of widespread habitat destruction and worldwide loss of biodiversity has demonstrated the pressing need for standardly collected data at all sites." He urges that procedural variations be reduced in the systematic gathering of information regarding the world's biota and that such efforts be expanded in both developed and developing countries.

Because amphibians are believed to be "potentially excellent indicators of health of the environment" even as their numbers are "declining sharply," McDiarmid advises that scientists can no longer afford procedural variations in assessing population. Thus, his manual is particularly aimed at establishing guidelines for monitoring and inventory programs for amphibians; it focuses on four components:

• A species list useful in comparing sites that could require years of fieldwork. "How closely the list approximates the species composition at a site depends on the complexity of the fauna, the duration of the inventory, and the availability of appropriate sampling techniques.

• "Documentation of different microhabitats at each site and the distribution of animals within and among these habitats," including breeding as well as nonbreeding habitats.

• Temporal variations, such as the inclusion of diurnal/nocturnal cycles and wet/dry seasonal data. "Precipitation and maximum and minimum temperatures should be recorded whenever possible."

• Individual abundance data, such as life history stage and sex ratios, allowing for "robust comparisons among habitats and sites than can be gained from comparison of species lists alone."

The manual provides a guide to research design, such as field observational methods and sampling techniques. Described are acoustic monitoring at fixed sites; radio and radioactive tracking, census taking in artificial habitats; automated data acquisition through dataloggers and recorded frog calls; geographic information systems (GIS) and remote sensing; and field trips—among other techniques.

McDiarmid is a consultant to the Species Survival Commission (IUCN/SSC) Amphibian Group and was recently made director of the American Association for Zoological Nomenclature.

Arctic National Wildlife Refuge (ANWR)
Fish and Wildlife Service
U.S. Department of Interior
1011 East Tudor Road
Anchorage, AK 99577 USA

Contact: **Bruce T. Batten,** Assistant Regional
     Director, Public Affairs
Phone: 1-907-786-3486 • *Fax:* 1-907-786-3495

Activities: Government Resource; Public Information • *Issues:*
Biodiversity/Species Preservation; Wildlife Management
Meetings: CITES

● **Organization**

Of the 16 wildlife refuges in Alaska, the Arctic National Wildlife Refuge is the most northern—encompassing "one of the spectacular assemblages of arctic plants, wildlife, and land forms in the world," according to the Fish and Wildlife Service. Its 19-million-acre tract is home to some 180,000 Porcupine caribou that migrate 1,000 miles from south of the Brooks Range to calving grounds on the northern coastal plains and Canada's Yukon Territory, as well as to 140 bird species including golden eagles, geese, swans, and ducks (some traveling as far as the Chesapeake Bay, Asia, Africa, and Antarctica). Free-roaming herds of muskoxen, Dall sheep, wolves, wolverines, and polar and grizzly bears also have their habitats there. Some 62 species of slow-growing fish in the northern Arctic waters are "easily affected by overharvest or changes in the environment." With permafrost within 1.5 feet of the surface, plants adapt and grow slowly; "it may take 300 years for a white spruce at tree line to reach a diameter of five inches; small willow shrubs may be 50–100 years old," continues the Service.

▲ **Projects and People**

In 1987, the Secretary of the Interior presented to the U.S. Congress a Coastal Plain Resource Assessment and Recommendation for a 1.5-million-acre portion known as the "1002 area," which refers to a section in the Alaska National Interest Lands Conservation Act (ANILCA) of 1980 that previously enlarged the refuge. The Recommendation stated, "The 1002 area is the nation's best single opportunity to increase significantly domestic oil production over the next 40 years. It is rated by geologists as the most outstanding petroleum exploration target in the onshore United States."

The document cited this area's advantageous proximity to Prudhoe Bay, where production was expected to decline over the next two decades and to the Trans-Alaska Pipeline System, which could carry "section 1002 area" oil to U.S. markets. The U.S. Department of Interior bolstered its appraisal with excerpts from a U.S. Department of Energy (DOE) *Energy Security* report that same year, which stated that domestic consumption of oil was expected to grow while "reduced U.S. oil exploration and production will increase U.S. reliance on oil from the unstable Persian Gulf region."

Referring to some 50 biological studies conducted in this area since 1980, the report stated that at the stages of exploration and development drilling "only minor or negligible effects on all wildlife resources" would be generated. Potentially "major"—not necessarily "adverse"—effects regarding oil production are "limited to the Porcupine caribou herd and the reintroduced muskox herd," stated

the assessment, which indicated that it was possible for such herds to "co-exist successfully with oil development" as seemed to be the case in the Prudhoe Bay development. Concern was expressed about the potential displacement of portions of the caribou herd as they seek calving areas.

The assessment concluded that "an orderly oil and gas leasing program for the entire 1002 area can be conducted in concert with America's environmental goals," and requested the authority and legislation to implement the Secretary's Recommendations for oil and gas leasings—superceding certain sections of ANILCA and also to open the Kaktovik Inupiat Corporation (KIC)/Arctic Slope Regional Corporation (ASRC) lands for such purposes.

As a result, environmental groups, such as the Sierra Club and the National Audubon Society, assert that the "Arctic Refuge is in grave danger" including the above-mentioned species and the icy coastal water's endangered bowhead whales and ringed seals that comprise 95 percent of the diet of some 1,800 Beaufort Sea polar bears. High-priority campaigns were organized to arouse public pressure and urge concerned citizens to contact members of the U.S. Congress in opposition to oil leasing and the desecration of wilderness values.

Fish and Wildlife Service literature also underscores the value of this refuge's biolgical heart—"a portion of one of the largest remaining areas of undisturbed land in the world"—that supports large animals as well as ground squirrels, shrews, arctic foxes, other wildlife, and plants that are part of a natural balance. "In the open tundra, one can easily observe the links in the chain of life. Any break in the chain can do irreparable and often unpredictable harm to the whole system." As part of the natural ecology, a community of about 210 Kaktovik natives and other villages subsist on the Porcupine herd and other resources.

According to the Audubon Society, "Interior admits there is only a one in five chance of finding any oil, and a much smaller chance of finding oil in commercially large quantities. The chances of finding an oilfield the size of Prudhoe Bay—9 billion barrels—are only one in a hundred." However, the Arctic Oil and Gas Association (AOGA) states the opposite view—describing the coastal plain as "one of the most promising areas for major discoveries of oil and gas of all untested onshore areas of the United States."

Environmental groups such as the Rocky Mountain Institute remind that oil use can be reduced considerably through improved efficiency of new cars and light trucks; and through superior electric lights, motors, and appliances. Worldwatch Institute calculates that as "much energy leaks through American windows each year as flows through the Alaska pipeline." George T. Frampton Jr., president of the Wilderness Society, was quoted in the *New York Times Magazine* ("Can Oil and Wilderness Mix?" by Timothy Egan, August 4, 1991) as, "We simply cannot drill ourselves to energy dependence" in light of the estimate that "65 percent of the world's known oil reserves are in the Middle East; the United States has only about four percent."

■ **Resources**

Free publications include a *Guide to National Wildlife Refuges of Alaska, Fish and Wildlife Resources of the Arctic Coastal Plain, Baseline Study Report Series,* and individual guides on each Alaskan national wildlife refuge. New technologies are needed for wildlife research; political support is needed for wildlife programs.

Endangered Species Program—Fairbanks
Fish and Wildlife Service
U.S. Department of Interior
1412 Airport Way
Fairbanks, AK 99701 USA

Contact: **Skip Ambrose**, Project Leader
  Phone: 1-907-456-0239 • *Fax:* 1-907-456-0346

Activities: Research • *Issues:* Biodiversity/Species Preservation
Meetings: CITES

● Organization

Fish and wildlife resources in interior and northern Alaska are identified, listed, protected, and recovered within the **Endangered Species Program**, according to the Endangered Species Act of 1973. Specifically, the Fish and Wildlife Service identifies species that have declined to "threatened or endangered levels" throughout all or a large portion of their range, prepares listings, and protects those listed species. A database is maintained on interior and northern Alaska's endangered, threatened, and candidate species. And the Service conducts surveys and recovery actions on listed species.

▲ Projects and People

Arctic and American peregrine falcons are the center of survey and recovery actions in interior and northern Alaska. These falcons began to decline in Alaska during the 1950s due to the use of the pesticide DDT, which affected adult behavior and reproduction. Although relatively little DDT was used in Alaska, falcons from this state migrate as far south as Argentina and were exposed to the pesticide during migration and in wintering areas. Peregrine falcons feed primarily on birds, and many of their prey are migratory and thus also exposed to DDT.

Because peregrine falcons are at the top of the food chain, the impact of DDT was much more severe to these birds than to their prey species. By the mid-1970s, "the peregrine falcons declined to approximately 25 percent of their historical levels," writes Endangered Species Program project leader Skip Ambrose. "The United States restricted the use of DDT in 1972, and peregrines began to increase in the late 1970s. Increases have continued through 1991, with some areas achieving numbers believed to be pre-DDT levels."

Surveys of the Arctic and American peregrine falcons include monitoring status, trends, and contaminant levels; and studying migration, dispersal, and mortality rates. Joining with other federal and state agencies in Alaska, the Fish and Wildlife Service surveys four index study areas and documents the number of occupied sites, the number of pairs successful in fledgling young, and the number of young per pair. The young are banded, as are the adult falcons, to assist in the dispersal and mortality studies. Unhatched eggs are also collected for pesticide analysis and prey remains are collected for prey studies.

The status of peregrine falcons is being reviewed to see if reclassification and/or delisting is appropriate.

■ Resources

Four publications are available: *Status of Peregrine Falcons in Alaska*; *Dispersal, Turnover and Migration of Peregrine Falcons in Alaska*; *Pesticides in Alaskan Peregrine Falcons*; and *Prey of Peregrine Falcons in Alaska*.

Grizzly Bear Recovery
Fish and Wildlife Service
U.S. Department of Interior
University of Montana
Natural Science 312
Missoula, MT 59812 USA

Contact: **Christopher Serveen, Ph.D.**, Coordinator
  Phone: 1-406-329-3223 • *Fax:* 1-406-329-3212

Activities: Development; Research • *Issues:* Biodiversity/Species Preservation; Deforestation

▲ Projects and People

A five-year effort is underway to augment the grizzly bear population in the Cabinet Mountains of Montana. The U.S. Fish and Wildlife Service research team is capturing young female grizzly bears in southeastern British Columbia and relocating them to the Cabinet Mountains. According to **Dr. Christopher Serveen**, "two or three more female bears will be relocated if and when they are captured."

Fragmentation of existing grizzly habitat in the USA is impacting on grizzly populations and may lead to "isolation of bear populations reducing the genetic viability of the populations," writes Dr. Serveen. Both Montana's Mission Mountains and the Bob Marshall Wilderness complex are home to grizzly populations. However, increased human settlements, logging, and road building are expanding into the Seeley-Swan River corridor between these two areas. A geographic information system (GIS) and remote sensing cameras are examining the impact of such development and activities on the grizzly bears and their habitat, particularly on fragmentation.

Research is also assessing the value of different habitat types to bears. The grizzly food biomass estimation per habitat and the weight gain of bears in varying habitats are under study in these bear energetics projects.

Dr. Serveen is responsible for coordinating all research and management activities involving grizzly bears in the continental USA. He is co-chair of the Species Survival Commission (IUCN/SSC) Bear Group.

■ Resources

Grizzly bear research papers and informational brochures are available.

Fish and Wildlife Service
Southeast Region
U.S. Department of Interior
330 Ridgefield Court
Asheville, NC 28806 USA

Contact: **V. Gary Henry**, Wildlife Biologist/Red Wolf
         Coordinator
  Phone: 1-704-665-1195 • *Fax:* 1-704-665-2782

Issues: Biodiversity/Species Preservation

▲ Projects and People

One of the most endangered species in North America, only 135 red wolves existed in 1990, and most of these were in captive projects. In America's pioneer days, three subspecies of wolves roamed from the eastern seaboard to the Ohio River valley and westward to what is now central Texas and Oklahoma. Two subspecies are extinct; only *C. r. gregori* survived. As people settled the region and land use changed, the coyote moved into the western portions of the red wolf's range, and eventually some hybridization occurred.

The effort to save the lanky, shy red wolf, *canus rufus*, is two-pronged: captive breeding programs and reintroduction into secured areas. The Point Defiance Zoological Garden, Metropolitan Park Board, Tacoma, Washington, established the first captive breeding program in 1973 with 40 wild-caught adult red wolves. Because of the productive vigor of red wolves in captivity, more U.S. zoos soon participated in captive breeding programs. Within 14 years, the number of red wolves in captivity increased to 80. In 1987, animals were introduced into the wild at the Alligator River National Wildlife Refuge, North Carolina.

The American Association of Zoological Parks and Aquariums (AAZPA) designed the strategy that maintains genetic vigor as red wolves are moved from one project to another. Three island propagation projects are on Bulls Island, Service's Cape Romain National Wildlife Refuge, South Carolina; St. Vincent National Wildlife Refuge, Florida; and Horn Island of the National Park Service's Gulf Islands National Seashore, Mississippi.

Red wolves actually vary in color from yellow to brown, gray to black. Females weigh about 50 to 60 pounds and males up to 80 pounds. It is thought that their long, slender legs are an adaptation to "long-distance running and pursuing prey in river-bottom swamps," says the Fish and Wildlife Service. Known to travel in family groups, they prefer small prey such as rodents, raccoons, muskrats, fish, insects, and plants—generally leaving alone livestock in fenced pastures. Secretive and nocturnal, red wolves do not threaten humans and their recovery in the wild is beginning to be appreciated.

Researchers are determining if the Great Smoky Mountains National Park is suitable for the reintroduction of red wolves. Park officials say, "If the first few animals displace coyotes and resist interbreeding, reintroduction of five or six adult pairs of wolves could begin in 1992." The goal is to sustain 50 to 75 red wolves in the park. The last wolf known to exist there was killed in 1905.

V. Gary Henry, with the Species Survival Commission/IUCN Wolf Group, succeeds Warren T. Parker as red wolf coordinator.

■ Resources

Henry says that "recapture collars to secure animals by remote control are needed," as are "adequate appropriated funds" to reintroduce red wolves into the wild. The public needs to know about the "value of predators," he says. "Unfounded fears of human or livestock predation" must be allayed.

Golden Gate National Recreation Area (NRA) ⊕
National Park Service
U.S. Department of Interior
Ft. Mason, 201
San Francisco, CA 94123 USA

Contact: **Judd A. Howell, Ph.D.**, Wildlife Ecologist
Phone: 1-415-556-9506

Activities: Research • *Issues:* Biodiversity/Species Preservation; Global Warming

▲ Projects and People

The **Coastal Scrub and Prairie Wildlife Inventory** was begun in mid-1991 at the Golden Gate National Recreation Area with the help of Earthwatch volunteers. In its first 8 weeks using various trapping and tracking methods, 25 species of terrestrial vertebrates were detected—including 17 mammals, 7 reptiles, and 1 amphibian. This study set the stage for more intensive inventories to research the effects of climate change and human activities on the diversity of vertebrates in this locality.

Biosphere Reserves, such as those in parks of the central California coast near San Francisco, CA, and D'Iroise near Brest, France, are examples of fragile terrestrial coastal ecosystems with similar offshore islands grouped in archipelago, grassland, heathland-shrub communities, and human use patterns. Here recreation and tourism have replaced agricultural uses and are impacting on the ecosystems; global warming may also pose a threat.

The two Reserves were recommended for comparative study sites to "examine the pattern and rates of ecosystem change after agricultural abandonment; examine the influences of recreational use following abandonment of these systems, and describe the mechanisms that influences the diversity of flora and fauna." A long-term goal is managing the biological diversity of these ecosystems while monitoring cultural uses. In the meantime, scientists share knowledge, experience, and technical expertise in computer science, remote sensing, and geographic information system (GIS) modeling.

The U.S. study site of plant and animal life is in the Marin Headlands District of Golden Gate National Recreation Area, Central California Coast Biosphere Reserve. The western France study site is the mer d'Iroise Biosphere Reserve, including the ile d' Ouessant. Similar methodologies of vegetation and wildlife sampling are being used by a team of plant and animal ecologists working with Earthwatch volunteers. **Dr. Judd A. Howell** and **Dr. Frederic Bioret** are chief scientists.

A recent **wildlife analysis** on **Alcatraz Island** in the midst of San Francisco Bay shows the need to provide alternative habitats when development reduces wetlands and other living sites. Dr. Howell describes Alcatraz as a "dynamic ecological system altered through its history by human use." For example, new horticultural species were introduced that differ from the original sparse mixture of grassland and coastal scrub. And increasing numbers of visitors—as many as 950,000 in 1990—impact sharply.

As Alcatraz is transformed from prison to park, disused island sections may be imperiled. These are important nesting and roosting areas especially for western gulls and black-crowned night herons, which are recognized as biological indicators of environmental poisoning. The deer mouse and the California slender salamander were also surveyed. And information compiled in databases, using

GIS modeling, revealed 108 bird species in a 10-year period but only 8 possible bird species as breeders and 9 vertebrates as breeders. Wildlife is especially vulnerable to human intrusion during breeding periods.

Environmentalists recommend that Alcatraz can be both managed for wildlife and still accommodate human use by identifying areas that can be restored to suitable habits, establishing shrub plantings for nesting, closing trails seasonally to keep nesters and potential bird colonizers from being disturbed, maintaining areas exclusively for wildlife, allowing visitors to observe wildlife without intruding into habitats, and educating the public about the value of island ecosystems.

U.S. Geological Survey ⊕
U.S. Department of Interior
345 Middlefield Road MS/975
Menlo Park, CA 94025 USA

Contact: **Howard G. Wilshire, Ph.D.**, Geologist
Phone: 1-415-329-4934 • *Fax:* 1-415-329-4936

Activities: Research • *Issues:* Energy; Geologic Processes; Global Warming; Health; Water Quality; Applications for Arid Lands • *Meetings:* USA–Japan Collaborative Research and Development

▲ Projects and People

Dr. Howard Wilshire recently worked on a report on surface geologic processes, which will provide information to policymakers about the importance of understanding these processes—all of which relate directly to global environmental issues.

With writer/producer **Doug Prose**, public television videos were made on recycling—*Reuse It or Lose It*—and threats to the natural integrity of the California deserts—*Desert Under Siege*. Proposed is a documentary on the environmental effects of the Mid-East Gulf War. Dr. Wilshire also documents the adverse effects of alternative energy sources, such as centralized wind- and solar-energy developments. He monitors the "desertifying" effects of road and utility corridor building, mining, grazing, deforestation, recreational use of off-road vehicles, military activities, agriculture, and urbanization in western U.S. arid lands. His goal is to create understanding of the magnitude of human impacts on arid lands that worsen the processes of desertification.

Working in controversial areas, Dr. Wilshire says he is hampered by lack of agency support and leadership that fails to recognize the importance of global environmental change. "My solution to this has been to do much of the work on my own, even follow-up of previously authorized studies," he says. "Since I do this on my own time and at my own expense, I feel free to advocate policies on the basis of my scientific findings."

He believes scientists and citizens ought to engage "both in dispassionate research and public advocacy"—an attitude denied by recent top levels of government, he says. Funding was turned down for studies of the combined effects of domestic animal grazing and deforestation in the western USA along with a proposal to produce public education videos on matters relating to global change/climate change.

Dr. Wilshire continues, "the opportunity to do good research is squelched by political appointees . . . in the guise of maintaining 'organizational credibility.'" Yet, he says, "products with scientifically questionable contents are encouraged if they fit a perceived political demand." As examples, Dr. Wilshire cites areas designated for mineral resource potential assessment reports because the may be targets of withdrawal for species preservation—such as the northern spotted owl—or for wilderness or national park status—such as the East Mojave National Scenic Area.

He writes that ". . . there is an enormous storehouse of scientific information relevant to environmental issues held by the federal government. This information, and the expertise of government scientists and engineers, needs to be made accessible to the public independently of the whims of changing administrations. . . . The presumption should be that scientific testimony is not for or against any party to a dispute, but rather represents an objective statement of facts and a dispassionate interpretation of those facts. When such changes are made, federal earth science can be coordinated with other approaches to addressing global environmental issues."

"The historical failure of nations that did not understand the consequences of their land-use practices can leave little doubt that land resources, as basic elements of the global system, profoundly affect the lives of every person on Earth," cautions Dr. Wilshire. "Earth science can and should play a central role in developing strategies for dealing with global environmental issues, and, in particular, federal earth science needs to be taken more aggressively to the public."

A fellow of the Geological Society of America and the American Association for the Advancement of Science, Dr. Wilshire was a consultant to the President's Council on Environmental Quality.

■ Resources

Dr. Wilshire stresses that greater political support is needed, such as encouragement from the U.S. Congress to bridge communications between scientists and the public. His hope is to "eliminate or minimize bureaucratic interference . . . and to lessen fears of scientists that going beyond simple publication of their (studies) will not jeopardize their careers."

Also needed, says Dr. Wilshire, is political support that will modify conflict-of-interest laws, which encourage government agencies to treat scientific information as proprietary to the government, "effectively denying public access to information for which it paid." Dr. Wilshire says this misuse frequently occurs in the environmental and health areas. Modifying the law would allow "public access to public testimony of government scientists in issues not a threat to national security."

Funding is a priority for the production of high-quality educational videos regarding global environmental matters. "Outside pressure is needed to encourage the U.S. Geological Survey to provide effective communication of earth science to the public as well as to policy makers," says Dr. Wilshire.

Available resources include three videos from GLOBAL VISION, 3790 El Camino Real, Suite 221, Palo Alto, CA 94306: *Desert Under Siege* about fragile ecosystems and human pressures; *Reuse It or Lose It* about municipal waste recycling; and *When the Bay Quakes* about San Francisco area earthquakes. Articles and maps on desertification and environmental effects of recreational and military vehicles can be obtained through the U.S. Geological Survey.

U.S. Man and the Biosphere Program (MAB) ⊕
OES/EGC/MAB
U.S. Department of State
Room 608 SA 37
Washington, DC 20522-3706 USA

Contact: **Roger E. Soles, Ph.D.,** Executive Director
Phone: 1-703-235-2946 • *Fax:* 1-703-235-3002

Activities: Research • *Issues:* Biodiversity/Species Preservation; Deforestation; Global Warming; Sustainable Development

● Organization

The **U.S. Man and Biosphere Program** describes itself as fostering "harmonious relationships between humans and the biosphere through an international program of policy-relevant research which integrates the social, physical, and biological sciences to address actual problems."

Housed at the U.S. Department of State, U.S. MAB is also supported by: Department of Agriculture (USDA)—Forest Service, Department of Energy (DOE), Department of Interior—National Park Service, Agency for International Development (USAID), Environmental Protection Agency (EPA), National Aeronautics and Space Administration (NASA), National Oceanic and Atmospheric Administration (NOAA), National Science Foundation (NSF), Peace Corps, Smithsonian Institution, and private-sector scientists.

U.S. MAB is a participant in the Man and Biosphere Program (MAB) that was set up in 1970 by the United Nations Educational, Scientific, and Cultural Organization (UNESCO). MAB brings 110 nations together to link natural and social sciences in conserving the Earth's biosphere of living organisms and improving the relationship between humans and the environment. Some 285 biosphere reserves are engaged in preserving ecosystems and species and managing land, water, and resources for sustainable development.

Directorates administrator U.S. MAB projects, and the directorate chairs make up the U.S. National Committee along with members from U.S. universities and private-sector groups. The State Department appoints the chairman; the Secretariat operates within State's Office of Global Change, Bureau of Oceans and International Environmental and Scientific Affairs.

▲ Projects and People

Project missions of the newest directorates emphasize the following:
• **High latitude ecosystems.** Here in Northern Hemisphere areas of permafrost and undeveloped lands where indigenous people practiced a subsistence lifestyle until recently—global warming, growing populations, and expanding resource use can have great impact. Activities and research feature managing sustainable resources, monitoring for global climate change, maintaining aquatic areas and wetlands, protecting biological diversity, and cooperating to recover lost or damaged ecosystems.
• **Human dominated systems.** Population growth is altering ecosystems that affect food production and resource extraction and cause air pollution in cities, soil degradation and tropical deforestation, and loss of beaches and coastal areas. This Directorate is quantifying the "intensity of activity than can be supported without causing the collapse of (essential) life support systems."

Activities also assess the role of population size and its effect on the environment, analyze causes of the greenhouse effect and costs to control; identify health and indirect effects of groundwater pollution, chemical runoff, and climate change; and define ways to integrate ecosystem management in human decision-making.
• **Marine and coastal ecosystems.** Because the oceans regulate this planet's ecological balance and moderate both world climate conditions and local weather patterns, ocean currents can carry pollutants that impair natural habitats across the globe. Damage is most apparent along coastal waters where population centers, industrial activities, and river inputs mix. This Directorate encourages activities to restore environmental quality for marine and coastal ecosystems including fisheries, enhance human welfare, and contribute to economic growth. Examined are sources and controls of marine pollution such as nutrient loading and eutrophication, siltation, and linkage to freshwater resources; sea level rise and coastal erosion; habitat loss and biological diversity; and red tides, harmful algal blooms, and results of intermingling between natural occurrences and human action.
• **Temperate ecosystems.** This zone is where two-thirds of the world's population lives and where most pollution and resource consumption occurs. Ecological conditions range from forests in wet regions to grasslands and deserts in dry regions and include aquatic ecosystems such as lakes, rivers, and freshwater wetlands. Research and activities focus on developing and applying environmental management practices that boost commodity production, protect biological diversity, adopt soil conservation measures in arid and semiarid regions, and adapt human populations to economically marginal or increasingly hazardous environments.
• **Tropical ecosystems.** Activities are undertaken to protect the planet's biological diversity and "biogeochemical cycles" while stemming losses of "precious genetic material," resources, and soil fertility. Focus is on restoring tropical forests, landscapes, and wetlands; improving communication among natural and social scientists working on tropical ecosystem conservation; and generating databases with solutions to tropical resource management.

Dr. Thomas E. Lovejoy (Smithsonian Institution) heads the U.S. National Committee. Members include executive director **Roger E. Soles** and **E.U. Curtis Bohlen** (Department of State), **Erich W. Bretthauer** (Environmental Protection Agency), **Dr. James M. Broadus** (Woods Hole Oceanographic Institution), **Dr. Stephen B. Brush** (University of California, Davis) **Dr. David Challinor** (Smithsonian Institution), **Dr. Arthur W. Cooper** (North Carolina State University), **Paul Coverdell** (Peace Corps), **Dr. Sylvia Earle** (National Oceanographic and Atmospheric Administration), **Dr. F. Eugene Hester** (Department of the Interior, National Park Service), **Dr. Michael A. Little** (State University of New York, Binghamton), **Dr. Ariel Lugo** and **Dr. Eldon W. Ross** (Department of Agriculture, Forest Service), **Dr. Helen C. McCammon** (Department of Energy), **Samuel McKee** (former U.S. MAB chairman), **Dr. Roberta B. Miller** (National Science Foundation), **Dr. Robert J. Naiman** (University of Washington), **Dr. Paul G. Risser** (University of New Mexico), **Dr. Shelby G. Tilford** (National Aeronautics and Space Administration), **Jack Vanderryn** (Agency for International Development), **Dr. Patrick J. Webber** (W.K. Kellogg Biological Station, Michigan State University), **Dr. Patricia A. Werner** (National Science Foundation), **Dr. Gilbert F. White** (University of Colorado), and **Dr. Robert G. Woodmansee** (Colorado State University).

■ Resources

The *U.S. MAB Bulletin*, published quarterly, is available from the U.S. MAB Secretariat, OES/EGC/MAB, Department of State, Washington, DC 20522-0508, USA, 1-202-632-2816 or 1-202-632-2786. Information on the international MAB program can be obtained from the MAB Secretariat, Division of Ecological Services, UNESCO, 7, place de Fontenoy, 75700 Paris, France.

Corvallis Environmental Research Laboratory
U.S. Environmental Protection Agency (EPA)
200 Southwest 35th Street
Corvallis, OR 97333 USA

Contact: **Robert T. Lackey,** Director
Phone: 1-503-757-4600 • *Fax:* 1-503-757-4799

Activities: Development; Research • *Issues:* Air Quality/Emission Control; Biodiversity/Species Preservation; Deforestation; Energy; Global Warming; Sustainable Development; Waste Management/Recycling; Water Quality

● Organization

The **Environmental Research Laboratory—Corvallis** focuses on terrestrial and watershed ecology; the ecological effects of global climate change; stratospheric ozone depletion; atmospheric pollutants, including acid rain; sustainable use of terrestrial ecosystems and loss of biological diversity; wetland and riparian ecosystems; ecological consequences of the use of genetically engineered organisms; effects of toxic chemicals on wildlife and terrestrial resources; and landscape and regional ecology.

The laboratory conducts research on the ecological effects of pollutants as they move through the air, soil, and water into the food chain. Research is targeted toward developing scientific information for regulations developed by the Environmental Protection Agency in response to existing legislation, international agreements, and treaties.

▲ Projects and People

Research is organized into the following three branches, with a total of approximately 350 employees:
- The **Terrestrial Branch** conducts research on effects of pollutants and other anthropogenic stressors on terrestrial ecosystems, such as agroecosystems, forested ecosystems, and rangelands, and on the effects of acidic deposition on soils, crops, and forests, and loss of biological diversity.
- The **Watershed Branch** conducts research on the effects of anthropogenically induced pollutants and other stressors on wetlands, lakes, streams, and their associated forests and watersheds. These include major fish populations in lakes and streams, criteria for maintenance of wetlands water quality and biota, cumulative losses and mitigation effectiveness for wetlands and riparian habitats, arctic tundra biomes, and the effectiveness of site-specific lake restoration practices relative to overall regional water quality.

- The **Ecotoxicology Branch** conducts research on effects of toxic chemicals and introduced organisms, particularly in terrestrial environments, which include wildlife, plants, and soil systems.

Microbial Ecology and Biotechnology Branch
Environmental Research Laboratory
U.S. Environmental Protection Agency (EPA)
Sabine Island, Building 7
Gulf Breeze, FL 32561 USA

Contact: **Parmely H. Pritchard, Ph.D.,** Research Microbiologist/Branch Chief
Phone: 1-904-934-9260 • *Fax:* 1-904-934-9201

Activities: Research • *Issues:* Bioremediation; Global Warming; Waste Management/Recycling; Water Quality

▲ Projects and People

The **Microbial Ecology and Biotechnology Branch** conducts research to determine the ways in which microbial communities interact to degrade toxic organic chemicals. Laboratory systems containing natural media and indigenous microorganisms attempt to simulate the delicate interactions of microorganisms as a function of natural ecosystems in order to compare the fate of chemicals in laboratory systems with natural systems.

Researchers also develop methods and data to assess the ecological risk of biotechnology. By applying molecular genetic techniques to ecological studies, they address questions on survival and growth of novel microorganisms.

Dr. Parmely H. Pritchard, branch chief, did postdoctoral work at the Woods Hole Oceanographic Institution. His primary research interests are the interactions of aquatic microbial communities in the degradation of toxic organic chemicals, the assessment of the validity of laboratory data in microcosms for predicting exposure concentrations in aquatic ecosystems, and the assessment of risk in the release of genetically engineered microorganisms to the environment.

Great Lakes National Program Office (GLNPO)
U.S. Environmental Protection Agency (EPA)
77 West Jackson
Chicago, IL 60604 USA

Contact: **Mary S. Setnicar,** Chief, Environmental Planning Staff
Phone: 1-312-886-3851, 1-312-353-2018

Activities: Interoffice, Interagency, International Coordination; Policy Development; Program Planning and Evaluation *Issues:* Biodiversity/Species Preservation; Health • *Meetings:* International Joint Commission, biennial; Soviet (former) Exchange Program

● Organization

The U.S. Environmental Protection Agency's Great Lakes National Program Office (GLNPO) was established in 1978 to focus attention on the natural resources of the Great Lakes. Most of GLNPO's functions are related to the United States–Canada Great Lakes Water Quality Agreement, as supported by the Clean Water Act of 1987, and to making progress in meeting its terms. GLNPO operates with a staff of 13.

GLNPO is mandated to "restore and maintain the chemical, physical, and biological integrity of the waters of the Great Lakes Basin Ecosystem." This means ensuring that the Great Lakes are safe for drinking as well as for swimming and other recreational activities, that biological populations are healthy, and that Great Lakes fish are safe to eat.

▲ Projects and People

The **Green Bay Mass Balance Study** is a pilot project to evaluate the feasibility of mass balance modeling for toxic substances as a basic planning and management tool in restoring Great Lakes water quality. If proven, the methodologies employed in the Green Bay model will offer an accurate process for pollution control and remedial action plans.

The **Great Lakes Fish Monitoring Program** is designed to support decisions on potential human exposure to pollutants and to provide indicators of the health of the Great Lakes ecosystem. The program monitors the presence of toxic contaminants in fish. Species are selected to represent a range of ages, locations, geographic dispersal, and sensitivity to pollutants.

The **Assessment and Remediation of Contaminated Sediments (ARCS)** program is a five-year study and demonstration program prompted by a growing awareness of the importance of sediment as a source for persistent toxic pollutants to the Great Lakes food chain, which includes a wide range of wildlife as well as people. Five Great Lakes areas of concern were chosen as demonstration sites for the project: Buffalo River, New York; Saginaw River, Michigan; Indiana Harbor Canal/Grand Calumet River, Indiana; Ashtabula River, Ohio; and Sheboygan River, Wisconsin.

■ Resources

GLNPO maintains a technical library of material pertaining mostly to the Great Lakes, their tributaries, and the surrounding watershed. Library materials are available to the public via interlibrary loan. Private citizens may visit the library by appointment. The library also provides a computer assisted literature search service. For information call the librarian: 1-312-353-7932.

Gulf Breeze Environmental Research Laboratory
U.S. Environmental Protection Agency (EPA)
Sabine Island
Gulf Breeze, FL 32561 USA

Contact: **William S. Fisher, Ph.D.,** Chief,
　　　　　Pathobiology Branch
　Phone: 1-904-934-9394 • *Fax:* 1-904-934-9201

Activities: Research • *Issues:* Estuarine and Marine Animal Health; Water Quality

● Organization

The **Pathology Branch** of the U.S. Environmental Protection Agency's (EPA) Office of Research and Development (ORD) and Office of Environmental Processes and Effects Research (OEPER) "conducts research on diseases of aquatic organisms caused or enhanced by chemical pollutants and biological agents in order to understand, predict and reverse their effects."

Following EPA's mandate to "provide research and monitoring support for regulation of effects of man's activities on the health status of flora and fauna in marine and estuarine ecosystems," the ORD issued an **Action Plan for the Implementation of the National Coastal and Marine Policy.** One of the stated goals is: "Greater understanding of the effects of pollution on complex coastal and marine ecosystems by expanding scientific research and monitoring programs, and the development of new technology."

▲ Projects and People

The Pathology Branch is attempting to meet this goal by conducting research on the "complex interactions of biological and environmental factors on a variety of aquatic systems." To accomplish this, the branch is continuously developing and evaluating methods for determining and predicting effects of biological and chemical agents. Objectives are to identify causes of disease and characterize host/disease agent interactions at molecular, cellular, tissue, and organismal levels; assess the impact of anthropogenic chemicals and other stressors on natural and introduced diseases in aquatic plants and animals; determine modes of action of toxic chemical agents in aquatic species and define endpoints to improve ecological risk assessments; establish realistic endpoints to assess aquatic animals as bioindicators of sublethal stress in the field; develop aquatic species as carcinogenesis models and potential indicators of environmental cancer risk; develop and employ state-of-the-art technology in pathobiology studies; and establish a Center for Marine and Estuarine Disease.

Dr. William S. Fisher is currently focusing his research on "the influence of natural environmental fluctuations and anthropogenic substances on the dynamic interactions of oyster immune mechanisms with oyster parasites and human bacterial pathogens transmitted by oysters." His other areas of expertise include crustacean pathology and aquaculture. He has written extensively on issues related to marine biology and serves on the editorial board of the *Journal of Invertebrate Pathology.* He is also a frequent reviewer of Sea Grant Programs at Texas A&M University and the University of Maryland.

Health Effects Research Laboratory (HERL) ⊕
U.S. Environmental Protection Agency (EPA),
　MD-58
UNC Campus Box 7315
Chapel Hill, NC 27599-7315 USA

Contact: **Timothy Gerrity, Ph.D.,** Chief, Clinical
　　　　　Research Branch
　Phone: 1-919-966-6202 • *Fax:* 1-919-966-6212

Activities: Research • *Issues:* Health

● Organization

The research staff of the Clinical Research Branch (CRB), under the direction of physiologist Dr. Timothy Gerrity, has expertise in the areas of physiology, epidemiology, physics, biophysics, neurotoxicology, virology, immunology, and cell and molecular biology. Collaborating with investigators in the Pulmonary Toxicology Branch, the staff has developed problem-oriented collaborative research projects, such as those described below.

▲ Projects and People

The CRB has conducted extensive ozone research since the last revision of the Criteria Document for Ozone in 1979. At that time, little health data were available to support the establishment of a National Ambient Air Quality Standard (NAAQS) for ozone. EPA established a one-hour standard at 0.12 ozone that was heavily dependent on margin of safety. EPA investigators, primarily in the CRB, designed, conducted and reported a series of experiments that today serve as the underpinnings of the database supporting a one-hour standard for ozone. Dr. Donald H. Hortsman and Dr. William Foskey McDonnell III received the EPA Silver Medal for their work on the one-hour ozone standard.

In response to the lack of strong health data in the Criteria Document for carbon monoxide, EPA investigators proposed to make physiological measurements of cardiac function that would provide more objective evidence of ventricular response to carbon monoxide. A research program was developed, using a digital gamma camera to measure the left ventricular ejection fraction, since there is evidence that, in patients with coronary artery disease, the ejection fraction changes measurably with the onset of angina, and this would provide a more objective measure of response in this patient population. The program has been conducted in collaboration with the Center for Environmental Medicine and Lung Biology of the University of North Carolina School of Medicine.

When the Criteria Document of Particulate Matter and Sulfur Oxides was being revised, CRB investigators were asked to evaluate the bronchoconstrictive response of asthmatics who were exposed to sulfur dioxide. Thus, Dr. Horstman, Dr. Lawrence J. Folinsbee, and others conceived, completed, and published a series of studies that addressed EPA's health research needs for sulfur dioxide.

Investigators from China and EPA are conducting multidisciplinary studies on indoor air pollution in China, specifically the effect of indoor coal combustion on lung cancer in Xuan Wei County. Interim results strongly suggest an etiological link between "smoky coal" use and lung cancer.

Dr. Susan Becker, a contract employee in the CRB Cell and Molecular Biology Section, is combining up-to-date technologies using recombinant growth factors to establish long-term cultures with human alveolar macrophages and pulmonary lymphocytes. These cultures can be used as models that, for the first time, allow investigators to examine the effects of *in vitro* exposures to these human pulmonary immune cells to environmental pollutants.

■ Resources

The EPA Health Effects Research Laboratory seeks ways to educate the public and all workers about the consequences of air pollution.

Environmental Toxicology Division
Health Effects Research Laboratory (HERL),
  ETD MD-66
U.S. Environmental Protection Agency (EPA)
Research Triangle Park, NC 27711 USA

Contact: Linda S. Birnbaum, Ph.D., Director
Phone: 1-919-541-2655 • *Fax:* 1-919-541-4324

Activities: Research • *Issues:* Air Quality/Emission Control; Health; Water Quality

▲ Projects and People

The Environmental Toxicology Division conducts research to determine the health effects of inhaled environmental pollutants. The research is designed to determine cause and effect relationships at pollutant concentrations which mimic those occurring in the environment. Particular emphasis is placed on the development and application of improved methods to enable advancement in the knowledge of the health effects of air pollutants.

The staff of 48, divided into multidisciplinary teams of scientists, is directed by Dr. Linda S. Birnbaum, whose expertise also extends to the areas of pharmacology and aging.

The Pulmonary Toxicology Branch investigates the inhalation of toxicology in environmental chemicals using laboratory animals. A major effort currently is underway to determine whether certain segments of the population might be more susceptible than others to air pollutants. Branch personnel investigate the health effects of a wide variety of gases and particles.

Within this branch, the Cardiopulmonary and Inhalation Exposure Section investigates the consequence of exposure to environmental pollutants on pulmonary and cardiovascular function. This section is also responsible for developing and monitoring state-of-the-art inhalation facilities in support of the branch's and the Laboratory's biological studies.

The Biochemical Toxicology Section conducts research to develop and evaluate quantitative biochemical markers for toxic chemical exposure, and to integrate these markers into current models of toxicity, adaptation, and repair.

The Immunotoxicology Branch investigates the effects of environmental chemicals on immune functions and on a variety of disease models. The goals of the research include the development of methods for assessing immune responses in the lung, improving methods for assessing the risk of increased disease, and the development of immunologic assays as biomarkers for allergenic reactions to inhaled and dermally applied compounds.

The Pharmacokinetics Branch focuses on developing a broad array of methods for studying the bioavailability, disposition kinetics, and metabolism of chemicals of interest to cross-media programs, including the volatile, semivolatile, and nonvolatile organics and inorganics.

Genetic Toxicology Division
Health Effects Research Laboratory (HERL)
U.S. Environmental Protection Agency (EPA), MD-68
Research Triangle Park, NC 27711 USA

Contact: **Michael D. Waters, Ph.D.**, Special Assistant to the Director
Phone: 1-919-541-2537 • *Fax:* 1-919-541-0694

Activities: Research • *Issues:* Health

● Organization
The genetic toxicology program of EPA's Health Effects Research Laboratory (HERL) explores the influences of environmental pollutants on genetic changes in somatic and germinal tissues (mutagenesis) and the conversion of normal cells to neoplastic cells (carcinogenesis).

Researchers analyze both direct and indirect interactions with genetic material for environmental chemicals, complex mixtures, and genetically engineered microorganisms. They perform under the direction of Dr. Michael D. Waters, former division head, who is also a professor of toxicology at the University of North Carolina at Chapel Hill.

The induction of cancer involves many potential mechanisms—such as gene or chromosomal mutation, changes in DNA transcription reflected in altered gene expression—and multiple stages in tumorigenesis. Researchers are focusing on understanding mechanisms of carcinogenesis and the factors that modulate neoplastic changes. Because certain substances seem to induce tumors without appearing to induce genotoxic effects, HERL must also develop methods and models for nongenotoxic carcinogens.

Monticello Ecological Research Station (MERS)
U.S. Environmental Protection Agency (EPA)
University of Minnesota
P.O. Box 500
Monticello, MN 55362 USA

Contact: **Michael C. Swift**, Research Associate/ Station Manager
Phone: 1-612-295-5145 • *Fax:* 1-612-270-2492

Activities: Education; Research • *Issues:* Waste Management/ Recycling; Water Quality • *Meetings:* INTECOL; Societas Internationalis Limnologiae

● Organization
The Monticello Ecological Research Station (MERS) is an aquatic biology field station operated jointly by the U.S. Environmental Protection Agency (EPA) and the University of Minnesota.

▲ Projects and People
MERS conducts chemical analyses, toxicity testing in ambient river or well water, and biological sample processing and identification.

The focus of this research is on eight outdoor experimental streams, each consisting of mud-bottom pools alternating with gravel riffles. The streams are supplied with ambient Mississippi River water which is considered unpolluted and classified for use as drinking water.

The mission of MERS is to develop an increased understanding of the effects of a variety of disturbances on experimental lotic ecosystems and to develop a predictive model of responses to disturbance.

Station manager **Michael C. Swift** operates through a steering committee to coordinate these efforts with a staff of six. Physical science technician **Kathleen Nordlie Allen** determines techniques for analyzing chemical toxicants in water, fish tissue, macrophytes, and sediments. **Roger O. Hermanutz**, a research aquatic biologist, designs and executes fish studies in MERS outdoor channels. **Thomas H. Roush**, an aquatic biologist, conducts research on the effect of pollutants on aquatic organisms.

■ Resources
Reprints of papers are available at no cost.

Neurotoxicology Division
U.S. Environmental Protection Agency (EPA)
MD-74B
Research Triangle Park, NC 27711 USA

Contact: **Hugh A. Tilson, Ph.D.**, Director
Phone: 1-919-541-2671 • *Fax:* 1-919-541-4849

Activities: Research • *Issues:* Health

● Organization
The Neurotoxicology Division plans, conducts, and evaluates a program to study the effects of physical and/or chemical agents on the nervous system. The objective is to provide the scientific basis and technological means to predict whether or not an environmental agent will produce neurotoxicity in humans—and to minimize the uncertainty of such predictions. The staff of 65 works in the following branches: Neurobehavioral Toxicology; Neurophysiological Toxicology; Cellular and Molecular; and Systems Development.

The division is headed by **Hugh A. Tilson**, who has also served as a consultant to the International Programme on Chemical Safety of the World Health Organization (WHO) and previously directed a behavioral toxicology and neurobiology laboratory at the National Institute of Environmental Health Sciences (NIEHS) in Research Triangle Park.

USEPA-ORP—National Air and Radiation
Environmental Laboratory (NAREL)
U.S. Environmental Protection Agency (EPA)
1504 Avenue A
Montgomery, AL 36115-2601 USA

Contact: **Edwin L. Sensintaffar**, Chief,
            Environmental Studies Branch
  Phone: 1-205-270-3420 • *Fax:* 1-205-270-3454
  E-Mail: EPA-6669

Activities: Development; Research • *Issues:* Air Quality/Emission Control; Global Warming; Water Quality • *Meetings:* Health Physics Society; International Symposium on Radon and Radon Reduction Technology; Natural Radiation Environment

● Organization

The **National Air and Radiation Environmental Laboratory (NAREL)** is a comprehensive environmental radiation laboratory managed by the U.S. Environmental Protection Agency and providing services to other EPA offices, federal and state agencies, and, in some cases, the private sector.

Edwin L. Sensintaffar directs a group of approximately 10 health physicists and engineers in environmental radiation measurements and analysis. He has been involved since 1975 in the direction of EPA laboratory and field studies of nuclear facilities or natural radiation exposure to the public.

▲ Projects and People

In the **Radon Program,** the NAREL provides analytical support to surveys of radon concentrations in homes, schools, and other public buildings across the United States. Measurements are made using a charcoal radon collector developed by the NAREL.

In its **National Environmental Monitoring Program** the NAREL publishes data collected by the Environmental Radiation Ambient Monitoring System (ERAMS). ERAMS is a national network of monitoring stations that regularly collect air, water, precipitation, and milk samples for analysis of radiation. ERAMS also tracks environmental releases of radioactivity from nuclear weapons tests and nuclear accidents.

In its **Superfund and Federal Facility Cleanup Program,** the NAREL assesses radiative contamination at Superfund and Federal facility sites and demonstrates remedial technology for minimizing the amount of radioactively contaminated soil that will ultimately require permanent disposal.

NAREL chemists work in the **Radiochemical Analysis Program** to analyze for radionuclides about 10,000 samples of air, water, soil, vegetation, human tissue, and food each year.

Using **Computer Dose Models,** the NAREL assesses the dispersion of radioactivity into the environment and the resulting risks to the public by measuring real-life disasters, such as the Chernobyl accident in the former Soviet Union.

■ Resources

The NAREL publishes numerous free publications about radon.

California

🏛 *State Government*

Chaffee Zoological Gardens
894 West Belmont Avenue
Fresno, CA 93728 USA

Contact: **Sean McKeown**, Curator of Reptiles
  Phone: 1-209-498-4860

Activities: Education; Research • *Issues:* Biodiversity/Species Preservation • *Meetings:* IUCN/SSC CBSG (Herpetology)

▲ Projects and People

Sean McKeown specializes in the management and breeding of reptiles and amphibians in captivity and has written extensively on the subject. He has a special interest in island ecosystems, particularly in the mid-Pacific and South Pacific and in the islands of the Indian Ocean.

McKeown's photographs of reptiles and amphibians have appeared in both popular and technical books and articles, and are utilized by the U.S. Forest Service, several universities and museums, the Chaffee Zoological Gardens, and the Honolulu Zoo Education Departments. He is a member of the Species Survival Commission (IUCN/SSC) Tortoise and Freshwater Turtle Group.

■ Resources

Publications are available upon request.

Colorado

🏛 *State Government*

Division of Wildlife
Colorado Department of Natural Resources
6060 Broadway
Denver, CO 80216 USA

Contact: **Ed Kochman**, Nongame Aquatic Program
  Phone: 1-303-291-7356

Activities: Education; Law Enforcement; Research • *Issues:* Biodiversity/Species Preservation; Wildlife Management

▲ Projects and People

Colorado's Division of Wildlife, Department of Natural Resources, initiated its nongame management effort in 1972. The program's goals are to reverse the decline of threatened and endangered species, to prevent other species from becoming threatened or endangered, and to encourage nonconsumptive (recreational/educational) uses of wildlife. Funds for the program are contributed through a Colorado income tax law that allows state residents to designate on their tax forms any amount they choose for the protection of nongame species. While the Colorado law was the first, similar laws now exist in 30 other states.

This **tax return checkoff** and individual contributions largely fund efforts to monitor and protect river otters, peregrine falcons, greenback cutthroat trout, greater prairie chickens, bald eagles, "big river" fishes, and other endangered species.

The Division has also undertaken several nonconsumptive use projects, including a **nature center** at Barr Lake (an isolated riparian zone on the eastern plains) and a series of **self-guided public tours** in North Park for the purpose of observing the courtship displays of sage grouse during the spring breeding months.

■ Resources

Publications include *Wild in the Streets* (an urban wildlife booklet), *Wildlife in Danger* (a status report on Colorado's threatened or endangered species), and *Amphibians and Reptiles in Colorado*.

Needed are political support and public relations.

Florida

🏛 *State Government*

Office of Protected Species Management
Division of Marine Resources
Florida Department of Natural Resources
3900 Commonwealth Boulevard
Room 321 MS 245
Tallahassee, FL 32399-3000 USA

Contact: **Patrick M. Rose**, Environmental
Administrator
Phone: 1-904-922-4330 • *Fax:* 1-904-922-4338

Activities: Education; Research • *Issues:* Biodiversity/Species Preservation

● Organization

The **Office of Protected Species Management**, Florida Department of Natural Resources, is responsible for the planning and implementation of management activities directed toward the protection of manatees, other marine mammals, and sea turtles, including their essential habitats. The Office also serves as the Department's liaison with federal, state, and local governments.

▲ Projects and People

Patrick M. Rose, environmental administrator, is responsible for all aspects of manatee, marine mammal, and sea turtle management activities, including habitat protection, permit review, and long-range planning. Rose belongs to the Species Survival Commission (IUCN/SSC) Sirenia Group.

The main research facility is the **Florida Marine Research Institute**, located at Bayboro Harbor, south of downtown St. Petersburg. The Institute is charged by a state statute to "conduct research necessary to develop marine resource management options, plans, and rules." Research includes studies on marine fisheries, marine

ecology, endangered and threatened species, and marine resource enhancement. **Dr. Karen Steidinger** is chief of marine research.

Research at the Institute encompasses studies in the following broad and interrelated program areas: **Fishery stock assessment studies** are done to determine population abundance, migration, and dispersal patterns, and monitor abundance and recruitment of juvenile and subadult fish and invertebrates in coastal and inshore areas. **Fisheries statistics studies** provide recreational and commercial fisheries data by species, area, gear, effort, and user. **Life history studies** help identify developmental stages of certain fishes and invertebrates, as well as spawning areas, age of reproduction and entry into a fishery, sex ratios, nursery areas, and feeding patterns. **Coastal hydrography and red tide studies** are performed to evaluate coastal processes such as hydrographical and biological features using satellite remote sensing, to forecast recruitment of larval fishes into nearshore waters, and to predict the occurrence of noxious plankton blooms. **Culture and rearing of marine animals studies** involve artificially spawning and rearing selected species for life history data to refine culture technologies and assess the cost/benefits of stocking selected estuarine species, particularly fishes, into Florida's estuaries and nearshore waters. **Benthic community studies** document assemblages of organisms and distribution patterns and provide data for an assessment of human influence on Florida's coral reefs. **Endangered and threatened species recovery studies** are conducted to determine life history, population abundance, movement or migration patterns, and causes of animal mortalities. **Marine animal health and contamination assessment studies** provide a reference collection of normal and diseased tissue to document and describe disease events, and to analyze animal disease in closed systems.

■ Resources

Staff scientists from the Institute serve as graduate advisors or guest lecturers at area universities and colleges. Staff members also serve on many international, national, regional, and state councils, committees, boards, and panels. Visiting scientists and training programs are being considered as part of the Institute's expansion.

Endangered and Threatened Species Management and Conservation Plan
Florida Game and Fresh Water Fish Commission (GFC)
620 South Meridian
Tallahassee, FL 32399 USA

Contact: **Don A. Wood**, Coordinator
Phone: 1-904-488-3831 • *Fax:* 1-904-488-6988

Activities: Development; Education; Law Enforcement; Political/Legislative; Research • *Issues:* Biodiversity/Species Preservation

● Organization

Since 1978, an update on the **Florida Endangered and Threatened Species Management and Conservation Plan** is presented to the legislature as required by state law. Coordination involves monitoring, facilitating, and correlating endangered species projects and studies; meeting federal and state requirements; and interacting with government agencies, conservation organizations, and others. Funding for coordination is 75 percent from the U.S. Fish and Wildlife Service and 25 percent from the Nongame Wildlife Trust Fund.

▲ Projects and People

A bald eagle reestablishment project is an ongoing cooperative state and governmental venture using Florida's eagle population as a donor source to bring back the species throughout the southeast USA. It is proven that when eggs are taken from Florida eagle nests early in the nesting season (December) for captive hatching, the adult eagles will lay new eggs; thus the Florida population is not adversely affected. Captive-hatched young can be successfully raised in captivity and then released into the wild at strategic sites in Alabama, Georgia, Mississippi, Oklahoma, and South Carolina. For example, 74 eggs collected from nests in December 1990 and transported to the George M. Sutton Avian Research Center, Oklahoma, produced 61 hatchlings for release in Alabama.

The U.S. Navy is funding a new management plan for marsh rabbits that inhabit Navy lands on the lower Keys marsh.

The Florida GFC and the Department of Natural Resources (DNR) are involved in important research projects on behalf of threatened and endangered species.

The crocodile research project is surveying the number, distribution, and reproductive status of the American crocodile in Florida through capturing, marking, and releasing. Of the likely several hundred, there are fewer than 20 known breeding females—all in the Everglades National Park and on northern Key Largo. Research biologist Paul Moler is principal investigator. Alligator populations are also being monitored statewide during annual night-light surveys in 17 wetlands areas, with Allan Woodward and Michael Delaney as project participants. Annual experimental alligator hunts at Orange, Lochloosa, and Newnans lakes are being conducted to determine impacts of sustained harvests. In a ranching study, wild alligator eggs are collected for hatchlings' captive rearing and harvest.

The panther is one of the USA's most endangered species; management strategies are being developed here based on habitat needs and the species status in Florida. Tracking with radio-telemetry devices shows that panthers prefer uplands over wetland forest cover; males sexually mature at age 3 and range 215 square miles, while females sexually mature at 1¹/₂ years, range 74 square miles, and are less likely to cross busy highways. Panther/deer/hog relationships are studied in the Big Cypress National Preserve. Sterilized mountain lions are being experimentally released and monitored as surrogate panthers in the Osceola National Forest area. Biomedical studies on panthers are also underway. Wildlife biologists include David Maehr, Bill Frankenberger, Chris Belden, Darrell Land, Jayde Roof, Walt McCown, and veterinarian Dr. Melody Roelke.

Snail kites are monitored each winter in southern Florida for movement trends and population numbers. Research biologist Jim Rodgers is in charge. Aerial brown pelican nesting surveys, undertaken every other year, confirm population stability. The 1990 aerial bald eagle nesting survey resulted in increasing numbers— 535 mated adult pairs, 366 of which successfully nested and produced 585 young. A new bald eagle project is assessing management guidlelines that establish primary and secondary protective zones around nests for up to one mile. Steve Nesbitt is investigator.

Habitats of gopher tortoises are being located and categorized to sustain introduced populations, with Joan Diemer as principal investigator. Reclaimed mining areas are being considered for properly managed habitats for gopher tortoises, now nearly depleted in certain areas by human overexploitation. Populations of the Sherman's fox squirrel and the threatened Big Cypress fox squirrel are being inventoried. John Wooding is principal investigator.

In nongame research, the white-crowned pigeon was the subject of a three-year ecological study relating its reproduction to the fruiting of tropical hardwood hammock trees. It was recommended that fruit sources be planted and maintained throughout the Keys to ensure the pigeons' survival. Florida scrub jays were relocated into suitable unoccupied habitats in a two-year experimental field study. In studies on the blackmouth shiner, it was determined that its quiet, backwaters habitats need immediate protection from encroaching development. The colorful Florida tree snail was found to reproduce partially by self-fertilization, "resulting in a high degree of genetic diversity among rather than within populations," writes Don Wood. Preservation will depend on "a large number of geographically isolated populations from throughout its range."

Ongoing studies include habitat evaluation of current and potential lands, within the Florida Key deer range, management of the Anastasia Island beach mouse and its protection from house mice, the reproductive biology of the Suwannee cooter turtle. GFC is waiting for grant approval to begin studies on the Florida gopher frog and striped newt, the Schaus' swallowtail butterfly, and the effects of fire on the endangered mint of the Florida Panhandle and the white birds-in-a-nest.

Wildlife Research Laboratory
Florida Game and Fresh Water Fish Commission (GFC)
4005 South Main Street
Gainesville, FL 32601-9009 USA

Contact: **Dennis N. David**, Wildlife Biologist
Phone: 1-904-336-2230 • *Fax:* 1-904-376-5359

Activities: Development; Law Enforcement; Research • *Issues:* Biodiversity/Species Preservation; Sustainable Development *Meetings:* IUCN/SSC

▲ Projects and People

Dennis David has taken the Statewide Alligator Management Program from its conception in the mid-1980s through its current full implementation. The goal of the program is to manage alligators on an optimum sustained-yield basis, recognizing these animals as an ecologically, aesthetically, and economically valuable natural resource. Developed with input from the public, the conservation community, and the crocodile industry, the program features regulations permitting the hunting of alligators for hides and meat and the collection of eggs and hatchlings for ranching alligators on both public waters and private lands.

The direction and focus of the program have been to base the establishment of harvest strategies and quotas on a sound biological basis, to encourage the conservation of the alligator and its wetland habitat, and to insist that a portion of the economic value of harvested alligators be directed to fund alligator management and research programs. According to David, "So long as this course is maintained, the program will continue to be a success and the alligator in Florida will be the benefactor."

The recovery of alligators from centuries of exploitation, along with a burgeoning human population in Florida, has resulted in an increasing number of alligator attacks. Dennis David is also responsible for the statewide coordination of the **Statewide Nuisance Alligator Program**, which has involved the development of procedures and regulations for dealing with 5,000 to 8,000 nuisance alligator complaints from the public annually, as well as a public education program to make people more aware of alligator habits. David belongs to the Species Survival Commission (IUCN/SSC) Crocodile Group.

■ Resources

A publications list featuring articles on alligator conservation and management is available.

Endangered Lands Acquisition Program— Volusia County

1005 North Dixie Freeway
New Smyrna Beach, FL 32168 USA

Contact: **Clay Henderson**, Attorney
  Phone: 1-904-427-2211 • *Fax:* 1-904-423-3334

Activities: Political/Legislative • *Issues:* Biodiversity/Species Preservation; Deforestation; Land Acquisition

▲ Projects and People

Florida local governments are leading the USA in acres purchased for open space protection. In recent years, 14 counties here have adopted such programs at more than $600 billion. Often they act in partnership with the state of Florida through **Preservation 2000**, a $300-million-a-year endeavor for 10 years beginning in 1990 that is described as America's "premier land acquisition program." Also, the newly formed **Florida Communities Trust** will make direct grants to local governments for land acquisition—another effort to help Florida communities lead a strong grassroots movement for this purpose.

Citizens pressured local governments to respond to Florida's rapid growth in this manner. The result is the **Local Government Comprehensive Planning and Land Development Regulation Act**, which imposes new requirements upon local governments to plan for conservation areas, open space, and recreation needs. Such land-use plans are enforced through "concurrency" and "land development regulations" to ensure that environmentally sensitive lands are so designated and land acquisition becomes part of the local government planning process.

A model for other local land acquisition programs, **Volusia County** in 1986 obtained 35,000 acres of environmentally sensitive lands to enhance a system of preserves totaling more than 150,000 acres. To keep the pro-active planning process free from political pressures, the Volusia County Council created a land acquisition selection committee, which adopted a manual of rules and began working on large project designs in cooperation with Preservation 2000 and the water management district. To save time, the committee meets monthly to review projects and contracts—pre-approving

appraisers and surveyors and doing tax planning in advance with the seller to bring down the ultimate purchase price. The Nature Conservancy also assists in structuring deals. "The county then acquires the property and negotiates with the water management district or the state for purchase by them," writes Clay Henderson. "The county has also used eminent domain and the threat of eminent domain. . . . Purchases cannot exceed appraised value. Through these techniques, local governments can acquire lands cheaper than the private sector."

The **Volusia Plan** is an illustration of the willingness of Florida voters to tax themselves to buy environmentally sensitive land. Other counties are following this example. Local land trusts are also expanding.

Volusia County councilman Clay Henderson is president of the Florida Land Trust Association and serves on Florida Communities Trust governing board. For his efforts, Henderson was honored by groups including The Nature Conservancy, Florida Environmental Regulators Association, Florida Audubon Society, Friends of Florida, and the Florida Parks and Recreation Association.

■ Resources

Volusia County's land acquisition plan and video are available.

Maryland

🏛 *State Government*

Maryland Environmental Trust (MET)

275 West Street, Suite 322
Annapolis, MD 21401 USA

Contact: **Thomas D. Saunders**, Director
  Phone: 1-301-974-5350 • *Fax:* 1-301-974-5340

Activities: Education • *Issues:* Biodiversity/Species Preservation; Deforestation; Land Conservation; Sustainable Development

▲ Projects and People

The Maryland Environmental Trust (MET) is cooperating with the Chesapeake Bay Foundation and other public, private, and academic organizations in its **land trust assistance program** to stimulate land formation and make it less complicated to protect them for open space, farmland, woodland, and other natural and historic areas. Land trusts are not-for-profit corporations or government-sponsored groups that protect land from development with the use of conservation easements, gifts, or land purchases, limited development, or promoting existing state easement programs. The Coastal Resources Division of Maryland's Department of Natural Resources provides funding.

Technical and legal assistance, training, funding, and cooperative land management services are provided—such as help with organizing a board of directors and showing how to operate such a group, set up conservation strategies, and apply for various grants. Land trusts can also participate in the **Maryland Land Trust Alliance**, an association of national, state, regional, and local land conservation organizations working there. One crucial local benefit of land trusts is protection of wetlands around the Chesapeake Bay to improve water quality.

The **conservation easements** program also helps preserve farmland, forestland, waterfront, rare or unique natural areas, scenic areas, endangered species habitat, historic properties, and other kinds of rural land. The easement document, the agreement between the landowner and MET, "makes it possible to merge environmental protection with substantial income, estate, and property tax benefits—while at the same time reserving for the landowner all rights of ownership, occupancy, and privacy," reports the Trust. Easements are one method whereby property owners, working together, can protect large areas in their communities without changes in local plans or zoning ordinances. MET is the country's third largest private easement holder by acreage.

The easements program assures permanent protection of natural resources and historic property, long-term monitoring, federal and state income tax deductions for the appraised value of the easement as a charitable gift, lower estate and inheritance taxes, timely processing of agreements, and a 15-year state and local property-tax credit on the unimproved land covered by a MET-held easement.

A third conservation measure is MET's **Rural Historic Village Protection Program**, which protects small towns and villages as well as the farmland, forests, and natural and historic open space that surround them with voluntary easements keeping housing subdivisions, strip commercial developments, or other industrial or commercial uses from disrupting the character of these communities. This is a cooperative venture with the Maryland Agricultural Land Preservation Foundation (MALPF), Maryland Historical Trust (MHT), Maryland Office of Planning (MOP), Chesapeake Bay Foundation (CBF), National Trust for Historic Preservation (NTHP), and Program Open Space, with funding from MHT, NTHP, and the J.M. Kaplan Fund, Inc.

Land-use pressures being exerted on Maryland are similar elsewhere in the USA. Here, "land is being converted from farmland to development at a rate that is about three times the rate of the state's population growth," writes former director H. Grant Dehart. Today, local governments are planning and zoning for "nearly 50 percent more land than was developed during the first 350 years of the state's settlement."

Ten historic rural villages are participating in this program. At Sharpsburg and the Antietam National Battlefield area, farmers and landowners are thrashing it out with historic preservation groups about who pays for the value of the land affected. "The lessons of the first year demonstrate the perils of top-down state or federal land preservation strategies, which would most likely fail because of the polarization of the local community and local suspicions about the motivations of governmental agencies," reports Dehart. "A collaborative local-state-private effort is likely to be the only way that the farmland around Sharpsburg and the Antietam Battlefield will be saved for future generations." Other areas include Barnesville, near Sugarloaf Mountain, a national natural landmark; Queenstown, near encroaching development on the Eastern Shore; Tunis Mills, surrounded by early waterfront plantations and rural farmlands; Claiborne, a waterfront fishing village; and seventeenth-century East New Market, now in the National Register of Historic Places.

Dehart shares pointers on other lessons learned as the village protection program got underway: Because each village community is unique, a variety of tools is needed to achieve conservation goals. Strong local leaders are essential and so are early efforts to preserve open space and forests before land is purchased for shopping centers or other developments. Property owners should be approached with

sensitivity and knowledge about their property. State or regional growth policies must exempt historic villages from concentrated development. Land preservation through easements cannot be clearly separated from land-use regulation. Local zoning can impact successfully on government programs designed to compensate landowners for *not* developing farmland. Participating communities should be honored and their pride bolstered. When conservation measures are planned, economic and social quality-of-life factors as well as physical and aesthetic factors should be considered.

■ Resources

Available publications are *Land Trust Assistance Manual, Conservation Easements*, and *Land Marks* newsletter.

MET seeks funding incentives for land conservation from the Maryland General Assembly and expanded media coverage to advance public knowledge of land conservation programs.

TREE-mendous Maryland
Maryland Department of Natural Resources
Tawes State Office Building
Annapolis, MD 21401 USA

Contact: **Wally Orlinsky,** Executive Director
Phone: 1-410-974-3776 • *Fax:* 1-410-974-5590

Activities: Education • *Issues:* Biodiversity/Species Preservation; Deforestation; Energy; Global Warming; Water Quality

▲ Projects and People

"We plant trees by the millions," says **Wally Orlinsky,** TREE-mendous Maryland executive director. In fact, in 1991, "36,000 people planted 1,260,000 trees and seedlings in 154 communities throughout the state." So far, 2,500,000 trees were planted in the program's initial 2-year period—with a government budget of $250,000 and a similar amount raised in matching funds.

Part of the program is modeled after the Jewish National Fund—giving Marylanders a chance to buy a tree or a grove in honor or memory of friends or family. Even without media advertising, some 12,000 persons have purchased trees for this purpose. Other trees are purchased for parks and public open spaces, with Maryland corporations and businesses involved in planting state highway clover leaves with trees—through the newly created **Clover Leaf Foundation.** The program also enlists the help of nurseries to promote the purchase of trees.

To augment TREE-mendous, Maryland's forestry division runs tree-planting seminars—developing a "cadre of some 400 tree planting supervisors to date," as U.S. Senator Paul Sarbanes reported in the Cong*ressional Record* (December 18, 1991). This effort recently expanded into **Maryland's Green Gorillas,** a "tree-care army of volunteer coordinators," who also receive planting materials and guides for their mission.

With "April as TREE-mendous month," school groups embark on planting as many as 10,000 seedlings in a day, such as in fields and along highways. Schools, community service organizations, scouts, and church groups can also take part in **Leaf by Leaf,** to encourage the planting of 40 trees at reduced costs. Species range from American hornbeam to Japanese dogwood, red maple to yellow poplar and green ash. Participants can also choose from disease-resistant liberty elm, white pine, crabapple, honeylocust, and pin oak, among others.

A main beneficiary is the **Chesapeake Bay**, as trees prevent soil erosion and, hence, sedimentation and pollutants in the nation's valuable estuary. TREE-mendous is the recipient of a Keep America Beautiful award for a government-sponsored environmental stewardship program.

🏛 *Local Government*

Farmland Preservation Programs

Montgomery County Office of Economic Development

101 Monroe Street, Suite 1500
Rockville, MD 20850 USA

Contact: **Timothy W. Warman**, Agricultural
             Program Manager
   Phone: 1-301-217-2358 • *Fax:* 1-301-217-2045

Activities: Farmland and Agricultural Industry Preservation
Issues: Farmland Preservation; Sustainable Development
Meetings: American Farmland Trust Conference; International Open Space Conference; Renew America Conference

▲ Projects and People

To compute the value of farmland preservation easements in most American localities, a real estate appraiser figures the full market value and the "agricultural value" and then determines the difference. Not so in Maryland's Montgomery County, where—with support from the local agricultural community and American Farmland Trust—a method was developed to value easements based on the relative importance and need to protect individual farms. Timothy W. Warman explains, "A formula was created that assigns points to factors such as size, soil productivity, road frontage, location, and ownership by an active farmer. The points have a dollar value which the county can adjust annually."

That is the basis of a county government program to target prime farms threatened with development and purchase farmland preservation easements. "The simplicity of the formula helps attract participation when farmers can see just what we are willing to pay." Warman recommends that "once a jurisdiction establishes priorities for preserving land, it is relatively easy to develop a formula that considers the factors that define the priority." Montgomery County purchases easements on about 1,000 acres of farmland annually.

To conserve farmland and foster the continuation of the agricultural industry, four preservation programs are in place: the county's Agricultural Easement Program (AEP); Maryland Agricultural Land Preservation Foundation (MALPF); Maryland Environmental Trust (MET) and other private trust organizations; and Transfer of Development Rights (TDR) Program. (*See also Maryland Environmental Trust in this section.*) MALPF purchases agricultural land preservation easements directly from the landowner for cash. Following sale of the easement, agricultural uses of the property are still permitted and are, in fact, encouraged. The TDR Program as part of an overall Master Plan for Preservation of Agricultural and Rural Open Space designates an agricultural reserve with rural density transfer zoning where preference is given to agriculture, forestry,

and open spaces, and some agriculturally related industry. Housing density is limited to one house per twenty-five acres on a minimum one-acre lot. Agricultural reserve properties also have transferable development rights (TDR's) at the rate of one per five acres which can be sold to developers who want to use them to construct houses in designated TDR receiving areas that allow greater density.

"The USA needs a national prime farmland protection policy," urges Warman. "Americans must be made aware of the importance of agricultural resources."

■ Resources

Publications include *Farmland Preservation Programs in Montgomery County*, *Preservation of Agriculture and Rural Open Space Master Plan*, and *Farmland Preservation Annual Report*.

Michigan

🏛 *Local Government*

Community Right to Know (CRTK)

Public Health Division, Environmental Health
  Bureau

Washtenaw County Services Department

4101 Washtenaw Avenue
P.O. Box 8645
Ann Arbor, MI 48107-8645 USA

Contact: **Robert Blake**, Director
   Phone: 1-313-971-7445 • *Fax:* 1-313-971-6947

Activities: Education; Law Enforcement • *Issues:* Health;
Waste Management/Recycling; Water Quality

▲ Projects and People

The **Community Right to Know (CRTK)** program grew from **Washtenaw County's** struggles to meet the requirements of Title III of the Superfund Amendments and Reauthorization Act (SARA)—as local and state governments are required to do by federal law. Facilities handling hazardous substances must reveal exceptional information about the types and quantities of chemicals they produce, use, and release into the environment. "This information must be reported to a Local Emergency Planning Committee (LEPC)," explains Robert Blake, "which then must formulate off-site response plans for each facility in its jurisdiction."

However, this community went a step beyond—by **combining disclosure measures with environmental protection laws.** Overcoming early obstacles, a regulation was eventually enacted that reduced the possibility of catastrophic releases of substances potentially lethal to surrounding populations; that would minimize the likelihood of chronic disease or acute injury from substances released into the air, groundwater, or surface water; and that would prevent adverse environmental affects by allowing emergency response personnel and the public access to information regarding chemicals in their communities. Thus, various environmental health programs were consolidated into one inspection package.

With broad community input, special reporting forms were designed, as were warning signs for workplace facilities which were effective but did not alarm the public. Local teams undertook hazardous materials training sessions of federal and state agencies. Emergency planning sessions with fire, police, elected officials, and coordinators were continuous.

Because of CRTK regulation, businesses must submit inventories of chemical substances located in their workplaces. In addition, all toxic or hazardous substances must be stored so as not to contaminate air, groundwater, and surface water and to prevent accidental discharge. Inspections in the first-year uncovered 272 out of 330 facilities without proper secondary containment. "An accidental spill at one of these facilities could have cost $50,000 to $100,000 to clean up," according to Blake. Findings included an estimated 10 years' accumulation of hazardous waste in 65 unmarked drums behind an occupied building, a refuse dumpster partially filled with flammable liquid waste and headed for a landfill, and a pool of flammable chemicals on the floor of a painting operation.

When state government agencies exempted themselves from county inspections, health department officials were hindered in their work. "Government entities can be severe polluters of the environment and should be subject to strict environmental controls," health inspectors believe. Some inspections at municipal facilities did turn up water and wastewater treatment plants without alarms or safety equipments and violations that could cause huge fish kills in a nearby river and other catastrophes.

Blake points to the benefits: Improvements are underway to reduce potential risks at facilities handling extremely hazardous substances. "Many facilities have installed secondary containment for chemicals and sealed off possible routes to the outside environment, while others have reduced their chemical inventories." Sensors for chemical releases and alarms for workplace and emergency response personnel are being installed. Less hazardous materials are sometimes substituted for more perilous ones. While environmental contamination is being cleaned up, air- and water-quality violations are referred to the appropriate agencies.

One unsettled matter relates to the filing and inspection fees paid by businesses which would prefer state government to bear these costs. County officials, however, believe that local needs will not be met through the "self-inspection philosophy of state environmental agencies."

■ Resources

Publications include reprints of *Journal of Environmental Health* "Right to Know" and Tufts University's *Center for Environmental Management* "Washtenaw County Community Right To Know Program"; and copies of Washtenaw County's CRTK Regulation and Status Sheet (Reporting Form).

New technologies to enable regulated facilities to send in computerized inventories are sought to reduce paperwork and to make data available to more interested parties.

Minnesota

🏛 *State Government*

Minnesota Natural Heritage Program
Minnesota Department of Natural Resources
500 Lafayette Road
St. Paul, MN 55155-4007 USA

Contact: **Robert Dana,** Coordinator
Phone: 1-612-297-2367 • *Fax:* 1-612-297-4961

Activities: Education; Research • *Issues:* Biodiversity/Species Preservation

● Organization

Minnesota is an "ecological crossroads," say environmentalists, so the Minnesota Natural Heritage Program originated an information system that locates geologic features, plants, animals, and natural communities needing special attention to allow citizens to "plan responsibly for future development and economic growth."

The eastern deciduous woodlands, western prairies, and northern conifer forests are home to many diverse and often rare plants and animals. But a growing population, complex technology, and dwindling wild lands are taking their toll. The Natural Heritage Program first identifies and locates those special plants and plant communities, animals, geologic features, aquatic environments, and other elements which are endangered, threatened, unique, or exemplary.

▲ Projects and People

Information is stored in a centralized, cost-effective, integrated data management system with map files, manual files, and computer files on 10,000 entries that is also compatible with the Minnesota Land Management Information System (MLMIS), an existing land-use planning database. Computer linkups are also planned with natural resources offices elsewhere. (In addition, The Nature Conservancy has developed a database network for the USA, eight South American countries, and two Canada provinces which provides a global perspective on managing endangered resources in this hemisphere.)

An essential planning key for public and private land purchases where the protection of endangered species and natural habitats are at stake, this cataloged information is also used to implement innovative land conservation programs, such as the **Reinvest in Minnesota Native Prairie Bank** and the **Prairie Landscape Reserve Bank.** Giving early notice of potential damage to biologically sensitive lands, it assists environmental impact reviews for development projects such as housing, highways, and utility corridors. The well-publicized program creates public awareness and encourages ecological research.

Among the natural assets it has helped save in conjunction with public land managers and special legislation are the sugar maple, basswood, and yellow birch forests of Moose Mountain Scientific and Natural Area; Magney Park; Yellow Birch Natural Area; Marble Lookout; and the mineral-rich wetlands, dry prairie, and the rare chestnut-collared longspur, Dakota skipper butterfly, and short-eared owls of Felton Prairie. Through CITES (Conference on the International Trade in Endangered Species of Fauna and Flora) and other laws, the exploited ginseng (*Panax quinquefolium*) is now protected. The Natural Heritage Program monitors and manages the mysterious ginseng along with the dwarf trout lily, prairie bush clover, grape fern, golden saxifrage, and other rare plants. Declared "state-threatened" are the Minnesota River granite outcrop community that harbors the rare pincushion cactus and five-lined skink, among others; northern hardwood-conifer forest; and the glacial till hill prairie.

Before the year 2000, the program intends to promote conservation at the landscape or watershed level, accelerate efforts to protect biological diversity, make its staff and information system more accessible to resource managers and the public, protect and manage more native plants before they become endangered, and complete the **Minnesota County Biological Survey (MCBS)**, began as a pilot project in 1987.

Information gathered from the county-by-county surveys is entered into the component databases and the **Geographic Information System (GIS)** of the Natural Heritage Program. Some early results include the discovery of new populations of the endangered bog bluegrass (*Poa paludigena*); protection of Falls Creek white pine and deciduous forests with Louisiana waterthrush, ginseng, red-shouldered hawk, and kitten-tails; and identification in six western counties of only 1.6 percent of the original prairie vegetation of the 1850s. The survey continues to find new locations of rare species, such as the bald eagle, upland sandpiper, snapping turtle, eastern and western hognose snakes, American water-pennywort, and lance-leaved violet; and imperiled natural communities, such as oak savannah, sand dune prairie, moist cliff, southern mesic pine-hardwood forest, black spruce bog, cattail marsh, and tamarack swamp.

Other inventories, such as **status surveys**, have put the prairie white-fringed orchid on the endangered species list.

On **relevé plot samples of vegetation**, biologists record the coverage of plant species and how they are distributed in vertical layers from canopy to ground; then the information is put into a **vegetation database** for managing.

■ Resources

Biological reports on natural heritage and nongame wildlife programs and brochures on the Natural Heritage Program and County Biological Surveys are available.

Conservation, Education, and Research ⊕
Minnesota Zoo
12101 Johnny Cake Ridge Road
Apple Valley, MN 55124 USA

Contact: **Ronald L. Tilson, Ph.D.,** Director of
Conservation
Phone: 1-612-431-9267 • *Fax:* 1-612-432-2757

Activities: Education; Research • *Issues:* Biodiversity/Species Preservation • *Meetings:* AAZPA; IUCN/SSC; IUCN/CBSG

● Organization

The **Conservation, Education, and Research** programs of the Minnesota Zoo encompass national and global conservation activities, including work on behalf of endangered species, with an operating budget of $1.1 million and 40 full-time and 24 seasonal employees.

This zoo hosts the IUCN's (World Conservation Union) **Species Survival Commission's (SSC) Captive Breeding Specialist Group**. It is also home to ISIS, the **International Species Information System**. International training is offered for conservationists in South America, India, and Southeast Asia.

With a National Science Foundation (NSF) grant, the Young Scholars Program—**Zooschool**—is conducted to encourage secondary student interest in conservation biology. **Teacher training workshops** in environmental education concentrate on vanishing animals, tropical forests, and conservation issues of the 1990s. Symposiums are offered on tropical forests, coral reefs and sharks, and Minnesota's prairie and marshes, in cooperation with the Bell Museum of Natural History. ZooArk is an educational outreach program that focuses on local and global conservation. "Steve Martin's World of Birds Show" features conservation education through entertainment. Public celebrations are held annually, such as Earth Day, Arbor Day, Conservation Day, and Southeast Asian Days. Supporting the zoo are the state of Minnesota, 14,000 members, and various grants such as from the Minnesota Zoo Foundation, Smithsonian Institution, and NSF.

▲ Projects and People

Ongoing conservation and education activities include coordinating the Siberian tiger breeding program of the American Association of Zoological Parks and Aquariums' (AAZPA) Tiger Species Survival Plan (SSP); and organizing the international symposium, "World Conservation Strategies for Tigers"; developing and coordinating the AAZPA Gibbon SSP; **Adopt-A-Park Program**, an *in situ* project to ensure ecological stability and save the world's last 50 Javan rhinoceroses in Ujung Kulon National Park, Java, Indonesia—in cooperation with the Indonesian Department of Nature Conservation; RELEAF Belize, an *in situ* student reforestation project; establishing North American regional **studbooks** for both **chimpanzee** and Nilgiri tahr; surveying rattlesnake population on Aruba Island; breeding and reintroducing the trumpeter swan—with the Minnesota Department of Natural Resources; and on-site eastern bluebird recovery project.

The first successful in *vitro* fertilization and embryo transfer with Siberian tigers (*Panthera tigris*) occurred here in cooperation with the National and Omaha zoos. Related activities are modeling neonatal growth for captive Siberian tiger cubs, hormonal events in male and female Siberian tigers, and scentmarking in Siberian tiger communication.

Other **research projects** include reproductive biology of Asian small-clawed otters (*Aonyx cinerea*), stallion behavior in a bachelor herd of Asian wild horses (*Equus przewalskii*), reproductive cycle and pregnancy in Malayan tapir (*Tapirus indicus*), communal suckling a herd of Bactrian camels (*Camelus bactrianus*), reproductive behavior in red panda (*Ailurus fulgens*), captive biology of an asocial ermine mustelid (*Mustela erminea*), aggressive behavior reduction in lion-tailed macaques (*Macaca silenus*), and first-time hatching and rearing Minnesota common loon (*Gavia immer*).

Headquartered here is ISIS, which provides global specimen and species catalogs and auxiliary information services for the purpose of preserving biotic diversity. Encompassing a network of 376 member zoos in 38 countries for recording standardized animal data, the electronic database includes 120,000 animals and another 300,000 of their ancestors. It is the only source of pedigree and demographic data for 97 percent of the species held by zoos and provides the basis for starting Species Survival Plans (SSPs) for the long-term survival of captive wildlife. ARKS and SPARKS are software programs developed here and used internationally for managing wildlife populations in captivity, as is MedARKS, a tool for specifically managing the medical health of captive wildlife.

Zoo research is linked with the IUCN/SSC Captive Breeding Specialist Group, which conducts **Population Viability Analyses** (PVA) to establish extinction risks for species and develop management methods to prevent extinction and win removal from endangered species lists. PVAs have been conducted for aridland antelope species, Asian wild horse, Bali mynah, black-footed ferret, Florida Key deer, Florida panther, golden lion tamarin, Mexican wolf, Puerto Rican parrot, and red wolf.

Dr. Ronald L. Tilson is also a member of the Species Survival Commission Asian Primate and Hyena groups.

■ Resources

The Minnesota Zoo publishes *Tiger Beat*, the Tiger SSP newsletter. A brochure is available on its wildlife conservation program.

Missouri

🏛 *Local Government*

St. Louis Zoological Park
1 Government Drive, Forest Park
St. Louis, MO 63139-1136 USA

Contact: R. Eric Miller, D.V.M., Associate
Veterinarian
Phone: 1-314-781-0900, ext. 277
Fax: 1-314-647-7969

Activities: Development; Education; Research • *Issues:* Biodiversity/Species Preservation; Waste Management/Recycling *Meetings:* AAZPA; American Association of Zoo Veterinarians; IUCN/SSC CBSG; IUDZG

● Organization

With 83 acres of animal exhibits amid bluffs, forests, and pampas, lecture and exhibit halls, and classrooms—the St. Louis Zoo shifted its emphasis in recent years to become a center both for propagation, conservation, and research and for general education, in addition to its traditional role as a family attraction in an aesthetic natural setting. The zoo is a subdistrict of the Metropolitan Zoological Park and Museum District that also includes the St. Louis Art Museum, St. Louis Science Center, and the Missouri Botanical Garden. It is supported by taxpayers of both the city and county of St. Louis.

▲ Projects and People

The St. Louis Zoo is committed to the Species Survival Programs (SSPs) of the American Association of Zoological Parks and Aquariums (AAZPA) in providing a refuge for healthy captive populations and encouraging breeding of rare and endangered species for possible release in the wild.

Although the Micronesian or Guam kingfisher and tuatara reptile are housed behind the scenes for breeding purposes, the species below can generally be viewed. Either protected by the U.S. Endangered Species Act and/or the Convention on International Trade of Endangered Species (CITES), not all of these species are SSP designated, however.

Mammals are black, ring-tailed, black-and-white ruffed, and red-ruffed lemurs; golden-headed lion and cottontop tamarins; lion-tailed macaque; siamang; orangutan; western lowland gorilla; chimpanzee; chinchilla; spectacled bear; red panda; Siberian tiger; snow and Amur leopards; cheetah; Asiatic elephant; Malayan tapir; black rhinoceros; pygmy hippopotamus; addax; bactrian camel; banteng; gaur or seladang; Eld's deer; Arabian oryx; Cuvier's gazelle; Grevy's zebra; and babirusa.

Dr. R. Eric Miller has coordinated a national veterinary research program evaluating diseases of captive black rhinoceroses (*Diceros bicornis*), including "studies into the causes of hemolytic anemia, fungal pneumonia, encephalomalacia, oral and skin ulcers, and the establishment of normal physiological values." Cooperative research institutions with the St. Louis Zoological Park have included Kansas State University, Mayo Clinic, National Animal Disease Center, National Zoological Park, New York State College of Veterinary Medicine, New York Zoological Society, University of California at Davis and Los Angeles, University of Miami, University of Missouri, University of Tennessee, Washington University of St. Louis, and the Zoological Society of San Diego.

Birds are Humboldt penguin; Aleutian Canada and ne-ne (Hawaiian) geese; white-winged wood duck; bald eagle; brown-eared and Elliot's pheasants; hooded crane; Mauritius pink pigeon; and Rothschild's or Bali mynah. The zoo strongly cautions visitors against purchasing wild birds as pets, as illegal importations combined with legal arrivals total more than one million birds in the USA annually and imperil parrots and other species. Rare birds are sometimes smuggled in spare tires and wrapped in tape or burlap and found dead on arrival. Habitat destruction also contributes to their demise. The zoo urges the purchase of captive-bred birds, such as budgerigars or parakeets, cockatiels, and canaries—which are generally tamer, less likely to carry diseases, less traumatized, and accustomed to being handled; consequently, these may live longer.

Reptiles are radiated and Galapogas tortoises; yellow-spotted side-necked and Coahuilan box turtle; American and dwarf crocodile; Chinese alligator; rhinoceros iguana; Jamaican and Brazilian boas; San Francisco garter snake; and Aruba Island rattlesnake. **Fish** are desert pupfish and Lake Victoria cichlids.

The overall "animal collection is an important resource for the practice of animal medicine and the study of animal nutrition," according to zoo literature. "The zoo cooperates with neighboring universities in animal behavior studies, a science of growing importance because of its relationship to human behavior."

The zoo's environmental education programs for children adults, and families focus on biodiversity, animal adaptations, and ecosystems such as tropical rainforests. To enhance school programs in Missouri, films, videocassettes, slide sets, and other teaching aids are available. The zoo's Library and Teacher Resource Center houses more than 2,000 reference books on zoology and conservation.

■ Resources

The Education Department provides a list of *Resource Materials for Educators*, including teacher's guides, audiovisuals and other loan material, informational leaflets, and kits on vanishing rainforests, mammals, and wildlife trade. The *Learning World* features ongoing zoo activities, such as exhibits and/or classes on bats, snakes, hoofed animals, and pygmy marmosets. The *What You Can Do* brochure explains how to recycle, conserve energy, and consume wisely.

New Hampshire

🏛 State Government

A Greener Workplace Project
New Hampshire Hospital (NHH)
105 Pleasant Street
Concord, NH 03301 USA

Contact: **Carolyn Mercer-McFadden, Ph.D.**
Phone: 1-603-271-5207 • *Fax:* 1-603-271-5395

Activities: Education; Research • *Issues:* Energy; Health; Transportation; Waste Management/Recycling • *Meetings:* UNEP: Global Assembly of Women and the Environment

▲ Projects and People

"We think of **A Greener Workplace** as something all employers should be doing, a moral imperative in the social leadership role of any corporation or social institution in the 1990s," writes **Dr. Carolyn Mercer-McFadden**. A Greener Workplace was established to save money and to advance the hospital's relationship with the environment. Now it is recognized as a model of comprehensive planning and action for corporate environmental responsibility.

At the state psychiatric hospital with 1,300 employees and patients in 9 facilities on 126 acres of open land in the center of New Hampshire's capital and an annual operating budget of $30 million, these project goals were set: **Reduce and recycle wastes** in all hospital operations. **Conserve energy** and use sustainable appropriate technologies throughout hospital operations. **Minimize toxic and hazardous materials** and protect people and the environment from risks of such materials. Train, educate, and inform employees and community members about conservation and environmental responsibility.

Specifically, the project targeted consumable supplies and disposals for waste reduction. In a year, they cut photocopying 70 percent and recycled 16 tons of office paper and 1.5 tons of aluminum paper. Washables are being substituted for disposables in dietary services and adult briefs. Workers were advised to "use electronic mail to send memos, instead of sending print-out memos on paper. Reuse manila folders and envelopes. Use printed stationery only for outside communications . . . instead of copying, route everything you can."

Conservation involved closing inefficient buildings and educating engineering staff and users on new energy management systems. Simple tips included turning off lights and equipment when not in use, dressing for levels of activity at work, letting the sun provide warmth and light when needed, and using carpools and public transportation.

NHH also recommends conservation steps an entire company can take: purchase in bulk to omit excess packaging; repair old furniture and equipment; buy materials and products made from recycled materials and that are energy efficient; get an energy audit; lower heating and raise air-conditioning temperatures; properly handle hazardous wastes; buy materials that are nonpollutants; sign onto the Valdez Principles; investigate the environmental records of vendors and other companies and suppliers; set long-term goals of environmental responsibility that begin today; and help the community in preservation measures.

■ Resources

Dr. Mercer-McFadden conducts workshops and presentations on motivating employees for waste reduction and recycling, targeting areas for waste reduction, analyzing the hospital wastestream, and preparing cost-benefit studies for product substitution. A *Handbook on Waste Reduction Campaigns in the Workplace* can be ordered.

New Jersey

🏛 Local Government

Berlin Township Public Works Department
170 Bate Avenue
West Berlin, NJ 08091 USA

Contact: **Mike McGee**, Director and Recycling Coordinator
Phone: 1-609-767-5052 • *Fax:* 1-609-767-4963

Activities: Public Works • *Issues:* Waste Management/Recycling • *Meetings:* League of Municipalities for State of New Jersey; NRC; Renew America Environmental Conference

● Organization

The Berlin Township Public Works Department—in a small suburban community of 1,800 households—began curbside collection of glass, newspaper, and cardboard in 1980. Mike McGee traces the history of this successful recycling program, which is a model for similar endeavors throughout the USA and Canada with its 60.4 percent recycling rate—the nation's highest.

Initially, a crew of 19 would "dig **glass** out of the hopper, or back of the trash truck, and fill burlap bags." A year later, scrap **metal, motor oil,** and car and truck **batteries** were added; a mandatory recycling ordinance was passed; and a local business was persuaded to purchase five-gallon buckets and give them to residents for curbside collection. **White goods** were next added to their pickup list. In 1985, they started chipping brush instead of landfilling. Next, they added bimetal cans to their pickup list for recycling. "We began recycling wood in 1987 and plastic containers and tires in 1988," writes McGee. Twenty-gallon cans soon replaced the five-gallon buckets. "In 1989, we opened our compost facility." In 1991, a grass recycling program was also started.

Wood and brush for mulching are picked up on Mondays, and all other recyclables are collected curbside on the weekly trash day. Leaves are collected twice yearly in the spring and fall and hauled to a local farm.

McGee says the "user-friendly" program has been both economically and environmentally sound. Proceeds are placed in a recycling trust fund "out of the regular budget and the hands of the politicians." Dump costs are avoided. Trash collection costs also receded as the "trash stream became more soft and compactible"; trucks could pack more trash per load and could go farther on each route, cutting down on the number of trips to the landfill and truck breakdowns. Thus, taxpayers' money is saved in a state with one of the nation's highest disposal or landfill costs, while pollution is curtailed and natural resources are conserved.

As the Berlin Township residents learned how and what to place at their curbs through meetings, media, and flyers, McGee believes the homeowners became closer with the Public Works Department.

"We make it simple. . . . Material can be put out commingled. We do the sorting. Recyclables are collected on the same day as the trash. This eliminates all confusion," advises McGee, who attributes one key to success to the use of the 20-gallon, brightly colored cans. The participation rate jumped from 60 percent to 95 percent with the use of these "very visible" yellow cans with peer pressure elements added, McGee continues. "Everyone knew who recycled and no one wanted to be pointed out for not having their recycling can at the curb."

Another key was strict adherence to a pickup schedule. "When the residents supply you with recyclables, they also demand that you pick them up on time."

■ Resources

As part of its public education program, the Department circulates a quarterly newsletter and an annual calendar. It will answer any requests about the recycling program.

New York

🏛 *State Government*

Industrial Energy Efficiency Programs
New York State Energy Research and
　Development Authority
Two Rockefeller Plaza
Albany, NY 12223-9998 USA

Contact: **Harry C. Howansky,** Program Manager
　Phone: 1-518-465-6251, ext. 214
　　Fax: 1-518-432-4630

Activities: Research • *Issues:* Energy

● Organization

Created in 1975, the **New York State Energy Research and Development Authority** develops and uses safe, dependable, renewable, and economic energy sources and conservation technologies. In two major program areas—energy efficiency and economic development, and energy resources and environmental research—it sponsors energy research, development and demonstration (RD&D) projects as well as finances programs to help utilities and other private companies fund certain energy-related undertakings.

The Energy Authority also constructs and operates facilities for disposing low-level radioactive wastes produced in New York State. "The generators of these wastes ultimately will bear the costs of the

construction of these facilities," according to chairman William D. Cotter, who is also State Energy Office commissioner. Irvin L. White, as president of the Energy Authority, manages program, staff, and facilities.

Revenues are mainly derived from an assessment on intrastate sales of the state's investor-owned electric and gas utilities. Investments from retained earnings as well as the New York State Power Authority's annual contribution are other funding sources.

▲ Projects and People

A railtruck developed through the Energy Authority's RD&D program is being hailed as a breakthrough in transportation technology. It enables cargo to be carried by both rail and highway without repackaging, thereby reducing fuel consumption by 25 to 50 percent. The RoadRailer, formerly called RailMaster, takes advantage of aerodynamic design and overcomes three hindrances to energy efficiency: excessive container weight, which adds to fuel costs; extra train height, which prohibits rail cars to navigate older and smaller tunnels; and severe wind drag, which requires more energy to move the freight.

The RoadRailer, with only two components—the modified truck trailer and an adapter to railroad wheels—eliminates 13 tons of dead weight. Costing about $17,000 a truck, about $8,500 less than a rail-compatible trailer, the 25,000 RoadRailer units are expected to be operable by the mid-1990s and will save 2 million barrels of oil and eliminate 800 tons of hydrocarbon emissions, 8,000 tons of carbon monoxide, 23,000 tons of nitrogen oxides, and 900 tons of particulate matters. Besides the fuel economy gain and pollution cutbacks, other energy and operating costs are reduced with the elimination of large gantry cranes and the complex loading and unloading of flat cars. Industries such as chemicals, iron, and steel are expected to benefit from this revolutionary design.

Some 1,000 units are now in operation in 10 states, and the technology is licensed in other countries. The United Nations is also interested in its application.

A recently completed project is the construction of an energy/emissions/economic model of New York State to evaluate the marginal costs of reducing carbon dioxide (CO_2) emissions here. Working with the MARKAL linear programming model, the evaluation includes all of the major energy-using sectors over 20- and 45-year time spans and analyzes all of the technologies that could reduce CO_2 emissions at the lowest cost. All agencies responsible for making state energy/environmental policy are receiving information; the U.S. Environmental Protection Agency (EPA) is also receiving information for inputs into federal policy in this area. Contractor is **Brookhaven National Laboratory,** Upton.

The **cullet color sorting** demonstration at Owens-Brockway, a New York State glass container manufacturer, can eventually save 500,000 to 1 million gallons of oil annually, cutting energy use by 7 to 15 percent, and boosting glass container recycling by 25 to 50 percent. The new sorter technology allows the purity of cullet or crushed glass to be increased by removing certain contaminants. "Higher purity cullet reduces the amount of finished glass container products which have to be rejected due to contamination," according to the researchers. A 10 percent increase in cullet use yields a 3 percent reduction in energy for heating the furnaces and gives off fewer emissions per pound of melted glass. Contractor is J. Busek and Company, Volney.

Researchers are experimenting with energy-efficient plastic bottle recycling in Yonkers, using the hydrocyclone separation technology. It is expected that some 12 million pounds of annually discarded plastic beverage bottles and other waste plastics will be recycled here. The recycled "fractions" or materials will be sold in domestic and foreign markets as a substitute for virgin plastics for value-added manufacturers who will produce fiber-fill garment stuffing, carpet, automobile bumper, and various nonfood containers. This will eliminate the need for about 144,000 barrels of crude oil used domestically for producing virgin plastic resins and for 38 million cubic feet of New York's landfill capacity. It will also create 10 permanent jobs and add about $1 million to the state's net economic product from the purchase of plastic wastes and resale of recovered fractions. **Amdak Corporation** is the contractor.

If a **solvent biofilter study** proves technically feasible and complies with environmental standards, it will "provide users in the printing industry and elsewhere with an economical alternative to more energy-intensive and expensive emission control technologies," report the researchers. They examined a prototype biofilter installation in Montreal, Canada, to see if it is applicable to the contractor's **Lucky Polyethylene Manufacturing Co.**, Brooklyn, and whether it can meet state and federal regulations.

Some New York City Transit Authority buses were equipped with experimental electric and fuel-burner regenerated filter traps developed earlier in this four-year study completed in 1992. With more than 3,000 tons of **diesel particulate emission** released into the city's atmosphere each year and contributing to lung-related illnesses such as lung cancer, reduced visibility, and dirty buildings and surroundings, the 5,000 diesel buses pollute as well as provide energy-efficient transportation. The filter, if successful, could help commercialize a technology to reduce harmful emissions of these diesel buses. Earlier results showed that one in-service bus had reduced particulates of about 80 percent. Contractor is **Ortech International.**

A **heating system environmental performance project** recently instructed manufacturers of various residential and commercial gas-fired systems in the fundamentals of pollution control and using laboratory testing equipment. The objective was to produce and test systems with lower levels of nitrogen dioxide (NO_x) and carbon monoxide emissions and to build a database of research knowledge about pollutant emission levels of many kinds of current heating systems. Manufacturers were to develop retrofit modifications to get compliance with local and environmental standards expected soon. The project is enabling an accelerated response to acid rain, photochemical smog, and global warming issues. **AGA Laboratories** is the contractor.

A high-speed, magnetically levitated train system based on **Maglev technology was assessed** according to its technical and economic application to New York State and certain potential industries. Domestic development of this technology is being encouraged to provide a fast, economical, energy-efficient, low- noise, and low-emission transportation mode for passengers in statewide and regional travel corridors and to have beneficial freight applications and positive environmental and energy impacts. "Maglev can bring depressed regions into an integrated national transport system, improving living standards and relieving over-crowding," reports the Maglev Technology Advisory Committee, who say that "Maglev is to diesel locomotives what *Star Trek* is to *Wagon Train.*"

Maglev is the suspension, guidance, and propulsion of a vehicle by magnetic forces, with no physical contact between the vehicle and its guideway. Maglev allows much greater speeds than wheeled suspensions and virtually eliminates roadbed maintenance. With

manufacturers outside the USA presently dominating this market, a New York consortium is being sought to pursue this technology. **Grumman Space Systems Division** is the contractor.

■ Resources

The Maglev Technology Advisory Committee prepared an *Executive Report on the Benefits of Magnetically Levitated High-Speed Transportation for the United States* for the U.S. Senate Committee on Public Works—which is published by Grumman Corporation, Bethpage, NY 11714-3580, and available through Harry Howansky of the Energy Research and Development Authority.

New York Flora Association (NYFA)
Biological Survey
New York State Museum
3132 C.E.C.
Albany, NY 12230 USA

Contact: **Richard S. Mitchell, Ph.D.**, State Botanist
Phone: 1-518-486-2027

Activities: Education; Research • *Issues:* Biodiversity/Species Preservation

● Organization

Dr. **Richard S. Mitchell** founded the **New York Flora Association** (NYFA) in 1990 under the auspices of the **New York State Museum** Institute and was its first co-director with **Robert E. Zaremba**, The Nature Conservancy. NYFA is a nonprofit, private group of 300 members and a 14-member advisory council with funding from dues, gifts, and grants.

An umbrella organization for field and herbarium botanists, NYFA promotes the study of the state's native and naturalized plant life, gathers voucher specimens and information for a complete atlas of state plant distribution, develops "ecological profiles" of native and introduced plant species, promotes conservation and wise management of native plant resources and their communities, fosters an information exchange among interested persons and clubs, and encourages the production of educational botanical publications.

Members hold meetings across the state and take field trips to interesting biological communities or floristic areas.

▲ Projects and People

The **Plant Distribution Atlas** is a major undertaking of the State Museum's Botany Office, with help from NYFA members. The 500-page preliminary atlas lists 3,500 known taxa of New York plants. Among the endangered species listings is the rare-hart's tongue fern, found in glacial plunge basins, cross channels or ravines, and growing best in central New York's steep rock crevices where organic matter accumulates and root anchorage is good. Trampling and habitat destruction by timber removal, quarrying, and residential or other developments are its enemies.

A New York State **Flora Database** with a geographic information system (GIS) that can create distribution maps and other documents is underway with plans to contain entries for some 500,000 plant specimens collected here. Voucher specimens will fully back up the database. Now botanists have the opportunity to pinpoint exact locations where specimens are found; the museum's older specimens were discovered in a less-precise manner.

Dr. Mitchell has undertaken numerous surveys of rare, endangered, and threatened plants of New York. In 1986, he produced a checklist of the state's 3,400 wild trees, shrubs, flowers, and ferns. He edits an ongoing series of monographs, known as *Contributions to a Flora of New York State*, and useful throughout the northeastern USA and eastern Canada.

■ Resources

A list of publications issued by the Botany Office is available, including *Ecosystem Management: Rare Species and Significant Habitats*.

Dr. Mitchell express concern about future funding as "our state is in a fiscal crisis. Political support for us in the past does little good if the money is withheld," he writes.

🏛 *Local Government*

Operation GreenThumb (OGT) 🌿
New York City Department of General Services
49 Chambers Street, Room 1020
New York, NY 10007 USA

Contact: **Jane Weissman,** Director
  Phone: 1-212-233-2926 • *Fax:* 1-212-233-2975

Activities: Development of Community Gardens and Meadows • *Issues:* Air Quality; Health; Land Reclamation

● Organization

Operation GreenThumb is New York City's community gardening program sponsored by the Department of General Services whereby city-owned vacant properties are leased to eligible, nonprofit organizations at one dollar a year and technical and material assistance is provided. For the benefit of thousands of New Yorkers, hundreds of community vegetable and flower gardens have sprouted up since 1978 in inner city neighborhoods, playgrounds, apartment projects, and residential blocks. It is the USA's largest municipality-run community garden project, with growing rows that stretch more than 60 miles if laid end to end. Funding comes from federal Community Development Block Grants.

Sponsors of this innovative landmark program include churches, schools, community centers, block associations, hospitals, senior citizen organizations, youth groups, day care centers, and drug rehabilitation centers. Besides bringing beauty, safety (by cleaning up hazardous areas of debris), and a sense of community to urban areas, gardens that are food producing grow $750,000 worth of vegetables each year. This is important to the gardeners who otherwise could not afford fresh produce. The urban gardening also provides an opportunity for many participants to return to the land after years of separation from their early farming roots.

After the lease is signed, the Department of Sanitation's Lot Cleaning Division clears the site of large debris. Garden sponsors attend workshops to learn about garden design, fence and raised-bed construction, and horticultural techniques. Operation GreenThumb provides fencing materials, tools, lumber, soil, ornamental and fruit trees, shrubs, seeds, and bulbs.

▲ Projects and People

Each March, the GreenThumb Growtogether conference for all gardeners opens the growing season. In August, participants bring their best fruits and vegetables to be judged at the **City Gardeners Harvest Fair,** Gateway National Recreation Area.

The **Land Reclamation Project** manages large tracts of city-owned vacant land within community development areas to thwart urban blight. Meadows of grasses and wildflowers replace dumping grounds, noxious weeds are controlled through mowing, property is made more presentable, neighborhood images are improved, and self-esteem is restored.

Other innovative ventures that are models nationwide include the **Urban Orchard Program,** with thousands of plantings of fruit trees, grape vines, and berries; **Artists in the Gardens,** where sculptures and murals have been installed in community gardens; **Education in the Gardens (EIG),** where school groups are invited to learn about and establish their own gardens; and the **Garden Preservation Program,** to expand long-term leases and create adjacent site long-term leases and preservation site designations.

In preparation for the documentary film "City Farmers" by Meryl Joseph about OGT community gardening, 100 local gardeners shared their experiences, anecdotes, and recipes. Here are two:

"When we first began our community garden, it meant changing an eyesore of a burnt-out building into something beautiful. Now, each morning I wake up to a dream come true. It also changed our mischievous teenagers to a positive junior block association, learning parliamentary procedures and conducting their own meetings instead of destroying the block."
  Ruth Fergus, Madison Community Garden

"This is the beauty. GreenThumb gave us a peach tree. Yearly, we got two or three bushels of peaches from the tree. People have come from near and far for Harlem-grown peaches from our garden tree."
  Bertha Jackson, 127th Street Block Association

■ Resources

Operation GreenThumb issues a quarterly newsletter and provides a *Planning Manual for Doing a Lot.*

North Carolina

🏛 *State Government*

North Carolina State Museum of Natural
  Sciences ⊕
North Carolina Department of Agriculture
P.O. Box 27647
Raleigh, NC 27611 USA

Contact: **David S. Lee,** Curator of Birds
  Phone: 1-919-733-7450 • *Fax:* 1-919-856-4139

Activities: Education; Research • *Issues:* Biodiversity/Species Preservation

● Organization

Established in 1879, the North Carolina State Museum of Natural Sciences, with Betsy Bennett as director, houses one of the three largest bird collections in the southeastern USA, including North American and global species—particularly seabirds. Working with this collection since 1975, curator David Lee has built it from 2,000 miscellaneous specimens to more than 17,000.

Emphasizing how plants and animals live together in ecological communities, the natural history collection of more than 250,000 examples also features mammals, reptiles, and amphibians as well as seashells, rocks, gems, and fossils. Serving students and teachers, residents and visitors, professional and amateur naturalists, and a magnet for attracting related industries to a revitalized urban core, the Museum is planning for a new building on Bicentennial Mall.

To accommodate extensive marine mammal exhibits, including the skeleton of the 55-foot sperm whale known as "Trouble" which once weighed 100,000 pounds, or as much as 10 elephants, a 200-foot-long hall with a 32-foot ceiling is being built. Dinosaurs and other large mammals from every continent will be placed in their proper astronomical, geological, and meteorological contexts. Also planned are an auditorium for educational lectures, a naturalist center for high school and college classes, a library to house 50,000 volumes, and facilities for expanded research—often in conjunction with the American Museum of Natural History, U.S. Fish and Wildlife Service, U.S. Department of the Navy, and other federal and state government agencies. Its staff also works with The Nature Conservancy, which established the Nags Head Woods Ecological Preserve and conducts ongoing studies of the Carolina bays.

According to the Museum, "No other agency in North Carolina has a zoological database that spans more than 100 years." A new greenhouse and animal care facility will also be provided. Native fishes are already on display in a new exhibit which emphasizes the dependence of fish on clean water, the unity of water resources from rain clouds over mountaintops to the saltwater of the oceans, and the effects on organisms of shore and stream activities. Support is generated through the nonprofit North Carolina State Museum of Natural Sciences Society.

▲ Projects and People

Early on, the Museum's bird collections and researchers who studied nesting habits helped to document that DDT and similar pesticides caused eggshell thinning of brown pelicans, ospreys, peregrine falcons, and other species whose reproduction dropped in the 1950s. The Carolina parakeet and passenger pigeon are already extinct, but there is hope for the red-cockaded woodpecker, on display here, which excavates its nest cavity in pine trees at least 80 years old.

Lee has conducted recent research projects for the National Oceanic and Atmospheric Administration on the outercontinental shelf of the pelagic Sargassum (seaweed) community; the U.S. Army Corps of Engineers on movements of striped bass through dam turbines and on the endangered piping plover on Topsail Island; the U.S. Department of the Navy on effects of electromagnetic pulse radiation environmental simulator on marine birds; and the North Carolina Wildlife Resources Commission on historical nesting areas of endangered and threatened birds.

Ornithologist John Gerwin is helping the Ecuadorean government research and compile conservation information about birds of a lowland tropical rainforest in a remote area accessible only by canoe as part of a team sponsored by the Academy of Natural Sciences of Philadelphia and funded by the MacArthur Foundation.

Ornithologists report that, to encourage bird reproduction, local gardeners should plant native and ornamental trees and shrubs, such as flowering dogwood, black gum, cedars, pines, oaks, cherries, hollies, pecans, and grapes, as well as vegetable, herb, or flower gardens. Dead limbs encourage woodpeckers and other cavity nesters. Cardinals and towhees are attracted to high spots furnished with rocks, logs, berms, or fences and low shrub and brush covers. A paste of yellow corn meal, peanut butter, and cooking oil or melted fat makes an excellent bird feed; seeds or suet are also good, and water—preferably moving—should be protected from freezing.

The Museum's Butterfly Garden, with 30 wild and cultivated plant species, encourages visitors to plant similar ones at home. Ironweed, hollyhock, butterfly weed, and Queen Anne's lace planted in a sunny area and screened with shrubs such as butterfly bush, lilac, or azalea are particularly recommended to attract butterflies such as American painted lady, common checkered skipper, monarch, and eastern black swallowtail.

Overall, Museum activities are geared to hands-on learning about living plants and animals in the Outdoor Science Program, in cooperation with the state's Department of Public Instruction. Ponds, school grounds, the Museum, and coastal, plateau, and mountainous regions are explored. Teachers also have an opportunity to participate in an annual Tropical Ecology Workshop in Belize.

Participating in a campaign to "recycle, reuse, and reduce waste and pollution," the Museum provides tips such as on saving gasoline while driving and making a household cleaner without chlorine bleach.

■ Resources

Books and posters appealing to young people are available on subjects including Native Americans, whales, snakes, and spiders. Reports are published on endangered species of North Carolinian mammals, marine and estuarine fishes, and birds. *Whalebones* is a monthly newsletter. The *Atlas of North American Freshwater Fishes* with supplement, various fishermen's guides, and documents on birds (including oil-spill effects) are also listed in the Museum's publications brochure. Films, classes, story hours, and information leaflets are regular features.

Natural Heritage Program (NHP)
North Carolina Department of Environment, Health, and Natural Resources
P.O. Box 27687
Raleigh, NC 27611-7687 USA

Contact: Leslie Pearsall, Coordinator
  Phone: 1-919-733-7701 • *Fax:* 1-919-733-2622

Activities: Education; Political/Legislative; Research • *Issues:* Biodiversity/Species Preservation; Park Planning; River Conservation

▲ Projects and People

The **Natural Heritage Program** of the **North Carolina Department of Natural Resources and Community Development** reports recent success in its program despite declining biological diversity in shrinking natural areas with tight budget constraints to cope with.

To its **inventory database,** nearly 2,000 new occurrences were added recently—including endangered and rare species, exceptional natural communities and ecosystems, special wildlife habitats, and exemplary geomorphic landforms. A new computer system that allows electronic information exchanges with other state and federal agencies and the national network of state heritage programs was implemented to manage biological and conservation data with technical assistance from The Nature Conservancy and financial aid from the Carolina Power and Light Company.

The Heritage Program helps the **Plant Conservation Program** to revise North Carolina's official list of endangered, threatened, and special concern species. In discovering new populations of 11 nationally endangered or threatened plant species, the Heritage Program scientists reported these and other population survey results to the U.S. Fish and Wildlife Service. Working with the **Animal Wildlife Resources Commission,** it is the **principal databank** for population records of endangered, threatened, and special concern animal species.

Natural areas were surveyed in a 10-county region along the Albemarle Sound and a 7-county region along the Pamlico Sound. A subsequent survey is being conducted in 18 counties in the upper drainage basins of the Roanoke, Tar, and Neuse rivers. Guidelines and specifications are prepared for the remnants of longleaf pine forest and mountain wetlands at the Fort Bragg Army Base, Camp Lejeune Marine Base, Croatan National Forest, Lumber River corridor, Yadkin River corridor, and other areas.

With the U.S. Forest Service, **national forest lands** are being inventoried. With the Wildlife Resources Commission's Nongame and Endangered Wildlife Program, species characterization abstracts for the state's endangered and threatened **bird species** were compiled. The state's first list of **rare butterfly species** was prepared, and populations are being monitored. Outstanding **water resources** such as rivers and lakes are nominated for preservation. A statewide list of **priority wetlands** and a wetlands conservation plan are other accomplishments.

With The Nature Conservancy, Trust for Public Land, and local land trusts, natural areas are designated for acquisition. Five new nature preserves were recently dedicated: Triangle Land Conservancy's **White Pine Preserve,** North Carolina State University's **Hill Forest Preserve,** and The Nature Conservancy's **Camassia Slopes, Devil's Gut,** and **Big Yellow Mountain Preserve.**

Training workshops on prescribed ecological burning in fire-dependent longleaf pine ecosystems, conferences on management of such forests, and sessions on local land trusts were held. NHF also convinced the Carolina Power and Light Company and the state's Department of Transportation to **maintain populations of endangered plant species in transmission line and highway right-of-ways.**

■ Resources

The *Natural Heritage Newsletter* is published. Also available are *Rare Plants List* and *Rare Animals List for North Carolina,* as well as literature relating to park management and critical assessments of the state's natural and scenic river systems.

Oregon

🏛 *State Government*

Oregon Department of Land Conservation and Development (DLCD)

1175 Court Street, NE
Salem, OR 97310 USA

Contact: **Mitch Rohse,** Communications Manager
Phone: 1-503-373-0064 • *Fax:* 1-503-362-6705

Activities: Land-Use Planning • *Issues:* Conservation

▲ Projects and People

Oregon's **Land-Use Planning Program** conserves farmland, forests, coastal resources, wetlands, and other natural resources vital to the economy and environment. Implementing a state growth management law of 1973, the program brings together government officials at all levels with citizens to implement local plans and enact zoning ordinances to keep urban development from encroaching onto forest land and other natural resources.

So far, 25 million acres of farm and forest lands are protected by tough conservation restrictions. All 241 cities have adopted **urban growth boundaries** effective over the next two decades which marks the limits of development.

A model program for similar endeavors in Florida, Georgia, Maine, and Vermont, it is assisted here by **1000 Friends of Oregon**—a private, nonprofit group that takes part in plan review, litigation, citizen assistance, research, and public education. As defenders of the urban growth boundaries ordinance, group members are watchdogs on government policymaking. While conserving farmland, forests, and coastal regions, they also work to improve economic opportunities in industrial areas. As a result of their efforts, 15 conservation groups from throughout the USA have formed the **National Growth Management Leadership Project.** *(See also 1000 Friends of Oregon in the USA NGO section.)*

■ Resources

A bibliography of publications relevant to Oregon's planning efforts is available from DLCD.

Natural Heritage Advisory Council (NHAC)
Oregon Division of State Lands

775 Summer Street
Salem, OR 97310 USA

Contact: **Caryn Throop,** Chairperson
Phone: 1-503-378-3805 • *Fax:* 1-503-378-4844

Activities: Conservation • *Issues:* Biodiversity/Species Preservation

● Organization

Oregon legislators put the **Natural Heritage Program** in place in 1979 to ensure that the state's remaining natural ecosystems and native species are preserved for future generations. Founded upon the natural area inventory of the Oregon chapter of The Nature Conservancy, it has two basic concepts: a scientific pursuit and listing of representative examples of the diverse ecosystem with rainforests, alpine meadows, high deserts, coastal marshes, lakes and streams; and cooperative and voluntary land conservation.

▲ Projects and People

The Heritage Plan brings together private landowners and public land management agencies for these purposes with the guidance of the 14-member Natural Heritage Advisory Council appointed by the governor. In addition to listing the resources of the state's terrestrial and aquatic ecosystems, conservation areas are designated to protect the most vulnerable plant and animal species. The Natural Heritage Data Base is the primary tool for selecting natural areas for conservation and is also used to assess potential environmental impacts of proposed development through land-use planning, federal environmental impact statements, and other private and public planning efforts. The Nature Conservancy first developed this computerized database for Oregon in 1974.

Terrestrial ecosystems include, as examples, the Oregon coast range with western hemlock and Sitka spruce zones; the conifer, hardwood, and riparian forests and grasslands of western Oregon's interior valley; the coniferous, tan oak, and redwood forests and ocean-front shrublands of the Siskiyou Mountains; the western and mountain hemlock, Pacific silver and shasta red fir zones, alpine community, and swamp forests of the Oregon Cascades' west slope and crest; grassland and shrubland steppe, ponderosa and lodgepole pine zones, and juniper and fir zones of the Oregon Cascades' east slope; sagebrush, conifer forests, and tundra of the Ochoco, Blue and Wallowa mountains; desert communities and mountain mahogany, cottonwood, and aspen groves of the basin and range; sand dunes, wheatlands, and pine forests of the Columbia Basin's high lava plains; and the black sagebrush, western juniper, and the valley bottom alkaline vegetation mosaic of the Owyhee Uplands.

Along the Oregon coast range, aquatic systems include estuaries, unvegetated sand beaches, kelp beds, waterfalls, and a Columbia River channel with marsh island vegetation. Ponds and cattail wetlands are found in western Oregon's interior valleys; eelgrass beds, coastal bogs, and large cold springs are among those water bodies in the Siskiyou Mountains. Glacier-fed streams, alpine and subalpine lakes, and hot and cold springs are in the Oregon Cascades with more stream systems, saltgrass ponds, marshes, grass wetlands, bogs, and hot springs in the Ochoco, Blue, and Wallowa mountains. Freshwater and saline ponds and alkaline wetland vegetation complex are in the basin and range, while a variety of lakes, ponds, flowing springs, and streams are in the Columbia Basin's high lava plains and the Owyhee Uplands.

Glaciers, lava floods, volcanoes, craters, lava and ice caves, sea stacks, tidal estuaries, and fossil locations are some of the unique geologic features in Oregon.

Endangered species include the shortnose sucker fish, spotted frog, peregrine falcon, Aleutian Canada goose, and California wolverine. Threatened animal species include the bald eagle, northern spotted owl, fringed bat, Oregon silverspot butterfly, and painted turtle. It is believed that the common loon, California condor, wood bison, gray wolf, southern sea otter, grizzly bear, and Columbia River tiger beetle are among the extinct species.

The Oregon Natural Heritage Plan also indicates endangered, threatened, and apparently extinct plants.

■ Resources

The Natural Heritage Plan is available and explains further how various state and federal agencies work together with the private sector to protect resources. It is being updated in 1993. The Division of State Lands office will provide information and fees on using the Natural Heritage computer database.

🏛 *Local Government*

Conservation Division
City of Ashland
City Hall, 20 East Main Street
Ashland, OR 97520 USA

Contact: **Dick Wanderscheid**, Conservation Manager
Phone: 1-503-488-5306 • *Fax:* 1-503-488-5311

Activities: Political/Legislative • *Issues:* Air Quality/Emission Control; Energy; Health; Water and Energy Conservation

▲ Projects and People

The **City of Ashland's Conservation Division** has four ongoing projects reaching into the community and one that begins in mid-1992.

The **weatherization program** provides cash grants of up to 60 percent of the costs to weatherize electrically heated houses. Low-income customers receive 100 percent grants. More than 1,500 residences have been completed in this program.

The **super good cents program** gives cash grants to builders or owners who construct energy-efficient new homes. More than 450 residential units have been certified as "super good cents" in this program.

The **energy-smart design program** offers free energy-efficient design assistance to major remodelers or builders of new commercial structures. Awards are presented to buildings which meet the standards recommended in the program. Cash grants to help recover the incremental costs of building to these standards were added in late 1991.

The **save our livability, view, and environment program** (SOLVE) provides zero-interest deferred payment loans to low- and moderate-income customers to weatherize and upgrade heating systems in homes presently heated with inefficient wood stoves. This program is helping to clean up Ashland's Rogue River valley particulate air quality problem.

In June 1992, customers of the city's water utility started getting free water conservation plumbing devices and landscape irrigation technical assistance. "The goal is to acquire cost-effective water conservation and delay or supplant the need to build expensive new water supply facilities," writes Richard J. **Wanderscheid**, division manager, who initiated all of the city's energy conservation programs.

For his personal residence, Wanderscheid coordinated the planning and construction of a solar home, which taught him the fundamentals of "siting and building techniques which are essential for an effective passive design," he indicates.

Rhode Island

🏛 *State Government*

Hazardous Waste Reduction Program
Office of Environmental Coordination
**Rhode Island Department of Environmental
  Management (DEM)**
83 Park Street
Providence, RI 02903 USA

Contact: **Janet Keller**, Division Chief
  Phone: 1-401-277-3434 • *Fax:* 1-401-277-2591

Activities: Development; Education; Research • *Issues:* Waste
Pollution Prevention; Management/Recycling

● Organization

The **Office of Environmental Coordination** helps Rhode Island
businesses reduce their use and disposal of toxic and hazardous
waste materials. Increased competitiveness and prosperity and decreased corporate and personal liability become the byproducts,
instead of dangerous wastes that threaten public health and natural
resources.

In taking a team approach to preventing pollution, the Department of Environmental Management cooperates with the University of Rhode Island, other government agencies and educational
institutions, and industry to establish manufacturing alternatives
that favor the environment.

▲ Projects and People

Elements of the **hazardous waste reduction program** include direct
technical assistance to waste reduction efforts; availability of $1.5
million in grants to establish industrial waste reduction programs at
their source—based on research, development, and demonstration
of promising technologies; the Hazardous Waste Reduction Information Center with technical and legal books and periodicals, and
access to a computerized database on waste disposal, treatment,
recycling, and reduction practices; state-sponsored training workshops and round-table forums where information and technology
transfers take place among industry representatives; publicity program to recognize outstanding achievements; annual Governor's
Award for excellence in hazardous waste management; and certificates of merit to other outstanding examples.

Technical assistance includes on-site information on waste-re-
duction practices relevant to each firm's operations; instructions on
how to handle wastes properly; help in developing waste reduction
and management plans; aid in setting up in-house waste reduction
teams; and details on applicable federal and state rules and regulations.

With the State Department of Economic Development, assessments are made of manufacturing processes within machine tooling
operations, electroplaters, photochemical etchers, polyester and plastic film coating firms, and an aluminum anodizing company. The
Manufacturing Jewelers and Silversmiths of America—whose member firms were also assessed—received a DEM grant and prepared a
Hazardous Waste *Reduction Technical Assistance Directory* for this
industry.

Local universities, electroplaters, and paper coaters have also
received grants. The Ocean State Cleanup and Recycling Program
(OSCAR), which implemented the USA's first mandatory recycling
effort, is also a DEM program administering grants to cities, private
entities, and nonprofit groups under the direction of Janet Keller.

With a grant from the U.S. Agency for International Development (USAID), former chief Victor Bell worked with pesticide
management and pollution prevention in Africa, particularly in
Benin and Zimbabwe.

■ Resources

The *Hazardous Waste Reduction Technical Assistance Directory* and
the newsletter *Options* are available free of charge.

South Carolina

🏛 *State Government*

Riverbanks Zoological Park ⊕
P.O. Box 1060
Columbia, SC 29202 USA

Contact: **Palmer Krantz**, Executive Director
  Alan H. Shoemaker, Curator of Mammals
  Phone: 1-803-779-8717 • *Fax:* 1-803-256-6463

Activities: Education; Research • *Issues:* Biodiversity/Species
Preservation

● Organization

One of the 15 most visited zoos in the USA, **Riverbanks Zoological
Park** is succeeding both as **South Carolina's** leading tourist attraction and as a wildlife conservation center where endangered species
are preserved and the natural history education of the audiences are
enhanced. With strong community backing, the Riverbanks Zoological Society is now the country's second largest per capita zoo
support organization.

Riverbanks, which is affiliated with the American Association of
Zoological Parks and Aquariums (AAZPA), is an example of today's
zoos that are becoming the "Noah's Arks of the 20th century and
beyond"—saving wildlife from modern catastrophes such as human
overpopulation, deforestation, habitat destruction, air pollution,
and illegal poaching.

Animals are housed in large, naturalistic exhibits where the barriers are moats, water, and light. They are exhibited in proper sex and
age ratios so that normal behavior and reproductive patterns will be
stimulated. The role of plants in ecology is also emphasized with
gardens such as the Woodlands Walk and the Lily Pond featuring
both indigenous and exotic plants. The Backyard Garden highlights
environmentally sound gardening techniques.

▲ Projects and People

Here, 14 endangered species—including the **Siberian tiger, clouded
leopard, Simang black ape, bald eagle, and Bali mynah**—are being
propagated to provide future breeding animals to other accredited
institutions and to reintroduce such animals back into the wild.

Riverbanks' three groups of black howler monkeys, the world's largest captive collection, have produced more than 25 viable births.

Participating in 16 AAZPA Species Survival Plans (SSP) to ensure that viable gene pools are maintained in captivity, the zoo is also helping black rhinos, spectacled bears, black and white ruffed lemurs, red-ruffed lemurs, lion-tailed macaques, golden lion tamarins, palm cockatoos, Guam kingfishers, and Chinese alligators. Dr. Terry Norton, zoo veterinarian, is leading several collaborative research projects on reproductive studies of African elephants, Baird's tapir, and radiated tortoises.

Since 1974, Alan H. Shoemaker, curator of mammals, is the International Studbook Keeper for Rare Leopards, including Amur, Persian, Chinese, and Ceylon leopards. The Studbook includes a country-by-country analysis on their status and is updated periodically. For example, information on Amur leopards (*Panthera pardus orientalis*) at the Moscow Zoo indicates that leopards obtained from the Pyongyang Zoo "are all wild born and unrelated to each other"— thus suitable for breeding. Promising too is the news that "a regional studbook for leopards in Chinese zoos is being developed and animals surplus to this program may become available for zoos outside that country." Generally, Amur and Persian leopard populations are increasing. Shoemaker indicates that "increasing numbers of owners are trying to minimize levels of inbreeding by taking care in the selection of potential mates."

The Studbook also points out that wildlife sanctuaries are aiding the survival of leopards in Thailand and South Africa; that urgent protection is being manifested in Saudi Arabia with research of the National Commission for Wildlife Conservation and Development (NCWCD) to study their diet of porcupines and feral dogs, hyraxes and baboons, while developing public awareness and initiating radio tagging and tracking; and that *Panthera pardus ciscaucasica (saxicolor)* is threatened with extinction in former Soviet Republics such as Azerbaijan, but that reserves, poaching decrease, and ungulates recovery are increasingly sighted.

Shoemaker is deputy chairman of the Species Survival Commission (IUCN/SSC) Cat Group and a member of Primate, Captive Breeding, Hyena, and Tapir groups.

Bob Seibels, curator of birds, coordinated an international effort to return zoo-bred mynahs to their native Indonesian habitat, where, reportedly, there were less than 40 remaining of these critically endangered species. The toco toucan and crimson seedcracker, never before bred in captivity, were hatched and raised at the zoo— which was also the first in the Western Hemisphere to breed milky eagle owls, blue-billed weavers, and cinereous vultures. Other notable bird breedings are those of pied hornbills, eclectus parrots, and Renauld's ground cuckoos. AAZPA has honored Riverbanks' various captive breeding programs.

The zoo's recent Aquarium Reptile Complex (ARC) houses thousands of fish and hundreds of reptiles, amphibians, and invertebrates from around the world. Species bred here include the king cobra and the endangered Aruba Island rattlesnake. The AAZPA named ARC one of the top three zoo exhibits in 1990.

A local wildlife rehabilitation program treats and releases injured eagles, hawks, owls, ospreys, kestrels, kites, and vultures. The zoo participates in the Wildlife Coalition of the Midlands, a nonprofit volunteer organization, by caring for orphaned or injured wild animals.

Week-long educational programs for school children and Saturday classes for all age groups focus on conservation, endangered species, animal behavior, and special animal adaptations.

■ Resources
Literature about the zoo and membership information is available.

Southeast Regional Climate Center (SERCC)
South Carolina Water Resources Commission
1201 Main Street, Suite 1100
Columbia, SC 29201 USA

Contact: **David J. Smith,** Director
Phone: 1-803-737-0800 • *Fax:* 1-803-765-9080

Activities: Development; Education; Research • *Issues:* Climatology; Global Warming • *Meetings:* AMS

● Organization
The Southeast Regional Climate Center (SERCC) was established in 1989 at the South Carolina Water Resources Commission with a plan approved by the National Climate Program Office (NCPO) of the National Oceanic and Atmospheric Administration (NOAA). SERCC provides funding and technical assistance to state climate programs in Alabama, Florida, Georgia, North Carolina, South Carolina, and Virginia.

▲ Projects and People
Computerized and other communication links are set up between the state climatology offices (SCO), the Climate Analysis Center (CAC), and the National Climatic Data Center (NCDC). Regional climate impact statements on agriculture, water resources, tourism, business, and industry are routinely reported; information is provided on hurricanes, tornadoes, and other storm warnings. With Mead Data Central, weather story coverage is provided across the southeastern USA and the Caribbean. Weather observations—including upper air data, solar radiation data, and hourly precipitation—are recorded in a regional database in cooperation with the University of South Carolina and Clemson University. Development was started on an enhanced Southeast Climate Information Service (SECIS).

Several projects on severe weather climatology include regional summaries of tornado and lightning statistics and a database with a comprehensive list of storm data in South Carolina since 1950. Surface and upper air daily weather guidance for agriculture, fire weather forecasting, and emergency planning operations is underway at the U.S. Department of Energy Savannah River Site. The Southern Region Agricultural Research Project is a database that facilitates validation of crop models and allows studies of temporal and spatial variation of climate across the South. In the Centennial Weather Station Project, pre-1948 data from the region's selected historical stations are being entered in cooperation with Auburn University, Georgia and South Carolina Agricultural Experiment stations, and North Carolina State University.

SCOs are prioritizing proposals for applied climate research within a five-year plan that ultimately leads to economic benefits for the region. Possibilities include updating a NOAA report on maximum rainfall probabilities, investigating climate trends, developing large-area empirical models specifically tailored to corn and soybean production, and analyzing pan evaporation and precipitation to "develop coefficients for irrigation system design and scheduling, drought indices, and water-use permitting." Research is expected to be concentrated on the coastal zone, an area of "economic importance, minimal climatological knowledge, and extreme susceptibility to climate variations and change," reports SERCC.

Global change is a key concern in the Southeast, where its projected impacts are magnified because of its geographic biodiversity, according to David Smith. "Changes in worldwide weather patterns are accompanied [here] by changes in the magnitude and distribution of key life sustaining elements—including precipitation, temperature, soil moisture, and air quality," he reports. "Such changes, particularly compounded by rapidly changing population, economic and agricultural activity, have had significant impacts on fragile ecosystems, the economy and quality of life in the southeast."

Climatologists recommend that resource managers, scientists, planners, policymakers, and students be educated as to the long-term impacts of global change; that the U.S. Global Change Program should be regionally specific; that the next-generation climate models adequately interface the atmosphere, biosphere, and oceans; that states mobilize to conduct their own global change studies on wetlands, coastal zone processes, human and natural resources; that public and private organizations work together on recycling and energy conservation programs; and that a Southeast regional program be designed to "identify and understand uncertainties and to develop more reasonable policy strategies which discourage further deterioration of fragile ecosystems."

■ Resources

The *Southeastern Climate Review* about current research and issues is published quarterly.

Texas

🏛 *State Government*

Organic Program
Texas Department of Agriculture (TDA)
P.O. Box 12847
Austin, TX 78711 USA

Contact: **Brent Wiseman**, Coordinator
 Phone: 1-512-475-1641 • *Fax:* 1-512-463-7643

Activities: Development; Education; Political/Legislative; Research • *Issues:* Sustainable Agriculture

▲ Projects and People

Under its **Organic Food Certification Program**, the Texas Department of Agriculture (TDA) inspects and certifies organic farms and other businesses that process, distribute, or retail organic food—which is defined as food grown without artificial chemical pesticides or fertilizers. The demand for organic food is rising among health-minded consumers who are concerned with pesticide residues. At the same time, farmers are showing an increased interest in produc-

ing crops without contaminating groundwater, eroding the land, or depleting soil fertility.

Participating farmers in this voluntary marketing and regulatory program use TDA's "Certified Organic" or "Transitional" logos in advertising and point-of-sale material and on labels, shipping boxes, and tags identifying their products. To qualify, farmers must not grow crops on land with synthetic pesticides applied to it for the last three years and synthetic fertilizers applied for the last two years. Such materials may not be used at any stage of production or handling of food. Since the program began in 1988, TDA has state-certified 210 farmers, 40 retail outlets, 5 processors, and 9 distributors. A "transitional" label indicates farmers are converting their operations from conventional to organic production.

Now a $3-billion-a-year growth industry in Texas—the USA's second largest farm state with its long growing season—organic food production helps farmers diversify their operations and capture a bigger share of a premium market. As the public and the environment are protected, Texas's agricultural economy gets a boost. A new alliance is fostered between consumers, farmers, and government. Because the TDA certification program has shifted regulation of organic food from industry to government, officials say consumers are better protected from cheating, and farmers are getting marketing help.

To overcome growers' skepticism of government regulation and to gain retailers' trust in the authenticity of the logo, TDA set up an 18-member task force of farmers, retailers, consumers, scientists, and processors who together devised the growing and handling standards of the certification program. TDA also regularly updates agricultural educators and students.

This is also the first such program to certify the entire distribution system from farmer to retailer with impartial inspections. Oklahoma, Colorado, and Kentucky have modeled their programs on TDA's concept; and now at least 20 states have organic labeling laws. The U.S. Congress in 1989 passed a national organics certification and definition bill that relies heavily on Texas' standards. These results of Texas' leadership are helping to move organic farming into the agricultural mainstream at a quick pace.

■ Resources

Fliers and brochures for farmers and consumers are available.

Division of Travel and Information
Texas Department of Highways and Public Transportation
11th and Brazos Streets
Austin, TX 78701-2483 USA

Contact: **J. Don Clark**, Director
 Phone: 1-512-463-8601 • *Fax:* 1-512-463-9896

Activities: Education • *Issues:* Highway Litter; Transportation; Waste Management/Recycling

▲ Projects and People

The message is bold: Don't Mess with Texas. In 1985, when this program received its first funding from the State Department of Highways and Public Transportation to fight roadside trash, the

UNITED STATES OF AMERICA

cost had reached $24 million for retrieving highway litter—an amount that had skyrocketed 15 to 20 percent annually in the previous six years. Slogans like "Pitch In" were having no effect on habitual litterers, mainly men 18 to 34, studies showed.

This tough campaign proved so effective that between its inception and 1991, visible litter in Texas decreased by 72 percent, according to the Institute for Applied Research, Sacramento, California. "The decrease achieved in Texas makes it the cleanest jurisdiction we have measured in the course of conducting or coordinating 50 surveys in 15 states. . . . It eclipses Hawaii's 58 percent litter reduction that took 10 years between 1978 and 1988 as well as Washington's 73 percent reduction . . . between 1975 and 1983," the Institute reports.

Litter reduction in Texas was the greatest along rural roads and highways at 75 percent reduction; rural and urban freeways were next at 61 percent, while urban streets achieved a 51 percent decline in the six-year period. The average 74 percent decline in littered beverage containers is *lower than* in measured states with deposit laws; in urban areas, the reduction was 81 percent.

The message is far-reaching. The Don't Mess with Texas program has kicked off aggressive campaigns throughout the USA.

The **Adopt-a-Highway** program was initiated in Texas where local groups "adopt" two-mile sections of highway which they agree to clean at least four times yearly. The Department provides trash bags and reflective vests, then hauls the trash; it also awards certificates of appreciation. At last count, more than 4,000 groups adopted some 8,000 miles of roadway; now the movement is spreading to local **Adopt-a-Country-Mile** and **Adopt-a-Beach** efforts. Moreover, similar programs are now underway in 47 states.

The **Great Texas Cleanup-Greenup** is an activity of the Adopt-a-Highway groups in partnership with Anheuser Busch, who donated $5 million to promoting and participating in litter removal and wildflower seed plantings along 3,000 miles of highway.

The **Great Texas Trash-Off** is an annual springtime occurrence where as many as 24,000 volunteers pick up 1.3 million gallons of trash along 3,500 miles of highway as the wildflowers begin to bloom. Corporate sponsors such as Coca-Cola provide trash bags, Sea World of Texas gives discount coupons, and television and radio stations, and newspapers promote the event. Through Texas' efforts, now there is a **Great American Trash-Off** with a **Don't Mess with U.S.** theme in June. As many as 40 states and the District of Columbia have participated with the endorsement of the U.S. Committee for the United Nations Environment Programme (UNEP) and the assistance of the Children's Association for Protection of the Environment (CAPE). **J. Don Clark** is the chairperson of the Great American Trash-Off Task Force. "The most visible way of working with, rather than against, the environment is to go out and pick up trash," he emphasizes.

Spread the Word . . . Not the Waste is an education program for grades seven and eight to make young people more aware of litter. Every middle school child and teacher in the public schools receive a kit including a video program and a handbook outlining individual and group activities. One year, a **Create-a-Commercial** contest resulted in student-produced radio and television public service announcements for the Don't Mess with Texas campaign.

Since 1990, Girl Scouts from Austin to Midland are sporting a new patch: **Girl Scouts of the Lone Star State Don't Mess with Texas** for participating in special antilitter activities, including visiting a landfill or recycling center.

About 40 corporate sponsors help support the campaign.

■ Resources
Don't Mess with Texas litterbags, bumperstickers, and decals can be ordered in bulk. Literature, kits, and other information are available on request.

🏛 *Local Government*
Alternative Energy
City of Austin Electric Utility
721 Barton Springs Road
Austin, TX 78704 USA

Contact: **John E. Hoffner**, Program Manager
Phone: 1-512-322-6284 • *Fax:* 1-512-322-6083

Activities: Development; Research • *Issues:* Electric Utility; Energy

● Organization
The citizens of Austin in 1983 voted in favor of revenue bonds to be used for the generation of electricity from renewable, alternative energy resources that cut down on harmful emissions related to the greenhouse effect. The new solar photovoltaic power plant (PV300) of the city's **Electric Utility Department** is a direct response to community interest in diversifying Austin's present fuel mix of gas, oil, and coal through the use of renewable energy resources. ARCO Solar, Inc., California—the nation's largest photovoltaic cell manufacturer—was selected in 1985 to design, install, and manage Austin's first solar power plant for $2.7 million.

Although it is expensive, solar photovoltaic technology is an interesting option for Austin's 450,000 population because the citywide peak demand for electricity occurs between 2 P.M. and 8 P.M. on hot summer days, when a solar installation performs well. A clean source of electric generation needing no fuel for operation, it requires minimal maintenance and thus keeps down operating and maintenance costs. Solar photovoltaic power is a result of the direct conversion of sunlight to electricity with no harmful byproducts.

▲ Projects and People
A number of applications are being developed through demonstration projects and test programs to gain hands-on experience, educate and interact with the local community, and evaluate technical concerns as the cost of photovoltaics decreases. The goal of these efforts, officials report, is to "pioneer cost-effective programs to utilize photovoltaics by the late 1990s in order to defer the need for additional fossil-fueled generating capacity in Austin."

Completed in 1986, the PV300 located at the Decker Power Plant is the largest solar plant in Texas and uses flat-plate photovoltaic cell technology and a unique sun-tracking system. It is also the nation's largest such project to be financed by a utility without federal subsidies. Austin also co-owns a **300-kilowatt plant** which uses a concentrating cell technology and is jointly funded by the city and state, U.S. Department of Energy, and 3M Company at the 3M/Austin Center. **Both large centralized plants with different photovoltaic technologies are being tested.** In addition, the Electric Utility is evaluating 1- and 2-kilowatt **rooftop systems** for residential use including low-income homes and an American Youth Hostel Facility, co-funded by the state, and a proposed 25-kilowatt commercial system at the new Austin Convention Center. The objective is "to develop a strategy in which the utility can cost-effectively participate with customers" on-site.

Stand-alone power applications not connected to the utility network are also being examined, such as cost-effective use of photovoltaic technology to replace diesel-powered **traffic control arrow boards**, reducing both air and urban noise pollution. Another project proposes **solar-generated lighting** for a 3.1-mile bicycle trail at Voloway Park, which could eliminate the need for electric utility line construction and reduce environmental impacts in a natural area.

Austin is addressing many pertinent issues before the widespread use of photovoltaics can occur: **increased public awareness of renewable** energy options and benefits through plant tours, demonstrations, and familiarity with the diversity of projects; **improved services and increased benefits for the community**, such as public safety through night lighting in recreational parks; **conservation of natural fossil fuel resource** as present demonstration projects are displacing about 1,000 megawatt hours of fossil-fueled generation equivalent to 11 million cubic feet of natural gas or 620 tons of coal each year; **air quality protection and waste reduction** by *not* contributing to ozone depletion or global warming, as a 5 percent penetration of photovoltaics will reduce annual carbon dioxide emissions by 100,000 to 200,000 tons; **water conservation** by using less water for electric generation; **reduction of environmental impact due to utility construction; evaluation of reliability and cost effectiveness**; and dissemination of results to national, state, and local audiences.

The city is helping other communities and electric utilities in the state to incorporate photovoltaics into their generation mix, such as through the Texas Renewable Energy Industries Association (TREIA), and is finding other means to transfer information from its projects. The Electric Utility is participating in two national programs on "Early Applications of Photovoltaics in the Electric Utility Industry" sponsored by the Electric Power Research Institute (EPRI) and Sandi National Laboratories.

■ Resources

Literature is available on the background and performance of the photovoltaic power plants and projects.

Washington

🏛 *Local Government*

Solid Waste Utility (SWU)
Seattle Engineering Department
710 2nd Avenue, Suite 505
Seattle, WA 98104 USA

Contact: **Nancy Glaser**, Director
   Phone: 1-206-684-7652 • *Fax:* 1-206-684-8529

Activities: Development; Education • *Issues:* Waste Management/Recycling

● Organization

A city of 500,000 people, Seattle turned a solid waste management crisis into an opportunity to bring together citizens, environmental groups, and public decisionmakers to revolutionize the **Solid Waste Utility (SWU)**, transforming it from a criticized public agency into a nationwide leader in recycling participation.

When the city landfills closed in the mid-1980s, Seattle was permitted to use the county landfill at high costs if it developed an alternate disposal option within two years; otherwise it would lose its waste stream authority to the county.

▲ Projects and People

Acting quickly, the following four planning processes were carried on simultaneously:

• **Comprehensive planning process.** Various options were examined to decide on managing the total wastestream over a 20-year period. This included a Recycling Potential Assessment (RPA) and a programmatic environmental impact statement (EIS) on all proposed disposal options including source-separated recycling, incineration, materials processing, and landfilling.

• **Curbside recycling planning.** A request for proposals (RFP) went out to contractors citywide for services providing free containers, frequent pickup, marketing plan for products collected; and collection of mixed paper, newspaper, glass, aluminum, and tin. Vendors were selected and contracts signed within the six-month process.

• **Garbage collection contracts.** An RFP also went out for new garbage collection contracts to move pickup from backyard to curbside and to initiate a yard waste collection and composting program as outlined in the RPA. Money saved would help fund the new recycling program.

• **Rate structure.** The purpose here was to initiate steeply escalating rates that would support the recycling goals and encourage waste production.

A goal of 60 percent recycling was set in the program's first year and the above recommendations were implemented. An environmentally conscious city, Seattle was already achieving about a 24 percent recovery rate from residential and commercial sectors.

Commingled pickup and segmented collection were offered in 1988 in different parts of the city for newspaper, mixed paper, glass, tin, aluminum, and plastic soft drink and liquor bottles. Special drop-off programs were started such as for household hazard wastes; milk, water, juice, shampoo, and detergent plastic bottles; and separated yard waste. Curbside pickups of stoves, refrigerators, and other large items yielded appliances for recycling. Special trash tags purchased at supermarkets were required for extra garbage. The other 40 percent, about 800 tons per day of residual, was hauled to a distant arid region landfill; plans were underway to ship by rail about 3,125 tons a week.

In addition, SWU had to completely reorganize and hire new staff to handle the garbage crisis. To convince the public to recycle and instill civic pride, communication and marketing programs were launched. These included direct mail packets, utility bill inserts, door hangers, information flyers, appealing and colorful brochures, community meetings, and intensified relations with the media, who viewed recycling as a hot topic.

Convincing outreach programs were begun with citizens and county and state officials who demanded immediate action, and this created a dilemma. "Taking more time to plan risked criticism for stalling or stonewalling," remembers **Diana Gale**, former director.

"Taking action immediately risked creating programs that might be inconsistent or perhaps not cost-effective. Striking the balance between action and deliberation was one of our biggest challenges."

SWU's strategy was to use its successful year-old curbside recycling as a base to win political and public support and engage other components. It also engaged in internal management techniques such as "team-work ethic," initiation of brainstorming "quality circles," and positive feedback on job performance so as to energize and keep up workers' morale as the utility system was overhauled. "Our customer service representatives usually answer about 1,000 calls per day. At peak periods we were getting as many as 35,000 calls a day," recalls Gale. "At one point, the telephone traffic for the Solid Waste Utility shut down the rest of the City's telephone system."

Latest surveys showed that 85 percent of customers were using the new garbage collection, more than 60 percent were using the yard waste pickup, and 63 percent of eligible households were signed up for curbside recycling.

Seattle is now addressing major challenges such as stable markets for newspaper, plastics, and compost products; keeping down costs on garbage rates; expanding public participation; maintaining public support; and achieving organizational stability as a customer-oriented service organization. At SWU's Cedar Grove Composting Facility, where 60,000 tons of Seattle's yard waste were processed in 1990, the final products are being tested in terms of trace organics, pathogens, metals, oil leakage from on-site vehicles, and other characteristics.

■ Resources

An SWU information kit and numerous brochures designed to peak public interest in recycling programs are available.

Bicycle/Pedestrian Program
Transportation Services Division
Seattle Engineering Department
708 Municipal Building
600 Fourth Avenue
Seattle, WA 98104-1879 USA

Contact: Peter A. Lagerwey, Coordinator
Phone: 1-206-684-7583 • *Fax:* 1-206-684-8581

Activities: Development • *Issues:* Transportation

▲ Projects and People

The city of Seattle started a comprehensive urban trails project in 1978 to promote nonmotorized models of transportation for recreational and practical purposes. With 3 major bike trails totaling 30 miles in length and 25 miles of secondary and loop trails completed or under construction, this system is serving more than 2 million people. In addition, 80 percent of bicycle use is on streets shared with motor vehicles, and efforts are underway to make all streets and bridges "bicycle friendly."

The urban trails program helps reduce street congestion, noise pollution, air pollution, and energy use. The 40 miles of linear greenways are providing wildlife habitat, open space, and shoreline protection while beautifying the urban areas. With so many bikers

using the trails along urban waterways, public support has been engendered to clean up pollution and debris in the water.

■ Resources

Program staff have developed a one-hour presentation for national to local conferences on "Institutionalization of Bicycle Transportation Planning and Development."

Wisconsin

🏛 *State Government*

Wisconsin Conservation Corps (WCC)
30 West Mifflin, Suite 406
Madison, WI 53703-2558 USA

Contact: Topf Wells, Executive Director
Phone: 1-608-266-7730

Activities: Development; Education • *Issues:* Biodiversity/Species Preservation; Deforestation; Energy; Water Quality

● Organization

Since it began in 1983, the **Wisconsin Conservation Corps (WCC)** has employed more than 2,000 jobless 18- to 25-year-old men and women across the state in lasting projects that conserve, develop, enhance, or maintain natural resources. The concept is based on the Civilian Conservation Corps (CCC) that President Franklin D. Roosevelt created in 1933 to put youth back to work in the country's forests and fields during the Depression.

About 320 local corps members and 55 crew leaders are selected annually for their backgrounds in conservation, environmental, agricultural, or youth issues. Screening is important because many corps members leave for other jobs after several months of compiling a good work record; others who are less motivated can be discharged. Those who remain for a year have the added bonus of an $1,800 college tuition voucher. Putting in a 40-hour week for minimum wage, corps members must be physically fit to endure varying work and weather conditions.

A staff of 12.5 full- and part-time persons is located at the Madison office. The Milwaukee office houses an Energy Team with five crews. Each crew of five to eight members has an individual project site sponsored by a federal, state, or local agency; tribal government; or nonprofit organization. The WWC Board of seven volunteers oversees the program. Regarded as "excellent" by community leaders, WWC is a model for a similar program in Arizona.

▲ Projects and People

Projects generally occur on lands owned by the public or nonprofit organizations and must benefit the general population. For example, the Ashland crew built **nesting islands for the endangered common tern** and made extensive improvements to the Lake Superior waterfront. In Trempealeau County, innovative **streambank structures** were installed to stop erosion and provide a superior habitat for trout throughout the Beaver Creek Watershed. A 12-mile spur was constructed to an "economically significant" bike trail in Juneau County. WCC helps citizens **plant hundreds of thou-**

sands of trees statewide. A crew worked with Rotary clubs to turn an abandoned quarry into a city park with a system of formal gardens and a nature center.

Windbreak planting, pheasant release, prairie seed collection, red pine thinning, trails and birdhouse construction, timber stand improvements, erosion control, wild rice survey, fish hatchery assistance, shoreline stabilization, well testing, and wild-prairie establishment are ongoing projects that build an esprit de corps and confidence among the participants.

The Energy Team concentrates its efforts on projects that install wall and ceiling insulation, weatherstripping, heat duct wrap, and lighting fixture upgrades. Foam and catalytic-set plaster materials were installed as foundation insulation in Department of Correction buildings. A planned project for park or recycling use involved the construction of items of plastic lumber made of recycled polyethylene milk bottles. The team cooperates with towns and villages on recycling programs. A woodworking shop at its headquarters allows recycling bins, compost bins, and related items to be mass produced.

The WCC has also joined forces with the Minnesota Conservation Corps (MCC) to create the Mississippi River Conservation Corps (MRCC) to improve water quality and outdoor recreational opportunities in the region.

■ Resources

WCC publishes the *On Corps!* newsletter and an *At Your Service* informational brochure.

Virgin Islands

 # *Government*

Bureau of Environmental Coordination
Division of Fish and Wildlife
**Department of Planning and Natural
   Resources**
101 Estate Nazareth
St. Thomas, VI 00802 USA

Contact: **Ralf H. Boulon, Jr.**, Chief, Endangered
             Species Coordinator
Phone: 1-809-775-6762 • *Fax:* 1-809-775-3972

Activities: Education; Natural Resources Management; Research • *Issues:* Biodiversity/Species Preservation; Natural Resources Survey/Inventory • *Meetings:* Caribbean Conservation Association; Gulf and Caribbean Fisheries Institute

● Organization

The **Division of Fish and Wildlife** oversees programs in environmental education, fisheries, wildlife, and protection of endangered species. **Ralf H. Boulon, Jr.**, is responsible for the development and distribution of environmental brochures, videos, newsletters, and other public information and educational activities. In the U.S. **Virgin Islands**, Sandy Point on St. Croix "supports the major concentration of nesting leatherback turtles" in the USA and the

northern Caribbean, says Boulon, who is the principal investigator in a 10-year study of the species' nesting biology and management.

The Department is also involved in collecting data from recreational fisheries landings, testing, and developing artificial habitats such as reefs to increase recreational fishery potential and to survey important species and habitats. Data is also collected on commercial fisheries programs with some effort directed at developing alternate fisheries. James P. Beets is in charge of both programs. The wildlife program entails research and surveys of nesting seabirds, wildlife, habitats, and the effects of artificial habitats on local wildlife. **David W. Nellis** is principal investigator.

Boulan writes that the Division of Fish and Wildlife actively seeks ways to act locally and think globally within the Virgin Islands and regionally. "We have been in contact with agencies and individuals in Puerto Rico, other Caribbean islands and countries in Central and South America, and have explored ways in which we can work cooperatively. . . . We also encourage global cooperation and are willing to be involved."

Education is the most important way to bring about the protection and recovery of natural ecosystems, he believes. "Only by providing the information in such a way as to make people want to learn about the problems our world is facing and to give them solutions will we see a significant change for the better. . . . Environmentalists must facilitate this endeavor."

▲ Projects and People

The **leatherback sea turtle** (*Dermochelys coriacea*), one of the seven known sea turtle species and the one thought to be the "most morphologically divergent," has been listed as an endangered species since 1970. Reportedly, there are only 13 significant nesting sites worldwide, and six of those are located in the Western Atlantic. Scientists estimate the mature adult female population to be about 115,000. In 1984, the U.S. Fish and Wildlife Service purchased Sandy Point for incorporation into the Caribbean Islands National Wildlife Refuge System. The National Marine Fisheries Service designated the surrounding waters as critical habitat for the leatherbacks. During the past decade, **Earthwatch** volunteers and the Center for Field Research, Watertown, Massachusetts, have joined in studies to assess the size, productivity, and management priorities of the Sandy Point population.

Recent Earthwatch teams formed nightly foot patrols in the spring, checking to see that turtles are tagged and recording information such as time, place, weather, and moon phase when turtles are encountered. Turtle length and width and nest depth were measured; yolked and yolkless eggs were counted, collected, and reburied in a safety zone if the nests were threatened by erosion, for instance. Follow-up continues throughout the year with weighing of adult leatherbacks; identifying them with reshaped titanium tags attached on the left front flipper and searching for and photographing deformities and injuries; observing hatchlings and guarding them from beach debris; and excavating and examining nests. In 1990, 22 leatherbacks were observed nesting on Sandy Point; 7,200 to 7,300 successful hatchlings were recorded; and no known adult female mortality was noted. Females generally produce five nests per season with an average of 80 yolked eggs per clutch.

Erosion is the greatest natural threat to leatherback eggs here, points out Boulan, who says that egg poaching has been nonexistent lately "due to Earthwatch patrols and the randomly-scheduled presence of the Environmental Enforcement officers." Boulan estimates that some 600 Earthwatch volunteers have contributed "more than

47,440 hours diligently patrolling over 35,820 miles of beach." In this process, they also guard against predators such as yellow-crowned night herons, ghost crabs, and mongooses that are attracted by increased garbage from the beach's greater public use. Volunteers clear away debris and common beach vines to avoid entanglement, and rescue hatchlings from automobile tire ruts in the sand.

Within its **environmental education program**, the Department advises on the area's greatest ecological threats: sedimentation from construction that smothers seagrass beds and coral reefs; habitat loss such as wetlands, salt ponds, and mangroves; improperly treated or discharged sewage that feeds algal growth and robs fish and marine animals of oxygen and room to grow; improperly operated landfills that host insects, rodents, and disease and leach toxic wastes into the environment; beach litter such as plastic that can cause death of fish and wildlife; overfishing; oil slicks; improper anchoring that can damage reefs and seagrass beds for centuries; poaching of seabird and turtle eggs; and auto pollution—particularly improperly disposed oil that gets into the drinking water or ocean.

■ Resources

The Department provides free helpful brochures such as *Plastic Pollution: A Local Solution, Recreational Fishing Facts, Snorkle Dive Tips*, and *Let's Clean Up Our Act!* A resource needed is satellite telemetry to track interseason movement of turtles.

 *NGO*

Adopt-A-Stream Foundation (AASF) ✹
P.O. Box 5558
Everett, WA 98206 USA

Contact: **Tom Murdoch**, Executive Director
Phone: 1-206-388-3487

Activities: Education • *Issues:* Biodiversity/Species Preservation; Sustainable Development; Waste Management/Recycling; Water Quality • *Meetings:* International Salmon Conference

● Organization

What started in 1981 as a Snohomish County, Washington, program was so successful that the **Adopt-A-Stream Foundation (AASF)** was established in 1985 to carry on the work of the original program on a wider basis, reaching the state, national, and international levels. A grassroots effort, the Foundation's staff of four has inspired and trained hundreds of individuals, both children and adults, to work toward restoring health and wildlife to streams and wetlands. These individuals, in turn, have spread the word and enthusiasm to thousands more.

▲ Projects and People

Tom Murdoch, founder of Adopt-A-Stream, believes: "If you can get people the right information and tell them how to use it, you can effect change." With this aim, the Foundation focuses on education and hands-on training. On the first day of a typical Adopt-A-Stream free workshop, several hundred people learn the essentials of planning, organizing, and implementing stream adoption groups, as well as how to get involved in environmental politics. The second day's lesson, limited to 20 committed participants who will go on to lead

stream adoption groups, focuses on actual, on-site stream monitoring procedures. Most participants are teachers, who later involve their students. One elementary school has managed to revive the salmon population of Pigeon Creek after 25 years without the fish. Other schools have, through Adopt-A-Stream, established "sister school" relationships with Japanese and Canadian schools that are also environmentally involved.

Adopt-A-Stream is working with Japan's **Come Back Salmon Society** and Canada's **Save the Salmon Society**, sharing experience and knowledge. This partnership has resulted in two international conferences in Japan with the goal of increasing children's involvement in environmental efforts. Schools as far away as Italy have asked for advice on how to begin their own environmental education programs.

The Foundation has also trained over 200 people to work as **volunteer data collectors** for the U.S. Environmental Protection Agency's (EPA) geographic information system (GIS). The information they will gather over two years of stream monitoring can then be used by state agencies and citizens to make informed decisions about actions with potential environmental impact.

■ Resources

Corporate and foundation financial support are needed to continue and expand educational and training activities.

The Foundation distributes two how-to manuals for those wishing to do their part for nature conservancy: *Adopting a Stream: A Northwest Handbook* and *Adopting a Wetland: A Northwest Guide*, both written by Steve Yates and illustrated by Sandra Noel. *Streamlines*, a quarterly environmental education newsletter, is available to members. *A Streamkeeper's Field Guide: Watershed Inventory and Stream Monitoring Methods* with a companion video is also available.

African Wildlife Foundation (AWF) ⊕
1717 Massachusetts Avenue, NW
Washington, DC 20036 USA

Contact: **Paul T. Schindler**, President and Chief
Executive Officer
Phone: 1-202-265-8394 • *Fax:* 1-202-265-2361

Activities: Education • *Issues:* Biodiversity/Species Preservation

● Organization

Founded in 1961, the independent, nonprofit **African Wildlife Foundation (AWF)** works with African governments and people to ensure the preservation of the continent's wildlife and protected areas. With the support of 100,000 members, as well as of various foundations, agencies and corporations, the Foundation employs 40 men and women, most of them based in its field office in Nairobi, Kenya. The AWF "remains the only such organization to work solely in Africa."

▲ Projects and People

The AWF spreads its conservation message through a variety of educational and other programs. Early on, the AWF established two wildlife management colleges, where hundreds of African rangers and wardens have received their training. The organization also funds many educational programs for Africa's children, working toward a more conservation-minded future generation.

In its efforts to protect African wildlife such as the mountain gorilla, black rhino, and elephant from poachers, the AWF supplies managers of African national parks and reserves with radios, vehicles, and other urgently needed equipment. This and other AWF projects help to improve enforcement of poaching laws. The rhinos and elephants are still at great risk; however, "only one gorilla has been lost to poachers since 1983" thanks to AWF-supported efforts at Rwanda's Parc des Volcans.

■ Resources

AWF's newsletter, *Wildlife News*, is published three times a year and is available to members.

Air and Waste Management Association (AWMA) ⊕

P.O. Box 2861
Pittsburgh, PA 15230 USA

Contact: **Martin E. Rivers**, Executive Vice President
Phone: 1-412-232-3444 • *Fax:* 1-412-232-3450

Activities: Education • *Issues:* Air Quality/Emission Control; Energy; Global Warming; Waste Management/Recycling

● Organization

Founded in 1907, the **Air and Waste Management Association** (AWMA) is a 12,000-member association for environmental professionals that promotes exchanges on environmental management issues and operates in more than 50 countries. This worldwide network includes physical and social scientists, engineers, lawyers, and managers. AWMA attracts government and agency decision makers, academics, and researchers who wish to exchange technical and managerial information about air pollution control and waste management.

AWMA's 15-member board evenly represents industry, government, and academic/research/consulting organizations. Four councils develop AWMA's programs: Technical, Sections, Education, and Communications and Marketing. Association sections and chapters also offer additional independent programs germane to their own geographical area.

The AWMA's annual meeting, one of the largest environmental conferences in North America and attended by thousands, features a five-day technical program, a three-day exhibition, and auxiliary meetings. Throughout the year, the association conducts hundreds of conferences, workshops, seminars, and section meetings.

Publishing a variety of periodicals, books, reprints of technical papers, and training manuals, AWMA also offers a widely quoted monthly technical journal. The association's awards program recognizes the accomplishments of environmental problem solvers.

▲ Projects and People

The Technical Council is divided into three groups: air, environmental management, and waste. Each group is made up of divisions.

The air group includes sources, emission control technology, basic sciences, and an intercommittee task force on ozone. Environmental management includes industrial processes, effects, measurements, and program administration. The waste division includes source control, physical/chemical treatment, thermal treatment, and land treatment and disposal.

The Sections Council serves as the communications link between AWMA and its local units. Formal education programs for professionals are the province of the Education Council, including classroom sessions, training manuals, computer software, videotape courses, and the U.S. Environmental Protection Agency's (EPA) Air Pollution Training Institute (APTI) manuals. The Communications and Marketing Council handles membership, international affairs, government affairs, publications, and marketing.

Every June, AWMA's international conference enables its members to present papers, meet in symposia, and interact informally. The annual meeting includes 180 technical sessions covering waste, air, and environmental management. AWMA also presents specialty conferences on critical topics, governmental affairs seminars, and workshops.

Martin E. Rivers, executive vice president since 1987, directs all AWMA activities. He previously served as director of environmental quality for the Tennessee Valley Authority (TVA) and Director General for the Air Pollution Control Directorate of Environment Canada.

■ Resources

Rivers publishes the *Journal of Air & Waste Management Association*, the Association's premier publication, which is listed in an available *Publications Catalog*. The monthly *Journal* carries peer review technical papers, and according to the Library of Congress, is one of the world's most frequently cited scientific journals. Every other month, members receive the association's newsletter featuring updates on association members, activities, and other timely information.

AWMA's *Transaction Series* compiles peer reviewed papers providing detailed technical analyses and advances. Members unable to attend the annual conference or specialty conferences can receive reprints from the *Specialty Proceedings Series*. AWMA also reprints various training manuals and an *Annual Directory* and *Resource Book*. Color videotapes on technology transfer for individual study supplement the educational program, which also features an *Environmental Resources Guide* designed as a teacher curriculum for grades six to eight to improve the "environmental literacy of our youth."

The Alan Guttmacher Institute (AGI) ⊕

111 Fifth Avenue
New York, NY 10003 USA

2010 Massachusetts Avenue, NW
Washington, DC 20036 USA

Contact: **Beth Fredrick**, Director of Communications and Development
Phone: 1-212-254-5656, 1-202-296-4012
Fax: 1-212-254-9891, 1-202-223-5756

Activities: Education; Legislative; Research • *Issues:* Population Planning

● Organization

Leonore Guttmacher incorporated AGI in 1977, named in honor of her late husband, **Alan F. Guttmacher**, a former president of Planned Parenthood Federation of America. AGI research, policy analysis, and public education, led by a staff of 55, stimulate the public debate about reproductive health and rights. AGI seeks to bridge the worlds of science and policymaking by focusing attention on health technologies and their relation to abortion politics. In addition to its data collection and analysis, AGI helps build coalitions that take its message to the U.S. Congress and state and local legislatures.

In the late 1980s, an AGI study of reasons underlying unwanted pregnancies in the Western world found that the proportion of sexually active women ages 18 to 44 had increased, despite growing awareness about the risk of HIV infection and sexually transmitted diseases. AGI examinations of the failure rates of condoms helped to explain why 43 percent of 3.4 million unintended pregnancies each year occur among couples who report using some type of birth control.

Another study reviewed state responses to teenage pregnancy and found most jurisdictions providing "after the fact" assistance to teens. The Institute's research, including legal queries, also supports the debate about freedom to terminate unwanted pregnancies. AGI aims to reduce the incidence of infant mortality in the USA by lowering financial barriers to quality maternity care for low-income women.

▲ Projects and People

AGI studies the **status of adolescents, family formation, and teenage pregnancy** in Brazil, Colombia, and Peru as part of its expanding international focus. Working with Latin American researchers, AGI found that rising educational levels and work-force participation and urbanization have not discouraged young women from early childbearing. The resulting study, *Adolescents of Today, Parents of Tomorrow*, published in Portuguese and Spanish, received widespread attention throughout the region.

What detailed information students receive about **sex education** comprises another major area of AGI inquiry. Studies such as *Risk and Responsibility: Teaching Sex Education in America's Schools Today* demonstrate how HIV/AIDS currently drives the type of information students receive, while basic information about sex education receives less emphasis.

AGI continues to gather information about who has abortions, under what circumstances, and in what settings. Of 1,900 women asked in AGI's first survey, two-thirds stated that they could not afford to raise a child. A second survey of 10,000 women closely linked reasons women gave for having an abortion with their socioeconomic status. A third study, *Abortion and Health*, analyzed available research concerning the link between abortion and women's health. Jeannie I. Rosoff, AGI's president, has the Institute gearing up for the major legal and legislative battles following in the wake of *Webster v. Reproduction Health Services* and pivotal reproductive rights cases now before the U.S. Supreme Court.

■ Resources

AGI publishes two professional journals, the bimonthly *Family Planning Perspectives* and the quarterly *International Family Planning Perspectives* with a worldwide circulation of 45,000. The Institute's public policy newsletter, *Washington Memo*, has a reader-

ship of over 2,000. Included in its publications list are periodicals on family planning; a *State Reproductive Health Monitor*, and books and other materials focusing on teenage pregnancy, sex education, abortion, family planning, maternal and child health, and public policy. A recently updated *Abortion Services in the United States, Each State and Metropolitan Area* presents articles and data from hospitals, clinics, and private physicians.

Alaska Center for the Environment (ACE) ❦
519 West 8th Avenue, Suite 201
Anchorage, AK 99501 USA

Contact: **Kevin Harun**, Executive Director
Phone: 1-907-274-3621 • *Fax:* 1-907-274-4145

Activities: Education • *Issues:* Air Quality/Emission Control; Deforestation; Energy; Transportation; Waste Management/Recycling; Water Quality

● Organization

Started in 1971, the **Alaska Center for the Environment (ACE)** is a nonprofit, grassroots organization which is committed to the preservation of Alaska's natural resources. ACE strives to heighten public consciousness and participation in environmental matters. To this end, ACE provides education, policy analysis, and policy development, mainly impacting on south-central Alaska—yet including statewide and federal issues.

ACE oversees policymaking in areas such as state land, rivers, and tidelands; state parks; forests; and wetland conservation. More than 104 million acres of uplands—a total area larger than the state of California—are included, as well as 50 million acres of tidelands. The Center educates the public and solicits help in passing choice legislation in such areas as deforestation and hazardous and toxic waste.

Volunteers and organization staff members actively contribute and serve other environmental or activist groups on a national and international level. Some ACE members are called upon to participate in **congressional advisory committee** meetings. After the Exxon *Valdez* 1989 oil spill in Prince William Sound, ACE became a part of the Oil Reform Alliance (ORA), a coalition of environmental groups united to assist the improvement of oil spill policies and oil industry. Concerns were expressed about Exxon's use of "bioremediation," which involves chemical fertilizers, its digging up of beaches in cleaning efforts, and lack of technology to fully complete its mission. ORA adopted measures that led to improved federal and state requirements for double hulled tankers and the formation of a **citizen oversight committee on oil and hazardous substances** to see that "hard-won reforms are implemented as intended."

▲ Projects and People

Following the oil spill, ACE took a central role in developing and petitioning for new legislation in spite of industry opposition—such as measures that allow "citizens to sue polluters directly if the state fails to take action." With ORA, the Center produced an **educational exhibit, Beyond the Spill**, informing the public of its details and ramifications.

With the 1990 opening of the Matanuska-Susitna Valley office, ACE is positioned to help protect the immense 15.7 million acres of land with parks, fisheries, and wildlife—especially grizzly bears, caribou, and wolves. ACE reports that local developers and multinational interests are threatening this area by demanding large scale clear-cutting of the forests. As part of a local coalition, ACE "helped to shape the management plan for six rivers in the Susitna Valley Recreation Rivers System" by calling for the prohibition of "logging within one half mile of these important fishing rivers and [designating] portions of these rivers as non-motorized."

ACE is the leader of the Trailside Discovery Camps (TDC) educational program for children, ages 4 to 13, and their families, which "teaches environmental concepts and awareness emphasizing hands-on learning and outdoor skills." These camps are located in Anchorage, the Matanuska-Susitna Valley, and Homer. The Alaskan Quest Programs are related environmental education experiences for teenagers.

In other pursuits, ACE is helping see that the Forest Practices Act is implemented as ACE and other grassroots conservation groups intended. ACE also supports broad citizen involvement in waste cleanup, pollution prevention, and waste management decisions.

■ Resources

ACE produced *Part of the Solution: A Guide to Greener Living in Southcentral Alaska*, a manual on how to achieve a cleaner, more efficient lifestyle—originally distributed to over 80,000 households on Earth Day 1990. ACE also publishes the *Center News* newsletter five times a year.

Alaska Craftsman Home Program (ACHP)
165 East Parks Highway, Suite 202
Wasilla, AK 99654 USA

Contact: **Tim Sullivan**, Executive Director
Phone: 1-907-373-2247 • *Fax:* 1-907-373-0793

Activities: Education • *Issues:* Energy; Health

● Organization

From wetlands to desert to permafrost, the climate in Alaska is significantly colder than the climate in the rest of the USA, but often home construction has followed the same patterns as in the lower 48 states. Moreover, energy and transportation costs in the nation's largest state are high. The **Alaska Craftsman Home Program (ACHP)** is a not-for-profit organization that is devoted to teaching Alaskans how to build homes that are much more energy-efficient and to changing attitudes within the industry and among the general public.

According to ACHP consultant **William Harvey Bowers**, "if all existing Alaskan homes were retrofitted to meet ACHP energy standards, the yearly savings would be 800 million dollars or greater . . . that equates to one half of the state budget . . . money that could be recycled back into the economy." ACHP points out that the U.S. government originally provided military housing designed for Guam.

▲ Projects and People

Through running workshops and teaching methods of building energy-efficient homes as well as retrofitting older homes to meet tougher energy standards, ACHP wants simultaneously to protect the Earth by saving fossil fuels and providing Alaskans with trouble-free housing that meets the health requirements of the occupants.

More than 2,000 people have participated in ACHP's home building workshops; more than 20 homes have been certified. Such homes use from 50 to 80 percent less energy than other homes of comparable sizes. Certified homes must meet conditions based on certain air-leakage tests, insulation standards, controlled ventilation, outside combustion air with no natural aspirating stoves or pot burners inside, energy-efficient lighting and appliances, and passive solar—whenever possible.

In addition, ACHP uses Canadian HOT-2000 computers to simulate desired insulation levels for houses in particular climates.

■ Resources

ACHP runs a number of two-day workshops on such topics as ventilation, airtightness, heating, and retrofitting. Information about schedules and costs can be obtained from ACHP, which also publishes a *Home Building Manual* and distributes assorted videos.

Alliance for the Chesapeake Bay (ACB)
6600 York Road, Suite 100
Baltimore, MD 21212 USA

Contact: **Kathleen K. Ellett**, Citizen Monitoring
Director
Phone: 1-410-377-6270 • *Fax:* 1-410-377-7144

Activities: Education; Research • *Issues:* Water Quality

● Organization

The **Alliance for the Chesapeake Bay (ACB)**, founded in 1971, is dedicated to building and maintaining a public-private alliance of business, civic, scientific, recreation, and conservation organizations that can work together to restore the Chesapeake Bay, one of the nation's most vital and complex estuaries.

▲ Projects and People

The ACB began a pilot water-quality testing project for volunteers in 1985 as one of the activities funded under its Chesapeake Bay Program public participation grant from the U.S. Environmental Protection Agency. **River Trend**, a citizen monitoring program, is a volunteer project that enables citizens to help monitor the quality of the water in their own backyards.

Volunteer-collected samples from near-shore sites show "some significant differences" from most government and university samples taken at mainstream sites, ACB reports. Volunteers weekly take a clean bucket and sampling kit to a designated spot along the river shores. "A sample is scooped from the river and the water temperature is measured; the bucket of water is then carried back to the house for further analysis." Filling vials, tubes, and bottles, the volunteers measure the pH (acidity or alkalinity), dissolved oxygen contents, and salinity of the water. Among other purposes, the data helps examine correlations between certain measured variables, such as low-dissolved oxygen, and the frequency of observed events, such as fish kills and algae blooms.

Approximately 75 River Trend monitors observe Virginia's Bay tributaries on five rivers as part of the Virginia Citizen Monitoring Program. About 100 monitors observe creeks and rivers on the Eastern Shore. In Pennsylvania, River Trend volunteers keep tabs on the water quality of Conestoga Creek.

■ Resources

An information brochure is available.

Alliance for Environmental Education
P.O. Box 368, 51 Main Street
The Plains, VA 22171 USA

Contact: **Thomas P. Benjamin**, Staff Director
Phone: 1-703-253-5812 • *Fax:* 1-703-253-5811
E-Mail: Econet: alliance

Activities: Education • *Issues:* Air Quality/Emission Control; Biodiversity/Species Preservation; Deforestation; Energy; Global Warming; Health; Population Planning; Sustainable Development; Transportation; Waste Management/Recycling; Water Quality

● Organization

Founded in 1972, the **Alliance for Environmental Education** is, according to its literature, "the largest advocate for environmental education in North America, representing over 50 million members through its 155 affiliate organizations." Made up of a coalition of business, labor, government agencies, and other nonprofit groups, the Alliance seeks to keep environmental concerns at the forefront of the national agenda, through the dissemination of materials to schools and to the public, and through advising the federal government on its role in environmental education.

Steven C. Kussman, chairman of the Alliance, states: "New values and attitudes toward the environment can only be established through a life-long learning process that goes beyond the school system and into the home, community, and workplace." By providing a neutral forum for exchange of information representing all shades of opinion, the Alliance works to build consensus for a national environmental policy.

▲ Projects and People

In 1989, in partnership with the Tennessee Valley Authority (TVA) and the U.S. Environmental Protection Agency (EPA), the Alliance created the **Network for Environmental Education**—100 U.S. and international centers for environmental education that operate on the campuses of colleges, universities, and other institutions and support each other with research, education programs, and ideas. New technologies link the centers electronically and provide instantaneous access to information and resources.

The centers work in the following four basic areas: **Professional teacher development**—offering teaching skills designed to help students learn to think for themselves in making educated and informed judgments about the environment. **Community outreach**—including issue forums, workshops and lectures, tours and technical assistance geared to specific community environmental problems. **Curriculum, program, and materials development and dissemination**—consisting of curriculum, programs and materials designed to address grassroots, regional, national, and international environmental issues and to reach citizens of all ages and geographic regions. **Research and evaluation**—in which the centers monitor and assess the success of programs and resource needs and actively pursue basic scientific research, especially the potential for environmental education through new technologies for learning.

An example of a new Alliance center is the **Resources Heritage Center at Deep Portage** in Hackensack, Minnesota, which is designed to be energy efficient with passive solar heating, super insulation, and partial Earth sheltering. A new energy center demonstration model will feature a dual system heat source, a four-day heat storage capacity, and an alternative electrical generating capability. Capital costs are being kept down in order to allow for "future retrofitting as technology evolves superior methods." Additional information can be obtained from Deep Portage, Route 1, Box 129, Hackensack, MN 56451, 1-218-682-2325.

Another recent network addition is the **Manomet Bird Observatory**, a center for long-term environmental research and education on Cape Cod Bay. As described in the Alliance's *Network Exchange* newsletter, the Observatory's research projects investigate bird population trends, behaviors, and migratory patterns; marine mammal adaptations to environmental change; and the dynamics of tropical forest regeneration. The Observatory is also a leader in environmental education in New England, providing science experiences for children in the field as well as in the classroom.

The Observatory's five long-term studies are on migrant land birds, with research being conducted in the Gaspe breeding grounds of **Canada** and wintering sites in **Belize**; shore migration and habitat use and the development of international conservation policy for wetlands and coasts, with work taking place from the Arctic to the Antarctic; breeding strategies and population dynamics of herons and gulls, based in New York City and Massachusetts; the structure and function of tropical forests and how their management can be improved, with work in progress in **Central America** and **Malaysia**; and seabirds and marine mammals—their population levels, food consumption, and possible conflicts with commercial fishing activity in the northwest Atlantic.

Another Alliance achievement was the signing into law in late 1990 of the **National Environmental Education Act**, marking the culmination of a two-year effort by the Alliance and other supporters. The Act established an Office of Environmental Education within the EPA, and a public/private National Environmental Education and Training Foundation. Both entities provide funding mechanisms for environmental education and training initiatives. "The grassroots support given the legislation by Alliance members was the key to its passage," said Kussman.

In its 1991 Washington, DC, conference, with the theme "Network Globally, Act Locally," the Alliance drew environmental experts from the former Soviet Union, Hungary, Czechoslovakia, Colombia, Brazil, and other countries, who demonstrated ways in which the world initiative for environmental education is being strengthened.

■ Resources

Members of the Alliance can participate in the Network for Environmental Education at a regional center; access the national electronic bulletin board, EcoNet, for information on environmental resource materials, speakers, meeting notices, and programs; receive the *Network Exchange,* a bimonthly newsletter; and may attend the Alliance's annual conference, which brings together worldwide environmental leaders and center directors.

The Manomet Bird Observatory can be reached for further information at P.O. Box 936, Manomet, MA 02345, 1-508-224-6521.

The Alliance to Save Energy
1725 K Street, NW, Suite 914
Washington, DC 20006-1401 USA

Contact: **William A. Nitze,** President
Phone: 1-202-857-0666 • *Fax:* 1-202-331-9588

Activities: Education; Political/Legislative; Research • *Issues:* Air Quality/Emission Control; Energy; Global Warming; Transportation • *Meetings:* International global warming negotiations

● Organization
The Alliance to Save Energy describes itself as "a nonprofit coalition of government, business, environmental and consumer leaders dedicated to increasing the efficiency of energy use." Funded by supporters ranging from philanthropic foundations and the federal government to individual consumers, the Alliance's efforts benefit all levels. Well thought out energy policies, the Alliance believes, will "improve the environment, strengthen the economy, enhance national security, and provide affordable and reliable energy to consumers."

After researching and analyzing given energy efficiency policies, programs, and technologies, the Alliance makes its findings known to policymakers, energy providers, the business world, the media, and consumers.

▲ Projects and People
On the federal level, the Alliance is a widely respected source of information and expertise, working closely with legislators and the Department of Energy to formulate a comprehensive National Energy Strategy, an ongoing endeavor. The Alliance has also worked to include energy efficiency provisions in the National Affordable Housing Act, and incentives for utilities to promote efficiency in the Clean Air Act Amendments of 1990. In addition, the Alliance has recommended changes in the federal government's own energy consumption patterns; if adopted, the new strategies could save taxpayers $864 million per year.

Individual consumers and homeowners benefit from the Alliance's activities to legislate more fuel efficiency for vehicles, more energy efficiency in homes, and mortgage breaks for homes that improve energy efficiency. They are also the targets of campaigns by the Alliance to educate the public on measures to promote efficient use of energy. Through one field test, the Alliance found that people who received "intensive education about home energy use lowered their energy consumption by 16.4 percent."

The president of the Alliance to Save Energy, William Nitze, was from 1987 to 1990, Deputy Assistant Secretary of State for Environment, Health and Natural Resources in the Bureau of Oceans and International Environmental and Scientific Affairs. In that capacity, he formulated policy and participated in international negotiations on global warming, the ozone problem, conservation of rainforests and wildlife, and international shipment of hazardous substances.

■ Resources
The Alliance to Save Energy publishes the findings of its research, educational, and policy advocacy projects in a variety of formats, including brochures, manuals, and computer software. Recent titles include *A Resource Guide for Exporting Energy-Efficient Products* and *Moving Consumers to Choose Energy Efficiency.*

American Association of Zoological Parks and Aquariums (AAZPA) ⊕
Oglebay Park
Wheeling, WV 26003-1698 USA

Contact: **Robert O. Wagner,** Chief Administrative Officer
Phone: 1-304-242-2160 • *Fax:* 1-304-242-2283

Activities: Education; Political/Legislative • *Issues:* Biodiversity/Species Preservation • *Meetings:* CITES

● Organization
AAZPA, a nonprofit organization founded in 1924, advances the concerns and interests of zoological parks and aquariums through conservation, education, scientific studies, and recreation. AAZPA has 18 staff members and represents nearly every zoological park, aquarium, wildlife park, and oceanarium in North America. AAZPA assists these organizations, their professionals and employees in acting as survival centers for endangered species through their animal exchange program as well as fosters animal behavior studies. AAZPA has approximately 3,800 members, including 220 nonprofit organizations which represent a combined membership of 350 citizens—offering national and regional conferences to members, including a professional zoo and aquarium management school.

AAZPA promulgates the theory that today's zoos are the Noah's Arks of the twentieth century; while zoo and aquarium professionals feel that working to preserve natural habitats is a conservation priority, many species teetering on the verge of extinction must have captive-breeding programs as survival insurance.

▲ Projects and People
The Species Survival Plan is AAZPA's strategy for the long-term survival of selected endangered species. Recognizing that propagation at zoos may be the last hope for fast-disappearing species, participating zoological institutions cooperate by sending endangered animals to zoos that are able to breed them. To further this effort, AAZPA operates an animal exchange which actively assists members in the distribution of surplus and unwanted animals for conservation, exhibition, and propagation.

International Species Inventory System (ISIS) is AAZPA's database for accumulating census and vital statistics on captive wildlife around the world.

■ Resources

AAZPA provides a monthly briefing of events and legislative and conservation activities to its members. Publications also include the directory *Zoological Parks & Aquariums in the Americas*, *Manual of Federal Wildlife Regulations*, *Zoological Park and Aquarium Fundamentals*, and several brochures.

American Cetacean Society (ACS) ⊕
National Headquarters
P.O. Box 2639
San Pedro, CA 90731 USA

Contact: **Paul Gold**, Executive Director
   Phone: 1-310-548-6279 • *Fax:* 1-310-548-6950

Activities: Education; Legislative; Research • *Issues:* Biodiversity/Species Preservation; Marine Mammals • *Meetings:* IWC

● Organization

The American Cetacean Society's (ACS) staff of two professionals serves 3,000 volunteer members concerned about the conservation of marine mammals, primarily whales and dolphins, and their natural environment. Founded in 1967, ACS calls itself "the oldest whale protection group in the world." Its members include marine scientists and educators as well as interested citizens.

ACS promotes an aggressive education, research, and conservation program—continually reminding the public that whales and dolphins are still endangered. For example, "3,000 pilot whales are killed in the [Danish] Faroe Islands drive fishery each year," ACS reports. Fourteen billion pounds of trash are discarded at sea each year, including 640,000 plastic containers daily—contributing to the deaths of many animals including sea birds, sea lions, sea otters, seals, and sea turtles. Lost fishing nets, oil spills, and other pollutants dumped in the ocean are other hazards. In 1988, U.S. tuna fishing cost the lives of 18,800 dolphins and porpoises.

Actively supporting marine mammal research, ACS maintains a cetacean research library at its national headquarters and holds biennial conferences and other occasions for information exchange. The Society's Washington, DC, representative monitors legislation, attends International Whaling Commission (IWC) meetings, and serves as liaison with other wildlife organizations. Conservation Committee members attend public hearings, prepare comments, and work with legislators and government officials to educate the public and influence world opinion. The Society's nine U.S. chapters are located in Galveston, Texas; Los Angeles, Monterey, San Diego, Santa Monica, Santa Barbara, Orange County, California; New York/New Jersey, and Seattle, Washington.

▲ Projects and People

The ACS biennial conference, entitled They're Not Yet Saved, serves both as a key opportunity for professionals to exchange information and views and as an educational and organizing tool. At the 1990 conference, participants heard that while the threat to large whales may be waning, small-toothed whales are in desperate trouble. In Mexico, the world's 200 to 500 remaining *vaquitas* porpoises find their way into the gillnets of totoaba fishermen who

have no alternative means of making a living. Nearly 80,000 dolphins die in the nets of tuna fishing boats each year in the eastern tropical Pacific. Driftnets kill thousands of dolphins annually in **Peru**. Entanglements also threaten baiji river dolphin in **China**, humpbacked and bottlenose dolphins in the Mediterranean, and harbor porpoises in the North Atlantic and off California.

Yet the greatest threat are habitats spoiled with toxins, sewage, and oil spills that poison mother whales' milk; contaminate calves with high levels of PCBs (polychlorinated biphenols) and DDT; and cause malignant tumors, infections, and other maladies.

ACS president **Steven Katona**, who says that education and the ability to strike compromises are the key to change, believes people are "in the process of giving up our historical attitude of taking what we wanted, when we wanted it, and as fast as we could." Meanwhile, ACS continues to be the voice for "whales [that] cannot speak for themselves; they depend on us for that."

■ Resources

Because it has never been easy to find good information about whales and dolphins, ACS developed the *Whale and Dolphin Fact Pack* that includes 11 dolphin species sheets, 6 how-to-draw dolphin sheets, 17 whale species sheets, 4 subject sheets, 5 how-to-draw a whale sheets, bibliography, and glossary.

Whalewatcher: Journal of the American Cetacean Society, and *WhaleNews*, the Society's newsletter, are the backbone of its educational efforts. The *Journal* publishes scientific, educational, and technical articles on ACS research and research germane to ACS interests. The newsletter provides updates, news alerts, reviews of international government policy changes, IWC activities, and descriptions of new threats to cetaceans.

Katona and Jon Lien, both respected scientists, have co-authored *A Guide to the Photographic Identification of Whales: Based on Their Natural & Acquired Markings*, a book to enable anyone with a camera who would like to assist in the urgently important work of whale research. Both a "how-to" book and a field guide for identifying whales, it describes how amateur photos can be used for research catalogs and identification processes. An ACS gift brochure is also available.

American Farmland Trust (AFT)
1920 N Street, NW, Suite 400
Washington, DC 20036 USA

Contact: **Ralph E. Grossi**, President
   Phone: 1-202-659-5170, 1-202-659-8339

Activities: Education; Direct Land Protection; Political/Legislative • *Issues:* Farmland Protection; Sustainable Development; Water Quality

● Organization

American farmers face an ever-increasing burden in the environmental age, says **Ralph E. Grossi**, president of the **American Farmland Trust** (AFT), established in 1980. "Not only are they being asked to provide a safe, abundant supply of food, but they are also being asked to be the keeper of the natural world around them. Farmers must seize this opportunity to show that they are careful stewards of the land and take the lead in promoting environmental action."

American Farmland Trust spearheads this effort through its stated mission of stopping the loss of productive farmland and promoting healthy farming practices. Through grassroots-type public education and technical assistance in policy development, AFT's coalition of 20,000 farmers, conservationists, educators, and business people assist landowners in meeting their farmland protection goals. Direct efforts have provided protection for 38,833 acres of land in 18 states through the use of conservation easements. These legally recorded agreements between the farmer and AFT ensure that land remains open for agricultural use.

▲ Projects and People

Tackling the environmental and economic concerns of farmers directly is American Farmland Trust's way. A current program offering small, **on-farm demonstrations of resource-conserving practices** shows farmers how they can slowly reduce the amount of pesticides, fertilizers, and fuel used on the farm without fear of lost revenues. With support from the state of Illinois and the Illinois Sustainable Agriculture Society, a dozen farms experimented with side-by-side plots to compare conventional and alternative farming practices. The AFT reports that three-quarters of the participants maintained or increased yields on the demonstration plots using fewer inputs. Similar or better results are expected from the now 150 farms involved in the demonstration projects.

"Never before have farmers had this level of encouragement and assistance from public and private agencies to actively protect natural resources," writes Grossi in the AFT's quarterly magazine. The 1990 farm bill includes over 10 conservation provisions, largely the work of AFT's lobbying efforts. Easement programs allow farmers to be financially compensated for voluntarily cooperating with environmental protection practices.

The United States annually loses 1.5 million acres of farmland to urbanization, according to the American Farmland Trust. The Seligson-Gum Farm, 339 acres of prime farm and forest land in Maryland, was nearly part of that grim statistic. The farm was in danger of being sold to developers and carved into 23 house lots when the AFT stepped in. It engineered a plan to purchase the land and keep it open for agricultural and recreational use. One-third was sold to the Maryland Department of Natural Resources for parkland and the remaining two-thirds leased to two farmers who have worked in the area for twenty years. The Seligson-Gum Farm is a favorite AFT success story.

For the 1990s, AFT has a six-point strategy to protect the USA's agricultural resources: "restructure farm programs to encourage stewardship, shift funding from traditional commodity programs to conservation programs, enforce existing conservation laws, implement conservation provisions under the 1990 Farm Bill, conduct a comprehensive study of land available for food and fiber production, and stimulate research on agricultural conservation issues."

Such a plan calls for the complete restructuring of federal farm programs so that long-term land productivity will not be sacrificed, indicates Grossi. "The $40 million we spend each day on farm support programs shortchanges conservation while encouraging production of crops that are in oversupply," he writes. "We're losing three acres of farmland each minute to encroaching urban sprawl, and three billion tons of topsoil each year to erosion."

A third-generation dairy and beef farmer, Grossi has received numerous agricultural and environmental awards.

■ Resources

American Farmland Trust prints a quarterly publication updating members on pending legislation, current projects, and opportunities for participation. Other public education materials are published by AFT; a publications list is available.

American Fisheries Society (AFS) ⊕
5410 Grosvenor Lane, Suite 110
Bethesda, MD 20814-2199 USA

Contact: **Paul Brouha**, Executive Director
   Phone: 1-301-897-8616 • *Fax:* 1-301-897-8096

Activities: Education; Political/Legislative; Research • *Issues:* Fisheries Resources and Management

● Organization

The **American Fisheries Society (AFS)** is an international, nonprofit organization controlled by its membership and dedicated to strengthening the fisheries profession, advancing fisheries science, and conserving fisheries resources. With a staff of 20 and more than 9,200 members, AFS offers career support for those in the fisheries field and professional development opportunities through annual meetings, symposia, workshops, and certification education as an Associate or Certified Fisheries Scientist. AFS influences legislative policy with respect to fisheries resource management and water pollution standards.

▲ Projects and People

AFS played a major role in the conception and passage of the Wallop-Breaux Amendment to the Sportfish Restoration Act, which has greatly increased state funding for **sportfish restoration**. Working with the USDA Forest Service, AFS initiated the **"Rise to the Future"** fisheries program in 1985, which has nearly tripled the number of federal fishery biologists and increased funding tenfold from $5 million to $50 million. AFS encourages opportunities for political activism through its broad membership, as part of the Society's efforts to protect and manage aquatic resources.

Paul Brouha, executive director, came to AFS from the U.S. Department of Agriculture (USDA), where he worked with fisheries programs. He writes publications on fisheries and lectures in the area of federal land management and the protection of aquatic species.

■ Resources

The following journals are published: *Transactions of American Fisheries Society* on original fishery biology research and related aquatic topics; *The Progressive Fish-Culturist* on aquaculture; *Northern American Journal of Fisheries Management; Journal of Aquatic Animal Health* on health research of fish and other aquatic species; and *Fisheries: A Bulletin of American Fisheries Society,* the official AFS magazine.

American Littoral Society (ALS)

Sandy Hook
Highlands, NJ 07732 USA

Contact: **D.W. Bennett**, Executive Director
Phone: 1-908-291-0055

Activities: Education; Legislative • *Issues:* Species Preservation; Tidal Wetland Preservation • *Meetings:* American Littoral Society

● Organization

A small group of divers, fishermen, and naturalists founded the **American Littoral Society (ALS)** in 1961. Members are both professionals and amateurs, including scientists, beachcombers, and other individuals interested in the enjoyment, study, and conservation of the coastal environments. Their purpose is to provide a "unified voice advocating protection of the delicate fabric of life along the shore." Executive director **D.W. Bennett** notes, "When we fill coastal wetlands, pollute estuaries, and dump toxic wastes in rivers and at sea, we threaten the ocean's future."

Devoting most of its energy to outdoor field experiences, the Society provides leaders for group explorations from Spruce Head, Maine, and the Olympic Peninsula in the Pacific Northwest to Cumberland Island, Georgia. The Society uses membership dues to watch and protect coastal beaches, wetlands, rivers, and estuaries through education and media coverage as well as lobbying efforts in Washington, DC, and state capitals. The emphasis is on preserving the entire ecosystem, threatened when juvenile fish and shrimp lose their wetland homes and waterfowl have no place to build shelter.

▲ Projects and People

Marine biologists increase their knowledge of fish migrations and growth rates with information supplied by the **ALS Fish Tag Program**. About 8,000 fish are tagged each year; the information is stored in the National Marine Fisheries Service computer system.

When the Society was founded in 1961, a group of divers established the Society's Divers' Section, the fastest growing division at ALS. Dive and study trips emphasize education and **monitoring the underwater environment**. A technical advisory group includes experts **Stan Waterman**, **Sylvia Earle**, and **Richard Ellis**. Expeditions to Bonaire in the Caribbean are led by Society members. Other field trips include oyster diving in the Chesapeake Bay, whale watching off the New England coast, visits to the Bermuda Biological Station, hiking in Washington State, and canoeing and rafting on various rivers.

■ Resources

Members receive the *Underwater Naturalist* quarterly magazine. The publications list includes a handbook for high school club advisors and catalogs of the California and New Jersey coastlands. Merchandise is offered through the Society's General Store.

American Lung Association (ALA)

1726 M Street, NW, Suite 902
Washington, DC 20036-4502 USA

Contact: **Ronald H. White**, Director of
Environmental Health
Phone: 1-202-785-3355 • *Fax:* 1-202-452-1805

Activities: Education; Legislative; Research • *Issues:* Air Quality/Emission Control; Health; Transportation

● Organization

For more than 40 years, the **American Lung Association (ALA)** has worked to eliminate air pollution and other environmental hazards that attack the lungs. After putting its considerable health research, grassroots education, and political clout behind the enactment of the Clean Air Act in 1990, the ALA has turned its attention indoors as well.

In addition to its efforts to combat ozone depletion, carbon monoxide, sulfur dioxide, nitrogen dioxide, particulate matter, lead, acid aerosols, and toxic chemicals, the ALA fights radon, secondhand smoke, asbestos, formaldehyde, toxic chemicals, combustion pollutants, and biological pollutants that attack individuals. The Association plans to advocate comprehensive laws to reduce exposure to indoor pollutants through safer consumer products, ventilation standards, and better building technologies.

▲ Projects and People

In cooperation with and with funding from the U.S. Environmental Protection Agency (EPA), ALA initiated the **Radon Public Information Project**. To mitigate the risks of high radon levels, and increase radon testing, the project provided 22 lung associations with $125,000 in grants.

ALA kicked off its eighth **Clean Air Week** with the **The Clean Air America Challenge** to stimulate personal action in addressing indoor and outdoor air pollution. As an example of how to take personal action, **Don't Drive Day** urged citizens to use alternative means of transportation to fight air pollution. Joining the EPA and the American Automobile Association (AAA), the ALA continues to lead the **National Car Care Program**, which highlights the role of car maintenance in fighting air pollution.

■ Resources

ALA publishes dozens of reports, articles, and pamphlets on air conservation including *The Health Effects of Ambient Air Pollution*, an 85-page review of animal, human, and population studies. In addition, the ALA has issued two special reports: *Pollution on Wheels II: The Car of the Future* analyzes the role of motor vehicles in contributing to air pollution and *The Health Costs of Air Pollution* concludes that health-related costs of air pollution are rising dramatically and may average $50 billion per year in the USA. A publications list is available.

American Methanol Institute (AMI)
815 Connecticut Avenue, NW, Suite 800
Washington, DC 20006 USA

Contact: **Raymond A. Lewis**, President
Phone: 1-202-467-5050 • *Fax:* 1-202-331-9055

Activities: Political/Legislative • *Issues:* Air Quality/Emission Control; Energy; Transportation

● Organization
The goal of the American Methanol Institute (AMI), founded in 1989, is to promote the use of methanol as an environmentally beneficial fuel in transportation and other applications. Methanol, a clear liquid usually made from natural gas, is increasingly being used as an alternative transportation fuel to reduce air pollution and meet air quality standards. It can be used either in its pure form (M100) or blended with 15 percent gasoline (M85).

According to AMI's co-founder and current president, Ray Lewis, major auto manufacturers are now producing flexible fuel vehicles (FFVs) that run on both methanol and gasoline. Studies have shown that methanol-powered vehicles create much less ozone, less nitrogen oxides, fewer particulates and fewer toxic compounds than gasoline and diesel fuel. In California, methanol fueling stations are now in use. In April 1992, Texas also approved methanol as an alternative transportation fuel.

■ Resources
General information, in the form of brochures and fact sheets, is available upon request. A biweekly news brief, *Fuel Line*, is distributed to members.

American Museum of Natural History (AMNH) ⊕
c/o 171 West 12th Street
New York, NY 10011 USA

Contact: **Arthur Greenhall**, Zoologist/Research
              Associate
Phone: 1-212-929-4353

Activities: Development; Education; Research • *Issues:* Biodiversity/Species Preservation; Deforestation; Health *Meetings:* International Bat Research Conference

▲ Projects and People
Retired from his federal government post for CITES—the Convention on International Trade in Endangered Species of Wild Fauna and Flora—Arthur Greenhall is a zoologist at the American Museum of Natural History's Department of Mammalogy, where he is currently undertaking the *Photographic Identification Field Guide for South American Mammals*. This *Field Guide*, which Greenhall is co-writing with colleague **Rexford D. Lord**, is prospectively classifying some 800 species of South American mammals.

Also a veterinary public health consultant to the Pan American Health Organization/World Health Organization, Greenhall is con-

sidered to be an expert in the field of rabies carried by bats, which is quickly becoming "a major public health problem," he indicates. His passion for mammals began more than 53 years ago in New York City, where he published his first paper on vampire bats. Earlier, as president of his high school's biology club, he enlisted the requisite 25 students so that a zoology course would be taught. Even in elementary school (P.S. 93), he was an avid reptile keeper.

Previously a bat ecologist with the UN's Food and Agriculture Organization (FAO) in Mexico City, Greenhall has developed "a long-term plan of action to deal with the rabies transmitted by vampire bats and other bats." Greenhall is currently compiling a guidebook which will "discuss methods and techniques for the control of vampire bats" and should aid governments and health organizations. He has written over 100 publications on the subject of bats, and edited with Uwe Schmidt *The Natural History of Vampire Bats*—a comprehensive book that describes the common *Desmodus rotundus*, the white-winged *Diaemus youngi*, and the hairy-legged *Diphylla ecaudata* and their social behavior, diseases, and care in captivity. Schmidt is with the Zoological Institute, University of Bonn, Germany.

As a zoologist, Greenhall has published children's books and guidebooks on certain animals in zoos where he was director, curator, or consultant. These include the Portland Zoo and Detroit Zoo, USA; Emperor Valley Zoo, Trinidad; Kingston Zoological Park, Jamaica; and the National Zoo (Parque Bolivar), Costa Rica. His first zoo assignment was in 1934 as assistant to the curator of mammals and reptiles at the Bronx Zoo, New York Zoological Park (NYZP). He presently belongs to the Species Survival Commission (IUCN/SSC) Chiroptera Group.

■ Resources
Greenhall has produced a list of his numerous publications on mammalogy and on specific animals. His books are used as textbooks and as study aids in related fields.

American Museum of Natural History (AMNH) ⊕
Central Park West at 79th Street
New York, NY 10024-5192 USA

Contact: **Michael Klemens, Ph.D.**, Scientist/
              Conservation Administrator
Phone: 1-212-769-5856 • *Fax:* 1-212-769-5031

Activities: Development; Education; Political/Legislative; Research • *Issues:* Biodiversity/Species Preservation; Sustainable Development • *Meetings:* CITES

▲ Projects and People
A herpetologist with the American Museum of Natural History's Department of Herpetology and Ichthyology, **Dr. Michael W. Klemens** is also director of the Global Turtle Recovery Program, which is a joint enterprise of AMNH and the World Conservation Union (IUCN). The program "encompasses 81 conservation projects located in 40 countries in tropical, subtropical, and temperate regions."

Alarmed by mounting extinction rates for tortoises and freshwater turtles, IUCN's Species Survival Commission (SSC) Tortoise and Freshwater Turtle Group—to which Dr. Klemens belongs—developed a Global Action Plan in 1989 to incorporate local people into the recovery program and to avoid as much natural resource damage as possible while actually rescuing the turtles. Of the 250 tortoise and freshwater turtle species in the world, more than 100 are "losing their habitat and facing irreversible population losses from human exploitation."

SSC's task force identified these as essential worldwide objectives: "Establishing sanctuaries; working with government and international organizations to curb trade in declining species; creating projects that conserve turtles while allowing them to remain a traditional subsistence resource; and initiating captive breeding programs to bolster depleted populations."

In Madagascar, scientists will design a tropical dry forest sanctuary in the country's northwest region, where the "world's rarest tortoise" (*Geochelone yniphora*), known locally as the *angonoka*, will live with a wide assortment of vegetation and animals. Jersey Wildlife Preservation Trust and Conservation International provide funding.

In Tanzania, "project scientists with faculty and students from the University of Dar es Salaam will initiate a monitoring and recovery program for the pancake tortoise (*Malacochersus tornieri*)," reports AMNH. "This flexible, crevice-dwelling species is restricted to *kopjes*, rock outcrops that dot the savannas of east Africa."

Surveys are underway in India to monitor the distribution and conservation status of freshwater turtles and tortoises there. The Wildlife Institute of India and the U.S. Fish and Wildlife Service are collaborating.

Other projects are being established "from New England to Namibia, from Viet Nam to Venezuela" in a concerted effort to save the turtles.

■ Resources

Funds are needed to reach the public about the Global Turtle Recovery Program; tax-deductible contributions are made payable to AMNH-IUCN-TURTLE. Desktop publishing software for educational materials is also needed.

The SSC/IUCN *Action Plan* is available to specialist group members and other project participants. AMNH has journal and newspaper articles describing their turtle work. Consultations are available on wetlands and other turtle habitats in New England and New York. Speakers will also discuss the turtle conservation action plan as well as amphibians and reptiles in the northeastern USA.

American Rivers
801 Pennsylvania Avenue, SE, Suite 400
Washington, DC 20003 USA

Contact: **Suzanne C. Wilkins**, Director of State
   Programs
Phone: 1-202-547-6900 • *Fax:* 1-202-543-6142

Activities: Education; Political/Legislative • *Issues:* Biodiversity/Species Preservation; Free-Flowing Rivers; Water Quality

● Organization

"American Rivers is the only nationwide charity whose mission is to preserve the nation's outstanding rivers and their landscapes," the organization reports. Since 1973, American Rivers and its conservation colleagues have helped preserve 9,000 miles of prime natural river, protect 7 million acres of adjacent lands, and stop scores of ecologically destructive dams, saving Americans $23 billion in taxes. It has approximately 17,000 members.

American Rivers, formerly American Rivers Conservation Council, strives to protect 180,000 miles of free-running rivers by the early part of the twenty-first century. The organization was formed because members believe "more rivers are being destroyed each year than are being saved for future generations." According to American Rivers, "600,000 of the nation's 3.5 million miles of river now lie motionless behind dams, never to be free-flowing rivers again. Countless additional stretches are drained nearly dry or choked brown with pollution. Fewer than 10,000 river miles—not even one quarter percent of all USA rivers—enjoy irreversible legal protection." For every mile of river preserved, 85 miles have been drowned behind concrete. The organization is trying to prevent the destruction of natural rivers.

Is is essential to save rivers in order to protect clean drinking water, replenish underground water supplies, preserve natural aquatic systems, provide fishing and water sports, save wildlife habitat including rare species, enhance open space, and reduce taxes through eliminating subsidized development.

▲ Projects and People

In the American Rivers conservation program, the U.S. Congress is persuaded to add rivers to the National Wild and Scenic Rivers System, and federal agencies are motivated to apply temporary legal protections for rivers and corridors awaiting preservation.

American Rivers brings together local organizations, state and federal agencies, and private citizens to inventory state rivers. The object is to identify which rivers to preserve and which to make available for judicious development. It then helps partners develop strong state laws and gain protection for rivers not eligible for federal safeguarding.

The American Rivers Hydro Center is "forging a new national policy that balances the need to develop some rivers and the equally legitimate need to preserve others." A staff lawyer directs dam builders away from rivers deserving protection, advises citizens and governments, and monitors the Federal Energy Regulatory Commission, the agency that licenses dams.

■ Resources

American Rivers has several manuals, listings, and books available for a charge, including *Rivers at Risk: The Concerned Citizen's Guide to Hydropower* by Echeverria, Barrow, and Roos-Collins and published by Island Press. *Two Decades of River Protection*, by Anne Watanabe, a 39-page overview of the history of the national wild and scenic river system, is free. A publications order form is available.

American Solar Energy Society (ASES)

2400 Central Avenue G-1
Boulder, CO 80301 USA

Contact: **Larry Sherwood**, Executive Director
   Phone: 1-303-443-3130

Activities: Education • *Issues:* Energy

● Organization

The **American Solar Energy Society (ASES)** promotes the efficient use of solar energy and energy efficiency in buildings, the use of solar energy technologies to generate electricity, the development and use of solar flux for heat-driven processes, and the development and use of solar radiation to produce or convert organic materials to fuels, chemicals, or other bio-based materials. With 19 state and regional chapters, the ASES addresses the measurement, modeling, and assessment of solar and wind resources and promotes and enhances solar energy education for students, professionals, and the general public.

▲ Projects and People

Speakers who are national leaders in their technical and professional fields come to ASES's annual **National Solar Energy Conference** for an exchange of information on advances in solar energy technologies. The conference typically includes over 200 paper presentations.

ASES sponsors a **Roundtable** in Washington, DC, providing a forum for congressional staffs, associations, policymakers, government officials, and the press on critical solar energy issues.

Critical analysis of technical, regulatory, and educational issues are addressed in periodic **White Papers**, which present critical analysis of important solar energy topics.

ASES educates the public and energy decisionmakers on the benefits of solar energy through a public relations campaign and information materials.

■ Resources

Advances in Solar Energy is a magazine published annually to inform on the latest research and developments in solar energy. *Solar Today* is a bimonthly magazine with more than 4,200 readers that highlights practical applications of solar energy, presents the latest results of solar energy research, covers developments in the nation's solar energy industry, and includes member discussion of solar-related issues. ASES has a publications catalog listing more than 40 books, a slide presentation series, back issues of conference proceedings, and a teacher's workbook.

American Wildlands (AWL)

6551 South Revere Parkway, Suite 160
Englewood, CO 80111 USA

Contact: **Sally A. Ranney**, President
   Phone: 1-303-649-9020 • *Fax:* 1-303-649-9017

Activities: Development; Education; Legislative; Recreation; Research • *Issues:* Deforestation; Land Management; Water Quality

● Organization

American Wildlands (AWL) promotes the protection and responsible management of U.S. and international wildland resources, including wilderness, wetlands, watersheds, free-flowing rivers, wildlife, fisheries, and forests. Founded in 1977 and formerly known as the American Wilderness Alliance, AWL maintains offices in Denver, Colorado; Bozeman, Montana; and Reno, Nevada, dividing its projects and functions among them. Field representatives are also assigned in Alaska and Idaho.

▲ Projects and People

In its fourth year, the **Timber Management Policy Reform Program (TMPRP)** promotes National Forest timber management and policies. Of 45 timber sale appeals filed, TMPRP has won 38, saving thousands of acres of timberland from damaging logging. Under TMPRP's auspices, AWL has developed an "informed activist timber management grassroots group" in each forest of the Greater Yellowstone Ecosystem as well as several forests outside the ecosystem. **Timber Management Leadership Clinics** have been held in almost all interior western states, training 200 citizens in three years.

AWL's new executive director, **Don Whittemore**, a Yale School of Forestry and Environmental Sciences graduate, previously managed AWL's Timber Management Policy Reform Program. Focusing its attention on the remnant ancient forests of the West and the old growth forests in the Northwest, which include virgin stands of now threatened timber, AWL took the lead in protecting such areas as New Mexico's Carrizo (Patos)–Tucson Mountains. Here, ancient junipers, 500 to 900 years old, are threatened by off-road vehicles, overgrazing, and the lack of a U.S. Forest Service fuelwood policy.

Filling the policy void in 1980, AWL's **Clifton R. Merritt** first proposed the criteria and concept for **Forest Service Special Management Area designations**. For such designations, AWL provides a host of nonlegislative land management tools to safeguard environmentally sensitive roadless lands as well as guidance and resource expertise to assist local citizens and ranchers. TMPRP's **Forest Watch Network** and **Technical Assistance Projects** provide communication and resource support within the western region.

AWL's **River Defense Fund** secures oversight hearings and legislative reforms for river protection and management, targeting the Grand Canyon below Glen Canyon Dam, the Colorado River, and the upper reaches of the Rio Grande—where excessive recreation use, overgrazing, and mining threaten raptor breeding. The agenda includes designating a **Birds of Prey Area** and **Wild and Scenic River status** for the upper reaches of the river and 16,000 acres of surrounding land.

AWL has also pushed for a **legislative river package** for river protection activities, including management of watersheds on national forest land, a river conservation stamp to generate revenues for river restoration, and grant-in-aid incentives to clean up urban rivers. In 1989, AWL took its proposals to the U.S. Fish and Wildlife Service and the U.S. Forest Service, and mounted another strong legislative initiative in Congress in support of the Tongass National Timber Reform Act.

Education campaigns are launched on wilderness protection and water rights issues. For example, AWL advised a group of landowners, the Dryland Farmers, to take action against the Bureau of Land Reclamation regarding water access; a settlement resulted that was favorable to both the farmers and to addressing environmental concerns, such as in-stream flow for fisheries, endangered species, and recreation.

AWL supports the proposed **Buffalo Commons**, a 30-year project and example of restoration ecology, to "rehabilitate a vast area to a near-original grassland ecosystem," according to **Sally Ranney**. Farmers would be paid for planting shortgrass instead of planting no crops. Drought-resistant shortgrass once covered the 11 million acres of prairie where 60 million buffalo, or American bison, once ranged, regenerating and fertilizing this sturdy groundcover before being diminished to 39 head by 1900. Ranney writes, "Unneeded agri-land would be added to existing national grassland preserves and state parks. In a 15,000 square mile area, 75,000 buffalo, 150,000 deer, 40,000 antelope and elk could again roam. . . . Hunting would be allowed. In two–three decades, vegetation would be restored. Suggested caretakers are Native Americans in coordination with state/federal agencies. Small cities remain 'urban islands in a sea of grass' and larger ones continue to prosper."

An ongoing initiative is the **Recreation-Conservation Connection**, which reaches the growing number of travelers and other public users of public lands about such matters. International projects include protection of the Tatshenshini River in **Canada**'s British Columbia, known as "North America's wildest river"; partnership with Island Foundation in Patagonia, **Argentina**, to establish a 7-million-acre roadless sanctuary, the Patagonia One World Earth preserve (POWER) for perhaps the planet's "largest, still intact biosphere reserve"; and assistance to **Dr. Richard Leakey** for his efforts to stop elephant poaching in **Kenya**.

AWL also co-sponsors the International Eco-Tourism and Adventure Travel Conference, which in 1991 drew 400 tour operators, environmentalists, educators, representatives of indigenous people, and ministers of tourism, commerce, and environment to discuss this controversial topic.

Ranney is coordinator of POWER in Argentina and a consultant to numerous organizations and projects, including National Center for the Western Horse, African NGOs and private industry, and Adventures International, Inc. Previously, she was the editor of *Wild America Magazine*. AWL water specialist is **Darcy Tickner**.

■ Resources

On the Wild Side is AWL's quarterly journal. The Timber Management Policy Reform Project publishes *Citizen Action Guides*.

American Wind Energy Association (AWEA) ⊕
777 North Capitol Street, NE, Suite 805
Washington, DC 20002 USA

Contact: **Laura J. Keelan**, Membership Director
Phone: 1-202-408-8988 • *Fax:* 1-202-408-8536

Activities: Education; Political/Legislative • *Issues:* Energy
Meetings: European Wind Energy Association Conference; UNCED

● Organization

"Wind energy, once considered a fanciful dream of environmentalists, has come of age," according to the **American Wind Energy Association** (AWEA). "After serving a decade-long apprenticeship in California's wind-buffeted mountain passes, wind energy is ready to deliver its share of the nation's electricity." Believing that wind energy is "economically and technically viable today," AWEA has been working since 1974 to further its development as a "clean and reliable energy alternative."

Headquartered in this nation's capital, AWEA acts as a lobbying and educating organization pressing for changes in environmental policy in the USA and abroad. Dealing with federal agencies such as the U.S. Departments of Commerce, Energy, and State, and the Agency for International Development (USAID)—AWEA is attempting to "level the playing field" of electricity generation, as policymakers grapple with "continuing dependence on oil, poor air quality, and growing evidence of global warming from fossil fuels."

Involved with supporting wind energy programs abroad, AWEA is a member of the U.S. Export Council for Renewable Energy and recommends strategies on implementing such programs.

▲ Projects and People

AWEA sponsors **trade missions** which send American workers and researchers abroad, and "reverse trade missions" in the USA for countries considering wind energy development. It provides information on this alternative energy source, including technology transfer, to multinational development banks.

AWEA is preparing "voluntary consensus standards" for wind turbine generators, sponsors an annual Research and Development meeting and a Conference and Trade Show. Media relations are ongoing to reach the public with current information about wind energy's capacity. For example, "modern wind turbines are reliable . . . and available for operation more than 95 percent of the time compared to only 50–60 percent in the early 1980s," according to AWEA, which also provides keynote speakers for audiences in the public and political sectors.

Scientific analyses and reports are prepared, which estimate costs involved in converting to wind energy. Outside California, wind turbines provide commercial bulk power in Hawaii, Denmark, Germany, Spain, Netherlands, and India, notes AWEA, which points to DOE studies identifying 37 states with "sufficient wind resources to support development of utility-scale wind power plants." In 1990, California's "wind power plants offset the emission of more than 2.5 billion pounds of carbon dioxide, and 15 million pounds of other pollutants that would have otherwise been produced," writes AWEA. "It would take a forest of 90 million to 175 million trees to provide the same air quality benefit."

According to AWEA, wind is "good for the economy," providing "more jobs per dollar invested than any other energy technology— more than five times that from coal or nuclear power." Like other forms of solar energy, wind plants are capital intensive, but "the fuel is free" and the plants are "less costly to operate, maintain, and fuel than conventional power plants," stresses AWEA.

■ Resources

The Association provides a monthly newsletter, *Windletter*, to its members on small wind turbines and related topics; and publishes *Wind Energy Weekly* covering "industry developments, legislative and regulatory issues, state and local decisions affecting the industry, windfarm projects, and international business opportunities." A publications list, including slide presentations, is available.

Americans for the Environment (AFE)
1400 16th Street, NW
Washington, DC 20036-2266 USA

Contact: **Roy Morgan**, President
  Phone: 1-202-797-6665 • *Fax:* 1-202-797-6646

Activities: Education; Political/Legislative • *Issues:* Electoral
Skills Training

● Organization
With its motto "each one teach one," **Americans for the Environment (AFE)**—a nonprofit, educational organization—engages voters to take on and solve environmental problems in the political arena.

Seeking "electoral action for a sound environment," AFE mobilizes citizen activists through national and regional electoral training workshops that teach campaign-related skills; heightened profiles of environmental issues among politicians, media, and general electorate; support for expanded local grassroots environmental networks; comprehensive training guidebooks; and interpretation of environmental polling trends that stir public policy debates and form political issues.

AFE's purpose is to harness and empower the green vote so that candidates and lawmakers will enact more environmental solutions such as the Clean Air Act, Clean Water Act, Wilderness Act, and Endangered Species Act. By doing so, it motivates more people to get involved in the political process.

▲ Projects and People
Training seminars and workshops in electoral skills cover campaign managing and planning, fundraising, organizing volunteers, voter contact, media relations and production, get-out-the-vote programs, and specialized skills for organizing initiatives and referendums.

Major publications are *The Power of the Green Vote*, a campaign training manual for environmental activists; *The Rising Tide: Public Opinion, Policy & Politics*, a comprehensive report about polling trends of voters to support candidates with strong positions on quality environmental issues; and *Taking the Initiative*, which analyzes the strengths and weaknesses of more than 60 environmental initiatives on local and statewide ballots in 1990 and also identifies the "resources that will be necessary to win environmental ballot propositions" in the near future.

AFE Board of Directors includes Chair **Randall Snodgrass**, National Audubon Society; Vice-chair **Betsy Loyless**, Sierra Club; Treasurer **Cindy Shogan**, Southern Utah Wilderness Alliance; and Secretary **Sharon Newsome**, National Wildlife Federation.

■ Resources
Membership and publication order forms are available.

America's Clean Water Foundation (ACWF) ⊕
750 First Street, NE, Suite 911
Washington, DC 20002 USA

Contact: **Roberta (Robbi) Savage**, Executive Director
  Phone: 1-202-898-0902 • *Fax:* 1-202-898-0929

Activities: Education; Legislative; Research • *Issues:* Water Quality

● Organization
Since 1989, **America's Clean Water Foundation (ACWF)** has "spearheaded a national campaign to reawaken American citizens' interest to the value of their water resources and reaffirm an international commitment to clean water." Foundation activities encourage citizen involvement, youth education, professional exchange of technology, documentation of water quality trends since 1972, and public participation in the Clean Water Act's twentieth anniversary commemorative celebration. Three advisory groups, a Board of Governors, a celebrity advisory council, and a steering committee provide direction for ACWF projects.

The Board of Governors is co-chaired by former U.S. **Senators Edmund Muskie** and **Howard Baker** and U.S. **Representatives John Blatnik** and **William Harsha**, and includes **Senators Bill Bradley, John Chafee, Claiborne Pell, Mark Hatfield**, and **Frank Lautenberg**, as well as other leaders from government and the environmental movement. Financial support comes from government and industry sources such as ARCO Foundation, Coca-Cola Company, E.I. Du Pont de Nemours & Co., World Resources Institute, U.S. Environmental Protection Agency (EPA), and states such as Alabama, California, Colorado, New York, and South Carolina.

Working with the U.S. Congress, states, and other national organizations under the banner of 1992 as the "Year of Clean Water," ACWF's staff of five relies upon a broad coalition of experts from government, public-interest groups, private citizens, professional organizations, and industry to accomplish its objectives. This partnership includes EPA, Keep America Beautiful, Tennessee Valley Authority (TVA), Take Pride in America, National Geographic Society, General Federation of Women's Clubs, and others.

▲ Projects and People
In 1991, ACWF sponsored a **World Water Summit** with a technology and trade exposition in Atlanta, Georgia, hosted by former President **Jimmy Carter**.

The Foundation's education program includes an environmental computer program for the learning disabled, environmental curricula for minority and rural schools, and environmental videos. **Pilot School Programs** to expand student learning opportunities have been established at the Lab School of Washington, DC; Lindberg Middle School in Long Beach, California (a program for inner city students); the Sound School in New Haven, Connecticut; and the **Rural Schools Program** for "one room school houses" in Georgia, Arizona, and Nevada.

To celebrate the twentieth anniversary of Earth Day, ACWF sponsored a 10-kilometer **Run for Clean Water** and hosted an **Environmental Expo** featuring a promotional video and portable exhibit. With Legacy International, it also designed and conducted a two-day YES (Youths for Environment and Service) Workshop to train international student leadership in encouraging community participation in environmental activities.

ACWF worked with EPA and the Izaak Walton League to create a compendium of organizations and contacts involved in **monitoring** activities in each state. Foundation experts organized and conducted a **water sampling program** for eight Washington, DC, area grade schools and high schools. ACWF president **Roberta Savage** formalized the foundation's working relationship with Take Pride in America.

ACWF is currently developing a **KIDS NETWORK Water Unit** based upon the National Geographic Society's acclaimed computer learning programs. The NETWORK enables elementary school students to learn about water issues and share information and ideas with students around the country. It is actively seeking funding for an **Environmental Science Institute** located at Prince George's Community College in Maryland. The Institute would train teachers during summer breaks in practical water related physics, chemistry, and natural resource protection.

The Foundation's **National Envirothon** tests high school students' knowledge of natural resources and environmental issues and brings students together to work on environmental problem solving. Teams compete in areas such as soils, wildlife, forestry, water, and ecology. A student information kit developed in connection with the TVA includes easy-to-read facts and terms about water issues. The National 4-H Council training facility in Chevy Chase, Maryland, has become the site for the **National Youth Water Summit**, where teams of young people from every state will participate in a multiday **Water Resources Workshop**. Participants practice leadership skills and decisionmaking in launching environmental initiatives.

National Kid Camp-Ins, in conjunction with the Association of Science-Technology Centers (ASTC), are planned for weekend explorations of water-related educational activities. To examine career opportunities, the first **National Young Women in Water Sciences Conference** was planned for 1992 in cooperation with the Girl Scouts of the USA.

The Foundation also hopes to establish a national campaign of **public service announcements** in support of **clean water strategies** under the auspices of the Advertising Council in New York.

■ Resources

ACWF publishes the *Gee Whiz Water Quiz*, an informative collection of water facts and questions to test public knowledge of water issues. Bimonthly newsletters highlight Foundation initiatives. *The Adventures of a Water Drop* computer-learning program, developed by ACWF and the Lab School of Washington, DC, provides a hypercard-based visual exploration of the numerous paths a water molecule can travel in the course of the Earth's biological and physical processes. At Prince George's County, Maryland, the Foundation jointly developed *A Year in the Life of a River*, a video documentary describing the process of community action needed to clean up a local body of water. A second video project, under an agreement with the Izaak Walton League of America, promotes citizen involvement in stream monitoring and training on the "how to's" of sampling and assessing water quality.

A governmental guide on conducting public involvement programs provides a step-by-step approach for enlisting public support for environmental and natural resource management projects. ACWF's first product, a *Citizens Brochure on the Clean Water Act*, provides tips to citizens on how to contribute to the protection of

water resources. *World Water Summit Proceedings* is among other publications.

Animal Welfare Institute (AWI) ⊕
P.O. Box 3650
Washington, DC 20007 USA

Contact: **Christine Stevens,** Founder and President
Phone: 1-202-337-2332 • *Fax:* 1-202-338-9478

Activities: Education; Research • *Issues:* Biodiversity/Species Preservation; Wildlife Management Reform; Wildlife Transportation • *Meetings:* CITES; IWC

● Organization

Founded in 1951 by Christine S. Stevens, who has served as president for more than four decades, the **Animal Welfare Institute (AWI)** addresses a wide range of animal rights concerns largely through the publication of educational materials. An extensive brochure of films, books, and magazines reveals the Institute's interest in programs ranging from treatment of laboratory animals to the trade of endangered species of wild birds. The goal is "to reduce the sum total of pain and fear inflicted on animals by man." The AWI presents the **Albert Schweitzer Medal** each year for outstanding achievement in the advancement of animal welfare. Dr. Schweitzer said of the award: "I would never have believed that my philosophy, which incorporates in our ethics a compassionate attitude toward all creatures, would be noted and recognized in my lifetime. Your medal celebrates this progress [of intervening] for animals when up till now man has shown so little interest in them."

▲ Projects and People

Direct observation and monitoring of **laboratories and farms** across the country is the most effective way to ensure compliance with regulations and the most humane treatment possible. AWI staff members use periodic site visits and analysis of U.S. Department of Agriculture inspection reports as research for documentation in books such as *Animals and Their Legal Rights* and *Beyond the Laboratory Door*. Specifically, AWI has actively opposed painful experiments on animals by high school students and intensive confinement in factory farms.

Continuous efforts to create a coalition with like groups has paid off for AWI. Investigations into the ivory trade with the Environmental Investigation Agency led to a campaign to **ban the international ivory trade** that began at the Convention on International Trade in Endangered Species (CITES) in 1989. That effort continues today. The **Save the Whales campaign** has been joined by environmentalists around the world since its launching in 1971. Boycotts of whaling nations and their products have proved effective in gaining public support for the permanent end to commercial whaling. AWI campaigns worldwide against **steel jaw leghold traps**, which indiscriminately maim yearly some 19 million fur-bearing animals in the USA alone.

Heading another coalition, AWI brought pressure to require regulations for the **humane and healthful transport of wild mammals and birds** particularly affecting the pet trade, which is regulated by the U.S. Department of Interior.

Three years after AWI was established, Christine Stevens founded the Society for Animal Protective Legislation. Twelve federal laws protecting animals have been passed through this Society's efforts. She is a member of the Board of Regents of Georgetown University and also serves on the boards of Monitor USA, Monitor International, Bat Conservation International, and the World Wildlife Fund/USA.

■ Resources

Materials for education purposes are available under the headings of Humane Education, Wildlife, Trapping, Factory Farming, Attitudes Toward Animals, Laboratory Animal, and Save the Whales. The Animal Welfare Institute publishes a quarterly report on cruel and inhumane treatment of animals, made available free of charge to libraries.

Appalachian Trail Conference (ATC) ❧
P.O. Box 807

Harper's Ferry, WV 25425 USA

Contact: **Brian B. King**, Director of Public Affairs
    Phone: 1-304-535-6331 • *Fax:* 1-304-535-2667

Activities: Education • *Issues:* Biodiversity/Species Preservation; Trail Maintenance, Management, and Promotion

● Organization

The Appalachian Trail Conference (ATC) is a nonprofit education organization representing the citizen interest in the Appalachian Trail and dedicated to the preservation, maintenance, and enjoyment of the Appalachian Trailway. Since 1925, the ATC and its member clubs have conceived, built, and maintained the Appalachian Trail in cooperation with federal and state agencies. It has approximately 23,500 members.

The Trail is a 2,144-mile footpath through the Appalachian Mountains from central Maine to the forests of northern Georgia; and its "natural greenway of backcountry wilderness" passes through some of the most populous and developed areas of the USA.

▲ Projects and People

Since 1937, the ATC has sought a protected trail extending along the entire range of the Trail in order to preserve not only a continuous footpath, but also a sampling of the natural and scenic resources along its route. Protection of these trails presents significant challenges to ATC from the threat of adjacent development or potential intrusions across the Trail corridor. Today, there are only 65 miles and 21,000 acres of the trail to be acquired—compared to 852 trail miles and 145,000 acres of surrounding, unprotected areas in 1978.

Some 4,600 year-round volunteers assisted by seasonal crews and ridge-runners repair privies, mark pathways, fix bridges, clean or repair campsites, put up information and safety signs, report threats from local development, and educate trail users.

ATC, through its natural-diversity inventory program, has identified more than 400 sites along the Trial in five of the 14 affected states where rare, threatened, or endangered species of flora and fauna reside. In addition, ATC monitors these resources and, in some cases, actively manages them to ensure their continual survival.

ATC also established a land-trust program in 1982. Initially, the focus of the trust was to assist federal agencies with acquisitions in the Trail-protection corridor. But, increasingly, the trust has broadened its focus to include lands and resources outside the basic corridor. Since its inception, the program has contributed to the purchase of more than 12,000 acres of land in every Trail state.

In addition to supporting federal and state land acquisition along the Trail and supplementing these acquisitions through the work of the trust, ATC plays a critical role in preserving the sensitive resources associated with the Trail by responding to many land-use and development proposals that might affect them.

The Conference is recognized internationally as a "model of successful public/private cooperation to preserve a rich cultural, recreational and natural resource."

■ Resources

The Conference publishes guidebooks and other educational literature about the Trail, the Trailway, and its facilities. Membership form and *Ultimate Trail Store* catalog are available.

Archbold Biological Station (ABS)
P.O. Box 2057

Lake Placid, FL 33852 USA

Contact: **John W. Fitzpatrick, Ph.D.**, Executive
    Director
    Phone: 1-813-465-2571 • *Fax:* 1-813-699-1927

Activities: Education; Land Management; Research • *Issues:* Biodiversity/Species Preservation; Deforestation; Ecosystem Preservation

● Organization

Archbold Biological Station (ABS), an independent ecological research facility located in south-central Florida, conducts long-term ecological research and conservation focusing on local and global concerns. For nearly 50 years, ABS has been a leader in the conservation of the unique scrub and sandhill habitats of the Lake Wales Ridge in Florida, which is known as the state's "water- and vegetation-rich spine"—attractive to both agriculture and housing developers.

Founded by aviator/explorer Richard Archbold, the Station was a gift from Richard Roebling. Station facilities include a library, laboratories, reference collections, a geographic information system (GIS), plus dining and housing facilities for visitors. Conferences are held here with the U.S. Fish and Wildlife Service, among other groups.

▲ Projects and People

ABS owns and manages 5,000 acres of pristine habitat containing "one of the highest concentrations of threatened and endangered organisms" in the USA; some exist nowhere else in the world. In 1987, the U.S. Department of Interior designated this preserve a National Natural Landmark. Elevations on ABS main property range from 110 to 213 feet and encompass one of the highest peaks in peninsular Florida.

The biota of the region is unusually diversified with xeric pine-oak habitats containing endangered plants and animals ranging from Florida panther and black bear to scarab beetles and funnel wolf spider to American alligator, Eastern indigo snake, gopher frog, and blue-tailed mole skink. The Florida jujube, once thought extinct, is said to be one of the world's rarest plants. The yellow scrub balm mint faces extinction from citrus growers. The scrub balm, protected only on ABS land, is believed to contain an ant and roach repellant. The sand skink is one of the planet's rarest reptiles. Abloom in the sand, the scrub-blazing star flower lures butterflies such as fairy sulphur.

ABS also operates the nearby 10,300-acre Buck Island Ranch as the John D. MacArthur Agro-Ecology Research Center (MAERC). The ranch enables scientists to study the ecological impact of Florida's two largest agricultural industries, cattle production and citrus cultivation.

As part of its land management preservation research, ABS studies the effects of fire, natural and prescribed, on the composition and structure of native elements and systems. A staff of 30 utilizes field laboratories to investigate population, biology, life history theory, and community ecology.

GIS stores maps—using points, lines, and polygons to represent the distribution of certain topography, soil types, vegetation zones, rare plant distribution, or animal territories and nest sites. Two software programs—ARC/INFO and ERDAS—operating on a sun workstation to provide baseline data sets of major variables (soils, surface water, land use) are being developed. GIS will also analyze historical fire patterns, biogeography of endemic scrub patterns, and how to design preserves for endangered species.

Staff biologists are working on 70 different research projects and several new initiatives that include research on Florida rare and endangered arthropods; gopher shell moth biology; Florida mouse; interactions between fire, trees, insects and woodpeckers; and demography and fragmentation of the Florida scrub jay with its threatened habitat. ABS developed a Conservation and Recreation Lands (CARL) program for acquiring 20 important scrub tracts on the Lake Wales Ridge, which is made up of relict dunes.

Environmental educational programs emphasize research internships for undergraduate and graduate students and formal environmental instruction for grade school children. Some 1,000 elementary and high school students from Highlands County become acquainted with "the real Florida" through ecological programs conducted by Nancy Derup, education specialist.

Staff also provides ecological advice to government agencies and helps The Nature Conservancy and other environmental organizations develop "biological priorities for land acquisition and habitat management." John W. Fitzpatrick, who leads this array of activities, first visited Archbold Station as a National Science Foundation (NSF) researcher from Harvard University. His field research has taken him to southeastern Peru to study flycatchers in the Manu National Park, and he has headed the Field Museum's Division of Birds in Chicago.

■ Resources

ABS publishes a *Florida Scrub Coloring Book* for elementary school children; an *Illustrated Guide to Endemic and Indication Plants and Animals of the Lake Wales Ridge, Florida*; and free, miscellaneous handouts on Florida scrub ecology and conservation. A Florida environmental education program for fourth graders is available.

ARISE Foundation
4001 Edmund F. Benson Boulevard
Miami, FL 33178 USA

Contact: **Edmund F. Benson**, Co-Founder
Phone: 1-305-592-7473 • *Fax:* 1-305-599-3750

Activities: Education; Research • *Issues:* Air Quality; Health; Nontoxic Pest Control; Waste Management/Recycling; Water Quality

● Organization

Former businessman and now citizen-activist Edmund F. Benson has spearheaded a multifaceted program to address environmental causes for degenerative diseases, including how to change personal lifestyle factors from food and tobacco to occupational chemicals, pesticides, and industrial pollutants.

ARISE offers more than 50 "self-help" strategies, including information on nutrition and environmental "self-defense" from sick buildings, formaldehyde exposure, contaminated ash, dioxin fallout, and pesticide drift. Founding ARISE with his wife, Susan, in 1986, Ed Benson, using galvanizing marketing strategies, has taken on the largest incinerator in southern Florida in a self-styled "Garbage War" waged with committees, petitions, information dissemination, and protest marches. ARISE even hired a helicopter to photograph the incinerator complex from the air. As a result, the State of Florida sued the county and the plant's operators in 1985 and cleaned up both plant management and the plant.

By 1989, ARISE helped convince McDonald's Corporation to institute its own recycling campaign, and from there went on to establish what many consider the "largest curbside recycling program in the country"—with Benson as chairman of the Dade County Recycling Task Force. Bringing youngsters into the recycling effort, ARISE collaborated with Montenay Power to establish the McAliley-Ruvin Environmental Awards, which offer $10,000 in recycling incentives to Dade County students.

▲ Projects and People

ARISE and Benson presently sponsor a new program known as ENVIRO-COPS for elementary school students. Students take a pledge to protect their own "in-vironment" by reading labels, investigating "corporate crimes" involving irresponsible television advertising claims, and patrolling at school and at home to reduce waste and promote recycling.

ENVIRO-COPS students participate in three major projects. In an Assault on Batteries, Dade County public schools collected 6,194 pounds of household batteries. Some were returned to the manufacturer along with letters asking for toxin-free batteries. The rest of the hazardous waste went to the landfill encased in 55-gallon drums. ENVIRO-COPS participate in their own recycling contests and learn more about the environment through GREAT EXPECTATIONS seminars.

Ed Benson, a Boston-born, former furniture executive, has received numerous awards for his work, including the 1990 Spirit of Excellence Award from the *Miami Herald*. Dade County, in fact, named a street after him in recognition for his efforts to clean up its 160-acre incinerator complex, known as the "Miami Monster." Currently, says Benson, "I'm working at making sure the county installs pollution control filters in incinerators . . . a must for medical waste incinerators." He also gleans research, for example, from Dr. Bill Wolverton, National Aeronautics and Space Administration (NASA), and circulates tips on using, at home and work, heart leaf, elephant ear, and finger leaf philodendron; green spider; Chinese evergreen; golden pothos; aloe vera; and mini schefflera, among others, to remove pollutants from the air.

He has also established and chairs the Dade County Nutritional Task Force. The Task Force's pilot program not only educates students and teachers about proper nutrition, but cites kitchen personnel with cash and recognition for enhancing food service preparation. A Friendly Foods Program urges people of all ages to avoid artery-clogging foods and empty calories, and promotes healthy doses of self-esteem. To get rid of the insecticide methyl bromide, an Integrated Pest Management program has been introduced into the school system for principals, custodians, and maintenance workers. With Benson's help, the public schools put together an environmental seminar for students in 1992.

■ Resources

ARISE offers a series of fact sheets entitled *Your Keys to Happiness* and dedicated "for your health—not for profit." Other informative brochures are also available. So is an ENVIRO-COP badge for those who arrest waste.

Ashoka: Innovators for the Public ⊕ ◉
1700 North Moore Street, Suite 1920
Arlington, VA 22209 USA

Contact: **Shawn McDonald**, Communications Director
Phone: 1-703-527-8300 • *Fax:* 1-703-527-8383

Activities: Development; Education; Law Enforcement; Political/Legislative; Research • *Issues:* Air Quality/Emission Control; Biodiversity/Species Preservation; Deforestation; Energy; Global Warming; Health; Sustainable Development; Transportation; Waste Management/Recycling; Water Quality

● Organization

Ashoka: Innovators for the Public is a worldwide fellowship program that brings social reform and democracy to communities on six continents while raising environmental awareness and impacting favorable ecological change.

Founded 10 years ago, its goal is to "help build mutual respect among the earth's peoples that is a prerequisite to world unity and peace." Fellows see themselves as path-breaking innovators and public service entrepreneurs. They have earned praise from members of the U.S. Congress, the media worldwide, and programs such as UNICEF's Action for Children, which heralds Ashoka for stepping in when the risks are greatest. "They seek to help individuals before they have succeeded—when no one else is ready to help and when a little help makes an enormous difference."

▲ Projects and People

Some 50 Ashoka Fellows are engaged in environmental fieldwork in countries such as Bangladesh, Brazil, India, Indonesia, Mexico,

Nepal, Pakistan, and Thailand. In addition, new fellowships were recently launched in Cameroon, Ghana, Nigeria, and South Africa. Profiles of selected Fellows follow.

Brazil: In 1988, the murder of Chico Mendes brought grief to colleague **Mary Allegretti** and the opportunity to carry on his work on behalf of 200,000 **Brazilian rubber tappers**, or *seringüeiros*, who live in harmony with the Amazonian rainforests. As cattle ranchers encroached upon this fragile ecology, deforestation, exhaustion of the thin soils, and displacement of the tappers and small-scale farmers were the results.

In the mid-1980s, Allegretti began organizing the tappers for political action, she reports. She established the **Institute for Amazon Studies** in 1986, to carry out the concept of "**extractive reserves**," where large tracts of rainforest are set aside for hunting, fishing, tapping rubber, gathering nuts, and harvesting herbs and other medicines—and ecological balance is maintained between man and nature. With the Brazilian government's adoption of this concept in 1987, she writes, more than 20 reserves have been set up for this purpose.

She is spreading the concept of extractive reserves to Malaysia and Indonesia and pressuring multilateral institutions in the USA to "end the financing of development projects that are neither economically viable nor environmentally sound." In addition, she works to improve the health, housing, and education of the *seringüeiros*.

The mayor of Cachoeira de Itapemirim, Brazil, hired agronomist **Nasser Youssef Nasr** several years ago to plant a garden on municipal land that would produce fruits and vegetables for school lunches. Using traditional methods, he cleared the land and applied an array of chemical herbicides, pesticides, and fertilizers; however, the yield was disappointingly low. Turning to **alternative agricultural techniques** and organic farming, Nasr quickly learned to leave native vegetation in place to keep the local insects from feeding on the food crops. Production costs dropped by almost half. Less water was required because the native vegetation and organic fertilizers helped the soil retain moisture; the ground cover also prevented soil erosion. Labor costs fell because there was little weeding and application of chemical products. Most important, the 10-hectare garden feeds the town's schools, hospitals, shelters, and day centers with "remarkably disease-resistant" citrus fruits, tomatoes, carrots, and other produce grown at a rate several times the national average. Moreover, ten different crops are planted where only two or three grew previously.

The municipal garden attracts thousands of annual visitors, and Nasr is now showing other farmers in Brazil how to become more ecologically minded and grow **toxin-free produce**, he reports.

Some farmers are forming cooperatives for this purpose. However, traditional agriculture is heavily chemically dependent, and agronomy schools still argue for this type of farming, Nasr indicates. Nasr, who produced a book and videotape to explain his farming methods in simple terms, co-founded an environmental group, the **Juqira Candiru Foundation**, to help disseminate these materials. Environment minister Jose Lutzemberger praises Nasr's techniques as "the birth of Brazilian agriculture and agronomy."

Edson Hiroshi Seo is another successful practitioner of alternative agriculture in Brazil. He demonstrates that both small holders and commercial farmers can employ **soil management** techniques appropriate to varying climates and can double and triple yields without using toxic chemicals; they also can tap hidden water resources without undue cost. He has experimented for years with alternative high-protein food preparations; with ground zeolite as a feed supplement; with "ferro-cement"—a simple construction material for houses, water tanks, and silos; and with bio-digestors that break down organic material and produce methane fuel and natural fertilizer.

His innovations and teaching across the country among agronomy students, agronomists, and local leaders led Seo to establish the **Center for Research and Training in Agriculture and Alternative Technology** on a university campus near Brasilia to attract students and facilitate his findings.

Ricardo Neves abandoned his plans to pursue a Ph.D. degree in engineering and instead formed the **Institute of Technology for the Citizen (ITC)** in 1988, with Ashoka support, to help Rio de Janeiro's multimillion population benefit from a model for development that was environmentally friendly. Neves first launched a **bike-riding campaign** as an alternative to the congestion and pollution of cars and buses and drew up plans for bike paths and a cycle park system at various train stations. ITC initiated a system of selective collection and recycling for hospital refuse and other waste. Through ITC, Neves aggressively publicizes such projects and gained the Portuguese language publishing rights to the public policy reports of the Worldwatch Institute and the Institute for Local Self-Reliance. He presently has funding from more than 20 national and international groups, he reports.

Veterinarian **Clovis Borges** of Parana created the nonprofit **Wildlife Research and Environmental Education Society (SPVS)** to address Brazil's mounting environmental problems as well as the loss of biologists and other environmental professionals to other work because of the lack of jobs existing outside government. With a small core staff and 40 active associates, these young environmental scientists and technicians are building an economic base by **helping institutions solve their environmental problems** on contract. Future profits are then reinvested in major environmental research, action, and education initiatives. Their goal is to expand across Brazil and into Uruguay and Paraguay as well as to inspire SPVS chapters elsewhere. Borges and his colleagues help groups develop conservation and resource management plans to put new or restored ecosystems in place; give technical assistance to zoos, animal reserves, and parks; document damage to wildlife and help reintroduce endangered species into natural habitats; prepare teaching materials for use in schools; and work with the local population to increase their sense of the value of the rainforests and a dense "herb matte" that is used for a local tea. A teacher training program and more intense and frequent media relations are envisioned.

Borges was much moved by the burning of the Amazon forests and believes "continued exclusive reliance on the government guarantees failure," reports Ashoka. His reaction is that "independent citizen groups with the highest level of technical competency must define the issue and demonstrate how society should—and can quite practically—respond."

Environmental chemist **Tania Mascarenhas Tavares** is a strong voice in the complex, volatile industrial pollution battles in the Reconcavo near a rich and vulnerable estuary that includes the point of land where the capital city of Bahia, Salvador, is located. Here there are many petrochemical, chemical, pharmaceutical, and heavy industries adding dangerous heavy metals and chemicals to the region's waters and air. Thousands of **industrial pollutants** are being combined that are dangerous to the human body, almost impossible to track, and overwhelming to the state regulatory agencies. Tavares' purpose is to "marry technical skill and environmental social action," according to Ashoka.

With some 30 professional colleagues, she is developing the **Interdisciplinary Environmental Center (NIMA)**, which is linked to, but independent of, Bahia's Federal University. With the goal of providing "technically competent, credible analysis for deliberations regarding industrial pollution" on a large scale, she is expanding in other parts of Brazil and working with other institutions to help this country "come to grips with its severe pollution problems." Tavares is setting up special training courses for professional environmental managers of private companies as well as those staffing government research and regulatory agencies. To build her reputation as a reliable and objective source of analysis, she sends her reports to the affected companies, interested government agencies, and those directly affected by the pollution. Ashoka also reports that Tavares "responds to the press but does not lead with it." Her plan is that effective public debate will produce decisions to help resolve the problems. Turkey has also sought her advice on dealing with heavy metal pollution problems.

India: Because aggressive fishing appears to be producing less fish such as cod and haddock from the North Sea to the Indian Ocean, **Nalina Nayak** knows that both the ocean and land must be managed in a sustainable way. After two decades of working with small-scale, traditional fishermen of Kerala and helping them develop marketing cooperatives and unions, the just-retired global convener of the **International Supporters of Fishermen** is now embarking on defining the harmful activities that affect the fisheries and are the likeliest to be resolved with political help. These include reducing harvesting as much as 50 percent to promote breeding and focusing on the interface between the land and the sea: rivers carrying more silt, destroyed mangroves wetlands, changing bird populations, dumping of chemicals, pollutants absorbed by rainfall that returns to the sea, and expanding populations.

After 10 years of public pressure, Nayak and her organized fishermen finally got the government to ban trawlers in the spawning months of June through August. She is investigating high-tech privatization of coastal wetlands and breeding groups such as for shrimp farms or other forms of intensive aquaculture. Nayak is also examining ways to encourage coastal forestry both to control erosion that destroys wetlands and to provide lumber for fishermen's boats. Her approach is to bring all interested groups together until they "edge their way towards new, mutually acceptable policies," reports Ashoka.

Nayak also founded the Self-Employed Women's Association of Trivandrum to help such individuals from fishing communities to find alternative jobs.

Engineer-ecologist **Dr. Dhrubajyoti Ghosh** demonstrates how to **treat municipal sewage, save wetlands, and raise fish safely and simultaneously**—using techniques of illiterate small farmers working wetlands to the east of Calcutta. Listening carefully to local unrecorded experience, Dr. Ghosh observed for years the practices of these farmers who use organically rich sewage. He devised a plan with economic and ecologic potential to turn low-yielding wetlands near cities into low-cost recovery schemes.

His goals are to enhance wetlands survival, lower sewage treatment costs, create jobs, increase production, and help clean India's "often chokingly polluted" waterways, according to Ashoka. Ghosh is developing a model that involves garbage gardens, waste water aquaculture, and paddy fields where he can observe "waste water loading, recruitment and harvesting of fish, cropping patterns on garbage, drainage diversion and desilting, pond management, fish disease and paddy cultivation using fish and effluent."

Dr. Ghosh received a United Nations Global 500 Roll of Honour award for outstanding environmental achievements. He is assistant chief/engineer of the Calcutta Metropolitan Water and Sanitation Authority (CMWSA).

Pandurang Hegde successfully worked with local groups to stop large-scale lumbering and other forms of environmental destruction. Now he is helping them restore the land and establish new patterns of sustainable agriculture fitted to each area's particular condition, starting with the region of his birth—India's west coast, once covered with tropical forests.

Hegde is a grassroots organizer and leader of the Appiko Andolan Movement that persuaded the government after a five-year struggle to end its deforestation policies in western India, where woodlands were reduced from 82 percent to 20 percent in 30 years. As described by Ashoka, "This was not a victory won in the drawing rooms of a distant capital. It was one of the first major examples of poor, seemingly powerless local people organizing in gradually widening circles in peaceful but very persistent defense of their environment. It began in Sirsi, Pandurang's home village, and eventually engulfed five hill districts and exerted an influence by example well beyond."

He intends to expand his mass movement to help thousands, perhaps millions, of villagers take up regionally adapted, sustainable agriculture and move away from current government policy, heavy dependence on chemicals, and high costs—using as models the Japanese One Straw Revolution and Australian Permaculture. A six-acre demonstration farm is central to his strategy so that he can test what will work, and what will not, before passing along these ideas to the community.

His plan is to build a Centre for Sustainable Development at the farm where local seed banks will be organized to collect, preserve, and make available locally adapted seeds likely to produce hardy, disease-resistant crops. The Centre would manage the farm and begin programs to reforest wastelands, help disseminate organic farming methods such as composting and green-leaf fertilization, and communicate these lessons throughout the country. Attached to the land where he grew up, it is said of Hegde, "Intentionally faceless, working not to be a banyan tree whose shadow would prevent others coming up and leading, he served as an activist wherever he was needed." Now he continues his leadership.

Indonesia: Before he effectively could advocate outlawing of chemical pesticides in a comprehensive nationwide program, Mr. Wiedjanarka embarked on an ambitious program with the help of leading rural private voluntary organizations (PVOs) to identify nonchemical alternatives such as biological and organic farming approaches. Once these alternatives were identified, they were explained to the rural population, such as through publications created with the Pesticide Action Network (PAN) to offset the advertising bombardment of chemical companies. A tri-annual journal of myths and facts about pesticides as well as posters and comic books are used in Wiedjanarka's campaign to enlighten farmers. Even though 57 pesticides were outlawed in 1986, according to Ashoka, farmers continued to purchase and use these readily available and illegal chemicals. Recently, Wiedjanarka helped prepare and edit a world report on "informed consent" as a principle to govern the international trade of dangerous chemical pesticides. He has documented the impact of the lack of disclosure in international trade in Indonesia as one major element of this work.

The coastal regions of eastern Indonesia are undergoing profound changes beyond the grasp of the local people and some of the newer institutions trying to help. Iwan Mucipto is putting together a union of all the widely differing groups trying to help fishermen and their coastal neighbors, reinforcing community groups and local organizations in the process. He is encouraging the PVOs to innovate, and he is instigating training and mutual help consultancy so that those who learn will share their expertise with others. Mucipto is helping this union to reach out regionally and nationally to make a greater impact and to seek funding and other support. With Mucipto's direction, the four most important coastal zone PVOs are joining together to create the core framework of this union to increase the quality of life for the citizens and save the environment.

The task is enormous because as the Indonesian government encourages large-scale fishing operations with far-sweeping trawling mechanisms that entrap vast numbers of the most valuable species, the small fishermen lose their livelihood and the ecology is endangered. Reef bombing is a result of fishermen "driven to the economic edge," according to Ashoka. Meantime, their competitors often ignore rules to abstain from fishing in certain areas during spawning season.

Mexico: Working the forest is the way to save the forest, believes Rodolfo Lopez, a skilled negotiator who is able to bridge the political bureaucracy of Mexico City with the myth-based society of the indigenous people of Oaxaca. Consequently, he is helping the Zapotec Indians reclaim their land from the Forestry Department's control, achieve independence, and take on long-term forest management through community-based companies that they own and operate. Already there are 12 such forestry companies within the Union of Forest Communities (UCEFO), which Lopez began, and this number is expected to grow to 300. The Zapotec Indians, who are mainly unschooled and lack management skills, face enormous challenges. According to Ashoka, Mexico has lost more than 95 percent of its tropical forest and more than two-thirds of its temperate forest. In addition, 80 percent of the country's land is seriously eroded. Only since 1982 has the government allowed local communities, rather than large timber companies, to develop their own resources; yet the movement away from slash-and-burn agriculture is difficult.

Lopez and UCEFO must help the native communities use and develop the limited forest resources in competitive markets and find nonforest-related economic activities such as vegetable and fruit growing to sustain them too. Forest company profits are being channeled back into the communities with the construction of schools and roads and the installation of drinking water and electricity.

How foul is the air of Mexico City? Luis Manuel Guerra, a top chemistry researcher, teacher, and innovator of new products and projects, knows. On a recent annual migration, thousands of flying birds choked on pollution and crashed to their deaths. In a city where the population will top 20 million by the year 2000, there is no hazardous waste treatment facility and water supply and protection are inadequate. Millions of vehicles spew out gasoline lead. Elsewhere, the tourist attraction of Acapulco Bay is at risk due to raw and inadequately treated sewage. Deforestation, soil loss and depletion, and drinking water contamination complicate the tremendous problems.

Guerra is dedicated to closing both the existing information gap that keeps the public and their leaders from understanding the causes and solutions to environmental degradation and the profound cooperation gap that keeps business, government, and foreign suppliers from working together because of suspicion and lack of faith in each other. He is building the **Instituto Autonomo de Investigaciones Ecologicas** to propose and pursue environmental solutions through competent design and implementation. To do so, Guerra is bringing together colleagues who are highly skilled technically, managerially, and institutionally, and who are respected and trusted in the community.

He gets results because he opens up communications and the possibility of collaboration between industry, government, environmentalists, and overseas technology companies. In his public education program, Guerra has a popular weekly radio talk program, "Ecocidio," that takes on pressing issues with guests and audience participation. He is publishing a magazine, *Ecologia/Politica/Cultura*, which targets policymakers and a selective nonscientific community. And he is working on a countrywide environmental impact study that calls for action on specific issues. As Guerra's environmental commitments grew, he left a private-sector job where he had developed new products. With help from Ashoka, Guerra is anchoring the policy debate with credibility, facilitating problem solving, and sparking practical reform measures. He and his colleagues are building a completely self-sufficient environmental community, where they will eventually live with their families.

Nepal: Badri Dahal is spearheading the introduction of new forms of smart, **environmentally and economically sustainable farming and forestry.** His prototype is a "family of six that must sustain itself on a half acre or less," because this is the situation of 60 percent of Nepal's population. Building a national network of small demonstration farms adapted to its regional ecosystem and social patterns of this diverse nation, Dahal is illustrating self-sufficiency in food, fodder, compost, fuelwood, and maintenance materials while also generating some surpluses for sale. Trees for fruit, fodder, and multiple uses grow on all unproductive land as well as in and around the vegetable and other agricultural fields. Experiments are ongoing on various low-input and no-minimum tillage ways or with mulching to raise vegetables, grains, pulses, and legumes. Beekeeping, zonation, companion planting, and shelter belts are added options.

Gradually a farm family will take charge of everything in its midst—land, crops, trees, buildings and other improvements, energy, and water. As the family becomes more responsible, the farm becomes more sustainable, self-renewing, and profitable. Dahal also champions mud-brick construction stabilized with one-seventeenth cement, and persuaded the Agricultural Development Bank of Nepal to finance a demonstration of a low-cost home built in this manner. He is also working on plant nurseries, seed exchanges, and other ways of preserving the diversity of the country's germ plasm.

Dahal takes his methods directly to the small-scale farmers, showing them on their land how to put in a segment of a living fence that will also produce fuelwood and fodder, how to check rain runoff and erosion, how to fix nitrogen in their soil, and how to take the initiative in finding new solutions. Relying on traditional farming and Western "green resolution" techniques have resulted in deforestation with fields climbing hills, the application of heavy and costly chemicals, nutrient-poor soils vulnerable to erosion, and health and livelihood risks. Dahal's **Institute for Sustainable Agriculture Nepal (INSAN)** is running and expanding the model farms so that they are commercially successful and within reach of many other farmers. Training courses are also taught, and Dahal is developing a sustainable agriculture school curriculum for students of all ages. Dahal is looking for support from mainstream organizations such as Secretaries of Agriculture and Planning and the World Bank representa-

tives as well as from the international alternative movement for whom he hosted a major Permaculture Congress in 1991.

Thailand: Tuenjai Deetes is working with tribal populations along the Thai/Burmese (Myanmar) border to get them to replace slash-and-burn techniques with sustainable agricultural practices, as migration and high birth rate pushes up the population and forests rapidly disappear.

With her **Hills Area Development Foundation,** she is encouraging valley contour cropping on eroded hillside fields where "sturdy, valuable, often nitrogen fixing bushes are planted in horizontal lines along these fields, breaking up run-off and encouraging absorption during rains." Crop rotation is also recommended so that one year tribes will plant cassava and the next soy beans in order to maintain soil nutrients. Keeping these fields fertile is a means to protect Thailand's remaining watershed forest from cultivation.

Deetes is able to make progress because she shows respect for the hill tribespeople; uses a "powerful, holistic model approach"; and encourages their ability to work out adaptations to a changing environment on their own terms. By nurturing relationships with universities, she also is a role model for students. She has started village schools serving 70 percent of the children. Here they prepare to enter government schools, learn about their environment and culture, and take part in a community seedling nursery. Adults take evening classes and learn the Thai language and how to access government services. As formerly nomadic people settle down and gain citizenship, security is maintained for the tribes and for Thailand.

The Mekong River—which has divided Indo-China and the Western-oriented economies for decades, often with bloodshed—is a central artery of Burma, Thailand, and Indo-China countries and must be kept healthy, believes Wiltoon Permpongsacharoen. He is creating a resource center to serve the region's future, especially the private voluntary organizations and other groups who will be responsible for advancing long-term environmental interests here and in the entire ecosystem of mainland Southeast Asia. To accomplish this, he must first build the human and institutional prerequisites for change. His **Project for Ecological Recovery (PER),** set up in 1986, is an outgrowth of a yearlong "Eco-forum" dialogue, or *niwes saywanna,* among civil servants, students, academics, artists, and journalists.

As a result, Permpongsacharoen and his colleagues are developing an action plan for each country. In Thailand, PER succeeded in its campaign against commercial logging projects and its blockage of the Nam Choan dam that would have flooded forests being proposed for World Heritage status. Instead villagers were bolstered to start their own environmental movement and take on major water development projects—keeping any new dams from being built for five years. PER's strategy is to spur the growth of Thailand's environmental movement by organizing a participatory, democratic framework for various diverse social groups with common interests. PER acts as a secretariat to coordinate the diverse interests.

Permpongsacharoen is also focusing on logging in **Burma (Myanmar)** and on the Mekong's development in Indo-China. His analyses are being forwarded to the World Bank and other major donors and investors to set new directions and build a broad network for the environment and cultures of mainland Southeast Asia.

Mistaken industrialization strategies have forced the most suffering on the poorest sectors, who rely on the disappearing forests for fuel, food, and livelihood, Permpongsacharoen believes. As social conflict and environmental crises worsen, Thailand's neighbors—

"countries with extensive resources and repressive political regime that stifle public debate—are becoming like shock absorbers for the Thai economy," he writes. "In short, we are exporting the environmental crisis."

■ Resources

Ashoka publishes the *Profile Book of Fellows* and a newsletter, *Ashoka Update*.

Bat Conservation International, Inc. (BCI) ⊕
P.O. Box 162603
Austin, TX 78716 USA

Contact: **Dr. Merlin D. Tuttle**, Founder and Executive Director
Phone: 1-512-327-9721 • *Fax:* 1-512-327-9724

Activities: Conservation; Education; Research • *Issues:* Biodiversity/Species Preservation; Deforestation; Health; Sustainable Development

● Organization

Bat Conservation International (BCI) was founded in 1982 because scientists around the world were witnessing alarming declines in these animals. BCI's purpose is to document and publicize the values and conservation needs of bats, to promote bat conservation projects, and to assist with management initiatives worldwide. It has approximately 12,000 members from more than 50 countries.

▲ Projects and People

BCI devotes a large portion of its resources to educational efforts. Current projects range from the production of new educator's packets and programs on American bats to the development of a bibliographic database to serve as a worldwide center for dissemination of information about bats.

"Worldwide, bats are the most important predators of night-flying insects," a conservation biologist explains. A single, endangered species gray bat can catch up to 3,000 insects, including mosquitoes, in a single night, and large colonies can eat billions. "In Texas, the Bracken Cave colony of 20 million Mexican free-tailed bats eats up to a half a million pounds of insects nightly, including countless crop pests," according to authority **Dr. Norman Myers**.

In tropical ecosystems, bats help the survival of rainforests through seed dispersal and pollination. Genes from wild varieties of bananas, for example, help in the development of disease-resistant strains and other crop improvements. Dependent on bat pollination in East Africa, the "tree of life," or the baobab, is crucial to the survival of many kinds of wildlife. Desert ecosystems in the southwestern USA will be threatened if the two remaining colonies of nectar-feeding bats that pollinate giant cacti and agaves disappear.

Bat droppings, known as guano, feed countless small invertebrates and other life forms in caves, are a choice agricultural fertilizer from the Caribbean to Africa and Southeast Asia, and have potential use to biotechnology and industry—for chemical waste detoxification and gasohol production.

BCI's active research program provides the documentation necessary to initiate conservation measures. Grants have supported field-work in Africa, Australia, Southeast Asia, Latin America, and the Pacific Islands. Currently, BCI is beginning a project to identify and prioritize the most important bat sites around the world, leading to management plans for the most critical places. BCI has received research grants from the National Geographic Society, New York Botanical Garden, World Wildlife Fund, Chapman Foundation, Laurel Foundation, and other sources.

BCI's founder, Dr. Merlin D. Tuttle, is an ecologist who has studied bats worldwide for the past 25 years. He is respected internationally as a research scientist, conservationist, and wildlife photographer, and is well known for his documentation of bat decline and its ecological and economic consequences. His book, *America's Neighborhood Bats*, was published by the University of Texas in 1988.

BCI publications provide education material for the general public, researchers, educators, and others. It also maintains and makes available the world's largest collection of bat photographs.

■ Resources

Members receive *BATS*, a quarterly publication that provides the primary source of conservation information on bats worldwide.

Bay Area Ridge Trail Council (BARTC) ❦
311 California Street, Suite 300
San Francisco, CA 94104 USA

Contact: **Barbara Rice**, Director
Phone: 1-415-391-0697 • *Fax:* 1-415-391-2649

Activities: Education; Development • *Issues:* Land Conservation; Recreation

● Organization

The Bay Area Ridge Trail Council (BARTC) is a volunteer organization of thousands of private citizens; local, state, and federal park agencies; recreational, environmental, and community groups; and business associates that advocates development of a 400-mile trail system accessible to more than 6 million area citizens.

The Bay Area Ridge Trail connects the Bay Area's parks and open space in a nine-county area in California lying on or just below the principal ridgeline closest to the San Francisco Bay with a bay view. More than 120 miles are already open. As this "dream takes shape," users can explore hidden beaches along rugged coastline; see springtime poppies, lupines, and other wildflowers on mountaintops; capture views of huge redwoods and Napa County vineyards; bird watch in the marshlands; discover Dinosaur Ridge with its ancient fossils from the ocean floor; and roam through sycamore-shaded canyons.

▲ Projects and People

Since launching the project to connect the 400-mile trail system, the BARTC has overseen the completion of more than one-third of the trail—200 miles. It hopes to realize its goal of completion of the trail by 1998.

Among its projects for the trail, the BARTC helps incorporate **multi-use connections** for hikers, equestrians, and mountain bicyclists of all abilities along single and parallel alignments within the

Ridge Trail corridor; helps develop trails in a manner that cultivates appreciation and protection of the Bay Area's natural, cultural, and historic resources; and follows existing trails where possible, connecting them by using public lands and lands on which a public access easement has been acquired, either through gift, dedication, or purchase.

The BARTC is operated through a partnership of agencies and organizations. It is administered through Greenbelt Alliance, a private, nonprofit organization. Chairing the Council are Brian O'Neill, general superintendent of the Golden Gate National Recreation Area, National Park Service, U.S. Department of Interior; and Marcia J. McNally, regional planner and principal in the firm Community Development by Design.

■ Resources

The BARTC needs volunteers with expertise in trail construction to build new or restore old trails. It also needs assistance in the area of landowner outreach to facilitate trail acquisition on private land. Information brochures and request forms for membership, merchandise, and trail guides are available.

Bicycle Federation of America (BFA)
1818 R Street, NW
Washington, DC 20009 USA

Contact: **Andy Clarke**, Project Manager
   Phone: 1-202-332-6986 • *Fax:* 1-202-332-6989

Activities: Development; Education; Political/Legislative; Research • *Issues:* Air Quality/Emission Control; Energy; Transportation • *Meetings:* Velo Mondiale

● Organization

The Bicycle Federation of America (BFA) is a national nonprofit organization established in 1977 to promote the increased, safe use of bicycles for transportation and recreation.

The BFA specializes in performing research, professional training, and planning projects for federal, state, and local agencies; organizing conferences and seminars; and serving as a clearinghouse for information on all aspects of bicycle policy and programs. BFA staff recently established the Pedestrian Federation of America along similar lines. The BFA works with Congress, federal agencies, and national associations to expand support for bicycling programs.

▲ Projects and People

The National Bicycle Policy Project (NBPP) was created in 1988 as a joint program of BFA and Bikecentennial, Inc. Its mission is to push for bicycle-friendly policies at the national level for the long-term benefit of all bicyclists.

To achieve this mission, the NBPP staff states that it works closely with members of Congress, federal agencies such as the U.S. Department of Transportation, and organizations such as the Institute of Transportation Engineers and the American Association of State Highway and Transportation Officials.

If bicycling is to continue to grow as a sport, a form of recreation, and a means of transportation, the NBPP says, "We need to develop more safe places to ride. Too often, transportation planners and engineers ignore bicyclists in the design of streets and highways." Rather than challenge every road and design individually, the NBPP seeks to change policies and guidelines used by the highway engineers and planners. "We want to make the few key policies and regulations that affect the design and operation of almost every highway in the nation bicycle friendly," says NBPP.

To date, the NBPP states that it has been successful in its efforts. For example, it has created a special task force of bicycle facility experts and worked with the American Association of Highway and Transportation Officials on their revision of the outdated 1981 *Guide to the Development of New Bicycle Facilities.* It also secured the appointment of a full-time national bicycle program manager in the Office of the Secretary of Transportation and encouraged the creation of a similar position in the Federal Highway Administration; published a four-page guide to *Improving Local Conditions for Bicycling* designed for state and local traffic engineers; and drafted legislative language for pro-bicycle bills introduced by U.S. Representatives Joe Kennedy (D-MA), Jim Oberstar (D-MN), and Peter DeFazio (D-OR) in the 101st and 102nd Congresses.

"The environmental community must recognize excessive automobile use as an environmental problem requiring more than a technical fix," writes Andy Clarke. "We are making this argument for years, and will continue to do so."

■ Resources

BFA has available publications, videos, and a newsletter that deal with pertinent biking issues. It also has a *Pro Bike Directory* which lists more than 1,000 experts involved in bicycling issues, and an education program with lesson plans and video on the basics of bicycling.

Pro Bike News is a monthly newsletter available by subscription that reports important issues to the community of traffic and engineering professionals, bicycle program specialists, bicycle advocates, and others—providing up-to-date information on subjects such as cycling promotions, mountain bikes and land access, jobs, education, technical details, facility design, and legislation.

Bio Integral Resource Center (BIRC) ⊕
P.O. Box 7414
Berkeley, CA 94707 USA
Contacts: **Sheila Daar**, Executive Director
          William Olkowski, Technical Director
   Phone: 1-510-524-2567 • *Fax:* 1-510-524-1758

Activities: Education; Library/Information Services; Research; Training

● Organization

Bio Integral Resource Center (BIRC) is a nonprofit organization founded in 1978 to provide practical information on the least-toxic methods for managing pests. With laboratory and offices based in Berkeley, California, BIRC also has a 60-acre field station located in Davis, California, for applied research. The goal of integrated pest management (IPM) is to reduce pesticide use and improve the quality of pest control. BIRC has an international membership of 3,000 and maintains a 10,000-volume library on pest management in English and Chinese.

▲ Projects and People

BIRC developed a pest program for the U.S. Department of Interior's National Park Service which resulted in a 70 percent reduction of pesticide use within the first three years of implementation and an IPM for the city trees of Berkeley, California, resulting in 90 percent reduced pesticide use and a $22,500 savings to the city in the first year. The China Program facilitates scientific exchanges between the USA and mainland China on pest management and sustainable agriculture. Mycological Studies Program presents workshops and consultations on the isolation and cultivation of edible and medicinal fungi. The IPM Program designs workshops and training programs to individual, community, or organizational needs.

■ Resources

Members supporting this educational and research effort are entitled to contact BIRC with pest problems by phone or in writing. Annual memberships begin at $25. *The IPM Practitioner*, published 10 times a year, monitors integrated pest management as applied to agricultural, landscape, structural, medical, range, veterinary, and forest settings. The *Common Pest Control Quarterly* describes in practical terms the nontoxic or least-toxic methods for pest management.

Biodiversity Resource Center ⊕
California Academy of Sciences
Golden Gate Park
San Francisco, CA 94118-4599 USA

Contact: **Thomas D. Moritz**, Academy Librarian
  Phone: 1-415-750-7101 • *Fax:* 1-415-750-7106
  E-Mail: Omnet: sciencenet:cas.library; Ontyme:class.
      caslib; Bitnet: caslib@cmsa.berkeley.edu;
      Econet: caos
Activities: Education; Research • *Issues:* Biodiversity/Species Preservation

● Organization

The Biodiversity Resource Center was opened at the California Academy of Sciences in 1991 to provide students, advanced researchers, and the public with a huge variety of information resources which describe biodiversity and human efforts to preserve it. Biodiversity refers to three levels of organisms: genetic variations, species, and the interdependence of ecosystems.

The Center assists in the training of scientists and researchers so they can undertake this task during a time of rapid environmental erosion. It estimates that approximately 1.4 million species of plants and animals have been described by scientists, and that there are at least 5 million remaining to be identified. Yet biodiversity is gravely threatened as 10,000 species become extinct each year—mainly through the destruction of natural habitats by humans, who in turn become vulnerable as sources of food, medicine, wood, and other products disappear.

"Many indigenous human cultures have also been driven to extinction by the same forces that have destroyed and continue to threaten non-human species," reports the Center. "Nearly every habitat on Earth is at risk: the rainforests and coral reefs of the tropics, the salt marshes and estuaries of our coastal regions, the tundra of the circumpolar north, the deserts of Asia and Australia, the temperate forests of North America and Europe, the savannahs of Africa and South America."

Staff researchers with the California Academy of Sciences study biodiversity worldwide and describe more than 100 new species each year. They do work in La Amistad Biosphere Reserve, Costa Rica; the Impenetrable Forest, Uganda; coral reefs of New Guinea and Madagascar; deserts of southwestern Asia; and Socorro Island off the west coast of Mexico.

The Center is affiliated with the Academy Library, a research library founded in 1853 and devoted to natural history and the natural sciences. It contains 170,000 volumes and 2,100 current journals.

■ Resources

The Center offers CD-ROM bibliographic databases as well as online access to remote databases, electronic mail, video tapes, optical laser disks, and extensive conventional print-form references sources in the areas of biodiversity and species preservation. Reference requests can be made by mail, phone, fax, or E-Mail.

The Cactus and Succulent Society of America (CSSA) ⊕
521 Sergeantsville Road
Flemington, NJ 08822 USA

Contact: **Gerald S. Barad, M.D.**, President
  Phone: 1-908-782-5571 • *Fax:* 1-908-782-7982

Activities: Education; Legislative; Research • *Issues:* Biodiversity/Species Preservation • *Meetings:* CITES

● Organization

The Cactus and Succulent Society of America (CCSA) is an international nonprofit organization founded in 1929 which supports 80 local and regional organizations with 3,300 members. Dedicated to the protection and preservation of cactus and succulent plants, CCSA's activities include research, international liaison, and conservation.

▲ Projects and People

Gerald Barad, active member since 1949, is also involved with the International Organization for Succulent Plant Study; New York Botanical Garden's Desert Plants Committee, which he's chaired; and the American Horticultural Society. He has written extensively on the preservation and propagation of succulent plant species. CSSA created the Baja plant display at the Wild Animal Park, San Diego, California.

■ Resources

Guest speakers and slide presentations are available. CSSA members have opportunities to obtain rare seeds at nominal costs.

California Institute of Public Affairs (CIPA) ⊕
517 19th Street
P.O. Box 189040
Sacramento, CA 95818 USA

Contact: **Thaddeus C. Trzyna, Ph.D.** , President
  Phone: 1-916-442-2472 • *Fax:* 1-916-442-2478
  E-Mail: Econet: cipa

Activities: Education; Research • *Issues:* Environmental Policy
Meetings: IUCN

● Organization
A forum for the discussion of global environmental issues, the **California Institute of Public Affairs (CIPA)** is what can be called a "think tank." Founded in 1969 as "an independent, nonpartisan, statewide organization," the Institute has grown into a far-reaching medium for opening up dialogue and collaboration "across professions, academic disciplines, governmental agencies, and other sectors of society." CIPA helps find common ground on concerns, through gathering and generating useful information.

CIPA is affiliated with the Claremont Graduate School, a part of the Claremont Consortium, which consists of Pomona, Scripps, Claremont-McKenna, Harvey Mudd, and Pitzer colleges. Staff and students of the Claremont Graduate School come to the Institute's workshops and projects conducted on the Claremont campus. The William and Flora Hewlett Foundation is a major source of funding for programs regarding environmental decisionmaking; other sources of support come from various foundation grants, corporations, Claremont Colleges, and sales of its publications. CIPA's policy, however, is to accept public funding only for projects it initiates.

▲ Projects and People
CIPA runs **Collaborative Policy Forums** to focus on fundamental policy decisions. These discussions are held with experts and leaders on issues such as energy, agricultural land protection, and hazardous materials policy. The Institute's role is to act as a catalyst and remain impartial while "bringing people together with disparate interests" and "bridging the gap between thought and action."

In 1989, CIPA initiated a program concerning **California Environmental Strategies** to examine "deficiencies and new policy approaches" regarding the state's environmental and natural resource problems and to make practical recommendations for their solutions. Participants from government, business, and public-interest groups began concentrating on sustainable agriculture, water management, and strategic planning and management of large bioregions.

With the California Senate Office of Research, a special CIPA group is exploring possibilities to set up an independent **California Research Council**—similar to the National Research Council, but on the state level.

The Institute examines public policy questions and policymaking, using social-science research. The Institute's research efforts have created studies such as *Breaking Political Gridlock: California's Experiment in Public-Private Cooperation for Hazardous Waste Policy.*

Dr. Thaddeus Trzyna, CIPA president, is also chairman of the World Conservation Union's (IUCN) **Commission on Environmental Strategy and Planning (CESP)** with its 150 members in 55 countries; consequently CIPA plays a prominent role in that organization. CESP has working groups on strategies and tools for sustain-

ability; population and the environment; and developing concepts of "red books" for threatened natural, seminatural, and agricultural landscapes. (IUCN prepares other Red Data Books for endangered species.) A business and environment study group is also underway. Dr. Trzyna chairs the Sierra Club's international committee and is a senior associate at the Center for Politics and Policy at the Claremont Graduate School.

■ Resources
CIPA produces reports and investigations on public policy, current issues, and leadership. A quarterly *Catalog* is prepared on *New Books on California Affairs and the Global Environment.* In cooperation with IUCN and the Sierra Club, CIPA publishes the *World Directory of Environmental Organizations.*

Caribbean Conservation Corporation (CCC) ⊕
P.O. Box 2866
Gainesville, FL 32602 USA

Contact: **Chris Starbird,** Director of Programs
  Phone: 1-904-373-6441 • *Fax:* 1-904-375-2449

Activities: Ecotourism; Education; Protected Area Development; Legislative/Political; Research • *Issues:* Biodiversity/Species Preservation; Deforestation; Sustainable Development
Meetings: International Sea Turtle Workshop; USAID; World Parks Congress; Annual

● Organization
Caribbean Conservation Corporation (CCC) was founded in 1959 to protect sea turtles and their habitats through research, conservation, and education. CCC's 11-member staff conducts regional programs in the Americas including marine and aquatic environmental research, habitat protection, community development, and training and environmental education.

In Tortuguero, **Costa Rica,** CCC has researched green turtle ecology and migration for nearly 40 years—including projects with the Archie Carr Center for Sea Turtle Research, University of Florida. It has also studied the Tortuguero River system and marine, coastal, and terrestrial ecosystems of the Miskito Coast Protected Area in **Nicaragua.** Construction of a year-round Environmental Interpretation and Extension Center (EIE) for research at Tortuguero was planned for 1992. CCC also offers short courses in marine turtle conservation and management in Tortuguero and has been heavily involved in Costa Rican zoning plans for sustainable economic growth, ecotourism development, and resource management for biodiversity preservation.

▲ Projects and People
Green turtle tagging is the mainstay of the CCC conservation program along the 22-mile nesting beach at Tortuguero. The largest nesting fleet of green turtles in the Western Hemisphere reproduce at the remote beach site where the endangered leatherback, loggerhead, and hawksbill sea turtles also nest. Says **Charles Luthin,** CCC director of programs, "A student can learn more about turtle biology in a summer at Tortuguero than in a year at a university." The founding of Tortuguero National Park in 1975 was a direct result of research in protecting turtles and their habitats.

The Environmental Interpretation and Education Center (EIE), with support from the U.S. Agency for International Development (USAID) and foundations, will draw more than 15,000 tourists yearly to Tortuguero and provide program outreach to local communities. The Center's tour guide training program for villagers helps bolster the community's economy and strengthens ecotourism in the region. Yet development and visitors will be closely managed to protect the turtle rookery and the area's rich biological diversity.

A training program in marine resource conservation is ongoing since the 1960s for marine resource managers from the Caribbean region and other developing nations. Courses in marine turtle biology and conservation are taught in English and Spanish.

A proposal to expand **Tortuguero National Park** from 40,000 acres northward to 160,000 acres is being promoted to create a major wildlife corridor and biologically viable protected area. The enlarged Tortuguero Park would play an even greater role in the region as CCC attempts to create a huge **Binational Park** with Nicaragua, "SI-A-PAZ." In all, the reserve would cover more than one million acres and would be the largest protected lowland rainforest in Central America.

Nicaragua's northeast coast includes beautiful cays, coral reefs, brackish lagoons, and extensive river systems in virtually untouched conditions. The shallow waters of the Miskito Cays serve as the foraging grounds for the largest resident population of green turtles in the Western Hemisphere—the same turtles that migrate to and nest on the protected Tortuguero beaches. CCC is actively working to establish a 5,000-square-mile **Miskito Coast Protected Area** for the region's major ecosystems. With full support of the Miskito Indians and the Ministry of Natural Resources (IRENA), CCC is developing a comprehensive research and conservation program. With a start-off grant from USAID of $150,000, CCC is seeking matching funds from donors.

The **Paseo Pantera**, or path of the panther, project is attempting to create a regional system of protected areas with a consortium of the Wildlife Conservation International, New York Zoological Society, and CCC. To begin, the group has selected four parks in four Central American countries to develop as models for ecotourism using education, buffer zone management, and research. Once again, USAID is providing a partial grant.

Sea turtle conservation is being urged for Florida's eastern coast, which supports the largest population of nesting loggerhead turtles in the Americas. As many as 12,000 nests are recorded from a single 20-mile stretch of beach in Brevard and Indian River counties south of Melbourne. Florida's entire coastline has suffered considerable alteration from residential development, and the turtle nesting population has suffered. CCC strongly supports efforts to establish an Archie Carr Wildlife Refuge to protect undeveloped beaches. It would be the first of its kind in the USA specifically for sea turtles. CCC has focused attention through media efforts, and early governmental cooperation yielded $15 million toward the $90 million goal needed to secure the remaining properties. Recent campaigns focused on state legislation to discourage further beach deterioration.

Marine turtle research projects in the Caribbean involving green and leatherback turtles are also being supported in Bermuda, Bahamas, and Costa Rica—with the help of the Archie Carr Center for Sea Turtle Research. International sea turtle tagging efforts are also being undertaken

Executive director David Carr combines a background in political science and biology. He was biologist and program director for the Florida Defenders of Wildlife and a natural resources legislative analyst in the Florida House of Representatives. **Chris Starbird** directs most of the CCC programs along with fundraising and developing new projects. **Susan Marynowski** is project coordinator for the Archie Carr National Wildlife Refuge.

■ Resources

Velador (meaning night watchman or hunter's assistant who watches for turtles to come ashore) is CCC's quarterly newsletter. *Turtle Log*, newsletter of the Archie Carr National Wildlife Refuge Project, is also published. An "Adopt-A-Turtle" program and an international short course on marine turtle conservation are offered.

Caribbean Natural Resources Institute ⊕ ❦ (CANARI)
1104 Strand Street, Suite 206
Christiansted, St. Croix
Virgin Islands 00820 USA

Contact: **Allen D. Putney**, President and Director
Phone: 1-809-773-9854 • *Fax:* 1-809-773-9854

Activities: Development; Education; Research • *Issues:* Biodiversity/Species Preservation; Deforestation; Sustainable Development • *Meetings:* World Parks Congress

● Organization

Working with governments and NGOs, the Caribbean Natural Resources Institute (CANARI) concerns itself with sustainable development and the quality of human life. CANARI seeks to develop programs that help both man and nature and that will also promote social equality. From 1977 to 1989, CANARI was known as the Eastern Caribbean Natural Area Management Programme (ECNAMP) and was a field program of the Caribbean Conservation Association and the University of Michigan. So far, CANARI has undertaken 6 regional projects, 16 field projects in 9 countries, major workshops, and short consultancies.

▲ Projects and People

According to CANARI president Allen D. Putney, CANARI focuses its efforts in two program areas: community-based management and parks and protected areas. Most projects range from 8 to 10 years; are situated in areas suitable for research, development, hands-on training, and extension; and are adaptable to other settings.

The **Community-Based Management Program** uses grassroots support to design and implement resource management projects that can also yield new economic opportunities. These projects have included a decade-long, southeast coastal zone management project in St. Lucia. Elements include mangrove and fuelwood management; community woodlots established for charcoal production; start of a seaweed mariculture commercial cottage industry; assistance to fishermen; development of local attractions; and ecological conservation. Elsewhere in St. Lucia, CANARI collects information on coral reef health and management, including sedimentation rates. In addition, a regional resource center has been set up for training with library and computer facilities.

In the ongoing **Seamoss Research and Development Programme**, in which seamoss farms are set up, "superior strains from wild populations [are selected] for cultivation," sites are monitored for the effects of environmental variables on productivity, and markets are identified.

In Dominica, a not-for-profit lumber purchasing and marketing corporation was set up for independent, small-scale sawers who adhere to this forest cottage industry's conservation measures. Using timber-stand improvement techniques, chain saws are used to fell mature trees and rip boards on site—"reducing disturbance to residual vegetation and soils," it is reported. Markets are guaranteed for lumber that is dried, dressed and sold, mostly for furniture.

CANARI documents its experiences and provides technical aid to assist students, resource managers, development workers, and others who could benefit from increased information about community-based initiatives.

The **Parks and Protected Areas Program** promotes the management of these areas for conservation and sustainable development. A key focus is helping to meet the training and information requirements of protected area managers. The program draws information and expertise about the Caribbean from within the region. One result is the creation of the **Caribbean Islands Park and Protected Area Network**, as technical and financial resources were pooled to learn, for example, about wildlife census, trail construction techniques, and marine park management.

Through the **Caribbean Heritage Programme**, CANARI is supporting Dominica Conservation Association revitalization; a National Trust in Anguilla; Barbuda nature tourism; Trust Fund for Jamaica's national park development; and other natural resource planning in St. Lucia's Soufriere region and in Virgin Gorda's North Sound, British Virgin Islands. Cooperating are local governments and the Caribbean Conservation Association with funding from the John D. and Catherine T. MacArthur Foundation. CANARI supports the **Consortium of Caribbean Universities for Natural Resource Management**, which it helped launch, and which provides educational opportunities for professionals.

Future plans include greater involvement in UNESCO's Man and Biosphere Preserves international program, particularly the Guànica Biosphere Reserve, Puerto Rico, and the Virgin Islands Biosphere Reserve, St. John. The principles of partnership, grassroots participation, and cooperation will continue to guide CANARI, it assures.

■ Resources

CANARI has technical reports and other publications from its work with community management. It also publishes the *Caribbean Park and Protected Area News*.

Carrying Capacity Network, Inc. (CCN)
1325 G Street, NW, Suite 1003
Washington, DC 20005-3104 USA

Contact: **John Sample**, Executive Director
Phone: 1-202-879-3044 • *Fax:* 1-202-879-3019
E-Mail: Econet conference: ccn.capacity

Activities: Education; Information Exchange; Outreach • *Issues:* Air Quality/Emission Control; Biodiversity/Species Pres-

ervation; Deforestation; Energy; Global Warming; Health; Population Planning; Sustainable Development; Transportation; Waste Management/Recycling; Water Quality

● Organization

Carrying Capacity Network (CCN) is a nonprofit, nonpartisan activist network that acts as an environmental information clearinghouse. CCN facilitates cooperation and information dissemination among organizations working on carrying capacity issues such as environmental protection, population stabilization, growth control, and resource conservation.

Carrying Capacity refers to the number of individuals who can be supported without degrading the physical, ecological, cultural, and social environments—without reducing the ability of the environments to sustain the desired quality of life over the long term. The board of directors includes **Dr. Virginia Abernethy**, Vanderbilt University; **Dr. Herman E. Daly**, World Bank; **David F. Durham**, attorney; **Tom McMahon**, computer writer involved in family planning and population issues for more than a quarter-century; and **Dr. Ieda Siqueira Wiarda**, U.S. Library of Congress and University of Massachusetts.

■ Resources

These include the publication of a quarterly journal, *Focus*, and a monthly newsletter, *Clearinghouse Bulletin*. *Focus* provides reprints of current and classic environmental essays; furnishes persuasive point-counterpoint discussions regarding controversial environmental issues such as the need for a national energy policy, global oil demand and supply, and "optimum population" for the USA; includes informative research articles; and features interviews with environmental personalities.

In a recent *Focus* issue, **Sandra Postel**, vice president for research, World Resources Institute, described how "pollution pays" in the calculation of the GNP. "The Alaskan oil spill of March 1989, the most environmentally damaging accident in U.S. history, actually created a rise in the GNP, since much of the $2 billion spent on labor and equipment was added to income," she wrote. "Equally perverse, much of the $40 billion in health care expenses and other damages incurred by U.S. citizens annually as a result of air pollution is counted on the plus side of the national income ledger."

Clearinghouse Bulletin is designed to succinctly highlight current environmental group initiatives, follow the most pressing environmental bills before Congress, and provide a fresh yet balanced perspective on environmental issues of interest, such as the impacts of "free trade," future of the Arctic National Wildlife Refuge (ANWR), dam relicensing, national flood insurance, and "auto-free cities." In addition, CCN keeps a list of speakers and writers available to environmental organizations.

In order to keep CCN and its participants aware of environmental organizations' activities, the group actively seeks submission of articles for its publications. Environmentalists or interested parties may call 1-800-466-4866.

Center for Environmental Information, Inc. (CEI)

46 Prince Street
Rochester, NY 14607-1016 USA

Contact: **Frances Gotcsik**, Manager, Program in
Environmental Risk Communication
Phone: 1-716-271-3550 • *Fax:* 1-716-271-0606

Activities: Collaborative Problem Solving; Development; Education; Environmental Mediation; Library/Information Services • *Issues:* Air Quality; Energy; Global Warming; Waste Management/Recycling; Water Quality

● Organization

The **Center for Environmental Information** (CEI) provides timely, accurate, and comprehensive environmental information to the public. The Rochester-based nonprofit organization, with a staff of 15, sustains its activities through membership dues, fees, contracts, grants, and contributions.

CEI sponsors educational programs, conferences, and seminars for decisionmakers, public officials, interest groups, business people, researchers, educators, students, and the public. The organization's information specialists access global databases through the Center's computer system, maintain an extensive library of materials available to the public, and handle information requests.

▲ Projects and People

A **Program in Environmental Risk Communication** (PERC) enhances decisionmaking by stimulating risk analysis from various perspectives, addressing "legitimate concerns voiced by a society seemingly besieged by environmentally risky situations," and enabling parties in an environment decision to gain credibility with one another.

A **survey short course in environmental law** is held annually in conjunction with the Monroe County Bar Association to provide an overview of environmental law at the local, state, and federal levels with focus on problem areas, unresolved issues, and future trends.

With the Center for Dispute Settlement, CEI offers sessions on **conducting environmental dialogue and making contact with the other side** to avoid paralyzing conflict and "environmental gridlock." Community and labor leaders, public officials, scientists, and environmental professionals participate.

■ Resources

CEI wants to expand its networking to "increase awareness of non-adversarial options to decisionmaking"; it also seeks help in marketing their services. CEI offers environmental information to callers and specialized services upon request. It publishes a newsletter, *Sphere.*

Center for Environmental Study ⊕

143 Bostwick NE
Grand Rapids, MI 49503 USA

Centro de Estudio Ambiental

APDO 347-2120
San Francisco de Guadalupe
295 Av. Calle 21-23
San Jose, Costa Rica

Contact: **Kay T. Dodge, Ph.D.**, President
Phone: 1-616-771-3935, 506 22 6608
Fax: 1-616-771-3907, 506 22 5052

Activities: Communications; Education; Research • *Issues:* Biodiversity/Species Preservation; Business Stewardship; Deforestation; Ecotourism; Global Warming; Health; Population Planning; Sustainable Development; Transportation; Water Management/Recycling; Water Quality • *Meetings:* CITES; IUCN; UNCED; Audubon; Great Lakes United; International Wildlife Congress

● Organization

The **Center for Environmental Study** (CES) has been a leader in environmental education, communications, and research in western Michigan since 1969. CES quickly became a focal point where citizens, industry, and government groups could come together to define and analyze problems and plan solutions for community needs. Today, the CES has its home on the campus of Grand Rapids Community College and also houses a collection of resource materials including books, magazines, clippings, videos, and slides.

CES's 11-member staff expanded its mission internationally in 1990 when it opened the **Centro De Estudio Ambiental** (CEA), San Jose, Costa Rica, to develop new programs for Latin America while effectively linking itself to North American conservation efforts. CES and CEA formed a partnership with a common mission and common goals.

▲ Projects and People

Among the Center's Tropical Forest Projects is the **Tree Amigos** program, which offers the people of the Americas an opportunity to form a common bond with the environment through tree-planting activities. Tree Amigos uses school and community programs in both the USA and Latin America to foster tree plantings, exchanges of materials, curriculum integration, teacher training, and translation. The program also recruits volunteers with special talents, such as educators, engineers, and media specialists who share their expertise in a professional talent exchange. Tree Amigos raises funds for the protection of tropical forests and encourages corporations to become partners in all phases of its efforts—including promoting environmentally safe products, educating employees, and conducting community relations campaigns.

CES has joined with other conservation groups to help the U.S. Department of Agriculture's (USDA) Forest Service improve its Threatened, Endangered, and Sensitive Species program. As part of a task force, the CES provided research, strategic planning, and services to develop an action plan for the Forest Service's **Every**

Species Counts national campaign. In Costa Rica, the Center organizes volunteers and corporate sponsors to encourage school children to clean up the country's beaches. To boost recycling, CEA has developed a system for identifying different types of plastics in commercial use and informing consumers about which types are reusable.

The guiding force in both CES's and CEA's efforts is **Kay T. Dodge**, the Center's president and executive director. Dr. Dodge produces television programs, such as *Warnings on the Wind* about the growing problem of air toxins and *Hot Potato* on solid waste management. Currently adjunct professor of ecology, world regional geography, and adult education theory at Grand Valley State University and Western Michigan University, Dodge has also taught innovative environmental education programs in the Grand Rapids public schools. A wildlife artist and photographer, she is a former Audubon Society state president.

■ Resources

CEA and CES offers speakers and workshop assistance to other groups; Tree Amigos teachers' resource book; and videos entitled *Birds of Michigan, Warnings on the Wind,* and *Hot Potato.* The partnership seeks resources on tropical subculture and reforestation projects. As it expands the Tree Amigos program, it welcomes new sponsors.

Center for Holistic Resource Management (CHRM) ⊕ ✹
5820 4th Street, NW
Albuquerque, NM 87107 USA

Contact: **Shannon A. Horst**, Chief Executive Officer
Phone: 1-505-344-3445 • *Fax:* 1-505-344-9709

Activities: Development; Education; Political/Legislative; Research • *Issues:* Biodiversity/Species Preservation; Community Development; Deforestation; Energy; Global Warming; Soil Erosion; Sustainable Development; Water Quality

● Organization

The **Center for Holistic Resource Management (CHRM)** was founded in 1984 to increase awareness and understanding of holistic management through education. A nonpolitical and nonprofit organization it was formed by a group of ranchers, farmers, researchers, and environmentalists to serve as a focal point for the exchange and dissemination of knowledge on holistic management. Membership includes approximately 1,500 families, individuals, and organizations. CHRM has 11 state, regional, and international branches, manned for the most part by volunteers.

Overall goals are to "produce stable environments with sound watersheds; restore profitability and sustainability to agricultural operations; improve the water resources of cities, industries, and agriculture; increase wildlife species . . . and the stability of wildlife populations; restore productivity and stability of riparian areas; prevent the waste of financial resources on faulty resource management practices; and increase citizen participation in sound resource management."

▲ Projects and People

A holistic resource management (HRM) model was developed by African-born ecologist **Allan Savory** as an alternative to conventional resource management approaches. As a wildlife biologist in Zambia in the early 1950s, Savory observed the effects of wild game on the land—some of which was lush and productive, and other areas which were degraded. Animal hooves "broke up and loosened crusted soils and trampled down old plant parts thereby creating mulch and an ideal seedbed," reports CHRM.

Land that appeared devastated would bounce back the next growing season, and Savory noted that "the extent to which the animals disturbed the land was directly related to the presence of predators which kept the herd bunched and excited." It also seemed that such areas tended to have erratic rainfall, whereas in jungles and certain grasslands, rainfall was more reliable—helping plant materials to break down rapidly.

It became obvious to Savory that plants, soils, herding prey, and predators needed each other—particularly in relation to the climate in which they co-existed. He also learned from a researcher in France, **Andre Voisin**, that huge herds roaming over large areas of land did not overgraze, while individual animals in a limited area did. Further studies of the 1920s holism philosophy of South African statesman and naturalist **Jan Smuts** led Savory to create the HRM planning model that takes into account many interacting variables of the ecosystems without losing sight of the "whole."

Rather than treating in isolation events such as the invasion or disappearance of a species, the holistic approach asks, "What quality of life can we expect? What can this place produce in order to support it? What will this landscape have to become to sustain this production indefinitely?"

Thus Savory developed a simple model that first addressed desertification, or the deterioriation of Africa's grasslands. It can be used by anyone, he says, to solve or make management decisions that are economically, socially, and environmentally sound.

Living in the USA since 1980, Savory practices his principles on his own farm and ranch. Also working with farmers, ranchers, and conservationists worldwide, Savory refined HRM with some outstanding results. According to CHRM literature, the Sonnleiten Ranch in Namibia increased both livestock and forage with this method and using dung beetles, termites, grasshoppers, and caterpillars in place of pesticides. Other "workers" are the wild pigs and antbears that dig up the soil.

Author and syndicated columnist **Alston Chase** writes, "The HRM model provides a kind of blueprint for conflict resolution: it aids the land manager's family or community in reaching a consensus on what they want from themselves, from others, and from the land."

■ Resources

CHRM offers 15 different courses on aspects of holistic management each year in various locations in North America and overseas, but mainly outside Albuquerque's international headquarters. Courses shape skills in holistic management and successful resource management policies, help families in business, devise long-range planning, and aid individual development.

The Center offers a degree program which was created to provide a worldwide core of practitioners with in-depth knowledge of HRM; the program is tailored to individual needs and time frame.

Other resources include the *HRM Newsletter, Holistic Resource Management* book and workbook, brochures and membership form, annual meetings, and international networking.

The Center needs additional persons interested in learning how to train others in HRM; as well as broader cooperation with groups, agencies, and institutions specializing in community development issues, such as conflict resolution, and leadership to help broaden the teaching base.

Available resources include training in holistic management and teaching for landowners/managers and others; and speakers, workshops, and private consultants on land sites.

Center for International Environmental Law— US (CIEL-US) ⊕
1621 Connecticut Avenue, NW, Suite 300
Washington, DC 20009-1076 USA

Contact: **Durwood J. Zaelke**, President and Founder
Phone: 1-202-332-4840 • *Fax:* 1-202-332-4865

Activities: Education; Law Enforcement; Political/Legislative; Research • *Issues:* Air Quality/Emission Control; Biodiversity/ Species Preservation; Deforestation; Energy; Global Warming; Health; Oceans Protection; Population Planning; Sustainable Development; Technology Transfer; Water Quality

● Organization
"CIEL-US was founded in 1989 to make environmental law and institutions more effective in solving global environmental problems and promoting sustainable development," according to the Center for International Environmental Law–US. "Through education and training, research and publications, and advocacy, CIEL-US seeks to broaden the role of citizens and NGOs in global environmental decisionmaking."

To carry out is purposes, CIEL-US forms partnerships with NGOs worldwide "to promote innovative and culturally appropriate solutions to environmental problems" that impact globally or regionally. These partnerships are characterized by a network of public-interest lawyers who "audit" implementation and compliance with environmental agreements and "focus the spotlight of public attention on progress towards environmental protection and sustainable development." These annual performance audits from various countries comprise the *State of Environmental Law* reports.

CIEL-US is pursuing a new level of cooperation among all the major "stakeholders"—NGOs, corporations, and governments—to search for "effective and efficient solutions [to] pollution prevention, total quality management, industrial ecology, and design for environment." What may be required, CIEL-US expects, are "new laws and institutional structures, with appropriate confidence-building mechanisms, so that the stakeholders can transcend their often adversarial relationships, shift away from the traditional command and control regulation that too often stifles innovative solutions, and move towards market-based incentives and disincentives."

▲ Projects and People
To help bring about "compliance," CIEL-US conducts independent research, case studies, and training workshops—with results disseminated through NGO partnerships and publications. Involv-

ing lawyers, law students, and scientists from the USA and abroad, it also offers a summer internship program, pro bono services, and Law Fellow and Law Associate positions that help launch careers in international and comparative law regarding global environment and development issues.

To help Central and Eastern Europe address the "staggering task of environmental reconstruction" as its countries move toward democracy and market economies, CIEL-US is assisting with several projects:

The European Bank for Reconstruction and Development (EBRD). With the encouragement of CIEL-US and others, the Bank's Charter requires that it promote democracy and "environmentally sound and sustainable development" in its activities. The lawyers' network is helping to see that implementation actually occurs, advancing environmental laws, and "facilitating loan applications for environmentally sustainable projects."

State of Environmental Law in Central and Eastern Europe. These annual reports, published in English and each official language, will measure the region's "evolving environmental law against those principles that should be found in any sound system of environmental laws and regulation." A regional lawyers' network will analyze progress; and the reports will help "mobilize citizen efforts to enforce compliance with the newly created environmental requirements and to assist industry in achieving voluntary compliance." The first in-country report was published in August 1991 regarding environmental decisions in the Czech and Slovak Federal Republic (CSFR).

Environmental Education and Assistance. Classroom and practical training in environmental protection strategies were held for public-interest lawyers, students, and environmental professionals, for example, in Budapest, Hungary, and at the Ecological University, Bucharest, Romania. Follow-up seminars are being undertaken and fellowships are being offered at the CIEL-US office, where permission for U.S. copyright is sought for foreign translations of appropriate titles. Under consideration is a CIEL-US office in Bucharest to help find technical assistance and to house an environmental law resource library, also containing materials on EBRD and World Bank activities. CIEL-US also sponsors and participates in public-interest exchanges and will assign an environmental lawyer to each country to develop legal capacities of environmental organizations in Central and Eastern Europe.

To ensure citizen participation, CIEL-US plans to organize task forces within certain countries "to develop culturally appropriate procedures" for broadening involvement in environmental decisionmaking. Legal assistance on specific environmental issues includes advising environmental organizations on oil development near Bulgaria's Black Sea coast and tourist development in CSFR's Tatra Mountains as well as analyzing the Slovak government legal position on construction of the Gabcikovo-Nagymoros dam. CIEL-US anticipates that it will provide assistance to "western investors and government ministries responsible for privatizing industry."

Other ongoing programs include the following:

Technology Cooperation between developed and developing nations, with attention to protecting "intellectual property rights" as technology is transferred for environmental purposes. CIEL-US also helps assess the role of multinational corporations in solving global environmental problems and ways they are able to succeed. It also is working to "integrate environmental protection and sustainable development into international trade policy" and assists NGOs that participate in global warming negotiations.

Oceans Conservation, whereby the "Freedom of the Seas" notion is being challenged and alternate principles of ocean governance are being explored. Being prepared for publication are 30 papers commissioned for a workshop on this topic held in conjunction with the Institute for Peace at the University of Hawaii Law School, the Peace Research Center of Canberra, Australia, and Greenpeace. With the National Oceanic and Atmospheric Administration (NOAA), CIEL–US is studying the "legal implications of wild ocean reserves for the high seas."

The Americas and Caribbean Program, whereby CIEL–US is working with World Wildlife Fund–USA and Brazil's Ecotropica Foundation to improve the enforcement of this country's environmental laws through training and education. This program particularly seeks to strengthen Brazil's *Curador do Meio Ambiente*, the "environmental guardians" within the Public Defender's office. CIEL–US is researching ways to protect indigenous peoples against deforestation, biodiversity loss, and other environmental and cultural threats. It is also exploring how "the human rights system—its legal standards, courts and commissions, network of public interest advocates, and successful strategies"—can serve as a model for the evolution of more democratic, and more effective international environmental laws.

Durwood J. Zaelke heads the CIEL–US staff of nine and is on the Board of Advisors for this Guide. Barbara L. Shaw is vice president. Attorney David B. Hunter directs the Central and Eastern Europe Program and law associate Chris A. Wold is assistant director. Hunter is also an adjunct professor of law at American University's Washington College of Law. Attorney Robert F. Houseman develops projects on environmental human rights and conducts general advocacy work.

■ Resources

CIEL–US's publications list includes reports and working papers on Central and Eastern Europe, Technology and Global Change, Oceans Conservation, and a *Press Relations Manual for NGOs*—among other documents.

Center for Plant Conservation (CPC)
P.O. Box 299
St. Louis, MO 63166-0299 USA

Contact: **Peggy Olwell**, Manager, Conservation Programs
Phone: 1-314-577-9450 • *Fax:* 1-314-664-0465

Activities: Development; Education; Research • *Issues:* Biodiversity/Species Preservation • *Meetings:* IUCN/SSC Specialist Groups; Kew Conference

● Organization

The Center for Plant Conservation (CPC) was founded in 1984 to prevent the extinction of the native flora of the United States. It is affiliated with the Missouri Botanical Garden in St. Louis. The heart of the CPC program lies in its national network of participating botanical gardens and arboreta.

These include the Arnold Arboretum of Harvard University, Berry Botanic Garden (Oregon), Bok Tower Gardens (Florida),

Denver Botanic Gardens (Colorado), Desert Botanical Garden (Arizona), Fairchild Tropical Garden (Florida), Arboretum at Flagstaff (Arizona), Garden in the Woods (Massachusetts), Holden Arboretum (Ohio), Mercer Arboretum (Texas), Missouri Botanical Garden, National Tropical Botanical Garden (Hawaii), Nebraska Statewide Arboretum, New York Botanical Garden, North Carolina Botanical Garden, Rancho Santa Ana Botanic Garden (California), Red Butte Gardens and Arboretum at the University of Utah, San Antonio Botanical Garden (Texas), University of California Botanical Garden, and Waimea Arboretum and Botanical Garden (Hawaii).

"Collectively, these gardens constitute an extraordinary resource for conservation and provide a world model," writes Don Falk, CPC director.

▲ Projects and People

A total of 20 institutions—all dedicated to plant conservation—are the essence of the CPC and participate in the National Collection of Endangered Plants that contains seeds, plants, and cuttings of 372 plant species native to the USA. Emphasis is on collecting "38 new species and to recollect 19 species in order to increase their genetic representation," CPC notes.

In one project, CPC is assisting various researchers around the world in their search for effective and safe biological control agents of leafy spurge, an exotic, introduced weed. Leafy spurge is an aggressive perennial which tends to displace other vegetation in pasture and rangeland habitats in places such as North Dakota. "Reductions of forage from 10 to 100 percent have been observed," according to CPC, which cooperates with the U.S. Department of Agriculture's (USDA) Animal and Plant Health Inspection Service (APHIS) in identifying an insect or disease pathogen that will be effective biocontrol—without harming economically important plants or endangered species.

To learn how to protect fragile habitats of certain rare plants CPC fellowship recipient Susan Wiser searched in the southern Appalachian Mountains for endangered species, such as Heller's blazing star, mountain avens, Blue Ridge goldenrod, Cain's reedgrass, and Roam Mt. bluet. She noted threats to their survival such as a planned ski resort, trampling by tourists on trails, acid rain which is causing red spruce and Fraser fir to decline on high peaks, higher acidity fogs, and rising temperatures due to global warming. However, Wiser was heartened to discover new populations of rare species.

■ Resources

The CPC publishes a quarterly publication, *Plant Conservation.*

Center for Population Options (CPO) ⊕
1025 Vermont Avenue, NW, Suite 210
Washington, DC 20005 USA

Contact: **Margaret Pruitt Clark**, Executive Director
Phone: 1-202-347-5700 • *Fax:* 1-202-347-2263

Activities: Development; Education; Political/Legislative • *Issues:* Health; Population Planning

● Organization

The **Center for Population Options (CPO)** is a nonprofit education organization founded in 1980 by **Judith Senderowitz** and dedicated to improving the quality of life for adolescents by preventing too-early childbearing.

CPO's national and international programs seek to improve adolescent decisionmaking through life planning and sexuality education programs. As part of its mission, CPO promotes access to comprehensive health care, including family planning through school-based and other community-based clinics and works to prevent the spread among adolescents of HIV and other sexually transmitted diseases.

CPO works to enhance opportunities for young people in key areas of their lives: continuing their education, planning their families, obtaining needed health and social services, and finding productive employment.

CPO places special emphasis upon the interconnections between family formation and other life actions. CPO believes that, in general, young people can more satisfactorily complete their education and prepare for employment if they delay parenthood. This, in turn, entails support for assistance in making decisions to delay sexual activity or to practice contraception; information about sexuality, birth control, and abortion; ready access to family planning and reproductive health services; and economic subsidy for these services, if needed.

▲ Projects and People

CPO works in partnership with existing organizations that already serve youth, such as the Salvation Army, National Youth Employment Coalition, and Boston Public Schools, to incorporate teen pregnancy and AIDS prevention into their programs.

The **High Risk Youth Demonstration Project** was started in 1989 with three organizations—Big Brothers/Big Sisters of America, the Salvation Army, and the Young Women's Christian Association of the USA. Along with CPO staff, these organizations selected three sites—Hartford/New Haven, Connecticut; Milwaukee/Racine, Wisconsin; and Atlanta, Georgia. The local affiliates have created a wide variety of HIV prevention programs. CPO has been working with the three organizations to help them devise plans for replicating the AIDS prevention projects in their affiliate chapters.

Since 1987, CPO has convened a group of 6 to 10 teenagers from high schools in the Washington, DC, areas to serve on its **Teen Council**. The teens receive training about teen pregnancy and AIDS prevention, as well as communication skills.

A **Teens for AIDS Prevention (TAP)** program trains teenagers to be advocates for and educators of their peers. The "Guide to Implementing TAP," supplies youth-serving professionals with information to set up a TAP program at their school or organization.

A **Support Center for School-Based Clinics** helps clinic practitioners design and refine their programs, offering technical assistance on fundraising, staff management, evaluation, and advocacy.

Who am I? Where am I going? How do I get there? These questions form the basis of CPO's sexuality education curriculum, **Life Planning Education: A Youth Development Program**.

CPO's research staff works with all of CPO's projects to help evaluate programs. The staff also is called upon to evaluate and assess promising models and programs developed by organizations throughout the country.

An **International Clearinghouse on Adolescent Fertility**, the only international program concerned exclusively with teen pregnancy prevention abroad, works to educate policymakers and the public about the problem of too-early childbearing in the developing world and to help local organizations based in Africa, Asia, and Latin America to start up teen pregnancy prevention programs.

■ Resources

Fact Sheets, *Teenage Pregnancy and Too-Early Childbearing: Public Costs, Personal Consequences* publication, and a list of reports and other materials are available.

Center for Reproduction of Endangered Species (CRES) ⊕

Zoological Society of San Diego
P.O. Box 551
San Diego, CA 92112 USA

Contacts: **Donald G. Lindburg, Ph.D.**, Head, Behavior Division
Phone: 1-619-557-3949;

Valentine A. Lance, Ph.D., Endocrinologist
Phone: 1-619-557-3944 • *Fax:* 1-619-557-3959

Activities: Education; Research • *Issues:* Biodiversity/Species Preservation • *Meetings:* IUCN

● Organization

To "combat a loss of wildlife and to assure the survival of imperiled animals for future generations," the **Zoological Society of San Diego** established the **Center for Reproduction of Endangered Species (CRES)**.

Founded in 1975 by **Dr. Kurt Benirschke**, CRES is an "intensive research program which applies the successes of modern technology to the world of endangered animals." Under the direction of **Dr. Werner Heschele**, CRES scientists develop a "complete profile for each species based on studies in the field of behavior, endocrinology, genetics, infectious diseases, pathology, and physiology."

Nationally and internationally known for their work, CRES scientists have achieved breakthroughs culminating in population increases among such rare species as the Indian rhinoceros, Chinese monal pheasant, Przewalski's wild horse, and others. CRES shares its advancements with zoos and wildlife organizations worldwide.

▲ Projects and People

CRES scientists, under the direction of **Dr. Donald G. Lindburg**, head of the behavior division, have been working on environmental projects that involve the reintroduction of endangered species to a remaining habitat after a period of captive breeding.

Examples include a project with the **California condor**, under the leadership of Noel Snyder and Bill Toone. CRES personnel took remaining birds from the wild and set up captive breeding. Release of the captive-born animals is being prepared. CRES behaviorists have been developing field and captive observational protocols and supervising the initial studies of the hatchlings from the first two years of captive incubation of wild harvested eggs.

Alan Lieberman, San Diego Zoo's curator of birds, led efforts to return 14 Andean condors, hatched and reared in California, to protected sites in the Colombian Andes. INDERENA, the Colombian Department of Interior; FES, a non-governmental conservation organization; and the Chiles Indian community assisted in relocating, even hand carrying, these birds to their new homes and continuing their care.

The reproductive biology and behavior of the lion-tailed macaque, an endangered monkey from India, is being studied with a reintroduction effort underway to return captive-born species to his native land. The cheetah is also the subject of successful captive-breeding experiments.

Dr. Lindburg believes that, "the single most important issue for environmental conservation and quality of life . . . is human population growth and growth-oriented economies in industrialized countries. My specialty is wildlife conservation, and I foresee an end to wildlife habitat as we know it today—if current human population trends continue." Dr. Lindburg recently organized a resolution effort within the American Society of Primatologists to call on "scientific organizations to join forces in shaping U.S. national policy on population growth."

■ Resources

The San Diego Zoo publishes the *CRES Report.*

Center for Reproduction of Endangered Wildlife (CREW)

Cincinnati Zoo and Botanical Garden
3400 Vine Street
Cincinnati, OH 45220 USA

Contact: **Betsy L. Dresser, Ph.D.,** Director of Research
Phone: 1-513-961-2739 • *Fax:* 1-513-569-8213

Activities: Research; Education; Development • *Issues:* Biodiversity/Species Preservation • *Meetings:* African Rhinoceros Workshop; Fifth World Conference on Breeding Endangered Species in Captivity; Reproductive Strategies for Endangered Wildlife

● Organization

The Cincinnati Zoo created a specific Research Department in 1981, with Dr. Betsy Dresser as its director. That same year, the Cincinnati Wildlife Research Federation was organized, as a joint effort of the zoo, the University of Cincinnati College of Medicine, and the Kings Island Wild Animal Habitat (KIWAH). The Center for Reproduction of Endangered Wildlife (CREW) supplanted previous research bodies in 1988. The CREW facility is being housed at the newly constructed Carl H. Lindner, Jr. Family Center for Reproduction of Endangered Wildlife—the world's first institution of its kind.

CREW has maintained its network of supporting organizations, among them KIWAH and the University of Cincinnati College of Medicine and Department of Biology.

▲ Projects and People

"Preserving wildlife . . . is far more complex than saving two of a species," CREW notes. An ongoing CREW project is the development of "universal surrogates." Through *in vitro* fertilization, more common animals, such as the domestic cat, have successfully been used as surrogate parents of closely related endangered species, such as the Indian desert cat.

The world's first birth of an exotic animal, the African eland antelope named "E.T." occurred here following a nonsurgical embryo transfer procedure in 1983; other elands subsequently became surrogate mothers to African bongo calves. In further research, a Holstein cow named "Alice" gave birth to a 70-pound male gaur calf—which is an endangered species of wild forest ox native to Malaysia. Meanwhile, "Louise," the calf's natural mother, also produced a female gaur the same year—illustrating the potential for boosting the birthrate of endangered species.

Through cryopreservation—freezing and storing genetic material such as embryos or seeds—CREW has created a "frozen ark" of threatened animals and plants. In suspended animation are the semen and embryos of cheetahs, black-footed cat, gaur, giant eland, Persian leopard, snow leopard, white tigers, white and black rhinos, even seeds from a rare species of wildflower called the royal catchfly. Researchers are able to thaw these samples and bring to life individual animals or plants using the surrogate mothers or a cloning process for plants.

CREW is studying techniques of cloning, via embryo splitting and plant tissue cultures. Much research centers on preserving tropical plants called recalcitrants whose seeds cannot normally withstand the drying and freezing involved in cyropreservation. Seeds of cacao, the source of chocolate, are the working models of successful experiments to produce a clonal copy of the original embryo which will, in turn, germinate into a healthy cacao plant. Such plants are ideal candidates for reforestation efforts in Central America. Nontropical recalcitrant plants being studied include wild rice of North American wetlands as well as walnut and horsechestnut trees.

An endangered species of the wilderness, *Trillium persistens,* is being regenerated in test tubes and then transplanted to cold frames at the Cincinnati Zoo; someday plant nurseries and wildflower gardeners may be supplied in this manner, boosting Trillium's survival in its natural habitat. Also, tissue culture research is leading to the propagation of a disease-resistant elm tree.

Beyond the walls of the laboratory, CREW sends teams of trained volunteers in its Outreach Program to explain and demonstrate the work of the research team, stressing the importance of linking conservation with technology. CREW is also expanding its overseas internship program, which permits the exchange of ideas and the sharing of techniques on an international level.

Dr. Betsy L. Dresser, who spearheaded the CREW programs, is also research associate professor, Department of Obstetrics and Gynecology, University of Cincinnati; member of the Species Survival Commission (IUCN/SSC) Asian Wild Cattle, Captive Breeding and Veterinary groups; serves on the editorial board of *Journal of Zoo Biology;* and is a leader of the Black Rhino Research Task Force.

■ Resources

CREW publishes a semiannual newsletter for members.

Center for Whale Research ⊕

1359 Smuggler's Cove
Friday Harbor, WA 98250 USA

Contact: **Kenneth C. Balcomb III**, Research Biologist
Phone: 1-206-378-5835 (seasonal)
Fax: 1-206-378-5954

Activities: Research • *Issues:* Biodiversity/Species Preservation
Meetings: IWC; Marine Mammal Society

● Organization

The Center for Whale Research, a nonprofit organization, was founded to "promote, support, and conduct scientific research on marine mammals of the Order Cetacea—the whales, dolphins, and porpoises." Governments, the public, and conservation organizations benefit from this knowledge.

A full-time volunteer director heads the Center staff, which includes up to five part-time or seasonal researchers as well as volunteers from **Earthwatch**, a Massachusetts-based volunteer environmental organization.

▲ Projects and People

The Center's principal studies are **Orca Survey**, a long-term photo-identification study of killer whales in the Pacific Northwest since 1976, and **Pacific Humpback Survey**, a photo-identification study since 1986 to "ascertain migratory patterns, destinations, and population status of humpback whales" in the northeastern Pacific Ocean.

Regarding the Orca Survey, researchers note that the "resident" population of killer whales in Greater Puget Sound was 90 in 1991, and they want to confirm whether this is the "current carrying capacity of the habitat" as well as other key factors, such as birth and mortality rates, and sociobiological parameters. According to Kenneth Balcomb, this research is vital to the future management of this ecosystem. "Monitoring the status and dynamics of the top predators is the most effective and realistic method of determining whether the activities are adversely affecting them," he reports.

In addition, a general survey was recently launched concerning marine mammals in the northern Bahamas and a possible reintroduction of monk seals, an endangered species close to extinction, in cooperation with the Bahamas National Trust and Earthwatch.

■ Resources

The Center publishes a variety of whale-related materials including the book *Killer Whales*, a killer and humpback whale poster, cards, and technical papers.

Citizen Petition ❦

34 Nathan Lord Road
Amherst, NH 03031 USA

Contact: **Mary Roy**
Phone: 1-603-673-7849 • *Fax:* 1-603-673-3111

Activities: Education; Law Enforcement; Political/Legislative
Issues: Health

● Organization

Mary Roy has worked since 1987 for labeling of produce in retail markets and formulated **Citizen Petition**, a nonprofit organization that petitions the government to require labeling of all ingredients added to food, including pesticides, waxes, colorants, and irradiants; that monitors compliance in grocery stores; and that takes legal action against violators.

▲ Projects and People

Despite food industry opposition—in 1990, New Hampshire law was amended to require detailed labeling of all ingredients in waxes on produce. However, rulemaking was delayed to allow grocers and produce associations to file a "citizen petition" that requested and received amendments of the rules.

Because so much importance had been given to a nonbinding citizen petition instead of a standing law, Roy presented a second citizen petition to the U.S. Food and Drug Administration (FDA).

Citizen Petition, with Mary Roy's leadership and sponsored this time by citizens not industry, requested that the FDA return to the original intent of Congress—which required counter signs in retail markets to detail all ingredients that were added postharvest to produce, including pesticides. Final rules have been adopted in 1991 which state that grocers in New Hampshire must post individual counter signs on waxed produce listing all ingredients, or face fines of $200 to $500.

Citizen's Clearinghouse for Hazardous Wastes (CCHW) ❦

P.O. Box 6806
Falls Church, VA 22040 USA

Contact: **Lois Marie Gibbs**, Executive Director
Phone: 1-703-237-2249 • *Fax:* 1-703-237-7449

Activities: Education; Organizing; Political Action; Research
Issues: Air Quality/Emission Control; Health; Sustainable Development; Waste Management/Recycling; Water Quality

● Organization

Started in 1981, Citizen's Clearinghouse for Hazardous Wastes (CCHW) is a result of founder Lois Gibbs' earlier efforts to combat Love Canal—a chemicals dump site in her town of Niagara Falls, New York—that was linked to birth defects, cancer, and other maladies among children and other community residents. Eventually, the federal government relocated Love Canal residents and enacted Superfund legislation to allow community leaders to hire technicians to investigate the impacts of hazardous sites.

Amid the controversy, Gibbs and CCHW—now with 10,000 members—continue to press for "environmental justice," stressing that "people have the right to be safe and secure from toxic exposure." Mainly in small, rural communities, CCHW helps organize local groups to "petition the government, to protest, to speak out and to hold industry and government accountable."

▲ Projects and People

More than a clearinghouse, CCHW provides guidebooks and information packages to concerned citizens and expects mutual involvement in return. Most persons who contact CCHW are asked to join or start community-action groups or "larger than local" groups—with the help of CCHW field office organizers. With established groups, CCHW helps strategize, plan workshops, research opponents, provide technical advice, and link up with like-minded organizations. The **Organizer Apprentice Program** trains grassroots leaders. Teaching others to speak for themselves, CCHW staff does not testify at public hearings.

Beyond Love Canal, CCHW is helping at least a dozen other communities nationwide evacuate from neighborhoods close to hazardous dump sites. **Forgotten Faces: The Children of Contaminated Communities** targets the "contaminated site of the month" through a media campaign to gain the attention of government officials such as the U.S. Surgeon General and U.S. Environmental Protection Agency (EPA) officials.

Mining is a project area in which residents, workers, and labor groups are brought together to deal with waste management, health risks, and groundwater pollution. The **Environmental Health Project** focuses on improving the "health assessments" of the Agency for Toxic Substance and Disease Registry (ATSDR), Centers for Disease Control, for the contaminated communities that it serves. An ongoing **Bad Boy Campaign** aims to highlight industry environmental track records and keep government agencies from entering into contracts with companies convicted of certain crimes.

The **Solid Waste Organizing Project** works to "stop unsafe and unsound disposal practices, such as landfilling and incineration; promote safe and effective alternatives, such as recycling and composting . . . and eliminate [industry] products that add to our solid waste problems." Disposal of medical, interstate, and radioactive waste are special concerns.

CCHW uncovers what it regards as industry-backed community groups that promote dumps in the guise of jobs. It notes that "People for the West" (PFW) and "Committee for a Constructive Tomorrow" (CFACT) are two such groups. CCHW publicizes **Operation SLAPP-Back** and has put together a **SLAPP-Back Fact Pack** to help keep citizens from being stifled by "Strategic Lawsuits Against Public Participation," which it says are "filed by polluters against grassroots activists."

■ Resources

CCHW publishes *Environmental Health Monthly* and *Everyone's Backyard* bimonthly magazine, and offers a list of some 65 guidebooks. "Wastebusters" and other T-shirts are available.

Citizens for a Better Environment (CBE) ❦
501 Second Street, Suite 305
San Francisco, CA 94107 USA

Contact: **Alan Ramo**, Legal Director
  Phone: 1-415-243-8373 • *Fax:* 1-415-243-8980

Activities: Development; Education; Law Enforcement; Research • *Issues:* Air Quality/Emission Control; Global Warming; Health; Ozone Depletion; Toxics; Transportation; Water Quality

● Organization

Citizens for a Better Environment (CBE), founded in 1978, is a nonprofit, California-based environmental group whose purpose is safeguarding the environment and public health from toxic pollution. CBE seeks to achieve its goal through technical research, litigation, policy advocacy, and public education.

CBE combines expert scientific research, effective legal action, sophisticated policy advocacy, and focused public education to bring about environmental reform for all Californians. CBE has approximately 30,000 members and a staff of 29.

"Our petroleum-based economy is a killer," writes CBE director **Michael Belliveau**. "Visible insults—oil-soaked birds or smoky black bus exhaust are just an indicator. In Los Angeles, where the air is unhealthful nearly every day, petroleum-based air pollution will cause premature death and respiratory disability in thousands of people. . . . In San Francisco Bay, petroleum hydrocarbons from spills and poisoned stormwater runoff have been linked with reproductive failure in striped bass."

Belliveau goes on to say that local hazards are minor compared to the "global disruption that's likely to result from the continued release of carbon dioxide (CO_2) from the burning of fossil fuel." CBE joins World Resources Institute in supporting efforts to "simultaneously raise oil and gasoline prices through 'green taxes' while lowering the income-tax burden for the poor and middle class."

▲ Projects and People

"Technical research is the cornerstone of CBE," reports CBE. With many years of experience, and dozens of reports published, CBE is "recognized and respected by government officials, industry executives, and the media." A CBE researcher exposed the fact that 3.5 million pounds of hidden toxic pollutants flow each year from Silicon Valley industries through public sewer systems into the Bay, for example.

CBE takes legal action when government agencies or industrial polluters will not negotiate. CBE initiates lawsuits and provides legal support to other groups. Through court action, **Alan Ramo** is responsible, for instance, in getting the appropriate California agencies to enforce the air quality laws.

Political advocacy is CBE's specialty. For example, in coalition with community organizations in the Los Angeles Basin, CBE convinced the Los Angeles City Council and County Board of Supervisors to stop trying to solve their garbage problems by building incinerators and landfills that may threaten the water and air quality of the region.

Public education and outreach stimulates action on the community level. Informed citizens got IBM's Silicon Valley plant to halt the use of a certain chlorofluorocarbon (CFC) by 1993, according to CBE.

■ Resources

CBE requests information on new technologies to argue for certain rules in the chemical and oil industries. CBE also requests findings from any economic research that may help combat the accusation that a clean environment always means lost jobs.

CBE publishes the *Environmental Review* and can provide a list of *Comments and Reports* regarding pesticide use, toxic effects, and air- and water-quality management.

Climate Institute ⊕

316 Pennsylvania Avenue, SE, Suite 402
Washington, DC 20003 USA

Contact: John C. Topping, Jr., President
   Phone: 1-202-547-0104 • *Fax:* 1-202-547-0111

Activities: Education • *Issues:* Global Warming

● Organization

Climate Institute was founded in 1986 by individuals active in the U.S. air pollution control movement. The organization describes itself as the "leading international environmental membership organization of climate scholars and international experts concerned with greenhouse warming and stratospheric ozone depletion." The Climate Institute has 5 staff members and a 43-member Board of Advisors.

▲ Projects and People

During its first three years, the Climate Institute organized regional conferences regarding implications of climate change for North America, Africa, and the Arctic as well as two symposia in 1988 in Washington, DC, and at the United Nations regarding implications of climate change for the Third World.

Since then, the Institute organized four major meetings that include an **international workshop** on a framework convention, a **North American Conference on Forestry Responses to Climate Change** in Washington which produced the first draft language for a climate convention and a forestry conservation protocol; the **Nagoya Asian Pacific Seminar on Climate Change**; and an **International Conference on Cities and Global Change.**

To further its objective of educating decisionmakers about the serious implications of climate change, the Institute has developed color graphic slide sets and texts in nine languages. In addition, the Institute has launched an active program for top decisionmakers in about 20 countries called the **Head of State and Ministerial Briefing Program.**

The United Nations has asked the Institute to organize **symposia** for UN diplomats, and the Institute holds several international, national, and regional **workshops** throughout the year.

Coalition for Environmentally Responsible Economies (CERES) ⊕

711 Atlantic Avenue
Boston, MA 02111 USA

Contact: Joan Bavaria, Co-chair
   Phone: 1-617-451-0927 • *Fax:* 1-617-482-2028

Activities: Environmental Disclosure • *Issues:* Air Quality/Emission Control; Biodiversity/Species Preservation; Deforestation; Energy; Global Warming; Health; Sustainable Development; Transportation; Waste Management/Recycling; Water Quality • *Meetings:* UNCED

● Organization

The Coalition for Environmentally Responsible Economies (CERES) began in 1988 as a Social Investment Forum (SIF) project uniting "environmental groups, churches, labor unions, concerned public and private entities entrusted with investment duties." *(See also Social Investment Forum in this section.)* After putting forth the Valdez Principles in 1989, CERES grew. Its Board of Directors now includes social investors and environmental group representatives; current co-chairs are Joan Bavaria and Denis Hayes. It has 6 staff members and represents "more than 10 million people and $150 billion in invested assets."

The "Coalition . . . believes that there is a natural alliance between people who are concerned about environmental degradation, whether they work for corporations, invest funds, or work for environmental organizations." It operates "by reaching directly into the corporate boardroom and influencing the policies that can and do cause irretrievable harm. Our highest goal is to forge a new direct dialogue among businesses, environmentalists and investors about appropriate corporate environmental performance and accountability to the public and the investment community."

To this end, CERES uses four strategies. One is "direct corporate dialogue." At least as important is the strategy of shareholder resolutions, often by proxy. In 1990 CERES sponsored 20 proxy resolutions, and in 1991, 40; these force dialogue on issues that corporations prefer to avoid. A main leader in the resolution strategy is the Interfaith Center on Corporate Responsibility (ICCR), a Coalition member. The third and fourth strategies involve measuring corporate environmental performance and public education. In the last area, CERES includes giving expert testimony in state and local legislatures, U.S. Senate hearings, and conferences regarding the 1992 World Environment Conference in Brazil.

▲ Projects and People

The keystone of CERES is the Valdez Principles, released in September 1989—10 principles for corporate environmental responsibility. The principles are "protection of the biosphere; sustainable use of natural resources; reduction and disposal of waste; wise use of energy; risk reduction; marketing of safe products and services; damage compensation; disclosure; environment directors and managers; [and] assessment and annual audit."

The goal of the Principles is to "encourage the development of positive programs designed to prevent environmental disasters and degradation"; they differ from existing corporate guidelines in that they require independent confirmation of compliance. CERES stresses that the Principles are in no way static; once a company signs on, constant dialogue ensues, and the guidelines are purposely flexible to fit individual companies' needs. The "intent is to create a voluntary mechanism of corporate self-governance that will maintain business practices consistent with the goals of sustaining our fragile environment for future generations, within a culture that respects all life and honors its interdependence." The eventual goal is "the elimination of all pollution and environmental degradation."

Each year, signatories must file a CERES Report, explaining and discussing their achievements in complying with each principle, where applicable, and any problems encountered. CERES reviews the reports, and "signatory status shall remain in effect unless and until a company resigns, fails to pay the annual service fee [ranging from $100 to $15,000; now consistent with the cost of managing the collection and dissemination of data], or it is determined by

CERES or the Corporate Advisory Committee that the company's performance is no longer in keeping with the spirit of the Valdez Principles."

According to CERES, "the Valdez Principles has just begun." At least 27 companies, most of them small, have signed; CERES is trying to get Fortune 500 companies to participate, but realizes that they "have more at stake and cannot move as quickly as smaller companies." The direct dialogue and shareholder resolution strategies may encourage more companies to agree to the Valdez Principles. Other potential sources of new members and support for shareholder resolutions are large state and university pension funds, which CERES is targeting.

Joan Bavaria, involved in social investing since 1975, was the founding president of the Social Investment Forum (SIF) and founder of CERES. President and CEO of Franklin Research and Development, a "socially responsible investment advisory firm," Bavaria is also a member of this Guide's Board of Advisors. Co-chair Denis Hayes, executive director of First Earth Day 1970 and International Chairman of Earth Day 1990, is president and CEO of Green Seal and an environmental lawyer. Hayes headed the federal Solar Energy Research Institute during the administration of President Jimmy Carter; he is author of the solar energy book *Rays of Hope*, and has received numerous environmental honors and awards.

Environmental directors are Jack Doyle, Friends of the Earth; Terry Gips, International Alliance for Sustainable Agriculture; Mike McCloskey, Sierra Club; Fran Spivey-Weber, National Audubon Society; and Joel Thomas, National Wildlife Federation. Social investment directors are Sherry Salway Black, First Nations Financial project; Gray Davis, State of California controller; Paul Freundlich, Fair Trade Foundation/Co-op America; John Guffey, Calvert Social Investment Fund/Calvert Foundation; Carol O'Cleireacain, New York City Department of Finance; and Andy Smith ICCR/American Baptist Churches USA. At-large directors are Carl Anthony, Earth Island Institute; Mindy Lubber, U.S. Public Interest Research Group (PIRG); and Howard Samuel, AFL-CIO.

■ Resources

Resources needed include Macintosh computers to assist with corporate outreach, newsletters, and environmental audits. CERES also needs the support of environmental organizations to join its coalition and help direct activities.

Committee for Conservation and Care of Chimpanzees (CCCC) ⊕

3819 48th Street, NW
Washington, DC 20016 USA

Contact: **Geza Teleki, Ph.D.**, Chairman, Primatologist
Phone: 1-202-362-1993 • *Fax:* 1-202-686-3402

Activities: Education; Research • *Issues:* Biodiversity/Species Preservation

● Organization

Chimpanzees share a remarkable close kinship with humans across a wide spectrum of biological traits. The purpose of the **Committee**

for Conservation and Care of Chimpanzees (CCCC) is to upgrade the long-term survival options of world chimpanzees through their natural ranges and in regions where remnant populations are severely threatened by expanding habitat loss, by hunting for local markets, and by exploitation for international trade.

Their goal is also to improve the physical, psychological, and social conditions of maintenance for captive chimpanzees in all types of holding facilities and particularly in those institutions which do not provide living conditions in keeping with the intrinsic needs of the species.

Founded in 1986, CCCC has a worldwide membership of over 200 professionals working in natural habitats, zoological parks, and research laboratories. Their collective knowledge provides a strong foundation for developing better conservation and care standards, influencing adoption of strict scientific ethics, promoting public awareness, reinforcing international legislation, and stimulating protective actions that measurably benefit chimpanzees in wild and captive settings. A 10-member Executive Council includes representatives from Gambia, Japan, Switzerland, Tanzania, Uganda, United Kingdom, USA, and Zaire.

▲ Projects and People

The CCCC has reported on remnant **wild populations** of chimpanzees, studied their habitat for damaging factors, and recommended places that would most enhance survival. They have assessed distributions of wild chimpanzees to stimulate international recognition of their endangered status.

Concerning the **international trade** of young wild-born chimpanzees, the CCCC has documented not only the numbers captured and destroyed but also the inhumane tactics used in taking and transporting infants for commercial purposes.

The CCCC has set guidelines for the proper care of captive chimpanzees and focuses on the physical and psychological enrichment of this highly social species.

■ Resources

Affiliated with the nonprofit Jane Goodall Institute, Tucson, Arizona, CCCC remains dependent on individual or institutional donors to support its activities. The CCCC is developing a budget for supporting urgent work on a limited scale.

Available resources include assorted reports and fact sheets on chimpanzee survival problems, such as population census and habitat assessment, trade analysis, captive care, sanctuary operation, and legislative action. Specific reports include *Action Plan for Chimpanzee Conservation in Africa*, *Current Threats to the Survival of Wild Chimpanzees*, *Review of Chimpanzee Trafficking Methods and Trade Trends*, and *Guidelines for Proper Care of Captive Chimpanzees*.

Community Conservation Consultants (CCC) ⊕ ✸

RD 1, Box 96
Gays Mills, WI 54631 USA

Contact: **Robert H. Horwich, Ph.D.**, Founder
Phone: 1-608-735-4717

Activities: Development; Research • *Issues:* Biodiversity/Species Preservation; Community Conservation; Deforestation;

Sustainable Development • *Meetings:* American Primatological Society; Conservation Biology

● Organization

Community Conservation Consultants (CCC) helps local people protect their environment while creating economic benefits for the people who safeguard their forests and wildlife. A community sanctuary is unlike a government sanctuary: "Since the land belongs to the landowners, nothing can occur which is not sanctioned by the people collectively or the sanctuary will fail," writes Dr. Robert Horwich.

He established the Community Baboon Sanctuary (CBS) in 1985 to protect a single endangered species, the black howler monkey, in a Belizean rainforest. Known locally as a baboon, the howler monkey eats fruit and leaves and lives in groups of two to ten in the canopy of tropical forests in a limited range of Belize, Mexico, and Guatemala. Its loud, lionlike roaring—often in chorus—accounts for the name. Lately, howler monkeys, with the help of expert Dr. Ken Glander, Duke University, have been marked with beaded ankle bracelets, blood tested, released, and tracked for knowledge that will help the species be reintroduced to other areas in Belize.

More than 100 landowners in an 8-village area signed voluntary pledges to abide by management plans. Results are that an education and ecotourism center has been established; employment is fostered; and villagers take responsibility for caring for the land and express pride in their riverine forests and wildlife—including mammals, reptiles, amphibians, and 170 bird species.

As an outgrowth of interest, the Community Conservation Consultants formed to help create a Manatee Community Reserve, Gales Point, Belize, with artist Chris Augusta based on both the sanctuary and biosphere reserve concepts; Sea Turtle Breeding Beach Sanctuary, Ambergris Cay, Belize, with Greg Smith; Kickapoo Valley Bio Refuge, Wisconsin, which is a soil sanctuary for organic vegetable growers; Turtle/Prairie Sanctuary, southwestern Wisconsin, with Gigi LaBudde and others to protect endangered and ornate box turtles; and the expanding Sauk-Prairie Community Eagle Sanctuary for the winter roosting area for bald eagles in southern Wisconsin along the Wisconsin River.

▲ Projects and People

Dr. Horwich chose a site in Belize for the sanctuary which contained a continuous high howler population and contacted the local people for ways the howler sanctuary would benefit them. In cooperation with private landowners, he developed the site by leaving a strip of forest along the riverbanks, between land boundaries, and around slash-and-burn cuttings and farm plots—leaving specific food trees, which are good for both howlers and livestock.

He retained a local manager, Fallet Young, to work in conservation, education, research, and tourism with the supervision of the Belize Audubon Society. Landowners who joined in the Community Baboon Sanctuary effort each received a vegetation map, conservation plan, signed pledge, World Wildlife T-shirt, and a certificate of participation.

In the *Journal of Medical Primatology,* Dr. Horwich and his colleague, Jonathan Lyon of the Institute for Research on Land and Water Resources, Pennsylvania State University, reported, "We have gone beyond tribal units to create a sanctuary based on cultural ties and ties to the land. By appealing to people's sense of stewardship of the land, we are attempting to instill or strengthen a proper land-use conservation ethic." They expect it will take 10 years to assess the success of the community sanctuary concept and its effects on population levels of plants and animals.

Meanwhile, small loans have been made available to local villagers to build house additions to accommodate tourists. A small museum was made to exhibit flora, fauna, history, and culture of the area. Eroded riverbanks have been replanted; hardwood tree species have been repropagated; the baboons are thriving; and the endangered, hunted river turtle (*Dermatemys mawii*) is now monitored. A three-mile labeled trail system is featured in the sanctuary project.

■ Resources

While the sanctuary has minuscule overhead cost, almost all funds are directed to the sanctuary. The sanctuary needs help for howler and wildlife conservation, education, tourism, and research.

A Belizean Rain Forest is a 420-page book developed on the Community Baboon Sanctuary. CCC gives lectures and provides consultations and assistance on setting up community sanctuaries. Reprints on materials such as *How to Develop a Community Sanctuary* are available on request.

Community Environmental Council (CEC)
Gildea Resource Center
930 Miramonte Drive
Santa Barbara, CA 93109 USA

Contact: Jon Clark, Executive Director
Phone: 1-805-963-0583 • *Fax:* 1-805-962-9080

Activities: Development; Education; Research • *Issues:* Sustainable Development; Waste Management/Recycling

● Organization

Did you know that recycling just one aluminum can saves enough energy to power a television for four hours? Or that 95 percent of the energy required to produce aluminum from its raw materials is saved when it's produced from a recycled can?

The Community Environmental Council (CEC) recycles for Santa Barbara County's south coast population of approximately 200,000. CEC was founded in 1970 as a direct citizen response to the catastrophic oil spill in the Santa Barbara Channel in 1969. CEC maintains an innovative hazardous waste collection program and conducts important testing to develop municipal composting plants. CEC provides expert testimony before legislative bodies and advises committees addressing recycling, source reduction, and land use at national, state, and local levels. Leaders in government, business, academia, and public interest fields are brought together for seminars on recycling technology, hazardous waste planning, municipal-scale composting, and resource and land use planning. Interns and scholars from around the world come to CEC to conduct research.

▲ Projects and People

The Gildea Resource Center serves as a community model for efficient energy design, drought-tolerant landscaping, water efficiency, nontoxic gardening, and as a center for environmental research and development. In the 1990s, it is aiming toward sustainability. "California faces the possibility of choking off real environmental progress in a tangle of needlessly expensive regulatory procedures," according to CEC literature. "Decisionmakers must now place more emphasis on eco-development, which means putting in place proven alternatives such as compost facilities, recycling facilities, water conservation and energy conservation."

CEC operates recycling buyback centers and coordinates a curbside recycling center in the Santa Barbara region. Special programs include office paper collection, such as at the University of California at Santa Barbara; school collections; hotel/bar glass recycling; compost testing; city parks recycling; and a multifamily-unit collection program. CEC also collects hazardous wastes on 22 designated days each year.

Pushing toward proper valuing of recycled materials so they are not regarded as "waste," CEC is encouraging "economically robust scrap-consuming industries in the coming decade."

Their Community Gardens Program has three gardens serving approximately 130 individuals and families. The Santa Barbara Safe Food Project is organizing a countywide consortium of growers, grocers, and consumers with the goal of converting 20 percent of county agriculture to organic growing by the year 2000.

■ Resources

CEC's recycling operations are self-supporting. Other funding sources are research grants, individual and corporate contributions, and advisory activities.

A publications list features print material regarding waste management, recycling and hazardous wastes, growth and sustainability, and seminar synopses such as on composting. The *Gildea Review* newsletter is also published.

The Conservation Agency ⊕
6 Swinburne Street
Jamestown, RI 02835 USA

Contact: **James D. Lazell, Ph.D.**, Biologist/President
Phone: 1-401-423-0866 • *Fax:* 1-401-423-0199

Activities: Research • *Issues:* Biodiversity/Species Preservation; Deforestation; Global Warming; Population Planning

● Organization

"Our over-riding goal is to preserve the diversity of life," writes Dr. James D. Lazell, president of The Conservation Agency, founded in 1980, which conducts research worldwide and gathers data to help save rare, endangered, and little-known species. Its field biologists publish natural history articles as well as scientific research in peer-reviewed technical journals that can be used by land preservation organizations and environmental lawmakers to maintain biotic diversity. The agency also specializes in management plans needed to carry out effective conservation. Through grants and volunteer opportunities, it has supported several hundred scientists and students working on projects related to the broadest aspects of biological conservation.

Working for expenses only, Dr. Lazell and biologist Dr. Numi C. Goodyear are the only two permanent staff scientists; The Conservation Agency pays no salaries. They are assisted by more than 20 associated scientists whose research is funded in part by the Agency. Scientists specialize in developing methods and carrying out studies

to determine a threatened species' critical resource requirements and to identify positive or negative impacts on its recovery or perpetuation. This information can be used to assess the value of land parcels to the threatened species.

The Agency also provides data to local government agencies that need information, but who have no research funds available.

▲ Projects and People

The Conservation Agency scientists have located and identified many rare and endangered animals and have done research to determine population status, life histories, and conservation needs. They also have rediscovered species thought extinct and populations thought destroyed completely. Altogether, its discoveries include more than 20 new species of mammals, reptiles, and amphibians such as Darlington's flat beetle, Walsingham's Virgin moth, Everett's flying dragon, and the silver rice rat—a small muskratlike mammal found by Goodyear. Through a special fund for tropical entomology, hundreds of new insect species have been uncovered.

Since 1986, significant research in South China Sea islands has turned up new species of land vertebrates. Now documented are life histories of threatened species such as Romer's frog, whiteheaded blindsnake, and the Chinese pangolin (a scaly ant-eating mammal).

With help from Universitas Sam Ratulangi, Sulawesi, Indonesia, and Silliman University, Philippines, Agency expeditions in these countries have resulted in the discovery of three new species of flying dragons (*Draco* lizards) and new data about the little-known anoa (a tiny forest ox), Jelesma's gecko, and the Sulawesi black racer. In Hawaii the rare and probably endemic rock wallaby is under study. In Fiji, the Agency is investigating rare reptiles, including the endemic mid-Pacific iguana.

In the Western Hemisphere, projects are ongoing from Newfoundland to Brazil. "In the Canadian Maritimes and New England, our subjects have been whales, marine and salt-marsh turtles, vernal pond wildlife, salamanders and acid rain, wetlands, protection, and coastal island biogeography," states Agency literature. A comparative study of Brazil's Outer Banks—"the world's largest sand barrier"—regarding worm lizards, insects, and iguanid lizards of the maritime forests was contrasted with surveys of North Carolina's Outer Banks. "In California, we have concentrated on the chaparral and old-growth forests of Mount Tamalpias. . . . We have challenged stereotypical views of controlled burning and other management techniques which are not appropriate in all cases."

In the West Indies, the Agency says it is "bringing back populations of lost birds such as flamingos, doves, ducks, and woodpeckers; restoring dry forest habitats; and making room for population expansion of iguanid lizards, skinks, ground geckos" as well as managing whole island ecosystems as wildlife sanctuaries. Such management plans are submitted to local governments at no cost.

Staff has dedicated two decades of research to the critically endangered Florida Keys wildlife from mice to miniature deer. In 1991, Agency and Sierra Club Legal Defense Fund efforts helped place the silver rice rat on the Federal List of Endangered Species. The U.S. Fish and Wildlife Service is able to use such research results in recovery plans for various species.

■ Resources

Individuals, numerous NGOs such as Earthwatch, and government agencies help support the Agency, which seeks greater political support and more publicity. A list of published works that is chronological and updated annually is available.

Conservation Council for Hawai'i (CCH) ✤
P.O. Box 2923
Honolulu, HI 96802 USA

Contact: **Bill Sager,** Chairman
    Phone: 1-808-247-2551 • *Fax:* 1-808-531-3050

Activities: Education; Political • *Issues:* Air Quality/Emission Control; Biodiversity/Species Preservation; Deforestation; Energy; Global Warming; Health; Population Planning; Sustainable Development; Waste Management/Recycling; Water Quality

● Organization
The Conservation Council for Hawai'i (CCH) is an all-volunteer organization promoting environmental health and education, conservation, and management of Hawaii's natural resources. Founded in 1950, CCH later became the Hawaii state affiliate of the National Wildlife Federation (NWF) and was honored in 1990 as NWF's "outstanding affiliate of the year."

With 500 members, active island chapters are on O'ahu, Kaua'i, and Hawai'i (Big Island). Affiliate organizations are the Kawai Nui Heritage Foundation, Hui Malama Pono O Lana'i, and Citizens Against Noise.

▲ Projects and People
CCH participants coordinate Hawaii Earth Day with a grant from the state legislator, prepare expert testimony for public hearings on issues of environmental concern, actively support environmental legislation, and sponsor forums to discuss the implications of public policies and to provide the technical background necessary for an informed citizenry. During the annual National Wildlife Week, they distribute more than 5,000 conservation learning kits to Hawaii's teachers. Networking with other environmental groups, they work on the conservation of the resources of the Kawai Nui marsh.

When necessary, CCH volunteers file appeals or go to court on behalf of Hawaii's irreplaceable natural resources to redress improper governmental and private actions. Apparent action of the Bush Administration to back down from the President's pledge of "no net loss of wetlands" is spurring the group to mobilize its members to lobby the White House and the Wetlands and Aquatic Resources Regulatory Branch, U.S. Environmental Protection Agency (EPA), against weakening revisions of the Federal Manual for Identifying and Delineating Wetlands.

CCH is concerned that valuable ecosystems such as the Florida Everglades, Louisiana Bayous, New Jersey Pine Barrens, Prairie Potholes, California Vernal Pools, and Hawaii's brackish lava coasts will be open to the development of condominiums, shopping malls, and parking lots. In addition to providing essential habitats and food for wildlife, these wetlands play a role in our economy, CCH points out. "They serve as nature's giant sponges, soaking up excess water after heavy rainfalls to protect our homes and communities from costly flood damage. And by filtering polluted runoff from farms and city streets, they keep our water clean and safe to drink." Coastal wetlands also provide nursery and spawning grounds for up to 90 percent of U.S. commercial fish catches.

Managing human population growth is a vital issue, and CCH adopted a resolution on international population assistance and family planning to encourage the USA to increase its funding levels to its "fair share" or $500 million. CCH also called on the USA to contribute to the United Nations Population Fund and the International Planned Parenthood Federation as well as to establish a Presidential Commission on Population Growth and Natural Resources Planning for formulating U.S. policy.

■ Resources
A membership form is available. CCH publishes *The Hawai'i Conserver* newsletter.

The Conservation Fund
1800 North Kent Street, Suite 1120
Arlington, VA 22209 USA

Contact: **Linda T. McKelvey,** Director of
               Administration
    Phone: 1-703-525-6300 • *Fax:* 1-703-525-4610

Issues: Biodiversity/Species Preservation; Land Conservation; Sustainable Development; Water Quality

● Organization
The Conservation Fund is a nonprofit nonmembership organization with a staff of 35 that encourages new means of land and water conservation. In its first five years the Conservation Fund has worked in 14 states on projects that are designed to merge financial return with conservation principles.

According to Fund president Patrick F. Noonan, learning from marketplace and private sector mechanisms and applying some of those results to conservation problems, the Conservation Fund seeks to establish connections between businesses and conservation interests. In serving as an organizer and investor, the Fund encourages experimental programs and new ideas.

▲ Projects and People
When people own land and want financial return from it, but also want the land use to be sustainable, two places they can go are the Fund's Land Advisory Service and the Water Advisory Service. Working with both individuals and corporations, the Fund develops site specific projects that are compatible with the economic goals of the owners as well as with conservation goals.

The Freshwater Institute, directed by Lawrence A. Selzer, is another project. The Institute runs a number of programs designed for the protection of water quality. Many of these programs are designed to help rural landowners use practical methods to increase conservation. The Institute is also involved in a water-quality project for the U.S. Department of Agriculture working to eliminate pollution from large-scale fish farms.

After years of coal mining damage to areas of the Potomac River, the Freshwater Institute—with the Interstate Commerce Commission, U.S. Army Corps of Engineers, and Maryland and West Virginia government—is studying the Potomac to see if there is any way to transform the area into a viable trout stream. This project, if successful, will provide environmental, economic, and recreational benefits to the region. It could serve as a model for restoring other rivers.

Other related projects concern **groundwater monitoring**; further studies on freshwater protection with the National Geographic Society; and the start of a **national freshwater conservation register** to recognize landowners, communities, corporations, and public agencies.

A **Public Conservation Partnership** is one way in which the Fund assists public acquisition of land. When land comes on the market it often must be purchased rapidly to prevent it from being sold to developers. By purchasing the land quickly and quietly and holding it until public funds become available, the Public Conservation Partnerships increase public space and save taxpayer money.

Management and communications among conservation groups are another priority. The Fund has completed the first phase of a study of the current and future **management needs of the conservation movement**.

Under the direction of **Keith Hays**, the **American Greenways** project is devoted to creating corridors of open space. These greenways are designed to protect the environment and also improve the quality of life of people living nearby. This program helps with design of new greenways, improvement of existing greenways, and the making of new partnerships between the diverse groups that are needed in the effort to create greenways.

■ Resources

The Conservation Fund publishes two newsletters, the biweekly *Land Letter* and the bimonthly *Common Ground*, and a brochure on *American Greenways*.

Conservation International (CI) ⊕

1015 18th Street, NW, Suite 1000
Washington, DC 20036 USA

Contacts: **Peter Seligmann,** Chairman and CEO
Russell A. Mittermeier, Ph.D., President
Phone: 1-202-429-5660
Fax: 1-202-887-0193, 1-202-887-5188
Telex: 910 240 9104 ciwdc
E-Mail: Econet: ciwash

Activities: Conservation; Policy; Scientific Understanding; Wildlands Management • *Issues:* Biodiversity/Species Preservation; Deforestation; Energy; Global Warming; Population Planning; Sustainable Development • *Meetings:* CITES; UNCED; World Parks Congress

● Organization

Conservation International (CI) is a nonprofit organization that is working to protect ecosystems worldwide. Currently active in Central and South America, Indonesia, Philippines, and Papua New Guinea, CI is also developing programs in other countries, including Japan, Zaire, Vanuatu, Venezuela, Guyana, New Caledonia, and Solomon Islands. The cornerstones of CI's work are "scientific understanding . . . ecosystem management . . . conservation-based development . . . [and] policy design." In its own words, "CI addresses the entire range of obstacles that stand in the way of a sustainable future." CI has some 30,000 members and a staff of 65.

▲ Projects and People

Realizing that heavy external debt often exacerbates pressures on a country's natural resource base, **Peter Seligmann** pioneered the now popular **debt-for-nature swaps** in Bolivia in 1987. Since then, CI has pioneered additional swaps in Africa and Latin America under the direction of **Steve Rubin**, CI director of conservation finance. Other environmental organizations like The Nature Conservancy and the World Wildlife Fund, as well as the Dutch and Swedish governments, have swapped debt for nature since 1987 totaling some $100 million of developing nations' debt.

New swaps involve several innovations. In **Madagascar**, a $5 million swap was the first to involve trade credits as well as bank credits. In 1991, the first-ever debt-for-nature swap was completed in **Mexico**, up to a limit of $5 million through the Salinas government term. In 1992, U.S. Agency for International Development (USAID) funds were swapped by CI to establish a trust fund to support conservation activities in the Petén region of **Guatemala**.

In 1991, BankAmerica announced it would donate $6 million in developing country loans through CI, World Wildlife Fund (WWF), and the Smithsonian Institution which will initially help rainforests in Mexico and Ecuador, such as Mexico's Selva Lacandona—"North America's largest and most threatened tropical rainforest," reports Seligmann. Dr. Mittermeier describes Mexico as a "megadiversity country . . . with more plant and animal species than all but three other nations." CI has also been involved in swaps in **Costa Rica**, and plans more in the future.

CI stresses that a debt-for-nature swap is not a panacea; before proceeding, participants must consider the cost, the economic viability, and the possibility of long-term funding and implementation. Above all, "those who have the greatest stake in the use of a country's natural resources must provide the leadership and direction to protect those resources."

Another program that combines economic and conservation issues is the **Sound Environmental Enterprise Development (SEED)** program. By identifying rainforest products other than timber and helping to market those products, CI is able to boost local economies by means other than deforestation. In **Ecuador**, a successful program called the **Tagua Initiative**, coordinated by **Karen Ziffer**, markets carved nuts which resemble ivory for use as buttons under a licensing agreement with two U.S. sportswear companies—Patagonia, Inc., and Smith & Hawken. The palm nut is harvested in what is considered a rainforest "hotspot" where this effort is an attempt to reverse a legacy of exploitation of local peoples and degradation of their environment—and, it is hoped, bring financial gain while saving the rainforest.

In **Propeten**, a project directed by CI's **Conrad Reining** in Guatemala, the goal is "to balance all the income sources of the region" to avoid overstressing any single resource. The inhabitants of the Peten area learn sustainable harvesting of chicle, allspice berries, herbs, oils, xate palm leaves, and other rainforest products. In addition, CI is involved with promoting **ecotourism** to encourage conservation and bring money into these communities.

One of CI's major projects is the **Rapid Assessment Program (RAP)**. In order to identify areas that are in need of immediate conservation attention, CI sends a RAP team composed of field scientists to make a study of an area. These teams mostly cover vast areas as quickly as possible to catalog the region's plant and animal life. By using top scientists and state-of-the-art technology, the RAP program succeeds in rapidly describing and cataloging the state of the rainforests such as in northern Bolivia along the Peruvian border and in northwest Ecuador.

RAP is a cornerstone to the **Rain Forest Imperative**, a 10-year plan to identify and launch conservation strategies to save exceptional areas, such as the South Pacific's Melanesian region, Brazil's Atlantic Forest, the Andes' eastern slope, and West and Central Africa's tropical forest belt. The urgency is clear as **Dr. Russell A. Mittermeier**, CI president, points out that one species of tree can harbor more than 1,000 beetle species, and "50 bird species are entirely dependent on the wanderings of army ants through the forest." The "conservation of rain forests is . . . a cornerstone of geopolitical stability," he says.

The **Plant Conservation and Ethnobotany** project is involved in a number of aspects of plant conservation. In addition to endangered species recovery and botanical inventories, CI—said to be the leading conservation organization in this field—is cataloging medicinal values of plants that have been used by local tribes. CI is also working to "develop models for sustainable harvesting of rainforest products."

In a race against time, researchers know that ecosystems in peril equally threaten the small American redstart bird that flies 7,000 miles yearly from New England to Brazil's shrinking rainforest and the disappearing shaman with their knowledge of healing plants that benefit local tribes. "Each time a medicine man dies, it is as if a library has burned down," CI ethnobotanist **Dr. Mark Plotkin** told the *New York Times* (June 11, 1991). CI is encouraging the passing of this wisdom to succeeding generations.

CI also funds research to design and implement **species recovery programs** for a variety of endangered animals under the direction of world-noted anthropologist **Dr. James D. Nations**. "The key to nature's creativity is variety [or] biological diversity," states CI. "Unfortunately, life on earth is losing its variety"—with present extinction rates "10,000 times" more than in recent geological history. A third of the Earth's species are presently in danger of vanishing. Yet, on a hopeful note, research recently uncovered a new species of black-faced lion tamarin (*Leontopithecus caissara*) on an island near Brazil. *(See also Instituto Pró-Ecologia [IPE] in the Brazil NGO section.)*

CI, the Organization of American States (OAS), and the Costa Rican government have worked together to create the **La Amistad Biosphere Reserve**. La Amistad is the first of what are called "Peace Parks" because they are often agreements between countries with border disputes. This park crosses the Costa Rican and Panamanian borders and will protect habitat and the indigenous population from large-scale mining, oil exploration, electric power plants, agribusinesses, and migrant farmers. Actually, the local Bribri and Cabecar Indian tribes took a bold new step in helping develop the conservation strategy. "It is the first time that the original residents of the region are being treated as equals in deciding how land and resources will be managed," noted **Hernan Seguro**, Bribri Indian and representative to the La Amistad Coordinating Commission.

The UNESCO classification "biosphere reserve" indicates that it is a region in which there are plans to protect threatened ecosystems, while still meeting economic needs of the communities. Half of Costa Rica's fresh water originates at La Amistad, the country's most biologically diverse region. Among the rare animals here are the jaguar, harpy eagle, and giant anteater. La Amistad is expected to be a conservation model in Central America, where at least three additional "Peace Parks" are being planned.

In the **Africa Program**, **Sharon Pitcairn** leads CI efforts to provide "technical assistance in protected areas development, conserva-

tion-based development, rapid biological assessment; geographic information systems (GIS), remote sensing analysis, sustainable enterprise development, [and] conservation finance."

Dr. Lee Hannah is program coordinator for the **Philippines**.

CI also has created **Ecotrust**, the first USA-based affiliate which works in the Pacific Northwest and in **Canada**'s British Columbia where it is expected to lead the conservation of "North America's greatest forests," points out Dr. Mittermeier. Director is **Spencer B. Beebe**, former president of The Nature Conservancy International Program, who writes that the "integration of ecological conservation and economic development is at least as urgent here at home as it is anywhere in tropical rainforests."

Initially, Ecotrust is working with the Willapa Bay Ecosystem Restoration and Development Program, Washington State; Clayoquot Sound Ecosystem Conservation and Development Program, Vancouver Island, British Columbia; Gardner Canal/Kitlope River Ecosystem Program, British Columbia; and Copper River/ Prince William Sound Watersheds/Northern Gulf of Alaska Ecosystem Program. A network of such CI affiliates is planned worldwide.

Russ Mittermeier is described as "one of a new generation of environmentalists who are as sophisticated in their methods as the industrial conglomerates they often confront." He is also called "Russell of the Apes"—being noted for efforts to save both the rainforests and some 200 endangered primate species by targeting 13 "megadiversity countries" containing 60 percent of the world's plant and animal species.

■ Resources

CI publishes the quarterly *TROPICUS Newsletter*, the *Orion Nature Quarterly*, *Debt-for-Nature Exchange: A Tool For Conservation*, and the *Rainforest Imperative* brochure and video. In addition, the organization offers the pamphlet *The Debt-for-Nature Exchange*, numerous Spanish language publications, and other collaborative efforts, such as *Conserving the World's Biological Diversity*, on which it worked along with the World Conservation Union (IUCN), World Resources Institute (WRI), World Wildlife Fund (WWF), and the World Bank. CI also helped produce McDonald's *Discover the Rain Forest* activity books for children.

Conservation and Research Foundation ⊕
Connecticut College
Call Box
New London, CT 06320 USA

Contact: **Richard H. Goodwin, Ph.D.**, President

Issues: Air Quality/Emission Control; Biodiversity/Species Preservation; Deforestation; Energy; Global Warming; Sustainable Development; Waste Management/Recycling; Water Quality

● Organization

This all-volunteer organization, founded in 1953, manages to keep its administrative costs to about 3 percent of total expenditures so that it can maximize the impact of its resources in the following ways: " . . . by direct grants to organizations to aid their conservation programs; by initiating studies, supporting activities, and publishing information that might have a catalytic impact upon the preservation or improvement of environmental quality; and by supporting biological research in neglected areas, especially those having environmental implications."

Previous grants of the Conservation and Research Foundation confronted the dangers of nuclear technology such as the dumping of radioactive waste into leaking landfills. Recommendations for regulating the hazardous waste management industry resulted from funding to the Council of Economic Priorities (CEP). Efforts to preserve Vermont farmlands and survey agricultural land resources were begun. Early support helped The Nature Conservancy survive to become one of the most prominent land preservation groups; lately, funds helped buy the Burnham Brook Preserve, Connecticut, introduced burning as a technique to manage woodland ecosystems, and successfully brought the Open Marsh Water Management mosquito approach to New England from the Chesapeake Bay area. Gray seals in Nantucket Sound, parakeets of Mauritius, and grizzly bears repelled from people contact by a harmless spray—these are beneficiaries of grants, also used to combat oil exploration in the Arctic National Wildlife Refuge.

In Latin America, the Foundation has aided Venezuela's BIOMA in publishing a vegetation map to identify ecosystems and foster biological diversity; Brazil's FUNATURA, which is conserving renewable resources and the SOS Mata Atlantica Foundation to save its Atlantic Coast tropical forest; Costa Rica's Finca La Selva biological research station for tropical rainforest preservation; the Latin American Natural Area Programs (LANAP); and Mexico in enhancing agricultural production without pesticides.

The Foundation supports efforts to protect Puerto Rico's Bahia de la Ballena and its teeming biological diversity from development; research to bring a simple "solar cooker" to Pakistan so save the Himalayan forests from destruction of fuelwoods; studying the Dom Doi wading bird colonies in southern Vietnam's Mekong Delta to strengthen their conservation; tracing the movements of the mud puppy salamander, threatened by chemicals used to control the sea lamprey in the Great Lakes; evaluating the effects of wildfires in national parks; reporting on hazardous waste sites on the Texas-Louisiana Gulf Coast; investigating illegal waste dumping in Nigeria and other nations; probing analyses of bird reproductive decline at the Point Reyes Bird Observatory, California, and the radioactive cloud fallout from the Chernobyl nuclear meltdown; supporting the Environmental Data Clearinghouse of the Council on Economic Priorities; helping the Yellowstone Conservancy initiate a waste recycling program at Yellowstone National Park; aiding the Lake Champlain Basin Project to plan ecosystem-based approaches since the 1989 designation of the International Biosphere Reserve by the United Nations Educational, Scientific, and Cultural Organization (UNESCO); and scrutinizing the effects of conventional weapons systems testing on soil, water, air, and buildings in New Mexico.

Dr. Goodwin is professor emeritus of botany at Connecticut College and was president of The Nature Conservancy. Other trustees include Wallace D. Bowman, World Resources Institute; Winslow R. Briggs, Carnegie Institution of Washington; Belton A. Copp, formerly of Copp, Berall and Hempstead law firm; Richard H. Goodwin, Jr., Neuro Probe, Inc.; Sarah M.B. Henry; Hubert W. Vogelmann, University of Vermont; Dr. Mary G. Wetzel, National Institutes of Health (NIH); and Alexander T. Wilson.

■ Resources

These and other projects are described in the Foundation's five-year report, to be published next in 1993.

Co-op America
2100 M Street, NW, Suite 403
Washington DC 20063 USA

Contact: **Alisa Gravitz,** Executive Director
  Phone: 1-202-872-5307 • *Fax:* 1-202-223-5821

Activities: Economic/Corporate Policy; Education • *Issues:* Corporate Watchdog; Deforestation; Energy; Global Warming; Marketplace Activism; Sustainable Business Development; Sustainable Development; Waste Management/Recycling

● Organization

If the environment is to survive, individuals must change their consumption patterns to a sustainable level. Co-op America is working toward that goal, and toward the goal of an alternative, environmentally friendly marketplace. A nonprofit member cooperative organization with a staff of 25, Co-op America is working to educate both consumers and businesses on how to "align buying and investing habits with values of peace, cooperation and environmental protection." Members include concerned individuals and organizations that have been screened and found responsible.

Education is key. Members are educated on a variety of issues and are offered concrete suggestions about what they personally can do to help the environment. Marketplace activism—shopping and investing to effect the environmental policies of businesses—is an important tool and one that Co-op America wants utilized. Combining economic and environmental issues through the slogan of "buy—invest—boycott—demand change" is the route that Co-op America is following toward a socially and environmentally responsible economy.

▲ Projects and People

Under the guidance of Alisa Gravitz, executive director, and with the goal of changing business practices and creating alternative environmentally friendly marketplaces, Co-op America focuses on educating its members and encouraging them to act in an environmentally responsible manner. Members are given the opportunity to shop from Co-op America's own catalog—a "marketplace for peace, cooperation, and a healthy planet"—or to patronize the organizations listed in its annual *Directory of Socially and Environmentally Responsible Businesses.* Catalog products include reusable grocery bags, recycling bins, solar battery chargers, non-electric household appliances, organic fertilizers, and toxic-free cleaners. In addition to shopping options, Co-op America publishes *Boycott Action News (BAN). BAN* provides information about products and companies with environmentally unsound policies and encourages boycotts until the companies reform their policies.

People often wonder how they can invest their money and receive a decent return without jeopardizing the environment. Co-op America wants people to have as much information as they need in order to make informed investment decisions. The *Socially Responsible Financial Planning Handbook* contains information to assist people with their investment plans, and give them socially responsible options. People are also referred to a financial network and given the option of using a Co-op America credit card.

Another publication is *Connections: Networking for a Sustainable Society* edited by Jyotsna Sreenivasan. This newsletter serves as a forum in which members can announce products and events, and also contains articles to help environmental entrepreneurs become successful.

The Recover the Earth program is concerned with rainforest deforestation. Their program seeks to inform members, encourages them to boycott products of companies involved in deforestation and informs the companies of the reasons behind the boycott. Co-op America also gives advice on ecotourism in areas of developing countries, such as Costa Rica's rainforest in the Carara Biological Preserve.

The Valdez Principles, to which Co-op America adheres, and to which others are encouraged to adhere, are designed to encourage sustainable and safe business practices.

■ Resources

Publications of Co-op America include *The Co-op America Quarterly, Socially Responsible Investing Handbook, Connections—Newsletter for Sustainable Businesses, Boycott Action News, Directory of Socially and Environmentally Responsible Businesses*, and *Co-op America Catalog.* Co-op America has a whole range of fact sheets on topics ranging from deforestation to energy efficient practices and *Living Green: 101 Green Things You Can Do.*

Council on Economic Priorities (CEP)
30 Irving Place
New York, NY 10003 USA

Contact: **Alice Tepper Marlin**, Executive Director
Phone: 1-212-420-1133 • *Fax:* 1-212-420-0988

Activities: Education; Information; Law Enforcement • *Issues:* Corporate Environmental Responsibility

● Organization

The **Council on Economic Priorities (CEP)**—an independent, nonprofit research organization—was established in 1969 to "inform and educate the American public and provide incentives for corporations to be good citizens responsive to the social concerns of all their stakeholders: employees, neighbors, investors, and consumers." When CEP began with a request for a "peace portfolio" of corporations not supplying weapons for the Vietnam war, it was able to obtain little information because "corporate social performance was essentially unmonitored," recalls CEP. This organization reports on a different environment today: "The withdrawal of some 190 corporations from South Africa, the rapid growth to over $450 billion in investment funds with social concerns, the dozens of cities and states which have adopted 'principled purchasing' programs, and the Exxon oil spill at *Valdez*, have all helped create a new climate awareness."

Priorities for the 1990s are to "stimulate environmental and corporate responsibility, and to redirect military spending to create a more productive, competitive economy." It is possible to do good and still make a profit, believes CEP. Private foundations, individual donors, and a nationwide membership of 7,000 support CEP.

▲ Projects and People

Overall, CEP has published more than 1,000 studies and reports on various environmental topics. The **Corporate Environmental Data Clearinghouse (CEDC)** is a major project monitoring corporate environmental behavior in the USA. Reports are available on more than 100 major U.S. companies—soon to be increased to 500 companies. These reports cover company-wide environmental issues, such as toxic releases, waste reduction programs, accidents or spills, recycling efforts, litigation, and compliance records. The varied and complex information culled from many sources is compiled in an easy-to-use format that provides a "one-stop environmental data resource." Its database will be a resource for individuals, grassroots groups, environmental NGOs, and government agencies.

In a recent report, for example, CEDC credited Amoco Corporation's "comprehensive and creative waste reduction programs in the petroleum industry," but also indicated there is much more to do among all corporations to "remedy past problems and prevent future negative impacts."

The new Institutional Investor Research Service (IIRS) provides ratings for several hundred companies and is regularly updated. *Shopping for a Better World* is a pocket-sized consumer guide, published annually, that lists more than 2,400 brand-name supermarket and whole-food products, and rates their makers according to factors such as environmental record, community outreach, animal testing, workplace issues, and promotion of women and minorities. For example, large companies with a positive environmental mark would use and encourage recycling, alternative energy sources, and waste reduction; small companies would be judged on using biodegradable and/or recyclable materials in packaging, disposal of waste in an environmentally sound manner; and using only natural ingredients and growing methods for food.

America's Corporate Conscience Awards are given annually to companies with outstanding work in environmental and other meritorious areas. Recent nominees for large companies included H. J. Heinz for its recycling programs and leadership in marketing "dolphin-safe tuna"; Herman Miller furniture manufacturer for surveying wood and veneer sources and excluding use of tropical woods, for buying back old furniture and reassembling parts, and for energy co-generation; and The Body Shop, for developing environmentally sound personal care products that are not animal tested. Recent small companies nominated were Aveda Corporation, first signer of the Valdez Principles, which produces shampoos from organically grown plants and flowers, uses no toxic pesticides, recycles plastics, and donates 2 percent of pre-tax earnings to environmental groups; Celestial Seasonings for eliminating teabag strings and tags, minimizing packaging, and replacing chlorine bleaches with oxygen bleaches; Real Goods Trading Company, a mail-order catalog for alternative energy and other sound environmental products, such as nontoxic paints; and Smith & Hawkin for its mail-order catalog for energy-efficient light bulbs, string bags, recycled household papers, and garden tools and furniture. *(See also Aveda Corporation in the USA Private section.)*

With the Committee for National Security (CNS), Washington, DC, the Council also studies national security issues and arms control, particularly regarding the USA/former USSR military expenditures, economic priorities, and productive transitions for the defense industry.

Volunteers in 42 states take part in the Letters to the Editor campaign to bring to readers' attention matters of corporate responsibility, military spending, and arms control.

■ Resources

Publications include the annual pocketbook *Shopping for a Better World* and *Better World Investment Guide* (Prentice Hall). A *Setting Priorities for the Nineties* brochure with membership form is available. CEP also has a speaker's bureau and internships.

Council on the Environment of New York City (CENYC) ❧

51 Chambers Street, Room 228
New York, NY 10007 USA

Contact: **Michael Zamm**, Director, Environmental
              Education/Training Student Organizers
   Phone: 1-212-788-7900 • *Fax:* 1-212-788-7913

Activities: Community Service; Education; Research • *Issues:* Energy; Health; Noise Pollution; Open Space Beautification and Preservation; Sustainable Development; Waste Management/Recycling; Water Quality

● Organization

The Council on the Environment of New York City (CENYC) is a privately funded citizens organization in the Office of the Mayor of New York. This is a local organization with an emphasis on practical ways to improve life and the environment in the challenging city of New York.

▲ Projects and People

Gerald Lordahl is director of the **Open Space Greening Program**. One of the many activities of this program is Plant-A-Lot. Between community gardens and the Lots for Tots which provide green playgrounds for children, the Plant-A-Lot program brings vegetation into a concrete city.

The Open Space Greening Program also runs a **Green Bank** providing matching funds for supplies for existing community gardens. More assistance to community gardeners comes in the form of the **Grow Truck**, a mobile unit which visits community garden sites throughout New York providing demonstrations of good gardening techniques, and lending gardening materials.

Greenmarket sites where local farmers can sell their produce now exist at 18 locations within New York. These give New Yorkers a place to purchase fresh produce, and also keeps a number of small farmers in business. Greenmarket encourages organic growers, and is involved in outreach efforts to bring more farmers into the program.

Education is an important part of the work of CENYC. The **Training Student Organizers (TSO) Program**, under the leadership of **Michael Zamm** teaches students the importance of civic participation and necessary organizational skills.

Students have undertaken projects involving water conservation, hazardous household substances and the preservation of natural areas. Workshops are run for both students and educators, to create a wide influence throughout New York City.

The **Office Recycling Service Plus (ORSP+)** works with organizations to encourage recycling and waste reduction. Aside from simple recycling, ORSP+ conducts feasibility studies, determines ways for organizations to reduce waste, and then designs and installs programs. Many organizations find that they actually make money through ORSP+ methods because they cut their costs and also market waste products.

■ Resources

CENYC needs access to mailing lists at little or no cost for outreach purposes. They are also in need of a computer modem and scanner in order to increase computer flexibility.

CENYC publications include *Training Student Organizers Curriculum*; *Environmental Bulletin*; *City Lot Fact Sheets*; *New York's Streets*; *Walking: A Realistic Approach to Environmental Education*; *Energy Conservation Action: An Action Approach*; *Greenmarket: The Rebirth of Farmer's Markets in New York City*; *From Farm to City: A Case Study*; *Neighborhood Composting in New York City*; *Office Waste Reduction*; *Sources of Recycled Paper*; and *Stepping Lightly on the Earth*.

The Cousteau Society ⊕

870 Greenbrier Circle, Suite 402
Chesapeake, VA 23320-2641 USA

Contact: **Jacques-Yves Cousteau**, Chairman of the
              Board/President
   Phone: 1-804-523-9335 • *Fax:* 1-804-523-2747

Activities: Education; Research • *Issues:* Biodiversity/Species Preservation; Sustainable Development

● Organization

The well-known explorer of oceans Jacques-Yves Cousteau created **The Cousteau Society** in 1973, thus formalizing his efforts to educate the world community about the environment. The Society is "dedicated to the protection and improvement of the quality of life for present and future generations."

The Society has three offices in the U.S., and one in France called **Fondation Cousteau**, altogether serving 350,000 members worldwide. Jean-Michel Cousteau, Jacques' son, is executive vice president and treasurer of the organization.

▲ Projects and People

Born in 1910, Jacques Cousteau spent his first 7 years battling chronic anemia and enteritis; at age 26 he was paralyzed for 8 months following an auto accident. Despite such adversity, his name has become synonymous with undersea exploration, and he has transformed a lifelong fascination with water into a crusade for the environment.

Films and television provide the Society's main source of income. Captain Cousteau made his first commercial film during the Nazi occupation of France, again overcoming steep obstacles in an era of chronic shortages: he bought dozens of rolls of 36 exposure film, and spliced them together at night under the blankets. With the results he made *18 Meters Below* (1942), a 10-minute film that was the first of many commercial successes. He later helped then 23-year-old **Louis Malle** launch a film career by allowing him to work on *The Silent World* (1957) despite a lack of experience.

Ironically, the man who produced films such as *Outrage at Valdez* (1990) once contracted with British Petroleum to find oil off the coast of Abu Dhabi. That success helped fund his next film, but also turned a nomad village into a modern city. It was not until 1958 that Cousteau "abruptly" realized the ecological damage caused by human activities and began to speak out in a series of campaigns. Central to the Society's activity is Cousteau's belief that "the most effective tool we have as citizens, as parents, is the sheer force of our numbers," and that, like dolphins, we must band together to face the threats to our environment, to "protect our progeny from [the] shortsightedness and ignorant abuse of awesome power."

With the Universal Bill of Rights for Future Generations, Cousteau seeks a formal United Nations Bill of Rights declaration, with five articles which guarantee "an uncontaminated and undamaged Earth." A Cousteau Society campaign garnered two million signatures for the Environmental Protection Protocol to the Antarctic Treaty, which "will preserve the ice continent as a natural reserve dedicated to peace and science."

An important element of Cousteau Society activities is research. Cousteau himself was instrumental in the creation of the Aqualung (with Emile Gagnan), deep-sea cameras, the small submarine, undersea habitats, the Turbosail (TM) (a patented wind-propulsion system), and the Sea Spider temperature probe. Vital to ongoing Society research efforts are the two research vessels: *Calypso*, acquired 1952, a World War II era wooden-hulled ship; and *Alcyone*, named for the daughter of Aeolus, God of the winds, launched in 1985 and featuring Turbosail.

In 1977, *Calypso* began circumnavigating the Mediterranean on a cruise whose findings led to the "Blue Plan" for Mediterranean Action; with this plan, Mediterranean countries began to view the sea as a *region* rather than a series of national waters. As Cousteau puts it, "there are no boundaries in the real Planet Earth." A more recent project is the Rediscovery of the World, a nine-year global circumnavigation during which both vessels will explore and film; the goal: 24 hours or more of quality film, much of it to be aired by Turner Broadcasting. The vessels will visit Haiti, Cuba, the tip of South America, the Sea of Cortez, Marquesas Islands, Channel Islands, Australia, Borneo, Indonesia, and Papua New Guinea, among other places. In Papua New Guinea, for example, they are investigating the long-term effects of deforestation on land and marine ecosystems.

Beyond research, the Cousteau Society seeks to educate the public and make them aware of its findings. In addition to the award-winning films, the Society publishes numerous books, articles, and teachers' guides. During two-week summer field study programs, participants can "explore a pristine tropical island ecosystem." The educational Parc Oceanique Cousteau opened in Paris in 1989.

■ Resources

The Cousteau Society publishes three periodicals: the bimonthly *Calypso Log* and *Dolphin Log*—the latter for children; and the *Calypso Dispatch*, which appears seven times each year. In addition, the Society's archives holds the approximately 100 underwater films produced since World War II by Cousteau teams, some 80 books, and 40 filmstrips. The Society also publishes a weekly syndicated newspaper column on the environment.

The Cousteau Society ⊕

Santa Monica Office
8440 Santa Monica Boulevard
Los Angeles, CA 90069-4221 USA

Contact: **Richard C. Murphy, Ph.D.**, Vice President, Science and Education
Phone: 1-213-656-4422 • *Fax:* 1-213-656-4891

Activities: Education; Research • *Issues:* Biodiversity/Species Preservation; Sustainable Development

● Organization

The Santa Monica Office of The Cousteau Society, in addition to pursuing the general objectives of the Society by working to make the world a better place for future generations, houses the editorial offices and produces the publications listed under Resources.

▲ Projects and People

Dr. Richard C. Murphy is based at the Santa Monica office in his capacity as Cousteau Society vice president for science and education—when he is not aboard one of the Society's two research vessels, the *Calypso* or the *Alcyone*. As a driving force behind the floating research programs, Dr. Murphy has organized and/or participated in numerous expeditions since he joined forces with Jacques Cousteau in 1969. In Project Ocean Search, Dr. Murphy participated in field studies in the islands of the South and North Pacific, the Atlantic, and off the coast of California. He has studied the possibilities for sustainable development and resource use along the Amazon River and in the Mississippi Basin; much of his recent work, however, has taken place in the Pacific Ocean.

In Papua New Guinea, he worked on studies of that nation's Motmot Island and on how the Sepik River contributed nutrients to the ocean. While circumnavigating the globe as part of the Society's Rediscovery of the World, he and his crew endured knee-deep mud to study and film the behavior of swiftlets, cave-dwelling birds on the island of Palawan, Philippines. A study of the coral reefs around the island nation of Nauru, 35 miles south of the equator, resulted in a report to the island's government which warned of the hazards of continued heavy phosphate mining there.

Dr. Murphy is also responsible for many of the Cousteau Society's education efforts, working on such endeavors as the Parc Oceanique Cousteau in Paris and similar centers in Montreal, Japan, and the USA. He developed a cartoon book intended to teach children in developing nations the basics of ecological awareness. In addition he has spoken at numerous conferences, including the Second Congress on Marine Sciences, Havana, Cuba, where he presented results of Cousteau Society studies on coral reefs, ocean productivity, and resource development strategies.

■ Resources

The Santa Monica editorial offices put out the Cousteau Society's bimonthly magazines *Calypso Log* and *Dolphin Log*, as well as the *Calypso Dispatch*, which comes out seven times each year.

Cultural Survival (CS) ⊕
215 First Street
Cambridge, MA 02142 USA

Contact: **Jason Clay**, Director
Phone: 1-617-621-3818 • *Fax:* 1-617-621-3814
E-Mail: Econet: cdp!cultsurv
Bitnet: survival@husc4
Internet: survival@husc4 harvard.edu

Activities: Advocacy; Development; Education; Research
Issues: Biodiversity/Species Preservation; Deforestation; Sustainable Development

● Organization

Concern for the fate of tribal peoples and ethnic minorities worldwide aroused a group of social scientists from Harvard to establish the nonprofit **Cultural Survival** in 1972. Today, its 20,000 members remain convinced that small, traditional societies can become "productive participants in multiethnic states" instead of "weak and tempting targets . . . destroyed in the name of development programs they are presumed to hinder, or in the name of the nation-state they are assumed to subvert."

Cultural Survival's mission is to "persuade governments and national or international agencies to stimulate action on behalf of traditional societies and to sponsor development programs that include traditional societies as participants and beneficiaries." Mindful that all societies are changing, the group does not insist societies retain their traditions; rather Cultural Survival assists them in maintaining aspects of their cultures that they consider important.

Empowerment of indigenous communities is gained through local institution building, resource managing and planning, and direct land use projects directed toward conserving biological diversity and working toward sustainable development in alliance with environmentalists. A consequence is the strengthening of human rights. "As struggles for resources become more acute worldwide," Cultural Survival says its programs will contribute "methods and analyses essential to national peace and security."

Cultural Survival, which is concentrating its efforts in rainforest areas of South America, is beginning to expand throughout Latin America and in the coastal and mainland communities and resources of South Pacific islands as well as Himalayan mountains and valleys—as funding becomes available. One regional activity sponsored with Colombian Fundación para la Educación Superior (FES) was a seminar/workshop for Indian resource management technicians from Colombia, Ecuador, Panama, and Peru, and including local Awá and Embera Indians.

▲ Projects and People

In **Ecuador,** CS supported efforts to bring about the 104,000-hectare Awá Ethnic-Forest Reserve in the environmental, rainforest "hot spot" known as the Southern Chocó. With the World Wildlife Fund (WWF), CS also provided technical assistance to establish a natural forest management program to stop deforestation.

Abutting the Ecuadorian lands, a similar reserve is planned for **Colombia** and will result in "one of the world's few, functioning binational Indian/environmental reserves." With project advisor **James Levy,** local Indian organizations are being strengthened. Two pilot projects along the Ecuadorian Reserve's most threatened frontiers will benefit from both Peru's Yanesha Forestry Cooperative (COFYAL) of the Palcazu Valley and CS technical assistance for natural forest management.

Also in Ecuador, CS staff is working with **Foster Brown,** Woods Hole Research Center; **Nicanor Gonzalez** and **Isaac Bastidas,** Kuna Indians from Panama's Project Pamasky; and Amuesha Indians from Peru's Yanesha Forestry Cooperative on **Project PUMAREN.** With support from the WWF-USA Tropical Forestry Program, a team is being developed which can design and implement a regional resource management program with the regional ethnic Federation for Indians of Napo Province (FOIN). An Indian-run community-based natural forest management program will be set up. COFYAL is also cooperating.

In **Bolivia,** as uncontrolled development—including logging and cattle raising—threaten the biologically diverse Chiman Forest and Isiboro-Secure Park, CS provided support to CIDOB, the Amazonian Indian organization, to help them defend their land and resource rights. As a result, according to CS, the government "promised title to approximately 1.2 million hectares." With ecologist Nicanor Gonzalez, CS is coordinating efforts to help the Beni Indians develop their conservation plan. As CS consultant and working with the Bolivian Coordinating Committee of Indian organizations, Gonzalez also initiated the International Indian Congress on Conservation and Resource Management. Theme of the second Congress in December 1991 was "the next 500 years: where are we going?"

In **Peru,** CS provided support to efforts of the Coordinating Group for Indigenous Organizations of the Amazon Basin (COICA) to link its concerns for land and resource rights with those of the global environmental community. Meetings set up in Iquitos, Peru, and Washington, DC, were a step in building mutual understanding and joint programming between Indians and environmental groups. Funding is presently being used to link COICA's work with local projects to demonstrate Indian resource management.

In the **Pacific Rim,** CS editor **John Cordell** began a survey in Papua New Guinea, Solomon Islands, Torres Straits, and north-central Australia to identify communities concerned with managing marine and upland resources and link them with the area's most active environmental/developmental organizations, national and global.

With **The Himalayas** being identified as another "hot spot" for biological diversity and deforestation, CS is working with regional specialist **Ann Forbes** to research perceptions and priorities of the area's residents and establish links with government agencies involved in resource management, forestry, and park planning. Her overview of local people and environmental issues, as well as recommendations on how CS can effectively work with local communities and NGOs in specific areas, is the basis for the CS developing program here.

In **Canada,** CS and *Ecologist Magazine* collaborated with the Canadian International Development Agency (CIDA) and the Canadian Aboriginal Forest Association (NAFA) in a series of meetings that brought together Indian "forest users" from North and South America who drafted a "Forest People's Charter" to define common interests and conservation concerns. Nicanor Gonzalez assists in circulating the draft charter among grassroots groups in Latin America through the World Rainforest Movement, which will "incorporate the various regional suggestions into a document and which reflects forest people's interests worldwide," CS reports. The Charter is expected to be included in the World Forest Convention agenda and presented for adoption at the 1992 United Nations Conference on the Environment and Development (UNCED) in Brazil.

With the Woodworkers Alliance for Rainforest Protection (WARP) formed in mid-1990 to set up social and ecological guidelines to define sustainable forest production and gain support of indigenous forest management in developing countries, CS is part of a consulting working group for "certifying" the extraction of wood in ways that incur the least forest damage, that respect the rights and needs of local people, and that inform wood consumers as to the nature and source of their purchases. Once these standards are established, the certification group will support marketing of timber products from indigenous forest management projects which appear the most socially and ecologically sustainable. This group also includes the UK's Ecological Trading Company (ETC), WWF-UK, WWF-USA, Green Cross, Rainforest Alliance "Smart Wood" Project, World Rainforest Movement, Greenpeace, and Rainforest Action Network (RAN).

■ Resources

Resources needed are new technologies for fruit processing and oil expelling as well as lab analyses of nontimber forest products, such as resins, essential oils, fruits, and flowers.

Defenders of Wildlife ⊕
1244 19th Street, NW
Washington, DC 20036 USA

Contact: **James K. Wyerman**, Director, International Wildlife Trade
Phone: 1-202-659-9510 • *Fax:* 1-202-833-3349

Activities: Development; Education; Law Enforcement; Political/Legislative; Research • *Issues:* Biodiversity/Species Preservation • *Meetings:* CITES

● Organization

The mission of Defenders of Wildlife is to protect wildlife and its habitat. An organization with a membership of 80,000 and a staff of 37, Defenders is able to work on a wide variety of projects to help protect wildlife.

▲ Projects and People

Refuge Management is one of the key projects of Defenders, which acts on behalf of current wildlife refuges in numerous ways including lobbying for better management. Believing that the refuges are full of activities that are harmful to the wildlife they are supposed to protect, Defenders has been working toward passage of an "organic act" which would clarify the purpose of the refuges and provide more protection for the wildlife. In 1990, Defenders convened its own **Commission on New Directions**—made up of biologists, wildlife managers, and academics—to stimulate public discussion of policies and practices for the National Wildlife Refuge System.

Defenders has also started organizing a number of **local refuge advocacy groups** and providing these groups with information and assistance. Examples are the Tinicum National Environmental Center in Pennsylvania, which sought to block development of wetlands

outside Philadelphia, and the Friends of the Stone Lakes Wildlife Refuge, which serves a similar purpose in Sacramento County, California.

Another issue of concern to Defenders is **international wildlife trade**. An estimated eight million birds are captured in the wild to be sold as pets, and more than half of them die before they are even sold. Approximately 500,000 are imported to the USA. Working toward a ban on wild bird imports, and convincing three major airlines to stop accepting shipments of wild birds are two of the approaches that Defenders have taken toward solving this problem.

Defenders has taken the U.S. government to court on several issues, including **violations of the Endangered Species Act**. Defenders also plans to use legal methods to force the government to **honor the Arctic National Wildlife Refuge** and keep it free from oil and gas exploration. In 1991, it joined with other conservation groups to urge the U.S. Congress not to authorize such drillings.

Driftnets, often called "curtains of death," continue to have a vast toll on dolphins and other marine animals. Defenders work to end marine entanglement in concert with other environmental groups.

Creating **Wildlife Viewing Guides** to provide a network of wildlife viewing sites throughout the USA is one of the first projects to combine tourism with wildlife appreciation. These guides are providing tourists with information on the best spots to see and photograph wildlife, while still protecting it. The first guides to be produced in this series are for Arizona, California, Colorado, Idaho, Indiana, Montana, New Jersey, New Mexico, Oregon, Texas, Utah, Washington, and Wisconsin.

Defenders also has a program devoted to the protection of **wolves** whereby it has taken numerous actions including suing the government to restore wolves to 2.2-million-acre Yellowstone National Park and establishing a fund to compensate farmers for the loss of livestock to some wolves that are now returning to their former habitat. Once a top predator in the Northern Rockies, the gray wolf disappeared 50 years ago. Now that wild populations of ungulates or hoofed mammals are so numerous there, conservationists believe the return of the wolf would help restore the natural balance to the ecosystem. Defenders also came to the aid of Alaskan wolves.

Other animals that Defenders recently championed include the California lion, the threatened Mojave desert tortoise, Florida Key deer, and the endangered Florida panther, peregrine falcon, and bald eagle.

In 1991, Defenders helped assess the **Persian Gulf War oil spill damage to wildlife and habitat**, advised on environmentally benign cleanup options, and assisted the President's Council on Environmental Quality with monitoring and developing a "Persian Gulf wildlife database to target species most imperiled"—among them the endangered dugong, Socotra cormorant, and endangered or threatened sea turtles, sea and shore birds, otters, dolphins, mollusks, fish, and whales.

The Defenders' John Fitzgerald is on the Board of Advisors of *Environmental Profiles*.

■ Resources

Defenders of Wildlife is in need of new technologies including online management for activist and membership databases and an internal LAN. They also are looking for activists interested in lobbying in their geographic areas and volunteers to help with outreach. To increase outreach, they would like public service announcements and press releases.

Publications include the bimonthly *Defenders Magazine*, action alerts, a wolf newsletter, fact sheets, and *Wildlife Viewing* guides. With the U.S. Bureau of Land Management, a *Watchable Wildlife* color brochure was produced to illustrate the localities of 270 million acres of federal land richly inhabited by starfish and sea lions, black bear and bighorn sheep, pine marten and marbled murrelet, antelope and leopard lizard, bison and yellow-billed cuckoo.

Ducks Unlimited (DU) ⊕
One Waterfowl Way
Memphis, TN 38120 USA

Contact: **Matthew B. Connolly, Jr.**, Executive Vice President
Phone: 1-901-758-3825 • *Fax:* 1-901-758-3850

Activities: Research; Education; Development; Political/Legislative • *Issues:* Air Quality/Emission Control; Biodiversity/Species Preservation; Deforestation; Population Planning; Sustainable Development; Water Quality; Wetlands Habitat
Meetings: CITES

● Organization

Ducks Unlimited (DU) is really two separate but complementary organizations: **Ducks Unlimited, Inc.** and the **Ducks Unlimited Foundation**, primarily a financing body which operates strictly to benefit DUI. Ducks Unlimited, Inc. was incorporated in 1937 with the purpose of restoring waterfowl populations to pre–Dust Bowl levels. With a staff of over 300, membership of over 515,000, and projects in all 50 states, Ducks Unlimited strives to protect the wetland ecosystems of North America, which lose nearly half a million acres to development and poor management annually.

John E. Walker is DU president and chief executive officer. Previous presidents Peter H. Coors, Coors Brewing Company, is DU Foundation President; Harry D. Knight is chairman of the Board; and Gaylord Donelley is a major supporter of wildlife conservation activities, including the ACE Basin Habitat Program in South Carolina.

▲ Projects and People

In conjunction with **Ducks Unlimited Canada (DUC)**, also organized in 1937, and **Ducks Unlimited de Mexico (DUMAC)**, in operation since 1974, the organization has helped conserve over 5.5 million acres of wildlife habitats in the 3 countries. Funds raised by DUI and its partners go to purchase tracts of land before they are damaged, as well as to rehabilitate threatened areas. Such wetland restoration efforts benefit not only ducks, but over 600 wildlife species, including whooping cranes, bald eagles, and other endangered fowl, fish, reptiles, amphibia, and plants.

DU reports that California has lost 91 percent of its wetlands and dramatic losses have also occurred "in the bottomland hardwoods of the Mississippi Valley, the coastal marshes of Louisiana, the prairie potholes of the northern Midwest, the rainwater basins of Nebraska, and the coastal plains of the southeastern United States." Wetlands are where almost one-third of the USA's endangered species live and where spawning grounds are located for "60 to 90 percent of the fish taken routinely by our commercial fishery."

For these reasons and to help save the breeding and wintering habitats of waterfowl and shorebirds, the organizations plays a leading role in implementing the North American Waterfowl Management Plan, signed by the USA and Canada in 1986 to conserve almost six million acres by the year 2000.

In addition, DUI directs efforts toward educating citizens and businesses, involving them in the future of wetlands conservation. Many DUI supporters are sportsmen and women, and an important message of the organization is that people must hunt responsibly, with "respect for the resource"; according to DUI, "true hunters . . . [are] this century's most effective conservationists."

Ducks Unlimited is currently working with The Nature Conservancy, the South Carolina Wildlife and Marine Resources Department, the U.S. Fish and Wildlife Service, and private landowners on the ACE Basin Task Force to protect that 350,000-acre region and the wildlife that inhabit it. Named for the 80 miles of 3 rivers winding through it—Ashepoo, Combahee and Edisto—ACE is home to species including wild turkey, bobcat, gray fox, white-tailed deer, loggerhead sea turtle, wood stork, green-winged teal, pintail, bald eagle, and osprey. The Edisto River is the longest blackwater river in North America—black from the tanic acids of the surrounding hardwood trees. Rice originating in Madagascar thrived in the area on large plantations for two centuries—aided by flocks of waterfowl who controlled unwanted plant pests by feeding on them. Although rice is no longer grown here, the ACE Basin provides about 20 percent of the state's commercial catch of fish, shrimp, oysters, and clams.

One major DU land acquisition in 1990 was the Mary's Island Reserve in the ACE Basin. Plans are underway to establish a National Estuarine Research Reserve among eight islands with saline, freshwater, and brackish tidal marshes and maritime forests; as well as a National Wildlife Refuge.

■ Resources

Ducks Unlimited publishes two bimonthly magazines, available with membership: *Ducks Unlimited Magazine*, containing articles on wetlands, wildlife, and hunting; and *Puddler: A Wetlands Wildlife Magazine for Children*. DU also offers workshops, speakers programs, and technical assistance regarding wetlands conservation.

Earth Island Institute ⊕ ❦
300 Broadway, Suite 28
San Francisco, CA 94133 USA

Contact: **Peter Fugazzotto**, Volunteer Coordinator
Phone: 1-415-788-3666 • *Fax:* 1-415-788-7324
E-Mail: Econet: earthisland

Activities: Research; Education; Political/Legislative • *Issues:* Biodiversity/Species Preservation; Deforestation; Energy; Global Warming; Sustainable Development; Transportation
Meetings: CITES; Conferences on the Fate of the Earth

● Organization

The Earth Island Institute was founded in 1982 by its current chairman of the board, David R. Brower, who was also first executive director of the Sierra Club and founder of Friends of the Earth. Seeing the need to provide a network for various environmental projects, he conceived of the Institute as a way "to forward the ideas of creative individuals and to provide organizational support for their work on ecologically linked issues." To this end, the Institute's staff of 25 coordinates various projects and groups, making office resources available for fundraising, outreach, and international environmental networking activities.

At the grassroots level are **Earth Island Centers**, local "organizing points" in the USA and abroad. A "sibling organization," the **Earth Island Action Group** focuses on legislative advocacy.

▲ Projects and People

True to its networking goal, the Earth Island Institute currently serves as umbrella to more than 20 different projects, both national and international in scope. The **Environmental Litigation Fund**, for example, "provides information to and holds conferences for citizen activists" who are contemplating lawsuits to enforce environmental laws. The **Urban Habitat Program** seeks to improve urban neighborhoods through enlisting "multicultural environmental leadership, . . . the creative use of public spaces and such techniques as 'urban barnraising.'"

International projects such as the **Conferences on the Fate of the Earth**, begun in 1982, stress the connections among "the issues of peace, environmental protection, human rights, and economic development." **Baikal Watch** works toward the preservation of Lake Baikal in the former USSR, and the **Rainforest Health Alliance** alerts health professionals to the risk of losing potentially useful plants as the tropical rainforests shrink. Other projects focus on animal protection and species preservation, such as the **International Marine Mammal Project**—working to save dolphins and whales from commercial slaughter—and the **Sea Turtle Restoration Project**.

■ Resources

Resources currently needed include Macintosh software training for staff, and people and organizations to work on Earth Island's legislative efforts.

The Institute publishes the *Earth Island Journal* quarterly for its approximately 35,000 members. Other materials include a video, *Where Have All the Dolphins Gone?*, *For Earth's Sake*, the memoirs of David Brower; and the *Bay Area Green Pages*, providing a comprehensive listing of Bay Area environmental organizations; a consumer guide to "green" products; and more information.

Earth Preservation Fund (EPF) ⊕ ❦
4011 Jackson Road
Ann Arbor, MI 48103 USA

Contact: **William Weber**, Director
Phone: 1-313-665-4407 • *Fax:* 1-313-665-2945

Activities: Development • *Issues:* Biodiversity/Species Preservation; Deforestation; Sustainable Development

● Organization

The **Earth Preservation Fund (EPF)** is a nonprofit organization through which travelers can assist and encourage better conservation and environmental practices at a local level worldwide.

EPF was founded in 1978 as an affiliate of Journeys International, a worldwide nature and culture travel company. It identifies and supports village and community-level projects that promote environmental or cultural preservation. Projects are carried out with volunteer assistance of Journeys' staff, travelers, and local officials and residents.

Culturally and environmentally responsible tourism can be a powerful force for conservation and community development, states EPF. As a result, the EPF is an ongoing experiment to discover and explore means by which group and individual travelers can contribute tangibly, directly, and significantly to global environmental conservation and human welfare in less developed regions of the world.

EPF's objective is to demonstrate small-scale, local-level models of preservation action which will inspire adoption by communities and wealthier funding agencies on a broader scale.

▲ Projects and People

Projects not only help local communities but also increase goodwill between hosts and guests and stimulate authentic cross-cultural interactions, says EDF. By educating travelers to be more sensitive and conscientious guests, EPF intends to provide an alternative to the exploitative and destructive elements of conventional tourism.

Projects supported by EPF fall into several classes. Some are designed and supervised by overseas guiding staffs in areas where they conduct trips. This is the case in **Nepal**, where EPF has ongoing **tree-planting and school programs**, and a beginning **Eagle Census** project as well as a **Medical Service Trek** bringing medical care to remote villages. In **Ladakh**, during the off-season, EPF has been involved in a variety of Buddhist heritage site restoration efforts. Previous projects include solar heating undertakings, bridge building, and vegetable seed distribution.

Some EPF projects involve direct participation of foreign travelers. Its **Inca Trail Preservation Trek** in **Peru** annually maintains the ancient Inca path to Machu Picchu. The **Human Warmth Project** is an ongoing project to collect warm clothing from friends and participants for free distribution to poor people in remote mountainous areas of Asia and South America. In **Costa Rica**, EPF has joined preservation efforts of the Monteverde Conservation League and the Caribbean Conservation Corporation.

Domestically, EPF offers travelers an opportunity to help **save whales in Hawaii**. Each winter humpback whales come to the shallow waters near Maui to mate and calve. Controversy over the effect of increased boat traffic on the whales is at a peak. According to EPF, there is evidence that boats may be displacing whales from the preferred nearshore habitat. Data on the impact of vessel traffic, which travelers with EPF will help collect, assist in the development of effective management strategies for this endangered population. The state of Hawaii and the National Marine Fisheries Service are supporting this research.

From shore-based observation stations, participants collect data on the movement and behavior of whales in response to approaching boats. They also assist in studying and documenting the underwater vocabulary and songs of the whales and record data on dive times, blow rates, aerial displays and other behaviors in response to boat traffic.

■ Resources

Trip Notes are available on destinations to upcoming journeys.

(See also Earth Preservation Fund [EPF] in the Madagascar NGO section.)

EarthSave
706 Frederick Street
Santa Cruz, CA 95062-2205 USA

Contact: **Patricia Carny**, Executive Director
Phone: 1-408-423-4069 • *Fax:* 1-408-458-0255

Activities: Development; Activity • *Issues:* Deforestation; Energy; Global Warming; Health; Sustainable Development; Water Quality

● Organization

In 1987, **John Robbins**, of Baskin-Robbins ice cream fame, published *Diet for a New America*, a book which urged Americans to change their meat-centered eating habits in order to save themselves and their planet. The massive public response to his book prompted him to found **EarthSave**, a nonprofit organization dedicated to educating people about the effects of their dietary choices on the Earth's ecology, and "to helping the Earth restore its delicate ecological balance."

To do this, EarthSave depends on the energies of a staff of 6, numerous volunteers, and 3,200 contributing members. Half the funds come from sales of products such as books, cassettes, and T-shirts; the rest come from membership dues, grants, and other contributions.

▲ Projects and People

EarthSave's primary activities are educational: informing "individuals, organizations, physicians, and legislators"; hosting seminars and workshops; and distributing literature. The head office provides advice and help to numerous **Local EarthSave Action Groups**, all-volunteer organizations that seek to spread EarthSave's message in their own communities. *Healthy People, Healthy Planet* is an EarthSave-sponsored curriculum package for schools, including lesson plans for teachers, posters, and other materials. These provide an alternative to the standard "four food groups" lesson, which EarthSave considers to be dangerous disinformation.

YES! (Youth for Environmental Sanity) is an EarthSave-sponsored troupe of six to eight young men and women (ages 16-20) who tour the nation's schools educating students about their role in improving the Earth's ecology. As a result of the efforts of YES!, students have begun demanding vegetarian alternatives in school cafeterias, composting projects for cafeteria organic garbage, and other environmental efforts. John Robbins, president of EarthSave, also tours the USA, speaking and leading workshops throughout the country.

Citing Americans' poor health, particularly their obesity, high blood pressure, and heart disease, and pointing to the environmental damage caused by current agricultural practices—wasted water and topsoil, methane gas emissions produced by cattle, aggravating global warming—EarthSave suggests: "A reduction in meat consumption is the most potent single act you can take to halt the destruction of our environment and preserve our precious natural resources. . . . With a national change toward a more plant-based diet, we will be able to grow ample food for the nation on approximately one-third of the land now used to produce an animal-based diet."

■ Resources

EarthSave offers *Diet for a New America*, an excerpted version entitled *Realities for the 90's*, and the quarterly *EarthSave Newsletter*. Resources needed include educational outreach to implement nutrition programs in public schools.

Earthwatch ⊕
680 Mt. Auburn Street
Box 403N
Watertown, MA 02272 USA

Contact: **Blue Magruder**, Public Affairs Director
Phone: 1-617-926-8200, 1-800-776-0188
Fax: 1-617-926-8532

Activities: Development; Education; Research • *Issues:* Air Quality/Emission Control; Biodiversity/Species Preservation; Deforestation; Energy; Global Warming; Health; Sustainable Development; Water Quality • *Meetings:* Principal Investigators Conference of Earthwatch

● Organization

Earthwatch is the brainchild of the late **Dr. Harold Edgerton**, MIT professor who founded the organization in 1971. Since 1972, **Brian A. Rosborough** has directed this "coalition of citizens and scientists working to sustain the world's environment . . . and foster world health and international cooperation."

In its Watertown headquarters, and in field offices in Los Angeles, Oxford, England, and Sydney, Australia, Earthwatch musters its 70 volunteer field representatives and 73,000 members to support for hundreds of projects around the world. Worldwatch's affiliate, the **Center for Field Research** (CFFR), screens and selects for Earthwatch involvement about 30 percent of more than 400 project proposals received annually. Over 28,000 **EarthCorps**-paying volunteers from more than 800 institutions have helped fund and staff many of these projects, continuing the search for solutions to environmental problems. More than 1,000 expeditions worldwide contributed nearly $15 million to such studies. The number of participating volunteers is growing yearly, with more than 3,000 volunteers having worked on more than 1 project; the record holder is an 81-year-old woman who has joined 30 projects.

▲ Projects and People

Recent Earthwatch-sponsored projects included 7 related to rainforest conservation; 47 in life sciences, covering conservation biology, restoration ecology, and wildlife management; and 21 in marine studies. The projects cover the globe, from the Amazon to Zimbabwe. *(Earthwatch projects are described throughout this Guide; see also Earthwatch, Europe in the UK NGO section.)*

Rosborough, Earthwatch president and CFFR chairman, received the U.S. Department of Interior Conservation Award in 1987 for Earthwatch's efforts to save the endangered Caribbean leatherback turtles. He has directed scientific investigations in Australia, China, Iceland, Mauritania, Nova Scotia, and Zaire and has funded and mobilized expeditions in more than 87 countries in support of the sciences and humanities.

■ Resources

Earthwatch provides teacher/student fellowships and scholarships for career training on Earthwatch research programs. The organization publishes a bimonthly magazine, *Earthwatch*.

Program on Environment ⊕
East-West Center
1777 East-West Road
Honolulu, HI 96848 USA

Contact: A. Terry Rambo, Ph.D., Director
Phone: 1-808-944-7265 • *Fax:* 1-808-944-7970

Activities: Education; Research • *Issues:* Air Quality/Emission Control; Biodiversity/Species Preservation; Energy; Global Warming; Sustainable Development; Transportation; Waste Management/Recycling; Water Quality

● Organization

The U.S. Congress established the **East-West Center** in 1960 "to promote better relations and understanding between the United States and the nations of Asia and the Pacific through cooperative study, training, and research." The Center's **Environment and Policy Institute (EAPI)** was set up in 1977 to further those aims in a more specific context.

Focusing on the Asia-Pacific region, EAPI is now the Program on Environment. The Program's goal is to use "interdisciplinary and multinational" study, training and research efforts for the development of "sustainable environmental management" programs. Its ultimate aim is "reconciling development with the environment," working with United Nations organizations, international banks, and both governmental and non-governmental organizations throughout the region. More than 27,000 men and women from the region—including some 2,000 research fellows, graduate students, and professionals in business and government annually—have participated in the Center's coooperative programs.

More than 20 Asian and Pacific governments, and private agencies and corporations join Congress in providing funding.

▲ Projects and People

In addition to sponsoring some 50 international graduate students each year for studies at the University of Hawaii, the Program has developed four core programs: **Habitat and Society** has projects studying urban environmental management in Asian cities, and focusing on sustainable rural development and resources management in Laos, Vietnam, and southwest China. **Land and Water** researches areas in which people, water, and land interact: biodiversity, lake and watershed management, tourism, and other issues important in the region. **Oceans and Atmosphere** researches current policies of ocean management and their effects on marine species. One project, for example, investigates the effects of global climate change on the sea level and on energy/technology policies of Asia-Pacific nations. **Risk and Development** focuses on the goal for many developing nations: "to develop effective control strategies to reduce environmental risk during economic development." Particular targets include air pollution, toxic wastes, and global warming.

■ Resources

Most of the Center's research is published by outside publishers as journal articles, chapters and books. One recent publication is *From Grave to Cradle: Trends in Hazardous Waste Management*, edited by Richard A. Carpenter and Richard R. Crillo.

The Center also publishes occasional papers and working papers available on request, Recent titles include *Human and Nonhuman Impacts on Pacific Island Environments*, by Patrick D. Nunn, and *Women's Livestock Production and Natural Resources in Pakistan*, by Carol Carpenter.

Eco-Home Network
4344 Russell Avenue
Los Angeles, CA 90027 USA

Contact: Julia Russell, Executive Director
Phone: 1-213-662-5207

Activities: Education • *Issues:* Ecological Living; Energy; Sustainable Development • *Meetings:* 1st International Ecological Cities Conference

● Organization

The **Eco-Home Network**, begun in 1985, became a nonprofit organization in 1989. Its 1,500 members work on information distribution, environmental projects, and support groups, with the goal of contributing to an earth-friendly lifestyle.

▲ Projects and People

The centerpiece of the Network's efforts is the **Eco-Home model habitat**. This actual house demonstrates to the over 3,000 people who have toured it the reality of "New City Living"—an urban lifestyle that improves the quality of life and at the same time protects the environment. Among the features of the Eco-Home and its surroundings are drought-resistant landscaping called a xeriscape that survives on natural rainfall once established, an ornamental vegetable garden and fruit trees watered by drip irrigation, a three-bin composting system, solar-powered light and solar-heated water, a low-flush toilet, and recycling containers. Also featured are walking seminars that have prompted many persons to change careers, revamp businesses, remodel homes, and generally modify lifestyles, according to Eco-Home.

The Eco-Home Network is also working to implement an alternative energy school curriculum, to educate children on steps they can take to improve the environment. In particular, the Network tries to educate consumers to use their buying power wisely, treating "economics as if people and the earth really mattered."

Founder Julia Russell, described by news media as "a modern day pioneer and environmental activist," brings her message to television audiences in the USA, Japan, Mexico, and Sweden. She and Eco-Home president Bob Walter work together on the television production, *The Land We Throw Away*.

■ Resources

Ecolution, Eco-Home Network's quarterly newsletter provides news, calendar activities, list of environmental businesses, and practical and philosophical articles related to the environment. It also shares with readers the activities of many other environmental groups in the Los Angeles area and beyond.

Eco-Home Network distributes pamphlets and information packets on specific home and lifestyle modifications that individuals can make to improve the environment, from how to get rid of ants without harmful chemicals to how to install water saving shower heads. The organization has also produced videotapes on sustainable living. Other resources include a library and study center that expects to be part of a Mnemodex database system for books, products, and people.

The Ecological Society of America (ESA) ⊕
2010 Massachusetts Avenue, NW, Suite 420
Washington, DC 20036 USA

Contact: **Marjorie M. Holland, Ph.D.,** Director, Public Affairs Office
Phone: 1-202-833-8773 • *Fax:* 1-202-833-8775

Activities: Education; Political/Legislative; Research • *Issues:* Biodiversity/Species Preservation; Global Warming; Sustainable Development; Sustainable Ecological Systems; Water Quality

● Organization

Founded in 1915, **The Ecological Society of America (ESA)** is a nonpartisan, nonprofit organization of scientists with 6,200 members in the USA, Canada, Mexico, and 62 other nations. ESA conducts research, teaches, and aids decisionmakers in government agencies, industry, and other organizations.

Its purpose is to "stimulate and publish research on the interrelations of organisms and their environment, to facilitate an exchange of ideas among those interested in ecology, and to instill ecological principles in the decisionmaking of society at large." It is known worldwide for its forums where these ideas are exchanged, for its scholarship research foundation in all branches of ecology, and for its high caliber journals.

With chapters in Washington, DC, the southeastern USA, and the central Rockies, the Society sponsors annual meetings, symposiums on timely topics, and field trips. Members also provide advice or testimony to governmental agencies, congressional subcommittees, and non-governmental bodies. In addition, through the work of its Public Affairs Office in Washington, DC, the Society may speak out on certain environmental issues and policies. The staff promotes policies that protect the environment and the welfare of ecology and ecologists and makes available the expertise of professional scientists on related matters. The Society maintains sections for ecologists with special needs and interests, such as paleoecology, aquatic, physiological, statistical, applied ecology, vegetation, education, and long-term studies.

▲ Projects and People

The Public Affairs Office has a "computerized databank of more than 1,500 ESA members who can provide expert scientific information to Congress and other groups on issues affecting domestic and international environmental quality." The **Ecological Information Network** provides rapid answers to questions about effects of human activities on animals, plants, and microorganisms in both natural and managed ecosystems. Network participants easily address questions about the effects of pesticides, oil spills, and radioactive substances, for example.

To promote wise management of the planet's resources, the ESA's **Sustainable Biosphere Initiative (SBI)** proposes a "multiyear, multi-disciplinary program which focuses on three priority areas: global change, biological diversity, and sustainable ecological systems." This is in response to a challenge of the National Academy of Sciences that "scientists set their own research priorities." SBI is a call to arms for "basic research for the acquisition of ecological knowledge," which is shared with the public and becomes the basis of sound policy and management decisions.

In a report entitled *The Planned Introduction of Genetically Engineered Organisms: Ecological Considerations and Recommendations,* ESA responded to questions about the ecological consequences of introducing such organisms into the environment by conducting an integrative scientific study. More than 100 ecologists reviewed the document before its release. Industry, government, and other scientific societies are using the final report.

ESA **awards and fellowships** recognize achievements in the field of ecology, including research. These include the Mercer Award, given for an outstanding paper published by a young ecologist; the Buell Award given for an outstanding talk by a graduate student at ESA's annual meeting; the MacArthur Award given for outstanding research contributions by an established ecologist; the Cooper Award given for the best paper in geobotany, physiographic ecology, and related subjects; the Whittaker Travel fellowship which brings a leading foreign scientist to this country; and the Forrest Shreve Sonoran Desert Research Fund which is used to support a young ecologist doing research there.

ESA also provides **professional certification** to ecologists who meet its standards.

■ Resources

The Society publishes four scientific journals distributed worldwide: *Ecology; Ecological Monographs;* and an applied ecology publication, *Ecological Applications. The Bulletin of the Ecological Society of America* is issued quarterly and includes articles which apply basic ecological principles to issues confronting mankind. Also, a pamphlet on *Careers in Ecology* describes present and future employment prospects. The report on *The Sustainable Biosphere Initiative: An Ecological Research Agenda* is available too.

EcoNet/Institute for Global Communications (EcoNet) ⊕
18 De Boom Street
San Francisco, CA 94107 USA

Contact: **Michael Stein,** Director
Phone: 1-415-442-0220 • *Fax:* 1-415-546-1794
E-Mail: Econet: bleland; Internet: bleland@igc.org

Activities: Computer Networking; Education • *Issues:* Air Quality/Emission Control; Biodiversity/Species Preservation; Deforestation; Energy; Global Warming; Health; Population Planning; Sustainable Development; Transportation; Waste Management/Recycling; Water Quality

● Organization

EcoNet is an international, computer-based communication system committed to serving organizations and individuals who are working for environmental preservation and sustainability.

EcoNet is a community of presently 2,500 members using the network for "information sharing and collaboration with the intent of enhancing the effectiveness of all environmentally-oriented programs." EcoNet connects persons with a personal computer and a modem with other people around the world. It offers easy-to-use tools for posting events on international bulletin boards, preparing joint projects through electronic conferences, and finding and discussing the latest information on environmental topics.

Resources on EcoNet include data such as the Environmental Grantmakers Association's directory of grantmakers, the Sierra Club National News Report, the National Wildlife Federation's Conservation Directory, Global Action Network's federal legislative information, and action alerts.

Local access to EcoNet is available from many countries. Partnership relationships are established with like-minded networks in Canada, England, Sweden, Australia, Nicaragua, and Brazil. Users also can exchange electronic mail with persons in Russia.

▲ Projects and People

To help coordinate Earth Day's twentieth anniversary celebration, offices all over the world used EcoNet to gather information on regional events, post summaries of activities, send online newsletters, and coordinate media outreach.

Calling themselves **Campus Earth**, university students around the USA use EcoNet to organize on-campus activities promoting public awareness of global warming.

EcoNet users received the story of the Exxon *Valdez* oil spill simultaneously through firsthand accounts of the devastation from local Alaskans who were online.

Network users frequently respond to an online letter-writing campaign request by immediately sending electronic mail, fax, and telex messages urging decisionmakers to take positive environmental action.

■ Resources

For a one-time modest sign-up fee, a user receives a manual, a private account and password, and one free hour of off-peak connect time each month. Additional connect time is charged at varying rates for both off-peak and peak hours. There is a small charge for users requiring large amounts of online storage.

Environmental Defense Fund (EDF) ⊕

National Headquarters
257 Park Avenue South
New York, NY 10010 USA

Contact: **Rodney M. Fujita, Ph.D.**, Staff Scientist
Phone: 1-212-505-2100 • *Fax:* 1-212-505-2375

Activities: Education; Law Enforcement; Political/Legislative; Research • *Issues:* Air Quality/Emission Control; Biodiversity/Species Preservation; Deforestation; Energy; Global Warming; Sustainable Development; Waste Management/Recycling; Water Quality

● Organization

The New York office of the **Environmental Defense Fund (EDF)**, like the other offices around the nation, has a highly trained staff of scientists, economists, and attorneys to carry on the work. Some of their projects are local in nature, but most are wider ranging.

▲ Projects and People

Staff scientist **Dr. Rodney M. Fujita** has been with the EDF since 1988. Much of his attention is directed at the effect of climate changes on marine life, especially coral reefs. He is co-founder and a board member of the Global Coral Reef Alliance, and has testified before the U.S. Congress and other decisionmaking bodies regarding the endangered status of reefs. Fujita has "developed a program to safeguard fragile coral reefs in U.S. and foreign waters."

The EDF's **Global Atmosphere Program** involves many staff members at the New York office. Astrophysicist **Dr. Michael Oppenheimer**, a scientist with the EDF since 1981, chairs the program dealing with air quality, climate change, acid rain, and ozone depletion. Oppenheimer worked on a joint EDF–United Nations study on limiting greenhouse gases, and co-authored *Dead Heat: The Race Against the Greenhouse Effect*. **Dr. Stephanie L. Pfirman** serves as scientific coordinator of the 1992 **Climate Change Exhibition** put on by the American Museum of Natural History with the EDF.

Scientist **Sarah L. Clark** and attorneys **D. Douglas Hopkins** and **Timothy D. Searchinger** are also involved in the Global Atmosphere Program. However, their focus is the **Eastern Water Program**, which Clark chairs. Under her direction, the EDF tackles issues such as pollution of the New York–New Jersey harbor, wetlands conservation, and protection of the Florida Everglades.

Senior economist **Dr. Daniel Dudek**, with the EDF since 1986, leads the organization's atmospheric policy work. Besides being instrumental in developing the acid rain control section of the Clean Air Act of 1990, he has testified before Congress and worked as consultant for the Polish, Hungarian, and former Soviet governments and the OECD (Organization for Economic Cooperation and Development) on ozone depletion, acid rain, and global warming. Dudek advocates "market-based solutions to environmental problems"—for example, an "emissions trading program" for sulfur dioxide (SO_2), carbon dioxide (CO_2), and chlorofluorocarbons (CFCs).

In *The Financial Times*, EDF staff economist **Alice LeBlanc** clarifies the relationship between economics and the environment. She worked on a 1991 program with the Chicago Board of Trade (CBOT) "to trade futures in U.S. pollution permits." Under this plan, the U.S. Environmental Protection Agency (EPA) issues permits to industries for a limited amount of SO_2 emissions; companies that quickly lower emissions with room to spare can sell or trade their excess permits to companies that emit excess SO_2. According to LeBlanc, this "will encourage polluters to control more than what is required by law."

■ Resources

Like other EDF offices, the New York branch carries the bimonthly *Environmental Defense Fund Letter*. Environmental historians might be interested in the book *Acorn Days*, by **Marion Lane Rogers**, which is a memoir of the early days of EDF and how it grew from the perspective of the author who was "sole support" staff at the beginning of the 1970s.

Environmental Defense Fund (EDF) ⊕
1616 P Street, NW
Washington, DC 20036 USA
Contacts: **Stephan Schwartzman, Ph.D.**, Senior
Scientist; **Korinna Horta**, Staff Economist
Phone: 1-202-387-3500 • *Fax:* 1-202-234-6049
E-Mail: Econet: edf

Activities: Education; Political/Legislative; Research • *Issues:*
Air Quality/Emission Control; Biodiversity/Species Preservation; Deforestation; Energy; Global Warming; Health;
Sustainable Development; Transportation; Waste Manage-
ment/Recycling; Water Quality • *Meetings:* CITES; UNCED
preparations; World Bank; World Climate Conference

● Organization

Volunteer conservationists angered by the dangers of widespread
DDT spraying founded the **Environmental Defense Fund (EDF)**
in 1967 in Stony Brook, New York. Early action halted the spraying
in Suffolk County. Now the EDF boasts a staff of over 100,
including scientists, engineers, lawyers, and economists, 200,000
members, and 7 offices from coast to coast. Regional offices focus
on issues of concern to their geographic areas, but all offices are
involved in international efforts as well.

Overall goals are to halt ozone depletion, stop acid rain, save
tropical rainforests, clean up toxic waste, preserve wetlands, control
global warming, achieve 50 percent recycling in the USA by the year
2000, protect Antarctica, guarantee clean water for the future, and
safeguard wildlife and habitats. For instance, the EDF was instrumental in 1991 in helping win a 90 percent reduction of pollution
shrouding the Grand Canyon in smog. A year earlier, EDF authored
an acid rain reduction plan that the U.S. Congress passed. Designed
to reduce air pollution at lower costs, this plan is attracting attention
from Eastern Europe. In the 1980s, EDF proved the link between
sulfur emissions and distant acid rain and achieved a legislated 90
percent reduction of lead additives in U.S. gasoline.

▲ Projects and People

The **International Program** is a joint effort of senior attorney **Bruce
Rich** with anthropologist **Dr. Stephan Schwartzman**, who works
with indigenous people's groups in the Amazon on behalf of sustainable forestry strategies; staff attorney **Lori Udall**, who is developing EDF's network with NGOs throughout Asia, Europe, and the
USA; staff scientist **Deborah Moore**, who also works with native
people's groups, farmers organizations, and government on advancing water conservation and watershed protection; and staff economist **Korinna Horta**, who takes on multilateral development organizations like the World Bank and the African Development Bank,
as well as numerous international companies involved in environmentally destructive activities such as logging and mining.

Through such activities, EDF works closely with grassroots groups
and non-governmental organizations in developing countries—
where most of the Earth's population, land mass, and natural resources are located—to promote sustainable alternatives, this
decade's most urgent need, according to EDF. One result is the
protection of more than "7.5 million acres of Amazon rainforest"
through the **establishment of special "extractive reserves" managed
by Brazilian rubber tappers**, who harvest Brazil nuts and latex

without harming the forest. In Africa and Indonesia, EDF is extending its rainforest work.

Two current trends are apparent to EDF: "An accelerating worsening of critical global environmental indicators and trends continues in areas such as deforestation, global warming, and growing
scarcity of water resources for human use." At the same time "public
pressure has increased on governments and international institutions to make sustainable development a priority," and some are
beginning to respond. EDF is part of the crucial effort to promote
environmental **reforms** in the "hitherto relatively unaccountable"
public **international financial institutions (PIFIs) and through
working with the "growing global network of NGOs** in developing
countries and in the industrialized world in collaborative research
and advocacy to analyze and change selected internationally financed development programs."

EDF chooses such projects as case studies based on local NGO
concerns and international policy implications for sustainable development. Consequently, it is helping to "close the circle" between
local community concerns, the highest decisionmaking levels of the
PIFIs, national parliaments, industrialized country environmental
NGOs, and the international media for news coverage and public
education.

For example, in 1990, because of opposition by EDF and its
Indian allies, the World Bank **halted** plans for the **Narmada Sagar
Dam** project in India, thus saving the area's tropical forest and
preventing displacement of 100,000 people. Environmentalists also
brought pressure to stop a **road project in Ivory Coast** that threatened coastal ecosystems and a rainforest that is home to the country's
last remaining elephant population and regarded as sacred by local
villagers. They pushed instead for a comprehensive program at the
grassroots level to protect the local area and livelihoods and foster
sustainable agriculture and forestry.

Current efforts center on protecting the rights of the **indigenous
Baka and Bakola peoples in Cameroon's rainforests**, who are
already suffering as a result of World Bank–condoned logging
operations, according to the EDF. Beset with economic woes, the
Cameroon government is turning to logging companies to provide
a source of foreign exchange by cutting down the rainforests and
severely threatening the biological wealth of this country, labeled as
"megadiversity" by the World Conservation Union (IUCN).

At the same time, current government policy "effectively denies
the forest people any legal rights to the natural resource base they
have depended on for thousands of years," writes Horta, who also
reports on criticism of the Tropical Forestry Action Plan for
Cameroon, drawn up by the UN Food and Agriculture Organization and the UN Development Programme.

"We view the respect for human rights and conservation of
natural resources as inextricably linked," continues Horta. "Through
a network of environmental organizations in all the major donor
countries, we advocate policy reform in international development
institutions to make them more transparent, participative, and accountable."

■ Resources

EDF publishes a bimonthly *Environmental Defense Fund Letter.* It
offers publications on a range of topics including acid rain, biotechnology, energy, environmental toxics, global atmosphere, solid waste
and recycling, degradable plastics, landfills, and wildlife.

Solid Waste Program
Toxic Chemicals Program
Environmental Defense Fund (EDF) ⊕
1616 P Street, NW, Suite 150
Washington, DC 20036 USA

Contact: **Richard A. Denison, Ph.D.**, Senior
Scientist; **Lois N. Epstein**, Toxic
Chemicals Specialist
Phone: 1-202-387-3500 • *Fax:* 1-202-234-6049
E-Mail: Econet: edf

Activities: Education; Political/Legislative; Research • *Issues:*
Air Quality/Emission Control; Biodiversity/Species Preservation; Deforestation; Energy; Global Warming; Health;
Sustainable Development; Transportation; Waste Manage-
ment/Recycling; Water Quality

● Organization

The **Solid Waste Program** makes up a significant part of the
operations of the **Environmental Defense Fund's (EDF)** Washington office. The program goals include ensuring "that economically
and environmentally sound waste reduction, reuse and recycling
programs are given priority . . . including full financial support";
reducing the risk to public health posed by such waste management
techniques as incineration and landfills; and reducing toxic materials in manufacturing processes and making products "more durable,
recyclable, and conserving of resources."

Strategies include litigation, scientific and economic research,
lobbying, intervention in regulatory and permit proceedings, testimony to government officials, technical assistance to agencies and
citizens' groups, public education programs, and collaborative
projects.

▲ Projects and People

Senior scientist on the Solid Waste Program **Dr. Richard A. Denison**
"specializes in hazardous and waste management issues ranging
from waste reduction and recycling to the health effects and regulatory requirements of landfilling and incineration." He is a former
fellow of the U.S. Congress' Office of Technology Assessment, and
is a frequent expert witness at congressional and other hearings. Dr.
Denison recommends "judicious use of both economic incentives
and government regulations" to effect changes necessary for the
environment.

A key recent success of the Solid Waste Program is fostering a
new **environmental consciousness** in **McDonald's Corporation**.
Because of the efforts of Dr. Denison and staff members **Jackie
Prince** and **John Ruston**, McDonald's has adopted EDF's recommendations, which "could divert 80 percent of McDonald's trash
from disposal in landfills and incinerators." By switching from
styrofoam to paper sandwich wrappings, buying recycled and recyclable packaging materials, composting, and other efforts,
McDonald's is setting an example for other businesses.

Scientist Prince researches recycling technologies and consumer
product life assessments; she has authored two books on hazardous
wastes. Economist Ruston works on identifying links between economic development and environmental quality, including develop-

ing a market for recycled materials and paper/plastic manufacturing
processes. Both have worked with Dr. Denison to take action
against "so-called degradable plastics," leading a consumer boycott
of bags and diapers with this label. As a result of the EDF's efforts,
the Federal Trade Commission began investigating those products,
and leading manufacturers dropped their packaging claims.

Regarding achievements in opposing toxic chemicals, World
Wildlife Fund–USA chairman **Russell E. Train** has commented,
"EDF's work in bringing about the banning of DDT was one of the
most important legal victories ever won for wildlife."

In such areas, attorney **Kevin P. Mills**, director of EDF's Environmental Information Exchange, is empowering state and local
groups on toxic chemicals and solid waste. **Lois N. Epstein**, toxic
chemicals specialist, has written citizens' guides on preventing leakage of underground storage tanks—thereby protecting drinking
water supplies from contamination. EDF agricultural chemist **Terry
F. Young** helped uncover toxic contamination at a wildlife refuge
and led the U.S. Environmental Protection Agency (EPA) to "reject
weak state standards for selenium and boron." Marine biologist and
senior scientist **Peter L. deFur** is protecting wildlife from toxic
discharges into Virginia's waterways and the Chesapeake Bay. Biologist **Rebecca J. Goldburg** draws national attention to "the misguided use of biotechnology to engineer crops with a higher tolerance for pesticides." Economist **Glen D. Anderson** aims to improve
funding mechanisms for hazardous waste sites cleanup. Attorney
David S. Bailey, who directs the EDF Virginia office, seeks "strong
regulation against dioxin pollution of rivers." Chemist **Diane C.
Fisher** is working on reducing human exposure to toxic chemicals.
Hydro-geochemist **Ann Maest** is striving to prevent pollution from
manufacturing and hard-rock mining.

Cancer research strides were made with the successful efforts to
win federal protection of the Pacific yew, whose bark is the source of
a promising drug treatment, by attorney/scientist **Bruce S.
Manheim, Jr.**, who also heads the Antarctic effort. With the Sierra
Club, EDF brought legislative action in California against 14 brands
of spot removers and paint strippers—"whose ingredients carry
some of the highest cancer risks found in common consumer products," according to EDF—which prompted the removal of such
products from some 2,000 stores and got certain manufacturers to
exclude carcinogenic ingredients from their formulas. Attorney
David Roe heads the toxic chemical program on the West Coast,
where the landmark toxics law "Proposition 65" was drafted, requiring companies to warn people when they are exposed to cancer-
causing chemicals.

In 1991, EDF released a report, *Legacy of Lead: America's Continuing Epidemic of Lead Poisoning*, which spurred U.S. congressional hearings within days to combat such poisoning in children.
Attorney **Karen L. Florini**, who helped prepare this report, is now
"working on legislative options and public education programs on
lead and other toxic substances." A campaign succeeded in keeping
a "toxic metal, manganese, as an octane booster" out of Ethyl
Corporation gasoline. Another longstanding effort is aimed against
"highly toxic dioxins, including those released into water from
papermaking plants during the bleaching process." Internationally
known expert **Ellen K. Silbergeld** testified on both issues.

EDF reports too that it stopped a proposed Colorado nuclear
weapons facility and instead had funds redirected to clean up the
site's radioactive waste; attorney **Melinda Kassen** was involved.
"EDF continues to press the [U.S.] Department of Energy (DOE)

to ensure that human health is not jeopardized by the transport or disposal of radioactive waste at the Waste Isolation Pilot Project near Carlsbad, New Mexico," states the organization.

Americans will be prohibited from exploring for oil, gas, and minerals in Antarctica because of a law that EDF helped draft and get passed with Greenpeace and others; also the U.S. State Department must seek international agreements to extend such provisions to foreign nationals to foster Antarctica as a world park. Coastal pollution is a target of innovative EDF reduction programs, such as in North Carolina, New Jersey wetlands, San Francisco Bay, Florida Keys National Marine Sanctuary, and the South Florida Water Management District, where polluted wastewater was being discharged into the Everglades. With the National Resources Defense Council, EDF spurred EPA to limit toxic discharges of mercury and copper from sewage treatment plants and other sources into the New York–New Jersey harbor. Among those working on such issues are scientist Sarah L. Clark, marine biologist Rodney M. Fujita, biologist Douglas N. Rader, and attorneys Allen E. Grimes, D. Douglas Hopkins, Timothy D. Searchinger, and Louisa C. Spencer.

■ Resources

The Solid Waste Program of the EDF produces the quarterly *Recycling World*, as well as a pamphlet entitled *Recycle*. *Recycling and Incineration: Evaluating the Choices* is a consumer-directed book encouraging environmentally sound waste management. A *Final Report* is available from the EDF-McDonald's Waste Reduction Task Force. A complete list of publications is available related to solid waste reduction, recycling, incineration, and landfilling; and a membership form can be provided.

Environmental Defense Fund (EDF) ⊕
California Office
5655 College Avenue
Oakland, CA 94618 USA

Contact: **Deborah Moore**, Staff Scientist
Thomas J. Graff, Senior Attorney
Phone: 1-510-658-8008 • *Fax:* 1-510-658-0630
E-Mail: Econet: edf

Activities: Education; Political/Legislative; Research • *Issues:* Air Quality/Emission Control; Biodiversity/Species Preservation; Deforestation; Energy; Global Warming; Health; Sustainable Development; Transportation; Waste Management/Recycling; Water Quality

● Organization

The California Office of the Environmental Defense Fund (EDF) carries out activities similar to those pursued by the EDF's other regional offices. Because of the nature of the western United States, much of the California office's energies focus on water resources, through their Rural Economy and Environment Program (REEP). The staff's expertise in this area has involved them in an international Global Water Resources Project as well, mainly seeking "sustainable water management strategies in developing countries."

▲ Projects and People

Dr. Zach Willey, senior economist, is the REEP program director. In his view, allocation of the resource is the key to environmentally sound water policies. Thus, REEP seeks to provide alternative water supplies to Los Angeles to protect the fragile Mono Lake from depletion; one way is by sale or lease of water from other California areas that can then use the revenue to improve their own water efficiency use.

REEP has other water rights agreements underway in the Truckee and Carson River basins in Nevada, the Verde River in Arizona, the Columbia River basin in Oregon and Washington, and the Wind River basin in Wyoming. In many of these areas, notes staff scientist Deborah Moore, Native American tribes and the environment have been the "traditional losers" in the battle for water resource control. The conflict diminishes when all sides concerned become more aware of the need to conserve and of alternative means of development.

Moore has worked closely with the American Indian Resources Institute, helping to organize an annual seminar to discuss issues such as tribal governance and conserving and managing water, mineral, fish, and wildlife resources on their lands.

The California EDF has carried out other West Coast programs to preserve or restore water quality and safety, in conjunction with state and municipal authorities, in areas such as the San Francisco Bay and Delta and the Central Valley. In addition, for more than ten years, the major California utility companies have used Elfin, EDF's "computer-based energy planning model." Elfin now has applications worldwide as electric utilities lower costs and convert to renewable energy and other conservation measures. Adolph S. "Spreck" Rosekrans is working with a consortium of environmental, agricultural, and urban interests to develop an "appropriate model to analyze potential policies (or projects) to make most efficient use of California's precious resource." After two years of study, EDF sees limitations in the California Department of Water Resources computer-based simulation model, DWRSIM, that is applied to both the State Water Project and Central Valley Project. According to Rosenkrans, " . . . the model assumes a much less flexible system than actually exists, and therefore its results indicate that the costs of environmental mitigation are exaggerated."

With the Regional Institute of southern California, EDF explores economic incentives to reduce emissions that cause smog and contribute to global warming. Michael Cameron manages EDF's air quality and transportation project in southern California.

In Washington State, water marketing incentives are examined to protect the threatened salmon and steelhead in the Yakima and Columbia rivers. In Oregon, a cooperative project with the Warm Springs Confederated Tribes is developing an environmental code especially to identify forestry and fisheries management options that are sustainable and economically sound.

In Nevada, the Pyramid Lake Paiute and Fallon Paiute-Shoshone Tribes as well as waterfowl, shorebirds, and endangered and threatened fisheries are benefitting from the permanent acquisition of water rights of the Stillwater National Wildlife Refuge wetlands—brought about through cooperative efforts of EDF, The Nature Conservancy, and other groups. Now, the EDF is following through with implementation of this far-reaching water allocation agreement.

In Arizona, EDF has legally taken on power plants whose emissions are polluting the Grand Canyon National Park and is analyzing patterns of electricity use and recommending changes in operation at the upstream Glen Canyon Dam, where peak power releases have caused severe habitat and beach erosion downstream in the Colorado River corridor, the organization reports. For a federally ordered Environmental Impact Statement (EIS), EDF is working to get reliable estimates on the impact of potential alternatives; and is also illustrating how "economic impacts [of the hydropower system] are only a fraction of what has been claimed." Along the Verde River, EDF is working with upstream farmers to develop and implement a water conservation program for Arizona's longest remaining stretch of a free-flowing river—following the recent enactment of the Fort McDowell Indian Community Water Rights Settlement Act.

Moore also focuses on water supply and management and wastershed protection in India, Indonesia, and Brazil; and has pursued related policy reform in the U.S. Congress, Bureau of Reclamation, World Bank, and United Nations.

Thomas Graff has written on federal water policy in the USA as an agenda for economic and environmental reform and related topics such as water marketing and public trust. He has "orchestrated western regional water and energy efforts since opening EDF's California office in 1971," reports EDF.

■ Resources

Moore's most recent publication is *Water in the American West: Institutional Evolution and Environmental Restoration in the 21st Century*. She also prepares articles, editorials, and testimonies on water conservation and water management.

Environmental Defense Fund (EDF)
Rocky Mountain Office
1405 Arapahoe Avenue
Boulder, CO 80302 USA

Contact: **Robert E. Yuhnke**, Attorney
   Phone: 1-303-440-4901 • *Fax:* 1-303-440-8052

Activities: Development; Legal • *Issues:* Air Quality/Emission Control; Biodiversity/Species Preservation; Energy; Global Warming; Health; Transportation; Waste Management/Recycling; Water Quality

▲ Projects and People

Since 1980, attorney **Robert E. Yuhnke** has directed the legal programs for this **Environmental Defense Fund (EDF)** regional office by identifying cases and forums where litigation can promote EDF's environmental goals, by maintaining media relations, and by building effective liaisons with members of U.S. Congress as well as with state legislatures and governors in the western USA.

Yuhnke was responsible for producing a sulfur pollution strategy for preventing acid pollution damage in a report entitled *Safeguarding Acid-Sensitive Waters in the Intermountain West* with Michael Oppenheimer. Yuhnke pointed out that most of the region's 10,000 lakes above 9,500 feet are sensitive to such pollution. The report was an effort to identify sources of sulfur dioxide (SO_2) such as smelters, new coal-fired power plants, and other potential pollutants in this air shed.

On another pressing issue, Yuhnke writes, "from all the evidence, **radon** in the home is the most deadly environmental hazard in America today." A radon expert, Yuhnke produced an EDF report on this "colorless, odorless, and tasteless radioactive gas" occurring naturally in all soil, and helped alert the nation to its serious health perils, including cancer.

After a decade of struggle, Rocky Mountain office environmental scientist **Dr. Daniel F. Luecke** helped defeat the proposed and "environmentally destructive water supply project" Two Forks Dam on the South Platte River southwest of Denver. "Luecke and the Colorado Environmental Caucus developed and advocated a set of cheaper, less harmful water supply options that were instrumental in EPA administrator William K. Reilly's ultimate veto of the Two Forks proposal," according to the EDF newsletter.

Global warming and transportation impacts on air pollution are also issues of concern.

■ Resources

In addition to the above-mentioned EDF reports, Yuhnke recommends *Reducing Greenhouse Gas Emissions with Alternative Transportation Fuels* and *Transportation Efficiency: Tackling Southern California's Air Pollution and Congestion*.

Environmental Defense Fund (EDF)
Texas Office
1800 Guadalupe
Austin, TX 78701 USA

Contact: **James Marston**, Director
   Phone: 1-512-478-5161 • *Fax:* 1-512-478-8140
E-Mail: Econet: edf

Activities: Development; Education; Political/Legislative; Research • *Issues:* Air Quality/Emission Control; Biodiversity/Species Preservation; Deforestation; Energy; Global Warming; Sustainable Development; Transportation; Waste Management/Recycling; Water Quality • *Meetings:* Group of Seven

● Organization

The recently established **Texas office** of the Environmental Defense Fund (EDF), with its staff of four aided by interns and volunteers, addresses regional environmental concerns.

▲ Projects and People

Led by director **James Marston**, the Texas EDF focuses on four main issues: "solid waste reduction, water resources preservation, global warming, and the Free Trade Agreement." The city of Houston is trying EDF-recommended waste reduction strategies, including source reduction, reuse, recycling, and composting." Once finalized, the EDF hopes this will serve as a "model solid waste plan" for other Texas cities.

Because Texas emits two times as much "greenhouse gas" as any other state, the EDF has developed options to help reduce those emissions—and lessen the threat of global warming.

The EDF sees the proposed North American Free Trade Agreement (NAFTA) as a chance to improve the environment. The USA-

Canada-Mexico treaty offers an opportunity to include provisions such as cleanup, protection of the Rio Grande, and limiting hazardous materials. For these purposes, a coalition was formed with the Natural Resources Defense Council (NRDC), National Wildlife Federation (NWF), Sierra Club, and Audubon Society.

During the 1990 Houston "Group of Seven" meeting, the EDF, in coalition with 150 other environmental organizations from the nations involved, forced global warming and other environmental issues onto the economic summit's agenda.

Typical activities of the EDF include working with the Texas legislature, offering expert testimony and recommending bills. The organization also makes appearances in the media, both print and television, with op/ed pieces, commentary, and interviews.

Environmental and Energy Study Institute (EESI)
122 C Street, NW, Suite 700
Washington, DC 20001-2109 USA

Contacts: **Ken Murphy,** Executive Director
Marilyn Linda Arnold, Program Associate
Phone: 1-202-628-1400 • *Fax:* 1-202-628-1825
E-Mail: EES

Activities: Education; Political/Legislative; Research • *Issues:* Air Quality/Emission Control; Deforestation; Energy; Global Warming; Population Planning; Sustainable Development; Transportation; Water Efficiency; Water Management/Recycling; Water Quality

● Organization
The Environmental and Energy Study Institute (EESI), established in 1984, is a cooperative enterprise with the leaders of the congressional Environmental and Energy Study Conference (EESC), the "largest legislative service" of the U.S. Congress. EESI, with a staff of 15, is the nation's only such independent organization set up by congressional leaders to "promote better informed national debate on environmental and energy issues and to generate innovative policy responses." In so doing, it provides information and options to Congress. EESI is known for its briefings, policy reports, blue-ribbon task force recommendations, testimony, national and regional conferences for key decisionmakers, and technical assistance.

Concerned with numerous local-to-global environmental issues, EESI focuses on leadership matters and sustainable development and technologies to encourage "cross-cutting," or new ways to use resources more efficiently and prevent pollution. "With some satisfaction, we have watched as ideas the Institute generated or disseminated showed up in legislation, in hearings . . . in administration proposals and elsewhere," reports EESI, whose chairman is James Gustave Speth, president, World Resources Institute. The bipartisan EESC is comprised of some 90 senators and 290 representatives.

▲ Projects and People
Worldwide projects include a **Task Force on International Cooperation for Sustainable Development,** a congressional working group on the United Nations Conference on Environment and Development (1992), and a **World Forest Agreement Working Group** to foster "appropriate congressional action" on global issues.

Addressing **water efficiency concerns** is a new domestic initiative. "As existing water supplies face ever-increasing demands," reports EESI, "conflicts over water, long an issue in the arid west, also have increased in the east." What's more, sites for large water projects such as reservoirs, dams, aqueducts, and pumping stations are lessening—as communities question environmental impact—and federal funding is decreasing. Yet the pressures for clean water mount among older cities grappling with aging infrastructure; agriculture, industry, thermoelectric power, and commercial facilities; and outdoor enthusiasts desiring to keep water in streams and lakes for endangered species and other wildlife, recreation, and tourism. Drought, global warming, and other climatic changes complicate these problems.

Moreover, the "mining" of groundwater can lead to overpumping that causes "land subsidence, higher pumping costs, and salt-water intrusions into underlying aquifers—tainting water supplies," according to EESI. At the same time, hazardous wastes are contaminating both ground and surface water.

The **Water Efficiency Project** stimulates rethinking of policy, identifies key issues, and helps develop new conservation and efficiency strategies through a series of regional workshops and Capitol Hill briefings with the assistance of water expert **Bruce Babbitt,** former Arizona governor, to help Congress determine the nation's direction. The Ford and Smith Richardson foundations help support the briefings.

A **working group** of congressional staff, environmentalists, educators, state and local officials, and water managers is being formed to explore policy actions in urban and rural communities; ultimately, a water policy paper will be prepared. A recent grant from the U.S. Environmental Protection Agency (EPA) will help EESI find opportunities for water conservation through federal agency programs and facilities; an options paper on water conservation for EPA will also be prepared.

Previously, EESI influenced a **Clean Air Act** amendment offered by **Congressmen Ed Markey** (D-MA) and **Carlos Moorhead** (R-CA) so that incentives are offered for electric utilities to use energy efficiency and renewable energy for halting acid rain. Also, inside the U.S. Capitol, lighting efficiency was begun. Energy education efforts continue to concentrate on the "relationship between greenhouse gases and conventional air pollutants and the potential of energy efficiency and renewables . . . in all economic sectors." EESI helped launch the **Surface Transportation Policy Committee** and serves on its steering committee.

■ Resources
EESI has an active intern program. Available EESI and Study Conference *Special Reports* are listed in the EESI annual report. A *Weekly Bulletin* is also published.

Environmental Hazards Management Institute (EHMI)
10 Newmarket Road
P.O. Box 932
Durham, NH 03824 USA

Contact: **Alan John Borner**, Executive Director
Phone: 1-603-868-1496 • *Fax:* 1-603-868-1547

Activities: Education; Research • *Issues:* Air Quality/Emission Control; Sustainable Development; Waste Management/Recycling; Water Quality

● Organization
With a new degree in international relations from LeHigh University, **Alan John Borner** traveled the world to reinforce his education. What he saw particularly in the Far East was unforgettable: "Rainbow-colored rivers were the results of pollutants dumped by unregulated, multi-national manufacturers hidden in the jungles upstream, and the results could be seen on the skin of natives diving for tourist coins." When he returned home to New Jersey and saw that "bad things were happening to good rivers as well," Borner sprung into action.

The Environmental Hazards Management Institute (EHMI) was formed in 1981 by Borner, its current director, as a result of the need for information, answers, and leadership expressed at the Northeast Conferences on Environmental Management. Headquartered in New Hampshire, with affiliate offices in Boston, San Francisco, Chicago, and Montreal, EHMI seeks to inform citizens, businesses, governmental and non-governmental agencies, and educators of the risks surrounding, laws controlling, and ways of minimizing environmentally hazardous materials and activities.

EHMI sees itself as "a trusted arbiter and mediator for opposing factions, often between environmental purists and industry."

▲ Projects and People
In working with government regulatory agencies, EHMI offers advice regarding "fair and effective" environmental legislation. Once laws are passed, EHMI explains to industries how to comply with the regulations, and informs citizens of ways to assess and control potential environmental threats.

EHMI's training programs include classroom instruction and simulated emergencies such as a toxic spill or chemical fire. The Institute also provides environmental audits to clients, highlighting areas in which companies need improved environmental safety measures and/or in which they are in violation of the law. These measures help prevent not only damaging fines and litigation, but also the very real possibility of an environmental accident.

■ Resources
EHMI seeks assistance in lobbying efforts and in curriculum development.

The Institute's regular publication, *HazMat World*, has a circulation of over 40,000. In addition, EHMI edits the *Environmental Managers Compliance Advisor* and the *OSHA Compliance Advisor*; these references provide the latest information on U.S. Environmental Protection Agency (EPA) and Occupational Safety and Health Administration (OSHA) regulations and on how to comply with them.

Additionally, EHMI produces *Wheels*, covering topics such as household hazardous waste, water sense, and recycling. These are interactive educational tools that can be used in both schools and businesses to increase awareness of environmental hazards.

(See also Environmental Investigation Agency in the UK NGO section.)

Environmental Law Institute (ELI)
1616 P Street, NW
Washington, DC 20036 USA

Contact: **James M. McElfish, Jr.**, Senior Attorney and Director, ELI Mining Center
Phone: 1-202-328-5150 • *Fax:* 1-202-328-5002

Activities: Education; Law Enforcement; Research • *Issues:* Biodiversity/Species Preservation; Deforestation; Energy; Global Warming; Health; Sustainable Development; Waste Management/Recycling; Water Quality

● Organization
Founded in 1969, the **Environmental Law Institute (ELI)** seeks to provide "reliable and objective information" regarding environmental issues, and to find "pragmatic solutions" to environmental problems. Financed by grants, federal and state contracts, foundations, and other sources, ELI's staff, which includes economists, scientists, journalists in addition to lawyers, serves as a link between various interest groups in the field, from industrialists to environmentalists.

▲ Projects and People
The main activity of ELI is education; they stress that they are not lobbyists or litigators. Since 1970, ELI has instructed more than 30,000 students and professionals on general and specific environmental law topics, negotiating skills, media and the environment, and the like.

For example, ELI's senior attorney and Mining Center director, **James M. McElfish, Jr.**, has developed and taught courses in cities across the USA in groundwater enforcement, hazardous waste inspection, cost-recovery for underground storage tank programs, and media coverage. Associates Programs update professionals, corporations, law firms, and other institutions in policy discussion forums and workshops. Special interdisciplinary undertakings include the **Center for State Environmental Programs, Eli Wetlands Program, Center for Surface Mining, Toxics and Groundwater Program,** and **International Projects**.

Specific research projects of the Institute include air and water quality, hazardous wastes, surface mining, coastal zone protection, and environmental management.

■ Resources
ELI's library of over 15,000 sources is available for use by staff and outside researchers. The Institute publishes reference texts, research reports, directories, monographs, and a list of publications. Its three bimonthly publications are *The Environmental Forum* provides a forum for professionals in the fields—from the U.S. Congress,

government agencies, industry, and environmental organizations—to exchange ideas creatively "in the search for effective solutions to complex environmental problems"; *Environmental Law Reporter*, published since 1970, gives analysis and updates on all law-related environmental issues: legislation, regulation of the Resource Conservation Recovery Act (RCRA), and litigation; and *National Wetlands Newsletter*, published since 1979, covers "legal, scientific, and regulatory developments affecting wetlands, the coastal zone and related sensitive lands."

Environmental Media Association (EMA)
10536 Culver Boulevard
Culver City, CA 90232 USA

Contact: **Lauren McMahon**, Executive Director
Phone: 1-310-559-9334 • *Fax:* 1-310-838-2367

Activities: Education; Research • *Issues:* Air Quality/Emission Control; Biodiversity/Species Preservation; Deforestation; Energy; Health; Population Planning; Sustainable Development; Transportation; Waste Management/Recycling; Water Quality

● Organization
Serving the entertainment industry as a "clearinghouse of environmental information and expertise," the Environmental Media Association (EMA) was founded in 1989 as a nonprofit organization. EMA reports that it "works with writers, producers, directors, and other members of the creative community to encourage the incorporation of environmental themes in entertainment productions." Since its beginning, EMA has worked with more than "60 primetime television shows, television specials, movies of the week and feature films and helped to produce home videos, public service announcements [PSAs], and music videos."

The board of directors includes chairmen and CEOs from major studios, three network television presidents, and industry leaders such as **Norman and Lyn Lear**, **Michael and Judy Ovitz**, and **Robert Redford**. The **Environmental Advisory Board** includes leaders from environmental organizations such as American Oceans Campaign (started by EMA directors **Ted and Casey Danson**), California League of Conservation Voters, Conservation International, Earth Island Institute, Environmental Defense Fund (EDF), Green Seal, Greenpeace USA, Natural Resources Defense Council (NRDC), National Audubon Society, National Center for Environmental Research (NCAR), National Wildlife Federation (NWF), Sierra Club Legal Defense Fund, Stanford University, TreePeople, United Nations Environment Programme (UNEP), Wilderness Society, World Wildlife Fund/Conservation Foundation, and WorldWatch Institute.

▲ Projects and People
EMA has seven goals to reach the public and work with government and industry leaders to help bring about sustainable development:
• **Engage the creative community**, through **organizing networks** among each entertainment specialty—"directors, producers, writers, actors, musicians who will then provide leadership, inspiration, and resources for their colleagues." EMA works through

the Directors Guild, Writers Guild, Screen Actors Guild, Academy of Motion Picture Arts and Sciences, American Film Institute, National Academy of Television Arts and Sciences, International Council of the Academy of Television Arts and Sciences, National Cable Television Academy, American Federation of Musicians, and the Caucus for Producers, Writers and Directors.
• **Sponsor educational forums** for the entertainment industry on current issues through briefings, seminars, interactive workshops, and presentations by scientists, activists, and politicians.
• **Provide written materials** in response to requests from the creative community and "match up producers and writers who need information with the individual or organization that can best provide it." EMA is establishing a **print and video information library** with materials from environmental groups. It is also assembling a **resource guidebook** to environmentally oriented programs, films, and documentaries and for use by schools, libraries, and other groups.
• **Encourage films, television programs, and other creative projects** to incorporate environmental themes. A quarterly newsletter that keeps the entertainment industry updated on environmental news also suggests such topics. Television examples include recycling highlights on *Murphy Brown*, a program that reaches an international audience, and *MacGyver* shows that present a range of issues from pesticides threats to ozone layer depletion.
• **Sponsor an Annual Awards program** to "give visibility to exemplary film and television productions" dealing responsibly with educational themes. At the inaugural ceremony, September 1991—with more than 100 entries—*MacGyver* and *Captain Planet* were cited for ongoing commitment. Other television award winners included *Chernobyl: The Final Warning* movie and *A User's Guide to Planet Earth: The American Environment Test* special; episodes from *Days of Our Lives*, *The Simpsons*, and *Shannon's Deal*; and children's programming *Earth to Kids: A Guide to Products for a Healthy Planet* and *Whales Tales*, *Tiny Toon Adventures*. *Dances with Wolves* won the feature film honor and *Yakety Yak, Take It Back* received EMA's music video award. EMA generates publicity for such programming—notifying sponsors, media, and public prior to their airing.
• **Be a source of communications expertise**, providing ideas for message development, volunteering publicity skills or video production expertise, or making talent referrals.
• **Inspire the creation of a global network of similar organizations**, "harnessing the power of the media for the environment." As a long-term effort, EMA will help develop similar groups in other countries and encourage the distribution of environmental programming worldwide.

■ Resources
In addition to the above-mentioned activities—EMA produces the quarterly *EMA News*; *Things Writers Can Do* poster; *30 Simple Things the Entertainment Community Can Do to Save Energy* book, among other educational materials. EMA also provides research assistance and fact verification to the entertainment industry.

Federation of Western Outdoor Clubs (FWOC)

512 Boylston Avenue, East #106
Seattle, WA 98102 USA

Contact: Hazel Wolf, Editor
Phone: 1-206-322-3041

Issues: Air Quality/Emission Control; Biodiversity/Species Preservation; Deforestation; Energy; Global Warming; Health; Population Planning; Sustainable Development; Transportation; Waste Management/Recycling; Water Quality

● Organization

"As the American frontier closed in the late 19th century and a land ethic emerged founded on enjoyment and preservation of natural beauty rather than economic exploitation, people throughout the West began to gather in local outdoor clubs."

So states the **Federation of Western Outdoor Clubs (FWOC)**, which was organized in 1932 in response to the growing need for the cooperation and coordination of these associations. Currently, 45 clubs make up FWOC.

The Federation's purposes include "the promotion of the proper use, enjoyment, and protection of scenic, wilderness, and outdoor recreation resources in western states; the cohesive organization of member clubs to further these objectives; the education and enlistment of the people and the government to support these objectives; the establishment of communication and the dissemination of information among member clubs; and the enrollment of individuals on a membership basis to strengthen the Federation."

FWOC works to "secure wilderness in state and national public lands, to acquire land for wildlife refuges, to protect wildlife and natural plants, to initiate and support legislation to preserve natural recreation resources (such as the quality of air, water, soils, and shorelands of rivers, lakes and coasts) and to promote wise forest practices on public and private lands."

▲ Projects and People

The Federation has representatives in the northwest USA and Washington, DC, who gather information, alert member clubs to needed action, and testify at national and state hearings.

Each year, delegates from member clubs gather in convention to decide Federation positions on issues of concern. **Resolutions**, submitted by clubs or individuals, provide an "early-warning system of new threats and provide plans for coping with them." Also, **long-range plans** are formulated to preserve the West's natural resources. Officers and clubs use such resolutions to urge legislators, public officials, and private bodies to take action needed to achieve conservation goals.

Hazel Wolf is a volunteer for both FWOC and the Seattle Audubon Society.

■ Resources

FWOC publishes *Outdoor West* twice a year. Topical bulletins appear on special issues.

Friends of the Earth (FOE-USA) ⊕

218 D Street, SE
Washington, DC 20003 USA

Contact: Public Information Coordinator
Phone: 1-202-544-2600 • *Fax:* 1-202-543-4710
E-Mail: Econet: foedc

Friends of the Earth—International ⊕

International Secretariat
Damrak 26
1012 LJ Amsterdam NET, Netherlands

Phone: (31) 20 221 366 • *Fax:* (31) 20 275 287
E-Mail: Econet: gn: foeintsecr.

Activities: Education; Political/Legislative; Research • *Issues:* Biodiversity/Species Preservation; Deforestation; Energy; Global Warming; Water Quality • *Meetings:* IMF; ITTO; LDC; Montreal Protocol; UNCED; World Bank

● Organization

Originally founded in 1969, **Friends of the Earth (FOE)** is dedicated to "protecting the planet from environmental disaster; preserving biological, cultural, and ethnic diversity; and empowering citizens to have a voice in decisions affecting their environment and lives." In 1990, FOE merged with two existing organizations, the **Environmental Policy Institute** and the **Oceanic Society**, expanding its network and resources. Currently the organization has 47 affiliates in over 40 countries; the U.S.-based group has a staff of 40 in its Washington, DC, headquarters; 4 persons in Seattle, Washington; and 1 staffer in Manila, Philippines. There are 50,000 U.S. members as well as numerous volunteers.

The International Secretariat in Amsterdam links FOE affiliates with a newsletter, electronic mail, executive committee meetings, and an annual gathering of representatives from each country. In addition to the special concerns of each FOE group, global campaigns are coordinated, such as "protecting the ozone layer"—a collaborative effort of FOE-USA and FOE-UK. *(See also Friends of the Earth [FOE] in country listings throughout this Guide.)*

FOE works with grassroots groups, lobbies Congress, holds workshops, and informs the public. Recently the group has focused on three key issues: "toxics and the involuntary poisoning of our populace; poverty, inequality and war; and controlling the development and use of new technologies." Many projects involve close cooperation with other environmental organizations, both national and international.

▲ Projects and People

A major project of FOE involves **ozone protection**. Under the direction of **Liz Cook**, FOE has worked with shareholders in Du Pont, which it considers to be one of largest producers of chlorofluorocarbons (CFCs) and halon—both damaging to the ozone layer—to force a new company policy, and has been in federal court challenging Du Pont's "attempt to stifle debate" as FOE reported at its 1991 annual meeting. FOE is concerned that the U.S. government is allowing the ozone problem to worsen by "delaying the elimination of ozone-causing chemicals" until beyond the year 2000.

It reports in the meantime "Germany and Sweden announced plans to phase out CFCs in 1995 and the European Community set 1997 as its target date." FOE-USA and FOE-UK continue to work together to strengthen the Montreal Protocol by pressuring countries to ban CFCs and other ozone-damaging products.

FOE's related **Tax Project** works toward abolishing tax incentives for destructive activities, replacing them with incentives for pro-environment efforts or, for example, a tax on ozone-depleting chemicals such as CFCs. Under FOE's related **Corporate Accountability Project** in 1990, current FOE acting president **Brent Blackwelder** with **Jack Doyle** played a "key role in developing shareholder resolutions aimed at forcing Exxon to adopt more environmentally responsible corporate policies and practices." Involved with the Coalition for Environmentally Responsible Economies (CERES), the organization supports the Valdez Principles—a code of conduct—noting that "in most cases, corporate policies that are good for the environment are good for the economy too." *(See also Coalition for Environmentally Responsible Economies in this section.)*

Another major interest of FOE is the waters of the world. With FOE–Netherlands, the U.S. group conducts an **Oceans Project**. In addition, Oceanic Society project director **Clifton E. Curtis**—on the scene soon after the Exxon *Valdez* incident—is working on a project in Chile with Greenpeace and the United Nations Economic Commission for Latin America to develop a **global program for confronting land-based marine pollution**. Various marine biodiversity projects confront seabed mining and seabed burial of wastes.

Since 1989, **Boyce Thorne-Miller**, FOE senior marine scientist has been working with an interdisciplinary group of scientists, economists, lawyers, and marine policy specialists convened by Woods Hole Oceanographic Institute's Marine Policy Center to solve various marine problems; Miller also works with the International Society for Reef Scientists to help protect coral reefs. She helped the U.S. Treasury Department develop guidelines to encourage the World Bank and other multilateral development institutions to "fund only projects which are protective of marine environments and the indigenous peoples which rely upon them."

Led by **Ann Shaughnessy**, FOE sponsored a summer "**Take Back the Coast!**" campaign, working together with Clean Water Action, 20 coastal organizations, and 30 grassroots groups along the Atlantic and Gulf coasts. The aim was to hold officials accountable for protecting beaches and coasts, particularly as more than half of all the people in the USA live and work in coastal counties.

Jennifer McAdoo and **Velma Smith** have both worked on FOE's **Groundwater Project**, crucial because of FOE's reported fact that Americans pump an alarming rate of "nearly 100 billion gallons of groundwater every day." Today, one-third of the USA's largest cities depend on groundwater for drinking water; farmers depend on this source for irrigation; so do ranchers for their livestock. In one recent effort, FOE, the Natural Resources Defense Council (NRDC), and the Sierra Club sued to force "release of long-overdue criteria to improve environmental protection at landfill sites."

At its annual **National Citizens Conference on Groundwater**, FOE provides a forum and information exchange for concerned citizens across the country. In 1990, FOE began working with citizens in North Carolina to fight the U.S. Environmental Protection Agency (EPA) over a proposed hazardous waste treatment plant on the Lumber River; the EPA preferred less lax federal restrictions over tighter state restrictions. FOE won the first round, but the court battle continues.

Through the Environmental Policy Institute, FOE has a significant history in battling strip mining and related abuses, now with the help of **Donald Barger**, formerly with Save Our Cumberland Mountains (SOCM) grassroots group, and **Jim Lyon**, coalfields project director. FOE's **Citizens' Mining Project** wants Congress "to reauthorize and expand the Abandoned Mine Land Fund" which expires in 1995, and is pushing for reform in at least nine states that would limit land degradation and pollution. For example, in West Virginia, FOE "went to court to force the state . . . to enforce laws aimed at controlling strip mining abuses"; the state eventually settled and agreed to more stringent controls.

On a wider level, FOE is working so that U.S. taxpayers' dollars, rather than being spent on nuclear weapons buildup, are spent cleaning up federal weapons facilities. It co-founded the **Plutonium Challenge Project** to clean up U.S. Department of Energy weapons sites. FOE's **David Albright**, as part of a gubernatorial-appointed health panel, began assessing toxicology and radiology at Rocky Flats nuclear weapons plant, Colorado, and their effects on the nearby population. FOE's **Appropriations Project**, launched in 1990 and directed by **Ralph De Gennaro** with the aid of policy analyst **Tom Zamora**, presses for more federal funding for environmental priorities, and has led to increased EPA and Defense Department cleanup project appropriations.

The **Toxic Chemicals Project**, directed by **Dr. Fred Millar**, provides advice to grassroots organizations like Texans United in communities threatened by explosions such as the 1989 disaster at Phillips Petroleum near Houston that killed 23 workers. The project also provides support for Local Emergency Planning Committees (LEPCs) established under federal law; builds coalitions with trade unions; and educates Congress and regulatory agencies.

For 10 years, FOE has been working with biotech analyst **Jack Doyle** to monitor **biotechnology and genetic engineering** and its impacts on public health, farming, and consumer products. FOE supported a congressional call for independent investigation of the Food and Drug Administration (FDA) review of bovine growth hormone (BGH), opposed federal spending for research on crops genetically tolerant to herbicides, and served as founder and principal advocate for the 1988 formation of the **Biotechnology Working Group**, a committee of 17 public-interest groups working on the issue. FOE, which supports sustainable agriculture, is concerned that biotech will "extend the pesticide era . . , rather than end it."

The organization's **Environmental Justice Program** was launched "to broaden and strengthen involvement with communities of color struggling with their own environmental concerns, and to involve hitherto excluded people within the national environmental movement through programs offering more access to information and leadership positions." Many of FOE's programs involve Native American communities and other minority groups.

FOE's **Northwest Office** opened in 1970 and is currently headed by **David Ortman**, focusing on regional issues such as deforestation, fisheries conservation, and the oil industry. The **Elwha Project**, under FOE's **Jim Baker**, is trying to force the removal of two large hydroelectric dams on the Elwha River, which have blocked salmon spawning for more than 60 years. The National Park Service, U.S. Fish and Wildlife Service, and National Marine Fisheries Service have joined FOE and others in the struggle, which will be decided either through negotiations or by litigation.

An international undertaking with which FOE-USA is involved is the **Central America Project**, led by **Margarita Suarez**, who is working to set up a biosphere reserve along the Costa Rican-

Nicaraguan border. This is a joint effort with the International System of Protected Areas for Peace (SI-A-PAZ), meaning "Yes-to-Peace"; FOE is also working on a regional pact banning all toxic waste import schemes. A major focus of FOE is international lending; the organization "successfully advocated the establishment of an environmental impact assessment process at the World Bank and environmental directives at the International Monetary Fund (IMF)." FOE monitors the activities of, among others, the World Bank, IMF, Asian Development Bank, International Tropical Timber Organization (ITTO), U.S. Agency for International Development (USAID), the Bank for European Reconstruction and Development (BERD), and the UN Tropical Forestry Action Plan (TFAP). Recently, FOE increased "international pressure on [TFAP] to reform—and to include indigenous peoples in planning and administering its new programs."

FOE sponsors a **Grants Program** for non-government organizations (NGOs) in developing nations for specific projects. Through its Appropriations Project, it worked to bar foreign aid funds from being used for development projects that destroy tropical rainforests in Panama and Nicaragua.

When **Paul McCartney** proposed a world concert tour to benefit Friends of the Earth, the results of his performances helped increase the annual budget, brought in new members, and enhanced the organization's public image. In gratitude, FOE gave McCartney its "Voice for the Planet" award on Earth Day 1990.

■ Resources

FOE offers student internships, many of them in Washington, DC, where a General Store with gift items is also located. The organization produces numerous publications including recent books and booklets on ozone depletion, pro-environment international aid projects, bottled water, and mining, such as the *Strip Mining Handbook*. Periodicals include *Friends of the Earth* monthly magazine, *Atmosphere* quarterly on climate change, the periodic newsletter *Groundwater News*, and *Community Plume* on chemical safety. The international organization publishes the newsletter *FOELink*. Interested volunteers can contact the DC office.

The Fund for Animals, Inc. ⊕
850 Sligo Avenue
Silver Spring, MD 20910 USA

Contact: **Wayne Pacelle**, National Director
Phone: 1-301-585-2591 • *Fax:* 1-301-585-2595

Activities: Education; Political/Legislative; Research • *Issues:* Biodiversity/Species Preservation; Deforestation

● Organization

The Fund for Animals, with more than 200,000 members, was founded by noted author **Cleveland Amory** to "promote the rights of all animals, wild and domestic, through education, litigation, and confrontation." On the legislative front, the Fund worked arduously for the passage and implementation of the U.S. Endangered Species Act and the Marine Mammal Protection Act.

It frequently takes to task federal and state governmental organizations mandated to protect wildlife, such as the U.S. Fish and

Wildlife Service, National Park Service, and the Florida Game and Fresh Water Fish Commission.

Three of its facilities are a rabbit sanctuary, wildlife rehabilitation center, and the **Black Beauty Ranch** in Texas, which rescues llamas, coyotes, elephants, chimpanzees, burros, and other animals that have abused and traumatized.

Wayne Pacelle, Fund for Animals national director, brings his animal-rights and antihunting messages to radio and television talk shows and on evening network news programs. He is often quoted in the print media, such as the *New York Times*, the *Los Angles Times*, *Time* magazine, and *U.S. News and World Report*. Particularly critical of pro-sport hunting groups—as the Fund for Animals describes Ducks Unlimited and Project Wild of the National Rifle Association—Pacelle asks the public to "put a stop to youth hunting and end this mayhem in the woods."

▲ Projects and People

The Fund for Animals protests against **laboratory animal testing**, wearing of fur coats, sport hunting, puppy mills, treatment of calves in rodeos, and circuses. Amory describes circuses as "a relic of a barbaric age, an anachronism . . . [not] the message children should have about saving animals in the wild."

A cause of great commitment is the **spaying and neutering of household pets**. The Fund for Animals backs legislation in local jurisdictions which would impose fines on cat and dog owners who do not spay or neuter their pets. To reduce the number of stray animals and those destroyed in shelters, mandatory sterilization was being considered in Montgomery County, Maryland; San Mateo County, California; and Kings County, Washington. In Mexico, the Fund started a program to help cut down the overpopulation of dogs and cats through the use of low-cost spaying and an estrus cycle inhibitor, Covinan.

The Fund helped get **Yellowstone National Park's buffalo** a reprieve from being slaughtered by hunters when wandering onto private lands in search of food. With pressure from the Fund and mounting concern about Montana's image and tourism industry, state legislators took action to prevent the killing. Also in Montana, the Fund for Animals helped bring a successful lawsuit to stop the 1991 fall season's **grizzly bear hunt**, which it said violated the federal Endangered Species Act.

"Grizzly bears are North America's slowest reproducing mammals, with many females often not becoming reproductively capable until seven or eight years old and then producing offspring every third year," reports the Fund. Moreover, "development, road building, timbering, and poaching are already taking a major toll on grizzly bears." The Fund is working to ensure the strengthening of the Endangered Species Act as it faces reauthorization in 1992.

The Fund is also urging supporters to lobby Florida's congressional delegation regarding the failure of the U.S. Fish and Wildlife Service to place the "imperiled" **Florida black bear** subspecies on the federal threatened species list. The fate of **Sherman's fox squirrel**, as a target for sport hunting and research shooting in Florida, is also a matter of concern. A declining longleaf pine forest habitat already threatens its existence.

Mountain goats of Olympic National Park, Washington, were the subject of another recent Funds for Animals campaign. According to the Fund, the National Park Service was considering hunting them to prevent the goats' damage to native alpine plant species.

For more than two decades, **Caroline Gilbert** has been salvaging abused rabbits on her 30-acre farm that is now the Fund's rabbit sanctuary. The rabbits arrive from laboratories, medical school research projects, animal shelters, and breeders cages. "The rabbits are allowed to do what rabbits do naturally, or almost—bucks are neutered," she told the *Charlotte Observer*.

Chuck and Cindy Traisi run the Fund's Wildlife Rehabilitation Center in Ramona, California. According to the *Ramona Sentinel* the Center cares for injured or orphaned animals from bobcats, opossums, and mule deer to hawks, eagles, and sparrows. In one year the rehabilitation center treated and released 582 animals to the wild, while others died or had to be euthanized. "Many of those deaths could have been prevented," write the Traisis. "Well-meaning people who find an injured animal will often try to treat it themselves—with disastrous results."

■ Resources

The Fund for Animals publishes free *Fund Facts* on topics as hunting, animal agriculture, and companion animals; *Action Alerts*; newsletter; and brochures such as the *Armchair Activist*; and reprints relevant articles such as "Should Hunting be Banned?" from *U.S. News and World Report. Jaws of Steel*, an antitrapping book by Thomas Eveland, can be ordered.

Global ReLeaf ❦
American Forestry Association
P.O. Box 2000
Washington, DC 20013 USA

Contact: **Deborah Gangloff**, Vice President for Programs
Phone: 1-202-667-3300 • *Fax:* 1-202-667-7751

Activities: Education; Development; Political/Legislative; Research • *Issues:* Deforestation; Global Warming; Sustainable Development • *Meetings:* World Forestry Conference

● Organization

Started in 1988, **Global ReLeaf** is a relatively new program of the 115-year-old **American Forestry Association** with the goal of reforestation and a call to action against the threats of climate change, global warming, and the greenhouse effect. According to Global ReLeaf, 50 million acres of tropical forests are being destroyed yearly—or "one-half million to one million trees per hour"; and the USA, with 5 percent of the world's population, uses about 22 percent of the Earth's fossil fuels that cause carbon dioxide.

With more than 112,000 members and the slogan "plant a tree, cool the globe," it seeks to help the global environment one tree at a time. In addition to the Washington, DC, office, there are coordinators in each state who are involved in organizing and educating. The Global ReLeaf umbrella network, **Global ReLeaf Coalition**, was begun in late 1990.

▲ Projects and People

Global ReLeaf has identified six areas for citizen action, and is encouraging individuals to help reach specific goals: **Urban Forests, Rural Forests, Tropical Forests, Environmental Research, Air Pollution,** and **Water Quantity** and **Quality**. Within each category, Global ReLeaf seeks to educate individuals as to problems, goals to increase the number of trees, and benefits that can result from planting trees.

For example, as urban trees die off or are removed more quickly than they are replaced, Global ReLeaf asks communities to plant twice as many trees as are lost during the next decade. Trees in cities reduce the "heat island" effect of air pollution, cut ozone and smog pollution, act as windbreaks, and can reduce air-conditioning costs and fuel consumption.

It encourages federal agencies such as the U.S. Department of Agriculture, Forest Service, and Bureau of Land Management to monitor and manage forest ecosystems more effectively. Regarding tropical forests, it looks to the strengthening of the Forest Service's Office of International Forestry and Institute of Tropical Forestry as well as intensified university research on sustainable uses.

Members are encouraged to plant and maintain healthy trees, survey existing parks and forests, start community tree-planting programs here and abroad, involve corporations and businesses, and write letters to legislators and key federal agencies, the World Bank, and other international organizations. McDonald's restaurants, for example, provided seedlings to school children. Aveda, Quintessence, and Ralston-Purina carry Global ReLeaf's messages on their products' packaging. Overseas, it inspired Hungary's Independent Ecological Center to begin tree-planting projects.

■ Resources

Global ReLeaf is in need of *pro bono* public relations assistance. Publications include several books, brochures, and the *Global ReLeaf Action Guide*. A merchandise list is available from the organization.

Global Tomorrow Coalition (GTC) ⊕ ❦
1325 G Street, NW, Suite 915
Washington DC 20005-3104 USA

Contact: **Jim Shepard**, Director, Management Services
Phone: 1-202-628-4016 • *Fax:* 1-202-628-4018

Activities: Education; Development; Research • *Issues:* Air Quality/Emission Control; Biodiversity/Species Preservation; Energy; Global Warming; Population Planning; Sustainable Development; Waste Management/Recycling; Water Quality • *Meetings:* CITES; Globescope Americas

● Organization

Founded in 1981, the **Global Tomorrow Coalition (GTC)** is composed of more than 100 U.S. organizations with more than 10 million Americans committed to a broad goal of sustainable development while confronting population growth, resource consumption, and environmental degradation. As a coalition, GTC is able to increase the effectiveness of its members, provide support, and expand activities on both national and international levels.

▲ Projects and People

Through the **Globescope Process** of information sharing and networking, GTC hopes to increase public awareness of environmental issues and stimulate patterns of sustainable development. It also encourages involvement of individuals and NGOs, and emphasizes U.S. responsibility to help find solutions to global problems.

Coalition Building is another program of GTC designed to increase membership and the range of organizations and individuals involved. Nationally, GTC wants to increase the effectiveness of membership organizations; internationally, it seeks to ensure that U.S. NGOs are represented and helping build international coalitions.

GTC's education activities are divided into **sustainable development education** and **formal educational services**. The formal educational services are designed to aid elementary and secondary teachers in teaching about global issues. Sustainable development education entails outreach through cooperative alliance forums, global town meetings, global issues resource centers, traveling exhibits, and a *Citizen's Guide* for community leaders.

Campaign 2000 seeks to make sustainable development the basis for every decision by the year 2000. This campaign includes leadership training, outreach, and the development of sustainability data at city and state levels.

GIANT, the **Global Issues Action Network Team,** is a separate but related nonprofit organization to "influence national legislation and policymaking on global issues."

The *Global Ecology Handbook* is prepared as a "stimulus to action" with overviews on topics ranging from biological diversity, tropical forests, and global security to hazardous substances, solid waste management, air quality, and climate change.

The GTC Board of Directors is a reflection of the wide and significant participation: **Dr. Lester Brown,** president, Worldwatch Institute; **Jim Fowler,** president, Fowler Center for Wildlife Education; **Kathryn S. Fuller,** president, World Wildlife Fund (WWF) and The Conservation Foundation; **Vera Gathright,** International Fund for Agricultural Development; **Dr. Jay D. Hair,** president, National Wildlife Federation (NWF); **Ladonna Harris,** president, Americans for Indian Opportunity Association; **Jan Hartke,** vice president, Humane Society of the United States; **John A. Hoyt,** president, Humane Society of the United States; **Dr. Bruce W. Karth,** vice president, E. I. Du Pont de Nemours & Company, Inc.; **Joan Martin-Brown,** United Nations Environment Programme; **Dr. Thomas W. Merrick,** president, Population Reference Bureau; **Dr. William J. Nagle; Dr. Russell W. Peterson,** president emeritus, National Audubon Society; **Paul C. Pritchard,** National Parks and Conservation Association; **Elizabeth Raisbeck,** senior vice president, National Audubon Society; **Dr. J. Joseph Speidel,** president, Population Crisis Committee; **Gus Speth,** president, World Resources Institute; **Charles Sykes,** vice president, CARE; **Dr. Konrad von Moltke,** Dartmouth University; and **Susan Weber,** executive director, Zero Population Growth.

Donald R. Lesh is GTC president; **Diane G. Lowrie** is vice president.

■ Resources

GTC has a number of publications including *The Global Ecology Handbook: What You Can Do About the Environmental Crisis; Global Issues Education Set; Sustainable Development: A New Path for Progress; Sustainable Development Bibliography,; Sustainable Development Issues in Education: A Status Report,* and a series of videotapes entitled *Race to Save the Planet.* In addition, GTC publishes a newsletter, *Interaction.*

Grand Isle Heritage Zoo ⊕
P.O. Box 1345
Grand Island, NE 68802 USA

Contact: **Gail E. Foreman, Ph.D.,** Zoo Director
Phone: 1-308-381-5416

Activities: Education; Research • *Issues:* Biodiversity/Species Preservation; Deforestation; Endangered Species • *Meetings:* International Snow Leopard Conference

▲ Projects and People

Pursuing the goal of conservation through education, preservation, and research, the **Grand Isle Heritage Zoo** is active in such environmental concerns. The zoo's **Conservation and Environmental Resource Center** serves as a library for books, magazines, journals, scientific papers, and audio/visual materials. These are available to both the community and the zoo staff.

The preservation of animals that are under threat in their native environment is a crucial aim. The **preservation program** of the Grand Isle Zoo focuses on the endangered animals of Nebraska, with a secondary focus on the animals of northern Eurasia. Local animals being considered for inclusion are the swift fox, beaver, mink, badger, cougar, bobcat, black bear, sandhill crane, whooping crane, black-footed ferret, bald eagle, golden eagle, goshawk, merlin. Reptiles, amphibians, and fish may also be included.

Animals from northern Europe and Asia that would adapt well to the cold Nebraska winters are being considered for the secondary focus. These are red panda, Pallas' cat, Scottish wildcat, Amur leopard cat, snow leopard, argyle, urial, Chinese water deer, musk deer, saiga antelope, Himalayan tahr, Mongolian wild horse, Chinese wolf, and pika, among others. With an active captive breeding program anticipated for both endangered and threatened animals, the zoo looks forward to participation in American Association of Zoological Parks and Aquariums (AAZPA) species survival plans.

Dr. Gail Foreman, zoo director, is closely involved in the third major aspect of the zoo's activities, and is also the Research Librarian for the Species Survival Commission (IUCN/SSC) Cat Group. IUCN, the World Conservation Union, is an international group that was founded by the United Nations and UNESCO to bring together governments, conservationists, and scientists. The Cat Group consists of 95 cat specialists throughout the USA and 28 other countries.

Foreman is co-authoring the first **IUCN Cat Action Recovery Plan,** which "will be the primary blueprint for saving the world's endangered cat species." The completed action plan will be distributed to governments, conservation organizations, and scientists. It will include a species-by-species report on the status of the world's cats, which will include information about threats, habitat, legal status, and conservation actions affecting them. It will also propose specific projects for cat conservation and an assessment of the priorities of these actions. In addition the group will create a combined database and publication. The database will link up with the Protected Areas Development Unit, World Conservation Monitoring Centre, England.

This plan is considered crucial; many small cats do not breed well in captivity, infant mortality is high, and there is also little research available. This action plan will establish needed conservation policies toward threatened cats.

■ Resources

Future research publications and conferences are planned as well as a volunteer program to aid in behavioral research.

Great Lakes United, Inc. (GLU) ⊕

State University College at Buffalo
Cassety Hall
1300 Elmwood Avenue
Buffalo, NY 14222 USA

Contact: **Terry Yonker**, Executive Director
   Phont: 1-716-886-0142 • *Fax:* 1-716-886-0303

Activities: Education; Political/Legislative; Research • *Issues:* Air Quality/Emission Control; Biodiversity/Species Preservation; Health; Waste Management/Recycling; Water Quality • *Meetings:* International Joint Commission

● Organization

Great Lakes United (GLU) is a binational coalition committed since 1982 to the protection of the Great Lakes and St. Lawrence River ecosystem. By coordinating the activities of more than 180 interest groups and numerous individuals, GLU is working to restore the Great Lakes Basin, where only 30 percent of the original wetlands remain.

Through "education, coordination, and action" GLU is having an effect. However, there are enormous challenges. According to GLU, some "90 percent of some toxic substances, such as PCBs, that enter the Lakes come from air pollutants." Toxic chemicals appear to be causing birth defects among cormorants; many of the 160 inactive landfill sites contain toxic chemicals and are nearby the Niagara River; and some "150 million tons of hazardous commodities" are transported on the Great Lakes each year—in a region where 25 percent of Canadian agriculture takes place and where the sport fishery industry is valued at $4 billion.

According to executive director Philip Weller, forming international environmental coalitions will result in a "cross fertilization" of ideas and activities. The benefits of international organizations and the need for international cooperation on environmental issues are the reasons for the GLU.

▲ Projects and People

GLU has a new policy for the Great Lakes: Zero Discharge or pollution prevention by eliminating the discharge of waste into the Basin. GLU is promoting programs that will help prevent the generation of toxic chemicals and developing educational materials that will teach individuals to eliminate toxic chemical discharge.

For regions of the Great Lakes Basin that already have been designated "areas of concern," GLU helps citizens with **Remedial Action Plans** (RAPs). Working with the RAPs on organization and cleanup efforts and aiding the RAP network are ways this organization helps citizens contribute to the Great Lakes ecosystem.

GLU also has a **Human Health Task Force** to investigate some of the human ramifications of the pollution in the Great Lakes region. GLU has held workshops on this area, and will be publishing a citizen's guide to environmental health standards.

Other issues of concern include **air quality, contaminated sediments, water levels and flows, fish and wildlife habitat, and shore-line protection.** GLU involves itself in a wide variety of issues all designed with the goal of conservation of the Great Lakes ecosystem.

■ Resources

GLU needs access to new pollution prevention techniques. Also needed are more national media coverage and increased political support to make the Great Lakes a national issue.

Publications include *Citizens' Guides* on how individuals can work on an effective cleanup plan for their contaminated area. These contain both general and site-specific advice. In addition, GLU publishes a range of reports on the Great Lakes region, and on political action that has been taken. A newsletter is published quarterly.

Green Cross Certification Company (GCCC)

1611 Telegraph Avenue, Suite 1111
Oakland, CA 94612-2113 USA

Contact: **Linda Brown**, Vice President of
         Communications
   Phone: 1-510-832-1415 • *Fax:* 1-510-832-0359

Activities: Education; Research • *Issues:* Environmental Claims; Evaluation; Food Testing and Safety

● Organization

The **Green Cross Certification Company** (GCCC), founded in 1989, is a "neutral scientific certification organization established to independently verify manufacturer claims of outstanding environmental achievement." It is described as "the first national certifying organization of its kind in the United States" and has certified specific environmental claims for more than 400 consumer products. Its focus is the education of consumers on environmental issues pertaining to marketplace claims, and of manufacturers and retailers in the development of sound environmental practices.

GCCC is a not-for-profit division of **Scientific Certification Systems** (SCS). SCS has specialized in the independent scientific evaluation of exceptional product claims since 1984.

NutriClean—the agricultural certification arm of SCS founded in 1984 as the nation's first independent, neutral certification system for fresh fruits and vegetables—applied for nonprofit status in December 1991 to highlight its joint "public mission" with Green Cross. Through its tough program of evaluation and verification on behalf of "clean food," pesticide use is being reduced on major commercial farms in the USA, and so far "more than 200,000 acres of farmland have been certified." NutriClean says its standards are up to 1,000 times stricter than government standards regarding pesticide residues and up to 100 times greater for current organic standards.

Because it has "no brokerage or ownership interest in the products it certifies and maintains complete financial independence from any special-interest organizations," GCCC says it is able to provide "neutral, third-party evaluation and certification service free from any conflict of interest."

According to GCCC, all certification work is performed in accordance with "strict scientific standards by a professional team of experienced scientists and engineers." Such records undergo a quarterly review. In addition, Green Cross provides full confidentiality assurances to participating manufacturers in the handling of any sensitive or proprietary information.

Within its **Environmental Literacy Campaign**, Green Cross addresses the growing confusion over environmental advertising claims in the marketplace. Some claims represent significant environmental achievement, but these can be lost amid the growing number of trivial and misleading claims. Green Cross provides a "unified, independent system for evaluating and identifying significant environmental claims across the full spectrum of consumer product categories."

▲ Projects and People

Green Cross has teamed with several other companies in verifying products in the marketplace. For example, a program to **identify and evaluate environmental marketing claims on labels of building materials**, home improvement products, and related packaging has been set up jointly by The Home Depot, "world's largest home center retailer," and Green Cross. The program requires vendors to submit for review any direct or implied environmental marketing claims such as "biodegradable," "environmentally safe," or "recycled content" for products and/or packages.

The Knoll Group, a leading designer and manufacturer of office furnishings, and Green Cross joined efforts to **identify timber products produced under stringent, carefully monitored sustainable forestry management methods**.

The Steel Can Recycling Institute recently launched with Green Cross a study to determine "the **optimum level of recycled steel content in cans** from an overall environmental perspective." Meanwhile, the tin mill products of six domestic integrated steel producers were certified to contain "a minimum of 10 percent postconsumer recycled content."

Green Cross also has designed the first comprehensive eco-labeling certification program based on life cycle or "cradle-to-grave" analysis. "Life cycle assessment is the only known methodology capable of accurately reflecting the complexities of modern industrial processes," explains **Dr. Stanley Rhodes**, GCCC president and chief executive officer. "Any eco-label which suggests to consumers that full environmental impact analysis has been conducted must, at a minimum, involve such an assessment."

A "life cycle" assessment involves an examination of the resources used, energy consumed, wastes produced, and emissions released as a result of the manufacture, distribution, use, and disposal of a product. Approved products receive the **Green Cross Environmental Seal of Achievement**. When there is an environmental trade-off, such as increasing the amount of energy required to boost the recycled content of a steel can in production, the Environmental Seal could be withheld. An **Environmental Report Card** features detailed information about the life cycle of a given consumer product, such as bathroom tissue, that is 100 percent recycled unbleached versus typical virgin-fiber bleached.

■ Resources

Green Cross has available brochures, background material, and a certification summary of its services.

The Green Guerillas 🌿
625 Broadway
New York, NY 10012 USA

Contact: **Barbara Earnest**, Executive Director
Phone: 1-212-674-8124

Activities: Education • *Issues:* Composting/Recycling; Gardening • *Meetings:* American Community Gardening Association

● Organization

The Green Guerillas is a nonprofit group with a 5-person staff, more than 250 volunteers, and 800 members. Determined to increase green spaces in New York City, the Guerillas are involved in community gardening.

The Guerillas provide technical support and assistance to community gardens. Volunteers assist a wide variety of organizations ranging from community groups to shelters for the homeless and schools. Each year volunteers devote more than 6,500 hours of their time to the goal of greening New York City.

▲ Projects and People

The first and longstanding project of the Guerillas is the **Liz Christy Bowery-Houston Community Garden** in the Bowery—containing vegetable, herb, and flower plantings. The Guerillas take the knowledge that they gained in creating this garden begun in 1973—on land that once belonged to Peter Stuyvesant's farm—and help other interested communities create gardens.

One important aspect of the Green Guerillas' philosophy is that they only create and assist with gardens when they have been approached by the community. They feel that with community support a garden can be extremely successful and be a major asset to a neighborhood. Without community support a new garden is likely to revert to its former state as soon as the Guerillas cease to tend it.

The Liz Christy garden is also the site of workshops and the distribution center for the plant material that the Guerillas collect. Donations of plant materials come from both individuals and businesses in the New York area. **Horticulture training** takes place at a men's shelter; a roof garden at an AIDS patients' residence and bushes at the Statue of Liberty are other examples of the Guerillas' work.

As executive director **Barbara Ernest** told *Town & Country* (June 1990), "A garden can turn a neighborhood around, give everyone a sense of enpowerment, of possibilities, of problems conquered. Most of all, a garden gives hope."

■ Resources

The Green Guerillas need volunteers for technical assistance in gardening.

Publications include a newsletter and a variety of fact sheets.

Green Library ⊕ ❧
2161 Shattuck Avenue, Suite 233
Berkeley, CA 94704 USA

Contact: **Jacek Purat,** Executive Director
  Phone: 1-510-841-9975 • *Fax:* 1-510-841-9996
E-Mail: MCI: 494-3789

Activities: Development; Education • *Issues:* Environmental Library Establishment • *Meetings:* American Library Association

● Organization
In 1986, **Green Library**—a nonprofit foundation—embarked on a mission of starting and equipping environmental science libraries in countries beset by ecological crises and lacking in related publications. "In providing books to the people of these ecologically at-risk nations, we are building public commitment to reverse such severe problems as the greenhouse effect, deforestation, and air, water and soil pollution," writes **Jacek Purat,** executive director and president.

▲ Projects and People
Green Library sends books to Eastern Europe, Southeast Asia, and Latin America. Publications are distributed to information centers and newly formed or existing libraries that are maintained by local public nonprofit organizations and are often located near universities.

   Bookraising campaigns are ongoing in Arizona, California, Idaho, and Washington, DC, as well as in Holland, Poland, and Sweden to gather books and journals on environmental sciences and related areas in agriculture, business, energy, law, and medicine. Since 1987, more than 160,000 volumes have been shipped to Latvia, Nepal, New Guinea, Peru, and Poland, among other countries.

   Those receiving these publications are part of an expanding **Green Library Network** which makes materials available to "everyone on an equal basis without censorship or pre-selection." Sponsoring organizations are requested to "catalog materials, translate major titles into the local language," and let the public know of their availability. Efforts are made to preserve and house the collections suitably; local fundraising drives are usually launched to support the building, furnishings, equipment, and "long-term financial stability."

   Green Library seeks independent grassroots groups to help establish similar facilities where books will reach "the most energetic social forces." Such groups must raise about 50 percent of their support in matching funds from local sources in their countries.

   In 1989, **a model library** was opened in Poznan, **Poland,** as a result of Green Library's joint efforts with the Polish Ecological Club in establishing **Fundacja Biblioteka Ekologiczna**—Poland's first independent ecological foundation. This Ecological Library is a collaborative effort with the University of Poznan's Life Sciences Department. A Polish language environmental science catalog of subject headings is being created; environmental translations and book distribution are also occurring.

   In an innovative step, Green Library and Biblioteka Ekologiczna are assisting the Environmental Commission of the new Sejm, or Polish Parliament, with **alternative energy development programs,** including plans for energy waste reduction and cost assessments of environmental protection.

   In Riga, **Latvia,** Green Library worked with VAK, the Latvian Environmental Protection Club, to establish **Zelo Bibloteka,** an Ecological Library providing "access to environmental literature for Latvia, Lithuania, and Estonia."

■ Resources
Published three times yearly, the *Green Library Journal—Environmental Topics in the Information World* was launched in 1992. With **Dr. Maria A. Jankowska** as editor, the *Journal* appeals to librarians, environmentalists, booksellers, publishers, regional planners, students, and researchers.

   Funds and books are needed for the Green Library Network.
   (See also Funacja Biblioteka Ekologiczna/Ecological Library Foundation in the Poland NGO section.)

Green Seal
1250 23rd Street, NW, Suite 275
Washington, DC 20037 USA

Contact: **Norman Dean,** President
  Phone: 1-202-331-7337 • *Fax:* 1-202-331-7533

Activities: Education; Research • *Issues:* Sustainable Development; Waste Management/Recycling • *Meetings:* ISO

● Organization
Green Seal is a national independent nonprofit environmental labeling and education organization that issues environmental seals of approval to consumer products meeting strict environmental standards. As described in the *New York Times* (July 14, 1991), Green Seal "intends to evaluate the environmental impact of products through their entire life cycles from the gathering of raw materials to disposal. For a fee, products that pass muster will be eligible for the Green Seal, which the company hopes will prompt manufacturers to protect the environment by giving their products a marketing advantage."

   National consumer and environmental groups are supporting the program, and Underwriters Laboratories (UL) is responsible for the "majority of Green Seal's product testing."

   Green Seal expects to give consumers "clear, expert advice about which products are less harmful to the environment." Green Seal states that it has no financial interest in the success or failure of any consumer product.

   When a manufacturer applies to Green Seal to have its product tested, it pays a fee to cover costs to both Green Seal and UL for time and materials. Fees vary for the specific product category and the manufacturing plant.

   Standards are reviewed at least every three years. And manufacturers of approved products pay a fee for continued monitoring. Subsequently, if the product fails to meet the criteria, the manufacturer is informed of the reasons and encouraged to apply again. All test results are kept confidential.

   With Denis Hayes as chairman, the Board of Directors also includes John Adams, Natural Resources Defense Council (NRDC); Joan Bavaria, Franklin Research and Development; Lester Brown, Worldwatch; Rev. Dr. Benjamin F. Chavis, Jr., Commission for Racial Justice of the United Church of Christ; Joan B.

Claybrook, Public Citizen; **David B. Crocker**, Shearson Lehman Bros.; **Norman L. Dean**, Green Seal; **Lynne Edgerton**; **Gretchen Long Glickman**; **Carol Goldberg**, Avcar Group, Ltd.; **Henry Hampton**, Blackside, Inc.; **Randall Hayes**, Rainforest Action Network; **Hubert H. Humphrey III**, Minnesota Attorney General; **Alan F. Kay**, Americans Talk Issues Foundation; **Frederick D. Krupp**, Environmental Defense Fund (EDF); **Dr. Joseph E. Lowery**, Southern Christian Leadership Conference; **Ira Magaziner**, SJS, Inc.; **Alice Tepper Marlin**, Council on Economic Priorities (CEP); **Michael McCloskey**, Sierra Club; **Thomas J. Peters**, The Tom Peters Group; **Douglas Phelps**, U.S. Public Interest Research Group (PIRG); **Rena M. Shulsky**, Galaxy Broadcasting; and **Stanley S. Welthorn**, Reid & Priest. (Organizations are listed for identification purposes only, and none is a product manufacturer.)

▲ Projects and People

Green Seal's premise is that no product is completely environmentally friendly; however, the objective is to **identify products that are less harmful than others in their category.** Environmental standards are set on a category-by-category basis. Once a category is selected, an **Environmental Impact Evaluation** identifies the characteristics of the product and the points in the manufacturing process that could cause significant harm.

"Based on this information, an **environmental standard** is developed that focuses on those factors that provide significant opportunities to reduce or eliminate pollution or other impacts on the Earth," reports Green Seal. Only the product's environmental impact is evaluated—not the company's environmental practices as a whole.

The public and industry can suggest product categories and comment on proposed standards which presently evaluate impacts as toxic chemical pollution; energy consumption; depletion and pollution of water resources; harm to wildlife, fish, and natural areas; waste of natural resources; destruction of the atmosphere, including ozone; and global warming.

Advisory panels, also including government and academia representatives, may be formed to develop certain standards. The **Environmental Standards Council**, comprised of independent experts, serves as an appeals board for manufacturers and others disagreeing with Green Seal's standards.

Its first proposed standard was issued in 1991 for **toilet tissue**, requiring, according to the *New York Times*, "100 percent recycled paper, no toxic solvents in removing ink from waste paper, and no dyes or fragrances in the final product; packaging would have to either be made from recycled paper or, to avoid waste, contain at least six rolls of tissue."

To appeal to consumers who want "environmentally friendly products," Green Seal's logo may appear on the product, packaging, and specific advertising and promotional materials; it may not be used in general corporate advertising.

With **media cooperation**, consumers are made aware of "environmentally friendly" products. Green Seal's **education campaign** also includes the distribution of fact sheets and brochures to the media, retailers, businesses, and schools.

Former director of the National Wildlife Federation's (NWF) environmental quality division, attorney Norman Dean heads Green Seal's efforts to "establish standards and criteria for evaluating the environmental impacts of consumer products." He authored *The Toxic 500* report on those industrial plants releasing the most toxic chemicals in the USA; *Danger on Tap* study about the "failure of the

government to enforce the federal safe drinking water law"; *The Acid Lakes Directory;* and the book *Energy Efficiency in Industry.*

■ Resources

Green Seal will publish a "description of the standard for each product category, a list of certified products, and an explanation of the environmental impacts reduced" by using products bearing the Green Seal. These materials will reach consumers through retailers, businesses, schools, and the media. A newsletter is being produced. Facts sheets and an information form are available.

Greenbelt Alliance ❦
116 New Montgomery Street, Suite 640
San Francisco, CA 94105 USA

Contact: **Jim Sayer**, Communication Director
    Phone: 1-415-543-4291 • *Fax:* 1-415-543-1093

Activities: Education; Political/Legislative; Research • *Issues:* Land Conservation; Sustainable Development

● Organization

The **Greenbelt Alliance** was founded in 1958 with the goal of preserving the "greenbelt" in the San Francisco Bay Area. By protecting the open areas, the Greenbelt Alliance seeks to ensure that the Bay Area remains a sustainable community, and to ensure the "economic, social and environmental well-being of the entire metropolitan region." The organization currently has a staff of 10 and a membership of 2,000.

▲ Projects and People

The Greenbelt Alliance is working to define and disseminate its vision for the Bay Area. It has published a number of reports containing views and research findings. One such publication is *At Risk*, a computer-generated map of the region showing urban areas, the Greenbelt, and areas that are at risk of development. This map serves to illustrate the pressures on undeveloped areas. Educating the region's people about the importance of the Greenbelt and its threats help garner support.

Public support is crucial for grassroots conservation groups, so the Greenbelt Alliance hopes to encourage as much as possible. Through increased media presence, volunteer outreach, and direct contact with local leaders the Alliance increased its visibility, and its support is growing as well.

Fieldwork with local citizen groups to protect the Greenbelt and change public policy regarding open spaces is another important aspect of the organization. Through **lobbying** local councils, the Alliance has managed to get a number of areas labeled as **open spaces** and to **limit zoning** changes on other areas.

To help mobilize the regional grassroots movement, the **Greenbelt Action Network** was founded in 1990. This network is designed to link people with specific opportunities to foster volunteerism.

■ Resources

The Greenbelt Alliance has more than 40 publications including a newsletter, public policy reports, case studies, and technical reports related to farmland and housing.

Greenpeace USA ⊕ ❦
1436 U Street, NW
Washington, DC 20009 USA

Contact: **Sanjay Mishra**, Public Information
                Coordinator
   Phone: 1-202-319-2444 • *Fax:* 1-202-462-4507

Activities: Education; Research; Political/Legislative • *Issues:*
Biodiversity; Deforestation; Energy; Global Warming; Transportation; Waste Management/Recycling; Water Quality

● Organization

With a staff of 250 and more than 4 million members, **Greenpeace**
is a large and highly visible international organization with offices in
23 countries and an international headquarters in Amsterdam. The
basic principle behind the work of Greenpeace is that members can
have a positive effect by simply "bearing witness"—drawing attention to the environmental abuse by being present. Originally formed
in 1971 from the Don't Make A Wave Committee in Vancouver,
Canada, one of its first ventures was to sail a small boat that year into
the atomic test zone near Amchitka, Alaska, and later help end this
activity.

Not allied with any political party, Greenpeace remains independent of the influence of any governments, groups, or individuals; it
has a policy of nonviolence or not attacking persons or property.
Its fleet includes a hot-air balloon, bus, and seven ships at sea: the
new *Rainbow Warrier, Sirius, Greenpeace, Gondwana, Moby Dick,
Beluga,* and *Vega.*

Greenpeace is best known for the courageous and well-publicized
protests its members have staged—sailing into nuclear testing zones,
hanging from bridges, and putting themselves in between harpoons
and the whales they are aimed at. In addition to these dramatic and
effective actions, the organization is involved in research, analysis,
informational briefings, and lobbying. It appears before the United
Nations, U.S. Congress, and town meetings worldwide.

▲ Projects and People

One of the major projects of Greenpeace is **Ocean Ecology**. Using
its fleet of ships and dinghies, Greenpeace is working to protect the
oceans, much of which falls outside of national territorial waters and
is therefore unregulated. Greenpeace action to protect whales and
dolphins and to stop waste dumping have met with some success.
Responsible food distributors, such as Heinz Company's StarKist,
now buy only "dolphin-free" tuna. The slaughter of whitecoat harp
seal pups has virtually ended. Greenpeace raises public awareness of
such problems and continues to monitor for illegal whaling.

Another aspect of the ocean ecology program is the campaign
against driftnets. These nets are killing as many as 100,000 dolphins
a year and "anything and everything that blunders into them."
Greenpeace has undertaken an international campaign to bring an
end to driftnet fishing. Florida sea turtles, such as Kemp's Ridley,
now are protected through the use of TED—turtle excluder devices—by shrimp vessels, for which Greenpeace campaigned.

Antarctica is another concern of Greenpeace. Working to keep
Antarctica in its natural and undamaged state, Greenpeace encountered an increase in groups and governments that were prepared to
exploit the natural resources in the region. Greenpeace established
the only non-governmental permanent base on the continent and
monitored the abuses of others. But in early 1992, Greenpeace was
preparing to pull out—as nations appeared to recognize Antarctica's
significance to the planet's survival.

The **Tropical Forests Campaign** aims to "counteract the destruction of the world's rainforests" from commercial logging, cattle
ranching, industrial development, nonsustainable agriculture, and
burning. In Guatemala, Greenpeace joined in a successful fight
against oil refinery construction in the Laguna del Tigre and the
Maya Biosphere Reserve.

Pressuring polluters is a large part of Greenpeace's **Toxic Pollution campaign.** By generating publicity about major polluters, and
by working to encourage the substitution of toxic materials with
nontoxic alternatives, Greenpeace hopes to limit the danger from
already existing toxins and prevent the creation of any more. Working against incineration, and also against the export of waste from
the First World to the Third, Greenpeace hopes to help create a
cleaner and safer environment. *(See also Greenpeace, International
Pesticide Campaign in this section.)*

Believing that nuclear devastation is the single greatest threat to
the planet, one of the major projects of Greenpeace is **Nuclear
Disarmament.** Greenpeace works to stop nuclear testing and weapons production and promote nuclear-free seas. After Greenpeace
protests of nuclear testing in the South Pacific, the French government went so far as to sink one of Greenpeace's ships, the *Rainbow
Warrior.* The Pacific Campaign draws attention to this fragile
environment and helps keep the London Dumping Convention
moratorium in place to prevent the dumping of radioactive wastes
there.

Greenpeace is also striving to **protect the atmosphere.** Working
against air pollution, and against the reliance on fossil fuels,
Greenpeace has an international campaign to help protect the environment beyond the limiting of chloroflurocarbons.

■ Resources

Greenpeace publishes several hundred magazine article reprints,
reports from their various campaigns, films on videocassette, and
books. A publications list is available. Greenpeace Stores are located
in San Francisco and Santa Cruz, California; Key West, Florida; and
Provincetown, Massachusetts. *(See P3 in the USA Private section.)*

International Pesticide Campaign (ITC) ⊕ ❦
Greenpeace
1436 U Street, NW
Washington DC 20009 USA

Contact: **Sandra Marquardt**, International
                Coordinator
   Phone: 1-202-462-2472 • *Fax:* 1-202-462-4507

Activities: Direct Action; Education; Information Dissemination; Political/Legislative; Research • *Issues:* Air Quality/Emis-
sion Control; Biodiversity/Species Preservation; Deforestation; Energy; Global Warming; Nuclear; Sustainable
Development; Transfer of Polluting Technology; Transportation; Waste Management/Recycling; Water Quality

▲ Projects and People

Touted as a "miracle of modern science" following World War II, pesticides were viewed as the end to insect pests, weeds, and hunger worldwide. According to Greenpeace, more than 45,000 pesticides have entered the marketplace since 1945; more than 6 billion pounds are produced yearly—"one pound for every person living today." But as Stanford University's **Paul Erkich** said in 1980, "Pesticides are an ideal product for big business: like heroin, they promise paradise and deliver addiction."

Under the coordination of **Sandra Marquardt**, the **Greenpeace International Pesticide Campaign** has several long-term objectives. These include **ending the production, trade, and use of particularly hazardous pesticides; halting the financing of chemically dependent agriculture,** and working with NGOs in developing countries to challenge development aid causing chemical dependency; and **promoting sustainable agriculture** by providing information on alternate methods and technologies.

To attain the goals, the pesticide campaign has staff in 12 countries and works in developing countries in Eastern Europe, Latin America, and North Africa with environmental, labor, consumer, farm, and development organizations. Promoting Greenpeace's agenda with governments, UN organizations, multilateral development banks, and regional organizations, the staff is also involved in lobbying internationally.

In the USA, Greenpeace led the push for pesticides export reform legislation in the U.S. Congress—the Circle of Poison Prevention Act—and published two reports *Never-Registered Pesticides: Rejected Toxics Join the Circle of Poison* and *Exporting Banned Pesticides: Fueling the Circle of Poison—A Case Study of Velsicol Chemical Corporation's Export of Chlordane and Heptachlor.* That year, Greenpeace also opposed the European Community directive on "harmonization of pesticide registration."

With other international NGOs of the **Pesticide Action Network,** Greenpeace worked to integrate Prior Informed Consent provisions in the Food and Agriculture Organization's (FAO) International Code of Conduct on the Distribution and Use of Pesticides. In preparing for the UN Conference on Environment and Development 1992 (UNCED), former coordinator Kay Treakle issued *A Call for Global Action to Promote Sustainable, Ecological Agriculture* that would preserve biological diversity of crops by protecting genetic resources, eliminate public health burdens, enhance livelihoods of small farmers and rural communities, and reduce "the severe soil, water, and marine pollution and nitrification from chemical inputs."

Greenpeace successfully lobbied the World Bank's Pest Management Advisory Group to improve their pest management guidelines and promote nonchemical alternatives. In the adopted Bamako Convention (1991) on trade of hazardous waste and hazardous products, Organization of African Unity (OAU), Greenpeace successfully included a ban on certain pesticides; also in **West Africa,** the use of incineration technology is being challenged for pesticide disposal.

In Italy, Greenpeace campaigned to stop the production and export of the toxic herbicide atrazine; in **Spain,** it opposed obsolete waste disposal technology in dealing with certain chemicals; in **Germany,** a drinking-water pollution campaign focused on pesticides and other agricultural chemicals; in **Belgium,** attention was drawn to Belgian corporate interests using "acutely toxic pesticides" paraquat and monocrotophos in Malaysian plantations; in the **Netherlands,** pesticides producers were targeted; in **France,** a campaign highlighted Rhone River pollution from pesticide production and use; in **Denmark,** organic farming was promoted; in **New Zealand,** pesticide problem research is underway; in **Australia,** dioxin contamination in harbor waters is being probed and a ban on organochlorine pesticides is sought; and in **Poland,** ecological agriculture is being promoted to counter the "Green Revolution blitzkrieg" by Western agribusiness concerns and international financial institutions—with pressure on the World Bank to include aid for small-scale ecological farm development. In **Argentina, Brazil, Mexico,** and **Central America,** pesticide/agriculture research is being conducted.

Greenbase is a computerized database on toxics, which includes all related reports and research papers. **Greenpeace QMW** is a laboratory in London where soil, and marine and fresh water are analyzed for pesticides. Among the chemicals researched to date are chlordane/heptachlor, lindane, DDT, paraquat and monocrotophos, and certain USA unregistered pesticides.

Issue projects of the ITC campaign include **Pulp and Paper, Incineration,** and **Waste Trade;** regional projects are **Great Lakes, North Sea,** and **Baltic Sea.**

Resistance from chemical companies and many organizations which the group lobbies, such as the multilateral development banks and UN organizations, is not surprising. The pesticide campaign is continuing its work as part of the **International Toxic Campaign** headquartered at the GPI Secretariat, Amsterdam, Netherlands, and plans to expand its activities and outreach.

"For the 1990s, Greenpeace will give greater attention to clean production as opposed to end of pipe solutions to toxics problems. Within this framework, the pesticide campaign has logically evolved to begin to address agricultural production."

■ Resources

This Greenpeace program is in need of financial assistance, information dissemination, political support, new technologies, and greater popular involvement. The pesticide campaign has produced briefings, fact sheets, videos, and much research.

Gulf Coast Tenants Organization (GCTO) ❦
1866 North Gayso Street
New Orleans, LA 70119 USA

Contact: **Pat Bryant,** Executive Director
Phone: 1-504-949-4919 • *Fax:* 1-504-949-0422

Activities: Education; Development; Political/Legislative; Research • *Issues:* Air Quality/Emission Control; Biodiversity/Species Preservation; Deforestation; Ecojustice; Global Warming; Health; Sustainable Development; Transportation; Waste Management/Recycling; Water Quality

● Organization

Started in 1982, the **Gulf Coast Tenants Organization (GCTO)** describes itself as an innovator in assisting community leaders in developing a vision of the future with "abundance of jobs paying a living wage, decent affordable housing, health care, quality education, wholesome recreation and a healthy environment." The staff of 6 professionals and 150 volunteers operate from offices in New Orleans and Baton Rouge, Louisiana, and Gulfport, Mississippi.

▲ Projects and People

GCTO redefines the meaning of the environment to include housing, health care, jobs, and physical resource conservation. The program has brought to light the "tremendous disproportionate poisoning of people of color in the USA, particularly African-Americans in 'Cancer Alley' and the deep South." GCTO joined with the SouthWest Organizing Project, The Panos Institute, and the Commission for Racial Justice of the United Church of Christ to develop the **National People of Color Environmental Leadership Summit.**

Cancer Alley is defined as the corridor along the Mississippi River between Baton Rouge and New Orleans. GCTO believes industrial and municipal poisons have been deposited disproportionately in communities where African-Americans, Native Americans, Hispanic-Americans, Asian-Americans, and poor Euro-Americans live. "Thousands of communities are on the verge of extinction from chemical poisonings," states GCTO.

According to its statistics, the risk of death, injury, and genetic changes from poisonings are 18 times greater in certain places along this corridor than the national average. Twenty-eight percent of the nation's petrochemical production also occurs in this 75-mile corridor, where there are more than 138 major petrochemical facilities dumping some 1 billion pounds of industrial poisons yearly. GCTO literature asserts this is a major factor leading to poor health in Louisiana and possibly the rest of the nation, yet industry leaders are quick to discount this.

GCTO is **training leaders** to fight back by "**utilizing mass protests, lawsuits, media** accounts that are unfavorable to poisoners, and the **ballot.**" Corporations retaliate by purchasing the land and houses of the people of the affected areas at low prices in exchange for release of liability from the poisonings, reports GCTO. Environmental educators also **teach community leaders how to discover chemicals** released locally, evaluate their health effects, and combat environmental racism.

The new **Municipal Incorporation Project** is helping small communities incorporate as towns enabling their leaders to "regulate and tax poisoning companies; clean up the air, land, and water; and create much needed job opportunities."

A **youth program** gets such people to organize for education, environment, and justice issues; phone counseling to youth is provided daily.

GCTO reaches people through **Workshops,** held six times yearly, and **personal visits** in the hearts of communities. Gulf Coast has more than 15,000 members from Baton Rouge to Birmingham, Alabama, situated in 37 communities along the Louisiana, Mississippi, and Alabama gulf coasts. Funding is raised from members, churches, individuals, and foundations.

Working with executive director **Pat Bryant** are **Rose Mary Smith**, president, and **Ruby Williams**, vice president.

■ Resources

GCTO produces a newsletter on its activities. It needs the following resources: air-monitoring equipment, political network with "progressive people and people of color internationally," media coverage, and funds for sustainable economic development.

The Holden Arboretum

9500 Sperry Road
Mentor, OH 44060 USA

Contact: **C.W. Eliot Paine**, Executive Director
Phone: 1-216-256-1110 • *Fax:* 1-216-256-1655

Activities: Education; Research • *Issues:* Biodiversity/Species Preservation

● Organization

The Holden Arboretum, "the nation's largest," spreads itself across 2,900 acres of horticultural collections, display gardens, and natural areas. Holden's mission is to "promote the knowledge and appreciation of plants for personal enjoyment, inspiration and recreation, for scientific research, and for educational and aesthetic purposes."

In the early 1900s, the Arboretum was the dream of mining engineer Albert Fairchild Holden, who was persuaded by his sister Roberta Holden Bole to make Cleveland the beneficiary for such an undertaking—rather than Harvard University. The Arboretum got underway in 1931, attracting many contributors, collections, and research and public programs on behalf of horticulture and natural environmental conservation. Primarily, Holden develops, maintains, and displays documented collections of woody plants, ornamentals, and other botanic specimens appropriate to the climatic zone of northeast Ohio.

But its mission is even greater. "In our magnificent Midwest, habitat destruction is the number one cause for plant rarity," it reports. "The vast forests, sweeping prairies, extensive wetlands, and miles of shoreline dune vegetation which graced the Great Lakes Region have been reduced to mere remnants."

▲ Projects and People

While thousands of acres have been set aside in the Midwest to protect such remnant plant communities, habitat protection alone cannot guarantee long-term species conservation. For many plant species, this may only be accomplished through the development of successful management regimes based upon a clear understanding of a species' life history and habitat requirements.

Unfortunately, for most of the region's rare flora, this information is unknown. The Arboretum has joined the Center for Plant Conservation (CPC) to help address unanswered questions regarding the **endangered flora** of Ohio, Indiana, Illinois, Iowa, Minnesota, Wisconsin, and Michigan.

Cooperatively with federal, state, and private conservation organizations, Holden **identifies the region's rarest plants and develops species biology research.** Such programs conducted in Holden's greenhouses use collected seeds, spores, or cuttings and produce valuable life history and cultural information as well as the young plants that make up Holden's additions to the CPC's National Living Collection of endangered species.

Working with its collections, Holden is intent on **developing superior landscape plants.** A particular focus might be using newly collected Asian material or improved shade and street trees for urban sites. In addition, floral and plant community surveys are ongoing.

With a grant from the Ohio Division of Natural Areas and Preserves, botanists study the **germination requirements and population viability** of selected monitored rare plant species in Ohio. At its David G. Leach Research Station, special programs include the hybridization of cold hardy ornamental **rhododendrons** and an investigation of the genetics of rhododendron, known as the haploid project.

A national crabapple (*Malus*) introduction program is underway, as are continual evaluations of horticultural collections. A cooperative project with the Ohio Division of Forestry concerns "plantation planting of high yield sugar maples for superior sap sugar production." With the Northern Nut Growers Association, black walnut (*Juglans nigra*), Persian walnut (*J. Regia*), and butternut (*J. Cinerea*) graftings and growing of hybrids are taking place. Being discussed is a project to "evaluate, select, introduce, and disseminate new clones" with the Lake County (Ohio) Nurserymen's Association.

The Holden Arboretum joined with the U.S. National Arboretum in a **cooperative plant exploration program** in China and Korea, sending staff on numerous trips; collecting sites are now being evaluated.

■ Resources

Holden publishes a quarterly class schedule and offers classes in art and nature, natural history, and horticulture. It also offers children's programs, hikes, and walks. The Arboretum has an intern program which draws groups of students from the USA and abroad.

Hoosier Environmental Council, Inc. (HEC)
1002 East Washington, Suite 300
Indianapolis, IN 46206 USA

Contact: **Jeffrey Stant**, Executive Director
Phone: 1-317-685-8800 • *Fax:* 1-317-686-4794

Activities: Education; Development; Political/Legislative; Research • *Issues:* Air Quality/Emission Control; Biodiversity/Species; Deforestation; Energy; Sustainable Development; Waste Management/Recycling

● Organization

The **Hoosier Environmental Council** (HEC) monitors state and federal government activities that affect Indiana's environment, including the policy and enforcement arms of the State Board of Health and Indiana Department of Natural Resources, Department of Environmental Management, State Legislative Study Committees, U.S. Fish and Wildlife Service, U.S. Army Corps of Engineers, Office of Surface Mining, U.S. Forest Service, and the U.S. Environmental Protection Agency (EPA).

▲ Projects and People

Statewide member groups are active in pursuing issues concerning Indiana's Management Plan of **Solid Waste Facilities**, a State Groundwater Protection Policy, and the Hoosier National Forest Land Management.

HEC confronted a plan in 1985 to dedicate more than four-fifths of the Hoosier National Forest to clearcutting and other developments. The 1990 climax of a subsequent campaign demonstrated massive public sentiment for protecting the wild lands, natural habitats, and native wildlife. HEC carried out the largest mass mail and phone campaigning in the history of the state's environmental movement, reaching over 100,000 Hoosiers and urging them to

write the Forest Service in support of the **Conservationist Alternative Management Plan**. Nearly 4,000 letters supported the alternative plan, nearly 10 times the response supporting predevelopment plans.

Today HEC is confronting the "assault on biodiversity" particularly in northwest Indiana. Battles it fought to **keep hazardous wastes from being injected underground**, especially by the steel industry, are being won through enforcement of a federal law, HEC reports in its literature. Incineration of hazardous wastes remain a great concern; and EPA has targeted the region to help abate its heavy industrial pollution in efforts to reduce risks to both human health and natural ecosystems, such as keeping toxic chemicals out of the food chain.

HEC is also focusing on efforts to clean up the "abused **Grand Calumet River**," once abundant with fish and wildlife and now beset with oil and sludge. "A century of industrial and municipal discharges have made the Grand Cal a waterway of almost unparalleled degradation," writes Dorreen Carey, Grand Cal Task Force executive director. "A 'dead' river in the 1960s, the Grand Cal could not support life because of its chemical poisons and thick layer of oil that floated on the surface. Today, the Grand Cal is said to be 90 percent wastewater due to the millions of gallons discharged daily by the steelmills and sanitary districts."

Yet Carey reports the river is slowly coming back to life. Pollution-resistant fishes and waterfowl species still survive. But the danger persists. "Heavy metals, arsenic, cyanide, PCBs, oil and grease, carcinogens from the coke-making process, and a long list of chemical constituents have created a poisonous stew which is impacting the drinking water supply of the Great Lakes."

Hoosiers are also fighting to **save the dunes and their rare and endangered species** at the National Lakeshore Park, where some 1,419 plants and animals have been identified. In addition to the dunes—wetlands, pristine oak savanna, northern paper birch, tall grass prairies, swale, and rich riverine forest abound.

Northwest Indiana is also afflicted with "the fourth worst ozone problem nationally" according to standards of the Clean Air Act of 1990, reports HEC literature. Industry and vehicles are the chief culprits. State and federal air quality monitoring programs are underway in attempts to resolve these long-range problems.

■ Resources

The Hoosier Environmental Council publishes the quarterly *HEC Monitor Newsletter*; the quarterly *In-Vironment Underground*; and the monthly *BoardWatch* for 16 member groups, HEC individual members, and state environmental leaders.

Human Environment Center (HEC)
1930 18th Street, NW, Suite 24
Washington, DC 20009 USA

Contact: **Hector-Ericksen-Mendoza**, Acting Executive Director
Phone: 1-202-588-8036 • *Fax:* 1-202-588-9422

Activities: Education; Research

● Organization

"Hungry people cannot conserve, unemployed people want jobs more than clean air or parks, and segregated people have little reason to trust majority movements. Yet the poor suffer the worst environments and benefit least from natural resources." Thus begins the mission statement of the **Human Environment Center (HEC)**. Because environmental ills disproportionately affect the poor and minorities, HEC assists minority participation in environmental concerns.

▲ Projects and People

The Center established the **Minority Environmental Science Internships** to expose high-school students to environmental careers. HEC recruits and places college and graduate students in such internships in collaboration with national and grassroots organizations. The Center staffs and houses the **National Association of Service and Conservation Corps (NASCC)**, which serves as a national clearinghouse and technical assistance organization for youth conservation and service corps to their communities.

More than 25 environmental organizations joined with the Human Environment Center to form the **Environmental Consortium for Minority Outreach**. The consortium formed two programs: the **Human Resources Program** and the Minority Environmental Internship Program (MEIP). The Human Resources Program recruits minorities from a data base of resumes to professional careers in the environment. The Minority Environmental Internship Program recruits undergraduate and graduate students from Asian, African, Hispanic, and Native American backgrounds and places interns with environmental organizations.

■ Resources

HEC uses the media, conferences, summits, and coalitions to define cooperative ventures between minorities and the environment. It publishes papers that effect equal or unequal participation in environmental concerns. HEC prepares a series of issue papers on water quality and other urban environmental problems disportionately affecting low-income and minority communities. An informational brochure is also available.

The Humane Society of the United States (HSUS) ⊕
2100 L Street, NW
Washington, DC 20037 USA

Contact: **John A. Hoyt,** Corporate Executive Officer
Phone: 1-202-452-1100 • *Fax:* 1-202-778-6132

Activities: Education • *Issues:* Biodiversity/Species Preservation; Deforestation; Energy; Global Warming; Health; Population Planning; Sustainable Development; Waste Management/Recycling • *Meetings:* CITES; International wildlife (birds and elephants); IUCN; IWC

● Organization

The Humane Society of the United States (HSUS) works on extensive information, legislation, investigation, and legal issues concerning the welfare of animals. The HSUS asks consumers to

pledge to not wear fur, buy ivory, or eat tuna; and to use cruelty-free cosmetics and eat lower on the food chain.

▲ Projects and People

The HSUS experienced two recent victories in dolphin-safe tuna policies; passing the **Dolphin Protection Consumer Information Act** and pressuring canneries to no longer purchase tuna caught in purse-seine nets which trap dolphins in with tuna catches. StarKist, Chicken of the Sea, and, more recently, Bumble Bee, Deep Sea, and Ocean Light have satisfied the dolphin-safe standards. However, HEC estimates more than seven million dolphins previously died in such nets.

According to HSUS, Mexico, Venezuela, and Vanuatu still engage in dolphin-deadly fishing practices—Mexico destroying 80,000 out of a total of 100,000 dolphins killed annually. The HSUS worked hard to pass the **Marine Mammal Protection Act** (1972) which slaps a U.S. trade embargo on these countries. It forbids the USA and foreign fleets to sell yellowfin tuna in the USA since they kill more dolphins than the law allows.

"In all the victories on behalf of the dolphins," HSUS reports, "consumers have had the most powerful voice. Consumers must continue to use that voice to save these unique, intelligent, and gentle-hearted creatures."

■ Resources

The HSUS publishes a catalog of informative books, tapes, videos, slides, and posters that deals with wild and domestic animal exploitation. HSUS is supported solely by charitable contributions.

INFORM, Inc. ⊕
381 Park Avenue South, Suite 1201
New York, NY 10016 USA

Contact: **Jerri K. McDermott,** Associate Director/ Communications
Phone: 1-212-689-4040 • *Fax:* 1-212-447-0689

Activities: Education; Law Enforcement; Political/Legislative; Research • *Issues:* Air Quality/Emission Control; Chemical Hazards Prevention; Energy; Transportation; Waste Management/Recycling

● Organization

INFORM was founded in 1974 to help combat the growing problem of air pollution. Supported by 30 foundations, 25 corporations and private donors, INFORM's major purpose is to educate the public about major health risks and natural resources.

Legislators, conservation groups, and business executives use INFORM research in areas of chemical hazard prevention, solid waste management, and land and water conservation. This information is published in books, abstracts, newsletters, and articles. Researchers appear before congressional briefings, community workshops, seminars, and business and environmental conferences.

Research entails "in-depth case studies of the practices of industries and individual companies that affect the environment" with deductions and evaluations. INFORM also has a communications branch which brings findings to those who request and/or benefit from them.

▲ Projects and People

INFORM's founder and president, Joanna D. Underwood, produced with two colleagues INFORM's first publication on **air pollution**. *A Clear View* showed local groups "how to investigate and combat industrial pollution," and helped establish the group's reputation as "a first-rate research organization."

According to INFORM, its studies on **chemical hazards prevention through source reduction** helped shape the first federal legislation on waste prevention and "spur the creation of EPA's [U.S. Environmental Protection Agency] Toxics Release Inventory, a nationwide computerized database of detailed information on toxic emissions." Using this database, INFORM documented the extent to which toxic wastes are transported across state boundaries.

Regarding **solid waste management**, INFORM documented an integrated Japanese model and started advising local and state planners, business leaders, and federal regulators on recycling.

Concerning **land and water conservation**, INFORM in 1984 began testing and promoting a "low-cost irrigation management method" that is growing in popularity among western farmers "because it saves them water, energy, labor, and money." Such farmers are able to reduce problems of "water waste, drainage, and runoff that deplete and degrade western water supplies." The method promises to help farmers "make the transition to sustainable agriculture," reports INFORM.

With studies in **urban air pollution control**, INFORM is pointing out the "environmental, economic, safety, and energy security benefits of natural gas as an alternative transportation fuel, especially in heavily urban areas." It is exploring "the role natural gas vehicles can play as a bridge to a hydrogen fuel economy."

■ Resources

INFORM publishes a quarterly newsletter and research reports on subjects ranging from toxic pollution to energy conservation. Planned publications are on successful natural gas vehicle programs, U.S. resource recovery plants, toxic chemicals in products, and office paper waste reduction—based on ongoing studies. A publications list is available.

Institute for Alternative Agriculture, Inc. (IAA) ❦

9200 Edmonston Road, Suite 117
Greenbelt, MD 20770 USA

Contact: **Garth Youngberg, Ph.D.**, Executive Director
Phone: 1-301-441-8777 • *Fax:* 1-301-220-0164

Activities: Education; Political/Legislative • *Issues:* Sustainable Agriculture

● Organization

Intensive use of farm chemicals is expensive and causes serious negative impact on ground and surface waters as well as affecting human and animal health and food quality and safety. Founded in 1983, the **Institute for Alternative Agriculture (IAA)** addresses low-input alternative farming methods. It works with producer groups, public research, education institutions, and government agencies to promote sustainable agriculture.

With 2,000 members, the Institute is governed by a grassroots board of directors that includes several commercial scale organic farmers. Its advisory groups include noted agricultural scientists from land grant universities, U.S. Department of Agriculture's (USDA) Agricultural Research Service (ARS) and Economic Research Service (ERS), nonprofit organizations, and industry.

▲ Projects and People

Institute representatives speak at universities and at conferences of producer associations and farm organizations. They testify at hearings in the U.S. Congress and provide scientific information to government agencies and congressional committees.

For example, IAA spokesperson **James Aidala** testified before the U.S. House of Representatives Committee on Agriculture on **new approaches to pest control** that reward and foster low-input agriculture. "If the pest problem were addressed as part of a total system of agricultural production . . . then alternatives such as crop rotation or changes in cultural methods might serve to keep the grower as an economically viable producer," he said. "Finding such answers requires regional, if not local, research . . . which takes time, research funds, and anticipation of regulatory problems."

IAA recently reported on the "bad news" that federal appropriations were shrinking for sustainable agriculture programs. It also provides leadership for the **Agricultural Task Force** of the Global Tomorrow Coalition.

Before joining the Institute, both **Dr. Garth Youngberg** and **Dr. Neill Schaller**, associate director, worked in organic farming and sustainable agriculture programs with the U.S. Department of Agriculture (USDA).

■ Resources

IAA promotes funding for the **Agricultural Productivity Act** in support of low-input agricultural research. IAA holds an annual scientific symposium. It is supported by memberships, donations, and grants from foundations, corporations, and individuals.

The Institute produces two monthly newsletters, the *Alternative Agriculture Resources Report* and the *Alternative Agriculture News*, which relates congressional and judicial developments as well as positions open and books on sustainable agriculture. The Institute publishes the quarterly *American Journal of Alternative Agriculture*, which reports on practical and scientific sustainable crop raising.

Institute for Development Anthropology (IDA) ⊕

99 Collier Street
P.O. Box 2207
Binghamton, NY 13902-2207 USA

Contact: **Vivian Carlip**, Editorial Associate
Phone: 1-607-772-6244 • *Fax:* 1-607-773-8993

Activities: Education; Development; Research • *Issues:* Biodiversity/Species Preservation; Deforestation; Energy; Fragile Lands; Health; River Basin Planning and Development; Settlement/Resettlement; Sustainable Development

● Organization

The **Institute for Development Anthropology (IDA)** seeks to apply anthropology to humane and compassionate development among disadvantaged people. In the 15 years of IDA's existence, much of the Third World's disadvantaged have increased because of civil strife, hunger, drought, AIDS, migrations, and refugees.

Anthropologists have become decisionmakers in multidisciplinary projects in host-country governments, actively coordinating river basin development; settlement and resettlement; natural resource management in such areas as forestry, range and livestock, fisheries, wildlife, and water use; regional development and planning; and sustainable and contract farming.

With its social justice goals, IDA aims to contribute to "equitable economic growth, environmental sustainability, resolution of ethnic conflict, and participatory government."

▲ Projects and People

IDA engages in research, training, and project design and evaluation in 50 countries. One such project centers on **rural Bolivia**; a combination of economic stagnation and environmental degradation has compelled rural people to migrate away from ruined farmlands for several decades. The growth in demand for cocaine in the USA and Europe has fueled a boom in coca leaf production in the Chapare region which in turn has resulted in the clearing of vast areas of tropical forest in areas unsuited for agriculture. IDA seeks to understand the driving forces behind the migrant workers' growing dependence on coca leaf to survive. Together with Bolivia's Alternative Development of Cochabamba, they are implementing **income-earning alternatives to cocaine production.**

"Sustainability cannot be achieved where impoverishment and the need for short-term survival result in environmental predation," according to IDA directors Dr. David W. Brokensha, University of California at Santa Barbara; Dr. Michael M. Horowitz, State University of New York at Binghamton; and Dr. Thayer Scudder, California Institute of Technology.

To help local populations and institutions manage and profit from their own resources, IDA concentrates on **arid and semiarid land and water management** as well as on **social and community forestry.** It has discovered that "those engaged in pastoral production systems . . . often prove to be effective managers of forage resources." IDA has participated in such research projects as **forestry in grazing systems**, White Nile Province, **Sudan**; Gestion de Pâturage in **Morocco, Algeria, Tunisia, Jordan, Syria, Iraq; Sudan** rangeland and livestock management; Masai range and livestock project evaluation, **Tanzania**; range and livestock appraisal, **Niger**; natural resource management in the Sebungwe Region, **Zimbabwe**; and range and livestock evaluations, **Mali and Niger.**

One IDA study regards the **economic activities of women in** pastoral and agropastoral societies in Africa, the Middle East, and Asia so that future productivity will be gender sensitive, cost effective, environmentally sound, and socioeconomically sustainable.

Other **natural resource management** projects took place in Ecuador, Indonesia, Malawi, Peru, and Zambia.

A study of **contract farming of small growers and cooperatives** was undertaken in seven African countries to determine how these could benefit from such arrangements with state or private organizations; IDA also helped formulate donor and government policies. Other **food and marketing projects** related to pesticide-residue reduction in the **Dominican Republic**, fisheries in Niger, and grain marketing in Kenya.

IDA provided research for the **Onchocerciasis (river blindness) Control Programme** of 11 West African countries. Since 1979, IDA provides technical assistance to **Sri Lanka** regarding the **Accelerated Mahaweli Programme** to bring about new government and donor policies that help generate income through diversified production on irrigated and home lot allotments as well as stimulate off-farm enterprise and employment. IDA aids in resettlement and development projects in **China, Pakistan, Bolivia, the Middle East,** the African River Basin, and Somalia.

In monitoring the **Middle Senegal Valley**, IDA with the Senegalese *Cellule Après-Barrages* is studying "impacts of changes in river flow on local production activities . . . and the relationships between production and income generation, off-farm employment, and regional development." Early findings from the four-year study seem to favor "a controlled release from the reservoir, replicating a natural flood (which) will allow the dam not only to produce hydropower and provide water for irrigation in the dry season, but also to enhance recession cultivation, fishing, and herding, and contribute to environmental sustainability."

Social and institutional studies are a large part of IDA's work. A current research and training **potable water project** supports a national strategy for water user associations in Tunisia to allow "equitable access and provide a means for users to contribute to the operation and maintenance" of 30 new drinking water sites. IDA helps host-country institutions set up computer-based information systems and encourages analyses of plans in developing nations worldwide.

■ Resources

IDA sponsors workshops, seminars, and conferences for governments, universities, and non-governmental organizations. IDA publishes *Development Anthropology Network* twice yearly, and *Working Papers* and *Monographs in Development Anthropology* in association with Westview Press. In the latter series, *Lands at Risk in the Third World* (by Drs. P.D. Little and M.M. Horowitz) is specifically about environmental issues.

IDA's library includes 14,000 computerized documents in socioeconomic and environmental dimensions of development. IDA maintains a computerized cartographic laboratory, with both fixed and field-based digitizing boards and an eight-color plotter, and provides geographic and land information system backstopping to overseas projects.

IDA maintains a roster of 1,000 professional associates in social sciences, agriculture, environment, health, and engineering.

Support for IDA comes from individuals, corporations, foundations, governments, and such multilateral agencies as the United Nations Development Programme (UNEP), the Food and Agriculture Organization (FAO), the United Nations Development Fund for Women, and the World Bank.

Institute for Local Self-Reliance (ILSR) ⊕
2425 18th Street, NW
Washington, DC 20009 USA

Contact: **Ingrid Komar**, Development Coordinator
Phone: 1-202-232-4108 • *Fax:* 1-202-332-0463

Activities: Research • *Issues:* Sustainable Development; Waste Management/Recycling

● Organization

Dedicated to "environmentally sound economic development," the Institute for Local Self-Reliance (ILSR) does research to develop a strategy for turning pollution into energy and economic opportunities. Some ideas include substituting agricultural materials for petroleum-based industrial products, whey-to-alcohol fermentation for fuel, and decentralized photovoltaics.

Co-founded in 1974 by economic development planners **Dr. David Morris** and **Dr. Neil Seldman**, ILSR gives conferences, workshops, and seminars for community groups and local officials on materials management in over 100 cities. Dr. Morris specializes in the impact of new technologies on development and has provided such assistance to numerous government agencies including the U.S. Department of Energy (DOE); U.S. Economic Development Administration; New York State Energy Research and Development Authority; Marin County, California; and the cities of St. Paul, Minnesota, and Philadelphia, Pennsylvania.

Dr. Seldman consults with city and state governments, citizen groups, and private industry on transforming "waste disposal from an increasingly costly service to a productive sector of the local economy." His view is that "materials recovery and reuse [is a] means to establish urban self-reliance."

▲ Projects and People

Research for New Jersey indicates the state can save over $2 billion, create thousands of new jobs, and reduce pollution dramatically. **Brenda Platt**, one of the authors of the report, says, "We hope to speed up the process whereby public policy catches up with both the new, environmentally conscious, regulatory climate and the many commercially viable technological innovations already available to meet them."

A 100-page report on the District of Columbia based on 8 months of research states that **DC can reduce its solid waste management costs**, reduce the flow of garbage to the Lorton, Virginia, landfill by almost two-thirds, and simultaneously create several hundred private sector entry-level and skilled manufacturing jobs.

ILSR has advised **Sesame Street** on ecological issues and assisted **Jolie** and **Quincy Jones' Take-It-Back Foundation** in the production of educational videos and pamphlets on recycling.

ILSR, in Minnesota, has helped form a coalition of 40 groups to advocate **energy self-reliance** in city councils and state agencies, in the state legislature, and before the general public to link issues of sustainable agriculture, recycling, energy efficiency, and alternative fuels to the goal of local self-reliance.

■ Resources

Platt, **Irshad Ahmed**, Morris, and Seldman together have written more than 20 books and technical publications. Among them are *Be Your Own Power Company; Alcohol Fuels from Whey: Novel Commercial Uses for a Waste Product; Proven Profits from Pollution Prevention;* and *Waste to Wealth: A Business Guide for Community Recycling Enterprises.* A recent manual is *Beyond 40 Percent: Record Setting Recycling and Composting Programs.* ILSR publishes a series of *Facts To Act On* news features for some 336 community groups on local aspects of sustainable economic development.

ILSR could use additional computers and media coverage.

The Institute of Scrap Recycling Industries, Inc. (ISRI) ⊕

1627 K Street, NW
Washington, DC 20006 USA

Contact: **Herschel Cutler**, Executive Director
Phone: 1-202-466-4050 • *Fax:* 1-202-775-9109

Activities: Education; Political/Legislative • *Issues:* Waste Management/Recycling • *Meetings:* BIR; NRC

● Organization

The Institute of Scrap Recycling Industries (ISRI) was formed in 1987 from a merger of the National Association of Recycling Industries, begun in 1913, and the Institute of Scrap Iron and Steel, formed in 1928. This trade association represents 1,800 companies that process, broker, and consume scrap commodities—including ferrous and nonferrous metals, paper, glass, plastics, and textiles. It is dedicated to "conserving the future by recycling the past."

According to ISRI, its members' expertise in resource recovery can save communities and organizations precious time and money in developing practical approaches. ISRI reports that in 1990 its members processed "more than 95 million tons of recyclables, including more than 9 million automobiles." The same year, exports of recycled materials totaled over 19 million tons and helped improve the U.S. trade balance. Moreover, says ISRI, "recycled aluminum scrap saves the nation 95 percent of the energy that would have been needed to make new aluminum from ore. Recycled iron and steel result in energy savings of 74 percent; recycled copper, 85 percent; recycled paper, 65 percent; and recycled plastic, over 80 percent." An environmental boon, such recycling also strengthens the U.S. economy.

In other statistical interpretations, 580 million refrigerators can be made from the 58 million tons of scrap iron and steel handled in 1990. At the same time, there are enormous savings in energy (74 percent), virgin materials use (90 percent), air pollution reduction (86 percent) water use reduction (40 percent), water pollution reduction (76 percent), mining wastes reduction (97 percent), and consumers wastes reduction (105 percent).

Moreover, 100 billion pizza boxes can be manufactured from 1990's 28 million tons of scrap paper and paperboard; siding for 8 million homes can be gotten from the 2.8 million tons of scrap aluminum; 360 trillion new pennies can be surfaced from the 1.8 million tons of scrap copper; 11 billion spoons can be made from the 900,000 tons of stainless steel scrap; and 4 billion quart-sized juice bottles can be produced from the 2 million tons of scrap beverage glass containers.

▲ Projects and People

ISRI supports legislation that encourages **removal of obsolete (pre-1980) autos** by supporting manufacturers to provide dealers incentives to trade them in for new and cleaner cars, that distinguishes recyclable or scrap material from solid waste or hazardous chemicals for export, that supports regulation of chlorofluorocarbons (CFCs) for **emptying CFC refrigerants** prior to delivery for scrap processing on the federal level, and that establishes a uniform procedure for the collection of undercharges on shipments by motor carriers. **Air bags** contain sodium azide, an explosive and poisonous substance. The recycling of autos containing airbags exposes employees to environmental health risks and the explosions causes injury and damages equipment. ISRI suggests the air bag containers should be removed from all autos just prior to delivery for recycling, just as is done with CFCs in auto air conditioners.

ISRI promotes **Design for Recycling,** a cooperative arrangement between manufacturers and recyclers, to make consumer products safely recyclable. It is recommended that small businesses particularly should be given economic and technical assistance during the transition into recyclable ability.

■ Resources

ISRI publishes safety guides, videotapes, commodity books, *ISRI Directory,* testing charts, industry magazines, children's literature, traveling exhibits, and brochures concerned with recycling. Publications and video lists are available.

Institute for 21st Century Studies (ITCS) ⊕
1611 North Kent Street, Suite 204
Arlington, VA 22209-2111 USA

Contact: **Gerald O. Barney, Ph.D.**, Executive
           Director, Founder
  Phone: 1-703-841-0048 • *Fax:* 1-703-841-0050

Activities: Education • *Issues:* Air Quality/Emission Control; Biodiversity/Species Preservation; Deforestation; Energy; Global Warming; Health; Population Planning; Sustainable Development; Transportation; Waste Management/Recycling; Water Quality

● Organization

The Institute for 21st Century Studies (ITCS) grew from The Global 2000 Report to President Jimmy Carter in 1980 which was directed by **Dr. Gerald O. Barney,** who founded this organization three years later as a result of the widespread interest in continuing such long-term studies in various countries.

Today, with the Institute's guidance, both industrialized and developing nations study alternative strategies for sustainable economic and ecologic development and national security. These studies are produced by a team of university, government, or independent institutes and provides detailed facts about the participants' countries, describes probable futures, and identifies possible strategic choices that lead to desired futures.

Some 31 countries in Africa, Asia, Europe, North America, Latin America, and the Caribbean are engaged in such studies. Both the World Council of Churches and His Holiness the Dalai Lama have projects in progress. **Dr. Martha J. Garrett** heads the European office in Mölndal, Sweden. Support comes from many sources including UNESCO, World Bank, Rockefeller Brothers Fund, individuals, corporations, religious institutions, grants, and fees for services.

According to the Institute's **draft definition,** a 21st Century Study "is led by the nation's own experts rather than by foreign consultants; examines multiple sectors, including trade foreign debt, demography, natural resources, environment, technology, health, education, security, and other key areas; integrates the sectoral analyses. focusing attention on the linkages among the sectors;

adopts a perspective of two or three decades rather than concentrating on short-term issues; reviews the adequacy and appropriateness of the political and other institutions; encourages participation by policymakers, community leaders, women, minorities, and young people; identifies national strategies that work well in the short-term without creating unwanted consequences in the long-term; and considers the impact the nation will have on global economy, environment, and security and the impact the global situation will have on the nation."

The Institute provides support for the growing international network of teams. It leads team training and provides information, strategies, design, and analysis. ITCS passes on microcomputer models and makes contact with those who could assist with particular aspects of the studies. ITCS leaders assist with publication and international distribution of their reports with assured national attention.

▲ Projects and People

Together with the Hunger and Peace Education Programs of the Evangelical Lutheran Church in America (ELCA), the Institute and six colleges and universities are studying how "hunger can be ended and peace established in countries where hunger and violence are common." Their purpose is to provide "blueprints for building a just, peaceful, and environmentally benign culture over the next few decades."

Basic core study materials are compiled in a **Tool Kit,** which includes leaders' handbook, generic computer model, data, background reports, and example studies. In 1992, the documentation for the **Standard National 21st Century Study Model** is being updated with the World Bank STARS data disks; *Justice, Peace, and the Integrity of Creation (JPIC) Primer;* summary of the *Brundtland Report* prepared by the Global Tomorrow Coalition; and the CEIP Fund's handbook, *Getting a Job in the Environmental Field.* Teams of professors plan research. Study trips are planned for students and professors. National Church officials select countries for study.

Training Workshops are presented, such as a recent program on **African Cultures—African Futures** led by Dr. Garrett and characterized by informality and lively exchanges on topics such as politics, women's issues, modern medicine, environment and public health, agriculture, sustainable development, education, energy, and computer models. Dr. Garrett recalls, "Most of the participants were used to thinking about the development of Africa and of their individual nations solely from a short-term perspective and solely by sector. Few were familiar with any kind of futures studies. . . . The workshop effectively changed this situation. . . . They also left knowing something about microcomputers, models, and databases and understanding their potential usefulness in futures studies."

For example, the need for a regional network to pursue future studies of sub-Sahara Africa became apparent. With ITCS, other recommended participants are Environment and Development Action in the Third World (ENDA), the Pan African Social Prospects Centre (CPPS), Analink computer consulting company based in West Africa, and the All-African Council of Churches (AACC), among NGOs and other international groups Dr. Garrett believes that West Africa can be the "ideal 'laboratory' for doing research on . . . alternative national futures."

In partnership with the Council for a Parliament of the World's Religions, the Institute is scheduling **The Second Parliament of the World's Religions** in late August 1993, Chicago, Illinois. Their purpose is to foster the values needed to sustain the world as it moves into the next century. Eight faiths—Baha'i, Buddhism, Christianity, Hinduism, Islam, Judaism, Theosophy, and Zoroastrianism—are represented by the Council. Both critical human community issues—such as environmental protection, population growth, economics and justice, consumerism, science and technology, and power/politics and liberation—as well as interfaith relations will be addressed.

In this way, the Institute encourages "cooperation between secular and sacred leaders" in solving the world's environmental, social, and justice problems.

■ Resources

A publications list includes *Managing a Nation: The Microcomputer Software Catalog, Peruvian 21st Century Study Model;* and *Studies for the 21st Century.* This list explains where publications can be ordered. The Institute offers a *Tool Kit for 21st Century Studies* as well as consultation, training, and technical assistance. Needed resources include microcomputer software for the third edition of *Managing a Nation;* and help in increasing the organization's visibility, in identifying teams to carry out 21st Century studies, and in acquiring floppy disks with accurate data on countries for use in studies.

New Forests Project (NFP) ⊕ ✺
International Center for Development Policy (ICDP)
731 Eighth Street, SE
Washington, DC 20003 USA

Contact: **Stuart N. Conway,** Director
Phone: 1-202-547-3800 • *Fax:* 1-202-546-4784
Telex: 5106017738

Activities: Development; Education; Research • *Issues:* Biodiversity/Species Preservation; Deforestation; Energy; Global Warming; Sustainable Development • *Meetings:* International Association for the Study of Common Properties; International Society of Tropical Foresters; UNCED

● Organization

The **New Forests Project (NFP)** was established in 1981 by the **International Center for Development Policy.** As a response to the destruction of natural resources, the New Forests Program is a direct-action program to help villagers in developing countries improve their quality of life by providing them with the means to plant trees.

With a staff of 8, 4 in the USA and 4 in **Guatemala,** and a membership of 20,000, the Project has aided more than 3,200 villages in 110 developing countries. Goals include "assisting villagers to initiate sustainable development projects," disseminating information about natural resources in the USA, and improving methods of technology transfers.

Tree by tree, country by country, NFP seeks to halt global warming through widespread plantings that can overtake the buildup of carbon dioxide. According to NFP, the four hottest years on record have all occurred in the 1980s. During 1988, "37 percent of the U.S. corn crop, 13 percent of the wheat, and 8 percent of the soybean crops were lost to heat and drought." Already, more than

25 percent of the Earth's land surface is in the advanced stages of desertification; "the Sahara moves steadily southward, famine in Africa is now chronic, and in our own southwest, 225 million acres of agricultural land is rapidly becoming a desert."

▲ Projects and People

The main activity of the New Forest Project is the **World Seed Program** providing tree seeds, technical assistance, and training materials to NGO's in developing countries. Under the direction of **Stuart Nelson Conway,** this program has expanded both in the number of countries involved, and also in the types of seeds distributed, which include *Leucaena leucocephala, Gliricidia sepium,* and *Robinia pseudoacacia.* The NFP plans to reach a still larger number of NGO's and farmers.

A preferred tree is the fast-growing **Leucaena,** or **Miracle Tree** (also known as ipil-ipil tree), which provides fuelwood and building materials, stops wind and water erosion, produces forage for livestock, controls grass fires, and brings rich organic fertilizer and moisture to the soil—preventing tropical forests from being cut down. The leaves and branches can be a source of food and medicine for humans.

In 1990, NFP helped to start 300 village-level projects. More than 80 percent of these new projects were in Africa including a major undertaking in the Tigray province of Ethiopia. In 1991, NFP sent more than 100 kilograms of seeds to Tigray alone. Senegal, Nigeria, and Ghana farmers also received leucaena, acacia, and blacklocust seeds.

Agronomist **Leonel Jarquin** serves as the project coordinator for the **Guatemala Reforestation Program.** With three community nurseries, NFP grew and distributed over 75,000 seedlings to more than 350 families. Citrus, coconut, papaya, and other fruit and nut trees are now being added at the request of the community. During the seasons when the nurseries are not fully occupied, they are used by local women's groups to grow vegetables to diversify their diets. Cooperating groups include the Guatemalan Forest Seed Bank (BANSEFOR), CARE, and Guatemalan Movement for Rural Reconstruction (MGRR). The NFP is also in the process of building a Regional Agroforestry Training Center in Guatemala.

A Sister School program for tree-planting has been established in the USA. Belize and the Caribbean are also sites for urban forestry projects.

The NFP provides assistance to in-service U.S. Peace Corps training, which is helping in this outreach program along with the Inter-American Foundation, International Society of Tropical Foresters, Development Innovations Network, African NGOs Environment Network, and other liaison and media groups.

In commenting on future global cooperation among environmental groups, Conway believes this will lead to "more efficient use of resources, less duplication of work, more effective conservation efforts, and better north-south relations." He adds that strengthening relations with grassroots NGO organizations in developing counties is a constructive means of protecting the environment.

Already, innovative techniques have been implemented, such as "buffer zones around nature preserves and national parks, sustainable tropical forest harvesting techniques, and extractive reserves." Conway believes these conservation techniques are "the best hope for saving the rainforests (and) combine tropical forest conservation with economic development."

As visiting senior fellow in 1991–1992, **Dr. M. Taghi Farvar** advised on ICDP's environment and sustainable development programs in Africa, Latin America, and Eastern Europe, while also serving as senior advisor to the secretary-general, United Nations Conference on Environment and Development (UNCED).

■ Resources

The NFP seeks new agroforestry and alternative agriculture technologies for both the training program and various projects. Volunteers to attend conferences and special events and to distribute NFP brochures are also needed. NFP is prepared to share its database of more than 3,000 contacts in 110 countries with other supportive environmental groups.

Publications include a newsletter, brochures (in English, French, and Spanish), posters, technical assistance, training materials, and filmstrips.

International Council for Bird Preservation (ICBP) ⊕

c/o World Wildlife Fund—USA
1250 24th Street, NW, Suite 500
Washington, DC 20037 USA

Contact: **Ron Naveen**, U.S. Representative
Phone: 1-202-778-9563 • *Fax:* 1-202-293-9342

George L. Shillinger, Acting Director
Phone: 1-202-778-9649

Activities: Education; Research • *Issues:* Biodiversity/Species Preservation • *Meetings:* CITES

● Organization

Founded in 1922, the International Council for Bird Preservation (ICBP) is devoted to the conservation of birds and their habitats—pioneering the cause of nature conservation worldwide and committed today to sustainable use of natural resources. ICBP is now a federation of more than 350 member organizations representing some 10 million people in 111 countries.

ICBP's founders, "prominent bird enthusiasts in Europe and America, were among the first to realize that only global efforts could protect birds along their migration routes." Today, a professional team of conservation experts and scientists carries out ICBP's mission in protecting endangered birds worldwide and promoting public awareness of their ecological importance.

According to ICBP, "200 species of birds have become extinct in the last 400 years. Today, more than 500 are in immediate danger of disappearing forever" from habitat destruction, pesticide poisoning, indiscriminate hunting and trapping, and other human pressures. Birds act as important indicators of environmental health. It is assumed that what harms them today—pollution and disruption of natural systems—will sooner or later be a threat to people.

▲ Projects and People

More than 100 field projects in ICBP's annual **Conservation Programme** target priorities. For example, the **Protect the Parrots** campaign lobbies legislators and decisionmakers to change public attitudes and to end importation of caged birds that menaces their existence. Meanwhile, in isolated parts of southeastern **Brazil**, the

red-tailed parrot remains severely endangered by tropical forest destruction; to save this species, ICBP supports local efforts to turn an island into a refuge. In **Dominica**, forest habitat of the imperial amazon parrot—down to 60 birds—is now a reserve.

Cameroon's Kilum Mountain Forest Project is an example of ICBP's new directions on behalf of **integrated land management** that helps local people earn their livelihood while practicing long-term conservation, reports Dr. **Christoph Imboden**, ICBP director-general. The **Biodiversity Project** identifies "hot spots" on a global map where 2,400 species—about one-quarter of all species—are restricted to small areas as an oceanic island, mountain range, or small forest patch. "It is vital that we know where species that are likely to come under threat occur—conserving them before they are critically endangered is cheaper and more effective than last-minute 'fire-brigade' action," writes Dr. Imboden. So far, **Indonesia** and the regions along the Andean chain in **South America** stand out—with deforestation a growing peril.

In **Mauritius**, ICBP helped save 10 endemic bird species, including the pink pigeon—but predators and loss of forest habitat still threaten. ICBP has joined with the government and the Jersey Wildlife Preservation Trust to benefit both the birds and the people. **Wings across Europe** is a program supporting conservation groups in Eastern Europe where many important bird areas (IBA) are located; ICBP assists with efforts to designate **Romania's Danube Delta** as a biosphere reserve.

ICBP's traditional concern for **migratory birds** continues. New field projects in African countries contribute important knowledge about this mystery, and educational programs promote the idea that migratory birds are a shared benefit. ICBP urges governments to enforce existing laws and draft new ones to protect these birds, their nesting areas, and crucial feeding sites along their flyways.

For instance, a detailed recovery plan was designed for the **lesser kestrel** (*Falco naumanni*); a report on the impact of **pesticides on migrants** in western Sahel was distributed to conservation agencies in Africa and the Mediterranean; in **Morocco**, ICBP trained reserve managers and wardens in **wetlands ecology** and helped develop a conservation plan; to safeguard the winter habitat of the **Palearctic wader**, nature reserves are being established in the **Republic of Guinea**; efforts in **Cyprus** are aimed at ending illegal hunting of migrants; conservation efforts are being triggered in **Turkey, Italy, Malta,** and **Yugoslavia**.

To save the **Seychelles brush warbler**, ICBP purchased Cousin Island—noted for its seabirds. Once reduced to only 30 birds, the population of this endemic species has increased to 10 times that level. In **Bali**, woodcutters and collectors pose a double threat to the snow-white **Rothschild's mynah**, where fewer than 250 birds remain on their lone island; an ICBP biologist is studying their little-known ecology and increasing local interest in their welfare. ICBP continues searching for nesting sites of the **slender-billed curlew** (*Numenius tenuirostris*), with only about 10 remaining worldwide. To create awareness, the Malagasy government distributed an ICBP poster, *Endemic Birds of Madagascar*, to the island's schools. A local artist painted the birds, and the educational message is in both Malagasy and French.

The old-grown evergreen and semideciduous forests in **Ivory Coast's Taï National Park** are habitats for **230 bird species** among wildlife such as pygmy hippopotamus, Liberian mongoose, and chimpanzee. ICBP undertook bird surveys and noted six threatened Upper Guinea endemics: white-breasted guineafowl, rufous fishing owl, western wattled cuckoo-shrike, white-necked picathartes, yellow-throated olive greenbul, and nimba flycatcher. ICBP, World Wildlife Fund (WWF), and other groups are working with the government to help develop management plans, curb poaching, and perhaps encourage ecotourism as a local "non-destructive" industry.

Among other recent projects—a **neotropical bird book series is a** joint venture with the U.S. Fish and Wildlife Service and WWF-USA for countries in the Americas lacking such reference material; an **environmental education** unit is being set up at Giza Zoo, Egypt; **seashore birds**, particularly the roseate tern, are getting protection in Ghana; conservation is underway for the **Hadejia/ Nguru wetlands** in Nigeria; **Arabuko-Sokoke Forest** management and research is at this key avian habitat in Kenya; efforts to halt the loss of **maleo** and exploitation of its eggs are focused in Salawesi, Indonesia; a management plan is being developed for the **forested Phulchowski Mountain** in Nepal; important sites are being surveyed and conservation biologists are being trained in the Philippines; support is provided for the **Forest Birds Working Group** in Vietnam and a pheasant survey is being undertaken.

A related organization—International Waterfowl and Wetlands Research Bureau (IWRB)—reports that in the USA the ingesting of lead shot results in the death of some 2.4 million wildfowl annually, mainly in wetland habitats. A movement is growing to replace lead gunshot with nontoxic cartridges through legislation.

To build membership and fundraising, ICBP has launched the World Environmental Partnership appealing to the business community. In a cooperative venture, ICBP and the Federation of British Zoos are raising funds to aid the Bali starling in Indonesia.

■ Resources

ICBP prints several publications. The *Bulletin* highlights conservation achievements. The quarterly newsletter, *Birdwatch*, keeps its worldwide network of members up to date. Technical publications examine major bird conservation issues in detail. ICBP study reports describe research undertaken and conclusions reached by project investigators. An award-winning series of books, *Save the Birds*, is published in 10 languages and tailored so far for 14 countries; Pro Natur in Germany is co-producer and main author is **Dr. Anthony W. Diamond.** A publications list is available.

As ICBP monitors the status of the world's threatened birds and their habitats, a computerized databank is constantly updated. ICBP, in cooperation with the World Conservation Union (IUCN), compiles the international *Bird Red Data Book*, which is used by government decisionmakers and conservationists.

ICBP also has a **World Bird Club** open to individuals who want to help protect birds.

(See also International Council for Bird Preservation in the UK NGO section.)

International Crane Foundation (ICF) ⊕
E-11376 Shady Land Road
Baraboo, WI 53913 USA

Contact: **George Archibald, Ph.D.,** Director
Phone: 1-608-356-9462 • *Fax:* 1-608-356-9465

Activities: Development; Education; Research • *Issues:* Biodiversity/Species Preservation • *Meetings:* Crane Workshops Worldwide

● Organization

The **International Crane Foundation (ICF)**, with 4,500 members and a staff of 19, breeds 15 species of cranes, including some of the largest, rarest, and most beautiful birds on Earth. Action by the ICF resulted in the conservation of more than five million hectares of wetlands in Asia, mostly in China and the former USSR. Conservation-education programs have been implemented among local people in remote regions of Africa, Australia, and Eurasia. ICF has a "species bank" of captive cranes and has successfully bred several endangered crane species.

▲ Projects and People

The total count of western flocks of the **Siberian Crane** is down to 14. The ICF went into the wilderness of western Siberia to contribute several eggs in hopes that the eggs would hatch, develop, and migrate with the rest of the flock on their 5,000-mile trek. "Our fear is that, if this flock is lost, the migration tradition will be lost, and then it will be too late to reintroduce cranes back into the western [former] USSR, Iran, or India, because they won't know where to go." The ICF plans to fit the fledglings with radio transmitters so they can be tracked by satellites.

To ICF's happy surprise, recent winter surveys in remote wetlands of Tibet turned up more than 2,800 **black-necked cranes**, least known among the 15 species. "Our findings more than double the known world population to 3,900 cranes," writes **Dr. Mary Anne Bishop,** ICF team leader. The government is planning a nature reserve in Linzhou County.

ICF designs school curriculums for children unable to visit the Baraboo site. One set, *Kids, Cranes, and Conservation,* is aimed at third, fourth, and fifth graders. The *Restoring the Land* curriculum, with experiments, exercises, and practical information on how to start or restore a prairie, is suitable for high school students as well as garden clubs and civic groups.

Dr. George Archibald has studied the ecology of eight species of cranes in Australia, China, Iran, India, Japan, Korea, and the USA. "I helped organize nine Working Groups on Cranes, including more than 900 researchers in 64 nations," he writes. Such groups meet every two to four years and publish research reports. Chairman of the Species Survival (IUCN/SSC) Crane Group, Dr. Archibald has appeared on the Global 500 honour list of the United Nations Environment Programme (UNEP) and was presented with the Order of the Golden Ark by Prince Bernhard of the Netherlands.

■ Resources

Publications include *Reflections: The Story of Cranes; Raising Crane, ICF Bugle* newsletter; and a video called *A Place for Whooping Cranes.* ICF has a new library and a visitors center complete with museum displays, slide shows, school packets, gift shop, and foot path. ICF needs assistance in purchasing video monitors to observe cranes for research purposes, with marketing a newly published book, and with mailing lists.

International Development Conference (IDC) ⊕

1401 New York Avenue, NW, Suite 1100
Washington, DC 20005 USA

Contact: **Andrew E. Rice,** Chairman Emeritus
Phone: 1-202-638-3111 • *Fax:* 1-202-638-1374

Activities: Education • *Issues:* Sustainable Development
Meetings: Society for International Development; UNCED

● Organization

The International Development Conference (IDC) describes itself as "a coalition of individuals associated with leading national organizations" primarily concerned with USA–Third World issues and desiring to educate Americans about these issues.

Supported by grants from individuals and organizations, IDC holds national meetings; co-produces the quarterly national newsletter *Ideas and Information about Development Education,* which deals with the theory and practice of educating Americans about sustainable development; promotes policy dialogues with official and private groups, such as on how to incorporate in a positive manner the Third World into the Bush Administration's "New World Order"; and promotes a series of **major global awards** to recognize outstanding individual and institutional contributions in international development. These awards are being given in memory of late President Harry S Truman, who established the USA's first major programs of international assistance.

A **biennial national conference** on world development issues is scheduled for 1993. In addition, IDC works closely with the U.S. Citizens Network on UNCED on disseminating information at the UN Conference on Environment and Development. The group is also involved with the Society for International Development, a worldwide professional association of persons engaged in sustainable development activities.

■ Resources

The above newsletter and reports from the biennial conferences are published; the most recent one is *From Cold Water to Cooperation: Dynamics of a New World Order.*

International Food Policy Research Institute (IFPRI) ⊕

1776 Massachusetts Avenue, NW
Washington, DC 20036 USA

Contact: **Barbara Rose,** Director of Information
Phone: 1-202-862-5600 • *Fax:* 1-202-467-4439

Activities: Research • *Issues:* Policy Analysis for agricultural development and economic growth of developing countries; Sustainable Development

● Organization

"As the policy center of the Consultative Group on International Agricultural Research (CGIAR) . . . the International Food Policy Research Institute (IFPRI) carries out research that promotes the use of, and enhances the returns on, the technological improvements developed by the other CGIAR centers," reports **Barbara Rose.**

With a staff of 140, IFPRI's research recognizes that the environmental matters associated with sustainable agriculture occupy a prominent place among policy issues facing the world, and that they pose particular challenges for developing countries, where "environment, degradation of the environment, and poverty are inseparable concerns." Rapid population growth, poor nutrition and hygiene, and low levels of education compound the difficulties of dealing with fragile ecosystems while technological change is being introduced.

▲ Projects and People

Research into **fertilizers and irrigation** reveals, for example, that quality of irrigation systems in the **Philippines** improves with the operation's financial autonomy, which government policies should encourage. It was also found that a balanced application of chemical fertilizers and organic manures stabilizes yield responses; and that "improved agricultural technologies in **India** was associated with declining fertility rates." One recent deforestation study in **Nepal** in the populated hill areas shows that over a decade the amount of time women spent collecting firewood and fodder increased by 1.13 hours a day. "This lowered agricultural productivity, which, in turn, reduced food consumption by about 100 calories per day."

As part of such research efforts, IFPRI works in information on natural resource bases—climate, soils, and topography relevant to agriculture—and uses this information to define agroecological zones by productivity, given agronomic management conditions, population densities, market linkages, and other socioeconomic conditions. In a project undertaken with the **International Center for Tropical Agriculture (CIAT),** IFPRI used these data to develop a system to identify priority regions for technological and policy intervention in the **Peruvian Amazon,** an area threatened by environmental degradation. The research also "identified the need for technological options in the short run to minimize agriculturally induced environmental degradation, while devising alternative production systems to enhance the comparative advantage of the Amazon in perennial and other crops."

Cooperating with the **International Center for Agricultural Research in the Dry Areas (ICARDA),** IFPRI identified measures to raise agricultural production while improving the deteriorating agricultural resource base in the **Middle East and North Africa**—where "waterlogging in irrigated areas, increased competition for water, and serious erosion and desertification of rainfed land through uncontrolled mechanization and extension of cultivation to marginal areas" has occurred.

With collaborators in **Brazil,** IFPRI identified and measured the extent to which **agroecological constraints determine the direction and speed of agricultural development** and how the policy environment affects this. The research suggests that in the early stages of agricultural development, "agroecological constraints are more important than the policy environment in determining which crops a farmer grows and which inputs he uses, but as development proceeds, technological change and the policy environment that promotes it become increasingly important and may eventually override ecological and climate constraints."

Recently, IFPRI has begun to identify **priorities in forestry and agroforestry policy** research for future consideration. These are expected to be the "effect of forest and upper watershed degradation on agricultural productivity; appropriate land use in wet tropical forest zones; options for reclamation and utilization of degraded forest lands in dry regions; the role of forestry and agroforestry in the generation of rural income and welfare; and the effect of macroeconomic policies on forest exploitation."

Key IFPRI staff members are the Environment and Production Technology Division's **Mark Rosegrant**, director; **Steve Vosti**, research fellow; and **Peter Oram**, research fellow emeritus.

■ Resources

A list of research reports and other publications is available. Some are printed in French and Spanish as well as English; a few abstracts appear in Arabic. Most IFPRI publications can be obtained at university libraries.

International Institute for Energy Conservation (IIEC) ⊕
750 First Street, NE, Suite 940
Washington, DC 20002 USA

Contact: **Deborah Lynn Bleviss**, Executive Director/ President
Phone: 1-202-842-3388 • *Fax:* 1-202-842-1565

Activities: Development • *Issues:* Air Quality/Emission Control; Energy; Global Warming; Sustainable Development; Transportation

● Organization

Founded in 1984, the **International Institute for Energy Conservation** (IIEC) seeks to benefit developing countries by increasing the efficiency of energy services needed for development so energy demand can be slow and stable while the economy thrives.

With a main office in Washington, DC, and a regional office in Bangkok, **Thailand**, IIEC has these thrusts: **information dissemination** among decisionmakers through workshops, conferences, and seminars; end-use analysis training programs for local experts; and the development of an energy efficiency information bank. IIEC gives **demonstrations** that illustrate both technological and financial efficiency and seeks projects that can serve as models for key industries or energy subsectors. The **development of local energy conservation business activity** is a third main goal. IIEC supports innovative business companies in this field and helps them gain access to others as well as to financing and essential technologies throughout the industrialized world.

IIEC recognizes that transfer of these technologies to the developing countries is very slow and is committed to generating projects that "simultaneously yield real energy savings in developing countries and serve as models for others to replicate."

▲ Projects and People

Under the sponsorship of the Chinese government, the IIEC has identified efficiency improvements at the Maanshan Iron and Steel Company and other steel businesses with the aid of six American companies in the **China-USA Metallurgical Energy Conservation Demonstration Project**.

A **Global Energy Efficiency Initiative** (GEEI) was launched in 1990 to develop a worldwide partnership among industrialized and developing countries to incorporate energy efficiency improvement consistent with national needs. Problems being addressed are oil vulnerability, global warming and other forms of environmental degradation, and capital constraints on development investment.

The first of several GEEI Working Groups—diverse informal coalitions—consists of U.S. government agencies, universities, national laboratories, private companies, environmental and energy organizations, and other non-government groups who work together to promote energy-saving investments worldwide. Members include the Environmental Defense Fund, American Public Power Association, and the U.S. Environmental Protection Agency (EPA). Efforts feature creating model programs, securing financing, promoting trade in needed energy efficiency and solar technologies, and building technical capacity within developing and Eastern European countries to design and implement policies.

Other projects focus on **strategies** and **end-use analysis training** for Thailand, a multilateral development bank guide to help such institutions alter their lending practices to spark energy efficiency, and **transportation energy growth** workshops.

■ Resources

IIEC and GEEI brochures are available. Conferences and workshops are held periodically.

International Marinelife Alliance, Inc.—USA (IMA) ⊕
201 West Stassney, Suite 408
Austin, TX 78745 USA

Contact: **Peter J. Rubec, Ph.D.**, President
Phone: 1-512-326-5265 • *Fax:* 1-512-326-4017

Activities: Education; Research • *Issues:* Biodiversity/Species Preservation; Sustainable Development

● Organization

The **International Marinelife Alliance** (IMA) is changing the way aquarium fish are collected from tropical coral reefs to protect the health of marine species and their natural habitats worldwide—and impact positively on local communities.

IMA is a nonprofit group with 200 members working through economic and technical cooperation, education, and research that is both environmental and humanitarian in its global concerns. It aims to conserve the diversity of marine life, protect and restore marine environments, promote the sustainable use of marine resources, enhance the quality of life and income of small-scale fisherfolk, foster participation of fisherfolk in managing their marine resources, provide enduring solutions through village-level education and training that encourages the use of environmentally sound harvesting methods, and sponsor and carry out relevant marine life research.

▲ Projects and People

With the highest diversity of colorful marine fishes of the world's reef areas, about 200 of the 2,177 species in the **Philippines** are exported for the pet fish industry. Since the early 1960s, the highly competitive enterprise has employed a quick, although toxic, method of collecting fish that is causing coral reefs to be destroyed and aquarium and food fishes to decline. In turn, the illegal practice of using squirts of sodium cyanide (NaCN) from plastic bottles to stun its catch ultimately robs local people of dietary protein that healthy coral reefs produce as well as coastal employment.

Deteriorating reefs produce less than five metric of fish yearly, and about 70 percent of the reef ecosystems are in this condition because of poisons such as NaCN, unlawful use of explosives, sedimentation and siltation results of deforestation, trawling, and muro-ami and kayakas fishing, according to the IMA. As a result, reef destruction affects local communities—leading to "malnutrition, poverty, and migration to city slums," writes Dr. Peter J. Rubec.

At first, it was thought that NaCN temporarily paralyzed marine fish so they would be easily collected among the complex coral configurations. But research shows that the metabolic process makes fish susceptible to stress and prone to fatal convulsions or Sudden Death Syndrome (SDS). The chemical cloud also kills fish trapped in the coral reefs. Cyanide is also used illegally to catch grouper and other food fish imported to Hong Kong, reports the IMA.

"More than 80 percent of the aquarium fishes captured die throughout the chain from the collector to the marine hobbyist," writes Rubec. According to the Coral Reef Research Section of the Philippine Bureau of Fisheries and Aquatic Resources (BFAR), the repeated use of NaCN kills coral heads, and there are 33 million coral heads sprayed with cyanide in the Philippines each year.

In aggressively promoting the use of environmentally sound methods to preserve the reefs and insure a livelihood for as many as 700,000 nearshore subsistence fishermen, IMA is training cyanide-using aquarium fish collectors and divers to use fine-mesh hand and barrier nets instead. The group is also developing a reliable cyanide detection test for screening tropical fish, before leaving the Philippines.

The "government under Corazon Aquino wishes to protect the marine environment through the creation of a viable marine conservation management plan, enforcement of existing laws, and through cooperation between government, non-government, and international agencies to provide education and research," Rubec points out. The Department of Agriculture, the authority for fisheries, designated the IMA as the lead NGO for fundraising, net-training, and finding alternatives to destructive fishing methods.

The IMA recommends village-based management plans which result from collaborative efforts between Philippine government agencies, and conservation and user groups—including fishermen, fish collectors, coral gatherers, and aquaculturists. A Coordinating Committee on Illegal Fishing was formed with this purpose. IMA also encourages the proper management of marine parks and reserves, marine research in fisheries in aquaculture, ongoing education programs, multiple use zoning, regulatory measures to ensure enforcement, and incorporation of a Marine Resource Management Plan into a larger Philippines National Conservation Strategy.

Dr. Rubec is also a coastal fisheries research specialist with the Texas Parks and Wildlife Department.

■ Resources

IMA seeks to intensify its net-training marine education efforts and related public information programs in the Philippines. *Sea Wind* is a quarterly bulletin for members; books, tapes, and T-shirts are available.

(See also Patricia C. Almada-Villela in the UK NGO section.)

International Oceanographic Foundation (IOF) ⊕

4600 Rickenbacker Causeway
Miami, FL 33149 USA

Contact: **Bonnie Gordon,** Editor, *Sea Frontiers*
   Phone: 1-305-361-4888 • *Fax:* 1-305-361-4131

Activities: Education • *Issues:* Air Quality/Emission Control; Biodiversity/Species Preservation; Energy; Global Warming; Waste Management/Recycling; Water Quality

● Organization

The International Oceanographic Foundation (IOF), with 45,000 members, has been dedicated to ocean ecology, education, and marine research for more than 35 years. It sponsors expeditions to foreign lands to experience sea life through special travel programs. With 40 destinations annually, these IOF Sea Safaris, led by a naturalist or scientist, go to some of the world's most pristine and fascinating areas. The University of Miami, National Aeronautics and Space Administration (NOAA), and scientific organizations worldwide offer technical assistance to IOF.

▲ Projects and People

Many mangrove creeks and channels riddle the shore around Biscayne Bay, Florida. Canal life may have a coffee-table-sized sea turtle, a three-foot nurse shark, a seven-foot lemon shark, or a sea horse or juvenile barracuda or queen angelfish swimming underwater, as described in IOF's publication *Sea Frontiers*. Mangrove leaves fall from the trees, decompose, then are eaten by larval crabs, shrimp, and molluscs, which are eaten by larger shrimp, clams, and lobsters, which are eaten by larger fishes. Seventy-five percent of Florida's game fish and 90 percent of its commercial fish, lobster, and shrimp grow fat at some point on food from mangroves, according to IOF.

Across the planet, the Great Barrier Reef of Australia excels in biological diversity with more than 300 different species of corals. Corals resemble "wheat sheaves, mushrooms, staghorns, cabbage leaves and a variety of other forms, glowing under water with vivid tints of every shade betwixt green, purple, brown and white," said Commander Matthew Flinders in the 1800s. Most corals are colonies of individually connected coral polyps. Each polyp secretes a limestone "cup" made of calcium carbonate from sea water. As the colony grows, the polyps increase in number and divide by asexual reproduction. Coral larvae also settles on dead corals, as described in *Sea Frontiers*. Corals grow very slowly. Some coral is as old as 20 million years, and coral in the central part of the Great Barrier Reef is "young," between 500,000 to 1 million years old. Fortunately, 98.5 percent of the Great Barrier Reef is protected as a marine park, created in 1975.

■ Resources

The IOF publishes bimonthly a full-color magazine, *Sea Frontiers*, which documents important scientific advancements, and calls to mind the beauty and value of the seas. It also includes a review of books, and offers two pages of gift items.

International Rivers Network (IRN) ⊕
1847 Berkeley Way
Berkeley, CA 94703 USA

Contact: **Owen Lammers,** Director
  Phone: 1-510-848-1155 • *Fax:* 1-510-848-1008
E-Mail: Econet: irn

Activities: Development; Education; Political/Legislative; Research • *Issues:* Biodiversity/Species Preservation; Dams/Irrigation; Deforestation; Energy; Health; Population Planning; Sustainable Development; Waste Management/Recycling; Water Quality • *Meetings:* ICOLD; UNCED; World Bank Directors

● Organization

The **International Rivers Network (IRN),** an affiliate of Friends of the Earth—International (FOE), is dedicated to preserving the world's rivers and watersheds. Founded in 1985, IRN has approximately 1,300 members including environmentalists, engineers, hydrologists, human rights activists, and academics who are committed to the study and defense of rivers and riverine communities. IRN says it has a network of citizens organizations and technical experts in more than 80 countries who "work to protect freshwater resources, endangered ecosystems, and the rights of indigenous peoples worldwide."

To IRN, rivers are among the planet's most important natural resources, with their waters sustaining and nourishing an incredible diversity of life. "Yet today, in the name of development, our rivers are dying as a direct result of reckless damming, channeling, and polluting," IRN reports. Problems are the severest in developing nations.

According to IRN, consequences include: "irreversible damage of natural treasures—free-flowing rivers and streams and the life they support; the forced displacement of millions of people; the loss of productive farmland to irrigation schemes that cause waterlogging and salinization; the destruction of fisheries, wetlands, estuaries, and the livelihood of people who depend on them, and an abrupt and often unpopular shift from sustainable farming practices to a cash-crop based agricultural economy."

IRN's underlying purpose is to update and foster development of water policy worldwide by providing a means of communication between members. IRN states that unless action is taken, most of the world's major rivers will be dammed within the next 25 years, and that modern dams precipitate environmental problems.

▲ Projects and People

With deep concern about large-scale development projects that are funded by the World Bank and other major lending institutions—particularly those containing large dams—IRN's World Bank Campaign exerts pressure through technical critiques, publications, and demonstrations, and by mobilizing popular opposition. Its goal is to convince the World Bank and other similar institutions to halt funding of ecologically destructive water projects.

In its literature, IRN describes villages adversely affected in India due to flooding resulting from the construction of Sardar Sarovar Dam in Gujarat. This submergence prior to resettlement plans, IRN says, is "in direct violation of the Narmada Tribunal." The group reports on local opposition to the Pangue Dam project on

Chile's Bio Bio River because of its perceived impact on the native Pehuenche people; on efforts to clean up Colombia's most significant river with the help of the newly formed Foundation for the Defense of the Magdalena River (FUNDEMAG); and on mosquitoes infesting communities near Brazil's Tucuruí reservoir, a hydroelectric project in the Amazon.

IRN also supports a USA/Canada coalition to create a huge international wilderness preserve that will prevent open-pit mining of copper, gold, silver, and cobalt in British Columbia's Windy Craggy Mountain and keep Tatshenshini River free from pollution as well as protect the region's grizzly bears, silver blue glacier bears, dall sheep, mountain goats, wolves, moose, gray falcons, and bald eagles. The river flows from the Yukon, through British Columbia and Alaska, and into the Gulf of Alaska.

IRN is disturbed about what it considers to be "unsafe designs, non-sustainable development policies, the forced relocation of indigenous communities, and the virtual obliteration of fragile river ecosystems." Consequently, IRN pressures associations such as the International Commission on Large Dams (ICOLD) to establish professional guidelines and standards.

IRN is building coalitions among conservation and river recreation organizations in the USA to save the world's river systems, and has established a river defense fund to provide direct financial support to pressing campaigns.

■ Resources

IRN publishes the *World Rivers Review,* a bimonthly newsletter about international water development issues; *Bankcheck Quarterly,* regarding World Bank activities; and special briefings about particular campaigns or threatened rivers. IRN has an information clearinghouse on freshwater issues, including a database on rivers and dams worldwide. It offers technical assistance to river protection organizations through an advisory panel of water professionals and in-house staff engineers, operates a speakers' bureau, and maintains a library on global water resources.

International Snow Leopard Trust ⊕
4649 Sunnyside Avenue North, Suite 342
Seattle, WA 98008 USA

Contact: **Helen Freeman,** President
  Phone: 1-206-632-2421 • *Fax:* 1-206-632-3967

Activities: Education; Research • *Issues:* Biodiversity/Species Preservation; Conservation Education; Sustainable Development • *Meetings:* International Snow Leopard Symposium; World Parks Congress

● Organization

"The demise of the snow leopard, regardless of how magnificent it is and how it may nourish our souls, may seem insignificant when compared to the environmental degradation threatening our entire planet," writes Helen Freeman, president, International Snow Leopard Trust. Yet because the snow leopard is at the top of the predator chain in much of the central Asian mountain region, it serves as an indicator species. "Its demise is a signal that something is going wrong in one of our natural ecosystems. The likelihood is great that other species will follow," she warns.

There are currently only about 1,500 wild snow leopards left in its historic range that includes Afghanistan, Bhutan, China, India, Mongolia, Pakistan, and the former USSR.

Organized in 1981 to combat ignorance about the snow leopard and its ecosystem, the International Snow Leopard Trust is comprised of more than 750 members, including scientists and concerned citizens. The Trust serves primarily to link related organizations in a common cause: to protect the snow leopard by combatting poaching and the loss of mountain wilderness areas.

▲ Projects and People

In order to reach its goals, the International Snow Leopard Trust works spread awareness of the snow leopard's plight; workshops and international conferences open to wildlife and reserve managers in snow leopard regions encourage information sharing. In cooperation with the Northwest Plateau Institute of Biology (under the Chinese Academy of Sciences) and the Qinghai Bureau of Agriculture and Forestry, the Trust is planning the Seventh International Snow Leopard Symposium in Qinghai, China; the symposium addresses the needs of local peoples, wildlife, and land relative to the snow leopard.

In addition, the Trust fulfills more concrete needs. It has provided local wildlife officers and field personnel with equipment and supplies to stop poachers. Working with local governments and through citizenry information centers, the Trust has helped establish wildlife reserves and funded existing ones. These reserves include **Hemis National Park** in India's Jammu and Kashmir, **Khitral Gol National Park** in Pakistan, and **Qomolangma (Everest) Nature Reserve** between Tibet and Nepal.

In her capacity as president, Freeman is currently co-chairperson of the 1992 Qinghai symposium, China. In addition, she directs the SLIMS project—the **Snow Leopard Information Management System**. This information network links the central Asian countries with current wild snow leopard populations.

A member of the Species Survival Commission (IUCN/SSC) Cat Group, Freeman also advises the American Association of Zoological Parks and Aquariums (AAZPA) in its Snow Leopard Species Survival Plan.

■ Resources

In addition to SLIMS, the Trust publishes a newsletter for members, *Snowline*, and has published proceedings from previous international symposia. Other publications include *A Review of the Status and Ecology of the Snow Leopard* and *An Annotated Bibliography of Literature on the Snow Leopard*, both by J.L. Fox.

International Society for the Preservation of the Tropical Rainforest (ISPTR) ⊕
3931 Camino de la Cumbre
Sherman Oaks, CA 91423 USA

Contact: **Arnold Newman**, President
Phone: 1-818-788-2002, 1-818-990-3333

Activities: Education; Political/Legislative; Research • *Issues:* Biodiversity/Species Preservation; Deforestation; Global Warming; Health; Population Planning; Sustainable Development • *Meetings:* UNCED

● Organization

The International Society for the Preservation of the Tropical Rainforest (ISPTR) was "founded in response to growing concerns for the cumulative effects of tropical deforestation, habitat destruction and the loss of countless species of plants and animals to unnatural extinction." The ISPTR works "at governmental and village levels, primarily in Peru and Brazil, to help the local people develop sustainable alternatives to various practices which threaten the rainforest ecosystem."

▲ Projects and People

Co-founder and president of ISPTR, **Arnold Newman** has had 25 years of experience working in tropical forests, in 31 countries worldwide. Newman, a nature photographer and television producer, wrote *Tropical Rainforest: A World Survey of Our Most Valuable and Endangered Habitat with a Blueprint for Its Survival* cited by environmentalists as a definitive work in the field. In addition, he served as chief of the U.S. delegation negotiating the U.S.-Soviet Cooperative Initiative on Tropical Deforestation in 1987, and has advised governments on and organized expeditions to various rainforests.

Specific projects of ISPTR focus both on the forests and on their indigenous peoples. **Project PARD—Preservation of the Amazonian River Dolphin**—helped establish the first river dolphin preserve on that river. The **Adopt an Acre of Rainforest** project has enabled ISPTR to preserve 77,000 acres in northern Peru, working with the Peruvian Ministry of Agriculture and three local villages. In addition, the Society has established a **species survival sanctuary** in Brazil, with plans for a children's museum, research station, and veterinary hospital.

Another project of ISPTR involves **educational conservation tourism**; through proper tourism development, the Society believes, the local area gains income, and both tourists and natives become more aware of the need to preserve the natural resources there. Although local Indians and river people also receive clothing, medical and educational supplies distributed by the Society, ISPTR considers that it is more important to develop "positive sustainable co-ops for river people without disrupting their indigenous cultures."

■ Resources

ISPTR publishes a newsletter. Arnold Newman's *Tropical Rainforest* book is also available.

International Society of Tropical Foresters (ISTF) ⊕

5400 Grosvenor Lane
Bethesda, MD 20814 USA

Contact: **Warren T. Doolittle**, President
Phone: 1-301-897-8720 • *Fax:* 1-301-897-3690

Activities: Education • *Issues:* Biodiversity/Species Preservation; Deforestation; Global Warming; Sustainable Development • *Meetings:* IUFRO Congress; Society of American Foresters; UNCED; World Forestry Congress

● Organization

Originally organized in 1950, the nonprofit **International Society of Tropical Foresters (ISTF)** is increasingly active as concern grows regarding rapid development of tropical forests. With some 2,000 members in 110 countries, its main objective is the transfer of technology and science to those concerned with the management, protection, and wise use of tropical forests.

With headquarters in Bethesda, Maryland, ISTF has three regional directors in Africa, Asia, and Latin America, and four directors-at-large. More than 60 countries have vice presidents and others are being sought. Financial support comes from membership dues, contributions, and grants. To date, the Tinker Foundation, World Wildlife Fund/Conservation Foundation, Foundation for Professional Forestry, and the U.S. Forest Service have made grants to ISTF.

▲ Projects and People

According to ISTF, most tropical forests are located in developing countries where the conservation and development of forests is often a controversial subject. "Deforestation and the interrelationship of agriculture, forestry, and industry are not well understood, so the prompt transfer of technical knowledge and its application is of prime importance to all people," states ISTF.

Projects are mainly designed to fill a need for a **communications network** among tropical foresters and others concerned with the fate of the forests.

Workshops and symposia are sponsored by ISTF in cooperation with other organizations. Local and regional chapters are continually being formed.

■ Resources

Available to members, *ISTF News*, the Society's quarterly newsletter, has editions in Spanish and English; a French version is upcoming. Members are encouraged to contribute news items on developments in tropical forestry, reviews of technical literature, and reports of accomplishments in the field. In addition, each issue contains a column for positions wanted. New members may place up to a fifty-word summary of their qualifications in one issue at no extra charge. A membership directory with names, addresses, and areas of technical interest is published periodically in the newsletter.

Individuals and organizations interested in tropical forests can contact ISTF headquarters to receive information about membership and a descriptive brochure in three languages. Funds to expand publications are sought.

International Water Resources Association (IWRA) ⊕

University of Illinois
205 North Mathews Avenue
Urbana, IL 61801 USA

Contact: **Glenn E. Stout**, Vice President and Executive Director
Phone: 1-217-333-0536 • *Fax:* 1-217-244-6633
E-Mail: stout@uiucvmd

Activities: Development; Education • *Issues:* Global Warming; Health; Sustainable Development; Water Quality • *Meetings:* IWRA Triennial World Congresses

● Organization

The **International Water Resources Association (IWRA)**, founded in 1972, describes itself as "an interdisciplinary worldwide organization for water managers, scientists, planners, manufacturers, administrators, educators, lawyers, physicians, and others concerned with the future of our water resources." Because there is an "urgent need for integrated long-term approaches to managing and developing the world's water resources," IWRA provides an "international forum and network "to promote . . . cooperation among professions in water resources fields."

▲ Projects and People

IWRA organizes/cosponsors more than 12 conferences, workshops, and symposia each year, with topics ranging from simple hand pumps to major hydroelectric projects. Its **triennial World Congresses**, lasting five days, draw water experts from around the world; the most recent was held in May 1991 in Rabat, Morocco.

To encourage excellence in water-related fields, the Association offers two **annual** and two **triennial awards**, to organizations or individuals who contribute to improving the world's water situation, to adapting water resources technology in a water management program, for outstanding papers, and for exemplary service to science and humanity.

IWRA encourages coordination with and support of international programs such as those sponsored by the United Nations and other international organizations.

■ Resources

Since 1975, IWRA has published a quarterly journal, *Water International*, addressing issues related to global water problems in both developed and developing nations. The Association also publishes congress, conference, and seminar proceedings.

International Wilderness Leadership Foundation (IWLF) (Wild Foundation) ⊕
211 West Magnolia
Fort Collins, CO 80521 USA

Contact: **Vance G. Martin**, President
Phone: 1-303-498-0303 • *Fax:* 1-303-498-0403

Activities: Direct Assistance; Education • *Issues:* Biodiversity/ Species Preservation; Wilderness • *Meetings:* World Wilderness Congress

● Organization
Established in 1974, the International Wilderness Leadership Foundation (IWLF) (Wild Foundation) describes itself as a "corporation working on projects which concern wilderness, wildlife and people." A small organization, with three full-time and eight volunteer staff members as well as other direct assistance volunteers, IWLF focuses its efforts on southern Africa—while concentrating on the "biological, economic, cultural, and spiritual values of wilderness."

▲ Projects and People
The IWLF has as its goals protecting the wilderness and wildlife; promoting the wise use of the Earth's wilderness areas; and providing education and training on the environment. To reach these ends, the Foundation works with various governments and organizations on many levels.

In southern Africa, the IWLF has ongoing projects in Zimbabwe, Botswana, Namibia, South Africa, and Mozambique. With the local governments, conservation groups and the private sector, IWLF has helped provide antipoaching equipment and other protective measures for wildlife, and has promoted ecotourism. In Namibia alone, IWLF-sponsored programs are at work to preserve the cheetah, rhino, and elephant.

Educational programs include the Wilderness Leadership Institute, which conducts wilderness experience programs in several countries leading to personal growth and an increased environmental awareness. Currently, the IWLF is working with the Bureau of Land Management, the U.S. Forest Service, the National Park Service, and the Fish and Wildlife Service to offer a course in wilderness management at Colorado State University.

Every three to five years, the IWLF conducts the World Wilderness Congress; the fifth is being held in Norway in 1993. At these congresses, participants "discuss—and *act* on—issues and projects involving wilderness, worldwide conservation, and sustainable development."

■ Resources
IWLF publishes proceedings of its World Wilderness Congresses. In addition, its newsletter *The Leaf* appears three times each year. The Foundation also offers the sourcebook *Wilderness Management*.

The Izaak Walton League of America (IWLA) ❧
1401 Wilson Boulevard, Level B
Arlington, VA 22209-2318 USA

Contact: **Maitland Sharpe**, Executive Director
Phone: 1-703-528-1818 • *Fax:* 1-703-528-1836

Activities: Education • *Issues:* Air Quality/Emission Control; Biodiversity/Species Preservation; Energy Efficiency; Stream Habitat Restoration; Sustainable Agriculture; Water Quality

● Organization
Named after the seventeenth-century English fisherman and author of *The Compleat Angler*, The Izaak Walton League of America (IWLA) was founded in 1922. It has since gathered a membership of 53,000, and boasts 21 regional and 400 local chapters. In the 1920s at the request of Calvin Coolidge, the League started a national campaign for clean water, which remains a top priority. Nevertheless, it has broadened its scope to cover other areas of environmental concern such as sustainable agriculture; urban open space, forestry, and transportation; U.S. energy policy; and stripmine controls on land and sea. Ultimately, the League's goal is to "promote citizen involvement in environmental protection efforts and educate the public about emerging national resource threats."

▲ Projects and People
In part because of the League's efforts, the U.S. Congress passed the Clean Water Act of 1972 and a newer version in 1990. Much of the League's work, however, takes place on the regional and local level, through programs like Save Our Streams; Wetlands Watch, which encourages people "to identify, adopt, and protect these wildlife nurseries"; Partners for Wetlands, which matches donors with individual wetland habitat projects on privately owned land and also gets matched funds from the U.S. Fish and Wildlife Service; the Uncle Ike Youth Education Program that teaches conservation to boys and girls; and the Outdoor Ethics Program to "reduce poaching, trespassing, and other illegal and inconsiderate activities" while promoting responsible fishing, hunting, and hiking. Funds raised through the ethics program paid for a helicopter donated to antipoaching officers on the Gulf of Mexico coast.

Regional programs, for example, emphasize saving the Chesapeake Bay, the USA's largest estuary; balancing competing uses of northwestern public forests and rangelands with vital water, fish, and wildlife resources; protecting species and advancing energy efficiency and clean-air issues in the Upper Mississippi River region; and combatting soil erosion and groundwater pollution while promoting sustainable agriculture and farm conservation in the Midwest.

Yet, there is distress about global climate change and its potential effects on wildlife in parks, refuges, and elsewhere. The League suggests a way to "buy time [by] driving fuel-efficient automobiles, planting trees, and encouraging global, national and local energy conservation policies." It also recommends urging "federal policymakers and wildlife managers to consider climate change impacts" when making decisions.

At the grassroots level, League volunteers—known as "Ikes"— have fought for the protection of the Everglades in Florida, defeated a landfill proposal in Tennessee, built fences, planted trees, organized cleanups, banded ducks—putting their energy to work for the environment.

Jack Lorenz, former executive director and creator of the League's Outdoor Ethics Program, has received much recognition for contributions to the environment. Co-author of the 1985 *Environmental Agenda for the Future*, Lorenz holds volunteer positions with the Wild Habitat Enhancement Council, the Alliance for Environmental Education, and the North American Wetlands Conservation Council, as well as with many other environmental organizations.

■ Resources

In addition to *A Grassroots Guide to Legal Action* and *Conservation Policies* among numerous educational pamphlets and brochures, the League produces four quarterly magazines: *Outdoor America*, for all members; *Splash*, for Save Our Stream members; *Outdoor Ethics*, directed toward resource managers and recreationists; and *League Leader*, a publication specifically for IWLA local and regional officers. *Make Peace with Nature*, a weekly environmental television program, airs on more than 50 PBS stations around the country.

Save Our Streams Program (SOS) ❦
The Izaak Walton League of America (IWLA)
1401 Wilson Boulevard, Level B
Arlington, VA 22209-2318 USA

Contact: **Karen Firehock**, Program Director
   Phone: 1-703-528-1818 • *Fax:* 1-703-528-1836

Activities: Conservation; Education; Political/Legislative; Research • *Issues:* Air Quality/Emission Control; Energy Efficiency; Global Warming; Public Lands Restoration; Stream Habitat Restoration; Sustainable Agriculture; Sustainable Development; Waste Management/Recycling; Water Quality; Wildlife

▲ Projects and People

Save Our Streams (SOS) is a major grassroots effort of The Izaak Walton League of America (IWLA) developed in 1969 and currently with projects in 37 states.

SOS works to educate volunteers, schools, organizations, and state and local government agencies on how to monitor and protect surface water areas. SOS cooperates with the U.S. Environmental Protection Agency's (EPA) Office of Water Regulations and Standards and the America's Clean Water Foundation, and is funded by grants from state and local agencies and major corporations.

Because federal and state agencies have resources to monitor only 30 percent of all U.S. surface water, there is a great need for volunteers to help in the effort. SOS helps government and private agencies organize citizen water monitoring programs. All the SOS projects help individuals and groups locate and adopt a healthy or damaged stream, then work to preserve, restore, and protect it. The organization sends monitoring packets and/or conducts hands-on workshops to teach people the methods of stream monitoring and restoration, emphasizing long-term solutions over stop-gap measures like restocking fish.

The Water Wagon is SOS's mobile laboratory—an educational vehicle that tours the country and assists in water monitoring projects.

Recently, SOS helped organize statewide water monitoring programs in Virginia, West Virginia, and Tennessee—working with state soil conservation and environment departments.

Other examples of the League's grassroots action help Lyme disease research as Minnesota bird banders gather deer ticks from low-flying birds in a forest research project; helped create the Patoka River Wildlife Refuge in Indiana; increased the local bluebird population in Illinois; boosted the mallard duck population in Colorado; and established a model Southern Maryland Outdoor Education Center on a historic farm site to teach people about contaminated runoff and careless land development.

■ Resources

SOS has developed MONITORS, a comprehensive, expanded database of all U.S. volunteer water-monitoring programs dealing with lakes, streams, wetlands, and estuaries. In addition, SOS publishes *Splash*, a quarterly newsletter, and produces fact sheets, brochures, stream monitoring kits, a training video, and a *Citizen's Directory for Water Quality Abuses*. The *Save Our Streambanks* guide is a survey of stabilization methods from improved vegetation to rock ripraps and log frame deflectors.

The Keystone Center ⊕
P.O. Box 8606
Keystone, CO 80435-7998 USA

Contact: **Michael T. Lesnick, Ph.D.**, Senior Vice
       President
   Phone: 1-303-468-5822 • *Fax:* 1-303-262-0152

Activities: Conflict Management; Dispute Resolution • *Issues:* Air Quality/Emission Control; Biodiversity/Species Preservation; Deforestation; Energy; Global Warming; Health; Population Planning; Sustainable Development; Transportation; Waste Management/Recycling; Water Quality

● Organization

"Like the course of heavenly bodies, harmony in national life is a resultant of the struggle between contending forces. In frank expression of conflicting opinion lies the greatest promise of wisdom in governmental action; and in suppression lies ordinarily the greatest peril."

These words of Louis D. Brandeis reflect the philosophy of The Keystone Center, a nonprofit organization founded in 1975, which focuses on science and public policy. Originally serving as a "neutral conflict management organization with a field science component for young people, the Center has expanded to provide "facilitation and mediation services on controversial environmental, health, natural resource, scientific, and technological issues and offers educational and professional programs to encourage scientific inquiry." For example, the Center was recently involved in a series of consultations and "innovative regulatory strategies" with the U.S. Environmental Protection Agency (EPA) on the implementation of the Clean Air Act of 1990. The Board of Trustees includes environmentalists, industry representatives, and media executives.

▲ Projects and People

The Center has three divisions offering distinct services.

• The **Keystone Science and Public Policy Program** is at the Center's heart. The staff attributes its success to what is called the "Keystone Process"—meaning neutrality and trust. Policy dialogues are structured problem-solving processes with goals that range from exploring diverse views and exchanging information to developing specific consensus recommendations on policy, regulations, or legislation.

Facilitation and mediation services are provided to government agencies, corporations, citizen and environmental groups, and labor unions regarding environmental, natural resources, and science-related issues. Training programs are offered to organizations to help them understand how conflict management strategies can be used in policy and regulatory decisionmaking. **Organizational development services** are available to groups desiring assistance in managing internal disputes more effectively.

Today, this international program manages more than 30 policy dialogues and conflict resolution projects with emphasis in six areas: environmental quality and health, biotechnology and genetic resources, natural resources, energy, food and agriculture, and science and technology.

U.S. Representative **John Dingell**, House Energy and Commerce Committee chairman, asked the Keystone Center to undertake a project regarding **AIDS vaccine liability** to examine both the possibilities and barriers of producing such a safe and effective vaccine. Dingell says, "Liability issues may constitute an additional impediment to research, development, and marketing for any such vaccine."

The Center's policy project on **national hazardous waste management strategies** provided input into the U.S. Resource Conservation and Recovery Act (RCRA).

Regarding genetic engineering, **biotechnology and genetic resources** studies are ongoing with representatives from developing and developed countries to "explore strategies for strengthening the international commitment to plant genetic resources." Dr. M.S. Swaminathan of India, first laureate of the World Food Prize and past president of the World Conservation Union (IUCN), guides this project.

Those involved in the project on **biological diversity on federal lands** are making recommendations for a "comprehensive national policy for conservation and sustainable use" as well as "guiding land managers in this policy's implementation."

In 1988, after years of negotiating, the Antarctic Treaty Parties agreed to the **Convention on the Regulation of Antarctic Mineral Resource Activities (CRAMRA)**. The Treaty never became ratified. Even so, several important papers were produced as a result of a Center facilitation process.

Food safety is the focus of the agriculture, food, and nutrition project—particularly since the nationwide publicity on alar, used on apples. The group examines pesticide residue on food as well as additives, biotechnology, and/or microbial contamination at later stages.

• The **Keystone Science School**, the Center's second component, targets students in kindergarten through twelfth grade. At 9,300 feet in the Rocky Mountains, the 16-acre campus nestles in the Snake River valley and operates under a special U.S. Forest Service permit. The programs last between 2 to 14 days, with emphasis on "the techniques and process of scientific inquiry, stewardship of the natural environment, and important problem solving skills." Students also learn how to care for themselves outdoors as well as work with others. The curriculum varies with the seasons, but focuses on the Rocky Mountain surroundings. Participating in **environmental conflict resolution simulations**, students "join together in a town meeting setting and through a role-playing process, debate, and attempt to create 'win-win' solutions to controversial problem situations."

An **Elderhostel** program is sponsored for people 60 years of age and older with week-long courses in wildlife biology, geology, mountain plants and wildflowers, photography, alpine ecology, snow physics, winter safety, and cross-country skiing.

The Science School also runs a **Discovery Camp** designed as a residential summer wildlife program for children ages 9 to 13. The camp combines "fun with learning, field studies focusing on alpine and forest ecology, aquatic studies, geology, wildlife biology, and the natural and broad cultural history of the Rocky Mountains."

The **Cancer Siblings Program** was founded in 1986 to support 11- to 17-year-old brothers and sisters of children with cancer as they learn to cope with this crisis. The program offers a weekend away from often difficult family situations and gives them a chance to meet friends who are going through the same experience. The oncology department of Children's Hospital, Denver, helps recruit the children and coordinates therapeutic discussions. The American Cancer Society funds this program.

• The Center's third component, the **Keystone Symposia on Molecular and Cellular Biology**, is international in scope, focusing on such emerging areas as they apply to basic biology, human medicine, and agriculture, reports its literature.

Dr. **Eric Lander**, Whitehead Institute for Biomedical Research, Massachusetts Institute of Technology (MIT), is Center chairman. **Robert W. Craig** is president; **John R. Ehrmann** and Dr. **Michael T. Lesnick** are senior vice presidents; **Abby P. Dilley** is senior mediator.

■ Resources

Consensus is a review of Center activities published three times a year. A list of publications includes reports on environmental quality and health, biotechnology and genetic resources, natural resources, energy, and science and technology policy.

The Land Institute
2440 East Water Well Road
Salina, KS 67401 USA

Contact: **Thomas O. Mulhern**, Director of Development
Phone: 1-913-823-5376 • *Fax:* 1-913-823-8728

Activities: Education; Research • *Issues:* Energy; Sustainable Agriculture; Sustainable Development

● Organization

In 1976, **Dana** and **Wes Jackson** started **The Land Institute**, with the desire "to help transform agriculture to protect and conserve the long-term ability of the earth to support a variety of life and culture." Because standard farming practices inevitably deplete the soil, destroy wilderness areas and rely far too heavily on chemicals and fossil fuels, the Jacksons sought to develop alternative methods. The ultimate goal is "an agriculture that mimics the native prairie."

To this end, the Institute relies primarily on the contributions of its 1,800 members and on foundations for funding.

▲ Projects and People

The development of new farming techniques requires research, which the Institute pursues in several different areas. Agroecologists work to breed plants that require less attention in the form of fertilizers, pesticides, and land preparation; they are currently focusing on perennial grains, which do not demand annual plowing.

Current fossil fuel use in world agriculture is seven times what it was in 1950; the Institute's **Sunshine Farm Project** is working to reverse that trend, by replacing such fuels with renewable energy sources like wind or solar power. The Institute also maintains an **Herbary** of prairie perennials, and three areas totaling 120 acres of native and restored prairie.

Education is another branch of the Institute's efforts. The **Intern Program** which began in 1983 allows up to ten graduate level interns to work on various ongoing projects or develop their own. Through publications, seminars, an annual Prairie Festival, and collaborative funding of research, The Land Institute encourages the public and other institutions to follow its example in combining ecology and agriculture.

■ Resources

The Land Institute publishes *The Land Report*, a 40-page journal, three times yearly as well as a newsletter called *The Grain Exchange*. Since 1985, the Institute has published an annual compilation of its research results under the title *The Land Institute Research Report*. In addition, books and articles by and about The Land Institute are numerous.

League of Conservation Voters (LCV) ✤

1707 L Street, NW, Suite 550
Washington, DC 20036 USA

Contact: **Jim Maddy**, Executive Director
Phone: 1-202-785-8683 • *Fax:* 1-202-835-0491

Activities: Education; Political/Legislative • *Issues:* Air Quality/Emission Control; Biodiversity/Species Preservation; Deforestation; Energy; Global Warming; Governmental Accountability; Health; Sustainable Development; Transportation; Waste Management/Recycling; Water Quality

● Organization

The Washington-based **League of Conservation Voters (LCV)**, founded in 1970, describes itself as "the political arm of the environmental movement in the United States." It fills the void left by laws prohibiting nonprofit organizations—in this case, most environmental groups—from contributing to political campaigns. With a staff of 10 at its head office and similar numbers at each of 3 New England regional offices, the nonpartisan League monitors elected officials and candidates for their environmental stances, and passes that information on to voters. By promoting environmental accountability, the League aims to "change the balance of power in Congress" in favor of the environment. LCV has 30,000 members.

▲ Projects and People

Since its start, the League of Conservation Voters has succeeded in swaying nominations and elections in favor of **pro-environment candidates** and against those with poor environmental records. League canvassers work with the public, discussing candidates, providing literature, and soliciting contributions. The League also organizes **grassroots lobbying campaigns**.

In the 1986 elections, League financial and educational contributions helped elect seven new pro-environment senators, the organization reports. In 1988, 60 of the LCV-supported incumbents won reelection. League members check to ensure that once in office the members of Congress uphold their campaign promises, and expose "false" environmental candidates.

For example, in 1990 the League **monitored congressional votes** on legislation related to elevating the U.S. Environmental Protection Agency (EPA) to Cabinet status and increasing its powers; closing irrigation law loopholes that encourage water wastage; creating national standards for "organically produced" foods; and the Clean Air Act amendments. Results of these and other votes appear in League publications.

■ Resources

Each year the League of Conservation Voters puts out a *National Environmental Scorecard* rating members of both houses of the U.S. Congress based on records on important environmental votes. Similar rating systems hold for the annual *Presidential Scorecard*, the *Election Report*, and the *Senate Energy Committee Scorecard*.

Committee on Environment
League of Women Voters (LWV) of New Castle, New York

P.O. Box 364
Chappaqua, NY 10514 USA

Contact: **Roberta Wiernik**, Chair
Phone: 1-914-241-7242

Activities: Education • *Issues:* Waste Management/Recycling

● Organization

The **League of Women Voters (LWV) of New Castle, New York**, began organizing **environmental shopping tours** in 1990 in an effort to combat the "American throwaway mentality" and desire for comfort at the expense of the environment. Bearing in mind that New York alone produces about two tons of garbage per person each year, and that one-third of all garbage is packaging, the goal of environmental shopping is to reduce the quantity and improve the safety level of garbage in landfills.

▲ Projects and People

Susan Schwarz, a member of New Castle's Environmental Control Commission who pushed for a local recycling program, designed the tours, which she describes as "common sense." A key element is to teach "pre-cycling." Admonishing citizens to "reduce, reuse, recycle, reject and react," the League's environmental shopping guides focus on the consumer's first step toward creating garbage—choosing products at the grocery store.

In the five-step plan, shoppers should buy products with the least packaging and buy in larger containers; buy reusable items such as cloth napkins and diapers rather than the disposable versions, reuse the containers of other items or donate them to school art classes, for example, and recirculate things like coat hangers by returning them to the dry cleaner; buy only products with recyclable packaging, and be aware of what kind of paper or plastic it is; not buy a product, if it does not meet these standards; and make suggestions to the store manager, or write the manufacturers of the worst offenders—if a store does not stock products which meet these standards.

A key element of environmental shopping, according the League members, is the work in conjunction with local programs and other community groups whenever possible. If it is incomplete or nonexistent, environmentally conscious shoppers should work to launch or improve a recycling program. This program also arouses awareness and action in local schools and energizes people to curb pesticide use and water pollution.

■ Resources

To those interested in conducting Environmental Shopping Tours in their own areas, the New Castle League of Women Voters will provide a starter kit including video, posters, brochures, resource lists, and other printed information necessary. *Alternative to Toxics in Your Home* is a popular publication. One poster is captioned, "The Second Most Important Walk Down An Aisle You Can Take."

Legal Environment Assistance Foundation, Inc. (LEAF) ❦
1115 North Gadsden Street
Tallahassee, FL 32303-6327 USA

Contact: B. Suzi Ruhl, President and General Counsel
Phone: 1-904-681-2591 • *Fax:* 1-904-224-1275

Activities: Education; Law Enforcement; Research • *Issues:* Health; Toxics; Waste Management/Recycling; Water Quality

● Organization

Among the seven staff members of the **Legal Environmental Assistance Foundation (LEAF)** are three lawyers and one law student. As its name implies, LEAF's purpose since current president and general counsel **B. Suzi Ruhl** founded it in 1979 has been provide legal advice and aid on environmental matters. Because it sees the poor and disenfranchised as the most common victims of corporate and governmental negligence, LEAF's legal services are free; funding comes from grants, membership fees, and donations. Working in three states in the Deep South—Alabama, Georgia, and Florida—LEAF focuses on community toxic pollution—particularly of ground and surface water. LEAF has 300 members.

▲ Projects and People

"To help citizens protect their health and environment," LEAF works on three levels: the permitting process, policy development, and advocacy. In the firs two areas, LEAF acts alone or with citizens and fellow grassroots organizations to influence state and local agencies toward a more responsible environmental role. As a result of their efforts, Georgia laws have changed to outlaw many of the more toxic chemicals previously allowed and to require better training for those using insecticides. LEAF activism also led to the removal of a dam in Chipola Basin, Florida.

As an advocate for citizens and the environment, LEAF's main weapon is the lawsuit. Through filing or threatening suits against both public and private organizations, LEAF combats their unwillingness to implement, enforce or comply with strict environmental regulations. A threatened suit against the U.S. Environmental Protection Agency (EPA) pushed the agency to process 23 permit hearing requests, some of which had been delayed as long as 10 years. Other such threats have led to the plugging of numerous injection wells which endanger drinking water in the region.

■ Resources

LEAF offers publications on toxic threats, Specific chemicals, citizens' guides to action and related topics. It seeks ways to maximize its outreach and publicity.

The David E. Luginbuhl Research Institute for Endangered Species, Inc.
P.O. Box 263
Ellington, CT 06029-9966 USA

Contacts: **Chris Luginbuhl**, Director
Phone: 1-203-871-6579

James Spotila, Ph.D., Research Scientist
Phone: 1-215-895-2627 • *Fax:* 1-215-895-1273

Activities: Education; Research • *Issues:* Biodiversity/Species Preservation

▲ Projects and People

"We must stop dumping non-biodegradable materials into our world's waterways. We are killing the leatherback turtle, the environment, and ourselves!" So proclaims literature of **The David E. Luginbuhl Research Institute for Endangered Species**, whose mission it is to educate people about the seriousness of marine pollution, especially the hazards of plastic bags and similar items that are fatally ingested when a sea turtle mistakes them for jellyfish, its major food source."

Since 1980, **Chris Luginbuhl**, son of the nonprofit group's namesake, has been tracking and tagging the large leatherbacks—some weighing as much as one ton—to study their migratory behavior from New England to Central America and the Caribbean. He is also fundraising for a center for endangered species in Newport, Rhode Island. The skeleton of a leatherback, caught by his father in 1951, was the first species of its kind to be displayed in the Yale Peabody Museum, according to an article in *Yankee* magazine (October 1989).

■ Resources

The Institute publishes a newsletter and makes video presentations to interested groups.

Maine Organic Farmers and Gardeners Association (MOFGA) ✹

P.O. Box 2176
Augusta, ME 04338 USA

Contact: **Nancy Ross**, Executive Director
Phone: 1-207-622-3118

Activities: Education; Political/Legislative; Research • *Issues:* Food Safety; Sustainable Agriculture • *Meetings:* IFOAM

● Organization

The eight staff members of the Maine Organic Farmers and Gardeners Association (MOFGA), along with several thousand members and other volunteers, make up, in their words, "the only organization dedicated to producing and consuming healthful food in Maine in a way that protects the environment and promotes farmers' economy and social well-being."

▲ Projects and People

Although many of MOFGA's efforts are directed at consumer education, the group has also played a role in legislative change. Two laws now on the books in Maine were the result of MOFGA work:

• The **Post-Harvest Treatment Labeling Law** requires retailers to post signs informing consumers that a product has or has not been chemically treated after harvest; if requested, stores must provide specific information on the treatment within two days of being asked.
• The **Country of Origin Labeling Law** requires labels on produce from countries which allow use of pesticides banned in the USA, and requires the Maine Department of Agriculture to inform consumers of these countries of the produce most likely to contain such toxins. According to MOFGA, imported produce can contain twice as much illegal residue from both pre- and post-harvest treatments.

MOFGA volunteers monitor stores for compliance with these laws and publicize the results.

An obvious element in many of MOFGA's ongoing programs is information sharing. Through the **Apprenticeship Program**, farmers who need extra help or who would like to pass on their expertise can find apprentices who want to learn organic farming techniques. In 1990 MOFGA placed more than 30 apprentices who worked in return for room, board, and hands-on experience at a sheep farm, dairy, organic farm, orchard, or other farm project.

MOFGA's annual **Farmer to Farmer Workshop**, launched in 1990, provides a forum at which organic farmers exchange ideas on soil testing, weed control, cooperative marketing and crop rotation, among many topics. In the **Grow Your Own** program, MOFGA organizes volunteers to advise beginning gardeners and help start community gardens, providing seeds, equipment and expertise as necessary. The **Common Ground Country Fair**, held each year

since 1976, is another MOFGA-sponsored opportunity for Maine's farmers and gardeners to share their knowledge.

Nancy Ross, MOFGA's executive director since 1987, has devoted much of her career to environmental concerns. Previously, she worked with the Audubon Society and with the Maine Department of Conservation. She is a member of the Maine League of Conservation Voters and is vice president of both the Agricultural Council of Maine and the Organic Farmers Associations Council.

■ Resources

MOFGA publishes a bimonthly newspaper, the *Maine Organic Farmer and Gardener*, which serves primarily as an information exchange. Other resources include the cookbook *A Bountiful Year*, fact sheets on specific crops, and technical bulletins for farmers and gardeners.

The Marine Mammal Center (MMC) ✹

Marin Headlands/Golden Gate National Recreation Area
Sausalito, CA 94965 USA

Contact: **Peigin Barrett**, Executive Director
Phone: 1-415-289-7325 • *Fax:* 1-415-289-7333

Activities: Education; Marine Mammal Medicine/Rehabilitation; Research • *Issues:* Biodiversity/Species Preservation; Health; Marine Mammal Science; Ocean Ecology; Water Quality • *Meetings:* International Association for Aquatic Animal Medicine; Ocean ecology meetings

● Organization

Located at an abandoned Nike missile base, **The Marine Mammal Center (MMC)** is largely a volunteer organization devoted to the rescuing, treating, and studying of distressed aquatic animals. With more than 28,000 members and a staff of 450 (95 percent unpaid), MMC provides care to seals, sea lions, sea otters, whales, dolphins, and porpoises which—before the Center was founded in 1975—were usually stranded and left to die. With modern rescue attempts, reports MMC, these animals have excellent chances for survival.

MMC is a member of the National Marine Fisheries Service's **Marine Mammal Stranding Network**, a group of scientific, governmental, and wildlife agencies dedicated to the maintenance of marine mammals and conservation. MMC is also affiliated with the U.S. Environmental Protection Agency (EPA), Marine Environmental Research Institute, universities, and other national organizations.

▲ Projects and People

MMC operates an animal help hotline (1-415-289-SEAL) to which the public can report animals in distress along the northern and central California coastline. Rescues in which the Center takes part are often "dangerous and frustrating because these wild animals, although weakened, struggle to avoid being captured." However difficult, rescue attempts have saved hundreds of marine mammals from death, reports MMC. Case histories on "Lepto" the California sea lion, "Spike" the harbor seal, "Valentino" the northern elephant seal, and "Humphrey" the humpback whale are available.

Rehabilitation, which is costly, can take up to eight months or more. In 1991, MMC was able to rehabilitate orphaned Steller sea lion pups on the Ano Nuevo Island breeding area near the coast at San Mateo at the request of the University of California at Santa Cruz. Two pups were returned to the wild and reported "in good health."

Near Sausalito on a seven-acre site in the Golden Gate National Recreation Area, the Center is open to visitors who wish to learn about aquatic animal life. Recently, the Center's Education Department produced a **Marine Mammal Curriculum Guide** to assist teachers with ocean ecology games, classroom assignments, and materials prepared in Spanish and English.

Regarding **scientific enterprises**, MMC provides "data and medical samples derived from routine medical treatment to scientific associates" worldwide. A current three-year study is evaluating the **immune status in marine mammals** "to determine possible connections between body pollutant levels and degree of immunosuppression." This study could be a major breakthrough in wildlife health, science, and rehabilitation as "the level of tolerance to toxins and pollutants in marine mammals has not yet been established," writes MMC. Results of studies are put into action in the wildlife field and/or utilized by scientific periodicals, such as the *Journal of Wildlife Diseases.*

Other **new studies** will investigate causes of a "mysterious skin disease suffered by northern elephant seals" and "seizures suffered by the **California sea lion**."

Staff members are executive director **Peigin Barrett**, who has a unique background in fine arts, film, and teaching; biologist **Lance E. Morgan**; and consulting veterinarian **Dr. Laurie Jean Gage**.

■ Resources

The Center is in need of medical diagnostic equipment and micro-laboratory equipment for use in animal rescues. MMC also needs paper and/or the ability to publish their educational materials. The Center also seeks advisors and medical, scientific, and natural history books to create a marine mammal library.

MMC offers its *Marine Mammal Curriculum and Activity Guide* at no cost to kindergarten to sixth grade educators and to the general public for a fee.

Marine Technology Society (MTS) ⊕
1825 K Street NW, Suite 203
Washington, DC 20006 USA

Contact: **Martin J. Finerty, Jr.**, General Manager
    Phone: 1-202-775-5966 • *Fax:* 1-202-429-9417

Activities: Education • *Issues:* Energy; Global Warming; Ocean and Atmosphere Technology; Water Quality • *Meetings:* Offshore Technology Conference

● Organization

The volunteer **Marine Technology Society (MTS)**, founded in 1963, describes itself as "the only professional society committed to serving the entire world ocean community—engineers, scientists, policymakers, educators, and every segment of the ocean business." With a staff of 4 and 2,800 members, the MTS serves as a network and clearinghouse for information related to all aspects of marine

technology. Its four professional divisions include advanced marine technology, marine resources, marine policy and education, and ocean and coastal engineering.

▲ Projects and People

MTS pursues its educational and information-sharing goals primarily through its **conferences, workshops, committee meetings, and resulting publications.** The Society is a founding participant in the annual Offshore Technology Conference which discusses ways to explore, develop, and protect the ocean resources and environment. A recent international conference highlighted the need for an "ocean cooperative made up of industry, government, and academia"; issues covered included oil spill prevention and cleanup, and endangered species preservation, with emphasis on sea turtles.

In chairing the conference, **Bob Howard**, Shell Offshore president, said, "The future of our domestic oil and gas business depends on continuing advances in ocean sciences and technology as we learn to develop the ocean's resources while protecting and preserving the oceans for society's myriad of other uses."

Martin Finerty, Jr., has a background in international environmental data collection. As an oceanographer with the U.S. Navy and National Oceanic and Atmospheric Administration (NOAA), Finerty supervised the development, production, and distribution of oceanographic, meteorologic, and sea ice products for polar regions. He was also with the National Academy of Sciences/National Research Council.

■ Resources

MTS is seeking outlets for its guide to *University Curricula in Oceanography and Related Fields*, sponsored by the National Science Foundation and the Office of Naval Research. The Society publishes a bimonthly newsletter, *Currents*, and the quarterly *Marine Technology Society Journal*, which recently focused on global change. The *Ocean Opportunities Career Guide* is directed toward high school students. A small library in the national office in Washington, DC, is available for limited use.

Mothers and Others for Safe Food (MOSF) ❦
P.O. Box 2975
Ventura, CA 93002 USA

Contact: **Cheryl Fletcher**, President
    Phone: 1-805-658-8915

Activities: Education • *Issues:* Health; Safe Food; Water Quality

● Organization

According to **Mothers and Others for Safe Food (MOSF)**, California agriculture uses about 200,000 tons of chemicals annually in crop production, many of them known carcinogens. Thousands of wells have been contaminated as a result, and the Natural Resources Defense Council (NRDC) cites agricultural chemical use as the single major cause of occupational illness in California.

Statistics like these led concerned citizens of Ventura County, California, to form MOSF in 1989 in order to "educate the public about harmful chemicals found in foods; provide . . . information on how to locate organic produce; ensure that safe produce is widely available; [and] inform the public about existing laws and pending legislation" related to pesticide use."

At monthly meetings—and more often for special projects—the more than 35 MOSF members gather to plan strategies for educating both consumers and producers.

▲ Projects and People

Education efforts of MOSF include holding public forums, manning booths at fairs and Earth Day celebrations, appearing on talk shows, and publishing articles in the press. An important arm of the organization is its **Speakers Bureau**; members volunteer for speaking engagements at schools, League of Women Voters meetings, local mothers clubs, and other venues. The current focus of such talks and articles includes combatting the application of fungicide-laden wax coatings on produce, seeking viable nontoxic solutions to the medfly problem, and providing information on pesticides and organic home gardening. The safety of bananas grown with aldicarb, a systemic pesticide, is another recent concern. So are aflatoxin in peanuts, other legumes, grains, and nuts; salmonella enteriditis in eggs; and rainwater with tracings of "atrazine and alachlor widely used on American corn and soybeans and both suspected human carcinogens," according to MOSF.

MOSF has worked with the **Ventura County Food Safety Study Group** since the latter's inception in 1990. Fueled by concerns at the time about alar on apples, the University of California Sustainable Agriculture Research and Education Program (SAREP) funded the group so that local farmers, retailers, and environmental groups could work together "to develop a systematic approach to growing safer produce in Ventura County."

MOSF reports it is succeeding in encouraging farmers to use less pesticides in their farming practices as well as creating consumer demand for nutritious, pesticide-free fruits and vegetables.

■ Resources

The organization publishes a newsletter for members; in addition, MOSF's *Organic Buying Guide for Ventura County* locates sources of organic food for interested consumers, including retail grocery stores, farmers markets, and private farms.

Current needs include office equipment, such as a Macintosh computer, and help with database development and public relations work.

National Aubudon Society ⊕
700 Broadway
New York, NY 10003 USA

Contact: Conservation Information
Phone: 1-212-979-3000

Activities: Development; Education; Law Enforcement; Political/Legislative; Research • *Issues:* Agriculture; Air Quality/Emission Control; Biodiversity/Species Preservation; Deforestation; Energy; Global Warming; Health; Population Planning; Public Lands Management/Protection; Sustainable Development; Transportation; Waste Management/Recycling; Water Quality

● Organization

Founded in 1890, the **National Audubon Society** is a long-established proponent of wildlife conservation and environmental protection. The organization prides itself on being considered "the voice of reason" on environmental issues—taken seriously by decisionmakers in government and industry.

Over the years, Audubon has grown into a large network—consisting of 10 regional offices, over 500 national chapters, and various divisions—through which information flows constantly on goals and activities relating to national and state environmental policies. Audubon is a nonprofit, tax-exempt organization supported by membership dues, gifts, and foundation and corporate grants. Scientists, educators, lobbyists, lawyers, and conservation professionals are among the staff of 250.

▲ Projects and People

Peter A.A. Berle became president of the 600,000-member National Audubon Society in 1985, following two decades of activism on a range of environmental and conservation issues in both the public and private sectors. In 1989, he was appointed by New York Governor Mario Cuomo to chair a special Commission on the Adirondacks in the 21st Century. He also serves on the Twentieth Century Fund, the U.S. Environmental Protection Agency's (EPA) Clean Air Act Advisory Committee, the National Commission on the Environment, the Keystone Center, and Clean Sites, Inc.

Brooks Yaeger, vice president for Government Relations, leads Audubon's highly successful campaign to protect the Arctic Refuge, heading a corps of activists and donors who have "committed themselves to the struggle to keep oil development out of a great wilderness refuge." At the core of the effort are 300 **Issue Leaders** across the country who encourage letter-writing campaigns, meetings with members of Congress, phone trees, and other means of persuading legislatures to designate the wilderness as a refuge and reject "unsound energy policies that rely on increased oil consumption." As one Issue Leader put it, "The Arctic is our template for the wild. It defines wilderness. . . . We must let the politicians know there are places that are sacred to us. The Arctic Refuge is one of those places."

The Audubon Society is gearing up for another intensive lobbying effort to prevent the weakening of the **Endangered Species Act**, up for reauthorization in 1992. **Randy Snodgrass**, Audubon's director of **Wildlife Programs** in Washington, DC, says, "By the year 2000, we're going to be losing 100 species a day from the face of this planet. How will this affect our own survival? What medicines won't we have? What foods won't we be able to produce? We have got to strengthen the act for the long-term well-being of society, the economy, and our quality of life."

Education plays an important part in Audubon's mission. In 1991, according to Audubon's Annual Report, nearly 450,000 children enrolled in **Audubon Adventures**, the Society's classroom program for upper-elementary children; a quarter of them belonged to minority groups. Audubon also has **educational facilities** in Connecticut, California, Ohio, and Wisconsin; **summer ecology camps** in Maine, Connecticut, Wisconsin, and Wyoming; and travel-study programs for high school and college students.

Scientific research is also a major component of Audubon's activities. Research is undertaken to aid endangered species and to develop ecologically sound management practices. In addition, Audubon convenes scientific conferences, workshops, and seminars for professionals in different disciplines. Audubon's Science Institute recently received a grant from the U.S. Department of Education (DOE) to develop a national strategy integrating environmental issues into science education.

The National Audubon Society manages a system of 75 **wildlife sanctuaries**, staffed by resident wardens; local chapters maintain an additional 100 sanctuaries. The primary objective of these sanctuaries is long-term protection of plants and animals, especially threatened and endangered species. For example, the 116-acre Tenmile Creed Sanctuary on the coast of Oregon preserves a key piece of the largest unlogged rainforest left in the United States, home to a significant population of the marbled murrelet (a seabird) and other threatened species. Some sanctuaries have limited public visitation; others are closed to all visitors because any human presence would cause harmful disturbance.

Among other recent Audubon accomplishments, **puffins** appear to be recolonizing Seal Island in Maine's Penobscot Bay; Long Island Sound citizens formed a **Watershed Alliance** to restore the region's health; in southwest USA, **Jane Lyons** leads a fight to save the habitat of the critically endangered **golden-cheeked warbler** songbird; the **Solid Waste Program** emphasizing reuse and recycling has expanded to 14 communities with the support of Fuji Photo Film, USA; Congress mandated 76 miles of the **Niobrara** as a "scenic river" in Nebraska; members lobbied to **defeat the Two Forks Dam project** along Colorado's South Platte River; activists are increasing to help save **wetlands** from the Everglades to Alaska.

Wetlands are defined as small freshwater or estuarine bogs, potholes, swamps, or marshes occurring where land and water meet; where plants are water-tolerant or water dependent; and where the soil is saturated at least part of the year. They are critical because they "filter out sediment and pollutants [as] wetland microorganisms break down waste and purify local groundwater supplies, are an essential source of food; protect from flooding by channeling excess water; prevent the erosion of beaches and inland soil"; and are a source of recreation. "More than one-third of U.S. endangered and threatened animal species (such as the wood stork and whooping crane) as well as several endangered plant species, are dependent on the continued existence of large areas of wetland habitat." So are millions of migratory birds along their flyways.

Audubon describes a few outcomes of the loss of more than half this nation's wetlands, or some 115 million acres: northern pintail populations are reduced by more than half as midwestern prairie potholes are drained for cropland; disappearing wetlands in Louisiana threaten $28 billion annually derived from commercial fishing, sport fishing, and hunting; agriculture has claimed more than 95 percent of Central Valley wetlands in California, a state "desperate for water"; and 75 percent of mitigation efforts required of developers by the U.S. Army Corps of Engineers failed to be carried out in Texas.

Office buildings produce 14 percent of the carbon dioxide emitted in the United States, contributing significantly to global warming. They also produce 15 percent of U.S. emissions of sulfur dioxide and nitrogen oxides, causes of acid rain. Under the leadership of President Berle, the Audubon Society acquired a century-old, eight-story building in lower Manhattan which is being converted into new, **environmentally sound headquarters** for the organization—a "living model" of energy efficiency, healthful living conditions, and waste reduction. As Mr. Berle explains, "We want to create a model that can be realistically replicated by anyone, from a multinational corporation building a new headquarters to a school board refurbishing an old school building." The architectural design emphasizes recycled building materials, energy efficiency, and fresh air ventilation.

■ Resources

Audubon members receive *Audubon*, a conservation and nature magazine; *Audubon Activist*, a news journal for those who want to be active on environmental issues; and *Audubon Policy Reports*. Brochures include a list of *Wildlife Sanctuaries*. The Society also produces the award-winning *National Audubon Society Specials* for televisions; teachers and libraries may obtain copies through PBS Video, Vestron Video, or local video stores. *Actionline*, a call-in service, provides a weekly recording of environmental news. *Audubon Wildlife Adventures* are educational computer software for children. *Audubon* field editor **Frank Graham, Jr.**, traces the organization's first 100 years in *The Audubon Ark: A History of the National Audubon Society*.

National Audubon Society ⊕
666 Pennsylvania Avenue, SE, Suite 301
Washington, DC 20003 USA

Contact: **Brock Evans**, Vice President for National Issues
Phone: 1-202-547-9009 • *Fax:* 1-202-547-9022

Activities: Education; Law Enforcement; Political/Legislative; Research • *Issues:* Agriculture; Air Quality/Emission Control; Biodiversity/Species Preservation; Deforestation; Energy; Global Warming; Population Planning; Public Lands Management/Protection; Sustainable Development; Waste Management/Recycling; Water Quality; Wetlands Protection *Meetings:* CITES

▲ Projects and People

As the **Audubon Society's** vice president for national issues, **Brock Evans** oversees a staff of 30 and has primary responsibility for the areas of public lands and forest management.

In this capacity, he manages the 100-member **Ancient Forest Alliance**, a major scientific/legislative campaign to rescue the remaining ancient forests of the USA on federal lands in Washington, Oregon, and California. According to the literature, "Forests older than Europe's cathedrals once stretched from northern California to the Alaskan panhandle. . . . Primordial groves sheltered wild creatures beyond number. Ancient forests like these still exist. But they are going fast: just fragments remain. And despite a national outcry, logging continues relentlessly."

The Audubon campaign stresses that if we lose these forests, we also lose clean, abundant water, farms, fisheries, endangered and rare plant and animal species, a living laboratory for understanding nature, a stabilizing influence on the global climate, and recreational opportunities. The Ancient Forest campaign is calling on the U.S. Congress to permanently protect these remaining tracts, connecting and restoring adjacent areas to complete the system. Audubon's maps are helping to lay the scientific groundwork for such a reserve.

Other activities in which Evans is engaged include **mining law reform**—working closely with the Mineral Policy Center and other environmentalists; **trails legislation**—seeking to weaken or change the Symms amendment, which would have created a special fund to assist motorized users in building new trails on public lands; **grazing reform** through raising such fees; and shaping a **New England forests land campaign** with David Miller.

Evans, an attorney, has received numerous awards for his environmental accomplishments, including Friends of the United Nations Environment Programme (UNEP) Award, League of Conservation Voters Outstanding Performance Award, and Sierra Club's John Muir Award. In 1990, he taught a course on politics and the environment as a fellow of the Institute of Politics, John F. Kennedy School of Government, Harvard University. He currently serves on the executive committee and as political committee chairman of the League of Conservation Voters, and on the board of directors of the Human Environment Center. He was chairman of the umbrella organization National Resources Council of America.

■ Resources

The Washington, DC, office provides extensive materials designed to help members organize and operate local chapters. Also available are guidebooks and videos on activist skills, federal agencies and programs, population and international programs, and teaching tools. Many of these materials may also be obtained from regional offices.

International Program ⊕
National Audubon Society
666 Pennsylvania Avenue, SE
Washington, DC 20003 USA

Contact: **Charlotte Fox**, Representative
   Phone: 1-202-547-9009 • *Fax:* 1-202-547-9022

Activities: Development; Education; Political/Legislative; Research • *Issues:* Agriculture; Air Quality/Emission Control; Biodiversity/Species Preservation; Deforestation; Energy; Global Warming; Population Planning; Sustainable Development; Waste Management/Recycling; Water Quality; Wetlands Protection • *Meetings:* IUCN; Ramsar; UNCED; World Parks Congress

● Organization

"International networking is an exciting way to work for policies that will meet the global needs of the future," states *Network News,* a publication distributed by **Audubon International**'s local chapters.

As the international component of the **Audubon Society,** Audubon International encourages **grassroots involvement in international conservation issues** to foster an appreciation of "the relationship between what is occurring abroad and what is happening in your community—similar problems, common disasters, natural heritage shared by the world, links with international finance, trade, and economy, international treaties and obligations, ethics and human compassion, news events."

▲ Projects and People

To get started on international projects and activities, local chapters are encouraged to make announcements at meetings about international issues; include articles in chapter newsletters on international problems and the work of conservation activists from other countries, particularly those working on problems similar to the ones with which the local chapter is involved; plan programs on the natural history of another country, including inviting Audubon International speakers to meetings; collect books, field guides, journals, and equipment to send abroad to partner conservation organizations or university libraries; extend the chapter's camp scholarship program to include attendance at an Audubon International ecology workshop; sponsor a foreign conservationist's attendance at an Audubon camp; conduct telephone and mail campaigns on international issues that relate to state or national legislation or regulatory policies; and develop a joint international project with other community organizations.

In a sense, it is difficult to differentiate "local," "national," and "international" Audubon programs. As the literature describes it, "every division of Audubon has initiated projects and campaigns that involve partnerships and/or cooperation with activists, organizations, and governments worldwide."

The **Beringia Conservation Program,** headed by Alaska regional vice president **David Cline,** sought mutual cooperation with the former USSR regarding preservation of the unique natural resources on land and in the Bering Sea and the establishment of a **Beringia International Marine Biosphere Reserve.** Cline, along with regional representative **Mary Core** and economist **John Tichotsky,** University of Alaska—Anchorage, traveled to the former USSR in 1990 as guests of the Soviet Academy of Science. The Alaska–Hawaii regional office worked throughout 1991 to promote the agreements achieved during the prior trip.

The **U.S. Citizens Network on the United Nations Conference on Environment and Development (UNCED)** was chaired by the former director of Audubon's International Program, **Frances Spivy-Weber.** The purpose of the Network was to organize the input of the non-governmental community at the conference in Brazil, June 1992. (The meeting was an "Earth Summit," bringing together leaders from all over the world to discuss and define strategies for promoting economic equality and environmentally sustainable development.) As a leader of the Network, Audubon worked to raise the Earth Summit's visibility with the American public. A central element of Audubon's work on the Summit is a campaign to change forest practices worldwide.

The Audubon Society also lobbyied for **reform of the Foreign Assistance Act,** which is being revised by the U.S. Congress. According to the literature, with the end of the Cold War, "there are many opportunities to improve the direction of foreign policy and to change the type of assistance given to developing nations." The Foreign Aid Reform project seeks to include environmental issues in U.S. global policy initiatives.

The **International Wetlands Project** devotes its efforts to the implementation of the **Ramsar Convention,** an international treaty established to protect the world's wetlands. This project also seeks to influence the policies of the World Bank, the InterAmerican Development Bank, and the U.S. Agency for International Development (USAID). According to the Audubon Society, "These banks dispense nearly $27 billion annually to developing countries and their activities often have enormous environmental impacts." Audubon seeks to halt projects and change policies that destroy wetlands while encouraging the redirection of other projects toward the conservation and wise use of wetlands. Recently, in large part because of Audubon's efforts, the World Bank has modified its **Bangladesh Flood Action Plan** to pay more attention to the environmental needs of that country. **Steve Parcells** directs the project and serves as an officer of the U.S. National Ramsar Committee.

In a "series of exchanges between Audubon Sanctuary managers and wildlife experts in developing countries," the **Population, Environment and Wildlife Project** evaluates the impact of human population pressures on wildlife and their habitat. The program was organized by **Patricia Baldi** in 1989 "to apply the expertise of Audubon's sanctuary managers . . . to the exploding world population problem and damaging environmental impacts." Managers from Audubon sanctuaries in Nebraska, North Dakota, Texas, South Carolina, Louisiana, and Florida visited "sister" sanctuaries in Central and South America, Asia, and Africa, and freely exchanged information and ideas. Later, the Americans hosted their foreign counterparts in their own sanctuaries. Baldi says that the program is intended to bring the population problem home to Americans who "need to realize that population growth is our problem, and it's our duty to preserve the natural world."

The **Western Hemisphere Shorebird Reserve Network** brings together wildlife agencies, private conservation groups, and other organizations to solve conservation challenges for migratory shorebirds and their irreplaceable habitats. These sites, which range from the Arctic to southern Chile, are vital stopover points for journeying shorebirds in fall and spring. The loss of even one site, due to an oil spill, development, or pollution, would have devastating effects on the entire bird population, the Society reports. The Network, headquartered in Manomet, Massachusetts, sponsors training and education programs, provides technical support, and promotes information sharing between geographically distant reserves.

The Audubon Society has vowed to end the **international wild bird trade**. The USA is the largest importer of wild birds, bringing in six to seven million in the last decade. The trade, while legal, involves "excessive mortality from exposure to smuggling, disease, dehydration, and starvation." The Audubon Society is pressing Congress to enact legislation restricting such trade and phasing out certain aspects altogether. It is seeking grassroots support for this effort.

The **Living Oceans Program** was started by **Dr. Carl Safina** to promote the same level of concern about oceans as exists for rainforests. Audubon's focus is on fisheries mismanagement, with other organizations working in different areas. The program's two major components are a **Marine Policy Center** and a **Marine Advocacy Network**. The strategy involves promoting "fish that are wildlife, not only commodities" and "heightened public awareness, especially among fishermen."

Stephen Kress, of the Maine Coast Sanctuaries, is in charge of the **Galapagos Project**, a joint effort between the Charles Darwin Research Center and the Galapagos National Park. The project seeks to develop techniques to encourage rare birds to live in a predator-free habitat.

George Powell's work on **neotropical migrants** involved a trip to Belize to "examine the impacts of shifting agriculture on wintering migrants in Central America." During the project's second year, research was completed in Monteverde, documenting the seasonal migration of Resplendent Quetzals. Powell helped in the planning of a $2.5 million **visitor-education-research complex** for the Monteverde Reserve that will absorb the burgeoning nature tourism in the area.

The **Save Our Songbirds Program** recognizes the importance of migratory songbirds as indicators of international biodiversity. According to Audubon literature, "Recent studies have shown alarming declines in many songbird species which depend on tropical forests in Latin and South America and temperate forests in the United States and Canada." The Audubon Society provides chapters and activists with the tools and techniques to reverse this decline. For example, a **Migratory Bird Information Kit**, published jointly with the Smithsonian Institution, contains materials on protecting migratory songbird populations at local, state, national, and international levels. **Susan Carlson** oversees the program.

Audubon's **Citizens Acid Rain Monitoring Network** is expanding its efforts to influence public policies on air pollution. Those who monitor acid rain and snow domestically are teaming up with monitors in places like San Jose, Costa Rica; Caracas, Venezuela; Tokyo, Japan; Budapest, Hungary; and Moscow, Russia, to exchange acid rain data and cultural information as well as illuminate the work of Audubon.

The **International Environmental Partnership Program** (IEPP) brings together high school students from different countries sharing similar natural habits, migratory species, environmental concerns, and interests. Four partnership programs were recently featured to study: acid rain by students in northeast USA and the former USSR; migratory birds by students in northeast USA and Trinidad; air quality, human population growth, water availability, and migratory birds by students in Los Angeles, California, and Mexico City, Mexico; and Beringia by students in Alaska and Siberia.

■ **Resources**

International Program members receive the *Audubon Activist*. The program also offers a *Conference-Call Workshop* series—a chance to share ideas about projects with other members across the country—as well as teachers' guides and other training materials.

Florida Audubon Society (FAS) 🌿
460 Highway 436, Suite 200
Casselberry, FL 32707 USA

Contact: **Terrie Diesbourg**, Executive Secretary
Phone: 1-407-260-8300 • *Fax:* 1-407-260-9652

Activities: Development; Education; Law Enforcement; Political/Legislative; Research • *Issues:* Air Quality/Emission Control; Biodiversity/Species Preservation; Deforestation; Energy; Global Warming; Population Planning; Sustainable Development; Transportation; Waste Management/Recycling; Water Quality

● **Organization**

Founded in 1900, the **Florida Audubon Society** (FAS) is a grassroots organization with a membership of 35,000 and a staff of 35, including **Dr. Bernard J. Yokel**, president; **Charles Lee**, senior vice president; **Dr. Peter C.H. Pritchard**, vice president, Florida International Wildlife Conservation; ornithologist **Dr. Herbert W. Kale II**, vice president; and environmental educator **Wendy Hale**. It consists of 47 local chapters, each with independent officers and directors, that work closely with the state office.

▲ Projects and People

Education is one of the primary activities of the Florida Audubon Society. Directed by Wendy Hale, the **Environmental Education** program reaches a diverse audience through various programs, newsletters, and activities aimed at both young people and educators. Teaching about such topics as the biodiversity of the Florida Keys, bird migrating habits, aquifers as "Florida's rain barrel," and wetlands, FAS hopes to have a wide impact.

The **Madalyn Baldwin Center for Birds of Prey**—such as bald eagles, hawks, and owls—is one of the Florida Audubon Society's important programs. The Center treats wounded birds, and when possible releases them after their recovery. Birds that cannot be released are kept at the Center or sent to other aviaries around the country. The Center is also a base for educational purposes, and for the **Adopt-a-Bird** program to encourage donations and to encourage people to promote increased interest in birds of prey.

Investigating and **recovering endangered species such as raptors** and manatees is another activity of the organizations. Working with the agencies and committees involved in conservation in Florida helps them to develop such strategies. This can include lobbying activities to gain environmental legislation.

The **Florida Breeding Bird Atlas** project, in sponsorship with the Florida Ornithological Society—and with funding from the Nongame Wildlife Trust Fund, Florida Game and Fresh Water Fish Commission—is an intensive volunteer effort to map the distribution of breeding birds and publish an atlas. One outcome is knowledge about new exotic species of plants and animals and their effects on native flora and fauna.

Recycle for Wildlife! is an ongoing campaign in cooperation with various businesses in central Florida.

Claiming to be the state's "leading conservation organization," FAS has many accomplishments including the "creation and expansion of Florida's Conservation and Recreation Lands Program providing over $400 million for environmental land acquisition; passage of landmark growth management legislation" to help local government achieve environmental protection; creation of the state's Nongame Wildlife Program; formation of the Everglades Coalition of state and national groups devoted to restoring the Everglades; and development of land acquisition program on North Key Largo to protect habitat of endangered species.

■ Resources

The Florida Audubon Society has a wide range of publications. These include *The Florida Naturalist* magazine; *The Young Naturalist*; *Florida Raptor News*; *Manatees: An Educator's Guide*; and the *Florida Breeding Bird Atlas* newsletter. In addition FAS has a speakers bureau and a collection of films that are available for rental to schools, libraries, clubs, and other organizations. A list is available.

Maine Audubon Society

118 Route One
Falmouth, ME 04105 USA

Contact: **Maureen Oates**, Director of Education
Phone: 1-207-781-2330

Activities: Education; Political/Legislative; Research • *Issues:* Air Quality/Emission Control; Biodiversity/Species Preservation; Deforestation; Energy; Pesticide Use; Waste Management/Recycling; Water Quality

● Organization

Established in 1843, the **Maine Audubon Society** aims to "promote and encourage understanding and appreciation of the natural environment, and to foster, through education, study, and advocacy, an awareness of the relationship between people and their environments, and of the problems caused by people."

As an independent, state-based organization, Maine Audubon focuses on Maine's natural resources and the environmental concerns of Maine residents. The staff of 25 consists of environmental educators, attorneys who are active environmental advocates, and a research and policy analysis office.

▲ Projects and People

Maine Audubon operates 15 sanctuaries; three of them are staffed—**Gilsand Farm sanctuary** in Falmouth, **Hamilton sanctuary** in West Bath, and **Mast Landing sanctuary** in Freeport.

Visitors to Mast Landing can explore trails that lead through natural communities, with a chance to observe a snowshoe hare, red fox, porcupine, raccoon, deer, great blue heron, broad-winged hawk, hairy woodpecker, or leopard frog. Since 1967, the sanctuary has been the setting for **Mast Landing Nature Camp**. It also serves as a program site for groups including schools, scouts, and the Freeport Historical Society.

The **Scarborough Marsh Nature Center** is the largest salt marsh in the state, comprising 3,100 acres of tidal marsh, salt creeks, fresh marsh, and uplands. The marsh is particularly significant for wildlife as a resting, breeding, and feeding ground. During the spring and summer months, Maine Audubon offers a regular schedule of nature education programs, including arts and crafts, day camp, nature clubs, and special events—such as full moon canoe tours, outings for senior citizens, foraging for edible and medicinal plants, observing the summer night sky, and classes in local history.

Maine Audubon sponsors a number of **nature walk programs.** One is the **Secrets of the Forest** program, during which groups are led through the forests and fields and, using all of the senses, "learn about adaptations and the interdependence of all living things, consider the season changes of autumn, and uncover signs of human history hidden within the landscape." Another is **Winter Ecology Walks** at Gilsand Farm, in which visitors explore how plants and animals prepare for winter and how their adaptations help them survive. The **Salt Marsh Ecology Walks** at Scarborough Marsh Nature Center highlight the sights and sounds of spring, from green shoots emerging from the earth to the fish, snails, and crabs hidden within the landscape. Finally, the **Bird Adventures** program teaches elementary school children about bird activities at habitats on the Orono Campus of the University of Maine.

Maine Audubon has a number of **environmental education** and **teacher enhancement** programs:

The **School Science and Natural History Enrichment Project** (SSNHEP) is a statewide program designed to provide teachers of grades K–12 with Maine-based science enrichment activities and materials; training in effective science teaching methods; access to, and knowledge of, local resources; and volunteer community partners to assist in science lessons. A cornerstone of the project is the four-volume curriculum supplement *Science and Natural History: A*

Maine Studies Sourcebook, which uses the Maine environment as a teaching resource and basis for scientific inquiry. Close to 350 teachers from the state participate in the project.

The **SSNHEP Leadership Institute** trains outstanding project participants to provide workshops and support services to teachers and volunteers in their own districts. The program involves a teacher/administrator team in developing a science improvement plan tailored to the needs of their own district. Also involved are the College of the Atlantic, University of Maine, and the Maine Departments of Education, Environmental Protection, and Fisheries and Wildlife.

Teaching programs are under the direction of Maureen Oates, a former science teacher and director of the Massachusetts Museum Network at the Boston Museum of Science. Oates is also a member of the Conservation Education Association, National Science Teachers Association, National Wildlife Federation (NWF), and National Marine Educators Association.

For the past 12 years, Maine Audubon has provided natural history tours, escorted by Audubon staff and experienced naturalist/guides, to special destinations through its World Tours program. 1991 tours encompassed Galapagos and Ecuador; Wassaw Island in the Georgia low country; the mountains, canyons, deserts, and forest habitats surrounding the Chiracahua Mountains in Arizona; and the Italian countryside.

■ Resources

Habitat, the Maine Audubon magazine, is available to members. The headquarters in Falmouth includes a reading library open to the public.

Maine Audubon is seeking a publisher to work with their organization on revising and updating the *Maine Studies Sourcebooks* to make them useful to a broader audience.

Audubon Society of New Hampshire (ASNH)
3 Silk Farm Road
P.O. Box 528-B
Concord, NH 03302-0516 USA

Contact: **Jeff Schwartz**, Education Director
   Phone: 1-603-224-9909 • *Fax:* 1-603-226-0902

Activities: Education; Political/Legislative; Research • *Issues:* Biodiversity/Species Preservation; Land Protection; Sustainable Development

● Organization

Established in 1914, the Audubon Society of New Hampshire (ASNH) is "dedicated to the conservation of wildlife and natural resources throughout the state." It is an independent, nonprofit organization, separate from the National Audubon Society. Its purpose is "working to ensure that wildlife needs and ecological values will be a lasting part of personal and public actions affecting New Hampshire's natural environments." ASNH has a membership of over 6,000 households.

▲ Projects and People

ASNH programs can be divided into four general areas:

Wildlife programs focus on nongame wildlife research and management in the state. Biologists and volunteers "conduct wildlife surveys, collect data on habitat needs, and study the breeding success of many species." Examples of programs include the **New Hampshire Endangered Species Program**, directed by ASNH in cooperation with the New Hampshire Fish and Game Department; the **Loon Preservation Committee**, sponsored by ASNH; and the **New Hampshire Breeding Bird Atlas**, cosponsored by ASNH.

Land protection involves diverse habitats in all 10 New Hampshire counties, which total over 4,000 acres. Audubon properties encompass freshwater and saltwater marshes, bogs, swamps, islands, and many different kinds of woodlands. ASNH maintains some of these ecosystems as natural areas, managing others for wildlife, public recreation, and education.

The purpose of ASNH **environmental affairs** activities is to provide a strong voice on "important local and national issues which affect New Hampshire's wildlife, habitats, and natural resources." ASNH representatives testify at hearings, communicate with public officials and legislators, and work with other environmental groups and resource agencies.

ASNH is committed to **environmental education** at all levels, "from preschoolers to families." Programs are offered year-round by the group's naturalist staff to over 20,000 people, through school programs, adult courses and workshops, youth programs, family programs, teacher workshops, field trips, and day camps. ASNH owns and operates two nature centers and participates in a cooperative coalition that operates the visitor center at Odiorne Point State Park.

■ Resources

The Northeast Field Guide to Environmental Education is a current directory of the Northeast, covering Connecticut, Massachusetts, Maine, New Hampshire, New Jersey, Pennsylvania, Rhode Island, and Vermont. It includes over 400 organizations in the region that offer environmental education programs, facilities, and resources. The guide provides information on nature centers; planetariums, aquariums, and zoos; state, regional, and national environmental organizations; national parks and wildlife refuges; conferences; and job listings. It also provides useful data on costs of programs, publications, internships, and available materials.

The book may be ordered from: *Northeast Field Guide*, Box 4, Antioch New England Graduate School, Roxbury Street, Keene, NH 03431.

National Audubon Society
Adopt-a-Forest
P.O. Box 462
Olympia, WA 98507 USA

Contact: **Jim Pissot**
   Phone: 1-206-786-8020 • *Fax:* 1-206-786-5054
   E-Mail: Dialcom: TCN2362

Activities: Education; Political/Legislative; Research • *Issues:* Biodiversity/Species Preservation; Deforestation; Sustainable Development; Water Quality

● Organization

Today, less than 10 percent of the original virgin forests remain in America, and the ones that do exist are only on federal lands. Despite the fact that the lands are administered by the U.S. Forest Service and Bureau of Land Management, nearly 100 square miles of forest are cut each year in the Pacific Northwest alone. The National Audubon Society's Adopt-a-Forest program is for those interested in getting involved "at the heart of what goes on."

Adopt-a-Forest volunteers work with Audubon maps to familiarize themselves with the critical features of their districts. They conduct surveys of proposed sales, work with Forest Service documentation, and raise their concerns with district personnel to make sure that when timber sales do occur they are consistent with sound environmental practices and federal law.

If necessary, Adopt-a-Forest groups make formal appeals, initiate litigation, solicit congressional intervention, or work with media to publicize problems. Negotiation has been found to be a preferable and more successful approach, however. According to the literature, "There are many people working within the Forest Service with a strong environmental ethos. By maintaining a strong and visible presence, an Adopt-a-Forest group supports those agency people who are our allies and helps them do the job of protecting the forest instead of harmfully destroying it."

Jim Pissot of Olympia, Washington, directs the program, coordinating activities with the National Audubon Society's Ancient Forest Alliance campaign, which is run by Brock Evans in Washington, DC.

■ Resources

Information on Pacific Northwest Adopt-a-Forest programs can be obtained from Audubon Society chapters in Olympia, Washington; Walla Walla, Washington; Eugene, Oregon; Mt. Shasta, California; and Polebridge, Montana. The *Adopt-a-Forest Manual* is available for purchase.

Environmental and Societal Impacts Group (ESIG) ⊕
National Center for Atmospheric Research (NCAR)
P.O. Box 3000
Boulder, CO 80307 USA

Contact: Michael H. Glantz, Ph.D., Program Director
Phone: 1-303-497-8119 • *Fax:* 1-303-497-8125

Activities: Education; Research • *Issues:* Climate/Society Interactions; Deforestation; Global Warming; Sustainable Development

● Organization

The National Center for Atmospheric Research (NCAR) established the Environmental and Societal Impacts Group (ESIG) in 1971 "to assess the societal aspects of the five-year National Hail Research Experiment." This in turn developed "a sub-field of research focused on climate-related impact assessment," studying "the

complex interactions between atmospheric processes and human activities . . . to identify the value and uses of meteorological information for economic development purposes." ESIG offers a rare opportunity for researchers in nonscientific disciplines to work closely with scientists and engineers, sharing information vital to the upkeep of the Earth.

ESIG researchers work within NCAR but often collaborate with researchers elsewhere, at universities, governmental and non-governmental research centers, and by other means. The group maintains an active outreach program with visiting scientists, and several study grants. Essential to the information sharing process are "multidisciplinary, multinational workshops and symposia on issues related to the human dimensions of environmental change." Much of the international networking is supported by the United Nations Environment Programme (UNEP). Other supporters include the National Science Foundation (NSF), National Oceanographic and Atmospheric Administration (NOAA), U.S. Environmental Protection Agency (EPA), U.S. Fish and Wildlife Service, and the National Marine Fisheries Service (NMFS).

▲ Projects and People

ESIG's work tends to focus "on issues related to the **impacts of human activities on the atmosphere, the impacts of the atmosphere on human activities, the use and value of meteorological information in decisionmaking processes, and on developing new methods for assessing societal responses to climate-related impacts.**" Future efforts will concentrate on expanding **education and training programs,** as interest grows "in the environment and in atmospheric processes."

Dr. Michael Howard Glantz, with NCAR since 1976, combines knowledge of science (metallurgical engineering, climate) with political science/philosophy. He believes that ESIG is "extremely beneficial for the implementation of multidisciplinary research and for building bridges between the social, biological, and physical sciences." Glantz has received numerous awards for articles and papers—including the UNEP's Global 500 Award—as well as fellowships and speaking engagements.

One of Dr. Glantz's main interests is in the interactions between climate and society, and the development process. He has pursued studies of climate change and desertification as related to droughts in Africa and elsewhere. Dr. Glantz has been a member of an international committee on the Conservation of the Aral Sea since 1991, also working on other projects in the former USSR. He has served as consultant and representative with UNEP, attending and participating in numerous workshops, conferences, and projects.

Dr. Richard W. Katz has been a scientist/statistician at ESIG periodically since 1976, and has worked under EPA Cooperative Agreements and National Science Foundation grants to study extreme meteorological events and the value of weather and climate information. He has written numerous articles, books, chapters, papers, and conference proceedings on those and related topics. Dr. **Kathleen A. Miller** has been a scientist with ESIG since 1985; her specialty is economics as it relates to the climate. In particular, she has investigated water rights issues and natural resource economics. **Dr. Steven L. Rhodes,** scientist with ESIG since 1989, had a postdoctoral fellowship at ESIG in 1982–1984 and was adjunct scientist there in 1985–1987. A political scientist, Dr. Rhodes researches the political and legislative connection to acid rain, nuclear regulation, and drought. Former University of Colorado law professor Dr. Daniel Magraw was an ESIG adjunct scientist and serves on this *Guide's* Board of Advisors.

■ Resources

ESIG scientists publish numerous articles and reports on their fields of study. NCAR has previously published a North American institutional directory on *Climate-Related Impacts Network.*

National Center for Environmental Health Strategies (NCEHS)
1100 Rural Avenue
Voorhees, NJ 08043 USA

Contact: **Mary Lamielle**, President
Phone: 1-609-429-5358

Activities: Education; Political/Legislative; Research • *Issues:* Air Quality/Emission Control; Health

● Organization

According to the statistics cited by the 2,000-member **National Center for Environmental Health Strategies (NCEHS)**, 15 percent of Americans suffer from multiple chemical sensitivity (MCS), a condition often brought on and then aggravated by the presence of low or high levels of chemicals in the surroundings. Although the government has worked to legislate improvements in the outside air quality, indoor air quality is often ignored, according to NCEHS. People who blame their headaches, disorientation, rashes, cardiovascular disturbances, and myriad other possible symptoms on their work environment too often meet with dismissive reactions. NCEHS is working to change attitudes and laws to take MCS into account, on behalf of the thousands of chemically sensitive individuals in the USA.

Culprits can be "emissions from building materials, furnishings, consumer and personal care products, and office equipment," the organization says. "Pesticides, particle board, plastics, tobacco smoke, carpeting and adhesives, deodorizers, cleaning agents, correction fluid, laser printers, copy machines, and fragrances and fragranced products are some of the many substances that contribute contaminants to indoor spaces. This chemical soup gets circulated in most buildings due to inadequate ventilation to dilute the mixture. Windows that do not open further impair attempts to introduce outdoor air. This is a prescription for disaster," according to **Mary Lamielle** and Dr. **Diane Reibel-Shinfeld**.

▲ Projects and People

Efforts of the Center focus on educating the public, employers, the medical community, and government institutions, emphasizing the need to improve the work environment as well as to research MCS in controlled studies. According to NCEHS president Mary Lamielle, herself a sufferer, employees with MCS "can be accommodated with a little creativity and common sense at minimal expense."

Representatives of NCEHS have testified before the House Subcommittee on Health and the Environment and before the Senate Subcommittee on Superfund, Ocean, and Water Protection regarding the Indoor Air Quality Act of 1991. Some of their suggestions,

such as using **integrated pest management** (IPM) rather than toxic pesticides, increasing office ventilation, and completely banning smoking, deodorizers, and fragranced products in the workplace, have been included in drafts of the Act.

In part because of NCEHS efforts to spread its message, the Department of Housing and Urban Development (HUD) in 1990 gave MCS disability status; the National Academy of Sciences/National Research Center recommended researching MCS at a recent U.S. Environmental Protection Agency (EPA)–sponsored workshop; and governmental organizations in New Jersey and California have recognized MCS as a problem that needs attention.

■ Resources

The Delicate Balance is a quarterly newsletter that keeps readers abreast of recent research, legislation, and legal action related to MCS. *Chemical Exposures: Low Levels and High Stakes* by Nicholas A. Ashford and Claudia S. Miller is an expanded version of the landmark study commissioned by the New Jersey Department of Health. The organization also publishes other materials on chemical sensitivity.

National Coalition Against the Misuse of Pesticides (NCAMP)
701 E Street, SE, Suite 200
Washington, DC 20003 USA

Contact: **Jay Feldman**, Executive Director
Phone: 1-202-543-5450

Activities: Education; Law Enforcement; Research • *Issues:* Health; Pesticides; Sustainable Development

● Organization

The **National Coalition Against the Misuse of Pesticides (NCAMP)** was formed in 1981 "to serve as a national network committed to pesticide safety and the adoption of alternative pest management strategies which reduce or eliminate a dependency on toxic chemicals." Emphasizing the power of the people's voice as a counter to chemical industry pressures, NCAMP targets both contact and systemic pesticides—those that lie on the surface and those that are absorbed into food. The staff of seven includes a toxicologist and an ecologist.

▲ Projects and People

NCAMP devotes its education and research efforts toward identifying the risks of pesticides and promoting alternatives, believing that knowledge is the key to pesticide control. According to NCAMP sources, perhaps 80 percent of currently used pesticides have not been completely tested. Between 1950 and 1983, the world's pesticide production increased from 100 tons to more than 1 million tons—but crop loss due to increasingly resistant insects doubled.

NCAMP's activities include testifying before congressional committees and state officials, handling thousands of requests each year for information, and providing expert advice. By making citizens and legislators aware of the alarming facts, NCAMP works to close legal loopholes that weaken the power of the U.S. Environmental Protection Agency (EPA) and of existing laws, and to broaden the scope of pesticide regulation. The group has brought the EPA to court over its methods of regulating pesticides.

Through its **Pesticides and Public Housing** project, NCAMP is working with major public housing administrators toward the development of a more responsible, safe pest management program such as integrated pest management (IPM), which reduces pests without toxic chemicals. NCAMP points out that a high proportion of public housing residents are the very old and the very young, both groups particularly vulnerable to pesticides.

NCAMP sifts through project proposals and using funds from its **Seed Grant Program**, begun in 1985, sponsors projects of local environmental groups.

A major ongoing effort of NCAMP is the **Chlordane/Heptachlor Litigation Clearinghouse**, in conjunction with Trial Lawyers for Public Justice. With this project, the two groups are creating a central information base to provide plaintiffs' attorneys with information useful to pursuing their cases. Successful litigation will lead to more corporate responsibility and accountability in the production of this termite insecticide and other pesticides.

■ Resources

NCAMP publishes three newsletters: *Pesticides and You*, five times yearly since 1981; *NCAMP's Technical Report*, monthly since 1986; and *chemicalWATCH*. Other publications include pamphlets and brochures on organic gardening, specific chemicals, and myths and facts about pesticides. The Chlordane/Heptachlor Litigation Clearinghouse has resulted in a 200-page *Materials Notebook*.

National Environmental Development Association (NEDA)

1440 New York Avenue, NW
Washington, DC 20005 USA

Contact: **George Eliades,** Executive Vice President
Phone: 1-202-638-1230 • *Fax:* 1-202-639-8685

Activities: Educationa; Political/Legislative • *Issues:* Air Quality/Emission Control; Waste Management/Recycling; Water Quality

● Organization

The **National Environmental Development Association (NEDA)** is a coalition of organized labor, agriculture, and industry that was started in 1973 to define public policy issues, to inform the media and general public, and to educate policymakers in the U.S. Congress and the Administration on the "importance of balancing environmental and economic considerations."

Forty-eight member groups include AFL-CIO's Building and Construction Trades, Anheuser-Busch Companies, Ashland Oil Company, AT&T, BP America, Campbell Soup Company, Chevron USA, Consolidation Coal Company, Corning, Digital Equipment Corporation, Dow Chemical Company, Du Pont, Eli Lilly, Exxon, Exxon Chemicals American, Florida Fruit and Vegetable Association, FMC Corporation, General Electric, General Motors Corporation, Geneva Steel, Giant Resource Recovery, Gulf Coast Waste Disposal, Hewlett-Packard, Hoechst-Celanese, and Honeywell.

Other members are International Brotherhood of Electrical Workers (IBEW), IBM Corporation, Kaiser Aluminum and Chemical Corporation, Kimberly-Clark Corporation, Laborers International Union, Lone Star Steel Company, Lukens, Maritrans Operating, Mobil Oil Corporation, Monsanto Company, Occidental Petroleum Corporation, Pacific Northwest Waterways, Phillips Petroleum Company, Procter & Gamble, Public Service Indiana Energy, Quantum Chemical Corporation, R.J. Reynolds Tobacco Company, Sugar Cane Growers, Sun Company, Texaco, United States Sugar Corporation, UNOCAL, and Velcon Filters.

NEDA is classified as a 501(c)6 organization, and contributions are not tax deductible. However, contributions to the nonprofit NEDA Foundation are tax deductible. The Foundation sponsors an annual awards dinner to recognize "key congressional, public sector, industry, and media leaders" striving for balanced environmental public policy. Beginning with Earth Day 1990, NEDA established an Honor Roll to recognize corporations for pollution prevention, environmental education, preservation of natural resources, and similar efforts. A Special Press Fund is used to encourage balanced environmental reporting through press briefings and media contact.

▲ Projects and People

Although it supported the basic goals of the Clean Air Act, NEDA established a **Clean Air Act Project (CAAP)** to recommend modifications that would streamline implementation, particularly regarding the U.S. Environmental Protection Agency's (EPA) rulemaking on state operating permit programs. Speaking for NEDA, **David A. Chittick,** AT&T corporate environment and safety vice president, urged that "U.S. manufacturing facilities be allowed fundamental operational flexibility to produce goods for competitive markets while proceeding through a permit program which provides timely implementation at a reasonable cost."

Chittick also noted AT&T's commitment, similar to other association members, to "phase out as rapidly as possible the use of materials which have been linked to the depletion of the stratospheric ozone layer." For example, by replacing methylene chloride and methyl chloroform with an aqueous solution for cleaning certain circuit boards, "emissions of hazardous air pollutant solvents were reduced from . . . about 185 tons per year of a hazardous emission to . . . about 1.7 tons per year of nonhazardous particulate."

NEDA's **Resource Conservation and Recovery Act (RCRA) Project** supports an expanded nonhazardous waste program and the view that a "truly viable national program must stress recovery and utilization of valuable resources from recyclable materials." It also supports a new RCRA reauthorization subtitle to address reuse and recycling of hazardous materials.

NEDA's **Groundwater Project** supports a "new national partnership between the states and federal government" to develop a balanced policy on groundwater availability for agriculture, industrial, and domestic uses. NEDA recommends "increased federal R&D [research and development] and technical assistance, while requiring states to develop comprehensive management programs tailored to the local physical conditions of ground water resources and their designated uses as aquifers."

NEDA occasionally profiles environmental successes such as AT&T's corporate policy to eliminate all emissions of chlorofluorocarbons (CFCs) from its manufacturing processes worldwide by 1994; it is sharing this technology and is a founding member of Industry Cooperative for Ozone Layer Protection (ICOLP).

Anheuser-Busch Companies are reclaiming and recycling processing residuals such as "spent grains, spent hops, yeast, and other byproducts" that are sold as animal feed. In New York, the company is composting byproducts. And at three breweries, "bio-energy recovery systems have been installed to handle brewery effluent; in closed chambers, wastewater is treated and methane gas recovered for use as fuel for production processes."

Proctor & Gamble plants in Japan and Venezuela are "role models," according to NEDA, for recycling all waste used in manufacturing. More than 80 percent of the company's paperboard packaging is made from recycled materials, and "within the laundry detergent category, all the paperboard cartons in the USA are made from recycled paper."

Quantum has launched a Waste and Release Reduction (WARR) Program whereby each USI Chemical Division facility is developing a baseline and annual inventories of amounts of waste generated and materials released to the environment to determine reduction priorities and implement such plans.

UNOCAL began the South Coast Recycled Auto Program (SCRAP), whereby the corporation pays $700 each for pre-1971 cars that "emit 15 to 30 times the exhaust pollutants of 1990 models [to] strike at the heart of air pollution" in the Los Angeles area. The cars are dismantled at a scrap yard, where some parts are recycled.

■ Resources

NEDA publishes the quarterly *Balance* newsletter and provides monthly insider updates to members.

National Environmental Law Center (NELC)
37 Temple Place
Boston, MA 02111 USA

Contacts: **Hillel Gray**, Policy Director
    Phone: 1-617-422-0880 • *Fax:* 1-617-422-0881

Activities: Law Enforcement; Political/Legislative; Research
Issues: Pollution Prevention; Toxic Waste Reduction; Waste Management/Recycling

● Organization

The nonprofit **National Environmental Law Center (NELC)** was founded in 1990 as a "litigation and policy development center" by the Public Interest Research Groups (PIRGs), a "national network of state-based consumer and environmental advocacy groups." NELC takes legal action against the polluters of the USA. It maintains a staff of attorneys and policy analysts who translate and enforce environmental policies throughout the country, usually on a state level, considered the "laboratories of democracy" in the federal system.

With a membership of 50,000 and staff of 13, NELC is particularly active in the fields of toxic pollution and solid waste pollution. For example, toxic specialists "promote policies to clean up past contamination, set strict limits on current emissions, and prevent future pollution by reducing the use of toxic chemicals." Solid waste specialists aim to "reduce waste, expand recycling markets, and stop unsafe disposal practices." Resource conservation is their main goal, especially lessening the use of virgin materials in products and packaging.

The Center runs a clearinghouse which keeps easily-accessible information on policy ideas and on other environmental groups who have expertise in certain areas of environmental law.

NELC also maintains the NELC Litigation Center, which actually takes action against environmental polluters. The Litigation Center gets court injunctions, secures fines, and arbitrates settlements that give money to other environmental efforts.

▲ Projects and People

The staff of NELC have strong backgrounds in environmental policy and in law. Recent policy director **Joel Ario**, Harvard Law School graduate, spent a decade working with PIRGs in Oregon, Massachusetts, and Wisconsin, and remains executive director of Oregon's PIRG (OSPIRG). He has directed many environmental and consumer crusades, recently fighting for Oregon's Toxic Use Reduction Law and for the upgrade of Oregon's recycling laws.

Strategies include **drafting model laws**, providing expert testimony, publishing case histories and other research in "easy-to-understand formats," and reaching the public through education materials and presentations.

Under the federal Clean Water Act, NELC attorneys have brought "more than 60 cases" and "won more than $29 million in fines and support for environmental projects in and around affected waterways." Working closely with the Rutgers University's Environmental Law Clinic, the Center reports that "by focusing on the worst polluters, NELC has never lost a case."

In one endeavor, a large citizen suit judgment of $4.2 million was ruled against Powell Duffryn Terminals, New Jersey, a company "refusing to build a treatment plant necessary to prevent toxic discharges into the Kill Van Kull and other waterways." The NELC succeeded again in a suit against Ohio's Wheeling-Pittsburgh Steel Corporation, which was ordered to stop contaminating the Ohio River. NELC reports similar victories in the states of Massachusetts, Washington, Illinois, Michigan, and Pennsylvania.

At least 20 cases concerning toxic, metals, oil, and other discharges are pending, from the Boston harbor to the Delaware and Schuylkill rivers to the Columbia River.

■ Resources

Aside from its clearinghouse of accessible information, NELC publications include the quarterly *National Environmental Law Center Report*, and *Unmasking Environmental Polluters: A Report on Chemical and Plastics Producers' Opposition to Measure 6*, a report on the five large producers of plastic resins: Union Carbide, Dow Chemical, Exxon, Chevron, and Occidental Chemical, who defeated an Oregon initiative advocating the use of recycled plastic.

Other reports look at feasibility of environmental technologies, assess data on toxic chemicals in the USA, and the findings of a group of scientists who were asked to draw up their ideal pollution prevention law. The *Introduction to the National Environmental Law Center* contains a publications list.

National Foundation for the Chemically Hypersensitive (NFCH) ❧
Route 649
Ophelia, VA 22530 USA

Contact: **Fred Nelson**, Director
Phone: 1-804-453-7538

Activities: Education; Research • *Issues:* Air Quality/Emission Control; Deforestation; Energy; Global Warming; Health; Sustainable Development; Toxic Exposure; Transportation; Waste Management/Recycling; Water Quality

● Organization
The National Foundation for the Chemically Hypersensitive (NFCH) is a nonprofit, tax-exempt, volunteer organization of 5,000 members devoted to research, education and dissemination of information including patient-to-doctor referrals; patient-to-attorney referrals; social security disability information; workers compensation information; and advice and resource assistance for the chemically injured and their relatives. Furthermore, it is a patient network for safe living, food and clothing referrals, and assistance with low-cost housing resources. NFCH compiles thousands of case histories and develops epidemiological studies.

According to NFCH, chemical hypersensitivity is a "chemically induced immune system disorder. It includes the classification of multiple chemical sensitivities, environmental illness, food intolerance, total allergy syndrome, candida, and chronic fatigue."

NFCH warns that chemicals remain in the body "long after use or exposure ends. Stored toxics can be released into body fluids leaving 'target' organs susceptible to dysfunction and disease," it reports. Also weakening the immune system are drugs and alcohol.

▲ Projects and People
NFCH identifies the common symptoms of chemical injury and the potential sources of exposure. For example, it describes the exposure and nature of an illness: "You may have had a low-level, long-term toxic substance exposure. Then you will have a suppressed immune system. If you had a short-term, acute exposure, your immune system will be highly activated. Sometimes, the immune system will be altered or imbalanced, and you will be very ill. If your system is impaired and altered, you usually will develop autoimmunity to one or more organ tissues."

If a person thinks he or she has had an adverse exposure, NFCH will counsel, make referrals, suggest a helpful reading list, take the case history to add to its database, help with chemical injury or dietary problem, and possibly have a member act as a big brother or big sister.

■ Resources
NFCH has a quarterly newsletter, *Cheers.* It publishes lists of common symptoms of chemical injury and potential sources of exposure.

National Recycling Coalition (NRC) ⊕
1101 30th Street, NW, Suite 305
Washington, DC 20007 USA

Contact: **Marsha Rhea**, Executive Director
Phone: 1-202-625-6406 • *Fax:* 1-202-625-6409

Activities: Education; Public Policy; Research • *Issues:* Sustainable Development; Waste Management/Recycling

● Organization
The National Recycling Coalition (NRC) is a nonprofit organization which is affiliated with numerous recycling groups including state recycling organizations (SROs) across the USA. NRC provides technical education on the plight of solid waste, helps to increase public awareness on issues, and looks for and advocates new laws which can help to solve the waste management problems around the world. NRC reminds the public, "Recycling is collection, processing, and reuse. Recycling is education, tax policy, and market development."

According to the NRC, in order for the recycling movement to grow and for the solid waste problem to diminish, there needs to be wide-scale public understanding. NRC believes that more input from a significantly large number of people can seriously help to shape national policies and find more solutions.

▲ Projects and People
To increase public awareness of the waste problem, NRC offers information on recycling and other resource-saving issues. Its 4,000 members receive an assortment of publications informing them of public policy and current events topical to the organization.

NRC's affiliation with nationwide state recycling organizations gives members many invaluable connections to successful recycling movements and numerous links with recycling experts around the country. Technical Councils are being established for the exchange of ideas and information among such members. So is a Peer Match program for hands-on assistance. The annual NRC-sponsored National Recycling Congress is the largest event of its kind.

■ Resources
To reach the general public, NRC gives discounts on national journals of recycling which discuss recycling and waste management. The magazines include *Resource Recycling, BioCycle, Garbage,* and *Recycling Times.* NRC also provides a wide range of publications on resource-conserving topics, available at member and non-member prices.

Environmental Health Center (EHC)
National Safety Council (NSC)
1019 19th Street, NW, Suite 401
Washington, DC 20036-5105 USA

Contact: **Bud Ward**, Executive Director
  Phone: 1-202-293-2270 • *Fax:* 1-202-293-0032
  E-Mail: CI$: 72730, 1035

Activities: Education; Journalism Continuing Education; Political/Legislative; Research • *Issues:* Air Quality/Emission Control; Energy; Global Warming; Health; Waste Management/Recycling; Water Quality

▲ Projects and People

The **Environmental Health Center (EHC)**, a division of the National Safety Council (NSC), is engaged primarily in continuing education projects for print and broadcast environmental journalists. With a staff of six, EHC publishes two periodicals—a monthly newsletter, *Environmental Writer*, for U.S. journalists, and a quarterly, *One Environment*, for journalists in Central and Eastern Europe. EHC has published several "peer-reviewed" reporters' guides, such as *Chemicals, The Press & The Public*, dealing with coverage of potential chemical risks in the community under the federal "right-to-know" law; and *Reporting on Radon*.

Companion guidebooks are being developed on municipal solid waste, coastal and on-shore development issues, air quality, and climate change. In addition, EHC is working to develop a semester-long **environmental journalism curriculum**, to be pilot-tested in four university journalism schools, and to serve as the basis for a two-week module in an existing science journalism curriculum.

Each year, EHC sponsors the **Central European Environmental Journalism Program**, under which reporters from these countries work and study in the USA for six weeks with American environmental journalists. EHC offers environmental journalism courses in Czechoslovakia, Hungary, Poland, and eastern Germany and is developing companion programs in other countries.

Beyond its environmental journalism projects, EHC sells and distributes emergency response planning software and provides technical support to users, and manages several environmental "hotlines" for providing citizens information on environmental issues.

With a 20-year background in environmental communications including teaching college courses in journalism and environmental issues, **Morris A. (Bud) Ward** founded EHC in 1988 and is a frequent public speaker in his field. As EHC executive director, Ward oversees a variety of foundation- and federally funded research and communications programs aimed at improving public understanding of pollution issues. He is the author of one book on the U.S. Clean Water Act and a co-author of a book on the Resource Conservation and Recovery Act (RCRA), as amended. In addition, he has assisted in managing the federally developed computer-based emergency planning and response software program known as CAMEO, Computer Aided Management of Emergency Operations.

■ Resources

EHC offers annual subscriptions to the *Environment Writer* newsletter. *Chemicals, The Press & The Public* can also be ordered.

National Tropical Botanical Garden
P.O. Box 340
Lawai Valley
Kauai, HI 96765 USA
Contacts: **W. Arthur Whistler, Ph.D.**, Botanist
  Phone: 1-808-956-3925 • *Fax:* 1-808-956-3923;

Diane Ragone, Ph.D., Program Botanist, Hawaii
       Plant Conservation Center
  Phone: 1-808-332-7324 • *Fax:* 1-808-332-9765

Activities: Education; Research • *Issues:* Biodiversity/Species Preservation • *Meetings:* AIBS; Society for Economic Botany

● Organization

An Act of Congress chartered the private, nonprofit **National Tropical Botanical Garden** in 1964 as "a scientific research center for tropical botany and horticulture, a comprehensive collection of plants from all parts of the tropical world, an educational center in tropical botany and horticulture, and a place of great natural beauty."

This national resource is headquartered on a 186-acre site in the Lawai Valley, where research, education, and the living collections—the Garden's "most important aspects"—are being developed. These special collections include plants of nutritional and medicinal value and ethnobotanical interest; rare and endangered species; plants of unexploited potential; tropical fruits; spices; and special groups such as palms, erythrinas, and gingers, and tropical ornamentals.

Three Satellite Gardens are the 180-acre Kahanu Gardens near Hana, Maui, which is a center for ethnobotanical plants and breadfruit, coconut, and loulu palm collections; the 1,000-acre Limahuli Gardens and Preserve on the northern coast of Kauai among rainforests and mixed mesophytic forest with areas set aside for native plants; and The Kampong, Florida, former home and garden of world-famous horticulturist Dr. David Fairchild, which contains an outstanding collection of tropical fruit trees. Rich with archaeological sites, Limahuli—which means "turning hands" in Hawaiian—is named for the early inhabitants who turned the soil to plant crops and turned over rocks to look for fish in streams. A laboratory and office is also located at the Department of Botany, University of Hawaii, Oahu. On preserves in Maui and the island of Hawaii, areas containing rare and endangered species are conserved.

Among the flora—the **baobab tree** has edible fruit, leaves used for cattle fodder, and a bark yielding gum, medicine, and fiber for rope, paper, and cloth. The **breadfruit** or *'ulu* provides a staple food with sources of vitamins A, B, and C; its wood is used for canoes, and the milky sap is a base for glue, caulk, and chewing gum. The **candlenut tree** or *kukui*—Hawaii's state tree—furnishes oily nut kernels for candles; at one time 10,000 gallons of oil were exported. The **coconut tree**, said to be one of the world's most beautiful palms, grows as high as 100 feet; all of its parts are usable—leaves as thatch, trunk into calabashes, drums, and small canoes; husk fibers for braided ropes; shells for bowls and utensils; nuts for nutrition, and oil from the meat as a hair and body lotion. Seeds of *Loulu* or *Pritchardia*, the only palm genus native to Hawaii, are thought to have arrived by pigeon from Fiji or afloat ocean currents as early as 40,000 years ago. Uniquely adapted to each Hawaiian island, *Pritchardia* are considered beautiful and select **ornamental palms**.

▲ Projects and People

Staff and visiting scientists are using the newly constructed, extensive laboratory, library, and herbarium complex. A consortium of U.S. universities is being formed to help establish the Garden as a true national center for tropical research.

The Hawaii Plant Conservation Center (HPCC) was set up in 1989 with a grant from the John D. and Catherine T. MacArthur Foundation as a "resource and information center dedicated to the collection, propagation, cultivation, and distribution of native Hawaiian plants." HPCC is compiling a reference file and bibliography on propagation of native Hawaiian species and related subjects; a master list of native species currently being propagated throughout the state; a calendar of flowering and fruiting times to facilitate seed collection; a slide and photograph collection; and fact sheets on how to grow and use native plants in cultivation.

HPCC distributes to the public at no charge native Hawaiian and Polynesian plants through an expanding service and is coordinating a statewide network of propagators to grow native species for cultivation and reintroduction into their natural habitats. A centralized seed storage and distribution center is being established as well as ongoing research to foster conservation and establishment of native species.

Xeroscaping, or gardening with less water, is being researched as Garden horticulturists expect that water rationing may be required on Oahu by the year 2000 as a result of increased population and development. Xeroscaping is based on "controlling weeds, using mulches, improving the soil, limiting lawn areas, practicing good maintenance, watering efficiently, having a good landscape design, and using low-water-use plants," writes Heide L. Bornhorst, HPCC education coordinator. "We have an abundance of native plants from dry coastal areas and lowland forests. These plants have evolved under extremely harsh conditions: hot, dry, windy, and often exposed to salt-laden breezes." Good candidates for testing in xeroscapes, indicates Bornhorst, are such trees and shrubs as *a'ali'i* (*Dodonaea viscosa*), *alahe'e* (*Canthium odoratum*), *koai'a* (*Acacia koaia*), *ma'o hau hele* (*Hibiscus brackenridgei*), *naio* (*Myoporum sandwicense*), *'ohi'a lehua* (*Metrosideros polymorpha*), and *wiliwili* (*Erythrina sandwicensis*).

The Growing Seed–*Ke ulu nei ke'ano* is an HPCC classroom series in which third to sixth graders learn about habitat, adaptation, native plants, life cycles, and conservation. Students are also encouraged to plant native species on their school grounds as educational display gardens.

A Professional Gardeners Training Program in Tropical Botany and Horticulture with instruction and paid work experience is being offered to high school graduates from throughout the USA and abroad. Special tours, classroom sessions, and field labs are held for school children in conjunction with the Hawaii State Department of Education's Environmental Awareness Program.

Na Lima Kokua or Helping Hands is the active volunteer organization that prepares recipe books, holds plant sales, assists in gardening, and conducts tours.

■ Resources

Aside from the many resources mentioned above, the Garden has an Internship Program for university students and also offers public lectures and guided tours. *The Bulletin* is published quarterly. A list of various publications, membership information, and a gift catalog are available.

National Wildlife Federation (NWF) ⊕
1400 16th Street, NW
Washington, DC 20036-2266 USA
Contacts: Jay Dee Hair, Ph.D., President,
            Sharon L. Newsome, Vice President,
            Resources Conservation Department
Phone: 1-202-797-6826 • *Fax:* 1-202-797-6646
E-Mail: Econet: nwf.enviroaction

Laurel Ridge Conservation Education Center
8925 Leesburg Pike
Vienna, VA 22184-0001 USA
Contacts: Gary J. San Julian, Ph.D., Vice President
            for Research and Education
Phone: 1-703-790-4495 • *Fax:* 1-703-442-7332;

            Barbara J. Pitman; Director, Educational
            Outreach
Phone: 1-703-790-4360

Activities: Education; Political/Legislative; Research • *Issues:* Air Quality/Emission Control; Biodiversity/Species Preservation; Deforestation; Energy; Global Warming; Population Planning; Sustainable Development; Transportation; Water Management/Recycling; Water Quality • *Meetings:* CITES; IAFW Agencies; IDB; ITTO; IUCN; World Bank

● Organization

Founded in 1936 by Jay "Ding" Darling, the National Wildlife Federation (NWF) has become, by its own description, "the largest citizens' conservation education organization in America"—expanding its interests from outdoor sports emphasis in its earlier years to areas of global concern today. Nearly 6 million members support a staff of 630 scientists, lawyers, lobbyists, resource specialists, and other staff; and 15 regional executives are spread among national headquarters, 7 natural resource centers, and 51 state and territorial offices.

The NWF's Natural Resource Centers, for example, work to combat oil and gas exploration in the northern and central Rocky Mountains and Alaska, deforestation in the Pacific Northwest, Great Lakes water degradation, and wetlands loss in North Dakota and Georgia.

Most recently, NWF dedicated itself to "making Earth Day every day," honoring former U.S. Senator and "Father of Earth Day" Gaylord Nelson as an exemplary conservationist. At annual meetings, delegates commit to major policies such as tackling global environmental problems; protecting wildlife habitat in U.S. legislative provisions; pushing for "no net loss" to protect wetlands; urging the U.S. Forest Service to protect such ecosystems and biological diversity and the U.S. Department of Agriculture (USDA) to protect soils, water, and species habitats; and calling on the U.S. Department of Energy (DOE) to "develop a national energy policy emphasizing energy efficiency" while generating alternative energy technologies.

As Dr. Jay D. Hair was quoted in the *Washington Post Magazine* (August 4, 1991), "The glue that binds us together is habitat. It's habitat where wild things live, it's habitat where humans live."

NWF also asked the Bureau of Land Management to see that millions of acres under its care be protected from oil and gas exploration and drilling; NWF is working toward population stability worldwide, and is part of the growing movement opposing "polluting facilities in minority or economically disadvantaged communities."

A subgroup, the **Resource Conservation Alliance (RCA)**, consists of 75,000 members whose main task involves writing letters to U.S. Congresspersons and other policymakers on behalf of the environment.

The **Laurel Ridge Conservation Education Center**, located on 43 acres in Virginia, provides the venue for NWF's **Conservation Summits** and other training programs.

▲ Projects and People

The NWF has sponsored an annual **National Wildlife Week** since 1938, and has as a long-term goal an **Environmental Quality Amendment** to the U.S. Constitution. Its Institute for Wildlife Research undertook a strategic plan for its Endangered Species Fund and conducts an annual midwinter bald eagle survey. In the meantime, NWF pursues other specific targets; for example, in 1989 the Federation filed suit against Exxon to force the set-up of a fund to restore the area damaged in the *Valdez* oil spill. That same year, the Federation sued the U.S. Department of Commerce to reinstate a rule requiring Turtle Excluder Devices (TEDs) on shrimp nets to protect sea turtles. NWF works with hundreds of grassroots groups on behalf of expanding the Coastal Barrier Resources System, for example. The Texas Coastal Cleanup, inspired by **Linda Maraniss**, is an annual event that motivated "at least one oil company to ban polystyrene cups from its drilling rigs in the Gulf of Mexico."

In an effort to involve members more deeply in the conservation effort, the NWF in 1973 began the **Backyard Wildlife Habitat Program**. To date, more than 9,000 homeowners have had their backyards certified as wildlife conservation areas, free of pesticides and other excessive treatment. Working with Native Americans such as the Shoshone and Arapahoe Indian tribes, NFW and its affiliate the Wyoming Wildlife Federation helped stock 10,000 rainbow trout in the Big Wind River, for example.

Other NWF affiliates' efforts include protecting the endangered Mount Graham red squirrel in Arizona, helping to achieve passage of a solid waste management bill and authorization for the Deep Fork River Wildlife Management Area and National Wildlife Refuge in Oklahoma, opposing a dam in Arkansas, gaining support for the Roanoke River National Wildlife Refuge in North Carolina, setting priorities on the Delaware Bay watershed, and stopping the commercializing of state parks in Pennsylvania. In a court battle, the Montana Wildlife Federation joined with NWF's Northern Rockies Natural Resource Center to limit livestock grazing at the Charles M. Russell National Wildlife Refuge; NWF's attorney in Missoula is **Tom Franke**.

With biologist Dr. Jay D. Hair as NWF president since 1981, the Federation has increased dialogue with major American business leaders through its **Corporate Conservation Council**—established in 1982. Members currently represent the following: Asea Brown Boveri, Inc.; AT&T; Browning-Ferris Industries; Ciba-Geigy Corporation; Dow Chemical USA; Duke Power Company; E.I. Du Pont de Nemours & Co.; General Motors Corporation; Johnson & Johnson; 3M Company (honored by NWF for its waste-prevention program); Monsanto Company; The Procter & Gamble Company;

Shell Oil Company; USX Corporation; Waste Management Inc.; and Weyerhauser Company. For graduate business students at Boston University, Loyola University, and the University of Minnesota, the Council initiated environmental management courses that are now a source of interest worldwide. The Council also co-sponsored the Pacific Rim Conference on Sustainable Development.

In 1989, the NWF created an **International Affairs Department**, administering conservation activities outside the USA, particularly in Latin America. Among other work, this department runs the **Chico Mendes Fund**, named to honor the rubber tapper killed for his efforts to preserve the Amazon rainforest; promotes debt-for-nature swaps; and pressures multilateral development banks to adopt environmental assessment procedures in their projects, for instance. Likewise, it convinced the Japanese government to "suspend its contributions for the first of a series of India's Narmada Valley dam projects" that lacked such assessments. This staff works particularly closely with local environmental groups in Bolivia, Brazil, Ecuador, Indonesia, and the Philippines.

The **Resources Conservation Department** is intent on increasing public awareness of toxic emissions from industrial facilities, tolerance of biotechnology to herbicides, and overgrazing of privately owned livestock on public rangelands, among other issues. In charge is **Sharon Newsome**, with NWF for 10 years and also a member of the board of directors of Americans for the Environment and the Coast Alliance.

The NWF's educational effort includes *CLASS Projects*, an environmental curriculum for middle schools. *NatureScope*, an 18-issue series of teaching materials on specific topics like insects, rainforests, and birds, edited by **Judy Braus**, received **Renew America** recognition. The Federation also produces the public television program *Conserving America* and offers **Environmental Conservation Fellowships** to graduate students. *(See also NWF COOL IT! in this section.)*

Each summer the NWF operates **Wildlife Camps** and **Teen Adventures** in North Carolina's Blue Ridge Mountains and the Colorado Rockies.

■ Resources

The Laurel Ridge Center houses the NWF resource library and hosts an annual Conservation Fair.

The National Wildlife Federation has published *National Wildlife* since 1962 and *International Wildlife* since 1971; both are bimonthly. Two children's publications, the award-winning *Ranger Rick* for 6- to 12-year-olds and *Your Big Backyard* for preschoolers, come out monthly, along with a newspaper, *The Leader*. The *Conservation Exchange* is the NWF's quarterly magazine for corporate leaders, and the bird-of-prey newsletter, *The Eyas*, is by the Federation's Institute for Wildlife Research. Books and other gift items are available through order forms or at the Laurel Ridge Center.

COOL IT!
National Wildlife Federation (NWF)
1400 16th Street, NW
Washington, DC 20036 USA

Contact: **Nick Keller**, Director, Campus Outreach
Division
Phone: 1-202-797-5435
E-Mail: Econet: nwfnatl

Activities: Education • *Issues:* Biodiversity/Species Preservation; Deforestation; Energy; Global Warming; Waste Management/Recycling

● Organization
The National Wildlife Federation (NWF) initiated COOL IT! in 1989 as "a nationwide challenge" to involve college communities—students, faculty, and administration—in the effort to halt global warming. The seven staff members, all recent college graduates, work to encourage college students to plan, organize, and implement projects such as recycling, tree planting, and energy conservation on their campuses.

▲ Projects and People
COOL IT! staff provides telephone consultations with students who need advice starting or continuing environmental projects. They also maintain a speakers bureau, hold workshops, and make visits to campuses when necessary to help tailor-plan projects to the specific needs of each college community. A job and internship information service is available for the use of active college COOL IT! participants, with résumé forwarding and job listings for careers in the environment.

COOL IT! has four criteria for accepting a project for publication in its national project directory:
• The project must deal with global warming.
• Students must take concrete action.
• The work must involve coalitions from different sectors of the campus community (the group especially encourages culturally diverse groups to cooperate).
• The project must be sustainable.

One example of a project that made it into the 1990–1991 COOL IT! directory is the effort of the student conservation group, College of William and Mary, Virginia. Students successfully worked with the college administration and the Marriott Corporation's campus food service to reject a proposed contract with the Dart Company which would have entailed switching from paper to polystyrene food and beverage containers. They also won a verbal commitment that the cafeteria would switch to permanent (glass and metal) dishes and utensils.

At Michigan's Lansing Community College, students made and distributed composting units of chicken wire and plastic tarp stapled to stakes. Profits from the sales will go to further pro-environment projects. Harvard University's undergraduate houses and dormitories took part in an ECOLYMPICS competition, achieving both a wider campus awareness of the need to conserve energy, and an actual savings to the college of $500,000 over six months—reducing heat, water, and electricity consumption by 15 percent.

In encouraging this type of student activity and providing guidance, COOL IT! hopes to produce students who have helped the environment and have learned a great deal about cooperation, planning, and management.

■ Resources
COOL IT! makes available to interested students the *EnviroAction* newsletter on congressional environment-related activity; another newsletter, the *COOL IT! Connection* and information packets on composting, recycling, energy efficiency, fundraising, and other topics of interest to fledgling student environmental groups. *Students Working for a Sustainable World* is a national project directory of campus environmental groups; the current listing includes 71 projects on colleges and campuses in 32 states. COOL IT! is also part of EcoNet, the environmental computer network.

National Wildlife Federation (NWF) ⊕ ⊌
Alaska Natural Resources Center (ANRC)
750 West Second Avenue, Suite 200
Anchorage, AK 99501-2168 USA

Contact: **S. Douglas Miller, Ph.D.**, Director
Phone: 1-907-258-4800 • *Fax:* 1-907-258-4811
E-Mail: MCI mail: dmiller 490-0571

Activities: Education; Research • *Issues:* Air Quality/Emission Control; Biodiversity/Species Preservation; Energy; Population Planning; Sustainable Development; Waste Management/Recycling; Water Quality

● Organization
The Alaska Natural Resources Center (ANRC) is one of seven such centers run by the National Wildlife Federation (NWF). Since the Alaska center opened in 1988, its staff of four has been very active—particularly focusing on oil and gas exploration efforts.

▲ Projects and People
The Exxon *Valdez* oil spill disaster in 1989 served as the ANRC's baptism by fire. The Center lobbied in the Alaska legislature for bills increasing the liability of oil companies for future spills. In 1990, the Center organized a series of citizens commission hearings on the effects of the disaster, which were eventually published in a report entitled *The Day the Water Died.*

Also in 1990, ANRC participated in an unprecedented contract signed by the Alyeska Pipeline Services Company and the Regional Citizens Advisory Committee. The contract effectively empowers citizens to become involved in monitoring TransAlaska Pipeline activity.

ANRC director **Dr. S. Douglas Miller** is a specialist in wildlife biology. He has worked with the NWF since 1981, and is also the Pacific Region regional executive for Alaska and Hawaii.

■ Resources
The Day the Water Died is available through the Center.

Native Plant Society of Oregon (NPSO) 🌱
393 Fulvue Drive
Eugene, OR 97405 USA

Contact: **Rhoda M. Love, Ph.D.**, Past President

Activities: Education; Political/Legislative; Research • *Issues:*
Biodiversity/Species Preservation; Deforestation

● Organization
Founded in 1961 by some 40 plant enthusiasts who wanted to enjoy
Oregon's flora, the **Native Plant Society of Oregon (NPSO)** has
grown in size and purpose. Currently, its more than 1,000 members
in local chapters are "dedicated to increasing the public's knowledge
of native plants and to habitat conservation in the Pacific Northwest."

▲ Projects and People
Working independently as well as with state and federal agencies
and other organizations, the Society has achieved protected status
for plants such as **Bradshaw's desert parsley** and has banned the sale
of the **cobra lily** at city markets in Oregon. Now NPSO is working
on protected status for other species, prairie and wetland conservation, and similar projects. Much of the work is on a volunteer basis,
with members donating time to do taxonomic research and collect
rare plant species for supervised breeding.

Sales of notecards depicting native Oregon plants, as well as
donations and membership fees, fund some of the Society's efforts.
With the proceeds, NPSO can provide **small research grants for
botanical field studies** and a yearly scholarship to a local university
student majoring in botany or ecology.

Dr. Rhoda M. Love, an ecologist, a botanist, and a biologist, has
served the Native Plant Society since 1982 in various capacities,
including president; she is currently vice president of the Eugene
chapter. In her years of educating and working for the environment,
Love has authored numerous articles and guidebooks on plant
ecology. In addition, she has served on various advisory committees,
working with the Oregon Department of Agriculture, the Mount
Pisgah Aboretum, and the Eugene Community.

■ Resources
The Society publishes a monthly *Bulletin*. A membership brochure
is available.

Natural Resources Council of America (NRCA)
801 Pennsylvania Avenue, SE, Suite 410
Washington, DC 20003 USA

Contact: **Andrea Yank**, Executive Director
   Phone: 1-202-547-7553

Activities: Education • *Issues:* Air Quality/Emission Control;
Biodiversity/Species Preservation; Deforestation; Energy;
Global Warming; Health; Population Planning; Sustainable
Development; Transportation; Waste Management/Recy-
cling; Water Quality • *Meetings:* CITES

● Organization
The **National Resources Council of America (NRCA)** is the "umbrella group" of 85 autonomous conservation organizations in the
USA. Since its founding in 1946 at Mammoth Cave National Park,
Kentucky, the Council has been serving as an information-sharing
network and a coalition for promoting "adoption of public policies
to further protect the environment" and to accomplish "the sound
management of the world's natural resources."

▲ Projects and People
The NRCA's **Conservation Round Table** is a monthly forum that
brings together representatives from the conservation movement,
including government and industry. Held at the National Press
Club in Washington, DC, the forum features keynote speakers who
are environmental leaders. Briefings and congressional breakfasts
are also scheduled regarding current topics such as oil exploration in
the Arctic National Wildlife Refuge, restoration of fisheries, disappearance of wetlands, and "the need for a national waste management policy." Other activities are **task forces and special projects**—
for example, on U.S. energy policy for the 1990s, and on increasing
minority participation in conservation efforts. NRCA **field trips**
explore areas surrounding Washington, DC, such as the Chesapeake Bay and Maryland's Patuxent Wildlife Research Center and
Jug Bay Natural Area, Patuxent River Park.

Former U.S. Secretary of the Interior and past **chairman Robert
L. Herbst** is now with the Tennessee Valley Authority (TVA). With
more than 500 awards for conservation and distinguished service, he
was executive director of both Trout Unlimited and the Izaak
Walton League of America—along with posts in national and state
government.

■ Resources
Among other publications, NRCA prepares a bimonthly newsletter,
NRCA NEWS, which discusses "current events, employment opportunities, state and local conservation issues, member activities, and
people in the conservation field." A list of NRCA member organizations is available.

Natural Resources Defense Council (NRDC) ⊕
40 West 20th Street
New York, NY 10011 USA

Contact: **John H. Adams**, Executive Director
   Phone: 1-212-727-2700 • *Fax:* 1-212-727-1773

1350 New York Avenue, NW
Washington, DC 20005 USA

Contacts: **Faith T. Campbell, Ph.D.**, Senior Research
      Associate
      David D. Doniger, Senior Attorney
      Janet S. Hathaway, Senior Attorney
      Daniel L. Lashof, Ph.D., Senior Scientist
   Phone: 1-202-783-7800 • *Fax:* 1-202-783-5917

617 South Olive Street
Los Angeles, CA 90014 USA

Contact: **Mary Nichols,** Director
Phone: 1-213-892-1500 • *Fax:* 1-213-629-5389

71 Stevenson Street
San Francisco, CA 94105 USA

Contact: **Laura B. King,** Scientist
Phone: 1-415-770-0990 • *Fax:* 1-415-495-5996

102 Merchant Street, Suite 203
Honolulu, HI 96813 USA

Contact: **Susan Miller,** Resource Specialist
Phone: 1-808-533-1075 • *Fax:* 1-808-521-6841

Activities: Education; Law Enforcement; Political/Legislative; Research • *Issues:* Agriculture; Air Quality/Emission Control; Biodiversity/Species Preservation; Deforestation; Energy; Global Warming; Health; Nuclear Waste and Weapons; Population Planning; Sustainable Development; Transportation; Waste Management/Recycling; Water Quality • *Meetings:* CFC Emissions; CITES; Endangered species; Energy efficiency; Global warming treaty negotiations; Montreal Protocol; SSC; UNEP

● Organization

After beginning in 1970 as a small group of environmentally concerned lawyers and others who saw the need for outside controls on both industry and the governmental agencies responsible for monitoring industry, the **Natural Resources Defense Council (NRDC)** now has a staff of about 150, including lawyers and scientists. Contributions from 170,000 members, as well as from foundations and grant programs, help the NRDC to work from its New York headquarters and four regional offices in Washington, San Francisco, Los Angeles, and Honolulu. NRDC describes its headquarters as the nation's "most energy-efficient remodeled office space." **Adrian W. DeWind,** who became chairman at the start of NRDC's second decade, leads multiple efforts to pursue the "increasingly compelling global issues that require global solutions." NRDC founding staff attorney **Gus Speth** recently was president of the World Resources Institute.

The NRDC is involved in litigation, advocacy, and research, and its staff lawyers testify in Congress and go to court on behalf of citizens and the environment, fighting for the enforcement of existing laws and the enactment of stricter legislation.

It seeks to fulfill two "ethical imperatives": promoting human health with pure air and water and safe food for every person; and upholding "the sanctity of the natural environment." NRDC's vision is a sustainable society whereby "human beings use our resources without fouling or depleting them. Such a world would not forego economic growth, especially that needed in poor nations and communities . . . but growth can only continue if it does not destroy the natural resources that fuel it." Moreover, NRDC is helping create new global environmental institutions to perpetuate this goal.

Morris Udall, former U.S. congressman and environmentalist, offers his opinion: "What distinguishes NRDC . . . is the ability of its high-caliber professional staff to get results—to change government policy for the better. Every American who values clean water and clean air, whether or not he or she realizes it, is in debt to NRDC."

▲ Projects and People

John H. Adams has been the executive director of the NRDC from its start. Under his guidance, the Council uses the policy process, the media, U.S. Congress, and the courts as tools to reach its goal of a better environment for all, from the local to the international level. Adams views the 1990s as the decade to leap "national boundaries and issue boundaries." He writes, "We have to translate our national work to the global arena [and] it's already happening . . . new work within the [former] Soviet Union [is helping] transform their energy system from one of the most wasteful in the world to one of the most efficient . . . our leadership on the issues of clean air and energy efficiency has led us to undertake our . . . **Atmospheric Protection Initiative,** which has [as] its goal . . . preserving the integrity of the global atmosphere."

Although active in hundreds of individual issues, the NRDC has more than 20 ongoing projects under 6 umbrella programs: **water and coastal; land; public health; urban; international and nuclear;** and **atmosphere and energy.** As senior project scientist for the Atmospheric Protection Initiative, **Dan Lashof** gives examples of what can be done to reduce carbon dioxide emissions, such as "halting the massive logging in our publicly-owned forests." Also, by improving the energy efficiency of three million refrigerators, for instance, we could save enough electricity to eliminate one $600 million coal fired power plant—thus saving hundreds of millions of dollars, while simultaneously preventing new carbon dioxide emissions." Lashof adds, "installing electricity-saving devices widely in homes and businesses could shut down more than 200 coal-burning power plants and would reduce carbon dioxide emissions from utilities by one-third." According to NRDC, coal power is one of the biggest causes of acid rain, global warming, and respiratory ailments.

David Doniger, senior staff attorney for the **Clean Air Project,** reports that NRDC is continuing to push for a "total phase-out of CFCs and related ozone-depleting chemicals." **David Hawkins,** also senior staff attorney, reiterates the need to confront air pollution problems in urban areas where the air is classified by the U.S. Environmental Protection Agency (EPA) as unsafe to breathe.

NRDC continually presses the EPA and other agencies to phase out toxic lead in gasoline, to stop auto and bus pollution, and to tackle the causes of acid rain, for example. In 1991, NRDC scored some successes with advancing federal proposed requirements for cleaner-burning gasoline in polluted cities, with gaining tougher state laws to curb auto pollution, with getting 1992 funding—with the help of **Deborah Sheiman**—for the Chemical Safety and Hazards Investigation Board to examine and prevent toxic chemical accidents, and with pushing to close loopholes in rules for emissions permits governing factories and other pollution sources in the Clean Air Act.

In litigation, the NRDC petitioned successfully in the 1970s for a ban on chlorofluorocarbons (CFCs) in aerosols, and pushed for a total ban on CFC emissions within 10 years. Although the EPA initially resisted, the Agency eventually called for a 95 percent ban for the same time frame. In 1977, the NRDC successfully sued the Tennessee Valley Authority (TVA), resulting in a reduction of sulfur dioxide emissions by more than one million tons.

What's wrong with our water? According to NRDC, "every year in the United States we dump five trillion gallons of industrial wastewater and 2.3 trillion gallons of sewage into our coastal waters . . . well over one billion pounds of dangerous toxic pollutants that contaminate our seafood and drinking water. . . . At least half of the nation's water pollution is caused by poison runoff from places like city streets, farms, building sites, and parking lots."

Goals of U.S. clean water legislation are not being met, as evidenced by EPA reports of some 17,000 bodies of water in the USA still "badly polluted," says NRDC. Thus, senior attorneys **Bob Adler** and **Jessica Landman**, NRDC Water and Coastal Program, spearheaded a clean water network of 60 national to local groups to lobby the U.S. Congress for an improved Clean Water Act. The message is: prevent water pollution for all sources such as underwater sediment, factories, sewage plants, farms, highways, parking lots, logging and mining operations; protect wetlands such as swamps, tidal marshes, bogs, and mangrove forests that habitats for fish and wildlife and natural controls for flooding and erosion; and enforce the law to improve water quality standards, ensure the public right to know, and adequately fund clean water programs. NRDC and its network want toxic discharges to be eliminated and groundwater and aquatic sediment to be cleaned up. It reports that nearly 30,000 acres of USA wetlands are lost yearly. To save the Northwest salmon, **Karen Garrison** got the Northwest Power Planning Council to "draft new rules for irrigation efficiency and water policy reform" so that spawning patterns are not disrupted; she also formed the "first coalition" of environmentalists and fishing interests to restore the Northwest salmon fishery.

Regarding NRDC's **Coastal Project**, Sarah Chasis, senior staff attorney, writes, "We have staved off oil development in environmentally sensitive ocean areas" like Bristol Bay—a vital habitat for salmon, birds, whales, and other marine mammals; Arctic Wildlife Refuge with its complete ecosystems and annual birthing grounds for the 180,000-head porcupine caribou; Georges Bank off the New England coast, Florida Keys, and California coast. Since the mid-1980s, with NRDC trustee Robert Redford's urging, the organization has negotiated with industry to delete "sensitive offshore areas in Alaska's Bering Sea from the government's oil and gas leasing program." Although the government initially balked at the proposal, NRDC continues to fight lease sales on an individual basis. In New Jersey, **Nina Sankovitch** spurred the end to sewage sludge dumping in the Atlantic Ocean, and she is working to get three communities in New York to stop a similar practice.

Johanna Wald is the NRDC attorney noted for **protecting federal public lands** from misuse by special interests, such as cattle-raising enterprises that lease some 175 million acres of public grazing lands from the Bureau of Land Management (BLM). In the *San Francisco Examiner* (September 12, 1990), writer Huey D. Johnson comments on lawsuits that Wald filed and won against BLM for permitting land to be overgrazed and lose its productivity. "As a result of her first big suit against the Bureau, the agency is now required to document the condition of the lands it is responsible for, prior to deciding on grazing permit renewals." Wald told Johnson, "I believe the public is becoming less and less tolerant of the abuse of public resources. The exploiters are on the way out. A hundred years from now the public lands will be in much better condition."

Other land achievements in 1991 include gaining with **David Edelson's** help the rejection of subsidized logging at Gallatin National Forest adjacent to Yellowstone National Park; protecting the Sierra Nevada Mountains' ancient forests harboring the California

spotted owl, among other wildlife, with leadership from Edelson, **Sami Yassa**, and **Nathaniel Lawrence**; influencing reduced logging in three Colorado National Forests with the help of **Kaid Benfield** and **Thomas Kuhnle**. Concerning agricultural land use, **Justin Ward** and others brought environmental gains to the 1990 U.S. Farm Bill such as new funding for wetlands conservation and water quality incentive programs and protecting crop rotation incentives that reduce dependence on pesticides and fertilizers.

"One of NRDC's chief accomplishments has been proving that we don't have to make a choice between environmental degradation and 'doing without,'" writes **Ralph Cavanagh**, senior staff attorney, Energy Project. "Instead, we're showing how go get more work from less energy." NRDC reports that the "amount of energy currently wasted by American cars, homes, appliances, and mass transit systems equal more than double the energy potential from oil and gas reserves in all of the Alaskan lands and the U.S. Outer Continental Shelf that have not been leased for energy development." In 1991, NRDC staff convinced major utilities in New York, California, and the Pacific Northwest to invest in energy efficiency and renewable energy rather than in new power plants; launched the **Golden Carrot program** to reward manufacturers of energy efficient household appliances and the **California Compact** to "spur production of higher-quality compact fluorescent lightbulbs"; persuaded the California Energy Commission to adopt the nation's most stringent standards for residential building construction that is expected to cut new homes' energy consumption by the equivalent of 80 million barrels of oil; and successfully lobbied with **Janet Hathaway** for U.S. Surface Transportation bills that earmark public transit funding. In Hawaii, NRDC tropics specialists **Laura King** and **Susan Miller** teamed with the Energy Project to get utilities to invest in energy efficiency programs.

About solid waste and recycling, NRDC worked hard to strengthen the U.S. Resource Conservation and Recovery Act (RCRA) with the leadership of senior scientist **Allen Hershkowitz** and attorneys **Linda Greer** and **Eric A. Goldstein**. NRDC points out that "unlike any other solution—even recycling—waste reduction consumes no natural resources and produces no pollution." It is concerned that RCRA leaves garbage management and recycling to state and local governments without "clear, stringent federal guidelines and incentives for reducing and recycling solid waste." NRDC also wants aggressive federal action on developing markets for recycled products.

Hershkowitz also helped draft legislation for 10 states to eliminate toxic heavy metals such as cadmium, lead, mercury, and hexavalent chromium from consumer goods packaging. When incorporated into plastic, these elements can "escape into the air when the plastic is incinerated, or leach into groundwater when the plastic or its incinerator ash is landfilled," according to NRDC.

Through Urban Projects, NRDC is helping forge "sound environmental policies in New York and Los Angeles that can be applied in urban areas" elsewhere, with "environmental justice" as a basis. New York City has a 1991 law requiring its fleet of cars, buses, and other vehicles to use cleaner-burning fuel, with Eric Goldstein's help. He and **Douglas Guevara** succeeded in getting a statewide regulation to reduce smog-causing vapors from aerosol cans and other consumer products. Goldstein, **Samuel Hartwell**, and **Renée Skelton**, through NRDC's WasteWatch Project, are monitoring the city's recycling program and advocating essential funding. With **Katherine Kennedy**, a new program is underway to prevent thousands of pounds of industrial toxic pollutants from being dumped in nearby waterways. With **Mitchell Bernard's** legal guidance, plans for a mammoth Trump City apartment and shopping mall project were changed to a residential complex proposal that would feature a public 23-acre waterfront park on the Hudson River. In southern California, **Veronica Kum** helped get an air pollution cleanup plan for transportation back on track. A toxic waste incinerator was canceled in a low-income Los Angeles community, due to efforts from **Joel Reynolds.** An ecological reserve is being created among the coastal sage scrub and other threatened species on the Santa Rosa Plateau to keep development out—with the help of senior attorney **Mary Nichols,** Clean Air Coalition award winner for citizen activism.

Regarding nuclear threats, **Dan Reicher,** senior staff attorney, writes, "One of the great ironies of our time is that the U.S. nuclear weapons program, which is supposed to assure our national security, is instead threatening the well-being of millions of Americans who live in the shadow of the bomb plants. It takes real courage for people to speak out against dangerous and dirty warhead production facilities . . . where the economies are so heavily dependent on bomb-making." NRDC reports that the U.S. government's "nuclear complex employs nearly 100,000 people at 17 major sites in 13 states [with] more than 3,000 radioactive, chemical, and solid waste sites, and hundreds of contaminated buildings. The cleanup costs could easily exceed $100 billion—which represents almost $2 million for each warhead produced, and more than what was spent to put a man on the moon." NRDC has led battles against a "plutonium economy" and has taken on the U.S. Department of Energy to obey hazardous waste laws.

In late 1991, **Dr. Thomas Cochran** of the NRDC Nuclear Project and cooperating with the Federation of American Scientists organized the American team's participation in an arms dismantlement conference in Moscow between arms designers and treaty experts from East and West. Also participating was NRDC senior researcher **Christopher Paine,** who told the *New York Times* (December 17, 1991), "We must work together with the new governments [**Commonwealth of Independent States**] to insure a complete inventory, secure storage and eventual dismantling of these warheads." To keep count of the warheads, NRDC suggested nuclear warheads bear an identification tag and a "seal" that would show tampering with storage containers. "These measures would insure a clear chain of custody for these weapons pending their destruction," Cochran described in the *Times.* Other key NRDC staff in this field are senior scientist **Dr. David Goldstein;** and coordinator **Kristen Suokko** and **Dan Reicher,** who set up a USA–Soviet environmental law exchange at the Leningrad State University Law School.

In the early 1970s, NRDC embarked on a global effort to save critical ecosystems and bring about environmentally responsible development with the evolving cooperation of U.S. Agency for International Development (USAID) and multilateral banks for "imposing environmental standards on projects affecting tropical forests, wetlands, and African grasslands." NRDC continues to build partnerships, for example, on energy efficiency planning in the Commonwealth and Sri Lanka and rainforest protection in Peru, Ecuador, and Chile. For example, **Jacob Scherr, Robert F. Kennedy, Jr.,** and **Lynn Fischer** "worked with the Peruvian Environmental Law Society to convince a Texas company to abandon its plans for oil drilling in a Peruvian rainforest reserve." **Glenn Prickett** helped secure a congressional ban on "all U.S. foreign aid for environmentally damaging logging in tropical forests and lobbied lawmakers with **Eugene Gibson** to see that local environmental groups help manage funds for environmental protection that become available through "debt-for-nature" swaps.

The fate of the tropical rainforests is the focus of the **Rescue the Rainforests Campaign.** In addition, the Council monitors such activities of the World Bank and AID projects in developing countries. NRDC staff also work to protect tropical forests under USA jurisdiction, including those in Hawaii, the Virgin Islands, and Puerto Rico. **Laura King,** senior staff scientist, comments, "We have no right asking Brazil and the rest of the world to stop the destruction of their rainforests if we aren't willing to protect our own." She reports that nearly a quarter of the endangered species in the USA are found in Hawaii alone—where **Susan Miller, Charles Clusen,** and **Dr. Faith Campbell** pushed for increased federal and state funding to preserve the state's tropical forests.

Also, Clusen and others got Congress to extend federal protection to rare coastal ecosystems in Puerto Rico and the U.S. Virgin Islands. A CITES participant, Dr. Campbell is particularly concerned with the conservation of tropical forests as well as endangered plants elsewhere; earlier, she fought for increased funding for the U.S. Fish and Wildlife Service to identify and recover endangered species. Recently, Campbell and **Johanna Wald** collaborated on brochures about plants of prairie ecosystems and helped gain critical habitat protection for the desert tortoise in Utah and California.

Other projects include **Mothers and Others for a Livable Planet,** a 1990 outgrowth of the more specific Mothers and Others for Pesticide Limits, begun the previous year. When the original group was successful in its campaign to ban the use of alar on apples—with the help of actress **Meryl Streep** who raised public consciousness through media exposure—the NRDC decided to expand this program's objectives. In 1991, this public health action project on behalf of children grew to 13,000 supporters; **Wendy Gordon** is co-chair. Senior scientist **Lawrie Mott,** with children's and health advocacy groups, got California to pass the nation's toughest lead poisoning prevention program. In another public health project, senior attorney **Albert Meyerhoff,** along with labor, consumer, and other environmental groups got the EPA to ban most uses of parathion, a widely used pesticide that has caused deaths and poisonings worldwide.

To fill the gap left by EPA inaction, the NRDC began a project to encourage citizen enforcement of existing laws and regulations. In 1991–1992, **Citizen Enforcement Projects** took on the oil industry in northern Alaska and won lawsuits to protect American rivers, such as the Potomac River and Pennsylvania's Conococheague and Brandywine creeks.

The Council works closely with grassroots and large environmental groups in the USA and abroad. Established in 1981, the **Duggan Fellowship** brings environmentalists from developing nations to the USA to work with local environmental groups. A significant international effort of the NRDC is its people-to-people exchange programs with the Soviet Academy of Sciences, in which NRDC and SAS scientists work together to monitor nuclear test ban treaty compliance.

■ Resources

The NRDC publishes *The Amicus Journal* quarterly and the *NRDC Newsline* five times a year as well as its *Annual Report* and an *EarthAction Guide* series. Mothers and Others for a Livable Planet also has a quarterly newsletter, *TLC.* In addition, the Council has publications on numerous areas of environmental interest, including agricultural policy, air quality, global warming, nuclear weapons, species conservation, and the urban environment. Examples are *Intolerable Risk: Pesticides in Our Children's Food; For Our Kids' Sake: How to Protect Your Child against Pesticides in Food;* and *Harvest of Hope* about the potential of alternative agricultural methods in reducing pesticide use.

In 1991, NRDC with other experts released a national energy efficiency plan *America's Energy Choices* and *Looking for Oil in All the Wrong Places* by **Robert Watson**, among other energy reports. *Tracking Arctic Oil* by senior scientist **Lisa Speer** is a study of oil industry pollution in Alaska, which won an award from the Natural Resources Council of America. *Testing the Waters* is a report on beach closings in 10 coastal states that is an effective tool in ending coastal sewage pollution. *The Right to Know More* by **Deborah Sheiman** about toxic polluters; *Going to the Source: A Case Study on Source Reduction of Toxic Industrial Waste;* and *The Lead Contamination Control Act: A Study in Non-Compliance* are reports used to spur tougher laws and enforcement. *Plant a Seedling, Cut a Forest* aids the fight to save ancient forests.

NRDC issued a summary of activities and accomplishments, *Twenty Years Defending the Environment,* in honor of its anniversary in 1990. Through Island Press, it published *The Challenge of Global Warming.* Through Living Planet Press, NRDC published in English and Spanish *The Amazing L.A. Environment* as a handbook for residents to "restore their environment and start living in greater harmony with their surroundings." "Rescue the Rainforest" T-shirts, audio tapes, children's books, and other products are available.

The Nature Conservancy (TNC) ⊕
1815 North Lynn Street
Arlington, VA 22209 USA

Contact: **John C. Sawhill**, President and Chief
            Executive Officer
Phone: 1-703-841-5330 • *Fax:* 1-703-841-1283

Activities: Education; Research • *Issues:* Biodiversity/Species Preservation; Deforestation; Global Warming; Sustainable Development

● Organization
Founded in 1951, **The Nature Conservancy (TNC)** is dedicated to the preservation of rare or endangered plant and animal species and ecosystems. The Conservancy controls the most extensive private system of nature preserves in the world. Practicing nonadversarial environmentalism, the group quietly buys land or otherwise arranges its protection, both within the United States and abroad. All 50 states have one or more chapter offices, and extra offices operate in the Tennessee Valley Authority (TVA) area, the Navajo Nation, and the Smoky Mountains. Outside the USA, the Conservancy works in Canada, Latin America, and the Pacific Islands. The head office in Arlington serves as an information and funding center, and encourages cooperation between or among the various local offices.

Funded by its nearly 600,000 members and by other contributions, TNC has safeguarded 5.5 million acres in the USA and has contributed to the preservation of more than 15 million acres outside the country. The organization currently manages 1.3 mil-

lion acres on 1,600 preserves. Most of TNC projects are at the recommendation of the **Natural Heritage Inventory Program** (NHIP), a data collection project developed by Conservancy vice president **Dr. Robert Jenkins** in 1974 and now generally run by individual states.

▲ Projects and People
The Conservancy's first **acquisition** project involved 60 acres of New York's **Mianus River Gorge**; that area has now spread over 555 acres, demonstrating typical TNC persistence in expanding the area of protected land. The early Conservancy emphasis on purchasing land it wishes to protect has undergone a gradual change, however. Much more important now are cooperative efforts with private landowners and with government and other conservation agencies.

Ducks Unlimited, for example, has been a frequent TNC partner in waterfowl and wetland conservation projects, as has the Environmental Defense Fund. **Conservancy collaborations** have also involved Dow Chemicals, General Motors, and other major corporations. For example, in 1990 the Tenneco Mineral Company donated more than 100,000 acres' worth of mineral rights to TNC, including 81,000 on the Conservancy's largest acquisition to date, the 500-square-mile Gray Ranch in New Mexico.

Because one-third of all U.S. land is federally owned, and 15 percent belongs to states, counties, or municipalities, the Conservancy, of necessity, works closely with governmental organizations in land protection activities. In 1989, TNC became the first private conservation organization to sign a land management agreement with the U.S. Department of Defense, which controls 25 million acres of national land; the following year, TNC finalized a partnership agreement with the Bureau of Land Management (BLM), which controls 200 million acres.

The Conservancy is also involved in **scientific and development research.** NHIPs are the most active and obvious examples of this; a recent joint effort of the Electric Power Research Institute (EPRI), the NHIP, and Biological and Conservation Data (BCD) studied the effects of global warming on 14,000 native vascular plant species north of Mexico. In addition, TNC actively seeks practical ways to change people's behavior to benefit the environment.

In conjunction with more than 100 public- and private-sector partners, the Conservancy unveiled in mid-1991 the **Last Great Places: An Alliance for People and the Environment**—12 working models or bioreserves for large-scale ecosystem conservation throughout the Western Hemisphere. At that time, **John C. Sawhill**, TNC president, remarked, "It is no longer sufficient to buy parcels of land, fence them off, and call them 'protected.' The solution is to focus on creative methods and partnerships to protect whole functioning ecosystems."

Evolving for a number of years, the concept is illustrated in Conservancy work at the Virginia Coast Reserve's 14 barrier islands along the Atlantic Ocean, where a buffer zone was created to protect these fragile ecosystems and their shorebirds and sea life from development, pollution, and other adverse conditions. In addition to the Virginia Coast Reserve's ecosystem preservation, the remaining dozen bioreserves of the "Last Great Places" are the following:
- **Big Darby Watershed, Ohio.** Runoff from agriculture and development as well as suburban expansion threaten a healthy aquatic system with more than 100 species of fish and mollusks. A coalition of some 20 public and private partners and computer modeling of land-use cause-and-effect relationships are meeting the challenges that also includes reforestation.

- **Tallgrass Prairie Preserve, Oklahoma.** "This project represents what may be the only opportunity to restore a functioning tallgrass ecosystem" that is now all but extinct. Bison and other native species will be reintroduced. The Osage Indian's oil extraction will be preserved, and a drive-through trail will allow low-impact tourism.
- **Nipomo Dunes, California.** The site of the 1923 silent movie version of Cecil B. DeMille's *The Ten Commandments*, these dunes along an 18-mile coastline provide nesting sites for rare shorebirds and also accommodates some industry, such as oil drilling and sand mining in a buffer zone.
- **Peconic/Block Island, New York/Rhode Island.** This watershed is home to endangered birds, plants, and the American burying beetle. An economy shift from farming and agriculture to tourism, with the cooperation of landowners, is expected to enhance protection of the core natural areas.
- **Rio Celestun/Rio Lagartos, Mexico.** Here, migratory waterfowl such as the greater flamingo depend upon 300,000 acres of estuary, marshlands, and tropical forest on the Yucatan peninsula that is also home to endangered sea turtles and jaguars in the scrub forests. Efforts will focus on building "staff and infrastructure to protect the area from encroachment and encourage sustainable fishing and ecotourism."
- **The Florida Keys.** Here, the human population has increased 65 percent in the last decade, and endangered Key deer and other natural communities are being squeezed out. The offshore coral reef also needs to be saved from mainland pollution and damage by boaters and divers.
- **Texas Hill Country.** Austin and San Antonio urban pressures are resulting in the drafting of a habitat conservation plan to "balance endangered species protection with economic growth."
- **Mbaracayu, Paraguay.** "This subtropical forest that shelters endangered species found nowhere else is currently held by the World Bank which has agreed to sell it to The Nature Conservancy and its NGO partner in Paraguay. Management plans call for the employment of local inhabitants to support sustainable agriculture, agroforestry, and other low-impact economic activities. Also to be tested are captive breeding of wild animals, beekeeping, fish farming, and nature tourism."
- **The Darien, Panama.** This park needs more rangers and stations and cooperation with indigenous communities to help protect the rivers and wetlands, forests, cloud forests, and rainforests along the Colombian border, and to bring about sustainable economic activities.
- **Southwest Ecosystems, Arizona/New Mexico.** Two somewhat dissimilar desert undertakings—the Gray Ranch and the San Pedro project—are expected to yield protection for rare plant and animal life in both mountain peaks and grasslands. Cattle grazing is expected to continue in the buffer zones of the Gray Ranch, which was "the largest single private acquisiiton in conservation history." Natural history tourism is expected to be incorporated into the San Pedro project.
- **Condor Reserve, Ecuador.** Considered to be one of the world's richest biological areas, the reserve spans the Amazon jungle to about 20,000 feet in the Andes—home of the condor. What is needed are infrastructure to protect the region and cottage industries to spur sustainable economic activities among local people.

Also in 1991, the Conservancy began providing editorial and promotional assistance to the award-winning PBS television series

Nature. In the same year, Chevron bestowed 3 of its 20 annual Conservation Awards on TNC individuals.

Chief botanist is **Dr. Larry E. Morse**, who points out that "to survive, species may need to shift their ranges considerable distances to keep up with the distribution of suitable habitat." Chief ecologist is **Dr. Dennis H. Grossman.**

■ Resources

The Conservancy publishes bimonthly the *Nature Conservancy* magazine In addition, it manages a video, film, and tape library including its own and others' productions outlining or detailing TNC activities.

The Nature Conservancy (TNC) ⊕
Latin America Division (LAD)
1815 North Lynn Street
Arlington, VA 22209 USA

Contact: **Geoffrey S. Barnard**, Vice President and Executive Director
Phone: 1-703-841-4861 • *Fax:* 1-703-841-4880

Activities: Education; Research • *Issues:* Biodiversity/Species Preservation; Deforestation; Sustainable Development • *Meetings:* World Parks Congress

● Organization

To protect biological diversity of tropical ecosystems, the **Latin America Division (LAD)** of The Nature Conservancy (TNC) covers four main areas: the Conservation Data Centers (CDCs) under the Latin America Science Program; Parks in Peril; general expansion of conservation leadership and training programs; and debt-for-nature swaps. The Conservancy works with more than 30 organizations covering 17 countries in Latin America and the Caribbean "to provide infrastructure, community development, professional training, and long-term funding for legally protected but underfunded areas" in the region.

▲ Projects and People

The Conservancy launched the **Parks in Peril** program in 1990, with the goal of bringing conservation management to 20 critical sites in Latin America and the Caribbean annually, to total 200 sites by the year 2000, or 100 million acres. Besides lacking money, many Latin American parks are "young and vulnerable"; TNC's program helps them to build a conservation infrastructure, integrate protected areas into local communities, and secure long-term funding. For example, with the help of the U.S. Agency for International Development (USAID), the government of Jamaica, and the Jamaica Conservation Development Trust, TNC's **Brad Northrop** is working to create a national park system in that country. Through the efforts of **Brian Houseal**, director of TNC's Latin America Stewardship Department, the Guatemalan Congress in 1990 created the **Maya Biosphere Reserve**, 500 square miles (3.5 million acres) of tropical forest with wetlands, where jaguar tracks indicate that the food chain is still functioning well.

Four of the Parks in Peril are included in TNC's **Last Great Places** program, launched in 1991. One of these is **Panama's Darien**

National Park; it covers 1.5 million acres of tropical rainforest, an area twice the size of Yosemite in the USA—but while Yosemite has a staff of 190 and 80 vehicles, Darien's staff of 14 has 1 jeep and 1 boat to combat deforestation and species extinction due to logging, mining, and the lack of economic opportunities for the native Embera and Kuna tribes.

Another is **Ecuador's Condor Reserve**, home to at least 40 Andean condors and the major source of Quito's drinking water. **Ria Lagartos** and **Ria Celestun** in Mexico's Yucatan Peninsula provide "critical refuge for wintering waterfowl" including the greater flamingo; TNC's work there is through the local group PRONATURA and the Mexican government. The **Mbaracayu**, 143,000 acres of subtropical forest in **Paraguay**, is the traditional hunting area for the 700-member Ache tribe, first "discovered" in 1976; TNC emphasizes the need to protect the Aches' economic base in its land protection plans.

Instead of forcing developing countries to exploit their resources in order to repay debts, **debt-for-nature swaps** allow the conversion of unpaid loans into conservation funding. Working through the central banks of the countries concerned, and with creditors, TNC—and other conservation organizations—"buys" debt, then converts the debt notes into local currency bonds to be channeled into environmental efforts. In 1991, for example, the Conservancy bought half a million dollars' worth of Argentina's debt from the American Express Bank at 16 cents per dollar (i.e., paying $80,000). **Costa Rica**, an aggressive debt-swapper, traded $80 million of debt to protect millions of acres and develop environmental research and training programs. According to TNC, there are billions of untapped dollars in debt that creditors would be happy to sell off at a low price. The tactic is gradually becoming popular.

The Conservancy encourages shared goals, even more important considering the linkage between tropical deforestation and dwindling U.S. species. The TNC **Campaign for the Delaware** program, a five-year, four-state cooperative effort begun in 1990, includes earmarking $1 million for the Conservancy's Latin American work. Indeed, that money will help ensure safe wintering sites for the more than 20 species of migratory birds that depend on the Delaware region in summer. The Oregon TNC chapter has a **sister forest program** with Costa Rica's Monte Verde National Park. Through TNC's **Adopt-An-Acre** program, launched in 1989, schools and other groups across the USA raise money for the Latin American rainforests; $30 preserves an acre. TNC's Latin America Program has also helped create a $1.5 million endowment for the Galapagos Archipelago.

Vice president and executive director of the Latin America Division since 1986, **Geoffrey S. Barnard** joined the Conservancy in 1972 after two years in Peru with the Peace Corps. He was the organization's first overseas advisor, to the National Parks Foundation of Costa Rica, in 1982.

■ Resources

The Latin America Division publishes brochures on The Nature Conservancy: International Programs, including *International Update*, *Directory of Latin American and Caribbean Conservation Data Centers (CDCs)*, *Biodiversity Network News*, and *Diversidata*. *Nature Conservancy* magazine also features articles on global issues.

The Nature Conservancy (TNC) ⊕
Latin America Science Program
1815 North Lynn Street
Alexandria, VA 22209 USA

Contact: **Roberto L. Roca, Ph.D.**, Chief Zoologist
Phone: 1-703-841-2712 • *Fax:* 1-703-841-1283

Activities: Research • *Issues:* Biodiversity/Species Preservation; Deforestation • *Meetings:* CITES; World Parks Congress

● Organization

The tropics of Latin America encompass one of the biologically richest areas on Earth; however, the land is in danger, with 40,000 square miles—the size of Ohio—of land lost each year to development. Globally, TNC estimates by the year 2000 "the world will lose one plant or animal species every hour of the day." Unfortunately, a lack of information hampers conservation efforts, particularly in this region.

The Nature Conservancy's **Latin America Science Program**, an important element of the overall **Latin America Program**, has as its primary task to establish and maintain **Conservation Data Centers (CDCs)** in Latin America and the Caribbean. Like the Natural Heritage programs of TNC in the USA, the CDCs gather the species and habitat information so vital to conservation and development planning. Thus, the Science Program with its CDCs works closely with multilateral and bilateral development banks and agencies, as well as with government organizations.

Staffed and operated by local scientists and conservationists, all CDCs use the same software and methodology, which the Conservancy developed to facilitate interaction between centers in different countries; as a TNC flier states, "distribution of plant and animal species do not respect political borders." *(See also TNC, Network of Natural Heritage Programs and Conservation Data Centers in this section.)*

▲ Projects and People

As chief zoologist for the Latin American Science Program, **Dr. Roberto L. Roca** coordinates the zoological work of the CDC network in Latin America and the Caribbean, assisting in training, technical, and scientific support. A recent addition to the Conservancy staff, Roca believes in wedding academic research with environmental activism.

In 1986, Roca began work on his dissertation topic, a study of the unique *guacharo*, or oilbird, of **Venezuela**. Borrowing techniques from "the true experts . . . the local poachers" to band the birds, he used radio telemetry to track the cave-dwelling, fruit-eating adults to learn why their numbers were shrinking. He discovered that the slash-and-burn land clearing activity of transient farmers (*conuqueros*) in the adjacent **Mata de Mango** forest region was affecting the species; but he also learned that the birds, because of their habit of eating fruit and then regurgitating the seeds during flight, were vital to the reforestation process—"better than any helicopter crew." In 1989, after Dr. Roca proposed enlarging **Guacharo National Park** by adding Mata de Mango, the Venezuelan government added 166,000 acres to its existing 38,750 acres of tropical forest.

■ Resources

(See TNC, Latin America Division. in this section.)

The Nature Conservancy (TNC) ⊕

Network of Natural Heritage Programs (NHPs)
  and Conservation Data Centers (CDCs)
1815 North Lynn Street
Arlington, VA 22209 USA

Contact: **Richard Warner**, Director, Central
           Conservation Databases
  Phone: 1-703-841-4888 • *Fax:* 1-703-841-1283

Activities: Education; Research • *Issues:* Biodiversity/Species
Preservation; Deforestation; Global Warming; Sustainable
Development

● Organization

"**Natural Heritage Programs (NHPs) and Conservation Data Centers (CDCs)** are continually updated, computer-assisted inventories of the biological and ecological features and biodiversity preservation of the country or region in which they are located," describes **The Nature Conservancy (TNC)**. These data centers assist in conservation planning, natural resource management, environmental impact assessment, and planning for sustainable development. Designed by **Dr. Robert Jenkins**, the network helps TNC prioritize sites for acquisition.

As of 1991, there were 82 data centers operating in the Western Hemisphere; NHPs operate in all 50 states in the USA, and CPCs operate in several U.S. bioreserves, national parks, Puerto Rico, 3 Canadian provinces, and 13 countries in Latin America and the Caribbean. TNC further describes the structure: "Each data center is established within a local institution, most frequently as part of a government agency responsible for natural resource management and protection."

Richard Warner, director of Central Conservation Databases, explains that "each program operates independently and the vast majority of them are now fully staffed by state government employees. Some of the data centers are in fact located in national government agencies." While the programs are not membership organizations, TNC—which is an NGO—is the "hub and centralized contact point" for these programs, with more than 300 biologists and computer specialists providing technical, scientific, and administrative support.

Operating within a network, the basic **Biological and Conservation Data (BCD) System** that has evolved over 16 years is a compilation of 30 interrelated computer files supported by extensive map and manual files, and a library. "A trained staff of biologists, natural resource specialists, and data managers interprets the data for use in local conservation and development planning, natural resource management, and environmental impact assessment," reports TNC.

▲ Projects and People

Richard Warner is the overall Central Conservation Database coordinator overseeing data exchanges with the Natural Heritage Programs and other cooperative agencies and institutions. Among the

Network's systematists, biogeographers, and field naturalists—Dr. Dennis H. Grossman, chief ecologist, manages the Heritage Ecology Network. Dr. Robert M. Chipley is director of the North American Heritage Operations. Dr. Larry Master directs the Eastern Heritage Task Force; Dr. Stephen Chaplin is science director at the Midwest Heritage Task Force (MHTF); Dr. Bennett Brown directs the Rocky Mountain Heritage Task Force (RMHTF). Susan R. Crispin is Canada's national coordinator of Conservation Data Centres.

■ Resources

A publication describes the *Natural Heritage Program and Conservation Data Center Network*, including a list of science division personnel at national headquarters as well as coordinators and technical and scientific staff of USA State Heritage Programs and Regional Heritage Centers; Latin American and Caribbean Conservation Data Centers (CDCs); Regional Support and Service Centers; and Data Network Cooperators, such as the Center for Plant Conservation, Flora of North America Project, Missouri Botanical Garden, International Plant Database, Rare Lichens Project, TNC of Canada, and Canadian Rare Plants Project.

The *Biodiversity Network News* describes biological and conservation data and applications. *Diversidata* newsletter is printed in Spanish and provides updates on the CDCs.

The Nature Conservancy (TNC) ⊕

Science Programs
1815 North Lynn Street
Arlington, VA 22209 USA

Contact: **Robert E. Jenkins, Ph.D.**, Vice President
           and Director
  Phone: 1-703-841-5320 • *Fax:* 1-703-841-1283

Activities: Education; Research • *Issues:* Biodiversity/Species
Preservation; Deforestation; Global Warming; Sustainable
Development

▲ Projects and People

Dr. Robert Ellsworth Jenkins, Jr., has directed TNC's science programs since 1970 and is vice president since 1972. He confesses to acquiring "biophilia or ecophilia" at an early age; in the eighth grade he gave a valedictory address on "endangered species and extinction—bison and passenger pigeons," according to a lengthy interview in *Natural Areas Journal* (1990).

All sources suggest that the arrival of Jenkins at TNC brought method and efficiency to what had been a fairly haphazard program of land acquisition. "By establishing a preserve we don't stop the destructive force in society, we just deflect it," says Jenkins; therefore, conservation efforts must focus on the most significant needs rather than wasting money and energy on relatively inconsequential tracts.

As an outgrowth of this belief, Jenkins conceived and established TNC's **Natural Heritage Inventory Program (NHIP)** in 1974, which has developed into "the world's most comprehensive and advanced inventory of plants, animals, and natural communities." Constantly thinking of improvements, Jenkins shifted the heritage ranking program from a single-species-based approach to a habitat-based approach after realizing that a "large habitat [is] a requirement for viable persistence of species."

Realizing that the best use of the program would require standardization, Jenkins conceived the "common standards of methodology and format," which make possible communication within a now international network of heritage programs. All use the same software and strategies. *(See also TNC, Network of Natural Heritage Programs and Conservation Data Centers in this section.)*

Another mark of Jenkins' efficiency-oriented mind is the **Rapid Ecological Assessment (REA)**. In 1987, he suggested combining satellite imagery and computer mapping with traditional ground-based field surveys, particularly in South American inventories. Because of the incredible diversity of species, the lack of groundwork, and the swift pace of habitat destruction in South America, there is rarely time for typical in-depth inventories. With REAs, conservationists can quickly see where a detailed inventory would be most effective, or can make the sometimes necessary snap judgments to preserve a given tract of land. An REA, for example, led to the inclusion of Paraguay's **Mbaracayu** subtropical forest in TNC's **Last Great Places** campaign.

Dr. Jenkins was also a founder of the Zero Population Growth movement in Massachusetts. **Dr. Hardy Wieting, Jr.**, is the Science Department's deputy director and Heritage Operations director.

■ Resources

Dr. Jenkins has published nearly 100 research reports, chapters, and articles on his specialties.

The Nature Conservancy (TNC) of Alabama ❦
2821C Second Avenue South
Birmingham, AL 35233 USA

Contact: **Kathy Stiles Cooley**, Executive Director
Phone: 1-205-251-1155 • *Fax:* 1-205-251-4444

Activities: Land Protection • *Issues:* Biodiversity/Species Preservation; Sustainable Development

● Organization

The state of Alabama is home to more than 75 rare or endangered species and natural communities. Like chapters of The Nature Conservancy (TNC) across America, The Nature Conservancy of Alabama has as its mission the safeguarding of species and areas in its state. With a staff of 5 and 3,000 members, the Alabama office follows standard TNC procedures of identifying, acquiring, and managing properties significant for the rarity and quality of the species and ecosystems they support.

The Alabama TNC office also administers the **Alabama Natural Heritage Inventory Program (NHIP)**, housed in the Department of Conservation and Natural Resources. Three Heritage biologists identify and locate the state's rare flora and fauna, using the information to influence development decisions, resource planning, and TNC acquisition schedules.

▲ Projects and People

Using NHIP data and its own or national TNC funds, the Alabama chapter chooses new protection projects or enlarges existing preserves. It then often resells the property to government or private conservation organizations, ensuring proper management and gaining income to use in further land acquisitions. Currently, the office maintains 4 preserves—the smallest, 3.3 acres, and the largest, 56 acres. Since opening, the chapter has preserved more than 34,000 acres through 14 acquisition projects, including the **Bon Secour National Wildlife Refuge**, **Chitwood and Pitcher Plant Bogs**, and **Shelta Cave**.

Kathy Stiles Cooley, executive director of the Alabama chapter since 1989 when the office was established, has a firm grounding in the Alabama environmental movement. She founded the Alabama Trails Association, which builds trails, and the Vulcan Trail Association, a backpacking group; she also founded and was executive director from 1979 to 1989 of Birmingham's Ruffner Mountain Nature Center, "a 538-acre inner-city environmental education center." A board member of the Alabama Forest Resource Center, with four years on the Alabama Department of Conservation's Natural Resources Committee, Cooley has been recognized as an outstanding conservationist by the Audubon Society, the Sierra Club, and the Alabama Conservancy.

■ Resources

The Alabama Conservancy has a slides program and distributes a quarterly newsletter to its members.

The Nature Conservancy (TNC) of Alaska
601 West 5th Avenue, Suite 550
Anchorage, AK 99501 USA

Contact: **Susan L. Ruddy**, Alaska Director
Phone: 1-907-276-3133 • *Fax:* 1-907-276-2584

Activities: Development; Education; Research • *Issues:* Biodiversity/Species Preservation; Deforestation; Sustainable Development

● Organization

"*When we try to pick out anything by itself, we find it hitched to everything else in the universe.*" Naturalist John Muir

Established in 1988, The Nature Conservancy's Alaska chapter is working with a staff of 8 and 2,200 members. Together they must safeguard one of the richest states in the USA in terms of minerals, fish, and timber—as Alaska is "critical to the life cycles of millions of migratory birds, mammals, and fishes." Although the huge state is sparsely populated—only half a million people live on an area equal to 20 percent of the continental USA—threats exist, as became clear with the Exxon *Valdez* disaster.

The typical TNC procedure is to acquire endangered habitats by purchasing them; however, the Alaska office must use different tactics. Less than 0.5 percent of Alaska's 44 million acres is privately owned; 88 percent is public, and 11.7 percent belongs to Native American tribal organizations.

As is the case with most TNC chapters, the local **Natural Heritage Program** plays a major role in gathering data, identifying systems and species most at risk, and recommending the best remaining locations for preservation efforts. In Alaska, this program receives primary funding from British Petroleum (BP)/Alaska and the state.

Because of Alaska's locality in Arctic and sub-Arctic regions, the Conservancy's mission is intensified in protecting biodiversity, such as millions of migratory birds that nest and breed in Alaska's uplands and wetlands, then return to the continental USA for the winter. Global climate changes are often first detected in the polar regions, providing early warnings. With considerable reductions of marine life recorded in the Bering Sea, scientists are watching for other disruptions in the Arctic waters containing the "critical links in the world's food chain." As in the tropics, plants of the Arctic may yield potential for medicine, science, education, and industry. The wilderness itself is a source of wonderment and renewal to those who experience it.

▲ Projects and People

Because there are so few private landowners, TNC–Alaska works intensively with governmental and Native American organizations. With the Bureau of Land Management (BLM), the Conservancy helped conduct a field inventory of the Black River, and worked on the purchase of 100 acres to add to Steese-White Mountain Recreational Area near Fairbanks. In 1990, the office completed its first acquisition: about 550 acres of wetland, home to hundreds of thousands of waterfowl, which it then transferred to the state of Alaska as an addition to the Palmer Hay Flats State Game Refuge.

Through its Conservation Joint Venture program to assist private landowners, the Conservancy acknowledges that it "often must consider economic development assistance to local people . . . to achieve its conservation mission." With the aid of a British Petroleum donation, the office has ongoing projects in Prince William Sound, the Pribilof Islands visited by some 2 million sea birds each summer from as far as Argentina and as near as Siberia, and the 1.5-million-acre Kenai River watershed.

Bioreserve conservation is being explored on St. Paul Island in the Pribilofs, where there are nearly 400 reindeer and Arctic blue fox; yet, fur and harbor seals and sea lions, pollock fish, and kittiwake chicks are declining to the bafflement of scientists.

To better understand its ecosystems, the Alaska Natural Heritage Program, with director Judy Sherborne, is building a data bank on rare and not-so-rare species such as the endangered Eskimo curlew shorebird, threatened Stellar sea lion, recovering peregrine falcon, and isolated plants such as the Aleutian shield fern.

Susan L. Ruddy established the Alaska field office in 1988, the Natural Heritage Program in 1989, and the field trips and lands programs in 1990. She did extensive early work with the Alaska Native Claims Settlement Act, mediating among state, private, and native organizations, and industry.

■ Resources

The Alaska chapter needs volunteers as tour guides, library aides, and general helpers, as well as office equipment, photographs, and hotel/rental car coupons.

The Nature Conservancy Alaska Newsletter is published quarterly. The organization offers a limited number of internships.

The Nature Conservancy (TNC) Connecticut Chapter
55 High Street
Middleton, CT 06457 USA

Contact: Leslie N. Corey, Jr., Executive Director
Phone: 1-203-344-0716 • *Fax:* 1-203-344-1334

Activities: Development; Education; Political/Legislative; Research • *Issues:* Biodiversity/Species Preservation; Habitat Protection

● Organization

The Nature Conservancy (TNC) has been at work in Connecticut since 1954, and since then has permanently protected more than 18,000 acres on 70 different preserves. Featured prominently in its brochure are Margaret Mead's words: "Never doubt that a small group of thoughtful, committed citizens can change the world; indeed, it's the only thing that ever has." The staff of 17 and 14,000 members take that message to heart.

▲ Projects and People

Educational programs feature prominently in the Connecticut chapter's activities. Besides conducting Summer Science and Stewardship programs, the organization recently began a Conservation Biology Research Program, awarding research grants for studies of Connecticut ecosystems. A 1991 agreement with Connecticut College gives that institution primary study and teaching access to the 437-acre Burnham Brook Preserve, established in 1960.

Canaan Mountain land purchases at a tax auction recently enabled the local TNC to obtain habitat for the rare timber rattlesnake (*Crotalus horridus*), yellow-eyed wood frog, blue heron, cliff swallows, other birds, and significant plants.

The Connecticut River Protection Program works to restore what used to be one of the world's best runs for shad and Atlantic salmon. In an unprecedented TNC coalition, Connecticut has been working since 1988 with three other states—Massachusetts, Vermont, and New Hampshire—to safeguard 100 key sites along the 407 miles of river. Thus far, they have secured more than 51,000 acres, working with the Connecticut State Recreation and Natural Heritage Trust Program, and with U.S. Congressmen to legislate National Fish and Wildlife refuges along parts of the river. TNC is also funding freshwater tidal wetlands research.

With the Connecticut River inventory underway, ornithologists such as Margarett Philbrick are searching for marsh birds—the American bittern, willet, least bittern, and northern harrier. Botanists such as Margaret Ardwin are looking for rare plants as sandplain gerardia, small whorled pogonia, green dragon, and twinflower. Entomologist Rich Packauskas is investigating insect populations. All information is being recorded in the Connecticut Department of Environmental Protection's Natural Diversity Data Base.

■ Resources

The Connecticut chapter seeks expertise for research in biology, zoology, ecology, and botany.

The organization publishes a quarterly newsletter, *From the Land*, and has available the guidebook *Country Walks in Connecticut* and the slide show *To Save a River*.

The Nature Conservancy (TNC)
Florida Natural Areas Inventory
1018 Thomasville Road, Suite 200-C
Tallahassee, FL 32303 USA

Contact: **James William Muller**, Coordinator
   Phone: 1-904-224-8207 • *Fax:* 1-904-681-9364

Activities: Education; Research • *Issues:* Biodiversity/Species
Preservation; Deforestation

● Organization

Established in 1981 as a cooperative effort of Florida's **Department
of Natural Resources** and **The Nature Conservancy (TNC)**, the
Florida Natural Areas Inventory (FNAI) is that state's natural
heritage program. Unlike most TNC-affiliated inventory programs,
however, it is not yet under state management; the Conservation
and Recreation Land (CARL) Trust Fund provides the necessary
income.

The FNAI serves as a statewide clearinghouse for environmental
data, which it collects from existing sources or generates from field
surveys. Although Florida has the largest land acquisition program
of any state—largely due to **Preservation 2000**, which pledges state
funding of $3 billion over 10 years for land purchase—it also has
one of the highest growth rates. In its inventories, FNAI concentrates on species that are rare, endangered, or of special concern.

▲ Projects and People

According the FNAI, its work is significant not for the amount of
data generated but for how that data "is used to guide the protection
and management of the state's important natural features." In the
first 10 years of operation, FNAI reviewed more than 425 CARL
land acquisition proposals, evaluating and then recommending action based on the quantity and quality of species and habitats
involved. Each year, the organization responds to about 1,400
requests for information from federal, state, local, and private agencies or organizations.

FNAI maintains **site tracking records** and **natural diversity
scorecards** with constantly updated information about given sites
and species. If a species has endangered status on the federal or state
level, it receives special attention; FNAI notes, however, that some
unlisted species are more in danger than some that are listed. For
example, there are more than 330 active bald eagle nesting sites in
Florida, but only 5 Orlando cave crayfish sites. Using its resources,
FNAI has successfully pushed for the listing of the rare Florida
skullcap and telephus spurge; it continues to work for other species.
James William Muller, a biological oceanographer, has served as
coordinator of FNAI since 1985.

■ Resources

The *Guide to the Natural Communities of Florida* is available through
the Department of Natural Resources. FNAI publications include
*Matrix of Habitats and Distribution by County of Rare/Endangered
Species in Florida* and *Element Lists* of special plants, lichens, vertebrates, invertebrates, and natural communities detailing 935 elements, including 465 that are rare or endangered.

The Nature Conservancy (TNC)
Georgia Field Office
1401 Peachtree Street, NE, Suite 136
Atlanta, GA 30309 USA

Contact: **Octavia C. McCuean**, Director
   Phone: 1-404-873-6946 • *Fax:* 1-404-873-6984

Activities: Development; Political/Legislative • *Issues:*
Biodiversity/Species Preservation; Deforestation

● Organization

According to **The Nature Conservancy's (TNC)** state field office,
Georgia ranks fourth in the USA (after California, Texas, and
Florida) in richness, uniqueness, and diversity of species and habitats. At the same time, at least 69 plant and 37 animal species in
Georgia are endangered; the 7 staff and 7,500 members of the
Georgia TNC chapter hope to ensure their survival. Although the
Conservancy has been working in the state since 1962, when it
began safeguarding the Ogeechee Swamp, the local office began
operations only in 1987. Since then, it has worked quickly to secure
more than 38 areas encompassing more than 145,000 acres, primarily in the southeast and northwest areas of the state, and including
coastal barrier islands as well as 15,000 acres in the Okefenokee
Swamp.

▲ Projects and People

A key endeavor of the Georgia field office is the **Altamaha River
Bioreserve Project**, launched in 1991 under the directorship of
Christi Lambert. The first stage involves a $600,000, two-year
biological study of the 137-mile river, funded by the Georgia Power
Company and the Woodruff Foundation. Much of the river, second
only to Chesapeake Bay in the amount of fresh water it pours into
the Atlantic, is in near-pristine condition. The Altamaha is home to
unique species such as the Altamaha spiny mussel and the day-flier
moth, and to 350-pound sturgeon carrying up to 80 pounds of
caviar. The Georgia office is negotiating with logging concerns and
owners of floating houses to reduce the damage to the river from the
cutting of hard woods and from the dumping of raw sewage.

A recent success has involved a unique cooperative effort between
the Georgia office, the Georgia Department of Natural Resources
(DNR), and the national Conservancy office. The Kerr-McGee
Corporation of Oklahoma gave 7,700 acres on **Little Tybee** and
Cabbage Islands to the Georgia office, which then sold the land to
the state to be managed by the DNR. The national office received
the proceeds, using the money to purchase more land on the Tallgrass
Prairie Preserve in eastern Oklahoma. "These islands are serving as
a model for [Georgia Governor Zell Miller's] new land-acquisition
program known as **Preservation 2000**."

Ongoing management programs include the **Marshall Forest
Preserve**, 250 acres of old growth forest within the Rome city limits;
the **Charles C. Harrold Preserve**, 73 acres protecting the gopher
tortoise and the Georgia Plume, a threatened tree found only in this
state; Heggie's Rock, said to be "the most outstanding National
Natural Landmark in the eastern piedmont" and home to 19 species
that live only where granitic flatrocks occur; and **Big Duck's Pond**,
swampland along the Carolina Bay that is a breeding ground for the
endangered wood stork.

The Georgia Natural Heritage Inventory is another cooperative effort with the DNR.

■ Resources

The Georgia chapter publishes its own quarterly newsletter. A brochure is available.

The Nature Conservancy (TNC)
Great Basin Field Office

P.O. Box 11486
Salt Lake City, UT 84147 USA

Contact: **David Livermore**, Vice President/Director
Phone: 1-801-531-0999 • *Fax:* 1-801-531-1003

Activities: Conservation • *Issues:* Biodiversity/Species Preservation

● Organization

The Nature Conservancy (TNC) ranks the state of Utah fifth "in terms of . . . ecological significance," and the 8 staff and 5,200 local members work in cooperation with state and private organizations and individuals to preserve lands vital to numerous species of the Great Basin, in both Utah and Nevada.

▲ Projects and People

The Great Basin field office began working in 1990 to help protect the Stillwater National Wildlife Refuge; designated in 1948, the wetlands area is a key way station for migrating waterfowl such as snow geese, dowitchers, and canvasbacks. Despite its status, however, the refuge is suffering. Since the early 1900s when a huge federal irrigation project dammed the Truckee and Carson rivers, water that used to supply the refuge has been diverted to cultivate the (now former) desert.

Farmers, cities, Native American groups, and the refuge have been vying over the same limited water supply, a situation that led some to dub the Lahontan Valley the "Middle East of the West." In cooperation with the U.S. Fish and Wildlife Service and the Environmental Defense Fund (EDF), the Conservancy has put more than $1 million into a program which pays farmers to stop cultivating portions of their lands and to sell the unused water at market prices to the refuge.

In 1988, the Great Basin office purchased from Dynamic American Corporation 1,730 acres along a 6.5-mile stretch of the Strawberry River in Duchesne County, Utah, an area rich in trout and raptors. By 1990, the office had acquired 13 miles of river land, with a long-range goal of 18 miles; eventually, the U.S. Bureau of Reclamation will buy the land from the Conservancy, to be managed by the Utah Division of Wildlife Resources.

Other recent ongoing acquisition projects include Ash Meadows, a 13,000-acre oasis near Death Valley; the 7,500-acre Cunningham Ranch, rich in elk and other wildlife; and a projected 700 acres of wetlands for the Moab Slough nature preserve along the Colorado River with its 140 species of waterfowl, passerines, shorebirds, and raptors. The Moab is also a "staging area for the rare Colordao squawfish and is visited by recently reintroduced river otters," writes David Livermore, vice president and director since 1989.

The Great Basin office has been working with the National Park Service since 1986 to catalog plant and animal species on the Colorado Plateau. With the Bureau of Land Management and the Utah Division of State Lands and Forestry, the Conservancy helped put up fencing to protect the state's endangered dwarf bear claw poppy from off-road vehicles (ORVs).

Conservancy publicity played a major role in the 1990 passage of Nevada's Question 5, which allocated millions of dollars of state funds to benefit parks and wildlife habitats, including the Stillwater and Lahontan Valley wetlands. In 1991, the office awarded three major research grants for the study of the endangered desert tortoise.

Livermore has worked with the Great Basin office since 1980. He started his nature-oriented career as a summer wrangler and guide at the Trial Creek Ranch in Wyoming while still a student. His key contribution to TNC activities in Nevada and Utah has been to apply to water the Conservancy's principle of buying land, with the belief that "capitalism works better than litigation."

■ Resources

The Great Basin office publishes a biannual newsletter.

The Nature Conservancy (TNC)
Illinois Field Office

79 West Monroe, Suite 900
Chicago, IL 60603 USA

Contact: **Adele Meyer**, Director of Development
Phone: 1-312-346-8166 • *Fax:* 1-312-346-5606

Activities: Conservation • *Issues:* Biodiversity/Species Preservation

● Organization

In operation since 1961, the Illinois Field Office, The Nature Conservancy (TNC), has protected more than 20,000 acres of valuable wildlife habitat. With 22,000 members, a staff of 15, and a 4,500-strong Volunteer Stewardship Network, TNC–Illinois is working to ensure that the state does not lose the small portion of untouched land remaining there (0.1 percent). In most cases, they follow the advice of Will Rogers: "Buy land. They ain't making it anymore."

▲ Projects and People

Like most TNC chapters, the Illinois office puts most of its effort into land acquisition. The Indian Boundary Prairies area, acquired in 1972 and now a National Natural Landmark, has grown to three times its original size. Habitat for the rare gray fox and Franklin's ground squirrel, the tract is only 30 minutes from Chicago. The Conservancy purchased the Nachusa Grasslands, part of the last remaining Illinois prairie, at a public auction, narrowly saving it from becoming a housing development. The 700-acre area is "large enough to sustain a full component of prairie wildlife," including badgers, upland sandpipers, rare prairie plants, and, eventually, bison herds. Until then, visitors are encouraged to do the beneficial trampling; there are no trails.

At the Nachusa Grasslands and on other TNC-managed properties in Illinois, prescribed burning is viewed "as essential . . . to a grassland's health"; some species like the bur oak depend on fire to propagate, while the action helps remove non-native plant species that threaten to usurp the land.

Fire was an essential part of Steve Packard's work on the **Miami Woods Restoration Project** in Northbrook. Although as a prairie restoration innovator Packard believes that the USA "needs at least one prairie big enough for the buffalo, the cougar, and for people too," at Miami Woods he actively dismantled what even he had considered prairie—because it was really savanna. The realization came after Packard discovered that prairie species he was trying to plant on the area kept dying. Prescribed burning destroyed intrusive plants, and research into nineteenth-century descriptions, along with diligent searching, provided seeds for the proper savanna species; gradually, native insects, and animals like the eastern bluebird, Cooper's hawk, and the Edwards hairstreak butterfly returned to the area.

Packard told the the *New York Times* (March 19, 1991) that the resurrected ecosystem is "so strikingly beautiful. It was like finding a Rembrandt covered with junk in somebody's attic."

The Cache River Wetlands project involves a 60,000-acre core area for a projected 475,000-acre bioreserve. The land is habitat for bobcat, river otter, and the gray bat, and contains what are possibly the oldest living organisms east of the Mississippi—1,000-year-old cypress trees. Cache River is also prime duck area, full of food plants like wild millet and smartweed, and is a key migratory stop for birds on the Mississippi flyway. Therefore, Ducks Unlimited is working with TNC, the Illinois Department of Conservation, and the U.S. Fish and Wildlife Service on the project, organized by TNC assistant director **Paul Dye**.

Not all of the Illinois chapter's activities involve purchasing land, however. In a move rivaling the UNESCO relocation of Abu Simbel, a force of staff and volunteers from the Conservancy and other organizations picked up the **10,000-year-old Healy Road Prairie**, a high-quality grassland, and put it down 5 miles south to save it from the bulldozers of its owner, a gravel mining company. In a three-tiered effort, teams first gathered as many insects and small animals as possible for later release, then dug and loaded pieces of the prairie onto truckbeds, and replanted them in the new location. Despite inevitable losses of some of the more fragile species, more than 75 percent of the original plant species were surviving a year later.

Al Pyott, the Illinois Field Office director, prefers to focus on larger tracts of land following the bioreserve idea; "we are only as good as our ability to think beyond the traditional concepts," he says.

■ Resources

The Illinois field office conducts burn workshops, gives lectures, and organizes field trips. The office also publishes a quarterly newsletter, *The Conservator*, as well as a *Steward's Handbook* and a compilation of *State of the Prairie Lectures*.

The Nature Conservancy (TNC)
Long Island Chapter
250 Lawrence Hill Road
Cold Spring Harbor, NY 11724 USA

Contact: **Donna M. Geluso**, Development Assistant
Phone: 1-516-367-3225

Activities: Development; Education • *Issues:* Biodiversity/Species Preservation

▲ Projects and People

Since 1980, a major undertaking of **The Nature Conservancy's** (TNC) **Long Island Chapter** is protection of Shelter Island's **Mashomack Preserve**, known as the "jewel of the Peconic," which features 10 miles of coastline, salt marshes, tidal creeks, woodlands, and fields. Visitors are cautioned to observe slowly and quietly, not to intrude or consume, in order to appreciate fully the wildlife such as spotted salamander, saw-whet owl, great blue heron, great and snowy egrets, harbor seal, river otter, golden eagle, osprey, and black duck, among others.

One unique area is a **pine swamp complex**, considered an important freshwater wetland. Elsewhere, 1,400 acres of "upland oak and beech forest are now being allowed to develop into an **old-growth forest**, a habitat scarce in the Northeast," reports TNC. The naturally nesting osprey colonies are rare in that part of the country; each spring, the osprey returns there from South America. Other sighted breeding species are red-tailed hawks, ruby-throated hummingbirds, and cedar waxwings. In the autumn, the deer mating season is evident by young saplings with bark rubbed completely away—an activity of "ardent bucks," notes TNC.

The Nature Conservancy (TNC)
Louisiana Nature Conservancy (LNC)
P.O. Box 4125
Baton Rouge, LA 70821 USA

Contact: **Nancy Jo Craig**, Executive Director/Vice
President
Phone: 1-504-338-1040

Activities: Conservation • *Issues:* Biodiversity/Species Preservation; Sustainable Development

● Organization

The **Louisiana Nature Conservancy** (LNC) currently employs a staff of 10 and is supported by 8,000 dues-paying members. Since opening in 1987, the office has worked to establish eight nature preserves, three new National Wildlife Refuges, and additional land for State Wildlife Management Areas.   Thus far, the Conservancy has protected more than 130,000 acres; however, because "forested wetlands in Louisiana's Mississippi Delta are disappearing at the alarming rate of 57,000 acres each year," efforts must intensify, and more organizations must cooperate.

In many of its projects, the Louisiana Nature Conservancy works to coordinate the efforts of the Fish and Wildlife Service (FWS), U.S. Forest Service, U.S. Army Corps of Engineers, U.S. Environmental Protection Agency (EPA), state land and wildlife agencies, private timber industry, farming community, and environmental groups such as the Louisiana Wildlife Federation and the Sierra Club.

▲ Projects and People

The lower Mississippi River valley is the largest wetland ecosystem in the USA, reports TNC. Still, it takes up less than half the area of the original 24-million-acre forested wetlands of the Alluvial Flood Plain. Intensive agriculture has been the main culprit; however, a decline in soybean prices in the 1980s resulted in less clearing activity, opening the way for conservation efforts. The first effort of the Louisiana office preserved 625 acres of White Kitchen Wetland, home to a bald eagle's nest in use for possibly 70 years. The size of the preserve has nearly doubled over the years.

In 1990, General Motors gave LNC 4,000 acres of rich hardwood forest at Bayou Cocodrie in the northeast; GM had owned the Fisher Lumber property since 1926, when the wood was used in manufacturing fine automobiles. The Conservancy then secured another 7,000 acres of the wetland, which hosts significant numbers of game and migratory waterfowl species as well as the endangered Louisiana black bear. Through the FWS, the Conservancy hopes to establish Bayou Cocodrie as a national wildlife refuge.

The 1,800-acre Little Pecan Island, part of Louisiana's Chenier Plain, is another of LNC's efforts. The 3,000-year-old ecosystem provides "critical habitat for neotropical migrating songbirds . . . [as the] only woodlands in a vast sea of coastal marsh and agriculture."

Nancy Jo Craig, an environmental biologist, has been working for Louisiana's natural resources since 1975 in various capacities. She has been executive director of the Conservancy chapter since she established it in 1987, and she became a TNC vice president in 1990. Focusing on wetland loss and conservation, she is an active member of numerous advisory boards and other conservation organizations, with many publications to her credit.

■ Resources

The Louisiana office publishes a quarterly newsletter, *Louisiana Nature Conservancy*.

The Nature Conservancy (TNC)
Lower Hudson Chapter ⊕
223 Katonah Avenue
Katonah, NY 10536 USA

Contact: **Dave Trynz Tobias**, Assistant Director
Phone: 1-914-232-9431

Activities: Development; Research • *Issues:* Biodiversity/Species Preservation; Deforestation

● Organization

Organized in 1976, the Lower Hudson Chapter of The Nature Conservancy (TNC) controls 25 preserves involving thousands of acres in 5 counties (Westchester, Putnam, Dutchess, Orange, and Rockland). With five million inhabitants, the area has the largest urban concentration in the USA. Nevertheless, the Hudson chapter's staff of 16 and 6,000 members pursue TNC goals of preserving the land by identifying, acquiring, and stewarding it.

TNC describes the Lower Hudson Valley as a "rich mixture of mountains and swamps, deep forests and swift rivers, smooth floodplains and sheer cliffs." The area's biological diversity—including its rare species—is recorded in the computerized database of the New York Natural Heritage Program, established with the state Department of Environmental Conservation. A Land Steward, working with volunteers, helps to protect each parcel of land; tasks range from posting of boundaries to monitoring of endangered and threatened species and communities.

▲ Projects and People

In addition to its local work, the Lower Hudson Chapter has donated training and technical assistance, as well as financial support, to TNC's international efforts in Latin America and the Caribbean.

■ Resources

The chapter publishes a quarterly newsletter.

The Nature Conservancy (TNC) ⊕
Maine Chapter
14 Maine Street, Suite 401
Brunswick, ME 04011 USA

Contact: **William A. McCue**, Development and
              Communications Director
Phone: 1-207-729-5181

Activities: Development; Education; Research • *Issues:* Biodiversity/Species Preservation

● Organization

The Maine Chapter of The Nature Conservancy (TNC) strives "to preserve plants, animals and natural communities that represent the diversity of life in Maine and on Earth by protecting the lands and water they need to survive." Fourteen staff members, supported by a network of 200 volunteers and by more than 14,000 dues-paying members, manage "the largest system of privately owned nature preserves" in Maine. Since its founding in 1956, the Chapter has protected more than 90,000 acres—nearly half of that area acquired in 1991 alone.

▲ Projects and People

Under Executive Director Kent Wommack, Maine's TNC Chapter in 1991 launched the $3.5 million Maine Legacy Campaign; money raised will go toward the purchase of at least 12 priority sites in Maine, including wetlands that shelter endangered plants and animals, Cobscook Bay estuary that has the Northeast's highest concentration of nesting bald eagles, seacoast sites that are bird sanctuaries for Leach's storm petrels and black guillemots, and Waterboro Barrens that is the Northeast's most threatened forest community. This campaign will help fund loans and stewardships and fulfill a $300,000 commitment to conservation partners in Bolivia, Mexico, and Panama for seven different projects. Further support for Latin America projects comes from TNC's Maine Forest to Rain Forest fund.

Also in 1991, the Chapter established the **Maine Natural Heritage Program** as part of the Department of Economic and Community Development. The Heritage Program processes daily requests for information, using its data on locations of endangered species and habitats to aid landowners and developers on land-use planning. TNC stewards help gather the data; in 1990, for example, they conducted inventories of the Maine mayfly, grasshopper sparrows, and northern blazing stars. Stewards also work to erect protective fences around piping plover and least tern nesting sites.

Overall, the Chapter manages 87 preserves and monitors 32 easements and 24 transferred lands; protected areas include the largest seabird nesting ground in the state, the largest old growth forest in New England, and 14 bald eagle nesting sites. Much of the work is in cooperation with state agencies, such as the Land for Maine's Future Board and the Maine Department of Inland Fisheries and Wildlife.

■ Resources

The Maine Chapter has published *Maine Forever: A Guide to The Nature Conservancy's Preserves in Maine*, containing more than 60 preserve maps, illustrations, and natural history information. The newsletter *Maine Legacy* appears quarterly.

The Nature Conservancy (TNC) Maryland Office

2 Wisconsin Circle, Suite 600
Chevy Chase, MD 20815 USA

Contact: **Wayne A. Klockner**, State Director
Phone: 1-301-656-8673

Activities: Conservation • *Issues:* Biodiversity/Species Preservation; Sustainable Development

● Organization

The Nature Conservancy's (TNC) Maryland Office, established in 1977, boasts 17,000 members. Its staff of nine uses their real estate and tax law expertise to pursue the Conservancy goal of protecting tracts of land from development. A corps of volunteers builds boardwalks, posts signs, pulls non-native weeds, and helps when needed on preservation and stewardship projects.

With wholly private funding, though frequently in cooperation with government agencies such as the state Department of Natural Resources (DNR), the Maryland Office has helped conserve more than 30,000 acres, 8,000 of which it currently owns and manages. The Natural Heritage Program under DNR auspices assists the Conservancy by identifying and ranking areas most worth saving.

▲ Projects and People

A major Maryland Conservancy project involves the **Western Pennsylvania Conservancy**; the two work together to preserve **Sideling Hill Creek**. From its beginning in south-central Pennsylvania, the 25-mile-long waterway joins the Potomac in western Maryland; home to diverse habitats and wildlife, the creek is relatively undeveloped and healthy. However, it faces threats from large-scale logging, the growing Potomac area population, flooding, toxins, and introduction of new, competing species. Conservancy efforts help pro-

tect the healthiest of 10 remaining populations of the endangered harperella wildflower, cousin to the carrot and to Queen Anne's lace, and conserve other rare endemic species like the cat's-paw ragwort and shale barren pussytoes.

In conjunction with the U.S. Fish and Wildlife Service, the Conservancy added 2,600 acres of marshlands and loblolly pine forest to the existing national wildlife refuge at **Blackwater**, a haven for nesting bald eagles, wintering Canadian geese, peregrine falcons, and muskrats. Other projects have preserved habitat and species at **Roundtop and Sugarloaf mountains, Finzel Swamp,** and **Nanjemoy Creek**, whose 288 acres support 500 to 600 pairs of blue heron, making the largest mid-Atlantic rookery. At an Eastern Shore TNC preserve, an ancient **cypress forest of Nassawango Creek** harbors river otters, pileated woodpeckers, 18 different warblers, 15 orchid species, and Maryland's "last large stand of Atlantic white cedar." The **Third Haven Woods** preserve is the only habitat of the endangered Delmarva fox squirrel.

Under the Maryland Office's **Natural Areas Registry Program**, more than 130 private landowners have volunteered to protect rare species on their land.

■ Resources

In addition to its quarterly newsletter, *The Nature Conservancy of Maryland*, the office has available the *Maryland Preserve Directory* and a guidebook, *Mountains to Marshes*. Lecturers are also available.

The Nature Conservancy (TNC) Minnesota Chapter ⊕

1313 5th Street, SE
Minneapolis, MN 55414-1524 USA

Contact: **Nelson T. French**, State Director
Phone: 1-612-331-0750 • *Fax:* 1-612-331-0770

Activities: Research • *Issues:* Biodiversity/Species Preservation

● Organization

The mission of the **Minnesota Chapter** of The Nature Conservancy (TNC) is "to preserve Minnesota's biological diversity by identifying, protecting, and managing the places that support rare or threatened plants, animals or natural communities." The **Minnesota Natural Heritage Program**, a data center run by the state Department of Natural Resources (DNR), was created in 1979 with the guidance and funding of the Conservancy. With information provided by the heritage program, the Minnesota Chapter has set more than 250 sites as top priorities for conservation efforts.

Depending on the situation, the Conservancy uses the "most appropriate mechanisms" to protect chosen sites, whether fee title purchases, conservation easements, land gifts, estate bequests, management agreements, or voluntary land registry.

Since its founding in 1958, the Minnesota Chapter has preserved more than 45,000 acres of prairie, peatlands, dunes, and woodlands. "One third of all the world's (endangered) dwarf trout lilies exist on Conservancy preserves and registered sites," TNC reports. Currently, 10 staff members and numerous volunteers help manage the more than 17,400 acres of the 65 preserves directly owned by the Conservancy.

▲ Projects and People

The Chapter introduced a **prescribed burning program** in 1975, enriching the soil and eliminating non-native plant species on 5,000 acres of prairie.

In 1991, the Conservancy led a drive that resulted in legislation designating 150,000 acres of **Minnesota peatlands**—comparable in quality only to those in Alaska—as **Scientific and Natural Areas** under Article 8 of the Wetland Conservation Act of 1991. In the recent **Minnesota Critical Area Campaign**, the Chapter raised $6 million for conservation efforts in the state and **Latin America**. In addition, TNC helped establish the **Minnesota Environment and Natural Resources Trust Fund** and the **Cannon River Watershed Partnership**, a seven-county coalition "committed to preserving water quality in south-eastern Minnesota." It is also involved with **tallgrass prairie management** and protection.

TNC supports the **Minnesota County Biological Survey**, a county-by-county inventory that classifies landscapes into "bioreserves" to foster "ecosystem-wide conservation approaches."

■ Resources

Materials needed include pro bono printing, art, and layout work for newsletters and brochures; microscope; and storage space. The Chapter publishes *Fact Sheets* and the *Minnesota Chapter Newsletter*, and, through the Natural Heritage Program, the *Minnesota County Biological Survey*. Seasonal internships are available, primarily for students.

The Nature Conservancy (TNC) Mississippi Field Office

P.O. Box 1028
Jackson, MS 39125-1028 USA

Contact: **Roger L. Jones, Jr.**, Director
  Phone: 1-601-355-5357 • *Fax:* 1-601-355-5360

Activities: Development • *Issues:* Biodiversity/Species Preservation

● Organization

Although **The Nature Conservancy (TNC)** has been working in Mississippi since 1975, the **Mississippi Field Office** did not open until 1989. The staff of four have as their mission "to find, protect and maintain the best examples of communities, ecosystems, and endangered species in Mississippi." They report, "We have already lost the passenger pigeon, ivory-billed woodpecker, and Florida panther. But we can still provide homes for the gopher tortoise, yellow ladies slipper, bayou darter, and snowy plover." They achieve these goals with the help of many volunteers, an elected Board of Trustees, appointed Corporate and Scientific Advisory Boards, and with the financial support of 1,600 members.

The office works in partnership with the **Mississippi Natural Heritage Inventory Program**, established in 1974 and administered by the state Department of Wildlife, Fisheries, and Parks. Heritage researchers use techniques developed by TNC, and have thus far identified more than 200 priority sites and 140 rare or endangered species and natural communities in the state. The Conservancy office acts on these findings, working to safeguard the best sites.

▲ Projects and People

Staff and volunteers from the Mississippi Conservancy monitor 13 **preserve areas** and more than 30 "registered natural areas" ranging in size from one to 33,000 acres, and totaling more than 94,000 protected acres.

In 1991, the office acquired 200 acres of wetland at **Grand Bay**, the "largest, best preserved wet savanna in the USA." With the cooperation of the Fish and Wildlife Service, the Conservancy hopes to expand that core into an 11,000-acre **national wildlife refuge** which will extend into Alabama and provide safe habitat for sandhill cranes, bald eagles, and other endangered species.

The Nature Conservancy purchased the 40,000-acre **Pascagoula River Wildlife Management Area**, described as Mississippi's finest natural habitat, in 1976; it later transferred ownership of the large and relatively untouched area, home to the Florida panther, to the state.

The Conservancy has enjoyed a profitable partnership with the Scott Paper Company, which not only donated 80 acres of **Black Creek Swamp**, location of the highest concentration of nesting ospreys in the state, but also participated in a tax-free exchange of land at **Dead Dog** and **State Line Bogs**. The resulting 360 acres, hosting lavender burrowing crayfish and the sweet pitcher plant, is now owned and managed by the state Department of Wildlife, Fish, and Parks.

TNC retains a conservation easement at **Buttercup Flats**, where forest habitats amble from long-leaf pine to savanna, bay swamp to bog; and rare orchid, bog button, Sarvis holly, and grass-pink appear. **Sweetbay Bogs Nature Area** is one of the state's last remaining examples of a quaking peatmoss bog and home to two globally endangered species: grass of Parnassus and bog spicebush. Other land-saving projects are **Clark Creek Natural Area** with some 40 waterfalls, black bear, rare land snail, and the threatened Carolina magnolia vine and southern red belly dace fish; **Dahomey National Wildlife Refuge**, a large stand of bottomland hardwood forest; **Hillside National Wildlife Refuge**, a critical site for migratory and wintering waterfowl; **McIntyre Scatters**, remnant of a once-vast delta bottomland forest ecosystem still inhabited by hoary bat coyote, Mississippi map turtle, and red-shouldered hawk, now encircled by farmland; **Mississippi Sandhill Crane National Wildlife Refuge**, home of the endangered sandhill crane; **Old River Wildlife Management Area** with more than 200 species of mammals, birds, and reptiles; **Panther Swamp National Wildlife Refuge**, where some 250,000 birds winter in the Yazoo Basin; **Plymouth Bush**; and **Shipland Wildlife Management Area**, bordering the Mississippi River and home to the rare scissor-tail flycatcher among many songbirds and waterfowl.

Roger L. Jones, Jr., the Mississippi office director, also established it. Holding degrees in biology and environmental management, Jones previously worked for six years with the Conservancy's South Carolina office.

■ Resources

The field office publishes a newsletter, *Mississippi: The Nature Conservancy*.

The Nature Conservancy (TNC)
South Carolina Office
P.O. Box 5475
Columbia, SC 29250 USA

Contact: **Patrick Morgan**, Executive Director
   Phone: 1-803-254-9049 • *Fax:* 1-803-252-7134

Activities: Conservation; Education; Research • *Issues:* Biodiversity/Species Preservation

● Organization
Beginning with the Francis Beidler Forest in 1970, the **South Carolina Office of The Nature Conservancy (TNC)** has helped to preserve more than 47 areas in the state. The staff of 7 and 6,500 members follow standard TNC procedures, working jointly with public and private sector groups to secure salvageable tracts of land, mainly through purchasing it.

▲ Projects and People
A major effort of the South Carolina Office has involved the 350,000-acre **ACE Basin** project to preserve the land where the Ashepoo, Combahee, and Edisto rivers converge at St. Helena Sound. *(See also Ducks Unlimited in this section.)* In cooperation with Ducks Unlimited, this is the South Carolina chapter's contribution to the **Atlantic Coast Joint Venture** portion of the **North American Waterfowl Management Plan (NAWMP)**, developed by the Department of Interior, U.S. Fish and Wildlife Service, and Canadian Wildlife Service. Among the land parcels involved is the 5,000-acre Hope Plantation, a conservation easement from **Ted Turner** of Turner Broadcasting; through the 1988 easement, Turner gives up his right to develop residential communities or pursue industrial and commercial activity on the land, though he retains ownership.

   In preserving places like the ACE Basin, Peachtree Rock, the Mountain Bridge Wilderness Area, and the Santee Coastal Reserve, the South Carolina Chapter also helps preserve native species such as the bald eagle, shortnose sturgeon, wood stork, and alligator.

■ Resources
Informational brochures and fact sheets listing local TNC-protected lands are available.

The Nature Conservancy
South Fork/Shelter Island Chapter
P.O. Box 2694
Sag Harbor, NY 11963 USA

Contact: **Kathleen Conrad**, Development and
          Communications Coordinator
   Phone: 1-516-725-2936 • *Fax:* 1-516-725-3720

Activities: Education; Research • *Issues:* Biodiversity/Species Preservation

● Organization
Supported by 5,000 members, the staff of 6 full-timers and 5 part-timers at **The Nature Conservancy's (TNC) South Fork/Shelter Island (SF/SI) Chapter** on eastern Long Island, New York, monitor 26 protected areas covering thousands of acres. Of the 96,000 acres of land available, all but 27,000 have been developed; thus, the Chapter is "committed . . . to preserving key areas of the East End's surviving maritime landscape and its wildlife." These include salt marshes, dunes, pine barrens, moorlands, maritime grasslands, and woodlands.

   Like most Conservancy regional offices, the SF/SI Chapter works in cooperation with the state's Natural Heritage Inventory (NHI) and Department of Environmental Conservation, as well as with local government councils and private groups or individuals.

▲ Projects and People
In the 1980s the Heritage Program identified 1,153 acres of the **Long Pond Greenbelt**, which runs from Sag Harbor to the Atlantic, as ecologically significant. The tract remains relatively intact, with more than 30 rare species and more endangered species than anywhere else in the state, including the tiger salamander, once thought extirpated.

   The Chapter purchased the 2,039-acre **Mashomack Preserve**, a salt marsh on Shelter Island, in 1980. Salt marshes are said to be "one of nature's most productive ecosystems and a fertile nursery for marine life." This habitat is home to such endangered species as osprey, oystercatcher, diamondback terrapin, and marsh fimbry sedge. Between the primary and secondary dune lines at the **Atlantic Double Dunes Preserve**, Amagansett is a wetland swale where diverse plants, some 130 bird species, and the spade-foot toad is found.

   In cooperation with the **New York State Diagnostic Laboratory** at Cornell University, TNC volunteers are experimenting with ways to preserve live ticks in the lab for a Lyme disease study. In 1986, the organization began a cooperative project, pooling its funds with those of Southampton Town's open space program and the Suffolk County Open Space Bond Act for a total of $9 million directed toward land acquisition. In 1990, with the Long Island Chapter and the New York State Office of Parks, Recreation and Historic Preservation, it signed a **protection advisory agreement** to aid in conserving Long Island's parks.

   Since 1986 the SF/SI Chapter has worked with the Long Island Chapter in a joint **Tern and Plover Protection Program**. Focusing on another bird species, volunteers annually erect or repair **osprey poles** in key nesting areas; because typical osprey nesting sites—large dead trees—are not as available as they used to be, 25-foot telephone poles, each with wire grating and a platform on top, go up in their stead. Osprey remain on the endangered list, but are making a clear comeback as a result of such efforts. Other volunteer activities include dune stabilization work, clearing and maintaining trails, erecting fencing, and posting signs.

   In the summer of 1991, Long Island property owners **Billy Joel** (accompanied by his wife **Christie Brinkley**), **Paul Simon**, and other celebrities performed at a Montauk **benefit concert**; 65 percent of the proceeds went to the SF/SI Chapter.

■ Resources
The South Fork Chapter needs help on a beach nesting bird protection program.

Available resources include a brochure and quarterly newsletter, *The Nature Conservancy South Fork/Shelter Island Chapter News*; William T. Griffith's *Preserve Guide* that is arranged geographically and describes each TNC-protected land parcel; and a handbook called *Healthy Households*. The Chapter offers a summer internship.

The Nature of Illinois Foundation (NIF)
208 South LaSalle Street, #1666
Chicago, IL 60604 USA

Contact: **John D. Schmitt**, Executive Director
Phone: 1-312-201-0650 • *Fax:* 1-312-201-0653

Activities: Education; Development • *Issues:* Biodiversity/Species Preservation; Energy; Water Quality

● Organization

The Nature of Illinois Foundation (NIF) is a nonprofit corporation, formed in 1983, to "foster an understanding of and appreciation for the natural resources of Illinois and to promote the activities of the three Illinois Scientific Surveys—Natural History, Water and Geological—and the Hazardous Waste Research and Information Center (HWRIC)." These partially state-funded agencies, under the Energy and Natural Resources Department, are directed by a governor-appointed board. NIF's mission is to support their programs and to provide citizens of Illinois with an enhanced understanding of the nature of their work.

NIF says it is also committed to fostering science literacy in both children and adults, and awareness and appreciation of Illinois' precious natural heritage.

The surveys were begun in the mid-1800s when scientists and civic and business leaders recognized the need to catalog the existence and location of Illinois resources, to understand their complex interdependencies, and to encourage their intelligent management. This leadership resulted in the early establishment of the Natural History Survey (1858), the Water Survey (1895), and the Geological Survey (1905). HWRIC was added in 1984.

▲ Projects and People

The Illinois Natural History Survey currently conducts more than 200 research projects in addition to maintaining collections that house millions of specimens of plants, insects, fish, birds, and mammals. One particular insect, the Asian Tiger mosquito, a container-inhabiting pest capable of carrying 26 viruses known to affect humans, is rapidly colonizing urban areas in Illinois. According to NIF, the survey took an active role in preparing legislation to control the storage of scrap tires that these and other mosquitoes use as larval habitat. Scientists currently are monitoring the spread of the deer tick and lone star tick, carriers of Lyme disease.

The Illinois State Water Survey investigates and monitors water and atmospheric resources—weather and climate, rivers, lakes, and underground aquifers. For example, NIF points out that fish and fowl have returned to a restored Lake Peoria backwater, once so murky it could not support aquatic life. This transformation, NIF says, is the result of a project to develop low-cost methods for reestablishing backwater fish habitats. The survey worked with the Illinois Department of Conservation to build artificial reefs.

The Illinois State Geological Survey provides maps and other information about the interconnections between geology and human culture so that "governments avoid hazards and costly errors when planning for resource development."

HWRIC is a multidisciplinary approach to better management and features a new research facility, the Hazardous Materials Laboratory, used also by the U.S. Environmental Protection Agency (EPA) and the U.S. Department of Defense for matters of national concern. With an EPA grant, HWRIC is getting results from selected industries in reducing wastes. For example, when a Chicago printer switched from volatile solvent-based to water-based inks, worker safety improved and toxic waste decreased. An electroplating company is attempting to reduce both waste volume and toxicity—and save costs too.

NIF distributed a natural history survey book, *The Natural Resources of Illinois*, and its educational videotape, *Baking Illinois*, to every school, 4-H Club and scout group in the state. The educational display **Biodiversity in Illinois** explains through photos, maps, narratives, and specimens the delicate balance that exists between the forces of nature and all living things. In addition, the Foundation mailed monthly "Science Feature Tip Sheets" to tell the media about programs and projects of the four scientific divisions.

NIF informs the legislature on issues of environmental and natural resource concerns; consults in school curriculum development; and co-sponsors workshops, conferences, and special events. It co-sponsored an hour-long television documentary, *Big River of the Heartland.*

■ Resources

Articles, monographs, maps, and pamphlets are available on an ongoing basis. Regular mailings keep members current on new publications and the progress of research projects. The award-winning magazine *The Nature of Illinois* has a circulation of approximately 12,000. A speakers bureau is maintained. NIF actively searches grants to support the surveys.

Nebraska Groundwater Foundation (NGF)
5561 South 48th Street, Suite 232
Lincoln, NE 68516 USA

Contact: **Susan S. Seacrest**, President
Phone: 1-402-423-7155

Activities: Education • *Issues:* Water Quality/Quantity

● Organization

The Nebraska Groundwater Foundation (NGF) was founded in 1985 by volunteers in order to educate Nebraskans about groundwater. According to Susan Seacrest, falling water tables and rising nitrate levels were threatening Nebraska's groundwater supply, which is the largest in America. Because NGF programs have an educational focus, rather than a political agenda or economic vested interest, divergent points of view and coalition building are encouraged.

In 1991, Seacrest was asked to join the U.S. Environmental Protection Agency's (EPA) national advisory group on water issues. The W.K. Kellogg Foundation asked NGF to chair the newly created National Groundwater Education Consortium.

▲ Projects and People

Each year NGF sponsors a fall symposium where industry leaders and environmental activists plan strategies to approach current environmental issues, such as making plans to protect the Platte River or creating climate response plans.

NGF also created the nation's first **Children's Groundwater Festival**, held annually. Seacrest reports that the festival gives 3,000 participating children a greater understanding of the importance of environmental issues affecting their lives.

■ Resources

The NGF publishes an award-winning quarterly journal, *The Aquifer*, for more than 2,500 readers. Typical stories include consumer tips, EPA updates, and interactive teaching activities. Other NGF resources include a lending library and a speakers bureau. A children's festival teaching packet which is groundwater oriented also is available, as are water games packaged in playing card decks, "Dripial Pursuit" and "Puddle Pictures." NGF requests assistance in indexing groundwater material into a computer file.

Institute of Systematic Botany (ISB) ⊕
Institute of Economic Botany (IEB) ⊕
The New York Botanical Garden (NYBG)
Bronx, NY 10458-5126

Contact: **Brian Boom, Ph.D.**, Vice President for
        Botanical Research
    Phone: 1-718-220-8628 • *Fax:* 1-718-220-6504

Activities: Development; Education; Research • *Issues:* Biodiversity/Species Preservation; Deforestation; Global Warming; Health; Sustainable Development

● Organization

Some 100 years ago, the Pierre Lorillard estate was dedicated as a scientific organization and opened to the public as The New York Botanical Garden, where today the treasury of plants ranges from 300-year-old native trees to rare orchids, ancient ferns, and exotics gathered worldwide. Facing the next century, Botanical Garden staff are mindful of four huge threats: ozone depletion, global warming, toxic pollution, and biodiversity loss through species extinction.

Together, with the Institute of Economic Botany (IEB) and the Institute of Ecosystem Studies, the combined staff is "larger than most universities, and their research is conducted throughout the world, with a focus on North and South America," writes Gregory Long, Botanical Garden president.

The Garden's ultimate aim is the "protection, creation, and interpretation of the living collections and the specimens in the Herbarium," Long adds. For example, the Rose Garden is graced with 2,700 roses of 270 different types; the Herbarium has some 5.4 million pressed and dried specimens in its highly studied collection. Also on the grounds are a Victorian glass conservatory, reflecting pools, Acid Rain Study Ponds—a grim reminder of the times—and a library with 200,000 volumes and 800,000 nonbook items that is said to be the "largest botanical/horticultural library under one roof

in North America." A living-plant collection computer database is a recent addition.

With environmental education programs that include horticulture, landscape design, botany, and ecology, some 500,000 people are reached annually at the 250-acre site. The Garden's education department also works closely with Bronx schools on gardening, exploring nature, and other special programming. Workshops, symposia on tropical forest medical resources, ecology walks, outdoor concerts, and college-level courses are also offered to adults. A lecture series highlights strategies addressing species loss, the greenhouse effect, and other global environment issues. **Bronx Green-Up**, a community project, helps transform hundreds of neighborhood eyesores into "green oases of flowers and vegetables that soften and beautify the urban landscape."

▲ Projects and People

Within the **Botanical Science Division**, research programs focus on two areas: systematic botany and economic botany—both supported by the Herbarium and the Harding and Lieberman Laboratories. Systematic botanists prepare monographs—"scholarly works summarizing all that is known about a particular group of plants"—and floras—"accounts of all the plants occurring in a particular geographic region." Important monographs produced here are about the blueberry family, mosses, gymnosperms, palms, Brazil nut, citrus, mimosoid legumes, and tree-of-heaven family. NYBG research mainly takes place in North America, Mexico and Central America, West Indies, Andes Mountains, Guayana region, Amazon region, and Planalto and eastern Brazil. This work is done by the scientists of the Institute of Systematic Botany, founded in 1991.

In North America, a major flora of the Intermountain West is being produced; **Dr. Arthur Cronquist** completed the second edition of *Manual of Vascular Plants of Northeastern United States and Adjacent Canada*; and **Dr. John Mickel**, in writing his monograph, discovered that "51 out of 121 species of ferns in the genus *Elaphoglossum* were new to science." Likewise, in **Guyana** and **Venezuela, Dr. Brian Boom** found various species new to science.

Collaborating with **Brazil's** Instituto Nacional de Pesquisas da Amazonia (INPA), Dr. Boom with NYBG botanists **Drs. Scott Mori and Andrew Henderson** surveyed plant diversity in the **Biological Dynamics of Forest Fragments Project.** Dr. Henderson is assessing economic uses of various palms. Also working with indigenous groups in **Bolivia** and **Venezuela,** Dr. Boom saw how rainforest tree species have the potential to produce new "foods, fuels, fibers, and medicines to Western society"—supporting the argument for more parks and reserves. Dr. Mori's research shows that the Brazil nut can also yield an oil for cooking, soap, and lamp fuel, and also can be a source of animal feed. Extractive reserves in southwestern Amazonian Brazil are the site of **Dr. Douglas Daly's** studies of plants in forests that promise long-term economic and ecological viability.

Dr. Gene E. Likens, director of the **Institute of Ecosystem Studies (IES),** founded the **Hubbard Brook Ecosystem Study** in 1963 while on the Dartmouth College faculty, which led to the "discovery of acid rain in North America," reports NYBG literature. Like Hubbard Brook, IES is in the forefront of long-term research projects such as investigating the ecology of the Hudson River ecosystem; probing the environmental stresses of a 40-acre forest of oak, hickory, and hemlock trees on the Garden grounds; determining ecologically sound management of vegetation beneath power lines; planning for outbreaks of the gypsy moth and other insects; and understanding the dynamics of small mammal populations and vegetation.

With **Dr. Gary M. Lovett's** supervision, data is being compiled on 12 forested sites in North America and Europe to determine the "magnitude of atmospheric deposition to forests and its possible impact on the natural nutrient cycling processes," which is essential to the forest ecosystem's health. In analyzing published data, **Drs. Jonathan J. Cole, Nina M. Caraco, and Michael L. Pace** demonstrated that "nitrate concentrations in the major rivers of the world are directly related to the human population density in the watersheds of those rivers." NYBG reports, "Understanding the sources of the nitrates, which include but are not limited to human sewage, will be a key to reducing nitrogen loading and curbing algal blooms and oxygen depletion."

Studies of water bodies of temperate zones, especially in the Northeast USA, examine "lake productivity and sedimentation, the liming of acidified Adirondack lakes to counter the effects of acid precipitation, and the investigation of the zebra mussel." **Dr. David L. Strayer** of IES reports that the inedible zebra mussels, originating in Europe and now appearing in the Great Lakes, reproduce swiftly, clog motorboats' cooling systems; and encrust drinking water intake pipes. With support from the Hudson River Foundation, Dr. Strayer is gathering baseline data among the bottom-living animal communities of the Hudson River to begin to measure ecological impact "before the mussels arrive," he says.

Founded in 1981, the **Institute of Economic Botany (IEB)** expresses the urgency to "find, collect, and document valuable plants before overexploitation of resources and destruction of habitats eliminate such treasures forever." With backgrounds in botany, anthropology, and ecology, IEB scientists are alarmed at the elimination of tropical rainforests at an "estimated rate of 100 acres per minute" or "twice the size of New York State every year."

A major intent is to **stem the loss of medicinal plants**, some of which treat cancer and cardiac patients, and to discover other plants with medicinal potential while learning how to use them from local cultures which are also disappearing as natural habitats are destroyed. IEB estimates that only about "one percent to five percent of the flora [in the Amazon Basin] has ever been analyzed for any pharmacological activity or chemical composition."

From 1986 to 1996, the National Cancer Institute (NCI) is funding IEB to **study plants that might yield chemical compounds to treat cancer and AIDS**. Principal investigators are IEB's philecology curator **Dr. Michael J. Balick** and Amazonian botany curator **Dr. Douglas Daly**. NCI tests some 1,500 samples collected yearly from neotropical rainforests by the IEB team. Lately, pharmaceutical companies are showing renewed interest in researching "previously unknown plant compounds" to help treat modern diseases. *(See also Rainforest Alliance in this section.)* IEB's strategy is to gain knowledge from local herbal healers, or *curanderos*, who are willing to share it—before their wisdom and the forests vanish.

In early 1992, with the Rainforest Alliance, IEB sponsored a symposium at Rockefeller University, New York, that revealed a new coalition of "conservationists, pharmaceutical companies, academic scientists, traditional healers and the government and entrepreneurs of developing countries," according to the *New York Times* (January 28, 1992).

For example, participants learned how "the world's largest drug company"—Merck & Company—and a small California company—Shaman Pharmaceuticals—are "both prospecting for prom-

ising plants," said the *Times*. Merck will share its proceeds with the National Institute of Biodiversity assisting in the project and the Costa Rican government. In turn, the Institute will train local people to learn about and collect the medicinal plant samples; and the government will set aside funds for conservation efforts. Shaman Pharmaceuticals, in working directly with its namesakes, traditional medicine men, says it is already testing a compound from a medical plant in South America for treatment of influenza and herpes virus, as reported in the *Times* article.

Dr. Balick was quoted: "It's out of the realm of an academic exercise now. You've got people whose lives depend on using the rainforest, and whose cultural traditions depend on having it available, and people whose businesses depend on new chemicals all sitting down to discuss these issues."

In Belize, Dr. Balick also works with *yierbateros*, who collect plants for sale in the marketplace as part of the primary health care system. His study shows that the earnings per hectare are "2 to 10 times the income of a farmer growing corn or beans"—which could provide an alternative livelihood to agriculture that destroys forests and depletes soils.

Seeking to **increase the stock of plants available as sources of nutrition**, economic botanists indicate that fewer than 20 major crop species provide most of the world's food. New food sources may be *camu-camu*, a "small red tropical fruit from the Peruvian Amazon with the highest known content of vitamin C in the plant kingdom—up to 30 times that of citrus." **Dr. Charles Peters** is undertaking studies of this sustainable resource that enables native people to conserve their habitats. So far, Dr. Peters has inventoried more than 120 types of edible fruits in the Iquitos region of Peru, including *sacha mangua* (*Grias peruviana*), rich in vitamin A. Peters has also begun ecological studies of three species of illipe nut in East Kalimantan, Borneo, whose oil is used in the preparation of cocoa butter and fine cosmetics. His colleague **Dr. Christine Padoch** searches for "promising native fruit species of the Peruvian Amazon" for potential propagation and marketing. Together, Peters and Padoch are also studying how tribal peoples manage forests in Indonesian Borneo.

Also found in Peru is the *cocona*, or **peach tomato** (*Solanum sessiliflorum*), which is rich in iron, vitamin A, and niacin. Honorary research associate **Dr. M. Jan Salick** is working on improved varieties. Easily grown native tubers of the upper Andes are being introduced to similar regions worldwide to improve local nutrition and increase diversity of local crops. The **peanut plant and its wild relatives** in Peru, Argentina, Bolivia, and Brazil are the source of germ plasm collections and research to breed improved crops. *Guaraná* is a native Amazonian plant that is a source of a commercial soft drink in Brazil. **Dr. Bradley Bennett** is completing a four-year survey of the useful plants of the Shuar Indians of Amazonian Ecuador.

With his colleagues, IEB director Dr. Balick is researching the properties of the oil of a palm, *patauá*, said to be comparable to olive oil and to provide treatment for asthma and bronchitis; the palm's fruit produces an "extremely nutritious" beverage; its protein source is similar to human milk and "40 percent higher than soybean in its biological value." A goal is to grow the palm more productively.

Confronting limitations on fossil fuels, the Institute's purpose is to "identify plants that can provide **renewable energy substitutes** and to develop their potential." Dr. Balick's long-term study of the *babassu* palm is aimed at "improved commercial and subsistence cultivation" of this Latin American plant that yields clean-burning charcoal and an oil suitable for cooking and fuel. Dr. Balick is on the Board of Advisors of this *Guide*.

■ Resources

NYBG publishes *Economic Botany*, the official journal of the Society for Economic Botany since 1959. In addition, *Advances in Economic Botany* is a volume series on ethnobotany and agroforestry of indigenous cultures and also includes proceedings of various symposia on useful plants. Offered are doctoral fellowships, short-term training courses for foreign students, and graduate courses in conjunction with Yale University and Lehman College of City University of New York. Regarding the Botanical Garden's on-site educational program, a registration booklet is available.

New York Zoological Society (NYZS) ⊕
Bronx, NY 10460 USA

Contact: **James Meeuwsen,** Director of Public Affairs
    Phone: 1-718-220-5090 • *Fax:* 1-718-364-7963

Activities: Education; Research • *Issues:* Biodiversity/Species Preservation

● Organization

"Founded in 1895, the New York Zoological Society (NYZS) consists of six divisions operating in the United States and worldwide: the **Bronx Zoo;** the **New York Aquarium;** the **Osborn Laboratories of Marine Sciences; Wildlife Conservation International (WCI);** the **Wildlife Survival Center;** and the **City Zoos** beginning with the new Central Park Zoo and soon to include Prospect Park and Flushing Meadows Zoos, and the Central Park Children's Zoo," according to NYZS literature, which continues:

"The Bronx Zoo, Central Park Zoo, and New York Aquarium, visited by nearly 4 million people each year, are in the forefront of wildlife exhibition, education, research, conservation and captive breeding. Wildlife Conservation International operates 113 field projects saving wild animals and wild places in 41 countries. The Osborn Laboratories of Marine Sciences concentrate on marine ecology and coastal conservation. The Wildlife Survival Center, off the coast of Georgia, is devoted to the propagation and study of endangered species."

The nation's largest urban zoo, the Bronx Zoo covers 265 acres and is home to more than 4,000 animals. Other city zoos are ongoing renovation. Some 2,500 marine animals inhabit the New York Aquarium. WCI has already helped to establish more than 60 parks and reserves worldwide. *(See also Wildlife Conservation International in this section.)*

▲ Projects and People

The New York Zoological Society has a long history in saving species—beginning with its efforts at the start of this century to rescue the American bison through the establishment of reserves in Oklahoma, South Dakota, and Montana that were stocked with animals from the Bronx Zoo. Today, many more vanishing species are being given a second chance at life through various NYZS breeding programs linked to research in ecology, wildlife behavior, nutrition, and reproduction, and through the WCI conservation

projects on five continents. With participation in the Species Survival Plans (SSPs) of the American Association of Zoological Parks and Aquariums (AAZPA), the Society is helping to preserve more than 50 endangered species. Through the ongoing, public-supported program **Sponsor-a-Species,** the zoo is now working to save the following and offers these descriptions:

Chinese Alligator. "Baby Chinese alligators are endearing creatures that make a sound much like a machine gun. The species is native to the lower reaches of the Yangtze Valley, where it has been hunted to near extinction for its skins. Being bred by the NYZS among seven crocodilian species, its SSP coordinator is the Bronx Zoo's herpetology curator. Life span is 75 years.

Palm Cockatoo. Distinguished by a large, powerful beak and its dark silvery gray color that is rare in tropical birds, the palm cockatoo of Indonesia is "most severely endangered by the illegal pet trade" as well as habitat destruction. NYZS helped convince Indonesia to demand the return of such birds. Life span is 50 years.

White Naped Crane. In eastern Mongolia and China, fewer than 2,000 of these "graceful birds remain in the wild, where they are threatened by habitat destruction and overhunting." SSP coordinator is located at the Bronx Zoo, whose captive-bred birds are in zoo populations as far away as Taiwan. Life span is 60 to 70 years.

Pere David's Deer. Extinct in the wild for some 1,800 years, this large Asian deer was native to the swampy marshlands of northeastern China. Zoos are breeding this species successfully and returning them to protected reserves in China. The Bronx Zoo recently hosted Pere David researchers from Beijing's Milu Reserve to learn about husbandry techniques. Life span is 20 years.

Indian Elephant. Only about 20,000 of these Asian elephants remain in the wild, mainly in the forests of India, south China, and Southeast Asia. Smaller than their African cousins, they face poaching and habitat destruction and fragmentation. Adult males are solitary and associate with females only during breeding periods; cows, adolescents, and calves "live in tight-knit herds led by a dominant female, or matriarch." Life span is 60 to 80 years.

Western Lowland Gorilla. At home in the tropical rainforests of Gabon, Congo, and Cameroon, fewer than 20,000 of these gorillas remain in the wild. These terrestrial vegetarians "live in cohesive family groups, led by adult males called silverbacks." The Bronx Zoo is engaged in a successful propagation program. Life span is 35 to 50 years.

Przewalski's Wild Horse. Formerly found throughout Asia, particularly in the open ranges of Mongolia and Sinkiang, this "powerfully built wild horse is believed to be extinct in the wild." Like most horses, young males generally band together in bachelor groups; an adult stallion controls a female group. The Bronx Zoo reared 8 young in 1989 as part of the SSP program that includes more than 70 collections. Life span is 25 to 35 years.

Snow Leopard. Secretive and beauteous, the remaining 5,000 to 10,000 of these leopards live in central Asia's mountainous regions, mainly above the timberline. "The Bronx Zoo has had tremendous success breeding this endangered species—more than 60 cubs have been produced in the past 20 years." The species' SSP coordinator is located here. Life span is 20 years.

Proboscis Monkey. Noted for the adult male's long pendulous nose, these monkeys are primarily leaf eaters and excellent swimmers at home in Borneo's lowland rainforests and mangrove swamps. Some 5,000 remain. "The Bronx Zoo's two breeding troops are the only proboscis monkeys in North America." Life span is 20 to 29 years.

Bali/Rothschild's Mynah. Captive breeding programs at various U.S. zoos have saved the Bali mynah from extinction; 40 of these beautiful blue-eyed, white birds were returned to Bali, Indonesia, in the mid-1980s, and the program continues. Life span is 20 to 25 years.

Arabian Oryx. A special trait is that this oryx of the central Arabian desert detects rainfall from considerable distances and migrates accordingly in search of fresh vegetation. Hunted to near extinction, the species has been revived through an "Operation Oryx" captive breeding herd in North America, which includes those at the Bronx Zoo. The young have been released into protected reserves in the Middle East. Life span is 10 to 15 years.

Oriental Small-Clawed Otter. Found throughout Southeast Asia, these tiny mammals that live in family groups of 10 to 12 animals are being used as models for captive breeding programs to save more endangered mustelids. Some 25 were born at the Bronx Zoo in recent years as part of the SSP program. Life span is 12 to 21 years.

Red Panda. At home in the mountainous bamboo forests of central Asia from The Himalayas to south China, this "shy creature" is busiest at dusk and dawn and behaves similarly to giant pandas. Two cubs were born to the breeding pair at the Bronx Zoo—which belongs to the red panda SSP. Life span is 14 years.

Mauritius Pink Pigeon. "Nearly extinct in the wild, between 200 and 300 of these beautiful pigeons exist in captivity. In addition to habitat destruction and hunting by man, nest-robbing monkeys threaten the pink pigeon in its precarious island habitat." The registered studbook is maintained at the Bronx Zoo, where these birds have been successfully hand-raised. Life span is 10 to 15 years.

Great Indian Rhinoceros. The largest of the Asian rhinos makes its home on the swampy grasslands of the Indian subcontinent, and only about 1,500 of this endangered species remain. "A relocation program for Indian rhinos is underway—moving animals from regions where they are numerous to under-utilized areas of their range." Two female calves were born at the Bronx Zoo in the past three years. Life span is 45 years.

Siberian Tiger. This largest of the tiger subspecies, weighing 500 pounds or more, has been reduced to fewer than 300 remaining animals in the northeast Asian forests. Imported from the former USSR, the Bronx Zoo's breeding female, Alisa, is helping to diversify the genetic makeup of the North American population as a member of the Siberian tiger SSP. So far, she has successfully reared six cubs in captivity. Life span is 15 to 20 years.

Radiated Tortoise. "Radiated tortoises, so named because of the lines that radiate from the center of each plate of their shells, live on fruits and fresh shoots in the cactus-filled parklands of Madagascar." Hunted for their tasty meat, they are now protected by the Malagasy government. Bred at the Bronx Zoo and the Wildlife Survival Center, the SSP species is coordinated by the Bronx Zoo herpetology supervisor.

Beluga Whale. Endangered—as are all large marine mammals—and covered by the U.S. Marine Mammal Protection Act, the beluga, or white, whale is now increasing in Alaska and the Hudson Bay; however, pesticides and toxic waste pollution threaten the St. Lawrence River population. This whale also makes its home in the Arctic Ocean. Life span is 20 to 30 years.

Grevy's Zebra. At home in Ethiopia, Somalia, and northern Kenya, this largest zebra species is endangered as a result of overhunting. Unlike other equids, adult animals form no strong bonds; stallions tolerate strange males in their midst. NYZS has successful breeding programs at the Bronx Zoo and the Wildlife Survival Center on St. Catherines Island.

■ Resources

NYZS's Education Department sponsors more than 20 programs for school children from kindergarten to twelfth grade. Families can also participate in special programs about crocodiles, elephants, and rhinos, for instance. Other adult courses focus on zoo medicine, endangered species, bird watching, and African wildlife. Literature is available on these **educational adventures** as well as on the Sponsor-a-Species program, Wildlife Conservation International, and other ecological education programs. The NYZS *Annual Report* lists current global environmental projects and publications.

Department of Ornithology ⊕
New York Zoological Society (NYZS)
Bronx, NY 10460 USA

Contact: **Donald F. Bruning, Ph.D.,** Curator and Chairman
Phone: 1-718-220-5159 • *Fax:* 1-718-220-7114

Activities: Development; Education; Research • *Issues:* Biodiversity/Species Preservation; Deforestation • *Meetings:* CITES; IUCN

▲ Projects and People

Dr. **Donald F. Bruning** describes worldwide projects with which he is involved as chairman of the **Department of Ornithology, New York Zoological Society (NYZS)**, which is supported by 40,000 members:

"We have been working with the American Association of Zoological Parks and Aquariums (AAZPA) and the International Council for Bird Preservation (ICBP) developing a comprehensive plan to protect the last few **wild Bali mynah and reintroduce captive-bred offspring** to augment the small wild population." Dr. Bruning adds that the release of captive-bred birds caused the population to double in 1991.

In **Papua New Guinea,** Dr. Bruning is working with the country's wildlife department and the Research Conservation Foundation (of which he is a founding board member) "to develop a series of **parks and reserves** to protect Papua New Guinea flora and fauna, especially **birds of paradise.**" He reports that "one reserve of over 900 square kilometers has been established. Efforts to secure other areas are underway. We are also assisting with training of wildlife officials and working with the Peace Corps to assist local people near the reserves."

In **Peninsular Malaysia and Sabah, East Malaysia,** Dr. Bruning is working with the wildlife departments to set up **captive breeding centers for pheasants.** "We are helping train personnel and hope that this combination will help assure the establishment and protection of reserves and parks," he continues.

In **Argentina,** "we continue to assist in the establishment of **parks and reserves** from the high Andes for **flamingos** to the coastal colonies of **penguins and cormorants.**"

In the **Caribbean,** Dr. Bruning and others are assisting the "lesser Antilles islands manage their **unique parrot species** by providing advice, training, and help while assisting them in managing a worldwide captive-breeding consortium for the St. Vincent Amazon," he reports.

With specific research interests in birds' ecology and behavior, embryology and reproductive biology, paleontology and zoogeography as well as in conservation of wildlife and habitats, Dr. Bruning has published extensively on rearing rheas, wild ducks, cranes, Malay peacock pheasants, Indian Pygmy geese, and other rare birds in captivity—mainly at the Bronx Zoo, NYZS.

Dr. Bruning represents CITES for AAZPA and is ICBP Parrot Specialist Group chairman, World Pheasant Association vice president, Bahamas National Trust Council member, American Pheasant and Wildfowl Society board member, and Species Survival Commission Trade and Parrot groups member—among his commitments to numerous professional organizations.

He also works closely with TRAFFIC on wildfowl trade issues and advises the governments of Mexico, Indonesia, and Malaysia on conservation matters. Dr. Bruning gives training courses and provides expert witness for the U.S. Fish and Wildlife Service, U.S. Customs Office, U.S. Department of Justice, and the New York Department of Environmental Conservation. He lectures regularly at several area universities.

■ Resources

Dr. Bruning reports that political support is needed in Indonesia, Malaysia, and Papua New Guinea to create parks and protected areas. Posters featuring the Amazon parrot, macaw, and cockatoo are available.

Northcoast Environmental Center (NEC) ⊕ ✍
879 Ninth Street
Arcata, CA 95521 USA

Contact: **Tim McKay, Ph.D.,** Executive Director
   Phone: 1-707-822-6918 • *Fax:* 1-707-822-0827

Activities: Development; Education; Law Enforcement; Political/Legislative; Research • *Issues:* Air Quality/Emission Control; Biodiversity/Species Preservation; Energy; Global Warming; Health; Population Planning; Reforestation; Sustainable Development; Transportation; Waste Management/ Recycling; Water Quality

● Organization

The **Northcoast Environmental Center (NEC),** with 5,000 members, is a grassroots organization founded in 1971 to unite the conservation groups of northwestern California. With a staff of 9 and some 100 volunteers, NEC focuses on educating the public on local and worldwide environmental issues.

The NEC "draws on people's experience to provide free advice, promoting the goals of citizenship through community involvement and greater public awareness of the environment." The diverse interests represented by the Center are "agriculture, land use and planning, energy development and alternatives, streams, rivers, and oceans, fish and wildlife, public lands, toxic substances, and cultural ecology."

▲ Projects and People

Center **outreach** begins with its **comprehensive, 8,000-volume library** that features "extensive files" on most environmental topics,

including 5,000 color slides, microfiche, videotapes, films, and 100 periodical titles. Center director for 16 years, **Tim McKay** runs two **weekly radio programs;** his *Econews* show is described as the "longest running public affairs program on KHSU-FM"—the area's national public radio station. NEC participates in community events through exhibits, displays, and workshops.

Accomplishments of NEC in northwestern California include **preserving local rivers** in the National Wild and Scenic Rivers System, initiating an **adopt-a-beach program** to help clean up the coastline's beaches, **adding 48,000 acres to the Redwood National Park** and **800,000 acres of wild forest land** in the Klamath-Siskiyou Mountains to the National Wilderness Preservation System, landscaping and revegetating park areas, and launching the successful **Arcata Community Recycling Center.** NEC reports that it tackles a range of pollution issues from toxic spraying to malodorous pulp mill emissions.

Former librarian **Dr. Barbara Vatter** is also associated with the College of the Redwoods and Humboldt State University. In 1987–1988, she was a foreign expert in American studies at Changsha Normal University of Water Resources and Electric Power, China.

■ Resources

NEC is in need of a computerized periodical index to help organize its library collections; a full-time librarian is also sought. Volunteers are welcome; work-study programs and internships are also available.

The Center maintains an "information clearinghouse" which, through a **phone referral service,** contacts members about needed action on environmental issues.

The NEC provides speakers and presentations free of charge to schools and organizations.

At its store, NEC sells books, posters, cards, T-shirts, and other items. A monthly newsletter, *Econews,* is distributed throughout the region and worldwide; its editor, **Sidney Dominitz,** was formerly with Reuters and United Press International (UPI).

Northeast Sustainable Energy Association (NESEA)
23 Ames Street
Greenfield, MA 01301 USA

Contact: **Nancy Hazard,** Associate Director
   Phone: 1-413-774-6051 • *Fax:* 1-413-774-6053

Activities: Education • *Issues:* Air Quality/Emission Control; Energy; Global Warming; Health; Sustainable Development; Transportation

● Organization

The **Northeast Sustainable Energy Association (NESEA),** with some 1,000 members, is a nonprofit membership organization founded in 1974 to foster the use of renewable and sustainable energy and to convey the value of these practices for the preservation of the environment.

In order to accomplish its goals, NESEA has established programs for professionals in the transportation, building, and energy management trades; advocacy through promoting positive solutions and NESEA's six-point energy plan; and public education through a network of local chapters, its magazine, and demonstration projects such as the American Tour de Sol.

▲ Projects and People

Attracting the attention of the U.S. Department of Energy, NESEA is pursuing a **Sustainable National Energy Policy** with these points as its cornerstone: increasing research and development funding for all renewable energy technologies; ensuring all real costs of energy be factored into the price of fuel and borne by the users; removing subsidies to polluting energy industries; reforming government and utility energy supply and production decisions to include long-term economic benefits of renewable resources; legislating increased fuel efficiency of vehicles, mass transit development, and bolstered funding for alternative fueled vehicles as solar and electric; and advocating building code changes that bring about energy efficient housing, healthy indoor air quality, and routine use of renewable energy.

NESEA develops programs of interest to people in the building trades. Its **Quality Building Council** hosts an annual conference for builders, architects, engineers, building suppliers, and educators. In addition, NESEA's **Energy Management Council** hosts a professional conference annually for architects, developers, energy consultants, utilities, engineers, facilities owners and managers, lighting specialists, and product suppliers.

NESEA fosters the development of practical nonpolluting solar and electric vehicles for everyday use, encourages young people to become involved in engineering, and demonstrates that solar energy can meet many energy needs through its sponsorship of its annual **American Tour de Sol** program, which includes a Solar and Electric Vehicle Symposium, Race Trials, and a five-day race held in May.

The program is designed to promote the development and use of practical nonpolluting electric and solar electric vehicles. The race was conceived by **Dr. Robert Wills**, who became interested in solar cars while working on his Ph.D. at Dartmouth. While involved in the solar car being built for the Swiss Tour de Sol (which was started in 1982), he decided that America should have a similar race. Both he and **Nancy Hazard** are co-directors of the race, which has grown from a 5-car event in 1989 to 260 cars in 1991.

NESEA reminds consumers that "transportation accounts for 63 percent of all the oil used in the USA, and 90 percent of the urban air pollution."

■ Resources

NESEA seeks any new technologies that reduce energy use, or produce energy in a sustainable, nonpolluting fashion. It also seeks political support with information on specific bills that relate to its work. In addition, NESEA needs assistance in the promotion of its events and in the development of fundraising efforts. NESEA's quarterly publication, *Northeast Sun*, is available to its members.

Northern Rockies Conservation Cooperative (NRCC) ⊕
Box 2705
Jackson, Wyoming 83001 USA

Contact: Tim W. Clark, Ph.D.
Phone: 1-307-733-6856

Activities: Education; Policy Development; Research • *Issues:* Biodiversity/Species Preservation; Sustainable Development

● Organization

The **Northern Rockies Conservation Cooperative (NRCC)** is a nonprofit corporation, founded in 1987, whose goal is creative, cooperative, practical problem-solving in the conservation of nature. Although its focus is on species and ecosystems in the northern Rocky Mountains, it seeks exemplary projects with national and international significance.

According to its mission statement, the Cooperative's work is "basic and applied ecological research, educational activities, organization and management development, research, and consulting, and policy research and analysis." It conducts active independent programs in these areas and offers services to existing agencies and conservation and business organizations.

Members include wildlife ecologists and conservationists, environmental educators, sociologists, organization and management development experts, natural resource policy experts, writers, and corporate executives.

▲ Projects and People

The Cooperative staff **collects, synthesizes, and analyzes data**, and has a history of working closely with government agencies, consulting firms, and conservation groups to enable the data to be used effectively.

For example, the 1976 National Forest Management Act calls for national forests to maintain species diversity, which they may do by managing for **indicator species**, sensitive species whose continued presence will indicate the health of the area. The **American marten** is listed as an indicator species in several northern Rockies forests, but according to the Cooperative, forest plans provide few management prescriptions for lack of data. The Cooperative proposes research and monitoring of the marten's role in forest ecology; comprehensive management plans will be developed in partnership with the U.S. Forest Service.

Other ongoing or planned projects include: behavior, ecology, and **conservation of mountain lions** in fragmented habitat; **endangered black-footed ferret recovery**; survey of endangered and rare plants and animals of the **Greater Yellowstone Ecosystem**; conservation of **hawks, eagles, and owls** in Jackson Hole, Wyoming, and northern Rocky Mountain **wolf recovery studies** in Montana, Idaho, and Wyoming.

According to the Cooperative, some conservation problems are the result of information remaining inaccessible to people who need it. Education programs can answer this need. **The Wolf Fund**, a Cooperative project, is one such program. The Greater Yellowstone Ecosystem is an incomplete ecosystem; the northern Rocky Mountain wolf, a major ecosystem regulator, has been missing since this century's early decades. Federal agencies appear ready to carry out wolf reintroductions to Yellowstone National Park, and the Wolf

Fund can help overcome one barrier to wolf reintroduction—misinformation and mythology concerning the behavior and ecology of wolves. The Wolf Fund seeks to distribute the facts and dispel the fears and myths about wolves.

The Cooperative says it locates, studies, and describes organized and managed natural resource programs that can be used as models for other programs, and assists existing programs through consultation, workshops, and education. Some projects in this area include case studies of such programs describing how to enhance organization and management performance; and consultation with groups on developing action plans, conflict management, and decision making.

■ Resources

The Cooperative publishes the *NRCC News* annually in the spring.

Northwest Coalition for Alternatives to Pesticides (NCAP) ❦

P.O. Box 1393
Eugene, OR 97440 USA

Contact: **Norma Grier,** Executive Director
Phone: 1-503-344-5044

Activities: Education • *Issues:* Pesticides

● Organization

The Northwest Coalition for Alternatives to Pesticides (NCAP) is a nonprofit, five-state, grassroots membership organization that describes its mission as educating the public and influencing public policy on pesticides. NCAP works to involve citizens in pest management decisions, focusing on prevention of pest problems and use of alternatives to pesticides.

NCAP is composed of individuals and groups from across the continent and around the world. Since 1977, it has worked to reduce pesticide use through policy reform and education about pesticide hazards and alternatives.

NCAP provides assistance in developing model policies to protect groundwater, food supply, and forest watersheds from pesticide contamination. It also provides model pest management policies for schoolgrounds, roadsides, and national forests, and information on hundreds of pesticides and alternatives for many pest problems.

NCAP also offers direct assistance and referrals for pesticide exposure victims and helps to organize assistance for citizens working for policy reform in their communities. The organization has approximately 1,400 members.

▲ Projects and People

According to NCAP, pesticides are broadly acting and often long-lasting poisons. Promoted as quick and easy solutions for pest problems, "they have been used casually and with little regard for the long term consequences. As a result, we now face an unprecedented environmental crisis."

NCAP states that toxic residues are found in human bodies, food supplies, and throughout the environment. Pesticides are used in agriculture, workplaces, parks, forests, schools, and many homes and gardens.

According to NCAP, in USA, 40 percent of the population depends on groundwater for their drinking water. The U.S. Environmental Protection Agency (EPA) has detected 46 pesticides in the groundwater of 23 states. At the same time, EPA's strategy for addressing pesticide pollution of groundwater places major responsibility for taking action with the states.

NCAP helps develop policies to protect groundwater, food supply, and forest watersheds from such contaminations. For example, in Oregon, NCAP took on the challenge of helping write, negotiate, and successfully lobby model legislation—the 1989 Oregon Groundwater Protection Act. The law is designed to both "prevent groundwater pollution from non-point sources [such as, applications of pesticides] and, where pollution is detected, involve local communities in solving groundwater problems before pollution reaches levels harmful to public health and the environment." The NCAP also worked at the implementation of the statute.

In addition to NCAP's successful effort to get a court injunction in 1984 that stopped all herbicide spraying by Northwest federal forest agencies, it has "convinced U.S. Forest Service managers to adopt a vegetation management policy that is based on sound, ecological forestry principles." The NCAP also continues to work for implementation of that policy and encourages the Bureau of Land Management, the other federal agency affected by the 1984 herbicide injunction, to adopt a similar policy.

NCAP recently supported the formation of the Northwest Action Center for Dioxin and Organochlorine Elimination to carry on this work, with staffer Mary O'Brien playing a major role—including researching pulp and paper technologies in Europe to learn about alternatives to paper chlorine-bleaching.

■ Resources

The NCAP provides a comprehensive information service on the hazards of pesticides and alternatives to their use. It maintains a computer-cataloged library of more than 5,000 articles, studies, books, government documents, videos, and other reference materials. It offers information packets and fact sheets and publishes a quarterly, *Journal of Pesticide Reform*. University of Oregon students, in a special work-study program, help answer questions to thousands of written requests for information.

The NCAP co-produced a documentary video, *Inert Alert: Secret Poisons in Pesticides.* Available from NCAP, the 17-minute video explains secret ingredients, human and environmental health problems that have resulted from their use, regulatory shortcomings, and ways for citizens to demand their right to know what toxic substances they are being exposed to when pesticides are used.

The NCAP requests information and reports about citizen efforts to stop pesticide use and promote alternatives.

Northwest Renewable Resources Center (NRRC)

1411 Fourth Avenue, Suite 1510
Seattle, WA 98101 USA

Contact: **Amy Solomon,** Executive Director
Phone: 1-206-623-7361 • *Fax:* 1-206-467-1640

Activities: Cross-Cultural Communications; Development; Dispute Resolution; Education; Facilitation; Mediation; Political/Legislative; Research • *Issues:* Biodiversity/Species Preservation; Deforestation; Fisheries; Forestry; Tribal Development; Tribal Relations; Natural Resource Use and Planning; Population Planning; Waste Management/Recycling; Sustainable Development; Water Quality

● Organization

The **Northwest Renewable Resources Center (NRRC)** is an organization that works to settle disputes arising over the use of resources. Founded in 1984, the Center now has a staff of five and a nationally respected reputation for effectiveness. Created by members of industry, Indian tribes, and environmental organizations, the Center provides assessment, mediation, facilitation, consultation, and training to help solve such resource disputes.

▲ Projects and People

The Center's projects are located throughout the Northwest. Two programs located in Idaho are the **Idaho Water Antidegradation Project** and the **Idaho Wildlife Depredation Project**. In the water antidegradation project, the Center accepted the role as mediator between mining, forest, and industrial representatives and environmental and tribal representatives. Despite the longstanding disagreements between the parties represented, an agreement was eventually reached. NRRC also served as a mediator in the depredation project, where it worked to reach consensus among farmers, environmentalists, and the Idaho Department of Fish and Game.

In Washington State, the Center facilitated 2 **water resources planning** retreats with more than 160 participants. Bringing together diverse people affected by this issue, a cooperative water planning agreement was reached.

NRRC has a number of projects working in conjunction with Native American tribes, including the **Tribes and Counties: Intergovernmental Cooperation** project and the **Indian Land Tenure and Economic Development** project. In the Tribes and Counties program, NRRC works to establish linkages between tribal governments and the counties within which the reservations are located. In the land tenure project, the Center works with individual tribes to help resolve the problems that have arisen through the unique status of Indian reservations.

One recent precedent-setting project resolved a conflict over prized coastal land use along Washington's Puget Sound between the Swinomish Tribal Community and the Skagit County Board of Commissioners, who signed a cooperative landmark agreement that recognized joint decisionmaking and the Native Americans' drive toward self-governance and a diverse economy.

Indian tribes are participants in many of the projects that are run by the Center. In the **Timber/Fish/Wildlife** project, negotiations over forest practice regulations included representatives from the timber industry, environmentalists, Indian tribes, and state agencies. The outcome means "better working relationships, better management of our resources, and a better way of life," according to Pam Crocker-Davis, Washington Environmental Council.

NRRC recently received a U.S. Environmental Protection Agency (EPA) grant to help facilitate **solid waste cooperation** between tribal and local governments on 38 reservations in Washington, Oregon, and Idaho.

In rural Alaska, NRRC is directing a three-year effort "to provide student and adult education and awareness of solid waste and home hazardous waste issues" for the EPA's regional offices. Four pilot villages "reflecting the state's geographic and cultural diversity" are being selected. **Frank L. Gaffney** is director of the Alaska program and has been project director of numerous other NRRC undertakings.

■ Resources

The Center is in need of computer support of all types, especially laptop computers as much of their work is off-site, and equipment which will increase their desktop publishing capabilities. Software and training would also be useful.

NRRC offers workshops, consultations, speaking programs, technical assistance, and presentations on cross-cultural communications. Publications include *Indian Land Tenure and Economic Development Project: Phase 1, Restoration of the Tribal Land Base, Living with Eagles: Status Report and Recommendations*, and a quarterly newsletter.

Oil, Chemical and Atomic Workers International Union (OCAW)
P.O. Box 2812
Denver, CO 80201 USA

Contact: **Anthony Mazzocchi,** Special Assistant to the President
Phone: 1-303-987-2229 • *Fax:* 1-303-987-1967

Activities: Collective Bargaining; Education; Political/Legislative • *Issues:* Energy; Global Warming; Health

▲ Projects and People

Concerned with health and safety issues for working people, **Anthony Mazzocchi** brought global warming to the attention of the Oil, Chemical and Atomic Workers International Union (OCAW).

In 1990, Mazzocchi spearheaded the **Global Warming Project**, a cooperative venture of OCAW and the Public Health Institute, which jointly publishes *Global Warming Watch* "to provide trade unionists and other interested individuals with a convenient way to keep abreast of the latest development . . . and its possible effects on working people around the world." Mazzocchi believes it is essential for labor to become part of this debate: "We cannot afford to be silent while corporate planners, economists, and politicians try to decide how they shall respond to the possibility of dramatic climate change. None of them will argue for our future interests unless we are prepared to do so."

The project's purpose is to provide a guide to this debate—particularly global warming effects on jobs, standard of living, and future economic development. It also provides "a unique perspective on climate change and its implications for human society" on behalf of working people, who, Mazzocchi points out, are generally omitted as topics of concern at such conferences.

"The interests of wage earners—who make up a growing proportion of Earth's population—need also to be taken into account," reports *Global Warming Watch*. "If we are to move from the present system of production which some say is threatening to turn the Earth into a hothouse, we need also to find a way both to preserve the high standard of living enjoyed today by the majority of working people in the industrial world and to extend it to the hundreds of millions of others who desire to emulate it."

The project views this debate as inseparable from "global industrial policy" and aims toward a global solution that includes analyses of economic and employment effects of alternative climate change scenarios. Mazzocchi believes this issue is "too important to be left to the scientists" who, along with politicians and corporate planners, "don't have all the answers." In addition to reporting on scientific community consensus, the project's publication is probing the issues for keener understanding and keys to solutions. It also illustrates the impact of legislation, such as the 1990 Clean Air Act, on jobs.

In a report from the first known labor-management conference on energy and the environment—co-sponsored in 1990 by the Energy and Chemical Workers Union (ECWU) and SaskPower private utility, Saskatoon, Saskatchewan, Canada—ECWU coordinator Russ Pratt urged ways to give trade unionists and community members more responsibility, such as setting up Union Environment Committees that are modeled on Union Safety and Health Committees to deal with environmental protection at the workplace; passing protective whistle-blower legislation for those reporting violations to proper authorities; selecting union or workforce representatives that are empowered to shut down operations that are environmentally unsound; and the right to refuse to perform environmentally unsafe work.

Unionists such as Pratt and Mazzocchi favor a Superfund for Workers, in place of defense expenditures, to help assure good jobs in industries that are not environmentally destructive as well as education to prepare for changing employment opportunities as a result of energy and environmental transitions.

Both past OCAW secretary-treasurer and president of OCAW Local 8-149 in New Jersey, Mazzocchi is noted for his negotiation skills worldwide. He helped bring into being the first dental insurance program in private industry; played a key role in the passage of the Occupational Safety and Health Act (OSHA); is visiting lecturer at Harvard University's School of Public Health; and holds awards from the American Public Health Association, Alice Hamilton organization, and *MS Magazine*, which in 1982 named him one of "40 Male Heroes of the Decade."

■ Resources

As described above, *Global Warming Watch* reports on climate change and the world economy from a trade union perspective and is obtainable through the Public Health Institute, New York City. Mazzocchi is publisher of the quarterly *New Solutions*, a journal of environmental and occupational health policy which fosters scholastic debate on issues intended to bridge the gap between labor and public health professionals, scientists, and the community. Interested journal contributors can submit articles to editor Dr. Charles Levenstein, Work Environment Program, University of Lowell, 1 University Avenue, Lowell, MA 01854 USA. OCAW publishes a *Resource Guide*.

Oklahoma Native Plant Society ❦
c/o Biology Department
Southeastern Oklahoma State University
Durant, OK 74701 USA

Contact: **Connie Taylor, Ph.D.**, Professor of Biology
Phone: 1-405-924-0121

Activities: Education; Political/Legislative; Sponsors Field Trips • *Issues:* Biodiversity/Species Preservation; Deforestation

● Organization

Founded in 1987, the **Oklahoma Native Plant Society** is a nonprofit organization whose purpose is to "encourage the study, protection, propagation, use and appreciation of the native plants of Oklahoma." With about 200 members, this organization is open to anyone interested in the flora of Oklahoma.

The goals of the Society include the beautification of Oklahoma by encouraging growth of wildflowers along roadsides and in other plantings throughout the state, efforts to encourage the preservation of rare and endangered species, and educational programs. The Society also sponsors field trips to various part of the state to give people the opportunity to view native plants and learn more about them.

The Society also runs the **Anne Long Fund**, which grants cash awards to groups and individuals doing work that furthers the Society's goals.

■ Resources

The Society publishes a newsletter which it distributes to members. They also have *An Annotated List of the Ferns, Fern Allies, Gymnosperms and Flowering Plants of Oklahoma; New, Rare and Infrequently Collected Plants in Oklahoma;* and *A Catalogue of Vascular Aquatic and Wetland Plants That Grow in Oklahoma.*

1000 Friends of Oregon
300 Willamette Building
534 SW Third Avenue
Portland, OR 97204 USA

Contact: **Henry R. Richmond**, Executive Director
Phone: 1-503-223-4396 • *Fax:* 1-503-223-0073

Activities: Education; Law Enforcement; Political/Legislative; Research • *Issues:* Growth Management; Land-Use Planning

● Organization

1000 Friends of Oregon is an independent watchdog of the nation's oldest land-use planning program in a state known for its natural beauty, economic stability, mountains and rugged coastland, "painted deserts and lush valleys, dense forests and rich farmland, small towns and big cities."

Enacted in 1973, this program focuses economic growth efficiently around urban centers, conserves important natural resources, and gives citizens a clear voice in the planning process. Two years later, 1000 Friends came into being to advocate protecting farms and forests from urban sprawl, to keeping housing at affordable prices, to protect coastal resources and wildlife habitats, and ensuring that transportation planning and land use are coordinated responsibly. Today, it has 3,600 members.

Accomplishments result from public/private interaction, such as designating 16 million acres of farmland for "exclusive farm use"; saving 9 million acres of private forest land from development; delineating urban growth boundaries for Oregon's 241 cities; speeding up the land-use appeals process; and increasing the supply of industrially zoned land by 79 percent in the state's 10 largest urban areas.

With a staff of 18 and a $1 million annual budget, 1000 Friends provides advice and referral services to citizens and organizations so they will be more effective in the planning process; reviews proposed changes to local land-use plans for consistency with state standards; and conducts and commissions research on current issues. "Staff attorneys initiate appeals when legal precedent is at stake," according to the nonprofit group. They also "recruit and train volunteer attorneys to participate in our Cooperating Attorneys Program (CAP)." Staff appear on radio and television, lead seminars, and produce educational materials. (See also Department of Land Conservation and Development [DLCD] in the USA Oregon State Government section.)

▲ Projects and People

Several model programs, developed as a result of 15 years' experience in land use, can have merit for other regions in the country that are attempting to reconcile conservation and growth issues.

The Nonindustrial Private Forestry Project (NIPF) focuses on improving management and incentives to "unlock" the value of immature timber. 1000 Friends is pushing for a bill that would create a market for small-scale owners to sell future cutting rights of immature timber.

A rural lands initiative intends to keep agriculture as the state's top economic industry. Broad efforts are underway to keep productive land under legislative protection.

A Grassroots Leadership Project (GLP) is training citizens and establishing strong networks of activists and county-level organizations to monitor land-use decisionmaking of the courts in targeted counties. The county organizations of the Council for the Protection of Rural England (CPRE), a 60-year-old group supporting its Town and Country Planning Act, are protypes for future efforts.

A Study of Managing Growth to Promote Affordable Housing examines the role of land-use planning in relation to moderate and low-income housing markets in urban areas. It will be a policy tool in preparing the state's growth management guidelines for Oregon's housing goals.

A National Demonstration Project known as LUTRAQ—for making the Land Use, Transportation, Air-Quality Connection—is looking at a 115-square-mile area of metropolitan Portland ultimately to see whether land-use changes can support transit alternatives that lessen dependence on automobiles, to determine how traffic congestion can be reduced and air pollution lessened, and to figure what the savings would be to taxpayers.

The National Growth Management Leadership Project (NGMLP) is a coalition in 17 states of conservation and growth management agencies which—during a five-year period—is building a resource network for exchanging information through conferences and educating members about the land-use reform movement in the USA.

Funding for these projects comes from members, foundations, business associations, and government agencies.

Attorney Henry R. Richmond has been executive director since the organization's start.

■ Resources

The 1000 Friends of Oregon Resource Guide lists available publications including newsletters, periodicals, studies, fact sheets, citizen-action guides, brochures, and membership forms. The group will also provide specific information about its projects and other program services upon request.

Oregon Natural Resources Council (ONRC) ❦
Yeon Building, Suite 1050
522 SW Fifth Avenue
Portland, Oregon 97204 USA

Contact: Andy Kerr, Conservation Director
Phone: 1-503-223-9001 • Fax: 1-503-223-9009

Activities: Education; Law Enforcement; Political/Legislative; Research • Issues: Air Quality/Emission Control; Biodiversity/Species Preservation; Deforestation; Water Quality

● Organization

The Oregon Natural Resources Council (ONRC), founded in 1972 as the Oregon Wilderness Coalition—and calling itself the "largest independent statewide conservation organization in the west"—has a mission "to protect Oregon's natural heritage through education, advocacy, and grassroots empowerment." The group's emphasis is on watershed integrity, wildlife and fisheries habitat, and wilderness, recreation, and regional economic diversity and stability.

With 6,000 members, a staff of 15, and a reputation as "a lightning rod" in preserving the remaining Northwest ancient forest ecosystems, ONRC is recognized as "consistently on the cutting edge of administrative, legal, grassroots organizing, educational and legislative strategies." As of 1991, ONRC said it won 41 percent of appeals of U.S. Forest Service and Bureau of Land Management ancient forest timber sales.

ONRC also worked to achieve listing of the northern spotted owl as a threatened species by the U.S. Fish and Wildlife Service; adding 11 rivers to Oregon's State Scenic Waterways System through an initiative campaign; establishing the 660,000-acre Hell's Canyon National Recreational Area; reduction of federal timber cutting levels closer to sustainable levels; passage of Oregon's 1984 Forest Wilderness Act, 1981 Riparian Area Protection Act, and 1979 Islands Wilderness Act, and enactment of the 1978 Endangered American Wilderness Act.

▲ Projects and People

With "less than 10 percent of Oregon's original ancient forests [remaining] unlogged" and so fragmented that habitat is inadequate for "hundreds of wildlife species," with beaches threatened by offshore drilling, with rangelands degraded by overgrazing, and with the decline of wild salmon runs in the Columbia basin by 97 percent since the mid-1800s, ONRC says its future work is evident.

One ongoing project, as Northwest field coordinator **Regna Merritt** reports, aims to stop a 3.2-mile stretch of "**Forest Service road 46**" that would bisect part of an 1,100-acre ancient forest grove and fill in 5 wetland areas which are potential habitat for **Cope's giant salamander** and the actual habitat for the red-legged frog. Also destroyed through logging would be prehistoric Native American cultural resources including cedar trees peeled by the Clackamas and Mollala Indians—used centuries ago for basket weaving. The government endeavor would muddy the waters of the Clackamas River where the salmon spawn, and which is also the principal drinking water source for the Clackamas Water District, asserts ONRC. Within the river corridor, deer and Roosevelt elk range in the winter and could be displaced as a result of building activity. The Forest Service responds that "widening Route 46 is essential reconstruction."

In southern Oregon, ONRC reports the decline of the endangered **Lost River and short-nosed sucker fish** in the dwindling Klamath basin wetland ecosystem apparently worsening by "damming of waterways, draining of marshes, and excessive irrigation withdrawals." In 1991, ONRC brought a lawsuit against the U.S. Bureau of Reclamation to require "the screening of the irrigation withdrawal and a halt to any operation of the Klamath Reclamation Project which endangers the fish." Believed to be an indicator species for the local marsh ecosystem, these sucker fish were once an important protein source for Native Americans in the region; under suitable conditions, Lost River suckers can grow to 30 inches and live more than 44 years. ONRC also says that the Upper Klamath Lake basin is dying from "massive amounts of nitrogen and phosphorous from fertilizers and cattle waste" as well as from excessive algal growth.

ONRC summarizes its interpretation of the intent of the Forest Service's first chief forester, Gifford Pinchot, who at the beginning of this century "developed the idea that national forest managers should harvest only the amount of timber which the forests produced new each year—managers were to retain the volume of standing timber in the forests as forest 'capital' for the future."

According to ONRC, "the Forest Service eventually developed a series of scientific innovations designed to increase forest production and the annual harvest. First, mixed species forests were replaced with monoculture plantations of Douglas fir or whatever was the most commerically valuable local tree species. Additionally, the Forest Service began applying pesticides and fertilizers and thinning forests to increase production. Finally, the agency began clearcutting old growth forests and replacing them with young, fast-growing stands. In their zeal to grow wood, they forgot to leave the forests."

Its handling of successful litigation, legislation, and timber sale appeals enables ONRC to set an example for similar activist groups in the USA. Regarding deforestation and other environmental problems in developing countries, ONRC writes, "We have a responsibility to help these nations through these ecological crises. Environmentalists can [educate] themselves on the struggles facing other peoples and countries, shifting [the] focus beyond the immediate problems close to home."

In *Time* magazine (June 25, 1990), Andy Kerr was described as "the Ralph Nader of the old-growth preservation movement," saying "he has spearheaded a guerrilla campaign in the courts, Congress, and the media to drive the old-growth timber industry out of business."

■ Resources

ONRC publishes *Wild Oregon*, a quarterly publication which provides updates in environmental issues and action alerts. Educational and recreational books are available on ancient forests, deserts, and hiking, as is the recently produced *A Walking Guide to Oregon's Ancient Forests* by **Wendell Wood**, which gives directions to more than 200 groves. A Natural Resources Conference is held annually at a wilderness site; an Oregon Coast and Ocean Conference is also sponsored yearly. Workshops are frequently conducted for members and other environmentalists. ONRC is in need of educational material on old-growth forests for high school and elementary school students.

Ouachita Watch League (OWL) ❦
Route 9, Box 95
Mena, AR 71953 USA

Contact: **Sherry Balkenhol**, Chair
Phone: 1-501-394-6593

Issues: Air Quality/Emission Control; Biodiversity/Species Preservation; Deforestation; Sustainable Development; Water Quality

● Organization

The Ouachita Watch League (OWL) consists of locally represented conservation, sporting, and outdoors groups, national forest inholders, horseback riders, environmentally sound logging operators, tourism and recreation organizations, and similar groups who share a long-range vision for responsible, true multiple-use management for the Ouachita National Forest.

OWL is made up of a coalition of 65 member organizations and 400 unaffiliated individual members. OWL has no paid staff; all activities are carried on by volunteers.

According to OWL's mission statement, their goals are to achieve responsible, sustainable, environmentally sensitive handling of public lands in the Forest. Consistent with the federal law requiring equal consideration for values other than timber such as recreation, wildlife, and watershed protection, OWL seeks multiple-use management. It believes that public land managers should give priority to providing benefits not supplied on private land.

To achieve these goals, OWL supports selection timber management, not clearcutting; native diversity, not pine farms; habitat for all wildlife; special protection for special areas; fewer roads, quality trails; a poison-free forest; and special protection for watersheds. Its literature points out that 8,000 miles of roads already criss-cross the Ouachita; more or rebuilt roads are expected within the next decade.

▲ Projects and People

OWL has a program, **Eyes of the Forest**, in which trained volunteers involve themselves in district decisionmaking by attending U.S. Forest Service scoping meetings, reviewing decisions, inspecting timber stands, making recommendations, and finally, if necessary, appealing individual timber sales. OWL trains volunteers who want to participate in this program.

In the summer of 1991, OWL, with the Sierra Club in Oklahoma, launched a campaign appealing to the U.S. Congress to end all forms of even-age management in the Ouachita—which OWL regards as pine farming. Conservationists and other forest lovers were asked to send hardwood tree leaves to their Senators and Representatives as reminders that the hardwoods need to be protected from clearcuts and herbicide use. According to OWL, the campaign has had a very positive effect.

OWL states that the campaign's purpose is to have congressional legislation passed that would allow single-tree selection logging as the only type of timber harvesting in the entire national forest. Since ending even-age management means no more pine plantations, native diversity of hardwoods and pines would be easily reached. According to OWL, no clearcutting also means no more dangerous chemical herbicides going into the soil and water. Presently, no national forest is managed entirely in such a manner.

In addition, OWL has been an intervenor in a Sierra Club lawsuit against the U.S. Forest Service. OWL also conducts legislative activities that include lobbying of congressional delegations, letter-writing campaigns, and support for forest reform legislation. OWL members sit on key Ouachita National Forest advisory committees.

Also at the grassroots level, OWL holds an annual weekend Fall Fest at a campgrounds where hundreds gather to "talk, listen, plan, and enjoy a fall forest together."

■ Resources

OWL needs help in both "influencing Congress for forest legislation" and opposing public relations efforts of the U.S. Forest Service and industry. According to OWL, details of issues are complex and hard for the public to grasp. OWL also requests legal and financial assistance for expenses of appeals and lawsuits.

Information is available on organizational and individual memberships, which include subscriptions to the *Hooter* quarterly newsletter. Speakers are available for a donation.

Environment and Development Program ⊕
Overseas Development Council (ODC)
1875 Connecticut Avenue, NW, Suite 1012
Washington, DC 20009 USA

Contact: **Patti L. Petesch**, Associate, Poverty and Environment Program
Phone: 1-202-234-8701 • *Fax:* 1-202-745-0067

Activities: Research • *Issues:* Biodiversity/Species Preservation; Deforestation; Energy; Global Warming; Sustainable Development

● Organization

The **Overseas Development Council (ODC)**, established in 1969, with its staff of 25 people, focuses on U.S. relations with developing countries in a post–Cold War era. ODC serves as a center for policy analysis and a resource for education, and provides opportunities for an exchange of ideas on development issues. Major concerns also include "international finance and easing the debt crisis," international trade, and environment and development.

▲ Projects and People

The Environment and Development program is divided into two projects, both run by Stuart Tucker and Patti Petesch. Seeing environmental degradation and rural poverty as continuing threats to regional peace and economic recovery, the Poverty, Natural Resources, and Public Policy in Central America project is working to develop strategies to combat these problems. ODC is working with experts in the USA and experts from Central America in this project, and is creating strategies that will be beneficial for both the poor and the environment.

The **Environmental Challenges to International Trade Policy** program "explores the linkage between the two issues and provides proposals to promote more environmentally sustainable development policies within the context of open trade." ODC sponsored a joint conference with the World Wildlife Fund to discuss the relationships between trade and the environment. The conference brought together "scientists, economists, policymakers, academics, industry leaders, activists, and many others." Topics of the conference included the "greening" of trade policy and the environmental impact of liberalized agricultural trade.

Global warming is a critical concern of ODC, which reports that this predicament "may seriously affect many developing countries by causing coastal flooding and large-scale population displacement in countries such as Bangladesh, India, China, Indonesia, and Egypt; devastation of some low-lying small island states; and serious disruption of agriculture and loss of energy, water, and wood in some nations." Global warming is a subject of one of its *Policy Focus* publications.

■ Resources

ODC has a range of publications which include the *Policy Focus Series*, 8- to 12-page papers providing background information and analysis on current USA–Third World issues. ODC also publishes several books and policy essays; a list can be obtained from the organization.

Pacific Energy and Resources Center (PERC) ⊕
Building 1055 Fort Cronkhite
Sausalito, CA 94965 USA

Contact: **Armin Rosencranz, Ph.D.**, President
Phone: 1-415-332-8200 • *Fax:* 1-415-331-2722
E-Mail: igc-perc

Activities: Education; Research • *Issues:* Air Quality/Emission Control; Biodiversity/Species Preservation; Deforestation; Energy; Global Warming; Ocean Pollution; Sustainable Development

● Organization

Started in 1987, the **Pacific Energy and Resources Center (PERC)** is a nonprofit organization with a staff of 13. Under the leadership of its president, Dr. Armin Rosencranz, PERC is involved with policy research and education about international resource management. PERC also serves as an umbrella organization for 10 nonprofit environmental groups.

▲ Projects and People

Environmental policy research is an important aspect of PERC's work. Current research includes a study of chlorofluorocarbon (CFC) usage in India and China and environmental law and policy in the former USSR and India. PERC also researches "international legal and regulatory aspects of global warming and acid rain."

Lisa Jacobson, education director, runs the environmental education program, which has developed and implemented interactive programs for elementary school children. In 1990, approximately 900 children participated in programs dealing with biodiversity, deforestation, global warming, and acid rain. PERC is also involved with teacher training on global environmental issues.

The Visiting Soviet Fellows Program has allowed Soviet environmental lawyers to meet with various San Francisco Bay Area environmental groups.

An international environmental law expert, Dr. Rosencranz has written on the Bhopal chemical plant disaster, deforestation in Siberia and India, and other global environmental issues. He has also co-authored books on *Acid Rain in Europe and North America*, *American Government*, and *Congress and the Public Trust*.

A conservation and resource studies lecturer at the University of California, Berkeley, Dr. Rosencranz has a 1992 Fulbright professorship to teach environmental law at the National University Law School, Bangalore, India; as a Fulbright senior scholar in 1987, he lectured on international and comparative environmental law in Australia. Previously, he was also a Fulbright professor at the Universities of Cochin and Poona, India.

■ Resources

PERC is in search of CFC substitutes for transfer to India and China. They also are hoping for the political support necessary to resist USA-Commonwealth joint ventures that could exploit the natural resources of Siberia.

PERC publishes a global environmental education set and a tropical deforestation teacher training set.

Pacific Institute for Studies in Development, Environment, and Security ⊕
1204 Preservation Park Way
Oakland, CA 94612 USA

Contact: Peter H. Gleick, Ph.D., Director, Global
Environment Program
Phone: 1-510-251-1600 • *Fax:* 1-510-251-2203
E-Mail: Econet: pacinst

Activities: Education; Research • *Issues:* Air Quality/Emission Control; Energy; Global Warming; Sustainable Development; Water Quality; Water Resources

● Organization

"The Earth is one but the world is not." So begins descriptive literature from the Pacific Institute, which sees the problems of environmental degradation, regional and global poverty, and political tension and conflict as interrelated, and believes long-term solutions to these problems will only be reached in an interdisciplinary manner. The Institute seeks to investigate and analyze the links between international security, the environment, and economic development. It also plans to facilitate communication among individuals and groups working in these policy areas, and to educate policymakers and the public about these problems.

▲ Projects and People

The Program on Sustainable Development takes varied approaches to developing different models of growth that are both ecologically and socially sustainable. Drs. Michael F. Maniates and Ronnie D. Lipschutz are working on a project entitled Social Organization and Sustainable Development: An Educational Challenge to create a college-level curriculum focusing on the "importance of organizational structure in an effort to foster sustainable development." A test implementation of this curriculum was planned in 1991 at the University of California, Santa Cruz, and Radford University, Virginia.

Working on another project in this program, Pablo Gutman focuses on Poverty Alleviation in Third World Countries Through Environmentally Sustainable Development. This project is reviewing the main approaches to poverty alleviation as well as evaluating several development programs.

The Program on the Global Environment focuses on the linkages between global environmental issues, changing agricultural production, growing populations, economic mismanagement, and international security issues. Dr. Peter H. Gleick, as director, is involved in most projects, including working with Linda Nash on Climactic Change, International Rivers, and International Security, and with Edwin P. Maurer on Costs of Adaptation to Climactic Change: Sea Level Rise. Gleick is also involved in research on Global Water Resources.

The Institute's third interest area is the Program on Global Security, which examines the connections between global security problems and environmental issues, and looks at the potential for security problems arising from sustainable development and environmental degradation. Lipschutz and Nash are working on a project entitled Sustainable Resource Management and Global Security.

In addition, the Institute is collaborating with the Transnational Foundation for Peace and Future Research in Sweden and the Institute of the World Economy and International Relations of the Soviet Academy of Sciences. Their joint project is preparing information on relationships between economics, security, values, and the environment.

■ Resources

The Institute has a large number of publications including papers, testimony, articles, books, and chapters written by its associates. A publications list is available on request.

Pinchot Institute for Conservation
6118 Hibbling Avenue
Springfield, VA 22150 USA

Contact: **James W. Giltmier**, Executive Director
Phone: 1-703-912-9535 • *Fax:* 1-703-912-9531

Activities: Education • *Issues:* Biodiversity/Species Preservation; Deforestation; Energy; Global Warming; Sustainable Development

● Organization

Gifford Pinchot, the Pinchot Institute for Conservation's name-sake, was the chief of the U.S. Forest Service 100 years ago and one of the first people to be a self-described "conservationist." Pinchot's forest policies on land management and ethics are still used today by the Forest Service. The Pinchot Institute is a nonprofit organization which aims to enhance the public's knowledge on the major issues of conservation. To this end, the Institute holds many conferences, workshops, and many other educational forums at **Grey Towers**, the old family residence of Gifford Pinchot, a national landmark in Pennsylvania.

The Pinchot Institute educates on the issues of urban development, land ethics, leadership in the conservation movement, environmental policy, and many other philosophical topics concerning the coexistence of humankind and the natural environment and land ethics and stewardship.

The Institute is dedicated to promoting good stewardship and land ethics. The basis for this principle is that "in caring for the land, the steward provides for the community—more than just for himself." This caring for the greater good, James W. Giltmier writes, leads the way to what Greek philosophers called a "decent life."

▲ Projects and People

The Institute, aside from its many workshops and seminars, offers a fellowship program, a liaison with leading corporations, and a scholar-in-residence program. The Institute plans to have an awards program for national journalists who have distinguished themselves in reporting. Future, more "ecumenical," seminars are planned which will explore the link between religion and nature.

The Pinchot Institute also runs and maintains a publishing arm called the **Grey Towers Press**. In its literary pursuits, the Grey Towers Press has donated the cumulative writings of Gifford Pinchot to the Library of Congress, which the Institute describes as the biggest personal collection donated to the library. These works are kept in excellent condition and have a catalog, making them more accessible to scholars who wish to utilize them.

In addition to hosting seminars, leaders of the Pinchot Institute also speak at national conferences on Institute concerns.

■ Resources

The Pinchot Institute offers a lecture series and a theatrical presentation about Gifford Pinchot which explains the namesake's philosophical and historical positions.

A members' newsletter, *The Conservation Legacy*, includes letters, speeches, and excerpts or full transcriptions of speeches which are topical to the Institute.

Planning and Conservation League (PCL) ❦
909 12th Street, Suite 203
Sacramento, CA 95814 USA

Contact: **Gerald H. Meral**, Executive Director
Phone: 1-916-444-8726 • *Fax:* 1-916-448-1789

Activities: Education; Political/Legislative; Research • *Issues:* Air Quality/Emission Control; Biodiversity/Species Preservation; Deforestation; Energy; Global Warming; Sustainable Development; Transportation; Waste Management/Recycling; Water Quality

● Organization

"There is only one California. Ours." So reminds the **Planning and Conservation League (PCL)**, a nonprofit organization dedicated to the education of the public and the passage of sound environmental laws and reforms in this state for over 25 years. PCL encourages Californians to participate in environmental protection through its handbooks, brochures, and pamphlets on topics such as conservation and petition writing. It also conducts research to help citizens find efficient, economical, and environmentally safe solutions to difficulties.

▲ Projects and People

Utilizing the initiative process, PCL draws up new propositions to be brought before the California State Legislature, builds coalitions, and lobbies elected officials to urge passage. PCL has successfully originated many bills and referendums including the **Rail Bond Act Initiative that launched a new mass transit system** that would reduce smog, conserve energy, and help reduce urban sprawl. PCL continues to work for implementation by keeping $1 billion rail bond acts on the state ballots in 1994.

With PCL's efforts, **toxic waste dumping remains a misdemeanor**, instead of being reduced to an "infraction" for which offenders pay small fines and serve no jail time. Regarding pesticides, PCL believed it made strides on behalf of public health when their regulation was moved from the California Department of Food and Agriculture to the new California Environmental Protection Agency (Cal-EPA). "The success of Cal-EPA will be measured by whether dangerous pesticides are phased out, industries reduce their use of toxic materials, and environmental protection laws are better enforced," reports PCL.

Two earlier initiatives proposed by PCL's **environmental campaigns program** that won statewide voter approval were the **California Wildlife, Coastal, and Parklands Initiative**, which protects and restores fish and wildlife habitat including more than "200,000 acres of potential parks and open space"; and the **Tobacco Tax and Health Protection Initiative**, which adds a 25-cent tax to a cigarette pack to benefit certain environmental and health programs. PCL also instructs developers on how to observe the landmark **California Environmental Quality Act (CEQA)**.

The **Planning and Conservation League Foundation** funds PCL, which is a member of the Environmental Federation of California.

■ Resources

PCL publications are obtainable for a small fee; a list is available. PCL also publishes a newsletter called *California Today*.

Pocono Environmental Education Center (PEEC)

RD 2, Box 1010

Dingmans Ferry, PA 18328 USA

Contact: **John J. Padalino, Ph.D.,** President
Phone: 1-717-828-2319 • *Fax:* 1-717-828-9695

Activities: Education • *Issues:* Biodiversity/Species Preservation; Energy; Global Warming; Transportation; Waste Management/Recycling; Water Quality

● Organization

The **Pocono Environmental Education Center** (PEEC), in cooperation with the **National Park Service**, describes itself as "the largest residential center in the Western Hemisphere for education about the environment." Converted in 1972 from a Pocono honeymoon resort, it is located 20 miles southwest of the meeting point of New York, New Jersey, and Pennsylvania in the **Delaware Water Gap National Recreation Area**. With PEEC's 38-acre campus and residential facilities plus 200,000 acres of public lands, there are 39 areas that The Nature Conservancy has designated suitable for preservation and study.

Such areas include the Delaware River and its banks, lowland and upland forests, scenic gorges, fields, a quarry with Devonian fossils, scrub oak barrens, ravines, talus slopes, reservoirs, upland lakes, streams, ponds, and acid bogs. In winter one can cross-country ski and see frozen waterfalls, an old Scottish game of curling on the ice, fresh animal tracks in the snow, and bald eagles. Summer provides an opportunity for canoeing, hiking, and exploring, and the fall and spring have extremely interesting natural wonders as well.

PEEC's mission is to inspire individuals, minorities, people with special needs, and social communities to better understand the complexity of natural and human-designed environments. PEEC offers teacher-education workshops, accredited graduate and undergraduate workshops, youth-at-risk workshops, and formal education programs in kindergarten through grade 12 schools. Working with scout, church, social, and special-interest groups, PEEC hosts conferences and retreats for professional organizations as well. Named Pennsylvania's "outstanding conservation organization," it is a charter member of the Alliance for Environmental Education.

▲ Projects and People

PEEC offers **workshops** in photography, environmental issues such as water quality monitoring, nature study, art and nature, science, birdwatching, and recreation; and family vacation camps with a combination of activities during holiday weekends and several weeklong vacations. The National Audubon Society helps sponsor an **annual bird count** in December and January to count the number of **bald eagles** in the area. Nature studies might include acid rain, water quality monitoring, weather monitoring, or beaver studies. Groups holding **conferences** include the American Nature Study Society and Conservation Education Association.

■ Resources

PEEC produces *PEEC SEASONS*, a quarterly newsletter containing natural history articles, poetry, educational resources, and a schedule of upcoming workshops. A bookstore is on the premises.

Elderhostel programs occur periodically. PEEC offers internships in environmental and outdoor education; scholarships are awarded to educators for certain workshops.

The Population Council ⊕

One Dag Hammarskjold Plaza

New York, NY 10017 USA

Contact: **Sandra Waldman,** Manager, Public Information
Phone: 1-212-339-0500 • *Fax:* 1-212-755-6052

Activities: Research • *Issues:* Health; Population Planning

● Organization

The **Population Council**, with a staff of some 300 persons from 62 countries, applies science and technology to the solution of population problems in 48 developing nations. Established in 1952, the Council's headquarters are in New York, and regional officers are in Bangkok, Cairo, Dakar, Mexico City, Nairobi, and nine more country offices.

▲ Projects and People

Collaborating with governments, non-governmental organizations, and private foundations in developing countries, the Council's **Programs and Research Divisions** work in social and health sciences on population policy, family planning, and fertility, reproductive health and child survival, women's roles and status, contraceptive introduction, and development of human and institutional resources.

The issue of **reproductive health and child survival** deals with the reduction and treatment of unsafe abortions, the prevention of sexually transmitted diseases including AIDS, and new approaches to postpartum care, including integration of breastfeeding with family planning strategies.

The Council's **Center for Biomedical Research**, based at Rockefeller University, investigates and develops **contraceptive technology** and studies male reproductive physiology. The Council developed two reversible long-acting contraceptive methods now in use worldwide—NORPLANT™ subdermal levonorgestrel implants and the copperbearing T-shaped IUDs. This achievement "justified the enormous faith of the governments, United Nations agencies, foundations, and individuals who helped fund our contraceptive development and introduction efforts" during the past two decades, Council president **George Zeidenstein** stated in his *Annual Report.* Among other research, new methods near final development include a hormone-releasing IUD, NORPLANT-2™ (using two instead of six implants), and contraceptive vaginal rings. With molecular and cellular biology techniques, the Center is advancing research in male contraceptives that block sperm production without reducing libido.

Researchers are furthering **AIDS knowledge** in several laboratories as they examine blood cells "known as lymphocytes and macrophages that can harbor the virus [and] are normally present in semen, breast milk, and blood." Collaboration between the National Institutes of Health (NIH) and the U.S. Agency for International Development (USAID) is fostering studies regarding bisexuality, for example, in about nine countries. The Council's AIDS prevention work is carried out through the **Robert H. Ebert Program on Critical Issues in Reproductive Health and Population.** Working in this program is physician **Dr. Christopher Elias,** who reports, "The Council's particular strength is in devising practical interventions . . . to prevent the transmission of AIDS."

What value do family planning programs have in reducing population? A demographic study conducted by **Dr. John Bongaarts,** Council vice president and Research Division director, with colleagues **W. Parker Mauldin** and **James E. Phillips,** concluded that the world's population "is now 412 million smaller than would have been the case without organized family planning programs"—upholding the Council's confidence in encouraging the availability of contraceptives and the encouraging of smaller families.

The **Research Division** is a small interdisciplinary staff of demographers, economists, sociologists, and anthropologists particularly concerned with population policy in China with 1.1 billion people, Kenya, Bangladesh, India, Pakistan, Nigeria, and the Sahel region of Africa. One pilot project, being conducted by **Cynthia Lloyd,** is on the consequences of high fertility. According to the Council, "fertility has declined more rapidly in Latin America than in any other continent." With the Institute for Resource Development, the Council was involved in 25 studies and analyses of **Demographic and Health Surveys (DHS)** in 14 Latin American, African, and Asian countries—reports of which are "collated in a series of monographs and distributed to universities, international agencies and foundations, research and training centers, and family service agencies." Investigations into child survival as well as into women's economic and family roles and status are other major undertakings of this division that correspond with the **Programs Division.**

The Council offers **workshops, briefings, and seminars** which are attended by professionals and students from developing countries and the United States. A **Distinguished Scholar** research agenda focuses on "economic and environmental consequences of rapid population growth." Fellowships are provided in social and health sciences to scholars from 90 countries as an integral part of the program in support of human and institutional resources. The Council gives research awards, and develops study groups and workshops to strengthen institutional capabilities in developing countries.

More than half the Council's funds come from governments and United Nations agencies. Important contributions come from nongovernmental organizations, foundations, and individuals, with a total budget exceeding $40 million.

McGeorge Bundy is board of trustees chairperson. **Robert H. Ebert** is chairperson emeritus.

■ Resources

The Council publishes two scholarly journals: the quarterly *Population and Development Review,* edited by **Paul Demeny,** which deals with the interrelationships between population and socioeconomic development; and *Studies in Family Planning,* a bimonthly journal concerned with all aspects of fertility regulation. They also produce working papers, pamphlets, newsletters, software, and books on population and fertility studies. One pamphlet series, *SEEDS,* documents innovative projects developed by and for low-income women that increased women's productivity and income. Another pamphlet series, *Quality/Calidad/Qualité,* shows efforts to improve family planning and related reproductive health services. A list of pub-

lications is available. The Office of Communications provides contraceptive information to other like groups, the media, and the general public.

Population Crisis Committee (PCC)
1120 19th Street, NW
Washington, DC 20036-3506 USA

Contact: **Robert Engelman,** Director, Population and Environment Program
Phone: 1-202-659-1833 • *Fax:* 1-202-293-1795

Activities: Education; Political/Legislative; Research • *Issues:* Family Planning; Population Planning; Sustainable Development; Women's Health • *Meetings:* Population and environment; UNCED

● Organization

The **Population Crisis Committee (PCC),** more than a quarter-century old, describes itself as a nonprofit organization with a staff of 38, one-fourth of whom are volunteers. Receiving no government funding, PCC is "directed by leading Americans from the fields of science, business, diplomacy, politics, and medicine who believe that rapid population growth is one of the most urgent problems facing the world community." In January 1993, PCC's name changes to Population Action International.

PCC is a source of world population information for legislators and policymakers. It builds coalitions with health, foreign policy, and other like-minded groups; encourages family-planning programs in more than 30 countries; and stimulates media coverage in the USA and abroad. PCC reminds that nearly a decade ago, the USA "withdrew its support for the United Nations Population Fund and the International Planned Parenthood Federation, which operate in over 130 countries."

▲ Projects and People

With director **Robert Engelman** and policy analyst **Stephanie L. Koontz,** PCC's new **Population and Environment Program** produced in 1992 a **Population Policy Information Kit** regarding such impacts in the developing world and advocating "universal access to high-quality, voluntary family planning services to . . . ultimately stabilize world population." An 11-minute videotape on this subject, narrated by journalist **Walter Cronkite,** will be available for distribution. Engelman, a founding member and secretary of the Society of Environmental Journalists, is also producing a reporter's guide to population issues as well as problem-specific reports. **Dr. Sharon L. Camp,** PCC senior vice president, speaks and writes on these issues.

■ Resources

Environmental publications include *Cities: Life in the World's 100 Largest Metro Areas* and a *Statistical Appendix* in disk or booklet format; *Why Population Matters: A Handbook for the Environmental Activist; Population Pressure, Poverty, and the Environment;* and an information sheet on *Population and the Environment: Impacts in the Developing World.* An order form is available. PCC seeks scientific databases to scan population/environment literature.

Population Reference Bureau, Inc. (PRB) ⊕
1875 Connecticut Avenue, NW, Suite 520
Washington, DC 20005 USA

Contact: **Barbara Torrey**, President
Phone: 1-202-483-1100 • *Fax:* 1-202-328-3937

Activities: Education; Policy Analysis • *Issues:* Population Planning; Population/Environment Connections; Sustainable Development

● Organization
The **Population Reference Bureau (PRB)** collects, interprets, and disseminates information on population trends and their public-policy implications. Its audiences include legislators and public officials, teachers and students, journalists, business leaders, and individuals concerned about world affairs. A nonadvocacy organization established in 1929, PRB believes that the facts on population speak best for themselves.

▲ Projects and People
The **International Program** provides technical support to organizations working in developing countries to communicate critical population information to policymakers in the Third World. Its previous five-year experimental program, **IMPACT** (Innovative Materials for Population Action), received funding from the Office of Population, U.S. Agency for International Development (USAID), for the production of materials on, for example, contraceptive safety distributed worldwide with sensitivity to various cultures.

To reach a wide audience, PRB collaborated with agencies such as the Inter-American Parliamentary Group on Population and Development: Institute for International Studies in Natural Family Planning, Georgetown University; PROFAMILIA, the International Planned Parenthood affiliate in Colombia; Demographic and Health Surveys project; Sahel Institute's Center for Applied Research on Population and Development in Mali; Tata Steel Family Welfare Programme in India; and the Zimbabwe National Family Planning Council.

IMPACT disseminated equipment, training, and technical assistance through established institutions in developing countries. PRB reaffirms, "No matter how important or well-presented information may be, it is better accepted when it comes from authorities who are known and trusted, rather than from 'outsiders.'" A "successful collaboration with colleagues at the University of Ghana has become a model for policy communications activities in Africa," according to PRB. Now IMPACT is the foundation of PRB's reorganized International Program as it prepares for "challenging global issues of the 1990s," including the United Nations Conference on Environment and Development (UNCED).

The **Policy Studies** program provides demographic information, analysis, and training in aspects of domestic social policy issues encompassing such concerns as ethnic diversity, economic well-being, and education.

Global Edition, which got its start under IMPACT, supports a network of Third World journalists reporting on the complex relationships between population and the environment. This project inspired **Connections**, which produces classroom materials, including 27 lesson plans for teaching students about population-environment links, such as deforestation, climate change, pollution, deser-tification, urbanization, farming, and food scarcities. These materials are being adapted for elementary level.

The **World Environment Data Sheet** lists statistical measures of environmental quality for each country and major region of the world. Other recent publications include sustainable development, environmental degradation, and resource management.

■ Resources
The **Publications Program** produces material using format and language suitable for nontechnical audiences, including booklets, wall charts, briefing materials, and a monthly newsletter, *Population Today*, published since 1945. The in-depth *Population Bulletin* is issued quarterly; topics range from America's children to Germany's uncertain future, to the global family planning movement. *World Youth and Environment Data Sheets* are publications. Series include *Population Trends and Public Policy* and *Population Handbook*. Order forms for publications and IMPACT services are available.

Population Resource Center ⊕
15 Roszel Road
Princeton, NJ 08540 USA

Contacts: **Jane S. De Lung**, President
Jeffrey N. Jordan, International Program Officer
Phone: 1-609-452-2822 • *Fax:* 1-609-452-0010

Activities: Research • *Issues:* Health; Population/Environment Interlinkages; Population Planning; Sustainable Development
Meetings: DHS Conference; Ditchley Conference; UN Population

● Organization
Java's population grew from 25 million to 95 million in this century, forcing landless families to clear forests for cropland. As a result, one million hectares (one hectare equals 2.74 acres) were depleted so that they can no longer support even basic agriculture. Though there are many unknowns in the population-environment equation, world population is estimated to double in the next hundred years with disastrous consequences for the environment.

The **Population Resource Center** shares such population statistics with policymakers in local, state, national, and international briefings. Its purpose is to bridge the gap between the demographic research community and policymakers in public and private sectors. "The Center is committed to improving policymaking through the provision of balanced, objective analyses," according to its literature. "The Center does not endorse particular policy outcomes; views expressed by experts participating in Center programs are their own."

Briefing topics focus on unmet needs for global family planning, urban poor, health consequences of women and children in the developing world in regards to contraception and reproduction, and the impact of immigration on today's jobs and tomorrow's workplace, among others. Briefings are organized for groups such as the World Bank, World Wildlife Fund, members of the U.S. Congress and congressional staff, and the U.S. Agency for International Development (USAID).

Founded in 1975, the Population Resource Center has 2,500 members and offices in San Francisco, California; Washington, DC; and Princeton, New Jersey. The international program focuses on Zambia and Mexico, with six briefings in Zambia during one year.

▲ Projects and People

The Population Resource Center puts together conferences such as America in the 21st Century and Asking the Right Questions: The Baby Boomers in Retirement.

The Center initiated a project with the African Development Bank to assist in the organization of a conference on Africa in the 21st Century. Jeffrey N. Jordan, the international program officer, works directly with Jane S. De Lung to support the program in Zambia as well as work in Latin America, Africa, and Asia. In this way, the Center calls attention to the United Nations statistic that population will reach 6 billion at the end of the next century with 90 percent of these people to be born into the developing world—such as in India, Nepal, and Bangladesh—where "the natural resource base is most fragile."

The Center brings forth data such as the loss of "two-thirds of the original habitat of species in Sub-Saharan Africa and Southeast Asia" through deforestation; Food and Agriculture Organization (FAO) estimates on the rate of fuelwood that is cut faster than it grows back and still cannot meet the needs of "125 million people in 23 developing countries"; the degradation of "60 percent of the world's rangelands"; and appearance of new deserts "at the rate of 6 million hectares a year." Meanwhile, UN sources say the industrialized countries "consume 75 percent of commercial fuels produced worldwide, 71 percent of the world's steel, 48 percent of the world's grain production, and 85 percent of the world's forest products. In total, approximately four-fifths of world trade goes to a quarter of the world's population."

■ Resources

The Center publishes an *Executive Summary* on topics such as population-environment linkages.

Programme for Belize (PFB) ⊕

P.O. Box 1088
Vineyard Haven, MA 02568 USA

Contact: A. Joy Grant, Managing Director
Phone: 1-508-693-0856

Activities: Development; Education; Political/Legislative; Research • *Issues:* Biodiversity/Species Preservation; Deforestation; Sustainable Development

● Organization

The Programme for Belize (PFB) is a nonprofit Belizean corporation which, in accordance with an agreement with the Belizean national government, buys, receives, and holds land in a trust for the Belizean people. This is an effort to conserve the country's vast natural resources. The PFB also acts as "a bridge between the private and public sectors," which both wish to protect the acres of rainforest and miles of rivers holding a great diversity of wildlife.

According to the PFB, Belize is largely covered with forests overflowing with hundreds of species of birds, animals, and plant life in a country the size of New Hampshire. With rainforests declining worldwide and "species extinction now 10,000 times greater than in normal geological history," it is expected that only a quarter of such rainforests will remain by the year 2000. Consequently, PFB is determined to preserve Belizean rainforests and their fauna and flora.

The PFB operates a land trust system which, with the help of the national government, the Belize Central Bank, private land holders, and major corporations (Coca Cola Foods, for example) acquires much land to be reserved for the people of Belize.

▲ Projects and People

The PFB has raised over $3 million for their campaigns to obtain land, manage that land, internationally educate people on biodiversity, and locally educate the people of Belize on their cultural heritage and on the importance of their natural surroundings.

PFB has acquired about 342,000 cumulative acres of land consisting of tropical rainforest in northwest Belize and tracts of savanna and wetlands to the east of the rainforests. To care for this land, the Programme operates the Rio Bravo Conservation and Management Area, a model for sustainable development and a research center. The PFB also has an Employment-Training Project which gives job opportunities in conservation and related fields.

A. Joy Grant is a member of the Conservation Advisory Board to the Belize Minister of Tourism and the Environment. Other Board members include: Chairman James V. Hyde, Belizean ambassador to the USA; Keith Arnold, financial secretary, government of Belize; James Baird, vice president, Massachusetts Audubon Society; Arnold K. Brown, president, Conservation Management Associates; L.M. Browne, managing director, Belize Sugar Industries, Inc.; Sir Edney Cain, governor, Belize Central Bank; David Gibson, permanent secretary, Belize Ministry of Natural Resources; and Therese Bowman Rath, director, Belize Audubon Society.

■ Resources

The Programme for Belize offers educational guides which help to increase public consciousness and awareness. The guides include a slide show on PFB and a Teacher Information Kit, both for free. The PFB also offers a newsletter to its members. Ongoing fundraising efforts are aimed at further activities.

Project Learning Tree (PLT) ⊕ ✿

1250 Connecticut Avenue, NW, Suite 320
Washington, DC 20036 USA

Contact: Andy Pasternak, Manager of Educational Programming
Phone: 1-202-463-2475 • *Fax:* 1-202-463-2461

Activities: Education • *Issues:* Air Quality/Emission Control; Biodiversity/Species Preservation; Deforestation; Economics; Energy; Forest Ecology; Global Warming; Health; Natural Resource Management; Sustainable Development; Transportation; Waste Management/Recycling; Water Quality
Meetings: NAAEE; NCSS; NSTA

● Organization

Project Learning Tree (PLT) is a grassroots, volunteer, environmental educational program created for educators working with students in kindergarten through grade 12. PLT attempts to make students "aware of their presence in the environment, their impact on it, and their responsibility for it, and to develop skills and knowledge to make informed decisions regarding the use and management of the environment—and the confidence to take action on their decisions."

PLT designs instructional activities and workshops to teach educators, foresters, park and nature center staff, and youth group leaders how to bring exciting, interdisciplinary methods to students. Since 1977, more than 200,000 educators and 10 million students have been reached in the USA, Canada, Sweden, and Finland through this program.

The PLT is co-sponsored by the American Forest Foundation (AFF) and the Western Regional Education Council (WREEC). Both are nonprofit institutions formed by various members of forestry-related organizations.

▲ Projects and People

To help teachers involve students in the learning process, PLT produces activities and curricula which are distributed at their workshops for educators. These activities are written with the cooperative efforts of classroom teachers and environmental workers, all who have different teaching styles and techniques. The hands-on activities are easily applicable to any curriculum and can be given to special-needs students. Overall, they are designed to help students "learn how to think, not what to think, about our complex environment."

PLT offers over 175 activities to enhance the teaching of a wide variety of subjects including science, math, social studies, and the humanities. The activities work with different teaching methods and student needs. PLT's activities are available at little or no cost and are handed out at PLT workshops in the USA and Canada.

■ Resources

In addition to the activities prepared by PLT, Project Learning Tree also produces *The Branch* newsletter filled with new ideas and activities, posters, guides, and computer simulation. PLT wants to expand testing of new environmental materials in urban, multicultural settings.

Rainforest Action Network (RAN) ⊕ ✿
450 Samsome Street
San Francisco, CA 94111 USA

Contact: **Randall Hayes**, Director
   Phone: 1-415-398-4404 • *Fax:* 1-415-398-2732

Activities: Education; Legislative/Political; Research • *Issues:* Biodiversity/Species Preservation; Deforestation; Indigenous Peoples; Tropical Timber Imports

● Organization

Saving the world's rainforests is the single, infinitely complex goal of the Rainforest Action Network (RAN). Since 1985, RAN has been collaborating with environmental and human rights organizations worldwide to increase protection for rapidly depleting forest lands. Grassroots organizing in the USA manifests itself in boycotts, demonstrations, consumer action, and letter-writing campaigns that put pressure on government agencies and corporations to change policies that endanger the rainforests. Facilitating communication among U.S. and Third World organizers is a stated priority of RAN. Affiliated Rainforest Action Groups in the USA and Europe team up with a twin group in the Third World to organize community actions and improve local preservation efforts.

▲ Projects and People

An international campaign to **stop the import of hardwoods from primeval tropical rainforests and promote ecologically sound nontimber forest products** has been a major focus of RAN resources for the last decade. By concentrating on public education about the effects of tropical hardwood logging, consumer alternatives, and economic development, RAN has succeeded in slowing down imports.

Burger King restaurants were the target of a recent boycott by RAN and its members worldwide. Due to economic and political pressure, Burger King announced it would **stop purchasing beef from rainforests**, a major victory for RAN. The next step is to enact a labeling law that would identify beef as coming from rainforest countries and deforested lands.

Other campaigns target an "unnecessary" power project on the island of Hawaii, said to be capable of destroying "the last extensive tract of intact lowland tropical rainforest in the USA"; Japanese corporations in the rainforest; Sarawak, Malaysian rainforests; World Bank and other multinational banks; oil exploration; and tribal crises.

■ Resources

Action Alerts are monthly one-page bulletins on issues needing immediate public action. "What to do" items and addresses are included for members' use in writing letters of protest or support. Other publications such as the *Rainforest Action Guide* and *Tropical Rainforest Press Brief* introduce laymen to the important issues and explain ways to get involved. Organizers' packets and educational materials for elementary school teachers are available also. A directory entitled *Amazonia: Voices from the Rainforest* lists over 250 organizations working on rainforest issues in the Amazon.

Rainforest Alliance ⊕
270 Lafayette Street, Suite 512
New York, NY 10012 USA

Contact: **Daniel Roger Katz**, Executive Director
   Phone: 1-212-941-1900 • *Fax:* 1-212-941-4986
E-Mail: canopy; mci: dkatz

Activities: Education; Development; Research • *Issues:* Biodiversity/Species Preservation; Deforestation; Sustainable Development • *Meetings:* ITTO; UNCED

● Organization

The Rainforest Alliance was established in 1986 as an international nonprofit organization "dedicated to the conservation of the world's endangered tropical forests" and whose primary mission is to "develop and promote sound alternatives to tropical deforestation." Today, with a staff of 20 and some 15,000 members and supporters, the Alliance works with local community groups to create stable income sources in harmony with conservation. With 50 to 100 acres of rainforests destroyed each minute, the consequences of erosion, flooding, and desertification are devastating. Only 2 percent of the Earth's surface is now covered with tropical rainforests.

▲ Projects and People

Rainforest Alliance ongoing projects include the Tropical Timber Project, Medicinal Plants Project, and Proyecto Banano Amigo (Friendly Banana Project). The Alliance also sponsors research and grant programs as well as a Tropical Conservation Newsbureau.

The Tropical Timber Project provides "economic incentives for the adoption of improved timber harvesting and management practices that do not degrade the integrity of the forests or its inhabitants," reports the Alliance. Under Ivan Ussach's direction, a special Smart Wood program uses the marketplace to promote better timber management. It creates guidelines for acceptable timber harvesting practices, identifies properly managed harvesting operations, and certifies socially and environmentally acceptable sources to encourage wood importers and consumers to purchase such tropical woods worldwide. The program also identifies Smart Wood businesses that buy tropical timber from certified Smart Wood sources.

The Medicinal Plants Project helps conserve tropical forests by "increasing the use of medicinal plants as an economic alternative to deforestation." The Alliance indicates that plants found in tropical forests supply modern medicines with drugs used to treat a range of maladies from heart disease and AIDS to menstrual cramps and diarrhea. For these reasons, the Alliance is completing an "intensive research program" that is producing a *Field Manual for Medicinal Plant Project Implementation* for use by other groups. The Alliance also intends to begin three "model medicinal plant projects" as community-based initiatives in Brazil, Peru, and Indonesia with technical training, financial support, workshops, and work plans.

Because all drugs cannot be synthesized effectively, efforts are underway by the National Cancer Institute among major medical research and pharmaceutical companies worldwide to search for and collect medicinal and herbal plant specimens. The New York Botanical Garden's Institute of Economic Botany is part of this undertaking and also joins the Alliance with Medicina da Terra and Nature's Way to advance ethnobotany. Early in 1992, the Alliance and the Institute sponsored a major symposium on tropical biodiversity conservation and health roles of such medicinal plants at Rockefeller University.

Local people benefit when medicinal materials are sold in regional, national, and international markets or when royalty fees are collected from pharmaceutical firms that successfully utilize the plants or the botanical knowledge of tribal doctors or shamans among indigenous cultures, reports the Alliance. Traditional healers such as shamans are estimated to provide "primary health care to 80 percent of the world's population—about four billion people." Yet only about 1 percent of tropical plants have been thoroughly researched in the Western world.

What has been discovered is that rosy periwinkle, found from Cuba to the Philippines to South America, contains two of the world's most powerful anti-cancer drugs, vincristine and vinblastine, that helped establish a basis for chemotherapy. These chemical agents are best known for treating childhood leukemia, Hodgkin's disease, and testicular cancer, according to the Alliance. The plant also treats eye inflammation, rheumatism, and diabetes.

The bark from the cinchona trees yields quinoline alkaloids such as quinine, a standard treatment for malaria that is also medicine for potentially fatal electrical rhythms, or dysrhythmias, of the heart.

A wild Mexican yam, diosgenin, produces a chemical from which oral contraceptives and sex hormones are synthesized that can aid in family planning, reports Alliance literature. The yam's other chemical products can be used in steroid drugs to treat skin diseases and forms of arthritis.

The compound pilocarpine used to treat glaucoma comes from a group of South American trees which are members of the citrus family. Other uses are expected to benefit the elderly. Coca, a shrub that grows on the eastern slope of the Andes Mountains and in the Amazon Valley, has leaves which yield cocaine and other derivatives including lidocaine—most commonly used in people who suffer irregularities of the heart. Other coca derivatives are used in local anesthetics in procedures like minor dental surgery. D tubocurarine yields derivatives that act as blocking agents during surgery to relax muscles and avoid spasms.

Psoralea corylifolia is a subtropical plant that produces psoralens, which are being tested by the National Cancer Institute to combat AIDS. In 1988, the Food and Drug Administration (FDA) approved one compound for the treatment of T-cell lymphoma. Clinical trials of psoralens are underway to treat chronic lymphocytic leukemia, scleroderma, and pemphigus. Castanospermine, another compound from the Moreton Bay chestnut tree of Australia, has been found to inhibit growth of the AIDS virus cell wall; however, this is still in the early test stages.

Susan Salas directs the Proyecto Banano Amigo (Friendly Banana Project) in Costa Rica whose goal is to "decrease the serious environmental damage caused by banana production in the tropics." When rainforests are cut down to make room for the banana plantations requiring large amounts of pesticides and fertilizers, chemicals are washed into the rivers and into neighboring forests—a huge price for bringing "low-cost unblemished fruits" to northern markets.

Proyecto Banano Amigo is a partnership effort of the Rainforest Alliance with the Asociacion Tsuli Tsuli/Audubon de Costa Rica and Fundacion Ambio to "explore potential solutions to the banana crisis." With representatives from environmental and scientific groups, fruit companies, and universities, 10 working groups were formed to investigate different facets of the problem.

Chris Wille and Diane Jukofsky head the Tropical Conservation Newsbureau in San Jose, Costa Rica, which leads media campaigns to sensitize the public, scientists, journalists, elected officials, and decisionmakers about the causes and consequences of tropical deforestation.

The Alliance sponsors the Periwinkle Project to gain public and professional support for funding medicinal plant research. Also provided are grants and fellowships for field research, such as Kleinhans Fellowship in Agroforestry Research for sustainable uses of tropical forests that benefit local people and the Eliane Souza Edelstein Fellowship for Medicinal Plant Research in Brazil for "sustainable use of medicinal plants as an economic alternative to rainforest destruction." The Small Grants Program gives up to

$10,000 for "carefully selected initiatives, where the infusion of a relatively modest amount of money can dramatically increase the effectiveness of a particular project," generally community based.

■ Resources

The Alliance says it needs tropical timber consultants for related projects. *The Canopy* is the quarterly newsletter for members. Publications such as *First Cut: A Primer on Tropical Wood Use and Conservation* and a bibliography of medicinal plants and ethnobotany are available, as is a *Teacher's Resource* pack and posters and brochures for use in medical offices. The Alliance also conducts conferences and workshops.

RARE Center ⊕ ✦

1529 Walnut Street
Philadelphia, PA 19102 USA

Contact: **Christine Psomiades**
Phone: 1-215-568-0420 • *Fax:* 1-215-568-0516

Activities: Education; Research • *Issues:* Biodiversity/Species Preservation; Deforestation; Sustainable Development

● Organization

The resplendent quetzal, national bird of Guatemala and sacred deity of ancient Aztec and Maya, is also the RARE Center's emblem and conservation catalyst. It is one of the species of endangered tropical birds RARE emphasizes in its efforts to ensure that the birds remain in their natural habitat for generations to come. The Center focuses on notable species of the Caribbean and Latin America in order to galvanize local government and public support to preserve the entire ecosystem on which the birds depend. Working with private organizations and government agencies in the host countries, RARE provides the education and training for conservation programs that continue to be effective after the Center has moved on to other regions.

▲ Projects and People

Forest reserve boundaries often are designed to satisfy political, not ecological, ends. This can mean disaster for ideologically neutral wildlife. RARE research at the world-famous **Monteverde Cloud Forest Reserve in Costa Rica** revealed that the quetzal bird makes a double migration to unprotected land on the Pacific slope never before thought important for quetzals—now known to be a "key indicator species of the cloud forest ecosystem." As a result, the Monteverde Conservation League acquired 5,000 acres of Atlantic slope to expand this Reserve. Ground-breaking tracking research, under the guidance of **Dr. George Powell**, shows how forest boundaries must be changed if the birds' natural habitat is to be preserved. Other projects have shown how the quetzal acts to distribute the seeds of a wild avocado tree species, once again emphasizing the interconnectedness of natural life.

Maintaining biological diversity, not political prosperity, is the main goal of RARE's project design of Montane Parks. For example, radio telemetry and other technology for tracking migratory birds are now helping to evaluate Mexico's El Triunfo Biosphere Reserve. Chief researcher is **Maria de Lourdes Alvila Hernandez.**

RARE will use the results to foster conservation action in the montane forest regions of Latin America.

RARE's education efforts in the Caribbean islands take a creative twist through the use of songs, music videos, church sermons, billboards, and bumper stickers. The goal is to involve the entire population in an effort to protect endangered national birds. As the birds become recognized as symbols of national pride and unity, public support for conservation increases.

One cooperative effort in **St. Lucia** involving RARE, the World Parrot Trust of Great Britain, and the U.S. Fish and Wildlife Service recently converted a bus into a **conservation classroom on wheels**, where messages on forestry and water quality are broadcast to cities and remote villages.

Caribbean program director **Paul Butler** leads the campaign in St. Lucia, Montserrat, St. Vincent, and Dominica. He prepared an **environmental education training manual** that appears to be adaptable to other larger Caribbean islands as well as to mainland Latin America. His work is part of the 10-year outreach endeavor **Conservation Education for the Caribbean (CEC)**, which is managed and financed at the local level. Attention is also being focused on **forest fragment conservation** and the stimulation of **ecotourism**.

Local governments have strengthened wildlife legislation and created key reserves in response to RARE's efforts. Involvement of local businesses has strengthened the educational efforts financially. A school visitation program on St. Lucia has reached over 16,000 children. Children on Dominica wrote and performed a musical "Song of the Sisserou" that is now sold on cassette with proceeds to fund continuing education. An environmental newspaper in Dominica and a conservation newsletter and wild parrot census in St. Vincent are other grassroots examples. Conservation education programs were extended to the Cayman Islands and the Bahamas in 1991.

■ Resources

The RARE Center publishes *RARE Center News* and an informational brochure. Conservation posters are available.

Renew America

1400 16th Street, NW
Washington, DC 20036 USA

Contact: **Tina Hobson**, President
Phone: 1-202-232-2252 • *Fax:* 1-202-232-2617

Activities: Education; Research • *Issues:* Air Quality/Emission Control; Biodiversity/Species Preservation; Deforestation; Energy; Global Warming; Health; Sustainable Development; Transportation; Waste Management/Recycling; Water Quality

● Organization

Renew America—a national, membership, environmental action organization—began its *Searching for Success* program in 1989 as a response to the "daily bombardment of negative environmental news." To focus on solutions and encourage cooperation among grassroots and government groups, individual activists and communities, Renew America conducts an annual search for programs that "protect, restore, and enhance the environment" and that can be used as models elsewhere in the nation.

Renew America acts as a "clearinghouse" for such programs and is the executive coordinator of the National Environmental Awards Council (NEAC), a coalition of the following 28 environmental organizations: Alliance to Save Energy, American Council for an Energy Efficient Economy, American Farmland Trust, California Planning and Conservation League, Center for Science in the Public Interest, The Conservation Foundation, Defenders of Wildlife, Environmental Action, Garden Club of America, Global Tomorrow Coalition, Humane Society of the USA, Institute for Local Self-Reliance, Izaak Walton League of America, National Audubon Society, National Wildlife Federation, Natural Resources Defense Council, National Parks and Conservation Association, The Nature Conservancy, Renew America, Rodale Institute, Sierra Club, Society of American Foresters, Soil and Water Conservation Society, Trout Unlimited, Trust for Public Land, Union of Concerned Scientists, The Wilderness Society, and Zero Population Growth.

The building of partnerships is a key element for the future of environmental movements, according to Renew America. **Tina Hobson**, executive director, points out, "We see our role as working within America's communities to identify environmental success at a local level, then promoting these 'success stories' so that they can be replicated. The programs that we recognize embody the concept of 'think globally, act locally' and are an important step in enhancing future global cooperation on environmental issues."

▲ Projects and People

Recognizing leaders in more than 20 policy areas, **National Environmental Achievement Awards** are presented each June on World Environment Day, Washington, DC. In 1991, the *Searching for Success* ceremony was accompanied with a two-day **Environmental Leadership Conference** in which participants prepared recommendations for presentation to policymakers and government officials on issues from air pollution reduction to range and wildlife conservation and including environmental beautification and education; energy and transportation efficiency; groundwater, surface water, drinking water, wetlands and open space protection; food safety; forestry; hazardous materials and solid waste reduction; and recycling.

Special Merit Certificates are also presented to those listed in the *Environmental Success Index*, published as a reference for public-interest groups, media, government, and policymakers, and interested individuals. With more than 1,600 "success stories" presently included, this list is expanded yearly. Each entry is reviewed, evaluated, and verified by NEAC representatives.

Renew America also prepares an annual **State of the States** "report card" that evaluates and compares programs and policies in every state.

The following Renew America award recipients appear in this *Guide*: Alaska Craftsman Home Program, Alliance for the Chesapeake Bay Citizen Monitoring Program, Appalachian Trail Conference, Arise Foundation ENVIRO-COPS, Ashland (Oregon) Conservation Division SOLVE program, Austin (Texas) Public Utility photovoltaic program, Bay Area (California) Ridge Trail Council, Berlin Township (New Jersey) Public Works Department, Bicycle Federation of America National Bicycle Policy Project, Billboard and Sign Control Project (Southern Environmental Law Center), CEHP Incorporated Save the Past for the Future, Center for Holistic Resource Management, *A Child's Organic Garden* (Earth Foods

Associates), Citizen Petition, Citizens for a Better Environment, Eco-Home Network, EcoNet, Flowerfield Enterprises, Global ReLeaf, Green Guerillas, Greenbelt Alliance, Institute for Local Self-Reliance, Interfaith Coalition on Energy, Izaak Walton League of America Save Our Streams, The Land Institute, Land Use Forums/Community Vision, Legal Environmental Assistance Foundation, League of Women Voters of New Castle (New York), Maine Organic Farmers and Gardeners Association, The Marine Mammal Center, Montgomery County (Maryland) Farmland Preservation Programs, Mothers and Others for Safe Food, National Audubon Society Adopt-a-Forest, Nebraska Groundwater Foundation, New Hampshire Hospital Greener Workplace Project, New York City Operation GreenThumb, New York State Energy Research and Development Authority RoadRailer, New York State Self-Help Support System (The Rensselaerville Institute), Northeast Sustainable Energy Association, 1000 Friends of Oregon, Oregon Department of Land Conservation and Development, Oregon Natural Resources Council, Ouachita Watch League, *P3 The Earth-Based Magazine for Kids*, Planning and Conservation League, Practical Farmers of Iowa, Project Learning Tree, Rhode Island Department of Environmental Management Hazardous Waste Reduction Program, Sacramento (California) Municipal Utility District, Salisbury (Connecticut) Land Trust, Santa Barbara (California) Community Environmental Council, Seattle (Washington) Bicycle/Pedestrian Program, Seattle (Washington) Solid Waste Utility, State University of New York Conserve UB Energy Program, Solviva, Texas Center for Policy Studies—Environmental Opportunities of the S&L Bailout, Texas Department of Highways and Public Transportation antilitter campaigns, Texas Organic Food Certification Program, Textbook Review Project (Texas), TREE-mendous Maryland, Total Resource Conservation Plan, TreePeople, Trees for Life, Tulane (Louisiana) Environmental Law Clinic, University of Michigan GREEN Project, University of Massachusetts Acid Rain Monitoring Project, Washtenaw County (Michigan) Services Department Community Right to Know, and Wisconsin Conservation Corps forest protection.

■ Resources

Renew America publishes the *Environmental Success Index*, a *Sharing Success* newsletter highlighting "success stories" and other solutions, information briefs, *State of the States* and other documents including *Sustainable Energy, The Emerging Environmental Consensus, The Oil Rollercoaster: A Call to Action; Communities at Risk: Environmental Dangers in Rural America*; among others appearing in a publications list. It offers a new *Kids Renew America* resource guide.

"It would be helpful if other organizations could assist in the distribution of applications for the *Searching for Success* program" to help identify and recognize those that could be replicated by other communities, requests Renew America.

The Rensselaerville Institute (TRI) ⊕
Pond Hill Road
Rensselaerville, NY 12147 USA

Contact: **Jane W. Schautz,** Director, Small Towns
Environment Program
Phone: 1-518-797-3783 • *Fax:* 1-518-797-3692

Activities: Community Development; Education; Water/
Wastewater • *Issues:* Sustainable Development; Water Qual-
ity • *Meetings:* Community Development Society; World
Water Summit

● Organization

"I have seen many educational programs and centers," wrote **Isaac
Asimov,** the late author and futurist. "For sheer energy, insight, and
capability, The Institute and its people stand out from the crowd."

The Rensselaerville Institute (TRI) is an "independent and nonprofit educational center which helps individuals and organizations
to become more inventive and effective in meeting change and
solving problems. In building and testing fresh solutions to social
and economic needs and issues, the Institute combines theory with
practice, scholar with practitioner."

Founded in 1963 and located in a nineteenth-century village 25
miles from Albany, New York, TRI owns and operates a residential
conference and training center. Known as **The Institute on Man
and Science** until 1983, it was a think tank that attracted leaders
from government, industry, science, and education to ponder how
technology could serve modern life. Reflection led to action; and
today, as TRI, the emphasis is on self-help, innovation, entrepre-
neurial capacity, divergent thinking, and focus on tangible results
for community and government betterment worldwide.

▲ Projects and People

The **Small Towns Environment Program** (STEP), one of TRI's
projects operating since 1990, is an outgrowth of the New York
State Self-Help Support System (SHSS)—in cooperation with the
Departments of Environmental Conservation, Health, and State—
by which 90 rural public water or **wastewater projects** are either
under construction or completed. Together these projects saved the
communities more than $13 million, and freed the state from the
expense, delay, and frustration of enforcement.

One of STEP's main principles, states TRI, is that the best way
for people and groups to solve problems is to take the initiative and
build on their own resources. The key to doing that is making and
finding opportunities which **empower people to act.** The STEP
approach recognizes that a state can also become an enabler along
with the more familiar role of regulator. The STEP approach allows
states to respond to increasing need despite shrinking resources.

The U.S. Environmental Protection Agency (EPA) supported the
start of several **demonstration states,** which include Maryland,
Arkansas, and North Carolina. For example, in Maryland, after 7
years of funding denials, the rural mountain community in Cash
Valley, with 64 households, voted overwhelmingly to solve its long-
term wastewater problems through STEP.

In its first three years of existence, **New York's Self-Help program reduced the cost of small community wastewater collection
and treatment facilities in the state by more than $1.6 million.**
According to TRI, Self-Help is an important alternative for many

small communities which can no longer anticipate federal or state
grant support for major repair or construction. In Arkansas, Governor Bill Clinton agreed this approach "makes sense [in assuring]
that all our people have access to clean, abundant water." "By self-
help, we mean any efforts made by a locality to do for itself some of
those things that traditionally were done by an outside firm," states
TRI.

These are the **strategies:** Calculate what households can afford,
then to make improvements, make all possible use of local resources
such as equipment, local work forces, and resident professionals.
Recognize how many tasks can be done by volunteers and what
outside resources—attorneys, engineers, contractors, bankers—will
be necessary. Divide large projects into smaller units which have end
points defining achievement.

Using the Self-Help concept in New York, the town of Mendon
in Monroe County provided water to a newly developed area where
residents had been forced to carry in all they used. Through cooperation between the town and Monroe Country Water Authority, a
12-inch water line was installed. In the town of Seward, Self-Help
sought to protect a stream from pollution by raw sewage from the
Hamlet of Seward. By having every possible task done by volunteers
and town employees, savings were achieved. Seward qualified for a
loan of $100,000 from the Self-Help loan fund.

■ Resources

The Institute holds invitational meetings on national and global
issues. For example, drought and desertification in the Sahel region
of Africa were the focus of one international conference.

TRI offers a *Self-Help Hand Book,* authored by **Jane W. Schautz,**
which focuses on improving or creating water and wastewater systems in small rural communities. The manual gives specific guidelines and techniques for establishing self-help projects.

TRI also offers *Minds On Workshops* for students and their teachers. The workshops, often designed as simulations, feature active
professionals who mentor 30 to 40 junior or high school youngsters,
sharing with them the tools of their trades which range from business to journalism, field biology to biography, history to theater. In
addition, TRI publishes quarterly newsletters for the Institute and
the small towns environment program. Both are available at no
charge. It also publishes a quarterly magazine, *Innovating.* Several
descriptive brochures are also produced. A publications list is available.

Resources for the Future (RFF) ⊕
1616 P Street, NW
Washington, DC 20036 USA

Contact: **Anne Jarrett,** Public Affairs
Phone: 1-202-328-5009 • *Fax:* 1-202-939-3460

Activities: Education; Research • *Issues:* Energy; Global Warming; Sustainable Development; Waste Management/Recy-
cling; Water Quality • *Meetings:* UNCED

● Organization

Resources for the Future (RFF), with a staff of 95, is an independent nonprofit organization that advances research and public edu-

cation in the development, conservation, and use of natural resources and in the quality of the environment.

Established in 1952 with the cooperation of the Ford Foundation, it is "supported by an endowment and by grants from foundations, government agencies, and corporations. Grants are accepted on the condition that RFF is solely responsible for the conduct of its research and the dissemination of its work to the public."

RFF describes its research as " . . . primarily social scientific, especially economic. It is concerned with the relationship of people to the natural environmental resources of land, water, and air; with the products and services derived from these basic resources; and with the effects of production and consumption on environmental quality and on human health and well-being.

"Grouped into four units—the Energy and Natural Resources Division, the Quality of the Environment Division, the National Center for Food and Agricultural Policy, and the Center for Risk Management—staff members pursue a wide variety of interests, including forest economics, natural gas policy, multiple use of public lands, mineral economics, air and water pollution, energy and national security, hazardous wastes, the economics of outer space, climate resources, and quantitative risk assessment."

▲ Projects and People

Research projects are underway within each division. In the **Quality of the Environment Division**, which examines the management of natural and environmental resources, RFF and the Institute of Economics at the Academia Sincia in Taiwan have begun a joint research effort to "examine **alternative regulatory strategies for urban air pollution problems** in Taipei and to develop methods for analyzing the impact of environmental regulations on economic growth in Taiwan." Project leaders are **Alan J. Krupnick** and **Maureen L. Cropper**.

In **China**, RFF's **Walter O. Spofford, Jr.**, is involved in working with the government on **air and water quality and waste management** for Beijing, Changzhou, the National Environmental Protection Agency (NEPA) and Research Academy of Environmental Sciences; participating in a pollution study with Krupnick; assessing **agricultural development** in the northeast; and preparing a planning guide regarding "**economic efficiency and sustainable development** and the need to integrate environmental considerations into regional economic planning" with the Asia Regional Office of the World Bank. Fifteen RFF books on environmental economics and management are being translated into Chinese at the request of the People's University of China.

In the USA, various **energy tax schemes** and their effects, or burdens, on households are being analyzed. So far, researchers show that "energy taxes are neither small nor evenly distributed, and that regional variation in the distribution of the taxes . . . can vary by as much as 60 percent," reports RFF.

The reliability of **electricity-generating technologies** is being estimated by Krupnick and **Karen J. Palmer**. A comparison of U.S. and European **hazardous waste** management and groundwater protection is being studied by **Allen V. Kneese** with Joanne Linnerooth, International Institute for Applied Systems Analysis (IIASA); a companion study focuses on reducing **Eastern Europe toxic pollutants** and environmental problems that are threats to public health.

What are the economic consequences of the development of **herbicide-tolerant weeds**? **Leonard P. Gianessi** and **Cynthia A. Puffer**, Weed Resistance Working Group, International Organization for Pest Resistance Management, are quantifying such possi-

bilities that can result from the over-reliance on herbicides. How does environmental costing affect electric utilities? RFF is undertaking such a study in Maryland. **Benefit-cost analyses** regarding "transnational environmental problems" such as carbon dioxide emissions control" are being conducted by **Dallas Burtraw** and **Raymond J. Kopp**, particularly the "obstacles to measuring the benefits side."

Plans are underway to develop a "comprehensive **model of environment and trade linkages**" that will enlighten efficient policies regarding global warming, hazardous wastes trade, and the export of "western-style" environmental rulemaking to developing countries. Researchers are Krupnick, Palmer, **Paul R. Portney**, and **Wallace E. Oates**. Other models continue to assess strategies for **mitigating environmental damage**, such as "hydroelectric and reservoir development on fish populations in the Columbia River Basin." Researchers are Kneese, Spofford, **Charles M. Paulsen**, **Jeffrey B. Hyman**, and **Kris Wernstedt**.

The **RFF Pesticide Usage Data Inventory** is being refined and similar herbicide, insecticide, and fungicide databases are also being developed. Researchers are Puffer and Gianessi, who are also involved with Kopp on studying the **economic impacts of pesticide removal** in southeastern states, where weeds, crop diseases, and insects from the Tropics abound. Gianessi is also researching how pesticide-use levels in the USA could be affected by four potential biotechnological developments: "insect-resistant cotton, insect-resistant walnuts, herbicide tolerant cotton, and herbicide-tolerant soybeans." How much does appearance influence the consumer's decision to buy produce and to what extent should pesticides be used to enhance cosmetics? Gianessi and Puffer are examining 13 crops with the help of the American Farm Bureau Federation.

The effects of undesirable facilities such as landfills and hazardous waste sites on local property values are being assessed by Cropper and **Winston Harrington**. With **Molly K. Macauley**, Harrington is engaged in studies on valuing urban open space and how local governments determine amounts of open space.

The **Energy and Natural Resources Division** mainly focuses its research on energy policy and the management of renewable resources, locally and globally. According to RFF, "vehicles that run on **compressed natural gas (CNG)** have many desirable characteristics from an environmental perspective and may be cost-competitive with vehicles that use gasoline." However, many factors may prevent them from enjoying wide consumer acceptance, such as trunk space and horsepower. **Margaret A. Walls** is using a statistical approach applied to data on gasoline-fueled vehicles and will make inferences about the costs to society of switching to vehicles running on CNG.

Walls, Krupnick, and **Carol T. Collins** are examining **alternative fuel vehicles** that run on methanol, CNG, and reformulated gasoline and comparing their emissions with those of conventional gasoline vehicles in the USA.

Peter M. Morrisette and **Norman J. Rosenberg** are assessing how greenhouse-induced **climate change** is influencing economic development and industrialization in developing countries. Overall economic policy responses to climate change are being researched.

Poland has been selected for a set of studies addressing the "dual problems of enhancing environmental protection and reforming natural resource markets" in Central Europe. **Michael A. Toman** is leading the effort.

Costs of recycling and landfilling in varying communities are being analyzed by Macauley and Walls, who are also studying the misuse of resources and potential markets for recycled materials, to help prevent a solid waste crisis in the USA.

Michael T. Bowes, Macauley, and Palmer are examining possible economic incentives under the Toxic Substances Control Act (TSCA) of 1976. As the Clean Air Act Amendments of 1990 require nearly a 50 percent reduction in the current level of sulfur dioxide emissions from electric-generating plants by the year 2000, **Douglas R. Bohi** and Burtraw are assessing "the efficiency of **emissions allowance trading** in an industry as heavily regulated as electricity." Numerous researchers, with Lawrence Berkeley Laboratory, are studying "how fuel cost adjustment mechanisms might be structured to avoid [biasing] investment preferences away from the least costly fuels." Walls is helping the U.S. Department of Energy's (DOE) Energy Information Administration modify its onshore oil and gas supply model.

Macauley is preparing an overview of "the potential contribution of economic methodology for assessing **space science research** priorities," such as "interplanetary missions, space-based research in the life and physical sciences, and data collection aimed at gauging global warming and other changes in the earth's environment." To help resolve key public policy issues regarding earth observation technology, such as remote sensing, Macauley and Toman are studying economic implications.

Pierre R. Corron is preparing a report on how to take into account all the economic and environmental costs of **sustainable agricultural production**. He and Toman are also investigating the "economic and philosophical questions underlying sustainability" and help clarify the term itself.

Bowes and **Roger A. Sedjo** are "developing a framework of a global system of tradable permits" to help capture the benefits of biodiversity and carbon sequestration in the **sustainable management of forests** worldwide. Regarding the sustainable management of tropical forestlands in the Asian-Pacific region, Sedjo is editing a book on this topic. He is also working with the Washington State Department of Natural Resources on "ecologically acceptable" timber harvests in varying forest management strategies. Timber supply, the effect of wastepaper recycling on the forest products industry, and the demand for forest resources are under study by Sedgo, **A. Clark Wiseman** of Gonzaga University, and **Kenneth S. Lyon** of Utah State University. With tropical forests as repositories of genetic resources, Bowes and Sedgo are probing the "evolving financial arrangements" between countries with such forests and pharmaceutical firms interested in using such resources. The value of access to tropical forests yielding medicinal plants is being assessed.

Kenneth D. Frederick is designing a research program for a better understanding of the "environmental values provided by water and the opportunities for accommodating these values in water use and investment decisions." RFF believes the USA has not given ample attention to preservation of fish and wildlife habitats, for example. **Mary J. McKenney** and Rosenberg are examining "the **impacts of a hotter and drier climate**, with and without carbon dioxide enrichment, on the amount of runoff from forestlands, rangelands, and croplands in the Missouri River basin." Frederick is investigating the effects of climate change on water supplies in this study area and socioeconomic implications.

The **National Center for Food and Agricultural Policy** examines policy issues related to the "linkages among agriculture, environmental quality, food safety, and health" both nationally and globally. **Dale E. Hathaway** is examining resource use "during periods of rising and of falling prices for agricultural products in the market and by comparing adjustments in resource use among countries

with different policy regimes, development levels, and economic systems."

Fred H. Sanderson is directing a project where East and West scholars examine and make recommendations about **agricultural and economic reforms in Eastern Europe** and their impact elsewhere. With the University of California at Berkeley, the University of Mexico, and the InterAmerican Institute for Cooperation on Agriculture, RFF researchers are examining the "potential exportation of health and environmental risks" as greater regulations are imposed on agriculture in developed countries—possibly prompting increased production activities in lesser developed countries that pose risks to farmers and local residents; also discussed will be policy options to help developing countries capitalize on the "revenue-enhancing windfall" of environmental controls in agriculture.

Carol S. Kramer is undertaking a big effort regarding the "management, policy, public perception, and communication issues associated with agricultural chemicals and pesticide substitutes." With consumers listing **pesticide residues on foods** as their most critical food safety concern for several years, according to RFF, the public and policymakers still face numerous uncertainties about the benefits and risks of the chemicals used in food production and processing.

RFF is releasing a report, compiled by visiting scholar **Lon C. Cesal**, regarding U.S. **trade growth with developing nations** and its impact on various segments on the agriculture and nonagriculture segments.

The **Center for Risk Management**, established in 1987, "carries out a comprehensive program of fundamental research, policy analysis, education, and outreach related to the management of risks to health and the environment in modern society."

Asking its respondents to choose between hypothetical government regulatory programs that save lives now or in the future, Portner, Cropper, and University of Maryland researcher **Sema K. Aydede** are conducting a nationwide **survey of attitudes** that may provide answers in allocating funds for cancer research that benefit older persons or for accident prevention affecting younger people, for example.

With **H. Keith Florig**'s leadership, RFF has three projects to help clarify scientific uncertainty about potential health risks from **electromagnetic fields (EMFs)**; outcomes are expected to improve "public and private polices for EMF risk management."

The Center recently provided peer review, with **Theodore S. Glickman**'s direction, of a joint pollution prevention and risk reduction study by the Amoco Corporation and U.S. Environmental Protection Agency (EPA) at the Amoco refinery, Yorktown, Virginia. Glickman is also investigating the transportation of hazardous materials, such as petroleum products, chemicals, and radioactive wastes. Glickman with **Christine A. Wnuk** and **Emily D. Silverman** are reporting on their investigation into natural and industrial disasters worldwide and the increasing rate at which certain catastrophes are taking place.

Various RFF researchers are helping to define national goals and explore national strategies in reducing health risks from environmental pollution. The Superfund law of 1980 and alternative approaches for funding hazardous waste site cleanups are being evaluated for these criteria: "speed of cleanup, incentives to clean up sites not on the EPA's National Priorities List (NPL), incentives for improved waste management practices, minimization of transactions costs, revenue adequacy, and equity considerations." Portney, **Katherine N. Probst**, and Wnuk are evaluating a trust twice the size of the existing fund, as one alternative. In a related study, Cropper and Shreekant Gupta are examining the "extent to which human health risks, ecological risks, and cleanup costs explain differences in the amount allocated to each site for cleanup." EPA visiting scholar Frederick W. Talcott is assessing pollution risks resulting from the release of chemicals at the point of production, during distribution and at the time of use—as well as clarifying the chemical flow analysis process.

■ Resources

RFF publishes a quarterly magazine, *Resources*, and an extensive publications list.

The Rockefeller Foundation ⊕
1133 Avenue of the Americas
New York, NY 10036 USA

Contact: **Robert W. Herdt,** Agricultural Sciences
          Division
Phone: 1-212-869-8500

Activities: Development; Education • *Issues:* Sustainable Development

● Organization

The Rockefeller Foundation is a philanthropic organization endowed by John D. Rockefeller and chartered in 1913 "to promote the well-being of mankind throughout the world." It is one of America's oldest private foundations and one of the few with strong international interests. From its beginning, its work has been directed toward identifying and attacking at their sources the underlying causes of human suffering and need.

The Foundation offers grants and fellowships in three principal areas: international science-based development, the arts and humanities, and equal opportunity for U.S. minorities. Within the science-based development, the emphases are on the global environment and on the agricultural, health, and population sciences. The Foundation also has smaller grant programs in international security and in school reform to improve public education for at-risk people.

The Foundation maintains the Bellagio Study and Conference Center in northern Italy for conferences of international scope and for residencies for artists and scholars.

With the John D. and Catherine T. MacArthur Foundation and the Pew Charitable Trusts, the Foundation created the **Energy Foundation**—an organization aimed at "making patterns of energy generation and consumption in the USA [where $450 billion a year is spent on energy] more efficient and sustainable."

▲ Projects and People

To help bring about "environmentally sustainable development," the Foundation embarked in 1990 on a collaborative effort to name each year 8 to 10 **fellows** from among scientists, policymakers, humanists, and entrepreneurs in 7 to 9 major developing countries. "Eventually, a worldwide class of 80 to 100 fellows will be added . . . to come together for a common curricular experience in global issues of environment and development."

In another joint effort including the Ford Foundation and the United Nations Development Programme (UNDP), the Foundation undertook an environmental initiative in China to help Hainan province design ecologically sound strategies for development.

For the 1992 UNCED, the Foundation produced background papers on options for accelerating environmentally sound technology transfer as well as briefings on global warming. The Foundation reports that the USA releases more than 40 percent of the world's CO_2 emissions.

Agroforestry projects of North Carolina State University and a Brazilian agricultural research agency being funded in the Western Amazon are emphasizing alternatives to slash-and-burn farming techniques and other deforestation solutions that include employment for local graduate students. Similar projects to aid small farmers with cash crops are ongoing in Malawi, Uganda, and Tanzania. In Southeast Asia, such projects assess shifts from timber exploitation to sustainable harvesting of fruits, nuts, organic gums, and oil.

Also in Latin America, the effects of **pesticides on potatoes,** a major Andean crop, are being assessed so that government policies can be modified. As one of 39 donors in the Consultative Group on International Agricultural Research, the Foundation supports collecting of seeds and other **germ plasm material** to be used in **plant breeding** and "biotechnologies to create improved plant varieties." Other **genetic engineering** studies, such as introducing "symbiotic **nitrogen fixation** into non-legume crops," could lead to a "significant contribution" that could "substantially reduce the need for fertilizer derived form non-renewable fossil fuels." This would also benefit millions of peasant farmers who cannot afford commercial fertilizer. With Foundation aid, such research is being conducted at England's University of Nottingham, Australian National University, the Chinese Academy of Science's Institute of Botany, and Mexico's Center for Nitrogen Fixation Research.

Other supported projects include creating **virus-resistant maize plants** in Africa and genetically engineered **chickpea** plants in Ethiopia, Pakistan, India and other developing countries; and starting a **plant gene bank** in China with 180,000 varieties of 294 different crop species that are expanded to 400,000 varieties within a 5-year period.

To promote nationally the neighborhood-based **Community Development Corporation (CDC)** concept in the USA, a consortium has been formed among seven foundations and The Prudential that will accelerate this movement, find new funders and lenders, and build a "national community development initiative" with substantial investment potential by about 1993.

The Foundation sponsors **The Working Group on Female Participation in Education,** which is a partnership with other international donors to help African governments formulate a large-scale, long-term strategy to increase female attendance and education performance. The group's work covers all levels of education, but pays particular attention to late primary and early secondary school—when female education has greatest impact on behaviors and attitudes that lead to strong future roles in national development. According to the Foundation, "only 10 percent of females attend secondary school in Africa, far lower than the percentage of males in Africa and also lower than the percentage of females in Asia and Latin America."

The Foundation launched a program in 1989 to improve **public education for children of poor families.** It targets some eight million children who are at risk of failure or dropping out because of an assortment of problems. Much of the Foundation's effort is geared to help parents and communities create the "conditions, climate, and demand for the reform of schools serving at-risk children, and to devise training and dissemination methods that will enable reform-minded educators to make the best practice the standard practice."

The Foundation states that a principal thrust of the program is to broaden the reach of **Dr. James Comer**'s philosophy and of his School Development Program (SDP), a tested approach stressing children's psychological preparation for school and emphasizing the collaboration of school staff and parents in children's academic and social development. The Foundation has produced a 14-part, how-to videocassette series on SDP.

The National Urban League, with a Foundation grant in 1990, selected and funded six local affiliates to begin mobilizing community support for school reform in selected cities in Louisiana, Michigan, Texas, and Washington State. In addition, a planning grant followed by a major school reform appropriation in 1990 underwrote a concerted effort in low-income Hispanic communities where language and cultural factors often create barriers to parental participation in their children's public schools.

■ Resources

As listed in its *Annual Report*, the Foundation offers several types of grants through its international program to support science-based development. They include agricultural sciences grants, health sciences grants, population sciences grants, global environmental program grants, special programming grants, fellowships, arts and humanities grants, equal opportunity grants, school reform grants, international security grants plus a very small number of projects that do not fall within established program guidelines. The *Report* also describes Foundation-funded programs.

Sacramento Municipal Utility District (SMUD)
6201 S Street
Sacramento, CA 95817 USA

Contact: **S. David Freeman**, General Manager
   Phone: 1-916-732-6160 • *Fax:* 1-916-732-6562

Activities: Development; Education • *Issues:* Air Quality/Emission Control; Energy; Global Warming; Transportation; Waste Management/Recycling

▲ Projects and People

With S. David Freeman's appointment in 1990, the Sacramento Municipal Utility District (SMUD) has launched an energy efficiency program—**Conservation Power**—designating new sources to serve its county's growth into the twenty-first century by cutting local energy use. Described as one of the nation's "most ambitious energy conservation programs," it "replaces" the potential of a 600-megawatt power plant that would offset projected growth by dispatching SMUD advisors into homes, businesses, public buildings, and schools—particularly all-electric and low-income neighborhoods—with free conservation kits including weatherstripping and caulking, low-flow showerheads, blankets for water

heaters, compact fluorescent light bulbs, and other energy-saving products.

SMUD also "provides **low-interest financing and cash rebates** to customers who replace older appliances with new, energy-efficient refrigerators and heat pumps," according to Freeman. "The customer makes finance payments for the new appliances along with their regular electric bill, with payments for electricity and improvements being less than the former electricity bill. . . . To ensure that the old, inefficient refrigerators do not enter the used appliance market, they are taken to a SMUD warehouse where they are broken down and disposed of in an environmentally sound manner."

This activity, including the recovery of freon and thousands of gallons of PCB-free oil sold for fuel, ties in to the vital **recycling program**—seen as a model by other utilities on the west coast. With the emphasis on rebuilding and reusing otherwise "disposable" items, both costs and solid waste are being reduced considerably. SMUD recycles car batteries, used motor oil, and other items from its fleet vehicles. Aluminum cans and paper are recycled in its offices; laser printer and copy machine toner cartridges are also refurbished and reused. Gone are throwaway plates, cups, and utensils. "Reduce, reuse, recycle" is the motto.

Also featured are entire **neighborhood audits** as well as a pilot **Total School Energy Management** package that includes teacher in-service training, student energy patrols, and community outreach.

Tree planting is a vital component of Conservation Power; SMUD is spearheading a community campaign to plant 500,000 trees in this decade to reduce the "urban heat island" effect and energy demands for air conditioning.

SMUD reports that its district's energy use was cut by more than 40 million kilowatt hours in about its first year. To convince the public of this program's merit, SMUD enlists **media cooperation** to report, for example, on "the recycling of the 10,000th refrigerator" or the "planting of the 10,000th tree." To ensure broad industry participation, SMUD asked power producers to propose projects to meet energy needs based on cost, environmental impact, and community benefits. The public as well participated in **workshops** where they helped to decide Sacramento's "energy future"—presently a recommendation of "a diverse generation mix of four local cogeneration plants, renewable resources, and power exchange agreements" that was endorsed by editorial boards as well as environmental, consumer, and business groups.

A cornerstone of policy to reduce the region's air pollution is to encourage the commercial and personal use of electric vehicles which SMUD is road-testing. SMUD also loans its 4 electric G-van vehicles to community and nonprofit groups, and more than 10,000 clean air miles were logged in a 6-month period. A network of thousands of "EV Pioneers" is being built to support the use of such commercial vehicles. Proposals for the **electrification of transportation** also include extending the region's light-rail system and replacing diesel buses with an electric trackless trolley system.

A number of new technologies are helping to provide cleaner air and energy self-sufficiency. These include two of the USA's largest **photovoltaic** or sun-powered generating plants; partnership in **Solar II**, a molten nitrate salt prototype generator; and **urban heat island** research with the California Institute for Energy Efficiency to study the "impact of shade trees and white surfaces on building peak loads and cooling energy savings." SMUD is also studying power quality impacts of new technologies to see what problems arise and how they can be resolved. Staff is also addressing the issue of electric and magnetic fields (EMF). The closed **Rancho Seco** nuclear plant is being **decommissioned. A nuclear spent-fuel cask project**, with the U.S. Department of Energy (DOE) is designing and testing such systems. **Solar energy application** through the **Solar Box Cooker** is being promoted at workshops, presentations, and fairs.

S. David Freeman is noted for "turning around troubled utilities." At the Tennessee Valley Authority (TVA), he "closed down eight nuclear reactors and set industry standards for the active promotion of energy conservation and environmentally sound energy practices." Working under Presidents Lyndon B. Johnson and Richard M. Nixon on energy policy, Freeman later helped develop the Carter Administration's energy/conservation plan; served on the U.S. Senate Commerce Committee which backed the Automobile Fuel Economy Act; and was head of the Lower Colorado River Authority. In 1974, he published *Energy: The New Era* about how renewable resources and conservation can move the USA toward "a more secure energy future." *A Time to Choose* is his report on how energy conservation can fuel a growing economy.

■ Resources

SMUD brings its Conservation Power program to the public through a volunteer employee speakers bureau; trade shows, exhibits, and community events; teachers' in-service workshops; classroom and special-interest group presentations; curriculum development; demonstrations; and contests. The SMUD library lends materials to schools, such as films; videos; and publications on electrical power and safety, generation resources, environment, and conservation. Brochures, fact sheets, and other conservation materials are available to customers.

Safe Energy Communication Council (SECC)
1717 Massachusetts Avenue, NW, Suite LL215
Washington, DC 20036 USA

Contact: **Scott Denman**, Director
   Phone: 1-202-483-8491 • *Fax:* 1-202-234-9194

Activities: Education; Media Training; Research • *Issues:* Energy; Global Warming; Sustainable Development; Transportation

● Organization

The **Safe Energy Communication Council (SECC)**, a national environmental coalition formed in 1980, helps "lead the fight for energy sources we can live with." SECC states that "energy efficiency and renewable energy resources are our nation's safest, cleanest and cheapest energy options." It cites polls taken which show Americans support of development of these environmentally safe energy choices. However, according to SECC, millions are being spent each year to sell the public on "costly, environmentally threatening energy sources, like nuclear power."

As the energy and environmental debate intensifies, SECC believes journalists, public officials, grassroots activists, and the public need reliable, timely information. Citizen groups also need to know how to work with the media to get their messages heard. To this end, SECC provides energy information programs, publications,

media workshops, strategy consultation, and technical assistance for grassroots and national organizations.

▲ Projects and People

With its **Information Dissemination Program**, SECC offers factual material to the press, public officials, and grassroots activists. This program meets the growing need for information about economically and environmentally sound energy options.

SECC has conducted media training workshops for hundreds of environmental, energy, consumer, and other public interest activists and volunteers in the USA. These "hands on" sessions teach participants how to work with the media and give them practical experience with interviews, press relations, media strategy, and creative media tactics. In addition, SECC provides team-building and coalition-building workshops.

SECC's **Media Strategy Consultation Program** responds to the many requests the organization receives each month from local, state, and national energy/environmental organizations seeking technical assistance and advice on media strategies and educational campaigns.

Countering nuclear energy advertising, SECC lists as accomplishments television and radio public service announcements (PSAs) including those by Robert Redford that aired on 500 radio stations; defeat of the Clinch River Breeder Reactor and the shutdown of the Rancho Seco nuclear power plant; and use of the Fairness Doctrine by local groups which received more than $8 million in broadcast time.

■ Resources

SECC offers media workshops—generally without charging for its "expertise and staff resources"—for energy, consumer, environmental, peace, agriculture, and other public interest activists and volunteers. These one- to two-day skills sessions have been held all over the country.

SECC also offers publications that include a *Mythbuster* series, a media skills manual and energy commentaries by experts on subjects such as nuclear power, energy efficiency, global warming, utility issues, communications, and alternative fuels. A publications list is available.

Salisbury Land Trust
Box 426
Salisbury, CT 06068 USA

Contact: **Mary Alice White, Ph.D.**, Chairperson
   Phone: 1-203-435-9121 • *Fax:* 1-203-435-8160

Activities: Conservation • *Issues:* Biodiversity/Species Preservation; Protection of Habitats, Open Space, and Wetlands; Sustainable Development; Water Quality • *Meetings:* Land Trust Alliance; TNC

● Organization

The **Salisbury Land Trust**, Connecticut, is an organization of 20 volunteers whose goal is the protection of open land, scenic views, farm land, wetlands, and special habitats through conservation easements, gifts of land, and purchases.

The Land Trust also plays an active role in land planning with its Planning and Zoning Commission, where it has developed maps and lists of protected land, habitats, and natural resources in need of preservation.

▲ Projects and People

According to chairperson Dr. Mary Alice White, "Land Trust's efforts have been successful due largely to a town population that is very conservation-minded." With 9 percent of town land under some form of protection, the Land Trust organized a drive to save Sages Ravine, a pristine and beautiful area that was to be sold. It raised enough money to ensure its purchase, now state-owned.

"We have collaborated with nearby land trusts on joint projects, the latest of which is helping the Sheffield (Massachusetts) Land Trust purchase and protect a 400-acre wetland preserve in Schenob Brook," reports White.

It also monitors land twice a year through its own easements and "land in fee," and gives conservation easements to other conservation organizations such as The Nature Conservancy and the American Farmlands Trust. "We feel that local monitoring by our volunteers is an effective way to maintain protection of these lands," writes White.

With the Land Trust's major interest in environmental education for the next generation, it funded for its local elementary school a National Geographic software program on acid rain, and plans similar efforts.

■ Resources

The Land Trust video, *Sages Ravine*, is available.

Scenic America (SA) ❧
21 Dupont Circle, NW
Washington, DC 20036 USA

Contact: **Sally Oldham**, President
Phone: 1-202-833-4300 • *Fax:* 1-202-833-4304

Activities: Education; Political/Legislative; Research • *Issues:* Transportation; Visual Quality and Visual Pollution

● Organization

Scenic America (SA), formerly the Coalition for Scenic Beauty, is a nonprofit organization founded in 1986 with the help of the American Conservation Association and dedicated to protecting America's scenic resources and to combatting toxic advertising. It has approximately 6,200 members and 15 state and local affiliate groups.

"The degradation of the scenic environment lowers our quality of life by destroying the uniqueness of each place and thereby robbing us of our sense of place," writes Sally Oldham, SA president. "Perhaps even more significant, the degradation of the scenic environment bears with it significant public health hazards." SA works with communities to combat these conditions.

SA is committed to conducting research which will be useful to citizen activists struggling to promote scenic resource protection agendas.

▲ Projects and People

SA educational efforts provide information and technical assistance on ways to protect and restore the visual quality of our country's natural beauty. In a single year, SA gave such help to more than 200 city and county government officials and citizen activists. Technical assistance requests typically cover billboard control, tree protection, scenic highway issues, and growth management.

Assistance relating to billboard control has led to the passage of new billboard laws in more than 100 communities including Boise, Idaho; Asheville, North Carolina; Columbia, Missouri; Roseburg, Oregon; and Philadelphia, Pennsylvania.

SA also conducts or supervises surveys which document the targeting of minority and low-income neighborhoods with tobacco and alcohol billboards. SA research indicates that the billboard industry chooses these areas to advertise alcohol and tobacco products. Through research, education, and advocacy, SA works with low-income and minority communities to "curb the intrusive and harmful presence of billboard advertising of these legal drugs." SA says its research has led to widespread press coverage on this issue and contributed to public policy debate and change in Detroit, Baltimore, Philadelphia, St. Louis, Chicago, and Berkeley.

Staff and board members speak and present research findings at national and local conferences and workshops. One attention-getting research paper explores national development of Heritage Areas—"large planning areas which combine recreation, natural and heritage resource preservation, public education, economic development, and tourism strategies."

As a result of its research on economic benefits of protecting scenic quality in urban, suburban, and rural areas, SA sponsored development of a model for determining the economic consequences of various aesthetic regulations. Statewide surveys detailing existing aesthetic regulations and identifying outstanding ordinances which could become models for other communities were recently conducted in New Jersey and California.

SA's role as the "primary advocacy group for the protection of the visual environment" has led it to pursue passage of legislation which will strengthen federal billboard controls and protect scenic highways. Such was the case with its coalition-building and other efforts to gain the passage in 1991 of the Visual Pollution Control Act to stop construction of new billboards along federal roads, halt destruction of publicly owned trees to improve billboard visibility, and restore local control over billboard removal.

Actively participating in the Surface Transportation Policy Project in 1991, as chair of the Environmental Issues Working Group, SA helped to define the "broad environmental agenda and progressive planning approach" acclaimed in the Senate-passed version of the reauthorization of the Surface Transportation Act, and continues actively to educate policymakers about these new approaches.

■ Resources

SA produces newsletters, videos, slide shows, and special publications. Two include, *The Economics of Community Character Preservation: An Annotated Bibliography*, and a *Survey of Aesthetic Regulations* for the state of New Jersey.

SA seeks to expand its network about issues of scenic quality and visual pollution in political action, and education.

Servicios Cientificos y Técnicos (SCT)/ Scientific and Technical Services
Condominio El Centro I, Oficina 607
Hato Rey
San Juan, Puerto Rico 00918 USA

Contact: **Neftalí García Martínez, Ph.D.**, Director
Phone: 1-809-759-8787 • *Fax:* 1-809-757-6757

Activities: Education; Research • *Issues:* Air Quality/Emission Control; Deforestation; Energy; Health; Waste Management/ Recycling; Water Quality • *Meetings:* First National People of Color Environmental Leadership Summit; Conference on Export-Oriented Industrialization in the Third World

● Organization
Servicios Cientificos y Técnicos (SCT) is a nonprofit organization founded in 1989 to work with the "environmental struggles in Puerto Rico." Its motto is: "A community that protects the health of its workers and the environment may aspire to true economic progress." Specializing in conservation and development of cultural and natural resources, environmental protection, health and economic development projects—SCT investigates, assesses, educates, and trains for communities, businesses, municipalities, and other public and private entities.

Its staff deals with pollution of air, water, soil, and noise—analyzing environmental impacts and looking for solutions to problems generated by waste that involve water treatment plants, garbage dumps, incinerators, hazardous waste dumps, emissions of toxic gases, sugar-processing plants, and agricultural and fishing activities.

▲ Projects and People
SCT has created the **Centro de Investigación, Información y Educación Social (CIIES)** as an autonomous unit to provide free scientific advisorship, education services, and technical and organizational assistance on environmental and natural resources issues to communities, workers, individuals, journalists, students, and professors. "We work as expert witnesses [and] attend public hearings, press conferences, and radio and TV programs," reports Dr. Neftalí García Martínez, director.

The Center offers books; periodicals; international publications; audiovisual programs; documentaries; computerized system with national and global organization hookups; seminars and workshops on topics such as community organizing, use of natural resources, and methods to combat environmental problems; and lectures on communication and legal action. SCT's public education program is motivated by this belief: "Only an informed, conscientious, and organized community can protect natural resources, the environment, and health."

SCT is examining and searching for solutions to the following: **Ciudad Cristiana**—hazardous waste and mercury pollution; **Vega Alta**—underground water pollution with polychlorinated solvents; **Hato Nuevo**—regional landfill of Browning Ferris Industries (BFI); **Cayey Municipal Landfill**—to improve operations; **Reforestation Project** with the Natural Resources Department; **Biomedical Waste Disposal**; **Mayaguez coal-fired electric plant** of Congentrix Company; and access to public beaches and to coastal regions.

Dr. Martinez notes, "A major obstacle is lack of economic resources. Our plans are to continue on-going investigations and support for local community organizations."

SCT's activities include preparing and analyzing environmental documents; conceptualizing project models; preparing proposals for funds; designing, administering, and analyzing questionnaires on occupational and environmental health; supervising special projects such as botanical gardens, historical museums, Museum of Natural History, and herbariums; controlling erosion and sedimentation; and developing green areas for tourism, recreation, nature study, and education.

■ Resources
SCT prepares materials such as bulletins, press releases, brochures, audiovisuals, documentaries, and films. Study trips and pleasure excursions are planned in addition to the activities described above.

New technologies that are needed concern "energy alternatives, recycling, treatment of landfill leachates." Political support is needed to "oppose and publicize the negative environmental impact of the BFI landfill project and Cogentrix project." Environmental outreach is needed to publish articles in English regarding the environmental situation in Puerto Rico . . . and to ask U.S. Congress to investigate U.S. Environmental Protection Agency's (EPA) inadequate management of the Ciudad Cristiana and Vega Alta Superfund Sites."

Dr. Martinez encourages an "exchange of publications on environmental and natural areas," development of lists of scientists by areas of expertise and of foundations that could provide economic resources, and "periodical meetings to exchange experiences and to development international strategies."

Shomrei Adamah ⊕
Church Road and Greenwood Avenue
Wynote, PA 19095 USA

Contacts: **Ellen Bernstein**, Director
Susan Mack, Associate Director
Phone: 1-215-887-3106

Activities: Education • *Meetings:* New England Conference on the Environment

● Organization
The mission of Shomrei Adamah/Keepers of the Earth is to "inspire environmental awareness and practice among Jews by unlocking the treasure of ancient Jewish ecological wisdom. Shomrei Adamah serves its members: rabbis, educators, youth, students, environmentalists, seminaries, and a network of affiliate groups across America with authentic traditional sources, curricula, publications, speakers, a newsletter and 'green synagogue' suggestions."

Founded nationally in 1988 when the Federation of Reconstructionist Congregations and Havurot joined with director **Ellen Bernstein's** Washington, DC–based chapter, the nondenominational Shomrei Adamah now has chapters throughout the USA and Canada. Several years earlier, what had inspired Bernstein—then a physical therapist with degrees in conservation biology and education, and with work experience as both a high school biology teacher and river trip guide—was a geology hike in the Grand Canyon on

Passover with several other students, where she improvised an open-air seder using wild foods and other natural elements. As Shomrei literature reports, "the experience impassioned her with Judaism's inherent feeling for nature and she began to seek ways to affirm the bond between her religion and her ecological sensibility."

In its statement of purpose, the Shomrei Adamah states that it hopes to better educate itself and its community through regular study and the introduction of environmental curricula into its congregational schools. It seeks to invest its observance of the Jewish holidays with greater environmental awareness. It intends to join hands with other local coalitions to enhance the quality of environmental health regionally.

▲ Projects and People

The organization has produced a Tu B'Shevat Haggadah book for the "New Year of the Trees" and other literature tying together Judaism and ecology with information on recycling, organic gardening, and pesticides, for example. Education programs are sponsored with other Jewish organizations; and leaders participate in nonsectarian national and global environmental groups. The newly developed high school level curricula is being offered countrywide; emphasis is on biodiversity, interconnectedness, conservation, life cycles, and rejuvenation. An institute for study is being planned.

The Shomrei Adamah's first national conference was held in 1989; a year later, at Earth Day, members developed a program that included a religious service to inspire participants to sustain life through environmental awareness. Environmental training programs for rabbis in Washington, DC, and elsewhere are being undertaken.

■ Resources

Shomrei Adamah has several available publications available, including the *Voice of the Trees* national newsletter and the *Roots & Branches* Washington, DC, newsletter; reprints of articles about Judaism and ecology, a universal seder book written by the organization director; a book of quotes from Jewish sources on the relationship with the natural world; a packet of speeches by the director; and a Jewish environment pack. Shomrei Adamah also offers a national speakers' bureau, recycling labels for envelope reuse, and note cards. The high school curriculum on Judaism and ecology can be ordered. Nature-oriented field trips are also sponsored.

Sierra Club ⊕
730 Polk Street
San Francisco, CA 94109 USA

Contact: **Michael Fischer,** Executive Director
   Phone: 1-415-776-2211 • *Fax:* 1-415-776-0350

Sierra Club ⊕
Legislative Office
408 C Street, NE
Washington, DC 20002 USA

Contacts: **Michael McCloskey,** Chairman
      David Gardiner, Legislative Director
   Phone: 1-202-547-1141 • *Fax:* 1-202-547-6009

Activities: Development; Education; Political/Legislative • *Issues:* Air Quality/Emission Control; Biodiversity/Species Preservation; Deforestation; Energy; Global Warming; Health; Population Planning; Sustainable Development; Transportation; Waste Management/Recycling; Water Quality

● Organization

After successfully campaigning to give Yosemite and the Grand Canyon national park status in 1890, poet and conservationist John Muir joined with a group of like-minded people to form the Sierra Club in 1892. Their early efforts safeguarded the Sierra Nevada mountains and forests; but a defeat over a San Francisco area dam in which the Club clashed with Gifford Pinchot, the USA's former chief forester, precipitated Muir's death in 1913 of a broken heart, environmental historians say. However, the battle elevated the Sierra Club to countrywide prominence, and it joined other conservation groups to convince the U.S. Congress to set up a bureau of national parks.

One hundred years later, the Club continues to work to protect and preserve wilderness areas, and at the same time encourages people to get out and experience the beauty of the world firsthand. Approximately 640,000 members spread among 57 chapters and 386 groups help the Sierra Club promote "conservation of the natural environment by influencing public policy decisions—legislative, administrative, legal, and electoral."

▲ Projects and People

The Sierra Club notes that the USA has about 91 million acres of National Wilderness Areas (more than half of which are in Alaska). The Club works for the continued protection of these areas, urging the Forest Service to adopt more responsible management policies, promoting the policy of "no net loss" of wetlands, and pressuring other government arms to implement existing legislation such as the Clean Air Act of 1990 and the Resource Conservation and Recovery Act (RCRA).

One such area is the Arctic National Wildlife Refuge wilderness expanse established by President Dwight Eisenhower in 1960; the 1.5 million acre area is home to the 180,000-head porcupine caribou herd, among hundreds of other species. *(See also Department of Interior, Fish and Wildlife Service, Arctic National Wildlife Refuge in the USA Government section.)* Oil companies are currently seeking rights to explore in the territory.

Of special interest to the Sierra Club is fuel efficiency; instead of increased environmentally harmful oil exploration, the Club advocates a standard of 45 miles per gallon for all American cars by the year 2000. Auto emission comparison studies and ozone research of the Club give added impetus to their campaign.

Related recent campaigns include the following:

Ancient Forest Protection. In the Pacific Northwest are some of the world's largest trees—hemlock, cedar, spruce, sequoia, and fir, as are chainsaws, reports the Sierra Club, which have wiped out about 90 percent of the forests. The Club is pushing for legislation that will "permanently protect the ancient forest ecosystem and defeat attacks on the Endangered Species Act and other environmental statutes being challenged by pro-logging interests."

Biological Diversity. Determined to protect the nation's dwindling species, the Sierra Club will push for various bills at the federal and state levels that range from comprehensive to specific forest practices acts and wildlife corridor protection efforts.

Global Warming/Energy Efficiency. Aggressive in its war against carbon dioxide buildup, chlorofluorocarbons (CFCs), and other greenhouse gases, the Sierra Club is calling for a "20 percent reduction in CO_2 emissions by the year 2000 through greater efficiency, federal and state regulations, and a carbon dioxide emissions tax." In addition to greater auto fuel efficiency, the Club supports accelerated research and development of solar and renewable energy-generating technologies "with multi-year authorizations" and tax incentives; domestic reforestation; and federal assistance for "least-cost state energy planning regulation." The Sierra Club is convinced that "current and emerging technology already exist that can reduce worldwide CO_2 emissions by 58 percent by 2050, even more with extensive substitution for coal-fired electricity."

Clean Air Act Implementation. Regarding the 1990 Clean Air Act, the Sierra Club is concentrating its efforts on state action. "In the most polluted areas, we will work for adoption of the strongest possible auto emission standards, the sale of cleaner fuels, and the development of transportation and land use plans designed to promote alternatives to the private car." This includes seeking increased funding for transit and passenger rail and other alternative transportation modes and reduced highway construction.

Solid Waste. Engaged in an intensive campaign being fought nationwide, the Sierra Club reports that each American generates an average of four pounds of garbage every day! In addition to tightening RCRA, Sierra has its own platform for Congress, which would permit states to enact tougher waste laws than federal statutes, would "dispose of waste close to its source of generation," would bolster the public's "right to know," would end the unfair burden on minority and low-income communities where facilities are often sited, and would "give Native American tribes the authority to administer waste laws." A priority is the cleanup of defense and energy facilities. Generally, at the state level, Sierra seeks "source reduction, recycling, and materials recovery"—with management plants to obtain such goals. It is also campaigning for bottle bills, restricted packaging, and improved marketing of recycled products.

Population. Sierra Club backs fully funded international family planning programs to achieve population stabilization.

International Development Lending and Tropical Forest Protection. Since 1971, the Sierra Club has reached beyond national boundaries in its International Program to extend wilderness and species protection to the entire globe. "Wild rivers don't stop at the border to show their passport, nor do clouds of radioactive steam," it emphasizes. Today, the Sierra Club encourages multilateral development banks (MDBs), such as the World Bank and the African Development Bank, as well as the U.S. Agency for International Development (USAID) to support projects that contribute to "sustainable and economically sound development in the Third World" in place of environmentally harmful loans for logging operations such as in Cameroon, road construction, or undertakings that clearcut tropical forests in the Ivory Coast, overgraze African grasslands, destroy wetlands, or flood wilderness reserves, for instance. It wants U.S. legislation that requires labeling for imported hardwoods to eliminate timber from unprotected forests.

The most recent focus of Sierra Club efforts has been U.S. election year 1992, with the possibility of up to 100 new congressional members in 1993. In 1990 elections, 86 percent of Sierra Club–endorsed candidates won congressional seats; the Club works to "separate the truth from the rhetoric" in political campaigns in an era when every politician wants to jump on the environmental bandwagon, it says. Liz Meyer chairs the Sierra Club Political

Action Committee, formed in 1976 specifically to help "environmental candidates" in election bids, and to educate the public. The Club's Activist Network consists of 84,000 who can quickly mobilize to write or telephone their senators or representatives when an important environmental issue is at stake.

The Club is also addressing the Mining Law of 1872, which is still on the books, though it demands no accountability for environmental damage from mining companies.

■ Resources

The Sierra Club has published 350 book titles, including a *Sourcebook* listing filmstrips, videos, posters, pamphlets, films, and other educational aids. The *Sierra* magazine has appeared bimonthly since 1893, and the *National News Report* is published biweekly. The International Program issues *Earthcare Appeals*. A *Mail-Order Service Guide* is available. Most chapters and groups have their own newsletters, in addition to which the Club publishes the following periodicals: *International Action Report*, quarterly; *Alaska Report*, quarterly; *Public Lands*, quarterly; *HazMat/Water Resources Newsletter*, quarterly; *Energy Report*, three times per year; *Population Report*, four to six times per year; *SierraEcology*, quarterly; and *Impacts*, focusing on the environmental impact of military development and wars.

The Club is known for its regular outings, including Inner City Outings, which allow youth, the disabled, and senior citizens in 29 cities across the country to experience nature. As John Muir wrote in an early plug for ecotourism: "If people in general could . . . hear the trees speak for themselves, all difficulties in the way of forest preservation would vanish."

Sierra Club (SC) ⊕
Midwest Regional Office ✔
214 North Henry, Suite 203
Madison, WI 53703 USA

Contact: **Carl A. Zichella**, Regional Staff Director
Phone: 1-608-257-4994 • *Fax:* 1-608-257-3513
E-Mail: dial com: sie004

Activities: Development; Education; Political/Legislative • *Issues:* Air Quality/Emission Control; Biodiversity/Species Preservation; Deforestation; Energy; Global Warming; Health; Population Planning; Sustainable Development; Transportation; Waste Management/Recycling; Water Quality

● Organization

The Midwest Regional Office of the Sierra Club (SC) operates over a nine-state region, including Minnesota, Wisconsin, Michigan, Illinois, Indiana, Ohio, Iowa, Missouri, and Kentucky. The staff of 4 focuses much effort on the Great Lakes Basin, a vital watershed of 295,000 square miles which encompasses the first 6 states named above, as well as Pennsylvania, New York, and part of Canada. Here in America's heartland, some 37 million people live.

Despite a great effort over the last decade or so to clean up the lakes—an effort resulting in substantially cleaner water—the Sierra Club office warns that much remains to be done. Food chain toxins provide one example, as a "fish can easily contain 50,000 times or more of a contaminant than the water in which it lives." The enemies are eutrophication; poisonous chemicals—sometimes DDT—carried by air from local smokestacks and other countries or discharged through pipes; "exotic parasites and other animal species arriving by ship from foreign ports of call"; nonpoint source pollution from the land such as pesticides, fertilizers, oil, grease, and raw sewage; storms; and tainted bottom sediments that cause deformities in terns, ducks, and fish.

▲ Projects and People

In its Great Lakes Program, the Midwest Office works to strengthen existing federal legislation, particularly RCRA (Resource Conservation and Recovery Act) hazardous waste provisions, the Clean Water Act, and the Clean Air Act of 1990. The Program is "dedicated to restoring environmental quality and biological diversity in the Great Lakes ecosystem"; the Chicago, Illinois, and Gary, Indiana, areas are at particular risk because of the local steel industry's coke oven emissions.

The Midwest Office is involved in the Blueprint for Zero program in cooperation with other local and national environmental groups, including the Natural Resources Defense Council (NRDC), the National Wildlife Federation (NWF), and Greenpeace. The goal is to eliminate completely toxic discharge into the Great Lakes.

Among public fears about communities becoming industrial sewers, the Sierra Club believes that solid waste disposal is possibly "the most explosive local environmental issue of the 1990s." In addition to pushing for tougher federal laws, Sierra calls to task the U.S. Environmental Protection Agency (EPA) for failing to "regulate large amounts of industrial waste that contain toxic chemicals; require effective management of wastes that are subject to regulation; ensure that wastes placed in land disposal facilities do not contaminate groundwater or migrate off-site; and enforce RCRA requirements for federal facilities."

According to the Sierra Club, "drilling wastes and mine tailings often contain radioactive and highly toxic chemical compounds. . . . Heaps of ore, through which acutely hazardous materials are leached to extract precious metals, are not considered wastes by EPA and thus escape regulation under RCRA." The "de-listing" of metals considered toxic, such as chromium, is an especially large concern of Sierra, as are the "recycling" of contaminated oils, the burning of hazardous wastes, and the failure to review groundwater monitoring systems. The military is "an enormous source of hazard wastes," Sierra asserts in its Blueprint. "The estimated cost of cleaning up existing military waste across the country is nearing $200 billion."

In 1991, the Office helped organize the Lake Superior Vision Conference, attended by Canadian and U.S. environmental organizations. The Great Lakes Wetlands Policy Consortium, made up of 21 U.S. and Canadian environmental groups, reports on the state of the wetlands in the region; 70 percent of the original wetlands area has been destroyed.

Regional staff director Carl A. Zichella writes, lobbies, and speaks to advance the cause of environmental protection, with a special focus on energy issues. While working for the Sierra Club in California, he was co-author of the 1980 Proposition B safe energy initiative in Humboldt County, which was overwhelmingly approved.

Hopeful about the "right direction" of certain regional environmental campaigns, Sierra's Midwest Office operates by the 1864 words of George Perkins Marsh: *"We have now felled forest enough everywhere, in many districts far too much. Let us restore this one*

element of material life to its normal proportions, and devise means for maintaining the permanence of its relations to the fields, the meadows, and the pastures, to the rain and the dews of heaven, to the springs and rivulets with which it waters the earth."

■ Resources

The Midwest Office publishes a bimonthly newsletter, *The Forest Networker*.

Sierra Club
Southeast Office
1330 21st Way South, Suite 100
Birmingham, AL 35205 USA

Contact: **James M. Price**, Staff Director
    Phone: 1-205-933-9111 • *Fax:* 1-205-939-1020

Activities: Political/Legislative • *Issues:* Air Quality/Emission Control; Biodiversity/Species Preservation; Deforestation; Energy; Global Warming; Health; Population Planning; Sustainable Development; Transportation; Waste Management/Recycling; Water Quality

▲ Projects and People

The Sierra Club's Southeast Office concentrates on wetlands protection and growth management and provides citizenship empowerment skill training—among other environmental activities crucial to this organization.

Sierra Club Legal Defense Fund, Inc. ⊕
180 Montgomery Street, Suite 1400
San Francisco, CA 94104 USA

Contact: **Vawter Parker**, Executive Director
    Phone: 1-415-627-6700 • *Fax:* 1-415-627-6740

Sierra Club Legal Defense Fund, Inc. (LDF)
Regional Office
1631 Glenarm Place, Suite 300
Denver, CO 80202 USA

Contact: **Lori Potter**, Managing Attorney/Regional
        Director
    Phone: 1-303-623-9466 • *Fax:* 1-303-623-8083

Activities: Law Enforcement; Litigation; Research • *Issues:* Air Quality/Emission Control; Biodiversity/Species Preservation; Deforestation; Energy; Global Warming; Health; Transportation; Waste Management/Recycling; Water Quality • *Meeting:* UN Subcommission on Prevention of Discrimination and Protection of Minorities

● Organization

Founded in 1971, the **Sierra Club Legal Defense Fund** (LDF) works with, but is not a part of, the Sierra Club. As a public interest law firm, the LDF, with a client list of more than 350 groups, "brings environmental litigation on behalf of the Sierra Club and other environmental organizations," which have included the Natural Resources Defense Council (NRDC), Greenpeace, and the Audubon Society. Typically, LDF battlefronts are herbicides, chlordane, radioactive wastes, incineration, and acid rain.

The Fund also represents Native American tribes and other individuals in its fight to protect wilderness areas from agriculture, mining, oil/gas exploration, real estate development, logging, toxic pollution, and other harmful activities. With headquarters in San Francisco and 7 other offices, the Fund employs a staff of 80, including 36 attorneys, and enjoys the support of 135,000 contributors as well as volunteer lawyers. It recently "served as midwife to the birth of the first public interest environmental law firm in Canada dedicated to using the courts, the Sierra Legal Defence Fund of Vancouver." SLDF will litigate logging in ancient forests, similar to its USA counterpart.

▲ Projects and People

Many of the LDF's activities involve suing, or challenging decisions and inaction of, government organizations such as the U.S. Fish and Wildlife Service (FWS), the Bureau of Land Management (BLM), the Forest Service, and the U.S. Environmental Protection Agency (EPA). Fund action led the FWS to add the northern spotted owl to the endangered species list. California staff attorney **Deborah Reames'** work led the BLM to cancel all further Barstow-to-Vegas motorcycle races, which in the past have destroyed vital desert tortoise habitat.

Lawyers in the Alaska regional office are still litigating to force Exxon to pay for environmental restoration following the *Valdez* oil spill disaster of 1989. Regarding the Tongass National Forest, LDF scored "interim victories" to block logging and road building in sensitive wildlife areas. Gold miners were kept away from mountain goats.

Ongoing action in California on the part of attorneys like Michael Sherwood and Stephen Volker concerns the threatened Sacramento/San Joaquin River delta area, inhabited by delta smelt and Chinook salmon. Air pollution in the San Francisco Bay area is bound to be reduced as a result of new regulations and recognition by transportation agencies of highway construction effects on air quality. The first suit against the Federal Deposit Insurance Corporation was filed to "block the sale of a valuable wetland near San Francisco Bay to a developer." The Castle Mountain gold mine in the desert is expected to be a standard for others as the world's most "environmentally benign heap-leach gold mine." Some 150 native California plant species are being added to the state's endangered species list.

LDF now has a permanent presence in Florida with its Tallahassee office's opening in 1990. Actions have already influenced the state and federal government to pursue "long-term research and monitoring" on behalf of the Everglades. In a lawsuit facing challenge, certain agricultural industries must treat and cleanse waters before they flow into the Everglades and Loxahatchie National Wildlife Refuge. Recycling programs are beginning to make inroads in place of incineration; the mainly high water table precludes landfills. The silver rice rat is now on the state's endangered species list.

In Hawaii, LDF won a court order for a thorough federal environmental impact study (EIS) concerning a plan to tap geothermal heat beneath Kilauea volcano on the Big Island to generate electricity by cable to Oahu. Another EIS is being prepared regarding a U.S. Army training center in a tropical dry forest, a scarce ecosystem. Water pollution lawsuits settled in 1991 set up funds for "environmental and cultural education, protection, and research"—and stopped the unlawful discharges. Parasails and jet skis must stay away from waters where humpback whales calve and nurse their young.

The Louisiana Office opened in 1991 and took the following action: filed suit to "block destruction of a wetland at the Stennis Space Center in Mississippi, . . . sent notices of their intent to sue the [U.S.] Department of the Interior for its failure to add the Louisiana black bear to the endangered species list [and] to sue several corporations that are violating the Clean Water Act, [worked] with a legal assistance organization on potential litigation that would combine civil rights and environmental issues, and . . . joined the state of Louisiana in a suit aimed at blocking permits for a new offshore drilling operation in the Gulf of Mexico until certain coastal-impact mitigation measures are put in place."

LDF attorneys in the Pacific Northwest are "vigorously" pursuing their campaign to save ancient forests, strengthened with a favorable court of appeals decision and an injunction to prohibit "new timber sales in spotted owl habitat in the national forests of Oregon, Washington and northern California" until the federal agencies "decide to obey the law." Protective measures continue on behalf of spotted owl habitat, marbled murrelet seabird, and restoration of sockeye salmon runs in the Columbia and Snake rivers.

The Rocky Mountain Office continues its efforts on behalf of the grizzly bear and its besieged habitat; to control pollution and commercial construction near the Grand Canyon; to stop a resort that "would have destroyed 8,000 acres of tidal wetlands"; to protect Colorado's water supply at the Holy Cross Wilderness; and to keep oil and gas exploration out of Arches National Monument.

Lori Potter has worked with the LDF since 1983, and has been regional director in the Denver Office since 1985. Managing a staff of 8 to 12, she litigates water, public lands, and environmental issues across a wide area. In 1985, for example, she helped protect the endangered brown pelican of Texas and Mexico, blocking high wire plans of a south Texas utility company 2,000 miles from her office. Also in the Denver Office, **Mark Hughes** threatened suit against Coors Brewing Company, leading Coors to clean a nearby creek, upgrade its plant facilities, and donate $160,000 for acquisition of protected land.

The Washington, DC, Office reports that although the Atlantic salmon is making a comeback, a proposed hydroelectric dam on Penobscot River, Maine, would have threatened spawning—had not LDF intervened successfully. A Resource Conservation and Recovery Act (RCRA) amendment was withdrawn that could have "hindered control of hazardous materials," and EPA was also required to complete and release "toxicological profiles" of a string of such chemicals.

Internationally, the Fund has represented the 1,500-strong Huaorani Indian tribe of the Ecuadorian Amazon in their case against Conoco, which wanted to build roads, wells, and pipelines throughout the tribe's hunting and gathering area; Conoco decided to pull out of the area in late 1991. In an effort to make environmental protection a basic human right, the Fund is working with the United Nations Sub-commission on the Prevention of Discrimination and Protection of Minorities as well as the Centre for Human Rights in Geneva to study the issue around the world.

■ Resources

The Legal Defense Fund publishes *In Brief*, a quarterly newsletter on environmental law; a *Docket*, or a sampling of active cases. The book *Wild by Law: The Sierra Club Legal Defense Fund and the Places It Has Saved*, text by Tom Turner and photographs by Carr Clifton, is available through mail order. In 1991, LDF established the Rick Sutherland Fund in memory of its late president.

Social Investment Forum (SIF)
430 First Avenue N, Suite 290
Minneapolis, MN 55401 USA

Contact: **Joan Kanavich**, Executive Director
Phone: 1-612-333-8338

Activities: Education • *Issues:* Ecojustice; Healthy Environment; Peace; Social Justice

● Organization

The Social Investment Forum (SIF) describes itself as a "national nonprofit professional association of individuals and investment professionals developing the concept and encouraging the practice of socially responsible investing." With a staff of 2, SIF is supported by 900 members (including more than 400 investment and research professionals); members must agree to abide by the Forum's Statement of Principles, upholding honor, integrity, and justice. Four officers, led by President John E. Schultz of Ethical Investments, Inc., and 13 board members contribute to the activities of the organization, which holds quarterly meetings across the USA.

SIF's goal is "to facilitate the beneficial use of economic power, providing information to help satisfy both financial and social concerns," and letting people know that "there are financially sound investments consistent with a commitment to a healthy environment, peace, and social and economic justice."

▲ Projects and People

The most environmentally significant move of SIF has been the initiation of the CERES (Coalition for Environmentally Responsible Economies) project in 1989 with that Coalition's **Valdez Principles**. *(See also Coalition for Environmentally Responsible Economies in this section.)*

Other activities are more general, such as offering support and advice while serving as a network and information clearinghouse for socially responsible investment. SIF determines good investments through both negative and positive screening; in the former, one avoids investments supporting oppressive governments, tobacco or alcohol producers, weapons manufacturers, or companies with substandard environmental records. In positive screening, one actively seeks companies with records of strong corporate citizenship, excellent employee relations, and safe, useful products.

SIF also participates in community investment—investing in low-income communities to improve conditions there. The Forum stresses that social investing is a growing, $625 billion industry, proving that "you can do well and do good at the same time."

■ Resources

The Forum quarterly newsletter tracks the performance of the socially responsible mutual funds; the *Social Investment Services Guide*, including 250 listings, indexed by region, describes community investments, financial institutions, financial professionals, information providers, mutual funds and money market funds, socially responsible organizations and services, and venture capital firms.

Society of American Foresters (SAF) ⊕
5400 Grosvenor Lane
Bethesda, MD 20814 USA

Contact: **William H. Banzhaf**, Executive Vice President
Phone: 1-301-897-8720 • *Fax:* 1-301-897-3690

Activities: Education; Public Affairs; Publications; Resource Policy • *Issues:* Biodiversity/Species Preservation; Deforestation; Global Warming; Sustainable Development

● Organization

The Society of American Foresters (SAF) describes its mission as "to advance the science, education, technology, and practice of forestry, to enhance the competency of its members, to establish professional excellence, and to use the knowledge, skills, and conservation ethic of the profession to ensure the continued health and use of forest ecosystems and the present and future availability of forest resources to benefit society." Founded in 1900 by **Gifford Pinchot**, SAF currently has 19,000 members throughout the USA.

▲ Projects and People

In one major national project, SAF is investigating the role of biological diversity in forest resource management, including the opportunities and tradeoffs involved with maintenance and enhancement. SAF identifies issues and problems about maintaining biological diversity on public and private lands and about forest ecosystems in general. It helps resource managers view issues of "timing and scale." Potential policy and program issues are also examined.

Another project underway relates to long-term forest health and productivity of forest ecosystems and biological diversity. It also assesses trends in demands being placed on forests and evaluates the adequacy of existing forestry knowledge, skills, and practices needed to meet them.

Proposed new approaches to forestry such as "ecosystem management" and "new forestry" are being considered for the "adequacy of their scientific base, . . . practicality for . . . implementation, and their effects on long-term forest productivity," SAF reports.

■ Resources

SAF publishes a brochure which lists books, resource policy series, proceedings of SAF-sponsored meetings and conventions, periodicals, and gift items.

Solid Waste Association of North America (SWANA)

P.O. Box 7219
Silver Spring, MD 20910 USA

Contacts: **H. Lanier Hickman, Jr.,** Executive Director
Eileen Crowe, Manager of Chapter
Programs
Phone: 1-301-585-2898 • *Fax:* 1-301-589-7068

Activities: Education • *Issues:* Waste Management/Recycling

● Organization

The Solid Waste Management Association of North America (SWANA) (formerly GRCDA), founded in 1961, "is a nonprofit education organization serving individuals and communities responsible for the management and operation of municipal solid waste management (MSWM). Dedicated to the advancement of professionalism in the fields, SWANA offers its members a variety of training programs, technical assistance, and educational opportunities to meet their evergrowing needs." A membership organization, SWANA consists mostly of "public-sector officials responsible for managing local government owned and/or operated MSWM systems." Other members are individuals and organizations responsible for corporate, commercial, and solid waste management. SWANA is organized with chapters throughout North America that tend to focus on issues concerning their local, state, or provincial governments. The staff at SWANA's headquarters helps evaluate and comments on local, state, provincial, and federal legislation and regulations in both the USA and Canada.

▲ Projects and People

SWANA runs a **Peer Match program** in conjunction with the U.S. Environmental Protection Agency (EPA) and the National Recycling Coalition (NRC). This program connects government officials looking for advice and information on solid waste management issues with those having direct experience with that problem. "Where possible, peers are used to assist their counterparts—mayors helping mayors, planners working with planners." The contact between the matched peers can involve either transfer information or be a direct on-site visit. Technical literature is also prepared and shared.

SWANA runs the Center for Regionalization of Municipal Solid Waste Management (MSWM). To foster such regionalization, the Center is available to provide both technical and policy assistance to local governments regarding merits and demerits of regional approaches; to help in the organizational structures for regional approaches; and to aid with the actual planning, formation, and operation of regional approaches.

The **Solid Waste Information Clearinghouse (SWICH)**, run by SWANA and funded by the EPA, is intended to provide assistance to government agencies, professional associations, industry, citizen groups, and other interested parties. With a library and an electronic bulletin board, SWICH contains information on source reduction, recycling, composting, planning, education and training, public participation, legislation and regulation, waste combustion, collection, transfer, disposal, landfill gas, and special wastes.

Working on the "determination of the incinerator language in the Clean Air Act reauthorization," SWANA wanted to ensure that the provisions of the act "while stringent, can be met with technology available today." SWANA was involved in the deletion of the 25 percent wastestream removal provision in the EPA-proposed New Source Performance Standards. It is now involved in a court case brought by the Natural Resources Defense Council (NRDC) to have the 25 percent provision reinstated.

The Solid Waste Action Coalition (SWAC), of which SWANA is a member, represents the interests of local government during the debate over the Resource Conservation and Recovery Act (RCRA) authorization.

The SWANA headquarters is also the site for the International Solid Wastes and Public Cleansing Association. ISWA studies the "management of solid waste worldwide and offers an opportunity for the exchange of information through journals, newsletters, various publications, international congresses, conferences, specialized seminars and exhibitions."

SWANA is also involved in the training and certification of waste management personnel. The **Manager of Landfill Operations (MOLO) Training and Certification Program** is such a program for managers of municipal and solid waste landfills and landfill enforcement officers. SWANA also has a **Professional Certification Program (PCP)** which "is available for managers, technicians, and designers of municipal solid waste management systems."

SWANA runs **workshops, symposia, technical tours, expositions,** and **training programs,** including the International Solid Waste Exposition. Courses are planned for monitoring groundwater and landfill gas, and managing household hazardous waste systems, among others.

■ Resources

SWANA publishes *Municipal Solid Waste News,* proceedings of meetings, and public policy statements ranging from resource recovery definitions to managing ash and biomedical wastes.

South and Meso American Indian Information Center (SAIIC) ⊕

1212 Broadway, #830
Oakland, CA 94618 USA

Contacts: **Nilo Cayuqueo,** Director
Karl Guevara Erb, Outreach and
Development Coordinator
Phone: 1-510-834-4263 • *Fax:* 1-510-834-4264
E-Mail: via Peacenet (cap: SAIIC)

Activities: Education • *Issues:* Biodiversity/Species Preservation; Deforestation; Indian Activities/Human Rights; Sustainable Development

● Organization

The South and Meso American Indian Information Center (SAIIC) promotes "peace and social justice for Indian people by providing information to the general public in North America, Europe, and to human rights and solidarity organizations regarding the struggles for survival and self-determination of Indian people of South and

Meso America." It also facilitates exchange and promotes direct communication and understanding between native peoples. In addition, SAIIC strives to disseminate the indigenous perspective to policy institutions whose work affects indigenous people.

SAIIC says that policy institutions, environmental groups, and human rights groups have little or no access to information regarding community-based Indian groups in Meso and South America. It contends that these communities and their lands face crises of survival on a daily basis. SAIIC monitors human rights violations, receives calls and notices of violations, and gets out alerts to call for support.

Indian people in the north need access to information regarding the struggles of their brothers and sisters in the south, and vice versa. SAIIC has found that the increased communication among Indian people has been empowering for all.

▲ Projects and People

SAIIC programs and projects include an Indian Visitor program that brings people together from various cultures; human rights advocacy; Environment Program; coordinating the 500 Years Resistance project; arranging the Indian Women's Project and the Community Organizing Project; public presentations; and a Resource Library for periodicals and videos. SAIIC is engaged in grassroots organizing and influencing policies of government agencies, NGOs, and finance institutions.

One of SAIIC's projects, The First Continental Meeting of Indigenous Peoples—500 Years of Indian Resistance, brought together more than 300 delegates and supporters from nearly every country in the hemisphere. SAIIC co-organized the five-day conference in Ecuador with the Confederation of Indian Nations of Ecuador and the National Indigenous Organization of Colombia. According to SAIIC, delegates debated, amended, and approved the work of eight commissions on issues of land rights, educational and cultural development, sovereignty, and legislative struggles.

The conference resolved to send a 500-member delegation to Spain in 1992 to "present demands for redress and reparations for the Spanish conquest on the occasion of the Olympic Games and the first congress of the European Community in Barcelona. Delegates also called for an international tribunal on continuing human rights abuses against Indian people, and for governments, churches, and multinational companies to honor indigenous rights to ancestral lands," as reported by *World* publication (August 31, 1990).

In the meantime, SAIIC helped coordinate the 1992 Native Network and the 1992 All Peoples Network Conference, both held in California. An alliance formed at the All Peoples Network Conference adopted the name Resistance 500. "Through these and other gatherings, a stronger, multicultural, multifaith network is being built for actions and support for indigenous struggles for self-determination," writes SAIIC.

■ Resources

SAIIC maintains Information Services, a program area that has available the *SAIIC Newsletter, Urgent Action Bulletins*, library, electronic communications, Women's Networking, research, and distribution of yearly campaign materials such as conference reports, news clippings, and schedules of events.

With Rainforest Action Network, Amazonia Film Project, and International Rivers Network, SAIIC co-authored *Amazonia—Voices from the Rainforest, A Resource and Action Guide*, designed to give added force to grassroots groups in the Amazon fighting in defense

of the rainforest and basic human rights for indigenous people there.

The *Quito Resolutions* from the Conference on 500 Years of Indian Resistance are also available for a fee. The SAIIC newsletter is provided free, in Spanish and English, to Indian organizations in South and Meso America.

Southern Environmental Law Center (SELC)
201 West Main Street, Suite 14
Charlottesville, VA 22901 USA

Contact: **Katherine E. Slaughter**, Staff Attorney
   Phone: 1-804-977-4090 • *Fax:* 1-804-977-1483

Activities: Development • *Issues:* Billboards and Sign Control; Energy; Public Lands; Water Quality; Wetlands and Coastal Areas

● Organization

"The Southern Environmental Law Center (SELC) is a regional, nonprofit, public interest law firm committed to providing effective legal representation, in court and before regulatory agencies, to environmental and conservation interests in the South." Concerned with protecting this region's natural resources and beauty, SELC divides its work into five areas: public lands, wetlands and coastal resources, toxics and water quality, billboards and visual pollution, and energy efficiency.

SELC was co-founded in 1986 by Rick Middleton, now executive director, and attorney Jim Dockery. Currently, SELC has eight attorneys plus a support staff of five. With an office in Charlottesville and another in Chapel Hill, North Carolina, SELC is involved in activities throughout the South. The North Carolina office is run by Lark Hayes who has a background in environmental advocacy. Legal services are provided free of charge.

▲ Projects and People

The Public Lands Project, led by David Carr, protects such lands from development. A recent success was the blocking of a proposed airport that was to be built on Georgia's scenic Chattooga River banks "that would have destroyed [its] natural character." Carr says that he has "seen SELC's legal work translate the energy of local conservationists into concrete results."

According to SELC, the Shenandoah National Park, Virginia, is the country's "most polluted" national park, with decreased visibility and sulfur pollution caused by acid rain. In 1991, SELC joined with the National Park Service and a coalition of other organizations to prevent further degradation by bringing action against Virginia utilities who propose to "increase electric-generating capacity in the state by 50 percent over the next decade." About 19 applications are pending for small, coal-burning power plants; as the state approves permits, SELC appeals them to the U.S. Environmental Protection Agency (EPA).

As a result of Hurricane Hugo, 70 percent of the Francis Marion National Forest outside Charleston, South Carolina, was destroyed; SELC assisted in a new management plan to restore the native ecosystems.

In Alabama, Georgia, and Tennessee, SELC is working with another coalition to assess the effect of potential wood chip mills planned along the Tennessee River. The chips would be prepared for pulp and paper manufacture mainly in Japan and Korea. Such mills pose a "major threat to forests, wildlife, water quality, and biodiversity in the region." It is expected that as much as seven million acres of hardwood forests would be harvested by clearcut means.

An undertaking of the North Carolina office is the Coastal and Wetlands Project with coastal issues specialist Derb Carter. While Americans love the coast, they have also done a good deal to threaten its existence, according to SELC. A current wetlands project is working to save the remaining 2,000 acres of the East Dismal Swamp that once comprised more than 100,000 acres. SELC with The Nature Conservancy helped save Georgia's Cumberland Island National Seashore from "detrimental dredging."

Recently, SELC stopped a condominium project pier on Okracoke Island, North Carolina, to prevent adverse impacts on shellfish beds and the area's unique environment. It joined the Sierra Club to protect a tract of wetlands near Atlanta, Georgia. SELC succeeded in postponing exploratory oil and gas well drilling off the North Carolina coast until 1992 so that impact studies on estuaries, wetlands, barrier islands, and offshore waters could be completed. It is helping the South Carolina Coastal Conservation League to halt highway expansion through ACE Basin and its "exceptional wetlands, water quality, and wildlife habitat."

The Clean Water Project provides free legal assistance to local environmental groups that are working to protect clean water. Under attorney Deborah Wassenaar's leadership, the project is working in a number of different communities. Representing the Cahaba River Society against further discharges from coalbed methane extraction wells into the Cahaba River is one example. Another is SELC action to ensure improved sewage effluent treatment elsewhere. SELC met recently with ALCOA to investigate their controls against discharging PCBs and other pollutants into surface waters near Knoxville, Tennessee.

The Billboard and Sign Control Project, with project head, Kay Slaughter, is helping towns pass ordinances that will limit billboards and is defending these laws in court if necessary. It also seeks to protect the character of southern communities and give them the legal help they need to stand up against the companies that want to put up billboards. SELC gives legal guidance to Scenic America and its fight to get the U.S. Congress to ban new billboards on federal and rural highways.

Jeff Gleason and Rick Parrish run the Energy Project. Believing that much of the future energy needs of the South can be obtained by employing energy-saving techniques instead of building new power plants, Gleason and Parrish are working to increase the efficiency of energy use. "We have initiated a region-wide regulatory reform effort to encourage and require electric utilities . . . to make direct investments in 'demand-side management programs,'" reports SELC. "Investments in demand-side programs [should be] on an equal footing with new power plant construction."

■ Resources

SELC publications include: *A Citizens' Guide to Protecting Wetlands in Georgia*; *A Citizens' Guide to Protecting Wetlands in Alabama*; *A Citizens' Guide to Protecting Wetlands in South Carolina*; *Visible Pollution and Sign Control: A Legal Handbook on Billboard Reform*; *A*
Road with a View: A Legal Handbook on Billboard Reform in North Carolina; and *Visible Pollution and Billboard Reform in Tennessee*.

Southern Women Against Toxics (SWAT) ❦
P.O. Drawer 1526
101 West Monroe Street
Livingston, AL 35470 USA

Contact: **Linda Wallace Campbell**, Director
Phone: 1-205-652-9854 • *Fax:* 1-205-652-9854

Activities: Education; Research; Organizing Communities
Issues: Waste Management; Stopping Garbage and Hazardous Waste Facilities

▲ Projects and People

Linda Wallace Campbell is an activist and a community organizer devoted to a wide variety of causes. She started her environmental activist career in 1985 with a letter to an editor; now she works in various capacities with grassroots organizations around the country. Her organization, **Southern Women Against Toxics (SWAT)**, is fighting garbage and hazardous waste facilities from its location which she says is "17 miles from the largest waste dump in the world" in Sumter County, Alabama.

Despite its name, the board of SWAT includes both men and women and is working against hazardous waste activity around the country. Campbell believes in community organizing, and will "go to the aid of the next community that calls me because we are being assaulted by pollution-for-profit organizations who poison the land, ruin the water and pollute the air, take the money and move on." She also believes in taking her fight to the street, organizing, and protesting to reach her goals. Campbell carries her message as well to the national media and speaks before universities and colleges in Alabama, California, Louisiana, Maryland, Mississippi, Pennsylvania, and Washington, DC; she has appeared before the American Civil Liberties Union and the United Nations.

Another of Campbell's major goals is for SWAT to teach other women how to carry on environmental protests. She means to empower these women by helping them gain the skills that they need to carry on this type of activity. She also wants them to learn from the experiences of "the first generation of women environmental activists."

SWAT is organizing African-American mayors of towns to prevent a well for drinking water from being constructed in close proximity to the 2,700-acre toxic waste dump at Emelle, Alabama. SWAT is also working in conjunction with Ala-Eco, another group founded by Campbell, to limit out-of-state waste dumping in Alabama. Campbell is also showing citizens of Mexico and New Zealand how to resist garbage and waste dumps in their respective countries.

"What will we leave for the seventh generation?" is a philosophical question raised by Campbell and based on the Native American's reverence for life.

SouthWest Organizing Project (SWOP) ❦

211 10th Street SW
Albuquerque, NM 87102 USA

Contact: **Jeanne Gauna,** Director
    Phone: 1-505-247-8832 • *Fax:* 1-505-247-9972

Activities: Education; Political/Legislative; Research • *Issues:* Air Quality/Emission Control; Energy; Health; Sustainable Development; Water Quality

● Organization

Since 1981, SouthWest Organizing Project (SWOP) has existed as a "multi-racial, multi-issue, grassroots, membership community organization." Devoted to community organizing, SWOP confronts a variety of different issues including the environment, racism, sexism, and discrimination by age or social class as it works for "social and economic justice."

▲ Projects and People

SWOP's **Community Environment Program** wants to protect communities from environmental hazards. Using its "Community Bill of Rights" as a platform, SWOP addresses environmental issues in conjunction with issues of "empowerment, information, and the responsibilities of industry and government."

This program has met with a number of successes: Ponderosa Products, Inc., was forced to clean up groundwater in Albuquerque's Sawmill neighborhood. In addition, SWOP forced a reduction of noise and sawdust emissions. The group also got a military base in Mountainview to "communicate directly with the community over its groundwater contamination of its drinking water."

Related community development and citizenship participation programs emphasize education on economic issues, housing stability, and voter registration to "promote greater accountability among politicians and officials." As a result, more than 30,000 ethnic Americans are new voters.

■ Resources

A descriptive brochure with membership form is available. SWOP makes presentations to schools and civic groups.

Species Survival Commission

(See World Conservation Union in this section.)

State and Territorial Air Pollution Program Administrators/ Association of Local Air Pollution Control Officials (STAPPA/ ALAPCO)

444 North Capitol Street, NW
Washington, DC 20001 USA

Contact: **S. William Becker,** Executive Director
    Phone: 1-202-624-7864 • *Fax:* 1-202-624-7863

Activities: Development; Political/Legislative; Research • *Issues:* Air Quality/Emission Control

● Organization

State and local governments must design plans that meet the air quality standards of the Clean Air Act. State and local officials play an important role in combatting problems such as acid rain, toxic air pollutants, and indoor air pollution.

The **State and Territorial Air Pollution Program Administrators (STAPPA)** is the "national association of state air quality officials in the 54 states and territories of the United States." The **Association of Local Air Pollution Control Officials (ALAPCO)** is the "national association representing air pollution control officials in over 150 major metropolitan areas across the United States." These two organizations share joint headquarters in Washington and work closely together.

These organizations help the air pollution controllers and managers exchange information, skills, and resources—thereby to improve air quality. In addition, the headquarters is a liaison office with the U.S. Environmental Protection Agency (EPA), U.S. Congress, environmentalists, and industrial organizations.

Issues that STAPPA/ALAPCO have worked on include "acid rain, long-range transportation of air pollution; toxic air pollutants; motor vehicle emissions, including lead in gasoline and emissions from refueling automobiles; and indoor air pollution in our homes and schools."

■ Resources

Publications of STAPPA/ALAPCO include: *Summary of the Clean Air Act Amendments of 1990; Highlights of Global Warming Control and Stratospheric Ozone Protection Activities in Selected States; Comments of STAPPA/ALAPCO on Conference Issues of the Clean Air Act; Summaries of S. 1630 & H.R. 3030; STAPPA/ALAPCO Membership Directory; Comments of STAPPA/ALAPCO on the Administration's Clean Air Proposal; Toxic Air Pollutants: State and Regulatory Strategies 1989; Summary of Selected Innovative Ozone and Carbon Monoxide Control Strategies; Air Permit and Emissions Fees;* and *Summary of State Acid Rain Laws.* A publications list is available.

Student Conservation Association (SCA)

P.O. Box 550
Charlestown, NH 03603 USA

Contact: **Scott D. Izzo,** President
    Phone: 1-603-543-1700 • *Fax:* 1-603-543-1828

Activities: Education • *Issues:* Biodiversity/Species Preservation; Conservation Services

● Organization

The Student Conservation Association (SCA) is a nonprofit organization with 15,000 members. It provides opportunities for student and adult volunteers to "assist in the stewardship and conservation of natural resources." SCA places volunteers in national parks, forests, wildlife refuges, and other public lands, where they have the opportunity to work with the rangers and get firsthand experience working with the environment.

▲ Projects and People

The Henry S. Francis, Jr. Wilderness Work Skills Program provides a chance for participants to work with forest professionals, conservation corps members, trail crew, and SCA supervisors. These are comprehensive courses, and the participants get to "try their hands at designing and maintaining trails, mending erosion damage, building rustic structures with timber and stone, and mastering the fine art of sharpening crosscut saws."

SCA runs a high school program giving students ages 16 to 18 "the opportunity to live and work for a month or more in magnificent outdoor settings." They live in backcountry camps while working on conservation projects, and backpack or canoe through the region. The work done by the crews includes trail construction and maintenance work, repairing damaged campsites, lakeshores and meadows, or building rustic bridges.

In addition, SCA runs a variety of special programs including several international exchanges bringing together young Americans with students from another country to complete conservation projects in both countries. Such successful exchange programs have occurred with the former Soviet Union and Germany. Other targeted programs have included those for inner-city children, and for deaf students.

The Resource Assistant (RA) Program "places volunteers with land management agencies for 12 weeks of intensive training and service alongside professional rangers, foresters, scientists, and other environmental specialists." RAs must be at least 18 years old. The RAs "conduct a wide range of recreation and interpretive programs for the public, safeguard remote backcountry, repair trails and meadows, and survey rare plants, mammals and seabirds." There are an increasing number of openings around the country, and opportunities exist for RAs to participate in a wide range of conservation activities.

The New Hampshire Conservation Corps is a seven-week employment and training program for boys and girls, many of whom are "at-risk or disadvantaged." While participants are involved in conservation projects, the curriculum also includes reading, writing, and math skills. This program is designed to teach the participants new skills, provide them with new opportunities, and help them to build their confidence and self-esteem.

The award-winning Greater Yellowstone Recovery Corps (GYRC) is involved in restoration and revegetation work in Yellowstone National Park. In 1988 fires swept through the region, and while many of the natural elements will recover on their own, GYRC is working to rebuild bridges, trails, and campsites. They are also working to encourage new growth in the areas that were cut or bulldozed by firefighters. GYRC directs volunteers in the rebuilding process

The Campground Administration and Maintenance Program (CAMP) is an opportunity for youth with developmental disabilities and other special needs to assist in managing the U.S. Forest Service Campgrounds. The participants live five days a week in "well-supervised rustic settings." They take part in the chores of the crew, and assist in important tasks. It is the first time many of the participants have been away from home, and "supervisors, parents and teachers have noted definite improvements in independence and self-esteem following a student's participation in the program." The CAMP Internship Program also provides college students on-site training for work with special populations.

Conservation Resources Incorporated (CRI) is a wholly owned for-profit subsidiary of SCA. "It is designed to take advantage of contracting opportunities not available to nonprofit organizations, including trail and resource management contracting, concession campground operations, and consulting." The profits generated by CRI will be used to support SCA programs.

■ Resources

SCA publishes *Earth Work*, a magazine that is designed to help people find work "to protect the land and the environment." It includes articles and job listings.

The Tarlton Foundation
50 Francisco Street, Suite 103
San Francisco, CA 94133 USA

Contact: **Dru Devlin**, Program Coordinator
Phone: 1-415-433-3163 • *Fax:* 1-415-989-2810

Activities: Education • *Issues:* Marine Science Conservation Education

● Organization

Formed in 1991, the Tarlton Foundation was named in memory of Kelly Tarlton, a New Zealand marine archaeologist and conservationist who developed the original Underwater World in Auckland, New Zealand. This aquarium concept lets the visitor walk through a tunnel surrounded by an aquarium with interpretive and interactive exhibits, thus simulating a "walk on the ocean floor."

The Tarlton Foundation took over the marine education programs of Ocean Alliance, a California-based, nonprofit organization. Affiliated with the Underwater World Aquarium, it developed a marine science education curriculum for children that is aligned with the host state's science framework and is adapted to the local region. At the Tarlton Center for Marine Education and Research, it also provides training for teachers and prepares materials.

▲ Projects and People

Early relationships between humans and gray whales were "violent and bloody," reports Tarlton's Adopt-a-Whale Program. The Atlantic gray whale population became extinct, and very few of the western Pacific population survive. Since 1972, the Marine Mammal Protection Act and international treaties are protecting the gray whale, which are making a comeback. But because whales live along the coast and migrate near populated areas, they are potentially threatened by offshore oil production, spills, and habitat destruction. Driftnets and ocean pollution also pose hazards for the gray whale.

Carrying on the work of the Whale Center and the Oceanic Society, the Adopt-a-Whale Program supports education and research. Each person who adopts a whale receives detailed information about gray whales and quarterly updates regarding their migration.

The Tarlton Foundation sponsors a WhaleBus, a classroom on wheels, which travels to over 20,000 children each year. The Foundation recruits students and other youth groups to care for an adopted area of seacoast or lake in an Adopt-a-Beach program. The Foundation has a Sea Camp offering marine science activities for children aged 6 to 14 in a summer residential and day camp. The program also emphasizes scholarship participation of underprivileged children.

Reminding youngsters that water covers more than 70 percent of this planet and that by the year 2000, some 80 percent of all Americans are expected to live along the coasts, Project OCEAN brings marine science into classrooms throughout the country and in Mexico and calls attention to the "unprecedented pressure on our marine and freshwater resources"—offering teacher training, educational materials, special "OCEANS weeks," and follow-up programs. Children have an opportunity to build three-dimensional tidepools, dissect squids and clams, explore a kelp forest, test water quality, and take field trips to lagoons, beaches, rivers, sewage treatment facilities, and aquariums.

■ Resources

The Tarlton Foundation offers *Project OCEAN Habitat Guides Adopt-a-Beach Curriculum*, and the Adopt-a-Whale program. They need interns for marine education programs, volunteers for fundraising to stimulate program support, and brochures and printing for publicity packets.

Terrene Institute
1717 K Street, NW, Suite 801
Washington, DC 20006 USA

Contact: **William Funk, Ph.D.**, President
Phone: 1-202-833-8317 • *Fax:* 1-202-296-4071

Activities: Education; Research • *Issues:* Air Quality/Emissions Control; Biodiversity/Species Preservation; Global Warming; Health; Sustainable Development; Waste Management/Recycling; Water Quality • *Meetings:* Alliance for Environmental Education Conference; International Association of Water Pollution Control; North American Lake Management Society Symposium; Water Pollution Control Federation Annual Conference

● Organization

The new Terrene Institute "is founded on the premise that maintaining a balanced environment is not a 'good guys vs. bad guys' issue—that human beings, when given the right information, will be inclined to act in ways that benefit the community." In the spirit of this philosophy, Terrene makes education, and the research to make its information credible, the cornerstones of the Institute. The Terrene Institute is a not-for-profit organization with a staff of 18 under its president **Dr. William Funk**.

▲ Projects and People

The effects and control of toxic pollution on lakes is a project that conducts "research and updating of technical procedures for identification and control of toxic substances in lakes. The **public outreach** project includes "the creation of educational programs, including posters, brochures, handbooks, technical documents, and seminars/meetings/workshops designed to educate both the citizen and the technician on environmental quality source issues." The **fisheries management** project is involved in the research and prepa-

ration for "scientifically-defensible" lake protection, management, and restoration procedures.

Terrene is also involved in the **facilitation of a national nonpoint source organization**. The Institute wants to organize and integrate the many organizations concerned with nonpoint source issues into a grassroots coalition. It is also publishing *Nonpoint Source NEWS-NOTES* for the U.S. Environmental Protection Agency (EPA). This is an "occasional bulletin dealing with the condition of the environment and the control of nonpoint sources of water pollution."

Terrene runs a **Clean Lake Clearinghouse** resource center for information on lake restoration, protection, and management. The Clearinghouse was initiated by the EPA's Clean Lakes Program to provide information to "federal staff, to state and local lake managers and associations, and to other academics and researchers." It includes a bibliographic database available to the public on EPA's Nonpoint Source Bulletin Board. Institute staff also maintains the database and assists users.

Former president Dr. Harvey Olem is the co-author of the recent *International Lake and Watershed Liming Practices, Diffuse Sources of Water Pollution*, and *Practical Guide to Managing Acid Surface Waters and Their Fisheries*. He has been a "professional volunteer" for the Alliance for Environmental Education; Federal Interagency Task Force on Acid Precipitation; North American Lake Management Society; Tennessee Academy of Scientists; *Water, Air, and Soil Pollution: An International Journal;* Water Pollution Control Federation; and Water Quality 2000, regarding urban and rural runoff.

■ Resources

The Terrene Institute publishes a number of publications on nonpoint source and clean lake issues; a list is available. Terrene has prepared a 10-minute educational video for lake associations that are about to begin volunteer lake monitoring programs. With EPA, Terrene has also produced glossy posters on issues such as pollution, urban water quality, and rural water quality. The Institute wants to extend its educational outreach to groups working with teacher training.

Texas Center for Policy Studies (TCPS)
P.O. Box 2618
Austin, TX 78768 USA

Contact: **Mary E. Kelly**, Executive Director
Phone: 1-512-474-0811

Activities: Education; Political/Legislative; Research • *Issues:* Deforestation; Environmental Implications of Economic and Trade Policy; Health; Sustainable Development

● Organization

The guiding principle of the **Texas Center for Policy Studies** (TCPS) is that "the adoption or alteration of economic development and financial policies *must* take into account the environmental and public health consequences." TCPS is a nonprofit organization that was established "to offer research, organizing, technical assistance, and policy development services on a variety of state, national and regional environmental issues."

▲ Projects and People

The TCPS **Free Trade Watch** project began in 1991 to monitor and begin documenting the potential environmental effects of a free trade pact between Mexico and the USA. With a border already suffering from environmental degradation, neither government anticipated the effects and the infrastructure needs of increased trade between these countries. TCPS is worried that if the free trade agreement goes through, the pace of environmental deterioration will accelerate. TCPS wants to develop policy alternatives that will protect the already threatened environment. One negative environmental impact of transnational corporations, or *maquiladoras*, relocating in Mexico is that U.S. industries are not required to provide "adequate sewers, water, or housing" for workers who live in nearby communities.

The Free Trade Watch project is related to the **Binational Project on the Environment.** TCPS and Bioconservación, A.C., a Mexican environmental group, are working together to establish a network of environmental and community organizations in Texas and the four bordering states of Mexico. They believe that they need to integrate sustainable development policies into the major economic projects in the region. Task forces are working on such issues as water supply, water quality, forestry development, and habitat destruction.

One successful example of work done by this project involved getting the World Bank to **stop loan disbursements** to Mexico for a **logging project.** The logging would have taken place in an area that contains over one-third of Mexico's remaining coniferous forests. The TCPS report showed this and other flaws in the World Bank project area, such as the Bank's lack of enforcement of its loan conditions. The disbursements have been halted until the full environmental impacts can be studied.

The **Rural Economic Policy Project** seeks to "survey and describe the current rural economic development environment in Texas with an eye toward developing a series of comprehensive policy recommendations for state leaders." This comes in response to the growing economic problems in rural Texas. TCPS wants to develop programs that will help the economy in rural Texas without harming the environment.

TCPS is also working on environmental issues that tend to affect people of color. In West Dallas, a predominately African-American area, the people have lived for years exposed to a wide variety of toxics. TCPS believes that communities of color have been marginalized in Texas both in terms of environmental issues and economic development. TCPS's **Toxics Organizing Project** is working to develop "a statewide network between African-American, Latino, and other communities of color to address these urgent environmental justice and economic development concerns."

TCPS also has **Savings and Loan Bailout–Related Projects.** These projects seek to move parcels of land that are now controlled by the Resolution Trust Corporation (RTC) or the Federal Deposit Insurance Corporation (FDIC) into the hands of county governments. This creates opportunities to acquire regions of land, many of which would have been slated for future development. These lands are now saved and protected for discounted prices. *(See also TCPS, Environmental Opportunities of the S&L Bailout in this section.)*

■ Resources

Publications of TCPS include *Texas Center for Policy Studies Newsletter; The Other Side of the Bailout: A Guide to Protecting Environmentally Significant RTC and FDIC Lands; Environmental Opportunities of the Crisis in the U.S. Financial Institutions; A Response to the Bush Administration's Environmental Action Plan for Free Trade Negotiation with Mexico; Mexico-U.S. Free Trade Negotiations and the Environment: Exploring the Issues; Evaluation of the Forestry Development Project of the World Bank in the Sierra Madre Occidental in Chihuahua and Durango, Mexico; Overview of Environmental Issues Associated with Maquiladora Development along the Texas Mexico Border; Lower Rio Grande Valley: Drinking Water Quality and Health Effects; Pesticide Regulation and the Texas Department of Agriculture;* and *The Pesticide Crisis: A Blueprint for the States.* A publications list is available.

Environmental Opportunities of the S&L Bailout
Texas Center for Policy Studies (TCPS)
c/o Henry, Kelly, & Lowerre
2103 Rio Grande
Austin, TX 78707 USA

Contact: **Richard W. Lowerre, Attorney**
Phone: 1-512-479-8125 • *Fax:* 1-512-479-8269

Activities: Legal; Political/Legislative; Research • *Issues:* Air Quality/Emission Control; Energy; Health; Sustainable Development; Waste Management/Recycling; Water Quality

▲ Projects and People

The Texas Center for Policy Studies (TCPS), a nonprofit organization, says it is the first conservationist group to raise environmental questions about the savings and loan and banking crises that are costing American citizens, billions, possibly one trillion dollars.

What does the S&L boom, bust, and bailout mean for the nation's environment? What wetlands and habitats were lost? How can domestic financial policy be integrated with the nation's environmental protection goals to avoid both fiscal and environmental bankruptcy?

A leader in researching the answers, the TCPS "bailout project" team is pioneering the effort to infuse federal banking agency actions with environmental protection. The goal is similar to that of global conservationists who seek to integrate environmental protection into international development banking through "debt-for-nature swaps."

Their conclusion is startling and hopeful: "The bust in banks and S&Ls and the resulting transfer of real estate assets into the hands of the federal deposit insurance agencies, coupled with the national slump in real estate markets, provides perhaps the best opportunity for protecting lands having special natural, cultural, or recreational values." This is a "rare opportunity," says the TCPS team, "to protect environmentally and culturally significant lands slated for the bulldozer, but now owned or controlled by the federal government."

With the support of the National Wildlife Federation, The Nature Conservancy, and Sierra Club as well as funding from the Rockefeller Family Fund, Education Fund of America, Schumann

Foundation, and National Fish and Wildlife Foundation, TCPS embarked on a countrywide program to help environmental groups identify significant lands that can be purchased for habitat protection, recreation, and open space.

For example, in a collaborative venture with the Resolution Trust Corporation (RTC) that manages the real property assets previously held by failed banks and savings and loans, the U.S. Fish and Wildlife Service, and local government, civic, business, and environmental leaders—a regional habitat conservation plan was designed to save nine endangered species in the Austin area. Other cities in Texas, California, Arizona, Colorado, and Florida are taking steps to acquire RTC- and Federal Deposit Insurance Corporation (FDIC)–held properties for parks and refuges.

Richard Lowerre, attorney with a private law firm, is the consultant with the Texas Center for Policy Studies working on environmental opportunities of the S&L bailout. His other current projects are on binational environmental cooperation between such groups in Texas and northern Mexico, and on forestry evaluation in northern Mexico that is funded by the World Bank.

The law firm also represents citizens groups and non-government organizations in a range of issues including pollution from oil and gas production in Ecuador, Peru, Panama, and Texas; and also in Texas—proposed hazardous waste disposal facilities and radioactive waste disposal sites, uranium and coal strip mining; and water-quality protection in lakes, rivers, and reservoirs.

Textbook Review Project ❧
3311 Bryker Drive
Austin, TX 78703 USA

Contact: Irene Pickhardt
Phone: 1-512-451-6820

Activities: Education • *Issues:* Air Quality/Emission Control; Biodiversity/Species Preservation; Deforestation; Energy; Global Warming; Health; Population Planning; Sustainable Development; Transportation; Waste Management/Recycling; Water Quality

● Organization

A Chinese proverb says, "If you are thinking a year ahead, sow a seed. If you are thinking 10 years ahead, plant a tree. If you are thinking 100 years ahead, educate the people." Such is the inspiration for the **Textbook Review Project** created by the **Lone Star Chapter of the Sierra Club** and the **Audubon Council of Texas**. The mission is to "monitor and review the environmental content of textbooks before they are purchased for student use." The group influences the specifications issued to publishers about the environmental content of textbooks in health, social studies, and science.

The impact of textbooks is enormous. While the Textbook Review Program only works in Texas, the change in one social studies textbook will reach 12 million students over the life of the book. Texas is also the second largest textbook market, so changes required by Texas are often reflected nationwide.

Textbooks are "teachers' most important teaching tool." The attitudes and information conveyed through textbooks shape the beliefs of future generations, reports program spokesperson Irene Prickardt, and any group "interested in creating environmentally responsible future citizens" can adopt this "simple, direct, and

inexpensive intervention." This program also "educates teachers, administrators, school-board members, and the students about the need for adequate information about air pollution reduction, drinking water protection, energy pollution control, and other areas of environmental concern."

Threshold, Inc.
International Center for Environmental
   Renewal ⊕
Drawer CU
Brisbee, AZ 85603 USA

Contact: **John P. Milton**, Chairman
Phone: 1-602-432-5814

Activities: Education; Research • *Issues:* Biodiversity/Species Preservation; Deforestation; Health; Sustainable Development; Water Quality • *Meetings:* UNCED

● Organization

Threshold was established in 1972 by **John P. Milton** as "an independent, non-governmental, nonprofit, and international center serving to improve mankind's understanding of and relationship to the environment. A major focus is on the development of ecologically sound alternatives for practical application in human society." Threshold's work over the years has been to create a global awareness of the destruction of tropical rainforests.

With field projects worldwide, Milton recently launched a related effort, **Sacred Passage**, which combines wilderness skills with "deep ecology meditation." These "passages" are being conducted with Milton in Arizona, Colorado, West Virginia, and Mexico.

▲ Projects and People

Threshold's early tropical rainforest projects were established to support the rubber tappers and the indigenous tribal peoples of the Amazon Basin.

The **Environmental Crisis Fund** was "created to help heal the major ecological threats now affecting planet Earth," with monies being dispatched "quickly and directly" to projects in environmentally critical areas.

Other projects include the production of **Seeds of Hope** video and traveling media exhibit on how local groups can preserve Asian tropical forests; and of a **Coral Reef** video and slide show on the uniqueness and fragility of the only "true" coral reef in the Gulf of Mexico. Milton describes the Pulmo coral reef as threatened by the increasing development of Baja California's Cape Region and associated tourist demands.

The **Organic Alternatives to Pesticides Project** in Baja California, Mexico, was designed to educate the Mexican people and government about pesticides such as malathion and their negative health and ecosystem effects.

To demonstrate **ecologically sound housing**, a video and slide show illustrates straw bale–based home construction which, according to Threshold, eliminates the impact on the forest, is highly energy efficient, very low cost, and can be implemented easily on a "do-it-yourself" basis.

Regarding his Sacred Passage, Milton prepares participants for solo retreats into the wilderness "to live as close to nature as possible, without artificial light, reading material, or unnecessary distractions." Milton believes that "for thousands of years, Nature has offered the space and time to plunge deep into new spiritual insight, face doubts, and fears, find healing and serenity."

■ Resources

Interested persons can order an information packet with cost details for a Sacred Passage. Brochures are available about both organizations.

TRAFFIC USA ⊕
World Wildlife Fund (WWF)
1250 24th Street, NW
Washington, DC 20037 USA

Contact: **Ginette Hemley**, Director
  Phone: 1-202-778-9605 • *Fax:* 1-202-293-9211

E-Mail: Econet: wwftraffic

  Telex: 64505 PANDA

Activities: Education; Research • *Issues:* Biodiversity/Species Preservation; Deforestation; Sustainable Development • *Meetings:* CITES

● Organization

TRAFFIC (Trade Records Analysis of Fauna and Flora in Commerce), is an international network that monitors international trade in wildlife and wildlife products. Its purpose is to stop illegal wildlife trade as well. A program of the **World Wildlife Fund** (WWF), TRAFFIC has 10 affiliated offices worldwide. The TRAFFIC network also works in conjunction with the World Conservation Union (IUCN) and the CITES Secretariat.

As the Convention on International Trade in Endangered Species of Wild Flora and Fauna, CITES is designed to prevent trade of such species. TRAFFIC is an objective source on information about world wildlife trade whose reports and analyses are used by other international agencies, governments, and NGOs. TRAFFIC also serves as a source of information on the wildlife trade for the media and public.

▲ Projects and People

In monitoring the U.S. bird trade, findings of TRAFFIC USA show that while imports of wild-caught birds have fallen, trade still is a major industry. TRAFFIC is a part of **The Cooperative Working Group on Bird Trade**, which has developed a set of recommendations including the elimination of wild-caught bird trade over the next five years and the strengthening of federal controls. Smuggling remains a serious problem, with the U.S. Department of Justice estimating that "some 150,000 exotic birds are smuggled into the United States from Mexico each year."

TRAFFIC also conducted a study which shows that **trade in bear parts** in several Asian countries is much more widespread than official government statistics show. In the Far East, bear gall is considered a medicinal treatment for acute human illnesses of the gall bladder, liver, spleen, and stomach, according to this report. Served in Asian restaurants, bear paw is considered a body-toning

delicacy. TRAFFIC research reveals that most Asian bear species are threatened with extinction, including Asiatic black bear (*Ursus thibetanus*), sloth bear (*Melursus ursinus*), sun bear (*Helarctos malayanus*), and certain brown bear (*U. arctos*) populations. TRAFFIC wants the American black bear (*U. americanus*) added to the CITES list of bear trade regulation to protect it from such slaughter. TRAFFIC contends that illegal trade is prevalent because illegal bear parts are labeled as legal.

TRAFFIC also monitors a variety of other wildlife trade that is covered by CITES, and publishes relevant information. **Ginette Hemley**, TRAFFIC USA director, is also in charge of the World Wildlife Fund's **Buyer Beware** program, which educates travelers and consumers about threatened wildlife and teaches them how to avoid purchasing endangered animal products.

■ Resources

The *TRAFFIC USA* newsletter is published monthly. A fact sheet with list of TRAFFIC offices is available. Also available is a brochure that lists products to avoid while traveling abroad, such as those made from reptile skins in Latin America, Nepal, India, and Pakistan; sea turtle products; most wild-bird feathers; large parrots such as macaws and cockatoos; and products made from elephant ivory and wild cats, such as jaguar, leopard, snow leopard, and tiger.

(See also TRAFFIC in the Argentina, Malaysia, UK, and Uruguay NGO sections.)

TreePeople ❦
12601 Mulholland Drive
Beverly Hills, CA 90210 USA
Contacts: **Andy and Katie Lipkis**
  Phone: 1-818-753-4600 • *Fax:* 1-818-753-4625
  E-Mail: MCI: TreePeople

Activities: Education • *Issues:* Urban Forestry

● Organization

In 1973, TreePeople founder Andy Lipkis organized his "fellow summer campers to pull up an old parking lot and plant trees [to] survive the smog creeping up the San Bernardino Mountains." Then he got the California Department of Forestry to contribute "8,000 saplings about to be destroyed" for a "California conservation project"—thus launching this new nonprofit organization, TreePeople.

Today, with a membership of 27,000 and a staff of 36, TreePeople promotes "personal involvement, community action, and global awareness as the solution to environmental problems." It encourages people to take steps as individuals to help the environment by planting trees in the Los Angeles area and to support similar programs in developing nations.

▲ Projects and People

The **Citizen Forester Program** has been so successful in training people to plant trees that some of these individuals are now running tree-planting programs throughout Los Angeles. Citizen foresters learn how to choose sites, get permission from the landowners or government, use community outreach skills, and understand urban tree care. They advise on selecting a tree species that will thrive in the site picked. This program does require a commitment from the participants to actually undertake their project and follow through with it.

Citizen foresters work with groups such as The Nature Conservancy, which planted 6,000 cottonwood and willow saplings to restore the natural habitat of the yellow-billed cuckoo. Children from the Jewish National Fund planted hundreds of seedlings in a park; Mothers Against Drunk Driving (MADD) established a memorial planting of 300 Canary Island pine seedlings; and on a 7-mile stretch of Martin Luther King, Jr. Boulevard, thousands planted similar pine trees in honor of Dr. King whose quote, "everybody can be great because anybody can serve," inspires this organization.

TreePeople believes that neighborhood projects can greatly enhance the localities in which the trees are planted in beautifying the region and helping the environment, such as acting as windbreaks, for example. These projects also provide a sense of community where members can meet and get to know one another. The group reminds their fellow citizens, "The best time to plant a tree was 20 years ago. The second best time is now."

The Environmental Leadership Program "takes TreePeople's message of individual power and responsibility straight to those responsible for the Earth—our young people." TreePeople's educators have worked with more than 100,000 children in the Los Angeles area, encouraging them to undertake the environmental activities. They learn about the need for trees, the need to save resources, and the need to participate and to encourage others such as their families and neighbors to participate as well.

TreePeople holds teacher training workshops to help educators utilize new materials and work environmental issues into the school curriculum.

TreePeople's global activities feature a Fruit Trees for Africa Program involving the airlifting of 5,000 trees to Africa and training in horticultural techniques. TreePeople also conducted a week-long workshop in Tanzania, where 26 project trainers were taught horticultural skills. Several trainers have established their own projects. Ethiopians were taught to use compost heaps for fertilizer and to graft more than 3,000 new trees. Kenyans learned technical horticultural aspects and how to cook fruits.

■ Resources

TreePeople runs workshops through its Citizen Forester Program. Environmental teaching materials are prepared, including a recycling video, *Tinka's Planet*; environmental coloring books; children's newsletters with special activities; and Spanish translations. Teachers' workshops provide further environmental training. Publications include *The Simple Act of Planting a Tree, Planter's Guide to the Urban Forest, Tree Boy*, and the bimonthly *Seedling News*.

Trees For Life (TFL) ⊕ ❦
1103 Jefferson
Wichita, KS 67203 USA

Contact: **Treva Mathur**, Administrator
Phone: 1-316-263-7294 • *Fax:* 1-316-263-5293

Activities: Development; Education • *Issues:* Deforestation; Health; Sustainable Development

● Organization

Balbir S. Mathur founded Trees For Life (TFL) in 1983 with the planting of food-bearing trees in India. Seeking to help people in developing countries plant and care for such trees, TFL works with voluntary organizations in the host countries. In addition to India, TFL now has programs in Brazil, Guatemala, Kenya, and Nepal. With more than 10 million trees already planted, TFL plans to plant 100 million fruit trees during the 1990s.

In the USA, TFL programs seek to educate school children by teaching them the value of trees and giving them the "hands-on" experience of growing a tree.

▲ Projects and People

Mathur, TFL executive director, decided to devote his life to fighting world hunger by planting fruit-bearing trees. In 1983 he visited his mother in India, planted some fruit trees, and convinced the villagers to plant more each year from seeds. After describing this experience to a class of eighth-grade students in Wichita, the students held a fundraising drive and managed to send another 103 fruit trees to India. While the scale of TFL has increased greatly since then, the principle of sending food-bearing trees to developing countries remains the same.

In Brazil's rainforests, TFL assisted the Amazon Basin Food Project (PABA) with an establishment of a fruit-tree nursery to bring better nourishment to the local people. Banana, coconut, citrus, passion fruit, cashew nut, and other native trees are examples; medicinal plants are also being raised. In Guatemala's San Lucas fruit-tree nursery, these plantings will soon nourish needy children with essential vitamins and minerals. Here, Balbir Mathur was told that "helping plant fruit trees give the same joy as listening to the birds or to fine poetry."

Through its USA educational program, TFL gives talks at schools—teaching the students about the environment and the importance of trees. The group also distributes kits which allow the children to plant and grow trees themselves. TFL recommends starting seeds soaked in boiling water in a coffee mug, temporarily planting swelled seeds in soil in a carton covered with plastic wrap, and transplanting the largest seedling outdoors when it's about one foot tall. Some one million children in this country have participated in the TFL program. TFL also connects "groups of children in the USA with children in other countries," believing that "helping children make the connection with those in developing countries will stimulate their awareness of the whole world, and not just their family, city and state."

■ Resources

TFL has a *Tree Kit* for children which provides seeds, a carton, an activity book, and information about trees. A workbook for teachers is available; and public presentations are also offered. In addition, TFL publishes *LifeLines*, a newsletter about TFL activities. Contributions from sales of "Let There Be Trees" buttons, bumperstickers, and "tree-shirts" help fund the program.

Trees Forever ❦
776 13th Street
Marion, IA 52302 USA

Contact: **Shannon Ramsay**, Program Director
   Phone: 1-319-373-0650

Activities: Development; Education • *Issues:* Urban and Rural Reforestation

● Organization
Founded in 1989 by two volunteers, **Trees Forever**, with a staff of seven, is now a special program of the **Iowa Natural Heritage Foundation**. The mission of Trees Forever is "to facilitate the planting of trees and the conservation and restoration of forests through action-oriented programs, education and public awareness." Trees Forever also cooperates with **Global ReLeaf**, American Forestry Association.

▲ Projects and People
In partnership with the Iowa Electric Light and Power Company, Trees Forever is involved in **IE Branching Out**, a five-year undertaking that is the first of its kind. This program offers matching funds and staff assistance to local steering committees for community urban forestry. The Trees Forever approach involves local volunteers in planning, fundraising, education, planting, and all other aspects of their mission. Participants plan to work with the existing networks to accomplish their goal of reforestation.

Other Trees Forever programs include **Adopt-a-Forest**, a program to "involve youth in tree planting and forest management on public lands"; the **Hardees Tree Challenge**, a "dollars-for-trees match program" in some 20 communities; the **Youth Tree Mini-Nurseries** for groups such as Future Farmers of America; and the **Annual Acorn Collection** that enables private and state nurseries to make more oak trees available.

■ Resources
A descriptive brochure with membership form is available.

Trout Unlimited (TU) ❦
800 Follin Lane, Suite 250
Vienna, VA 22180 USA

Contact: **Charles F. Gauvin**, Executive Director
   Phone: 1-703-281-1100 • *Fax:* 1-703-281-1825

Activities: Conservation; Education; Research • *Issues:* Air Quality/Emission Control; Biodiversity/Species Preservation; Deforestation; Energy; Global Warming; Health; Transportation; Waste Management/Recycling; Water Quality

● Organization
Trout Unlimited (TU) was founded in 1959 when a group of friends became worried about the state of public trout management and decided that they had to take some action. While large numbers of hatchery trout were being put into Michigan trout streams, little

or no attention was being paid to the habitat, or to the wild fish population.

Trout Unlimited takes political action toward protecting the trout. TU believes that as trout fishing is sport fishing—fishing for the challenge and not to capture the greatest number possible—an important element is "communing with nature." TU also says, "what's good for the trout is good for the fisherman, and managing trout for the trout rather than for the trout fisherman is fundamental for the solution to our trout problems."

Trout Unlimited is growing; it now has a staff of 20 and 66,000 members associated with 400 local chapters throughout the USA. **Robert Herbst** is executive director. The group continues to work toward managing trout as a resource, protecting trout habitats, preventing the release of hatchery fish that will satisfy fishermen while damaging the wild trout population, and lobbying for the protection of trout and other fish.

▲ Projects and People
Trout Unlimited's most important objective is to "enhance its ability to serve as an effective coldwater fisheries advocate on the federal level." The Clean Water Act and wetlands protection are two issues of particular importance to TU. The group currently is concerned that the U.S. Congress will repeal or weaken legislation instrumental in convincing the U.S. Environmental Protection Agency (EPA) to stop Colorado's Two Forks Dam project. TU is also a part of the **Clean Water Coalition**, a group lobbying on clean water issues. Trout Unlimited staff have also testified at congressional hearings regarding the addition of Michigan's rivers to the Wild and Scenic Rivers Act, the Connecticut River National Wildlife Refuge Act, and the proposed Fish Habitat Conservation Act which would bring fishery habitats into the National Wildlife Refuge System. TU also testified at the National Marine Fisheries Service (NMFS) about the need to protect the Snake River sockeye salmon. NMFS now proposes to classify the salmon as an endangered species. TU seeks to be involved in **salmon recovery programs** for this region.

At the council and chapter levels, volunteers are involved in a variety of projects ranging from political work to **hands-on stream improvement work**. Examples include a major cleanup project on Paulins Kill, New Jersey, where the initial effort collected six tons of garbage, and a Utah program called **Hooked on Fishing, Not on Drugs** which provides educational programs to teach youth about recreational fishing and discourage them from drug use. **Embrace-A-Stream** (EAS) is TU's vehicle for restoring and preserving coldwater resources at the community levels.

The Trust for Public Land (TPL)
116 New Montgomery, 4th Floor
San Francisco, CA 94105 USA

Contact: **Martin Rosen**, President
   Phone: 1-415-495-4014 • *Fax:* 1-415-495-4013

Activities: Land Preservation • *Issues:* Biodiversity/Species Preservation; Land Conservation; Open Space; Sustainable Development

● Organization

As Will Rogers is quoted in **The Trust for Public Land** literature, "Land, they ain't making it any more." TPL is a national, private nonprofit, nonmember organization with a staff of 185 that seeks to protect land. It works to help transfer lands to public agencies or nonprofit conservation groups for permanent protection. Since 1972, TPL has protected over half a million acres of land valued at over $600 million with projects conducted in 39 states and Canada.

In addition to its headquarters, regional and field offices are located in Washington, DC; Tallahassee, Florida; New York City; Boston, Massachusetts; Minneapolis, Minnesota; Morristown, New Jersey; Santa Fe, New Mexico; Portland, Oregon; and Seattle, Washington.

▲ Projects and People

The Trust can **work with landowners** who are interested in conservation and serves as an intermediary between these landowners and public agencies that can purchase and protect the land. The Trust can also move quickly to **acquire land** when it comes on the market, and then hold onto that land until state and local governments can gather up the funds to purchase the lands from the Trust. By moving quickly, the Trust can secure an interest in land and manage to protect it. "By holding property with a small downpayment, the Trust can often protect up to $30 in actual land value for every dollar invested."

The Trust **protects parcels of all sizes from small urban gardens to large rural tracts** of land. They work with historic sites from preserving Martin Luther King's boyhood home in Atlanta to bringing additional acreage into California's Toiyabe National Forest. TPL's aim is to protect land of "environmental, recreational, historic, or cultural significance."

TPL has contributed to recent key projects, such as Walden Woods, Massachusetts; Denali National Park and Preserve, Alaska; Grand Island Recreation Area, Michigan; Columbia River Gorge National Scenic Area, Oregon and Washington; and Lower East Side Community Gardens, New York City.

It has also helped protect the Oregon Ridge stream valley in the Baltimore metropolitan area; Prairie Creek in Florida; bear habitat along Black River in Michigan; Ramapo Mountains in New Jersey; Monte Sano Mountain in Alabama; and the Conway Summit corridor (with its migrating mule deer herds) to Mono Lake in California—among others.

TPL also works to help establish local land trusts. These are community-based nonprofit organizations that manage and protect local land. TPL offers "training and technical assistance in land trust incorporation, real estate negotiation, tax incentives for potential donors of land, and joint venture land acquisition."

■ Resources

TPL publishes *Land and People* magazine three times a year, and an Annual Report. Members of the TPL staff have collaborated on the following publications: *Conservation Easement Handbook*, and *Tools and Strategies, Protecting the Landscape and Spacing Growth*.

Trustees for Alaska
725 Christensen Drive, Suite 4
Anchorage, AK 99501-2101 USA

Contact: **Michael Wenig**, Acting Director
  Phone: 1-907-276-4244 • *Fax:* 1-907-258-6688

Activities: Law Enforcement • *Issues:* Air Quality/Emission Control; Energy; Health; Sustainable Development; Waste Management/Recycling; Water Quality

● Organization

Trustees for Alaska describes itself as a "nonprofit, public interest, environmental law firm." Since 1974, it has been working for a balanced approach to the preservation and development of Alaska's natural resources. Former director **Randall M. Weiner** sums up the organization's mission as follows: "In response to national and international events, Alaska has seen a recent proliferation of oil and gas activity. . . . Generally neglected in the flurry of oil wells and drill ships are the voiceless whales, walrus, polar bear, and other unique mammals that inhabit the Arctic waters and their environs. These incredible species, as well as the interests of the native communities who rely on them, can and must be protected despite the nation's demand for oil."

▲ Projects and People

Recent activities resulted in the following court victories:

Alaska Center for the Environment v. *Denali State Park Resort (DNR)*, in which Trustees challenged the decision of the state's Department of Natural Resources (DNR) to build a resort on a "largely pristine wilderness." The superior court held that such a resort had little to do with improving visitor access and understanding of the park, but was intended as an attraction by and of itself. DNR has abandoned the project.

Trustees for Alaska v. *DNR*, in which the Alaska Supreme Court ruled in favor of the Trustees' challenge to an oil and gas lease sale just off the coast of the Arctic National Wildlife Refuge, determining that the state should have first studied the riskiness of transport by oil companies of any crude oil from ice-clogged Camden Bay to market. Now, offshore oil and gas lands may not be leased until these issues are addressed.

Trustees for Alaska v. *DEC*, in which Trustees challenged a decision by the State Department of Environmental Conservation (DEC) approving the effluent limit for settleable solids (SS) in federal permits for several hundred Alaskan placer miners. The Alaska Supreme Court established a more stringent SS limit for those gold mines with insufficient dilution of wastewater.

Trustees for Alaska v. *DNR*, in which Trustees successfully brought suit against the state, claiming that the mineral leasing system violated the Alaska Statehood Act by failing to require the payment of rents and royalties on minerals extracted from certain state lands. The Alaska State Legislature recently enacted legislation implementing a new court-ordered leasing system, including a requirement that miners reclaim the lands they mine.

Trustees for Alaska v. *Hodel*, in which the federal district court ruled that the U.S. Fish and Wildlife Service violated the National Environmental Policy Act by failing to allow public comment on a report to the U.S. Congress on whether to permit oil and gas activities on the coastal plain of the Arctic National Wildlife Refuge

(ANWR)—thus vindicating the public's right to participate in important government decisions.

Trustees for Alaska v. *Horn*, in which Trustees, acting on behalf of seven environmental groups, challenged Department of Interior negotiations to trade subsurface oil and gas rights in the ANWR coastal plain in return for Native Corporation lands elsewhere in the state. Trustees argued that these negotiations were "premature and highly inappropriate" in light of the fact that Congress had not yet decided whether to open the region for oil and gas development. While awaiting a federal district court decision, Congress passed a law shelving the project pending congressional action on the Arctic Refuge.

Trustees for Alaska v. *Fink*, in which Trustees sued the EPA under the Freedom of Information Act to obtain certain information, including "Hazard Ranking System" scores relating to hazardous wastes sites in Alaska. EPA "grudgingly released the information" as the lawsuit progressed.

Rybacheck, et al. v. *Reilly*, in which Trustees intervened on behalf of EPA and against a challenge by miners (recently dismissed) to EPA regulations that would substantially reduce water pollution from placer mines in Alaska. Among other things, the regulations require the miners to recirculate all their wastewater for use in sluicing.

Tuliksarmute Native Community School v. *BLM*, in which Trustees challenged the Bureau of Land Management's (BLM) approval of plans for a mining dredge in an Eskimo village in southwestern Alaska which "severely interfered with the villagers' reliance upon the Tuluksak River for subsistence uses such as fishing and drinking water." BLM must now reassess the impacts of the dredging operation and conduct further proceedings.

ARCO v. *Coastal Policy Council*, in which Trustees represented the native village of Kaktovik off the coast of the ANWR in an attempt to impose limits on expected oil development activities in the Beaufort Sea. ARCO Alaska sued the Alaska Coastal Policy Council after the Council agreed to hear the villagers' concerns. Procedures were implemented allowing coastal residents to challenge agency and district decisions that violate the Alaska Coastal Management Program.

Trustees for Alaska v. *Brady*, in which Trustees sought to block the sale proposed in 1980 by the Department of Natural Resources of 14,000 acres of a wetlands area known as Potlatch Pond for "agricultural purposes." In a final February 1991 decision, the superior court ruled against the sale of the land because it had never been proven fit for agricultural activity.

■ Resources

Trustees for Alaska produced, along with Natural Resources Defense Council (NRDC) and National Wildlife Federation (NWF), a 36-page report entitled *Tracking Arctic Oil: The Environmental Price of Drilling the National Wildlife Refuge*, which is available for a small fee.

United Earth ⊕
700 East Daisy Lane
Milwaukee, WI 53217 USA

Contact: **Marcus Nobel**, Development
Phone: 1-414-351-2737 • *Fax:* 1-414-351-4493

Activities: Educational; Promotion • *Issues:* Air Quality/Emission Control; Biodiversity/Species Preservation; Deforestation; Energy; Global Warming; Health; Population Planning; Sustainable Development; Transportation; Waste Management/Recycling; Water Quality • *Meetings:* UNCED

● Organization

Founded in 1974 by Nobel Prize family member **Claes Nobel**, Earth Aid Society preceded United Earth—so named in 1987. In 1975, the organization drafted a document entitled **The Nobel Laureates Declaration on Human Survival**—a "symbolic endorsement" of humanity's need to address critical issues of the planet's survival, which was signed by 78 Nobel Laureates. One of its first activities was to sponsor and co-fund with the Bank of Sweden an international Nobel Symposium in Stockholm on "Nitrogen—An Essential Life Factor and Growing Environmental Hazards," which was jointly conducted by the Royal Swedish Academy of Science and the United Nations Environment Programme (UNEP).

Bringing together the Audubon Society, International Oceanographic Foundation, National Wildlife Federation (NWF), and World Wildlife Fund (WWF), United Earth established a Joint Membership Program to place environmental literature in waiting rooms of physicians, dentists, and other professionals as well as in employee lounges. United Earth also inspired NWF to publish its *International Wildlife Magazine* in Japanese for distribution in that nation.

United Earth says it has brought substantial financial support to Earth Day in New York as well as to the International Federation of Institutes of Advanced Study, Linus Pauling Institute of Science and Medicine, Global Forum of Spiritual and Parliamentary Leaders on Human Survival, among other groups dedicated to "working for the common cause of the Earth's future."

▲ Projects and People

Claes Nobel's belief that "at this point of the history of mankind, our environmental life support systems are in the process of collapsing" led him to link like-minded individuals and create "global environmental awards, or Earth Prizes, that could help publicize all of the important environmental issues and, in turn, prompt others to do what they could do to help the planet survive the ecological crisis."

According to United Earth, "inspired by his early studies and by the world renowned Nobel Prizes and the related vision and dedication of his illustrious relative Dr. Alfred Nobel, Claes Nobel established the Earth Prizes for Environmental Leadership and Humanitarian Excellence."

Demonstrating respect for both humanity and nature, "the Earth Prizes will give international recognition and financial support to the most outstanding and inspiring achievements related to the creation of a sustainable environment," he writes. Fundamentally, the awards would inspire commitments from individuals worldwide to change values, priorities, and lifestyles to stem environmental decay. Cooperating in the program are UNEP, Dayton Hudson Corporation and its Target chain of retail outlets, the Pulsar Group of Mexico, and the Hearst Corporation with *Good Housekeeping Magazine*.

In 1991, in conjunction with World Earth Day, five **Inaugural Earth Prize Laureates** were announced at the United Nations: H.E. Carlos Salinas de Gortari, president of Mexico; Dr. Gro Harlem

Brundtland, prime minister of Norway; Tenzin Gyatso, XIV Dalai Lama of Tibet; Ted Turner, chairman and founder of Turner Broadcasting Systems, Inc.; and James Grant, executive director for UNICEF.

Recipient of the **environmental statesmanship** award, **Carlos Salinas** remarks, "Ecology is a fundamental consideration of our general strategy. We want a clean Mexico because we are concerned not only about our own generation but also for our children and grandchildren." United Earth recognized President Salinas for implementing measures to improve air quality in urban centers such as introducing unleaded gasoline and less polluting fuels and closing a Mexico City oil refinery. He also took action to preserve the Lacandona forest and create newly protected natural areas, ban the hunting of marine turtles and whales, and preserve the habitat of the Monarka butterfly.

Recognized for **environmental leadership** is **Dr. Gro Harlem Brundtland**, who believes, "A new era of sustainable growth cannot take place without an intimate linking of environmental and economic activities. We need political reform and real options.... Those responsible for activities that have an impact on a nation's environment and resource base must assume responsibility and accountability." As chairperson of the World Commission on the Environment and Development, former environment minister Dr. Brundtland authored *Our Common Future*, the "groundbreaking report" on the biosphere's decline.

Honored for his outstanding contribution to **ethics and the environment**, the **Dalai Lama** is also a 1989 Nobel Peace Prize recipient for "constructive and forward-looking proposals for the solution of international conflicts, human rights issues, and global environmental problems." He affirms, "It is not difficult to forgive destruction in the past that resulted from ignorance. Today, however, we have access to more information. It is essential that we reexamine ethically what we have inherited, what we are responsible for, and what we will pass on to coming generations." This Earth Prize category is permanent.

As a contributor to **environmental communications**, **Ted Turner** comments, "The world will heal itself; it's the people I'm concerned about. The world's been here four billion years; life's been here three billion years.... The real question is, what happens to us? All I really want is to see myself and my species survive, and all the other things that live here during our time. I'd hate to see it all just waste away." Among his environmental programming on cable television is *Captain Planet*, a children's show. Turner also launched a notable corporate recycling program in the USA.

Speaking on behalf of UNICEF, **youth and the environment** recipient **James Grant** says, "We in UNICEF worked for the World Summit for Children—for the Declaration and Plan of Action which it adopted, and for the Convention on the Rights of the Child which it advanced . . . because we believed it also could be the first collective step toward a new ethic for humanity."

The 1992 Earth Prize Awards ceremony was held in December in Washington, DC, and featured seven categories: conservation of nature and natural resources, elimination of hunger and poverty, energy, pollution abatement, population stabilization, Earth security, and Earth ethics. In addition to the inaugural "summit awards," future Earth Prizes will recognize children, adults, and organizations on the grassroots level for setting exceptional and practical examples in saving the planet.

United Earth has already recognized posthumously Clinton Hill, the 11-year-old founder of **Kids for Saving Earth (KSE)** and cancer victim, who started an environmental club that has grown to become "a global network including more than 7,000 individual clubs with more than 200,000 members" in North America, Europe, the Middle East, Eastern Europe, and the Philippines. Target Stores of Dayton Hudson Corporation is the international sponsor of this program to "educate and empower kids to act on behalf of the environment."

United Earth is also designing an **Earth-Ethics educational program** to help solve a "current severe lack in the world of ethics and courageous 'right action' that are necessary to guide human behavior, attitudes, and priorities in addressing environmental problems," writes Nobel. Such a subject would be introduced into school curriculum worldwide. A **United Earth Corps** is also being forged, as a worldwide volunteer effort in service to the environment for both youth and adults.

Claes Nobel is a director of **Earth Day USA** and also author of **The Universal Declaration of Earth Ethics.** A Milwaukee, Wisconsin, resident, he founded a recreational homestead in his state for wildlife and nature preservation; and is also president of Alternative Combustion Engineering, a Florida-based company developing technology to use less fuel and create more horsepower with less pollution; and vice president of Bedminster Bioconversion Corporation, New Jersey, which is concerned with municipal solid waste and sewage disposal.

■ Resources

Descriptive information about United Earth is available from the Milwaukee office.

U.S. Citizens Network on UNCED
300 Broadway, Suite 39
San Francisco, CA 94133 USA

Contact: **Catherine J. Porter**, Executive Director
Phone: 1-415-956-6162; 1-212-373-4200
E-Mail: Econet: citizensnet

Issues: Air Quality/Emission Control; Biodiversity/Species Preservation; Deforestation; Energy; Global Warming; Health; Population Planning; Sustainable Development; Transportation; Waste Management/Recycling; Water Quality • *Meetings:* EcoForum North America; Globescope Americas; UNCED

● Organization
Citizen groups were invited to participate in the process for the 1992 **United Nations Conference on Environment and Development** (UNCED). "To facilitate the efforts of U.S. organizations trying to affect the Earth Summit process, 160 organizations created the **U.S. Citizens Network** in October 1990. Now, with more than 250 organizations as members, the Citizens Network is organized into committees and Issue Working Groups."

The Network also encouraged citizens to get involved in the process, and provided them with concrete advice of how they could do so. This advice included having elected officials hold hearings on the U.S. National Report, and offering and publicizing alternative goals and actions, collaborating with citizens groups. The Network remains in existence following UNCED, and a New York office is being opened in 1993.

▲ Projects and People

The Network is divided into committees. The **Outreach Committee** worked to increase public understanding of the UNCED and UNCED issues. Network members held meetings on the town and regional levels, and hearings, with the goal of educating the public. The **Clearinghouse Committee** worked with EcoNet to help disseminate information about the Conference. While this was designed to be an EcoNet-based service, Citizens Network provided hard copy materials to those who do not have access to EcoNet. **Issue Working Groups** served as a forum for work on environmental and development options. The Commentary from the Issue Working Groups is published as the *U.S. Citizen's National Report*, as an alternative to the U.S. Government's *National Report to the Conference*. "The **Policy Committee** [supported] the members of the Network seeking to develop or influence policies for the UN Conference and on the issues affected by the Conference." The Network encouraged its members to pursue a wide range of policy actions, but only takes a position if there is a consensus among its membership.

The Network works on **Media Coverage**. This involves creating a media campaign and a message. The message is "Join the Network's diverse membership to make a difference worldwide. Learn how to leapfrog an unresponsive federal government to change outdated domestic environment and development practices through international peer pressure."

The **International Committee** looks for opportunities for U.S. groups to work with citizen groups in other countries, and with other networks. It also sent people to preparatory committee meetings to represent the Network at UNCED.

■ Resources

The Network publishes a newsletter, *Introductory Guide to UNCED*, and *U.S. Citizens Report*.

Office for Interdisciplinary Earth Studies (OIES) ⊕

University Corporation for Atmospheric Research (UCAR)

P.O. Box 3000
Boulder, CO 80307 USA

Contact: **George William Curtis, Ph.D.**, Interim Director
Phone: 1-303-497-1680 • *Fax:* 1-303-497-1679
E-Mail: Omnet:w.curtis.bc

Activities: Education; Research • *Issues:* Global Warming

● Organization

We shall not cease from exploration
And the end of all our exploring
Will be to arrive where we started
And know the place for the first time. T.S. Eliot

The **Office for Interdisciplinary Earth Studies (OIES)** was established by the University Corporation for Atmospheric Research

(UCAR) in 1986 "on behalf of all disciplines that study the Earth as an interconnected system." This organization is designed to assist the U.S. Global Change Program and the International Geosphere-Biosphere Programme (IGBP) through "advocacy, the dissemination of information, and the definition and instigation of scientific research." OIES former director **Dr. John A. Eddy** serves on the Scientific Committee of the IGBP. Dr. Tom M.L. Wigley will become OIES director in April 1993.

UCAR literature indicates that program sponsors include the National Aeronautic and Space Administration (NASA) Earth Science and Applications Division, National Science Foundation (NSF), National Oceanic and Atmospheric Administration (NOAA), U.S. Environmental Protection Agency (EPA), and the U.S. Department of Energy (DOE).

▲ Projects and People

Each year, OIES conducts a **Global Change Institute**. The themes of these Institutes change yearly. Past topics have included trace gases and the biosphere, understanding global changes of the past, modeling the geosphere-biosphere system, and future impacts of global change.

OIES seeks to encourage and assist with **interdisciplinary research** on the grounds that all of the Earth's processes are interconnected. "The staff and associates of OIES are centrally involved in the formulation of national and international programs in interdisciplinary Earth Science," it reports.

Indeed, OIES points out that "more than 30 percent of the land area of the Earth is now under active management for the purposes of mankind." Results include the greenhouse effect, acid rain, tropical deforestation, desertification, and global changes in soils and groundwater from intensive agriculture and industry. Yet, with man's still "limited understanding" of natural processes, it is "difficult if not impossible" to predict the outcome.

Dr. Eddy writes, "We must keep reminding ourselves . . . that global change is more . . . than climate change or global warming. Most would agree that the global changes that will strike the hardest blows on people and other living things in the next 20 or 50 or 100 years will not be those of climate change. If the world is in environmental extremis today, it is more through the rapidly changing chemistry of the air and soils and water, and through the inexorable and wide-reaching pressures of urbanization and intensive agriculture and land use."

Research specialties of Dr. Eddy are "solar physics, history of solar behavior of climate, and the history of astronomy including archaeoastronomy." A fellow of the American Association for the Advancement of Science, Dr. Eddy indicates he is "best known for his historical work on the Maunder Minimum of solar activity and its association with the Little Ice Age, and for his work, in archaeology, on the astronomical significance of stone medicine wheels left by the prehistoric Indians of the western plains."

■ Resources

OIES has a large number of publications including *Global Changes of the Past*; *Reports to the Nation on Our Changing Planet*, which explains the impact of climatic change; and the quarterly journal *EarthQuest*. A publications list is available.

Whale and Dolphin Conservation Society USA (WDCS)

191 Weston Road
Lincoln, MA 01773 USA

Contact: **Charlotte Reynders**, Director
    Phone: 1-617-259-0423 • *Fax:* 1-617-259-0288

Activities: Development; Education; Research • *Issues:* Biodiversity/Species Preservation; Waste Management/Recycling; Water Quality • *Meetings:* IWC

● Organization

The **Whale and Dolphin Conservation Society USA (WDCS)** is the sister office to the Whale and Dolphin Conservation Society UK. Formerly known as the Long Term Research Institute, WDCS is devoted to "conduct[ing] and promot[ing] worldwide, research, conservation, and education activities needed to support the conservation of cetaceans and their habitats." The organization has a staff of four and—including the USA and UK organizations—a membership of over 30,000.

▲ Projects and People

The commercial whaling moratorium is not in itself enough to protect whales. Deaths still occur from pollution and traditional hunts. Deaths also occur in fishery-related incidents. Work is needed to protect cetaceans—whales, dolphins, and porpoises—and that is the work of WDCS.

One WDCS USA project that received much attention was an **opinion poll in Japan** which showed that the overwhelming majority of Japanese do not believe eating whale meat is an important part of Japanese culture, and feel that whales should be protected.

Current WDCS programs include "**right whale population analysis work in Argentina** (the world's longest continual study using photo-identification techniques), participation or attendance at meetings with **cetacean conservation** relevance, the **funding and training of young biologists** in developing countries in benign research techniques, the investigation of **toxic marine pollutants** and the effect on cetacean fitness (with funding from the W. Alton Jones Foundation), and **mobilizing public opinion** on the need to protect marine mammals and their habitats." WDCS is working to build support for their projects within the New England area.

WDCS president **Dr. Roger Payne**, also World Wildlife Fund (WWF) senior scientist wrote and narrated a two-hour Discovery Channel television special, *In the Company of Whales*, which is also a book.

■ Resources

WDCS resources include speakers, a whale adoption program, the *SONAR* magazine, and two cassette tapes of whale songs from among the "world's largest whale song collection," says WDCS.

The Wilderness Society

900 17th Street, NW
Washington, DC 20006-2596 USA

Contact: **George T. Frampton, Jr.**, President
    Phone: 1-202-833-2300 • *Fax:* 1-202-429-3958

Activities: Education; Political/Legislative; Research • *Issues:* Biodiversity/Species Preservation; Deforestation; Energy; Sustainable Development

● Organization

Founded in 1935, the **Wilderness Society** has grown to encompass Washington, DC, headquarters, 16 regional offices, a staff of 130, and a membership of 400,000. In addition to Society President **George T. Frampton, Jr.**, the staff includes foresters, lawyers, wildlife biologists, resource managers, economists, ecologists, writers, and media specialists.

The Society's activities mostly relate to federal public lands, including the provisions of accurate information and analysis on issues regarding the U.S. Department of Interior. To meet its commitment to the preservation and expansion of public lands, the Society conducts research, then works to gain publicity for its findings. It both educates the public and lobbies the U.S. Congress. "The Wilderness Society employs three primary tools: research, public education, and advocacy."

▲ Projects and People

At both national and regional levels, the Wilderness Society is involved in a large number of programs relating to public lands. One current major goal is to protect **Ancient Forests of the Pacific Northwest**. Only 10 to 15 percent of the old-growth forests in the Pacific Northwest remain today, and only 923,000 acres are protected in national parks or as federally designated wilderness; 1.4 million acres are open for logging. Such forests are "being liquidated at a rate of nearly 70,000 acres every year," writes the Society.

These forests are also home to thousands of species, including the endangered northern spotted owl, pine marten, marbled murrelet, and fisher. It is estimated that some of the trees—cedar, Douglas fir, western hemlock, and Sitka spruce—were seedlings when "Marco Polo first traveled to China some 700 years ago" and are now 250 feet tall.

In asking Congress to designate an **Ancient Forest Reserve System** by 1993, the Society established a five-point **plan of action:** cease logging or road building in crucial, old-growth stands of 200 acres or more, or spotted owl habitat, or low-elevation stands; reduce federal timber sales by about 40 percent; authorize a state export tax on raw logs to be dedicated to economic redevelopment projects in timber communities and grant a tax incentive to millers who upgrade facilities; improve forest management on private lands; and diversify rural economies with technical assistance, loan programs, and tax credits that fund job retraining and reemployment.

The Society is also involved in blocking the Bush Administration's plan to open the Arctic National Wildlife Refuge to oil and gas development. The Society played a role in building a **coalition of 16 energy and conservation groups** to work on this project in conjunction with a national energy policy. They also called for a reduction in U.S. dependence on oil on the anniversary of the Exxon *Valdez* spill when the Society issued a report on the 100 worst oil spills in U.S. history.

"A national energy strategy based on conservation, efficiency, renewable fuels, alternative energy sources, and development of a non-petroleum-driven automobile would be faster, cheaper, and more secure than one based primarily on producing a little more domestic oil," writes Frampton.

In 1990, the Congress passed the **Tongass Timber Reform Act,** for which the Society labored many years. In the 1980s, the Society's Resource Planning and Economics Department published a study which showed that the policies toward the Tongass National Forest were causing environmental damage and costing "taxpayers millions of dollars in lost revenues." With this information publicized, the timber reform legislation resulted.

The Society supports additions to **federal land acquisition** that is already made possible through the Land and Water Conservation Fund, Migratory Bird Conservation Fund, and the North American Wetlands Conservation Fund. Other means of acquisition include state bond action, purchases by private groups such as the Trust for Public Land or The Nature Conservancy who may later sell to the government, and the Pittman–Robertson Fund for wildlife habitat and the Dingell–Johnson Fund for fisheries, both which provide federal monies to states for those purposes.

In 1990, the work of the Wilderness Society resulted in the creation or **expansion of national parks** in Florida, Alaska, Arizona, Illinois, and Maine.

Among its **regional activities,** the Society backs the Walden Woods Project, Massachusetts, to save the historic site from development; coordinates a pioneer ecosystem study of six southern Appalachian national forests; works for federal protection of the Florida Key Deer habitat; inspired a recommendation to add some 1.6 million acres of new, designated wilderness in Colorado; opened a new office in Santa Fe, New Mexico, where a priority is to establish the Animas National Wildlife Refuge; helped achieve passage of the Arizona Desert Wilderness Act of 1990, the first statewide Bureau of Land Management (BLM) wilderness bill affecting nearly 4.5 million acres; in San Francisco, California, gained with its coalition "major backfill, hydrological monitoring, a citizen advisory council, and creation of $4 million environmental trust fund" in settlement with Viceroy Mining Company; and helped forge a "singular wilderness agreement" with timber and mining officials and the timber workers' union for the Kootenai and Lolo forests, Montana.

The Wilderness Society has also worked in conjunction with The University of Arizona School for Renewable Natural Resources, the U.S. Forest Service, and KUAT-TV public television to create the **Green Scene.** This is an environmental teaching tool aimed at middle-school children and consisting of six freestanding lessons, activity sheets, a map, and a video.

■ Resources

The Wilderness Society has a variety of publications including fact sheets, books, and brochures on national forests, energy and minerals, public education, and video tapes—listed on its order form. A needed resource is computer graphics technology.

Wildlife Conservation International (WCI) ⊕
New York Zoological Society (NYZS)
Bronx, NY 10460 USA

Contact: **John G. Robinson, Ph.D.,** Director
Phone: 1-718-220-6864 • *Fax:* 1-718-220-7114

Activities: Education; Research • *Issues:* Biodiversity/Species Preservation • *Meetings:* CITES; Society of Conservation Biology

● Organization

A division of the New York Zoological Society, **Wildlife Conservation International (WCI),** with 60,000 members, is conducting more than 150 projects in 44 countries on behalf of biological diversity and ecological processes. With a full-time staff of ecologists and wildlife biologists, WCI concentrates its efforts mainly in developing countries where "biological diversity is greatest and threats posed by population and development are the most intense." Research fellows, in-country wildlife biologists, and volunteer field assistants also take part in exploring interactions of humans and wildlife, gathering scientific information, developing locally sustainable solutions, and implementing conservation plans. *(See also New York Zoological Society in this section.)*

Because training of in-country professionals is crucial, all WCI projects have such a component. WCI maintains productive working relationships worldwide with governments and local NGOs. "The success of our approach is reflected in the increasing numbers of in-country conservation professionals and the growing list of protected areas resulting from our recommendations," reports WCI.

During this century, WCI has contributed to hundreds of landmarks in nurturing biodiversity. These include discovery and disclosure of whales' songs; early protection of fur seals in Alaska; American bison's return to the western plains; recovery of endangered North American black-footed ferret; first studies of mountain gorillas in their habitats; first main study of rainforest primates; leading studies of African lions, Himalayan wild sheep and goats, and snow leopards; scientific leadership in saving the giant panda, Komodo lizard, white rhino, and Formosan pheasant; first U.S. projects to protect East African wildlife; global coordination to preserve African elephants and rhinos; global effort to ban ivory imports, causing the ivory market's collapse; legislation in the USA and elsewhere to curb destructive hunting practices and wildlife products' commerce; U.S. regulation of imported skins of spotted cats, crocodilians, and other endangered species; U.S. ban on importing bird plumes; establishing 70 wildlife sanctuaries worldwide, including 17 North American reserves; establishing Brazilian rainforest sanctuary of nearly three million acres; establishing Arctic National Wildlife Refuge; establishing the Peruvian rainforest Tambopata-Candamo Reserve; creating Crater Mountain Refuge for rare birds in Papua New Guinea; creating a refuge for Magellanic penguins along Argentina's Patagonian coast; creating Tibetan reserve for the snow leopard, yak, and others; and in Belize, creating the world's first jaguar reserve and leading efforts to save the Western Hemisphere's largest coral reef.

▲ Projects and People

Current projects are focused in Central America and the Caribbean with Dr. Archie Carr III as regional coordinator and Kathleen Williams as assistant coordinator; in tropical South America with Latin American programs assistant director Dr. Stuart Strahl and officer Dr. Alejandro Grajal and research zoologists Dr. Marcio Ayres and Dr. Charles Munn III; in temperate South America with regional coordinator Dr. William Conway and resident conservationist Dr. Guillermo Harris; in East African Savannas with regional coordinator Dr. David Western, zoologists Dr. Patricia Moehlman and Dr. Jesse C. Hillman, and conservation officer Dr. Christopher Gakahu of Nairobi; in African forests with African programs assistant director Dr. William Weber, research zoologists Drs. Terese and John Hart, and biodiversity coordinator Dr. Amy Vedder; in temperate Asia with regional coordinator Dr. George B. Schaller; and in tropical Asia and the Pacific with Asian programs assistant director Dr. Mary Pearl and research zoologists Dr. Alan Rabinowitz of Thailand, Dr. Bruce Beehler of Papua New Guinea, and Dr. Elizabeth Bennett of Malaysia. In North America, the focus is primarily on black-footed ferret conservation and wildlife training courses on Navajo lands with Brian Miller of Navajo Nations.

Regarding worldwide education and training, Dr. Pearl is in charge of the research fellowship program; Dr. Weber directs the Pew Charitable Trust field training grants; and Dr. James Connor oversees the environmental education outreach programs.

Subjects of WCI-supported conservation projects in Central American and Caribbean countries include: Belize—tropical forest reserve planning, Barrier Reef management; Costa Rica—tarpon, park planning; Guatemala—Morelet's crocodile, El Peton planning; Honduras—environmental education; Mexico—Calakmul Biosphere Reserve, Lacandona forest; horned guan, white-lipped peccary and habitat, Highland guan, Chinchorro Banks reef, and tropical mammal community disturbances; Panama—marine turtles, Bastimentos National Marine Park management, ethnic groups' use of natural resources in Bocas del Toro National Park.

With the daily leveling of more than 100,000 acres of tropical rainforests worldwide, says WCI, research projects are crucial to save these rainforests providing "at least 25 percent of all prescription drugs," an ecosystem on which 200 million people depend for their livelihood, and commodities such as perfume, vanilla and coffee beans, rubber, and exotic wood. In Central America's El Peten tropical forest, which spans Guatemala, Mexico, and Belize—Howard Quigley, Milton Cabrera, and María José González are working to create the region's first trinational reserve. Chuck Carr III is working on a cooperative Paseo Pantera project among WCI, Caribbean Conservation Corporation, and U.S. Agency for International Development (USAID), which includes protected areas in Belize, Costa Rica, Guatemala, Honduras, and Nicaragua. *(See also Wildlife Conservation International in the Guatemala NGO section.)*

Regional programs in tropical South America include short courses in conservation biology, zoos conservation, and student grants. The Species Survival Commission (IUCN/SSC) Parrot and Cracidae Group is supporting Dr. Strahl and ornithologist Dr. Donald Bruning. Specific conservation projects focus on the following: Bolivia—cracids, hunting effects, peccary reserve planning; Brazil—establishing Brazil's new Mamirauá Reserve in the Amazon, primate studies in flooded forests; hyacinth macaw, white-lipped peccary, western Amazonian biogeography, Pantanal mam-

mals, black-fronted piping guan; Colombia—avifauna survey, animals hunted by indigenous communities in Utría National Natural Park of the diverse Chocó; Ecuador—cloud forest, protecting Podocarpus National Park from gold mining, curassow surveys, mountain tapir, Pinzon Island giant tortoise; Paraguay—Chaco peccary; Peru—hunting and the Manu National Park, Brazil nut trees, Tambopata reserve planning, Amazon parrots; and Venezuela—creation of Orinoco-Casiquiare Biosphere, Orinoco crocodile and turtle, breeding and reintroduction of rare Orinoco caiman, cracids and parrots related to wildlife trade, Rio Nichare rainforest, Margarita Island parrot ecology, wildlife/hunter education, herbivorous birds in Henri Pittier National Park, private landowner initiatives, Imataca Reserve timber exploitation, spider monkey, freshwater turtles, jaguar management, biocide impacts, student education programs, and general parrot and crocodile preservation.

In temperate South America, the following conservation is ongoing: Argentina—sea lion ecology and coastal management, support for Chubut Province's Department of Conservation, Punta León seabird and mammal colonies, oiled penguins, flamingo and seabird surveys, Magellanic penguins at Punta Tombo, Valdes Research Station; Chile—Flamingo Center and habitat preservation endangered by mining practices, Humboldt penguins; and Peru—fur seals and Punta San Juan protection.

In 1991, WCI researcher Dr. Patricia Majluf was victorious in halting Peruvian government action that would have allowed the hunting of sea lions. With the majority of the population crowding the coastlines due to "poverty and terrorist activities in the mountains," according to WCI, overfishing and a cholera epidemic cut the income of fishermen who blamed their losses on marine mammals. Media and education campaigns along with the support of IMARPE, Peruvian National Oceanographic Institute, underscored Majluf's position. In Argentina, Alejandro Grajal reports that a Punta León tourist development was halted so that seabird nesting would not be disrupted.

Conservation projects in the East African Savannas relate to the following: Botswana—Okavango Delta wildlife survey; Ethiopia—Awash National Park management plant, education, Simien jackal ecology, wildlife government advisor; Kenya—ecological monitoring, Kitengela ecological corridor design, Nakuru National Park and rhino sanctuary, elephant genetics, elephant impacts outside parks, rhino rescue fund, coral reef ecology, and government capital improvements; Namibia—rhino horn; Somalia—Somali wild ass; Tanzania—jackal ecology, Ngorongoro Crater monitoring and training, conservation biology graduate training, Ruaha National Park, Lake Manyara National Park, Tarangire National Park; and Zimbabwe—wild dog. The African Elephant Action plan is underway, and so is support for the IUCN/SSC African Elephant and Rhino Group.

In Kenya, for example, WCI's David Western reports he is successfully working with Ogulului ranchers of Amboseli and Richard Leakey, Kenya Wildlife Services (KWI), to keep this area open to wildlife and communal livestock grazing and ostensibly overturns a national policy that did not benefit wildlife outside parks. A Kitengela Landowners Association was formed to draw up a regional land-use plan, he adds.

Also in Kenya, the Laikipia Project with principal investigator John M. Waithaka is assessing whether elephants change animal diversity by affecting habitats as well as their economic impacts on commercial forestry and damage to human dwellings and farms. Noting elephant overcrowding in Kenya's Amboseli and Aberdare

National Parks is causing deforestation, Waithaka is extending his project to arid Tsavo National Park as well as to wetter areas for further studies. With poaching on the decline, "controlled elephant viewing and other wildlife tourism" is bringing profits to local peoples.

"The role of elephants as agents of seed dispersal has been investigated by germinating the seeds found in their dung," Waithaka writes. "Dung was collected and irrigated periodically in confined plots within the study areas." So far 17 plant species have germinated, and more studies are underway. "This aspect demonstrates the ecological consequences of eradicating elephants from their natural ranges."

African Forests conservation work is concerned with: Cameroon—elephant surveys, Korup Forest primates and other mammals; **Congo**—Congo Forest, reserve planning, elephant surveys; **Central African Republic**—forest elephant monitoring and genetics, rainforests; **Gabon**—logging impact, human impacts on wildlife and reserve planning; **Ivory Coast**—manatees and coastal conservation; **Madagascar**—butterflies as biodiversity indicators; **Rwanda**—Nyungwe Forest's animal seed dispersers, forest conservation, mountain gorilla surveys, and forest dynamics; **Sierra Leone**—Tiwai Island primates, secondary forests' primates; **Tanzania**—Uzungwa Mountains' forest birds; **Uganda**—wild coffee ecology and use, chimpanzees, community forests; **Zaire**—okapi and duiker ecology and small carnivores of Ituri Forest, rainforest dynamics, training with USAID, Grauer's gorilla survey, and Maiko National Park surveys. Together, the Congo, Cameroon, and Central African Republic are doing trinational conservation planning. These countries, with Gabon, Equatorial Guinea, and Zaire, are also undertaking forest elephant surveys and management plans.

A five-year **Congo Forest Conservation** project regards lowland tropical forests that are believed to contain some of the most biologically diverse sites in Africa. However, intensified commercial logging and lack of a clear management policy threaten the forest in the short term while providing a low economic return. With forest resources second to petroleum in importance, the government asked WCI to help "establish and develop the Nouabale-Ndoki National Park" and other areas while providing a forest conservation policy and hands-on training to local Congolese foresters and wildlife managers, according to the WCI proposal to the USAID. Loss of the tall moist forest ecosystems imperil the habitats of endangered and threatened species such as forest elephant (*Loxodonta africana cyclotis*), gorilla (*Gorilla gorilla gorilla*), chimpanzee (*Pan troglodytes*), bongo (*Tragelaphus euryceros*), and leopard (*Panthera pardus*). The park site was selected because of its high populations of such mammals. **Dr. J. Michael Fay** is director of WCI's Congo program.

The virgin forests are also believed to affect climate changes; with headwaters of many river systems, the region's hydrological system is ecologically regulated; and the area is potentially suited for low-impact activities such as ecotourism. In the meantime, attention to long-term development and its economic potential is being focused here through the Tropical Forest Action Plan of the Food and Agricultural Organization (FAO) and a Natural Resource Management project financed by the World Bank.

Richard Barnes, who surveys Asian elephants, believes that domesticated elephants can be assets to the logging industries of village cooperatives in Gabon and Congo—saying that they do less damage than machines and "require little maintenance or fuel because their food is supplied by nature." Such action would give local people "a stake in the future of their forests," which now yield profits to foreign companies. Early reactions to this concept were mixed.

An **elephant census** with satellite tracking (PTT) is also ongoing in **Cameroon's Korup Forest**, where geographical information system (GIS) maps are also being used for the first time, reports James A. Powell, chief investigator. Elephant darting is tricky. Powell writes, "We developed a strategy to build a blind . . . near fruiting trees which attract elephants; the tagging team can safely hide and dart elephants without fear of being charged." To reduce poaching, WCI has initiated innovative programs in vocational job training for young village men with promising results. "Hunters . . . have voluntarily turned in their guns and snares. Once common, gunshots are now rarely heard in the research area." Also, an avifauna study in Korup is expected to yield a field guide to the birds.

In **Zaire**, John and Terese Hart report on the progress of the USAID-funded and WCI-directed Centre de Formation et Recherches Forestieres (CEFREF), where University of Kisangani (UNIKIS) field botany courses are taught and foreign researchers are hosted. Here a scientific basis is being provided for management and protection of forest parks and protected areas, including park personnel training. The Harts' **Ituri Forest Project** includes the study of okapi (a ruminant similar to a small giraffe with a shorter neck) and of the rainforest duiker (frugivorous forest antelope) which wore their first radio collars in 1991; tree inventories; and leopard data collection. In the Ituri, which is known for its mix of swamp and other forest types, the duiker ungulate is a main source of protein; its study is being enhanced by a traditional net-hunting technique of the Mbuti pygmies. Researchers are seeking to find out how "so many species with similar diets coexist in the same forest."

In **temperate Asia and tropical Asia/Pacific** conservation projects concern: China—Guizhou golden monkey, gibbon and other wildlife in Yunnan; India—Nagarahold National Park tigers and other carnivores; Indonesia—tropical ecology and training at Kalimantan, training at Sumatra, primate conservation at Sulawesi, economic fruits regeneration; Laos—wildlife managers' training; Malaysia—wildlife surveys and logging impact at Sarawak, proboscis monkey at Sabah, logging effects, wildlife surveys, training; Mongolia—wildlife research; Papua New Guinea—specialized frugivory, bird behavior at flower and fruit trees, dwarf cassowary, Crater Mountain Reserve planning, comprehensive surveys, short training courses, national conservation organization support; Thailand—Project Tiger, ungulate distribution, Thai national parks' wildlife surveys, conservation biology research training; and Tibet—wildlife surveys and reserve planning.

WCI partnerships are vital in addressing the huge biogeographic zones of the Indian subcontinent, continental Asia, Indochina, biologically diverse Indonesian archipelago, and the South Pacific Islands. Here is where pressures of an enlarging, richly cultured human population coexist with commercial exploitation of forests and other resources. Opportunities for field studies are sought and investigations generally begin in Borneo and radiate outward.

In **Indonesia**, WCI supports training programs at the **Cabang Panti research station** in Gunung Palung National Park with UNESCO and the country's Protected Areas Department (PHPA) and National Academy of Sciences (LIPI). One study is "documenting the success of illipe nut collection in logged and unlogged forests and the effect of logging nearby on the production of this economically important seed." At Gunung Leuser National Park in northern Sumatra, Indonesian forest ecologists are trained in censusing rainforest animals and estimating fruits—the resource base. Dr. Pearl is consulting with Dr. Syamsuni Arman, Universitas Tanjungpura, and Dr. Christine Padoch, New York Botanical Garden, on nontimber forest products studies, including forest vertebrates. Veterinarian **Dr. W.B. Karesh**, New York Zoological Society (NYZS); **Harmony Frazier-Taylor**, Woodland Park Zoo; and **Dondin Sajuthi**, Institut Pertanian Bogor's primate studies, are studying free-ranging orangutans and the diversity within and between geographically isolated populations.

The most biological diverse habitats are considered to be in Malaysia's Sarawak and Sabah, where WCI is conducting intensive training and research. To restructure the Wildlife Department of Sabah, **Dr. Alan Rabinowitz** is working with its director, **Mahedi Andau**. Dr. Elizabeth Bennett is documenting the impact of commercial logging, hunting, shifting cultivation, and botany and mineral levels on wildlife. With **Ramesh Boonratana**, she is also surveying the Kinabatangan area with its orangutans, elephants, Sumatran rhinos, and proboscis monkeys—and urges the **conservation of the coastal mangrove habitat** for these monkeys. With support from the John D. and Catherine T. MacArthur Foundation, WCI is involved in research on the abundance of large vertebrates to recommend on game hunting management.

Carnivores are being radio collared and monitored at Nagarahole National Park, India, to find out why a high biomass of wildlife exists here, how it can be preserved, and how other reserves in southern India may benefit. **Ullas Karanth**, Center for Wildlife Studies of Mysore, and **Mel Sunquist**, University of Florida, are principal investigators, often working closely with the Bombay Natural History Society. With at least 19 tiger reserves in India, the team is studying this big cat's needs for space, prey, and social interaction. Researchers are also investigating the Indian leopard and *dhole*, or wild dog. A tiger management plan and a visitors' education center are to follow, and so is a joint endeavor with local villagers to protect humans and livestock from tiger attacks.

In 1991, a breakthrough **memorandum of understanding** was entered into by WCI and the **Laotian government** for wildlife surveys and training of wildlife management personnel.

WCI's science director, **Dr. George Schaller**, had a rare opportunity to work with Vietnamese conservation scientists and, while completing a cooperative wildlife survey in 1988, rediscovered the Javan rhino and noted the decline of species such as Tonkin and Douc langurs. At the same time, WCI helped the Vietnamese launch an education **campaign to save the Javan rhino**. Drs. Rabinowitz and Bennett are presently conducting a field research training course.

With the world's fastest growing economy—and one that is based on agribusiness—Thailand is benefitting from WCI-generated assessments of wildlife in key protected areas. Mahidol University professor **Warren Brockelman** and **Dr. Sompoad Srikosamatara**, both WCI Fellows, are cooperating, and the Sub-Minister for Science expresses a commitment to transpose results into government policy.

In Mongolia, Dr. Schaller has launched WCI projects on the **Gobi brown bear and snow leopard** and conducts his own research on **wild camels** in the Gobi desert "with the goal of eventual reserve designation." Most attention is given to Mongolian researchers through "training, information, and moral support." In China, Dr. Schaller surveys large mammals such as the Tibetan antelope, wild yak, and wild ass—with the hope of establishing a 100,000-square-mile reserve, which would be the planet's largest. **Dr. William Bleisch** and **Xie Jiahua** of Guiyang Normal University, who are co-principal investigators of the Guizhou golden monkey studies, are producing a management plan to preserve the species' habitat that also includes rare flora, such as a fir tree that exists nowhere else.

Urgency is expressed about the Yunnan attempts to improve the "rapidly deteriorating tropical and sub-tropical habitats," where more than half of China's endemic animal and plant species are found; Drs. Pearl and Bleisch are working with Ji Weizhi, Kumming Institute of Zoology, and others.

WCI's Papua New Guinea program includes "research on forest regeneration, reserve protection through rural development and ecotourism, and public education through support for [the local] Research Conservation Foundation of Papua New Guinea (RCFPNG)." The Foundation for the Peoples of the South Pacific is also participating with WCI in designating and strengthening reserves in this country which is three-quarters tropical forest and "home to fabulously beautiful birds of paradise, endemic bowerbirds, fruit pigeons, tree kangaroos and other wildlife," according to WCI literature. *(See also Foundation for the People of the South Pacific and Research Conservation Foundation of PNG in the Papua New Guinea NGO section.)*

Local tribesmen are setting aside part of their lands for biodiversity protection in the Crater Mountain Wildlife Management Area in exchange for nature tourism benefits resulting from a tourist lodge created by Pacific Helicopters Ltd. of Goroka and local village investors. International wildlife authority Dr. Bruce Beehler leads WCI's applied research program. Other researchers at Crater Mountain are **Andrew Mack** and **Debra Wright**, who are working on the role of the dwarf cassowary in forest regeneration; **Lawong Balun**, who is developing seed banks at Bulolo Valley; **Miles Hopkins**, of the University of PNG; and **Eleanor Brown**, of the Smithsonian Institution, who are investigating the rainforest birds and flower and fruit trees.

Plans are to create a model conservation/rural development program including "studies on the effects of tourism on local people and wildlife," as well as efforts to improve "local people's lives through garden intensification, cooperative manufacturing and marketing of artifacts and carefully selected nontimber forest products." Meanwhile, new WCI field projects are being initiated worldwide from Madagascar to Zaire to Venezuela and Paraguay.

■ Resources

WCI publishes a list of current projects as well as brochures on *The Challenge of Conserving Wildlife in Asia, Protecting the World's Tropical Forests, WCI Bulletin* newsletter; and membership information. Dr. Alan Rabinowitz is the author of *Chasing the Dragon's Tail: The Struggle to Save Thailand's Wild Cats* (Doubleday) as well as *Jaguar: One Man's Battle to Establish the World's First Jaguar Preserve* (Anchor Book). Dr. Mary Pearl, assistant director, expresses WCI's need to be "better known within the USA." The *Bulletin* also lists these needs: camera equipment, binoculars, portable computers with software, altimeters, calculators, boats with inboard or outboard motors, four-wheel-drive vehicles, and field clothing.

(Wildlife Conservation International [WCI] projects and people appear throughout this Guide.)

Wildlife Conservation International (WCI) ⊕
New York Zoological Society
c/o P.O. Box 456
Minden, NV 89423 USA

Contact: **Craig C. Downer**, Wildlife Ecologist
Phone: 1-702-267-3484

Activities: Education; Research • *Issues:* Biodiversity/Species Preservation; Deforestation • *Meetings:* CITES

▲ Projects and People
A research fellow with the Wildlife Conservation International (WCI), Craig C. Downer has been tracking and studying the endangered mountain tapirs in South America since 1977 when he discovered the boarlike animal as a U.S. Peace Corps volunteer in Colombia.

Downer believes he is the first wildlife biologist to successfully capture, radio collar, and track four tapirs in Ecuador's Sangay National Park. After years of searching for tapirs, such as in the Andes, Downer is on a two-year expedition as a WCI fellow in national parks and reserves of Ecuador, Venezuela, and Colombia to help stem their extinction.

With only about 2,000 tapirs left in the world, according to Downer, those remaining "are falling prey to the methodical destruction of the lush natural environment." The endangered tapir is unprotected by national law and therefore difficult to save from killings by farmers and other private citizens who generally value animals for "meat and pelt and whatever utilitarian benefit can be derived from it," Downer told the *Reno Gazette-Journal* (June 4, 1990).

Downer hopes to change attitudes in South America toward the tapir and other natural resources. In so doing, he must discourage the "burning of forests, overgrazing of Andes, and hunting within national parks." Rainforests and hillsides are frequently set afire as a means to clear land for cattle and crops. As Downer told his home newspaper, "people can learn to farm on lower, flatter land. . . . It would not only preserve the forests and wildlife but makes good sense . . . since the rugged sometimes nearly vertical terrain is a critical watershed to the towns below."

His present goal is to capture an additional nine tapirs to begin a tapir reservation. The tapir weighs 400 to 600 pounds; has a "unique snout for eating, drinking, bathing, and self defense"; and is related to a horse and rhinoceros.

Before devoting himself to the tapir's rescue, Downer studied the black bear in Great Smokies National Park, Tennessee, and worked with other conservation groups on behalf of wild horses and bighorn sheep. One way to encourage global cooperation, he says, is "by being a good example of harmonious and respectful living with other life forms on this planet."

■ Resources
Downer needs adequate funding for his tapir capture missions as well as radio telemetry tracking for large mammal capture. In seeking protection of tapir habitats, the biologist needs support in encouraging public education on the plight of endangered species and in pressuring authorities to create new parks and habitats. He gives talks on this subject. WCI can provide additional information on the tapir and other endangered species.

Wildlife Preservation Trust International (WPTI) ⊕
3400 West Girard Avenue
Philadelphia, PA 19104 USA

Contact: **Jerry Eberhart, Ph.D.**, Executive Director
Phone: 1-215-222-3636 • *Fax:* 1-215-222-2191

Activities: Education; Research • *Issues:* Biodiversity/Species Preservation

● Organization
"The world is as delicate and as complicated as a spider's web. If you touch one thread you send shudders through all the other threads. We're not just touching the web, we're tearing great holes in it." Gerald Durrell

The Wildlife Preservation Trust International (WPTI) was founded in 1963 by noted naturalist, wildlife conservationist, and author Gerald Durrell, to concentrate on saving endangered species worldwide. Its symbol is the dodo, "a flightless and defenseless bird hunted to extinction soon after it was first encountered by sailors stopping at the island of Mauritius," writes Durrell. "It reminds us of the consequences of mankind's thoughtless exploitation of the natural world."

The origin of WPTI is rooted in the Jersey Wildlife Preservation Trust (JWPT), which was established in 1959 on the English Channel island, also by Durrell. JWPT "founded a small zoo committed to the preservation and propagation of species disappearing in the wild."

A zoo must "fulfill three functions in order to justify its existence," according to Durrell: "help people learn and appreciate the importance of other life forms; study animal behavior so that successful conservation plans can be drafted; and be a sanctuary for threatened species, keeping and breeding them so that they will not vanish as the dodo, the guagga, and the passenger pigeon have," as reported in *Animals Magazine* (July/August 1989).

In 1963, Durrell took his operation "a great leap forward" and instituted the WPTI, which brought captive breeding and species perpetuation to a global level. Now composed of 2 united preservation trusts in the USA and Canada, JWPT in the UK, and more than 3,000 worldwide contributors, WPTI has been operating as a consolidated group since 1985. *(See also Jersey Wildlife Preservation Trust [JWPT] in the UK NGO section.)*

The Trust's International Training Center for Conservation and Captive Breeding of Endangered Species (ITC) prepares "New Noahs"—a "new generation of front-line conservatives"—to take leadership in sustaining animal populations worldwide. So far, zoologists from more than 60 countries have participated, many with scholarships. With projects on almost every continent, WPTI funds and sends workers, strategists, and experts to help previously established programs needing financial and physical assistance. WPTI often works in association with the World Conservation Union's Species Survival Commission (IUCN/SSC).

▲ Projects and People
In Africa's Madagascar, projects at Zoo Ivoloina in conjunction with Andrea Katz and Charles Welch, Duke University Primate Center, include the creation of an island in this port town to allow the pre-release of captive-bred black-and-white ruffed lemurs to

prepare them for reintroduction. On behalf of **Parc Tsimbazaza**, WPTI is collaborating with the Missouri Botanical Garden and the Madagascar Faunal Interest Group on a **redevelopment plan for this national zoo**. A survey at the Betampona Reserve turned up **aye-aye, lemur, and diademed sifaka**; lemurs were also studied at Berenty Reserve to determine ecotourism effects, with **Dr. Alison Jolly** and **Jennifer Davidson**. A successful breeding project for the **angonoka, or plowshare tortoise**, is combined with community education. In **Mauritius**, the kestrel, pink pigeon, olive white-eye, cuckoo-shrike, paradise flycatcher, mascarene cave swiftlet, and other endangered birds will be assisted through the approval of the Black River Gorges as a national park, with support from the World Bank. With only 18 remaining, the echo parakeet may be the world's "rarest bird"; a special feeding program is underway and a captive birth is reported.

In the Americas, WPTI is involved with the thick-billed parrot reintroduction, black-footed ferret recovery, golden lion tamarin, black lion tamarin, golden-headed lion tamarin, woolly spider monkey, Baird's tapir, and Guatemalan manatee projects. Projects are ongoing in the USA, Brazil, Central America, and the Caribbean. The new Tropical Education Center at the Belize Zoo is being named in Durrell's honor.

By setting up "stationary arks" worldwide, such as for endangered marmosets and tamarins, Durrell—whose passion for wildlife started in early childhood—hopes that someday the ITC will no longer be necessary. Dr. Jerry Andrew Eberhart joined WPTI in 1991.

■ Resources

WPTI needs computers for research site use. WPTI produces a quarterly newsletter, *On the Edge*, and sponsors a **Run for the Rarest** in Philadelphia. A catalog lists educational publications, posters, note cards, and other gift items. WPTI also offers travel tours, which help support conservation efforts.

Wildlife Science Center
P.O. Box 190
Cedar, MN 55011 USA

Contact: **Terry John Kreeger, D.V.M., Ph.D.,**
Executive Director
Phone: 1-612-464-3993

Activities: Education; Research • *Issues:* Biodiversity/Species Preservation • *Meetings:* Wildlife Disease Association; Midwest Fish and Wildlife Conference; CBSG

▲ Projects and People

Under the directorship of Dr. Terry John Kreeger, the Wildlife Science Center is involved in scientific research on gray wolf, red fox, black bear, white-tailed deer, and other wildlife. It seeks "to advance our understanding of the biology of wild animals through the long-term, humane, scientific studies on captive populations in order to contribute to the conservation and maintenance of such species both in the wild and in captivity."

The Center, which was begun in 1976 by biologists from the University of Minnesota and the U.S. Fish and Wildlife Service, also seeks to serve as an educational resource. It wants to reach all ages, and give individuals some exposure to wild animals and conserva-

tion information. The Center provides an opportunity for collaborative scientific research, and for technical training for wildlife agencies, educational institutions, and conservation organizations. By disseminating information, the Center hopes it can influence people to make informed decisions about conservation and the protection of wild animals.

The Center is determined to conduct **humane research** and maintain a quality lifestyle for the animals. The animals are cared for around the clock by wildlife biologists. It is approved by the American Association for Accreditation of Laboratory Animal Care, and it exceeds the U.S. government standards for animal health and welfare. The research that has been conducted at the Center is published, for example, in *The Journal of Wildlife Management*, *The Journal of the American Veterinary Medical Association*, *The Canadian Journal of Zoology*, *The Journal of Mammalogy*, and other publications.

Dr. Kreeger is newsletter editor for the Species Survival Commission (IUCN/SSC) Captive Breeding Group. With a recent grant from Advanced Telemetry Systems, Minnesota, he and J.R. Tester sampled and analyzed radio-transmitted heart rates and body temperatures in gray wolves.

■ Resources

The Center is always seeking collaborative scientific research on captive wolves, bears, foxes, and deer.

Publications available from the Wildlife Center include a *Gray Wolf Bibliography* and a *Black Bear Bibliography*. Both of these are available in either computerized form or in hard copy. Published also is *Chemical Immobilization of Wildlife with Computerized Bibliography*.

Winrock International Institute for Agricultural Development ⊕
Route 3, Box 376
Morrilton, AR 72110-9537 USA

Contact: **Earl D. Kellogg, Ph.D.,** Senior Vice President
Phone: 1-501-727-5435 • *Fax:* 1-501-727-5242
E-Mail: dialcom 41:tcn400
Telex: 910-720-6616 WI HQ UD

Activities: Development; Education; Research • *Issues:* Biodiversity/Species Preservation; Deforestation; Energy; Sustainable Development

● Organization

Winrock International Institute for Agricultural Development was formed in 1985 by the merger of Winrock International Livestock Research and Training Center, Agricultural Development Council, and International Agricultural Development Service. Headquartered in Arkansas and working with a staff of 225 people from nearly 20 countries, Winrock is involved in agricultural research, education, development, and training. Winrock has fully staffed and equipped offices in Washington, DC; Abidjan, Ivory Coast; and Manila, Philippines. It has another 25 field offices in 4 world regions to "coordinate on-site contact and project implementation."

Winrock has working agreements and memoranda of understanding with more than 100 organizations including consortia, universities, private companies, and international centers. It implements and manages programs for such groups as the Asian Development Bank, Food and Agriculture Organization of the UN (FAO), Ford Foundation, International Fund for Agricultural Development (IFAD), Rockefeller Foundation, United Nations Development Programme (UNDP), U.S. Agency for International Development (USAID), and the World Bank.

In 1989, Winrock staff worked in 34 countries including the USA to provide technical expertise in areas such as agro-enterprise development, irrigation and watershed management, natural resource policy and planning, information systems design and maintenance, soil conservation, agroforestry, and forestry.

▲ Projects and People

Winrock has four "program themes" that govern its project: "strengthening agricultural institutions, developing human resources, designing and implementing sustainable forestry and livestock systems, and encouraging policies that enhance environmental resources and rural development."

The **Development Studies Center** employs natural and social scientists to work on integrated analysis and training. The center yields "concept papers on development issues, data to support outlook and situation analysis, and literature searches" on a variety of issues related to **agricultural development and resource management.**

Winrock implements contracts for a variety of organizations listed earlier in this entry. They currently have approximately 45 projects underway in Asia, Latin America and the Caribbean, Africa, and the USA. These projects have a total annual funding of more than $27 million and involve a multidisciplinary staff that includes agricultural scientists, foresters, economists, social scientists, policy analysts, communications specialists, and computer personnel.

Because of ties with national and community leaders in many nations, Winrock is able to mobilize people to meet the specific needs of the host country, and implement programs for donor agencies to benefit the host country. Because Winrock is politically independent, it says it can undertake any project that **sustainably improves the lives of the rural poor.** Winrock is also well suited for involvement in multicountry or regional projects.

Two long-term global programs that Winrock is running for USAID are: **Economic Analyses of Small Ruminant Production and Marketing Systems** (Brazil, Indonesia, Kenya, and Peru), and **Biomass Energy Systems and Technology.** The Economic Analyses program is "evaluating the economic feasibility of efforts to improve the productivity of small ruminants (such as sheep and goats) and thereby raise farmers' incomes and improving the host country's abilities to conduct similar economic analyses." This program began in 1979 and is scheduled to continue through 1995. The Biomass Energy program is involved in "promoting environmentally sound production of renewable fuels from agriculture and forestry, encouraging commerce in products associated with renewable energy systems, and facilitating the participation of business in renewable energy activities."

A few of the approximately 45 local or regional programs include: **Comparative Analysis of Agricultural Development in Africa and Asia, National Agricultural Research in Egypt, Agroforestry and Participatory Forestry in Bangladesh, Farming Systems Research** in Indonesia, **Irrigation Development in Honduras,** and **Low Input Sustainable Agriculture in the USA.**

■ Resources

Winrock publishes over 50 publications; in addition it runs the **Winrock International Agribookstore,** which is a catalog bookstore of books, maps, slides, and computer software on agricultural science, development, and natural resources. Lists of Winrock publications and the Agribookstore *Catalog* can be obtained from: Winrock International, 1611 North Kent Street, Arlington, VA 22209-2134 USA.

World Association of Soil and Water Conservation (WASWC) ⊕
7515 NE Ankeny Road
Ankeny, IA 50021-9764 USA

Contact: **William C. Moldenhauer,** Executive Secretary
Phone: 1-515-289-2331 • *Fax:* 1-515-289-1227

Activities: Education ; Development; Soil and Water Conservation • *Issues:* Sustainable Development (agriculture)

● Organization

In January 1983, at the International Conference on Soil Erosion and Conservation in Honolulu, Hawaii, delegates from more than 30 countries agreed to form a **World Association of Soil and Water Conservation (WASWC).** Its purpose, according to WASWC, is to serve and inform its members in the soil and water conservation profession and to champion the cause of resource conservation worldwide. Today, more than 500 members represent 63 countries. Current president is Hans Hurni, University of Berne Geography Institute, Switzerland.

According to WASWC, although "unprecedented cooperation among governments, international agencies, and private and public institutions" has occurred since World War II on behalf of agricultural production—lesser efforts are noted among soil and water scientists for such resource conservation, even though this need today is "more apparent than ever." Population growth, the need for more food and fiber, more intensive use of land, more land in production, and deforestation all have combined to increase land and water degradation at an alarming rate, notes WASWC.

Among its activities, WASWC monitors soil and water conservation practices, assesses needs, and supports conservation action in each nation, and maintains a worldwide information exchange for members for awareness of one another's problems and possible solutions to those problems. WASWC also serves as a forum for scientists, conservation professionals, and policymakers and works for the adoption of sound, long-term policies regarding soil and water conservation.

▲ Projects and People

WASWC co-sponsors workshops worldwide, such as the **Steep Lands Workshop** in Puerto Rico that addresses soil erosion which undermines food, feed, and fiber production and a **Soil Manage-**

ment **Workshop** in Edmonton, Canada. Similar workshops are being held in **Kenya, Australia,** and **Hawaii.** Proceedings of these and six international conferences are published for distribution.

A 1991 international workshop in **Indonesia** regarding **conservation policies for sustainable hillslope farming** explored the socio-economic, legal, and institutional issues relating to the formulation of soil conservation policies; considered alternative policy strategies applicable to soil conservation in such areas, and provided a forum for the exchange of ideas and experiences among participants from developing countries. Sponsors with WASWC were Indonesian government ministries, U.S. Agency for International Development (USAID), the Asia Soil Conservation Network for the Humid Tropics (ASOCON), United Nations Development Programme (UNDP), and Food and Agriculture Organization (FAO).

Also in 1991, **Environment and the Poor: An International Workshop on Soil and Water Management for Sustainable Smallholder Development** was held in Kenya and Tanzania and emphasized on-farm research and development, as well as sustainable soil and water management practices that can be transformed into smallholder production systems and assure environmental protection. Methodological development and the appraisal of each region's needs—such as information, training, and networking—were highlighted.

WASWC plans a 1993 training workshop in Manila, Philippines, and is continuing its regional workshops in Africa with emphasis on restoration of degraded and desertified lands, watershed management, and sustainable resource management.

■ Resources

WASWC requests information on soil and water conservation practices that may improve production and reduce sediment production.

WASWC has available several books and conference and workshop proceedings that include *Development of Conservation Farming on Hillslopes, Conservation Farming on Steep Slopes,* and *Land Husbandry: A Framework for Soil and Water conservation.* A quarterly newsletter is also published.

(See also World Association of Soil and Water Conservation—Europe in the UK NGO section.)

The World Conservation Union (IUCN) ⊕
USA Office
1400 16th Street, NW
Washington, DC 20036 USA

Contact: **Byron Swift,** Executive Director
Phone: 1-202-797-5454 • *Fax:* 1-202-797-5461

Activities: Education; Research • *Issues:* Biodiversity/Species Preservation

● Organization
The **World Conservation Union (IUCN),** founded in 1948, provides the means by which governments, government agencies, and the non-governmental environment movement can debate analyze and coordinate their actions to deal with the world's crucial environmental problems. Over 60 sovereign states, more than 100

government agencies, and over 500 non-governmental bodies are linked in IUCN membership, which spreads over 120 countries.

Some 3,000 individuals, drawn from member organizations and professions, form ICUN's network of commissions, expert groups, and task forces. Their collective interaction and debate is orchestrated by a staff of about 75 at ICUN's Swiss headquarters and 100 at expert centers and regional and national offices around the world.

The World Conservation Union's mission is "to harness the insights and skills of the world conservation movement so as to promote the sustainable and equitable use of nature and natural resources and so build harmony between humanity and nature." *(See also World Conservation Union [IUCN] in the Kenya, Niger, Pakistan, Switzerland, and Zimbabwe NGO sections.)*

▲ Projects and People
IUCN embarked on a new plan in 1991–1993: **Caring for the World: A Strategy for Sustainability.** The plan supplements and extends the **World Conservation Strategy** which IUCN, United Nations Environment Programme (UNEP), and World Wildlife Fund (WWF) published in 1980. It worked with the secretariat preparing the United Nations Conference on Environment and Development (UNCED) in 1992.

IUCN works closely with the United Nations system and has consultative status with ECOSOC. With UNEP, Food and Agricultural Organization (FAO), UN Educational, Scientific, and Cultural Organization (UNESCO), the World Bank, and World Wildlife Fund (WWF), it is a member of the **Ecosystems Conservation Group,** formed to coordinate the work of those bodies in the environmental field.

Species Survival Commission (SSC) ⊕
The World Conservation Union (IUCN)
c/o Chicago Zoological Society
Brookfield, IL 60513 USA

Contact: **George B. Rabb, Ph.D.,** Chairman
Phone: 1-708-387-0269 • *Fax:* 1-708-485-3532

Activities: Research • *Issues:* Biodiversity/Species Preservation
Meetings: CITES; IUCN ; IUDZG; SSC

● Organization
"Species and their habitats are under increasing pressure everywhere from humankind. To conserve biodiversity from species level to ecosystem requires management based on understanding of biological sciences as well as cultures, environmental economics, and governmental dynamics."

So states the **Species Survival Commission (SSC),** the largest commission of the **World Conservation Union (IUCN),** which was formed in 1949 to provide leadership to species conservation efforts. The SSC network encompasses 3,400 volunteer member scientists, field researchers, governmental officials, and conservation leaders in 137 countries. SSC members provide technical and scientific counsel for biodiversity conservation projects throughout the world and also serve as resources to governments, international conventions, and conservation organizations.

▲ Projects and People

SSC works principally through some 100 Specialist Groups, most of which represent particular plant or animal groups that are threatened with extinction, or are of special importance to human welfare. A few groups are disciplinary—veterinary medicine, captive breeding, reintroductions, sustainable utilization of wildlife, and trade.

Each taxonomic group is charged to assess the conservation status of the chosen species and their habitats, to develop a **Conservation Action Plan** that specifies conservation priorities, and, finally, to implement required activities outlined in the Plan. Developing a Conservation Action Plan may take several months; full implementation may span decades. Action Plans now underway cover a range of species from tortoises to whales. *(Conservation work of SSC Group members appears throughout this Guide).*

Dr. George B. Rabb, SSC chairman, has been director of the Chicago Zoological Park and president of the Chicago Zoological Society since 1976. He also serves on the University of Chicago's Committee on Evolutionary Biology; as a research associate for the Field Museum of Natural History; on the Biological Parks Advisory Committee of the Santa Fe Community College in Gainesville, Florida; as chairman of the American Committee for International Conservation; and as chairman of the board of the International Species Inventory System.

■ Resources

Funding is sought to assist in the Action Plan program and to enable SSC to respond to urgent situations, including the phenomenon of declining amphibian populations, dramatic worldwide decimation of sharks, and rescue of imperiled rhinoceros populations in Asia.

Captive Breeding Specialist Group (CBSG) ⊕
Species Survival Commission (SSC)
The World Conservation Union (IUCN)
12101 Johnny Cake Ridge Road
Apple Valley, MN 55124 USA

Contact: **Ulysses Seal, Ph.D.,** Chairman
    Phone: 1-612-431-9325 • *Fax:* 1-612-432-2757

Activities: Education; Research • *Issues:* Biodiversity/Species Preservation • *Meetings:* CITES

● Organization

CBSG is a specialist group of the Species Survival Commission, World Conservation Union (IUCN/SSC) operating in 50 countries through 250 members. A network of volunteers and a permanent staff of four serves CBSG and the other groups. These organizations operate captive propagation programs that foster and coordinate international conservation strategies for preserving threatened species.

CBSG helps formulate viable population and conservation action plans involving intensive protection and management of small and fragmented populations in the wild as well as captive propagation programs. Within SSC, CBSG provides a major link between the captive community and the field conservationists in other IUCN specialist groups. Sixty-one zoos and six zoo associations provide most of the funding for an annual operating budget of $358,000.

▲ Projects and People

Captive Action Plans (CAPs) provide a strategic framework by reinforcing viable species populations both in the wild and in captivity. CAPs make recommendations about captive propagation programs and intensive management of species in the wild, hold workshops for Population Viability Analyses, and conduct *in situ* and *ex situ* research. For species in captivity, optimal conservation management programs are designed, priority uses for limited resources are established, and captive breeding habitats are assessed. Using statistical analytical techniques, CBSG categorizes concern about species as "critical," "endangered," "vulnerable," or "high anxiety." Expectations are that species within the "high anxiety" group could move into the first three groups in one to ten years.

CBSG is developing a **Primate Global Captive Action Plan** as a prototype for other CAPs involved in the vertebrate groups. A taxon advisory group, similar to those active in North America, was recommended for all regions in order to improve coordination of efforts.

The **Golden Lion Tamarin (GLT) Management Committee** helped engineer the transfer of ownership of the entire world zoo population of 500 golden lion tamarins to the government of Brazil—the only country where the GLT is indigenous. The action represented a breakthrough in international cooperation among zoos and the first such transfer of its kind of any species. With a world population of about 1,000 animals, some zoos are reintroducing GLTs to protected areas in Brazil; most others remain in zoos and the GLT Management Committee will manage their reintroduction to their natural habitat. Only 2 percent of the GLT's original habitat in Brazil's unique Atlantic coastal rainforest remains.

Zoos are also actively cooperating on **tiger breeding programs** for the Siberian, Sumatran, Bengal, and Indochinese tigers. Breeding sessions are being scheduled. The National Zoo (Washington, DC) and the Minnesota and Omaha zoos conduct research into *in vitro* fertilization and embryo transfer techniques for tigers. The USA and former USSR agreed to a joint study of Siberian tigers in the wild.

Two large projects are underway to **reintroduce the Przewalski's horse into the Mongolian People's Republic** at Hustain nuuru, Ulan Bator, and Great Gobi National Park. The Hustain nuuru projects involve the Foundation Reserves' Przewalski's horse (FRPH), based in the Netherlands, and the Academy of Sciences, former Soviet Union, where the State Committee for the Reintroduction of the Takhi (Przewalski's) is leading the effort.

The **Indonesia-Malaya Faunal Group,** works with the following projects: the Komodo Monitor Consortium of the American Association of Zoological Parks and Aquariums (AAZPA), Bali Starling Conservation Project, Orangutan Research Project (New York Zoological Society and the Harmony Frazier-Taylor, Woodland Park Zoo, Seattle, Washington), Mentawai Island Ecology Study (located in Sumatra, the world's leading island with more species of endemic primates per area), and training programs for zoo biologists in Indonesia. Also, NGOs in Washington, DC, are banding together to pursue greater interagency cooperation for Indonesia's environmental conservation, land planning, and a national biodiversity database proposed by **Dr. Marty Fujuita** of the National Zoo's Conservation Research Center.

In recent years, the National Zoological Park (Washington, DC) has worked closely with Malaysia's Department of Wildlife and National Parks (DWNP) to create a **Malaysian International Center for Wildlife Conservation** in order to establish training programs in wildlife management for the Association of Southeast Asian Nations (ASEAN). **Dr. Rudy Rudran** is spearheading the program; and the Smithsonian Institution, which promotes biodiversity in the region, will be involved. Malaysia's national university has also agreed to establish a degree program to support conservation training.

Other major CBSG programs include the **Asian Rhino Specialist Group**; the **European Breeding of Endangered Species Program (EEP)**, which also has a coordinating committee for the clouded leopard; and the **Invertebrate Working Group**, which prepared a resolution for the IUCN on "Conservation of Insects and Other Invertebrates."

■ Resources

CBSG publishes a newsletter distributed to 5,000 individuals in 150 countries. Biodiversity journals and other publications are also available.

Species Survival Commission (SSC)/Crocodile Specialist Group (CSG) ⊕
The World Conservation Union (IUCN)
Florida Museum of Natural History
University of Florida
Gainesville, FL 32611-2035 USA

Contact: **James Perran Ross, Ph.D.**, Department of
Natural Sciences
Phone: 1-904-392-1721 • *Fax:* 1-904-392-9367

Activities: Education; Political/Legislative; Research • *Issues:* Biodiversity/Species Preservation; Sustainable Development *Meetings:* CITES; IUCN/SSC Crocodile Specialist Group

● Organization

The Florida Museum of Natural History is home to two leaders of the Crocodile Specialist Group (CSG), Species Survival Commission, World Conservation Union (IUCN/SSC). They are **Prof. F. Wayne King**, CSG deputy chairman, and **Dr. James Perran Ross**, CSG executive officer. CSG is a worldwide network of individuals and organizations involved in the conservation of crocodiles, alligators, caimans, and gharials (together called crocodilians) within the IUCN—the world's largest consortium of conservation organizations. CSG's 300 members include scientists, researchers, management authorities, crocodile farmers, hide traders and leather tanners, manufacturers and importers of crocodile products, and retail trade representatives. With other volunteers, they give expert advice to governments, do surveys, estimate population, and provide technical information and training. Private donors raise about $65,000 yearly to fund CSG. At the international level, CSG works closely with the Convention on Trade in Endangered Species (CITES) and other groups espousing sustainable use as a conservation measure. "Economic benefits derived from carefully regulated harvest, farms and ranches, and trade provide a powerful incentive for the preservation of wild crocodilians," writes CSG. So do management plans and the protection and study of certain species.

Although crocodilians are the largest predators in their habitats—with some preying on humans and their livestock—habitat loss and pollution of aquatic habitats take their toll. "Several species are exploited for their valuable skin, which supports an international trade worth over $200 million US annually," reports CSG. Yet crocodilians benefit wetland ecosystems by selectively consuming fish species, recycling nutrients, and maintaining wet refuges during droughts.

Surveys in **Gambia** reveal crocodiles are declining due to habitat loss, such as canopy cover removal, which causes hydrological changes that keep essential wet season forest pools from forming. At the Abuko Nature Reserve, however, a series of **plastic pools** irrigated with plastic piping were buried in the forest floor and attracted a colony of the nocturnal forest dwelling *Osteolaemus* for breeding.

With 34 crocodile farms operating in **South Africa**, two commercial rearing units were recently established as extensions of established breeding operations in Transvaal and Natal. An estimated 4,400 crocodiles are in the wild and game reserves at Natal. "Widespread habitat destruction, poaching, incompatibility with livestock production, and competition with people for resources ensure that the Nile crocodile has little chance of survival outside protected areas in Natal," reports CSG. Also in Transvaal some 4,000 crocodiles are adversely affected by the damming and drying up of rivers, water removal for irrigation, and water pollution. Problem animals and hundreds of hatchlings from doomed nests are transferred to farms. In **Kenya**, the Mamba Village Crocodile Farm is experimenting with feeding space and its effects on growth rate and mortality and also testing how the rearing of young crocodiles in different sex groups affects their growth rate. **Zambia** researchers are seeking funds for aerial surveys and field biology research.

In **Malaysia**, the Sandakan Crocodile Farm is attempting to improve the fertility of laid eggs and build its breeding stock. The Crocodile Farming Institute in the **Philippines** is researching the diets of wild crocodiles; breeding successes are reported. On mainland **China**, a new crocodile farm called **Crocodilian Island is being built** in the Guangdong Province.

In **India**, a saltwater crocodile conservation program is operating successfully at Orissa Research Center. About 200 young crocodiles from the Bhitarkanika Wildlife Sanctuary are expected to be released into the mangroves area of Sundarban in west Bengal, where few crocodiles remain. Captive breeding is proceeding for three Indian species at the Nandanakanan Zoological Park.

Researchers in **Belize** are examining how certain turtle species and crocodiles share habitats. Conversion of private land up for sale to protected wildlife area is being promoted. Long-term surveys of populations of Morelet's crocodile in **Mexico** show that the species is struggling with reproduction, particularly at Catemaco Lake, where there is increased human activity.

In **Argentina**, the caiman breeding program at the El Bagual Ecological Reserve produced their first nests. Brazilian researchers are observing the social fishing behavior of Paraguayan caiman when the shallow-pond habitat offers limited space during the dry season. The outlook is pessimistic for broad-snouted caiman in São Paulo State, where the construction of a hydroelectric station is threatening a key habitat in the Paraña River marsh.

In **Colombia**, caiman and iguana are being ranched in a new project funded by Promatlantico and Acopi—two investment and industrial associations—to produce salted caiman skins and iguanas for the pet trade. For the first time at the Estacion de Biologia

Tropical Roberto Franco, eggs were successfully hatched from a pair of captive Orinoco crocodiles. In Venezuela, captive Orinoco crocodiles are being released into the wild at a ranch close to a refuge. To prevent the consumption of crocodile eggs, FUDENA reports that local Indians are being paid to locate eggs and are offered chicken eggs at a bargain rate instead.

In the USA, crocodile breeding projects are underway at Florida's Gator Jungle and nature exhibits of Silver Springs. A nutritious alligator food has been developed by Gold Kist of Georgia, with reports that "floating food gives more timid alligators equal access to food. . . . Alligators (so fed) grow faster and more uniformly than animals grown in the wild." Recent research shows that alligators lack the enzyme sucrase for hydrolyzing plant sucrose into usable blood glucose; treatments for improper hygiene, nutrition, and crowding are also being studied. To encourage egg fertility, the Ellen Trout Zoo in Texas built new enclosures for Siamese crocodiles.

■ Resources

A list of source materials on crocodiles and alligators is available; a newsletter is published quarterly. A commercial farming guide can be obtained from the American Alligator Farmers Association, P.O. Drawer 1208, Keystone Heights, FL 32656 USA. Dr. James P. Ross also reports the drafting of a worldwide "Action Plan for Crocodilian Conservation." International meetings are held every other year.

Species Survival Commission (SSC)/Sirenia Specialist Group ⊕
The World Conservation Union (IUCN)
Howard University
Department of Anatomy
Washington, DC 20059 USA

Contact: **Daryl P. Domning, Ph.D.,** *Sirenews* Editor
Phone: 1-202-806-6026 • *Fax:* 1-202-265-7055

Activities: Education; Research • *Issues:* Biodiversity/Species Preservation • *Meetings:* International Theriological Congress

● Organization

Dugong or manatee, sea pig or river cow—whichever the name, these are Sirenians, explains biologist and paleontologist Dr. Daryl P. Domning, *Sirenews* editor for the Sirenia Specialist Group of the Species Survival Commission (IUCN/SSC). At Domning's request, group members from Australia, Canada, Ecuador, Indonesia, Japan, Mexico, Netherlands, Palau, and the USA are reconsidering names that "better reflect both classification and major niche differences among living and recently extinct sirenians."

Domning says the "strictly marine" dugong with front flippers and tail that feeds on a variety of seagrasses could be called a sea pig because, among other reasons, it is a "somewhat omnivorous rooter into the bottom. . . . The dugong's 'pigness' is recognized by indigenous peoples who know it well, as indicated by the Sinhalese and Tamil names 'cudalpandi' and 'kadalpani,' both of which translate literally as sea pig." Domning suspects the Malay word "dugong" is similarly translated.

Because the manatee is a grazer and browser in fresh water, "river cow" might be an appropriate name, Domning says, whereas the marine *Hydrodamalis gigas* that grazes on marine algae is a "sea cow."

Formal classification places the dugong and the manatee in a relationship similar to that existing between canids and felids, Domning points out.

▲ Projects and People

In Australia, the tracking of 13 dugongs at Moreton Bay, Queensland, revealed females maintained larger home ranges than males and both were more active in daytime. During winter, dugongs move from the feeding areas on the Bay's seagrass banks to a warm-water refuge in the oceanic waters—synchronizing their movements with the tidal currents.

In Brazil, the Manatee Study and Conservation Center of IBAMA—Instituto Brasileiro do Meio Ambiente e dos Recursos Naturais Renováveis (Brazilian Institute of the Environment and Renewable Natural Resources)—enforces protective laws that have reduced illegal hunting of manatees; however, the vastness and inaccessibility of the Amazonian region create difficulties. According to Iona Colares, IBAMA inspects areas near large lakes in the dry, low-water season to stop massacre of the vulnerable animals. Equipment is confiscated from lawbreaking hunters, fines are imposed, and illegal captures are recorded publicly. The Center is also educating fishermen and school children about preserving manatee and a similar information project, in cooperation with the Aquatic Mammals Laboratory of INPA, is underway in Amazonia. With support from the World Wildlife Fund—USA, Salvatore Siciliano conducted a marine mammal survey along the coast of northeastern Brazil, where the presence and hunting of West Indian manatees (*T. managus*) were discovered.

In Cameroon, James Powell conducted a preliminary survey of manatees (*T. senegalensis*) with support from WWF and Wildlife Conservation International. Information was obtained for four regions, using land and boat surveys of specific watersheds and in conjunction with standard village interviews. For example, fishermen reported seeing as many as five manatees a day during the rainy season in the Korup area's Ndian River and its tributaries. In the Edea region, it is believed the Sanaga River hydroelectric dam is adversely affecting manatee populations and could potentially harm habitats by "amplifying tidal changes downstream and hence altering food availability," reports Melissa M. Grigione. Results of the three-month study are leading to a manatee conservation strategy plan that could include an extension of the Korup National Park borders to provide an aquatic sanctuary for manatees.

In Indonesia, an aerial survey was carried out in keeping with the Dugong Management and Conservation Project. Encouragingly, 15 dugongs were sited, and aerial surveys will continue.

In Ivory Coast, it is reported that road construction through a tropical rainforest is threatening coastal ecosystems, including the mangrove forests around lagoons that support West African manatees.

In Jamaica, it is being decided that four captive Antillean manatees at the Alligator Hole River nature reserve—segregated for years from other manatees—be allowed to roam the river freely while an environmental study monitors the manatees and assesses their impact on the river system. A team led by **Antonio A. Mignucci Giannoni**, acting with United Nations Environment Programme (UNEP) funding under the auspices of the Jamaican Natural Resources Conservation Department (NRCD), stated that "no justification exists, not even for education, to hold four, possibly reproducing female manatees apart (from) the rest of the Jamaican manatee gene pool."

In **Mozambique**, dolphin and dugong occurrence is being studied "in relation to increasing gillnet use and environmental degradation through massive population increase and demographic changes," writes **Vic Cockcroft**, Center for Dolphin Studies, Port Elizabeth Museum, South Africa.

The **Palauan waters** support "what is probably the most isolated dugong population in the world," reports Dr. Helene Marsh, Sirenia Group chairperson. Elsewhere in the Micronesian region, dugongs have been exterminated from several archipelagoes by human exploitation. With The Nature Conservancy's sponsorship, Dr. Marsh led a team to resurvey Palau by seaplane and concluded there are "fewer than 200 dugongs in Palau and their numbers are probably decreasing." Poaching for sport, meat, and even jewelry from rib continues to occur even though it is illegal; penalties are regarded as trivial and few hunters are caught. "Dugongs will become extinct" here if the poaching continues, believes Dr. Marsh.

In the **Persian Gulf**, the six million barrels of oil spilled during the war took four weeks to reach Abu Ali, an island almost halfway down the Saudi coast, preventing the further spread of the main slick. According to **Tony Preen**, 93 marine mammals representing four species were killed in the western Gulf. However, chronic pollution also may be a factor in a series of die-offs of marine mammals and deserves investigation, he believes.

In **Peru's** Pacaya-Samiria rainforest wildlife reserve, oil exploration is threatening the Amazonian manatees and a violation of law, according to environmentalists.

In the **USA**, Florida studies of manatee mortality show that hazards from watercraft impacts are increasing faster than overall threats, according to the Florida Department of Natural Resources. Floodgate and canal-lock deaths are also rising.

World Nature Association, Inc. (WNA) ⊕ 🌿
P.O. Box 673, Woodmoor Station
Silver Spring, MD 20901 USA

Contact: **Donald H. Messersmith, Ph.D.**, President
Phone: 1-301-593-2522 • *Fax:* 1-301-593-2522

Activities: Education; Research • *Issues:* Biodiversity/Species Preservation

● Organization

The World Nature Association (WNA) describes itself as "a small, low-overhead, membership-controlled organization of concerned naturalists [including] amateur bird watchers, self-trained wildflower experts, professors of the natural sciences, international tour guides, professional wildlife managers, research scientists, and nature photographers" from several countries.

WNA was started in 1969 as a World Nature Club by the "ardent" **Orville Wright Crowder**, who five years earlier created one of the world's first nature tour organizations. The all-volunteer

association—under the leadership of **Dr. Donald Messersmith**, professor emeritus of the University of Maryland—adheres to Crowder's concept of bringing together both amateur and professional naturalists from various countries to support and work in small, vital conservation projects worldwide.

Through members' dues and contributions, WNA funds projects that are usually too small to attract support from international conservation groups. WNA's most important function, Dr. Messersmith reports, is to "provide help to individuals who are working on saving rare and endangered habitats and wildlife of all types outside the United States where financial help is not nearly as readily available. . . . We especially like projects that include educating the local inhabitants in conservation matters and sustainable yield of their resources."

▲ Projects and People

Several projects have been undertaken in **mainland China**, where equipment and educational materials were furnished to a nature preserve in the northeast, and assistance with educational materials for Love of the Birds Week and crane research was extended in cooperation with the International Crane Foundation and the Chinese Ministry of Forestry. Funding is provided for *The Call of the Crane*, a Chinese publication. Nanjing Normal University received copies of *The Field Guide to the Birds of Japan* for ornithology classes. A six-volume set of *The Birds in the Soviet Union* was given to the Beijing Natural History Museum.

A bird-banding project was undertaken in **Cyprus** with publication of results. Andy P. Damalas reported here that "time is running out for the birds of Cyprus. Hunting is one of the most popular pastimes. Over 40,000 hunters shoot more than one million birds every year. Most hunters shoot birds whether they are protected or not. . . . The government is seeking ways and means to educate sportsmen."

Audio equipment was provided to a national park visitor center in **Costa Rica**. Craig Downer studied the mountain tapir in Ecuador to prescribe a rescue program for the endangered species which he says was well received by the government and pursued by conservation organizations in the cloud forest region where "a very exuberant mountain tapir population" makes its home. In the **Dominican Republic**, Susan Miller studied Armando Bermudez National Park management and the interface between ecological processes and local uses. She says the park—"established within a subsistence agricultural economy" which may circumvent the park's preservation and functional objectives—is viewed by the local residents as an imprisonment of utilizable natural resources. Her findings are being shared with the Dominican government.

Studies of Mayan Gardens, Wildlife Density, and Subsistence Hunting in **Mexico** helped Jeffrey P. Jorgenson understand that such centuries-old practices of Mayan farmers "may represent a sustainable use of their wildlife resource. Development is threatening the forest," he writes. "If the forest is lost, the wildlife will likely disappear with serious implications for the local Mayan residents."

Rosemarie Gnan's work with Breeding Biology of the Bahama Parrot on Abaco led her to conclude that "feral cat predation is less in barren nesting areas than in areas with vegetation." Her recommendations for conservation of this parrot, which nests underground in limestone cavities, is being submitted to the Bahamian government and international conservation groups.

In observing vicuna and alpaca habitats in the National Fauna Reserve of Ulla Ulla, **Bolivia**, Maria Lilian Villalba Murillo concluded that "a sustainable industry using these animals can be developed with proper management."

In **Papua New Guinea**, Andrew and Debra Mack began a research station to study the effects of cassowary foraging and seed dispersal on the flora and to promote conservation incentives through work and income for local people. One reported success is that local landowners placed a moratorium on all hunting and clearing within several kilometers of the research station.

WNA literature warns against the threat of the greenhouse effect, and states that birds, which traditionally warn of ecological disasters such as gasses in mines and water pollution on farms, may now predict the hazards of rising temperatures. For example, snow geese who migrate from the southern USA to the northern end of Hudson Bay every spring are responding to a heavy cloud cover that delays snow melt.

"Now the birds halt their migration 90 to 150 miles south of where they used to nest," WNA reports. "As a result, they may have become salt marsh menaces, their taste for young plants turning productive marshes into mud flats. Arctic wading birds are following the geese in lower numbers since the insects and worms they consume are almost gone from the marsh. All the species dependent upon the marsh for food and nest sites will be affected. Man, too will pay the price as the marsh loses its ability to detoxify runoff, stock bodies of water with fish, and provide places of aesthetics and wonder."

As an educator, Dr. Messersmith's areas of expertise are entomology, ornithology, and environmental science.

■ Resources

WNA publishes a biannual newsletter for members; maintains a library of books and periodicals on nature study, conservation, and world travel (some of which are rare volumes); and holds meetings alternately in the USA and other countries. An annual scholarship is offered to a natural sciences educator in the USA to attend a summer Audubon nature camp.

World Resources Institute (WRI) ⊕
1709 New York Avenue, NW
Washington, DC 20006 USA

Contact: **Wallace Borman**, Secretary/Treasurer
   Phone: 1-202-638-6300 • *Fax:* 1-202-638-0036

Activities: Development; Research • *Issues:* Biodiversity/Species Preservation; Deforestation; Energy; Global Warming; Sustainable Development; Transportation; Environmental and Resource Information • *Meetings:* IUCN; TFAP; UNCED

● Organization

"How can societies meet human needs and nurture economic growth without destroying the natural resources and environmental integrity that make prosperity possible?" The **World Resources Institute** (WRI), through policy research and technical assistance, helps governments, the private sector, environmental and development organizations, and others address this central question. The new president is Jonathan Lash, formerly Environmental Law Center direcor, Vermont Law School.

The aim of WRI's policy studies is to "generate accurate information about global resources and environmental conditions, analyze emerging issues, provide early warning of tomorrow's questions, and develop creative responses to both problems and opportunities." WRI publishes a variety of books, papers, and reports, holds briefings, seminars and conferences, and provides the media with new perspectives and materials on environmental issues. Through such communication, WRI seeks to "build bridges between scholarship and action."

WRI's programs reach developing countries, where the organization provides technical support as well as policy analysis and other services to both government and non-government entities.

Created in 1982, WRI has a 95-member interdisciplinary staff, augmented by a network of advisors, collaborators, international fellows, and cooperating institutions in more than 50 countries.

WRI president (through 1992) James Gustave Speth, writes, "A series of large-scale transitions, redirecting growth, are essential if human society is to approach sustainability." WRI is adopting six "mega-trends" by which to measure environmental leadership in the 1990s. These are:
- "A demographic transition toward stable populations both in nations where growth is explosive and on a global basis before the world's population doubles again."
- "A technological transition away from today's wasteful and polluting technologies, particularly in energy, to a new generation of technologies that sharply reduce environmental impact per unit of prosperity."
- "An economic transition to a world economy based on reliance on nature's 'income,' not depletion of its 'capital.'"
- "A social transition to a more equitable sharing of environmental and economic benefits both within and among nations."
- "A transition in understanding and consciousness to a far more profound appreciation of the requirements of global sustainability."
- "An institutional transition to new arrangements among governments and peoples."

▲ Projects and People

WRI's activities are divided into three broad areas: Policy Research, the Center for International Development and Environment, and Policy Affairs and Publications.

Under **Policy Research**, WRI is currently carrying out research projects relating to forests and biodiversity; economics and institutions; climate, energy, and pollution; resource and environmental information; technology; and special initiatives in institutions and governance.

The **Center for International Development and Environment** "fields technical experts at the request of developing countries that want help assessing or better managing natural resources." The Center's six programs are: Natural Resources Management Strategies and Assessments; Natural Resources Data; Biological Diversity and Sustainable Agriculture; Forestry and Land Use; NGO Support Services; and From the Ground Up.

Responsible for general management of the center and its budget is Deputy Director **Walter Arensberg**, who is also directly responsible for the Center's Cooperative Agreement with the U.S. Agency for International Development (USAID), entitled the **Environmental Planning and Management Project**. This project supports the Center's program throughout the world and facilitates its work with USAID missions in developing countries.

Examples of projects undertaken under Arensberg's direction include assistance to **Bolivian NGOs** in the formation and development of an umbrella NGO to strengthen Bolivian NGOs' capacities to address environment and development issues; a participatory **rural appraisal workshop in Ecuador**, which succeeded in identifying problems, resources, and lines of activities to improve agroforestry; support for the participation of 80 grassroots groups in the Ecuadorian **Tropical Forest Action Plan (PAFE)**, which demonstrated that "forest dependent peoples can provide constructive alternatives to national problems and become part of global environmental initiatives when provided with the opportunity"; defining a strategy for USAID's **environmental program in Chile**; and country environmental profiles (CEPs) of **Colombia** and **Paraguay**.

Aaron Zazueta is manager of the **NGO Support Services Program (NGOSS)** which acts as WRI's "field practitioner for sustainable development, both through policy research and technical assistance." NGOSS is frequently included in advisory teams for official planning processes in Africa and Latin America. Within this context, the goal of NGOSS is to "enhance the capacities of NGOs and grassroots organizations to contribute to the search for environmentally sound policy options that meet the needs of the poor." This entails developing and disseminating methods and tools in the form of handbooks, videos, and other guides for improved natural resource management; conducting case studies of local initiatives; and organizing regional forums to discuss recommendations for successful natural resource management.

Director of **Sustainable Agriculture** is **Lori Ann Thrupp**, who organizes and administers projects and activities in Latin America. Her work involves developing strategies, participatory rural appraisal methods, planning program budgets, and support of grassroots agroecology projects.

It has been discovered that many of the most effective, rural-based environmental efforts are tied to centuries-old practices. **From the Ground Up** is a project that "identifies communities already pursuing a course of ecologically sound self-development, and analyzes the root causes and key relationships of their efforts." From the Ground Up is underway in more than 10 nations in sub-Saharan Africa and in two countries in Latin America. Contacts with Asian organizations are being initiated. Each country program represents a joint effort involving government and NGOs, universities, development assistance agencies, and concerned grassroots organizations. **Peter Veit** is manager of the Africa NGO Policy Impact Program.

A case study in Waddaa community in **Sudan**, for example, shows how farmers practice sound land-use management based on soil characteristics. Erodible sandy soils are planted with tree crops. On clay soils, farmers build small ridges along slope contours to capture water.

By promoting sharing of information about such practices—both among villages and among decisionmakers at all levels—policy changes can begin to favor "decentralized, small-scale natural resource management methods," which in turn leads to a "people-oriented approach to development."

The purpose of the **Policy Affairs and Publications** division is to put WRI's recommendations into the hands of those who can act on them. **Donna Wise** is policy affairs director. **Jessica Matthews**, who writes regularly for *The Washington Post* and various journals, is vice president. **Sarah Paula Burns**, NGO liaison, oversees collaboration with other non-governmental organizations. **Rafe Pomerance**, **Peter Thacher**, and **Robert Blake** handle interactions with U.S.,

United Nations, and other official agencies and the U.S. Congress. **Miwako Kurosaka** is liaison with Japanese officials and non-governmental bodies. **Nina Kogan** and **Bruce Smart** are in charge of initiatives with private businesses in the USA and abroad. **Shirley Greer** handles contact with the communications media.

■ Resources

A current catalog of WRI's publications is available. In addition to extensive printed materials, WRI's *World Resources 1992–1993* is now available on IBM-compatible diskette, providing the latest data on economic, population, natural resource, and environmental conditions for 147 countries.

For further information, contact: WRS Publications, P.O. Box 4852, Hampden Station, Baltimore, MD 21211; phone 1-800-822-0504 (USA only); fax 1-410-516-6998.

World Wildlife Fund (WWF) ⊕
1250 24th Street, NW
Washington, DC 20037 USA

Contact: **Susan Krug**, Program Assistant for Public Affairs
Phone: 1-202-293-4800 • *Fax:* 1-202-293-9211

Activities: Development; Political/Legislative; Research • *Issues:* Biodiversity/Species Preservation; Deforestation; Energy; Global Warming; Sustainable Development • *Meetings:* CITES

● Organization

In July 1990, **World Wildlife Fund (WWF)** and **The Conservation Foundation (CF)** officially merged to form a single, legally incorporated organization. This formalized a relationship that dated back to 1985. WWF was started in 1961, with the launching of organizations in the USA, United Kingdom, Netherlands, and Switzerland. CF was created over 40 years ago for the purpose of bringing objective, interdisciplinary analysis to a broad range of environmental issues. Today, WWF is the largest private organization in the USA working worldwide to protect endangered wildlife and wildlands. The international WWF family consists of national organizations or representatives in nearly 40 countries.

The organization is "committed to reversing the degradation of our planet's natural environment, and to building a future in which human needs are met in harmony with nature. We recognize the critical relevance of human numbers, poverty, and consumption patterns to meeting these goals."

▲ Projects and People

In the *Annual Report*, Chairman of the Board **Russell E. Train** writes: "The rise of our multi-faceted organization, which is now poised to confront the compelling conservation challenges of the new decade, has been interwoven with one of the great movements of human history—the growth of environmental awareness and action in the United States and around the world."

President **Kathryn S. Fuller** adds: "Given the pressures of human need, economic development, and industrial pollution, WWF seeks to exert more influence than ever before. Creating nature reserves is not enough where people in neighboring communities lack adequate land, seeds, credit, and training to farm successfully, and the capital and skills to serve tourists. So long as this is the case, the cycle of poaching, overuse of resources, and poverty can only continue, until once protected reserves are destroyed. In pursuit of our mission, therefore, we must necessarily become more deeply engaged in socio-economic issues: not just identifying species and their habitats, but also addressing the human context in which they must survive."

Highlights of current programs follow:

Despite the diverse wildlife and lush plant cover, the **Amazon** rainforest is described as a "fragile tapestry." Only 10 percent of the land is suitable for long-term agriculture, and if extensive plant cover is destroyed, the odds are against recovery of the natural forest and animal species. A meeting in Manaus, Brazil, supported by WWF and other organizations, brought together for the first time more than 100 prominent scientists working on biodiversity in the Amazon. The findings of this meeting enabled WWF to set new priorities to accelerate conservation throughout the region.

The United States is the world's largest importer of wild birds, accounting for more than 600,000 of the estimated 3 million birds entering the trade annually. The **Cooperative Working Group on Bird Trade**, convened by WWF and staffed by TRAFFIC USA, WWF's trade monitoring arm, brought together representatives of the pet industry, aviculturists, zoos, animal welfare organizations, and conservationists. The group reached a "precedent-setting agreement" in March 1990 that the USA "reduce and ultimately end within five years the import of wild-caught birds for sale specifically as pets." The policy will also facilitate captive breeding, ideally in countries of origin.

TRAFFIC has also undertaken a comprehensive review of U.S. implementation and compliance with the Convention on International Trade in Endangered Species of Wild Fauna and Flora (CITES).

The island nation of **Madagascar** contains an exceptional variety of plants and animals: 200,000 species exist in an area only about 40 percent larger than California, and an estimated 150,000 of these are unique—for example, insectivorous mammals called tenrecs; hundreds of endemic reptiles and amphibians; 7 of the world's 9 species of baobab trees; and the rosy periwinkle, source of a drug used to combat Hodgkin's disease and childhood leukemia.

Yet, by some estimates, more than 80 percent of the natural forest cover is gone; less than 2 percent of the island is protected. WWF's efforts in Madagascar began over 20 years ago. Debt-for-nature swaps in 1989 and 1990 have provided the funds needed to begin meeting urgent conservation priorities to help assure basic protection for the most sensitive habitats throughout Madagascar.

"Although they encompass only one-twentieth of the world's land area," according to WWF, "wetlands are under pressure to accommodate competing natural and human demands." Fortunately, however, "perhaps just in time, the public and policymakers are beginning to understand the special role such places can play in nature." WWF wetlands projects are underway in Latin America—including the world's largest wetland, the Pantanal, covering 55,000 square miles of central Brazil, Paraguay, and Bolivia—and the Caribbean. In addition, WWF helped create the Convention on Wetlands of International Importance, called the **Ramsar Convention** for the Iranian town where it was signed in 1971, and has since encouraged greater participation among developing nations.

WWF assisted the U.S. Environmental Protection Agency (EPA) in convening the **National Wetlands Policy Forum**, which endorsed a nationwide policy of "no overall net loss" of wetland resources and long-term actions to increase the quantity and improve the quality of the nation's wetlands. The language was adopted by Congress in 1989 in the North American Wetlands Conservation Act.

WWF, in conjunction with the Institute for Research on Public Policy, released a report in 1989 called *Great Lakes, Great Legacy?*, addressing ecosystem contamination in the Great Lakes region. This contamination is most dramatically symbolized by the serious decline in the number of bald eagles that are successfully reproducing along the Great Lakes shores. In response to the publication, the United States and Canada reached a landmark agreement to work to restore and maintain the "chemical, physical, and biological integrity of the waters of the Great Lakes Basin." Yet only limited headway has been made toward achieving these broad goals. WWF staff are continuing to work with Congress on legislation to speed the cleanup.

■ Resources

WWF has an extensive publications catalog. Information is available from: World Wildlife Fund Publications, P.O. Box 4866, Hampden Post Office, Baltimore, MD 21211, phone 1-410-516-6951, fax 1-410-516-6998. The WWF *Annual Report* lists specific projects worldwide with names of investigators.

Members receive a newsletter with reports on projects, invitations to visit project sites and attend lectures, and *Wildlife Alerts* on endangered species and WWF efforts.

(World Wildlife Fund and World Wide Fund for Nature [WWF] projects and people appear throughout the Guide.)

Worldwatch Institute ⊕
1776 Massachusetts Avenue, NW
Washington, DC 20036 USA

Contact: **Lester Brown**, President
Phone: 1-202-452-1999 • *Fax:* 1-202-296-7365

Activities: Education; Research • *Issues:* Air Quality/Emission Control; Biodiversity/Species Preservation; Deforestation; Energy; Health; Population Planning; Sustainable Development; Transportation; Water Quality; Waste Management/Recycling

● Organization

Since 1975, Worldwatch has worked to educate policymakers and the general public about "the interdependence of the world economy and its environmental support systems. The Worldwatch research staff analyzes issues from a global perspective and within an integrated, interdisciplinary framework." Much of Worldwatch's research is public policy related. Worldwatch is a nonprofit organization with a staff of 30.

▲ Projects and People

Worldwatch president **Lester Brown** is involved in many aspects of the organization's activities. A co-author of the yearly series *State of* *the World* since 1984, Brown is the editor of *World Watch* bimonthly magazine. He has written more than 18 Worldwatch papers on topics ranging from population trends and policies to redefining "national security."

In addition to his work with Worldwatch, Brown has written 10 books, including *By Bread Alone* and *The Twenty-Ninth Day* about world food problems, that were cited with literary awards. A recipient of at least 12 honorary doctorates in law, humane letters, and science, he was also named MacArthur fellow by the John D. and Catherine T. MacArthur Foundation in 1986. Brown serves on the boards of a number of organizations including: Overseas Development Council, Better World Society, Green Seal, Renew America, Global Studies Center, and the Global Tomorrow Coalition.

In addition to co-authoring *State of the World*, Brown serves as its project director. Continuing to grow in popularity, the book is now available worldwide, and has been translated into more than 17 languages. Worldwatch feels that "since neither the United Nations nor any national government attempts to produce a comprehensive, annual analysis of changing global environmental conditions, our *State of the World* report is acquiring semi-official status, widely used by national governments, U.N. agencies, the international development community, and concerned individuals everywhere."

State of the World contains information on both the current state of the environment—such as sustainable energy, solid waste, urban transportation pollution, forestry reform, global economy, emissions in Eastern Europe, and military impacts—and different public policy responses. This can include descriptions of different policies that are being used, and an analysis of the effectiveness of those policies.

Worldwatch conducts its own research, and also has an information gathering network through subscriptions to approximately 200 periodicals and publication exchange agreements with some 90 research institutes and public interest research groups.

■ Resources

In addition to the annual *State of the World*, Worldwatch has published nearly 100 papers on specific issues. *World Watch Magazine* is a bimonthly publication aimed at a more general audience. A complete publications list is available.

WorldWIDE Network ⊕
1331 H Street, NW, Suite 903
Washington, DC 20005 USA

Contact: **Waafas Ofosu-Amaah**, Managing Director
Phone: 1-202-347-1514 • *Fax:* 1-202-347-1524

Activities: Education; Research • *Issues:* Sustainable Development

● Organization

Recognizing that women can have very different views of environmental problems than their male counterparts, WorldWIDE is devoted to creating networks among women environmentalists and providing them with support. Founded in 1982, WorldWIDE is a nonprofit, membership organization headquartered in Washington, DC, with an International Advisory Council with representatives from 17 different countries.

WorldWIDE goals are "to establish a worldwide network of women concerned about environmental management and protection; to educate the public and its policy makers about the vital linkages between women, natural resources, and sustainable development; to promote the inclusion of women and their environmental perceptions in the design and implementation of development problems; and to mobilize and support women, individually and in organizations, in environmental and natural resource programs."

▲ Projects and People

A major ongoing project is the *WorldWIDE Directory of Women in the Environment* coordinated by Helen Freeman. With entries from more than 70 countries, this directory continues to expand each year. The directory includes women who are involved in environmental and resource management issues at the grassroots level, women whose professions relate to environmental issues, and other women committed to environmental solutions. Women who are interested in being listed in this directory should obtain an application from WorldWIDE. The only requirement is that people submit the application on their own behalf to demonstrate their interest in joining the network.

WorldWIDE mobilizes the **Global Assembly of Women and the Environment**, which is convened under the sponsorship of the Senior Women's Advisory Group on Sustainable Development (SWAGSD) with the executive director of the United Nations Environment Programme (UNEP). The conference is an opportunity for women to demonstrate their abilities in environmental management. Waafas Ofosu-Amaah serves as WorldWIDE's project director for the conference.

At the conference, there are presentations of environmental success stories in areas of "environmentally friendly systems, products and technologies, water, energy as it relates to climate change, and waste." Those projects selected are "repeatable, affordable sustainable, and visible." "New generation leaders," people between the ages of 18 and 25, are invited to participate in the assembly to gain exposure to the issues involved. **Regional Assemblies** precede the Global Assembly.

WorldWIDE Forums are groups that are formed by women on the local or national level. These are autonomous forums that are affiliated with WorldWIDE and are designed to deal with environmental and natural resource issues. Currently there are women in more than 10 countries organizing or planning Forums.

■ Resources

Available resources include the *Directory of Women in the Environment* and *WorldWIDE News*, an international newsletter published six times a year.

WorldWIDE also developed the **WorldWIDE Information Service**, a "computerized database of sources of information and materials on environmental and development topics." Materials are prepared in Spanish and English, with plans to include other languages in the future.

The Xerces Society ⊕
10 SW Ash Street
Portland, OR 97204 USA

Contact: **Melody Mackey Allen**, Executive Director
   Phone: 1-503-222-2788 • *Fax:* 1-503-222-2763

Activities: Education; Research • *Issues:* Biodiversity/Species
Preservation

● Organization
The Xerces Society was founded in 1971 to fight "human caused
extinctions of rare invertebrate populations and their habitats." The
organization's name comes from the name of the first North American butterfly to become extinct, the Xerces blue butterfly. The scope
of the organization has grown, and "since 1985 the society has
committed itself to protecting invertebrates as the major component
of global biological diversity." Xerces now has a staff of 4 and a
membership of 3,000.

▲ Projects and People
Xerces has three program areas: **Conservation Science, Education,**
and **Public Policy.**

One project is **Biodiversity Research and Training in Madagascar.** Xerces is working in collaboration with other organizations to
encourage the protection of endangered resources and areas. In
addition to providing invertebrate biodiversity data for use in conservation planning, it also works to increase public awareness of
conservation issues and train Malagasies in conservation techniques.

In **Jamaica,** Xerces is working in conjunction with The Nature
Conservancy to "[focus] on ecosystem requirements for a spectacular **butterfly** (*Papilio homerus*) and its rapidly vanishing habitat. The
homerus is the largest and most endangered swallowtail in the
Western Hemisphere."

Xerces also runs **Invertebrate Indicator Workshops.** These workshops are designed to bring together ecologists and invertebrate
biologists to work on practical conservation strategies and methods
for assessing biodiversity and monitoring long-term stability of
ecosystems.

In the USA, Xerces is working on **Science Based Conservation
Strategies for Old-Growth Forests.** These programs are designed to
shift the focus from conservation aimed at a single species to programs aimed at preserving the forest ecosystem and structure. In the
Pacific Northwest, where the old-growth forests can have as many
200 to 250 species of invertebrates per square meter of soil, there has
been more focus on preserving the habitat of the spotted owl than
on careful preservation of the whole ecosystem, believes Xerces.

In Clatsop County, Oregon, Xerces is involved in the **Silverspot
Butterfly Habitat Preservation Project.** In collaboration with the
U.S. Fish and Wildlife Service and The Nature Conservancy, Xerces
is working to develop a recovery plan for the silverspot butterfly,
which is on the brink of extinction. Xerces is also publishing a
brochure on the silverspot which will be distributed in public parks
and museums throughout the butterfly's range.

In California, Xerces is working to protect the overwintering
habits of the **monarch butterfly,** including establishing a grassroots
network to protect the overwintering sites. Xerces is working with
state and local governments, scientists, land trusts, and landowners
in a bid to establish protection for monarch habitats.

For the last 15 years, Xerces has sponsored its **Annual Xerces
Society North American Butterfly Count.** Taking place on or close
to the Fourth of July, Xerces publishes the results annually. This
gives people a chance to monitor butterfly populations in their own
area and around the country.

Xerces also sponsors the **International Register of Invertebrate
Specialists,** which is a database supplying information to encourage
"biodiversity conservation planning in the Western Hemisphere."

The Society includes **Dr. Stephen R. Kellert,** president; **Dr. Paul
Opler,** vice president; **Ed Grosswiler,** secretary; **Katherine Janeway,**
treasurer; and **Melody Mackey Allen,** executive director.

As **Dr. E.O. Wilson** (Harvard University), Scientific Advisory
Committee chair, reminds, "Quite simply, the terrestrial world is
turned by insects and a few other invertebrate groups; the living
world would probably survive the demise of all vertebrates, in
greatly altered form of course, but life on land and in the sea would
collapse down to a few simple plants and micro-organisms without
invertebrates."

■ Resources
Xerces has a large number of publications, and a full publication list
can be obtained from the organization. These publications include
*The Common Names of North American Butterflies; Butterfly Garden-
ing: Creating Summer Magic in Your Garden; Wings: Essays on Invertebrate Conservation; Atala;* and the *Fourth of July Annual Butterfly
Count Reports.*

Zero Population Growth (ZPG)
1400 16th Street, NW, Suite 320
Washington DC 20036 USA

Contact: **Susan Weber,** Executive Director
   Phone: 1-202-332-2200 • *Fax:* 1-202-332-2302

Activities: Education; Political/Legislative; Research • *Issues:*
Air Quality/Emission Control; Biodiversity/Species Preservation; Deforestation; Energy; Global Warming; Health;
Population Planning; Sustainable Development; Transportation; Waste Management/Recycling; Water Quality

● Organization
Since 1968, **Zero Population Growth (ZPG)** has been working "to
achieve a sustainable balance of population, resources and the environment—both in the United States and worldwide." ZPG is a
nonprofit organization with a membership of 40,000 and a staff of
20.

▲ Projects and People
Susan Weber, executive director, is responsible for a number of
projects, such as the *Urban Stress Test,* a guide that ranks U.S. cities
based on how well they are withstanding the stresses of population
growth. The Stress Test addresses issues, including population density, population change, crowding, education, violent crime, community economics, individual economics, birth, air quality, hazardous wastes, water, and sewage. Each of these topics was rated on a
scale of 1 to 5, and the results included how each city fared on each

topic and a ranking by averages. Cedar Rapids, Iowa, had the best score while Gary, Indiana, had the worst.

Weber is also the author of *USA by Numbers: A Statistical Portrait of the United States*, published by ZPG. Treasurer for the Global Tomorrow Coalition, she has served on the boards of Natural Resources Council of America and Earth Day 1990.

ZPG runs a **Population Education Program** that provides "quality population education training and teaching materials to educators." These materials are "hands-on" and cross-disciplinary. They require little teacher preparation. There are a wide variety of curricula, and while many are suitable for high school students, there are also many programs suitable for students at all levels. ZPG also runs teacher training programs.

ZPG has prepared a report entitled *Planning the Ideal Family: The Small Family Option* written by Pamela Wasserman and edited by Dianne Sherman. This booklet provides information designed to help people with decisions about whether to have children, and how many children to have. This guide encourages people to have small families by discussing economic, social, and environmental impacts that are associated with children, and with a large population growth. It also analyzes the trends in family sizes in the USA. This booklet also works to debunk social myths about childless marriages and only children.

ZPG also is involved in citizen action efforts to "build public support for domestic and international family planning, and other key population programs." This includes keeping ZPG members informed of congressional activities that can affect population planning through the quarterly *ZPG Activist*.

■ Resources

ZPG publishes a wide range of information. This includes books, fact sheets, *The ZPG Reporter* bimonthly, the *Urban Stress Test*, resources for educators, and more. A complete list is available from ZPG.

Zoo Atlanta ⊕
800 Cherokee Avenue, SE
Atlanta, GA 30315 USA

Contact: C. Dietrich Schaaf, Ph.D., Assistant Director/General Curator
Phone: 1-404-624-5619 • *Fax:* 1-404-627-7514

Activities: Education; Research • *Issues:* Animal Population Planning; Biodiversity/Species Preservation; Conservation; Recreation; • *Meetings:* AAZPA; American Association of Zoo Vets; IUDZG

● Organization

Zoo Atlanta is a zoo with a broad scope—involved in research and conservation work worldwide. It has a membership of 176,000 and a staff of 105 including Zoo Atlanta director Dr. Terry L. Marple, assistant director/general curator Dr. C. Dietrich Schaaf, veterinarian Dr. Rita McManamon, Dr. Debra L. Forthman, Dr. Elizabeth Franke Stevens, and Jeffery S. Swanagan.

According to Schaaf, "We see education as being key to promoting environmental awareness at all levels. The public is often ignorant of environmental matters, and if information is available to them it is frequently misunderstood or misinterpreted. Consequently, environmentalists and conservation biologists need to become good educators and populizers of their work. . . . Equally important is the education of political leaders at all levels."

He continues, "In terms of getting the message across to a large segment of the public, zoos and similar institutions (aquariums, museums, botanical gardens) have the potential to play a major role in environmental education. The mission of today's zoo in the USA and many other countries involves conservation, research, and education. . . . There is a need to take a more holistic approach to environmental education."

▲ Projects and People

Zoo Atlanta runs a variety of field research and conservation projects. They provide financial assistance and technical expertise to the field research, and are involved in conservation projects in natural ecosystems on four continents.

In Akagera National Park, Rwanda, Debra Forthman and Wayne Esarove are studying the behavior and ecology of orphaned elephants to learn the effects of early loss of the "adult kin group on reproductive fitness." Comparisons will be made with studies by Cynthia Moss, African Wildlife Foundation, among populations with "more normal demographics" at Kenya's Ambolesi National Park and by Douglas-Hamilton at Tanzania's Lake Manyara. The Rwandan government has also asked for Zoo Atlanta's assistance in running their national park and encouraging tourism.

Schaaf, Thomas M. Butynski, and Gail W. Hearn are engaged in a project to study drills in Bioko, West Africa. Drills and their mandrill cousins are forest-dwelling baboons that are endangered by habitat destruction and illegal hunting. With less than 60 drills in captivity worldwide, including 3 at Zoo Atlanta, such studies will help improve captive-breeding programs.

In *ZOOM* magazine, Schaaf describes his first encounter with drills in the unpopulated island's "forested interior of the Gran Caldera Volcanica De Luba, a vast, collapsed volcanic cone that broods over the southern shores of Bioko," where the rare Preuss's monkey was also sighted: "We were able to watch as [drills] ate wild figs and as the youngsters cavorted in the dappled shadows of their forest world.

"We found evidence of their feeding on a variety of plants and invertebrates, and our guides showed us how to imitate the distressed calls of duikers—small forest antelopes—that people use to lure drills when hunting them." This confirmed for Schaaf that "drills are at least occasional predators, as are savannah baboons and chimpanzees, and that they would take disabled or vulnerable duikers if the opportunity arose."

Because of the remoteness of the Caldera, Schaaf is searching for another site in Bioko suitable for long-term study. He and Butynski are also "proposing to guide the rehabilitation of the Entebbe Zoo and its development into a conservation education center."

Elizabeth Stevens is working with the Charles Darwin Research Station in the Galapagos Islands to study the reproductive behavior of flamingos, which tend to breed poorly in captivity. Her observations of group courtship display reveal, "It seems to take more than just . . . a male and a female. . . . There seems to be a minimum number of birds needed to provide an adequate level of stimulation for successful breeding." Stevens is hoping to determine the optimum flock size; she is also establishing a national studbook for Chilean flamingos "to coordinate the breeding efforts of zoos across the country."

Stevens is also bringing some of Zoo Atlanta's expertise in breeding giant tortoises to help enhance the Galapagos breeding program.

Lorraine Perkins and Dr. McManamon are undertaking a project in Indonesia to enhance communication between field researchers about orangutans. To overcome the organutan's lethargy in captivity, Zoo Atlanta experimented with "holzrugels" or small raisin-filled logs—as shown by Dr. Christian Schmidt of Zoo Zurich, Switzerland. Increased activity, it is believed, will help stimulate successful reproduction. Perkins is international and North American Regional Studbook keeper for orangutans.

In Belize, Zoo Atlanta is assisting with the renovation of the Belize Zoo and Tropical Education Center and offers the native Belizean staff help with horticulture, environmental education, and animal management.

Zoo Atlanta and the San Diego Zoo are "developing a program for assisting Chinese biologists in developing captive propagation programs for endangered Asian monkeys." Zoo Atlanta also expects to help develop ways to protect the giant panda in its habitat.

In the USA, Zoo Atlanta has ongoing projects in Georgia, including peregrine falcon release project, a study of alligator in the Okefenokee Swamp, a gopher tortoise program whereby zoo-bred hatchlings are released near a wild colony, work to protect indigo snakes, breeding a bald eagle pair that have so far produced eggs but no chicks, and Dennis Herman's monitoring of bog turtles in the wild.

One goal is to "repopulate the skies" with peregrine falcons that will prey on Atlanta's starlings and pigeons. A bird released in 1989 roosts at the Marriott Marquis Hotel. Reptile curator Howard Hunt's studies focus on nesting ecology of American alligators; his videotapes show bears destroying such nests.

In addition to research, Zoo Atlanta is involved in educational activities—drawing up activity packets for teachers on such topics as endangered species and dinosaurs. They also have educational packets designed to enhance the educational experience of a trip to the Zoo, and run a series of educational programs at the Zoo.

Schaaf strongly holds that "conservationists need to communicate more effectively with the public at large." To reach beyond "the converted," he recommends placing articles in popular news magazines or specialty "airline" periodicals; encouraging television coverage on *60 Minutes* and similar programs; writing columns in local newspapers; and developing new school science curricula emphasizing ecological understanding and responsibility.

Zoos, he believes, can be used to "draw the public into programs that show the interrelatedness of a variety of environmental problems—air and water pollution, solid and toxic waste disposal, energy and non-renewable resource conservation, global warming, and the thorniest of all issues, human population growth." Schaaf reminds, "Unless we begin to make more progress on all environmental fronts, wildlife conservation efforts on their own have little chance for successful outcomes in the long run."

■ Resources

Zoo Atlanta publishes *ZOOM* magazine; *Field Guide to the African Rainforest* exhibit produced with Yerkes Regional Primate Research Center of Emory University and Ford Motor Company; *Against Extinction*—a resource manual for teachers on endangered and extinct animals; *Zoo Atlanta Encounters: A Teacher's Guide to Rainforest Extinction*; and various brochures.

◊ *Private*

American Board of Environmental Medicine
International Board of Environmental Medicine ⊕
4205 McAuley Boulevard, Suite 385
Oklahoma City, OK 73120 USA

Contact: **Clifton Rowland Brooks, M.D.,** Executive Director
Phone: 1-405-749-0193 • *Fax:* 1-405-751-5168

Activities: Development; Education • *Issues:* Accreditation

● Organization

The **American Board of Environmental Medicine** examines and evaluates the qualifications of physicians "previously certified by an American Specialty Board in a field of Clinical Medicine or Surgery, who have, and wish to be recognized for, special interest, training, experience, and knowledge in the discipline of Environmental Medicine," according to their literature. This Board also certifies "such physicians to be experts in the field after examination by their peers, and periodically producing a *Register of Experts* in the field to be made available to the public, as well as the profession."

Training programs in this field are also evaluated, as are hospital units such as environmental control units, emergency care facilities and biodetoxification units, and rehabilitation centers. Recommendations are made to the medical and allied health science professions on establishing and maintaining the adequacy of and improving the quality of care in this discipline.

Practicing physicians who make voluntary application for examination to the American Board generally have at least five years of practice in the field, a residency, fellowship, or postgraduate training from an institution approved for such training, and are "thoroughly knowledgeable of the work."

The USA is divided into 10 geographic areas by U.S. Department of Labor's Occupational Safety and Health Administration (OSHA) regions and states. Regional offices are in Boston, Massachusetts; New York, New York; Philadelphia, Pennsylvania; Atlanta, Georgia; Chicago, Illinois; Dallas, Texas; Kansas City, Missouri; Denver, Colorado; San Francisco, California; and Seattle, Washington.

The **International Board of Environmental Medicine** is established as an independent accreditation agency. Members are from the USA, Puerto Rico, Canada, Japan, and United Kingdom.

■ Resources

An annual **Register** of the American and International Boards of Environmental Medicine is available.

Applied Biology, Inc. (ABI)
Environmental Consultants
P.O. Box 974
Jensen Beach, FL 34958-0974 USA

Contact: **R. Erik Martin,** Senior Scientist, Marine
Biologist
Phone: 1-407-334-3729

Activities: Research • *Issues:* Biodiversity/Species Preservation; Water Quality • *Meetings:* Annual Workshop on Sea Turtle Conservation and Biology

● Organization

From its Jensen Beach field office, **Applied Biology, Inc. (ABI)**, a woman-owned small business, does environmental impact assessments of aquatic and terrestrial ecosystems, biological and water quality monitoring programs from southern Florida to the Great Lakes, and endangered species surveys for government and industry. Corporate headquarters are in Decatur, Georgia, and founder is **Dr. Nancy W. Walls**, whose previous experience includes expeditions to Antarctica to investigate system dynamics in the Antarctic Ocean. ABI field and laboratory work involves water chemistry; microbiology; bioassay; terrestrial, wetland, and aquatic macrophytes; plankton, fish, and macroinvertebrates; birds and small mammals; and aerial photo mapping and interpretation.

In a sampling of more than 200 studies, ABI has determined when landfill runoff contaminates water and coral reefs; how thermal discharges of utility plants impact lagoons and seagrasses; how industrial effluents and drilling affect fish abundance and health; what heavy metals, oils and greases, and pesticides do to sediments; and how to protect sea turtles and their habitats.

▲ Projects and People

R. Erik Martin, manager of ABI's sea turtle programs, has been studying the effects of coastal construction and power plant operation on the nesting of loggerhead and green sea turtles since the early 1970s.

Research began on Hutchinson Island, a then undeveloped barrier island near the southeast coast of Florida, recognized as an important rookery for loggerhead turtles. Initially, nine survey areas were monitored during daylight hours each weekday and every other year. A decade later, the survey area increased to 36 stations along the island's entire Atlantic coastline; monitoring was daily and annually. By 1988, it was shown that high beachfront lighting and conspicuous human activity in areas marked by coastal development kept turtles from emerging and females from reproducing. Although nighttime construction at a power plant also deterred turtles, sea turtle nesting was not affected by the power plant's operation as potential disturbances to nesting behavior were eliminated. In other island areas where development is buffered by vegetated dunes and beachfront lighting is at a minimum, emerging and nesting rates appeared high—although subject to change. Erosion control measures, more human beach activity at night, and denser residential development could impact on nesting. "It is imperative that systematic sea turtle nesting surveys continue on Hutchinson Island," writes Martin.

The **St. Lucie Power Plant** on Hutchinson Island has been a key player in turtle conservation. "The plant draws its condenser cooling water through an enclosed intake canal connected with the Atlantic Ocean via submerged pipes," reports Martin. "Sea turtles, which are apparently attracted to the offshore structures housing the intake pipes, are often entrained with cooling water and become trapped in the intake canal. Since the plant began operating, entrapped sea turtles have been systematically captured, measured, weighed, tagged, and returned to the ocean."

In a 10-year period, 1,322 loggerhead and 192 green turtles were removed from the canal. Young turtles were captured mostly in the winter; adult loggerheads were more abundant in the summer and mostly female en route to nesting beaches. When these captured turtles were compared with those at other locations along the east coast of Florida, Georgia, and South Carolina, it was shown that there is a relatively uniform population of loggerheads in the coastal waters of the southeastern USA; that loggerheads do not enter coastal waters until they are at least 40 cm in length; and that Florida coastal waters are an intermediate habitat for green turtles leaving the pelagic environment to enter lagoonal feeding grounds, but maturing sub-adult green turtles are generally absent there.

ABI also helped draft the comprehensive sea turtle protection ordinance for St. Lucie County to ensure that beachfront lighting, beach access, beach/dune stabilization and restoration projects, mechanical beach cleaning, and other coastal activities do not disturb sea turtles and their nesting habitats.

Martin is a member of the Species Survival Commission (IUCN/SSC) Marine Turtle Group.

■ Resources

Facilities include chemistry laboratories, water-quality sampling equipment, boats and gear for biological studies, sediment-testing devices. For the Florida Power & Light Company, ABI prepared a public information booklet, *Florida's Sea Turtles,* used by schools and civic groups.

Aveda Corporation
4000 Pheasant Ridge Drive
Minneapolis, MN 55434 USA

Contact: Horst Rechelbacher, Chairman
Phone: 1-612-783-4000 or 1-800-283-3224
Fax: 1-612-783-4110

Activities: Manufacturer of plant- and flower-based beauty products • *Issues:* Health; Sustainable Development; Waste Management/Recycling • *Meetings:* CEP; CERES; Social Venture Network

● Organization

Aveda Corporation makes beauty and home products from organically grown plants and flowers. Horst Rechelbacher started the company in 1978 with the purpose of providing consumers an alternative to the synthetic, petroleum-based products that dominated the beauty industry. From its shampoos to dishwashing liquids to lipsticks, Aveda's products are blended from all natural substances. No animals are used for product testing. At the same time, Aveda is dedicated to promoting environmental action worldwide through sponsorship of "green causes." Rechelbacher maintains, "We strive to set an *example* for environmental leadership and responsibility, and we hope the rest of the world will follow that example."

▲ Projects and People

The company recently instituted a corporate environmental plan to "conserve waste and energy to the maximum extent possible," including fully recycling product packagings. Customers can return such packaging to Aveda's 20,000 authorized salons and distributors as they participate in this program known as Aveda Earth Action: Agenda for Recycling. "The program has the potential of reducing the salon's waste by up to 70 percent," predicts Aveda, while costs to the salons are minimal. Rechelbacher acknowledges, "This is the first program of its kind in the beauty industry."

Working out standards with the Society of Plastics Institute, which denotes the types of plastic for recycling with a number indicating density level, Aveda uses HDPE (high density polyethylene) bottles and LDPE (low density polyethylene) tubes. The company is working with Waste Management, Inc., which turns the plastics into secondary-use products.

In 1991, the company underwrote the U.S. Aveda Environmental Film Festival; sponsored an Earth Day Cut-A-Thon to benefit Global ReLeaf, the American Forestry Association's tree-planting project; and inaugurated a Student Catalyst Program to "help our country's youth share and act upon their concerns for the environment." A year earlier, Aveda raised more than $130,000 for Global ReLeaf in the USA; was selected as an official sponsor of Earth Day; and became the first corporation to sign the Valdez Principles that was created by an investors' coalition as an advanced set of guidelines for corporate responsibility. *(See also Coalition for Environmentally Responsible Economies and Council on Economic Priorities in the USA NGO section.)*

■ Resources

Aveda's environmental affairs manager is available for presentations.

Libby Bassett
Environmental Consultant and Writer
521 East 14th Street, #4F
New York, NY 10009 USA

Phone: 1-212-533-3082 • *Fax:* 1-212-533-3082
E-Mail: Econet: lbassett

Issues: Air Quality/Emission Control; Biodiversity/Species Preservation; Deforestation; Energy; Global Warming; Health; Population Planning; Sustainable Development; Transportation; Waste Management/Recycling; Water Quality • *Meetings:* UNCED; World Women's Congress for a Healthy Planet

▲ Projects and People

Libby Bassett's career as writer, editor, and consultant on international environment and development issues has taken her to Africa, the Americas, Asia, Europe, and the Middle East. Her current projects include writing, editing, designing, and producing the United Nations Environment Programme's (UNEP) *North America News and Environmental Sabbath Newsletter.*

Bassett was communications coordinator for the World Women's Congress for a Healthy Planet, which has as its goal the full participation of women in policies which were to be determined by the 1992 United Nations Conference on Environment and Development (UNCED) in Brazil. The World Women's Congress met in Miami, Florida, in November 1991 to organize for a more active policymaking role and to strengthen and expand networks to influence the UN Conference through an Action Agenda to be used over the next decade and beyond.

Bassett was also communications coordinator for the Women's Environment & Development Organization (WEDO) of the Women's Foreign Policy Council, whose mission is also to enable women to become full partners in global environmental decisionmaking. She also writes, designs, and produces the Council's newsletter, *News & Views.* In addition, Bassett served as press officer for the Earth Summit in Rio de Janeiro, Brazil, for the Global Forum of Spiritual and Parliamentary Leaders on Human Survival. The Global Forum started in October 1985, when a core group of religious and political leaders, representing five continents, met for the first time in Tarrytown, New York. The purpose of their meeting was to explore the possibility of a dialogue that intermingled their perspectives. According to Global Forum's literature: "The lawmakers tended toward practical solutions while the spiritual leaders emphasized morality and ethics. From their differing viewpoints new possibilities emerged. In the end, the religious and political leaders . . . called upon colleagues worldwide to join them in a conference on global survival."

The first Global Survival Conference in April 1988 drew nearly 200 spiritual and legislative leaders to Oxford, England. Among the participants were cabinet members, speakers of parliaments, and the Dalai Lama, Mother Teresa, the Archbishop of Canterbury, High Priest of Togo's Sacred Forest, Cardinal Koenig of Vienna, and Native American Spiritual Chief Oren Lyons. They conferred with (among others) astronomer Carl Sagan, scientist Evguenij Velikhov of the former Soviet Union, Gaia scientist James Lovelock, Kenyan environmentalist Wangari Maathai, and Cosmonaut Valentina Tereshkova.

Then, in 1990, the Global Forum on Environment and Development took place in Moscow. More than 1,000 spiritual and parliamentary leaders, artists, journalists, business persons and young people from 83 countries came to the Moscow Forum, which then President Mikhail Gorbachev called "a major step toward the ecological consciousness of humanity."

The literature summarizes the importance of the Global Forum as follows: "Throughout much of human history, spiritual and political leaders consulted regularly. Only in modern times have the two been separated. Today it has become increasingly clear to the leadership of these two great constituencies that only together can they confront the complex, interconnected threats to the survival of all life on Earth."

Whitman Bassow & Associates, Inc. ⊕
655 Third Avenue
New York, NY 10017 USA

Contact: **Whitman Bassow, Ph.D.**, President
Phone: 1-212-867-6356 • *Fax:* 1-212-697-6554

● Organization
Dr. Bassow founded Whitman Bassow & Associates, a consulting service on management and communications policy for international environmental affairs, in 1989 after he retired from the World Environment Center (WEC) as president emeritus.

▲ Projects and People
With funding from the United Nations Environment Programme (UNEP), Dr. Bassow in 1974 founded the **World Environment Center**, as "a nonprofit, non-advocacy organization concerned with international environment and development issues." WEC was "the first environmental organization to link multinational corporations with developing countries to provide volunteer assistance in environmental training and management," writes Dr. Bassow.

Dr. Bassow was also instrumental in originating two other programs: The **International Environment Forum** (1977) is "a meeting place for industry executives and senior environmental officials from around the world, now comprising 60 multinational corporations." In 1984, Dr. Bassow also created the **Gold Medal Award for International Corporate Environmental Achievement** "which provides public recognition for outstanding industry accomplishments," he says.

An environmentalist, journalist with *Newsweek* and United Press International (UPI), and CBS News correspondent, Dr. Bassow was also a United Nations senior public affairs officer in charge of media and public information. He participated in the landmark UN Conference on the Human Environment held in Stockholm in 1972, which inspired his creation of the World Environment Center.

Business Publishers, Inc. (BPI)
951 Pershing Drive
Silver Spring, MD 20910-4464 USA

Contact: **Kevin Adler**, Marketing Director
Phone: 1-301-587-6300 • *Fax:* 1-301-587-1081

Activities: Environmental Publications • *Issues:* Air Quality/Emission Control; Energy; Health; Transportation; Waste Management/Recycling; Water Quality

● Organization
Predating the U.S. Environmental Protection Agency (EPA), it was 1963 when **Business Publishers, Inc.** (BPI) produced its first *Air/Water Pollution Report*. Since then, it publishes some 59 "weekly, biweekly, and monthly newsletters that address issues of global importance." According to their literature, these publications have one goal: "to bring you the news you need—on target and on time—from EPA, Congress, the courts, federal agencies, state and local governments, industry, academia, and international sources."

The *World Environmental Directory* (WED), described as eleven directories in one, first appeared in 1974 and is now in its seventh edition for the USA and Canada. It contains some 1,000 pages of listings of manufacturers of pollution control equipment "classified by air, asbestos, hazardous waste, noise, radiation, sludge, solid waste, toxic substances, and water"; consultants, designers, labs, and researchers providing professional services; EPA, independent agencies, and other federal and state/provincial government environmental offices; and non-governmental organizations, including United Nations' INFOTERRA.

Newly released is the *Directory of Radioactive Waste Officials*, including both federal government and corporate sections in the USA.

Among others, BPI's publications/price list contains the weekly *Toxic Materials News* and biweeklies, such as *International Solar Energy Intelligence Report, Land Use Law Report*, and *Hazmat Transport News*.

CEHP Incorporated
1133 20th Street, NW, Suite 200
Washington, DC 20036 USA

Contact: **Loretta Neumann**, President
Phone: 1-202-293-1774 • *Fax:* 1-202-293-1782

Activities: Education; Political/Legislative; Research • *Issues:* Biodiversity/Species Preservation; Deforestation; Energy; Global Warming; Transportation; Water Management/Recycling; Water Quality

● Organization
CEHP Incorporated provides government relations, research, planning and information services in the fields of conservation, environment, and historic preservation for educational and nonprofit groups, professional societies, and government agencies. (CEHP is an acronym for conservation, environment, and historic preservation.)

▲ Projects and People
CEHP tracks activities of federal agencies, such as the endangered species program of the U.S. Fish and Wildlife Service, the effect on cultural resources of coal-mining programs under the Office of Surface Mining, and the implementation of the Native Grave Protection and Repatriation Act for the National Park Service.

Regarding the U.S. Congress, CEHP tracks and analyzes legislation, prepares testimony and position papers, drafts proposed amendments, and arranges meetings with lawmakers and staff. Issues include clean air and water, pesticides, biodiversity, global warming, Alaska conservation, hazardous and solid waste management, and a national energy policy. Coalition building and grassroots constituency development help get complex legislation enacted. Databases are developed. Research and policy analysis is conducted regarding environmental protection and cultural resources; CEHP recently assessed the Clean Water Act and the Wild and Scenic Rivers Act.

CEHP manages national projects, including administering the **Save the Past for the Future** project of the Society for American Archaeology that addresses archaeological looting and vandalism. **Passport in Time** is a nationwide initiative of the U.S. Forest Service and CEHP which enlists volunteers to reroute mountain trails, work with Indians in archaeological excavations, inventory ancient cultural sites, survey rock art, catalog photographs, take oral histories, and make maps. **Training in government relations, professional development, and conference management** are also provided. **Jill Schaefer** is program administrator. **Kathleen Reinburg** is editor.

■ Resources

CEHP Notes is a newsletter for clients and subscribers that highlights congressional action. The *Archaeology and Historic Preservation Federal Affairs Notebook* is a subscription series for educators, professionals, and practitioners in these fields.

Community Vision, Inc. ⊕
60 East Hanover Avenue
Morris Plains, NJ 07950 USA

Contact: **Karl Kehde**, Land Use Consultant
Phone: 1-201-267-3244 • *Fax:* 1-201-267-9816

Activities: Education; Research • *Issues:* Sustainable Development

● Organization

Community Vision, Inc. is an umbrella organization which fosters **Land Use Forums (LUFs)** throughout the USA. Land Use Forums are a union of citizens, landowners, and officials who work together to create plans for "diversified communities with affordable housing, common recreation facilities, open spaces, and pedestrian connections to adjoining areas."

Land Use Forums are created in order to emphasize cooperation between all strata of people concerned with growth and planning and citizens wishing to have a say in their expanding communities. They are an alternative to traditional methods of official decisionmaking regarding land development which frequently takes place "in an adversarial setting that pits developer against planning board and planning board against the citizen it represents."

Community Vision helps to get these Forums started and functioning on their own in an effort to assist the members of the community-at-large in becoming a direct part of their ever-changing, ever-growing surroundings. The Forums also help these citizens substantially influence the formation of community plans.

▲ Projects and People

The organization sees the ideal LUF as being made up of 5 to 25 members who can be town council legislators, members of the local planning board, environmental officers, and representatives of groups interested in preservation and community betterment, such as historical societies, League of Women Voters, or Parent-Teachers Association (PTA).

The LUF should be an "informal place to debate and discuss plans to solve environmental and social problems and apply these solutions to a specific site, therefore enhancing the sense of community," says **Karl Kehde**. Ways of handling waste, roads, solar access, agricultural lands, and habitats for endangered species are the kinds of issues that are addressed. Generally, LUFs produce plans which support "diversified communities with mixed land uses, affordable

housing, common recreation facilities, substantial protected parkland, and a variety of pedestrian, open space, and architectural connections to adjoining areas," according to Kehde. The atmosphere of trust that is engendered during this process helps generate public support for the project during the approval process.

To help with the formation of LUFs, Community Vision sometimes can advance money from its revolving fund to cover expenses that the Forum needs, such as for aerial photos, topographic surveys, and other land planning needs. Mostly, landowners fund the effort, hoping to gain a superior development plan that will win official approval. Usually the process eliminates duplicate engineering and legal fees and saves money and other resources. The community also holds **LUF Advisors workshops** to help the Forums function as cohesive groups.

Three New Jersey residential projects that Kehde designed and developed include Heron Park, with 20 acres donated to the NJ Natural Lands Trust to protect critical environmental areas; Forest Park, Boonton Township, where wetlands, steep slopes, and stream corridors are protected from development; and Highfields at Hardyston, which donated a 67-acre permanent wildlife sanctuary to the state's Lands Trust.

■ Resources

Community Visions, Inc. offers brochures which introduce the LUF process, the LUF *Guidebook*, training for advisors, scale model buildings, and a *This Place, Our Place Workbook* for LUF participants.

Cumberland Science Museum
800 Ridley Boulevard
Nashville, TN 37203-4899 USA

Contact: **Celeste Hauser**, Marketing Director
Phone: 1-615-862-5160, ext. 56
Fax: 1-615-862-5178

Activities: Development; Education; Research • *Issues:* Air Quality/Emission Control; Biodiversity/Species Preservation; Deforestation; Energy; Global Warming; Health; Waste Management/Recycling; Water Quality • *Meetings:* American Association of Museums; Association of Science and Technology Centers

● Organization

Originating in 1944 as the Nashville Children's Museum, today the Cumberland Science Museum is a major educational and cultural attraction in this city—serving almost 400,000 people in 1991 and supported jointly by private corporations, state and metropolitan government, and some 8,500 members. Cumberland Museums manages both the science museum and Grassmere Wildlife Park.

▲ Projects and People

Cooperating with the Tennessee Department of Education and WSMV–Channel 4 television, the Cumberland Science Museum has developed **Tennessee Environmental Issues Teaching Kits** with an approach known as the **Science, Technology, and Society (STS)** method that is supported by the National Science Teachers Association. Science is learned "in context of an issue, so it is more relevant, more student-oriented, and more fun," reports the museum. Specific to the region, topics for grades 4 through 10 feature landfill problems; whether to clearcut Tennessee forests; and endangered species such as the peregrine falcon, bald eagle, Tennessee coneflower, and mussels. In a unit designed for high school students, oil prices resulting from Middle East situations, energy policy, and the state's solar power potential are examined together.

In conjunction with Steiner Liff and the *Nashville Banner*, a permanent recycling exhibit shows students how to "reduce waste, reuse items, and recycle materials." Traveling displays on clean air, rainforests, and endangered animals are offered through the Museum Mobile educational programs presented at local schools.

Supported in part by the National Science Foundation (NSF), a Science-by-Mail program for grade levels 4–6 and 7–9 gives children, teachers, parents, and scientists an opportunity to solve science problems with learning kits and pen pals.

Charles A. Howell III, president and CEO of Cumberland Museums and former Tennessee Commissioner of Conservation, also heads Trust for the Future—an environmental education foundation.

■ Resources

The Cumberland Science Museum has brochures describing its various environmental programs.

Earon S. Davis, J.D., M.P.H.
Environmental Health Consultant
2530 Crawford Avenue, #115
Evanston, IL 60201 USA

Phone: 1-708-475-8620 • *Fax:* 1-708-475-8520

Activities: Education; Political/Legislative; Research • *Issues:* Air Quality/Emission Control; Health • *Meetings:* American Academy of Environmental Medicine; American Public Health Association

▲ Projects and People

Earon S. Davis is an environmental health law consultant, a writer, and a lecturer who concentrates on indoor pollution, multiple chemical sensitivities, and the rights and responsibilities of consumers and industry—including various compensation systems, regulatory systems, and litigation. "The recognition of the subtle health effects of environmental contamination, indoors, outdoors, and at work, are central to our efforts to motivate change. Man is also an endangered species," writes Davis.

Described as a crusader against ecological illness, Davis sees the perception of environment "moving from the national parks and scenic areas to the neighborhoods . . . from the great outdoors and into our homes and workplaces . . . from an engineering and legal focus to that of public health professionals and 'victims' organizations."

In the 1970s, he says the environment movement emphasized caring for nature—the planet's animals and plants—with the growing prominence of groups such as the Sierra Club, National Wildlife Federation (NWF), and the Natural Resources Defense Council

(NRDC). Gradually, the focus turned to toxic chemicals and hazardous wastes. Organized labor and urban groups got involved, and people were talking about "Love Canal," asbestos in the schools, Kepone, "agent orange," dioxin, and the "workers' right to know" about toxic exposures in the workplace. New measuring devices began detecting pollutants in neighborhoods, human bloodstreams, and breast milk.

Davis says that "psychological denial" of chemical contaminants' threats to human life is a main reason why individuals and groups fail to act with resolve, creativity, and speed in meeting such challenges.

People responded to the energy crisis by insulating their homes with materials containing formaldehyde. Hermetically sealed office buildings are associated with recorded "decreases in worker productivity and increases in absenteeism," he says.

Davis also disputes the federal government's "irrational approaches" to compartmentalizing people into units,—outdoor, occupational, housing, traveling, for examples—regulating and controlling these units separately as if they were not interrelated, thus ignoring the cumulative effect of exposure to toxic chemicals. Davis asks citizens to think about the names of national and state agencies. "The Environmental Protection Agency [EPA] seems focused on protecting the environment, not people," he writes.

A positive step is that environmental workers are beginning to be trained in human health matters such as toxicology, and epidemiology, along with law and engineering. Davis recommends a better government/industry/public approach to the human environment that emphasizes the "'whole person' and not merely segments of people depending upon when and where they work . . . spend time indoors, and . . . live in urban or rural areas."

In presentations to government, legal and health workshops and forums, Davis stresses, "It is time to expand our notion of the human environment and begin to investigate the larger social issue involving potentially serious health problems that may be related to the overall chemical 'overload' facing our society." Davis is on the Advisory Board of the National Center for Environmental Health Strategies.

■ Resources

Needed are political support, educational outreach, and public relations for the "recognition of multiple chemical sensitivity as an environmental illness."

Development Strategies for Fragile Lands (DESFIL) ⊕
605 Ray Avenue
Silver Spring, MD 20910 USA

Contact: Dennis V. Johnson, Consultant
Phone: 1-301-587-2840

Activities: Research • *Issues:* Biodiversity/Species Preservation; Deforestation; Sustainable Development • *Meetings:* CITES

● Organization

Development Strategies for Fragile Lands (DESFIL) is a private for-profit technical resource of the Office of Rural and Institutional

Development, Bureau for Science and Technology, and the Office of Rural Development, Bureau for Latin America and the Caribbean, U.S. Agency for International Development (USAID). DESFIL is a consortium of Development Alternatives, Inc. (DAI), Tropical Research and Development, Inc. (TR&D), Earth Satellite Corporation (EarthSat), and Social Consultants International (SCI), which provides technical assistance to USAID missions and national governments in Latin America and the Caribbean. Its goal is to "arrest the degradation of natural resources while encouraging the increased production of food and fuel for income generation on steep hillsides and human tropical lowlands."

▲ Projects and People

According to consultant Dennis V. Johnson, DESFIL undertakes the following assignments: raising public awareness and involvement in sustainable use and management of fragile areas while analyzing and formulating related policies; analyzing appropriate technologies for managing fragile area resources and incentives to stimulate the use of those technologies while developing institutional arrangements that promote and facilitate sustainable development; developing geographic, management, and natural resource information systems, and sharing that information; formulating strategies and plans concerning natural resources for host country clients and AID missions including environmental assessments, land-use planning, natural resource strategies, action plans, and project designs; and promoting collaboration among international donors, including USAID, and public and private development organizations.

Recent DESFIL projects include natural resource management in Central America; workshop on sustainable uses for steep slopes in Latin America and the Caribbean; agricultural strategy in the Andean region; tropical forestry action plan in Belize; land-use planning in the high valleys of Bolivia; northern zone consolidation in Costa Rica; on-farm water management in the Dominican Republic; environmental workshops, Sumaco land-use planning, and forestry sector development in Ecuador; natural resources strategy in El Salvador; geographic information system for highland agriculture development, tropical forestry action plan, and Maya resources management in Guatemala; national agroforestry program and watershed management in Haiti; strategy planning and environmental education in Honduras; hillside agriculture in Jamaica; biological diversity in Peru; and southeast peninsula land-use management in St. Kitts.

Dennis Johnson was a member of DESFIL's core team of specialists in agronomy, forestry, geography, animal science, institutional anthropology, and agricultural and resource economics. He has managed the Guatemala Tropical Forestry Action Plan and organized an international conference on the humid tropical lowlands, among other activities.

■ Resources

DESFIL publishes working papers, project summaries, technical reports, newsletter, and brochure with background on the consortium members and information on how to acquire services. A list of publications is available.

Earth Foods Associates ⊕
11221 Markwood Drive
Wheaton, MD 20902 USA

Contact: Lee Fryer, President
Phone: 1-301-649-6212

Activities: Development; Education • *Issues:* Abate Food Pollution; Energy; Global Warming; Health; Sustainable Development; Waste Management/Recycling; Water Quality

▲ Projects and People

Lee Fryer has been in food production from most every angle and through most of this century—as migrant farmer, trucker, New Deal rural-assistance program administrator, fertilizer manufacturer, technologist, federal agricultural bureaucrat, and entrepreneur. For two decades, as head of Earth Foods Associates, Fryer shows how to take the best from both worlds—organic and mainstream high-technology—as he develops model farms across the country.

His enemies are the pollutants of land, water, and foods that are the staples of "petroagriculture"—pesticides and herbicides that enter ecosystems and food chains. Fryer's message is to "use resources of both science and nature to attain high yields of safe, nutritious farm and garden foods." "Eco-agriculture" is his method. Otherwise, traditional organic methods cannot feed millions of people, he believes.

For fertilizers, he recommends "ecologically sound" humanates, humic acids, biological supplements, and sprays for crop feeding through leaves. Because he recognizes that few farm districts have ample and affordable manure and compost to fertilize crops, he believes innovation is the key. Slowly releasing timed nutrients into the soil and adding minerals through "seaweed sprays," for instance, are the kinds of new ideas he is advocating. His firm formulates energy and cost-saving fertilizers and plant protectants, using seaweed and fishery materials as main ingredients.

Fryer is also vice-chairman of Acres Foundation for Foods and Agriculture so that "safe, nutritious foods can be readily available in supermarkets at reasonable prices." Plans for this new group are to reassess existing fertilizer and pest control technologies and products to identify those that are ecologically sound; to build success models of ecologically sound systems and demonstrate these in varied soil and climate zones in the USA; to educate the market as to sound farming systems and classes of products; to stimulate federal and state agencies to update themselves on new approaches; to involve school children in environmental projects; and to keep the media and public informed about successful techniques.

Author of books including *The Bio-Gardener's Bible*, *Earth Foods*, and *Whole Foods for You*, Fryer wrote *A Child's Organic Garden* with his grandchild Leigh Bradford at age nine.

Encycle/Texas, Inc.
5500 Up River Road
Corpus Christi, TX 78407 USA

Contact: **John K. Likarish**, General Manager
    Phone: 1-512-289-0300 • *Fax:* 1-512-289-7415

Activities: Recycling and Resource Recovery; Research • *Issues:* Transportation; Waste Management/Recycling; Water Quality

● Organization
Encycle/Texas, with a staff of 100, is the largest hydrometalurgical waste recycling and reclamation facility in North America, it claims. This operation includes a 100-acre production complex which recovers, reclaims, and processes hazardous and nonhazardous wastes with an annual capacity of 175,000 tons. It extracts beneficial metals and other valuable inorganics from waste materials which are now lost through landfilling and other less desirable disposal methods. So far, Encycle has recovered more than 15,000 tons of copper, lead, zinc, nickel, silver, and other elements from wastes which otherwise would have been lost as natural resources. The marketing and sales program is national.

Established in 1988, Encycle has developed and/or utilized more than 50 different treatment and processing steps for recovering natural resources from waste. "The results are optimistic in the application of these steps and technologies," writes **Lee Deets**, former CEO. "The results are somewhat less optimistic from an economic standpoint while companies can landfill these types of waste rather than chemically process."

The corporation's philosophy is "to use environmentally sound technologies in managing industrial waste and to recycle natural resources." Encycle is a subsidiary of ASARCO, which has provided more than $20 million in funding for research and development, equipment, and expenses.

▲ Projects and People
Encycle recently undertook a feasibility study evaluating the technologies and corresponding products manufactured from byproducts following the extraction of select metals. The byproducts include silica, iron, calcium, magnesium, and certain hydroxides. Current research and development projects are in technologies that "enhance optimization of recycling and recovery."

With 20 years in the environmental services industry, Deets says, "Environmentalists must recognize that many segments of industry are working on processes which reduce adverse impacts on the environment. We must work with these industries proactively." Deets is also a founding member of the State of the Environment Committee, Boston, Massachusetts, and has been active in the Environmental Hazards Management Institute. He wrote on "Recycling Hazardous Metal Plating Waste" in the *Journal of Metals*, 1991.

Espey, Huston & Associates, Inc. (EH&A)
916 Capital of Texas Highway South
P.O. Box 519
Austin, TX 78767 USA

Contact: **Derek Green**, Staff Ecologist
    Phone: 1-512-327-6840 • *Fax:* 1-512-327-2453

Activities: Environmental Consultation • *Issues:* Air Quality/Emission Control; Biodiversity/Species Preservation; Deforestation; Waste Management/Recycling; Water Quality

● Organization
Founded in 1972, **Espey, Huston & Associates, Inc.** (EH&A), Engineering and Environmental Consultants, with a staff of 450, devotes itself to modern technological problem solving. EH&A has "consistently adapted to new conditions, requirements, and technology [to help] confront the critical challenges of the future," it reports.

With expertise in more than 30 professional and scientific disciplines, EH&A provides innovative, cost-effective solutions to organizations in such diverse fields as energy and mining, government, development, petrochemicals, manufacturing, and other scientific disciplines including: "ecology, biology, archaeology, geology, hydrology, socio-economics, meteorology, and oceanography."

▲ Projects and People
EH&A provides **environmental impact assessment (EIA) services** with its staff, "trained and experienced in pertinent environmental fields" to study the consequences of such forces as dredging, drilling, mining, and building on certain areas, including wetlands and coastal regions. Analyses are prepared regarding air, water, and soil pollution, and impacts of transportation corridors, industrial facilities, and residential and commercial developments.

The company surveys **endangered species** and prepares **Habitat Conservation Plans** under the guidelines of the Endangered Species Act of 1973. EH&A also prepares biological assessments, identifies potential habitat, interprets aerial photos and mapping, and develops public awareness and education programs.

With **wetland protection** project experience ranging from "the North Slope of Alaska to the islands of the Caribbean and from the Pacific Basin to eastern Africa," EH&A conducts wetlands investigations—including soil surveys, hydrology studies, and groundwater well design, installation and monitoring—environmental assessments, regulatory services, habitat preserve creation and planning, and mitigation activities. In the USA, "concern for the protection of our nation's wetlands is a national priority," says EH&A, whose personnel "serve as instructors in the U.S. Army Corps of Engineers (USACE) in-house regulatory training program and serve on national committees developing certification procedures for wetland professionals." EH&A works with wetlands "ranging from a few isolated acres associated with small developments to tens of thousands of acres associated with multi-state mining or timber projects."

Regarding **hazardous waste management**, the firm's services range from the "investigation and interpretation of environmental substances to the design and management of contamination remediation programs." Working closely with federal, state, and local governments and under such recent legislative measures as the Resource

Conservation and Recovery Act (RCRA) and the Comprehensive Environmental Response, Compensation, and Liability Act (CERCLA), the corporation addresses management, disposal, and remediation of environmental contaminants.

EH&A also provides other services regarding ecology; air quality; groundwater, wastewater, stormwater, and water quality; bioassay, biomonitoring, and bioaccumulation testing; marine cultural resources; construction and building design; aviation; surveys; advanced technology industrial facilities; industrial facilities engineering; cultural resources; highway engineering and transportation planning; preacquisition site assessment; municipalities; and oil, gas, and petrochemical industries regarding environmental site assessment, oil spills, distribution, storage, and regulatory compliance.

Staff ecologist **Derek Green** is also a recognized authority in **sea turtles**, particularly the east Pacific green turtle, having studied them in mainland Ecuador and at the Charles Darwin Research Station, Galapagos. While involved in EH&A terrestrial and aquatic field surveys inland and along the Texas coast, he continues expeditions to the Galapagos, Peru, and north Yemen for surveys of sea turtles and other coastal resources.

■ Resources

Among EH&A's resources are biological, bioassay, and archaeological laboratories; state-of-the-art computer center; extensive field survey equipment; and approximately 137,000 square feet of workspace, including 4,500 square feet for equipment maintenance. Its technical library contains about 10,000 volumes of research reports, monographs, and periodicals. An EH&A information kit is available. Derek Green offers a list of publications and technical reports ranging from sea turtles to environmental assessments.

Flowerfield Enterprises
10332 Shaver Road
Kalamazoo, MI 49002 USA

Contact: **Mary Appelhof**, President and Owner
  Phone: 1-616-327-0108 • *Fax:* 1-616-343-4505

Activities: Education; Research • *Issues:* Waste Management/ Recycling • *Meetings:* Global Assembly of Women and the Environment; International Society of Earthworm Ecology

● Organization

Through **Flowerfield Enterprises** and its publishing entity, **Flower Press**, Mary Appelhof is committed to "spreading the word about the value of earthworms in converting organic residues to useful soil amendments."

For 19 years, Appelhof, a biologist, experimented with small-scale composting of organic kitchen waste. She tested various containers, beddings, types of worm, and maintenance procedures. Her research led to a "vermicomposting system design that is simple, effective, and convenient."

Start-up materials are one pound of redworms in a wood, metal, or plastic box with aeration holes and properly moistened bedding such as shredded newspaper. "Food waste—lettuce leaves, citrus rinds, apple peels, and plate scrapings—is buried in the worm bin," describes Appelhof. "Worms and bacteria consume this waste, converting it in about four months to a dark, crumbly humus which homeowners use to fertilize plants."

Appelhof says she has had to overcome negative public attitudes about worms and garbage. The rewards of home worm boxes, however, benefit the environment in significant ways: organic waste is converted on site to a nutrient-rich, sweet-smelling product that makes plants grow better. This eliminates the need to buy commercial fertilizers and cuts down on consumption of fossil fuel and natural resources. Vermicomposting protects groundwater and saves water through lessened use of garbage disposals. When organic wastes are kept out of landfills, there is less toxic leachate that results from chemical reactions in unsorted buried waste, and there is more space for burial of nonrecyclable materials.

Appelhof's innovative research with this ancient concept intensified when she coordinated a 1980 workshop about earthworms and the stabilization of organic residues at Western Michigan University, Kalamazoo, Michigan, with grants from the National Science Foundation and the Small Business Innovation and Research (SBIR). Four international conferences followed, as did the publication of *Worms Eat My Garbage*, which is used worldwide in home and school projects. Appelhof estimates that a classroom vermicomposting system can handle more than 150 pounds of lunchroom waste per year. In Seattle, Washington, high school students are teaching others how to maintain their worm bins. Cornell University Cooperative Extension teaches worm-composting workshops; elsewhere there are 4-H Club projects, children's garden programs in East Harlem, New York, and a bin at Seattle's Woodlawn Zoo.

■ Resources

Books and other materials available through Appelhof are *Worms Eat My Garbage*; *Worms Eat Our Garbage: Classroom Activities for a Better Environment*; the Worm-A-Way™ Vermicomposting Bin; and printed Recycling Labels for source separation. She also conducts workshops and presentations.

Mary Appelhof wants to "identify and correspond with teachers using live worms in classrooms" and is seeking information about viable worm composting projects worldwide.

Global Environment Fund (GEF)
1250 24th Street, NW
Washington, DC 20037 USA

Contacts: **H. Jeffrey Leonard, Ph.D.**, President
  Sumner Pingree, Vice President
  Phone: 1-202-466-0529 • *Fax:* 1-202-466-6454

Activities: Environmental Investing/Financial Management *Issues:* Air Quality/Emission Control; Energy; Health; Transportation; Waste Management/Recycling; Water Quality

● Organization

The **Global Environment Fund (GEF)** describes itself as the "first investment fund conceived and managed by professional environmentalists" who are committed to realizing "superior investment performance" through applying sound environmental and financial principles.

The Fund is designed for investors whose programs or charitable donations support environmental causes; their investments are used to promote environmental goals. Combining environmental and financial objectives, the Fund seeks to expand opportunities for businesses to "do well by doing good" and expects that the "increasing economic demand for environmental protection, cleaner forms of industrial and energy production, and more efficient utilization of energy and natural resources will provide dynamic business opportunities in coming decades."

Working with government agencies, environmental groups, and the scientific community, companies are identified that "actively engage in solving environmental problems—not merely those that provide products and services labeled as environmental." In the screening process, both positive and negative impacts of a company's environmental activities are fully scrutinized. Any company is eliminated when current practices result in pollution, waste, or other environmental damage or risk. If it passes this first test, a company is next evaluated according to the Fund's conservative financial criteria.

Typical portfolio companies engage in energy and water conservation; pollution reduction; improved air and water pollution control; alternative forms of energy or substitute fuels that reduce environmentally harmful side effects of energy production and use; energy production from renewable resources, as waste-to-energy, geothermal, solar, and wind; materials reuse or recycling; safe handling of toxic materials; and methods of cultivating, harvesting, using, and processing natural resources under environmentally sound and sustainable conditions.

In its first 18 months (January 1, 1990–June 30, 1991), GEF reported its assets quadrupled and investment partners grew from 5 to 30—including individuals, trusts, partnerships, foundations, and other institutional investors. At the end of that period, there were 57 companies in its broad-based portfolio; and the Fund said it "substantially outperformed" Standard & Poor's 500 Index, Lipper Index of Environmental Mutual Funds, and the O'Neill Index of Pollution Control Stocks. In June 1992, GEF reported "substantial growth" throughout 1991, with assets expanding to nearly $25 million that year—well above the "$20 million target" set for 1991.

Investment capital has helped commercialize innovative technologies such as **Aurora Flight Sciences'** development of high-altitude unmanned aircraft that monitor atmospheric conditions; **Investor Responsibility Research Center's** subscription service that enables institutional investors to review environmental performances of large investors; and environmental magazine *Buzzworm*, with expanded circulation and printing equipment.

President H. Jeffrey Leonard, former vice president of the World Wildlife Fund and The Conservation Foundation, also worked with the Princeton Center for Energy and Environmental Studies and the Investor Responsibility Research Center. He was a consultant for the World Bank and U.S. Environmental Protection Agency (EPA). He is a director of the Center for International Tropical Agriculture and the Biomass Users Network, a nonprofit corporation promoting economic development in the Third World through efficient use of renewable natural resources.

Vice President Sumner Pingree is also chairman of the Chesapeake Bay Foundation and director of the School for Field Studies and Americans for the Environment. He was vice president of the Wilderness Society and a founder of Conservation International.

Global Environmental Management Initiative (GEMI) ⊕

2000 L Street, NW, Suite 710
Washington, DC 20036 USA

Contact: **Andrew Mastrandonas,** Program Director
Phone: 1-202-296-7449 • *Fax:* 1-202-296-7442

Activities: Development; Education • *Issues:* Sustainable Development • *Meetings:* UNCED

● Organization

The **Global Environmental Management Initiative (GEMI)** was formed in 1990 by an executive workgroup of the Business Roundtable as a corporate partnership to improve business's environmental performance worldwide. Its intent is to take charge of its environmental destiny by spurring change from within. Administered by the U.S. Council Foundation as a nonprofit activity, GEMI is strengthening working relations with the International Chamber of Commerce (ICC), Paris, France, and the International Environmental Bureau (IEB), Geneva, Switzerland. The U.S. Council Foundation is the educational arm of the U.S. Council for International Business—ICC's U.S. national committee.

Member companies are Allied Signal, AT&T, The Boeing Company, Browning-Ferris Industries, Digital Equipment Corporation, Dow Chemical Company, Duke Power Company, Eastman Kodak Company, E.I. DuPont de Nemours & Company, Florida Power & Light, ICI Americas, Merck & Company, Occidental Petroleum, Procter & Gamble, The Southern Company, Tenneco, Union Carbide Corporation, USX Corporation, and W.R. Grace and Co. **George Carpenter,** Procter & Gamble, is GEMI chairman.

The Environmental Forum, the Environmental Law Institute's policy journal, called GEMI a "jewel in the rough" (September/October 1990). The article summed up, "[GEMI] does not profess to be the panacea for industry's environmental problems. What it can do, however, is help the world realize that every facet of society has its part to play in the cleanup and prevention of pollution. GEMI, representing a pooling of the resources and insight of a number of prominent companies and industries to foster excellence in environmental management, is the type of collaborative effort that should be encouraged and supported."

▲ Projects and People

GEMI is establishing projects to carry out its goals which are "to stimulate worldwide critical thinking on environmental management, to improve environmental performance, . . . to promote a worldwide business ethic for environmental management and sustainable development, to enhance the dialogue between business [and NGOs], governments, and academia," and to forge global partnerships, such as with the United Nations Environment Programme (UNEP). GEMI participated with ICC in the development of a global **Business Charter for Sustainable Development** to serve as a guideline.

At its first annual **conference on corporate quality and environmental management,** GEMI brought together 240 representatives of industry, academia, and government to launch its concept of **Total Quality Management (TQM)** principles, which were later presented at the Second World Industry Conference on Environmental Management (WICEM II) Rotterdam, Netherlands. The

TQM handbook, detailing waste management, recycling, product package design, models as effective measuring tools, and other aspects of environmental application was to be distributed to 1992 UNCED participants. A TQM workgroup is ongoing.

A stakeholder communications workgroup is underway to further communications on environmental matters with investors, public interest groups, regulators, employees, media, and the public. Through its outreach workgroup, GEMI is participating in conferences and distributing reports and proceedings to expand its message. With other industry groups and international organizations, GEMI's data measurement workgroup is "exploring the development of a database" to track business efforts to implement ICC's sustainable development charter. The database would also be used to "spread leading-edge thinking on environmental management" as practiced by corporations worldwide.

■ Resources

GEMI publishes a quarterly newsletter, *GEMI-NEWS*. Proceedings from the first Corporate Quality/Environmental Management Conference can be ordered. An information packet is available.

Greenwire
American Political Network, Inc.
282 North Washington Street
Falls Church, VA 22046 USA

Contact: **Philip Shabecoff,** Executive Publisher
  Phone: 1-703-237-5130 • *Fax:* 1-703-237-5148
Modem: 1-703-237-5130

Activities: Media Publishing • *Issues:* Air Quality/Emission Control; Biodiversity/Species Preservation; Deforestation; Energy; Global Warming; Health; Population Planning; Sustainable Development; Transportation; Waste Management/ Recycling; Water Quality

▲ Projects and People

Executive publisher **Philip Shabecoff** describes *Greenwire* as a "daily executive briefing on the environment." This news service of the American Political Network is an "electronic 12-page daily summary of the last 24 hours of news coverage of environmental issues—50 states and worldwide," accessed Monday through Friday at 10 A.M. EST with a personal computer modem and an 800 number.

Greenwire covers environmental protection, law and the environment, energy/natural resources, solid waste, business science, marketplace battles, local news, worldwide headlines, Capitol Hill, and television highlights. It includes a daily calendar, focus interviews, and a spotlight story.

Other publishers include: **Philip Angell,** former chief of staff, U.S. Environmental Protection Agency (EPA) and now with Browning-Ferris Industries; **Jim Maddy,** executive director, League of Conservation Voters; and **Douglas Bailey,** *HOTLINE* executive publisher. In addition, a 27-member board of analysts includes environmentalists in government, non-government organizations, universities, industry, and global groups.

A sample edition reports on wetlands definitions by the Bush Administration, EPA awards for ozone protection, environmental monitoring in China, logging ban rally in the Philippines, U.S. Department of Energy (DOE) comments on energy taxes that could more than double oil costs as the price for curbing carbon dioxide emissions, environmental progress in Sweden, and a new household water and energy saver—the "ShowerCOCOON" invented by Eric Reeves.

With the *New York Times* for three decades, Shabecoff was the newspaper's environmental correspondent to its Washington Bureau for 14 years until 1991. He is on the Board of Advisors for *Environmental Profiles.*

■ Resources

Interested subscribers can contact **Donald Tighe** for information on annual rates.

Interfaith Coalition on Energy (ICE)
P.O. Box 26577
Philadelphia, PA 19141 USA

Contact: **Andrew Rudin,** Project Coordinator
  Phone: 1-215-635-1122 • *Fax:* 1-215-635-1903

Issues: Energy • *Meetings:* American Society of Heating, Refrigerating, and Air Conditioning Engineers

● Organization

Pressures on urban congregations are many: aging buildings in decaying neighborhoods; dwindling memberships; and social problems such as drugs, AIDS, hunger, and homelessness. When congregations control energy costs, they have more money to confront these other problems. In 1980, the **Interfaith Coalition on Energy** (ICE) started the movement to reduce energy costs in congregations in Philadelphia where these expenses represent about 14 percent of the total operating budget—or $2.40 per family each week for heating, cooling, and lighting. Smaller congregations are hurt even more. ICE's view is that a dollar saved in cutting energy costs puts an extra dollar in the collection plate—without the efforts of fundraising.

ICE is a nonprofit venture made up of clergy, lay leaders, utility representatives, engineers, and business persons who work together to reduce energy costs. Participants learn when and how much indoor temperatures can be lowered in unoccupied buildings without harming woodwork, pipe organs, paint, plaster, and fibrous materials. Through meters and its database, ICE knows how much electricity is used by organs or elevators and the amount of energy required for efficient heating and cooking.

▲ Projects and People

Congregations are advised to set temperatures as low as 45° F when buildings are not in use; to install freeze-protection thermostats set at 37° to keep water in pipes from freezing; to turn off all motorized systems when not in use; to replace broken or worn systems with more energy-efficient ones.

ICE recounts many success stories including the Palethorp/Zion Presbyterian Church, which reduced its energy use by 45 percent and energy cost by 40 percent by shutting down the boiler during the week. One parish saved $8,000 a year by lowering its thermostat. Another congregation repaired its roof, windows, and heating system and began to make savings.

In Philadelphia, ICE hires a consultant, Andrew Rudin, who compiles data and provides energy surveys, workshops, newsletters, and audiovisual aids to the 4,200 member congregations in its area. (Rudin has also analyzed energy use for congregations in New York City, Chicago, and Phoenix, Arizona.) Foundations, such as The Pew Charitable Trusts, provide funding along with the Philadelphia Electric Company, corporations, fees from energy surveys, and other sources.

ICE helps similar groups get underway in other cities with kits of materials, but urges that each city put together its own coalition. "With the support of the Solar Energy Research Institute, ICE did research which showed that congregations do not readily accept information from sources outside of their local area."

Lowry Park Zoological Garden ⊕
7530 North Boulevard
Tampa, FL 33604 USA

Contact: **Lex Salisbury**, General Curator
   Phone: 1-813-935-8552 • *Fax:* 1-813-935-9486

Activities: Education; Research • *Issues:* Biodiversity/Species Preservation • *Meetings:* AAZPA

● Organization
The metropolitan **Lowry Park Zoological Garden** is an example of a modern zoo where conservation is a major objective. In addition to recreational and educational activities—as well as research conducted by a staff of 33, volunteers, and university faculty and students—emphasis is on captive breeding of animals for eventual reintroduction into the wild and on creating public awareness to curb the destruction of natural habitats and the loss of species. The zoo takes part in the decade-old, international Species Survival Plan (SSP)—today's "Noah's Ark"—of the American Association of Zoological Parks and Aquariums (AAZPA) to "reinforce natural populations which have been reduced by human activity, natural catastrophe, or epidemic diseases." To participate in such a plan, zoos must have suitable facilities, size, and economic support to sustain breeding sites for endangered species.

▲ Projects and People
The Lowry Park Zoo is currently working with 14 species listed with SSP. These are Arabian oryx, Asian elephant, Bali mynah, Bornean orangutan, chimpanzee, Dumeril's ground boa, golden lion tamarin, Guam rail, Indian rhinoceros, radiated tortoise, red-ruffed lemur, red wolf, siamang, and Sumatran tiger.

Among the zoo's other threatened and endangered animals are Africa's mandrill; Asia's Bactrian camel, Malayan tapir, Palawan peacock pheasant, and Persian leopard; Madagascar's collared and ring-tailed lemurs; North America's alligator and crocodile, bald eagle, brown bear, plains bison, Florida sandhill crane, fox squirrel, and West Indian manatee; and South America's golden-headed lion tamarin and woolly monkey. In addition, the zoo houses 84 reptile species, 19 amphibian species, 138 bird species, and numerous fish species.

Recent research studies include **Lex Salisbury's Bioenergetics of Two Temperate Parakeets** as a Griffith University thesis, Australia. The veterinary staff is working on **Evaluation of Male Fertility in Indian Rhino** and **Correlation of Fecal Steroid and Blood Level Analyses in Manatees**. Other projects include **Mandrill Social Group Time/Activity Study** and **Behavioral Patterns of the Manatee in Captivity**, in conjunction with Eckard College, Florida; **Hearing Capabilities of the Manatee** for Mote Marine Lab, and Florida Atlantic University, Florida; and **Infrasonic Vocalizations in Manatees** with Old Dominion University, Virginia. Current studies include **Manatee Learning and Problem Solving, Experiments in Manatee Navigation**, and **Manatee Local Behavior** with University of Windsor, Ontario, Canada; **Assimilation Efficiency, Water Balance, and Body Condition of the Manatee** and **Metabolism, Thermoregulation, and Lactation Energetics of the Manatee** with Texas A&M University; and **Dietary Requirements of the Manatee** with Michigan State University.

"To facilitate global conservation, collaborative programs have been instituted that include to date three aerial surveys of the Manatee in Belize and Guatemala," reports Salisbury, who maintains the AAZPA regional studbook for the West Indian manatee. "We have funded computers and software to better compile, access, and use medical data necessary for management of endangered species."

The zoo's $3.3 million **Manatee and Aquatic Center** opened in 1990 and includes three rehabilitation pools for injured manatees, an underwater office, and laboratory facilities for staff and visiting researchers. The research office is adjacent to the manatee underwater pools and is on view to the public, as are poster abstracts of ongoing studies. "The zoo's commitment to the conservation of this endangered species is more profound than a classical species maintained in a public zoo," according to staff. Research results are published and presented at seminars and professional meetings; final results are maintained in the zoo library.

In its education programs, the Lowry Zoo teaches the many reasons to conserve wild animals and their habits: "Their basic right to exist, their beauty, economics and the fact that they are all a part of a vital ecosystem . . . preservation is a contribution to maintaining biological diversity."

The zoo asks the public to "think globally" and "become better informed" through publications, the news media, and classes; "make views known to newspapers, politicians, corporations, and those in power"; support conservation group efforts; recycle reusable materials; conserve fuel and energy resources; refrain from buying plants or animals that have been taken from the wild or products made from endangered species; and also protect wildlife close to home.

■ Resources
Materials needed include electrochemistry oxygen analyzer to determine metabolism and thermoregulation of manatee as well as weighing meter and hanging load cell with 1,100 kg capacity (Western scales). Zoo staff make public presentations on the zoo's role in species preservation and environmental education for a donation to Lowry Park Zoo's conservation account. Natural habitat zoo/biopark planning is available for a negotiable fee. Literature about conservation and research programs is published.

M.B. Services for International Development Corp. (MBSID)
3 La Costa Court
Towson, MD 21204 USA

Contact: **Menajem M. Bessalel**, President
Phone: 1-410-321-1979 • *Fax:* 1-410-494-0594

Activities: Development • *Issues:* Transportation; Waste Management/Recycling

● Organization
Established in 1988, MBSID provides management consulting services and technical assistance on a worldwide scale in the fields of transportation, urban planning, and environmental studies. The company's areas of expertise include analysis of environmental and infrastructure problems, industrial pollution control, and energy recovery.

▲ Projects and People
MBSID president, Menajem (Mito) Bessalel, of Argentina, with a background in international and domestic engineering and municipal and industrial environmental projects, has evaluated the urban infrastructure components of more than 100 cities worldwide. Projects he has managed or participated in include the following:

A comprehensive study, conducted for the Societe d'Equipement des Terrain Urbains (SETU), for 56 cities of Cote d'Ivoire, where gradual deterioration of the storm sewer system, erosion of existing roads, and inadequate garbage collection were creating unhealthy environmental conditions, and where problems were compounded by rapid population growth. Master plans were developed for each city to improve the urban infrastructure and subsequently the environmental conditions.

A project, for the U.S. Agency for International Development (USAID), that had as its objective the reduction of **industrial pollutants in the Philippines**, where most industries were reported to be out of compliance with air or water control regulations. Surveys were conducted, industries ranked, and waste management techniques recommended.

An **urban environmental evaluation in Chihuahua, Mexico**, conducted for the U.S. Trade and Development Program, to evaluate the municipal services for garbage treatment, traffic and transportation, storm drainage, collection and treatment of wastewater, and water supply. The city's rapid population growth (from 277,000 people in 1977 to over 622,500 people in 1990), along with rapid industrial development, created difficulties in delivering municipal services. MBSID made recommendations relating to more efficient garbage collection, improved public transportation management, better maintenance of streets and buses, flood damage control, and new channel construction.

An assessment of **solid waste management in Port-au-Prince, Haiti**, was conducted for the U.S. Agency for International Services. Collection and treatment of solid waste in Port-au-Prince has been a problem for many years, with garbage remaining uncollected for as long as months on the streets or open fields, creating health hazards. MBSID assessed the environmental impact of the conditions and made recommendations for utilization of new and existing equipment and landfill rehabilitation.

■ Resources
MBSID conducts workshops in English, Spanish, and French.

ManTech Environmental Technology, Inc.
200 Southwest 35th Street
Corvallis, OR 97330 USA

Contact: **George King**, Research Scientist
Phone: 1-503-757-4486 • *Fax:* 1-503-757-4338

Activities: Research • *Issues:* Air Quality/Emission Control; Biotechnology; Biodiversity/Species Preservation; Deforestation; Global Warming; Sustainable Development; Water Quality • *Meetings:* Technical symposia in ecology, geophysics, global change, ecophysiology, resource management

▲ Projects and People
George King specializes in the interdisciplinary study of large-scale interactions of the biosphere with the atmosphere and paleological processes. He has a particular interest in the effects of climate change on the carbon cycle, biogenic emission of trace gases fluxes in terrestrial ecosystems, and biophysical processes that are mediated by terrestrial water and energy balance.

He currently is involved with the **Global Change Research program** at the U.S. Environmental Protection Agency's (EPA) Corvallis Environmental Research Laboratory. The program focuses on predicting the response of terrestrial ecosystems to climate change, and subsequent feedbacks to the climate system; and assessing the extent to which terrestrial ecosystems, particularly forests, affect the global carbon cycle. *(See also Environmental Protection Agency, Corvallis Environmental Research Laboratory in the USA Government section.)*

One aspect of the program, **global processes and effects research**, assesses the impacts of global climate change on the terrestrial biosphere. The research supports the following policy issues:
- What are the probable magnitudes, rates, and variations of human-induced change in ecological systems; how can these changes be distinguished from natural fluctuations?
- In which ecological systems and species are significant changes most likely to occur; what attributes of importance to humans (such as harvestable resources, arable land, and species diversity) will be at risk?
- How will predicted changes in terrestrial ecosystems alter biospheric CO_2 fluxes, radiation, heat, or water fluxes in the climate system and how will these change with time?

Another aspect of the program, **terrestrial carbon sequestration research**, assesses the degree to which forests and agroforest ecosystems conserve and sequester carbon and, consequently, reduce the accumulation of CO_2 in the atmosphere. The primary policy questions driving this research effort include the following:
- Sequestration—Do forested ecosystems and the associated land management practices substantially affect the Earth's carbon cycle and the concentration of carbon dioxide in the atmosphere?
- Conservation—If rapid climate change occurs as projected, how can forest technology and management practices help adapt forests to a changing environment so that their health and productivity are maintained?

• Global Forestry Agreement—As a key element for international cooperation, how can forest systems in the tropical, temperate, and boreal regions be managed for sustainable human health and prosperity?

N-Viro Energy Systems Ltd. ⊕
3450 West Central Avenue, Suite 250
Toledo, OH 43606 USA
Contacts: **J. Patrick Nicholson**, Chief Executive
Sally J. Robinson, Director of Marketing and Communications
Phone: 1-419-535-6374, 1-800-66-NVIRO
Fax: 1-419-535-7008

Activities: Research • *Issues:* Health; Sustainable Development; Waste Management/Recycling; Water Quality

▲ Projects and People

To cut back on ocean dumping, landfills, and incineration, **N-Viro Energy Systems Ltd.**—a sludge (decomposed sewage) treatment company—worked with Ohio universities in the mid-1980s to develop N-Viro Soil. According to its inventor J. Patrick Nicholson and co-principal investigator/microbiologist Dr. Jeffrey Burnham, Medical College of Ohio, N-Viro is a "pasteurization and chemical fixation process that achieves the highest level of disinfection and stabilization recognized by the U.S. EPA [Environmental and Protection Agency]."

In municipal sludge management—such as in Toledo, Ohio's award-winning compost and wastewater treatment plant, N-Viro describes its seven-day process of converting sludge to soil through the use of low-cost and highly available cement kiln dust (CKD) or lime (containing calcium, potassium, and sulfur) to "immobilize heavy metals" and to help create a "valuable organic fertilizer" when balanced with nitrogen and phosphorous.

"Dewatered sludge from the treatment plant is mixed with N-Viro alkaline admixture. Immediate and significant odor reduction occurs. Material is dispensed and stored in containers. Pathogen kill begins. The temperature is raised to 52°C or above for 12 hours and the pH is raised to greater than 12 to 72 hours. . . . Material is moved to a drying pad . . . and turned every other day for six days. Greater than 50 percent solids is achieved and the process is completed."

The "pathogen-free and nearly odorless" N-Viro Soil with a "natural earth color and granularity" that "looks, smells, and performs like natural soil" is used worldwide in "daily landfill cover material, land and mine reclamation, land remineralization, highway and median construction, landscaping and turf material, [and] aglime/soil conditioner." Capital costs are said to be lower than that for "compost, incineration or pelletization." The company markets both its N-Viro Soil, by bag or in bulk, and technology.

Nicholson believes that the "by-product resources of industrialized nations are desperately needed" to fertilize soil "until Third World countries can treat and develop their own organic residuals resources."

P3 The Earth-Based Magazine for Kids
c/o Greenpeace
1436 U Street, NW
Washington, DC 20009 USA

Contact: **Elke Martin**, National Campaign Coordinator
Phone: 1-202-462-1177 • *Fax:* 1-202-462-4507

Activities: Education • *Issues:* Air Quality/Emission Control; Biodiversity/Species Preservation; Deforestation; Energy; Global Warming; Health; Population Planning; Sustainable Development; Transportation; Waste Management/Recycling; Water Quality

▲ Projects and People

Now teamed with Greenpeace, the eco-journal *Planet 3* emerged to fill a void in the environmental education of American children. A colorful magazine printed on recycled paper and written for ages 7 to 12, its purpose is to "show kids that Earth is the coolest, hippest planet to live on and empowers them to help keep it that way." Its editors say that the environment is not a required unit in the elementary school curriculum. "That means we're raising yet another generation of children who are *not* learning how to care for Earth and will continue to contribute to the monumental environmental problems facing our planet." Their prior experience in communicating with children through *The Electric Company Magazine*, *Sports Illustrated for Kids*, *ALF*, *Muppet Magazine*, and *The Big Picture* tells them that children are scared. "They hear about the ozone hole, the greenhouse effect, acid rain, elephant slaughter, and the garbage crisis but they have no current, consistent source of information
to help them understand what's happening and act to change it." The editors are worried too that problems adults fail to handle today, such as fossil fuel dependency and chlorofluorocarbon (CFC) pollution, will shortly be the responsibilities of the younger generation.

Based in Vermont, **Randi Hacker** and **Jackie Kaufman** conceive and write *P3* contents that are reviewed before publication by a panel parents, teachers, children, editors, and environmentalists—including **Paul T. Schindler**, president of the African Wildlife Foundation. To get input from their readers after each issue is mailed, the editorial staff visits subscribing classrooms and conducts focus groups with students and "find out what works, what doesn't, and what they want to read more about." Readers also serve as advisors on a rotating basis so that the points of view of many children nationwide can be reflected in the content.

A typical issue explains acid rain; how prehistoric dinosaurs really are fossil fuel; the Great Plains' value to buffalo, topsoil, and tourists; the peril of snow leopards in The Himalayas; and folklore of Native Americans. Featured are cartoons, puzzles, games, recognition for special readers—called the Earth Patrol—and letters to the editor. The magazine urges children to get involved by writing to legislators and environmental leaders.

Beginning as a private endeavor with a board that included Schindler, **Jane Bloom** of the Natural Resources Defense Council, author John Javna, Wendy Gordon Rockefeller, Lauren Hutton, and Sigourney Weaver, *P3* is now published by the nonprofit Greenpeace.

■ **Resources**

Yearly subscriptions are available.

Pavich Family Farms
Route 2, Box 291
Delano, CA 93215 USA

Contacts: Stephen Paul Pavich; Thomas Pavich; Tonya Pavich
Phone: 1-805-725-1046 • *Fax:* 1-805-725-8223

Activities: Development; Education; Research • *Issues:* Sustainable Agriculture; Sustainable Development

● **Organization**

The **Pavich Family Farms**—an environmentally conscious enterprise using biological and ecological farming methods—started out differently in the 1950s when chemical-based agriculture depleted the soil and diminished fruit production.

Shortly after **Stephen Pavich** returned to the family farm from college in 1971, he was overcome by pesticide fumes. "I knew right then we were dealing with substances that were not of benefit. I questioned worker safety, the pollution of the ground water, and overall quality of the crops." Immediately, the conversion began to an **organic farm free of synthetic pesticides and chemicals** that is now the "world's largest grower and shipper of organic table grapes."

On 2,300 acres, soil nutrition strengthens plant immunology to control plant pests; predatory insects and parasites are encouraged to attack unwanted pests; tillage and the planting of nitrogen-building cover crops control weeds; natural minerals and nutrients obtained from fossilized lake beds or nutrient-rich rock quarries worldwide replace synthetically compounded fertilizers.

Pavich Farms makes **compost from raw manure and plant wastes** that return nutrients to the soil, saving rivers and water from contamination with nitrates, preventing soil erosion, and eliminating solid waste from landfills. Some farming costs have been lowered and yields match those in a commercial vineyard. In 1991, they produced 1,400,000 boxes of grapes.

Research and development teams are working on producing a 100 percent recycled packing system to replace virgin paper and wood product use, helping conserve forests and reduce the greenhouse effect.

Pavich Family Farms meets the standards of the California Certified Organic Farmers Association (CCOF). An independent laboratory, *NutriClean*, tests their crops, soil, and water for any chemical pesticide residues.

Motivated by concern for his land's future, Stephen Pavich says, "When I leave, I want this spot on the planet to be clean. I don't want my child or any other child to worry about the side effects of the soil we work." **Tom Pavich** and his wife **Tonya Pavich** are partners in the enterprise—and vision.

■ **Resources**

A Farm Tour Day is held for educators, students, consumers. A newsletter is available. Grapes, melons, vegetables, and raisins are marketed worldwide.

Resource Recycling
P.O. Box 10540
Portland, OR 97210 USA

Contact: **Jerry Powell,** Editor-in-Chief
Phone: 1-503-227-1319 • *Fax:* 1-503-227-6135

Activities: Publishing • *Issues:* Waste Management/Recycling

▲ **Projects and People**

Resource Recycling is a comprehensive, monthly magazine on municipal waste recycling, such as paper, glass, metals, plastics, and yard waste. According to its editor **Jerry Powell**, the publication is a "leading analytical journal in recycling covering collection processing and end use of recyclables; public and private sector activities; legislation; and equipment." Distribution is worldwide.

■ **Resources**

Twelve-month subscriptions are available for *Resource Recycling* as well as for *Plastics Recycling Update* and *Bottle/Can Update.*

Barry Roth, Ph.D.
Environmental Science Consultant
745 Cole Street
San Francisco, CA 94117 USA
Phone: 1-415-387-8538

Activities: Education; Research • *Issues:* Biodiversity/Species Preservation; Invertebrate Ecology • *Meetings:* International Malacological Congress

▲ **Projects and People**

Dr. **Barry Roth** is an independent consultant in environmental science, malacology, and paleobiology. Formerly affiliated with both the California Academy of Sciences and the Western Society of Malacologists, he specializes in the ecology and taxonomy of land snails and slugs as well as in the history of global climate.

Current projects include the analysis of the survival status of rare and endangered species of land mollusks in the western USA and a comprehensive examination of biogeography and the distribution of the snails and slugs of California, particularly relating to environmental threats. For the State Departments of Food and Agriculture, and Fish and Game, he is preparing a *Manual of the Land Snails and Slugs of California.*

Dr. Roth is also involved in a descriptive study of the fossil land mollusks of western North America and is developing a methodology for their use as indicators of the history of global climate through the past 65 million years. Related research interests are the "evolution of marine faunas in the North Pacific-Beringian-Arctic regions" and "problems and perquisites of simultaneous hermaphroditism."

■ **Resources**

Dr. Roth can provide copies of his scientific papers and others indexed in *Biological Abstracts.* He also offers services as an expert witness.

Scharlin/Taylor Associates ⊕
10 West Drive
Sag Harbor, NY 11963 USA

Contact: **Patricia J. Scharlin**, President
   Phone: 1-516-725-7055 • *Fax:* 1-516-725-4723

Activities: Development; Education; Policy Analysis; Publishing • *Issues:* Sustainable Development; Waste Management/Recycling • *Meetings:* UNCED

● Organization
Scharlin/Taylor Associates are consultants in environment and development, policy planning, and communications. They produce information materials adapted from technical studies for broader audiences; coordinate international meetings, conferences, and seminars; design and implement communications and education projects for international agencies and organizations; produce public outreach strategies; and conduct media relations.

▲ Projects and People
Patricia J. Scharlin, a principal, is international editor for *Environment, Health and Safety Management* newsletter for senior executives in business and government. *EHS Management* is compiling a directory of environmental consultants. She is also a senior international representative for the **Environmental Law Institute**, where she conducts programs on environmental policy and law for professionals from other countries. At Tufts University, she coordinates the **Talloires Seminar series** on sustainable development and institutional change for senior officials from the private sector, international agencies, and NGOs. Scharlin is a special advisor to the New York Office of the United Nations Environment Programme, where she is media liaison for North America activities. Also at the UN, she is IUCN (World Conservation Union) advisor and serves on the IUCN Commission on Environmental Policy and Law. Earlier, she was director of the Sierra Club International Office and Program.

Solviva ⊕
Box 582 RFD
Vineyard Haven, MA 02568 USA

Contact: **Anna Edey**, Director
   Phone: 1-508-693-3341 • *Fax:* 1-508-693-3341

Activities: Development; Education; Research • *Issues:* Energy; Global Warming; Health; Sustainable Development; Waste Management/Recycling; Water Quality

▲ Projects and People
Snow-bound, bitter-cold Martha's Vineyard, Massachusetts, in the wintertime is where **Anna Edey** lives and grows sweet tomatoes, raddichio, swiss chard, arugola, nasturtiums, orchids, jasmine, sweetpeas, and more in her **Solviva Solar Greenhome**. For more than 10 years, Edey has been living a dream that she hopes will set a real example for others to emulate. "The solutions are far more achievable than almost anyone realizes. The technology is all here right now . . . in ways that are not only environmentally, but also economically vastly superior to our present standard ways."

In her asymmetrical A-frame greenhouse heated by sunlight and the presence of chickens and angora rabbits and powered by photovoltaic cells, the growing season is September to May—unassisted by fossil fuels or chemicals and generating no pollution—even though the outdoor temperature may be below zero.

Built mostly by volunteers in 1983, the Solviva greenhouse is 3,000 square feet, energy self-sufficient, and proves that vigorous seedlings can germinate "in the darkest, coldest times with no added heat or light" and the **highest quality organically grown greens, herbs and edible flowers** can grow "throughout even the coldest nights and shortest, cloudiest days of mid-winter," writes Edey. These and the lower-cholesterol eggs from relaxed hens are marketed to restaurants at premium prices.

Thirty square feet of photovoltaic panels provide enough electricity to circulate solar-heated air and water in winter and pump water for irrigation in the summer. The south-facing wall is glazed with four layers of polyester sheeting. Edey burns compacted waste newspaper in her Franklin stove in an alcove of heat-absorbing concrete and bricks that retain the "wonderful heat" hours after the fire has gone out. She says newspaper is superior to wood because it "produces less ashes, makes less mess, and the chimney never needs cleaning."

She uses **compostoilets**. "The compost resulting from my toilets and kitchen wastes is great as is for landscaping, but if I wish to use it for food plants, I first bake it to 170 fahrenheit in a simple solar hotbox, in order to eliminate any pathogens. These composting toilet systems could bring low-cost, reliable sanitation to all nations, wasting no water, electricity or nutrients."

Edey estimates that a "single-family solar greenhome could save over $5,000 per year on heating, water, electricity, solid waste, and sewage bills. Such a home would prevent the emission of over 40,000 pounds per year of CO_2 global warming gas, as compared with a standard home, as well as prevent the waste and pollution of many natural resources."

Edey sees solar adaptations to community swimming pools, schools, hospitals, day-care centers, elder housing; and she imagines retrofitting city skyscrapers, such as the southern wall of the United Nations building, New York, to create "solar heat equivalent to 60,000 gallons of fuel oil and 120,000 kilowatt-hours (kwh) of electricity per year, plenty of hot water, clean air, and fresh salad greens and herbs."

To correct the algae infestation of Vineyard waterways on the island, Edey recommended that the community "plant an 'emerald necklace' around the Lagoon, trees and bushes which thrive in wetland areas, like beetlebung, red maple, swamp azalea, and in the water's edge plant reeds and hedges and other aquatic plants . . . to absorb as much nitrate as possible before it has a chance to hit the Lagoon water and feed the algae." She believes that the excessive nitrogen had leeches into the ground from septic systems, yet is "beyond the reach of roots that could benefit from it."

■ Resources
Edey consults with individuals, speaks with groups, and conducts seminars and workshops. Resources needed include "manufacturing of Solviva design for mass distribution"; and educational outreach for lectures, seminars, and internships.

Total Resource Conservation Plan 🌿
3500 W. Bohman Lane
Morgan, UT 84050 USA

Contact: **Frank W. Bohman**, Farmer
Phone: 1-801-876-3039

Issues: Sustainable Development; Water Quality

● Organization

A Total Resource Conservation Plan is how Frank Bohman has been managing his farm for the past 35 years. Of his 4,155 acres, 120 are used for crop production and 35 for meadow hay; the remaining 4,000 are for range. When Bohman first bought this land, it was "mostly covered with sagebrush, mule's ear dock, and some oak, maple, and aspen. . . . Sheet erosion was in evidence nearly everywhere and some gully erosion was beginning. . . . Water and wildlife populations were decreasing. . . . It didn't take long to see that a person couldn't stay in business unless some major changes took place."

Bohman began reading about the American pioneers of the 1840s. "The grass was up to the saddle stirrups, very little sagebrush and clear flowing streams of water. I decided there must be some way to restore the land . . . but where to start? What are the costs? What might the benefits be [to] protecting the soil, water, wildlife, and other natural resources?"

With the help of the **Morgan Soil Conservation District** and the **Soil Conservation Service (SCS)**, Bohman prepared a long-term plan for both his cropland and rangeland. His first step was to gain a thorough understanding of his soil, rainfall, topography, and other resources. Next, he appraised the soil, water, plant, timber, minerals, livestock, and wildlife resources. "We identified problems and opportunities for improvement." The plan marked areas for reseeding, spraying, or rest and rotation. Every possible site to develop water was marked. Fencing was planned to manage a rotation system of grazing and protect young seedlings. Grass seed was selected for the soil types, rainfall, and grazing to get the most production and keep the soil from eroding.

The ranch's restoration was launched the following year with a fireguard and burning—the "cheapest way" to clear sagebrush and because ashes provide a good seedbed. Seeding followed the next spring and within several years, stock ponds began filling. "Water began to flow and seeps showed up where I had never before seen water. . . . The spring that serves the home and corrals has increased from a trickle to an inch-and-a-quarter pipe full of water. . . . Now we have enough water to supply the house, a large lawn, a small garden, and a steel tank to raise trout" for his and the neighbors' stock ponds.

His soil and water conservation techniques include rest-rotation grazing, crop rotation, brush control, range seeding, wildlife habitat improvement, and water development. Where 100 cows once grazed, the land now supports 200 cattle for 3 months longer each year. However, to protect the land from overgrazing, he has cut his stock to 150 head and his sheep from 370 to 230. By not disturbing the oak, quaking aspen, and other shrubs, wildlife habitats are protected and their population has also doubled; beaver ponds, geese, elk, moose, and deer abound; and the land is enjoyed for camping, hiking, and horseback riding. Vehicles are not allowed off the roads.

Bohman experiments with various grasses to keep down erosion, testing them for the SCS Plant Materials Center. Utah State University also has an experimental grass plot on the Bohman ranch, which hosts state and national conservation groups and agricultural leaders from Africa. "I believe that each of us has an obligation to treat our land with respect and to try and leave it a little better than we found it," writes Bohman.

Winthrop, Stimson, Putnam & Roberts ⊕
1133 Connecticut Avenue, NW
Washington, DC 20036 USA

Contact: **Kenneth Berlin**, Managing Partner, Washington Office
Phone: 1-202-775-9800 • *Fax:* 1-202-822-0825

Activities: Legal Representation • *Issues:* Biodiversity/Species Preservation; Environmental Audits; Waste Management/Recycling

● Organization

Winthrop, Stimson, Putnam & Roberts is a law firm dealing with environmental issues covering litigation, regulatory counseling and agency advocacy, audits and compliance, and "monitoring and predicting the evolving environmental regulatory scheme." Winthrop, Stimson's environmental group is comprised of 12 attorneys headed by Kenneth Berlin, Donald Carr, and Howard Goldman.

▲ Projects and People

Ken Berlin, director of the Washington office, supervised over 150 industrial facility environmental audits while head of the U.S. Department of Justice's Environment and Natural Resources Division. Among his cases of "national environmental policy importance" was the landmark *National Audubon Society* v. *Watt, et al.* concerning Alaskan oils and gas development. Berlin has also done work in litigating Superfund cases and has represented utilities and nuclear fuel processing facilities.

Don Carr also came to the law firm from the U.S. Department of Justice, where he represented the federal government in litigation regarding "the environment, oil and gas, minerals management, timber and public lands." Carr tries all aspects of major issues including the reauthorization of the Clean Air Act, wetlands conservation, Superfund, toxic tort, and environmental crime issues.

In the firm's New York office, **Howard Goldman** is the specialist in land use, development, and environmental matters. He previously was involved in Clean Air Act litigation regarding the proposed sale of the New York Coliseum site; earlier, he was counsel to the Alaska Coastal Management Program and local governments in that state.

The law firm analyzes "environmental liabilities that could affect proposed merges, acquisitions, and financing," it reports. The purpose of **environmental audits** is to investigate "existing environmental liabilities," which is difficult because these may result from activities of prior owners or actions "discontinued decades earlier." Audit teams include environmental engineers and attorneys who "visit the relevant site, analyze documents, and interview employees." Testing and a review of government records can also be involved.

"We attempt to **identify environmental cleanup costs** associated with the sites involved in the transaction, sites adjacent to properties involved in the transaction, and off-site cleanup under the Superfund statute," the law firm reports. "We also analyze compliance with laws and regulations since inattention to environmental requirements ultimately can lead to very substantial remediation costs as well as monetary penalties."

Audits also attempt to assess future costs that can result as environmental laws are changed or implemented, such as the 1990 amendments to the Clean Air Act that relate to hazardous air pollutants, "utility emissions of acid rain precursors," and new technology standards. According to Winthrop, Stimson, the detailed auditing process is cost-effective in enabling clients to understand environmental liabilities and how to protect against them.

The law firm expects that federal legislation and international gatherings will soon address anticipated climate changes linked to fossil fuel emissions. It is also watching for the "environmental consequences" of any legislation that could result from a new national energy strategy.

■ Resources

The law firm has prepared *Emergency Reporting Requirements for Environmental Spills and Releases, Allocating Liability Under Superfund,* and *Environmental Liabilities in Mergers and Acquisitions*—among other reports.

🎓 Universities

Department of Zoology and Wildlife Science
Auburn University
331 Funchess Hall
Auburn, AL 36849-5414 USA

Contact: **Robert H. Mount, Ph.D.,** Professor Emeritus
Phone: 1-205-844-4850 • *Fax:* 1-205-844-9234

Activities: Biological Consulting; Education; Research • *Issues:* Biodiversity/Species Preservation

▲ Projects and People

Dr. Robert H. Mount is principally involved in biological consulting; recent projects in Alabama and Mississippi are biotic surveys and impact assessments. "One with which I am currently involved as principal investigator is an assessment of the status of the **flattened musk turtle,** a threatened species, in the Locust Fork of the Black Warrior River in northern Alabama," he writes.

He is also directing a **faunal and floral survey** of the Fort Rucker Military Reservation in southeastern Alabama; and he serves as **herpetological consultant** to the Alabama Highway Department "when projects it proposes to undertake might impact a threatened or endangered species." Previous endangered species surveys have been conducted for the U.S. Fish and Wildlife Service and the Butler County Water Authority.

This Alabama Conservancy board member says he is most interested in herpetology, general ecology, and biogeography of the southeastern USA. He belongs to the Species Survival Commission (IUCN/SSC) Tortoise and Freshwater Turtle Group. Subjects of

his publications include salamanders, water snakes, worm snakes, rattlesnakes, greater sirens, pine barrens treefrogs, fire ants, and lizards among various reptiles and amphibians at risk.

Center for Remote Sensing ⊕
Boston University
725 Commonwealth Avenue
Boston, MA 02215 USA

Contact: **Farouk El-Baz, Ph.D.,** Director
Phone: 1-617-353-5081 • *Fax:* 1-617-353-3200
E-Mail: paula@bucisb.bu.edu

Activities: Research • *Issues:* Energy; Sustainable Development

● Organization

Three departments at **Boston University** sponsor the **Center for Remote Sensing:** archaeology, geography, and geology. Its purpose is to provide "interdisciplinary research that uses remotely-sensed data, undergraduate and graduate education in remote sensing, and training for professionals from several fields in the application of remote-sensing techniques." The W.M. Keck Foundation, Los Angeles, California, contributed the initial grant for equipment and program support. In a newly renovated facility, the Center is comprised of research areas, offices for faculty and support staff, teaching laboratory, computer room, library, and map and drafting room.

The Center describes its present, powerful hardware and software for image processing and geographic information system (GIS) work: "Sun 3/160 file server and time-sharing system with one attached Sun 3/50 workstation, a VAX 11/750 computer running under VMS with over 1.4 GBytes of storage and a 1600/6250 bpi tape drive, two DIPIX Aries II image processing and display workstations, a Comtal display monitor, a Model 3000 Matrix Camera, and a 36 x 48-inch Altek coordinate digitizer" with Aries Software and ELAS—Earth Resources Laboratory Applications Software.

▲ Projects and People

Within the Department of Archaeology, the Center offers a series of **annual training seminars** and is developing a **national archive** of information on remotely sensed data.

Remote-sensing applications in the Department of Geography include population estimation, land-use mapping as an aid to planning in developing countries, and "mapping the extent and location of **tropical deforestation** in order to study the social, economic, and political factors that influence deforestation rates." Other studies use remote sensing to map and inventory forests and other natural environments, evaluate the **degradation of arid lands,** and assess the effects of **air pollution and acid rain** on forest environments.

Within the Department of Geology, remote sensing is used in **arctic and periglacial areas** to "'map' sea ice and study surging glacier dynamics and glacial landform evolution in order to model fjord and glacio-marine processes." **Paleodune monitoring** in the Sahel region of northern West Africa recreates climates from more than 720 million years ago. Nearby, water and mineral resources relating to large-scale "craton-penetrating lineaments" are being assessed. Elsewhere, **water resources** in deserts and arid lands are being analyzed, as is the coliform pollution of marine embayments and estuaries.

Dr. Farouk El-Baz is known for his "pioneering work in the applications of space photography to the understanding of arid terrain, particularly the location of groundwater resources." An investigator of the world's major deserts, his analyses led to the finding of groundwater resources in the Sinai Peninsula, Western Desert of Egypt, northern Somalia, and the Red Sea Province of Eastern Sudan. Altogether he has 30 years' experience in Earth and planetary geology and the application of remote-sensing techniques in many disciplines.

Dr. El-Baz participated in the landing site selection for the *Apollo* lunar missions for six years and was the founding director of the Center for Earth and Planetary Studies, National Air and Space Museum, Smithsonian Institution, Washington, DC. In 1975, the National Aeronautics and Space Administration (NASA) selected Dr. El-Baz to be principal investigator for Earth observations and photography on the Apollo-Soyuz Test Project, the joint American-Soviet space mission. The recipient of many awards for extraordinary accomplishments, Dr. El-Baz is a member of the Third World Academy of Sciences—which he represents on the UN Economic and Social Council.

Department of Biology
California State University
Dominquez Hills, CA 90034 USA

Contact: **David J. Morafka, Ph.D.**, Professor of Biology
Phone: 1-213-516-3407, 1-213-516-3381
Fax: 1-213-516-3449

Activities: Education; Research • *Issues:* Biodiversity/Species Preservation • *Meetings:* IUCN/SSC Tortoise and Freshwater Turtle Specialist Group

▲ Projects and People
Dr. David J. Morafka of California State University describes his current proposal for a continuation of research reconstructing the "ecological requirements of the eggs and young of the Mojave desert tortoise (*Gopherus agassizii*)."

"This project would bring together behavioral, anatomical, physiological, ecological, and genetic information in order to resolve the special environmental needs and vulnerabilities of desert tortoise eggs and neonates (hatchlings, the first year of life)," the biologist writes. "These stages of tortoise life history are the least studied, yet the most vulnerable to loss, and potentially, the most powerful components for accelerating recovery in a population."

Elements of Dr. Morafka's research, in which he and his colleagues approach "eggs and neonate tortoises as if they were a species separate and distinct from adults," include isolating the mechanisms of environmentally influenced sex determination (ESD); and "delineating ecological requirements of tortoise eggs, and the role of nest placement as an ESD factor." Seasonal and daily patterns in movement, feeding, and thermoregulation are documented with field telemetric radio tracking and captive choice trails; egg incubation is documented in both field and controlled labora-

tory settings. Other elements of tortoise health are monitored in both settings.

In related studies with veterinarians **Rebecca A. Yates**, Department of Biology, and **Michael L. Christianson**, United States Army, **blood cell and serum chemistry values** are compared for free ranging and captive neonatal California desert tortoises. It was discovered that captive neonate turtles kept within an "environmental chamber" at the university had "significantly lower" white blood cell counts and "absolute heterophil, lymphocyte, and basophil counts" than the Mohave desert tortoises maintained in "predator proof enclosures" at Fort Irwin.

As principal investigator, Dr. Morafka with consultant Dr. Yates is identifying, comparing, and contrasting the "**sex determination systems** of two different groups of reptiles—pit vipers (subfamily *Crotalinae*) and **tortoises** (family *Testudinidae*)." Studies of pit viper phylogenetics are expected to aid toxicologists and immunologists "in their search for **antibodies** effective **against the venom** components of close relatives as well as pharmacologists in their search for related sources of venom fractions valuable to medicine." Dr. Morafka, who describes pit vipers as a "widespread and extremely dangerous group of poisonous snakes," previously studied rattlesnake phylogenetics regarding "improved antivenin specificity." A research grant is from Minority Biomedical Research Support (MBRS), a division of the National Institutes of Health.

Dr. Morafka was a participant in the U.S. Department of State's Man and the Biosphere Program to negotiate USA–Mexican joint research in the Chihuahuan Desert. He belongs to the Species Survival Commission (IUCN/SSC) Tortoise and Freshwater Turtle Group.

■ Resources
Radio-telemetry equipment is needed to monitor neonate turtles.

Engineering and Public Policy (EPP)
Carnegie Institute of Technology
Carnegie-Mellon University
Pittsburgh, PA 15213 USA

Contact: **M. Granger Morgan, Ph.D**, Professor
Phone: 1-412-268-2672 • *Fax:* 1-412-268-3757

Activities: Education; Research • *Issues:* Air Quality/Emission Control; Energy; Global Warming; Health; Transportation; Water Quality

● Organization
Carnegie-Mellon University was established in 1967 when the Carnegie Institute of Technology—which grew from the Carnegie Technical Schools founded by steel industrialist Andrew Carnegie in 1900—merged with the Mellon Institute, the "prototype of independent applied research institutes" in the USA. Today, Carnegie-Mellon is considered one of the nation's top 20 in total research funding.

Engineering and Public Policy (EPP) addresses problems in technology and policy, including environmental matters that relate to energy, health, and safety. Research frequently leads to solutions of specific problems, while developing better analytical tools and insights that broaden knowledge and standards and lead to improved future policymaking.

▲ Projects and People

Since 1969, undergraduates take on a final project involving a "real client and a review panel of outside experts." Some of their work assessed Pennsylvania's periodic motor vehicle inspection system; examined Pentachlorophenol as a case study in pesticide regulation; investigated the regulation and disposal of used motor oil and household batteries; and developed strategies to reduce population exposure to certain electric and magnetic fields. Extensive research is undertaken in the graduate program, such as estimating the "supply curve" for CO_2 emissions reduction, modeling climate change, developing environmental/energy policies, and environmental control technologies for electric-power generation.

Much of EPP's work involves applying energy and material-balance techniques to analyzing large process industries. Work done here and at the Resources for the Future influenced the U.S. Environmental Protection Agency (EPA) to adopt a "multi-media approach to environmental controls." Recently, similar techniques were applied to integrated modeling of acid rain problems whereby a system was developed to assess the impacts of alternative acid rain abatement strategies for the northeast USA and Canada. Another set of engineering-economic models built by Professor Ed Rubin and colleagues is being applied to conventional and advanced coal-to-electric conversion systems.

Dynamic modeling of atmospheric chemical systems, such as for the Los Angeles basin, is helping that area improve its air quality as well as increase understanding of the impacts of alternative transportation fuels such as methanol. Professors Greg McRae and Ted Russell are key investigators. With "quantitative models of uncertainty," Professor Mitchell Small is working on problems of groundwater contamination, indoor radon, and cleanup and control of toxic and hazardous substances. Professor Cliff Davidson, with expertise in aerosol physics, is observing the distribution and fate of heavy metals in the environment. Together, Professor Granger Morgan and Dr. Indira Nair assess possible risks associated with human exposure to power-frequency electric and magnetic fields and help develop appropriate policy responses to this issue. Professor Scott Farrow takes on oil pollution at sea and extraction of depletable resources at the Woods Hole Oceanographic Institute.

Future environmental research will largely focus on assessing uncertainties of global climate changes. Researchers have already begun analyzing time dynamics of climate change, sea level rise, and economic impacts in the developed and developing worlds.

Risk communication and risk analysis research also has environmental bearings. Professor Baruch Fischhoff, pioneer in "behavioral decisionmaking and the psychology of risk perception," is conducting studies on people's willingness to pay for environmental protection, and communicating with people on risks such as radon and AIDS, as examples. Professor Lester Lave studies risks of air pollution and environmental controls; he is improving screening techniques for chemical carcinogens and studying risks of nuclear power sources in space vehicles.

■ Resources

Descriptions of EPP research and graduate studies are available in English, Chinese, Japanese, and Russian in its departmental brochure.

Department of Biology
Central College
Pella, IA 50219 USA

Contact: **John B. Bowles, Ph.D.**, Professor of Biology
Phone: 1-515-628-5204 • *Fax:* 1-515-628-5316

Activities: Education • *Issues:* Biodiversity/Species Preservation

▲ Projects and People

At Central College, an undergraduate liberal arts college, Dr. John B. Bowles is conducting research in the following areas:

Spiders. According to Dr. Bowles, "Less is known about Iowa's spider fauna than in other midwestern states." Searches for verified Iowa species, limited collecting, and information from nearby states form the basis for his preliminary checklist. Providing an extensive spider fauna are the Iowa landscape ranging from grasslands and open sand to caves, algific slopes, and relict or remnant forests. Spiders have ecological importance, reminds Dr. Bowles.

Bats. Research is determining the summer distribution of bats in Iowa, particularly summer nursery sites of the federally endangered Indiana bat (*Myotis sodalis*) and the evening bat (*Nycticeius humeralis*). "One recent work with the latter involves the use of radio transmitters to determine the roost trees," describes Dr. Bowles. "Use of dead or dying trees are necessary components of the summer habitats for both species."

Dr. Bowles has concluded studies on **eastern screech owls** (*Otus asio*) in which, during several surveys, 18 screech owls were located often near riparian woodland. With screech owl population densities comparable to those in other regions, the data is serving as a baseline from which to determine population trends in the future. "During the survey, we studied screech owl responses to the introduction of nest boxes," Dr. Bowles recalls. "If screech owls continue to decline in the region, they may respond to appropriately placed nest boxes the same way that kestrels have" in the past. Earlier studies were conducted with grants from the Iowa Conservation Commission.

Dr. Bowles has published notes, guides, checklists, and articles on his research.

Environmental Psychology
The Graduate School and University Center
City University of New York (CUNY)
33 West 42nd Street
New York, NY 10036-8099 USA

Contact: **Setha M. Low, Ph.D.**, Professor
Phone: 1-718-768-8156 • *Fax:* 1-718-768-6806

Activities: Education • *Issues:* Culturally Appropriate Sustainable Development • *Meetings:* American Anthropological Association; Environmental Design Research Association

● Organization

Within the USA's largest urban university system, **The Graduate School and University Center** houses a number of research institutes including a subprogram in **Environmental Psychology** that was established in 1968 as the first Ph.D. program in this field. Attracting students and visiting scholars from throughout the world, the subprogram provides interdisciplinary doctoral training from fields such as urban planning, landscape architecture, architecture, geography, anthropology, sociology, and psychology which are concerned with the relationships between the physical environment and behavior.

Environmental psychology is defined as "the study of the relationship between person and place" and is concerned with the developing individual as well as with social meaning, political power and control, group interactions, cultural history, technological change, and human possibilities.

▲ Projects and People

The scope of recent Environmental Psychology Subprogram research projects ranges from evaluating low-income cooperative housing and medical facilities to child rearing in Japanese and U.S. homes to studies of open space and urban policy space to the meaning of home and homelessness.

The **Center for Human Environments** is a nonprofit organization of social scientists, psychologists, designers, planners, and educators with the subprogram. A major undertaking is the **Children's Environments Research Group** which develops designs, policies, and programs to improve the quality of children's environments and enhance children's interactions with them. For example, the Group provides assistance to educators on environmental curriculum for all grade levels; it develops behavior-based specifications for children's environments such as day-care centers, schools, hospitals, psychiatric facilities, and playgrounds. One desired result is to "foster more dynamic and empowering relationships between children and the environment."

Similarly, the **Housing Environments Research Group (HERG)** takes on projects regarding the housing needs of residents including the currently and potentially homeless—with special attention to gender differences, children's needs, concerns of the elderly, low-income cooperatives, and community participation. The impact of residential crowding on children and an analysis of intergenerational housing are sample projects.

Dr. Setha Low directs both the **Public Space Research Group** and **Cultural Aspects of Design** with a **Network** of more than 200 design professionals, educators, anthropologists, environmental psychologists, sociologists and others from North and South America, Australia, Western and Eastern Europe, the Middle East, and Asia. "Research interests (of the Network group) include the study of cross-cultural design variations, vernacular architecture, symbolism, and meaning in the built environment; design in the Third World; and the socio-cultural context of design," she writes. Design in hot and humid places, "nest-building of higher apes," low-energy housing, biosphere creation, and traffic impacts are examples of specific pursuits.

Network members are able to exchange views at meetings of the Environmental Design Research Association, Society for Applied Anthropology, American Anthropological Association, Council of Educators in Landscape Architecture, and Built Form and Culture Research Conference.

■ Resources

CUNY literature is available on the Environmental Psychology Subprogram and the Children's and Housing Environments Re-

search Groups. The *Children's Environments Quarterly* and the *Children's Environments Monograph* series are also published. Interested persons may subscribe to the Cultural Aspects of Design Network and receive a worldwide directory and newsletter.

Archbold Tropical Research Center (ATRC) ⊕
Clemson University
126 Lehotsky Hall
Clemson, SC 29634-1019 USA

Contact: **Thomas E. Lacher, Jr., Ph.D.,** Director
Phone: 1-803-656-0457 • *Fax:* 1-803-656-0231

E-Mail: lachert@clemson.edu

Activities: Education; Research • *Issues:* Biodiversity/Species Preservation; Deforestation; Sustainable Development • *Meetings:* International Theriological Congress

● Organization

Clemson University founded the **Archbold Tropical Research Center (ATRC)** in Dominica after receiving **John D. Archbold's** donation of his 92-hectare Springfield plantation, which was then opened as a Field Station in late 1989. A consortium of institutions from the USA, Latin America, and the Caribbean run ATRC with **Dr. Thomas Lacher** as director.

The consortium includes the Smithsonian Institution, National Park Service, U.S. Forest Service, Medical University of South Carolina, Penn State University, State University of New York College of Environmental Science and Forestry, Syracuse University, Texas A&M University, University of Georgia, Yale University, and University of the West Indies at Barbados. In addition, affiliates include universities and agencies from Argentina, Barbados, Brazil, Colombia, Costa Rica, and Mexico.

The Center's purpose is to "provide an opportunity for students, faculty members, scientists, and other professionals to participate in field experiences in education, training, research, and public service in a unique moist tropical setting," writes Dr. Lacher. It is expected to play an important role in coordinated teaching and research activities related to sustainable development in tropical nations.

Dominica is known as the "nature island" of the Caribbean, and the plantation is a "living laboratory" from which to explore rainforests, fertile volcanic soils, abundant fresh water, and the many species of plants and animals including the endangered imperial and red-necked parrots.

ATRC is promoting undergraduate and graduate education and research in tropical ecology, conservation biology, natural resources management, nature tourism, island culture, anthropology, health, and sustainable economic development. For example, plant biologist **Constance S. Stubbs,** University of Maine, is surveying **lichen** diversity and its relationship to invertebrate associate diversity. Other researchers are discovering **land ownership** is complex here, and coordination between government agencies is difficult.

An early project with **Cleber J.R. Alho,** World Wildlife Fund—Brazil, surveyed **mammals in the Pantanal of Brazil** and examined threats to their survival.

Liliana Madrigal and **Manuel Ramirez,** Conservation International, are cooperating with ATRC on a five- to seven-year project on conservation and reforestation in Costa Rica and Panama—including economic recovery of degraded watersheds and with funding from the McDonald's Corporation.

Lacher is encouraging networking on biodiversity and conservation among Western Hemisphere colleagues and will use the ATRC *Tropicalia* newsletter to publish articles in English, Spanish, or Portuguese for "information but also as a source of ideas and inspiration." At Clemson University, he is also associate professor, Department of Aquaculture, Fisheries and Wildlife, and has an adjunct appointment with the Institute of Wildlife and Environmental Toxicology. Lacher belongs to the Species Survival Commission (IUCN/SSC) New World Marsupial Group.

Institute of Arctic Studies ⊕
Dartmouth College
6182 Murdough Center
Hanover, NH 03755 USA

Contact: **Gail Osherenko, J.D.**, Senior Fellow
  Phone: 1-603-646-1278 • *Fax:* 1-603-646-1279
  E-Mail: Bitnet: gail.osherenko@dartmouth.edu

Activities: Education; Research • *Issues:* Arctic Environment; Biodiversity/Species Preservation; Cultural Diversity; Global Warming; Sustainable Development

● Organization
The **Institute of Arctic Studies** is a unit of Dartmouth College's John Sloan Dickey Endowment for International Understanding. The Institute "uses its resources to strengthen and expand the Arctic community centered at Dartmouth and to support innovative and broad-ranging research on Arctic topics."

The Institute reports ". . . a common thread running through its research programs is a concern for the design and implementation of institutional arrangements that allow peaceful development of the Arctic as a distinctive region, while acknowledging the links between the region and the rest of the world." The Circumpolar North is an unusually attractive setting in which to conduct research on institutions dealing with human/environment relationships or "resource regimes." Under study are divergent regimes of indigenous or aboriginal cultures alongside modern Western systems and ways to meld "contrasting institutional arrangements" through "co-management regimes."

No longer perceived as "a frozen wasteland" of concern to a few "explorers, traders, missionaries, scientists, and indigenous people, the Arctic has recently emerged as a rapidly changing international region of global importance in strategic, economic, cultural, and ecological terms," writes the Institute. For examples, some two-thirds of the former Soviet Union's oil and gas is produced in northwestern Siberia and almost one-fourth of the USA's crude oil comes from Alaska's North Slope. Its unique cultural and biological diversity requires preservation. And global climate change in the Arctic—a "carbon and methane sink"—is expected to intensify the greenhouse effect as temperatures could raise "2–3 times those experienced in the temperate zones."

▲ Projects and People
The **Arctic Peace and Cooperation Program** analyzes the Arctic as an international region in its own right. The Institute has sponsored a circumpolar **Working Group on Arctic International Relations** and also organized research "on international cooperation both to encourage the development of an Arctic zone of peace and to contribute to our understanding of international cooperation more generally."

The **USA/Russia Arctic Collaboration Program** involves issues surrounding international cooperation, environmental protection, sustainable development, and cultural survival.

The **Arctic Indigenous Peoples Program** is in collaboration with the Northern Native governments and non-governmental organizations "bringing indigenous students to Dartmouth and conducting research on topics of mutual interest."

Current research on **global environmental issues** focuses on what makes international environmental regimes "effective." Gail Osherenko, an environmental lawyer, writes, "Case studies focus on operational discharge of pollutants in the oceans, long-range transboundary air pollution (LRTAP), and fisheries in the Barents Sea."

The recently completed **International Regime Formation Project** studied the conditions necessary to bring about international cooperation in a specific issue area.

Osherenko, senior fellow, is on the Board of Advisors of this *Guide.*

■ Resources
Cornell University Press is publishing results of the International Regime Formation Project, *The Politics of Regime Formations: Lessons from Arctic Cases.* University Press of New England, Hanover, New Hampshire, publishes an Arctic book series entitled *Arctic Visions* by Osherenko and **Oran R. Young** that emphasizes the human dimension of the Arctic and northern regions. The Institute produces occasional papers and *Northern Notes.*

Center for Tropical Conservation (CTC) ⊕
Duke University
3705 Erwin Road
Durham, NC 27705 USA

Contact: **John W. Terborgh, Ph.D.**, Director
  Phone: 1-919-490-6834 • *Fax:* 1-919-490-6718

Activities: Education; Research • *Issues:* Air Quality/Emission Control; Biodiversity/Species Preservation; Deforestation; Global Warming; Sustainable Development; Waste Management/Recycling

● Organization
The **Center for Tropical Conservation (CTC)** directs its activities toward "improvement of the scientific, informational, and institutional basis for reform of environmental as well as non-environmental policies affecting sustainable development in the Tropics and . . . Eastern Europe."

In 1990, its first year, researchers from Duke University, North Carolina State University (NCSU), University of North Carolina

(UNC), University of Florida, and University of Washington provided expertise in the biological sciences, resource management, economics, and public policy with experience in tropical and nontropical developing countries. Additional faculty and consultants from the USA and other countries are expected to participate, according to the CTC's five-year plan.

▲ Projects and People

The first projects centered on Tropical Silviculture and Forest Management Alternatives, Enhancement and Expansion of Parks and Reserves for Tourism and Conservation, the Role of Institutions, and Eastern European Pilot Studies.

With fieldwork in Ecuador, Costa Rica, and Mexico, a research team began analyzing natural forest management programs in the American tropics—including government policy environment, socioeconomic conditions under which the programs operate, the ecological basis for silvicultural approaches employed, and the sustainability of different forest management methods.

Dr. Frances E. Putz, University of Florida, has prepared a paper on an alternative practice in which "commercially valuable products are extracted from forested areas while retaining substantial canopy cover." Tracing the process from seed production to forest product yields, he also analyzes the sustainability of various harvesting and extractive systems. CTC says conservation groups view this approach as preferable to the "more destructive practices of replacing forest cover with agricultural crops, pastures, or plantations." Because entire management systems may not transfer from one forest to another, Dr. Putz is preparing a parallel work plan based on Southeast Asian case studies, such as in Malaysia, Indonesia, and the Philippines.

Duke researchers Carel van Schaik and Randall Kramer are focusing on tropical rainforests as they review both economic and management methods for preserving biodiversity. "The interests of sustainable development, which often stress ecosystem function, are not sufficient to protect biodiversity," they say. Their findings are being combined into a series of management recommendations for U.S. Agency for International Development (USAID) project design.

In Guatemala and Indonesia, Dr. John Terborgh compared "existing non-timber forest production extraction systems." His findings concluded that "although extractive reserves can play a significant role in preserving tropical forests, their effectiveness is highly dependent on local circumstances. [They] will not by themselves save the rainforest, and must be a component of a broader land use pattern." Research is continuing at the Manau National Park field station, Peru, where Terborgh has conducted various projects with more than 100 investigators since 1973. Here, ecosystems are being studied and understood so that the economics of extractive reserves can be analyzed.

In reviewing tax policy and tree species selection in Latin America, Indonesia, Malaysia, and Malawi, Duke researchers Robert Conrad and Evan Mercer are examining whether taxes should be sensitive to the tree species harvested. For example, they are reviewing how tax policies affect harvesting decisions—such as clearcutting versus high-grading—wood mass volumes, and waste or use of low-value stems.

Regarding enhancement and expansion of parks and reserves, studies of ecotourism and forest resource industries initially were focused in Eastern Europe, particularly Poland, where discussions with officials from government, universities, and environmental

NGOs centered on local environmental planning, land-use regulation, tourism, and sustainable development. Dr. Robert Healy, Duke, is continuing research to examine how ecotourism potential can be part of assessments of overall development strategies in both Eastern Europe and Latin America; specific parks and reserves are being evaluated.

In Czechoslovakia, economic approaches to environmental policy are being researched by Richard N. Andrews, UNC. Edward M. Bergman, UNC, began a study of opportunities for joint research projects here and in Hungary regarding democractization and privatization of regional economies and environmental effects, such as pollution and recycling needs.

Honduras and Costa Rica were selected for case studies to determine how "forestry institutions can acquire and maintain the capabilities to formulate sound forestry strategies." Duke researchers William Ascher and Marie Lynn Miranda concentrated on standing timber fee systems and state forest agency's changing role in Honduras and reforestation incentive schemes and natural forest management in Costa Rica. Similar case studies are continuing in Indonesia and Malaysia to identify factors that "facilitate forestry institutions to use state-of-the-art forest management science." They are also determining whether earlier findings hold "despite the shift in regional and cultural contexts."

Miranda is also studying the role of NGOs in improving natural resource policy particularly in the ASEAN (Association of Southeast Asian Nations) region and "to pressure governments toward a proper balance of revenue generation, environmental protection, and social objectives." The differences between international and local NGOs are being reviewed "in light of the growing skepticism toward the motives of the international non-governmental organizations."

Guidelines for minimizing waste and preventing pollution are being developed for AID projects in its recipient countries. Duke's Michael Regan is outlining several alternative strategies designed to meet USAID environmental management objectives.

In global warming studies, Duke researchers Wesley Magat and Kip Viscusi are developing a framework for adjusting energy taxes according to environmental damage, particularly air pollution, caused by each fuel type. Results "could be used either directly to calculate 'no-regrets' energy taxes or to guide policymakers in energy strategy—even if such taxes cannot be fully implemented," they report.

Ranomafana National Park (RNP) Project ⊕
Duke University
Wheeler Building, 3705-B Erwin Road
Durham, NC 27705 USA

Contact: **Patricia C. Wright, Ph.D.,** Assistant
        Professor of Biological Anthropology
Phone: 1-516-632-7425, 1-919-490-6286
  Fax: 1-516-632-9165, 1-919-490-4718

Activities: Development; Education; Research • *Issues:* Biodiversity/Species Preservation; Conservation Education; Health; Sustainable Development; Water Quality

● Organization

As international director of the **Ranomafana National Park (RNP) Project** in Madagascar, **Dr. Patricia Wright** brings the resources of Duke University and other North American educational institutions. She is also affiliated with the State University of New York at Stony Brook.

Situated in the southeastern region of the world's fourth largest island, Ranomafana is a naturalist's treasure, with 12 species of playful lemurs, the white-water river Namorona surging toward the Indian Ocean, orchid-covered mountains, and rare birds, such as the groundroller, iridescent coas, and jewel-like kingfishers.

The park was actually established to protect two newly discovered species—the golden bamboo lemur and the greater bamboo lemur, found nowhere else. Chameleons and lizards are profuse. Romping about are elegant carnivores and forest-dwelling fossa cats. Altogether, 12 species of primates coexist in this tropical cloud forest.

The U.S. Agency for International Development (USAID) awarded Duke University more than $3 million for a three-year, first phase of Ranomafana village development—in keeping with sustainable development and integrated rainforest conservation in Madagascar. The MacArthur Foundation is contributing $150,000 for biodiversity studies and training of Malagasies. Offices are set up in both Ranomafana and Duke University.

▲ Projects and People

RNP has a wide range of projects with participation of scientists and researchers from various universities.

A first **health and parasite survey** was undertaken in 1991–1992 of about 525 families living in 18 communities near the National Park. The information gathered is serving as a "baseline from which to monitor the effects of changing forest use on the parasite burden, nutrition, and general health of the local human populations; and will provide information to help assess the current health needs," writes **Lon Kightlinger**, School of Public Health, University of North Carolina. A team of 11 researchers with 14 facilitators and collaborators are testing their theory that individuals with a certain "deficient cellular immune response" are predisposed to heavy worm infections because they lack the genetic constitution or "immunosuppression" to resist the parasite. Severe malnutrition was found in about 10 percent of the children with worms.

Earlier, the Malagasy Lutheran Church (SA.FA.FI.) joined with RNP on a **socioeconomic study** examining leadership, ethnic background, diet, agriculture, animal husbandry, forestry, drinking water, sanitation, education, and stores and markets in six pilot villages. It was recommended that each village buy tools, seeds, building and school supplies, and build dams to increase rice production.

A study of **land management and sustainable development** showed that the acid and "surprisingly deep" soils of Ranomafana yield low crop production—particularly rice. Deforestation and *alang-alang*, a nonpalatable, non-native grass are present. Researchers **Pedro Sanchez, Bob Kellison, Greg Minnick,** and **Joe Peters** recommend further soil surveys; improving, expanding, and irrigating rice paddies; stopping erosion; and planting food-producing trees. "Crop production diversification along with alternative food and livestock production systems are necessary to break the dependency on rice and cassava," they say. Slash-and-burn agriculture (*tavy*) on shortened rotations and uncontrolled harvest of forest products are the Park's greatest threat. To reduce pressure on the native forests, the team recommends long-term forestry develop-

ment in the degraded lowland areas and in the pine and eucalyptus plantations of the buffer zone.

In evaluating surveys of the **potential for ecotourism, Joe Peters,** North Carolina State University, is identifying opportunities for research and future development projects that will promote the growth of this nature-oriented industry.

Researchers are studying the habitats and diets of **rare lemurs** to increase their chances of survival. Dr. Wright is searching for the aye-aye (*Daubentonia madagascariensis*), and studying three species of bamboo lemurs and two groups of "marked Milne Edward's diademed sifakas" (*Propithecus diadema edwardsi*). **Deborah Overdorff** is concerned with the red-fronted lemur and the red-bellied lemur. **Adina Merenlender** and **Andy Dobson,** Princeton University, are collecting information on population genetics on which to base better management plans for the endangered primates. **Marion Dagosta,** Northwestern University, is comparing the locomotion and positional behavior of the "vertical clinging and leaping sifakas" and the "more quadrupedal" black-and-white ruffed lemur (*Varecia variegata*). It is believed that selective logging is eliminating the *Varecia* because it feeds on tree species selected for timber; because of its locomotor behavior, it relies on big trees with horizontal crowns. In a seed dispersal study of primates, **Larry Drew** and **Nigel Asquith** discovered that "seeds passed through a lemur digestive tract sprouted three months earlier than those not eaten. "Primates can have an important positive effect on plant distribution in the Malagasy rain forest," they say, although some lemurs are seed predators.

In radio collaring **red forest rats,** it was discovered that these rodents are also seed dispersers for plants and an important part of the ecosystem. Recent researchers include **James Ryan,** Hobart and William Smith Colleges, and **Ernestine Ralamavo.**

Steve Zack, Yale University, is undertaking a long-term study of seven species of **vanga shrikes** and also surveys **mixed species flocks** of birds in different habitats within the park. His observations show that there are "six times the number of mixed species flocks in the pristine rain forest region . . . than were in the selectively logged habitats." Such information enriches understanding of the rainforest ecosystem and gives insights into how to best preserve the rarest species.

Crayfish are a main source of cash income for villagers. **Paul Ferraro** and **Alex Dehgan** worked on studies to determine the effects of increased exploitation and to see if crayfish can be harvested more efficiently or grown in rice paddies. They may also be a good indication of water quality in watershed areas.

Plants collected during the rainy season yielded a rare species of monomaciae. Botanist **David Seigler,** University of Illinois, discovered there were heavy doses of cyanide in the golden bamboo lemur's diet of bamboo. The five species of tree fern in the park are a cash crop for villagers who sell them in flower pots along the highway outside the park. Plants used by a village "medicine man" and eaten by lemurs are being chemically analyzed, notes Ferraro.

Six species of **snails** were noted at RNP. Life-history studies of the giant tree snail show that they "gorge themselves on mushrooms and the females produce a single ping-pong ball size egg per year." **Ken Emberton,** Academy of Natural Sciences, Philadelphia, PA, conducted the search.

Unusual forms of **frog reproduction** were observed, including species that breed inside bamboos, tree ferns, and other tree cavities; species that attach their eggs to aerial roots suspended within the splash zone of waterfalls; and species that attached their eggs to vegetation over quiet water. Samplings by John Cadle, Academy of Natural Sciences, also show at least 7 species of chameleons, 8 species of other lizards, and 15 species of snakes—which are secretive and difficult to locate. The rare water tenrec is also under study.

■ Resources

A bibliography of recent publications resulting from research at Ranomafana is available.

School of the Environment
Duke University
Durham, NC 27706 USA

Contact: **Lynn A. Maguire, Ph.D.,** Assistant
      Professor of Ecology
Phone: 1-919-684-2619, 1-919-549-4033
Fax: 1-919-684-8741

Activities: Education; Research • *Issues:* Sustainable Development

▲ Projects and People

Dr. Lynn A. Maguire describes her research interests as focusing on "the application of quantitative tools, such as decision analysis, statistical analysis, and simulation modeling" and relating these to the management of natural resources.

She uses "stochastic simulation and decisionmaking under uncertainty to analyze management strategies for a variety of endangered species," such as grizzly bears, tigers, and black-footed ferrets. Dr. Maguire—who has given short courses and workshops for universities, the U.S. Forest Service, and zoos—is also interested in landscape scale effects and forest fragmentation.

Her current research uses a "combination of decision analysis and environmental dispute resolution to develop consensus plans" for management of both endangered species populations and national forests in southern Appalachiana. Dr. Maguire belongs to the Species Survival Commission (IUCN/SSC) Captive Breeding Specialist Group.

■ Resources

Dr. Maguire offers consulting on decision analysis, conflict resolution, and population modeling for endangered species management and multiple-use land planning.

Department of Biological Sciences
Florida Atlantic University
Boca Raton, FL 33461-0991 USA

Contact: **Peter L. Lutz, Ph.D.,** Chair
Phone: 1-407-367-2886 • *Fax:* 1-407-367-2749

Activities: Education • Research • *Issues:* Biodiversity/Species Preservation; Marine Pollution

▲ Projects and People

Sea turtles are highly vulnerable to pollutants, such as oil and nonbiodegradable debris, including plastic ingestion—topics which biologist Dr. Peter L. Lutz researches at Florida Atlantic University.

Green and loggerhead sea turtles do not seem to detect and avoid oil slicks or distinguish tar balls from food. Laboratory studies indicate that chronic crude oil exposure can cause "cell abnormalities of the skin, alteration of respiratory patterns, and blood cell dysfunctions." Ingested oil sometimes appears in the feces, and the salt gland's regulatory ability can be reduced or delayed—a failure that could prove fatal at sea. Tests also showed that exposure to oil decreased turtles' oxygen consumption and that ingestion of oil resulted in lower red blood cell counts.

"In an oil slick, the turtles' habit of surfacing to breath ensures a continued contact with the oil," writes Dr. Lutz with Dr. Molly Lutcavage, University of Wisconsin. And in the Gulf of Mexico—where five of the seven living turtle species are found today—oil slicks pose a real damage to Kemp's ridley (*Lepidochelys kempi*), loggerhead (*Caretta caretta*), green (*Chelonia mydas*), hawksbill (*Eretmochelys imbricata*), and leatherback (*Dermochelys coriacea*). Above all, Kemp's ridley turtles—already suffering from reduced populations and restricted nesting habitats—face greatest risks. However, field studies of some sea turtle eggs incubating in oiled sand showed reduced survival rates, while others—including Kemp's ridley eggs—did not. What needs to be critically assessed, writes Lutz, is how "oil exposure disrupts behavior and physiological functions." Because sea turtles are known for their "high tolerance" to severe physical damage, some of these results surprised scientists. Charles W. Caillouet, Jr., National Marine Fisheries Service, and André M. Landry, Jr., Texas A&M University at Galveston, are updating findings.

Dr. Lutz cautions about the hazards of plastics and latex ocean debris. Not only can larger pieces of ingested plastic block the gut and cause death, green and loggerhead sea turtles will "seek out" and consume smaller pieces that can collect, build up, bind together, and decompose in the intestine for a period of days to months. Sea turtles also show a color preference—with clear plastic having the lowest acceptance rate. Further investigations are needed, believes Lutz, of the effects of sea turtles' ingestion of "higher, more realistic levels of nonbiodegradable materials."

Dr. Lutz is an Explorers Club Fellow and member of the Species Survival Commission (IUCN/SSC) Marine Turtle Group.

Department of Biology ⊕
Frostburg State University
Frostburg, MD 21532-1099 USA

Contact: **Howard Quigley, Ph.D.,** Assistant Professor
Phone: 1-301-689-4166 • *Fax:* 1-301-689-4737

Activities: Research • *Issues:* Biodiversity/Species Preservation; Deforestation; Sustainable Development

● Organization

Frostburg State University, located within 150 miles of Washington, DC, Baltimore, Maryland, and Pittsburgh, Pennsylvania, offers a Master of Science program in Applied Ecology and Conservation

Biology with global dimensions through the collaboration of universities in Malawi, Zimbabwe, and Nicaragua. Students address those issues pertaining to ecosystem fragmentation, conservation and development conflicts, and integrated resource management on domestic and international levels. The program trains people for vital environmental roles in maintaining biodiversity.

▲ Projects and People

Dr. Howard Quigley, who teaches in this program, is involved in several research projects—including the **Vertebrate Populations Project, El Peten, Guatemala**, with Wildlife Conservation International; and the **Ecology and Conservation of the Siberian Tiger in the former Soviet Union.**

Guatemala: The Peten is largely in a government-established Biosphere Reserve in northern Guatemala. The United Nations is considering this area's inclusion for the worldwide Biosphere Reserve program—the Peten being the second largest expanse of undisturbed tropical forest in Central America and widely significant in conserving Latin American resources. Because of these reasons, human and financial resources are attracted here mainly at the planning level. Addressing the need for data collection, the **vertebrate populations project** in Tikal National Park is focusing on mammals and large birds in the beginning; a census design for vertebrates in tropical tall forest will allow comparisons to be made with vertebrates under different forms of resource management and use; its framework allows Guatemalan biologists and students to conduct wildlife research in the field; and information gathered will benefit science, public awareness, and lasting conservation.

Getting into the field during the dry season (April 1991), the number of species encountered were initially low: trapped and released spiny pocket mouse and big-eared climbing rat, sightings of birds, and sound-detected white-tailed deer. **Kathy Quigley, DVM,** is investigating **leishmaniasis protozoa** by taking samples of trapped animals; lab analysis in Guatemala City is determining the disease's incidence in wild mammals. Project managers are **María José González** and **Milton Cabrera.**

Former Soviet Union: A study of the **ecology and conservation of the Siberian tiger in the Sikhote-Alin Biosphere Reserve**, Primorye, is underway with **Dr. Maurice Hornocker**, Wildlife Research Institute, Moscow, Idaho, USA. Without effective conservation, Siberian tigers could be the fifth subspecies to become extinct. Quigley's research is aimed at scientifically gathering information to help save the world's largest cat, which could disappear in 40 years or less.

"Beyond the preservation of a single subspecies, the project can bring further examination of the conflicts and compromises involved in dealing with the conservation of a critical component of natural communities: large predators," describes Dr. Quigley. "Strategies which assure the future of large predators—which require large areas and a stable prey base—normally accommodate a number of other species through the 'coattail effect.'"

Within the Sikhote-Alin Biosphere Reserve in the former Soviet Far East, all captured tigers are being fitted with radio collars, then released. Their locations will be monitored several times weekly to determine what types of habitat they prefer, and how their movements are associated with other tigers and their prey. While the animals are immobilized, biological materials will be taken for nutritional and genetic analyses. Research results will be the basis for the tigers' conservation strategy in the Primorye province.

The National Geographic Society and the National Fish and Wildlife Foundation are helping to finance the project. Dr. Quigley has 15 years' experience in wildlife ecology, with special emphasis on carnivore ecology and research experience in China and Brazil; he has worked with jaguars and mountain lions. Dr. Hornocker has more than 30 years' experience in carnivore wildlife ecology, and experience in India, China, Africa, and South America.

Department of Biology ⊕
George Mason University
Room 3019, David King Hall
Fairfax, VA 22030 USA

Contact: **Luther Brown, Ph.D.**, Associate Professor
Phone: 1-703-993-1050 • *Fax:* 1-703-993-1046

Activities: Education; Research • *Issues:* Biodiversity • *Meetings:* ESA

▲ Projects and People

Dr. Luther Brown is involved with the ongoing **Earthwatch** project, **Behavioral and Ecological Investigations of Freshwater Bahamian Fish Communities.** This Bahamas' Blue Holes project is a detailed study of the demography and population biology of freshwater fishes, especially the endemic mosquito fish *Gambusia manni* that lives in the blue holes, lakes, and ponds of Andros Island. "These sites provide an ideal location for the analysis of microevolution," writes Brown, whose co-director is **Jerry Downhower**, The Ohio State University. They supervise teams of 8 to 12 volunteer researchers each year.

Brown and Downhower are also trying to establish a **teaching and research institute** on Andros Island in the fields of biology, geology, and anthropology/sociology of the Bahamas. Their plan is to promote ecotourism or the "sustained, environmentally sound touring of the coral reef and terrestrial habitats" of Andros Island.

At George Mason University, Brown teaches and directs undergraduate research in animal behavior and field biology. He writes laboratory manuals for studies in ecology. He also produced slide shows and video tapes on Andros Island, the Bahamas, and the Western Atlantic region—featuring native fish and birds, terrestrial and marine invertebrates, coral reef, and bush medicine and agriculture.

■ Resources

Needed are vehicles and materials for the establishment and support of the Blanket Sound research station, Andros Island; and overall political support for conservation issues in the Bahamas.

Harvard Environmental Law Review (HELR)
Harvard Law School
Hastings Hall Publication Center
Cambridge, MA 02138 USA

Contact: Editor-in-Chief
   Phone: 1-617-495-3110 • *Fax:* 1-617-495-1110

Activities: Publishing • *Issues:* Environmental Law

▲ Projects and People

Harvard Law School students comprise the staff of the *Harvard Environmental Law Review (HELR)*, which is published semiannually in winter and summer. Students founded the scholarly legal journal in 1976 to provide a forum for "in-depth technical and legal analysis of complex environmental problems." According to the 1992 editor-in-chief, "HELR presents all sides of the environmental debate but maintains a commitment to the ideals of environmental preservation and enhancement."

Issues covered include land use; air, water, and noise regulation; toxic substances control; radiation control; energy use; workplace pollution; science and technology control; and resource use and regulation with interests on local, federal, and global levels. Submitted articles are reviewed and critiqued by an editorial team of up to six persons as well as four to 12 subciters. "Organization, style, and substance are all carefully scrutinized."

■ Resources

HELR subscriptions are available.

MIT-Harvard Public Disputes Program (PDP) ⊕
Harvard Law School
512 Pound Hall
Cambridge, MA 02138 USA

Contact: **Bruce J. Stedman,** Associate Director
   Phone: 1-617-495-1684 • *Fax:* 1-617-495-7818

Activities: Education; Research • *Issues:* Negotiation and Alternative Dispute Resolution; Sustainable Development
Meetings: IUCN General Assembly; IWC; UNCED

● Organization

Established in 1979, the **MIT-Harvard Public Disputes Program (PDP)** is dedicated to enhancing the "fairness, efficiency, stability, and wisdom of public sector decisionmaking." An international action-research center, it seeks to replace "'win-lose' outcomes with 'all-gain' solutions in the highly charged, complex arena of public policymaking." PDP cornerstones are "hands-on mediation, research, teaching, and theory building." Neutral facilitators assist the process.

A decade ago, it was a revolutionary idea to attempt balancing scientific and political considerations through negotiating environmental decisionmaking. Today, this new approach brings about creative resolution to multiparty, multi-issue conflicts involving government, citizens, and business on issues such as locations of nuclear power plants and hazardous waste disposal sites; land development; preservation and use of natural resources; and establishment of public health and safety standards.

▲ Projects and People

Early collective dispute settlements resolved the expansion of mass transit facilities in a wetlands area, allocated costs of a pollution cleanup, set rules for the national park system, and mediated offshore oil exploration.

Based on real situations, **simulations** are produced as teaching aids to show, for example, how to settle fishing rights among Native Americans, fishing interests, and government agencies; how to develop cleanup policies for hazardous wastes or dispose of low-level radioactive wastes; or how to develop a statewide energy assistance plan for low-income residents.

National workshops are being sponsored to formulate "a diagnosis of the most important conceptual problems and to develop public policy prescriptions" regarding **facility sitings.** With the American Energy Assurance Council and support from the National Institute of Dispute Resolution and the Hewlett Foundation, PDP is holding mediated work sessions to reach **consensus on a national energy policy;** other workshops are assessing ways that EPA, industry, and environmental groups can bring about Superfund settlements.

PDP is currently launching a series of programs on international environmental negotiations through the **Salzburg Initiative and the International Environmental Negotiation Network (IENN).** Director of both PDP and the IENN Secretariat Dr. Lawrence E. Susskind invites "senior government officials, grassroots leaders, journalists, business executives, and scholars" to join the Network and help "resolve global and transboundary environmental conflicts through more efficient negotiation and consensus building" particularly among representatives of government and non-government organizations (NGOs) in environmental treaty making. IENN also intends to "build support for a variety of institutional reforms [ensuring] that global and regional environmental negotiations" will be conducted more efficiently and fairly while achieving scientifically informed outcomes.

With more than 300 members, the Network is beginning the following activities: **Environmental Negotiators Database; Negotiation Advisory Service** to assist individuals, organizations, or countries; **negotiation training;** follow-up to the **United Nations Conference on Environment and Development (UNCED);** promoting the Salzburg Initiative; publishing the quarterly IENN Bulletin, **Environmental Diplomacy booklets,** and an **International Compendium of National Environment and Development Priorities; prenegotiation training;** and **regional prenegotiation sessions** aimed at reducing confrontations between wealthier countries in temperate areas and developing nations from tropical regions.

The **Salzburg Initiative** is a result of ongoing conversations among politicians, executives, journalists, and environmentalists from 20 countries to improve environmental diplomacy. Its declaration supports 10 strategies: a decentralized United Nations Environment Programme (UNEP) support structure to reorganize nontraditional clusters of countries; prenegotiation assistance to individual countries; new approaches to treaty drafting; expanded roles for NGO interests; recategorizing countries for purposes of prescribing action, such as according to ways environmental problems are caused or responded to; ongoing scientific involvement in political decisions that negotiate contingent agreements with "continuous monitoring and periodic recalibration of standards and responses"; encouraging creative issues linkage that generate economic incentives for environmental protection; removing penalties for constructive unilateral action; involving an informed mass media; and reforming or upgrading international institutions.

■ Resources

Workshops, seminars, and short courses are taught; environmental dispute resolution courses at MIT are now models for similar courses at other universities. Videotapes, case studies, working papers, books, and curriculum materials are available to interested parties and listed in *Publications and Teaching Materials*. PDP publishes the quarterly *Consensus* which highlights successful dispute resolution cases. IENN welcomes inquiries about membership.

PDP seeks opportunities to serve as "third-party neutrals in environmental disputes involving nongovernmental organizations."

Botanical Museum ⊕
Harvard University
26 Oxford Street
Cambridge, MA 02138 USA

Contact: **Richard Evans Schultes, Ph.D.**, Director
(Emeritus)
Phone: 1-617-495-2326

Activities: Research • *Issues:* Biodiversity/Species Preservation; Ethnobotanical Conservation

▲ Projects and People

Dr. Richard Evans Schultes, Jeffrey Professor of Biology and Harvard Botanical Museum director (emeritus), carries out ethnobotany and conservation with **Dr. Mark J. Plotkin**, Conservation International (CI) vice president and Botanical Museum research associate.

The **Ethnobotanical Conservation** program, under the aegis of CI, aims at "stressing the urgency of preserving as much of the knowledge of plant utilization as possible by field work." Dr. Schultes writes, "This knowledge, much of which could be of great value to humanity as a whole and which has been amassed over millennia is today in grave danger of annihilation as a result of the rapidity of acculturation in many parts of the world." Knowledge of plant uses among primitive societies is particularly fast disappearing, notes Dr. Schultes.

This program's purpose is "to educate the public through bulletins, articles, lectures, and primarily through direct correspondence in the practical value of salvaging this native knowledge before it disappears forever."

Dr. Schultes has conducted seminars on the plant kingdom as a source of new medicines; and on plant hallucinogens from Native American shamanism to modern medicine. His research takes him into Amazonian Colombia and Malaysia almost yearly; other study and lecture trips are worldwide. He and Dr. Plotkin are chairman and co-chairman, respectively, of the Species Survival Commission (IUCN/SSC) Ethnobotany Group.

■ Resources

Dr. Schultes encourages more research in the field. A newsletter is published on occasion.

Howard University
(See World Conservation Union [IUCN], Species Survival Commission/Sirenia Specialist Group in the USA NGO section.)

Department of Anthropology ⊕
Hunter College of the City University of New York (CUNY)
695 Park Avenue
New York, NY 10021 USA

Contact: **John F. Oates, Ph.D.**, Professor
Phone: 1-212-772-5410 • *Fax:* 1-212-772-5423

Activities: Research • *Issues:* Biodiversity/Species Preservation; Deforestation

▲ Projects and People

In Sierra Leone, West Africa, Dr. John Oates started the ongoing Tiwai Island Project in 1982 on the ecology of a rainforest primate community. Tiwai Island is 12 square kilometers in the Moa River close to the edge of the Gola Forest Reserves. His research led to the development of a field research station run by a consortium of Njala University College (Sierra Leone), Hunter College, and the University of Miami (UM). Working with local chiefdom authorities, Sierra Leone's government and Conservation Society, and the U.S. Peace Corps, a wildlife sanctuary and visitors' center has also been established.

Efforts are now being made to extend conservation activities to downstream islands, and to the Gola West Forest Reserve in coordination with the Gola Rainforest Project. The National Science Foundation, Wildlife Conservation International (New York Zoological Society), World Wildlife Fund (WWF), African Wildlife Foundation, and the Federation of British Zoos have joined Hunter, CUNY, and UM in funding this research. Research and conservation efforts were disturbed in mid-1991 by a "rebel insurgency spilling over from Liberia; however the Sierra Leone government was able to neutralize rebel activity and the project began rebuilding," according to Oates. **Emmanuel Alieu**, secretary of the Conservation Society, is coordinator in Sierra Leone. A television documentary entitled "Tiwai: Island of the Apes" was made for British viewers by an organization known as Survival Anglia.

In Nigeria, also in 1982, Dr. Oates began working with the Nigerian Conservation Foundation (NCF), university colleagues in Nigeria, and other researchers on various forest primate research and conservation projects. Results of some of these activities are: the establishment of NCF's Okomu Forest Project and the creation of a wildlife sanctuary in the Okomu Forest Reserve in Bendel State; surveys to assess the status of the endangered Sclater's guenon monkey in eastern states and the initiation of a study directed by Oates on the behavioral ecology of the monkey in Enugu; proposals for wildlife sanctuaries or reserves at Stubbs Creek and Taylor

Creek; and a management study of the proposed Boshi-Okwangwo sector of the Cross River National Park, with a special focus on the conservation of gorillas. Dr. Pius Anadu, NCF executive director, collaborates with Dr. Oates on the Nigerian projects. Funding is from Wildlife Conservation International, National Geographic Society, WWF-USA, and WWF-UK.

Oates, who is coordinator of the African Section, Species Survival Commission (IUCN/SSC) Primate Group, is currently updating an Action Plan for African Primate Conservation that will review population changes and recommend future actions.

Department of Geosciences ⊕
Indiana/Purdue University at Fort Wayne (IPFW)
Fort Wayne, IN 46805-1499 USA

Contact: **Solomon A. Isiorho, Ph.D.,** Assistant Professor
Phone: 1-219-481-6249 • *Fax:* 1-219-481-6880

Activities: Education; Research • *Issues:* Energy; Global Warming; Health; Sustainable Development; Waste Management/ Recycling; Water Quality

▲ Projects and People
Water shortage problems in the Sahel, Nigeria, region of Africa are a long-term interest of **Dr. Solomon A. Isiorho, Indiana/Purdue University at Fort Wayne.**

An Earthwatch study in 1991 continued his five-year exploration of the Chad Basin, which is important to 10 million people in 4 countries: Nigeria, Niger, Cameroon, and Chad. Located in a semiarid region, **Lake Chad** has the world's largest drainage basin of any lake. It is a shallow closed-basin lake with surprisingly fresh, reasonably salt-free open waters. "It is possible that the lake remains relatively fresh due to significant removal of water and solutes by seepage into the groundwater beneath the lake," writes Isiorho. This seepage along with a series of droughts and the introduction of large-scale irrigation have complicated the region's water problems. Isiorho studies the volume of water "lost" from the lake to help bring about better water resources management and planning.

In the field, Isiorho uses seepage meters and core tracer experiments, slug tests, borehole dilution tests, electrical resistivity surveys, and lakebed analysis. Laboratory experiments evaluate the "aquifer properties of porosity, diffusion coefficient, and diversity" as well as analyze sediment with scanning electron miscroscope (SEM) and x-ray diffraction techniques.

Seepage rates from the lake into the ground are confirmed at various sites and indicate the level of groundwater recharge. Research results show that the amount of water "lost" from the lake is as much as 30 percent of the total water input to the lake each year. This amount represents 15 times more than water used annually for irrigation in the southwest Chad Basin. Oil and gas exploration bring the threat of pollution, which adversely affects groundwater quality. As water demands are growing for the region's domestic, agricultural, and industrial needs, Dr. Isiorho urges coordinated geophysical, hydrological, and geological efforts to address these matters. He also recommends regional rules, regulations, and laws to plan for and manage these crucial waters cooperatively.

The National Geographic Society and Indiana/Purdue University also provide grants for these studies.

Dr. Isiorho teaches environmental geology, climatology, remote sensing, hydrogeology, and related earth sciences at IPFW; his research often utilizes remote sensing for groundwater and geologic exploration in developing countries.

Practical Farmers of Iowa (PFI) ⍦
Iowa State University
2104 Agronomy Hall
Ames, IA 50011 USA

Contact: **Rick Exner,** On-Farm Trials Coordinator
Phone: 1-515-294-1923 • *Fax:* 1-515-294-3163

Activities: Education; Research • *Issues:* Energy; Sustainable Agriculture; Waste Management/Recycling

● Organization
In search of "sustainable agriculture," a small group of farmers started **Practical Farmers of Iowa (PFI)** in 1985 to find new ways to benefit both themselves and the environment. With a vision that is holding and a growing membership of 350, PFI is determined to lower input costs, do less harm, and increase net returns. As founding member Sharon Thompson says, "Profitable and environmentally sound farming—pure and simple. It's got to sustain the land, the soil, the people, the communities, and the pocketbook. It all has to fit together."

▲ Projects and People
With help from Iowa's Agriculture Energy Management Program, Extension Service, and industry, PFI's **on-farm research demonstrations** are changing and improving farming practices throughout the upper Midwest. Generally, a customary practice is paired with an alternative method as on-farm trials are carried out and data is collected. For example, ridge-till farmers compared chemical weed control to nonchemical weed control in soybean and corn demonstrations and found they could maintain yields and save more than more than $11 an acre by eliminating herbicides and substituting cultivation. In other tests, corn growers found they could reduce nitrogen fertilizer without impairing yields, or control weeds by cross cultivation.

PFI farmers are also **testing practices,** such as planting winter cover crops, comparing weed counts between tillage systems, improving pasture land through intensive rotational grazing, and using biological controls. They also record manure and weed management, and rates of seeding, potash, and phosphate. One PFI farmer obtained his highest oat yields by growing crops in narrow strips; some claim good results from planting seeds in elevated rows, such as the ridge-till method, which also lowers machinery and labor costs compared to conventional practices.

Farmers find new uses for **recyclable products,** such as Tom Franzen, who grinds newspaper in a converted forage harvester for hog bedding. In encouraging diversification, a cooperative is experimenting with growing chickens without feed additives; other farmers are adding new enterprises such as production of hothouse tomatoes, herbs, and exotic grains. Some farmers avoid commercial pesticides in favor of natural control methods and others cut back on chemicals but do not eliminate them. A program of the Institute for Agricultural Medicine, is monitoring farmers for exposure to pesticides.

The Educational Foundation of America and the federal program "Agriculture in Concert with the Environment (ACE) fund a PFI education project in sustainable agriculture, primarily aimed at school children and 4-H Club members.

As farmers apply trial practices to the whole farm, they will further evaluate economic and environmental impacts. Agriculturists from throughout the USA and other countries express interest in the PFI program.

■ Resources

Members receive quarterly newsletters and on-farm trial information and summaries. The public is invited to view the demonstrations at farm field days.

Environmental Law
Lewis and Clark College
10015 SW Terwilliger Boulevard
Portland, OR 97219 USA

Contact: **Leslie Carlough**, Editor-in-Chief
    Phone: 1-503-768-6700 • *Fax:* 1-503-246-8542

Activities: Education; Research • *Issues:* Legal Research • *Meetings:* CITES

▲ Projects and People

Environmental Law is the "country's oldest journal dedicated solely to environmental issues," it asserts. Published quarterly since 1970, the legal journal includes articles ranging from analyses of recent cases to abstract discussions of pollution prevention theories and practical and helpful pieces that "lead the way in environmental thinking."

As part of its third issue, *Environmental Law* publishes an annual *Ninth Circuit Review* covering the nation's biggest federal circuit that hears a "large percentage of the environmental cases decided each year and therefore, is an important influence on environmental law," reports **Sarah Himmelhoch**. This might include decisions involving federal environmental statutes, Native American rights, common-law environmental protections, procedural aspects of citizen suits, and any other environmental topic.

The "letters to the editor" section, **Clear the Air**, gives "critics and skeptics a chance to respond to essays and stimulates answers and information from subscribers that can benefit researchers and practitioners.

Environmental Law is presently printed on recycled paper. The staff participates in Northwestern School of Law's recycling program so that paper for office use is continually recycled.

■ Resources

Essays and articles on environmental and legal issues are sought.

Center for Technology, Policy, and Industrial Development ⊕
Massachusetts Institute of Technology (MIT)
MIT Room E40-239
Cambridge, MA 02139 USA

Contact: **Nicholas A. Ashford, Ph.D., J.D.**, Associate Professor
    Phone: 1-617-253-1664 • *Fax:* 1-617-253-7140

Activities: Development; Political/Legislative; Research • *Issues:* Air Quality/Emission Control; Energy; Health; Sustainable Development; Waste Management/Recycling; Water Quality

▲ Projects and People

In a three-year multidisciplinary study, **Dr. Nicholas A. Ashford**, associate professor of **Massachusetts Institute of Technology's (MIT) Center for Technology, Policy, and Industrial Development**, heads a research team which investigates the monitoring of people facing health risks as a result of toxic substance exposure in certain communities.

According to the MIT Center, "In addition to the technical and political issues involved in either environmental or human monitoring, complex ethical questions arise which, if unresolved, may lead to distrust or even conflict among communities, scientists, and agency professionals.

"Monitoring is ethically and legally problematic, in part, because the data collected may be used for a variety of purposes, some of which conflict." The Center continues, "For example, results of monitoring may be used: to contribute to problem solving and decisionmaking; to promote health by identifying environmental threats; to encourage compliance with existing laws; or to provide the basis for compensation."

The study is supported jointly by the Centers for Disease Control (CDC), Agency for Toxic Substance and Disease Registry (ATSDR) that is financed by the U.S. Environmental Protection Agency (EPA) Superfund, and the National Institute for Occupational Safety and Health (NIOSH).

CDC's role is to coordinate with health care providers to give medical care and testing of exposed individuals, to collect and analyze laboratory specimens, to perform periodic screening for diseases, to determine "the extent of danger to the public health from a release or threatened release of a hazardous substance," to maintain registries of those exposed, and to "evaluate associations between human exposure to environmental toxicants and adverse health effects."

Human monitoring begins, reports the MIT Center, "when a potential hazard to human health is officially acknowledged . . . continues through planning, conduct, and evaluation of testing and communication of the results, and ends ideally with decisions and actions taken to reduce the risk or degree of harm to those exposed."

Three case studies are being used "to illustrate the fundamentally different sets of circumstances that may result in a decision to monitor," points out Dr. Ashford. The famed **Love Canal disaster** is one example where leaking barrels of toxic waste were identified as the hazard before systematic monitoring of health effects began. In a different situation, examinations of **leukemia victims and their families** preceded a search in a town called Woford for chemical contamination. In Michigan, a flame retardant containing poly-brominated biphenyls (PBBs) was mistakenly combined with a fumigant distributed statewide for use on grain which was directly fed to dairy farm animals—creating serious deformities in the cattle. The chemical was conveyed to humans through cows' milk, also causing abnormalities. Monitoring in this case began after the substance was isolated in affected cattle, notes the Center.

The investigatory team's objective is to "establish some guidelines for communities and agencies to follow when a health hazard is suspected." One observation is that "problems encountered during human monitoring are more difficult to resolve if there is antago-nism due to earlier confrontations over monitoring of the community's air, water, or soil."

Professor Ashford is also involved in a one-year research project with the EPA's Office of Chemical Emergency Preparedness and Prevention to **investigate long-term strategies for encouraging tech-nological change to prevent chemical accidents and spills**—par-ticularly those that are released suddenly rather than gradually, involve "significant quantities of acutely toxic chemicals," and occur at a site of chemical manufacture or use. The study will focus on "prerelease accident prevention rather than postrelease accident miti-gation" and stimulate fundamental technological changes. Impetus for this research is the deadly release in 1984 of methyl isocyanate from a chemical plant in Bhopal, India, followed by a lethal petro-leum/plastics plant explosion in Pasadena, Texas.

Among his honors, Dr. Ashford received the Macedo Award (established by the Pan American Health Organization) from the American Association for World Health on behalf of the New Jersey Department of Health for "promoting greater awareness of environ-mental health issues through its report on chemical sensitivity." *The Washington Post* cited the report, co-written by Ashford and **Dr. Claudia S. Miller**, University of Texas Health Science Center, as the "most comprehensive study" of its kind, and so did "public interest crusader" **Ralph Nader** who summoned clinical ecologists, aller-gists, and health departments to work together on "this fast-growing affliction."

The report linked **chemical sensitivity to certain synthetic or-ganic chemicals** in consumer products, furnishings, and construc-tion materials as well as restricted air circulation in buildings erected since 1970. Indoor pollutants can also emit from carpeting, office equipment, perfumes, and insecticides, according to the study, and immune, nervous, and hormonal systems can be affected at varying levels. Particularly vulnerable are industrial workers exposed to chemicals on the job and people who live near "toxic waste sites, sources of diesel fumes, or pesticide spraying." Among recommen-dations made to the state of New Jersey were "setting standards for dealing with low-level chemical exposure," educating primary care physicians on how to identify chemical sensitivity, and preparing guidelines for and a registry of chemically sensitive persons.

Dr. Ashford is also adjunct faculty, Harvard School of Public Health; adjunct associate professor, Boston University School of Public Health; and EPA consultant on toxic substances. At MIT, he teaches courses on regulating toxic substances, monitoring expo-sure, public policy, and tort law and insurance in preventing and compensating damage from such exposure.

Institute for Environmental Toxicology ⊕
Michigan State University (MSU)
10C Natural Resources Building
East Lansing, MI 48824 USA

Contact: **John Paul Giesy, Ph.D.**, Professor
Phone: 1-517-353-2000 • *Fax:* 1-517-336-1699
E-Mail: Bitnet: 16990gny@msu

Activities: Education; Research • *Issues:* Air Quality/Emission Control; Biodiversity/Species Preservation; Energy; Global Warming; Health; Sustainable Development; Waste Man-agement/Recycling; Water Quality • *Meetings:* American Association for the Advancement of Science; American Chemical Society; American Fisheries Society; American So-ciety for Limnology and Oceanography; American Society of Zoologists; CITES; ESA; International Association for Great Lakes Research; SETAC

● Organization

A heavily industrialized state covered with forests and surrounded by the world's largest bodies of fresh water, Michigan is a leader in environmental legislation and efforts—enacting the nation's first Environmental Protection Act in 1969.

As a land-grant institution, **Michigan State University (MSU)** is at the forefront in pursuing environmental research, education, and outreach endeavors with the coordination of the **Institute for Envi-ronmental Toxicology**. National programs here focus on hazardous substance management, food toxicology and safety, microbial ecol-ogy, plant biology, and analytical chemistry. The Cooperative Ex-tension Service helps disseminate information to the public.

As MSU persists in its search for excellence and answers to problems of contamination and hazardous chemicals, it is aggres-sively recruiting both faculty and graduate students. Some 100 faculty members in about 30 departments of 7 colleges are affiliated with the Institute.

MSU support for environmental studies also comes from the following units: **Pesticide Research Center** concerned with impacts of contaminants, biotechnology, pesticide toxicology, and integrated pest management; **Institute of Water Research**, which studies the effects of natural and manmade chemicals on surface and ground-water systems; **U.S. Environmental Protection Agency Hazardous Substance Research Center** which is investigating ways to remediate soil and groundwater contamination by organic chemicals; **Na-tional Science Foundation Center for Microbial Ecology**, which is using microorganisms for similar purposes; **Carcinogenesis Labora-tory**, which researches molecular mechanisms of cancer; three **Medi-cal Schools** focused on protecting health from contaminants; **Na-tional Institutes of Health Regional Mass Spectrometry Facility** with its state-of-the-art analytical instrumentation; and the new **National Food Safety and Toxicology Center**.

▲ Projects and People

Among its activities, the Institute conducts **workshops for media reporters** dealing with environmental issues such as risk assessment, water quality, and pesticides. A **Community Assistance Program in Environmental Toxicology** helps small communities solve their own environmental contamination problems. It coordinates an ongoing U.S. Department of Agriculture (USDA) grant program finding new ways to protect the food chain from environmental contaminants. The Institute participates in a Hazardous Materials Management Consortium with the University of Michigan, which researches the movement of chemicals in the environment, their potential health effects, and methods of degrading pollutants. National and global seminars and conferences for scientists and students are held periodically.

Research emphasis is in the following fields:

Biochemical Toxicology, including studies on the receptor responsible for dioxin toxicity, lung cell changes characteristic of pyrrolizidine alkaloid toxicity, biochemical modifications in cells undergoing carcinogenic changes, role of calcium ions in heavy metal neurotoxicity, mode of action of pesticides in insects, and mechanisms of chemically induced diabetes.

Analytical Toxicology, including development of improved instrumentation such as mass spectrometers; more rapid and sensitive immunological assay techniques, particularly for mycotoxins; and techniques that assess contaminants in the environment and in foods.

Aquatic and Wildlife Toxicology, including studies on sediments and surface waters and investigations of chemicals from simple inorganic metals to complex chlorinated organics, such as PCBs. Proximity to the Great Lakes provides "abundant natural laboratories" to study effects of contaminants on wildlife and aquatic organisms.

Degradation of Environmental Contaminants, including studies of sediment microorganisms that can degrade complex chlorinated organics, modified clays that can perform similar steps, natural soil processes that can degrade agricultural chemicals, and physical processes that can remove volatile organics in the subsurface environment.

Epidemiology, which is newly investigating occupational factors responsible for cancer in workers; epidemiology of environmental disease in agricultural and rural settings; and breast cancer etiology, epidemiology, and control.

Food Toxicology, including studies of the effects of PCBs, PBBs, and other contaminants in food on human health; food additives; migration of chemicals from packaging materials into foods; persistence of certain contaminants in food products; and reducing food contaminants through improved preparation and cooking methods.

Regulatory Toxicology, examining the interactions between risk management, risk assessment, and risk communication—particularly relating to pesticides in various environmental agents, such as food.

At the Pesticide Research Center, William Wesley Bowerman IV is researching bald eagles and how biocontaminants affect reproduction. Working with Earthwatch volunteers, Bowerman is viewing the eagle as an environmental indicator and is using satellite telemetry to study seasonal movements in relation to food habits and basic ecology. Methods include banding nestlings and adults, attaching radio and satellite transmitters, and collecting blood, feathers, and eggs for contaminant analysis.

President of the Society of Environmental Toxicology and Chemistry (SETAC) which is affiliated with SETAC–Europe in Brussels, Dr. Giesy has conducted numerous research projects concerning bald eagles as well as effects of fish toxicity on river otters and birds, PCBs and trace metals in the Great Lakes, swine fed soil contaminated with lead sulfide, sediment toxicity, and others. Funding sources include the EPA, U.S. Fish and Wildlife Service, USDA,

Michigan Department of Natural Resources, and General Electric Co.

■ Resources

Brochures are available about the Institute and the multidisciplinary graduate program in environmental toxicology. Expressed needs are for new technologies and more educational outreach.

School of Forestry and Wood Products
Michigan Technological University
Houghton, MI 49931 USA

Contact: **Rolf O. Peterson, Ph.D.**, Professor
   Phone: 1-906-487-2179 • *Fax:* 1-906-487-2915

Activities: Research • *Issues:* Wildlife Ecology

▲ Projects and People

Dr. Rolf O. Peterson pursues research on various wildlife and their habitats in the northern USA, such as beaver, deer, moose, muskrat, fisher, and the endangered Isle Royale gray wolf. His studies explore interactions between the beaver, moose, and wolf and concentrate on areas such as moose osteoarthritis and calf mortality, the importance of natural mineral licks to moose, and management of large carnivores in natural ecosystems. A member of the Species Survival Commission (IUCN/SSC) Canid Group, Peterson was recognized for his efforts in moose management on this continent by the North American Moose Conference.

■ Resources

Dr. Peterson publishes the annual *Ecological Studies of Wolves on Isle Royale*, available from the Isle Royale Natural History Association, 87 North Ripley, Houghton, MI 49931.

Department of Fishery and Wildlife Sciences
New Mexico State University (NMSU)
Box 4901
Las Cruces, NM 88003 USA

Contact: **Sanford D. Schemnitz, Ph.D.**, Professor, Wildlife Sciences
   Phone: 1-505-646-1136 • *Fax:* 1-505-646-5975

Activities: Education • *Issues:* Biodiversity/Species Preservation; Wildlife Management • *Meetings:* North American Wildlife and Natural Resources Annual Conference

● Organization

Classrooms and laboratories extend to New Mexico's mesas and mountains, valleys and rivers in the Department of Fishery and Wildlife Sciences, NMSU, where students prepare for careers in conservation and management of renewable animal resources and their habitats—as competition intensifies for land and water among recreational, agricultural, industrial, and municipal interests.

Nongame and game wild animals and their habitats are studied in the field. Research includes behavior, population dynamics, terrestrial and aquatic ecology, experimental management, and influences of environmental disturbance. The Gila trout, once almost extinct, is one species that has been restored in the Gila National Forest—with help from this Department and cooperating environmental agencies. Dr. Schemnitz's recent grants were for research on Gould's turkey ecology, bighorn sheep/wild horse interactions, and reclamation treatments of coal-mine lands on wildlife.

Students learn about protecting rare or endangered species, managing animals for commercial and recreational uses, and controlling rodents, blackbirds, starlings, and other wildlife that cause economic losses to crops or other property. Career possibilities explored are in management of wildlife and their habitats; research in land-use changes, pollution, ecology, or population dynamics; education; public relations; law enforcement; and related fields such as recreation and land-use planning. The faculty urges potential students to gain a strong background in mathematics, applied and basic sciences, and communication.

The Department also conducts ecosystem short courses for non-wildlife professionals. Subjects covered include wildlife habitat management in the riparian and wetland ecosystem, and landscape ecology in the southwest USA.

The Acid Rain Foundation, Inc. ⊕
SCI-LINK
North Carolina State University (NCSU)
1410 Varsity Drive
Raleigh, NC 27606-2010 USA

Contact: **Harriett S. Stubbs, Ph.D.**, Executive Director
Phone: 1-919-828-9443 • *Fax:* 1-919-515-3593

Activities: Education • *Issues:* Air Quality/Emission Control; Global Warming

● Organization
The Acid Rain Foundation is a publicly supported, tax-exempt, non-advocacy organization that fosters a broader understanding of global atmospheric issues. Its goals are to develop and raise the level of public awareness, supply educational resources and materials to a wide range of audiences, support research, and help bring about resolution of air pollution issues.

The Foundation describes the pollutants contributing to air pollution and acid rain as primary—sulfur dioxide, nitrogen oxides, volatile organic compounds, and heavy metals; and secondary—photochemical oxidants and acid deposition.

▲ Projects and People
GLOBE-NET: Changes in the Global Environment is a curriculum project for grades 4–12 being developed with a National Science Foundation grant. Principal investigator Dr. Harriett Stubbs is designing new instructional materials that incorporate current scientific research on global environmental changes into ongoing biology, chemistry, physics, environmental sciences, and other studies.

With 10 years' experience in developing, publishing, and disseminating materials on air-quality issues, GLOBE-NET will "create, evaluate, and prepare materials for publication which are bias-free and scientifically and educationally sound," writes Dr. Stubbs. It will translate and transfer the most current information on global environmental changes to teachers, students, and parents so that citizens will be more informed about scientific issues of major international concern.

Walt Heck and Steve Businger are co-investigators; the materials will meet the standards of the National Diffusion Network; Carolina Academic Press will publish; and Carolina Biological Supply Company and others will disseminate. GLOBE-NET is a joint project of North Carolina State University's Colleges of Agriculture and Life Sciences, Physical and Mathematical Sciences, Forest Resources, and Education and Psychology in cooperation with the University of Minnesota's Colleges of Education and Ecology and Behavioral Biology.

So that teachers can bring the most current knowledge on air and water quality topics to the classroom, an innovative two-year project—SCI-LINK—is underway. Middle and junior/senior high school teachers in North Carolina and Minnesota are working together with leading scientists with environmental research backgrounds from two major universities and federal and state agencies to produce new curricula on acid rain, ozone, carbon dioxide warming, and global climate change. As content evolves, new teaching practices are expected to be developed. Dr. Ellis B. Cowling and Dr. Stubbs are principal investigators. NCSU, University of Minnesota, and the National Science Foundation are cooperating.

■ Resources
The Foundation is an information source for media, museums, groups, and individuals. Resources include visual aids, conferences, speaker's bureau, curricular materials for teachers and students from kindergarten through college, library, intern and volunteer programs, and databases.

Department of Zoology ⊕
North Carolina State University (NCSU)
Box 7617
Raleigh, NC 27695-7617 USA

Contact: **Roger A. Powell, Ph.D.**, Associate Professor
Phone: 1-919-515-2741
E-Mail: Bitnet: newf@ncsumvs

Activities: Education; Research • *Issues:* Air Quality/Emission Control; Biodiversity/Species Preservation; Deforestation; Energy; Global Warming; Health; Population Planning; Sustainable Development; Transportation; Water Management/Recycling; Water Quality

▲ Projects and People
A wildlife biologist, Dr. Roger A. Powell widely researches animals—many of which are vanishing species—and predator-prey relationships. He particularly is interested in mustelids and viverrids—such as the elusive fisher, endangered black-footed ferret, captive-raised Siberian ferret, American marten, and the weasel's hunting behavior. He also has conducted studies on pine and red-

backed voles, prairie dog, coyote, porcupine, Siberian polecat, and the American black bear.

For several years, Powell has been a principal investigator for Earthwatch concerning the "social organization, sexual dimorphism, and energetics" of the black bear. Working with students in his state's Southern Appalachian Mountains, he explores how "predictability (or lack of predictability) and irregularities in the distribution of food affect animal spacing." He writes, "Black bears have been live-trapped, uniquely tagged, and some have been outfitted with radio transmitters or miniature computers connected to radio transponder units." His research shows that "the difference in spacing is due to differences in food supply and it pays to defend a territory only when the productivity of food is limited and therefore food must be defended." Powell's studies also reveal that poaching and other human-caused mortality influence population levels of the black bear—an indicator species of the U.S. Forest Service—in a way that may elude knowledge of whether forest management changes are helpful or harmful to its survival.

In other studies, Dr. Powell has "developed mathematical modes of territorial behavior, mating systems, and dispersal in mammals." When applied to weasel family members trapped for fur, his models indicate that "depending on conditions [including age], trapping may improve or decrease a population's ability to respond to increases in food supply.

Powell belongs to the Species Survival Commission (IUCN/SSC) Mustelid and Viverrid Group and has done consultant work for the National Audubon Society, National Geographic Society Books, U.S. Forest Service, U.S. National Park Service, Wisconsin Department of Natural Resources, and Young Naturalist Foundation. As a National Aeronautics and Space Administration (NASA) workteam member, he helped develop an automated telemetry system for wildlife. At North Carolina State University, Powell is on the graduate faculty in ecology, wildlife, and zoology, and also teaches undergraduate courses.

Conserve UB Energy Program
State University of New York (SUNY) at Buffalo

120 John Beane Center
Amherst, NY 14260 USA

Contact: **Walter Simpson**, Director
 Phone: 1-716-645-3636 • *Fax:* 1-716-645-2034

Activities: Education; Energy Conservation • *Issues:* Energy; Global Warming; Waste Management/Recycling

▲ Projects and People

For 10 years, the State University of New York at Buffalo (UB) has committed itself to an energy conservation program that is now a model in the statewide SUNY system and for universities across the country.

Despite the nation's apathy about energy issues in the 1980s, **Walter Simpson, Conserve UB** director, mobilized two large campuses that are the academic home of 25,000 students with more than 100 buildings containing 8 million gross square feet of office space where annual energy costs are more than $15 million.

With administrative, student, and maintenance support, Simpson initiated more than 300 conservation measures and retrofit projects that save over $3 million a year as well as cut down on environmental damage. In tracking its effectiveness, Conserve UB says it is using 7 million *fewer* kilowatt hours of electricity than it did in 1983—even though 6 new buildings have expanded the campus by 20 percent. Through instituting low- or no-cost operational measures such as lower thermostat settings, reduced fan run-times, and vacation shutdowns, an additional $500,000 is saved. Energy-related carbon dioxide emissions are reduced by 50,000 tons each year.

One large-scale construction effort to renovate energy-inefficient fume hood exhaust systems in three laboratory buildings is expected to save up to $1 million a year in energy costs and reduce annual carbon dioxide emissions by another 18,000 tons.

In a *Recipe for an Effective Campus Energy Conservation Program,* a report written by Simpson for the Union of Concerned Scientists, he outlines 21 steps in undertaking a successful campus energy conservation program. These steps include securing top-level support; developing effective leadership and support network; adjusting to changed heating and cooling temperatures; establishing a high-profile energy-awareness campaign that seizes the moral high ground; being fair to faculty and students; nurturing a creative staff; employing low-tech solutions such as *delamping,* turning off equipment not in use, and installing light switches, time clocks, insulation, weatherstripping, and low-flow showerheads; expediting large energy and money savers, such as converting to cheaper and cleaner fuels; testing new products, such as compact fluorescent lights and reflective window film; obtaining expert technical services; securing sufficient funds; taking advantage of vendor competition; using utility rebates and demand-side management programs; piggybacking conservation on repair projects; considering both fast-payback and long-term payback projects; designing for conservation; documenting savings and accomplishments; and exploring new measures that cut waste and inefficiency.

Both Conserve UB and the Union of Concerned Scientists encourage similar programs at other schools with distribution of this report to 1,500 colleges and universities.

Together, they promote a "billion-pound, carbon-dioxide diet." Simpson also lectures at regional and national conferences and meets one-on-one with managers of facilities.

The community at large benefits. "By conserving electricity, natural gas and coal, we are significantly reducing carbon dioxide and acid rain emissions as well as conserving energy resources for future energy," writes Simpson, who teaches environmental courses and reaches young children through public television programs. Taxpayers save too. The UB Environmental Task Force, on which Simpson serves, seeks additional ways to keep the university from impacting negatively on the environment.

"We need to work with both the political leadership and grassroots, urges Simpson, who also co-founded (with wife Nan) the Animal Rights Advocates of Western New York, which espouses vegetarianism as another significant conservation measure. "Individual and group efforts *do* make a difference. We need to reach beyond the converted and bring the environmental message to new people. We need to set an example with dedication, hard work, and love of life."

■ Resources

Publications include *Recipe for an Effective Campus Energy Conservation Program; Greenhouse Blues: Energy Choices and Global Warming,*

The Unresolved Crisis: Why Energy Conservation Remains a Priority; and *Easy on the Spaceship: Energy Saving Tips for a Sustainable Future.*

Conserve UB needs "innovative financing for conservation projects" and seeks greater political support, public involvement and new technologies.

International Center for Arid and Semiarid Land Studies (ICASALS) and the Association for Arid Land Studies (AALS) ⊕
Texas Tech University
Box 41036
Lubbock, TX 74909-1036 USA

Contact: **Idris Rhea Traylor, Jr.,** Director
Phone: 1-806-742-2218 • *Fax:* 1-806-742-1954

Activities: Development; Education; Research • *Issues:* Deforestation; Desertification; Energy; Sustainable Development; Water Quality • *Meetings:* AALS

● Organization
Idris Rhea Traylor directs both the International Center for Arid and Semiarid Land Studies (ICASALS) and the Association for Arid Land Studies (AALS).

ICASALS is dedicated to the study of arid and semiarid environments and the relationship of people to those environments. In these ecologically sensitive water-short areas covering about one-third of the Earth's land surface live an estimated 700,000 million people—or 14 percent of the world's population. Such regions where human life began continue to challenge the ability to survive, especially as populations grow and demand even more from the Earth's marginal lands.

Created in 1966, ICASALS combines teaching, research, and public service activities at **Texas Tech University** about all aspects of the world's arid and semiarid lands, their people, and problems. It is suitably situated at Lubbock on the high plains adjacent to the desert regions of the USA and Mexico and within a state which is more than half arid and semiarid. More than 150 faculty are designated "ICASALS Associates"; international research and developmental programs are ongoing; visitors from 100 countries have sought research data and technical assistance on the campus.

ICASALS has formal affiliations with universities and research centers in Turkey, Egypt, Colombia, and mainland China, among other countries. It has conducted technical assistance projects for the U.S. Agency for International Development (USAID); through Texas Tech as an institutional representative to the Consortium for International Development (CID), it submits proposals to CID for overseas consultancies, development contracts, and long-term assistance projects. The Center also offers a Women in Development (WID) Committee, workshops, conferences, and symposia. Consultancies have been provided for Egypt, Yemen, Saudi Arabia, Kenya, Senegal, Zaire, Niger, Chad, Botswana, and Mexico—among others.

AALS was formed in 1977 and now represents most of the USA and more than 20 other nations. Headquartered at ICASALS, it is also affiliated with the Western Social Science Association and the Southwestern and Rocky Mountain Divisions of the American Association for the Advancement of Science. Its purposes include: encouraging an increased awareness of the problems and potentials of arid and semiarid lands and human adjustment to and impact upon them; stimulating research and teaching; promoting contact with others having similar specializations; and maintaining communication through periodicals, abstracts, and annual meetings.

■ Resources
An order form is available for publications on such subjects as urbanization and plant resources in arid lands, weather modification research, and water-recharge projects.

Environmental Citizenship Program ⊕
The Lincoln Filene Center
Tufts University
Medford, MA 02155 USA

Contact: **Nancy W. Anderson,** Director
Phone: 1-617-627-3451 • *Fax:* 1-617-627-3401

Activities: Development; Education; Political/Legislative; Research • *Issues:* Air Quality/Emission Control; Biodiversity/Species Preservation; Deforestation; Energy; Global Warming; Health; Population Planning; Sustainable Development; Transportation; Waste Management/Recycling; Water Quality • *Meetings:* Eastern Africa Environmental Network; ECO World; Global Tomorrow Coalition Forum; IUCN; Second Global Structures Convocation; UNCED

● Organization
The Lincoln Filene Center is a partnership between **Tufts University** and the Civic Education Foundation started in 1945 to promote American democratic values. It draws upon faculty resources from throughout Tufts University for public policy research and training. It also incorporates graduate students from Tuft's Department of Urban and Environmental Policy in field projects and internships.

▲ Projects and People
Preparing people to shape environmental policies is a main function of the Center's Environmental Citizenship Program. It sponsors a Growth Management Forum, Environmental Leadership Training Institute, Globescope Forum, and other environmental courses. Citizens, corporate officials, and public leaders come together to debate environmental issues and forge improved policies.

At the community level, the New England Environmental Network invites nonprofits, businesses, government agencies, and private citizens to participate in the annual New England Environmental Conference to address local and global challenges. Featured are speakers from the most prominent environmental groups and 50 workshops on pressing topics including safe drinking water, birds as sensitive environmental indicators, the Arctic oil/wildlife dilemma, Antarctica's protection, neglect of oceans, and elected leaders as environmentalists. Recognized as a "model of regional collaboration," the Network is inspiring similar groups in the USA and east Africa.

The *New England Environmental Network News* carries legislative updates, practical tips for a greener Earth, messages from key spokespersons on vital issues, and reports on studies. In an editorial, Director **Nancy Anderson** urged readers to write for a World Conservation Union–IUCN Briefing Statement on "Environmental Consequences of the (Persian) Gulf War." Said Anderson, "IUCN already has good baseline information on the marine environment of the Gulf, having conducted comprehensive and detailed coastal studies on behalf of Saudi Arabia's Meteorological and Environmental Protection Agency [MEPA]. . . . The World Conservation Union is therefore well prepared . . . to plan for the restoration and future protection of an environment which is of international economic and ecological importance."

Worldwide environmental information on the **Global Action Network (GAN)** computerized communication system, housed within EcoNet, is available to individuals and groups. Issues range from developing a safe national energy strategy to controlling exports of dangerous chemicals, from supporting funding for international population assistance to sustainable use of tropical and temperate forests.

■ Resources

Available are New England Environmental Conference registration forms and a publications list/order form, including subscriptions to the *New England Environmental Network Newsletter*. The Global Action Network and EcoNet hotline is 1-800-669-4246 or 1-617-381-3423.

Tulane Environmental Law Clinic
Tulane Law School
7039 Freret Street
New Orleans, LA 70018 USA

Contact: **Robert Kuehn**, Director
Phone: 1-504-865-5789 • *Fax:* 1-504-862-8721

Activities: Education • *Issues:* Air Quality/Emission Control; Waste Management/Recycling; Water Quality

● Organization

The **Tulane Environmental Law Clinic** is designed to train students to be "effective environmental lawyers with sensitivity to the natural environment and the needs of the public." Through its student attorneys, the Clinic provides free legal assistance to individuals and more than 30 community organizations seeking to protect and restore the environment for the benefit of the public—when those individuals and organizations are not otherwise able to obtain representation from the private bar. The Clinic pursues cases and issues that present an educational opportunity for the students as well as an opportunity to establish important precedents for the protection of the environment.

▲ Projects and People

In its first three years, the Clinic has been dealing with issues such as hazardous and solid waste landfills, toxic wastewater and toxic air emissions, hazardous and medical waste incinerators, tax breaks for polluters, regulations of pesticide use, scenic river protection, and development in wetlands and coastal areas. For example, the Clinic prevented a landfill from being situated next to a national wildlife refuge and near residential neighborhoods in New Orleans. A permit was blocked—preventing some 2,200 acres of lowland wetlands from being filled for farming.

When Renew America honored the environmental law training program, it particularly praised efforts to force the state to tighten its regulation and tax treatment of hazardous waste incinerators. "This is a crucial program for Louisiana," reported this clearinghouse for environmental solutions, "a state with serious pollution problems, badly underfunded state agencies, and no legal representatives of national environmental groups."

■ Resources

The Clinic publishes a *Citizen's Guide to Environmental Activism in Louisiana*.

Department of Biology
Universidad Interamericana de Puerto Rico
Recinto de San Germán
Call Box 5100
San Germán, Puerto Rico 00683 USA

Contact: **Héctor E. Quintero, Ph.D.**, Associate
          Professor
Phone: 1-809-264-1912, ext. 297

Activities: Education; Research • *Issues:* Biodiversity/Species Preservation; Water Quality • *Meetings:* International Conference on Tropical Rainforests

▲ Projects and People

Within the Interamerican University of Puerto Rico's **Department of Biology**, Dr. Héctor E. Quintero heads the **Manatee Project** to save this endangered species from humans who foster conditions that threaten their survival. Quintero, who researches the behavior and density of the West Indian manatee found mostly on the eastern and southern coasts, says there are only 300 to 400 of these large, gray-brown, aquatic mammals known as sea cows remaining in Puerto Rico and that each year many more die than are born.

The Department of Biology describes *Trichechus manatus* with "a seal-like body that tapers to a flat, paddle-shaped tail. Their two small forelimbs on the upper body are shaped as flippers and have three or four nails on each flipper. The head and face are wrinkled and the snout has stiff whiskers . . . [and] body hair is almost absent. Adults normally weigh 1,000 pounds and measure nine feet, but they can reach 13 feet and weigh 3,000 pounds. At birth, manatees are approximately four feet long and weigh between 60 to 70 pounds."

Because manatees are slow and gentle creatures that feed on sea grasses such as *thalassia* in calm, shallow coastal waters, they are mostly found on the water surface where they breathe every four minutes or less. They have poor eyesight and tend to communicate by sound particularly when playing or breeding, or if they are scared. Having a low reproduction rate, they generally bear one calf in intervals between three to five years and the gestation period is approximately 13 months—one of the longest among mammals.

"Mothers nurse their young for a long period and a calf may remain dependent for up to two years," reports the Biology Department. "Manatees do not become sexually mature until five to nine years of age."

Even though they can live more than 40 years, tragically, their life habits make them vulnerable to boat collisions—the major cause of human-related deaths. "Small boats with outboard engines often cause wounds, scarring and even death. Large power vessels and barges moving through shallow water can also run over manatees, crushing them against the bottom," according to the Department. Other causes of death include fishing nets, in which manatees can get tangled and drown; debris, such as discarded lines, hook, plastics, and other objects; pollution, harassment, and vandalism.

The Manatee Project educates the public on preserving these tame and harmless animals. Persons are advised to operate boats at low speeds where manatees are sighted, not to frighten them when diving or wearing scuba gear, to avoid discarding monofilament lines and hooks in the ocean where they can tangle in the sea grasses, and to educate others about protecting this species from extinction.

Dr. Quintero is a member of the Species Survival Commission (IUCN/SSC) Sirenia Group.

■ Resources

The brochure, *West Indian Manatee in Puerto Rico*, and a "Save the Manatee" poster are free to the public.

Biology Department ⊕
University of Akron
Akron, OH 44325-3908 USA

Contact: **Warren P. Stoutamire, Ph.D.,** Professor
           Emeritus
Phone: 1-216-972-5864

Activities: Education; Research • *Issues:* Biodiversity/Species Preservation

▲ Projects and People

Orchid researcher since 1972, Dr. Warren Stoutamire today studies the biology of rare and endangered plants, including terrestrial orchids. As university greenhouse and herbarium manager, Dr. Stoutamire has four interest areas:

Sexual and vegetative reproduction of Celastraceae (*Paxistima canbyi*), an endangered plant in Ohio that consists of two populations that are "probably single clones," writes Dr. Stoutamire. "The self-sterile plants are incapable of producing fruits, and I have worked in conjunction with the Ohio Natural Heritage Program investigating the biology of the species and its management.

Biology of Orchidaceae (*Cypripedium candidum*) in an Ohio prairie. Dr. Stoutamire is investigating the development of the underground organs to better understand the species' biology and preserve its threatened habitat. He is also researching its seed production and seedling biology.

Seed germination and seedling development of Orchidaceae (*Cypripedium reginae*). Also in cooperation with the Ohio Natural Heritage Program, Dr. Stoutamire is producing laboratory-reared seedlings in flask and reintroducing these to a natural habitat. "We are monitoring seedling survival and development. The goals are to

determine the best management practices for encouraging growth and survival of this orchid," he reports.

Seed viability studies of species of terrestrial orchids from North America, Australia, and South Africa. "Some species of seasonally dry habitats appear to have long survival as dormant seeds and this has implications for the disappearance and reappearance of orchid populations over long intervals of time."

Dr. Stoutamire is a member of the Species Survival Commission (IUCN/SSC) Orchid Group, Australasian Native Orchid Society, Royal Horticultural Society, and American Association of Plant Taxonomists, among others.

Geophysical Institute ⊕
University of Alaska
Fairbanks, AK 99775-0800 USA

Contact: **Daniel Jaffe, Ph.D.,** Assistant Professor of
           Chemistry
Phone: 1-907-474-7910 • *Fax:* 1-907-474-7290
E-Mail: Bitnet: ffdaj@alaska

Activities: Education; Research • *Issues:* Air Quality/Emission Control; Energy; Global Warming

▲ Projects and People

At the University of Alaska's Geophysical Institute, which was established by the U.S. Congress to maintain geophysical research regarding the Arctic regions, Dr. Daniel Jaffe researches global air pollution. His principal areas of interest are "Arctic air pollution, especially nitrogen oxides and tropospheric (lower atmospheric) ozone and acid precipitation."

Dr. Jaffe also studies "air pollution impacts of several large industries in the Arctic, most notably the smelters in the [former] Soviet Union and oil and gas industry." With a group of five graduate students and postdoctoral researchers, he builds chemical instrumentation and conducts field measurements "in some of the most remote areas of the planet."

He recently worked with scientists from the former Soviet Union to study "some of the largest Soviet pollution sources located on the Kola Peninsula in the Soviet Arctic."

Dr. Jaffe, who teaches chemistry and environmental chemistry in the Department of Chemistry, is principal investigator on projects with both the National Science Foundation and Earthwatch. A recent Earthwatch endeavor investigated acid snows on the Kola Peninsula.

Previously, Jaffe participated in a National Center for Atmospheric Research colloquium. He recently wrote *The Global Nitrogen Cycle in Biogeochemical Cycles*, among other publications; and has presented conference papers on the influence of Prudhoe Bay's industrial area on the Barrow air chemistry record and on "All the Wrong Places: How Mankind's Activities Create and Destroy Ozone."

Alaska Cooperative Wildlife Research Unit ⊕
Department of Biology and Wildlife
University of Alaska—Fairbanks
211 Irving Building
Fairbanks, AK 99775 USA

Contact: **Merav Ben-David,** Teaching Assistant
   Phone: 1-907-474-7671 • *Fax:* 1-907-474-6716
   E-Mail: etmb1@alaska

Activities: Research • *Issues:* Biodiversity/Species Preservation

▲ Projects and People

Zookeeper, marbled polecat researcher, hedgehog tracker, and veterinary assistant for dairy cows in Israel; safari guide in Kenya; Ph.D.-candidate researcher on river otters and mink in southeast Alaska—Merav Ben-David proposed a project to search for the Colombian weasel in the cloud forests of Colombia and Ecuador at the time of publication.

As a member of the Species Survival Commission (IUCN/SSC) Mustelid and Viverrid Group, Ben-David is involved in a conservation action plan that "aims to be different. "In addition to demonstrating what we would lose in terms of beauty, cultural, scientific, and also economic values if mustelids and viverrids disappeared, we emphasize what can be done to reverse the present negative trend," this plan states.

The group wants to involve all interested persons in their **mission to save Mustelids and Viverrids**—two extraordinarily diverse families of mammals such as weasels, civets, and mongooses that include species living on the ground, in the water, in trees, and in burrows. Some are carnivores, some are earthworm feeders; others eat fruit or insects. Weasels feed on rats and mice; mongooses favor venomous snakes. Viverrids mostly live in the Old World tropics; they were the only carnivores to reach Madagascar; their meat is a protein source in African villages. Civets secrete musk, used in the French perfume industry. Mustelids appear everywhere but Australia and Antarctica and include valuable fur-bearers as ermine, mink, sable, and marten. Conservationists believe "carefully controlled culling of wild fur-bearers can be a viable alternative to other forms of land use." Unfortunately, some mongoose and weasels earned a bad reputation by threatening other species when introduced into island ecosystems.

Probably the least known carnivores, their evolution and behavior are intriguing to scientists. Some communicate with scent, some use "tools" to "open hard-shelled food such as eggs by skillfully throwing them against stones," and some can withstand high doses of rattlesnake poison—of potential interest to medical researchers. As some species continue to be discovered, such as the giant striped mongoose from southwest Madagascar, others are disappearing—thus the urgency to investigate them in their natural habitats.

Ben-David chose to search for the Colombian weasel (*Mustela felipei*) in the cloud forests of Colombia and Ecuador because it is "the rarest, most recently described (1978) and the least known of all new world carnivores . . . and was never seen alive by scientists," she proposed.

Surmising that the animal is semiaquatic and possibly threatened by pollution, her goal is to pursue it in several habitats in the Ecuadorian Andes and Colombia—collaborating with local authorities, scientists, and conservation groups. The survey is based on phototrapping, live trapping, and mark and recapture methods with the involvement of local students to "enhance the development of a long-term study," her proposal continued. "It is hoped that these local students will later be involved with conservation issues of the cloud forests and their endemic species."

Ben-David also belongs to the Society for Protection of Nature in Israel.

■ Resources

Funding is needed for such studies. *Mustelid & Viverrid Conservation Newsletter* subscriptions are available from H. Van Rompaey, Jan Verbertlei 15, 2520 Edegem, Belgium.

California Insect Survey
University of California
201 Wellman Hall
Berkeley, CA 94720 USA

Contact: **Jerry A. Powell,** Professor of Entomology
   Phone: 1-510-642-3207

Activities: Education; Research • *Issues:* Biodiversity/Species Preservation

---

Department of Ecology and Evolutionary
  Biology ⊕
University of California
Irvine, CA 92717 USA

Contact: **F. Lynn Carpenter, Ph.D.,** Professor
   Phone: 1-714-856-6006

Activities: Education; Research • *Issues:* Biodiversity/Species Preservation; Deforestation; Ecology • *Meetings:* International Ornithological Congress

▲ Projects and People

University of California's Dr. F.L. Carpenter is the principal investigator of a recent Earthwatch project regarding **floral neighborhood, pollination, and herbivore damage.**

Dr. Carpenter chooses research areas where she observes "how interactions at the individual level affect ecological phenomena at the population and community levels" and specifically focuses on "plant foraging behavior and competition." She narrows this down further to "systems in which plants supply nectar to their avian pollinators, which in turn may be territorial, develop specialized foraging behaviors, and migrate at least in part due to competition."

With long-term studies of **Rufous hummingbirds** that migrate in the summer south through the California Sierra Nevada, Dr. Carpenter reports that these birds "use most of their fat store flying from their Pacific Northwest breeding grounds on their way to Mexico to winter, and must stop in the Sierras to refuel." With her colleagues, she has gathered data in **bird energetics,** including bird weight which is monitored on artificial perch-balances and the size of a single meal during a foraging bout.

Because nectar is a food resource with relatively easy-to-determine production and availability, this factor enhances the studies of bird behavior and pollinator effectiveness as it relates to food limitation. According to Dr. Carpenter, these food limitations include territoriality, meal size, intra- and interspecific interactions between the nectarivores, body masses at the beginning and end of an individual's migratory stopover, rate of mass gain, stopover duration, pollen grains deposited on plant stigmas, and number of seeds produced.

The Earthwatch project is a study of hummingbird-pollinated epiphyte in the Costa Rican rainforests. Epiphytes are plants growing above the ground, deriving nutrients from water, air, dust—such as air plants. Dr. Carpenter and a research volunteer team are discovering the "biotic factors that determine seed output in this native species by experimentally manipulating plant density and species composition of the floral neighborhood surrounding a central epiphyte." The effects of local herbivores "as a function of plant neighborhood" are measured as well, and so is seed output to assess the "ultimate effect of the interplay between pollination and herbivory in determining reproductive output of the plant."

By examining these individual interactions, Dr. Carpenter believes researchers will gain knowledge that will translate to "important phenomena" at population and community levels.

Departmental and Cell Biology ⊕
University of California
Irvine, CA 92717 USA

Contact: **Joseph Arditti, Ph.D.,** Professor
Phone: 1-714-856-5221 • *Fax:* 1-714-856-4709
E-Mail: jarditti@uci

Activities: Education; Research • *Issues:* Biodiversity/Species Preservation

▲ Projects and People
Dr. Joseph Arditti has conducted orchid and taro studies worldwide from the Arctic Ocean to Latin America to Australia and Malaysia. Taro (*Colocasia esculenta*) is an edible, stemless plant mainly cultivated in tropical regions. One current effort is to develop salt-tolerant taro in brackish soils.

Concerned with these plants' physiology and development, Dr. Arditti, **Departmental and Cell Biology,** University of California, describes some of his research as follows: "tissue and cell culture of orchids and aroids; post-pollination phenomena in orchid flowers; biological effects of surfactants; uptake and metabolism of sugars by orchid seedlings; resupination; and effects of simulated acid rain on tropical epiphytes."

Completed projects involve "phytoalexin production and effects; chemotaxonomy using mostly anthocyanins and sugars; sugar content of floral and extrafloral exudates; nutritional requirements of orchid mycorrhizal fungi; culture media for orchids which do not require sterilization; tissue culture of *Parthenium*; and niacin biosynthesis in plants."

One of his published articles describes orchids as contraceptives; others tell how orchids use carbon dioxide, whether tomato juice makes a good media for seedling cultures, and how clonal propagation takes place *in vitro*. Dr. Arditti has also written about orchids

"in mystery, adventure and science-fiction novels as well as in "music, parables, quotes, [and] secrets." His scientific reports have been translated into Chinese, Japanese, German, Hebrew, Italian, Portuguese, and Spanish.

Dr. Arditti's work has received support from the American Orchid Society, U.S. Department of Interior, U.S. Department of Agriculture (USDA), U.S. Environmental Protection Agency (EPA), and National Science Foundation (NSF), among others. He is active in the Botanical Society of America, various orchid societies, and the Species Survival Commission (IUCN/SSC) Orchid Group. On occasion, he consults with a Tahitian group on vanilla.

Marine Science Institute
University of California
c/o 3372 Martin Road
Carmel, CA 93923 USA

Contact: **John A. Laurmann, Ph.D.,** Research
        Scientist
Phone: 1-408-625-6696

Activities: Research • *Issues:* Global Warming

▲ Projects and People
Dr. John A. Laurmann is both a research scientist with the Marine Science Institute, University of California, Santa Barbara, and consulting professor with Stanford University, Department of Civil Engineering.

In commenting on public misperceptions of the greenhouse gas/climate change environmental issues, Dr. Laurmann observes that "misconceptions concerning the state of scientific understanding are a direct responsibility of the scientific community to correct"; otherwise the credibility of science is endangered, he believes.

"The large uncertainty that is indigenous to the greenhouse gas issue, and its long term, globally heterogeneous nature, means that we cannot today expect universality in attitude towards its seriousness, and hence towards methods for dealing with it," he writes. For instance, the world community does not presently accept lowering energy use to reduce greenhouse gas emissions, and so a transformation of attitudes of large segments within the world community would have to accompany this effort—if reduced energy consumption is the goal.

He writes further, "Because of the large uncertainties that abound in the greenhouse question, biased and one-sided depictions of the state of scientific understanding and of the policy options that flow from them are easy to pass undetected. This means that opposite conclusions can be drawn, depending on which extremes of the ranges of uncertainty are chosen. Even actions based on the median 'best guess' estimates have a good chance of being found inappropriate as information and understanding are updated and revised. Too narrow a view will inevitably lead to erroneous ranking of research and remediative technology priorities, as well as policy choices; the probabilistic nature of the decisionmaking process should not be ignored."

A "scientifically based" prescription on greenhouse gas policy needs to contain "valuations" concerning "uncertainty, future discounting, and risk taking." He also reminds that costs for reaching global-warming reduction goals will be borne differently by groups dissimilarly impacted by the prospective climate changes, which also leads to conflicting perspectives on the problem.

Former executive scientist with the Gas Research Institute, Dr. Laurmann has been a consultant for the National Commission on Air Quality, National Oceanic and Atmospheric Administration (NOAA), U.S. Environmental Protection Agency, Oak Ridge National Laboratory, National Center for Atmospheric Research, and the United Nations Conference on Human Settlements, among others. His published works include books on wing theory and rarefied gas dynamics and papers on carbon dioxide emission reduction, energy policy, global warming, and countering climatic change.

UCI Arboretum and Gene Bank ⊕
Ecology and Evolutionary Biology
University of California
Irvine, CA 92717 USA

Contact: **Harold Koopowitz, Ph.D.**, Director
Phone: 1-714-856-5833 • *Fax:* 1-714-856-8511

Activities: Education; Research • *Issues:* Biodiversity/Species Preservation

● Organization

"If you have two pennies, use one to buy bread and the other to buy flowers," says the quote from an unknown source in **UCI Arboretum and Gene Bank**'s literature. The Arboretum has an urgent mission: to quell the rate of plant extinction that has already reached one species per day and threatens to climb to one per hour during the next decade. A global pioneer in this struggle to rescue valuable plants, its scientists are beginning to sense a change in public attitude as a result of greater awareness of the greenhouse effect and what plant losses can mean to ecosystems, medicines, and mankind.

Founded in 1965, the Arboretum and Gene Bank got underway when **Dr. Harold Koopowitz** was named director in 1976. It was the first facility in the Western Hemisphere to house a gene bank devoted to wild plant species. Here seeds are carefully packaged and stored at subfreezing temperatures to keep them alive for hundreds of years so that they may be thawed and planted in case such species vanish. Pollen is also frozen. There are more than 7,500 entries from about 800 plant species, mostly from Africa, in cryogenic storage. About 200 of these species are considered endangered, including 76 species grown only at UCI. Research with a living plant collection, mostly from southern Africa and Chile, is geared toward conservation.

The Arboretum explains that plants living today are the result of five billion years of experimenting by nature. As "phytochemists," plants and their genes produce countless natural chemicals with applications from food production to pharmaceuticals, petroleum to horticulture. Conservation is crucial, especially as biotechnologists learn to transfer genes from one organism to another.

▲ Projects and People

Scientists are working on **models for predicting rates of biodiversity collapse** in relation to various forms of land conversion, indicates Dr. Koopowitz.

Plans are under way to expand the Arboretum's staff and facility in the next decade. A new building to house the expanded gene bank will include "rooms where seed and pollen can be cleaned, processed and stored; a laboratory to test seed viability prior to storage; walk-in freezers (-40°C) to store the seed and pollen; a computer-aided research library; a propagation facility to utilize modern tissue culture techniques, and administrative office space."

Orchids and bulbous and cormous plants, such as gladiolus, will be added to the existing African collection. With two-thirds of the world's 15,000 tropical orchid species considered endangered, the gene bank will become vital to their survival. The Arboretum reports, "Cryogenic preservation is considered by many to be the best hedge against extinction." Millions of orchid seeds will be stored within a few cubic centimeters at much less space and cost than maintaining living plants which are susceptible to disease and other nursery hazards. Also, seeds can be preserved under the same conditions; different species require different growing conditions.

In the future, the Arboretum hopes to create special gardens and lawns with geographic or plant family themes. A Cymbidium living plant collection is also planned.

Dr. Koopowitz is assistant director of the International Bulb Society. A member of the Species Survival Commission (IUCN/SSC) Orchid Group, he is on the conservation and committees of both the International Orchid Commission and the American Orchid Society. His book *Plant Extinction: A Global Crisis* is in its third printing.

Department of Environmental, Population, and Organismic (EPO) Biology ⊕
University of Colorado
Campus Box B-334
Boulder, CO 80309-0334 USA

Contact: **Charles H. Southwick, Ph.D.**, Professor of Biology
Phone: 1-303-492-5468

Activities: Education; Research • *Issues:* Air Quality/Emission Control; Biodiversity/Species Preservation; Deforestation; Energy; Global Warming; Health; Sustainable Development; Water Quality • *Meetings:* International Congress of Primatology; UNCED

▲ Projects and People

Within five years, **Dr. Charles H. Southwick, Department of EPO Biology, University of Colorado**, has undertaken primate studies for the National Institutes of Health (NIH), National Geographic Society, National Science Foundation (NSF), and World Wildlife Fund (WWF), mostly involving macaques in Asia. According to Dr. Southwick, the rhesus monkey has the broadest geographical and ecological distribution of any nonhuman primate—ranging from Afghanistan to China and including parts of India, Thailand, and Vietnam and inhabiting deserts to tropical forests to The Himalayas.

Having also served on faculties at Hamilton College, Ohio University, and Johns Hopkins University, and formerly holding postdoctoral fellowships at Oxford University, Stanford, and India's Aligarh University, Dr. Southwick's primary research areas are animal population ecology and primatology. Since 1951, he has been a leader or member of more than 75 research expeditions throughout Central and South America, Asia, and Africa—including 32 years of field study in India.

He and his colleagues focus on "commensal animals which live in close associations with human populations" involving "problems ranging from the conservation of endangered species to the control of vertebrate pest problems." His research programs also extend into broader areas of agriculture, public health, human ecology, and sustainable development.

Working with biologists at China's Institute for Endangered Animals, Guangzhou, and Institute of Geography, Academy of Sciences, Beijing, Dr. Southwick observes that the highly endangered Hainan gibbon (*Hylobates concolor hainanus*), with a known population of 21 individuals in Bawanglin Nature Reserve, and the rhesus macaque (*Macaca mulatta brachyurus*) at the Nanwan Nature Reserve are responding well to "dedicated conservation efforts by Chinese scientists and officials." He reports that the gibbon population declined from 2,000 animals 30 years ago to less than 50 in 1978 with only 7 or 8 at Bawanglin a decade ago. When the Nanwan Reserve was founded in 1965, there were only about 100 macaques; now these have an "average birth rate of 78 percent and an annual population growth rate of 12.7 percent," reports Dr. Southwick in the *American Journal of Primatology*.

Dr. Southwick describes comparative ecology studies of rhesus monkeys from China's tropical monsoon forests to temperate woodlands and snow-covered mountains, illustrating the "remarkable" adaptability of the species. It is noted that macaques in tropical China have "higher population densities, smaller home ranges, higher birth rates, and higher rates of population growth than populations in temperate regions."

In Kathmandu, Nepal, rhesus macaques—studied here between 1974 and 1985—live close to humans as "fostered by religious and philosophical beliefs," he notes. For example, as many as 300 monkeys in 5 to 7 social groups live at 2,000-year-old Buddhist and Hindu temple site, Swayambhu, where populations remained stable throughout this study. Southwick points out benefits to the monkeys—"supplemental feeding, protection from trapping, freedom from predation, and a complex and varied habitat" as well as detriments—"restricted home ranges, crowding, competition, harassment from people, and the potentials of disease and parasite exchange from people, livestock, and feral animals." The Kathmandu rhesus showed high levels of social interaction and aggression.

In 1988, Dr. Southwick, with Dr. M. Farooq Siddiqi, Aligarh Muslim University, published results in the *American Journal of Primatology* of a field study in India begun in 1959 which showed a rhesus population decline of more than 90 percent "from an estimated two million animals in 1960 to approximately 180,000 by 1980," due mainly to increased agricultural pressures, loss of habitat, less monkey protection, and high levels of trapping. As conservation efforts are undertaken, populations are beginning to rise; the estimated rhesus population in 1985 was about 410,000–460,000 individuals.

Similarly in northern Sulawesi, Indonesia, studies showed that significant population declines among Celebes crested or black macaque (*Macaca nigra*), Gorontalo macaque (*M. nigrescens*), and Heck's macaque (*M. hecki*) result from "habitat shrinkage, increasing human population pressure, and drought conditions." Dr. Southwick also discovered "a shortage of juveniles and infants" and that "group sizes were smaller" than in previous studies.

Dr. Southwick's most recent study for NSF regarded development of computer-operated remote tracking system (CORTS) for monitoring radio-tagged animals. He has also been a consultant to the Pan American Health Organization, National Academy of Sciences, Rockefeller Foundation, and Ford Foundation, among others. Other affiliations include American Institute of Biological Sciences (AIBS), Wisconsin Regional Primate Research Center, and Species Survival Commission (IUCN/SSC) Primate Group. He is the author or editor of seven books including two which deal with worldwide environmental issues: *Ecology and the Quality of Our Environment* and *Global Ecology*.

■ Resources

Greater funding is needed "at state and national levels for projects in international primatology and biodiversity research and conservation," writes Dr. Southwick.

Geography and Environmental Studies
University of Colorado
Box 7150
Colorado Springs, CO 80933 USA

Contact: **Eve Gruntfest, Ph.D.,** Associate Professor
Phone: 1-719-593-3513 • *Fax:* 1-719-593-3662
E-Mail: Bitnet: ecgruntfest@coldspgs.

Activities: Education; Political/Legislative; Research • *Issues:* Deforestation; Energy; Flash Flood Mitigation; Flood Plain Management; Global Warming; Hazard Warning Systems; Population Planning; Sustainable Development; Transportation; Waste Management/Recycling; Water Quality

▲ Projects and People

Dr. Eve Gruntfest has been investigating flash floods in the USA since the Big Thompson Canyon disaster in 1976 killed 140 people in Colorado. It was the second major flash flood in 4 years, following the deaths of 237 people in Rapid City, South Dakota. Occurring within minutes or hours of a downpour, levee failure or a sudden release of water escaping an ice jam—flash floods are marked by high-velocity flows, extremely short warning and response time, and extreme danger of loss of life. A harrowing combination of flash floods and slow-rise floods is also possible.

Gruntfest was principal investigator of several research projects, the most recent being a four-year "comprehensive assessment of flash flood hazard mitigation" funded by the National Science Foundation. The effectiveness of warning systems is a priority. Researchers recommend that warnings be assessed with coordinated audits, that false alarms be addressed, that response systems be made and marketed as competently as detection systems, and that standards be developed. Positive findings are that large cities will have the means to lower flash-flood losses; however, smaller communities are expected to have fewer resources.

One post-audit held 10 years after the Big Thompson Canyon flood brought together forecasters, hydrologists, sociologists, civil defense officials, water engineers, insurance brokers, lawyers, local residents, geographers, and geomorphologists. They concluded that a definition of "flash floods" was essential for effective warning and response, improving data collection, and resolving conflicts among the experts.

Researchers now recognize three types of flash floods: "those resulting from precipitation in basically natural drainages . . . [or] in catchments altered by humans or those resulting from the sudden release of water impounded by natural or human-made dams." But there are regional differences too. In the eastern USA, "flash foods often result from hurricane landfalls, providing several hours warning lead time," writes Dr. Gruntfest. In the west—with arid conditions, steep topography, and intense thunderstorms—"flash floods can offer less than one hour advance warning." Aging dams where breaks can occur are a threat. And in urban areas, other hazards are lack of vegetation, "greatly expanded impermeable surfaces" of buildings and paving, and bridges and culverts that constrict flows. Rainfall intensity greatly affects severity.

Regarding warning systems, the "Automated Local Evaluation in Real Time (ALERT) developed by the National Weather Service in the California-Nevada River Forecast Center and available from vendors" is the most common, observes Dr. Gruntfest. Most mentioned problems include frequent battery burnout and unexpected damage by weather, wildlife, and vandalism; remote location of gages in areas accessed mainly by helicopters or four-wheel-drive vehicles; and lack of manufacturing standards in designing, implementing, and pricing.

Communities need to be responsible for operating and responding to the systems, urge scientists who predict that "vulnerability is increasing." "We will never have warning systems in all hazardous areas. We need to educate the public about environmental cues . . . because in many cases they will be the only warning."

Dr. Gruntfest is co-chair, Commission on the Status of Women in Geography, Association of American Geographers; and consultant to *P3 The Earth-Based Magazine for Kids.*

World Data Center A for Glaciology/National Snow and Ice Data Center (WDC/NSIDC) ⊕
CIRES
University of Colorado
Box 449
Boulder, CO 80309 USA

Contact: **Roger G. Barry,** Ph.D., Director
Phone: 1-303-492-5171 • *Fax:* 1-303-492-2468

Activities: Research • *Issues:* Snow and Ice Data Management

● Organization
The National Snow and Ice Data Center (NSIDC), established by the National Oceanic and Atmospheric Administration (NOAA) in 1982, operates as a "national information and referral center for the snow and ice community." The NSIDC shares its task and operating space with the World Data Center A for Glaciology (WDC), one of three such international data centers; the other two are in Cambridge, England, and Moscow. The WDC performs the same function as the NSIDC, but on an international level. Both operate through CIRES, the **Cooperative Institute for Research in the Environmental Sciences** at the University of Colorado at Boulder, in association with the National Geophysical Data Center and the NOAA.

▲ Projects and People
The WDC/NSIDC exists for its resources—the data. Scientists from throughout the USA and the world use the center on working visits, and the data is gathered in cooperation with universities, government laboratories, and individual researchers in the USA and abroad. Topics covered include avalanche research, freshwater ice, glacier fluctuations, glacier mass balance, ground ice or permafrost, paleoglaciology, polar ice sheets, sea ice, and seasonal snow cover.

The archives include digital data on snow and ice research as well as published material on snow and ice. Among the most significant are the snow cover data sets, the sea ice data sets, charts of snow and ice limits for the Northern Hemisphere, and a collection of 80,000 aerial photos of Alaska, both historic and current.

■ Resources
As a service to those who wish to use the archives, the WDC/NSIDC publishes a *New Accessions List,* a quarterly listing of recent acquisitions. The *Glaciological Data* report series appears twice each year.

Natural Resources Law Center ⊕
University of Colorado School of Law
Campus Box 401
Boulder, CO 80309 USA

Contact: **Sarah Bates,** Assistant Director
Phone: 1-303-492-1286 • *Fax:* 1-303-492-1297

Activities: Education; Research • *Issues:* Deforestation; Energy; Global Warming; Land Use; Mining; Water Resources

● Organization
The Natural Resources Law Center is a program of the University of Colorado School of Law to improve public understanding of issues relating to environmental and natural resources law and policy through conferences, research, publications, and visitors.

▲ Projects and People
Founded in 1982, the Center has attracted more than 3,000 participants to various conferences on topics from the ethics of representing environmental law clients to western water law and management and for international groups such as environmental attorneys from Brazil and China.

Funded studies result in the publication of books and papers that are mostly about water use and policy. With a grant from the U.S. Environmental Protection Agency (EPA), the Center studied the legal protection for water supporting a wetlands area and prepared *Wetlands Protection and Water Rights.* The effects of depletion, degradation, physical alteration, pollution, and migration incident on water quality are described in *Controlling Water Use: The Unfinished Business of Water Quality Protection.* This report also recommends how to integrate water quality protection into water allocation decisions. Conserving irrigation water and allocating water during drought in California and Arizona are subjects of other studies.

Board of advisors includes attorney and former Arizona governor Bruce Babbitt; Dr. John W. Firor, National Center for Atmospheric Research; and Karin P. Sheldon, The Wilderness Society.

■ **Resources**

Associates enrollment and publications list/order forms are available.

Archie Carr Center for Sea Turtle Research ⊕
University of Florida
Zoology Department, Bartram Hall
Gainesville, FL 32611 USA

Contact: **Jeanne A. Mortimer, Ph.D.**, Adjunct
Assistant Professor
Phone: 1-904-373-4480 • *Fax:* 1-904-392-9166

Activities: Development; Education; Legislative; Political; Research • *Issues:* Biodiversity/Species Preservation; Sustainable Development

▲ Projects and People

Dr. Jeanne A. Mortimer recently was project leader for the **Save the Sea Turtle campaign**, a three-year endeavor of the **World Wide Fund for Nature (WWF)—Malaysia**. To implement a national strategy, the campaign raised funds for various projects to "provide professional advice and assistance to the Department of Fisheries and the state authorities in their efforts to conserve Malaysia's sea turtle resources." Specific problems were addressed in Peninsular Malaysia, Sabah, and Sarawak.

As a consultant with **WWF-International**, Dr. Mortimer conducted **education and training workshops** in marine turtle biology and conservation for Fisheries personnel and tour guides. She produced some 12,000 **educational charts** in both the Bahasa Malaysia and English languages which were distributed free of charge throughout the school systems.

Other marine turtle projects included: assessing **nesting populations and critical habitats**, identifying sites where new nesting sanctuaries and/or hatchery programs are needed, providing advice and assistance in managing nesting sanctuaries and **artificial hatcheries**, and coordinating of management-oriented research in the sanctuaries and hatcheries.

Dr. Mortimer worked with WWF-Malaysia staff **Dr. Mikaail Kavanagh**, executive director; **Ken Scriven**, former executive director; **Sarala Aikanathan**, scientific officer; **Lee Kup Jip**, education director; and **Encik Dzuhari bin Daud** and **Dionysius Sharma**, both scientific trainees.

Also a research fellow of the Caribbean Conservation Corporation and a member of the Species Survival Commission (IUCN/SSC) Marine Turtle Group, Dr. Mortimer specializes in the "conservation of marine and coastal ecosystems especially in the tropics," she writes. Other research has focused on hawksbill turtles of Seychelles, green turtles of the Caribbean, and Amazonian freshwater turtles.

■ **Resources**

A collection of 30 unpublished reports on *Marine Turtle Conservation—National Planning* (469 pages) is available from the WWF-Malaysia.

Program for Studies in Tropical Conservation (PSTC) ⊕
Center for Latin American Studies
University of Florida (UF)
319 Grinter Hall
Gainesville, FL 32611 USA

Contact: **Kent H. Redford, Ph.D.**, Director
Phone: 1-904-392-0083 • *Fax:* 1-904-392-0085

Activities: Education; Research • *Issues:* Biodiversity/Species Preservation; Deforestation; Population Planning; Sustainable Development

● Organization

The **Program for Studies in Tropical Conservation (PSTC)** is a companion to the **Tropical Conservation and Development Program (TCD)** and the **Amazon Research and Training Program (ARTP)**—all housed at the **Center for Latin American Studies, University of Florida (UF)**. This Center, the oldest continuously funded one of its kind in the USA, is internationally known for tropical research. TCD conducts interdisciplinary research and training to address problems of biological conservation and the livelihoods of Latin America's rural poor. ARTP offers graduate and undergraduate studies and seminars emphasizing sustainable development in collaborative programs with the Núcleo de Altos Estudos Amazonicos of the Federal University of Pará, the Museu Paraense Emílio Goeldi in Belém, and the Federal University of Acre in Rio Branco.

PSTC's goals are to "promote the rational management, sustainable use, and conservation of tropical wildlife." Representatives from developing tropical countries are provided scientific training so that they can make "critically important decisions affecting their natural resources." PSTC got underway 10 years ago with help from the Smithsonian Institution's National Zoological Park, the National Institute of Mental Health (NIMH), and UF. With the Federal University of Minas Gerais, Brazil, it developed South America's first advanced degree in wildlife management and conservation. PSTC presently is affiliated with 13 Latin American institutions in Argentina, Bolivia, Brazil, Colombia, Costa Rica, Mexico, Peru, and Venezuela.

▲ Projects and People

Coursework is individually tailored to the interests of students and their country's conservation needs. Field courses are held in conjunction with the Smithsonian Institution's National Zoological Park and the University of Florida's Department of Wildlife and Range Sciences.

Wildlife conservation and management fieldwork is carried out in the home countries, where it addresses problems such as social, economic, and legal influences resulting in deforestation of eastern

Brazilian Atlantic ecosystems; forest fragmentation and changes in forest structure on mammal population in the eastern Brazilian Atlantic forests chicle extraction in Mexico; Colombian Amazon colonists' attitudes toward wildlife use; surveying primate populations in the Paraguayan Chaco; and a land-use plan for a cloud-forest habitat in northern Peru that harbors the endemic yellow-tailed woolly monkey.

Several students including **Rodrigo A. Medellin, Ph.D.** candidate, are pursuing research in the **Selva Lacandona** tropical rainforest of Chiapas, Mexico—exploited between the early 1800s and late 1940s by European and U.S. companies for its then abundant mahogany and red cedar wood. It is an area with a rich cultural heritage and Mayan roots.

With settlements stimulated there as a result of conflicts and turbulence elsewhere, humans are exerting a "powerful pressure on the ecosystem," which contains about 25 percent of total Mexican biodiversity including 112 species of nonmarine mammals, over 300 species of birds, 800 species of diurnal butterflies, and about 800 species of vascular plants, according to Medellin. Rapid change and degradation include a fast-growing population, slash-and-burn agricultural activities, heavy pesticides use, health hazards, oil exploration and extraction, poaching of endangered species, growing of marijuana and other drugs, discrepancies in land ownership, and lack of organized production.

As director of the Chajul Tropical Biology Station, Medellin is encouraging continued conservation and research interests of the PSTC and other organizations. "Rising official concern in Mexico and around the world make the present time the best and last opportunity to save one of the greatest structural, cultural and biological complexities," writes Medellin.

In **Brazil**, PSTC student **Suzana M. Padua** describes her efforts to begin an **environmental education** (EE) outreach program to stem deforestation and save the endangered ecosystem of the Morro do Diabo (Devil's Hill) State Park.

Wildlife surveys were begun in the Rio Bravo Conservation and Management Area, Belize, with the direction of PSTC's **Dr. Susan Jacobson.** Cooperating on the surveys is the Programme for Belize, which plans to make the area self-supporting with income earned through tourism and sustainable extraction of forest resources, such as timber and chicle, according to PSTC student **Susan Walker.**

In **Argentina,** PSTC graduate students are encouraging **conservation research in a newly established biological station** in chacoan habitats rich with grasslands, palm savannahs, lagoons, varieties of wetland birds, and mammals such as howler and night monkeys, tapir, anteater, puma, river otter, pampas deer, maned wolf, jaguarundi, foxes, and white-lipped peccaries.

Dr. Kent H. Redford is co-authoring the book *Mammals of Southern South America*; other publications are on conservation and human resource use, insects and insectivory, and mammalian biology and zoogeography.

■ Resources

A *Tropical Conservation Development* brochure describes these programs. The *TCD Newsletter* is published in English, Spanish, and Portuguese.

Department of Wildlife and Range Sciences ⊕
University of Florida
118 Newins-Ziegler Hall
Gainesville, FL 32611 USA

Contact: **Thomas T. Struhsaker, Ph.D.,** Adjunct Professor
Phone: 1-904-392-1791 • *Fax:* 1-904-392-1707

Activities: Education; Research • *Issues:* Biodiversity/Species Preservation; Deforestation • *Meetings:* International Primatological Society

● Organization

The **Department of Wildlife and Range Sciences** graduate program with which **Dr. Thomas T. Struhsaker, University of Florida** is affiliated, concentrates on conservation biology and includes a Program for Studies in Tropical Conservation. Other emphases are on urban wildlife management (the first such program of its kind in the USA, according to university literature); forest/wildlife relations; wetlands ecology/restoration, applicable to Florida, where about half of the original wetlands have been lost or greatly altered; and range ecology.

▲ Projects and People

Studies of the endangered **woolly-spider monkeys** in coastal Brazil are among the research pursuits of Dr. Struhsaker. His worldwide scientific interests center on **tropical rainforest ecology and conservation,** the impact of **logging** on tropical forest regeneration and wildlife, and **mammalian behavior and ecology.**

For the World Conservation Union (IUCN), Struhsaker evaluated **forest conservation and management** issues in the Ngorongoro Conservation Area, **Tanzania.** Recently ending his six-year post as Carter Chair of Rain Forest Biology, Wildlife Conservation International (WCI), New York Zoological Society (NYZS)—Struhsaker conducts research in **Kenya** where he is a member of that country's Forest Working Group of East African Wildlife Society. Originating the **Kibale Forest Project** (KFP) in **West Uganda,** he was its director for 17 years—supervising doctorate, postdoctorate, and other students in this highly researched African rainforest, where KFP encourages tree plantings in the local community and field trips to the forest by local students and teachers.

Reproductive successes of certain African forest monkeys, the "impact of selective logging on effective population size in **red colobus and redtail monkeys**" in Kibale, and "prey selectivity by crowned hawk-eagles on monkeys" in Kibale are subjects of recent papers.

Struhsaker belongs to the Species Survival Commission (IUCN/SSC) Primate Group.

■ Resources

Funds are sought for conservation research and training overseas, as are policy changes in national and international lender/donor agencies, which Dr. Struhsaker did not specify.

Florida Museum of Natural History
University of Florida
(See World Conservation Union [IUCN], Species Survival Commission/Crocodile Specialist Group in the USA NGO section.)

Savannah River Ecology Laboratory (SREL)
University of Georgia
Drawer E
Aiken, SC 29802 USA

Contacts: **J. Whitfield Gibbons, Ph.D.**, Professor of
Zoology
Anton D. Tucker, Herpetology Lab
Coordinator
Phone: 1-803-725-2472 • *Fax:* 1-803-725-3309

Activities: Education; Research • *Issues:* Biodiversity/Species
Preservation; Energy

● Organization

The **University of Georgia** founded the **Savannah River Ecology Laboratory (SREL)** in 1951 to do ecological inventories at what was then an Atomic Energy Commission site. Here, scientists conducted "pre-installation" surveys of biological populations to assess the future impacts of nuclear production facilities; they used radioactive tracers to chart food chains and devoted much effort to radiation ecology. A permanent ecology laboratory, called the Laboratory of Radiation Ecology, was set up 10 years later—also enabling the studies of basic principles, such as competition in animal and plant communities. In 1964, it was renamed Savannah River Ecology Laboratory to emphasize the broad spectrum of ecological pursuits. SREL is now under contract with the U.S. Department of Energy (DOE), which oversees the Savannah River Site (SRS) operation.

SRS is a protected outdoor laboratory covering more than 77,000 hectares that became the USA's first National Environmental Research Park. With limited public access, SRS attracts research scientists who can easily study environmental impacts of energy technology. Permanent set-aside areas for ecological research are among the bottomland hardwood and upland oak-hickory forests, cypress-tupelo swamps, blackwater streams, Carolina bays, and man-influenced habitats such as pine plantations and thermally-altered lakes and streams. Both natural and perturbed ecosystems are investigated.

▲ Projects and People

Recognized worldwide, SREL research programs are in three areas: biogeochemical ecology, stress and wildlife ecology, and wetlands ecology.

Biogeochemical principles are investigated in terrestrial and aquatic ecosystems emphasizing "interactions among biological and chemical cycling processes and how they affect the transport, bioavailability, fate, and effects of potential contaminants." Research deals with cycling of radionuclides, metals, and organics in the environment; environmental chemistry and toxicology; and modeling of the transport and fate of environmental contaminants.

Stress and wildlife ecology research with deer, waterfowl, fish, reptiles, amphibians, and nongame small mammals and vertebrates emphasizes human-caused disturbances. Current research is on the effects of reactor effluents on aquatic organisms, effectiveness of offering alternative habitats to populations whose primary habitat was destroyed, and endangered species' ecology and response to perturbations.

Wetlands ecology research is concerned with factors affecting biological community development in natural and disturbed wetlands. Sample studies are on regeneration of the cypress/tupelo forest, population dynamics of zooplankton, organic matter processing in stream floodplains, and wetland creation or rehabilitation.

A collaborative project on diamondback terrapins is conducted with Earthwatch during the summer on Kiawah Island, South Carolina. Volunteers and staff work together to find out how to best protect the terrapin as well as properly manage and save the turtles' habitat—the salt marsh. Once overexploited as a "gourmet delicacy," today's terrapin faces major threats of salt marsh destruction from development and pollution along the Atlantic and Gulf of Mexico coasts. In exploring how turtles mate among the sand dunes, grow, and adjust their diets to the changing tides, researchers capture, weigh, measure, and tag them—returning them to the marsh once their eggs have hatched. Marked or coded terrapins are recognizable for 25 years or more.

One finding was that "terrapins are virtually unaffected by certain natural phenomena that can be devastating to humans," Whit Gibbons wrote in the *Tuscaloosa News*. "After Hurricane Hugo slammed into the Charleston coastline, many terrestrial habitats were destroyed. Yet the coded terrapins from previous years were found in the same tidal creeks they had lived in before the hurricane. They had weathered the storm and within days were going about their business as though nothing had ever happened."

Dr. Gibbons is the co-author of the book *Poisonous Plants and Venomous Animals of Alabama and Adjoining States* and *Guide to the Reptiles and Amphibians of the Savannah River Site*, among others. Anton Tucker, is co-investigator on numerous turtle projects and endangered species studies.

■ Resources

Herpetological reprints and information on Environmental Outreach Programs are available.

Hawaii Natural Energy Institute (HNEI) ⊕
University of Hawaii
2540 Dole Street
Honolulu, HI 96822 USA

Contact: **Patrick K. Takahashi, Ph.D.**, College of
Engineering
Phone: 1-808-956-8346 • *Fax:* 1-808-956-2336

Activities: Development; Education; Research • *Issues:* Energy; Global Warming; Sustainable Development

● Organization

From a decade of reliance on abundant oil to rekindled confidence in renewable energy and ocean resources—that is the direction the Hawaii Natural Energy Institute (HNEI) moves determinedly, even when public support lags behind. With backing from the U.S. Department of Energy (DOE), University of Hawaii and its School of Ocean and Earth Science and Technology, state officials, and the private sector, HNEI has pushed forward on working with hydrogen, geothermal, biomass, wind, solar, and ocean thermal energy conversion; HNEI is solidifying Hawaii's reputation as "the preeminent site in the world for renewable energy and ocean research." Director Patrick K. Takahashi is certain "the combination of marine resource and ocean thermal energy conversion technologies will present untold opportunities for economic progress in Hawaii and elsewhere."

▲ Projects and People

The Hawaii Integrated Biofuels Research Program assesses the fastest-growing, highest-yielding tree and grass crops for biomass conversion to liquid and gas transportation fuels, such as methane, methanol, or ethanol. It examines land suitability for biomass plantation sites, tests species such as "energy cane" hybrids and sterile banagrass, designs and evaluates harvest and post-harvest processing systems for cost-effectiveness, improves yields of glucose in certain thermochemical conversion processes, and sets up models to test the effects of feedstock composition on biomass gasification. Its analyses determined that the "pressurized, oxygen-blown, fluidized-bed biomass gasification system" works best for methanol production in Hawaii.

The Hawaii gasifier scale-up project is a cooperative venture between HNEI, the Pacific International Center for High Technology Research (PICHTR), Hawaiian Commercial and Sugar Company, Institute of Gas Technology, and Ralph M. Parsons Company to design, build, and test a 100-ton per day pressurized biomass gasification plant at the sugar company. As technical advisor, HNEI says this is an "ideal case whereby the University of Hawaii conducted the basic research and transferred the technology. PICHTR, as lead, is serving as the technology transfer bridge to commerce." DOE is funding half the amount for the $10 million project; the state of Hawaii is appropriating $4 million, and industry is providing the balance.

A significant research program is the production of hydrogen from renewable sources: pyrolytic gasification, photoelectrochemistry, photobiology where genetic engineering plays a role in the potential of cyanobacteria to produce this fuel using only sunlight and water; and novel storage materials to inhibit losses of hydrogen.

Renewable transportation alternatives are advanced with the testing of an electric-power car that can travel 40 to 50 miles on a single charge at speeds up to 60 mph, and battery systems that can offer longer life and range at lower cost. Scientists are experimenting with a two-kilowatt photovoltaic system to charge the vehicle as well as a "range extender" which adds electrical energy for trips extending beyond a full battery charge. HNEI researchers say they are seeking to "address and overcome the negative aspects of methanol: its corrosiveness, poor visibility flame, toxicity, and explosive nature." Mainland China and HNEI are cooperatively testing hydrogen as a fuel—one of several collaborative renewable energy projects. "An 11-kilowatt electrolyzer will split water into hydrogen and oxygen, with the hydrogen stored in a metal hydride tank. The

hydrogen fuel will be used in a spark ignition engine and a fuel cell to provide supplemental power to an electric vehicle to double its range."

Sugarcane research can help strengthen the economies in developing nations as ways are sought to diversify markets into electricity and ethanol production. The U.S. Agency for International Development (USAID), Tennessee Valley Authority (TVA), and the Hawaiian Sugar Planters' Association are helping.

Can algae remove all or part of the carbon dioxide from power plant stack gases? Scientists are experimenting with ways for the environment to benefit from the ability of algae to convert carbon dioxide into organic carbon.

Scientific Observation Holes (SOH) are being drilled to "confirm the existence of geothermal resources on the Big Island and Maui and stimulate resource development . . . for electricity, heating, refrigeration, foods and materials processing, animal husbandry, agriculture, aquaculture, and geothermal spas. A technique known as slim-hole drilling is being employed at lower cost and less environmental impact. This technique can also replace the expensive drilling of wells as electric utility companies test property before committing to buy options for the purchase of power.

HNEI is testing wind-powered sources of energy and water supply for Maui through computer modeling. Water heaters wrapped with a jacket can reduce energy consumption and astrofoil is almost as effective as fiberglass as an insulator: that is what HNEI with the Hawaii Department of Labor and Industrial Relations determined under the federal government's Weatherization Assistance Program. Hawaii says it is the only state in the USA doing this testing. A University of Hawaii library is the site of a space cooling project that seeks to reduce air conditioning energy consumption in commercial buildings. HNEI is pursuing cold fusion research with innovative elevated-temperature theories featuring a molten-salt electrochemical system which could help increase the efficiency of power generation with less expensive materials.

Scanning Tunneling Microscopy, currently the world's finest microscope system, is being used to gain information on the structural and electronic properties of materials—such as today's silicon and transition-metal compounds of the future—to improve solar energy devices. Several other projects focus on materials testing to enhance energy storage and discover other advantages.

The Marine Minerals Technology Center/Ocean Basins Division conducts seabed resource research projects under the umbrella of HNEI's Center for Ocean Resource Technology and as assigned by the U.S. Department of Interior. The following is a sampling:

The Sand Deposit Characterization and Environmental Assessment project is examining Oahu beach erosion and replenishment, and determining the potential impact of mining operations on the coastal environment.

Computer-Aided Design Methodology for Power, Communication, and Strength Umbilicals is concerned with finding a practical and economic means to connect surface vessels with mobile seabed mining platforms by way of umbilical cables.

High-resolution topographic models of the sea floor are being developed from 35mm photographs of cobalt-rich manganese crust deposits as a prerequisite step in designing suitable exploration equipment. The study is called Microtopography of Manganese Crusts.

Collecting large numbers of mineral specimens from specific sea floor sites in a minimum amount of time is being aided with the development of an "untethered, free-fall corer that uses percussion coring technology to recover precisely-located samples of manganese crusts and other deposits."

HNEI researchers are seeking funding for oceanic research to help stem global warming. The studies would evaluate "enhanced carbon uptake via nutrient subsidy to marine algae and subsequent deposition in marine sediments," and "enhanced dimethyl sulfide production via marine algae to increase cloud formation and albedo." Other proposals expected to be underwritten involve the "solution, dispersal, and fate of carbon dioxide delivered at depth with application to ocean thermal energy conversion processes" and the construction and evaluation of a high-resolution open ocean model of the carbon cycle that will show the flux and fate of greenhouse gases. Another model will measure carbon dioxide concentrations in Hawaiian waters. An inexpensive artificial reef that attracts marine life has been built, deployed, and successfully tested; the Atlantis Submarines company—a research co-sponsor with Sea Grant College—is installing these reefs off Waikiki and Kona.

The effects of waves are being tested on exposed submarine pipes in open trenches as an option to burying pipelines for carrying fresh or waste water, petroleum or natural gas—which is more costly and time-consuming. Fiber-optic sensors are being developed for use in the marine environment.

Dr. Takahashi, professor of the College of Engineering, is the principal investigator of a number of HNEI research projects. As an energy special assistant to the late U.S. Senator Spark Matsunaga, he helped introduce and gain passage of various environmental measures dealing with ocean energy, windpower, geothermal energy, biomass, fusion, fossil fuels, advanced batteries, hydrogen, and deep seabed mining.

■ Resources

Available publications and reports are listed in the Hawaii Natural Energy Institute Annual Report.

Department of Botany
University of Hawaii at Manoa
3190 Maile Way
Honolulu, HI 96822 USA

Contact: **Charles H. Lamoureux, Ph.D.**, Professor of Botany
Activities: Education; Research • *Issues:* Biodiversity/Species Preservation

▲ Projects and People

Dr. Charles H. Lamoureux is associate dean for academic affairs, College of Arts and Sciences, at the University of Hawaii at Manoa, where he is concurrently professor of botany.

This former president of both the Hawaiian Botanical Society and Hawaii Audubon Society has conducted botanical field studies worldwide and is especially interested in conservation of biological diversity—especially endangered species in Hawaii, Pacific islands, and Southeast Asia. Specific pursuits are pteridophytes, tropical forest ecology, and wood anatomy and seasonality of growth of tropical woody plants.

Hawaii Research Center for Future Studies (HRCFS) ⊕
Social Science Research Institute
University of Hawaii
2424 Maile Way, Porteus 720
Honolulu, HI 66822 USA

Contact: **James A. Dator, Ph.D.**
  Phone: 1-808-956-6601 • *Fax:* 1-808-956-2884
  E-Mail: dator@uhccvx.uhcc.hawaii.edu

Activities: Education; Research • *Issues:* Alternative Futures; Global Warming

● Organization

The Hawaii Research Center for Future Studies (HRCFS) is a member of the Policy and Planning Group on Global Climate Change (PPGGCC) housed at the Social Science Research Institute at the University of Hawaii.

▲ Projects and People

With funding from the University of Hawaii, the PPGGCC first undertook a research project to develop state and regional sea level rise/climate change proposals. Results assessed the scientific debate on sea-level rise and global climate change. A paper was prepared on the development of policy and planning for climate change and sea-level rise in the Pacific Islands.

A second endeavor was a regional pilot project on policy and planning for global-climate change in the Marshall Islands. With funding from the U.S. Environmental Protection Agency and the U.S. Department of Interior, this project involved workshops on the environment and climate change for government leaders and high school and college students.

A third project set up a workshop and brought together leading natural scientists in fields relating to sea-level rise and climate change. The project explored methodologies which might be used to identify probabilities that scientists can assign to different scenarios of climate change and sea-level rise.

Key researchers are Dr. James A. Dator, also president of the World Futures Studies Federation, and professor, Department of Political Science; research associate Dr. Christopher B. Jones, who also coordinated a project for the Pacific Basin Development Council (PBDC) to assess feasibility and implications of recycling municipal solid waste for USA Pacific Islands; and Dr. Michael P. Hammett, coordinator of both the Ocean Resources Management Program (PBDC) and the Center for Development Studies, Social Science Research Institute, where he develops training and policy research projects to meet the development needs of countries and territories in the Asia–Pacific region.

Department of Biological Sciences
University of Idaho
Moscow, ID 83843 USA

Contact: **Rodney A. Mead, Ph.D.**, Professor of
         Zoology
Phone: 1-208-885-6185 • *Fax:* 1-208-885-7905

Activities: Education; Research • *Issues:* Biodiversity/Species
Preservation • *Meetings:* Society for Study of Reproduction

● Organization

Within the Department of Biological Sciences, University of Idaho,
Dr. Rodney A. Mead heads a team of laboratory researchers who are
studying methods for the captive propagation of four species of
mustelid carnivores, including the ferret (*Mustela putorius*), steppe
polecat (*Mustela eversmanni*), wolverine (*Gulo gulo*), and the spotted skunk (*Spilogale putorius*). The ferret and steppe polecat projects
are performed for the U.S. Fish and Wildlife Service to develop
methods for the more rapid propagation of the highly endangered
black-footed ferret (*Mustela nigripes*).

▲ Projects and People

Researchers are experimenting with a hormonal regimen to induce
estrus in sexually inactive female ferrets so that they would be
prone to breeding and successful pregnancy.

Steppe polecats that inhabit the plains of the former USSR,
Romania, Hungary, Czechoslovakia, Yugoslavia, and eastern China
are in decline. They are closely related to black-footed ferrets, which
once ranged from Saskatchewan, Canada, to Mexico but now are
nearly extinct because of government programs to eradicate prairie
dogs on which they depended. According to Mead, the goal of the
U.S. Fish and Wildlife Service is to "reestablish a prebreeding
population of 1,500 free-ranging adult black-footed ferrets in 10 or
more locations by 2010. This is a large and expensive task when one
considers that litter size ranges from one to six and averages 3.4 kits
a year." Ideal surrogate species, 20 steppe polecats from Russia—age
3 months—were the subjects of experiments to manipulate the
photoperiod (amount of light in a 24-hour period) and induce 2
reproductive cycles within a year. "We demonstrated that it was
possible to obtain two litters during each breeding season by removing the young shortly after birth and fostering the kits to domestic
ferrets," writes Dr. Mead.

Reproduction studies of wolverines and spotted skunks are focal
points of these captive breeding programs.

Dr. Mead is a member of the Species Survival Commission
(IUCN/SSC) Mustelid and Viverrid Group.

■ Resources

Needed are additional funds to support research.

School of Biological Sciences
University of Kentucky
T.H. Morgan Building
Lexington, KY 40506-0225 USA

Contact: **Wayne H. Davis, Ph.D.**, Professor
Phone: 1-606-257-1828 • *Fax:* 1-606-257-5889

Activities: Education; Research • *Issues:* Biodiversity/Species
Preservation

▲ Projects and People

Current research of **Dr. Wayne H. Davis** concentrates on saving
the eastern bluebird (*Sialia sialis*) from decline. "Competition from
the imported house sparrow (*Passer domesticos*), which evicts bluebirds, kills young and adults, and takes over the nesting site, is a
major factor," he writes. "Based upon subtle differences in nest site
preferences of the two species, my research is an effort to develop a
nest box that bluebirds like and house sparrows won't take."

At the University of Kentucky, where his overall research interests concern the natural history of mammals and birds, Dr. Davis
reports he is "making a lot of progress" with his studies partially
funded by the North American Bluebird Society.

Dr. Davis has extensively studied the **bats of New England** with
Harold Hitchcock for the National Science Foundation and the
toxicity of pesticides to wild native mice with Mark Luckens for
the National Institutes of Health—among other projects. He has
also written about Kentucky's endangered, threatened, and rare
animals and plants, specifically about the red bat, short-eared owls,
ants and snakes in bird boxes, sassafras herb of the Appalachians,
stripmining, and pollution. With **Roger W. Barbour**, he co-authored
The Bats of the United States.

■ Resources

Dr. Davis builds sizable bat houses to order, and has various bat
research publications available. Resources needed include new technologies to build a better bluebird house and an expansion of
educational workshops on bluebird conservation.

Center for Global Change (CGC) ⊕
University of Maryland (UM)
7100 Baltimore Avenue, Suite 401
College Park, MD 20740 USA

Contact: **Alan Miller**, Executive Director
Phone: 1-301-403-4165 • *Fax:* 1-301-403-4292
E-Mail: Bitnet: amiller@umdacc.umd.edu

Activities: Development; Political/Legislative; Research • *Issues:* Air Quality/Emission Control; Deforestation; Energy;
Global Warming; Transportation

● Organization

The **Center for Global Change (CGC)** at the University of Maryland (UM), College Park, is an "interdisciplinary research group
that studies environmental quality issues and their relationship to

energy use and economic growth." In cooperation with faculty from throughout the University, the Center conducts research assessing the risks of environmental degradation and investigates technologies and policies to reduce it, according to **Alan Miller**, executive director. "The Center's initial priorities include programs dealing with ozone depletion, global climate change, and sea level rise."

According to CGC literature, "The Center disseminates its research to the policy, science, and business communities through publications, conferences, and other educational activities. The Center neither litigates nor lobbies."

The mission of the CGC is based on the belief that environmental problems are part of a larger cluster of related issues, including energy consumption, economic growth, resource allocation, and development; therefore, there is a need for a program to bring together diverse disciplines and perspectives.

▲ Projects and People

CGC staff is interdisciplinary, holding degrees in subjects including business, economics, energy and environmental studies, engineering, law, and public policy. The Center's projects are as diverse as its staff. They include the compilation of a **database of state and municipal policy proposals to address global climate change**, the compilation of a directory for environmental technologies and services in Maryland, and the identification and analysis of environmental implications of energy use for Vermont's comprehensive energy plan.

The Center **briefs corporations and teaches courses** on subjects such as management of the planet, international environmental law, environment and public policy, and global change. It collaborates with related UM research programs, such as the Laboratory for Coastal Research and the East-West Science and Technology Center.

In one report, **Alan Miller** and co-author **Curtis Moore** present perspectives on **Japan's relationship to the global environment**—concluding that although the country is strong in pollution reduction and energy efficiency technologies, these systems have not been readily applied at home. As public concern is mounting, however, the government has begun to "reconsider its position on many environmental issues." Recent actions include suspending the widespread use of driftnets until "regulatory measures are established" and showing some sensitivity to endangered species through banning the ivory trade and imports of green sea turtle and to global atmospheric pollution through stabilizing CO_2 emissions by the year 2000 "at the lowest possible level" and acceptance of the Montreal Protocol on Substances That Deplete the Ozone Layer.

A report on *Environmental Stress and National Security*, written by **Dr. Sara Hoagland** of American University and **Susan Combere** of CGC, examines how environmental problems affect the conduct of international relations and how increased global cooperation can "prevent crises on issues such as tropical deforestation, greenhouse gas emissions, ozone and water depletion, acid rain, and photochemical smog." The authors underscore today's opportunity "as the world order is changing to meet this challenge with new vigor."

■ Resources

A list of the Center's selected books, reports, and case studies is available.

Water Resources Research Center
University of Massachusetts
Blaisdell House
Amherst, MA 01003 USA

Contact: **Paul J. Godfrey, Ph.D.**, Director
Phone: 1-413-545-2842 • *Fax:* 1-413-545-2304

Activities: Research • *Issues:* Water Quality

● Organization

Water Resources Research Institutes at land-grant universities throughout the USA are operating under a 1984 federal law and administrated by the U.S. Geological Survey's Water Resources Division. In the late 1980s, 249 research projects were ongoing at 96 universities. One example of an innovation developed in Texas is a low-energy precision application (LEPA) irrigation system where water is applied at lower pressures and near the soil surface. Savings are expected to amount to over $1 billion from 1980–2020 in the Texas high plains. Water professionals receive science and engineer career training at the Institutes, which also reach out to the community with conferences, seminars, databases, libraries, and publications. The Institutes collaborate with each other as well as with other universities and colleges to confront regional water and land-related problems.

▲ Projects and People

The **Massachusetts Acid Rain Monitoring (ARM) Project** at the **Water Resources Research Center, University of Massachusetts,** is a cooperative venture between staff and "citizen volunteers" to assess the effects of acid deposition on local surface waters. Citizens collect seasonal water samples from lakes and streams statewide, which undergo analyses in more than 70 professional laboratories volunteering their services. Since 1983, a coordinated corps of some 1,000 citizens have sampled 85 percent of the state's surface waters. Collectors report to district coordinators in each county; county coordinators and the volunteer laboratories report to the Research Center, where samples are further analyzed. Volunteers receive feedback as rapidly as possible; committed local leaders and media support undergird the program's success.

Among other uses, data is submitted to the state legislature and the U.S. Congress to support pollution reduction bills. With funding from the Massachusetts Division of Fisheries and Wildlife, these groups also participate: Massachusetts Audubon Society, Trout Unlimited, Massachusetts Sportsmen's Council, Appalachian Mountain Club, Massachusetts Grange, and League of Women Voters. ARM has inspired similar volunteer efforts in Rhode Island, Vermont, New Jersey, and Maryland. An **Acid Rain Working Group** with environmental, industry, and state representatives and a five-year acid rain research program has also been created. ARM is also influencing a national program for emission control.

Recent research shows that 5.5 percent of Massachusetts surface waters are presently acidified and 63 percent are vulnerable—with lakes and ponds being "significantly more sensitive" than streams. Acidity is crucial because it can injure and kill fish as well as other vital aquatic life forms such as insects, clams, and microscopic organisms.

The **Massachusetts Water Watch Partnership (MassWWP)** is a statewide collaboration of scientists, educators, industry, and volunteer groups working to involve more citizens in collecting water quality information. Among its outreach programs are a school program "aimed at teaching young scientists the importance of healthy aquatic ecosystems" and an annual Massachusetts Citizen Monitoring Congress.

■ Resources

ARM provides a list of research and other available publications with how-to-order information.

Rosenstiel School of Marine and Atmospheric Science ⊕
University of Miami
4600 Rickenbacker Causeway
Miami, FL 33176 USA

Contact: **Samuel H. Gruber, Ph.D.**, Professor, Marine Biology and Fisheries
Phone: 1-305-361-4146 • *Fax:* 1-305-361-4600

Activities: Education; Research • *Issues:* Biodiversity/Species Preservation • *Meetings:* American Elasmobranch Society; American Society of Ichthyologists and Herpetologists

▲ Projects and People

As a pre-med student, Samuel H. "Sonny" Gruber turned to fisheries biology, it is said, after a near strike by a hammerhead shark in the ocean. Ever since, Dr. Gruber devotes his professional life and personal interests to the study and conservation of elasmobranch fishes, particularly large predatory sharks.

Beyond the university campus, Gruber takes the story of shark survival and behavior into the community to speak before school children and civic, water-sports, and environmental group audiences. Fascinated with sharks' evolutionary adaptations that "place these creatures in a 'favored position' in the tropical marine ecosystem" and sensitive to their nasty reputation that evokes fear and hate among people, Gruber is a recognized spokesman and is often interviewed by the news media.

Active globally—in 1990, Gruber was asked to form and chair a **Shark Specialist Group** within the **Species Survival Commission** (IUCN/SSC) of the World Conservation Union. Previously, he was a visiting scientist at Germany's Max-Planck Institut für Verhaltenphysiologie, Israel's Heinz Steinitz Marine Biological Laboratory, and Egypt's Al Ghardaqua Marine Station.

With **Earthwatch**, he conducts a series of studies with **lemon sharks in the Bimini Islands, Bahamas**. Here, team members capture, tag, and monitor the movements of lemon sharks in Bimini Lagoon, where electronic tracking devices are planted on its bottom. They also observe lemon shark interactions among each other and with other species. In **Bimini**, he also established a **biological field station** for research and teaching.

In his research, Gruber notes sharks have more in common with mammals than with bony fish. Sharks grow, mature, and reproduce slowly—having a few young at a time. But contrary to popular myth, sharks eat comparatively less than humans pound for pound. Young lemon sharks can burn 1,000 calories per hour, helping to fuel their constant motion.

Particularly interested in **shark vision**, Gruber was among the scientists who found that shark corneas can be transplanted to the human eye with success. With support from the National Science Foundation, he is producing a "**conceptual model of biological production** in the lemon shark as a basis for understanding energy flow at the top level of the tropical marine ecosystem." Fifty days are spent at sea each year in this study involving bioenergetics, population dynamics, ethology, and classical physiology. With a contract from the Office of Naval Research, he is developing an **effective chemical shark repellent** and documenting the behavioral and medical aspects of shark attacks. Gruber also is interested in researching **shark oil as a treatment for cancer tumors**.

His fight for the survival of sharks includes changing people's attitudes and convincing sport fishermen to tag and release their catch, and putting in place a sound management plan for sharks under the U.S. Magnuson Fishery Conservation and Management Act of 1976. Growing Asian markets for shark fins (used in soup stock) and American markets for shark flesh are depleting these species.

In his laboratories and at sea, Gruber also trains high school seniors in the **Navy Science Student Program**. He founded the American Elasmobranch Society and produced a conservation book for the American Littoral Society.

Center for Political Studies ⊕
Institute for Social Research (ISR)
University of Michigan
P.O. Box 1248
Ann Arbor, MI 48106-1248 USA

Contact: **Harold K. Jacobson, Ph.D.**, Director
Phone: 1-313-763-1348 • *Fax:* 1-313-662-4785
E-Mail: Bitnet: hkj@um.cc.umich.edu

Activities: Research • *Issues:* Biodiversity/Species Preservation; Deforestation; Energy; Global Warming; Population Planning; Sustainable Development

● Organization

As chair of both **Center for Political Studies, University of Michigan** and the Standing Committee on the Human Dimensions of Global Change, International Social Sciences Council (ISSC), **Dr. Harold K. Jacobson** emphasizes the need for further research on the "human dimensions of global environmental change."

With Dr. Martin Price, Canadian geographer, Dr. Jacobson prepared a framework for such research which recommends international and interdisciplinary programs to complement natural science research carried out by the International Council of Scientific Unions' Geosphere-Biosphere Programme on changes in Earth's physical, chemical, and biological systems. The ISSC encourages its member associations, which cover all social sciences, to establish research committees or working groups on global change.

▲ Projects and People

Research must consider "human activities both as they contribute to and as they are affected by global environmental change," writes Dr. Jacobson, who identifies three broad research topics to be investigated both separately and in combination: "the social dimensions of resource use; the perception and assessment of global environmental conditions and change; and the impacts of local, national, and international social, economic, and political structures and institutions on the global environment."

He proposes studies on those topics as well as land use, energy production and consumption, industrial growth, and environmental security and sustainable development.

The need is crucial for collaboration among social scientists and between social and natural scientists and so is understanding of worldwide interconnections among the phenomena, Dr. Jacobson points out. The world's population has doubled since 1950 while the demand for energy has quadrupled—yet economic growth has not been uniform. Moreover, greenhouse gas concentrations are rising, ozone depletion is evident, acid deposition is damaging aquatic ecosystems and causing forest dieback, deforestation and agricultural activities are contributing to climate change, biodiversity losses are resulting from changes in land use and rainforest ecosystems, and halocarbons including chlorofluorocarbons (CFCs) continue to be released.

"The challenges of understanding the human dimensions are daunting, but the topic demands that the social sciences take up and meet these challenges," urges Jacobson, a *Jesse S. Reeves* political science professor, who is affiliated with numerous organizations, including the American Academy of Arts and Sciences, American and International Political Science Associations, National Academy of Sciences, Commission to Study the Organization of Peace, United Nations Association of the USA, and the *American Journal of International Law*.

School of Natural Resources ⊕
University of Michigan
86 Museums Building
Ann Arbor, MI 48109 USA

Contact: **John B. Burch, Ph.D.,** Professor
Phone: 1-313-764-0470
E-Mail: jbburch

Activities: Education; Research • *Issues:* Biodiversity/Species Preservation; Health

▲ Projects and People

At the University of Michigan's School of Natural Resources, Dr. John B. Burch is both professor of zoology and of natural resources. Since 1963, he has served as curator of mollusks at the university's Museum of Zoology—a post he also held at the Australian Museum, where he remains affiliated. Burch is also associated with Mahidol University, Thailand, and Bishop Museum, Hawaii; he serves on the World Health Organization's Expert Advisory Panel on Parasitic Diseases such as schistosomiasis.

Some specific interests are freshwater snails of North America (particularly the Great Lakes), Jordan, Korea, Philippines; land snails and other terrestrial mollusks; "medically important mollusks" of Thailand; and bulinine snails from Lake Nassar, Egypt.

Dr. Burch is on the editorial boards of *American Malacological Bulletin* and *The Korean Journal of Malacology*; is editor of *Malacological Review*, among other publications; and holds leadership posi-

tions in the International Society for Medical and Applied Malacology, Society for Experimental and Descriptive Malacology, and Institute of Malacology.

Endangered Species UPDATE
School of Natural Resources
University of Michigan
Dana Building, 430 East University
Ann Arbor, MI 48109-1115 USA

Contact: **Alice Clarke**
   Joel Heinen, Co-Editors
Phone: 1-313-763-3243 • *Fax:* 1-313-936-2195

Activities: Publication • *Issues:* Biodiversity/Species Preservation

● Organization

The *Endangered Species UPDATE* is a monthly forum for information exchange on such issues for professionals in both scientific and policy fields—from corporations, zoos and botanical gardens to universities and nonprofit organizations. It includes a reprint of the latest issue of the U.S. Fish and Wildlife Service's *Endangered Species Technical Bulletin* along with complementary articles and information about species conservation efforts outside the federal program.

Feature articles include topics such as the Pacific Northwest spotted owl controversy, the effectiveness of international debt-for-nature swaps, and the impact of the savings and loan bailout on habitat protection for endangered species.

In turn, *UPDATE* welcomes articles related to species protection in areas including, but not limited to, research and management for specific endangered or threatened species, theoretical approaches to species conservation, and strategies for habitat protection and design of preserves. Book reviews, editorial comments, and announcements of current events and publications are also welcome.

UPDATE was first published in 1983, and copies of back issues are available.

■ Resources

Contact the editors for annual subscription rates and instructions to authors.

The GREEN Project ⊕
School of Natural Resources
University of Michigan
216 South State Street, Suite 4
Ann Arbor, MI 48104 USA

Contact: **William B. Stapp, Ph.D.,** Director
Phone: 1-313-761-8142 • *Fax:* 1-313-936-2195

Activities: Education; Research • *Issues:* Air Quality/Emission Control; Health; Water Quality • *Meetings:* North American Association for Environmental Education

● Organization

Uncertainty about the world's water quality prompted the creation of the **Global Rivers Environmental Education Network (GREEN)**—an international network which seeks to bring students, teachers, and communities closer together through the bond of learning about and improving our common river systems. It began with a water-quality monitoring project involving high school students in the Great Lakes region. Today it is composed of programs involving thousands of students on every continent. GREEN's goals are to acquaint students with the environmental problems and characteristics of their local watershed, giving them hands-on experience in theory and practice of chemical, biological, and sociological research; empower students through community problem-solving strategies, thereby enabling them to see the relevance of school subjects to the real world; and promote intercultural communication and understanding, thus fostering awareness of the global context of local environmental issues and of the significance of cultural differences in choosing effective solutions.

▲ Projects and People

It was health risks of wind surfing on a river near Ann Arbor that motivated a local high school to monitor water. With the help of the **University of Michigan's School of Natural Resources**, tests showed that the river should not be used for body contact after heavy rain. Questions about quality both upstream and downstream led to the monitoring program's expansion to three schools the following year. From this grew the **Rouge River project**, in which students from diverse socioeconomic classes exchanged information, using computers, about the deteriorating water quality of the river that connected them—learning about each other as well. Within a few years, similar school projects were set up in watersheds in all 50 states.

In 1989, **Dr. William B. Stapp** and his seminar of graduate students originated the concept of GREEN. Later that year, teams of student interns, educators, and consultants conducted seminars on water monitoring and networking in 18 nations. Since then, several countries have started school-based water monitoring programs, such as Taiwan with river monitoring programs, an environmental monitoring network in German schools, and river studies in Israel's new senior high school curriculum. More than 35 nations are participating in GREEN.

River studies can be incorporated into courses in the humanities, social studies, geography, and science to make students more environmentally aware and capable of developing action plans. Students study local rivers chemically, biologically and physically, then use the data to identify issues and determine how personal behavior affects water quality. The global connection expands educational opportunities as students are linked with others in different regions and cultures with a common mission to end river pollution and improve water quality.

Elements of GREEN include the Sister School/Watershed, where schools from different regions are matched so that students can jointly advance water quality while getting to know the geography and local cultures of others. The first participants were schools in Australia, Canada, Hungary, Mexico, New Zealand, Taiwan, and USA. **International Workshops and Congresses** bring together teachers and students to share monitoring data, develop action strategies, and design future cooperative projects. GREEN plans to establish **regional offices** worldwide as a "decentralized infrastructure" to identify each area's needs and resources, based upon student research. The GREEN **computer network**, based at the University of Michigan, is shared by students from Australia, Canada, Germany, Japan, and the USA. The EcoNet system for environmentalists enables both international and local watershed computer conferences to be held. The semiannual *GREEN Newsletter* updates participants about water quality projects and research findings.

■ Resources

Available publications include a *Field Manual for Water Quality Monitoring*, a handbook on *Establishing a Cross-Cultural Sister Watershed Program*, *Directory of GREEN Participants*, and a curriculum of 10 activities. Dr. Stapp will provide information about membership for schools, educators, and resource persons.

Needs include low-tech water monitoring equipment to facilitate understanding of water pollution health risks, and environmental educator consortia and networks to disseminate materials.

Center for International Food and Agricultural Policy ⊕
Department of Agricultural and Applied Economics
University of Minnesota
33 H Classroom–Office Building
1994 Buford Avenue
St. Paul, MN 55108 USA

Contact: **Harald von Witzke, Ph.D.**, Director
Phone: 1-612-625-1712 • *Fax:* 1-612-625-6245

Activities: Development; Education; Research • *Issues:* Agricultural Development; Air Quality/Emission Control; Deforestation; Sustainable Development; Water Quality

● Organization

The Center for International Food and Agricultural Policy provides leadership in research to "improve our knowledge of the international aspects of food, agriculture, natural and human resources, and the environment." It encompasses four major program areas: commodity and trade policy, research policy, development assistance and policy, and natural resource and environmental policy.

▲ Projects and People

In addition to its graduate-level course on research planning and priority setting, the Center conducts **training programs on agricultural research** for U.S. and international researchers. Program members work with national and global organizations including the Consultative Group on International Agricultural Research (CGIAR), United Nations Food and Agriculture Organization (FAO), U.S. Agency for International Development (USAID), International Service for National Agricultural Research (ISNAR), and U.S. Department of Agriculture (USDA).

The **natural resource and environmental policy** program area involves research concerning irrigation and water management, farming systems and biological nitrogen fixation, soil erosion, watershed management, deforestation and the decline of upland areas in development countries, environmental effects of export-oriented crop production, forestry for sustainable development, and management of international common-property resources.

Chemical contamination of groundwater, groundwater depletion, accelerated soil erosion, or compaction are relevant to the Center, particularly when they result from international trade or development policies. Projects in Thailand and India have dealt with small-scale irrigation and upland watershed management. "Work in the USA has evaluated the Conservation Reserve Program and the effects of new laws regulating use of water from the Great Lakes," reports the Center.

Recent work includes a global research strategy for tropical forestry, impacts of bans on pesticides, common-property resources, watershed planning and management, natural resource management policy relating to recreation use and resource degradation, and off-site economic benefits of erosion control.

A collaborative agreement with the University of Padova, Italy, is focusing on the links between agricultural and trade policies and environmental quality. According to the Center, Minnesota's "primary focus is on pollution of subsurface aquifers, while Padova's major problem is contamination of streams flowing into the lagoons and estuary surrounding Venice." With the University of Guelph, Canada, a joint agreement is concerned with environmental impacts on agricultural policies on groundwater and the Great Lakes as well as the effects of the USA/Canada Free Trade Agreement.

The Center's literature states that environmental research is expected to result in "development, investment, and management strategies which are consistent with the economic and sustainable development of the world's natural resources."

Humphrey Institute of Public Affairs
University of Minnesota
301 19th Avenue South
Minneapolis, MN 55455 USA

Contact: **Dean Abrahamson, Ph.D.**, Professor
Phone: 1-612-625-2338 • *Fax:* 1-612-625-6351

Activities: Education; Research • *Issues:* Energy; Global Warming

▲ Projects and People
Recently returned from Iceland, Dr. Dean Abrahamson, Humphrey Institute of Public Affairs, University of Minnesota, is engaged in "writing and analysis with energy policy responses to global climatic change." He is currently working on a "summary of the comparative advantages of the several non-fossil primary energy sources."

■ Resources
Dr. Abrahamson can furnish reprints of articles in his field. He is also available for speaking, consulting, and lectures, with varying charges for these services depending upon the contracting organization's ability to pay.

Freshwater Biology Program
Department of Biology
University of Mississippi
University, MS 38677 USA

Contact: **James A. Kushlan, Ph.D.**, Chairman and Professor
Phone: 1-601-232-7203 • *Fax:* 1-601-232-5144

Activities: Education; Research • *Issues:* Biodiversity/Species Preservation; Water Quality

● Organization
"Fresh water is the most vital natural resource in our world," according to the University of Mississippi's Department of Biology, which is home to the Freshwater Biology Program—one of the most extensive programs of its kind in the USA. Here undergraduate and graduate courses are offered in watershed studies, major freshwater ecosystems, diversity of freshwater organisms, and analyses of complex aquatic systems. Students and researchers have the benefit of a Biological Field Station on 710 acres with more than 140 regulated ponds and wetlands, fish hatchery building, artificial streams, and many aquatic and terrestrial habitats.

Programs are collaborative with other Ole Miss departments and the U.S. Department of Agriculture's (USDA) Agricultural Research Service's National Sedimentation Laboratory; U.S. Forest Service Hydrology Laboratory; Mississippi Department of Wildlife, Fisheries, and Parks; and Mississippi Department of Environmental Quality. Other resources nearby are Sardis Lake, a U.S. Army Corps of Engineers flood-control reservoir; Sardis Waterfowl Refuge; Yazoo–Little Tallahatchie Drainage and Mississippi River, where research is conducted with forestry practices on streams in the Tallahatchie Experimental Forest, microbes, and stream energetics; University of Mississippi forest lands on 23,000 acres; and another 275 acres of estaurine-wetlands habitat.

▲ Projects and People
Research areas include the following: water quality; ecology of ponds, lakes, and reservoirs; wetlands as transformers of agricultural chemicals; nutrient cycling; ecology and conservation of aquatic birds and aquatic plants; responses of plants and animals to water management; effects of toxicants on aquatic systems and sediments; environmental microbiology; biotransformation of pollutants; and natural products in aquatic plants.

Dr. James A. Kushlan is also director and curator of ornithology, University of Mississippi Biological Museum and the chairman of the Species Survival Commission (IUCN/SSC) Heron Group. He is the co-author of books on herons; Southern Florida's freshwater fishes; and storks, ibises, and spoonbills.

University of Nebraska State Museum
University of Nebraska
307 Morrill Hall
Lincoln, NE 68588-0338 USA

Contact: **Hugh H. Genoways, Ph.D.,** Director
    Phone: 1-402-472-3779 • *Fax:* 1-402-472-8899

Activities: Education; Research • *Issues:* Biodiversity/Species
Preservation

▲ Projects and People

With eight scientific research curators at the **University of Nebraska
State Museum,** projects include **freshwater clams of Nebraska's
Platte River system; mammals of the Platte River valley; systematics
of neotropical beetles; survey for the endangered American burying
beetle** on the northern Great Plains; **pollen surveys in Lincoln;
surveys of parasites of Nebraskan fishes;** and systematics and biogeography of **Pacific fishes.**

In 1991, the Ashfall Fossil Beds State Historical Park, a cooperative project with the state Game and Parks Commission, was dedicated, along with a rhino barn museum.

Research interests of Dr. Hugh H. Genoways are "systematics,
biogeography, and ecology of New World mammals, especially
rodents and bats," along with computer application to data analysis
and retrieval. A member of the Species Survival Commission (IUCN/
SSC) Chiroptera Group, he has investigated bats, shrews, and other
mammals in field studies throughout the Americas. Dr. Genoways
is past president of the American Society of Mammalogists and has
published widely in his field.

Department of Biology ⊕
University of Redlands
P.O. Box 3080
Redlands, CA 92373-0999 USA

Contact: **James S. Malcolm, Ph.D.,** Associate
             Professor
    Phone: 1-714-793-2121, ext. 2423
     Fax: 1-714-793-2029

Activities: Education; Research • *Issues:* Biodiversity/Species
Preservation • *Meetings:* International Theriological Congress

▲ Projects and People

Dr. **James S. Malcolm** is pursuing the survival of a small scaleless
freshwater fish, the **unarmored threespine stickleback,** which is
presently restricted to two small streams in southern California
under the protection of the U.S. Endangered Species Act. Named
for Lt. R.S. Williamson who discovered *Gasterosteus aculeatus
williamsoni* in the mid-1800s while surveying for a western railroad
route, the greenish and silvery fish is believed to lack protective
armor because "predatory fishes are scarce in the upper reaches of
the river system where it made its home," report researchers.

This subspecies was once abundant in the Los Angeles, San
Gabriel, and Santa Ana river systems of the Los Angeles basin. With
these rivers nearly drained, the last self-sustaining and somewhat
abundant populations are confined to the Soledad Canyon section
of the Santa Clara River and a few small tributaries as well as in San
Antonio Creek on Vandenburg Air Force Base, Santa Barbara
County.

Whatever threatens these river ecosystems—such as agricultural,
industrial, and municipal water pollution; channelization; and other
urban intrusions on habitat—can harm the continued existence of
this stickleback. So, conservationists are introducing the subspecies
into San Filipe Creek, San Diego County, and Hondo Creek on
Vandenburg Air Force Base, where native vegetation and other
conditions help the unarmored threespine stickleback thrive and
breed. "Sticklebacks cannot survive in concrete-lined channels,"
reports Dr. Malcolm.

Those sensitive to the stickleback's survival caution that its depletion is an "indicator to us that something is wrong, and that special
effort is required to protect and preserve the habitat. . . . This means
maintaining the wild habitat and not just removing a few fish to an
aquarium or pond so the last habitat can be destroyed." Malcolm
currently advises the recovery team of biologists appointed by the
Fish and Wildlife Service to restore the habitat of the subspecies so
that it can be reclassified from endangered to threatened status.

Loss of species is also an indicator that humans may be left
without habitats for their own survival and become threatened as
well.

Dr. Malcolm is also a member of the Species Survival Commission (IUCN/SSC) Canid Group, which **monitors 36 species of
foxes, jackals, wolves, dogs, and their relatives** and supports research worldwide such as for simien jackal conservation in Ethiopia
and on behalf of African wild dogs in Tanzania.

■ Resources

Information on the unarmored threespine stickleback is available
from Dr. Malcolm. The canid survival action plan is available from
IUCN.

The Herpetologists' League (HL) ⊕
University of Richmond
Department of Biology
Richmond, VA 23173 USA

Contact: **Joseph C. Mitchell, Ph.D.,** Research
             Biologist/Ecologist
    Phone: 1-804-289-8234 • *Fax:* 1-804-289-8482

Activities: Research • *Issues:* Biodiversity/Species Preservation; Biology of Amphibians and Reptiles • *Meetings:* World
Congress of Herpetology

● Organization

The Herpetologists' League (HL) is an international organization
founded in 1936 for people interested in studying the biology of
amphibians and reptiles. Its 1,150 members are active in various
committees, including conservation.

▲ Projects and People

HL encourages the publication of books, such as the *Venomous Reptiles of Latin America*. Future projects are focusing on the herpetofauna of Baja California and continuing the amphibian checklists of the Association of Systematics Collections. At HL's annual meeting, herpetological research achievements are recognized with awards for best student paper and poster; and a Distinguished Herpetologist Lecture Series is offered.

At the University of Richmond, Dr. Joseph C. Mitchell, a visiting associate professor, teaches ecology, environmental biology, and botany classes and is particularly interested in the "historical effects of landscape alteration on the ecology of animal populations." He is pursuing the following research projects:

Conservation biology of the Peaks of Otter salamander (*Plethodon hubrichti*). "This small, terrestrial amphibian is endemic to the Peaks of Otter region of the Blue Ridge Mountains in Virginia," writes Dr. Mitchell. "The project focuses on determining the effects of logging practices on population size and life history characteristics of this salamander. Long-term monitoring will be conducted."

Structure and dynamics of amphibian communities in eastern deciduous forests. Species richness and diversity as well as seasonal variations within such habitats are being examined. "The objective is to determine how amphibian communities vary in different environments and through time," describes Dr. Mitchell.

Population ecology of the tiger salamander (*Ambystoma tigrinum*). Dr. Mitchell is monitoring Virginia's endangered species in a national forest to determine population size and dynamics in order to develop management plans that focus on long-term survival.

Population ecology of freshwater turtles. By monitoring several such groups, Dr. Mitchell is evaluating geographic variation in life-history characters, sexual dimorphism, and population structure.

Reproductive cycles in snakes and freshwater turtles. Ongoing projects are focused on variations in the male and female reproductive cycles of several species, such as copperhead and snapping turtle.

Biogeography of the amphibians and reptiles of the Virginia barrier islands. "I am nearing completion of a project that examines the causes of the diversity of the herpetofauna on these islands," he reports.

Dr. Mitchell has consulted with Virginia Power and other corporations and institutions on terrestrial vertebrate ecology studies. At the Quantico Marine Corps Base and Shenandoah National Park, he conducted biological and rare species inventories.

With C.A. Pague, Dr. Mitchell is co-author of the book *The Amphibians and Reptiles of Virginia*; he has published extensively in his field.

■ Resources

HL publishes the quarterly scholarly journal *Herpetologica* and its annual supplement *Herpetological Monographs* with research articles, syntheses, and reviews. A membership form is available.

Department of Biology
University of South Carolina at Aiken (USCA)
171 University Parkway
Aiken, SC 29801 USA

Contact: **Garriet W. Smith, Ph.D.**, Associate Professor
Phone: 1-803-648-6851, ext. 3427

Activities: Education; Research • *Issues:* Biodiversity/Species Preservation; Marine Pollution; Water Quality

▲ Projects and People

Where Christopher Columbus first landed in the New World 500 years ago is precisely the locality of Dr. Garriet Smith's search for ocean pollution in the Caribbean—and for similar reasons. While easterly San Salvador Island was the first land sighted for Columbus, it is also the farthest from populated and industrial areas of the remaining Bahamas.

Smith and his Earthwatch teams are uncovering the effects of pollution in seagrass. There are "great undersea meadows of seagrass" in tropical shallows worldwide that influence the food chains of near-shore ecosystems. Smith believes too that "seagrasses can serve as sensitive indicators of potential environmental change caused by human-made pollutants, such as pesticides, petroleum compounds, and heavy metals." And he is attempting to answer some critical questions about seaweed biology: "Do seagrasses exhibit seasonal variations in distribution or biomass, and do these variations change over the years? Are the three species of San Salvador seagrass—*Halodule wrightii*, *Syringodium filiforme*, and *Thalassia testudinum*—distributed differently at various sites? Does above- and below-sediment biomass vary according to location?" Smith knows that the sediments and water masses are very nutrient-poor in carbonate islands like San Salvador.

The Earthwatch volunteers are based at the Bahamian Field Station on San Salvador. Snorkeling team members collect and bag seagrass samples along 20-meter transects in water generally less than 3 meters deep. At the lab, the grasses are washed and sorted by species, then separated, dried, weighed, and packaged.

Smith, who received the USCA excellence in research award in 1991, is interested in all types of microbial-plant interactions, marine microbiology, nitrogen cycling, and pathogens in the environment. He is also research associate with the Belle W. Baruch Institute of Marine Biology and Coastal Research. With four other investigators, Smith participated in a recent National Aeronautics and Space Administration (NASA) grant for testing the use of "remote sensing to evaluate ecological and environmental impacts of commercial development."

Department of Zoology ⊕
University of Texas at Austin
Austin, TX 78712 USA

Contact: **Clark Hubbs, Ph.D.**, Regents Professor of
Zoology Emeritus
Phone: 1-512-471-1176 • *Fax:* 1-512-471-9651

Activities: Education • *Issues:* Biodiversity/Species Preservation; Energy; Water Quality

▲ Projects and People
Research under **Dr. Clark Hubbs'** supervision at the **University of Texas** is "designed to answer two sequential questions," he writes: "Why are animals (fishes) able to live in their environments? And how did they become adapted to succeed there?"

During Dr. Hubbs notable career, he has investigated and written about snails as a food source for the endangered *Gambusia*, declining fishes of the Chihuahuan Desert, pupfish recovery plan, stream fish ecology, Texas freshwater fish, Nile perch survival, black bass, greenthroat darter, and sperm viability in frequency of fish hybridization—among other subjects.

He is a director and research committee chairman of the Hubbs Sea World Research Institute (H-SWRI), a nonprofit foundation established in 1963 to "serve the community by returning to the sea some measure of the benefits derived from it." Scientists here are examining worldwide issues to conserve water resources, grow fishes for the future, and protect marine mammals and birds. Other studies relate to human health concerns. Such research is the foundation for "sound legislation, environmental management decisions, and conservation programs that protect coastal resources," according to H-SWRI.

For example, biologists are studying the habits of marine mammals around the California Channel Islands and trying find ways to lessen their contact with marine fishes. Another goal is to develop hatcheries for fishes whose populations have been depleted, so that they can be returned to their natural environments. Working with the Southeastern Stranding Network, researchers are rehabilitating injured manatees and probing the ecology of this endangered species. A program is underway to raise white sea bass as a sport fish. Bioacoustics studies with whales and dolphins may soon help resolve conflicts between marine mammals and fish. The effects of water diversion from the tributaries of Mono Lake, California will have impact on migratory birds as a future staging area and on surrounding populations as an important water source.

Curator of Ichthyology at Texas Memorial Museum, Dr. Hubbs is also on the Glen Canyon Environmental Studies Committee, National Research Council; leader of the Rio Grande Fishes Recovery Team, U.S. Department of Interior; trustee of the Texas Nature Conservancy; and Environmental Advisory Committee member, Texas Parks and Wildlife. In recent years he was honored by the Southwestern Association of Naturalists, Texas Organization for Endangered Species, and the American Fisheries Society.

■ Resources
A descriptive H-SWRI brochure with membership application is available.

Department of Botany ⊕ ✿
University of Washington
420 Hitchcock Hall, KB–15
Seattle, WA 98195 USA

Contact: **Arthur R. Kruckeberg, Ph.D.**, Professor
Emeritus
Phone: 1-206-543-1976

Activities: Education; Research • *Issues:* Biodiversity/Species Preservation

▲ Projects and People
Born into a horticultural printing and publishing family, **Dr. Arthur R. Kruckeberg**, former Botany Department chairman, **University of Washington**, is researching the flora and vegetation of Pacific Coast serpentine soils in addition to the various influences of geology on the biology of higher plants. As well as teaching botany and biology courses on campus, he instructs adult education classes in wildflowers of the Pacific Northwest, native and woody ornamental plants, conservation and ecological reclamation with plants, and rare and endangered species. Dr. Kruckeberg has authored three books on plants and gardening, including *Natural History of Puget Sound Country* published in 1991.

Co-founder of the both the Washington Native Plant Society (WNPS) and the Washington State Natural Heritage Program, he conducts rare plant surveys and currently oversees a study of the impact of livestock grazing on the Pasayten Wilderness of the North Cascades.

The 1,000-member Society introduces people to the state's unique flora, such as salmonberry, mountain hemlock, and *Lewisia tweedyi*, to encourage preservation of their botanical heritage. Dr. Kruckeberg edits the group's quarterly newsletter *Douglasia*. He belongs to the Species Survival Commission (IUCN/SSC) North American Plant Group.

■ Resources
A WNPS membership brochure is available.

Institute for Environmental Studies (IES) ⊕ ✿
University of Washington
Engineering Annex FM-12
Seattle, WA 98195 USA

Contact: **P. Dee Boersma, Ph.D.**, Professor
Phone: 1-206-543-1812 • *Fax:* 1-206-543-2025

Activities: Education; Research • *Issues:* Biodiversity/Species Preservation

● Organization
"This we know: the Earth does not belong to man, man belongs to the Earth. All things are connected like the blood that unites us all. Man did not weave the web of life, he is merely a strand in it. Whatever he does to the web, he does to himself." Chief Seattle, 1852

Such is the introduction to the **Institute for Environmental Studies (IES)**, established in 1972 to develop environmental curricula for **University of Washington** students, to encourage interdisciplinary research, and to develop public services programs—including continuous education.

Environmental courses are offered to both undergraduate and graduate students. The **Continuing Environmental Education Program** conducts conferences, seminars, public forums, and short courses on environmental topics of importance to government, business, citizens, educators, and the general public.

▲ Projects and People

Dr. P. Dee Boersma, a zoologist, is engaged studies of seabirds as environmental indicators, population dynamics and reproductive strategies of Magellanic penguins and other seabirds, forest structure in the Hoh rainforest, and conservation biology. **Dr. Robert J. Charlson**, atmospheric scientist, is researching the cycling of elemental carbon in the atmosphere, cloudwater and rainwater chemistry, and the influence of volcanic and/or anthropogenic particles on climate—among other studies. **Dr. David J. Eaton** is investigating biochemical mechanisms of toxicity and carcinogenicity of aflatoxins disposition of organic pollutants in the liver as well as assessing health risks of environmental and occupational exposures to toxic chemicals. **Dr. Kai N. Lee**, political scientist, is examining public policy issues such as on global climate change, adaptive management, and Pacific Northwest resources. **Dr. Estella B. Leopold**, botanist, is researching forest history, including wetlands restoration; climate and vegetation patterns since glaciation, using pollen analysis; prairie history; co-evolution of grasslands and grazing ungulates; and Pacific Northwest forest associations' development. **Dr. Conway B. Leovy**, atmospheric scientist, is probing global circulation of Earth's atmosphere, atmospheric remote sensing, transport of trace substances in the stratosphere, and planetary atmospheres. **Dr. Gordon H. Orians**, zoologist, is surveying the ecological aspects of vertebrate social organization, environmental aesthetics, structures of ecological communities, and conservation biology.

■ Resources

IES publishes a monthly newsletter, *Environmental Outlook*, describing environmental courses, conferences, and hearings; featuring updates on faculty activities; and listing publications and videos of interest. It also produces the *Northwest Environmental Journal* to broaden public understanding of regional environmental concerns. Available brochures list courses and study options.

Primate Center Library ⊕
Wisconsin Regional Primate Research Center
University of Wisconsin—Madison
1220 Capitol Court
Madison, WI 53715-1299 USA

Contact: **Lawrence Jacobsen**, Librarian
  Phone: 1-608-263-3512 • *Fax:* 1-608-263-4031
  E-Mail: jacobsen@primate.wisc.edu

Activities: Education; Research • *Issues:* Biodiversity/Species Preservation; Health

● Organization

Research and total budgets at the **University of Wisconsin** are first among public universities in the USA; in 1990, totals were $300 million and $1 billion, respectively. There are more than 40,000 graduate and undergraduate students, 2,350 faculty, and 4,300 academic staff involved in multidisciplinary interaction and collaboration.

The **Wisconsin Regional Primate Research Center**, based in the Graduate School, has strong research and teaching links in the Schools or Colleges of Medicine, Letters and Science, Agriculture and Life Science, and Veterinary Medicine. One of seven Primate Centers in the USA with grants from the National Institutes of Health (NIH), it is concerned with fundamental research in primate biology that is relevant to human and animal health. Studies are at the molecular, cellular systemic, whole animal and environmental levels and include projects in conservation biology.

"The Center is self-sufficient in breeding colonies of primates and does not import from the wild," reports **Dr. John P. Hearn**, director. Improving the breeding, health, and management of primates in captivity and in the wild is a conservation goal. Other objectives include advancing biomedicine to benefit human health and serving as the Midwest's research, development, and information center nationally and globally.

▲ Projects and People

Regarded as first-rate, the **Primate Center Library** was established in 1973 and serves university needs as well as primatologists worldwide. Material on prosimians, monkeys, and apes appears in more than 15,000 volumes and special resources—including rare books, slides, sound recordings and tapes, annual reports of the NIH Primate Centers, and some 300 periodicals, indexes, newsletters, and bulletins. Video has become an "invaluable tool for behavioral research," enabling scientists such as **Dr. Maxine Biben** of NIH to record play behavior in squirrel monkeys, for instance.

A new electronic mail listserver called **PRIMATE-TALK** is available as a worldwide, open forum for the discussion of primatology and related subjects. Persons with Internet, BITNET, or UUCP addresses can communicate about news, research, meetings, information requests, veterinary/husbandry topics, book reviews, or jobs, for example. Interested users can send a message to PRIMATE-TALK-REQUEST@PRIMATE.WISC.EDU and request to sign on.

■ Resources

The *Primate Library Report* is published bimonthly, and a catalog of audiovisual resources is also available. The Center's first international directory of primatology is being published in 1992. Users can order library materials by phone, fax, or electronic mail. The library welcomes copies of research papers and pertinent books, book collections, slides, tapes, sound recordings, photographs, and ephemera.

Flora of North America Project
c/o Department of Biology ⊕
Utah State University
Logan, UT 84321 USA

Contact: **Leila M. Shultz, Ph.D.,** Curator,
Intermountain Herbarium
Phone: 1-801-750-1586 • *Fax:* 1-801-750-1575

Activities: Education; Research • *Issues:* Biodiversity/Species
Preservation • *Meetings:* AIBS; IOPB

▲ Projects and People

With a five-year grant from the National Science Foundation (NSF)
and Dr. **Leila M. Shultz** as co-principal investigator with **Dr.
Nancy R. Morin,** Missouri Botanical Garden, and **Dr. Richard W.
Spellenberg,** New Mexico State University, the **Flora of North
America Project** is developing the standard identification resource
catalog of the botanical diversity of North America north of
Mexico—the first of its kind. All that is known about plants, their
characteristics, relationships, and distributions will be included in
this inventory. As a result, "wildlife management, forestry, range
management, horticulture, and research will all benefit," assert the
project leaders. Without such an accessible research tool, conservation efforts—including biodiversity and species preservation—are
thwarted.

Scientists from more than 20 major botanical institutions in the
USA and Canada are together producing the *Flora of North America.*
The editorial committee members work from their home institutions to consolidate data from "hundreds of scattered sources into a
single series of 12 volumes covering vascular plants north of
Mexico." Some 17,000 species of vascular plants will be included—
representing about 7 percent of the world's total. The Hunt Institute for Botanical Documentation is the bibliographic center where
a comprehensive bibliographic database is being designed; the **Missouri Botanical Garden is the organizational and floristic database
center.**

According to the researchers, "The description of each species
will be written and reviewed by experts from the systematic botanical community worldwide and will be based on original observations of living and herbarium specimens supplemented by a critical
review of the literature." The volumes are being published over a 10-
year period through 1999, with the final book containing a comprehensive bibliography and complete text. "Each treatment will include scientific and vernacular names, taxonomic descriptions,
identification keys, distribution maps, illustrations, summaries of
habitat and geographic ranges, pertinent synonymies, descriptions,
chromosome numbers, phenological information, and other important biological observations," the researchers write.

The computerized database will allow easy access, sorting, and
comparison of large amounts of floristic information. The database
will even store materials not appearing in the volumes, such as
detailed morphological descriptions and geographical ranges.
Housed at the Missouri Botanical Garden, it will allow users to
access information through hard copy, magnetic tapes, and other
media. Most important, "Researchers can quickly sort species by
characteristics or habitat or correlate them with their physical parameters, creating an exciting new era for botanical research," the
project leaders report.

Also helping to support the Flora Project are the Pew Charitable
Trusts, the Robert and Lucile Packard Foundation, the National
Fish and Wildlife Foundation, and the Hewlett Foundation. An
authority on flora and endangered species of the intermountain
west, Dr. Shultz is co-author of *The Atlas of the Vascular Plants of
Utah,* based upon the examination of 400,000 specimens.

■ Resources

New technologies needed for specimen information are database/
museum management programs. Monetary support from user groups
is also requested. The *Atlas* and the free publication *Gardening for
Wildlife in Utah* are available.

Beza Mahafaly Reserve Project
Department of Anthropology ⊕
Washington University
St. Louis, MO 63130 USA

Contact: **Robert W. Sussman, Ph.D.,** Professor
Phone: 1-314-889-5264 • *Fax:* 1-314-889-5799

Activities: Education; Research • *Issues:* Biodiversity/Species
Preservation; Deforestation; Energy; Global Warming;
Health; Population Planning; Sustainable Development
Meetings: International Primatological Society

● Organization

Washington University began collaborating with Yale University
and the University of Madagascar in 1977 and the World Wildlife
Fund—USA a year later to create a natural reserve in Madagascar
for practical conservation measures and a field station for related
research and teaching; a field training program in conservation
biology for Malagasy students; and training for students and technical staff of École Superieur des Sciences Agronomiques, E.S.S.A.
(School of Agronomy), University of Madagascar. Consequently,
the **Beza Mahafaly Reserve** was established in 1978 and became a
Special Government Reserve in 1985. At that time, it was decided to
expand both the Beza Mahafaly Special Reserve from 600 to 1,400
hectares and include people bordering the reserve in the education
and development programs as well as to strengthen the protection of
the Andohahela Special Reserve in southeastern Madagascar—also
incorporating education and research programs.

Overall, local involvement has been the key to the success of this
international venture. Researchers recognize E.S.S.A. as playing a
central role in establishing the Beza Mahafaly Reserve and its place
as a training ground for students who move onto positions of
responsibility in ministries directly concerned with managing the
country's resources. Other Malagasy institutions and employees also
contribute to this long-term project, where the initial accomplishments appear below.

▲ Projects and People

As "Madagascar is biologically one of the richest areas on Earth, and
its plants and animals are among the most endangered," principal
investigators Dr. **Robert W. Sussman,** Dr. **Alison F. Richard** of
Yale University, and Dr. **Andrianasolo Ranaivoson** of Universite de
Madagascar set four interrelated project goals: conservation, educa-

tion, sustainable development, and research in addressing the threat of extinction of forest species resulting from "legitimate economic needs and demands of the growing human population," they report.

At the Beza Mahafaly Reserve, **field courses** were offered for local students; the E.S.S.A. Forestry Library was enlarged; a seminar/conference was held for directors of natural reserves; primary populations were studied; ecological and behavioral research continued with vertebrate populations; regeneration of vegetation was monitored and density and distribution were described both inside and outside the reserve; **ethnobotany and ethnography studies** were conducted of people bordering the reserves; plants were collected and identified for compiling a plant list and florula for the general area; and satellite imagery was used to define and monitor endemic plant communities in the south and compare them with those in the Beza Mahafaly region.

Closely investigated are **sifakas and ring-tailed lemurs**—the region's largest diurnal primates—and long-term research results are expected to give an indication of the general health of the forest over time. With the lemurs almost all marked, some interesting observations are emerging: termite feeding is a daily occurrence and seems important to the diet year-round; and females stay in a tree "only long enough to get prime ripe fruit and then [move] off, whereas males and juveniles stay in the same tree much longer," according to researcher Michelle Sauther.

Seasonal, long-term monitoring of the 69 species and 44 genera of Malagasy plant communities of riverine and dry forests lead to understanding forest dynamics and the effects of reduced grazing. A **floristic inventory** collection will be housed at the Herbariums of the Parc de Tsimbazaza, Antananarivo, and the Missouri Botanical Garden, St. Louis. "It will serve as a database for future studies of plant systematics, general ecology, nutrition, ethnobotany, ethnopharmacology, agronomy, and other related disciplines," relates Dr. Sussman. Pressed and dried **herbarium specimens** were made of all species found in fruit or flower, and this collection will result in a published plant list for the reserve. **Medicinal plant species** were also collected, pressed and dried, and verified with most of the villagers who reported using them. Documented sets are being deposited at the Missouri Botanical Garden, Parc de Tsimbazaza, and possibly, the Centre National de Recherches Pharmaceutiques.

As building maintenance, fence improvements, school construction, road repairs, surveys, and mapmaking was begun at both Beza and Andohahela, a plan was initiated to **improve sustainable agriculture and fuelwood production** in areas surrounding the two reserves. According to Dr. Sussman, "a major priority in the near future is to greatly improve our local education and sensitization programs so that local inhabitants understand the importance of conservation for their own future and feel that they are an integral part of the Beza Mahafaly project, and not victims of it."

With **Dr. G.M. Green,** Department of Earth and Planetary Sciences and the McDonnell Center for the Space Sciences, Washington University, Dr. Sussman participated in a **satellite imaging** project to detail the **"deforestation history of the eastern rain forests of Madagascar"**—described in *Science* (April 13, 1990). These satellite images and vegetation maps, compared with earlier aerial photographs, monitored the rate of deforestation over a 35-year period between 1950 and 1985. It was shown that the rate of deforestation averaged 111,000 hectares annually—cutting in half 7.6 million hectares of rainforest in 1950 to 3.8 million in 1985. And originally this rainforest was 11.2 million hectares.

"Deforestation was most rapid in areas with low topographic relief and high population density," report the researchers, who conclude that "if cutting of forests continues at the same pace, only forests on the steepest slopes will survive in the next 35 years." Using LANDSAT Multispectral Scanner imaging, recent deforestation was noted in established reserves. These forest losses along with intense rainstorms and frequent hurricanes contribute to high soil erosion rates. Dr. Sussman says that many Landsat images from the past two decades exist of most of the world's tropical forests, providing "a remarkable but essentially unused database."

To preserve tropical rainforests effectively, what is essential, say these researchers, is "sustainable agriculture and agroforesty to provide local inhabitants with needed food and fuel, accompanied by reduction of population growth." Detailed "ethnographic studies" are needed as well to address the "social and economic needs of local peoples." With the guidance of R. Barbour and R. Hagen, new crops and agricultural practices were being tested and farmers were being trained with new tools to increase production of rice, for example.

Over a 10-month period, interviews were conducted of local households to determine their use of noncultivated plants for food, medicine, and construction. The impacts of harvesting wild plants and of injurious grazing by goats and cattle on new forest growths are under study, and so is the chemical analysis of certain species which are used for treating illness, feeding newborns, and for postpartum care of mothers. The researchers report that "the residents of the area purchase little. Most of their essential resources other than the foods they grow come from the forests—for medicine, for housing, tool handles, oxcarts, toys, baskets, etc. The people—men, women, old, young—are very knowledgeable about their environment, especially the flora. They appreciate its importance to them and the importance of not misusing it. For example, they know how to harvest most medicinal plant species—branches or bark or roots, etc.—without injuring or killing the individual plant. . . . The Mahafaly of the region have a great pride in their traditional way of life and . . . are aware of the importance of preserving the land of their ancestors.

In preliminary findings of **health studies,** it was found that the Mahafaly in the Analafaly region have a great knowledge of the forest and medicinal plants which is passed down through the generations. There are also healers who tend to specialize in treating particular illnesses, which—according to the Mahafaly—might be caused by natural causes, displeasure of the ancestors as a result of breaches in the moral order, sorcery, and spirit possession. While researchers say the villagers are generally in good health, improvements can be made in sanitation and accessible care for pregnant women and newborns.

Dr. Sussman and his colleagues are **encouraging** more scientists and students to conduct both **long-term and short-term research** at the reserve. Plans are to continue with the studies of the sifaka and ring-tailed lemur, vegetation, satellite monitoring, and ethnographic and agricultural research on local people.

Dr. Sussman is receiving funding from the National Science Foundation (NSF) for studies of the black lemur (*Lemur macaco macaco*). The National Geographic Society, Fulbright Scholar Program, National Institutes of Health (NIH), and WWF have also supported his recent studies in Madagascar. He is a member of the Species Survival Commission (IUCN/SSC) Primate Group and of the executive committee of the American Association of Physical Anthropologists.

Department of Biology ⊕
Washington University
Campus Box 1137
St. Louis, MO 63130 USA

Contact: **Nicholas Georgiadis, Ph.D.,** Conservation
Biologist
Phone: 1-314-935-6867 • *Fax:* 1-314-935-4432

Activities: Research • *Issues:* Biodiversity/Species Preservation; Population Genetics

▲ Projects and People

Dr. Nicholas Georgiadis is a research fellow with **Wildlife Conservation International, New York Zoological Society,** as well as research associate with both **Washington University's Department of
Biology,** and the **Serengeti Wildlife Research Institute, Arusha,
Tanzania.**

Based on current investigations, he is now preparing several
papers on the genetic structure of elephant populations in Africa.

Research experience includes the "evolutionary and population
genetics of the African Bovidae and of African elephants using . . .
restriction fragment and sequencing methods"; carbon and nutrient
cycling in the Serengeti Ecosystem; and immobilization of wild
bovids.

With 15 years' study of the African savanna ecosystems, Dr.
Georgiadis undertakes ecological monitoring, aerial animal and
vegetation surveys, elephant population biology, herbivore nutrition and energetics, plant ecology, physiology and chemical de-
fenses, soil analysis, ecosystem-level nutrient cycling, and quantifying impacts of poaching.

Dr. Georgiadis is a member of the Species Survival Commission
(IUCN/SSC) Antelope Group, Ecological Society of America, and
other conservation groups.

Awash National Park Baboon Project ⊕
Department of Anatomy & Neurobiology
Washington University, School of Medicine
Box 8108, 660 South Euclid Avenue
St. Louis, MO 63110 USA

Contact: **Jane Phillips-Conroy, Ph.D.,** Associate
Professor
Phone: 1-314-362-3396 • *Fax:* 1-314-362-3446
E-Mail: baboon@wums

Activities: Education; Research • *Issues:* Biodiversity/Species
Preservation • *Meetings:* American Association of Physical
Anthropologists; American Association of Primatologists;
CITES; International Society of Primatologists

▲ Projects and People

Even though baboons are not endangered, they are "a yardstick of
sorts" in appreciating the potential loss of genetic diversity in species
that are. Conservation is still a vital factor in two baboon subspecies–the olive baboon (*Papio hamadryas anubis*) and the sacred or

hamadryas baboon (*Papio hamadryas hamadryas*) which interbreed
in Ethiopia where their ranges intersect. The sacred baboon particularly is "potentially quite vulnerable" as it is hunted extensively for
biomedical research purposes. Yet in sub-Saharan Africa, these baboons are mainly regarded as "a pest and a nuisance," writes Dr.
Jane Phillips-Conroy.

During 10 years of annual and semiannual fieldwork in the
Awash National Park, Shoa province, Ethiopia, she and **Dr. Clifford
Jolly, New York University,** have been capturing, identifying with
eartags and sometimes palmprints, photographing, and collecting
biological data on seven groups of baboons to "explore how the
structure of the [Awash National Park] hybrid zone has changed
over time; document the extent of gene flow into and out of the
parental groups; and determine the behavioral mechanisms of the
hybridizing groups that have contributed to hybridization and that
may be limiting its spread."

Body measurements are taken; blood, saliva, and hair samples are
examined; and fecal samples are collected for parasite analysis. One
subset of "extraordinarily healthy" Awash baboons were studied for
periodontal disease and compared with baboons living in other
habitats and with different diets, such as those in Amboseli, Kenya.

Following long-term behavioral studies, Phillips-Conroy and researchers—including **Earthwatch** volunteers and Ph.D. students—
report that "the hybrid zone is formed primarily through immigration and long-term residence of hamadryas males into these formerly
anubis, now 'mixed' social groups." These studies are a continuation
of those begun 20 years ago.

More recently, Phillips-Conroy initiated a study on **SIV (simian
immunodeficiency virus or monkey AIDS) in sympatric green
monkeys** (*Cercopithecus aethiops aethiops*), including its incidence
and mode of transmission. Using genetic studies of the early 1970s,
researchers are able to compare genetic changes in two primate
populations with fundamentally similar social organizations.

In the mid-1980s, a study of the **yellow baboons** (*Papio hamadryas cynocephalus*) in Mikumi National Park, Tanzania, with **Dr. Jeff
Rogers,** now of the Southwest Foundation for Research and Education (SWFRE), San Antonio, Texas, focused on the effects of demographics on the population's genetic structure.

Dr. Phillips-Conroy is an associate professor in the Department
of Anatomy & Neurobiology, where she teaches human anatomy
courses in the Medical School, and an associate professor of anthropology in the Department of Anthropology, where she teaches,
advises, and supervises Ph.D. students.   She belongs to the Species
Survival Commission (IUCN/SSC) Primate Group.

■ Resources

The Harry Frank Guggenheim Foundation and Earthwatch/Center
for Field Research provide support for the Awash baboon project
with supplementary funding from National Institutes of Health
(NIH) for the monkey AIDS studies. Slides and lectures are available for public purposes.

Department of Geography ⊕
Western Michigan University
Kalamazoo, MI 49008 USA

Contact: **Vernon K. Jones, Ph.D.**, Instructor
  Phone: 1-517-543-7251

Activities: Research • *Issues:* Climatology; Global Warming

▲ Projects and People
With a background in agricultural climatology, Dr. Vernon K. Jones has expertise in the field of global warming. He gives seminars and talks on the impacts of carbon dioxide and climate change on agriculture; on nuclear winter and scientific models; on mountain meteorology, glaciers, and geoclimatic research; and on general areas regarding agriculture, food, geography, hunger, environment, and their impacts by and upon people. His focus is on resource management education and societal impacts as well as on energy problems.

Dr. Jones is a co-author of *The Impact of Climate Change from Increased Atmospheric Carbon Dioxide on American Agriculture* for the U.S. Department of Energy and of a similar publication for Lawrence Berkeley Laboratory. He is visiting professor, Social Studies Department, Ferris State University, Big Rapids, Michigan, 49307.

Department of Anthropology ⊕
Yale University
Box 2114
New Haven, CT 06520 USA

Contact: **Arthur H. Mitchell**, Researcher
  Phone: 1-203-432-3692

Activities: Research • *Issues:* Biodiversity/Species Preservation

▲ Projects and People
Arthur H. Mitchell recently completed an 18-month study of the ecology of a native colobine monkey, Hose's langur (*Presbytis hosei*), in a lowland rainforest of Sabah, Malaysia, in northeast Borneo. Described as the first such intensive study, the research is expected to contribute to "our understanding of the consequences of habitat alteration on the resources of a forest primate and has implications for management of primate populations in altered forests," points out Mitchell, whose doctoral dissertation for Yale University's joint degree program of Biological Anthropology/Forestry and Environmental Studies was being completed in 1992.

In a range of forest microhabitats, from undisturbed to highly disturbed, three main types were selected—unlogged forest, logged forest adjacent to unlogged forest, and logged forest distant from unlogged forest—where Mitchell collected ecological and behavioral data of the Hose's langur.

Mitchell describes his methods: "The forest habitat's structure, species composition, and phenology were examined to determine differences among the study area's forest microhabitats and to assess the relative availability and quality of langur resources within them." The phytochemistry of plant materials was also analyzed to help evaluate how forest disturbance affects resource distribution which, in turn, influences langur feeding and ranging.

In Indonesia, where Mitchell was previously a U.S. Peace Corps volunteer, he planned an environmental education center and began a program of orangutan rehabilitation at Sepilok. He was also a project leader for two World Wildlife Fund (WWF) conservation programs that involved management planning and implementation of reserves and parks, such as on Siberut Island. A member of the Species Survival Commission (IUCN/SSC) Primate Group, Mitchell was a recipient of a Fulbright-Hays Doctoral Dissertation Research Abroad Fellowship.

School of Forestry and Environmental Studies ⊕
Yale University
205 Prospect Street
New Haven, CT 06511-2189 USA

Contact: **John C. Gordon, Ph.D.**, Dean and
          Professor
  Phone: 1-203-432-5107 • *Fax:* 1-203-432-5942

Activities: Education; Research • *Issues:* Air Quality; Biodiversity/Species Preservation; Deforestation; Global Warming; Sustainable Development; Water Quality

● Organization
The School of Forestry and Environmental Studies educates professionals in managing the environment and renewable natural resources worldwide. Research contributes to scientific understanding of the biosphere, both its ecological and social systems. Students are encouraged to design and manage solutions to problems. Faculty members serve on state and federal environmental commissions and on boards of national and global environmental groups. In collaborative efforts, the School is involved with other colleges, universities, government agencies, and private organizations in the USA, Latin America, Europe, and Asia.

Yale's Forestry School, the nation's first, was established in 1900 by Gifford Pinchot and Henry S. Graves, the first Americans to receive professional forestry training in Europe—in that order. Pinchot later created and served as first chief of the U.S. Forest Service, originated the phrase "conservation of natural resources," and defined conservation as "the wise use of the earth for the good of present and future generations." He was also an advisor to President Theodore Roosevelt. Graves became the School's first dean. Since its founding, 2,000 men and women have graduated as natural resource managers and environmental scientists.

▲ Projects and People
Global ecology that emphasizes understanding of ecosystems, management technology of sustainable resource systems, and policy analysis of the ways society allocates economic, social, political, and legal resources are the unifying themes of research and teaching programs. In a cooperative effort, the School participates in the Hubbard Brook Experimental Forest, New Hampshire, which is said to be the world's model for ecosystem studies. The Tropical Resources Institute, the School's largest and fastest-growing program, is described as having developed "one of the most substantial

social and policy components of any resource management program in the world."

A sampling of research projects of faculty and students include: transport and consequences of pollutants in the environment; scientific and social bases for agroforestry; chemical regulatory policy and implications for human diets; genetic, biochemical, and ecological fundamentals of nitrogen fixation by trees; interactions between humans and animals; nutrient cycling in managed and unmanaged forests; ecology of food sources for endangered species; and financial and policy aspects of forest land ownership.

Dr. Dwight Baker, associate research scientist, is studying symbiotic interactions of forest plants and soil microorganisms—including the initiation, compatibility, and physiology of mutualistic, nitrogen-fixing symbioses of trees. In the laboratory and field, methods for measuring or estimating the contribution of nitrogen-fixing trees to natural forest ecosystems or agroforesty plantations are being tested. Much emphasis is on the isolation, characterization, and utilization of the symbiotic actinomycete, *Frankia*, and other forest soil microsymbionts.

Dr. Paul K. Barten, assistant professor of water resources, is researching the integration of field monitoring and computer modeling to quantify the rates and pathways of water movement through upland, wetland, and estaurine ecosystems. He recently completed studies of upland–peatland watersheds in Minnesota and the hydrologic impact of climate variability on the Hubbard Brook experimental watersheds. He also established a regional cooperative network of four experimental watersheds in Connecticut, Massachusetts, and New Hampshire (Hubbard Brook) to refine and verify a computer algorithm that links a water balance model with a geographic information system (GIS). Another study will quantify the water balance of a tidal freshwater marsh and the adjacent shallow embayment in the Tivoli Bays National Estaurine Research Reserve on the Hudson River. Ongoing applied research includes the delineation of critical streamflow and sediment source areas on the 126-square-kilometer watershed that drains into the Tivoli Bays and a second watershed that drains into the Ashokan Reservoir—part of New York City's Catskill Water Supply system.

Dr. Steven R. Beissinger, associate professor of wildlife ecology, is concerned with the behavioral ecology and life history strategies of animals—particularly birds, small mammals, and aquatic snails—and with using this information to manage endangered species or commercially valuable wildlife. His studies of the snail kite examined how "sexual differences in the allocation of reproductive effort and demographic traits may affect the evolution of mating systems and mating desertion." In Florida's Everglades, where this snail-eating neotropical hawk is endangered, models of the impact of water levels on snail kite population viability are progressing. In Venezuela, snail kites eat crabs in addition to snails, and researchers are determining why. Other studies examine the impact of the bird trade on neotropical parrots. Small Venezuelan parrots are nesting in artificial boxes as a model for understanding hatching evolution and for sustained management of parrots for the pet trade.

Dr. Gaboury Benoit, assistant professor of environmental chemistry, is focusing on "elucidating metal speciation in natural waters and sediments," emphasizing the identity and characteristics of organic and inorganic colloids and their influence on metal speciation and behavior. Naturally occurring radionuclides are being used as tracers.

Dr. William R. Bentley, senior research associate in forestry, examines forests as development assets, leadership and organizational development in forestry and environmental institutions, and the economics of agroforestry. His latest books are on agroforestry in South Asia and on problem solving for rural resource managers.

Dr. Graeme P. Berlyn, professor of anatomy and physiology of trees, is director of the School's Forest Biotechnology Program—a field he helped pioneer. He is interested in tissue culture regeneration of trees, DNA characterization of cell and tree populations, wood formation, effects of toxic and radioactive materials on tree growth and development, light processing by tree leaves in relation to physiology and adaptation, and trees in alpine and stressed environments. His research team developed organic biostimulants that enhance growth, yield, and stress resistance while reducing fertilizer use.

Dr. F. Herbert Bormann, professor of forest ecology, is the co-founder of the internationally recognized Hubbard Brook Ecosystem Study; for 22 years Dr. Bormann has studied nutrient cycling and the effects of acid rain on the forest environment in the Hubbard Brook and the surrounding 7,500 acres of the White Mountain National Forest, New Hampshire. He is an authority on ecosystem development, biogeochemistry, land-water interaction, air pollution stress, and environmental policy.

Dr. Garry D. Brewer is professor of natural resource policy and management, organization and management, public health, political science, and at the Institution for Social and Policy Research. He develops and applies various theories and methods associated with policy sciences to improve understanding and management of large-scale and complex social/environmental/political systems. In a five-year study with the National Academy of Sciences, he assessed environmental science and studies done in conjunction with drilling for oil and gas offshore. With the International Management Center and the Regional Center for the Environment in Budapest, Hungary, he has designed and executed professional training programs for environmental managers in Eastern Europe.

Dr. William R. Burch, professor of natural resource management and at the Institution for Social and Policy Studies, pursues research in leisure and the nature of social bond, energy and social structure, and social change and land use. Each area has application in forecasting future patterns of behavior for planning or in identifying the social dimensions of management decision. Burch is developing a research program on public participation and management in biosphere reserves of the Caribbean Basin.

Dr. William J. Cronon, professor of history and of environmental studies, is an environmental historian who studies the way human communities modify the landscapes in which they live and, in turn, are affected by changing geological, climatological, epidemiological, and ecological conditions. President of the American Society for Environmental History, he writes books on how the New England landscape changed as control of the region shifted from the Indians to the European colonists, on Chicago's relationship to environmental and economic change in its rural hinterland during the second half of the nineteenth century, and on environmental "prophets of doom" from Malthus to the Club of Rome.

Dr. Robert E. Evenson, professor of economics, has done empirical studies on farm households in India, Brazil, Philippines, and Thailand. His global studies on the development and adoption of improved agricultural technology focus on public sector research institutions, invention in both public and private enterprises, and intellectual property rights.

Dr. George M. Furnival, professor of forest management, researches the application of biometrical procedures to forestry and ecological problems. He is working on forest yield prediction and more efficient techniques for sampling biomass.

Dr. Gordon T. Geballe, assistant dean, is studying cities and using concepts of ecosystem ecology such as chemical cycling, energy flow, and succession. A forest ecologist, he is researching the development of community organizations that are trying to identify and solve urban environmental problems.

Dr. John C. Gordon, dean, is currently refining a general model to estimate carbon fixation and allocation in forests. He is strongly interested in the physiological basis of agroforestry and intensive culture systems.

Dr. Stephen R. Kellert, associate dean, is a social ecologist who researches policy and management issues relating to the interaction of people and natural resources—mainly wildlife and biological diversity. Current projects include a three-year study in sub-Saharan Africa on basic values and perceptions relating to the conservation of biological biodiversity, endangered species conservation, and management projects; and a study of methods and concepts for valuing natural resources and the environment. Xerces Society president and a recipient of the Society for Conservation Biology's distinguished individual achievement award, he is co-authoring several books on wildlife policy as well as *Ecology, Economics, Ethics: The Broken Circle* with Dr. Bormann.

Dr. Bruce C. Larson, director of school forests, quantifies the development patterns in forest stands and investigates the effect on growth and yield. Current research includes size/density/growth relationships in single species stands, bole growth effects of asymmetric tree crowns, use of computer simulation models to predict the effect of forest management decisions on changes in forest growth resulting from climate change, and stand management option modeling in upstate New York.

Dr. Robert O. Mendelsohn, professor of forestry policy and of economics, researches various facets of natural resource economics—including solving nonmarket commodities such as air pollution, wildlife, and recreation sites; and pursuing dynamic models of forestry, forest taxation, and common property fisheries.

Dr. Joseph A. Miller, forest history lecturer and librarian, is compiling a reference guide to sources of environmental information to be published by Island Press. Items in the "sourcebook" are arranged by a classification he has developed using brief, number-letter codes. These codes are used in the library's microcomputer database of current environmental information—some 15,000 items drawn from journal articles, books, and technical reports published since 1981.

Dr. Florencia Montagnini, assistant professor of tropical forestry, researches nutrient cycling processes in natural and man-managed ecosystems—particularly the influence of tree species on soil fertility and nutrient cycling as a key aspect when designing and managing agroforesty and reforestation projects. Current research is centered on native tree species with potential value for agroforestry and tree plantations in Costa Rica, northeast Argentina, and Brazil. Future research will be expanded to include more species of economic value in other tropical regions and certain aspects of tree-soil interactions. Her goal is to develop alternatives to deforestation and other tropical forest resource losses—such as properly managed tree plantations and agroforesty systems.

Dr. Charles L. Remington, professor of biology, forest entomology, and museology, concentrates studies in linked areas of evolutionary processes and their genetic control—mostly in insects with some comparisons with birds, mammals, and woody plants mainly in the northeastern USA, California islands, and Colorado Rockies. Research includes measuring levels of sterility; insect-foodplant and

pollination interactions; evolutionary ecology of extinction-prone animals; and mimicry and insect palatability to predators.

Dr. Alison F. Richard, professor of anthropology, of environmental studies and Peabody Museum director, has studied living primates in Central America, tropical Africa, and the Himalayan foothills, with most work focused on the lemurs of Madagascar. *(See also Washington University, Department of Anthropology in this section.)* Investigating life history patterns, demography, and social dynamics in a wild population of 120 captured and marked sifakas *(Propithecus verrauxi)* in southern Madagascar, she is concentrating on fluctuations in the population's sex ratio through simulation modeling; research is beginning on the energetic costs of reproduction. One observation: in some years, most animals born are male and survival among reproductive females appears to drop sharply. This contrasts with most other primate populations where adult females outnumber male adults—a result attributed to differential survival.

Dr. Thomas G. Siccama, forest ecology lecturer and field studies director, researches trace element cycling in the terrestrial ecosystems. Cooperating with the University of Pennsylvania, he is establishing baseline data on the accumulation of trace metals in the forest floor of the northeastern USA. Siccama is also working on the suggested effects of environmental pollution on the growth of forest trees especially in relation to pitch pine and red spruce, which are declining. He is involved with the Hubbard Brook Experimental Watershed Ecosystem project as well as land-use planning, natural areas documentation, and inventory studies.

Dr. William H. Smith, professor of forest biology, investigates fungal ecology and physiology, tree root exudation and associated rhizosphere ecology and biochemistry, and the relationships between air pollution and woody plants and microorganisms. Certain studies are concentrating on aluminum and heavy metals in northern hardwood forest rhizospheres. The interaction of regional air pollutants, including ozone, trace metals, and acid deposition as well as biotic stress factors, such as microbial and insect pests, of trees are current research interests. Another ongoing program involves documentation of the ability of forest soils to sequester and detoxify trace organic air pollutants. Dr. Smith began a major research effort in 1987 to better understand the interactive ecology of fungi belonging to the *Armillaria* and *Trichoderma* genera in northeastern forest soils.

Dr. Bruce B. Stowe, professor of biology and of forestry, researches the physiology and biochemistry of hormone action in plants such as the effects of lipids augmenting hormone action in peas and fig trees and the relation of these to membrane structure. Involved are computer modeling from bioassay data and analytical biochemistry. Other pursuits are hormonal regulation of plant developments, especially in fruit ripening and rooting of cuttings, and the biochemistry of secondary products in plants.

Dr. Kristiina A. Vogt, associate professor of forest ecology and below ground ecology director, is concerned with ecosystem ecology and studies in forest productivity, mycorrhizal ecology, tree physiology, nutrient cycling, plant development, and environmental interactions. Her research seeks to answer how plants respond to disturbances by shifting allocation between above and below ground components and to secondary defensive chemicals. Vogt is investigating how disturbances such as acid rain, aluminum toxicity, fragmentation of ecosystems, gaps, and hurricanes affect ecosystem sustainability in tropical and temperate zones.

Dr. John F. Wargo, assistant professor of forestry and environmental studies and of political science, explores the **relationship between scientific uncertainty and regulatory policy**. He is presently researching the scientific basis of toxic substance regulations. Using microcomputers, he is designing simulation and gaming strategies to estimate ecological, public health, and economic effects of alternative regulatory strategies.

Located in the southeast part of the continent, Uruguay is South America's smallest country. It is bordered by Brazil, the estuary of the Rio de la Plata, and the Atlantic Ocean. The entire western border is defined by the Uruguay River, which separates the country from Argentina.

The topography consists mainly of low plateaus and low hilly regions. Forests cover less than 5 percent of the total area. Indigenous fauna include puma, jaguar, fox, deer, and wildcat. Uruguay has the lowest birth rate of any South American country and a mixed public/private economy. Pastures cover four-fifths of the land area and support large herds of livestock. The European Community is the main market for the country's meat exports.

🏛 *Government*

Museo Nacional de Historia Natural
National Museum of Natural History
Casilla Correo 399
11000 Montevideo, Uruguay

Contact: **Ricardo Praderi Gonzalez**, Curator of Cetaceans

Activities: Research • *Issues:* Biodiversity/Species Preservation • *Meetings:* IUCN/SSC Cetacean Group

▲ Projects and People

Ricardo Praderi is the curator of cetaceans at the **Museo Nacional de Historia Natural (National Museum of Natural History)** and a founding member of the Sociedad Taguató (Taguató Society), devoted to the natural sciences, and the Sociedad Zoológica de Uruguay (Uruguay Zoological Society). As a member of the Species Survival Commission (IUCN/SSC) Cetacean Group, he has developed numerous projects for the World Conservation Union's **Cetacean Action Plan**. These include the continued monitoring of the incidental kill of Franciscanas in Uruguay, and a regional plan to coordinate research on this species of whale in Brazil, Uruguay, and Argentina.

■ Resources

Economic support is needed for developing new cetacean projects. His publications list contains more than 35 scientific articles on the cetaceans of South America.

 *NGO*

Centro Interdisciplinario de Estudios sobre el Desarrollo (CIEDUR)
Interdisciplinary Center for Development Studies
Joaquín Requena 1375
11200 Montevideo, Uruguay

Contact: **Carlos Perez Arrarte**, Forest Project Director
Phone: (598) 2 484674 • *Fax:* (598) 2 480908
E-Mail: Geonet:chasque

Activities: Education; Research • *Issues:* Energy; Forest Management; Sustainable Development; Wetlands • *Meetings:* UNCED

● Organization

The **Centro Interdisciplinario de Estudios sobre el Desarrollo (CIEDUR) (Interdisciplinary Center for Development Studies, Uruguay)** was founded in 1977 to promote research in the social sciences at the national level, publicize the results of this work, and provide assistance to social organizations such as cooperatives, trade unions, craft and merchants' guilds, and farming associations. With 25 members, CIEDUR provides training and new contacts for researchers through special courses and seminars and also executes its own research projects. These projects link environmental issues with the economic and social development of Uruguay because they are "complementary aspects of the same theme: improving the quality of life for present and future generations."

▲ Projects and People

In charge of CIEDUR's forest management projects, agricultural engineer **Carlos Perez Arrarte** focuses on alternative uses of Uruguay's natural resources and the probable environmental effects of these uses, taking into account the technical, social and political factors which determine how natural resources are managed. This comprehensive study includes an analysis of the current agroindustrial forestry complex and changes of the biophysical environment associated with its activities as well as a plan for dealing with the economic and social changes that accompany forest development at the local and regional levels.

■ Resources

CIEDUR's Technical, Economic and Social Advisory Department (DATES) produces a monthly newsletter for social organizations and a quarterly newsletter focusing on economic issues and their relationship to employment and salaries; in addition to this, a library and documentation center are available for use. CIEDUR publishes books, research and statistical reports, notes from seminars and workshops, and brochures and pamphlets, including roughly 15 publications dealing with forest management and the environment. A publications list is available, printed in Spanish.

Red de Ecología Social (REDES-AT) ⊕
Amigos de la Tierra/Friends of the Earth (FOE)
Avda. Millán 4113
Montevideo 12900, Uruguay

Contact: **Karin Nansen,** Ex-Secretary
  Phone: (598) 2 356265 • *Fax:* (598) 2 381640
  E-Mail: Ax: Redesur

Activities: Development; Education; Research • *Issues:* Agroecology; Appropriate Technologies; Biodiversity/Species Preservation; Deforestation; Energy; Health; Sustainable Development; Transportation; Waste Management/Recycling; Water Quality • *Meetings:* FOEI; UNCED

● Organization

REDES-AT, Social Ecology Network, Friends of the Earth (FOE), was founded in 1988 to confront natural and sociocultural environmental problems brought about by current political and economic

systems and the predominant models of development based on the exploitation of nature and of large sectors of the population. It is committed to campaigns and actions directed toward hindering the effects of acid rain, global warming, disappearing ozone layer, growing desertification, spread of radioactive and toxic wastes, and extinction of species.

With 100 members and a staff of 3, REDES-AT promotes meetings and the exchange of ideas and experiences between people and groups striving to create an ecologically and socially responsible ethic. It proposes research and the promotion of development models on a human scale, which pay attention to the needs of everybody, and each body, based on the principles of social ecology. Links are forged at the regional and international level with organizations and people that share these concerns and goals, such as Friends of the Earth International. Research teams are organized on topics such as appropriate technology, health and the environment, renewable energy, agroecology, and social ecology, in collaboration with similar organizations, regional as well as global, and universities. The latter includes various departments of the University of Uruguay and its Environment Commission and UNISINOS, a Brazilian university.

▲ Projects and People

REDES-AT holds **seminars and conferences** about twice a month at its *Casaencuentro* (meeting house), on subjects such as women in development, the use of poisons in agriculture, water pollution, and social movements. The cooperative and educational ecocommunity project is an experiment for alternative development, using appropriate technologies in construction, ecological agriculture, solar energy, and sanitation systems. Working with "peasants and inhabitants of marginal zones," REDES-AT brings them information obtained from the experimental center.

The training center was established in 1989 on a 30-acre farm, with facilities to house up to 90 people including staff and consultants. Its purpose is to meet the needs of "large numbers of people all over Latin America [who] are fleeing from the poverty of farms only to find overcrowding, social alienation, and even more poverty in the cities." REDES-AT believes that "sustainable development, understood as a global strategy to ensure a sustainable future for the present and coming generations, needs to come down to the ground in the form of concrete actions, programs, and policies."

The ecocommunity helps people either "stay on or return to farms, but with better ecological practice and economic return; or to stay in the city and do better there." Programs in **agroecology** include organic family gardening, soil restoration and conservation, crop rotations, biological pest control, ecological animal husbandry, vermiculture, eco-apiculture, composting, and organic waste recycling. **Energy** programs focus on such renewable sources as wind, solar, and biogas, for the household, farm, and cooperative; and energy efficiency and conservation. **Housing demonstrations** show how to build with adobe and other local materials, design housing cooperatives and ecocommunities, clean and recycle water. **Health** programs in herbal medicine, reproductive care, conflict resolution, and preventive measures are also developed.

Among other groups that REDES-AT works with are the Pantanoso Stream Neighbourhood Commission and the Miguelete Stream Neighbourhood Commission, "formed in response to the pollution and problems provoked by these montevidean streams."

■ Resources

REDES-AT publishes *Tierra Amiga* (Friendly earth) magazine about grassroots programs and features articles from other Latin American environmental organizations. More publications are planned.

TRAFFIC Sudamérica/South America ⊕
Carlos Roxlo 1496/301
Montevideo 11200, Uruguay

Contact: **Juan Sebastian Villalba-Macias,** Director
Phone: (598) 2 493384 • *Fax:* (598) 2 493384

Activities: Education; Research • *Issues:* Biodiversity/Species Preservation • *Meetings:* CITES; IUCN; TRAFFIC

● Organization

At the CITES (Convention on International Trade in Endangered Species of Wild Flora and Fauna) Conference in Buenos Aires, Argentina, in 1985, it was agreed that the **TRAFFIC** (Trade Records Analysis of Fauna and Flora in Commerce) Sudamérica (South America) office should be established in Montevideo later that year. The only developing nation among 11 offices in the TRAFFIC network, Uruguay was selected because of its political stability and location in the "heart" of the region comprising Argentina, Chile, Paraguay, Bolivia, Brazil, and Uruguay, according to the World Wildlife Fund—USA (WWF)—supporter of TRAFFIC, along with the World Conservation Union (IUCN). Montevideo is also a principal trading port and headquarters of the Latin-American Organization of Integration (ALADI) that coordinates commerce in wildlife products. *(See also TRAFFIC in the Argentina, Malaysia, UK, and USA NGO sections.)*

In its earliest days, TRAFFIC Sudamérica investigated illegal trading of peccary skins, sales of endangered species, and hunting of Pampas deer. As the *WWF Yearbook* (1987/88) reports, TRAFFIC rescued "live animals from the hands of illegal dealers in Argentina and Paraguay" in 1987, including the extremely rare Spix's macaw (*Cyanopsitta spixii*), and later a group of hyacinth macaw. More recently, TRAFFIC uncovered illegal wildlife traffickers, including members of the Paraguay government and army who were involved in a transaction of 35,000 caiman skins and nearly 3,500 greater rhea skins, according to *WWF News* (March–April 1991).

Today, TRAFFIC Sudamérica remains a watchdog on wildlife trade and conservation. Its purpose is to "monitor and report trade in wild animals and plants and wildlife products from and to all countries in South America." Nonetheless, poaching, illegal hunting, and smuggling of endangered plants and animals continues.

▲ Projects and People

TRAFFIC Sudamérica reports that "George"—the "ultimate loner"—is the last surviving giant tortoise of its kind of Isla Pinta in the Galapagos Islands. Unless a female of George's subspecies is found, he will have the dubious distinction of being the last of his ancient family line. George is just one of many species endangered by destruction of natural habitats and commercial exploitation. Often extinction has negative consequences for human communities as well.

In Uruguay, several years ago, fox hunting was legalized because farmers blamed foxes for killing off their sheep. In reality, however, although the foxes did kill some young sheep every springtime, the rest of the year they ate rats, mice, and rabbits, and so helped to preserve the farmer's crops. When the farmers killed thousands of foxes in a few months and sold their pelts, the rodent population increased alarmingly. The next year, the farmers asked the Ministry of Agriculture and Fishing to help them control the rabbits and other pests threatening their crops.

Juan S. Villalba-Macias is also representative for South American, Central American, and Caribbean countries at the CITES Animal Committee; and vice chair of the Species Survival Commission (IUCN/SSC) Crocodile Group and member of its Cat Group. Author of books and other publications, he presented Uruguay president Dr. Luis Alberto Lacalle Herrera with a WWF/TRAFFIC award in 1990 for creating one Ministry to deal with environmental issues and for his support.

■ Resources

Alerta magazine is published about three or four times a year and includes news of important meetings, major arrests of animal and plant traffickers, successful reintroduction of seized animals, new sightings of rare species, news of environmental education campaigns, and articles on restaurants and stores in South America that sell such illegal products as grilled caiman and jaguar furs. *Alerta* also reprints articles from South American newspapers relating to trafficking in endangered species and scholarly essays on conservation and preservation of endangered species.

Uzbekistan is the third most populous republic of the former Soviet Union, comprising almost 7 percent of the Soviet population. Between 1979 and 1989, the population increased 30 percent, a rate more than three times the average for the USSR. Uzbekistan also accounts for about two-thirds of Central Asia's irrigated land.

Soviet planners built the Uzbek economy almost exclusively around cotton, which led to an unbalanced economy and a serious water shortage due to the massive irrigation required.

A major casualty of this practice has been the Aral Sea, which over the past 20 years dropped roughly 11 meters and doubled in salinity. The heavy use of agrochemicals also has taken a toll, polluting the air, degrading the soil, and causing devastating health effects.

The situation is slowly improving, however, with some cotton production being replaced by feed grains and other crops, and more efficient irrigation methods. Unfortunately, the damage to the Aral Sea may be permanent.

NGO

Museum of Peace and Solidarity
P.O. Box 76
SU-703000 Samarkand, Uzbekistan

Contact: **Anatoly Ionesov**, Director
Phone: (7) 33 1753 (operator assistance)

Activities: Citizen Diplomacy; Education; Research • *Issues:* Air Quality/Emission Control; Biodiversity/Species Preservation; Deforestation; Energy; Global Warming; Health; Human Rights; Peace; Population Planning; Sustainable Development; Transportation; Waste Management/Recycling; Water Quality

● Organization

The **Museum of Peace and Solidarity**, located in the 2,500-year-old city of Samarkand, was set up by local Esperantists in 1986, in honor of the International Year of Peace and inside the new political climate of the (then) USSR.

With over 10,000 exhibits from 100 countries, these include "posters, streamers, paintings, children's drawings, photographs, literature, badges, stickers, T-shirts, films, slides, balloons, medals, original handmade items, . . . everything depicting the most vital global problems: peace, disarmament, development, protection of human rights." It also houses "international large-scale environmental and healthy living exhibitions with the motto 'Our Home—Earth.'"

Among its activities, the Museum pursues a "nuclear-free and non-violent world that can ensure the survival of the human race."

According to the Museum, which seeks to involve itself in all types of global issues and encourage "citizen diplomacy," these exhibits represent the political and environmental realities of the contemporary world. With a staff of three and volunteers, the Museum is "ready to work together for the humanization of international relations. . . . There is no more important task today than to make men of all creeds and countries, of all ranks of life, conscious of the danger right now threatening our very life on earth," writes director **Anatoly Ionesov**.

One Museum department that takes its name from the words of author Dostoyevsky—"beauty will save the world"—displays such objects of nature and symbols of friendship and peace that are gifts from writers and artists "who, through their creative work, make their plea for a better world."

Museum staffers organize permanent and traveling exhibits, collect signatures for various campaigns, and sponsor meetings and discussions with foreign guests—among its activities. They welcome the opportunity "to attend international meetings on the invitation of the organizations."

■ Resources

The Museum is in need of a computer and printer to catalog its collection, VCR to show videos about world problems, office equipment such as copy and fax machines, and resources on global problems. It does not have a formal membership, but rather asks people to join by "offering your help in whatever form you are able."

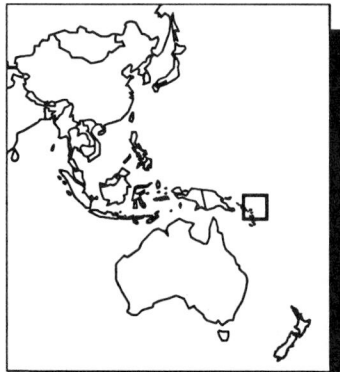

Formerly New Hebrides, Vanuatu is a chain of 13 principal and many smaller islands in the Southwest Pacific, east of Australia. The diverse terrain ranges from rugged mountains and high plateaus to rolling hills and low plateaus with coastal terraces and offshore reefs. Sedimentary and coastal limestones and volcanic rock predominate. Natural disasters, including active volcanoes and frequent earthquakes, are a continuing threat. In 1987 Vanuatu was ravaged by the worst cyclone in history. All the islands are well forested.

Vanuatu has very high population growth and accompanying agricultural pressures. Major threats to the country's biodiversity include severe soil destruction, deforestation, overgrazing, and mining; major timber harvesting; indiscriminate use of pesticides; and illegal poisoning of fish.

🏛 *Government*

Department of Agriculture, Livestock, and Horticulture
Private Mail Bag
Port Vila, Vanuatu

Contact: **Philip J. Dovo,** Director
Phone: (678) 22525 • *Fax:* (678) 23185

Activities: Development; Research • *Issues:* Deforestation; Sustainable Development

■ Resources

With a background in tropical agriculture, **Philip J. Dovo, Department of Agriculture, Livestock, and Horticulture,** seeks new technologies in slope agriculture for soil conservation.

Department of Geology, Mines, and Rural Water Supply
Private Mail Bag 01
Port Vila, Vanuatu

Contact: **Jules Temakon Stanallison,** Director Designate
Phone: (678) 22423 • *Fax:* (678) 22213

Activities: Development; Law Enforcement • *Issues:* Energy; Health; Supervision of Prospect Mining; Sustainable Development; Waste Management/Recycling; Water Quality • *Meetings:* SOPAC

■ Resources

Jules Temakon Stanallison, director designate, **Department of Geology, Mines, and Rural Water Supply,** can provide a publications list, including geology maps, legislation, and regional and general reports. Stanallison, recently with the South Pacific Applied Geoscience Commission (SOPAC)—where he did geological mapping, evaluated industrial minerals, surveyed volcanoes, and investigated geothermal energy—seeks staff for his government department "from international source agencies in hydrology, geology, and chemistry."

Vanuatu

Ports and Marine Department ⊕
P.M.B. 046
Port Vila, Vanuatu

Contact: **Captain Paul Peter**, Harbour Master
  Phone: (678) 22339 (operator assistance)
    Fax: (678) 22475

Activities: Development • *Issues:* Transportation • *Meetings:*
CITES; Marine Pollution Seminar

■ Resources
Captain Paul Peter, Ports and Marine Department, seeks new
technologies and materials on preventing oil spills and related magazines and pamphlets for teaching and public outreach.

 NGO

Vanuatu Natural Science Society (VNSS)
P.O.Box 944
Port Vila, Vanuatu

Contact: **Thorkil Casse**, Secretary
  Phone: (678) 2 2605 • *Fax:* (678) 2 3087

Activities: Education; Research • *Issues:* Biodiversity/Species
Preservation • *Meetings:* ICBP

▲ Projects and People
The Vanuatu Natural Science Society (VNSS) organizes nature
walks and evening events on environmental issues, with the help of
secretary **Thorkil Casse** and Martin Horrocks, former editor of
Naika, the VNSS bulletin.

During recent expeditions to Laika Island in the Shepherd group
of Vanuatu, members observed the wedge-tailed shearwater colony,
which VNSS is trying to protect from harvesting by the thousands
during the breeding season. "We have had discussions with the chief
of the village Kuramambe, and the villagers are willing to protect the
island from harvesting if the government can pass a legislation
converting the island into a natural reserve," reports Casse. "The
Environment Unit, Ministry of Home Affairs, is now pursuing the
issue with the Minister. If agreed upon, Laika Island will be the first
natural reserve in Vanuatu."

The purpose of an expedition to the "mist forests on Santo, the
biggest island in the archipelago," was to determine the status of the
Santo Mountain starling (*Aplonis santovestris*), first recorded in
1933 and last seen in 1934, according to VNSS. "Today, only one
stuffed specimen exists in the British Museum in London." Yet, the
expedition discovered the starling at an altitude of 1,700 meters,
and "took the first photos ever of the bird alive. "In the 1930s, most
ornithological expeditions were executed with a shotgun. The very
rare bird is not in any immediate threat, but once more the issue is
being raised with the environment unit." The Australia and Pacific
Science Foundation helped fund this expedition, which was also
attended by Jim Reside and Peter Montgomery, ornithologists
from Australia and New Zealand.

VNSS writes that the country is presently formulating its first
environment strategy. In-depth studies began in 1992 with support
from the South Pacific Regional Environmental Programme
(SPREP).

■ Resources
Naika is published quarterly for members.

Venezuela

Located on the northern coast of South America, along the Caribbean Sea, Venezuela enjoys vast energy, mineral, and other natural resources, an educated populace, and the highest per capita income on the continent. Most of the population resides along the coast and in major cities; the interior consists largely of undeveloped flatlands and rainforest.

Since the 1970s, Venezuela has been a modern urban country; however, falling oil prices, corruption, and political unrest brought economic decline in the 1980s. In addition to oil reserves, the country has large undeveloped coal fields, and is looking for ways to convert its reserve of bitumen into an inexpensive fuel. In terms of environmental problems, Venezuela lost an average of 1,000 square feet of forest a year in the 1980s, and the flatlands are being overgrazed and degraded. Urban and industrial pollution is most severe along the coast, and air and freshwater are both negatively affected.

🏛 *Government*

National Parks Institute (INPARQUES)
Avenida Romulo Gallegos, APDO 76471
1071-A Caracas, Venezuela

Contact: Jorge Luis Pérez Emán, Biologist
Phone: (58) 2 2854106

Activities: Research • *Issues:* Biodiversity/Species Preservation; Deforestation; Sustainable Development

▲ Projects and People

Pérez Emán reports that he has recently completed studies on river turtles, which, he says, constitute an important alimentary resource of the regional people from the Venezuelan Amazonia. He has investigated the biology, ecology, and use of the turtle in the Territorio Federal Amazonas, Venezuela. The project received financial support from Wildlife Conservation International (WCI) and the EcoNatura.

Currently, Pérez Emán is a researcher for the National Parks Institute (INPARQUES) at the Duida-Marhuaca National Park, also located in the Territorio Federal Amazonas, Venezuela, where he is collecting data to develop a management plan. Working with biologist Maricela Sosa, Los Andes University, he is investigating the frugivory of bats and birds in the rainforest. Such information is useful for future park management. Funding is being sought from WCI.

NGO

Asociación Educativa para la Conservación de la Naturaleza (EcoNatura)
APDO 63.109
1067-A Caracas, Venezuela

Contact: Isabel Novo Torres, Ph.D., General Manager
Phone: (58) 2 923268 • *Fax:* (58) 2 910716

Activities: Education; Research • *Issues:* Biodiversity/Species Preservation; Deforestation; Grants and Courses; Sustainable Development • *Meetings:* World Parks Congress

● Organization

The nucleus of EcoNatura, founded in 1989, is a training program started by Wildlife Conservation International (WCI) and funded by the Jessie Smith Noyes Foundation. Providing an organized national base for both student grants and other associated training programs, EcoNatura states, "We are the only Venezuelan NGO dedicated exclusively to university-level training programs in conservation education and to funding applied student projects in related fields."

Known also for cooperating with other national and international NGOs, EcoNatura is "dedicated to improving the status of conservation on a national level through the promotion of student research, materials in conservation topics, and coursework, while encouraging the further development of appropriate curricula at leading universities [thereby] promoting a broad-based educational plan for Venezuelan conservation."

▲ Projects and People

EcoNatura funds various doctoral and master's theses and postgraduate research projects. Examples are evaluating the "Tendedor" fishing technique and its effects on the fish *Caranx latus*, in Los Roques National Park; ecology and social organization of the spider monkey in the Nichare River Basin; ecological impact of forestry practices on mammal communities in the Imataca Forest reserve; and feeding ecology of the yellow-shouldered parrot on the Island of Margarita. Studies also focus on crocodile, wildlife commercialization, alternative agriculture, bats, and the socioeconomic use of the *Cabezón* turtle in the Amazon.

■ Resources

Currently developing a directory of Venezuelan conservation organizations, EcoNatura presently offers its *Bulletin Communiction Network* publication to a network of Venezuelan NGOs known as the ARA. It plans to begin publishing biannually a short volume of expanded abstracts of student projects.

EcoNatura needs periodicals for its library as well as information on grant organizations.

Fundación para la Defensa de la Naturaleza (FUDENA)
Foundation for the Defense of Nature
APDO 70376
1071-A Caracas, Venezuela

Contact: **Glenda Medina-Cuervo**, Director for Institutional Planning
Phone: (58) 2 238 2930 • *Fax:* (58) 2 239 6547

Activities: Education; Management of Resources; Research
Issues: Biodiversity/Species Preservation; Deforestation; Sustainable Development; Water Quality • *Meetings:* CITES

● Organization

FUDENA, the **Foundation for the Defense of Nature**, is a Venezuelan conservation organization with a dynamic conservation concept based on the maintenance of the essential ecological process, preservation of genetic diversity, and sustainable use of the species and ecosystems as the means for attaining permanent development. FUDENA's action is especially oriented toward the conservation of endangered animal and plant species. Founded in 1975, the nonprofit FUDENA aims to carry out is purposes by "sponsoring, financing, or executing scientific research and conservation projects."

▲ Projects and People

The **Orinoco Crocodile Project** intends to conserve this endangered species that lives only in Venezuela and Colombia and is considered by the World Conservation Union (IUCN) to be among the planet's 12 most endangered species, mainly because of overexploitation for its hide.

Since 1978, FUDENA has been consolidating an integral program to reestablish the species. One phase consists of **Captive Rearing of the Orinoco Crocodile** with co-sponsors Fundo Percario Masaguaral WCI (World Conservation International, New York Zoological Society), World Wildlife Fund (WWF) International, and Smithsonian Institution. Another phase concentrates on the **Status of the Wild Population of the Orinoco Crocodile**, which examines the remnant populations within areas subject to human intervention.

Population Ecology of the Spectacled Caiman in Venezuelan Llanos is a program experiencing a successful recovery, due in part to experimental harvest seasons allowed by the Ministry of the Environment and Renewable Natural Resources.

Fourteen species of the **cracid bird family** of guans, curassows, and chacalacas inhabit Venezuela and are legally hunted and comprise the diet of rural and indigenous people. Susceptible to human disturbance and to habitat alterations, they are recognized as good indicators of habitat quality. WCI is co-sponsoring this conservation project.

With five **sea turtle species** nesting in insular and continental coasts of Venezuela, they are threatened both as adults and as eggs, which are part of the human diet, and their shells are used to manufacture artcrafts—even though the turtles are protected by international and national laws and agreements. FUDENA reports this situation as "critical." It is one reason why FUDENA defines "environmental education" as its most crucial task.

■ Resources

FUDENA seeks funding and equipment for resource management, environmental education, and community development; and assistance with marketing, development planning, membership, and lobbying.

Programa Hortalizas/Vegetable Programs
**Fundación Servicio para el Agricultor
(FUSAGRI)**
Service Foundation for Agriculture
Km 3, via La Segundera
2122 Cagua, Venezuela

Contact: **Jorge Manuel Gonzalez,** Coordinator
Phone: (58) 44 79184 • *Fax:* (58) 44 75607

Activities: Development; Education; Research • *Issues:* Sustainable Development

▲ Projects and People

With a 40-year history of agricultural services, **Fundación Servicio para el Agricultor (FUSAGRI)** is shifting its focus to "more natural and ecological solutions" to crop production and especially in curbing the use of pesticides that pollute water and soil. Entomologist and agricultural engineer Jorge Manuel Gonzalez, who coordinates vegetable programs for FUSAGRI, lists the following current projects to benefit small farmers, which are getting good results:

Usage of plastics in tomato and cantaloupe production to diminish the use of herbicides. "We are actually using and evaluating two kinds of plastic sheets as mulch—black and co-extruded. Even though mulch covers are known and used in developed countries . . . they haven't been successful [in Venezuela]."

Use of alternatives to pesticides to control sweet potato whitefly. "We are evaluating pathogens, oils, neem extracts, specific insecticides, cultural practices, [and] fertilization to use in appropriate integrated pest management (IPM)."

Protection of water sources in Sierra San Luis through low-impact agriculture, soil protection, and nonuse of pesticides. Crops, such as coffee, are being managed to integrate with the forest, reports Gonzalez.

In addition, biotechnology is being developed both to **reproduce orchids in danger of extinction** and to be made available to orchid growers.

Program staff include agricultural engineers **Kinido Gomez Pereira, Dario Rafael Boscan Odor, Hector Honorio Rodriguez Orellana, Pedro Ivan Valenzuela Gonzalez,** and **Alexis Antonio Infante Viloria;** agroindustrial engineer **Enrique C. Avila;** and biologist **Ariadne L. Vegas Garcia.**

■ Resources

All projects, particularly orchid growing, need financial support. A publications list of interest to farmers is available in Spanish. Gonzalez has co-authored numerous articles in English on parasitic wasps and other insects.

**La Fundación Venezolana para la Conservación
de la Diversidad Biológica (BIOMA)**
**Venezuelan Foundation for Biodiversity
Conservation**
Avenida Este 2, Edif Camara De Comercio De
Caracas
Piso 4, Los Caobos
1010-A Caracas DF, Venezuela

Contact: **Aldemaro Romero Diaz,** Executive Director
Phone: (58) 2 571 8831 • *Fax:* (58) 2 571 1412

Activities: Education; Research • *Issues:* Biodiversity/Species Preservation; Deforestation; Environmental Education *Meetings:* IUCN

■ Resources

BIOMA, with membership of 800 and a staff of 40, offers a Venezuela Vegetation Map, *Venezuela State of the Environment Report,* and BIOMA's *Bulletin,* published bimonthly. The conservation organization needs funding.

PROVITA
APDO 47552
1041-A Caracas, Venezuela

Contact: **Christopher J. Sharpe,** Director of
Fundraising and Promotion
Phone: (58) 2 576 2828 • *Fax:* (58) 2 576 1579

Activities: Communication; Education; Research • *Issues:* Biodiversity/Species Preservation; Deforestation; Sustainable Development • *Meetings:* World Parks Congress

● Organization

PROVITA was formed in 1987 by a group of university students who wanted to bridge the gap between the scientific community and rural populations regarding conservation and environmental issues. Their goal is the "conservation of Venezuelan wildlife and the habitats on which it depends, with special emphasis on those species whose survival is in jeopardy." Now with 80 members, aims are achieved through "investigation, publication, and environmental education in both rural and urban settings."

Support is from members and sponsors within Venezuela and grants from foreign institutions: Wildlife Conservation International (WCI), New York Zoological Society; Jessie Smith Noyes Foundation, La Fundación para la Defensa de la Naturaleza (FUDENA), Lincoln Park Zoo; British and Australian embassies in Caracas; and Maraven, a Venezuelan oil company.

▲ Projects and People

The **Spectacled Bear Program** raises the public awareness of this animal among the communities where the animal occurs, and gathers information, in conjunction with the National Parks Institute (INPARQUES) and other local groups and funding from FUDENA and WCI. Presently, the project aims to build an **ethnozoological inventory** of the Sierra Nevada National Park to serve as a base for a comprehensive program of environmental education.

Regarding the **Marine Turtles Program**, PROVITA reports that preparatory work has been carried out in the Paraguaná Peninsula in collaboration with FUDENA. Once funds are secured work will begin on an environmental education campaign for artesanal fishing communities in the peninsula and more widely along the Caribbean coast.

In the **Psittacids Program**, PROVITA is producing educational materials to conserve the parrot endemic to Margarita Island, with help from WCI and Profauna, the Venezuealan Wildlife Service. PROVITA also leads the effort to protect the **oilbird**, to coincide with the recent expansion of El Guáchara National Park. In the **Cracids Program**, particularly regarding the **northern helmetted curassow**, environmental materials are being prepared; and ethnozoological information is being gathered in Sanare.

The **Environmental Education Workshop for Preschool Children**, operating in Caracas, designs educational materials and methods with a conservation/environment theme for use at children's events and workshops. PROVITA maintains close relations with such institutions as the Children's Museum in Caracas.

With the National Parks Institute, PROVITA is lobbying the government to create Turuépana National Park, "a huge mangrove island," harboring **manatee, giant river otter,** and **crocodile**. Agricultural alternatives will be offered to rural communities practicing unsustainable slash-and-burn agriculture in the Paria Peninsula National Park. A study of **tiger beetles** as indicators of habitat quality and forest type in tropical forests is being planned.

The new commission on a **Latin American Conservation Network** is establishing a communication linkup of voluntary conservation groups throughout Latin America to facilitate information exchanges and work on joint projects. The **Threatened Species Commission** is compiling a database, with results to be published in a **Red Data Book** for Venezuela, in conjunction with the World Conservation Union.

■ Resources

PROVITA produces posters, stickers, information flyers, children's games, displays, and audiovisuals, mostly for local grassroots groups. It publishes the quarterly *Bulletin.*

Sociedad Conservacionista Audubon de Venezuela (SCAV)
Venezuelan Audubon Society
APDO 80450
80450-A Caracas, Venezuela

Contact: **Jose Ochoa G., Ph.D.,** Project Manager/ Ecologist
Phone: (58) 2 922812 • *Fax:* (58) 2 910716

Activities: Education; Research • *Issues:* Air Quality/Emission Control; Biodiversity/Species Preservation; Deforestation; Sustainable Development; Wetlands • *Meetings:* American Bat Research Symposium; Audubon Congress; ITTO; IUCN; Latin-American Zoo Congress; World Parks Congress

● Organization

Driven by a growing concern about the progressive deterioration of wildlife and natural habitats, in 1970, a group of Venezuelan and foreign residents interested in nature founded the **Sociedad Conservacionista Audubon de Venezuela (SCAV) (Venezuelan Audubon Society)**, which is modeled after those in the USA.

"Present generations have the obligation to guarantee the survival of future ones; thus, policies affecting natural resource and land use should be based on thoroughly comparing economic benefits with long-term environmental costs," writes the Society, whose activism follows this pattern: "once a problem arises, the volunteers and the staff gather the information required to define and evaluate the situation and prepare a report, then try to give the widest publicity to their analysis, recommendations and suggestions for possible alternatives." With 600 members and a staff of 7, support is from Jessie Smith Noyes Foundation, New York Zoological Society (NYZS), and H. John Heinz III Charitable Trust, among others.

▲ Projects and People

Among its lobbying efforts, Audubon won a victory to keep the salt industry out of the Los Olivitos estuary, a feeding area for **flamingoes** and migratory shorebirds and breeding area for **fisheries** on which local communities depend. With PROFAUNA, the Venezuelan Fish and Wildlife Service, the Society stopped deforestation for a coffee field in an area adjacent to Sierra Nevada National Park, gained the land for the park, and saved the habitat for 80 species of birds, including the "spectacular" **Andean cock-of-the-rock** and **orange-eared tanager**. In other issues, the Society prompted a campaign to return illegally traded Venezuelan parrots to a national park.

The Society does not always win, such as regarding a government "management program" for the spectacled caiman, which it criticized—fearing that the **caiman** population was being imperiled, while the agency issuing the licenses benefits. Audubon is asking for census and status report of this species in the wild. PROFAUNA also licenses the hunting of **jaguars**, targeted for "supposedly preying on cattle"; Audubon opposes their hunting. Presently, SCAV is fighting the destruction of a limestone cave system from quarry development, the construction of a superhighway and industrial salt complex near the Lagoon of Píritu, and the reduction of the Cuare Wildlife Refuge—Venezuela's only Ramsar site—where a golf course, "five-star resort," and a possible sanitary landfill are planned.

Some 14 percent of the territorial surface of Venezuela has been set aside for the System of Forest Reserves and Wooded Lots, which defines area of natural forests set aside for lumber extraction, reports the Society. Eighty-nine percent of this land is south of the Orinoco River, which includes primary forests that have some of the richest levels of biodiversity in the country. Government policies for these reserves foresee intensive development of the lumber industry, and management plans include selected extraction of trees and government-controlled silvicultural methods.

Conflicts are arising over the use of the lands, such as mining, agriculture, the establishment of plantations and urban sprawl. The Society feels that, "where exploitation has occurred, the forests have been substituted by secondary vegetation communities. This, in turn, has brought about the local extinction of animal species."

The Society continues, "The present situation of the Forest Reserves north of the Orinoco have been subjected to exploitation for more than 20 years. The Forest Reserve of Turen has totally disappeared. Its lands are now occupied by farmers and devoted to farming and cattle breeding. Tipcoporo, Caparo, San Camilo, and Rio Tocuyo . . . have suffered extensive invasions by peasants, who slashed and burned significant areas of forest within the reserves."

The **Forest Project** aims to change public policy, increase public awareness about the reserves and the importance of adequate management methods, evaluate existing legislation, and develop a "well-balanced research program based on sustainable methods."

Worried about the declining trends of many **migratory northern birds**, SCAV has launched their study at the Pass of Portachuelo, Henri Pittier National Park. With some causes believed to be pressures in either the wintering or the breeding areas, researchers agree on the need for solid data, especially information gathered where it is most scarce, at the wintering areas in Central and South America. The Society states, "The **Bird Migration Project** developed from . . . the critical need for more information and the availability of an ideal site for data gathering, within a short distance from Caracas and in a national park." Scientists are working from the Rancho Grande Research Station.

Observation of local species are also part of this project. Evidence suggests the patterns of migration of local species may range from 1,000 to 4,000 kilometers within the neotropical region and that short-distance migrations may be connected to weather conditions and/or the availability of food, such as fruit trees and insects.

Ornithologist **Miguel Lentino** is in the process of netting and banding northern species at Portachuelo (using U.S. Fish and Wildlife Service bands donated by the NYZS). The Society reports that "approximately 3,000 records have been collected, representing 220 species. A database has been set up and the data is frequently reviewed with statistical methods."

The Society is cooperating with the U.S. National Audubon on its **Citizens Acid Rain Monitoring Network.**

■ Resources

Monthly bulletins in Spanish, a monthly lecture, and weekly excursions are ongoing resources. The book *Birding in Venezuela* assists foreign birders. In preparation are a *Field Guide to the Mammals of Venezuela*, checklist of birds of Venezuelan national parks, studies of the **Venezuelan coastal wetlands**, and an analysis of **commercial logging** of the lowland tropical forests in southern Venezuela.

The Society requests the following technological and scientific journals, which they cannot afford or obtain in Venezuela: *Journal of Field Ornithology, The Ecologist, Conservation Biology, Journal of Applied Ecology,* and *Biological Conservation.*

 *Universities*

Jardin Botanico/Botanical Garden
Instituto de Botanica/Botanical Institute
Facultad de Agronomia/Agronomy Faculty
Universidad Central de Venezuela
APDO 4579, UCV/Maracay 2101
UJ 79 Aragua, Venezuela

Contact: **Baltasar Trujillo**, Director, Agronomist
            Engineer
   Phone: (58) 43 24565 • *Fax:* (58) 43 453224

Activities: Education; Research • *Issues:* Biodiversity/Species Preservation • *Meetings:* SSC; World Parks Congress

▲ Projects and People

With "approximately 90 percent of the world's plants found in the tropics," according to **Baltasar Trujillo**, only 10 percent of these have been studied, which leaves much work to be done, such as the discovery and classification of new species, studies of their habitats, and the search for new sources of food, medicinal drugs, and plants with other economic uses. The ultimate aim is to "plan for sustainable development where health, water quality, recycling [and other environmental issues] play an important role."

Trujillo supports *in situ* seed banks for endangered species, introduction of desirable species, and improving species. He is concerned that Venezuela's natural botanical resources are being destroyed by progress in industry and agriculture—even though important discoveries in medicine, wood, fibers, resins, rubber, and dyes have yet to be made. He points out that the country needs more botanists to classify collections of flora at herbariums. "The government should help support botanical exploration and create nature reserves in each of Venezuela's different habitats," he writes. The country's 34 native tribes and their territory and cultural values must also be respected.

As the Fairchild Botanical Garden of USA points out, says Trujillo, "A knowledge of tropical plants and their ecology is basic to the development of a rational use of the Earth [in order to] provide a satisfactory quality of life for the growing populations in the tropical lands. . . . The need for conservation is urgent."

The Botanical Garden's activities include **environmental education** for primary, secondary, and university students through exhibitions and promotional campaigns. Research is carried out in the areas of taxonomy, ecology, plant biology, and agricultural botany. **Ecological research** focuses on studying Venezuela's biodiversity, particularly in **aquatic plants**, which are food for many animal species. There is also extensive research being done on weeds, including the systematic collection of their seeds, which is hoped will lead to the development of alternatives to herbicides in agriculture.

With the Botanical Garden bordering the Pozo del Diablo (Devil's Pit) Reserve, the World Conservation Union (IUCN) has a role in its development. Here are found pigeons, partridges, toads, frogs, a huge number of iguanas, chameleons, snakes, bats, mice, rabbits, and deer, as well as aquatic wildlife, such as herons and ornamental fish.

Trujillo recently completed *Important Plant Species in the Venzuelan Amazonas* and is working on a **global inventory** of the specific **composition of dry vegetation types** in Venezuela—the study of which is a main purpose of the Botanical Garden. While researching endangered species, Trujillo is also representing his university in a new project regarding a federation of institutions in favor of a future for the Orinoquia-Amazonia.

He also proposes that environmentalists, especially in the tropics, work with indigenous groups on a concrete conservation endeavor, such as Venezuela's *Unuma* project—meaning "everybody together" in Jiwi language—which is succeeding in the Amazon. A "rapid exchange of knowledge" and meaningful communications at the grassroots level create awareness of the "conservation labor of the people correctly called the 'keepers of the forests,'" writes Trujillo, who belongs to the Species Survival Commission (IUCN/SSC) Cacti and Succulent Group.

■ Resources
The Botanical Garden requests the exchange of scientific literature, seeds, photographs, and related materials.

Programa de Recursos Naturales/Natural Resources Program
Universidad de los Llanos Ezequiel Zamora (UNELLEZ)
Antiguo Convento San Francisco, Carr. 3
3310 Guanare, Portuguesa, Venezuela

Contact: **Luis Enrique Rengel-Aviles**, Field Coordinator
Phone: (58) 57 680006 • *Fax:* (58) 57 511690

Activities: Education; Research • *Issues:* Air Quality/Emission Control; Biodiversity/Species Preservation; Population Planning; Soil Research; Sustainable Development • *Meetings:* CITES; International Botanical Congress; International Congress of Americanism; Latin American Soil Science Congress

▲ Projects and People
UNELLEZ is a regional university system, located in the western plains of Venezuela and dedicated, on the one hand, to studying the particular problems of the region and avoiding, on the other hand, concentrating on a specific urban center. Thus, its contribution to national development is accomplished through the social, scientific, economic, and cultural transformation of Apure, Barinas, Cojedes, and Portuguesa states.

To accomplish its aims, UNELLEZ, with a staff of fifty, is four universities in one—with an office coordinated by a vice rector in each state capital and diverse centers of experimentation in different zones of the region. Areas of activities are agricultural production, infrastructure and industrial processes, regional planning and development, and social planning and development.

Just outside Guanare, in Mesa de Cavacas, are teaching and research facilities—including a herbarium, zoological and Earth science museum, and mapping center—for the agricultural production program as well as an agricultural mechanization workshop, where cows, rabbits, birds, bees, and horses are kept, along with a plant nursery. With preserved forest, savanna, and chaparral areas that are representative of the western Andean piedmont and links to high mountains, high and low plains, and important rivers, the strategic location enhances teaching in real situations.

Once students have completed foundation courses in Barinas and Apure, they may take professional courses in agricultural plant production, agricultural animal production, or renewable natural resources, and graduate with the respective titles of agronomical engineer, zoo technician, or renewable natural resources engineer.

Through teaching, research, and extension, the program aims to contribute to the "rational utilization of renewable natural resources promoting harmony between man and the environment to assure adequate biological support for the human community." The new professionals are capable of understanding environmental and conservation problems and confronting them from a global perspective. Thus, they are able to play "a role of major importance in the future development of the country."

Work opportunities include, for example, the inventory of renewable natural resources, planning and management of river basins, environmental impact studies, planning and executing soil and water conservation, and designing and managing recreational parks and wildlife refuges in both the public and private sectors.

■ Resources
UNELLEZ seeks technologies in tropical willdife research, support from the U.S. Fish and Wildlife Service and Smithsonian Institution, and a conference for Latin American geographers.

Available publications are *Biollania Magazine* and *RNR Technical Bulletin*, printed in Spanish, as well as information on postgraduate courses and master's degree programs in environmental education, natural resource planning, soil and water assessment, and wildlife resources.

Herbario/Herbarium
Programa de Recursos Naturales/Natural Resources Program
Universidad de los Llanos Ezequiel Zamora (UNELLEZ)
3310 Guanare, Portuguesa, Venezuela

Contact: **Francisco J. Ortega M., Ph.D.**, Professor
Phone: (58) 57 68007 • *Fax:* (58) 57 519121

Activities: Development; Education; Research • *Issues:* Biodiversity/Species Preservation • *Meetings:* CITES; International Botanical Congress

▲ Projects and People
Founded in 1980, the Herbarium is a project within the UNELLEZ Museum of Natural Sciences subprogram, which maintains links with the Missouri Botanical Garden and New York Botanical Garden, for example. Research is on environmental impacts on Venezuelan flora and featured is a collection of over 65,000 specimens of flora from Venezuela's Portuguesa, Apure, Barinas, and Cojedes states. Specimens are on public display. Recent projects include studies of the taxonomy and geographical distribution of the fern *Pteridium* and of arthropod communities found in bracken in the Venezuelan Andes.

Prof. Francisco J. Ortega, who also belongs to the Species Survival Commission (IUCN/SSC) Northwest South American Plant and Pteridophyte groups, has authored numerous scientific articles and three books, including *Helechos del Estado Portuguesa* (Ferns of Portuguesa State).

■ Resources

The Herbarium offers issues of *Biollania*, the scientific newsletter *Noti Port* (News from Portuguesa) that features articles on ethnobotany, and scientific monographs. Prof. Ortega seeks a microcomputer for a biodiversity database, outreach, and exchanges with other professors, and herbarium cases.

Vietnam

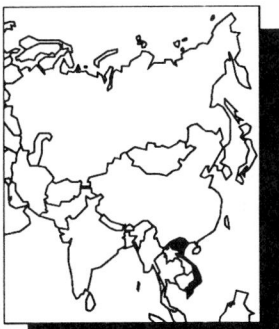

Located in Southeast Asia, Vietnam borders China, Laos, Kampuchea, and the South China Sea. Monsoon rains occur during both the hot summer and the cooler winter. The southern region is extremely fertile for rice cultivation because of its is low, flat, marshy composition; most of Vietnam is hilly, however, and covered by thick jungle. The northern lowlands are heavily populated and more industrialized.

The country still bears the scars of the bulldozing, bombing, and chemical defoliation of the Vietnam war. Forests have never recovered, fisheries lost their variety and productivity, and the decimation of mangrove swamps led to a serious loss of wildlife habitat. In addition, the population doubled in the postwar decade, compounding environmental problems. A national conservation strategy, including massive reforestation, will be required to reverse current trends.

🎓 *Universities*

Centre for Natural Resources Management and Environmental Studies
University of Hanoi
19 Le Thanh Tong
Hanoi, Vietnam

Contact: **Ha Dinh Duc, Ph.D.**, Professor of Zoology
Phone: 42 53506 (operator assistance)
Telex: 411556 TNMT VT

Activities: Education; Research • *Issues:* Biodiversity/Species Preservation • *Meetings:* World Parks Congress

● Organization

With the **University of Hanoi** since 1963, **Dr. Ha Dinh Duc** has conservation interests ranging from the endangered **kouprey** (wild ox) and **rhinoceros** to **primates**, particularly the **snub-nose monkey**, and Vietnamese **catfish**. Much of his literature stresses the need for wildlife to make a comeback and a love of nature. "Do Not [Let] Nature Shun Us," is the title of one recent article.

Dr. Dinh Duc has collaborated on kouprey surveys with both **Roger Cox**, World Wildlife Fund (WWF), and the University of Cambridge, UK; as well as with Cox and Poland's **Radoslaw Ratajsnczak** on primate surveys in North Vietnam. He participated in the Smithsonian Institution's National Zoological Park's Wildlife Conservation and Management Training Program, USA. A member of the Species Survival Commission (IUCN/SSC) Asian Wild Cattle Group, Dr. Dinh Duc wrote his country's reports for the SSC's Caprinae and Cat Action Recovery Plans.

Environmentalists can enhance global cooperation, he believes, through the publication of directories, "networking globally and acting locally," international organizations' support of environmental projects in developing countries, and "coordinating and collaborating studies between environmentalists."

Department of Vertebrate Zoology ⊕
University of Hanoi
90 Nguyen Trei Strees
Hanoi, Vietnam

Contact: **Le Vu Khoi, Ph.D.**, Chairman, Professor
Phone: 42 53506 (operator assistance)

Activities: Education; Research • *Issues:* Biodiversity/Species Preservation; Sustainable Development • *Meetings:* CITES; International Otter Colloquium

● Organization

Dr. Le Vu Khoi earned his doctorate at the University of Leningrad, former USSR, while a biology lecturer with the University of Hanoi, where he has taught and conducted research for over 30 years and currently chairs the Department of Vertebrate Zoology.

A member of the Species Survival Commission (IUCN/SSC) Asian Elephant and Asian Wild Cattle groups, Dr. Khoi has been coordinating a project on researching and conserving the elephant in Vietnam. He also belongs to the Society of Zoologists of Vietnam and the Society of Environmental Conservation of Vietnam. Among his 50 papers and reports on mammology, endangered species, and wildlife conservation—including otter, kouprey (wild ox), and elephant—is literature for children on *The Life of Birth* and *The Life of Mammals.*

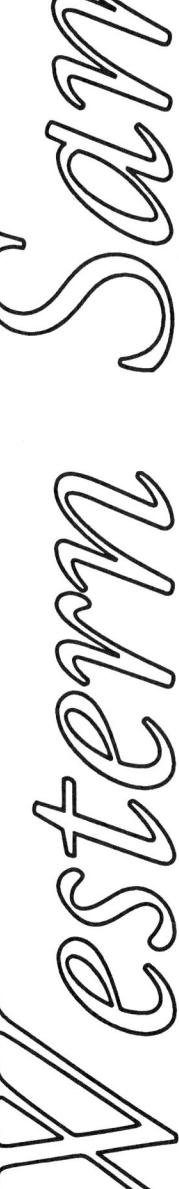

Western Samoa

Western Samoa is a small island group in the South-Central Pacific. It comprises the western half of the Samoa Islands. Both major islands of Western Samoa—Upolu and Savai'i—have numerous flowing rivers with rapid waterfalls. The volcanic soils are rich but porous and easily exhausted. The island centers are forested with tall evergreens. Mangroves thrive in the lower elevations. Animal life is sparse with only flying foxes, small bats, lizards, and snakes. Birds are more common. The economy is based mainly on agriculture, fishing, lumber, and tourism.

Major threats to the country's biodiversity are timber harvesting, standardization of garden and agricultural crops, indiscriminate use of pesticides and illegal poisoning of fish. In addition to these human activities, natural disasters such as tropical cyclones and floods have threatened some rare species.

🏛 *Government*

Department of Lands and Environment
Private Bag
Apia, Western Samoa

Contact: **Pati K. Liu Sailimaio**, Environmental Planning Officer
Phone: (685) 224822 • *Fax:* (685) 23176

Activities: Development; Education; Law Enforcement; Political/Legislative; Research
Issues: Air Quality/Emission Control; Biodiversity/Species Preservation; Deforestation; Energy; Global Warming; Health; Population Planning; Sustainable Development; Transportation; Waste Management/Recycling; Water Quality • *Meetings:* SPREP; UNCED

■ Resources
Pamphlets on national parks and reserves and posters featuring bird, fish, and other wildlife can be ordered from the Department of Lands and Environment, staff of 40, which encourages "the creation of a global association of environmentalists." Pati Sailimaio reports that educational resources are needed to teach environmental topics on primary and high school levels. Technologies on sewage/waste disposal are sought for new urban facilities.

Located directly on the equator in eastern central Africa, Zaire is the third largest country on the continent. The landscape varies from tropical rainforests (one-third of the total area) to mountains, plateaus, savannahs, and grasslands. The primary forests are home to over 100 different tree species. The native wildlife are among the country's richest resources, and include lions, leopards, and mountain gorillas.

Most of the population is in the cities, where water contamination is a major problem. Despite the fact that Zaire has abundant forest resources—including the second largest area of tropical forest in the world—poor management and land-use practices have led to permanent tree loss. Wildlife too are endangered, because of poaching. While the government has indicated a commitment to conservation of forests and biological diversity, it is hampered by lack of economic resources.

NGO

Societe Africaine d'Etudes sur l'Environnement et la Conservation de la Nature (AGRIBO)
African Society for the Study of the Environment and Conservation of Nature
B.P. 5698
Kinshasa Gombe 10, Zaire

Contact: **Longo Lond Efengu Isa**, Director-General, Ecologist

Activities: Education; Research • *Issues:* Agroforestry; Air Quality; Soil Conservation; Sustainable Development

▲ Projects and People

A member of the World Conservation Union (IUCN), **AGRIBO** collaborates with Zaire's Ministry of the Environment and National Reforestation Service on the **Ngafula** project to benefit agroforestry in Kinshasa. The project aims to fight soil erosion, create "green spaces" to generate oxygen; contribute to the local economies by planting fruit trees; and prevent future "anarchical construction." AGRIBO combines traditional techniques with modern science to achieve its results, such as planting in antierosion bands, planting to follow curves and levels, planting hedgerows, and improving fallow forests. AGRIBO grows watermelons and avocado, orange, and mango trees.

Director-General **Longo Lond Efengu Isa** has researched the environment in Africa and Europe—participating in conferences and colloquia in Moscow, Paris, San Jose (Costa Rica), and Brazzaville (Congo). He edits a scientific review on African environmental issues, and has several ongoing studies.

■ Resources

In its densely populated country, AGRIBO encounters many difficulties: lack of money, materials, collaboration with public and international institutions, qualified personnel, and has difficulty obtaining credit. In order to accomplish its mission, AGRIBO says a moral, financial, and technical framework is desperately needed to guide and support work in Zaire. At a minimum, AGRIBO reports that it needs $15,000 for its current projects.

Zambia

Zambia, a landlocked country with comparatively low population and little industrial development, is the home of Victoria Falls, one of the world's greatest waterfalls. The wildlife is among the richest in Africa, and consists of herds of elephants and impalas as well as 4,600 species of plants. Nineteen national parks have been established since the country gained independence in 1964.

Zambia has been forced to diversify its economy due to declining prices for copper and cobalt, which accounted for most of its export earnings. Agriculture and tourism have been expanded. Environmental problems include soil erosion; deforestation, resulting from slash-and-burn agriculture, firewood gathering, and coal production; range degradation from overgrazing; and health problems from contaminated water and inadequate sanitation.

🏛 *Government*

Administrative Management Design for Game Areas of Zambia (ADMADE)
P.O. Box 82
Mfuwe, Zambia

Contact: Dale M. Lewis, Ph.D., Technical Advisor/Training Coordinator

Activities: Education; Development; Law Enforcement; Publications; Research • *Issues:* Biodiversity/Species Preservation; Deforestation; Population Planning; Sustainable Development

● Organization

"ADMADE is a blessing for Zambia," writes Luke Daka, permanent secretary for tourism. "Without it I am sure our great wildlife estate would be imperiled beyond any reasonable way of safeguarding it. Today, unlike earlier years of unabated wildlife slaughter, we find local residents assisting National Parks and Wildlife Services [NPWS] in the protection and management of wildlife throughout the nation. We also find more qualified personnel undertaking these responsibilities. . . . As a result, wildlife is better protected and illegal, wasteful uses are declining. Only in this way will there be a long-term future for tourist development in Zambia."

Representing the official policy of NPWS, the Administrative Management Design for Game Areas of Zambia, staff of 450, involves "local community participation in the management of wildlife resources and the distribution of its economic benefits." ADMADE funding comes from safari hunting concessions and a 50 percent retention of all licenses issued for sustained yield uses of wildlife. The U.S. Agency for International Development (USAID) and World Wide Fund for Nature (WWF) provide support and technical assistance.

As Dr. Patrick Chipungu, NPWS director, advises, "ADMADE was conceived through scientific methods in close consultation with traditional village communities in the game management areas."

Led by NPWS trained leaders, here is how the administrative wildlife management units are organized, according to ADMADE. If the unit is "sufficiently stocked with wildlife to support its management costs, a Wildlife Management Authority is established to help formulate policies for the planning and direction of that unit. Chaired by the District Governor, members include local chiefs, ward chairmen, local MPs [military police], managing director of the safari hunting company operating in the unit, wildlife warden, and unit leader. Wildlife Management Sub-authorities are created to approve local resident hunting permits, resolve wildlife management problems within chiefdoms, and approve community development projects." Local chiefs chair Wildlife Management Sub-authorities, engage local residents in the planning and decisionmaking process for local wildlife, and represent the views of their constituencies.

ADMADE's involvement in community government and its giving communities a share in the wildlife resources creates trust among the communities, ADMADE, and NPWS. It is also leading to a decrease in poaching, and an increase in arrests of poachers, reports ADMADE. By involving the community, it is creating allies where they have long been needed.

Dr. Dale Lewis, who serves as a technical advisor, training coordinator, and WWF-USA project executant, says, "A good conservationist . . . will establish strong and diplomatically positive links with those government offices which regulate resource use and management. People who are able to make the bridge between real world field problems and the hallways of government need the encouragement and support from those organizations professing to serve the cause of conservation."

Among its endeavors, the ADMADE's Luana Wildlife Management Unit was the first to succeed in launching a major ecotourism project. A "multiple use approach" is being adopted to attract tourists to "river safaris, walking photo safaris, game ranching, and exclusive hunting safaris."

Sixteen grinding mills are being constructed throughout the country using ADMADE funds. Schoolrooms, senior staff house, tanning tanks for processing hides, and a health clinic are being built with the community development funds. Mapping, correlated with a computerized data system, is underway to facilitate land-use planning. Leathercraft courses are offered. Staff have been sent to the University of Zimbabwe and the Mweka College of African Wildlife Management in Tanzania to enhance wildlife management skills. NPWS is hiring a development officer to help facilitate such projects.

New programs, such as those beginning in Namibia, are based on the ADMADE concept.

Dr. Lewis is also interested in "sustainable tree yields and local market economies based on indigenous tree uses of tropical deciduous species, large mammal ecology, fire ecology . . . and appropriate technology transfers to Sub-saharan Africa."

He also comments on professional researchers in the field. "They have little if any job security, are rarely communicated with by their support agency, and generally find themselves very much out of the mainstream of conservation thinking. This is sad because I feel the mainstream thinking could be more enhanced with the real-world insights by those who live and work close to the problems."

■ Resources

The *Zambian Wilderness and Human Needs Newsletter* is published. Overseas volunteers are needed as ADMADE facilitators.

Species Protection Department (SPD)
Anti-Corruption Commission
P.O. Box 50486
15102 Lusaka, Zambia

Contact: **Paul Russell**, Director of Operations
Phone: (260) 1 229377 • *Fax:* (260) 1 251397

Activities: Law Enforcement • *Issues:* Species Preservation

● Organization

The **Species Protection Department (SPD)** of the **Anti-Corruption Commission** believes that conservation organizations alone cannot stop animal trafficking. The capture and sale of endangered species are frequently masterminded by people who are "very sophisticated and wealthy, but criminal, entrepreneurs who finance these criminal activities." Unfortunately, corruption also plays a role in commercial poaching, but it is these corrupt officials who are also targeted by SDP. People apprehended by the SPD face a mandatory minimum prison term of five years. This successful approach is gaining worldwide attention.

Working with the World Wide Fund for Nature (WWF), SPD reminds, "Poachers are ruthless in their illegal war. They are heavily armed with weapons that kill not only their prey, but also the legal protectors of our wildlife." To stop the "slaughter of man and beast," SPD sets four tasks: "learning about planned poaching operations; tracing the ownership of weapons used in poaching activities; apprehending the people responsible for the illegal trade in wildlife products and securing their prosecution; [and] promoting public cooperation through mass-media awareness and information campaigns."

With the African elephant and black rhino among the victims, SPD reminds the public that these animals with their horns and tusks—and live birds as well—belong to all the people of Zambia. "We can earn more money for our nation by keeping these animals alive and free. Tourists pay a lot of money to come and see our wildlife. That money helps to build clinics and schools. It helps us all towards a better standard of living for the future."

Luangwa Integrated Resource Development Project (LIRDP) ❦
P.O. Box 510249
Chipata, Zambia

Contact: **Dr. Richard H.V. Bell**, Co-Director, Technical
Phone: (260) 62 21126 • *Fax:* (260) 62 21092

Activities: Development; Law Enforcement; Research • *Issues:* Biodiversity/Species Preservation; Deforestation; Population Planning; Sustainable Development; Tourism; Transportation; Wildlife Management

▲ Projects and People

Established in 1986, the **Luangwa Integrated Resource Development Project (LIRDP)** is located in the central part of the Luangwa Valley, in an area known as Malambo, covering an area of about 14,000 square kilometers with a population of some 40,000. Here are important natural resources, especially wildlife, for which the Luangwa Valley is famous. Because poaching has considerably reduced the wildlife population, the program aims to improve the standard of living through sustainable development of resources, to ensure long-term self-sustainability, and to transfer control of the resource use to the local communities.

Getting underway in 1988, LIRDP reports that it has reduced famine, poaching, and misuse of wildlife—especially elephants; and increased the number of usable roads, park revenues, and community involvement. The Zambian president chairs the steering committee for LIRDP, which resembles the CAMPFIRE project in Zimbabwe and Zambia's ADMADE and Wetlands Programmes. **Fidelis B. Lungu** is LIRDP management co-director.

LIRDP has established several **resource-use businesses** as subsidiaries, such as Malambo Safaris, a safari **hunting** operation; Malambo Trails, **a photographic walking safari** company; Malambo Milling, a **maize production** plant; and Malambo Transport, a **public transport and heavy haulage** utility.

Co-director **Dr. Richard Hugh Vincent Bell**, with a background in zoology at Manchester University (UK), early ecological fieldwork on **grazing ungulates** and their use of the herb layer in the Serengeti National Park (Tanzania), and research on the black lechwe antelope and Bangweulu wetland system in northern Zambia, belongs to the Species Survival Commission (IUCN/SSC) African Elephant and Rhino, Equid, and Antelope groups. Authored and co-authored articles include "Conservation with a Human Face; Conflict and Reconciliation in African Land-Use Planning," and "Tracing Ivory to Its Origin: Microchemical Evidence."

NGO

Chimfunshi Wildlife Orphanage
P.O. Box 11190
Chingola, Zambia

Contacts: **David Siddle**
          Sheila Siddle

Activities: Education; Law Enforcement; Research • *Issues:* Biodiversity/Species Preservation

● Organization
Even with a CITES ban on the trade of **chimpanzees**, poachers hunt the endangered species for food or to capture infants to supply the local trade or to be sold to dealers for the pharmaceutical or entertainment industries. (CITES is the Convention on International Trade in Endangered Species of Wild Fauna and Flora.)

When the Zambian government confiscates chimps from illegal traders, it sends them to the **Chimfunshi Wildlife Orphanage,** where **David and Sheila Siddle** either care for them in captivity or release the animals on their 10,000-acre ranch. The Siddles also aid **baboons, vervet monkeys,** and several species of **birds.**

"The first chimpanzee arrived at Chimfunshi in 1983," writes Carole Noon, Ph.D. candidate, University of Florida (*Animal Keepers' Forum, 1991*). "The chimpanzee was suffering from a severe facial wound and several of his teeth had been smashed. The Siddles' success in caring for this orphan prompted Zambian game rangers to send more confiscated chimpanzees to Chimfunshi."

For the chimps that cannot be released into the wild, the Siddles provide an environment as close to their natural environment as possible. Some 2,000 acres of the ranch have been designated as a chimpanzee sanctuary. The first step was the construction of a seven-acre walled compound where the chimps could establish themselves as a social group before being transferred to the new sanctuary.

When Noon returned to Chimfunshi in 1990, with funding from the Jane Goodall Institute, she assisted in the care of 10 newly arrived chimps. At that time, "six of the new arrivals are less than 1.5 years old. A seventh was brought to the ranch during my stay. The cruel and crude capture methods that I had read about became more real to me. Mothers, older siblings, and other protective group members, sometimes pursued by dog teams, are shot or poisoned so poachers can remove the infants which continue to cling to their dead or dying mothers. The orphans arrived at Chimfunshi traumatized and often [were] injured themselves. Receiving little care on their journey from the bush, many arrived dehydrated, sick, and malnourished."

Noon also reports that new chimps cannot be integrated into the existing group, and a new compound with feeding stations must be built to accommodate a second social group. At the same time, inflation rates are high, money is tight, and funds are desperately needed, particularly for the long term. "Captive chimpanzees can live to be 50 years old. The newly arrived orphans can still be expected to require care in the year 2039," writes Noon.

With an estimated 5 chimps killed for each chimp that is rescued, as of late 1991, the 34 chimps enjoying refuge at Chimfunshi statistically represented some 170 dead members of this species "that is rapidly losing numbers." Chimfunshi is helping protect the remaining wild populations.

■ Resources
Funding is sought to care for the orphaned chimps, as is help with educational outreach.

Chongololo and Conservation Club Leaders Association (CCCLA) ❧
Wildlife Conservation Society of Zambia (WCSZ)
P.O. Box 30255
Lusaka, Zambia

Contact: **Wasamunu Charles Akashambatwa**
Phone: (260) 1 254226

Activities: Education • *Issues:* Biodiversity/Species Preservation; Deforestation; Energy; Global Warming; Sustainable Development

● Organization
The **Chongololo and Conservation Club Leaders Association** (CCCLA) runs Chongololo clubs (Choco Clubs for short) for primary school students, and Conservation Clubs for members in secondary school and college. Chongololo is the name of a Zambian millipede that curls up into a ball when it is disturbed, a funny sight to young children who often call it the "wheel-worm." Prominent during the rainy season, it is also the source of names for local tribes.

CCCLA was founded in 1981 by **Marrianthy Noble,** who at that time was an education officer for the Lusaka branch of the Wildlife Conservation Society of Zambia (WCSZ). She formed the Association to create a forum where teachers could meet and discuss club management and share ideas. This brought teachers from all over the country together in ways that had not ben possible before. The number of clubs also grew from 50 to 286—with at least 50 children in each. Noble is now trustee for the David Shepherd Conservation Foundation for Wildlife and the Environment, UK, which helps support CCCLA's publications. These describe, for example, the value of local wetlands such as Kafue Flats, an acacia habitat and its wildlife, birds of Zambia such as the orange-breasted bush shrike, and *The Adventures of Mr. Chongololo.*

▲ Projects and People

Teacher, education officer, and "trying conservationist" **Wasamunu Charles Akashambatwa** coordinates club activities in schools, works with club leaders, and runs educational meetings nine times a year. He also visits clubs, where he gives talks on environmental issues, organizes antipollution and fruit tree–planting campaigns, introduces environmental education leaders from other countries, shows conservation films, and inspires ecological songs, debates, and dramas. In 1989 he arranged a visit for club leaders to Malawi's Wildlife Clubs.

■ Resources

CCCLA distributes *Chipembele: Environmental Science Magazine for Secondary Schools* and *Bata Chongololo: The Children's Wildlife Magazine* with the help of WCSZ and the World Wide Fund for Nature International (WWF). Additional funding is sought.

International Council for Research in Agroforestry (ICRAF)
SADCC/ICRAF Agroforestry Project
Chalimbana Agricultural Station
Lusaka, Zambia

Contact: **Cherrnor Sullay Kamara, Ph.D.,** Soil Scientist
Phone: (260) 1 293080 • *Fax:* (260) 1 293080

Activities: Research • *Issues:* Biodiversity/Species Preservation: Deforestation; Energy; Global Warming; Sustainable Development; Waste Management/Recycling

■ Resources

Dr. Cherrnor Sullay Kamara, ICRAF soil scientist, recently co-authored a series of papers on "Intercropping Maize and Forage Type Cowpeas in the Ethiopian Highlands"; "Concentration, Macronutrients Yields, and Crop Residue Quality"; "Irrigation Effects on Lowland Arid Soils and Cotton Growth and Yield"; and "Soil Moisture and Maize Productivity under Alley Cropping with *Leucaena* and *Flemingia* Hedgerows in Semi-Arid Conditions in Chalimbana, Zambia," among others.

He seeks soil and plant analysis research equipment.

(See also International Council for Research in Agroforestry [ICRAF] in the Kenya NGO and Uganda Government sections.)

Wildlife Conservation Society of Zambia (WCSZ)
P.O. Box 30255
10101 Lusaka, Zambia

Contact: **Nikki Ashley,** Executive Director
Phone: (260) 1 254226

Activities: Advisory; Education • *Issues:* Biodiversity/Species Preservation; Deforestation; Sustainable Development; Waste Management/Recycling

● Organization

The Wildlife Conservation Society of Zambia (WCSZ) is a nationwide NGO with a staff of 4 and a membership of 2,000. The oldest NGO in Zambia, its advice is often sought "by governmental and non-governmental organizations alike." The Society has links with a number of other conservation groups including Chongololo and Conservation Club Leaders Association of Zambia, Commercial Farmers Bureau, Wildlife Producers' Association, Curriculum Development Centre, Zambia Environmental Education Programme, National Conservation Committee, Forestry Association of Zambia, Zambian Ornithological Society, and Environment and Population Centre. Among other global groups, WCSZ is affiliated with the World Conservation Union (IUCN) and the World Wide Fund for Nature (WWF).

▲ Projects and People

Hit hard by the Zambian "economic situation," the school system has a shortage of educational materials. Consequently, the Society welcomes the opportunity to create environmentally sound teaching materials for the schools. "A set of books [are] based on the national character, 'Mr. Chongololo,' who is the spokesman for conservation issues," writes Nikki Ashley, WCSZ executive director. The series includes *The Adventures of Mr. Chongololo*—books of questions and answers, puzzles, songs, common trees, favorite stories, and bird migration.

WCSZ conducts one-day mini-workshops in select centers to reach teachers with skills and activities that foster environmental education. The Society anticipates that teachers who participate will not only teach their students, but will also help other teachers to develop skills in environmental education. It also holds local and national exhibits on conservation education. Also with WWF sponsorship, the Society produces *Chimpembele Magazine*, an environmental periodical for schools.

With the Zambian Ornithological Society, WCSZ launched the Owls Want Loving awareness campaign to change people's attitudes. "The end result is a primary school English reader written and illustrated entirely by the children," reports Ashley. She adds that the book will be used in all primary and secondary schools.

Among recent actions are WCSZ's role in holding a national seminar that helped clear the way for game ranching in Zambia. Although the government was "initially apprehensive, . . . especially as it would involve significant changes in the laws with regard to ownership of wildlife, . . . there are now 13 registered game ranches," writes Ashley.

She believes that environmentalists can enhance future global cooperation through realistic lobbying at the national and international levels, and "by supporting attempts to develop within social frameworks [and] by allowing environmentalists to work within their own regions and national boundaries, then giving them the opportunity to share their experiences with others at international fora." Ashley thinks that the proceedings of international conferences should be distributed more widely, so groups that do not have the funding to attend can still benefit from the knowledge.

An observation that Ashley makes about global environmental activities is that "in general Third-World countries are not credited with having the knowledge to prepare and substantiate their own project proposals. Too often, a country's nationals are not given the credibility they warrant and are therefore ignored in favor of overseas 'consultants/experts.' This often leads to inappropriate advice and actions."

Former owner of a recording studio, Ashley writes and produces conservation scripts, conferring with "societies in other African countries, including Liberia and Botswana, [which are] now designing their own radio-education conservation programs." Ashley is also involved in a nationwide **adopt-a-log campaign** "to encourage low-cost micro-observation in any available habitat by anyone possessing a note pad and pencil."

■ Resources
WCSZ seeks educational outreach support and materials in reaching children and adults in Zambia.

Zambia Country Office
World Wide Fund for Nature (WWF)
P.O. Box 50551
Lusaka, Zambia

Contact: Richard Cleveland Vanson Jeffery, Country
Representative, Biologist
Phone: (260) 1 253749 • *Fax:* (260) 1 250658

Activities: Development; Education; Research • *Issues:* Biodiversity/Species Preservation; Sustainable Development • *Meetings:* CITES

▲ Projects and People
In Zambia, the World Wide Fund for Nature (WWF) works with the Administrative Management Design for Game Areas of Zambia (ADMADE), Species Preservation Department of the Anti-Corruption Commission, and the Wildlife Conservation Society of Zambia (WCSZ), among other groups.

▮ *Private*
City Investments Ltd.
P.O. Box 30093
Lusaka, Zambia

Contact: Peter T.S. Miller, Conservation Farmer,
Civil Engineer
Phone: (260) 1 228682

Activities: Development; Wildlife Ranching • *Issues:* Sustainable Development

Gabriel Ellison
Artist/Sculptor
Box 320122
Lusaka, Zambia

Phone: (260) 1 261378

Activities: Conservation Education Through Art • *Issues:* Biodiversity/Species Preservation

▲ Projects and People
Gabriel Ellison is a Zambian artist who uses her talents to aid the conservation movement. She has donated the proceeds of her shows to organizations such as the Wildlife Conservation Society of Zambia (WCSZ) and the Zambia Ornithological Society. Ellison contributed artwork for a 1992 wildlife calendar that raised money for wildlife organizations. She also has illustrated such books as *Common Birds of Zambia*, with text compiled and donated by Wildlife/Ornithological Society members Dylan Aspinwall, Terry Taylor, Grace Conacher, Jim Grant, and Alstair Scott; *Common Mammals of Zambia*, similarly compiled by Richard Jefferres, Leo O'Keefe, Rory Nefdt, and Robert Monroe; and *118 Rare Birds of Zambia*. These books are primarily aimed at young people as an educational tool.

Among her work for the Zambian government, she designs the country's flags, coins, Parliamentary Mace, medals, and a number of postage stamps. She recently designed an important stamp of another sort, the First International Non-Resident Safari Stamp (with limited edition prints). Upon entering, all non-resident visitors to Zambia's national parks will be required to purchase and sign these stamps. The money raised will go to the Wildlife Conservation Revolving Fund of the Zambian National Parks and Wildlife Service for explicit use to protect endangered wildlife, such as the elephant, and their habitats.

For example, less than 600,000 elephants are in Africa today, compared with over 1.5 million a decade earlier—victims of poachers who sell their ivory tusks for "everything from jewelry and trinkets, to statues, piano keys, billiard balls, and chopsticks," according to WCSZ.

For her service to the arts, Gabriel is a Grand Officer of the Order of Distinguished Service, presented by the Zambian government, and a member of the Order of the British Empire, bestowed by the British government. A fellow of the Royal Society of Arts, Gabriel exhibits her paintings of landscapes, birds, and animals worldwide.

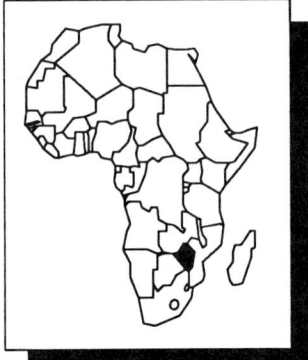

Zimbabwe is a landlocked country bordered by South Africa, Zambia, and Mozambique. The terrain consists of a central plateau, grasslands, and mountains in the eastern region. It shares the world's largest waterfall, Victoria Falls, with Zambia.

Agriculture is the chief industry (especially tobacco, corn, tea, sugar, and cotton), but the manufacturing sector and infrastructure are also strong. Wood is the population's primary energy source. The country has an effective conservation program, with game reserves that contain over 50,000 elephants and the largest black rhinoceros population in Africa. Many of Zimbabwe's environmental problems were exacerbated by the years of civil war and international economic sanctions. Land degradation due to overuse, deforestation caused by slash-and-burn agriculture, and pollution caused by industrialization are continuing concerns.

🏛 *Government*

Agricultural Development Authority (ADA)
Box 8439, Causeway
Harare, Zimbabwe

Contact: **Liberty Mhlanga**, General Manager
Phone: (263) 4 700099 • *Fax:* (263) 4 705847

Activities: Development; Research • *Issues:* Agricultural Development; Sustainable Development • *Meetings:* International Policy Council on Agriculture and Trade; IUCN; UNCED; WRI

● Organization

A quasi-government or "parastatal" organization (run similar to a public utility), the current Agricultural Development Authority (ADA) is a merger of the Sabi-Limpopo Authority, the Tribal Trust Land Development Corporation, and ADA—organizations formed "during the pre-independence era." Today the ADA offers consultancy services to international, regional, and local clients in fields including agricultural planning, land-use planning, rural development, farm management, finance management, administration and human resources management. ADA employs agronomists, planners, cost and management accountants, hydrologists, ecologists, and economists. Among services offered are remote sensing, aerial photography, soil mapping, irrigation planning, soil coding, marketing, and manpower training.

Since its inception, the ADA has grown from 13 estates in 1982 to 28 in 1990. The estates grow coffee, produce milk, and maintain livestock.

▲ Projects and People

The ADA advises a great number of projects throughout Zimbabwe. Several are **irrigation schemes**, like the Middle Sabi Stage 111b Expansion project, Manicaland Province, which involves several thousand hectares of irrigated cotton, wheat, and soybean crops. The project seeks to add 1,000 hectares. Other examples are the Jotsholo Irrigation Expansion, the Amend Tsovane Irrigation Scheme, the Exchange Block Irrigation Scheme, the Fair Acres Irrigation Projects, and the Chisumbanje Irrigation Project—together covering thousands of hectares of cropland that are being developed and managed by ADA.

The ADA manages several projects with external funding. The **Smallholder Coffee, Fruit, Irrigated Food Crop and Vegetable Development Programme**, with support from the European Community (EC), is designed to assist and improve production as well as marketing for over 1,800 small-scale farmers. The program also focuses on post-harvest handling of the product, maintenance of roads and transport of the product, and training in general.

ADA also aids agricultural projects in communal areas to increase agricultural output and efficient resource use.

■ Resources

ADA offers "studies, teaching tools, and consultations."

Department of National Parks and Wildlife Management

Box 8365, Causeway
Harare, Zimbabwe

Contact: **Willie Kusezweni Nduku, Ph.D.,** Director
Phone: (263) 4 792783 • *Fax:* (263) 4 724914

Activities: Development; Education; Law Enforcement; Political/Legislative; Research • *Issues:* Biodiversity/Species Preservation; Deforestation; Sustainable Development; Wildlife and Fisheries Development • *Meetings:* CITES; IUCN; SADCC; UNESCO

▲ Projects and People

Dr. Willie Kusezweni Nduku, director, Department of National Parks and Wildlife Management, reports on three major conservation projects on behalf of rhinoceros, elephant, and aquaculture.

To protect *Diceros bicornis* and *Ceratotherium simum* from poachers, Dr. Nduku is **researching and monitoring wild rhino populations;** translocating "wild rhinoceros from the border areas to safe farmers within Zimbabwe, dehorning wild populations to render them safe from poachers who require them for their horns, and captive breeding [involving] cooperative collaborative programmes with *bona fide* zoos of the world."

According to Dr. Nduku, "Vegetation management is the key for the survival of **elephants.**" Ecosystems are protected in order to ensure adequate food and water supply for the elephant (*Loxodonta africana*). This programme also takes into account the survival of other animal species.

In charge of **fisheries management and research,** the Department also investigates **water pollution and quality.** "Being a land-locked country, management of water becomes a priority," writes Dr. Nduku. "In order to increase fish production, we have seriously entered a period of aquacultural development using indigenous fish species with breeding centres for fry" at various research stations. "However, the major constraints are droughts causing rivers and water supplies to dry."

Dr. Nduku is a member and rapporteur (reporter) of UNESCO's Man and the Biosphere (MAB) Program, regional member of the World Conservation Union's (IUCN) Species Survival Commission, member of IUCN/SSC's Ethnozoology Group, and co-author with G.F.T. Child of the recent report *Wildlife and Human Welfare in Zimbabwe,* Food and Agriculture Organization (FAO).

■ Resources

The Department seeks new technologies for scientific research and development and educational outreach and political support for conservation awareness at local to global levels. Publications are available from the librarian.

 ## NGO

Regional Office for Southern Africa ⊕
The World Conservation Union (IUCN)
P.O. Box 745
Harare, Zimbabwe

Contact: **India Musokotwane,** Regional Director
Phone: (263) 4 728266 • *Fax:* (263) 4 720738

Activities: Development • *Issues:* Biodiversity/Species Preservation; Deforestation; Energy; Global Warming; Population Planning; Sustainable Development; Water Quality; Wildlife Conservation • *Meetings:* CITES; IUCN

▲ Projects and People

India Musokotwane, who heads the **Regional Office for Southern Africa** of the World Conservation Union (IUCN), has a staff of six concentrating on conservation, environment, and sustainable development programs, attending global meetings in these fields, and "looking for international expertise in our major areas of activity." *(See also World Conservation Union [IUCN] in the France, Kenya, Niger, Pakistan, Switzerland, and USA NGO sections and throughout this Guide.)*

Zimbabwe National Conservation Trust

P.O. Box 8575, Causeway
Harare, Zimbabwe

Contact: **John A. Pile,** Executive Director
Phone: (263) 446105 • *Fax:* (263) 446105

Activities: Development; Education; Research • *Issues:* Biodiversity/Species Preservation; Sustainable Development

● Organization

The **Zimbabwe National Conservation Trust** was inaugurated in 1974 by a group of dedicated Zimbabwean conservationists. A group of environmentalists acts as a board of trustees drawn from all segments of the community, industry, and government. The nonprofit Trust raises and dispenses money for environmental conservation.

The Trust acts on behalf of environmental conservation and education as well as scientific research. Among other activities, it finances students in environmental programs, captures and relocates animals, proposes legislation and policy, and assists any organization with similar objectives with money or resources.

Early on, the Trust started a campaign to purchase **game capture units** for the Department of National Parks and Wildlife Management. The project purchased several units and now uses them to capture game for transport and sale. The captured game is sold to farmers who pay an agreed fee plus transport and handling costs. The farmers include the game in their own herds of domestic animals. The Trust uses the funds generated by these sales to reinvest in the Department of National Parks and Wildlife Management. The project safeguards against poaching in remote areas and assures species survival, reports the Trust.

In 1976, the Trust began breeding **nonendangered Gaboon vipers** in captivity—believing that this program will lead to the successful breeding of endangered species. The snakes are housed in three large outdoor enclosures, and detailed records are kept concerning general health, weight, length, and breeding activities. All births are documented and if possible recorded on video tape and film. Trust funds enabled the project to construct a third enclosure and winter quarters.

The **Zimbabwe Rhino Survival Campaign**, in cooperation with the Department of National Parks and Wildlife Management, has helped make possible the capture and translocation of more than 300 black rhinos, "from poacher-danger zones to more protected areas." The Trust also helped fund **Quiet Waters**, an outdoor classroom at Falcon College, where students partake of educational tours and wilderness research and birds and other wildlife benefit from an improved water source.

The Trust supports Zimbabwe's **National Conservation Strategy**, which is being carried out by an Inter-Ministerial Committee. It also works with the Natural Resources Board (public trustee of all natural resources); Boy Scouts; Agricultural Development Authority (ADA); Aloe, Cactus and Succulent Society of Zimbabwe; Association of Women of Zimbabwe; Association of Tanhaca Herbal Medicine, and Zimbabwe Trust, among other groups.

(Note: The Zimbabwe National Conservation Trust and Zimbabwe Trust are separate organizations.)

Zimbabwe Trust

4 Lanark Road, Box 4027
Harare, Zimbabwe
The Old Lodge, Christchurch Road
Epsom, Surrey KT19 8NE, UK

Contact: **Robert Monro**, General Secretary
Phone: (263) 4 722957 • *Fax:* (263) 4 795150

Activities: Development • *Issues:* Biodiversity/Species Preservation; Deforestation; Sustainable Development • *Meetings:* CITES; UNCED

● Organization

The **Zimbabwe Trust** was established in 1980 in the United Kingdom, upon Zimbabwe's attainment of political independence. At the UK head office, The Lady Soames, DBE, is president. A governing board of Trustees is comprised of five residents of the UK and two from Zimbabwe.

The Trust acts as the local agent for the Oak Foundation, a private, funding foundation based in the Cayman Islands.

Emphasis is on the "development of representative community-based institutions," particularly in rural or "marginal" areas deprived in terms of "socioeconomic infrastructure, skills, resources, services, and access to markets." Some 56 percent of Zimbabwe's population lives on communal land, which comprises 42 percent of the country's total land area.

▲ Projects and People

Trust activities are categorized into three general groups: programs, projects, or microprojects/small grants. The criteria for these categories are based on the nature and extent of Trust involvement, potential for duplication, and scale.

CAMPFIRE (Communal Areas Management Programme for Indigenous Resources) was started in 1989 as a collaborative project among the Zimbabwe Trust, the World Wide Fund for Nature, University of Zimbabwe, and the Centre for Applied Social Science. Conceived by the Department of National Parks and Wildlife Management, the CAMPFIRE concept was formed to help communal areas develop their capacities to acquire authority from the central government and manage their own local wildlife and natural resources.

According to CAMPFIRE literature, "For several decades, a number of far-sighted wildlife biologists and rural development specialists have believed that indigenous wildlife—and in particular the large mammals for which Africa is renowned—is one of the continent's greatest natural resources, and that it can and should be used by rural communities to generate income and improve the quality of their lives." *(See also Centre for Our Common Future in the Switzerland NGO section.)*

The largest CAMPFIRE project is in **Nyaminyami**, described as "starving in the midst of plenty," where some income is now being generated by regulated, safari sport hunting; fishing and other marine resource development is being considered for development. In areas such as Mzarabani, local people are learning to rely on tourism, instead of sport hunting.

The Trust is involved in the **Rusitu Valley Smallholder Tea Pilot Program** designed to allow peasant farmers in the lower valley's Ngorima Communal Area to plant tea and have it processed by a neighboring commercial tea estate. Implemented in 1989, it will be expanded to other areas—if successful.

The Trust helped launch **Mutare South Cooperative Development Programme (MSDP)** so that "pre-cooperative agricultural groups" could secure group credit and other economic benefits through joint purchases.

The Trust has supported numerous projects, notably the AIDS Counselling Trust, *Action Magazine*, Citizen's Advice Bureau, Zimbabwe National Association for Mental Health, Zimbabwe Integration Through the Arts, and Zimbabwe National Traditional Healers Association (ZINATHA), among others.

■ Resources

The Zimbabwe Trust makes available field and general reports, case studies, occasional papers, articles, and papers for oral presentations, including a booklet on *People, Wildlife, and Natural Resources—the Campfire Approach to Rural Development in Zimbabwe*. Robert Monro, general secretary, seeks to promote the Trust's ideas in the USA and Europe. Needed are field equipment for community-based environmental initiatives, information technology for project monitoring, resource materials, and political support for consumptive resource utilization.

◣ *Private*

Jonathan Hutton, Ph.D.
Consultant, Wildlife Utilization
P.O.Box HG 690, Highlands
Harare, Zimbabwe

Phone: (263) 4 739163 • *Fax:* (263) 4 739163

Activities: Political/Legislative; Research • *Issues:* Sustainable Development; Wildlife Utilization • *Meetings:* CITES

■ Resources

Dr. Jonathan Hutton, a consultant who is also Species Survival Commission (IUCN/SSC) Africa vice chair of the Crocodile Group and member of the SSC Trade Group, conducts "workshops on wildlife utilization and the conservation effects of CITES," particularly in countries belonging to the Southern Africa Development Coordination Committee (SADCC). Presently, he directs both The Ostrich Producers Association of Zimbabwe and the Crocodile Farmers' Association of Zimbabwe and is a consultant on crocodile farming in Mauritius and Tanzania.

He also provides input on environmental impact assessments (EIAs) and baseline ecological surveys regarding roads, hydroelectric power dams, pipelines, agricultural development, fuelwood use, and deforestation. Dr. Hutton has produced tourism management plans for Zimbabwe and Botswana and provides services in park, town, and land-use planning; community-based agroforestry; and ecological health among other issues.

 *Universities*

Department of Biological Sciences
University of Zimbabwe
P.O. Box MP 167, Mount Pleasant
Harare, Zimbabwe

Contact: **John Péri Loveridge**, Professor of Zoology
  Phone: (263) 4 303211, ext. 1364
  Fax: (263) 4 732828

Activities: Education; Research • *Issues:* Biodiversity/Species Preservation; Deforestation; Energy; Health; Sustainable Development; Water Quality

● Organization

The University of Zimbabwe has been undertaking crocodile research since 1968, with the amount of funding increasing steadily since 1981. The **Department of Biological Sciences** collaborates with other University departments as well as the Department of National Parks and Wildlife Management, and other universities. The Department sponsors many research studies on population ecology, feeding ecology, and crocodiles.

▲ Projects and People

Physiological research objectives are to provide basic information about how changes in the environment affect the crocodiles. Environmental variables include temperature, water (or lack of water), and food. Research ranges from the simple capture and measurement (of weight, length, sex) to the use of radioactive markers to measure drinking rates. The Nile crocodiles range in size from a 100-gram hatchling to a full-grown adult of 1,000 kilograms. The size and incredible strength of full-grown adult crocodiles prevents their use on a large scale, but sedatives and anesthesia are being developed. Currently, most research in the laboratory utilizes small crocodiles of less than one meter in length and under two kilograms. The largest crocodile studied weighed 423 kilograms.

In his studies, **Dr. Jon Hutton** showed that the sex of crocodiles is determined by temperature as the eggs incubate. It has been known for years that other animals, such as Australian crocodiles, alligators, and many species of turtles and terrapins, have their sex determined in the incubation stage. Dr. Hutton indicated that "eggs incubated at 31°C or below produce only females, and eggs incubated at 31–34°C produce males, depending on the clutch of eggs." The way in which temperatures determine sex is not clear, and further studies are beginning.

Dr. Chris Kofron has conducted research on the seasonality of reproduction in crocodiles in the Runde River, Gonarezhou National Park. Dr. Kofron captured crocodiles, measured and weighed them, determined their sex, and took blood samples. The blood samples were analyzed by radio-immunoassay at Henderson Research Station and the University's Department of Animal Science to determine the presence and levels of reproductive hormones, calcium, magnesium, and iron. A calendar of crocodile reproduction was prepared, noting that peak mating occurred from late June to early August.

Dr. John Péri Loveridge, entomologist and zoology professor, has worked at the Universities of Rhodesia, Cape Town, and Zimbabwe. He is particularly interested in how animals survive extremes of heat and desiccation, as well as in the conservation of vegetation and habitat for amphibians and reptiles. A member of the Species Survival Commission (IUCN/SSC) Crocodile Group and African Reptile and Amphibian Group, Dr. Loveridge has authored, co-authored, and edited dozens of research and scientific papers.

■ Resources

A bibliography is available on literature that includes research results. Funding is needed for research.

Department of Geography
University of Zimbabwe
P.O. Box MP 167
Harare, Zimbabwe

Contact: **Daniel Silas Tevera, Ph.D.**, Lecturer
  Phone: (263) 4 303211 • *Fax:* (263) 4 732828

Activities: Development; Education; Research • *Issues:* Sustainable Development; Waste Management/Recycling

▲ Projects and People

Dr. Daniel Silas Tevera, whose specialty is ecological geography at the University of Zimbabwe, researched "municipal solid waste management in sub-Saharan Africa" as a World Bank African-American fellow in 1991. Also that year, Dr. Tevera prepared an "assessment of environmental impact of the urban waste management system in Zimbabwe," with funding from the African-American Institute.

At the University, Dr. Tevera lectures on "regional and urban development, economic geography, and the environment and growth"; and has been researching the "development of the waste-recycling industry in Zimbabwe" and, with funding from the Swedish Agency for Research Cooperation with Developing Countries (SAREC), "a survey of rural-urban food distribution and population migration in Zimbabwe." Recent subjects of papers include the role of trade in African development, environmental effects of solid waste disposal in Harare, "The Changing Geography of Africa," and "Industrial Transnational Corporations and Their Impacts on Developing Countries."

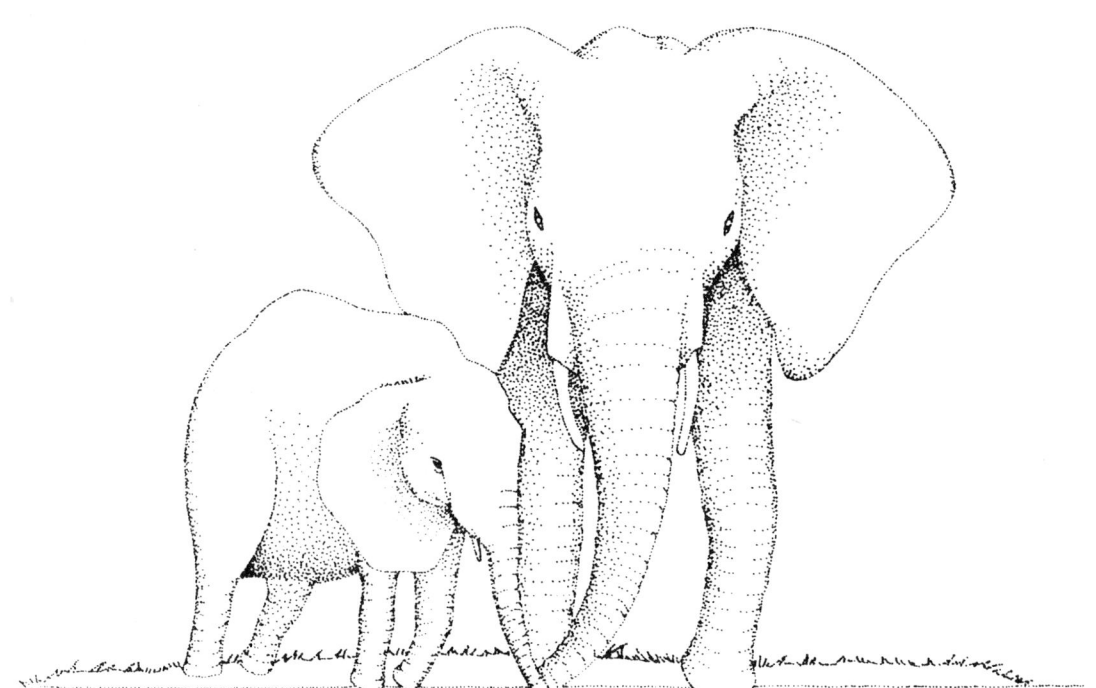

Appendix

The listings in the appendix will guide you to the organization, project, or individual involved in the designated issues. For further information, please consult the index and/or the country sections.

Air Quality/Emission Control

Biodiversity/Species Preservation

Deforestation

Energy

Global Warming

Health

Population Planning

Recycling/Waste Management

Sustainable Development

Transportation

Water Quality

Index

B

Brazil/Peru, border communities, 6
Brazilian Atlantic ecosystems, 848
Brazilian Foundation for Nature Conservation, 91
Brazilian Hunting and Conservation Association, **86–87**
Brazilian Institute for the Environment and Natural
  Resources, 84, 87, 89, 92
Brazilian Movement for Defense of Life, 91
Brazilian pine, 90
Brazilian Primatological Society, 93
Brazzaville University of Science, 145
breadfruit, 232, 714
breastfeeding, 750
breeding, captive. *See* captive breeding
Breeding Bird Atlas, New Hampshire, 708
Breen, Charles, 456
Brehm, Wolf W., 211
Brehm Funds for International Bird Conservation/Brehm-
  Fonds für Internationalen Vogelschutz, 172, **210–
  211**
Breitner, Paul, 216
Bretthauer, Erich W., 580
brevipalpis, 430
Brewer, Garry D., 866
brewery byproducts, 712
Breymeyer, A., 197
Bribri Indian tribe, 649
Bridges-Goodall, Clarita M., 26
Brieva, Lila M., 129
Briggs, Winslow R., 650
Brinckman and Associates, Canada, 334
Brindley, Harry P., 75
Brinkley, Christie, 734
British Association for the Advancement of Science, 526
British Association of Nature Conservation, 539
British Broadcasting Company, 537, 542, 544
British Chelonia Group, 545
British Council, 91
British Ecological Society, 345, 523, 539, 542, 546, 547
British Herpetological Society, 124, **508–509**
British Institute of Management, 519
British Museum, 875
British Museum of Natural History, 69
British National Space Centre, 524, 542
British Organic Farmers/Organic Growers Association, 512
British Organic Standards Committee, 512
British Ornithologists Union, 518, 539
British Petroleum (BP), 145, 711, 727
British Telecom, 528
British Trust for Ornithology, 518, 529
Brittany, 193
Brklacich, M., 102
Broad, Steven, 527
Broadbent, Peter, 520
Broadus, James M., 580
Brockelman, Warren Y., 488, 489, 790
Broggi, Mario, 308
Brokensha, David W., 684
bromelia, 3

bromeliads, 86, 170
bronchitis, 737
Bronx Green-Up, 736
Bronx Zoo, NY, 534
  Education Department, 166
  New York Zoological Park, 187, 617, 738–739
Brookhaven National Laboratory, 595
Brooks, Clifton Roland, 805
Brooks, P. Martin, 447, 448
Brooks, Ralph H., 558
Brouha, Paul, 615
Broun, Richard H., 573
Broussalis, P., 224
Brousset B., Jorge, 397
Brower, David R., 656
Brown, Bennett, 726
Brown, David A., 501
Brown, Eleanor, 393, 790
Brown, Enrique, 147
Brown, Foster, 654
Brown, Keith S., Jr., 92, 93
Brown, Lester, 11, 673, 676, 801
Brown, Linda, 674
Brown, Luther, 830
Browne, L.M., 753
Browning-Ferris Industries, 716, 765, 814, 815
Bruce, Maxwell, 511
bruchids, 359
Brundtland, Gro Harlem, 376, 474, 783–784
Brundtland Report, 686
Brundtland Report: Our Common Future, 114, 116
Brunei, 314, 317
Bruning, Donald F., 739–740, 788
Brunner, Karla, 92
Brush, Stephen B., 580
Bryant, Pat, 679, 680
Buchan Countryside Group, 528
Buck, Pearl S., 466
Buckler, Ernest, 100
Budapest Technical University, 241
Buddhism, 236–237, 378, 466, 657
  Mahayana, 236–237
  Theravada, 236–237
Buddhist Perception of Nature, **236–237**
Buddhist Perspectives on the Ecocrisis, 237
Buddhist Temple, Mt. Emei, China, 137
budgerigar. *See* parakeet
buffalo, 28, 226, 304, 339, 447, 450, 454, 671
  African, 351
  Asian, 72
  Cape, 447, 456
  dwarfed, 125
  wild, 247, 264
  wild water, 342
Buffalo Commons, 620
buffaloberry, 99
buffer zones, 152, 260, 317
Built Form and Culture Research Conference, 825

J

L

Landscape Institute, The, **520–521**, 528
landslides, 280, 366. *See also* erosion
Landvernd, 243
Landvernd—Icelandic Environmental Union, **244**
Landwise Glasgow, 528
Lane, Leonard J., 563
Lang, Brian, 523
Lange, Jürgen, 209, 210
Lange, Manfred, 206
Langer, Julia, 111
Langkawi Declaration on the Environment, 314
Langley Research Center, VA, 553
langoustine, 452
Langston, John, 518
Langtang Valley, Nepal, 343
Langton, Thomas E.S., 538
langur, 247, 249, 342, 790, 865
Lansing (MI) Community College, 718
Laona Project, **156**
Laos, xiv, 248, 317, 659, 789
Lapalala Wilderness School, 454–455
LaPierre, Louis, 104
Lapka, Miroslav, 159
lapwing, 348
larch, 239
larouman plant, 170
Larrondobuno, Alberto, 23
Larson, Bruce C., 867
Larson, Susan G., 54
Las Cruces Tropical Botanical Garden, Costa Rica, xi
laser-modified surfaces, 472
Lashof, Daniel L., 718
Last Great Places: An Alliance for People and the Environment, 722–723, 723–724, 725
Last Interglacial in the Arctic, 121
Laster, Richard, 274
Laster and Gouldman, **274**
Latin America, xiv, 19, 111, 352, 557, 565, 695, 716, 723–724, 724–725, 731, 731, 733, 793
Latin America Science Program, 723
Latin American and Caribbean Conservation Data Centers, 726
Latin American Committee of National Parks, Ecuador, 226
Latin American Conservation Network, 879
Latin American ecosystems, 21
Latin American Federation of Young Environmentalists, 151
Latin American Natural Area Programs, 650
Latin-American Organization of Integration, 870
Latour, P., 102
latrine, 180, 181, 230, 406, 468
Latvia, 676
Lau, Peter C.K., 105, 106
Laurel Foundation, 629
Laurel Ridge Conservation Education Center, 715, 716
Laurentian University, 108
Laurie, Andrew, 541, 542
Laurmann, John A., 843, 844

Laursen, Karsten, 163
Lautenbach, William E., 108
Lautenberg, Frank, 621
Lavalle de Vignaroli, Vivian, 24
Lave, Lester, 823
law, 405, 416. *See also names of individual laws*
    environmental, xii, xiii, 53, 148, 210, 213, 215, 216, 268, 274, 333, 404, 405, 469, 461, 549, 635, 700, 712, 748, 782–783, 810, 821–822, 831, 834, 840, 846–847, 852. *See also* legislation, environmental
    international, xiv, xv
    toxic tort, 821
Lawford, Richard (Rick) G., 103
Lawrence, Nathaniel, 720
Lawrence Berkeley Laboratory, California, 368, 760, 865. *See also* US Department of Energy
Lawrence Livermore National Laboratory, Atmospheric and Geophysical Sciences Division, **569–570**
lawsuits, environmental, 700. *See also* law, environmental
Lazcano-Barrero, Marco A., 330
Lazell, James D., 646
leachates, 765. *See also* groundwater: seepage
lead, 113, 186, 193, 201, 240, 241, 376, 518, 522, 616, 663, 719, 774
leafhopper, 503
League of Conservation Voters, **699**, 705, 815
League of Women Voters, **699–700**, 757, 853
Leakey, Richard E., 297, 620, 788
Lear, Norman and Lyn, 668
Learned Society, ZSL, 534
leathercraft, 888. *See also* crafts
Lebanon, 480
LeBlanc, Alice, 661
lechwe, 339, 454, 889
LeCroy, Mary, 392
Ledesma, Fr. Antonio, 409
Ledger, John A., 450
Lee, C.A., 109
Lee, Charles, 706
Lee, David S., 597, 598
Lee, Kai N., 861
Lee, Wo Yen, 13
Leeb, Hermann, 215
Lega Italiana Protezione Uccelli, **280**
Lega per L'Abolizione della Caccia, **279–280**
Legacy International, 621
Legal Environmental Assistance Foundation, 700, 757
legislation, environmental, 64, 74, 89, 402, 428, 471–472, 808. *See also* law, environmental
legume(s), 409, 481
    chick pea, 481, 761
    cowpea, 358, 890
    guar, 640
    honey locust, 590
    lentil, 364, 481
    mimosoid, 736
    pulse, 504

ℳ

P

$\mathcal{V}$

Species Survival Commission, 791, **794-795**
USA, **794**
World Coral Fish Programme, Species Survival Commission, 508
World Council for the Biosphere, 475
World Council of Churches, 686
World Data Center A for Glaciology, 846
World Data Centre "C" for Glaciology, 543
World Earth Day, 783
World Ecologists Foundation, **411-412**
World Energy Council, UK, 258
World Environment Center, 116, 551, 808
World Environment Data Sheet, 752
World Environment Day, 14, 229, 254
World Environmental Week, 283
World Forest Agreement Working Group, 666
World Forest Convention, 654
World Futures Studies Federation, 851
World Health Organization (WHO), 7, 13, 14, 24, 64, 167, 273, 294, 345, 379, 383, 435, 479, 571
Expert Advisory Panel on Parasitic Diseases, 855
International Programme on Chemical Safety, 584
World Heritage Convention, 164
World Heritage List, 30, 32, 37, 477
World Heritage Site, 342, 387
World Life Zone System of Ecological Classification, 150
World Meteorological Organization, 10, 13, 429, 470, **478**, 481, 507, 526, 542, 794
World Nature Association, **798-799**
World Ocean Circulation Experiment, 478, **507-508**, 526
International Planning Office, 501
"World Ocean" program, 427
World Parrot Trust, UK, 756
World Pheasant Association, 65, 740
World Plan of Action on the Ozone Layer, 10
World Rainforest Movement, 655
World Resources Institute, xii, xv, 41, 106, 116, 147, 258, 290, 540, 555, 621, 634, 642, 673, 719, 794, **799-800**
World Seed Program, 687
World Soils and Terrain Digital Database, 347
World Soils Policy, 11
World Summit for Children, 784
World Underwater Federation, **192-193**
World University Service of Canada, 297
World War II, 239, 418, 447
World Who's Who in Environment and Conservation, 475
World Wide Fund for Nature, 82, 182, 205, 215, 218, 223, 251, 252, 253, 263, 303, 315, 317, 342, 355, 356, 365, 373, 453, 474, 477, 477, 484, 488, 509, 515, 518, 527, 533, 536, 802, 887, 888, 890, 894
Australia, **38-39**, 43
India, **260**, 261
International, 340, 545
Japan, 292
Malaysia, 315, 319, 543, 847
Netherlands, 351

New Zealand, **367-368**
Pakistan, 381, **382-383**
United Kingdom, 542
United States, 88
Zambia Country Office, **891**
World Wilderness Congress, 696
World Wildlife Fund, xii, 11, 58, 69, 73, 80, 87, 92, 99, 131, 137, 140, 146, 148, 165, 172, 174, 178, 192, 199, 202, 226, 229, 230, 233, 238, 264, 283, 300, 329, 333, 343, 360, 390, 398, 405, 407, 413, 424, 460, 428, 450, 465, 475, 524, 525, 529, 530, 535, 536, 539, 555, 557, 629, 648, 668, 673, 688, 695, 747, 752, 779, 783, 794, **800-801**, 814, 832, 844, 865, 883
Australia, 11, 38
Belgium, **65-66**
Brazil, 825
Canada, 104, 111, 114
Denmark, **169**
Hong Kong, 237
India, 254, 262, 546
Indonesia, 316
International, 62, 451, 470, 847, 877
Japan, 288, 289
Uganda, **497-498**
United Kingdom, 542, 655, 832, 833
United States, 55, 76, 89, 125, **152**, 237, 261, 316, 331, 433, 539, 540, 546, 623, 655, 663, 797, 832, 833, 863, 870
World Women's Congress for a Healthy Planet, 807
Worldwatch Institute, 11, 576, 626, 668, 673, 676, **801-802**
WorldWIDE Directory of Women, 802
WorldWIDE Network, **802**
worm, 431, 813
Wright, Anne, 260
Wright, Debra, 790
Wright, Janice C., 368
Wright, Patricia C., 827, 828
Wright, Tim, 453
Wrist, Peter E., 106
Writers Guild, 668
Wróbel, Stanisław, 417
Wrocław University, Museum of Natural History, **419**
Wu, Changhua, 131
Wu, Su-gong, 136
Wuebbles, Donald J., 570
Wuhan University, 133
Wünschmann, Arnd, 218
Wuorenrinne, Heikki, 185
WWF. *See* World Wide Fund for Nature; World Wildlife Fund
Wyerman, James K., 655
Wyoming Wildlife Federation, 716

XYZ